Stereochemistry of
Organic Compounds

Stereochemistry of Organic Compounds

ERNEST L. ELIEL
Department of Chemistry
The University of North Carolina at Chapel Hill
Chapel Hill, North Carolina

SAMUEL H. WILEN
Department of Chemistry
The City College of the City University of New York
New York, New York

With a Chapter on Stereoselective Synthesis by

LEWIS N. MANDER
Research School of Chemistry
Australian National University
Canberra, Australia

A Wiley-Interscience Publication

JOHN WILEY & SONS, INC.

New York • Chichester • Brisbane • Toronto • Singapore

Library of Congress Cataloging in Publication Data:

Eliel, Ernest Ludwig, 1921-
    Stereochemistry of organic compounds / Ernest L. Eliel, Samuel H.
Wilen, Lewis N. Mander.
        p.   cm.
    "A Wiley-Interscience publication."
    Includes index.
    ISBN 0-471-01670-5
    1. Stereochemistry. 2. Organic compounds. I. Wilen, Samuel H.
II. Mander, Lewis N. III. Title.
QD481.E52115   1993
547.1'223--dc20   93-12476

Contents

Preface

It is now over 30 years since the last comprehensive text on organic stereochemistry (*Stereochemistry of Carbon Compounds*, McGraw-Hill, 1962) was published by one of us. Since then there has been enormous interest and activity in the field and entirely new concepts have sprung up. To give just a few examples: Conformational analysis has come of age. Nuclear magnetic resonance has become ubiquitous in chemistry and NMR spectra very often require consideration of configuration and conformation before they can be interpreted and, in turn, permit inferences about the configuration and conformation of the compounds whose spectra are recorded. Molecular mechanics, in its infancy in 1962, is now a widely used tool. The concepts of prostereoisomerism (commonly known under the more restrictive term *prochirality*), which were barely understood in 1962 are now broadly disseminated and utilized. The preparation of pure enantiomers has become of consuming interest in the pharmaceutical, flavor, and agricultural industries, not to speak of university laboratories. As a result, there has been enormous progress in the area of enantioselective synthesis, in the techniques for separating enantiomers, and in the methods for analysis of enantiomeric purity, the latter spurred by instrumental developments in spectroscopy and in chromatography. There has also been significant development in the conceptual and mathematical foundations of stereochemistry, especially in the application of symmetry concepts.

Contemporary elementary textbooks in organic chemistry have taken cognizance of these developments by increasingly incorporating stereochemical principles with the result that stereochemical ideas and concepts are introduced to undergraduate students to a greater extent than was the case 30 years ago. Within the last 15 years a number of briefer stereochemistry books have appeared. While these books elaborate on this greater awareness of stereochemistry, we felt that the time had come for the preparation of an up-to-date comprehensive text. This book is intended to serve as a textbook for graduate students and advanced undergraduate students for whom it may serve as a guide to subsequent studies and research. We have endeavored to prepare a book that may also serve as a comprehensive guide for research workers who might wish to have stereochemical principles explained and illustrated under one cover.

The task of writing such a text has been rather daunting, and, as a result, has extended over more years than we would have liked. As implied in the first paragraph, the field has grown enormously since the last comprehensive text was put together; the fact that this book has three authors rather than a single one is only one token of this development. Even three authors cannot be conversant with all the ramifications of the subject as it exists in 1994. We must therefore ask the reader's forbearance if, on occasions, discussion of a given subject is not as authoritative or as detailed as one might wish. As with the 1962 book, this one is limited to the stereochemistry of organic compounds, but unlike in the earlier work, we were forced to exclude certain subjects even though they are parts of organic stereochemistry. Thus limitations of space and time have forced us to exclude polymer stereochemistry. The stereochemistry of reaction mechanism has been touched on but peripherally and this has led to the exclusion of discussion of the Woodward-Hoffmann rules. The enormously large area of stereoselective synthesis could only be covered in overview form; we are greatly indebted to Professor Lewis N. Mander of the Australian National University for contributing the chapter on this topic.

Even with these omissions, the book is more extensive than we would have liked. In order to make it somewhat easier for the reader to cover the essential material, we have used smaller print for subject matter that may not be in the mainstream of the argument, but may be of interest to some of our readers.

Along with the burgeoning literature on stereochemistry has come a proliferation of new and modified terminology that befuddles novices and sometimes experienced scientists as well. We have been conservative in our usage of new terms by limiting ourselves, as much as possible, to established terminology. In addition to defining terms at appropriate places in the text, we have gathered the more important definitions in a glossary appearing at the end of the book. The glossary anticipates, but does not duplicate, a listing of IUPAC terminology on stereochemistry that has been in gestation for quite a few years.

Stereochemistry is an old subject and we have tried to pay some attention to its historical development. One important ingredient in understanding the history of any subject is to know who did what when. This cannot be accomplished solely by listing references at the ends of chapters, since readers are apt to turn to such references only when they have an interest in a specific topic. Listing the references at the bottom of each page would unfortunately have added appreciably to the production cost, and hence to the price of this book. We have, therefore, included authors' names and the year of publication of the work as the reference citation within the text with the name of the senior author included even if not the first of several listed in the ordinary style of citations as "et al." In addition, to facilitate entry into the literature of sterochemistry, we have included the titles of review articles in the reference list at the end of each chapter.

One of the authors (ELE) is grateful to the John Simon Guggenheim Foundation for fellowships during the academic years 1975–1976 and 1982–1983. Without this help, supplemented by academic leaves provided by the University of North Carolina at Chapel Hill, this book could not have been launched. ELE also acknowledges the hospitality of Stanford, Princeton, and Duke Universities during these leaves, as well as many stimulating conversations with Harry S. Mosher, Hans Gerlach, Kurt Mislow, James G. Nourse, and Jack D. Dunitz. Another author (SHW) is pleased to acknowledge a Fellowship leave granted in 1983–1984 by the City College, City University of New York, and the hospitality of the Chemistry Department of the University of North Carolina at Chapel Hill during several visits occasioned by the preparation of this book as well as for a visiting professorship in the Spring of 1984.

A number of colleagues have read and made suggestions for the improvement of various versions of the manuscript of this book. Professors James H. Brewster and Michael P. Doyle read the entire text and Professors William A. Bonner, André Collet, Jack D. Dunitz, Mark M. Green, Jean Jacques, Henri B. Kagan, Meir Lahav, Kurt Mislow, Laurence A. Nafie, Vladimir Prelog, Hans-Jürg Schneider, George Severne, Roger A. Sheldon, Grant Gill Smith, Dr. Jeffrey I. Seeman, and the late Günther Snatzke commented on entire chapters or sections. To all these colleagues we are grateful. Nevertheless we ourselves take full responsibility for the contents. We also wish to express our gratitude to Eva Eliel and to Rosamond Wilen for assistance in the checking of proof.

We appreciate the release, on the part of McGraw-Hill, Inc., of the rights to *Stereochemistry of Carbon Compounds* (Eliel, 1962), which has enabled us to use several figures from that book. We are also grateful to Springer-Verlag GmbH & Co., Heidelberg, for permission to use the text and figures of the chapter "Prostereoisomerism (Prochirality)" by E.L. Eliel, which appeared in *Topics in Current Chemistry*, Vol. 105. A substantial amount of material in Chapter 8 has been taken from this earlier work.

ERNEST L. ELIEL
SAMUEL H. WILEN

Stereochemistry of Organic Compounds

1

Introduction

1-1. SCOPE

Stereochemistry (from the Greek *stereos*, meaning solid) refers to chemistry in three dimensions. Since most molecules are three-dimensional (3D), stereochemistry, in fact, pervades all of chemistry. It is not so much a branch of the subject as a point of view, and whether one chooses to take this point of view in any given situation depends on the problem one wants to solve and on the tools one has available to solve it.

In the evolution of chemical thought, the stereochemical point of view came relatively late; much of the often excellent chemistry of the nineteenth century ignores it. By the same token, some important contemporaneous developments, such as the computer design of synthesis (Wipke, 1974; Wipke et al., 1977) and the computer-assisted elucidation of chemical structure (Carhart, Djerassi, et al., 1975), legitimately *started out* by disregarding the third dimension; however, this shortcoming has since been remedied (e.g., Djerassi et al., 1982; Corey et al., 1985).

Nevertheless, there is little question that, at least in the last 25 years, the third dimension has become all-important in the understanding of problems not only in organic, but in physical, inorganic, and analytical chemistry as well as biochemistry, so that no chemist can afford to be without a reasonably detailed knowledge of the subject.

It has become customary to factorize stereochemistry into its static and dynamic aspects. *Static stereochemistry* (perhaps better called stereochemistry of molecules) deals with the counting of stereoisomers, with their structure (i.e., molecular architecture), with their energy, and with their physical and most of their spectral properties. *Dynamic stereochemistry* (or stereochemistry of reactions) deals with the stereochemical requirements and the stereochemical outcome of chemical reactions, including interconversion of conformational isomers or topomers (cf. Chapter 2); this topic is deeply interwoven with the study and understanding of reaction mechanisms. Like most categorizations, this one is not truly dichotomous and some subjects fall in between; for example, quantum mechanical treatments of stereochemistry may deal with either its structural or its mechanistic aspects; spectroscopic measurements may fathom reaction rate as well as molecular structure.

Limitations of space, plus the sheer vastness to which the subject of stereochemistry has grown, force us to make choices in what will be discussed in this book. Static aspects, including spectroscopic ones, will be covered in detail. Dynamic stereochemistry and reference to calculations will be woven into the text, but reaction mechanisms will not be described in a systematic fashion. In particular, there will be no specific coverage of electrocyclic reactions and orbital symmetry. While we regret these omissions, we take note of the fact that there are a number of excellent books on reaction mechanism that include stereochemical aspects (Ingold, 1969; Deslongchamps, 1983; Lowry and Richardson, 1987; Carey and Sundberg, 1991; March, 1992) and several detailed treatments of the Woodward–Hoffmann rules (Woodward and Hoffmann, 1969, 1970; Anh, 1970; Lehr and Marchand, 1972; Fleming, 1976; Marchand and Lehr, 1977; Gilchrist and Storr, 1979). With equal regret (because we consider the division between inorganic and organic chemistry to be artificial) we had to omit inorganic compounds (cf. Geoffroy, 1981; Kepert, 1982), both coordination compounds (cf. Sokolov, 1990) and compounds of the main group elements, such as silicon (cf. Corriu, 1984), sulfur (Mikołajczyk and Drabowicz, 1982; Mikołajczyk, 1987), and phosphorus (Gallagher and Jenkins, 1968; Quin, 1981; Verkade and Quin, 1987). Only the stereochemistry of nitrogen will be marginally touched on. We have also had to forgo a treatment of the stereochemistry of polymers (Bovey, 1969, 1982; Farina, 1987). Nor will we deal, in this book, with the mathematical foundations of stereochemistry, but we draw attention to a recent book (Mezey, 1991) and review article (Buda, Mislow, et al., 1992) in this area.

1-2. HISTORY

Only a very abbreviated history of stereochemistry will be given here, since two authoritative books (Bykov, 1966; Ramsay, 1981) plus a collection of pertinent essays (Ramsay, 1975) are available (see also Mason, 1976).

Historically, the origins of sterochemistry stem from the discovery of plane-polarized light by the French physicist Malus (1809). In 1812 another French scientist, Biot (q.v.), following an earlier observation of his colleague Arago (1811), discovered that a quartz plate, cut at right angles to its crystal axis, rotates the plane of polarized light through an angle proportional to the thickness of the plate; this constitutes the phenomenon of optical rotation. Some quartz crystals turn the plane of polarization to the right, while others turn it to the left. Three years later, Biot (1815) extended these observations to organic substances—both liquids (such as turpentine) and solutions of solids (such as sucrose, camphor, and tartaric acid). Biot recognized the difference between the rotation produced by quartz and that produced by the organic substances he studied: The former is a property of the crystal; it is observed only in the solid state and depends on the direction in which the crystal is viewed, whereas the latter is a property of the individual molecules, and may therefore be observed not only in the solid, but in the liquid and gaseous states, as well as in solution.

With respect to the question of the cause of optical rotation, the French mineralogist Haüy (q.v.) had already noticed in 1801 that quartz crystals exhibit the phenomenon of hemihedrism. Hemihedrism (cf. Section 6-4.c) implies inter alia that certain facets of the crystal are so disposed as to produce nonsuperposable species (Fig. 1.1, **A** and **B**), which are related as an object to its mirror image. (Such mirror-image crystals are called "enantiomorphous," from the Greek *enantios* meaning opposite and *morphe* form.) In 1822, Sir John Herschel (q.v.), a British astronomer, observed that there was a relation between hemihedrism and optical rotation: All the quartz crystals having the odd faces inclined in one direction rotate the plane of polarized light in one and the same sense, whereas the enantiomorphous crystals rotate polarized light in the opposite sense.

It was, however, left to the genius of Pasteur to extend this correlation from the realm of crystals, such as quartz, which rotate polarized light only in the solid state, to the realm of molecules, such as *dextro*-tartaric acid, which rotate both as the solid and in solution. [*dextro*-Tartaric acid, henceforth denoted as (+)-tartaric acid, rotates the plane of polarized light to the right, see Section 1-3.] In 1848 Pasteur (q.v.) had succeeded in separating crystals of the sodium ammonium salts of (+)- and (−)-tartaric acid from the racemic (nonrotating) mixture. When the salt of the mixed (racemic) acid, which is found in wine caskets, was crystallized by slow evaporation of its aqueous solution, large crystals formed which, to Pasteur's surprise and delight, displayed hemihedric crystals similar to those found in quartz (Fig. 1.1). By looking at these crystals with a lens, Pasteur was able to separate the two types (with their dissymmetric facets inclined to the right or left) by means of a pair of tweezers. When he then separately redissolved the two kinds of crystals, he found that one solution rotated polarized light to the right [the crystals being identical with those of the salt of the natural (+)-acid], whereas the other rotated to the left. [(−)-Tartaric acid had never been encountered up to that time.]

> The relationship between crystal morphology and molecular structure (in particular, molecular configuration) has presented a challenge to chemists and crystallographers ever since. However, it was only in 1982 (cf. Addadi et al., 1986) that a definite relationship was established, allowing one to deduce configuration from crystal habit (cf. Chapter 5).

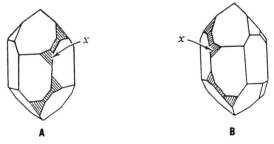

A   **B**

Figure 1.1. Hemihedrism of quartz crystals. [Reprinted with permission from L. F. Fieser and M. Fieser (1956), *Organic Chemistry*, 3rd ed., Heath, Lexington, MA.]

Pasteur (1860) soon came to realize the analogy between crystals and molecules: In both cases the power to rotate polarized light was caused by dissymmetry, that is the nonidentity of the crystal or molecule with its mirror image, expressed in the case of the ammonium sodium tartrate crystal by the presence of the hemihedric faces. Similarly, Pasteur postulated, the molecular structures of (+)- and (−)-tartaric acids must be related as an object to its mirror image. The two acids are thus enantiomorphous at the molecular level; we call them enantiomers. [The ending -mer (as in isomer, polymer, and oligomer from the Greek meros meaning part) usually refers to a molecular species.]

By the time Pasteur had arrived at this insight, his interests had shifted from chemistry to microbiology, and he never couched the mirror-image relationship in unequivocal geometric terms, even though the structural theory of organic chemistry, which must form the basis of any such precise specification, was beginning to unfold at that time thanks to the publications of Kekulé (1858), Couper (1858) and Butlerov (1861). It was not until 1874 that van't Hoff (1874, 1875) in Utrecht, the Netherlands and Le Bel (1874) in Paris, France independently and almost simultaneously proposed the case for enantiomerism in a substance of the type Cabcd: the four substituents are arranged tetrahedrally around the central carbon atom to which they are linked. van't Hoff, who had worked with Kekulé and whose views were based on structural theory, specified the 3D arrangement quite precisely: The four linkages to a carbon atom point toward the corners of a regular tetrahedron (Fig. 1.2) and two nonsuperposable arrangements (enantiomers) are thus possible.

We call the model corresponding to a given enantiomer (e.g., Fig. 1.2, **A**) and the molecule that it represents "chiral" (meaning handed, from Greek cheir, hand) because, like hands, the molecules are not superposable with their mirror images. The term chiral was first used by Thomson (later elevated to the peerage as Lord Kelvin) in 1884 (Kelvin, 1904), was rediscovered by Whyte (1957, 1958), and was firmly reintroduced into the stereochemical literature by Mislow (1965) and by Cahn, Ingold, and Prelog (1966) who define a model as chiral when it has no element of symmetry (plane, center, alternating axis; cf. Chapter 4) except at most an axis of rotation.

A certain amount of confusion or ambiguity has arisen in the use of the term. When a *molecule* is chiral, it must be either "right-handed" or "left-handed." But if a *substance or sample* is said to be chiral, this merely means that it is made up of

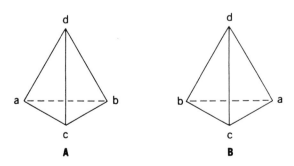

Figure 1.2. Tetrahedral carbon.

chiral molecules; it does not necessarily imply that all the constituent molecules have the same "sense of chirality" (R or S, or M or P; cf. Chapter 5). We may distinguish two extreme situations (plus an infinite number of intermediate ones): (a) The sample is made up of molecules that all have the same sense of chirality (homochiral molecules). In that case the sample is said to be chiral and "non-racemic." This serves to distinguish this case from the opposite situation where (b) the sample is made up of equal (or very nearly equal) numbers of molecules of opposite sense of chirality (heterochiral molecules), in which case the sample is chiral but racemic. Thus the statement that a macroscopic sample (as distinct from an individual molecule) is chiral is ambiguous and therefore sometimes insufficient; it may need to be further stated if the sample is racemic or nonracemic. Lack of precision on this point has led to some confusion, for example, in the titles of articles where the synthesis of a chiral natural product is claimed, but it is not clear whether the investigator simply wishes to draw attention to the chirality of the pertinent structure or whether the product has actually been synthesized as a single enantiomer (i.e., an assembly of homochiral molecules, which should not, however, be called a homochiral sample).

The situation is even slightly more complex than so far implied. There is little ambiguity about the meaning of "chiral, racemic": Chiral, racemic means that (within the limits of normal stochastic fluctuations) the sample is made up of equal numbers of molecules of opposite sense of chirality. But in a "chiral, nonracemic" sample there can be some molecules of a sense of chirality opposite to that of the majority; that is, the sample may not be enantiomerically pure (or *enantiopure*). Experimental tests as to whether a sample is enantiopure or merely *enantioenriched* will be discussed in Section 6-5.

> In consequence of these definitions, use of the word "chiral" should be restricted to molecules (or models thereof) and substances as in chiral substrate, chiral catalyst, chiral stationary phase, and so on. However, we strongly discourage the application of the word to processes, as in chiral synthesis, chiral catalysis, chiral recognition, chiral chromatography, and so on.

It immediately follows from van't Hoff's hypothesis that in an alkene, where the tetrahedra are linked along one edge, cis–trans isomerism is possible (see Chapter 9) and already in 1875 van't Hoff (q.v.) predicted the stereoisomerism of allenes, not actually observed in the laboratory until 1935 (cf. Chapter 14; Fig. 1.3).

In contrast, Le Bel, whose ideas were based on Pasteur's analogy between crystals and molecules, admitted the possibility of asymmetric (or, as we would now say, chiral) arrangements other than tetrahedral in Cabcd (cf. Snelders, 1975). As a consequence he left open the question as to whether alkenes were planar and he actually looked for enantiomeric alkenes, a search that he abandoned only 20 years later (Le Bel, 1894).

Figure 1.3. Tetrahedral representation of alkenes and allenes.

The hypothesis of van't Hoff and Le Bel has stood with but minor modifications until today (cf. Eliel, 1974 for an account of the centennial of the theory). Both the visualization of molecules by X-ray and electron diffraction and the interpretation of vibrational [infrared (IR) and Raman] spectra have confirmed that carbon is, indeed, tetrahedral. Quantum mechanical calculations concur in predicting a much lower energy for tetrahedral methane than for (hypothetical) methane of *planar geometry* (Monkhorst, 1968; Hoffmann et al., 1970).

van't Hoff (1874) had already pointed out that if CX_2Y_2 were planar (or, for that matter, square pyramidal), two isomers should exist but only one is found. For a detailed discussion see Wheland (1960).

1-3. POLARIMETRY AND OPTICAL ROTATION

It was mentioned in Section 1-2 that the discoveries of polarized light and optical rotation led to the concept of molecular chirality which, in turn, is basic to the field of stereochemistry. Polarized light and optical rotation are therefore usually given considerable play in elementary treatments of stereochemistry. In the present text we take the view that the central theme of stereochemistry is molecular architecture, notably including chirality, and the resultant fits (as of a right hand with a right glove or of an enzyme with its natural substrate) or misfits (as of a right hand with a left glove or of an enzyme with the enantiomer of its natural substrate). In this theme, polarimetry and optical rotation are but epiphenomena (side issues), which are important, indeed, as diagnostic tools for chirality but not central to its existence. We shall therefore treat polarimetry only briefly at this point, assuming that the nature of polarized light and the workings of a polarimeter are already familiar to the reader.

Methods of palpating chirality by optical tools [polarimetry, optical rotatory dispersion (ORD), and circular dichroism (CD)] have been called "chiral–optical" methods by Weiss and Dreiding (Weiss, 1968) later contracted to "chiroptical" methods (Prelog, 1968; Henson and Mislow, 1969; see also Kelvin, 1904, p. 461), a term that will be used in this book. These methods will be discussed in detail in Chapter 13.

The observed angle of rotation of the plane of polarization by an optically active liquid, solution, or (more rarely) gas or solid is usually denoted by the symbol α. The angle may be either positive ($+$) or negative ($-$) depending on whether the rotation is clockwise, that is, to the right (*dextro*) or counterclockwise, that is, to the left (*levo*) as seen by an observer *towards whom* the beam of polarized light travels. (This is opposite from the direction of rotation viewed *along* the light beam.) It may be noted that no immediate distinction can be made between rotations of $\alpha \pm 180 \, n°$ ($n =$ integer), for if the plane of polarization is rotated in the field of the polarimeter by $\pm 180°$, the new plane will coincide with the old one. In fact α, as measured, is always recorded as being between $-90°$ and $+90°$. Thus, for example, no difference appears between rotations of $+50°$, $+230°$, $+410°$, or $-130°$. To make the distinction, one must measure the rotation at least at one other concentration. Since optical rotation is proportional to concentration (see below), if solutions of the above rotations were diluted to one-tenth of their original concentrations, their rotations would become $+5°$, $+23°$, $+41°$, and $-13°$, values that are all clearly distinct. Readings taken at two different concentrations almost always determine α unequivocally. An alternative for solutions and the method of choice for pure liquids is to measure the rotation in a shorter tube. In the above cases, if a tube of a quarter of the original length [e.g., 0.25 decimeters (dm) instead of 1 dm] is used, the rotations as recorded become $+12.5°$, $+57.5°$, $-77.5°$ (equivalent to $+102.5°$), and $-32.5°$, again all clearly distinguishable. [Note that halving the tube length (e.g., from 1 to 0.5 dm) would have left the ambiguity between the first and third observation ($+25°$ vs. $+205° = 180° + 25°$) and between the fourth and second ($-65°$ and $+115° = 180° - 65°$).]

Biot discovered that the observed rotation is proportional to the length ℓ of the cell or tube containing the optically active liquid or solution and the concentration c (or density in the case of a pure liquid): $\alpha = [\alpha] c \ell$ (Biot's law). The value of the proportionality constant $[\alpha]$ depends on the units chosen; in polarimetry it is customary to express ℓ in decimeters, because the cells are usually 0.25, 0.5, 1, or 2 dm in length, and c in grams per milliliter ($g\,mL^{-1}$) or (and this is preferred for solutions) in $g\,100\,mL^{-1}$. Thus,

$$[\alpha] = \frac{\alpha}{\ell(\text{dm})\, c(\text{g mL}^{-1})} = \frac{100\,\alpha}{\ell(\text{dm})\, c'(\text{g 100 mL}^{-1})} \tag{1.1}$$

The value of $[\alpha]$, the so-called *specific rotation*, depends on wavelength and temperature which are usually indicated as subscripts and superscripts, respectively; thus $[\alpha]_D^{25}$ denotes the specific rotation for light of the wavelength of the sodium D-line (589 nm) at 25°C. (Evidently, a polarimeter requires a source of monochromatic light as well as a thermostatted cell; moreover, the solution must be made up in a thermostatted volumetric flask at the same temperature as that of the measurement, or else the volume correction must be applied.) In addition, $[\alpha]$ also depends on the solvent and to some extent on the concentration (in a fashion not taken into account by the concentration term in Biot's law), which must thus also be specified. This is usually done by adding such information in parentheses, thus $[\alpha]_{546}^{20} - 10.8 \pm 0.1$ (c 5.77, 95% ethanol) denotes the specific rotation at 20°C for light of wavelength 546 nm in 95% ethanol solution at a concentration of $5.77\,g\,100\,mL^{-1}$. The importance of solvent and concentration is occasioned by association phenomena, which will be discussed in more detail below and in Chapter 13.

The dimensions of $[\alpha]$ are $\deg\,cm^2\,g^{-1}$ *not* degrees (Snatzke, 1989/90). In this book $[\alpha]$ will always be given without the units (understood to be $10^{-1}\,\deg\,cm^2\,g^{-1}$) and (in contrast to the observed rotation α) will *not* be given in degrees.

For pure liquids, since the density is fixed at a given temperature, one may simply state the observed rotation, along with the cell length, such as $\alpha_D^{25} + 44°$

(neat, $\ell = 1$ dm), the word "neat" (or sometimes "homog" for homogeneous) denoting that the rotation refers to the undiluted (pure) liquid. However, even here it is preferable to give the specific rotation. Thus, if the density of the liquid in question at 25°C is 1.1, the specific rotation is $[\alpha]_D^{25} + 40$ (neat) $(44/1 \times 1.1)$.

Since optical rotation is proportional to the number of molecules encountered by the beam of polarized light, if two substances have unequal molecular weights but are alike with respect to their power of rotating polarized light, the substance of smaller molecular weight will have the larger specific rotation simply by virtue of having more molecules per unit weight. In order to compensate for this effect and to put rotation on a per-mole basis, one defines the term "molar rotation" as the product of specific rotation and molecular weight divided by 100. (The divisor serves to keep the numerical value of molar rotation on the same approximate scale as that of specific rotation. For a substance of molecular weight (MW) 100, molar and specific rotation are the same.) Thus, denoting molar rotation by $[M]$ or $[\Phi]$, the latter symbol being preferred

$$[M] = [\Phi] = \frac{[\alpha].\text{MW}}{100} = \frac{\alpha}{\ell(\text{dm}).c''(\text{mol } 100 \text{ mL}^{-1})} \tag{1.2}$$

The choice of solvent particularly affects the rotation of polar compounds because of its intervention in solvation and association phenomena (cf. Chapter 13). Substantial changes of specific rotation with solvent are not uncommon; reversals of sign are less frequent but have been explicitly reported in a number of instances (Chapter 13). A pH dependence of rotation is also common in the case of acids and bases and reversals are recorded, for example, for (S)-$(+)$-lactic acid, dextrorotatory in water, whose sodium salt is levorotatory (Borsook et al., 1933) and for L-leucine, which is levorotatory in water but dextrorotatory in aqueous hydrochloric acid (Stoddard and Dunn, 1942).

An even more remarkable change in rotation, from positive to negative, is seen in 2-methyl-2-ethylsuccinic acid (Krow and Hill, 1968) as its solution in chloroform (containing 0.7% ethanol) is diluted, with a reversal of sign (corresponding to null rotation) occurring at a concentration of 6.3%. The phenomenon (presumably due to association) is confined to solvents of low polarity ($CHCl_3$ or CH_2Cl_2); no reversal is seen in alcohol solvents, pyridine, diglyme, or acetonitrile. 2-Methyl-2-ethylsuccinic acid is also a case where the presence of one enantiomer affects the rotation of the other beyond the obvious way of partially canceling it (Horeau, 1969; Horeau and Guetté, 1974). These points will be returned to later (Chapter 13).

REFERENCES

Addadi, L., Berkovitch-Yellin, Z., Weissbuch, I., Lahav, M., and Leiserowitz, L. (1986), "A Link between Macroscopic Phenomena and Molecular Chirality: Crystals as Probes for the Direct Assignment of Absolute Configuration of Chiral Molecules," *Top. Stereochem.*, **16**, 1.

Alembic Club Reprint (Engl. Transl.) No. 14, Edinburgh, UK, 1905.

Anh, N. T. (1970), *Les Règles de Woodward–Hoffmann*, Ediscience, Paris.

Arago, D. F. (1811), *Mem. Cl. Sci. Math. Phys. Inst. Imp. Fr.*, **12**, 93, 115.

Benfey, O. T., Ed. (1963), "Classics in the Theory of Chemical Combination," *Classics of Science*, Vol. 1, Dover Publications, New York; reprinted by Krieger, Malabar, FL, 1981.

Biot, J.B. (1812), *Mem. Cl. Sci. Math. Phys. Inst. Imp. Fr.*, **13**, 1.

Biot, J. B. (1815), *Bull. Soc. Philomath. Paris*, 190.

Borsook, H., Huffman, H. M., and Liu, Y.-P. (1933), *J. Biol. Chem.*, **102**, 449.

Bovey, F.A. (1969), *Polymer Conformation and Configuration*, Academic, New York.

Bovey, F.A. (1982), *Chain Structure and Conformation of Macromolecules*, Academic, New York.

Buda, A. B., Auf der Heyde, T., and Mislow, K. (1992), "On Quantifying Chirality," *Angew. Chem. Int. Ed. Engl.*, **31**, 989.

Butlerov, A. M. (1861), *Z. Chem. Pharm.*, **4**, 546. Translation by Kluge, F. F. and Larder, D. F. (1971), *J. Chem. Educ.*, **48**, 289.

Cahn, R. S., Ingold, Sir C., and Prelog, V. (1966), "Specification of Molecular Chirality," *Angew. Chem. Int. Ed. Engl.*, **5**, 385.

Carey, F.A. and Sundberg, R. J. (1990), *Advanced Organic Chemistry. Part A: Structure and Mechanism*, 3rd ed., Plenum, New York.

Carhart, R. E., Smith, D. H., Brown, H., and Djerassi, C. (1975), *J. Am. Chem. Soc.*, **97**, 5755.

Corey, E. J., Long, A. K., and Rubenstein, S. D. (1985), *Science*, **228**, 408.

Corriu, R. J. P., Guérin, C., and Moreau, J. J. E. (1984), "Stereochemistry at Silicon," *Top. Stereochem.*, **15**, 43.

Couper, A. S. (1858), *Philos. Mag.*, [4], **16**, 104; *C. R. Acad. Sci.*, **46**, 1157; see also Benfey (1963), p. 132.

Deslongchamps, P. (1983), *Stereoelectronic Effects in Organic Chemistry*, Pergamon, New York.

Djerassi, C., Smith, D. H., Crandell, C. W., Gray, N. A. B., Nourse, J. G., and Lindley, M. R. (1982), "The Dendral Project: Computational Aids to Natural Products Structure Elucidation," *Pure Appl. Chem.*, **54**, 2425.

Eliel, E. L. (1974), *CHEMTECH*, 758.

Farina, M. (1987), "The Stereochemistry of Linear Macromolecules," *Top. Stereochem.*, **17**, 1.

Fleming, I. (1976), *Frontier Orbitals and Organic Chemical Reactions*, Wiley, New York.

Gallagher, M. J. and Jenkins, I. D. (1968), "Stereochemical Aspects of Phosphorus Chemistry," *Top. Stereochem.*, **3**, 1.

Geoffroy, G. L., Ed. (1981), "Inorganic and Organometalic Stereochemistry," in *Top. Stereochemistry*, Vol. 12, contains several pertinent chapters. Wiley, New York.

Gilchrist, T. L. and Storr, R. C. (1979), *Organic Reactions and Orbital Symmetry*, 2nd ed., Cambridge University Press, New York.

Haüy, R. J. (1801), *Traité de Mineralogie*, Chez Louis, Paris.

Henson, P. D. and Mislow, K. (1969), *J. Chem. Soc. D*, 413.

Herschel, J. F. W. (1822), *Trans. Cambridge Philos. Soc.*, **1**, 43.

Hoffmann, R., Alder, R. W., and Wilcox, C. F. (1970), *J. Am. Chem. Soc.*, **92**, 4992.

Horeau, A. (1969), *Tetrahedron Lett.*, 3121.

Horeau, A. and Guetté, J. P. (1974), *Tetrahedron*, **30**, 1923.

Ingold, C. K. (1969), *Structure and Mechanism in Organic Chemistry*, 2nd ed., G. Bell and Sons, London and Cornell University Press, Ithaca, NY.

Jacques, J. (1986), *Sur la Dissymétrie Moléculaire*, Christian Bourgeois, Paris.

Kekulé, A. (1858), *Justus Liebigs Ann. Chem.*, **106**, 129; cf. Benfey (1963) 109 (Engl. transl.).

Kelvin, Lord (W. Thomson) (1904), *Baltimore Lectures on Molecular Dynamics and the Wave Theory of Light*, C. J. Clay & Sons, London. The lectures were given in 1884 and 1893.

Kepert, D.L. (1982), *Inorganic Stereochemistry*, Springer-Verlag, New York.

Krow, G. and Hill, R. K. (1968), *Chem. Commun.*, 430.

Le Bel, J. A. (1874), *Bull. Soc. Chim. Fr.*, [2], **22**, 337; see also Richardson, G. M. (1901).

Le Bel, J. A. (1894), *Bull. Soc. Chim. Fr.*, [3], **11**, 295.

Lehr, R. and Marchand, A. (1971), *Orbital Symmetry: A Problem Solving Approach*, Academic, New York.

Lowry, T. H. and Richardson, K. S. (1987), *Mechanism and Theory in Organic Chemistry*, 3rd ed., Harper & Row, New York.

Malus, E. L. (1809) *Mem. Soc. d'Arcueil*, **2**, 143.

March, J. (1992), *Advanced Organic Chemistry. Reactions, Mechanisms and Structure*, 4th ed. Wiley, New York.

Marchand, A. P. and Lehr, R. E., Eds. (1977), *Pericyclic Reactions* (2 vols.), Academic, New York.

Mason, S. F. (1976), "The Foundations of Classical Stereochemistry," *Top. Stereochem.*, **9**, 1.

Mezey, P. G., Ed. (1991), *New Developments in Molecular Chirality*, Kluwer Academic, Boston, MA.

Mikołajczyk, M. and Drabowitz, J. (1982), "Chiral Organosulfur Compounds," *Top. Stereochem.*, **13**, 333.

Mikołajczyk, M. (1987), "Sulfur Stereochemistry—Old Problems and New Results," Zwanenburg, B. and Klunder, A. J. H., Eds., *Perspectives in the Organic Chemistry of Sulfur*, Elsevier, New York, p. 23.

Mislow, K. (1965), *Introduction to Stereochemistry*, Benjamin, New York, p. 52.

Monkhorst, H. J. (1968), *Chem. Commun.*, 1111 has calculated that planar methane would be 250 kcal mol^{-1} (1046 kJ mol^{-1}) less stable than tetrahedral.

Pasteur, L. (1860), Two lectures delivered before the Societé Chimique de France, Jan. 20 and Feb. 3. cf. Jacques, 1986. For English translation see references to Richardson, 1901 and Alembic Club Reprint, No. 14.

Prelog, V. (1968), *Proc. Koninkl. Ned. Akad. Wetenschap.*, **B71**, 108.

Quin, L. D. (1981), *The Hetereocyclic Chemistry of Phosphorus*, Wiley-Interscience, New York.

Ramsay, O.B., Ed. (1975), *van't Hoff–Le Bel Centennial*, ACS Symposium Series 12, American Chemical Society, Washington, DC.

Ramsay, O. B. (1981), *Stereochemistry*, Heyden & Son, Philadelphia.

Richardson, G. M., Ed. (1901), *The Foundations of Stereochemistry*, American Book Co., New York.

Snatzke, G. (1989/90), personal communication to SHW; see also Section 13-5.a.

Snelders, H. A. M. (1975), "J. A. Le Bel's Stereochemical Ideas Compared with those of J. H. van't Hoff (1974)" in *van't Hoff–Le Bel Centennial*, Ramsay, O. B., Ed., ACS Symposium Series 12, American Chemical Society, Washington, DC., p. 66.

Sokolov, V. I. (1990), *Chirality and Optical Activity in Organometallic Compounds*, Gordon and Breach, New York.

Stoddard, M. P. and Dunn, M. S. (1942), *J. Biol. Chem.*, **142**, 329.

Thomson, W. – see Kelvin.

van't Hoff, J. H. (1874), *Arch. Neerl. Sci. Exactes Nat.*, **9**, 445. The Dutch version of this seminal article was published simultaneously in the form of a pamphlet in September, 1874 and has been reprinted in the Netherlands on the occasion of the van't Hoff Centennial in 1974. The French version was republished in abbreviated and revised form in *Bull. Soc. Chim. Fr.*, [2], **23**, 295 (1875). English translations are available in Richardson (1901), p. 37 and in Benfey (1963), p. 151.

van't Hoff, J. H. (1875), *La Chimie dans L'Espace*, Bazendijk, Rotterdam, The Netherlands, pp. 13–14.

Verkade, J. G. and Quin, L. D., Eds. (1987), *Phosphorus-31 NMR Spectroscopy in Stereochemical Analysis*, VCH Publishers, New York.

Wheland, G. A. (1960), *Advanced Organic Chemistry*, 3rd ed., Wiley, New York.

Whyte, L. L. (1957), *Nature* (*London*), **180**, 513.

Whyte, L. L. (1958), *Nature* (*London*), **182**, 198.

Weiss, U. (1968), *Experientia*, **24**, 1088.

Wipke, W. T. (1974), "Computer-Assisted Three-Dimensional Synthetic Analysis" in Wipke, W. T., Heller, S. R., Feldman, R. J., and Hyde, E., Eds., *Computer Representation and Manipulation of Chemical Information*, Wiley, New York, p. 147.

Wipke, W. T. and Howe, W. J., Eds. (1977), *Computer Assisted Organic Synthesis*, ACS Symposium Series 61, American Chemical Society, Washington, DC.

Woodward, R. B. and Hoffmann, R. (1969), *The Conservation of Orbital Symmetry*, Angew. Chem. Int. Ed. Engl., **8**, 781; id. (1970) Academic, New York.

2

Structure

2-1. MEANING, FACTORIZATION. INTERNAL COORDINATES. ISOMERS

We have seen (Section 1-1) that static stereochemistry deals with the shape of molecules (molecular architecture or molecular structure). The nomenclature rules of IUPAC (International Union of Pure and Applied Chemistry) (Cross and Klyne, 1976) do not provide an unequivocal definition of "structure"; we shall use the term in the sense of the crystallographer as denoting the position in space of all the atoms constituting a molecule. Molecular structure may thus be defined in terms of the Cartesian coordinates of the atoms, or the oblique coordinates that crystallographers often use for crystals belonging to the monoclinic, triclinic, and trigonal hexagonal systems.

For many purposes (cf. Section 2-7), it is more convenient to employ so-called "internal coordinates": bond lengths (or distances) r, bond angles θ, and torsion angles ω or τ [polymer chemists, unfortunately, use τ for the bond angle and θ for the torsion angle (cf. Jenkins, 1981)]. Since the absolute position and orientation of a molecule is of no structural significance, only relative positions of atoms within the molecule need to be specified. For a diatomic molecule A–B (Fig. 2.1), structure is thus completely defined by the nature of the nuclei A and B and the bond distance r between their centers. For a triatomic molecule ABC (Fig. 2.1), besides the nature of A, B, and C, their connectivity (i.e., which atom is connected to which) and the bond lengths A–B (r_1) and B–C (r_2), we must specify the bond angle θ. With a tetratomic molecule ABCD the situation is slightly more complicated: In addition to the nature and connectivity of the atoms and the bond distances r_1, r_2, and r_3 and bond angles θ_1 and θ_2, one must now specify the torsion angle ω (Fig. 2.1) in order to fix the position of all four atoms A, B, C, and D. The torsion angle ω is defined as the angle between the planes ABC and BCD (Fig. 2.2); this angle has sign as well as magnitude. If the turn from the ABC to the BCD plane (front to back) is clockwise (Fig. 2.2), ω is

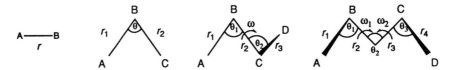

Figure 2.1. Internal coordinates.

positive, if it is counterclockwise (Fig. 2.2), ω is negative. In determining the sign of ω it is immaterial whether one views the ABCD array from AB looking toward CD or from CD looking toward AB.

For each additional atom (e.g., E in ABCD–E) three more coordinates need to be specified: the bond length D–E, the bond angle θ(CDE), and the torsion angle ω(BCDE). Thus the total number of independent coordinates for a nonlinear n-atomic molecule is $3n - 6$; for a chain of n atoms these may be taken as $n - 1$ bond distances (the first atom defines no such distance), $n - 2$ bond angles (not defined for the first two atoms), and $n - 3$ torsion angles (defined only for atoms after the third). Figure 2.1 (right) displays an example (5 atoms, 9 coordinates).

The situation is more complicated for branched molecules, rings, or molecules with three or more collinear atoms. For example, the nine independent coordinates for CHFClBr (Fig. 2.8) are four bond distances, five bond angles, and no torsion angle. Actually, the four bonds form six bond angles which, however, are not independent but are interrelated by an equation of constraint. For the five-atom carbon skeleton of 2-methylbutane (isopentane) we can choose four bond distances, four bond angles, and one torsion angle. In these cases, the total number of independent coordinates remains $3n - 6$; but when three or more atoms are collinear, as in methylallene (H_3C–CH=C=CH_2) or butatriene (H_2C=C=C=CH_2) this number is less (five instead of six for the carbon skeleton in these two cases: three bond distances and two bond angles.)

Whereas torsion angles change widely from one molecule to another and even bond angles can vary appreciably from their standard magnitudes (e.g., the normal CCC angle of 112° in propane is reduced to 88° in cyclobutane), bond distances are usually quite constant (see, however, Sections 2-2 and 2-7). Standard bond distances (Sutton, 1958, 1965) for common bonds to carbon are given in Table 2.1.

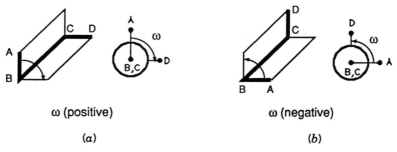

ω (positive)   ω (negative)

(a)   (b)

Figure 2.2. Torsion angle.

TABLE 2.1. Standard Bond Distances.

Units	C–C	C=C	C≡C	C–H	C–F	C–Cl	C–Br	C–N	C–O	C=O	C–S
Å	1.53	1.34	1.20	1.09	1.38	1.77	1.94	1.47	1.43	1.22	1.82
pm	153	134	120	109	138	177	194	147	143	122	182

Computer programs are in existence for transformation of internal to Cartesian coordinates and vice versa. Formulas are also available to calculate the distance between any two given atoms in a molecule; this distance is needed for the computation of nonbonded interaction. For atoms separated by more than a few bonds these formulas become unmanageably complicated; fortunately, most molecular mechanics programs (cf. Section 2-6) yield the distance, as well as the nonbonded interaction energy, between specified atoms directly.

Although structure may thus be completely described by a system of coordinates (Cartesian, internal, or other type), we shall find it convenient, from the conceptual point of view, to subdivide (factorize) structure into a constitutional and stereochemical part. The stereochemical aspect may be further factorized [if with some difficulty (see below)] into configuration and conformation. Structure thus embraces constitution (connectivity), configuration, and conformation.

By "isomers" we mean substances that have the same composition and molecular weight but differ in properties. On the molecular level, such substances have the same number and kind of atoms, but differ in structure. Again, the structural difference may be in constitution (constitutional isomers) or stereochemical arrangement (stereoisomers). Stereoisomers, in turn, may differ (see Sections 2-3 and 2-4) in configuration (configurational isomers) or, having the same configuration, may differ in conformation (conformational isomers or conformers). A different subdivision of stereoisomers will be considered in Chapter 3.

In a crystal the position of the molecules and their constituent atoms is generally well defined. Disregarding the generally small atomic motions, molecules in crystals may thus be considered rigid objects with well-defined shape, symmetry, dimensions, and may be appropriately represented by rigid models. However, this rigidity does not persist in the fluid (liquid, solution, or gaseous) state. Molecular vibrations become larger, leading to less well defined bond distances and bond angles, and in some cases, as in the inversion or "umbrella motion" of tertiary amines (NRR'R'') or rotations around single bonds (torsions), for example, around the central bond in $ClH_2C–CH_2Cl$, may actually lead to different structures. Such changes, which frequently occur rapidly at room temperature, put in question what is actually meant by a "molecule." Commonly, the term has been defined as "the smallest entity of a chemical species." It is implicit in this definition that the entities are all the same, but this may not be true. Thus the species chlorocyclohexane (Fig. 2.3), is well known to consist of molecules with equatorial and others with axial chlorine atoms. In the IR spectrum of the species, the two types of molecules can be clearly discerned, for example, by their different C–Cl stretching frequencies. Yet in most respects, for example, in distillation, chromatography, or chemical reaction, chlorocyclohexane appears to be a homogeneous substance. Even that appearance depends on the

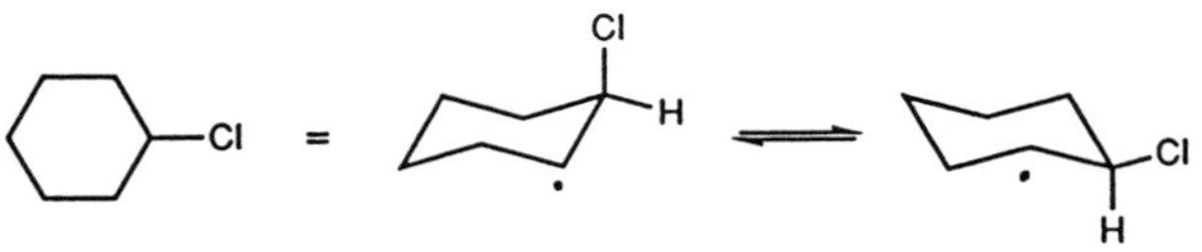

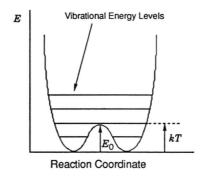

Figure 2.3. Chlorocyclohexane.

temperature of observation. At room temperature chlorocyclohexane displays a simple NMR spectrum; thus there are only four different ^{13}C signals. But at $-100°C$, well below the "coalescence temperature" (cf. Section 8-4.d), separate spectra of the two structures shown in Figure 2.3 are seen and the number of carbon signals doubles to eight. Thus chlorocyclohexane is clearly made up of two kinds of molecules, but whether these molecules can be separately discerned is a matter of both the technique of observation (isolation, NMR, or IR) and the temperature: At $-150°C$, the two different chlorocyclohexane isomers can actually be isolated (Jensen and Bushweller, 1966, 1969; cf. Section 11-4).

The barrier between the two isomeric chlorocyclohexanes is actually quite high, between 10 and 11 kcal mol^{-1} (42 and 46 kJ mol^{-1}). Let us now make the conceptual experiment of lowering the barrier gradually until it disappears. At what point shall we cease to speak of two molecules and say that the molecules making up the macroscopic species are all the same? It was suggested at one time (Eliel, 1976/77) that rapidly interconverting structures be considered as two (or more) different molecules only if the barrier between them exceeds kT, whereas interconverting structures separated by barriers of less than kT be considered single molecules (cf. Fig. 2.4). The rationale for this suggestion was that, in classical theory, kT is the energy of the lowest vibrational energy level so that, if the barrier is less than kT, one molecular vibration will lead to the barrier being crossed. However, it is probably not desirable to introduce a sharp dividing line into what is in actual fact a continuum, especially since quite small barriers can be measured by microwave spectroscopy [e.g., $CH_3–C_6H_5$ rotation in toluene, 14 cal mol^{-1} (59 J mol^{-1}), see Chapter 10; ring inversion in (nonplanar) trimethylene oxide, 100 cal mol^{-1}, (420 J mol^{-1}), Moriarty, 1974]. Moreover, because of the quantization of vibrational energy levels and the existence of zero-point energy, the classical value of kT for the lowest vibrational energy level is in any case not meaningful. What remains important, however, is the concept that, whether in a given situation

E | Vibrational Energy Levels

E_0 | kT

Reaction Coordinate

Figure 2.4. Double potential minimum. If $E_0 < kT$, the first vibrational energy level will lie above the barrier separating the two energy minima.

a substance is palpably homogeneous or not (i.e., made up of seemingly identical or of differing though rapidly interconverting molecules) may, especially in the fluid state, depend on the temperature and the time scale of the observation (see also Section 3-1.b).

Despite these complications, identifying a molecule as an isolable entity (under whatever conditions it is observed) has merit on thermodynamic grounds, for the number of components C in the phase rule $F = C - P + 2$ (where F is the number of degrees of freedom of the system and P is the number of phases present) is equal to the number of *isolable* species (for further discussion, see Chapter 3).

2-2. CONSTITUTION

The term "constitution" connotes the number, kind, and connectivity of the atoms in a molecule. Constitution may be represented by a two-dimensional (2D) graph in which the atoms linked to each other are connected by a bond (single, double, or triple). An alternative representation is an n^2 matrix (for a molecule with n atoms) in which the elements are zero for nonbonded atoms, or 1, 2, or 3, respectively, for atoms linked by single, double, or triple bonds. Thus the isomeric molecules ethyl alcohol (ethanol) and dimethyl ether (oxybismethane) (both C_2H_6O) may be represented by the graphical formulas **A** and **B** (cf. Rouvray, 1975; King, 1983) or by the matrices **C** and **D** below (cf. Wheland,

$$
\begin{array}{c}
\quad\;\; H_4\;\; H_2 \\
\quad\;\; |\quad\; | \\
H_5 - C_2 - C_1 - O - H_1 \\
\quad\;\; |\quad\; | \\
\quad\;\; H_6\;\; H_3
\end{array}
\qquad \text{and} \qquad
\begin{array}{c}
\quad\;\; H_3\;\;\;\; H_4 \\
\quad\;\; |\quad\quad\; | \\
H_2 - C_1 - O - C_2 - H_5 \\
\quad\;\; |\quad\quad\; | \\
\quad\;\; H_1\;\;\;\; H_6
\end{array}
$$

$$\textbf{A} \qquad\qquad\qquad\qquad\qquad\qquad \textbf{B}$$

	C_1	C_2	O	H_1	H_2	H_3	H_4	H_5	H_6
C_1		1	1	0	1	1	0	0	0
C_2	1		0	0	0	0	1	1	1
O	1	0		1	0	0	0	0	0
H_1	0	0	1		0	0	0	0	0
H_2	1	0	0	0		0	0	0	0
H_3	1	0	0	0	0		0	0	0
H_4	0	1	0	0	0	0		0	0
H_5	0	1	0	0	0	0	0		0
H_6	0	1	0	0	0	0	0	0	

$$\textbf{C}$$

	C_1	C_2	O	H_1	H_2	H_3	H_4	H_5	H_6
C_1		0	1	1	1	1	0	0	0
C_2	0		1	0	0	0	1	1	1
O	1	1		0	0	0	0	0	0
H_1	1	0	0		0	0	0	0	0
H_2	1	0	0	0		0	0	0	0
H_3	1	0	0	0	0		0	0	0
H_4	0	1	0	0	0	0		0	0
H_5	0	1	0	0	0	0	0		0
H_6	0	1	0	0	0	0	0	0	

$$\textbf{D}$$

1960; such matrices are particularly useful for computer manipulation). The two molecules clearly differ in connectivity and are constitutional isomers. These isomers may be distinguished by diffraction experiments, but simpler criteria, such as acidity, volatility (reflecting the presence or absence of hydrogen bonding), or spectroscopic criteria for the OH group in the alcohol may be used more conveniently.

To draw a constitutional formula or to construct a connectivity matrix, one must decide whether or not a bond exists between two given atoms within a molecule of known structure. Since it is difficult to determine the electron density between the atoms in question experimentally (see also Section 2-5), bonds are usually drawn in such a way as to (a) satisfy the normal valencies of the atoms and (b) take into account what is known about bonded and nonbonded distances between atoms. Thus two carbon atoms may normally be considered bonded if their internuclear distance is less than 160 pm (1.6 Å) [the sum of the single-bond radii is 153 pm (1.53 Å); cf. Table 2-1] but not bonded if their distance exceeds 270 pm (2.7 Å) [the sum of the van der Waals radii is 340 pm (3.4 Å)]. Distances in the intermediate range are encountered rarely (but see below concerning long bonds; see also Ōsawa and Kanematsu, 1986).

A more pragmatic definition has been given by Pauling (1960, p. 6): "... there is a chemical bond between two atoms or groups of atoms in case that the forces acting between them are such as to lead to the formation of an aggregate with sufficient stability to make it convenient for the chemist to consider it as an independent molecular species." This definition, which is in terms of energy rather than geometry, primarily applies to intermolecular bonds. Thus we may consider the hydrogen bridges in the acetic acid dimer (Fig. 2.5, **A**) as true bonds, since the dimer persists even in the vapor phase and affects the boiling point and the IR spectrum; in contrast, we may not wish to consider the hydrogen bridge between two molecules of methyl mercaptan, CH_3SH (Fig. 2.5, **B**) as a bond, since it scarcely manifests itself in the properties of the substance. In an intramolecular analogy, we may wish to pay no attention to the very weak hydrogen bond that links the two ends of an ethylenediamine molecule (Fig. 2.5, **C**) but we cannot disregard the hydrogen bonds between peptide units in a polypeptide (Fig. 2.5, **D**) since these bonds are responsible for the important secondary structure of proteins.

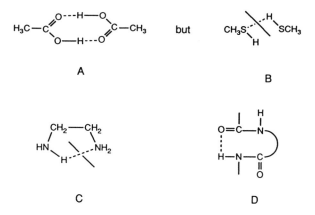

Figure 2.5. Hydrogen bonds.

In Section 2-1 we discussed the question of structural homogeneity. Constitutional isomerism provides examples involving a variety of time scales. For example, the keto **1** and enol **2** forms of acetoacetic ester, which coexist in equilibrium, display distinct NMR spectra and are separable by low-pressure

distillation

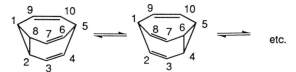

$$CH_3\!-\!\underset{\underset{O}{\|}}{C}\!-\!CH_2\!-\!CO_2C_2H_5 \rightleftharpoons CH_3\!-\!\underset{\underset{OH}{|}}{C}\!=\!CH\!-\!CO_2C_2H_5$$

1   **2**

in quartz equipment. But when warmed in the presence of a trace of base they become "tautomeric," that is, rapidly equilibrating; under those circumstances the two species are no longer separable and even their NMR spectra may coalesce. Nevertheless, two molecular species still coexist, as can be readily verified by IR spectroscopy or by careful separation following neutralization of the base.

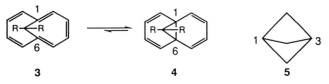

Figure 2.6. Bullvalene.

Another example of tautomerism called "valence tautomerism" is represented by bullvalene (Fig. 2.6). This molecule exists in 1,209,600 (10!/3) structures because there is possibility of bond migration, as shown in Figure 2.6. The structures are degenerate (i.e., superposable except for the numbering of the atoms) but, at room temperature, the carbon atoms may be distinguished in a ^{13}C NMR spectrum as being vinylic, allylic, cyclopropanoid, and so on. Upon heating to 100°C, however, bond migration or reorganization (fluxion) becomes so fast that the carbon atoms lose their distinctive character on the NMR time scale and only a single ^{13}C signal is seen. Nevertheless, the greater than 10^6 fluxional structures still coexist; the NMR spectrum corresponds to an average of these structures, not to an actual, unique molecule (cf. Saunders, 1963).

Compounds capable of valence tautomerism, such as **3** ⇌ **4** (Fig. 2.7) sometimes display unusual bond lengths in the crystal. Thus the fluorinated bridged [10]annulene **3** shown in Figure 2.7 (R = F) has the unusually short nonbonded C(1)–C(6) distance of 225 pm (2.25 Å). The corresponding cyclopropanoid dimethyl compound **4** (R = CH$_3$), in contrast, does not appear to be aromatic and is nonplanar; it has an unusually long *bonded* C(1)–C(6) distance of 181 pm (1.81 Å) (Simonetta, 1974; Bürgi, Dunitz, et al., 1975). This distance is barely less than the *nonbonded* C(1)–C(3) distance in the strained bicyclo[1.1.1]pentane **5** (Fig. 2.7), which amounts to 186 pm (1.86 Å) (average of values reported by Chiang and Bauer, 1970 and by Almenningen et al., 1971). Apparently, the valence bond tautomerism **3** ⇌ **4** is somehow foreshadowed even in the crystal.

Figure 2.7. Compounds with unusual bond distances.

2-3. CONFIGURATION

Molecules of identical constitution may yet differ in structure. Thus fluoro-chlorobromomethane (CHFClBr) has a unique connectivity but there are two enantiomers (nonsuperposable mirror-image structures; Fig. 2.8; Hargreaves and Modarai, 1969, 1971; Wilen et al., 1985; Doyle and Vogl, 1989). These molecules are said to differ in configuration. Again, when the substance 1,2-dichloroethane is viewed by IR spectroscopy (cf. Mizushima, 1954), it is found that at least two different isomers coexist, gauche and anti (Fig. 2.9); these isomeric structures are said to differ in conformation. Figure 2.9 shows the *three* staggered conformational isomers of 1,2-dichloroethane. Structures **A** and **C** are enantiomeric, and therefore indistinguishable by IR spectroscopy (or any other scalar observation; cf. p. 59).

In contrast to constitution, which can be represented by a 2D graph of atoms linked by bonds, configuration, and conformation embody the three-dimensional (3D) or stereochemical part of structure.

> The distinction between configuration and conformation is subtle and is, in fact, not universally agreed upon (cf. Cross and Klyne, 1976). Some authors, especially physical chemists, use the term configuration synonymous with "arrangement in space of the atoms in a molecule of defined constitution," that is, comprising conformation as well. A tentative postulate that configurational distinctions among molecules of identical constitution come about by reflection or by an interchange of ligands, whereas differences in conformation come about by rotation about single bonds is clearly not tenable: Isomers **A** and **C** in Figure 2.9 can be interconverted *either* by rotation about the carbon–carbon bond *or* by reflection (or by an appropriate interchange of H and Cl at one of the carbon centers).

Perhaps the most fundamental distinction one can make between configuration and conformation is to say that configurational differences imply differences in bond angles, whereas conformational differences involve differences in torsion angles (including, in both cases, differences that are exclusively in sign). If one considers an assembly of four atoms, A, B, C, and D, one can envisage two

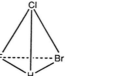

Figure 2.8.   Enantiomers of CHFClBr.

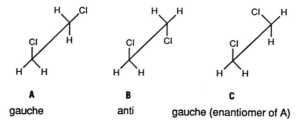

Figure 2.9.   1,2-Dichloroethane.

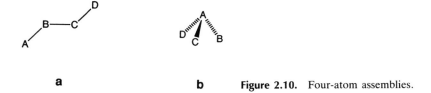

a   **b**   **Figure 2.10.** Four-atom assemblies.

different types of connectivity (constitution) shown in Figure 2.10. In array **a**, assuming constant bond distances A–B, B–C, and C–D and constant bond angles ABC and BCD, one can generate an infinite number of structures by rotating around the BC bond, that is, by changing the torsion angle ABC/BCD, ω(ABCD). These structures are said to differ in conformation. The conformational differences (cf. also Figs. 2.2 and 2.9) are independent of the nature of the atoms A, B, C, and D; conformation for the array can be defined even if all the atoms are the same (i.e., in A–A–A–A) as long as no three of them are collinear. Array **b**, in contrast, presents no conformational variability but if B $\neq$ C $\neq$ D, that is, if the atom A is chiral [or (see later discussion) stereogenic], there are two possible arrangements of the type shown in Figure 2.8. These two arrangements are said to differ in configuration.

   If one views the molecule **b** and its mirror image (enantiomer) from the side of the B–C–D "tripod" (with A at the base of the tripod), one sees the two different arrangements shown in Figure 2.11. In the former, the sequence B–C–D describes a clockwise array, in the latter a counterclockwise array. [The former array might be described as *R* in the Cahn–Ingold–Prelog system (cf. Chapter 5) and the latter as *S*.] This difference implies that the 3D angle subtended by the ligands B, C, and D at the pivot atom A is of opposite sign in the two (*R* or *S*) configurations.

   Configuration and conformation as defined above are *not* delineated from each other by considerations of energy barriers. The energy barrier between molecules of opposite configuration may be quite low, as in NMeEtPr [ethylmethylpropylamine, ca. 8 kcal mol^{-1} (33.5 kJ mol^{-1})], moderate, as in PMePrPh

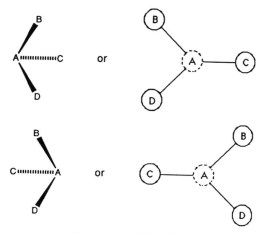

Figure 2.11. Tripodal array.

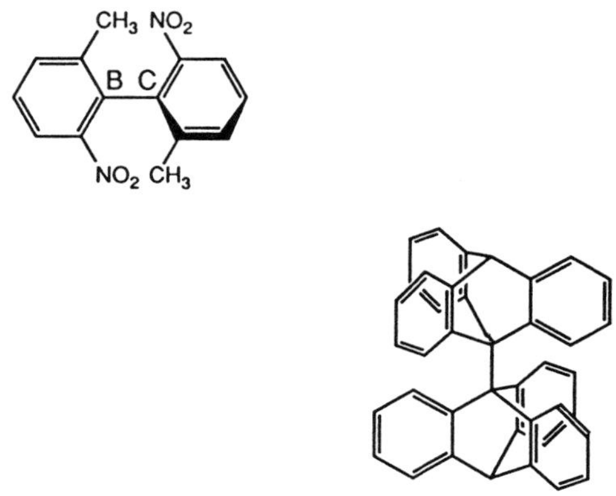

Figure 2.12. Molecules with "hindered rotation."

[methylphenylpropylphosphine, 32.1 kcal mol^{-1} (134 kJ mol^{-1}) (cf. Lehn, 1970; Lambert, 1971)], or large, as in a tetrasubstituted methane [where the structure shown in Figure 2.10, **b** has been modified by addition of a fourth ligand]. In all three cases the enantiomers are considered to differ in configuration.

> However, in view of the thermodynamic importance of the isolability criterion (p. 15) an argument can be made to consider the enantiomers of ethylmethyl-propylamine as conformational isomers. The same argument would not apply to chiral phosphines that are configurationally stable at room temperature. In effect, amine inversion may be considered either a configurational or a conformational change, depending on the predilection of the individual describing it (see also Section 5-1).

Whereas molecules of type **b** are usually considered to fall under the definition of configurational isomerism, the picture with molecules of type **a** is less clear-cut. This type covers a wide range of species. Even when the central bond B–C is "single," type **a** ranges from simple, rapidly rotating ethanes, such as those depicted in Figure 2.9 to biphenyl and bitriptycyls (Fig. 2.12; cf. Section 14-5) where rotation about the central C–C bond may be slow enough to permit the isolation of perfectly stable stereoisomers that nonetheless differ only by the rotational arrangement about this bond. There is no agreement as to whether these isomers should be called configurational or conformational isomers, with the isolability criterion favoring the former alternative. The situation is aggravated when the central bond (B–C in Fig. 2.10, **a**) is a double bond or when it has a bond order between single and double. We shall return to this point in Section 2-4. Configuration is considered in more detail in Chapter 5.

2-4. CONFORMATION

Reference to Figure 2.1 discloses that definition of constitution and configuration does not suffice to locate the atoms of a molecule in space, since structural differences may be engendered by change of torsion angle ω, as exemplified in

TABLE 2.2. Specification of Torsion Angle (Klyne–Prelog).

Angle of Torsion (ω)	Designation	Symbol
-30 to $+30°$	synperiplanar	sp^a
$+30$ to $+90°$	+ synclinal	$+sc^b$
$+90$ to $+150°$	+ anticlinal	$+ac$
$+150$ to $-150°$	antiperiplanar	ap^c
-150 to $-90°$	− anticlinal	$-ac$
-90 to $-30°$	− synclinal	$-sc^b$

[a] The designation syn or eclipsed are often used for $\omega \approx 0°$.
[b] The designation gauche is frequently used for $\omega \approx 60°$.
[c] The designation anti (or, less properly, trans) is often used for $\omega \approx 180°$.

Figure 2.9. Thus to complete the description of structure one must specify the torsion angles. By "conformation" of a molecule of given constitution and configuration is meant the rotational arrangement about all bonds as defined by the magnitude and sign of all pertinent torsion angles. Conformation may be described exactly, by specifying the exact magnitude and sign of the torsion angles (cf. Fig. 2.2) or it may be described approximately, by a range of these angles. The latter approach is often preferable, since the exact values of the torsion angles, especially for molecules in the liquid and gaseous states, are frequently not known. An appropriate system of classification (Klyne and Prelog, 1960) is summarized in Table 2.2 and Figure 2.13.

Many conformations, such as the eclipsed conformations in ethane, do not correspond to energy minima. Those that do, such as the conformations of $ClCH_2CH_2Cl$ shown in Figure 2.9, may be called "conformational isomers" or "conformers."

We have already pointed out that the above description of structure in terms of torsion angle may shift the stereoisomerism of molecules such as biphenyls into the conformational category. More disturbingly, this is also true if the B–C bond in structure **a**, Figure 2.10 is a double bond. Thus one might argue that the cis Z and trans E (cf. Chapter 9) isomers of 1,2-dichloroethene (Fig. 2.14) should be considered as differing in conformation rather than configuration. While such a view is permitted under IUPAC rules (Cross and Klyne, 1976), cis–trans isomers are usually said to differ in configuration and we shall (with some discomfort) conform to the convention of considering them configurational isomers. The same applies to stereoisomeric allenes, for example CHCl=C=CHCl (cf. Chapter 14).

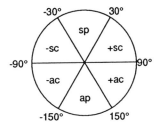

Figure 2.13. Specification of torsion angle (Klyne–Prelog).

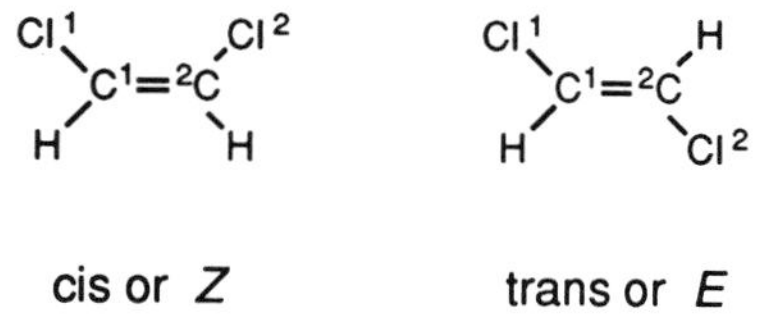

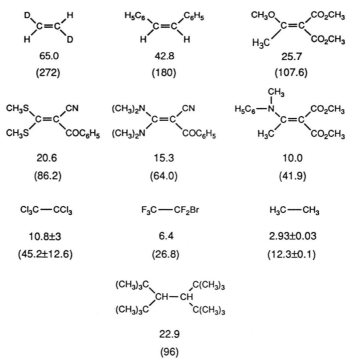

Figure 2.14. cis-trans Isomers.

The two isomers shown in Figure 2.14 also differ in the relative sense of the Cl(1)–C(1)–C(2) and Cl(2)–C(2)–C(1) bond angles; these angles are of equal sign (both positive or both negative, depending on how the molecule is written) in the trans but of opposite sign (one positive, one negative) in the cis isomer. (Here a clockwise order of the three sequential atoms is taken to define a positive angle, a counterclockwise order a negative one.) Thus, analogy with the 3D case shown in Figure 2.10, **b** and 2.11 provides some logic for considering the difference between the cis–trans isomers as being in their configuration.

The problem is further compounded because the bond order of the central bond may be intermediate between one and two; thus it is not always clear whether the central B–C bond in array **a**, Figure 2.10, is double or single. Figure 2.15 (Kalinowski and Kessler, 1973; Sandström, 1983; see also Chapter 10)

Figure 2.15. Barriers about carbon–carbon bonds. $\Delta G^{\ddagger}$ or E_a is in kcal mol^{-1}. The values in parentheses are in kilojoules per mole (kJ mol^{-1}). [Data from Kalinowski and Kessler, 1973; Sandström, 1983; Umemoto and Ouchi, 1985; see also Chapter 9.]

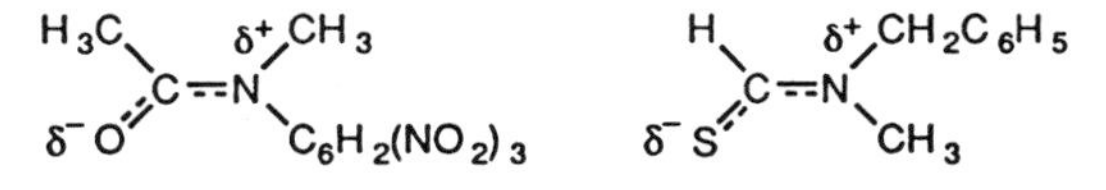

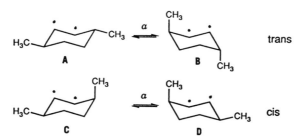

Figure 2.16. Compounds with intermediate barriers.

displays examples in which the rotational barrier about double bonds drops from 65 kcal mol^{-1} (272 kJ mol^{-1}) in the parent ethylene to as low as 10.0 kcal mol^{-1} (41.9 kJ mol^{-1}) in an ethylene of the "push–pull" type (cf. Chapter 9). On the other hand, the single-bond barrier of 2.93 kcal mol^{-1} (12.3 kJ mol^{-1}) in ethane is shown to rise to 10.8 kcal mol^{-1} (45.2 kJ mol^{-1}) in hexachloroethane and to 22.9 kcal mol^{-1} (96 kJ mol^{-1}) in tetra-*tert*-butylethane, so that there is an overlap in the range of barrier height between single-bonded and double-bonded species. The overlap would be much more extensive, were we to include examples of biphenyl and related isomerism shown in Figure 2.12. Barrier height thus appears unsuitable as a criterion to delineate configurational from conformational differences. It is better to speak about configurational isomerism (cis–trans isomerism) in the case of formal double bonds (first two cases in Fig. 2.15; it happens that in the subsequent four cases, the substitution pattern is such that cis–trans isomerism is absent; for the means of determining the barrier in these cases, see Chapter 8) and about conformational isomers in the case of formal single bonds (cf. the last four cases: Note that isomerism is present in but one of these cases, which were selected only to show the variation in barrier height). An important case where the choice of integral bond order is arbitrary is that of amides and thioamides in Figure 2.16. The amide barrier is 21.0 kcal mol^{-1} (88 kJ mol^{-1}) (Weil et al., 1967) and is easily measured by NMR spectroscopy (cf. Chapter 10); the thioamide barrier is even higher, 25.1 kcal mol^{-1} (105 kJ mol^{-1}); in this case the two stereoisomers can be separated (Walter et al., 1966). While the amide or thioamide isomers shown in Figure 2.16 are generally implied to differ in conformation (by rotation about the C–N bond) the E/Z nomenclature of double-bonded species is commonly applied, demonstrating a certain degree of ambivalence in these cases!

Conformation in acyclic molecules is considered in more detail in Chapter 10.

We shall end this discussion with a few more subtle cases. The ring reversal of chlorocyclohexane (Fig. 2.3) is a conformational change; the two isomers differ in the sense of the torsion angles but not in the sense of the bond angles (although some change of bond angles must occur during the flipping process). The trans and cis isomers of 1,4-dimethylcyclohexane (Fig. 2.17) differ in configuration but

Figure 2.17. 1,4-Dimethylcyclohexanes. *a*: Conformational change. The cis and trans isomers differ in configuration.

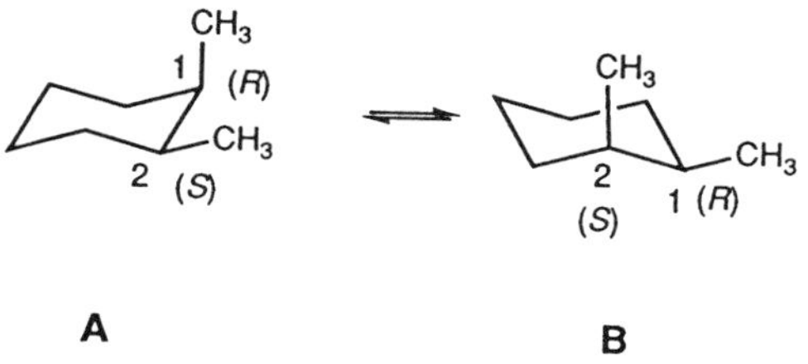

Figure 2.18. *cis*-1,2-dimethylcyclohexane.

the ring-reversed forms (**A/B** or **C/D**, respectively) of each individual configurational isomer differ from each other in conformation only. The ring-reversed forms of *cis*-1,2-dimethylcyclohexane (Fig. 2.18, **E**, **F**) also differ in conformation, even though these two forms are enantiomers (cf. Section 11-4.c and Fig. 11.27).

That the configurations of chlorocyclohexane and of *cis*- as well as of *trans*-1,4-dimethylcyclohexane do not change upon ring inversion can best be seen by arbitrarily considering one of the C(2) atoms labeled (cf. the dots in Figs. 2.3 and 2.17). If the dotted carbon precedes the undotted one, both conformers of chlorocyclohexane shown in Figure 2.3 have the *R* configuration, both conformers (**A**, **B**) of the trans isomers in Figure 2.17 are *R*, *R* and both conformers of the cis isomer (**C**, **D**) are *R*, *S*. In *cis*-1,2-dimethylcyclohexane (Fig. 2.18), both conformers (**E**, **F**) are 1*R*, 2*S*, but in the first one the *R* center is axial (and the *S* center equatorial), whereas in the second one this situation is reversed. This case demonstrates that the existence of enantiomers does not necessarily imply a configurational difference; here, as in 1,2-dichloroethane (Fig. 2.9, structures **A** and **C**), we encounter conformational enantiomers.

2-5. DETERMINATION OF STRUCTURE

The structure of a molecule may be established by separate investigation of constitution, configuration, and conformation, or it may be determined all at once. We shall not deal in this book with the separate investigation of constitution; the determination of configuration will be treated in Chapter 5 and that of conformation in Chapters 10 and 11. Here we shall discuss briefly the most important methods for the integral determination of structure: X-ray and neutron diffraction analysis, electron diffraction, and microwave spectroscopy. The former two methods are applied in the solid state, the latter two in the gas phase. There are few methods for the direct determination of structure in the liquid or solution phase; IR spectroscopy combined with normal coordinate analysis is of very limited application and the method of lanthanide shift reagents in NMR, once thought to be quite powerful, is probably only of limited usefulness. Nuclear magnetic resonance in the nematic phase has also been used; the best method is probably 2D NMR with quantitative evaluation of nuclear Overhauser effects (NOESY, see p. 30)

X-ray diffraction (Glusker and Trueblood, 1985; Dunitz, 1979; Lipscomb and Jacobson, 1972; Stewart and Hall, 1971; Mills and Speakman, 1969) is by far the most powerful technique for the structure determination of crystalline materials—from the smallest molecules that can be conveniently crystallized (the analysis may be carried out at low temperature if necessary) to species as large as proteins. In this method, X-rays of suitable wavelength are allowed to impinge on a single crystal, 0.1–1 mm in length, of the material to be investigated. The X-rays are scattered by the atoms (by their electrons rather than by the nuclei) and the interference of the scattered radiation is recorded as a diffraction pattern.

If one knows the structure of a compound, one can unequivocally calculate the diffraction pattern, but the desired reverse process is not so straightforward. The so-called "structure factor" which is a function (the Fourier transform) of the scattering density and which is crucial in the determination of the structure, has both magnitude and phase. The scattering pattern provides the magnitude of the structure factors but not (at least not in a straightforward way) their phases. It is necessary to obtain the missing phase information before the structure can be solved.

In a classical (if no longer much used) approach (Patterson, 1935) the phase problem is solved by the introduction of a heavy atom (usually from the second complete row of the periodic system or heavier) in the molecule. The position of the heavy atom in the unit cell can usually be ascertained directly from the diffraction pattern and the phases of a number of the scattering amplitudes can then be determined. (The unit cell is the smallest geometric unit from which the complete 3D array of molecules is generated by translation along the axes. It usually contains more than one molecule.) This procedure leads to the establishment of a "trial structure" from which improved atomic coordinates and the phases of the remaining scattering amplitudes are obtained by a process of least squares "refinement." Refinement involves calculation of the diffraction pattern from the trial structure followed by adjustment of the latter until good agreement in terms of the R factor (see below) is attained. The introduction of the heavy atom usually involves the formation of a derivative, such as p-bromobenzoate, but it suffices even to have heavy atoms in solvent of crystallization (Akimoto et al., 1968).

The usefulness of the heavy atom method diminishes as the heavy atom constitutes a smaller and smaller portion of the total molecule. For very large molecules, such as proteins, a modification of the method called the "method of isomorphous replacement" (Green, Perutz, et al., 1954) is used. Here one looks at two or more derivatives with different heavy atoms placed in otherwise identical (isomorphous) crystals (Blundell and Johnson, 1976).

As a result of the pioneering work of Karle and Hauptmann (1956; cf. Karle, 1986), recognized by the 1985 Nobel Prize in Chemistry, it has become possible to solve the phase problem by so-called "direct methods," which involve a mathematical manipulation of the intensity data (symbolic addition). The method (Karle and Karle, 1972; Ladd and Palmer, 1980) is now used routinely for molecules of moderate size, including quite large organic structures (concerning potential applications to macromolecules see Hendrickson, 1991); it produces more accurate structural data than the heavy atom method because the potentially disturbing effect of the heavy atom is absent.

The availability of good crystals is a prerequisite for X-ray analysis. The crystals need not be large but must be single, not twinned. Since molecular vibrations occur in the crystal, diffraction analysis establishes an average position of the atoms. The thermal motion of the atoms can be assessed in the analysis; the "temperature factors" along the three axes measure the mean square amplitudes of the thermal vibration. In the general case, these vibrations are anisotropic (i.e., differ in magnitude along the three direction axes) and may be included in the drawings of the atomic positions in molecules as ellipsoids or "footballs." Sometimes molecular vibrations in certain parts of a molecule are so large that the positions of the atoms are not well defined; in that case one speaks of *dynamic disorder* in that part of the structure. More commonly, disorder (*static disorder*) is due to the coexistence of two different structures (e.g., two different conformations of the same molecule, Dunitz et al., 1967) in the crystal. It may lead to misinterpretations of structure (see Ermer, 1983). Uncertainties in atomic position due to thermal motion can be minimized by performing the crystallography at low temperature.

The goodness of fit of an X-ray structure determination (important during refinement, see above) is expressed as the R factor (agreement factor), which measures the agreement between the observed structure amplitudes and those calculated from an intermediate or the final structure. The smaller R, the better the fit.

$$R = \left[\frac{\sum (|F_{obs}|^2 - |F_{calc}|^2)}{\sum |F_{obs}|^2} \right]^{1/2}$$

where the F values are the absolute values of the structure factors.

Recently, it has become possible to achieve sufficiently high resolution in X-ray crystallography to map not only hydrogen atoms (which normally, because of their low electron density, are not seen well) but also the electron density between atoms (cf. Angermund et al., 1985). Some of the results have been surprising. Thus, if the "difference density" (difference of atomic electron density and that in the molecule) between electronegative atoms, such as two oxygen atoms, is determined, it turns out to be negative (Dunitz and Seiler, 1983)! The primitive notion that one can always tell a bond from the finding of high electron density between atoms is therefore not necessarily correct (see also Cremer and Kraka, 1984; Spackman and Masler, 1985; Schwarz, Ruedenberg, et al., 1985).

Neutron diffraction analysis (Hastings and Hamilton, 1972; Bacon, 1975; Speakman, 1978) is sometimes used as a complement to X-ray diffraction. It requires the availability of a strong flux of neutrons from an atomic reactor and the crystals must be somewhat larger than for X-ray analysis. However, there are several advantages, of which the principal one is the much greater scattering power for neutrons (compared to X-rays) of light atoms, such as hydrogen, deuterium or lithium. The location of hydrogen atoms from X-ray diffraction patterns is inaccurate, being usually based on the difference of the observed electron density with that calculated disregarding the hydrogen atoms. In some instances the hydrogen atoms are located simply on the basis of model considerations (but see above). In contrast, neutron diffraction permits accurate location of hydrogen or deuterium atoms.

Deuterated compounds are preferred because hydrogen absorbs neutrons strongly and thereby seriously reduces the intensity of the diffraction pattern.

Neutron diffraction also permits a more precise analysis of thermal motion than is possible by X-rays, because the scattering is mainly by the nuclei rather than the electrons. This, in turn, makes possible an interesting combination of X-rays and neutron diffraction analysis that provides an alternative approach to pinpointing electron density between atoms, thereby gaining information on the shapes of chemical bonds and location of unshared electron pairs. The existence of a bent bond between the ring carbon atoms in tetracyanoethylene oxide has thus been demonstrated (Mathews and Stucky, 1973; cf. Coppens, 1974; Fig. 2.19).

Electron diffraction (cf. Bartell, 1972; Karle, 1973; Zeil, 1974; Schäfer, 1987; Hargittai and Hargittai, 1988, part A) is a means for determining structure in the vapor phase. The scattering of the impinging electrons is caused mainly by the atomic nuclei and the method yields internuclear distances between both bonded and nonbonded atoms. Bond distances are obtained with greater precision than by X-ray diffraction but the method is limited to molecules of relatively small size, not only because of the requirement for appreciable volatility but also because the difficulty of interpreting the diffraction pattern increases as the square of the number of atoms. Its scope may be enhanced by combination with other techniques, such as microwave or vibrational spectroscopy (see below) or a priori calculations (cf. Section 2-6; Hargittai and Hargittai, 1988). The method, as we shall see in Chapter 11, is useful for assessing conformational isomerism.

Microwave spectroscopy (cf. Flygare, 1972; Gordy and Cook, 1984) is an alternative method of structure determination in the vapor phase, applicable to small molecules that have dipole moments. This method primarily yields the dipole moment and the moments of inertia of the molecular species investigated. To obtain structural data, it is generally necessary to investigate several isotopically substituted species and combine the information obtained from them. Because of the need to use several different species, because of the anharmonicity of the potential function, and for other reasons, the bond distances obtained by microwave spectroscopy differ slightly but systematically from those obtained by diffraction methods; this problem has been discussed extensively (Kuchitsu, 1968 and references cited therein; Bartell, 1972, p. 15; Robiette, 1972). Structure determinations by microwave spectroscopy, just as those by electron diffraction, may be enhanced through combination with theoretical calculations (see below: Schäfer et al., 1987).

Normal coordinate analysis is a means to deduce structure from IR and Raman spectroscopic data by complete theoretical interpretation of all the vibrational frequencies. The technique (cf. Wilson et al., 1955) is not often used to determine total structure and is limited to relative small molecules for which

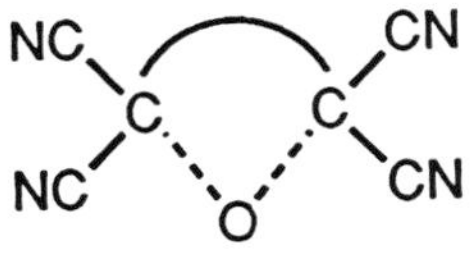

Figure 2.19. Tetracyanoethylene oxide bonding.

extensive spectroscopic information is available (see also the series edited by Durig, 1972–1991).

Another technique used occasionally is that of NMR spectroscopy in liquid crystals (Meiboom and Snyder, 1968; Saupe, 1968; Luckhurst, 1968; Emsley and Lindon, 1975; Emsley, 1985; Diehl and Jokisaari, 1986). When a (small) molecule is dissolved in a nematic or cholesteric liquid crystal (partially ordered), the solute will become oriented as well, and its geometry can then be inferred from the ^{1}H NMR spectrum. Thus the puckered conformation of cyclobutane has been confirmed in this fashion through NMR analysis in the nematic solvent

$$\overset{\displaystyle O}{\overset{\displaystyle |}{}}$$

p-(n-C$_6$H$_{13}$O)C$_6$H$_4$N=NC$_6$H$_4$(OC$_6$H$_{13}$-n)-p (Meiboom and Snyder, 1970). Since cholesteric mesophases (liquid crystals) are chiral, distinct signals for dissolved enantiomeric materials can be seen in such phases (Luckhurst, 1968; cf. Chapter 6).

A once popular approach for determining structure in solution is the lanthanide shift method in NMR (cf. Mayo, 1973; Cockerill et al., 1973; Willcott and Davis, 1975; Hofer, 1976; Morrill, 1986). As a method for determining total structure it is marginal; nonetheless, it is of some usefulness in the determination of conformation and configuration.

Only the basic principles of the method can be given here. To the substrate to be investigated in a suitable solvent is added a chelate of a suitable rare earth element (lanthanide), most commonly europium (Eu), praseodymium (Pr), or ytterbium (Yb). Suitable chelate complexes of dipivaloylmethane (dpm),

$$\underset{\overset{\displaystyle ||}{\displaystyle O}}{(CH_3)_3CC}\underset{}{CH_2}\underset{\overset{\displaystyle ||}{\displaystyle O}}{CC(CH_3)_3} \, ,$$

"Eu(dpm)$_3$" or of pivaloylheptafluorobutyrylmethane (fod) "Eu(fod)$_3$" (CH$_3$)$_3$CCOCH$_2$COCF$_2$CF$_2$CF$_3$, are commercially available. The latter reagent is the more soluble, and therefore is often more useful, since it can be added in greater concentration. The substrate must have a coordination site for the lanthanide shift reagent (LSR), for example a carbonyl, ether, amine, or sulfox-

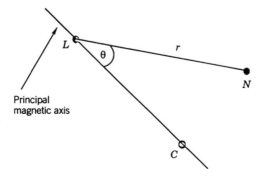

Figure 2.20. Lanthanide induced shifts: geometry. In this figure L is lanthanide, C is the coordination site, and N is the nucleus whose LIS is observed.

ide group, so that a (loose) complex can be formed. This method cannot be used for noncoordinating substrates, such as saturated hydrocarbons. The effect of adding the LSRs is to enhance the chemical shifts of the nuclei observed (e.g., protons) so that these shifts are spread over a much wider range of the spectrum than in the absence of the lanthanide. The extent of the shifting depends on the distance of the nucleus observed from the lanthanide (r) and the angle θ between the line joining the nucleus to the lanthanide and the direction of the magnetic dipole passing through the (paramagnetic) lanthanide and the atom coordinated with it (cf. Fig. 2.20). As a result, some quantitative structural information can be obtained from a lanthanide induced shift (LIS) experiment.

If, as is often the case in ^{1}H NMR spectroscopy, the LIS Δ is principally a "pseudocontact shift," which is due to a magnetic-dipolar type of interaction similar, in principle, to that causing other chemical shifts, it may be expressed by the McConnell–Robertson equation (cf. Fig. 2.20).

$$\Delta = \frac{3\cos\theta - 1}{r^3}$$

For a method of factoring out other contributions to Δ by the use of several lanthanides, see Reilley et al. (1975) and Hofer (1976).

Unfortunately, even if Δ is determined for a number of different nuclei in a molecule, the LIS experiment does not give enough independent pieces of information to allow the complete structure to be solved in the absence of simplifying assumptions (e.g., of bond lengths and angles). This shortage of information is in contrast to the X-ray experiment in which there is a redundancy of data. Even at best, refinement of the LIS data may not yield an unequivocal structure (Sullivan, 1976).

Lanthanide induced shifts may, however, provide a method for determining relative configuration (cf. Chapter 5) or for determining conformation in solution. The example shown in Figure 2.21 relates to the configuration of oximes (cf. Wolkowski, 1971). This example is a relatively simple case where the answer can be obtained even in a purely qualitative way, since it turns out that all protons on the side of the OH group, which complexes the lanthanide, undergo larger shifts than corresponding groups on the opposite side. A more subtle case, shown in Figure 2.22, concerns the conformation of β-adenosine 3′,5′-phosphate (3′, 5′-AMP) (Lavallee and Zeltman, 1974). This species, in the crystalline state, shows two conformational isomers coexisting in the unit cell (Watenpaugh et al., 1968). The left (extended) conformation (Fig. 2.22) has a torsion angle (O_1–C_1–N_9–C_8) of $-50°$; for the right (compact) conformation this angle is $+102°$. The LIS determination with Pr^{3+} in aqueous solution indicates a conformation close to the compact one with a torsion angle of $86 \pm 22°$ ($R = 0.048$).

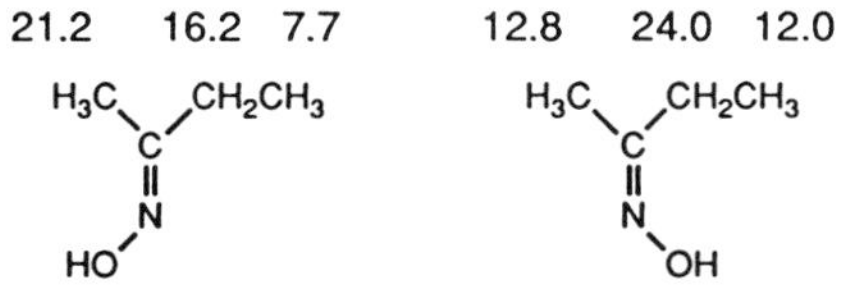

21.2   16.2   7.7   12.8   24.0   12.0

Figure 2.21. Oxime configuration by LIS. The numbers refer to europium induced shifts (in ppm) at the protons indicated.

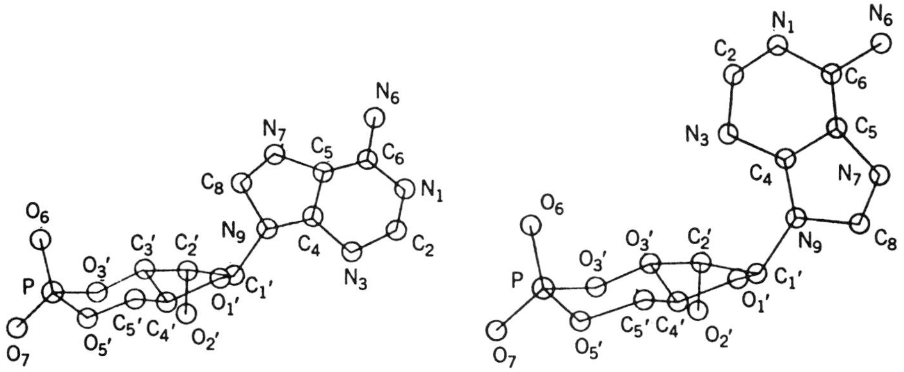

Figure 2.22. Conformations of β-adenosine 3',5'-phosphate. [Reprinted with permission from Lavalle, D. K. and Zeltman A. H. (1974), *J. Am. Chem. Soc.*, **96**, 5552. Copyright © (1974) American Chemical Society.]

One of the advantageous aspects of this case is that the complexation site (phosphate ester) is far from the conformationally mobile part of the molecule (purine). When this is not the case, the conformation of the (uncomplexed) molecule may sometimes be altered by the complexation with the lanthanide and the conformation (or mixture of conformations) determined by LIS may thus not correspond to that of the uncomplexed substrate (e.g., Kessler and Molter, 1973).

The case depicted in Figure 2.22, where there are two conformers in the crystal and only one in aqueous solution, is atypical. More often, in the case of substances that are not conformationally homogeneous, the crystal has a unique conformation and the conformational heterogeneity appears in fluid phases (liquid, vapor, or solution). This problem, and the related one as to whether conformation in solution, if unique, corresponds to conformation in the crystal has been discussed by Jeffrey (1971) with particular reference to carbohydrates.

It would appear at this time that although LIS can be a useful tool for ascertaining conformation in solution (and, in some cases, to distinguish between two possible diastereomeric configurations), it is best used in conjunction with other methods or when some structural information is already available.

The best currently available method for determining total structure in solution is undoubtedly 2D NMR (Wüthrich, 1986). The principal (though not exclusive) tools are proton correlated spectroscopy (COSY) and determination of nuclear Overhauser effects by nuclear Overhauser and exchange spectroscopy (NOESY) (Neuhaus and Williamson, 1989). The most fruitful application of these tools has been in the structure analysis of peptides and small proteins (up to ca. 100 residues) and nucleic acids. Only a brief discussion of the method can be given here.

A combination of COSY and NOESY is used to assign proton resonances to the protons belonging to individual amino acids in a peptide chain. Only geminal and vicinal protons will normally couple strongly enough to give cross-peaks in proton–proton COSY, though the method can be extended to certain more long-range couplings. Thus the low-field $-CH(NH_2)-$ proton of alanine can be uniquely recognized because it is the only such proton in a naturally occurring amino acid that is coupled to a methyl group: $CH_3-CH(NH_2)-CO_2H$, though coupling between an even lower field proton and a methyl group also occurs in threonine, $CH_3-CHOH-CH(NH_2)-CO_2H$. Nuclear Overhauser effects (NOEs) can be exploited to determine protons that are close to each other within one residue, such as the ortho and alpha protons in aromatic (Ar) residues $Ar-CH_2-$ and protons that are located 1,3 to each other ($H-C-C-C-H$). Its greater usefulness, however, lies in identifying protons that are close to each other in distinct residues. Thus NOEs are seen for protons in sequential residues, especially for CH/NH protons, as in $-CH-CO-NH-$, and will thus help in the establishment of sequence. The NOEs can also be found, however, in residues that are distant in sequence but close in space, as in α-helices and β-pleated sheets; thus they provide a tool for inferring conformation. Because the NOE falls off with the inverse sixth power of distance, protons must be within approximately 500 pm (5 Å) of each other for an NOE between them to be observed. Conclusions are beginning to be drawn as to the distance between NOE active protons on the basis of the intensity of the NOE between them.

> It must be remembered, however, that the lack of an NOE is not evidence that two protons are farther than 500 pm (5 Å) apart, because NOE strength is modulated by the dynamics of the two protons. On the other hand, the presence of an NOE does not constitute absolute proof that two protons are close, because of spin diffusion effects.

For peptides of moderate size, complete structures in solution can be inferred from the presence or absence of NOEs between assigned protons by what is known as the "distance geometry algorithm," provided the sequence is known from routine chemical determination. A somewhat easier task is to determine if the structure of a protein that has been determined in the crystalline state is maintained in solution, that is, if it is compatible with the observed NOEs.

Additional information can be obtained from NMR studies of peptides that are specifically ^{15}N or ^{13}C labeled, nitrogen labeling being particularly useful since it is the easier to effect, and since it is particularly helpful in heteronuclear COSY and NOESY. Sometimes information can be obtained from exchange rates of amide protons; while these are normally very fast, they can become slow on the NMR time scale for amide protons that are strongly hydrogen bonded in the secondary structure of peptides or are buried within the protein. Last, but not least, NOESY can also be used to demonstrate intermolecular association of peptides (as in the quaternary structure of proteins), of nucleic acids (as in double helices), of nucleic acids with proteins, and of nucleic acids with drugs.

Because it is often important for the chemist to ascertain the structure either of the molecules they are working with or, if that structure is not available, of chemically similar molecules, it is fortunate that a number of tabulations of molecules for which accurate structures have been determined are available (Sutton, 1958, 1965; Sim and Sutton, 1973, 1974; Kennard et al., 1970–1984; Duax et al., 1976; Duax and Norton, 1975; see also Allen, Kennard et al., 1983; Allen, 1987 concerning the Cambridge Crystallographic Database.)

2-6. A PRIORI CALCULATION OF STRUCTURE

For an overview, see Clark, 1985.

When, hypothetically, one throws together an appropriate array of atoms to form a molecule, one should be able to calculate, by the principles of quantum mechanics (cf. Náray-Szabó et al., 1987), which arrangement of the atoms corresponds to the absolute minimum in the total energy of the resulting molecule. This arrangement should then correspond to the actual structure.

In practice, there are severe impediments to such an approach. The accuracy of the calculations is limited by their complexity in terms of the number of electron orbitals that must be considered for molecules of even modest size and the resultant amount of computational effort (computer time) that is required. In the so-called ab initio method (Radom and Pople, 1972; Lathan, Pople et al., 1974; Hehre et al., 1986; see also Schaeffer, 1986) no input of experimental information of any kind is required. Nevertheless, except for first-row diatomic molecules, such as H_2, a complete solution of the Schrödinger equation is not, at present, feasible. Restrictions as to the number and type of atomic orbitals (basis set) considered must be introduced and the correlated motion of the electrons with respect to each other (electron correlation) is sometimes disregarded. As a result of such simplifications, the results of the calculations tend to be imperfect; extensive testing on molecules where the answer is known from experiment is required before any reliance can be placed on calculations of molecular properties as yet unknown. For relatively simple molecules, such as propene, extensive information can be obtained from calculation, whereas in the case of more complex species, such as acetophenone, often only the energy and a single property (e.g., dipole moment or a rotation barrier) are computed (cf. Parr, 1975; Hehre, Schleyer, Pople et al., 1986).

For more complex molecules, so-called semiempirical quantum mechanical methods have been used (Sadlej, 1985). In these methods, some of the parameters entering the calculations are obtained from experimental results in a limited number of molecules and then used for additional species (e.g., Klopman and O'Leary, 1970; Pullman, 1972; Dewar, 1975). The dependability of such methods in calculations of global molecular geometry and energy has been questioned (Pople, 1975; Hehre, 1975).

In summary, it appears that the quantum mechanical approach to the prediction of molecular structure is limited to molecules of modest size, but improvements in the calculations are continually forthcoming (cf. Schaeffer, 1986), and the situation is also improving as the result of the increased availability of supercomputers allowing the use of larger basis sets as well as their applications to larger molecules.

It must be recognized that calculations refer to isolated molecules, that is, to the gas phase. Structure in the solid may be similar because the relative magnitude of packing forces is often small (Jeffery, 1971) except when intermolecular hydrogen bonding or other strong electrostatic interaction occurs, but calculations from the vapor phase cannot usually be transferred to the liquid phase because of sizeable solvation as well as association energies. Efforts are being made to take the solvent into account either by considering specific interactions (e.g., Pullman et al., 1974; Tomasi et al., 1987) or by assessing solvation energy (cavitation, electrostatic interaction, or dispersion) in a general way (Beveridge et al., 1974).

A different approach, which can be applied with much lesser restrictions as to the complexity of the molecules considered, was suggested by Westheimer and Mayer (1946; Westheimer, 1956) and by Hill (1946). In this method (Hendrickson, 1961; Altona and Faber, 1974; Osawa and Musso, 1982; Burkert and Allinger, 1982; Boyd and Lipkowitz, 1982; Allinger, 1992) one computes, by a "mechanical" method, the excess energy of a given array of atoms in an (as yet hypothetical) molecule over the minimum energy that the array would possess if certain kinds of interactions (see below) were "turned off."

> Various names have been coined for this excess energy, for example, "conformational energy," "strain energy," and "steric energy." We shall use the term steric energy (for our use of conformational energy, and of strain energy, see Chapter 11).

One expresses this energy in terms of a number of contributions: bond stretching or compression strain, bond angle and torsion angle strain, nonbonded interaction, electrostatic interaction, and (in the liquid phase) solvation energy. One then changes these parameters in concert (using computer programs) so as to minimize the total energy. The geometry at the energy minimum is taken to be the actual predicted structure of the molecule in question, and the sum of the absolute minimum energy plus the residual steric energy at the calculated minimum is taken to be the heat of formation. Corrections must be made for zero point and thermal energies. Fortunately, often one is interested in calculating energy *differences* between isomeric configurations or conformations. In that case the minimum energy term cancels out in the difference of energy between the two isomers.

The method permits the a priori calculation of both molecular structure and energy. It is sometimes called the "method of molecular mechanics," the "energy optimization method," or the "force field method." The first and preferred designation stems from the fact that the strain parameters are calculated in essentially a classically mechanical way. The force field designation comes from the idea that the array of atoms finds itself in a field of interatomic forces whose resultants must vanish for the equilibrium structure.

In detail, the total strain energy of the molecule V_{tot} may be set equal to

$$V_{tot} = V_r + V_\theta + V_\omega + V_{UB} + V_{nb} + V_E - V_S \qquad (2.1)$$

where V_r is the energy due to bond stretching or compression (summed for all bonds), V_θ is the energy increment (summed for all angles) for bond angle deformations, V_ω is the sum total of the excess energy due to changes of torsion angles from their (energetic) optima, V_{UB} stands for the (Urey–Bradley) nonbonded energy of atoms 1,3 to each other (this term is often subsumed in V_{nb}), V_{nb} is the sum total of the remaining nonbonded energy within the molecule (involving more distant atoms), V_E is the sum total of the intramolecular electrostatic energy, and V_S is the solvation energy of the molecule.

If one wants to find the structure of a molecule using Eq. 2.1, one begins by assuming a trial structure (e.g., one obtained from a molecular model; cf. Section 2-7) and calculates the energy from Eq. 2.1 using an appropriate computer program. One then allows the computer to change the coordinates and to recompute the energy. The computer thus explores the energies of all structures slightly deformed from the original trial structure; from these it selects the one that has the lowest energy and then repeats the exploration in the same way, always along the path of "steepest descent" in energy (McCammon and Harvey, 1987, p. 47. Other computational approaches have been used). The process continues until the structure found is at an energy minimum such that all further deformations lead to a higher energy. At this point the energy and the atomic coordinates (structure) are computed. A problem that arises not infrequently is that the structure exploration ends in a local minimum that may still lie well above the most stable structure (cf. Kollman and Merz, 1990). Often this pitfall can be avoided by starting the search with different trial structures. Thus, if one started with a trial structure near the gauche form of butane (Section 10-1.a), one would probably find a slightly deformed version of the gauche form as minimum, but if one then restarted near the anti form, the latter would emerge as the true minimum in energy for butane.

An alternative approach to avoiding local minima is to deliberately "drive" the computer to perform a rotation about a chosen bond after the apparent minimum has been reached and thereby to find a deeper minimum. For example, if, in *gauche*-butane, one drives the rotation about the C(2)–C(3) bond, the energy will increase for small increments of the torsion angle (cf. Chapter 10, Fig. 10.3) but will eventually decrease again and, when the angle is changed by 120° the true (anti) minimum will be found. Yet other approaches have been suggested (Scheraga, 1986).

Before we discuss the individual energy terms, the following points should be noted: (a) Equation 2.1 is based on the assumption that the energy terms are separable, that is, that there are no cross terms, this is an oversimplification. Some high-quality force field treatments do include cross-terms at the cost of increasing the number of parameters in Eq. 2.1. (b) Specific intermolecular interactions are disregarded; thus Eq. 2.1 best approximates the situation in the vapor phase or (with an appropriate V_S term) in dilute solution. However, intermolecular forces of a specific nature can be taken into account by considering, as the "molecule", a dimer or even higher aggregate (supermolecule). This approach is similar to that taken in the quantum mechanical calculation (see p. 32).

In the assessment of the individual terms in Eq. 2.1, two approaches may be distinguished in principle. In one of these (the consistent force field method), one tries to parametrize the force field by using data from other experimental areas, for example, stretching and bending force constants from vibrational spectra, torsional force constants from rotational spectra, and nonbonded interactions from the theory of liquids (Warshel and Lifson, 1970). The other (and more common) approach (empirical force field method) is to select a few molecules whose structure and energy (heat of formation or heat of atomization) are well known and to parametrize Eq. 2.1 in such a way that the experimental energy for such molecules in the experimentally found structure is well reproduced (cf. Allinger et al., 1968). This technique has been called the Hendrickson–Wiberg–Allinger approach after its principal practioners.

The bond stretching term V_r is assumed to follow a Hooke's law expression (harmonic potential):

$$V_r = \tfrac{1}{2} k_r (r - r_0)^2 = \tfrac{1}{2} k_r \Delta r^2 \tag{2.2}$$

where Δr is the stretching or compression deformation from the "equilibrium" bond length. The equation has two adjustable parameters, r_0 and k_r. Although the standard C–C single-bond distance is near 153 pm (1.53 Å), smaller values are usually chosen for r_0, for example: $r_0 = 152$ pm (1.52 Å), $k_r = 0.2654$ kJ mol^{-1} pm^{-2} (634 kcal mol^{-1} Å^{-2}) (Engler, Schleyer et al., 1973). It is clear that bond stretching or compression is expensive in terms of energy: To stretch or compress a C–C bond by 10 pm (0.1 Å) costs 3.2 kcal mol^{-1} (13.4 kJ mol^{-1}). This result is presumably the reason for the general near constancy of bond distances mentioned earlier.

Angle bending is similarly assumed to follow a Hooke's law relation

$$V_\theta = \tfrac{1}{2} k_\theta (\theta - \theta_0)^2 = \tfrac{1}{2} k_\theta \Delta \theta^2 \tag{2.3}$$

where $\Delta \theta$ is the change in angle from the "equilibrium" bond angle that may be taken (Engler, Schleyer, et al., 1973) as the tetrahedral angle 109.5°.

> This choice is again somewhat arbitrary. For example, the CCC valency angle in propane or a longer straight-chain alkane is 112.5°. This result may be due to opening of the "normal" angle by 1,3-repulsion (CH_3/CH_3) from 109.5° to 112.5° or it may be the natural angle of a
>
> $$\begin{array}{c} H \qquad\qquad C \\ \diagdown \qquad \diagup \\ C \\ \diagup \qquad \diagdown \\ H \qquad\qquad C \end{array}$$
>
> fragment. In the former case, choice of an 109.5° angle for the unstrained situation is justified, in the latter it is artificial.

A typical value for k_θ is 82.0 kcal rad^{-2} mol^{-1} (343 kJ rad^{-2} mol^{-1}) or 0.025 kcal mol^{-1} deg^{-2} (0.105 kJ mol^{-1} deg^{-2}) for CCC. Bending angles (contrary to what appears from a Dreiding or similar molecular model!) is thus facile: The energy required for a 5° deformation is only 0.3 kcal mol^{-1} (1.3 kJ mol^{-1}); for a 10° deformation, 1.25 kcal mol^{-1} (5.2 kJ mol^{-1}).

> Eqs. 2.2 and 2.3 do not apply to large deformations (as in cyclopropane or other highly strained systems) for which softer potentials must be employed since the parabolic Hooke's law function approximates the real (Morse) function only near the energy minimum; at distances or angles far from the minimum it exaggerates the strain (Williams, Schleyer, et al., 1968).

The torsional potential about a C–C single bond is generally assumed to follow a simple cosine function ("Pitzer potential" after its originator):

$$V_\omega = \tfrac{1}{2} V_0 (1 + \cos 3\omega) \tag{2.4}$$

Qualitative consideration of this expression discloses that the torsional potential is at a maximum for eclipsed (synperiplanar) substituents ($\omega = 0°$; $V = V_0$) and at a minimum for staggered (gauche, synclinal) substituents ($\omega = 60°$; $V = 0$). The value recommended (Engler, Schleyer, et al., 1973) for V_0 for a C–C–C–C bond sequence is 3.1 kcal mol^{-1} (13.0 kJ mol^{-1}).

The single-term equation 2.4 is limited to simple ethanoid bonds, as in molecules of the type X_3C–CY_3. In other molecules, such as biphenyls, a twofold (so-called V_2) potential is found and in yet others, such as CH_3O–OCH_3 (if one disregards the unshared pairs), a onefold one (V_1) is found. [The numbers one, two, three here refer to the number of energy maxima (or the equivalent number of energy minima) found in a 360° rotation about the bond in question. Thus nitromethane, H_3C–NO_2, considering that the two oxygen ligands on nitrogen are equivalent, has a sixfold (V_6) barrier, in as much as maxima (or minima, as the case may be) occur at every 60° interval of torsion angle.] In a more complex case, several terms need to be considered:

$$V = \tfrac{1}{2}V_1(1 + \cos \omega) + \tfrac{1}{2}V_2(1 - \cos 2\omega) + \tfrac{1}{2}V_3(1 + \cos 3\omega) + \cdots$$

For 1,3 interactions at a distance r_{ij}, the Urey–Bradley expression

$$V_{UB} = \tfrac{1}{2}F_{ij}(r_{ij} - r_{ij}^0)^2 \tag{2.5}$$

is used, where r_{ij}^0, the equilibrium 1,3 distance, is taken to be 250 pm (2.5 Å) for two carbon atoms and F_{ij} for this case is 55.0 kcal mol^{-1} Å^{-2} (0.023 kJ mol^{-1} pm^{-2}) (Warshel and Lifson, 1970).

This leaves the nonbonded interaction V_{nb} in Eq. 2.1 applied to nonpolar molecules. Experimental information for this term is scarce and a variety of expressions have been used for V_{nb} (cf. Dunitz and Bürgi, 1975). One aspect all of them have in common is an attractive term that predominates at larger distances and a repulsive term that becomes dominant on close approach. The situation is depicted in Figure 2.23. At infinite distance, two nonbonded atoms do not interact ($V_{nb} = 0$). As the atoms approach each other, an attractive force, the

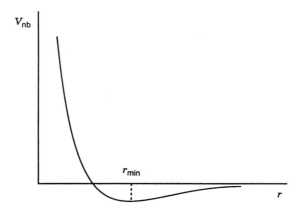

Figure 2.23.   Nonbonded interaction.

so-called London or dispersion force, comes into play leading to a lowering of the energy. The energy corresponding to this force is commonly taken as $-ar^{-6}$, where a is a constant and r is the internuclear distance. The coefficient a is proportional to the polarizability of the two atoms that approach each other, thus the attractive interaction will be greatest (other things being equal) for atoms in the lower right of the periodic table, such as S, Br, or I.

At a still closer distance of approach, a repulsive force, (due to closed-shell repulsion) is encountered. The repulsive energy term may be expressed as an inverse twelfth power of the internuclear distance (br^{-12}, b being a constant). In that case the overall nonbonded potential V_{nb} is expressed by Eq. 2.6,

$$V_{nb} = -ar^{-6} + br^{-12} \qquad (2.6)$$

called a 6–12 or Lennard–Jones potential (after its first proponent); it is plotted in Figure 2.23. Equation 2.6 implies an energy minimum at an internuclear distance that turns out to be an additive property of the two atoms A and B that approach each other. It is convenient to assign to these atoms radii (called van der Waals radii) r_A and r_B whose sum $r^* = r_A + r_B$ is approximately equal to the internuclear distance r_{min} at which the nonbonded energy is at a minimum. It should be realized that V_{nb} is negative (attractive) not zero, at that distance (in fact the net attraction is maximal at r_{min}). When $r < r_{min}$, V_{nb} increases, crosses the zero energy axis and, at still shorter internuclear distances, becomes repulsive. Thus, nonbonded interactions may be either attractive or repulsive, depending on the internuclear distance.

As an alternative to the 6–12 potential, the so-called Buckingham potential in which the repulsive term is exponential: $V_{nb} = -ar^{-6} + be^{-cr}$ has been used, where a, b, and c are constants. Or, in the Lennard–Jones potential, the exponents may be modified on grounds of better empirical fit, thus an inverse ninth-power term (br^{-9}) has been used in lieu of the inverse twelfth-power term.

Some authors prefer to use an expression due to Hill, equivalent to the Lennard–Jones or Buckingham potentials; thus, for example, for the 6–12 potential (Hill, 1948)

$$V_{nb} = -2.0\eta(r^*/r)^6 + \eta(r^*/r)^{12} \qquad (2.7)$$

In Eq. 2.7, η is a measure of the "softness" or "hardness" of the potential; the harder the potential, the steeper the curve (Fig. 2.23) and the deeper the minimum; η increases with the size of the interacting atoms A and B. Equation 2.7 unlike Eq. 2.6 does not have separate coefficients for the attractive and repulsive parts of the potential, and thus has one less disposable parameter. From Eq. 2.7 one may derive $V_{min} = -\eta$ at $r = r^*$. The crossing point occurs at $r/r^* = 0.89$, that is, one can reduce the internuclear distance 11% below the optimum (r_{min}) before repulsion sets in.

The electrostatic term V_E is a Coulombic term

$$V_E = e_A e_B r^{-1} \varepsilon^{-1} \qquad (2.8)$$

where e_A and e_B are the charges on the interacting atoms, r is the distance between them, as before, and ε is the dielectric constant. Assessing the magnitude of these charges is not straightforward; it has been customary to compute the charges from bond dipoles and locate them at the center of the pertinent atomic nuclei. This approach has been taken even for the C–H bond ($\mu = 0.3$ D) in some calculations (Warshel and Lifson, 1970). A better procedure is to calculate the charges from quantum mechanics. Yet another alternative is to compute the energy from the electrostatic energy between two dipoles (cf. Eliel et al., 1965):

$$V_E = (\mu_1 \mu_2 / \varepsilon r^3)(\cos \chi - 3 \cos \alpha_1 \cos \alpha_2) \qquad (2.9)$$

In Eq. (2.9) μ_1, μ_2 are the magnitude of the two dipoles, r is the distance between their centers, α_1 and α_2 are the angles between the dipoles and a line connecting their centers, and χ is the dihedral angle between the two planes containing the two dipoles and the line joining their centers. Equation 2.9 strictly applies to dipoles that are at large distances from each other relative to their own extent; it cannot properly be applied to dipoles that are close together, as in small molecules, such as $BrCH_2CH_2Br$. The parameter ε in Eqs. 2.8 and 2.9 is the effective dielectric constant whose value also presents problems. Since, in a molecule, part of the space between the atomic point charges is occupied by the molecule itself and part by the solvent, ε corresponds neither to the dielectric constant of the solvent nor to that of the solute, but is somewhere in between (Kirkwood and Westheimer, 1938). It is customary to use an empirical value for ε somewhere between 2 and 5; this value may, in fact, be treated as a disposable parameter in model calculations on molecules where the experimental results are known.

That the V_E term is of considerable importance in dipolar molecules is seen, for example, in the conformational equilibrium of *trans*-1,2-dibromocyclohexane (Fig. 2.24). On purely steric grounds this molecule should prefer the diequatorial conformation but, in fact, 75% of the molecules are in the diaxial conformation in carbon tetrachloride ($\varepsilon = 2.2$), whereas the preference is for the diequatorial form (64%) in acetonitrile, CH_3CN ($\varepsilon = 36$) at room temperature (cf. Abraham and Bretschneider, 1974). The diaxial preference in CCl_4 clearly has its origin in the electrostatic repulsion of the dipoles when they are equatorially oriented (see also Chapter 11).

Because the change in *effective* dielectric constant is relatively small, the large change in ΔG^0 with solvent (Fig. 2.24) cannot have its main source in the V_E term. In fact, most of the change is due to V_S, the solvation energy. Solvation of a dipole leads to stabilization and one might think that the results in Figure 2.24 could be explained straightforwardly in this fashion. However, *trans*-1,4-dibromocyclohexane (Fig. 2.25) shows a similar change in conformational equilibrium with solvent even though the dipole moments of both the diequatorial and

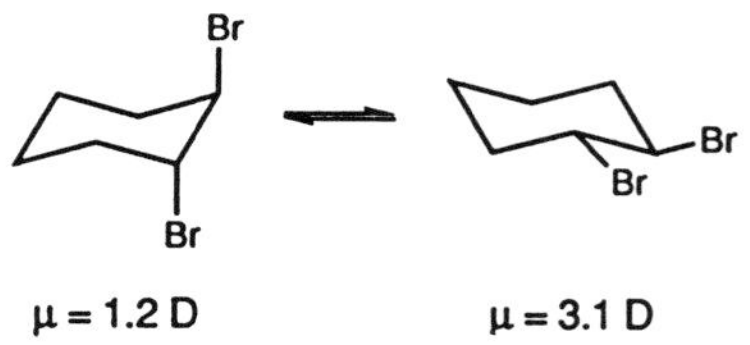

Figure 2.24. *trans*-1,2-Dibromocyclohexane, where $\Delta G^0 = 0.65$ kcal mol^{-1} (2.7 kJ mol^{-1}) for CCl_4 and $\Delta G^0 = -0.30$ kcal mol^{-1} (-1.3 kJ mol^{-1}) for CH_3CN.

$\mu = 1.2$ D $\mu = 3.1$ D

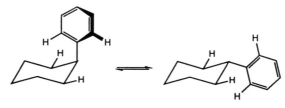

Br
μ = 0 D

Br
Br
μ = 0 D

Figure 2.25. *trans*-1,4-Dibromocyclohexane, where $\Delta G^0 = 0.21\,\text{kcal mol}^{-1}$ $(0.88\,\text{kJ mol}^{-1})$ for CCl_4 and $\Delta G^0 = -0.41\,\text{kcal mol}^{-1}$ $(-1.7\,\text{kJ mol}^{-1})$ for CH_3CN.

diaxial conformations are zero. The reason for the solvent dependence here is that this molecule has a quadrupole moment and, in assessing the V_S term, both dipole and quadrupole solvation must be taken into account (cf. Abraham and Bretschneider, 1974). An equation has been derived

$$V_S = \frac{kx}{1 - lx} + \frac{3hx}{5 - x} \qquad (2.10)$$

where $k = m^2/a^3$ (m being the dipole moment of the conformer in question, and a the radius of the solute cavity, assumed to be spherical), $x = (\varepsilon - 1)/(2\varepsilon + 1)$, $l = 2\alpha/a^3$ [α being the molecular polarizability of the solute given by $\alpha = 2(n^2 - 1)/(n^2 + 2)$, where n is the refractive index of the solute] and $h = 3q^2/2a^5$ where q is the molecular quadrupole moment of the conformer in question.

The calculated $\Delta\Delta G^0$ in going from CCl_4 to acetonitrile, CH_3CN, in the case of *trans*-1,4-dibromocyclohexane (Fig. 2.25) is $0.62\,\text{kcal mol}^{-1}$ $(2.6\,\text{kJ mol}^{-1})$. The diequatorial isomer is favored in the more polar solvent, since it has the larger quadrupole moment because of the shorter perpendicular distance between the dipoles. The calculated difference for *trans*-1,2-dibromocyclohexane (Fig. 2.24) is $0.92\,\text{kcal mol}^{-1}$ $(3.85\,\text{kJ mol}^{-1})$ favoring the equatorial isomer versus $0.95–1.01\,\text{kcal mol}^{-1}$ $(4.0–4.2\,\text{kJ mol}^{-1})$ observed experimentally. In this case, however, it is the dipole interaction that dominates; the quadrupole interaction actually opposes the dipole interaction in that it (slightly) favors the diaxial conformation in more polar solvents (Abraham and Bretschneider, 1974).

The molecular mechanics method as described can be applied only to systems that do not possess resonance energy (aromatics, conjugated alkenes, etc.) unless consideration is confined to a difference in energy between two structures for which delocalization energy remains constant, for example, the equatorial and axial forms of phenylcyclohexane (Allinger and Tribble, 1971). However, it has been possible to calculate the energy of conjugated and aromatic systems by combining quantum mechanical and molecular mechanical calculations (the so-called MMP methods of Allinger and Sprague, 1973).

Perpendicular phenyl   Bisecting phenyl

Figure 2.26. Phenylcyclohexane.

One of the attractive aspects of molecular mechanics calculations is that they provide insight as to the source of experimentally observed steric interactions. Thus, for example, Allinger and Tribble (1971; see also Hodgson et al., 1985) showed that the calculated difference of $3\,kcal\,mol^{-1}$ ($12.6\,kJ\,mol^{-1}$) between the energy of equatorial and axial phenylcyclohexane [later found, experimentally, to amount to $2.87\,kcal\,mol^{-1}$ ($12.0\,kJ\,mol^{-1}$) (Eliel and Manoharan, 1981)] is mainly due to the interaction of ortho hydrogen and carbon atoms of the benzene ring with the adjacent equatorial hydrogen atoms on the cyclohexane ring when the phenyl is axial (Fig. 2.26), rather than to the compression of the synaxial hydrogen atoms, which is normally principally responsible for the instability of the axial conformers (cf. Chapter 11). However, some caution must be exercised in the dissection of steric interactions as among angle bending, torsion, and nonbonded interactions; it must be remembered that in the Hendrickson–Wiberg–Allinger approach the parameters ascribed to these various interaction modes are picked empirically and somewhat arbitrarily. Therefore, two different types of calculation using one or the other of the several available force fields can both give the right answers in terms of energy and structure, but they are not likely to be both correct as to the exact physical origin of whatever interaction energies are found. By the same token, it is not surprising that force fields are constantly being modified and improved.

Only a brief, and no doubt incomplete, enumeration of the many existing force fields can be given here. For small molecules, the MM (molecular mechanics) force fields developed by Allinger and his group (MMP2, MM2/82, MM2/85, see Sprague, Allinger et al., 1987; MM3, see Allinger et al., 1989; Lii and Allinger, 1989a,b; 1990; Schmitz and Allinger, 1990; Allinger et al., 1991; Chen and Allinger, 1991; Lii and Allinger, 1992; Allinger et al., 1992a; 1992b and references cited therein; Aped and Allinger, 1992; Fox, Allinger, et al., 1992) seem to be the best as indicated by comparison with experimental data (Gundertofte et al., 1991; Clark et al., 1989). Other force fields are used, however, in many current studies of peptides, proteins, and nucleic acids (McCammon and Harvey, 1987); among these are AMBER (Weiner and Kollman, 1981; Weiner, Kollman et al., 1984); CEDAR (Carson and Hermans, 1985); ChemX (cf. Gundertofte et al., 1991); CHARMM (Brooks, Karplus, et al., 1983); GROMOS (van Gunsteren et al., 1983); OPAL (Jorgensen and Tirado-Rives, 1988); SYBIL (Clark et al., 1989) and force fields developed by Lifson (Lifson et al., 1979) and by Scheraga (Momany, Scheraga, et al., 1975).

2-7. MOLECULAR MODELS

Only a very brief discussion of molecular models will be given here (for a detailed discussion, see Walton, 1978).

There are essentially three types of models:

1. Those that simply help to visualize the 3D architecture and stereochemistry but are not to scale.
2. Framework-type models that indicate correct bond distances and bond angles and can be used to measure distances between nonbonded atoms in molecules but do not show the atoms as such.
3. So-called space-filling models that provide a fairly realistic 3D representation of what the molecule actually looks like.

The first kind of models are the so-called "ball-and-stick" kind, though the "balls" (wooden or plastic) are sometimes polyhedra; the "sticks" may be made of wood, of hard rubber, or of other polymeric material. These models are generally inexpensive and help beginning students visualize molecules in three dimensions, manipulate projection formulas, and count stereoisomers, but they do not show the dimensions of the molecule.

The second (framework-type) kind of model is exemplified by the well-known "Dreiding models," which are machined so that when the atomic models are put together, interatomic distances as well as bond angles are nearly correct. In these models, if the conformation is arranged properly (and this is a rather important condition, not easily fulfilled in flexible structures), distances between nonbonded atoms can be measured and nonbonded repulsions leading to strain can be sensed and even estimated as to their magnitude. The Kendrew Skeletal Molecular Models are of this type.

The third kind, space-filling models, give a better picture of the true molecular shape and dimensions but are too congested to allow measurements of interatomic distances. The so-called Fisher–Hirschfelder–Taylor, Stuart–Briegleb, C–P–K (Cory–Pauling–Koltun), Catalin, Courtauld, and Godfrey models are examples of this type.

A serious shortcoming, common to virtually all molecular models, is that they have fixed bond angles and that rotation about single bonds is excessively facile, especially in the first two types of models. This is in contrast to the real situation (cf. the discussion of molecular mechanics, Section 2-6) of relatively easily deformable bond angles and substantial barriers to rotation about single bonds. Moreover, if one wants to measure an intramolecular bond distance between two atoms in a Dreiding model, one must fix the model in its actual conformation. For molecules having a number of single bonds, this may be quite inconvenient (even though mechanical devices to stop bond rotation are available) and it may be difficult to set the actual torsion angles with any kind of precision.

As a result of all these difficulties, molecular modeling in suitable computers in conjunction with appropriate displays has become a superior substitute for the use of mechanical models in situations where quantitative (as distinct from qualitative or semiquantitative) information about exact molecular shapes and intra- or intermolecular interactions is desired. The molecule can be input with standard coordinates (bond lengths, bond angles, and torsion angles) or may even be drawn on the screen by a "mouse" or other device, and is then manipulated in the computer by stretching or compressing bonds, opening or closing bond angles, and changing torsion angles so as to minimize the energy (or in any other way desired); interatomic distances may then be read off directly on screen. The energy minimization is performed by a built-in molecular mechanics program, as described in Section 2-6.

Molecular modeling of this type has been used quite extensively to study the fit of enzymes with their substrates or of drug receptors with drugs. Assuming that an X-ray structure of the enzyme is available, and that the conformation in solution is close to that in the crystal, one can model both the enzyme and the (small) substrate and then try to "dock" the substrate in the active site of the enzyme. This may, of course, require some conformational changes, whose energetic penalty can, in turn, be readily assessed. The approach is useful if one tries to devise enzyme inhibitors that must fit into the active site but should not undergo the subsequent chemical transformations that the natural substrate will undergo (cf. Kollman, 1985; Kollman and Merz, 1990).

If the structure of the enzyme is not known, or in the case of a drug receptor (whose shape is usually not accessible by X-ray crystallography), one can actually try to model the active site or receptor if one knows the structure of a number of substrates that interact with it, since there must be a fit or induced fit in each case. Then one can try to devise new substrates (drugs) that will fit the enzyme cavity or receptor as modeled. The finding of pharmacological activity in such cases would suggest that the model of the active site or receptor is valid (cf. Ondetti and Cushman, 1981). Studies of this type are being pursued very actively at the time of writing.

REFERENCES

Allinger, N. L. (1992), "Molecular Mechanics," in Domenicano, A. and Hargittai, I., Eds., *Accurate Molecular Structures*, Oxford University Press, New York, p. 336.

Allinger, N. L., Chen, K. Rahman, M. Pathiaseril, A. (1991) *J. Am. Chem. Soc.*, **113**, 4505.

Allinger, N. L., Hirsch, J. A., Miller, M. A., Tyminski, I. J., and Van-Catledge, F. A. (1968), *J. Am. Chem. Soc.*, **90**, 1199.

Allinger, N. L., Kuang, J., and Thomas, H. D. (1990), *THEOCHEM*, **68**, 125.

Allinger, N. L., Li, F. and Yan, L. (1990), *J. Comput. Chem.*, **11**, 848.

Allinger, N. L., Li, F., Yan, L., and Tai, J. C. (1990), *J. Comput. Chem.*, **11**, 868.

Allinger, N. L., Quinn, M., Raman, M., and Chen, K. (1991), *J. Phys. Org. Chem.*, **4**, 647.

Allinger, N. L., Rahman, M., and Lii, J.-H. (1990), *J. Am. Chem. Soc.*, **112**, 8293.

Allinger, N. L., Rodriguez, S., and Chen, K. (1992b), *THEOCHEM*, **92**, 161.

Allinger, N. L. and Sprague, J. T. (1973), *J. Am. Chem. Soc.*, **95**, 3893.

Allinger, N. L. and Tribble, M. T. (1971), *Tetrahedron Lett.*, 3259.

Allinger, N. L., Yuh, Y. H., and Lii, J.-H. (1989), *J. Am. Chem. Soc.*, **111**, 8551.

Allinger, N. L., Zhu, Z.-q. S., and Chen, K. (1992a), *J. Am. Chem. Soc.*, **114**, 6120.

Abraham, R. J. and Bretschneider, E. (1974), "Medium Effects on Rotational and Conformational Equilibria," in Orville-Thomas, W. J., Ed., *Internal Rotation in Molecules*, J. Wiley, New York, pp. 481, 555, 556.

Akimoto, H., Shioiri, T., Iitaka, Y., and Yamada, S. (1968), *Tetrahedron Lett.*, 97.

Allen, F. H. (1987), "The Cambridge Structural Data Base as a Research Tool in Chemistry," in Maksić, Z. B., Ed., *Modelling of Structure and Properties of Molecules*, Wiley, New York, p. 51.

Allen, F. H., Kennard, O., and Taylor, R., "Systematic Analysis of Structural Data as a Research Technique in Organic Chemistry" (1983), *Acc. Chem. Res.*, **16**, 146.

Almenningen, A., Andersen, B., and Nyhus, B. A. (1971), *Acta Chem. Scand.*, **25**, 1217.

Altona, C. and Faber, D. H. (1974), "Empirical Force Field Calculations. A Tool in Structural Organic Chemistry," *Top. Curr. Chem.*, **45**, 1.

Angermund, K., Claus, K. H., Goddard, R., and Krüger, C. (1985), "High Resolution X-ray Crystallography—An Experimental Method for the Description of Chemical Bonds," *Angew. Chem. Int. Ed. Engl.*, **24**, 237.

Aped, P. and Allinger, N. L. (1992), *J. Am. Chem. Soc.*, **114**, 1.

Bacon, G. E. (1975), *Neutron Diffraction*, 3rd ed., Oxford University Press, Oxford, UK.

Bartell, L. S. (1972), "Electron Diffraction by Gases," in Weissberger, A. and Rossiter, B. W. Eds., *Techniques of Chemistry*, Vol. 1, Part IIId, Wiley, New York, p. 125.

Beveridge, D. L., Kelly, M. M., and Radna, R. J. (1974), *J. Am. Chem. Soc.*, **96**, 3769.

Blundell, T. L. and Johnson, L. N. (1976), *Protein Crystallography*, Academic, New York, chapter 6.

Boyd, D. and Lipkowitz, K. B. (1982), "Molecular Mechanics: The Method and its Underlying Philosophy," *J. Chem. Educ.*, **59**, 269.

Brooks, B. R., Bruccoleri, R. E., Olafson, B. D., States, D. J., Swaminathan, S., and Karplus, M. (1983), *J. Comp. Chem.*, **4**, 187.

Bürgi, H. B., Schefter, E., and Dunitz, J. D. (1975), *Tetrahedron*, **31**. 3089.

Burkert, U. and Allinger, N. L., (1982), *Molecular Mechanics*, ACS Monograph 177, American Chemical Society, Washington, DC.

Chen, K. and Allinger, N. L. (1991), *J. Phys. Ong. Chem.*, **4**, 659.

Chiang, J. F. and Bauer, S. H. (1970), *J. Am. Chem. Soc.*, **92**, 1614.

Clark, T. (1985), *A Handbook of Computational Chemistry*, Wiley-Interscience, New York.

Clark, M., Cramer, R. D., and Van Opdenbosch, N. (1989), *J. Comp. Chem.*, **10**, 982.

Cockerill, A. F., Davies, G. L. O., Harden, R. C., and Rackham, D. M. (1973), "Lanthanide Shift Reagents for Nuclear Magnetic Resonance Spectroscopy," *Chem. Rev.* **73**, 553.

Coppens, P. (1974), *Acta Cryst.*, **B30**, 255.

Cremer, D. and Kraka, E. (1984), *Angew. Chem. Int. Ed. Engl.*, **23**, 627.

Cross, L. I. and Klyne, W. (1976), "Rules for the Nomenclature of Organic Chemistry-Section E: Stereochemistry" *Pure Appl. Chem.*, **45**, 11; see also (1970), "IUPAC Tentative Rules for the Nomenclature of Organic Chemistry: Fundamental Stereochemistry," *J. Org. Chem.*, **35**, 2849.

Dewar, M. J. S. (1975), "Quantum Organic Chemistry," *Science*, **187**, 1037.

Diehl, P. and Jokisaari, J. (1986), "NMR Studies in Liquid Crystals: Determination of Solute Molecular Geometry," in Takeuchi, Y. and Marchand, A. P., Eds., *Applications of NMR Spectroscopy to Problems of Stereochemistry and Conformational Analysis*, VCH, Deerfield Beach, FL., p. 41.

Doyle, T. R. and Vogl, O. (1989), *J. Am. Chem. Soc.*, **111**, 8510.

Duax, W. L., Weeks, C. M., and Rohrer, D. C. (1976), *Top. Stereochem.*, **9**, 271.

Dunitz, J. D. (1979), *X-ray Analysis and the Structure of Organic Molecules*, Cornell University Press, Ithaca, NY.

Dunitz, J. D. and Bürgi, H. B. (1975), "Non-bonded Interaction in Organic Molecules," MTP International Reviews of Science, Physical Chemistry Society, 2, Vol. 11, Butterworths, Boston, MA, p. 81.

Dunitz, J. D., Eser, H., Bixon, M., and Lifson, S. (1967), *Helv. Chim. Acta.*, **50**, 1572.

Dunitz, J. D. and Seiler, P. (1983), *J. Am. Chem. Soc.*, **105**, 7056.

Durig, J. R., Ed. (1972–1991), *Vibrational Spectra and Structure*, Vols. 1–19, Marcel Dekker, New York.

Eliel, E. L. (1976/77), "On the Concept of Isomerism", *Isr. J. Chem.*, **15**, 7.

Eliel, E. L., Allinger, N. L., Angyal, S. J., and Morrison, G. A. (1965), *Conformational Analysis*, Interscience-Wiley, New York, reprinted by American Chemical Society, Washington (1981), Chapter 7 and p. 461.

Eliel, E. L. and Manoharan, M. (1981), *J. Org. Chem.*, **46**, 1959.

Emsley, J. W., Ed. (1985), *Nuclear Magnetic Resonance in Liquid Crystals*, Reidel, Boston, MA.

Emsley, J. W. and Lindon, J. C. (1975), *NMR Spectroscopy Using Liquid Crystal Solvents*, Pergamon Press, New York.

Engler, E. M., Andose, J. D., and Schleyer, P. v. R. (1973), *J. Am. Chem. Soc.*, **95**, 8005.

Ermer, O. (1983), *Angew. Chem. Int. Ed. Engl.*, **22**, 251.

Flygare, W. H. (1972), "Microwave Spectroscopy," in Weissberger, A. and Rossiter, B. W., Eds., *Techniques of Chemistry* Vol. 1, Part IIIa, Wiley, New York, p. 439.

Fox, P. C., Bowen, J. P., and Allinger, N. L. (1992), *J. Am. Chem. Soc.*, **114**, 8536.

Glusker, J. P. and Trueblood, K. N. (1985), *Crystal Structure Analysis, A Primer*, 2nd ed., Oxford University Press, New York.

Gordy, W. and Cook, R. L. (1984), *Microwave Molecular Spectra*, Wiley, New York.

Green, D. W., Ingram, V. M., and Perutz, M. F. (1954), *Proc. R. Soc. (London)*, A225, 287.

Gundertofte, K., Palm, J., Petterson, I., and Stamvik, A. (1991), *J. Comp. Chem.*, **12**, 200.

Hargittai, I. and Hargittai, M. (1988), *Stereochemical Applications of Gas-Phase Electron Diffraction* Part A, "The Electron Diffraction Technique"; Part B, "Structural Information for Selected Classes of Compounds," VCH Publishers, New York.

Hargreaves, M. K. and Modarai, B. (1969), *Chem. Commun.*, 16; (1971), *J. Chem. Soc.*, 1013.

Hastings, J. M. and Hamilton, W. C. (1972), "Neutron Scattering," in Weissberger, A. and Rossiter, B. W., Eds., *Techniques of Chemistry*, Vol. 1, Part IIId, Wiley, New York, p. 159.

Hehre, W. J. (1975), *J. Am. Chem. Soc.*, **97**, 5308.

Hehre, W. J., Radom, L., Schleyer, P. V. R., and Pople, J. A. (1986), *Ab Initio Molecular Orbital Theory*, Wiley, New York.

Hendrickson, J. B. (1961), *J. Am. Chem. Soc.*, **83**, 4537.

Hendrickson, W. A. (1991), "Determination of Macromolecular Structures from Anomalous Diffraction of Synchrotron Radiation," *Science*, **254**, 51.

Hill, T. L. (1946), *J. Chem. Phys.*, **14**, 465.

Hill, T. L. (1948), *J. Chem. Phys.*, **16**, 399.

Hodgson, D. J., Rychlewska, U., Eliel, E. L., Manoharan, M., Knox, D. E., and Olefirowicz, E. M. (1985), *J. Org. Chem.*, **50**, 4838.

Hofer, O. (1976), "The Lanthanide Induced Shift Technique: Applications in Conformational Analysis," *Top. Stereochem.*, **9**, 111.

Inagaki, F. and Miyazawa, T. (1981), "NMR Analyses of Molecular Conformation and Conformational Equilibria with the Lanthanide Probe Method," *Prog. Nucl. Magn. Reson. Spectrosc.*, **14**, 67.

Jeffrey, G. A. (1973), "Conformational Studies in the Solid State: Extrapolation to Molecules in Solution," in Isbell, H. S., Ed., *Carbohydrates in Solution*, Advances in Chemistry Series No. 117, American Chemical Society, Washington, DC.

Jenkins, A. D. (1981), "Stereochemical Definitions and Notions Relating to Polymers," *Pure Appl. Chem.*, **53**, 733.

Jensen, F. R. and Bushweller, C. H. (1966), *J. Am. Chem. Soc.*, **88**, 4279.

Jensen, F. R. and Bushweller, C. H. (1969), *J. Am. Chem. Soc.*, **91**, 3223.

Jorgensen, W. J. and Tirado-Rives, J. (1988), *J. Am. Chem. Soc.*, **110**, 1657.

Kalinowski, H. O. and Kessler, H. (1973), "Fast Isomerization about Double Bonds," *Top. Stereochem.*, **7**, 295.

Karle, J. (1973), "Electron Diffraction" in Nachod, F. C. and Zuckerman, J. J., Eds., *Determination of Organic Structure by Physical Methods*, Vol. 5, Academic, New York, p. 1.

Karle, J. (1986), "Recovering Phase Information from Intensity Data," *Science*, **232**, 837.

Karle, J. and Hauptman, H. (1956), *Acta Cryst.*, **9**, 635.

Karle, J. and Karle, I. (1972), "Application of Direct methods in X-ray Crystallography," in Robertson, J. M., Ed., MTP International Review of Science, Physical Chemistry Series, 1, Vol. 11, Butterworths, London, p. 247.

Kennard, O., and colleagues (1970–1984), *Molecular Structure and Dimensions*, Crystallographic Data Center, Cambridge, UK, Vol. 1–15.

Kessler, H. and Molter, M. (1973), *Angew. Chem. Int. Ed. Engl.* **12**, 1011.

King, R. B., Ed. (1983), *Chemical Applications of Topology and Graph Theory*, Elsevier, New York.

Kirkwood, J. G. and Westheimer, F. H. (1938), *J. Chem. Phys.*, **6**, 506.

Klopman, G. and O'Leary, B. (1970), "All Valence Electrons S.C.F. Calculations of Large Organic Molecules," *Top. Curr. Chem.*, **15**, 445.

Klyne, W. and Prelog, V. (1960), "Description of Steric Relationships Across Single Bonds," *Experientia*, **16**, 521.

Kollman, P. (1985), "Theory of Complex Molecular Interactions: Computer Graphics, Distance Geometry, Molecular Mechanics and Quantum Mechanics," *Acc. Chem. Res.*, **18**, 105.

Kollman, P. A. and Merz, K. M. (1990), "Computer Modeling of the Interactions of Complex Molecules," *Acc. Chem. Res.*, **23**, 246.

Kuchitsu, K. (1968), *J. Chem. Phys.*, **49**, 4456.

Ladd, M. F. C. and Palmer, P. A. (1980), *Theory and Practice of Direct Methods in Crystallography*, Plenum Press, New York.

Lambert, J. B. (1971), "Pyramidal Inversion," *Top. Stereochem.*, **6**, 19.

Lathan, W. A., Curtiss, L. A., Hehre, W. J., Lisle, J. B., and Pople, J. A. (1974), "Molecular Orbital Structures for Small Organic Molecules and Cations," *Prog. Phys. Org. Chem.*, **11**, 175.

Lavallee, D. K. and Zeltmann, A. H. (1974), *J. Am. Chem. Soc.*, **96**, 5552.

Lehn, J.-M. (1970), "Nitrogen Inversion," *Top. Curr. Chem.*, **15**, 311.

Lifson, S., Hagler, A. T., and Dauber, P. (1979), *J. Am. Chem. Soc.*, **101**, 5111.

Lii, J.-H. and Allinger, N. L. (1989a), *J. Am. Chem. Soc.*, **111**, 8566.

Lii, J.-H. and Allinger, N. L. (1989b), *J. Am. Chem. Soc.*, **111**, 8576.

Lii, J.-H. and Allinger, N. L. (1991), *J. Comput. Chem.* **12**, 186.

Lii, J.-H. and Allinger, N. L. (1992), *J. Comput. Chem.* **13**, 1138.

Lipscomb, W. N. and Jacobson, R. A. (1972), "X-Ray Crystal Structure Analysis," in Weissberger, A. and Rossiter, B. W., Eds., *Techniques of Chemistry*, Vol. 1, Part IIId, Wiley, New York, p. 1.

Luckhurst, G. R. (1968), "Liquid Crystals as Solvents in Nuclear Magnetic Resonance," *Q. Rev.*, **22**, 179.

Mathews, D. A. and Stucky, G. D. (1973), *J. Am. Chem.*, **93**, 5954.

Mayo, B. C. (1973), "Lanthanide Shift Reagents in Nuclear Magnetic Resonance Spectroscopy," *Chem. Soc. Rev.* **2**, 49.

McCammon, J. A. and Harvey, S. C. (1987), *Dynamics of Proteins and Nucleic Acids*, Cambridge University Press, New York.

Meiboom, S. and Snyder, L. C. (1968), "Nuclear Magnetic Resonance in Liquid Crystals," *Science*, **162**, 1337.

Meiboom, S. and Snyder, L. C. (1970), *J. Chem. Phys.*, **52**, 3857.

Mills, H. H. and Speakman, J. C. (1969), "Crystallography and Stereochemistry," *Prog. Stereochem.*, **4**, 273.

Mizushima, S.-I. (1954), *Structure of Molecules and Internal Rotation*, Academic, New York, p. 7 ff.

Momany, F. A., McGuire, R. F., Burgess, A. W., and Scheraga, H. A. (1975), *J. Phys. Chem.*, **79**, 2361.

Moriarty, R. M. (1974), "Stereochemistry of Cyclobutane and Heterocyclic Analogs," *Top. Stereochem.*, **8**, 271.

Morrill, T. C., Ed. (1986), *Lanthanide Shift Reagents in Stereochemical Analysis*, VCH Publishers, Deerfield Beach, FL.

Náray-Szabó, G., Surján, P. R., and Ángyán, S. G. (1987), *Applied Quantum Chemistry*, Reidel, Boston, MA.

Neuhaus, D. and Williamson, M. P. (1989), *The Nuclear Overhauser Effect in Structural and Conformational Analysis*, VCH Publishers, New York.

Ondetti, M. A. and Cushman, D. W. (1981), *Biopolymers*, **20**, 2001.

Ōsawa, E. and Kanematsu, K. (1986), "Generation of Long Carbon–Carbon Single Bonds in Strained Molecules by Through-Bond Interaction," Liebman, J. E. and Greenberg, A., Eds. *Molecular Structure and Energetics*, Vol. 3, VCH Publishers, Deerfield Beach, FL, p. 329.

Ōsawa, E. and Musso, H. (1982), "Application of Molecular Mechanics Calculations to Organic Chemistry," *Top. Stereochem.*, **13**, 117.

Parr, R. G. (1975), *Proc. Natl. Acad. Sci. USA*, **72**, 763.

Patterson, A. L. (1935), *Z. Kryst.*, **90**, 517.

Pauling, L. (1960), *The Nature of the Chemical Bond*, 3rd ed., Cornell University Press, Ithaca, NY.

Pople, J. A. (1975), *J. Am. Chem. Soc.*, **97**, 5306.

Pullman, A. (1972), "Quantum Biochemistry at the All- or Quasi-All-Electrons Level," *Top. Curr. Chem.*, **31**, 45.

Pullman, B., Courrière, P. C., and Berthod, H. (1974), *J. Med. Chem.*, **17**, 439.

Radom, L. and Pople, J. A. (1972), "Ab initio Molecular Orbital Theory of Organic Molecules," in Brown, W. B., *MTP International Review of Science*, Physical Chemistry Series, 1, Vol. 1, Butterworths, London, p. 71.

Reilley, C. N., Good, B. W., and Desreux, J. F. (1975), *Anal. Chem.*, **47**, 2110.

Robiette, A. G. (1972), "The Interplay between Spectroscopy and Electron Diffraction" in *Molecular Structure by Diffraction Methods*, A Specialist Periodical Report, Chemical Society, London, Vol. 1, p. 161.

Rouvray, D. H. (1975), *J. Chem. Educ.*, **52**, 768.

Sadlej, J. (1985), *Semi-Empirical Methods of Quantum Chemistry*, Wiley, New York.

Sandström, J. (1983), "Static and Dynamic Stereochemistry of Push–pull and Strained Ethylenes," *Top. Stereochem.*, **14**, 83.

Saunders, M. (1963), *Tetrahedron Lett.*, 1699.

Saupe, A. (1968), "Recent Results in the Field of Liquid Crystals," *Angew. Chem. Int. Ed. Engl.*, **7**, 97.

Schäfer, L., Siam, K., Ewbank, J. D., Caminati, W. and Fantoni, A. (1987) "Ab initio Studies of Structural Features not easily amenable to Experiment: Some Surprising Applications of Ab initio Geometries in Microwave Spectroscopic Conformational Analyses," in Maksić, Z. B., Ed., *Modeling of Structure and Properties of Molecules*, Wiley, New York, p. 79.

Schaeffer, H. F. (1986), "Methylene: A Paradigm for Computational Quantum Chemistry" *Science*, **231**, 1100.

Scheraga, H. (1968), "Calculation of Conformations of Polypeptides," *Adv. Phys. Org. Chem.*, **6**, 103.

Schmitz, L. R. and Allinger, N. L. (1990), *J. Am. Chem. Soc.*, **112**, 8307.

Schwarz, W. H. E., Valtazanos, P., and Ruedenberg, K. (1985), *Theor. Chim Acta*, **68**, 471.

Simonetta, M. (1974), "Structural Investigations in Organic Molecules and Crystals by means of Molecular Mechanics and X-ray Diffraction," *Acc. Chem. Res.*, **7**, 345.

Spackman, M. A. and Masler, E. N. (1985), *Acta Cryst. A.*, **39**, 1259.

Speakman, J. C. (1978), "Neutron Diffraction" in *Molecular Structure by Diffraction Methods*, A Specialist Periodical Report, Chemical Society, London, Vol. 6, p. 117.

Sprague, J. T., Tai, J. C., Yuh, Y., and Allinger, N. L. (1987), *J. Comp. Chem.*, **8**, 581.

Stewart, R. F. and Hall, S. R. (1971), "X-Ray Diffraction" in *Determinations of Organic Structure by Physical Methods*, Vol. 3, Nachod, F. C. and Zuckerman, J. J., Eds., Academic, New York, p. 73.

Sullivan, G. R. (1976), *J. Am. Chem. Soc.*, **98**, 7162.

Sutton, L. E., Ed. (1958, 1965), *Tables of Interatomic Distances and Configuration in Molecules and Ions*, Special Publication No. 11, and Supplement No. 18, The Chemical Society, London.

Tomasi, G. J., Alagona, G., Bonaccorsi, R. and Ghio, C. (1987), "A Theoretical Model for Solvation—Some Application to Biological Systems," in Maksić, Z. B., Ed., *Modelling of Structure and Properties of Molecules*, Wiley, New York, p. 330.

Umemoto, K. and Ouchi, K. (1985), "Hindered Internal Rotation and Intermolecular Interactions," in Orville-Thomas, W. J., Ratajczak, H., and Rao, C. N. R., Eds., *Topics in Molecular Interaction*, Elsevier, New York, p. 1.

Walton, A. (1978), *Molecular and Crystal Structure Models* John Wiley & Sons, New York.

Walter, W., Maerten, G. and Rose, H. (1966), *Justus Liebigs Ann. Chem.*, **691**, 25.

Warshel, A. and Lifson, S. (1970), *J. Chem. Phys.*, **53**, 582.

Watenpaugh, K., Dow, J., Jensen, L. H., and Furberg, S. (1968), *Science*, **159**, 206.

Weil, J. A., Blum, A., Heiss, A. H., and Kinnaird, J. K. (1967), *J. Chem. Phys.*, **46**, 3132.

Weiner, P. K. and Kollman, P. A. (1981), *J. Comp. Chem.*, **2**, 287.

Weiner, S. J., Kollman, P. A., Case, D. A., Singh, U. C., Ghio, C., Alagona, G., Profeta, S., and Weiner, P. (1984), *J. Am. Chem. Soc.*, **106**, 765.

Westheimer, F. H. (1956), "Calculation of the Magnitude of Steric Effects," in M. S. Newman, Ed., *Steric Effects in Organic Chemistry*, Wiley, New York, p. 523.

Westheimer, F. H. and Kirkwood, J. G. (1938), *J. Chem. Phys.*, **6**, 513.

Wheland, G. W. (1960), *Advanced Organic Chemistry*, 3rd ed., Wiley, New York, p. 41.

Wilen, S. H., Bunding, K. A., Kascheres, C. M., and Wieder, M. J. (1985), *J. Am. Chem. Soc.*, **107**, 6997.

Willcott, M. R. and Davis, R. E. (1975), "Determination of Molecular Conformation in Solution," *Science*, **190**, 850.

Williams, J. E., Stang, P. J. and Schleyer, P. v. R. (1968), "Physical Organic Chemistry: Quantitative Conformational Analysis; calculation methods," *Ann. Rev. Phys. Chem.*, **19**, 531.

Wilson, E. B., Decius, J. C. and Cross, P. C. (1955), *Molecular Vibrations*, McGraw-Hill Book Co., New York.

Wolkowski, Z. W. (1971), *Tetrahedron Lett.*, 825.

Wüthrich, K. (1986), *NMR of Proteins and Nucleic Acids*, Wiley-Interscience, New York.

Zeil, W. (1974), "Molecular Structure Determination by Electron Diffraction of Gases. Progress and Results," in Hoppe, W. and Mason, R., Eds., *Advances in Structure Research by Diffraction Methods*, Pergamon Press, New York.

3

Stereoisomers

3-1. NATURE OF STEREOISOMERS

a. General

In Chapter 2 isomers were defined as compounds having the same molecular formula but differing in structure. The isomers were subdivided according to whether they differ in constitution, in configuration, and/or in conformation. Isomers differing only in configuration and/or conformation were recognized as stereoisomers.

In this chapter we consider an alternative subdivision of stereoisomers, namely, into enantiomers and diastereomers. Unlike the somewhat fuzzy (cf. Section 2-3) division between configuration and conformation, the dichotomy between enantiomers and diastereomers is unequivocal. Moreover, the two classifications are orthogonal: Enantiomers may differ in configuration or only in conformation; the same is true of diastereomers (diastereoisomers).

Enantiomers are pairs of isomers related as an object is to its mirror image. This relationship may stem from a configurational difference, as in CHFClBr (Fig. 2.8) or (depending on definition, see Chapter 2) in ethylmethylbenzylamine (Fig. 3.1) or from a conformational (sign of torsion angle) difference, as in a tetra-*o*-substituted biphenyl (Fig. 3.2), the gauche forms of 1,2-dichloroethane (Fig. 2.9, **A** and **C**), or the two chair forms of *cis*-1,2-dimethylcyclohexane (Fig. 2.18). The stability to interconversion of the enantiomers, which is high in the case of CHFClBr and appropriately tetra-*o*-substituted biphenyls (Chapter 14) but fleeting in the other three examples, is of no concern in the definition; conceptually, the structures that are compared are considered as being rigid.

> The synonyms *antimer*, *optical antipode*, and *enantiomorph* for enantiomer are found in the older literature. The first term is now archaic and the second rarely used. The term enantiomorph (and the adjective enantiomorphous or enantiomorphic) properly applies to mirror-image objects (e.g., crystals, Fig. 1.1), including chemical ligands.

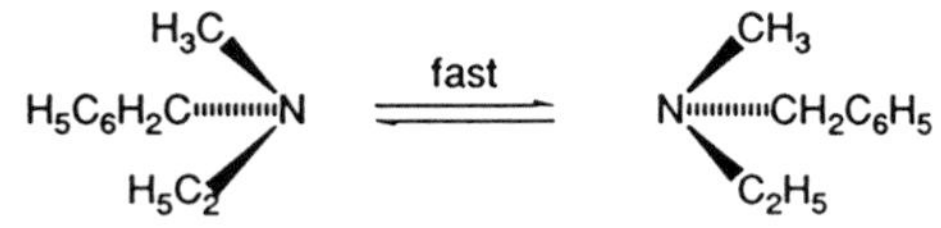

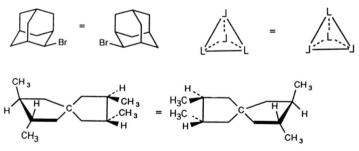

Figure 3.1.   Enantiomers of ethylmethylbenzylamine.

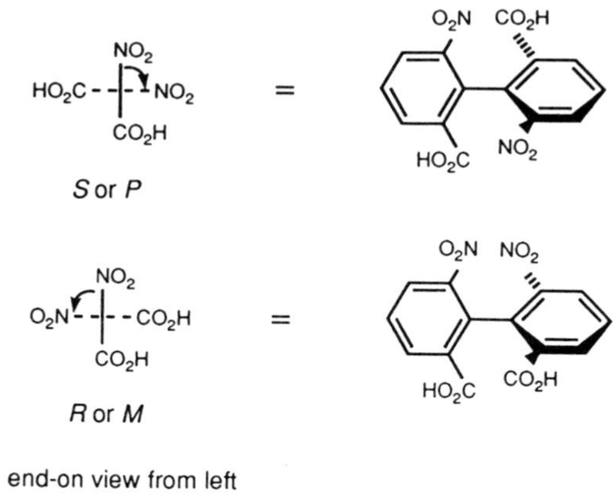

Figure 3.2.   Enantiomers of 6,6′-dinitro-2,2′-diphenic acid.

Enantiomers must be isomers as well as mirror images, that is, they must not be superposable. One can draw many structures (some exemplified in Fig. 3.3) which bear a mirror-image relationship but, upon rotation of the entire model (rigid rotation) around an appropriate axis, turn out to be superposable; such structures are identical (or homomeric, see below) not enantiomeric.

Diastereomers (or diastereoisomers) are stereoisomers (i.e., isomers of identical constitution but differing three-dimensional architecture) which do not bear a mirror-image relation to each other. Diastereoisomerism may be due to differences in configuration (or conformation) at several sites in the molecule, as in the tartaric acids (Fig. 3.4) or in appropriately substituted terphenyls (Fig. 3.5, **A**) or biphenyls (Fig. 3.5, **B**); we are dealing in the first case with configurational differences, in the second case with differences in conformation between the isomers, and in the third case with a combination of the two types. In these particular cases the barriers between the diastereomers are high and each isomer is stable, at least at room temperature. However, this is not essential: The rapidly

Figure 3.3.   Identical mirror-image structures.

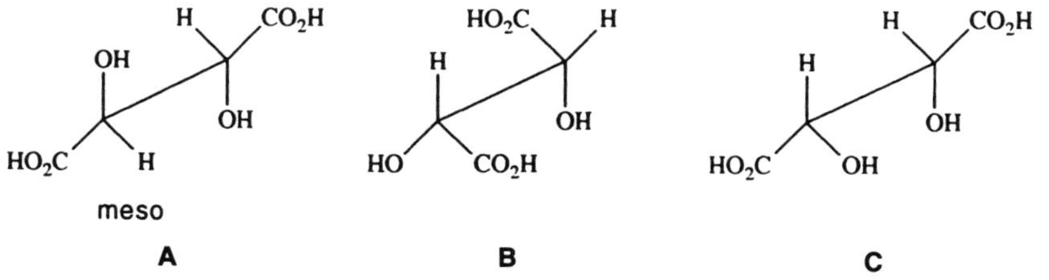

Figure 3.4.   The tartaric acids. Enantiomers are **B**, **C** diastereomers are **A**, **B** and **A**, **C**.

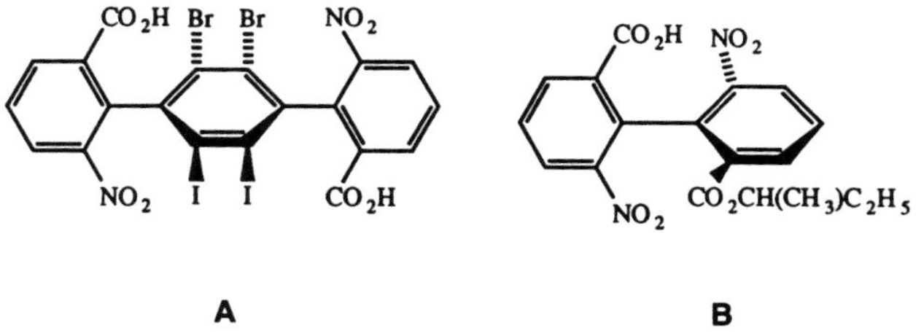

Figure 3.5.   Diastereoisomerism in terphenyls and biphenyls. **A**: Terphenyl capable of existing in diastereomeric forms. **B**: Diastereoisomers generated by combination of a chiral center and a chiral axis.

interconverting isomers of tertiary amines shown in Figure 3.6 and the gauche and anti form of 1,2-dichloroethane (Fig. 2.9, **A** and **B**) are also diastereoisomers.

The latter case is notable in that the difference between the diastereoisomers is in the magnitude of the torsion angle (60° vs. 180°) rather than in a difference of sign of several torsion or bond angles.

Gross differences in the magnitude of bond angles leading to diastereoisomers are seemingly unknown in purely organic compounds but are found in such inorganic species as dichlorodiaminoplatinum (Fig. 3.7) and PCl_2F_3 (Fig. 3.8; Muetterties et al., 1964); in the former case the diastereomers are quite stable, in the latter case they are readily interconverted. Rosenberg et al. (1967; see also Rosenberg, 1971) discovered that the cis but not the trans isomer of $PtCl_2(NH_3)_2$ is a potent anticancer agent (cf. Sherman and Lippard, 1987).

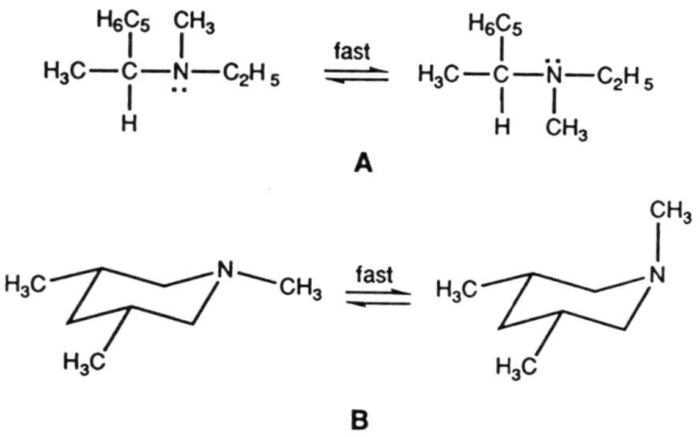

Figure 3.6.   Rapidly interconverting diastereomers.

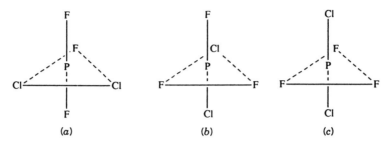

Figure 3.7. Diastereomers of dichlorodiaminoplatinum.

F
|
.F.
P
Cl⟋‾‾‾‾‾‾Cl
|
F
(a)

F
|
Cl
P
F⟋‾‾‾‾‾‾F
|
Cl
(b)

Cl
|
F.
P
F⟋‾‾‾‾‾‾F
|
Cl
(c)

Figure 3.8. Diastereomers of PCl_2F_3.

The terms enantiomer and diastereomer relate to molecules as a whole. Thus, if two molecules have the same constitution (connectivity) but different spatial arrangements of the atoms (i.e., if they are stereoisomeric), they must either be related as mirror images or not: In the former case they are enantiomers, in the latter case they are diastereomers. The differentiation can be made without considering any particular part of the molecule. Nonetheless, in viewing and specifying (cf. Chapter 5) enantiomers and diastereoisomers one often focuses on particular sites in the molecule, such as the carbon atom in CHFClBr or the torsion axis containing the phenyl–phenyl bond in the biphenyls and terphenyls shown in Figures 3.2 and 3.5. In the tradition of van't Hoff, one "factorizes" (cf. Cahn, Ingold, and Prelog, 1966a) stereoisomerism by attributing it to a "chiral center" (cf. Chapter 1) or a torsion axis.

A chiral center (or, more properly, center of chirality – Chiralitätszentrum in German, cf. Cahn, Ingold, and Prelog, 1966b) is a focus of chirality; in the case of carbon, at least, it corresponds to the asymmetric tetrahedral atom of van't Hoff (1874) as shown in Figure 3.9. The existence of enantiomers is usually, but not invariably, associated with at least one chiral center or chiral torsion axis (axis of chirality) as seen in the above examples (exceptions will be discussed in Chapter 14).

Diastereomers, as we have seen, often contain two or more chiral centers (Fig. 3.4), chiral (torsion) axes (Fig. 3.5, **A**), or a combination thereof (Fig. 3.5, **B**). However, this is not necessarily the case; Figures 3.6b and 3.10 illustrate cases where diastereomers are neither chiral nor contain chiral centers; alkene diastereomers (Fig. 3.11) are also of this type. Thus, diastereoisomerism is not

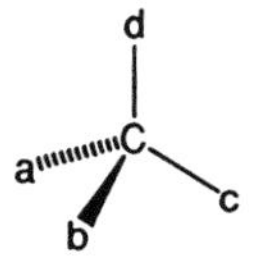

Figure 3.9. Center of chirality.

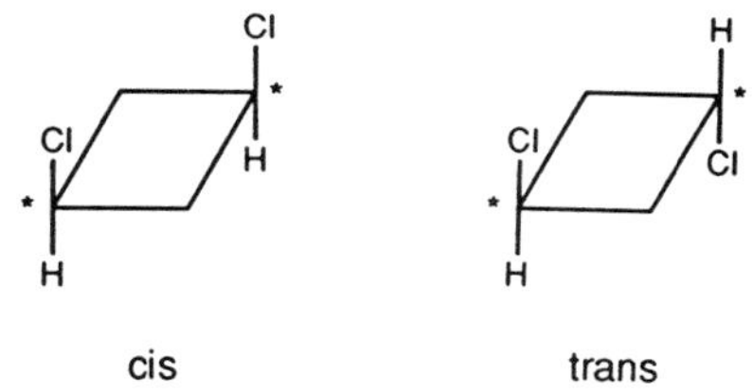

Figure 3.10. Diastereomeric 1,3-dichloro-cyclobutanes.

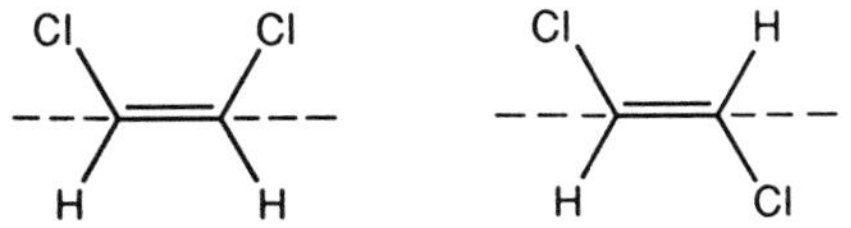

Figure 3.11. Diastereomeric alkenes.

necessarily associated with chiral centers or torsion axes; a general scheme for factorizing diastereoisomerism must go beyond consideration of these chiral elements.

The starred carbon atoms (C*) in Figure 3.10 (which are the foci of the diastereoisomerism but are not chiral, since each bears two identical ligands) have been called "centers of stereoisomerism" (Hirschmann and Hanson, 1971) or "stereogenic centers" (McCasland, 1950; Mislow and Siegel, 1984; cf. also Eliel, 1982). Interchange of two ligands at a stereogenic center leads to a stereoisomer. By extension one might call the axis containing the olefinic double bond (dashed in Fig. 3.11) a "stereogenic axis." The axis containing the phenyl–phenyl bonds in the terphenyl shown in Figure 3.5, **A** is also a stereogenic axis.

Stereogenic centers thus may be or may not be chiral (i.e., centers of chirality). Conversely, however, *all* chiral centers are stereogenic. Both terms (chiral and stereogenic) will be used in this book as appropriate. Mislow and Siegel (1984), who have discussed in depth the subject of stereogenicity as being distinct from chirality have coined another term, *chirotopic.* A point in a molecule (it need not be a material point coincident with an atom) is chirotopic if it is located in a chiral environment. *Any* point in a chiral molecule is chirotopic, but even in an achiral molecule there may be many chirotopic points or atoms. For example, in *meso*-tartaric acid (Fig. 3.4, **A**) all atoms are chirotopic; only the center of symmetry [midway between C(2) and C(3)] is "achirotopic" (i.e., not chirotopic). In general, "all points in a model that remain invariant under a rotation–reflection operation (cf. Chapter 4) are achirotopic"; thus all points on a plane of symmetry in a molecule are achirotopic. Although the concept of chirotopicity may be of limited practical applicability, it serves to stress the fact that van't Hoff's asymmetric atom has two separate aspects: one, that it is a focus of dissymmetry or chirality (chirotopic), the other, that the exchange of two ligands at the asymmetric atom gives rise to stereoisomers (it is stereogenic). With carbon, these two aspects are usually linked, although, as seen in Figure 3.10 and again later in Figure 3.27, this is not always the case. With atoms of other than tetrahedral coordination geometry stereogenicity and chirotopicity are usually not linked at all. Thus the square planar platinum atom in Figure 3.7 and the trigonal bipyramidal phosphorus atom in Figure 3.8 are stereogenic, since an exchange of appropriate ligands (Cl and NH_3 or Cl and F, respectively) leads to diastereoisomers. Yet the central atoms are achirotopic, since they lie on symmetry planes. No enantiomers exist in either $PtCl_2(NH_3)_2$ or PF_3Cl_2 and all the heavy atoms in these structures are achirotopic. The hydrogen atoms in the NH_3 groups may be chirotopic, depending on conformation and time scale of Pt–NH_3 rotation (cf. Section 4-5).

b. Barriers between Stereoisomers. Residual Stereoisomers

In the discussions so far, the question as to whether stereoisomers (of whatever type) can or cannot be isolated or otherwise observed was considered immaterial. Defining stereoisomers in that fashion has advantages and disadvantages. The advantage is that the number of stereoisomers so counted is independent of the method of observation and, generally, of the temperature. Thus chlorocyclohexane has two dominant stereoisomers (Fig. 2.3), one with equatorial chlorine, the other with axial chlorine (the concentration of twist forms is small enough to be disregarded; cf. Chapter 11). With this insight one knows what one might expect upon physical examination of chlorocyclohexane by methods that give an "instantaneous" picture, such as IR spectroscopy or electron diffraction. The disadvantage is that the classification does not correspond to the experience of the preparative chemist who, at or near room temperature, encounters only a single substance called chlorocyclohexane in the laboratory. [This occurs because the rate of interconversion of the isomers (Fig. 2.3) is very high; the isomers are separated by energy barriers of much less than ca. 20 kcal mol^{-1} (84 kJ mol^{-1}) and thus are interconverted too rapidly for isolation at room temperature.]

> A barrier ($\Delta G^{\ddagger}$) of 20 kcal mol^{-1} (84 kJ mol^{-1}) corresponds to an interconversion rate of 1.3×10^{-2} s^{-1} at 25°C, that is, a half-life ($t_{1/2}$) of about 1 min; a barrier of 25 kcal mol^{-1} (105 kJ mol^{-1}) indicates an interconversion rate of 2.9×10^{-6} s^{-1}; $t_{1/2} = 66$ h at 25°C. (In general, $k = 2.084 \times 10^{10} T\, e^{-\Delta G^{\ddagger}/1.986T}$, where $\Delta G^{\ddagger}$ is in calories pre mole; the exponent is $-\Delta G^{\ddagger}/8.315T$ when $\Delta G^{\ddagger}$ is in joules per mole.)

Even in NMR spectroscopy, the question of whether one sees one or two spectra of a substituted cyclohexane depends on temperature and instrumental frequency. At -100°C, the spectra of the axial and equatorial conformers would appear separately, but at room temperature a single (averaged) spectrum would be seen (cf. Chapter 11).

Thus in deciding whether structures are effectively enantiomeric (or diastereomeric) or not, one must make allowance for those stereochemical changes (rotations or inversions) that occur rapidly on the scale of the experiment under consideration (chemical transformation, spectroscopic observation, determination of a physical property). The resulting need for introducing a time scale presents an area of conflict between the definition of stereoisomers given above, which relates to rigid structures, and the more practical view relating to substances as they are encountered in the laboratory. The latter must necessarily include energy considerations, and this affects the isomer count. Several illustrations follow.

Structures **A** and **B** in Figure 3.12 represent stereoisomers of 2-bromobutane. These structures are diastereomeric (i.e., they are neither superposable nor have an object to mirror-image relationship.) However, rotation around the C(2)–C(3) bond, which occurs very rapidly, converts **B** into a diastereomer **C**, which is enantiomeric with **A**. Similarly, **A** can, by rotation, be converted into diastereomer **D**, which is enantiomeric with **B**. Further rotation of **A** leads to another diastereomer **F**, which is enantiomeric with **E**, another diastereomer of **B**.

Thus in 2-bromobutane there are six stereoisomers (**A–F**), which may be grouped in two triads **A–D–F/B–C–E** of enantiomeric pairs. A spectroscopist

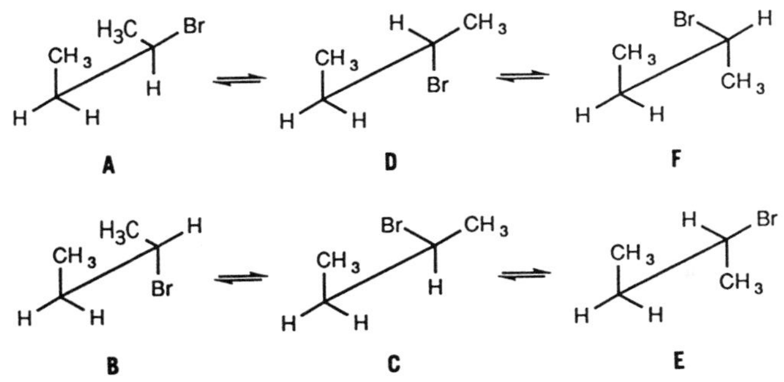

Figure 3.12. 2-Bromobutanes: A complete view.

might, indeed, observe the three diastereomers in these triads spectrally or, in a chiral medium (see Chapter 6) might even see all six stereoisomers. But a preparative chemist, working on the ordinary laboratory time scale at room temperature or slighty above, will find the triads averaged by rotation (just as the diastereoisomers of chlorocyclohexane are averaged by ring reversal) and so encounters only two species, (+)- and (−)-2-bromobutane, which can be separated. Thus, normally 2-bromobutane is considered as existing in only two enantiomeric forms and the planar projection formulas often used (Fig. 3.13) reinforce this view by sweeping the conformational details under the rug. One may say that on the laboratory time scale there are only two so-called residual entantiomers (see below), as shown in Figure 3.13.

   Just as fast rotation reduces the number of isolable species below that actually existing, so does fast inversion. Thus, although there are two enantiomers of ethylmethylbenzylamine (Fig. 3.1), their interconversion is too fast to permit resolution. By the same token the four configurational isomers (Fig. 3.14, **A′**, **A″**, **B′**, and **B″**) of N-methyl-1-phenylethylamine (Fig. 3.14) are reduced to two observable "residual enantiomer" species (**A** and **B**, Fig. 3.14) by rapid inversion at nitrogen.

   A very useful concept introduced by Finocchiaro, Mislow et al. (1973), namely, that of "residual stereoisomerism," has been generalized (Eliel, 1976/77) to cover the above cases. We define as "residual stereoisomers" those subsets of the total set of stereoisomers that can be distinguished under specified conditions by a given technique. Thus 2-bromobutane (Fig. 3.12) has six stereoisomers but [in the limit of fast rotation around the C(2)–C(3) bond that applies to chemical

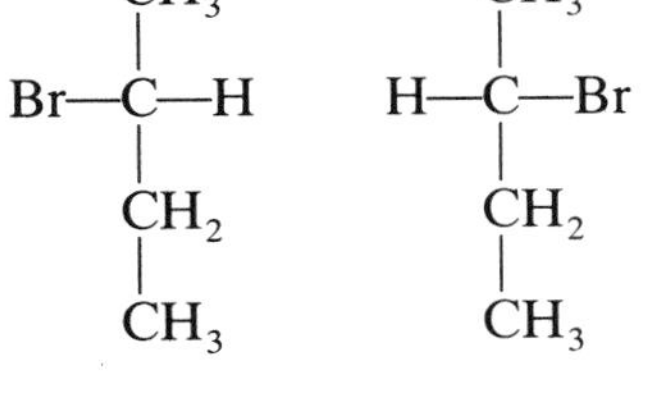

Figure 3.13. 2-Bromobutanes. A laboratory view.

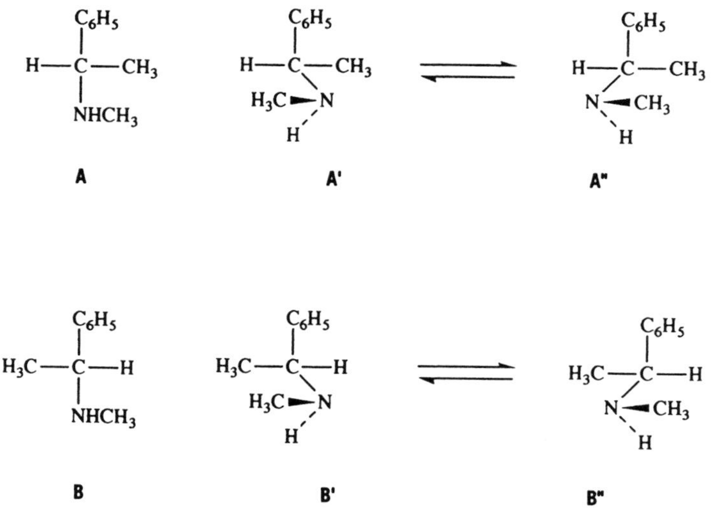

Figure 3.14. N-Methyl-1-phenylethylamine.

manipulations in the laboratory at room temperature] only two residual stereo-isomers with respect to isolation: the (+) and (−) enantiomers (Fig. 3.13). Similarly, N-methyl-1-phenylethylamine (Fig. 3.14) has two such residual stereo-isomers (the two enantiomers differing in configuration at the benzylic carbon). Chlorocyclohexane (Fig. 2.3) displays two stereoisomers (with equatorial and axial chlorine) when viewed by IR or Raman spectroscopy, when handled in the laboratory at −150°C, or when viewed by NMR spectroscopy at −100°C (see Chapter 11). At room temperature, the two stereoisomers can still be seen by vibrational spectroscopy, but as far as NMR or isolation is concerned, there is only a single substance, that is, a single residual species (since there is only one, it cannot be called an isomer).

Thus, unlike stereoisomerism, which is a structural concept independent of the means of detection, residual stereoisomerism depends on the circumstances under which it is determined. As a further example, consider the case of 2,5-dimethylpiperidine (Fig. 3.15). The total number of stereoisomers is 16 (the eight A–H shown in Figure 3.15 and their eight enantiomers); four by virtue of the fact that C(2) and C(5) are chiral centers, each of them giving rise to two conformational isomers through ring inversion or "reversal," and finally a dou-bling again due to the stereochemistry at nitrogen (axial or equatorial hydrogen). To a vibrational spectroscopist making observations in a chiral medium all 16 isomers might have real existence. Eight racemates (Fig. 3.15, rac-A–H) might be detected in a normal coordinate analysis of the IR–Raman spectrum in the vapor, or in an achiral solvent, or in an analysis of the microwave spectrum of the mixture. Nuclear magnetic resonance spectroscopy in a chiral solvent at low temperature would show eight but different residual stereoisomers (I–L and enantiomers) or four racemates (rac-I–L) in an achiral medium, since nitrogen (NH) inversion is so fast as to be averaged out on the NMR time scale. Finally, a chemist working in the laboratory would see two or four (M, N, and enantiomers) residual stereoisomers, depending on whether or not he or she had the means to

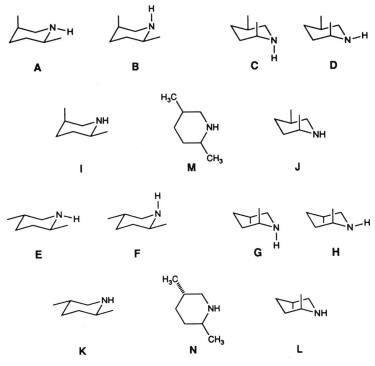

Figure 3.15 Stereoisomers of 2,5-dimethylpiperidine.

resolve enantiomers and observe optical activity. Clearly the number of residual stereoisomers depends on the observational technique used, whereas the total number of stereoisomers does not.

> As discussed elsewhere (Sections 8-4.d and 11-4.g) the barriers to nitrogen inversion are usually far too low to permit isolation of the two stereoisomers produced by such inversion. However, since the inversion of the sp^3 hybridized nitrogen pyramid involves a planar (sp^2 hybridized) transition state, increase of p character of the bonds to nitrogen slows down the inversion. Such an increase may be brought about by a constraint of the angles at nitrogen (as in azetidine and, a fortiori, aziridine where the internuclear angles are near 90° and 60°, respectively) and/or by electronegative substituents on nitrogen, such as chlorine (cf. Lehn, 1970; Jennings and Boyd, 1992). When these two factors cooperate, the barrier may be so high that the two stereoisomers can be isolated. An example is *N*-chloro-2-methylaziridine in which cis and trans isomers (with respect to CH_3 and Cl) have been isolated (Brois, 1968) with a barrier between them of about 27 kcal mol^{-1} (113 kJ mol^{-1}) (Kostyanovsky et al., 1969; Lehn, 1970; Jennings and Boyd, 1992).

Returning to the 2-bromobutane case (Fig. 3.12), we may represent the complete set of six isomers as shown in Figure 3.16. Only under special circumstances (e.g., by IR spectroscopy in a chiral medium) will it be possible to distinguish all six members of the set. Under polarimetric investigation one may discern two subsets: A/D/F and C/B/E (separated by the horizontal line in Figure 3.16)

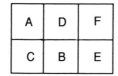

Figure 3.16. 2-Bromobutane isomers.

Figure 3.17. Cycloheptatriene-norcaradiene valence isomerism.

corresponding to the (R)-$(+)$ and (S)-$(-)$ isomers, respectively (cf. Fig. 3.13). These are the two residual stereoisomers, since the conformational isomers cannot be distinguished under the conditions of the experiment. On the other hand, spectroscopic investigation (e.g., by IR) in an achiral medium might disclose the presence of three residual stereoisomers separated, in Figure 3.16, by the vertical lines: conformers A/C, D/B, and F/E. Finally, chemical manipulation in an achiral medium would put into evidence only a single residual 2-bromobutane species (comprising the whole box in Fig. 3.16). This case shows not only that the number of residual isomers depends on the method of observation, but also that different submultiples (3 or 2 as well as 1) of the total number of isomers may appear as residual isomers or species under different circumstances.

In conclusion of this section we point out that the concept of "residual isomerism" is not restricted to stereoisomerism; it applies to constitutional isomerism as well, notably in the context of tautomerism as in the keto and enol forms of ethyl acetoacetate and of fluxional isomerism (p. 17), as exemplified in Figure 3.17 (Ciganek, 1971).

3-2. ENANTIOMERS

Enantiomers are characterized by nonsuperposability and a mirror-image relationship. The absence of superposability may be tested directly in the model or, as further explained in Chapter 4, it may be recognized from symmetry considerations. Molecules having planes, centers, or alternating axes of symmetry are superposable with their mirror images, those lacking all such elements of symmetry are not.

As explained in Section 3-1, one is usually interested in residual enantiomerism, and therefore disregards events that are rapid on the time scale of the experiment. Thus, 1,2-dichloroethane ($ClCH_2CH_2Cl$) is generally considered a single achiral substance: Even though, as shown in Figure 2.9, there are, in principle, two enantiomers and a diastereomer, there is a single residual species when one considers chemical manipulation at room temperature.

As indicated in Section 2-3, the simplest source of chirality in an organic molecule is a carbon atom with four different ligands. A historically much studied example is lactic acid (Fig. 3.18). Although a number of stereoisomers exist by virtue of rotation around the C–OH and C–CO$_2$H bonds there are only two residual enantiomers, which are depicted in Figure 3.18. Both occur in nature. The (+)-lactic acid is found in muscle fluid and results from the biochemical processes occurring when a muscle does work. It is sometimes called "sarcolactic acid" in the old literature (*sarcos* meaning muscle in Greek), melts at 25–26°, and has a specific rotation $[\alpha]_D^{15}$ 3.8 in 10% aqueous solution.

It is perhaps unfortunate that lactic acid was one of the earliest studied chiral substances. In many respects it is badly behaved: it is unstable to heat; its specific rotation diminishes appreciably on dilution and on heating, the salts rotate in the opposite direction from the free acid, it is low melting, and it is difficult to obtain free of water. No wonder that its early history (cf. Fisher, 1975) is turbid!

The (−)-lactic acid is found in certain fermentations of sugars, but when so formed it is usually contaminated with its enantiomer. However, anaerobic fermentation of glucose with *Leishmania brazilensis panamensis* gives enantiomerically pure (−)-lactic acid among other products (Darling, Blum, et al., 1987); the acid melts at 26–27°C, and has the same magnitude of rotation, but in the opposite sense, as the (+) isomer.

Since most properties of matter are invariant to reflection ("scalar properties"), it is to be expected that (+)- and (−)-lactic acid will be identical in many respects. Not only do they have the same melting point, but they also have the same solubility, density, refractive index, IR, Raman, UV, and NMR spectra and, at least very nearly, the same X-ray diffraction pattern. It is only with respect to properties or manipulations that change sign, but not magnitude, upon reflection ("pseudoscalar properties"), that (+)- and (−)-lactic acid differ. This is true for chiroptical properties (cf. Chapter 1), such as optical rotation, optical rotatory dispersion (ORD), circular dichroism (CD), topics to be further pursued in Chapter 13.

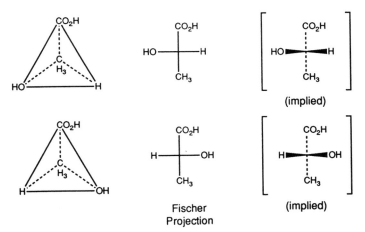

Fischer Projection

Figure 3.18. Lactic acids.

Some fermentative processes generate a mixture of (+)- and (−)-lactic acid in equal amounts, that is, they yield the so-called (±)- or racemic lactic acid that does not rotate polarized light and melts at 18°C (see Chapter 6 for properties of racemates).

The production of racemic modifications (cf. Chapters 5 and 6) in nature is rather unusual. Most chiral natural products are either single enantiomers or, at least, predominantly in one enantiomeric form.

Various representations of the lactic acids are shown in Figure 3.18. [The question as to which model corresponds to the (+) and which to the (−) enantiomer will be taken up in Chapter 5.] The representation of the tetrahedron (with the chiral carbon atom at the center not shown) is cumbersome. The model is depicted by way of an invitation to the readers to build their own. A common representation is the planar projection formula, first proposed by Fischer (1891). It is important to perform the Fischer projection correctly: The atoms pointing sideways must project forward in the model but those pointing up and down in the projection must extend toward the rear. There are 24 (= 4! this being the number of permutations of 4 ligands among 4 sites) ways of writing the projection formula; 12 correspond to one enantiomer and 12 to the other. Twelve of these, corresponding to the (+) enantiomer, are depicted in Figure 3.19; the 12 mirror images of those shown would represent the (−) isomer.

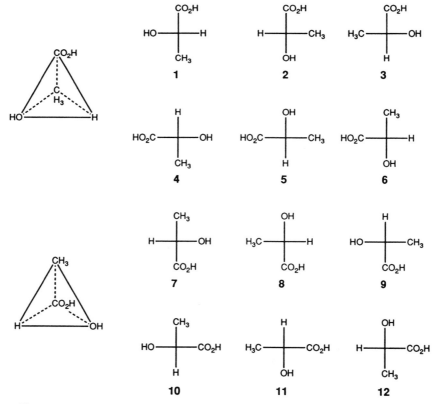

Figure 3.19. Twelve representations of (+)-lactic acid (Fischer projections).

The 12 permutations shown in Figure 3.19 can be generated rather simply, either by permuting groups in threes (e.g., $1 \rightarrow 2 \rightarrow 3$, $3 \rightarrow 4 \rightarrow 7$, or $4 \rightarrow 5 \rightarrow 6$) or by turning the formula by 180° ($1 \rightarrow 7$, $6 \rightarrow 12$, etc.) The former permutation corresponds to holding the model at one of the ligands ($1 \rightarrow 2 \rightarrow 3$ and $4 \rightarrow 5 \rightarrow 6$, CO_2H; $3 \rightarrow 4 \rightarrow 7$, OH) and turning 120° around an axis linking that ligand to the chiral center. The latter change corresponds to turning the model upside down. (The reader is invited to carry out these operations with a model.)

Not permitted are interchanges of two groups (such interchanges clearly lead from the molecule to its enantiomer) and 90° rotations of the formula. (Rotating the model 90° does not yield a projection formula rotated by 90°, since the sideways groups no longer point forward nor do those on top and bottom point backward.) However, two successive exchanges of two groups are permitted as converting the model to its enantiomer and then back to the original stereo-isomer; such double exchanges (e.g., $1 \rightarrow 5$) are, in all cases, equivalent to permutations already shown in Figure 3.19.

If the configuration (or sense of chirality) at the chiral (stereogenic) center is indeed responsible for the direction of optical rotation, then reversing the sense of chirality by chemical manipulation should reverse the rotation. This classical experiment was carried out by Fischer and Brauns in 1914 (q.v.) and is shown in Figure 3.20.

The configurations of the chiral compounds shown in Figure 3.20 are unknown; that is, it is not known which arrangement corresponds to the (+) and (−) isomer. The representations given here represent an enlightened guess (cf. Eliel, 1962, p. 18).

In a modern modification of the Fischer–Brauns experiment, (S)-(mono-methyl 3-hydroxy-3-methylglutarate) (obtained from the dimethyl ester by selec-tive hydrolysis catalyzed by pig liver esterase; see Section 12-4.e) has been converted to both (R)-mevalolactone and (S)-mevalolactone by reducing either the esterified carboxy group (using borohydride or Na/NH_3) or the unesterified carboxyl group (using diborane) to a primary alcohol group that then sponta-neously lactonizes (Huang, Sih, Caspi, et al., 1975). The sequence is shown in Figure 3.21; the diborane reduction requires protection of the tertiary alcohol by acetylation or else the esterified rather than the free carboxyl group is reduced. A somewhat analogous, purely chemical synthesis of the mevalolactone enantiomers from a common chiral, nonracemic precursor has been described by Frye and Eliel (1985).

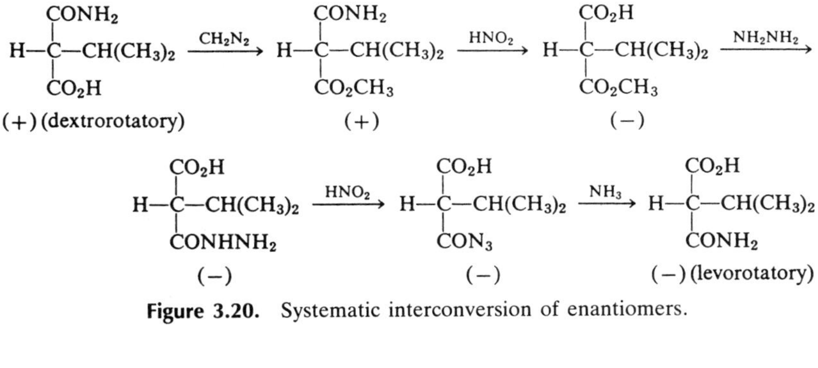

Figure 3.20. Systematic interconversion of enantiomers.

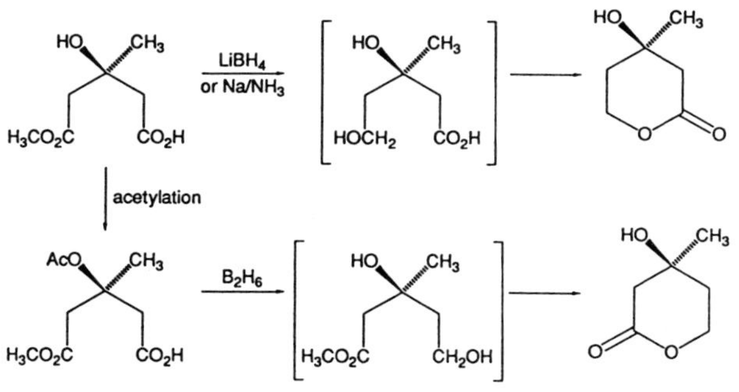

Figure 3.21. Reduction of (S)-$(+)$-(methyl 3-methyl-3-hydroxyglutarate) to (R)-$(-)$- and (S)-$(+)$-mevalolactone.

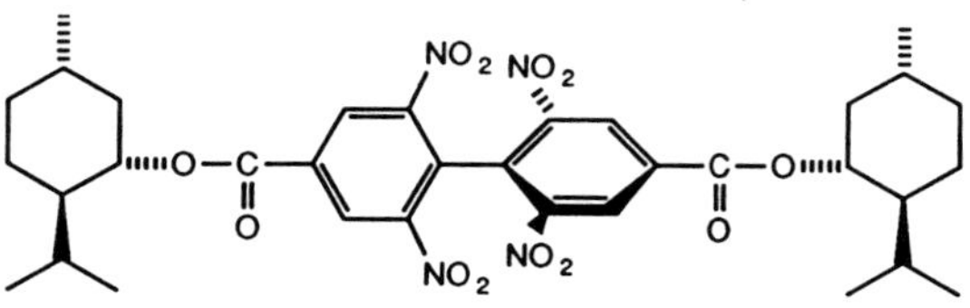

Figure 3.22. $(-)$-Methyl $(+)$-menthyl 2,2′,6,6′-tetranitro-4,4′-diphenate.

On rare occasions one encounters a compound that lacks the usual symmetry elements and is yet achiral, that is, superposable with its mirror image. The $(-)$-menthyl $(+)$-menthyl 2,2′,6,6′-tetranitro-4,4′-diphenate (Fig. 3.22) is such a substance (Mislow and Bolstad, 1955). Since the biphenyl moiety swivels easily between the carboxyl groups (even though the two phenyl groups cannot rotate freely with respect to each other), it is clear that the molecule is superposable with its mirror image on the usual laboratory time scale, that is, there is no residual enantiomerism. Yet there are no discernible symmetry elements in the molecule in any possible conformation. This molecule thus presents a case where the symmetry of the interconverting (rotating) system is higher than that of any of the contributing static arrangements (cf. Trindle and Bouman, 1974). We shall return to this point in Chapter 4.

3-3. DIASTEREOMERS

a. General Cases

Diastereomers are stereoisomers that are not related as object and mirror image. Whereas a set of enantiomers can contain only two members, there is no such limitation for diastereomers. Perusal of Section 3-1 indicates that the numbers of diastereoisomers corresponding to any given constitution may be quite large, since rotation about *each* bond in a molecule is apt to give rise to several energy minima. Thus, if there are n bonds and m minima for each, the total number of

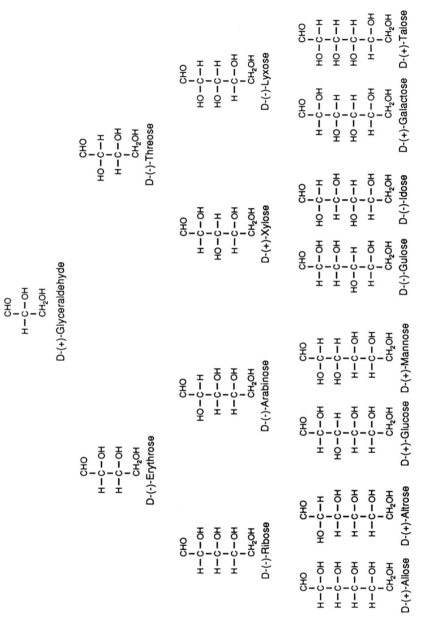

Figure 3.23. The aldose sugars.

diastereomers may be as large as $n \cdot m$, even if there are no stereogenic centers at all. (In practice, the matter is complicated because the number of minima may vary from bond to bond, degeneracies may occur, some conformations may be excluded on steric grounds and because, in overall achiral structures, some conformational isomers are enantiomeric rather than diastereomeric (cf. Fig. 2.9).

Although in polymer chemistry one has to cope with situations of this kind (cf. Flory, 1969) we shall restrict ourselves here to the residual diastereoisomerism on a time scale where most bond rotations are fast. Under such circumstances, the sources of stereoisomerism are usually either stereogenic centers or stereogenic axes, including double bonds. Diastereoisomerism of alkenes will be dealt with in Chapter 9; here we discuss molecules containing more than one stereogenic center. (In most cases, these will be chiral centers; see however, Fig. 3.10.)

Since each stereogenic center can exist in one or the other of two configurations, there will be two stereoisomers for a single stereogenic center, $2 \cdot 2$ or 2^2 combinations for two such centers, $2 \cdot 2 \cdot 2$ or 2^3 for three centers and, in general, 2^n for n centers. Since each stereoisomer has its enantiomer, the 2^n stereoisomers will exist as 2^{n-1} pairs of enantiomers, with each pair of enantiomers being diastereomeric with any other pair.

The aldose sugars (open-chain forms) illustrate this progression ($n = 1-4$). Glyceraldehyde, the simplest sugar, exists as a pair of enantiomers: for the tetroses, there are two enantiomer pairs; for the pentoses, four; for the hexoses, eight; and so on. Sugar chemists assign different names to the (diastereomeric) enantiomer pairs; within each pair they distinguish the enantiomers by giving the prefix D to the isomer that has the hydroxyl group on the right in the Fischer projection formula at the highest numbered chiral carbon atom. (The other enantiomer is given the prefix L.) In Figure 3.23 the D isomers of the aldoses are represented; the L enantiomers (not shown) would be generated by mirror imaging each structure (or, equivalently, exchanging H and OH at each chiral center). In interpreting the Fischer projection formulas the reader must keep in mind the points made in Section 3-2; in particular, the sideways groups all point forward, whereas the "backbone" of the sugars (the sequence of carbon atoms

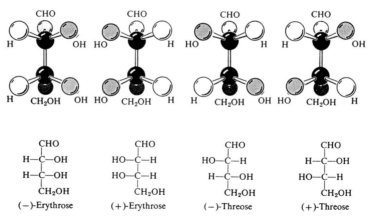

Figure 3.24. Projection formulas of the aldotetroses.

written vertically) curves backward into the plane of the paper (in horseshoe fashion) both at the top and at the bottom of the structure. The situation for the aldotetroses is shown in both three dimensions and Fischer projection formulas in Figure 3.24; it is customary to write the most highly oxidized (CHO) carbon atom (No. 1 in IUPAC numbering) at the top.

It was pointed out in Section 3-2 that enantiomers are identical in all scalar properties (unless examined in a chiral environment; cf. Chapter 6). In contrast, diastereomers differ in most, if not all, physical and chemical properties; in fact, diastereomers tend to be as different from each other as many constitutional isomers. The basic reason for this difference is that enantiomers are "isometric" (Mislow, 1977); that is, for each distance between two given atoms (whether bonded or not) in one isomer there is a corresponding identical distance in the other. No such "isometry" exists in diastereomers or in constitutional isomers.

> An *isometry* is a transformation that preserves the lengths of all line segments between all pairs of points, that is, a transformation that preserves the size and shape of a figure (Coxeter, 1989) but not necessarily its handedness. Mislow (1977) calls two figures "isometric" if they can be related by an isometry; "anisometric" if they cannot be so related. Enantiomers may thus be defined as isometric nonsuperposable molecules; isometric structures are either superposable or enantiomeric.

The practical consequences of the presence or absence of isometry are best seen in an example, the cyclopentane-1,2-diols shown in Figure 3.25. The two trans enantiomers (**A, B**) are isometric; as a result, the distances between the hydroxyl groups in the two are exactly the same. Neither isomer undergoes intramolecular hydrogen bonding (Kuhn, 1952), since the hydroxyl groups are too far from each other, and both are oxidized by lead tetraacetate at the same slow rate (Criegee et al., 1940; cf. Chapter 11). In contrast, the cis diastereomer **C**, in which the hydroxyl groups are much closer to each other, and which is thus not isometric with **A** and **B** forms an intramolecular hydrogen bond and is oxidized by lead tetraacetate over 3000 times as fast as **A** or **B**.

b. Degenerate Cases

In the instances discussed so far, the chiral elements were distinct. Degeneracies occur when this is not the case. The cyclopentanediols (Fig. 3.25) constitute an example: diastereomer **C** is (on the laboratory time scale, when rotation about the C–O bonds is fast) achiral; hence, instead of four stereoisomers (2^2) there are only three. A similar example involving two identical chiral centers is tartaric acid

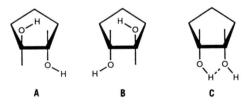

Figure 3.25. The cyclopentane-1,2-diols.

(Fig. 3.26). In this case also one of the diastereoisomeric sets contains two enantiomers, whereas the other contains a single, achiral, isomer. Instead of four isomers there are again only three. The achiral diastereomer is called a meso form, defined as an achiral member of a set of diastereomers that also contains chiral members. [Thus the term "meso form" is not applied to the 1,3-dichlorocyclobutanes shown in Figure 3.10, nor to the 1,2-dichloroethylenes in Figure 3.11, since there are no chiral members in these sets.]

> It might appear (Fig. 3.26) that *meso*-tartaric acid has a plane of symmetry. However, the (eclipsed) Fischer formulas do not ordinarily represent stable conformations; rather, *meso*-tartaric acid exists in the staggered conformations shown in Figure 3.4. Thus, in one of the conformers, there is a center of symmetry; this conformer is achiral. [The other staggered conformers probably also contribute, but since they are enantiomeric and equally populated, they constitute a racemate: though each is chiral, there is no residual chirality just as there is no residual chirality in 1,2-dichloroethane (cf. Figure 2-9 and earlier discussions).]

As expected, *meso*-tartaric acid differs form its chiral diastereomers in physical properties; it melts at 140°C [the (+) and (−) acids melt at 170°C] and is less dense, less soluble in water, and is a weaker acid than the active forms. [The (+) and (−) isomers are, of course, identical to each other in all physical properties, other than chiroptical ones such as specific rotation (cf. Chapter 13)].

The acids obtained by oxidation of the aldopentose sugars (trihydroxyglutaric acids) constitute a somewhat more complex case (see Fig. 3.27). In this set of isomers there are two meso forms and one pair of enantiomers. Carbon (3) in these compounds is of particular interest. In the chiral members of the set, C(3) is not a stereogenic center since its two chiral ligands ($CHOHCO_2H$) are homomorphic (identical). [The reader will realize that transposing H and OH at C(3) in either enantiomer leads to an identical structure; this is best seen by turning the new structure by 180°, which is an allowed operation (see Fig. 3.19).] These structures are of the type CL^+L^+XY. In the achiral members of the set (type CL^+L^-XY or $CЈLXY$), permuting the ligands at C(3) leads from one diastereomer to the other, both diastereomers being meso forms (but see Nourse,

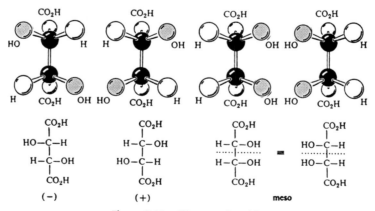

Figure 3.26. The tartaric acids.

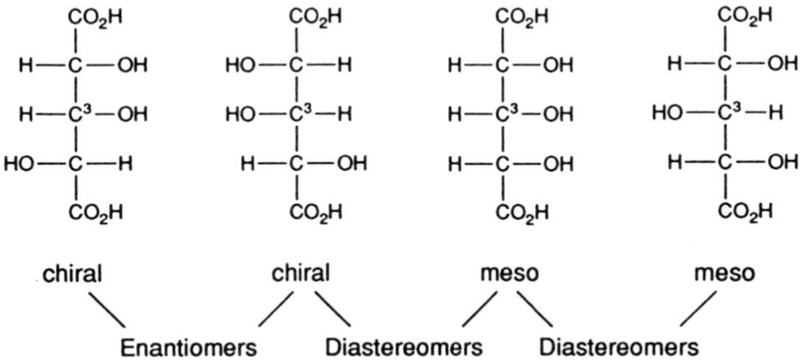

Figure 3.27. The trihydroxyglutaric acids.

1975). Carbon (3) is a center of stereoisomerism (stereogenic center), but it is not a chiral center. It has been called a "pseudoasymmetric" center.

> The concept of pseudoasymmetry in CXYLJ has been discussed at length by Landolt (1899); unfortunately the English translation, in contrast to the original German, (cf. Prelog and Helmchen, 1972) does not contain the term as such. The term was again used by Werner (1904) and has been extensively discussed in the recent literature (Prelog and Helmchen, 1972; Hirschmann and Hanson, 1971, 1974; Nourse, 1975; Prelog and Helmchen, 1982); use of the term "pseudochirality" (Nourse, 1975) is not recommended. Mislow and Siegel (1984) have pointed out that the pseudoasymmetric C(3) in the achiral diastereomers in Figure 3.27 is stereogenic (interchange of two of its ligands lead from one meso form to the other) but *not* chirotopic (since it lies on a local symmetry plane).

The case of the tetrahydroxyadipic or hexaric acids is depicted in Figure 3.28. Carbon (2) is constitutionally equivalent to C(5) and C(3) is constitutionally equivalent to C(4). This case is therefore sometimes said to be of the "ABBA" type (the trihydroxyglutaric or pentaric acid case, by the same notation, is "ABA," the tartaric acid case "AA;" the use of achiral letters here implies achiral ligands but not necessarily absence of chirality). The reader may work out that the hexaric acids exist as four racemic pairs and two meso forms; there are no pseudoasymmetric carbon atoms. The next higher homologue (heptaric acid) is of the "ABCBA" type and presents six racemic pairs and four meso isomers; here, in two of the six isomers that generate racemic pairs C(4) is not stereogenic, but chirotopic, whereas in all the meso isomers it is pseudoasymmetric (i.e., achirotopic, yet stereogenic). In general, in the degenerate cases with an even number n of chiral centers, the number of chiral stereoisomers is 2^{n-1} (i.e., the number of enantiomer pairs is 2^{n-2}) and the number of meso forms is $2^{(n-2)/2}$ (Landolt, 1899). When the number of like chiral centers m is odd, the number of chiral

$$HO_2C\text{---}CHOH\text{---}CHOH\text{---}CHOH\text{---}CHOH\text{---}CO_2H$$

Carbon No.   2   3   4   5

Figure 3.28. Tetrahydroxyadipic or hexaric acids.

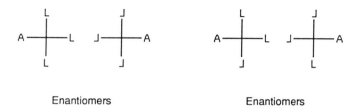

Figure 3.29. The CAL$_3$ case.

stereoisomers is $2^{(m-1)} - 2^{(m-1)/2}$ (corresponding to $2^{m-2} - 2^{(m-3)/2}$ enantiomer pairs) and the number of meso forms $2^{(m-1)/2}$ (Landolt, 1899).

The above formulas apply only to cases where the chiral centers are arranged in a straight chain. The more general cases with chiral centers in branches have been treated by Senior (1927; see also Nourse, 1975). A few specific cases are shown here for illustration. The CAL$_3$ case (A achiral group, L chiral substituent) shown in Figure 3.29 has two pairs of enantiomers, the CAL$_2$G case (G is also a chiral ligand) in Figure 3.30 has four pairs of enantiomers and the CL$_4$ case (Fig. 3.31) has two pairs of enantiomers and one meso form. The CAL$_2$G case is interesting because in the two pairs of enantiomers (G, L chiral ligands) (Fig. 3.30, **E, F** and **G, H**), the central carbon is stereogenic (Hirschmann and Hanson, 1971). Interchange of L and ⌐ leads from **E** to **G**, which is neither identical nor enantiomeric (but diastereomeric) to **E**. The same is true, of course, for the corresponding enantiomers **F** and **H**. A controversy has arisen as to whether such a center should be called chiral (Prelog and Helmchen, 1972) or pseudo-asymmetric (Hirschmann and Hanson, 1971, 1974); there is, of course, no doubt that the structure as a whole is chiral and that the central carbon atom is both stereogenic and chirotopic which, in our view, qualifies it as a chiral center.

The CL$_4$ case encompasses a meso form (Fig. 3.31, **S$_4$**) in the set of diastereomers. By exchanging two pairs of ligands (cf. Fig. 3.19 and accompanying discussion) the reader can easily discern that the projection of molecule **E** shown is superposable with the projection of the mirror image. Yet the molecule

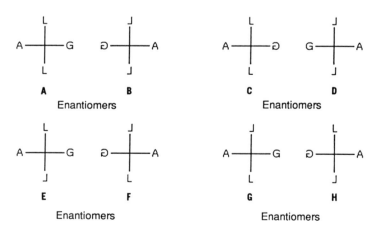

Figure 3.30. The CAL$_2$G case.

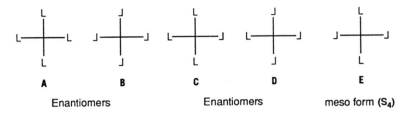

Figure 3.31. The CL_4 case.

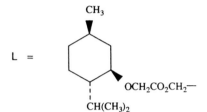

L =

$CH(CH_3)_2$ **Figure 3.32.** Menthoxyacetylcarbinyl ligands.

possesses no plane of symmetry; it does, however, have a fourfold alternating axis of symmetry (S_4), which is a sufficient condition for the absence of chirality (cf. Chapter 4). An actual example of a molecule of the type shown in Figure 3.31 (and which, indeed, was found to be achiral) is the di-(+)-menthoxyacetyl-di-(−)-menthoxyacetyl derivative of pentaerythritol (McCasland et al., 1959); the ligands L are those shown in Figure 3.32.

We shall return to the stereoisomerism of cyclanes (cf. Figs. 3.10 and 3.25) in Chapter 11 (Section 11-1).

REFERENCES

Brois, S. J. (1968), *J. Am. Chem. Soc.*, **90**, 508 1680.

Cahn, R. S., Ingold, C. K., and Prelog, V. (1966a), *Angew. Chem. Int. Ed. Engl.*, **5**, 385.

Cahn, R. S., Ingold, C. K., and Prelog, V. (1966b), *Angew. Chem.*, **78**, 413.

Ciganek, E. (1971), *J. Am. Chem. Soc.*, **93**, 2207.

Coxeter, H. S. (1989), *Introduction to Geometry*, 2nd ed., Wiley, New York.

Criegee, R., Büchner, E., and Walther, W. (1940), *Ber. Dtsch. Chem. Ges.* **73B**, 571.

Darling, T. N., Davis, D. G., London, R. E., and Blum, J. J. (1987), *Proc. Natl. Acad. Sci. USA*, **84**, 7129.

Eliel, E. L. (1962), *Stereochemistry of Carbon Compounds*, McGraw-Hill Book Co., New York.

Eliel, E. L. (1976/77), "On the concept of isomerism," *Isr. J. Chem.*, **15**, 7.

Eliel, E. L. (1982), "Prostereoisomerism (Prochirality)," *Top. Curr. Chem.*, **105**, 1.

Finocchiaro, P., Gust, D., and Mislow, K. (1973), *J. Am. Chem. Soc.*, **95**, 8172.

Fischer, E. (1891), *Ber. Dtsch. Chem. Ges.*, **24**, 2683.

Fischer, E. and Brauns, F. (1914), *Ber. Dtsch. Chem. Ges.*, **47**, 3181.

Fisher, N.W. (1975), "Wislicenus and Lactic Acid: The Chemical Background to van't Hoff's Hypothesis," in *van't Hoff–Le Bel Centennial*, Ramsay, B., Ed., ACS Symposium Series 12, American Chemical Society, Washington, DC.

Flory, P. J. (1969), *Statistical Mechanics of Chain Molecules*, Interscience-Wiley, New York.

Frye, S. V. and Eliel, E. L. (1985), *J. Org. Chem.*, **50**, 3402.

Hirschmann, H. and Hanson, K. R. (1971), *J. Org. Chem.*, **36**, 3293.

Hirschmann, H. and Hanson, K. R. (1974), *Tetrahedron*, **30**, 3649.

Huang, F.-C., Lee, L. F. H., Mittal, R. S. D., Ravikumar, P. R., Chan, J. A., Sih, C. J., Caspi, E., and Eck, C. R., (1975), *J. Am. Chem. Soc.*, **97**, 4144.

Jennings, W. B. and Boyd, D. R. (1992), "Strained Rings," in Lambert, J. B. and Takeuchi, Y., Eds., *Cyclic Organonitrogen Stereodynamics*, VCH Publishers, New York, p. 105.

Kostyanovsky, R.G., Samojlova, Z. E., and Tchervin, I. I. (1969), *Tetrahedron Lett.*, 719.

Kuhn, L. P. (1952), *J. Am. Chem. Soc.*, **74**, 2492.

Landolt, H. (1899), *Optical Activity and Chemical Composition* (Engl. transl.), Whittaker & Co., New York.

Lehn, J. M. (1970), "Nitrogen Inversion," *Top. Curr. Chem.*, **15**, 311.

McCasland, G. E. (1950), *A New General System for the Naming of Stereoisomers*, Chemical Abstracts Service, Columbus, OH.

McCasland, G. E., Horvat, R., and Roth, M. R. (1959), *J. Am. Chem. Soc.*, **81**, 2399.

Mislow, K. (1977), *Bull. Soc. Chim. Belg.*, **86**, 595.

Mislow, K. and Bolstad, R. (1955), *J. Am. Chem. Soc.* **77**, 6712.

Mislow, K. and Siegel, J. (1984), *J. Am. Chem. Soc.*, **106**, 3319.

Muetterties, E. L., Mahler, W., Packer, K. J., and Schmutzler, R. (1964), *Inorg. Chem.*, **3**, 1298.

Nourse, J. G. (1975), *J. Am. Chem. Soc.*, **97**, 4594.

Prelog, V. and Helmchen, G. (1972), *Helv. Chim. Acta*, **55**, 2581.

Prelog, V. and Helmchen, G. (1982), "Basic Principles of the CIP System and Proposals for a Revision," *Angew. Chem. Int. Ed. Engl.*, **21**, 567.

Rosenberg, B. (1971), *Platinum Metals Rev.*, **15**, 42.

Rosenberg, B., Van Camp, L., Grimley, E. B., and Thomson, A. J. (1967), *J. Biol. Chem.*, **242**, 1347.

Senior, J. K. (1927), *Ber. Dtsch. Chem. Ges.*, **60B**, 73.

Sherman, S. E. and Lippard, S. J. (1987), *Chem. Rev.*, **87**, 1153.

Trindle, C. and Bouman, T. D. (1974), "A Group Theory for Fluxional Systems," in Bergmann, E. D., and Pullman, B., Eds., *Chemical and Biochemical Reactivity*, Reidel, Boston, MA, p. 51 ff.

van't Hoff, J. H. (1874), *Arch. Neerl. Sci. Exactes Nat.*, **9**, 445; (Engl. Transl.) Richardson, G. M. (1901), *The Foundation of Stereochemistry*, American Book Co., New York, p. 37.

Werner, A. (1904), *Lehrbuch der Stereochemie*, Verl. Gustav Fischer, Jena, Germany.

4

Symmetry

4-1. INTRODUCTION

Symmetry is an esthetically pleasing attribute of objects found in architecture and in various forms of art as well as in the realm of nature. It also plays an important part in science: in molecular spectroscopy, in quantum mechanics as well as (in the present context) in the determination of structure and the understanding of stereochemistry. The essence of symmetry is the regular recurrence of certain patterns within an object or structure (see Orchin and Jaffe, 1970, 1971; Donaldson and Ross, 1972; Shubnikov and Koptsik, 1974; Kettle, 1985; Hargittai and Hargittai, 1987; Heilbronner and Dunitz, 1993).

As pointed out in Chapter 2, it is often convenient to think of molecules as idealized static entities that can be represented by rigid mechanical models. The symmetry relationships to be considered in this chapter will generally refer to such ideal molecules or molecular models (but see Section 4-5). One must also keep in mind that, since molecules are three-dimensional (3D), in general only 3D representations will be completely adequate as models. Thus the tetrahedron representing (+)-lactic acid or its perspective drawing (Fig. 3.19) is a proper model, but the Fischer projection (without the appropriate specifications as to which groups are in front and which are in the back) is not; a novice contemplating the formulas of (+)- and (−)-lactic acids in their Fischer projections might come to the erroneous conclusion that these structures have a plane of symmetry and are superposable (as they would be if carbon were square planar rather than tetrahedral).

4-2. SYMMETRY ELEMENTS

Symmetry elements (not to be confused with the elements in a set) are the operators that generate the repeat pattern of symmetry. In finite objects they are the simple or proper axis of symmetry C_n, the plane of symmetry σ, the center of

71

symmetry i, and the alternating or improper axis of symmetry S_n (Orchin and Jaffe, 1970a).

In objects of essentially infinite extent (on the molecular scale), such as crystals and polymer molecules, there is an additional element of translational symmetry which, in conjunction with reflection, may give rise to glide planes and, in conjunction with rotation, to screw axes (cf. Donaldson and Ross, 1972, pp. 80–82).

The axis (or simple axis) of symmetry, of multiplicity n, also called the "n-fold axis" and denoted by the symbol C_n is an axis such that if one rotates the model (or molecule) around the axis by $360°/n$, the new position of the model is superposable with the original one. Examples are *cis*-($1R,3S$)-*sec*-butylcyclobutane (Fig. 4.1, **A**), which has a twofold (C_2) axis and *r*-1,*c*-2,*c*-3,*c*-4-($1R,2R,3R,4R$)-tetra-*sec*-butylcyclobutane (Fig. 4.1, **B**), which has a fourfold axis of symmetry (C_4). (Regarding nomenclature, cf. Chapters 5 and 10: *c* stands for *cis*, *t* for *trans*, and *r* for reference group.)

It is clear that any figure or model, turned 360° around any axis will be superposable upon itself. The C_1 symmetry axis is thus a universal (and therefore trivial) symmetry element; it is equivalent to the "identity operation," (E or I) in group theory.

Rotation is a "real" operation in the sense that the points that are brought into superposition are actual, material points. Such an operation is called a "symmetry operation of the first kind" or "proper operation" (in contrast to the rotation–reflection operations discussed below) (cf. Kettle, 1985, p. 10). The presence of a symmetry axis does not preclude chirality; indeed, both of the molecules shown in Figure 4.1 are chiral. Hence, chirality cannot be equated with asymmetry (i.e., the total absence of symmetry); in the older literature the word "dissymmetry" is often used as a synonym for what we now call chirality (cf. Chapter 1).

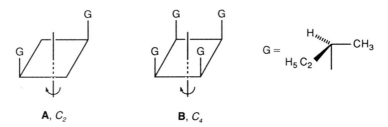

The (chiral) letter G stands for a chiral ligand, in this case a *sec*-butyl group. The letter Ɔ will be used in Figures 4.2–4.5 as being enantiomorphous to G.

Figure 4.1. Examples of molecules with C_2 and C_4 symmetry axes.

Pasteur was well aware of the difference between dissymmetry and asymmetry as evidenced by the French title of his 1860 lecture "Recherches sur la Dissymmétrie Moléculaire des Produits Organiques Naturels" (Jacques, 1986, p. 47). Unfortunately, this was translated into English as "Researches on the Molecular

Asymmetry of Natural Organic Products" (Pasteur, 1905) and the word dis-symmetry in the sense of what we now call chirality seems to have been lost to the English language until it was revived by Jaeger over 50 years later (cf. Jaeger, 1920, pp. 108, 245; see also Wheland, 1960, p. 216; O'Loane, 1980).

Planes, centers, and alternating axes of symmetry are elements corresponding to "symmetry operations of the second kind" or "improper operations" because, rather than bringing into coincidence material points, they bring into coincidence one material point with the reflection of another, a coincidence that might be termed virtual rather than real.

A plane of symmetry σ is a reflection plane by the operation of which each part of a model (each atom of a molecule) is brought into coincidence with a like part (or atom), located elsewhere in the model (or molecule). The compound *cis*-(1*R*,3*S*)-di-*sec*-butylcyclobutane (Fig. 4.2) provides an example; the plane shown is a plane of symmetry.

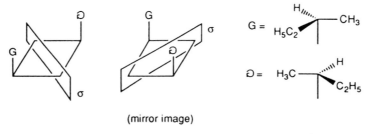

(mirror image)

Figure 4.2. Example of a molecule with a symmetry plane.

The center of symmetry i is a point such that if a line is drawn from any part of the model (or atom in a molecule) to that point and extended an equal distance on the other side, a like part (or atom) is encountered. An example of a molecule with a center of symmetry is *trans*-(1*R*,3*S*)-di-*sec*-butylcyclobutane (Fig. 4.3): The center of the four-membered ring is the center of symmetry, sometimes also called "point of inversion." If the point of inversion is taken as the origin of a Cartesian coordinate system, the effect of the inversion operation is to transpose a point of coordinates x, y, z into a point of coordinates $-x, -y, -z$.

Simple reflection (σ) changes one coordinate but not the other two (e.g., x, y, z to $-x, y, z$). Twofold rotation (C_2) changes two of the three coordinates (e.g., x, y, z to $-x, -y, z$).

The remaining symmetry element is the rotation–reflection or alternating axis of symmetry of order n, S_n. This is an axis such that if the model is turned $360°/n$

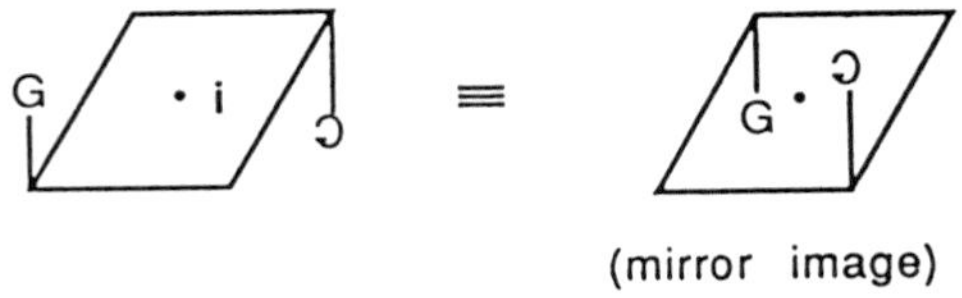

(mirror image)

Figure 4.3. Example of a molecule with a center of symmetry. See Figures 4.1, 4.2 for explanation of G and Ɔ.

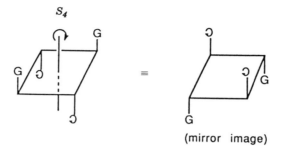

(mirror image)

Figure 4.4. Example of a molecule with a fourfold alternating axis of symmetry S_4. See Figures 4.1, 4.2 for explanation of G and Ɔ.

about the axis and then reflected across a plane perpendicular to that axis, each part of the original model encounters an equivalent part in the rotated–reflected model (and likewise for atoms in a molecule). An example with an S_4 axis [r-1,t-2,c-3,t-4-($1R,2S,3R,4S$)-tetra-*sec*-butylcyclobutane] is displayed in Figure 4.4. Another example of a molecule with an S_4 axis is shown in Figure 3.31.

A 360° rotation–reflection operation about an S_n axis is necessarily equivalent to the identity (i.e., leads back to the original model) only if n is even. In that case, there is also a simple $C_{n/2}$ axis contained within the S_n axis. When n is odd, a single 360° turn will not necessarily lead back to the original molecule, but it may do so in special circumstances, for example, in the structure of PCl_3F_2 shown in Figure 3-8c. However, this molecule contains other symmetry elements (planes and simple axes) and in fact the symmetry element S_n (where n is odd) can exist only in conjunction with such other elements (cf. Section 4-3): S_n (n = odd) cannot be the sole symmetry element in a group.

An S_2 axis corresponds to a point of inversion ($S_2 = i$) located at the intersection of the axis and reflection plane and an S_1 axis is equivalent to a plane of symmetry ($S_1 = \sigma$). It can be proved in group theory that any structure that possesses a plane, center, or alternating axis of symmetry can be superposed with its mirror image, that is, such a structure is achiral. (By way of an exercise, the reader should note that structures shown in Figures 4.2–4.4 are superposable with their mirror images. Some of the structures may have to be rotated in space to make this evident.) It follows, as a corollary, that the presence of an S_n axis in a molecule is a necessary and sufficient condition for absence of chirality.

> However, molecules whose *average* symmetry is greater than the symmetry of any of their rigid representations may be achiral even in the absence of an S_n axis. The structure shown in Figure 3.22 is an example. See the discussion in Section 4-5.

4-3. SYMMETRY OPERATORS. SYMMETRY POINT GROUPS

A symmetry operation is an operation that carries a molecule (or a model, or, for that matter, any figure or set of points) into a position indistinguishable from (or equivalent to) the original position. (For reviews, see Mislow, 1965; Salthouse and Ware, 1972; Orchin and Jaffé, 1970b.) It is brought about by the operation of one or several of the earlier mentioned symmetry elements that thus function as "symmetry operators." For example, successive operation of the C_4 axis in the molecule in Figure 4.1, **B** starting with the original molecule, gives rise to three additional, superposable species rotated around the C_4 axis by 90°, 180°, and 270°, respectively; rotation by 360° leads back to the original position. These symmetry operations so performed are called E, C_4^1, C_4^2, C_4^3; E being the identity operation.

The sum total of all possible symmetry operations defines a group. The criteria for a set of operations to constitute a group is that: (a) The combination of any two operations must be another operation in the group, for example, $E \times C_4^1, = C_4^1$, $C_4^1 \times C_4^2 = C_4^3$. The reader should note that the combination of the first two operations is equivalent to the third in each of these cases. (b) The associative law must hold, that is, the combination of two operations followed by a third must be equal to the first operation followed by the combination of the second and third: $(A \times B) \times C = A \times (B \times C)$. *Example*: $(C_4^1 \times C_4^2) \times C_4^1 = C_4^3 \times C_4^1 = E$. $C_4^1 \times (C_4^2 \times C_4^1) = C_4^1 \times C_4^3 = E$. (c) One of the operations of the group must be the identity operation E, which commutes with all the other operations and leaves them unchanged. This condition clearly holds: $E \times C_4^3 = C_4^3 = C_4^3 \times E$. (d) Each member of the group must have an inverse that is also a member of the group, thus $A \times A_{inv} = E = A_{inv} \times A$. In the above example C_4^1 is the inverse of C_4^3, and C_4^2 is its own inverse. (The operation C_4^2 corresponds to the C_2 axis that necessarily accompanies the C_4 axis).

The number of different operations that can be performed in a group is called the *order* of the group (cf. Orchin and Jaffe, 1970c). In the case of C_4, there are four operations: E, C_4^1, C_4^2, C_4^3, so the order of the group generated by the C_4 operator acting alone is four. (The orders of other point groups will be taken up as they are discussed.)

It quickly becomes clear that one cannot combine symmetry operators at random in a group. The existence of certain symmetry elements implies the simultaneous existence of certain others. For example, the existence of a C_2 axis and a σ plane perpendicular to it necessarily implies the presence of an S_2 axis (i.e., a center of symmetry, i). [The reverse is not true, however, i.e., i (or S_2) can be present in the absence of C_2 and σ as exemplified in Fig. 4.3.] Thus, $C_2 \times \sigma = S_2$ or, in words, C_2 followed by σ is equivalent to S_2. The point is exemplified by the molecule shown in Figure 4.5. It is also true that the existence or combination of certain symmetry elements excludes certain others. For example, a species that has a C_3 axis cannot have a collinear C_2 axis unless it also has a C_6 axis ($C_2 \times C_3 = C_6$). Nor can it have a single C_2 axis at right angles to the C_3; there must be either three such axes, at 60° angles to each other, or none. Thus

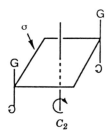

Figure 4.5. Molecule with C_2 and σ (implying $i = S_2$). See Figures 4.1, 4.2 for explanation of G and Ɔ.

we cannot combine symmetry elements indiscriminately but only according to the rules applicable to groups (see above). A proper combination of symmetry elements gives rise to what is known as a "symmetry point group," (as distinct from the "space groups" of the crystallographer, which include glide planes as symmetry operators, in addition to proper and improper axes.) There is a finite number of point groups as will be shown in the sequel.

Chiral molecules necessarily belong to point groups C_1, C_n, or D_n (or rarely to T, O, or I), that is, groups that have *only* proper axes. All other point groups, that is, those containing alternating (improper) axes, including reflection planes and inversion centers, are associated with achiral molecules. In these latter point groups, the number n of rotation operations for C_n (including $C_1 = E$) is always equal to the number of reflection operations (σ or S_n).

The notation for symmetry point groups used here is that due to Schoenflies. Crystallographers often use a different system called the Hermann–Maugin notation (cf. Donaldson and Ross, 1972, p. 34). A translation table between the two systems may be found elsewhere (Donaldson and Ross, 1972, p. 30).

a. Point Groups Containing Chiral Molecules

Point Group C_1

This point group has the lowest degree of symmetry. It is represented by a molecule of the type Cabcd (e.g., CHFClBr, Fig. 2.8); such a molecule has no symmetry at all and is thus truly "asymmetric." The only symmetry element is the identity E (always present) or the equivalent onefold axis C_1 and this point group is therefore denoted as C_1; its order is 1.

Point Groups C_n

In the C_n point groups, the only symmetry element is the C_n axis.

Point group C_2 is of fairly common occurrence. Examples are (+)- or (−)-tartaric acid (Fig. 3.26), chiral biphenyls in which both rings bear the same substituents (e.g., Fig. 3.2), 1,3-dichloroallene (Fig. 4.6; the axis passes diagonally through the central carbon atom) and the cyclobutane derivative shown in Figure 4.1, **A**, as well as the gauche forms of 1,2-dichloroethane (Fig. 2-9).

Point group C_3 is quite rare. Tri-o-thymotide (Fig. 4.7, **A**) constitutes an example in two of the four possible conformations attainable by tilting the rings. (The other two conformations are C_1.) The optically active compound racemizes with an activation energy of about $22\,\text{kcal mol}^{-1}$ ($92\,\text{kJ mol}^{-1}$) (cf. Farina and Morandi, 1974) by a flipping of the rings. The similar cyclotriveratrylene deriva-

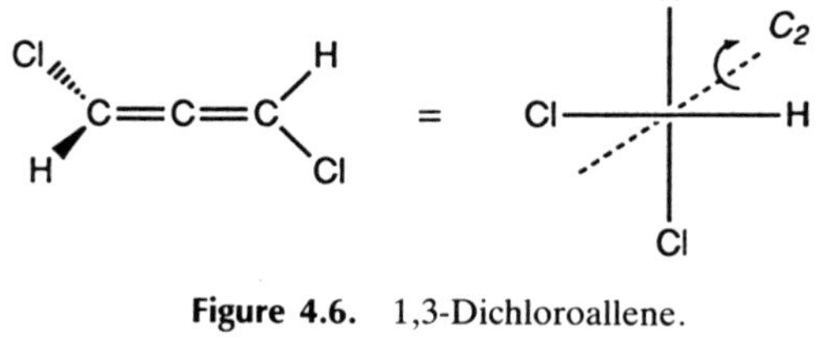

Figure 4.6. 1,3-Dichloroallene.

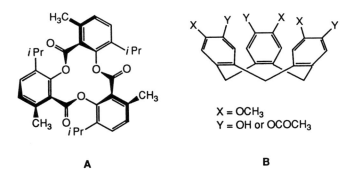

Figure 4.7. Tri-*o*-thymotide **A** and the cyclotriveratrylene derivative **B**.

tive (Fig. 4.7, **B**) is optically more stable; the barrier to racemization is 26.5 kcal/mol^{-1} (111 kJ mol^{-1}) (cf. Collet and Jacques, 1978; for related examples see Canceill et al., 1985).

Another chiral **C$_3$** compound is *trans,trans,trans*-3,7,11-trimethyl-cyclododeca-1,5,9-triene (Fig. 4.8), which is obtained along with its double-bond isomers and some head-to-head constitutional isomers upon trimerization of 1,3-pentadiene (Furukawa et al., 1968). The compound may be obtained optically active either by asymmetric synthesis with a chiral titanium methoxide–Et$_2$AlCl catalyst or by partial asymmetric destruction with tetrapinanyldiborane (cf. Chapter 7).

A tripodal ligand (cf. Fig. 11.133) with **C$_3$** symmetry has been described (Adolfsson et al., 1992).

Figure 4.1, **B** displays a (hypothetical) example of a species of symmetry point group **C$_4$**.

Point group **C$_6$** is found (Hybl et al., 1965) in cyclohexaamylose (Fig. 4.9), also called α-cyclodextrin.

Extensive lists of chiral compounds in the less common symmetry groups (**C$_n$**, $n > 2$; **D$_n$**) have been compiled by Farina and Morandi (1974) and by Nakazaki (1984). These compounds have been called "high-symmetry chiral molecules" by these authors. (The term "gyrochiral molecules" also used in the literature is not recommended.) As we have already seen, the order of a **C$_n$** point group is n.

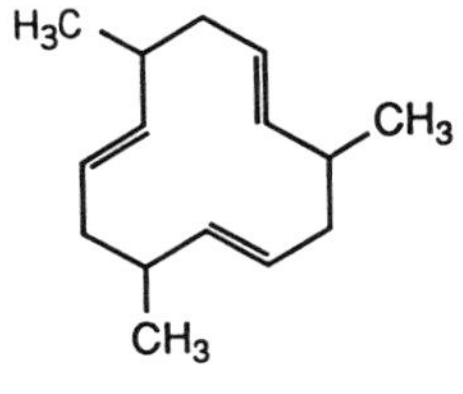

Figure 4.8. *trans,trans,trans*-3,7,11-Trimethyl-cyclododeca-1,5,9-triene.

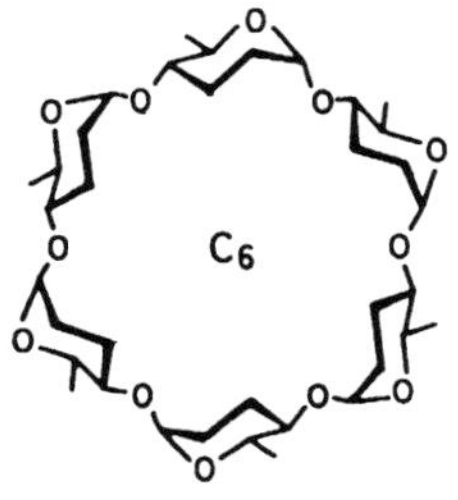

Figure 4.9. Cyclohexaamylose (α-cyclodextrin).

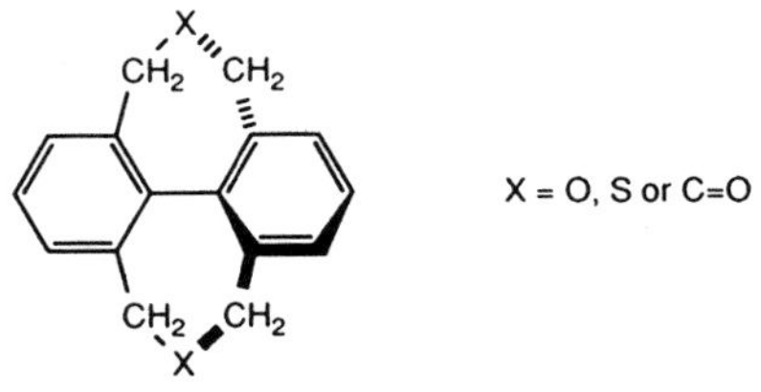

X = O, S or C=O

Figure 4.10. o,o'-Bridged biphenyls, where X = O, S, or C=O.

Point Groups D_n

These are the so-called "dihedral" point groups. These point groups are characterized by n C_2 axes perpendicular to the main C_n axis. The symmetry of these point groups is thus quite high; nevertheless they are chiral.

> The axis of highest multiplicity is taken as the "main" axis and, in the customary representation, is oriented along the z axis of a right-handed coordinate system. When all axes are C_2, the one passing through the most atoms is taken as the "main" axis.

The **D$_2$** [sometimes called **V** (for German *Vierergruppe*)] point group (cf. Farina and Morandi, 1974; Nakazaki, 1984) displays three mutually perpendicular C_2 axes. Among examples are the tetra-$(-)$-menthoxyacetyl derivative of pentaerythritol (McCasland et al., 1959), similar to the molecules shown in Figures 3.31 and 3.32 but with all four menthyl groups of the same configuration, the bridged biphenyls shown in Figure 4.10. (Mislow and Glass, 1961; Mislow et al., 1964), and the interesting molecule twistane (Whitlock, 1962; Adachi, Nakazaki, et al., 1968) shown in Figure 4.11. A number of additional examples of molecules in the **D$_2$** point group (allenes, spiranes, and biphenyls) will be found in Chapter 14.

The first organic **D$_3$** compound to be obtained in optically active form is *trans-transoid-trans-transoid-trans*-perhydrotriphenylene (Fig. 4.12); (Farina and Audisio, 1970). Other interesting examples are trishomocubane (Fig. 4.13, **A**) (Helmchen and Staiger, 1977; Eaton and Leipzig, 1978; Nakazaki et al., 1978) and a dimer of the previously mentioned cyclotriveratrylene (Gabard and Collet, 1981; Fig. 4.13, **B**).

A common example of **D$_3$** symmetry is the (ephemeral) skew form of ethane (cf. Chapter 10), in a conformation that is neither staggered nor eclipsed.

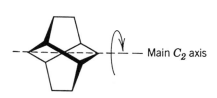

— Main C_2 axis

Figure 4.11. Twistane.

Figure 4.12. *trans-transoid-trans-transoid-trans*-Perhydrotriphenylene.

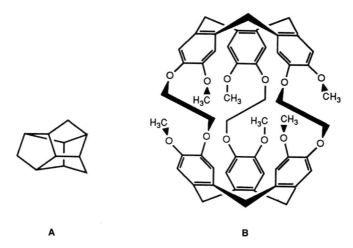

Figure 4.13. Trishomocubane **A** and cyclotriveratrylene dimer **B**. Reprinted with permission from Gabard and Collet, *J. Chem. Soc. Chem. Commun.* 1137 (1981). Copyright Royal Society of Chemistry, London, 1981.

No organic molecules of $\mathbf{D_4}$ symmetry have yet been synthesized but two inorganic niobium(IV) compounds of this symmetry have been obtained as racemates (Pinnavaia et al., 1975; Peterson, Brown et al., 1976).

The order of a $\mathbf{D_n}$ point group is $2n$. Thus, for $\mathbf{D_2}$ the possible symmetry operations are E, C_2, C_2', and C_2'' (there are three C_2 axes, unprimed, primed, and double primed for distinction). For $\mathbf{D_4}$ the symmetry operations would be E, C_4^1, C_4^2 (or C_2), C_4^3, $C_2{}'$, $C_2{}''$, $C_2{}'''$, and $C_2{}''''$ (for the C_4 and four perpendicular C_2 axes, respectively), thus the order is 8.

b. Point Groups Containing Only Achiral Molecules

Point groups other than $\mathbf{C_n}$ and $\mathbf{D_n}$ generally have planes, a center or an alternating axis of symmetry, and are therefore achiral. These point groups will be discussed here in order of increasing number of symmetry elements.

Point Group C_s (or C_{1h})

This point group has a symmetry plane σ *only* (no C_n). Examples are common among properly desymmetrized (cf. Section 4-4) alkenes, aromatics, and heterocyclic compounds. Chloroethylene, $CHCl{=}CH_2$, *m*-chlorobromobenzene, and furfural (furan-2-carboxaldehyde (cf. Abraham and Bretschneider, 1974) may be cited as examples. The compound *m*-chlorotoluene also has $\mathbf{C_s}$ symmetry provided either the conformation of the methyl group is such that one of the C–H bonds lies in the plane of the benzene ring, which then bisects the H–C–H angle of the other two, or the methyl group rotates fast enough on the time scale of the observation that it may be considered, *on the average*, to have a symmetry plane coincident with the plane of the benzene ring (cf. Section 4-5). With a corresponding proviso, methanol (CH_3OH) also belongs in this group. Molecules of the type CH_2XY or CR_2XY and aldehydes (RHC=O) are other common examples of C_s provided the R groups have a plane of symmetry either intrinsic (e.g., for R = Cl) or on the average (i.e., through fast rotation on the experimental time scale, see above). The order of the C_s point group is 2 (operations E, σ).

Point Group S_n

Molecules in point group S_n have an n-fold alternating axis of symmetry. When n is even there is the additional condition that there must be no symmetry planes, but there will necessarily be a proper rotation axis $C_{n/2}$ coextensive with S_n. When $n = 4m + 2$ ($m = 0, 1, 2, \ldots$) there is also a center of symmetry, but when $n = 4m$ there is no center of symmetry. When n is odd, the S_n axis cannot exist by itself but must coexist with C_n and σ_h. Point groups in this category are customarily called C_{nh} (see below) rather than S_n (odd n).

An S_2 axis corresponds to a center of symmetry and so the point group S_2 may also be called C_i (cf. Fig. 4.3). Examples (provided one properly orients the methyl groups) are the anti conformation of *meso*-2,3-dichlorobutane (**A**), the *trans*-diketopiperazine derived from 1 mol of D-alanine and 1 mol of L-alanine (**B**), and the dibromo[2.2]paracyclophane (**C**) (Cram et al., 1974; cf. Chapter 14), all shown in Figure 4.14.

> Dimerization of two alanine molecules gives either meso (C_i) dimer **B**, if the two alanine molecules are of opposite configuration (heterochiral) or a chiral (C_2) dimer (*cis*-diketopiperazine, diastereomeric to **B**) if they are of the same configuration (homochiral). One must not, however, conclude that dimerization of two identical homochiral molecules *necessarily* gives rise to a chiral dimer; this has been shown not to be the case (Anet, Mislow, et al., 1983; see also Mislow, 1985). One example where dimerization of homochiral molecules gives an achiral dimer is shown in Figure 4.15 (cf. Nouaille and Horeau, 1985).
>
> The reverse process—"cutting" an achiral molecule into two homochiral fragments, the so-called "coupe du roi" (for the history of this intriguing operation, see Anet, Mislow, et al., 1983)—has been realized by Cinquini, Cozzi, et al. (1988).
>
> It is worth noting that the stereochemical result of dimerizing pure enantiomers is substantially different from that of dimerizing the corresponding racemate. Thus, in the case of alanine, dimerization of the S enantiomer gives only the chiral (S,S)-diketopiperazine, but dimerization of (racemic) (RS)-alanine gives not only the racemic (RS,RS)-diketopiperazine but also the diastereomeric RS,SR meso isomer (Fig. 4.14, **B**). Similarly, in Figure 4.15, dimerization of the pure $1R, 2S$, enantiomer gives only the meso (C_{2v}) dimer shown (RS,SR), but dimerization of the corresponding racemic half-ester would give a mixture of the meso (C_{2v}) and meso (RS,RS, C_{2h}) dimers.

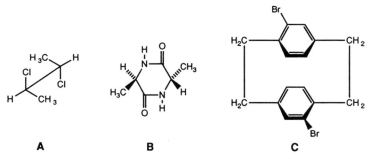

A   **B**   **C**

Figure 4.14. Molecules in S_2 point group (C_i).

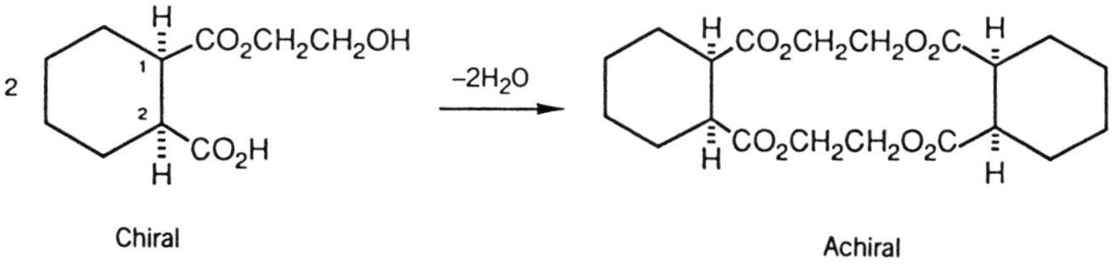

Figure 4.15. An example of dimerization of homochiral molecules yielding an achiral dimer.

The order of the S_2 point group is 2, operators E and i.

A hypothetical example of the relatively rare S_4 point group is shown in Figure 4.4 (again one must properly orient the alkyl groups) and a real one in Figures 3.31 and 3.32. Other examples (**A**, McCasland and Proskow, 1955, 1956; **B**, Helmchen, Prelog, et al., 1973) are shown in Figure 4.16.

It is an interesting exercise to assign the symmetry point group to the isomer of spirane shown in Figure 4.16, **A**, wherein the methyl groups on the right ring are switched to the alternative trans arrangement (upper methyl backward, lower methyl forward) (cf. Farina and Morandi, 1974) and to ascertain the symmetry of the isomer of the biphenyl **B**, wherein all α-phenethyl groups are of the same configuration (Helmchen, Prelog, et al., 1973).

The order of S_4 is 4, operators E, S_4^1, C_2, and S_4^3.

Molecules belonging to the S_6, S_8, and S_{10} point groups are exemplified by the class of cyclopeptides (Prelog and Gerlach, 1964) and will be discussed in Chapter 14.

The order of S_6 is 6, operators E, S_6^1, C_3^1, i, C_3^2, and S_6^5.

The remaining symmetry point groups have both axes and planes of symmetry. The planes are distinguished as either σ_v, σ_d symmetry planes that contain the main axis, or σ_h symmetry planes that are perpendicular to the main axis. Various combinations of these planes with axes of the C_n or D_n type generate most of the following groups, with the exception of some highly symmetric ones to be mentioned at the end.

Point Groups C_{nv}

This point group has a single C_n axis and n symmetry planes σ_v, which all contain the axis and intersect at it. If the axis is assumed to be vertical, so are the planes, hence the symbol v.

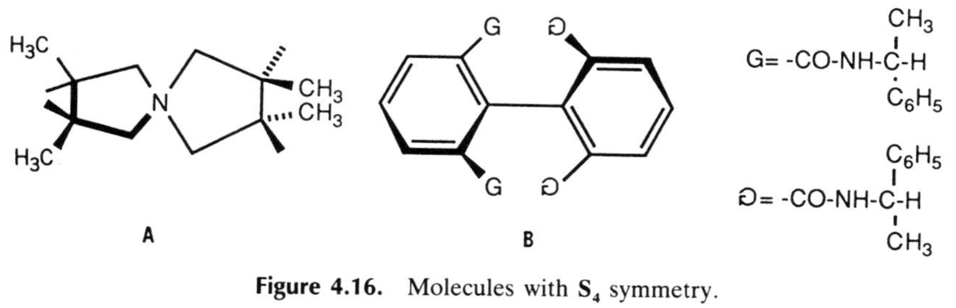

Figure 4.16. Molecules with S_4 symmetry.

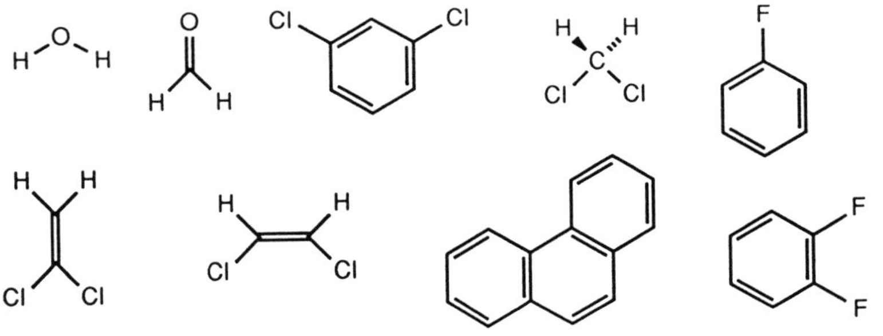

Figure 4.17. Molecules belonging to the C_{2v} point group.

A number of common planar molecules belonging to the C_{2v} point group are depicted in Figure 4.17. Any planar molecule also having a twofold axis in the plane will necessarily have a second plane of symmetry at right angles to the first plane and will belong to this group.

The order of C_{2v} is 4, operators E, C_2, and $2\sigma_v$.

The point group C_{3v} is also common, with such representatives as $CHCl_3$, NH_3, eclipsed CH_3CCl_3, and $C_6H_6Cr(CO)_3$ (benzenechromium tricarbonyl).

The order of C_{3v} is 6, operators E, C_3^1, C_3^2, and $3\sigma_v$.

Point groups C_{4v}, C_{5v}, and so on are rare, though not unknown. The all-*cis*-1,2,3,4-tetrachlorocyclobutane would have C_{4v} symmetry and the corresponding all-*cis*-1,2,3,4,5-pentachlorocyclopentane C_{5v}, if their rings were planar. As will be discussed in Chapter 11, these molecules are in fact not planar; however, due to rapid vibrational motion of the atoms, they may be considered to be planar *on the average* on the laboratory time scale and their average symmetry (cf. Section 4-5) will be that given above. Similarly, C_{nv} would be the average symmetry of an *n*-numbered ring with *n* equal, all-cis substituents. The octahedral molecule SF_5Cl is a rigid representative of the C_{4v} group.

The order of C_{nv} is 2*n*, as may be readily extrapolated from the specific cases above.

A C_∞ symmetry axis is an axis about which rotation by *any* angle (no matter how small) gives a molecule superposable with the original one. The point group $C_{\infty v}$ contains such an axis along with an infinite number of symmetry planes intersecting the axis, but no other symmetry elements. Hydrogen chloride (H–Cl), carbon monoxide (C=O), and chloroacetylene (H–C≡C–Cl) belong to the $C_{\infty v}$ point group. This type of symmetry is that displayed by a cone, and is therefore often called "conical symmetry." The order of this group is infinite.

Point Groups C_{nh}

The C_{nh} point groups have a C_n axis but only a single symmetry plane σ_h, which is perpendicular to the axis (i.e., *horizontal*, assuming the axis to be vertical).

Point group C_{2h} is the symmetry point group of *trans*-1,2-dichloroethylene and similarly substituted alkenes and of the *s*-trans forms of 1,3-butadiene (**A**) (cf. Chapter 9) and glyoxal (**B**) (Fig. 4.18). 1,4-Dichloro-2,5-dibromobenzene (**C**) also belongs in this group. The order of this group is 4; operators E, C_2, σ, and i.

Higher C_{nh} groups are rare and generally limited to specific conformations of the molecules in question. Phloroglucinol (Fig. 4.18, **D**) with all hydroxyl groups

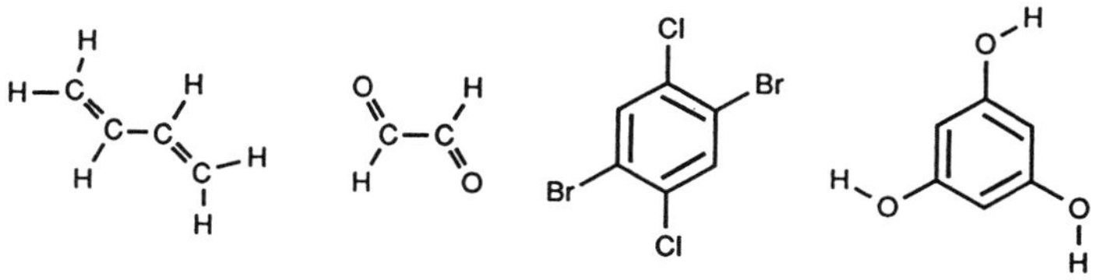

Figure 4.18. Molecules in point groups C_{2h} and C_{3h}.

pointing in the same direction belongs to point group C_{3h} and an appropriate conformer of hexahydroxybenzene would similarly belong to C_{6h}. The order of C_{3h} is 6: operators E, C_3^1, C_3^2, σ, S_3^1, and S_3^2; in general the order of C_{nh} is $2n$. An actual example of C_{6h} is found in hexaisopropylbenzene (Siegel, Mislow, et al., 1986), where "gearing" holds the isopropyl group in conformations in which the methine hydrogen atoms all point in the same direction (cf. Section 14-6.b).

Point Groups D_{nd}

We have already seen that the dihedral point groups contain a C_n axis and n C_2 axes perpendicular to it. When, in addition, there are n symmetry planes intersecting in the principal axis, the symmetry point group is D_{nd}. [The symmetry planes are considered to be *diagonal* (σ_d), since they do not contain the horizontal axes but rather the bisectors between two such axes.]

Point group D_{2d}, also called V_d (*Vierergruppe*), is found in allene, certain spiranes, and the perpendicular conformation of biphenyl shown in Figure 4.19. As depicted these molecules have a vertical C_2 symmetry axis, two vertical symmetry planes, and two perpendicular horizontal (diagonal) symmetry axes not contained in the σ_d planes. There is also an S_4 axis. The order of D_{2d} is 8: operators E, C_2, $2C_2'$, $2\sigma_d$, S_4^1, and S_4^3.

Point group D_{3d} occurs in the staggered form of ethane and the chair form of cyclohexane as the most important representatives of this group. In addition to C_3, nC_2, and $3\sigma_v$ there is an S_6 axis in this group. The order is 12, operators E, C_3^1, C_3^2, $3C_2$, $3\sigma_v$, i, S_6^1, and S_6^5.

Molecules in higher D_{nd} groups are rare. Ferrocene in the staggered conformation is D_{5d} (Fig. 4.20), whereas in the eclipsed conformation it is D_{5h} (see below). Dibenzenechromium and uranocene (Fig. 4.20) may be considered as D_{6h} and D_{8h}, or D_{6d} and D_{8d}, respectively, according to whether they are eclipsed or staggered. The order of D_{nd} is $4n$, as exemplified above.

Figure 4.19. Molecules in point groups D_{2d}.

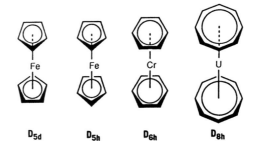

Figure 4.20. Ferrocene, dibenzenechromium, and uranocene.

D_{5d} D_{5h} D_{6h} D_{8h}

Point Group D_{nh}

This symmetry point group has symmetry elements similar to those of D_{nd} (except that σ_v planes, containing the horizontal axes, take the place of σ_d planes) *plus* a horizontal plane. It is more common than D_{nd}.

Common representatives of the point group D_{2h} are ethylene, 1,4-dichlorobenzene, naphthalene, dibenzocyclobutadiene, and anthracene. Its order is 8, operators E, C_2, $2C_2'$, $2\sigma_v$, σ_h, and i.

Cyclopropane, 1,3,5-trichlorobenzene, boron trifluoride (BF_3), and triphenylene (Fig. 4.21), as well as the trigonal bipyramidal PF_3Cl_2 (Fig. 3.8c) are found in the D_{3h} point group. Its order is 12, operators E, C_3^1, C_3^2, $3C_2$, $3\sigma_v$, σ_h, S_3^1, and S_3^2.

Point group D_{4h} represents the symmetry of the (average) planar form of cyclobutane. Square planar metal compounds with four equal ligands, such as $PtCl_4$, are also in this group. The order of this group is 16, operators E, C_4^1, C_2, C_4^3, $4C_2'$, $4\sigma_v$, σ_h, S_4^1, i, and S_4^3.

The symmetry point group D_{5h} represents the planar (or average planar) cyclopentane. It is also seen in the eclipsed form of ferrocene (Fig. 4.20) and in the cyclopentadienyl anion.

The symmetry point group D_{6h} is common; it is represented by benzene, hexachlorobenzene, and coronene (Fig. 4.21). Hexachlorocoronene, however, is twisted because of crowding of the chlorine substituents and belongs to the D_{3d} point groups (Baird, MacNicol, et al., 1988). The interesting molecule kekulene (Diederich and Staab, 1978; Fig. 4.22) has D_{6h} symmetry.

The symmetry point groups D_{7h}, D_{8h}, and D_{9h} are found in the tropylium cation (D_{7h}); in uranocene (Fig. 4.20); (Streitwieser and Müller-Westerhoff, 1968; Zalkin and Raymond, 1969; Streitwieser et al., 1973), in the cyclooctatetraenyl

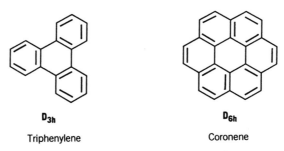

D_{3h} D_{6h}

Triphenylene   Coronene

Figure 4.21. Structure of triphenylene and coronene.

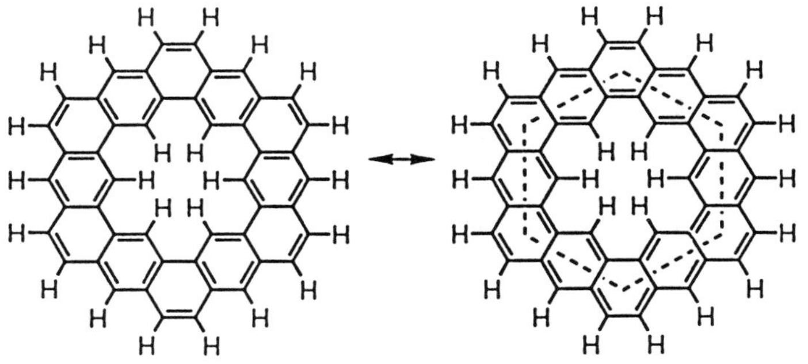

Figure 4.22. Kekulene.

dianion ($\mathbf{D_{8h}}$) (Katz, 1960), and in the cyclononatetraenyl anion ($\mathbf{D_{9h}}$) (Katz and Garratt, 1963, 1964; LaLancette and Benson, 1963, 1965). The general class of corannulenes (cf. Agranat et al., 1980), of which coronene (Fig. 4.21) is an example, belong to the $\mathbf{D_{nh}}$ point groups.

The order of $\mathbf{D_{nh}}$ is $4n$, as exemplified for specific cases above.

The point group $\mathbf{D_{\infty h}}$, in addition to a C_∞ axis, contains an infinite number of symmetry planes intersecting in it and an infinite number of C_2 axes perpendicular to the main axis, and has a symmetry plane perpendicular to the C_∞ axis. This type of symmetry is called "cylindrical" since a cylinder displays it. Molecules in the $\mathbf{D_{\infty h}}$ group must be linear and end-over-end symmetrical, such as ethyne (H–C≡C–H), carbon dioxide (O=C=O), and diatomic molecules such as di-hydrogen (H_2). The order of this group, as that of $\mathbf{C_{\infty v}}$, is infinite.

Point Groups Corresponding to the Platonic Solids: T_d, O_h, I_h

We now come to the point groups of the most highly symmetric bodies, the tetrahedron, cube, octahedron, dodecahedron (12 faces), and icosahedron (20 faces), the so-called "Platonic solids" (mentioned in Plato's dialogue *Timaeus*). Examples of all of these bodies are now represented in the molecular realm.

The Tetrahedral Point Group (T_d). The point group of the regular tetra-hedron is $\mathbf{T_d}$. It has four C_3 axes passing through each apex and the center of the opposite face, three C_2 axes passing through pairs of opposed edges (i.e., edges ending in noncommon vertices), and six σ_d planes, each containing one edge and bisecting the opposite one. Examples of molecules in group $\mathbf{T_d}$ are methane (**A**) and adamantane (**B**) (Fig. 4.23). The basic tetrahedral skeleton of tetrahedrane (Fig. 4.23, **C**; R = H) has not yet been obtained, but the corresponding tetra-*tert*-butyl derivative [**C**, R = C(CH$_3$)$_3$] has been synthesized by Maier et al. (1978, 1981, see Section 11-6.e).

Desymmetrization by distortion (cf. Section 4-4) destroys the symmetry planes but not necessarily the axes and thus leads to the chiral point group $\mathbf{T}$. Force field calculations on tetra-*tert*-butyltetrahedrane in fact suggest (Hounshell and Mislow, 1979) that the *tert*-butyl groups are so disposed in their most stable conformation as to place this molecule in point group $\mathbf{T}$. However, since rotation about the (CH$_3$)$_3$C– bond is rapid, the *average* symmetry (cf. Section 4-5) of

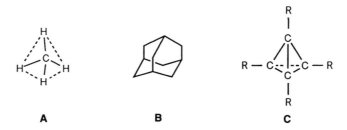

Figure 4.23. Molecules in point group T_d.

tetra-*tert*-butyltetrahedrane is T_d. A stable chiral molecule in point group **T** has been synthesized (Nakazaki and Naemura, 1981; Nakazaki et al., 1982) by desymmetrizing (cf. Section 4-4) adamantane (T_d; Fig. 4.23, **B**) in such a way that the planes of symmetry are abolished but the axes remain. This synthesis was achieved by attaching a substituent of local C_3 symmetry, trishomocubane (Fig. 4.13), to each of the four apices of adamantane. In the initial work of Nakazaki and Naemura (1981), the linking group was $-CH_2OCOCH_2$ (linking the tertiary carbon in adamantane to the tertiary carbon in trishomocubane) but this was considered inappropriate, since the link itself lacks local C_3 symmetry (Mislow, 1981). In subsequent work (Nakazaki et al., 1982) a $-C{\equiv}C-C{\equiv}C-$ (local symmetry $C_{\infty v}$) was used and the product thus can have **T** symmetry and is optically active having been synthesized from an optically active precursor.

> Another, rare, modification of the tetrahedral group is T_h in which, instead of the σ_d planes of T_d, there are four σ_v planes and a center of symmetry; this point group has been found only in inorganic complexes $[Co(NO_2)_6]^{3-}$ (with opposite nitro groups eclipsed) and $W[N(CH_3)_2]$ (Bradley, Chisholm, et al., 1969).

The order of **T** is 12 (identity, 2 operations for each of the 4 C_3 axes and one for each of the C_2 axes) and that of T_d is 24 (the additional 12 operations stem from the 6 σ_v planes and 3 S_4 axes).

The Cubic Point Group O_h. The cube and the octahedron belong to the octahedral point group O_h. This group has three C_4 axes (passing through the centers of opposite faces of the cube or opposite apices of the octahedron), four C_3 axes (the space diagonals of the cube or axes passing through the centers of opposite faces in the octahedron), and six C_2 axes (passing through the centers of opposite edges). In addition there are nine σ planes, three passing through the middle of opposite faces (bisecting the edges) and six passing diagonally through opposite faces. Octahedral symmetry is found in cubane (C_8H_8), first synthesized by Eaton and Cole (1964; see also Section 11-6.e), sulfur hexafluoride (SF_6), and octahedral coordination compounds with six equal ligands. The (hypothetical) point group **O** would be obtained by removing the symmetry planes.

The order of O_h is 48 (E, 9 stemming from the 3 C_4 axes, 8 from the 4 C_3 axes, and 6 stemming from the C_2 axes. The resulting total of 24 is doubled by the existence of the symmetry planes).

The cube and octahedron are called "reciprocal solids" in as much as, if the faces of the cube are replaced by vertices and the vertices by faces, an octahedron is obtained, and vice versa. (Note that the tetrahedron is its own reciprocal.)

The Icosahedral Point Group (I_h). The remaining regular polyhedra are the dodecahedron and the icosahedron, the former having 12 faces in the shape of regular pentagons, the latter 20 in the shape of equilateral triangles. (The icosahedron has 12 vertices, the dodecahedron has 20.) These two (reciprocal) solids belong to the symmetry point group I_h, which has 6 C_5, 10 C_3, and 15 C_2 axes as well as 15 σ planes. The substance dodecahedrane ($C_{12}H_{12}$, Fig. 4.24), which has the shape of a regular dodecahedron has been synthesized by Paquette and co-workers (Ternansky, Paquette, et al., 1982; see also Paquette, 1984 and Section 11-6.e); the icosahedron is found in the dodecaborane dianion $B_{12}H_{12}^{2-}$ and the corresponding halides $B_{12}X_{12}^{2-}$ (X = Cl, Br, or I, cf. Muetterties and Knoth, 1968). The order of the icosahedral (I_h) point group is 120 (cf. Kettle, 1985, p. 274) and the group also has six S_{10} axes. The substance buckminsterfullerene (Section 11-6.e) corresponds to a truncated icosahedron and also has I_h symmetry.

The highest symmetry, spherical symmetry (K_h; K for German *Kugel*, meaning sphere), for obvious reasons is not represented by a finite point group.

The symmetry point group of a given molecule or model can be inferred systematically by an appropriate dichotomous tree (Donaldson and Ross, 1972, p. 38; Kettle, 1985, p. 167; Orchin and Jaffe, 1970b). However, the result can be arrived at faster by proceeding in nondichotomous fashion: Does the molecule have a finite symmetry axis, an infinite symmetry axis, or no symmetry axis at all? If the symmetry axis is infinite, the point group is $C_{\infty v}$ or $D_{\infty h}$, depending on the absence or presence of a perpendicular symmetry plane. If there is no symmetry axis, the possible point groups are C_1 (asymmetric), C_i (point of symmetry only), or C_s (plane of symmetry only). If there is at least one symmetry axis, what is its order, and are there other axes (C_2) perpendicular to it? If there are no other symmetry axes, we are dealing with the C_{nx} family, if there are other axes, we are dealing with D_{nx}, n being the order of the principal axis. If the family is C_{nx}, are there no planes of symmetry, n planes of symmetry containing the axis, or one plane of symmetry perpendicular to the axis? In the first case, the point group is C_n, unless there is also an alternating axis of order $2n$, in which case the point group is S_{2n}. In the second case the group is C_{nv}, in the third case C_{nh}. If the molecule belongs to the D_{nx} family, is it devoid of symmetry planes, does it have diagonal symmetry planes, or does it also have a symmetry plane perpendicular to the principal axis? In the first case, the point group is D_n, in the second case it is D_{nd}, and in the third case D_{nh}. This leaves only the highly symmetric point groups T_d, O_h, and I_h, which are rare and readily recognized.

Figure 4.24. Dodecahedrane.

4-4. DESYMMETRIZATION

Symmetry point groups may be arranged in a hierarchy by starting with a highly symmetrical one, such as O_h or I_h, and then systematically removing elements of symmetry. This process is called desymmetrization. As already explained in Section 4-3, symmetry elements cannot be removed arbitrarily but only in certain combinations, in as much as the removal of one symmetry element may automatically imply the removal of others, lest the remaining set left no longer constitute a group (Section 4-3).

One way of desymmetrizing a model is by distorting it (Donaldson and Ross, 1972, p. 50). For example, if one distorts (elongates or squashes) a cube (O_h) along a C_4 axis, one obtains a rectangular parallelepiped (square prism) whose symmetry is D_{4h}, the same as that of a square. The square, in turn, may be distorted into a rectangle (D_{2h}) or it may be given a sense of turn; (C_{4h}; Fig. 4.25). Again it may be distorted out of the plane into a square pyramid (C_{4v}) or folded along a diagonal to give a wing-shaped figure (D_{2d}). Finally, the top square face of the square prism may be twisted with respect to the bottom face by an angle of less than 90° to give a figure of D_4 symmetry. Any one of these figures may be distorted further, resulting in more reduced symmetry; a complete "hierarchy of distortions" of O_h as well as T_d and D_{6h} has been given by Donaldson and Ross (1972, pp. 59–61). Distortions of molecular frameworks are common in chemistry. For example, if one looks at the formula for corannulene (Fig. 4.26, **A**) one might think that the molecule is planar, D_{5h}. In fact, however, synthesis followed by X-ray study has disclosed (Barth and Lawton, 1971; Hanson and Nordman, 1976) that it is distorted out of plane and, therefore, C_{5v}; C_{5v} necessarily corresponds to a subsymmetry of D_{5h}. A similar argument applies to peristylane (Fig.4.26, **B**; Eaton et al., 1977).

von Baeyer thought cyclohexane was a planar hexagon but conformational analysis (Chapter 11) has taught us that the hexagon (D_{6h}) is distorted into a chair whose actual symmetry D_{3d} is a subsymmetry of D_{6h} (see also Section 4-5 concerning conformational averaging that may lead from the lesser to the greater symmetry). As a final example consider tetraphenylsilane, $(C_6H_5)_4Si$. Since the phenyl groups lack C_3 axes, this molecule cannot have the full tetrahedral symmetry T_d of the parent SiH_4; the highest symmetry possible is D_{2d} (Fig. 4.27). This would require, however, that one of the symmetry planes of the aryl rings (either the ring plane or the plane perpendicular to it) coincide with the C–Si–C

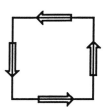

Figure 4.25. Directed square.

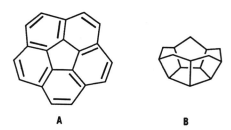

A   **B**

Figure 4.26. Structures of corannulene ([5]circulene) and peristylane.

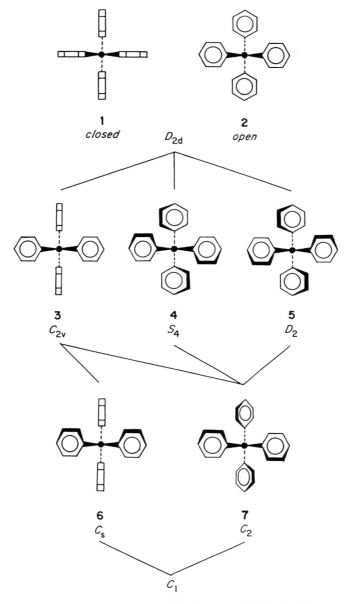

Figure 4.27. Possible point group symmetries of $(C_6H_5)_4C$ and $(C_6H_5)_4Si$. [Reprinted with permission from Hutchings et al., *J. Am. Chem. Soc.* (1975) **97**, 4553. Copyright © (1975) American Chemical Society.

plane. In fact the rings are tilted in such a way as to generate $\mathbf{S_4}$ symmetry (Hutchings, Mislow, et al., 1975). In contrast, in tetraphenylmethane, $(C_6H_5)_4C$, the maximal $\mathbf{D_{2d}}$ symmetry is preserved (Hutchings, Mislow et al., 1975). Figure 4.27 shows other possible lower symmetries.

More commonly, desymmetrization in chemistry results from substitution on a symmetrical lattice framework rather than from distortion of such a framework. For example, the $\mathbf{T_d}$ of CH_4 can be desymmetrized to $\mathbf{C_{3v}}$ in CH_3Cl or $\mathbf{C_{2v}}$ in

CH_2Cl_2. Further desymmetrization of C_{3v} or C_{2v} to C_s and finally to C_1 is exemplified by CH_2ClBr and $CHFClBr$. It must be noted, however, that distortion of the framework is a necessary consequence of desymmetrization by substitution; thus in CH_3Cl and CH_2Cl_2 the bond angles will no longer be exactly tetrahedral (109.5°). In fact, in CH_2Cl_2, the H–C–H angle is 113°, the Cl–C–Cl angle 111.8° and, hence, the H–C–Cl angles are 108°. (In CA_2B_2 the relation between the A–C–A, B–C–B, and A–C–B bond angles is $\cos A$–C–$B = -\cos\frac{1}{2}A$–C–$A \cdot \cos\frac{1}{2}B$–C–B.) These distortions are small but real and are demanded by the symmetry of the system; it would be a pure coincidence if the bond angles in CH_3Cl or CH_2Cl_2 were all equal (109.5°).

By distortional desymmetrization, we can convert the T_d of methane into the D_{2d} of allene $H_2C=C=CH_2$ (Fig. 4.19), which may be further desymmetrized by substitution to $Cl_2C=C=CH_2$ (C_{2v}), $ClCH=C=CHCl$ (C_2; cf. Fig. 4.6) $CHCl=C=CH_2$ (C_s), and finally $CHCl=C=CHBr$, which is C_1.

Finally, consider ethylene, which is D_{2h}. Desymmetrization of $H_2C=CH_2$ to $Cl_2C=CH_2$ or *cis*-$ClHC=CHCl$ leads to C_{2v}; desymmetrization to *trans*-$ClHC=CHCl$ leads to C_{2h}; and further desymmetrization to $H_2C=CHCl$ leads to C_s (see also Prelog, 1968 for additional examples).

A more difficult problem arises when one deliberately wants to follow a given desymmetrization path. For example (cf. Farina and Morandi, 1974) how could one convert T_d to T, destroying the symmetry planes but retaining all the axes? This problem has been elegantly solved by Nakazaki et al. (1982) (cf. p. 86). If substituents of local C_3 symmetry are attached to each carbon atom in cubane, one obtains a molecule belonging to symmetry point group O.

An interesting hypothetical desymmetrization leading to a chiral molecule of C_3 symmetry is shown in Figure 4.28. (Farina and Morandi, 1974). The keto form of phloroglucinol (cyclohexane-1,3,5-trione, **A**; D_{3h}) may (at least hypothetically)

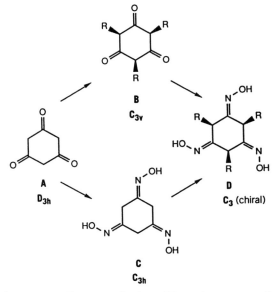

Figure 4.28.   Desymmetrization of D_{3h} point group to C_3.

be alkylated in all-cis fashion to give the molecule shown in **B**, which belongs to C_{3v} or it may be oximated to the oxime **C** in which all oxime groups point in the same direction (C_{3h}). Combination of the two transformations in either sequence gives the chiral molecule **D** of symmetry C_3. Chiral ligands are not required in this desymmetrization.

This last example shows rather nicely that symmetry is a "subtractive property." Pierre Curie stated already in 1894 (q.v.): "When several natural phenomena are superimposed in the same system, the reductions in symmetry are additive. There remain in the system only those symmetry elements that are common to each phenomenon taken separately."

4-5. AVERAGED SYMMETRY

In the previous section we discussed desymmetrization. Here we shall look at the opposite process, namely, that of a molecule that simulates higher symmetry than the symmetry present in any of its contributing structures because of a rapid interconversion of these structures on a time scale that is appreciably faster than that of the experiment designed to determine the symmetry. As discussed in Chapter 2, this situation is not uncommon in real molecules.

> Unfortunately, only a superficial treatment of the subject of averaged or nonrigid symmetry is possible here, since a rigorous approach (cf. Longuet-Higgins, 1963) requires a knowledge of permutation group theory and the concept of isomorphism of groups that goes beyond the scope of this book.

By way of examples that are intuitively easy to grasp, let us consider cyclohexane. Cyclohexane is known to exist in the chair form of symmetry D_{3d} and at −100°C the NMR spectrum of the compound (cf. Chapter 11) is indeed appropriate for a molecule of that symmetry. At room temperature, however, due to rapid inversion of the chair, the ^{1}H NMR spectrum shows a single signal due to the equivalent averaged hydrogen atoms. This observation is what would be expected for planar cyclohexane (D_{6h}) and it is therefore reasonable to assume that the average symmetry of cyclohexane is indeed D_{6h}, even though the planar form is of very high energy and does not even correspond to the transition state for chair inversion (Chapter 11). A rigorous demonstration of this intuitive conclusion has been given (Leonard, Hammond, and Simmons, 1975).

However, not all cases of "nonrigid symmetry" can be treated in this intuitive fashion; in fact, most cannot (cf. Bauder, Günthard, et al., 1974). For example, while we were able to grasp intuitively that the molecule shown in Figure 3.22 was superposable with its mirror image due to swiveling of the central portion, we may find it difficult to assign this structure to an achiral symmetry point group by intuition, even if we try to take the averaging into account (but see below). Again, take the molecule toluene (Fig. 4.29). The symmetry of any of the conformations shown is C_s; however, in fact, there are six such equivalent conformations (rotamers) and we might therefore guess that the nonrigid symmetry of toluene is C_{6v}. [Indeed, the nonrigid symmetry group is isomorphous (cf. Kettle, 1985, p. 63) to C_{6v}.] An even less obvious case is presented by ethane in

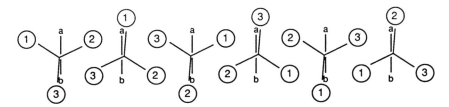

Figure 4.29. Toluene.

which the rigid point group of the staggered conformation is $\mathbf{D_{3d}}$, and there are three equivalent conformations of this type. The resulting nonrigid symmetry group is not even isomorphous to one of the point groups discussed earlier.

What conclusions, then, can one draw on an elementary level? Perhaps the most useful result from more in-depth treatments is that the order of the nonrigid symmetry group is equal to $n \times o$ where n is the number of distinct conformations in which the (rigid) symmetry point group, which generates the nonrigid group, can appear and o is the order of that symmetry point group. Thus for toluene the order (of $\mathbf{C_s}$) is 2, and the number of conformations is 6 (Fig. 4.29), so the actual order is $2 \times 6 = 12$, which is also the order of $\mathbf{C_{6v}}$. For ethane, $n = 3$ (three staggered conformations) and $o = 12$ (for $\mathbf{D_{3d}}$), so $n \times o = 36$.

It is instructive to check this computation against the "intuitive result" in those cases where intuition seems to provide the answer. For cyclohexane, the order of $\mathbf{D_{3d}}$ (the chair form) is 12 and the number of conformations is 2 (chair and inverted chair), so the order of the nonrigid point group is $2 \times 12 = 24$, which is indeed the order of the intuitively inferred $\mathbf{D_{6h}}$ point group of the average planar hexagon.

Finally, we may consider the Mislow–Bolstad molecule (Fig. 3.22). The molecule as written is asymmetric ($\mathbf{C_1}$, order 1). There appear to be four pertinent conformations, generated by rotating the biphenyl moiety through successive angles of $90°$. The order of the group is thus $1 \times 4 = 4$, which corresponds to the order of $\mathbf{S_4}$ and, indeed, it seems intuitively reasonable that the nonrigid point group of the molecule would be $\mathbf{S_4}$ (see Frei and Günthard, 1976 for an in-depth treatment of this case).

4-6. SYMMETRY AND MOLECULAR PROPERTIES

In general, physical properties of matter are dependent on molecular structure and among these are properties that are predictable on the basis of symmetry arguments. We shall deal here with three such properties: the ability to rotate the plane of polarized light, the ability to display a permanent dipole moment, and the thermodynamically important entity called symmetry number. (We will take up the relation of symmetry properties to NMR spectra in Chapter 8, and their relation to chiroptical properties in Chapter 13.)

a. Rotation of Polarized Light

The relation of optical rotation (cf. Chapter 1) to symmetry is simple: only substances whose molecules belong to chiral point groups can rotate polarized light. The presence of an S_n axis (including $S_1 = \sigma$ and $S_2 = i$) in a molecule precludes its displaying optical activity. Optical activity is a *pseudoscalar property*, which means the property remains invariant under a proper operation (rotation) but changes sign under an improper operation (reflection). In contrast, *scalar properties* are properties that are invariant to both proper and improper operations. Since an improper operation (reflection) transforms one enantiomorph into another, enantiomorphs (including enantiomeric substances) have identical scalar properties as already indicated in Chapter 3.

It follows immediately that only molecules in point groups $\mathbf{C_n}$ (including $\mathbf{C_1}$) and $\mathbf{D_n}$, as well as the less common groups $\mathbf{T}$, $\mathbf{O}$, and $\mathbf{I}$, can give rise to optical rotation.

The reverse does not necessarily follow. Molecules in a chiral point group will normally display optical rotation, but they may occasionally fail to do so because of what is called "accidental degeneracy" (Mislow and Bickart, 1976/77). We already saw (Chapter 1) that change of temperature, solvent, or pH may carry optical rotation from a positive to a negative value through a null, and the same is true of changes in wavelength (Chapter 13). Moreover, a mixture of two different chiral substances of opposite rotation (e.g., in an extract from a natural product) may accidentally display zero rotation by fortuitous compensation. Finally, the rotation of a chiral substance may, in some cases, be so small as to be unobservable, as a consequence of instrumental limitations. Examples are *n*-butylethyl-*n*-hexyl-*n*-propylmethane (Wynberg et al., 1965; Wynberg and Hulshof, 1974; see also Section 13-5.a and Fig. 13.71), which has a specific rotation at 280 nm of less than 0.04, and neopentyl-1-*d* alcohol, $(CH_3)_3CCHDOH$, which displays no rotation above 300 nm in 80% acetone solution (Mosher, 1974), but does display a plain negative ORD curve for the *S* isomer in hexane solvent with palpable rotations between 350 and 220 nm (Anderson, Mosher, et al., 1974). Such molecules have been called *cryptochiral* (Mislow and Bickart, 1976/77).

A more trivial case of zero rotation in chiral molecules, concerned with racemates, is discussed in Chapter 13.

Mention of neopentyl-1-*d* alcohol brings up the general question of optical activity of chiral compounds, two of whose ligands differ only in isotopic composition. Clearly, compounds of the general type R_1R_2CXX', where X and X' are isotopically distinct, are chiral; the question is whether, if nonracemic, they display palpable optical activity. The earliest candidates for investigation were deuterated compounds of the type R_1R_2CHD and $R_1R_2CX^DX^H$ (X^D and X^H being otherwise identical ligands that differ in being substituted by one or more deuterium atoms taking the place of one or more hydrogen atoms). The reason for this was twofold since both the early availability of deuterium and the fact that the relative difference between hydrogen and deuterium is greater than that of any other two stable isotopes made it easy for optical activity, if any, to be detected in this case.

Surprisingly, the early investigations (1936–1942) led to no observation of unequivocal optical activity (cf. Arigoni and Eliel, 1969; Verbit, 1970). In light of

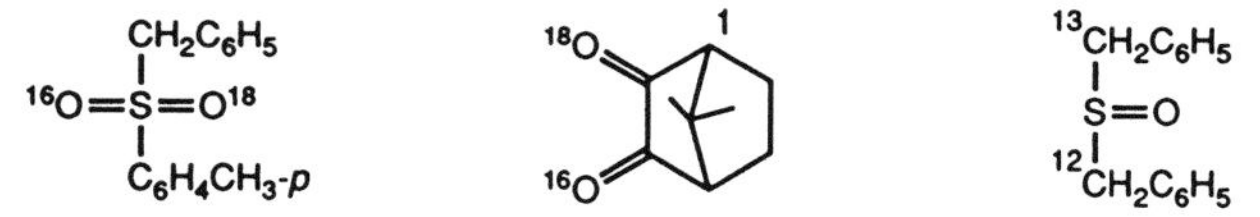

$$^{16}O=\overset{\overset{\displaystyle CH_2C_6H_5}{|}}{\underset{\underset{\displaystyle C_6H_4CH_3\text{-}p}{|}}{S}}=O^{18}$$

$$^{13}CH_2C_6H_5$$
$$\overset{|}{S}=O$$
$$^{12}CH_2C_6H_5$$

(S)-(−), $[\alpha]_D^{25}$ -0.16
(c 7.8, CHCl$_3$)

1-(R), positive CD

(R)-(+), $[\alpha]_{280}^{25}$ +0.71
(c 5.6, CHCl$_3$)

A   **B**   **C**

Figure 4.30. Optical activity due to isotopic differences other than H or D.

later findings (e.g., Makino, Mosher et al., 1985) this was clearly due to accidental causes, for when the investigations were resumed in more systematic fashion, a number of optically active compounds were found beginning in 1949 (Alexander and Pinkus), including $C_6H_5CHDCH_3$ (Eliel, 1949), and $CH_3CHOHCD_3$ (Mislow et al., 1960), which exemplify the two types of chiral deuterio compounds mentioned above. The configuration and maximum optical rotation of (+)-$C_6H_5CHDCH_3$ have been determined unequivocally by Elsenbaumer and Mosher (1979); the dextrorotating hydrocarbon is S (i.e., the earlier studied LiAlD$_4$ reduction of $C_6H_5CHClCH_3$ proceeds with nearly complete inversion of configuration; cf. Eliel, 1949) and the specific rotation $[\alpha]_D^{20}$ of neat, enantiomerically pure material is 0.81 ± 0.01. The compound (+)-$CH_3CHOHCD_3$ is S and its $[\alpha]_D^{27}$ is 0.2 (c 4, CHCl$_3$). The compound $C_6H_5CHOHC_6D_5$ has also been obtained optically active (Makino, Mosher, et al., 1985); $[\alpha]_D^{20}$ (c 16, CHCl$_3$) for enantiomerically pure material was calculated to be 1.00.

> Other interesting cases, in which the deuterium is remote from the chiral center are nonadeuterated cyclotriveratrylene (Fig. 4.7, **B**, X = OCH$_3$, Y = OCD$_3$), $[\alpha]_D^{25}$ 3.0 (c 2, CHCl$_3$) (Collet and Gabard, 1980) and a corresponding tribenzylene (Fig. 4.7, **B**, X = H, Y = D), $[\alpha]_D^{25}$ −2.3 (c 2.6, CHCl$_3$) (Canceill, Collet, and Gottarelli, 1984). Another example, altogether devoid of a chiral center, is ring monodeuterated paracyclophane (cf. Chapter 14-8.b), $[\alpha]_{546}^{25}$ −4.0 (c 0.87, CHCl$_3$) (Hoffman, Nugent, et al., 1973).
>
> For historical accounts, see Arigoni and Eliel (1969), Verbit (1970) and Makino, Mosher, et al. (1985).

Cases of optical activity due to isotopes of other elements are few; compounds investigated with their respective rotations are shown in Figure 4.30, **A** (Stirling, 1963), **B** (Kokke and Oosterhoff, 1972; Kokke et al., 1973), and **C** (Andersen, Stirling, et al., 1973).

b. Dipole Moment

The (permanent) dipole moment of a chemical compound is due to an unbalanced charge distribution in its molecules. In principle, any two distinct atoms in a molecule (say a grouping −X−Y−) will give rise to a local dipole, since X and Y will, to a greater or lesser extent, differ in electronegativity. However, the molecule as a whole will have a dipole moment only if the local dipole of an

–X–Y– grouping in one region is not compensated by an opposite local dipole elsewhere in the molecule. Symmetry properties dictate whether this will or will not be the case. Specifically, if a molecule has a center of symmetry, to each local dipole there will correspond an equal and opposite one, and the overall dipole moment will be zero. Thus molecules in point groups S_n with even n (including C_i), among other point groups, cannot have a permanent dipole moment. If a molecule has a C_n axis, the component of the dipole moment vector at right angles to the axis averages to zero and the molecule can, at best, have a moment along the direction of the axis. However, this component of the moment will also vanish if (a) there is a symmetry plane perpendicular to the axis or (b) there is at least one other axis perpendicular to the first. Condition (a) eliminates $\mathbf{C_{nh}}$ and condition (b) eliminates all dihedral classes: $\mathbf{D_n}$, $\mathbf{D_{nd}}$, and $\mathbf{D_{nh}}$. It follows that only molecules in symmetry point groups $\mathbf{C_n}$ (including $\mathbf{C_1}$), $\mathbf{C_s}$, and $\mathbf{C_{nv}}$ can have permanent dipole moments.

By way of an example we may consider the two chair conformations of *trans*-1,4- and *trans*-1,2-dibromocyclohexane, respectively (Fig. 4.31). Both conformational isomers of the *trans*-1,4 compound belong to symmetry point group $\mathbf{C_{2h}}$ and, therefore, necessarily have zero dipole moments. In contrast, *both* conformations of the *trans*-1,2 compound belong to symmetry point group $\mathbf{C_2}$ and thus have dipole moments (the dipole of the diaxial conformer is erroneously claimed to be zero in some textbooks). From model studies [dipole measurements of 4-*tert*-butyl substituted analogues (cf. Chapter 11)] the moment of the diequatorial conformer is 3.3 D and that of the diaxial conformer 1.2 D. The latter moment is surprisingly large, the calculated value being only 0.37 D (cf. Abraham and Bretschneider, 1974; Eliel, et al., 1965, 1981).

Molecular symmetry must *not* be averaged (Section 4-5) in the decision whether a molecule does or does not have a dipole moment. Thus, 1,2-dichloroethane (Fig. 2.9) has a dipole moment (1.12 D at 32°C in the gas phase) even though $\mu = 0$ for the centrosymmetric anti form and it might be thought (erroneously!) that the dipole moments of the equally populated gauche forms cancel each other. The reason they do not is that the polarizations (i.e., squares of the dipole moments) average, not the moments themselves.

Provided a molecule is in the proper symmetry point group, even the C–H bond moments may be sufficient to impart a dipole moment to it. Thus small but finite dipole moments were found in cycloheptane and cyclooctane (Dowd, Klemperer, et al., 1970).

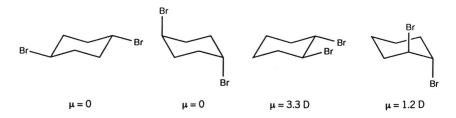

$\mu = 0$ $\mu = 0$ $\mu = 3.3\,D$ $\mu = 1.2\,D$

Figure 4.31. Dipole moments of *trans*-1,4- and *trans*-1,2-dibromocyclohexane.

c. Symmetry Number

The symmetry number σ of a molecule is defined as the number of indistinguishable but nonidentical positions into which the molecule can be turned by rigid rotation. It is important in the computation of entropy. Symmetry reduces entropy, because it implies that arrangements of a molecule obtainable by virtue of rigid rotation (i.e., rotation of the molecule as a whole), and that would otherwise contribute to the rotational entropy, are superposable, and are therefore thermodynamically indistinguishable. Thus the number of distinct rotational arrays is diminished and the rotational entropy is thereby reduced.

To attain the appropriate indistinguishable positions (indistinguishable except for conceptual labeling that avails nothing in thermodynamics), it is easiest to operate the various symmetry axes in a molecule. Thus for water, operation of the C_2 axis (E, C_2) will interchange the position of the hydrogen atoms: the symmetry number is thus 2. A more complex case, that of 1,3,5-trichlorobenzene (symmetry point group D_{3h}, $\sigma = 6$) is shown in Figure 4.32. The top three arrangements are generated by rotation about the threefold axis and the bottom three from the first (E) by rotation about each of the twofold axes. If we remember the concept of the order of a symmetry point group (Section 4-3), we recognize that, for a chiral point group, the symmetry number is equal to the order of the groups, whereas for a finite achiral point group it is one-half the order (since only one-half of the symmetry operations are rotations, whereas the other one-half are operations of the second kind, which are not pertinent to the concept of symmetry number in that they do not produce a *real* coincidence of atoms). Exceptions are $C_{\infty v}$ and $D_{\infty h}$, whose orders are infinite, but whose symmetry numbers are 1 and 2, respectively.

Table 4.1 lists the order of various point groups and the corresponding symmetry numbers.

We shall return to the use of symmetry numbers in computing entropy differences between stereoisomers in Chapters 10 and 11.

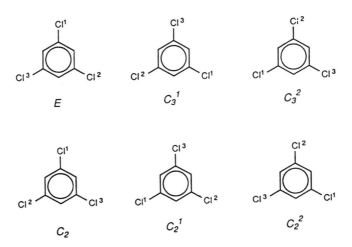

Figure 4.32. Rigid rotation of 1,3,5-trichlorobenzene.

TABLE 4-1. Point Groups, Their Order, and Their Symmetry Number.

Group	C_1	C_n	D_n	C_s	$S_n{}^a$	C_{nv}; C_{nh}	$C_{\infty v}$	D_{nd}; D_{nh}	$D_{\infty h}$	T_d	O_h	I_h
Order	1	n	$2n$	2	n	$2n$	∞	$4n$	∞	24	48	120
σ	1	n	$2n$	1	$n/2$	n	1	$2n$	2	12	24	60

^{a} Includes $C_i = S_2$, order 2, $\sigma = 1$.

REFERENCES

Adolfsson, H., Warnmark, K., and Moberg, C. (1992), *J. Chem. Soc. Chem. Commun.*, 1054.

Abraham, R. J. and Bretschneider, E. (1974), "Medium Effects on Rotational and Conformational Equilibria," in *Internal Rotation in Molecules*, Orville-Thomas, W. J., Ed., Wiley, New York. pp. 481, 567.

Adachi, K., Naemura, K., and Nakazaki, M. (1968), *Tetrahedron Lett.*, 5467.

Agranat, I., Hess, B. A., and Schaad, L. J. (1980), *Pure Appl. Chem.*, **52**, 1399.

Alexander, E. R. and Pinkus, A.G. (1949), *J. Am. Chem. Soc.*, **71**, 1786.

Andersen, K. K., Colonna, S., and Stirling, C. J. M. (1973) *J. Chem. Soc. Chem. Commun.*, 645.

Anderson, P. H., Stephenson, B., and Mosher, H. S. (1974), *J. Am. Chem. Soc.*, **96**, 3171.

Anet, F. A. L., Miura, S. S., Siegel, J., and Mislow, K. (1983), *J. Am. Chem. Soc.*, **105**, 1419.

Arigoni, D. and Eliel, E. L. (1969), "Chirality Due to the Presence of Hydrogen Isotopes at Non-Cyclic Positions," *Top. Stereochem.*, **4**, 127.

Baird, T., Gall, J. H., MacNicol, D. D., Mallinson, P. R., and Mitchie, C. R. (1988), *J. Chem. Soc. Chem. Commun.*, 1471.

Barth, W. E. and Lawton, R. G. (1971), *J. Am. Chem. Soc.*, **93**, 1730.

Bauder, A., Meyer, R., and Günthard, Hs. H. (1974), *Mol. Phys.* **28**, 1305.

Bradley, D. C., Chisholm, M. H., Heath, C. E., and Hursthouse, M. B. (1969), *J. Chem. Soc. D Chem. Commun.*, 1261.

Canceill, J., Collet, A., and Gottarelli, G. (1984), *J. Am. Chem. Soc.*, **106**, 5997.

Canceill, J., Lacombe, L., and Collet, A. (1985), *J. Am. Chem. Soc.*, **107**, 6993.

Cinquini, M., Cozzi, F., Sannicolo, F., and Sironi, A. (1988), *J. Am. Chem. Soc.*, **110**, 4363.

Collet, A. and Gabard, J. (1980), *J. Org. Chem.*, **45**, 5400.

Collet, A. and Jacques, J. (1978), *Tetrahedron Lett.*, 1265.

Cram, D. J., Hornby, R. B., Truesdale, E. A., Reich, H. J., Delton, M. H., and Cram, J. M. (1974), *Tetrahedron*, **30**, 1757.

Curie, P. (1894), *J. Phys. (Paris)*, III, **3**, 393.

Diederich, F. and Staab, H. A. (1978), *Angew. Chem. Int. Ed. Engl.*, **17**, 372.

Donaldson, J. D. and Ross, S. D. (1972), *Symmetry and Stereochemistry*, Halsted, Wiley, New York.

Dowd, P., Dyke, T., Neumann, R. M., and Klemperer, W. (1970), *J. Am. Chem. Soc.*, **92**, 6325.

Eaton, P. E. and Cole, T. W. (1964), *J. Am. Chem. Soc.*, **86**, 963, 3157.

Eaton, P. E. and Leipzig, B. (1978), *J. Org. Chem.*, **43**, 2483.

Eaton, P. E., Mueller, R. H., Carlson, G. R., Cullison, D. A., Cooper, G. F., Chou, T.-C., and Krebs, E. P. (1977), *J. Am. Chem. Soc.*, **99**, 2751.

Eliel, E. L. (1949), *J. Am. Chem. Soc.*, **71**, 3970.

Eliel, E. L., Allinger, N. L., Angyal, S. J., and Morrison, G. A. (1965), *Conformational Analysis*, Wiley-Interscience, New York, reprinted by American Chemical Society, 1981.

Elsenbaumer, R. L. and Mosher, H. S. (1979) *J. Org. Chem.*, **44**, 600.

Farina, M. and Audisio, G. (1970), *Tetrahedron*, **26**, 1827, 1839.

Farina, M. and Morandi, C. (1974), "High Symmetry Chiral Molecules," *Tetrahedron*, **30**, 1819.

Frei, H. and Günthard, Hs. H. (1976), *Chem. Phys.* **15**, 155.

Furukawa, J., Kakuzen, T., Morikawa, H., Yamamoto, R., and Okuno, O. (1968), *Bull. Chem. Soc. Jpn.*, **41**, 155.

Gabard, J. and Collet, A. (1981), *J. Chem. Soc. Chem. Comm.*, 1137.

Hanson, J. C. and Nordman, C. E. (1976), *Acta Cryst.*, **B32**, 1147.

Hargittai, I. and Hargittai, M. (1987), *Symmetry through the Eyes of a Chemist*, VCH Publishers, New York.

Heilbronner, E. and Dunitz, J.D. (1993), *Reflections on Symmetry*, VCH Publishers, New York.

Helmchen, G., Haas, G., and Prelog, V. (1973), *Helv. Chim. Acta*, **56**, 2255.

Helmchen, G. and Staiger, G. (1977), *Angew. Chem. Int. Ed. Engl.*, **16**, 116.

Hoffman, P. H., Ong, E. C., Weigang, O. E., and Nugent, M. J. (1974), *J. Am. Chem. Soc.*, **96**, 2620.

Hounshell, W. D. and Mislow, K. (1979), *Tetrahedron Lett.*, 1205.

Hutchings, M. G., Andose, J. D., and Mislow, K. (1975), *J. Am. Chem. Soc.*, **97**, 4553.

Hybl, A., Rundle, R. E., and Williams, D. E. (1965), *J. Am. Chem. Soc.*, **87**, 2779.

Jacques, J., Ed. (1986), *Sur la Dissymétrie Moléculaire*, Christian Bourgois, Paris, France.

Jaeger, F. M. (1920), *Lectures on the Principle of Symmetry*, 2nd ed., Elsevier, Amsterdam, The Netherlands.

Katz, T. J. (1960), *J. Am. Chem. Soc.*, **82**, 3784.

Katz, T. J. and Garratt, P. J. (1963), *J. Am. Chem. Soc.*, **85**, 2852; id. (1964); *ibid.* **86**, 5194.

Kettle, S. F. A. (1985), *Symmetry and Structure*, Wiley, New York.

Kokke, W. C. M. C. (1973), *J. Org. Chem.*, **38**, 2989.

Kokke, W. C. M. C. and Oosterhoff, L. J. (1972), *J. Am. Chem. Soc.*, **94**, 7583.

LaLancette, E. A. and Benson, R. E. (1963), *J. Am. Chem. Soc.*, **85**, 2853; id. (1965); *ibid.* **87**, 1941.

Leonard, J. E., Hammond, G. S., and Simmons, H. E. (1975), *J. Am. Chem. Soc.*, **97**, 5052.

Longuet-Higgins, H. C. (1963), *Mol. Phys.*, **6**, 445.

Maier, G., Pfriem, S., Schäfer, U., Malsch, K. D., and Matusch, R. (1981), *Chem. Ber.*, **114**, 3965.

Maier, G., Pfriem, S., Schäfer, U., and Matusch, R. (1978), *Angew. Chem. Int. Ed. Engl.*, **17**, 520.

Makino, T., Orfanopoulos, M., You, T.-P., Wu, B., Mosher, C. W., and Mosher, H. S. (1985), *J. Org. Chem.*, **50**, 5357.

McCasland, G. E., Horvat, R., and Roth, M. R. (1959), *J. Am. Chem. Soc.*, **81**, 2399.

McCasland, G. E. and Proskow, S. (1955), *J. Am. Chem. Soc.*, **77**, 4688; (1956) **78**, 5646.

Mislow, K. (1965), *Introduction to Stereochemistry*, Benjamin, New York.

Mislow, K. (1981), *J. Chem. Soc. Chem. Commun.*, 234.

Mislow, K. (1985), *Croatica Chem. Acta*, **58** (353).

Mislow, K. and Bickart, P. (1976/77), "An Epistemological Note on Chirality," *Isr. J. Chem.*, **15**, 1.

Mislow, K. and Glass, M. A. W., (1961), *J. Am. Chem. Soc.*, **83**, 2780.

Mislow, K., Glass, M. A. W., Hopps, H.B., Simon, E., and Wahl, G. H. (1964), *J. Am. Chem. Soc.*, **86**, 1710.

Mislow, K., O'Brien, R. E., and Schaefer, H. (1960), *J. Am. Chem. Soc.*, **82**, 5512.

Mosher, H. S. (1974), *Tetrahedron*, **30**, 1733.

Muetterties, E. L. and Knoth, W. H. (1968), *Polyhedral Boranes*, Marcel-Dekker, New York, p. 21 ff.

Nakazaki, M. (1984), "The Synthesis and Stereochemistry of Chiral Organic Molecules with High Symmetry," *Top. Stereochem.*, **15**, 199.

Nakazaki, M. and Naemura, K. (1981), *J. Org. Chem.*, **46**, 106.

Nakazaki, M., Naemura, K., and Arashiba, N. (1978), *J. Org. Chem.*, **43**, 689.

Nakazaki, M., Naemura, K., and Hokura, Y. (1982), *J. Chem. Soc. Chem. Commun.*, 1245.

Nouaille, A. and Horeau, A. (1985), *C.R. Séances Acad. Sc. Ser. II.*, 335.

O'Loane, K. (1980), "Optical Activity in Small Molecules, Non-enantiomorphous Crystals and Nematic Liquid Crystals," *Chem. Rev.*, **80**, 41.

Orchin, M. and Jaffé, H. H. (1970), "Symmetry, Point Groups, and Character Tables," *J. Chem. Educ.* **47**, 246, 372, 510.

Orchin, M. and Jaffé, H. H. (1971), *Symmetry, Orbitals and Spectra*, Wiley-Interscience, New York.

Paquette, L. (1984), *Strategies and Tactics of Organic Synthesis*, Lindberg, T., Ed., Academic, New York, p. 175.

Pasteur, L. (1905), *Researches on the Molecular Asymmetry of Natural Organic Products*, (English translation); Alembic Club Reprint No. 14, Alemic Club, Edinburgh, UK.

Peterson, E. J., Von Dreele, R. B., and Brown, T. M. (1976), *Inorg. Chem.*, **15**, 309.

Pinnavaia, T. J., Barnett, B. L., Podolsky, G., and Tulinsky, A. (1975), *J. Am. Chem. Soc.*, **97**, 2712.

Prelog, V. and Gerlach, H. (1964), *Helv. Chim. Acta.*, **47**, 2288.

Prelog, V. (1968), "Problems in Chemical Topology," *Chem. Br.*, **4**, 382.

Salthouse, J. A. and Ware, M. J. (1972), *Point Groups, Character Tables and Related Data*, Cambridge University Press, Cambridge, UK.

Shubnikov, A. V. and Koptsik, V. A. (1974), *Symmetry in Science and Art*, Plenum Press, New York.

Siegel, J., Gutierrez, A., Schweizer, W. B., Ermer, O., and Mislow, K. (1986), *J. Am. Chem. Soc.*, **108**, 1569.

Stirling, C. J. M. (1963), *J. Chem. Soc.*, 5741.

Streitwieser, A. and Müller-Westerhoff, U. (1968), *J. Am. Chem. Soc.*, **90**, 7364.

Streitwieser, A., Müller-Westerhoff, U., Sonnichsen, G., Mares, F., Morrell, D. G., Hodgson, K. O., and Harmon, C. A. (1973), *J. Am. Chem. Soc.*, **95**, 8644.

Ternansky, R. J., Balogh, D. W., and Paquette, L. A. (1982), *J. Am. Chem. Soc.*, **104**, 4503.

Verbit, L. (1970), "Optically Active Deuterium Compounds," *Progr. Phys. Org. Chem.*, **7**, 51.

Wheland, G. W. (1960), *Advanced Organic Chemistry*, 3rd ed., Wiley, New York.

Whitlock, H. W. (1962), *J. Am. Chem. Soc.*, **84**, 3412.

Wynberg, H., Hekkert, G. L., Houbiers, J. P. M., and Bosch, H. W. (1965), *J. Am. Chem. Soc.*, **87**, 2635.

Wynberg, H. and Hulshof, L. A. (1974), *Tetrahedron*, **30**, 1775.

Zalkin, A. and Raymond, K. N. (1969), *J. Am. Chem. Soc.*, **91**, 5667.

5

Configuration

5-1. DEFINITIONS: RELATIVE AND ABSOLUTE CONFIGURATION

Configuration has been defined (Cross and Klyne, 1976; IUPAC, 1979) as "The arrangement of the atoms in space of a molecule of defined constitution without regard to arrangements that differ only as after rotation about one or more single bonds."

> According to Noël (1987) the term *configuration* was introduced by Wunderlich in 1886 (see also Auwers and Meyer, 1888). Victor Meyer (loc. cit., p. 789), one of the pioneers of stereochemistry, has defined it as the "geometrical" aspect of structure.

The IUPAC definition presents some difficulties and is, in fact, not universally agreed on for reasons already touched on in Chapters 2 and 3. One difficulty, mentioned in Chapter 2, is that bond order may vary continuously, and it is therefore awkward, if not impossible, to single out rotation about a *single bond*. Are the C–N bonds in the amides and thioamides depicted in Figure 2.16 to be considered single or partial double bonds, and are the cis and trans isomers of these compounds thus to be considered as differing in configuration, or are they simply conformational isomers of compounds of unique configuration? The same problem arises in biphenyls (Fig. 3.2; and Chapter 14) and in a number of other structures. As detailed in Chapter 3 in connection with the 2-bromobutanes (Fig. 3.12) the problem is related to the technique of observation. Physical chemists, whose methodology (e.g., vibrational spectroscopy) is capable of detecting species separated by quite low barriers, have, in fact, tended to define configuration with*out* the exemption regarding rotation about single bonds. If this is done, configuration simply refers to the spatial arrangements of atoms in a molecule of given constitution, that is, to the stereochemical aspect of structure, as originally envisaged by Victor Meyer (see above) who, however, believed rotation about single bonds to be free. The gauche and anti forms of butane then become configurational isomers. Few organic chemists use the term configuration in this sense. At the opposite extreme, it has been suggested that, in the definition of configuration (see above), no regard should be given to rotation about bonds of any order, including double bonds. Under this definition, *cis-* and *trans*-2-butene would be conformers, not configurational isomers; it is understandable that such a view has not become popular either.

In this book we shall use configuration in the sense of "the arrangement of the atoms in space of a molecule of defined constitution without regard to arrangements that differ only by rotation about one or more *single* bonds, provided that such rotation is so fast as not to allow isolation of the species so differing." Under this definition stereoisomers of appropriately substituted biphenyls (*atropisomers*, cf. Chapter 14) are considered configurational isomers as long as they can be isolated, and so, of course, are cis–trans isomers of alkenes (the question as to the delineation of single and double bonds is swept under the rug).

It is recognized that this definition is quite imperfect for several reasons, one being the just mentioned uncertainty as to what constitutes a single bond. Another is that it arbitrarily takes isolation (rather than some other factor, such as spectroscopic observability by one or other technique) as the criterion of existence of configurational isomers, that is, it deals with "residual isomers" employing the criterion of isolability (cf. Chapter 3). A more serious imperfection is that "isolability" is itself not a clearly defined term, in that it lacks reference to both the temperature of the experiment and the lifetime required to consider a species "isolable." Nonetheless, the criterion has merit in a thermodynamic sense since, as already mentioned in Chapter 2, in application of the phase rule, only isolable species count as separate components.

> Yet another potential problem connected with the above definition (also mentioned in Chapter 2) is the emphasis on fast *rotation* as a bar to configurational distinction. There are other rapid processes precluding isolation of certain isomeric species, notably inversion in amines RR'R"N: (cf. Fig. 3.14), which is extremely rapid (Saunders and Yamada, 1963; Lehn, 1970; Lambert, 1971). Under the above definition the two rapidly inverting stereoisomers **A'** and **A"** (or **B'** and **B"**) (Fig. 3.14) must be considered configurationally different. Some authors (e.g., Mosher, 1980) have limited the term "configuration" to spatial arrangements that can be interconverted only by breaking and reforming of chemical bonds. This limitation would exclude not only atropisomers but presumably also such species as the enantiomers of phosphines of the type RR'R"P: which can be resolved and isolated but are interconverted readily (through inversion at phosphorus) by heating (Horner et al., 1961; Horner, 1964; McEwen, 1965; Gallagher and Jenkins, 1968). An argument based on the phase rule (see above) could be made here also, namely, that species isolable under the conditions of the experiment, such as enantiomeric phosphines at room temperature, should be considered to differ in configuration, but those not isolable, such as enantiomeric amines, should be considered to differ in conformation.

Despite these shortcomings, chemists usually agree as to what constitutes a difference in configuration; in those few cases where there is doubt, the meaning of the term as used by a given author should be clearly defined.

The determination and specification of configuration is an essential part of the structure determination of molecules displaying stereoisomerism, including chiral ones. Later in this chapter we shall explore how configuration is determined experimentally and how it is specified, either graphically or by descriptors attached to the chemical names (see also Chapters 9 and 11). Before this is done, however, it is desirable to distinguish between relative and absolute configuration. We shall illustrate this distinction with an analogy.

Suppose that a child who has not yet learned to distinguish right from left is given a right and a left shoe. The child will not be able to tell which of the shoes is right and which is left (i.e., their absolute configuration), but the child may well be able to tell that they are different in that one fits one of his or her feet (foot A) but not the other foot (B) and vice versa for the other shoe. Moreover, when provided with an assembly of shoes the child may (with some experimentation) be able to divide them into two kinds: one kind that fits foot A, the other that fits foot B. In other words, the child will be able to determine the configuration of the shoes relative to the feet and relative to each other (i.e., their relative configuration), even without knowing which shoe (or foot) is "right" and which is "left," i.e., their absolute configuration.

> In the terminology of Ruch (1968, 1972) the child can tell whether two shoes (or a foot and a shoe) are *homochiral* or *heterochiral* (referring to the same or opposite relative configuration) even though the child does not know the actual *sense of chirality* (absolute configuration) of either object (see also Kelvin, 1904.)

Prior to 1951, chemists were in the position of the child: The configurational correlation (i.e., relative configuration) of a large number of chiral substances with respect to others was known (cf. Section 5-4), but the absolute configuration or sense of chirality was not. [It was usual to correlate configuration with that of (+)- or (−)-glyceraldehyde ($HOCH_2CHOHCHO$).] Also known was the relative configuration of the chiral centers within numerous molecules having more than one chiral center; this correlation may, for example, be established by X-ray structural determination. However (cf. Chapter 2), X-ray diffraction was thought to be unable to differentiate between enantiomers and thus unable to reveal which enantiomer is which. This situation was changed only in 1951, thanks to the ingenious application of anomalous X-ray scattering to the determination of absolute configuration, to be described in Section 5-3 (Bijvoet et al., 1951).

5-2. ABSOLUTE CONFIGURATION AND NOTATION

Just as constitutional isomers are characterized by names as well as formulas, it is desirable to give differentiating symbols (so-called descriptors) to enantiomers, such as those of the lactic acids shown, in three-dimensional (3D) or projection formulas, in Figure 3.18. Of course one can characterize the two acids experimentally by the observable signs of their rotation but one also wants to specify their 3D structure without resorting to a formula drawing. It is for the purpose of such specification that systems of configurational notation have been devised.

There is no 1:1 relation between the sign of optical rotation (+ or −) and configurational notation (e.g., *R* or *S*). In fact, the sign of rotation may change not only with wavelength λ (cf. Chapter 13) but also with solvent, or even with concentration (Chapter 13). The conditions of the polarimetric measurement (λ, temperature, solvent, and concentration) must thus always be specified (see also Lowry, 1935). Even so, the presence of impurities, especially those with high specific rotation in a substance of low specific rotation, may give rise to errors (e.g., McKenzie, 1906; Canet, 1973) which can be avoided by careful purification.

Confusion is likely to ensue if one tries to determine the configuration of a chiral substance that is not properly characterized by its optical rotation (e.g., Brewster, 1982), optical rotatory dispersion (ORD), circular dichroism (CD), or in some other reproducible manner, such as behavior toward a given enzyme (see Chapter 7) or order of elution on a given enantioselective chromatographic column (Chapter 6).

Perhaps by coincidence, the now universally used Cahn–Ingold–Prelog (CIP) system (Cahn et al., 1956, 1966; Prelog and Helmchen, 1982) for the specification of molecular configuration originated in the same year (Cahn and Ingold, 1951) in which absolute configuration was first experimentally determined by Bijvoet. In this system, the descriptors *R* and *S*, which are italicized, placed in parentheses, and connected with a hyphen only when used as prefixes (*configurational descriptors*) of chemical names, are used to denote the configurations of enantiomers and diastereomers. To determine the proper descriptor for a chiral center, the *CIP chirality rule* (Cahn, Ingold, and Prelog, 1966) is employed. In general, the four ligands attached to the chiral center can be arranged in a unique sequence; for the moment let us call these ligands A, B, D, and E with the proviso that the sequence (or priority) is A precedes B precedes D precedes E (A > B > D > E or A→B→D→E). The CIP chirality rule requires that the model be viewed from the side opposite to that occupied by the ligand E of lowest priority (Fig. 5.1). The remaining three ligands then present a tripodal array, with the legs extending toward the viewer. If the sequential arrangement or sense of direction of these three ligands (A→B→D) is then clockwise (as in Fig. 5.1), the configurational descriptor is *R* (for Latin *rectus* meaning right), if it is counterclockwise, it is *S* (for Latin *sinister* meaning left).

Other rules apply to torsional stereoisomers, such as biphenyls (Fig. 3.1), and allenes and to stereoisomers possessing "chiral planes," such as paracyclophanes; these cases will be discussed in Chapter 14.

Figure 5.1. Chirality rule.

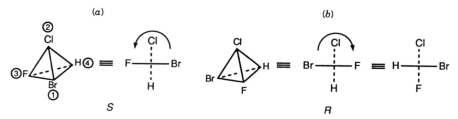

Figure 5.2. Configuration of CHFClBr.

We next discuss the *sequence rules* (Cahn et al., 1966, Prelog and Helmchen, 1982), that is, the rules for arranging groups A, B, D, and E in order of priority. In the first instance, this is done on the basis of atomic number: Ligands of higher atomic number precede ligands of lower atomic number. Figure 5.2 shows the designation of one of the simplest chiral molecules known, CHFClBr (Chapter 2).

The reader should refer to Figure 3.19 regarding the conversion of the 3D into a Fischer projection formula: If the projection formula is written with the atom of lowest priority at the bottom or at the top, clockwise array of the remaining three corresponds to the *R* configuration and counterclockwise array corresponds to the *S* configuration. This correspondence does not hold when the lowest sequence group is on the side of the Fischer projection (see Fig. 5.2*b*, *R* configuration); in that case the opposite assignment is correct (see also p. 111).

To include pyramidal triligant (three-coordinate) atoms, such as N or P, in the scheme, the "absent" ligand (usually a lone pair) is considered to have atomic number zero and will thus automatically have the lowest precedence.

Examples shown in Figure 5.3 are methylphenyl-*n*-propylphosphine (Horner et al., 1961; regarding its configuration, see McEwen, 1965; Gallagher and Jenkins, 1968) and *o*-carboxylphenyl methyl sulfoxide (Dahlén, 1974); the corresponding meta isomer, whose configuration does not appear to be known, was one of the first two sulfoxides to be resolved, by Harrison, Kenyon, and Phillips (1926). (Regarding chirality at sulfur, see Mikołajczyk and Drabowicz, 1982.)

When two or more atoms directly attached to the chiral center are the same, as in the case of lactic acid (Fig. 5.4), a principle of outward exploration is applied: If the proximal atom attached to the chiral center does not provide a decision, one proceeds outwardly to the second atom, then, if needed, to the third

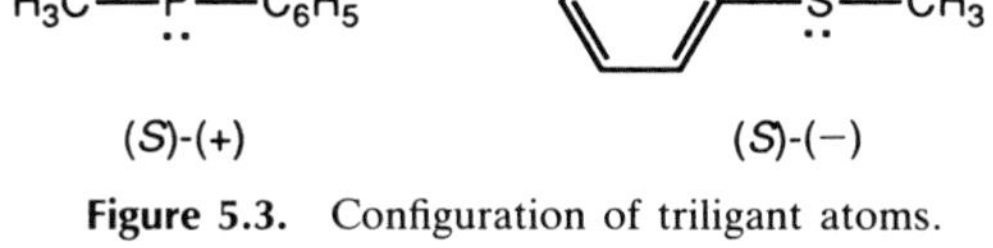

$$CH_2CH_2CH_3$$
$$H_3C-\overset{..}{P}-C_6H_5$$

(*S*)-(+)

$$HO_2C \qquad O$$
$$\overset{\|}{\underset{..}{S}}-CH_3$$

(*S*)-(−)

Figure 5.3. Configuration of triligant atoms.

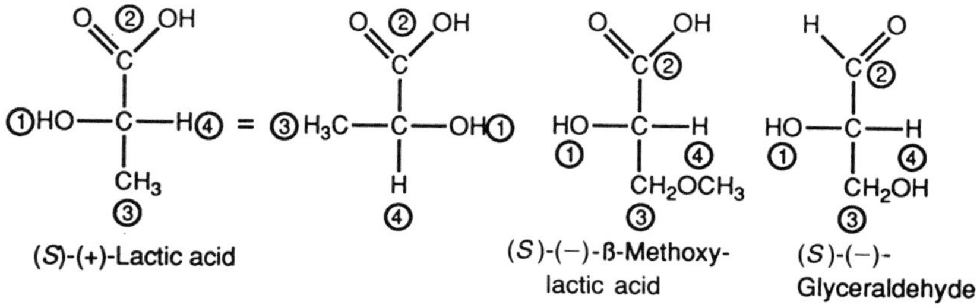

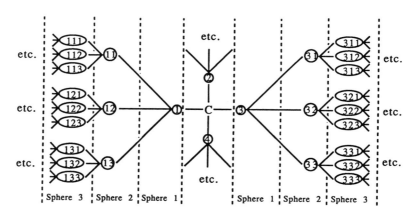

Figure 5.4. Applications of the sequence rule.

and so on. In the case of the lactic acids, since the carbon atoms of CO_2H and CH_3 provide no decision, one proceeds to the next attached atom and notes that $O > H$. The proper execution of this tree-graph exploration (Fig. 5.5) requires a well-defined hierarchy (Prelog and Helmchen, 1982). The most important principles are (a) *all* ligands in a given sphere (Fig. 5.5) must be explored before one proceeds to the next sphere. Thus in β-methoxylactic acid (Fig. 5.4), after one finds one oxygen atom in the second sphere attached to each carbon, one must not proceed to the third sphere, but one must explore the second sphere (Fig. 5.5) further and note that there is a second oxygen atom attached to the carbon in CO_2H but not in CH_2OCH_3 and since $O > H$, $CO_2H > CH_2OCH_3$. (b) In contrast, however, once a precedence of one path of exploration over another has been established in one sphere, that precedence carries to the next sphere. Suppose one wants to determine the sequence of the two ligands shown in Figure 5.6. Clearly, no decision is reached at the first (C), second (C, C, H), or third (C, H, F, C, H, Br) sphere. One therefore has to proceed to the fourth sphere, and one does so using the path of higher precedence at the third sphere (Br > F). The decision that ligand B precedes ligand A is therefore made on the basis that Br > Cl in the fourth sphere of the preferred branch rather than on the basis of the observation that I is the highest atomic number atom in the fourth sphere (of the less preferred branch).

Figure 5.5. Tree-graph.

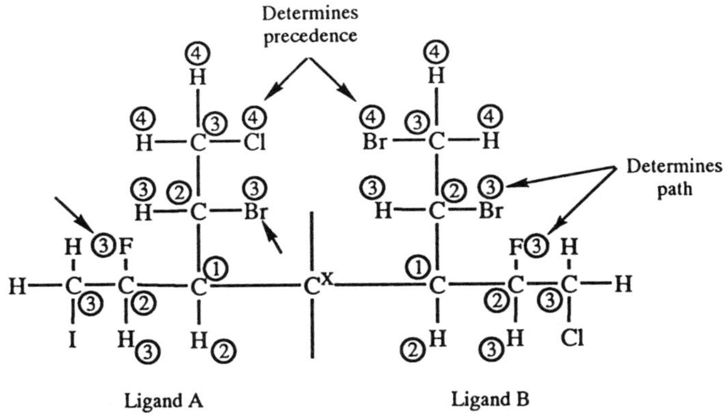

Figure 5.6. Hierarchy of paths.

We next consider the case of glyceraldehyde (Fig. 5.4). Clearly, $OH > C > H$, but there is a question as to the priority of CH_2OH and $CH{=}O$. This question is resolved by the convention of regarding each atom at a multiple bond as being associated with a "phantom" (duplicate) atom or atoms of the species at the other end of the multiple bond. Thus

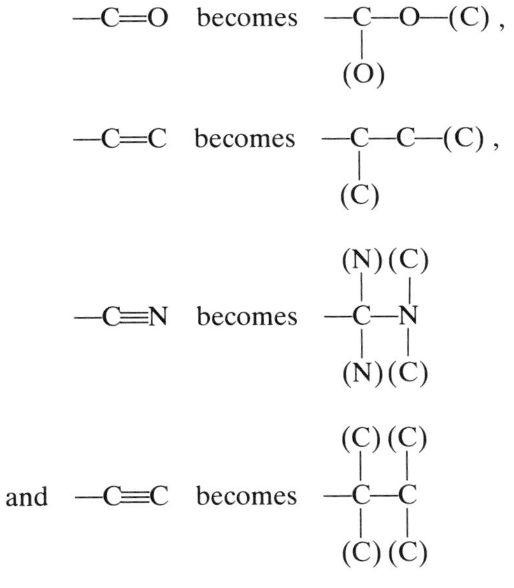

$$-C{=}O \quad \text{becomes} \quad -\underset{(O)}{C}-O-(C),$$

$$-C{=}C \quad \text{becomes} \quad -\underset{(C)}{C}-C-(C),$$

$$-C{\equiv}N \quad \text{becomes} \quad -\underset{(N)(C)}{\overset{(N)(C)}{C}}-N$$

$$\text{and} \quad -C{\equiv}C \quad \text{becomes} \quad -\underset{(C)(C)}{\overset{(C)(C)}{C}}-C$$

[The complemented (duplicate or phantom) atoms are enclosed in parentheses and considered to bear no further substituents.] It follows that in glyceraldehyde C=O has precedence over C–O (Fig. 5.4).

Ligand complementation is also required when a ligand is bidentate (or tri- or tetradentate) or when there is a cyclic or bicyclic component to a ligand (Prelog and Helmchen, 1982; see also Eliel, 1985). In either case, each branch of the cyclic structure is severed at the branch point (where it doubles onto itself) and

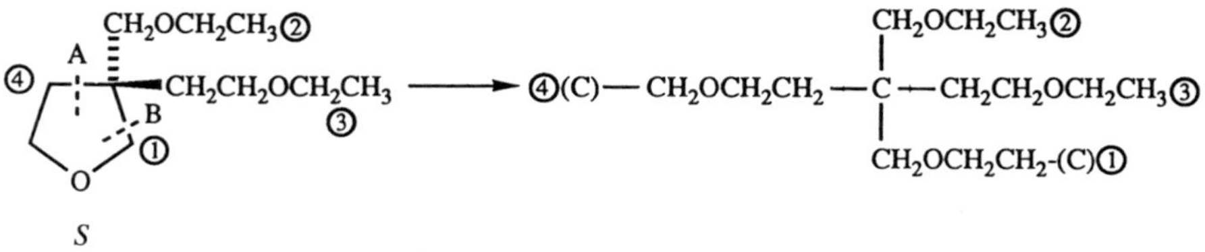

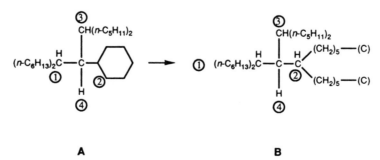

Figure 5.7. Stereochemical nomenclature for bidentate ligands.

the atom at the branch point is then complemented at the end of the chain resulting from the disconnection. The case of a bidentate ligand is exemplified by the tetrahydrofuran (THF) derivative shown in Figure 5.7. The two branches of the five-membered ring are (separately) severed at the chiral center, which is then complemented at the end of each of the two chains so created. It is seen that in disjuncture A the hypothetical ligand produced, $-CH_2OCH_2CH_2-(C)$, has precedence over the (real) acyclic ligand $CH_2OCH_2CH_3$ because of the presence of the phantom C at the end of the former ligand. In contrast, with disjuncture B the hypothetical ligand created, $-CH_2CH_2OCH_2-(C)$, falls behind the real one, $-CH_2CH_2OCH_2CH_3$, since the latter has three hydrogen atoms attached to the terminal carbon and the former has none. The configurational symbol is therefore S.

The related case of cutting a ring substituent is exemplified by the structure in Figure 5.8. Figure 5.8, **B** illustrates the treatment of the cyclohexyl ring. The proper sequence is therefore di-n-hexylcarbinyl > cyclohexyl > di-n-pentyl-carbinyl > H and the configuration is S.

We are now ready to consider phenyl. Since each of the six carbon atoms is doubly bonded (in either Kekulé structure) to another carbon, each ring carbon bears a duplicate carbon as a substituent. The ring so complemented (Fig. 5.9, **B**) is then opened according to the rules for cyclic systems to yield the representation in Figure 5.9, **C**.

It is evident that phenyl has precedence over most other tertiary carbinyl groups. Not only does each carbon of the benzene ring bear a phantom carbon atom, but in proceeding around the phenyl ring, and keeping in mind that real carbon atoms precede phantom carbon atoms in the path of exploration, one encounters a chain of a total of six tertiary carbon atoms before truncation occurs. This situation is in

Figure 5.8. Cyclic ligands in stereochemical nomenclature.

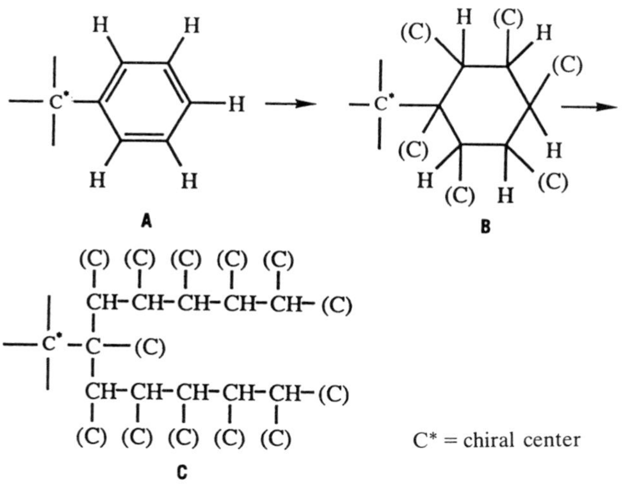

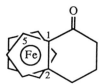

Figure 5.9. Treatment of benzene in the sequence rule.

contrast to that in the ethynyl group $-C\equiv CH$, which (see above) has precedence over $-C(CH_3)_3$ but yields to $(H_3C)_2C-CH(CH_3)_2$. The smallest saturated alkyl group that has precedence over phenyl is $(H_3C)_2C-C(CH_3)_3$.

Of the other rules applicable to constitutional differences between ligands (Cahn et al., 1966), most of which are rare in their applications, we mention only two. One has to do with π complexes: A multiple carbon bond π complexed to a metal is complemented at both ends by the metal (plus the metal's other ligands, if any) and correspondingly has its multiplicity reduced by one. Thus a silver complex of olefinic substituent $-CR=CR'R''$ would be considered $-\overset{\mathrm{Ag}}{\underset{|}{C}}R-\overset{\mathrm{Ag}}{\underset{|}{C}}R'R''$. An interesting example is the chiral ferrocene shown in Figure 5.10 (Schlögl and Falk, 1964; Falk and Schlögl, 1965; Schlögl, 1967). The configuration of the entire molecule is given by that of C(1), which, in view of the sequence $Fe > C(2) > C(5) > C=O$ is S (see also Chapter 14).

The last rule relating to material or constitutional differences (as distinct from configurational ones) to be mentioned in this chapter applies to molecules in which ligands differ *only* as a result of isotopic substitution. The sequence rule here is "higher atomic mass precedes lower atomic mass isotope," for example, $D > H$ and $^{13}C > {}^{12}C$. Applications to $C_6H_5CHDCH_3$, $CD_3CHOHCH_3$, benzyl p-tolyl sulfone-${}^{16}O$, ${}^{18}O$ and L-valine-4-${}^{13}C$ (cf. Chapter 4) are shown in Figure 5.11. On the other hand, 3-hexanol-2-d (Fig. 5.11) is an example where this rule

Figure 5.10. Configuration of a chiral ferrocene.

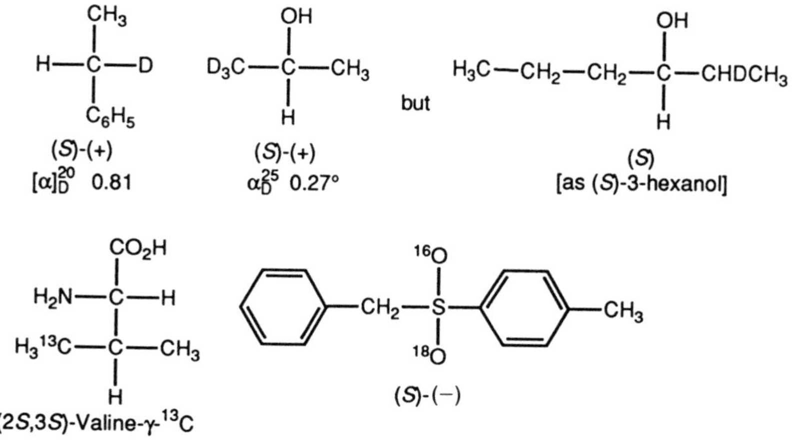

Figure 5.11. Configuration of chirally labeled compounds.

does not apply because the propyl group has precedence over the ethyl group in the absence of isotopic substitution; the symbol is therefore the same as it would be for unlabeled 3-hexanol of corresponding configuration.

We consider now chiral compounds in which the differences between ligands are not material or constitutional but configurational. Most such compounds contain more than one chiral center and will be considered in Section 5-4. Here we are considering ligands that differ in (olefinic) cis–trans isomerism. According to Prelog and Helmchen (1982) the olefinic ligand in which the substituent of higher sequence is on the *same* side of the olefinic double bond as the chiral center has precedence over the ligand in which the substituent of higher sequence is trans to the chiral center. Examples are shown in Figure 5.12.

This precedence is not related either to the classical cis–trans designation of double-bond configuration or to the *E–Z* designation; in compound **A** in Figure 5.12 the cis group (here also *Z*) happens to have precedence over the trans (*E*) group, but in **B** the opposite is true. In the latter case, the symbol R_n (*n* for new) is used because the 1982 notation differs from that proposed (Cahn, Ingold, and Prelog) in 1966.

One frequently encounters cases where a formula is depicted with the lowest sequence group in front rather than in back or with some group other than the

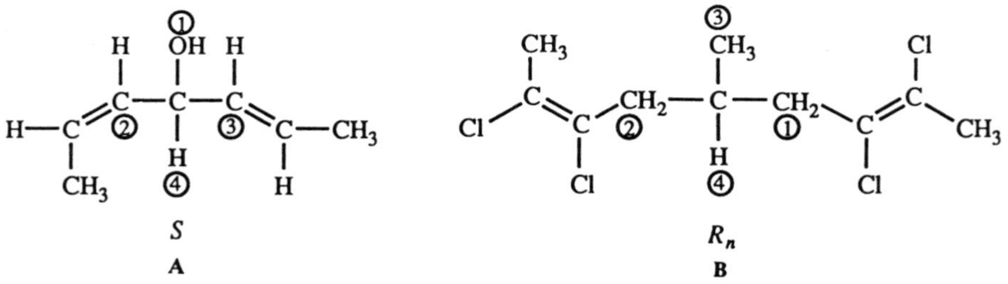

Figure 5.12. Configuration of chiral centers with cis–trans isomeric ligands.

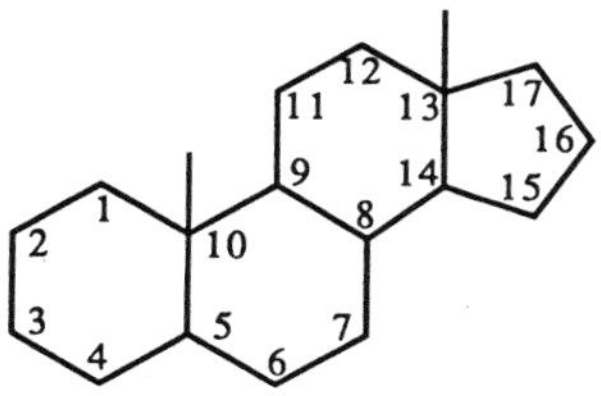

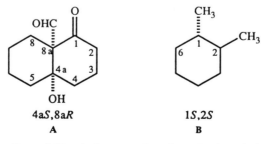

Figure 5.13. Steroid skeleton.

In front:	+	−	+	−
Ligand No:	1	2	3	4
In back:	−	+	−	+

Figure. 5.14. Ligand permutation scheme.

lowest sequence group in back. The steroid shown in Figure 5.13 (methyl groups above the plane of the paper) will illustrate the former case; let us consider its configuration at C(10). The sequence of the attached atoms is C(9) > C(5) > C(1) > CH_3. It should be evident by now that $CH > CH_2 > CH_3$; the sequence C(9) > C(5) follows from the precedence of the attached atoms [C(8) vs. C(4) or C(6).] Since the lowest sequence group CH_3 is in front, the representation is opposite to that demanded in Figure 5.1 (p. 104). Therefore, the descriptor R, which would be obtained by considering the (clockwise) array of the remaining three groups C(9), C(5), and C(1), must be reversed; the correct descriptor is S. One can, of course, arrive at the same conclusion by placing oneself *behind* the plane of the paper.

In general, there are eight possibilities that are summarized schematically in Figure 5.14 (Eliel, 1985). The middle row indicates the CIP priority of the ligands that are in front (upper row) or in back (lower row) in the 3D formula. The signs in the upper and lower rows indicate whether the apparent array (clockwise, R; counterclockwise, S) of the remaining three ligands gives the correct (indicated by +) or reversed (−) descriptor. Clearly, when the No. 4 ligand is in the back (Fig. 5.1), the descriptor obtained by observing the array of the remaining three is correct, so the corresponding entry in the scheme is +. For other ligands in front or in back the signs alternate, as shown in Figure 5.14, so the scheme can be reconstructed at a moment's notice when needed.

Additional applications are shown in Figure 5.15. In the decalone derivative **A** the sequence at C(4a) is OH > C(8a) > C(4) > C(5). Since the highest priority group OH is in back, the clockwise array of C(8a) → C(4) → C(5) does not give the correct descriptor (cf. Fig. 5.14, lower row, minus sign for ligand No. 1); the correct descriptor is therefore S. At C(8a), however, where the sequence is C(1) > CHO > C(4a) > C(8), the group of second priority (CHO) is in the back. This situation corresponds to a plus sign in Figure 5.14 (lower row), so the

Figure 5.15. Assignment of configurational symbol.

clockwise array $C(1) \rightarrow C(4a) \rightarrow C(8)$ gives rise to the correct descriptor R. Compound **A** is therefore $4aS, 8aR$. Finally, in compound **B** at $C(1)$ the sequence is $C(2) > C(6) > CH_3 > H$ (H in front, not shown). The group at the rear (CH_3) is third in priority (Fig. 5.14, lower row, minus sign) and so the clockwise array $C(2) \rightarrow C(6) \rightarrow H$ corresponds to S. At $C(2)$ the configuration is obviously S also (H is implied in the rear), and, indeed, in the trans isomer the configurations of $C(1)$ and $C(2)$ must be the same, $1S, 2S$.

We conclude with a listing of some common substituents in order of priority in the CIP sequence (for a more extensive list, see Cross and Klyne, 1976; IUPAC, 1979):

—I, —Br, —Cl, —PR$_2$, —SO$_3$H, —SO$_2$R, —SOR, —SR, —SH, —F, —OTs,

—OCOCH$_3$, —OC$_6$H$_5$, —OCH$_3$, —OH, —NO$_2$, —N$^+$(CH$_3$)$_3$, N(C$_2$H$_5$)$_2$, —N(CH$_3$)$_2$, —NHCOC$_6$H$_5$,

—NHR, —NH$_2$, —CO$_2$R, —CO$_2$H, —COC$_6$H$_5$, —COCH$_3$, —CHO, —CH$_2$OR, —CH$_2$OH,

—CN, —CH$_2$NH$_2$, —C$_6$H$_5$, —C≡CH, (CH$_3$)$_3$C—, cyclohexyl, *sec*-butyl, —CH=CH$_2$,

isopropyl, —CH$_2$C$_6$H$_5$, —CH$_2$CH=CH$_2$, isobutyl, —C$_2$H$_5$, —CH$_3$, —D, —H.

> The CIP system, though applicable to the vast majority of chiral compounds, is not perfect. The system has some inconsistencies, especially when the chiral element is a chiral axis or plane (see Chapter 14) or when chirality is due to the molecular framework as a whole and cannot be readily factorized (e.g., twistane, Fig. 4.11). For a criticism of the system see Dodziuk and Mirowicz (1990) and references cited therein.

Prior to the 1960s, it was common to denote the configuration at chiral centers with references to planar projection (Fischer) rather than 3D formulas; the descriptors used to this end are D and L (cf. Eliel, 1962). At the present time, the DL system is used only for amino acids and carbohydrates, as shown in Figure 5.16 (see also Figs. 3.23 and 3.24). In α-amino acids, the configuration is L if the amino (or ammonio) group is on the left of the Fischer projection formula with the carboxylate group written on top, D for the enantiomer. For sugars, the configurational notation is based on the highest numbered (farthest from the carbonyl end) chiral CHOH group. If in this group (with the Fischer projection written so that the C=O is on top and the CH$_2$OH is on the bottom) OH is on the right, the configuration is D, if OH is on the left, the configuration is L.

A number of other configurational names and symbols are used in compounds having more than one chiral center, for example, in ring compounds, and alkenes. These names and symbols will be introduced in later sections or chapters, where the appropriate compounds are treated.

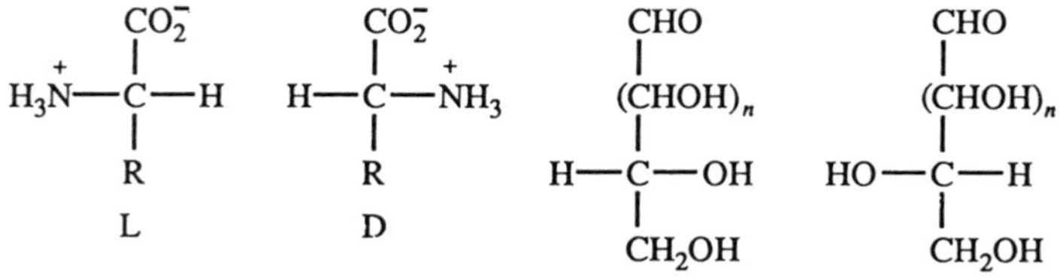

Figure 5.16. Configuration of α-amino acids and monosaccharides.

5-3. DETERMINATION OF ABSOLUTE CONFIGURATION

a. Bijvoet Method

It was only in 1951 that the first experimental determination of the absolute configuration of any chiral molecule was achieved (Bijvoet et al., 1951; Bijvoet, 1955). In ordinary X-ray crystallography the intensities of the diffracted beams depend on the distances between the atoms but not on the absolute spatial orientation of the structure. This comes about because the phase change due to scattering of the incident radiation is (nearly) the same for all atoms. Thus a chiral crystal and its enantiomorph produce the same X-ray patterns and cannot be distinguished from one another. As a simple example, the one-dimensional (1D) chiral array A–B cannot be distinguished from B–A because the interference pattern at any point I in space will be the same for the two 1D enantiomers since it depends only on the absolute value of the path length difference $|AI - BI| = |BI - AI|$ (Fig. 5.17). Thus there will be no change in interference pattern when atoms A and B are interchanged. The same is true in the more complex case of a 3D chiral crystal: Since there is nothing to distinguish propagation of X-rays in a direction A–B from that in the opposite direction B–A (Friedel's law), X-ray diffraction patterns are centrosymmetric, whether the crystal under investigation has a center of symmetry or not (see Dunitz, 1979).

The solution to this dilemma was, in principle, found in 1930 when Coster et al. (q.v.) used so-called anomalous X-ray scattering to determine the sequence of planes of zinc and sulfur atoms in a crystal of zinc blende (ZnS). The method involves using X-rays of a wavelength near the absorption edge of one of the atoms, in this case zinc. This results in a small *phase change* of the X-rays scattered by the zinc atoms relative to the sulfur atoms, which does *not* depend on their relative positions. If B in Figure 5.17 denotes the zinc atom, the effect of retarding, say, the phase of the diffracted X-ray is the same as if the path B–I were longer than A–I, by a hypothetical extra path length p, simulated by the phase change. Thus the scattered intensities for A–B and B–A are no longer the same, since the phase change is in opposite directions in the two cases. (In terms of the hypothetical quantity p, it appears as if the path length for A–B is $|AI - BI| - p$, but for B–A is $|AI - BI| + p$.) As a result, the diffraction pattern is no longer centrosymmetric; pairs of spots in the pattern that are related by the center of symmetry (cf. Chapter 4), which are now called *Bijvoet pairs* (see below), become unequal in intensity. Provided that the structure is known except

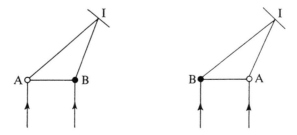

Figure 5.17. Determination of absolute configuration.

for the absolute configuration, one can then calculate the relative intensities of the Bijvoet pairs for the R and S isomers, and by comparison with experiment one can tell which is which. Thus one can determine the sense of chirality by using X-rays on the absorption edge of one of the atoms in the structure. Usually, this is a relatively heavy atom (sulfur or larger) since the phase change generally increases with atomic mass.

This principle was applied by Bijvoet et al. in 1951 (q.v.) by using zirconium $K\alpha$ X-rays in the X-ray crystallographic study of sodium rubidium tartrate (Fig. 5.18). The (+)-tartrate anion was found to have the R, R configuration. Since (+)-tartaric acid had been correlated with many other chiral compounds, notably the sugars (Section 5-5), the determination of its configuration marked a major milestone in stereochemistry. Soon afterwards the absolute configuration of the hydrobromide of the amino acid D-(−)-isoleucine (Fig. 5.18) was determined utilizing the phase lag at bromine introduced by the use of uranium $L\alpha$ radiation (Trommel and Bijvoet, 1954).

> Isoleucine hydrobromide has two chiral centers, C(2) and C(3). Ordinary X-ray structure determination discloses their *relative* configuration. The Bijvoet method leads to the *absolute* configuration (handedness, sense of chirality) of a given molecule; being an X-ray method, it also gives the relative spatial arrangement of all the atoms, and thus the absolute configuration at *each* chiral center.
>
> Listings of compounds whose absolute configuration has been determined by anomalous X-ray scattering are available (Allen and Rogers, 1966; Allen et al., 1968–1970; Klyne and Buckingham 1978, Buckingham and Hill, 1986).

An interesting variation of the Bijvoet method has been used in the determination of the absolute configuration of (−)-glycolic-*d* acid (−)-CHDOH–CO_2H, obtained by enzymatic reduction of glyoxylic-*d* acid CDO–CO_2H with NADH (nicotinamide adenine dinucleotide, reduced form) in the presence of liver alcohol dehydrogenase (Johnson et al., 1965). The optically active acid was converted to its ^6Li salt, which was then subjected to X-ray and neutron diffraction analysis. The X-ray analysis not unexpectedly failed to distinguish hydrogen and deuterium (but see Seiler, Dunitz et al., 1984 for an attempt to make the distinction on the basis of the difference in the vibrational behavior of H and D). However, neutron diffraction does discriminate between the two isotopes. Moreover, neutrons of wavelength 107.8 pm (1.078 Å) produce anomalous scattering from ^6Li so that the configuration could be deduced as being S from the

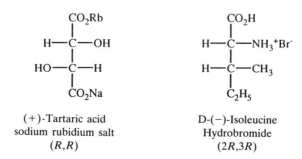

Figure 5.18. Absolute configuration as determined by anomalous X-ray scattering.

neutron diffraction analysis. This result is important inasmuch as the configurations of numerous other chiral deuterium compounds (RCHDR′) have been correlated with that of (S)-(−)-glycolic-d acid (cf. Section 5-6). Moreover, chiral glycolic-t acid has been correlated, configurationally, with the deuterium analogue by correspondence in enzymatic degradation (cf. Chapter 8) and can thus serve as a standard of reference for chiral RCHTR′ species.

b. Theoretical Approaches

A second (theoretical) approach to the determination of absolute configuration that actually antedated the experimental approach depends on the comparison of measured optical rotations with values computed by theory, an approach pioneered by Kuhn (1935, 1952) and Kirkwood (1937) (see also Wood et al., 1952; Caldwell and Eyring, 1971). More recent extensions include the comparison of experimental and theoretically computed ORD and CD curves (Kauzmann et al., 1940; Moffitt, 1956; reviewed by Bosnich, 1969) as well as the exciton chirality method (cf. Chapter 13-4.d).

An apparent contradiction between theoretical and experimental methods seemed to develop in 1972 when the work of Tanaka and co-workers (Tanaka et al., 1972, 1973a,b) on 2,7-disubstituted triptycenes purported to demonstrate that the configurations of these compounds as determined by the Bijvoet method (see above) were incorrect and the Bijvoet method thereby flawed. However, it turned out (Mason, 1973; Hezemans and Groenewege, 1973) that Tanaka's result rested on an erroneous application of exciton theory.

The 1972 challenge to Bijvoet's result created quite a flurry of excitement at the time, since it suggested that the assigned absolute configuration of *every* chiral compound should be reversed: This concern led to a careful reexamination of the basis of Bijvoet's experiment. As was mentioned earlier, anomalous X-ray scattering had originally been used to determine the sequence of the layers of zinc and sulfur atoms in a polar crystal of zinc sulfide. A supposedly independent assessment of this order was made on the basis of appearance (one face of the crystal is shiny, the other is dull), crystal growth (the shiny face is well developed, the dull one is not), and etching properties. If this assessment were in error, the interpretation of the Coster anomalous scattering experiment of ZnS as well as the interpretation of Bijvoet's experiment would have to be reversed. Fortunately, by 1972 it was possible to make an independent determination of the sequence of the layers in the ZnS crystal (i.e., of the sense of polarity of the crystal) by using noble gas ion scattering. This type of scattering depends on the mass number of the scattering atoms and thus unequivocally distinguishes a layer of zinc atoms from one of sulfur atoms. Application of the technique using Ne^+ ions led to the conclusion that the polarity of the ZnS crystal assumed on the basis of its etching pattern was correct (Brongersma and Mul, 1973) and that the basis of the interpretation of Bijvoet's original experiment was therefore sound.

In general, the result of an *individual* determination of absolute configuration by the Bijvoet method is more prone to error than other structure determinations by X-ray diffraction, since it depends on small differences in the intensity of pairs of related reflections (the *Bijvoet pairs*). The probability of error increases in the absence of a heavy atom (e.g., Hulshof, Wynberg et al., 1976; Shirahata and Hirayama, 1983).

c. Modification of Crystal Morphology in the Presence of Additives

The concept of polarity in crystals has not only contributed to the confirmation of the correctness of Bijvoet's method, but has also given rise to another, independent method of determining absolute configuration experimentally (Berkovitch-Yellin, Lahav, et al., 1982; cf. Addadi, Lahav, et al., 1985; Addadi, Leiserowitz, et al., 1986). Suppose that one examines a crystal of a given enantiomer. For the sake of simplicity we shall again assume 1D chirality, as in a molecule W–Y (enantiomer: Y–W). Let us further suppose that the disposition of the molecules in the crystal has been determined by ordinary X-ray diffraction and that they are oriented along the crystal axis as shown in Figure 5.19. As already pointed out (cf. Fig. 5.17) ordinary X-ray diffraction cannot distinguish between W–Y and Y–W arrangements. Thus, even though the two ends of the polar crystal (along the W–Y or Y–W axis) may be visually distinct (cf. the case of ZnS above), it is not known which direction in the crystal corresponds to which of the constituent molecules (W–Y or Y–W) and thus the configuration of the latter remains unknown. The breaking of this impasse rests on an observation (Addadi, Leiserowitz, et al., 1982a) that impurities or additives adsorbed on growing crystal faces tend to selectively stunt the growth of the crystal in the direction orthogonal to the faces in question. Thus, impurity W–A in Figure 5.19, being adsorbed on the face exposing Y, would impede growth in the direction +b (i.e., along faces f_1 and f_2), whereas additive Z–Y will impede growth along −b by being adsorbed on exposed W at faces f_3, f_4, f_5. It remains to find additives which, in a defined way, attach themselves to the W or Y ends of the molecules (or rather, cf. Fig. 5.19, to the faces where W or Y groups, respectively, are exposed). Provided a consistent set of such additives can be found and if the stunting of growth in one or other direction along the crystal axis can be observed (this is relatively easy, since substantial changes in crystal habit are often seen), then the orientation of W–Y relative to the macroscopic crystal axis can be defined and the configuration of W–Y thus determined.

A 3D example (Berkovitch-Yellin, Lahav, et al., 1982) is provided by L-lysine hydrochloride dihydrate (Fig. 5.20, **A**). In this crystal the carbon chain as written orients itself more or less along one of the crystal axes. Impurities tend to interfere with growth along directions where there are constitutional or configurational differences between the substrate and the impurity. Thus the first two extraneous substances (**B** and **C**) shown in Figure 5.20 are identical with L-lysine hydrochloride in the chain-end NH_3^+ but differ at the carboxylic amino acid end. These compounds will, in fact, inhibit growth of the L-lysine crystal in one and the

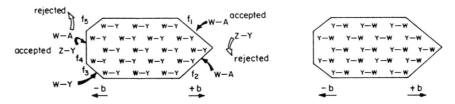

Figure 5.19. Polar crystal W–Y in the presence of impurities W–A or Z–Y. [Reprinted with permission from Berkovitch-Yellin, et al. (1982), *Nature (London)*, **296**, 27.]

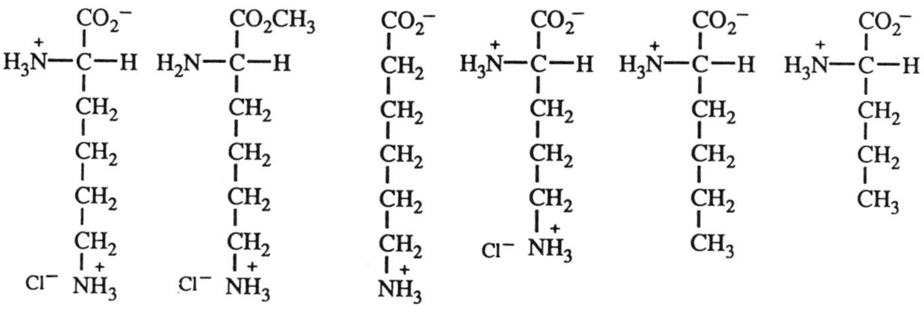

Figure structures:

A — L-Lysine hydrochloride
CO_2^- — H_3N^+—C—H — CH_2 — CH_2 — CH_2 — CH_2 — $Cl^-\ NH_3^+$

B — L-Lysine methyl ester hydrochloride
CO_2CH_3 — H_2N—C—H — CH_2 — CH_2 — CH_2 — CH_2 — $Cl^-\ NH_3^+$

C — ε-Amino caproic acid
CO_2^- — CH_2 — CH_2 — CH_2 — CH_2 — CH_2 — NH_3^+

D — L-Ornithine hydrochloride
CO_2^- — H_3N^+—C—H — CH_2 — CH_2 — CH_2 — $Cl^-\ NH_3^+$

E — α-Amino caproic acid
CO_2^- — H_3N^+—C—H — CH_2 — CH_2 — CH_2 — CH_3

F — α-Amino valeric acid
CO_2^- — H_3N^+—C—H — CH_2 — CH_2 — CH_3

Figure 5.20. L-Lysine hydrochloride and addends used in the determination of its configuration.

same direction; this direction is assumed to be that along the CO_2^- end of the chain. The latter three impurities (**D–F**), in contrast, are all α-amino acid zwitterions, but differ at the other end of the chain, either by chain length or by end group. These additives impede crystal growth in the opposite direction from the first two; this is taken to be the direction along the $-CH_2-NH_3^+$ moiety. The direction of the lysine molecule along the crystal axis, and hence its absolute configuration as L, are thus determined. It should be noted that one of the addends (ε-aminocaproic acid, **C**) is achiral; in this method the addends need not be chiral or, if they are chiral, their absolute configuration need not be known.

Other methods for distinguishing the two ends of a polar crystal have been described (see, e.g., the case of zinc blende, p. 113). Thus the two ends of crystals of chiral nonracemic gluconamides $H(CH_2)_nNHCO(CHOH)_4CH_2OH$ ($n = 7$–10), can be distinguished by the wettability of the crystal face containing the CH_2OH groups (hydrophilic) and the lack of wettability of the faces containing the terminal CH_3 groups (hydrophobic) (Wang, Lahav, and Leiserowitz, 1991; see also references cited therein for other, related methods, notably the probing of crystal polarity by pyroelectric effects: Pennington, Paul, Curtin, et al., 1988; Curtin and Paul, 1981).

It is even possible to use crystals of racemic compounds (Chapter 6) in the determination of absolute configuration; this has been described by Addadi, Leiserowitz, et al. (1982b) whose paper should be consulted for details.

5-4. RELATIVE CONFIGURATION AND NOTATION

As was mentioned earlier, the term "relative configuration" may be used in two contexts: relative configuration of a chiral center of one compound with respect to that of another, or relative configuration of several chiral centers within one compound with respect to each other. We shall postpone the comparison of

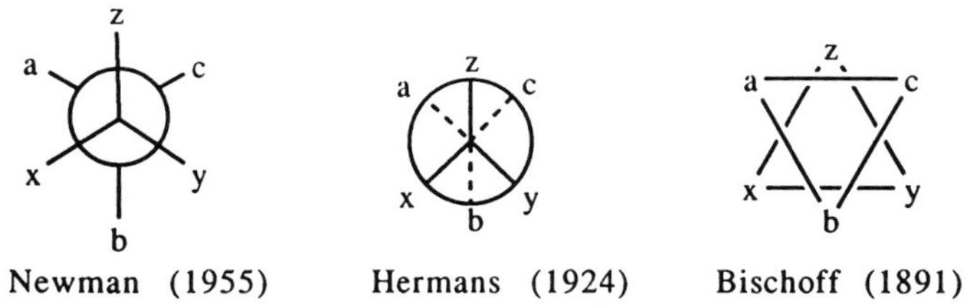

Newman (1955)   Hermans (1924)   Bischoff (1891)

Figure 5.21. Front-on projection formulas.

configuration of distinct compounds until Section 5-5. In this section we will deal with single compounds containing two or more chiral centers.

Fischer projection formulas for representation of compounds with more than one chiral center have already been discussed in Chapter 3 (e.g., Figs. 3.24 and 3.26). These formulas have the drawback, however, that they depict the molecule in the unrealistic eclipsed conformation (cf. Chapter 10). Two other systems, the so-called Newman projection formulas (Fig. 5.21; Newman, 1955) and the so-called "sawhorse" formulas (Fig. 5.22) show the molecules in their staggered conformations. "Front-on view" formulas very similar to the Newman projections have been used much earlier (Bischoff, 1891; Hermans, 1924) and are also shown in Figure 5.21. Figure 5.22 also indicates the interconversion of the various perspectives. Figure 5.23 shows the so-called "flying wedge" and "zigzag" formulas that are useful to show the actual conformation (as well as configuration) of compounds having two or more chiral centers. (Newman and sawhorse formulas are most useful for compounds having no more than two chiral centers.) The flying wedge and zigzag formulas have the advantage of showing the molecular backbone in the (true) staggered conformation and entirely in the plane of the paper. The zigzag formula, although simpler, can be used only for chiral centers bearing a hydrogen atom, since this atom is assumed and not drawn. The flying wedge formula does not suffer from this limitation and is convenient for the representation of any number of chiral centers in a straight chain. In a molecule having two or more identical chiral centers, such as tartaric acid (Fig. 5.23), the fact that two corresponding groups (here OH) are on the same side of the molecular framework does not imply the presence of a symmetry plane; this is in contrast to the situation in the Fischer projection (Fig. 3.26).

The notation for compounds having two or more chiral centers should, in principle, be simple: Each center is given its proper CIP descriptor along with the

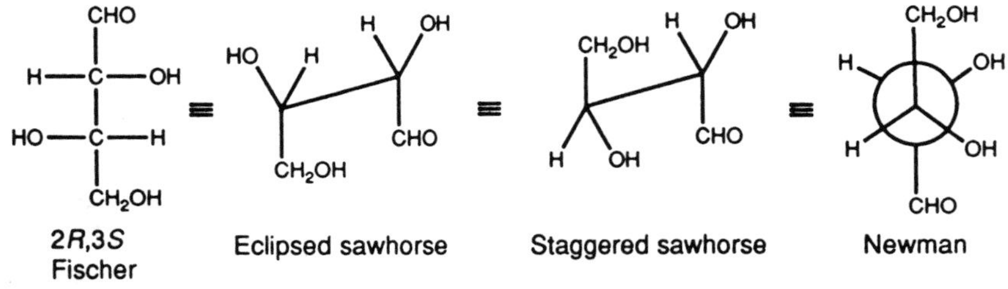

2*R*,3*S*   Eclipsed sawhorse   Staggered sawhorse   Newman
Fischer

Figure 5.22. Fischer, sawhorse, and Newman projections of L-threose.

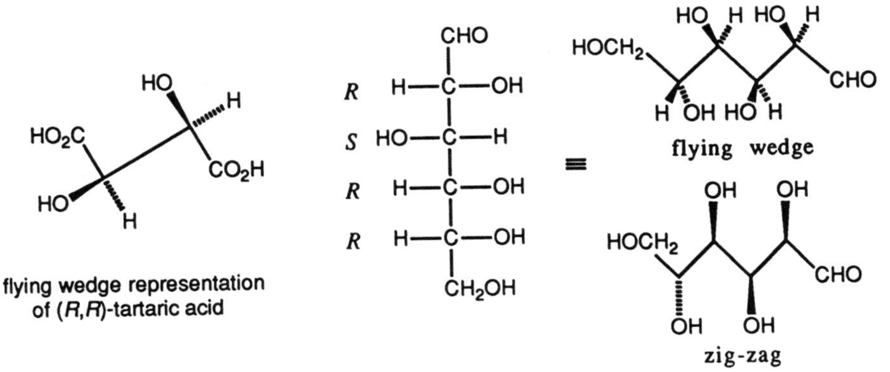

Figure 5.23. Flying wedge and zigzag formulas.

usual positional number (locant). Thus L-threose (Fig. 5.22) is (2R,3S)-2,3,4-trihydroxybutanal and D-glucose (Fig. 5.23) is (2R,3S,4R,5R)-2,3,4,5,6-pentahydroxyhexanal. This system works well when the compounds are resolved (nonracemic) and their absolute configurations at each chiral center (i.e., their absolute as well as relative configurations) are known. But if, as frequently happens, one deals with racemates, or if only the relative but not the absolute configuration is known, the system requires modification.

For racemates, the symbols RS or SR are used for the chiral centers. The lowest numbered (or sole) chiral center is automatically given the symbol RS, which thus serves as a reference; let us call this chiral center No. 1. The next chiral center No. 2 is then denoted by RS if, in the enantiomer where the No. 1 chiral center is R, No. 2 is R also (and, likewise, in the enantiomer where No. 1 is S, No. 2 is S also). Conversely, if the relative configuration is such that, when No. 1 is R, No. 2 is S (and, therefore, if No. 1 is S, No. 2 is R), the configuration of the second center is denoted by SR. The two racemic diastereomers are therefore 1RS,2RS and 1RS,2SR, the former denoting a racemic mixture of 1R,2R and 1S,2S, the latter of 1R,2S and 1S,2R. By way of examples, let us consider the racemates of the open-chain forms of threose depicted in Figure 5.22 and glucose shown in Figure 5.23. The former would be (2RS,3SR)-2,3,4-trihydroxybutanal and the latter (2RS,3SR,4RS,5RS)-2,3,4,5,6-pentahydroxyhexanal. Had we started from the descriptor of the enantiomeric D-threose, 2S,3R, we should still have arrived at the symbol 2RS,3SR (and not 2SR,3RS) for the racemate, since, by convention, the lowest numbered chiral center is always RS.

On occasion, only the relative configuration of several chiral centers, not the absolute configuration of any one of them, is known. Since the RS/SR system only implies relative configuration, it may be used in this case also. Alternatively, a system used by *Chemical Abstracts* (CA) may be employed: The lowest numbered chiral center is arbitrarily assumed to be R and the configuration of the others is denoted as R^* or S^*, relative to it. In order to signal that the configurations are relative, not absolute (i.e., the configuration of the first chiral center is assumed arbitrarily), all symbols are starred. Thus, prior to 1951, when the absolute configuration of glucose was not known (though its relative configuration had been determined by Fischer; cf. Section 5-6) its configurational descriptor would have been $2R^*,3S^*,4R^*,5R^*$. Beilstein uses $rel\text{-}(2R,3S,4R,5R)$ for the same case of unknown absolute configuration. It might be noted here that CA uses separate symbols for absolute and relative configuration; thus the configurational notations of CA for the tartaric acids (Fig. 3.26) are $[S\text{-}(R^*,R^*)]\text{-}(-)$, $[R\text{-}(R^*,R^*)]\text{-}(+)$ and (R^*,S^*) (meso). The descriptor for racemic tartaric acid is $(R^*,R^*)\text{-}(\pm)$.

The use of very similar symbols R^*,S^* for relative and R,S for absolute configuration is somewhat confusing, for, whereas the symbols for relative configuration are reflection invariant, those for absolute configuration change with reflection. Brewster (1986) proposed a modified use of the $l\text{-}u$ (like-unlike) nomenclature of Seebach and Prelog (1982, see next paragraph) by referring each chiral center to the lowest numbered one, which is arbitrarily given the symbol l (in place of R^*). Then, if with this chiral center as R, another is also R, the latter is given the descriptor l (like), if it is S, it is given the descriptor u (unlike). The absolute configuration, if known, is denoted separately by indicating the actual configuration of the lowest numbered chiral center as R or S. Thus the descriptor for D-glucose (Fig. 5.23) would be $(R)\text{-}(2l,3u,4l,5l)$, that for $(+)$-tartaric acid (Fig. 3.26) would be $(R)\text{-}(2l,3l)$, that for racemic tartaric acid $(2l,3l)$, that for meso-tartaric acid $(2l,3u)$.

Because of the very frequent occurrence of compounds with two chiral centers, chemists have wanted to have words as well as symbols for the relative chirality of the two centers involved without individually specifying the configuration of either one. It is, of course, possible to do this on the basis of the CIP system; thus Seebach and Prelog (1982) use the symbol l (for like) and u (for unlike) when the CIP descriptors for the two chiral centers are like (R,R or S,S or RS,RS) or unlike (R,S or S,R or RS,SR), respectively. While this notation is as unambiguous as the CIP system itself, it is not particularly convenient, because it requires ascertaining the CIP descriptors for both chiral centers before the l or u descriptor can be assigned. What most chemists want is an "at a glance" designation for a system of two chiral centers.

Such an "at a glance" nomenclature was coined many years ago; it is based on the names of the four-carbon sugars erythrose and threose (Fig. 3.23). As shown in Figure 5.24, **A**, and **B** ($R \neq R'$), isomers with the two identical ligands attached to a carbon chain on the same side of the Fischer projection formula (as in erythrose) were called erythro and those with the identical ligands on opposite sides (as in threose) were called threo. (Although individual enantiomers are depicted in Fig. 5.24, the same nomenclature applies to diastereomeric racemates.)

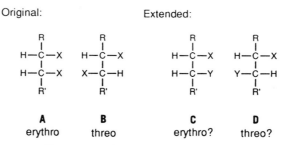

Figure 5.24. Erythro–threo nomenclature.

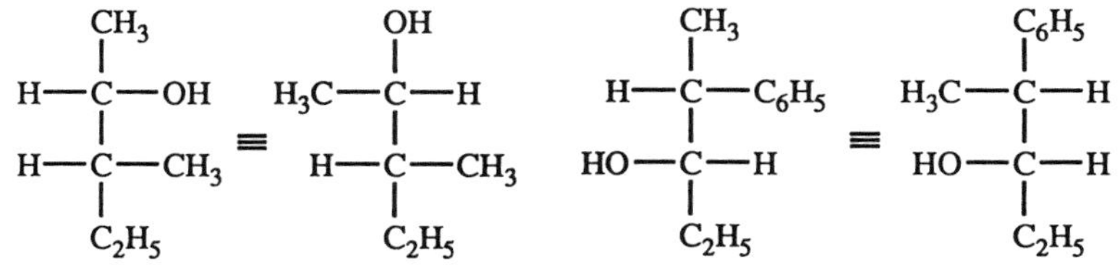

Figure 5.25. Conflicts in erythro–threo nomenclature.

An alternative way of defining threo and erythro is in terms of the configuration of the fragments R–CHX–Cβ and R'–CHX–Cβ: If the configurations of these two fragments in isolation correspond (R' taking the place of R), the compound is threo, if they are opposite, the configuration is erythro (Gielen, 1977). This nomenclature would appear simple enough and is quite unequivocal. However, complications set in when attempts were made to extend it to more complex cases (Fig. 5.24, **C** and **D**). Here, identical (or opposite) configurations of R–CHX–Cβ and R'–CHY–Cβ can no longer be unequivocally defined, because now *two* substitutions or comparisons, R/R' and X/Y, must be made and it is not always clear which group is to be substituted for which. The nomenclature still works when X and Y are heteroatoms, such as OR, NRR', halogen, and when R and R' are alkyl or aryl. However, when X or Y, also, are alkyl or aryl, there may be ambiguity. The R and R' groups have been defined as being part of the *main chain* (Eliel, 1962). Unfortunately, this may lead to a conflict with the definition given earlier, as shown in Figure 5.25, **A**. There may also be a problem with defining the main chain, for example, in case **B**, Figure 5.25 (Cram, 1952). The root of the problem is that erythro/threo, like D/L, is based on Fischer projection formulas and that, moreover, there is no clear definition of ligand precedence.

Heathcock et al. (1981) and Hoffmann and Zeiß (1981) tried to solve the former problem by basing the threo–erythro nomenclature on the more realistic staggered (zigzag or flying wedge) formulas (Fig. 5.23) but this only caused more confusion, since (apart from the continuing need for defining a main chain in this system also) the new definition was frequently in conflict with the old. Nonetheless, it is generally realized that new nomenclature should be based on staggered conformational representation, that it should be applicable to chains bearing more than two chiral centers, and that an unequivocal ligand precedence (preferably CIP) is essential.

One can simplify the system by realizing that the bond joining the two chiral centers occupies a special position and can thus be factored out. (There may be complications if the path joining the two chiral centers is not unique, i.e., if one or both of them form part of a ring: see Carey and Kuehne, 1982.) Then it is necessary only to order three ligands at each chiral center (Fig. 5.26, $a > b > c$ and $a' > b' > c'$). One does this by viewing the ligands from the side of the bond joining them. The CIP sequence rules will then put the three ligands either in a

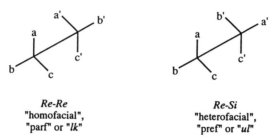

Re-Re
"homofacial",
"parf" or "*lk*"

Re-Si
"heterofacial",
"pref" or "*ul*"

Figure 5.26.   Combinations of two chiral centers.

clockwise (*Re*, cf. Chapter 8) or counterclockwise (*Si*) order (Fig. 5.26). The relative configuration may then be specified as *Re/Re* (or *Si/Si*) for one diastereomer and *Re/Si* (or *Si/Re*) for the other. Noyori et al., (1981) used threo to denote the former case and erythro for the latter; in most cases, this conserves correspondence of the new nomenclature to the old. Nevertheless the redefinition of old terms in this fashion is clearly undesirable, since one can readily construct cases [e.g., $C_2H_5CHClCH(CH_3)C_2H_5$] where the correspondence breaks down. Kreiser (1981) therefore used different terms: "homofacial" for *Re/Re–Si/Si* and "heterofacial" for *Re/Si–Si/Re* (but see Section 5-5.b), and Carey and Kuehne (1982) used *pref* (priority reflective) for the latter and *parf* (priority antireflective) for the former. Presumably one could also use the Seebach–Prelog system (1982); since prochirality (Chapter 8) rather than chirality is involved, one would have to use the symbols *ul* and *lk*, (rather than *u* and *l*) for *Re/Si–Si/Re* and *Re/Re–Si/Si*, respectively (cf. Fig. 5.26). None of these systems have, as of this writing, been generally embraced.

The above systems are unequivocal but still require assignment of priority to at least three ligands; the systems are therefore not the "at a glance" type that the synthetic chemist desires. If one can define a main chain in a molecule (and this may sometimes be troublesome), it then requires ordering of only two substituents at each of the chiral centers to define relative configuration; such ordering can usually be performed at a glance. An example is shown in Figure 5.27 for a compound with three chiral centers; the designation syn–anti proposed by Masamune et al. (1980) is now preferred for such compounds. Carey and Kuehne (1982) proposed the more lengthy descriptors "syncat" and "ancat" to incidate that the syn–anti nomenclature is here applied to the specific case of a hydrocarbon chain (Latin *catena*, meaning backbone). The ordering of the two substituents at each chiral center of the chain is best done by the CIP system, (thus OH > H and OH > CH_3 in Fig. 5.27) though other suggestions have been made (Carey and Kuehne, 1982).

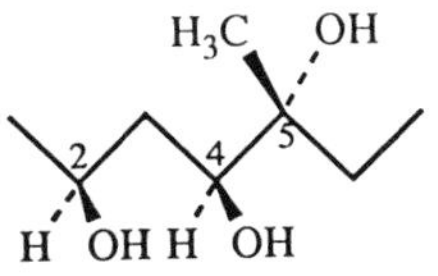

H₃C   OH

2   4 5

H OH H OH

2,4-syn,2,5-anti

Figure 5.27.   The syn–anti nomenclature for diastereomers.

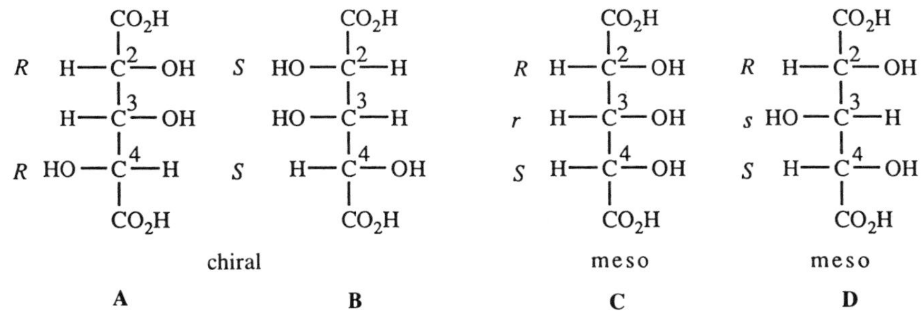

$$
\begin{array}{cccc}
& \text{CO}_2\text{H} & \text{CO}_2\text{H} & \text{CO}_2\text{H} & \text{CO}_2\text{H} \\
R\ \ \text{H—C}^2\text{—OH} & S\ \ \text{HO—C}^2\text{—H} & R\ \ \text{H—C}^2\text{—OH} & R\ \ \text{H—C}^2\text{—OH} \\
\text{H—C}^3\text{—OH} & \text{HO—C}^3\text{—H} & r\ \ \text{H—C}^3\text{—OH} & s\ \ \text{HO—C}^3\text{—H} \\
R\ \ \text{HO—C}^4\text{—H} & S\ \ \text{H—C}^4\text{—OH} & S\ \ \text{H—C}^4\text{—OH} & S\ \ \text{H—C}^4\text{—OH} \\
& \text{CO}_2\text{H} & \text{CO}_2\text{H} & \text{CO}_2\text{H} & \text{CO}_2\text{H} \\
& \text{chiral} & & \text{meso} & \text{meso} \\
& \textbf{A} & \textbf{B} & \textbf{C} & \textbf{D}
\end{array}
$$

Figure 5.28. 2,3,4-Trihydroxyglutaric acids.

Nomenclature for Pseudoasymmetric Centers

Figure 5.28 represents the four stereoisomers of 2,3,4-trihydroxyglutaric acid, which were discussed earlier in Section 3-3.b (Fig. 3.27). Structures **A** and **B** are mirror images and thus represent a pair of enantiomers $2R,4R$ and $2S,4S$. Carbon (3) though chirotopic (p. 53) is not a stereogenic center in **A** and **B** because it bears two identical ligands; the structure generated by interchange of H and OH at C(3) can be converted into the original structure by a 180° rotation. In contrast, C(3) in the meso isomers **C** and **D** *is* stereogenic (cf. Mislow and Siegel, 1984). In fact, **C** and **D** differ only in configuration at that center. Since reflection of **C** or of **D** converts it into a superposable structure, neither **C** nor **D** are chiral and C(3), being achirotopic, is not appropriately called a chiral center and so cannot be described as being R or S. [Chiral centers change configuration and configurational descriptors upon reflection of the molecule as a whole. The C(3) atom in **C** and **D** is reflection invariant.] The name "pseudoasymmetric center" has historically been used for such an achirotopic stereogenic center (cf. Chapter 3); the notation for such centers is lower case r or s (Cahn et al., 1966) to distinguish them from chiral centers. Figure 5.28 (isomers **C** and **D**) also implies that in the ordering of the enantiomorphous CHOHCO$_2$H ligands, the sequence $R > S$ is followed.

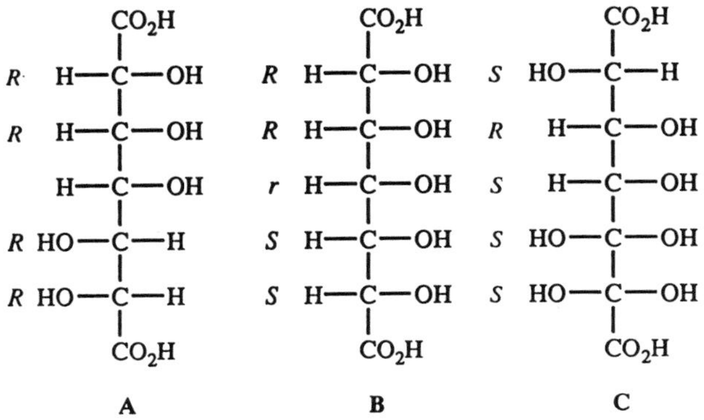

$$
\begin{array}{ccc}
\text{CO}_2\text{H} & \text{CO}_2\text{H} & \text{CO}_2\text{H} \\
R\ \ \text{H—C—OH} & R\ \ \text{H—C—OH} & S\ \ \text{HO—C—H} \\
R\ \ \text{H—C—OH} & R\ \ \text{H—C—OH} & R\ \ \text{H—C—OH} \\
\text{H—C—OH} & r\ \ \text{H—C—OH} & S\ \ \text{H—C—OH} \\
R\ \ \text{HO—C—H} & S\ \ \text{H—C—OH} & S\ \ \text{HO—C—OH} \\
R\ \ \text{HO—C—H} & S\ \ \text{H—C—OH} & S\ \ \text{HO—C—OH} \\
\text{CO}_2\text{H} & \text{CO}_2\text{H} & \text{CO}_2\text{H} \\
\textbf{A} & \textbf{B} & \textbf{C}
\end{array}
$$

Figure 5.29. Representative pentahydroxypimelic acids.

No particular nomenclature problem is presented by compounds with four, pairwise corresponding chiral centers, such as the tetrahydroxyadipic acids (Fig. 3.28). However, the next higher homologue, of which three diastereomers are shown in Figure 5.29, presents six racemic pairs and four meso forms. In two of the racemic pairs C(4) is achiral (type **A**), and in all of the meso forms C(4) is pseudoasymmetric (type **B**). Structure **C** is representative of the remaining four pairs of enantiomers in which C(4) is chiral. To assign it a descriptor (here S), the rule *like precedes unlike* (Prelog and Helmchen, 1982) is used; in the case of **C**, $S,S > S,R$. (This rule, where applicable, has precedence over $R > S$.)

5-5. DETERMINATION OF RELATIVE CONFIGURATION OF SATURATED ALIPHATIC COMPOUNDS

There are many methods for correlating configurations, that is, for determining the configuration of one chiral center relative to another either in the same molecule or another (Klyne and Scopes, 1969; Brewster, 1972). This is fortunate, because the determination of absolute configuration by the Bijvoet method is limited to substances that are crystalline and possess an appropriate "heavy" atom; only a small percentage of all chiral organic compounds has had its configuration determined either in this way or by the examination of crystal morphology in the presence of tailor-made additives (Section 5-3). On the other hand, a very large number of correlations has been effected and catalogued (Klyne and Buckingham, 1978; Buckingham and Hill, 1986), and the configuration of compounds with a single chiral center at carbon has been tabulated (Jacques et al., 1977).

Methods used to determine relative configuration are (a) X-ray structure analysis. (b) Chemical interconversion not affecting bonds to the chiral atom. (c) Methods based on symmetry properties. (d) Correlation via diastereomers ("confrontation correlation"). (e) Correlation via quasi-racemates. (f) Chemical correlations affecting bonds to a chiral atom in a "known" way. (g) Correlations by asymmetric synthesis of "known" course. (h) Various spectroscopic and other physical methods, among which ORD–CD and NMR techniques should be singled out. We shall deal with the first seven methods in this chapter, though methods b, e, and f are now largely of historic interest. The application of asymmetric synthesis (g), which is a somewhat uncertain method for configurational assignment at best, will be discussed in more detail in Chapter 12. Spectroscopic methods (h) will also be dealt with elsewhere in this book, principally in Chapters 10 and 11 (NMR) and 13 (ORD–CD), as will be the configuration of alkenes in Chapter 9, and of compounds with axial and planar chirality or helicity in Chapter 14.

a. X-Ray Structure Analysis

X-ray structure determination yields the relative configuration of all chiral centers in a molecule, though, in the absence of Bijvoet anomalous scattering, it does not yield the absolute configuration of the molecule as a whole. Nevertheless, if the absolute configuration of one chiral center in the molecule is known, that of all the others follows. Thus the structure determination of the natural product (+)-S-methylcysteine sulfoxide (Fig. 5.30, **A**; Hine and Rogers, 1956) revealed that the configuration at sulfur is S, given that the product can be related to natural L- or (R)-cysteine by methylation and oxidation of the latter (so that the configuration at carbon must be R) and that the relative configuration of the two chiral centers is evident from the X-ray analysis.

> It is sometimes convenient, in a molecule where chirality at an atom other than carbon is present, to use a subscript with the descriptor indicating the heteroatom to which the descriptor refers. Thus sulfoxide **A** (Fig. 5.30) is S_S and one can use this statement without referring to the configuration at carbon R_C. Other symbols of this type are S_P (e.g., Fig. 5.3), R_{Si}, R_{Ge}, and so on. The configuration at sulfur in **A** is understood to be represented by the Fischer convention (lone pair on the right).

Mathieson (1956) subsequently pointed out that the X-ray correlation method can be used to determine the absolute configuration of a chiral compound of unknown configuration by linking it chemically (covalently or otherwise) to another of known absolute configuration, and then subjecting it to X-ray analysis. An example is the determination of the absolute configuration of (R)-(−)-1,1'-dimethylferrocene-3-carboxylic acid (Fig. 5.30, **B**; regarding the descriptor, cf. Fig. 5.10) by X-ray structure analysis of its quinidine salt (Carter, Sim, et al., 1967). The absolute configuration of quinidine is known (cf. Prelog and Zalan, 1944) and that of the ferrocene moiety was thus deduced to be R as shown in Figure 5.30. In this case the result was confirmed by an independent Bijvoet analysis of the salt which yielded the absolute configuration directly. Another example is the determination of configuration of Tröger's base (Fig. 7.39) through its (−)-1,1'-binaphthyl-2,2'-diyl hydrogen phosphate salt (Wilen et al., 1991). The dextrorotatory base has the S,S configuration, contrary to what had been deduced from an exciton chirality calculation.

> An interesting variation of this method is the determination of the configuration of $(CH_3)_3CSOCHDC_6H_5$ at the deuterated chiral center by a combination of X-ray and neutron diffraction methods (Iitaka et al., 1986).

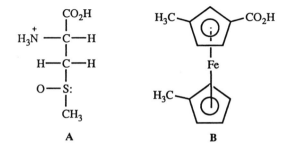

A   B

Figure 5.30. Compounds studied by the X-ray correlation method.

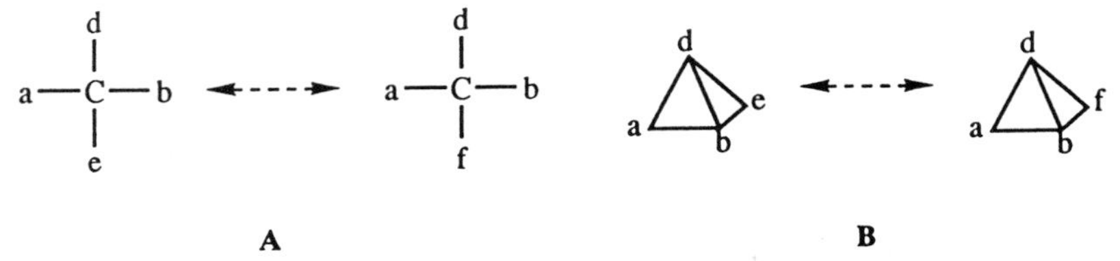

A   **B**

Figure 5.31. Chemical interconversion not affecting bonds to the chiral atom: Principle.

b. Chemical Interconversion Not Affecting Bonds to the Stereogenic Atom

If a compound Cabde can be converted to another, Cabdf, without severing the C–e (or C–f) bonds, the configurations of the two compounds are unequivocally correlated: Figure 5.31, **A**. If we join the three common ligands a, b, and d by a triangle (Fig. 5.31, **B**) we note that the new ligand f is located on the same (back) face of the triangle as the original one e: We may say that Cabde and Cabdf are "homofacial" molecules. Stepwise correlation of molecules of this type is exemplified in Figure 5.32 (cf. Klyne and Buckingham, 1978) for the correlation of (R,R)-$(+)$-tartaric acid [whose absolute configuration is known by Bijvoet's original experiment (see above)] with $(+)$-glyceraldehyde and $(-)$-lactic acid whose configurations are thus determined to be R in both cases. As shown in the correlation of (S)-$(+)$-lactic acid with (S)-$(-)$-phenylmethylcarbinol and (R)-$(-)$-mandelic acid (Fig. 5.33; Mislow, 1951), the method is not confined to homofacial molecules; in fact, in the correlation of the two acids, it involves molecules that are heterofacial.

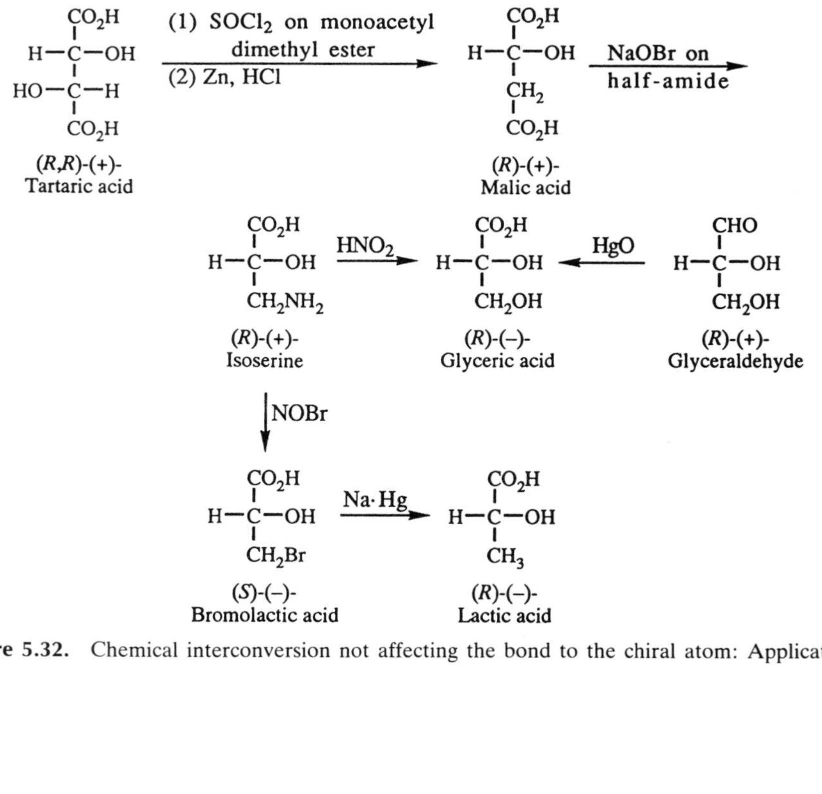

Figure 5.32. Chemical interconversion not affecting the bond to the chiral atom: Application.

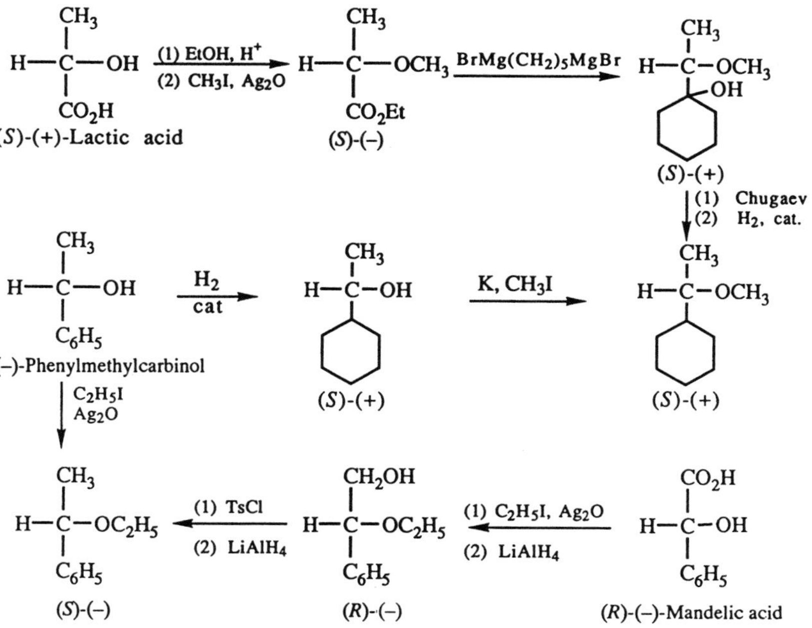

$$
\begin{array}{c}
\text{CH}_3 \\
| \\
\text{H}-\text{C}-\text{OH} \\
| \\
\text{CO}_2\text{H} \\
(S)\text{-}(+)\text{-Lactic acid}
\end{array}
\xrightarrow[\text{(2) CH}_3\text{I, Ag}_2\text{O}]{\text{(1) EtOH, H}^+}
\begin{array}{c}
\text{CH}_3 \\
| \\
\text{H}-\text{C}-\text{OCH}_3 \\
| \\
\text{CO}_2\text{Et} \\
(S)\text{-}(-)
\end{array}
\xrightarrow{\text{BrMg(CH}_2)_5\text{MgBr}}
\begin{array}{c}
\text{CH}_3 \\
| \\
\text{H}-\text{C}-\text{OCH}_3 \\
\qquad \text{OH} \\
(S)\text{-}(+)
\end{array}
$$

(with cyclohexane ring)   (1) Chugaev (2) H₂, cat.

$$
\begin{array}{c}
\text{CH}_3 \\
| \\
\text{H}-\text{C}-\text{OH} \\
| \\
\text{C}_6\text{H}_5 \\
(S)\text{-}(-)\text{-Phenylmethylcarbinol}
\end{array}
\xrightarrow[\text{cat}]{\text{H}_2}
\begin{array}{c}
\text{CH}_3 \\
| \\
\text{H}-\text{C}-\text{OH} \\
(S)\text{-}(+)
\end{array}
\xrightarrow{\text{K, CH}_3\text{I}}
\begin{array}{c}
\text{CH}_3 \\
| \\
\text{H}-\text{C}-\text{OCH}_3 \\
(S)\text{-}(+)
\end{array}
$$

C₂H₅I / Ag₂O ↓

$$
\begin{array}{c}
\text{CH}_3 \\
| \\
\text{H}-\text{C}-\text{OC}_2\text{H}_5 \\
| \\
\text{C}_6\text{H}_5 \\
(S)\text{-}(-)
\end{array}
\xleftarrow[\text{(2) LiAlH}_4]{\text{(1) TsCl}}
\begin{array}{c}
\text{CH}_2\text{OH} \\
| \\
\text{H}-\text{C}-\text{OC}_2\text{H}_5 \\
| \\
\text{C}_6\text{H}_5 \\
(R)\text{-}(-)
\end{array}
\xleftarrow[\text{(2) LiAlH}_4]{\text{(1) C}_2\text{H}_5\text{I, Ag}_2\text{O}}
\begin{array}{c}
\text{CO}_2\text{H} \\
| \\
\text{H}-\text{C}-\text{OH} \\
| \\
\text{C}_6\text{H}_5 \\
(R)\text{-}(-)\text{-Mandelic acid}
\end{array}
$$

Figure 5.33. Chemical interconversion not affecting the bond to the chiral center: Lactic and mandelic acids.

The terms *homofacial* and *heterofacial* were suggested by Ruch (communicated to Prelog and Helmchen, 1972); these terms require that three of the four ligands at the chiral centers of the molecules compared be identical (cf. Fig. 5.31). These terms must not be confused with the terms *homochiral* and *heterochiral* (p. 103), which imply that chiral molecules have either the same (homo) or the opposite (hetero) sense of chirality and strictly require comparison of isometric molecules (see p. 65), although Ruch (1972) used the terms in a slightly broader sense (e.g., all right shoes are homochiral).

It is common to say that homofacial molecules *have the same configuration,* whereas heterofacial molecules *have opposite configuration.* Indeed, (S)-(+)-lactic acid and (R)-(−)-mandelic acid have opposite configurations (Fig. 5.33); these acids are, of course, heterofacial. Alternatively in the case of the (S)-(−)-bromolactic acid and (R)-(−)-lactic acid shown in Figure 5.32, while these acids are clearly homofacial, their configurational descriptors are opposite. Thus the statement that these two acids have the same configuration might prove confusing, whereas to say that they are homofacial is not. It might be noted that, in both correlations (Figs. 5.32 and 5.33), bonds to the chiral center are never broken!

A correlation involving chiral deuterated compounds is shown in Figure 5.34: The configuration of (−)-ethanol-1-*d* was shown to be *S* by chemical correlation (not affecting the chiral center) with (S)-(−)-glycolic-*d* acid (cf. Weber, 1965, cited in Arigoni and Eliel, 1968; see also p. 115).

In actual fact, the deuterated alcohol was characterized not by its rotation but by the fact that in enzymatic dehydrogenation with NAD⁺ (nicotinamide adenine dinucleotide, oxidized form) and yeast alcohol dehydrogenase it yielded largely

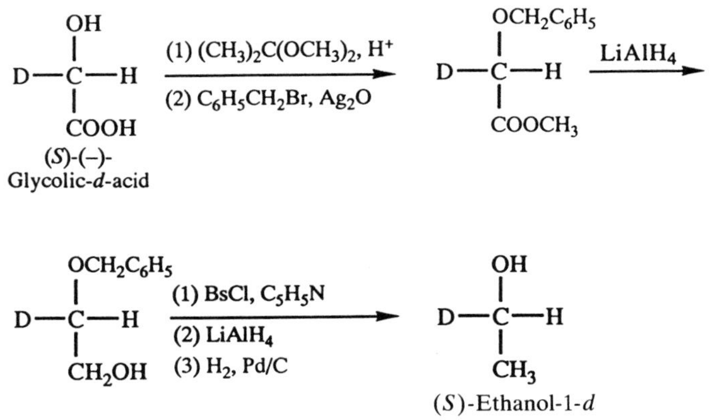

Figure 5.34. Correlation of configuration of deuterated molecules. BsCl represents *p*-bromo-benzenesulfonyl chloride.

deuterated acetaldehyde (CH_3CDO), that is, the hydrogen removed in the dehydrogenation is protium rather than deuterium. It is known from other experiments that this is characteristic of the levoratatory isomer (cf. Chapter 8).

c. Methods Based on Symmetry Considerations

Emil Fischer's famous proof of the configuration of glucose, summarized in many undergraduate textbooks of organic chemistry, relies heavily on symmetry properties.

For example, the diastereomeric tartaric acids [(+), (−) and meso] shown in Figure 3.26 may be distinguished in that the (+) or (−) compounds are optically active (or can be resolved if encountered as a racemate), whereas the meso form is inactive. The relative configuration of the CHOH groups follows immediately as being R^*,R^* in the active or resolvable species and R^*,S^* in the meso or nonresolvable species. The relative configuration of the tetroses shown in Figure 3.24 can then be established by oxidation (HNO_3) to the corresponding tartaric acids: meso from erythrose, (+) or (−) from threose.

> The absolute configuration of (−)-threose ($2S,3R$, or D) can thus also be deduced since its oxidation leads to (−)-tartaric acid, the enantiomer of the acid found to be R,R by the Bijvoet method (cf. Figs. 5.18 and 3.26). Obviously, the absolute configuration of (−)-erythrose cannot be determined similarly, since its oxidation gives the achiral meso acid; but it can be correlated with (−)-threose through their common origin in the homologation of (R)-(+)-glyceraldehyde (cf. Fig. 3.23).

Similar principles can be applied to the trihydroxyglutaric acids (Fig. 3.27): The chiral acid must be $2R^*,4R^*$. Referring to Figure 3.23, one notes that arabinose and lyxose will be oxidized to the chiral acid, whereas ribose and xylose give rise to one or the other of the meso acids, which cannot be distinguished by symmetry arguments alone.

Arabinose and lyxose can be distinguished in that degradation (i.e., removal of the aldehyde carbon and conversion of the adjacent carbon to an aldehyde group) of arabinose gives erythrose, whereas lyxose similarly gives rise to threose (Fig. 3.23); these two carbon sugars are distinguished as explained above. To complete the assignment of relative configuration of glucose [the absolute configuration of (+)-glucose was, of course, not known in Fischer's time but was arbitrarily assumed to be D], it was synthesized by homologation of arabinose; this gave rise to a mixture of glucose and mannose (Fig. 3.23). These two hexoses, in turn, were distinguished by symmetry arguments applied to the corresponding dicarboxylic acids (Fig. 5.35): The diacid **A** obtained from mannose has C_2 symmetry but **B**, which is derived from glucose, belongs to the C_1 point group. Both acids are chiral but can be distinguished because **A** can *only* be obtained by oxidation of D-mannose, whereas **B** can be obtained *not only* by oxidation of D-glucose *but also* by oxidation of L-gulose (Fig. 5.35).

An alternative way of distinguishing hexaric acids **A** and **B** (and, thereby, identifying the structure of glucose and mannose) would have been to prepare their monoesters (e.g., monomethyl esters) either by partial esterification or by partial saponification of the corresponding diesters. Because of the existence of the C_2 axis in **A**, the two carboxyl groups are equivalent (homotopic, cf. Chapter 8), and therefore, regardless of which one is esterified, one and the same monoester results. In contrast, in **B** there is no C_2 axis and the two carboxyl groups are different (heterotopic), hence two monoesters can be formed. It might also be noted here that **A**, because of pairwise equivalence of C(1,2,3) with C(4,5,6), will show only three signals in the ^{13}C NMR spectrum, whereas **B** will (at least in principle) display six signals (a test obviously not available to Fischer!).

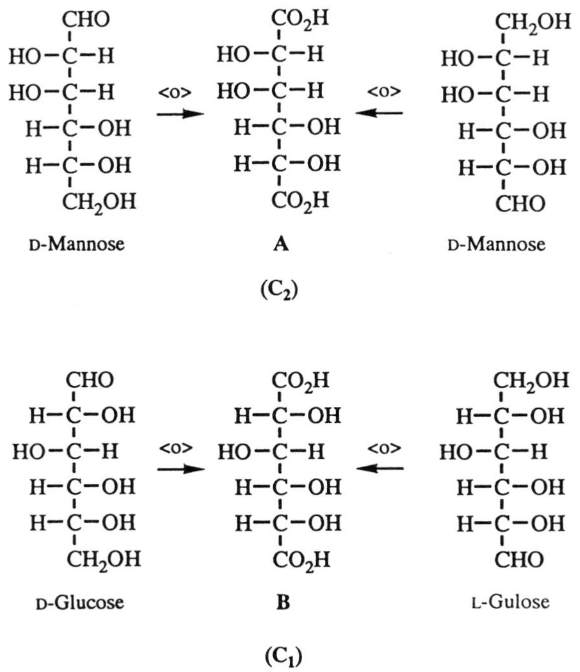

Figure 5.35. Hexaric acids from mannose and glucose.

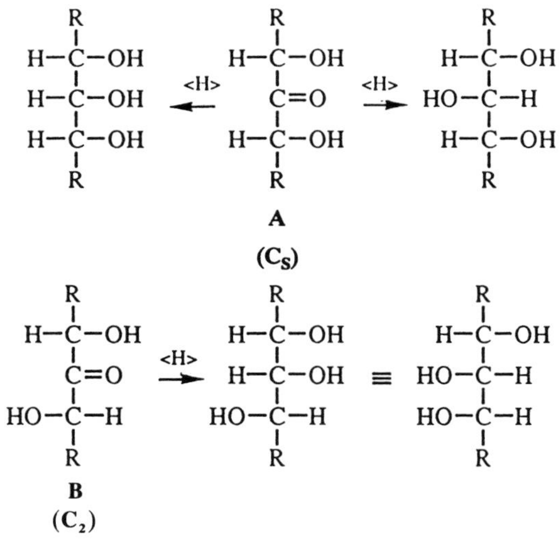

$$
\begin{array}{c}
\text{R} \\
| \\
\text{H—C—OH} \\
| \\
\text{H—C—OH} \\
| \\
\text{H—C—OH} \\
| \\
\text{R}
\end{array}
\xleftarrow{\langle H \rangle}
\begin{array}{c}
\text{R} \\
| \\
\text{H—C—OH} \\
| \\
\text{C=O} \\
| \\
\text{H—C—OH} \\
| \\
\text{R}
\end{array}
\xrightarrow{\langle H \rangle}
\begin{array}{c}
\text{R} \\
| \\
\text{H—C—OH} \\
| \\
\text{HO—C—H} \\
| \\
\text{H—C—OH} \\
| \\
\text{R}
\end{array}
$$

A

(C$_S$)

$$
\begin{array}{c}
\text{R} \\
| \\
\text{H—C—OH} \\
| \\
\text{C=O} \\
| \\
\text{HO—C—H} \\
| \\
\text{R}
\end{array}
\xrightarrow{\langle H \rangle}
\begin{array}{c}
\text{R} \\
| \\
\text{H—C—OH} \\
| \\
\text{H—C—OH} \\
| \\
\text{HO—C—H} \\
| \\
\text{R}
\end{array}
\equiv
\begin{array}{c}
\text{R} \\
| \\
\text{H—C—OH} \\
| \\
\text{HO—C—H} \\
| \\
\text{HO—C—H} \\
| \\
\text{R}
\end{array}
$$

B

(C$_2$)

Figure 5.36. Differentiation between isomers of C$_S$ and C$_2$ symmetry.

In summary, symmetry properties may be utilized to distinguish diastereomers that have C_2, C_s, or higher symmetry from those that lack axes or planes of symmetry (for a more detailed discussion, see Chapter 8). The presence of symmetry planes leads to lack of chirality and resolvability; axes lead to a reduction in the number of possible derivatives (such as esters, see above); both axes and planes lead to degeneracies in ^{1}H and ^{13}C NMR spectra. Compounds possessing C_2 axes of symmetry cannot usually be distinguished by NMR spectroscopy from those possessing symmetry planes (see also Chapter 8) but sometimes other means of distinction are available. Thus (Fig. 5.36) ketone **A** of C_s symmetry gives rise to two triols on reduction, whereas **B** (C_2) can give rise to only one because the two faces of the carbonyl group are equivalent (homotopic) (Chapter 8).

> Differences in derivatives, spectra, and so on, expected for compounds lacking symmetry may, through accidental degeneracies, fail to be observed in a particular case (e.g., two distinct carbon nuclei may be accidentally isochronous or only one of two possible monoesters may be formed for steric reasons). Therefore, such criteria must be used with some caution.

d. Correlation via Compounds with Chiral Centers of Two Types

The term "confrontation analysis" has been coined by Brewster (1972) for this type of analysis. It involves comparing one type of chiral center (say RCabR′) with another (say RCxR′R″, where the R's represents alkyl groups and a,b,x represent hydrogen or heteroatoms or groups) and determining their relative configuration after building the two chiral centers into one and the same compound (confronting them). Then, if both centers can be carved out without loss of chirality, and the configuration of one of them (say RCabR′) is known, that of the other (say RCxR′R″) can be deduced. We have already seen this method in operation in cases where X-ray structure analysis was used to correlate the relative configuration of two chiral centers (cf. Fig. 5.30). However, other correlation methods may, in principle, be used.

A number of examples have been presented by Brewster (1972) and by Klyne and Scopes (1969). Here we shall exemplify the method with the determination of the absolute configuration of (+)-ethanol-1-d, (+)-CH$_3$CHDOH (Lemieux and Howard, 1963). (The significance of this determination will be returned to in Chapter 8.) The pertinent series of transformations is shown in Figure 5.37. The first three steps involve the synthesis of D-xylose-5-d (**D**) from the glucofuranose derivative **A**. Since the aldehyde intermediate **B** is chiral, deuteride approaches the two faces of the carbonyl at unequal rates and one diastereomer of **D** is formed in preference to the other (ca. 2 : 1 ratio). The stereochemistry of **D** may be analyzed as follows: This isomer is the deuterated analogue of D-xylopyranose

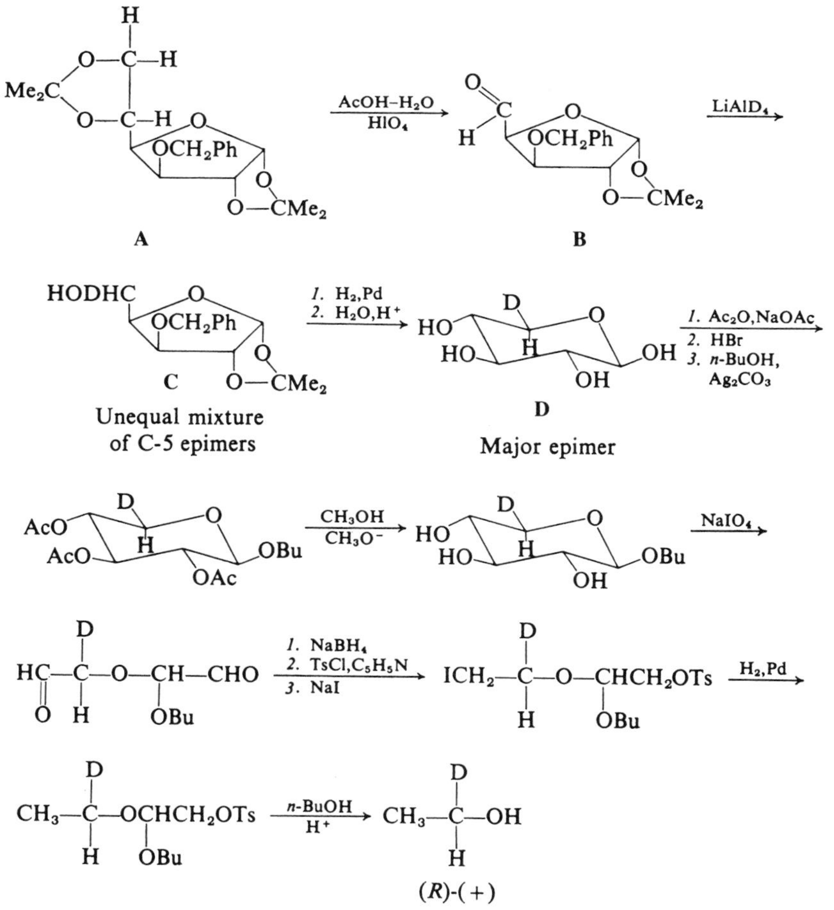

Figure 5.37. Proof of configuration of (+)-ethanol-1-d as R. Reprinted with permission from Arigoni and Eliel, *Topics Stereochem.*, **4**, 160 (1968). Copyright John Wiley & Sons, New York, 1968.

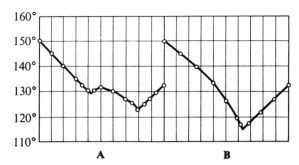

Figure 5.38. Phase diagrams for a mixture of the dextrorotatory ethyl xanthate derivative of malic acid with the (**A**) dextrorotatory and (**B**) levorotatory ethyl thionecarbonate derivative of thiolsuccinic acid. (From "The Svedberg" Memorial Volume, Almquist & Wiksells, Uppsala, Sweden, 1944. By permission of the publishers.)

of known absolute configuration (see above). The configuration of chiral centers C(1)–C(4) is thus known and the conformation of the chair will be such that the four hydroxyl groups occupy equatorial positions. The configuration at C(5) is now determined by ^{1}H NMR spectroscopy which readily distinguishes axial and equatorial hydrogen atoms (cf. Chapter 11): It turns out that the major product has axial H, and hence equatorial D, that is, the configuration at C(5) is R. The remainder of the procedure (Fig. 5.37) serves to carve out C(5) from the xylose molecule, to remove all other chiral centers, and to convert C(5) to the chiral center CH_3–CHD–OH of ethanol-1-d without ever severing any of the bonds to this chiral center and without racemizing it. These transformations were accomplished as shown; the CH_3CHDOH so obtained was dextrorotatory. Contemplation of the correlation (Fig. 5.38) shows that the –CHD–O– moiety of the D-xylose-5-d becomes the corresponding moiety of CH_3–CHD–OH and that the –CHOH–C part of the xylose becomes the CH_3 of the ethanol-d. Hence, the ethanol-1-d obtained as shown has the R configuration; it follows that (R)-ethanol-1-d is dextrorotatory. This result agrees with the conclusion reached earlier (Fig. 5.34) that the S isomer is levorotatory.

> The four methods discussed so far are compelling: If they are properly carried out, and if a meaningful result is obtained, the answer is not in doubt. In contrast, the four remaining methods do not have the same certitude: Since they depend on chemically less well established principles, they yield probable (sometimes highly probable) but not compelling answers (e.g., Anderson and Fraser-Reid, 1985 for disproof of an earlier correlation). It must be pointed out here that, since a random guess at configuration has a 50% chance of being correct, even an unreliable method can repeatedly give the correct result.

e. The Method of Quasi-racemates

This method, introduced by Timmermans (1928) based on earlier observations by Centnerszwer (1899), has been used most extensively by Fredga who has reviewed it in detail (Fredga, 1960, 1973; see also Brewster, 1972). It rests on the phase behavior of conglomerates and racemic compounds (Chapters 6 and 7).

When two *chemically very similar* chiral substances, such as (+)-chloro-succinic and (−)-bromosuccinic acid, which are heterofacial (see above), are mixed, a phase behavior similar to that seen in racemic compounds may be observed, that is, a molecular compound in the solid state may be formed (Centnerszwer, 1899). Such a compound is called a *quasi-racemate* and shows typical compound behavior in the phase diagram. In contrast, the homofacial enantiomers (+)-chlorosuccinic acid and (+)-bromosuccinic acid will show the eutectic behavior similar to that of a conglomerate. Figure 5.38 shows melting point phase diagrams for such a pair. This particular phase diagram refers to the case of the (+)-ethyl xanthate derivative of malic acid and the (+)-ethyl thionecarbonate derivative of thiolsuccinic acid (Fig. 5.39). As shown in Figure 5.39, this correlation (which showed the two dextrorotatory compounds to be heterofacial, since they formed a quasi-racemic compound, whereas the opposite pair gave rise to a eutectic mixture) is part of the configurational correlation of (S)-(−)-malic acid with (−)-methylsuccinic acid. Direct correlation was not possible, because the acids were too different to form a compound for either combination of enantiomers. For the same reason it was not possible to correlate malic acid and thiolsuccinic acid directly; it was necessary to first convert them to chemically very similar derivatives. On the other hand, correlation of (+)-thiolsuccinic acid with (−)-methylsuccinic acid by the quasi-racemate method was successful and showed the two enantiomers in question to be heterofacial inasmuch as they formed a compound; in this case the opposite pairing gave rise not to a mixture but to a solid solution. In other cases, one combination gives rise to a solid solution and the other to a mixture; in such cases it is held (Mislow and Heffler, 1952) that the pair forming the solid solution is homofacial and that forming the mixture is heterofacial.

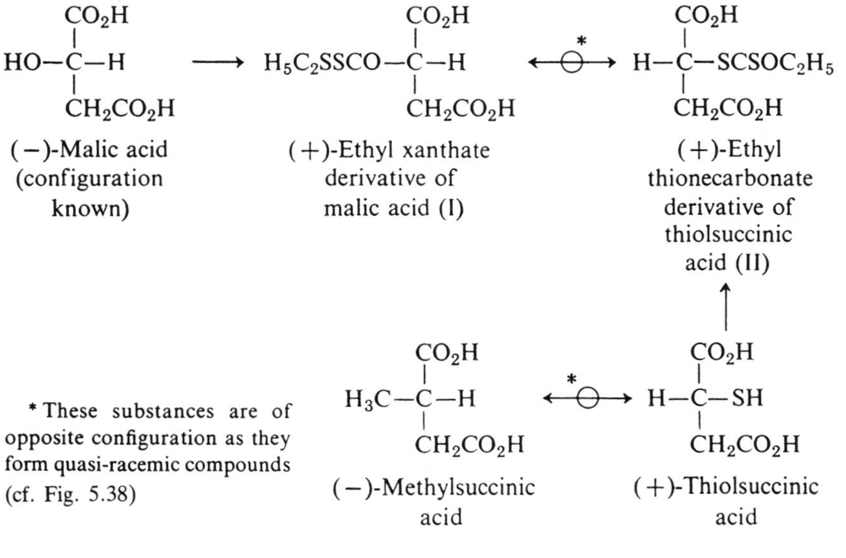

Figure 5.39. Correlation of the configuration of (S)-(−)-malic and (−)-methylsuccinic acid.

The method would seem to fail when both combinations give rise to mixtures or if both yield compounds (the latter situation is rare: Fredga and Gamstedt, 1976). However, it has been pointed out (Ricci, 1962; Raznikiewicz, 1962; see also Seebach et al., 1984) that it is really the deviation from ideality in phase behavior that counts: The combination of heterofacial species almost invariably deviates more than that of the corresponding homofacial compounds. Detailed analysis of the melting point or solubility phase diagram will therefore often give an answer even when there is no gross difference, such as that between a compound and a mixture. In addition, compound formation may sometimes be deduced from IR spectra (Rosenberg and Schotte, 1955; Gronowitz, 1957) or from X-ray powder diagrams (Patterson, 1954; Gronowitz and Larsson, 1955; Gronowitz, 1957). Nevertheless, the quasi-racemate method has been little used in recent years (but see Ionov et al., 1976; Tambute and Collet, 1984); it is, of course, limited to crystalline solids and many other procedures applicable to both liquids and solids are now available.

f. Chemical Correlations Affecting Bonds to a Chiral Atom in a "Known" Way

For an overview, see Brewster, 1972.

Nucleophilic Displacement Reactions

In principle, a transformation Cabde + f → Cabdf + e (where e and f are ligands *directly* attached to the chiral carbon atom or other chiral center) could be used to establish the absolute configuration of Cabdf given that of Cabde, provided the stereochemical course (homofacial or heterofacial, cf. p. 127) of the replacement of e by f is *reliably* known. Thus, if the replacement of e by f proceeds homofacially, (i.e., with *retention* of configuration), Cabde and Cabdf will be *homofacial* (i.e., have the *same* configuration), whereas they will be *heterofacial* (i.e. have *opposite* configuration) if the reaction proceeds heterofacially (with *inversion* of configuration). Indeed these may be considered to be the definitions of "retention" and "inversion." The problem, which will be addressed in this section, is to *know* whether a given reaction proceeds with retention or inversion and also to assess the reliability of that knowledge.

The primitive notion that all displacement reactions proceed homofacially was dispelled by Walden (1896, 1897) who, at the end of the nineteenth century, performed the sequence of reactions shown in Figure 5.40. The enantiomeric (−)-chlorosuccinic and (+)-chlorosuccinic acids obviously have opposite configurations; thus if the AgOH reaction proceeds homofacially with retention, the KOH and PCl$_5$ reactions must proceed heterofacially with inversion. Conversely, if the AgOH reaction involves inversion, the other two reactions must proceed with retention. Unfortunately, it was not known in Walden's time which of the two possibilities was the correct one and it was only a quarter of a century later (Phillips, 1923, 1925) that the stereochemical course of *any* chemical reaction proceeding at a stereogenic center was elucidated, as shown in Figure 5.41.

Of the four reactions involved in Phillips' correlation, three (acetylation, *p*-toluenesulfinylation, and oxidation of the sulfinate to the sulfonate) do not

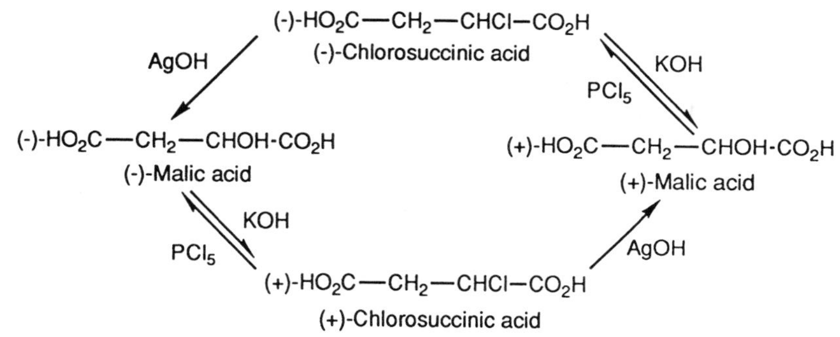

Figure 5.40. Walden inversion.

involve breaking of bonds to the chiral center and, therefore, must proceed homofacially (with retention of configuration). It follows that $(-)$-2-octyl acetate and $(-)$-2-octyl p-toluenesulfonate (tosylate) have the same configuration (i.e., are homofacial) and that the reaction of the latter with acetate to give $(+)$-2-octyl acetate therefore proceeds with inversion of configuration (heterofacially; indicated by a looped arrow in Fig. 5.41). The conclusion that nucleophilic displacement reactions of the type shown in Figure 5.41 proceed with inversion of configuration was later extended, by analogy, to the reaction of 2-octyl tosylate with halide ions (Houssa, Phillips, et al., 1929) and the configurations of 2-halooctanes were thus assigned. The correctness of the argument was put on a firmer basis by the work of Hughes et al. (1935), shown in Figure 5.42. Optically active 2-iodooctane, when treated with iodide ion, is racemized because inversion of configuration occurs in the displacement and eventually one-half of the molecules will have a configuration opposite to that of the other one-half. At this point the system is in equilibrium. Treatment of 2-iodooctane with radioiodide (I^{*-}) leads to exchange and incorporation of radioiodine into the organic material. The rate of racemization can be measured polarimetrically and the rate

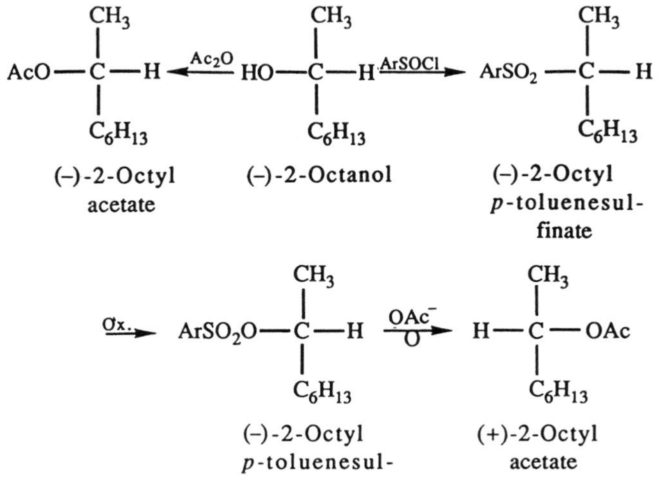

Figure 5.41. Inversion in the reaction of $(-)$-2-octyl p-toluenesulfonate with acetate.

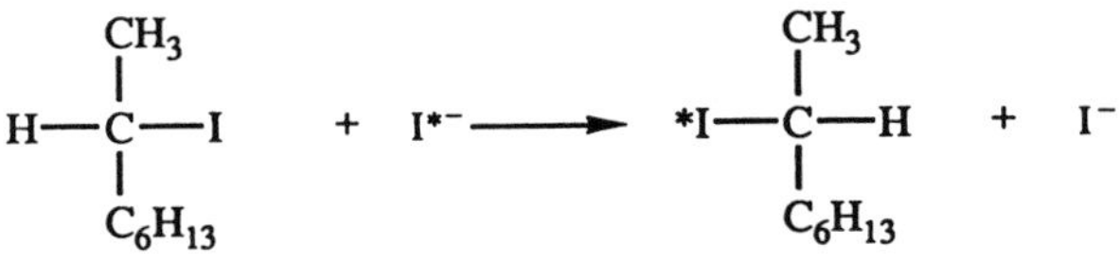

$$
\begin{array}{ccc}
& CH_3 & \\
& | & \\
H\!-\!C\!-\!I & & *I\!-\!C\!-\!H \quad + \quad I^- \\
& | & \\
& C_6H_{13} & C_6H_{13}
\end{array}
$$

H—C—I + I*⁻ ⟶ *I—C—H + I⁻

Figure 5.42. Reaction of optically active 2-iodooctane with radioactive iodide ion. (In actual fact, the racemization and the radioactive exchange were carried out in two separate experiments and the rates were then compared. This result does not affect the argument.)

of radioiodine incorporation by assessing the radioactivity of the recovered organic material. It turns out that the initial rate of inversion (or one-half the initial rate of racemization, since for each molecule inverted, two are racemized) is equal to the initial rate of radioiodide displacement. In other words, each act of nucleophilic displacement involves inversion.

However, this finding is clearly not the whole story. In Walden's scheme (Fig. 5.41) some reactions are evidently heterofacial, (i.e., they involve inversion), whereas others are homofacial (that is, they proceed with retention). If the nucleophilic displacement reaction is to be used as a tool for assigning configuration, one must know which reaction course is followed in a specific instance. This problem was addressed by the group of Hughes and Ingold in further studies of the mechanism of nucleophilic displacement reactions. It was concluded that inversion occurs reliably only in bimolecular nucleophilic displacement (S_N2) reactions (Ingold, 1969); this postulate is sometimes called the S_N2 rule. Thus to assure that a reaction of the type $R\!-\!X + Y^-$ (or $R\!-\!X^+ + Y^-$, or $R\!-\!X + Y$, or $R\!-\!X^+ + Y) \rightarrow R\!-\!Y + X^-$ (or $R\!-\!Y + X$, or $R\!-\!Y^+ + X^-$, or $R\!-\!Y^+ + X$) proceeds heterofacially, one must also study its kinetics; inversion of configuration is assured only if the reaction is bimolecular (generally second order in substrate and nucleophile).

Figure 5.43 shows the overall picture of the S_N2 reaction, in which the incoming nucleophile engages the σ^* orbital of the atom that is being inverted (Meer and Polanyi, 1932; Olson, 1933). (For an extensive discussion of the detailed course of the S_N2 reaction, both in solution and in the gas phase, see Shaik, Schlegel, and Wolfe, 1992.)

It might appear from the above that the S_N2 rule rests on a very limited experimental basis. In fact, however, quite apart from the theoretical underpinnings (e.g., Olson, 1933), there are a number of other instances where the

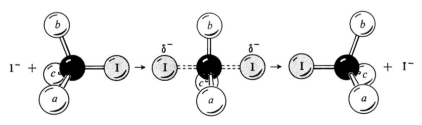

Transition state

Figure 5.43. Mechanism of the S_N2 reaction.

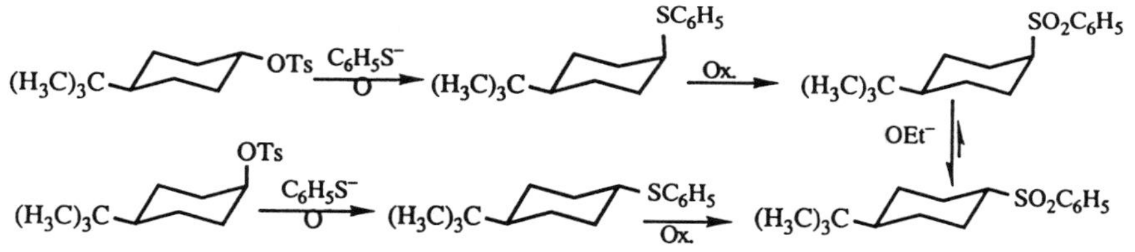

Figure 5.44. The S_N2 reaction of *trans-* and *cis-4-tert*-butylcyclohexyl *p*-toluenesufonates.

configuration of the starting material and that of the product are independently known to be opposite, that is, where starting material and product are heterofacially related. In some of these cases the configurations have been determined by independent methods discussed elsewhere in this Chapter. In others, the substrate had two (or more) stereogenic centers of which only one was inverted in the nucleophilic displacement. In such cases correlation of the relative configuration of the pertinent stereogenic center with that of other stereogenic centers in the same molecule (whose configuration is not affected) will document the occurrence of inversion. An example is the reaction (shown to be second-order kinetically) of *trans-4-tert*-butylcyclohexyl *p*-toluenesulfonate with thiophenolate to give, as the major product, *cis-4-tert*-butylcyclohexyl phenyl sulfide (Eliel and Ro, 1957; cf. Fig. 5.44). The cis configuration of the product was inferred from the S_N2 rule but was independently confirmed by oxidizing the sulfide to the corresponding sulfone and demonstrating that the presumed axial sulfone could be epimerized, by base, to its more stable equatorial isomer (Fig. 5.44); it was later corroborated (Eliel and Gianni, 1962) by ^{1}H NMR spectroscopy.

An application of the S_N2 rule (Brewster, Ingold, et al., 1950) to the configurational correlation of $(+)$-α-methylbenzyl chloride and $(-)$-α-methylbenzylamine via the azide is shown in Figure 5.45. The kinetics of the azide displacement shows it to be bimolecular, so inversion occurs and the chloride and amine shown have opposite configurations. Displacement by azide followed by reduction rather than displacement by amide or ammonia was chosen, because the azide reaction is cleaner and more easily studied kinetically.

Reactions of this type have been most useful to correlate amines with halides and with alcohols (via the displacement of their sulfonates), of halides with alcohols (via displacement of halide by acetate or benzoate followed by saponification, or via displacement of a sulfonate of the alcohol by halide), and of halides $RR'CHX$ with acids $RR'CH(CH_2)_nCO_2H$ ($n = 0$ or 1) by displacement with cyanide or malonate.

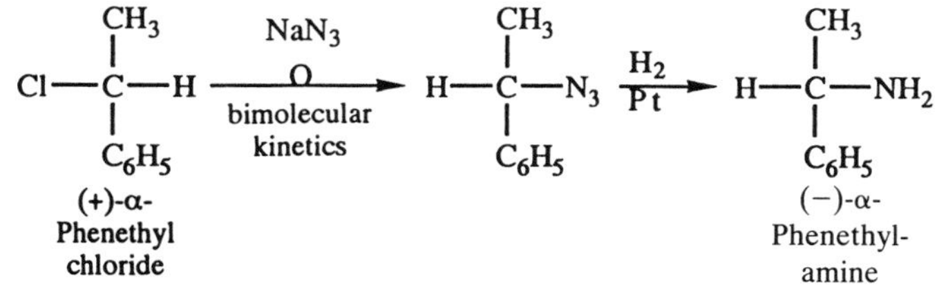

Figure 5.45. Correlation of configuration by application of the S_N2 rule.

$$\text{zyxC—C—X} \longrightarrow \text{C—X—Cxyz}$$

Figure 5.46. Intramolecular 1,2-rearrangements.

Rearrangement Reactions at Saturated Carbon Atoms

The reactions to be discussed here are of the general type shown in Figure 5.46. A chiral carbon atom, zyxC–, migrates from carbon to a heteroatom X [in the case of the Stevens rearrangement (see below) the migration is from nitrogen to carbon]; thus if the configuration of the starting chiral center, zyxC*–C– is known, that of the product zyxC*–X, where X may be nitrogen or oxygen, can be inferred, provided the steric course of the migration (homofacial or heterofacial with respect to the migrating group –Cxyz) is known. Both on theoretical grounds and on the basis of many experiments it has been concluded that most reactions of this type proceed homofacially, that is, with retention of configuration at the migrating group. A number of examples, involving migration from carbon to nitrogen [Hofmann bromamide, Curtius, Schmidt, Lossen, and Beckmann (cf. p. 562) rearrangements], from carbon to oxygen (Baeyer–Villiger rearrangement), and the similar Stevens rearrangement (nitrogen to carbon) have been summarized elsewhere (Eliel, 1962, pp. 120–121; Brewster, 1972). The steric course of these rearrangements was first elucidated in cases where the configuration of both the starting materials and products was known. The reactions were subsequently used to correlate configurations of products and starting materials where only one of these configurations was known (see also Fiaud and Kagan, 1977). A less than routine example is shown in Figure 5.47 (Weber, Arigoni, et al., 1966); it provides one of three configurational correlations of the (−)-ethanol-1-*d* obtained by enzymatic reduction of CH_3CDO (see earlier discussions). The overall correlation involves three different principles: the confrontation method (see above), correlation of chiral centers by nucleophilic displacement, and the Baeyer–

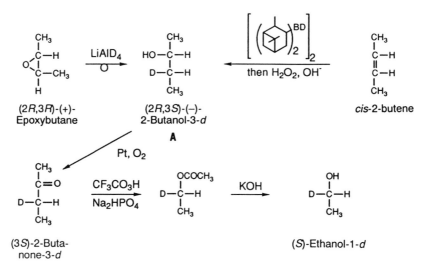

Figure 5.47. Configuration of (*S*)-(−)-ethanol-1-*d* by Baeyer–Villiger reaction.

Villiger reaction. Deuteride reduction of (+)-2,3-epoxybutane gives levorotatory 2-butanol-3-d (Skell et al., 1960). Since C(3), chiral by virtue of the presence of deuterium, contributes little to the rotation of the alcohol, it is safe to assume that the latter's configuration at C(2) is R, the same as that of (−)-2-butanol, which has been correlated with tartaric acid (cf. Klyne and Buckingham, 1978). The configuration of the starting epoxide is therefore unequivocally shown to be 2R,3R. (It cannot be 2R,3S, because that would constitute an inactive meso form.) Since opening of aliphatic secondary epoxides with nucleophiles invariably involves inversion of configuration* (Wohl, 1974), the configuration of **A** at C(3) must thus be S. The carbinol C(2) and CHD C(3) chiral centers in **A** are thus "confronted" as shown in Figure 5.47. The carbinol center is subsequently destroyed and the CHD center incorporated in ethanol-1-d by Baeyer–Villiger reaction and saponification as shown. An additional feature of the correlation is that the final product, (S)-ethanol-1-d, was characterized not by optical rotation (which is small, and therefore subject to falsification by small amounts of chiral impurities), but by enzymatic oxidation which resulted largely in retention of the deuterium atom in the acetaldehyde (CH$_3$CDO) so formed (Weber, Arigoni, et al., 1966). This behavior is characteristic of levoratory ethanol-1-d (Chapter 8), which is thus shown to be S.

Figure 5.47 also shows an alternative synthesis of **A** (Weber, Arigoni, et al., 1966) by asymmetric hydroboration of cis-2-butene with tetrapinanyldiborane (Brown and Zweifel, 1961; cf. Chapter 12).

g. Correlation by Stereoselective Synthesis of "Known" Stereochemical Course

A stereoselective synthesis (Chapter 12) involves introducing a stereogenic center in a molecule in such a way that one stereoisomer of the product is formed in preference to the other or others. The synthesis may be either enantioselective or diastereoselective; in the former case one of two enantiomers is formed preferentially, in the latter case one of two or more diastereomers is formed. Both types will be discussed here, though in the case of diastereoselective syntheses we shall confine ourselves to cases where the newly introduced stereogenic center is chiral. (Cases involving achiral stereogenic centers, such as the reduction of 4-tert-butylcyclohexanone to cis- or trans-4-tert-butylcyclohexanol, will be taken up in Chapter 11.) In a typical case, reaction of an achiral precursor with a chiral reagent (or in the presence of a chiral auxiliary substance, called a *chiral adjuvant*) will give a chiral product. If the configuration of the reagent (or chiral adjuvant) is known, and if there is a unique configurational correlation between the configuration of the reagent or adjuvant and that of the newly created chiral center, then the configuration of the latter is correlated with that of the former. The main problem with this type of correlation is that the steric course of stereoselective "asymmetric" syntheses is even more uncertain than that of reactions at chiral centers discussed in Section 5-5.f. Although an asymmetric

---

*This is not necessarily true for tertiary or aromatically substituted epoxides (cf. Eliel, 1962, p. 102).

synthesis may have taken the same steric course in n known cases (where n is usually a relatively small number), it cannot be guaranteed to take the same course in the $n + 1$st case, perhaps because some steric factor in the substrate alters the course of the reaction. Therefore correlations of this type can never be considered entirely secure, especially in the (frequent) absence of detailed mechanistic information. Nevertheless, we shall present here four examples of correlations of this kind, the first two of which have been frequently used in assigning configuration: Horeau's, Prelog's, Cram's, and Sharpless' rule.

Horeau's Rule

This topic has been reviewed by Brewster (1972) and Horeau (1977). If an optically active secondary alcohol (RR'CHOH) is esterified with a racemic acid (R^*-CO_2H) the transition states $[(-)\text{-}RR'CHOH.(-)\text{-}R^*\text{-}CO_2H]^{\ddagger}$ and $[(-)\text{-}RR'CHOH.(+)\text{-}R^*\text{-}CO_2H]^{\ddagger}$ are diastereomeric and thus unequal in energy. Therefore the activation energies and hence reaction rates of esterification with the $(+)$ and $(-)$ acids will be unequal (the energies of the initial states are the same, since the alcohol is one and the same in the two combinations and the acids are enantiomeric, and therefore equienergetic). Thus, if an excess of acid is used, one enantiomer of the acid will be preferentially incorporated in the ester and the other will be predominantly left behind (kinetic resolution, see Chapter 7).

Horeau (1961) introduced inactive 2-phenylbutyric anhydride (or sometimes the corresponding chloride) as the reagent for an optically active alcohol of unknown configuration. It matters not that the anhydride is a mixture of RR/SS (racemic) and RS (meso) species: One of the chiral $C_6H_5CH(C_2H_5)CO-$ moieties will react faster with the alcohol than the enantiomeric one. The residual acid (after hydrolysis of the left-over anhydride) is submitted to polarimetry: If it is levorotatory, the configuration of the optically active secondary alcohol RR'CHOH is such that, in its Fischer projection, the hydroxy group is down, the hydrogen atom up, and the larger of the two remaining substituents is on the right (Fig. 5.48, **A**). The reverse obviously applies if the recovered acid is dextrorotatory (Fig. 5.48, **B**). Evidently the $(-)$ acid reacts more slowly with alcohol **A** and vice versa for the $(+)$ acid.

> It might be noted that in many cases (though *not* invariably) the CIP sequence will be $OH > L > M > H$, in which case formation of a levorotatory residual acid implies the S configuration of the starting alcohol, whereas dextrorotatory residual acid implies (R)-RR'CHOH; however, this correlation must not be used by rote.

The method and its limitations have been discussed in detail in a review by Horeau (1977). It is possible to start with racemic alcohol and use optically active

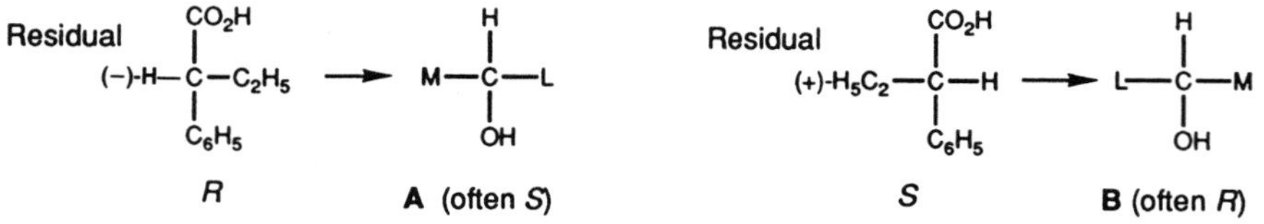

Figure 5.48. Horeau's rule.

2-phenylbutyric anhydride to lead, at the same time, to kinetic resolution (partial) of the alcohol and to determination of the configuration of the alcohol so resolved (Weidmann and Horeau, 1967). Thus if (S)-(+)-2-phenylbutyric anhydride is used with an excess of racemic RR'CHOH, the left-over alcohol (whose sign of rotation needs to be determined) will have configuration **B** in Figure 5.48.

The configuration of alcohols that are chiral merely through the presence of deuterium, such as RCHDOH (Horeau and Nouaille, 1966) and even $CH_3CHOHCD_3$ (Horeau et al., 1965), has been successfully determined by this method. Deuterium is appreciably smaller than hydrogen because of its lesser zero-point vibration and although the optical yields of the residual 2-phenyl-butyric acid in the determination of the configuration of the above deuterated alcohols are only about 0.5%, the high specific rotation of the acid (nearly 100) makes it possible to measure its rotation even at such low enantiomeric purity. It should be noted that since D is smaller than H (but see Chapter 11), the rule embodied in Figure 5.48 must be modified: Isolation of residual levorotatory acid points to

$$
\begin{array}{c}
D \\
| \\
H-C-R, \\
| \\
OH
\end{array}
$$

that is, R configuration of the alcohol.

Modifications of Horeau's procedure allow determination of configuration of optically active alcohols on a microscale, where measurement of the optical rotation of the recovered acid is not feasible. In one method (Weidmann and Horeau, 1973) the alcohol is esterified with an excess of racemic 2-phenylbutyric anhydride and the diastereomer excess of the product ester determined by gas chromatography (GC) or high-performance liquid chromatography (HPLC) or ^{1}H NMR spectroscopy. As explained above, the major peak must be due to either the R,R or S,S isomer, depending on the configuration of the starting material. In a second experiment, the same alcohol (again on a microscale) is esterified with nonracemic (say S) anhydride; the product will be R,S if the alcohol was R, but S,S if it was S; the two possibilities are again diagnosed by chromatography or NMR. The peaks are identified by comparing them with the position of the major peak in the first experiment.

In another method (Horeau and Nouaille, 1990) one of the enantiomers of 2-phenylbutyric anhydride (say S) is deuterium labeled (either at the α or the γ position) and mixed with the other, unlabeled R enantiomer. This enantiospecifical-ly half-labeled "racemate" is then allowed to react with a very small amount of the alcohol of unknown configuration. As explained earlier, if the alcohol has configuration **A** (Fig. 5.48) the S (deuterated) acid will be predominantly incorporated into the ester formed [and vice versa, if it is **B**, the R (undeuterated) acid will predominate as part of the ester formed]. The purified ester is then submitted to mass spectrometry (MS). The major fragment ion is $C_6H_5CHCH_2CH_3^+$, $m/z = 119$ or the corresponding α- or β-deuterated species, $m/z = 120$. Thus, if the original alcohol had configuration **A**, the ratio of the 120:119 peaks will be in excess of unity, but if it was **B**, that ratio will be less than unity.

In yet another modification of the method (Kato, 1990, Kato et al., 1991), the esterifying agent for the alcohol whose configuration is to be determined is 1,1'-binaphthyl-2,2'-diyl phosphoryl chloride; in general, the *S* chloride reacts preferentially with the *S* alcohol (and vice versa). The acid predominantly incorporated in the ester is diagnosed by CD; because of the occurrence of exciton chirality (Davidov splitting, see Section 13-4.d), the CD amplitude of the ester spectrum is very large and the method should therefore be applicable on a small scale.

In summary, the Horeau method can be used for microscale determinations provided a highly sensitive method (GC, HPLC, MS, or CD) is available to determine the composition of the diastereomeric esters formed; the detection method must be chosen so as to be independent of the chemical nature of the alcohol moiety of the esters.

Prelog's Rule

This rule (Fiaud, 1977a) relates to the course of asymmetric synthesis when a Grignard reagent is added to an α-ketoester (frequently a pyruvate) of a chiral secondary or tertiary alcohol, SMLC–OH (S, M, and L stand for the small, medium-sized, and large substituents, respectively; S may be H). Schematically, the reaction generally follows the stereochemical course outlined in Figure 5.49 (Prelog, 1953): If the ketoester is in the conformation in which the two carbonyl groups are antiperiplanar and in which the L group occupies the same plane as the two carbonyl groups and the alkyl oxygen, the Grignard reagent (RMgX) will approach the keto carbonyl function from the side of the S group of the two remaining alcohol substituents. This rule correlates the configuration of the α-hydroxyester moiety (and therefore of the α-hydroxyacid obtained by hydrolysis thereof, Fig. 5.49) with that of the starting alcohol SMLCOH incorporated in the α-ketoester.

Contemplation of Figure 5.49 indicates that if the CIP sequence in the alcohol is OH > L > M > S and the sequence in the acid is OH > CO_2H > R' > R the ester of the *S* alcohol will give the *R* acid. However, this type of symbolic correlation is not generally valid, for while the CIP sequence given for the alcohol is usually correct, that for the acid has only a 50% chance of being so (in one-half of the cases R will precede R'). A more useful generalization (Seebach and Prelog, 1982) is that if the alcohol is *S* (on the above premise of CIP sequence), the Grignard reagent will approach the keto carbonyl from the *Re* face (the face above the paper in Fig. 5.49, cf. Chapter 8). Since *S* and *Re* are "opposite," the reaction is of the *ul* (unlike) rather than of the *lk* (like) type. That the reaction is *ul* holds, of course, even if the starting alcohol is of the *R* configuration (as long as the CIP sequence is as indicated above); in that case the picture in Figure 5.49 must be mirrored and the reagent will attack from the rear or *Si* face of the carbonyl.

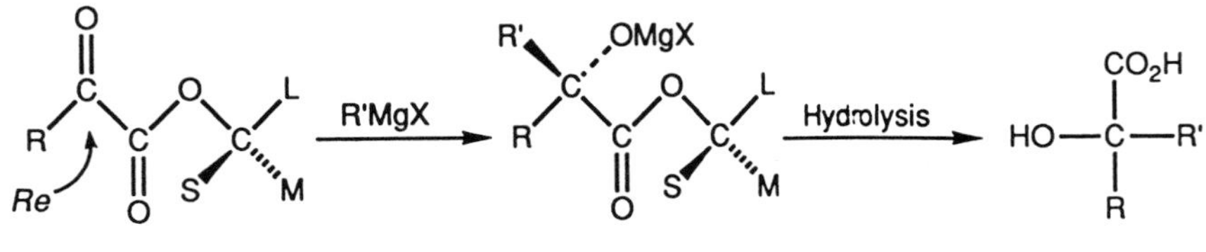

Figure 5.49. Prelog's rule.

Prelog's rule has been used to assign the configuration of numerous alcohols, generally secondary ones (S = H) (Fiaud, 1977). In most cases, the addition studied is that of phenylmagnesium bromide to the pyruvate of the alcohol under study; this produces an atrolactate. The atrolactic acid obtained after saponification is identified polarimetrically as (S)-$(+)$- or (R)-$(-)$-$C_6H_5(CH_3)COHCO_2H$ and the configuration of the alcohol is deduced therefrom (S acid → R alcohol, and vice versa). Some precautions need to be taken: It is important that the saponification of the atrolactate be quantitative, otherwise the results may be falsified by kinetic resolution (Chapter 7). Thus, if the predominant ester product is that of the $(+)$ acid, but the (diastereomeric) ester of the $(-)$ acid is saponified more rapidly, it might appear, if saponification is incomplete, that the $(-)$ acid predominated in the product; a false conclusion would thus be drawn. This problem is avoided by saponifying the ester completely.

Erroneous conclusions may also be drawn if the diastereomer excess is very small and measurement of the rotation therefore problematic. [In many cases, because of the large distance between the inducing (alcohol) and the induced (hydroxy acid) chiral center, the asymmetric induction or diastereomer excess in the reaction shown in Figure 5.49 is not large.] There may also be a problem as to which of the two alkyl substituents in the secondary alcohol LMCHOH is the larger, and there may be difficulties if one of the substituents can complex with the Grignard reagent. Problems resulting from kinetic resolution can also arise if too much Grignard reagent is used and the initially formed α-hydroxy ester is converted into a glycol by secondary reaction at the ester function.

It must be emphasized that the picture shown in Figure 5.49 is schematic, but not necessarily mechanistic. Indeed in the reactions of α-aldehydo and α-keto esters of 7-phenylmenthol (cf. Fig. 5.50), which are much more stereoselective than corresponding reactions with menthol esters, there is evidence that the carbonyl groups in the transition state are syn- rather than antiperiplanar (Whitesell et al., 1982, 1983; Whitesell, 1985). Reactions of this type are useful both for the asymmetric synthesis and for configurational assignment of secondary and tertiary α-hydroxy acids, $RR''COHCO_2H$ ($R' = H$ or alkyl).

Cram's Rule; Sharpless' Rule

Cram's rule refers to the diastereomer ratio formed in the addition of organometallic reagents to chiral ketones of the type R′COCHXR, whereas Sharpless' rule deals with the ratio of enantiomers formed in epoxidation of RR′C=CR″CH₂OH with *tert*-butyl hydroperoxide and titanium tetraalkoxide in

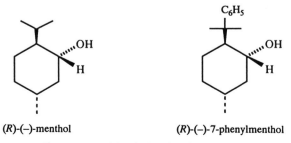

(R)-(–)-menthol   (R)-(–)-7-phenylmenthol

Figure 5.50.   Menthol and 7-phenylmenthol.

the presence of an R,R or S,S dialkyl tartrate. Since the major application of these rules is in stereoselective synthesis, they will be taken up in Chapter 12. It might be noted, however, that both rules (along with Prelog's rule) have been used in the assignment of the absolute configuration of $(CH_3)_3C–COH(CH_3)–CO_2H$ (Eliel and Lynch, 1987; see also Inch et al., 1968).

Other Asymmetric Syntheses

Other asymmetric syntheses and kinetic resolutions can be used for configurational assignment (Fiaud, 1977b) but since the basis of such methods is empirical, their reliability is often questionable. Among the more reliable methods are enzymatic ones which, however, depend on having available an enzyme that is not only very stereoselective (this is true of many *pure* enzymes) but which is also quite *un*selective with respect to substrate so that it can be used for a variety of compounds. An example is hog kidney acylase which mediates the hydrolysis of many N-acyl-L-amino acids. If this enzyme functions in the case of an N-acyl derivative of an optically active α-amino acid of unknown configuration, it demonstrates that the amino acid in question is L (cf. Fig. 5.16, p. 112); if only the N-acyl derivative of the racemate is available and hydrolysis is catalyzed, the L-acylamino acid will be hydrolyzed to the free L-amino acid, whereas the acyl derivative of the D isomer will remain unchanged. However, a negative result (i.e., no hydrolysis) is not conclusive for the D configuration of the substrate, since there are L-acylamino acids for which the reaction fails.

h. Chiroptical, Spectroscopic, and Other Physical Methods

By chiroptical (from chiral and optical) methods we mean optical methods of assessing chirality. The term thus refers to optical rotation, ORD, CD (for all three methods see Legrand and Rougier, 1977), vibrational CD (Nafie and Diem, 1979; Freedman and Nafie, 1987), and Raman optical activity (Barron, 1980). Since the application of these methods to the determination of configuration requires a detailed background on the methods themselves, consideration will be deferred to Chapter 13 (Sections 4 and 5b).

A number of correlation methods depend in NMR spectroscopy as a tool (Gaudemer, 1977); some of these have already been taken up in Chapter 2. Most of the others depend in one way or another on conformation and the way it affects chemical shifts and especially coupling constants. (Regarding a caveat in the use of such applications, cf. Cha, Kishi, et al., 1982). These methods will be discussed in Chapters 10 and 11. A variation of the NMR method is to deduce configuration from the sense of shift produced in corresponding nuclei (^1H, ^{13}C, and ^{19}F) of enantiomers under the influence of chiral solvating agents (CSA) (Pirkle and Hoover, 1982; Weisman, 1983) or chiral shift reagents (Sullivan, 1978; Fraser, 1983) or upon conversion to one of a pair of diastereomers by combination with a chiral, enantiomerically pure reagent (e.g, Helmchen, 1974; Yamaguchi, 1983). These methods have been reviewed elsewhere (Rinaldi, 1982; see also Chapter 8) and, because of their empirical nature, will not be discussed further here; use of these same methods in the determination of enantiomeric purity will be discussed in Chapter 6. Particularly interesting results are obtained by combining chiral derivatizing reagents (CDA) with achiral shift reagents (Yamaguchi et al., 1976).

> Since regularities in the sense of shift can only be established on the basis of the known absolute configuration of a number of test compounds, the method is, in fact, a correlative one, even though this point is often misstated in the literature.

Among other physical methods for the determination of relative configuration, X-ray structure analysis has already been mentioned prominently in this chapter and in Chapter 2. Infrared spectroscopy (Golfier, 1977) is sometimes useful in cyclic (especially six-membered) compounds and in compounds capable of forming intramolecular hydrogen bonds; examples will be given in Chapters 10 and 11. Chapters 10 and 11 will also address the use of dipole moments (Minkin, 1977), which have been used in configurational assignment of compounds having more than one polar substituent. Mass spectra (Green, 1976; Mandelbaum, 1977) are of limited application in assignment of relative configuration (but see Splitter and Tureček, 1994); Figure 5.51 shows one of the rare examples (Audier et al.,

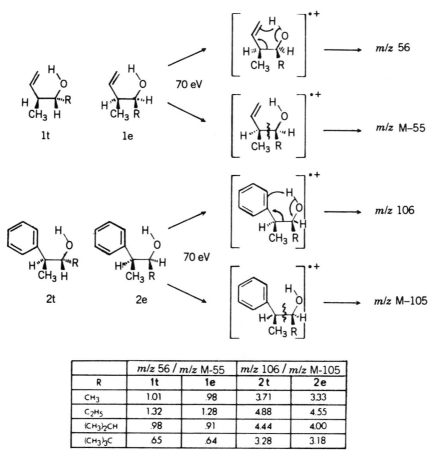

	m/z 56 / m/z M-55		m/z 106 / m/z M-105	
R	1t	1e	2t	2e
CH$_3$	1.01	.98	3.71	3.33
C$_2$H$_5$	1.32	1.28	4.88	4.55
(CH$_3$)$_2$CH	.98	.91	4.44	4.00
(CH$_3$)$_3$C	.65	.64	3.28	3.18

Figure 5.51. Configurational assignment by mass spectrometry. Reprinted with permission from Green, M. M., *Topics in Stereochemistry*, **9**, 41 (1976). Copyright John Wiley & Sons, New York, 1976.

1965) where they have served for assignment of configuration in an acyclic system. The electrocyclic process leading to formation of the m/z 56 ion is more favorable when it does not require eclipsing of CH_3 and R; therefore the 56 ion is more abundant in the spectrum of the R^*,S^* isomer than in the spectrum of the R^*,R^*. (Even though only one enantiomer of each is shown in Fig. 5.51, the experiment was carried out with racemic diastereomers.)

Chromatography often serves to separate diastereomers either analytically or preparatively. If the order of elution can be firmly correlated with the relative configuration of the chiral centers (either mechanistically, or empirically, or both) a method of configurational assignment is at hand. This correlation has been used for amides of chiral acids with chiral amines by Helmchen et al. (1972); in this way, the configuration of the acid can be determined by converting it into an amide with an amine of known configuration and vice versa for an amine whose configuration is unknown. Unfortunately, the method fails in a number of instances (Bergot et al., 1978, 1979; Kasai, Ziffer et al., 1983; Kasai and Ziffer, 1983).

Nematic liquid crystals are converted into cholesteric ones upon addition of small amounts of chiral addenda. The sign of the CD of the resulting cholesteric mesophase is sometimes systematically related to the configuration of the added chiral material (Gottarelli et al., 1975; Rinaldi et al., 1982); the method and its limitations will be discussed in Section 13-4.e. Cases of systematic color changes between enantiomers included in crown ethers covalently linked to steroids have also been reported (Nishi, Shinkai, et al., 1991); these changes are due to a differential change in helical pitch of the cholesteric liquid crystals brought about by the two opposite enantiomers.

Two other techniques will be mentioned here. One is related to the determination of absolute configuration by observing the growth of polar crystals in the presence of impurities (Section 5-3). Crystallization of a racemic substance that forms a conglomerate (i.e., a mixture of crystals of the two enantiomers; cf. Chapter 6) in the presence of a resolved added chiral impurity will lead to the preferential crystallization of the enantiomer whose configuration is opposite to that of the impurity (Addadi et al., 1982c, 1986). For example, solutions of DL-α-amino acids that form conglomerates, when seeded with crystals of a different L-amino acid, will deposit crystals of the D enantiomer. The configuration of the material that crystallizes can thus be correlated with that of the seed material.

The second technique to be mentioned involves chemical reactions in chiral crystals of achiral molecules leading to chiral products. Examples include photochemical reactions in the solid state, such as the formation of chiral cyclobutane derivatives by light-induced $2+2$ dimerization of alkenes. If the chirality of the crystal can be established crystallographically, that of the reaction product can be deduced mechanistically (Green et al., 1979, 1986).

Again it will be realized that many of these methods involve the confrontation principle (Section 5-5.d): The configuration of a chiral unknown is determined by combining it, in some way or other, with a chiral compound of known configuration; if then the relative configurations of the known and unknown can be determined by whatever method, that of the unknown can be deduced.

5-6. CONCLUSION: NETWORK ARGUMENTS

In concluding this chapter, it must be stressed that many chiral compounds have been correlated with others by multiple pathways (cf. Klyne and Buckingham, 1978; Buckingham and Hill, 1986). Changing the configurational assignment of one compound in such a network means changing the configuration assignment of all the others correlated with it (unless it can also be shown that *all* the individual correlations with the compound under scrutiny are wrong). Thus it is incumbent on anyone who claims a change in configurational assignment of a given compound to investigate all the other changes that would entail and rationalize or explain all the concomitant changes (for an example, see Brewster, 1982). Of course, the configurations of some compounds are more firmly anchored than those of others because they are more tightly tied into the network by multiple correlations, whereas others may have been correlated by only a single method and possibly not a very reliable one. This point must be kept in mind when configurational assignments are discussed.

REFERENCES

Addadi, L., Berkovitch-Yellin, Z., Domb, N., Gati, E., Lahav, M., and Leiserowitz, L. (1982a), *Nature (London)*, **296**, 21.

Addadi, L., Weinstein, S., Gati, E., Weissbuch, I., and Lahav, M. (1982c), *J. Am. Chem. Soc.*, **104**, 4610.

Addadi, L., Berkovitch-Yellin, Z., Weissbuch, I., Lahav, M., and Leiserowitz, L. (1982b), *J. Am. Chem. Soc.*, **104**, 2075.

Addadi, L., Berkovitch-Yellin, Z., Weissbuch, I., Lahav, M., and Leiserowitz, L. (1986), "A Link between Macroscopic Phenomena and Molecular Chirality: Crystals as Probes for the Direct Assignment of Absolute Configuration of Chiral Molecules," *Top. Stereochem.*, **16**, 1.

Addadi, L., Berkovitch-Yellin, Z., Weissbuch. I., van Mil, J., Shimon, L. J. W., Lahav, M., and Leiserowitz, L. (1985), "Growth and Dissolution of Organic Crystals with "Taylor-Made" Inhibitors—Implications in Stereochemistry and Materials Science," *Angew. Chem. Int. Ed. Engl.*, **24**, 466.

Allen, F. H., Neidle, S., and Rogers, D. (1968), *Chem. Commun.*, 308.

Allen, F. H., Neidle, S., and Rogers, D. (1969), *J. Chem. Soc. Chem. Commun.*, 452.

Allen, F. H. and Rogers, D. (1966), *Chem. Commun.*, 838.

Anderson, R. C. and Fraser-Reid, B. (1985), *J. Org. Chem.*, **50**, 4781.

Arigoni, D. and Eliel, E. L. (1968), "Chirality Due to the Presence of Hydrogen Isotopes at Noncyclic Positions," *Top. Stereochem.*, **4**, 160.

Audier, H. E., Felkin, H., Fetizon, M., and Vetter, W. (1965), *Bull. Soc. Chim. Fr.*, 3236.

Auwers, K. and Meyer, V. (1888), *Ber. Dtsch. Chem. Ges.*, **21**, 784.

Barron, L. D. (1980), "Raman Optical Activity: A New Probe of Stereochemistry and Magnetic Structure," *Acc. Chem. Res.*, **13**, 90.

Bergot, B. J., Anderson, R. J., Schooley, D. A., and Henrick, C. A. (1978), *J. Chromatogr.*, **155**, 97.

Bergot, B. J., Baker, F. C., Lee, E. and Schooley, D. A. (1979), *J. Am. Chem. Soc.*, **101**, 7432.

Berkovitch-Yellin, Z., Addadi, L., Idelson, M., Leiserowitz, L., and Lahav, M. (1982) *Nature (London)*, **296**, 27.

Bijvoet, J. M. (1955), "Determination of the Absolute Configuration of Optical Antipodes," *Endeavour*, **14**, 71.

Bijvoet, J. M., Peerdeman, A. F., and van Bommel, A. J. (1951), *Nature (London)*, **168**, 271.

Bischoff, C. A. (1891), *Ber. Dtsch. Chem. Ges.*, **24**, 1085.

Bosnich, B. (1969), "The Application of Exciton Theory to the Determination of the Absolute Configurations of Inorganic Complexes," *Acc. Chem. Res.*, **2**, 266.

Brewster, J. H. (1972), "Assignment of Stereochemical Configuration by Chemical Methods," in Bentley, K. W. and Kirby, G. W., Eds., *Elucidation of Organic Structures by Physical and Chemical Methods*, Vol. IV, Part III, 2nd ed., Wiley-Interscience New York, pp. 1–249.

Brewster, J. H. (1982), *Helv. Chim. Acta*, **65**, 317.

Brewster, J. H. (1986), *J. Org. Chem.*, **51**, 4751.

Brewster, P., Hiron, F., Hughes, E. D., Ingold, C. K., and Rao, P. A. D. S. (1950), *Nature (London)*, **166**, 179.

Brongersma, H. H. and Mul, P. M. (1973), *Chem. Phys. Lett.*, **19**, 217.

Brown, H. C. and Zweifel, G. (1961), *J. Am. Chem. Soc.*, **83**, 486.

Buckingham, J. and Hill, R. A. (1986), *Atlas of Stereochemistry—Supplement*, Chapman & Hall, New York.

Cahn, R. S. and Ingold, C. K. (1951), *J. Chem. Soc.*, 612.

Cahn, R. S., Ingold, C. K., and Prelog, V. (1956), *Experientia*, **12**, 81.

Cahn, R. S., Ingold, Sir C., and Prelog, V. (1966), "Specification of Molecular Chirality," *Angew. Chem. Int. Ed. Engl.*, **5** 385.

Caldwell, D. J. and Eyring, H. (1971), *The Theory of Optical Activity*, Wiley-Interscience New York.

Canet, D. (1973) *C. R. Séances Acad. Sc. Ser. C*, **276**, 315.

Carey, F. A. and Kuehne, M. E. (1982), *J. Org. Chem.*, **47**, 3811.

Carter, O. L., McPhail, A. T., and Sim, G. A. (1967), *J. Chem. Soc. A.*, 365.

Centnerszwer, M. (1899), *Z. Phys. Chem.*, **29**, 715.

Cha, J. K., Christ, W. J., Finan, J. M., Fujioka, H., Kishi, Y., Klein, L. L., Ko. S. S., Leder, J., McWhorter, W. W., Pfaff, K.-P., and Yonaga, M. (1982), *J. Am. Chem. Soc.*, **104**, 7369.

Coster, D., Knol, K. S., and Prins, J. A. (1930), *Z. Phys*, **63**, 345.

Cram, D. J. (1952), *J. Am. Chem. Soc.*, **74**, 2152.

Cross, L. C. and Klyne, W. (1976), "Rules for the Nomenclature of Organic Chemistry, Section E: Stereochemistry," *Pure Appl. Chem.*, **45**, 13; see also *J. Org. Chem.* (1970), **35**, 2849.

Curtin, D. Y. and Paul, I. C. (1981), "Chemical Consequences of the Polar Axis in Organic Solid-State Chemistry," *Chem. Rev.*, **81**, 525.

Dahlén, B. (1974), *Acta Cryst.*, **B30**, 642.

Dodziuk, H. and Mirowicz, M. (1990), *Tetrahedron: Asym.*, **1**, 171.

Dunitz, J. D. (1979), *X-ray Analysis and the Structure of Organic Molecules*, Cornell University Press, Ithaca, New York, pp. 129–148.

Eliel, E. L. (1962), *Stereochemistry of Carbon Compounds*, McGraw-Hill Book Co., New York, pp. 22, 88–92.

Eliel, E. L. (1985), *J. Chem. Educ.*, **62**, 223.

Eliel, E. L. and Gianni, M. H. (1962), *Tetrahedron Lett.*, 97.

Eliel, E. L. and Lynch, J. E. (1987), *Tetrahedron Lett.*, **28**, 4813.

Eliel, E. L. and Ro, R. S. (1957), *J. Am. Chem. Soc.*, **79**, 5995.

Falk, H. and Schlögl, K. (1965), *Monatsh. Chem.*, **96**, 266.

Fiaud, J. C. (1977a), "Prelog's Methods," in Kagan, H. B., ed., *Stereochemistry, Fundamentals and Methods*, Vol. 3, Georg Thieme Publishers, Stuttgart, Germany, p. 19.

Fiaud, J. C. (1977b), "Determination of Absolute Configurations of Organic Compounds by Asymmetric Synthesis, by Resolution and by Enzymatic Methods," in Kagan, H. B., Ed., *Stereochemistry, Fundamentals and Methods*, Vol. 3, Georg Thieme Publishers, Stuttgart, Germany, p. 95.

Fiaud, J. C. and Kagan, H. B. (1977), "Determination of Stereochemistry by Chemical Correlation Methods," in Kagan, H. B., Ed., *Stereochemistry, Fundamentals and Methods*, Vol. 3, Georg Thieme Publishers, Stuttgart, Germany, p. 1.

Fraser, R. R. (1983), "Nuclear Magnetic Resonance Analysis Using Chiral Shift Reagents." in Morrison, J. D., Ed., *Asymmetric Synthesis*, Vol. 1, Academic, New York, p. 173.

Fredga, A. (1960), "Steric Correlations by the Quasi-Racemate Method," *Tetrahedron*, **8**, 126.

Fredga, A. (1973), *Bull. Soc. Chim. Fr.*, 173.

Fredga, A. and Gamstedt, E. (1976), *Chem. Scr.* **9**, 5.

Freedman, T. B. and Nafie, L. A. (1987) "Stereochemical Aspects of Vibrational Optical Activity," *Top. Stereochem.*, **17**, 113.

Gallagher, M. J. and Jenkins, I. D. (1968), "Stereochemical Aspects of Phosphorus Chemistry," *Top. Stereochem.*, **3**, 1.

Gaudemer, A. (1977), "Determination of Relative Configuration by NMR Spectroscopy," in Kagan, H. B., Ed., *Stereochemistry, Fundamentals and Methods*, Vol. 1, Georg Thieme Publishers, Stuttgart, Germany, p. 44.

Gielen, M. (1977), *J. Chem. Educ.*, **54**, 673.

Golfier, M. (1977), "Determination of Configuration by Infrared Spectroscopy," in Kagan, H. B., Ed., *Stereochemistry, Fundamentals and Methods*, Vol. 1, Georg Thieme Publishers, Stuttgart, Germany, p. 1.

Gottarelli, G., Samorì, B., Marzochhi, S., and Stremmenos, C. (1975) *Tetrahedon Lett.*, 1981.

Green, M. M. (1976), "Mass Spectrometry and the Stereochemistry of Organic Molecules," *Top. Stereochem.*, **9**, 35.

Green, B. S., Arad-Yellin, R., and Cohen, M. D. (1986), "Stereochemistry and Organic Solid-State Reactions," *Top. Stereochem.*, **16**, 131.

Green, B. S., Lahav, M., and Rabinovich, D. (1979), "Asymmetric Synthesis via Reactions in Chiral Crystals," *Acc. Chem. Res.*, **12**, 191.

Gronowitz, S. (1957) *Ark. Kemi*, **11**, 361.

Gronowitz, S. and Larsson, S. (1955), *Ark. Kemi*, **8**, 567.

Harrison, P. W. B., Kenyon, J., and Phillips, H. (1926), *J. Chem. Soc.*, 2079.

Heathcock, C. H., White, C. H., Morrison, J. J. and VanDerveer, D. (1981), *J. Org. Chem.*, **46**, 1296.

Helmchen, G. (1974), *Tetrahedron Lett.*, 1527.

Helmchen, G., Ott, R., and Sauber, K. (1972), *Tetrahedron Lett.*, 3873.

Hermans, P. H. (1924), *Z. Phys. Chem.*, **113**, 337.

Hezemans, A. M. F. and Groenewege, M. P. (1973), *Tetrahedron*, **29**, 1223.

Hine, R. and Rogers, D. (1956), *Chem. Ind.*, 1428.

Hoffmann, R. W. and Zeiß, H.-J. (1981), *J. Org. Chem.*, **46**, 1309.

Horeau, A. (1961), *Tetrahedron Lett.*, 506.

Horeau, A. (1977), "Determination of the Configuration of Secondary Alcohols by Partial Resolution," Kagan, H. B., Ed., *Stereochemistry, Fundamentals and Methods*, Vol. 3, Georg Thieme Publishers, Stuttgart, Germany, Chap. 3.

Horeau, A. and Nouaille, A. (1966), *Tetrahedron Lett.*, 3953.

Horeau, A. and Nouaille, A. (1990), *Tetrahedron Lett.*, **31**, 2707.

Horeau, A. and Nouaille, A., and Mislow, K. (1965), *J. Am. Chem. Soc.*, **87**, 4957.

Horner, L. (1964), "Darstellung und Eigenschaften Optisch Aktiver, Tertiärer Phosphine," *Pure Appl. Chem.*, **9**, 225.

Horner, L., Winkler, H., Rapp, A., Mentrup, A., Hoffmann, H., and Beck, P. (1961), *Tetrahedron Lett.*, 161.

Houssa, A. J. H., Kenyon, J., and Phillips, H. (1929), *J. Chem. Soc.*, 1700.

Hughes, E. D., Juliusburger, F., Masterman, S., Topley, B., and Weiss, J. (1935), *J. Chem. Soc.*, 1525.

Hulshof, L. A., Wynberg, H., van Dijk, B., and De Boer, J. L. (1976), *J. Am. Chem. Soc.*, **98**, 2733.

Iitaka, Y., Itai, A., Tomioka, N., Kodama, Y., Ichikawa, K., Nishihata, K., Nishio, M., Izumi, M., and Doi, K. (1986), *Bull. Chem. Soc. Jpn.*, **59**, 2801.

Inch, T. D., Ley, R. V., and Rich, P. (1968), *J. Chem. Soc. C*, 1683, 1693.

Ingold, C. K. (1969), *Structure and Mechanism in Organic Chemistry*, 2nd ed., Cornell University Press, Ithaca, New York, Chap. VII.

Ionov, L. B., Kunitskaya, L. A., Mukanov, I. P., and Gatilov, Y. F. (1976), *J. Gen. Chem. USSR*, **46**, 68.

IUPAC (1979), *Nomenclature of Organic Chemistry*, Pergamon Press, New York, p. 473.

Jacques, J., Gros, C., and Bourcier, S. (1977), "Absolute Configurations of 6000 Selected Compounds with One Asymmetric Carbon Atom," in Kagan, H. B., Ed., *Stereochemistry, Fundamentals and Methods*, Vol. 4, Georg Thieme Publishers, Stuttgart, Germany.

Johnson, C. K., Gabe, E. J., Taylor, M. R., and Rose, I. A. (1965), *J. Am. Chem. Soc.*, **87**, 1802.

Kasai, M., Froussios, C., and Ziffer, H. (1983), *J. Org. Chem.*, **48**, 459.

Kasai, M. and Ziffer, H. (1983), *J. Org. Chem.*, **48**, 712.

Kato, N. (1990), *J. Am. Chem. Soc.*, **112**, 254.

Kato, N., Iguchi, M., Kato, Y. (1991) *Tetrahedron Asym.*, **2**, 763.

Kauzmann, W. J., Walter, J. E., and Eyring, H. (1940), "Theories of Optical Rotatory Power," *Chem. Rev.*, **26**, 339.

Kelvin, Lord (W. Thomson) (1904), *Baltimore Lectures on Molecular Dynamics and the Wave Theory of Light*, C. J. Clay & Sons, London. The lectures were given in 1884 and 1893.

Kirkwood, J. G. (1937), *J. Chem. Phys.*, **5**, 479.

Klyne, W. and Buckingham, J. (1978), *Atlas of Stereochemistry*, Vols. I and II, 2nd ed., Chapman & Hall, London.

Klyne, W. and Scopes, P. M. (1969), Aylett, B. J. and Harris, M. M., Eds., "Stereochemical Correlations," *Progress in Stereochemistry*, Vol. 4, Butterworths, London.

Kreiser, W. (1981), *Nachr. Chem. Tech. Lab.*, **29**, 555.

Kuhn, W. (1935), *Z. Phys. Chem.*, **B31**, 23.

Kuhn, W. (1952), *Z. Elektrochem.*, **56**, 506.

Lambert, J. B. (1971), "Pyramidal Atomic Inversion," *Top. Stereochem.*, **6**, 19.

Legrand, M. and Rougier, M. J. (1977), "Application of the Optical Activity to Stereochemical Determinations," in Kagan, H. B., Ed., *Stereochemistry, Fundamentals and Methods*, Vol. 2, Georg Thieme Publishers, Stuttgart, Germany, Chap. 2.

Lehn, J. M. (1970), "Nitrogen Inversion," *Top. Curr. Chem.*, **15**, 311.

Lemieux, R. U. and Howard, J. (1963), *Can. J. Chem.*, **41**, 308.

Lowry, T. M. (1935), *Optical Rotary Power*, Longmans Green & Co., New York, Chap. 7.

Mandelbaum, A. (1977), "Application of Mass Spectrometry to Stereochemical Problems," in Kagan, H. B., Ed., *Stereochemistry, Fundamentals and Methods*, Vol. 1, Georg Thieme Publishers, Stuttgart, Germany, p. 137.

Masamune, S., Ali, Sk. A., Snitman, D. L., and Garvey, D. S. (1980), *Angew Chem. Int. Ed. Engl.*, **19**, 557.

Mason, S. F. (1973), *J. Chem. Soc. Chem. Commun.*, 239.

Mathieson, A. McL. (1956) *Acta Cryst.*, **9**, 317.

McEwen, W. E. (1965), "Stereochemistry of Reactions of Organophosphorus Compounds," *Top. Phosphorus Chem.*, **2**, 1.

McKenzie, A. (1906), *J. Chem. Soc.*, **89**, 365.

Meer, N. and Polanyi, M. (1932), *Z. Phys. Chem.*, **B19**, 164.

Mikołajczyk, M. and Drabowicz, J. (1982), "Chiral Organosulfur Compounds," *Top. Stereochem.*, **13**, 333.

Minkin, V. I. (1977), "Dipole Moments and Stereochemistry of Organic Compounds. Selected Applications," in Kagan, H. B., Ed., *Stereochemistry, Fundamentals and Methods*, Vol. 2, Georg Thieme Publishers, Stuttgart, Germany, p. 1.

Mislow, K. (1951), *J. Am. Chem. Soc.*, **73**, 3954.

Mislow, K. and Heffler, M. (1952), *J. Am. Chem. Soc.*, **74**, 3668.

Mislow, K. and Siegel, J. (1984), *J. Am. Chem. Soc.*, **106**, 3319.

Moffitt, W. (1956), *J. Chem. Phys.*, **25**, 467.

Mosher, H. S. (1980), *Glossary to Audio Course "Stereochemistry,"* American Chemical Society, Washington, DC.

Nafie, L. A. and Diem, M. (1979), "Optical Activity in Vibrational Transitions: Vibrational Circular Dichroism and Raman Optical Activity," *Acc. Chem. Res.*, **12**, 296.

Neidle, S., Rogers, D., and Allen, F. H. (1970), *J. Chem. Soc. C*, 2340.

Newman, M. S. (1955), *J. Chem. Educ.*, **32**, 344.

Nishi, T., Ikeda, A., Matsuda, T., and Shinkai, S. (1991), *J. Chem. Soc. Chem. Commun.*, 339.

Noël, Y. (1987), *New J. Chem.*, **11**, 211.

Noyori, R., Nishida, T., and Sakata, J. (1981), *J. Am. Chem. Soc.*, **103**, 2106.

Olson, A. R. (1933), *J. Chem. Phys.* **1**, 418.

Patterson, K. (1954), *Ark. Kemi*, **7**, 347.

Pennington, W. T., Chakraborty, S., Paul, I. C., and Curtin, D. Y. (1988), *J. Am. Chem. Soc.*, **110**, 6498.

Phillips, H. (1923), *J. Chem. Soc.*, **123**, 44.

Phillips, H. (1925), *J. Chem. Soc.*, **127**, 2552.

Pirkle, W. H. and Hoover, D. J. (1982), "NMR Chiral Solvating Agents," *Top. Stereochem.*, **13**, 263.

Prelog, V. (1953), *Helv. Chim. Acta*, **36**, 308.

Prelog, V. and Helmchen, G. (1972), *Helv. Chim. Acta*, **55**, 2581.

Prelog, V. and Helmchem, G. (1982), *Angew. Chem. Int. Ed. Engl.*, **21**, 567.

Prelog, V. and Zalán (1944), *Helv. Chim. Acta*, **27**, 535, 545.

Raznikiewicz, T. (1962), *Acta Chem. Scand.*, **16**, 1097.

Ricci, J. E. (1962), *Tetrahedron*, **18**, 605.

Rinaldi, P. L. (1982), "The Determination of Absolute Configuration Using Nuclear Magnetic Resonance Techniques," *Prog. Nucl. Magn. Spectrosc.* **15**, 291.

Rinaldi, P. L., Naidu, M. S. R., and Conaway, W. E. (1982), *J. Org. Chem.*, **47**, 3987.

Rosenberg, A. and Schotte, L. (1955), *Ark. Kemi*, **8**, 143.

Ruch, E. (1968), *Theor. Chim. Acta*, **11**, 183, 462.

Ruch, E. (1972), "Algebraic Aspects of the Chirality Phenomenon in Chemistry," *Acc. Chem. Res.*, **5**, 49.

Saunders, M. and Yamada, F. (1963), *J. Am. Chem. Soc.*, **85**, 1882.

Schlögl, K. (1967), "Stereochemisty of Metallocenes," *Top. Stereochem.*, **1**, 39.

Schlögl, K. and Falk, H. (1964), *Angew. Chem. Int. Ed. Engl.*, **3**, 512.

Seebach, D. and Prelog, V. (1982), "The Unambiguous Specification of the Steric Course of Asymmetric Syntheses," *Angew. Chem. Int. Ed. Engl.*, **21**, 654.

Seebach, D., Renaud, P., Schweizer, W. B., Züger, M. F., and Brienne, M.-J. (1984), *Helv. Chim. Acta*, **67**, 1843.

Seiler, P., Martinoni, B., and Dunitz, J. D. (1984), *Nature (London)*, **309**, 435.

Shaik, S. S., Schlegel, H. B., and Wolfe, S. (1992), *Theoretical Aspects of Physical Organic Chemistry. The S_N2 Mechanism*, Wiley New York.

Shirahata, K. and Hirayama, N. (1983), *J. Am. Chem. Soc.*, **105**, 7199.

Skell, P. S., Allen, R. G., and Helmkamp, G. K. (1960), *J. Am. Chem. Soc.*, **82**, 410.

Splitter, J. S. and Tureček, F. (1994), *Application of Mass Spectrometry to Organic Stereochemistry*, VCH, New York.

Sullivan, G. R. (1978), "Chiral Lanthanide Shift Reagents," *Top. Stereochem.*, **10**, 287.

Tambute, A. and Collet, A. (1984), *Bull. Soc. Chim. Fr.*, II-77.

Tanaka, J., Katayama, C., Ogura, F., Tatemitsu, H., and Nakagawa, M. (1973a), *J. Chem. Soc. Chem. Commun.*, 21.

Tanaka, J., Ogura, F., Kuritani, M., and Nakagawa, M. (1972), *Chimia*, **26**, 471.

Tanaka, J., Ozeki-Minakata, K., Ogura, F., and Nakagawa, M. (1973b), *Spectrochim. Acta*, **A29**, 897.

Thomson, W. See Kelvin (Lord).

Timmermans, J. (1928), *Recl. Trav. Chim. Pays-Bas*, **48**, 890.

Trommel, J. and Bijvoet, J.M. (1954), *Acta Cryst.*, **7**, 703.

Walden, P. (1896), *Ber. Dtsch. Chem. Ges.*, **29**, 133.

Walden, P. (1897), *Ber. Dtsch. Chem. Ges.*, **30**, 3146.

Wang, J.-L., Lahav, M., and Leiserowitz, L. (1991), *Angew. Chem. Int. Ed. Engl.*, **30**, 696.

Weber, H. (1965), Ph.D. Dissertation No. 3591, Eidgenössische Technische Hochschule, Zurich, Switzerland.

Weber, H., Seibl, J., and Arigoni, D. (1966), *Helv. Chim. Acta*, **49**, 741.

Weidmann, R. and Horeau, A. (1967), *Bull. Soc. Chim. Fr.*, 117.

Weidmann, R. and Horeau, A. (1973), *Tetrahedron Lett.*, 2979.

Weisman, G. R. (1983), "Nuclear Magnetic Resonance Analysis Using Chiral Solvating Agents," in Morrison, J. D., Ed., *Asymmetric Synthesis*, Vol. 1, Academic, New York, p. 153.

Whitesell, J. K. (1985), "New Perspectives in Asymmetric Induction," *Acc. Chem. Res.*, **18**, 280.

Whitesell, J. K., Bhattacharya, A., and Henke, K. (1982), *J. Chem. Soc. Chem. Commun.*, 988.

Whitesell, J. K., Deyo, D., and Bhattacharya, A. (1983), *J. Chem. Soc. Chem. Commun.*, 802.

Wilen, S. H., Qi, J. Z., and Willard, P. G. (1991), *J. Org. Chem.*, **56**, 485.

Wohl, R. A. (1974), *Chimia*, **28**, 1.

Wood, W. W., Fickett, W., and Kirkwood, J. G. (1952), *J. Chem. Phys.*, **20**, 561.

Wunderlich, A. E. (1886), cited by Noël (q.v.).

Yamaguchi, S. (1983), "Nuclear Magnetic Resonance Analysis Using Chiral Derivatives" in Morrison, J.D., Ed., *Asymmetric Synthesis*, Vol. 1, Academic, New York, p. 125.

Yamaguchi, S., Yasuhara, F., and Kabuto, K. (1976), *Tetrahedron*, **32**, 1363.

6

Properties of Stereoisomers. Stereoisomer Discrimination

6-1. INTRODUCTION

This chapter will focus on the physical properties of enantiomer pairs and on methods for the determination of enantiomer composition. We begin by examining some of the ways in which the enantiomers of a chiral substance interact with one another. The properties of enantiomerically pure compounds are contrasted with those of the corresponding racemates, and also with those of unequal mixtures of the enantiomers. We shall see that the properties of chiral substances are affected by the proportion of the enantiomers in the mixture and by the state of the system. Knowledge of such properties and of their differences is essential to the design of efficient separation methods (Chapter 7). In addition, the study of these properties (and those of diastereomers) is useful in its own right in regard to such diverse applications as the determination of enantiomeric and diastereomeric purity and the generation of enantiomeric bias in Nature.

6-2. STEREOISOMER DISCRIMINATION

The word racemate describes an equimolar mixture of a pair of enantiomers. An important question that will concern us through the first one-half of this chapter is whether such a mixture behaves ideally, that is, does this mixture behave in the same manner as the individual constituent enantiomers. Therefore, the question is whether homochiral and heterochiral interactions do or do not differ significantly. Once it was believed that, in the liquid and in the gaseous phase, these interactions do not differ, but we shall see that this is not true in general.

    The terms homochiral and heterochiral, which describe the relatedness of chiral classes of compounds, were first used by William Thomson (Lord Kelvin) a century ago (Kelvin, 1904; see also Ruch, 1968, 1972). "Homochiral interactions," interactions between homochiral molecules (cf. Chapter 5), are defined as intermolecular nonbonded attractions or repulsions present in assemblies of molecules having like chirality sense. For the R enantiomers of a given substance and *in the absence of chemical reaction*, homochiral interactions might be represented as in Eq. 6.1:

$$R + R \rightleftharpoons R \cdots R \tag{6.1}$$

Homochiral interactions must be identical for either enantiomer. The phrase "heterochiral interactions" is a shorthand term for interactions between molecules of unlike chirality sense (heterochiral molecules), again in the absence of a chemical reaction, for example,

$$R + S \rightleftharpoons R \cdots S \tag{6.2}$$

The nonbonded interactions in question (Eqs. 6.1 and 6.2) comprise van der Waals and electrostatic interactions, hydrogen bonding, π-complex formation, and other forms of electron donation and acceptance that are readily reversible. The forces involve long-range dispersion (London forces), and have inductive, and permanent multipolar components (Craig and Schipper, 1975). The nature and magnitude of these forces is partially dependent on the symmetry of the molecules (see Section 4-6). Although Eqs. 6.1 and 6.2 describe only dimeric interactions, the possible formation of higher aggregates is not excluded.

The surrounding medium, not unexpectedly, affects the interactions implicit in Eqs. 6.1 and 6.2, for example, an alcohol solvent often overwhelms, and thereby destroys, solute–solute hydrogen-bonding interactions and a high dielectric solvent diminishes dipolar interactions (cf. Craig and Elsum, 1982).

Homochiral (Eq. 6.1) and heterochiral (Eq. 6.2) interactions among molecules of like constitution are unlikely ever to be exactly equal in magnitude ($\Delta G_{\text{homo}} \neq \Delta G_{\text{hetero}}$) because the two types of aggregates are anisometric (diastereomeric). The difference between these interactions was heretofore generally assumed to be nonexistent in solution or the liquid or gaseous state, or at least too small to be measurable, and of no significant consequence. However, as will presently be made clear, the difference $\Delta\Delta G$ between homochiral and heterochiral interactions ($R \cdots R$ vs. $R \cdots S$) is responsible for measureable differences in physical properties exhibited by racemates on one hand, and by the corresponding enantiomerically pure compounds on the other. This difference is a manifestation of *enantiomer discrimination*.

Interactions between a given enantiomer of one compound and the two enantiomers of another species give rise to diastereomeric pairs, such as $R_{\text{I}} \cdots R_{\text{II}}$ and $R_{\text{I}} \cdots S_{\text{II}}$; the difference between these interactions is called *diastereomer discrimination* (Craig and Mellor, 1976; Stewart and Arnett, 1982). A comprehensive term for the two types of discrimination is *stereoisomer discrimination*.

If enantiomer discrimination results from palpable dimeric or oligomeric aggregates (Eqs. 6.1 and 6.2) that bear diastereomeric relationships to one another, then the distinction between enantiomer discrimination and diastereomer discrimination is blurred. The distinction made here is more one of convenience than of principle; all stereoisomer discrimination is diastereomeric in nature.

We prefer the term stereoisomer discrimination to the more widely used expression "chiral discrimination" (or chiral recognition) to emphasize its nature. There is nothing chiral about the discrimination per se; and while it is exhibited by chiral substances, it is caused by diastereomer, not enantiomer differences. The alternative term chirodiastaltic discrimination, proposed by Craig and Elsum (1982), has found little favor.

As pointed out by Gillard (1979), stereoisomer discrimination is one of the major themes of contemporary stereochemical studies.

Figure 6.1 is a scheme describing the relationships between enantiomeric interactions and the diastereomeric interactions that exist among the enantiomers of two different compounds, **I** (R_I and S_I) and **II** (R_{II} and S_{II}). Only the horizontally related enantiomeric interactions are necessarily equal in magnitude (by symmetry) whether the systems behave ideally or not [if $M(X \cdots Y)$ represents the numerical magnitude of any physical property measured in a bulk sample, then $M(R_I \cdots R_{II}) = M(S_{II} \cdots S_I)$ and $M(R_I \cdots R_{II}) - M(R_I \cdots S_{II}) = M(S_{II} \cdots S_I) - M(R_{II} \cdots S_I)$ and, of course, $M(R_I \cdots R_I) = M(S_I \cdots S_I)$] (Stewart and Arnett, 1982).

Stereoisomer discrimination is strongly phase dependent. At one extreme, enantiomer discrimination in the solid state is responsible for the occurrence of well-defined racemic compounds having properties that are significantly different from those of enantiomerically pure samples of the same substance. Moreover, the physical properties of a given chiral substance in bulk evidently depend on the proportion of the enantiomers, that is, on the *enantiomer composition*, of the sample. At the other extreme, stereoisomer discrimination in the gaseous state (Section 6-4.o) is rarely observed. The situation in liquids and solutions is intermediate.

The magnitude of enantiomer discrimination in aqueous solution may be estimated from calorimetric data for the mixing of solutions of the enantiomers of a given chiral compound. For the tartaric acid enantiomers, the heat of mixing, $\Delta H^m = 0.48$ cal mol^{-1} (2.0 J mol^{-1}), that is, the homochiral combination is enthalpically preferred; for the threonine enantiomers: $\Delta H^m = -1.3$ cal mol^{-1} (-5.5 J mol^{-1}) at 25.6°C. The two substances exhibit opposite senses of homochiral versus heterochiral enthalpic preference (Takagi, Fujishiro, and Amaya, 1968). For fenchone, dilution of one enantiomer with the other gives $\Delta H^m =$

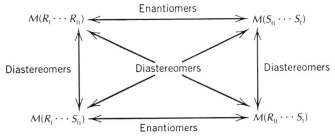

Figure 6.1 Schematic relationship of enantiomeric and diasteromeric interactions for two pairs of enantiomers. [Reprinted with permission from Stewart, M. V. and Arnett, E. M. (1982), *Top. Stereochem.*, **13**, 195. Copyright © 1982 John Wiley & Sons, Inc.]

$-1.1\,\text{cal mol}^{-1}$ $(-4.5\,\text{J mol}^{-1})$; the corresponding values for α-methylbenzylamine are $\Delta H^{m} = +1.7\,\text{cal mol}^{-1}$ $(+7.3\,\text{J mol}^{-1})$ at 30°C. It appears that, for the amine, ΔH^{m} changes sign in the vicinity of 67°C (Atik, Ewing, and McGlashan, 1983). Thus the nonbonded attractive interactions in the homochiral versus the heterochiral aggregates responsible for enantiomer discrimination in solution may be either greater or smaller, according to conditions, and these interactions are quite small.

Calorimetric measurements of the heats of solution, ΔH^{soln}, of crystalline racemates and of the corresponding crystalline enantiomer components (in water) illustrate the much greater magnitude of enantiomer discrimination in the solid state relative to that in solution. For tartaric acid, $\Delta H^{\text{soln}}_{\text{rac}} - \Delta H^{\text{soln}}_{\text{enant}} = \Delta\Delta H^{\text{soln}} = 2.2 \pm 0.07\,\text{kcal mol}^{-1}$ $(9.4 \pm 0.3\,\text{kJ mol}^{-1})$ and for threonine $\Delta\Delta H^{\text{soln}} = 0.04 \pm 0.03\,\text{kcal mol}^{-1}$ $(0.17 \pm 0.13\,\text{kJ mol}^{-1})$ at 25°C (Matsumoto and Amaya, 1980).

The importance of these enthalpies is best appreciated from the thermodynamic cycle illustrated in Figure 6.2. In this cycle, ΔH^{form} is the enthalpy of formation of the racemic compound from the crystalline enantiomers.

From the cycle, we see that $\Delta H^{\text{soln}}_{\text{rac}} - \Delta H^{\text{soln}}_{\text{enant}} = \Delta H^{m}_{\text{enant}} - \Delta H^{\text{form}}_{\text{rac}}$. For a conglomerate, $\Delta H^{\text{form}}_{\text{rac}} = 0$, hence the equation reduces to

$$\Delta\Delta H^{\text{soln}} = \Delta H^{\text{soln}}_{\text{rac}} - \Delta H^{\text{soln}}_{\text{enant}} = \Delta H^{m}_{\text{enant}} \qquad (6.3)$$

Applying this relationship to threonine (a conglomerate in the solid state), we see that the experimental values are not consistent: $\Delta H^{m} = -5.5\,\text{J mol}^{-1}$, a value that should be equal to $\Delta\Delta H^{\text{soln}}$, whereas the latter is $170\,\text{J mol}^{-1}$ (see above; also cf. Table 6.1). It would seem that the latter value must be very imprecise, being a small difference between two large numbers.

Numerous studies have been carried out over the past 100 years in attempts to demonstrate the occurrence of "racemic compounds" in the liquid state (see, e.g., Wheland, 1951, p. 160). Such investigations have been summarized by Horeau and Guetté (1974) who concluded that all studies purporting to demonstrate the

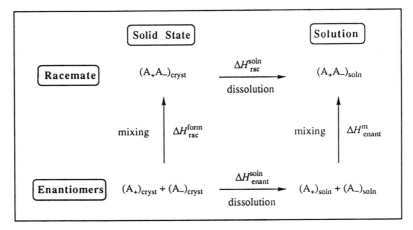

Figure 6.2. Thermodynamic cycle for the mixing of enantiomers.

formation of liquid racemic compounds are flawed principally by insufficient purity of the optically active or racemic samples utilized, and by the insensitivity of the measurements relative to the small difference in magnitude of the observed effects (see also Dunlop, Gillard, Wilkinson, et al., 1966; Al-Ani and Olin, 1983).

> The measurement of properties of chiral systems in general, and of the constituent enantiomers versus the racemates in particular, is often subject to errors (due to achiral as well as chiral impurities) and to instrumental limitations (sensitivity and artifacts). Great care is required in carrying out such measurements and in interpreting them. Stewart and Arnett (1982) have described an *absolute* method for reducing to a minimum the effect of impurities on properties of chiral systems.
>
> In the absolute method, each of the enantiomers of a compound is individually purified rigorously until the properties of the two are identical. Enantiomer discrimination may then be probed by carrying out similar measurements individually on the pure enantiomers and on mixtures of the two enantiomers (including the racemate) prepared from the purified samples. For example, the heat of solution of solid (+)-tartaric acid should match that of (−)-tartaric acid. Only when this is confirmed should comparison be made with ΔH^{soln} of racemic tartaric acid prepared from the individually purified enantiomers.
>
> Similarly, diastereomer discrimination is revealed with confidence if a chiral cross-check is undertaken (comparison of properties involving individual enantiomers of two different substances, e.g., $R_{\text{I}} + S_{\text{II}}$ with those of the enantiomeric combination, $R_{\text{II}} + S_{\text{I}}$). The equivalent pairs of measurements are summarized in Figure 6.1 (e.g., Arnett and Zingg, 1981).

While enantiomer discrimination may not be measureable in the gas phase or in dilute solution (ideal behavior obtains), differences in physical properties reflecting enantiomer discrimination will be increasingly manifest as the concentration of solutions is increased (see above), as measurements are compared in the neat liquid state and, a fortiori, as they are carried out in liquid crystals and in the solid state (cf. Wynberg, 1976; Wynberg and Feringa, 1976).

Ideal behavior is understood to mean that the enthalpy of mixing of enantiomer pairs in the liquid (or gaseous) state is equal to zero for any mixture of enantiomers, and the corresponding entropy of mixing is equal to $R \ln 2 = 1.38 \text{ cal mol}^{-1} \text{ K}^{-1}$ ($5.77 \text{ J mol}^{-1} \text{ K}^{-1}$). In fact, the enthalpy of mixing ΔH^{m} is often nonzero in the liquid state and in solution (for a summary, see Jacques, Collet, and Wilen, 1981a, p. 47ff). The values are typically quite small [ca. 0.5–50 cal mol^{-1} (ca. 2–200 J mol^{-1})] and they vary over quite a range depending on the substances that are mixed.

Enantiomer discrimination in the solid state can also be assessed by comparing heats of sublimation ΔH^{s} of enantiomerically pure and racemic samples of a given substance. The magnitude of this energy difference is about 1000 times greater than that found for the liquid state (see, e.g., Chickos, Garin, et al., 1981). The origin, sense, and theoretical basis of these energy discriminations have been discussed by Craig and Mellor (1976), and by Mason (1982, Chapter 11).

A major consequence of the nearly ideal behavior of neat liquid mixtures of enantiomers is that enantiomeric enrichment by distillation is, in general, not practical. It has been suggested that such separation should be thermodynamically impossible, the vapor pressures of a single enantiomer and of mixtures of the two enantiomers of a substance *in the liquid state* being equal (Mauser, 1957). However, since the behavior of chiral compounds in the liquid state is often not quite ideal, a small but finite difference in vapor pressure must exist. If we assume that the vapor phase behaves ideally, then the boiling point difference between a racemate and the corresponding pure enantiomer may be estimated by means of Trouton's rule ($\Delta T^v = \Delta H_\ell^m / 20$, where ΔH_ℓ^m is the heat of mixing of the enantiomers in the liquid state in calories per mole). For 2-(p-nitrophenyl)butane $\Delta H^m = 0.45$ cal mol^{-1}; the boiling point difference thus amounts to 2.2×10^{-2}°C, too small to allow separation by distillation. However, for 2-octanol ($\Delta H^m = 3.1$ cal mol^{-1}, at 25°C), the calculation gives a ΔT^v value of about 0.15°C and for α-methylbenzylamine ($\Delta H^m = 2.3$ cal mol^{-1}, at 25°C), $\Delta T^v = 0.12$°C (Horeau and Guetté, 1974; Jacques et al., 1981a, p. 165). One must conclude that separation of a racemate from one of its enantiomer constituents by distillation (or by other methods based on vapor pressure differences), while difficult, should be possible in some cases (e.g., Koppenhoefer and Trettin, 1989).

We emphasize that this statement does *not* imply that a racemate can be resolved into its enantiomer constituents by distillation. However, an unequal mixture of enantiomers ($x_R \neq x_S$, where x is the mole fraction) could, in principle, be enriched with respect to one of the enantiomers. Such separations of enantiomers from racemates have, in fact, been achieved by liquid chromatography on *achiral* columns (see Section 6-4.n).

> The above discussion relates to stereoisomer discrimination in a three-dimensional (3D) system. Recently reported calculations (Craig and Elsum, 1982) suggest that, in a two-dimensional (2D) system, stereoisomer discrimination factors for comparable substances should be about two orders of magnitude larger than the value given above. This prediction calls to mind a study by Arnett and Thompson (1981) demonstrating significant enantiomer discrimination in chiral monolayers (see also Arnett et al., 1989).

Enantiomer enrichment based on manipulation of solid samples of chiral substances, in contrast, is entirely practical. The substrate at hand may be racemic ($x_R = x_S$) or it may already be partially enriched ($x_R \neq x_S$). In the latter case, further enrichment, which may ultimately lead to enantiomeric purity, is possible (in fact, may be unavoidable!) by conventional (i.e., excluding the use of optically active solvents or reagents) crystallization, sublimation, extraction (even washing), or by combinations of these processes (Horeau, 1972). In a racemic sample, separation of the enantiomers (resolution) is possible without undertaking chemical reactions or conventional chromatography provided only that the solid racemate is a conglomerate (see below).

To summarize, stereoisomer separations are possible without the use of chiral reagents or chiral solvents. Diastereomers are more or less easy to separate from one another by conventional crystallization, distillation, or sublimation. To some extent, this is also true for enantiomer pairs: For chiral substances, such separations depend on the difference between homochiral and heterochiral interactions. In general, separation procedures that depend on the crossing of phase boundaries rather than on chemical transformations are likely to succeed if the entities involved (enantiomers, racemates, and diastereomers) differ by energies of the order of kilocalories per mole (kilojoules per mole) and to fail if these energies are only of the order of small calories per mole (or joules per mole).

6-3. THE NATURE OF RACEMATES

Three types of crystalline racemates are known (Roozeboom, 1899).

1. The racemate is simply a 1:1 mechanical mixture or *conglomerate* of crystals of the two enantiomers, each crystal being made up of homochiral molecules.

The term racemic mixture has often been used to describe this type of racemate in the literature. Since "racemic mixture" has also been used to describe 1:1 mixtures of enantiomers of unspecified type (in this book called racemate), we will not use the term here.

2. The racemate consists of crystals in each of which the (+) and (−) enantiomers are present in a 1:1 ratio down to the unit cell level. This corresponds to the formation of a solid compound, called a *racemic compound*.

A few cases, called *anomalous racemates*, are known in which enantiomers form addition compounds in ratios other than 1:1 (Jacques et al., 1981a, p. 147).
   The term true racemate, or simply racemate, has occasionally been used to describe a racemic compound. We prefer the latter term for reasons of clarity; we use "racemate" for any 1:1 mixture of enantiomers.

3. The racemate consists of a solid solution of the two enantiomers, that is, a single homogeneous phase in which a 1:1 stoichiometric mixture of the two enantiomers is present unordered in the solid phase. This type is called *pseudoracemate*.

Inspection of Figure 6.3c, in particular, makes clear our inability to identify any composition differing sufficiently from others as to warrant identification as a "racemate" [referring to 1:1 (+)/(−) composition] in the solid solution case.
   Such solid solutions are occasionally called mixed crystals. The misuse of the term pseudoracemate to describe a 1:1 mixture of (+)- and (−)-enantiomers, one of whose components is isotopically labeled (and hence has a different mass) (Testa and Jenner, 1978), is to be discouraged. Care should also be taken not to confuse pseudoracemates with *quasi-racemates* (see Chapter 5).

Samples of chiral compounds having no observable optical activity are generally assumed to comprise 1:1 mixtures of the enantiomers, that is, to be racemates. However, one must not ignore the possibility that such a measurement may be accidental. At a different temperature, wavelength, or in a different solvent, optical activity could be revealed (cf. Chapter 13), and consequently the sample in question is not racemic; it is said to be *cryptochiral*. The definition of a racemate also disregards statistical fluctuations. Thus, the likelihood that a racemate (such as the 2-butanol produced by hydrogenation of 2-butanone in the absence of any chiral influence) consists precisely of a 50:50 mixture of the enantiomers is statistically improbable. Yet it is not possible to demonstrate the low statistical preponderance of one enantiomer in a racemic sample with any tool presently available; the term cryptochiral has also been applied to such systems (Mislow and Bickart, 1976/77).

One of the simplest ways of recognizing a racemate type is by inspection of binary phase diagrams. Such characterization was first undertaken by Roozeboom (1899) as an application of the phase rule. Typical phase diagrams that relate the composition to the melting point for the three racemate classes are shown in Figure 6.3.

The most common of these racemate types is seen in Figure 6.3b; that is, the majority of racemates of chiral organic compounds (ca. 90%) exist in the form of racemic compounds, only about 10% as conglomerates (Section 7-2.b). The third type (Fig. 6.3c) is relatively rare. While conglomerate behavior is less common than racemic compound formation, the simplicity of enantiomer separation in the case of conglomerates gives this category special importance. The utility of the above phase diagrams will be taken up in the following section.

> Only about 60 examples of racemates exhibiting solid solution properties are listed in the book by Jacques et al. (1981a, pp. 104–113).

The above classification is not meant to imply that the racemate of a given chiral substance may exist in only one crystalline form. Many compounds are known whose racemate may crystallize in either of two (racemic compound or conglomerate) of the three principal forms depending on conditions. When this occurs, one of the forms is metastable (within a given range of temperature and pressure). In the solid state, the metastable form may often be maintained for a long time without change.

Whenever two crystalline forms of a given substance can be isolated, there is also a fair chance that they can be interconverted. So too with racemate types. For example, 1,1'-binaphthyl (Fig. 7.11, **15**) crystallizes as a racemic compound (mp 154°C) that is converted to the thermodynamically stabler conglomerate (mp 159°C) on heating (Wilson and Pincock, 1975; Kress, Paul, Curtin, et al., 1980). Numerous other examples of such transformations are known (e.g., Jacques et al., 1981a, pp. 134–144 and Section 7-2.c).

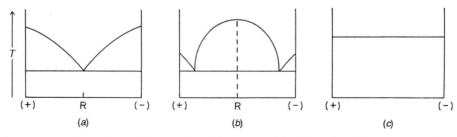

Figure 6.3. Binary phase diagrams describing the melting behavior of the common types of racemates. (a) Conglomerate, (b) racemic compound and (c) ideal solid solution (pseudoracemate).

The existence of a given substance in more than one crystalline form is called polymorphism. Polymorphs of the same compound differ in melting point and in crystal structure; polymorphs can sometimes be transformed into one another by changing the temperature or pressure. The incidence of two crystalline forms of racemates is but a special case of polymorphism. Differently solvated forms of a given compound have sometimes loosely been called polymorphs as well, even though this is incorrect.

The rather complex factors permitting the prediction of conditions under which racemic compound $\rightleftharpoons$ conglomerate transformations may occur as a function of temperature, pressure, and solvation have been probed (Leclerq, Collet, and Jacques, 1976; Chickos, Garin, et al., 1981; see also Jacques et al., 1981a, p. 140). Polymorphism of enantiomers and of changes in the extent of solvation have often been observed in cases of racemate type interconversions (Jacques et al., 1981a, pp. 131, 203; Mason, 1982, p. 152).

The two pure enantiomers of a chiral substance always form enantiomorphous (oppositely handed) crystals in the solid state, though this may not be visually discernible. For a given enantiomer, the sample may well be enantiopure, that is, consisting of molecules of only one chirality sense (within experimental limits). In this connection we use the following terms: (a) the older adjective *enantiomorphous* is used to describe appropriate chiral crystals; for molecules, we use *enantiomeric*; (b) a molecule, or an object, is either chiral or it is achiral. In contrast, a macroscopic chiral sample (loosely, a substance) is either racemic or *nonracemic*. The term nonracemic conveys the concept of a sample being made up mainly of homochiral molecules (rather than being racemic); nonracemic is not synonymous with "enantiomerically pure" (abbreviated to enantiopure), however. The latter expression relates to an experimentally determined fact [implying 100% enantiomeric excess (ee); see also Section 6-5.a].

> In the description of the enantiomer composition of macroscopic samples, the terms nonracemic and/or enantiopure are to be used in preference to "homochiral," a term meaning "of the same chirality sense," as in (all) right-handed shoes; use of the term homochiral should be reserved for the description of a chiral class (Ruch, 1972).

It is possible, in principle, to create racemates by mixing enantiomorphous crystals of a chiral substance in the absence of solvent. Such artificial racemates would, by definition, be conglomerates; hence conglomerate formation is possible for *all* chiral compounds (for an application of an artificial conglomerate, see Walker, 1985). Most such conglomerates would likely be metastable, however. The latter statement follows from the statistical analysis of racemate types.

> Mixing of enantiomorphous crystals is not to be confused with the preparation of a racemate by mixing of *solutions* of the individual enantiomers, for example, in the preparation of a racemic rubredoxin protein (Zawadzke and Berg, 1992). Crystallization from such racemic solutions may lead to formation of racemic compounds, in this instance in an attempt to simplify the crystallographic phase problem.

The low frequency of formation of conglomerates as against racemic compounds is not dependent on the availability of crystallographic space groups permitting the creation of homochiral crystals (the preferred space groups occupied by either conglomerates or racemic compounds are limited to a few [Jacques et al., 1981a, p. 8]), but it has much to do with optimal packing of molecules in a 3D array (geometric packing model), that is, packing in so-called racemic space groups may be more favorable than in chiral crystal space groups (Kitaigorodskii, 1973; Brock, Schweizer, and Dunitz, 1991); this, in turn, is dependent on the symmetry of the constituent molecules. However, the relationships are complex and are insufficiently understood to permit prediction of formation of conglomerates as against racemic compounds as a function of structure (Mason, 1982, pp. 164–174; see also Section 6-4.d). Kinetic factors operating during crystal growth in solution or from the melt may also play a role in favoring racemic compound formation. One of the enantiomers may act as an inhibitor during the crystallization of the other in a chiral space group, but not in the crystallization of racemic crystals (see Section 7-2.d; Brock, Schweizer, and Dunitz, 1991).

A recent useful finding in this connection is that the frequency of conglomerate formation among salts appears to be two to three times as great as for covalent chiral compounds (Jacques, Leclercq, and Brienne, 1981b). The consequences of this fact are taken up in Chapter 7. The oppositely charged counterions in ion pairs, for example, α-methylbenzylammonium cinnamate, form helical columns of hydrogen-bonded ion pairs that appear to be incompatible with centrosymmetric space groups; such salts exhibit conglomerate behavior (Saigo et al., 1987; see also Section 7-2.c).

6-4. PROPERTIES OF RACEMATES AND OF THEIR ENANTIOMER COMPONENTS

a. Introduction

Elementary textbooks in organic chemistry uniformly stress the fact that pairs of enantiomers have identical physical properties, save the sign of their optical rotation. This emphasis is misleading: It tends to imply that optical activity is the only significant measurement available for characterizing chirality-related properties of substances; and, it ignores the fact that *all* properties of chiral substances differ according to their enantiomeric composition. The textbook emphasis is on comparison of pure (+) with pure (−) enantiomers. However, many and perhaps most samples of optically active materials encountered in the laboratory are not enantiomerically pure.

> Naturally occurring optically active substances isolated from their original source and purified conventionally are generally regarded as being enantiomerically pure. However, many examples are known where this is not the case (see, e.g., Section 7-3.a); moreover, the slow racemization to which many chiral substances (e.g., amino acids) are subject over time causes different samples of the same substance isolated from nature or prepared in the laboratory to vary in enantiomeric purity according to their history as well as their origin.

Since the optically active materials that one works with in the laboratory (or in the factory) are often not enantiopure, their properties should be routinely contrasted not just with those of the pure enantiomers, but also with those of the corresponding racemate. Indeed, it is often helpful to think of enantiomerically impure samples as mixtures of single enantiomers and of their racemate.

A comparison of the properties of racemates and of their enantiomer components in the solid state is equivalent to the comparison of properties of enantiomorphous crystals (of pure enantiomers as well as of conglomerates) with those of crystalline racemic compounds. This section examines such properties as densities, melting points, and various types of spectra of pure enantiomers and of enantiomer mixtures. Toward the end of this section, we shall compare the properties of racemates with those of their enantiomer components in the liquid state and in solution. We conclude with a brief discussion of the biological properties of enantiomer pairs and of the origin of the chiral homogeneity in Nature.

b. Optical Activity

Given a sample of a chiral substance, one often wants to know whether the sample is optically active. The answer to this question is a function of the sampling as well as of the nature of the racemate and the state of the system. Provided that the sensitivity of measurement is adequate and that the sample is not fortuitously cryptochiral (cf. pages 93 and 159), a clearcut answer is possible for substances in the liquid or gaseous states. Not so for crystalline solids, however. Thus, measurement of the optical rotation of a single well-formed crystal of a conglomerate (either in solution or on the crystal itself) would reveal optical activity even in a sample that was globally racemic as established by a rotation measurement carried out on a larger, homogeneous, portion.

An example demonstrating the limitations of optical activity (or of chiroptical measurements in general) in revealing the enantiomer composition of a crystalline sample is provided by the following study. Solid state photodimerization of (E)-cinnamamide **1** and its 2-thienyl analogue **2** in a mixed crystal (solid solution) of **1** + **2** grown from the melt gives rise to a single heterodimeric product **3** (Eq. 6.4).

$$\text{(6.4)}$$

Th = 2–Thienyl

The composition of individual crystals of dimer **3** is governed by the topochemical principle (Schmidt, 1971), that is, in the solid state, substrate reactivity and product structure are controlled by the crystal lattice (or crystal structure) of the substrate. Product stereochemistry is determined by the "contact geometry" provided that the C=C bonds of the two interacting molecules (**1** and **2**) are within 400 pm (4 Å) of one another in the lattice.

It has been demonstrated that crystal pieces taken from opposite ends of a single crystal of **3** isolated from a globally racemic sample prepared as described above exhibit optical activity and circular dichroism (CD) with one end of the crystal being enriched in one enantiomer and the other end enriched in the second enantiomer (Vaida, Lahav, et al., 1988; Weissbuch et al., 1991).

c. Crystal Shape

Crystallography played a prominent role in the early phases of stereochemical experimentation. Ever since the description of the first experiments of Pasteur, it has been recognized that individual enantiomers can sometimes be recognized from the outward appearance (morphology or habit) of their crystals.

All enantiomerically pure solid samples inhabit enantiomorphous crystals, that is, they inhabit one of the 11 noncentrosymmetric (chiral) crystal classes. Should it not be possible to differentiate individual crystals of the two enantiomers of *any* chiral substance by inspection of their outward shape? To put it another way, can one tell whether two crystals of the same chiral substance have the same or opposite handedness? In fact, such recognition is possible only when the crystals are hemihedral and possess hemihedral faces (hemihedry).

> Crystals derived from optically active samples may be holohedral or hemihedral. Holohedral crystals have the highest symmetry within their crystal class; each face is accompanied by a corresponding parallel face on the other side of such crystals. In hemihedral crystals, parallel faces are absent and the crystals possess only one-half the number of faces required by the symmetry of the crystal system (see, e.g., Kauffman and Myers, 1975). In addition, hemihedral faces are themselves in a chiral environment; those "turned" in the *same* way, as described by Pasteur, are outward manifestations of homochiral crystals. Hemihedral faces turned in the *opposite* way identify enantiomorphous crystals.

The presence of hemihedral faces is related to the crystal class and space group of the crystals being investigated. The 65 space groups comprising the 11 enantiomorphous crystal classes are unequally represented in the set of chiral substances. Two of these space groups account for about 80% of chiral crystals among organic compounds, namely, space groups $P2_1$ and $P2_12_12_1$ (Mighell, Ondik, and Molino, 1977; Jacques et al., 1981a, p. 8). The probability of finding hemihedral faces is good in $P2_1$ and poor in $P2_12_12_1$ (Kress, Paul, Curtin, et al., 1980). Unfortunately, hemihedral faces are present less often in the crystal classes typically found among chiral organic compounds than one might like. Hemihedry is a sufficient but not a necessary condition for enantiomorphism. Thus, most crystals of conglomerates do not exhibit hemihedrism, and the sense of chirality of individual crystals cannot be ascertained from their morphology. We will return to this problem in Chapter 7.

> The morphology of crystals can be predicted, that is, calculated, either by empirical methods requiring information about the space group and the unit cell, or by molecular mechanics methods based on the assumption that the energy released when a new layer of molecules is deposited on the face of a growing crystal

(attachment energy) determines the rate of growth of that crystal face. For an example of such calculations, see Black, Davey, et al. (1990).

The utility of all of this lies in the possibility of carrying out a resolution by mechanical separation (triage) of crystals from a conglomerate à la Pasteur (see Section 7-2.a). It becomes evident that the rarity of this type of resolution relates not just to the probability of finding chiral crystals (conglomerate formation) but on the more severe requirement of observing hemihedry. The existence of hemihedry is the significant and fortuitous circumstance that made possible the famous experiment wherein Pasteur separated crystals of (+)- from (−)-sodium ammonium tartrate (Pasteur, 1848; Kauffman and Myers, 1975).

There are other, no less fortuitous, circumstances that made this first resolution possible: The racemic double salt that Pasteur prepared was a conglomerate, and it crystallized as such rather than as a racemic compound because the crystallization happened to take place at a temperature lower than 28°C. Above this temperature, the conglomerate (a tetrahydrate) is converted to a monohydrate, which is a racemic compound. Crystals of the latter, called Scacchi's salt, are neither enantiomorphous nor hemihedral.

d. Density and Racemate Type

Solid racemic compounds differ significantly in density from the corresponding pure enantiomers. The difference can be as large as 5%, for example, the densities of racemic and enantiomerically pure *trans*-1,2-cyclohexanecarboxylic acid are 1.43 and 1.38, respectively ($\Delta d = 3.6\%$) (Jacques et al., 1981a, p. 28). This difference stems from the enthalpy change associated with the following reaction, that is, the formation of a racemic compound from the enantiomers in the solid state:

$$A_+ + A_- \rightarrow A_+ A_-$$

The difference in density is reflected in the magnitude of the melting point difference between that of the racemate ($A_+ A_-$) and that of the pure enantiomer A_+ (or A_-) as well as in the shape of the melting point phase diagram (Section 6-3 and 6-4.e).

Contrary to what was once believed, namely, that *on average* racemic compounds in the solid state are denser than the corresponding enantiomers (*Wallach's rule*; Wallach, 1895), the cases are about equally divided between those in which the racemate is denser and those in which the enantiomer is denser. The available data upon which the classical rule is based come from measurements carried out on 24 compounds most of which are quite polar (Jacques et al., 1981a, p. 28).

Hydrogen bonding is an important feature of the crystal networks of polar compounds. Differences in density may well depend on the number and geometry of the hydrogen bonds that may in turn differ significantly according to the nature of the intermolecular interactions: homochiral versus heterochiral.

A new statistical study of 129 compounds culled from the Cambridge Structural Database has confirmed the fact that the average difference in density between racemic compounds and enantiopure samples of the same compound is very small. This study demonstrates that, on the average, the set is nearly equally divided between compounds the density of whose racemic compounds is greater than that of the individual enantiomers, and those for which the opposite is true (Brock, Schweizer, and Dunitz, 1991). Yet, a small but significant increase in the average density of racemic compounds over that of the pure enantiomers was found for a subset composed of stable enantiomers, that is, those that are resolvable, i.e., not in rapid equilibrium with one another.

> It has been suggested that the slight increase in density is accounted for by a statistical bias in the subset: The subset can contain pairs (racemic compounds and pure enantiomers) in which the racemic compound is significantly stabler than the enantiomers, and pairs in which the two forms are about of equal stability but it cannot contain pairs in which the racemic compound is much less stable than the pure enantiomers. If the latter situation obtains, the racemic compound "disappears" from the phase diagram and the crystalline racemate obtained (on crystallization or evaporation) from a racemic solution must be the eutectic, that is, the conglomerate (Brock, Schweizer, and Dunitz, 1991).

There remains a widespread notion that enantiomers can be packed into crystals more tightly in a heterochiral manner than homochirally, in other words, that racemic compounds are more stable than the corresponding enantiomers. There may indeed be a "genuine tendency" for racemic compounds to be more stable (and therefore be slightly denser) than the corresponding enantiomers. This would account, in part, for the greater incidence of racemic compounds over conglomerates among racemates (Jacques et al., 1981a, p. 81). However, it has also been suggested that this tendency is not thermodynamic in origin but rather that it reflects either kinetic factors dealing with nucleation and growth of crystals from racemic solution and/or packing arrangements in crystallographic space groups that favor racemic molecular crystals over those of pure enantiomers (Brock, Schweizer, and Dunitz, 1991; see also Mason, 1982, p. 164ff).

> In principle, a racemic solution can crystallize either as a racemic compound or as a conglomerate. Conglomerates necessarily crystallize in one of the 65 chiral space groups (Sections 6-4.c and 6-4.k) whereas racemic compounds crystallize (with rare exception) in one of the 165 racemic (centrosymmetric) space groups. The increase in packing possibilities afforded by the larger number of centrosymmetric space groups may account, in part, for the greater number of racemates crystallizing as racemic compounds than as conglomerates (Heilbronner and Dunitz, 1993, p. 114).

Beyond this, can one identify any structural factors that determine the shape of the racemate binary phase diagram (Section 6-3)? The shape of the diagram, in particular the location of the eutectics, has significant practical consequence with respect to the ease with which a nonracemic sample can be enantiomerically enriched (Section 7-4). This question has been analyzed by Collet who suggested on empirical grounds that one should consider two subsets of chiral compounds (Gabard and Collet, 1986; Collet, 1990). The first subset comprises chiral compounds that have a strong tendency to pair up about a center of symmetry. Simple carboxylic acids devoid of hydrogen-bonding functional groups, such as OH and SH, and whose chiral center is located close to the carboxyl group have a great tendency to form very stable racemic compounds, that is, to form crystals made up of heterochiral molecules.

A second subset consists of molecules that are prevented for various reasons from pairing up about a center of symmetry and in which the tendency for heterochiral packing is thus reduced. It is in this subset that the tendency toward formation of conglomerates rather than racemic compounds during the crystallization of racemic samples is somewhat increased. Carboxylic acids bearing one or more hydroxyl groups, carboxylic acid salts (cf. Section 7-4), and possibly some molecules in which a C_2 symmetry axis is present are structural types belonging to the second subset. These generalizations are supported by crystal structure arguments that often depend on the symmetry of hydrogen-bonding networks and on the consequence of these on the efficiency of packing in appropriate space groups (for the details of this analysis, see Collet, 1990).

e. Melting Point

The melting point of a solid chiral substance is a highly revealing physical property. For the purpose of extracting stereochemically relevant information, the racemate type must be taken into account, and both the beginning and the termination of melting must be known.

> Conventional melting point determination usually does not suffice when both of these temperatures must be determined. The preferred techniques that provide the necessary data are either differential thermal analysis (DTA) or differential scanning calorimetry (DSC).

The necessary information is conveniently summarized by means of symmetrical binary phase diagrams that relate the composition to the melting point. The first of the three basic types (identified in Section 6-3), that of conglomerate behavior, is illustrated in Figure 6.4. This system implies no mutual solubility of the

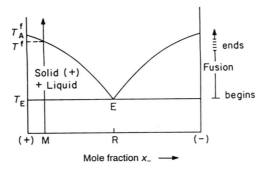

Figure 6.4.   Melting of a mixture M in a conglomerate system (E = eutectic; R = racemate). [Adapted with permission from Jacques, J., Collet, A. and Wilen, S. H. (1981), *Enantiomers, Racemates and Resolutions*, p. 44. Copyright © 1981 John Wiley & Sons, Inc.]

enantiomers in the solid state. Composition E (eutectic) consists of a mixture of two kinds of crystals: those of the $(+)$ enantiomer and those of the $(-)$ enantiomer. The melting point of the pure enantiomer is given by T_A^f, while that of the racemate, corresponding to the eutectic in the melting point diagram, is T_E (or T_R). It is characteristic of conglomerate systems that the racemates *always* melt below the corresponding pure enantiomers.

Mixtures of intermediate composition, for example, M [rich in, say, the $(+)$ enantiomer], begin melting at T_E. Crystals of both enantiomers disappear during the melting process so as to produce a liquid that is racemic (the eutectic disappears first) and the temperature stays constant at T_E. After all the racemate has melted [i.e., no $(-)$ enantiomer is left in the solid phase], the remaining $(+)$ enantiomer gradually melts in a liquid of varying composition with the last bit of solid disappearing at T^f. The same diagram also describes the reverse process, namely, crystallization from the melt; the key point is that it is $(+)$ solid alone [the pure $(+)$ enantiomer] that crystallizes beginning at temperature T^f when a liquid mixture of composition M is cooled. This is true also when solvent is present.

The composition of mixture M (Fig. 6.4) is given by the ratio $MR/(+)R$ and the $ee = 2x_+ - 1$, where $x_+ =$ mole fraction of the dominant enantiomer (see Section 6-6). An alternative way of describing mixture M is as a mixture consisting of $2x - 1$ mol of the $(+)$ enantiomer and $2(1 - x)$ mol of racemate ($=$ eutectic). That such a description is not a figment of the imagination is revealed by examination of a mixture of enantiomers by DSC. In DSC, one measures the enthalpy absorbed by a sample as a function of temperature (Fig. 6.5). The DSC trace for a mixture of composition M actually shows separately the

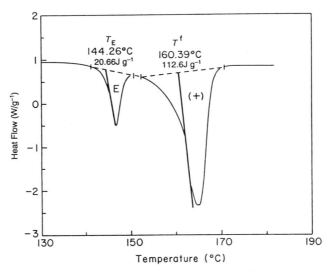

Figure 6.5. Differential scanning calorimetry trace of the melting of an enantiomer mixture (approximating composition M in Figure 6.4) of α-methylbenzylammonium cinnamate (3.1-mg sample) exhibiting conglomerate behavior. The "peak" at left represents the enthalpy of melting of the eutectic E (racemate) while the larger "peak" to the right is the enthalpy of melting of the $(+)$ enantiomer in the mixture. The temperatures shown correspond to those in Figure 6.4 (Wilen and Toporovsky, 1992).

absorption of energy during the melting of the eutectic E and of the (+) enantiomer with increasing temperature. The area of the first event (peak) is directly proportional to the heat necessary to melt the racemate present in the mixture. In the case of a conglomerate, if one knows the molar heat of fusion of the racemate (ΔH_R) and the total weight of the sample, the proportion of racemate and of pure enantiomer, hence also the enantiomer composition, are easily determined from the DSC trace provided that the peaks are reasonably well separated.

> The molar heats of fusion of the racemate ΔH_R and of the pure enantiomers ΔH_A are unequal. That this is so even in a conglomerate system follows from the fact that the melting points of the two forms differ while the specific heats do not. Consequently, the relative amounts of the racemate and the enantiomers are not directly given by the areas of the DSC peaks (Jacques et al., 1981a, pp. 48, 151).

Some examples of compounds that behave in the way just described are ($\pm$)-hydrobenzoin **4** ($T_A^f = 147.5°C$ and $T_R^f = 121°C$; $\Delta T = 27.5°C$), 1-phenyl-1-butanol **5** ($T_A^f = 50°C$ and $T_R^f = 16°C$; $\Delta T = 34°C$), and hexahelicene **6** ($T_A^f = 265–267°C$ and $T_R^f = 231–233°C$; $\Delta T = 34°C$; Fig. 6.6). Phenylglycine **7** (dec $> 200°C$) also behaves in this way but precise temperatures cannot be obtained due to decomposition during melting. This result is a significant limitation of the DSC technique. The examples reveal an important characteristic of conglomerate systems, namely, that the racemate melts at a much lower temperature than the pure enantiomer. The difference ranges from 25 to 35°C (cf. Table 7.2) (Jacques et al., 1981a, p. 80).

The relationship between the composition (as mole fraction) and the melting point of a mixture of enantiomers exhibiting conglomerate behavior (Fig. 6.3a) is given by the Schröder–van Laar equation:

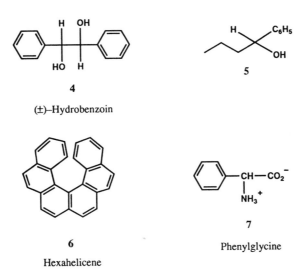

4

(±)–Hydrobenzoin

5

6

Hexahelicene

7

Phenylglycine

Figure 6.6. Compounds exhibiting conglomerate behavior.

$$\ln x = \frac{(\Delta H_A^f)}{R}\left(\frac{1}{T_A^f} - \frac{1}{T^f}\right) \tag{6.5}$$

That is, the mole fraction x of the dominant enantiomer of the mixture is related to the molar heat (enthalpy) of fusion of the enantiomer (ΔH_A^f) as well as to the melting temperature, that of the pure enantiomer T_A^f and that of termination of fusion of the mixture T^f (Schröder, 1893; van Laar, 1903). With two of these values known, typically ΔH_A^f and T_A^f, the composition, hence enantiomeric purity of the sample, may be ascertained by measuring T^f. The equation, in which R represents the gas constant, is useful for the construction of phase diagrams as well.

There are three basic assumptions upon which the validity of the Schröder–van Laar equation rests. First, the relationship assumes that the enantiomers are immiscible in the solid state; they are fully miscible in the liquid state. Second, it is assumed that the enantiomer mixture behaves ideally in the liquid state (see Section 6-2). Third, it is further assumed that the enthalpy of fusion is independent of T in the range of temperature considered, this being equivalent to neglect of the heat capacity difference between solid and liquid enantiomers. The fact that the Schröder–van Laar equation, in general, accurately describes real systems, is testimony to the validity of the stated assumptions.

> Heat capacity terms do not appear in the abridged form of the Schröder–van Laar equation (Eq. 6.5), which is adequate for most purposes (Jacques et al., 1981a, p. 46).

A second type of racemate (Fig. 6.3b) corresponds to the case in which the two enantiomers of a given compound coexist in the same unit cell, that is, the enantiomer pair forms a well-defined racemic compound. Figure 6.7a and b describes such cases, which correspond to the vast majority of organic racemates. The eutectics in racemic compounds (Fig. 6.7a) may be treated as mixtures of crystalline enantiomer and of a racemic compound. Thus, E represents a mixture of $2x_E - 1$ mol of crystalline (+) enantiomer and $2(1 - x_E)$ mol of racemic compound R.

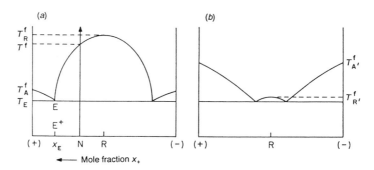

Figure 6.7. Racemic compound formation. The melting point of the racemic compound is T_R^f (or $T_{R'}^f$). [Adapted with permission from Jacques, J., Collet, A. and Wilen, S. H. (1981), *Enantiomers, Racemates and Resolutions*, p. 90. Copyright © 1981 John Wiley & Sons, Inc.]

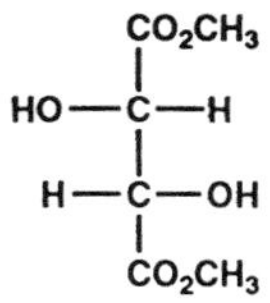

8

Dimethyl tartrate

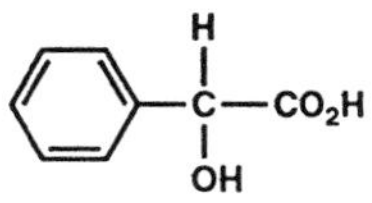

9

Mandelic acid

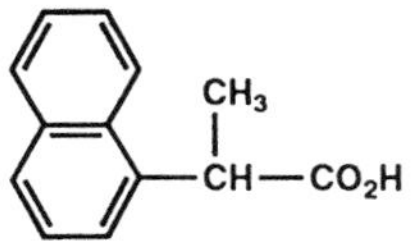

10

2-(1-Naphthyl)propanoic acid

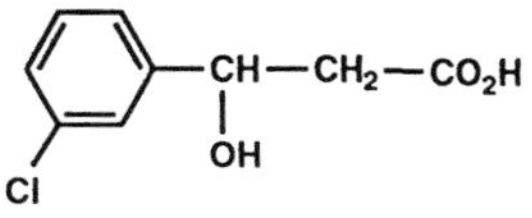

11

3-(3-Chlorophenyl)-3-hydroxy-propanic acid

Figure 6.8. Chiral substances forming racemic compounds.

Racemic compound systems are characterized by the fact that addition of a small amount of pure enantiomer to the racemate invariably *lowers* the melting point of the racemate (contrary to what obtains in conglomerate systems). On heating a mixture of composition N [poorer in the (+) enantiomer than the eutectic E^+; Fig. 6.7*a*], melting begins at T_E and it continues until all the eutectic in the sample becomes liquid. The solid phase remaining above T_E is the racemic compound. To ascertain whether a given mixture lies to the right or the left of the eutectic, it suffices to add a small amount of racemate to the mixture. Redetermination of the melting point of the mixture shows where the composition lies (on the racemate side of the eutectic if the melting point rises and on the pure enantiomer side of the eutectic if it drops).

Racemic compound formation may be illustrated by dimethyl tartrate, **8** (Fig. 6.8) ($T_A^f = 43.3°C$ and $T_R^f = 86.4°C$; $\Delta T = -46.1°C$) (see Fig. 6.7*a*), and by mandelic acid, **9** (Fig. 6.8) ($T_A^f = 132.8°C$ and $T_R^f = 118.0°C$; $\Delta T = 14.8°C$) (see Fig. 6.7*b*). Occasionally, extreme cases are observed, such as 2-(1-naphthyl)-propanoic acid **10** (Fig. 6.8) (see Fig. 6.7*a*), where the eutectic is so close to the pure enantiomer that it is hardly detectable (Sjöberg, 1957) and 3-(*m*-chlorophenyl)-3-hydroxypropanoic acid **11** (Fig. 6.8) (see Fig. 6.7*b*), where the racemic compound is difficult to detect by inspection of the phase diagram (Collet and Jacques, 1972).

The Schröder–van Laar equation (Eq. 6.5) relates the composition of that part of the phase diagram lying between the pure enantiomer and the adjacent eutectic. An analogous expression, the Prigogine–Defay equation (Eq. 6.6), governs the relation between composition and melting temperature in the region between the two eutectics:

$$\ln 4x(1-x) = \frac{(2\,\Delta H_R^f)}{R}\left(\frac{1}{T_R^f} - \frac{1}{T^f}\right) \tag{6.6}$$

Equation 6.6 relates the mole fraction x of the dominant enantiomer in a given mixture to the molar heat (enthalpy) of fusion of the racemate (ΔH_R^f) and to the melting temperatures (that of the racemate T_R^f and that of the mixture T^f).

Only a few racemates exhibit solid solution behavior (Figure 6.3c) in which the two enantiomers are miscible in the solid state (p. 160).

Ideal solid solution behavior is exemplified by camphor whose racemate and pure enantiomers each melt at $T = $ ca. 178°C. The phase diagram describing this type of behavior is shown in Figure 6.9a. Solid solution behavior is particularly prevalent among molecules forming plastic crystals and among those substances forming rotationally disordered crystals made up of molecules of spheroidal shape (Mason, 1982, p. 161).

> Few cases of solid solution behavior have been studied in detail. However, whenever the racemate and one of the pure enantiomers are reported to melt at the same temperature, existence of a solid solution may be suspected.

Figure 6.9b, which describes positive deviations from ideality and Figure 6.9c, which describes negative deviations from ideality, illustrate even rarer cases than solid solutions behaving ideally; only 19 examples of these rare cases are listed in the inventory of Jacques et al. (1981a, pp. 110–13). Note that there is no danger of confusing a pseudoracemate system of the type illustrated by Figure 6.9c with a conglomerate system. Mixtures of composition P (Figure 6.9c) exhibit only one peak in the DCS trace (no eutectic peak is present; e.g., see Pella and Restelli, 1983).

> Partial miscibility, in the solid state, of enantiomers with one another, or of enantiomers with racemic compounds, is more prevalent than the full miscibility that characterizes pseudoracemates (Fig. 6.9). Figure 6.10 illustrates a situation (for a conglomerate system) in which *in the vicinity* of pure (−) enantiomer, for example, a small amount of the (+) enantiomer may dissolve to form a miscible phase. Thus, on cooling a mixture of composition O, the solid that begins to deposit at temperature T_o is not pure (−), but rather a solid solution of (+) in (−) having composition O′. Figure 6.10 thus illustrates a case that is hybrid between a conglomerate and a pseudoracemate. A similar situation can arise in compounds exhibiting racemic compound behavior. The nearly pure dominant enantiomer can dissolve some of the racemate in the solid state and, in addition, the racemic compound (behaving as an independent substance) can form a solid solution in the vicinity of $x = 0.5$ by dissolving some of the dominant enantiomer. The consequences of such partial miscibility will be examined in Chapter 7.

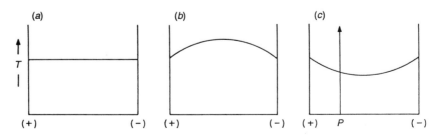

Figure 6.9. Melting point phase diagrams of solid solutions of enantiomers.

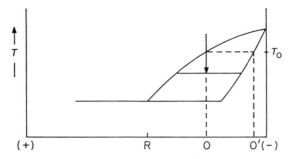

Figure 6.10. Partial miscibility of enantiomers in the solid state (hybrid of conglomerate and pseudoracemate systems, Fig. 6.3 *a* and *c*, respectively); only one-half of the phase diagram is shown.

f. Solubility

Stereoisomer discrimination is most obviously manifested in the fusion and solubility properties of chiral compounds. The former properties were examined in Section 6-4.e immediately preceding. The latter are revealed in the measurement of heats of solution and in solubility properties of the enantiomers relative to those of the racemates. Differences in the heats of solution, corresponding to the enthalphy of mixing of the enantiomers of a given compound in the solid state, are summarized for amino and hydroxy acids (in water at 25°C) in Table 6.1 (Matsumoto and Amaya, 1980). The most revealing aspect of these data are their large magnitude relative to that found for comparable mixing in solution (Section 6-2), where averaging of interactions obtains (Takagi, Fujishiro, and Amaya, 1968).

Most chiral compounds exhibit significantly different solubilities for the racemate and the corresponding pure enantiomer. This fact forms the basis of a relatively simple enantiomeric enrichment process that may be applied when a nonracemic but enantiomerically impure sample is available. However, rational application of such a process requires a knowledge of the solubility behavior of the racemate–enantiomer system, that is, the ternary phase diagram (at least its essential features) must be known.

TABLE 6.1. Differences in Heats of Solution between Enantiopure Solid Chiral Compounds and the Corresponding Racemates[a,b]

Compound	(kcal mol^{-1})	(kJ mol^{-1})
Alanine	0.268	2.00 ± 0.21
Glutamic acid	0.98	4.1 ± 0.02
Histidine	0.36	1.5 ± 0.2
Threonine	0.041	0.17 ± 0.13
Valine	0.547	2.29 ± 0.01
Tartaric acid	2.2	9.4 ± 0.3

[a] At 298.15 K.
[b] Matsumoto and Amaya (1980).

Knowledge of the solubility of enantiomers in solution furnishes information about the crystallization of the enantiomers from solution. This fact is entirely analogous to the use of binary phase diagrams in understanding both the melting of a sample and the crystallization of a chiral nonracemic sample from the melt. Evaluation of solution properties requires consideration of an additional variable, namely, the solvent, leading to the use of ternary phase diagrams. Solvent presents a constraint in that changes in concentration as well as in temperatures can no longer be represented all at once in two dimensions. Representation of temperature as well as three concentrations would require a 3D diagram (a triangular prism), which is awkward to say the least. For most purposes, it suffices to examine an isothermal slice of the triangular prism corresponding to a ternary diagram representing the solubility of (+) and (−) enantiomers in solvent S at a given temperature T_o.

> For a detailed treatment of ternary phase diagrams see Findlay (1951) and Ricci (1951). Abbreviated treatments are given by Jacques et al. (1981a, Chapter 3) and by MacCarthy (1983).

Conglomerate systems are illustrated by Figure 6.11. In such a diagram, the two sides of the equilateral triangle represent the concentrations of (+) and (−) enantiomers in solvent, while the bottom describes the mole fraction of the enantiomer mixture, just as in the case of the binary phase diagram. In this example, composition E corresponds to the solubility of the racemate (eutectic) at T_o (the concentrations of the two enantiomers in solvent may be expressed as mole fractions, or conveniently as weight/weight (w/w) percentages; these two concentration modes do not give rise to identical phase diagrams).

The proportion of solvent in E is given by segment (−)a, that of (+) by Sb, and that of (−) by (+)c; with respect to E, each of these segments is measured in directions parallel to the sides of the triangle. Figure 6.11 represents the behavior of a conglomerate in solution under equilibrium conditions. Note that the

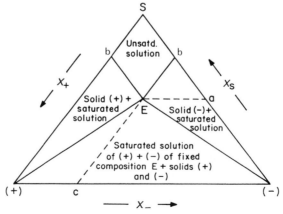

Figure 6.11. Solubility diagram for a conglomerate system at fixed temperature T_o. The saturated solutions in equilibrium with pure (+) and (−) enantiomers, respectively, have variable composition. The dashed lines show that point E has a composition given by concentrations a, b and c measured along the sides.

solubility "curve" bEb looks remarkably like the melting curve in the corresponding binary phase diagram (Fig. 6.4). The solubility of each enantiomer in solvent S is given by segment Sb for the (+) enantiomer and by segment (−)b for the mirror-image isomer (note the direction of the arrows that specify increasing concentrations). These solubilities are, of course, equal; the values are for temperature T_o, the temperature of an isothermal slice of the triangular prism alluded to above. The solubility of the racemate is given by E (segment Sa, measured on either side of the diagram).

From simple thermodynamic considerations, it is evident that the solubility of the conglomerate is greater than that of the individual enantiomers [this is equally true of any mixture of (+) and (−)]. This situation brings to mind the old empirical double-solubility rule rationalized by Meyerhoffer (1904), that is, that the racemate is twice as soluble as the enantiomers. This rule is applicable to neutral (nondissociable) organic compounds having the same thermodynamic constants and behaving in an ideal way (implying that solvent properties do not enter into the calculation of solubilities). The same double solubility conclusion may be reached in a more precise way by application of the Schröder–van Laar equation. Effectively, this relationship requires that, for an ideal solution, liquidus curve $T_A^f E$ (Fig. 6.4) is unchanged when (−) is replaced by (−) plus solvent.

The solubility–concentration relations are given by straight lines that are parallel to the solvent–enantiomer sides of the diagram provided that the composition is expressed as mole fraction. Experimental results are in accord with the foregoing statements (Jacques and Gabard, 1972) and the solubility ratio α, defined as α = solubility of racemate/solubility of enantiomer, is indeed approximately equal to 2 for those covalent compounds whose solubilities have been measured quantitatively (Collet, Brienne, and Jacques, 1980, p. 182). The precise value 2 is found if (and only if) solubilities are expressed as mole fractions x: $\alpha = 2$, where $\alpha_x = x_d/x_\ell$. By way of example, the solubilities of the racemate and one of the pure enantiomers of α-methylbenzyl 3,5-dinitrobenzoate are 27.4 g% (g 100 g^{-1} solution) and 13.2 g%, respectively (Brienne, Collet and Jacques, 1983).

The solubility ratio for fully dissociated solutes (+1/−1 salts) is $\alpha = \sqrt{2}$ and the lines describing the solubility (and the supersaturation) in the ternary phase diagram are no longer straight and parallel to the sides due to the common ion effect of the counterion (Fig. 6.12).

To see what the utility of a ternary phase diagram may be, consider what happens when a mixture of enantiomers of composition M ($x_+ \neq x_-$) exhibiting conglomerate behavior is added to a small amount of pure solvent. This combination is represented by point P in Figure 6.13. At P, some (but not all) of the solid dissolves, there being insufficient solvent to form an unsaturated solution. The dissolution process may take a very long time since equilibrium is established only slowly at constant temperature. When equilibrium is attained, the solid and the solvent each have a new composition: The solute has composition N (by extension of tie line EP), and the solvent that now contains some solute, has composition E, (racemic).

The result of this simple dissolution process (of part of a chiral but non-racemic solid sample) is that a new solid is obtained that of necessity is enriched

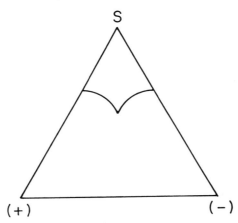

Figure 6.12. Solubility of a fully dissociated conglomerate ($\alpha = \sqrt{2}$).

in the major enantiomer (M→N). This result holds true for all conglomerate systems (Fig. 6.3a). Even washing a solid mixture with solvent to remove impurities affects the enantiomer ratio; for conglomerates, it always increases the enantiomeric purity. The phase diagram requires that if the same mixture M is added to a larger amount of pure solvent so as to yield system Q, then the remaining solid is pure (+) surrounded by a solution of composition U.

When the temperature is changed, a new phase diagram is obtained (Fig. 6.14). Typically, as the temperature increases the solubility increases, as does the area of unsaturation of the diagram. In spite of this change in solubility, it is entirely possible to describe a recrystallization by means of a single ternary diagram: A partially resolved conglomerate in the presence of a given amount of solvent is warmed and the system is allowed to cool until temperature T_o is restored (equilibrium is reestablished). Indeed this is a way of speeding up the slow equilibration described above.

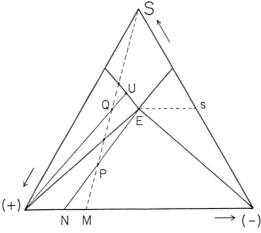

Figure 6.13. Crystallization of an enantiomer mixture of composition M from solution (conglomerate system). [Adapted with permission from Jacques, J., Collet, A. and Wilen, S. H. (1981), *Enantiomers, Racemates and Resolutions*, p. 178. Copyright © 1981 John Wiley & Sons, Inc.]

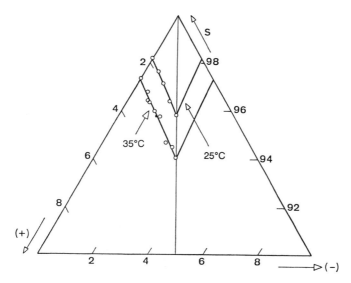

Figure 6.14. Effect of temperature on the solubility of a conglomerate. Solubility diagram of the (+)-and (−)-N-acetylleucines in acetone at 25 and 35°C. Only the upper part of the diagram is shown. Concentrations are in g $100\,g^{-1}$ of mixture. [Jacques, J. and Gabard, J. (1972) *Bull. Soc. Chim. Fr.*, 344. Reprinted with permission of the Société Française de Chimie.]

Effectively, the same increase in enantiomer purity may be achieved by isothermal evaporation of an unsaturated solution (shaded region in Fig. 6.15) containing a mixture of (+) and (−) $(x_+ \neq x_-)$. When $x_+ > x_-$, the first crystals to be deposited are those of pure (+). Beyond point z, both enantiomers deposit together from a solution that remains racemic (solid compositions N′, N″, etc.). When evaporation takes place concomitant with cooling, a similar crystallization process takes place; however, the diagram changes, with the solubility curves moving upward due to reduced solubility of (+) and (−) enantiomers, and of the racemate. Under such conditions, it would be more difficult to obtain one of the enantiomers pure than under conditions of constant temperature.

For ternary diagrams illustrating the solubilities of racemic compound systems analogous to the binary diagrams shown in Figure 6.7, see Jacques et al. (1981a,

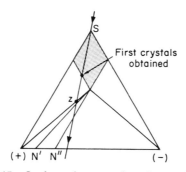

Figure 6.15. Isothermal evaporation of a conglomerate.

p. 197). For racemic compound systems, the following general statements may be made: (a) The solubility of the racemates is less clearly related to that of the corresponding enantiomers (unlike the case of conglomerates). The racemic compound solubility may be either greater or lower than that of the enantiomer. (b) Eutectic compositions apparently are close to those found in binary phase diagrams for the corresponding $(+)$ plus $(-)$ mixture. (c) Even in unfavorable cases, enrichment is possible. Consider that crystallization of a mixture of low enantiomeric purity yields a precipitate of racemic compound. Enrichment takes place in the mother liquor.

> The correspondence of eutectic compositions in binary and ternary phase diagrams tends to enhance the utility of binary phase diagrams. After all, binary phase diagrams are easier to construct than the ternary ones, so that one would rather use the former type in predicting or rationalizing solubility behavior (Wilen, Collet, and Jacques, 1977).
> For examples of "reverse" enrichment (enrichment in the mother liquor), see Ohfune and Tomita (1982); ten Hoeve and Wynberg (1979); Leclercq, Jacques, et al. (1976); and Downer and Kenyon (1939).

Figure 6.16 illustrates the three types of pseudoracemate solubility behavior. The similarity to the melting point diagrams (Fig. 6.9) is again evident. It suffices to point out for this relatively rare type that solubility is little affected by the enantiomer composition. For the type illustrated by Figure 6.16a (the common type), the solubility is completely independent of enantiomer composition. Consequently, purification by recrystallization of a mixture with $x_+ \neq x_-$ exhibiting this type of behavior is effectively precluded.

Consider also the possibility that a given compound may exhibit more than one kind of solubility behavior even without changing the solvent. This possibility is reminiscent of the transformation of one crystalline racemate type into another as a function of temperature and pressure discussed in Section 6-3.

Figure 6.17a illustrates the solubility behavior of histidine·HCl in H_2O and Figure 6.17b illustrates the solubility behavior of α-phenylglycine sulfate [(α-Phegly)$_2$·SO$_4$] in 30% H_2SO_4 as a function of temperature. We see that histidine·HCl crystallizes as a racemic compound at room temperature and slightly above but as a conglomerate at 45°C and that, conversely, (α-Phegly)$_2$·SO$_4$ crystallizes as a conglomerate at 5°C and below but as a racemic compound at 10°C and above. The transition temperature above which the histidine·HCl racemic compound decomposes lies in the vicinity of 45°C (Jacques and Gabard, 1972); the conglomerate–racemic compound transition temperature

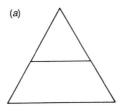
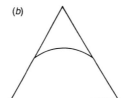
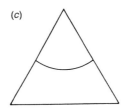

Figure 6.16. Solubility diagrams of pseudoracemate systems.

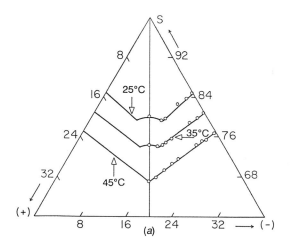

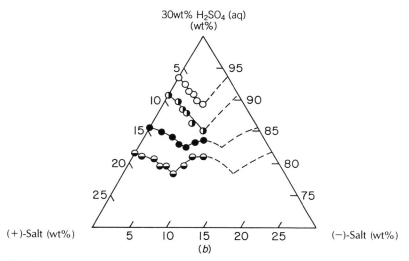

Figure 6.17. Solubility diagram (*a*) for histidine·HCl in H$_2$O. [Jacques, J. and Gabard, J. (1972), *Bull. Soc. Chim. Fr.*, 344. Reproduced by permission of the Société Française de Chimie.]; and (*b*) for α-phenylglycine sulfate in 30% aqueous H$_2$SO$_4$. For the latter, temperature (°C): ○ 0; ◑ 5; ● 10; ◕ 20. [Shiraiwa, T., Ikawa, A., Fujimoto, K., Iwafugi, K. and Kurokawa, H. (1984), *Nippon Kagaku Kaishi*, 766. Reproduced by permission of the Chemical Society of Japan.]

of $(\alpha\text{-Phegly})_2 \cdot SO_4$ lies in the vicinity of 5°C (Shiraiwa et al., 1984). Few transformations among racemate types *in solution* have been reported. However, it is known that such transformations are usually associated with changes in solvation (Jacques et al., 1981a, p. 204).

g. Vapor Pressure

Just as the melting points and the solubilities of racemates differ from those of their enantiomer constituents, so too do the vapor pressures. This statement refers specifically to the vapor pressures of solids; for differences in the vapor pressure of liquid enantiomers and the corresponding racemates, see Section 6-2. Such a difference was first suggested in the case of dimethyl tartrate by Adriani (1900).

Differences between the heats of sublimation (ΔH_{sublim}) of enantiopure samples and of the corresponding racemic compounds (Eq. 6.7) have been measured in the case of just a few compounds (Chickos, Garin, et al., 1981) (Table 6.2).

$$\text{versus} \quad \left.\begin{array}{l} (+)A_{solid} \rightarrow (+)A_{vapor} \\ (\pm)A_{solid} \rightarrow (\pm)A_{vapor} \end{array}\right\} \Delta\Delta H_{sublim}^{25} \qquad (6.7)$$

The fact that both negative and positive values of $\Delta\Delta H_{sublim}$ have been found implies that heterochiral interactions between the solid enantiomers (Section 6-2) can be either stronger or weaker than the corresponding homochiral interactions.

The first application of differences in vapor pressure is due to Pracejus (1959) who observed that fractional sublimation of an amino acid derivative led to increased optical activity. Kwart and Hoster (1967) correctly inferred that enantiomeric enrichment of partially resolved α-ethylbenzyl phenyl sulfide (Fig. 6.18, 12) by sublimation as well as by recrystallization could be ascribed to the same "intercrystalline force." The preferential vaporization of an enantiomer relative to that of the racemate was demonstrated mass spectrometrically by differential isotopic labeling of the two samples (Zahorsky and Musso, 1973; cf. Section 6-4.o).

A systematic study of the sublimation of mandelic acid [$C_6H_5CH(OH)CO_2H$]

TABLE 6.2 Differences between Heats of Sublimation of Enantiopure Compounds and the Corresponding Racemates[a,b]

	$\Delta\Delta H_{sublim}$		
	(kcal mol^{-1})	(kJ mol^{-1})	mp (°C)[c]
Carvone oxime	2.6	10.9	71(91)
	8.4	35.1	49(87)
menthol	−4.1	−17.2	43(28)

[a] At 25°C.
[b] Chickos, Garin, et al. (1981).
[c] mp of (+) or (−) enantiomer (mp of racemate).

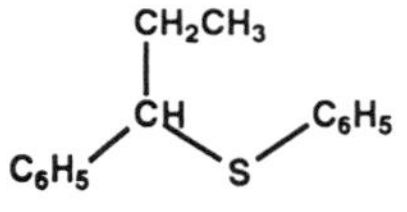

12

α–Ethylbenzyl phenyl sulfide

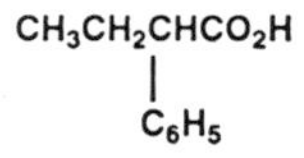

13

α–Phenylbutyric acid

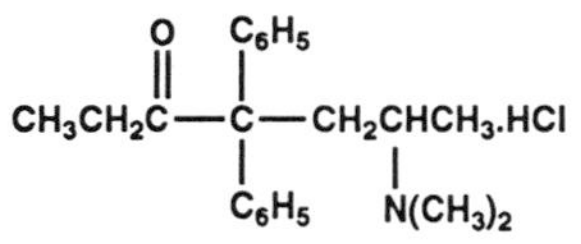

14

Methadone.HCl

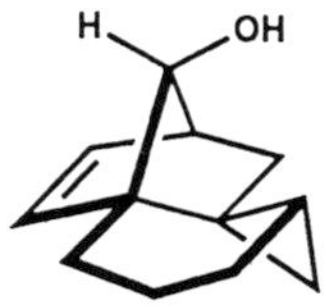

15

Figure 6.18.   Structures **12–15**.

was carried out by Garin et al. (1977). The latter group of investigators as well as Kwart and Hoster (see above) call attention to the high efficiency of separation by sublimation. Indeed, it appears that sublimation may be a significantly more efficient process than recrystallization in effecting enantiomer enrichment in enantiomer mixtures.

The principal limitations are that the samples need to have reasonable volatility at the operating temperatures, which need to be below the melting points of the lowest melting crystalline form, and that these temperatures not be so high as to lead to thermal decomposition. Zahorsky and Musso (1973) realized that the extent of enrichment could be rationalized by means of solid–gas phase diagrams reminiscent of the well-known melting point composition curves (though they did not elaborate the precise relationship between these two types of diagrams). Intuitively, there must exist a fairly close relationship between the phase diagrams describing melting and the sublimation of enantiomer mixtures. After all, the heat necessary to pass from the solid state to the gas state ΔH_{sublim}, corresponds to the heat of fusion plus the heat of vaporization (the latter being the heat required to pass from the liquid to the vapor state) and we have already seen (Section 6-2) that the latter is practically the same for enantiomers and for their mixtures. This analogy must, however, take into account the fact that, in the case of sublimation, a new variable can be introduced, namely, pressure.

Calculated (pressure–temperature) phase diagrams have since been reported (Farina, 1987a); the vapor pressure of a conglomerate is exactly twice that of the corresponding pure enantiomers. On the other hand, the vapor pressure of a racemic compound can be either higher or lower than that of the corresponding enantiomers; it depends chiefly on the enthalpy of decomposition of the racemic compound and on the temperature at which this decomposition takes place (Farina, 1987a).

The sublimation of enantiomer mixtures exhibiting racemic compound behavior is illustrated by means of a phase diagram (Fig. 6.19). Only the case of separation at constant pressure will be considered; a different phase diagram is required for the analogous separation at constant temperature, e.g., see Jacques et al., 1981a, p. 161.

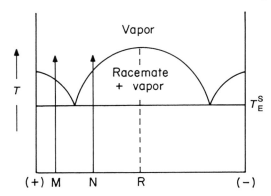

Figure 6.19. Sublimation (vapor pressure) diagram for a racemic compound system at constant pressure, P_0. [Adapted with permission from Jacques, J., Collet, A. and Wilen, S. H. (1981), *Enantiomers, Racemates and Resolutions*, p. 161. Copyright © 1981 John Wiley & Sons, Inc.]

The constant pressure sublimation diagram resembles that of the fusion phase diagram (Fig. 6.7), with the vapor and liquid eutectic compositions in the same regions of the respective diagrams (Jacques et al., 1981a, p. 161). Consider first what happens when a mixture of composition M (enantiomeric purity greater than that of the eutectic) is sublimed by raising the temperature. Sublimation begins at T_E^S and the vapor (corresponding to the first sublimate, if provision is made to condense the vapor) has the composition of the eutectic. Consequently, as a result of partial sublimation, the residue is enriched in the (+) enantiomer.

Sublimation of a less enantiomerically pure mixture, for example, of composition N, at constant pressure, leads to accumulation of more (+) enantiomer in the first sublimate than was present in the initial mixture. The residue is enriched in racemate. For example, a sample of α-phenylbutyric acid (Fig. 6.18, **13**) [$T_A^f = 41°C$; the racemate is a liquid at room temperature], having an enantiomeric composition of 7% (ee) is "sublimed" at 40°C by evacuation to 0.5 mm. The sublimate has ee = 17.2%. Sublimation at a lower temperature yields a product having an enantiomeric excess approaching 70% (Horeau and Nouaille, 1972). For a system that behaves as shown in Figure 6.19, it is expected that sublimation of the pure enantiomer will take place at a lower temperature than will the racemate. This phenomenon has actually been observed, for example, with methadone·HCl (Fig. 6.18, **14**; Kassau, 1966). It is important to note that it is not always the lowest melting form that sublimes preferentially; the behavior of mixture N (Fig. 6.19) is independent of the relative magnitudes of the melting points of racemate and of enantiomer. What matters is the composition of the eutectic relative to that of the composition of the mixture to be sublimed.

Conglomerate and pseudoracemate systems exhibiting maxima or minima would be expected to behave analogously, but examples of such processes on enantiomerically impure samples have yet to be described.

Contrary to the inference of Kwart and Hoster (1967), a partially enriched sample of a conglomerate could be purified by sublimation. It would behave just as a racemic compound system of composition M (Fig. 6.19). The enantiomer composition of an ideal solid solution of enantiomers would be expected to be unchanged on sublimation, however.

A different situation is found in the sublimation of racemic **15** (Fig. 6.18), which gives rise to crystal clusters, some of which are dextrorotatory and others levorotatory (rotation measured in ethanol solution). This finding, corresponding to a spontaneous resolution, requires that *rac*-**15** (a noncrystallizable solid) is a conglomerate whose enantiomers are independently nucleated on the walls of the sublimation vessel (Paquette and Lau, 1987; see also Section 7-2.a). One wonders whether preferential crystallization (by seeding; cf. Section 7-2.c) from the vapor state might be possible.

h. Infrared Spectra

While differences in solid state IR spectra of enantiomers and the corresponding racemates were observed long ago (Wright, 1937, 1939; Eliel and Kofron, 1953), the origin of these differences was sometimes ignored (Rao, 1963; Avram and Mateescu, 1972). It is now clear that such differences are not to be expected between spectra (measured either as KBr disks or as mulls) of either enantiomer and of the corresponding racemate when the latter is a conglomerate. On the other hand, IR spectra of racemic compounds usually differ significantly from those of the corresponding enantiomer (Jacques et al., 1981a, p. 18). Infrared spectroscopy is often the diagnostic procedure of choice in ascertaining whether or not a racemate is a conglomerate. However, in case of doubt (see below), it would seem best to utilize more than one criterion to determine the racemate type (for a discussion of the several methods available, see Section 7-2.b).

The observed differences in IR spectra reveal that certain molecular interactions are modified. One sees differences, even remarkable ones, in the spectra of the several crystalline (polymorphic) forms of one and the same substance, for example, the classic case of estradiol (Rao, 1963, p. 23). In carboxylic acids, for example, strong hydrogen bonds are usually present. Eliel and Kofron (1953) observed that such hydrogen bonding was much more important in the racemate than in the individual enantiomers of the hydrogen phthalate ester **16** that they studied (Fig. 6.20). The hydrogen bonds are responsible for dimer formation and the IR spectrum reveals that the dimer is a racemic compound. Conversely, in the enantiomer, intermolecular hydrogen bonding, as revealed by the position of the OH and C=O stretching frequencies, is less important. This difference is then responsible for the contrast in IR spectra between the two stereoisomer forms.

Differentiation between racemates and the corresponding enantiomers in the case of hydrogen-bonded carboxylic acids (mandelic, lactic, and tartaric acids), carbohydrates, for example, arabinose and salts (sodium ammonium tartrate), is readily effected by IR spectroscopy (Wirzing, 1973). Brockmann and Musso (1956) observed differences in IR spectra of racemic amino acid derivatives and the corresponding enantiomers and have correlated them with differences in solubility.

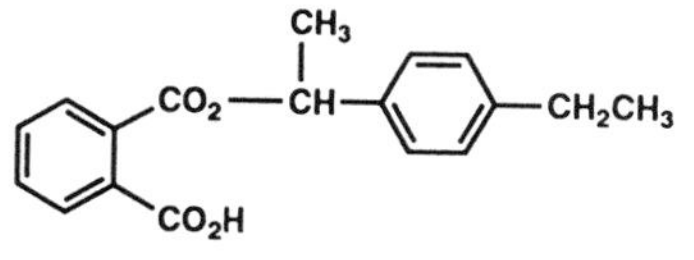

16

Figure 6.20. α-Methyl-(p-ethylbenzyl) hydrogen phthalate.

When no intermolecular hydrogen bonding is present, the IR spectral differences between enantiomers and racemates may be quite small. For example, in the case of chiral sulfide **12** (Fig. 6.18), the two spectra were superposable even though the racemate is clearly a racemic compound (Kwart and Hoster, 1967).

> The differences are sometimes quite subtle. Though ($\pm$)-alanine is not a conglomerate, spectra of racemic and resolved alanine have been reported to be identical (Wright, 1937, 1939; Brockmann and Musso, 1956). Yet, in an extensive study of IR spectra of amino acids, Tsuboi and Takenishi (1959) chose to place alanine in a class exhibiting "small but clear" differences between racemate and enantiomer spectra. However, asparagine (known to be a conglomerate since it is spontaneously resolved) appears in the same class!

Solid state Raman spectra may also exhibit the differences just described (Mathieu and Gouteron-Vaisserman, 1972; Mathieu, 1973). Stereoisomer discrimination is also observed in IR spectra of solutions (see Section 6-4.m).

i. Electronic Spectra

(1*S*)-Dicarbonylrhodium(I) 3-trifluoroacetylcamphorate (Fig. 6.21, **17**), a chiral planar d^8 metal complex, is a yellow solid, mp 134°C. Admixture of this substance

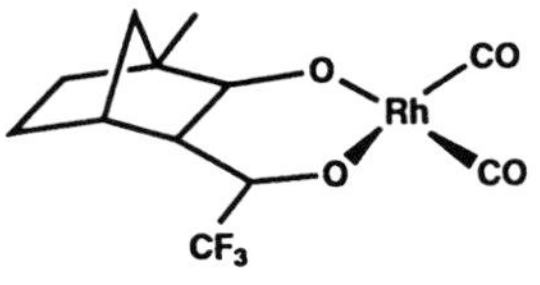

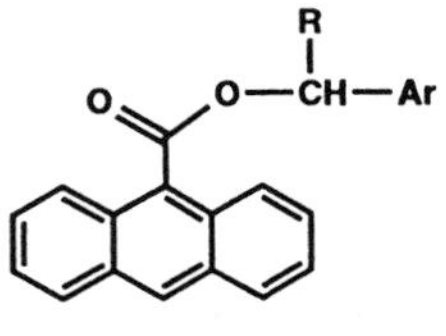

e.g. R = CH$_3$, Ar =

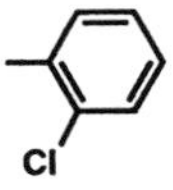

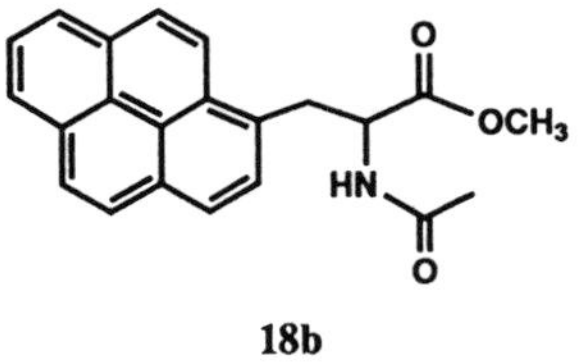

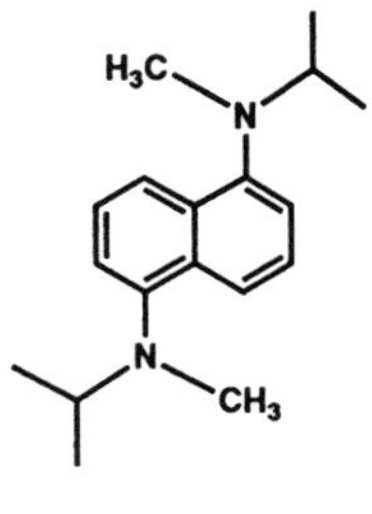

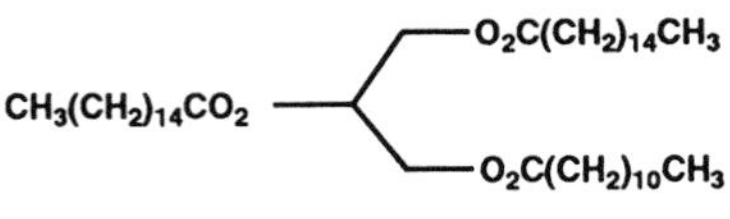

Figure 6.21. Structures **17–20**.

with the independently prepared ($1R$) enantiomer in solution and evaporation of the solvent leaves a red-green (dichroic) solid, mp 130.5°C, which can be easily identified as a racemic compound (a mixture melting point with a small amount of the enantiomer leads to melting point lowering) (Schurig, 1981). This striking color change constitutes unequivocal visual proof that electronic absorption spectra of racemic compounds and of the corresponding enantiomers in the solid state can differ significantly. Diffuse reflectance spectra of the two forms also differ markedly. In contrast, the colors of the corresponding melts were alike (brown).

Analogous differential behavior in solid state emission spectra (fluorescence), as well as differences in color, has been observed in several typical organic compounds, for example, **18**, (Fig. 6.21) whose racemates had been shown to be racemic compounds by X-ray analysis (Lahav et al., 1976; Ludmer et al., 1982). The racemic compounds exhibited characteristic α-type structureless excimer emission while the enantiomers have γ-type monomer emission. The utility of this differential behavior in the determination of enantiomeric purity has been demonstrated (Ludmer et al., 1982). Enantiomer discrimination is observed (by fluorescence) in the dissociation of intermolecular homochiral and heterochiral excimers of methyl N-acetyl-1-pyrenylalaninate (Fig. 6.21, **18b**), but not in the formation of the excimers. Differences in hydrogen bonding favor the D–L (heterochiral) excimer more than the L–L (homochiral) excimer in CH_3CN but not in N,N'-dimethylformamide (DMF), the latter solvent being a hydrogen-bond acceptor able to interfere with excimer formation (López-Arbeloa, De Schryver, et al., 1987).

Stereoisomer discrimination in the fluorescence of the enantiomers of 1,1'-binaphthyl (Fig. 7.11, **15**) has been observed in solution as monitored by quenching with (S)-(−)- and (R)-(+)-N,N-dimethyl-α-methylbenzylamine at 22°C. The difference in the rate of quenching was ascribed to the formation of a contact pair of an excited binaphthyl molecule and a quencher molecule as the result of a preferred orientation of the emitter and quencher molecules (Yorozu, Hayashi, and Irie, 1981). A related system wherein a surface [silica derivatized with (R)-(−)-1,1'-binaphthyl-2,2'-dihydrogen phosphate (Fig. 7.21, **65**)] has been endowed with chirality has also been probed by fluorescence quenching with the same pair of enantiomeric amines. Diastereomer discrimination on the surface is responsible for differential quenching (Avnir, Wellner, and Ottolenghi, 1989).

j. Nuclear Magnetic Resonance Spectra

Differences in solid state ^{13}C NMR spectra of the enantiomers and the racemic compound have been demonstrated for tartaric acid by Hill, Zens, and Jacobus (1979). Instrumental facilities for measuring high-resolution solid state NMR spectra by a combination of cross-polarization (CP) and magic angle spinning (MAS) are becoming widely available, hence such measurements are likely to be increasingly important in identifying racemate types. Since enantiopure and racemic samples of the same substance in the solid state can exhibit different chemical shifts (as in tartaric acid), when conglomerates are not formed, solid state NMR may serve to determine the enantiomer purity of a solid sample (see Section 6-6). Enantiomer discrimination in solution is taken up in Section 6-4.m.

Solid state ^{13}C NMR spectra of inclusion compounds obtained when solutions of tri-o-thymotide (TOT; Fig. 7.10, **13**) in rac-2-halobutanes ($CH_3CHXCH_2CH_3$) or 2-butanol are allowed to evaporate demonstrate stereoisomer discrimination. Virtually all the carbon signals observed, of host as well as guest molecules, exhibit anisochrony. All signals due to host carbon atoms become triplets (TOT in the solid state loses its threefold symmetry) and signals due to guest molecules (when not obscured by peaks arising from the host) are doublets that reveal the diastereomeric environments of the two guest enantiomers. Solid state NMR provides a way of measuring the enantiomer composition of the guest; however, allowance must be made for the CP time-dependent line intensities.

In the case of X = Cl and Br, the doublets exhibit unequal intensities since (S)-(+) guest molecules prefer (or enforce) the (P)-(+) configuration of host molecules (diastereomer discrimination). Application of dipolar dephasing techniques reveals differences in guest molecule dynamics where TOT preferentially encloses one guest enantiomer; the major enclosed enantiomer is held rigidly in the TOT "cage," whereas the minor enantiomer is mobile (Ripmeester and Burlinson, 1985).

Application of the CP–MAS technique sometimes permits the differentiation of meso from racemic *chiral* diastereomers. In the solid state, the ^{13}C NMR spectrum of 2,3-dimethylsuccinic acid yields two signals (for the methyl and the methine carbon atoms) in the case of the meso diastereomer, whereas in the ($\pm$) diastereomer each of these signals is split into a doublet since the pairs of CH and CH_3 carbon atoms are not related by a plane or center of symmetry. This finding is equivalent to stating that, in contrast to what obtains in solution (cf. Chapter 8), the solid chiral diastereomer exists as two noninterconverting conformers whose nuclei are externally diastereotopic. It must be noted that this result is possible because the meso acid adopts a single symmetric (anti) conformation in the solid state, a situation that does not universally apply (e.g., it does not in meso-tartaric acid). Solid state CP–MAS NMR spectra may also be useful in differentiating meso and chiral conformers that cannot be physically separated, for example, **19** (Fig. 6.21), in which the methyl and isopropyl groups are twisted with the methyl groups on the same or opposite sides of the naphthalene plane yielding two diastereomeric conformations (Casarini, Lunazzi and Macciantelli, 1988).

k. X-Ray Spectra

We have already seen that X-ray diffraction furnishes two types of stereochemical information: relative configuration of chiral centers, both intra- and intermolecular, can be determined (cf. Section 5-5), and absolute configuration can be established when the wavelength of the X-rays is chosen to be close to the absorption wavelength of an inner-shell electron of one of the atoms in the crystal (Section 5-3). There is yet more stereochemical information extractable from X-ray spectra.

Diffraction experiments begin with the assignment of the space group of the crystal (which reflects the symmetry of its constituent molecules) by examination of the diffraction pattern (Dunitz, 1979, p. 96). This assignment is important because enantiopure samples of chiral molecules necessarily crystallize in noncentrosymmetric space groups. The space group [together with information about the number of molecules (Z) per unit cell] reveals the presence or absence of a center of symmetry, that is, the space group (together with the value of Z) tells us whether the crystal is centrosymmetric or noncentrosymmetric. The latter category (65 out of 230 space groups, cf. Section 6-4.c) comprises all enantiomorphous crystals. In other words, determination of the space group (and of Z) reveals the presence or absence of enantiopure crystals; for racemic samples, this is equivalent to determining whether or not the compound exists as a conglomerate (for an early example, see Shoemaker, Corey, et al., 1950). Racemic compounds crystallize mostly in centrosymmetric space groups. A few examples of racemic compounds crystallizing in chiral space groups are known (Jacques et al., 1981a, p. 7 and Brock, Schweizer, and Dunitz, 1991), whereas conglomerates necessarily crystallize in noncentrosymmetric space groups.

The space group also reveals the presence or absence of a polar axis, a crystal attribute that is in turn responsible for interesting and useful chemical and physical properties of numerous common chemicals (Paul and Curtin, 1975; Curtin and Paul, 1981; cf. Section 6-4.l). The utility of the polar axis in the determination of absolute configuration through crystal habit modification was described in Chapter 5.

> A crystallographic polar axis represents a linear dipolar alignment of molecules within a crystal lacking a center of symmetry. Properties at one end of a polar axis will consequently differ from those at the other end. The presence of a major polar axis, which is often visually (i.e., morphologically) evident, defines a polar crystal.

Finally, in a strictly empirical manner, comparison of X-ray powder diffraction spectra (Debye–Scherrer diagrams) measured on powders rather than on single crystals (e.g., Glusker and Trueblood, 1985) show distinct bands for enantiopure samples and for the corresponding racemates but only when the latter are racemic compounds. Thus, X-ray powder spectra can reveal the nonracemic character of cryptochiral samples, such as the triglyceride 1-laurodipalmitin (Fig. 6.21, **20**) (Schlenck, Jr., 1965a,b).

l. Other Physical Properties

The development of static electricity induced by mechanical action (e.g., compression or torsion), that is, development of charges on opposite faces of crystals, is called piezoelectricity. A related effect called pyroelectricity is the generation of static electricity by heat. Both effects require that the crystals be noncentrosymmetric, that is, a polar axis must be present (Curtin and Paul, 1981; Paul and Curtin, 1987). Many chiral substances exhibit these effects, as well as some achiral substances crystallizing in chiral space groups.

The pyroelectric effect has been applied (on single crystals) to the determination of the absolute configuration of chiral hydrobenzoin (Pennington, Paul, and Curtin, 1988). Natural triglycerides with very low or zero optical activity (cryptochiral compounds) may be characterized through use of their piezoelectric properties. The synthetic racemic triglycerides exhibit no piezoelectricity while the naturally occurring ones do. Therefore, the conclusion, independently confirmed (see Section 6-4.m), is that the latter are nonracemic (Schlenk, Jr., 1965a,b).

A related property, triboluminescence, is the emission of light under mechanical constraint (Zink, 1978). This effect, discovered centuries ago, is often associated with the generation of electric fields created when crystals fracture. Most well-known triboluminescent crystals are noncentrosymmetric. The tartaric acid enantiomers (but not the centrosymmetric racemate) and many sugars are triboluminescent (Hardy, Zink, et al., 1981).

There is a coupling between chirality and electromechanical properties of crystals. The piezoelectric effect is occasionally used to assist in the assignment of the space group (centrosymmetric vs. noncentrosymmetric) when crystallographic methods give ambiguous results (Glusker and Trueblood, 1985, p. 100). Applications of such properties (piezoelectricity, pyroelectricity, triboluminescence, and ferroelectricity) that relate to chirality are scarce; see, however, Addadi et al. (1986); Goodby (1986).

> By way of example, nonracemic samples of rodlike compounds such as **21** (Fig. 6.22) exhibit liquid-crystal (cholesteric) behavior with the sign of the mesophase optical activity alternating according to the configuration of the stereocenter (indicated by *), for example (−) if the latter is of S configuration and the number n of $(CH_2)_n$ intervening between the stereocenter and the biphenyl moiety is odd; the sign is (+) if n is even. Both the sign and the magnitude of $[\alpha]$ are determined by the helical ordering between layers of the liquid crystal (Section 13-4.e). If the helix pitch is close to that of the wavelength of visible light, the liquid crystal can selectively reflect light and it is temperature and electrically sensitive (ferroelectric); such a compound is then useful as an indicator in thermometers and thermochromic devices, such as battery testers (Goodby, 1986).

Dielectric relaxation phenomena, which directly probe the rotational tum-

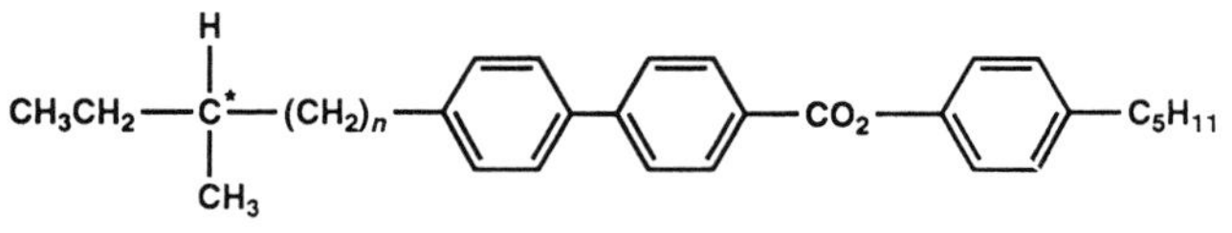

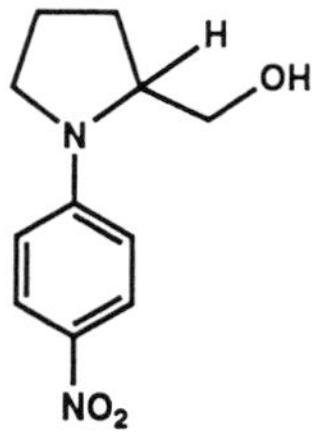

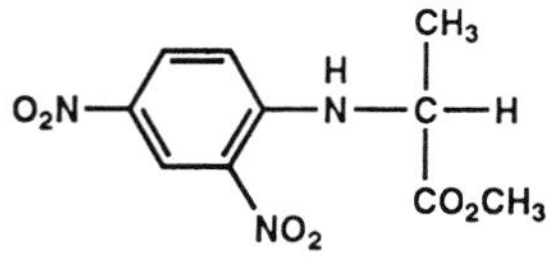

Figure 6.22. Structures **21–23**.

bling of molecules, provide another type of experimental measurement for probing enantiomer discrimination: The pure enantiomers of 3-methylcyclo-hexanone and of menthone relax faster than do mixtures of the enantiomers (Stockhausen and Daschwitz, 1985).

Noncentrosymmetric crystals [both of enantiopure samples of chiral compounds and of (racemic) conglomerates] are presently the focus of intensive research with respect to their optical properties that have large so-called second-order nonlinear components. Properties (such as second harmonic generation) that is, the generation of a second beam having twice the frequency of the incident beam, on passage of light through a crystal, are only found in non-centrosymmetric crystals. These properties have important practical applications in the field of electronics, communications, and optics technology, that is, in the construction of electrooptic devices, such as amplifiers, switches, and modulators (Williams, 1984; Chemla and Zyss, 1987). N-(4-Nitrophenyl)-L-prolinol (NPP), 22, and methyl N-(2,4-dinitrophenyl)alaninate (MAP), 23 (Fig. 6.22), are examples of conjugated molecules endowed with polar functionalities and polar axes (cf. Section 6-4.k; Nicoud and Twieg, 1987). These features give rise to large (= hyper)polarizabilities that are the basis of the aforementioned nonlinear properties.

m. Liquid State and Interfacial Properties

It should now be quite clear that the properties, both physical and chemical (though the latter have yet to be described), of enantiomeric systems will often be unequal whenever one compares two samples differing in enantiomeric purity [ranging from total (= pure enantiomer) to zero (= racemate)]. Such differences, which manifest themselves most strongly in the solid state, will be evident to some extent in other states of matter as well.

Among the most striking illustrations of enantiomer discrimination in the liquid state are those found in solution NMR spectra. In solution, enantiopure and racemic samples of chiral compounds generally exhibit identical NMR spectra (small differences in the chemical shifts are sometimes observed, e.g., Dobashi, Hara, et al., 1986). However, nonracemic but not enantiopure samples of compounds that strongly self-associate, for example, hydroxyamines, amides, and carboxylic acids (either as neat liquids or in solutions of *achiral* nonpolar solvents) sometimes exhibit anisochrony (split signals) reflecting the enantiomer composition even in the absence of chiral inducing agents, such as nonracemic chiral solvents or shift reagents (Review: Kabachnik et al., 1978).

Williams, Uskokovic, et al. discovered (in 1969) that a racemic solution of dihydroquinine (Fig. 6.23, 24; ca. $0.3\,M$ in $CDCl_3$) exhibits an ^1H NMR spectrum different from that of the optically active compound (at 100 MHz). Chemical shifts were affected throughout the spectrum, and peak intensities of partially resolved material depended on the enantiomeric purity of the sample. (The spectra of the racemate and of the optically active sample tended to become identical at high dilution.) Accordingly, the enantiomer composition of the mixture may be deduced without resorting to a chiral solvent or to a shift reagent. That the observed effect is a manifestation of enantiomer discrimination is made

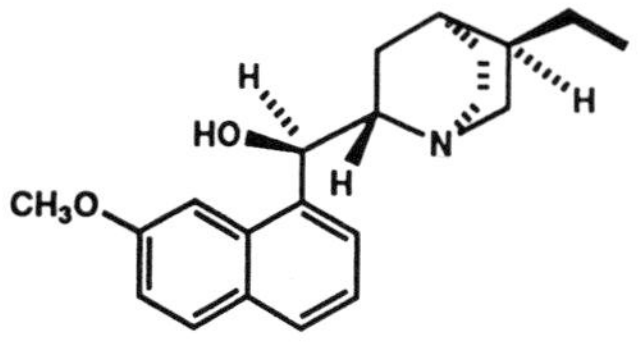

24

Dihydroquinine

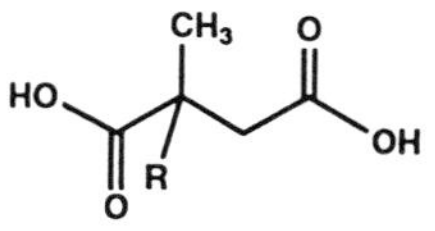

25

R = Et, *i*-Pro

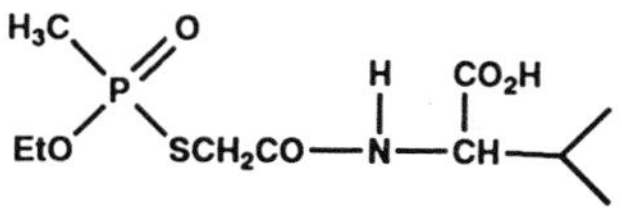

26

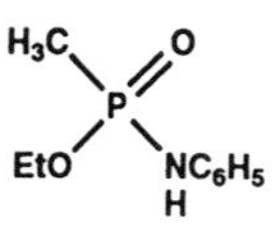

27

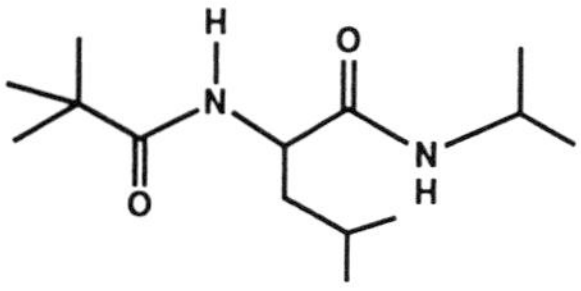

28

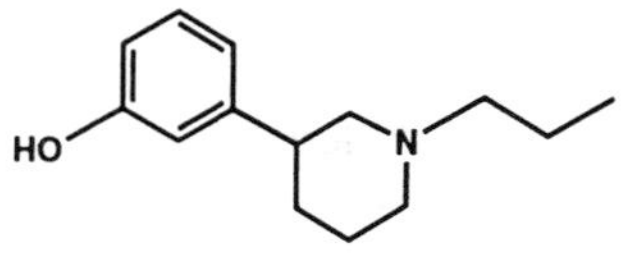

29a

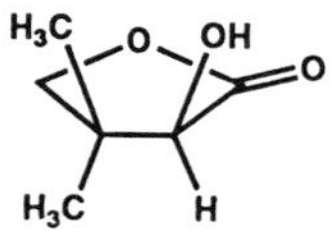

29b

Pantolactone

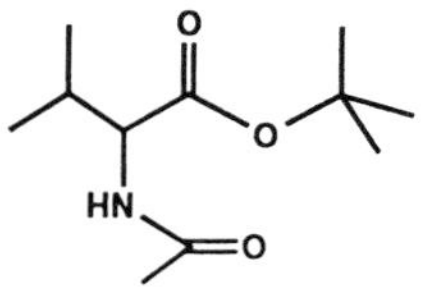

30a

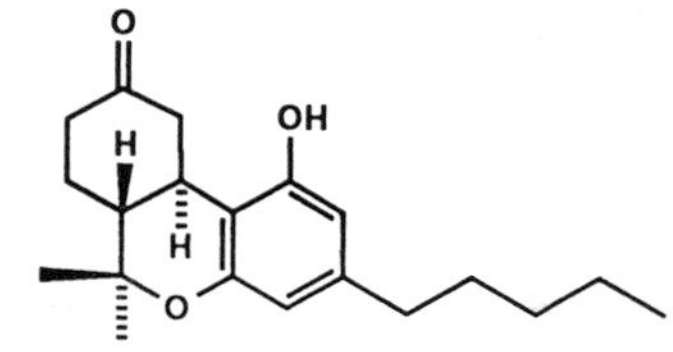

30b

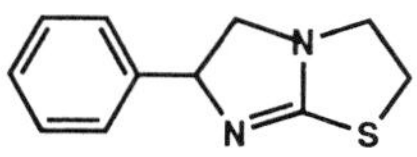

30c

Figure 6.23. Compounds exhibiting enantiomer discrimination in their NMR spectra.

evident by the following additional facts: (a) the additional peaks disappear on acetylation of the hydroxyl group; and (b) replacement of the aprotic solvent by CH_3OH leads to identical spectra for enantiopure and racemic **24** as well as for all mixtures of (+)- and (−)-**24**, presumably by destroying the hydrogen-bonded dimeric or oligomeric "associates" responsible for the effect.

The self-induced anisochrony requires formation of diastereomeric associates and consequently one would expect to observe this otherwise unexpected phenomenon only when hydrogen bonding or other strong association is present. The time-average environments of [(−)-**24** · (−)-**24**, (+)-**24** · (+)-**24**], and [(−)-**24** · (+)-**24**] dimeric associates (as well as oligomeric associates) are different in the absence of extraneous influences such as high temperature and polar solvents able to interfere with the self-association (Weisman, 1983). The anisochrony occurs under conditions of fast exchange and is favored by low temperature and high concentration. A linear relationship exists between $\Delta\delta$ and the proportion of the two enantiomers present indicating that the proportion of homochiral and heterochiral aggregates formed is under statistical control in this case (Kabachnik et al., 1976; Dobashi, Hara, et al., 1986).

Other compounds have been reported to display this type of enantiomer discrimination in NMR spectra: substituted succinic acids **25** (Horeau and Guetté, 1974); organophosphorus compounds with phosphorus and carbon chiral centers **26** (Kabachnik et al., 1976) and simple phosphinic amides (e.g., **27**; Harger, 1977; see also Harger, 1978); *N*-acylamino acid amides **28** (e.g., Cung, Neel, et al., 1978); substituted piperidines **29a**, where the anisochrony was observable even with a 98:2 enantiomer ratio (Arnold, Imhof, et al., 1983); pantolactone **29b** (Nakao et al., 1985); *N*-acylamino acid ester **30a** (Dobashi, Hara, et al., 1986); a hydroxy ketone cannabinoid intermediate **30b** (Hui et al., 1991); and several carboxamides, all bearing a hydrogen atom on the carboxamide nitrogen atom (Jursic and Goldberg, 1992). Surprisingly, the fused imidazo[2,1*b*]thiazole **30c** is claimed to exhibit anisochrony of signals in CD_3OD but not in $CDCl_3$ (Ács, 1990). Enantiomer discrimination of this type is likely to be observed increasingly often as ever higher field NMR spectrometers become available.

The search for solutes exhibiting enantiomer discrimination in NMR spectra has been influenced by the observation of chelated diastereomeric "solvates" that are formed between chiral solvating agents (CSA) and solutes (Horeau and Guetté, 1974; Dobashi and Hara, 1983). Such solvates are responsible for anisochrony in NMR spectra (Section 6-5.c, p. 231) and for separation in high-performance liquid chromatography (HPLC) when the CSA are added to chromatographic mobile phases (Section 6-5.d, p. 246).

In the case of **30a**, evidence for NH···O=C hydrogen bonding (involving both amide and ester groups) between solute molecules was obtained from NMR chemical shifts and from IR spectra in solution. Consequently, enantiomer discrimination was ascribed to the formation of diastereomeric dimers as illustrated in Figure 6.24. Dimer **A** results from homochiral interaction (Eq. 6.1) whereas dimer **B** results from heterochiral interaction (Eq. 6.2; Dobashi, Hara, et al., 1986). In contrast, in the case of carboxamides lacking additional hydrogen-bonding acceptor sites, the observed enantiomer discrimination cannot be ascribed to cyclic dimers in view of the preferred *Z* conformation (CO and NH anti) of the typical carboxamide functionality (Section 10-2.a). Instead, the observed discrimination is ascribed to linear hydrogen-bonded associates (Jursic and Goldberg, 1992).

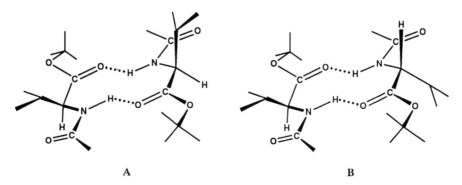

            A   B

Figure 6.24.   Proposed diastereomeric dimers (**A** = homochiral; **B** = heterochiral) formed from **30a** [Adapted with permission from Dobashi, A., Saito, N., Motoyama, Y., and Hara, S. (1986), *J. Am. Chem. Soc.*, **108**, 307. Copyright © 1986 American Chemical Society.]

If the requirement for observable enantiomer discrimination is aggregation giving rise to homochiral and heterochiral associates, then reaction with achiral reagents promoting such aggregation may increase the tendency for the development of anisochrony in NMR spectra. This possibility has been demonstrated with nonracemic mixtures of chiral 1,2-diols whose enantiomer composition can be measured by ^{13}C NMR spectroscopy following their quantitative conversion to dioxastannolanes (Luchinat and Roelens, 1986) by reaction with (achiral) dibutyltin(IV) oxide in $CDCl_3$ or in other nonpolar solvents (Fig. 6.25). In the case of 1,2-propanediol, anisochrony ($\Delta\delta$) of the methine carbon was observable on mixtures ranging from 20 to 89% ee with the intensities of the pairs of signals closely reflecting the proportion of R and S diol enantiomers present in the analyte. Similar analysis is also possible with dioxastannolanes formed in situ (with Bu_2SnCl_2 in $CDCl_3$).

Critical analysis of "self-discrimination" by Luchinat and Roelens (1986) reveals that (a) systems unable to form heterochiral dimers (heterodimers; Eq. 6.2) do not give rise to enantiomer discrimination, (b) maximal enantiomer discrimination is exhibited by systems favoring formation of heterochiral over homochiral dimers, and (c) calculation of $\Delta\delta$ values must take into account the K_{equil} of the $SS + RR \rightleftharpoons 2RS$ process; $\Delta\delta$ is a linear function of the enantiomeric excess only for the case of $K = 4$ (corresponding to the measured value in the case of dioxastannolane formation with 1,2-propanediol and 1-phenyl-1,2-ethanediol). The value 4 is, in fact, statistically expected:

$$SS + RR \rightleftharpoons 2RS$$

$$K = \frac{[RS]^2}{[RR][SS]} = \frac{[RS]^2}{[SS]^2} = \frac{2^2}{1^2} = 4$$

Pasquier and Marty (1985) translated the enantiomer discrimination potentially demonstrable in a nonracemic sample into diastereomer discrimination in order to increase the possibility of observing the former. The diastereomer discrimination has been achieved by use of the coupling (duplication) method of Horeau (Vigneron, Dhaenens, and Horeau, 1973; see also Section 6-5.c and Section 7-4). Reaction of (±)-1-diphenylphosphino-2-propanethiol, $(C_6H_5)_2PCH_2CH(CH_3)SH$, with nickel nitrate in solution leads to formation of a mixture of meso and chiral diastereomeric trans-Ni(thiol)$_2$ complexes whose composition is easily analyzable by ^{31}P NMR spectroscopy. Provided that the

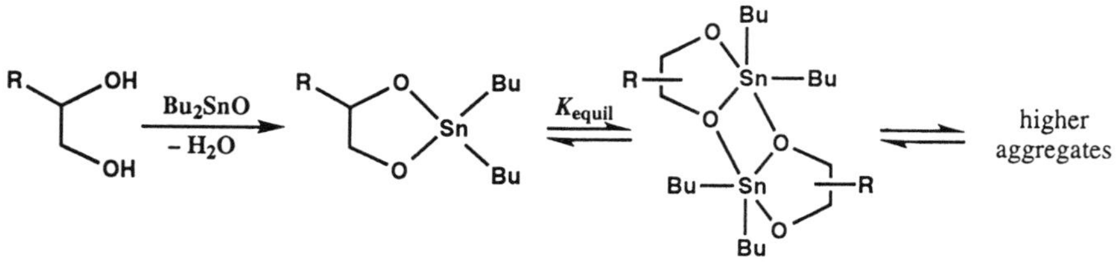

Figure 6.25. Formation of dioxastannolanes ($R = CH_3, C_6H_5$) for the enantiomeric purity determination of chiral 1,2-diols without chiral auxiliaries [Adapted with permission from Luchinat, C., and Roelens, S. (1986), *J. Am. Chem. Soc.*, **108**, 4873. Copyright © 1986 American Chemical Society.]

diastereoselectivity of the reaction is thermodynamic in origin (reversible reaction) and its extent $\{s = ([m]/[e])_{racemate}$ where $[m]$ is the concentration of the achiral meso species and $[e]$ is the concentration of the chiral ($\pm$) complex$\}$ is known from reaction of the bifunctional Ni(II) with *racemic* substrate then, if no free (uncomplexed) analyte remains, the enantiomer composition of an enantiomerically enriched sample, the phosphinopropanethiol, can be directly calculated by means of Eq. 6.8

$$\%ee \approx \frac{\sqrt{K^2 - s^2}}{K + 1} \times 100 \tag{6.8}$$

where K is the measured $[m]/[e]$ ratio of the product (see also p. 231). Another instance of the enhancement of magnetic nonequivalence by coupling has been observed on partially resolved amines, such as α-methylbenzylamine, by addition of Eu(fod)$_3$ (Fig. 6.41), an achiral lanthanide shift reagent (Ajisaka and Kainosho, 1975; for additional examples, see Jacques et al., 1981a, p. 410). For a general treatment of the coupling of enantiomers, see Marty, Pasquier, and Gampp (1987) and Kagan and Fiaud (1988, p. 278).

The involvement of dimeric and oligomeric species in enantiomer discrimination is demonstrated even in aqueous solution by "mere addition" of lanthanide ions. Thus, the methyl resonance of a $80(S):20(R)$ mixture of sodium lactate was resolved in the presence of EuCl$_3$ (3:1 or 2:1 ligand to EuCl$_3$ ratio; Reuben, 1980). Anisochrony in the NMR spectrum of *racemic* lactate was observed in the presence of another chiral (nonracemic) ligand, for example, (R)-($+$)-malate or (S)-($+$)-citramalate; here the discrimination is more obviously diastereomeric in nature.

In the case of α-hydroxycarboxylates, anisochrony resulting from lanthanide:mixed ligand complexation occurs when the stoichiometry (ligand–lanthanide) is 3:1 and is absent in 2:1 complexes. This finding suggests that crowding about the lanthanide ion forces the ligands into stereochemically constrained environments that maximize the discrimination leading to spectral resolution (Reuben, 1980).

In any event, anisochrony resulting from "self-discrimination" can no longer be considered exceptional; the statement often made that enantiomers and racemates (or mixtures of intermediate enantiomer composition) have identical NMR spectra in solution is not universally correct.

It has been demonstrated that the enantiomeric purity of nonracemic samples isolated from nature, for example, α-pinene, may be estimated from $^2H/^1H$ isotope ratios in ^2H NMR spectra at the natural abundance level. This estimation is made possible by the finding that (as a consequence of the operation of isotope effects in biogenesis that differ for the two enantiomers) deuterium is not statistically distributed in the pinene molecule; the $^2H/^1H$ ratio varies with the specific (carbon) site examined (Martin, Jurczak, et al., 1986).

Stereoisomer discrimination has also been observed on examination of IR spectra in solution. Frequencies and intensities of group vibration bands that are sensitive to association (e.g., OH) are the ones subject to enantiomer discrimination (Walden, Pracejus, et al., 1977; see also Cung, Neel, et al., 1978). 2-Butanol is known to form dimers and oligomers in nonpolar solvents, such as CCl$_4$. Equilibrium constants for dimer and for cyclic pentamer formation differ according to whether enantiopure or racemic 2-butanol is being examined (Herndon and Vincenti, 1983); a similar stability difference has been observed in the IR spectra (hydrogen-bonded ν_{NH}) of enantiopure and racemic N-acetylvaline $tert$-butyl ester (Dobashi, Hara, et al., 1986).

Evans and Evans (1983) found evidence of enantiomer discrimination in the far IR spectrum of 3-methylcyclohexanone. This finding is ascribed to molecular rotation–translation coupling.

> Other manifestations of enantiomer recognition are less well known: differential rates of electron-transfer reactions (Chang and Weissman, 1967); and rate and equilibrium differences in excimer formation (fluorescence) from association of ground and excited state monomers (Tran and Fendler, 1980).

Stereoisomer discrimination may also be responsible for differences in reaction rates and product distribution according to whether chiral reactants are pure enantiomers or racemates (Wynberg and Feringa, 1976) . Examples of such reactions are given in Section 12-1.c.

Horeau and Guetté (1974) summarized attempts to demonstrate enantiomer discrimination in solution through measurements of classical physical properties: surface tension, index of refraction, viscosity, and so on. The effects are uniformly very small, barely, if at all, detectable. However, careful experimentation coupled with improved instrumentation is beginning to reveal such effects, for example, IR spectra of alcohols and of amino acid derivatives in solution give clear evidence of intermolecular association (hydrogen-bonded dimers and oligomers) that differentiates enantiomers from racemates (see above).

Highly oriented monolayers of chiral surfactants exhibit differences in properties reminiscent of those described in melting point and solubility diagrams of enantiomers. For example, force–area ($\pi - A$) curves (variation of the surface area A of a monolayer with surface pressure π as measured in a Langmuir film balance) of (S)-$(+)$- and rac-2-tetracosanyl acetate clearly show enantiomer discrimination that is additionally affected differentially by temperature (Lundquist, 1961). This discrimination is interpreted as being due to differences in molecular packing in an approximately 2D liquid. Hydrogen bonding as well as hydrophobic interactions play significant roles in determining the area occupied by the monolayers (Stewart and Arnett, 1982).

Other measures of stereoisomer discrimination in chiral monolayers are surface viscosity and hysteresis (timing of relaxation from the compressed monolayer state). Stereoisomer discrimination is even visually palpable by (epifluorescence) microscopy. All of these methods reveal a much greater degree of order and close packing in the quasicrystalline homochiral films of chiral surfactants, such as N-stearoylserine methyl ester, than in the corresponding heterochiral films (Review: Arnett, Harvey, and Rose, 1989).

n. Chromatography

It has generally been assumed that, in order for enantiomers to be separated (or for the enantiomer composition of a nonracemic sample to be modified) by chromatography, either the stationary phase or the mobile phase must be nonracemic (Jacques et al., 1981a, p. 413). The following results make it clear that the preceding assumption is not generally justified.

High-performance liquid chromatography of ^{14}C radiolabeled racemic nicotine on an ordinary reversed phase column gives rise, as expected, to but a single peak in the chromatogram. However, in the presence of varying amounts of unlabeled (S)-(−)-nicotine (Fig. 6.26, **31**), the isotopically labeled nicotine exhibits two radioactive peaks in the chromatogram as monitored by a radioactive flow-through detector. The peaks are ascribed to the two enantiomers of the radiolabeled nicotine; but see p. 196 (Cundy and Crooks, 1983). A similar study has demonstrated the chromatographic separation (on ordinary silica gel) of ^{14}C radiolabeled racemic N-acetylvaline tert-butyl ester (Fig. 6.23, **30a**) diluted with (−)-**30a** into two peaks (Dobashi, Hara, et al., 1987). The order of emergence of the two radiolabeled enantiomers was determined by coinjection experiments with nonradioactive (−)-**30a** and (+)-**30a**.

Chromatography of nonracemic (but enantiomerically impure) samples of N-lauroylvaline tert-butylamide (Fig. 6.26, **32**) on silica gel eluted with the usual achiral solvents (hexane + ethyl acetate mixtures) furnished fractions differing in their melting points. Careful chromatography on Kieselgur 60 of a sample having an initial enantiomer composition L/D = 87:13 (74% ee) gave a fraction (ca. 30% of the sample) of lower enantiomeric purity (46% ee) than that of the analyte. Later fractions exhibited enrichment (as high as 97% ee) in the L enantiomer; this is consistent with the easier elution of the racemate relative to the predominant enantiomer. Similar results were obtained with acylated dipeptide esters. As required by the mass balance, the initial reduction in enantiomeric excess matches the enrichment in the later fractions (Charles and Gil-Av, 1984).

Similarly, fractionation of a sample of Wieland–Miescher ketone (Fig. 6.26, **33a**), of 65% ee by chromatography on silica gel led to 10 fractions differing significantly in enantiomer composition: The initial fraction had ee = 84% and the final fraction ee = 51%. Control experiments confirmed the fact that the observed effect was not due to decomposition of the sample, to racemization, or to impurities in the sample or on the column. The observation of enantiomer

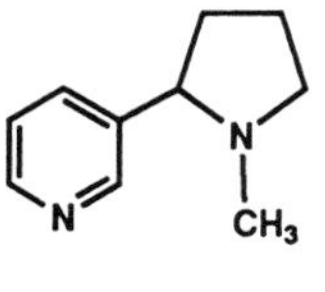

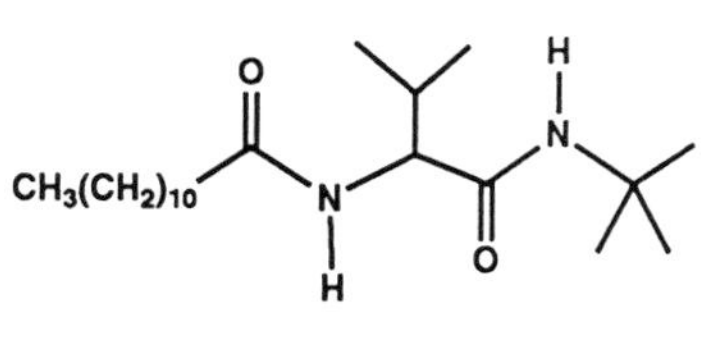

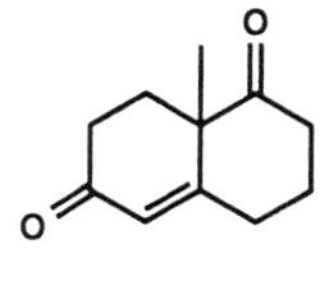

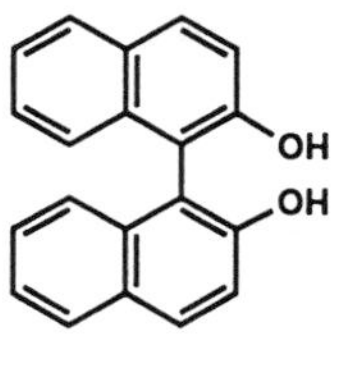

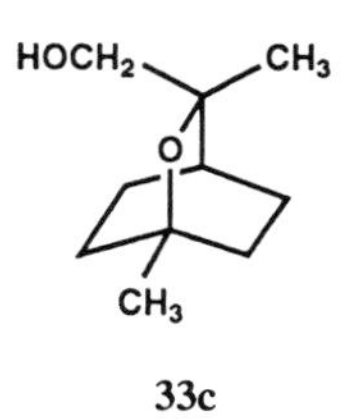

Figure 6.26. Compounds exhibiting enantiomer discrimination in chromatography on achiral stationary phases.

discrimination in ketone **33a** supports the contention that this effect is not limited to molecules that are able to form hydrogen bonds (Tsai, Dreiding, et al., 1985).

Enantiomeric enrichment has been observed also on chromatography of nonracemic samples of binaphthol (Fig. 6.26, **33b**; Matusch and Coors, 1989) specifically on aminopropyl silica gel (and not on silica gel itself), and most recently in cineole metabolites (Fig. 6.26, **33c**) isolated from the urine of female Australian brushtail possum (Carman and Klika, 1991).

It is evident from the foregoing results that chromatography of enantiomerically impure (but nonracemic) samples, including natural products in their native state, on achiral stationary phases and with achiral mobile phases can lead to fractions in which enantiomeric enrichment is observed, sometimes in the early chromatographic fractions and sometimes in the later ones (Review: Martens and Bhushan, 1992). While explanations for such chromatographic enrichment remain speculative, it has been pointed out that dilution of the analyte sample leads to less effective separation (Tsai, Dreiding, et al., 1985; Carman and Klika, 1991). This finding is consistent with the hypothesis that separation is due to diastereomeric aggregates of analyte molecules (cf. Section 6-2) that are concentrated on the surface of the achiral stationary phase (Matusch and Coors, 1989; Martens and Bhushan, 1992).

> The attribution of separation to the formation of diastereomeric associates on the surface of the stationary phase is consistent with the observation that, when well-resolved peaks are obtained, one peak contains optically active material whereas the other represents inactive material as measured polarimetrically (Matusch and Coors, 1989). Thus, when chromatography of nonracemic samples on strongly bonding achiral stationary phases (aminopropyl silica gel) leads to separation, the fractions are the dominant enantiomer and the racemate and not the enantiomers themselves. In contrast, reports of chromatography of nonracemic samples on ordinary silica gel indicate that the separated fractions are those of the enantiomers, e.g., Dobashi, Hara, et al. (1987); Dobashi, et al. (1991). The latter have proposed that separation may be due to differences in stability of the aggregates in solution (compare Sec. 6-4.m). Yet, it is hard to see how chromatography of a mixture of diastereomeric (homochiral and heterochiral) aggregates with an achiral stationary phase can give rise to *separated* fractions consisting of the enantiomers themselves. The separation of nonracemic samples of chiral compounds on achiral stationary phases clearly merits further study.

Two additional sets of chromatographic separations have been reported in which neither a chiral mobile phase nor a chiral stationary phase are employed. Chromatography of several *racemic* amino acids (e.g., tyrosine and histidine) on an ion exchanger containing apparently achiral iminodi(methanephosphonic acid) ligands bonded to $[Cu(NH_3)_2]^{2+}$ with aqueous ammonia as the mobile phase is claimed to lead to separation of the enantiomers (Szczepaniak and Ciszewska, 1982). Also, racemic lactic acid is claimed to be resolved into its enantiomers by thin-layer chromatography (TLC) on silica gel plates impregnated with copper(II) acetate. The mobile phase is aqueous dioxane (Cecchi and Malaspina, 1991). These reports are in contrast to those described above in which the samples chromatographed are enantiomerically impure *and* nonracemic. These two reports are almost certainly spurious.

o. Mass Spectrometry

The preferential vaporization of a single solid enantiomer relative to its solid racemate discovered by mass spectrometry (MS) was one of the first bits of evidence of potential enantiomer enrichment based on sublimation (Section 6-4.g). The separation was revealed in differences in the relative abundance of the two molecular ions produced in the electron impact (EI) mass spectrum of an enantiomer mixture as a function of time as compared with that of the isotopically labeled, and thus distinct, pure enantiomer (Zahorsky and Musso, 1973). The same effect has been observed in the chemical ionization (CI) mass spectrum (presumably with CH_5^+ ions) of mixtures of unequal amounts of dimethyl-$d_6(2S,3S)$-tartrate (M_{d_6}) and dimethyl-d_0 $(2R,3R)$-tartrate (M_{d_0}) in the MH^+ peaks, where $m/z = 185$ and 179, respectively (Fales and Wright, 1977).

An additional and very interesting finding in the work of Fales and Wright, made on inspection of the protonated dimer $(2M + H)^+$ peaks, is that the combined abundance of $(2M_{d_6} + H)^+$ and $(2M_{d_0} + H)^+$ homodimers exceeds that of the heterodimer $(M_{d_6} + M_{d_0} + H)^+$. Instead of finding equal intensities for the sum of two homodimer $(2M + H)^+$ peaks and for the heterodimer peak [as expected for a 50:50 mixture of the quasi-enantiomeric ions, and as observed in the case of a 1:1 d_0-S + d_6-S mixture (1:2:1 triplet), thus ruling out isotope effects in the ionization], the central peak of the triplet ($m/z = 363$ due to the heterodimer) has an intensity only 78% of that calculated. The heterochiral $(2M + H)^+$ ion is seen to be destabilized relative to the homochiral ions. The same result is found for an analogous 1:1 mixture of differentially labeled diisopropyl tartrate enantiomers; in this case the central peak of the corresponding triplet ($m/z = 483$) has only 48% of the calculated intensity (see also Winkler and Splitter, 1994). Incidentally, these results make it quite clear that enantiomer discrimination may be observed in the gas phase.

The relative stability of the homochiral versus the heterochiral protonated dimer ions of diisopropyl tartrate has been measured by CI mass spectrometry (with $C_4H_9^+$ ions) on a mixture of d_0-S + d_{14}-R isotopically labeled quasi-enantiomers. The ratio of virtual equilibrium constants K_{SS}/K_{SR} ($= K_{RR}/K_{SR}$) ≈ 1.6 corresponding to $-\Delta\Delta G \approx 0.29$ kcal mol^{-1}(1.2 kJ mol^{-1}), a value similar to those found for enantiomer discrimination of strongly hydrogen-bonded compounds in solution (Winkler, Stahl, and Maquin, 1986; see also Baldwin, Winkler, et al., 1988). A similar result has been obtained by Fourier transform ion cyclotron resonance mass spectrometry (Nikolaev et al., 1988).

p. Interaction with Other Chiral Substances

Sections 6-4.a–o dealt mostly with properties resulting from enantiomer discrimination. Much better known, however, are properties that reflect diastereomer discrimination, that is, those involving reversible interaction of a chiral substance, whether racemic or enantiomerically enriched, with a second chiral substance in which one enantiomer predominates (Section 6-2). Such interactions are responsible for the NMR anisochronies and chromatographic separations that permit the determination of enantiomer composition without conversion of enantiomer mixtures into diastereomer mixtures. These applications of diastereomer discrimination are described in detail in Section 6-5. Diastereomer discrimination in biological systems is treated in Section 6-4.q.

Diastereomer discrimination in the solid state is manifested by significant differences in heats of fusion, in melting points, in heats of solution, and in solubilities of diastereomers; for example, $\Delta\Delta H_{fus}$ of the diastereomeric α-methylbenzylammonium and ephedrinium mandelates R,R or S,S versus R,S exceed $5\,kcal\,mol^{-1}$ ($\geq 20\,kJ\,mol^{-1}$) (Zingg, Arnett, et al., 1988; for other examples, see van der Haest, Wynberg, et al., 1990; Valente et al., 1992). These differences between crystalline diastereomers form the basis for their separation in the classical Pasteurian resolutions, as well as for resolution by inclusion compound formation (Section 7-3.c).

Diastereomer discrimination in the liquid state is several orders of magnitude larger than liquid state enantiomer discrimination (Mason, 1982, p. 227). Titration (neutralization) of either α-methylbenzylamine (b) enantiomer with each mandelic acid (a) enantiomer in water produced an equal amount of heat (ΔH_{neut}). However, in dioxane or in dimethyl sulfoxide (DMSO), thermometric titrations reveal differences of the order of $0.25\,kcal\,mol^{-1}$ ($1\,kJ\,mol^{-1}$) between the two nonenantiomeric pairs of reactants (R)-a, (R)-b and (R)-a, (S)-b; application of the absolute method of chiral cross-checks between enantiomeric ion-pairs (S)-a, (S)-b and (S)-a, (R)-b assured the validity of the data (cf. Section 6-2). Similar results have been obtained in the reactions of the mandelic acid enantiomers with the enantiomers of ephedrine and of pseudoephedrine (Fig. 6.27), respectively. Diastereomer discrimination is observed also in the measurement of enthalpies of solution and of dissociation [average values of the several processes measured in solution for the three above-mentioned sets of salts lie between 0.20 and $0.70\,kcal\,mol^{-1}$ (0.84–$2.9\,kJ\,mol^{-1}$)] and in NMR spectra (in chemical shifts as well as in vicinal spin–spin coupling constants) of the salts (Arnett and Zingg, 1981; Zingg, Arnett, et al., 1988). Detailed interpretation of the results in terms of the nature of ion aggregation awaits further study.

The measurement of equilibrium constants of analogous acid–base reactions (by analysis of IR spectra in CCl_4) clearly reveals thermodynamic diastereomer discrimination. Substituent effects in the acid–base reactions of *para*-substituted (S)-α-methylbenzylamines plus (S)- or (R)-indane-1-carboxylic acids (no chiral cross-checks performed) are said to be inductive and conjugative in nature rather than steric (Takenaka and Koden, 1978). The results of conductimetric titration of amino acid enantiomers, for example, of L-glutamic (Glu) acid with L-histidine

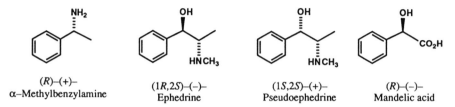

| (R)–$(+)$– | $(1R,2S)$–$(-)$– | $(1S,2S)$–$(+)$– | (R)–$(-)$– |
| α–Methylbenzylamine | Ephedrine | Pseudoephedrine | Mandelic acid |

Figure 6.27. Compounds exhibiting diastereomer discrimination in acid-base reactions (in the liquid state).

(His), when compared with the corresponding titration of L-Glu with D-His (all taken as hydrochlorides; no chiral cross-checks performed), suggests stronger interaction (ion pair formation) between the homochiral ions than between the heterochiral ions (Akabue and Hemmes, 1982).

Diastereomer discrimination may also be revealed through differences in solubility of racemate components in chiral solvents. Jacques et al. (1981a, p. 245) concluded that measurements reported up to the 1930s reveal no solubility differences between the enantiomers of stable organic compounds in optically active solvents. Differences have been clearly observed in the case of ionic organometallic complexes, however. But the interpretation of such solubility differences remains tenuous. Moreover, the operation of kinetic phenomena in these cases of heterogeneous equilibrium cannot be completely ruled out (Jacques et al., 1981a, p. 245; Mason, 1982, p. 210).

Differences in solubility between the enantiomers of chiral organic compounds do nevertheless exist, as is readily demonstrated by the chromatographic separation of enantiomers on chiral stationary phases. Such differences in retention behavior of the enantiomers may be ascribed to differences in their solubilities in chiral stationary phases. Use of moderately high temperatures (as in GC) is no impediment to such separation. Even modest diastereomer discrimination is subject to amplification in efficient chromatographic columns as the partition process is repeated numerous times (see Section 6-5.d). Diastereomer discrimination also operates in monolayers (Stewart and Arnett, 1982, p. 249).

Evidence for diastereomer discrimination in the gas phase is provided by chemical ionization MS of amino acids enantiomers (e.g., phenylalanine and methione) and of mandelic acid, when the ionizing gas (gaseous $CH_4 + H_2O$) is modified by addition of $(-)$-2-methyl-1-butanol. Relative abundances of ions produced in ion–molecule reactions of each enantiomer within a given pair with the same ionizing reagent differ with the R^+/S^+ ratio being as high as 13 in the case of $m/z = 311$ for mandelic acid $[C_6H_5CH(OH)COOH \cdot C_5H_{11} \cdot C_5H_{11}OH]^+$ ion (Suming, Yaozu, et al., 1986; cf. Section 6-4.o). No chiral cross-checks were performed; it has been reported that the effect reported by Suming, Yaozu, et al. (1986) could not be reproduced (Baldwin, Winkler, et al., 1988).

Yet another manifestation of diastereomer discrimination is the difference in solid–gas reaction rate that is observed when mixtures of crystalline chiral carboxylic acid enantiomers (mandelic, tartaric, and 2,2-diphenylcyclopropane-1-carboxylic acids) are exposed to $(+)$- or $(-)$-α-methylbenzylamine vapor. Based on the observed rate at which the crystals become opaque, dextrorotatory and levorotatory acid crystals clearly react at different rates with one of the enantiomeric gaseous amines (Lin, Curtin, and Paul, 1974; Paul and Curtin, 1987).

Diastereomer discrimination in solution also is revealed by the differential shielding or deshielding of magnetic nuclei from large external magnetic fields when such nuclei (e.g., ^1H, ^{13}C, and ^{31}P) are contained in chiral molecules that are dissolved in optically active solvents; these solvents are specifically chosen for their ability to strongly interact intermolecularly. This type of NMR experiment is described in Section 6-5.c in connection with the determination of enantiomer purity. Information about the configuration of chiral solute molecules may also be obtained from such experiments (see Chapter 5).

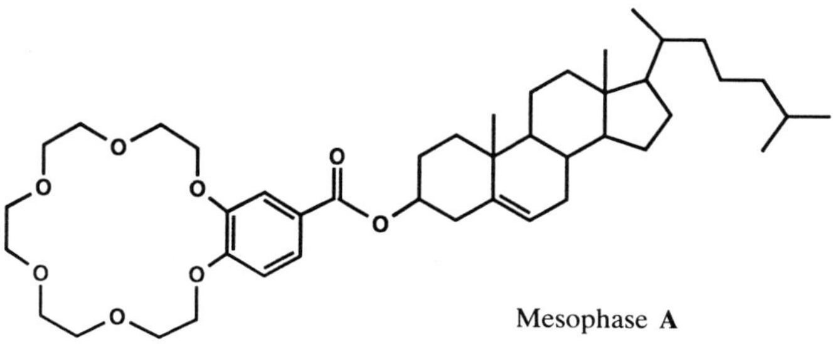

Mesophase **A**

Figure 6.28. Mesophase **A**.

The application of diastereomer discrimination has made possible the visual distinction between the enantiomers of chiral compounds (Review: Vögtle and Knops, 1991). The idea was to combine a chiral crown ether (host) capable of exhibiting enantioselective complexation (Section 7-7) with a chromophore so as to elicit a change in color that would differ according to the absolute configuration of the chiral guest. Success was achieved by linking an achiral monobenzo-18-crown-6 to cholesterol (Fig. 6.28) thereby endowing the crown ether simultaneously with chirality and with liquid crystal properties (cf. Section 13-4.e).

Addition of alkali metal mandelates to the cholesteric mesophase shown in Figure 6.28, leads to enantioselective complexation with an attendant change in the helical pitch of the liquid crystals. In the case of potassium mandelate, the difference in wavelength of maximum reflection for incident light, $\Delta\lambda_{Refl}$ [$\lambda_{Refl} = nP$, where n is the mean index of reflection and P is the helical pitch of the cholesteric mesophase (Section 13-4.e)], amounts to 61 nm (for a given ratio of potassium mandelate to mesophase **A**, in $CHCl_3$), which is detectable as a change in color, blue in the case of the S enantiomer and green in the case of the R enantiomer (Shinkai et al., 1991). The principle underlying this remarkable result is that the organization of the liquid crystal system (Section 13-4.e) serves to amplify the small diastereomer discrimination that attends the enantioselective complexation of the guest by the chiral host and translates the discrimination into one detectable with the naked eye (for another example, see Nishi, Shinkai, et al., 1991).

A visual distinction (color difference) has also been observed in the case of diastereomeric salts (ion pairs): One diastereomer of the brucine salt of phthaloyl-*threo*-β-hydroxyleucine is yellow while the other is white (Kuwata et al., 1989).

Diastereomer discrimination may also be responsible for stereoselectivity in chemical reactions. Consider, for example, the chlorination (at $-60°C$) of 2,2-diphenylaziridine with *tert*-butyl hypochlorite. When the reaction is carried out in a chiral solvent [CH_2Cl_2 containing (S)-$(+)$-2,2,2-trifluoro-1-(6-anthryl)-ethanol], optically active 1-chloro-2,2-diphenylaziridine (optical purity >29%) is produced (Bruckner et al., 1982; Eq. 6.9; see also p. 57):

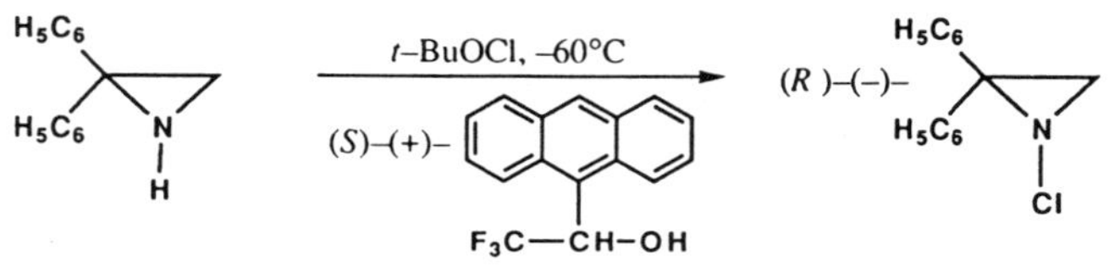

(6.9)

The chiral solvating agent responsible for the asymmetric induction observed is apparently recovered unchanged. Other chiral solvating agents and other reaction types, for example, oxidation, exhibit similar behavior (Forni, Moretti, et al., 1981; Bucciarelli, Torre, et al., 1980. For a review of the earlier literature, see Morrison and Mosher, 1976, Chapter 10). Enantioselectivity is also exhibited in reactions taking place in the solid state, that is, in crystal lattices of enantiopure chiral hosts (Review: Toda, 1991).

q. Biological Properties

Diastereomer discrimination is especially striking in biological systems (biological recognition or biodiscrimination) where it is responsible for the differences in taste, in odor, and in other physiological responses to the individual enantiomers of a given substrate, and to racemates as compared to their corresponding pure enantiomers (Craig and Mellor, 1976; for a brief historical account, see Holmstedt, 1990).

To the extent that these stereoisomers interact with chiral receptors, biodiscrimination is diastereomer discrimination. One of the first reports of biodiscrimination is that of Piutti (1886) who reported the isolation of dextrorotatory asparagine, $HO_2CCH(NH_2)CH_2CONH_2$, as having a sweet taste, whereas the naturally occurring levorotatory asparagine is tasteless. In his presentation of Piutti's work before the Académie des Sciences de Paris, Pasteur (1886) interpreted this difference in taste as arising from differential interaction of the two enantiomers with dissymmetric nerve tissue (matière nerveuse).

Such differential physiological response was unknown up to that point (Piutti states, "les isomères physiques connus sont doués de la même saveur"). Piutti considered the possibility that the two asparagines, though endowed with identical chemical properties and having identical optical rotation save the sign thereof, were constitutional isomers (de véritables isomères chimiques) having structural formulas **a** and **b**:

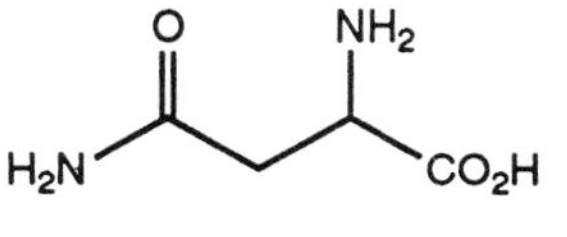

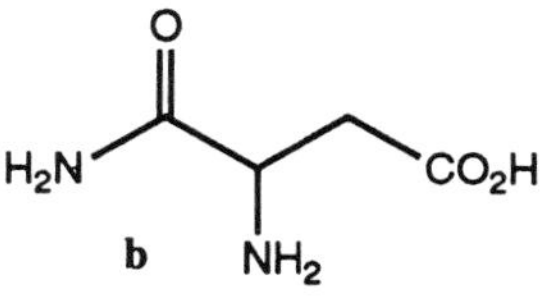

Piutti eliminated this possibility by showing that all derivatives of the two (enantiomeric) asparagines possess identical chemical properties and display equal rotations but opposite signs.

Stereochemical differences affecting the human senses are quite common, as in the case of many amino acids (Greenstein and Winitz, 1961, p. 150; Solms et al., 1965), but are by no means universal. For example, it has been reported that for some of the monosaccharides, both enantiomers have virtually the same sweetness (Schallenberger et al., 1969). In contrast, of the four stereoisomers of N-aspartylphenylalanine methyl ester, it is the L,L isomer (Fig. 6.29, **34a**) that is marketed as a synthetic sweetening agent (under the name aspartame; it is more than 100 times as sweet as sucrose); the L,D diastereomer, for example, is bitter (Mazur et al., 1969).

> These findings must be tempered by the understanding that sensory physiological response to the same stereoisomer may vary among different individuals (Bentley, 1969, p. 284; see also Theimer and McDaniel, 1971).

Stereoisomer discrimination in odor perception is also well recognized (Bentley, 1969, p. 286; Ohloff, 1986; Holmstedt, 1990). It has become evident that chirality plays a role in the olfactory properties of perfumes and fragrances; cases are known in which the two enantiomers of a pair possess significantly different olfactory properties (Ohloff et al., 1980; Ohloff, 1990). The experimental results on the carvone and limonene enantiomers (Fig. 6.29, **34b** and **35**, respectively) are particularly striking (Russell and Hills, 1971; Friedman and Miller, 1971; Leitereg et al., 1971): (S)-$(+)$-carvone possesses the odor perception of caraway while (R)-$(-)$-carvone has a spearmint odor; (R)-$(+)$-limonene has an orange odor while its enantiomer has that of lemons. Clearly, differences of this kind (stereoisomer discrimination) have commercial consequences. A specific example is that only the $(-)$-menthol enantiomer (Fig. 6.29, **36a**; cf. Section 7-5.b) exhibits the desirable cooling effect in tobacco smoke, as well as the lower threshold (concentration at which the effect is perceived; Emberger and Hopp, 1987); these facts have prompted the continued incorporation of $(-)$-menthol in cigarettes.

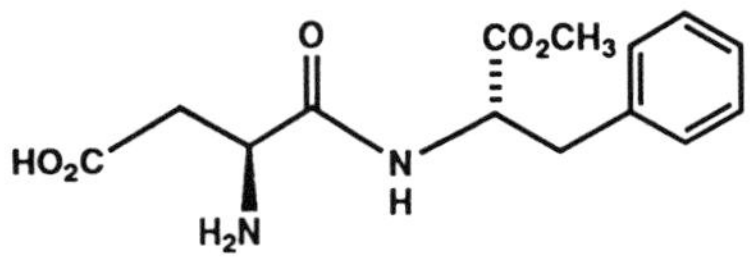

34a

L,L-(-)-Aspartame

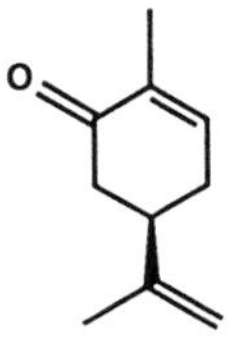

34b

(R)–$(-)$–Carvone

35

(R)–$(+)$–Limonene

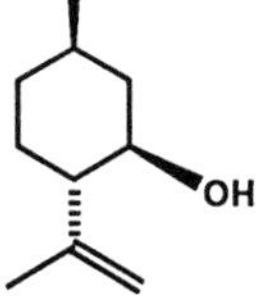

36a

(1R,3R,4S)–
(-)–Menthol

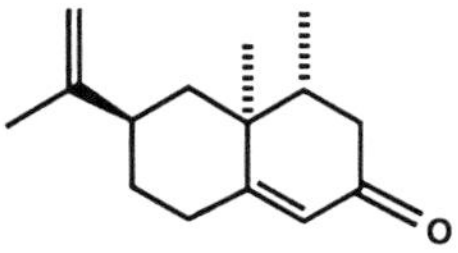

36b

(+)–Nootkatone

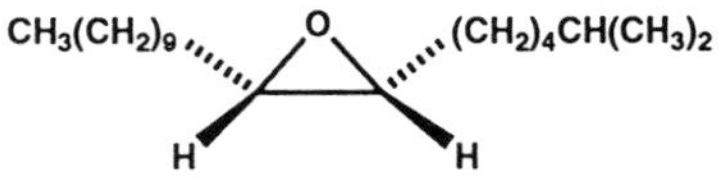

37

Disparlure

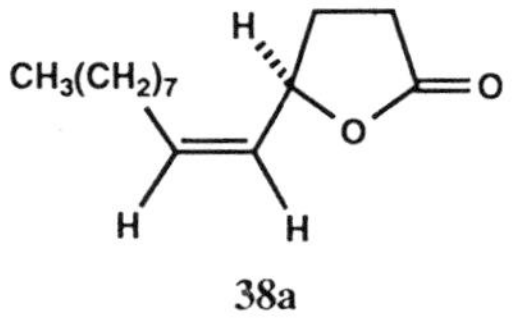

38a

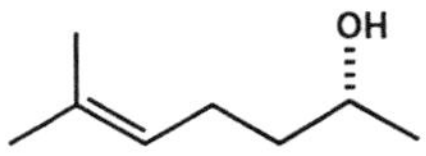

38b

(+)–Sulcatol

Figure 6.29. Stereoisomers exhibiting taste or odor discrimination.

Not only does odor quality show considerable differences between many, but not all, enantiomeric compounds but so does their potency. Thus, the odor threshold of (+)-nootkatone (Fig. 6.29, **36b**; 0.8 ppm), which is responsible for the aroma of grapefruit, is some 750 times lower than that of its enantiomer (600 ppm; Ohloff et al., 1980). There is also evidence that, with respect to some substances, anosmia (loss of the sense of smell) may be stereochemistry dependent (Theimer and McDaniel, 1971).

Numerous studies on insect pheromones revealed that "olfactory" communication among insects is subject to stereoisomer discrimination, for example, in disparlure (Fig. 6.29, **37**), the sex attractant of the gypsy moth (Beroza, 1970). As little as 1% of the "wrong" enantiomer of lactone **38a** (Fig. 6.29), the Japanese beetle pheromone, can significantly reduce the biological activity (Tumlinson et al., 1977); and, in the case of sulcatol (Fig. 6.29, **38b**), the aggregation pheromone of the ambrosia beetle (a timber pest), the racemate is more active than either enantiomer, that is, the response to the enantiomers is synergistic (Borden et al., 1976). In some cases, the "wrong" enantiomer can even be repellent or counterproductive (Silverstein, 1979, p. 133).

Naturally occurring chiral foodstuffs all have the "correct" stereochemistry relative to those of the enzymes that catalyze the conversion of polymeric nutrients to monomeric constituents of living cells and that burn up nutrients as fuel for energy production. The issue of stereoisomer discrimination does not arise in the use of naturally occurring foodstuffs. On the other hand, it has been found that while L-glucose is comparable in sweetness to the natural D-enantiomer (Schallenberger et al., 1969) microorganisms do not metabolize L-glucose (Brunton et al., 1967). Taking advantage of the inability of enzymes to metabolize L-sugars, the production of nonnutritive sweeteners (L-hexoses having sweetness comparable to D-hexoses or to sucrose) has been patented (Levin, 1981).

L-Amino acids, such as lysine, are food supplements (in cereals) both for human beings and for poultry and cattle. The quantities of amino acids required for this purpose are large enough that some must be produced synthetically. The matter of stereoisomer discrimination becomes relevant since only the L-enantiomer is usable; some of the strategies used in the economical synthesis of enantiomerically pure amino acids on a large scale are dealt with in Chapter 7. A large and specialized industry has evolved in Japan to provide for the production of amino acids for this purpose, for example, L-lysine ($>10^4$ tons per year) (Kaneko et al., 1974, Chapter 6).

D-Amino acids do occur in free and peptide-bound form in Nature (Corrigan, 1969). Cell walls and capsules of bacteria and fungi contain peptides constructed partially with "unnatural" D-amino acids, for example, the pathogenic anthrax bacillus capsule consists entirely of poly-D-glutamate (Zwartouw and Smith, 1956; Glwysen et al., 1968). The deleterious effect (virulence) of these microorganisms on human beings and on farm animals may stem from the inability of phagocytes to digest bacteria containing D-amino acids in their cell walls and capsules (Brubaker, 1985).

On the other hand, D-amino acid oxidase (found in human neutrophilic leucocytes), one of the few enzymes able to process D-amino acids, catalyzes the oxidation of D-amino acids derived from ingested bacteria in the presence of myeloperoxidase. The byproduct of the oxidation, H_2O_2, is what actually kills the invading bacteria (Cline and Lehrer, 1969). This byproduct is an instance of a specific defense mechanism that relies on diastereomer discrimination for its effect.

D-Amino acids also are found in oligopeptide antibiotics, such as Gramicidin S (Abraham, 1963). D-Amino acids resulting from the racemization of the naturally occurring L-enantiomers have been found, for example, in aged wine (Chaves das Neves et al., 1990) and in processed foodstuffs (Brückner and Hausch, 1990). The effect of the incorporation of D-amino acids in peptide hormones has been studied (Geiger and König, 1990).

More recently, an enzyme consisting of a 99 amino acid polypeptide chain, HIV-1 protease, synthesized from D-amino acids only, has been shown to cleave only D-amino acid peptides, whereas the analogous enzyme synthesized from L-amino acids cleaves only L-amino acid peptides (Milton, Milton, and Kent, 1992; Petsko, 1992). The two enantiomers of a chiral inhibitor exhibit similar specificity toward the corresponding enzyme forms. The D-form of the enzyme exhibited equal and opposite CD to that of the L-form consistent with the mirror-image folding of the two enantiomeric enzyme forms (cf. Section 13-4.g).

In recent years, numerous meetings, articles, and reviews have addressed the pharmacological implications of stereoisomerism, especially those of chirality (Ariëns et al., 1983; Wainer and Drayer, 1988; Borman, 1990; Stinson, 1992, 1993). Prior to the 1960s, differences in the biological activity of racemates versus single enantiomers were largely ignored. This in spite of the fact that attention to differences in pharmacological properties [e.g., between (±)- and (−)-atropine **39** and between (±)- and (−)-epinephrine (adrenalin) **40a** (Fig. 6.30)] was drawn more than 80 years ago by Cushny (Cushny, 1926; Smith and Caldwell, 1988; Holmstedt, 1990). Differences in potency between the enantiomers of sex hormones (e.g., equilenin, Fig. 6.30, **40b**) have been known since 1940 (Bachmann et al., 1940).

The increased interest has been fueled by several factors among which are the present availability of sensitive analytical methods permitting the monitoring of the enantiomer composition of chiral medicinal agents and their metabolites at therapeutic concentrations in physiological fluids, and the increased facility in synthesizing enantiomerically pure organic compounds.

The one significant incident that is alleged to have had a major impact on the revision of the U.S. Pure Food and Drug Act of 1906 (the 1962 Kefauver–Harris Drug Amendments) was the introduction in 1961 (in Europe) of racemic thalidomide (Fig. 6.30, **41**) as a sedative and antinausea agent for use especially during early pregnancy. Unfortunately, it was soon found that thalidomide is a very potent teratogen (causing fetal abnormalities); children born to the women who used thalidomide had a high incidence of deformities in their limbs (phocomelia). The teratogenicity was eventually traced to the (S)-(−)-thalidomide enantiomer (Blaschke et al., 1980), whereas the (R)-(+) enantiomer is claimed not to cause deformities in animals even in high doses. The tragedy is claimed to have been entirely avoidable had the physiological properties of the individual thalidomide enantiomers (and of the racemate) been tested prior to commercialization [Scott, 1989; for contrary views, see De Camp (1989) and Crossley (1992)]. In any event, it was not until 1988 that the Food and Drug Administration (FDA) explicitly required the submission of information about the

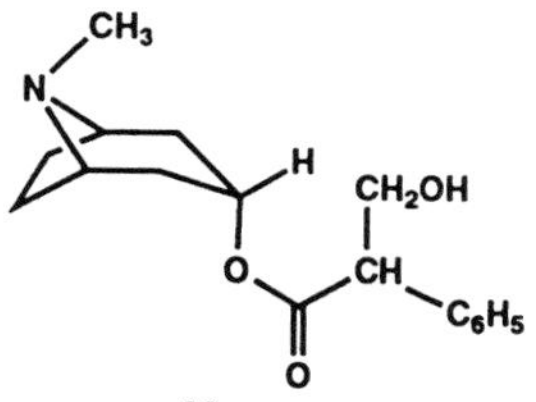

39

Atropine

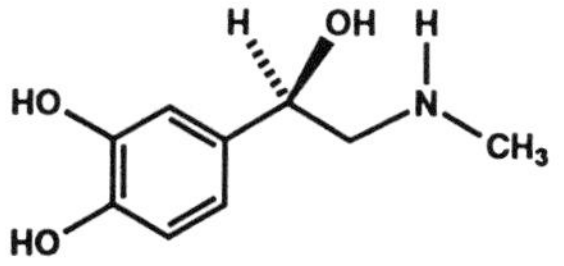

40a

(*R*)-(–)-Epinephrine

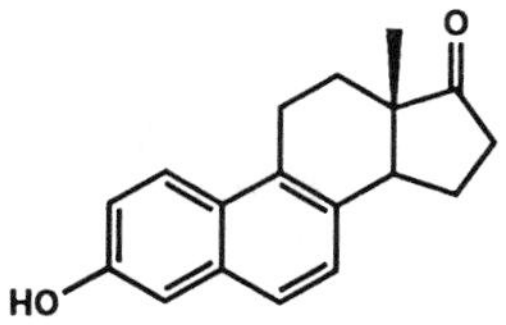

40b

(+)-Equilenin

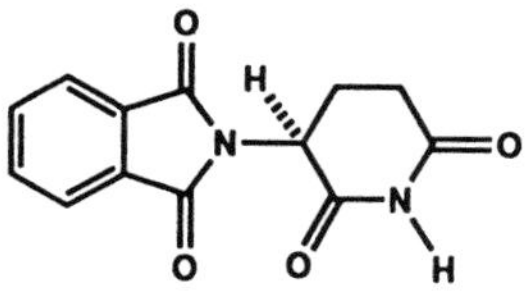

(*S*)-(–)-Thalidomide, **41**

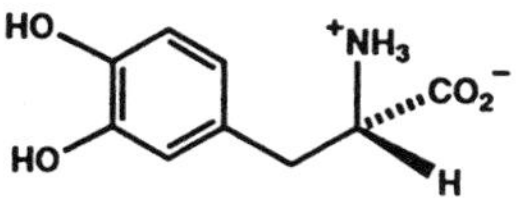

L-DOPA, **42**

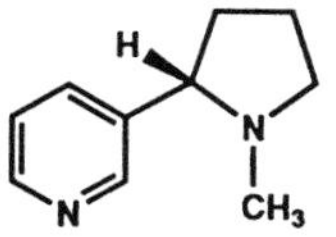

(*S*)-(–)-Nicotine, **43**

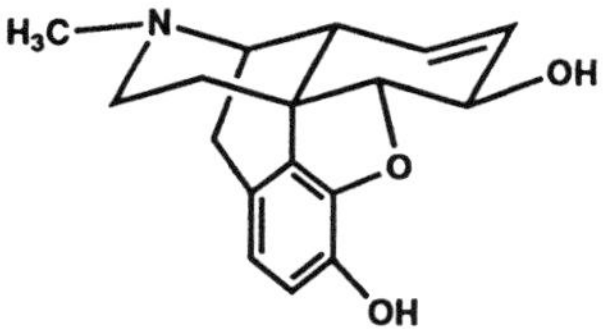

Morphine, **44**

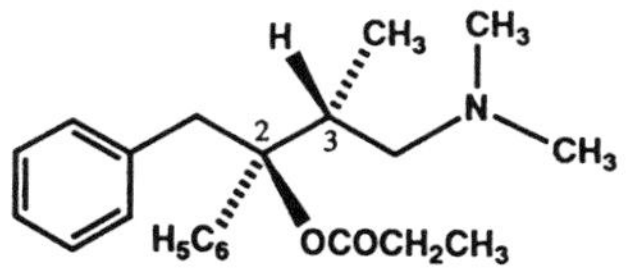

(2*S*,3*R*)-(+)-Propoxyphene, **45**

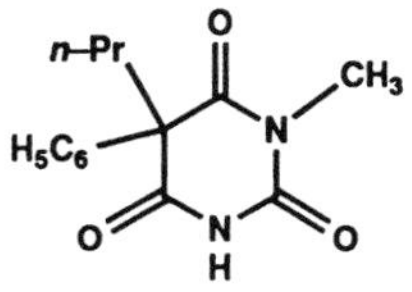

46

1-Methyl-5-phenyl-5-propylbarbituric acid

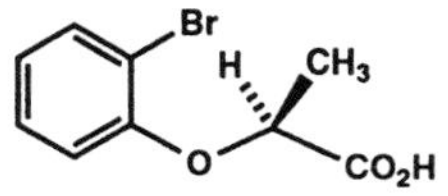

(*S*)-(+)-α–(2-Bromophenoxy)-propionic acid, **47**

Figure 6.30.   Biodiscriminating stereoisomers.

enantiomer composition of chiral substances in new drug applications (De Camp, 1989).

It is by now quite clear that diastereomer discrimination abounds in the domain of medicinal chemistry and pharmacology (for reviews of this subject, see Ariens et al., 1964; Sastry, 1973; Patil et al., 1975; Lehmann, Ariëns et al., 1976; Witiak and Inbasekaran, 1982; Ariëns, 1988; Wainer and Drayer, 1988; Holmstedt et al., 1990). A few additional examples must suffice here. Only (*S*)-(−)-3-(3,4-dihydroxyphenyl)-alanine (L-DOPA) (Fig. 6.30, **42**) is active chemotherapeutically in Parkinson's disease. The toxicity of naturally occurring (−)-nicotine

(Fig. 6.30, **43**) is much greater than that of the dextrorotatory enantiomer. (−)-Morphine (Fig. 6.30, **44**), and not the synthetic (+) enantiomer, is the one with the analgesic activity.

> The biologically more active isomer of a stereoisomeric pair has been named *eutomer*; the corresponding less potent or inactive isomer is then called the *distomer* (Lehmann, Rodrigues de Miranda, and Ariëns, 1976). The ratio of the activities (eutomer:distomer), the so-called eudismic ratio, is a measure of the degree of stereoselectivity of the biological activity. Eudismic ratios greater than 100 are not uncommon among chiral medicinal agents (Ariëns, 1986).

It must not be thought that enantiomers behave only in the manner of one having biological activity and the other serving only as "stereochemical ballast" when a drug is marketed and consumed as a racemate (Ariëns, 1986). In the case of drugs, the distomer can be a eutomer with respect to a different type of activity. Propoxyphene (Fig. 6.30, **45**), a synthetic compound structurally related to morphine and to methadone, exemplifies this phenomenon. The (+) enantiomer is an analgesic while the (−) enantiomer is an antitussive. Both enantiomers are marketed (individually). Barbiturate enantiomers, such as those of 1-methyl-5-phenyl-5-propylbarbituric acid (Fig. 6.30, **46**), exhibit opposite effects on the central nervous system (CNS). One enantiomer is a (useful) sedative (hypnotic or narcotic, according to the dosage); the other is an (undesirable) convulsant (Knabe et al., 1978).

An interesting feature of some stereoisomeric pairs of biologically active compounds is the possibility that their activities will cancel one another (Ariëns et al., 1964). An example is α-(2-bromophenoxy)propionic acid, whose (*S*)-(+) enantiomer (Fig. 6.30, **47**) is a plant growth stimulant (auxin) while the (−) enantiomer is an antagonist thereof (anti-auxin; Draber and Stetter, 1979). On the other hand, in some cases, both enantiomers must be present for physiological activity to be manifest (Barfknecht and Nichols, 1972).

With respect to chiral compounds in general, four different types of behavior may obtain: (a) the desired biological activity resides entirely in one of the enantiomers, whereas the other is essentially without effect; (b) the enantiomers have identical (or nearly identical) qualitative and quantitative pharmacological activity; (c) the activity is qualitatively identical but quantitatively different between the enantiomers; and (d) the activities of the enantiomers are qualitatively different (Powell, Ambre, and Ruo, 1988). For additional examples and summaries of the types of relationship possible between biological activity and the two enantiomers of a given medicinal agent, see Patil et al. (1974); Ariëns, (1986). In spite of the high degree of biodiscrimination exhibited by stereoisomers, many chiral synthetic medicinal agents are still marketed as racemates (Ariëns, 1988). On the other hand, the possibility of patenting an enantiopure drug even when the corresponding racemate is already patented is rapidly changing this situation (Stinson, 1993). One commercial advantage is that patent protection on the drug is indirectly extended.

Biodiscrimination occurs when a chiral compound having a "messenger" function binds to a specific site (or sites) in a receptor molecule thus being activated so as to elicit a response. In the case of chiral compounds, discrimination has tended to be explained in terms of three-point interaction models. These models [pioneered by Easson and Stedman, 1933; see also Bergmann, 1934; Ogston, 1948; Dalgliesh, 1952; and Chapter 8] stipulate that binding eliciting optimal recognition of the stereochemistry of the reagent (the eutomer) requires interaction at three complementary sites in the messenger and in the receptor molecules (Fig. 8.8). A stereoisomer of the messenger (the distomer) might well bind with the eutomer receptor at two or at only one site with the result that the response relative to that of the eutomer is decreased or annihilated. On the other hand, the distomer may bind very well (better than the cited eutomer) with a different receptor with the consequence that the eutomer–distomer roles are reversed at the latter receptor (Ariëns, 1988). The greater the affinity of a biological agent toward its receptor the greater its stereoselectivity. This common relationship is known as Pfeiffer's rule (Pfeiffer, 1956).

An interesting aspect of biodiscrimination is the possibility that a given bioactive enantiomer (eutomer) may racemize in solution. An example is (S)-$(-)$-hyoscyamine (an anticholinergic agent; eudismic ratio $(-)/(+) = 200$). In view of its rapid racemization, it is administered as the racemate (atropine, Fig. 6.30, **39**; Ariëns, 1988). The antiinflammatory agent ibuprofen (Fig. 6.31, **48**), is also administered as the racemate because the distomer [the (R)-$(-)$ enantiomer, as measured in vitro] undergoes inversion to the eutomer [(S)-$(+)$ enantiomer] in vivo (Ariëns, 1988; Kumkumian, 1988).

So as not to mislead the reader, we should state that the above treatment has been somewhat oversimplified. Differences in the physiological activity of stereoisomers (of diastereomers as well as enantiomers) do not rest solely on the chirality of receptor molecules and associated diastereomer interactions (pharmacodynamics). Additional (pharmacokinetic) factors responsible for these differences include: rate differences in the transport across membranes and cell walls; rate differences between stereoisomers binding to plasma proteins; and rate differences in the metabolism and clearance (elimination from the body) of stereoisomers (Drayer,

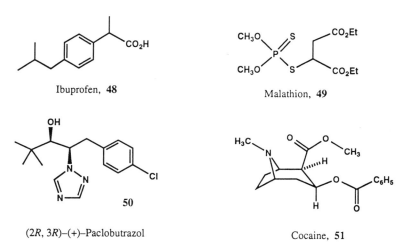

Ibuprofen, **48**

Malathion, **49**

50

(2R, 3R)–(+)–Paclobutrazol

Cocaine, **51**

Figure 6.31. Structures **48–51**.

1988; Jamali et al., 1989 and Crossley, 1992). Thus, although a medicinal agent may be administered as a racemate, when it reaches the target site, the agent may no longer be racemic, it may be enriched with respect to one of its enantiomers. Which stereochemical discrimination factor is dominant in any given case may not yet be known; in due course, one hopes that the basis of many of these stereospecific interactions will be understood.

Chiral herbicides, pesticides, and plant growth regulators, chemicals widely used in agriculture, are also subject to biodiscrimination. Examples are the insecticide malathion (Fig. 6.31, **49**) which is applied as a racemate and paclobutrazol (Fig. 6.31, **50**). In the case of paclobutrazol, both enantiomers are biologically active, as fungicides active against cereal mildew and rust and as plant growth regulators (e.g., in apple seedlings). However, the $(2S,3S)$-$(−)$ enantiomer has the greater activity as a growth regulator, whereas the $(2R,3R)$-$(+)$ enantiomer is the eutomer with respect to the fungicidal activity (Ariëns, 1988). In contrast, of the eight possible stereoisomers of the insecticide deltamethrin (Fig. 7.72), the only two having insecticidal activity are the $(1R,3R,\alpha S)$- and $(1R,3S,\alpha S)$-diastereomers (see Section 7-5.a).

Biodiscrimination is, of course, not limited to compounds having carbon stereocenters. The four major nerve gases, including sarin, tabun, and soman, which are potential chemical warfare agents, all are endowed with a stereogenic phosphorus atom having the general structure $R(R'O)P(=O)X$. Although the four have been produced as racemates, it has been demonstrated that the enantiomers (diastereomers in the case of soman) differ in their anticholinesterase activity and lethality (Review: Benschop and De Jong, 1988).

Stereoisomer discrimination even has significant legal implications. By way of example, forensic chemists employed by law enforcement agencies have had to defend in court analytical work leading to the identification of controlled substances, such as cocaine (Fig. 6.31, **51**). Lawyers for defendants have argued that the identification of the CNS active substance was not carried out "beyond a reasonable doubt" when the analysis, for example, by TLC or by nonchiroptical spectroscopic techniques, could not be expected to differentiate between $(−)$-cocaine (the naturally occurring and psychoactive isomer) and $(+)$-cocaine (presumed to be inactive and apparently not a controlled substance) (People vs. Aston, 1984).

> Parenthetically, the report of the legal proceedings refers to "an isomeric form of cocaine, *dextro*-cocaine, also known as pseudococaine . . . and which could not be distinguished from the controlled form of cocaine, *levo*-cocaine, the naturally occurring form . . . ." Pseudococaine is actually a *diastereomer* of cocaine! (Allen et al., 1981).

This problem may have arisen because many forensic laboratories are not equipped with polarimeters permitting the differentiation of the two enantiomers or, more likely, because insufficient sample was available for polarimetric analysis. Microcrystalline and IR tests are now available involving, for example, the conversion of the cocaine enantiomers to the easily distinguishable diastereomeric O,O'-ditoluoyltartrate salts that obviate the need to carry out polarimetric analysis on small samples of confiscated cocaine (Ruybal, 1982; Sorgen, 1983).

r. Origins of Enantiomeric Homogeneity in Nature

While nonracemic and enantiomerically pure compounds found in Nature are understood to arise mainly from chemical reactions catalyzed by enantioselective catalysts (i.e., enzymes), the original source of the latter and of their components, enantiomerically pure amino acids, eludes us. This mystery is a fascinating aspect of science that lends itself to much speculation and indirect experimentation; yet by its very nature the experimentation has led to few definitive conclusions. Part of the fascination stems from the question of the origin of life on earth that is invariably linked to that of the origin of enantiomerically pure compounds.

A recent and very thorough review of this subject is that by Bonner (1988; see also Bonner, 1972; Mason, 1984, 1989). Bonner classifies all theories bearing on the origin of chiral homogeneity as being *biotic* or *abiotic*. Biotic theories presuppose that life originated at an advanced stage of chemical evolution in the presence of numerous racemic building blocks. These theories are consistent with the notion that competing life forms gradually *selected* one enantiomer (the L-amino acids and D-sugars) as being more efficient to survival than their enantiomers. A corollary of such theories is that prebiotic enantiomeric homogeneity is not a *prerequisite* for the origin of life. Bonner classifies biotic theories as speculative and probably impossible to verify experimentally.

Abiotic theories, on the contrary, presuppose that life originated *after* the initial establishment of enantiomeric excess, that is, that the molecules that characterize life processes (e.g., RNA, proteins, and/or DNA) could not have originated or evolved in the absence of the predominance (albeit small) of one of the enantiomeric forms of the requisite precursor molecules. The "abiotic" establishment of an enantiomerically enriched chemical environment could have taken place by chance or in a determinate way. Mechanisms conforming to the chance hypothesis (in any process, the probability of forming either enantiomer of a pair is equal) include the spontaneous crystallization of conglomerates (Section 7-2) and the occurrence of asymmetric transformations, that is, the spontaneous crystallization of conglomerates of easily racemized compounds (Section 7-2.e). Other chance mechanisms include the occurrence of chemical reactions in chiral crystals under lattice control, in cholesteric (liquid crystal) phases, and adsorption or catalysis on chiral solid supports, such as quartz crystals of a given configuration. Although the operation of chance mechanisms is quite plausible, Bonner (1988) concluded on the basis of statistical considerations that the random estabishment of chiral homogeneity at several sites on earth would not likely have led to the dominance of one of these over time.

Determinate mechanisms assume that some chiral physical force acting on (or during formation of) racemates is responsible for establishment of an initial enantiomer excess. The latter is subsequently amplified (e.g., by polymerization mechanisms; see below) until enantiomeric homogeneity obtains and the resulting enantiopure compounds are converted to those molecules that we recognize as being essential to life processes. Experimental efforts aimed at the verification of determinate mechanisms have been the subject of numerous and fruitful studies over the past three decades. The principal mechanisms studied involve the violation of parity and the interaction of matter with chiral radiation.

The notion of universal dissymmetric forces that pervade the world and are responsible for optically active natural products stems from the 1850s and was pioneered by Pasteur (Mason, 1986). Many wrong (i.e., theoretically impossible) forces have been suggested based on dubious experiments (see page 211).

In 1956 Lee and Yang (q.v.) proposed that parity is not conserved in the weak interactions (e.g., in β decay). Experimental verification of their proposal was provided shortly thereafter; Wu et al. (1957) demonstrated that electrons emitted during the decay of ^{60}Co nuclei were longitudinally polarized in a left-handed way to a greater extent than in a right-handed way. These results constituted the first experimental evidence that the parity principle (that the laws of Nature are invariant under spatial reflection) is not conserved for weak interactions. Since the parity principle also requires that elementary particles be present in mirror-image forms (Ulbricht, 1959), the violation of parity bespeaks an imbalance between the quantities of matter and antimatter (Mason, 1982, p. 240; Mislow and Bickart, 1976/77).

A mechanism for linking the asymmetry of β decay to the chirality of biomolecules was proposed by Vester and Ulbricht (V–U hypothesis) (Ulbricht, 1959; Ulbricht and Vester, 1962). During the passage of the longitudinally polarized β-decay electrons through matter, the electrons slow down while emitting γ rays (so-called bremsstrahlung). The γ radiation has been shown to be circularly polarized. It was proposed that, on absorption of the bremsstrahlung by organic compounds, stereoselective photochemical processes might take place that would lead to accumulation of enantiomerically enriched compounds (Ulbricht and Vester, 1962).

Positive results in accord with the V–U hypothesis were reported in 1968 (Garay, 1968). The latter reported that separate β irradiations of D- and L-tyrosine solutions (by ^{90}SrCl$_2$) led, over an 18-month period, to greater decomposition of the D than the L enantiomer. This finding was based solely on reduction in the intensity of UV absorption by the amino acid samples, that is, not a method directly probing the enantiomer composition of the samples. A very thorough reinvestigation of this type of asymmetric destruction using more sensitive analytical methods (GC analysis on enantioselective columns), including chiroptical methods, failed to substantiate the validity of the V–U hypothesis (Bonner, 1974; Bonner and Liang, 1984). It has also been pointed out that the occurrence of radioracemization of amino acids (Section 7-8.b) might tend to reduce the efficacy of the V–U mechanism if the relative rates of enantioselective radiolysis and of racemization were similar or skewed in the direction of faster racemization. As of 1988, essentially all of the numerous tests of the V-U hypothesis led to negative or nonreproducible results (Bonner, 1988).

Some tantalizing results have also been obtained when spin-polarized electrons, protons, muons, or positrons have been allowed to interact with racemates leading in some cases to apparent stereoselective degradation. Yet, most of the experiments of this type have eluded efforts to duplicate them (Bonner, 1988).

An additional consequence of the nonconservation of parity is the prediction that the two enantiomers of a chiral compound are of slightly unequal energy (Rein, 1974). This prediction arises from the parity-violating neutral current interaction between the two enantiomers of a compound. The parity nonconserving (pnc) (or violating) energy difference ΔE_{pnc} is approximated by Eq. 6.10

$$\Delta E_{pnc} \sim (\eta Z^5) \times 10^{-8}\,eV \qquad (6.10)$$

where η is a chirality factor estimated to be of the order of 10^{-2} and Z is the atomic number, for example, that of a stereogenic atom (Mason, 1982, p. 244). One estimate of the energy difference, for molecules containing heavy atoms, is $10^{-12}\,eV$ (Zel'dovich et al., 1977). In the case of a peptide containing chiral glycine conformations, the parity-violating stabilization (L > D peptide) either in the α-helix or the β-sheet conformations, is calculated to be ca. $-2.10^{-14}\,J\,mol^{-1}$ per amino acid residue, corresponding to $10^{-15}\%$ ee (Mason and Tranter, 1983). Such small theoretical enantiomeric excesses are far beyond the abilities of detection by present day experimental techniques. The evaluation of parity-violating energy differences as a possible source of chiral homogeneity in Nature by different authors has been reviewed by Bonner (1988); see also MacDermott and Tranter (1989).

While a detailed quantum mechanical explanation of the parity-violating neutral current interaction between enantiomers is beyond the scope of this book, an intuitively useful and equally fundamental appreciation of its origin may be gained by consideration of the concept of "true" and "false" chirality proposed by Barron (1986). Barron has given the following definition of these terms: "True chirality is possessed by systems that exist in two distinct enantiomeric states that are interconverted by space inversion but not by time reversal combined with any proper spatial rotation." This definition leads to the conclusion that the space-inverted enantiomers of a chiral compound are not *true* enantiomers. A pair of true enantiomers would be a chiral molecule having a given configuration and another having the identical constitution, the mirror-image configuration but composed of antimatter particles. The nondegeneracy of ordinary enantiomers (see above) is a consequence of their not being true enantiomers (see also Bonner, 1988).

Attempts have been made to produce enantiomeric excesses by the intervention of electric, magnetic, and gravitational fields on stereoselective syntheses; the results have been highly controversial (for a summary, see Bonner, 1988). In this connection, it has been pointed out that the possibility of effecting absolute asymmetric synthesis, that is, the generation of enantiomeric excess in the absence of nonracemic chiral catalysts or reagents (Eliel, 1962, p. 79), under the influence of the above-named fields, can be assessed on the basis of the true and false chirality criterion; it was concluded that only "a truly chiral influence can induce absolute asymmetric synthesis in a reaction mixture which is isotropic in the absence of the influence and which has been allowed to reach thermodynamic equilibrium. But for reactions under kinetic control, false chirality might suffice" (Barron, 1986).

In contrast to the above very small and highly controversial effects, absolute asymmetric synthesis under the influence of circularly polarized light (cpl, a "truly chiral" physical force) has been observed numerous times. The most successful experiments are of two types: photochemical asymmetric synthesis and asymmetric photolysis. The former type is exemplified by the cyclization of diarylethylenes

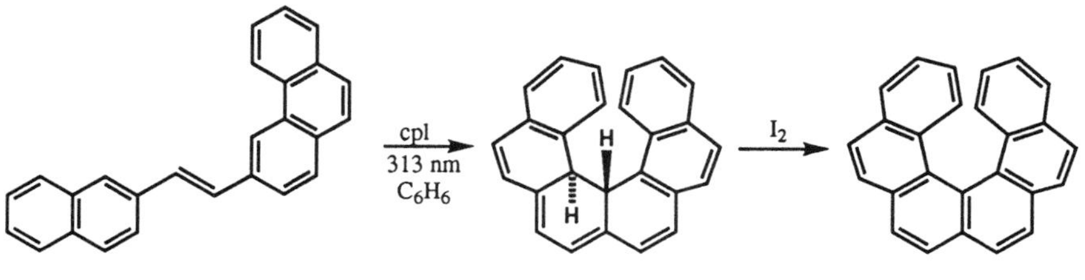

Figure 6.32a. Photochemical asymmetric synthesis of hexahelicene.

to helicenes (e.g., Fig. 6.32a). Optically active hexahelicene (Fig. 6.6, **6**), $[\alpha]_{436}^{23} - 30.0$ (CHCl$_3$) was obtained by irradiation of compound **52** with right circularly polarized light (rcpl), while **6**, having $[\alpha]_{436}^{23} + 30.5$ (CHCl$_3$) resulted on irradiation with left cpl (ee $<$ ca. 0.2%) (Moradpour, Kagan, et al., 1971). Similar results were obtained by Bernstein, Calvin, and Buchardt (1972; review: Kagan and Fiaud, 1978). In such syntheses, the enantiomeric purity of the product is independent of the conversion and is related to the differential absorbance of the cpl: $\Delta\varepsilon/2\varepsilon = g/2$ (for a definition of the γ number, see Section 13-4.a; Buchardt, 1974); see also Chapter 14.

The second of the mentioned processes, asymmetric photodestruction, is the more important one from the standpoint of its import to the abiotic generation of enantiomerically enriched compounds. Three reasons for the importance attached to asymmetric photodestruction are that (a) the process is widely applicable to virtually any type of racemic organic compound possessing a chromophore; (b) the enantiomeric excess attainable can be significantly higher and can approach 100% if the γ number (Section 13-4.a) is relatively high and the conversion is appreciable (cf. Section 7-6); and (c) a source of cpl is available on earth since cpl is known to be generated in the sky by reflection and scattering from aerosols (Bonner, 1988). The asymmetric photolysis of a biologically significant compound, DL-leucine, with rcpl (at 212.8 nm; the light was obtained from a laser source) generated samples of 2% ee (L-leucine $>$ D-leucine) at a conversion of 59% and of 2.5% ee (D $>$ L) on 75% conversion with lcpl without concurrent photoracemization (Flores, Bonner, and Massey, 1977).

In view of the negative outcome of virtually all of the other experiments proposed to validate determinate mechanisms for the origin of the observed enantiomeric bias on Earth, those involving the stereoselective photodestruction of racemates mediated by cpl remain among the most likely (Bonner, 1988); for an alternative assessment, see MacDermott and Tranter (1989). A crucial question on which this possibility depends is whether there is a necessary terrestrial predominance of one sense of cpl over the other. While there is no unambiguous answer to this question, experimental and theoretical assessments of the matter have not ruled out either spatial or temporal excesses of one sense of cpl on earth. According to Bonner (1988), "cpl-mediated reactions, particularly stereoselective photolyses, may provide the most likely determinate mechanisms for the origin of the enantiomeric bias in nature."

The operation of such determinate mechanisms provides for formation of small enantiomer excesses of chiral compounds. Amplification processes must be operative if the latter are to yield, ultimately, enantiomerically pure compounds.

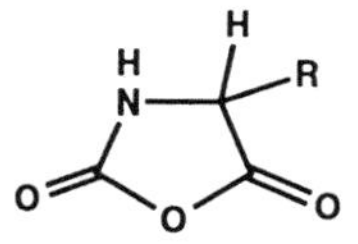

53

Figure 6.32b. Alanine N-carboxyanhydride ($R = CH_3$).

A number of such processes are described in Chapter 7, namely, total spontaneous resolution (Section 7-2.e) and amplification during incomplete reaction (Section 7-4). However, since these processes are not general ones, one could hardly claim that they are the likely ones to have led to the enantiomeric homogeneity that characterizes the important biomolecules found nowadays in living systems.

One set of amplification experiments that is quite pertinent to the origin of biomolecular chiral homogeneity is that which occurs during polymerization. Incomplete polymerization of enantioenriched alanine N-carboxyanhydride (Fig. 6.32b, **53**, $R = CH_3$; L > D) leads to preferential incorporation of the predominant enantiomer at the beginning of the polymerization (Matsuura et al., 1965). The selectivity has been attributed to the configuration of the amino acid at the growing end of the polymer chain as well as by a cooperative amplification resulting from formation of an α-helix conformation in accord with an earlier proposal by Wald (1957). The latter rationale is reminiscent of the formation of partially isotactic polymers by *stereoelective* polymerization of nonracemic, enantiomerically pure monomers (Farina, 1987b).

Analogous results were found for leucine N-carboxyanhydride (Fig. 6.32b, **53**, $R = i$-Bu) but not for valine (Fig. 6.32b, **53**, $R = i$-Pr), the latter being unable to form an α-helix due to steric hindrance (Akaike et al., 1975; Blair and Bonner, 1980). Subsequently, Blair, Dirbas, and Bonner (1981) demonstrated that enantiomeric enrichment also took place during partial hydrolysis of polypeptides, for example, with poly(DL-leucine) hydrolyzing faster than poly(L-Leu) or poly(D-Leu). Consequently, enantiomeric enrichment takes place during partial hydrolysis of poly(Leu) prepared from nonracemic leucine, e.g., 45.4% ee, with the recovered peptide containing leucine having as much as 54.9% ee.

Also pertinent is the synthesis of nonracemic poly(triphenylmethyl methacrylate), an isotactic linear polymer whose chiroptical properties are due solely to the generation of a preferred helicity sense (cf. Section 6-5.d). The latter arises during the polymerization of (achiral) triphenylmethyl methacrylate monomer by butyllithium in the presence of ($-$)-sparteine (monomer/initiator ratio $\geq 40:1$). It is evident that one chiral molecule in the anionic polymerization initiator (a molecule that is not even incorporated in the polymer) is responsible for the generation of a preferred helicity sense in a polymer molecule incorporating as many as 200 monomers (at least during the initial phases of the polymerization) (Yuki, Okamoto, and Okamoto, 1980).

Cooperative effects responsible for the amplification of enantiomeric bias is also seen during the copolymerization of achiral hexyl isocyanate with as little as 0.12% of nonracemic chiral isocyanate. The resulting polyisocyanate copolymer consists of a 56:44 mixture of mirror-image helices (at $-20°C$) (Green et al., 1989). The even more dramatic influence of minor chiral perturbations is seen in the polymerization of (R)-1-deuterio-1-hexyl isocyanate $(n\text{-}C_5H_{11}CHD\text{-}NCO$ giving rise to a helical polymer having a large optical rotation, $[\alpha]_D^{10} - 450$ $(CHCl_3)$. This rotation has been interpreted as arising from an excess of one of the helical senses and not from a structural or conformational perturbation. The amplification mechanism giving rise to the helical sense excess is due to a conformational equilibrium isotope effect in which the energy difference per deuterium, hence per polymer residue, is very small [ca. $1 \, cal \, mol^{-1}$ (ca. $4 \, J \, mol^{-1}$)] (Green et al., 1988; Green et al., 1991).

These model experiments suggest that the secondary structure of polymers is implicated in the amplification of enantiomeric bias during polymer synthesis and that it can protect the stereochemical integrity of the polymer once formed during subsequent partial degradation. The relevance of model experiments demonstrating amplification of low enantiomeric enrichment in polymers to the broader question of the origin of enantiomeric bias in Nature is evident because of the important role that chiral biopolymers (polysaccharides, polynucleotides, and proteins) play in the maintenance of the enantiomeric homogeneity in Nature (Bonner, 1988; see also Mason, 1986).

6-5. DETERMINATION OF ENANTIOMER AND DIASTEREOMER COMPOSITION

a. Introduction

Let us review the terms that are used to describe the enantiomer composition (i.e., the proportion of enantiomers) of chiral samples. Although the adjective *chiral* has often been used descriptively to imply that a sample is nonracemic (commonly optically active) (e.g., Nogradi, 1981, p. 116), it should be reserved to characterize, in a conceptual way, molecules (and by extension, compounds), crystals, or objects that are not superposable with their mirror images (cf. Section 1-2).

The enantiomer composition of macroscopic samples of chiral compounds requires its own descriptive adjectives. The term racemic (1:1 proportion of enantiomers) is time-honored, unequivocal, and universally accepted. Note that racemic and chiral are not incompatible adjectives: only a chiral substance can give rise to a racemate. A term is clearly needed to describe the composition of samples that until now have been called optically active. The latter expression suffers from being too closely linked to the measurement of chiroptical properties that are gradually being deemphasized for the purpose of enantiomer purity determination. Moreover, it is now much better understood than heretofore that enantiomerically enriched (even pure) samples need not be optically active (at a given concentration, temperature, wavelength, or in a given solvent; see Section 13-2). In this book, we have used the adjective nonracemic to describe samples of chiral compounds whose enantiomer composition is somewhere between 50:50 and 100:0 (Halevi, 1992). It is thus correct, albeit imprecise, to speak of a nonracemic sample even if it is enantiomerically pure.

We observe that some authors have taken to using the adjective "homochiral" to describe chirally homogeneous, that is, enantiomerically pure samples. However, "homochiral" has been previously defined to mean "of the same chirality sense," as in two right hands (Ruch, 1972). Homochiral thus does not denote chiral homogeneity. Its continued use in that sense will degrade the original definition and lead to confusion (Halevi, 1992). We recommend instead that the unequivocal expression "enantiomerically pure" (or the contraction enantiopure) be used to describe chirally homogeneous samples (see also Sections 1-2 and 6-3).

The enantiomer composition of a sample may be described as a dimensionless mole ratio (or as mole percent of the major enantiomer) and this is, in fact, the most generally useful way to describe the composition of all types of stereoisomer mixtures. A second and very common term used in this connection is *enantiomer excess*, usually expressed as a percentage. The latter expression describes the excess of one enantiomer over the other. The percentage enantiomer excess, $ee = 100(x_R - x_S)/(x_R + x_S)$, where $x_R > x_S$. Alternatively, $ee = 100(2x - 1)$, where x is the mole fraction of the dominant enantiomer in a mixture. For a mixture that is rich in one enantiomer, for example, 80:20, $x = 0.8$ but the enantiomer excess is only 60%. The converse relation is $x_R = (ee_R + 100)/200$, where the dominant enantiomer is R.

The analogous term enantiomeric purity (ep) that is sometimes seen in the literature is unfortunately not unequivocally defined. One definition, $ep = 100x_R/(x_R + x_S)$, corresponding to the mole percent of one enantiomer in a nonracemic sample, is preferred by some (Horeau, 1972; Martin, 1973; Testa, 1979, p. 162). In contrast, other sources have equated enantiomeric purity to enantiomer excess (Raban and Mislow, 1967; Jacques et al., 1981a, p. 33).

Enantiomer excess is defined so as to correspond to the older expression, *optical purity*: $op = ([\alpha]_{obs}/[\alpha]_{max}) \times 100\%$. The maximum or absolute rotation $[\alpha]_{max}$ (Farmer and Hamer, 1966) is that of an enantiomerically pure sample.

For the use of an analogous expression, diastereomer excess, see Thaisrivongs and Seebach (1983).

Although in the late 1980s the use of rotation measurements to determine stereoisomer composition has decreased relative to other methods (see above), the expression optical purity continues to be used even when rotation is not the basis of the determination. This usage is not only anachronistic but it is occasionally inappropriate, since it stems from the generally held belief that enantiomer excess and optical purity are numerically equal (Raban and Mislow, 1967), whereas experimental studies (to be described) have shown that this is not necessarily the case (Horeau, 1969). In consequence, we do not use the two expressions synonymously in this book.

The principal methods for the determination of enantiomer composition are summarized in Table 6.3. The table gives the basis of the measurement and its experimental nature. It indicates whether the measurement is undertaken on the intact mixture or whether pretreatment of the original mixture has taken place, that is, quantitative conversion of an enantiomer mixture to a mixture of diastereomers, or quantitative conversion to a different enantiomer mixture to allow a particular analysis to be carried out (e.g., conversion of a liquid to a solid

TABLE 6.3. Methods for the Determination of Enantiomer Composition

Basis of Measurement	Nature of Measurement	Treatment[a]	Species Examined[b]
1. Chiroptical	A. α, ϕ, or $\Delta\varepsilon$	I[c]	E or D
	B. Circular polarization of emission	I	E
2. Diastereotopicity (external comparison) (see Chap. 8)	A. NMR of diastereomers in achiral solvents[d]	Der	D
	B. NMR in chiral solvents (chiral solvating agents)	I	E
	C. NMR with chiral shift reagents	I	E
3. Diastereomeric interactions (separation)	A. Chromatography on diastereoselective stationary phases		
	i. GC	Der	D
	ii. HPLC	Der	D
	iii. HPLC with chiral solvent	I	E
	iv. TLC	Der	D
	B. Chromatography on enantioselective stationary phases	I	E
	i. GC		
	ii. HPLC		
	iii. TLC		
	C. Electrophoresis with enantioselective supporting electrolyte	I	E
4. Kinetics	Product composition	I or Der	E or D
5. Enzyme specificity	Quantitative enzyme-catalyzed reaction	I	E
6. Fusion properties	Differential scanning calorimetry	I or Der	E or D
7. Isotope dilution	Isotope analysis	I[e]	E
8. Potentiometry	Potential of an electrochemical cell	I	E

[a] I = intact mixture; Der = analysis on a diastereomeric derivative.
[b] E = original enantiomer mixture; D = diastereomer mixture prepared from enantiomer mixture to be examined.
[c] A derivative prepared with an achiral, chromophoric reagent may be used.
[d] Also, NMR in the solid state.
[e] This method requires reisolation of a sample following the dilution procedure.

mixture). Note that complete separation of the stereoisomers is required only in the case of the chromatographic methods.

The classification of Table 6.3 is convenient if somewhat arbitrary. It is evident that direct determination of enantiomer composition by NMR spectroscopy (categories 2.B and C) is a consequence of diastereomeric interactions between solutes and solvents or shift reagents and thus, in principle, is not different from category 2.A, which requires preparation of covalent diastereomeric derivatives. We have chosen to focus attention particularly on the topic (structural) relationships in the case of analyses that do not require the separation of stereoisomers and to reserve the category of diasteromeric interactions (category 3) for analyses that depend on separations.

Also, although the category enzyme specificity (category 5) is an analytical technique based on differences in rates of reaction exhibited by enantiomers, that is, kinetics (cf. category 4), the two categories are kept separate for convenience and because of the substantial difference in technique.

Few functional group types are not now amenable to analysis by one of the methods shown in Table 6.3. Even chiral hydrocarbons (at least some of them) respond to one or more of these methods. The principal limitation in methodology is that a very low enantiomer excess (below ca. 2%) or a very high one (above ca. 98%) remains difficult to measure with precision, that is, reproducibly in different laboratories.

The choice of method to be used depends on a variety of factors, not the least of which is convenience and the availability of the necessary instrumentation. Moreover, it is necessary to take into account the purpose of the measurement in choosing the method to be used. Very precise analyses would not normally be required while monitoring the progress of a resolution.

Great care must be taken not to modify the enantiomer composition of a mixture prior to the analysis lest the results be invalidated. Chemical purification, as in the workup of a reaction mixture, involving washing, crystallization, or sublimation of solids will change the enantiomer ratio of the mixture. Even chromatography of a partially resolved chiral sample on an achiral stationary phase may occasionally modify the enantiomer composition (Section 6-4.n) especially if carried out carefully, that is, when small fractions are individually analyzed. This possibility follows from the principles outlined in Section 6-4.

It is fair to say that no one method for the determination of enantiomer composition, whether spectroscopic, chromatographic, or other technique is universally applicable. Each new case requiring such analysis must be examined individually and a choice of method made based on the structure of the chiral analyte, the state of the art at the moment, local resources and experience, and the required precision of measurement.

It is remarkable that in 1962 few methods were available for assessing enantiomer composition and none were considered completely reliable (Eliel, 1962, p. 83). The first comprehensive review of modern methods appeared in 1967 (Raban and Mislow). During the following decade, an avalanche of papers describing discoveries and applications of enantiomeric purity determinations found their way into the literature. General surveys of the literature may be found in the books by Jacques et al., 1981a, p. 405; Potapov, 1979, p. 156; and Izumi and Tai, 1977, p. 218; and a comprehensive survey of the better known and most useful methods comprises an entire volume (Morrison, 1983). Reviews dealing with individual methods are cited in the appropriate sections that follow.

b. Chiroptical Methods

These methods, involving the measurement of optical rotation typically at a single wavelength, provide results (optical purity) rather fast but the information is often not very precise, nor is it necessarily very accurate. The measurement of optical rotation is the traditional method for assessing enantiomer composition; it requires a knowledge of the specific rotation, or molar rotation, of the pure compound under investigation, $op = 100[\alpha]/[\alpha]_{max}\%$. Numerous examples are known wherein the enantiomer composition determined in this way in earlier years has been found to be incorrect. Such errors arose principally because the assumption of complete resolution of the sample used for determining $[\alpha]_{max}$ was invalid (e.g., Guetté et al., 1974) or because the sample examined contained solvent residues or other impurities.

We must also remember that the optical rotation, and hence the optical purity determination, may be affected by numerous variables: wavelength of light used, presence or absence of solvent, and the nature of the solvent used in the rotation measurement, concentration of the solution (even though this variable is factored into the calculation of the specific rotation, Eq. 1.1 and Section 13-5.a), temperature, and presence of impurities; the effect of these variables will be analyzed in detail in Chapter 13. This is a long list of variables that potentially modify the specific rotation to be compared to $[\alpha]_{max}$. The latter, whether obtained directly or indirectly by calculation, may have been determined in another laboratory and under circumstances that are not sufficiently well specified; for optimal results, the rotations to be compared should, of course, be measured under precisely identical conditions (see also Section 1-3). Since this is difficult to do, if not impractical in most circumstances, and since the precision of polarimetric measurements is typically no better than 1–2%, the optical purity will, in many cases, give only a rough estimate of the enantiomer composition of a mixture. New developments in polarimetry are taken up in Section 13-5.a.

Consider the following examples illustrating the problems associated with the use of optical purity. Reduction of (presumably) enantiopure L-leucine to leucinol (Fig. 6.33, **54**) was found to afford a product whose rotation varied with the reducing agent used. Reduction of leucine with borane–dimethyl sulfide led to a sample having $[\alpha]_D^{20} + 4.89$ (neat) while reduction of leucine ethyl ester hydrochloride with $NaBH_4$ or of the ester with $LiAlH_4$ afforded leucinol having $[\alpha]_D^{20} + 1.22$–1.23 (neat) (Poindexter and Meyers, 1977). Originally this discrepancy was considered to be due to possible racemization under the second and third reduction conditions. However, independent analysis of the leucinol samples by ^{19}F NMR analysis of the Mosher amide derivatives (see below) showed the samples from all three reductions to be enantiopure. The discrepancies were ascribed to the presence of trace amounts of strongly dextrorotating impurities (removable by recrystallization of the leucinol as its hydrochloride salt) in the sample having high rotation.

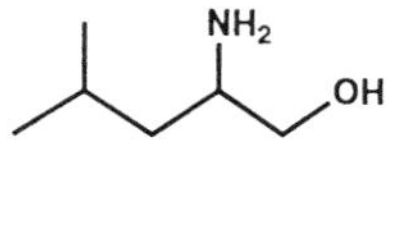

| **54** | **55** |
| Leucinol | 2-Phenylpropanal |

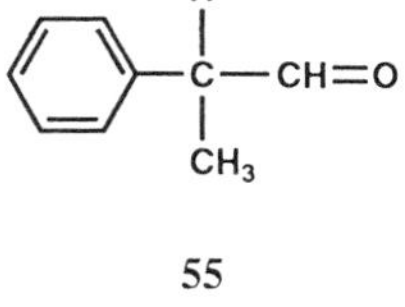

Figure 6.33. Leucinol (**54**) and 2-phenylpropanal (**55**).

Asymmetric hydroformylation of styrene leads to optically active 2-phenyl-propanal (Fig. 6.33, **55**) with an optical purity of about 95% (max) as determined by comparison of the rotation of purified product $[\alpha]_D^{21} + 224.8$ measured in benzene (c 1.5–20) with the maximum specific rotation reported earlier $[\alpha]_D^{25} + 238$ (neat) (Pittman et al., 1982). Subsequent reevaluation of the optical purity revealed a significant difference in the specific rotation of the product as measured in benzene and in the absence of solvent (neat) as well as a concentration effect, $[\alpha]_D^{21} + 214.7$ (c 1.5, benzene) versus $[\alpha]_D^{21} + 182.2$ (c 46.4 benzene) with all measurements being made using a sample of fixed enantiomer composition (op 68%) (see page 7 for units of c). Note the substantial increase in specific rotation as the concentration is decreased. Consequently, the actual optical purity of the product obtained as described above needed to be recalculated: op ca. 73% instead of 95% (Consiglio, Pino, Pittman, et al., 1983).

The concentration effect arises as a consequence of association phenomena that vary according to the functional groups that are present and the solvent that is used. It is pronounced when polar molecules (especially alcohols and carboxylic acids) are measured in nonpolar solvents. Polyfunctional molecules, for example, diols and hydroxy acids, are also prone to exhibit concentration effects as intramolecular aggregates give way to intermolecular ones when fewer solvent molecules are present.

These data suggest that the measured optical purity may not always be numerically equal to the enantiomer excess. The discrepancies discussed so far occur because the two specific rotations being compared were not measured under strictly identical conditions. However, there is another possible source of nonequivalence of enantiomeric excess with optical purity, namely, enantiomer discrimination in mixtures of enantiomers of composition intermediate between 50–50 and 100%. Such discrimination would undermine the assumption of strict additivity of specific rotations that is the basis of the optical purity determination.

The nonequivalence of enantiomer and optical purities was first observed by Horeau (1969) in α-ethyl-α-methylsuccinic acid (Fig. 6.34, **56A**). For the pure acid, $[\alpha]_D^{22} + 4.4$ (c 15, CHCl$_3$), hence Horeau calculated $[\alpha]_D^{22} + 2.2$ for a 75:25 enantiomer mixture (ee = 50%). He actually observed $[\alpha]_D^{22} + 1.6$ (c 15, CHCl$_3$), corresponding to op ca. 36%. The effect was only observable in weakly polar solvents (CH$_2$Cl$_2$, CHCl$_3$, and C$_6$H$_6$); the discrepancy between enantiomeric excess and optical purity was found to disappear in polar solvents (ethanol, pyridine, diglyme, acetonitrile). The structurally related α-isopropyl-α-methyl-succinic acid (Fig. 6.34, **56B**) has also been shown to exhibit the *Horeau effect* as

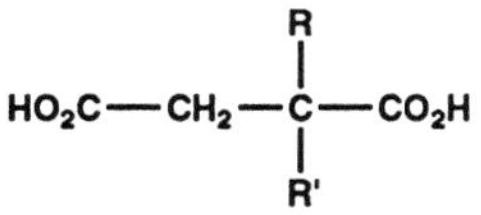

HO$_2$C—CH$_2$—C—CO$_2$H (with R above and R' below the central C)

56

A : R = Et ; R' = CH$_3$

B : R = i-Pr ; R' = CH$_3$

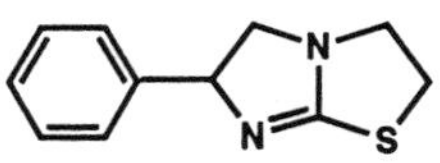

56 C

Figure 6.34. Compounds exhibiting the Horeau effect.

this lack of linearity between specific rotation and enantiomer purity has come to be known. It shows an even larger divergence of measured $[\alpha]_D^{22} + 7.3$ (c 0.8, CHCl$_3$) for an enantiomerically pure sample from that calculated from the rotation of sample of op ca. 25%: $[\alpha]_D^{22} + 17$. The discrepancy amounts to 118% (Horeau, 1972; Horeau and Guetté, 1974).

> Though the source of the nonequivalence is enantiomer discrimination, it is not evident how this is translated into a change in specific rotation. There are two possibilities: (a) preferential heterochiral association or repulsion can affect the *effective* concentration of oligomeric species contributing to the specific rotation, or (b) the magnitudes and wavelengths of the Cotton effects that are utimately responsible for the optical activity (Chapter 13) differ for homochiral and heterochiral assemblages of molecules. This difference may be sufficiently large so that the specific rotations of enantiomer mixtures measured at wavelengths far removed from the Cotton effects may differ significantly from those calculated as if they were derived from Cotton effects of enantiomerically pure materials.

The departure from linearity is influenced by factors such as the wavelength at which the specific rotation is measured and by the concentration (note that for enantiopure acid **56A**, whose rotation is strongly concentration dependent in chloroform, $[\alpha]_D^{22} + 0$ at c 6.3, CHCl$_3$). The effect is the more important the smaller the enantiomer excess of the sample; it vanishes in the vicinity of 100% ee and as the ee tends to 0%. A striking demonstration of the Horeau effect is given by the finding that for acid **56A** the pure R enantiomer is dextrorotatory, $[\alpha]_D^{22} + 0.6$ for c 7.5, while a 75:25 enantiomer mixture is levorotatory, $[\alpha]_D^{22}$ -1.5 at the same concentration (Horeau and Guetté, 1974). In the case of **56C** (Fig. 6.34), the observed Horeau effect has also been shown to depend on temperature (Ács, 1990).

The Horeau effect appears not to have been observed in other covalent organic compounds. It has been observed in ionic complexes where it occurs even in aqueous solution [mandelic acid in the presence of ammonium molybdate (Guetté, Boucherot, and Horeau, 1974), and mixtures of cupric and potassium tartrates (Morozov et al., 1979)]. That some studies have confirmed the accuracy of polarimetric determinations of enantiomer composition (Seebach and Langer, 1979; Sugimoto et al., 1978; Webster, Zeng, and Silverstein, 1982) does not represent any sort of contradiction since these findings were for enantiomerically pure (or nearly so) samples.

The Horeau effect, being a manifestation of enantiomer discrimination, is likely to be significant in strongly associated substances, for example, more so in carboxylic acids than in alcohols (Seebach et al., 1977), and more so in polyfunctional compounds than in monofunctional ones. Nonracemic enantiomer mixtures exhibiting anisochrony of NMR signals in solution in the absence of chiral shift reagents or chiral solvating agents (Section 6-4.m) are good candidates for operation of the Horeau effect since the underlying explanation is common to both. But many such candidate compounds have been found not to exhibit the effect (Horeau and Guetté, 1974; Lang and Hansen, 1979). Although lack of constancy of specific rotation as a function of concentration (Chapters 1 and 13) is suggestive of intermolecular association, such a finding gives no assurance that the system under observation will manifest the Horeau effect. Nor does the observation that optical rotation is linear with respect to concentration assure the equivalence of optical purity and enantiomer excess (i.e., linearity of optical rotation with enantiomer composition).

Some indirect chiroptical methods have been used in the estimation of enantiomer and diastereomer composition. For example, the specific rotation of a diastereomeric salt (e.g., one formed from reaction of a chiral acid with a chiral base) measured in dilute solution may be estimated by the approximate additivity of $[\Phi]_D$ values of the constituent ions (Walden's rule) (Jacques et al., 1981a, p. 318; cf. Section 13-5.b). Also, the composition of optically active diastereomer mixtures, as in the construction of a ternary (solubility) phase diagram, may be estimated by assuming the additivity of the specific rotations of the constituent compounds. The latter estimate assumes that diastereomer composition is linear with specific rotation; however, this linearity does not always obtain (Guetté and Guetté, 1977).

During the purification (enrichment) of a diastereomer mixture (or of an enantiomer mixture) by crystallization, comparison of the rotation of the solid isolated with that of the residue recovered from the mother liquor may indicate the approximate stereoisomeric composition (Jacques et al., 1981a, p. 408).

> This comparison is possible because the solid that crystallizes and the mother liquor bathing it bear a relationship determined by the phase diagram of that system, when operating under equilibrium conditions. If the crystallized solid has a low enantiomeric (or diastereomeric) purity, then the residue recovered from the mother liquor has a composition tending to the nearest eutectic. In Figure 6.13, recrystallization of solid P yields enriched solid N in mother liquor of composition tending to E. With further recrystallizations (beginning with solid N), the re-crystallized solid will tend to pure (+) and the rotation of solid recovered from the mother liquor will jump to match the rotation of (+). When the two rotations are equal, the purification is complete; an example is given by Martin and Libert (1980).

For compounds having measurable fluorescence, the circular polarization of emission (CPE) provides an absolute means of determining enantiomeric purity. CPE involves the evaluation of the g number, g_{em}, for spontaneous emission of radiation induced when chiral molecules are excited by radiation. Comparison of g_{em} with the g number (corresponding to the ratio of CD to isotropic absorbance at a given wavelength, $\Delta\varepsilon/\varepsilon$) permits the calculation of the enantiomer excess (Section 13-7). The method, which gives best results with samples of low enantiomeric purity, is limited by the paucity of CPE spectrometers (not available commercially as of 1990) (Eaton, 1971; Kokke, 1974; Schippers and Dekkers, 1982).

c. NMR Methods Based on Diastereotopicity

NMR of Diastereomers. Chiral Derivatizing Agents

The first NMR technique to be applied in the determination of enantiomer composition was the analysis of covalent diastereomer mixtures (Reviews: Gaudemer, 1977; Rinaldi, 1982; Yamaguchi, 1983; Parker, 1991; Parker and

Taylor, 1992). This approach is exemplified by the derivatization of chiral alcohols and amines with optically active acids.

In the pioneering work of Raban and Mislow (1965; see also Jacobus, Raban, and Mislow, 1968), alcohols and amines were converted to esters and amides, respectively, by reaction with the *chiral derivatizing agent* (CDA) (R)-(−)-O-methylmandeloyl chloride (acid chloride of **57**, Fig. 6.35; Jacobus and Raban, 1969; Jacobus and Jones, 1970) (Eq. 6.11):

$$(6.11)$$

[For the application of (−)-menthoxyacetic acid (Fig. 6.35, **58**) to such determinations, see Galpin and Huitric (1968) and Cochran and Huitric (1971)]. Raban and Mislow pointed out the advantage of having easily identifiable diastereotopic nuclei in the CDA which, in some instances, might serve as a resolving agent (see Chapter 7 for the latter usage) as well as serving to monitor the resolution by NMR spectroscopy. If the NMR resonances contributed by the diastereomer moiety originating in the analyte are not sharp and well resolved, so as to permit accurate integration, then diastereotopic nuclei contributed by the CDA, for example, the isopropyl CH_3 groups or the side chain acetate CH_2 groups in the menthoxyacetates or menthoxyacetamides might serve this purpose.

Subsequent developments were prompted, in part, by the observation that some O-methylmandelate esters and O-methylmandelamide derivatives are subject to epimerization at the hydrogen alpha to the carbonyl. This factor leads to incorrectly low values of the enantiomer composition (Dale and Mosher, 1968). α-Methoxy-α-trifluoromethylphenylacetic acid, *Mosher's reagent* (MTPA reagent; Fig. 6.35, **59**) was designed to avoid this problem by eliminating an α-hydrogen atom in the CDA (Dale, Dull, and Mosher, 1969). At the same time, incorporation of fluorine by way of a CF_3 group made possible analysis of the ester and amide derivatives by means of ^{19}F NMR spectroscopy. The latter often simplifies the analysis since there is much less likelihood of finding overlapping peaks; the number of peaks is smaller and they are better separated than are comparable peaks in ^{1}H NMR (the chemical shift dispersion of ^{19}F NMR is greater than that of ^{1}H). For an example of the application of MTPA, see Guerrier, Husson, et al. (1983).

Parallel application of such diastereomer derivatives to enantiomer composition determination by GC (see Section 6-5.d) led to the development of other CDAs including MMPA (Fig. 6.35, **60**; Pohl and Trager, 1973). In a similar way, application of the *Anderson–Shapiro reagent* (Fig. 6.35, **61**) to the analysis of

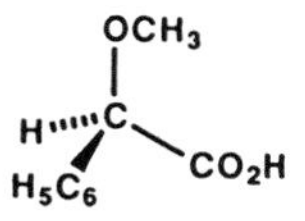

(R)–(–)–O–Methyl–
mandelic acid, **57**

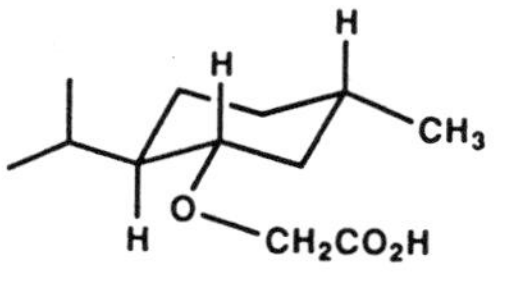

Menthoxyacetic
acid, **58**

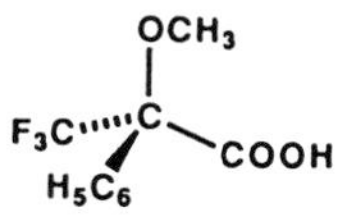

(R)–(+)–α–Methoxy–
α–trifluoromethyl–
phenylacetic acid
(MTPA; Mosher's reagent)
59

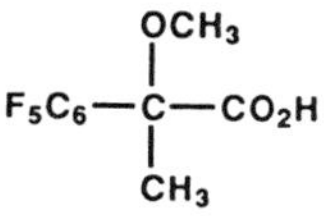

α–Methoxy–α–methyl–(pentafluoro-
phenyl) acetic acid (MMPA), **60**

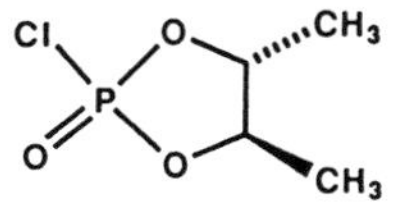

61

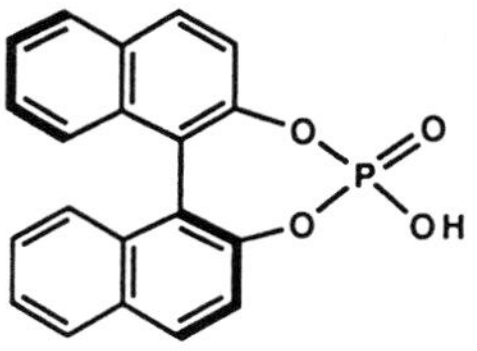

62a

(S)-(+)-Binaphthyl-2,2'-diyl
hydrogen phosphate

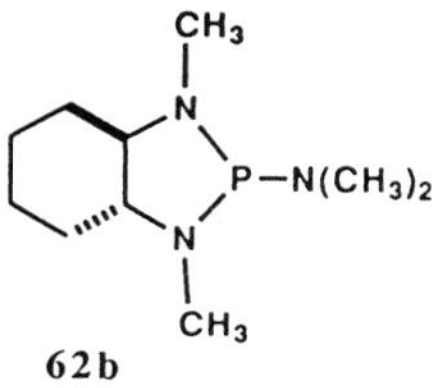

62b

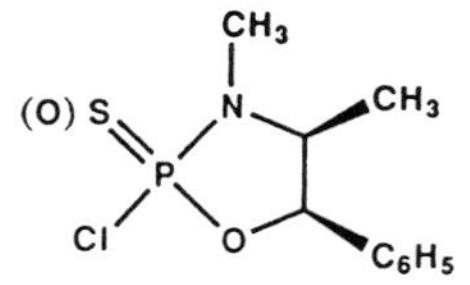

62c

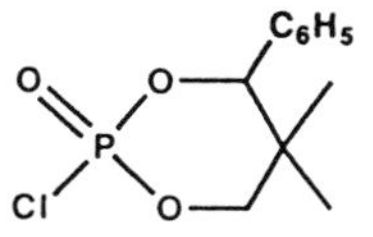

63

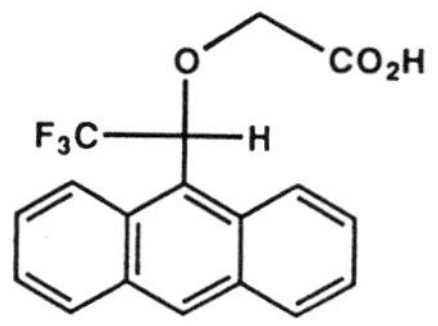

α–[1–(9–Anthryl)–
2, 2, 2–trifluoroethoxy]-
acetic acid (ATEA), **64a**

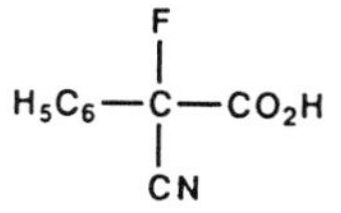

α–Cyano–α–fluorophenylacetic
acid (CFPA), **64b**

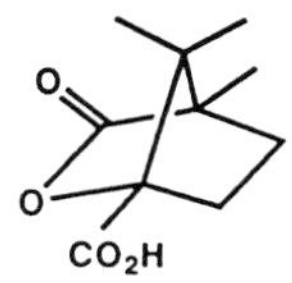

(1S,4R)–(–)–ω–
Camphanic acid, **65**

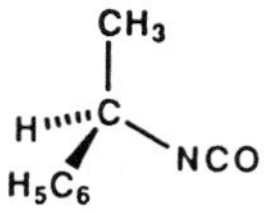

(R)–α–Methylbenzyl
isocyanate, **66**

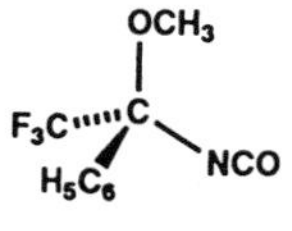

(S)–(–)–α–Methoxy–
α–(trifluoromethyl)–
benzyl isocyanate, **67**

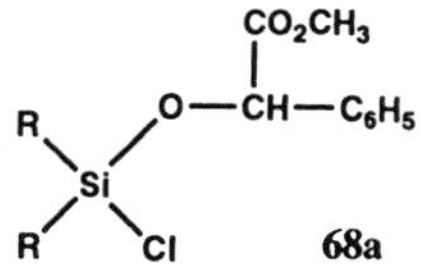

R = CH₃, C₆H₅

68a

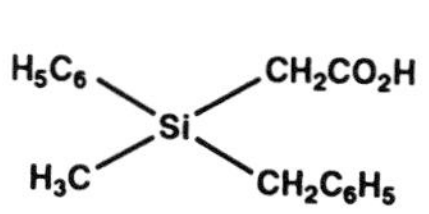

68b

Figure 6.35. Chiral derivatizing agents for NMR analysis of diastereomers.

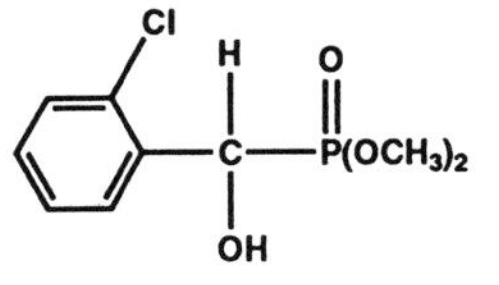

α–Hydroxy–(o–chlorobenzyl)–
O,O–dimethylphosphonate, **69**

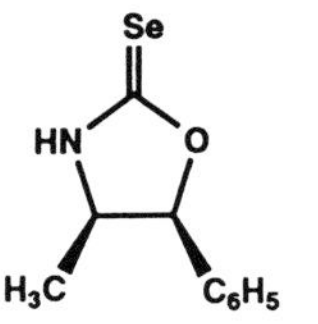

70

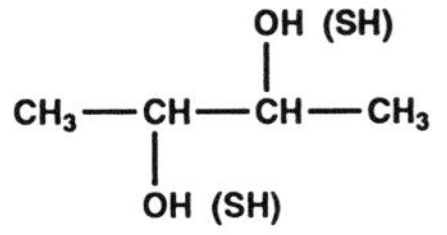

chiral–2,3–Butanediol
and dithiol, **71**

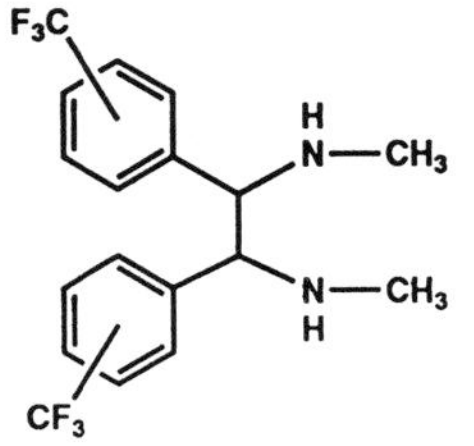

(o, m, p)   **72**

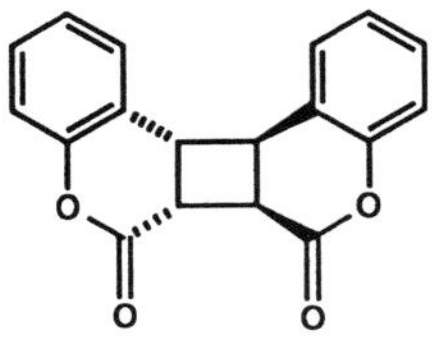

73

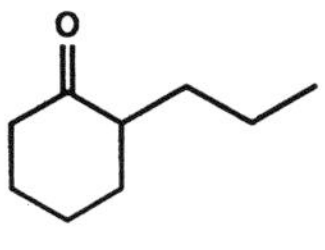

74

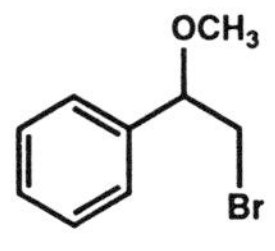

75

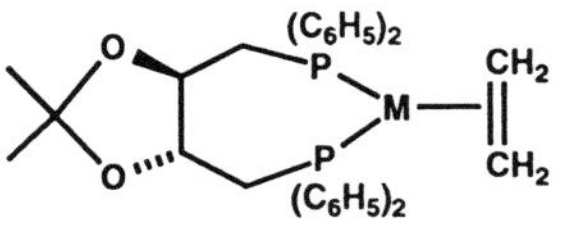

Metal (M) = Pt, Pd   **76a**

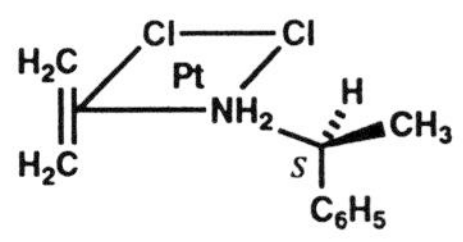

76b

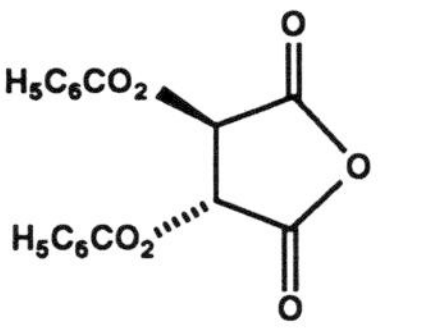

77

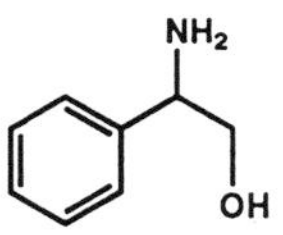

78

Phenylglycinol

Figure 6.35   (*Continued*)

nonracemic samples of primary and secondary alcohols by ^{31}P NMR spectroscopy is simple in view of the absence of interfering bands in the spectrum of the phosphorus ester mixture (Anderson and Shapiro, 1984). Since the phosphorus atom is nonstereogenic as a consequence of the C_2 symmetry axis of the parent glycol moiety, derivatization of a given analyte enantiomer with **61** gives rise to but one diastereomer irrespective of the mechanism of bond formation to phosphorus, retention, or inversion. The binaphthylphosphoric acid **62a** (Fig. 6.35), a C_2 symmetric reagent, may be a useful CDA for chiral alcohols (Kato, 1990; compare Johnson et al., 1984). More recently, a C_2 symmetric diazaphospholidine (Fig. 6.35, **62b**) has been shown to be a powerful and easy to use CDA for the determination of enantiomeric purity of a wide range of alcohols and thiols, even in the presence of other functionalities (Alexakis et al., 1992). Completion of the reaction is checked by conversion of the trivalent phosphorus product **A** to an air-stable and chromatographable tetravalent thio derivative **B** (Eq. 6.12):

$$(6.12)$$

Several 1,3,2-oxazaphospholidine sulfides and oxides derived from (−)-ephedrine (Fig. 6.35, **62c**) have been applied as CDAs in the determination of the enantiomeric purity of alcohols and amines (Johnson et al., 1984). Here the phosphorus atom is chiral and, in contrast to C_2-symmetric CDAs (above), an enantiopure alcohol can give rise to more than one diastereomeric phosphorus ester according to the mechanism of reaction. In fact, mixed mechanisms have been observed in reactions of **62c** (especially in the presence of pyridine) leading to loss of stereochemical integrity and this reduces the utility of this CDA (Cullis et al., 1987). A 1,3,2-dioxaphosphorinane oxide (Fig. 6.35, **63**; ten Hoeve and Wynberg, 1985) is an excellent CDA for the determination of the enantiomer composition of amines, even though it suffers from the same problem.

Dale and Mosher (1968, 1973) proposed that the spectral anisochrony (chemical shift difference, $\Delta\delta$) observed between the two diastereomers produced from enantiomeric substrate molecules is dependent on a combination of steric and electronic nonbonded interactions enhanced by the anisotropy of the aromatic ring in the CDA. These interactions tend to populate conformations that are quite distinct for the two diastereomers. In consequence, selected nuclei in these conformations exhibit significant chemical shift differences for the two diastereomers.

Another development is the application of the ATEA reagent (Fig. 6.35, **64a**) to the determination of enantiomer composition of amines, alcohols, and thiols (Pirkle and Simmons, 1981). The design of this CDA is based on the hypothesis that conformationally more rigid diastereomers would engender greater anisochrony. More specifically, the observed anisochronies are weighted averages of effects caused by all conformations present in the diastereomer mixtures, both those that exhibit nonequivalence of chemical shifts and those that do not. Reduction of conformational mobility might increase the population of the former (see below for another way of achieving this end).

In order that analysis of the amide (or other diastereomer) mixture accurately reflect the substrate enantiomer composition, it is essential that reaction be complete. The CFPA reagent, α-cyano-α-fluorophenylacetic acid (Fig. 6.35, **64b**), has recently been proposed as an alternative CDA to the widely used MTPA reagent (see above) particularly for hindered substrates, such as 3,3-dimethyl-2-butanol, whose reaction with MTPA chloride is relatively slow and may be incomplete under a given set of conditions (Takeuchi et al., 1992). In cases tested thus far, diastereomers prepared with the CFPA reagent also have been found to have greater $\Delta\delta$ values than for the corresponding MTPA diastereomers in both ^{19}F and ^1H NMR spectra (Takeuchi et al., 1993). In addition, neither reaction partner must be prone to racemization during reaction and neither equilibration nor alternation of the diastereomer ratio must be allowed to take place following reaction.

> Equilibration of the diastereomers *during* the derivatization reaction would be equivalent to an asymmetric transformation of the second kind (Section 7-3.e). Alteration of the diastereomer ratio from that expected for a given enantiomer composition may result if insufficient CDA reagent is used or if reaction is incomplete. In that event, the analysis could be "falsified" as a consequence of kinetic resolution (Section 7-6). This possibility argues for application of a control reaction on racemic substrate whenever an analysis with a CDA is carried out (Jacobus, Raban, and Mislow, 1968). Such a control reaction is essential also to prove that the two diastereomers can give rise to separate signals under the conditions of measurement.
>
> The finding that a racemic substrate yields diastereomeric derivatives in other than a 1:1 proportion under some conditions may also be interpreted as indicating that kinetic resolution (sometimes described as asymmetric induction) has taken place (Dutcher, Macmillan, and Heathcock, 1976). In contrast, reaction of racemic substrate with *racemic* CDA leads to a diastereomer mixture whose ratio reflects the relative rates of the two reactions taking place (Jacobus, Raban, and Mislow, 1968), even if the two reactants are in equimolar proportions and the reaction is allowed to go to completion.

A significant limitation of the method is that the nonracemic CDA *must* be enantiomerically pure (see Section 6-5.d for a justification); it is wise to check this before using the reagent, for example, by combining it with a known enantiomerically pure substrate and ascertaining that only one product diastereomer results. In spite of this disadvantage relative to some of the other analytical methods, diastereomer analysis by NMR is very popular.

An additional feature of this technique is that, with synthetic CDAs, either enantiomer of the reagent can be used thereby reducing possible analytical ambiguities. Confirmation of the validity of the method (by observation of anisochronous signals) must always be undertaken by reaction of the CDA with a racemic sample of the analyte. If the racemate is unavailable (as in the case of natural products), then the validity of the method could be ascertained by analysis of a derivative obtained on reaction of the analyte with the enantiomeric CDA. An additional real problem is that in which a small second signal appears following derivatization of a sample of unknown enantiomer composition: Is the signal that of the other isomer or is it an artifact? Here too, reaction of the analyte with the enantiomeric CDA would resolve the ambiguity. Most of the early applications involved analysis by ^{19}F NMR and by ^1H NMR even at low magnetic field strengths. However, contemporary methodology is characterized by application of higher magnetic fields and/or a wider range of probes: ^2H NMR (e.g., Brown and Parker, 1981), ^{13}C (e.g., Bordeaux and Gagnaire, 1982), ^{19}F (Kawa and Ishikawa, 1980), ^{29}Si for alcohols (Chan et al., 1987), ^{31}P (see above), ^{77}Se for carboxylic acids (see below; Silks, III et al., 1991; see also Michelsen, Annby, and Gronowitz, 1984), ^{195}Pt for allylic compounds (Salvadori et al., 1988) and for allenes (Salvadori et al., 1990; see below).

The CDA–NMR method, for example, with $(1S,4R)$-$(-)$-ω-camphanic acid as CDA (Fig. 6.35, **65**) has also been applied to the enantiomer purity determination of compounds whose chirality depends on deuterium substitution (Gerlach, 1966; Raban and Mislow, 1966; Gerlach and Zagalak, 1973; see also Chapter 8). In addition, one may take advantage of two other means of enhancing the anisochrony between diastereomers: solvent effects and addition of achiral shift reagents. Determination of NMR spectra in benzene (also pyridine, trifluoromethylbenzene, and halogen-substituted aromatics; see Dale and Mosher, 1968) instead of in CCl_4 or $CDCl_3$ often leads to increased peak separation (Aromatic Solvent Induced Shifts, ASIS) that have facilitated configurational assignments. The improved resolution may also permit or simplify the quantitative analysis of diastereomer mixtures (Laszlo, 1967; Kalyanam, 1983).

The second mentioned enhancement process is an application of *achiral* shift reagent techniques (Review: Yamaguchi, 1983). For example, addition of $Eu(dpm)_3$ (dpm = dipivaloylmethane; see Fig. 6.41) to the camphanates of chiral α-deuterated primary alcohols leads to a chemical shift difference of as much as 0.5 ppm between the diastereotopic α-methylene protons (Gerlach and Zagalak, 1973; Armarego, Milloy, and Pendergas, 1976). An example involving enhanced anisochrony in ^{19}F spectra is given by Merckx (1983). It has been postulated that achiral lanthanide shift reagents [e.g., $Eu(fod)_3$; Fig. 6.41] enhance the separation ($\Delta\delta$) of specific peaks, such as the OCH_3 in the MPTA moiety (see **59** in Fig. 6.35) of MTPA esters by reducing conformational mobility within the diastereomeric derivatives (Yamaguchi et al., 1976).

The structures of the most useful CDA reagents are shown in Figure 6.35. In addition to the CDA already mentioned, α-methylbenzyl isocyanate (Fig. 6.35, **66**) has been applied to the quantitative analysis of chiral amines by formation of diastereomeric ureas (Rice and Brossi, 1980; Hauser et al., 1984); an analogous reagent is **67** (Fig. 6.35; Nabeya and Endo, 1988). The chiral silyl reagent **68a** (Fig. 6.35) has been applied to the analysis of chiral alcohols (Chan et al., 1987). A second silyl reagent (Fig. 6.35, **68b**) incorporating a stereogenic silicon atom has been applied as CDA for alcohols and amines (Terunuma et al., 1985); for an example of its application, see Saigo et al. (1986). Diastereomeric esters or amides prepared with **68b** generate signals in the Si-CH_3 region of the ^1H NMR spectrum that are generally free of interference by other signals thus simplifying spectral interpretation.

Chiral carboxylic acids can often be analyzed as amides with optically active α-methylbenzylamine (Fig. 6.27; e.g., Mamlok, Marquet, and Lacombe, 1973; Paquette et al., 1974; and Rosen, Watanabe, and Heathcock, 1984). Compound **69** has been proposed as a reagent for the determination of the enantiomer composition of carboxylic acids (Smaardijk, 1986). 2-Chloropropionic acid has been applied to the enantiomeric purity analysis of amino acids (Kruizinga, Kellogg, and Kamphuis, 1988). More recently, a selenium reagent, oxazolidine-2-selone **70** (Fig. 6.35) having a very high ^{77}Se chemical shift sensitivity, has been shown to be a useful reagent for the measurement of enantiomeric purity of carboxylic acids even when the latter bear stereocenters remote from the carboxyl group (Silks, III et al., 1991).

Ketones have been analyzed by ^{13}C NMR following conversion to diastereomeric ketals with *threo*-2,3-butanediol, or with *threo*-2,3-butanedithiol (Fig. 6.35, **71**) to thioketals (Hiemstra and Wynberg, 1977; ten Hoeve and

Wynberg, 1979; Meyers et al., 1981) and aldehydes by conversion to imidazolidines with reagent **72** (Fig. 6.35; Cuvinot, Mangeney, et al., 1989). The
latter three CDA possess a C_2 axis (see also Fig. 6.35, **73**, a CDA useful in the
analysis of amines; Saigo et al., 1985). In absence of this symmetry element, an
additional stereogenic center (hence more than two diastereomers) would be
created during formation of the derivative, unless the ketone itself possesses a C_2
axis. The reciprocal process has been applied to the analysis of glycols as ketals
formed with optically active 2-propylcyclohexanone (Fig. 6.35, **74**; Meyers,
White, and Fuentes, 1983).

The enantiomer composition of phosphines may be determined following
their conversion to diastereomeric phosphonium salts by reaction with 2-phenyl-2-
methoxyethyl bromide (Fig. 6.35, **75**; Casey, Lewis and Mislow, 1969). The
enantiomer composition of alkenes and allenes has been measured by ^{31}P NMR
following ligand exchange on metal complex **76a** (Fig. 6.35; Parker and Taylor,
1987) and on **76b**, *cis*-dichloro[(S)-α-methylbenzylamine](ethylene)platinum(II)
(Fig. 6.35) for allyl derivatives (Salvadori et al., 1988), and for allenes, on the
corresponding trans complex, both by ^{195}Pt NMR (Salvadori et al., 1990).

One of the early applications of CDAs was in the assessment of the degree of
racemization attending the coupling of protected amino acids in peptide synthesis.
The ^{1}H NMR spectra at 60 MHz of *N*-acylamino acids derivatized either with
L-Phe–OCH$_3$ or L-Ala–OCH$_3$ give cleanly separated peaks for the diastereomeric
peptides (Halpern et al., 1967). A later study demonstrates the convenience and
sensitivity of ^{13}C NMR spectroscopy in racemization studies, for example, on
threonine-containing glycosyl amino acids and glycopeptides (Pavia and
Lacombe, 1983). Anhydride **77** (Fig. 6.35) is an example of a CDA applied to the
determination of the enantiomer composition of amino acids and peptides, as well
as amines and amino alcohols (Kolasa and Miller, 1986).

Covalent diastereomers prepared in connection with resolutions, either by
crystallization or by chromatography, are often analyzable directly by NMR. The
direct determination of enantiomer composition is a valuable concomitant of such
separations. For an outstanding example, see the application of phenylglycinol
(Fig. 6.35, **78**) as a resolving agent and CDA (Helmchen et al., 1979). Additional
examples are camphanate (Briaucourt, Guetté, and Horeau, 1972; Gerlach and
Zagalak, 1973) and chiral carbamate esters (Section 7-3.a; Nicoud and Kagan,
1976/77; Pirkle and Hoekstra, 1974).

While the CDAs thus far described have all involved formation of covalent
derivatives, it was observed by Horeau and Guetté (1968) and Guetté et al.,
(1968) that ionic derivatives could also serve as probes for the determination of
enantiomer composition. The induction of nonequivalence in NMR spectra of
diastereomeric salts, that is, chiral ions in the presence of chiral counterions, is
illustrated in Figure 6.36.

When racemic acids and bases are combined to form diastereomeric salt
mixtures, each ion in solution is surrounded by both of the enantiomeric
counterions. Since the ions change partners at a rate rapid on the NMR time
scale, chemical shift differences of groups are averaged out and the mixture of the
four salts (two sets of enantiomeric pairs; Fig. 6.36a) behaves as if it were a single
substance. However, when one of the counterions is present in only one enantiomeric form, as in the resolution of *rac*-2-chloropropanoic acid with $(+)$-α-

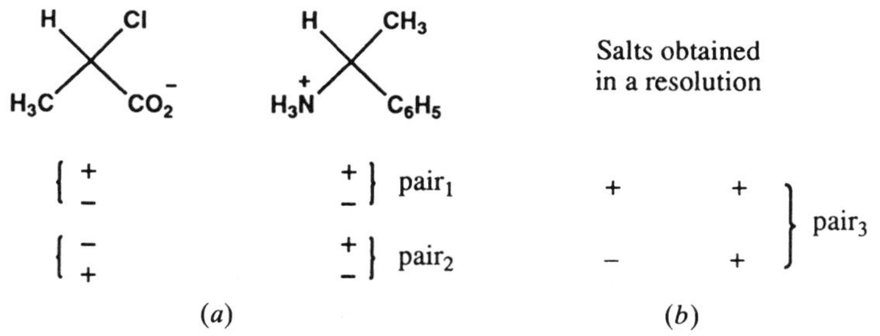

Figure 6.36. Nuclear magnetic resonance nonequivalence in diastereomeric salts.

methylbenzylamine, diastereomeric ion pairs are obtained (Fig. 6.36b) and persist regardless of the rate of counterion exchange. One might consider that the amine present in only one enantiomeric form [that generated from the (+) amine in the example, Fig. 6.36b], serves as a chiral solvating agent (CSA; see p. 231). The resulting anisochrony of the diastereomeric ion pairs provides a very simple means for determining the enantiomer composition during resolutions by diastereomeric salt formation. The same considerations and limitations listed for CSA apply (Section 6-5.c, p. 231); for the effect of temperature, concentration, and stoichiometry, see Fulwood and Parker, 1992.

The nonequivalence is observable in nonpolar aprotic solvents (CDCl$_3$, C$_6$D$_6$, and especially in pyridine), and even in DMSO, in which ion pairs or aggregates are present. Use of protic solvents, such as CH$_3$OH, destroys the effect presumably by disaggregating the ion pairs. Typical diastereomeric salts have sufficient solubility in one or more of the solvents cited so that lack of solubility is not a major limitation.

In connection with the analysis of nonracemic cyclic phosphoric acids (e.g., Fig. 7.21, **66**) with (−)-ephedrine, it has been observed that reciprocal experiments (Section 7-3.a) do not lead to anisochrony. Whereas (−)-ephedrine acts as a CDA toward the racemic acids, the enantiopure cyclic phosphoric acid cannot be used for the determination of enantiomer composition of ephedrine (van der Haest, 1992).

Nonequivalence in diastereomeric salts has been observed in ^{1}H, ^{31}P, and ^{13}C NMR spectra (Mamlok et al., 1971, 1973; Mikołajczyk et al., 1971, 1978; Mikołajczyk and Omelanczuk, 1972; Ejchart and Jurczak, 1970, 1971; Ejchart et al., 1971; Kabachnik et al., 1976; Baxter and Richards, 1972; Kuchen and Kutter, 1979; Anet, Mislow et al., 1983; Villani, Jr. et al., 1986; Björkling, Norin, et al., 1987; van der Haest, 1992). Application of salts to the determination of enantiomer composition of acids and bases is a direct and simple process that merits wider use than it has so far received. For an example involving the use of quinine as CDA, see Webster, Zeng, and Silverstein (1982). Mandelic acid, C$_6$H$_5$CH(OH)CO$_2$H; MTPA (Fig. 6.35, **59**); and N-(3,5-dinitrobenzoyl)phenylglycine (Fig. 6.37, **79**) have also been used in this way (Benson, Snyder, et al., 1988) as has binaphthylphosphoric acid (Fig. 6.35, **62a**; Shapiro et al., 1989) and 1,2-diphenyl-1,2-diaminoethane (e.g., for analysis of 2-arylpropanoic acids; Fulwood and Parker, 1992).

Finally, Feringa, Smaardijk, and Wynberg (1985; Smaardijk, 1986) emphasize the utility of the "duplication" method (Section 7-4) as an analytical technique for

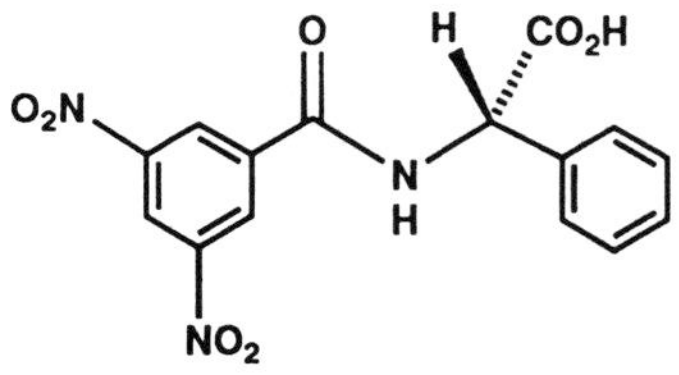

Figure 6.37. *N*-(3,5-Dinitrobenzoyl)phenylglycine.　　　　　　　　　(S)–$(+)$–**79**

determining the enantiomer composition of alcohols. This method eschews the need for a nonracemic chiral reagent. By way of an example, alcohols, such as 2-octanol, are quantitatively converted to mixtures of diastereomeric O,O-di-alkylphosphonates by reaction with PCl_3 (Eq. 6.13) whose ratio must be directly

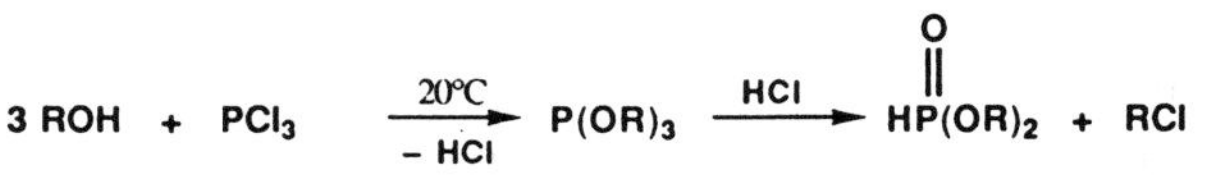

$$3\ \text{ROH} + \text{PCl}_3 \quad \xrightarrow[-\text{HCl}]{20^\circ\text{C}} \quad \text{P(OR)}_3 \quad \xrightarrow{\text{HCl}} \quad \overset{\overset{\text{O}}{\|}}{\text{HP(OR)}_2} + \text{RCl} \qquad (6.13\text{a})$$

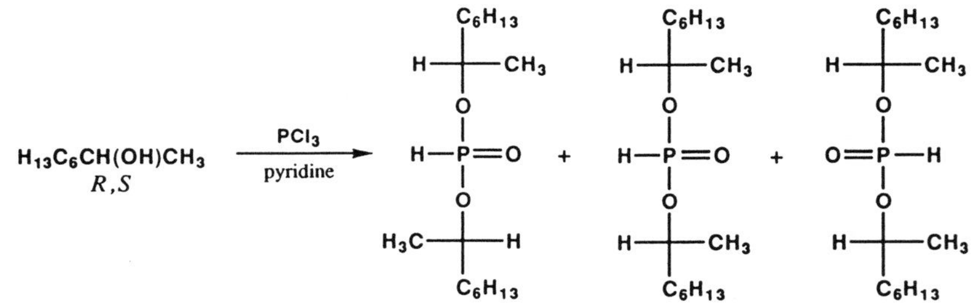

$$(6.13\text{b})$$

related to the proportion of enantiomers in the substrate. Analysis of these diastereomers by ^{31}P NMR gives rise to but three peaks in a ratio of $2:1:1$ in the case of racemic 2-octanol, whereas enantiomerically pure (S)-$(-)$-2-octanol exhibits only one peak (that for the S,S diastereomer).

The duplication process was first conceived by Vigneron, Dhaenens, and Horeau (1973) and applied by them especially to the *purification* of enantiomer mixtures (see Section 7-4). In the case of a racemate, the proportion of the chiral and meso diastereomers is perforce $1:1$ if the reaction is not subject to stereisomer (kinetic) discrimination. This assumption [**s** = 1, where **s** is the stereoselectivity factor (p. 396)] is usually, but not always valid.

If the mole fraction of the major enantiomer, say R, is x and that of the minor enantiomer, S, is $1 - x$, then the product diastereomer mixture will be proportional to x^2 and $(1 - x)^2$ for R,R and S,S (both chiral), respectively, and to $2x(1 - x)$ for R,S [= S,R (meso)]. When expressed in terms of the enantiomeric purity (p = ee) of the starting material, the proportion of the products of duplication are given by $(1 + p)^2$, $(1 - p)^2$, and $2(1 - p^2)$ for R,R, S,S, and R,S, respectively, that is, formation of the diastereomeric products is statistically driven. The analysis of the diastereomer mixture can be easily carried out by quantitative chromatography (GC or HPLC) or by NMR. In the example given above, the ratio of the peak areas of the chiral and (sum of the) meso diastereomers (K, determined by chromatography or NMR) is related to that of the enantiomers in the mixture: $K = (1 + p^2)/(1 - p^2)$ (Kagan and Fiaud, 1988, p. 282). Equation 6.14 is obtained on rearranging terms:

$$p^2 = \frac{K - 1}{K + 1} \qquad (6.14)$$

The principal advantage of this method is that no chiral auxiliary reagent is required for the enantiomeric purity determination. Thus it is also independent of the, sometimes unknown, enantiomeric purity of such auxiliary reagents. The enantiomer purity of cis,trans-1,5-cyclooctadiene has been determined in this way by taking advantage of the spontaneous thermal dimerization of the diene (Leitich, 1978). For an example in which the achiral polyfunctional reagent is a divalent cation, see p. 192 (Pasquier and Marty, 1985). In the more general case in which $s \neq 1$ and the observed diastereoselectivity may be thermodynamic in origin (reversible reaction), Eq. 6.15 (equivalent to Eq. 6.8, p. 193) obtains

$$p \approx \frac{\sqrt{K^2 - s^2}}{K + 1} \qquad (6.15)$$

when the amount of unreacted substrate is low (Kagan and Fiaud, 1988).

The duplication method has been applied to the analysis of the enantiomer composition of chiral amines and amino acid esters [with $CH_3P(=S)Cl_2$ as duplicating agent; Feringa, Strijtveen, and Kellogg, 1986], and to the analysis of thiols using methylphosphonic dichloride [$CH_3P(=O)Cl_2$] as the duplication agent (Strijtveen, Feringa, and Kellogg, 1987). The enantiomeric purity of chiral carboxylic acids has similarly been determined in the presence of tris(tetraphenylimidodiphosphinato)praseodymium [$Pr(tpip)_3$] that converts carboxylate salts into complexes containing two molecules of the carboxylate ligand (Alvarez, Platzer, et al., 1989).

NMR in Chiral Solvents. Chiral Solvating Agents

Following the prediction of anisochrony in NMR spectroscopy with the use of nonracemic chiral solvents (Raban and Mislow, 1965), Pirkle (1966) promptly demonstrated the possibility of distinguishing enantiomers by recording the ^{19}F NMR spectrum of racemic 2,2,2-trifluoro-1-phenylethanol, TFPE (Fig. 6.38, **80**) in (−)-α-methylbenzylamine (PEA, Fig. 6.38). The trifluoromethyl groups of the two enantiomers were nonequivalent with a chemical shift difference of 2 Hz (at 56 MHz). In the same year, Burlingame and Pirkle (1966) found that such nonequivalence was exhibited also in ^{1}H spectra. The observation that the two enantiomers of a compound may exhibit different NMR spectra when the measurement is carried out in nonracemic solvents is one of the most direct manifestations of diastereomer discrimination (Section 6-2). It is not necessary to use a chiral solvent; a nonracemic CSA (i.e., a chiral reagent that complexes, at

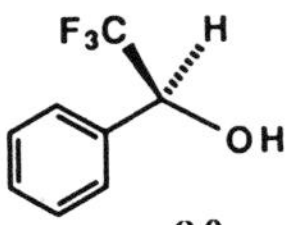

Figure 6.38. Chiral solvating agents (CSA) (Pirkle and Hoover, 1982; Desmukh and Kagan, et al., 1984; Dunach and Kagan, 1985; Toda, Mori and Sato, 1988a and 1988b; Rosini, Salvadori, et al., 1988; Wilen et al., 1991; Jursic and Goldberg, 1992b).

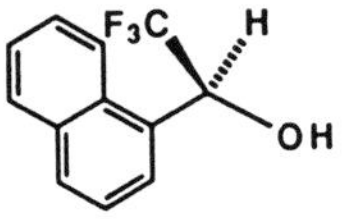

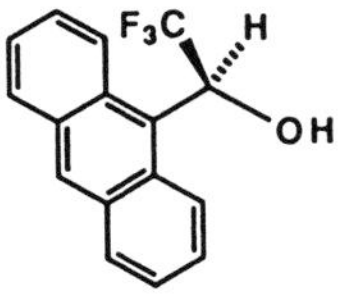

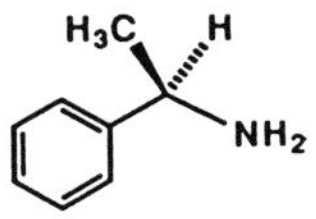

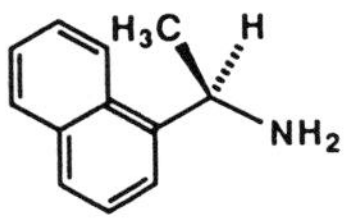

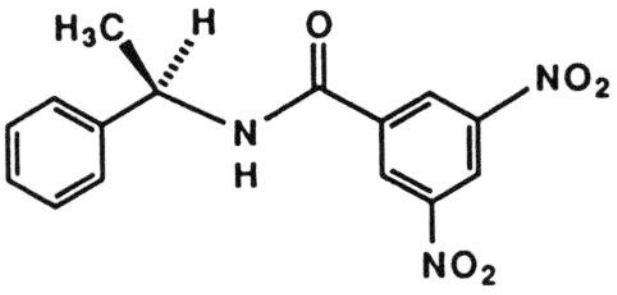

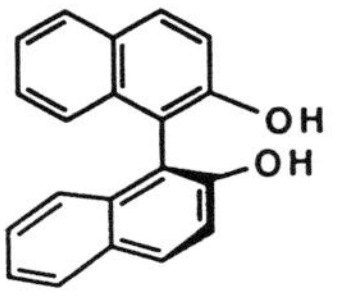

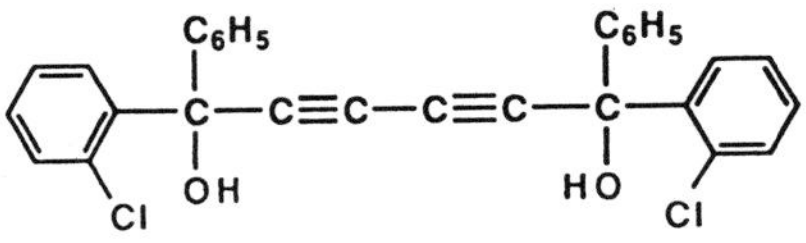

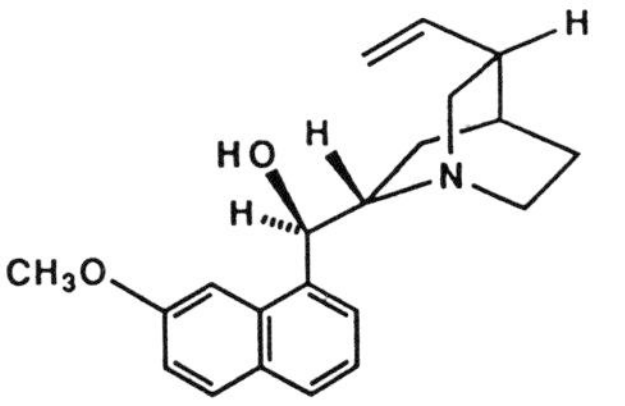

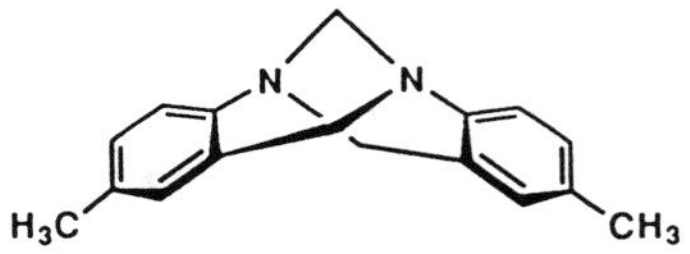

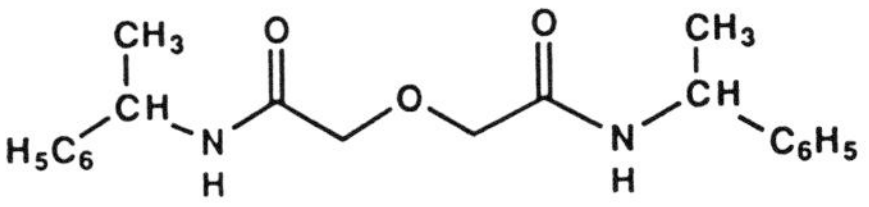

least to some extent, with the substrate) in an achiral solvent is an often used alternative (see below). The application of chiral solvating agents in the determination of enantiomer composition has been reviewed by Pirkle and Hoover (1982), by Weisman (1983) and by Parker (1991).

The structures of the most frequently used CSAs are shown in Figure 6.38. A wide variety of solutes (alcohols, amines, amino acids, ketones, carboxylic acids, lactones, ethers, oxaziridines, sulfoxides, and amine and phosphine oxides) have been analyzed by NMR spectroscopy in the presence of CSAs for the determination of enantiomer purity. Enantiopure 2,2,2-trifluoro-1-(anthryl)-ethanol, TFAE (Fig. 6.38, **81a**; Pirkle and Beare, 1969) is one of the most widely used CSAs in the determination of enantiomer composition, for example, of lactones (see, e.g., Strekowski, Visnick, and Battiste, 1986). Compound **81b** (Fig. 6.38) has been applied to the analysis of chiral methyl sulfoxides (Deshmukh, Kagan, et al., 1984; Toda et al., 1990) and phosphine oxides (Duñach and Kagan, 1985). Based on an earlier finding that carboxylic acids complex strongly with sulfoxides (Nishio, 1969), the determination of enantiomeric composition of chiral sulfoxides (including quasi-symmetrical dialkyl sulfoxides) has been effected with (S)-$(+)$-O-methylmandelic acid (Fig. 6.35, **57**; Buist and Marecak, 1992).

Compounds **82**, **83** (Toda, Mori, et al., 1988a; Toda, Mori and Sato, 1988b), **84a** (quinine, Fig. 6.38; Rosini, Salvadori, et al., 1988), and Tröger's base (Fig. 6.38; Wilen et al., 1991) are more recently introduced CSAs, the latter two for the analysis of alcohols. Compound **84b**, a dicarboxamide CSA, has been applied to the determination of the enantiomer composition of carboxamides and hence of their precursors, amines and acids (Jursic and Goldberg, 1992b). The application of quinine as CSA (also for the analysis of binaphthyl derivatives and for N-(3,5-dinitrobenzoyl)-β-hydroxy-β-phenethylamines) is noteworthy because of its ready availability and low cost (Salvadori et al., 1992). The utility of β- and γ-cyclodextrins (Section 7-3.c) as CSA in the determination of the enantiomeric composition of amine salts (e.g., propranolol hydrochloride) *in water* has been described (Greatbanks and Pickford, 1987). Spectral nonequivalence generated by cyclodextrins is due to formation of diastereomeric inclusion compounds *in solution* (Section 7-3.c).

It was first recognized by Mislow and Raban (1965) that interaction of chiral solute molecules with a chiral nonracemic environment transforms the previous enantiotopic relationships of specific nuclei into diastereotopic ones (both by internal and by external comparison; cf. Chapter 8). The differences in magnetic environment of nuclei are generated by diastereomeric solvates that form and dissociate rapidly on the NMR time scale [Eq. 6.16 where (R)-A and (S)-A stand for the enantiomers of the substrate (analyte) and (R)-X stands for a chiral solvating or shift reagent]. The NMR shifts observed are thus average shifts of the solvated (right-hand side of Eqs. 6.16a and 6.16b) and unsolvated species (left side of the equations), and since the shifts for the former, though not the latter, differ for the enantiomers, so does the average.

$$(R)\text{-A} + (R)\text{-X} \underset{}{\overset{K_{RR}}{\rightleftharpoons}} (R)\text{-A} \cdot (R)\text{-X} \tag{6.16a}$$

$$(S)\text{-A} + (R)\text{-X} \underset{}{\overset{K_{SR}}{\rightleftharpoons}} (S)\text{-A} \cdot (R)\text{-X} \tag{6.16b}$$

There are two reasons why the equilibria in Eqs. 6.16a and 6.16b taking place in the presence of CSA lead to differently shifted signals: (a) Intrinsic: (R)-A $\cdot (R)$-X and (S)-A$\cdot (R)$-X are diastereomers and therefore have, at least in principle, different chemical shifts. Of course, (R)-A and (S)-A, being enantiomers, have the same shift, so the existence of rapid equilibrium diminishes the observed shift difference (by averaging) but does not obliterate it. (b) Differential stability of diastereomeric solvates: $K_{RR} \neq K_{SR}$, since the species on the right-hand side of the two equations are diastereomers, and hence may be unequally stable (the species on the left-hand side of Eqs. 6.16a and 6.16b, in contrast, have equal stability). Thus complexing may not be the same for the two enantiomers; other things being equal, the more complexed enantiomer is likely to present the larger chemical shift change by the CSA or lanthanide shift reagent (LSR). However, complexing to unequal extents cannot be the only reason for the shift differences between the two enantiomers. Explanation (a) is also significant; this follows from the observation of shift differences of enantiotopic nuclei *within one molecule* (by use of CASs or LSRs; for the latter cf. page 237. See also section 8-4.a). Here differential solvation or complexation cannot come into play.

The observed chemical shift differences ($\Delta\Delta\delta$) vary with spectrometer frequency but are typically quite small (0–10 Hz for ^{1}H NMR at 100 MHz). Sharp, well-resolved signals are required particularly for quantitative work. The limitation of inconveniently small shift differences can be overcome in several ways. One way is by increasing the spectrometer frequency. Another way is to lower the temperature thereby increasing solvate formation; however, this possibility is subject to solubility limitations. The magnitude of the NMR spectral nonequivalence can also be increased by addition of achiral shift reagents, for example, Eu(fod)$_3$ (Fig. 6.41) (mixed CSA-shift reagent system: Jennison and Mackay, 1973). The effects observed upon such addition have been interpreted as arising from the differential stability of the solute-CSA solvates (Wenzel, 1987, p. 135); shift reagent molecules can, in some cases, preferentially displace solute molecules from the least stable solvate (Pirkle and Sikkenga, 1975). The enhancement has been observed especially in the case of analytes that associate with lanthanides independently of the presence of CSAs, such as **81b** (Fig. 6.38), that themselves associate weakly, if at all, with lanthanide ions (Wenzel et al., 1992). The enhancement is also seen with cyclodextrins acting as CSAs (see above).

The following additional considerations apply: (a) the effect has been observed on ^{31}P, ^{13}C, and ^{15}N as well as ^{1}H and ^{19}F nuclei; (b) the CSA need not be used as neat liquids; achiral diluents (CCl$_4$ or CS$_2$) are tolerated and solid CSAs, such as TFAE (Fig. 6.38, **81a**), can be used; (c) the association between solute molecules and the CSA, and hence the $\Delta\Delta\delta$ are maximized in nonpolar solvents. Polar solvents, such as CH$_3$OH or DMSO often, but not invariably, interfere with diastereomer solvate formation. There is also a significant concentration dependence of the CSA (at a fixed solute concentration) on the magnitude of $\Delta\Delta\delta$; (d) the CSA need *not* be enantiomerically pure. The last of these considerations is noteworthy for it conveys an absolute character to this enantiomeric purity determination.

In general, analytical methods that involve formation of transient diastereomers or dynamic diastereomer systems (ion pairs, charge-transfer complexes, solvates, and complexes with shift reagents) in which the diastereomeric partners and the enantiomer substrates undergo rapid exchange do not require that the chiral detecting agent (solvent, shift reagent, or stationary phase; Section 6-5.d) be enantiomerically pure. In the absence of complete enantiomer purity of the reagent, the observed anisochrony or separation is simply attenuated. With racemic CSA, the chemical shift differences disappear altogether, as a consequence of the rapid statistical averaging of solvate formation (Sullivan, 1978, p. 320; Jacques et al., 1981a, p. 393).

Enantiomerically pure solutes in CSA will evidently not exhibit the multiple signals just described. However, since there is no a priori way of insuring that lack of peak splitting is not accidental in such a case, that is, that the theoretically demanded nonequivalence might just be insufficient to produce additional peaks, the determination of enantiomer purity on samples of unknown enantiomer composition must be preceded by a comparable analysis carried out on a racemic or partly racemic sample where signals for both enantiomers must be seen so as to validate the method. In the absence of such a sample, the magnitude of the nonequivalence could still be assessed by observing the signals obtained when the spectrum is measured first with one enantiomer of the CSA and then with the other, under rigorously identical conditions, (Jochims et al., 1967; cf. p. 226 for a similar situation with chiral derivatizing agents).

Analysis in the presence of CSA has also been carried out for the assignment of absolute configuration. The latter use follows from the possibility of relating the sense of chemical shift nonequivalence (upfield vs. downfield shift of a signal) for one enantiomer of a compound dissolved in a given CSA, to the configuration of that enantiomer. This configurational correlation depends on the application of models that specify the number and kind of interactions possible between solute and solvent (e.g., see Fig. 6.39); the correlation is subject to the limitation of mechanistic understanding of all such models (cf. Chapter 5).

The dibasic solute model (see also Fig. 8.40; Pirkle and Hoover, 1982, p. 28) requires that CSA and solute molecules have complementary functionalities permitting formation of chelate-like solvates (Fig. 6.39). In appropriate solutes (e.g., amines), there is simultaneous interaction of two basic sites (B_1 and B_2) in the solute with acidic sites in the CSA. This interaction accounts for the widespread use of TFPE and of TFAE (Fig. 6.38, **80** and **81a**, respectively) with their acidic hydroxyl and carbinyl hydrogen atoms both of which may participate in hydrogen bonding. The aromatic rings serve as anisotropic shift perturbers (enhancers), thus providing yet a third site for interaction between CSA and solute. A size difference between sites B_1 and B_2 is not essential. At 100 MHz, the CH_3 protons of the RS solvate (Fig. 6.39) are shifted to lower field (2.8 Hz) while the carbinyl hydrogen and the CH_3' (ester) group of this solvate are both shifted upfield (by 2.3 and 1.3 Hz, respectively; Pirkle and Beare, 1969).

Compounds that act as inclusion hosts, e.g., **82** and **83** (Fig. 6.38) and **115** (Fig. 7.36) also serve as CSAs. The observation of ^{1}H NMR differential shifts,

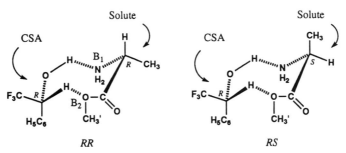

Figure 6.39. Dibasic solute model for diastereomeric solvates. Effect of (R)-$(-)$-TFPE on ^{1}H NMR of methyl alanate enantiomers (100 MHz, 29°C, $CFCl_3$ diluent); B_1 = hydrogen bond receptor and B_2 = carbinyl hydrogen bond receptor (Pirkle and Beare, 1969).

e.g., of α-methylbenzylamine (Fig. 6.38, PEA) with these CSAs has been attributed to formation of 2:1 (e.g., **115**: PEA) molecular complexes (clathrates; Sec. 7-3.c) in solution. This interpretation is supported by the X-ray crystal structure of the crystalline diastereomeric host-guest inclusion compound formed between (R, R)-$(-)$-**115** (Fig. 7.36) and (S)-$(-)$-PEA (Toda, Hirotsu, et al., 1993).

Chemical shift differences between nonracemic chiral diastereomers observed in achiral solvents may also be enhanced in the presence of CSAs (e.g., Thomas, 1966 and Kaehler and Rehse, 1968). Moreover, since *achiral* molecules possessing enantiotopic nuclei exhibit nonequivalence of peaks for these nuclei in non-racemic chiral solvent, use of CSAs permits the differentiation, assignment of configuration, and quantitative estimation of many meso and chiral diastereomer mixtures. For example, the methine hydrogen atoms H_a of *rac*-dimethyl 2,3-diaminosuccinate (Fig. 6.40) are homotopic (singlet), whereas the corresponding hydrogen atoms $H_{a'}$ in the meso diastereomer are enantiotopic (see also Section 8-4.b). In the presence of TFPE (Fig. 6.38), the latter become diastereotopic and exhibit coupling. An AB quartet, $J_{AB} = 3.7$ Hz is discerned. In contrast, the racemic diastereomers exhibit two singlets for the methine hydrogen atoms, one within each enantiomer (Kainosho, Pirkle, et al., 1972) so there is no coupling. Other cases are described in Chapter 8 (Section 8-4.b) and by Bovey (1969).

A second, less common but more dramatic, way of measuring enantiomer composition is possible by NMR spectroscopy with molecules that strongly self-associate. Partially racemic samples, for example, of some amines and carboxylic acids exhibit split signals ascribable to the enantiomer constituents in neat liquid samples or in solutions of *achiral* nonpolar solvents. In this case, the bulk sample itself serves as CSA.

The self-induced nonequivalence still requires formation of diastereomeric species (associates), that is, *RR* (or *SS*) and *RS* dimers (cf. Eqs. 6.1 and 6.2), which differ in their NMR properties. Consequently, one would expect to observe this otherwise unexpected phenomenon only when hydrogen bonding or other strong association is present. Unlike the typical CSA experiment, the self-induction of diastereotopicity observed in achiral solvents (or in absence of any solvent) is a manifestation of enantiomer discrimination. This possibility has been described earlier (see Section 6-4.m).

Finally, the possibility of applying liquid crystals as CSA has been explored. A homogeneously oriented cholesteric solvent mixture (composed of nematic compounds together with cholesteryl propionate) has been found in which every ^{1}H NMR signal of *rac*-1,2-epoxy-3,3,3-trichloropropane is doubled with the origin of the anisotropy ascribed to dipolar coupling (Lafontaine, Bayle, and Courtieu, 1989). Such solvent mixtures have been applied to the determination of enantiomer composition (Lafontaine et al., 1990).

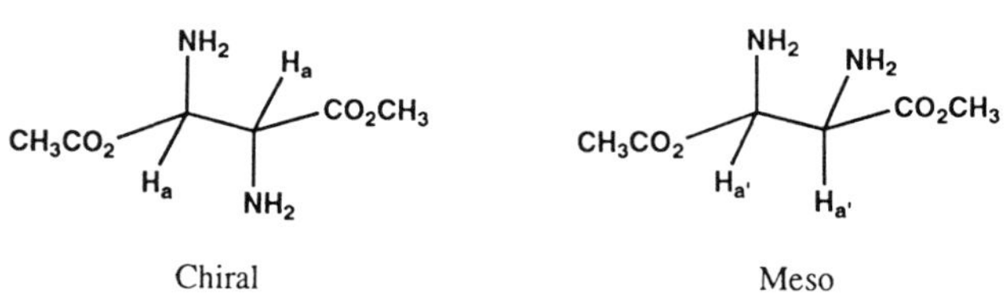

Figure 6.40. Nuclear magnetic resonance differentiation of diastereomers in CSA.

NMR with Chiral Shift and Chiral Relaxation Reagents

Shift Reagents. Lanthanide shift reagents (LSRs) are compounds prepared by reaction of certain transition metal (Eu, Pr, or Yb) salts with β-diketones. The LSRs themselves are tris-chelates of coordination number 6 that behave as weak Lewis acids. In nonpolar solvents (e.g., $CDCl_3$, CCl_4, or CS_2), these paramagnetic salts are able to bind Lewis bases, especially amides, amines, esters, ketones, and sulfoxides, by expansion of the coordination sphere of the metal. The organic solutes rapidly exchange between bound and unbound states.

As a result, protons, carbon atoms, and other nuclei have averaged magnetic environments and are chemically shifted (usually deshielded relative to their positions in the uncomplexed substrates) to an extent that depends on the strength of the complex and how far the nuclei are from the paramagnetic metal atom. Thus, different nuclei are shifted to different extents, thereby enhancing spectral dispersion (much as a higher magnetic field would) and leading to spectral simplification.

Achiral LSRs (Fig. 6.41) are occasionally used to enhance anisochrony in the spectra of *diastereomer mixtures*, facilitating quantitative analysis of such mixtures (see Corfield and Trippett, 1971).

When the permanent coordination sphere of the metal in the LSR is chiral (through complexation with nonracemic β-diketones, such as 3-acylcamphor or dicampholylmethane), chiral shift reagents (CSRs) are created. Addition of such reagents to racemic organic solutes in solution gives rise to diastereomeric complexes. Individual groups in the organic solute are differentially shielded from the external magnetic field just as they are when achiral LSRs are used; the peaks are spread over a wide range of chemical shifts leading to spectral simplification. Superimposed upon this is the anisochrony of externally enantiotopic groups that have become diastereotopic under the influence of the chiral LSR (cf. Chapter 8). If peaks due to the two enantiomers of a racemic solute are well separated, quantitative analysis of enantiomer composition becomes feasible.

> The term CSR is limited to chiral paramagnetic compounds containing transition elements that produce so-called pseudocontact shifts (Morrill, 1986a; Wenzel, 1987). Chiral diamagnetic substances not containing metals that effect NMR spectral nonequivalence in chiral compounds are considered to be CSAs and have been treated in the preceding subsection (page 231). The two phenomena have, however, a number of features in common.

Lanthanide induced shifts (LISs) due to chiral LSRs were first observed by Whitesides and Lewis in 1970 [Fig. 6.41, Eu(*t*-cam)$_3$]. Shortly thereafter, Fraser et al. (1971) and Goering et al. (1971, 1974) described CSRs that had a much greater range of applications. The application of chiral LSRs to the determination of enantiomer composition has been reviewed by Sullivan (1978), Reuben and Elgavish (1979), Fraser (1983); Schurig (1985b), Wenzel (1987, Chapter 3) and Parker (1991). The structures of the most frequently used CSRs are given in Figure 6.41.

The following particulars apply to the use of CSRs. Substrate concentrations range from 0.1 to 0.25 M while the CSR/substrate ratio generally is in the range of 0.5–1. Observed ^{1}H chemical shift differences $\Delta\Delta\delta$ between corresponding groups in the two enantiomers are in the $0.1 \rightarrow 0.5$-ppm range (occasionally as

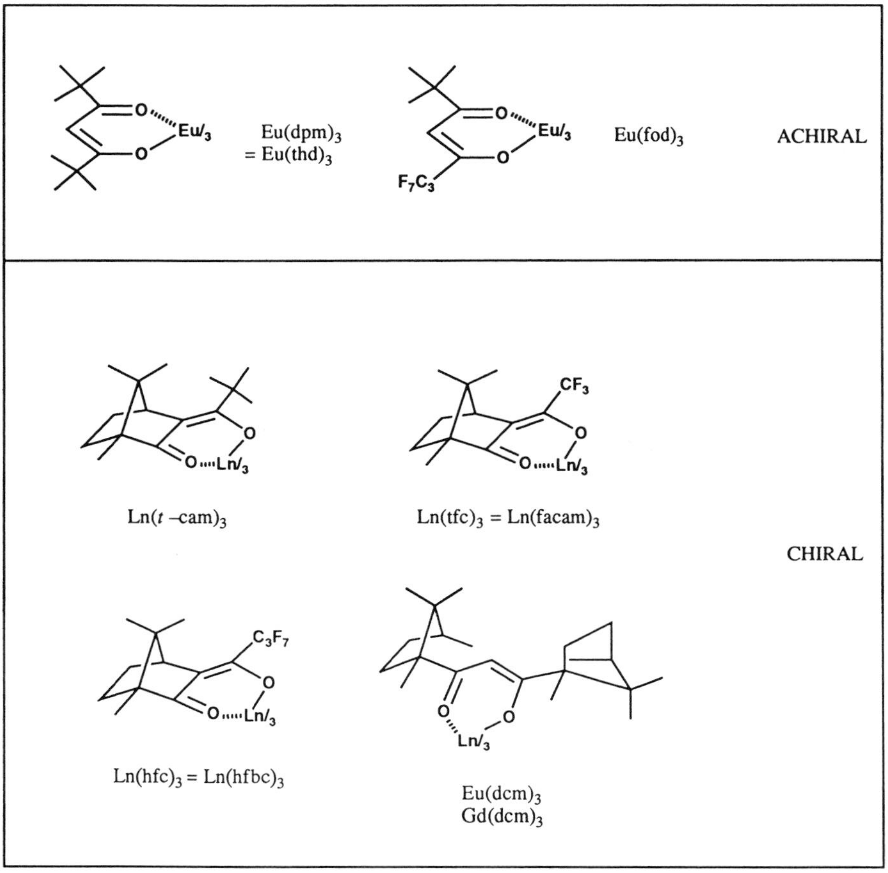

Figure 6.41. Lanthanide shift and relaxation reagents (Ln = a lanthanide metal). Eu(dpm)$_3$ = Eu(thd)$_3$: tris(dipivaloylmethanato)europium(III); Eu(fod)$_3$: tris(6,6,7,7,8,8,8-heptafluoro-2,2-dimethyl-3,5-octanedionato-O,O')europium(III); Eu(t-cam)$_3$ (Ln = Eu): tris(3-*tert*-butylhydroxy-methylene-d-camphorato)-europium(III); Eu(tfc)$_3$ = Eu(facam)$_3$ (Ln = Eu): tris(3-trifluoroacetyl-d-camphorato)-europium(III); Eu(hfc)$_3$ = Eu(hfbc)$_3$ (Ln = Eu): tris(3-heptafluorobutyryl-d-camphor-ato)-europium(III); Eu(dcm)$_3$: tris(d,d-dicampholylmethanato)-europium(III).

large as 4 ppm), that is, 10–50 times as large as the nonequivalences observed with CSAs. CSRs are usable with ^{1}H, ^{13}C, and ^{19}F probe nuclei and the accuracy of enantiomer compositions are unaffected if the CSRs are not enantiomerically pure, though the resolution of the enantiotopic signals decreases with decreasing enantiomeric purity of the CSR. The choice of metal for the CSR determines the magnitude of the LIS, the direction (upfield or downfield) of the LIS, and the extent of peak broadening resulting from the paramagnetic character of the metal. The specific CSR selected for a particular use is often a compromise of several factors including accessibility and cost of the reagent (Fraser, 1983; Wenzel, 1987, Chapter 2).

The CSRs have many advantages over CSAs or the use of CDAs in the determination of enantiomer composition by spectroscopy. Among these are the convenience of the method and the ease of interpretation and manipulation of the spectroscopic data; and, in comparison with CDAs, the absence of the need for enantiomerically pure reagents and of the need for quantitative derivatization. The CSRs are applicable to the analysis of most "hard" organic bases. One of the most effective CSRs in differentiating nuclei that are enantiotopic by external comparison is $Eu(dcm)_3$ (Fig. 6.41; McCreary, Whitesides, et al., 1974).

However, the application of CSRs is not always successful. In most instances only a few nuclei in the analyte exhibit stereoisomer discrimination in their shifts. Insufficient peak resolution, signal broadening, chemical degradation of the CSR, and inattention to instrumental conditions may thwart success in the use of this technique. For an example (determination of the enantiomer composition of mevalonolactone), see Wilson, Scallen, and Morrow (1982). Computer processing of NMR spectra (rephasing, base-line correction and Gaussian line narrowing) has been shown to increase the precision of the analyses (Peterson and Stepanian, 1988). The use of high-field (even 300 MHz) NMR spectrometers is detrimental to the application of LSRs since, under the typically prevailing fast exchange conditions, signal broadening is proportional to the square of the magnetic field strength (Parker, 1991). This problem is particularly acute with compounds exhibiting large LISs (e.g., alcohols). The broadening can be reduced and useful anisochrony restored on warming the sample (e.g., see Wenzel et al., 1992) or on application of NMR spin-echo techniques (Bulsing, Sanders, and Hall, 1981).

Chemical exchange is the origin of the signal broadening effect. The two principal variables affecting the signals due to species undergoing exchange are field strength, which affects line separation, and temperature, which affects the exchange rate. It is the interplay between these variables that is responsible for both the line broadening at high field and its reduction on warming. The fast exchange approximation $\delta\nu = \pi\nu_{AB}^2/2k$, where k is the rate constant for exchange, shows that line width ($\delta\nu$) is proportional to the square of the separation (chemical shift difference ν_{AB}) between two lines; this separation is, in turn, proportional to the square of the field (Anet and Bourn, 1967).

Combination of CSRs with soft acids (Ho, 1977; e.g., $AgNO_3$ or $AgOCOCF_3$) or with an achiral LSR [e.g., Ag(fod)] gives rise to so-called binuclear lanthanide(III)–silver(I) shift reagents. The latter sometimes permit the determination of enantiomer composition of weak or "softer" bases, such as alkenes, arenes, and halogen compounds (Offermann and Mannschreck, 1981; Wenzel and Sievers, 1981, 1982; Wenzel and Lalonde, 1983) and allenes (Mannschreck et al., 1986; Peterson and Jensen, 1984) with which they complex [Reviews: Wenzel, 1986; Wenzel, 1987, Chapter 4].

Enantiomer mixtures may need to be converted to derivatives to obtain reasonable peak dispersion with CSRs. Optically active glycols, for example, have been converted to mixtures of epimeric 1,3-dioxolanes on condensation with benzaldehyde. The analysis of the mixture by NMR [by addition of $Eu(hfc)_3$; Fig. 6.41] has been described by Eliel and Ko (1983). Subsequently, determination of the enantiomer composition of polar substrates (diols, triols, or glycidol) was shown to be possible in polar solvents, such as CD_3CN, acetone-d_6, and $CDCl_3$, with $Eu(facam)_3$ and with $Eu(hfbc)_3$ (Fig. 6.41; Sweeting, Crans, and Whitesides, 1987).

The reagent Eu(hfc)$_3$ has been applied to the determination of the enantiomer composition of compounds whose chirality depends only on the presence of H and D or CH$_3$ and CD$_3$ at the stereogenic atom (Eliel et al., 1986). Chiral shift reagents permitting enantiomer purity analysis of amino and hydroxy acids and of carboxylic acids in aqueous solutions have also been described (Reuben, 1980; Peters, van Bekkum, et al., 1983; Kabuto and Sasaki, 1984, 1987). Chiral shift reagents can also help to differentiate *meso-* from *rac-*diastereomers (Goe, 1973; cf. Chapter 8) and to induce chemical shift differences between internally enantiotopic nuclei such as the two benzylic protons of benzyl alcohol (Fraser et al., 1972).

An achiral dinuclear lanthanide shift reagent, Pr(tpip)$_3$, has been applied to the enantiomeric purity determination of carboxylic acids (Alvarez, Platzer, et al., 1989). The complexes formed in the presence of carboxylic acid salts contain two carboxylate ligands hence, when racemic salts are analyzed, two diastereomeric complexes are formed giving widely separated signals. This finding is yet another example of the duplication method described in Section 6-5.c, p. 229 (see also Section 7-4).

Two possible and probably not independent explanations for the spectral nonequivalence observed in the presence of CSRs have been advanced. These explanations have much in common with the anisochrony observed with CSAs (Section 6-5.c, p. 231): (a) In the presence of CSRs, transient time-averaged diastereomeric complexes of different stabilities are formed that equilibrate with the unbound enantiomers of the substrate (alcohol, amine, etc.); see Eq. 6.16 (Fraser, 1983). (b) The geometries of the CSR–chiral substrate diastereomeric complexes will differ, and hence the magnetic environments of selected nuclei in the complexes may differ sufficiently to lead to anisochrony (Goering et al., 1971; see also McCreary, Whitesides, et al., 1974 and Peterson and Stepanian, 1988). The CSRs may also generate transient diastereomeric species of more than one type of stoichiometry (Fraser et al., 1971; Goering et al., 1974), which further affects the position of equilibrium of complexed and uncomplexed species as well as the intrinsic shifts in the complexed species.

Relaxation Reagents. It has been observed that Gd(dcm)$_3$ (Fig. 6.41) forms diastereomeric complexes with chiral substrates (e.g., methylephedrine or camphor) in which the two enantiomers of the base do not exhibit different chemical shifts. However, the ^{13}C relaxation rates of selected nuclei differ significantly for the two enantiomers. The reagent Gd(dcm)$_3$ is a substance that promotes the differential relaxation of nuclei from their excited to the ground state, that is, it is a chiral relaxation reagent (CRR). This difference in properties of the nuclei has been adapted to the quantitative determination of enantiomer composition. A significant limitation of CRRs is that the NMR inversion recovery experiments required to measure relaxation rates are quite time consuming (Hofer et al., 1984).

d. Chromatographic and Related Separation Methods Based on Diastereomeric Interactions

Chromatographic methods, methods that depend on the total separation (direct or indirect) of the enantiomers of a chiral substance, are among the most powerful available for the determination of enantiomer composition. The earliest determinations of this type required prior conversion of enantiomer mixtures to diastereomeric derivatives. The latter were then analyzed by GC on achiral stationary phases. Subsequently, HPLC has been applied to such analyses.

A second type of chromatographic process applied to enantiomer purity determinations is the separation of enantiomers on chiral stationary phases (CSPs). Both GC and HPLC techniques have been used in this way. Most recently, enantiomeric composition analyses have been carried out by HPLC on achiral stationary phases with the aid of chiral mobile phases.

All of the separation processes mentioned rely on the intervention of stable or transient diastereomeric species whose different solubilities, stabilities, or adsorption characteristics are responsible for the separation of the stereoisomers.

Although the focus of this section is on separations (resolutions) effected for analytical purposes, it will be evident that small scale preparative chromatographic resolutions partake of identical methodology and even conditions. Although preparative chromatographic resolutions are described in Chapter 7, much of what follows here is pertinent to such macroscopic separations. Indeed, many articles in the literature dealing with chromatographic resolutions leave unclear the principal intent of the work, analytical or preparative.

Chromatography on Diastereoselective Stationary Phases

Gas Chromatography. Since the report by Casanova, Jr., and Corey (1961) that the resolution of camphor could be effected by means of GC of its dioxolane derivatives with (−)-2,3-butanediol, it has been apparent that GC of diastereomer mixtures offers a simple route to the assessment of enantiomer composition of chiral substances. An even earlier report describes the analysis of *rac*-phenylalanine by GC of its L-alanyl derivatives [as the *N*-trifluoroacetyl (TFA) methyl esters; Weygand et al., 1960]. The development of this method was carried out during the early 1960s by several groups, notably those of Gil-Av and Nurok (1962), Halpern and Westley (1965a,b), Guetté and Horeau (1965), Weygand et al. (1963), Feigl and Mosher (1965), and Sanz-Burata et al. (1965) with the focus at first being very much on the analysis of amino acids. (For a brief historical account, see Feibush and Grinberg, 1988. For reviews of the early literature, see Raban and Mislow, 1967; Karger, 1967; and Rogozhin and Davankov, 1968. For more recent reviews, see Gil-Av and Nurok, 1975; Halpern, 1977; Knapp, 1979; Drozd, 1981 and Souter, 1985. Additional reviews are cited below).

Gas chromatographic analysis of diastereomer mixtures has the same limitations as NMR analysis does, including the requirement that the CDA must be enantiomerically pure. With an enantiomerically impure CDA, the minor enantiomer generates a diastereomer mixture that is enantiomeric with the diastereomer mixture produced by reaction of the dominant CDA enantiomer with the analyte. Since the two (enantiomeric) diastereomer mixtures are not separable in an achiral stationary phase (just as they do not give rise to separate signals on NMR analysis in achiral solvents), the analysis is falsified.

By way of example, consider an analyte **A** consisting of 99.5% (+) and 0.5% (−) enantiomers (99% ee) that is derivatized with enantiomerically impure CDA (+)-**B** [the latter consisting of 99% (+)-**B** and 1% (−)-**B**]. Following quantitative derivatization, four diastereomeric products are present: I (+)-**A** · (+)-**B** (98.5%), II (−)-**A** · (+)-**B** (0.5%), III (+)-**A** · (−)-**B** (1.0%), IV (−)-**A** · (−)-**B** (0.0%). Chromatographic analysis of the mixture on an achiral stationary phase yields but two peaks: I + IV (these being enantiomers), and II + III (also enantiomers) with the mixture having an apparent enantiomer composition of 98.5% (+) and 1.5% (−), hence 97% ee (Allenmark, 1991, p. 60).

Care must be taken during the preparation and isolation of the derivative to avoid accidental fractionation lest the diastereomer ratio not be equal to that of the enantiomers prior to derivatization (Schurig, 1983). In addition, the derivatives must be thermally stable and reasonably volatile if the analysis is not to be excessively time consuming due to long retention times. Here, as in NMR, a control experiment on derivatives of the racemate or of a synthetic mixture of the two enantiomers of the analyte must be carried out to insure that absence of a second GC peak reflects enantiomeric homogeneity and not the inability of the column to resolve the diastereomeric derivatives.

As pointed out by Karger (1967), the degree of separation of the diastereomeric derivatives, and hence the efficiency of the separation, depends principally on the CDAs rather than on the (achiral) stationary phase (see also Rose, Stern and Karger, 1966). In consequence, many CDAs have been examined for their potential in the determination of enantiomer composition. More recently, capillary columns have been increasingly applied to such analyses thereby further enhancing the effectiveness of the method. The number and variety of CDAs used with GC (and analogous HPLC techniques, to be described below) is so large that only a limited number of examples can be cited here (see Fig. 6.42). For a summary, see Allenmark, 1991, Chapter 4.

Chiral derivatizing agents derived from amino acids have been applied to the GC analysis of chiral secondary alcohols (Ayers, 1970), of amines (Halpern and Westley, 1966), and of amino acids (Halpern and Westley, 1965a). Natural product derived CDAs (other than amino acids) have been used for the GC determination of enantiomer purity of alcohols, such as borneol and menthol (e.g., as their tetra-*O*-acetyl glucosides; Sakata and Koshimizu, 1979). In turn, menthol, as the chloroformate (Fig. 6.42), has been applied to the enantiomer purity determination of amino acids (e.g., as the *N*-TFA derivatives) as well as of alcohols and α-hydroxy acids (Halpern and Westley, 1965b; Westley and Halpern, 1968; Hauser et al., 1974; and Hirota, Izumi, et al., 1976). A chromanecarboxylic acid CDA derived (following resolution) from the racemic commercial antioxidant Trolox™ (Fig. 6.42) has been applied to the enantiomeric purity determination of chiral 1° and 2° alcohols (Walther, Netscher, et al., 1991). As an alternative, 1° alcohols have been analyzed following chromic acid oxidation to carboxylic acids; the oxidation is known to proceed without racemization. The latter are derivatized to amides with α-methylbenzylamine (Sonnet, 1987; Högberg et al., 1992).

Lactones have been analyzed as ortho esters following their derivatization with chiral nonracemic 2,3-butanediol (Saucy, Jones, et al., 1977). Mono- and sesquiterpenoids (chrysanthemic and drimanoic acids) have been applied to the analytical resolution of alcohols and amines by GC (Brooks et al., 1973). These

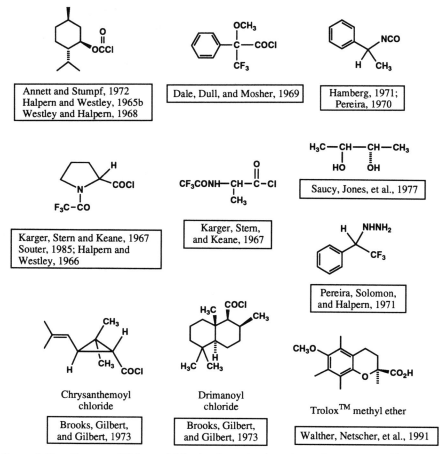

Figure 6.42. Common CDAs useful in the determination of enantiomer composition by GC.

CDAs (Fig. 6.42) were chosen because the stereocenters of the reagent are embedded in conformationally rigid molecular skeletons (see also, Rose, Stern, and Karger, 1966). These reagents led to a better separation than that observed with menthoxyacetic acid and menthoxychloroformate reagents when applied to amine and alcohol substrates. A chiral hydrazine reagent (Fig. 6.42) provides a means for the enantiomer purity determination of ketones via conversion of the latter to hydrazones followed by GC resolution (Pereira, Solomon, and Halpern, 1971). Diels-Alder adducts of dienes with α,β-unsaturated aldehydes have been analyzed for their enantiomer content as acetals following derivatization with chiral nonracemic 2,4-pentanediol (Furuta, Yamamoto, et al., 1989). Stereo-chemical analysis of open-chain terpenoid carbonyl compounds (e.g., hexahydro-farnesylacetone) has been carried out by GC on dioxolane derivatives prepared by reaction with $(2R,3R)$-$(+)$-diisopropyl tartrate. The diastereomeric deriva-tives derived from racemic analyte exhibit no anisochrony in their ^{1}H or ^{13}C NMR spectra (Knierzinger, Walther, et al., 1990).

High-Performance Liquid Chromatography. While the analysis of diastereo-mer mixtures by GC will most likely continue to be a useful method for a long time to come, it is limited to compounds that are volatile and reasonably thermally stable. In recent years, it has been complemented, and sometimes superseded, by the analogous HPLC methods and by methods that employ chiral stationary phases. The versatility of HPLC, its wider range of application, and the wider choice of process variables (adsorption vs. partition mode, choice of mobile as well as stationary phase, etc.) is responsible for the increase in the use of this modern technique.

Among the earliest applications of HPLC to the determination of enantiomer composition is the report of the chromatographic separation of lactates and mandelates of alcohols (Leitch, Rothbart, and Rieman, 1968). Application of the HPLC technique became more common in the mid-1970s when HPLC instruments and columns filled with microparticulate stationary phases became routinely available.

Some of the very same CDAs that have been applied to the determination of enantiomer composition by NMR spectroscopy (e.g., Mosher's reagent, MTPA; Fig. 6.35) have also served in the analysis of stereoisomer mixtures by chromatography (e.g., Koreeda, Weiss, and Nakanishi, 1973; Valentine, Saucy et al., 1976).

The principal types of chiral derivatizing agents used in the preparation of diastereomeric derivatives used in HPLC are acylating agents, amines, isocyanates and isothiocyanates, and alkylating agents (Fig. 6.43) [Reviews: Allenmark, 1991, Chapter 4; Lindner, 1988; Ahnoff and Einarsson, 1989]. Many of these have figured in drug metabolism studies of enantiomer mixtures (Testa, 1986). Some of these reagents have been developed specifically to resist racemization prior to or during the derivatization reaction (Cook, 1985). [The problem of racemization of the reagent may be illustrated with the commercial derivatizing agent (−)-N-TFA-1-prolyl chloride which, at least on one occasion, was discovered to be contaminated to the extent of 4-15% with the (+) enantiomer, presumably due to racemization on storage (Wainer and Doyle, 1984).] It is fair to say that *all* CDAs should be checked for enantiomer purity before use.

The formation and chromatographic resolution of stable diastereomeric platinum complexes has been described in connection with the determination of enantiomer composition of chiral alkenes and sulfoxides by HPLC (Goldman, Gil-Av, et al., 1982; Köhler and Schomburg, 1981, 1983). Selection of CDAs for liquid chromatographic resolution of acids has been influenced by the obervation that diastereoisomeric amides are more strongly hydrogen bonded to silica gel and to alumina than are diastereomeric esters. Diastereoselectivity is enhanced by the presence of suitably located aromatic and polar functional groups, such as hydroxyl (Helmchen et al., 1979a,b,c; Pirkle and Finn, 1983). The determination of the enantiomer composition of both carboxylic acids and of amines can readily be effected by prior conversion to diastereomeric amides. Hydroxyamides, such as those derived from phenylglycinol or from β-phenylbutyrolactone (Fig. 6.43) have been shown by Helmchen to be especially good CDAs for both analytical and preparative purposes (Helmchen et al., 1972, 1977, 1979a,b; Helmchen and Strubert, 1974; Helmchen and Nill, 1979c). Separation factors, α, which are a measure of the ease of separation of the diastereomers (Section 7-3.d), are quite large ($\alpha > 2.5$) for diastereomeric amide pairs derived from phenylglycinol (Fig. 6.43), whereas α values greater than 2 are rarely found in other typical diastereomer systems.

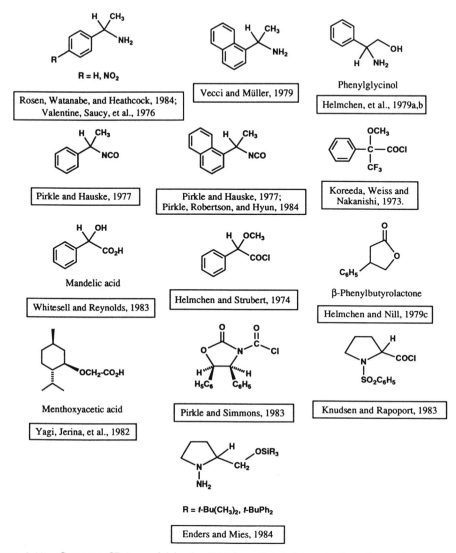

Figure 6.43. Common CDAs useful in the HPLC mediated determination of enantiomer composition.

Diastereomeric carbamates, ureas, and allophanates derived (e.g., from nonracemic isocyanate, 2-oxazolidone, and ureide CDAs) have been applied to the chromatographic resolution and assessment of enantiomer purity of alcohols, amines, and lactams. The preferential population of conformations possessing approximately rigid planar backbones permits such molecules to lie flat on the surface of polar adsorbents (Fig. 6.44). The selective bulkiness of the R_1 to R_4 substituent groups that protrude from this planar backbone contribute to the ease of approach and preferential "anchoring" of individual diastereomers to the surface. This effect is mitigated by polar effects in the substituents that may repel or "ward off" molecules from the surface. The relative attractive and repulsive effects contribute to the elution order of the diastereomers. Correlation of

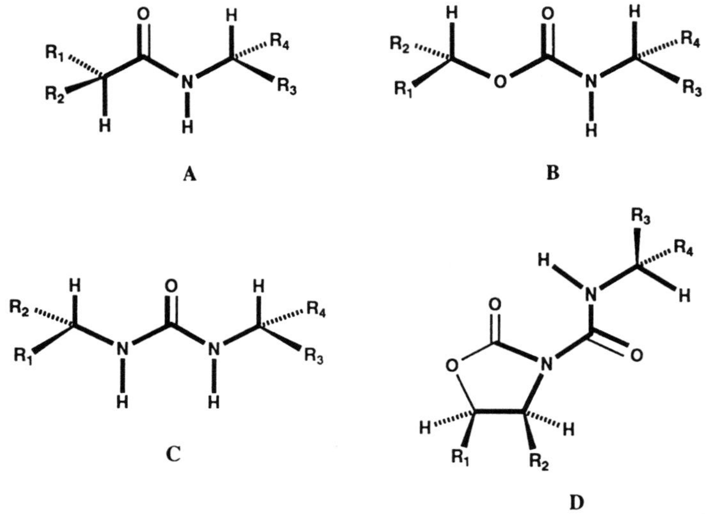

Figure 6.44. Common solution conformations of diastereomeric amides (**A**), carbamates (**B**), ureas (**C**), and allophanates (**D**) (after Pirkle and Hauske, 1977 and Pirkle and Simmons, 1983). The planar backbone is emphasized.

stereochemistry and elution order is now possible in many cases on the basis of models such as that described above (Pirkle and Hauske, 1977; Pirkle and Simmons, 1983). In this connection, Sonnet (1984) proposed that, in the absence of specific solute–solvent interactions, increased retention of diastereomeric alkenes on nonpolar stationary phases should be associated with the ability of some of the diastereomers to assume conformations better able to align themselves with polymeric stationary phase molecules.

The possibility of assigning configurations from the chromatographic elution order (see Chapter 5) has been the impetus of many studies of chromatographic resolutions. However, exceptions to the elution order models are not infrequent (e.g., Bergot, Schooley, et al., 1979; Kasai, Froussios, and Ziffer, 1983; Kasai and Ziffer, 1983), and skepticism is necessary in the application of HPLC for configurational assignments. For reviews of the extensive literature dealing with resolution and determination of ee by liquid chromatography (LC) especially on achiral stationary phases, see Lochmüller and Souter (1975); Lindner (1981, 1988); Pirkle and Finn (1983); Wainer and Doyle (1984); Souter (1985); Testa (1986); Gal (1987); Görög (1990).

High-Performance Liquid Chromatography with Chiral Solvents. One of the more recent developments in the liquid chromatographic determination of enantiomer purity is the application of chiral mobile phases (eluents) together with achiral stationary phases. (Reviews: Szepesi, 1989; Pettersson, 1989.)

The possibility of effecting such analyses was first demonstrated by Pirkle and Sikkenga in 1976 when they partially resolved a racemic sulfoxide by LC on silica gel. The mobile phase consisted of CCl_4 to which nonracemic 2,2,2-trifluoro-1-(9-anthryl)ethanol (Fig. 6.38, **81a**) had been added. Other simple chiral mobile phase additives, such as (+)-N,N-diisopropyltartramide, have been applied to analytical chromatographic resolutions of amino alcohols, glycols, hydroxy ketones, and hydroxy carboxylic and amino acids (Dobashi, Dobashi, and Hara, 1984; Dobashi and Hara, 1983a, 1985a). In these instances, separation may depend on the differential stability of diastereomeric intermolecularly hydrogen-bonded dimers.

Analyses of chiral amines (β-blocking drugs), such as propranolol (Fig. 6.45), have been effected by addition of (+)-10-camphorsulfonic acid to a mobile phase of low polarity. Formation of diastereomeric ion pairs that are bound both by electrostatic forces and by hydrogen bonds, and their differential migration accounts for the resolution (Pettersson and Schill, 1981). Chiral acids (e.g., naproxen, Fig. 6.45) have similarly been analyzed with solvents containing alkaloids (Pettersson, 1984).

Chiral reagents capable of transiently forming inclusion compounds (e.g., β-cyclodextrin, Fig. 6.46; see also Section 7-3.d), have been used as mobile phase modifiers in the analytical resolution of mandelic acid (Debowski, Sybilska, and Jurczak, 1982). Macromolecular complexing agents, for example, α_1-acid glyco-protein (α_1-AGP), added to the mobile phase have served to resolve and analyze chiral tertiary amines (Hermansson, 1984).

Most reports dealing with this type of enantiomer purity determination describe resolutions that are dependent on ligand-exchange chromatography (LEC). This process requires that a nonracemic metal complex be present in the mobile phase whose coordination sphere contains two or more ligand molecules at least one of which is chiral (see p. 257).

Replacement of one of the achiral ligands by a chiral molecule of the analyte [e.g., (+)] gives rise to a diastereomeric complex different from that formed by inclusion of a (−) analyte molecule in the coordination sphere. The stereoselec-tivity induced by the chiral mobile phase has been ascribed to differences in stability of the transient diastereomeric complexes in solution. However, there are indications that the diastereomer recognition takes place in the stationary phase (Grushka et al., 1983). The principal applications are to analyses of amino acids and their derivatives: Cu^{2+}-proline additives (Hare and Gil-Av, 1979; Gil-Av, Tishbee, and Hare, 1980); Zn^{2+}-2-alkyl-4-octyldiethylenetriamine additive Le Page, Karger, et al., 1979); Cu^{2+}- or Zn^{2+}-aspartame additive (a dipeptide: L-aspartyl-L-phenylalanine methyl ester) (Gilon, Grushka, et al., 1979; review: Hare, 1988).

The particular simplicity and sensitivity of such systems is appealing. Since switching the chirality sense of chiral eluents is far easier than switching the sense of chirality of enantioselective columns, this technique easily lends itself to the distinction of achiral artifacts from true enantiomers (Hare and Gil-Av, 1979).

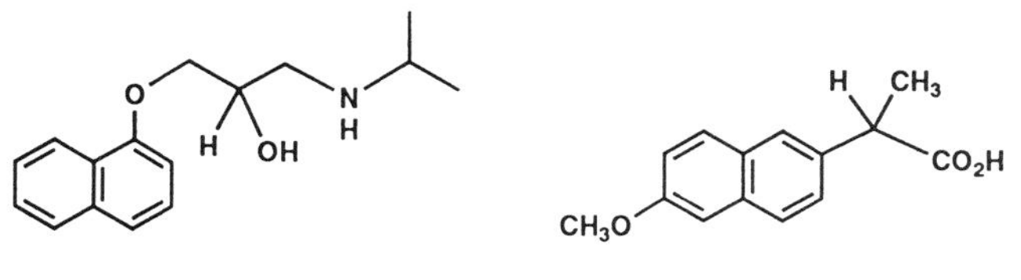

Propranolol   Naproxen

Figure 6.45. Propranolol and naproxen.

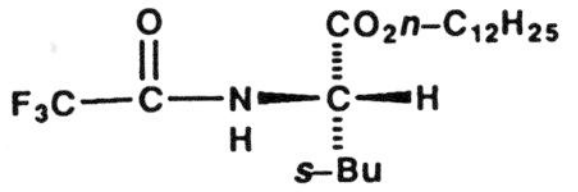

N–TFA–L–isoleucine lauryl
ester (Gil–Av and Feibush, 1967)

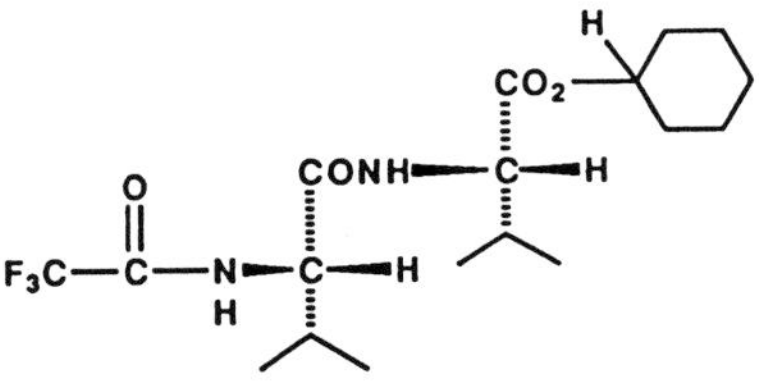

A typical dipeptide phase
(Gil–Av, 1985)

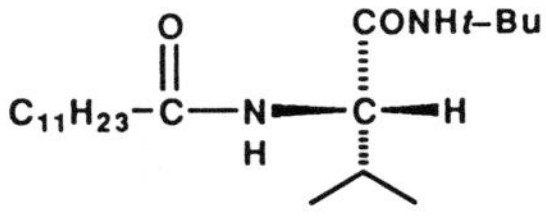

A typical diamide phase

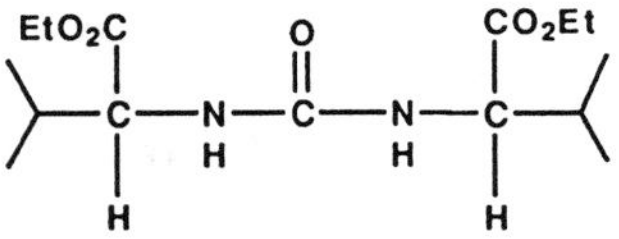

Carbonyl–bis–(L–Valine
ethyl ester). A typical
chiral mesophase

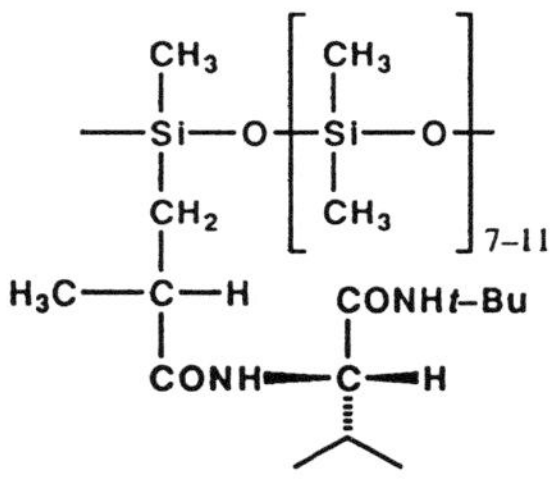

Chirasil–Val™
(a chiral polysiloxane)

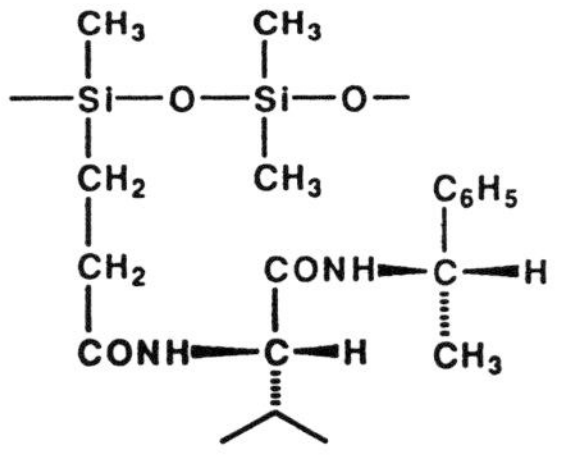

Chrompack™
(a chiral phase based on
polysiloxane XE–60)

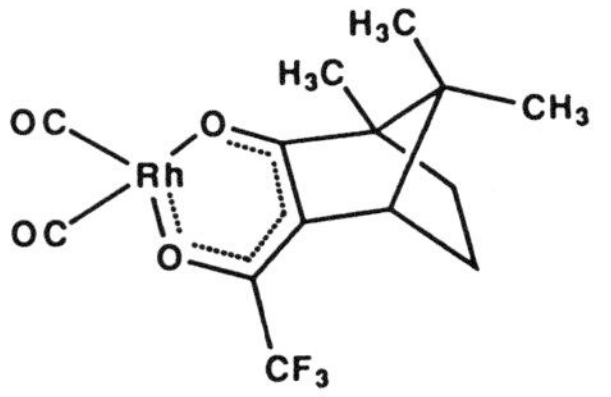

Dicarbonylrhodium (I)
trifluoroacetyl–(1R)–camphorate

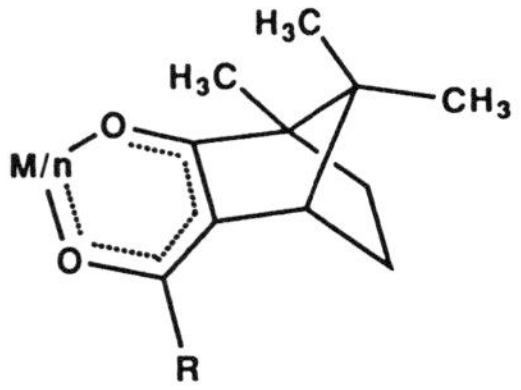

e.g., R = CF₃CF₂CF₂
and Metal (M)/3 = Eu/3, Ni/2
Metal bis– or tris–[(1R)–3–
(heptafluorobutyryl)camphorate]

Figure 6.46. Basic structural types of gas chromatographic enantioselective stationary phases.

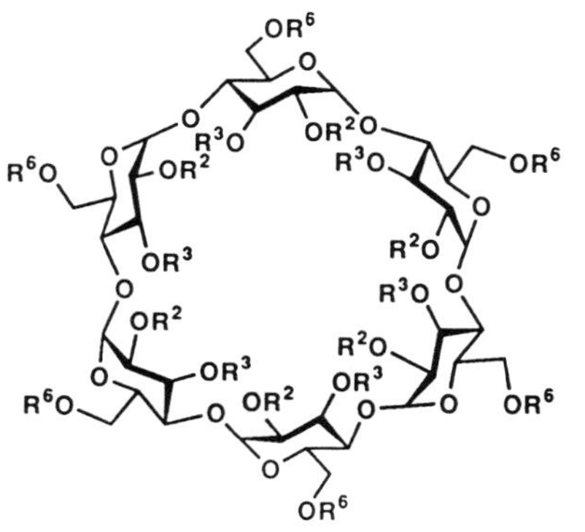

α–Cyclodextrin = cyclohexaamylose
OR^2, OR^3 and OR^6 = OH

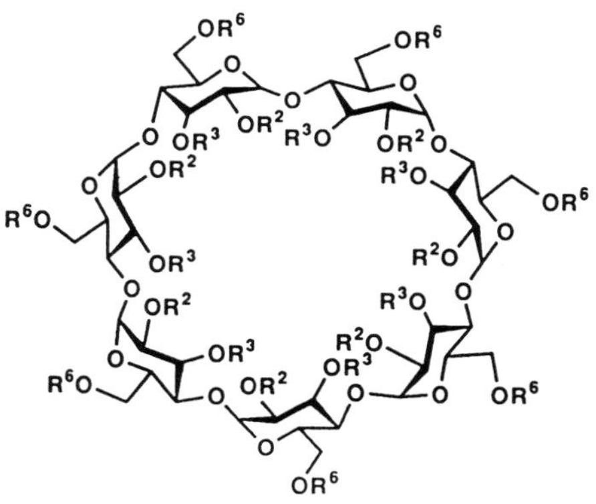

β–Cyclodextrin = cycloheptaamylose
OR^2, OR^3 and OR^6 = OH

Lipodex^{TM}
OR^2 and OR^6 = O–pentyl
OR^3 = O–pentyl *or* O–acyl

Peroctylated α-CD
OR^2, OR^3 and OR^6 = O–octyl

Figure 6.46   (*Continued*)

Nevertheless, the number of applications of chiral eluents has been limited possibly due to the required quantities and costs of the reagents together with the impracticality of recovering them (Testa, 1986).

Chromatography on Enantioselective Stationary Phases

The direct resolution of enantiomers by chromatography on nonracemic chiral stationary phases is a goal that has tantalized chemists virtually since chromatography was recognized as a potentially valuable separation technique. Early trials with such columns were oriented to preparative resolution (see Section 7-3.d) and were not very successful. No resolution, or only partial resolution, was achieved. Some of the early trials are described in Eliel (1962, p. 61) and in Feibush and Grinberg (1988). The advent of instrumental chromatographic techniques (GC and HPLC) with their high-resolution columns and sensitive detectors made analytical resolutions with enantioselective columns and their application to enantiomer purity determinations possible.

Numerous enantioselective stationary phases have been developed over a 25-year period (1966–1990); the end of this development is nowhere in sight. A summary of the various types is given in Table 6.4 (cf. Armstrong, 1992) along with the preferred modes of use (GC, HPLC, and TLC). While the principles of separation are also given for most cases, there is no absolute certainty about these. It is likely that in many instances more than one mechanism applies. [Books: Souter, 1985; Krstulovic, 1989; Allenmark, 1991. Reviews: Lochmüller and Souter, 1975; Allenmark, 1984; Gil-Av, 1985.]

TABLE 6.4. Principal Types of Enantioselective Chromatographic Stationary Phases

	Type	Principle of Separation	Mode
I[a]	Amide	Attractive interactions, hydrogen bonding, π–interaction,	GC, HPLC
I	Fluoro alcohol	Dipole attraction, charge transfer	HPLC
I	π Acid		
II	Carbohydrate	Attractive interactions + inclusion	HPLC
III	Cyclic hexose oligomers	Inclusion compound formation	GC, HPLC
III	Crown ether	Inclusion	HPLC
IV	Metal chelates	Ligand exchange	GC, HPLC, TLC
V	Proteins	Hydrophobic and polar interactions in a protein	HPLC
	Ureide	Interaction with mesophases	GC[b]

[a] Categories I–V as defined by Wainer (1987, 1989).
[b] Below selector melting point.

Gas Chromatography. The earliest nonracemic chiral stationary phases were patterned after structural elements present in enzymes; they contained –CO– and –NH– moieties present in the vicinity of stereocenters. It was anticipated that hydrogen-bond formation between the stationary phase and chiral amino acid derivatives would provide a small degree of enantioselectivity that might be sufficient for quantitative analysis of the enantiomer composition provided that the effect could be amplified. The use of long capillary columns provided for the amplification (Gil-Av, 1985). The first successful CSP incorporating these elements was *N*-trifluoroacetyl-L-isoleucine dodecyl ester (Gil-Av, Feibush, and Charles-Sigler, 1966). This phase was able to resolve simultaneously the enantiomers of several TFA-α-amino acid isopropyl esters by GC.

The following terms, used in the sequel, have been suggested for describing the two chromatographic partners in the stereoisomer discrimination phenomenon: *selector* (CSP = resolving agent) and *selectand* (chiral solute = analyte) (Mikes, Boshart, and Gil-Av, 1976).

The structural similarity and complementary functionalities of the analyte and of the chiral component of the stationary phase reveals the basis of the column design, formation of transient diastereomeric complexes (Feibush and Gil-Av, 1970). Subsequent developments in analytical GC included the application of dipeptide, diamide, and similar phases permitting multiple contact points between selectand and selector molecules (Fig. 6.46) (Lochmüller and Souter, 1975). Enantioselective capillary GC columns are capable of separating enantiomers with separation factors α [$\alpha = (t_2 - t_1)/(t_1 - t_0)$, where t_2 and t_1 are the retention times of the second and first enantiomers eluted, respectively, and t_0 is that of an unretained substance; see also Section 7-3.d] of 1.05 (or even less) corresponding to a difference of only $\Delta\Delta G^0 = 29$ cal mol^{-1}(121 J mol^{-1}) between the transient diastereomeric solvates (Allenmark, 1991, Chapter 5). [Book: König, 1987. Reviews: König, 1982; Schurig, 1983, 1984, 1986; Allenmark, 1991, Chapter 6].

Enantioselectivity is ascribed to differences in stability, as manifested by stability constants K_{RR} and K_{SR}, of the transient solvates formed between the selectand enantiomers and the nonracemic selector molecules (Eq. 6.16). Since these stability constants are thermodynamic quantities, they should be temperature dependent. Indeed, a linear temperature dependence exists between ln α (where $\alpha = K_{RR}/K_{SR}$ with, arbitrarily, $K_{RR} > K_{SR}$) and $1/T$, and enantioselectivity (α) decreases with increasing temperature (the two stability constants become equal at a characteristic temperature, the isoenantioselective temperature T_{iso}).

Inversion of the sense of stereoselectivity (i.e., of the elution order of the enantiomers) of an enantioselective chromatographic (selector–selectand) system as a function of temperature has been predicted on the basis of the Gibbs–Helmholtz relationship (Koppenhöfer and Bayer, 1984; Schurig, 1984). While such inversion is not often observed (T_{iso} typically is too high), there have been at least two reports of GC enantioselectivity reversal, that is, reversal of peak elution order along with increasing chromatographic resolution with increasing temperature (above T_{iso}), one of amino acid derivatives on a hydrogen-bonding stationary phase (Watabe, Charles, and Gil-Av, 1989) and a second of complexation on a stationary phase containing a chiral metal chelate selector (see below) (Schurig et al., 1989; Schurig and Betschinger, 1992).

Stationary phases of related structure exhibiting liquid crystalline behavior in appropriate temperature ranges (chiral mesophases; Fig. 6.46) have been found to be useful in the separation of enantiomers (Lochmüller and Souter, 1973, 1974).

Higher thermal stability and lower volatility than is available with columns containing amino acid or dipeptide derivatives was achieved by linking such compounds covalently to polymer backbones. An example of such enantioselective stationary phases usable to at least 220°C in the form of an open tubular column is *N*-propionyl-L-valine *t*-butylamide bound to a copolymer of dimethylsiloxane and carboxyalkylmethylsiloxane units (Chirasil-Val™, Fig. 6.46; Frank, Nicholson, and Bayer, 1977). Dipeptides linked to polysiloxanes (e.g., Chrompack™) exhibit a wider range of utility. Analytical GC resolution of sugar derivatives and of ketones (as their oximes) is possible on such a column (König, 1982). It is now recognized that the precise nature of the functional group and other polar sites present in the derivatized analytes strongly affects the degree of separation of the enantiomers. In the case of chiral ketones, neither the free ketones nor the less polar *O*-methyloximes or *O*-trimethylsilyloximes could be separated (König, 1982). While hydrogen bonding between analyte and chiral selector molecules is very likely to play a role in the enantioselectivity of the columns described (Schurig, 1984), other types of intermolecular forces (e.g., dipole–dipole interactions or van der Waals forces) must be involved since dipeptide chiral selectors lacking NH can fully separate racemic *N*-TFA-proline isopropyl ester. There are no sites for hydrogen bonding in such systems (Stölting and König, 1976).

A complementary technique applicable to the determination of enantiomer composition by GC is complexation GC. The possibility of separating enantiomers by the selective and reversible incorporation of gaseous chiral molecules in the coordination sphere of metal complexes appears to have been first suggested by Feibush et al. (1972). They postulated that enantiomer mixtures could be separated by GC on columns containing a chiral LSR in the stationary phase. This idea in somewhat modified form was first reduced to practice by Schurig in 1977 in a study of the enantiomeric resolution of a chiral alkene (3-methylcyclopentene) on a metal capillary column containing an optically active rhodium(I) chelate dissolved in squalane $C_{30}H_{62}$ (Schurig and Gil-Av, 1976/77; Schurig, 1977; Fig. 6.46). Subsequently, stationary phases containing europium and nickel complexes have been applied to the analytical resolution of chiral molecules (e.g., oxiranes; Schurig and Betschinger, 1992), including those devoid of functional groups; an example is shown in Fig. 7.16.

1-Chloro-2,2-dimethylaziridine (Section 3-1.b) is an example of a compound observed to undergo enantiomerization (inversion of configuration at the chiral center of both enantiomers; cf. Section 7-9.a) during complexation GC (Schurig and Bürkle, 1982; Schurig, 1983; Allenmark, 1991, Chapter 5). The metal chelate may participate in facilitating the inversion of nitrogen in the aziridine (cf. Section 7-2.e and Fig. 7.16).

Methylated cyclodextrins (*CD*s) dissolved in achiral liquid supports, liquid nonpolar alkyl derivatives [e.g., hexakis-(2,3,6-tri-*O*-methyl)-β-cyclodextrin (Cyclodex-B^{TM})] and mixed regioselectively alkylated and acylated α- and β-cyclodextrins [e.g., Lipodex^{TM}, Fig. 6.46; Lipodex A^{TM} is hexakis-(2,3,6-tri-*O*-pentyl)-α-cyclodextrin] have been applied, since about 1987, as stationary phases in the capillary GC determination of enantiomer purity (König, et al., 1988, 1989. Reviews: König, 1989; König and Lutz, 1990; Schurig and Nowotny, 1990; Armstrong, 1992). These commercially available phases are remarkably thermally stable (to 220°C). A wide range of structural types are base-line separated on these columns consistent with the presumed operation of an inclusion mechanism: alcohols, alkenes, alkyl halides [e.g., the chiral inhalation anesthetic halothane, $CF_3CHBrCl$, is fully resolved on a Lipodex A^{TM} column at 30°C (Meinwald, et al., 1991)], amines, carbohydrates (e.g., as *O*-trifluoro-acetates), ethers, ketones, lactones, and even saturated hydrocarbons. However, other mechanisms involving interaction of the analytes with the chirotopic periphery of the derivatized cyclodextrins cannot be excluded.

Polar hydrophilic *CD* derivatives (e.g., permethyl-(*R*)-2-hydroxypropyl-β-cyclodextrin) are also useful in the resolution of a wide range of compounds including many that are devoid of aromatic groups. The configuration of the chiral 2-methoxypropyl side chains does not affect the resolution order of analyte enantiomers (Armstrong et al., 1990b). On the other hand, it has been demonstrated in the case of 2-amino-1-propanol as the TFA derivative that the elution order of its enantiomers can be reversed on shifting between derivatives in the same β-*CD* series, for example, from the less polar trifluoroacetylated 2,6-di-*O*-pentyl-β-cyclodextrin to the above mentioned polar *CD* stationary phase (Armstrong et al., 1990a).

An interesting modification of enantiomer purity determinations by GC involves the use of the enantiomer of the dominant chiral species to spike the analyte in the role of an "internal standard" (Schurig, 1984). Thus, for the enantiomeric purity determination of L-amino acids the D-amino acids are used in lieu of a true internal standard. The procedure, known as "enantiomer labeling," is recommended quite generally for checking the reliability and accuracy of chromatographic (and NMR) enantiomer purity determinations, especially in those cases in which the proportion of the dominant enantiomer is greater than about 95% (Bonner, 1973; Frank, Nicholson, and Bayer, 1978). It serves to indicate qualitatively that the minor peak is, in fact, due to the enantiomer and not to something else. Quantitatively, if one adds 2.5% (by weight) of the minor enantiomer to a weighed sample of a compound having 95% ee (enantiomer ratio = 97.5:2.5), the resulting mixture should show an enantiomeric excess of about 90% (ca. 95:5). If this is not found, then something is wrong.

High-Performance Liquid Chromatography. High-performance liquid chromatography is a second instrumental chromatographic technique permitting enantiomer purity determinations to be carried out relatively simply. Here a much wider range of enantioselective columns is available (in part because thermal stability is not a limitation) and, moreover, the liquid (mobile) phase can play an important role in facilitating resolution (Books: Souter, 1985, Chapter 3; Allenmark, 1991, Chapter 7; Zief and Crane, 1988; Krstulovic, 1989; Lough, 1989; Ahuja, 1991. Reviews: Krull, 1978; Audebert, 1979; Blaschke, 1980; Lindner, 1981; Davankov et al., 1983; Pirkle and Finn, 1983; Wainer and Doyle, 1984; Wainer, 1986; Pirkle and Pochapsky, 1989). As of 1992, more than 50 enantioselective LC stationary phases appear to be available (Armstrong, 1992).

Enantiomer purity determinations by either GC or HPLC with chiral stationary phases have the advantage of being absolute, that is, both techniques give correct enantiomer purity information even with stationary phases that are not enantiomerically pure as recognized by Davankov (1966). With reduced enantiomer purity, the separation factor α is reduced.

$$\alpha^* = \frac{\alpha(P + 100) + (P - 100)}{(P + 100) + \alpha(P - 100)} \tag{6.17}$$

Equation 6.17 permits the calculation of α^*, the separation factor that would be obtained if P [the enantiomeric purity (%ee)] of the stationary phase were 100% (Beitler and Feibush, 1976; Davankov, 1989).

> As pointed out by Davankov (1980), a difference in free energy of interaction ($\Delta\Delta G^0$) of the two selectand enantiomers with the stationary phase as small as $10\,\text{cal}\,\text{mol}^{-1}$ ($42\,\text{J}\,\text{mol}^{-1}$), corresponding to $\alpha = 1.01$, suffices to achieve a complete analytical separation with columns having a theoretical plate number of 10^5.

Unlike CSPs designed for GC resolution, most selectors studied in the LC mode, particularly until the mid-1970s, were chosen or designed with preparative goals in mind. Some of these are described in Chapter 7. Instrumental advances in pumps and detectors and the increased availability of these components gradually made analytical HPLC more accessible. Coincidentally, the design of new CSPs for HPLC increased the use of the latter technique for direct enantiomer purity determinations. One of the first reports of a rationally designed CSP was that of a column incorporating L-arginine covalently bonded to Sephadex G-25 (a carbohydrate polymer) via a cyanuric chloride linkage. The analytical capabilities of this stationary phase were limited in scope, but excellent toward the anti-Parkinsonian agent 3-(3,4-dihydroxyphenyl)alanine (DOPA) (Baczuk et al., 1971).

In the late 1970s, Pirkle and co-workers began to adapt NMR chiral solvating agents that exhibit stereoisomer discrimination in solution (Section 6-5.c, p. 231) to chromatographic resolutions (Reviews: Pirkle and Finn, 1983; Finn, 1988; Doyle, 1989; Macaudière, Tambuté, et al., 1989; Pirkle and Pochapsky, 1990; Perrin and Pirkle, 1991). A CSA analogous to TFAE (Fig. 6.38; 10-methyl-

TFAE) was anchored to γ-mercaptopropyl silanized silica to produce CSP **A** (Fig. 6.47), useful in the enantiomer purity determination of sulfoxides, amines, alcohols, thiols, amino and hydroxy acids, and lactones (Pirkle and House, 1979; Pirkle et al., 1980). A later development also based on the general structure **E** (Fig. 6.47) and incorporating N-(2-naphthyl)-α-amino acids (alanine and valine) as the chirotopic elements in the CSP is exemplified by **D** (Fig. 6.47). Use of the latter requires that the selectands (amines, alcohols, etc.) be converted to their 3,5-dinitrobenzoyl derivatives prior to analysis (Pirkle and Pochapsky, 1986; Pirkle et al., 1986).

This and other CSPs were designed on the basis of the three-point recognition model (cf. p. 207 and Fig. 8.8). Since the above cited functional groups in the analyte provide only two interaction sites, for example, a basic site (NH, OH, etc.) and a carbinyl hydrogen atom (the latter being postulated as being sufficiently acidic to interact with a basic site in the selectand), the compounds are derivatized to provide the required third site. This process typically involves incorporation of a π-acid, for example, a 3,5-dinitrobenzoyl group that is able to form a charge-transfer complex with the aryl group of the CSP. The three interactions, postulated as giving rise to the more stable of the two possible diastereomeric solvates, are illustrated in Figure 6.48 and Figure 8.40.

Additional CSPs have been designed on the basis that reciprocity obtains: If racemate **A** can be resolved by nonracemic **B**, then racemate **B** should be resolvable by nonracemic **A** (see Section 7-3.a). Linking N-(3,5-dinitrobenzoyl)

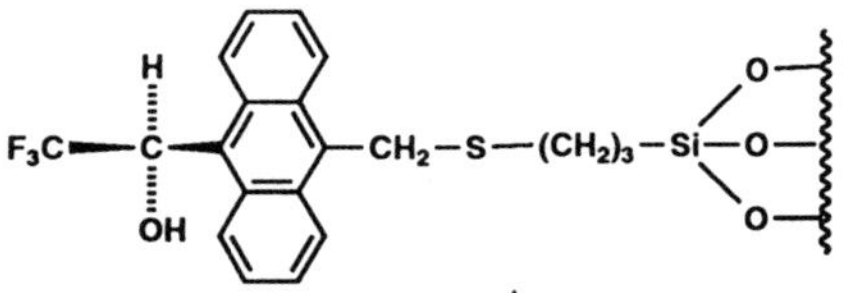

(R)–2,2,2–Trifluoro–1–[9–(10–α–thio–
methyl) anthryl]ethanol–derived CSP.
Covalently bonded (1979)

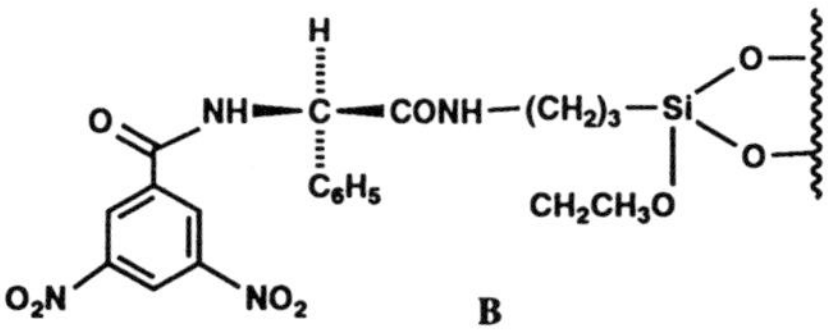

A

(R)–N –(3,5–Dinitrobenzoyl)phenyl–
glycine–derived CSP.
Covalently bonded (1980)

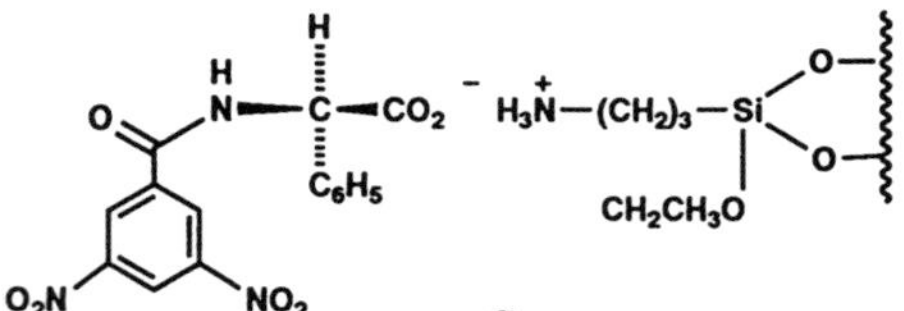

B

(R)–N –(3, 5–Dinitrobenzoyl) phenyl–
glycine–derived CSP.
Ionically bonded (1981)

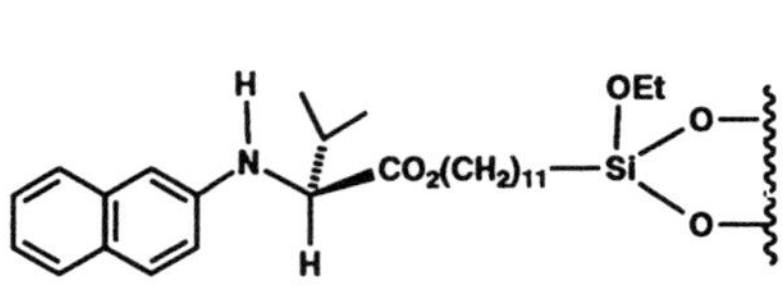

C

Covalently bonded CSP derived
from N –(2-naphthyl)–α–amino
acids (1986)

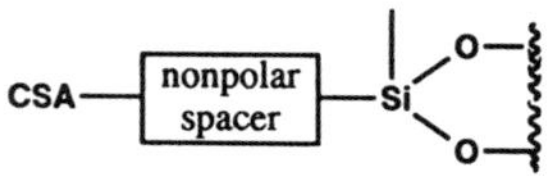

D

Generalized structure of
covalently bonded CSP
based on silica gel

E

Figure 6.47. Chiral stationary phases for enantiomeric purity determination by HPLC: Pirkle columns.

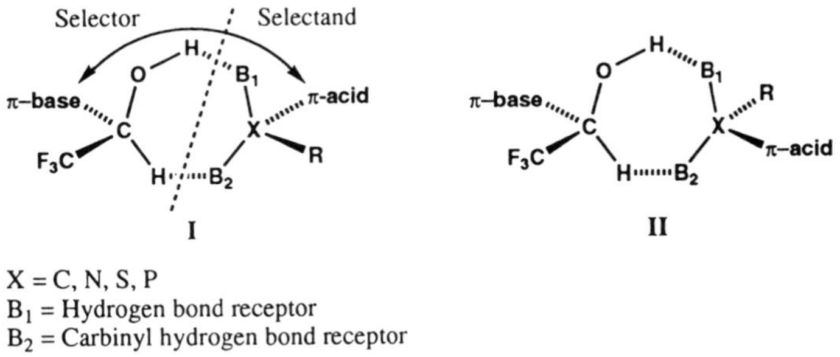
Selector
Selectand
π-base
π-acid
F3C
R
I
II
X = C, N, S, P
B1 = Hydrogen bond receptor
B2 = Carbinyl hydrogen bond receptor

Figure 6.48. Three point interaction model responsible for the preferential relative retention of solutes on chiral stationary phases. Fluoroalcohol model. Diastereomeric solvates **I** and **II**.

phenylglycine to γ-aminopropyl-derivatized silica gel covalently gives rise to CSP **B** (Fig. 6.47; Pirkle et al., 1980) that can resolve alcohols. A similar but ionically bonded CSP (Fig. 6.47, **C**) has overlapping but not identical applications (Pirkle and Finn, 1981; Wainer and Doyle, 1984). Numerous chiral heterocyclic systems are amenable to resolution with the ionic Pirkle CSP (Pirkle et al., 1982) as are underivatized alcohols possessing at least one aryl group able to serve as a π-base (Pirkle et al., 1981) and atropisomeric (cf. Chapter 14) binaphthols and analogues (Pirkle and Schreiner, 1981). The diastereomeric solvate responsible for preferential retention of the enantiomer having the longest retention time is illustrated in Figure 6.49.

Both the covalent and the ionic Pirkle CSPs are commercially available. Amines may be analyzed on the covalent Pirkle CSP (Fig. 6.47, **B**) following acylation, for example, with α-naphthoyl chloride (Pirkle et al., 1983; Pirkle and Welch, 1984; Pirkle, Meyers, et al., 1984). The CSPs of this type have been applied to the assessment of enantiomer composition of chiral diastereomer mixtures whose enantiomer purity determination on achiral stationary phases was unsuccessful (e.g., Evans et al., 1985).

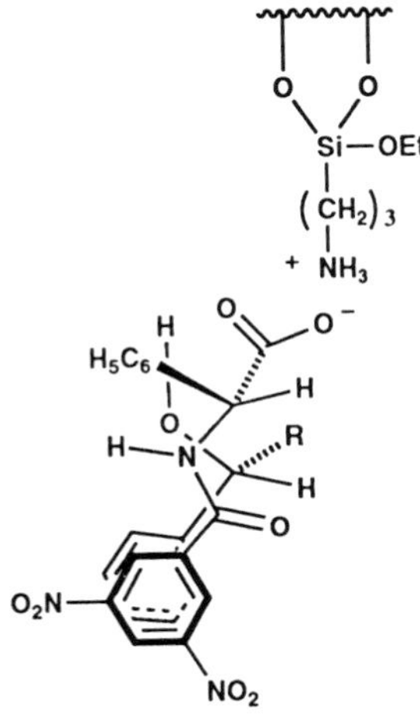
Si—OEt
(CH2)3
NH3
H5C6
O2N
NO2

Figure 6.49. Application of the general "recognition" model to HPLC of a chiral alkyl aryl carbinol on CSP **C** (Figure 6.47). The model shows the interaction of (*R*)-CSP with the more strongly retained (*R*) enantiomer of the alcohol (corresponding to the stabler diastereomeric solvate). [Adapted with permission from Pirkle, W.H., Finn, J.M. Hamper, B.C., Schreiner, J., and Pribish, J.R. (1982), *Asymmetric Reactions and Processes in Chemistry*, A.C.S. Symposium Series, No. 185. Copyright © 1982 American Chemical Society.]

The Pirkle CSPs (Fig. 6.47, **B** and **C**) depend on a combination of hydrogen bonding, $\pi-\pi$ (charge-transfer) interactions, dipole–dipole interactions (dipole stacking; Pirkle, 1983), and steric interactions to achieve both absolute and enantioselective retention of analyte molecules. The likely interaction sites in a typical selector are identified in Figure 6.50.

Inversion of the elution order for corresponding enantiomers in homologous series of analytes suggests that the several interaction and recognition mechanisms can operate independently and occasionally in opposite senses. Such heuristic models are especially useful in the fine tuning of chiral selector design (Pirkle, Hyun, and Bank, 1984). It has been pointed out that so-called chiral recognition models make some assumptions about the conformation of the selector molecules. Calculations of the molecular mechanics (MM2) type are not always in accord with those invoked by the models (Lipkowitz et al., 1987).

> The selector–selectand interaction mechanisms and models proposed by Pirkle and co-workers principally served as very effective heuristics in the development of CSPs. However, they should not be accepted as being proven. Three contact point interaction models have been challenged as being incompatible with recent findings (Allenmark, 1991, Chapter 6; see also Topiol and Sabio, 1989). The $\pi-\pi$-interaction in *SR* as compared to *SS* (or *RR*) solvates do not appear to contribute significantly to chromatographic separation suggesting that derivatization of the selectand and incorporation of π-acid or π-base moieties in selector molecules may be unnecessary for enantiomer discrimination to be observed (Wainer and Alembik, 1986; Topiol et al., 1988).

Yet other CSPs have been developed based on the foregoing considerations. These selectors incorporate chiral ureas (*Supelco Reporter*, 1985); chiral amino acid derivatives, such as *N*-formyl-L-valine (Dobashi, Oka, and Hara, 1980); tartramide (Dobashi and Hara, 1985; review: Dobashi et al., 1991); α-(1-naphthyl)alkylamine analogues (Pirkle and Hyun, 1984); and hydantoins (Pirkle and Hyun, 1985) (see the generalized CSP structure **E** in Fig. 6.47).

In some instances, chemoselective derivatization of the analyte may be necessary to promote adequate enantioselectivity. In amino acid analysis on (*N*-formyl-L-valylamino)propyl (FVA) silica gel, *O*-alkylation by bulky groups (as in *tert*-butyl esters) is more helpful for increasing the selectivity than are changes in the *N*-acyl substituent (Dobashi, Oka, and Hara, 1980).

Cinchona alkaloids (e.g., quinine, Fig. 6.38, **84a**), incorporated in CSPs by reaction with a mercaptopropyl silanized silica gel (by free radical addition of the thiol group to the quinine vinyl group), are effective in the resolution of alcohols

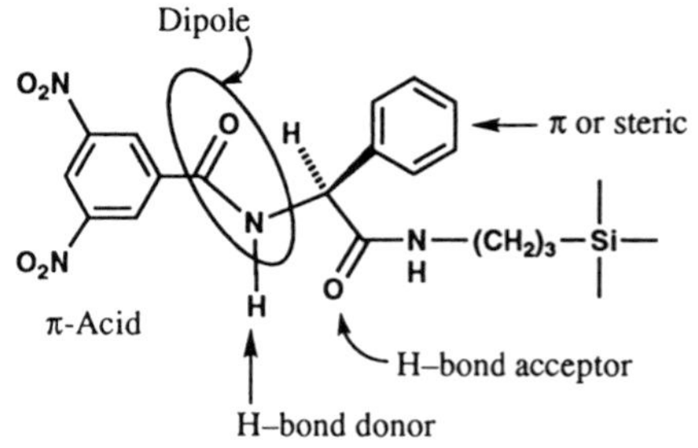

Figure 6.50. Enantioselective Pirkle Type **B** CSP showing interaction sites and types [Adapted with permission from Cook, C.E. (1985) *Pharm. Int.*, **6**, 302.]

and binaphthol derivatives and acylated amines (Rosini, Salvadori, et al., 1985; Salvadori et al., 1992). It is noteworthy that, with the analogous selector based on quinidine, inversion of the elution order of selectands is observed [C(8) and C(9) in quinidine are "quasi-enantiomeric" with the corresponding chiral centers in quinine; Salvadori et al., 1992].

While most chiral selectors presently in use are anchored to the support by covalent attachment, typically to microparticulate silica gel, the earliest stationary phases were often simply coated mechanically. An example of such a stationary phase is TAPA [2-(2,4,5,7-tetranitrofluoren-9-ylideneaminooxy)-propionic acid, Fig. 6.51*a*, **85**]. This charge-transfer agent is a powerful resolving agent for helicenes and heterohelicenes (Mikes, Boshart, and Gil-Av, 1976; Numan, Helder and Wynberg, 1976). Subsequently, another chiral selector that separates helicene enantiomers principally by charge-transfer interaction was introduced: 1,1′-binaphthyl-2,2′-diyl hydrogen phosphate (Fig. 6.35, **62a**) covalently bonded to an achiral support polymer (Fig. 6.51*b*; Mikes and Boshart, 1978).

Chiral selectors based on proteins have been developed based on accumulated knowledge of protein binding (multiple bonding sites and enantioselective binding; Allenmark, 1991, Chapter 7). One of these stationary phases consists of bovine serum albumin (BSA) bonded to silica gel (Allenmark et al., 1983; Allenmark, 1985). More recently, an enantioselective column based on α_1-acid glycoprotein (AGP), another serum protein, has been investigated (Hermansson, 1983, 1984, 1985. Review: Hermansson and Schill, 1988; see also Allenmark, 1991, p. 129). Both protein columns are commercially available (Resolvosil[TM] and Enantiopac[TM], respectively).

In consequence of the mixed interactions (hydrophobic, electrostatic, or steric) that are likely to obtain between proteins and typical analytes and the high sensitivity of the binding sites to small changes in the mobile phase, especially pH and chemical modifiers (acids, bases, or alcohols), these columns can be used to analyze a very wide range of compounds, for example, amino acid derivatives, amines, alcohols, sulfoxides, and carboxylic acids including numerous medicinal agents (Wainer, Barkan, and Schill, 1986).

Ligand-exchange chromatography (LEC) has been investigated as an alternative mode for the analysis of optically active compounds, especially amino

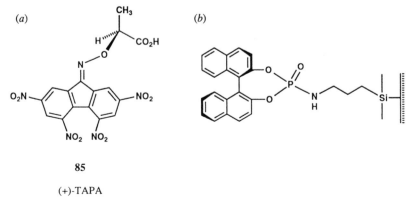

(*a*)   (*b*)

85

(+)-TAPA

Figure 6.51. Chiral charge-transfer selectors.

acids. The method was first proposed in 1968 by Rogozhin and Davankov (1970) and by Bernauer (1970) and exploited particularly by the Russian group (Davankov, 1980; Davankov et al., 1983). In LEC, resins are designed to hold and retain transition metal ions whose coordination spheres contain chiral chelating ligands. In a typical situation, chiral analyte molecules reversibly displace some but not all of these ligand molecules thereby forming diastereomeric complexes of unequal stability (Review: Lam, 1989).

Ligand-exchange chromatography differs from complexation GC (see above) in that analyte molecules act as competitors toward chiral ligands attached to metal-containing polymeric backbones in the former technique, whereas in the latter, chiral nonpolymeric metal complexes similar (or identical) to LSRs, dissolved in squalane, interact selectively with analyte molecules by *expanding* their coordination sphere.

Figure 6.52 illustrates the analysis of tyrosine by LEC on mixed copper

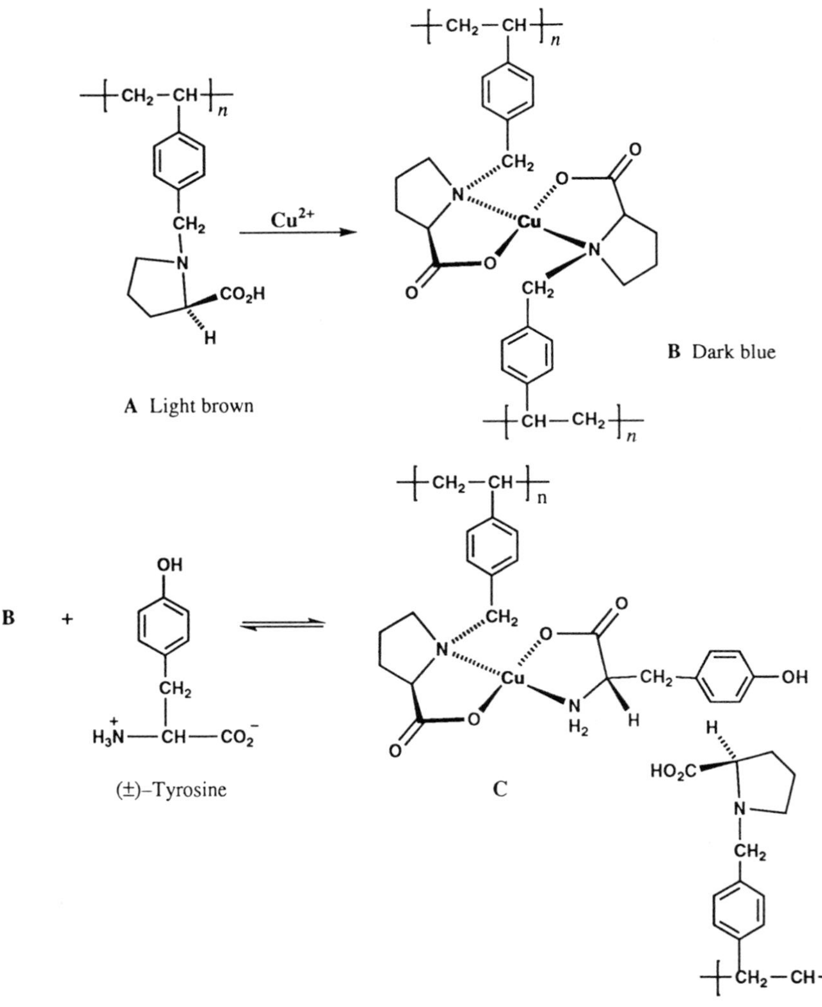

Figure 6.52. Ligand exchange chromatography of tyrosine. (**A**) Polystyrene-bound L-proline. (**B**) Copper chelate complex. (**C**) Mixed copper chelate complex formed when a tyrosine molecule displaces one bound proline from attachment to copper.

chelate-L-proline complex. Two labile diastereomeric complexes of structure **C** are formed that, in the case of tyrosine, lead to a separation exhibiting $\alpha = 2.46$. Such complexes differ in energy content by up to $825\,\text{cal}\,\text{mol}^{-1}$ ($3.45\,\text{kJ}\,\text{mol}^{-1}$) (Blaschke, 1980) with the less stable one regenerating free amino acid (typically the L-enantiomer) more rapidly so that the latter elutes first. In one model used to explain the elution order, coordination of solvent (e.g., H_2O molecules) in the axial position of a five-coordinate complex, is subject to steric effects that vary according to the configuration of ligands derived from the analyte (Davankov et al., 1983). Columns for LEC are commercially available (Daicel: Chiralpak W series) and may be used in the analysis of compounds other than amino acids by derivatization of the chiral analyte with achiral reagents that convert poor ligands into better ones, for example, amino alcohols converted into Schiff's bases with salicylaldehyde (Gelber, Karger, et al., 1984).

Optically active ion-exchange resins, originally thought to be promising CSPs for chromatographic resolutions because of the many successful resolutions involving the crystallization of diastereomeric salts, have been found to be only marginally effective in the separation of enantiomers (Davankov, 1980).

Numerous other CSPs that have utility in analytical or preparative resolutions have been studied over the years. These include the following types of column packings:

(a) Chiral stationary phases prepared by polymerization of optically active unsaturated amides, studied particularly by Blaschke et al., are used in the analysis of amino acids (Blaschke, 1980; Krull, 1978). Chiral crown ethers bonded to polystyrene (Sogah and Cram, 1976) or to silica gel (Sousa, Cram, et al., 1978) are useful in the separation of chiral compounds bearing primary amino groups (see Section 7-7). Poly[(S)-(−)-p-(p-tolylsulfinyl)styrene] bearing chirotopic sulfur atoms is effective in the chromatographic resolution of chiral aromatic alcohols and phenols (Kunieda et al., 1990).

(b) *Microcrystalline* cellulose triacetate (triacetylcellulose, TAC) is a very useful type of optically active polymer used in the analysis of mixtures of racemates including some that are devoid of functional groups (Hesse and Hagel, 1973, 1976a,b; Shibata, Okamoto, and Ishii, 1986. Reviews: Ichida and Shibata, 1988; Johns, 1989a; Allenmark, 1991, Chapter 7). When prepared by heterogeneous acetylation of cellulose in benzene suspension, TAC retains regions of crystallinity. Cavities within these regions allow enantioselective inclusion of solutes, especially those incorporating substituent-free phenyl groups. The resulting "inclusion chromatography" is effective with a wide variety of chiral compounds: hydrocarbons including atropisomers, heterocycles, and medicinal agents (Blaschke, 1980; Frejd and Klingstedt, 1983; Petterson and Berg, 1984). A proposed model for this inclusion is shown in Figure 6.53. A comparative study of TAC and poly[ethyl (S)-N-acryloyl phenylalaninate] shows that polyamide and TAC stationary phases have complementary utility (Blaschke et al., 1983).

> In connection with the chromatographic resolution of bridged biphenyls on TAC, Isaksson, Wennerström, and Wennerström (1988) showed experimentally and theoretically, using statistical methods, that C_2 symmetric

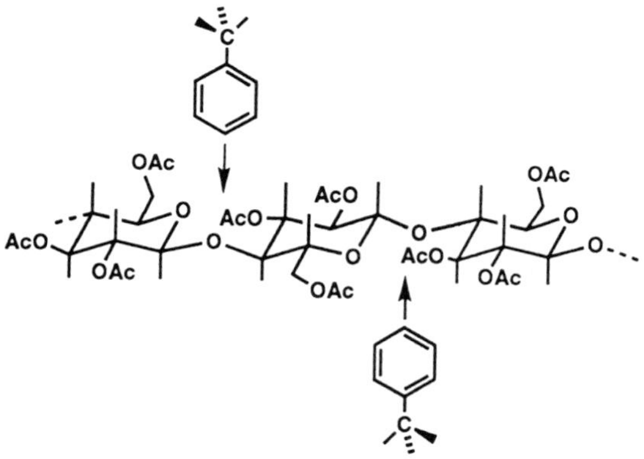

Figure 6.53. Model for inclusion chromatography of chiral analytes in microcrystalline cellulose triacetate [Adapted with permission from Blaschke, G. (1980), *Angew. Chem. Int. Ed. Engl.*, **19,** 113.]

molecules are better separated than less symmetric ones, that is, those with C_1 symmetry. Selectand symmetry is an important factor in the resolution of enantiomers (at least on TAC columns). It is suggested that symmetrical molecules may permit more interactions (binding) to take place with the selector relative to less symmetrical analyte molecules with the attending consequence of amplification of separation effects and improved resolution.

It has been observed that when TAC is solubilized and reprecipitated, its resolving power is largely lost due to a breakdown of the original crystalline structure (Hesse and Hagel, 1976a). However, TAC, cellulose tribenzoate, and the phenylcarbamates of cellulose, amylose, and other polysaccharides exhibit good resolving power when adsorbed from solution onto silica gel. Here too, it was found that none of the cited CSPs was effective for all solute types tried (Okamoto et al., 1984a,b). Cellulose ester, carbamate, and ether columns are commercially available [Daicel: Chiralcel^{TM} OA, OB (esters); OC (carbamate); OE, OK (ether); see Fig. 6.54].

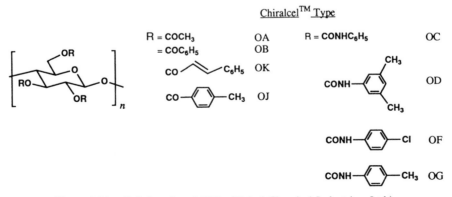

Figure 6.54. Cellulose based CSPs (Daicel Chemical Industries, Ltd.).

(c) Liquid chromatography chiral stationary phases have been prepared by a process called *molecular imprinting* (review: Wulff, 1986; Wulff and Minárik, 1988; see also Allenmark, 1991, p. 125). The process consists of three steps: (1) An achiral monomer is covalently bonded to a non-racemic compound serving as a template (also called a print molecule); (2) copolymerization of the modified monomer in the presence of a cross-linking agent gives rise to a rigid polymeric network; and (3) the rigid polymer is freed of the chiral template moiety by hydrolysis. There results a polymer that retains the ability to recognize the template molecule if the latter is passed in solution through a bed containing the rigid polymer. The stationary phase evidently contains chiral cavities that are retained over long periods of time in spite of repeated washing with solvent. These cavities are responsible for the stereoisomer discrimination by the column in which either the template molecule enantiomer, an analogue thereof, or the enantiomer of the template molecule is pref-erentially retained during chromatography (e.g., Sellergren and Anders-son, 1990; Wulff and Schauhoff, 1991; Fischer, Mosbach, et al., 1991). The process is illustrated in Figure 6.55 with L-*N*-propionyl-2-amino-3-(4-hydroxyphenyl)-1-propanol **A** serving as the template molecule.

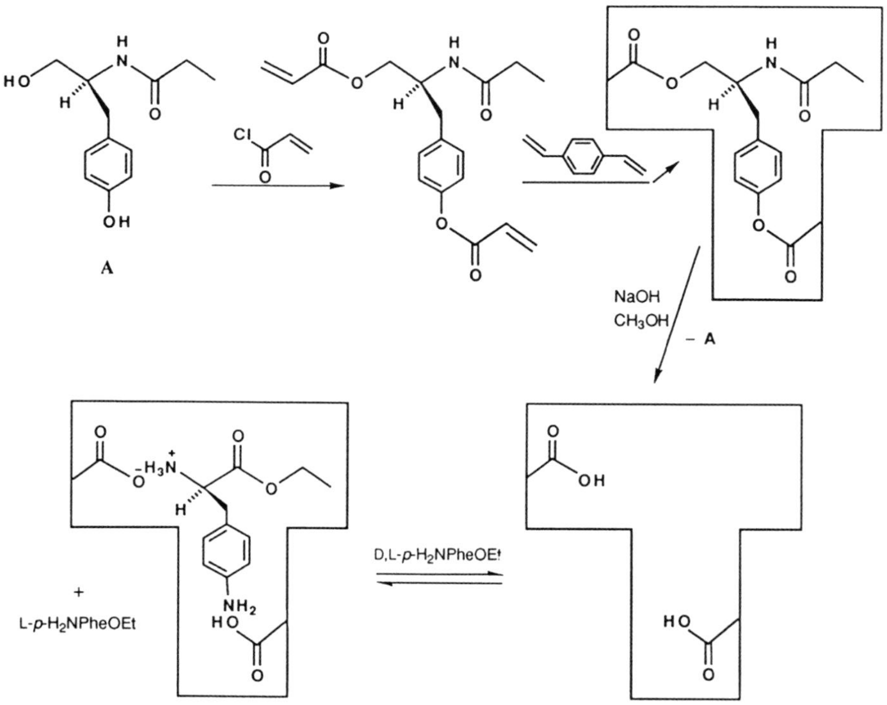

Figure 6.55. Model for chromatographic resolution of enantiomer mixtures by molecular imprinting. In the example, L-*N*-propionyl-2-amino-3-(4-hydroxyphenyl)-1-propanol (**A**) serves as the template (or print) molecule. The imprinted polymer discriminates between the enantiomers of *p*-amino-phenylalanine ethyl ester [*p*-H$_2$NPheOEt] [Adapted with permission from Sellergren, B. and Anders-son, L., (1990), *J. Org. Chem.*, **55**, 3381. Copyright © 1990 American Chemical Society.]

(d) Inclusion complex formation with β-cyclodextrin (Fig. 6.46) bonded to silica gel by means of a covalently bonded spacer (Armstrong and DeMond, 1984; Armstrong, 1985; Armstrong et al., 1985a. Review: Ward and Armstrong, 1988) is widely applicable to the determination of the enantiomer composition of, for example, chiral amines, amino alcohols (medicinal agents), and binaphthyl derived crown ethers by HPLC (Armstrong et al., 1985a,b; Hinze, Armstrong, et al., 1985).

Computer-assisted molecular modeling (3D projections and graphic images) based on X-ray crystallographic data assist in providing an understanding of the factors that promote enantioselective inclusion complex formation: for example, (+)-propanolol > (−)-propanolol (Fig. 6.45) both with β-cyclodextrin. Hydrogen bonds between the (+)-propanolol and the 2- and 3-OH groups of immobilized β-cyclodextrin are postulated to be shorter, and hence stronger than is possible for the (−) enantiomer (Fig. 6.56). This preferential interaction of the analyte with the chiral stationary phase leads to the prediction of a longer retention time for the (+) enantiomer (Armstrong et al., 1986).

While inclusion of analyte molecules in the selector is required for diastereomer discrimination to take place, this mechanism is not the only factor contributing to differential chromatographic retention; the order of capacity factors ($k' = A_s/A_m$, where A_s and A_m represent the amounts of analyte in the stationary and mobile phases, respectively) does not always mimic that of binding constants (K_b) in a series of measurements (Armstrong et al., 1985a).

(e) Chiral isotactic poly(triphenylmethyl methacrylate) (PTrMA) is the preeminent example of a cooperative CSP, one in which the stereoisomer discrimination does not arise from interaction between analyte residues and specific functional moieties in the chiral selector acting independently (Blaschke, 1980; Pirkle and Finn, 1983). In the latter, the polymer

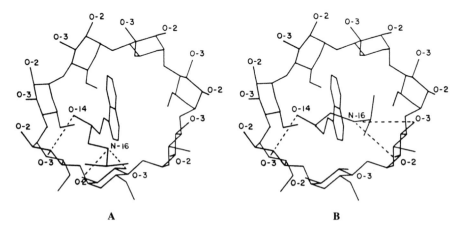

Figure 6.56. Computer projections of inclusion complexes of (**A**) (+)-propanolol, and (**B**) (−)-propanolol both in β-cyclodextrin. The dotted lines represent potential hydrogen bonds [Reprinted with permission from Armstrong, D.W. et al. (1986), *Science* (*Washington, D.C.*), **232**, 1132. Copyright 1986 by the American Association for the Advancement of Science.]

molecules as a whole are chiral without possessing side chains bearing stereocenters; such polymers have a chiral helical backbone. Cooperative phenomena (those involving numerous chirotopic sites in the selector acting simultaneously) may account for the stereoisomer recognition and resolution that is observed in this type of enantioselective stationary phase.

Poly(triphenylmethyl methacrylate) is prepared by enantioselective polymerization of triphenylmethyl methacrylate monomer by a non-racemic anionic catalyst [(−)-sparteine/n-butyl lithium, Fig. 6.57]. The chirality of the resulting isotactic polymer is due solely to its helicity (Okamoto, Yuki, et al., 1979; Yuki, Okamoto, and Okamoto, 1980. Reviews: Okamoto, 1987; Okamoto and Hatada, 1988; Johns, 1989b; cf. Section 13-4.g). The polymer is unusual in that the helices are stable in solution even at high temperature. Conformational transitions involving the polymer backbone that would "unravel" the helix, and hence cause a loss in chirality, are prevented by the bulkiness of the trityl appendages (Wulff et al., 1986; Farina, 1987b).

The chromatographic enantioselectivity of PTrMA may be due to intercalation of aryl substituents in analyte molecules into chirotopic channels present in the polymer (Pirkle and Finn, 1983). Crystalline regions in the polymer may play a role in the separations (as in separations effected in TAC). Two variants of PTrMA are commercially available (Chiralpak^{TM} OT(+) and OP(+), the latter being coated on silica gel; Okamoto et al., 1981).

Many hydrophobic substances including hexahelicene (Fig. 6.6, **6**), tetramesitylethylene (Gur, Biali, Rapoport, et al., 1992), and Tröger's base (Fig. 6.38) are resolved on PTrMA columns with separation factors α ranging from 1.1 to >2 ($\alpha > 13$ for hexahelicene). Virtually all have structures bearing one or more aromatic or heteroaromatic rings. The utility of PTrMA is comparable (in some cases superior) to TAC.

In enantiomeric purity determinations by HPLC, it is sometimes desirable to use a detector capable of measuring chiroptical properties (optical rotation or

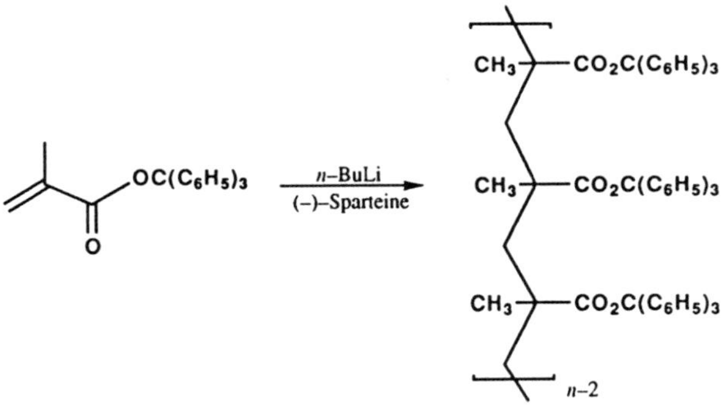

Figure 6.57. Preparation of optically active, isotactic PTrMA. The degree of polymerization (DP = n) is as high as 220 for the insoluble PTrMA and for the soluble PTrMA that is adsorbed on silanized silica gel.

circular dichroism) instead of, or in addition to, the usual absorbance (or refractivity) detector. Clearly, such a detector can confirm the success of a resolution, or enantiomeric purity determination, while the chromatography is in progress (Hesse and Hagel, 1976b); cf. Sections 13-4.e and 13-5.a. A third simultaneous measurement (beyond, e.g., CD and absorbance), that of the g number $= \Delta A/A$, where A = absorbance and ΔA = differential absorbance of lcpl and rcpl radiation (cf. Section 13-4.a) permits the assessment of the enantiomeric purity (Drake, Gould, and Mason, 1980; Mason, 1982, page 28).

The possibility of determining the enantiomer purity by HPLC in spite of peak overlap or of peak fusion has been demonstrated when absorbance and chiroptical detectors are simultaneously used (Mannschreck et al., 1980, 1982).

Thin-Layer Chromatography. Application of TLC to enantiomer purity determination via separation of diastereomeric derivatives is a relatively straightforward though seemingly little used technique; an example is the analysis of a nonracemic carboxylic acid synthetic intermediate as the amide formed with 1-(α-naphthyl)ethylamine (Fig. 6.38, NEA; Fujisawa et al., 1981. Reviews: Souter, 1985, Chapter 4; Günther, 1988).

More recently, TLC plates impregnated with nonracemic chiral selectors, or coated with the latter bonded to achiral stationary phases, such as silica gel, have been developed for qualitative and quantitative analysis especially of multicomponent mixtures containing enantiomer pairs. The first report of such an analysis appears to be that of Wainer et al. (1983) who demonstrated that the chiral alcohol TFAE (Fig. 6.38, **81a**) could be resolved chromatographically on plates coated with (R)-N-(3,5-dinitrobenzoyl)phenylglycine covalently bonded to γ-aminopropyl silanized silica gel (Fig. 6.47, **B**) with an estimated separation factor $\alpha = 1.50$.

Thin-layer chromatography plates coated with chiral ligand-exchange media have been applied to the analysis of amino acid mixtures, either after derivatization (Weinstein, 1984; Grinberg and Weinstein, 1984) or as such (Günther et al., 1984; 1985); β-cyclodextrin-coated plates have been shown to fully resolve dansyl amino acids (Ward and Armstrong, 1986).

Quantitative enantiomer purity determination by TLC is possible using densitometry or by measurement of fluorescence or UV absorbance following extraction of the spots. Visual estimation of even trace amounts of one enantiomer (at the level of <1% in some cases) is also possible. The potential of enantiomer purity determinations by TLC is evident from the report that such an analysis could be carried out in as little as 5 min on commercial plates (Chiralplate^{TM}; Brinkman and Kamminga, 1985).

Electrophoresis with Enantioselective Supporting Electrolytes

Formation of diastereomeric complexes by ligand exchange is the basis of a highly sensitive analytical method for the determination of enantiomeric purity by high-voltage electrophoresis in capillary columns (Wallingford and Ewing, 1989; Camilleri et al., 1992). The analyte, for example, a chiral amino acid [derivatized with 5-(dimethylamino)naphthalene-1-sulfonyl (dansyl, DNS) chloride] is dissolved in a support electrolyte consisting of copper(II)-L-histidine (or copper(II)–

aspartame) complex. The sample migrates by a combination of electroosmosis and electrophoretic action under the influence of a strong (300 V cm^{-1}) electric field. Analysis of the separated DNS–amino acids is carried out with a laser–fluorescence detector.

The principle of the method is analogous to that which obtains in ligand-exchange chromatographic resolutions with chiral mobile phases where diastereomeric complexes are formed that migrate at different rates (Section 6-5.d, p. 247). In the present method, diastereomeric complexes are also formed. Differences in complexation constants cause these transient charged species to acquire different mobilities under the influence of the applied electric field (electrokinetic separation). Note that there is no mobile phase per se in electrophoresis. The method permits rapid (ca. 10 min) analysis of the enantiomer composition of femtomolar amounts of analyte (Gassmann, Kuo, and Zare, 1985; Gozel, Zare, et al., 1987). Diastereomer discrimination is also observed in differences in the intensity of the fluorescence signals of the diastereomeric DNS–amino acid complexes.

Dansylated amino acids have also been electrophoretically separated in the presence of ionic cyclodextrin derivatives (Terabe, 1989). Selective inclusion and inclusion complex migration is responsible for the separation. While typically electrophoresis is limited to ionic analytes (in this context mixtures of enantiomeric ions and of diastereomeric ions), neutral analytes can also be electophoretically analyzed by this method in the presence of surfactants that convert conventional electrolytes to micellar ones with ionic micelles acting as carrier (Terabe, 1989; Nishi and Terabe, 1990). The enantiomers of 1,1'-binaphthalene-2,2'-diol (Fig. 6.26, **33b**) and of the related binaphthylphosphoric acid (Fig. 6.35, **62a**) are resolved by capillary electrophoresis in the presence of bile salts, for example, sodium deoxycholate.

e. Kinetic Methods

Several analytical methods are based on differences in rates of reaction of enantiomers as their mixtures react with nonracemic chiral reagents. Preparative methods based on this principle are called kinetic resolutions (Section 7-6). The rate differences stem from the fact that diastereomeric transition states are formed at rates that reflect the differing free energies of activation (Mislow, 1965, p. 122 ff.).

Methods Based on Enzyme Specificity

Reactions of chiral substrates brought about by enzymes are often subject to enormous rate differences for the two enantiomers so that, to all intents and purposes, only one of them reacts. The enzyme, e.g., hog kidney acylase (Price and Greenstein, 1948), may be said to be "specific" to one of the enantiomeric substrates. The rate difference may be made the basis of a powerful analytical method for the determination of enantiomer composition (Bergmeyer, 1984–1985). Moreover, in many cases, pairs of enzymes of opposite stereospecificity are known, so enantiomerically impure samples may be analyzed with either enzyme with one analysis serving as a check on the other since the two results must sum to

100%. For example, the proportion of enantiomers in α-hydroxy acids may be determined by oxidation of the acid with β-nicotinamide adenine dinucleotide (NAD$^+$) in the presence of either D- or L-lactic acid dehydrogenase (LDH) depending on the information and precision desired (Gawehn, 1984; Matos, Smith, and Wong, 1985; Wong and Matos, 1985).

A particularly precise way of determining enantiomer purities employs enzymes that catalyze the reactions of the *minor* enantiomer present in an enantiomerically impure sample. In the following example, the enantiomeric purity of (*R*)-lactate is determined enzymatically by oxidation of the minor (*S*)-lactate enantiomer with (*S*)-lactate dehydrogenase in the presence of the stoichiometric cofactor NAD$^+$, which is concommitantly converted to its reduced form NADH as shown in Eq. 6.18:

$$
\underset{\substack{\text{OH}\\|}}{H_3C-CH-CO_2H} + NAD^+ \longrightarrow H_3C-\overset{\substack{O\\||}}{C}-CO_2H + NADH + H^+
$$

$$(6.18)$$

The formation of NADH is monitored by UV spectroscopy at 340 nm, the wavelength at which it absorbs (Wong and Whitesides, 1981). In such analyses, care must be taken to insure that the reaction goes to completion, or else a correction must be applied that requires knowledge of the equilibrium constant measured under identical conditions.

While enzymatic methods have been applied mostly to the analysis of amino acids (Greenstein and Winitz, 1961; Hinkkanen and Decker, 1985), other functional groups lend themselves to this approach: alcohols (e.g., Caspi and Eck, 1977), halogenated carboxylic acids (e.g., Motusugi, Esaki, and Soda, 1983), and carbohydrates (e.g., galactose; Whyte and Englar, 1977).

Enzymatic methods are the principal ones available for the determination of enantiomeric purity of compounds containing chiral methyl groups (cf. Section 8-6.a).

Nonenzymatic Methods

Kinetic resolutions are governed by the relative rates of two competing reactions having rate constants k_R and k_S (i.e., those of the two enantiomers of the substrate; Section 7-6). The enantiomeric purity attainable in such a reaction is dependent on the conversion C. The relation between C, the relative rate k_R/k_S (the stereoselectivity factor **s**) and the ee of the unreacted substrate is given by Eq. 6.19.

$$\mathbf{s} = \frac{\ln[(1-C)(1-\text{ee})]}{\ln[(1-C)(1+\text{ee})]} \qquad (6.19)$$

It is evident that the enantiomeric purity of the resolved sample (unreacted kinetic resolution substrate) may be determined if **s** and C ($C < 1$) are known. An alternative equation (Eq. 6.20),

$$[S] - [R] = 0.5(e^{-k_S t} - e^{-k_R t}) \qquad (6.20)$$

where [S] and [R] refer to the concentrations of the enantiomers shows that the enantiomer composition of the substrate may be determined at a given *time t* after the start of the reaction from a knowledge of the relative rate constants (Kagan and Fiaud, 1988). In either of these approaches, the enantiomeric purity is not measured directly, rather it is calculated from the conversion or time. Applications of these equations for the purpose of determining enantiomer compositions are few in number (e.g., to hydrocarbons, Fig. 6.58, **86**), which were kinetically resolved by enantioselective rearrangement to achiral indenes **87** (Fig. 6.58) in the presence of a chiral catalyst (Meurling and Bergson, 1974; Meurling, Bergson, and Obenius, 1976). Equations 6.19 and 6.20 serve mostly to guide preparative kinetic resolutions (see Section 7-6).

In this connection, Kagan, Bergson, and their co-workers explicitly pointed out that two separate kinetic resolutions on the same substrate, when carried out to known conversions, suffice to determine the degree of stereoselectivity (**s**) of the reaction and the absolute rotation of the substrate, $[\alpha]_{max}$, and hence, the enantiomer composition of the substrate (assuming that the ratio $ee_1/ee_2 = [\alpha]_1/[\alpha]_2$ for the two reactions with $[\alpha]$ being that of the residual substrates) (Balavoine, Moradpour, and Kagan, 1974; Meurling, Bergson, and Obenius, 1976; see also Schoofs and Guetté, 1983). For a recent application, see Hawkins and Mayer (1993).

In 1964, Horeau described an elegant but elaborate kinetic method for determining enantiomer compositions that requires two partial consecutive resolutions to be carried out. This method, which also permits the determination of configuration in some cases, is relative in that it correlates an unknown enantiomer composition with that of another substance whose enantiomer composition is known (Reviews: Raban and Mislow, 1967; Schoofs and Guetté, 1983).

Subsequently, Horeau (1975) described a second kinetic method permitting both the enantiomeric enrichment of a chiral sample and the assessment of its composition just as in an enzymatic (kinetic) resolution (see above).

A racemic sample of compound A, the specific rotation of whose enantiomers is unknown, is allowed to react with an insufficient amount of optically pure B_+. Two reactions ensue but at different rates. The unreacted A will be optically active, that is, it will be enriched in one of the enantiomers. In a second reaction, an excess of racemic B is allowed to react with an insufficient amount of optically active A, whereupon optically active B accumulates in the residue. Correlation of the rotation of residual A with that of residual B, based on the equality of the reaction rate of A_+ with B_+ to that of A_- with B_- (and similarly for A_+ with B_- vs. A_- with B_+), permits the determination of the absolute rotation of A. For applications of this method, see Horeau and Nouaille (1966); Horeau et al. (1966); Christol et al. (1968); and Brugidou et al. (1974).

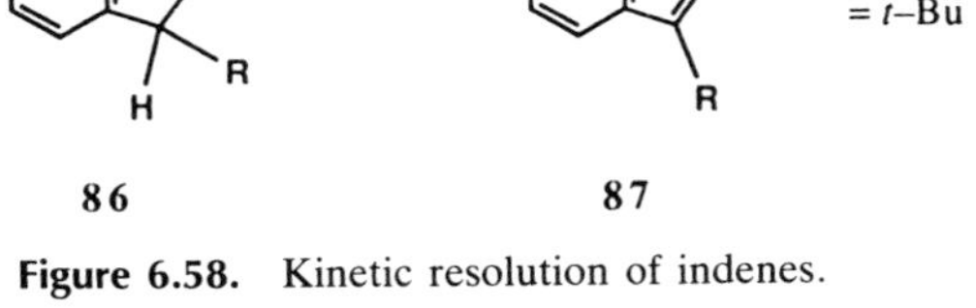

Figure 6.58.   Kinetic resolution of indenes.

When two racemic chiral substances R,S and D,L, respectively, react with one another in a stereoselective process, the diastereomeric products RD + SL and RL + SD (**A** and **B**, respectively) are formed at different rates, $k_A/k_B = K$ (when $k_A > k_B$). This ratio may be easily measured by carrying out the reaction with racemic substrates and reagents, and measuring the ratio of the racemic diastereomers formed. At the same time, when the reactant samples are nonracemic, knowledge of K and of the conversion allows one to calculate the maximum rotatory power of the remaining (slower reacting) substrate enantiomer. With this information, the optical purity of the original substrate mixture may be calculated.

Application of the formula for "homocompetitive reactions" $[(R/R_0 = S/S_0)^K$, where R_0 and R and S_0 and S refer to the reactant enantiomers before (R_0/S_0) and after (R/S) reaction] allows one to derive the quantitative expression (Eq. 6.21):

$$(1 - F)^{K-1} = \frac{1 - p'}{1 - p}\left(\frac{1 + p}{1 + p'}\right) \tag{6.21}$$

where F is the fraction of substrate that has reacted, p is the enantiomeric purity (ee) of the substrate prior to reaction (L > D) and p' is the enantiomeric purity of the substrate after reaction ($p' > p$ when $k_D > k_L$) (Kagan and Fiaud, 1988, p. 262).

Equation 6.21 allows one to calculate the extent of reaction F of the substrate necessary to raise its enantiomeric purity from p to p' for a given value of K (Horeau, 1975; Briaucourt and Horeau, 1979). If p of the substrate is reasonably high, then it may be possible to attain $p' = 99.9\%$ in one step with a reasonable conversion. For example, reaction of an enantiomerically impure sample of $(-)$-$C_6H_5CH(OH)CH_3$, $\alpha_D^{22} - 42.75°$ (neat, $\ell = 1$) with insufficient $(+)$-α-phenylbutyric anhydride [setting $F = 0.75$ (75% conversion) and $K = 4.7$], leads to the accumulation of alcohol, $\alpha_D^{22} - 44.0°$ (neat, $\ell = 1$), having $p' = 99.9\%$ in the residue. Equation 6.21 requires that the substrate (original alcohol sample) had $p = 90.0\%$.

Horeau is responsible for yet another analytical method permitting the determination of enantiomer compositions, the duplication method. The method is described in Section 6-5.c.

f. Calorimetric Methods

The measurement of the melting point of enantiomer mixtures provides sufficient information for the determination of enantiomeric composition as was pointed out in Section 6-4.e. The composition of such mixtures can, in principle, be directly read off the binary (melting point) phase diagram. Construction of the entire diagram can be avoided, however, and the precision of the measurement increased, if one knows the enthalpy of fusion of the racemate and/or of the enantiomerically pure substance. Obviously, a calorimeter of some sort is required, the most convenient of which is a differential scanning microcalorimeter. Such instruments can measure heats of fusion fairly accurately on samples of the order of 1 mg while at the same time measuring the melting point lowering of samples (relative to an enantiomerically pure compound or a racemic compound) with precision (Jacques et al., 1981a, pp. 39, 151).

In the case of compounds exhibiting conglomerate behavior (Fig. 6.3a) and for mixtures whose enantiomeric purity is roughly in the range 30–90%, the measurement of the enantiomer composition by differential scanning calorimetry (DSC) is *direct*, that is, the proportion of enantiomers may be directly taken from the area of the eutectic (racemate) peak in the DSC (melting) trace (see, e.g., Fig. 6.5) provided that the enthalpy of fusion of the pure enantiomer and the weight of the sample be known.

For racemic compounds (Fig. 6.3b), the composition may be calculated by means of the equation of Prigogine and Defay (Eq. 6.6, Section 6-4.e), if the sample composition lies on the racemate side of the eutectic (e.g., N in Fig. 6.7a). This calculation requires a knowledge of the enthalpy of fusion of the racemate as well as the melting point lowering of the sample vs. that of the racemate. The composition of samples richer in pure enantiomer than the eutectic may be determined by means of the Schröder-van Laar equation (Eq. 6.5) as for conglomerates (Fig. 6.3a).

An *indirect method* is applicable to samples that are either very pure or very impure such that no eutectic is visible in the DSC trace. This method involves the analysis of the melting scan according to procedures applicable to purity determinations in general that are based on accurate determination of melting point lowering (McNaughton and Mortimer, 1975). The method thus makes no distinction between "chemical" impurities and enantiomeric impurities (Jacques et al., 1981a, p. 151; Wildmann and Sommerauer, 1988).

Differential scanning calorimetry has been applied to the determination of the enantiomer composition of norleucine. The method involves comparison of the heat of transition ΔH^{DL} from the α- to the γ-(polymorphic) forms of racemic norleucine with the measured ΔH^{obs} of nonracemic samples of the amino acid (Matsumoto et al., 1987).

Calorimetric methods for the determination of the enantiomer composition are broadly applicable and deserve to be used more frequently than they have been. A significant limitation is the fact that DSC instruments are not often found in organic chemistry laboratories. For a detailed analysis of the method and discussion of their limitations, see Jacques et al., 1981a, pp. 151 ff, 416.

g. Isotope Dilution

Isotope dilution analysis is an analytical technique useful in the determination of a single substance, say **A**, that is present in a mixture consisting of chemically similar substances (**A**, **B**, or **C**) from which **A** cannot be quantitatively separated (Tolgyessy, Braun, and Kyrs, 1972). In this method, only a small sample of **A** need be isolated, and large losses during its purification are allowed (Waller, 1972).

The principle of the process, as applied to enantiomeric purity determinations, involves the dilution of a substance of unknown enantiomer composition by the corresponding isotopically labeled compound of *known* enantiomer composition. Measurement of the isotope concentration following the commingling of the labeled and unlabeled samples and reisolation of either a pure enantiomer or of the pure racemate permits calculation of a dilution factor that can be related to the enantiomer composition of the unknown mixture. Isotope dilution was first applied to the analysis of organic compounds by Rittenberg and Foster (1940) and Graff, Rittenberg and Foster (1940). The first application to the determination of enantiomeric purity is due to Berson and Ben-Efraim (1959).

Originally, the procedure called for the dilution of the optically active analytical sample (of unknown enantiomer composition) with isotopically labeled racemate. Recrystallization of such mixtures often leads to reisolation of labeled racemate whose isotope content is reduced by the extent to which the original (optically active) sample was "contaminated" by racemate. The crux of the process is that a *chemically* pure sample must be isolated. In the context of enantiomeric purity analysis, a purification step is required that permits either the pure racemate *or* the pure enantiomer to be isolated.

For purposes of calculation, a sample consisting of unequal amounts of the enantiomers of a given compound is considered to be a mixture of one enantiomer and the racemate, for example, a sample consisting of 0.6 mol of one enantiomer and 0.4 mol of the other (ee = 0.2) is considered to be comprised of 20% enantiomer and 80% racemate (cf. Fig. 6.4). The relation between the fraction of isotope label before and after reisolation, assuming that there is no isotope effect during the crystallization, is given by Eq. 6.22

$$xC_0 = (x + y)C \qquad (6.22)$$

where x and y are the amounts of isotope labeled racemate (or other mixture of known enantiomer composition) and of substance to be determined, respectively. The parameters C_0 and C are the measured isotope contents of x and y, respectively. The isotope analysis itself may be carried out by NMR or, especially, by MS in the case of stable isotopes, and by scintillation counting if the isotope is radioactive. The determination of the enantiomer composition of methionine by isotope dilution using pure ^{14}C-labeled L-methionine as a tracer is described by Elvidge and Sammes (1966).

Equation 6.22 can be elaborated to

$$(ee)_y = \frac{C(x + y)^2 - C_0 x(x + y)}{y^2 C} \qquad (6.23)$$

If the specific rotations of the sample before and after isotope dilution have been measured, then Eq. 6.23 may be elaborated to Eq. 6.24

$$[\alpha_A] = \frac{Cy^2[\alpha_1]^2}{C(x + y)^2 - C_0 x(x + y)} \qquad (6.24)$$

assuming that the reisolated sample is racemic (C is the isotope content of the reisolated racemate, $[\alpha_A]$ is the specific rotation of enantiomerically pure material and $[\alpha_1]$ is the specific rotation of the mixture before the isotope dilution experiment is begun) (Andersen, Gash, and Robertson, 1983).

As we have seen in Section 6-4.f, recrystallization of an enantiomerically impure sample may give rise to an enantiomerically pure sample, to a racemate, or to mixtures of intermediate enantiomer composition according to the nature of the phase relationship exhibited by the substance, the original composition of the mixture being recrystallized, and experimental conditions. Accordingly, a more general treatment is desirable, one that requires only that the reisolated material differ in enantiomer composition from the original.

The following more general equation permits the analysis of mixtures that do not yield a racemate on reisolation following the isotope dilution experiment (the isotope tracer used is still a racemate).

$$[\alpha_A]^2 = \frac{Cy^2[\alpha_1]^2 - C_0xy[\alpha_1][\alpha_2]}{C(x+y)^2 - C_0x(x+y)} \qquad (6.25)$$

In Eq. 6.25, $[\alpha_2]$ is the specific rotation of the reisolated mixture (following the isotope dilution experiment); the enantiomeric purity of the original (chemically but not enantiomerically pure) sample equals $[\alpha_1]/[\alpha_A]$. Note that when the reisolated mixture is a racemate, $[\alpha_2] = 0$ and Eq. 6.25 reduces to Eq. 6.24.

Three compounds whose enantiomer compositions were determined by isotope dilution are **88** (Bruzik and Stec, 1981) and **89a** and **b** (Hulsholf and Wynberg, 1979; Fig. 6.59). A stable isotope (deuterium) was used in the case of **88** ($4,4,6,6\text{-}d_4$) and a radioactive isotope (^{14}C) was used in the case of **89a** ($-^{14}CO_2H$). Neither of these compounds lent themselves readily to enantiomer purity determination by alternative methods, that is, by methods available at the time when the work was done. In particular, HPLC analysis of some diastereomeric esters and amides of **89a** failed due to lack of separation of the diastereomers; nor was the NMR analysis of alcohol **89b** in the presence of CSRs successful. It was inferred that the stereocenters of these compounds are too far removed from the sites of interaction of the molecules with the stationary phases or with the shift reagent to induce differences either in chromatographic retention or in chemical shifts.

The method is quite general in applicability but it suffers somewhat from being time consuming. The review by Raban and Mislow (1967) gives a list of examples. A more recent review by Andersen, Gash, and Robertson (1983) gives a very detailed derivation of the quantitative expressions used in isotope dilution analysis.

Limitations in sensitivity makes the "ordinary" isotope dilution method unsuitable for samples having high enantiomeric purities. Such samples include synthetic peptides prepared from optically active amino acid derivatives by nominally nonracemizing coupling procedures. Analysis of peptides for racemate

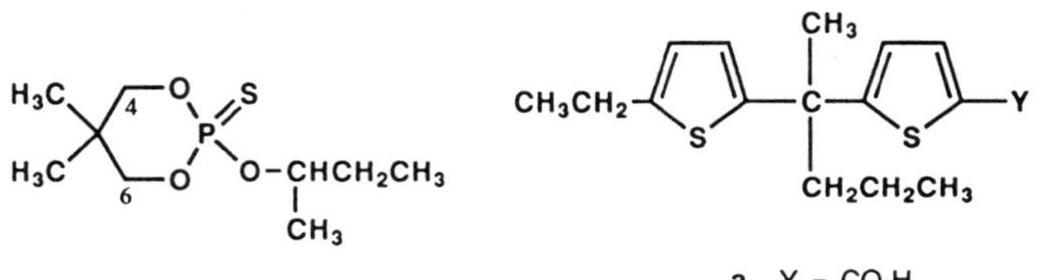

Figure 6.59. Compounds whose enantiomer composition was determined by isotope dilution.

content in the 0.001–1% concentration range can, however, be effected by successive isotope dilution. The latter process is based on the original Berson and Ben-Efraim method (1959) much as fractional crystallization is related to crystallization. The principle is that isotopically labeled and enantiomerically enriched samples on crystallization in the presence of unlabeled racemate (with the racemate being recovered) have the label slightly diluted with respect to the dominant enantiomer but significantly reduced with respect to the minor enantiomer. Repetition of the crystallization process makes it possible to attain labeled mixtures having known enantiomer ratios of the order of 10^6–10^7. Such labeled mixtures have been used as probes in assessing the extent of racemization in several peptide coupling procedures (Kemp et al., 1970).

An alternative method for the determination of enantiomer composition in mixtures having high enantiomer purity values has been called the "radioactive tracer method" (Jacques et al., 1981a, p. 418). In this method, a small amount of an isotopically labeled enantiomer (the tracer) is added to a sample of the (unlabeled) other enantiomer, both being of high enantiomeric purity. The mixture is subjected to purification and the enrichment is assessed by following the decrease in radioactivity of the sample. For an application involving purification by single-pass zone refining of the analgesic agent (+)-α-propoxyphene **45** (Fig. 6.30), see Tensmeyer, Landis, and Marshall (1967).

h. Miscellaneous Methods

The enantiomer composition of chiral ions (e.g., in ephedrinium salts) may be measured potentiometrically in an electrochemical cell fitted with two liquid polyvinyl chloride (PVC) membranes each containing one enantiomer of an electrically neutral chiral ionophore, for example, enantiopure (R,R and S,S)-5-nonyl tartrate (Fig. 6.60; Bussmann and Simon, 1981).

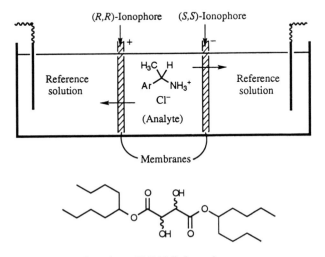

Ionophore: (R,R / S,S)-5-nonyl tartrate

Figure 6.60. Electrochemical cell for the potentiometric determination of the enantiomer composition of chiral ionic compounds.

Each of the membranes selectively extracts one enantiomer of the analyte forming diastereomeric complexes that formally permeate across the membrane. An electric potential difference is established between the analyte solution and the reference solution; strictly speaking, this potential difference is the sum of two phase boundary potentials and a potential within the membrane. The potential difference, which is also affected by the analyte concentration and by the necessarily unequal analyte enantiomer ratio, differs for the left-hand side and right-hand side of the electrochemical cell (Fig. 6.60). Calibration is effected separately with solutions of fixed concentration of each analyte enantiomer (Bussman and Simon, 1981). Along similar lines, potentiometric ion-selective electrodes incorporating peroctylated α-cyclodextrin (Fig. 6.46) [e.g., in a membrane containing PVC and bis(butylpentyl)adipate] have been shown to be useful in the determination of the enantiomer composition of ephedrine (Fig. 6.27) in the presence of serum cations (Kataky, Bates, and Parker, 1992; Bates, Kataky, and Parker, 1992).

The differential interaction of analyte enantiomers with membranes containing chiral ionophores has recently been applied to the determination of enantiomer composition by a spectrophotometric method that remarkably eschews the need for polarized light (Holy, Simon, et al., 1990). The principle of the method lies in the nature of the so-called optode membranes that contain a lipophilic chiral ionophore as well as a "chronoionophore," that is, one that drastically changes its optical properties (absorption intensity as a function of wavelength) on ion complexation.

A promising analytical method for the determination of enantiomer composition, especially in complex biological fluids and at very low concentrations, is radioimmunoassay (RIA). The possibility of applying RIA methodology derives from the finding (ca. 1929) that serum reactions are enantioselective (Landsteiner, 1962), that is, specific antibodies (antisera) may be produced in living organisms "against practically any type of organic compound, including enantiomers of any chiral molecule" (Huhtikangas et al., 1982. Reviews: Cook, 1988; Porter, 1991).

The process requires that typically small molecules, *haptens* (from the Greek $\alpha\pi\tau\epsilon\iota\nu$, to fasten), radioisotopically labeled for easy detection, be conjugated, that is, covalently bound to a macromolecule (e.g., a protein), to form immunogens capable of stimulating the formation of antibodies.

Immunization of rabbits with (−)-propranolol (Fig. 6.45) conjugated with BSA generates antisera that have very low affinity (Cook, 1988) for the enantiomer of the hapten, that is (+)-propranolol. Analyses consist of the addition of biological fluids containing an unknown amount of (−)-propranolol to the labeled (−)-propranolol–antibody complex from which labeled (−)-propranolol is quantitatively displaced. Following separation of free analyte from the complex (by electrophoresis, precipitation, or chromatography), the concentration of the radiolabeled material is measured in a scintillation counter. As little as 10 pg of (−)-propranolol was recognized by the antiserum (Kawashima et al., 1976).

Immunoassays are simple to run on multiple samples (Cook, 1985). On the other hand, the procedure requires the generation of stereoselective antisera that is certainly not routine practice and may take several weeks (Vermeulen and Breimer, 1983). In addition to propranolol, the RIA method has been applied to the enantiomer analysis of warfarin (Fig. 6.61) (Cook et al., 1979), to the alkaloid atropine (Fig. 6.30, **39**), which relatively easily racemizes (Huhtikangas et al., 1982), to the enantiomers of ephedrine (Fig. 6.27; Midha et al., 1983) and of pentobarbital (Cook et al., 1987).

Yet another analytical method, one involving isotope labeling of one of the enantiomers of a pair, permits the quantitative assessment of in vivo enantioselective effects. Administration to a living system of a 1:1 mixture of (+) and (−) enantiomers, one of whose components is labeled with a stable isotope [the mixture is regrettably misnamed pseudoracemate in the original paper (see Section 6-3)], permits the determination of the enantiomer composition of the recovered substrate by mass spectrometric analysis of isotope ratios. An essential requirement of the method is that isotope effects be known not to bias the results of the analysis (Cook, 1985). For applications, see McMahon and Sullivan (1976), Weinkam et al., (1976), and Howald et al. (1980).

The method has also been used in the assessment of the enantioselectivity of chemical reactions, for example, in the reaction of (R,S)-methyloxirane {quasi-racemic mixture of (R)-methyloxirane with (S)-$[2,2\text{-}^2H_2]$methyloxirane} with L-valyl-L-leucine (Golding, 1988). An analogous assessment of the enantioselectivity of transport across a chiral membrane (Section 7-8.b) has been made for α-methylbenzylammonium ions by measuring the $^3H/^{14}C$ ratio of electrodialyzed solutions with a scintillation counter, using a quasi-racemic mixture of ^3H-(R)- and ^{14}C-(S)-ammonium ions as substrate (Thoma, Simon, et al., 1975).

It is not uncommon to find that the racemate of a chiral compound and the corresponding pure enantiomers crystallize with different amounts of solvent of crystallization (Jacques et al., 1981a, p. 204). If the solvent of crystallization is easily removed from the racemate leaving a solvent-free enantiomer, then such removal makes it possible, by measuring the resulting weight loss, to assess the enantiomer composition of a sensibly enantiomerically impure nonracemic sample.

An example of this method is given by Jacques and Horeau (1949) in the case of 2,2-dimethyl-3-(6′-methoxy-2′-naphthyl)-pentanoic acid. Demethylation of the 6′-methoxy group gives rise to a derivative whose racemate crystallizes (from benzene) with a molecule of solvent, whereas the enantiomers crystallize solvent-free. A control experiment demonstrates that at 120°C, the racemate easily loses the theoretical amount of benzene, whereas under identical conditions, the enantiomer does not lose any weight. The method was applied in demonstrating the enantiopure character of the sample tested.

Finally, just as it is possible to determine the enantiomer composition of chiral samples by analysis of the fusion phase diagram (Section 6-5.f) so is it possible to do the same, though far less rapidly and easily, from the analysis of solubility diagrams. For details, see Jacques et al. (1981a, p. 207ff).

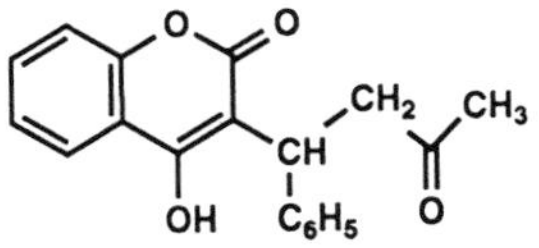

Figure 6.61. Warfarin.

REFERENCES

Abraham, E. P. (1963), "The Antibiotics," in Florkin, M. and Stotz, E. H., Eds., *Comprehensive Biochemistry*, Vol. 11, Elsevier, Amsterdam, The Netherlands, p. 181.

Ács, M. (1990), "Chiral Recognition in the Light of Molecular Associations," in Simonyi, M., Ed., *Problems and Wonders of Chiral Molecules*, Akadémiai Kiadó, Budapest, Hungary, pp. 111–123.

Addadi, L., Berkovitch-Yellin, Z., Weissbuch, I., Lahav, M., and Leiserowitz, L. (1986), "A Link Between Macroscopic Phenomena and Molecular Chirality: Crystals as Probes for the Direct Assignment of Absolute Configuration of Chiral Molecules," *Top. Stereochem.*, **16**, 1 (see p. 71).

Adriani, J. H. (1900), *Z. Phys. Chem.*, **33**, 453.

Ahnoff, M. and Einarsson, S. (1989), "Chiral Derivatization," in Lough, W. J., Ed., *Chiral Liquid Chromatography*, Blackie, Glasgow, UK. Chap. 4.

Ahuja, S., Ed. (1991), *Chiral Separations by Liquid Chromatography*, ACS Symposium Series 471, American Chemical Society, Washington, DC.

Ajisaka, K. and Kainosho, M. (1975), *J. Am. Chem. Soc.*, **97**, 1761.

Akabue, B. E. and Hemmes, P. (1982), *Adv. Mol. Relaxation Interact. Processes*, **22**, 11.

Akaike, T., Aogaki, Y., and Inoue, S. (1975) *Biopolymers*, **14**, 2577.

Al-Ani, N. and Olin, A. (1983), *Chem. Scr.* **22**, 105.

Alexakis, A., Mutti, S., and Mangeney, P. (1992), *J. Org. Chem.*, **57**, 1224.

Allen, A. C., Cooper, D. A., Kiser, W. O., and Cottrell, R. C. (1981), *J. Forensic Sci.*, **26**, 12.

Allenmark, S. (1984), "Recent Advances in Methods of Direct Optical Resolution," *J. Biochem, Biophys. Methods*, **9**, 1.

Allenmark, S. (1985), *LC, Liq. Chromatogr. HPLC Mag.*, **3**, 348, 352.

Allenmark, S. G. (1991), *Chromatographic Enantioseparation: Methods and Applications*, 2nd ed., Ellis Horwood, New York.

Allenmark, S., Bomgren, B., and Borén, H. (1983), *J. Chromatogr.*, **264**, 63.

Alvarez, C., Barkaoui, L., Goasdoue, N., Daran, J.C., Platzer, N., Rudler, H., and Vaissermann, J. (1989), *J. Chem. Soc. Chem. Commun.*, 1507.

Andersen, K. K., Gash, D. M., and Robertson, J. D. (1983), "Isotope-Dilution Techniques," in Morrison, J. D., Ed., *Asymmetric Synthesis*, Vol. 1, Academic, New York, Chap. 4.

Anderson, R. C. and Shapiro, M. J. (1984), *J. Org. Chem.*, **49**, 1304.

Anet, F. A. L., and Bourn, A. J. R. (1967), *J. Am. Chem. Soc.*, **89**, 760.

Anet, F. A. L., Miura, S. S., Siegel, J., and Mislow, K. (1983), *J. Am. Chem. Soc.*, **105**, 1419.

Annett, R. G. and Stumpf, P. K. (1972), *Anal. Biochem.*, **47**, 638.

Ariens, E. J., Simonis, A. M., and van Rossum, J. M. (1964), "Drug-Receptor Interaction: Interaction of One or More Drugs with One Receptor System," in Ariens, E. J., Ed., *Molecular Pharmacology. The Mode of Action of Biologically Active Compounds*, Vol. 1, Academic, New York, p. 119.

Ariëns, E. J. (1983), "Stereoselectivity of Bioactive Agents: General Aspects," in Ariëns, E. J., Soudijn, W., and Timmermans, P. B. M. W. M., Eds., *Stereochemistry and Biological Activity of Drugs*, Blackwell, Oxford, UK, p. 11.

Ariëns, E. J. (1986), "Chirality in bioactive agents and its pitfalls," *Trends Pharmacol. Sci.*, **7**, 200.

Ariëns, E. J. (1988), "Stereospecificity of Bioactive Agents," in Ariëns, E. J., van Rensen, J. J. S., and Welling, W., Eds., *Stereoselectivity of Pesticides: Biological and Chemical Problems*, Elsevier, Amsterdam, The Netherlands, Chap. 3.

Ariëns, E. J., Soudijn, W., and Timmermans, P. B. M. W. M., Eds. (1983), *Stereochemistry and Biological Activity of Drugs*, Blackwell, Oxford, UK.

Armarego, W. L. F., Milloy, B. A., and Pendergast, W. (1976), *J. Chem. Soc., Perkin Trans. 1*, 2229.

Armstrong, D. W. (1985), *U.S. Patent* 4 539 399, Sep. 3, 1985; *Chem. Abstr.*, **103**, 226754f (1985).

Armstrong, D. W. (1992), "Chiral Separations," *LC.GC*, **10**, 249.

Armstrong, D. W. and DeMond, W. (1984), *J. Chromatogr. Sci.*, **22**, 411.

Armstrong, D. W., DeMond, W., and Czech, B. P. (1985b), *Anal. Chem.*, **57**, 481.

Armstrong, D. W., Li, W., Chang, C.-D., and Pitha, J. (1990b), *Anal. Chem.*, **62**, 914.

Armstrong, D. W., Li, W., and Pitha, J. (1990a), *Anal. Chem.*, **62**, 214.

Armstrong, D. W., Ward, T. J., Armstrong, R. D., and Beesley, T. E. (1986), *Science*, **232**, 1132.

Armstrong, D. W., Ward, T. J., Czech, A., Czech, B. P., and Bartsch, R. A. (1985a), *J. Org. Chem.*, **50**, 5556.

Arnett, E. M., Harvey, N. G., and Rose, P. L. (1989), "Stereochemistry and Molecular Recognition in 'Two Dimensions'," *Acc. Chem. Res.*, **22**, 131.

Arnett, E. M. and Thompson, O. (1981), *J. Am. Chem. Soc.*, **103**, 968.

Arnett, E. M. and Zingg, S. P. (1981), *J. Am. Chem. Soc.*, **103**, 1221.

Arnold, W., Daly, J. J., Imhof, R., and Kyburz, E. (1983), *Tetrahedron Lett.*, **24**, 343.

Atik, Z., Ewing, M. B., and McGlashan, M. L. (1983), *J. Chem. Thermodyn.*, **15**, 159.

Audebert, R. (1979), "Direct Resolution of Enantiomers in Column Liquid Chromatography," *J. Liq. Chrom.*, **2**, 1063.

Avnir, D., Wellner, E., and Ottolenghi, M. (1989), *J. Am. Chem. Soc.*, **111**, 2001.

Avram, M. and Mateescu, Gh. D. (1972), *Infrared Spectroscopy*, Wiley, New York, pp. 482–483.

Ayers, G. S., Mossholder, J. H., and Monroe, R. E. (1970), *J. Chromatogr.*, **51**, 407.

Bachmann, W. E., Cole, W., and Wilds, A. L. (1940), *J. Am. Chem. Soc.*, **62**, 824; see p. 830.

Baczuk, R. J., Landram, G. K., Dubois, R. J., and Dehm, H. C. (1971), *J. Chromatogr.*, **60**, 351.

Balavoine, G., Moradpour, A., and Kagan, H. B. (1974), *J. Am. Chem. Soc.*, **96**, 5152.

Baldwin, M. A., Howell, S. A., Welham, K. J., and Winkler, F. J. (1988), *Biomed. Environ. Mass Spectrom.*, **16**, 357.

Barfknecht, C. F. and Nichols, D. E. (1972), *J. Med. Chem.*, **15**, 109.

Barron, L. D. (1986), *J. Am. Chem. Soc.*, **108**, 5539.

Bates, P. S., Kataky, R., and Parker, D. (1992), *J. Chem. Soc. Chem. Commun.*, 153.

Baxter, C. A. R. and Richards, H. C. (1972), *Tetrahedron Lett.*, 3357.

Beitler, U. and Feibush, B. (1976), *J. Chromatogr.*, **123**, 149.

Benschop, H. P. and De Jong, L. P. A. (1988), "Nerve Agent Stereosiomers: Analysis, Isolation and Toxicology," *Acc. Chem. Res.*, **21**, 368.

Benson, S. C., Cai, P., Colon, M., Haiza, M. A., Tokles, M., and Snyder, J. K. (1988), *J. Org. Chem.*, **53**, 5335.

Bentley, R. (1969), *Molecular Asymmetry in Biology*, Vol. 1, Academic, New York.

Bergmann, M. (1934), *Science*, **79**, 439.

Bergmeyer, H. U., Ed. (1984–1985), *Methods of Enzymatic Analysis*, 3rd ed., Vols. VI–VIII, VCH, Weinheim, Germany.

Bergot, B. J., Baker, F. C., Lee, E., and Schooley, D. A. (1979), *J. Am. Chem. Soc.*, **101**, 7432.

Bernauer, K. (1970), *Swiss Patent* 490292, March 6, 1970; *Chem. Abstr.*, 1971, **74**, 64395t.

Bernstein, W. J., Calvin, M., and Buchardt, O. (1972), *J. Am. Chem. Soc.*, **94**, 494.

Beroza, M., Ed. (1970), *Chemicals Controlling Insect Behavior*, Academic, New York.

Berson, J. A. and Ben-Efraim D. A. (1959), *J. Am. Chem. Soc.*, **81**, 4083.

Björkling, F., Boutelje, J., Hjalmarsson, M., Hult, K., and Norin, T. (1987), *J. Chem. Soc. Chem. Commun.*, 1041.

Black, S. N., Williams, L. J., Davey, R. J., Moffatt, F., McEwan, D. M., Sadler, D. E., Docherty, R., and Williams, D. J. (1990), *J. Phys. Chem.*, **94**, 3223.

Blair, N. E. and Bonner, W. A. (1980), *Origins Life*, **10**, 255.

Blair, N. E., Dirbas, F. M., and Bonner, W. A. (1981), *Tetrahedron*, **37**, 27.

Blaschke, G. (1980), "Chromatographic Resolution of Racemates," *Angew. Chem. Int. Ed. Engl.*, **19**, 13.

Blaschke, G., Kraft, H. P., and Markgraf, H. (1980), *Chem. Ber.*, **113**, 2318.

Blaschke, G., Kraft, H. P., and Markgraf, H. (1983), *Chem. Ber.*, **116**, 3611.

Bonner, W. A. (1972), "Origins of Molecular Chirality," in Ponnamperuma, C., Ed., *Exobiology*, North-Holland, Amsterdam, The Netherlands, p. 170.

Bonner, W. A. (1973), *J. Chromatogr. Sci.*, **11**, 101.

Bonner, W. A. (1974), *J. Mol. Evol.*, **4**, 23.

Bonner, W. A. (1988), "Origins of Chiral Homogeneity in Nature," *Top. Stereochem.*, **18**, 1.

Bonner, W. A. and Liang, Y. (1984), *J. Mol. Evol.*, **21**, 84.

Bordeaux, D. and Gagnaire, G. (1982), *Tetrahedron Lett.*, **23**, 3353.

Borden, J. H., Chong, L., McLean, J. A., Slessor, K. N., and Mori, K. (1976), *Science*, **192**, 894.

Borman, S. (1990), "Chirality Emerges as Key Issue in Pharmaceutical Research," *Chem. Eng. News*, **68**, July 9, 1990, 9.

Bovey, F. A. (1969), *Polymer Conformation and Configuration*, Academic, New York, p. 5ff.

Briaucourt, P., Guetté, J.-P., and Horeau, A. (1972), *C. R. Seances Acad. Sci. Ser. C*, **274**, 1203.

Briaucourt, P. and Horeau, A., (1979), *C. R. Seances Acad. Sci. Ser. C*, **289**, 49.

Brienne, M.-J., Collet, A., and Jacques, J. (1983), *Synthesis*, 704.

Brinkman, U. A. Th. and Kamminga, D. (1985), *J. Chromatogr.*, **330**, 375.

Brock, C. P., Schweizer, W. B., and Dunitz, J. D. (1991), *J. Am. Chem. Soc.*, **113**, 9811.

Brockmann, Jr., H. and Musso, H. (1956), *Chem. Ber.*, **89**, 241.

Brooks, C. J. W., Gilbert, M. T., and Gilbert, J. D. (1973), *Anal. Chem.*, **45**, 896.

Brown, J. M. and Parker, D. (1981), *Tetrahedron Lett.*, **22**, 2815.

Brubaker, R. R. (1985), "Mechanisms of Bacterial Virulence," *Annu. Rev. Microbiol.*, **39**, 21.

Brückner, S., Forni, A., Moretti, I., and Torre, G. (1982), *J. Chem. Soc. Chem. Commun.*, 1218.

Brückner, H. and Hausch, M. (1990), "D-Amino Acids in Food: Detection and Nutritional Aspects," in Holmstedt, B., Frank, H., and Testa, B., Eds., *Chirality and Biological Activity*, Liss, New York, Chap. 11.

Brugidou, J., Christol, H., and Sales, R. (1974), *C. R. Seances Acad. Sci. Ser. C*, **278**, 725.

Bruton, J., Horner, W. H., and Russ, G. A. (1967), *J. Biol. Chem.*, **242**, 813.

Bruzik, K. and Stec. W. J. (1981), *J. Org. Chem.*, **46**, 1618.

Bucciarelli, M., Forni, A., Moretti, I., and Torre, G. (1980), *J. Chem. Soc. Perkin Trans 1*, 2152.

Buchardt, O. (1974), "Photochemistry with Polarized Light," *Angew. Chem. Int. Ed. Engl.*, **13**, 179.

Buist, P. H. and Marecak, D. M. (1992), *J. Am. Chem. Soc.*, **114**, 5073.

Bulsing, J. M., Sanders, J. K. M., and Hall, L. D. (1981), *J. Chem. Soc. Chem. Commun.*, 1201.

Burlingame, T. G. and Pirkle, W. H. (1966), *J. Am. Chem. Soc.*, **88** 4294.

Bussmann, W. and Simon, W. (1981), *Helv. Chim. Acta*, **64**, 2101.

Camilleri, P., Brown, R., and Okafo, G. (1992), "The Shape of Things to Come," *Chem. Brit.*, **28**, 800.

Carman, R. M. and Klika, K. D. (1991), *Aust. J. Chem.*, **44**, 895.

Casanova, Jr., J. and Corey, E. J. (1961), *Chem. Ind. (London)*, 1664.

Casarini, D., Lunazzi, L., and Macciantelli, D. (1988), *J. Org. Chem.*, **53**, 177.

Casey, J. P., Lewis, R. A., and Mislow, K. (1969), *J. Am. Chem. Soc.*, **91**, 2789.

Caspi, E. and Eck, C. R. (1977), *J. Org. Chem.*, **42**, 767.

Cecchi, L. and Malaspina, P. (1991), *Anal. Biochem.*, **192**, 219.

Chan, T. H., Peng, Q.-J., Wang, D., and Guo, J. A. (1987), *J. Chem. Soc. Chem. Commun.*, 325.

Chang, R. and Weissman, S. I. (1967), *J. Am. Chem. Soc.*, **89**, 5968.

Charles, R. and Gil-Av, E. (1984), *J. Chromatogr.*, **298**, 516.

Chaves das Neves, H. J., Vasconcelos, A. M. P., and Costa, M. L. (1990), "Racemization of Wine Free Amino Acids as a Function of Bottling Age" in Holmstedt, B., Frank, H., and Testa, B., Eds., *Chirality and Biological Activity*, Liss, New York, Chap. 12.

Chemla, D. S. and Zyss, J., Eds. (1987), *Nonlinear Optical Properties of Organic Molecules and Crystals*, Vol. 1, Academic, Orlando, FL.

Chickos, J. S., Garin, D. L., Hitt, M., and Schilling, G. (1981), *Tetrahedron*, **37**, 2255.

Christol, H., Duval, D., and Solladié, G. (1968), *Bull. Soc. Chim. Fr.*, 4151.

Cline, M. J. and Lehrer, R. I. (1969), *Proc. Natl. Acad. Sci. USA*, **62**, 756.

Cochran, T. G. and Huitric, A. C. (1971), *J. Org. Chem.*, **36**, 3046.

Collet, A. (1990), "The Homochiral versus Heterochiral Packing Dilemma," in Simonyi, M., Ed., *Problems and Wonders of Chiral Molecules*, Akadémiai Kiadó, Budapest, Hungary, p. 91.

Collet, A., Brienne, M.-J., and Jacques, J. (1980), "Optical Resolution by Direct Crystallization of Enantiomer Mixtures," *Chem. Rev.*, **80**, 215.

Collet, A. and Jacques, J. (1972), *Bull. Soc. Chim. Fr.*, 3857.

Consiglio, G., Pino, P., Flowers, L. I., and Pittmann, Jr., C. U. (1983), *J. Chem. Soc. Chem. Commun.*, 612.

Cook, C. E. (1985), "Enantioselective drug analysis," *Pharm. Int.*, **6**, 302.

Cook, C. E. (1988), "Enantiomer Analysis by Competitive Binding Methods," in Wainer, I. W. and Drayer, D. E., *Drug Stereochemistry. Analytical Methods and Pharmacology*, Marcel Dekker, New York, pp. 45–76.

Cook, C. E., Ballentine, N. H., Seltzman, T. B., and Tallent, C. R. (1979), *J. Pharmacol. Exp. Ther.*, **210**, 391.

Cook, C. E., Seltzman, T. B., Tallent, C. R., Lorenzo, B., and Drayer, D. E. (1987), *J. Pharmacol. Exp. Ther.*, **241**, 779.

Corfield, J. R. and Trippett, S. (1971), *J. Chem. Soc. D*, 721.

Corrigan J. J. (1969), "D-Amino acids in Animals," *Science*, **164**, 142.

Craig, D. P. and Elsum, I. R. (1982), *Chem. Phys.*, **73**, 349.

Craig, D. P. and Mellor, D. P. (1976), "Discriminating Interactions between Chiral Molecules," *Top. Curr. Chem.*, **63**, 1.

Craig, D. P. and Schipper, P. E. (1975), *Proc. R. Soc. London Ser. A*, **342**, 19.

Crossley, R. (1992), "The Relevance of Chirality to the Study of Biological Activity," *Tetrahedron*, **48**, 8155.

Cullis, P. M., Iagrossi, A., Rous, A. J., and Schilling, M. B. (1987), *J. Chem. Soc. Chem. Commun.*, 996.

Cundy, K. C. and Crooks, P. A. (1983), *J. Chromatogr*, **281**, 17.

Cung, M. T., Marraud, M., Neel, J., and Aubry, A. (1978), *Biopolymers*, **17**, 1149.

Curtin, D. Y. and Paul, I. C. (1981), "Chemical Consequences of the Polar Axis in Organic Solid-State Chemistry," *Chem. Rev.* **81**, 525.

Cushny, A. R. (1926), *The Biological Relations of Optically Isomeric Substances*, Williams & Wilkins, Baltimore.

Cuvinot, D., Mangeney, P., Alexakis, A., Normant, J.-F., and Lellouche, J.-P. (1989), *J. Org. Chem.*, **54**, 2420.

Dale, J. A., Dull, D. L., and Mosher, H. S. (1969), *J. Org. Chem.*, **34**, 2543.

Dale, J. A. and Mosher, H. S. (1968), *J. Am. Chem. Soc.*, **90**, 3732; see also Dale, J. A. and Mosher, H. S. (1973), *ibid.*, **95**, 512.

Dalgliesh, C. E. (1952), *J. Chem. Soc.*, 3940.

Davankov, V. A. (1966), Thesis, Moscow [cited in Davankov, V. A. (1980)].

Davankov, V. A. (1980), "Resolution of Racemates by Ligand Exchange–Exchange Chromatography," *Adv. Chromatogr.*, **18**, 139.

Davankov, V. A. (1989), "Separation of Enantiomeric Compounds Using Chiral HPLC Systems. A Brief Review of General Principles, Advances and Development Trends," *Chromatographia*, **27**, 475.

Davankov, V. A., Kurganov, A. A., and Bochkov, A. S. (1983), "Resolution of Racemates by High-Performance Liquid Chromatography," *Adv. Chromatogr.*, **22**, 71.

De Camp, W. H. (1989), "The FDA Perspective on the Development of Stereoisomers," *Chirality*, **1**, 2.

Debowski, J., Sybilska, D., and Jurczak, J. (1982), *J. Chromatogr.*, **237**, 303.

Deshmukh, M., Duñach, E., Juge, S., and Kagan, H. B. (1984), *Tetrahedron Lett.*, **25**, 3467.

Dobashi, A. and Hara, S. (1983a), *Anal. Chem.*, **55**, 1805.

Dobashi, A. and Hara, S. (1983b), *Tetrahedron Lett.*, **24**, 1509.

Dobashi, A., Dobashi, Y., and Hara, S. (1991), "Liquid Chromatographic Separation of Enantiomers by Hydrogen-Bonding Association," in Ahuja, S., Ed., *Chiral Separations by Liquid Chromatography*, ACS Symposium Series 471, American Chemical Society, Washington, DC, Chap. 10.

Dobashi, A., Motoyama, Y., Kinoshita, K., Hara, S., and Fukasaku, N. (1987), *Anal. Chem*, **59**, 2209.

Dobashi, A., Oka, K., and Hara, S. (1980), *J. Am. Chem. Soc.*, **102**, 7122.

Dobashi, A., Saito, N., Motoyama, Y., and Hara, S. (1986), *J. Am. Chem. Soc.*, **108**, 307.

Dobashi, Y. and Hara, S. (1985a), *J. Am. Chem. Soc.*, **107**, 3406.

Dobashi, Y. and Hara, S. (1985b), *Tetrahedron Lett.*, **26**, 4217.

Dobashi, Y., Dobashi, A., and Hara, S. (1984), *Tetrahedron Lett.*, **25**, 329.

Downer, E. and Kenyon, J. (1939), *J. Chem. Soc.*, 1156.

Doyle, T. D. (1989), "Synthetic Multiple-Interaction Chiral Bonded Phases," Lough, W. J., Ed., *Chiral Liquid Chromatography*, Blackie, Glasgow, UK., Chap. 6.

Draber, W. and Stetter, J. (1979), "Plant Growth Regulators," in *Chemistry and Agriculture*, Spec. Publ. No. 36, The Chemical Society, London, p. 128.

Drake, A. F., Gould, J. M., and Mason, S. F. (1980), *J. Chromatogr.*, **202**, 239.

Drayer, D. E. (1988), "Pharmacokinetic Differences between Drug Enantiomers in Man," in Wainer, I. W. and Drayer, D. E., Eds., *Drug Stereochemistry. Analytical Methods and Pharmacology*, Marcel Dekker, New York.

Drozd, J. (1981), *Chemical derivatization in gas chromatography*, Journal of Chromatography Library, Vol. 19, Elsevier, Amsterdam, The Netherlands.

Duñach, E. and Kagan, H. B. (1985), *Tetrahedron Lett.*, **26**, 2649.

Dunitz, J. D. (1979), *X-Ray Analysis and the Structure of Organic Molecules*, Cornell University Press, Ithaca, NY.

Dunlop, J. H., Evans, D. F., Gillard, R. D., and Wilkinson, G. (1966), *J. Chem. Soc. A*, 1260.

Dutcher, J. S., Macmillan, J. G., and Heathcock, C. H. (1976), *J. Org. Chem.*, **41**, 2663.

Easson, L. H. and Stedman, E. (1933), *Biochem. J.*, **27**, 1257.

Eaton, S. S. (1971), *Chem. Phys. Lett.* **8**, 251.

Ejchart, A., and Jurczak, J. (1970), *Bull. Acad. Polon. Sci. Ser. Sci. Chim.*, **18**, 445; Mikołajczyk, M., Para, M., Ejchart, A., and Jurczak, J. (1970), *J. Chem. Soc. D*, 654.

Ejchart, A. and Jurczak, J. (1971), *Bull. Acad. Polon. Sci. Ser. Sci. Chim.*, **19**, 725.

Ejchart, A., Jurczak, J., and Bankowski, K. (1971), *Bull. Acad. Polon. Sci. Ser. Sci. Chim.*, **19**, 731.

Eliel, E.L. (1962), *Stereochemistry of Carbon Compounds*, McGraw-Hill, New York.

Eliel, E. L., Alvarez, M. T., and Lynch, J. E. (1986), *Nouv. J. Chim.*, **10**, 749.

Eliel, E. L. and Ko, K.-Y. (1983), *Tetrahedron Lett.*, **24**, 3547.

Eliel, E. L. and Kofron, J. T. (1953), *J. Am. Chem. Soc.*, **75**, 4585.

Elvidge, J. A., and Sammes, P. G. (1966), *A Course in Modern Techniques of Organic Chemistry*, 2nd ed., Butterworths, London, p. 218.

Emberger, R. and Hopp, R. (1987), *Spec. Chem.*, **7**, 193.

Enders, D. and Mies, W. (1984), *J. Chem. Soc. Chem. Commun.*, 1221.

Evans, D. A., Mathre, D. J., and Scott, W. L. (1985), *J. Org. Chem.*, **50**, 1830.

Evans, G. J. and Evans, M. W. (1983), *Chem. Phys. Lett.*, **96**, 416.

Fales, H. M. and Wright, G. J. (1977), *J. Am. Chem. Soc.*, **99**, 2339.

Farina, M. (1987a), *J. Chem. Soc. Chem. Commun.*, 1121.

Farina, M. (1987b), "The Stereochemistry of Linear Macromolecules," *Top. Stereochem.*, **17**, 1.

Farmer, R. F. and Hamer, J. (1966), *J. Org. Chem.*, **31**, 2418.

Feibush, B. and Gil-Av, E. (1970), *Tetrahedron*, **26**, 1361.

Feibush, B. and Grinberg, N. (1988), "The History of Enantiomeric Resolution," in Zief, M. and Crane, L. J., Eds., *Chromatographic Chiral Separations*, Marcel Dekker, New York, Chap. 1.

Feibush, B., Richardson, M. F., Sievers, R. E., and Springer, Jr., C. S. (1972), *J. Am. Chem. Soc.*, **94**, 6717.

Feigl, D. M. and Mosher, H. S. (1965), *Chem. Commun.*, 615.

Feringa, B. L., Smaardijk, A., and Wynberg, H. (1985), *J. Am. Chem. Soc.*, **107**, 4798.

Feringa, B. L., Strijtveen, B., and Kellogg, R. M. (1986), *J. Org. Chem.*, **51**, 5484.

Findlay, A. (1951), *The Phase Rule*, 9th ed., revised by Campbell, A. N. and Smith, N. O., Dover, New York.

Finn, J. M. (1988), "Rational Design of Pirkle-Type Chiral Stationary Phases," in Zief, M. and Crane, L. J., Eds., *Chromatographic Chiral Separations*, Marcel Dekker, New York, Chap. 3.

Fischer, L, Müller, R., Ekberg, B., and Mosbach, K. (1991), *J. Am. Chem. Soc.*, **113**, 9358.

Flores, J. J., Bonner, W. A., and Massey, G. A. (1977), *J. Am. Chem. Soc.*, **99**, 3622.

Forni, A., Moretti, I., Prosyanik, A. V., and Torre, G. (1981), *J. Chem. Soc. Chem. Commun.*, 588.

Frank, H., Nicholson, G. J., and Bayer, E. (1977), *J. Chromatogr. Sci.*, **15**, 174.

Frank, H., Nicholson, G. J., and Bayer, E. (1978), *J. Chromatogr.*, **167**, 187.

Fraser, R. R. (1983), "Nuclear Magnetic Resonance Analysis Using Chiral Shift Reagents," in Morrison, J. D., Ed., *Asymmetric Synthesis*, Vol. 1, Academic, New York, Chap. 9.

Fraser, R. R., Petit, M. A., and Miskow, M. (1972), *J. Am. Chem. Soc.*, **94**, 3253.

Fraser, R. R., Petit, M. A., and Saunders, J. K. (1971), *J. Chem. Soc. Chem. Commun.*, 1450.

Frejd, T. and Klingstedt (1983), *J. Chem. Soc. Chem. Commun.*, 1021.

Friedman, L. and Miller, J. G. (1971), *Science*, **172**, 1044.

Fujisawa, T., Sato, T., Kawara, T., and Ohashi, K. (1981), *Tetrahedron Lett.*, **22**, 4823.

Fulwood, R. and Parker, D. (1992), *Tetrahedron: Asymmetry*, **3**, 25.

Furuta, K., Shimizu, S., Miwa, Y., and Yamamoto, H. (1989), *J. Org. Chem.*, **54**, 1481.

Gabard, J. and Collet, A. (1986), *Nouv. J. Chim.*, **10**, 685.

Gal, J. (1987), "Stereoisomer Separations via Derivatization with Optically Active Reagents," *LC.GC*, **5**, 106.

Galpin, D. R. and Huitric, A. C. (1968), *J. Org. Chem.*, **33**, 921.

Garay, A. S. (1968), *Nature (London)*, **219**, 338.

Garin, D. L., Greco, D. J. C., and Kelley, L. (1977), *J. Org. Chem.*, **42**, 1249.

Gassmann, E., Kuo, J. E., and Zare, R. N. (1985), *Science*, **230**, 813.

Gaudemer, A. (1977), "Determination of Configurations by NMR Spectroscopy Methods," in Kagan, H. B., Ed., *Stereochemistry. Fundamentals and Methods*, Vol. 1, Thieme, Stuttgart, Germany, p. 117.

Gawehn, K. (1984), "D-(−)-Lactate," in Bergmeyer, H. U., Ed., *Methods of Enzymatic Analysis*, 3rd ed., Vol. 6, VCH, Weinheim, Germany, pp. 588.

Geiger, R. and König, W. (1990), "Configurational Modification of Peptide Hormones," in Holmstedt, B., Frank, H., and Testa, B., Eds., *Chirality and Biological Activity*, Liss, New York, Chap. 21.

Gelber, L. R., Karger, B. L., Neumeyer, J. L., and Feibush, B. (1984), *J. Am. Chem. Soc.*, **106**, 7729.

Gerlach, H. (1966), *Helv. Chim. Acta*, **49**, 2481.

Gerlach, H. and Zagalak, B. (1973), *J. Chem. Soc. Chem. Commun.*, 274.

Gil-Av, E. (1985), "Selectors for Chiral Recognition in Chromatography," *J. Chromatogr. Libr.*, **32**, 111.

Gil-Av, E., and Feibush, B. (1967), *Tetrahedron Lett.*, 3345.

Gil-Av, E., Feibush, B., and Charles-Sigler, R. (1966), *Tetrahedron Lett.*, 1009.

Gil-Av, E. and Nurok, D. (1962), *Proc. Chem. Soc. London*, 146.

Gil-Av, E. and Nurok, D. (1974), "Resolution of Optical Isomers by Gas Chromatography of Diastereomers," *Adv. Chromatogr.*, **10**, 99.

Gil-Av, E., Tishbee, A., and Hare, P. E. (1980), *J. Am. Chem. Soc.*, **102**, 5115.

Gillard, R. D. (1979), "The Origin of the Pfeiffer Effect," in Mason, S. F., Ed., *Optical Activity and Chiral Discrimination*, Reidel, Dordrecht, The Netherlands, p. 353.

Gilon, C., Leshem, R., Tapuhi, Y., and Grushka, E. (1979), *J. Am. Chem. Soc.*, **101**, 7612.

Glusker, J. P. and Trueblood, K. N. (1985), *Crystal Structure Analysis, A Primer*, 2nd ed., Oxford, New York.

Ghuysen, J. M., Strominger, J. L., and D. J. Tipper (1968), "Bacterial Cell Walls," in Florkin, M. and Stotz, E. H., Eds., *Comprehensive Biochemistry*, Vol. 26A, Elsevier, Amsterdam, The Netherlands, p. 53.

Goe, G. L. (1973), *J. Org. Chem.*, **38**, 4285.

Goering, H. L., Eikenberry, J. N., and Koermer, G. S. (1971), *J. Am. Chem. Soc.*, **93**, 5913.

Goering, H. L., Eikenberry, J. N., Koermer, G. S., and Lattimer, C. J. (1974), *J. Am. Chem. Soc.*, **96**, 1493.

Golding, B. T. (1988), "Synthesis and Reactions of Chiral C_3-Units," *Chem. Ind. (London)*, 617.

Goldman, M., Kustanovich, Z., Weinstein, S., Tishbee, A., and Gil-Av, E. (1982), *J. Am. Chem. Soc.*, **104**, 1093.

Goodby, J. W. (1986), "Optical Activity and Ferroelectricity in Liquid Crystals," *Science*, **231**, 350.

Görög, S. (1990), "Enantiomeric Derivatization," in Lingeman, H. and Underberg, W. J. M., Eds., *Detection-Oriented Derivatization Techniques in Liquid Chromatography*, Marcel Dekker, New York, Chap. 5.

Gozel, P., Gassmann, E., Michelsen, H., and Zare, R. N. (1987), *Anal. Chem.*, **59**, 44.

Graff, S., Rittenberg, D., and Foster, G. L. (1940), *J. Biol. Chem.*, **133**, 745.

Greatbanks, D. and Pickford, R. (1987), *Magn. Reson. Chem.*, **25**, 208.

Green, M. M., Andreola, C., Muñoz, B., Reidy, M. P., and Zero, K. (1988) *J. Am. Chem. Soc.*, **110**, 4063.

Green, M. M., Lifson, S., and Teramoto, A. (1991), "Cooperation in a Deep Helical Energy Well," *Chirality*, **3**, 285.

Green, M. M., Reidy, M. P., Johnson, R. J., Darling, G., O'Leary, D. J., and Wilson, G. (1989), *J. Am. Chem. Soc.*, **111**, 6452.

Greenstein, J. P. and Winitz, M. (1961), *Chemistry of the Amino Acids*, Vols. 1 and 2, Wiley, New York.

Grinberg, N. and Weinstein, S. (1984), *J. Chromatogr.*, **303**, 251.

Grushka, E., Leshem, R., and Gilon, C. (1983), *J. Chromatogr.*, **255**, 41.

Guerrier, L., Royer, J., Grierson, D. S., and Husson, H.-P. (1983), *J. Am. Chem. Soc.*, **105**, 7754.

Guetté, J.-P., Boucherot, D., and Horeau, A. (1974), *C.R. Seances Acad. Sci. Ser. C*, **278**, 1243.

Guetté, M. and Guetté, J.-P. (1977), *Bull. Soc. Chim. Fr.*, 769.

Guetté, J.-P. and Horeau, A. (1965), *Tetrahedron Lett.*, 3049.

Guetté, J.-P., Lacombe, L., and Horeau, A. (1968), *C.R. Seances Acad. Sci. Ser. C*, **267**, 166.

Guetté, J.-P., Perlat, M., Capillon, J., and Boucherot, D. (1974), *Tetrahedron Lett.*, 2411.

Günther, K. (1988), "Thin-Layer Chromatographic Enantiomeric Resolution via Ligand Exchange," *J. Chromatogr.*, **448**, 11.

Günther, K., Martens, J., and Schickedanz, M. (1984), *Angew. Chem. Int. Ed. Engl.*, **23**, 506.

Günther, K., Schickedanz, M., and Martens, J. (1985), *Naturwissenschaften*, **72**, 149.

Gur, E., Kaida, Y., Okamoto, Y., Biali, S. E., and Rappoport, Z. (1992), *J. Org. Chem.*, **57**, 3689.

Halevi, E. A. (1992), *Chem. Eng. News*, **70**, Oct. 26, 1992, 2.

Halpern, B. (1978), "Derivatives for Chromatographic Resolution of Optically Active Compounds," in Blau, K. and King, G. S., Eds., *Handbook of Derivatives for Chromatography*, Heyden, London, p. 457.

Halpern, B., Chew, L. F., and Weinstein, B. (1967), *J. Am. Chem. Soc.*, **89**, 5051.

Halpern, B. and Westley, J. W. (1965a), *Chem. Commun.*, 246.

Halpern, B. and Westley, J. W. (1965b), *Chem. Commun.*, 421.

Halpern, B. and Westley, J. W. (1966), *Chem. Commun.*, 34.

Hamberg, M. (1971), *Chem. Phys. Lipids*, **6**, 152.

Hardy, G. E., Kaska, W. C., Chandra, B. P., and Zink, J. I. (1981), *J. Am. Chem. Soc.*, **103**, 1074.

Hare, P. E. (1988), "Chiral Mobile Phases for the Enantiomeric Resolution of Amino Acids," *Chromatogr. Sci.*, **40** (Chromatogr. Chiral Sep.), 165.

Hare, P. E. and Gil-Av, E. (1979), *Science*, **204**, 1226.

Harger, M. J. P. (1977), *J. Chem. Soc. Perkin Trans.* **2**, 1882.

Harger, M. J. P. (1978), *J. Chem. Soc., Perkin Trans.* **2**, 326.

Hauser, F. M., Coleman, M. L., Huffman, R. C., and Carroll, F. I. (1974), *J. Org. Chem.*, **39**, 3426.

Hauser, F. M., Rhee, R. P., and Ellenberger, S. R. (1984), *J. Org. Chem.*, **49**, 2236.

Hawkins, J. M. and Meyer, A. (1993), *Science*, **260**, 1918.

Heilbronner, E. and Dunitz, J. D. (1993), *Reflections on Symmetry in Chemistry . . . and Elsewhere*, VHCA, Verlag Helvetica Chimica Acta, Basel, Switzerland.

Helmchen, G., Nill, G., Flockerzi, D., Schühle, W., and Youssef, M. S. K. (1979a), *Angew. Chem. Int. Ed. Engl.*, **18**, 62. Helmchen, G., Nill, G., Flockerzi, D., and Youssef, M. S. K. (1979b), *ibid.*, **91**, 65 and **18**, 63. Helmchen, G. and Nill, G. (1979c), *ibid.*, **91**, 66 and **18**, 65.

Helmchen, G., Ott, R., and Sauber, K. (1972), *Tetrahedron Lett.*, 3873.

Helmchen, G. and Strubert, W. (1974), *Chromatographia*, **7**, 713.

Helmchen, G., Völter, H., and Schühle, W. (1977), *Tetrahedron Lett.*, 1417.

Hermansson, J. (1983), *J. Chromatogr.*, **269**, 71.

Hermansson, J. (1984), *J. Chromatogr.*, **298**, 67.

Hermansson, J. (1985), *J. Chromatogr.*, **325**, 379.

Hermansson, J. and Schill, G. (1988), "Resolution of Enantiomeric Compounds by Silica-Bonded α_1-Acid Glycoprotein," in Zief, M. and Crane, L. J., Eds., *Chromatographic Chiral Separations*, Marcel Dekker, New York, Chap. 10.

Herndon, W. C. and Vincenti, S. P. (1983), *J. Am. Chem. Soc.*, **105**, 6174.

Hesse, G. and Hagel, R. (1973), *Chromatographia*, **6**, 277.

Hesse, G. and Hagel, R. (1976a), *Chromatographia*, **9**, 62.

Hesse, G. and Hagel, R. (1976b), *Justus Liebigs Ann. Chem.*, 996.

Hiemstra, H. and Wynberg, H. (1977), *Tetrahedron Lett.*, 2183; see also H. Hiemstra, Ph.D. Dissertation, University of Groningen, Groningen, The Netherlands, 1980, p. 62.

Hill, H. D. W., Zens, A. P., and Jacobus, J. (1979), *J. Am. Chem. Soc.*, **101**, 7090.

Hinkkanen, A. and Decker, A. (1985), "D-Amino Acids," in Bergmeyer, H. U., Ed., *Methods of Enzymatic Analysis*, 3rd ed., Vol. 8, VCH, Weinheim, Germany, p. 329.

Hinze, W. L., Riehl, T. E., Armstrong, D. W., DeMond, W., Alak, A., and Ward, T. (1985), *Anal. Chem.*, 57, 237.

Hirota, K., Koizumi, H., Hironaka, Y., and Izumi, Y. (1976), *Bull. Chem. Soc. Jpn.*, **49**, 289.

Ho, T.-L. (1977), *Hard and Soft Acids and Bases Principle in Organic Chemistry*, Academic, New York.

Hofer, E., Keuper, R., and Renken, H. (1984), *Tetrahedron Lett.*, **25**, 1141; Hofer, E. and Keuper, R. (1984), ibid., **25**, 5631.

Högberg, H.-E., Hedenström, E., Fägerhag, J., and Servi, S. (1992), *J. Org. Chem.*, **57**, 2052.

Holmstedt, B. (1990), "The Use of Enantiomers in Biological Studies: An Historical Review," in Holmstedt, B., Frank, H., and Testa, B., Eds., *Chirality and Biological Activity*, Liss, New York, Chap. 1.

Holmstedt, B., Frank, H., and Testa, B., Eds. (1990), *Chirality and Biological Activity*, Liss, New York.

Holý, P., Morf, W. E., Seiler, K., Simon, W., and Vigneron, J.-P. (1990), *Helv. Chim. Acta*, **73**, 1171.

Horeau, A. (1964), *J. Am. Chem. Soc.*, **86**, 3171.

Horeau, A. (1969), *Tetrahedron Lett.*, 3121.

Horeau, A. (1972), "Safety on the Routes to Asymmetric Syntheses," Lecture presented at La Baule, France.

Horeau, A. (1975), *Tetrahedron*, **31**, 1307.

Horeau, A. and Guetté, J.-P. (1968), *C.R. Seances Acad. Sci. Ser. C*, **267**, 257.

Horeau, A., and Guetté, J.-P. (1974), "Interaction Diastéréoisomères d'Antipodes en phase liquide," *Tetrahedron*, **30**, 1923.

Horeau, A., Guetté, J.-P., and Weidmann, R. (1966), *Bull. Soc. Chim. Fr.*, 3513.

Horeau, A. and Nouaille, A. (1966), *Tetrahedron Lett.*, 3953.

Horeau, A. and Nouaille, A. (1972), unpublished observations cited in Horeau (1972).

Howald, W. N., Bush, E. D., Trager, W. F., O'Reilly, R. A., and Motley, C. H. (1980), *Biomed. Mass Spectrom.*, **7**, 35.

Huhtikangas, A., Lehtola, T., Virtanen, R., and Peura, P. (1982), *Finn. Chem. Lett.*, 63.

Hui, R. A. H. F., Salamone, S. and Williams, T. H. (1991), *Pharmacol. Biochem. Behav.*, **40**, 491.

Hulshof, L. A. and Wynberg, H. (1979), *Stud. Org. Chem. (Amsterdam)*, **1979**, 3 (New Trends Heterocyl. Chem.), 373; *Chem. Abstr.*, 1980, **92**, 75738e.

Ichida, A. and Shibata, T. (1988), "Cellulose Derivatives as Stationary Chiral Phases," in Zief, M. and Crane, L. J., Eds., *Chromatographic Chiral Separations*, Marcel Dekker, New York, Chap 9.

Isaksson, R., Wennerström, H., and Wennerström, O. (1988), *Tetrahedron*, **44**, 1697.

Izumi, Y. and Tai, A. (1977), *Stereo-differentiating Reactions: The Nature of Asymmetric Reactions*, Kodansha, Tokyo and Academic, New York, p. 218ff.

Jacobus, J. and Jones, T. B. (1970), *J. Am. Chem. Soc.*, **92**, 4583.

Jacobus, J. and Raban, M. (1969), *J. Chem. Educ.*, **46**, 351.

Jacobus, J., Raban, M., and Mislow, K. (1968), *J. Org. Chem.*, **33**, 1142.

Jacques, J. and Gabard, J. (1972), *Bull. Soc. Chim. Fr.*, 342.

Jacques, J. and Horeau, A. (1949), *Bull. Soc. Chim. Fr.*, **16**, 301.

Jacques, J., Collet, A., and Wilen, S. H. (1981a), *Enantiomers, Racemates and Resolutions*, Wiley-Interscience, New York.

Jacques, J., Leclercq, M., and Brienne, M.-J. (1981b), *Tetrahedron*, **37**, 1727.

Jamali, F., Mehvar, R., and Pasutto, F. M. (1989), "Enantioselective Aspects of Drug Action and Disposition: Therapeutic Pitfalls," *J. Pharm. Sci.*, **78**, 695.

Jennison, C. P. R. and Mackay, D. (1973), *Can. J. Chem.*, **51**, 3726.

Jochims, J. C., Taigel, G., and Seeliger, A. (1967), *Tetrahedron Lett.*, 1901.

Johns, D. M. (1989a), "Binding to Cellulose Derivatives," in Lough, W. J., Ed., *Chiral Liquid Chromatography*, Blackie, Glasgow, UK., Chap. 9.

Johns, D. M. (1989b), "Binding to Synthetic Polymers," in Lough, W. J., Ed., *Chiral Liquid Chromatography*, Blackie, Glasgow, UK., Chap. 10.

Johnson, C. R., Elliott, R. C., and Penning, T. D. (1984), *J. Am. Chem. Soc.*, **106**, 5019.

Jursic, B. S. and Goldberg, S. I. (1992a), *J. Org. Chem.*, **57**, 7172.

Jursic, B. S. and Goldberg, S. I. (1992b), *J. Org. Chem.*, **57**, 7370.

Kabachnik, M. I., Mastryukova, T. A., Fedin, E. I., Vaisberg, M. S., Morozov, L. L., Petrovskii, P. V., and Shipov, A. E. (1976), *Tetrahedron*, **32**, 1719.

Kabachnik, M. I., Mastryukova, T. A., Fedin, E. I., Vaisberg, M. S., Morozov, L. L., Petrovskii, P. V., and Shipov, A. E. (1978), "Optical Isomers in Solution Investigated by Nuclear Magnetic Resonance," *Usp. Khim.*, **47**, 1541; *Russ. Chem. Rev.*, **47**, 821.

Kabuto, K. and Sasaki, Y. (1984), *J. Chem. Soc. Chem. Commun.*, 316.

Kabuto, K. and Sasaki, Y. (1987), *J. Chem. Soc. Chem. Commun.*, 670.

Kaehler, H. and Rehse, K. (1968), *Tetrahedron Lett.*, 5019.

Kagan, H. B. and Fiaud, J. C. (1978), "New Approaches to Asymmetric Synthesis," *Top. Stereochem.*, **10**, 175.

Kagan, H. B. and Fiaud, J. C. (1988), "Kinetic Resolution," *Top. Stereochem.*, **18**, 249.

Kainosho, M., Ajisaka, K., Pirkle, W. H., and Beare, S. D. (1972), *J. Am. Chem. Soc.*, **94**, 5924.

Kalyanam, N. (1983), "Application of Aromatic Solvent Induced Shifts in Organic Chemistry," *J. Chem. Educ.*, **60**, 635.

Kaneko, T., Izumi, Y., Chibata, I., and Itoh, T., Eds. (1974), *Synthetic Production and Utilization of Amino Acids*, Kodansha, Tokyo and Wiley, New York.

Karger, B. L. (1967), "New Developments in Chemical Selectivity in Gas–Liquid Chromatography," *Anal. Chem.*, **39**(8), 24A.

Karger, B. L., Stern, R. L., and Keane, W. (1967), *Anal. Chem.*, **39**, 228.

Kasai, M., Froussios, C., and Ziffer, H. (1983), *J. Org. Chem.*, **48**, 459.

Kasai, M. and Ziffer, H. (1983), *J. Org. Chem.*, **48**, 712.

Kassau, E. (1966), *Dtsch. Apoth. Ztg.*, **106**, 1455.

Kataky, R., Bates, P. S., and Parker, D. (1992), *Analyst*, **117**, 1313.

Kato, N. (1990), *J. Am. Chem. Soc.*, **112**, 254.

Kauffman, G. B. and Myers, R. D. (1975), *J. Chem. Educ.*, **52**, 777.

Kawa, H. and Ishikawa, N. (1980), *Chem. Lett.*, 843.

Kawashima, K., Levy, A., and Spector, S. (1976), *J. Pharmacol, Exp. Ther.*, **196**, 517.

Kelvin, Lord (W. Thomson)(1904), *Baltimore Lectures on Molecular Dynamics and the Wave Theory of Light*, C. J. Clay & Sons, London, p. 619. The lectures were given in 1884 and 1893 at Johns Hopkins Univ., Baltimore, MD.

Kemp, D. S., Wang, S. W., Busby III, G., and Hugel, G. (1970), *J. Am. Chem. Soc.*, **92**, 1043.

Kitaigorodsky, A. I. (1973), *Molecular Crystals and Molecules*, Academic, New York and London.

Knabe, J., Rummel, W., Buech, H. P., and Franz, N. (1978), "Optically active barbiturates. Synthesis, configuration and pharmacological effects," *Arzneim.-Forsch./Drug Res.*, **28**(II), 1048.

Knapp, D. R. (1979), *Handbook of Analytical Derivatization Reactions*, Wiley, New York, p. 405.

Knierzinger, A., Walther, W., Weber, B., Müller, R. K., and Netscher, T. (1990), *Helv. Chim. Acta.*, **73**, 1087.

Knudsen, C. G. and Rappoport, H. (1983), *J. Org. Chem.*, **48**, 2260.

Köhler, J. and Schomburg, G. (1981), *Chromatographia*, **14**, 559.

Köhler, J. and Schomburg, G. (1983), *J. Chromatogr.*, **255**, 311.

Kokke, W. C. M. C. (1974), *J. Am. Chem. Soc.*, **96**, 2627.

Kolasa, T. and Miller, M. J. (1986), *J. Org. Chem.*, **51**, 3055.

König, W. A. (1982), "Separation of Enantiomers by Capillary Gas Chromatography with Chiral Stationary Phases," *HRC CC, J. High Resolut. Chromatogr. Chromatogr. Commun.*, **5**, 588.

König, W. A. (1987), *The Practice of Enantiomer Separation by Capillary Gas Chromatography*, Hüthig., Heidelberg, Germany.

König, W. A. (1989), "Eine neue Generation chiraler Trennphasen für die Gaschromatographie," *Nachr. Chem. Tech. Lab.*, **37**, 471.

König, W. A., and Lutz, S. (1990), "Gas Chromatographic Enantiomer Separation with Modified Cyclodextrins," in Holmstedt, B., Frank, H., and Testa, B., Eds., *Chirality and Biological Activity*, Liss, New York, Chap. 4.

König, W. A., Lutz, S., and Wenz, G. (1988), *Angew. Chem. Int. Ed. Engl.*, **27**, 979.

König, W. A., Lutz, S., Wenz, G., Görgen, G., Neumann, C., Gäbler, A., and Boland, W. (1989), *Angew. Chem. Int. Ed. Engl.*, **28**, 178.

Koppenhoefer, B. and Bayer, E. (1984), *Chromatographia*, **19**, 123.

Koppenhoefer, B. and Trettin, U. (1989), *Fresenius Z. Anal., Chem.*, **333**, 750.

Koreeda, M., Weiss, G., and Nakanishi, K. (1973), *J. Am. Chem. Soc.*, **95**, 239.

Kress, R. B., Duesler, E. N., Etter, M. C., Paul, I. C., and Curtin, D. Y. (1980), *J. Am. Chem. Soc.*, **102**, 7709.

Krstulovic, A. M., Ed. (1989), *Chiral Separations by HPLC. Applications to Pharmaceutical Compounds*, Horwood, Chichester, UK.

Kruizinga, W. H., Bolster, J., Kellogg, R. M., Kamphuis, J., Boesten, W. H. J., Meijer, E. M., and Schoemaker, H. E (1988), *J. Org. Chem.*, **53**, 1826.

Krull, I. S. (1978), "The Liquid Chromatographic Resolution of Enantiomers," *Adv. Chromatogr.*, **16**, 175.

Kuchen, W. and Kutter, J. (1979), *Z. Naturforsch.*, **34B**, 1332.

Kumkumian, C. S. (1988), "The Use of Stereochemically Pure Pharmaceuticals: A Regulatory Point of View," in Wainer, I. W. and Drayer, D. S., Eds., *Drug Stereochemistry. Analytical Methods and Pharmacology*, Marcel Dekker, New York, Chap. 12.

Kunieda, N., Chakihara, H., and Kinoshita, M. (1990), *Chem. Lett.*, 317.

Kuwata, S., Tanaka, J., Sakamoto, Y., Onda, N., Yamada, T., and Miyazawa, T. (1989), *Chem. Lett.*, 2031.

Kwart, H. and Hoster, D. P. (1967), *J. Org. Chem.*, **32**, 1867.

Lafontaine, E., Bayle, J. P., and Courtieu, J. (1989), *J. Am. Chem. Soc.*, **111**, 8294.

Lafontaine, E., Pechine, J. M., Courtieu, J., and Mayne, C. L. (1990), *Liq. Cryst.*, **7**, 293.

Lahav, M., Laub, F., Gati, E., Leiserowitz, L., and Ludmer, Z. (1976), *J. Am. Chem. Soc.*, **98**, 1620.

Lam, S. (1989), "Chiral Ligand Exchange Chromatography," in Lough, W. J., Ed., *Chiral Liquid Chromatography*, Blackie, Glasgow, UK, Chap. 5.

Landsteiner, K. (1962), *The Specificity of Serological Reaction*, Rev. ed., Dover, New York, p. 172.

Lang, R. W. and Hansen, H.-J. (1979), *Helv. Chim. Acta*, **62**, 1025.

Laszlo, P. (1967), "Solvent Effects and Nuclear Magnetic Resonance," *Prog. NMR Spectrosc.*, **3**, 231.

Leclercq, M., Collet, A., and Jacques, J. (1976), *Tetrahedron*, **32**, 821.

Lee, T. D. and Yang, C. N. (1956), *Phys. Rev.*, **104**, 254.

Lehmann F., P. A., Rodrigues de Miranda, J. F., and Ariëns, E. J. (1976), "Stereoselectivity and Affinity in Molecular Pharmacology," *Prog. Drug Res.*, **20**, 101.

Leitich, J. (1978), *Tetrahedron Lett.*, 3589.

Leitch, R. E., Rothbart, H. L., and Rieman, III, W. M. (1968), *Talanta*, **15**, 213.

Leitereg, T. J., Guadagni, D. G., Harris, J., Mon, T. R., and Teranishi, R. (1971), *Nature (London)*, **230**, 455.

LePage, J. N., Lindner, W., Davies, G., Seitz, D. E., and Karger, B. L. (1979), *Anal. Chem.*, **51**, 433.

Levin, G. V. (1981), US Patent 4 262 032, Apr. 14, 1981; *Chem. Abstr.*, **95**, 78771h (1981).

Lin, C.-T., Curtin, D. Y., and Paul, I. C. (1974), *J. Am. Chem. Soc.*, **96**, 6199.

Lindner, W. (1981), "Trennung von Enantiomeren mittels moderner Flussigkeits-Chromatographie," *Chimia*, **35**, 294.

Lindner, W. (1988), "Indirect Separation of Enantiomers by Liquid Chromatography," in Zief, M. and Crane, L. J., Eds., *Chromatographic Chiral Separations*, Marcel Dekker, New York, Chap. 4.

Lipkowitz, K. B., Demeter, D. A., Parish, C. A., and Darden, T. (1987), *Anal. Chem.*, **59**, 1731.

Lipkowitz, K., Landwer, J. M., and Darden, T. (1986), *Anal. Chem.*, **58**, 1611.

Lochmüller, C. H. and Souter, R. W. (1973), *J. Chromatogr.*, **87**, 243.

Lochmüller, C. H. and Souter, R. W. (1974), *J. Chromatogr.*, **88**, 41.

Lochmüller, C. H. and Souter, R. W. (1975), "Chromatographic Resolution of Enantiomers. Selective Review," *J. Chromatogr.*, **113**, 283.

López-Arbeloa, F., Goedeweeck, R., Ruttens, F., De Schryver, F. C., and Sisido, M. (1987), *J. Am. Chem. Soc.*, **109**, 3068.

Lough, W. J., Ed. (1989), *Chiral Liquid Chromatography*, Blackie, Glasgow, UK.

Luchinat, C. and Roelens, S. (1986), *J. Am. Chem. Soc.*, **108**, 4873.

Ludmer, Z., Lahav, M., Leiserowitz, L. and Roitman, L. (1982), *J. Chem. Soc. Chem. Commun.*, 326.

Lundquist, M. (1961), *Ark. Kemi*, **17**, 183.

Macaudière, P., Lienne, M., Tambuté, A., and Caude, M. (1989), "Pirkle-type and related chiral stationary phases for enantiomeric resolutions," in Krstulovic, A. M., Ed., *Chiral Separations by HPLC. Applications to Pharmaceutical Compounds*, Horwood, Chichester, UK, Chap. 14.

MacCarthy, P. (1983), "Ternary and Quaternary Composition Diagrams," *J. Chem. Educ.*, **60**, 922.

MacDermott, A. J. and Tranter, G. E. (1989), "Electroweak Bioenantioselection," *Croat. Chem. Acta*, **62**, 165.

Mamlok, L., Marquet, A., and Lacombe, L. (1971), *Tetrahedron Lett.*, 1039.

Mamlok, L., Marquet, A., and Lacombe, L. (1973), *Bull. Soc. Chem. Fr.*, 1524.

Mannschreck, A., Eiglsperger, A., and Stühler, G. (1982), *Chem. Ber.*, **115**, 1568.

Mannschreck, A., Mintas, M., Becher, G., and Stühler, G. (1980), *Angew. Chem. Int. Ed. Engl.*, **19**, 469.

Mannschreck, A., Munninger, W., Burgemeister, T., Gore, J., and Cazes, B. (1986), *Tetrahedron*, **42**, 399.

Martens, J. and Bhushan, R. (1992), "Resolution of Enantiomers with Achiral Phase Chromatography," *J. Liq. Chromatog.*, **15**, 1.

Martin, G. J., Janvier, P., Akoka, S., Mabon, F., and Jurczak, J. (1986), *Tetrahedron Lett.*, **27**, 2855.

Martin, R. H. (1973), *Problémes de Chimie Organique liés à la notion de chiralité* (Syllabus), Faculté des Sciences, Université Libre de Bruxelles, Brussels.

Martin, R. H. and Libert, V. (1980), *J. Chem. Res. Synop.*, 130 and *Miniprint*, 1940.

Marty, W., Pasquier, M. L., and Gampp, H. (1987), *Helv. Chim. Acta*, **70**, 1774.

Mason, S. F. (1982), *Molecular Optical Activity and the Chiral Discriminations*, Cambridge University Press, Cambridge, UK.

Mason, S. (1984), "Origins of biomolecular handedness," *Nature*, (*London*), **311**, 19.

Mason, S. (1986), "Biomolecular Handedness from Pasteur to Parity Non-Conservation," *Nouv. J. Chem.*, **10**, 739.

Mason, S. F. (1989a), "The Origin of Biomolecular Chirality in Nature," in Krstulovic, A. M., Ed., *Chiral Separations by HPLC. Applications to Pharmaceutical Compounds*, Horwood, Chichester, UK, Chap. 1.

Mason, S. F. (1989b), "The Development of Concepts of Chiral Discrimination," *Chirality*, **1**, 183.

Mason, S. F. and Tranter, G. E. (1983), *J. Chem. Soc. Chem. Commun.*, 117.

Mathieu, J. P. (1973), *J. Raman Spectrosc.*, **1**, 47.

Mathieu, J. P. and Gouteron-Vaisserman, J. (1972), *C.R. Acad, Sci. Ser. B*, **274**, 880.

Matos, J. R., Smith, M. B., and Wong, C.-H. (1985), *Bioorg. Chem.*, **13**, 121.

Matsumoto, M. and Amaya, K. (1980), *Bull. Chem. Soc. Jpn.*, **53**, 3510.

Matsumoto, M., Yajima, H., and Endo, R. (1987), *Bull. Chem. Soc. Jpn.*, **60**, 4139.

Matsuura, K., Inoue, S., and Tsuruta, T. (1965), *Makromol. Chem.*, **85**, 284.

Matusch, R. and Coors, C. (1989), *Angew. Chem. Int. Ed. Engl.*, **28**, 626.

Mauser, H. (1957), *Chem. Ber.*, **90**, 299.

Mazur, R. H., Schlatter, J. M., and Goldkamp, A. H. (1969), *J. Am. Chem. Soc.*, **91**, 2684.

McCreary, M. D., Lewis, D. W., Wernick, D. L., and Whitesides, G. M. (1974), *J. Am. Chem. Soc.*, **96**, 1038.

McMahon, R. E. and Sullivan, H. R. (1976), *Res. Commun. Chem. Pathol. Pharmacol.*, **14**, 631.

McNaughton, J. L. and Mortimer, C. T. (1975), "Differential Scanning Calorimetry," in Buckingham, A. D., Ed., *IRS* (*International Review of Science*), Physical Chemistry Series 2, 1975, Vol. 10, Thermochemistry and Thermodynamics, Skinner, H. A., Ed., Butterworths, London.

Meinwald, J., Thompson, W. R., Pearson, D. L., König, W. A., Runge, T., and Francke, W. (1991), *Science*, **251**, 560.

Merckx, E. M., Lepoivre, J. A., Lemière, G. L., and Alderweireldt, F.-C. (1983), *Org. Magn. Reson.*, **21**, 380.

Meurling, L. and Bergson, G. (1974), *Chem. Scr.*, **6**, 104.

Meurling, L., Bergson, G., and Obenius, U. (1976), *Chem. Scr.*, **9**, 9.

Meyerhoffer, W. (1904), *Ber. Deutsch. Chem. Ges.*, **37**, 2604.

Meyers, A. I., White, S. K., and Fuentes, L. M. (1983), *Tetrahedron Lett.*, **24**, 3551.

Meyers, A. I., Williams, D. R., Erickson, G. W., White, S., and Druelinger, M. (1981), *J. Am. Chem. Soc.*, **103**, 3081.

Michelsen, P., Annby, U., and Gronowitz, S. (1984), *Chem. Scr.*, **24**, 251.

Midha, K. K., Hubbard, J. W., Cooper, J. K., and Mackonka, C. (1983), *J. Pharm. Sci.*, **72**, 736.

Mighell, A. D., Ondik, H. M., and Molino, B. B. (1977), *J. Phys. Chem. Ref. Data*, **6**, 675.

Mikeš, F. and Boshart, G. (1978), *J. Chem. Soc. Chem. Commun.*, 173.

Mikeš, F., Boshart, G., and Gil-Av, E. (1976), *J. Chem. Soc. Chem. Commun.*, 99; *J. Chromatogr.*, **122**, 205.

Mikołajczyk, M., Ejchart, A., and Jurczak, J. (1971), *Bull. Acad. Polon. Sci. Ser. Sci. Chim.*, **19**, 721.

Mikołajczyk, M. and Omelańczúk, J. (1972), *Tetrahedron Lett.*, 1539.

Mikołajczyk, M., Omelańczúk, J., Leitloff, M., Drabowicz, J., Ejchart, A., and Jurczak, J. (1978), *J. Am. Chem. Soc.*, **100**, 7003.

Milton, R. C. deL., Milton, S. C. F., and Kent, S. B. H. (1992), *Science*, **256**, 1445.

Mislow, K. (1965), *Intoduction to Stereochemistry*, Benjamin, New York.

Mislow, K. and Bickart, P. (1976/77), "An Epistemological Note on Chirality," *Isr. J. Chem.*, **15**, 1.

Moradpour, A., Nicoud, J. F., Balavoine, G., Kagan, H., and Tsoucaris, G. (1971), *J. Am. Chem. Soc.*, **93**, 2353.

Morozov, L. L., Vetrov, A. A., Vaisberg, M. S., and Kuz'min, V. V. (1979), *Dokl. Akad. Nauk. SSSR*, **247**, 875, [Engl. Transl. (1980), p. 655].

Morrill, T. C. (1986), "An Introduction to Lanthanide Shift Reagents" in Morrill, T. C., Ed., *Lanthanide Shift Reagents in Stereochemical Analysis*, VCH, Deerfield Beach, FL., Chap. 1.

Morrison, J. D. and Mosher, H. S. (1976), *Asymmetric Organic Reactions*, Prentice-Hall, Englewood Cliffs, NJ, 1971, and American Chemical Society (reprint), Washington, DC.

Morrison, J. D., Ed. (1983), *Asymmetric Synthesis*, Vol. 1: *Analytical Methods*, Academic, New York.

Motosugi, K., Esaki, N., and Soda, K. (1983), *Anal. Lett.*, **16**, 509.

Nabeya, A. and Endo, T. (1988), *J. Org. Chem.*, **53**, 3358.

Nakao, Y., Sugeta, H., and Kyogoku, Y. (1985), *Bull. Chem. Soc. Jpn.*, **58**, 1767.

Nicoud, J. F. and Kagan, H. B. (1976/77), *Isr. J. Chem.*, **15**, 78.

Nicoud, J. F. and Twieg, R. J. (1987), "Design and Synthesis of Organic Molecular Compounds for Efficient Second-Harmonic Generation," in Chemla, D. S. and Zyss, J., Eds., *Nonlinear Optical Properties of Organic Molecules and Crystals*, Academic, Orlando, FL, Chap. 11–3.

Nikolaev, E. N., Goginashvili, G. T., Tal'rose, V. L., and Kostyanovsky, R. G. (1988), *Int. J. Mass Spectrom. Ion Processes.*, **86**, 249.

Nishi, H. and Terabe, S. (1990), *Electrophoresis (Weinheim, Germany)*, **11**, 691.

Nishi, T., Ikeda, A., Matsuda, T., and Shinkai, S. (1991), *J. Chem. Soc. Chem. Commun.*, 339.

Nishio, M. (1969), *Chem. Pharm. Bull.*, **17**, 262.

Nógrádi, M. (1981), *Stereochemistry*, Pergamon, Oxford, UK.

Numan, H., Helder, R., and Wynberg, H. (1976), *Recl. Trav. Chim. Pays-Bas*, **95**, 211.

Offermann, W. and Mannschreck, A. (1981), *Tetrahedron Lett.*, **22**, 3227.

Ogston, A. G. (1948), *Nature (London)*, **162**, 963.

Ohfune, Y. and Tomita, M. (1982), *J. Am. Chem. Soc.*, **104**, 3511 (see note 7).

Ohloff, G. (1986), "Chemistry of Odor Stimuli," *Experientia*, **42**, 271.

Ohloff, G. (1990), *Riechstoffe und Geruchssinn. Die molekulare Welt der Düfte*. Springer, Berlin, Germany, Chap. 2.3.7.

Ohloff, G., Vial, C., Wolf, H. R., Job, K., Jégou, E., Polonsky, J., and Lederer, E. (1980), *Helv. Chim. Acta*, **63**, 1932.

Okamoto, Y. (1987), "Separate Optical Isomers by Chiral HPLC," *CHEMTECH*, **17**, 176.

Okamoto, Y. and Hatada, K. (1988), "Optically Active Poly(Triphenylmethyl Methacrylate) as a Chiral Stationary Phase," in Zief. M. and Crane, L. J., Eds., *Chromatographic Chiral Separations*, Marcel Dekker, New York, Chap. 8.

Okamoto, Y., Honda, S., Okamoto, I., Yuki, H., Murata, S., Noyori, R., and Takaya, H. (1981), *J. Am. Chem. Soc.*, **103**, 6971.

Okamoto, Y., Kawashima, M., and Hatada, K. (1984a), *J. Am. Chem. Soc.*, **106**, 5357.

Okamoto, Y., Kawashima, M., Yamamoto, K., and Hatada, K. (1984b), *Chem. Lett.*, 739.

Okamoto, Y., Suzuki, K., Ohta, K., Hatada, K., and Yuki, H. (1979), *J. Am. Chem. Soc.*, **101**, 4763.

Paquette, L. A., and Lau, C. J. (1987), *J. Org. Chem.*, **52**, 1634.

Paquette, L. A., and Ley, S. V., and Farnham, W. B. (1974), *J. Am. Chem. Soc.*, **96**, 312.

Parker, D. (1991), "NMR Determination of Enantiomeric Purity," *Chem. Rev.*, **91**, 1441.

Parker, D. and Taylor, R. J. (1987), *J. Chem. Soc. Chem. Commun.*, 1781.

Parker, D. and Taylor, R. J. (1992), "Analytical methods: determination of enantiomeric purity," in Aitken, R. A. and Kilényi, S. N., Eds., *Asymmetric Synthesis*, Blackie, London, Chap. 3.

Pasquier, M. L. and Marty, W. (1985), *Angew. Chem. Int. Ed. Engl.*, **24**, 315.

Pasteur, L. (1848), *C.R. Hebd. Seances Acad. Sci.*, **26**, 535.

Pasteur, L. (1886), *C.R. Hebd. Seances Acad. Sci.*, **103**, 138.

Patil, P. N., Miller, D. D., and Trendelenburg, U. (1975), "Molecular Geometry and Adrenergic Drug Activity," *Pharmacol. Rev.*, **26**, 323.

Paul, I. C. and Curtin, D. Y. (1975), "Reactions of Organic Crystals with Gases," *Science*, **187**, 19.

Paul, I. C. and Curtin, D. Y. (1987), "Gas-Solid Reactions and Polar Crystals," in Desiraju, G. R., Ed., *Organic Solid State Chemistry*, (Studies in Organic Chemistry, Vol. 32), Elsevier, Amsterdam, The Netherlands, Chap. 9.

Pavia, A. A. and Lacombe, J. M. (1983), *J. Org. Chem.*, **48**, 2564; Lacombe, J. M. and Pavia, A. A. (1983), *ibid.*, **48**, 2557.

Pella, E. and Restelli, R. (1983), *Microchim. Acta* **1**, 65.

Pennington, W. T., Chakraborty, S., Paul, I. C., and Curtin, D. Y. (1988), *J. Am. Chem. Soc.*, **110**, 6498.

People vs. Aston, J. G. (1984), *California Appellate Reports, Third Dist.*, **154**, 818.

Pereira, W., Bacon, V. A., Patton, W., Halpern, B., and Pollock, G. E. (1970), *Anal. Lett.*, **3**, 23.

Pereira, Jr., W. E., Solomon, M., and Halpern, B. (1971), *Aust. J. Chem.*, **24**, 1103.

Perrin, S. R. and Pirkle, W. H. (1991), "Commercially Available Brush-Type Chiral Selectors for the Direct Resolution of Enantiomers," in Ahuja, S., Ed., *Chiral Separations by Liquid Chromatography*, ACS Symposium Series 471, American Chemical Society, Washington, DC., Chap. 3.

Peters, J. A., Vijverberg, C. A. M., Kieboom, A. P. G., and van Bekkum, H. (1983), *Tetrahedron Lett.*, **24**, 3141.

Peterson, P. E. and Jensen, B. L. (1984), *Tetrahedron Lett.*, **25**, 5711.

Peterson, P. E. and Stepanian, M. (1988), *J. Org. Chem.*, **53**, 1907.

Petsko, G. A. (1992), "On the Other Hand . . . ," *Science*, **256**, 1403.

Pettersson, I. and Berg, U. (1984), *J. Chem. Res. Synop.*, 208.

Pettersson, C. (1984), *J. Chromatogr.*, **316**, 553.

Pettersson, C. (1989), "Formation of Diastereomeric Ion-Pairs," in Krstulovic, A. M., ed., *Chiral Separations by HPLC. Applications to Pharmaceutical Compounds*, Horwood, Chichester, UK, Chap. 6.

Pettersson, C. and Schill, G. (1981), *J Chromatogr.*, **204**, 179.

Pfeiffer, C. C. (1956), *Science*, **124**, 29.

Pirkle, W. H., House, D. W., and Finn, J. M. (1980), *J. Chromatogr.*, **192**, 143.

Pirkle, W. H. (1966), *J. Am. Chem. Soc.*, **88**, 1837.

Pirkle, W. H. (1983), *Tetrahedron Lett.*, **24**, 5707.

Pirkle, W. H. and Beare, S. D. (1969), *J. Am. Chem. Soc.*, **91**, 5150.

Pirkle, W. H. and Finn, J. M. (1981), *J. Org. Chem.*, **46**, 2935.

Pirkle, W. H. and Finn, J. (1983), "Separations of Enantiomers by Liquid Chromatographic Methods," in Morrison, J. D., Ed., *Asymmetric Synthesis*, Vol. 1, Academic, New York, Chap. 6.

Pirkle, W. H., Finn, J. M., Hamper, B. C., Schreiner, J., and Pribish, J. R. (1982), "A Useful and Conveniently Accessible Chiral Stationary Phase for the Liquid Chromatographic Separation of

Enantiomers," in Eliel, E. L. and Otsuka, S., Eds., *Asymmetric Reactions and Processes in Chemistry*, ACS Symposium Series, No. 185, American Chemical Society, Washington, DC, 1982, Chap. 18.

Pirkle, W. H., Finn, J. M., Schreiner, J. L., and Hamper, B. C. (1981), *J. Am. Chem. Soc.*, **103**, 3964.

Pirkle, W. H. and Hauske, J. R. (1977), *J. Org. Chem.*, **42**, 1839.

Pirkle, W. H. and Hoekstra, M. S. (1974), *J. Org. Chem.*, **39**, 3904.

Pirkle, W. H. and Hoover, D. J. (1982), "NMR Chiral Solvating Agents," *Top. Stereochem.*, **13**, 263.

Pirkle, W. H. and House, D. W. (1979), *J. Org. Chem.*, **44**, 1957.

Pirkle, W. H. and Hyun, M. H. (1984), *J. Org. Chem.*, **49**, 3043.

Pirkle, W. H. and Hyun, M. H. (1985), *J. Chromatogr.*, **322**, 309.

Pirkle, W. H., Hyun, M. H., and Bank, B. (1984), *J. Chromatogr.*, **316**, 585.

Pirkle, W. H. and Pochapsky, T. C. (1986), *J. Am. Chem. Soc.*, **108**, 352.

Pirkle, W. H. and Pochapsky, T. C. (1989), "Considerations of Chiral Recognition Relevant to Liquid Chromatographic Separation of Enantiomers," *Chem. Rev.*, **89**, 347.

Pirkle, W. H. and Pochapsky, T. C. (1990), "Theory and Design of Chiral Stationary Phases for the Direct Chromatographic Separation of Enantiomers," *Chromatogr. Sci.*, **47**, 783.

Pirkle, W. H., Pochapsky, T. C., Mahler, G. S., Corey, D. E., Reno, D. S., and Alessi, D. M. (1986), *J. Org. Chem.*, **51**, 4991.

Pirkle, W. H., Robertson, M. R., and Hyun, M. H. (1984), *J. Org. Chem.*, **49**, 2433.

Pirkle, W. H. and Schreiner, J. L. (1981), *J. Org. Chem.*, **46**, 4988.

Pirkle, W. H. and Sikkenga, D. L. (1975), *J. Org. Chem.*, **40**, 3430.

Pirkle, W. H. and Sikkenga, D. L. (1976), *J. Chromatogr.*, **123**, 400.

Pirkle, W. H. and Simmons, K. A. (1981), *J. Org. Chem.*, **46**, 3239.

Pirkle, W. H. and Simmons, K. A. (1983), *J. Org. Chem.*, **48**, 2520.

Pirkle, W. H. and Welch, C. J. (1984), *J. Org. Chem.*, **49**, 138.

Pirkle, W. H., Welch, C. J., and Hyun, M. H. (1983), *J. Org. Chem.*, **48**, 5022.

Pirkle, W. H., Welch, C. J., Mahler, G. S., Meyers, A. I., Fuentes, L. M., and Boes, M. (1984), *J. Org. Chem.*, **49**, 2504.

Pittman, Jr., C. U., Kawabata, Y., and Flowers, L. I. (1982), *J. Chem. Soc. Chem. Commun.*, 473.

Piutti, A. (1886), *C.R. Hebd. Seances Acad. Sci.*, **103**, 134.

Pohl, L. R. and Trager, W. F. (1973), *J. Med. Chem.*, **16**, 475.

Poindexter, G. S. and Meyers, A. I. (1977), *Tetrahedron Lett.*, 3527.

Porter, W. H. (1991), "Resolution of Chiral Drugs," *Pure Appl. Chem.*, **63**, 1119.

Potapov, V. M. (1979), *Stereochemistry*, translated by A. Beknazorov, Mir Publishers, Moscow.

Powell, J. R., Ambre, J. J., and Ruo, T. I. (1988), "The Efficacy and Toxicity of Drug Stereoisomers" in Wainer, I. W. and Drayer, D. E., Eds. in *Drug Stereochemistry. Analytical Methods and Pharmacology*, Marcel Dekker, New York, p. 245.

Pracejus, G. (1959), *Justus Liebigs Ann. Chem.*, **622**, 10.

Price, V. E. and Greenstein, J. P. (1948), *J. Biol. Chem.*, **175**, 969.

Raban, M. and Mislow, K. (1965), *Tetrahedron Lett.*, 4249.

Raban, M. and Mislow, K. (1966), *Tetrahedron Lett.*, 3961.

Raban, M. and Mislow, K. (1967), "Modern Methods for the Determination of Optical Purity," *Top. Stereochem.*, **2**, 199.

Rao, C. N. R. (1963), *Chemical Application of Infrared Spectroscopy*, Academic, New York, pp. 103, 105, 381.

Reid, E., Scales, B., and Wilson, I. D. (1986), *Methodological Surveys in Biochemistry and Analysis*, Vol. 16. *Bioactive Analytes, Including CNS Drugs, Peptides, and Enantiomers*, Plenum, NY.

Rein, D. W. (1974), *J. Mol. Evol.*, **4**, 15.

Reuben, J. (1980), *J. Am. Chem. Soc.*, **102**, 2232.

Reuben, J. and Elgavish, G. A. (1979), "Shift Reagents and NMR of Paramagnetic Lanthanide Complexes" in Gschneidner, Jr. K. A. and Eyring, L., Eds., *Handbook on the Physics and Chemistry of Rare Earths*, Vol. 4, Elsevier, Amsterdam, The Netherlands, p. 483.

Ricci, J. E. (1951), *The Phase Rule and Heterogeneous Equilibrium*, Van Nostrand, New York.

Rice, K. and Brossi, A. (1980), *J. Org. Chem.*, **45**, 592.

Rinaldi, P. L. (1982), "The Determination of Absolute Configuration Using Nuclear Magnetic Resonance Techniques," *Prog. Nucl. Magn. Res. Spectrosc.*, **15**, 291.

Ripmeester, J. A. and Burlinson, N. E. (1985), *J. Am. Chem. Soc.*, **107**, 3713.

Rittenberg, D. and Foster, G. L. (1940), *J. Biol. Chem.*, **133**, 737.

Rogozhin, S. V. and Davankov, V. A. (1968), "Chromatographic Resolution of Racemates on Dissymmetric Sorbents," *Usp. Khim.*, **37**, 1327; *Russ. Chem. Rev.*, **37**, 565.

Rogozhin, S. V. and Davankov, V. A. (1970), *German Patent*, 1 932 190, Jan. 8, 1970; *Chem. Abstr.*, 1970, **72**, 90875c. *Idem.*, *J. Chem. Soc. D*, 1971, 490.

Roozeboom, H. W. B. (1899), *Z. Physik, Chem.*, **28**, 494.

Rose, H. C., Stern, R. L., and Karger, B. L. (1966), *Anal. Chem.*, **38**, 469.

Rosen, T., Watanabe, M., and Heathcock, C. H. (1984), *J. Org. Chem.*, **49**, 3657.

Rosini, C., Bertucci, C., Pini, D., Altemura, P., and Salvadori, P. (1985), *Tetrahedron Lett.*, **26**, 3361.

Rosini, C., Uccello-Barretta, G., Pini, D., Abete, C., and Salvadori, P. (1988), *J. Org. Chem.*, **53**, 4579.

Ruch, E. (1968), *Theor. Chim. Acta*, **11**, 183.

Ruch, E. (1972), "Algebraic Aspects of the Chirality Phenomenon in Chemistry," *Acc. Chem. Res.*, **5**, 49.

Russell, G. F. and Hills, J. I. (1971), *Science*, **172**, 1043.

Ruybal, R. (1982), *Microgram*, **15**, 160.

Saigo, K., Kimoto, H., Nohira, H., Yanagi, K., and Hasegawa, M. (1987), *Bull. Chem. Soc. Jpn.*, **60**, 3655.

Saigo, K., Sekimoto, K., Yonezawa, N., Ishii, F., and Hasegawa, M. (1985), *Bull. Chem. Soc. Jpn.*, **58**, 1006.

Saigo, K., Sugiura, I., Shida, I., Tachibana, K., and Hasegawa, M. (1986), *Bull. Chem. Soc. Jpn.*, **59**, 2915.

Salvadori, P., Pini, D., Rosini, C., Bertucci, C., and Uccello-Barretta, G. (1992), *Chirality*, **4**, 43.

Salvadori, P., Uccello-Barretta, G., Bertozzi, S., Settambolo, R., and Lazzaroni, R. (1988), *J. Org. Chem.*, **53**, 5768.

Salvadori, P., Uccello-Barretta, G., Lazzaroni, R., and Caporusso, A. M. (1990), *J. Chem. Soc. Chem. Commun.*, 1121.

Sanz-Burata, M., Julia, S., and Irurre, J. (1965), *Afinidad*, **22**, 259.

Sastry, B. V. R. (1973), "Stereoisomerism and drug action in the nervous system," *Annu. Rev. Pharmacol.*, **13**, 253.

Saucy, G., Borer, R., Trullinger, D. P., Jones, J. B., and Lok, K. P. (1977), *J. Org. Chem.*, **42**, 3206.

Shallenberger, R. S., Acree, T. E., and Lee, C. Y. (1969), *Nature (London)*, **221**, 555.

Schippers, P. H. and Dekkers, H. P. J. M. (1982), *Tetrahedron*, **38**, 2089.

Schlenk, Jr., W. (1965a), "Recent Results of Configurational Research," *Angew. Chem. Int. Ed. Engl.*, **4**, 139.

Schlenk, Jr., W. (1965b), *J. Am. Oil Chem. Soc.*, **42**, 945.

Schmidt, G. M. J. (1971), "Photodimerization in the Solid State", *Pure Appl. Chem.*, **27**, 647.

Schoofs, A. R. and Guetté, J.-P. (1983), "Competitive Reaction Methods for the Determination of Maximum Specific Rotations," Morrison, J. D., Ed., *Asymmetric Synthesis*, Vol. 1, Academic, NY, Chap. 3.

Schröder, I. (1893), *Z. Phys. Chem.*, **11**, 449.

Schurig, V. (1977), *Angew. Chem. Int. Ed. Engl.*, **16**, 110.

Schurig, V. (1981), *Angew. Chem. Int. Ed. Engl.*, **20**, 807.

Schurig, V. (1983), "Gas Chromatographic Methods," in Morrison, J. D., Ed., *Asymmetric Synthesis*, Vol. 1, Academic, New York, Chap. 5.

Schurig, V. (1984), "Gas Chromatographic Separation of Enantiomers on Optically Active Metal-Complex-Free Stationary Phases," *Angew. Chem. Int. Ed. Engl.*, **23**, 747.

Schurig, V. (1985), "Current Methods for Determination of Enantiomeric Compositions (a) Part 1: Definitions, Polarimetry, *Kontakte* (*Darmstadt*) (1) 54; (b) Part 2: NMR Spectroscopy with Chiral Lanthanide Shift Reagents," *Kontakte* (*Darmstadt*), (2) 22.

Schurig, V. (1986), "Current Methods for Determination of Enantiomeric Compositions, Part 3: Gas Chromatography on Chiral Stationary Phases," *Kontakte* (*Darmstadt*) (1) 3.

Schurig, V. and Betschinger, F. (1992), "Metal-mediated enantioselective access to unfunctionalized aliphatic oxiranes: prochiral and chiral recognition," *Chem. Rev.*, **92**, 873.

Schurig, V. and Bürkle, W. (1982), *J. Am. Chem. Soc.*, **104**, 7573.

Schurig, V. and Gil-Av, E. (1976/77), *Isr. J. Chem.*, **15**, 96.

Schurig, V. and Nowotny, H.-P. (1990), "Gas Chromatographic Separation of Enantiomers on Cyclodextrin Derivatives," *Angew. Chem. Int. Ed. Engl.*, **29**, 939.

Schurig, V., Ossig, J., and Link, R. (1989), *Angew. Chem. Int. Ed. Engl.*, **28**, 194.

Scott, J. W. (1989), "Enantioselective Synthesis of Non-Racemic Chiral Molecules on an Industrial Scale," *Top. Stereochem.*, **19**, 209.

Seebach, D. and Langer, W. (1979), *Helv. Chim. Acta*, **62**, 1701.

Seebach, D., et al., (1977), *Helv. Chim. Acta*, **60**, 301.

Sellergren, B. and Andersson, L. (1990), *J. Org. Chem.*, **55**, 3381.

Shapiro, M. J., Archinal, A. E., and Jarema, M. A. (1989), *J. Org. Chem.*, **54**, 5826.

Shibata, T., Okamoto, I., and Ishii, K. (1986), "Chromatographic Optical Resolution on Polysaccharides and their Derivatives," *J. Liq. Chromatogr.*, **9**, 313.

Shinkai, S., Nishi, T., and Matsuda, T. (1991), *Chem. Lett.*, 437.

Shiraiwa, T., Ikawa, A., Fujimoto, K., Iwafugi, K., and Kurokawa, H. (1984), *Nippon Kagaku Kaishi*, 764.

Shoemaker, D. P., Donohue, J., Schomaker, V., and Corey, R. B. (1950), *J. Am. Chem. Soc.*, **72**, 2328.

Silks, L. A., III, Peng, J., Odom, J. D., and Dunlap, R. B. (1991), *J. Org. Chem.*, **56**, 6733.

Silverstein, R. M. (1979), "Enantiomer composition and bioactivity of chiral semiochemicals in insects," in Ritter, F. J., Ed., *Chemical Ecology: Odour Communication in Animals*, Elsevier, Amsterdam, The Netherlands.

Sjöberg, B. (1957), *Ark. Kemi*, **11**, 439.

Smaardijk, A. D. (1986), Ph.D. Dissertation, University of Groningen, The Netherlands, Chap. 4.

Smith, R. L. and Caldwell, J. (1988), "Racemates: Towards a New Year Resolution?" *Trends Pharmacol. Sci.*, **9**, 75.

Sogah, G. D. Y. and Cram, D. J. (1979), *J. Am. Chem. Soc.*, **101**, 3035.

Solms, J., Vuataz, L., and Egli, R. H. (1965), *Experientia*, **21**, 692.

Sonnet, P. E. (1984), *J. Chromatogr.*, **292**, 295.

Sonnet, P. E. (1987), *J. Org. Chem.*, **52**, 3477.

Sorgen, G. J. (1983), *Microgram*, **16**, 126.

Sousa, L. R., Sogah, G. D. Y., Hoffman, D. H., and Cram, D. J. (1978), *J. Am. Chem. Soc*, **100**, 4569.

Souter, R. (1985), *Chromatographic Separations of Stereoisomers*, CRC Press, Boca Raton, FL.

Stewart, M. V. and Arnett, E. M. (1982), "Chiral Monolayers at the Air–Water Interface," *Top. Stereochem.*, **13**, 195.

Stinson, S. C. (1992), "Chiral Drugs," *Chem. Eng. News*, **70**, Sept. 28, 1992, 46.

Stinson, S. C. (1993), "Chiral Drugs," *Chem. Eng. News*, **71**, Sept. 27, 1993, 38.

Stockhausen, M. and Dachwitz, E. (1985), *Chem. Phys. Lett.*, **121**, 77.

Stölting, K. and König, W. A. (1976), *Chromatographia*, **9**, 331.

Strekowski, L., Visnick, M., and Battiste, M. A. (1986), *J. Org. Chem.*, **51**, 4836.

Strijtveen, B., Feringa, B. L., and Kellogg, R. M. (1987), *Tetrahedron*, **43**, 123.

Sugimoto, T., Matsumura, Y., Tanimoto, S., and Okano, M. (1978), *J. Chem. Soc. Chem. Commun.*, 926.

Sullivan, G. R. (1978), "Chiral Lanthanide Shift Reagents," *Top. Stereochem.*, **10**, 287.

Suming, H., Yaozu, C., Longfei, J., and Shuman, X. (1986), *Org. Mass. Spectrom.*, **21**, 7.

Supelco Reporter (1985), IV (2) 1 [Supelco, Inc., Bellefonte, PA].

Sweeting, L. M., Crans, D. C., and Whitesides, G. M. (1987), *J. Org. Chem.*, **52**, 2273.

Szczepaniak, W. and Ciszewska, W. (1982), *Chromatographia*, **15**, 38.

Szepesi, G. (1989), "Ion-Pairing," in Lough, W. J., Ed., *Chiral Liquid Chromatography*, Blackie, Glasgow, UK, Chap. 11.

Takagi, S., Fujishiro, R., and Amaya, K. (1968), *Chem. Commun.*, 480.

Takenaka, S. and Koden, M. (1978), *J. Chem. Soc. Chem. Commun.*, 830.

Takeuchi, Y., Itoh, N., and Koizumi, T. (1992), *J. Chem. Soc. Chem. Commun.*, 1514.

Takeuchi, Y., Itoh, N., Satoh, T., Koizumi, T., and Yamaguchi, K. (1993), *J. Org. Chem.*, **58**, 1812.

ten Hoeve, W. and Wynberg, H. (1979), *J. Org. Chem.*, **44**, 1508.

ten Hoeve, W. and Wynberg, H. (1985), *J. Org. Chem.*, **50**, 4508.

Tensmeyer, L. G., Landis, P. W., and Marshall, F. J. (1967), *J. Org. Chem.*, **32**, 2901.

Terabe, S. (1989), "Electrokinetic Chromatography: An Interface between Electrophoresis and Chromatography," *Trends Anal. Chem.*, **8**, 129.

Terunuma, D., Kato, M., Kamei, M., Uchida, H., and Nohira, H. (1985), *Chem. Lett.*, 13.

Testa, B. (1979), *Principles of Organic Stereochemistry*, Marcel Dekker, New York.

Testa, B. (1986), "The Chromatographic Analysis of Enantiomers in Drug Metabolism Studies," *Xenobiotica*, **16**, 265.

Testa, B. and Jenner, P. (1978), "Stereochemical Methodology," in Garrett, E. R. and Hirtz, J. L., Eds., *Drug Fate and Metabolism: Methods and Techniques*, Vol. 2, Marcel Dekker, New York, p. 143.

Thaisrivongs, S. and Seebach, D. (1983), *J. Am. Chem. Soc.*, **105**, 7407.

Theimer, E. T. and McDaniel, M. R. (1971), *J. Soc. Cosmet. Chem*, **22**, 15.

Thoma, A. P., Cimerman, Z., Fiedler, U., Bedeković, D., Güggi, M., Jordan, P., May, K., Pretsch, E., Prelog, V., and Simon, W. (1975), *Chimia*, **29**, 344.

Thomas, E. W. (1966), *Biochem. Biophys. Res. Commun.*, **24**, 611.

Thomson, W. See Kelvin (Lord).

Toda, F. (1991), "Molecular Recognition," *Bioorg. Chem.*, **19**, 157.

Toda, F., Mori, K., and Sato, A. (1988b), *Bull. Chem. Soc., Jpn.*, **61**, 4167.

Toda, F., Mori, K., Okada, J., Node, M., Itoh, A., Oomine, K., and Fuji, K. (1988a), *Chem. Lett.*, 131.

Toda, F., Toyotaka, R., and Fukuda, H. (1990), *Tetrahedron: Asymmetry*, **1**, 303.

Tölgyessy, J., Braun, T., and Kyrš, M. (1972), *Isotope Dilution Analysis*, Pergamon, Oxford, UK.

Topiol, S. and Sabio, M. (1989), *J. Am. Chem. Soc.*, **111**, 4109.

Topiol, S., Sabio, M., Moroz, J., and Caldwell, W. B. (1988), *J. Am. Chem. Soc.*, **110**, 8367.

Tran, C. D. and Fendler, J. H. (1980), *J. Am. Chem. Soc.*, **102**, 2923.

Tsai, W.-L., Hermann, K., Hug, E., Rohde, B., and Dreiding, A. S. (1985), *Helv. Chim. Acta*, **68**, 2238.

Tsuboi, M. and Takenishi, T. (1959), *Bull. Chem. Soc. Jpn.*, **32**, 726.

Tumlinson, J. H., Klein, M. G., Doolittle, R. E., Ladd, T. L., and Proveaux, A. T. (1977), *Science*, **197**, 789.

Ulbricht, T. L. V. (1959), "Asymmetry: The Non-Conservation of Parity and Optical Activity," *Q. Rev. Chem. Soc.*, **13**, 48.

Ulbricht, T. L. V. and Vester, F. (1962), *Tetrahedron*, **18**, 629.

Vaida, M., Shimon, L. J. W., Weisinger-Lewin, Y., Frolow, F., Lahav, M., Leiserowitz, L., and McMullan, R. K. (1988), *Science*, **241**, 1475.

Valente, E. J., Zubrowski, J., and Eggleston, D. S. (1992), *Chirality*, **4**, 494.

Valentine, D., Chan, K. K., Scott, C. G., Johnson, K. K., Toth, K., and Saucy, G. (1976), *J. Org. Chem.*, **41**, 62.

van der Haest, A. D. (1992), Ph.D. Dissertation, University of Groningen, The Netherlands, Chap. 4.

van der Haest, A. D., Wynberg, H., Leusen, F. J. J., and Bruggink, A. (1990), *Recl. Trav. Chim. Pays-Bas*, **109**, 523.

van Laar, J. J. (1903), *Arch. Neerl. Sci. Exactes Nat.*, [2] **8**, 264.

Vecci, M. and Mueller, R. K. (1979), *HRC&CC J. High Resolut. Chromatogr. Chromatogr. Commun.*, **2**, 195.

Vermeulen, N. P. E. and Breimer, D. D. (1983), "Stereoselectivity in Drug and Xenobiotic Metabolism," in Ariens, E. J., Soudijn, W., and Timmermans, P. B. M. W. M., Eds., *Stereochemistry and Biological Activity of Drugs*, Blackwell, Oxford, UK, p. 33.

Vigneron, J. P., Dhaenens, M., and Horeau, A. (1973), *Tetrahedron*, **29**, 1055.

Villani, Jr., F. J., Costanzo, M. J., Inners, R. R., Mutter, M. S., and McClure, D. E. (1986), *J. Org. Chem.*, **51**, 3715.

Vögtle, F. and Knops, P. (1991), "Dyes for Visual Distinction between Enantiomers: Crown Ethers as Optical Sensors for Chiral Compounds," *Angew. Chem. Int. Ed. Engl.*, **30**, 958.

Wainer, I. W. (1986), "Comparison of the Liquid Chromatographic Approaches to the Resolution of Enantiomeric Compounds," *Chromatographic Forum*, **1**, 55.

Wainer, I. W. (1987), "Proposal for the classification of high-performance liquid chromatographic stationary phases: how to choose the right column." *Trends Anal. Chem.*, **6**, 125.

Wainer, I. W. (1989), "Some Observations on Choosing an HPLC Chiral Stationary Phase," *LC-GC*, **7**, 378.

Wainer, I. W. and Alembik, M. C. (1986), *J. Chromatogr.*, **367**, 59.

Wainer, I. W., Barkan, S. A., and Schill, G. (1986), *LC, Liq. Chromatogr. HPLC Mag.*, **4**, 422.

Wainer, I. W., Brunner, C. A., and Doyle, T. D. (1983), *J. Chromatogr.*, **264**, 154.

Wainer, I. W. and Doyle, T. D. (1984), "Stereoisomeric separations: Use of chiral stationary phases to resolve molecules of pharmacological interest," *LC, Liq. Chromatogr. HPLC Mag.*, **2**, 88.

Wainer, I. W. and Drayer, D. E., Eds., (1988), *Drug Stereochemistry. Analytical Methods and Pharmacology*, Marcel Dekker, New York.

Wald, G. (1957), *Ann. N.Y. Acad. Sci.*, **69**, 352.

Walden, W., Zimmerman, C., Kolbe, A., and Pracejus, H. (1977), *Tetrahedron*, **33**, 419.

Walker, D. C. (1985), "Leptons in Chemistry," *Acc. Chem. Res.*, **18**, 167.

Wallach, O. (1895), *Justus Liebigs Ann. Chem.*, **286**, 90.

Waller, G. R. (1972), *Biochemical Applications of Mass Spectrometry*, Wiley, New York.

Wallingford, R. A. and Ewing, A. G. (1989), "Capillary Electrophoresis," in Giddings, J. C., Grushka, E., and Brown, P. R., Eds., *Advances in Chromatography*, Vol. 29, Marcel Dekker, NY, Chap. 1.

Walther, W., Vetter, W., Vecchi, M., Schneider, H., Müller, R. K., and Netscher, T. (1991), *Chimia*, **45**, 121.

Ward, T. J. and Armstrong, D. W. (1986), *J. Liquid Chromatogr.*, **9**, 407.

Ward, T. J. and Armstrong, D. W. (1988), "Cyclodextrin-Stationary Phases," in Zief, M. and Crane, L. J., Eds., *Chromatographic Chiral Separations*, Marcel Dekker, New York, Chap. 5.

Watabe, K., Charles, R., and Gil-Av, E. (1989), *Angew. Chem. Int. Ed. Engl.*, **28**, 192.

Webster, F. X., Zeng, X-n., and Silverstein, R. M. (1982), *J. Org. Chem.*, **47**, 5225.

Weinkam, R. J., Gal, J., Callery, P., and Castagnoli, Jr., N. (1976), *Anal. Chem.*, **48**, 203.

Weinstein, S. (1984), *Tetrahedron Lett.*, **25**, 985.

Weisman, G. R. (1983), "Nuclear Magnetic Resonance Analysis Using Chiral Solvating Agents," in Morrison, J. D., Ed., *Asymmetric Synthesis*, Vol. 1, Academic, New York, Chap. 8.

Weissbuch, I., Addadi, L., Lahav, M., and Leiserowitz, L. (1991), "Molecular Recognition at Crystal Interfaces," *Science*, **253**, 637.

Wenzel, T. J. (1986), "Binuclear Lanthanide (III)–Silver (I) NMR Shift Reagents," in Morrill, T. C., Ed., *Lanthanide Shift Reagents in Stereochemical Analysis*, VCH, Deerfield Beach, FL, Chap. 5.

Wenzel, T. J. (1987), *NMR Shift Reagents*, CRC Press, Boca Raton, FL.

Wenzel, T. J. and Lalonde, Jr., D. R. (1983), *J. Org. Chem.*, **48**, 1951.

Wenzel, T. J., Morin, C. A., and Brechting, A. A. (1992), *J. Org. Chem.*, **57**, 3594.

Wenzel, T. J. and Sievers, R. E. (1981), *Anal. Chem.*, **53**, 393.

Wenzel, T. J. and Sievers, R. E. (1982), *J. Am. Chem. Soc.*, **104**, 382.

Westley, J. W. and Halpern, B. (1968), *J. Org. Chem.*, **33**, 3978.

Weygand, F., Kolb, B., Prox, A., Tilak, M., and Tomida, I. (1960), *Z. Physiol. Chem.*, **322**, 38.

Weygand, F., Prox, A., Schmidhammer, L., and König, W. (1963), "Gas Chromatographic Investigation of Racemization during Peptide Synthesis," *Angew. Chem. Int. Ed. Engl.*, **2**, 183.

Wheland, G. W. (1951), *Advanced Organic Chemistry*, 2nd ed., Wiley, New York.

Whitesell, J. K. and Reynolds, D. (1983), *J. Org. Chem.*, **48**, 3548.

Whitesides, G. M. and Lewis, D. W. (1970), *J. Am. Chem. Soc.*, **92**, 6979; Whitesides, G. M. and Lewis, D. W. (1971), *ibid.* **93**, 5914.

Whyte, J. N. C. and Englar, J. R. (1977), *Carbohydr. Res.*, **57**, 273.

Wildmann, G. and Sommerauer, H. (1986), "Applications of DSC Purity Analysis," *Am. Lab.*, **20**, 107.

Wilen, S. H., Collet, A. and Jacques, J. (1977), "Strategies in Optical Resolutions," *Tetrahedron*, **33**, 2725.

Wilen, S. H. and Toporovsky, I. (1992), unpublished data.

Wilen, S. H., Qi, J. Z., and Williard, P. G. (1991), *J. Org. Chem.*, **56**, 485.

Williams, D. J. (1984), "Organic Polymeric and Non-Polymeric Materials with Large Optical Nonlinearities," *Angew. Chem. Int. Ed. Engl.*, **23**, 690.

Williams, T., Pitcher, R. G., Bommer, P., Gutzwiller, J., and Uskoković, M. (1969), *J. Am. Chem. Soc.*, **91**, 1871.

Wilson, K. R. and Pincock, R. E. (1975), *J. Am. Chem. Soc.*, **97**, 1474.

Wilson, W. K., Scallen, T. J., and Morrow, C. J. (1982), *J. Lipid. Res.*, **23**, 645.

Winkler, F. J. and Splitter, J. S. (1994), "Stereochemical Effects in the Positive- and Negative-Ion Chemical-Ionization Mass Spectra of Stereoisomeric Molecules," Chap. 16 in Splitter, J. S. and Turecek, F., *Applications of Mass Spectrometry to Organic Stereochemistry*, VCH, New York. See, especially, pages 365–367.

Winkler, F. J., Stahl, D., and Maquin, F. (1986), *Tetrahedron Lett.*, **27**, 335.

Wirzing, G. (1973), *Z. Anal. Chem.*, **267**, 1.

Witiak, D. T. and Inbasekaran, M. N. (1982), "Optically Active Pharmaceuticals," in Grayson, M., Ed., *Kirk–Othmer Encyclopedia of Chemical Technology*, 3rd ed., Vol. 17, Wiley, New York, p. 311.

Witter, A. (1983), "Stereoselectivity in Peptides," in Ariens, E. J., Soudijn, W., and Timmermans, P. B. M. W. M., Eds., *Stereochemistry and Biological Activity of Drugs*, Blackwell, Oxford, UK, p. 151.

Wong, C.-H. and Matos, J. R. (1985), *J. Org. Chem.*, **50**, 1992.

Wong, C.-H. and Whitesides, G. M. (1981), *J. Am. Chem. Soc.*, **103**, 4890.

Wright, N. (1937), *J. Biol. Chem.*, **120**, 641; Wright, N., *ibid.*, (1939), **127**, 137.

Wu, C. S., Ambler, E., Hayward, R. W., Hoppes, D. D., and Hudson, R. P. (1957), *Phys. Rev.*, **105**, 1413.

Wulff, G. (1986), "Molecular Recognition in Polymers Prepared by Imprinting with Templates," in *Polymeric Reagents and Catalysts*, Ford, W. T., Ed., ACS Symposium Series 308, American Chemical Society, Washington, DC, Chap. 8.

Wulff, G. and Minárik, M. (1988), "Tailor-Made Sorbents: A Modular Approach to Chiral Separations," in Zief, M. and Crane, L. J., Eds., *Chromatographic Chiral Separations*, Marcel Dekker, New York, Chap. 2.

Wulff, G. and Schauhoff, S. (1991), *J. Org. Chem.*, **56**, 395.

Wulff, G., Sczepan, R., and Steigel, A. (1986), *Tetrahedron Lett.*, **27**, 1991.

Wynberg, H. (1976), "Asymmetric Catalysis in Oxidation Reactions. The Consequences of Interactions between Enantiomers," *Chimia*, **30**, 445.

Wynberg, H. and Feringa, B. (1976), "Enantiomeric Recognition and Interactions," *Tetrahedron*, **32**, 2831.

Yagi, H., Vyas, K. P., Tada, M., Thakker, D. R., and Jerina, D. M. (1982), *J. Org. Chem.*, **47**, 1110.

Yamaguchi, S. (1983), "Nuclear Magnetic Resonance Analysis Using Chiral Derivatives," in Morrison, J. D., Ed., *Asymmetric Synthesis*, Vol. 1, Academic, New York, Chap. 7.

Yamaguchi, S., Yasuhara, F., and Kabuto, K. (1976), *Tetrahedron*, **32**, 1363.

Yorozu, T., Hayashi, K., and Irie, M. (1981), *J. Am. Chem. Soc.*, **103**, 5480.

Yuki, H., Okamoto, Y., and Okamoto, I. (1980), *J. Am. Chem. Soc.*, **102**, 6356.

Záhorsky, U.-I. and Musso, H. (1973), *Chem. Ber.*, **106**, 3608.

Zawadzke, L. E. and Berg, J. M. (1992), *J. Am. Chem. Soc.*, **114**, 4002.

Zel'dovich, B. Ya., Saakyan, D. B., and Sobel'man, I. I. (1977), *Sov. Phys. JETP Lett.*, **25**, 95.

Zief, M. and Crane, L. J., Eds. (1988), *Chromatographic Chiral Separations*, Marcel Dekker, New York.

Zingg, S. P., Arnett, E. M., McPhail, A. T., Bothner-By, A. A., and Gilkerson, W. R. (1988), *J. Am. Chem. Soc.*, **110**, 1565.

Zink, J. I. (1978), "Triboluminescence," *Acc. Chem. Res.*, **11**, 289.

Zwartouw, H. T. and Smith, H. (1956), *Biochem. J.*, **63**, 437.

7

Separation of Stereoisomers. Resolution. Racemization

7-1. INTRODUCTION

The production of nonracemic samples of chiral organic compounds from achiral or from racemic precursors by whatever means is called *optical activation*.

> Optical activation is a rather unsatisfactory expression. It obviously stands for the generation of a physical property (optical activity), whereas the description of the process ought properly to emphasize the enantiomerically enriched *composition* of the desired product quite apart from the use of rotation as a probe. Moreover, the term suffers from the limitation already identified in Chapter 6 (cf. also Chapter 13) that bulk samples of enantiomerically enriched chiral substances may have low or even zero rotation. Unfortunately no accepted alternative expression exists.

If the isolation of chiral substances from plant or animal sources and their transformations is excluded, there are two broad categories of optical activation methods: resolutions and stereoselective syntheses (Table 7.1). A resolution is the separation of a racemate into its two enantiomer constituents. In resolution methods, the point of departure is a racemate; therefore, the maximum yield of each enantiomer is 50%. Resolution methods may involve (a) physical processes only (see Section 7-2), or (b) chemical reactions. In resolutions mediated by chemical reactions, diastereomeric transition states or diastereomeric products usually intervene. In using these methods, one may take advantage of the operation of either thermodynamic or kinetic control.

It is implied that whatever be the nature of the chemical transformation that takes place in connection with a resolution, the reaction is either reversible or, in some other way, the process ultimately leads back to the starting material separated into its enantiomer components. Hybrid processes in which the resolved products are derivatives of the starting racemates are also known and, occasion-

TABLE 7.1   Optical Activation Methods

Resolutions
   Crystallization of enantiomer mixtures. Mechanical
   separation of enantiomers
      Processes under thermodynamic control
      Processes under kinetic control
   Chemical separation
      Conversion to diastereomers
         Thermodynamic control
         Kinetic control
      Intervention of diastereomeric transition states
      or excited states
         Kinetic (including enzymatic) resolutions
         Asymmetric destruction or transformation
         with circularly polarized light
Stereoselective synthesis
   Reaction of achiral substrates with chiral reagents
   Reaction at prochiral centers of chiral substrates with either
   chiral or achiral reagents

ally, only one of the enantiomers is recovered from the separation. These processes do not neatly fit the above classification but they are nonetheless called resolutions.

The focus of this chapter is on resolution methods. We begin our survey by examining one of the oldest and most fascinating resolution methods: the crystallization of enantiomers from solutions of racemates in the absence of resolving agents (Section 7-2). Classical diastereomer-mediated resolutions, resolutions mediated by inclusion compound formation and chromatographic resolutions are taken up in Section 7-3. This section concludes with a description of general separation methods applicable to mixtures of achiral as well as chiral diastereomers (Section 7-3.f). These and related topics are followed by a section that addresses the general question of enantiomeric enrichment in nonracemic samples of any provenance (Section 7-4). Section 7-5 deals with large scale resolutions, a topic of considerable practical importance rarely treated in stereochemistry texts. Kinetic resolutions are surveyed in Section 7-6. Section 7-7 illustrates less traditional yet promising resolution methods involving such principles as selective extraction and transport. The chapter concludes with a survey of racemization (Section 7-8).

An asymmetric synthesis in its most common form is the conversion of an achiral starting material into a chiral product by stereoselective reaction with a chiral reagent or catalyst. These and other processes that fit the stereoselective synthesis category are elaborated in Chapter 12.

7-2. SEPARATION OF ENANTIOMERS BY CRYSTALLIZATION

a. Crystal Picking. Triage

When crystallization of a racemate leads to the formation of a conglomerate then, by definition, the substance is said to be spontaneously resolved. This means that during the crystallization of the racemate under equilibrium conditions, whether this takes place spontaneously at a slow rate or more rapidly when induced with seed crystals, both enantiomers of the substance deposit in equal quantities as enantiomorphous crystals.

The manual sorting of conglomerate crystals into fractions whose solutions are dextrorotatory and levorotatory (as measured by polarimetry or circular dichroism, CD) is called triage. Palpation of the crystal handedness can speed up the process and this is possible when the crystals are well formed and endowed with hemihedral faces. Two examples of such resolutions are that described by Pasteur (1848; Kauffman and Myers, 1975; cf. Chapter 6) and that of asparagine (Piutti, 1886). Various stratagems for distinguishing crystals giving rise to oppositely signed solutions without relying on the presence of hemihedral faces (Section 6-4.c) have been discussed by Jacques, Collet and Wilen (1981a, p. 217) and Kaneko et al. (1974, p. 19); see also Sections 7-2.b and 7-2.d.

When a crystal that is, say, 2 mm long is available, a small piece may be cut off with a razor blade. The handedness of the crystal may be determined chiroptically or chromatographically after the small piece is dissolved in a suitable solvent. Solutions of different crystals of the same sense of chirality may then be combined for recovery of the crystal pieces and for combination with the corresponding original crystals. This process is obviously tedious. When information about the chirality sense is available from crystal shapes, resolution by triage becomes more efficient (see Section 6-4.c and 7-2.d).

While this type of resolution is rare [a well-known example is *rac*-hydrobenzoin (Fieser, 1964)], examples of resolution by triage continue to be recorded in the literature, even if only to obtain just enough nonracemic sample for configurational studies (Allen, Green, et al., 1983; Asaoka et al., 1988; Dossena et al., 1983; Ireland, Johnson, et al., 1970; Overberger and Labianca, 1970).

> If one remains dubious that the mere occurrence of a conglomerate should properly be called a "spontaneous resolution," consider the fact that sublimation of a *racemic* sample of a tetracyclic alcohol (Fig. 6.18, **15**) gave rise to crystal clusters that were optically active, some dextrorotatory and some levorotatory (Paquette and Lau, 1987). While the racemate should have exhibited this spontaneous resolution just as well prior to sublimation, the absence of crystals in the crude product prevented establishment of the fact that the product was a conglomerate. Spontaneous resolution has also been observed in 1,1'-binaphthyl (Fig. 7.11, **15a**) when samples are allowed to crystallize from the melt (Pincock and Wilson, 1971 and 1973; Kuroda and Mason, 1981). However, in this case the entire mass was found to have crystallized in one or the other enantiomeric form. This behavior results from the operation of an asymmetric transformation (cf. Section 7-2.e).

One must not underestimate the possibility of resolving small amounts of a racemate in this very direct and economical way that completely eschews the use of chiral auxiliary reagents. Neither hemihedral facets nor the use of a hand lens is required for success in carrying out resolutions by direct crystallization of enantiomer mixtures.

b. Conglomerates

We emphasize that, in order for resolutions of enantiomer mixtures to be possible without the use of chiral reagents or of enantioselective chromatographic columns, the racemate must be a conglomerate under the conditions of the crystallization. While only several hundred conglomerates have been explicitly described (Jacques et al., 1981a, p. 58), many more must exist in the domain of chiral organic compounds. A sampling of 1308 neutral compounds drawn from the Beilstein Handbook reveals that conglomerates constitute between 5 and 10% of the totality of chiral organic solids (Jacques, Leclercq, and Brienne, 1981b).

Even if a given racemate does not crystallize as a conglomerate, it may be possible to reversibly convert it to a derivative that does. For example, while few of the common amino acids are spontaneously resolved, many derivatives are known that are conglomerates (cf. Section 7-2.c). Alanine, leucine, and tryptophan crystallize as racemic compounds, but their benzenesulfonate salts are conglomerates; histidine hydrochloride is a conglomerate, and there are many other cases (Collet, Brienne, and Jacques, 1980). In general, it has been shown that the frequency with which such derivatization leads to a conglomerate is significantly increased when the derivative is a salt rather than a covalent compound (Jacques, Leclercq, and Brienne, 1981b). Thus while ($\pm$)-tartaric acid crystallizes as a racemic compound, the first substance ever resolved, Pasteur's salt (racemic $NaNH_4$ tartrate·$4H_2O$), is a conglomerate below 28°C as is "Seignette's salt" (racemic NaK tartrate·$4H_2O$) below −6°C; these double salts are spontaneously resolved. These two cases also illustrate the possibility of transforming some conglomerates into less highly solvated racemic compounds as the temperature is raised (Jacques et al., 1981a, pp. 131, 201) and vice versa.

Two further examples will demonstrate the utility in resolutions of derivatives that exhibit conglomerate behavior (Fig. 7.1). The amine **1** and the alcohol **2** are both liquids at room temperature. The conventional resolutions of these substances via crystalline diastereomeric derivatives is not particularly difficult, although a certain amount of tedium may be associated with the required recrystallizations. However, solid derivatives **3–5** are conglomerates. The implications of this fact are as follows: Crystallization of each of the racemates (**3–5**) under equilibrium conditions (slowly, with cooling or by evaporation even in the presence of *racemic* seeds) gives rise to a globally racemic precipitate from which it may be possible to retrieve (or separate) reasonably well-formed homochiral crystals of *each* chirality sense (see below). Such crystallization corresponds to a classic Pasteurian resolution of the first kind (see above).

Seeding with homochiral crystals of one of the enantiomers moves the process into the domain of nonequilibrium crystallizations. Here, a preponderance of one of the enantiomers of the substance may be deposited in a mother liquor whose optical rotation changes during the crystallization. *Preferential crystallization* is a practical variant of such nonequilibrium crystallizations carried out under carefully specified conditions designed to prevent the spontaneous crystallization of the undesired enantiomer during the controlled crystallization of its chiral partner (Section 7-2.c).

If one has a sample of enantiomerically enriched but impure **1** or **2**, one of the simplest ways of transforming it into an enantiomerically pure one is to convert the substance to one of the derivatives (**3**, **4**, or **5**). One crystallization should suffice to remove the minor enantiomer (in the mother liquor), leaving an enantiomerically pure crystalline sample of the major constituent. This result is possible *no matter what* the initial enantiomeric composition is. It is for this reason, in particular, that conglomerate-forming derivatives are so useful.

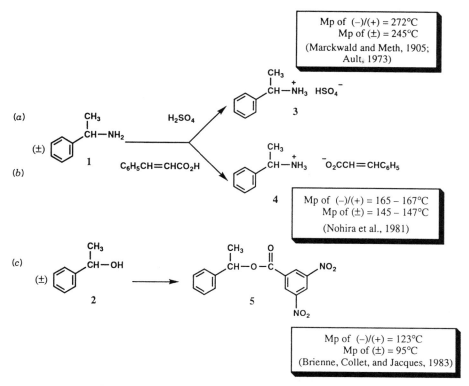

Figure 7.1. Derivatives exhibiting conglomerate behavior.

Since conglomerate systems are so much easier to resolve and to enrich than are racemic compound systems, it is important to be able to identify conglomerates with confidence. In general, when a racemate that must ultimately be resolved is produced in the laboratory, knowledge of the racemate type should simplify subsequent manipulation. The more common methods for identifying conglomerates are summarized in Table 7.2 (Jacques et al., 1981a, p. 53).

TABLE 7.2   Methods for the Identification of Conglomerates

1. First screening: If the melting point of the enantiomer exceeds that of the racemate by 25°C or more, the probability of conglomerate formation is high (the difference in melting point is generally in the range 25–35° though differences of 20° or even less are not uncommon).

2. Demonstration of spontaneous resolution via measurement of a finite optical rotation on a solution prepared from a single crystal of the racemate.

3. Solution in a nematic phase of a single crystal taken from a racemate; conversion of the nematic phase to a cholesteric liquid crystal.

4. a. Analysis of the binary phase diagram of the enantiomer mixture.
   b. Application of the contact method of Kofler.

5. Solubility behavior of one of the enantiomers in the saturated solution of the racemate. Insolubility is indicative of conglomerate behavior.

6. Determination of the crystal space group and of Z (number of asymmetric units per unit cell) by X-ray diffraction.

7. Comparison of the *solid state* IR spectra (also of solid state NMR spectra or of the X-ray powder diffraction patterns) of the racemate with that of one of the enantiomers; identity of the spectra is characteristic of conglomerate behavior.

Method 1 serves for a first screening only; it must be followed by one of the other tests. However, in the absence of a significant melting temperature difference, or if the racemate melting point is *higher* than that of one of the pure enantiomers, conglomerate behavior is precluded. A very useful test of conglomerate character is the calculation of the melting point of the racemate by means of the Schröder–van Laar equation (Eq. 6.5); agreement (within 1°C) between the melting point calculated for a conglomerate and the experimental melting point of the racemate is generally observed if conglomerate behavior actually obtains (e.g., Coquerel et al., 1988a).

Method 2 requires the measurement of optical activity of a solution prepared from a single crystal of the racemate. This method will be successful except when twinning occurs or when the crystals are too small to make it practical. (For a rare example of conglomerate single crystals that do not uniformly give rise to optically active solutions, see Davey et al., 1990; Black, Davey, et al., 1990; see also Section 7-2.c). Thus, absence of optical activity as measured on a single crystal is not a sufficient criterion for excluding conglomerate behavior. Of course, with a small crystal, the magnitude of rotation may be too small to be observed.

> Twinning is the formation of heterochiral crystals containing both enantiomers. Homochiral blocks of molecules alternating in chirality sense may be found along a growth direction (ordinary twinning) or in alternating layers (lamellar twinning). The latter phenomenon is responsible for the lack of optical activity of individual crystals of racemic hexahelicene (a conglomerate) in spite of the fact that these crystals are homochiral based on the crystal space group criterion (Green and Knossow, 1981; cf. Section 6-4.b).

Method 3 is effectively a variant of Method 2. The induced cholesteric behavior is an alternative and effective way of revealing optical activity in an extremely small sample (e.g., a single crystal) (Jacques et al., 1981a, p. 54). In fact, such behavior would be revealed even on a cryptochiral sample (i.e., in the absence of optical activity). A variant of this method, called liquid-crystal induced circular dichroism (LCICD) is described in Chapter 13.

Analysis of binary phase diagrams (Method 4) has been taken up in Section 6-4. The contact method of Kofler is a simpler way of obtaining the essential information conveyed by the phase diagram, namely, whether the diagram contains one or two eutectics (see Fig. 6.3). The method of Kofler can be carried out on the hot stage of a polarizing microscope (Jacques et al., 1981a, p. 40).

Method 5 deals with the solubility behavior of one of the enantiomers in a saturated solution of the racemate. An excess of one of the enantiomers is insoluble in such a solution if the racemate is a conglomerate. This phenomenon follows as a consequence of the shape of the solubility diagram (see Fig. 6.11).

Method 6 is a direct method, applicable if a well-formed crystal is available. If the crystal *space group* found is one of the noncentrosymmetric kinds, then (with rare exception; see Sec. 6-4.k) the sample is a conglomerate. *This analysis does not require a full structure determination by X-ray diffraction.*

Perhaps the simplest of the methods is the comparison of the IR spectra in the solid state of one of the enantiomers and of the corresponding racemate (Method

7). If the IR spectra of the two forms, as measured by the KBr pellet method or in a mull, are completely superposable, then the racemate is a conglomerate. If there is a significant difference between the spectra, then the racemic sample is most probably a racemic compound.

Three examples illustrating conglomerate behavior are shown in Figure 7.1 (compounds **3–5** with their melting points). Conglomerate behavior is confirmed (for **4** and **5**) by complete superposability of the IR spectra of the nonracemic and racemic forms.

Methods 1, 4, 5, and 7 require samples of both racemate and enantiomer (for Method 1, literature values of melting points would suffice) while Methods 2, 3, and 6 can be carried out with samples of racemate alone.

In order for a spontaneous resolution to have practical significance, the crystals must be separated. We say "mechanically separated" though what we really mean in most cases is manually separated, for no mechanical device is yet available that will unerringly pick all dextrorotatory crystals out of the crystalline matrix of the racemate and segregate them from the remaining levorotatory ones (however, see below for separation by sifting). Even for manual separation, there remains the question of how to identify the sense of chirality of any one crystal.

For some purposes, this requirement is irrelevant. For example, in the determination of absolute configuration, all that is required is to hand pick one crystal, to carry out the X-ray analysis (Bijvoet method; Section 5-3) and, following the conclusion of the analysis, to characterize the crystal, that is, to link it to other observables, by measuring its optical rotation or CD spectrum (e.g., Allen, Green, et al., 1983).

> When a crystal whose X-ray analysis is undertaken is damaged by prolonged radiation, or if its size is too small, chiroptical measurements may be precluded. However, in that instance, the absolute configuration would not have been ascertained in a useful way. Any attempt to reproduce the X-ray experiment would founder since one could not correlate the configuration of the crystal newly studied with another observable property of the previously studied crystal. For an example, see Dossena et al., 1983.

When the crystals are well formed, their external or morphological characteristics, in particular the occurrence of hemihedrism, may permit the differentiation of the enantiomers. However, as pointed out in Section 6-4.c, this possibility is dependent on the presence and the recognition of hemihedral facets. There are two troublesome aspects in connection with this requirement: Only a fraction of chiral crystals exhibit hemihedral facets as a consequence of inherent symmetry characteristics of the crystal and, even when present, such facets may be difficult to recognize. Recognition of morphological characteristics is thus not a generally useful way of identifying crystals of like chirality sense.

The sign of rotation of individual crystals may be measured conventionally, following which solutions of like sign may be combined and the solid enantiomers recovered from the respective solutions. Alternatively, fragments of isolated crystals of measurable size may be broken off (e.g., cut off with a razor blade) and melted in the presence of similar fragments cut from other crystals. Melting point lowering or absence thereof will indicate the relative sense of chirality of the pair being tested. Application of the contact method of Kofler (see above) is suggested as a simple way of doing this. The small sample may also be analyzed chromatographically (e.g., Addadi et al., 1982b; Arad-Yellin et al., 1983) or by Method 3.

Enantiomorphous crystals may be cleanly differentiated by stereoselective etching of just one of the enantiomers during the partial dissolution of the racemate (conglomerate) in the presence of a homochiral (see pp. 103 and 127) additive. Thus, racemic asparagine is partially dissolved in water in the presence of (R)-aspartic acid (20% by weight). Etch pits are observed only on the {010} faces of undissolved (R)-asparagine crystals; etch pits are not observed on the crystals of (S)-asparagine. Following the Pasteur-type sorting of more than 100 etched crystals, their rotation was measured. All had been correctly sorted (Shimon, Lahav, and Leiserowitz, 1986).

In spite of the dramatic results just described, resolution by manual separation is unlikely to be used frequently. In addition to the difficulties and pitfalls apparent from the preceding discussion there are others, such as the occasional occurrence of microtwinning, which may completely preclude resolution by crystallization (see above). Yet it is the most direct resolution method and it can quickly settle issues such as the relative configuration (meso or $\pm$) of a symmetrically constituted compound containing two or more stereocenters since only a single crystal is needed for this purpose (for an example, see Wittig and Rümpler, 1971). A single crystal may also suffice for the construction of a binary phase diagram of a conglomerate system or to serve as a seed crystal in the first step of a nonequilibrium crystallization.

Two variants can transform the simultaneous crystallization of the two enantiomers of a compound into practical resolution methods. The first consists in the deposition of the two enantiomers at different places in the racemic solution under the influence of localized seeds, the $(+)$-seeds forming large homochiral $(+)$-crystals while the $(-)$-seeds induce formation of homochiral $(-)$-crystals elsewhere in the solution whose liquid-phase homogeneity may be maintained by stirring or by use of a circulating system. Several devices have been described to facilitate the localization of crystallization (Collet, Brienne, and Jacques, 1980).

In the second variant, one encourages the formation of large crystals of one of the enantiomers while the other enantiomer spontaneously deposits in the form of small crystals. Relatively large seeds of the first enantiomer are required to initiate what has been called "differentiated" crystallization. The two sizes of enantiomeric crystals may then be separated by sifting (Collet, Brienne, and Jacques, 1980).

c. Preferential Crystallization

When the resolution of a racemate is initiated by inoculation of a saturated or supersaturated solution of the racemate with seeds (crystals) of one of the enantiomers, crystallization ensues in a nonequilibrium process called *preferential crystallization*, since it is only the seeded enantiomer that deposits.

Crystallization experiments undertaken on aqueous solutions of sodium chlorate $(NaClO_3)$, an achiral substance crystallizing in the chiral cubic space group $P2_13$, have dramatically demonstrated the distinction between crystallization under equilibrium and nonequilibrium conditions. Undisturbed crystallization (equilibrium conditions) gives rise to equal numbers of enantiomeric crystals, that is, to a racemic, or nearly racemic, solid. On the other hand, if crystallization is carried out with constant stirring (kinetic control) without seeding, the solid product is nearly enantiopure (ee > 99%) with a distribution of 18 predominantly (+) and 14 predominantly (−) crystal batches found in 32 crystallization trials (Kondepudi et al., 1990; see also McBride and Carter, 1991).

It must be emphasized that preferential crystallization works only for substances that are conglomerates (Inagaki, 1977). More precisely, it may work also with compounds existing either as conglomerates or as racemic compounds within an accessible temperature range, provided that the racemic compound does not crystallize during the operation. The latter qualification is to allow for the possible existence of metastable racemic compounds that may impede a preferential crystallization (Collet, Brienne, and Jacques, 1980). Thus, histidine·HCl is resolvable by preferential crystallization above 45°C consistent with the phase diagram examined in Section 6-4.f (Fig. 6-17a).

The resolutions of fenfluramine **6**, an anorectic (appetite suppressing) medicinal agent, and of its precursor norfenfluramine **7** illustrate some of the limitations of the preferential crystallization procedure (Fig. 7.2). Numerous salts (50) of **6** and **7** with achiral organic acids were tested to identify conglomerate character. Five salts of **6** and three of **7** were found to be conglomerates. Only two of these could not be preferentially crystallized, namely, the salt of **6** with phenoxyacetic acid and the dichloroacetate salt **8** of **7** (Coquerel, Bouaziz, and Brienne, 1988a).

The phenoxyacetate salt of **6** crystallizes simultaneously as an unstable conglomerate and a stable racemic compound at room temperature. In the case of salt **8**, it appears that some of the crystal faces of homochiral crystals of **8** may act as seeds for the other enantiomer. In the presence of racemic solution, crystals of one enantiomer grow onto the faces of the other (heterochiral growth) leading to

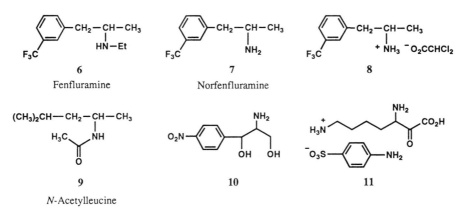

Figure 7.2.   Structures **6–11.**

composite crystals of very low optical purity (Coquerel, Perez, and Hartman, 1988b).

For similar analyses of covalent derivatives of **7** exhibiting conglomerate behavior, see Coquerel, Bouaziz, and Brienne (1990), and of which substituted benzoate salts of α-methylbenzylamine are conglomerates, see Shiraiwa et al. (1985b) and Nakamura et al. (1986).

Other examples of compounds resolvable by preferential crystallization above or below a known transition temperature, including the classic Pasteur case, are shown in Figure 7.3. The common case is that of a racemic compound that gives rise to a conglomerate on heating; the converse case (Pasteur's salt; phenylglycine sulfate; 2,2′-diamino-1,1′-biphenyl), though rarer, is also known (Jacques et al., 1981a, pp. 131, 201).

Conglomerates are sometimes obtainable in metastable form outside the temperature range in which they are normally stable, for example, 1,1′-binaphthyl by crystallization at room temperature (Kuroda and Mason, 1981). The preferential crystallization of a conglomerate as a metastable polymorphic form in the presence of additives, for example, that of histidine·HCl at 25°C, has been described (see Section 7-2.d).

In a preferred procedure, the excess enantiomer [e.g., (+)] crystallizes (either spontaneously or with seeds) from a solution of a conglomerate in which some of the (+) enantiomer has been dissolved by heating. When the solution is cooled, it

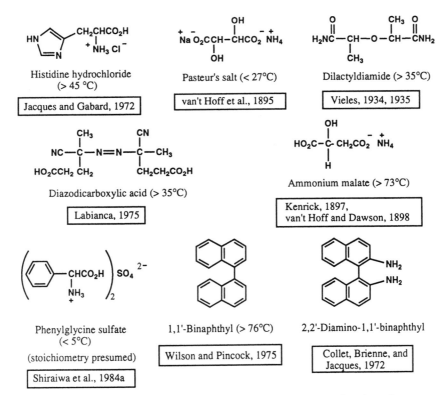

Figure 7.3. Compounds resolvable by preferential crystallization as a function of temperature.

becomes supersaturated with respect to the (+) enantiomer and the latter crystallizes. Moreover, the crystalline (+) enantiomer *entrains* the crystallization of that enantiomer so that more of this enantiomer crystallizes than was originally added to the racemate at the beginning of the experiment (Kaneko et al., 1974, p. 19; Collet, Brienne, and Jacques, 1980).

This concept gives rise to the name *entrainment* that has been used as an alternative term for the process. Preferential crystallization was developed by Amiard, Velluz, and co-workers at Roussel-Uclaf, Paris (Amiard, 1956, 1959; Velluz et al., 1953a,b) and applied with verve by Japanese investigators. It is based upon a discovery by Gernez (1866) who was a student of Pasteur's. The efficacy of the process was recognized by Duschinsky (1934, 1936) to whom we owe the first practical resolution by preferential crystallization, that of histidine·HCl.

> One of the earliest preferential crystallizations described in the literature, that of epinephrine (Fig. 6.30, **40a**) is found in a patent (Calzavara, 1934). However, it was not until the appearance of the review by Secor (1963), an industrial chemist, that the potential for this type of resolution began to become known in academic circles. Credit is due to Secor for reintroducing the use of phase diagrams in the analysis of resolution processes.

The process is best understood with the aid of a solubility diagram (Fig. 7.4*a*) (cf. Section 6-4.f). Spontaneous crystallization of a mixture of composition M (enantiomer composition x) gives rise to crystals of composition x′ with enrichment (x→x′) in enantiomer D. The solution that was originally dextrorotatory (the sign was arbitrarily chosen) gradually becomes racemic as the crystals deposit (solution composition tends toward E).

Crystallization initiated by seeding with *pure* D, on the other hand, leads to deposition of *pure* D initially along the line MD (×××) as a consequence of the unequal *rate of crystallization* of the two enantiomers. The rotation of the crystallization mother liquor drops to zero and then changes sign as the composition of the solution crosses the SR line. Indeed, measuring the rotation of the mother liquor is the most direct way of monitoring such a kinetically controlled

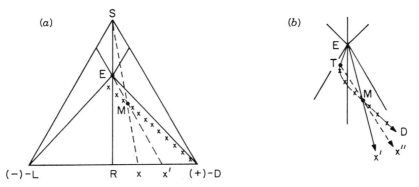

Figure 7.4. Solubility diagram for preferential crystallization. [Adapted with permission from Collet, A., Brienne, M.-J. and Jacques, J. (1980), *Chem. Rev.*, **80**, 215. © 1980 American Chemical Society.]

resolution. To emphasize the distinction between preferential crystallization and the spontaneous resolutions described in Sections 7-2.a and 7-2.b, the reader is reminded that in the latter (equilibrium processes), the rotation of the mother liquor remains zero all the while that crystallization takes place.

Consider the following example involving preferential crystallization of α-methyl-benzylamine cinnamate (Fig. 7.1b, **4**; Nohira et al., 1981). Racemic cinnamate (**4**, 15.5 g) is dissolved in aqueous CH_3OH by heating it along with 12.1 g ($\pm$)-α-methylbenzylamine and aq. HCl (12 M). The cooled racemic solution is inoculated with 20 mg of pure (+) salt. After 1 h the white crystals obtained are washed and dried. Yield: 3.4 g of (+)-**4** salt ($3.4/7.75 \times 100 = 44\%$ of theory); ee = 75%. The mother liquor is levorotatory. Racemic salt is now added in a quantity roughly equal to the solid collected (3.5 g), dissolved by heating, cooled and seeded with 20 mg of pure (−) salt. After 1 h, 5.4 g (57% of theory) of precipitated (−)-**4** salt is collected; ee = 85%.

The procedure is efficient, simple and, moreover, rapid. The speed may seem surprising but is at the heart of this kinetically controlled process. Longer crystallization times would improve the yield, but the enantiomer composition would tend toward 50:50, which is what recrystallization of a racemate would produce under normal (i.e., equilibrium) conditions.

If enantiomerically pure products are required, one recrystallization of each of the enantiomers (in the above example, from CH_3OH) suffices to produce pure (+) and (−) salt, respectively. The second part of the process leading to the (−) salt suggests that, if the beginning solution is enriched in one enantiomer, the yield of crystals depositing within a fixed period of time will be higher than when the starting solution is racemic [after the second batch of racemic salt is added, the solution from which (−) salt deposits is more highly supersaturated with respect to this salt]. Note that the yield of (−) salt (5.4 g) is larger than the excess present at the onset of its preferential crystallization (3.4 g). This enhancement in yield is the essence of the entrainment. In any event, many resolutions by preferential crystallization are initiated with nonracemic solutions. The readily soluble racemic salts formed by reaction of the ($\pm$) amine with HCl (or HOAc) act as "a kind of buffer" stabilizing the preferential crystallization of one of the enantiomers significantly (Saigo, Nohira, et al., 1981). While this is not further explained, it is reminiscent of the effect of pH changes that can take place in nonstoichiometric classical resolutions involving diastereomers (Section 7.4.a).

Addition of an achiral acid or base to an ionizable racemate leads to the recovery of a higher yield of the desired enantiomer by controlling the degree of super-saturation of the enantiomer remaining in solution (Asai and Ikegami, 1982; Asai, 1983; see also Asai, 1985).

The reasons for the incomplete enantiomeric purity of the crystals collected bear further scrutiny (Fig. 7.4b). At the onset of the preferential crystallization, pure D crystallizes alone (rate of deposition D > L). However, when the mother liquor composition becomes richer in L enantiomer, deposition of the latter becomes competitive with that of the now depleted D. In other words, the kinetically controlled process tends to revert to an equilibrium one as crys-

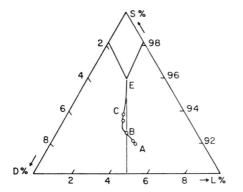

Figure 7.5 (*a*). Crystallization curve of *N*-acetylleucine in acetone at 25°C. At **A** (initial optical purity 11%) the solution was seeded with a very small amount of the pure crystalline L enantiomer. Point **B** (change of sign of the solution) occurs at the end of about 20 min, and point **C** is reached at the end of 40 min. Only the upper part of the ternary diagram is shown. [Reprinted with permission from Collet, A., Brienne, M.-J. and Jacques, J. (1980), *Chem. Rev.*, **80**, 215. Copyright © 1980 American Chemical Society.]

tallization slows down. The tie line DM then curves toward E (Fig. 7.4*b*) (as explicitly demonstrated for the cases of *N*-acetyleucine (Fig. 7.2, **9**; see Fig. 7.5*a*) and of chiral hydrobenzoin, C₆H₅CH(OH)CH(OH)C₆H₅ (Collet, Brienne, and Jacques, 1980). Solid depositing when the global composition is given by T has composition x″. This is still more enriched in D than is the solid obtained in a standard crystallization of mixture x (Fig. 7.4*a*) under equilibrium conditions, namely, x′. A key advantage of preferential crystallization is that it allows the enantiomer purity of the crystallization product to exceed that obtained in a normal (equilibrium controlled) crystallization.

Preferential crystallization may be carried out in a cyclic, repetitive, manner as illustrated in Figure 7.5*b* for hydrobenzoin. Initial crystallization takes place from a stirred solution containing a slight excess of one of the enantiomers [0.37 g of (−) and 11.0 g of racemate in 85 g of 95% ethanol]. In the example, the solution was cooled to 15°C, seeded with 10 mg of pure (−)-hydrobenzoin, and crystallization was allowed to proceed for 20 min. The crystals collected (0.87 g) were about double in weight with respect to the excess levorotatory isomer taken. *rac*-Hydrobenzoin was added to the mother liquor in an amount (0.9 g) equal to the (−) crystals collected and the solution heated to dissolve the solid. The composition of the solution, which had been N after removal of the first crop of crystals, is given by P after addition of the *rac*-hydrobenzoin. Composition P is symmetrical with M, the starting composition. Cooling the solution and seeding it

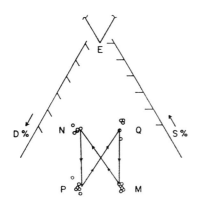

Figure 7.5 (*b*). Successive cycles of preferential crystallization of hydrobenzoin (experimental conditions are given in the text) [Reprinted with permission from Collet, A., Brienne, M.-J. and Jacques, J. (1980), *Chem. Rev.*, **80**, 215. Copyright © 1980 American Chemical Society.]

with 10 mg of pure (+)-hydrobenzoin gave 0.9 g of (+) crystals after 20 min. After removal of the latter, addition of 0.9 g of racemate to mother liquor Q and warming reconstituted solution M whose preferential crystallization was repeated (8 cycles) as just described. The products [6.5 g of (−)- and 5.7 g of (+)-hydrobenzoin] had an enantiomeric composition approaching 99:1. A similar cyclic resolution process has been described for 1-phenylethyl 3,5-dinitrobenzoate (Fig. 7.1, **5**) on a larger scale (Brienne, Collet, and Jacques, 1983). Examples of even larger scale preferential crystallizations (kilogram lots) carried out for 30–40 cycles may be found in the literature (e.g., Jommi and Teatini, 1962).

A protocol for the optimization of such cyclic preferential crystallizations has been developed by Amiard and co-workers (Velluz and Amiard, 1953; Velluz, Amiard, and Joly, 1953) at Roussel-Uclaf in connection with the industrial scale resolution of *threo*-1-(*p*-nitrophenyl)-2-amino-1,3-propanediol (Fig. 7.2, **10**), whose (−) enantiomer is an intermediate in the synthesis of the antibiotic chloramphenicol. The protocol seeks to control as much as possible the relative rates of crystallization of the two enantiomers within a fixed time period and in a given set of experimental conditions with the aim of totally preventing the crystallization of the unseeded isomer during the crystallization of the seeded one. The key variables here are concentrations and temperatures (Collet, Brienne, and Jacques, 1980). For an example of such control involving lysine *p*-amino-benzenesulfonate (Fig. 7.2, **11**), see Yamada et al. (1973a).

The efficiency of preferential crystallization is expressed by a resolution index (RI) defined as the ratio of resolution product weight (calculated as pure enantiomer) to that of the initial excess of this enantiomer taken, that is, $[W_{recovered} \times ep - W_{seeds}]/E_{excess\ taken}$, where ep (enantiomeric purity) is equivalent to ee. Since RI = 1 corresponds just to recovery of the excess enantiomer taken, a value of RI ≥ 2 should obtain if the alternate crystallization of each enantiomer by preferential crystallization is to be considered efficient (Coquerel, Bouaziz, and Brienne, 1988a). The value of RI calculated for the first cycle of the hydrobenzoin resolution (p. 309) described above (assuming a product ee of 97%) is $[(0.87)(0.97) - 0.010]/0.37 = 2.2$.

The factors favoring preferential crystallization (Collet, Brienne, and Jacques, 1980) are as follows: (a) racemic salts (hydrochlorides, sulfonates, etc.) are more prone to resolution by preferential crystallization than are covalent racemates (Yamada et al., 1973b; Kimoto, Saigo, et al., 1989); (b) a solubility ratio, $\alpha_x = S_R/S_A$ (where S_R and S_A are the solubilities of the racemate and of one of the enantiomers, respectively, expressed as mole fractions) less than 2 is preferable to >2. Contrary to intuition, the reduced solubility of the racemate relative to that of the enantiomers gives rise to a more favorable situation by enlarging the part of the solubility diagram usable for preferential crystallization; (c) the stirring rate can be optimized to increase the rate of crystal growth as well as to promote the growth of homochiral crystals (Kondepudi et al., 1990; McBride and Carter, 1991). However, increasing the stirring rate indiscriminately is counterproductive since spontaneous nucleation of the wrong enantiomer may ensue; (d) it is desirable to use seeds of uniform small size and composition; (e) sterilization of the solution by boiling prior to the cooling and inoculation step may eliminate unwanted small seeds (nuclei) invisible to the naked eye; and (f) it has been suggested that generation of microcrystalline seeds effected by ultrasonic irradiation (10–100 kHz) may reproducibly lead to a very rapid crystallization (Anon., 1965a).

Additional examples of compounds whose resolution by preferential crystallization has been described (in addition to others cited elsewhere in the book) are α-amino-ε-caprolactam, for example, as the hydrochloride (Watase et al., 1975); N-acetylnorleucine as the ammonium salt (Shiraiwa et al., 1986c); phenylalanine, as the salt of the N-acetyl derivative (Shiraiwa et al., 1984b) and of the N-formyl derivative (Shiraiwa et al., 1986a) with various alkylamines; phenylglycine, as the sulfate (Shiraiwa et al., 1984a) and as the 1,1-3,3-tetramethylbutylammonium salt of the N-formyl derivative (Shiraiwa et al., 1987a); carnitine nitrile, as the oxalate (Jacques et al., 1984); mandelic acid, as the 4-methylpiperidinium salt at 10°C (Shiraiwa et al., 1986b); 2-aryl-3-methylbutanoic acids as the diethylamine salts (Nohira et al., 1986); tyrosine, as the dicyclohexylammonium salt of the N-formyl derivative (Shiraiwa et al., 1987b); cysteine, as the salt with benzenesulfonic acid (Shiraiwa et al., 1987c); thiazolidine-4-carboxylic acid (Shiraiwa et al., 1987d); valine as the hydrochloride (Feldnere et al., 1988; Nohira et al., 1988); naproxen as the salt with ethylamine (Piselli, 1989); and a triazolyl ketone intermediate in the synthesis of the pesticide paclobutrazol (Fig. 6.31, **50**; Black et al., 1989).

Preferential crystallization in a supercooled melt has also been described (see Collet, Brienne, and Jacques, 1980). The cyclic, repetitive, preferential crystallization of a mixture of *diastereomers* has been described in the case of the mandelate salts of *erythro*-2-amino-1,2-diphenylethanol (Saigo et al., 1986a). The resolution of the analgesic ibuprofen (Fig. 6.31, **48**) by preferential crystallization of its salts with (S)-lysine has also been reported (Tung et al., 1991).

Numerous papers and patents attest to the potential and actual economic importance of preferential crystallization as a technique for resolution on a large scale. Examples of these and some key references are given in Section 7-5.

d. Preferential Crystallization in the Presence of Additives

Numerous experiments are described in the chemical literature in which the crystallization of optically active materials has been encouraged by use of foreign seed crystals (either achiral or chiral but racemic) or by use of chiral solvents. The results of these experiments which, for the most part, are not of any practical significance, have been summarized by Jacques et al. (1981a, p. 245). Successful cases are likely to involve the operation of kinetic phenomena (rates of crystallization or dissolution) rather than changes in equilibrium (differences in solubility).

Insoluble additives favor the growth of crystals that are identical or isomorphous with the seed crystal. In contrast, the effect of soluble additives is inverse to this. Among the early experiments described, (R,R)-$(+)$-sodium ammonium tartrate preferentially crystallizes from an aqueous solution of the racemate containing either dissolved (S)-$(-)$-sodium ammonium malate, $Na(NH_4)[O_2C-CH(OH)CH_2CO_2]$, or (S)-$(-)$-asparagine, $H_2NCOCH_2CH(NH_2)CO_2H$ (Ostromisslenski, 1908). Similarly, crystallization of racemic aspartic acid in the presence of dissolved L-$(+)$-glutamic acid leads to precipitation of D-$(-)$-aspartic acid of configuration opposite to that of the chiral cosolute (Purvis, 1957).

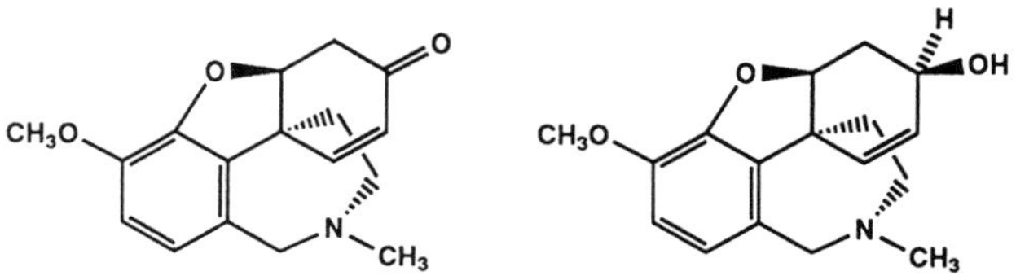

Narwedine   Galanthamine

(b)

Figure 7.6. Narwedine (left) and galanthamine (right).

The definitive rationalization and application of this process is due to a group at the Weizmann Institute of Science in Israel (Addadi, van Mil, and Lahav, 1981; Addadi, Lahav, et al., 1986; see also Chapter 5). Adsorption of the additive (cosolute) on the surfaces of growing crystals of one of the solute enantiomers inhibits that enantiomer from crystallizing normally. Preferential crystallization of the *other* enantiomer of the solute ensues (rule of reversal) (Addadi, Lahav, et al., 1982c). This explanation was originally suggested by Barton and Kirby (1962) in connection with the crystallization of the alkaloid *rac*-narwedine (Fig. 7.6) in the presence of (−)-galanthamine (Fig. 7.6) [(+)-narwedine crystallized] and is supported by several findings that are consistent with the adsorption argument.

Preferential crystallization with reversal takes place only with conglomerates and is interpreted as a kinetic phenomenon on crystal growth. Allowing the crystallization to continue for a long time eventually allows both the affected and the unaffected enantiomers to precipitate, that is, the precipitate becomes globally racemic.

The "tailor made" additives that are effective in promoting this type of preferential crystallization (called "replacing" crystallization by Kaneko et al., 1974, p. 24) are configurationally and constitutionally related to the crystallizing solute (cf. Fig. 5.20). This relationship is required if the additive is to fit into the growing crystal lattice, replacing the normal lattice constituents in one of the enantiomers and thereby stunting its growth. Normal crystallization of the other enantiomer thus proceeds preferentially.

For example, crystallization of racemic threonine from an aqueous supersaturated solution containing 10% of (S)-(+)-glutamic acid (L-glutamic acid) yields a crop of crystals identified as (2R,3S)-(+)-threonine (D-threonine) of enantiomer composition 97:3. The (+)-threonine that precipitates fastest is configurationally related to (R)-glutamic acid (D-glutamic acid) (Fig. 7.7). The L-glutamic acid additive therefore inhibits the crystallization of the configurationally related L-threonine. A second crop of crystals obtained is identified as L-(−)-threonine. Deposition of (−)-threonine in the presence of D-(−)-glutamic acid exemplifies reversal of the configurations of the two successive crops.

A corollary of the explanation requires that if crystalline *rac*-threonine is added to a solution of L-(+)-glutamic acid (10%), D-(+)-threonine preferentially dissolves while the remaining crystals are enriched in (−)-threonine whose dissolution is inhibited. This result is actually found. When approximately one-half of the racemate has dissolved, the remaining crystals are preponderantly

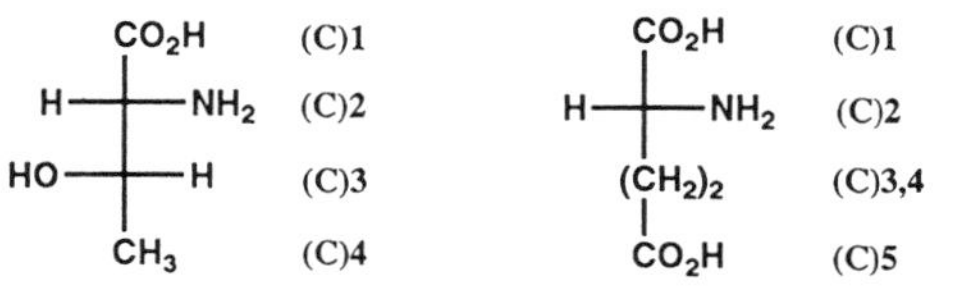

$$
\begin{array}{lll}
& \text{CO}_2\text{H} & \text{(C)1} \\
\text{H}\!-\!\!\!-\!\!\!-\text{NH}_2 & & \text{(C)2} \\
\text{HO}\!-\!\!\!-\!\!\!-\text{H} & & \text{(C)3} \\
& \text{CH}_3 & \text{(C)4}
\end{array}
\qquad
\begin{array}{lll}
& \text{CO}_2\text{H} & \text{(C)1} \\
\text{H}\!-\!\!\!-\!\!\!-\text{NH}_2 & & \text{(C)2} \\
(\text{CH}_2)_2 & & \text{(C)3,4} \\
& \text{CO}_2\text{H} & \text{(C)5}
\end{array}
$$

(2*R*,3*S*)-(+)-Threonine (D)   (*R*)(−)Glutamic acid (D)

Figure 7.7. Configurational relationship in the preferential crystallization of (2*R*,3*S*)-threonine in the presence of a soluble additive (for an explanation see the text).

L-(−)-threonine (82% ee). To demystify this finding, it is worthwhile to emphasize that we are dealing with *kinetic* phenomena: The fastest crystallizing enantiomer is also the fastest dissolving.

The preferential crystallization of thiazolidine-4-carboxylic acid (THC, Fig. 7.21, **72a**) has been carried out in the presence of various amino acids (as additives). L-4-Hydroxyproline was found to give the best results with (*S*)-THC crystallizing first followed by the *R* enantiomer (Shiraiwa et al., 1992).

Preferential crystallization in the presence of added soluble additives may be coupled with seeding by using crystals of the desired enantiomer as seeds while, at the same time, the dissolved additive inhibits the crystallization of the unwanted enantiomer.

Small amounts of the additive (0.03–2%) are found in the crystallized products. Preferential inclusion of the additive having the same chirality sense as the crystallized substrate is demonstrated in the slower crystallizing product crop that has the same configuration as the additive responsible for the rate phenomenon. Thus, crystallization of L-(−)-threonine in the presence of racemic glutamic acid leads to incorporation of the glutamic acid additive in a ratio of 5:2 (L′:D′) in the crystals. But the distribution of the glutamic acid enantiomers in the crystals of (−)-threonine is not uniform. Progressive dissolution of the crystals in pure water and analysis of the various fractions showed that the D′ additive was present only in the first (outer surface) fractions (being the last to crystallize) while the L′ additive was evenly distributed throughout the crystal.

Since the rates of growth at different faces of a crystal are unequal, the presence of incorporated additives will differentially affect growth directions as shown in Figure 7.8. Morphological changes are predicted and observed in that crystal whose constituent molecules are configurationally related to the dissolved impurity. The unaffected enantiomorphous chiral crystal will have essentially unmodified morphology. For example, when a supersaturated solution of racemic asparagine is preferentially crystallized in the presence of approximately 10% of dissolved (*S*)-(+)-aspartic acid, $HO_2CCH_2CH(NH_2)CO_2H$, spontaneous crystallization takes place. After 2–4 days, the crystals that are collected are seen to be of two distinct morphologies. As expected, the major crop consists of (*R*)-asparagine whose morphology is effectively unchanged from that of (*R*)-asparagine crystallized independently. The other crystals, (*S*)-asparagine, are present in significantly smaller amount and are sufficiently different in appearance from the major enantiomer as to make manual separation facile (Addadi, Gati, and Lahav, 1981). In some cases, the crystals of one enantiomer are well formed while those whose component molecules are configurationally related to the additive are powders (Addadi, Lahav, Leiserowitz, et al., 1982a; 1986).

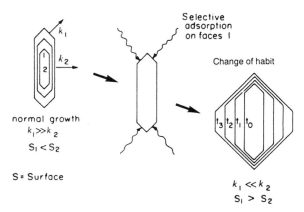

Figure 7.8. Morphological changes during crystal growth in the presence of additives [Reprinted with permission from Addadi, L., Berkovitch-Yellin, Z., Weissbuch, I., Lahav, M. and Leiserowitz, L. (1986), *Top. Stereochem.*, **16**, 1. Copyright © 1986 John Wiley & Sons, Inc.]

Preferential crystallization in the presence of additives constitutes a new method for visual separation of conglomerates that is more broadly applicable than Pasteur's first resolution method in that it does not depend on identification of hemihedral facets or even on their very presence (see Section 7-2.b).

Resolution of α-amino acids, such as glutamic acid, p-hydroxyphenylglycine, and methionine (all considered to be nonhydrophobic), by occlusion inside enantiomorphous glycine crystals has been demonstrated when the crystallization takes place in the presence of 1% (by weight) of (S)-leucine in aqueous solution (Weissbuch et al., 1988). The hydrophobic amino acid leucine promotes the orientation of growing *chiral* glycine crystals at the surface of the solution, that is, at the water–air interface, such that only one of the enantiomers of the nonhydrophobic amino acid additives is incorporated from solution into the enantiotopic glycine crystal faces pointing toward the solution. Only the R enantiomers of the additives are found within the glycine crystals. The mechanism of this type of resolution (hydrophobic orientation effect) is distinct from the kinetic effect on crystal growth described above.

Additives may also inhibit the crystallization of metastable polymorphs when the accessible polymorphic phases are relatively close to one another in energy. *rac*-Histidine·HCl (Fig. 7.3) was crystallized at 25°C, 20°C lower than the transition temperature between His·HCl·2H$_2$O (racemic compound, α-polymorph, stable <45°C) and His·HCl·H$_2$O (conglomerate, β-polymorph, stable >45°C) in the presence of 1–3% of a chiral polymeric inhibitor, for example, poly(p-acrylamido)-(R)-phenylalanine, $-CH_2-CH[CONH-C_6H_4-CH_2-CH(CO_2^-)NH_3^+\}_n$. The ($S$)-histidine·HCl preferentially crystallized (100% ee) in the presence of (S)-His·HCl seeds, though even sand could be used as seed. Thus, the polymer stereospecifically inhibited the nucleation and crystal growth of both the α-(racemic compound) and the β-(R)-polymorphs (Weissbuch, Lahav, et al., 1987).

e. Asymmetric Transformation of Racemates. Total Spontaneous Resolution

The yield of a resolution is necessarily limited to 50%, that is, 100% of one enantiomer of the pair being resolved. If the compound being resolved is to be used in a synthesis, 50% of the resolution substrate may be wasted. There are, however, two ways in which this limitation may be circumvented: (a) both enantiomers may conceivably be used in an enantioconvergent synthesis (Trost, 1982; cf. Chapter 12, Section 4.7), and (b) the undesired enantiomer may conceivably be racemized and subjected again to resolution. If the resolution could be coupled to the racemization step (Sec. 7-8), then a yield exceeding 50% of one enantiomer could be achieved efficiently essentially in one step. Such a process, corresponding to an asymmetric transformation (see below), actually has been described in a number of cases (Arai, 1986).

Many chiral substances are known that are configurationally labile (i.e., their enantiomers interconvert) in solution. Displacement of the equilibrium in solution as a result of (a reversible) interaction with some external chiral entity corresponds to an *asymmetric transformation of the first kind* (Eliel, 1962, p. 40 ff) a process that is more commonly observed in diastereomeric systems (Section 7-3.e). In the more recent literature, the term *enantiomerization*, applied to the interconversion of enantiomeric conformations by torsional motion or by inversion (e.g., of nitrogen compounds) in nonplanar environments (Mislow et al., 1974) has often come to replace the older expression asymmetric transformation of the first kind (see below; Craig and Mellor, 1976). This type of stereoisomerization is manifested principally in NMR spectra where diastereotopic nuclei become enantiotopic (and consequently become isochronous; cf. Sec. 8-4.d) as the temperature is raised (e.g., Johnson, Guenzi, and Mislow, 1981; Casarini, Placucci, et al., 1987). A few examples of photochemical enantiomerizations, either under the influence of circularly polarized light (cpl) or of chiral photosensitizers have been described (Review: Inoue, 1992).

A beautiful example of an enantiomerization, that of a chiral polymer, has recently been described by Khatri, Green, et al. (1992). Polyisocyanates such as poly(n-hexyl isocyanate) $-(N(n-C_6H_{13})-C(=O)-)_n$, are polymers devoid of chiral centers whose chirality is due to formation of helical conformations of the polyisocyanate backbone (Bur and Fetters, 1976). The absence of optical activity in such polymers stems from the equal probability of forming right (P) and left (M) handed helical forms during typical syntheses. The dynamic equilibrium existing in solution between the two enantiomeric conformations of the polymer has been both revealed and perturbed by dissolving the racemic polymer in a chiral (nonracemic) solvent, (R)-2-chlorobutane. Induction of positive CD (see Section 13-4.e), at 250 nm, a spectral region in which the solvent is transparent, indicates that (R)-2-chlorobutane (91% ee) displaces the (1:1)$P \rightleftharpoons M$ equilibrium in the direction of excess P helix (i.e., the polymer undergoes enantiomerization).

From studies of the molecular weight and temperature dependence of the CD magnitude, the free energy bias per residue arising from interaction of the polymer enantiomers with the chiral solvent (an interaction of as yet unknown nature) has been calculated to be $0.04 \, \text{cal} \, \text{mol}^{-1}$ ($0.2 \, \text{J} \, \text{mol}^{-1}$) (this energy being a measure of diastereomer discrimination). The observed CD has been proposed to result from a sort of "cooperation" in which a small perturbation per unit residue favoring one helical sense is multiplied over many backbone residues thus leading to a large amplification (Green et al., 1991).

When configurationally labile racemates are crystallized rapidly, the enantiomeric composition of the liquid-phase equilibrium is reproduced in the solid state: the crystalline mass is racemic. This solid racemate may exist either as a racemic compound or it may be a conglomerate just as is found with configurationally stable chiral compounds. With conglomerate systems, however, *slow* crystallization occurring spontaneously, or induced by means of enantiopure seeds, may result in exclusive deposition of only one of the enantiomers. This phenomenon, that traditionally (but not very appropriately) has been called *asymmetric transformation of the second kind*, has given rise to the term *total spontaneous resolution*. Such asymmetric transformations were first observed in diastereomeric systems and they are more common in such systems (Section 7-3.e).

For systems consisting of racemates, the more descriptive term "crystallization-induced asymmetric disequilibration" has been suggested to describe this process. An even better term, applicable to both enantiomers and to diastereomers (see Section 7-3.d) is *crystallization-induced asymmetric transformation* (Jacques et al., 1981a, Chapter 6). In a system containing enantiomeric molecules, a crystallization-induced asymmetric transformation is a process for converting a racemate into a single (pure) enantiomer of the substrate (Fig. 7.9). In that sense, the process is the converse of racemization. The hallmark of such a transformation is the separation of one enantiomer from a racemate in a yield greater than 50%.

Two early discovered examples of compounds resolved by crystallization-induced asymmetric transformations are shown in Figure 7.10. When *rac-N,N,N*-allylethylmethylanilinium iodide **12** is allowed to crystallize slowly from chloroform it forms enantiopure crystals. A characteristic of total spontaneous resolutions is that the mother liquor remains racemic because the enantiomer that crystallizes is constantly being replenished by a $(+) \rightleftarrows (-)$ equilibration (50:50) in solution (Havinga, 1954).

Crystallization of tri-*o*-thymotide (TOT, Fig. 7.10, **13**) leads to formation of clathrate inclusion complexes by inclusion of solvent molecules. These inclusion complexes are conglomerates (Baker, Gilbert, and Ollis, 1952; Newman and Powell, 1952). Individual crystals of the conglomerate form containing either achiral or chiral guest molecules surrounded by several TOT molecules are homochiral. However, dissolution of these crystals regenerates racemic TOT as a result of interconversion of the TOT enantiomers by bond rotations. The application of TOT to resolutions (when the guest solvent molecules are replaced by chiral molecules) is elaborated in Section 7-3.b.

Figure 7.9. Crystallization-induced asymmetric transformation of a racemate.

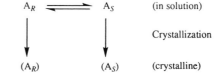

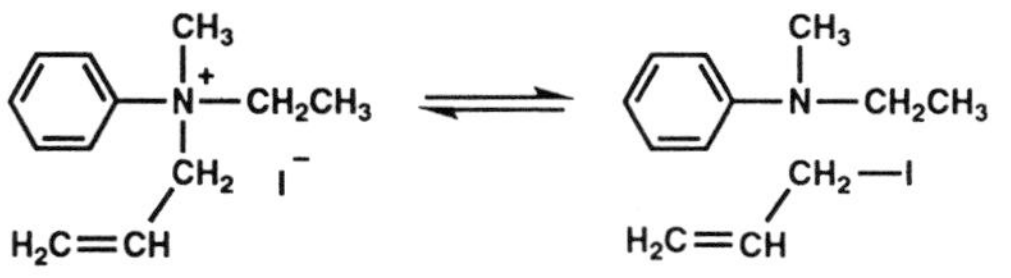

12

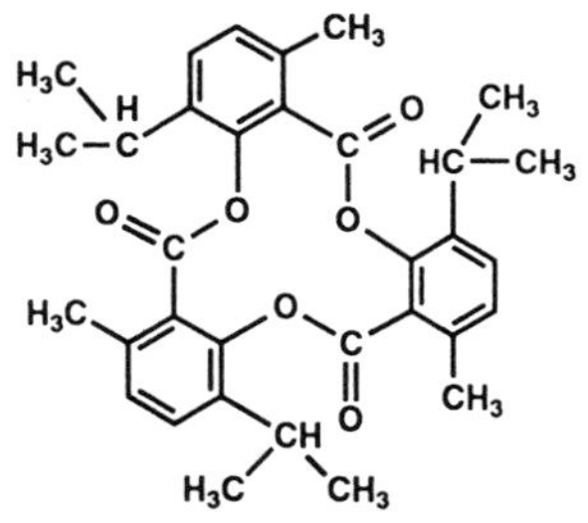

13

Tri-*o*-thymotide (TOT)

Figure 7.10. Compounds that undergo crystallization-induced asymmetric transformations. The enantiomers of **12** interconvert by dissociation into the achiral amine and alkyl halide precursors of the salt; those of TOT (**13**) are interconverted by bond rotations.

The dithia-hexahelicene **14** (Fig. 7.11) undergoes spontaneous resolution on crystallization in benzene. When optically active **14** is dissolved in chloroform it racemizes at room temperature with $t = 230$ min ($t = t_{1/2}$; see Section 7-8); this implies that crystallization of **14** is attended by asymmetric disequilibration (Wynberg and Groen, 1968).

Spontaneous resolution has been observed in *rac*-1,1′-binaphthyl (Fig. 7.11, **15a**) on crystallization either from solution or from the melt. Optical activity as

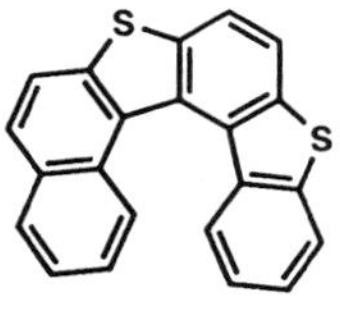

14

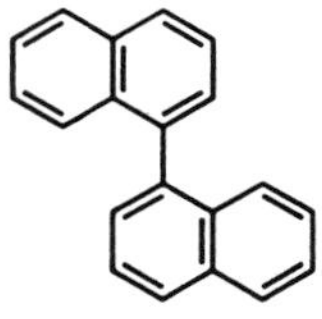

15a

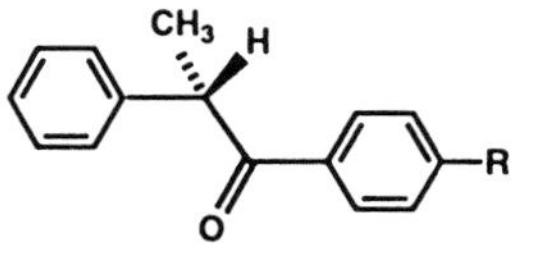

15b R = OCH_3
15c R = SCH_3

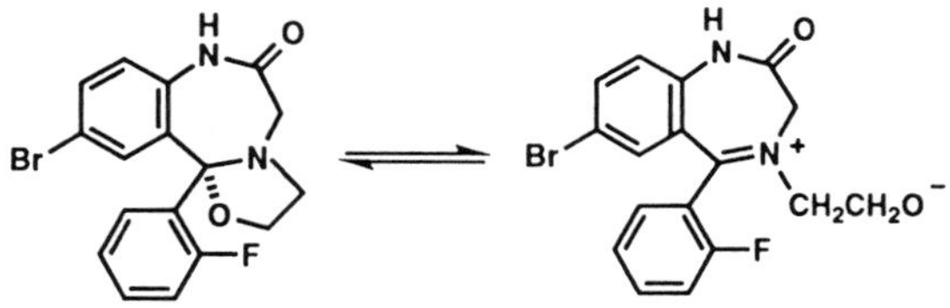

16a 16b

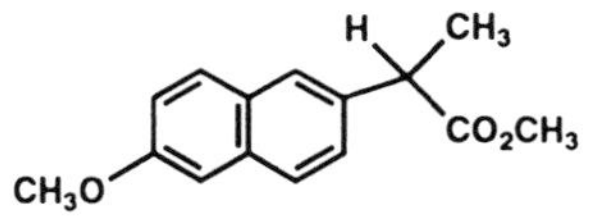

16c

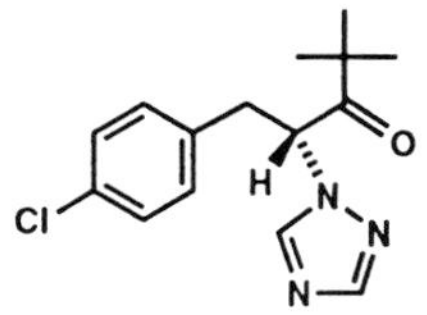

16d

Figure 7.11. Additional examples of compounds that exhibit crystallization-induced asymmetric transformation.

high as $[\alpha]_D \pm 233$ (presumably in benzene) corresponding to 95% ee has been measured in individual samples (Wilson and Pincock, 1975; Kuroda and Mason, 1981). The extent of resolution is very sensitive to the length and temperature of storage of the sample, to the extent of grinding of the crystals, and to inadvertent seeding. These facts have been rationalized by the finding that the racemic compound (mp 145°C) and the conglomerate (mp 158°C) polymorphic forms of **15a** can coexist over a wide range of temperatures. The conglomerate is more stable than the racemic compound at temperatures greater than 76° but a metastable conglomerate can easily be maintained even at room temperature. In solution, optical activity is gradually lost; $t_{1/2}$ for racemization is about 10 h (25°C) in several solvents and in the melt $t_{1/2}$ = about 0.5 s at 150°C with interconversion of the enantiomers taking place by rotation about the 1,1' bond. Spontaneous resolution is thus understood to follow from the ease of racemization and from the presence of the conglomerate form that allows one of the enantiomers to grow at the expense of the racemate. This growth of nonracemic crystals can, in the case of **15a**, also take place in the solid state, albeit more slowly than in solution (Wilson and Pincock, 1975).

The oxazolobenzodiazepinone **16a** (Fig. 7.11) deposits optically active crystals when crystallized slowly from methanol containing acetic acid. Specific rotations, which can exceed $[\alpha]_D^{25}$ 300 (c 0.2% in dioxane), are sometimes (+) and at other times (−), while that of the mother liquor remains $[\alpha]$ 0. When optically active crystals of **16a** are dissolved in methanol, the solutions are racemic. The racemization is believed to take place via an achiral quaternary iminium salt (Fig. 7.11, **16b**; Okada et al., 1983, 1989).

Chiral ketones with a stereogenic atom alpha to the carbonyl group undergo crystallization-induced asymmetric transformations in basic medium provided that the stereocenter bears a hydrogen atom. Examples are *p*-anisyl α-methylbenzyl ketone (Fig. 7.11, **15b**; Chandrasekhar and Ravindranath, 1989), and the ketone precursor **16d** (Fig. 7.11) of the plant growth regulator paclobutrazol (Black et al., 1989).

A sample of (±)-methyl α-(6-methoxy-2-naphthyl)propionate (naproxen methyl ester; Section 7-3.a; Fig. 7.11, **16c**), was fused at 70°C, and sodium methoxide was added. The mixture was supercooled (to 67°C), seeded with (+)-**16c**, then allowed to crystallize; (+)-**16c** was obtained in 87% yield (Arai et al., 1986). The interpretation of this experiment is that ester **16c** is a conglomerate; crystallization of one enantiomer takes place simultaneously with racemization of the other. The result is a crystallization-induced asymmetric transformation from the melt. A similar transformation, with a yield in excess of 90%, takes place in solution on the ethylamine salt of naproxen (Piselli, 1989).

α-Amino-ε-caprolactam (ACL, Fig. 7.12), a precursor of the amino acid lysine, is a configurationally stable compound whose crystallization-induced asymmetric transformation is made possible by formation of a metal coordination complex (e.g., with $NiCl_2$) that is a conglomerate (Kubanek et al., 1974). Preferential crystallization of either enantiomer of the latter with seeding is coupled to racemization of the complex (presumably via a carbanion generated at the α-position of the chelated ACL) in the presence of excess ACL and a catalytic amount of ethoxide ion (Sifniades et al., 1976; Boyle et al., 1979; Sifniades, 1992). What is remarkable about this asymmetric transformation is the fact that

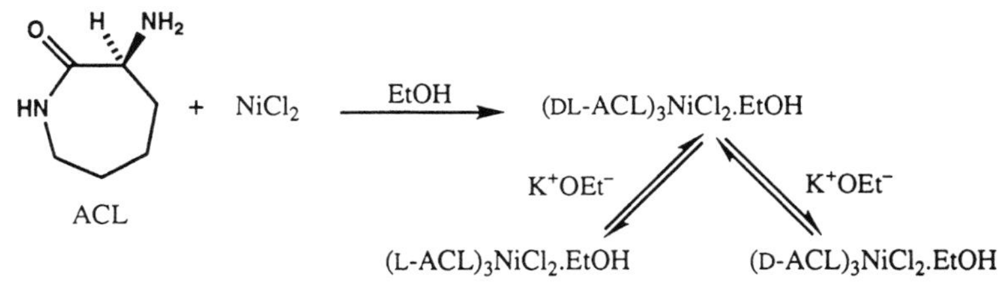

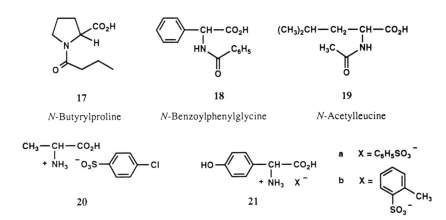

Figure 7.12. Crystallization-induced asymmetric transformation of α-amino-ϵ-caprolactam (ACL) (only the R enantiomer is shown) when complexed with nickel(II) chloride. The dissolved mixture of diastereomers remains racemic,

only one conglomerate salt (Fig. 7.12) crystallizes even though numerous other diastereomers of the complex may exist in solution. Coupling of fast ligand exchange with fast racemization under the conditions of the experiment must account for the asymmetric transformation observed (see also van Mil, Lahav, et al., 1987).

Among amino acid derivatives, examples, particularly from the patent literature, reveal the practical significance of asymmetric transformations in the design of economically viable syntheses of optically active compounds without having to resort to separate racemization and recycling steps (Fig. 7.13): *N*-butyrylproline **17** and *N*-benzoylphenylglycine **18** (Hongo, Yamada, and Chibata, 1981a); *N*-acetylleucine **19** (Hongo, Yamada, and Chibata, 1981b); alanine as the *p*-chlorobenzenesulfonate **20** (Hongo et al., 1983); and *p*-hydroxyphenylglycine as its benzenesulfonate salt **21a** (Chibata et al., 1982) or its *o*-toluenesulfonate salt **21b** (Hongo et al., 1985).

In all of these cases the conglomerate form is stable under the conditions of the experiment and crystallization of one enantiomer from the supersaturated solution is promoted by seeding with the desired enantiomer. The *N*-acylamino acids **17–19** (Fig. 7.13) suffer concomitant racemization either in the molten state or in solution in the presence of a catalytic amount of acetic anhydride. Racemization of **20**, **21a**, and **21b** (Fig. 7.13) is promoted by aldehydes, such as butanal or salicylaldehyde (for a discussion of racemization mechanisms of amino acids, see Section 7-9.g).

17

N-Butyrylproline

18

N-Benzoylphenylglycine

19

N-Acetylleucine

20

21

Figure 7.13. Amino acid derivatives that undergo crystallization-induced asymmetric transformation.

While the enantiomer purity of the recovered amino acid derivative is not necessarily 100%, the substrate recovered after filtration of the optically active product is racemic. Hence, the process is not a simple preferential crystallization; it is attended by an asymmetric transformation.

The enzymatic hydrolysis of racemic amino acid derivatives is a fairly common kinetic resolution method wherein one enantiomer of the racemate is recovered as the unhydrolyzed derivative while the other may be recovered from the hydrolysate as a pure amino acid or as a new amino acid derivative (Section 7-6). In a number of instances, the yield of the latter exceeds 50% signaling concomitant racemization and the occurrence of an asymmetric transformation (Sih and Wu, 1989). Since no crystallization is involved and a chiral catalyst is required to effect the process, we are not dealing here with a crystallization-induced asymmetric transformation. The process has much in common with asymmetric transformations of diastereomers in that diastereomer discrimination is operative during the hydrolysis. Nevertheless, since the substrate (both starting and recovered) is a racemate and the product is a single enantiomer, it is appropriate to describe examples of such processes in this section.

Hydantoins substituted in the 5 position are well-known precursors of amino acids. For example, 5-phenylhydantoin **22** is readily hydrolyzed in the presence of hydropyrimidine hydrase (a hydantoinase) isolated from calf liver. If the reaction medium is slightly alkaline (pH ca. 8), the hydantoins undergo racemization. Under the conditions of the reaction, racemic **22** is completely converted to (R)-carbamoylphenylglycine **23**, a phenylglycine precursor. No (S)-**23** is found in the reaction mixture since the unreacted (S)-**22** undergoes spontaneous racemization during the hydrolysis of the R enantiomer for which the enzyme is specific (Fig. 7.14; Cecere et al., 1975). In the presence of N-carbamoyl-D-amino acid amidohydrolase (or of the organism *Agrobacterium radiobacter* from which the enzyme can be isolated) *rac*-**22** yields 100% D-phenylglycine directly (Olivieri et al., 1979, 1981). Similar enzymatic hydrolysis of 5-(4-hydroxyphenyl)-hydantion has been reported (Anon., 1980; Yokozeki et al., 1987). Analogous processes have been reported requiring the intervention of two enzymes, one of which is a racemase (Fig. 7.15) (see the review by Sih and Wu, 1989).

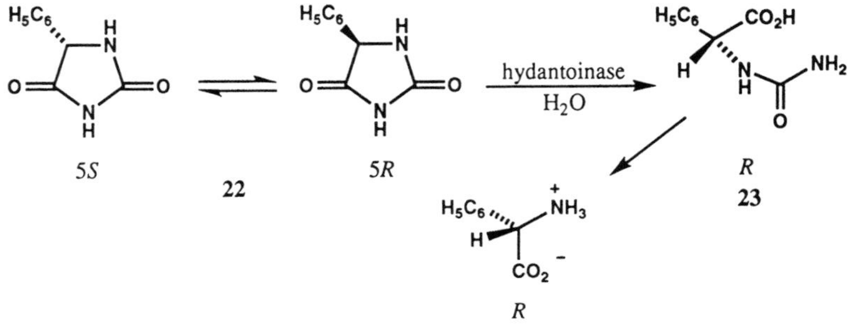

Figure 7.14. Enzyme catalyzed kinetic resolution of *rac*-5-phenylhydantoin **22** with concomitant racemization of (5*S*)-**22**.

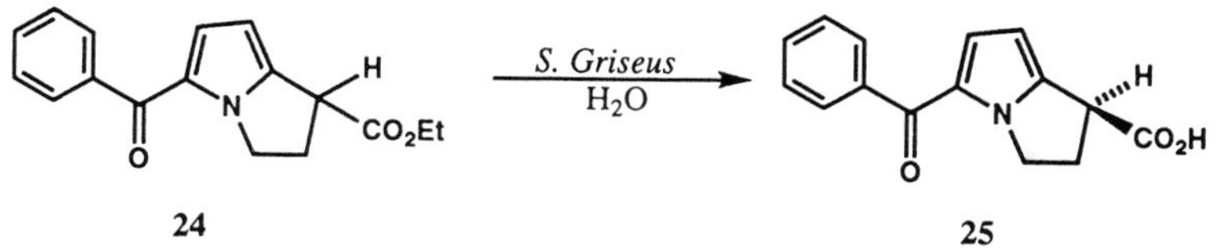

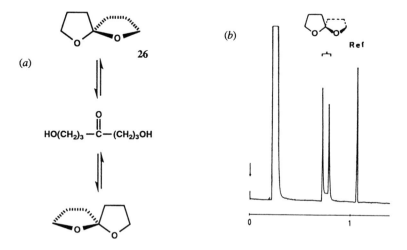

Figure 7.15. Kinetic resolution/asymmetric transformation leading to (S)-$(-)$-ketorolac **25**.

The hydrolysis of a racemic carboxylic ester **24** is catalyzed by the protease of *Streptomyces griseus* at pH 9.7 (24 h at 22°C). Under these conditions, kinetic resolution and racemization take place simultaneously in view of the intervention of a highly stabilized ester enolate. (S)-$(-)$-Ketorolac, **25**, a potent antiinflammatory and analgesic agent, is isolated in 92% yield (ee = 85%) (Fig. 7.15) (Fülling and Sih, 1987).

Asymmetric transformations may also take place in other contexts. The inversion of configuration of the enantiomers of 2,6-dioxaspiro[4.4]nonane (Fig. 7.16a, **26**) has been demonstrated during the chromatographic resolution of the racemate on an enantioselective stationary phase, where it is presumed that the process takes place via a didydroxyketone intermediate (Fig. 7.16a) (Schurig and Bürkle, 1982). The enantiomerization is revealed by a distinctive signature in the chromatogram (Fig. 7.16b). The "plateau" bridging the peaks due to the two enantiomers is caused by molecules that have inverted their configuration *in the column* and travel with velocities approaching those of their enantiomers. The process results in tailing of the leading peak and fronting of the peak due to the second appearing enantiomer (Allenmark, 1991, pp. 84–86). Since the horizontal equilibration of Figure 7.9 now takes place in a nonracemic chiral environment,

Figure 7.16. (*a*) Enantiomerization of spiroketal **26**; (*b*) Peak coalescence of **26** during its GC analysis on nickel(II) bis[[(1R)-3-(heptafluorobutyryl)camphorate] (0.1*M* in squalane) at 70°C [Reprinted with permission from Schurig, V. and Bürkle, W. (1982), *J. Am. Chem. Soc.*, **104**, 7573. Copyright © 1982 American Chemical Society.]

this type of enantiomerization has much in common with asymmetric transformations of diastereomers. The latter are described in Section 7-3.e.

7-3. CHEMICAL SEPARATION OF ENANTIOMERS VIA DIASTEREOMERS

a. Formation and Separation of Diastereomers. Resolving Agents

The largest number of recorded resolutions has been effected by conversion of a racemate to a mixture of diastereomers. In this type of reaction, the substrate to be resolved is treated with one enantiomer of a chiral substance (the resolving agent). The first such resolution, described by Pasteur in 1853, is outlined in Figure 7.17 (Jacques et al., 1981a, pp. 253, 257). Diastereomer pairs prepared in connection with resolutions may be ionic (diastereomeric salts), covalent, charge-transfer complexes, or inclusion compounds. The latter two types of diastereomers are discussed in Section 7-3.c.

The vast majority of resolutions mediated by diastereomers (diastereomeric salt mixtures, in particular) have been based on solubility differences of solids; however, in the contemporary literature, covalent diastereomer separations based on chromatography in all of its variants are used with great frequency. Chromatography has freed resolutions from the constraint of dependency on crystallization as the technique on which diastereomer separation has traditionally depended. As a result, resolutions in general are much more successful at present than they were in the past.

Not infrequently oily covalent diastereomer mixtures eventually crystallize and their resolution may then be performed in the more traditional way by taking advantage of solubility differences. Separation of diastereomeric salt mixtures by chromatography is also now possible (Section 7-3.d). Because of this interplay between ionic and covalent structure and the several ways of separating diastereomer mixtures, we have chosen in this section not to treat resolving agents separately according to whether they form covalent or ionic diastereomer mixtures.

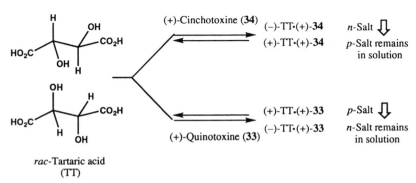

Figure 7.17. First resolution via diastereomers. Tartaric acid resolution with cinchotoxine and quinotoxine (Fig. 7.19) as resolving agents (Pasteur, 1853). The n and p symbols are defined on p. 326.

Details of chromatographic resolutions are examined in Section 7-3.d. The differential stability of diastereomers is another basis for their separation. Transient diastereomeric species are formed in chromatographic columns as flowing racemate samples interact with chiral stationary phases or with achiral stationary phases in the presence of nonracemic chiral mobile phases. This type of resolution is also dealt with in Section 7-3.d. Section 7-3 concludes with an examination of the asymmetric transformation of diastereomers (Section 7-3.e).

In this section, our analysis focuses on resolving agents with emphasis on the recent literature and with examples of their use. The desirable characteristics of a good resolving agent are (Wilen, 1971):

(a) Ready availability
(b) Stability of supply
(c) Stability in use and in storage
(d) Low price or ease of preparation
(e) Ease of recovery and reuse
(f) Low molecular weight
(g) Availability in high enantiomeric purity
(h) Availability of both enantiomers
(i) Low toxicity
(j) Reasonable solubility

α-Methyl-β-phenylethylamine, $C_6H_5CH_2CH(NH_2)CH_3$, illustrates the application of feature (a). This amine is a potentially useful resolving agent (for a recent application to the resolution of gossypol, see Kai, Liang, et al., 1985). However, the amine (amphetamine) is a central nervous system (CNS) active compound, and accordingly it is a controlled substance. Like all such substances (e.g., deoxyephedrine and morphine) it is difficult to obtain. The acquisition of controlled substances for use as resolving agents is so complicated and time consuming (at least in the United States) that their use for this purpose is essentially precluded.

The supply of resolving agents that are derived from natural sources, such as brucine and 10-camphorsulfonic acid, may be shut off by economic or political problems that impede access to the sources (feature b).

Some resolving agents are awkward to use and to store without precaution. Liquid primary amines, such as α-methylbenzylamine and dehydroabietylamine (Fig. 7.19), readily form solid carbamates on exposure to air (Rosan, 1989). It may consequently be desirable to store these amines as salts [feature (c)]; if so, one may profitably choose salts that are conglomerates, since enantiomer purification of such salts would be concomitant with chemical purification (e.g., during recovery). α-Methylbenzylamine hydrogen sulfate (Fig. 7.1) and α-(1-naphthyl)-ethylamine phenylacetate are examples of salts that are conglomerates (Jacques et al., 1981a). All other things being equal, high expense is a negative feature in the choice of a resolving agent, although this feature (d) may be mitigated by the possibility of recovery and reuse (feature e). When preparation of a resolving agent is required, the yield and complexity of the synthesis is likely to be a consideration.

Since resolving agents are purchased by weight but are used on a molar basis, low molecular weight (feature f) is an advantage. This is a significant consideration especially in resolutions carried out on an industrial scale. Unfortunately, many naturally occurring resolving agents, notably alkaloids, have high molecular weights (e.g., brucine, MW 394.4); this is less likely to be the case for synthetic ones (for lists of resolving agents giving molecular weights, see Jacques et al., 1981a, pp. 255–256). Moreover, synthetic resolving agents are usually obtainable in both enantiomeric forms and this feature (h) is advantageous, since it permits the preparation of both enantiomers of a compound by means of mirror-image resolutions (*Marckwald principle*, Marckwald, 1896; type a in Table 7.3). A fair number of such pairs of enantiomers are available commercially (e.g., α-methylbenzylamine, ephedrine, tartaric acid, or 10-camphorsulfonic acid). Some synthetic resolving agents have been designed that explicitly incorporate many of the features listed above (e.g., ten Hoeve and Wynberg, 1985).

Use of synthetic resolving agents requires their prior resolution. This requirement leads us to discuss the possibility of effecting *reciprocal resolutions*: If *rac*-N-benzyloxycarbonylalanine [($\pm$)-Z-Ala] is resolvable with ($-$)-ephedrine [($-$)-Eph], then, as is often (but not invariably) the case, the resolving agent ($\pm$)-Eph will be resolvable with either ($+$)- or ($-$)-Z-Ala (type b in Table 7.3; Overby and Ingersoll, 1960; Jacques et al., 1981a, p. 306).

TABLE 7.3 Types of Diastereomer-mediated Resolutions[a]

Type of Resolution[a]	Resolution Substrate	Resolving Agent	Diastereomeric Products	
			Less Soluble	More Soluble
a. Normal Marckwald	($\pm$)-**Z-Ala** + ($-$)-Eph $\longrightarrow$		($-$)-**Z-Ala**·($-$)-Eph + ($+$)-**Z-Ala**·($-$)-Eph	
	($\pm$)-**Z-Ala** + ($+$)-Eph $\longrightarrow$		($+$)-**Z-Ala**·($+$)-Eph + ($-$)-**Z-Ala**·($+$)-Eph	
b. Normal Reciprocal	($\pm$)-**Z-Ala** + ($-$)-Eph $\longrightarrow$		($-$)-**Z-Ala**·($-$)-Eph + ($+$)-**Z-Ala**·($-$)-Eph	
	($\pm$)-**Eph** + ($+$)-Z-Ala $\longrightarrow$		($+$)-Z-Ala·($+$)-**Eph** + ($+$)-Z-Ala·($-$)-**Eph**	
c. Mutual	($+$)-**Z-Ala**[b] + ($\pm$)-Eph $\longrightarrow$		($+$)-**Z-Ala**·($+$)-**Eph** + ($+$)-Z-Ala·($-$)-**Eph**	
d. Mutual	($\pm$)-**Z-Ala** + ($\pm$)-**Eph** $\longrightarrow$		$\begin{cases}(+)\text{-}\mathbf{Z\text{-}Ala}\cdot(+)\text{-}\mathbf{Eph}^c + (-)\text{-}Z\text{-}Ala\cdot(-)\text{-}Eph^e \\ (-)\text{-}\mathbf{Z\text{-}Ala}\cdot(-)\text{-}\mathbf{Eph}^d + (+)\text{-}Z\text{-}Ala\cdot(+)\text{-}Eph^e\end{cases}$	

[a] Types of resolutions: (a) Normal and Marckwald resolutions. (b) Normal and reciprocal resolutions (Overby and Ingersoll, 1960); (c) Mutual resolution; see text (Ingersoll, 1925); (d) Mutual resolutions (Wong and Wang, 1978). Resolution substrates and products are in boldface.
[b] This resolution was carried out on partially resolved Ala enriched in ($+$)-Z-Ala.
[c] On seeding with ($+$, $+$) salt.
[d] On seeding with ($-$, $-$) salt.
[e] The ($+$)-Z-Ala·($-$)-Eph and ($-$)-Z-Ala·($+$)-Eph diastereomeric salts did not crystallize.

Z–Ala = (structure of N-Benzyloxycarbonylalanine)

N-Benzyloxycarbonylalanine

Eph = (structure of ephedrine)

($1R,2S$)-($-$)-Ephedrine

The reciprocal resolutions shown in Table 7.3 (type b) lead to diastereomeric combinations of salt pairs (the two sets of resolution products). It follows that separability of the diastereomeric salts in one case does not guarantee such separability in the other (Mislow, 1962). However, it does not preclude it either; reciprocal resolutions very often are successful. Moreover, the likelihood of success in reciprocal resolutions has served as a guide in the design of new resolvings agents (including the design of new enantioselective stationary phases for chromatography; Section 7-3.c). Factors leading to the prediction of success in reciprocal resolutions have been evaluated (Fogassy et al., 1981).

Another potentially useful approach to the preparation of synthetic resolving agents is the application of *mutual resolution*. The idea of effecting the mutual resolution of a racemic acid and of a racemic base was first advanced by Ingersoll (1925) in connection with the resolution of phenylglycine with (+)-10-camphorsulfonic acid. The compound (−)-phenylglycine was recovered from the less soluble diastereomeric salt and (+)-phenylglycine (ee = 63%) was recovered from the mother liquor. Reaction of the latter with *rac*-10-camphorsulfonic acid led to formation of a precipitate from which pure (+)-phenylglycine *and* (−)-10-camphorsulfonic were recovered. This process is schematically illustrated in Table 7.3 (type c) for partially resolved (+)-Z-Ala (admixed with *rac*-Z-Ala) recovered from the more soluble product (+)-Z-Ala·(−)-Eph (method a, top line). The less soluble product isolated from reaction with (±)-Eph contains both pure (+)-Z-Ala and resolved "resolving agent" (+)-Eph. Although the method is attractive for the isolation of the substrate enantiomer incorporated in the more soluble diastereomeric product (Table 7.3), we are unaware of any application of this process other than the cases described by Ingersoll (1925).

Although it is not implicit in Table 7.3 (type a), only *one* enantiomer of the substrate is readily obtained in conventional (hence, also in reciprocal) diastereomer-mediated resolutions, for example, (−)-Z-Ala in the resolution of (±)-Z-Ala with (−)-Eph, and (+)-Eph in the reciprocal resolution of (±)-Eph with (+)-Z-Ala (a and b in Table 7.3). In either case, it is only the enantiomer incorporated in the less soluble product that is readily obtained. A change in resolving agent or use of the enantiomeric resolving agent [(+)-Eph as in Table 7.3 (type a), second line (Marckwald principle)] is usually required to obtain the other enantiomer of the substrate to be resolved (see, e.g., Saigo et al., 1986b). On the other hand, crystallization of solutions containing equivalent amounts of *racemic* substrate and *racemic* "resolving agent" may permit the isolation of either enantiomer of the material to be resolved and simultaneously either enantiomer of the "resolving agent" provided that the racemic salt is a conglomerate. Wong and Wang (1978) demonstrated the possibility of effecting such mutual resolutions by alternately seeding racemic solutions containing the four possible salts with crystals of one of the less soluble salts and then with crystals of the enantiomeric salt. In each case, the salt that precipitated had the same composition as that of the seeds; the enantiomers of only one of two possible diastereomeric salts crystallized (d in Table 7.3). As expected, on admixture, the two precipitated enantiomers formed a conglomerate. Hence, mutual resolution, though performed on diastereomeric salts, has the attributes of preferential crystallization. The mutual resolution of (±)-malic acid and of (±)-α-methylbenzylamine has been reported (Shiraiwa et al., 1983) as has that of ($\pm$)-α-phenylglycine and ($\pm$)-10-camphorsulfonic acid (Nohira et al., 1982); for more recent examples, see Nohira et al., 1988b; Shiraiwa et al., 1989b. The incidence of such conglomerate versus addition compound formation in combinations of racemic chiral acids and bases has been studied by Jacques et al. (1981b). The data of the latter predict the possibility of a mutual resolution being effected in the case of malic acid with α-methylbenzylamine for the 1:1 (acid) salt with the *n* salt (see below), being the one isolated as was subsequently observed (with seeding) by Shiraiwa et al. (1983).

> Mutual resolution as defined here is not to be confused with "mutual kinetic resolution," a term originally used by Heathcock et al. (1979, 1981), to describe diastereoselective reactions (e.g., Michael additions) between chiral donors (nucleophiles) and chiral acceptors (electrophiles) that are both racemic (Kagan and Fiaud, 1988, p. 273). In order not to give the impression that a resolution in the traditional sense is involved in such diastereoselective reactions, the expression "mutual kinetic resolution" has been replaced by "mutual kinetic enantioselection" (Oare and Heathcock, 1989, p. 241).

The thermodynamic properties of pairs of diastereomeric salts formed from racemic acid and racemic base, corresponding to "*rac*-resolution substrate" plus "*rac*-resolving agent," [formation of such pairs is a special case of "reciprocal salt-pairs" studied a century ago by Meyerhoffer (see Findlay, 1951, p. 409ff)] have been studied by Leclercq, Jacques, and Cohen-Adad (1982). They observed that two kinds of behavior are possible: (a) formation of a simple eutectic as exemplified by the cases cited above, and by atrolactic acid, $CH_3C(C_6H_5)(OH)CO_2H$, with α-methylbenzylamine, permitting mutual resolution to take place under conditions of kinetic control (seeding) and (b) formation of a double salt (addition compound; phase diagram as in Fig. 6.3*b*) in which mutual resolution is precluded. More specifically, in the former case, one salt pair (the *p* salt-pair in the case of atrolactic acid plus α-methylbenzylamine) is more stable and is the one precipitated on seeding; the less stable *n* salt-pair (that can be independently prepared) is not manifested. Mandelic acid plus deoxyephedrine is an example of the second type of behavior; here, both *p* and *n* salt-pairs are addition compounds (with eutectics having differing composition).

> Diastereomeric salts are designated *p* if the acid and base reaction partners giving rise to the salts have the same sign of rotation (p_+ if both are + and p_- if both are −) and *n* if the rotation signs of the diastereomer constituents are unlike (Jacques et al., 1981a, p. 251; see below).

In the case of resolutions, does it matter whether one uses enantiomerically pure resolving agents? Provided that the resolving agent is nearly, even if not completely, enantiopure, the answer is no.

> In this connection, it is generally assumed that naturally occurring chiral compounds, including resolving agents, are enantiomerically homogeneous. Such an assumption is, in fact, dangerous to make. A notorious case is α-pinene (Fig. 7.18),

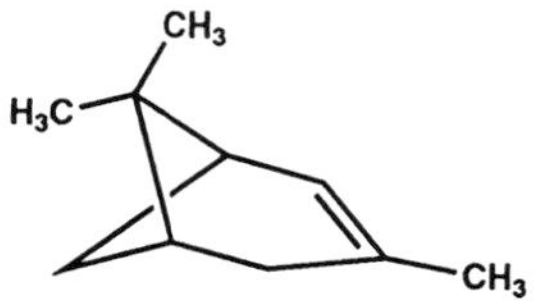

Figure 7.18 α-Pinene and camphor.

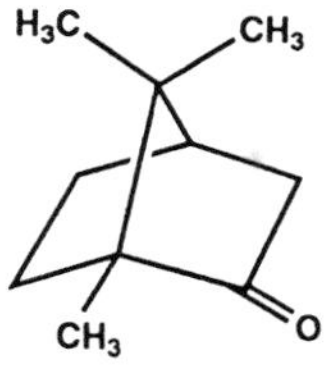

a chiral auxiliary useful in enantioselective hydroborations, whose enantiomeric purity is only about 92% (Scott, 1984, p. 8). To this one must add another caveat: A substance that appears to be a naturally occurring resolving agent or one derived from a naturally occurring precursor may in fact be totally synthetic. At one time, resolving agents such as (+)-10-camphorsulfonic acid (Fig. 7.21), were prepared from naturally occurring and optically active starting materials (in this case, camphor, Fig. 7.18). In the 1970s, natural camphor, from Japan, Taiwan, and Korea, became scarce and the price shot up. It is conjectured that suppliers turned to the cheaper synthetic (racemic) camphor as a source of the sulfonic acid, which then had to be resolved. The optical rotation of 10-camphorsulfonic acid (stated to be 99%) cited in the catalog of a leading supplier is $[\alpha]_D^{20} + 19.9$ (H$_2$O), whereas the maximum rotation appears to be $[\alpha]_D^{20} + 24.0$ (H$_2$O) (Hilditch, 1912). This discrepancy suggests that the optical purity of their 10-camphorsulfonic acid is far from optimal. In contrast, it is obviously not possible to assume anything about the enantiomeric composition of synthetic resolving agents. Their enantiomer composition needs to be independently determined prior to their use in stereochemically relevant studies.

Contrary to what was earlier believed (Eliel, 1962, p. 51), the enantiomer purity of a substance resolved via crystallization of diasteromeric derivatives may *exceed* that of the resolving agent. This result is evident from inspection of the relevant phase diagram provided only that the enantiomer purity of the relevant less soluble crystalline diastereomer is greater than that of the eutectic.

Consider the resolution of compound ($\pm$)-A with resolving agent ($-$)-B that is contaminated by a small amount of (+)-B:

$$(\pm)\text{-A} + \text{ca. } (-)\text{-B} \rightarrow [(+)\text{-A}\cdot(-)\text{-B}] \downarrow + (-)\text{-A}\cdot(-)\text{-B}_{\text{diss}}$$

$$[(-)\text{-A}\cdot(+)\text{-B}] \downarrow$$

$$n \qquad\qquad\qquad\qquad p$$

The enantiomer composition of the less soluble (n) salt (upper line n_+; lower line n_-) is approximately given by that of the resolving agent ($-$)-B. However, the composition of the salt that *crystallizes* from solution depends on the nature of the phase diagram for system n_+/n_- and the position of the eutectic(s) relative to the initial composition of the less soluble resolution product. This explanation ignores the possible coprecipitation of the more soluble p diastereomeric salt and focuses only on the precipitation of the less soluble n_+ salt together with its enantiomeric n_- contaminant. See Figure 6.13 with the enantiomer composition $(+)/(-)$ replaced by n_+ and n_- (or, assuming only a modest enantiomer impurity, the enantiomer branch of the corresponding diagram for a racemic compound between n_+ and n_-). Whereas the composition of the solution with respect to n_+ and n_- is given by point P (reflecting the enantiomer purity of the resolving agent), the composition of the crystals in equilibrium with the solution is N (whose ee exceeds that of P) and, if sufficient solvent is taken, crystallization from solution of composition Q (still along tie line SM) would lead to enantiopure n_+.

Nevertheless, use of resolving agents of low enantiomeric purity is not desirable since the number of recrystallization steps necessary to attain stereochemical homogeneity, hence also the yield, is dependent on the original enantiomer composition of the resolving agent (Jacques et al., 1981a, pp. 305 and 393).

In contrast, when the diastereomeric derivatives are separated chromatographically, on an achiral stationary phase, use of enantiomerically impure resolving agents leads to enantiomerically impure products even when the diastereomers are totally separated (see page 241).

What structural features are desirable for resolving agents? Polyfunctional compounds are generally preferred over monofunctional ones and aromatic compounds over aliphatic ones. Thus, 2-amino-1-butanol and α-methylbenzylamine are more frequently used as resolving agents than is *sec*-butylamine. It would seem that multiple sites of (nonbonding) interaction between the counterions of diastereomeric salts or between different groups in covalent diastereomers enhance solubility differences between diastereomer pairs and thus favor their separation (Wilen, 1971). Multiple interaction sites in covalent diastereomers are also responsible for "anchoring" the latter selectively to chromatographic adsorbents or for "dissolving" them in chromatographic liquid stationary phases in such a manner as to permit differential elution. The constitutional and conformational rigidity of diastereomer molecules may be responsible for selectively "fending off" one of the diastereomers from the adsorbent thereby contributing to separation (Pirkle and Simmons, 1983). Rigidity may also play a role in the separation of diastereomeric salts on the basis of solubility differences though the reasons for this are less evident. The alkaloid brucine has a rigid heptacyclic skeleton (Fig. 7.19) that is remarkably effective in "conferring" low solubility on carboxylate salts; brucine is one of the most widely used basic resolving agents (Jacques et al., 1981a, p. 254); however, see below.

Resolutions mediated by salts are more likely to succeed when the acidic and basic functional groups whose interaction result in salt formation are closer "to those factors that render each asymmetric", for example, the chiral centers (Woodward et al., 1963). Strongly acidic and basic resolving agents are normally preferred over weak ones. The latter may yield salts of low stability relative to dissociation, or salt formation may be altogether precluded. High acid or base strength is one of the advantages of synthetic resolving agents, of amines in particular, which are often stronger than the typical alkaloidal resolving agents; for example, α-methylbenzylamine, pK_b 4.5 versus brucine, pK_b 5.9 in 95% ethanol (10°C) (Leclercq and Jacques, 1979). A number of relatively new synthetic acidic resolving agents designed with this structural feature in mind are described below (e.g., in Fig. 7.21).

The creation of an additional chiral center during a resolution, that is, incidental to the conversion of a racemate to a diastereomer mixture, is normally to be avoided. This situation may arise when a trigonal atom is converted to a tetrahedral one, for example, formation of an amine bisulfite derivative of a ketone. The possible formation of additional stereoisomers in such a process is not always detrimental, however (Wallis, 1983).

The structures of the principal resolving agents of modern usage are illustrated in Figures 7.19–7.25 and 7.27, arranged according to resolution substrate type, together with references (mostly after 1979) that illustrate their use. No attempt has been made to present an exhaustive list of resolving agents from the large literature on resolutions but the compounds whose structures are given are representative (for comments on the magnitude of this literature, see Section 7-5). More extensive lists of resolving agents, as well as their properties and methods of preparation, may be found in the compilations of Newman (1978–1984) and of Wilen (1972). These compilations also contain descriptions of numerous resolutions. For applications, see also the reviews by Boyle (1971), Wilen (1971), Potapov (1979, Chapter 2), and Jacques et al. (1981a).

Many of the applications are to be found in the recent patent literature. This fact is evidence for the significant effort expended in the search for better, cheaper, and more specific resolving agents for chiral organic compounds of commercial interest (Section 7-5). Resolution principles and practice are summarized in Section 7-3.b.

Resolving Agents for Acids and Lactones

These agents are listed in Figures 7.19 and 7.20. Figure 7.19 brings together a wide variety of naturally occurring bases and compounds derived from them, alkaloids (**27–35**), and terpene derivatives in particular (**41**, **42**, **48a**, **48b**), as well as synthetic amines (**38–40b**, **43–47**). In this figure, no distinction is made between resolutions whose diastereomeric products are separated by crystallization or by chromatography. Few carbohydrate derivatives have seen use as resolving agents. A significant exception is the use of inexpensive N-alkyl-D-glucamines, for example, R = CH_3 (**36**), C_8H_{17} (**37**) (Fig. 7.19) in the resolution of naproxen **83** (Fig. 7.23).

> Medicinal agents, agricultural agents (e.g., pesticides), and intermediates that figure in their synthesis, are increasingly being used as resolving agents. This use stems principally from the availability of nonracemic samples of these substances as commercial materials and from the feasibility of reciprocal resolutions; quinine (**31**) is an obvious example. More recently introduced resolving agents that fit this category are ephedrine **38**, compound **40b**, phenylglycine **51a**, and p-hydroxyphenylglycine **51b**. Even codeine, a morphine analogue widely used as an antitussive, has seen use as a resolving agent (Knabe et al., 1987). Among acidic resolving agents, one may cite compounds **67** (Fig. 7.21b) and **74** (Fig. 7.21d); in the list of resolving agents for alcohols, a notable example is naproxen **83** (Fig. 7.23a), a widely used antiinflammatory and analgesic agent.

Amino acids and derived bases figure prominently as resolving agents for carboxylic acids. These acids are exemplified by compounds **50–55** (Fig. 7.20).

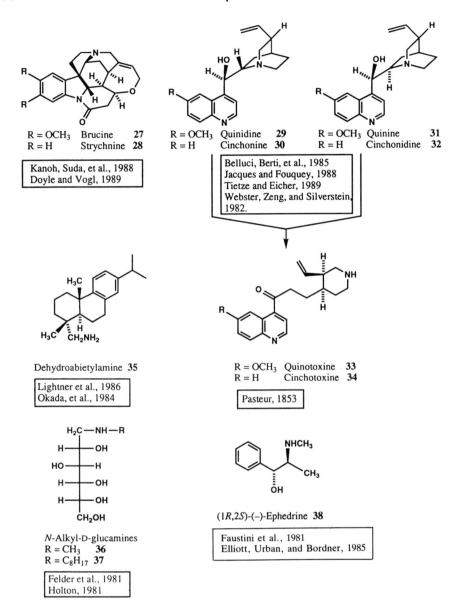

Figure 7.19 (*a*). Resolving agents for acids. In the case of the naturally occurring agents, the absolute configurations shown are in agreement with those given in the *Atlas of Stereochemistry* (Klyne and Buckingham, 1978). In the case of synthetic resolving agents, though both enantiomers may have been used in resolutions, only one is shown.

Basic resolving agents have generally been used to transform racemic covalent substrates (carboxylic, sulfonic, and a variety of phosphorus acids) into diastereomeric salts that are separated by crystallization. Increasingly, resolutions of acids and of lactones have involved formation of covalent diastereomers, notably amides, for example, by compounds **54** and **55** (Fig. 7.20), and esters (for the latter, see menthol **49** in Fig. 7.19*c*) that are separated by either crystallization or

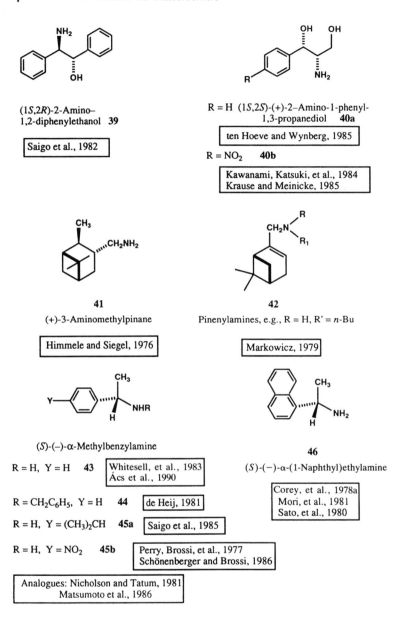

Figure 7.19 (b).

by chromatography (see Jacques et al., 1981a, p. 330). Resolving agents that have been applied to the resolution of lactones via diastereomeric amide formation are exemplified by **41**, **43**, and **48a**. Amides are generally more easily separated chromatographically than are esters. The major limitation in amide-mediated resolutions has been the harsh conditions often required to hydrolyze the separated amide diastereomers. For examples of resolutions in which such hydrolysis may be facilitated (e.g., by neighboring group participation), say by a

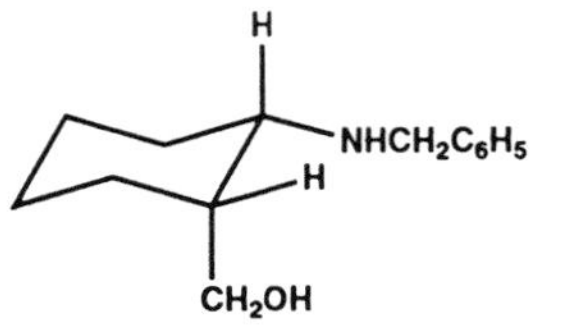

47

cis-N-Benzyl-2-(hydroxymethyl)-
cyclohexylamine

Nohira, 1981
Nishikawa, Nohira, et al., 1979

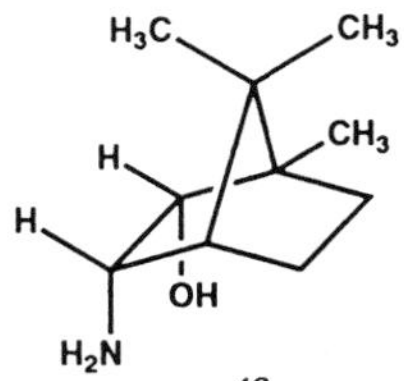

48a

(1R)-3-*endo*-Aminoborneol

Fizet, 1986

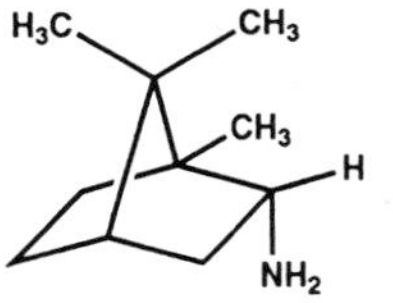

48b

endo-Bornylamine

Paquette and Gardlik, 1980

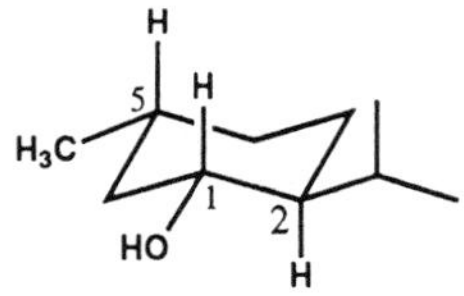

49

(1R,2S,5R)-(–)-Menthol

Yodo, Harada, et al., 1988

Figure 7.19 (*c*).

suitably placed hydroxyl group (e.g., Fig. 7.19*c*, **48a** and Fig. 7.38), when permitted by the structure of either the resolution substrate or the resolving agent, or with the latter expressly modified (labilized) for the purpose following the resolution, see Helmchen et al. (1979); Sonnet (1982); Sonnet et al. (1984); Vercesi and Azzolina (1985); Fizet (1986); Webster, Millar, and Silverstein (1986); Wani, Nicholas, and Wall (1987); and Jacques et al. (1981a, pp. 329–330).

Resolving Agents for Bases

Tartaric acid **56** and its acyl derivatives (e.g., dibenzoyl- and di-*p*-toluyltartaric acids **57**; R = H and CH_3, respectively), continue to be widely used in resolutions (Fig. 7.21*a*). The latter are more acidic than the parent acid and the presence of aroyl groups may enhance diastereomer discrimination by providing additional anchoring points for the resolution substrate. One of the first applications of dibenzoyltartaric acid as a resolving agent (in the 1930s) was the resolution of *rac*-lobeline, which has been used as a deterrent to smoking (Levy, 1977). Resolutions effected with polybasic acids, such as **56** and **57**, are generally carried out with a 1:1 stoichiometry with respect to the resolution substrate; the diastereomeric salts isolated are acid salts (Jacques et al., 1981a, p. 387).

Mandelic acid **60**, *O*-acetylmandelic acid **61**, and *O*-methylmandelic acid **62** (Fig. 7.21*a*), figure prominently as resolving agents among resolutions of amines (1° and 2°) reported in the last decade. To a lesser extent, this generalization applies also to the analogous Mosher's acid (MTPA acid, Fig. 7.21*a*, **63**). The 1,1'-binaphthylphosphoric acid **65** (Fig. 7.21*b*) ($pK_a = 2.50$) has proven to be a

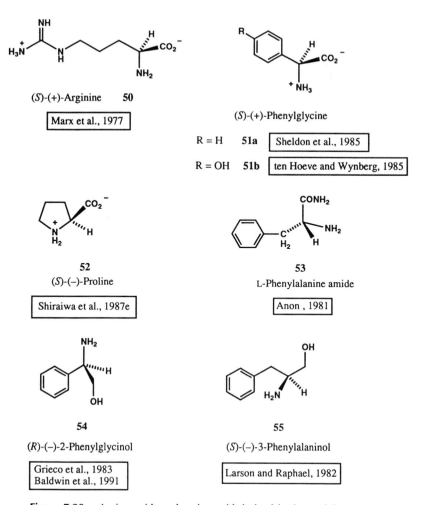

Figure 7.20. Amino acids and amino acid derived basic resolving agents.

very powerful resolving agent, even for the difficult resolution of 3° amines (Imhof, Kyburz, and Daly, 1984). A second synthetic and strongly acidic resolving agent (introduced by ten Hoeve and Wynberg, 1985) that has a wide range of utility is the cyclic phosphoric acid **66** (Fig. 7.21*b*; R = *o*-Cl). Deoxycholic acid **68** has been applied to the resolution of a highly water-soluble amine (Kelly, Wierenga, et al., 1979).

For the most part, the cited acidic resolving agents (and others whose structures are given in Fig. 7.21) are involved in diastereomeric salt-mediated resolutions. In contrast, compounds **62**, **72b** and **73** (Fig. 7.21) have been applied to amide-mediated resolutions; isocyanate **75** (Fig. 7.21*c*) has been applied to the resolution of amines by urea formation (diastereomeric ureas are formed even by chiral hydroxyamines; see Rozwadowska and Brossi, 1989), and menthol-derived **76** (Fig. 7.21*c*) to resolution of amines via carbamate formation. Acid **64b** (Fig. 7.21*a*) has the interesting advantage of not bearing a hydrogen atom at the stereocenter, hence avoiding the danger of racemization during its recovery and recycling particularly if these processes are carried out under basic conditions.

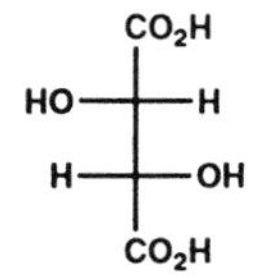

56

(2R,3R)-(+)-Tartaric acid

Corey et al., 1989
Geue, McCarthy, and Sargeson, 1984
Suda et al., 1979.

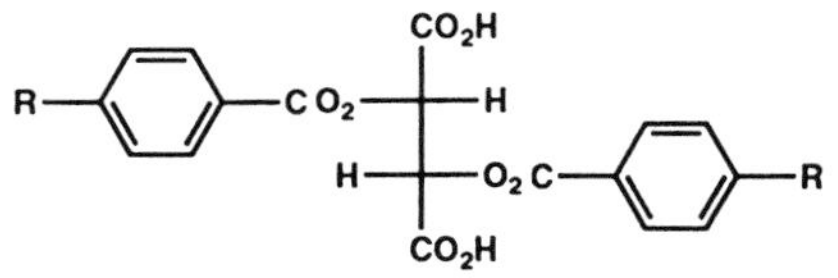

57

R = H   O,O'-Dibenzoyltartaric acid
R = CH₃ O,O'-di-p-Toluoyltartaric acid

Blaschke and Walther, 1987
Dumont, Brossi, and Silverton, 1986
Abu Zarga and Shamma, 1980; Smith et al., 1983

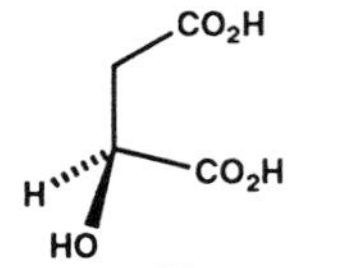

58

(S)-(–)-Malic acid

Anon., 1983

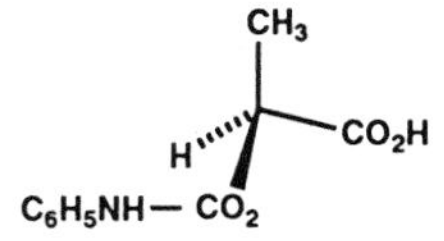

59

(S)-(–)-Carbamalactic acid

Brown, Viot, and Le Floc'h, 1985a

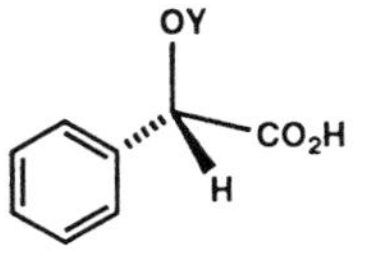

60

Y = H (S)-(+)-Mandelic acid

Fitzi and Seebach, 1988
Ohgi, Kondo, and Goto, 1979
Whitesell et al., 1988

61   Y = COCH₃ O-Acetylmandelic acid

Corey et al., 1987

62   Y = CH₃ O-Methylmandelic acid

Hecker and Heathcock, 1986
Nilsson and Hacksell, 1988

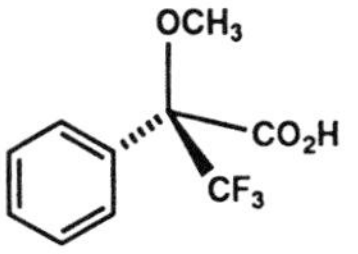

63

α-Methoxy-α-(trifluoromethyl)-
phenylacetic acid [Mosher's acid; MTPA]

Jacob, III, 1982

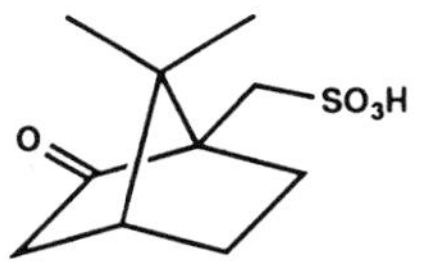

64a

(+)-10-Camphorsulfonic acid

Brown, Berry, and Murdoch, 1985b
Dumont, Brossi, and Silverton, 1986
Smith et al., 1983b

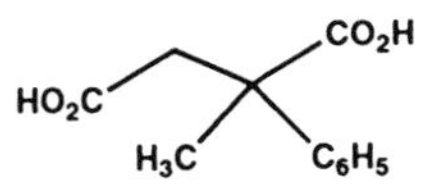

64b

2-Methyl-2-phenyl-
butanedioic acid

Gharpure and Rao, 1988a,b

Figure 7.21 (a).   Resolving agents for bases. In the case of synthetic resolving agents, though both enantiomers may have been used in resolutions, only one is shown.

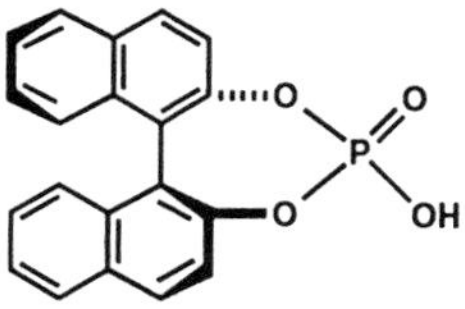

65

(S)-(+)-1,1'-Binaphthyl-2,2'-diyl
hydrogen phosphate

Bey et al., 1979
Imhof, Kyburz, and Daly, 1984
Jacques and Fouquey, 1988
Wilen et al., 1991

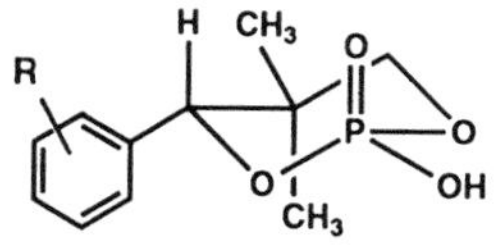

66

4-Aryl-5,5-dimethyl–2–
hydroxy-1,3,2-dioxaphos-
phorinane 2-oxide

ten Hoeve and Wynberg, 1985
Vriesema, Wynberg, et al., 1986

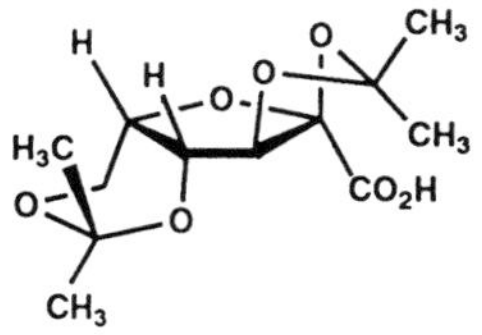

67

(−)-Diisopropylidene-
2-keto-L-gulonic acid

Brossi and Teitel, 1970
Fitzi and Seebach, 1988

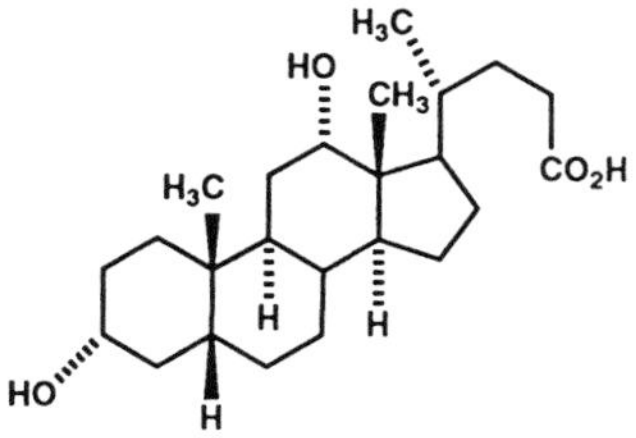

68

(+)-Deoxycholic acid

Kelly, Wierenga, et al., 1979

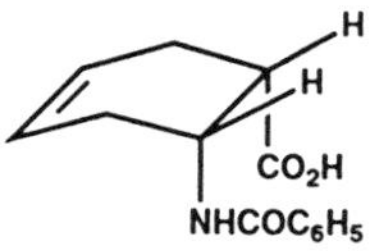

69

3-*endo*-Benzamido-5-norbornene-
2-*endo*-carboxylic acid

Saigo, Nohira, et al., 1981

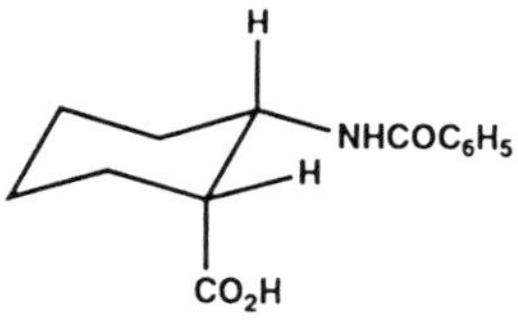

70

cis-2-Benzamidocyclohexane-
carboxylic acid

Nohira et al., 1988b

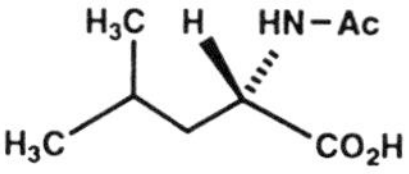

71

(−)-*N*-Acetyl-L-leucine

Gold et al., 1982
Smith et al., 1977; 1983b

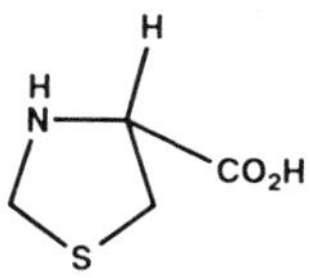

72a

Thiazolidine-4-
carboxylic acid

Tomuro et al., 1987
Shiraiwa et al., 1989a

Figure 7.21 (b).

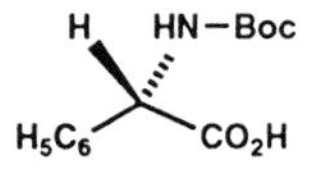

72b

(−)-*N*-Butyloxycarbonyl-L-phenylglycine

Rittle et al., 1987

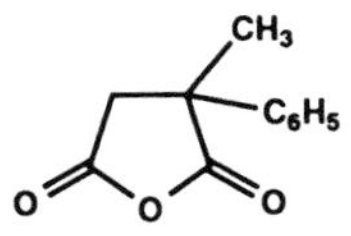

73

(*S*)-(−)-α-Methyl-α-
phenylsuccinic anhydride

Gharpure and Rao, 1988b

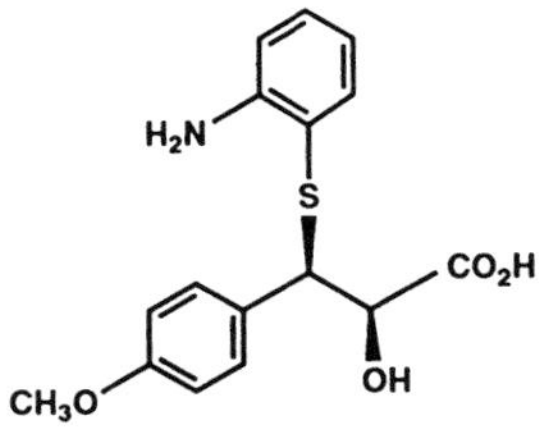

74

(2*S*,3*S*)-(+)-3-[(2–Aminophenyl)thio]-
2-hydroxy-3-(4-methoxyphenyl)
propionic acid

Nohira and Yamamoto, 1986

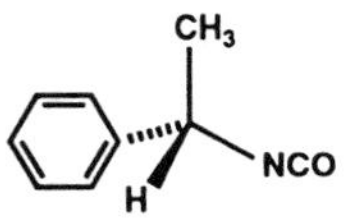

75

α-Methylbenzyl isocyanate

Schönenberger and Brossi, 1986

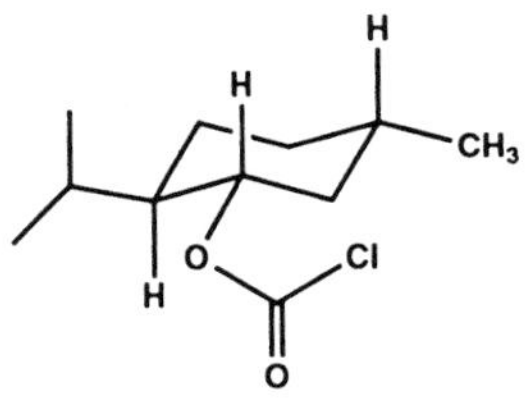

76

Menthyl chlorocarbonate

Pirkle and Hauske, 1977

Figure 7.21 (*c*).

Resolving Agents for Amino Acids

Amino acids are most often resolved in "protected" form, mainly via derivatization of the amino group (Wilen, 1971; Jacques et al., 1981a). The groups that figure prominently in the methodology are *N*-acetyl, *N*-formyl, *N*-benzoyl, *N*-tosyl, *N*-phthalyl, *N*-carbobenzyloxy, and *N*-(*p*-nitrophenyl)sulfenyl. These amino acids may be resolved by diastereomeric salt formation with basic resolving agents, such as brucine **27**, quinine **31**, ephedrine **38** (Fig. 7.19*a*) and pseudo-ephedrine, and chloramphenicol base **40a** (Fig. 7.19*b*) among others and with tyrosine hydrazide (Fig. 7.22, **77**). For a representative procedure using **77**, see Yamada and Okawa (1985). Lactone **79** (a protected sugar alcohol) has been applied to the resolution of the *N*-phthalimido derivatives of γ-aminobutyric acid analogues via ester formation (Allan and Fong, 1986).

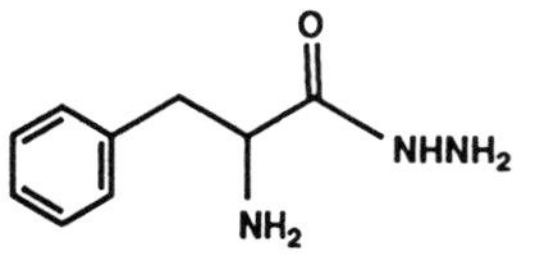

77

Tyrosine hydrazide

Boggs III, Hiskey, et al., 1979
Yamada and Okawa, 1985

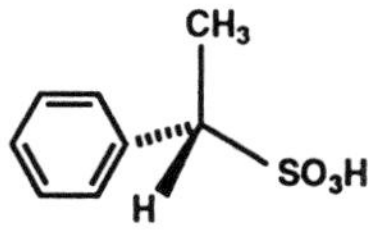

78

α-Phenylethanesulfonic acid

Chibata et al., 1984
Yoshioka et al., 1987

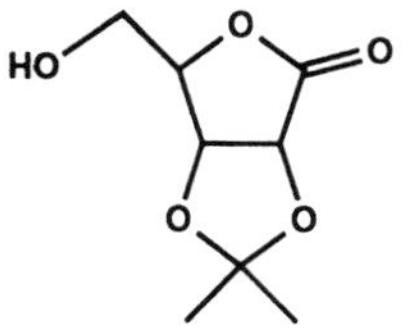

79

Ribonolactone

Allan and Fong, 1986

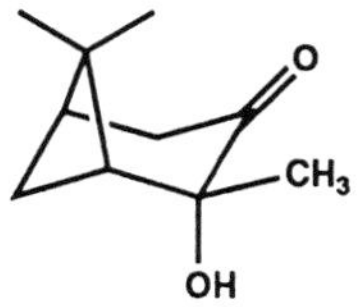

80

2-Hydroxypinane-3-one

Bajgrowicz, Jacquier, et al., 1984

Figure 7.22.   Resolving agents for amino acids.

A widely applicable and attractive method that is apparently specific to amino acids is the simultaneous derivatization and salt formation exemplified by Eq. 7.1 (Gal et al., 1977):

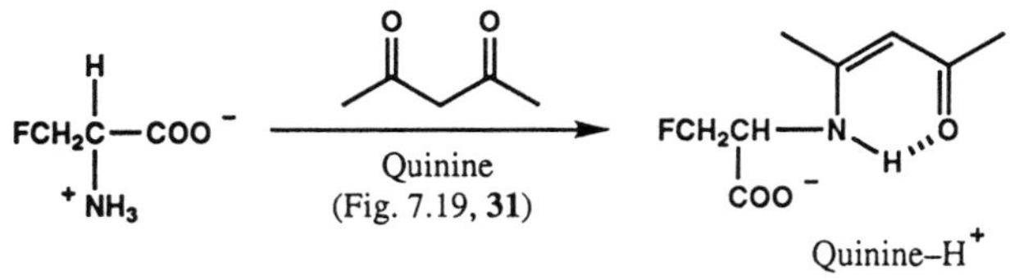

$$\text{FCH}_2\overset{\overset{\displaystyle H}{|}}{\underset{\underset{\displaystyle +NH_3}{|}}{C}}-\text{COO}^- \xrightarrow[\substack{\text{Quinine}\\ \text{(Fig. 7.19, 31)}}]{} \text{FCH}_2\underset{\underset{\displaystyle \text{COO}^-}{|}}{\text{CH}}-N \cdots \quad (7.1)$$

Quinine–H$^+$

Resolution of amino acids with acidic resolving agents is fairly common. Amino acids in unprotected form may be resolved: with mandelic acid **60** via amide formation (Baldwin et al., 1985) or via salt formation (Tashiro and Aoki, 1985; Nohira and Ueda, 1983). Other representative examples are terleucine $(CH_3)_3CCH(NH_3^+)COO^-$ with 10-camphorsulfonic acid **64a** (Fig. 7.21*a*) (Viret, Patzelt, and Collet, 1986); *o*-tyrosine (Garnier-Suillerot, Collet, et al., 1981) and homomethionine (Vriesema, Wynberg, et al., 1986) with 1,1'-binaphthylphosphoric acid **65** (Fig. 7.21*b*); and arginine (Chibata et al., 1983) and *p*-hydroxyphenylglycine (Yoshioka et al., 1987) with acid **78** (Fig. 7.22).

The resolution of amino acid esters as Schiff's bases, with ketone **80** (Fig. 7.22), an α-pinene derivative, has been reported (Bajgrowicz, Jacquier, et al., 1984).

Resolving Agents for Alcohols, Diols, Thiols, Dithiols, and Phenols

Twenty-five years ago most alcohols requiring resolution were transformed to hydrogen phthalate derivatives following which these acidic derivatives were converted to diastereomeric salts by reaction with alkaloidal bases (Reviews: Klyashchitskii and Shvets, 1972; Wilen, 1972; Jacques et al., 1981a; see also Givens et al., 1984; Mori et al., 1981; and for an application to a dihydroxyadipate ester, see Burns, Martin, and Sharpless, 1989). For an application of this approach to the resolution of glycols mediated by 4-boronobenzoic acid (Fig. 7.23*b*, **91**), see Fry and Britton (1973); Brown and Fuller (1986).

Nowadays, the phthalate method is rarely used, rather alcohols are most often resolved by conversion to diastereomeric esters; the latter are separated by crystallization or chromatography. What has changed most significantly is the widespread application of synthetic and semisynthetic acidic resolving agents to this process and the simplicity with which diastereomeric ester mixtures may be separated chromatographically. Resolving agents applied to the resolution of alcohols and diols as diastereomeric esters (in the case of diols often as the mono ester) are mandelic and *O*-acetylmandelic acids **60** and **61**, respectively (Fig.

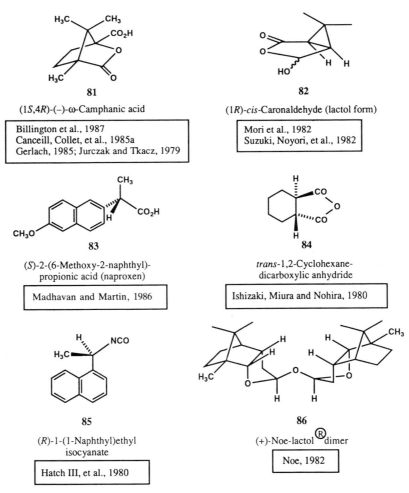

81

(1*S*,4*R*)-(–)-ω-Camphanic acid

Billington et al., 1987
Canceill, Collet, et al., 1985a
Gerlach, 1985; Jurczak and Tkacz, 1979

82

(1*R*)-*cis*-Caronaldehyde (lactol form)

Mori et al., 1982
Suzuki, Noyori, et al., 1982

83

(*S*)-2-(6-Methoxy-2-naphthyl)-propionic acid (naproxen)

Madhavan and Martin, 1986

84

trans-1,2-Cyclohexane-dicarboxylic anhydride

Ishizaki, Miura and Nohira, 1980

85

(*R*)-1-(1-Naphthyl)ethyl isocyanate

Hatch III, et al., 1980

86

(+)-Noe-lactol®dimer

Noe, 1982

Figure 7.23 (*a*). Resolving agents and resolution adjuvants for alcohols, diols, thiols, dithiols, and phenols.

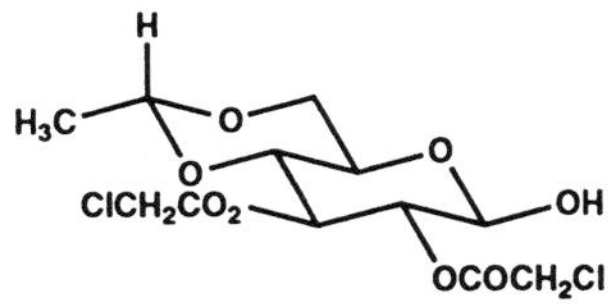

87

Saito, Nishimura, et al., 1987

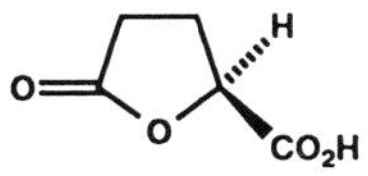

88

(S)-(+)-5-Oxo-2-tetrahydro-
furancarboxylic acid

Doolittle and Heath, 1984

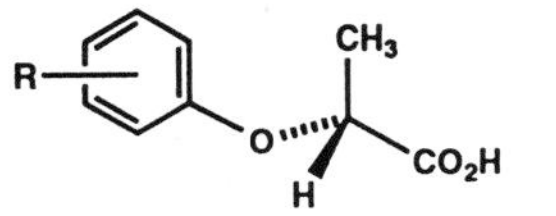

89a

(R)-(+)-2-Phenoxy-
propionic acid (R = H)

Canceill, Collet, et al., 1985a
Canceill and Collet, 1986

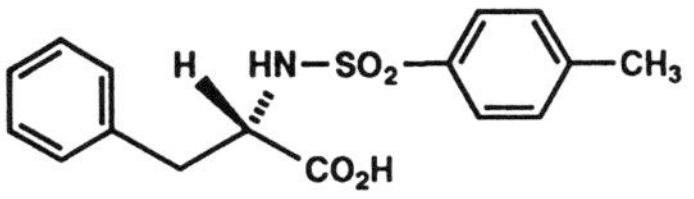

89b

N-(p-Toluenesulfonyl)-(S)-
phenylalanine

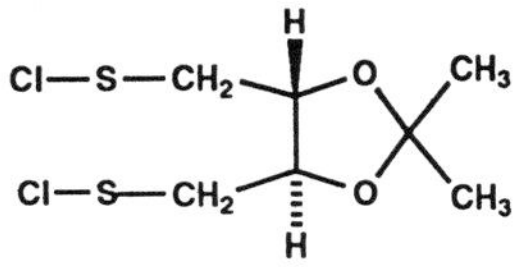

90

Ottenheijm et al., 1977

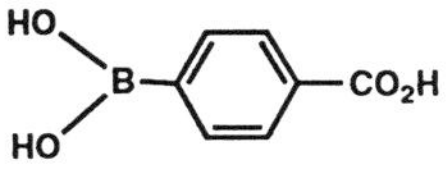

91 4-Boronobenzoic acid

Fry and Britton, 1973
Brown and Fuller, 1986

Figure 7.23 (b).

7.21a, e.g., Whitesell and Reynolds, 1983), O-methylmandelic acid **62** e.g., (Corey et al., 1979; Smith and Konopelski, 1984; Trost et al., 1986), Mosher's acid **63** (Fig. 7.21a; e.g., Koreeda and Yoshihara, 1981), cis-2-benzamidocy-clohexanecarboxylic acid **70** (Fig. 7.21b, e.g., Saigo et al., 1979), ω-camphanic acid **81** (Fig. 7.23a) [also for the resolution of chiral phenols (e.g., Canceill, Collet, et al., 1985a) and myo-inositol derivatives (e.g., Desai, Gigg, et al., 1990; review: Desai, Gigg, et al., 1991); the latter review makes clear the possibility of recovering **81** following resolution, contrary to what is stated in Jacques et al., 1981a, p. 332], cis-caronaldehyde **82**; naproxen **83**; trans-1,2-cyclohexanedicar-boxylic anhydride **84**; and tetrahydro-5-oxofurancarboxylic acid **88**. Alcohols may be resolved also as 10-camphorsulfonate esters (Lauricella et al., 1987). 2-Phenoxypropionic acid (Fig. 7.23b, **89a**), is useful for the resolution of phenols. The N-tosyl derivative of phenylalanine (Fig. 7.23b, **89b**), as the acyl chloride, has been applied to the resolution of 1-octyn-3-ol (Hashimoto, Ikegami, et al., 1991).

Urethane (carbamate) esters have seen repeated application to the resolution of alcohols (even 3° alcohols, e.g., Corey et al., 1978b). Specific resolving agents for this purpose are α-methylbenzyl isocyanate (Fig. 7.21c, **75**; e.g., Donaldson,

Fuchs, et al., 1983 and Whitesell, Minton, and Felman, 1983) and 1-(1-naph-thyl)ethyl isocyanate (Fig. 7.23a, **85**).

Hemiacetal dimer **86** (derived from camphor) has been shown to be a versatile resolving agent for alcohols, thiols, and cyanohydrins (Noe, 1982). Reaction of **86** with racemic substrates gives rise to diastereomeric acetals that are separable by crystallization or chromatography. The derivatives may be cleaved by reaction with methanol.

D-Glucose (and even L-glucose) in a protected form (Fig. 7.23b, **87**), has finally been applied as a resolving agent in the resolution of podophyllotoxin, the aglycone alcohol of an antineoplastic glucoside. The diastereomeric glycosides were separated chromatographically. The analogous 2-amino-2-deoxyglucopyran-oses have also served as resolving agents (Saito, Nishimura, et al., 1987).

Chiral thiols RSH may be resolved in an indirect way. The chloride pre-cursors RCl are converted to isothiouronium salts $RS-C(=NH_2)NH_2^+Cl^-$ by reaction with thiourea. Anion exchange with nonracemic potassium mandelate, for example, converts the enantiomeric isothiouronium salts into diastereomeric salts that may be separated by crystallization. The desired thiol enantiomers may be obtained by hydrolysis in base (Golmohammadi, 1982).

Compound **90** (Fig. 7.23b), an analogue of the key ligand (−)-DIOP (Fig. 12.172, **283**) figuring in optically active transition metal catalysts for enantioselec-tive synthesis (Brunner, 1988) is a resolving agent for the resolution of dithiols (Ottenheijm et al., 1977).

Resolving Agents for Aldehydes and Ketones

Formulas of resolving agents for the direct resolution of aldehydes and ketones are shown in Figure 7.24. Among the more obvious types are those leading to

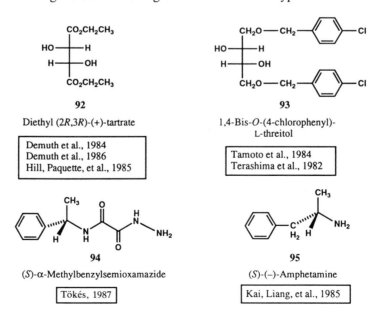

Figure 7.24 (a). Resolving agents for aldehydes and ketones. Resolution adjuvants for carbonyl compounds.

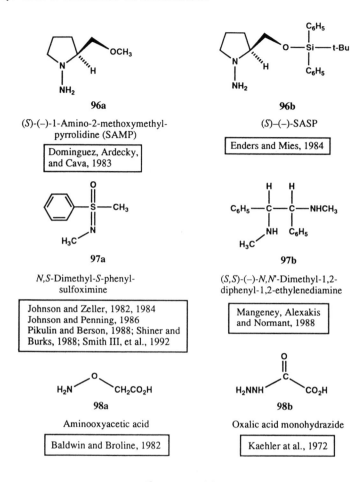

96a

(S)-(–)-1-Amino-2-methoxymethyl-
pyrrolidine (SAMP)

Dominguez, Ardecky,
and Cava, 1983

96b

(S)–(–)-SASP

Enders and Mies, 1984

97a

N,S-Dimethyl-S-phenyl-
sulfoximine

Johnson and Zeller, 1982, 1984
Johnson and Penning, 1986
Pikulin and Berson, 1988; Shiner and
Burks, 1988; Smith III, et al., 1992

97b

(S,S)-(–)-N,N'-Dimethyl-1,2-
diphenyl-1,2-ethylenediamine

Mangeney, Alexakis
and Normant, 1988

98a

Aminooxyacetic acid

Baldwin and Broline, 1982

98b

Oxalic acid monohydrazide

Kaehler at al., 1972

Figure 7.24 (b).

diastereomeric acetals and ketals, namely, tartaric acid derivatives **92** and **93** (Fig. 7.24a) and (R,R)-2,3-butanediol (Fessner and Prinzbach, 1986). α-Methylbenzylsemioxamazide (Fig. 7.24a, **94**) is a carbonyl derivatizing agent that has been applied to the resolution of ketones. Schiff's bases of carbonyl compounds prepared with a variety of amines or amino acid esters (e.g., **95**) are separable by crystallization (Arcamone et al., 1976) or thin-layer or high-performance liquid chromatography (HPLC). The potential antifertility agent gossypol (Fig. 7.25), a dialdehyde, has been resolved in this way (Si et al., 1987; Tyson, 1988).

(R)- and (S)-1-Amino-2-methoxymethylpyrrolidine (RAMP and SAMP reagents, Fig. 7.24b, **96a**), derived from (S)-proline and from (R)-glutamic acid, respectively (Enders et al., 1987), and applied principally as chiral adjuvants in diastereoselective alkylations, aldol condensations, and Michael additions (cf. Chapter 12; Oare and Heathcock, 1989, p. 326), have also seen use as resolving agents via formation of diastereomeric hydrazones. The analogous SASP reagent **96b** (Fig. 7.24b) has been advanced as a specific reagent for the resolution of aldehydes.

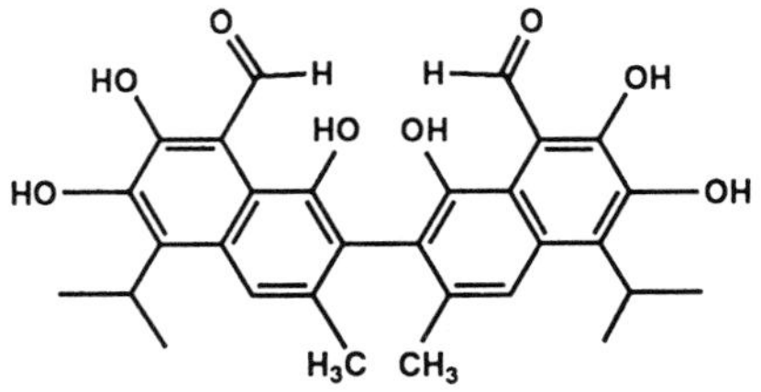

Figure 7.25. Gossypol.

N,S-Dimethyl-*S*-phenylsulfoximine **97a** is a versatile reagent for the resolution of ketones (Johnson, 1985); the example shown in Eq. 7.2 reveals the fact that application of reagent **97a** represents an exception to the usual rule that formation of additional stereocenters during a resolution is to be avoided (a new stereocenter is formed during the lithium salt addition, hence four diastereomeric products are possible). Since only two diastereomers have been observed in some cases, the addition often takes place with high diastereoselectivity. Reciprocal resolution of **97a**, for example, with (−)-menthone has been reported (Johnson and Zeller, 1982).

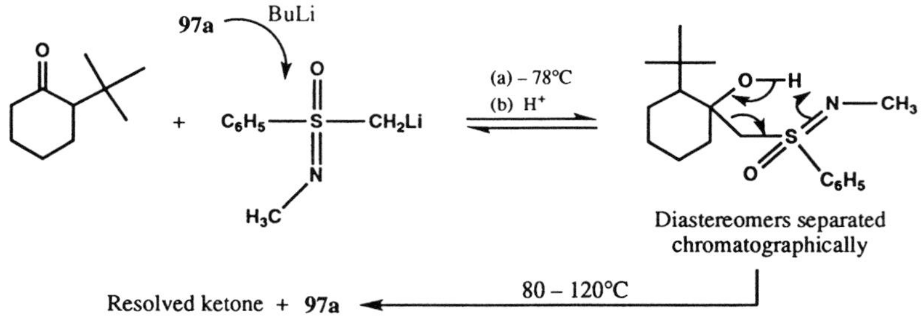

$$(7.2)$$

Additional stereocenters are also generated when aldehydes and ketones are resolved through the separation of diastereomeric oxazolidines, which are prepared by reaction with ephedrine **38** (Fig. 7.19*a*; Eaton and Leipzig, 1978; Just et al., 1982), and through bisulfite addition compounds formed by reaction of chiral ketones with SO_2 and, for example, **43** (Fig. 7.19*b*; Wallis, 1983). An example is given in Eq. 7.3. The diastereomeric salt mixtures may be separated by crystallization as well as, remarkably, by chromatography on silica gel.

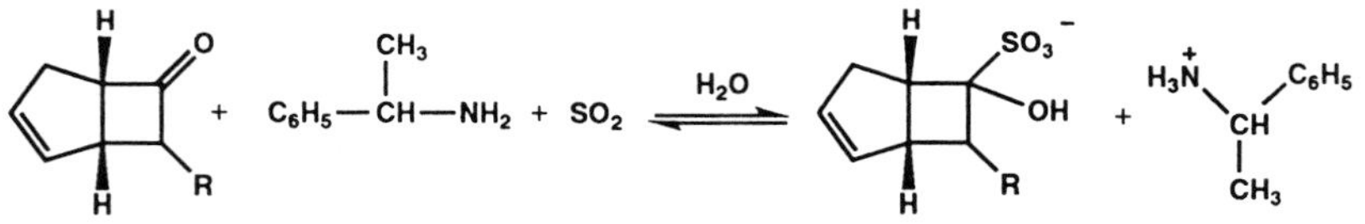

$$(7.3)$$

Compound **97b** (Fig. 7.24*b*) has been applied specifically to the resolution of aldehydes via formation and chromatographic separation of diastereomeric imidazolidines (Eq. 7.4). The absence of additional stereocenters in the imidazolidines illustrates the advantage of building C_2 symmetry into the structure of a resolving agent (Mangeney, Alexakis, and Normant, 1988).

$$R-CH=O \qquad (7.4)$$

Reagent **98a** (Fig. 7.24*b*), aminooxyacetic acid (Anker and Clarke, 1955), illustrates a type of auxiliary bifunctional reagent that still finds use in the resolution of carbonyl compounds. Conversion of a ketone, for example, into an oximino derivative bearing a carboxyl group permits resolution of the derivative by means of basic resolving agents (e.g., Fig. 7.19*b*, **43**) in the example cited in Figure 7.24*b*. Oxalic acid monohydrazide (Fig. 7.24*b*, **98b**), another such auxiliary bifunctional reagent, has been applied to the resolution of 3-methylcyclohexanone with **44** (Fig. 7.19*b*). Finally, the indirect resolution of ketones by prior reduction to the corresponding alcohol, resolution by standard procedures (see above) followed by reoxidation, is still occasionally practiced. For a recent example, see Wood, Smith III, et al. (1989).

Miscellaneous Resolving Agents

Chiral triazolidinediones, such as **99** (Fig. 7.26), derived from *endo*-bornylamine (Fig. 7.19*c*, **48b**; Paquette and Doehner, Jr., 1980) undergo facile Diels–Alder reaction with a variety of dienes. Separation of the diastereomeric urazole adducts followed by Diels–Alder reversal by hydrolysis–oxidation has been made the basis of the resolution of chiral dienes, such as substituted cyclooctatetraenes.

The resolution of an α,β-unsaturated ketone (4-*tert*-butoxycyclopent-2-one) by a novel reversible Michael addition–elimination is noteworthy (Eq. 7.5). The resolving agent, derived from 10-camphorsulfonic acid, is (−)-10-mercaptoisobor-neol (Fig. 7.26, **100**). The success of this resolution hinges on several factors all of which tend to keep the increase in the number of stereocenters generated both in

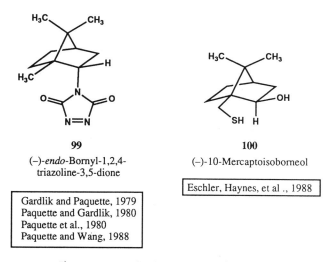

99

(−)-*endo*-Bornyl-1,2,4-
triazoline-3,5-dione

Gardlik and Paquette, 1979
Paquette and Gardlik, 1980
Paquette et al., 1980
Paquette and Wang, 1988

100

(−)-10-Mercaptoisoborneol

Eschler, Haynes, et al., 1988

Figure 7.26. Miscellaneous resolving agents.

the sulfide and the sulfoxide adducts from producing intractable mixtures of diastereomers. The Michael addition occurs exclusively in trans fashion and the oxidation of the sulfide adducts is highly stereoselective (Eschler, Haynes, et al., 1988).

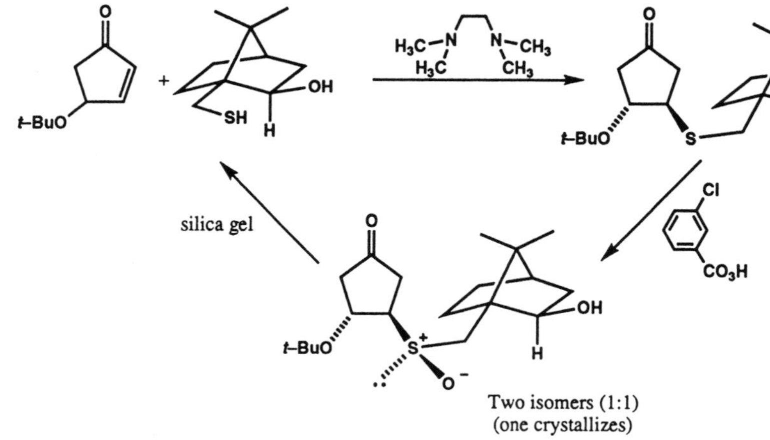

$$(7.5)$$

The optical activation of *rac*-menthone **101** illustrates a type of combined resolution and stereoselective synthesis that is not unique. Conversion of the ketone into a 1,3-dithiolane is followed by enantioselective oxidation under Sharpless conditions with diethyl (R,R)-tartrate (DET) (cf. Section 7-6 and Chapter 12). The mixture of diastereomeric 1,3-dithiolane-S-oxides is separable by low-pressure chromatography. Deoxygenation of the separated oxides and cleavage of the nonracemic dithiolanes affords menthone, ee = 93% (Bortolini, Furia, et al., 1986); (Fig. 7.27). The analogy to the resolution shown in Eq. 7.5 is obvious; however, note that no resolving agent is used in the process illustrated by Figure 7.27.

b. Resolution Principles and Practice

It is fair to say that few chiral compounds are totally resistant to resolution. Yet guidelines for successfully choosing resolving agents and for the selection of the most appropriate technique are mostly still qualitative in nature. In this section,

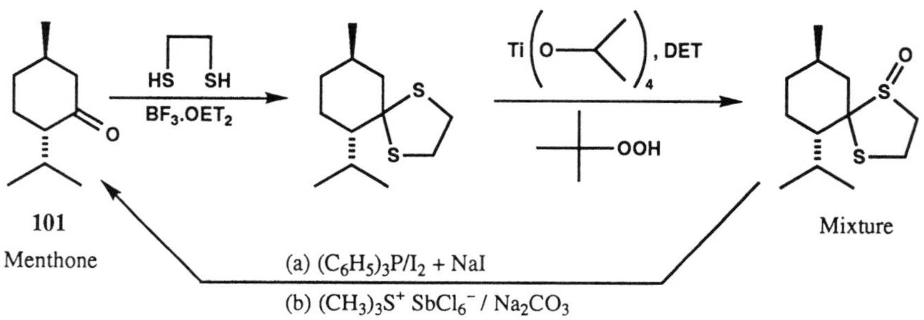

Figure 7.27. Optical activation of menthone.

we identify efforts that have been made to explain why one resolving agent may be better than another; features of resolutions that are known to impede success; and some studies aimed at understanding why a specific resolving agent forms a more stable (or less soluble) diastereomeric salt with one enantiomer of the resolution substrate than with the other. We also describe some recent studies designed to show how diastereomeric-salt mediated resolutions may be optimized with respect to yield and the enantiomer purity of the desired enantiomer.

It has long been recognized, in a qualitative way, that multiple interactions between the resolution substrate and the resolving agent are beneficial, if not essential, to efficient resolutions (Wilen, 1971; see above). This application of the three-point interaction model (Section 6-4.q; Fig. 8.8) may be responsible, in part, for the effectiveness of resolving agents incorporating aromatic rings and additional functional groups beyond the acid or basic functionality, as in O,O'-dibenzoyltartaric acid **57**, and in mandelic acid **60** (Fig. 7.21a). The model also has served implicitly as a guide in the design of new resolving agents.

A statistical study of the resolution of phenylglycine (Fig. 7.20, **51a**) derivatives with tartaric acid has shown that the primary influence on resolution results (yield and enantiomer composition of the resolution product) is the structure of the racemate (resolution substrate), notably the nature of the substituent on the benzene ring (Fogassy, Töke, et al., 1980). This type of study is suggestive and, if refined, may be of real use in the future.

Resolutions that depend on the formation and chromatographic separation of covalent diastereomers depend to a large extent on the nature and degree of interaction between stationary phases and mobile phase constituents. Some of these factors are described in Section 7-3.d. It must be emphasized that in the formation of these covalent diastereomers either an excess of resolving agent or a very long reaction time is required, or else reaction will be incomplete and a kinetic resolution may ensue as evidenced by the formation of diastereomer mixtures in proportions other than 1:1; for an example, see Smith and Konopelski (1984).

Resolutions mediated by diastereomeric salts depend principally on solubility differences. This fact is very clearly illustrated by a study of the reaction between mandelic acid (Fig. 7.21a, **60**) and α-methylbenzylamine (Fig. 7.19b, **43**). When carried out in aqueous solution, the first precipitate is practically pure n salt (Section 7-3.a; Ingersoll, Babcock, and Burns, 1933). This fact is consistent with the later finding that the p salt formed between **60** and **43** is much more soluble in water than the n salt, at 10 and 30°C (Leclercq and Jacques, 1975). In contrast, when the same two reaction partners are allowed to react in the absence of solvent (crystalline **60** with **43** in the vapor phase), the p salt is found to form much more rapidly than the n salt, by a factor of the order of 10^2 (Patil, Paul, and Curtin, et al., 1987). The latter study has more to do with the stereoselectivity expressed by molecules of amine **43** as they attempt to penetrate the crystal lattice of **60** (i.e., kinetic factors) than with the intrinsic relative energies of the n and p diastereomers. Kinetic factors also intervene when, occasionally, the "wrong" (i.e., the more soluble) diastereomer precipitates first presumably as a result of adventitious seeding.

Quantitative studies have aimed at explaining the relative diastereomer discrimination between the above reaction partners: What is the origin of the large melting point difference and of the large $\Delta\Delta H_{\text{fus}}$ between the salts in the above example, and why is the n salt less soluble and more stable than the p salt in the same example? Such studies are few in number; the study by Zingg, Arnett, et al. (1988) is the most comprehensive to date. The most intriguing studies are those that have probed the crystal structures of the salts by X-ray crystallography. These studies revealed extensive hydrogen-bonding networks, not just between ion pairs in a unit cell, but also in columns exhibiting helical networks of hydrogen bonds involving many ions (Jacques et al., 1981a, p. 283). The solid state conformations of the salt pairs are controlled by hydrogen bonding and to a lesser extent by van der Waals attraction between the benzene rings of the cation and anion. Yet the authors of the above-mentioned comprehensive study conclude with the statement "We are unable to provide a complete structural interpretation of the ion-pairing behavior, even for this carefully chosen system" (Zingg, Arnett, et al., 1988).

The diastereomeric α-methylbenzylamine salts of *cis*-2-hydroxy-4-cyclopentenylacetic acid exhibit significant differences in the respective crystals: The two components of the $(1R,2S)$-$(+)$ acid:(R)-$(+)$ base pair (p) form a well-ordered crystal by assuming a rodlike (linear) orientation via two-dimensional (2D) hydrogen bonding between the cation and the anion. In contrast, the $(1R,2S)$ acid: (S) base (n) diastereomer forms a distinctly more globular aggregate and is disordered in the crystal (the presence of two anion conformers in a 59:41 ratio is evident). The large difference in solubility between the p and n diastereomers (11.4 and 48.3 g cm^{-3} in H_2O at 298 K, respectively) and the difference in dissociation constants (0.021 and 0.012 mol dm^{-3} in H_2O at 298 K, respectively) are accounted for by the very different geometries of the salts in the lattice. Moreover, in order to solvate the p salt, the salt bridges must be broken, that is, dissociation must occur simultaneously with dissolution (Czugler, Ács, et al., 1989). In another X-ray crystallographic study, the benzene rings of the cation and anion counterions in the less soluble diastereomeric p salt were shown not to interact (ten Hoeve and Wynberg, 1985).

The resolving ability of a resolving agent may be measured in terms of a parameter **S** (which ranges from 0 for no resolution to 1 when 100% of one enantiomer having 100% ee is obtained):

$$\mathbf{S} = k \times \mathrm{p} = \frac{k_p - k_n}{0.5 C_0} \tag{7.6}$$

where k = the chemical (diastereomer) yield (with $k = 2$ for a 100% yield; however, in principle, **S** cannot exceed 1 since the maximum yield in a resolution, absent an asymmetric transformation, does not exceed 50% of enantiopure material) and p = the enantiomer purity ($\mathrm{p} = 1$ for 100% ee) of the recovered resolution substrate (Fogassy et al., 1980). As shown in Eq. 7.6, **S** is related to the solubilities k_p and k_n of the p and n salts, respectively, as well as to the initial concentration of the resolution substrate C_0. Equation 7.6 permits one to calculate the maximum yield in a resolution from the measured solubilities (van der Haest, Wynberg, et al., 1990). Figure 7.28a demonstrates that the maximum value of **S** is attained when $0.5 C_0 = k_p$ (saturation of the p salt) in the case where the p salt is the more soluble diastereomer.

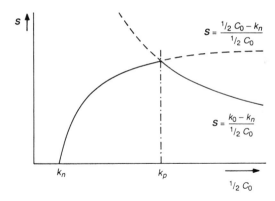

Figure 7.28 (a). Idealized plot of resolving ability **S** as a function of the initial concentration C_0 of product when the p-salt is more soluble than the n-salt (for definitions of p and n, see page 326) [Adapted from van der Haest, A.D., Wynberg, H., Lausen, F.J.J. and Bruggink, A. (1990), *Recl. Trav. Chim. Pays-Bas*, **109**, 523 with permission of the Royal Netherlands Chemical Society.]

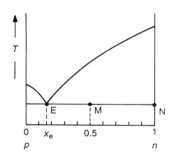

Figure 7.28 (b). Typical phase diagram for the melting of a mixture of two diastereomers (E = eutectic).

The resolving abilities **S** of a series of chiral phosphoric acids (Fig. 7.21b, **66**, and analogues) on *rac*-ephedrine were correlated with differences in enthalpies of fusion between the diastereomer pairs (van der Haest, Wynberg, et al., 1990; van der Haest, 1992). The greater the difference, the better the resolving ability. These authors observed that, in the resolution of ephedrine with acid **66** and five of its analogues, **S** is almost independent of the alcohol solvent used; the resolving ability is, however, highly dependent on the difference in lattice energy of the crystalline diastereomeric salts.

Binary mixtures of diastereomers obtained in a resolution often form simple eutectics, that is, their melting point behavior is summarized by unsymmetrical phase diagrams, such as that shown in Figure 7.28b (Jacques et al., 1981a, p. 289). Since the maximum yield of a pure diastereomer (e.g., n) is given by EM/EN, the "best" systems are those in which the eutectic lies far from composition M (the 50:50 mixture obtained in a diastereomer-mediated resolution; Collet, 1989). The efficiency of a diastereomer-mediated resolution, as reflected in parameter **S**, can be calculated from the eutectic composition x_e of the binary phase diagram (Fig. 7.28b). If crystallization (under equilibrium conditions) is stopped when the mother liquor reaches the eutectic composition, then pure diastereomer is obtained (p = 1, Eq. 7.6) and

$$S = \frac{1 - 2x_e}{1 - x_e} \qquad (7.6a)$$

In the case of seven pairs of diastereomeric salts, it has been shown that the eutectic composition, hence also the shape of the binary phase diagram, can be calculated from the differential scanning calorimetry (DSC) melting trace of a single mixture of the diastereomers, namely, the 1:1 mixture of the resolution substrate (racemate) with the putative resolving agent (Kozma, Pokol, and Ács, 1992). The foregoing analysis is predicated on the assumption that the system behaves ideally or nearly so, in which case the composition of the eutectic is independent of the nature of the solvent used (for an example illustrating such behavior see above; van der Haest, 1992, Chapter 3). Comparison of calculated **S** values obtained from 1:1 mixtures of a resolution substrate with several test resolving agents facilitates selection of the preferred resolving agent in the set.

In an analysis of hydrogen tartrate salts and complexes found in the Cambridge Crystallographic Database, at least one helical 2_1 axis (analogous to those mentioned in the case of the mandelate salts above) was found to be present in all structures examined as a result of head-to-tail hydrogen bonding (Fogassy et al., 1986). Superimposed on this primary feature of the solid state structure are weaker bonds of the van der Waals type that play a role in binding the helical networks to one another. It has been proposed—and some evidence has been presented in support of the hypothesis—that these weaker (second-order) interactions are responsible also for the diastereomer discrimination observed in resolutions (Fogassy et al., 1985; Fogassy et al., 1986). In the case of α-methylbenzylamine hydratropate, analysis of the crystal structures of the *p* and *n* salts (Brianso, 1976) reveals that conformational differences involving the benzene ring side chains prevents the cocrystallization (formation of solid solutions) of the diastereomeric salts and in that sense makes possible their resolution by crystallization (Jacques et al., 1981a, p. 284).

Comparison of the crystal structures of the *N*-benzoylalanine (Bz-Ala) salts of brucine **27** and of strychnine **28** (Fig. 7.19*a*) revealed the source of the different abilities of these very similar basic resolving agents to resolve acids. Brucine, which is much more widely used as a resolving agent than is strychnine (Wilen, 1971), forms a corrugated monolayer sheet structure (evidently the result of packing of the wedge-shaped molecular structure induced by the methoxy groups) leaving substantial room for the D-Bz-Ala anions (see Sec. 7-3.c). The channels between the monolayer sheets are variable in size thereby having the ability of accommodating a great variety of acids while retaining the corrugated sheet packing arrangement. In contrast, strychnine crystals are composed of bilayers of strychnine molecules in which there is much less room for the L-Bz-Ala (and presumably other) anions. The preference for the L-Bz-Ala anion enantiomer in strychnine versus the D-Bz-Ala anion in brucine presumably is a consequence of specific shape differences of the cavities (Gould and Walkinshaw, 1984).

Empirical formulas based on ab initio self-consistent field (SCF) calculations have provided the binding energies of cations derived from α-methylbenzylamine

(Fig. 7.19b, **43**) and amine **102** toward Lasalocid A, an ionophore previously shown to resolve amines (Fig. 7.29). Preferential "complexation," that is, diastereomer salt formation of (R)- versus (S)-**43**, and of (S)- versus (R)-**102** (Fig. 7.29), the latter preference being influenced by conformational factors, toward Lasalocid were of the order of less than 2.5 kcal mol^{-1} (Gresh, 1987).

For resolutions that depend on solubility differences between diastereomers, there are two essential requirements for success (Leclercq and Jacques, 1975): (a) at least one of the pair of diastereomers formed must be crystalline, (b) the solubility difference between the two must be significant. The solvent plays a significant role in these requirements though often it does so in unpredictable ways (Jacques et al., 1981a, pp. 292, 383; Wilen, Collet, and Jacques, 1977).

> The resolution of 2,2'-dimethyl-6,6'-biphenylcarboxylic acid with brucine **27** (Fig. 7.19a) as resolving agent (in an unspecified solvent) was previously reported to fail (Bell, 1934); repetition of the resolution in methanol–acetone (7:3 v/v) provided the (S)-(+) isomer with about 100% ee after one crystallization; crystallization in acetone of the salt recovered from the mother liquor afforded the (R)-(−)-isomer (ca. 99% ee) (Kanoh et al., 1987). For an illustration of the wide difference in solubility that occasionally permits diastereomer pairs to be separated simply by washing with warm solvent (without recrystallization) or by simple trituration, see Fizet (1986).

Two potential problems may forestall success: (c) the two diastereomers may interact to form a double salt or addition compound. This result leads both diastereomers to precipitate together. Recovery of the resolution substrate yields the racemate, and (d) the two diastereomers may be partially miscible in one another in the solid state forming solid solutions. The inability of one of the diastereomeric counterions to fit in the crystal lattice of the other diastereomer tends to minimize this possibility (see above and Jacques et al., 1981a, p. 299).

We shall see later (Section 7-4) how one may circumvent these difficulties whose unforseen incursion is largely responsible for the unpredictability of success in resolutions. In any event, none of the above limitations need impede a resolution if separation is effected by chromatography. We are obliged to point out, however, that separation of diastereomers by chromatography, though very common, is not universal and some resolutions are "rescued" when one of the diastereomers in a chromatographically unseparable pair unpredictably crystallizes (see, e.g., Jonas and Wurziger, 1987).

On a more practical level, in addition to choice of resolving agent (see above), the outcome of a diastereomeric salt-mediated resolution, that is, the

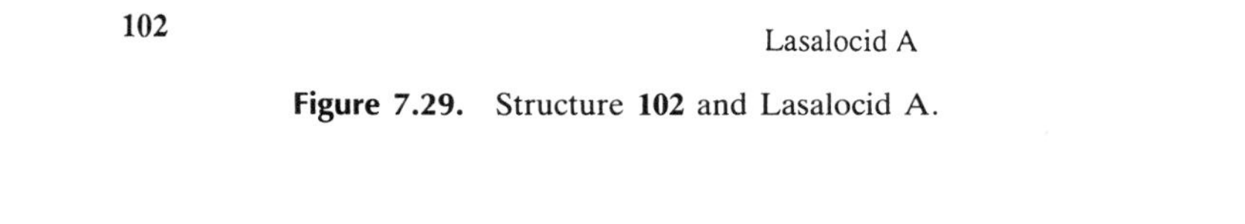

Figure 7.29. Structure **102** and Lasalocid A.

yield and the enantiomeric enrichment of the initial precipitate, are significantly affected by variables, such as the nature and volume of solvent, stoichiometry, temperature (e.g., Shiraiwa et al., 1987e), and pH. One of the reasons for the sometimes dramatic effect of the nature of the solvent on a resolution is that the solvent can solvate the two *crystalline* diastereomeric species differently. The consequence is that the relative solubilities of the two species is changed. Examples of the effect of solvent, and of the effect of water in particular, on the relative solubilities of *p* and *n* salts are given in Jacques et al. (1981a, pp. 293, 383) and Wilen et al. (1977). The crystal structure of the solvated salt can sometimes provide insight into the otherwise mysterious role that water plays in facilitating resolutions (Gould and Walkinshaw, 1984).

The increasingly common use of nonstoichiometric amounts of resolving agent (e.g., 1 : 0.5; resolution substrate/resolving agent) is principally a way of manipulating the pH of a system of dissociable diastereomeric salts (Leclercq and Jacques, 1979; Fogassy et al., 1981). For some examples from the recent literature, see Arnold et al. (1983); Schwab and Lin (1985); Snatzke and Meese (1987); and Whitesell et al. (1983). The main advantages of such manipulations are the economy in amount of resolving agent and solvent required and in the yield of resolved product obtained (Jacques et al., 1981a, p. 307). When the acid–base ratio in a "nonstoichiometric" resolution is restored to 1 : 1 by addition of an achiral acid or base (Pope and Peachey, 1899), additional salts are evidently present in the reaction mixture. This situation has been analyzed in detail in the case of the camphorate salts of fenfluramine (Fig. 7.2, **6**) and a methodology described that permits the yield of desired diastereomer to be optimized as a function of the substrate/resolving agent ratio (Mofaddel and Bouaziz, 1991).

The pH dependence on the yield and on the enantiomer purity of the diastereomeric salts formed in the resolution of *cis*-permethrinic acid (Fig. 7.30), an intermediate in the synthesis of the insecticide permethrin, with (*S*)-(+)-2-(*N*-benzylamino)-1-butanol, has been investigated (Fogassy et al., 1988). Observed yields, enantiomeric purities, and optical yields could be correlated with thermodynamic parameters (solubility and dissociation constants, heats of fusion) by means of a previously proposed thermodynamic equilibrium model (Fogassy et al., 1981) that is valid if the diastereomeric salts do not form aggregates. This condition is fulfilled if these salts exhibit simple eutectic behavior [i.e., if they do not form an addition compound (see above)].

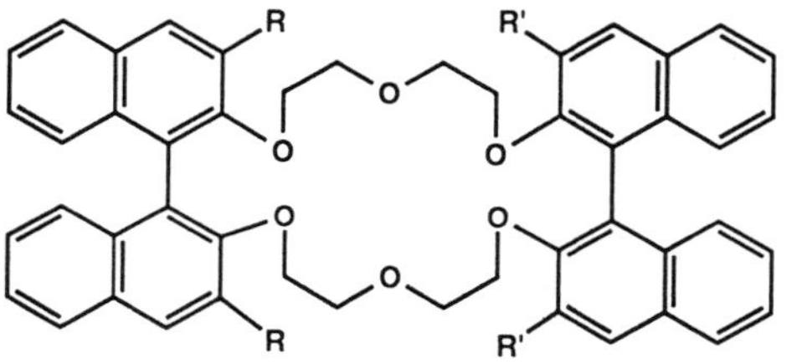

103

R, R' = H, CH₃

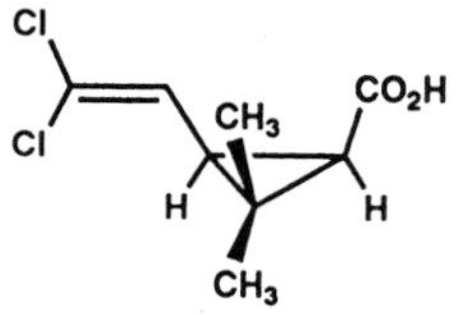

Figure 7.30.   Structure **103** and *cis*-permethrinic acid.

In view of the nearly equal solubilities of the diastereomeric salts (solubility constants K_{sR} and $K_{sS} = 1.5$ and $1.2 \, \text{mol dm}^{-3}$ in H_2O at 25°C for the p and n diastereomers, respectively), it could be postulated that resolution of *cis*-permethrinic acid (Fig. 7.30) with 2-benzylaminobutanol is practically impossible. However, adjustment of the pH by addition of NaOH (optimally 0.45 equivalent) affects the yield [49% maximum; $S = k \times p = 0.49 \times 0.98 = 0.48$ (see Eq. 7.6)] *provided* that the dissociation constants of the two salts differs significantly, as was observed in this case (Fogassy et al., 1988). A similar study permitted the optimization of the resolution of *trans*-permethrinic acid with the same resolving agent (Simon, Fogassy, et al., 1990).

The rational design of reaction partners illustrating "optimal" resolution by crystallization may be found in the resolution of chiral multiheteromacrocyclic ethers **103** (R = CH$_3$, R' = H; Fig. 7.30) by complexation with enantiomerically pure phenylglycine (Fig. 7.20, **51a**) perchlorate in ethyl acetate. Only 0.5 equivalent of resolving agent was required and crystallization of the thermodynamically favored and diastereomerically homogeneous complex (R,R)-**103**-(D-**51**·HClO$_4$) was essentially complete. The enantiomer discrimination implicit in this resolution is of the order of $\Delta(\Delta G°) = -1.5 \, \text{kcal mol}^{-1}$ $(-6.3 \, \text{kJ mol}^{-1})$. The less favored (S,S)(D) complex could not be induced to crystallize even when enantiomerically pure components were mixed (Peacock, Cram, et al., 1980; see also Lingenfelter, Helgeson, and Cram, 1981).

The advantages of carrying out diastereomeric salt-mediated resolutions in mixtures of immiscible solvents have been explored (Ács, Fogassy, and Faigl, 1985).

While the resolution of racemates by distillation of diastereomeric esters has been previously reported and patented (Bailey and Hass, 1941; Hass, 1945), this approach has, to our knowledge, rarely been applied. However, the resolution of carboxylic acids, e.g., tetrahydrofuran-2-carboxylic acid, by fractional distillation of the diastereomeric amides formed with, e.g., methyl esters of (S)-valine or (S)-leucine on a kilogram scale has been described (Fritz-Langhals, 1993). The resolved acids as well as the amino acid resolving agents were recovered following acid hydrolysis of the separated amides.

c. Separation via Complexes and Inclusion Compounds

Compounds devoid of conventional functionality (e.g., chiral alkenes and arenes, sulfoxides, and phosphines), for which alternative resolution routes are limited (see also Section 7-6 on kinetic resolution) may be resolved by incorporation into diastereomeric metal complexes or by reaction with chiral π-acids (Jacques et al., 1981a, pp. 273, 339).

Cope *et al.* first described the resolution of *trans*-cyclooctene by incorporation of the alkene enantiomers into diastereomeric trans platinum(0) square planar complexes (Cope et al., 1962, 1963) that are separable by low-temperature crystallization (Fig. 7.31).

Other alkenes, dienes, for example, spiro[3.3]hepta-1,5-diene (Hulshof, McKervey, and Wynberg, 1974), allenes, arsines, phosphines and sulfoxides have been analogously resolved. 2-Vinyltetrahydropyran **104** has been resolved by

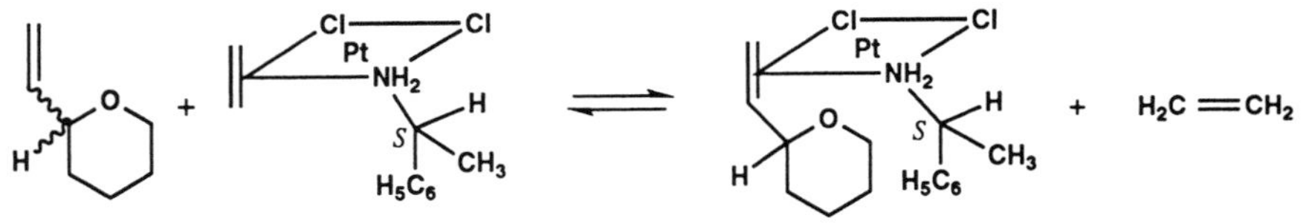

Figure 7.31. Resolution of chiral alkenes via neutral platinum coordination compounds (only one diastereomeric complex is shown).

formation of a cis platinum complex (Fig. 7.31). Displacement of the pyran from the complex with excess ethylene gave (S)-(−)-**104** (92% ee; Lazzaroni, Salvadori, et al., 1983). For a recent example of the resolution of an atropisomeric phosphinamine by separation of diastereomeric palladium complexes, see Alcock, Brown, and Hulmes (1993). A chromatographic variant, called ligand exchange chromatography, wherein one of the chiral ligands of the complex is incorporated into a polymer, has been applied mainly to the resolution of amino acids (cf. Section 6-5.d, pp. 247 and 257).

Diastereomeric π- (or charge-transfer) complexes, especially between the π-acid   α-(2,4,5,7-tetranitro-9-fluorenylideneaminoöxy)propionic   acid   **105** (TAPA; Fig. 7.32; Block, Jr., and Newman, 1968) and chiral Lewis bases, may be reasonably strong. Such complexes may be separated by crystallization (Newman, Lutz, and Lednicer, 1955; Newman and Lutz, 1956). Chiral aromatic hydrocarbons, heteroaromatic compounds, and aromatic amines whose basicity is too low to permit resolution with conventional acidic resolving agents have been resolved with TAPA.

In cases of borderline stability, one of the diastereomeric π-complexes crystallizes while the other remains in solution largely dissociated. The converse situation is illustrated by the classic resolution of hexahelicene (Fig. 6.6, **6**) with (−)-TAPA; (−)-hexahelicene crystallizes whereas the diastereomeric π-complex containing the (+)-hexahelicene remains in solution (Newman et al., 1955; Wynberg and Lammertsma, 1973). For specific applications of TAPA, see Jacques et al., 1981a, p. 273. Chromatographic resolutions based on TAPA are described on p. 257 and in Section 7-3.d. An analogous resolution method that depends on

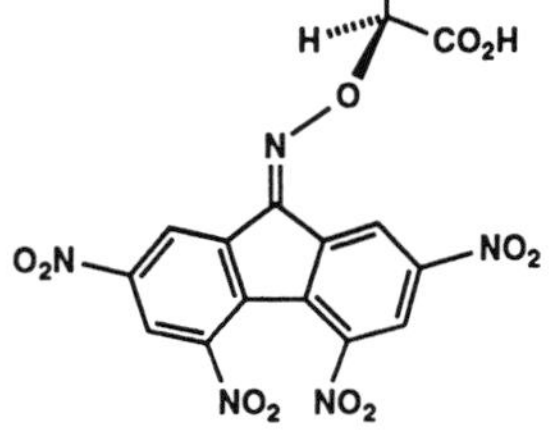

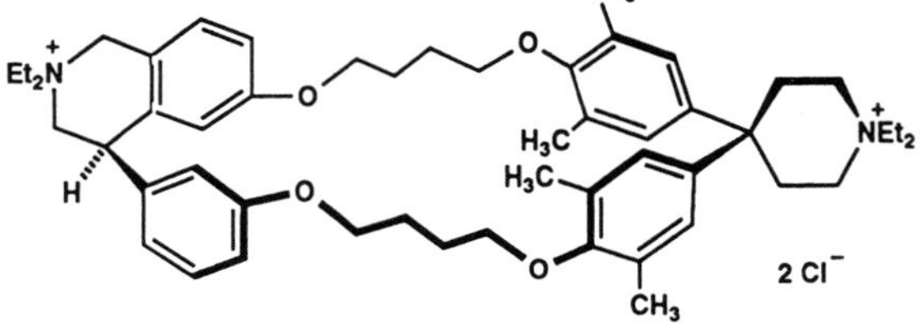

Figure 7.32. TAPA and structure **106**. [Structure **106** is reprinted with permission from Dharanipregada, R., Ferguson, S.B. and Diederich, F. (1988), *J. Am. Chem. Soc.*, **110**, 1679. Copyright © 1988 American Chemical Society.]

coordination of chiral Lewis bases to nonracemic organometallic compounds (complexation GC) is described in Section 6-5.d, p. 251.

Inclusion of molecules within others is a widespread phenomenon and the application of the "inclusion method" to resolutions, pioneered by Schlenk (1965), can no longer be considered to be a mere curiosity. We distinguish two broad classes of inclusion compounds: (a) the *cavitates* (Cram, 1983) in which the chiral substrate to be resolved (the guest molecule) is partially or entirely enclosed within a second chiral substance (the resolving agent and host molecule) that is endowed with a chiral cavity (e.g., Fig. 7.87, **220**), and (b) those in which the guest molecules are surrounded by several molecules of the resolving agent forming a cage or channel; these are often called *clathrates* (Weber, 1987, Chapter 1). For reviews of resolution by inclusion, see Arad-Yellin, Green, et al. (1984); Worsch and Vögtle (1987); Toda (1987); Tsoucaris (1987).

Preferential inclusion of one substrate enantiomer is the result of hydrogen bonding and van der Waals attraction between guest and host molecules. However, in addition, there must be a match between the size and shape of the guest molecule and a corresponding size and shape of the host cavity. While the size and shape of guest molecules may vary if they are conformationally flexible, there may be more play in the space available in the host molecules particularly in the case of lattice inclusion compounds. Separation of the diastereomeric inclusion compounds may depend on differential stability or solubility. Recovery of the resolution substrate may involve melting, dissolution, or extraction of the enantiomerically enriched guest.

In the first category (cavitates), inclusion of one enantiomer of amino acid derivatives in the cavity of racemic cyclic polyethers (chiral crown ethers; Stoddart, 1987) leads to crystallization of only the more stable diastereomeric complex in enantiomerically pure form (Peacock, Cram, et al., 1980; Fig. 7.30). Geometric differences are responsible for the greater ability of the chiral (*RR*) crown ether, **103** R,R′ = H (Fig. 7.30) whose cavity is more open to form inclusion complexes as compared to its meso (*RS*) diastereomer (Brienne and Jacques, 1975); for a crystallographic study of cavitate complex **103**, R,R′ = H plus methyl phenylglycinate·HPF_6, see Goldberg (1977); for the analogous complex **103**, R = CH_3, R′ = H plus phenylglycine (Fig. 7.20, **51a**) perchlorate, see Knobler, Gaeta, and Cram (1988). A similar resolution of a chiral crown ether comprising only one stereogenic element has been described (Lingenfelter, Helgeson, and Cram, 1981). The reciprocal resolution of racemic phenylglycine (as the perchlorate) by complex formation with one of the crown ether enantiomers is also possible (see Section 7-7).

A recent study of complexation in a macrocyclic host (Fig. 7.32, **106**) designed to form a cavitate with naproxen (Fig. 7.23a, **83**) has shown that while both naproxen and its methyl ester are included in the host cavity, the complex with the ester is more stable than that with the acid. Consequently, ion pairing between CO_2^- of the guest molecule and N^+ in the host is not as effective in binding and in diastereomer discrimination as is $\pi-\pi$ interaction, at least in aqueous solution. The nature of the medium may contribute to this fact by requiring desolvation of ionic sites in guest and host molecules during complexation. The desolvation may be energetically more costly in naproxen than in its ester (Dharanipragada, Ferguson and Diederich, 1988).

The cyclodextrins, water-soluble macrocyclic glucose oligomers having six or more glucopyranose units linked α-1,4, form crystalline 1:1 cavitate inclusion complexes in solution. Alternatively, the complexes can be formed by grinding together (or sonicating) solid guest and host reaction partners. Partial resolution of a variety of functionalized compounds (carboxylic acids, esters, and alcohols) by inclusion in β-cyclodextrin (cycloheptaamylose; Fig. 7.33a, **107**) was first described by Cramer and Dietsche (1959) (cf. also Cramer and Hettler, 1967). Though the enantiomeric enrichment was low ($<$11% ee), this is not inherent in the process. Enrichment of the order of 60–70% ee is feasible in specific cases, for example, for phosphinates (Benschop and Van den Berg, 1970). Moreover, the inclusion process may be repeated to enhance the enantiomeric enrichment. Derivatized cyclodextrins form the basis of enantioselective GC stationary phases that are able to chromatographically resolve numerous racemates (see Section 6-5.d, p. 252).

Partial resolution of sulfinyl compounds by crystallization with β- or α-CD (Fig. 6.46) has also been described (Mikołajczyk and Drabowicz, 1978; Drabowicz and Mikołajczyk, 1982). A more recent study of the resolution of chiral 1,4-benzoquinones in β-CD demonstrates the importance of hydrogen bonding in enhancing the stereoselectivity (Jurczak and Gankowski, 1982). A model that accounts for the effect of hydrogen bonding and for variations in the sense of stereoselectivity of inclusion is shown in Figure 7.33b. Model b accounts for the reversal of the sense of preferential inclusion of (R)-(+)-*tert*-butyl benzyl sulfoxide having opposite relative configuration to (R)-(−)-methyl benzyl sulfoxide.

The resolution of dansyl derivatives of amino acids (e.g., compound **108**, Fig. 7.34), by formation of crystalline inclusion complexes with β- and γ-CD (cyclooctaamylose) has been reported (Jin, Stalcup, and Armstrong, 1989). The crystal structures of the complexes of β-CD with the antiinflammatory, antipyretic, and

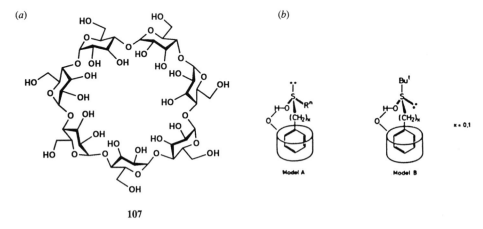

107

β-Cyclodextrin (β-CD; cycloheptaamylose)

Figure 7.33. (a) β-Cyclodextrin. (b) Model for the preferential inclusion of chiral sulfoxides in cyclodextrins [Reprinted with permission from Mikołajczyk, M. and Drabowicz, J. (1978), *J. Am. Chem. Soc.*, **100**, 2510. Copyright © 1978 American Chemical Society.]

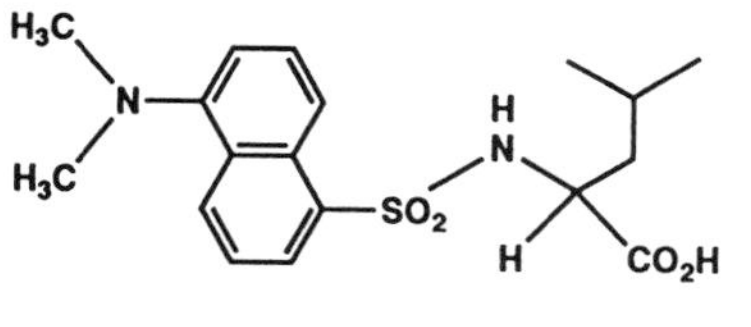

Figure 7.34. Structure **108** and fenoprofen.

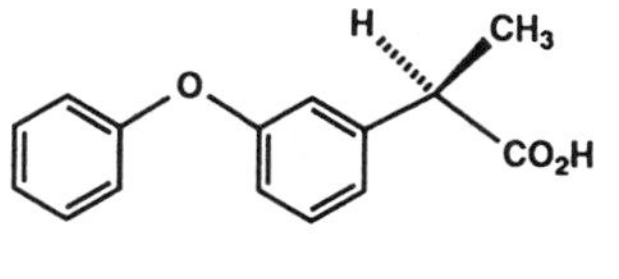

analgesic agent fenoprofen **109a** (Fig. 7.34), have been determined. One molecule of fenoprofen is included in the cavity of each β-CD molecule but in the crystal each host–guest pair exists as a dimer with the alignment being head to head in the case of the (S)-**109a** enantiomer and head to tail in the case of the (R)-**109a** enantiomer (Hamilton and Chen, 1988).

In the second category of inclusion complexes (clathrates, in which guest molecules are surrounded by many host molecules) it is sometimes the very presence of guest molecules that induces the formation of channels within the host crystal lattice that perforce contains the included guest molecules. Remarkably, lattice inclusion compounds useful in resolutions are formed with urea, an achiral compound, and with TOT (Fig. 7.10, **13**), a chiral compound that easily racemizes in solution (Section 7-2.e; Addadi et al., 1985; Worsch and Vögtle, 1987). While free urea crystallizes in a tetragonal lattice, in the presence of various guest substances, it is induced to crystallize in an hexagonal lattice in which the guest molecules are included in tubular channels having either right-handed or left-handed helical configurations (Schlenk, 1952; Asselineau and Asselineau, 1964). Crystallization of urea from methanol in the presence of unbranched chiral hydrocarbons, acids, or esters bearing small substituents (e.g., NH_2, OH, or Cl) may be induced by seeding or in a variety of other ways leading to preferential deposition of one of the diastereomers (Schlenk, 1973). Even alkanes, such as 3-methyloctane, have been partially resolved in this way (Klabunovskii et al., 1960).

The most useful of these types of hosts thus far employed for resolution by inclusion compound formation is TOT (Fig. 7.10, **13**; Tam et al., 1989; Gnaim, Green, et al., 1991). Though the TOT crystallizes in guest-free form as a racemic compound, in the presence of any of numerous guest molecules, it crystallizes as a conglomerate. The latter crystals may contain TOT in either the (P)-$(+)$ (right-handed propeller) or (M)-$(-)$ (left-handed propeller) configuration and the cavities wherein reside the guest molecules are correspondingly chiral. On dissolution of the separate guest-containing crystal batches at room temperature, the optical rotation falls nearly to zero as the TOT molecules undergo enantiomerization [interconversion $P \rightleftharpoons M$, $\Delta H^{\ddagger}$ ca. 21 kcal mol^{-1} (ca. 88 kJ mol^{-1})]. Cage clathrates (space group $P3_121$; host-guest ratio 2:1) are formed with small molecules (up to six nonhydrogen atoms; guest molecule length <9 Å) and the enantiomeric enrichment observed ranges from less than 1% ee to at least 83% ee (Arad-Yellin, Green, et al., 1983, and Gerdil and Allemand, 1980). Although 3.5 kcal mol^{-1} (15 kJ mol^{-1}) separate the two diastereomeric complexes in the case of the 2-halobutanes [X = Br and Cl; the (P)-$(+)$-TOT/(S)-2-halobutanes are the preferred stereoisomers], twofold disorder of the guest molecules in the lattice may be responsible for the limited stereoselectivity observed; there are two *R* and one *S* guest molecules in each of three cages comprising one unit cell (32–45% ee observed for 2-chloro- and 2-bromobutanes). The very low enantiomeric enrichment observed with 2-butanol (<5% ee) is associated with an additional (torsional) degree of freedom of the guest molecule.

Similarly, the channel clathrates formed by interaction between TOT and large molecules, such as 2- and 3-halooctanes (space groups $P6_1$, $P6_2$, and $P6_3$), exhibit low but measurable diastereomer discrimination as a consequence of the higher degree of disorder within the larger chiral cavities. Indeed, it was considered remarkable that any degree of recognition was observed at all in view of the many sources of disorder present. Repeated enclathration of partially enriched guest via channel clathrates yields large enhancements of enantiomeric purity (e.g., for 2-bromooctane). Consequently, resolutions of this type are practical for preparative (though small scale) purposes (Gerdil, 1987).

Lattice inclusion compounds are also formed with other chiral hosts. Among the better known examples are the alkaloids brucine and sparteine (Worsch and Vögtle, 1987; Toda, 1987). In 1981, it was observed that chiral tertiary acetylenic alcohols, e.g., **109b**, form stable 1:1 molecular complexes with brucine (Fig. 7.19*a*, **27**) (Toda, Tanaka, and Ueda, 1981). The alcohols recovered from the crystalline complexes formed in acetone exhibited high enantiomeric purity (Eq. 7.7).

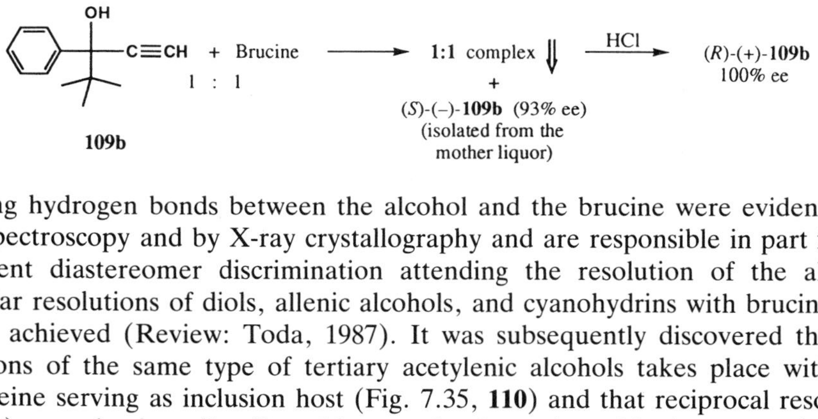

$$(7.7)$$

Strong hydrogen bonds between the alcohol and the brucine were evidenced by IR spectroscopy and by X-ray crystallography and are responsible in part for the efficient diastereomer discrimination attending the resolution of the alcohol. Similar resolutions of diols, allenic alcohols, and cyanohydrins with brucine have been achieved (Review: Toda, 1987). It was subsequently discovered that resolutions of the same type of tertiary acetylenic alcohols takes place with (−)-sparteine serving as inclusion host (Fig. 7.35, **110**) and that reciprocal resolution of (±)-sparteine is easily effected by complex formation with alcohol **111** (X = Br; Toda et al., 1983b). It has been observed that the nonracemic methosulfate salt of

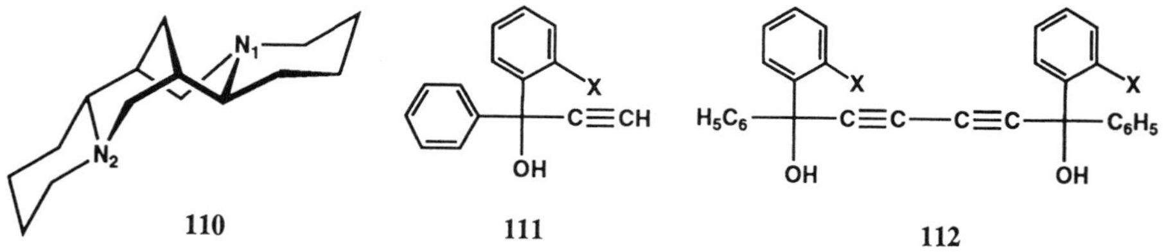

Figure 7.35. Sparteine and structures **111–112**.

Tröger's base (Fig. 7.39, **121**), can serve as a resolving agent for alcohols (e.g., 2-phenyl-1-propanol) by inclusion compound formation (Wilen and Qi, 1991).

Dimers obtained by oxidative coupling of alcohol **111** (Fig. 7.35, **112**, X = Br, Cl, or F) form 1:2 crystalline complexes with numerous small chiral molecules, for example, cyclanones and lactones and with epoxycyclohexanones (Toda et al., 1983b, 1983c; Tanaka and Toda, 1983; Toda, 1987). Isolation of the optically active carbonyl compounds was effected by thermal decomposition of the complexes and distillation of the resolved compounds. It is significant that only enantiopure **112** (X = Cl) and its meso isomer form inclusion compounds; the racemate does not (Leclercq and Jacques, 1985). This finding serves to underscore the suggestion made by Toda, Ward and Hart (1981) and by Hart et al. (1984) that, in addition to specific hydrogen bonds involving the hydroxy groups and π interactions between benzene rings, the overall shape of the host (having spacer and axle character in the case of **112**) contributes in an important way to its ability to include guest molecules. The finding of Leclercq and Jacques calls attention to the importance of the overall geometry and symmetry of the host–guest partners to their association in the solid state.

2,2'-Dihydroxy-1,1'-binaphthyl (Fig. 7.36, **113**) has been applied as an inclusion compound host to the resolution of sulfoxides (Toda et al., 1984a,b), phosphine oxides and phosphinates [Toda, Goldberg, et al., 1988; this study includes the X-ray crystallographic comparison of the two diastereomeric complexes of **113** and $C_6H_5P{=}O(CH_3)(CH_3O)$], and amine oxides (Toda, Goldberg, et al., 1989). The resolution of 3-methylpiperidine by clathration in 2,2'-dihydroxy-9,9'-spirobifluorene (Fig. 7.36, **114a**) has been described (Toda, 1986). A chiral host derived from lactic acid (Fig. 7.36, **114b**) effects the resolution, inter alia, of ketones, alcohols, sulfoxides, and lactones by recrystallization of enantiopure host from excess guest as well as by vapor sorption, that is, exposing the solid host to vapor of racemic guest compounds (Weber et al., 1992a).

With the exception of brucine and of the cyclodextrins, both of which are natural products, all of the cited host compounds have required resolution prior to their use in resolutions. This limitation has now been circumvented by

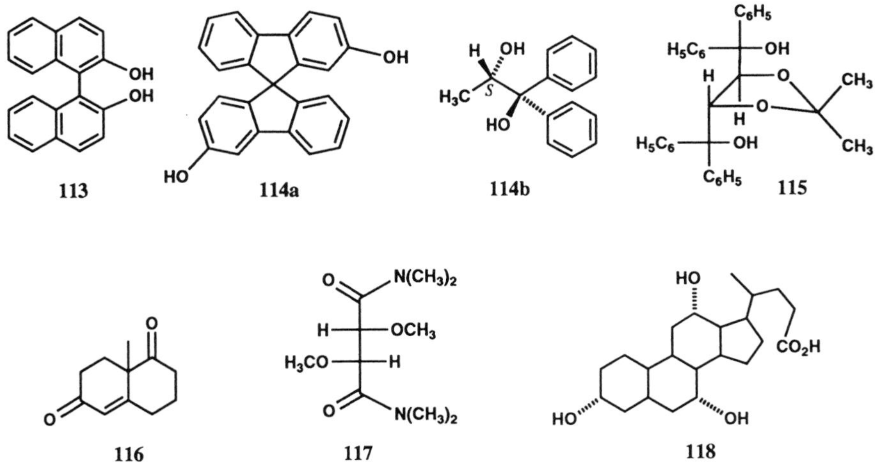

Figure 7.36. Structures **113–118**.

application of compound **115** (Fig. 7.36), derived from diethyl (2*R*,3*R*)-tartrate (made from natural tartaric acid), to the resolution of Wieland–Miescher ketone **116** (Toda and Tanaka, 1988a); compound **117** (and analogues), also derived from diethyl (2*R*,3*R*)-tartrate, have been shown to be useful hosts for the resolution of chiral compounds with C_2 symmetry, such as **113** (Toda and Tanaka, 1988b). Another host not requiring prior resolution is cholic acid **118** (Fig. 7.36); the latter has been shown to resolve lactones by inclusion (Miyata et al., 1988).

The high efficiency and simplicity of these resolutions is as remarkable as is the fact that they are effective with neutral compounds that are otherwise difficult to resolve. It has become apparent that the diastereomer discrimination responsible for the resolution is due to favorable packing of guest and host molecules in channel inclusion complexes. In some cases, the discrimination can be attributed to specific hydrogen bonding between guest and host (Toda et al., 1993).

The heterogeneous brucine-mediated resolution of alkyl halides (1,2-dibromopropane and 3-chloro-2-butanol) was discovered by Lucas and Gould (1942) and applied also to (±)-2,3-dibromobutane (Pavlis, Skell, et al., 1973). Since the recognition that this type of resolution does not involve a kinetic resolution, that is, dehydrohalogenation (Pavlis, Skell, et al., 1973; Skell and Pavlis, 1983), but instead is due to complexation (the stoichiometry appears to be 1:1), the alcohol resolutions of Toda et al., by inclusion in brucine (see above) and those of the haloalkanes have been rationalized by the finding that the brucine crystal lattice possesses large channels of variable height that are capable of including solvent molecules, guest ions as well as neutral guest molecules (Gould and Walkinshaw, 1984), that is, all of these resolutions with brucine (even resolutions of chiral acids by salt formation) are due to inclusion compound formation (cf. Section 7-3.b). The incursion of stereoselective inclusion of the halogen compound in the brucine lattice in the resolution of α-hexachlorocyclohexane with brucine (Cristol, 1949; cf. Section 7-4.c) must be considered a possibility.

Partial resolution of the anesthetic halothane $CF_3CHBrCl$ and of the chiral methane CHBrClF has been achieved by complexation with brucine. Both enantiomers of the latter may be recovered, the more strongly bound one by dissolution of the complex in acid (Wilen et al., 1985). The possibility of resolving a chiral methane by interaction with a chiral amine had been suggested by Hassel in his Nobel Prize address (1970) because of the relative stability of CH $\cdots$ N bonding. The enantiomeric purity of the partially resolved CHBrClF was determined by formation of a cavitate complex with a bis(cyclotriveratrylyl) macrocage compound (Canceill, Lacombe, and Collet, 1985b).

Inclusion compound formation is evidently a very powerful separation technique whose application is in active development. A less conventional application is the enrichment of a conformationally mobile system in that conformation that is normally disfavored (Gerdil, 1987; Thomas and Harris, 1987). When included in the channels of their inclusion compound with thiourea, bromo- and chlorocyclohexane are found by IR and Raman spectroscopy to be present largely in the axial conformation (Nishikawa, 1963; Fukushima, 1976; Allen et al., 1976; Gustavsen et al., 1978); axial chlorocyclohexane is also isolated as a clathrate with TOT (Fig. 7.10, **13**; Gerdil and Frew, 1985). Hexa-host **119** (Fig. 7.37) preferentially forms a crystalline inclusion complex with the normally disfavored chair form of 3,3,6,6-tetramethyl-*s*-tetrathiane **120** (MacNicol and Murphy, 1981).

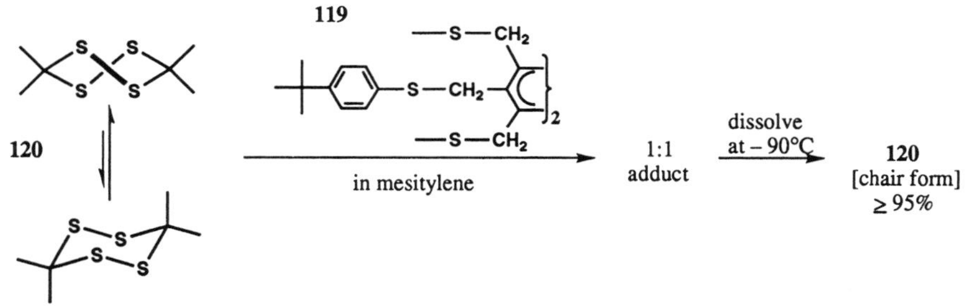

Figure 7.37. Conformational selection by clathration.

d. Chromatographic Resolution

The vast majority of the numerous chromatographic resolutions reported in the literature (cf. Fig. 7.69) have been analytical in nature, that is, their purpose has been the determination of stereoisomer and, in particular, enantiomer composition. These have been surveyed in Chapter 6, Section 5.d. The present section deals with preparative chromatographic resolutions.

Preparative chromatographic resolutions are often performed on covalent diastereomer mixtures (Section 7-3.a). In numerous cases, chromatographic resolution of diastereomers on silica gel succeeds even with the rapid and simple flash chromatography technique (Still et al., 1978) with the guidance of preliminary thin-layer chromatographic (TLC) experiments in which solvent composition is the only variable (e.g., Comber and Brouillette, 1987). The resolution of gossypol (Fig. 7.25) as Schiff's bases with various amino acid esters (e.g., methyl L-phenylalaninate) by TLC on silica gel could be scaled up to separate 5 g (and even larger) samples on silica gel columns (Tyson, 1988). The chromatographic resolution of alcohols as diastereomeric carbamates (prepared with isocyanate 85, Fig. 7.23a) in an automated liquid chromatograph (on 1-g samples) has been described (Pirkle and Hoekstra, 1974).

The ease with which covalent diastereomers are separated chromatographically serves to emphasize the anisometric character of such isomers. Failures of such chromatographic resolutions are probably the result of not building into the structures a sufficient number of intermolecular binding sites (or sites permitting sufficiently strong intermolecular binding) to permit selectivity to manifest itself as the diastereomeric molecules travel along the stationary phase. For reviews of chromatographic resolutions involving covalent diastereomers, see Gal (1987, 1988); Jacques et al. (1981a, p. 348); Allenmark (1991, p. 60); Lindner (1988); Ahnoff and Einarsson (1989). The reader is reminded of an important limitation of such resolutions: separation via covalent diastereomers will necessarily be incomplete if the resolving agent is not enantiomerically pure (see p. 241).

When carried out on chromatographic columns of high efficiency ($>10,000$ theoretical plates), baseline separations are easily achieved when separation factors attain or exceed $\alpha = 1.2$ [$\alpha = (t_2 - t_1)/(t_1 - t_0)$ where t_2 and t_1 are the retention times of the second and first diastereomers eluted, respectively, and t_0 is that of an unretained substance, e.g., pentane]. In contrast, when $\alpha = 2.0$ [corresponding to $-\Delta(\Delta G)$ of $0.4 \, \text{kcal mol}^{-1}$ ($1.7 \, \text{kJ mol}^{-1}$) between the diastereomers], as is possible with some diastereomeric amides, the separations can

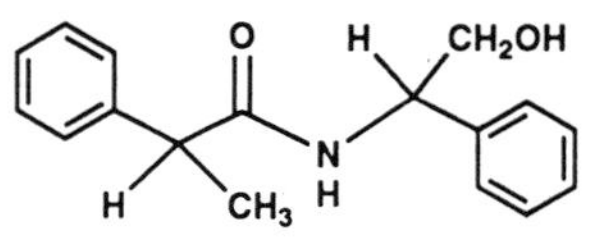

Figure 7.38. Diastereomeric amides separated by chromatography on "primitive" columns; $\alpha = 2.56$ (column = silica gel 60; mobile phase = hexane/EtOAc, 1:1).

be carried out even on relatively "primitive" columns, for example, the resolution of 2-phenylpropionic acid with phenylglycinol (Fig. 7.20, **54**) by chromatography of the diastereomeric amides (Fig. 7.38; Helmchen et al., 1979). The choice of resolving agent was also dictated by the presence of a suitably placed hydroxyl group that facilitates the hydrolysis of the diastereomeric amides following their separation, presumably by neighboring group participation.

Although, in principle, preformed ionic diastereomer mixtures are separable by chromatography (see, e.g., Knox and Jurand, 1982), in practice ion-pair chromatographic resolutions have involved the addition of nonracemic counterions to the mobile phase. (+)-10-Camphorsulfonic acid (Fig. 7.21a, **64a**) has served as the mobile phase additive in the resolution of chiral amino alcohols and (−)-quinine·HCl (Fig. 7.19a, **31**) has been used as additive in the resolution of (±)-10-camphorsulfonic acid. These separations have for the most part been analytical in scale (Reviews: Pettersson and Schill, 1986, 1988; Szepesi, 1989).

The most versatile and powerful chromatographic resolutions are those that take place on enantioselective stationary phases; with such columns, it has been suggested that preparative resolutions are feasible when $\alpha \geq 1.4$ (Davankov, 1989). The various types of enantioselective columns have been described in Section 6-5.d. Numerous examples of such resolutions on a scale of $10–10^2$ mg, but fewer examples of resolutions of 1 g or more of a racemate by one of these methods, have been described in the literature (Reviews: Pirkle and Hamper, 1987; Zief, 1988; Allenmark, 1991, Chapter 9; Taylor, 1989; Francotte and Junker-Buchheit, 1992).

The very earliest chromatographic resolutions (e.g., of Tröger's base, Fig. 7.39, **121**) were carried out on readily available natural products (lactose in the case of **121**; Prelog and Wieland, 1944) and potato starch is still being used for the purpose (Hess, Burger, and Musso, 1978; Konrad and Musso, 1986; for applications of cellulose derivatives, see below). Later, synthetic resolving agents impregnated on achiral supports, such as silica gel, were applied. The π-acid TAPA (Fig. 7.32, **105**) was found to be effective in the resolution of chiral hydrocarbons devoid of functional groups, such as the helicenes (Klemm and Reed, 1960; Mikes, Boshart, and Gil-Av, 1976; Numan, Helder, and Wynberg, 1976). Subsequently, TAPA covalently bonded to silica gel and, similarly, binaphthylphosphoric acid (Fig. 7.21b, **65**) bound to silica gel were found to be effective in the resolution of helicenes (Kim, Tishbee, and Gil-Av, 1981 and Mikes, Boshart, 1978, respectively).

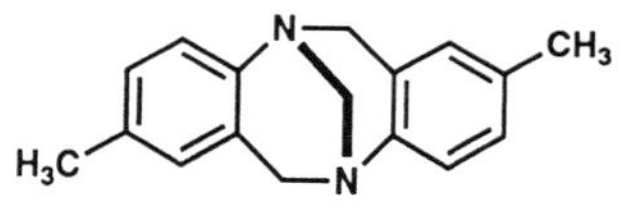

121

Figure 7.39. Tröger's base.

The first of the chiral stationary phases (CSP) incorporating a π-base allowing for charge-transfer interaction as well as having hydrogen bonding capabilities is illustrated in Figure 7.40. This CSP was based on Pirkle's TFAE chiral solvating agent (Fig. 6.38, **81a**) covalently bonded to mercaptopropyl silica. The structure of a generalized resolution substrate derivatized for optimal interaction with this stationary phase is given in Figure 7.40 (**122**). Chiral amines (Q = NH), amino acid esters (Q = NH, B = CO_2CH_3), alcohols (Q = O), sulfoxides and lactones, among others, could be resolved with this stationary phase following derivatization with 3,5-dinitrobenzoyl chloride, the latter to provide a π-acid bonding site complementary to the π-base incorporated in the stationary phase molecules (Pirkle and House, 1979).

On the assumption that reciprocal resolution would be successful, a structure analogous to **122** (Fig. 7.40) was incorporated in a new stationary phase that is ionically bonded to 3-aminopropyl silica (as in Fig. 6.47, **C**). This CSP is effective in resolving chiral arylalkylcarbinols (Pirkle and Finn, 1981) and many other types of organic compounds as well (sulfoxides, amides, and heterocycles) all possessing at least one aryl group. Subsequently, CSPs derived from *N*-3,5-dinitrobenzoylphenylglycine and *N*-3,5-dinitrobenzoylleucine covalently bonded to 3-aminopropyl silica were applied to chromatographic resolutions (Fig. 6.47, **B**; Pirkle and Welch, 1984). For a discussion of applications of Pirkle enantioselective stationary phases and stereoisomer discrimination models rationalizing their use, see Section 6-5.d, p. 253).

Resolutions on a gram scale (up to 8 g in some cases) were reported with large columns (30 in. $\times$ 2 in.) filled with ionic phase **C** (Fig. 6.47; Pirkle et al., 1982) and with the covalently bonded phenylglycine phase (Pirkle and Finn, 1982). Preparative resolution by flash chromatography on the two covalently bonded enantioselective phases has also been reported (Pirkle, Tsipouras, and Sowin, 1985). Even larger scale resolutions (up to 125 g of racemate in a single run) with Pirkle enantioselective phases have been carried out (Pirkle, 1987a,b). More recently, an enantioselective stationary phase derived from *N*-(1-naphthyl)leucine having separation factors as large as $\alpha = 60$ toward some analytes has been described (Pirkle et al., 1991).

Among the synthetic polymeric CSPs, the poly(acrylamide) and poly-(methacrylamide) stationary phases (Fig. 7.41) developed by Blaschke and co-workers have also been applied to preparative chromatography. These are

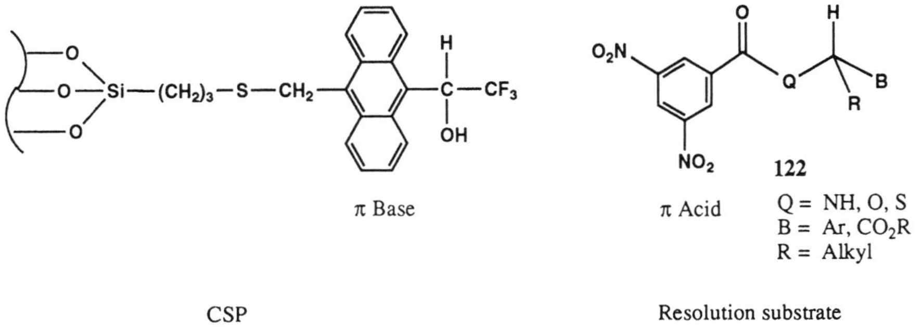

π Base π Acid   **122**

Q = NH, O, S
B = Ar, CO_2R
R = Alkyl

CSP   Resolution substrate

Figure 7.40. Complementary enantioselective stationary phase (CSP) and resolution substrates incorporating, respectively, π-base and π-acid moieties.

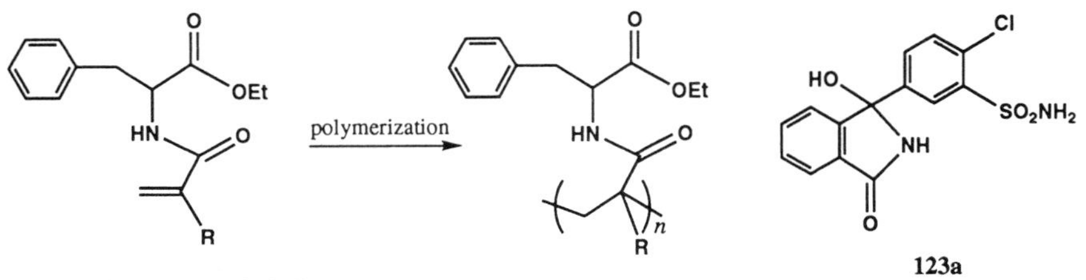

R = H   Poly(acrylamide) phase
R = CH$_3$ Poly(methacrylamide) phase

123a
Chlorthalidone

Figure 7.41. Chlorthalidone **123a** and the CSP on which it was resolved.

typically prepared by reaction of nonracemic amino acid esters with unsaturated acylating agents followed by polymerization (Blaschke, 1988). Racemic chlorthalidone **123a** (a diuretic–antihypertensive agent), was preparatively resolved (0.530 g in one pass) on nonracemic poly(ethyl *N*-acryloylphenylalaninate) (Blaschke and Markgraf, 1980). The analogous poly(menthylacrylamide) and poly[menthyl(meth)acrylamide] phases either covalently bonded to silica gel or in the form of cross-linked polymer beads, are effective stationary phases for preparative HPLC (Bömer, Arlt, et al., 1988). Acid **123b** (Fig. 7.42) (50 g in one pass) was resolved on 3 kg of poly(menthylacrylamide) beads.

Among the least expensive and most widely used chiral polymeric stationary phases are those based on cellulose and amylose (Section 6-5.d, p. 259). Numerous preparative resolutions on the 100-mg scale have been reported with microcrystalline cellulose triacetate (TAC; Fig. 6.53). An advantage of such phases is that they lend themselves to resolution of nonfunctionalized (or non-derivatizable) racemates (e.g., perchlorotriphenylamine, **124**), whose partial resolution on TAC has been reported (Hayes, Mislow, et al., 1980). Stable atropisomers (e.g., Fig. 7.42, **125a**) have been resolved on TAC (Roussel and Chemlal, 1988) as has the strained hydrocarbon **125b** (Fig. 7.42) that adopts a chiral *syn*-conformation (Agranat et al., 1987).

Underivatized carboxylic acids (e.g., mandelic acid) have been chromatographically resolved on both cellulose and amylose tris(3,5-dimethylphenyl-carbamates) (**126** and **127**, respectively; Fig. 7.43) analogous to TAC (Okamoto et al., 1988). Compound **128** (Fig. 7.42, in which the CHCl$_2$ groups are geared, cf. Sec. 14-6.b) that slowly undergoes enantiomerization has been preparatively resolved on stationary phase **126** (Biali, Mislow, et al., 1988). For other examples of resolutions carried out on polysaccharide stationary phases, see Ichida and Shibata (1988); Zief (1988); Allenmark (1991); Johns (1989). In spite of the low efficiency of TAC, resolution of gram and multigram samples [one report mentions a separation of up to 200 g per pass on TAC (Francotte, Lang, and Winkler, 1991)] with this CSP have been reported, for example, oxapadol, **129a** (Blaschke, 1986); and 2-phenyl-1,3-dioxin-4-ones **129b** (Fig. 7.42) (Seebach et al., 1991). Cellulose tribenzoate beads, that is, polymer free of inorganic carrier, and having a high loading capacity, have been applied to the preparative resolution (1 g) of α-methylbenzyl alcohol (Francotte and Wolf, 1991).

Another polymeric CSP reported to have been used in preparative chromatographic resolutions is poly(triphenylmethyl methacrylate) (PTrMA, Section 6-5.d, p. 262). Complete (baseline) resolution of amine **124** (Fig. 7.42), with

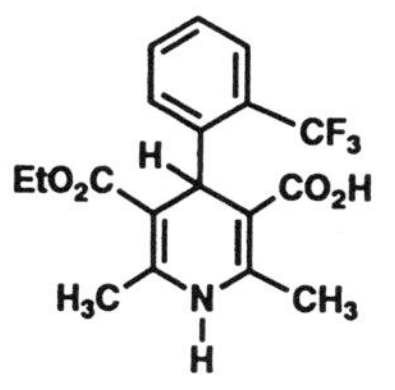

123b

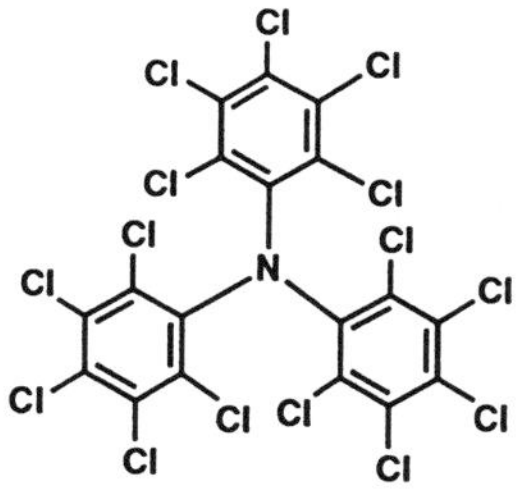

124

Perchlorotriphenylamine

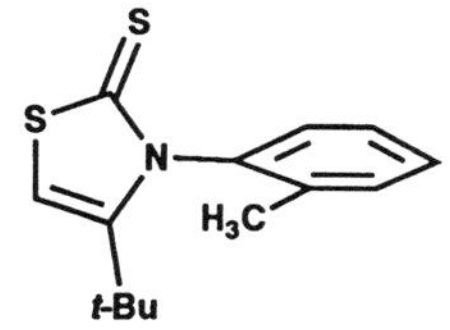

125a

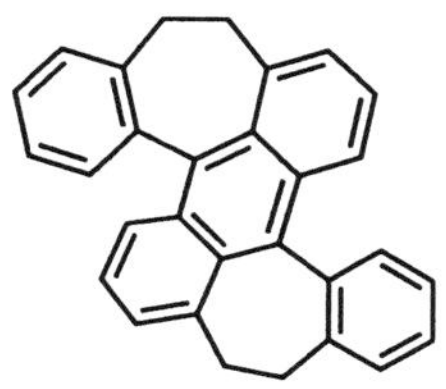

125b

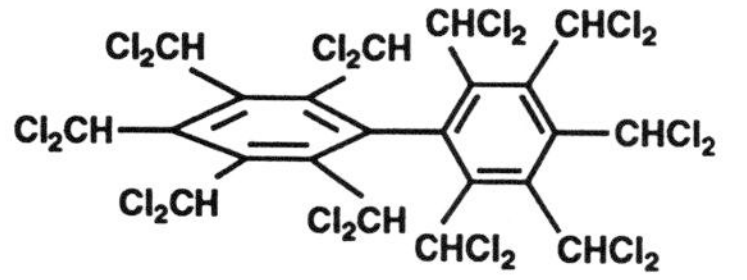

128

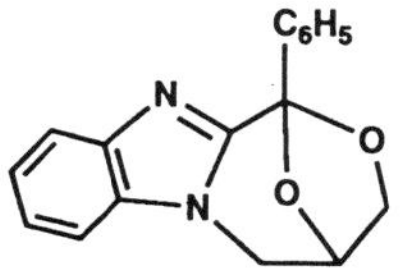

129a

Oxapadol

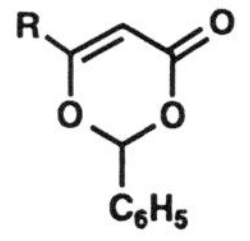

129b

R = CH$_3$, H

Figure 7.42. Structures **123b–125b** and **128–129b**.

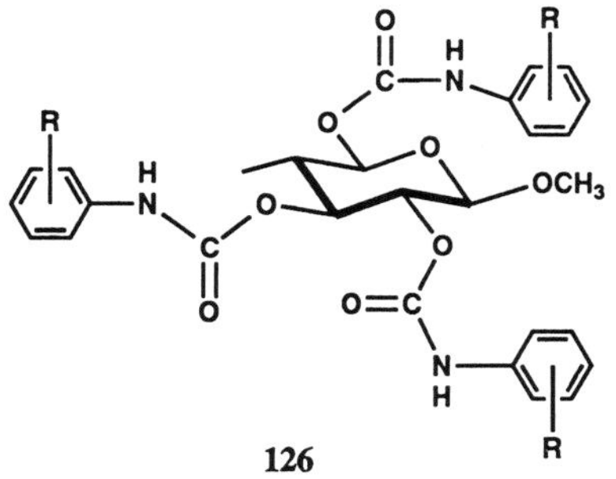

126

R = 3, 5-(CH$_3$)$_2$

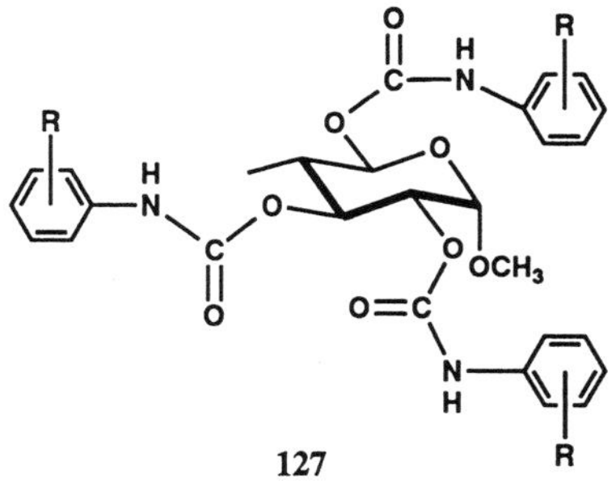

127

R = 3, 5-(CH$_3$)$_2$

Figure 7.43. Cellulose (**126**) and amylose (**127**) based enantioselective stationary phases (left and right, respectively).

separation factor $\alpha = 2.9$ (albeit only on ca. 1-mg scale) has been reported on PTrMA (Okamoto, Mislow, et al., 1984) as well as the resolution of Tröger's base (Fig. 7.39; Okamoto and Hatada, 1988). α-Methylbenzyl alcohol was partially resolved on PTrMA on a 219-mg scale (Yuki, Okamoto, and Okamoto, 1980).

Finally, preparative resolutions have been reported also by means of ligand-exchange chromatography (Section 6-5.d, pp. 247 and 257, Zief, 1988). Even gram amounts of *rac*-proline and *rac*-threonine were resolved on an enantioselective stationary phase of the ligand exchange chromatography type (see Fig. 6.52); (Davankov, Zolotarev, and Kurganov, 1979).

e. Asymmetric Transformations of Diastereomers

We have already seen that it is sometimes possible to obtain more than 50% of one enantiomer of a substance by a process whose departure point is a racemate (Section 7-2.e). Asymmetric transformation, the most general name applied to such processes, is also possible with diastereomers. Indeed, asymmetric transformations of diastereomers, either diastereomeric salts or covalent diastereomers, are more common than are transformations of enantiomeric species.

Figure 7.44 summarizes the process that obtains when components of diastereomeric mixtures, such as those formed in resolutions, are equilibrated. When equilibration, in solution or in the molten state, is rapid, mixtures are obtained that reflect the relative thermodynamic stability of the diastereomers. This spontaneous equilibration of stereoisomers in solution is often called an *asymmetric transformation of the first kind* (for a discussion of the history of the term, see Eliel, 1962, p. 42).

Asymmetric transformations of the first kind are the result of epimerization (a change in configuration at just one of several chiral centers present in a molecule) of both diastereomers present in a mixture. When a single diastereomer crystallizes from a solution containing several equilibrating diastereomers in a yield exceeding that given by its solution concentration, the process is termed *asymmetric transformation of the second kind* (Eliel, 1962, p. 63; Mason, 1982, p. 208). Here again, the term "crystallization-induced asymmetric transformation" has

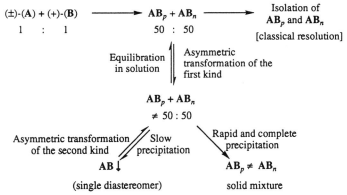

Figure 7.44. Asymmetric transformations of chiral diastereomers; **A** represents a racemate and **B** a resolving agent. The *p* and *n* subscripts are defined on page 326.

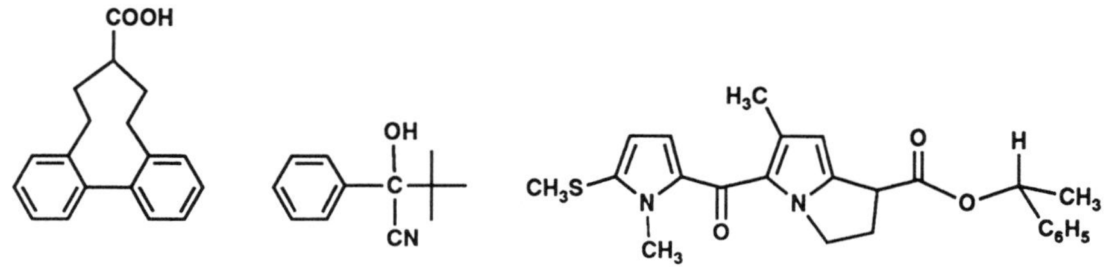

Figure 7.45. Compounds that undergo crystallization-induced asymmetric transformations.

been suggested as a more descriptive alternative (Jacques et al., 1981a, Chapter 6).

Asymmetric transformations are usually observed during resolutions when more than 50% of one diastereomer is isolated. This finding is illustrated by the optical activation of the bridged biphenylcarboxylic acid (Fig. 7.45, **130**) with quinidine (Fig. 7.19a, **29**). Both the yield of quinidine salt (theoretical yield of one diastereomeric salt 50%; found 79%) and the fact that the acid recovered from the salt is optically labile ($t_{1/2} = 53$ min in o-xylene at 50°C) are indicative of the operation of an asymmetric transformation (Mislow et al., 1962).

A number of investigators have observed that, in crystallization-induced asymmetric transformations, it is the minor (hence less stable) diastereomer present in solution that crystallizes preferentially (Buchanan and Graham, 1950; Harris, 1958; Jacques et al., 1981a, p. 374). These findings have been generalized into a "rule," the so-called "van't Hoff-Dimroth rule," (Buchanan and Graham, 1950; Smrčina, Kočovský, et al., 1993), yet as pointed out by Harris (1958), "there seems to be no valid reason why this should be a general rule" and the generality of such findings remains to be proven.

In contrast to epimerization, the term asymmetric transformation is limited to processes in which the ultimate product is necessarily enantiomerically enriched (Eq. 7.8 in which the chiral acid H-R is nonracemic). However, the epimerization of *rac-cis*-1-decalone **131** to *trans*-1-decalone under the influence of an achiral catalyst or reagent is not considered an asymmetric transformation.

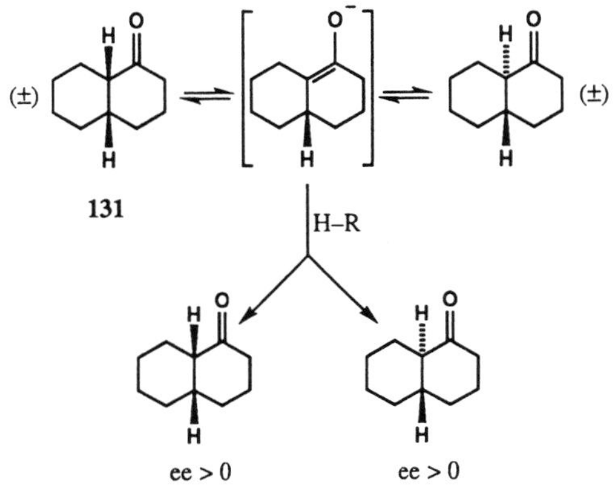

$$(7.8)$$

The occurrence of asymmetric transformations among mixtures of diastereomers is much more common than is supposed. These occurrences may arise
in diastereomeric salts whenever one of the ions, or the molecule that the ion can
dissociate into, is labile (i.e., the ion can racemize). Such racemization can be
spontaneous or induced, for example, by heating or on changing the pH.
Asymmetric transformations are also observed in covalent diastereomeric systems. In both cases, the change in configuration at the labile stereocenter is
induced by a chiral agent (by the counterion in the case of ionic diastereomers,
and by the resolving agent in the case of covalent diastereomers). Asymmetric
transformations less commonly take place when racemates are crystallized in
optically active solvents. Epimerization of α- and β-anomers, representing diastereomer equilibration in aldohexoses, is dealt with in Chapters 10 and 11. In this
section we principally survey those asymmetric transformations that have appeared in the literature since about 1979 (Review: Arai, 1986); for earlier
examples, see Jacques et al. (1981a, Chapter 6; see also Kagan and Fiaud, 1978,
p. 229 where, inadvertently, they are called asymmetric "conversions").

> Asymmetric transformations are often confused with kinetic resolutions and with
> stereoselective syntheses. For example, the thermodynamically controlled asym-
> metric transformations discussed by Morrison and Mosher in the classic monograph
> "Asymmetric Organic Reactions" (1976; p. 15) are of the type described above.
> However, the processes that they call "kinetically controlled asymmetric trans-
> formations" are in fact kinetic resolutions and enantioselective syntheses (see also
> below).

A beautiful example of a crystallization-induced asymmetric transformation is
that of the 3-amino-1,4-benzodiazepin-2-one (Fig. 7.46, **132**), an intermediate in
the synthesis of a selective antagonist to the gastrointestinal hormone cholecys-
tokinin (CCK). Resolution of **132** was effected with 0.5 equiv of (1S)-(+)-10-

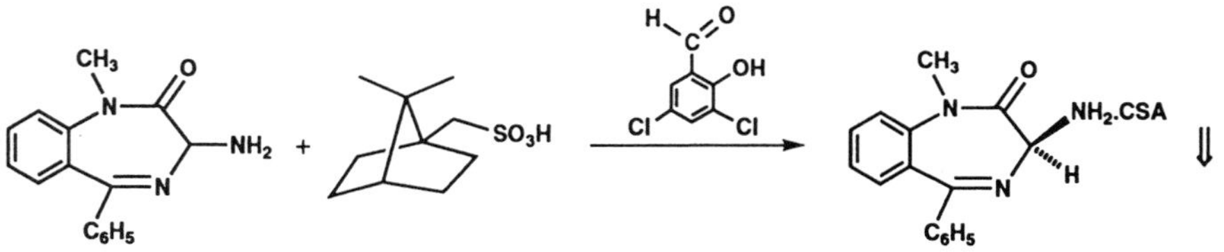

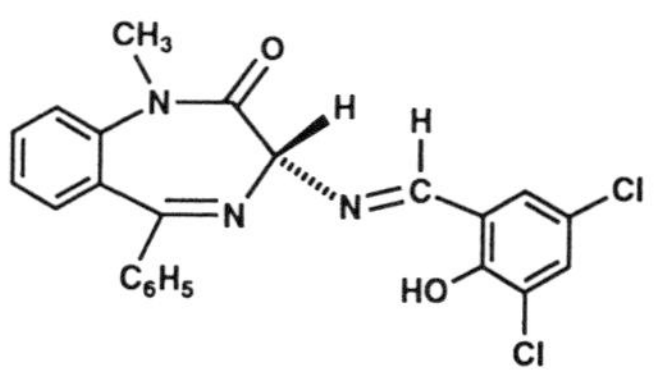

Figure 7.46. Asymmetric transformation of a benzodiazepinone.

camphorsulfonic acid (CSA; Fig. 7.21, **64a**) in isopropyl acetate to afford about 40% of the (3S)-amine-CSA salt (>99.5% ee). From the observation that the undesired (3R)-**132** racemizes at 90°C in the presence of the resolving agent (this process may, alternatively, be viewed as an epimerization), a system was devised to enhance the acidity of the α-hydrogen atom of **132** so that the epimerization could take place under mild conditions.

The epimerization was achieved by including a catalytic amount of an aromatic aldehyde (e.g., 3,5-dichlorosalicylaldehyde, 3 mol%) in the resolution reaction mixture, with the result that epimerization of the **132**·CSA diastereomers became possible at room temperature. The enhancement in the speed of the epimerization is attributed to formation of imine **133** (in low concentration). The ensuing crystallization-induced asymmetric transformation is driven by the insolubility of the (3S)-**132**·(1S)-CSA diastereomer. With all conditions optimized, there results a one-pot process (equivalent to a combined resolution-racemization): (±)-**132** (23 mol) treated with (1S)-(+)-CSA (21.16 mol) and (3S)-**132**·CSA seed (10 g) in isopropyl acetate/CH$_3$CN followed by addition of 3,5-dichlorosalicylaldehyde (0.69 mol). In this way, greater than 90% of the racemic substrate is converted to enantiomerically pure (3S)-**132**·(1S)-CSA. In the reaction, epimerization is actually effected by a small amount of unprotonated base **132** (present as a result of an 8 mol% deficiency of resolving acid) generating an achiral enolate from **133** (Reider et al., 1987; Fig. 7.46).

During the resolution of cyanohydrin **134** (Fig. 7.45) by inclusion in brucine, it was observed that the yield of the precipitated 1:1 complex exceeded 50%. Moreover, the cyanohydrin recovered from the filtrate was found to be optically inactive. (*Note*: This is another hallmark of an asymmetric transformation of the second kind.) On slow crystallization, the yield of cyanohydrin could be raised to 100%. In this crystallization-induced asymmetric transformation, the stereocenter is labilized by reversible dissociation of the cyanohydrin into a ketone and HCN, two achiral precursors (Toda et al., 1983a; Tanaka and Toda, 1987).

Treatment (at room temperature) of *rac*-2,2′-dihydroxy-1,1′-binaphthyl (Fig. 7.36, **113**) with a stoichiometric amount of a complex generated from CuCl$_2$ and (−)-sparteine (Fig. 7.35, **110**) affords 94% of (−)-diol (80% ee) (Smrčina, Kočovský, et al., 1992). Sparteine is known to be an enantioselective host in inclusion compound formation (e.g., of **114a**, Fig. 7.36; Toda et al., 1983b; Toda, 1987); it is likely that this asymmetric transformation also involves formation of an inclusion compound (see above).

In the presence of 1,5-diazabicyclo[4.3.0]non-5-ene (DBN), ester **135** (Fig. 7.45) prepared by esterification of the (±) acid with (−)-α-methylbenzyl alcohol crystallizes to give 88% of the (−) ester (Hagmann, 1986).

Asymmetric transformations resulting from the lability of atoms other than hydrogen are known. The product in Eq. 7.9, where DCC = dicyclohexyl-carbodimide, is produced as a mixture of diastereomers (**136a**:**136b**) (each of the two diastereomers being present as a mixture of rotamers, i.e., cis–trans amides) that slowly converges to 50:50 composition. Evaporation of the solvent leads to the deposition of pure **136a**. Since equilibration of the diastereomers **136** in CH$_3$OD did not lead to incorporation of deuterium, it was inferred that the epimerization was the result of C–Br dissociation at the exocyclic stereocenter (Zeegers, 1989).

DCC = dicyclohexyl–
carbodiimide

136a   136b

(7.9)

The resolution of Tröger's base (TB, Fig. 7.39) with binaphthylphosphoric acid (Fig. 7.21*b*, **65**) is attended by a crystallization-induced asymmetric transformation; the yield of enantiopure TB obtained is 93% (Wilen et al., 1991). The attending acid-catalyzed racemization of TB is described in Section 7-8.a.

Numerous reports exist of the asymmetric transformation of amino acids and their derivatives. These findings stem from the economic importance of amino acids (cf. Section 7-5). *p*-Hydroxyphenylglycine (Fig. 7.20, **51b**) is resolved with α-phenylethanesulfonic acid (Fig. 7.22, **78**). During the crystallization of the less soluble D-**51a**·(+)-**78**, the more soluble L-**51a**·(+)-**78** diastereomer selectively and simultaneously undergoes epimerization. Epimerization is accelerated by heating to 140°C and by acid (e.g., sulfuric acid). As a result, both of the diastereomers initially present are converted into the less soluble one (Chibata et al., 1983; Yoshioka et al., 1987b).

Resolution of ($\pm$)-phenylglycinate esters [e.g., $C_6H_5CH(NH_2)CO_2C_2H_5$] with (+)-(2R,3R)-tartaric acid in ethanol containing 10% dimethyl sulfoxide (DMSO) is attended by an asymmetric transformation (the yield of the less soluble ethyl (2R)-phenylglycinate diastereomer is 74%; >90% ee; Clark et al., 1976a). Resolution of ($\pm$)-methyl phenylglycinate and substituted phenylglycine esters $ArCH(NH_2)CO_2CH_3$ with (+)-(2R,3R)-tartaric acid (1 equiv) in the presence of 1 equiv of benzaldehyde speeds up the asymmetric transformation process; (2R)-$ArCH(NH_3^+)CO_2CH_3$ hydrogen (2R,3R)-tartrate ($Ar = C_6H_5$; 99% ee) is formed in 85% yield (reuse of the filtrate raises the yield of D salt to 95%; Clark et al., 1976b). Epimerization has also been carried out on L-**51a**·(+)-**78** by heating in acetic acid in the presence of salicylaldehyde (see Section 7-9 for a discussion of the function of aldehydes in the asymmetric transformations of amino acid derivatives; Yoshioka et al., 1987a). Salicylaldehyde also facilitates the asymmetric transformation of ($\pm$)-thiazolidine-4-carboxylic acid (Fig. 7.21*b*, **72a**) with tartaric acid (yields: 68–78%; Shiraiwa, Kataoka, and Kurokawa, 1987f). ($\pm$)-Proline undergoes asymmetric transformation (in 80% yield) and so does ($\pm$)-2-piperidinecarboxylic acid (in 70% yield) on crystallization of the salt formed with (2R,3R)-tartaric acid in the presence of butanal (in butanoic acid solvent) (Shiraiwa, Shingo, and Kurokawa, 1991). The asymmetric transformation of substituted phenylglycine esters with tartaric acid has been investigated by correlation analysis (Lopata et al., 1984).

More commonly, asymmetric transformations of amino acids require prior derivatization of the resolution substrate. Derivatization permits a wider range of resolving agents or resolution methods to be used. That of phenylglycine (Fig. 7.20, **51a**) may be carried out on the *N*-benzoyl derivative concomitant with the resolution [resolving agent: excess $(-)$-α-methylbenzylamine (Fig. 7.19b, **43**)]. The yield of D-N-benzoylphenylglycine is 95–98% (Shiraiwa et al., 1985a). A similar asymmetric transformation has been reported for N-acetylphenylglycine (Shiraiwa et al., 1988).

Asymmetric transformations of α-amino acids may further be "promoted" by formation of N-acyl derivatives and complexation with transition metal ions. In a system designed to mimic the mechanism of action of a racemase enzyme (i.e., labilization of the α-hydrogen atom), the Schiff's bases of amino acids [e.g., $(\pm)$-alanine with $(-)$-menthyl 3-(2-hydroxybenzoyl)-propionate], was treated with cobalt(II) acetate. A mixture of diastereomeric salts was formed from which one precipitated. D-Alanine·HCl (22% ee; L-enantiomers in the case of leucine and valine) was isolated from the precipitate (Fig. 7.47; Numata et al., 1979).

An unusual, perhaps unique, example involving epimerization at two dissimilar stereocenters is that shown in Figure 7.48. On heating with $(-)$-10-camphorsulfonic acid (Fig. 7.21a, **64a**) in EtOAc for 22.5 h, a racemic diastereomer mixture (**137**) yields a single dextrorotatory isomer in 80% yield (Openshaw and Whittaker, 1963). The resolving agent catalyzes the equilibration of **138** at C(11b) (by reversible Mannich reaction) and at C(3) (by reversible enolization) and selectively causes precipitation of the least soluble salt (Oppolzer, 1987).

Even when one of the stereogenic elements is so labile that isolation of a pure stereoisomer is precluded, an asymmetric transformation can nevertheless be a useful process. Recrystallization of chiral sulfonium malate (Fig. 7.49a, **139**) followed by conversion of the resulting salt to an optically active fluoborate demonstrates that resolution has taken place. It has been suggested that, since chiral sulfonium salts epimerize in the vicinity of room temperature, the resolution is attended by an asymmetric transformation, that is, the diastereomer mixture is entirely converted to one diastereomer (Trost and Hammen, 1973). The chiral bias of the stereochemically labile sulfonium cation could be transferred to a carbon atom. Sulfonium salt **140** (Fig. 7.49b) $[\alpha]^{25}_{365} + 22.2$ (c 1.10, EtOH), resolved as the dibenzoyl hydrogen tartrate salt, is converted to (R)-1-

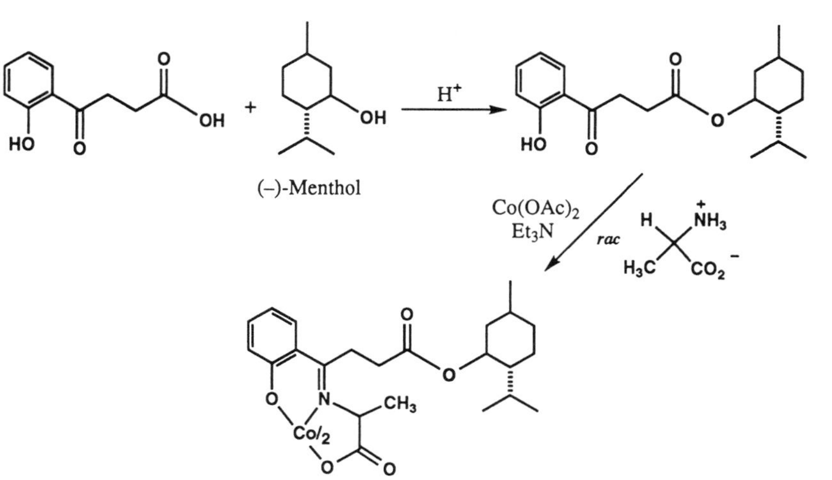

Figure 7.47. Synthesis of a Schiff's base derivative of alanine complexed with Co(II) to facilitate the asymmetric transformation of the amino acid.

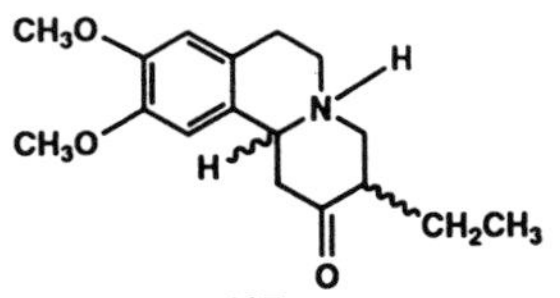

Figure 7.48. Asymmetric transformation involving epimerization at two stereocenters (Reprinted from Oppolzer, W. (1987), *Tetrahedron*, **43**, 1969. Copyright 1987, with permission from Pergamon Press, Oxford, UK).

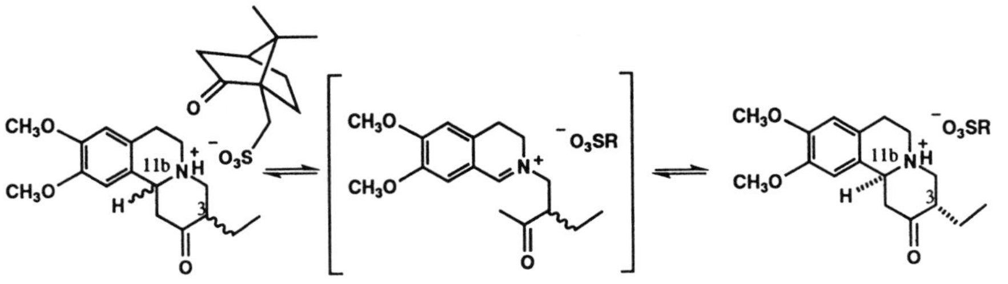

adamantyl 2-pent-4-enyl sulfide on reaction with potassium *tert*-butoxide (via [2,3]sigmatropic rearrangement; 94% ee). A stereochemically labile thia-heterohelicene has been totally converted to a single diastereomer by crystalliza-tion-induced asymmetric transformation with (*S*)-TAPA (Fig. 7.32, **105**), a reagent that induces charge-transfer complexation (Sec. 7-3.c) (Nakagawa, Yamada, and Kawazura, 1989).

The above examples of crystallization-induced asymmetric transformations all take place under thermodynamic control. Analogous processes, also called asym-metric transformations or "deracemizations" (see below), have been described that take place under kinetic control (Review: Duhamel et al., 1984). Two approaches may be envisioned. In one, the racemic substrate is treated with a nonracemic basic derivatizing agent to produce a chiral intermediate lacking the original stereocenter. Enantioselective protonation of the intermediate (under the influence of the covalently bonded chiral auxiliary agent) releases the derivatizing

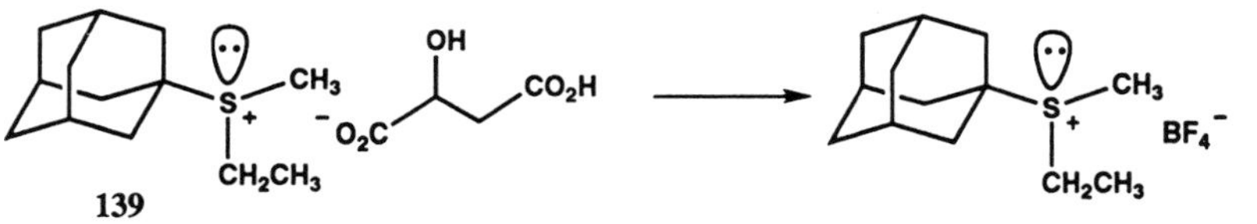

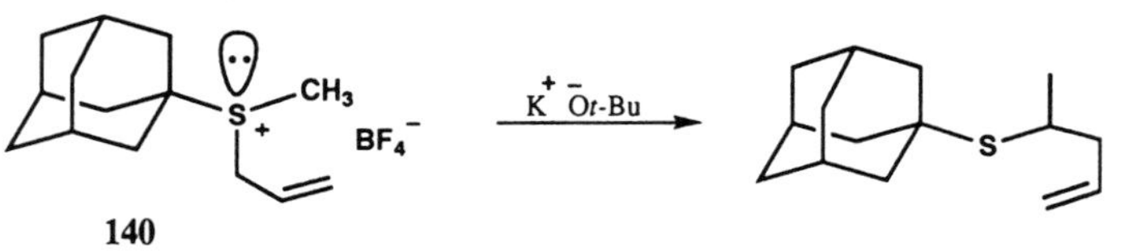

Figure 7.49. (*a*) Asymmetric transformation of a labile chiral sulfonium salt; (*b*) application of a chiral sulfonium salt.

agent and simultaneously restores the original stereocenter in enantiomerically enriched form (e.g., Eq. 7.10).

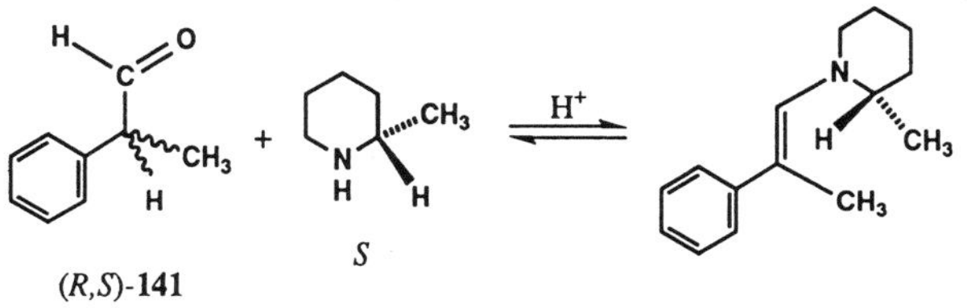

$$(7.10)$$

(R,S)-**141**

Atropaldehyde

Atropaldehyde **141** is (re)isolated in optically active form (up to 33% ee; Matsushita, Noguchi, and Yoshikawa, 1976). Alternatively, on hydrolysis in the presence of optically active acids, enamines of racemic aldehydes regenerate the aldehyde in nonracemic form (ep < 33%) (Matsushita, Yoshikawa, et al., 1978).

In the second approach, a racemic substrate is converted to an *achiral* intermediate or derivative (e.g., an enolate) and the latter is enantioselectively protonated by an optically active acid as shown for benzoin in Eq. 7.11.

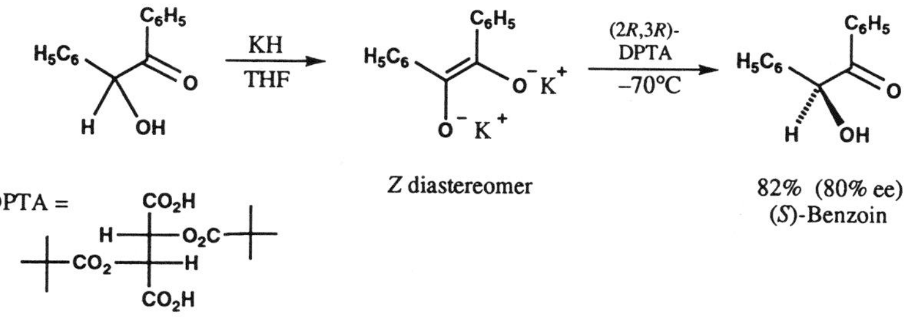

$$(7.11)$$

On isolation, the benzoin crystallizes in enantiomerically pure (100% ee) form (Duhamel and Launay, 1983). While the "deracemization" of (±)-benzoin, a kinetically controlled enantioselective protonation, has features in common with crystallization-induced asymmetric transformations, the process is more usefully viewed as an asymmetric (i.e., enantioselective) synthesis.

It should be evident that enantioselective protonations, such as that described in Eq. 7.11, are kindred to other enantioselective reactions, for example, enantioselective alkylations, of compounds bearing prostereogenic centers. Enantioselective reactions that transform a racemate into the nonracemic form of the same substance have been termed *deracemizations* (Duhamel et al., 1984). Like crystallization-induced asymmetric transformations taking place under equilibrium control, deracemizations (a somewhat unfortunate term since the converse, racemization, *is* an equilibrium process, whereas deracemization as originally defined is not) also have the potential for converting a racemate entirely into either of its enantiomers. Readers are advised that some authors have chosen to apply the term deracemization as a general synonym for asymmetric transformations (e.g., Pirkle and Reno, 1987; Smrčina, Kočovský, et al., 1992).

When the enolate of ketone **142**, generated with lithium diisopropylamide (LDA) is protonated with (−)-mandelic acid, with ephedrine, or with (−)-

menthol, the recovered **142** is not enantiomerically enriched. However, if the enolate is generated with lithium (S,S)-α,α'-dimethyldibenzylamide, hydrolysis regenerates **142** with 48% ee (Eq. 7.12; Hogeveen and Zwart, 1982).

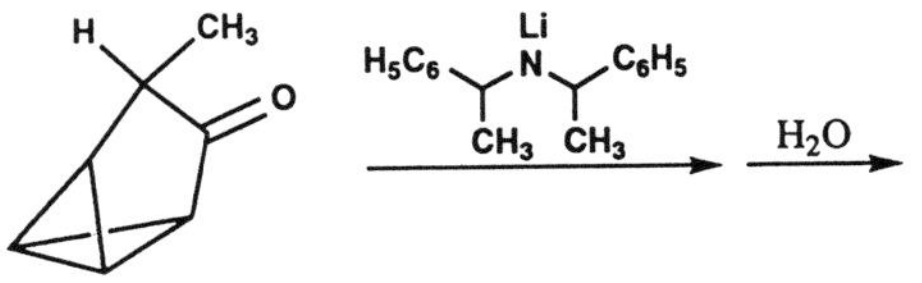

$$(7.12)$$

142   48% ee

Schiff's bases of amino acid esters undergo deracemization on sequential deprotonation (with LDA) and enantioselective protonation of the intermediate lithium enolate with optically active acids (Duhamel and Plaquevent, 1978, 1982). Ligand exchange within the enolate intermediate increases the enantiomeric enrichment of the recovered amino acid (Duhamel, Fouquay, and Plaquevent, 1986).

The enantioselective protonations are complicated by the presence of more than one enamine or enolate stereoisomer, and by equilibration between stereoisomeric iminium salts (Fig. 7.50). Available evidence suggests that the principal driving force in these reactions is the relative rate of protonation (i.e., enantioselective protonation), whether R in the iminium salts or A in HA are stereogenic (Duhamel, 1976; Matsushita, Yoshikawa, et al., 1978; Duhamel and Plaquevent, 1982). Optical activation of carboxylic acids (e.g., hydratropic acid, $C_6H_5CH(CH_3)CO_2H$) by protonation or alkylation of 2-substituted 4-benzyloxazoline derivatives [the latter being generated from (S)-(−)-3-phenylalaninol **55**, (Fig. 7.20)] is also mainly kinetically controlled (e.g., Shibata et al., 1982a,b; Shibata et al., 1984).

An asymmetric transformation of the first kind has been described that is mediated by the formation of diastereomeric complexes. N-(±)-(3,5-Dinitrobenzoyl)leucine butyl thioester **143** (0.045 M) was allowed to stand in mixed cy-

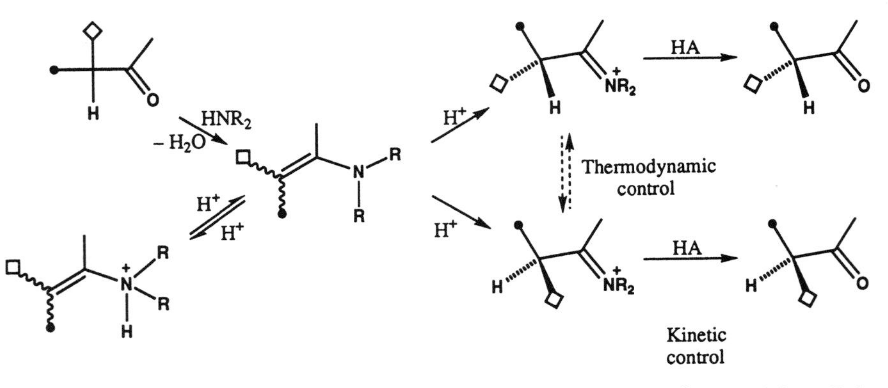

Figure 7.50. Mechanism of deracemization of chiral carbonyl compounds. [Adapted from Duhamel, L., Duhamel, P., Launay, J.-C., and Plaquevent, J.-C. (1984), *Bull. Soc. Chim. Fr.*, II-421. Reprinted with permission of the Société Francaise de Chime.]

clohexane/CH$_2$Cl$_2$ solvent containing triethylamine (0.18 M) and 10-undecenyl (R)-N-(1-naphthyl)alaninate (Fig. 7.51, **144a**; 0.20 M). Periodic chromatographic analysis on an enantioselective stationary phase analogous to **144a** (cf. also Fig. 6.47, **D**) revealed that thioester **143** was slowly undergoing an asymmetric transformation, with the enantiomeric enrichment reaching 78% ee of (R)-**143** after 28 days (Pirkle and Reno, 1987). The choice of a thioester rather than an ordinary ester serves to increase the acidity of the α-hydrogen atom so as to enhance the rate of the Et$_3$N-catalyzed epimerization.

The diastereomer discrimination that provides the driving force for the asymmetric transformation of the thioester is similar to that operative in chromatographic resolutions with enantioselective stationary phases **B** and **C** (Fig. 6.47). Preferential formation of the R,R' transient complex as the result of one additional hydrogen bond (Fig. 7.51) is supported by nuclear Overhauser effect (NOE) measurements (Pirkle and Pochapsky, 1987).

The enantiomers of chiral hemiacetals (e.g. **144b**) are in rapid equilibrium with one another (Fig. 7.52). Acetylation of the hemiacetals with formation of **144c** serves to "fix" their configuration. If the acylation is carried out in a nonracemic chiral solvent **144d**, optical activation ensues and the product is enantiomerically enriched (yield = 92%; 20% ee; Potapov et al., 1986).

Since all participants involved in the reaction are liquids, the process is not a

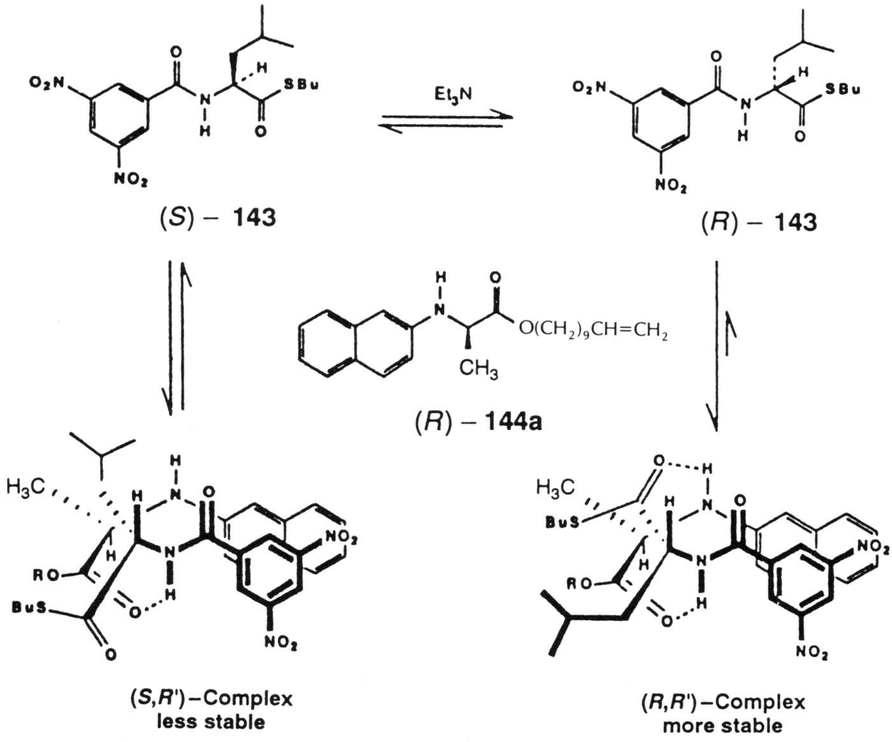

Figure 7.51. Asymmetric transformation on an amino acid derivative. Adapted with permission from Roush, W.R., (1988) *Chemtracts: Org. Chem.*, **1**, 136. Copyright © by Data Trace Chemistry Publishers, Inc.

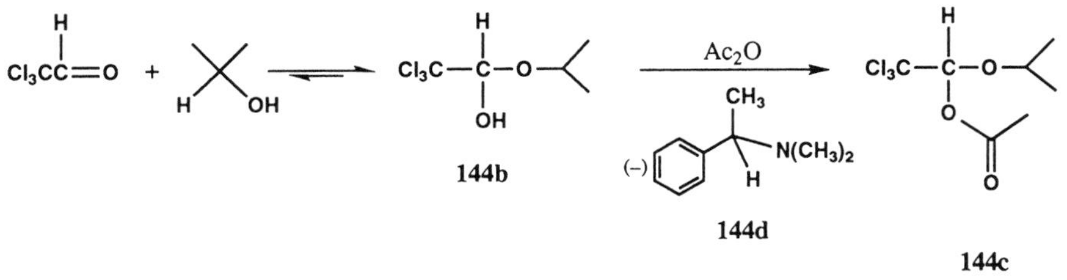

Figure 7.52. Optical activation of hemiacetals.

crystallization-induced asymmetric transformation. It is very similar to processes described in Section 7-2.e as well as to the just-mentioned epimerization of the diastereomeric **143:144** complex. Potapov et al. (1986), correctly pointed out that the acylation of the hemiacetals most likely takes place at a much slower rate than that of the enantiomerization of **144b**. The observed enrichment reflects not the enantiomer composition of the aldehyde–hemiacetal equilibrium (the latter would not necessarily be expected to be 50:50 in a nonracemic solvent) but, as required by the Curtin–Hammett principle (cf. Chapter 10), the outcome is governed by the inherent stereoselectivity s of acylation (Section 7-6), that is, by the relative rates of reaction (or $\Delta\Delta G^{\ddagger}$) of the two **144b** enantiomers in the nonracemic medium.

The optical activation of hemiacetals (Fig. 7.52) is thus not properly an asymmetric transformation nor strictly is it a resolution; it is more accurately designated as an enantioselective synthesis. This example serves to emphasize the fact that the demarcation lines between enantioselective syntheses, asymmetric transformations, and kinetic resolutions can, under appropriate circumstances, be very fuzzy indeed.

f. General Methods for the Separation of Diastereomers

Diastereoisomer mixtures other than those obtained in resolutions frequently need to be separated. These are typically covalent compound mixtures of epimers, anomers, meso and chiral diastereomers and stereoisomers differentiated by permutations at several stereocenters. Mixtures of compounds differing in geometry at double bonds require the same treatment for their separation as do those above-mentioned in which chirality (though often present) is not at issue; separation of chiral (and racemic) or achiral diastereomers from their mixtures is probably more common in organic laboratory practice than is the separation of optically active diastereomers during a resolution.

The components of diastereomer mixtures, like mixtures of constitutional isomers, bear anisometric relationships to one another. Such mixtures may be separated by whatever method seems most expedient. Yet some of the most common and obvious separation methods are of little use in the separation of stereoisomers, to wit, extractive separation based upon acid–base properties. This fact follows from the very character of diastereomers: All members of the set possess the same kind and number of functional groups. While the acid–base properties do differ somewhat, the differences are rarely as large as those, say, between fumaric and maleic acids ($K_{a_1} = 1.0 \times 10^{-3}$ and 1.5×10^{-2}, respectively) and even this difference is difficult to exploit for separation.

Distillation on the other hand, though unexciting and relatively slow, is a perfectly useful and satisfying technique for separating stable stereoisomers differing in boiling point by about 5°C or more (even less if a good spinning band column is at hand) or at least for purifying them. The pivotal demonstration of the existence of two decalin stereoisomers that helped to convince organic chemists that cyclohexane is not planar (cf. Eliel, 1962, p. 279) depended on the ability of Hückel et al. (1925) to separate *cis*- from *trans*-decalin by distillation (the boiling point difference is 8°C; see also Seyer and Walker, 1938). Some additional examples of such separations are that of *cis*- and *trans*-3-isopropyl-cyclohexanols (Eliel and Biros, 1966), and the separation of *meso*- from *rac*-2,4-pentanediol by fractional distillation of the cyclic sulfite esters, bp 72 and 82°C/12 torr, respectively (Pritchard and Vollmer, 1963).

Diastereomers may differ significantly in solubility and this difference may be exploited in the separation of their mixtures. For example, the *meso*-2,4-pentane-diamine·2HCl **145** (Fig. 7.53), has a solubility of 3.3 g/100 mL in boiling EtOH. In contrast, the chiral diastereomer **146**, dissolves in the same medium only to the extent of 0.1 g/100 mL. This large solubility difference has been exploited in a separation described by Bosnich and Harrowfield (1972). Electrooxidative tri-fluoromethylation of methyl acrylate furnishes a 1:1 mixture of *meso*- and *rac*-dimethyl 2,3-bis(2,2,2-trifluoroethyl)succinate **147**. The meso isomer was isolated in pure form on repeated recrystallization of the mixture from hexane. Analogous recrystallization of the mixture from pentane afforded the chiral diastereomer (Uneyama et al., 1989).

The isolation on an industrial scale of the γ-isomer of hexachlorocyclohexane (the insecticide lindane) from a mixture with its diastereomers is based on solubility differences (Colson, 1979).

Even trituration may suffice if the solubility difference is large enough, for example, in the case of the epimeric pair **148** (Fig. 7.54) with boiling toluene (Sheehan and Whitney, 1963). In the same report there is a description of the separation of the *rac-trans*- and *cis*-3-methoxyprolines **149** by conversion to Cu(II) salts. Following the crystallization of their salts from ethanol, the *cis*-**149** compound is regenerated from the less soluble copper salt by passage through an ion exchange column. Washing with hot solvent suffices to separate amides **150** (Fig. 7.53) with a configurational difference at the carbinol carbon (C*) [solubility (g L^{-1}) ratio (S/R) = 12 at room temperature, in CH$_2$Cl$_2$](Fizet, 1986).

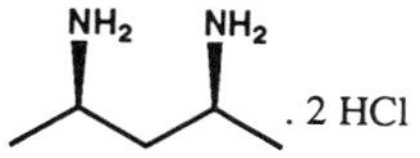

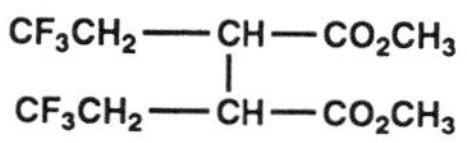

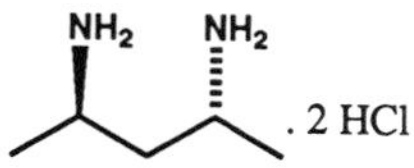

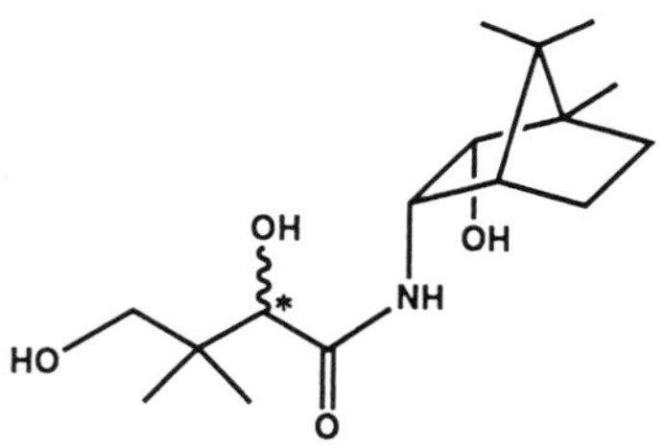

Figure 7.53. Structures **145–147** and **150**.

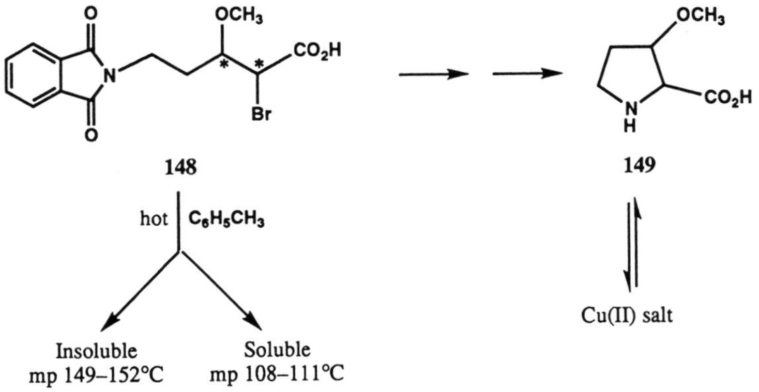

Figure 7.54. Separation of diastereomers (starred stereocenters) by trituration and on formation of Cu(II) salts.

Separation of relatively soluble substances, such as polycarboxylic acids, is facilitated by their conversion to derivatives, such as amine salts (cf. Section 7-4). The *threo-* and *erythro*-2-fluorocitric acids (Fig. 7.55, **151**) are separated by fractional crystallization of the cyclohexylammonium salts of the mixed diethyl 2-fluorocitrates. The erythro isomer is the less soluble one (Dummel and Kun, 1969).

Very efficient and rapid separation of stereoisomeric alcohols from mixtures may be achieved by selective complexation with inorganic cations. For example, either $CaCl_2$ or $MnCl_2$ form alcoholates with *trans-* but not *cis*-4-*tert*-butyl-cyclohexanol, with *erythro-* but not *threo*-3-phenyl-2-butanol, and with only one of four possible decahydro-1-naphthols. Ethanol serves as a catalyst in the complexation (Sharpless et al., 1975). In the partial separation *meso-* and *rac*-2,5-hexanediols, much of the achiral isomer is extracted from a chloroform solution of the two isomers by contacting it with anhydrous powdered $CaCl_2$ (Whitesell and Reynolds, 1983). Several Lasalocid A stereoisomers (Fig. 7.29) are complexed and isomerized concomitantly in the presence of barium hydroxide (Still et al., 1987).

Complexation is a powerful and general method that is applicable to separation of regioisomers as well as to stereoisomers. Mixtures of catechol and hydroquinone, or of the latter with resorcinol, and *m-* and *p*-cresols (the boiling point difference between these cresols is ca. 1°C) yield to separation by treatment with $CaBr_2$ (Leston, 1984).

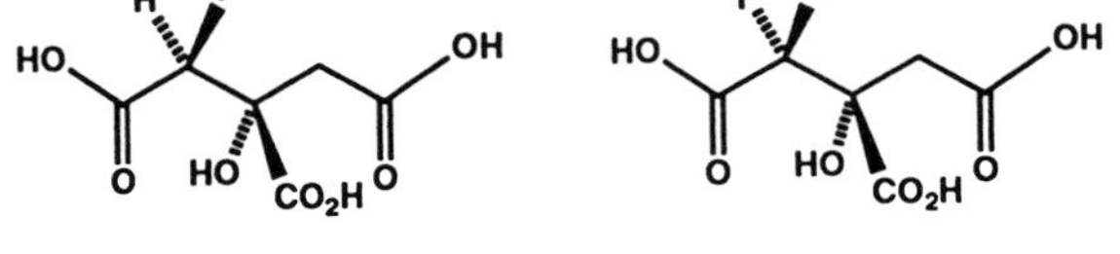

Figure 7.55. Separation of diastereomeric acids by fractional crystallization of their cyclohexyl-ammonium salts: *threo*-**151** (2*S*,3*R* enantiomer shown) and *erythro*-**151** (2*R*,3*R* enantiomer shown).

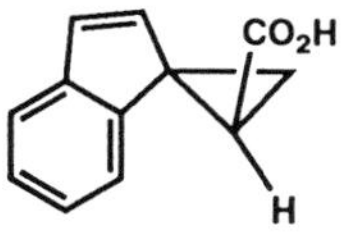

Figure 7.56. Epimeric 2-spiro[cyclopropane-1,1-indene]carboxylic acids.

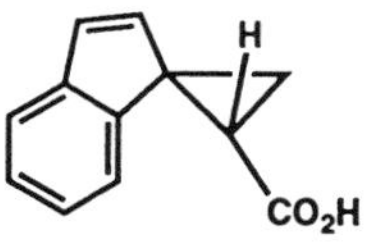

Cooperative separation by serial application of two techniques is not infrequent. A 2:1 mixture of epimeric 2-spiro[cyclopropane-1,1-indene]carboxylic acids (**152** and **153**, Fig. 7.56) was separated by crystallization from hexanes containing ethyl acetate (10:1). The major isomer (trans) was isolated in reasonably pure form (97% trans + 3% cis). Resolution of the trans acid (with quinine) freed it of the cis isomer. The latter accumulated in the resolution mother liquor (Baldwin and Black, 1984).

It is not uncommon to find that one stereoisomer in a mixture crystallizes spontaneously from solution or on standing of the solvent-free mixture. Many classic isolation experiments depend on such fortuitous occurrences. The selective crystallization of one sugar anomer from a mixture consisting of several anomers and alternative ring types (furanoses and pyranoses) exemplifies such separations. Ordinary D-glucose crystallized from aqueous solution is a single species, α-D-glucopyranose (see below).

However, the modern literature is replete with examples of diastereomer mixtures that do not crystallize at all, even following separation. Such mixtures are nowadays separated invariably by chromatography. Just a few examples of such separations follow.

There are numerous reports of the application of flash chromatography to stereoisomer separation including the example given in the original report (Fig. 7.57, **154**; Still et al., 1978). Two fairly representative additional examples are the adamantanone isomers **155** that could not be readily separated and in any event epimerized upon chromatography on silica gel. These isomers were separated as ketals **156** (Fig. 7.57) by flash chromatography on basic alumina (Henkel and Spector, 1983). The *meso-* and *rac-*2,4-pentanediol diastereomers (see above) can also be separated by chromatography of their benzaldehyde acetals (Denmark and Almstead, 1991).

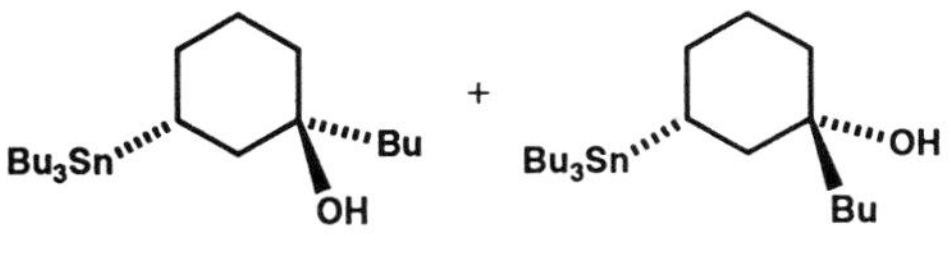

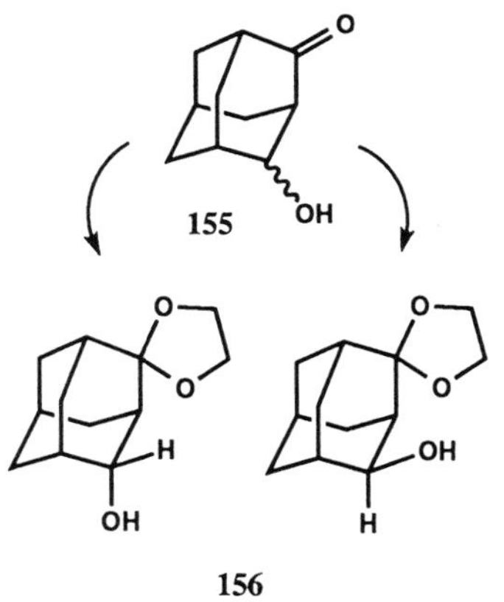

Figure 7.57. Structures **154–156**.

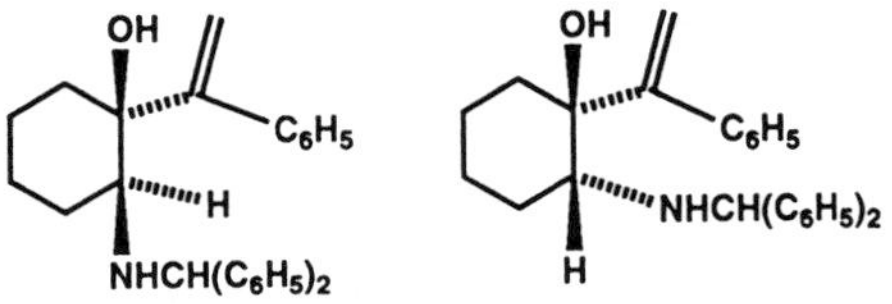

157

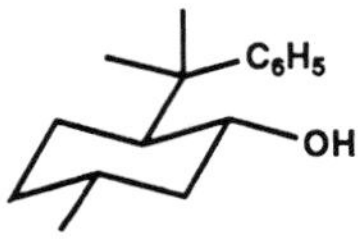

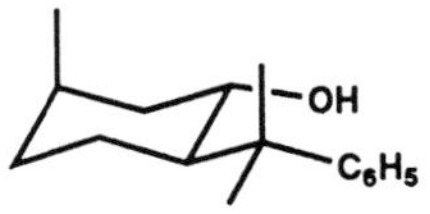

158 Phenylmenthol

159 Epientphenylmenthol

Figure 7.58. Structures **157–159**.

Though gel permeation chromatography is a particularly powerful tool for the separation of macromolecules, it is capable of separation of small molecules as well. Diastereomers of structure **157** (Fig. 7.58) separated in a polystyrene gel column with recycling. The molar volume of the isomers was affected by hydrogen bonding to the solvent mixture (tetrahydrofuran/diisopropyl ether; Lesec et al., 1974).

A difficult diastereomer separation, necessitating the use of an efficient preparative HPLC system (as well as recycling), is that of "phenylmenthol" **158** from "epientphenylmenthol" **159** (Fig. 7.58; Whitesell et al., 1986).

Thin-layer chromatography is a popular technique that has seen much use in stereoisomer separation. An example of a remarkable TLC separation is that of the racemates of all four possible flexible diastereomeric α-methylbenzylamides of biphenyl-2,2′,6,6′-tetracarboxylic acid **160** (Helmchen, Haas, and Prelog, 1973; Fig. 7.59).

Specific separation methods for stereoisomers take advantage of molecular symmetry or proximity of functional groups. The two diastereomeric 1-methoxybicyclo[2.2.2]oct-5-ene-2-carboxylic acids **161** are separated by taking advantage of the ease with which the endo isomer forms a lactone (by iodolactonization). Unreacted exo acid **161** is separated from lactone **162** by extraction with base (Fig. 7.60; Elliott, Urban, and Bordner, 1985).

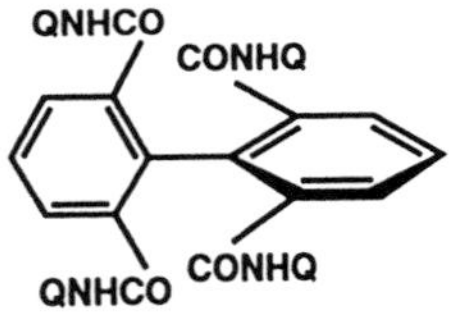

160

$Q = C_6H_5{-}\underset{|}{C}HCH_3$

Isomer*	Point Symmetry	Rf
S, S, S, S	D_2	0.15
2*R*, 6*R*, 2′*S*, 6′*S*	S_4	0.255
2*R*, 6*S*, 2′*S*, 6′*R*	C_2	0.277
2*S*, 6*R*, 2′*R*, 6′*R*	C_1	0.231

Figure 7.59. TLC separation of diastereomeric atropisomers on silica gel (eluent = C_6H_6/CH_2Cl_2/AcOEt 2:2:1); only one enantiomer of each pair is listed.

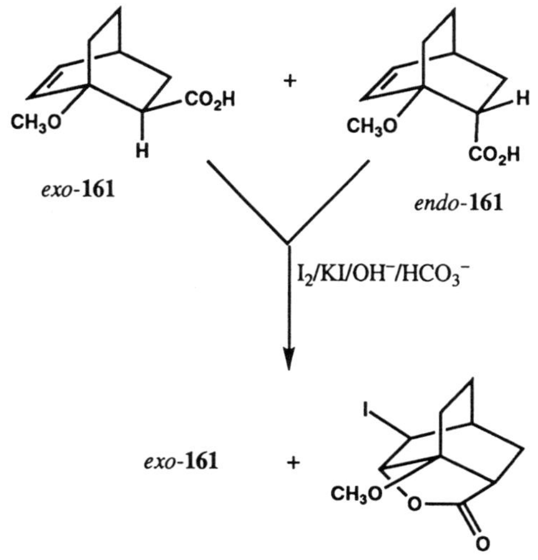

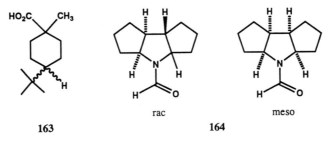

Figure 7.60. Separation of the diastereomeric 1-methoxybicyclo[2.2.2]oct-5-ene-2-carboxylic acids.

Differences in reaction rates may also permit diastereomers to be separated. As an example, 1-methyl-*trans*-4-*tert*-butylcyclohexanecarboxylic acid (Fig. 7.61, **163**) is esterified completely with 10% BF$_3$ in CH$_3$OH in 5 min (steam bath); the cis stereoisomer was similarly esterified only on refluxing for 1.5 h (Krapcho and Dundulis, 1980); for an example involving amide hydrolysis (of **164**) at different rates, see Whitesell et al., 1988.

Derivatization of diols or diol ethers with formation of volatile products (Fig. 7.62, **165**; Brown and Zweifel, 1962) or of insoluble ones (Fig. 7.63, **166**; Eliel and Brett, 1963) similarly permits facile isolation of relatively homogeneous stereoisomers and recovery of the less volatile or less reactive epimers, respectively.

Separation is built into a system through the operation of a crystallization-induced asymmetric transformation (Section 7-3.e). As a consequence of rapid equilibrium among the several possible diastereomers in solution, crystallization of one of these as triggered adventitiously or purposefully by seeding can lead, in principle, to total precipitation of one of the components. One can imagine that other diastereomers could be crystallized instead by changing conditions (e.g.,

Figure 7.61. Diastereomer separation based on rate differences.

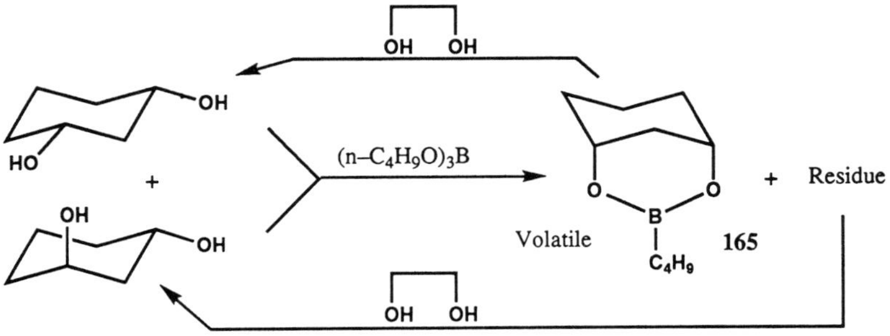

Figure 7.62. Separation of the diastereomeric 1,3-cyclohexanediols.

changing the solvent and/or using seeds of the other diastereomers). Changing the solvent need not result in a change in the equilibrium composition. Only the relative solubilities of the diastereomers need be affected. The two equilibria involved, one homogeneous and the other heterogeneous, are not equally affected by changes in conditions. The sugar series has numerous examples of such separations (e.g., glucose), which is present in solution as an equilibrium mixture of α and β forms. The relative concentrations are little affected by the solvent. However, crystallization from ethanol affords the α anomer while crystallization from warm pyridine leads to the β-anomer.

Separation of diastereomers may also be effected by inclusion compound formation. The *cis*- and *trans*-4-*tert*-butylcyclohexanecarboxylic acids are separable by "complex formation" with thiourea (van Bekkum et al., 1970). *cis*-3,5-Dimethylcyclohexanone **167** (bp 179°C) forms an inclusion compound with 1,1,6,6-tetraphenyl-2,4-hexadiyne-1,6-diol (Fig. 7.35, **112**, X = H) while the trans stereoisomer **168** (bp 179°C) does not (Fig. 7.64; Toda, 1987, p. 55). Similarly, racemic *cis*-permethrinic acid (Fig. 7.30) readily forms inclusion compounds on crystallization from simple aromatic compounds (e.g., thiophene), whereas the trans diastereomer does not (Dvořák, Loehlin, et al., 1992).

Water-soluble diastereomers may be separated by clathrate formation with polyhydroxy macrocycles. Extraction of a 1,4-cyclohexanediol mixture (cis/trans = 53:47) with **169** (Fig. 7.64, R = $(CH_2)_{10}CH_3$) in CCl_4 yields a solution containing diol in a cis/trans = 83:17 ratio; the complexation of diol with the cyclotetramer is stereoselective. Aldohexose diastereomers are also selectively extracted from H_2O into CCl_4 with **169** (Aoyama, Tanaka, and Sugahara, 1989).

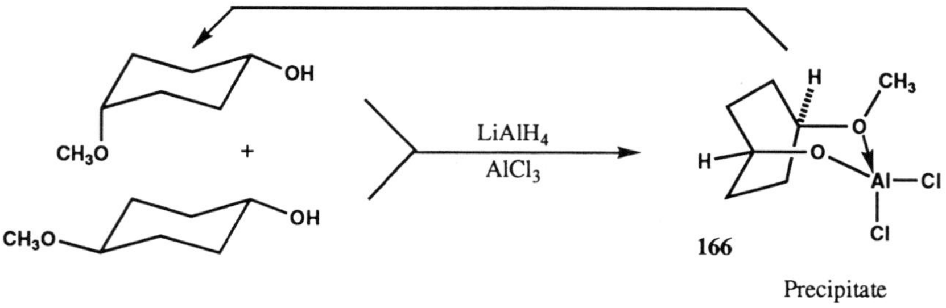

Figure 7.63. Separation of the diastereomeric 4-methoxycyclohexanols. Decomposition of the crystalline chelate with H_2SO_4 affords the cis isomer (>95% cis).

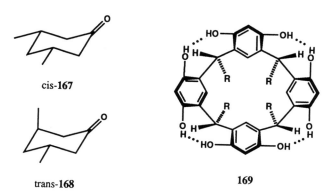

cis-**167**

trans-**168**

169

Figure 7.64. Diastereomer separation by inclusion compound formation and by clathration. [Structure **169** is reproduced with permission from Aoyama, Y., Tanaka, Y. and Sugahara, S. (1989), *J. Am. Chem. Soc.*, **111**, 5397. Copyright © 1989 American Chemical Society.]

7-4. ENANTIOMERIC ENRICHMENT. RESOLUTION STRATEGY

Numerous statements may be found in the literature that describe the enantiomeric enrichment of nonracemic and enantiomerically impure samples. Enantiomerically impure naturally occurring compounds as well as nonracemic samples prepared by resolution or stereoselective synthesis may be subjected to enantiomeric enrichment by crystallization, by sublimation (Section 6-4.g), by chromatography on enantioselective stationary phases (Section 7-3.d), and in some cases even by chromatography on achiral stationary phases (Section 6-4.n). Alternatively the nonracemic samples may be converted to diastereomer mixtures and the latter may be separated by the methods described in Sections 7-3 and 7-7. In this section we address the basis on which enantiomeric enrichment by crystallization takes place (Collet, 1989).

We begin with the finding that vitamin K_3 epoxide **170** prepared by epoxidation of the unsaturated quinone precursor in the presence of a catalytic amount of the chiral phase-transfer agent, benzylquininium chloride **171**, is optically active. The product of this reaction (Eq. 7.13) had $[\alpha]_{587}^{21}$ 0; $[\alpha]_{436}^{21}$ − 6.8 ($CHCl_3$ or acetone). Recrystallization of the product (4–10 times) consistently produced enantiopure material $[\alpha]_{436}^{21}$ − 124 ± 5. It was discovered only after the purification efforts that the enantiomer excess of the product originally obtained from the epoxidation reaction was only 5–10%! Prior estimation of the enantiomer composition "might well have discouraged any but the most optimistic chemists from attempting purification by crystallization" (Snatzke, Wynberg, et al., 1980).

$$\text{(7.13)}$$

170

The enantiomeric enrichment (to ee ca. 100%) of other products with initial enantiomeric purity as low as 50% (or less) by means of a few recrystallizations is perhaps rare (e.g., Pluim and Wynberg, 1979; Bucciarelli et al., 1983; Rossiter and Sharpless, 1984; for other, less dramatic examples, see Fryzuk and Bosnich, 1978; Regan and Staunton, 1987). It is significant that the enantiomeric enrichment by means of a few recrystallizations remains sufficiently remarkable that authors explicitly describe such findings in wonderment.

Enantiomeric enrichment by crystallization from the melt has been demonstrated in the case of the so-called Corey-lactone II (Fig. 7.65) whose pure enantiomer melts at 46°C. The solid product obtained on cooling a molten sample with ee = 29.2% at 5°C was enriched to 72.5% ee. Cooling the melt derived from the same sample to 10°C led to recovery of a sample with ee = 90.7% ee (Ács et al., 1988).

The enantiomeric enrichment of liquids or oily products is easily effected by their conversion to solid derivatives, for example, 1,2-propanediol as the ditosylate (Fryzuk and Bosnich, 1978); arylalkylcarbinols as 3,5-dinitrobenzoates or phenylcarbamates (Červinka, Fábryová, and Sablukova, 1986); see also Fig. 7.1. A noteworthy instance of enantiomeric enrichment of a liquid is that of α-pinene (Fig. 7.18). This enrichment is effected most simply by crystallization from pentane at −120°C (Bir and Kaufmann, 1987); for another method, see below.

In other cases, it has been reported that recrystallization of nonracemic samples leads to a *decrease* in enantiomeric purity. The enriched enantiomer could then be recovered from the mother liquor (often as an oil), for example, Domagala and Bach (1979); Hashimoto, Komeshima and Koga (1979); O'Donnell, Bennett, and Wu (1989). In some instances, it was verified that the apparent reversal in the enrichment process is due to the fact that the racemate is higher melting than are the individual enantiomers.

During the recrystallization of the Wieland–Miescher ketone (Fig. 7.36, **116**) prepared by enantioselective Robinson annulation, it was observed that enantiomeric enrichment took place if the initial enantiomer composition exceeded 75:25 (50% ee), whereas if the initial ee < 50%, the recrystallized product had a lower ee. In the latter case, product isolated from the mother liquor was found to be enantiomerically enriched (Gutzwiller, Buchschacher, and Fürst, 1977).

All of the above observations may be rationalized by reference to phase diagrams for the racemate–enantiomer system under study (Wilen et al., 1977). Ideally, a ternary diagram including the solvent should be used (e.g., Fig. 6.13). Ternary diagrams are rarely available, however. A binary phase diagram (e.g., Fig. 6.7) is often sufficient since the composition of the eutectics found in both the binary and the ternary diagrams are close to one another (Jacques et al., 1981a, pp. 194, 290). Even melting point information often suffices to establish whether

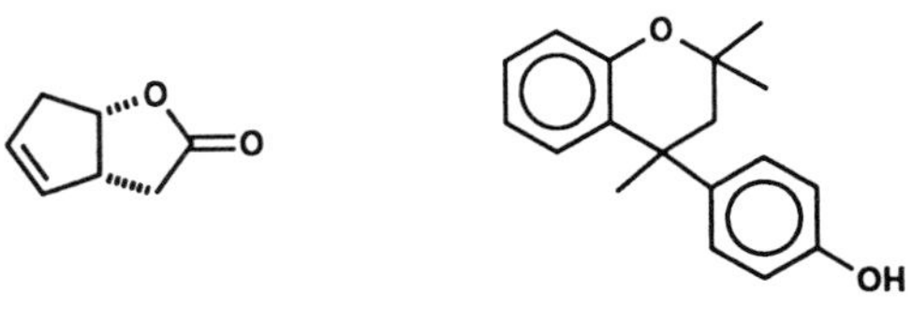

Figure 7.65. "Corey lactone II" (left) and Dianin's compound (right).

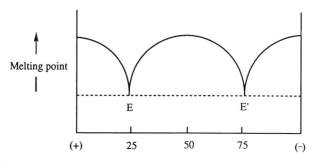

Figure 7.66. Proposed phase diagram of compound **116** (Fig. 7.36).

the racemate is a conglomerate or a racemic compound and, if the latter, whether the racemic compound is very stable (high melting point relative to that of the pure enantiomer; cf. Fig. 6.7*a*). These facts are indicative of the general form of the binary (and by implication also the ternary) phase diagram and roughly of the location of the eutectics; for example, the fact that (+)-ketone **116** (Fig. 7.36) has mp 50–51°C while its racemate has an almost identical melting point (mp 49–50°C) suggests that the phase diagram may have eutectics in the vicinity of enantiomeric compositions 75:25 and 25:75, that is, 50% ee, with respect to either enantiomer (Fig. 7.66), which is precisely the break point where the outcome of its recrystallization undergoes reversal (see above) (Gutzwiller, Buchschacher, and Fürst, 1977).

The following generalizations account for the results observed in enrichment by recrystallization: (a) If the racemate is a conglomerate, then enrichment is observed for *any* initial enantiomer composition (Section 6-4.f; e.g., see Brewster and Prudence, 1973). (b) More frequently, the racemate is a racemic compound. In such cases, if the initial purity is low, with an enantiomer composition less enriched than that of the eutectic, that is, lying on the *racemate* branch of the phase diagram (EE′ in Fig. 7.66), then recrystallization *reduces* the enantiomeric purity of the initial product. A mixture lying on the enantiomer branch [i.e., (+)E or E′(−)], *can* be enriched during recrystallization. If sufficient solvent is taken, then *one* recrystallization suffices to reach 100% ee (Jacques et al., 1981a, p. 192). The latter statement suggests that there is necessarily a trade-off between yield and enantiomer purity; moreover, during the enrichment process it is essential to monitor the enantiomer composition (preferably by a method independent of $[\alpha]_D$).

Purification of carboxylic acids by recrystallization of their ammonium salts has the advantage of furnishing highly crystalline derivatives. Acids that are enantiomerically impure are often incidentally enriched during the recrystallization process (e.g., Hengartner, Valentine, et al., 1979); for the use of dicyclohexyl- and dibenzylammonium salts in the enantiomeric enrichment of carboxylic acids see Kikukawa, Tai, et al. (1987). Whereas nonracemic 2-phenoxypropionic acids (Fig. 7.23*b*, **89a**, R = H; ep 75–88%) cannot be purified by recrystallization, their *n*-propyl-, cyclohexyl-, or dicyclohexylammonium salts can. Note that all of these are salts of chiral carboxylic acids with *achiral* amines, that is, the salts are those of the enantiomers. Comparison of the binary phase diagrams of the acids with those of the ammonium salts reveals the reason for the difference in recrystallization outcome. For **89a** (R = H, and *o*-, *m*-, *p*-Cl, or NO$_2$) the composition of the eutectics shifts from a range encompassing 10:90 and 1:99 (79–97% ee) in the case of the acids to one encompassing 20:80 and 50:50 (61–0% ee; i.e., one of the salts is a conglomerate) for the *n*-propyl- or cyclohexylammonium salts studied (Gabard and Collet, 1986). The results and their import are best appreciated by examination of the phase diagrams (Fig. 7.67), where the typical shape of the fusion behavior of the acids as a function of their enantiomer composition is that of the dotted line (*a*), whereas that of most of the salts is represented by the solid line (*b*). For an acid whose initial composition is represented by M, it is evident from the diagram that recrystallization of the acid would furnish the racemate, whereas recrystallization of a salt of the acid might well yield enantiopure product (eventually the acid as well) with the maximum yield of enantiopure material (in the form of either acid or salt) given by ME/DE. Analogously, nonracemic ibuprofen (Fig. 6.31, **48**) and naproxen (Fig. 7.23*a*, **83**) have been enantiomerically enriched by crystallization of their sodium salts (Manimaran and Stahly, 1993).

> Recrystallization of nonracemic *n*-propylammonium 2-phenoxypropionate of composition M (75% ee) can give a product with 100% ee with the theoretical yield given by ME/DE (60%, Fig. 7.67). The maximum yield obtainable in the crystallization of a nonracemic sample is found when the composition of the eutectic E is 50% (i.e., ME/DE is largest when E = 0.5). The phase diagram then corresponds to that of a conglomerate (Fig. 6.3*a*). Hence, when the yield of recovered enantiopure material *must* be maximized, a search for a derivative whose racemate exhibits conglomerate behavior is worthwhile (see Section 7-2.b).

If an enantiomer mixture has a composition corresponding to that of the eutectic (E in Fig. 7.66), then enrichment by crystallization becomes quite difficult. Such a situation is encountered in the asymmetric transformation of 2,2'-dihydroxy-1,1'-binaphthyl (Fig. 7.36, **113**; Section 7-3.e). Recrystallization of enantioenriched samples leads to a product that does not exceed about 82% ee (evidently, the composition of the eutectic). During the crystallization trials it was observed that two types of crystals were formed, one being the enantiopure diol (faster crystallizing), the other the racemic diol (the latter, crystallizing more slowly, is a racemic compound). By taking advantage of the different rates of crystallization of the two types of crystals on recrystallization of the eutectic mixture (kinetic control, with or without seeding with enantiopure diol), it was

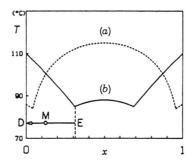

Figure 7.67. Binary phase diagram (*a*) of 2-phenoxy-propionic acid (Fig. 7.23*b*, **89a**, R = H) and (*b*) of its *n*-propylammonium salt (*x* is the mole fraction of one of the enantiomers). [Adapted from Gabard, J. and Collet, A. (1986), *New J. Chem.*, **10**, 685. Reprinted with permission of Editions Gauthier-Villars.]

possible to surmount the "thermodynamic obstacle" and to obtain enantiopure **113** (Smrčina, Kočovský, et al., 1992).

Enantiomeric enrichment to 100% ee of nonracemic samples of certain 2° and 3° alcohols, e.g., Fig. 7.35, **111**, X = Cl, has been effected by crystallization of a 1:2 complex (amine:2 ROH) with 1,4-diazabicyclo[3.3.3]octane (DABCO), a rigid achiral amine (Toda and Tanaka, 1986). It was subsequently established that the enrichment is due to crystallization of the much stronger homochiral (−)-**111**·DABCO·(−)-**111** (X = Cl) complex relative to the corresponding heterochiral (+)-**111**·DABCO·(−)-**111** complex (Yasui, Kasai, Toda, et al., 1989).

When the racemate of a chiral compound (e.g., *cis*-permethrinic acid, Fig. 7.30) forms a centrosymmetric inclusion compound on crystallization from a given solvent, whereas the individual enantiomers cannot (this is also true of Dianin's compound, Fig. 7.65; Brienne and Jacques, 1975), enrichment of an enantiomerically impure sample is possible. Recrystallization of a sample of *cis*-permethrinic acid of 37% ee from thiophene led to crystallization of the racemic thiophene inclusion compound. The acid recovered from the mother liquor had ee = 87% (Dvořák and Loehlin, et al., 1992).

A process has also been described in which formation of a solid solution between constitutionally analogous compounds, *p*-anisyl α-methylbenzyl ketone (Fig. 7.11, **15b**) and its thio analog **15c** (Fig. 7.11), forms the basis of a resolution of the latter. Crystallization of (+)-**15b** (400 mg) in the presence of (±)-**15c** (200 mg) at room temperature gives rise to a large crystal (223 mg) containing 92% of (+)-**15b** and 8% of **15c**. Analysis of the crystal by DSC revealed that the crystal consists of but a single-crystal phase (a solid solution). Chromatographic separation of **15b** from **15c** led to recovery of 13.4 mg of (+)-**15c** (98% ee; Garcia and Collet, 1992).

Purification of mixtures of diastereomers (by chromatography or by recrystallization) is the alternative to the above process, that is, diastereomer enrichment leading ultimately to enantiomer enrichment when the diastereomers are cleaved. When chromatography is employed, the result is straightforward; this process has been examined in Section 7-3.d. Diastereomer enrichment by recrystallization is usually effective because the shape of the unsymmetrical phase diagrams (either binary or ternary) is most often analogous to those of conglomerate systems, that is, there is but a single eutectic present (as in Fig. 7.28 (*b*); Jacques et al., 1981a, pp. 289, 342).

> Complications arise mainly as a result (a) of solubility differences between the diastereomers that are too small, and (b) the occurrence of solid solutions especially at the extremes of the phase diagrams (corresponding to partial mutual solubility of diastereomers, salts in particular, in their respective crystal lattices); for an example, see Viret, Patzelt, and Collet (1986). While these complications may reduce the efficiency of the enrichment process, they do not often completely prevent it (Collet, 1989).
>
> For an analysis of the question: Is it best to continue recrystallizing a mixture of diastereomers or to cleave the latter and push on with the enantiomeric enrichment of the recovered enantiomer mixture, see Jacques et al. (1981a, p. 426).

While enantiomeric enrichment may indeed be carried out by conversion of a nonracemic sample into a mixture of diastereomers, e.g., of α-methylbenzylamine into its tartrate followed by recrystallization of the latter (see Leeson et al., 1990), enrichment by crystallization of conglomerate salts formed from the amine and *achiral* acids (e.g., cinnamic acid; see the example in Section 7-2.b) or *o*-chlorobenzoic acid (Shiraiwa et al., 1985b), should be just as effective and likely be less expensive. Enantiomeric enrichment has also been observed on sublimation of mixtures of diastereomeric salts (Ács, von dem Bussche, and Seebach, 1990).

Two special procedures involving chemical reactions exist for the attainment of enantiomeric purity from enantiomerically impure samples. The first, called "duplication," requires that two molecules of the impure sample be chemically combined into one new molecule by whatever process seems best for the case at hand, for example, two amine molecules having opposite chirality sense may be incorporated in a neutral *achiral* nickel complex. Following the removal of the complex, enrichment is observed in the uncomplexed amine (Hanotier-Bridoux et al., 1967). This type of process has been generalized by Horeau especially for the enantiomeric enrichment of enantiomerically impure alcohols (through formation of carbonate, malonate, or phthalate esters, e.g., Eq. 7.14; for a recent application, see Fleming and Ghosh, 1994), and of carboxylic acids, through formation of the anhydride (Vigneron, Dhaenens, and Horeau, 1973).

$$RS \quad \text{meso (liquid) +}$$
$$RR/SS \quad \text{chiral (solid)}$$

The enantiomeric purity p' (ee) of the alcohol recovered from the chiral RR/SS diastereomer following separation (by chromatography or by crystallization) of the latter from the concomitantly formed meso RS diastereomer is given by Eq. 7.15 where p is the enantiomeric purity of the starting nonracemic alcohol (Jacques et al., 1981a, p. 430; Kagan and Fiaud, 1988):

$$p' = \frac{2p}{1 + p^2} \qquad (7.15)$$

Equation 7.15 is derived as follows: Let the concentrations of the nonracemic alcohol to be enriched be x and $x - 1$ for the R and S enantiomers, respectively. If it be assumed that the duplication (Eq. 7.14) be bimolecular and first order with respect to each reactant, and that no kinetic discrimination occurs, then the quantities of chiral RR and SS diastereomer formed will be proportional to x^2 and $(1 - x)^2$ and that of the meso RS diastereomer proportional to $2x(1 - x)$. Expressing these quantities in terms of the enantiomeric purity p of the starting alcohol, the quantity of the dominant enantiomer being $(p + 1)/2$ (Section 6-5.a), then those of RR, SS and RS are proportional to $(p + 1)^2$, $(p - 1)^2$ and $2(p - 1)^2$, respectively. Following the removal of the meso diastereomer, the enantiomer purity p' of the remaining chiral diastereomer is

$$p' = \frac{(p + 1)^2 - (p - 1)^2}{(p + 1)^2 + (p - 1)^2} = \frac{2p}{1 + p^2} \qquad (7.15a)$$

One of the more noteworthy applications of this enrichment process is to the purification of α-pinene (Fig. 7.18). The liquid terpene (bp 155–156°C; 91% ee) was hydroborated to crystalline tetra-3-pinanyldiborane [commonly called diisopinocampheylborane (Ipc$_2$BH) ignoring the dimeric character of the borane] in the presence of 25% excess of α-pinene (Fig. 7.68). Liberation of the α-pinene by reaction of the crystalline [Ipc$_2$BH]$_2$ with benzaldehyde is catalyzed by BF$_3$·OEt$_2$. The isolated α-pinene, $[\alpha]_D^{23} + 51.4$ (neat) has ee = 99.6% (Brown and Joshi, 1988; see also Brown, Jadhav, and Desai, 1982; Brown and Singaram, 1984; application to longifolene and to 3-carene: Jadhav, Prasad, and Brown, 1985).

This improvement in enantiomeric purity of α-pinene is actually due to the equilibration and selective crystallization of the chiral tetra-3-pinanyldiborane diastereomer from a mixture containing the latter in admixture with the meso diastereomer (Jacques et al., 1981a, p. 430). In other words, the observed enantiomeric enrichment is an instance of the duplication process of Horeau (calculation of p' by means of Eq. 7.15 gives a value of 99.6%, precisely the value found by Brown and Joshi).

The other special enrichment process is essentially a kinetic resolution carried out under rigorously specified conditions (Horeau, 1975). It involves the stereoselective reaction, for example, of nonracemic α-methylbenzyl alcohol, C$_6$H$_5$CH(OH)CH$_3$, with an insufficient but specified quantity of (−)-α-phenyl-butyric anhydride, [C$_6$H$_5$CH(C$_3$H$_7$)CO]$_2$O. The required conversion may be calculated by means of Eq. 7.16 where p and p' have the same significance as in Eq. 7.15 and F is the specified conversion of the reaction.

$$\frac{1}{(1-F)^{(k_D/k_L)}} = \frac{1-p'}{1-p}\left[\frac{1+p}{1+p'}\right]^{(k_D/k_L)} \tag{7.16}$$

If the relative rate $k_D/k_L = 4$, then an enrichment from 95 to 99% ee is calculated to require a conversion of 47% of the alcohol to ester. The enantiomeric purity of the residual alcohol would be 99% ee. The required relative rate may be readily obtained from the ratio of diastereomers formed in the reaction of rac-(−)-α-phenylbutyric anhydride with rac-α-methylbenzyl alcohol as measured chromatographically or by NMR spectroscopy (see also Brandt, Ugi, et al., 1977, and Kagan and Fiaud, 1988, p. 262).

The amplification of small enantiomer excesses by the foregoing mechanism is one possible route for the prebiotic generation of nonracemic compounds in nature. Briaucourt and Horeau (1979) demonstrated experimentally that starting from reactants (an alcohol and an acyl chloride) having an enantiomer purity not exceeding 0.1% each, successive incomplete esterifications generated residual nonracemic products whose enantiomeric purity reached 98%.

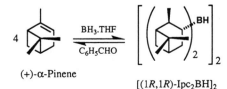

(+)-α-Pinene   [(1R,1R)-Ipc$_2$BH]$_2$

Figure 7.68. Purification of α-pinene by duplication to tetra-3-pinanyldiborane.

7-5. LARGE SCALE RESOLUTION

As a rule, the resolution of a racemate yields only 50% of its mass in the form of the desired enantiomer. This fact may lead to the misconception that resolutions are impractical on a large scale. The economic consequences of the 50% maximal yield are real enough, so much so that present-day efforts in the development of enantioselective syntheses, where such a limitation does not exist (cf. Chapter 12), are outstripping the output of research work on resolutions; for surveys of industrial scale enantioselective syntheses, see Scott (1989) and Collins et al. (1992). The limitation in the yield of resolutions may, of course, be mitigated if the undesired enantiomer can be racemized, or if the resolution and racemization can be combined in an asymmetric transformation. These tactics have actually been built into commercial resolutions. Some statistics bearing on resolutions are summarized in Figure 7.69. The numbers in the figure (ordinate) refer to numbers of articles, reports, and patents dealing specifically with resolutions and do not reflect the much larger number of resolutions reported in the literature.

The numbers, and the trends, shown in Figure 7.69 suggest that resolution remains a viable and practical unit process to nonracemic compounds. Moreover, one can gain an idea of the potential use of resolutions in the preparation of optically active compounds on a large scale from the number of patents that are granted for new resolution processes. By one count, 1527 patents dealing with resolutions were granted from 1967 to 1991, inclusive. This number represents 25% of all publications dealing with resolutions entered into the *Chemical Abstracts* data base (CA File) over the 25-year period. Moreover, thus far, the number of patents granted for resolutions continues to significantly outstrip those granted for "asymmetric syntheses." The former group of patents comprises all of

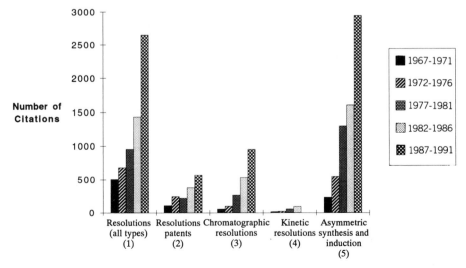

Figure 7.69. Literature survey of resolutions (1967–1991, inclusive): left to right, (1) the first set of bars refers to the number of articles, reports, and patents dealing with resolutions of all types; (2) number of resolution patents; (3) number of chromatographic resolution articles and patents; (4) number of kinetic resolution articles and patents; (5) number of asymmetric synthesis and asymmetric induction articles and patents (CAS Online search). Bar sets (2)–(4) are subsets of the first.

the principal types of resolution, namely, those involving diastereomer separation, those effected by preferential crystallization of enantiomer mixtures (Section 7-2.c), by kinetic resolution (Section 7-6) and by chromatographic resolution (Section 7-3.d).

While this large number of patents gives a very clear idea of the interest in commercializing resolutions, it tells one nothing about the actual application of resolutions to large-scale commercial processes. Yet the enormous investment in research and development required to produce the above-mentioned patents would hardly have been made were there not a reasonable expectation that some of the processes developed would be economically viable.

In fact, there are strong indications that all but the last type of resolution (chromatographic resolution) are presently being used industrially and on a large scale in the production of medicinal, agricultural, and food products (Books: Collins et al., 1992; Sheldon, 1993. Reviews: Sheldon et al., 1985; Sheldon, 1990; Sheldon et al., 1990; Crosby, 1991; Stinson, 1992, 1993). Since details and even essential information about such applications are hard to find in the open literature (reports of large scale resolutions, i.e., kilogram scale and up, are scattered in the literature, but most often are not described at all), only a few examples will be given whose authenticity appears not to be in doubt.

A point that needs to be made explicitly is that, on the industrial scale, the selection among known optical activation methods is strictly made on the basis of economics. That is why a "perfectly good" existing commercial resolution method (one with high chemical and enantiomer yields, with allowance being made for possible racemization and reuse of the undesired substrate enantiomer) may be abandoned in favor of another, for example, it may be replaced by a more economical kinetic resolution process, or a resolution may be entirely replaced by a stereoselective synthesis.

a. Diastereomer-Mediated Resolution

(R)-(−)-Phenylglycine (Fig. 7.20, **51a**) is an intermediate in the synthesis of semisynthetic penicillins (e.g., ampicillin and cephalosporin antibiotics). A key

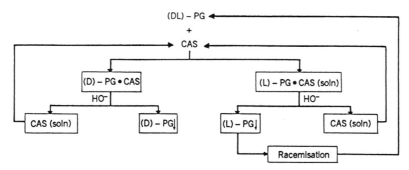

Figure 7.70. Commercial resolution of phenylglycine (PG) with (+)-10-camphorsulfonic acid (CAS) [Reproduced with permission from Sheldon, R.A., Porskamp, P.A. and ten Hoeve, W. (1985) in *Biocatalysts in Organic Syntheses*, Tramper, J., van der Plas, H.C. and Linko, P., Eds., Elsevier Science Publishers, p. 67.]

step in the DSM/Andeno process for the manufacture of **51a** is a conventional resolution involving the selective crystallization of the salt with (+)-10-camphorsulfonic acid (Fig. 7.21*a*, **64a**). The resolution is a batch process carried out in vessels larger than 5000 L; annual production is in excess of 1000 metric tons per year (Sheldon, 1990). The crystallization is rapid and the process parameters can be adjusted so as to obtain the desired diastereomeric salt in high yield (>40%, with 50% being the maximum attainable) and a high diastereomer ratio ($n:p >$ 95:5).

Two requirements must be met for such a resolution to be economically profitable or even viable. The resolving agent must be easily recoverable and the unwanted enantiomer of the substrate (remaining in the mother liquor) must be readily racemized and recycled. Both of these requirements are met in the phenylglycine resolutions that produces on the order of 100 kg per batch. The process is illustrated schematically in Figure 7.70.

Similar large-scale resolutions are also presently carried out on *p*-hydroxyphenylglycine (Fig. 7.20, **51b**; see also below), which is an intermediate in the production of the antibiotic amoxicillin and on (*S*)-acetyl-β-mercaptoisobutyric acid (Houbiers, 1980; de Heij, 1981), an intermediate in the production of captopril (Fig. 7.71), which is a widely used antihypertensive agent that operates via inhibition of the angiotensin converting enzyme (ACE). The manufacture of the adrenergic (beta) blocker timolol also involves a resolution, with an acidic resolving agent (Wasson et al., 1972; Lago, 1984; Fig. 7.71).

The photostable pyrethroid deltamethrin (Fig. 7.72) is one of eight possible stereoisomers having the same constitution. Only two of these (1*R*,3*R*,αS and 1*R*,3*S*,αS) have any insecticidal activity (Tessier, 1982). The former, having the highest activity, is prepared on an industrial scale at Roussel-Uclaf (Davies et al., 1989) in a process that incorporates a resolution, namely, that of *trans*-chrysanthemic acid with (1*R*,2*R*)-(−)-*threo*-2-(*N*,*N*-dimethylamino)-1-(4-nitrophenyl)-1,3-propanediol (Fig. 7.72, **172**) serving as resolving agent, the latter being derived from (−)-**40b** (the enantiomer of which is shown in Fig. 7.19*b*; see below), an intermediate in the synthesis of chloramphenicol (Martel, 1970).

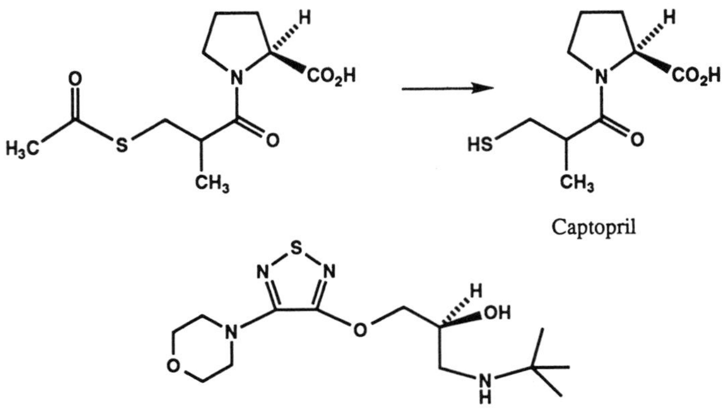

Captopril

Timolol

Figure 7.71.   Products of large-scale diastereomer-mediated resolutions.

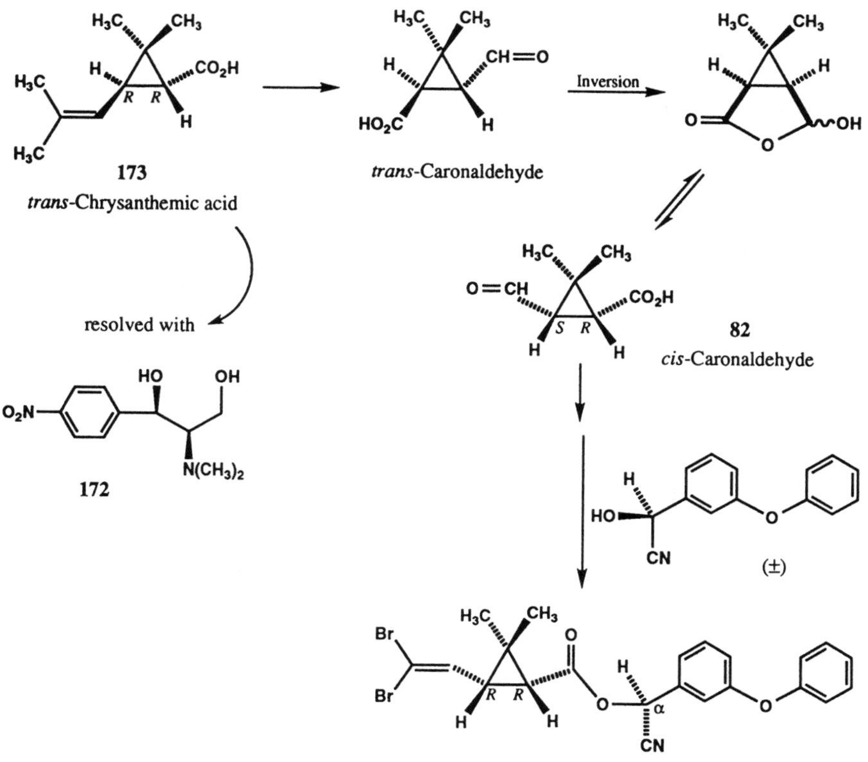

Figure 7.72. Resolution and asymmetric transformation in the commercial synthesis of deltamethrin.

Ozonolysis of $(1R,3R)$-$(+)$-*trans*-chrysanthemic acid (Fig. 7.72, **173**), leads to *trans*-caronaldehyde that epimerizes via formation of a cyclic *cis*-lactol derivative. The same cyclic $(1R,3S)$-*cis* lactol is accessible from the enantiomeric $(1S,3S)$-$(-)$-*trans*-chrysanthemic acid by prior epimerization at C(1) to the $(1R,3S)$-*cis*-chrysanthemic acid, thus insuring that both enantiomers of the resolved *trans*-chrysanthemic acid are incorporated into the most active deltamethrin enantiomer (Reviews: Tessier, 1982; Martel, 1992). Thus, this comprises a stereoconvergent synthesis of $(1R,3S)$-*cis*-caronaldehyde or its lactol. The resulting resolved *cis*-caronaldehyde (Fig. 7.72, **82**) is converted to a dibromovinyl cyclopropanecarboxylic acid that is coupled with a *racemic* diphenyl ether cyanohydrin. The desired deltamethrin diastereomer is isolated as the result of a crystallization-induced asymmetric transformation involving epimerization at $C(\alpha)$ (Warnant et al., 1977; Fig. 7.72).

Both naturally occurring and synthetic resolving agents are used in industrial resolutions with the advantage of the latter being the possibility of insuring a supply of the resolving agent by carrying out reciprocal resolutions of the resolving agent on demand.

b. Resolution by Preferential Crystallization

Demand for the broad-spectrum antibiotic chloramphenicol (Fig. 7.73*a*) led, in the early 1950s, to the development of a resolution process permitting synthesis of the (+) enantiomer (rotation measured in ethanol) on a large scale. The pioneering work of Amiard (1956, 1959) and Velluz, et al. (1953a) led to the application of the preferential crystallization process to the intermediate (−)-*threo*-2-amino-1-(*p*-nitrophenyl)-1,3-propanediol (Fig. 7.19*b*, **40b**). Tens of metric tons of chloramphenicol were produced at Roussel-Uclaf in 1958 by this method (Amiard, 1959). The synthesis of the analogous antibacterial agent thiamphenicol (Fig. 7.73*a*) involves a resolution by preferential crystallization of the thiomethyl amino alcohol precursor (Fig. 7.73*b*). Since there are two stereocenters in the amino diol molecule, racemization of the undesired *S,S* enantiomer is not straightforward. One approach is to convert the amine to a phenyloxazoline derivative thereby protecting the amino group; oxidation of the underivatized alcohol functionality (α-carbon atom) affords a racemic ketone. Diastereoselective reduction and deprotection returns the carbon skeleton to the racemate pool for resolution (Fig. 7.73*b*; Giordano et al., 1988).

In an apparently independent development, widespread use of the flavor enhancing agent monosodium glutamate (MSG) in Japan was the impetus for the commercialization of a preferential crystallization process for glutamic acid (Wakamatsu, 1968). As much as 13,000 metric tons per year of (*S*)-glutamic acid (Fig. 7.73*a*) was produced by the Ajinomoto Company during the years 1963–1973 in a completely automated plant situated near Nagoya. The process incorporated a racemization step with recycling thereby avoiding the waste of the undesired enantiomer. (*S*)-Glutamic acid is now commercially produced by microbiological fermentation of molasses or starch (Kawakita, 1992).

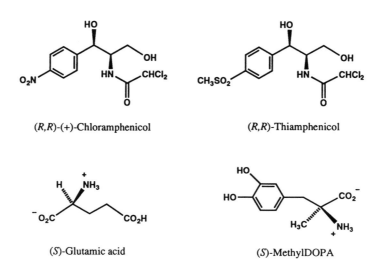

(*R,R*)-(+)-Chloramphenicol   (*R,R*)-Thiamphenicol

(*S*)-Glutamic acid   (*S*)-MethylDOPA

Figure 7.73 (*a*). Products of large-scale resolutions by preferential crystallization.

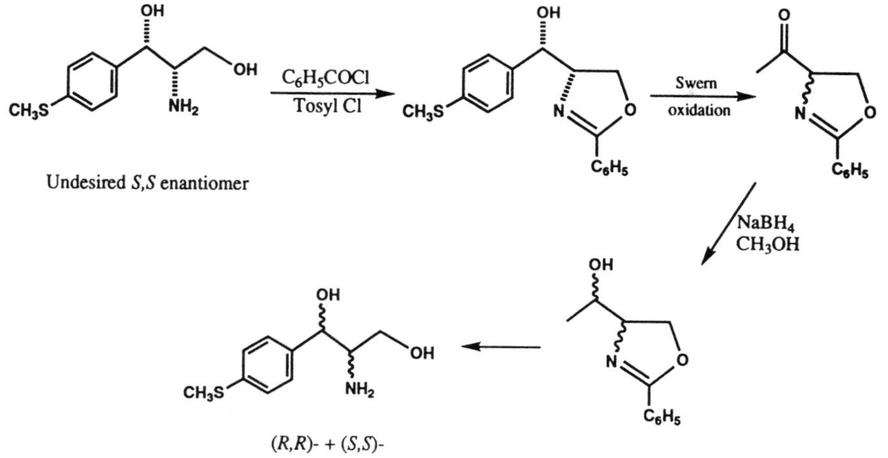

Figure 7.73 (*b*).   Racemization of a thiamphenicol intermediate.

The antihypertensive agent L-3-(3,4-dihydroxyphenyl)-2-methylalanine (3-hydroxytyrosine; (*S*)-methylDOPA; Fig. 7.73*a*) is also resolved by preferential crystallization on a large scale (Lago, 1984). A plant capable of producing 3×10^5 lb/year of methylDOPA by preferential crystallization of the bisulfite salt was constructed and operated by Merck & Co., in Albany, Georgia for this purpose (Anon., 1965b). A limitation of the resolution of methylDOPA is that it cannot be racemized without destroying the molecule. On the other hand, the resolution can be performed (also by preferential crystallization) on α-acetylamino-α-vanillylpropionitrile (Fig. 7.73*c*, **A**; Reinhold et al., 1968; Krieger et al., 1968), an intermediate that is easily convertible into methylDOPA. The α-acetamidonitrile intermediate can be racemized by reaction with NaCN in DMSO (Fig. 7.73*c*), hence circumventing the 50% limitation on the yield (Firestone et al., 1968).

Since neither racemic nor (+)-menthol serves the purpose, (−)-menthol (Fig. 7.19*c*, **49**), a flavoring and "cooling" agent in cigarettes, has been obtained by resolution on a large scale by preferential crystallization of the benzoate (Fleischer et al., 1976; Davis, 1978). Much (or all) of the optical activation of menthol by preferential crystallization has given way to a more efficient synthesis involving an enantioselective alkene isomerization (Tani, Otsuka, et al., 1984. Reviews: Scott, 1989; Akutagawa, 1992).

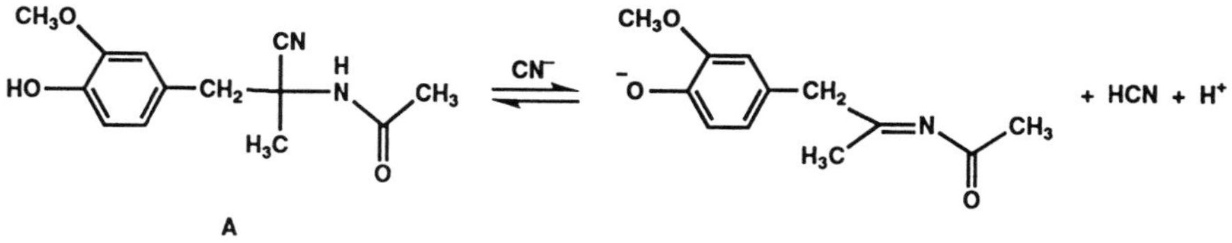

Figure 7.73 (*c*).   Mechanism of racemization of α-methylDOPA intermediate **A** (Firestone, et al., 1968).

Industrial scale resolution of several other amino acids by preferential crystallization of salts (sulfonates or hydrochlorides) was claimed to be "promising" (Yamada et al., 1973a). A semicontinuous batch process and a fully continuous process for resolution of *N*-acetylleucine (Fig. 7.21*b*, **71**), incorporating a total spontaneous resolution (Section 7-2.e) has been described by Hongo, Yamada, and Chibata (1981b).

c. Kinetic Resolution

The selective hydrolysis of *N*-acyl derivatives of amino acids mediated by aminoacylase enzymes is a kinetic resolution process that is highly efficient (Section 7-6). It has been widely adopted for the commercial production of optically active amino acids (Kaneko et al., 1974, p. 33; Izumi et al., 1978). Among the compounds resolved in this way are L-phenylalanine, L-tryptophan, L-valine, L-methonine, and L-DOPA. The economically most attractive variant of this enzymatic resolution method is that in which the enzyme is immobilized on a water-insoluble carrier (Review: Bommarius, et al., 1992), for example, ionic binding to a polysaccharide (DEAE Sephadex) support. Immobilized aminoacylases have increased stability such that they may be used without degradation for long periods of time making continuous production feasible (Tosa, Chibata et al., 1969a,b; Kaneko et al., 1974; Izumi et al., 1978; Chibata, 1982; Chibata, et al., 1992). A more recent development is illustrated by resolution of ethyl *N*-benzoyl-(±)-tyrosinate with a lipase immobilized in a hollow-fiber membrane reactor (Stinson, 1992).

D-*p*-Hydroxyphenylglycine (D-HPG) is kinetically resolved microbiologically by hydrolysis of *rac-p*-hydroxyphenylhydantoin, that is, the hydrolysis is catalyzed with immobilized cells of *Bacillus brevis*. A second process uses *Agrobacterium radiobacter* to catalyze the hydrolysis of the hydantoin (Fig. 7.74; Sheldon et al., 1985). Both processes benefit from the fact that the substituted hydantoin spontaneously racemizes during the enzymatic hydrolysis (cf. Section 7-3.e).

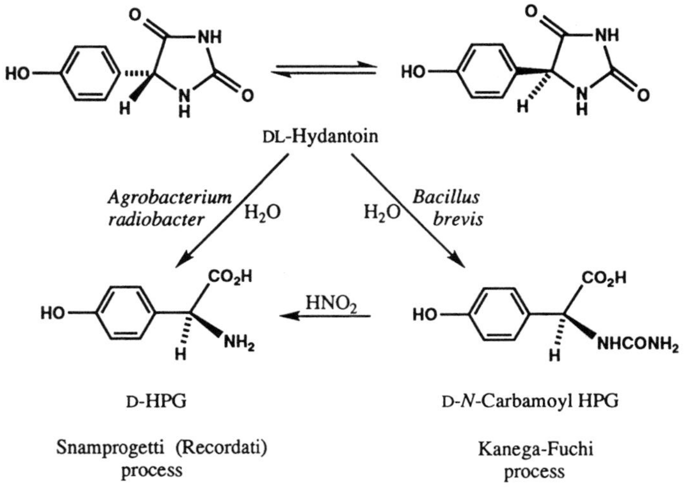

Figure 7.74. Enzymatic resolution of *p*-hydroxyphenylglycine (HPG). (Sheldon, et al., 1985).

Other commercial resolution processes for amino acids are discussed by Kaneko et al. (1974); Izumi et al. (1978); Sheldon et al. (1985); and Collins et al. (1992).

Ibuprofen (Fig. 6.31, **48**), an important antiinflammatory medicinal agent, has been resolved by kinetic resolution on a pilot plant scale in a bioreactor containing an immobilized enzyme (Brandt, 1989).

7-6. KINETIC RESOLUTION

A kinetic resolution is a chemical reaction of a racemate in which one of the enantiomers forms a product more rapidly than the other. The rate difference arises from a difference in E_a required to reach the transition states (for the respective enantiomers of the substrate) of the reaction (Eq. 7.17). Recovery of the unreacted enantiomer, (S)-A, in nonracemic form constitutes a resolution. The other enantiomer (R)-A may often be recovered as well (from product B) in nonracemic form by reversing the original enantioselective reaction in a nonselective manner (Eq. 7.17; k_R and k_S are rate constants).

$$(R,S)\text{-A} \xrightarrow[\substack{\text{Chiral} \\ \text{reagent}}]{k_R > k_S} \text{B} + (S)\text{-A} \qquad (7.17)$$
$$\big\downarrow$$
$$(R)\text{-A}$$

The following conditions apply to kinetic resolutions:

1. If A and the chiral reagent are in stoichiometric ratio and sufficient time is allowed, both enantiomers of A are converted to product B and no resolution is effected. In order to be of practical use, it is essential that the reaction be stopped at some point short of 100% conversion. Either adjustment of the reaction stoichiometry or of the reaction time (together with monitoring of the enantiomeric composition of the unreacted resolution substrate) may be used to control the extent of conversion.

2. Product B may be either chiral or achiral. If B is achiral, the reaction still constitutes a kinetic resolution but then only one enantiomer of the original racemate can be recovered in the resolution.

3. A chiral and nonracemic reagent, a chiral solvent, or a chiral physical force, must be present in order for a kinetic resolution to take place, but the reagent need not be present in stoichiometric amount, that is, it can be a chiral catalyst.

For an extensive and general review of kinetic resolutions, see Kagan and Fiaud (1988); see, also, Eliel (1962; p. 65); Morrison and Mosher (1976); Schoofs and Guetté (1983).

a. Theory. Stoichiometric and Abiotic Catalytic Kinetic Resolution

In this section, we describe the mathematical relationships that govern kinetic resolutions and then examine examples of stoichiometric and catalytic kinetic resolutions. Enzymatic resolutions, a subset of the latter type, are described in Section 7.6.b.

> Kinetic resolutions are closely related to stereoselective (asymmetric) syntheses (cf. Chapter 12) in that diastereomeric transition states govern the ratio of products that are formed. It has been pointed out that the two processes are often "inseparable if distinct aspects of the same phenomenon" (Mislow, 1965, p. 139; for illustrations of this point, see below).
>
> Both types of reactions are sometimes grouped together under the heading "kinetically controlled asymmetric transformations" (see the comment in Section 7-3.d; Morrison and Mosher, 1976, p. 28) or "enantiodifferentiating reactions" (Izumi and Tai, 1977, p. 77). However, kinetic resolution substrates are invariably chiral and usually racemic thus insuring that their free energies are identical ($\Delta\Delta G° = 0$, assuming lack of interaction, e.g., solvation or hydrogen bonding, between the substrate and the chiral reagent prior to the beginning of the reaction) and that the differential rate of reaction depends only on the free energy difference ($\Delta\Delta G^{\ddagger}$) between the competing diastereomeric pathways.
>
> No distinction is warranted between kinetic resolution proper and reactions called "asymmetric destruction" in the earlier literature. For the meaning of the term "mutual kinetic resolution," see p. 326.

From the conditions listed above, it follows that the efficiency of kinetic resolutions depends on the conversion (C) $(0 < C < 1)$ and the rate constants of the two competing reactions, k_R and k_S. More precisely, it is the relative rate of reaction of the two enantiomers ($k_R/k_S = s$, the stereoselectivity factor) that is governing. The efficiency is measured by the enantiomer excess of the reaction substrate [unreacted starting material A (Eq. 7.17)] following the resolution. The fundamental relationship between these three variables is given by Eq. 7.18 (for a derivation, see Kagan and Fiaud, 1988).

$$s = \frac{\ln[(1 - C)(1 - \text{ee})]}{\ln[(1 - C)(1 + \text{ee})]} \qquad (7.18)$$

This equation is valid when the competing reactions giving rise to B and unreacted A (Eq. 7.17) are first order with respect to A_R and A_S and regardless of the order in the reagent (chiral or achiral but catalyzed by a chiral catalyst). A related equation (Eq. 7.19) gives the enantiomer composition as a function of the rate constants and of time (t):

$$[S] - [R] = 0.5(e^{-k_S t} - e^{-k_R t}) \qquad (7.19)$$

Such an equation had been derived and experimentally verified as early as 1958 by Newman, Rutkin, and Mislow (1958).

The evolution of enantiomerically enriched product resulting from a kinetic resolution is most clearly appreciated from plots of data calculated by means of Eq. 7.18 (Fig. 7.75). Since Eq. 7.18 is applicable to all types of reactions (see, however, Section 7-6.b), kinetic resolution appears to be a practical and general route to the optical activation of chiral compounds; this would seem to be true especially for reactions having a stereoselectivity factor $s > 10$. Stated in words, in order to obtain unreacted resolution substrate having 99% ee, a kinetic resolution

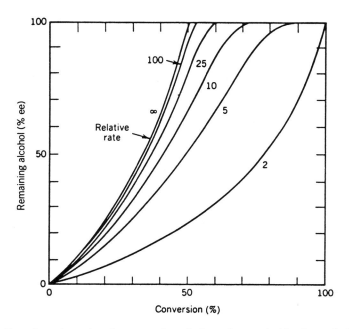

Figure 7.75. Enantiomeric purity of unreacted resolution substrate in kinetic resolutions. Effect of stereoselectivity factor, **s**, and of conversion (C) on the efficiency of resolution [Adapted with permission from Martin, V.S., Woodard, S.S., Katsuki, T., Yamada, Y., Ikeda, M., and Sharpless, K.B. (1981), *J. Am. Chem. Soc.*, **103**, 6237. Copyright © 1981 American Chemical Society.]

having a relative rate ratio of 10 would have to be taken to 72.1% conversion, that is, the yield of the unreacted substrate would be 27.9% (the latter value must, of course, be compared to the maximum yield of one enantiomer obtainable in any resolution, i.e., 50%).

Equation 7.20 relates C and **s** to the enantiomeric purity, ee′, of the product B (Eq. 7.17) of a kinetic resolution when B is chiral and the fast reacting enantiomer (R)-A gives rise to product ($R′$)-B:

$$\mathbf{s} = \frac{\ln[1 - C(1 + \text{ee}')]}{\ln[1 - C(1 - \text{ee}')]} \tag{7.20}$$

Combination of Eqs. 7.18 and 7.20 gives Eq. 7.21.

$$\frac{\text{ee}}{\text{ee}'} = \frac{C}{1 - C} \tag{7.21}$$

The latter equation illustrates the fact that the enantiomeric purities of the unreacted substrate and chiral product of a kinetic resolution are necessarily related and independent of the stereoselectivity factor. As the enantiomeric purity of the starting material goes up, so must that of the product go down. From Eq. 7.21 it also follows that it is impossible to maximize both the enantiomeric purity of the unreacted substrate *and* its yield (Kagan and Fiaud, 1988). These facts are also appreciated best in a graphical presentation (Fig. 7.76).

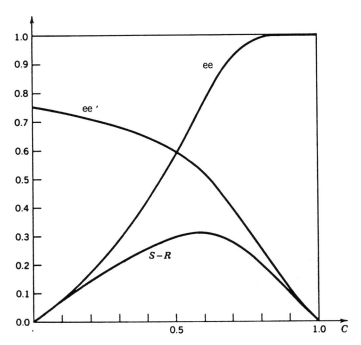

Figure 7.76. Relationship between the extent of reaction (conversion, C) and the enantiomeric enrichment of unreacted starting material (ee) and of chiral product (ee') in a kinetic resolution. Computed curve ($\mathbf{s} = 7$) for a kinetic resolution in which the reactants exhibit pseudo-first order kinetics [Reprinted with permission from Kagan, H.B. and Fiaud, J.C. (1988), *Top. Stereochem.*, **18**, 249. Copyright © 1988 John Wiley & Sons, Inc.]

In a kinetic resolution, the enantiomer composition of chiral products is governed by **s** even when the enantiomers of the resolution substrate are in equilibrium with one another (in accord with the Curtin–Hammett principle; cf. Chapter 9). For an example, see Section 7-3.e, Figure 7.52.

In kinetic resolutions that are second order with respect to reactant (S)- or (R)-A, Eq. 7.17 no longer applies (as is true in stereo*elective* polymerizations; Farina, 1987). In the second-order cases, the unreacted starting material does not generally attain as high an enantiomeric excess for comparable conversions as is true of the more common first-order (or pseudo-first-order) cases (Kagan and Fiaud, 1988).

A special case of kinetic resolutions is that of photochemical destruction of a racemate during irradiation with cpl; this is an instance in which the chiral "reagent" responsible for the enantioselectivity is cpl (Review: Inoue, 1992). It was suggested as early as 1874 by LeBel (q.v.) and again by van't Hoff (1894) that such reactions might account for the prebiotic origin of optically active molecules in nature. The molar absorption coefficients of the enantiomers of a resolution substrate toward left and right cpl would be expected to differ, $\varepsilon_R \neq \varepsilon_S$, and the ee_{max} value depends on the g number [$\Delta\varepsilon/\varepsilon$, where $\Delta\varepsilon = |\varepsilon_R - \varepsilon_S|$; see Section 13-4.a]. In photochemical resolutions, the stereoselectivity factor is given by g with $g = 2(\mathbf{s} - 1)/(\mathbf{s} + 1)$.

Optical activation by photochemical resolution was demonstrated in 1929–1930 by Kuhn et al. (Kuhn, 1936). Enantiomeric purities obtained in photochemical kinetic resolution with cpl are typically of the order of 1–2% ee or less (Izumi and Tai, 1977, p. 158; Zandomeneghi et al., 1983, 1984). The highest enantiomeric purities thus far obtained in such resolutions are 20% ee (for camphor at 99% conversion; $g = 0.08$; Balavoine, Moradpour, and Kagan, 1974) and 30% ee (*trans*-2-hydrindanone; cited in Kagan and Fiaud, 1988).

Kinetic resolutions have been known for nearly a century (Marckwald and McKenzie, 1899) yet, except in the area of enzymatic resolutions (cf. Sections 7-5 and 7-6.b), they had not been considered really useful optical activation methods (Boyle, 1971). The reasons for this are twofold: The reactivity ratios of most of the cases described were small ($s < 10$) and the underlying theory of the method was not widely known. Kinetic resolution also suffered from the same factors limiting the early development of efficient stereoselective synthesis, for example, the inability until the 1960s of accurately measuring enantiomer compositions by nonchiroptical methods. A significant discovery by Sharpless and co-workers in 1981, that of the highly efficient enantioselective epoxidation of allylic alcohols coupled to the efficient kinetic resolution of the unreacted starting material, also helped to change the perception that kinetic resolution was not a very practical optical activation route (Martin, Sharpless, et al., 1981; Fig. 7.77). The unreacted (R)-**174** isolated ($C_{estimated} = 0.55$; $s_{experim.} = 104$) has ee $> 96\%$. The epoxide products of the reaction (erythro/threo ratio 97 : 3) are also optically active ($> 96\%$ ee for **175**, when $C = 0.52$). The epoxide products have been shown to be convertible to additional (R)-**174** by reaction of the methanesulfonate ester with telluride ion (Te^{2-}) (Discordia and Dittmer, 1990).

The foregoing reaction (Fig. 7.77) is an example of a kinetic resolution coupled to an enantioselective synthesis during which one or more additional stereocenters are created. The yields and enantiomeric purities of the epoxy alcohol products obtained in this reaction are shown in Figure 7.78 for the limiting case in which one of the diastereomeric products formed is present in too small an amount to be detected (Fig. 7.78, $C = 50\%$). For a discussion of the mathematical relations giving rise to the curves in Figure 7.78, see Kagan and Fiaud (1988; the more general case has been treated by El-Baba, Poulin, and Kagan, 1984).

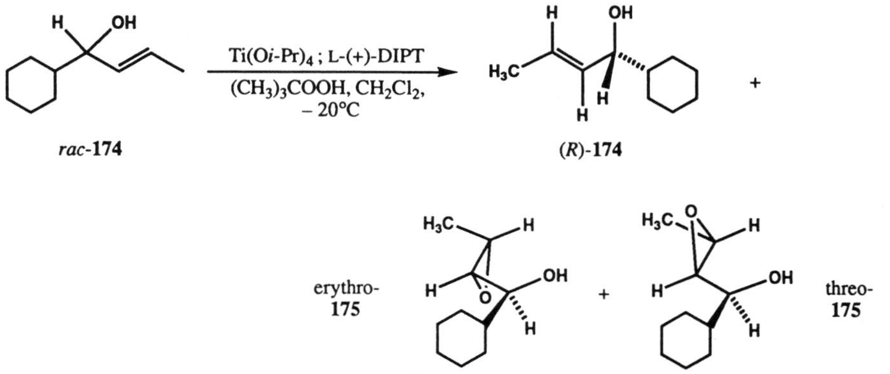

Figure 7.77. Kinetic resolution of a racemic allylic alcohol. Epoxidation of racemic *E*-**174** with titanium tartrate catalyst prepared from diisopropyl L-tartrate (DIPT).

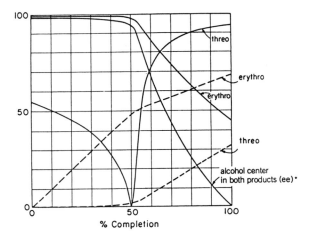

Figure 7.78. Enantioselective epoxidation of *rac-E*-$C_6H_{11}CH(OH)CH=CHCH_3$ (**174**). Yields, % (---) and ee, % (—) of epoxy alcohol products *erythro-* and *threo*-**175** as a function of conversion, *C* (completion). ee* is that of the allylic alcohol obtained on complete deoxygenation of the mixture of epoxy alcohol products. (Sharpless, 1981, reproduced with permission).

An analogous process begins with an enantioselective synthesis whose substrate is *achiral* but endowed with prostereoisomeric groups or faces. In the presence of a nonracemic catalyst, one or more stereoisomers of the product begin to accumulate (rate constants = $k_1, k_2, k_3, \ldots$). Since this product is partially racemic, reaction of the *product* (rate constants = $\beta_1, \beta_2, \beta_3, \ldots$) under the influence of the same catalyst constitutes a kinetic resolution.

Such a system of consecutive reactions, one enantioselective synthesis and the other a kinetic resolution, has been analyzed in detail by Schreiber et al. (1987); see also Kagan and Fiaud (1988, p. 266).

The achiral divinylcarbinol (Fig. 7.79, **176**), for example, is epoxidized with Sharpless reagent (Fig. 7.77) to give two of four possible monoepoxidation

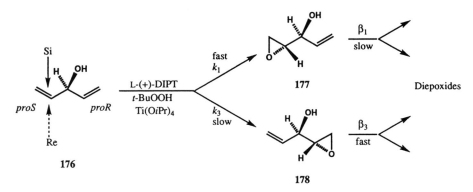

Figure 7.79. Epoxidation of an achiral divinyl carbinol with Sharpless L-(+)-DIPT reagent. Consecutive enantioselective synthesis and kinetic resolution. The minor diastereomeric monoepoxide products formed and the diepoxides subsequently formed from them are not shown (Schreiber, et al., 1987).

products, **177** and **178**; the major product **177** is formed as a result of a combination of selectivity for one enantiotopic group (*pro-S* vinyl group) and one diastereotopic (*Re*) face ($k_1 > k_3$).

Products **177** and **178** are further oxidized under the influence of the already present Sharpless catalyst. If the epoxy groups have little influence on the rates of addition to the remaining alkenyl groups, then $\beta_1 \sim \beta_2 \sim (k_3 + k_4)$ and **178** is oxidized faster to a diepoxide than **177** ($\beta_3 > \beta_1$). The ratio **177**:**178**, hence the enantiomeric purity of the monoepoxide, increases as the reaction proceeds and it can become as large as desired (with a correspondingly large conversion). For the case in which an excess of reagent (*tert*-butyl hydroperoxide) is present, an expression relating the enantiomer ratio to the rate constants at a given conversion was developed

$$\frac{X_1}{X_3} = \left[\frac{\delta_1(\delta_3 + \delta_4)}{\delta_3(\delta_1 + \delta_2)} \right] \left[\frac{s^{-(\delta_1+\delta_2)} - 1}{s^{-(\delta_3+\delta_4)} - 1} \right] \tag{7.22}$$

where $\delta_i = k_i / \Sigma \, k_i$ (fractional rate constant) and $s = [S]/[S]_{init}$ (fractional substrate concentration). Reaction of **176** with 4.8 equivalents of *tert*-butyl hydroperoxide in the presence of Sharpless catalyst derived from $(2R,3R)-(+)$-DIPT (Fig. 7.77) for 24 h ($-25°C$) afforded **177** with 93% ee over **178** and a diastereomer ratio of 99.9:0.01 (Schreiber et al., 1987).

Another kinetic resolution of great interest is that of the simplest of the known chiral fullerenes, C_{76}, having $\mathbf{D}_2$ symmetry (Section 11-6.e). Enantioselective osmylation of C_{76} has provided the first sample of an optically active allotrope of the element carbon (Hawkins and Meyer, 1993).

The resolution was achieved by adapting the protocol of Sharpless for the enantioselective dihydroxylation of alkenes (Sharpless et al., 1991 and 1992; see Section 12-4.c) to reaction with the fullerene. The slower reacting C_{76} enantiomer recovered from a two-stage osmylation (to a conversion of 95%) exhibited $[\alpha]_D - 4000$ (c 0.0034, toluene). The other enantiomer was isolated following reduction of the osmylation product $[(C_{76}(OsO_4L_2)_n$ where L is ligand] by $SnCl_2$. The enantiomer purity of the slower reacting enantiomer (>97% ee) could be calculated from its $[\alpha]_D$ measured at two different conversions in the kinetic resolution (see Section 6-5.e).

A significant feature of this resolution is the high regioselectivity obtained; only two isomeric osmylation products were observed (by HPLC) even though 30 different types of carbon-carbon bonds are present in the C_{76} substrate. Osmylation at different reaction sites might have taken place with opposing, hence countervailing, enantioselectivities. In the event, the osmylation product was reduced without chromatographic separation to afford a product having a CD spectrum enantiomeric with (though not of equal enantiomer purity with) that of the slower reacting C_{76} enantiomer.

Kinetic resolution may evidently be applied to the further enantiomeric enrichment of nonracemic samples. Two stratagems for doing this have been described in Section 7-4. Kinetic resolution has also been applied to the determination of configuration (Section 5-5.g, Horeau's method) and to the determination of enantiomeric purity (ee) (Section 6-5.e); for an analysis of kinetic resolutions occurring during polymerizations, see Kagan and Fiaud, 1988, p. 255).

Kinetic resolutions of preparative interest are of two types: stoichiometric and catalytic (including enzymatic). Examples of both types, excluding cases leading to low enantiomeric enrichments, follow (for additional examples, see Kagan and Fiaud, 1988).

Stoichiometric Reactions

The simplicity of kinetic resolutions as a general approach to optical activation is illustrated by application to alcohols, such as α-methylbenzyl alcohol, by way of esterification with a deficiency of a nonracemic acid. Dicyclohexylcarbodiimide-mediated formation of the ester with one-half equivalent of (R)-2,4-dichlorophenoxypropanoic acid yields the corresponding ester (75:25 diastereomer ratio) and unreacted (S) alcohol having 43% ee (Chinchilla, Heumann, et al., 1990). For this reaction, one calculates a stereoselectivity factor $s = 3.6$ (Eq. 7.18).

A useful kinetic resolution of secondary alcohols consists of reaction with a deficiency of L-valine, $(CH_3)_2CHCH_2CH(NH_3^+)CO_2^-$, in the presence of p-toluenesulfonic acid. The products, tosylate salts of (+) alkyl esters of L-valine (obtained in 60–80% yields with respect to one alcohol enantiomer), were recrystallized and hydrolyzed to afford enantiomerically homogeneous alcohols (Halpern and Westley, 1966; Jermyn, 1967).

> The foregoing alcohol resolution illustrates the fact that occasionally kinetic and thermodynamic control may be combined to optimize a resolution, that is, the initial enrichment obtained under kinetic control was followed by enantiomer enrichment based on solubility differences. In the case described, the occurrence of a kinetic resolution was not mentioned (the unreacted alcohol was characterized as being racemic, but its rotation does not appear to have been measured).

In a method patterned after Horeau's method (Section 5-5.g), kinetic resolution of alcohols by incomplete reaction with, e.g., O,O'-dibenzoyltartaric acid (Fig. 7.21, **57**), affords alcohols of ee up to 48% (Bell, 1979). The reciprocal process, the kinetic resolution of acids (e.g., Fig. 7.23a, **83**), by reaction of its racemic anhydrides (without prior separation of the meso and threo diastereomers) with alcohols, e.g., with nonracemic 1-(4-pyridyl)ethanol, has also been observed (Franck and Rüchardt, 1984; Salz and Rüchardt, 1984).

An even simpler kinetic resolution of alcohols takes place on reaction of two equivalents of a racemic chiral alcohol, e.g., α-methylbenzyl alcohol, with one equivalent of an achiral acyl halide (e.g., acetyl chloride) in the presence of one equivalent of a synthetic "inductor base" (e.g., (R)- or (S)-N,N-dimethyl-α-methylbenzylamine; Weidert, Geyer, and Horner, 1989). Enantiomeric purities as high as 60–70% ee (of ester product and/or unreacted alcohol) are easily obtained.

Glycols, such as 3-chloro-1,2-propanediol **179** (Fig. 7.80a) may be resolved by reaction with D-camphorquinone **180**. Four monoketals are formed under kinetic control (excess of diol) in the proportion 27:45:17:12. Crystallization of the dominant isomer and recovery of the diol from the latter affords enantiomerically pure (R)-(−)-**179**. The latter is a precursor of nonracemic epichlorohydrin (Ellis, Golding, and Watson, 1984).

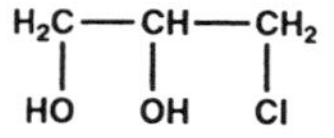

179

resolved with

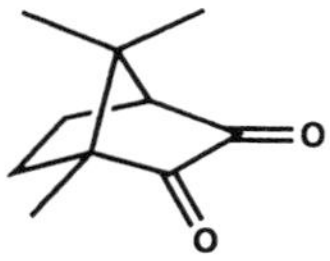

180

D-Camphorquinone

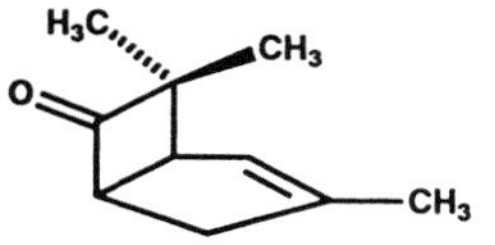

181

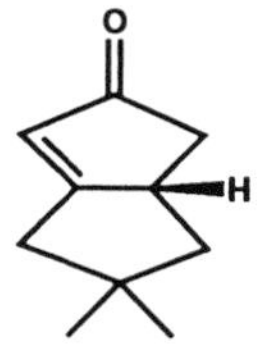

183

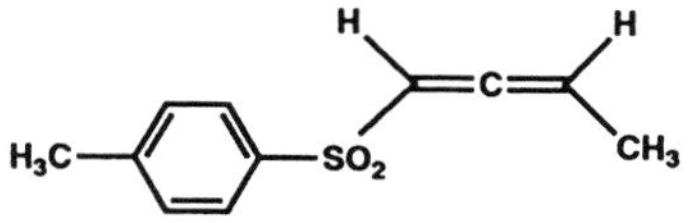

184

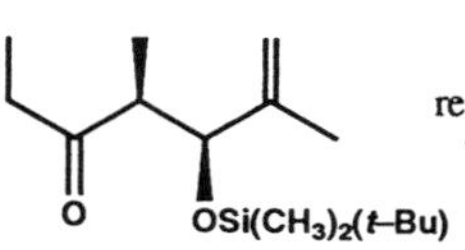

185

resolved with

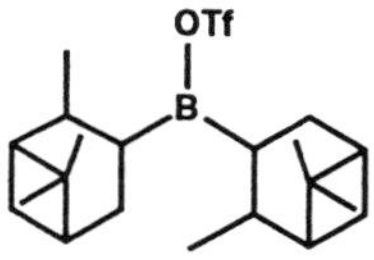

186a

(Tf = triflate)

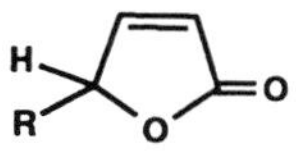

186b

$R = CH_3, (CH_3)_2CH$

resolved with

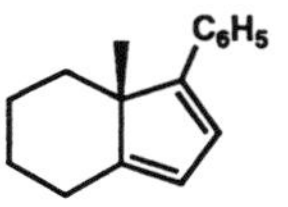

186c

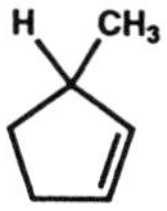

187

Figure 7.80(*a*). Compounds resolved by stoichiometric kinetic resolution.

The kinetic resolution of amines is occasionally practiced, for example, by partial reaction with D-10-camphorsulfonyl chloride (Wiesner et al., 1972) and by reaction of 1° amines with (*S*)-2-phenylbutyric anhydride (Hiraki and Tai, 1982).

Kinetic resolution of ketones (e.g., **181**, Fig. 7.80*a*) by reaction with an insufficient amount of nonracemic chiral primary amine, such as dehydroabietylamine (Fig. 7.19*a*, **35**), has been described; the product is a Schiff's base. Hydrolysis of the Schiff's base product following distillation of the optically active unreacted ketone gave the enantiomeric ketone (Huber and Dreiding, 1970). Ketones, such as 2-methylcyclohexanone, (±)-**182** (Eq. 7.23), can be resolved via selective cleavage of their diastereomeric acetals [formed by acid-catalyzed reaction with (2*R*,4*R*)-(−)-2,4-pentanediol] by triisobutylaluminum (Eq. 7.23); the kinetically controlled step (a diastereoselective elimination) is the acetal cleavage to an enol ether. The (2*R*)-(−)-ketone **182** is obtained at −20°C (5 h; >95% ee); at 0°C (3 h), (+)-**182** is recovered from unreacted acetal (Mori and Yamamoto, 1985).

(7.23)

Kinetic resolution of α,β-unsaturated ketone **183** (Fig. 7.80*a*) takes place on addition of a nonracemic sulfoxide enolate (Hua, 1986). Allenic sulfones (e.g., Fig. 7.80*a*, **184**) having a double bond highly activated toward nucleophilic addition, may be efficiently resolved by partial reaction with chiral amines, such as **43** (Fig. 7.19*b*) (Cinquini et al., 1978).

The aldol condensation has been adapted to kinetic resolutions with fore-knowledge of the degree of diastereofacial selectivity of a given system. *rac*-Ketone **185** is converted to its enolate by reaction with nonracemic di-(3-pinanyl)borane triflate (Fig. 7.80*a*, **186a**). Ketone (+)-**185** is isolated (ee > 95% with *C* ca. 75%) following partial diastereoselective reaction of the enolate mixture with methacrolein. Kinetic resolution also attends the enolization but with a lower efficiency (Paterson et al., 1989).

($\pm$)-α,β-Unsaturated lactones, e.g., **186b** (Fig. 7.80*a*) can be kinetically resolved when they react as dienophiles in enantioselective Diels-Alder reactions with chiral nonracemic diene **186c**. Both the unreacted lactone and its enantiomer isolated by thermal retro-Diels-Alder reaction were found to be enantiomerically enriched (Wegener, Hansen, and Winterfeldt, 1993).

Among the most useful applications of stoichiometric kinetic resolutions are the preparation of nonracemic alkenes and allenes whose resolution by other means is often problematic. Hydroboration of racemic alkenes with a deficiency of tetra-3-pinanyldiborane (derived from α-pinene) leaves unreacted alkenes, such as **187** (Fig. 7.80*a*), in an enantiomerically enriched state (up to 65% ee; Brown, Ayyangar and Zweifel, 1964). Allenes may be similarly resolved (Review: Brown and Jadhav, 1982). Racemic boranes may themselves be kinetically resolved by reaction with a deficiency of a nonracemic Lewis base (Masamune et al., 1985).

Resolution of *trans*-1,2-dimethylcyclopropane (Eq. 7.24, **188**), is achieved by conversion of the racemate to the platinacycle **189a** that is treated with 0.5 equivalent of a chiral phosphine or a chiral alkene to yield one of the enantiomers of **188**. Reaction of the residual platinacyclobutane with $(C_6H_5)_3P$ leads to the enantiomeric (*S*, *S*) cyclopropane (Johnson et al., 1979).

$$(7.24)$$

Resolutions of halogen compounds effected by reaction with brucine (Lucas and Gould, 1942) have heretofore been considered to be kinetic resolutions involving enantioselective elimination (e.g., α-hexachlorocyclohexane, Section 7-3.c). More recent studies demonstrate that when carried out under mild and heterogeneous conditions such resolutions are nondestructive. Formation of 1:1 solid adducts between brucine and the halide as well as recovery of both enantiomers (one from the filtrate following removal of the adduct, the other on acidification of the adduct) have shown that the resulting optical activations are most likely due to inclusion compound formation (cf. Section 7-3.c) (Pavlis, Skell, et al., 1973; Wilen et al., 1985). When the reaction time is extended (Tanner et al., 1977; Pavlis and Skell, 1983; Tanner et al., 1985) or at high temperatures (Singler and Cram, 1972) elimination products are observed and kinetic resolution proper does compete with inclusion compound formation. Secondary halides are kinetically resolved (up to 49% ee for $C \approx 50\%$) at low temperature (-60 to $-70°C$) on reaction with nonracemic lithium salts of oxazolines (Meyers and Kamata, 1976).

Kinetic resolutions have not always been monitored by examination of unreacted resolution substrate (see above), nor have nonracemic samples of the latter always been the objectives of such resolutions. Mixtures of covalent diastereomers obtained from racemates under kinetically controlled conditions have in many cases been directly converted to products of interest rather than to enantiomerically enriched starting materials. Although not strictly meeting the definition of resolution, the processes nevertheless are often considered to be examples of kinetic resolutions, for example, application of the sulfoximine reagent to kinetic resolution (Johnson and Meanwell, 1981) and application of the ene reaction of N-sulfinylcarbamates (Whitesell and Carpenter, 1987).

Catalytic Reactions

Organic and organometallic catalysts have increasingly been studied as reagents for effecting kinetic resolutions (Reviews: Brown, 1988; Brown, 1989); preparative procedures have lagged somewhat behind the mechanistic studies. A more stringent limitation of some preparative catalytic kinetic resolutions relative to the stoichiometric ones is that separation of unreacted starting materials from products may be more difficult to carry out, for example, unsaturated alcohol starting materials from saturated alcohol products. Nevertheless, observations of selectivities s of the order of 10^1–10^2 in some kinetic resolutions catalyzed by "abiotic"

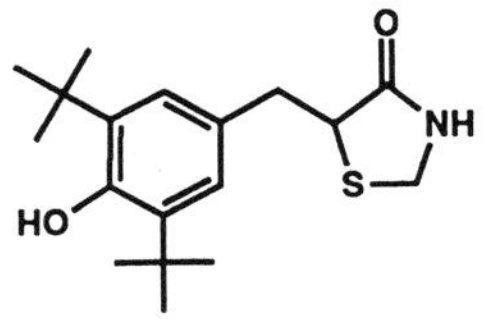

189b

Figure 7.80(*b*). Structure **189b**.

catalysts (with recovered starting material >99% ee) make these types of kinetic resolutions very attractive.

The principal application of nonracemic catalysts to kinetic resolution has been that of the Sharpless reagent (see above) applied to the resolution of allylic alcohols (Reviews: Finn and Sharpless, 1985; Rossiter, 1985). The process, as originally described in 1981 (see p. 399), required the use of a stoichiometric amount of catalyst. However, in the presence of molecular sieves, the process becomes truly catalytic, for example with 5–10 mol% of the Ti$^{(IV)}$ reagent in the presence of (+)-dicyclododecyl tartrate (recovered alcohols: >98% ee, $C = 0.52$– 0.66; Gao, Sharpless, et al., 1987). With bulky β substituents on the double bond, selectivities as high as $s = 700$ have been found (Carlier, Sharpless, et al., 1988). β-Amino alcohols are susceptible to kinetic resolution with the Sharpless reagent, as the result of their enantioselective conversion to *N*-amine oxides (Miyano, Sharpless, et al., 1985) and so are 2-furylcarbinols (Kobayashi, Sato, et al., 1988). Kinetic resolution of a chiral sulfide, the antioxidant **189b** (Fig. 7.80*b*), by enantioselective oxidation to the sulfoxide (catalyzed by a water-modified Sharpless reagent) has been described (Phillips et al., 1992).

Other applications of organometallic catalysts to kinetic resolutions are summarized and illustrated in Table 7.4 and Figure 7.81.

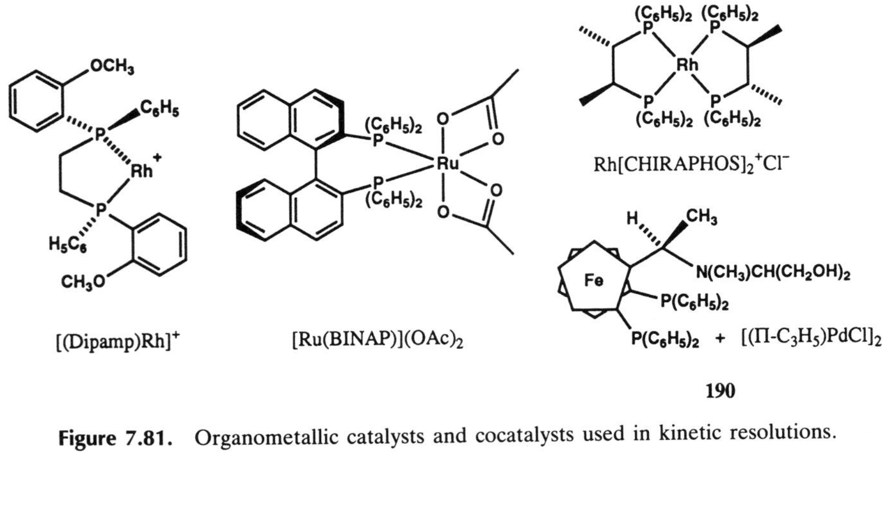

190

Figure 7.81. Organometallic catalysts and cocatalysts used in kinetic resolutions.

TABLE 7.4 Examples of Abiotic Catalytic Kinetic Resolutions

Reaction Type	Examples	Catalyst[a]	s	ee (%)[b]	References
Homogeneous hydrogenation		$[(Dipamp)Rh]^+$	4.5	>90% ($c = 0.70$)	Brown and Cutting, 1985
		$Ru(BINAP)(OAc)_2$	74	>99% ($c \approx 0.55$)	Kitamura, Noyori, et al., 1988
Isomerization		$Rh(BINAP)(OCH_3)_2^+$	5	?	Kitamura, Noyori, et al., 1987
Cyclization		$Rh(CHIRAPHOS)^+Cl^-$	?	?	James and Young, 1985
Allylic alkylation		"Ferrocenylbiphosphine" + "π-allyl Pd" **190**	14	>99% ($c = 0.80$)	Hayashi, Ito, et al., 1986

[a] Catalyst structures are given in Figure 7.81.
[b] Recovered starting mterial.

A number of kinetic resolutions are catalyzed by alkaloids. Quinidine **29** (Fig. 7.19*a*), catalyzes the isomerization of *rac*-1-methylindene **191**, to 3-methylindene (Fig. 7.82); the former is recovered (by GC) with up to 73% ee ($C = 0.65$; s $\cong$ 5; Meurling, 1974). Kinetic resolution also attends the Wittig rearrangement of the atropisomeric ether **192** triggered by butyllithium. (−)-Sparteine (Fig. 7.35, **110**), is the chiral inductor. Unreacted ether (Fig. 7.82, **192**), both product alcohols, and byproduct pentahelicene all exhibit high rotations (Mazaleyrat and Welvart, 1983).

The chiral cyclohexenone **193** (Fig. 7.82) reacts with a deficiency of *p-tert*-butylthiophenol in a Michael reaction (1.5:1) catalyzed by cinchonidine (Fig. 7.19*a*, **32**). Unreacted (+)-ketone was separated from the *cis*- and *trans*-sulfide adducts by careful distillation. Retrograde Michael reaction of the trans adduct afforded (−) ketone (Pluim, 1982, p. 138). Reduction of sulfimide **194** (Fig. 7.82)

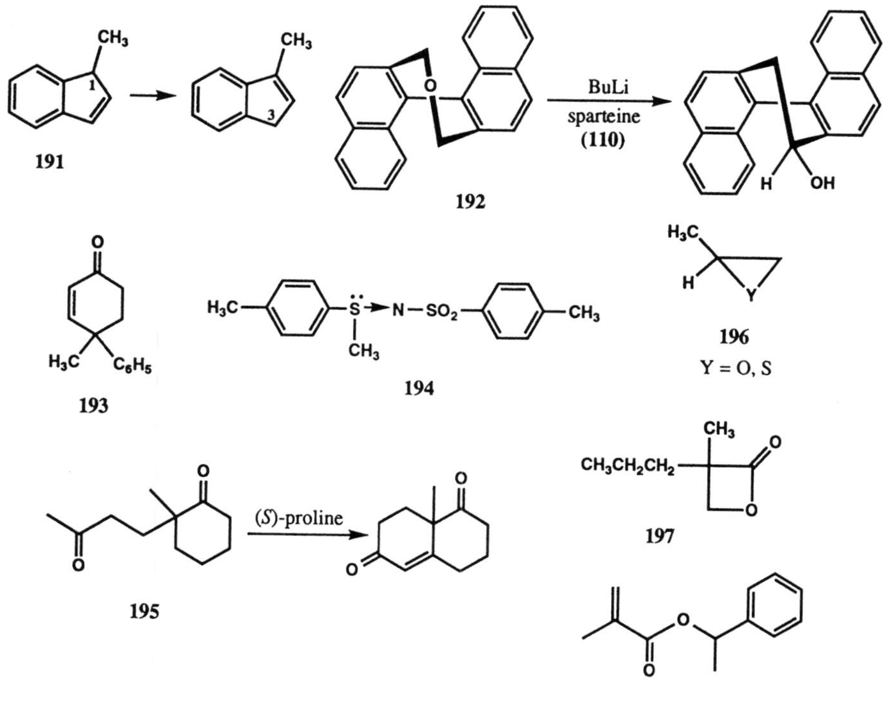

Figure 7.82. Compounds that undergo kinetic resolution with alkaloid and amino acid catalysts.

by mesitylenethiol in the presence of quinine (Fig. 7.19a, **31**), led to recovery of unreacted sulfimide highly enriched in one enantiomer (Dell'Erba et al., 1983). Diketone **195** is resolved concomitant with its (S)-proline catalyzed annulation to an octalone (Fig. 7.82; Agami et al., 1984). A catalytic version of the kinetic resolution of allene sulfone (Fig. 7.80, **184**) has also been described (Cinquini et al., 1978).

A nonracemic solvent (2R,3R)-(+)-2,3-dimethoxybutane has been shown to induce the kinetic resolution (4.9% ee) of a racemic Grignard reagent $CH_3CH_2CH(C_6H_5)CH_2MgCl$ during reaction of the latter with iso-butyrophenone; hydrolysis of the unreacted Grignard reagent affords optically active 2-phenylbutane (Morrison and Ridgway, 1969).

Polymerization reactions may also be used to effect kinetically controlled resolutions (Kagan and Fiaud, 1988, p. 285). Diethylzinc catalyzed polymerization of (±)-propylene oxide and of (±)-propylene sulfide (Fig. 7.82, **196**, Y = S), in the presence of (−)-3,3-dimethyl-1,2-butanediol gives rise to optically active polymers [stereo*elective* polymerization (Farina, 1987)]. The unreacted monomers are enantiomerically enriched (Sépulchre, Spassky, and Sigwalt, 1972; Spassky, Leborgne, and Sépulchre, 1981). Kinetic resolution of chiral lactone (Fig. 7.82, **197**) attends its polymerization to a nonracemic polyester in the presence of a chiral initiator (Leborgne, Spassky, and Sigwalt, 1979). *rac*-α-Methylbenzyl methacrylate (Fig. 7.82, **198**) undergoes anionic polymerization (to a mostly isotactic polymethacrylate) in the presence of achiral Grignard reagents modified by $(-)$-sparteine (Fig. 7.35, **110**). Unreacted **198** monomer exhibited up to 83% ee with cyclohexylmagnesium bromide-**110** initiator (Okamoto, Yuki, et al., 1982).

b. Enzymatic Resolution

Resolutions catalyzed by enzymes have been exploited to a larger degree than any other kind of kinetic resolution. The very first kinetic resolution discovered was of this type (Pasteur's third method in terms of chronology; Pasteur, 1858), namely, the resolution of tartaric acid by fermenting yeast, an experiment that marks the introduction of stereochemistry into "physiological principles" (i.e., into biochemistry; Pasteur, 1860). Consequently, there exists a larger body of data about enzymatic resolutions than about any other kind of kinetic resolution.

An enzymatically catalyzed reaction, such as an ester hydrolysis, may be described by a simplified mechanism (Eq. 7.25)

$$
\begin{array}{l}
\text{E} + R \underset{k_{-1}}{\overset{k_1}{\rightleftharpoons}} \text{E-}R \underset{k_{-2}.\text{R'OH}}{\overset{k_2}{\rightleftharpoons}} \text{E-}R^* \underset{k_{-3}}{\overset{k_3.\text{H}_2\text{O}}{\rightleftharpoons}} \text{E-P} \underset{k_{-4}}{\overset{k_4}{\rightleftharpoons}} \text{E} + \text{P} \\[2mm]
\text{E} + S \underset{k'_{-1}}{\overset{k'_1}{\rightleftharpoons}} \text{E-}S \underset{k'_{-2}.\text{R'OH}}{\overset{k'_2}{\rightleftharpoons}} \text{E-}S^* \underset{k'_{-3}}{\overset{k'_3.\text{H}_2\text{O}}{\rightleftharpoons}} \text{E-Q} \underset{k'_{-4}}{\overset{k'_4}{\rightleftharpoons}} \text{E} + \text{Q}
\end{array}
\qquad (7.25)
$$

where $\text{E} = \text{enzyme}$, R and S represent the fast and slow reacting substrate enantiomers, respectively, and $\text{E-}R$ (and $\text{E-}S$) represent the enzyme–substrate complexes. Further reaction leads to the acyl–enzymes $\text{E-}R^*$ and $\text{E-}S^*$, to the enzyme-product complexes E–P and E–Q and finally to the products, P and Q, respectively, with liberation of the enzyme. Application of steady-state kinetics to the rate constants of Eq. 7.25 with the assumption that the steps leading to $\text{E-}R^*$ (and $\text{E-}S^*$) are essentially irreversible (and that products P and Q do not inhibit the reaction) leads to Eq. 7.26 (Chen, Sih, et al., 1982; Sih and Wu, 1989)

$$
\mathbf{E} = \frac{\ln[(1 - C)(1 - \text{ee})]}{\ln[(1 - C)(1 + \text{ee})]}
\qquad (7.26)
$$

where **E** is the biochemical stereoselectivity factor (corresponding to s). The parameter **E** may also be defined in terms of the rates of reaction of the competing enantiomer substrates, k_2 and k'_2; for cases in which $k_{-1} \gg k_2$ and $k'_{-1} \gg k'_2$ (Eq. 7.25) the following relation holds:

$$
\frac{\left(\dfrac{k_{\text{cat}}}{K_{\text{m}}}\right)_R}{\left(\dfrac{k_{\text{cat}}}{K_{\text{m}}}\right)_S} = e^{-\Delta\Delta G^{\ddagger}/RT} = \mathbf{E}
\qquad (7.27)
$$

where k_{cat} and K_{m} denote the turnover number and Michaelis constant, respectively. While Eqs. 7.18 and 7.26 are formally identical and equally predictive, the latter equation is applicable only to irreversible reactions (see below). Equation 7.26 is also dependent on the ratio of enzymatic specificity constants V/K, where

V and K are the maximal velocities and Michaelis constants of the substrate enantiomers, and it is independent of substrate concentrations (Chen, Sih, et al., 1982). Comparison of enzyme efficiencies must be carried out at equal conversions; values of C are conveniently obtained from the enantiomer compositions of unreacted resolution substrates (ee) and of resolution products (ee′) by means of Eq. 7.21. However, under conditions such that the substrate can be racemized in situ during the kinetic resolution, that is, an asymmetric transformation takes place (cf. Section 7-2.e), the enantiomeric purity of the product becomes independent of the extent of conversion; moreover, the reaction (e.g., hydrolysis) is apparently more enantiospecific than when concomitant racemization does not take place (Fülling and Sih, 1987).

Enzymatic resolutions that take place in two-phase (aqueous + organic) systems are reversible, and hence are not governed by Eq. 7.26. For such systems, a new expression relating $\mathbf{E}$, C, and ee must be employed that incorporates K, the equilibrium constant for the reactions

$$E + R \rightleftharpoons E + P$$
$$E + S \rightleftharpoons E + Q$$

$$(7.28)$$

where $K = k_2/k_1 = k_4/k_3$ and

$$\mathbf{E} = \frac{\ln[1 - (1 + K)(C + \text{ee}\{1 - C\})]}{\ln[1 - (1 + K)(C - \text{ee}\{1 - C\})]} \qquad (7.29)$$

Note that K, being a thermodynamic parameter unlike $\mathbf{E}$ (a kinetic parameter that varies from enzyme to enzyme), is independent of the nature of the enzyme (Sih and Wu, 1989).

A value of $\mathbf{E} > 100$ is necessary if *both* enantiomers of a racemate are to be obtained with high enantiomeric purity in a single kinetic resolution. With lower $\mathbf{E}$ values, one enantiomer (the unreacted one) may be obtained with high enantiomeric enrichment (at high conversion); to obtain the other one with high enantiomeric purity requires recycling (Laumen and Schneider, 1988).

It is surprising to find that in quite a few instances only modest $\mathbf{E}$ values (1–10) obtain when commercial enzyme preparations or intact microorganisms are used in kinetic resolutions in vitro. The reasons for this are not easy to establish; the enzyme may lack stereoselectivity or the preparation in use may contain several enzymes, some of which may have countervailing enantioselectivities. In order to raise the enantioselectivity, it is not uncommon to screen enzymes against a given substrate, to modify the substrate, to change reaction conditions, or to recycle the product, that is, to repeat the resolution on the product of an enzymatic resolution (Sih and Wu, 1989).

Treatment with sodium deoxycholate and precipitation with ether and ethanol (1:1) leads to a significant increase in the enantioselectivity of lipase (of *Candida cylindracea*) catalyzed ester hydrolysis (e.g., on arylpropionate esters). This result has been ascribed to a noncovalent modification of the native enzyme protein, that is, to the generation of a more stable conformer (Wu, Guo, and Sih, 1990). Following a pragmatic and often empirical selection of enzyme, substrate, and reaction conditions, enantiomeric enrichments greater than 90% ee in enzymatic resolutions are not difficult to attain.

An alternative approach to raising the enantioselectivity that is likely to be increasingly useful in the future is the development of reliable active-site models that permit the interpretation and prediction of an enzyme's stereospecificity, that is, will a given enzyme catalyze a reaction on a specific substrate and if so what will be the configuration of the product? An example is the cubic-space section model encompassing the active site region of horse liver alcohol dehydrogenase (HLADH) for predicting the enzyme's specificity in oxidoreductions (Jones and Jakovac, 1982). This model (inspired by the diamond lattice section model developed by Prelog, 1964), is empirical in that it incorporates specific X-ray and kinetic data about the enzyme. The model has features very reminiscent of a chiroptical sector rule (Chapter 13); elaboration of the model by eventual application of computer graphics is evident.

> Active site mapping means devising a geometric model of the active site of an enzyme that incorporates information about the specific amino acids participating in the enzyme catalyzed reaction (where are these amino acids in the primary structure, how far apart are they, and what is their angular relationship) leading to a hypothesis of how the substrate binds to these amino acids during the reaction and what does the transition state of the reaction look like.

A significant development in enzymatic catalysis that may also reduce some of the extant empiricism in enzymatic resolutions is the discovery that antibody proteins catalyze chemical reactions (Pollack, Jacobs, and Schultz, 1986; Tramontano, Janda, and Lerner, 1987). Since antibodies can be made toward many chemical species, this finding opens up the possibility of "enzymes," that is, catalytic *anti*bodies (so-called abzymes), being designed to catalyze specific chemical reactions, including resolutions. The possibility of designing catalytic antibodies with enzymatic activity may circumvent the need to adjust substrate structures and conditions to overcome the limitations of naturally occurring enzymes (Reviews: Schultz, 1989; Schultz and Lerner, 1993).

A stable phosphonate ester **199** (Fig. 7.83) that mimics the transition state of a transesterification reaction has (after conjugation with a carrier protein) elicited the formation of a monoclonal antibody that catalyzes a kinetic resolution. Compound **199** acts as a hapten (Section 6-5. h). Synthesis of a single diastereomer of the hapten bearing a suitably placed stereogenic atom (albeit as the racemate) insured that the antibody would behave in a stereoselective manner. Intramolecular transesterification of *rac*-**200** (Fig. 7.83) in the presence of antibody 24B11 spontaneously terminated at $C = 0.5$ suggesting that the reaction was indeed a kinetic resolution. Confirmation of this fact was obtained by NMR analysis of the product lactone **201**, with a chiral shift reagent (ee = $94 \pm 8\%$) (Napper, Benkovic, et al., 1987). In this study, only antibodies preferring one enantiomer of the substrate **200** were found (Benkovic, 1991). However, in a later study also using a racemic hapten, two classes of antibodies were found: some antibodies that catalyzed reaction of the *S* enantiomer of the substrate and others that catalyzed reaction of the *R* enantiomer (Janda, Benkovic, et al., 1989).

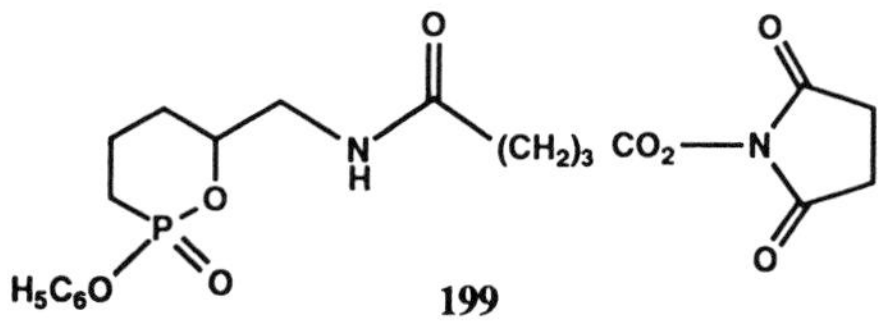

Figure 7.83. Kinetic resolution in a transesterification catalyzed by a monoclonal antibody. Compound **199** is a transition state mimic designed to elicit formation of the antibody.

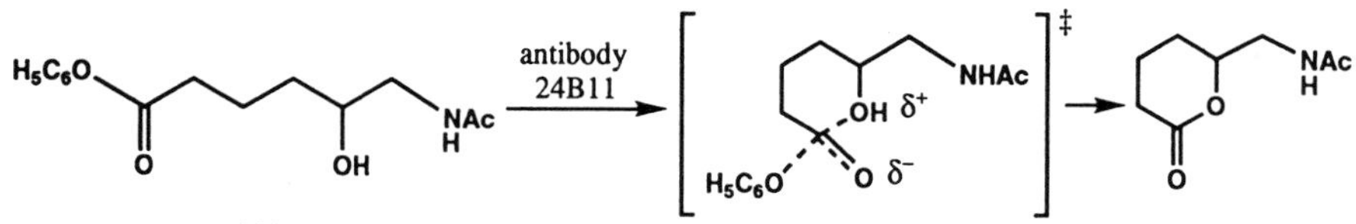

In general, enzymes are versatile, highly efficient, and very selective in their activity, that is, selective with respect to functionality, regioselectivity, and stereoselectivity. Two significant limitations in the use of enzymes are that (a) they are easily denatured and (b) they often require stoichiometric amounts of cofactors [coenzymes, such as adenosine triphosphate (ATP)]. These limitations have been addressed, respectively, by (a) immobilization (attachment of the enzyme to a support without affecting its active site); this tactic has been especially important in connection with enzymatic resolutions carried out on a commercial scale (Section 7-5), and (b) incorporation of cofactor regeneration schemes into the enzymatic resolution process (Reviews: Whitesides and Wong, 1985; Wong, 1989).

Of the six main classes of enzymes (Section 12-4.e), the first three (oxidoreductases, transferases, and hydrolases) have been the most useful in kinetic resolutions. Both intact cells, for example, microbial preparations or yeast (crude yet effective mixtures containing enzymes) and cell-free enzyme "preparations" have been used in enzymatic resolutions. A significant increase in the application of enzymes to organic synthesis in recent years may be attributed in large measure to new purification methods that have significantly lowered the price of commercial enzyme preparations. In light of the very large number of enzymatic resolutions in the literature and the availability of excellent reviews (Books: Davies, et al., 1989; Abramowicz, 1990. Reviews: Jones and Beck, 1976; Fischli, 1980; Svedas and Galaev, 1983; Jones, 1985; Whitesides and Wong, 1985; Jones, 1986; Kagan and Fiaud, 1988; Sih and Wu, 1989; Santaniello et al., 1992), we limit ourselves to some examples that illustrate significant features of enzymatic resolutions.

Numerous carboxylic acids [including amino acids (Review: Verkhovskaya and Yamskov, 1991), e.g., arylglycines (cf. Williams and Hendrix, 1992)], alcohols and amines have been enzymatically resolved by hydrolysis catalyzed by hydrolases (acylases). These were among the first enzymatic resolutions to be exploited. Substrates for such reactions include esters, amides, carbamates, and hydantoins. An example follows:

On a 2-g scale, hydrolysis of *rac-N*-acetylphenylalanine methyl ester (Fig. 7.84, **202**) in the presence of a commercial serine proteinase (Subilopeptidase A; Alcalase™) is complete in 45 min with the *R* ester being quantitatively recovered by extraction into CH_2Cl_2: The reaction comes to a near standstill at $C \approx 50\%$ as would be expected for a resolution occurring with a high value of **E**. The *S* acid is isolated from the reaction mixture following acidification (96%; 98% ee after one recrystallization) (Roper and Bauer, 1983).

Microbial resolution of Mosher's acid (Fig. 7.21a, **63**) widely used as a reagent for the determination of enantiomer composition (Section 6-5), has been effected by hydrolysis of cyanohydrin acetate (Fig. 7.84, **203**) with dry cells of *B. coagulans*. The recovered (*R*)-**203** (100% ee at $C = 0.7$) was converted to **63** in three steps without loss of enantiomeric integrity (Ohta, Miyame, and Kimura, 1989).

The discovery that lipases catalyze reactions very well in nearly anhydrous organic solvents has been a stimulus to the application of such enzymes to esterification and transesterification reactions. These are reactions that normally are suppressed when carried out in water as the result of competing hydrolysis (Zaks and Klibanov, 1984. Reviews: Klibanov, 1986, 1990). Esterification of 2-bromopropionic acid with 1-butanol in hexane in the presence of yeast lipase (the latter is insoluble in organic media), for example, afforded optically active butyl (*R*)-(+)-2-bromopropionate (96% ee at $C = 0.45$) and unreacted (*S*)-(−) acid (99.6% ee at $C = 0.78$). The catalyst could be repeatedly used in resolutions with little loss in activity (Kirchner, Scollar, and Klibanov, 1985).

Menthol (Fig. 7.19c, **49**) may be enzymatically resolved by ester hydrolysis, by ester formation, or by ester interchange in the presence of lipase from *C. cylindracea*. These reactions are carried out in water (hydrolysis) or in heptane (ester formation or interchange) as appropriate (Langrand, Triantaphylides, et al., 1986). The aggregation pheromone of ambrosia beetle (a pest that attacks North American timber) is alcohol **204** (Sulcatol, Fig. 7.84). The corresponding racemate has been resolved, for example, with pig pancreatic lipase (PPL) by

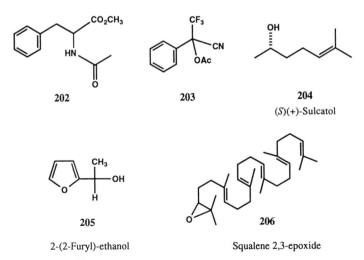

202	**203**	**204**
		(*S*)(+)-Sulcatol

205	**206**
2-(2-Furyl)-ethanol	Squalene 2,3-epoxide

Figure 7.84. Enzymatic resolution substrates.

transesterificaton of laurate from trifluoroethyl laurate in diethyl ether ($E = 100$) (Stokes and Oehlschlager, 1986; Belan, Veschambre, et al., 1987).

In enzymatic esterification or transesterification a given enzyme catalyzes both the forward and the reverse reaction. Consider the hypothetical reaction described in Eq. 7.30 in which $k_R > k_S$.

$$(7.30)$$

The R acetate accumulates in the reaction along with unreacted S alcohol (together with minor amounts of the enantiomeric ester and alcohol). As a consequence of microscopic reversibility, it is the major acetate R enantiomer that reacts faster with methanol in the reverse reaction, because [R acetate] > [S acetate] together with the rates k_R (acetate) > k_S (acetate). Consequently, in such reversible reactions, as the conversion increases, the enantiomeric purity of both products gradually decreases. This difficulty has been circumvented, as first described by Degueil-Castaing, De Jeso, et al. (1987), by causing the reaction to become irreversible by application of enol esters, for example, vinyl acetate (the byproduct is acetaldehyde), as esterification (or transesterification) agents (Wang, Bergbreiter, Wong, et al., 1988; Laumen, Breitgoff and Schneider, 1988).

A two-step enantioselective synthesis of chiral cyanohydrin acetates RCH(OAc)CN, e.g., R = p-CH$_3$C$_6$H$_4$, is effected by (a) base-catalyzed formation of rac-cyanohydrin from aldehydes in the presence of acetone cyanohydrin, $(CH_3)_2C(OH)CN$, and (b) enantioselective and irreversible acetylation of the unstable racemic cyanohydrin by isopropenyl acetate $CH_2=C(CH_3)OAc$ in diisopropyl ether in the presence of $P.$ $cepacia$. The yield of a single enantiomer (63–100%; ee up to 94%) exceeds 50% in view of the continuous racemization of the slower reacting cyanohydrin enantiomer that equilibrates with the precursor aldehyde. Thus the entire process incorporates the elements of a kinetic resolution coupled to an asymmetric transformation (Inagaki, Oda, et al., 1992).

Enzymatic double (or sequential) kinetic resolutions represents one strategy for enhancing the enantiomeric enrichment of one of the substrate enantiomers. As an example, ($\pm$)-2,4-pentanediol is resolved by sequential esterification with hexanoic acid in anhydrous isooctane. The enantiomer yield is increased during esterification of the initial nonracemic monoester product (30% ee) to the diester (>98% ee at $C = 0.24$) both in the presence of the lipase $Pseudomonas$ sp. (AK) (Guo, Sih, et al., 1990). In another example, (2-furyl)-ethanol (Fig. 7.84, **205**) is resolved by esterification with octanoic acid with $Candida$ lipase. The predominant ester formed was the R enantiomer (52% ee at $C = 0.5$). Hydrolysis of this ester under the influence of the same enzyme, with R ester hydrolyzing faster than its enantiomer (see above), led to R alcohol of 94% ee; $C = 0.25$ (Drueckhammer, Wong, et al., 1988).

Enzymatic resolution of 2,2'-dihydroxy-1,1'-binaphthyl (Fig. 7.36, **113**) by hydrolysis of diesters was effected with several strains of soil microorganisms. With *Bacillus* sp. LP-75, the hydrolysis rate, the efficiency (ee) and the sense of enantioselectivity depended strongly on the nature of the R group of the acid. The diacetate furnished the *S* diol (50% ee) directly while the (±) dibutyrate preferentially hydrolyzed to the *R* diol (97% ee; Fujimoto, Iwadate, and Ikekawa, 1985). In other cases, enzymatic resolution of diesters furnish monoesters (Ganey, Berchtold, et al., 1989). Subsequently, a much faster resolution of diol **113** (hydrolysis of the divalerate ester) has been described that is catalyzed by an enzyme (PPL) preparation affording (*S*)-diol (95% ee) at 46% conversion (Miyano et al., 1987).

The enzymatic resolution of simple 2° alcohols [e.g., $C_6H_5CH(CH_3)OH$] by lipase catalyzed hydrolysis of the acetate from *Pseudomonas* sp. affords both enantiomers of the alcohol (after hydrolysis of the unreacted one) in nearly quantitative yields (ee > 99% for each enantiomer; **E** > 1000) on a molar scale. The high yield, simplicity, and efficiency of this type of optical activation process may lead it to displace resolution methods mediated by covalent diastereomers as well as enantioselective ketone reductions (Laumen and Schneider, 1988).

The enzymatic hydrolysis of numerous meso diacetates to the nonracemic monoacetate has been reported with the same enzymes as those used in the enzymatic resolution of alcohol acetates (e.g., Xie, Suemune, and Sakai, 1987; cf. Section 12-4.e). The unreacted diacetate is, of course, achiral, hence optically inactive. Nevertheless, such enantioselective syntheses are sometimes erroneously called enzymatic resolutions.

The resolution of amino alcohols (e.g., 2-amino-1-butanol) is effected by PPL catalyzed acylation of the amino and hydroxyl groups; ethyl acetate serves both as acylating agent and as solvent in the reaction (Gotor et al., 1988); alternatively, resolution can be effected by lipase catalyzed hydrolysis of the *N,O*-diacetyl derivative (Bevinakatti and Newadkar, 1990). The enzymatic resolution of oxaziridines whose chirality is solely due to the presence of a stereogenic nitrogen atom has been reported (Bucciarelli, Moretti, et al., 1988).

Lactic acid and other 2-hydroxycarboxylates are resolved enzymatically, the former to enantiomerically pure (*S*)-lactate, by an oxidoreductase contained in crude extracts of *Proteus vulgaris*. The dehydrogenation requires a cofactor, carbamoylmethylviologen, that is regenerated electrochemically during the course of the resolution (Skopan, Günther, and Simon, 1987).

The enzymatic resolution of steroidal ketones by oxidoreductase-containing microorganisms (*Hansenula capsulata*) has been reported (Neef et al., 1985). *rac*-Squalene oxide (Fig. 7.84, **206**) is enzymatically resolved in the course of its cyclization to lanosterol under the influence of a sterol cyclase contained in bakers' yeast and stimulated by ultrasonic irradiation (Bujons, Guajardo, and Kyler, 1988).

Metal containing substrates, for example, metallocene aldehydes (Top, Jaouen, et al., 1988) and ferrocene aldehydes (Izumi et al., 1989) are also resolved by bakers' yeast.

7-7. MISCELLANEOUS SEPARATION METHODS

Two possible ways of exploiting diastereomer discrimination between the enantiomers of a solute and a chiral solvent are (a) separation based on a solubility difference between the enantiomers in chiral solvents, and (b) differential partition of the enantiomers between immiscible solvents one of which is chiral. The first of these methods leads to no practical separation.

> Attempts have been made to resolve chiral compounds by crystallization from optically active solvents. It was reasoned that the two enantiomers of a given substance must have different solubilities in optically active solvents, hence separation on this basis should be feasible. For compounds whose enantiomers are thermally stable, the experimental results have generally been negative, however. Solubility differences within the limited range of solvents tried are usually insufficient to effect enrichment. With some ionic organometallic complexes and organic compounds that are able to invert their configuration easily, that is, having limited configurational stability (see Section 7-2.e), enantiomeric enrichment is found. It has been suggested that the enrichment reflects not the interplay of heterogeneous equilibria but rather differences in rates of crystallization (or dissolution), that is, the observations may depend on kinetic phenomena akin to preferential crystallization (Jacques et al., 1981a, p. 245).

Experiments of the second kind carried out intensively since the early 1970s have been quite successful. These experiments take the form of resolution by extraction, by transport across a chiral membrane, or by some combination of these techniques.

The greater magnitude of diastereomer discrimination observed by application of the techniques just mentioned relative to those involving solubility differences stem from the larger number of interactions (multiple binding sites) available, that is, from a better choice of chiral "reagent" or solvent, as well as from the additional selectivity derived from the need of substrate molecules to cross a physical barrier (phase boundary or membrane). The chiral reagents that are responsible for the diastereomer discrimination, typically solutes diluted by solvents, such as chloroform, toluene, or 1,2-dichloroethane, are called hosts, ligands, carriers, or ionophores, the latter terms emphasizing the ability of the reagents to bind and transport ions. While resolutions by extraction and transport have some resemblance to other resolution processes, notably chromatography, they are conveniently discussed separately in this section.

a. Partition in Heterogeneous Solvent Mixtures

The first quantitative study demonstrating the possibility of partially extracting one of the enantiomers of a racemate preferentially from water into an optically active solvent [(+)-diisoamyl tartrate (Fig. 7.85, **207a**)] is that of Bowman et al., (1968). These investigators proposed that the preferential extraction of (±)-camphoric acid and of (±)-hydrobenzoin (Fig. 7.86, **208** and **209**) is related to the formation of hydrogen bonds between substrate and chiral solute molecules; two such hydrogen bonds are necessary for resolution to take place. The efficiency,

207

R =
207a $(CH_3)_2CHCH_2CH_2$
207b CH_3CH_2
207c $(CH_3CH_2CH_2CH_2)_2CH$

210

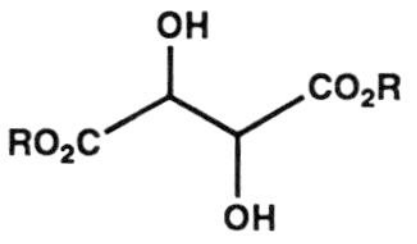

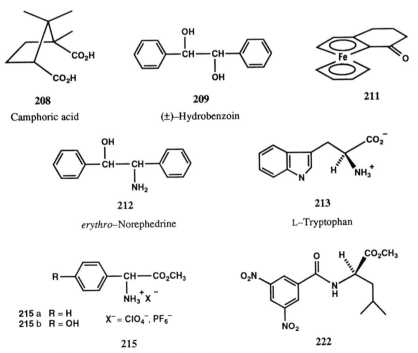

Figure 7.85. Chiral solvents and lipophilic carriers used in resolution by partition.

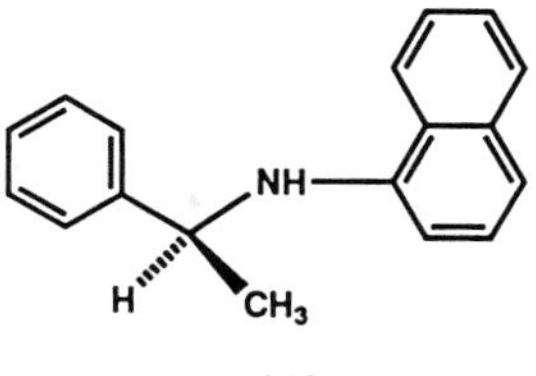

which is quite low (1–2% at best), can be increased with multiple extractions. Romano, Rieman, et al. (1969) described the extraction of sodium mandelate, $C_6H_5CH(OH)CO_2Na$, from H_2O into $CHCl_3$ containing the chiral amine **210** (Fig. 7.85) present as an ion pair.

At about the same time, Bauer, Falk and Schlögl (1968) described the partial resolution of 1,2-(2-oxytetramethylene)ferrocene (Fig. 7.86, **211**) by countercurrent distribution. The partitioning liquids were (+)-diethyl tartrate (Fig. 7.85, **207b**) and cyclohexane. The Craig countercurrent technique has also been applied to the resolution of sodium mandelate (Romano, Rieman, et al., 1969).

More recently, Prelog et al. (1982) found that chiral α-amino alcohols (e.g., *rac-erythro*-norephedrine, Fig. 7.86, **212**) can be resolved by partition of their salts with lipophilic anions (e.g., PF_6^-) between an aqueous phase (aq) and an immiscible lipophilic phase (lp) (1,2-dichloroethane) containing an ester of tartaric acid, for example, (−)-di-5-nonyl-(S,S)-tartrate (Fig. 7.85, **207c**). The

208
Camphoric acid

209
(±)-Hydrobenzoin

211

212
erythro-Norephedrine

213
L-Tryptophan

215 a R = H
215 b R = OH

$X^- = ClO_4^-, PF_6^-$

215

222

Figure 7.86. Compounds resolved by partition or transport.

selectivity rose with decreasing temperature, reaching an enantiomer excess ratio $ee_{lp}/ee_{aq} = 7.1$ [23.3 $(1R)$–3.3 $(1S)$] at 4°C in the two phases. No selectivity was exhibited by α-methylbenzylamine salts, thus demonstrating again that at least two hydrogen bonds between the substrate and the chiral extractant molecules are necessary for differentiation (three such bonds according to Prelog). This partition experiment was translated into a preparative chromatographic resolution.

The reverse process, that is, translation of partition chromatographic resolutions into a liquid–liquid partition (or a countercurrent distribution) system has been explored in order to increase the scale of separation. Chiral phases (aqueous yet immiscible), such as dextran (a low molecular weight polysaccharide) and polyethylene glycol (PEG, a polyalcohol) saturated with bovine serum albumin (BSA, a protein) partition *rac*-tryptophan (Fig. 7.86, **213**) between the two layers. More L- than D-tryptophan is retained in the lower (aq PEG/BSA) layer at 4°C; no separation takes place in the absence of the BSA. A separation factor (reflecting the enantioselectivity of the upper dextran phase) $[G_D/G_L] = 3.1$ was found following countercurrent distribution of the two phases in a 60 tube system (Sellergren et al., 1987).

Donald J. Cram and co-workers at UCLA (Los Angeles) have been particularly successful in the design of synthetic chiral "solvents" to effect extraction with high selectivity. In their experiments, the racemic solute and solvent (a solution of an optically active compound diluted by, e.g., $CHCl_3$) are often called guest and host, respectively, and the interaction is so specific and so tight that the diastereomeric complexes thus formed may even be isolated (Section 7-3.c). The terminology reflects the formation of cavitates between solute and "solvent" molecules.

Extraction of aqueous *rac*-α-methylbenzylammonium hexafluorophosphate $C_6H_5CH(CH_3)NH_3^+PF_6^-$, with a solution of (−)-crown ether **214** (Fig. 7.87) dissolved in $CHCl_3$, leads to recovery of (+) amine having ee = 24% (Kyba, Cram, et al., 1973). Subsequently, partial resolution by extraction (one equilibrium = one plate distribution) was achieved with phenylglycine methyl ester (Fig. 7.86, **215a**) and other amino acid derivatives using host compound **216a** (Fig. 7.87). The selectivity was determined by measuring enantiomer distribution constants, EDC = D_A/D_B, where D_A is the distribution constant for $CDCl_3$ versus D_2O of the enantiomer more tightly bound to the chiral crown ether, while D_B is the corresponding coefficient for the less tightly complexed enantiomer. The EDC values are directly related to the diastereomer discrimination energy: $\Delta(\Delta G°) = -RT\ln(EDC)$ (Cram and Cram, 1978).

It was found (Helgeson, Cram, et al., 1974) that the highest selectivity was achieved with compounds **215a** and **215b** (Fig. 7.86) for which EDC = 12 and 18, respectively, using host **216a** (Fig. 7.87). By modifying the nature and the amount of inorganic salt in the water layer ($LiPF_6$ or $LiClO_4$) and of CH_3CN in the organic layer, the amount of guest transferred and the selectivity could be adjusted. Amino acids, such as phenylglycine, $C_6H_5CH(NH_3^+)CO_2H$ ClO_4^-, could be extracted with EDC = 52 leading to the possibility of obtaining guest with an enantiomer purity of 96% ee through just one equilibration (Peacock and Cram, 1976).

The fact that compounds such as **214** and **216a** (Fig. 7.87) exhibit enantiomer selectivity toward α-methylbenzylamine salts requires that the nature of the

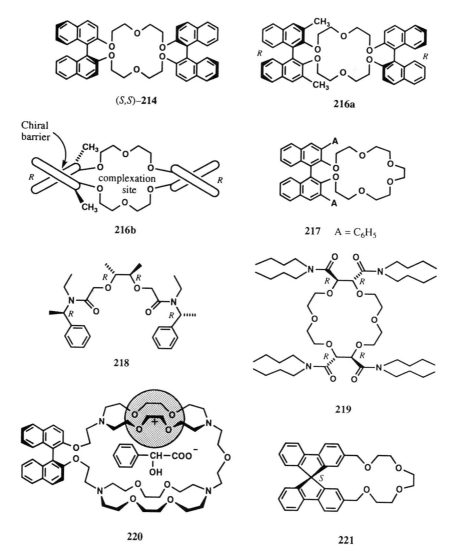

Figure 7.87. Ion-selective ligands (ionophores). [Structure **216** is reprinted with permission from Cram, D.J. and Cram, J.M. (1978), *Acc. Chem. Res.*, **11**, 8. Copyright © 1978 American Chemical Society; structure **220** is reprinted from Lehn, J.-M. (1979), *Pure Appl. Chem.*, **51**, 979 with permission of the International Union of Pure and Applied Chemistry.]

hydrogen bonding in these cases differ from that which obtains with the tartrate ester (see above). This selectivity is understandable in light of the rigid shape of the chiral host molecule and the necessity of the substrate to fit *inside* the host (ligand) cavity wherein the hydrogen bonding is manifest. As cogently stated by Cram and Cram (1978), in macrocyclic chiral ligands the binding sites *converge* while those of the substrate molecule (three hydrogen bonds involving $-NH_3^+$) *diverge*. The resultant complementary stereoelectronic arrangement of binding sites and steric fit may account for the greater selectivity of these ligands relative to that possible with open ligands, such as "simple" tartrate esters, although the latter are capable of a greater variety of hydrogen-bonding modes with multifunctional substrates.

The model of guest–host complexation that emerges is pictured in Figure 7.87, **216b**. This general model has been elaborated on the basis of ^{1}H NMR evidence as well as from X-ray spectra of both diastereomeric complexes of a given guest–host pair (Goldberg, 1977; Kyba, Cram, et al., 1978). Hosts, such as **214** (called dilocular since they have two stereogenic units), have homotopic faces [C_2 or higher (e.g., D_2) symmetry] such that complexation may take place with equal ease from either face to provide identical diastereomeric complexes. Substituent groups, such as CH_3, on the naphthalene rings in **216a** (Fig. 7.87) extend the chiral barrier and are responsible for the dramatically higher diastereomer discrimination, for example, to EDC = 52, where $\Delta\Delta G° = -2.15$ kcal mol^{-1} (9.0 kJ mol^{-1}). Figure 7.88 is a model describing the interaction of crown ether **216a** with guest molecules (L and M = large and small substituents, respectively) that is responsible for the observed diastereomer discrimination and recognition. The model also predicts the configuration of the guest giving rise to the stabler complex (Cram and Cram, 1978).

Chromatographic resolutions based on the chiral crown ethers have been of two types: (a) preformed diastereomeric complexes (salts of racemic amines or amino acid esters plus chiral host mixed in the mobile phase) separated on silica gel columns and (b) separation of the same racemates on chiral crown ethers covalently bound to silica gel (enantioselective columns) (Sousa, Cram, et al., 1978; Sogah and Cram, 1979). Baseline separation of the enantiomers was achieved in many cases. A significant aspect of these resolutions is that the order of appearance of the enantiomers was predictable on the basis of the general model described above.

Partition experiments have been carried out with monolocular hosts (chiral hosts having just one stereogenic element) that retain the desirable C_2 axis permitting complexation from either face of the macrocyclic ring. Host **217** (A = C$_6$H$_5$, Fig. 7.87) was the most discriminating toward **215a**$^+$ ClO$_4^-$, which shows an EDC = 20 (Lingenfelter, Helgeson, and Cram, 1981). A detailed analysis of host and guest structures, and of the environment of the complex, on the selectivity (diastereomer discrimination) and binding affinities has appeared (Peacock, Cram, et al., 1978).

Similar extraction experiments (potential or actual) have been carried out with configurationally chiral macrocyclic polyether hosts incorporating sugars (mannitol and threitol derivatives; Stoddart, 1979, 1987), tartaric acid (Lehn, 1979), a 1,1'-binaphthyl unit (Lehn et al., 1978), and a 9,9-spirobifluorene moiety

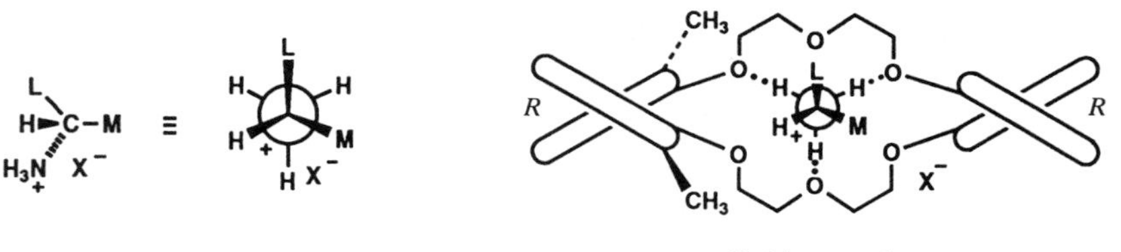

Stabler complex

Figure 7.88. Complexation model for dilocular hosts [Reprinted with permission from Cram, D.J. and Cram, J.M. (1978), *Acc. Chem. Res.*, **11**, 8. Copyright © 1978 American Chemical Society.]

(Prelog, 1978; see also Jolley, Bradshaw, and Izatt, 1982). Finally, it was found that some naturally occurring ionophores, for example, the desvalino–boromycin (DVB) anion $[C_{40}H_{64}BO_{14}]^- X^+$, a degradation product of the antibiotic boromycin, also exhibit diastereomer discrimination. A solution of $(DVB)^- N(CH_3)_4^+$ in 1,2-dichloroethane selectively extracts ester and amide derivatives of phenylglycine from water into the chiral lipophilic phase with preference for the R enantiomers (EDC > 2) (Prelog, 1978).

b. Transport across Membranes

Interest in the transport of amino acids, and of other substances, across membranes in biological systems has stimulated experiments in which synthetic lipophilic substances of the type described in Figures 7.85 and 7.87, and others, might serve as carriers, that is, ferry ions across an otherwise repelling barrier (Behr and Lehn, 1973). The idea was to create synthetic membranes consisting of water immiscible solvents into which one of the lipophilic chiral compounds might be dissolved. Since these "membranes" are chiral and nonracemic, it was hoped that enantioselective transport might be observed as a consequence of the operation of diastereomer discrimination within the membrane. The results of partition experiments (see above) provided guidance in the design of several types of transport experiments that are summarized schematically in Figure 7.89. Bulk toluene and chloroform liquid phases served as representative membranes. One of the goals of such transport experiments was the development of practical resolution methods.

One of the first reports of enantiomer discrimination in transport is that of Newcomb, Helgeson, and Cram (1974). The driving force for transport in this experiment is provided by entropy of dilution and "salting out" of the chiral organic salt from its reservoir by an achiral inorganic salt (Fig. 7.89a, U-tube at left). The enantiomers of methyl phenylglycinate·HPF$_6$ (Fig. 7.86, 215a) were shown to migrate (from the α aqueous reservoir toward the β aqueous reservoir) across a chloroform membrane containing R,R ligand 218 (Fig. 7.87) (the carrier C) at differing rates (rate ratio $\cong$ 10). The faster moving (R)-215a (obtained with ee = 78%) is that shown to form a tighter complex with R,R ligand 218 in one-plate partition experiments. Similar experiments carried out in W-shaped tubes (Fig. 7.89a, right) permit simultaneous and competitive transport of both enantiomers of the same glycine ester toward separate reservoirs thereby accomplishing a virtually complete resolution of the racemate. Newcomb, Cram, et al., (1979) dubbed such a system a *resolving machine* (Fig. 7.90). Resolution of 215a (Fig. 7.86) with hosts (R,R)- and (S,S)-216a (97 and 98% ee, respectively) in the W-tube at 0°C [EDC $\cong$ 30 for this system] led to delivery of (R)-215a (ee = 90%) and of (S)-215a (ee = 86%), respectively, in the two tubes. This nonoptimized finding may be contrasted to an expected 94% ee (ideal value) for each enantiomer based on the EDC value.

For other enantioselective crown ether transport experiments (Fig. 7.89a systems), see Naemura, Fukunaga, and Yamanaka (1985); Yamamoto, Noda, and Okamoto (1985); Yamamoto, Kitsuki, and Okamoto (1986).

Lehn et al., (1975) modulated the transport experiment by allowing an achiral

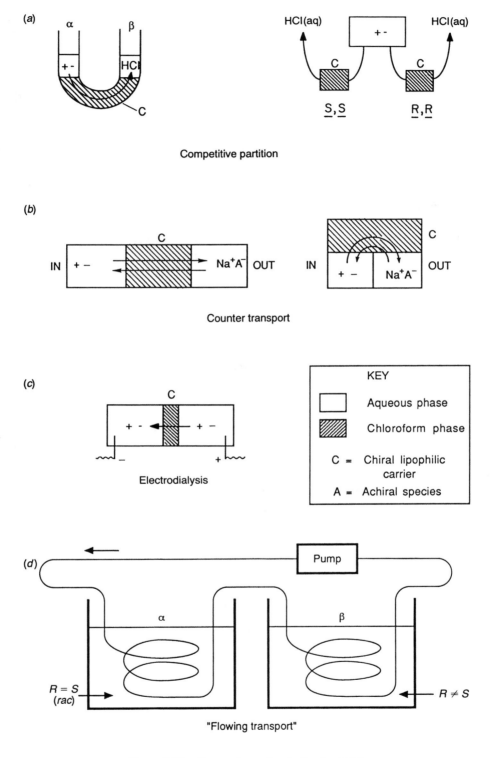

Figure 7.89. Types of enantioselective transport.

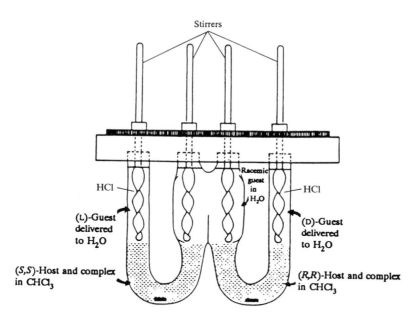

Figure 7.90. Resolving machine [Adapted with permission from Cram, D. J. (1990), *From Design to Discovery*," American Chemical Society, Washington, DC, p. 67. Copyright © 1990 American Chemical Society.]

species to be transported against the flux of chiral ions migrating through the chloroform phase (equivalent to a liquid membrane) containing the chiral lipophilic carrier **219** (Fig. 7.87). The racemic salt solution, for example, of sodium mandelate, $C_6H_5CH(OH)CO_2^-Na^+$, initially on the IN (left) side, of Figure 7.89*b*, becomes optically active (<10% ep) in all three phases. Both the rate and the enantioselectivity of the transport of mandelate ion are affected by the nature of the countertransported (called antiport) achiral species (Na^+A^- = NaCl, sodium propionate, etc.).

The evolution of optical activity in the IN and OUT phases resembles quantitatively the changes in enantiomeric enrichment observed in kinetic resolution experiments (Section 7-6) when Na^+A^- is sodium propionate. However, as the extent of transport increases, back-transport of mandelate becomes important and the enrichment begins to decrease. This type of experiment may be a reasonable model (cf. Behr and Lehn, 1973) to describe how selective transport of a variety of substances takes place across a membrane in biological systems.

Transport of mandelate ion through a membrane containing **220** (Fig. 7.87) leads to resolution whose extent is cation dependent. Ion pair formation of mandelate ion with K^+ (shaded circle) with both ions apparently penetrating the cavity of the crown ether is more enantioselective than with Na^+ or Cs^+ (Lehn, 1979).

In a third type of transport experiment, electrodialysis provides the driving force for migration across the thin (0.2 mm) liquid membrane (Fig. 7.89*c*). The membrane consists of the ionophore **218** (Fig. 7.87) dissolved in *o*-nitrophenyl octyl ether containing PVC. Migration of α-methylbenzylammonium [MBA, $C_6H_5CH(CH_3)NH_3^+$] ions ($10^{-3} M$) under a potential difference, affords a selectivity of 8% for the two enantiomers (Thoma, Prelog, Simon, et al., 1975; Thoma, Simon, et al., 1977). The results were subjected to a chiral cross-check (Section 6-2). In these experiments, the enantiomeric compositions of the very dilute solutions were measured radiochemically by means of a novel double-labeling technique (^{3}H label for the R enantiomer of MBA and ^{14}C label for the S) thus avoiding the need for measuring optical rotations (Section 6-5.h).

A membrane containing a chiral ionophore (Fig. 7.87, **219**) exhibits a selectivity of 2.6 for the two enantiomeric MBA ions (where the preference is $R > S$) (Bussmann, Lehn, Simon, et al., 1981). The order of selectivity (2.65 for MBA and 1.19 for **215a**, Fig. 7.86) was found to differ from that found with a membrane containing the ligand **216a** (Fig. 7.87). Chiral ionophores based on the spirobifluorene nucleus (Fig. 7.87, **221**) also exhibit selectivity toward MBA (Thoma, Prelog, Simon, et al., 1979).

Permeable silicone rubber tubing serves as an enantioselective membrane when impregnated by octadecyl (S)-N-(1-naphthyl)leucinate dissolved in dodecane circulating through the tubing (Fig. 7.89d). Racemic analyte, for example, methyl N-(3,5-dinitrobenzoyl)leucinate (Fig. 7.86, **222**) dissolved in container α (CH$_3$OH/H$_2$O 4:1) slowly diffuses through the tubing and appears in container β with concomitant enantioselection. After several hours, the enantiomer compositions of **222** in containers α and β were significantly unequal (Pirkle and Doherty, 1989).

Hollow fiber membrane reactors (analogous to Fig. 7.89d, but with opposing flow systems) have been adapted to the large-scale kinetic resolution of esters (e.g., of ibuprofen) with the lipase enzyme immobilized in the membrane (Young and Bratzler, 1990; Stinson, 1992). Amino acid derivatives are also resolved in a hollow fiber membrane system in which solutions of the racemate and the chiral selector, for example, derivatives of N-(1-naphthyl)leucine, flow in opposite directions (Pirkle and Bowen, 1991).

A *water-based* liquid membrane containing cyclodextrin carriers (e.g., Fig. 7.33, **107**) has been applied to the enantioselective transport of several chiral organic compounds (Armstrong and Jin, 1987).

7-8. RACEMIZATION

Racemization is the formation of a racemate from a pure enantiomer having the same constitution (Eq. 7.31). While the constitution of the compound is unaffected, the configurational integrity of the sample is lost, that is, an enantiopure (or nonracemic) sample becomes racemic during the process. Racemization is an irreversible process arising from the reversible interconversion of enantiomers; it is always associated with the disappearance of optical activity (Mislow, 1965, p. 69).

Racemization was discovered by Pasteur in 1853 (q.v.) when he sought purposefully to prepare *racemic* tartaric acid, since sources of this acid had been depleted. Pasteur discovered that on heating the cinchonine salt of (+)-tartaric to 170°C, the latter was converted largely to (the salt of) the inactive *rac*-tartaric acid. For this discovery, Pasteur was decorated by the French Government and he received a prize from the Paris Pharmaceutical Society (Holmes, 1924; Jacques, 1992).

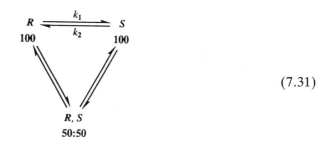

$$(7.31)$$

Epimerization is a related term that we use to describe a (reversible) change in configuration at one or more (but not all) stereocenters in a molecule possessing several stereocenters. Diastereomers differing in configuration at just one stereocenter are often called *epimers* (Eliel, 1962, p. 40). Because at least one stereocenter remains unaffected in the molecule, epimerization of a non-racemic substance does not lead to the loss of optical activity. However, the magnitude and sign of the sample's rotation may, and generally will, change over time. This change in rotation is called mutarotation (Lowry, 1899).

> Epimerization of chiral diastereomers is a common process. In certain cases it has significant commercial importance. Two examples will suffice to make the point: (a) hydrogenation of thymol is one route to the menthol (Fig. 7.19c, **49**)/isomenthol/ neomenthol/neoisomenthol diastereomer manifold ($2^3 = 8$ stereoisomers); epimeri-zation of the stereoisomer mixture (at C(1) and C(2)) can lead to a mixture rich in the most stable and most valuable component (menthol; Derfer and Derfer, 1983), and (b) (R^*,S^*)-ephedrine (Fig. 7.19a, **38**) and (R^*,R^*)-pseudoephedrine are both medicinal agents, but they are of unequal value. The former is the more costly. Epimerization of the latter under acidic conditions increases the value.

Racemization is an energetically favored process since the concentration of the starting enantiomer is reduced by one-half during its operation. But, it is favored energetically only to the extent of $\Delta G° = RT \ln \frac{1}{2} = -RT \ln 2$, corre-sponding to $-0.41 \text{ kcal mol}^{-1}$ (-1.7 kJ mol^{-1}) at 25°C. Since $\Delta H° = 0$ for the enantiomers undergoing racemization (assuming no differential intermolecular interactions; cf. Chapter 6), it follows that the driving force for racemization is entirely entropic.

Racemization is generally quite slow unless a suitable mechanistic pathway is available. Yet, because numerous mechanisms exist by which the configuration at a stereocenter may be "lost," racemization is a ubiquitous phenomenon which is described and discussed in numerous papers incidental to other work. Racemiza-tion has the distinction of being at times something to avoid (as during the synthesis of a polypeptide) and, at other times, something to encourage (to avoid having to discard the unwanted enantiomer after a resolution). For two contrast-ing studies of racemization mechanisms that illustrate these opposite goals, see Smith and Sivakua (1983) and Yamada et al. (1983).

Racemization has also been defined operationally as the loss in optical activity of a substance over a period of time (Mislow, 1965, p. 69; Potapov, 1979, p. 110). While in most cases the loss in optical activity parallels the inversion of *all* stereocenters in one-half the molecules of an enantiopure sample, the change in rotation may not be easily measurable (cf. Chapter 13) and it is preferable to monitor the process by other methods for the determination of enantiomeric composition. In spite of this, the old expression "optically labile," meaning easily racemized, is still being used.

From the rate expression for reversible first-order reactions, it can be shown (assuming that rate constants $k_1 = k_2 = k$) that after a given time t

or

$$\ln\left[\frac{1 + [S]/[R]}{1 - [S]/[R]}\right] = 2kt \tag{7.32}$$

$$\frac{1}{t}\ln\frac{a}{a - 2x} = 2k \tag{7.33}$$

where a is the initial concentration of enantiomer R, x is the concentration of enantiomer R at time t and $a = 1$ if no S is present prior to racemization. If some S is present at onset, then

$$\ln\left[\frac{1 + [S]/[R]}{1 - [S]/[R]}\right] = 2kt + C \tag{7.34}$$

where C, the constant of integration, is no longer equal to zero (Williams and Smith, 1977).

If racemization is considered to be an irreversible transformation of R into a racemate R,S (Eq. 7.31, lower part), then the rate expression becomes that of an irreversible first-order reaction

$$\frac{1}{t}\ln\frac{a}{a - y} = k' \tag{7.35}$$

where y is the concentration of racemate R,S at time t. The reader is cautioned to note that $k' = 2k$, that is, the rate of racemization (Eq. 7.35) is equal to twice the rate of interconversion of the enantiomers (Eq. 7.33). In reading the literature, it is important to note which rate constant is being used (Eliel, 1962, p. 33).

For a system undergoing racemization, a useful quantity is the racemization half-life, τ (more accurately, τ is the inversion half-life in view of its definition). This quantity may be calculated from a more general form of Eq. 7.34 (applicable also to epimerizations in which $k_1 \neq k_2$):

$$\tau = \frac{\ln 2}{k_1 + k_2} = \frac{\ln 2}{(1 + K')k_1} \tag{7.36}$$

for the time when the enantiomeric purity of a chiral sample has been reduced from 100 to 50% ee. In Eq. 7.36, $K' = 1/K = k_2/k_1$ and, when $k_1 = k_2$, as is true in racemizations, Eq. 7.36 reduces to $\tau = \ln 2/2k$. (Williams and Smith, 1977). Another quite useful and practical quantity, particularly in the study of racemization of compounds having large rotations, is $t_{1/100}$, the time necessary to lose 1% of the rotation (Canceill, Collet, and Gottarelli, 1984).

a. Racemization Processes

There are three general ways of making racemates. (a) Most frequently, a racemate is formed whenever a synthesis results in the generation of a stereocenter in the absence of a chiral (intramolecular or extramolecular) influence. By way of example, addition of a nucleophilic reagent, such as methyllithium, to either of the heterotopic faces of an achiral but unsymmetrical aldehyde (e.g., benzaldehyde) gives rise to a racemic carbinol. (b) Racemates may be formed as the result of chemical transformations on nonracemic samples. Typically, this requires the cleavage of bonds (usually just one bond) at a stereocenter or the stretching of or rotation about bonds. The cleavage can take place via formation of free radicals, carbocations, carbanions, or excited states. The basis of racemization mediated by these intermediates and transition states is that they are planar (or that they easily attain a planar geometry). Examples of this approach to racemization are given below. (c) The most direct way of making racemates is by mixing equal parts of the two enantiomers of a given compound. While this process may seem trivial, it is occasionally practiced when a sample of a racemate (say of a natural product) is needed that is otherwise unavailable, for example, dimers of camphor (Huffman and Wallace, 1989), α-pinene (Weber et al., 1992b) and a protein (Zawadzke and Berg, 1992). A special application of the mixing approach is the preparation of 50:50 mixtures of enantiomers wherein one enantiomer is isotopically labeled (or both enantiomers are labeled, each with a different isotope) so as to permit facile subsequent analysis of the enantiomer composition (e.g., by mass spectrometry; cf. Section 6-6.h).

The second of the general types of racemization processes (see above) may take place by mechanisms involving the intervention of discrete intermediates (Review: Henderson, 1973) or of planar (or near-planar) transition states. In this section we limit ourselves to citing some examples of the several types from the large literature of racemization studies. Leading references to the earlier literature may be found in Wheland (1960, p. 334); Eliel (1962, p. 33); and in Morrison and Mosher (1976, p. 19). Racemization of amino acids is dealt with in Section 7-8.b.

Thermal Methods

Racemization without the breaking of covalent bonds may take place if thermal molecular deformation (of bond angles and lengths) can be achieved without demolition of the structure. Attainment of a truly planar transition state in thermal racemization is less important than is the facile interconversion of the enantiomers (Eq. 7.31, top line). The total spontaneous resolution of configurationally labile compounds (e.g., TOT **13** and 1,1'-binaphthyl **15**; Figs. 7.10 and 7.11, respectively), implies their easy racemization without breaking of covalent bonds; for a discussion of this point, see Section 7-2.e and Chapter 14.

The pioneering studies of thermal racemizations were those carried out on the biaryls. It is well worth pointing out that the calculation of activation energies for the racemization of biaryls are among the seminal studies leading to the development of the powerful empirical force field calculation (molecular mechanics, MM) method (Westheimer, 1956; cf. Sec. 2-6). In contrast to tetra-o-alkylsubstituted biphenyls, single (**223**) and doubly bridged biphenyls (Fig. 7.91, **224**) undergo thermal racemization. The E_a values calculated by the fledgling MM method were in remarkably good accord with the experimental ones (Mislow et al., 1964a; cf. Section 14-5). The measurement of racemization rates of nonracemic **223** and of

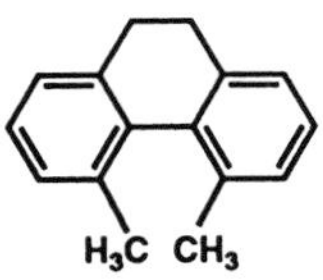

	E_a	
	(kcal mol^{-1})	(kJ mol^{-1})
X = Y = C=O	31.2	130.5
X = Y = O, S	24.0; 35.0	100.4; 146.4
X = O, Y = S	30.6	128.0

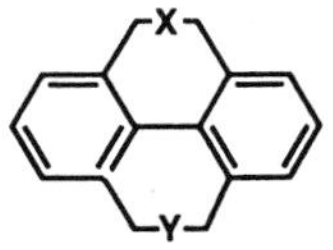

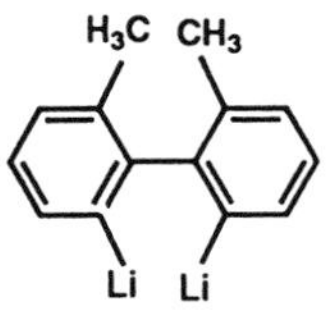

223   **224**

225 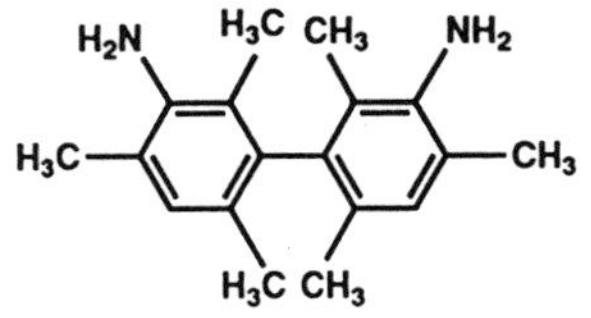   **226**

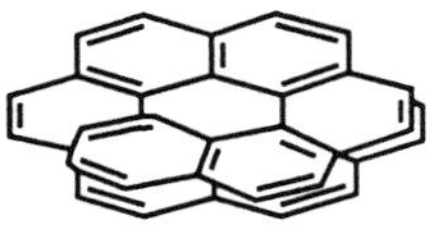

227

[9]Helicene

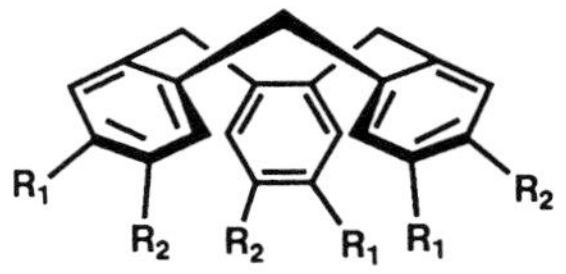

232

$R_1 = OCH_3$ $R_2 = OCD_3$

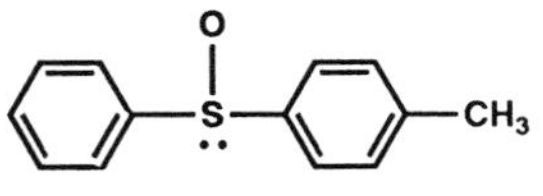

235 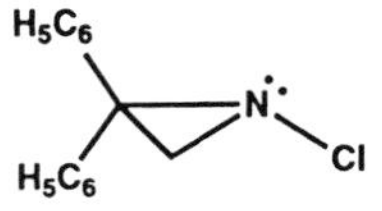   **236** 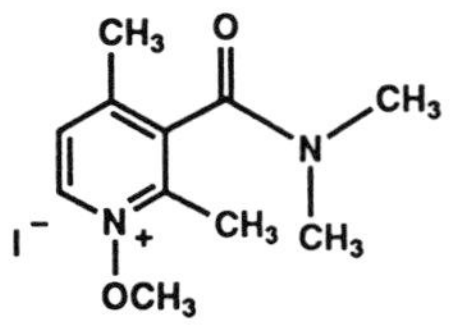   **237**

Figure 7.91.   Thermally racemizable compounds.

its dimethyl-d_6 analogue (with the finding that $k_D < k_H$) was the first demonstration of the operation of a steric kinetic isotope effect (Mislow et al., 1964b). 2,2'-Dilithio-6,6'-dimethylbiphenyl **225** (Fig. 7.91) is stable to racemization at −10°C (Fredj and Klingstedt, 1983). Racemization of **128** (Fig. 7.42) where $t = 7.8$ h (25°C) takes place not by rotation about the central phenyl–phenyl bond but by "internal" rotation of the CHCl$_2$ groups; the barrier to racemization of **128** $[\Delta G^{\ddagger} = 23.7$ kcal mol^{-1} (99.2 kJ mol^{-1})] is substantially smaller than that of 3,3'-diaminobimesityl **226** $[\Delta G^{\ddagger} = 29.8$ kcal mol^{-1} (124.7 kJ mol^{-1})] having four significantly less bulky methyl groups at the ortho positions (Biali, Mislow, et al., 1988).

The helicenes (e.g., hexahelicene, Fig. 6.6), are prototypes of overcrowded molecules adopting helical configurations. The fact that helicenes as large as [9]helicene (Fig. 7.91, **227**) have been found to racemize thermally [the latter is

Figure 7.92. Thermal racemization of annulated cyclooctatetraenes.

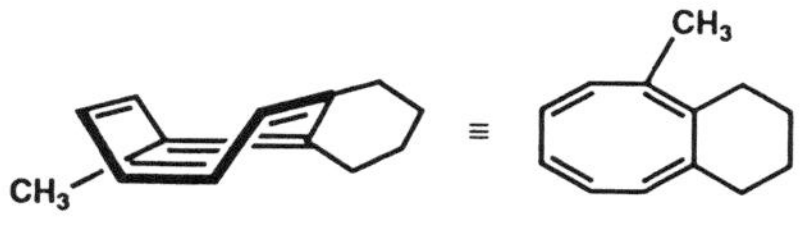

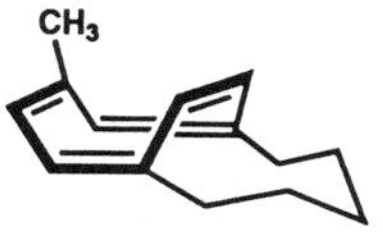

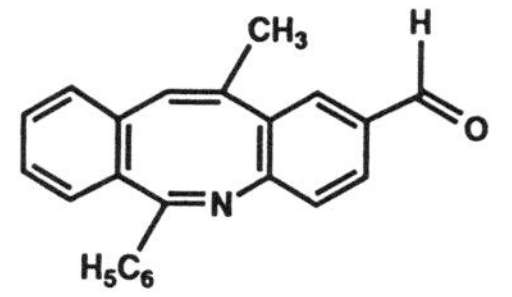

completely racemized in 10 min at 380°C with little decomposition; $\Delta G^{\ddagger} =$ 43.5 kcal mol^{-1} (182 kJ mol^{-1}) in naphthalene] is remarkable (Martin and Marchant, 1974). Even [11]helicene has been found to undergo thermal racemization (at 400–410°C) (Martin and Libert, 1980). The ingeneous suggestion by J. Nasielski, that racemization of helicenes might conceivably take place by an internal double Diels–Alder reaction, was discarded by means of an isotope label study. The most reasonable explanation for the racemization remains that it takes place by a "conformational pathway," that is, the helicenes are significantly more flexible than our prejudices and examination of space-filling models previously led us to believe, with the necessary bond deformations being spread over much of the molecular skeleton (Martin, 1974). For analogous studies of the racemization of thiaheterohelicenes, see Yamada, Kawazura, et al. (1986).

Chiral substituted cyclooctatetraenes (COTs; e.g., **228**, Fig. 7.92), thermally racemize principally by ring inversion rather than by bond shifting of **228** to isomer **229** (Paquette and Wang, 1988). Thermal racemization is completely inhibited in a cyclooctatetraene that is 1,4-annulated (Fig. 7.92, **230**; Paquette and Trova, 1988). The racemization rate of COT analogue **231** was measured by following changes in the pitch of the cholesteric mesophase generated when **231** was added to a nematic liquid crystal (Ruxer, Solladié, and Candau, 1978; cf. Section 13-4.3); for an analogous study of the racemization of 1,1'-binaphthyl in nematic solvents, see Naciri, Weiss, et al. (1987); see also Sec. 14-7.

Cyclotriveratrylene (Fig. 7.91, **232**), a cup-shaped chiral compound, racemizes fairly easily [$\Delta G^{\ddagger} = 26.5$ kcal mol^{-1} (111 kJ mol^{-1}) at 25°C] by a "crown-to-crown" interconversion (Collet and Gabard, 1980; Canceill, Collet, and Gottarelli, 1984).

Allyl *p*-tolyl sulfoxide **233** cleanly racemizes at modest temperatures (50–70°C) that are much lower than those required for the racemization of other sulfoxides (see below); $\Delta G^{\ddagger} = 24.7$ kcal mol^{-1} (103 kJ mol^{-1}) at 51°C in C_6H_6. Based on available thermodynamic and kinetic data together with isotope labeling experiments it was concluded that racemization of **233** (and of sulfoxides such as allyl methyl sulfoxide) takes place by a [2,3] sigmatropic (concerted) rearrangement involving the reversible formation of an achiral intermediate, allyl *p*-toluenesulfenate **234** (Eq. 7.37) (Bickart, Mislow, et al., 1968).

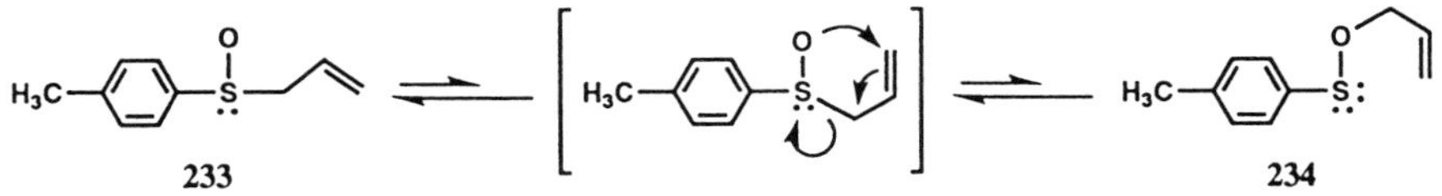

$$(7.37)$$

Racemization of the alkaloid (−)-vincadifformine occurs by a sequential retro Diels–Alder and Diels–Alder cycloaddition process. Racemization is assured by the fact that the intermediate (cycloreversion) product is achiral (Fig. 7.93; Takano, et al., 1989).

Perchlorotriphenylamine (Fig. 7.42, **124**), a compound whose chirality is due to its shape as a molecular propeller (the nitrogen is chirotopic but non-stereogenic), undergoes thermal racemization at 120°C [$\Delta G^{\ddagger} = 31.4$ kcal mol^{-1} (131 kJ mol^{-1})] presumably by a "two-ring flip" mechanism (Hayes, Mislow, et al., 1980).

Some chiral compounds with chirotopic and stereogenic heteroatoms are able to racemize by a pyramidal inversion mechanism at the heteroatom (equivalent to quantum mechanical tunneling). Two examples of compounds subject to this type of racemization mechanism are phenyl p-tolyl sulfoxide (Fig. 7.91, **235**) with inversion at sulfur [$\Delta G^{\ddagger} = 38.6$ kcal mol^{-1} (162 kJ mol^{-1}) at 200°C in p-xylene] and 2,2-diphenyl-N-chloroaziridine (Fig. 7.91, **236**) (with inversion at nitrogen [$\Delta G^{\ddagger} = 24.4$ kcal mol^{-1} (102 kJ mol^{-1}) in CCl$_4$] (Rayner, Mislow, et al., 1968 and Forni, Moretti, et al., 1981, respectively). Note the much lower inversion barrier at nitrogen.

The nicotinamide analogue **237** (Fig. 7.91) is an example of a type of compound whose chirality is due to an out-of-plane orientation of the carbox-amide group that is enforced by the two flanking methyl groups on the pyrimidinium ring (atropisomerism); the three-dimensional structure may be thought of as a sort of tetrasubstituted chiral allene in which the carboxamide is perpendicular to the unsymmetrical pyrimidinium moiety (see also p. 552). Thermal racemization of **237** is much faster in nonpolar solvents (e.g., hexane) than it is in water [$\Delta G^{\ddagger} = 27.1$ kcal mol^{-1} (113 kJ mol^{-1}) at 27°C in H$_2$O]; the decreased solvent polarity would be expected to decrease the double-bond (dipolar) character of the amide thereby increasing k (van Lier et al., 1983).

Racemization via Stable Achiral Intermediates, Excited States, and Free Radicals

Partial racemization of **238** (Fig. 7.94) takes place on catalytic hydrogenation in the presence of palladium; the enantiomeric purity is reduced (by >10% ee)

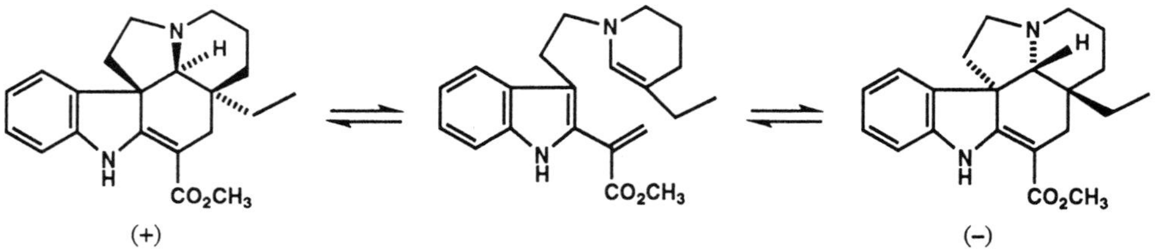

Figure 7.93. Racemization of (−)-vincadifformine.

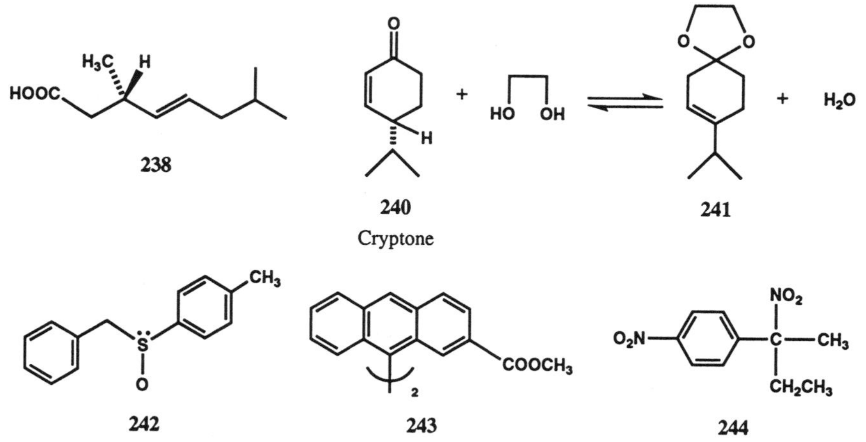

238 240 Cryptone 241 242 243 244

Figure 7.94. Racemization via stable achiral intermediates and radicals.

during the reduction. It is believed that the racemization is most likely due to double-bond isomerization to an achiral alkene occurring concurrently with the reduction (Chan et al., 1976; cf. also Chapter 9).

In contrast to other biaryls, it has been suggested that the thermal racemization of heterobiaryls **239** is best accounted for by a ring opening–ring closure reaction (Eq. 7.38; Roussel et al., 1988a).

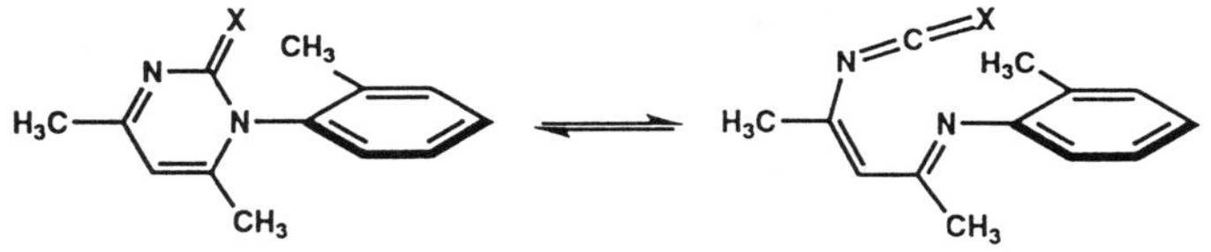 (7.38)

239
X = O, S

achiral

Acyl halides with a stereocenter at the α-carbon atom are prone to racemization provided that at least one α-hydrogen atom is present in the structure. There is evidence that elimination of and readdition of HCl to a ketene intermediate is responsible for this racemization (Sutliff, 1966).

rac-Cryptone (Fig. 7.94, **240**) is obtained from the more abundant (−)-isomer on ketalization with ethylene glycol. Derivatization is attended by double-bond migration to the achiral Δ^3-dioxolane **241**. Regeneration of the ketone on acid hydrolysis returns the double bond to its original location (Soffer and Günay, 1965) but leaves the product racemic.

Racemization has been observed to take place during the course of the Wittig reaction on a β-hydroxyaldehyde. It is presumed that this is a consequence of the operation of an intramolecular equilibration (with a β-ketocarbinol) of the Meerwein–Ponndorff–Verley/Oppenauer type catalyzed by sodium ion rather than by formation of a carbanion intermediate (by deprotonation of the α-carbon atom) (Lubell and Rapoport, 1989).

Palladium black catalyzes the racemization and concomitant alkylation of nonracemic α-methylbenzylamine (Murahashi et al., 1983). Deuterium label studies have demonstrated the intermediacy of achiral enamine and/or imine palladium complexes in the racemization (Eq. 7.39):

$$
\underset{\substack{\text{H}}}{\overset{\substack{\text{CH}_3}}{C_6H_5-\overset{|}{\underset{|}{C}}-NH_2}} \; \underset{\rightleftharpoons}{\overset{Pd}{}} \; \underset{\underset{PdH_2}{\uparrow}}{\underset{C_6H_5}{\overset{CH_3}{>}}C=NH} \; \rightleftharpoons \; \underset{\underset{\substack{Pd \\ H_2}}{}}{\overset{C_6H_5}{\underset{H_2C}{\diagdown}\overset{|}{\underset{C}{\diagup}}\diagdown_{NH_2}}} \qquad (7.39)
$$

Along similar lines, 2-amino-1-butanol, $CH_3CH_2CH(NH_2)CH_2OH$, racemizes under hydrogenation conditions in the presence of a cobalt catalyst (Ichikawa et al., 1976); this racemization may proceed reversibly via a configurationally labile α-aminoaldehyde. A different mechanism must apply to the partial racemization attending the hydrogenolysis of 2-isopropylaziridine to 3-methyl-2-butanamine (Rubinstein, Feibush, and Gil-Av, 1973; see also Schurig and Bürkle, 1982).

Racemization mediated by an excited state takes place when methyl p-tolyl sulfoxide is irradiated by light of wavelength longer than 285 nm either in the absence or presence of a photosensitizer (naphthalene). This reaction may be considered as a light-induced pyramidal inversion of a chiral sulfur atom. The presence of an arenesulfinyl chromophore is essential for this racemization to take place (Mislow, Hammond, et al., 1965). For additional examples of photoracemization, see Tétreau et al. (1982) and Okada, Samizo, and Oda (1986).

Racemization of [(R)-1-cyanoethyl](piperidine)cobaloxime has been observed to occur without decomposition upon irradiation of a crystalline sample with X-rays. This type of racemization is an apparently general phenomenon that occurs if the reaction cavity, i.e., the free space available to the stereogenic group in the lattice, is of sufficient size (Ohashi, 1988).

The X-ray irradiation was repeated on the corresponding racemic cobaloxime crystallized in a chiral space group ($P2_12_12_1$) containing two heterochiral molecules per unit cell. In such a crystal, the stereogenic 1-cyanoethyl groups of the two enantiomers reside in diastereomeric and not enantiomeric environments. On irradiation, it was observed that the 1-cyanoethyl group was racemized in only one of the two molecules in the unit cell. The resulting irradiated crystal contained (R)- and (S)-1-cyanoethyl groups in a 3:1 ratio, that is, the crystal was enantiomerically enriched. The difference in reactivity of the enantiomeric 1-cyanoethyl groups was attributed to differences in sizes in the reaction cavities surrounding the cyanoethyl groups (Osano, Ushida, and Ohashi, 1991).

Benzyl p-tolyl sulfoxide (Fig. 7.94, **242**) thermally racemizes faster by a factor of 10^3 than, for example, alkyl aryl sulfoxides (see above) [for **242**, $\Delta G^{\ddagger} = 32.7\ \text{kcal mol}^{-1}$ ($137\ \text{kJ mol}^{-1}$) at 145°C]; this racemization is accompanied by significant formation of byproducts including bibenzyl ($C_6H_5CH_2CH_2C_6H_5$), a product of the coupling of benzyl radicals. The free energy of activation of the racemization ($\Delta G^{\ddagger}$) is significantly smaller (at comparable temperatures) than that of sulfoxide racemization by pyramidal inversion (see above) and the energy of activation [$E_a = 44\ \text{kcal mol}^{-1}$ ($184\ \text{kJ mol}^{-1}$)] is within the range of expected benzyl C–S bond dissociation energy values. Accordingly, the mechanism proposed for racemization of **242** is one of homolytic cleavage and recombination of the radical pair (Miller, Mislow, et al., 1968). The large increase in the rate of racemization of **242** (Fig. 7.94) over that, for example, of **235** (Fig. 7.91) is accounted for by a much more positive entropy of activation ($\Delta S^{\ddagger}$), 24.6 vs. -5.1 eu for **242** and **235**, respectively (Rayner, Mislow, et al., 1968).

Addition of phenyl radicals to 2,2'-dicarbomethoxy-9,9'-bianthryl (Fig. 7.94, **243**) causes racemization. It is proposed that reversible addition of the radical to an anthracene ring of **243** generates a σ-complex having a lower C(9)–C(9') rotational energy barrier than is present in the bianthryl (Schwartz et al., 1989).

Substitution of the aliphatic nitro group in 2-(4-nitrophenyl)-2-nitrobutane (Fig. 7.94, **244**) by nucleophiles (e.g., azide or thiophenoxide) leads to completely racemized products. The racemization takes place concurrently with displacement since, at 50% completion, unreacted **244** retains all of its enantiomeric purity. The results are consistent with the operation of an electron-transfer chain mechanism mediated by free radicals (Kornblum and Wade, 1987).

trans-1,2-Disubstituted cyclopropanes enantiomerize, that is, racemize thermally without competing epimerization to the cis isomers, via ring-opened 1,3-disubstituted trimethylene diradicals. This insight stems from the fact that experimental $\Delta G^{\ddagger}_{\text{enant}}$ values correlate linearly with the sum of substituent radical stabilization energy terms (Baldwin, 1988).

Acid-Catalyzed Processes

Acid catalysts have the potential of generating carbocations at chiral centers provided that suitable leaving groups are present there. Since the resulting charged intermediates are typically planar (for exceptions, see March, 1992, p. 172), hence achiral, the topographic criterion would seem to demand that the products of reactions of carbocations generated at chiral centers be invariably racemic. However, due to the intervention of ion pairs (especially in solvolytic reactions), or of solvated (hence potentially chiral) carbocations, matters are significantly less clear-cut than is implied by the preceding statement. (For a summary of results bearing on this point see Eliel, 1962, p. 372; March, 1992, p. 302.)

Carbocations that cannot achieve planarity in the vicinity of the positively charged carbon may be able to maintain their configuration for long periods of time. When ion **246**, prepared by dissolving the ($+$) alcohol **245** in H_2SO_4 at $-20°C$, is quenched by pouring it into H_2O, ($-$) alcohol is recovered (with 91% ee; Fig. 7.95). Since rotation of the α-naphthyl group would lead to interconversion of the chiral cation to its enantiomer, the maintenance of optical activity forces the conclusion that such rotation is prevented. It was suggested that ionization occurs from the syn conformer of the alcohol, and that water attacks the more accessible rear face of the cation (with inversion). In contrast, ionization of ($+$)-**245** with BF_3 in CH_2Cl_2 leads to ($+$)-**245** on quenching. These results were explained by assuming that both ionization and quenching steps involved inversion due to increased steric requirements of the $-OH \cdot BF_3$ leaving group (Murr and Feller, 1968; Murr and Santiago, 1968).

Protonation of phenols on the benzene ring is suggested as being responsible for the racemization of 2,2'-dihydroxy-1,1'-binaphthyl (Fig. 7.36, **113**) on reflux-

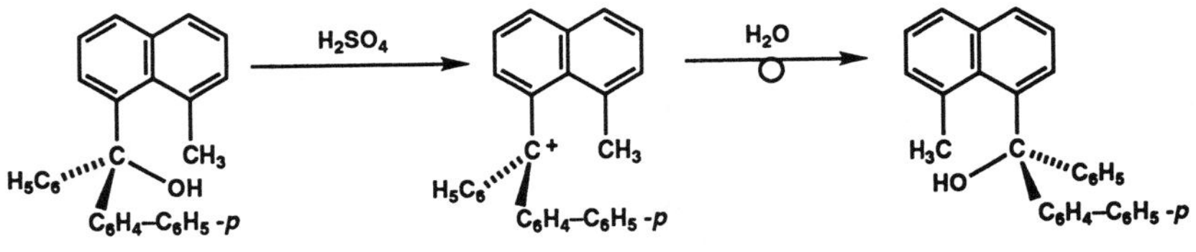

Figure 7.95. Hydration of a chiral carbocation.

ing of an acidic solution. It is suggested that protonation occurs at the peri position; one of the biaryl pivot carbon atoms becomes tetrahedral in the intermediate with the result that rotation about the aryl–aryl bond is facilitated. Racemization of **113** in basic medium was also observed (Kyba, Cram., et al., 1977).

Dissolution in nonprotonic polar solvents such as liquid SO_2 suffices to racemize α-methylbenzyl alcohol ($t = 18$ h at 0°C; Tokura and Akiyama, 1966). Esterification of nonracemic lactic acid, $CH_3CH(OH)CO_2H$, when catalyzed by H_2SO_4 or by $BF_3 \cdot Et_2O$, is accompanied by a small degree of racemization (Massad, Hawkins, and Baker, 1983).

Cases previously termed *autoracemization* (spontaneous racemization on standing in the absence of added reagents) are now believed to be due to the effect of traces of catalysts, including the glass of the container (Potapov, 1979, p. 110). α-Methylbenzyl chloride racemizes during chromatography on silica gel (and on acidic, but not basic, alumina); formation of styrene as a minor byproduct of the chromatography was noted (Denney and DiLeone, 1961). Because of the extremely high polarity of silica gel (Flowers and Leffler, 1989) this racemization may reasonably be attributed to the intervention of the $C_6H_5CH(CH_3)^+$ cation.

Addition of $(CF_3CO_2)_2Hg$ to **247** refluxing in THF or dioxane leads to complete racemization but without the accumulation of the trans isomer. It was concluded that the symmetrical carbocation **249** must account for the racemization (Fig. 7.96). On equilibration with $(CF_3CO_2)_2Hg$, the analogous urethane **248** affords a 1:2 mixture of diastereomers (asymmetric transformation) from which the enantiomerically enriched methyl 5-hydroxy-3-cyclohexenecarboxylate ester may be recovered (Trost, 1982).

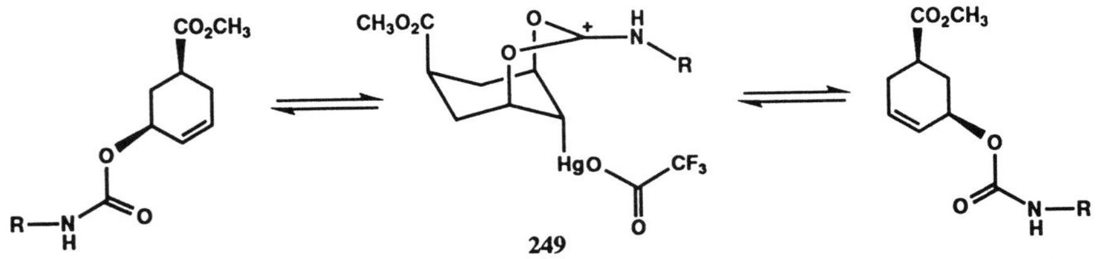

(S,S) **247** R = C_6H_5–CH_2
(S,S,S) **248** R = α–(1–naphthyl)methyl

(R,R) **248**
(R,R,S) **248**

Figure 7.96. Acid-catalyzed racemization via a symmetrical intermediate.

Racemization of Tröger's base (Fig. 7.39, **121**) with acid has been ascribed to the reversible formation of a ring-opened methyleneiminium intermediate (Prelog and Wieland, 1944). However, a spectroscopic study (UV and ^{13}C NMR) has failed to turn up any evidence for such an intermediate (Greenberg, Molinaro, and Lang, 1984). Nevertheless, formation of a methyleneiminium ion in lower than detectable concentrations remains the most likely acid-catalyzed racemization mechanism for **121**.

Racemization of compounds with two or more stereocenters without formation of mixtures of diastereomers (in contradistinction to epimerization; see above) is evidently an unlikely prospect since configurational inversion at *every* stereocenter is required to effect the racemization. Surprisingly, such racemization is not unknown. In addition to the case described in Figure 7.96, and that of racemization of Tröger's base (above), we cite three examples, all involving inversion at two stereocenters:

1. Racemization of camphor (by $AlCl_3$ at 80–85°C) was discovered nearly a century ago (Debierne, 1899). The operation of the Nametkin rearrangement (migration of a methyl group in a carbocation) in the racemization of camphor derivatives was recognized by J. Bredt over 60 years ago (Kauffman, 1983; see also Finch, Jr., and Vaughan, 1969).

2. (−)-*trans*-Chrysanthemic acid (Fig. 7.72, **173**) is racemized (as its anhydride) to (±)-*trans* acid by elemental iodine (Suzukamo and Fukao, 1982). A concerted process accelerated by Lewis acid catalysis may be postulated to account for the inversion of both stereocenters without loss of relative configurational integrity; see also Martel, 1992 .

3. Racemization of thiamphenicol by a novel functional group protection and activation scheme has been described in Section 7-5 (Fig. 7.73*b*).

Base Catalyzed Processes

Removal of α-hydrogen atoms from carbonyl compounds and nitriles gives rise to carbanions. This property plays a central role in the methodology of synthetic organic chemistry. If the α-carbon atom is a stereogenic atom, then as a rule, stereochemical integrity is not maintained when carbanion formation initiates with a nonracemic sample, with the rate of racemization equaling that of enolization and of deuterium incorporation (in protonic solvents containing −OD) (March, 1992, p. 574).

Numerous studies dealing with carbanion racemization have appeared. In general, carbanions have potentially less tendency to racemize quickly than do carbocations (for summaries bearing on this point, see Cram, 1965, p. 138 and March, 1992, p. 574). Two instances in which stereochemical integrity is partially or fully retained during carbanion formation are

1. formation of nonracemic CHBrClF in the haloform reaction of CHBrClC(=O)CH$_3$ (Hargreaves and Modarai, 1971; see Wilen et al., 1985) and in the decarboxylation of the strychnine salt of CFClBrCO$_2$H (Doyle and Vogl, 1989).

2. *N*-Pivaloylphenylalanine dimethylamide (Fig. 7.97, **250**) has been found to exchange its α-hydrogen atom for deuterium (k_c) at a rate faster than the loss of optical activity (k_a)[$k_c/k_a = 2.4$ with *t*-BuOK in *t*-BuOD at 30°C] (Guthrie and Nicolas, 1981).

Although simple ketones bearing an α-hydrogen atom are remarkably resistant to racemization on standing and distillation (Enders and Eichenauer, 1976), in the presence of base, rapid racemization ensues. The rate of racemization is highly sensitive to structure; whereas ketone **251** (Fig. 7.97) racemizes with $t_{1/2} = 18.4$ min, replacement of α-CH$_3$ by α-*tert*-butyl leads to $t_{1/2} = 13,680$ min (both with EtONa in EtOH at 25°C; Mills and Smith, 1960).

Carbanionic intermediates have been implicated also in the racemization of mandelic acid by mandelate racemase enzyme from *Pseudomonas putida* (Kenyon and Hegeman, 1970). The racemization requires a metal cation (Mg^{2+} is most effective); it is inferred that the cation increases the acidity of the mandelate substrate in the active site (Fee, Hegeman, and Kenyon, 1974). Based on a double isotope labeling study, it has been demonstrated that proton transfer by the enzyme is strictly intramolecular and heterofacial, that is, the proton abstracted from one face of the substrate is returned to the opposite face of the same molecule (Sharp, Hegeman, and Kenyon, 1977).

b. Racemization of Amino Acids

There is a remarkably large literature dealing with the racemization of amino acids and peptides (Benoiton, 1983). There are at least four reasons for the interest: (a) efforts to minimize racemization incidental to peptide and protein synthesis; (b) attempts to find ways of promoting racemization in connection with the large scale resolution and asymmetric transformation of amino acids so as to avoid the need to discard 50% of the starting material (see Section 7-5); (c) the relation of amino acid racemization to the age of fossils requires clarification of the mechanism of racemization; and (d) the study of racemization contributes to the search for the origins of the chiral homogeneity of natural products and, by extension, for the origins of life. For reviews of the subject, see Kaneko, et al. (1974); Williams and Smith (1977); Bada (1985); Bodanszky (1988). It is worthwhile to emphasize the fact that the modern study of racemization of amino acids has benefited greatly from the very sensitive techniques now available for the determination of their enantiomer composition by nonchiroptical methods, principally those involving GC analysis of volative derivatives on enantioselective GC columns (cf. Section 6-5.d, p. 250), e.g., Smith, Khatib and Reddy (1983a).

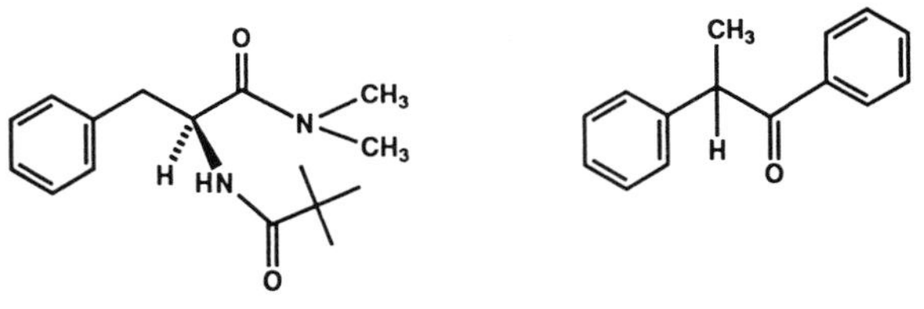

Figure 7.97. Structures **250–251**.

It is well known that free amino acids are relatively difficult to racemize in aqueous medium. Racemization is catalyzed by both acids and bases, especially the latter. A mechanism for racemization was proposed as early as 1910 by Dakin (1910). The proposal of a mechanism at the time is remarkable since it antedated the work of Robert Robinson (1932) by some decades. The contemporary accepted mechanism, a modification of that of Dakin (a) in basic medium and (b) in acidic medium, is that of Neuberger (1948; Eq. 7.40).

$$R-\underset{\underset{NH_2}{|}}{\overset{\overset{H}{|}}{C}}-CO_2^- \quad \rightleftharpoons \quad R-\underset{\underset{NH_2}{|}}{\overset{..}{C}}-C\overset{\overset{O}{\diagup}}{\underset{O^-}{\diagdown}} \tag{7.40a}$$

$$R-\underset{\underset{^+NH_3}{|}}{\overset{\overset{H}{|}}{C}}-CO_2H \quad \rightleftharpoons \quad R-\underset{\underset{^+NH_3}{|}}{\overset{\overset{H}{|}}{C}}-C\overset{\overset{+}{\diagup}O-H}{\underset{OH}{\diagdown}} \quad \rightleftharpoons \quad R-\underset{\underset{^+NH_3}{|}}{\overset{..}{C}}-C\overset{\overset{+}{\diagup}O-H}{\underset{OH}{\diagdown}} \tag{7.40b}$$

The racemization rate is enhanced by a number of factors: an increase in the electronegativity of R; a decrease in the negative charge on the carboxylate group; substitution of hydrogen on the amino group by electronegative atoms or groups, such as acyl and by transition metal ions (e.g., Smith, Khatib, and Reddy, 1983a). A detailed modern study (of the racemization of phenylglycine as a model compound) indicates that the racemization is in accord with the S_E1 mechanism (Smith and Sivakua, 1983). The mechanism is consistent with the observation that amino acid derivatives (N-acyl, esters) and peptides racemize faster than do free amino acids. In addition, racemization of N-acyl amino acids is facilitated by intramolecular formation of and deprotonation of azlactones (oxazolinones); see, for example, Fryzuk and Bosnich (1977), and of hydantoins, for example, Lazarus (1990). Racemization of dipeptides is influenced by neighboring groups and by formation of diketopiperazines (Smith and Baum, 1987). The role of hydantoins in racemization and the labilization of amino acids and their derivatives to racemization by transition metal ions have been dealt with in Section 7-2.e.

> Readers are cautioned that, in the literature, the racemization of free amino acids and of "bound" amino acids (e.g., peptides and proteins) is often treated collectively without distinction. The latter process is, of course, more accurately considered as epimerization and is subject to quite different forces as the aforementioned more rapid racemization of peptides than amino acids makes clear (see Wehmiller and Hare, 1971).

Efforts to suppress racemization during peptide synthesis (coupling activated by dicyclohexylcarbodiimide) have principally taken the form of adding weak acids (e.g., 1-hydroxybenzotriazole) that suppress proton abstraction from the α-carbon atom of amino acid intermediates and have additional benefits as well (Bodanszky, 1988).

The avoidance of racemization in the context of stereoselective synthesis is subject to both thermodynamic and kinetic control. A strategy for alkylation of amino acids without loss of stereochemical integrity is illustrated by the electrophilic substitution of (S)-proline at the α-carbon atom. Formation of the N,O-acetal **252** of proline and pivalaldehyde (one diastereomer only) takes place under thermodynamic control while the subsequent alkylation of the acetal enolate (e.g., with methyl iodide) affords, after hydrolysis, enantiomerically pure (S)-2-methylproline under kinetic control (steered by the induced chiral center; Eq. 7.41; Seebach et al., 1983).

$$(7.41)$$

Readers are reminded that the racemization of amino acids has been coupled to the preferential crystallization of derivatives crystallizing as conglomerates giving rise to highly efficient asymmetric transformations (Section 7-2.e). Efforts to simplify their racemization have led to the finding that a wide range of amino acids (15 out of the 17 amino acids studied) are substantially racemized on heating in acetic acid in 1 h (at 80–100°C) in the presence of aldehydes, such as salicylaldehyde (Yamada et al., 1983). The proposed mechanism of this racemization involves an imine (Schiff's base). Proton abstraction from the α-carbon atom of the protonated imine yields a stabilized zwitterion (Eq. 7.42).

$$(7.42)$$

Cupric ion had a negligible effect on the rate of racemization in spite of its ability to chelate the intermediate Schiff's base (see also, Smith et al., 1986). The beneficial effect of aldehydes is modeled after the proposed mechanism by which amino acid racemase enzymes act; these enzymes require pyridoxal phosphate as a coenzyme (Metzler, Ikawa, and Snell, 1954).

Careful examination of aging crystalline samples of ^{14}C labeled amino acids led to the discovery that β radiolysis (destruction) of the samples might be accompanied by slight racemization (Bonner, Lemmon, and Noyes, 1978; Bonner and Lemmon, 1978). Racemization induced by ionizing radiation (*radioracemization*) had been observed in other substances (Bonner, 1988). Radioracemization of amino acids (effected by γ-rays from a ^{60}Co source) was clearly observed in solution; the effect was greater (up to 13.2% racemization) in basic medium (sodium salts of amino acids) than on comparable acidic solutions (amino acid hydrochlorides). These results were accounted for in terms of the relative stabilities of radical ions formed when HO · radicals (from the radiolysis of water) attack the predominant amino acid species present. Radical anion **253** (Fig. 7.98), formed in basic medium, is significantly stabler than the comparable radical cation **254** former in acidic solution due to the possibility of delocalization of the odd electron (Bonner and Lemmon, 1978).

It was also observed that mineral surfaces (silica or clays) increased the tendency of amino acids to undergo radioracemization (Bonner and Lemmon, 1978, 1981; Bonner et al., 1985). Radioracemization of amino acids in contact with such materials was intended to more closely approximate conditions under which geologic samples (either terrestrial or extraterrestrial) were irradiated (e.g., by cosmic rays). Radioracemization tends to reduce the significance of the finding that amino acids in meteorites are racemic in regard to their biogenic or abiogenic origin (Bonner, 1988).

Finally, we consider the application of amino acid racemization to the dating of fossil samples. Abelson (1955) suggested that the decomposition of protein as a function of time might provide a dating method. After noting that amino acid samples in fossil shell and bone were significantly racemized, a dating method based on measurement of the enantiomer composition was developed (Hare and Mitterer, 1967–1968). This dating method (often called *aminostratigraphy*) is based on the following premises: (a) biosynthesis of amino acids in most living cells (bacteria excepted) produces material having the L-configuration. After the death of the individual, replenishment of amino acids ceases and accumulation of increasing amounts of the D-enantiomers takes place as a result of racemization; (b) the racemization rate is governed by first-order kinetics; (c) the rate of racemization is dependent on environmental conditions in which the amino acids are stored, the main variable (along with pH, bound and free metals, and surfaces) being the average temperature of the sample t_0 (called the diagenetic

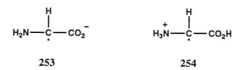

Figure 7.98. Radical ions formed on radiolysis of amino acids.

temperature). Application of the method requires the determination of k_1 and K values for the individual amino acids (in the laboratory), and calibration by an independent method of the environment in which a given sample of amino acid is found (e.g., ^{14}C dating). Samples as old as 10^5–10^6 years can be dated by this method, this being a time range complementary to those accessible by other dating methods, for example, ^{14}C dating that is only useful for 30–70×10^3 years if activation analysis methods are used. Numerous examples, evaluation of assumptions, and description of the pitfalls of this method are given in the detailed review by Williams and Smith (1977); see also Sykes (1988) and Meyer (1991; 1992).

The amino acid dating method is applicable even to living humans (this might be used, e.g., if their age is not well documented) provided that the analysis is carried out on a sample derived from metabolically inert tissue (e.g., tooth enamel or the eye lens). The analysis focuses on the enantiomer composition of aspartic acid, one of the fastest racemizing amino acids (Bada, 1985).

REFERENCES

Abelson, P. H. (1955), *Yearbook–Carnegie Inst. Washington*, **54**, 107.

Abramowicz, D. A., Ed. (1990), *Biocatalysis*, Van Nostrand Reinhold, New York.

Abu Zarga, M. H. and Shamma, M. (1980), *Tetrahedron Lett.*, **21**, 3739.

Ács, M., Fogassy, E., and Faigl, F. (1985), *Tetrahedron*, **41**, 2465.

Ács, M., Pokol, G., Faigl, F., and Fogassy, E. (1988), *J. Thermal Anal.*, **33**, 1241.

Ács, M., von dem Bussche, C., and Seebach, D. (1990), *Chimia*, **44**, 90.

Addadi, L., Berkovitch-Yellin, Z., Domb, N., Gati, E., Lahav, M., and Leiserowitz, L. (1982a), *Nature (London)*, **296**, 21.

Addadi, L., Berkovitch-Yellin, Z., Weissbuch, I., Lahav, M., and Leiserowitz, L. (1986), "A Link Between Macroscopic Phenomena and Molecular Chirality: Crystals as Probes for the Direct Assignment of Absolute Configuration of Chiral Molecules," *Top. Stereochem.*, **16**, 1.

Addadi, L., Berkovitch-Yellin, Z., Weissbuch, I., Lahav, M., Leiserowitz, L., and Weinstein, S. (1982b), *J. Am. Chem. Soc.*, **104**, 2075.

Addadi, L., Berkovitch-Yellin, Z., Weissbuch, I., van Mil, J., Shimon, L. J. W., Lahav, M., and Leiserowitz, L. (1985), *Angew. Chem. Int. Ed. Engl.*, **24**, 466.

Addadi, L., Gati, E., and Lahav, M. (1981), *J. Am. Chem. Soc.*, **103**, 1251.

Addadi, L., van Mil, J., and Lahav, M. (1981), *J. Am. Chem. Soc.*, **103**, 1249.

Addadi, L., Weinstein, S., Gati, E., Weissbuch, I., and Lahav, M. (1982c), *J. Am. Chem. Soc.*, **104**, 4610.

Agami, C., Levisalles, J., and Sevestre, H. (1984), *J. Chem. Soc. Chem. Commun.*, 418.

Agranat, I., Suissa, M. R., Cohen, S., Isaksson, R., Sandström, J., Dale, J., and Grace, D. (1987), *J. Chem. Soc. Chem. Commun.*, 381.

Ahnoff, M. and Einarsson, S. (1989), "Chiral Derivatization," in Lough, W. J., Ed., *Chiral Liquid Chromatography*, Blackie, Glasgow, UK, Chap. 4.

Akutagawa, S. (1992), "A Practical Synthesis of (−)-Menthol with the Rh–BINAP Catalyst," in Collins, A. N., Sheldrake, G. N., and Crosby, J., Eds., *Chirality in Industry: The Commercial Manufacture and Applications of Optically Active Compounds*, Wiley, Chichester, UK, Chap. 16.

Allan, R. D. and Fong, J. (1986), *Aust. J. Chem.*, **39**, 855.

Alcock, N. W., Brown, J. M., and Hulmes, D. I. (1993), *Tetrahedron: Asymmetry*, **4**, 743.

Allen, A., Fawcett, V., and Long, D. A. (1976), *J. Raman Specrosc.*, **4**, 285.

Allen, D. L., Gibson, V. C., Green, M. L. H., Skinner, J. F., Bashkin, J., and Grebenik, P. D. (1983), *J. Chem. Soc. Chem. Commun.*, 895.

Allenmark, S. G. (1991), *Chromatographic Enantioseparation: Methods and Applications*, 2nd ed., Ellis Horwood, Chichester, UK.

Amiard, G. (1956), *Bull. Soc. Chim. Fr.* 447.

Amiard, G. (1959), *Experientia*, **15**, 38.

Anker, H. S. and Clarke, H. T. (1955), *Organic Syntheses, Collective Volume III*, Horning, E. C., Ed., Wiley, New York, p. 172.

Anon. (1965a), *Fr. Patent* 1 389 840, Feb. 19, 1965 (to the Noguchi Research Foundation); *Chem. Abstr.*, **63**, 5740f (1965).

Anon. (1965b), *Chem. Eng. (N.Y.)*, **72**, Nov. 8, 1965, p. 247.

Anon. (1980), *Jpn. Kokai Tokkyo Koho JP 80 104 890*, Aug. 11, 1980 (to Kanegafuchi Chemical Industry Co.); *Chem. Abstr.*, **93**, 236942e (1980).

Anon. (1981), *Jpn. Kokai Tokkyo Koho JP 57 188 563* [82 188, 563] Nov. 19, 1982; *Chem. Abstr.*, **98**, 160423s (1983).

Anon. (1983), *Jpn. Kokai Tokkyo Koho JP 58 41 847* [83 41847], Mar. 11, 1983; *Chem. Abstr.*, **99**, 104949b (1983).

Aoyama, Y., Tanaka, Y., and Sugahara, S. (1989), *J. Am. Chem. Soc.*, **111**, 5397.

Arad-Yellin, R., Green, B. S., Knossow, M., and Tsoucaris, G. (1983), *J. Am. Chem. Soc.*, **105**, 4561.

Arad-Yellin, R., Green, B. S., Knossow, M., and Tsoucaris, G. (1984), "Enantiomeric Selectivity of Host Lattices," in Atwood, J. L., Davies, J. E. D., and MacNicol, D. D., Eds., *Inclusion Compounds*, Vol. 3, Academic, London, Chap. 9.

Arai, K. (1986), "Isomerization–Crystallization Method in Optical Resolution," (in Japanese) *Yuki Gosei Kagaku Kyokaishi*, **44**, 486.

Arai, K., Obara, Y., Takahashi, Y., and Takakuwa, Y. (1986), *Jpn. Kokai Tokkyo Koho JP 61 238 734*, Oct. 24, 1986; *Chem. Abstr.*, **106**, 196063x (1987).

Arcamone, F., Bernardi, L., Patelli, B., and Di Marco, A. (1976), *Ger. Offen.* 2 601 785, July 29, 1976; *Chem. Abstr.*, **85**, 142918 (1976).

Armstrong, D. W. and Jin, H. L. (1987), *Anal. Chem.*, **59**, 2237.

Arnold, W., Daly, J. J., Imhof, R., and Kyburz, E. (1983), *Tetrahedron Lett.*, **24**, 343.

Asai, S. (1983), *Ind. Eng. Chem. Process Des. Dev.*, **22**, 429.

Asai, S. (1985), *Ind. Eng. Chem. Process Des. Dev.*, **24**, 1105.

Asai, S. and Ikegami, S. (1982), *Ind. Eng. Chem. Fundam.*, **21**, 181.

Asaoka, M., Shima, K., and Takei, H. (1988), *J. Chem. Soc. Chem. Commun.*, 430.

Asselineau, C. and Asselineau, J. (1964), "Utilisation de quelques composés d'inclusion en chimie organique," *Ann. Chim. (Paris)*, **9**, 461.

Ault, A. (1973), *Organic Syntheses, Collective Volume V*, Baumgarten, H. E., Ed., Wiley, New York, p. 932.

Bada, J. L. (1985), "Racemization of Amino Acids" in Barrett, G. C., Ed., *Chemistry and Biochemistry of Amino Acids*, Chapman & Hall, London, Chap. 13.

Bailey, M. E. and Hass, H. B. (1941), *J. Am. Chem. Soc.*, **63**, 1969.

Bajgrowicz, J. A., Cossec, B., Pigière, Ch., Jacquier, R., and Viallefont, P. (1984), *Tetrahedron Lett.*, **25**, 1789.

Baker, W., Gilbert, B., and Ollis, W. D. (1952), *J. Chem. Soc.*, 1443.

Balavoine, G., Moradpour, A., and Kagan, H. B. (1974), *J. Am. Chem. Soc.*, **96**, 5152.

Baldwin, J. E. (1988), *J. Chem. Soc., Chem. Commun.*, 31.

Baldwin, J. E., Adlington, R. M., Rawlings, B. J., and Jones, R. H. (1985), *Tetrahedron Lett.*, **26**, 485.

Baldwin, J. E. and Black, K. A. (1984), *J. Am. Chem. Soc.*, **106**, 1029.

Baldwin, J. E. and Broline, B. M. (1982), *J. Org. Chem.*, **47**, 1385.

Baldwin, J. E., Ostrander, R. L., Simon, C. D., and Widdison, W. C. (1991), *J. Am. Chem. Soc.*, **112**, 2021.

Barton, D. H. R. and Kirby, G. W. (1962), *J. Chem. Soc.*, 806.

Bauer, K., Falk, H., and Schlögl, K. (1968), *Monatsh. Chem.*, **99**, 2186.

Behr, J.-P. and Lehn, J.-M. (1973), *J. Am. Chem. Soc.*, **95**, 6108.

Belan, A., Bolte, J., Fauve, A., Gourcy, J. G., and Veschambre, H. (1987), *J. Org. Chem.*, **52**, 256.

Bell, F. (1934), *J. Chem. Soc.*, 835.

Bell, K. H. (1979), *Aust. J. Chem.*, **32**, 65.

Bellucci, G., Berti, G., Ferretti, M., Mastrorilli, E., and Silvestri, L. (1985), *J. Org. Chem.*, **50**, 1471.

Benkovic, S. J. (1991), personal communication to SHW.

Benoiton, N. L. (1983), "Quantitation and Sequence Dependence of Racemization in Peptide Synthesis" in Gross, E. and Meienhofer, J., Eds., *The Peptides: Analysis, Synthesis. Biology.* Vol. 5. *Special Methods in Peptide Synthesis*, Part B, Academic, New York, p. 217.

Benschop, H. P. and Van den Berg, G. R. (1970), *J. Chem. Soc. D*, 1431.

Bevinakatti, H. S. and Newadkar, R. V. (1990), *Tetrahedron: Asymmetry*, **1**, 583.

Bey, P., Vevert, J.-P., Van Dorsselaer, V., and Kolb, M. (1979), *J. Org. Chem.*, **44**, 2732.

Biali, S. E., Kahr, B., Okamoto, Y., Aburatani, R., and Mislow, K. (1988), *J. Am. Chem. Soc.*, **110**, 1917.

Bickart, P., Carson, F. W., Jacobus, J., Miller, E. G., and Mislow, K. (1968), *J. Am. Chem. Soc.*, **90**, 4869.

Billington, D. C., Baker, R., Kulagowski, J. J., and Mawer, I. M. (1987), *J. Chem. Soc., Chem. Commun.*, 314.

Bir, G. and Kaufmann, D. (1987), *Tetrahedron Lett.*, **28**, 777.

Black, S. N., Williams, L. J., Davey, R. J., Moffatt, F., Jones, R. V. H., McEwan, D. M., and Sadler, D. E. (1989), *Tetrahedron*, **45**, 2677.

Black, S. N., Williams, L. J., Davey, R. J., Moffatt, F., McEwan, D. M., Sadler, D. E., Docherty, R., and Williams, D. J. (1990), *J. Phys. Chem.*, **94**, 3223.

Blaschke, G. (1986), *J. Liq. Chromatog.*, **9**, 341.

Blaschke, G. (1988), "Substituted Polyacrylamides as Chiral Phases for the Resolution of Drugs," in Zief, M. and Crane, L. J., Eds., *Chromatographic Chiral Separations*, Marcel Dekker, New York, Chap. 7.

Blaschke, G. and Markgraf, H. (1980), *Chem. Ber.*, **113**, 2031.

Blaschke, G. and Walther, B. (1987), *Liebigs Ann. Chem.*, 561.

Block, Jr., P. and Newman, M. S. (1968), *Org. Synth.*, **48**, 120; *Organic Syntheses, Collective Volume V*, Baumgarten, H. E., Ed., Wiley, New York, p. 1031 (1973).

Bodanszky, M. (1988), *Peptide Chemistry*, Springer, Berlin, Chap. 8.

Boggs III, N. T., Goldsmith, B., Gawley, R. E., Koehler, K. A., and Hiskey, R. G. (1979), *J. Org. Chem.*, **44**, 2262.

Bömer, B., Grosser, R., Schwartz, U., Arlt, D., and Piejko, K.-E. (1988), *Ger. Offen. DE 3 706 890*, Sep. 21, 1988; *Chem. Abstr.*, **110**, 76300h (1989).

Bommarius, A. S., Drauz, K., Groeger, U., and Wandrey, C. (1992), "Membrane Bioreactors for the Production of Enantiomerically Pure α-Amino Acids," in Collins, A. N., Sheldrake, G. N., and Crosby, J., Eds., *Chirality in Industry: The Commercial Manufacture and Applications of Optically Active Compounds*, Wiley, Chichester, UK, Chap. 20.

Bonner, W. A. (1988), "Origins of Chiral Homogeneity in Nature," *Top. Stereochem.*, **18**, 1.

Bonner, W. A. and Lemmon, R. M. (1978), *Bioorg. Chem.*, **7**, 175.

Bonner, W. A. and Lemmon, R. M. (1981), *Origins Life Evol. Biosphere*, **11**, 321.

Bonner, W. A., Hall, H., Chow, G., Liang, Y., and Lemmon, R. M. (1985), *Origins Life Evol. Biosphere*, **15**, 103.

Bonner, W. A., Lemmon, R. M., and Noyes, H. P. (1978), *J. Org. Chem.*, **43**, 522.

Bortolini, O., Di Furia, F., Licini, G., Modena, G., and Rossi, M. (1986), *Tetrahedron Lett.*, **27**, 6257.

Bosnich, B. and Harrowfield, J. M. (1972), *J. Am. Chem. Soc.*, **94**, 3425.

Bowman, N. S., McCloud, G. T., and Schweitzer, G. K. (1968), *J. Am. Chem. Soc.*, **90**, 3848.

Boyle, P. H. (1971), "Methods of Optical Resolution," *Q. Rev. Chem. Soc.*, **25**, 323.

Boyle, W. J., Jr., Sifniades, S., and Van Peppen, J. F. (1979), *J. Org. Chem.*, **44**, 4841.

Brandt, J., Jochum, C., Ugi, I., and Jochum, P. (1977), *Tetrahedron*, **33**, 1353.

Brandt, S. (1989), Personal communication to SHW. See also *Performance Chemicals*, May 1987.

Brewster, J. H. and Prudence, R. T. (1973), *J. Am. Chem. Soc.*, **95**, 1217.

Brianso, M. C. (1976), *Acta Crystallogr.*, **B32**, 3040.

Briaucourt, P. and Horeau, A. (1979) *C.R. Seances Acad. Sci., Ser. C*, **289**, 49.

Brienne, M.-J., Collet, A., and Jacques, J. (1983), *Synthesis*, 704.

Brienne, M.-J. and Jacques, J. (1975), *Tetrahedron Lett.*, 2349.

Brossi, A. and Teitel, S. (1970), *J. Org. Chem.*, **35**, 3559.

Brown, B. R. and Fuller, M. J. (1986), *J. Chem. Res., Synop.*, 140.

Brown, E., Viot, F., and Le Floc'h, Y. (1985a), *Tetrahedron Lett.*, **26**, 4451.

Brown, H. C., Ayyangar, N. R., and Zweifel, G. (1964), *J. Am. Chem. Soc.*, **86**, 397.

Brown, H. C. and Jadhav, P. K. (1982), "Asymmetric Hydroboration," in Morrison, J. D., Ed., *Asymmetric Synthesis*, Vol. 2, Academic, New York, Chap. 1.

Brown, H. C., Jadhav, P. K., and Desai, M. C. (1982), *J. Org. Chem.*, **47**, 4583.

Brown, H. C. and Joshi, N. N. (1988), *J. Org. Chem.*, **53**, 4059.

Brown, H. C. and Singaram, B. (1984), *J. Org. Chem.*, **49**, 945.

Brown, H. C. and Zweifel, G. (1962), *J. Org. Chem.*, **27**, 4708.

Brown, J. M. (1988), "Catalytic Kinetic Resolution," *Chem. Ind. London*, 612.

Brown, J. M. (1989), "Asymmetric Homogeneous Catalysis," *Chem. Br.*, **25**, 276.

Brown, J. M. and Cutting, I. (1985), *J. Chem. Soc. Chem. Commun.*, 578.

Brown, K. J., Berry, M. S., and Murdoch, J. R. (1985b), *J. Org. Chem.*, **50**, 4345.

Brunner, H. (1988), "Enantioselective Synthesis of Organic Compounds with Optically Active Transition Metal Catalysts in Substoichiometric Quantities," *Top. Stereochem.*, **18**, 129.

Bucciarelli, M., Forni, A., Marcaccioli, S., Moretti, I., and Torre, G. (1983), *Tetrahedron*, **39**, 187.

Bucciarelli, M., Forni, A., Moretti, I., and Prati, F. (1988), *J. Chem. Soc. Chem. Commun.*, 1614.

Buchanan, C. and Graham, S. H. (1950), *J. Chem. Soc.*, 500.

Buckingham, J., Ed. (1982), *Dictionary of Organic Compounds*, 5th ed., Chapman & Hall, New York.

Bujons, J., Guajardo, R., and Kyler, K. S. (1988), *J. Am. Chem. Soc.*, **110**, 604.

Bur, A. J. and Fetters, L. J. (1976), "The Chain Structure, Polymerization, and Conformation of Polyisocyanates," *Chem. Rev.*, **76**, 727.

Burns, C. J., Martin, C. A., and Sharpless, K. B. (1989), *J. Org. Chem.*, **54**, 2826.

Bussmann, W., Lehn, J.-M., Oesch, U., Plumeré, P., and Simon, W. (1981), *Helv. Chim. Acta*, **64**, 657.

Calzavara, E. (1934), *Fr. Patent* 763 374, Apr. 30, 1934; *Chem. Abstr.*, 1934, **28**, 5472.

Canceill, J. and Collet, A. (1986), *Nouv. J. Chim.*, **10**, 17.

Canceill, J., Collet, A., Gabard, J., Gottarelli, G., and Spada, G. P. (1985a), *J. Am. Chem. Soc.*, **107**, 1299.

Canceill, J., Collet, A., and Gottarelli, G. (1984), *J. Am. Chem. Soc.*, **106**, 5997.

Canceill, J., Lacombe, L., and Collet, A. (1985b), *J. Am. Chem. Soc.*, **107**, 6993.

Carlier, P. R., Mungall, W. S., Schröder, G., and Sharpless, K. B. (1988), *J. Am. Chem. Soc.*, **110**, 2978.

Casarini, D., Lunazzi, L., Placucci, G., and Macciantelli, D. (1987), *J. Org. Chem.*, **52**, 4721.

Cecere, F., Galli, G., and Morisi, F. (1975), *FEBS Lett.*, **57**, 192.

Červinka, O., Fábryová, A., and Sablukova, I. (1986), *Collect. Czech. Chem. Commun.*, **51**, 401.

Chan, K.-K., Cohen, N., De Noble, J. P., Specian, A. C., and Saucy, G. (1976), *J. Org. Chem.*, **41**, 3497.

Chandrasekhar, S. and M. Ravindranath (1989), *Tetrahedron Lett.*, **30**, 6207.

Chen, C.-S., Fujimoto, Y., Girdaukas, G., and Sih, C. J. (1982), *J. Am. Chem. Soc.*, **104**, 7294.

Chibata, I. (1982), "Application of Immobilized Enzymes for Asymmetric Reactions," in Eliel, E. L. and Otsuka, S., Eds., *Asymmetric Reactions and Processes in Chemistry*, ACS Symposium Series 185, American Chemical Society, Washington, D.C., p. 195.

Chibata, I., Tosa, T., and Shibatani, T. (1992), "The Industrial Production of Optically Active Compounds by Immobilized Catalysts," in Collins, A. N., Sheldrake G. N., and Crosby, J., Eds., *Chirality in Industry: The Commercial Manufacture and Applications of Optically Active Compounds*, Wiley, Chichester, UK, Chap. 19.

Chibata, I., Yamada, S., Hongo, C., and Yoshioka, R. (1982), *Eur. Pat. Appl.*, EP 70 114, June 25, 1982; *Chem. Abstr.*, **99**, 6039k (1983).

Chibata, I., Yamada, S., Hongo, C., and Yoshioka, R. (1983), *Eur. Pat. Appl.*, EP 75 318, Mar. 30, 1983; *Chem. Abstr.*, **99**, 105702 (1983).

Chibata, I., Yamada, S., Hongo, C., and Yoshioka, R. (1984), *Eur. Pat. Appl.*, EP 119 804, Sept. 26, 1984; *Chem. Abstr.*, **101**, 231031f (1984).

Chinchilla, R., Nájera, C., Yus, M., and Heumann, A. (1990), *Tetrahedron: Asymmetry.*, **1**, 851.

Cinquini, M., Colonna, S., and Cozzi, F. (1978), *J. Chem. Soc. Perkin Trans. 1*, 247.

Clark, J. C., Phillipps, G. H., and Steer, M. R. (1976b), *J. Chem. Soc. Perkin 1*, 475.

Clark, J. C., Phillipps, G. H., Steer, M. R., Stephenson, L., and Cooksey, A. R. (1976a), *J. Chem. Soc., Perkin 1*, 471.

Collet, A. (1989), "Optical resolution by crystallization methods," in Krstulović, A. M., Ed., *Chiral Separations by HPLC. Applications to Pharmaceutical Compounds*, Ellis Horwood, Chichester, UK, Chap. 4.

Collet, A., Brienne, M.-J., and Jacques, J. (1972), *Bull. Soc. Chim. Fr.*, 127.

Collet, A., Brienne, M.-J., and Jacques, J. (1980), "Optical Resolution by Direct Crystallization of Enantiomer Mixtures," *Chem. Rev.*, **80**, 215.

Collet, A. and Gabard, J. (1980), *J. Org. Chem.*, **45**, 5400.

Collins, A. N., Sheldrake, G. N., and Crosby, J., Eds. (1992), *Chirality in Industry: The Commercial Manufacture and Applications of Optically Active Compounds*, Wiley, Chichester, UK.

Colson, J. G. (1979), Grayson, M., Ed., in Kirk–Othmer *Encyclopedia of Chemical Technology*, 3rd ed., Vol. 5, Wiley, New York, p. 808.

Comber, R. N. and Brouillette, W. J. (1987), *J. Org. Chem.*, **52**, 2311.

Cope, A. C., Ganellin, C. R., and Johnson, Jr., H. W. (1962), *J. Am. Chem. Soc.*, **84**, 3191.

Cope, A. C., Ganellin, C. R., Johnson, Jr., H. W., Van Auken, T. V., and Winkler, H. J. S. (1963), *J. Am. Chem. Soc.*, **85**, 3276.

Coquerel, G., Bouaziz, R., and Brienne, M.-J. (1988a), *Chem. Lett.*, 1081.

Coquerel, G., Perez, G., and Hartman, P. (1988b), *J. Cryst. Growth*, **88**, 511.

Coquerel, G., Bouaziz, R., and Brienne, M.-J. (1990), *Tetrahedron Lett.*, **31**, 2143.

Corey, E. J., Bakshi, R. K., Shibata, S., Chen, C.-P., and Singh, V. K. (1987), *J. Am. Chem. Soc.*, **109**, 7925.

Corey, E. J., Danheiser, R. L., Chandrasekaran, S., Keck, G. E., Gopalan, B., Larsen, S. D., Sizer, P., and Gras, J.-L. (1978b), *J. Am. Chem. Soc.*, **100**, 8034.

Corey, E. J., Hopkins, P. B., Kim, S., Yoo, S.-e., Nambiar, K. P., and Falck, J. R. (1979), *J. Am. Chem. Soc.*, **101**, 7131.

Corey, E. J., Imwinkelried, R., Pikul, S., and Xiang, Y. B. (1989), *J. Am. Chem. Soc.*, **111**, 5493.

Corey, E. J., Trybulski, E. J., Melvin, Jr., L. S., Nicolaou, K. C., Secrist, J. A., Lett, R., Sheldrake, P. W., Falck, J. R., Brunelle, D. J., Haslanger, M. F., Kim, S., and Yoo, S.-e., (1978a), *J. Am. Chem. Soc.*, **100**, 4618.

Craig, D. P. and Mellor, D. P. (1976), "Discriminating interactions between chiral molecules," *Top. Curr. Chem.*, **63**, 1.

Cram, D. J. (1965), *Fundamentals of Carbanion Chemistry*, Academic, New York.

Cram, D. J. (1983), *Science*, **219**, 1177.

Cram, D. J. (1990), *From Design to Discovery*, American Chemical Society, Washington, DC, p. 67.

Cram, D. J. and Cram, J. M. (1978), "Design of Complexes between Synthetic Hosts and Organic Guests," *Acc. Chem. Res.*, **11**, 8.

Cramer, F. and Dietsche, W. (1959), *Chem. Ber.*, **92**, 378.

Cramer, F. and Hettler, H. (1967), "Inclusion Compounds of Cyclodextrins," *Naturwissenschaften*, **54**, 625.

Cristol, S. J. (1949), *J. Am. Chem. Soc.*, **71**, 1894.

Crosby, J. (1991), "Synthesis of Optically Active Compounds: A Large Scale Perspective," *Tetrahedron*, **47**, 4789.

Czugler, M., Csöregh, I., Kálmán, A., Faigl, F., and Ács, M. (1989), *J. Mol. Struct.*, **196**, 157.

Dakin, H. D. (1910), *Am. Chem. J.*, **44**, 48.

Davankov, V. A. (1989), "Separation of Enantiomeric Compounds Using Chiral HPLC Systems," *Chromatographia*, **27**, 475.

Davankov, V. A., Zolotarev, Y. A., and Kurganov, A. A. (1979), *J. Liq. Chromatogr.*, **2**, 119.

Davey, R. J., Black, S. N., Williams, L. J., McEwan, D., and Sadler, D. E. (1990), *J. Cryst. Growth.*, **102**, 97.

Davies, H. G., Green, R. H., Kelly, D. R., and Roberts, S. N. (1989), *Biotransformations in Preparative Organic Chemistry*, Academic, San Diego, CA.

Davies, S. G., Brown, J. M., Pratt, A. J., and Fleet, G. (1989), "Asymmetric Synthesis–Meeting the Challenge," *Chem. Br.*, **25**, 259.

Davis, J. C. (1978), *Chem. Eng. (N.Y.)*, **85** (12), May 22, 62.

de Heij, N. A. (1981) *Eur. Pat. Appl.*, EP 35 811, Sept. 16, 1981; *Chem. Abstr.*, **96**, 52011f (1982).

Debierne, A. (1899), *C.R. Hebd. Séances Acad. Sci.*, **128**, 1110.

Degueil-Castaing, M., De Jeso, B., Drouillard, S., and Maillard, B. (1987), *Tetrahedron Lett.*, **28**, 953.

Dell'Erba, C., Novi, M., Garbarino, G., and Corallo, G. P. (1983), *Tetrahedron Lett.*, **24**, 1191.

Demuth, M., Chandrasekhar, S., and Schaffner, K. (1984), *J. Am. Chem. Soc.*, **106**, 1092.

Demuth, M., Ritterskamp, P., Weigt, E., and Schaffner, K. (1986), *J. Am. Chem. Soc.*, **108**, 4149.

Denmark, S. E. and Almstead, N. G. (1991), *J. Am. Chem. Soc.*, **113**, 8089.

Denney, D. B. and DiLeone, R. (1961), *J. Org. Chem.*, **26**, 984.

Derfer, J. M. and Derfer, M. M. (1983), "Terpenoids," in Grayson, M., Ed., *Kirk-Othmer Encyclopedia of Chemical Technology*, Vol. 22, Wiley-Interscience, New York, p. 742.

Desai, T., Fernandez-Mayoralas, A., Gigg, J., Gigg, R., and Payne, S. (1990) *Carbohydr. Res.*, **205**, 105.

Desai, T., Fernandez-Mayoralas, A., Gigg, J., Gigg, R., Jaramillo, C., Payne, S., Penades, S., and Schnetz, N. (1991), "Preparation of Optically Active *myo*-Inositol Derivatives as Intermediates for the Synthesis of Inositol Phosphates," in Reitz, A. B., Ed., *Inositol Phosphates and Derivatives: Synthesis, Biochemistry and Therapeutic Potential*, ACS Symposium Series 463, American Chemical Society, Washington, DC, 1991, Chap. 6.

Dharanipragada, R., Ferguson, S. B., and Diederich, F. (1988), *J. Am. Chem. Soc.*, **110**, 1679.

Discordia, R. P. and Dittmer, D. C. (1990), *J. Org. Chem.*, **55**, 1414.

Domagala, J. M. and Bach, R. D. (1979), *J. Org. Chem.*, **44**, 3168.

Domínguez, D., Ardecky, R. J., and Cava, M. P. (1983), *J. Am. Chem. Soc.*, **105**, 1608.

Donaldson, R. E., Saddler, J. C., Byrn, S., McKenzie, A. T., and Fuchs, P. L. (1983), *J. Org. Chem.*, **48**, 2167.

Doolittle, R. E. and Heath, R. R. (1984), *J. Org. Chem.*, **49**, 5041.

Dossena, A., Marchelli, R., Armani, E., Fava, G. G., and Belicchi, M. F. (1983), *J. Chem. Soc. Chem. Commun.*, 1196.

Doyle, T. R. and Vogl, O. (1989), *J. Am. Chem. Soc.*, **111**, 8510.

Drabowicz, J. and Mikołajczyk, M. (1982), in Szejtli, J., Ed., *Proceedings of the First International Symposium of Cyclodextrins*, 1981, Reidel, Dordrecht, The Netherlands, pp. 205–216.

Drueckhammer, D. G., Barbas III, C. F., Nozaki, K., Wong, C.-H., Wood, C. Y. and Ciufolini, M. A. (1988), *J. Org. Chem.*, **53**, 1607.

Duhamel, L. (1976), *C.R. Séances Acad. Sci.*, Ser. C, **282**, 125.

Duhamel, L., Duhamel, P., Launay, J.-C., and Plaquevent, J.-C. (1984), "Asymmetric Protonation," *Bull. Soc. Chim. Fr.*, II-421.

Duhamel, L., Fouquay, S., and Plaquevent, J.-C. (1986), *Tetrahedron Lett.*, **27**, 4975.

Duhamel, L. and Launay, J.-C. (1983), *Tetrahedron Lett.*, **24**, 4209.

Duhamel, L. and Plaquevent, J.-C. (1978), *J. Am. Chem. Soc.*, **100**, 7415.

Duhamel, L. and Plaquevent, J.-C. (1982), *Bull. Soc. Chim. Fr.*, II-69, II-75.

Dummel, R. J. and Kun, E. (1969), *J. Biol. Chem.*, **244**, 2966.

Dumont, R., Brossi, A., and Silverton, J. V. (1986), *J. Org. Chem.*, **51**, 2515.

Duschinsky, R. (1934), *Chem. Ind. (London)*, 10.

Duschinsky, R. (1936), in Wüest, H. M., Ed., *Festschrift Emil Barell*, F. Reinhardt Verlag, Basel, Switzerland, p. 375.

Dvořák, D., Závada, J., Etter, M. C., and Loehlin, J. H. (1992), *J. Org. Chem.*, **57**, 4839.

Eaton, P. E. and Leipzig, B. (1978), *J. Org. Chem.*, **43**, 2483.

El-Baba, S., Poulin, J.-C., and Kagan, H. B. (1984), *Tetrahedron*, **40**, 4275.

Eliel, E. (1962), *Stereochemisty of Carbon Compounds*, McGraw-Hill, New York.

Eliel, E. L. and Biros, F. J. (1966), *J. Am. Chem. Soc.*, **88**, 3334.

Eliel, E. L. and Brett, T. J. (1963), *J. Org. Chem.*, **28**, 1923.

Elliott, M. L., Urban, F. J., and Bordner, J. (1985), *J. Org. Chem.*, **50**, 1752.

Ellis, M. K., Golding, B. T., and Watson, W. P. (1984), *J. Chem. Soc., Chem. Commun.*, 1600.

Enders, D. and Eichenauer, H. (1976), *Angew. Chem. Int. Ed. Engl.*, **15**, 549.

Enders, D., Fey, P., and Kipphardt, H. (1987), *Org. Synth.*, **65**, 173; *Organic Syntheses, Collective Volume VIII*, Freeman, J. P., Ed., Wiley, New York, p. 26 (1993).

Enders, D. and Mies, W. (1984), *J. Chem. Soc. Chem. Commun.*, 1221.

Eschler, B. M., Haynes, R. K., Kremmydas, S., and D. D. Ridley (1988), *J. Chem. Soc. Chem. Commun.*, 137.

Farina, M. (1987), "The Stereochemistry of Linear Macromolecules," *Top. Stereochem.*, **17**, 1.

Faustini, F., DeMunari, S., Panzeri, A., Villa, V., and Gandolfi, C. A. (1981), *Tetrahedron Lett.*, **22**, 4533.

Fee, J. A., Hegeman, G. D., and Kenyon, G. L. (1974), *Biochemistry*, **13**, 2528.

Felder, E., Pitre D., and Zutter, H. (1981), *U.S. Patent* 4 246 164, Jan. 20, 1981; *Chem. Abstr.*, **93**, 132285b (1980).

Feldnere, V., Peica, D., and Viksna, A. (1988), *Latv. PSR Zinat. Akad. Vestis, Kim. Ser.*, 216; *Chem. Abstr.*, **109**, 211424y (1988).

Fessner, W.-D. and Prinzbach, H. (1986), *Tetrahedron*, **42**, 1797.

Fieser, L. F. (1964), *Organic Experiments*, Heath, Boston, MA, p. 229.

Finch, Jr., A. M. T. and Vaughan, W. R. (1969), *J. Am. Chem. Soc.*, **91**, 1416.

Findlay, A. (1951), *The Phase Rule and its Applications*, 9th ed., by Campbell, A. N. and Smith, N. O., Dover, New York, p. 409ff.

Finn, M. G. and Sharpless, K. B. (1985), "On the Mechanism of Asymmetric Epoxidation with Titanium-Tartrate Catalysts," Morrison, J. D., Ed., *Asymmetric Synthesis*, Vol. 5, Academic, New York, Chap. 8.

Firestone, R. A., Reinhold, D. F., Gaines, W. A., Chemerda, J. M., and Sletzinger, M. (1968), *J. Org. Chem.*, **33**, 1213.

Fischli, A. (1980), "Chiral Building Blocks in Enantiomer Synthesis Using Enzymatic Transformations," Scheffold, R., Ed., *Modern Synthetic Methods*, Vol. 2, Salle and Sauerländer, Frankfurt, Germany, and Aarau, Switzerland, p. 269.

Fitzi, R. and Seebach, D. (1988), *Tetrahedron*, **44**, 5277.

Fizet, C. (1986), *Helv. Chem. Acta*, **69**, 404.

Fleischer, J., Bauer, K., and Hopp, R. (1976), *U.S. Patent* 3,943,181, Mar. 9, 1976; *Ger. Offen.* 2,109,456, Sep. 14, 1972; *Chem. Abstr.*, **77**, 152393h (1972).

Fleming, I. and Ghosh, S. K. (1994), *J. Chem. Soc. Chem. Commun.*, 99.

Flowers, G. C. and Leffler, J. E. (1989), *J. Org. Chem.*, **54**, 3995.

Fogassy, E., Ács, M., Faigl, F., Simon, K., Rohonczy, J., and Ecsery, Z. (1986), *J. Chem. Soc. Perkin Trans.* 2, 1881.

Fogassy, E., Faigl, F., and Ács, M. (1985), *Tetrahedron*, **41**, 2837.

Fogassy, E., Faigl, F., Ács, M., and Grofcsik, A. (1981), *J. Chem. Res., Synop.* 346; *Miniprint*, 3981.

Fogassy, E., Faigl, F., Ács, M., Simon, K., Kozsda, É., Podányi, B., Czugler, M., and Reck, G. (1988), *J. Chem. Soc. Perkin Trans.* 2, 1385.

Fogassy, E., Lopata, A., Faigl, F., Darvas, F., Ács, M., and Töke, L. (1980), *Tetrahedron Lett.*, **21**, 647.

Forni, A., I., Moretti, I., Prosyanik, A. V., and Torre, G. (1981), *J. Chem. Soc. Chem. Commun.*, 588.

Franck, A. and Rüchardt, C. (1984), *Chem. Lett.*, 1431.

Francotte, E. and Junker-Buchheit, A. (1992), "Preparative chromatographic separation of enantiomers," *J. Chromatogr.*, **576**, 1.

Francotte, E. and Wolf, R. M. (1991), *Chirality*, **3**, 43.

Francotte, E., Lang, R. W., and Winkler, T. (1991), *Chirality*, **3**, 177.

Frejd, T. and Klingstedt, T. (1983), *J. Chem. Soc. Chem. Commun.*, 1021.

Fritz-Langhals, E. (1993), *Angew. Chem. Int. Ed. Engl.*, **32**, 753.

Fry, A. J. and Britton, W. E. (1973), *J. Org. Chem.*, **38**, 4016.

Fryzuk, M. D. and Bosnich, B. (1977), *J. Am. Chem. Soc.*, **99**, 6262.

Fryzuk, M. D. and Bosnich, B. (1978), *J. Am. Chem. Soc.*, **100**, 5491.

Fujimoto, Y., Iwadate, H. and Ikekawa, N. (1985), *J. Chem. Soc. Chem. Commun.*, 1333.

Fukushima, K. (1976), *J. Mol. Struct.*, **34**, 67.

Fülling, G. and Sih, C. J. (1987), *J. Am. Chem. Soc.*, **109**, 2845.

Gabard, J. and Collet, A. (1986), *Nouv. J. Chim.*, **10**, 685.

Gal, G., Chemerda, J. M., Reinhold, D. F., and Purick, R. M. (1977), *J. Org. Chem.*, **42**, 142.

Gal, J. (1987), "Stereoisomer Separations via Derivatization with Optically Active Reagents," *LC. GC*, **5**, 106.

Gal, J. (1988), "Indirect Chromatographic Methods for Resolution of Drug Enantiomers—Synthesis and Separation of Diastereomeric Derivatives" in Wainer, I. W. and Drayer, D. E., Eds., *Drug Stereochemistry. Analytical Methods and Pharmacology*, Marcel Dekker, New York, Chap. 4. See also, Gal. J. (1993) "Indirect Methods for the Chromatographic Resolution of Drug Enantiomers: Synthesis and Separation of Diastereomeric Derivatives" in Wainer, I. W., Ed., *Drug Stereochemistry. Analytical Methods and Pharmacology*, 2nd ed., Marcel Dekker, New York, Chap. 4.

Ganey, M. V., Padykula, R. E., Berchtold, G. A., and Braun, A. G. (1989), *J. Org. Chem.*, **54**, 2787.

Gao, Y., Hanson, R. M., Klunder, J. M., Ko, S. Y., Masamune, H., and Sharpless, K. B. (1987), *J. Am. Chem. Soc.*, **109**, 5765.

Garcia, C. and Collet, A. (1992), *Tetrahedron: Asymmetry*, **3**, 361.

Gardlik, J. M. and Paquette, L. A. (1979), *Tetrahedron Lett.*, 3597.

Garnier-Suillerot, A., Albertini, J. P., Collet, A., Faury, L., Pastor, J.-M., and Tosi, L. (1981), *J. Chem. Soc. Dalton Trans.*, 2544.

Gerdil, R. (1987), "Tri-*o*-Thymotide Clathrates," *Top. Curr. Chem.*, **140**, 72.

Gerdil, R. and Allemand, J. (1980), *Helv. Chim. Acta*, **63**, 1750.

Gerdil, R. and Frew, A. (1985), *J. Incl. Phenom.*, **3**, 335.

Gerlach, H. (1985), *Helv. Chim. Acta*, **68**, 1815.

Gernez, D. (1866), *C.R. Acad. Sci.*, **63**, 843.

Geue, R. J., McCarthy, M. G., and Sargeson, A. M. (1984), *J. Am. Chem. Soc.*, **106**, 8282.

Gharpure, M. M. and Rao, A. S. (1988a), *Synthesis*, 410.

Gharpure, M. M. and Rao, A. S. (1988b), *Synth. Commun.*, **18**, 1833.

Giordano, C., Cavicchioli, S. Levi, S., and Villa, M. (1988), *Tetrahedron Lett.*, **29**, 5561.

Givens, R. S., Hrinczenko, B., Liu, J. H.-S., Matuszewski, B., and Tholen-Collison, J. (1984), *J. Am. Chem. Soc.*, **106**, 1779.

Gnaim, J. M., Green, B. S., Arad-Yellin, R., and Keehn, P. M. (1991), *J. Org. Chem.*, **56**, 4525.

Gold, E. H., Chang, W., Cohen, M., Baum, T., Ehrreich, S., Johnson, G., Prioli, N., and Sybertz, E. J. (1982), *J. Med. Chem.*, **25**, 1363.

Goldberg, I. (1977), *J. Am. Chem. Soc.*, **99**, 6049.

Golmohammadi, R. (1982), *Chem. Scr.*, **20**, 32.

Gotor, V., Brieva, R., and Rebolledo, F. (1988), *J. Chem. Soc. Chem. Commun.*, 957.

Gould, R. O. and Walkinshaw, M. D. (1984), *J. Am. Chem. Soc.*, **106**, 7840.

Green, B. S. and Knossow, M. (1981), *Science*, **214**, 795.

Green, M. M., Lifson, S., and Teramoto, A. (1991), "Cooperation in a Deep Helical Energy Well," *Chirality*, **3**, 285.

Greenberg, A., Molinaro, N., and Lang, M. (1984), *J. Org. Chem.*, **49**, 1127.

Gresh, N. (1987), *New J. Chem.*, **11**, 61.

Grieco, P. A., Zelle, R. E., Lis, R., and Finn, J., (1983), *J. Am. Chem. Soc.*, **105**, 1403.

Guo, Z.-W., Wu, S.-H., Chen, C.-S., Girdaukas, G., and Sih, C. J. (1990), *J. Am. Chem. Soc.*, **112**, 4942.

Gustavsen, J. E., Klæboe, P., and Kvila, H. (1978), *Acta Chem. Scand.*, **A32**, 25.

Guthrie, R. D. and Nicolas, E. C. (1981), *J. Am. Chem. Soc.*, **103**, 4637.

Gutzwiller, J., Buchschacher, P., and Fürst, A. (1977), *Synthesis*, 167.

Hagmann, W. K. (1986), *Synth. Commun.*, **16**, 437.

Halpern, B. and Westley, J. W. (1966), *Aust. J. Chem.* **19**, 1533.

Hamilton, J. A. and Chen, L. (1988), *J. Am. Chem. Soc.*, **110**, 4379.

Hanotier-Bridoux, M., Hanotier, J., and De Radzitzky, P. (1967), *Nature*, (*London*), **215**, 502.

Hare, P. E. and Mitterer, R. M. (1967–68), *Carnegie Inst. Wash. Yearb.*, **67**, 205.

Hargreaves, M. K. and Modarai, B. (1971), *J. Chem. Soc. C*, 1013.

Harris, M. M. (1958), "The Study of Optically Labile Compounds," *Progress Stereochem.*, **2**, 157.

Hart, H., Lin, L.-T. W., and Ward, D. L. (1984), *J. Am. Chem. Soc.*, **106**, 4043.

Hashimoto, S.-i., Kase, S., Suzuki, A., Yanagiya, Y., and Ikegami, S. (1991), *Synth. Commun.*, **21**, 833.

Hashimoto, S.-i., Komeshima, N., and Koga, K. (1979), *J. Chem. Soc. Chem. Commun.*, 437.

Hass, H. B. (1945), *U.S. Patent* 2,388,688, Nov. 13, 1945; *Chem. Abstr.*, **40**, 1538 (1946).

Hassel, O. (1970), "Structural Aspects of Interatomic Charge-Transfer Bonding," *Science*, **170**, 497.

Hatch III, C. E., Baum, J. S., Takashima, T., and Kondo, K. (1980), *J. Org. Chem.*, **45**, 3281.

Havinga, E. (1954), *Biochim. Biophys. Acta*, **13**, 171.

Hawkins, J. M. and Meyer, A. (1993), *Science*, **260**, 1918.

Hayashi, T., Yamamoto, A., and Ito, Y. (1986), *J. Chem. Soc. Chem. Commun.*, 1090.

Hayes, K. S., Nagumo, M., Blount, J. F., and Mislow, K. (1980), *J. Am. Chem. Soc.*, **102**, 2773.

Heathcock, C. H., Pirrung, M. C., Buse, C. T., Hagen, J. P., Young, S. D., and Sohn, J. E. (1979), *J. Am. Chem. Soc.*, **101**, 7076.

Heathcock, C. H., Pirrung, M. C., Lampe, J., Buse, C. T., and Young, S. D. (1981), *J. Org. Chem.*, **46**, 2290.

Hecker, S. J. and Heathcock, C. H. (1986), *J. Am. Chem. Soc.*, **108**, 4586.

Helgeson, R. C., Timko, J. M., Moreau, P., Peacock, S. C., Mayer, J. M., and Cram, D. J. (1974), *J. Am. Chem. Soc.*, **96**, 6762.

Helmchen, G., Haas, G., and Prelog, V. (1973), *Helv. Chim. Acta*, **56**, 2255.

Helmchen, G., Nill, G., Flockerzi, D., Schühle, W., and Youssef, M. S. K. (1979), *Angew. Chem. Int. Ed. Engl.*, **18**, 62; Helmchen, G., Nill, G., Flockerzi, D., and Youssef, M. S. K., *Angew. Chem. Int. Ed. Engl.*, **18**, 63; Helmchen, G. and Nill, G., *Angew. Chem. Int. Ed. Engl.*, **18**, 65.

Henderson, J. W. (1973), "Chirality in Carbonium Ions, Carbanions and Radicals," *Chem. Soc. Rev.*, **2**, 397.

Hengartner, U., Valentine, Jr., D., Johnson, K. K., Larscheid, M. E., Pigott, F., Scheidl, F., Scott, J. W., Sun, R. C., Townsend, J. M., and Williams, T. H. (1979), *J. Org. Chem.*, **44**, 3741.

Henkel, J. G. and Spector, J. H. (1983), *J. Org. Chem.*, **48**, 3657.

Hess, H., Burger, G., and Musso, H. (1978), *Angew. Chem. Int. Ed. Engl.*, **17**, 612.

Hilditch, T. (1912), *J. Chem. Soc.*, **101**, 192.

Hill, R. K., Morton, G. H., Peterson, J. R., Walsh, J. A., and Paquette, L. A. (1985), *J. Org. Chem.*, **50**, 5528.

Himmele, W. and Siegel, H. (1976), *Tetrahedron Lett.*, 907, 911.

Hiraki, Y. and Tai, A. (1982), *Chem. Lett.*, 341.

Hogeveen, H. and Zwart, L. (1982), *Tetrahedron Lett.*, **23**, 105.

Holmes, S. J. (1924), *Louis Pasteur*, Harcourt, Brace and Co., New York; Dover reprint, New York, 1961, p. 20.

Holton, P. G. (1981), *U.S.Patent* 4 246 193, Jan. 20, 1981; *Chem. Abstr.*, **95**, 61852n (1981).

Hongo, C., Tohyama, M., Yoshioka, R., Yamada, S., and Chibata, I. (1985), *Bull. Chem. Soc. Jpn.*, **58**, 433.

Hongo, C., Yamada, S., and Chibata, I. (1981a), *Bull. Chem. Soc. Jpn.*, **54**, 3286.

Hongo, C., Yamada, S., and Chibata, I. (1981b), *Bull. Chem. Soc. Jpn.*, **54**, 3291.

Hongo, C., Yoshioka, R., Tohyama, M., Yamada, S., and Chibata, I. (1983), *Bull. Chem. Soc. Jpn.*, **56**, 3744.

Horeau, A. (1975), *Tetrahedron*, **31**, 1307.

Houbiers, J. P. M. (1980), *Eur. Pat. Appl.*, EP 8833, Mar. 19, 1980, *Chem. Abstr.*, **93**, 71003j (1980).

Hua, D. H. (1986), *J. Am. Chem. Soc.*, **108**, 3835.

Huber, U. A. and Dreiding, A. S. (1970), *Helv. Chim. Acta.*, **53**, 495.

Hückel, W., Mentzel, R., Brinkmann, W., and Goth, E. (1925), *Justus Liebigs Ann. Chem.*, **441**, 1.

Huffman, J. W., and Wallace, R. H. (1989), *J. Am. Chem. Soc.*, **111**, 8691.

Hulshof, L. A., McKervey, M. A., and Wynberg, H. (1974), *J. Am. Chem. Soc.*, **96**, 3906.

Ichida, A. and Shibata, T. (1988), "Cellulose Derivatives as Stationary Chiral Phases," in Zeif, M. and Crane, L. J., Eds., *Chromatographic Chiral Separations*, Marcel Dekker, New York, Chap. 9.

Ichikawa, Y., Nakagawa, K., and Yoshisato, E. (1976), *Jpn. Kokai* 76 06 911, Jan. 20, 1976; *Chem. Abstr.*, **84**, 164151k (1976).

Imhof, R., Kyburz, E., and Daly, J. J. (1984), *J. Med. Chem.*, **27**, 165.

Inagaki, M. (1977), *Chem. Pharm. Bull.*, **25**, 2497.

Inagaki, M., Hiratake, J., Nishioka, T., and Oda, J. (1992), *J. Org. Chem.*, **57**, 5643.

Ingersoll, A. W. (1925), *J. Am. Chem. Soc.*, **47**, 1168.

Ingersoll, A. W., Babcock, S. H., and Burns, F. B. (1933), *J. Am. Chem. Soc.*, **55**, 411.

Inoue, Y. (1992), "Asymmetric Photochemical Reactions in Solution," *Chem. Rev.*, **92**, 741.

Ireland, R. E., Baldwin, S. W., Dawson, D. J., Dawson, M. I., Dolfini, J. E., Newbould, J., Johnson, W. S., Brown, M., Crawford, R. J., Hudrlik, P. F., Rasmussen, G. H., and Schmiegel, K. K. (1970), *J. Am. Chem. Soc.*, **92**, 5743.

Ishizaki, T., Miura, H., and Nohira, H. (1980), *Nippon Kagaku Kaishi*, 1381; *Chem. Abstr.*, **94**, 14600i (1981).

Izumi, T., Hino, T., Shoji, K., Sasaki, K., and Kasahara, A. (1989), *Chem. Ind. (London)*, 457.

Izumi, Y. and Tai, A. (1977), *Stereo-differentiating Reactions: The Nature of Asymmetric Reactions*, Kodansha, Tokyo and Academic, New York.

Izumi, Y., Chibata, I., and Itoh, T. (1978), "Production and Utilization of Amino Acids," *Angew. Chem. Int. Ed. Engl.*, **17**, 176.

Jacob III, P. (1982), *J. Org. Chem.*, **47**, 4165.

Jacques, J. (1992), *La Molécule et son Double*, Hachette, Paris, France, p. 41; *The Molecule and its Double* (English translation by Lee Scanlon), McGraw-Hill, NY, 1993, p. 41.

Jacques, J., Brienne, M.-J., Collet, A., and Cier, A. (1984), *Fr. Demande FR* 2 536 391, May 25, 1984; *Chem. Abstr.*, **101**, 231026h, (1984).

Jacques, J., Collet, A., and Wilen, S. H. (1981a), *Enantiomers, Racemates and Resolutions*, Wiley, New York.

Jacques, J. and Fouquey, C. (1989), *Org. Syn.*, **67**, 1.

Jacques, J. and Gabard, J. (1972), *Bull. Soc. Chim. Fr.*, 342.

Jacques, J., Leclercq, M., and Brienne, M.-J. (1981b), *Tetrahedron*, **37**, 1727.

Jadhav, P. K., Vara Prasad, J. V. N., and Brown, H. C. (1985), *J. Org. Chem.*, **50**, 3203.

James, B. R. and Young, C. G. (1985), *J. Organomet. Chem.*, **285**, 321.

Janda, K. D., Benkovic, S. J., and Lerner, R. A. (1989), *Science*, **244**, 437.

Jermyn, M. A. (1967), *Aust. J. Chem.*, **20**, 2283.

Jin, H. L., Stalcup, A., and Armstrong, D. W. (1989), *Chirality*, **1**, 137.

Johns, D. M. (1989), "Binding to cellulose derivatives," in Lough, W. J., Ed., *Chiral Liquid Chromatography*, Blackie, Glasgow, UK, p. 166.

Johnson, C. A., Guenzi, A., and Mislow, K. (1981), *J. Am. Chem. Soc.*, **103**, 6240.

Johnson, C. R. (1985), "Applications of Sulfoximines in Synthesis," *Aldrichim. Acta*, **18**, 3.

Johnson, C. R. and Meanwell, N. A. (1981), *J. Am. Chem. Soc.*, **103**, 7667.

Johnson, C. R. and Penning, T. D. (1986), *J. Am. Chem. Soc.*, **108**, 5655.

Johnson, C. R. and Zeller, J. R. (1982), *J. Am. Chem. Soc.*, **104**, 4021.

Johnson, C. R. and Zeller, J. R. (1984), *Tetrahedron*, **40**, 1225.

Johnson, T. H., Baldwin, T. F., and Klein, K. C. (1979), *Tetrahedron Lett.*, 1191.

Jolley, S. T., Bradshaw, J. S., and Izatt, R. M. (1982), "Synthetic Chiral Macrocyclic Crown Ligands: A Short Review," *J. Heterocyclic Chem.*, **19**, 3.

Jommi, G. and Teatini, A. (1962), *Chim. Ind. (Milan)* **44**, 29, cited in the Newman (1978) compendium, Vol. 1, p. 134.

Jonas, R. and Wurziger, H. (1987), *Tetrahedron*, **43**, 4539.

Jones, J. B. (1985), "Enzymes as Chiral Catalysts" in Morrison, J. D., Ed., *Asymmetric Synthesis*, Vol. 5, Academic, New York, Chap. 9.

Jones, J. B. (1986), "Enzymes in Organic Synthesis," *Tetrahedron*, **42**, 3351.

Jones, J. B. and Beck, J. F. (1976), "Asymmetric Syntheses and Resolutions Using Enzymes," in Jones, J. B., Sih, C. J., and Perlman, D., Eds., *Applications of Biochemical Systems in Organic Synthesis, Technique of Chemistry*, Vol. **10**, Part 1, Wiley-Interscience, New York, Chap. 4.

Jones, J. B. and Jakovac, I. J. (1982), *Can. J. Chem.*, **60**, 19.

Jurczak, J. and Gankowski, B. (1982), *Pol. J. Chem.*, **56**, 411.

Jurczak, J. and Tkacz, M. (1979), *J. Org. Chem.*, **44**, 3347.

Just, G., Luthe, C., and Potvin, P. (1982), *Tetrahedron Lett.*, **23**, 2285.

Kaehler, H., Nerdel, F., Engemann, G., and Schwerin, K. (1972), *Justus Liebigs Ann. Chem.*, **757**, 15.

Kagan, H. B. and Fiaud, J. C. (1978), "New Approaches in Asymmetric Synthesis," *Top. Stereochem.*, **10**, 175.

Kagan, H. B. and Fiaud, J. C. (1988), "Kinetic Resolution," *Top. Stereochem.*, **18**, 249.

Kai, Z. D., Kang, S. Y., Ke, M. J., Jin, Z., and Liang, H. (1985), *J. Chem. Soc., Chem. Commun.*, 168.

Kaneko, T., Izumi, Y., Chibata, I., and Itoh, T., Eds. (1974), *Synthetic Production and Utilization of Amino Acids*, Kodansha, Tokyo and Wiley, New York.

Kanoh, S., Hongoh, Y., Motoi, M., and Suda, H. (1988), *Bull. Chem. Soc. Jpn.*, **61**, 1032.

Kanoh, S., Muramoto, H., Kobayashi, N., Motoi, M., and Suda, H. (1987), *Bull. Chem. Soc. Jpn.*, **60**, 3659.

Kauffman, G. B. (1983), *J. Chem. Educ.*, **60**, 341.

Kauffman, G. B. and Myers, R. D. (1975), *J. Chem. Educ.*, **52**, 777.

Kawakita, T. (1992), "L-Monosodium glutamate," in Kroschwitz, J. I., Ed., *Kirk-Othmer Encyclopedia of Chemical Technology*, 4th ed., Vol. 2, Wiley-Interscience, New York, p. 571.

Kawanami, Y., Ito, Y., Kitagawa, T., Taniguchi, Y., Katsuki, T., and Yamaguchi, M. (1984), *Tetrahedron Lett.*, **25**, 857.

Kelly, R. C., Schletter, I., Stein, S. J., and Wierenga, W. (1979), *J. Am. Chem. Soc.*, **101**, 1054.

Kenrick, F. B. (1897), *Ber. Dtsch. Chem. Ges.*, **30**, 1749.

Kenyon, G. L. and Hegeman, G. D. (1970), *Biochemistry*, **9**, 4036.

Khatri, C. A., Andreola, C., Peterson, N. C., and Green, M. M. (1992), Polymer Preprint, Polymer Division, American Chemical Society, Washington, DC.

Kikukawa, T., Iizuka, Y., Sugimura, T., Harada, T., and Tai, A. (1987), *Chem. Lett.*, 1267.

Kim, Y. H., Tishbee, A., and Gil-Av, E. (1981), *J. Chem. Soc. Chem. Commun.*, 75.

Kimoto, H., Saigo, K., Ohashi, Y., and Hasegawa, M. (1989), *Bull. Chem. Soc. Jpn.*, **62**, 2189.

Kirchner, G., Scollar, M. P., and Klibanov, A. M. (1985), *J. Am. Chem. Soc.*, **107**, 7072.

Kitamura, M., Kasahara, I., Manabe, K., Noyori, R., and Takaya, H. (1988), *J. Org. Chem.*, **53**, 708.

Kitamura, M., Manabe, K., Noyori, R., and Takaya, H. (1987), *Tetrahedron Lett.*, **28**, 4719.

Klabunovskii, E. I., Patrikeev, V. V., and Balandin, A. A. (1960), *Izv. Akad. Nauk. SSSR*, 552; Engl. transl.: *Bull. Acad. Sci. USSR*, **1960**, 521.

Klemm, L. H. and Reed, D. (1960), *J. Chromatogr.*, **3**, 364.

Klibanov, A. M. (1986), "Enzymes that work in organic solvents," *CHEMTECH*, **16**, 354.

Klibanov, A. M. (1990), "Asymmetric Transformations Catalyzed by Enzymes in Organic Solvents," *Acc. Chem. Res.*, **23**, 114.

Klyashchitskii, B. A. and Shvets, V. I. (1972), "Resolution of Racemic Alcohols into Optical Isomers," *Russ. Chem. Rev.*, **41**, 592; *Uspekhi Khim.*, **41**, 1315 (1972).

Klyne, W. and Buckingham, J. (1978), *Atlas of Stereochemistry*, 2nd ed., Vols. I and II, Oxford, New York.

Knabe, J., Buech, H. P., and Kirsch, G. A. (1987), *Arch. Pharm. (Weinheim, Ger.)*, **320**, 323; *Chem. Abstr.*, **108**, 94484j (1988).

Knobler, C. B., Gaeta, F. C. A., and Cram, D. J. (1988), *J. Chem. Soc. Chem. Commun.*, 330.

Knox, J. H. and Jurand, J. (1982), *J. Chromatogr.*, **234**, 222.

Kobayashi, Y., Kusakabe, M., Kitano, Y., and Sato, F. (1988), *J. Org. Chem.*, **53**, 1586.

Kondepudi, D. K., Kaufman, R. J., and Singh, N. (1990), *Science*, **250**, 975.

Konrad, G. and Musso, H. (1986), *Liebigs Ann. Chem.*, 1956.

Koreeda, M. and Yoshihara, M. (1981), *J. Chem. Soc. Chem. Commun.*, 974.

Kornblum, N. and Wade, P. A. (1987), *J. Org. Chem.*, **52**, 5301.

Kozma, D., Pokol, G., and Ács, M. (1992), *J. Chem. Soc. Perkin Trans. 2*, 435.

Krapcho, A. P. and Dundulis, E. A. (1980), *J. Org. Chem.*, **45**, 3236.

Krause, H. W. and Meinicke, C. (1985), *J. Prakt. Chem.*, **327**, 1023.

Krieger, K. H., Lago, L., and Wantuck, J. A. (1968), *U.S. Patent*, 3 405 159, Oct. 8, 1968 and *Neth. Appl.* 6 514 950, May 18, 1966; *Chem. Abstr.*, **65**, 14557b (1966).

Kubanek, A. M., Sifniades, S., and Fuhrmann, R. (1974), *U.S. Patent* 3 824 231, July 16, 1974; *Chem. Abstr.*, **81**, 135484d (1974).

Kuhn, W. (1936), *Angew. Chem.*, **49**, 215.

Kuroda, R. and Mason, S. F. (1981), *J. Chem. Soc. Perkin 2*, 167.

Kyba, E. P., Gokel, G. W., de Jong, F., Koga, K., Sousa, L. R., Siegel, M. G., Kaplan, L., Sogah, G. D. Y., and Cram, D. J. (1977), *J. Org. Chem.* **42**, 4173.

Kyba, E. P., Koga, K., Sousa, L. R., Siegel, M. G., and Cram, D. J. (1973), *J. Am. Chem. Soc.*, **95**, 2692.

Kyba, E. P., Timko, J. M., Kaplan, L. J., de Jong, F., Gokel, G. W., and Cram, D. J. (1978), *J. Am. Chem. Soc.*, **100**, 4555.

Labianca, D. A. (1975), *J. Chem. Educ.*, **52**, 156.

Lago, J. (1984), personal communication to SHW.

Langrand, G., Baratti, J., Buono, G., and Triantaphylides, C. (1986), *Tetrahedron Lett.*, **27**, 29.

Larson, E. R. and Raphael, R. A. (1982), *J. Chem. Soc.*, *Perkin 1*, 521.

Laumen, K. and Schneider, M. P. (1988), *J. Chem. Soc. Chem. Commun.*, 598.

Laumen, K., Breitgoff, D., and Schneider, M. P. (1988), *J. Chem. Soc. Chem. Commun.*, 1459.

Lauricella, R., Kéchayan, J., and Bodot, H. (1987), *J. Org. Chem.*, **52**, 1577.

Lazarus, R. A. (1990), *J. Org. Chem.*, **55**, 4755.

Lazzaroni, R., Uccello-Barretta, G., Pini, D., Pucci, S., and Salvadori, P. (1983), *J. Chem. Res.*, *Synop.*, 286.

Le Bel, J.-A. (1874), *Bull. Soc. Chim. Fr.*, [2] **22**, 337.

Leborgne, A., Spassky, N., and Sigwalt, P. (1979), *Polym. Bull. (Berlin)*, **1**, 825.

Leclercq, M. and Jacques, J. (1975), *Bull. Soc. Chim. Fr.*, 2052.

Leclercq, M. and Jacques, J. (1979), *Nouv. J. Chim.*, **3**, 629.

Leclercq, M. and Jacques, J. (1985), *C.R. Séances Acad. Sci. Ser. 2*, **301**, 1231.

Leclercq, M., Jacques, J., and Cohen-Adad, R. (1982), *Bull. Soc. Chim. Fr.*, I-388.

Leeson, P. D., Williams, B. J., Baker, R., Ladduwahetty, T., Moore, K. W., and Rowley, M. (1990), *J. Chem. Soc. Chem. Commun.*, 1578.

Lehn, J.-M. (1979), "Macrocyclic Receptor Molecules: Aspects of Chemical Reactivity. Investigations into Molecular Catalysis and Transport Processes," *Pure Appl. Chem.*, **51**, 979.

Lehn, J.-M., Moradpour, A., and Behr, J. P. (1975), *J. Am. Chem. Soc.*, **97**, 2532.

Lehn, J.-M., Simon, J., and Moradpour, A. (1978), *Helv. Chim. Acta*, **61**, 2407.

Lesec, J., Lafuma, F., and Quivoron, C. (1974), *J. Chromatogr. Sci.*, **12**, 683.

Leston, G. (1984), 187th American Chemical Society National Meeting, St. Louis, MO, April 12, 1984, Abstracts ORGN 236 and 237; *Chem. Eng. News.*, **62**, April 30, 1984, p. 31.

Levy, J. (1977) personal communication to SHW.

Lightner, D. A., Bouman, T. D., Wijekoon, W. M. D., and Hansen, A. E. (1986), *J. Am. Chem. Soc.*, **108**, 4484.

Lindner, W. (1988), "Indirect Separation of Enantiomers by Liquid Chromatography," in Zief, M. and Crane L. J., Eds., *Chromatographic Chiral Separations*, Marcel Dekker, New York, Chap. 4.

Lingenfelter, D. S., Helgeson, R. C., and Cram, D. J. (1981), *J. Org. Chem.*, **46**, 393.

Lopata, A., Faigl, F., Fogassy, E., and Darvas, F. (1984), *J. Chem. Res. Synop.*, 322; *Miniprint*, 2953.

Lowry, T. M. (1899), *J. Chem. Soc.*, **75**, 211.

Lubell, W. and Rapoport, H. (1989), *J. Org. Chem.*, **54**, 3824.

Lucas, H. J. and Gould, Jr., C. W. (1942), *J. Am. Chem. Soc.*, **64**, 601.

MacNicol, D. D. and Murphy, A. (1981), *Tetrahedron Lett.*, **22**, 1131.

Madhavan, G. V. B. and Martin, J. C. (1986), *J. Org. Chem.*, **51**, 1287.

Mangeney, P., Alexakis, A., and Normant, J. F. (1988), *Tetrahedron Lett.*, **29**, 2677.

Manimaran, T. and Stahly, G. P. (1993), *Tetrahedron: Asymmetry*, **4**, 1949.

March, J. (1992), *Advanced Organic Chemistry*, 4th ed., Wiley, New York.

Marckwald, W. (1896), *Ber. Dtsch. Chem. Ges.*, **29**, 42, 43.

Marckwald, W. and McKenzie, A. (1899), *Ber. Dtsch. Chem. Ges.*, **32**, 2130.

Marckwald, W. and Meth, R. (1905), *Ber. Dtsch. Chem. Ges.*, **38**, 801.

Markowicz, S. W. (1979), *Pol. J. Chem.*, **53**, 157.

Martel, J. (1970), *Ger. Offen.* 1 935 320, Jan. 15, 1970; *Chem. Abstr.*, **72**, 121078b (1970).

Martel, J. (1992), "The Development and Manufacture of Pyrethroid Insecticides," in Collins, A. N., Sheldrake, G.N., and Crosby, J., Eds., *Chirality in Industry: The Commercial Manufacture and Applications of Optically Active Compounds*, Wiley, Chichester, UK, Chap. 4.

Martin, R. H. (1974), The Helicenes, *Angew. Chem. Int. Ed. Engl.*, **13**, 649.

Martin, R. H. and Libert, V. (1980), *J. Chem. Res., Synop.*, 130; *Miniprint*, 1940.

Martin, R. H. and Marchant, M. J. (1974), *Tetrahedron*, **30**, 347.

Martin, V. S., Woodard, S. S., Katsuki, T., Yamada, Y., Ikeda, M., and Sharpless, K. B. (1981), *J. Am. Chem. Soc.*, **103**, 6237.

Marx, M., Marti, F., Reisdorff, J., Sandmeier, R., and Clark, S. (1977), *J. Am. Chem. Soc.*, **99**, 6754.

Masamune, S., Kim, B., Petersen, J. S., Sato, T., Veenstra, S. J., and Imai, T. (1985), *J. Am. Chem. Soc.*, **107**, 4549.

Mason, S. F. (1982), *Molecular Optical Activity and the Chiral Discriminations*, Cambridge University Press, Cambridge, UK.

Massad, S. K., Hawkins, L. D., and Baker, D. C. (1983), *J. Org. Chem.*, **48**, 5180.

Matsumoto, H., Obara, Y., Arai, K., and Tsuchiya, S. (1986), *Jpn. Kokai Tokkyo Koho* JP 61 83,144 [86, 83,144], Apr. 20, 1986; *Chem. Abstr.*, **105**, 225789a (1986).

Matsushita, H., Noguchi, M., and Yoshikawa, S. (1976), *Bull. Chem. Soc. Jpn.*, **49**, 1928.

Matsushita, H., Tsujino, Y., Noguchi, M., Saburi, M., and Yoshikawa, S. (1978), *Bull. Chem. Soc. Jpn.*, **51**, 862.

Mazaleyrat, J. P. and Welvart, Z. (1983), *Nouv. J. Chim.*, **7**, 491.

McBride, J. M. and Carter, R. L. (1991), *Angew. Chem. Int. Ed. Engl.*, **30**, 293.

Metzler, D. E., Ikawa, M., and Snell, E. E. (1954), *J. Am. Chem. Soc.*, **76**, 648.

Meurling, L. (1974), *Chem. Scr.*, **6**, 92.

Meyer, V. R. (1991), "Amino Acid Racemization. A Tool for Dating?", in Ahuja, S., Ed., *Chiral Separations by Liquid Chromatography*, ACS Symposium Series 471, American Chemical Society, Washington, D.C., Chap. 13.

Meyer, V. R. (1992), "Amino acid racemization: a tool for fossil dating," *CHEMTECH*, **22**, 412.

Meyers, A. I. and Kamata, K. (1976), *J. Am. Chem. Soc.*, **98**, 2290.

Mikes, F. and Boshart, G. (1978), *J. Chem. Soc. Chem. Commun.*, 173; *J. Chromatogr.*, **149**, 455.

Mikes, F., Boshart, G., and Gil-Av, E. (1976), *J. Chromatogr.*, **122**, 205.

Mikołajczyk, M. and Drabowicz, J. (1978), *J. Am. Chem. Soc.*, **100**, 2510.

Miller, E. G., Rayner, D. R., Thomas, H. T., and Mislow, K. (1968), *J. Am. Chem. Soc.*, **90**, 4861.

Mills, A. K. and Smith, A. E. W. (1960), *Helv. Chim. Acta*, **43**, 1915.

Mislow, K. (1962), "Stereoisomerism" in Florkin, M. and Stotz, E. H., Eds., *Comprehensive Biochemistry*, Vol. 1, Elsevier, Amsterdam, The Netherlands, p. 223.

Mislow, K. (1965), *Introduction to Stereochemistry*, Benjamin, New York.

Mislow, K., Axelrod, M., Rayner, D. R., Gotthardt, H., Coyne, L. M., and Hammond, G. S. (1965), *J. Am. Chem. Soc.*, **87**, 4958.

Mislow, K., Glass, M. A. W., Hopps, H. B., Simon, E., and Wahl, Jr., G. H. (1964a), *J. Am. Chem. Soc.*, **86**, 1710.

Mislow, K., Graeve, R., Gordon, A. J., and Wahl, Jr., G. H. (1964b), *J. Am. Chem. Soc.*, **86**, 1733.

Mislow, K., Gust, D., Finocchiaro, P., and Boettcher, R. J. (1974), "Stereochemical Correspondence Among Molecular Propellers," *Top. Curr. Chem.*, **47**, 1.

Mislow, K., Hyden, S., and Schaefer, H. (1962), *J. Am. Chem. Soc.*, **84**, 1449.

Miyano, S., Kawahara, K., Inoue, Y., and Hashimoto, H. (1987), *Chem. Lett.*, 355.

Miyano, S., Lu, L. D.-L., Viti, S. M., and Sharpless, K. B. (1985), *J. Org. Chem.*, **50**, 4350.

Miyata, M., Shibakami, M., and Takemoto, K. (1988), *J. Chem. Soc. Chem. Commun.*, 655.

Mofaddel, N. and Bouaziz, R. (1991), *Bull. Soc. Chim. Fr.*, 773.

Mori, A. and Yamamoto, H. (1985), *J. Org. Chem.*, **50**, 5444.

Mori, K., Nukada, T., and Ebata, T. (1981), *Tetrahedron*, **37**, 1343.

Mori, K., Uematsu, T., Minobe, M., and Yanagi, K. (1982), *Tetrahedron Lett.*, **23**, 1921.

Morrison, J. D. and Mosher, H. S. (1976), *Asymmetric Organic Reactions*, Prentice-Hall, Englewood Cliffs, NJ., 1971; American Chemical Society, Washington, DC, corrected reprint, 1976.

Morrison, J. D. and Ridgway, R. W. (1969), *Tetrahedron Lett.*, 569.

Murahashi, S.-I., Yoshimura, N., Tsumiyama, T., and Kojima, T. (1983), *J. Am. Chem. Soc.*, **105**, 5002.

Murr, B. L. and Feller, L. W. (1968), *J. Am. Chem. Soc.*, **90**, 2966.

Murr, B. L. and Santiago, C. (1968), *J. Am. Chem. Soc.*, **90**, 2964.

Naciri, J., Spada, G. P., Gottarelli, G., and Weiss, R. G. (1987), *J. Am. Chem. Soc.*, **109**, 4352.

Naemura, K., Fukunaga, R., and Yamanaka, M. (1985), *J. Chem. Soc. Chem. Commun.*, 1560.

Nakagawa, H., Yamada, K.-i., and Kawazura, H. (1989), *J. Chem. Soc. Chem. Commun.*, 1378.

Nakamura, M., Shiraiwa, T., and Kurokawa, H. (1986), *Technol. Rep. Kansai Univ.*, **27**, 111; *Chem. Abstr.*, **108**, 5648a (1988).

Napper, A. D., Benkovic, S. J., Tramontano, A., and Lerner, R. A. (1987), *Science*, **237**, 1041.

Neef, G., Petzoldt, K., Wieglepp, H., and Wiechert, R. (1985), *Tetrahedron Lett.*, **26**, 5033.

Neuberger, A. (1948), "Stereochemistry of Amino Acids," *Adv. Protein Chem.*, **4**, 297.

Newcomb, M., Helgeson, R. C., and Cram, D. J. (1974), *J. Am. Chem. Soc.*, **96**, 7367.

Newcomb, M., Toner, J. L., Helgeson, R. C., and Cram, D. J. (1979), *J. Am. Chem. Soc.*, **101**, 4941.

Newman, A. C. D. and Powell, H. M. (1952), *J. Chem. Soc.*, 3747.

Newman, M. S. and Lutz, W. B. (1956), *J. Am. Chem. Soc.*, **78**, 2469.

Newman, M. S., Lutz, W. B., and Lednicer, D. (1955), *J. Am. Chem. Soc.*, **77**, 3420.

Newman, P. (1978–1993), *Optical Resolution Procedures for Chemical Compounds*, Optical Resolution Information Center, Manhattan College, New York, Vols. 1, 2A, 2B, 3, and 4 (Parts I and II).

Newman, P., Rutkin, P., and Mislow, K. (1958), *J. Am. Chem. Soc.*, **80**, 465.

Nicholson, J. S. and Tatum, J. G. (1981), *U.S. Patent* 4 209 638, June 24, 1980; *Chem. Abstr.*, **95**, 6831e (1981).

Nilsson, B. M. and Hacksell, U. (1988), *Acta Chem. Scand. Ser. B*, **42**, 55.

Nishikawa, J., Ishizaki, T., Nakayama, F., Kawa, H., Saigo, K., and Nohira, H. (1979), *Nippon Kagaku Kaishi*, 754; *Chem. Abstr.*, **93**, 7767j (1980).

Nishikawa, M. (1963), *Chem. Pharm. Bull.*, **11**, 977.

Noe, C. R. (1982), *Chem. Ber.*, **115**, 1576, 1591, 1607; see *Aldrichim. Acta*, **16**, 10 (1983).

Nohira, H. (1981), *Jpn. Kokai Tokkyo Koho* JP 81 133 244 (Oct. 13, 1981); *Chem. Abstr.*, **96**, 105549j (1982).

Nohira, H., Fujii, H., Yajima, M., and Fujimura, R. (1982), *Eur. Patent Appl.*, EP 30 871, June 24, 1981; *Chem. Abstr.*, **96**, 34875m (1982).

Nohira, H., Kai, M., Nohira, M., Nishikawa, J., Hoshiko, T., and Saigo, K. (1981), *Chem. Lett.*, 951.

Nohira, H., Miura, H., Otani, H., Saito, K., and Yajima, M. (1988a), *Jpn. Kokai Tokkyo Koho* JP 63 57 560 [88 57,560] March 12, 1988; *Chem. Abstr.*, **109**, 231541n (1988).

Nohira, H., Nohira, M., Yoshida, S.-i., Osada, A., and Terunuma, D. (1988b), *Bull. Chem. Soc. Jpn.*, **61**, 1395.

Nohira, H., Terunuma, D., Kobe, S., Asakura, I., Miyashita, A., and Ito, T. (1986), *Agric. Biol. Chem.*, **50**, 675.

Nohira, H. and Ueda, K. (1983), *Eur. Patent Appl.*, EP 65 867; *Chem. Abstr.*, **98**, 161164v (1983).

Nohira, H. and Yamamoto, M. (1986), *Jpn. Kokai Tokkyo Koho* JP 61 260, 044 [86 260,044] Nov. 18, 1986; *Chem. Abstr.*, **107**, 6953n (1987).

Numan, H., Helder, R., and Wynberg, H. (1976), *Recl. Trav. Chim. Pays-Bas*, **95**, 211.

Numata, Y., Okawa, H., and Kida, S. (1979), *Chem. Lett.*, 293.

O'Donnell, M. J., Bennett, W. D., and Wu, S. (1989), *J. Am. Chem. Soc.*, **111**, 2353.

Oare, D. A. and Heathcock, C. H. (1989), "Stereochemistry of the Base-Promoted Micheal Addition Reaction," *Top. Stereochem*, **19**, 227.

Ohashi, Y. (1988), "Dynamical Structure Analysis of Crystalline-State Racemization," *Acc. Chem. Res.*, **21**, 268.

Ohgi, T., Kondo, T., and Goto, T. (1979), *J. Am. Chem. Soc.*, **101**, 3629.

Ohta, H., Miyamae, Y., and Kimura, Y. (1989), *Chem. Lett.*, 379.

Okada, K., Samizo, F., and Oda, M. (1986), *J. Chem. Soc. Chem. Commun.*, 1044.

Okada, M., Sumitomo, H., and Atsumi, M. (1984), *J. Am. Chem. Soc.*, **106**, 2101.

Okada, Y., Takebayashi, T., and Sato, D. (1989), *Chem. Pharm. Bull.*, **37**, 5.

Okada, Y., Takebayashi, T., Hashimoto, M., Kasuga, S., Sato, S., and Tamura, C. (1983), *J. Chem. Soc. Chem. Commun.*, 784.

Okamoto, Y., Aburatani, R., Kaida, Y., and Hatada, K. (1988), *Chem. Lett.*, 1125.

Okamoto, Y. and Hatada, K. (1988), "Optically Active Poly(Triphenylmethyl Methacrylate) as a Chiral Stationary Phase," in Zief, M. and Crane L.J., Eds., *Chromatographic Chiral Separations*, Marcel Dekker, New York, Chap. 8.

Okamoto, Y., Suzuki, K., Kitayama, T., Yuki, H., Kageyama, H., Miki, K., Tanaka, N., and Kasai, N. (1982), *J. Am. Chem. Soc.*, **104**, 4618.

Okamoto, Y., Yashima, E., Hatada, K., and Mislow, K. (1984), *J. Org. Chem.*, **49**, 557.

Olivieri, R., Fascetti, E., Angelini, L, and Degen, L. (1979), *Enzyme Microb. Technol.*, **1**, 201; *idem.*, *Biotechnol. Bioeng.*, **23**, 2173 (1981).

Openshaw, H. T. and Whittaker, N. (1963), *J. Chem. Soc.*, 1461.

Oppolzer, W. (1987), "Camphor Derivatives as Chiral Auxiliaries in Asymmetric Synthesis," *Tetrahedron*, **43**, 1969.

Osano, Y. T., Uchida, A., and Ohashi, Y. (1991), *Nature (London)*, **352**, 510; see Lahav, M., "Optical Enrichment of a Racemic Chiral Crystal by X-ray Irradiation," *Chemtracts-Organic Chemistry*, **4**, 440 (1991).

Ostromisslenskii, I. (1908), *Ber. Dtsch. Chem. Ges.*, **41**, 3035.

Ottenheijm, H. C. J., Herscheid, J. D. M., and Nivard, R. J. F. (1977), *J. Org. Chem.*, **42**, 924.

Overberger, C. G. and Labianca, D. A. (1970), *J. Org. Chem.*, **35**, 1762.

Overby, L. R. and Ingersoll, A. W. (1960), *J. Am. Chem. Soc.*, **82**, 2067.

Paquette, L. A. and Doehner, Jr., R. F. (1980), *J. Org. Chem.*, **45**, 5105.

Paquette, L. A. and Gardlik, J. M. (1980), *J. Am. Chem. Soc.*, **102**, 5016.

Paquette, L. A., Gardlik, J. M., Johnson, L. K., and McCullough, K. J. (1980), *J. Am. Chem. Soc.*, **102**, 5026.

Paquette, L. A. and Lau, C. J. (1987), *J. Org. Chem.* **52**, 1634.

Paquette, L. A. and Trova, M. P. (1988), *J. Am. Chem. Soc.*, **110**, 8197.

Paquette, L. A. and Wang, T.-Z. (1988), *J. Am. Chem. Soc.*, **110**, 8192.

Pasteur, L. (1848), *Ann. Chim. Phys.*, [3] **24**, 442.

Pasteur, L. (1853), *C.R. Acad. Sci.*, **37**, 162.

Pasteur, L. (1858), *C.R. Acad. Sci.*, **46**, 615.

Pasteur, L. (1860), *Researches on the Molecular Asymmetry* (sic) *of Natural Organic Products*, Alembic Club Reprint No. 14, Clay, W. F., Edinburgh, UK, p. 43.

Paterson, I., McClure, C. K., and Schumann, R. C. (1989), *Tetrahedron Lett.*, **30**, 1293.

Patil, A. O., Pennington, W. T., Paul, I. C., Curtin, D. Y., and Dykstra, C. E., (1987), *J. Am. Chem. Soc.*, **109**, 1529.

Pavlis, R. R. and Skell, P. S. (1983), *J. Org. Chem.*, **48**, 1901.

Pavlis, R. R., Skell, P. S., Lewis, D. C. and Shea, K. J. (1973), *J. Am. Chem. Soc.*, **95**, 6735.

Peacock, S. C. and Cram, D. J. (1976), *J. Chem. Soc. Chem. Commun.*, 282.

Peacock, S. C., Domeier, L. A., Gaeta, F. C. A., Helgeson, R. C., Timko, J. M., and Cram, D. J. (1978), *J. Am. Chem. Soc.*, **100**, 8190.

Peacock, S. S. (sic), Walba, D. M., Gaeta, F. C. A., Helgeson, R. C., and Cram, D. J. (1980), *J. Am. Chem. Soc.*, **102**, 2043.

Pearson, D. E. and Rosenberg, A. A. (1975), *J. Med. Chem.*, **18**, 523.

Perry, C. W., Brossi, A., Deitcher, K. H., Tautz, W., and Teitel, S. (1977) *Synthesis*, 492.

Petterson, C. and Schill, G. (1986), *J. Liq. Chromatog.*, **9**, 269.

Petterson, C. and Schill, G. (1988), "Enantiomer Separation in Ion-Pairing Systems," in Zief, M. and Crane, L.J., Eds., *Chromatographic Chiral Separations*, Marcel Dekker, New York, Chap. 11.

Phillips, M. L., Berry, D. M., and Panetta, J. A. (1992), *J. Org. Chem.*, **57**, 4047.

Pikulin, S. and Berson, J. A. (1988), *J. Am. Chem. Soc.*, **110**, 8500.

Pincock, R. E. and Wilson, K. R. (1971) *J. Am. Chem. Soc.*, **93**, 1291.

Pincock, R. E. and Wilson, K. R. (1973), *J. Chem. Educ.*, **50**, 455.

Pirkle, W. H. (1987a) cited in Allenmark, S. G. (1991).

Pirkle, W. H. (1987b), "New Developments in Chiral Stationary Phases for HPLC," *11th Int. Symp. on Column Liquid Chromatography*, Amsterdam, The Netherlands, cited in Taylor, D. R. (1989).

Pirkle, W. H. and Bowen, W. E. (1991), Abstracts of the *2nd Int. Symp. on Chiral Discrimination*, Rome, May 27–31, 1991, p. 3.

Pirkle, W. H., Deming, K. C., and Burke III, J. A. (1991), *Chirality*, **3**, 183.

Pirkle, W. H. and Doherty, E. M. (1989), *J. Am. Chem. Soc.*, **111**, 4113.

Pirkle, W. H. and Finn, J. M. (1981), *J. Org. Chem.*, **46**, 2935.

Pirkle, W. H. and Finn, J. M. (1982), *J. Org. Chem.*, **47**, 4037.

Pirkle, W. H., Finn, J. M., Hamper, B. C., Schreiner, J., and Pribish, J. R. (1982), "A Useful and Conveniently Accessible Chiral Stationary Phase for the Liquid Chromatographic Separation of Enantiomers," in Eliel, E. L. and Otsuka, S., Eds., *Asymmetric Reactions and Processes in Chemistry*, ACS Symposium Series 185, American Chemical Society, Washington, DC, p. 245.

Pirkle, W. H. and Hamper, B. C. (1987), "The Direct Preparative Resolution of Enantiomers by Liquid Chromatography on Chiral Stationary Phases," in Bidlingmeyer, B. A., Ed., *Preparative Liquid Chromatography*, Elsevier, Amsterdam, The Netherlands, p. 235.

Pirkle, W. H. and Hauske, J. R. (1977), *J. Org. Chem.*, **42**, 2436.

Pirkle, W. H. and Hoekstra, M. S. (1974), *J. Org. Chem.*, **39**, 3904.

Pirkle, W. H. and House, D. W. (1979), *J. Org. Chem.*, **44**, 1957.

Pirkle, W. H. and Pochapsky, T. C. (1987), *J. Am. Chem. Soc.*, **109**, 5975.

Pirkle, W. H. and Reno, D. S. (1987), *J. Am. Chem. Soc.*, **109**, 7189.

Pirkle, W. H. and Simmons, K. A. (1983), *J. Org. Chem.*, **48**, 2520.

Pirkle, W. H., Tsipouras, A. and Sowin, T. J. (1985), *J. Chromatog.*, **319**, 392.

Pirkle, W. H. and Welch, C. J. (1984), *J. Org. Chem.*, **49**, 138.

Piselli, F. L. (1989), *Eur. Patent Appl.* EP 298 395, Jan. 11, 1989; *Chem. Abstr.*, **111**, 7085a (1989).

Piutti, A. (1886), *C.R. Hebd. Séances Acad. Sci.*, **103**, 134.

Pluim, H. (1982), Ph.D. Dissertation, University of Groningen, The Netherlands.

Pluim, H and Wynberg, H. (1979), *Tetrahedron Lett.*, 1251.

Pollack, S. J., Jacobs, J. W., and Schultz, P. G. (1986), *Science*, **234**, 1570.

Pope, W. J. and Peachey, S. J. (1899), *J. Chem. Soc.*, **75**, 1066.

Potapov, V. M. (1979), *Stereochemistry*, translated by A. Beknazarov, Mir, Moscow.

Potapov, V. M., Dem'yanovich, V. M., and Khlebnikov, V. A. (1986), *Dokl. Akad. Nauk SSSR*, **289**, 117; Engl. Transl. (Doklady Chemistry) **289**, 254.

Prelog, V. (1964), *Pure Appl. Chem.*, **9**, 119.

Prelog, V. (1978), "Chiral Ionophores," *Pure Appl. Chem.* **50**, 893.

Prelog, V. and Wieland, P. (1944), *Helv. Chim. Acta.*, **27**, 1127.

Prelog, V., Stojanac, Z., and Kovačević (1982), *Helv. Chim. Acta*, **65**, 377.

Pritchard, J. G. and Vollmer, R. L. (1963), *J. Org. Chem.*, **28**, 1545.

Purvis, J. L. (1957), *U.S. Patent* 2 790 001, April 23, 1957; *Chem. Abstr.*, **51**, 13910i, (1957).

Rayner, D. R., Gordon, A. J., and Mislow, K. (1968), *J. Am. Chem. Soc.*, **90**, 4854.

Regan, A. C. and Staunton, J. (1987), *J. Chem. Soc. Chem. Commun.*, 520.

Reider, P. J., Davis, P., Hughes, D. L., and Grabowski, E. J. J. (1987), *J. Org. Chem.* **52**, 955.

Reinhold, D. F., Firestone, R. A., Gaines, W. A., Chemerda, J. M., and Sletzinger, M. (1968), *J. Org. Chem.*, **33**, 1209.

Rittle, K. E., Evans, B. E., Bock, M. G., DiPardo, R. M., Whitter, W. L., Homnick, C. F., Veber, D. F., and Freidinger, R. M. (1987), *Tetrahedron Lett.*, **28**, 521.

Robinson, R. (1932), *Outline of an Electrochemical (Electronic) Theory of the Course of Organic Reactions*, The Institute of Chemistry of Great Britain and Ireland, London.

Romano, S. J., Wells, K. H., Rothbard, H. L., and Rieman III, W. (1969), *Talanta*, **16**, 581.

Roper, J. M. and Bauer, D. P. (1983), *Synthesis*, 1041.

Rosan, A. M. (1989), *J. Chem. Educ.*, **66**, 608.

Rossiter, B. E. (1985), "Synthetic Aspects and Applications of Asymmetric Epoxidation" in Morrison, J. D., Ed., *Asymmetric Synthesis*, Vol. 5, Academic, New York, Chap. 7.

Rossiter, B. E., and Sharpless, K. B. (1984), *J. Org. Chem.*, **49**, 3707.

Roush, W. R. (1988), "Extension of Chromatographically Derived Molecular Recognition Concepts to First-Order Asymmetric Transformations," *Chemtracts: Org. Chem.*, **1**, 136.

Roussel, C., Adjimi, M., Chemlal, A., and Djafri, A. (1988a), *J. Org. Chem.*, **53**, 5076.

Roussel, C. and Chemlal, A. (1988), *New J. Chem.*, **12**, 947.

Rozwadowska, M. D. and Brossi, A. (1989), *J. Org. Chem.*, **54**, 3202.

Rubinstein, H., Feibush, B., and Gil-Av, E. (1973), *J. Chem. Soc. Perkin Trans. 2*, 2094.

Ruxer, J.-M., Solladié, G., and Candau, S. (1978), *J. Chem. Res., Synopsis*, 82.

Saigo, K., Kai, M., Yonezawa, N., and Hasegawa, M. (1985), *Synthesis*, 214.

Saigo, K., Kimoto, H., Nohira, H., Yanagi, K. and Hasegawa, M. (1987), *Bull. Chem. Soc. Jpn.*, **60**, 3655.

Saigo, K., Kubota, N., Takebayashi, S., and Hasegawa, M. (1986b), *Bull, Chem. Soc. Jpn.*, **59**, 931.

Saigo, K., Ogawa, S., Kikuchi, S., Kasahara, A., and Nohira, H. (1982), *Bull. Chem. Soc. Jpn.*, **55**, 1568.

Saigo, K., Okuda, Y., Wakabayashi, S., Hoshiko, T., and Nohira, H. (1981), *Chem. Lett.*, 857.

Saigo, K., Sugiura, I., Shida, I., Tachibana, K., and Hasegawa, M. (1986a), *Bull. Chem. Soc. Jpn.*, **59**, 2915.

Saito, H., Nishimura, Y., Kondo, S., and Umezawa, H. (1987), *Chem. Lett.*, 799.

Salz, U. and Rüchardt, C. (1984), *Chem. Ber.*, **117**, 3457; see also Rüchardt, C., Gärtner, H., and Salz, U., *Angew. Chem. Int. Ed. Engl.*, **23**, 162 (1984).

Sano, K., Eguchi, C., Yasuda, N. and Mitsugi, K. (1979), *Agric. Biol. Chem.* **43**, 2373.

Santaniello, E., Ferraboschi, P., Grisenti, P., and Manzocchi, A. (1992), "The Biocatalytic Approach to the Preparation of Enantiomerically Pure Chiral Building Blocks," *Chem. Rev.*, **92**, 1071.

Sato, T., Kawara, T., Nishizawa, A., and Fujisawa, T. (1980), *Tetrahedron Lett.*, **21**, 3377.

Schlenk, Jr., W. (1952), *Experientia*, **8**, 337.

Schlenk, Jr., W. (1965), Recent Results of Configurational Research, *Angew. Chem. Int. Ed. Engl.*, **4**, 139.

Schlenk, Jr., W. (1973), *Justus Liebigs Ann. Chem.*, 1145, 1156, 1179, 1195.

Schönenberger, B. and Brossi, A. (1986), *Helv. Chim Acta*, **69**, 1486.

Schoofs, A. R. and Guetté, J.-P. (1983), "Competitive Reaction Methods for the Determination of Maximum Specific Rotations," in Morrison, J. D., Ed., *Asymmetric Synthesis*, Vol. 1, Academic, New York, Chap. 3.

Schreiber, S. L., Schreiber, T. S., and Smith, D. B. (1987), *J. Am. Chem. Soc.*, **109**, 1525.

Schultz, P. G. (1989), "Catalytic Antibodies," *Acc. Chem. Res.*, **22**, 287.

Schultz, P. G. and Lerner, R. A. (1993), "Antibody Catalysis of Difficult Chemical Transformations," *Acc. Chem. Res.*, **26**, 391.

Schurig, V. and Bürkle, W. (1982), *J. Am. Chem. Soc.*, **104**, 7573.

Schwab, J. M. and Lin, D. C. T. (1985), *J. Am. Chem. Soc.*, **107**, 6046.

Schwartz, L. H., Aria, P. S., and Henschel, R. F. (1989), 197th American Chemical Society National Meeting, Dallas, TX, April 12, 1989, Abstract ORGN 203.

Scott, J. W. (1984), "Readily Available Chiral Carbon Fragments and their Use in Synthesis," in Morrison, J. D. and Scott, J. W., Eds., *Asymmetric Synthesis*, Vol. 4, Academic, New York, Chap. 1.

Scott, J. W. (1989), "Enantioselective Synthesis of Non-racemic Chiral Molecules on an Industrial Scale," *Top. Stereochem.*, **19**, 209.

Secor, R. M. (1963), "Resolution of Optical Isomers by Crystallization Procedures," *Chem. Rev.*, **63**, 297.

Seebach, D., Boes, M., Naef, R., and Schweizer, W. B. (1983), *J. Am. Chem. Soc.*, **105**, 5390.

Seebach, D., Gysel, U., and Kinkel, J. N. (1991), *Chimia*, **45**, 114.

Sellergren, B., Ekberg, B., Albertsson, P. A., and Mosbach, K. (1987), *PCT Int. Appl.* WO 87 00165 (Jan. 15, 1987); *Chem. Abstr.*, **107** 133602x (1987).

Sépulchre, M., Spassky, N., and Sigwalt, P. (1972), *Macromolecules*, **5**, 92.

Seyer, W. F. and Walker, R. D. (1938), *J. Am. Chem. Soc.*, **60**, 2125.

Sharp, T. R., Hegeman, G. D., and Kenyon, G. L. (1977), *Biochemistry*, **16**, 1123.

Sharpless, K. B. (1981), personal communication to SHW.

Sharpless, K. B., Amberg, W., Beller, M., Chen, H., Hartung, J., Kawanami, Y., Lübben, D., Manoury, E., Ogino, Y., Shibata, T., and Ukita, T. (1991), *J. Org. Chem.*, **56**, 4585.

Sharpless, K. B., Amberg, W., Bennani, Y. L., Crispino, G. A., Hartung, J., Jeong, K.-S., Kwong, H.-L., Morikawa, K., Wang, Z.-M., Xu, D., and Zhang, X.-L. (1992), *J. Org. Chem.*, **57**, 2768.

Sharpless, K. B., Chong, A. O., and Scott, J. A. (1975), *J. Org. Chem.*, **40**, 1252.

Sheehan, J. C. and Whitney, J. G. (1963), *J. Am. Chem. Soc.*, **85**, 3863.

Sheldon, R. (1990), "Industrial Synthesis of Optically Active Compounds," *Chem. Ind. (London)*, 212.

Sheldon, R. A. (1993), *Chirotechnology, Industrial Synthesis of Optically Active Compounds*, Marcel Dekker, New York.

Sheldon, R. A., Hulshof, L. A., Bruggink, A., Leusen, F. J. J., van der Haest, A. D., and Wynberg, H. (1990), "Crystallization techniques for the industrial synthesis of pure enantiomers" in *Proceedings of "Chiral 90" Symposium*, University of Manchester, UK, p. 101.

Sheldon, R. A., Porskamp, P. A., and ten Hoeve, W. (1985), "Advantages and Limitations of Chemical Optical Resolution," in Tramper, J., van der Plas, H. C. and Linko, P., Eds., *Biocatalysts in Organic Syntheses*, Elsevier, Amsterdam, The Netherlands, p. 59.

Shibata, S., Matsushita, H., Kaneko, H., Noguchi, M., Saburi, M., and Yoshikawa, S. (1982a), *Chem. Lett.*, 1983.

Shibata, S., Matsushita, H., Kaneko, H., Noguchi, M., Saburi, M., and Yoshikawa, S. (1982b), *Bull. Chem. Soc. Jpn.*, **55**, 3546.

Shibata, S., Matsushita, H., Kaneko, H., Noguchi, M., Sakurai, T., Saburi, M., and Yoshikawa, S. (1984), *Bull. Chem. Soc. Jpn.*, **57**, 3531.

Shimon, L. J. W., Lahav, M., and Leiserowitz, L. (1986), *Nouv. J. Chim.*, **10**, 723.

Shiner, C. S. and Berks, A. H. (1988), *J. Org. Chem.*, **53**, 5542.

Shiraiwa, T., Chatani, T., Matsushita, T., and Kurokawa, H. (1985a), *Technol. Rep. Kansai Univ.*, **26**, 103; *Chem. Abstr.*, **104**, 149365w (1986).

Shiraiwa, T., Ikawa, A., Fujimoto, K., Iwafuji, K., and Kurokawa, H. (1984a), *Nippon Kagaku Kaishi*, 764; *Chem. Abstr.*, **101**, 73070m (1984).

Shiraiwa, T., Kataoka, K., and Kurokawa, H. (1987f), *Chem. Lett.*, 2041.

Shiraiwa, T., Miyazaki, H., Ikawa, A., and Kurokawa, H. (1984b) *Nippon Kagaku Kaishi*, 1425; *Chem. Abstr.*, **102**, 62560r (1985).

Shiraiwa, T., Miyazaki, H., Imai, T., Sunami, M., and Kurokawa, H. (1987a), *Bull. Chem. Soc. Jpn.*, **60**, 661.

Shiraiwa, T., Miyazaki, H., Sakamoto, Y., and Kurokawa, H. (1986a), *Bull. Chem. Soc. Jpn.*, **59**, 2331.

Shiraiwa, T., Miyazaki, H., Uramoto, A., Sunami, M., and Kurokawa, H. (1987b), *Bull. Chem. Soc. Jpn.*, **60**, 1645.

Shiraiwa, T., Morita, M., Iwafuji, K., and Kurokawa, H. (1983), *Nippon Kagaku Kaishi*, 1743; *Chem. Abstr.*, **100**, 175181v (1984).

Shiraiwa, T., Nagata, M., Kataoka, K., Sado, Y., and Kurokawa, H. (1989b), *Nippon Kagaku Kaishi*, 84: *Chem. Abstr.*, **110**, 212673m (1989).

Shiraiwa, T., Nakamura, M., Nobeoka, M., and Kurokawa, H. (1986b), *Bull. Chem. Soc. Jpn.*, **59**, 2669.

Shiraiwa, T., Nakamura, M. Taniguchi, S., and Kurokawa, H. (1985b), *Nippon Kagaku Kaishi*, 43; *Chem. Abstr.*, **103**, 37160y (1985).

Shiraiwa, T., Sado, Y., Fujii, S., Nakamura, M., and Kurokawa, H. (1987e), *Bull. Chem. Soc. Jpn.*, **60**, 824.

Shiraiwa, T., Sakata, S., and Kurokawa, H. (1988), *Chem. Express*, **3**, 415; *Chem. Abstr.*, **110**, 95749c (1989).

Shiraiwa, T., Sado, Y., Komure, M., and Kurokawa, H. (1987d), *Chem. Lett.*, 621.

Shiraiwa, T., Shinjo, K., and Kurokawa, H. (1991), *Bull. Chem. Soc. Jpn.*, **64**, 3251.

Shiraiwa, T., Tazoh. H., Sunami, M., Sado, Y., and Kurokawa, H. (1987c), *Bull. Chem. Soc. Jpn.*, **60**, 3985.

Shiraiwa, T., Yamauchi, M., Kataoka, K., and Kurokawa, H. (1989a), *Nippon Kagaku Kaishi*, 1161; *Chem. Abstr.*, **112**, 55161q (1990).

Shiraiwa, T., Yamauchi, M., Tatsumi, T., and Kurokawa, H. (1992), *Bull. Chem. Soc. Jpn.*, **65**, 267.

Shiraiwa, T., Yoshida, H., Mashima, K., and Kurokawa, H. (1986c), *Nippon Kagaku Kaishi*, (2) 177; *Chem. Abstr.*, **106**, 196744v (1987).

Si, Y., Zhou, J., and Huang, L. (1987), *Sci. Sin.*, *Ser. B* (*Engl. Ed.*), **30**, 297; *Chem. Abstr.*, **108**, 5775q (1988).

Sifniades, S., Boyle, Jr., W. J., and Van Peppen, F. (1976), *U.S. Patent* 3 988 320, Oct. 26, 1976 (to Allied Chemical Corp.); *Chem. Abstr.*, **86**, 43216t (1976).

Sifniades, S. (1992), "L-Lysine via Asymmetric Transformation of α-Amino-ε-caprolactam," in Collins, A. N., Sheldrake, G.N., and Crosby, J., Eds., *Chirality in Industry: The Commercial Manufacture and Applications of Optically Active Compounds*, Wiley, Chichester, UK, Chap. 3.

Sih, C. J. and Wu, S.-H. (1989), "Resolution of Enantiomers via Biocatalysis," *Top. Stereochem.*, **19**, 63.

Simon, K., Kozsda, E., Böcskei, Z., Faigl, F., Fogassy, E., and Reck, G. (1990), *J. Chem. Soc. Perkin Trans. 2*, 1395.

Singler, R. E. and Cram, D. J. (1972), *J. Am. Chem. Soc.*, **94**, 3512.

Skell, P. S. and Pavlis, R. R. (1983), *J. Org. Chem.*, **48**, 1901.

Skopan, H., Günther, H., and Simon, H. (1987), *Angew. Chem. Int. Ed. Engl.*, **26**, 128.

Smith III, A. B. and Konopelski, J. P. (1984), *J. Org. Chem.*, **49**, 4094.

Smith III, A. B., Rano, T. A., Chida, N., Sulikowski, G. A., and Wood, J. L. (1992), *J. Am. Chem. Soc.*, **114**, 8008.

Smith, G. G. and Baum, R. (1987), *J. Org. Chem.*, **52**, 2248.

Smith, G. G., Evans, R. C., and Baum, R. (1986), *J. Am. Chem. Soc.*, **108**, 7327.

Smith, G. G., Khatib, A., and Reddy, G. S. (1983a), *J. Am. Chem. Soc.*, **105**, 293.

Smith, G. G. and Sivakua, T. (1983), *J. Org. Chem.*, **48**, 627.

Smith, H. E., Burrows, E. P., Marks, M. J., Lynch, R. D., and Chen, F.-M. (1977), *J. Am. Chem. Soc.*, **99**, 707.

Smith, H. E., Neergaard, J. R., de Paulis, T., and Chen, F.-M. (1983b), *J. Am. Chem. Soc.*, **105**, 1578.

Smrčina, M., Lorenc, M., Hanuš, V., Sedmera, P., and Kočovský, P. (1992), *J. Org. Chem.*, **57**, 1917.

Smrčina, M., Poláková, J., Vyskočil, S., and Kočovský, P. (1993), *J. Org. Chem.*, **58**, 4534.

Snatzke, G., and Meese, C. O. (1987), *Liebigs Ann. Chem.*, 81.

Snatzke, G., Wynberg, H., Feringa, B., Marsman, B. G., Greydanus, B., and Pluim, H. (1980), *J. Org. Chem.*, **45**, 4094.

Soffer, M. D. and Günay, G. E. (1965), *Tetrahedron Lett.*, 1355.

Sogah, G. D. Y. and Cram, D. J. (1979), *J. Am. Chem. Soc.*, **101**, 3035.

Sonnet, P. E. (1982), *J. Org. Chem.*, **47**, 3793.

Sonnet, P. E., McGovern, T. P., and Cunningham, R. T. (1984), *J. Org. Chem.* **49**, 4639.

Sousa, L. R., Sogah, G. D. Y., Hoffman, D. H., and Cram. D. J. (1978), *J. Am. Chem. Soc.*, **100**, 4569.

Souter, R. W. (1985), *Chromatographic Separations of Stereoisomers*, CRC, Boca Raton, FL.

Spassky, N., Leborgne, A., and Sépulchre, M. (1981), *Pure Appl. Chem.*, **53**, 1735.

Still, W. C., Hauck, P., and Kempf, D. (1987), *Tetrahedron Lett.*, **28**, 2817.

Still, W. C., Kahn, M., and Mitra, A. (1978), *J. Org. Chem.*, **43**, 2923.

Stinson, S. C. (1992), "Chiral Drugs," *Chem. Eng. News*, **70**, Sept. 28, 1992, p. 46.

Stinson, S. C. (1993), "Chiral Drugs," *Chem. Eng. News*, **71**, Sept. 27, 1993, p. 38.

Stoddart, J. F. (1979), "From Carbohydrates to Enzyme Analogues," *Chem. Soc. Rev.*, **8**, 85.

Stoddart, J. F. (1987), "Chiral Crown Ethers," *Top. Stereochem*, **17**, 207.

Stokes, T. M. and Oehlschlager, A. C. (1987), *Tetrahedron Lett.*, **28**, 2091.

Suda, H., Motoi, M., Fujii, M., Kanoh, S., and Yoshida, H. (1979), *Tetrahedron Lett.*, 4565.

Sutliff, T. (1966), M. S. Thesis, Ohio State University; cited in Newman, M.S., *An Advanced Organic Laboratory Course*, Macmillan, New York, 1972, p. 41.

Suzukamo, G. and Fukao, M. (1982), *Eur. Patent App.* EP 61 880, Oct. 6, 1982; *Chem. Abstr.*, 1983, **98**, 126416g.

Suzuki, M., Kawagishi, T., Suzuki, T., and Noyori, R. (1982), *Tetrahedron Lett.*, **23**, 4057.

Švedas, V. and Galaev, I. U. (1983), "Enzymic Conversion of Racemates into Aminoacid Enantiomers," *Usp. Khim.*, **52**, 2039; *Russ. Chem. Rev.*, **52**, 1184.

Sykes, G. A. (1988), "Amino Acids on Ice," *Chem. Br.*, **24**, 235.

Szepesi, G. (1989), "Ion-pairing," in Lough, W. J., Ed., *Chiral Liquid Chromatography*, Blackie, Glasgow, UK, p. 185.

Takano, S., Kijima, A., Sugihara, T., Satoh, S., and Ogasawara, K. (1989), *Chem. Lett.*, 87.

Tam, W., Eaton, D.F., Calabrese, J. C., Williams, I. D., Wang, Y., and Anderson, A. G. (1989), *Chem. Mat.*, **1**, 128.

Tamoto, K., Sugimoro, M., and Terashima, S. (1984), *Tetrahedron*, **40**, 4617.

Tanaka, K. and Toda, F. (1983), *J. Chem. Soc. Chem. Commun.*, 1513.

Tanaka, K. and Toda, F. (1987), *Nippon Kagaku Kaishi*, (3) 456; *Chem. Abstr.*, **107**, 197525g (1987).

Tani, K., Yamagata, T., Akutagawa, S., Kumobayashi, H., Taketomi, T., Takaya, H., Miyashita, A., Noyori, R., and Otsuka, S. (1984), *J. Am. Chem. Soc.*, **106**, 5208.

Tanner, D. D., Blackburn, E. V., Kosugi, Y., and Ruo, T. C. S. (1977), *J. Am. Chem. Soc.*, **99**, 2714.

Tanner, D. D., Ruo, T. C. S., and Meintzer, C. P. (1985), *J. Org. Chem.*, **50**, 2573.

Tashiro, Y. and Aoki, S. (1985), *Eur. Patent. Appl.*, EP 133 053, Feb. 13, 1985; *Chem. Abstr.*, **103**, 37734p (1985).

Taylor, D. R. (1989), "Future trends and requirements," in Lough, W. J., Ed., *Chiral Liquid Chromatography*, Blackie, Glasgow, UK, p. 287.

ten Hoeve, W. and Wynberg, H. (1985), *J. Org. Chem.*, **50**, 4508.

Terashima, S., Tamoto, K., and Sugimori, M. (1982), *Tetrahedron Lett.*, **23**, 4107.

Tessier, J. R. (1982), in Lhoste, J., Ed., *Deltamethrin*, Roussel-Uclaf, Paris, France, Chap. 2.

Tétreau, C., Lavalette, D., Cabaret, D., Geraghty, N., and Welvart, Z. (1982), *Nouv. J. Chim.*, **6**, 461.

Thoma, A. P., Cimerman, Z., Fiedler, U., Bedeković, D., Güggi, M., Jordan, P., May, K., Pretsch, E., Prelog, V., and Simon, W. (1975), *Chimia*, **29**, 344.

Thoma, A. P., Pretsch, E., Horvai, G., and Simon, W. (1977), in Semenza, G. and Cafaroli, E., Eds., *Biochemistry of Membrane Transport*, FEBS Symposium No. 42, Springer, Berlin, p. 116.

Thoma, A. P., Viviani-Nauer, A., Schellenberg, K. H., Bedeković, D., Pretsch, E., Prelog, V., and Simon, W. (1979), *Helv. Chim. Acta.*, **62**, 2303.

Thomas, J. M. and Harris, K. D. M. (1987), "Organic Molecules in Constrained Environments," in Desiraju, G. R., Ed., *Organic Solid State Chemistry*, Elsevier, Amsterdam, The Netherlands, p. 179.

Tietze, L.-F. and Eicher, T. (1989), *Reactions and Syntheses in the Organic Chemistry Laboratory*, University Science Books, Mill Valley, CA, p. 412.

Toda, F. (1986), *Jpn. Kokai Tokkyo Koho* JP 61 268 633 [86, 268,633], Nov. 28, 1986; *Chem. Abstr.*, **106**, 195559b (1987).

Toda, F. (1987), "Isolation and Optical Resolution of Materials Utilizing Inclusion Crystallization," *Top. Curr. Chem.*, **140**, 43.

Toda, F., Mori, K., Stein, Z., and Goldberg, I. (1988), *J. Org. Chem.*, **53**, 308.

Toda, F., Mori, K., Stein, Z., and Goldberg, I. (1989), *Tetrahedron Lett.*, **30**, 1841.

Toda, F. and Tanaka, K. (1986), *Chem. Lett.*, 1905.

Toda, F. and Tanaka, K. (1988a), *Tetrahedron Lett.*, **29**, 551.

Toda, F. and Tanaka, K. (1988b), *J. Org. Chem.*, **53**, 3607.

Toda, F., Tanaka, K., and Mak, T. C. W. (1984a) *Chem. Lett.*, 2085.

Toda, F., Tanaka, K., and Nagamatsu, S. (1984b), *Tetrahedron Lett.*, **25**, 4929.

Toda, F., Tanaka, K., Omata, T., Nakamura, K., and Ōshima, T. (1983c), *J. Am. Chem. Soc.*, **105**, 5151.

Toda, F., Tanaka, K., Ootani, M., Hayashi, A., Miyahara, I., and Hirotsu, K. (1993), *J. Chem. Soc. Chem. Commun.*, 1413.

Toda, F., Tanaka, K., and Ueda, H. (1981), *Tetrahedron Lett.*, **22**, 4669.

Toda, F., and Tanaka, K., (1983a), *Chem. Lett.*, 661.

Toda, F., Tanaka, K., Ueda, H., and Ōshima, T. (1983b), *J. Chem. Soc. Chem. Commun.*, 743.

Toda, F., Ward, D. L., and Hart, H. (1981), *Tetrahedron Lett.*, **22**, 3865.

Tökés, A. L. (1987), *Liebigs Ann. Chem.*, 1007.

Tokura, N. and Akiyama, F. (1966), *Bull. Chem. Soc. Jpn.*, **39**, 838.

Tomuro, K., Tamura, Y., Morimoto, Y., and Katoh, S. (1987), *Eur. Patent Appl.*, EP 213 785, Mar. 11, 1987; *Chem. Abstr.*, **108**, 6421h (1988).

Top, S., Jaouen, G., Gillois, J., Baldoli, C., and Maiorana, S. (1988), *J. Chem. Soc. Chem. Commun.*, 1284.

Tosa, T., Mori, T., and Chibata, I. (1969b), *Agric. Biol. Chem.*, **33**, 1053.

Tosa, T., Mori, T., Fuse, N., and Chibata, I. (1969a), *Agric, Biol. Chem.*, **33**, 1047.

Tramontano, A., Janda, K. D., and Lerner, R. A. (1986), *Science*, **234**, 1566.

Trost, B. M (1982), "Approaches for Asymmetric Synthesis as Directed Toward Natural Products," in Eliel, E. L. and Otsuka, S., Eds., *Asymmetric Reactions and Processes in Chemistry*, ACS Symposium Series 185, American Chemical Society, Washington, DC, p. 3.

Trost, B. M., Belletire, J. L., Godleski, S., McDougal, P. G., Balkovec, J. M., Baldwin, J. J., Christy, M. E., Ponticello, G. S., Varga, S. L., and Springer, J. P. (1986), *J. Org. Chem.*, **51**, 2370.

Trost, B. M. and Hammen, R. F. (1973), *J. Am. Chem. Soc.*, **95**, 962.

Tsoucaris, G. (1987), "Clathrates," in Desiraju, G. R., Ed., *Organic Solid State Chemistry*, Elsevier, Amsterdam, The Netherlands, Chap. 7.

Tung, H. H., Waterson, S., and Reynolds, S. D. (1991), *U.S. Patent* 4 994 604 (to Merck & Co.,

Inc.), Feb. 19, 1991; *Chem. Abstr.*, 1991, **115**, 15586n; see also in Ramanarayanan, R. et al., Eds., *Particle Design via Crystallization*, A. I. Ch. E. Symposium Series, No. 284, Vol. 87, American Institute of Chemical Engineers, New York, 1991.

Tyson, R. (1988), *Chem. Ind. (London)*, 118.

Uneyama, K., Makio, S., and Nanbu, H. (1989), *J. Org. Chem.*, **54**, 872.

van Bekkum, H., van De Graaf, B., van Minnen-Pathius, G., Peters, J. A., and Wepster, B. M. (1970), *Recl. Trav. Chim. Pays-Bas*, **89**, 521.

van der Haest, A. D. (1992), Ph.D. Dissertation, University of Groningen, The Netherlands.

van der Haest, A. D., Wynberg, H., Leusen, F. J. J., and Bruggink, A. (1990), *Recl. Trav. Chim. Pays-Bas*, **109**, 523.

van Lier, P. M., Meulendijks, G. H. W. M., and Buck, H. M. (1983), *Recl. Trav. Chim. Pays-Bas*, **102**, 337.

van Mil, J., Addadi, L., Lahav, M., Boyle, Jr., W. J., and Sifniades, S., (1987), *Tetrahedron*, **43**, 1281.

van't Hoff, J. H. (1894), *Die Lagerung der Atome im Raume*, 2nd ed., Vieweg, Germany, p. 30.

van't Hoff, J. H. and Dawson, H. M. (1898), *Ber. Dtsch. Chem. Ges.*, **31**, 528.

van't Hoff, J. H., Goldschmidt, H., and Tomssen, W. P. (1895), *Z. Phys Chem.*, **17**, 49.

Velluz, L., Amiard, G., and Joly, R. (1953), *Bull. Soc. Chim. Fr.*, 342.

Velluz, L. and Amiard, G. (1953), *Bull. Soc. Chim. Fr.*, 903.

Vercesi, D. and Azzolina, O. (1985), *Farmaco, Ed. Prat.*, **40**, 396; *Chem. Abstr.*, **105**, 114681q (1986).

Verkhovskaya, M. A. and Yamskov, I. A. (1991), "Enzymic methods of resolving racemates of amino acids and their derivatives," *Usp. Khim.*, 2250; *Russ. Chem. Rev.*, **60**, 1163.

Vièles, P. (1934), *C.R. Hebd. Séances Acad. Sci.*, **198**, 2102.

Vièles, P. (1935), *Ann. Chim. (Paris)*, **3**, 143.

Vigneron, J. P., Dhaenens, M., and Horeau, A. (1973), *Tetrahedron*, **29**, 1055.

Viret, J., Patzelt, H., and Collet, A. (1986), *Tetrahedron Lett.*, **27**, 5865.

Vriesema, B. K., ten Hoeve, W., Wynberg, H., Kellogg, R. M., Boesten, W. H. J., Meijer, E. M., and Schoemaker, H. E. (1986), *Tetrahedron Lett.*, **27**, 2045.

Wakamatsu, H. (1968), "MSG by Synthesis," *Food Eng.*, **40**, (11) 92.

Walker, D. C., Ed. (1979), *Origins of Optical Activity in Nature*, Elsevier, Amsterdam, The Netherlands.

Wallis, C. J. (1983), *Eur. Patent App.* EP 74 856, Mar. 23, 1983; *Chem. Abstr.*, 1983, **99**, 139627x.

Wang, Y.-F., Lalonde, J. J., Momongan, M., Bergbreiter, D. E., and Wong, C.-H. (1988), *J. Am. Chem. Soc.*, **110**, 7200.

Wani, M. C., Nicholas, A. W., and Wall, M. E. (1987), *J. Med. Chem.*, **30**, 2317.

Warnant, J., Prost-Marechal, J., and Cosquer, P. (1977), *Ger. Offen.*, 2 718 039, Nov. 17, 1977; *Chem. Abstr.*, **88**, 104929t (1978).

Wasson, B. K., Gibson, W. K., Stuart, R. S., Williams, H. W. R., and Yates, C. H. (1972), *J. Med. Chem.*, **15**, 651.

Watase, H., Ohne, Y., Nakamura, T., Okada, T., and Takeshita, T. (1975), *U.S. Patent* 3 879 382; see *Chem. Abstr.*, **81**, 13146z, (1974).

Weber, E., Ed. (1987), *Top. Curr. Chem.*, **140**, (Molecular Inclusion and Molecular Recognition-Clathrates I), Springer, Berlin.

Weber, E., Wimmer, C., Llamas-Saiz, A. L., and Foces-Foces, C. (1992a), *J. Chem. Soc. Chem. Commun.*, 733.

Weber, L., Imiolczyk, I., Haufe, G., Rehorek, D., and Hennig, H. (1992b), *J. Chem. Soc. Chem. Commun.*, 301.

Webster, F. X., Millar, J. G., and Silverstein, R. M. (1986), *Tetrahedron Lett.*, **27**, 4941.

Webster, F. X., Zeng, X.-n., and Silverstein, R. M. (1982), *J. Org. Chem.*, **47**, 5225.

Wegener, B., Hansen, M., and Winterfeldt, E. (1993), *Tetrahedron: Asymmetry*, **4**, 345.

Wehmiller, J. and Hare, P. E. (1971), *Science*, **173**, 907.

Weidert, P. J., Geyer, E., and Horner, L. (1989), *Liebigs Ann. Chem.*, 533.

Weissbuch, I., Addadi, L., Leiserowitz, L., and Lahav, M. (1988), *J. Am. Chem. Soc.*, 110, 561.

Weissbuch, I., Zbaida, D., Addadi, L., Leiserowitz, L., and Lahav, M. (1987), *J. Am. Chem. Soc.*, 109, 1869.

Westheimer, F. H. (1956), "Calculation of the Magnitude of Steric Effects," in Newman, M. S., Ed., *Steric Effects in Organic Chemistry*, Wiley, New York, Chap. 12.

Wheland, G. W. (1960), *Advanced Organic Chemistry*, 3rd. ed., Wiley, New York.

Whitesell, J. K. and Carpenter, J. F. (1987), *J. Am. Chem. Soc.*, 109, 2839.

Whitesell, J. K., Liu, C.-L., Buchanan, C. M., Chen, H.-H., and Minton, M. A. (1986), *J. Org. Chem.*, 51, 551.

Whitesell, J. K., Minton, M. A., and Chen, K.-M. (1988), *J. Org. Chem.*, 53, 5383.

Whitesell, J. K., Minton, M. A., and Felman, S. W. (1983), *J. Org. Chem.*, 48, 2193.

Whitesell, J. K. and Reynolds, D. (1983), *J. Org. Chem.*, 48, 3548.

Whitesides, G. M. and Wong, C.-H. (1985), "Enzymes as Catalysts in Synthetic Organic Chemistry," *Angew. Chem. Int. Ed. Engl.*, 24, 617.

Wiesner, K., Jay, E. W. K., Tsai, T. Y. R., Demerson, C., Jay, L., Kanno, T., Krepinsky, J., Vilím, A., and Wu., C. S. (1972), *Can. J. Chem.*, 50, 1925.

Wilen, S. H. (1971), "Resolving Agents and Resolutions in Organic Chemistry," *Top. Stereochem.*, 6, 107.

Wilen, S. H. (1972), *Tables of Resolving Agents and Optical Resolutions*, Eliel, E. L., Ed., University of Notre Dame Press, Notre Dame, Indiana.

Wilen, S. H., Bunding, K. A., Kascheres, C. M., and Wieder, M. J. (1985), *J. Am. Chem. Soc.*, 107, 6997.

Wilen, S. H., Collet, A., and Jacques, J., (1977), "Strategies in Optical Resolutions," *Tetrahedron*, 33, 2725.

Wilen, S. H. and J. Z. Qi (1991), 4th Chemical Congress of North America and 202nd ACS National Meeting, New York, NY, August 25, 1991, Abstract ORGN 41.

Wilen, S. H., Qi, J.-Z., and Williard, P. G. (1991), *J. Org. Chem.*, 56, 485.

Williams, R. M. and Hendrix, J. A. (1992), "Asymmetric Synthesis of Arylglycines," *Chem. Rev.*, 92, 889.

Williams, K. M. and Smith, G. G. (1977), "A Critical Evaluation of the Application of Amino Acid Racemization to Geochronology and Geothermometry, *Origins of Life*," 8, 91.

Wilson, K. R. and Pincock, R. E. (1975), *J. Am. Chem. Soc.*, 97, 1474.

Wittig, G. and Rümpler, K.-D. (1971), *Justus Liebigs Ann. Chem.*, 751, 1.

Wong, C.-H. (1989), "Enzymatic Catalysts in Organic Synthesis," *Science*, 244, 1145.

Wong, C.-H. and Wang, K.-T. (1978), *Tetrahedron Lett.*, 3813.

Wood, J. L., Liverton, N. J., Visnick, M., and Smith, III, A. B. (1989), *J. Am. Chem. Soc.*, 111, 4530.

Woodward, R. B., Cava, M. P., Ollis, W. D., Hunger, A., Daeniker, H. V., and Schenker, K. (1963), *Tetrahedron*, 19, 247; see, in particular, the footnotes on p. 259.

Worsch, D. and Vögtle, F. (1987), "Separation of Enantiomers by Clathrate Formation," *Top. Curr. Chem.*, 140, 21.

Wu, S.-H., Guo, Z.-W., and Sih, C. J. (1990), *J. Am. Chem. Soc.*, 112, 1990.

Wynberg, H. and Groen, M. B. (1968), *J. Am. Chem. Soc.*, 90, 5339.

Wynberg, H. and Lammertsma, K. (1973), *J. Am. Chem. Soc.*, 95, 7913.

Xie, Z.-F., Suemune, H., and Sakai, K. (1987), *J. Chem. Soc. Chem. Commun.*, 838.

Yamada, K.-i., Nakagawa, H., and Kawazura, H. (1986), *Bull. Chem. Soc. Jpn.*, 59, 2429.

Yamada, M. and Okawa, K. (1985), *Bull. Chem. Soc. Jpn.*, 58, 2889.

Yamada, S., Hongo, C., Yoshioka, R., and Chibata, I. (1983), *J. Org. Chem.*, 48, 843.

Yamada, S., Yamamoto, M., and Chibata, I. (1973a), *J. Agr. Food Chem.*, 21, 889.

Yamada, S., Yamamoto, M., and Chibata, I. (1973b), *J. Org. Chem.*, 38, 4408.

Yamamoto, K., Kitsuki, T., and Okamoto, Y. (1986), *Bull. Chem. Soc. Jpn.*, **59**, 1269.

Yamamoto, K., Noda, K., and Okamoto, Y. (1985), *J. Chem. Soc. Chem. Commun.*, 1065.

Yasui, M, Yabuki, T., Takama, M., Harada, S., Kasai, N., Tanaka, K., and Toda, F. (1989), *Bull. Chem. Soc. Jpn.*, **62**, 1436.

Yodo, M., Matsushita, Y., Ohsugi, E., and Harada, H. (1988), *Chem. Pharm. Bull.*, **36**, 902.

Yokozeki, K., Nakamori, S., Eguchi, C., Yamada, K., and Mitsugi, K. (1987), *Agric. Biol. Chem.*, **51**, 355.

Yoshioka, R., Tohyama, M., Ohtsuki, O., Yamada, S., and Chibata, I. (1987a), *Bull. Chem. Soc. Jpn.*, **60**, 649.

Yoshioka, R., Tohyama, M., Yamada, S., Ohtsuki, O., and Chibata, I. (1987b), *Bull. Chem. Soc. Jpn.*, **60**, 4321.

Young, J. W. and Bratzler, R. L. (1990), "Membrane Reactors for the Biocatalytic Production of Chiral Compounds," in *Proceedings of "Chiral 90" Symposium*, University of Manchester, UK, p. 23.

Yuki, H., Okamoto, Y., and Okamoto, I. (1980), *J. Am. Chem. Soc.*, **102**, 6356.

Zaks, A. and Klibanov, A. M. (1984), *Science*, **224**, 1249.

Zandomeneghi, M., Cavazza, M., Festa, C., and Pietra, F. (1983), *J. Am. Chem. Soc.*, **105**, 1839.

Zawadzke, L. E. and Berg, J. M. (1992), *J. Am. Chem. Soc.*, **114**, 4002.

Zeegers, H. J. M. (1989), Ph.D. Dissertation, Catholic University of Nijmegen, the Netherlands, p. 40.

Zief, M. (1988), "Preparative Enantiomeric Separation," in Zief, M. and Crane, L. J., Eds., *Chromatographic Chiral Separations*, Marcel Dekker, New York, Chap. 13.

Zingg, S. P., Arnett, E. M., McPhail, A., Bothner-By, A. A., and Gilkerson, W. R. (1988), *J. Am. Chem. Soc.*, **110**, 1565.

8

Heterotopic Ligands and Faces (Prostereoisomerism, Prochirality)

8-1. INTRODUCTION. TERMINOLOGY

Often (e.g., in stereoselective synthesis) one is interested in the fact that in certain molecules, such as propionic acid (Fig. 8.1, **1**), a nonstereogenic center (here C_α) can be transformed into a stereogenic center by replacement of one or other of two apparently identical ligands by a different one. Such ligands are called "homomorphic" from Greek *homos* meaning same and *morphe* meaning form (Hirschmann and Hanson, 1974); they are identical only when separated from the rest of the molecule. Thus the replacement of H_A at C_α in propionic acid by OH generates the chiral center of (*S*)-lactic acid (Fig. 8.1, **2**), whereas the analogous replacement of H_B gives rise to the enantiomeric (*R*)-lactic acid. The C_α center in propionic acid has therefore been called a "prochiral center" (Hanson, 1966; Mislow and Siegel, 1984); H_A and H_B at such a center are called "heterotopic ligands" (Mislow and Raban, 1967; Hirschmann and Hanson, 1971a,b; cf. Eliel, 1980, 1982) from Greek *heteros* meaning different and *topos* meaning place (see also Section 8-3). Prochiral axes and planes may similarly be defined in relation to chiral axes and planes (see below).

Substitution is one of the common ways of interconverting organic molecules, another is addition. The chiral center in lactic acid (Fig. 8.1, **2**) can also be generated by the addition of hydride (e.g., from sodium borohydride) to the carbonyl group of pyruvic acid (Fig. 8.1, **3**). Depending on the face of the keto acid the hydride adds to, either (*S*)- or (*R*)-lactic acid is obtained. (Readers should convince themselves that addition to the rear face of the keto acid as depicted in Fig. 8.1 will give rise to (*S*)-lactic acid **2**, whereas (*R*)-lactic acid is obtained by addition to the front face.) Thus the carbonyl group in pyruvic acid is also said to be prochiral and to present two heterotopic faces.

Although the term prochirality is properly used in relation to prochiral centers, faces, axes, and so on (cf. Bentley, 1969), it suffers from a limitation that arises from a corresponding limitation in the definition of chirality. We have already seen that molecules, such as cis–trans isomers of alkenes and certain

465

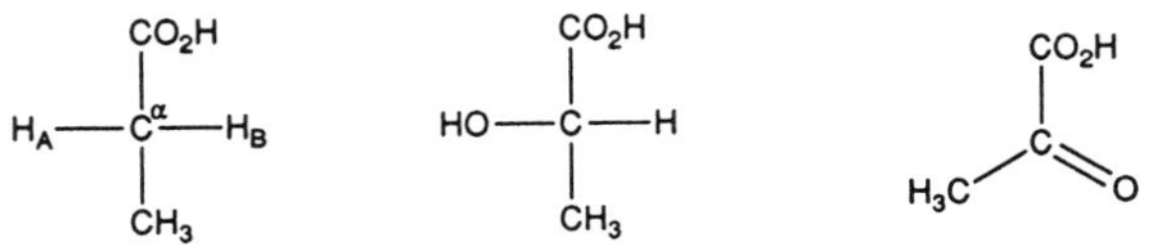

Figure 8.1. Chiral and prochiral molecules.

cis–trans isomers of cyclanes, may display stereochemical differences without being chiral. Thus (Z)- and (E)-1,2-dichloroethylene (**4, 5**) are achiral diastereomers, as are cis- and trans-1,3-dibromocyclobutanes (**6, 7**) (Fig. 8.2); being devoid of chirality these compounds have no chiral centers (or other chiral elements). Thus, just as it is inappropriate to associate stereoisomerism only with the occurrence of chiral elements (cf. Chapter 3) the concept of prochirality needs to be generalized to one of prostereoisomerism (Hirschmann and Hanson, 1974). It is exemplified by chloroethylene (Fig. 8.2, **8**) and bromocyclobutane (**9**); these molecules display prostereoisomerism inasmuch as replacement of the homomorphic atoms H_A and H_B in **8** by chlorine gives rise to stereogenic centers in the achiral stereoisomers **5** and **4**, respectively. Similarly, replacement of H_A and H_B, respectively, in **9** by bromine gives rise to the stereoisomers **6** and **7**. Thus **9** has a prostereogenic (but not prochiral) center (center of prostereoisomerism) at C(3) and **8** may be said to have a prostereogenic axis (axis of prostereoisomerism) coinciding with the axis of the double bond. The atoms H_A and H_B in both **8** and **9** are heterotopic.

Cases of a prochiral axis (in allene **10**, convertible by replacement of H_A by Cl into chiral allene **11**) and a prochiral plane (in paracyclophane **12** which can be converted into the chiral structure **13** by replacement of H_A by CO_2H) are shown in Figure 8.3 (see also Chapter 14).

Some authors (e.g., Mislow and Siegel, 1984) use the terms "prochiral" (or prostereogenic) as referring to a molecule as a whole, that is, to an achiral molecule that becomes chiral by substitution of one of a pair of heterotopic ligands or by addition to one or other of a pair of heterotopic faces. This definition avoids the factorization into prochiral centers, axes, planes, and so on (cf. Chapter 5 for the corresponding factorization of chirality) but it has the disadvantage of trivializing— the concept (most molecules are prochiral). Under this definition, bromo-cyclo-butane **9** is both prostereogenic [with respect to both C(2) and C(3)] and prochiral [with respect to C(2) only], whereas chloroethene **8** is prostereogenic but not prochiral.

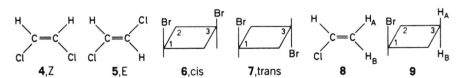

Figure 8.2. Stereogenic and prostereogenic elements.

10 X = H_A
11 X = Cl

12 X = H_A
13 X = CO_2H

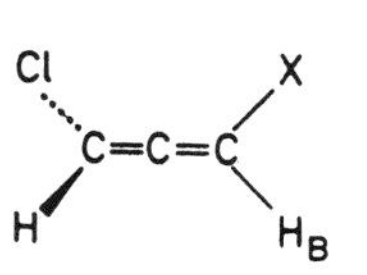

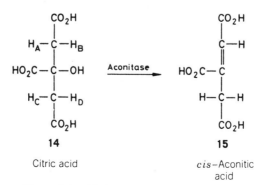

Figure 8.3. Chiral and prochiral axes and planes.

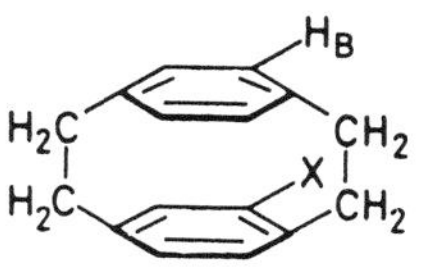

8-2. SIGNIFICANCE. HISTORY

The most significant aspect of the present subject of prostereoisomerism lies in the possibility of differentiating heterotopic ligands or heterotopic faces. The concept of heterotopic ligands and the possibility of their differentiation, in suitable instances, by NMR spectroscopy was first presented in a pioneering article by Mislow and Raban (1967; see also Mislow, 1965). Differentiation of heterotopic ligands or faces may be chemical or biochemical (as in stereoselective synthesis, including transformations by enzymes) or spectroscopic (notably by NMR spectroscopy). Before entering upon these topics in detail, we pose here a challenge to illustrate the utility of the concepts: In citric acid (Fig. 8.4, **14**), can the four methylene hydrogen atoms H_A, H_B, H_C, and H_D be distinguished by NMR spectroscopy, or by virtue of their involvement in the enzymatic dehydration of citric acid to *cis*-aconitic acid **15**, or both? This question can easily be answered once the tenets of prostereoisomerism are understood: all the hydrogen atoms can be distinguished by appropriate enzymatic reactions and H_A and H_B (as well as H_C and H_D) can give rise to distinct signals in the ^1H NMR spectrum, whereas H_A and H_C (or H_B and H_D) give rise to coincident signals, except in chiral media, where all four protons differ in chemical shift (Anet and Park, 1992).

Historically, the first significant observation involving prochirality (though not recognized as such) concerned the decarboxylation of ethylmethylmalonic acid **16** to α-methylbutyric acid **17** in the presence of brucine (Fig. 8.5; Marckwald, 1904). Removal of one or other of the heterotopic (enantiotopic; see below) carboxyl groups of **16** gives rise to one or other of the enantiomers of **17**. In fact, the product (Fig. 8.5) is optically active; indeed, this is one of the first recorded

14 Citric acid

15 *cis*-Aconitic acid

Figure 8.4. Citric acid and *cis*-aconitic acid.

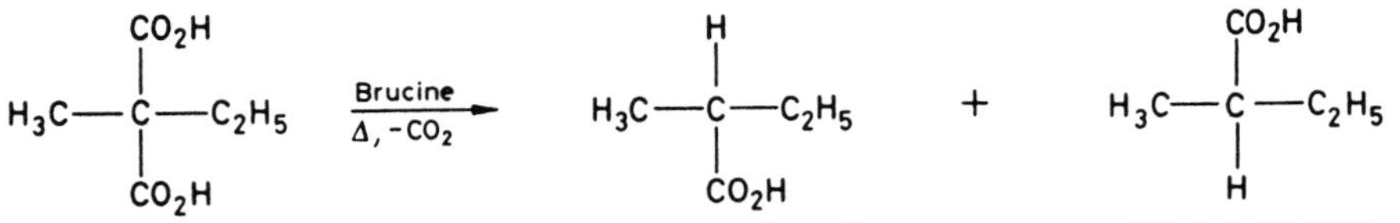

Figure 8.5. Asymmetric decarboxylation of methylethylmalonic acid.

enantioselective (asymmetric) syntheses. A better documented case is that of citric acid (Fig. 8.6, **14**). It was long known (Evans and Slotin, 1941; Wood et al., 1942) that when oxaloacetic acid **18** labeled at C(4) is taken through the Krebs cycle, the α-ketoglutaric acid **19** formed is labeled exclusively at C(1) (next to the keto group), and not at all at C(5). This finding seemed to throw doubt on the theretofore assumed intermediacy of citric acid **14** in the cycle since, it was argued, the two ends of citric acid $-CH_2CO_2H$ are "equivalent" and, therefore, the α-ketoglutaric acid formed through this intermediate should be labeled equally at C(1) and C(5). However, it is now clear (see Sections 8-3 and 8-5) that the experiment in no way eliminates citric acid as a potential intermediate in the oxaloacetic acid–α-ketoglutaric acid transformation, since the two CH_2CO_2H branches are, in fact, distinct and distinguishable by enzymes because they are enantiotopic (see below). Similarly, since phosphorylation of glycerol **20** with adenosine triphosphate (ATP) in the presence of the enzyme glycerokinase gives exclusively (R)-$(-)$-(glycerol 1-phosphate) (Bublitz and Kennedy, 1954; Fig. 8.7, **21**), it is clear that the enzyme can distinguish between the two enantiotopic primary alcohol groups of glycerol.

The first glimpse of understanding of this type of differentiation came when Ogston (1948; see Bentley, 1978 for a historical review) pointed out that an

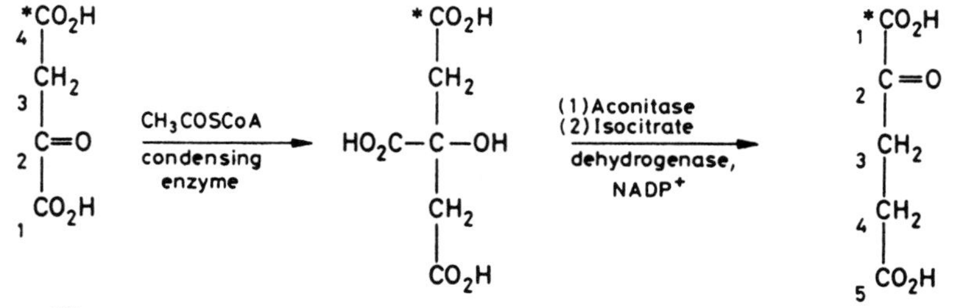

Figure 8.6. Part of citric acid cycle.

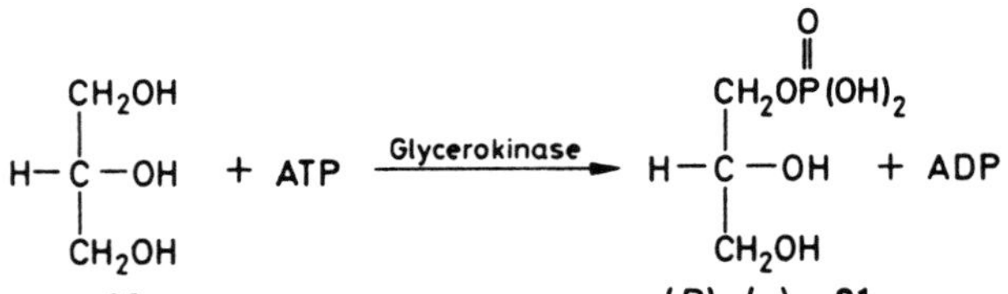

Figure 8.7. Enzymatic phosphorylation of glycerol.

attachment of a substrate Caa'bc (a = a') to an enzyme at three sites (so-called three-point contact) could lead to the observed distinction between the homomorphic (as we would now say) groups a and a', as shown in Figure 8.8. If A is a catalytically active site on the enzyme and B and C are binding sites, Figure 8.8 shows that only a but not a' can be brought into juxtaposition with the active site A when b and c are bound to B and C. Therefore a but not a' may be enzymatically transformed; a and a' are clearly distinguishable.

> The first mention of a three-point contact (between a chiral drug and its receptor) is actually found in an article by Easson and Stedman published in 1933 (q.v.), and a year later, Max Bergmann (1934) postulated a three-point contact (involving CO_2H, H_2N, and the dipeptide linkage) between peptidases and the dipeptides hydrolyzed by them.
>   The "contact" need not be covalent and may be repulsive as well as attractive.
>   For similar considerations in the use of chiral solvating agents for the determination of configuration, see p. 495 and Pirkle and Hoover (1982).

It was subsequently recognized that, whereas Ogston's picture provides a mechanistic rationale for the observed distinction of *apparently* equivalent groups (for a more detailed picture, see Fig. 8.54) his rationale is not unique. Indeed, the distinguishability of the a and a' groups in Caa'bc (a prochiral center), is a consequence of symmetry properties and is independent of any mechanistic principles. This point was probably first recognized by Schwartz and Carter (1954) who called such carbon atoms (Caabc) "meso carbon atoms" (now supplanted by prochiral carbon atoms). Several reviews concerning the nature of prochiral centers (and other prostereogenic elements) are available (Hirschmann, 1964; Mislow and Raban, 1967; Jones, 1976; Eliel, 1982).

At this point, we must pick up one other, at the beginning apparently unrelated, historical thread. In 1957 two groups of investigators (Drysdale and Phillips, 1957; Nair and Roberts, 1957) discovered that in molecules of the type CX_2YCabc [e.g., $CF_2BrCHBrC_6H_5$ or $CH_2BrC(CH_3)BrCO_2CH_3$] the X nuclei (F in the first example, H in the second) display distinct NMR signals. Although the phenomenon was not clearly understood until some time later (Waugh and Cotton, 1961; Gutowsky, 1962), it is now clear that the nonequivalence of such X nuclei in NMR rests on the same symmetry principles (Mislow and Raban, 1967) as the earlier-mentioned nonequivalence in enzymatic reactions and other reactions involving chiral reagents. Sections 8-3 – 8-5 will deal with the explanation of these nonequivalencies and their chemical and spectral consequences.

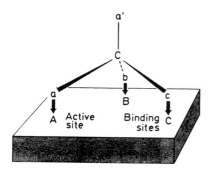

Figure 8.8. Ogston's three-point contact model. From Florkin, M. and Stotz, G., eds. *Comprehensive Biochemistry*, Elsevier, Vol. 12 (1964), p. 237.

8-3. HOMOTOPIC AND HETEROTOPIC LIGANDS AND FACES

Ligands and faces may be homotopic or heterotopic (Hirschmann, 1964; Mislow and Raban, 1967; Eliel, 1980, 1982). Heterotopic ligands and faces may be either enantiotopic or diastereotopic. In the following sections we shall define these terms and explain how the pertinent nature of ligands or faces is recognized. (For a preview of the terminology, see Fig. 8.21.)

a. Homotopic Ligands and Faces

We indicated in Section 8-2 that some apparently alike (homomorphic) ligands are, in fact, not equivalent towards enzymes or in their NMR signals. How does one decide, then, whether such ligands are equivalent (homotopic) or not? There are two alternative criteria: a substitution and a symmetry criterion (Mislow and Raban, 1967). Similar criteria (addition or symmetry) serve to test the equivalency (homotopicity) of faces.

> Unfortunately, the commonly used term "equivalent" is equivocal. Thus the methylene hydrogen atoms in propionic acid (Fig. 8.1) are equivalent when detached (i.e., they are homomorphic) but, as already explained, they are not equivalent in the $CH_3CH_2CO_2H$ molecules because of their placement (i.e., they are heterotopic). Ligands that are equivalent by the criteria to be described below are called "homotopic" from Greek "homos" meaning same and "topos" meaning place (Hirschmann and Hanson, 1971), those that are not are called "heterotopic" (see p. 465).

Substitution and Addition Criteria

Two homorphic ligands (see p. 465) are homotopic if replacement of first one and then the other by a different ligand leads to the same structure. (The replacement ligand must be different not only from the original one but also from all other ligands attached to the same atom.) Thus, as shown in Figure 8.9, the two hydrogen atoms in methylene chloride **22** are homotopic because replacement of either by, say, bromine gives the same $CHBrCl_2$ molecule **23**; the three methyl hydrogen atoms in acetic acid **24** are homotopic because replacement of any one of them by, say, chlorine gives one and the same chloroacetic acid **25**; the two methine hydrogen atoms in (R)-$(+)$-tartaric acid **26** are homotopic because replacement of either of them, for example by deuterium, gives the same $(2R,3R)$-tartaric-2-d acid **27**.

Two corresponding faces of a molecule (usually, but not invariably, faces of a double bond) are homotopic when addition of the same reagent to either face gives the same product. For example, addition of HCN to acetone **28** will give the same cyanohydrin **29**, no matter to which face addition occurs (Fig. 8.10) and addition of bromine to ethylene similarly gives $BrCH_2CH_2Br$ regardless of the face of approach. The two faces of the C=O double bond of acetone and of the C=C double bond of ethylene are thus homotopic.

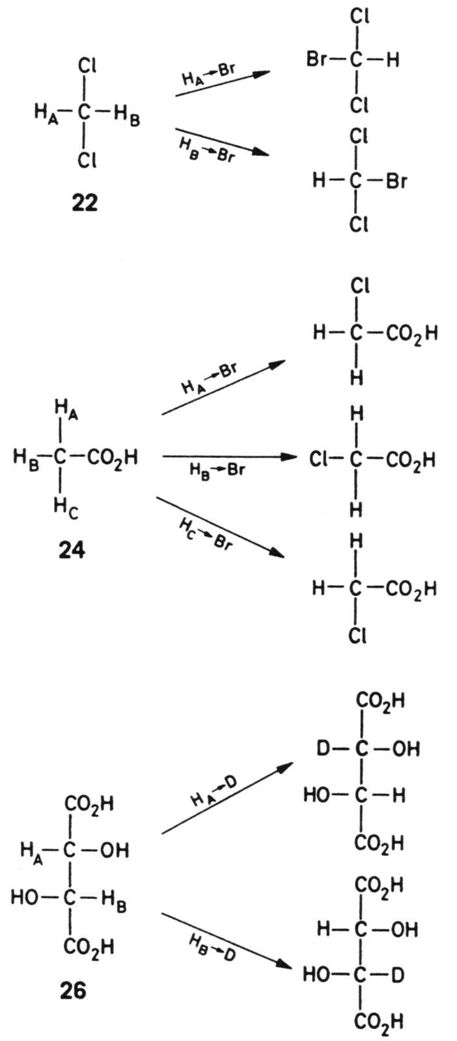

22 → 23, 23 — Superposable (by 180° turn)

24 → 25, 25, 25 — Superposable molecules (by rotation of CH_2Cl groups, see text)

26 → 27, 27 — Superposable (by 180° turn)

Figure 8.9. Homotopic ligands.

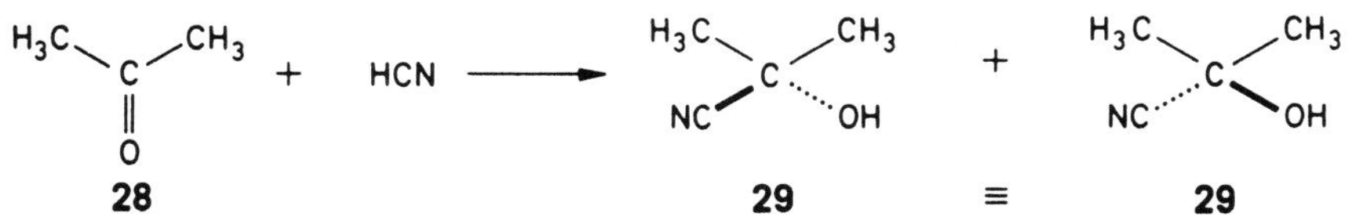

28 29 ≡ 29

Front attack Rear attack

Figure 8.10. Homotopic faces: Addition of HCN to acetone.

Symmetry Criterion

Ligands are homotopic if they can interchange places through operation of a C_n symmetry axis (cf. Chapter 4). Thus the chorine atoms in methylene chloride (symmetry point group $\mathbf{C_{2v}}$) are homotopic since they exchange places through a 180° turn around the C_2 axis (C_2^1). Similarly, the methine hydrogen atoms of (+)-tartaric acid (Fig. 8.9, **26**) are interchanged by operation of the C_2 axis (the molecule belongs to point group $\mathbf{C_2}$). The situation in acetic acid is somewhat more complicated. If we depict this molecule as stationary in one of its eclipsed conformations, we see (Fig. 8.11) that the hydrogen atoms are heterotopic. However, rotation around the $H_3C–CO_2H$ axis is rapid on the time scale of most experiments. We are therefore dealing with a case of averaged symmetry (cf. Section 4-5) leading to interchange of the three methyl hydrogen atoms of CH_3CO_2H, which are thus homotopic when rotation is fast on the time scale of whatever experiment is being considered.

The presence of a symmetry axis in a molecule does not guarantee that homomorphic ligands will be homotopic. It is necessary that operation of the symmetry axis make the nuclei in question interchange places. Thus in 1,3-dioxolane (Fig. 8.12), in its average planar conformation, the hydrogen atoms at C(2) are homotopic, since they are interchanged by operation of the C_2 axis (the symmetry point group of the molecule is $\mathbf{C_{2v}}$). On the other hand, the geminal hydrogen atoms at C(4), or C(5), are not interconverted by the C_2 symmetry operation and are therefore heterotopic (H_A with respect to H_B and H_C with respect to H_D). However, H_A and H_D are homotopic (as are H_B and H_C), being interchanged once again by the C_2 axis.

Faces of double bonds are similarly homotopic when they can be interchanged by operation of a symmetry axis. (Since there are only two such faces, the pertinent axis must, of necessity, be of even multiplicity so as to contain C_2.) Thus the two faces of acetone (Fig. 8.10) are interchanged by the operation of the C_2 axis (the molecule is of symmetry $\mathbf{C_{2v}}$); the two faces of ethylene ($\mathbf{D_{2h}}$) are interchanged by operation of two of the three C_2 axes (either the one containing the C=C segment or the axis at right angles to the first one and in the plane of the double bond).

Acetic acid   1, 3 – Dioxolane

Figure 8.11. Eclipsed conformation of acetic acid.   **Figure 8.12.** 1,3-Dioxolane.

By way of an exercise, the reader may be convinced that the two hydrogen atoms in each of the three dichloroethylenes (1,1-, *cis*-1,2-, *trans*-1,2-) and the four hydrogen atoms in methane (CH_4), allene ($H_2C=C=CH_2$), and ethylene ($H_2C=CH_2$) are homotopic. It might be noted that, in a rigid molecule, the number of homotopic ligands in a set cannot be greater than the symmetry number (Section 4-6) of the molecule in question. Thus the four hydrogen atoms H_{A-D} in 1,3-dioxolane (Fig. 8.12) cannot possibly all be homotopic, since the symmetry number of the molecule is only 2. Similarly, rigid molecules in the nonaxial point groups C_1, C_s, and C_i cannot display homotopic ligands because $\sigma = 1$ for these groups; the same is true of $C_{\infty v}$. This limitation does not apply to cases of averaged symmetry as in acetic acid, which has homotopic methyl hydrogen atoms even though its nonaveraged symmetry point group is C_s.

Returning to the above-mentioned unsaturated compounds, the reader might also note that the faces in four of the five are homotopic, the exception being *trans*-1,2-dichloroethylene, which has heterotopic faces.

The addition criterion tends to be confusing when applied to a molecule like ethylene where addition occurs at both ends of the double bond. The reader is advised, in such cases, either to use the symmetry criterion or to choose epoxidation as the test reaction for the addition criterion. For additional examples involving the heterotopic faces of not only alkenes and carbonyl compounds but also species of the oxime or hydrazone type ($RR'C=NX$), trigonal planar species, such as carbonium ions ($RR'R''C^+$), and even bent disubstituted atoms bearing unshared electron pairs, such as sulfides ($RR'S:$) and related compounds, see Kaloustian and Kaloustian (1975). In this reference, a double bond of an alkene is considered as a single entity, whereas other authors consider each end of the double bond separately. The two views lead to different designations, for example, in *cis*-2-butene where the two faces of the double bond are homotopic overall, but the two faces of each $CH_3CH=$ moiety taken separately are heterotopic. This point will be discussed further below.

b. Enantiotopic Ligands and Faces

Just as one divides stereoisomers into two sets, enantiomers and diastereomers, it is convenient to divide heterotopic (nonequivalent) ligands or faces into enantiotopic and diastereotopic moieties (cf. Fig. 8.21). Enantiotopic ligands are ligands in mirror-image positions, whereas diastereotopic ligands are in stereochemically distinct positions not related in mirror-image fashion; similar considerations relate to faces of double bonds.

Substitution–Addition Criterion

Two ligands are enantiotopic if replacement of either one of them by a different achiral ligand, which is also different from other ligands to the prochiral element, gives rise to enantiomeric products. Examples are shown in Figure 8.13. The marked hydrogen atoms (H_A, H_B) in CH_2ClBr (**30**), *meso*-tartaric acid (**32**), cyclobutanone (**34**) [at C(2) and C(4) but not at C(3)], and chloroallene (**36**) [at C(3)] are enantiotopic as are the methyl carbon atoms in isopropyl alcohol (**38**). *meso*-Tartaric acid exemplifies one of the rare instances of a molecule with heterotopic ligands but no discernible prochiral atom or other element of prochirality.

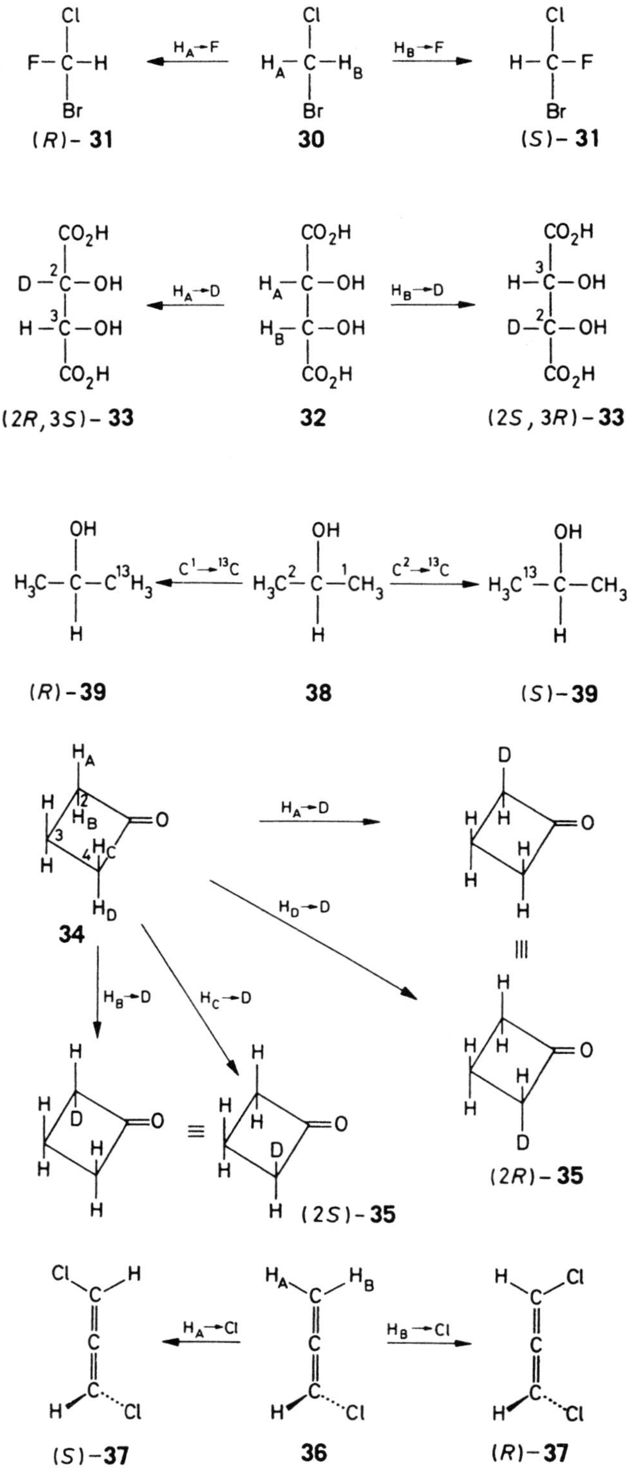

Figure 8.13. Enantiotopic ligands.

(viewed from H_A) **a** (*Si*)   (viewed from H_B) **b** (*Re*)

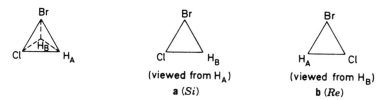

Figure 8.14. The compound CH_2ClBr. A view of the rest of the molecule from each enantiopic ligand.

Those encountering the phenomenon of enantiotopic ligands for the first time are sometimes puzzled by the nature of the difference between such ligands. One way of explaining the difference is by the very substitution criterion: If replacement of two ligands, in turn, by a third ligand gives rise to different (enantiomeric) products, then the ligands can, *by definition*, not be equivalent (homotopic). A perhaps more satisfying view of the matter (Schwartz and Carter, 1954; cf. Mislow, 1965) is shown in Figure 8.14. If one views the rest of the CH_AH_BClBr molecule from the vantage point of H_A, one perceives the atoms Br–Cl–H_B in a counterclockwise direction (**a**), that is, one sees the *Si* face (cf. Section 8-3.d) of the Br–Cl–H remainder. Conversely, if one views the remainder of the molecule from H_B, Br–Cl–H_A are seen in a clockwise (*Re*) sequence **b**. Therefore the environment of H_B is the mirror image of the environment of H_A. We shall return to this view in Section 8-3.d.

Similar criteria, but of addition, can be established for enantiotopic faces. Faces are enantiotopic if addition of the same achiral reagent to either one or the other will give rise to enantiomeric products. Thus (Fig. 8.15) addition of HCN to the two enantiotopic faces of acetaldehyde gives rise to the two enantiomers of lactonitrile. [Here, as in the case of substitution, the added group must be different from any group already there. Thus we cannot test the enantiotopic nature of the two faces of the C=O function of acetaldehyde **40** by addition of CH_3MgI, since the group added (CH_3–) is the same as one of the existing groups.]

Symmetry Criterion

Enantiotopic ligands and faces are not interchangeable by operation of a symmetry element of the first kind (C_n, simple axis of symmetry) but must be interchangeable by operation of a symmetry element of the second kind (σ, plane of symmetry; i, center of symmetry; or S_n, alternating axis of symmetry). It follows that, since chiral molecules cannot contain a symmetry element of the second kind, there can be no enantiotopic ligands or faces in chiral molecules. (Nor, for different reasons, can such ligands or faces occur in linear molecules, $\mathbf{C_{\infty v}}$ or $\mathbf{D_{\infty h}}$.)

40   (*R*) – **41** Front addition   (*S*) – **41** Rear addition

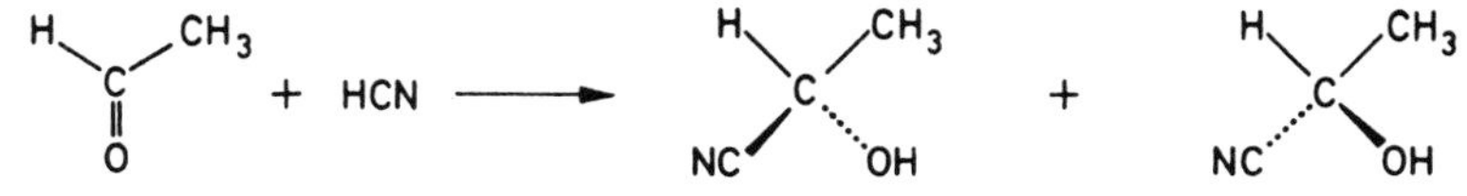

Figure 8.15. Addition of HCN to acetaldehyde.

The symmetry planes σ in molecules **30**, **32**, **34**, **36**, and **38** in Figure 8.13 should be readily evident. It is possible to have both homotopic and enantiotopic ligands in the same set, as exemplified by the case of cyclobutanone **34**: H_A and H_D are homotopic as are H_B and H_C. The ligand H_A is enantiotopic with H_B and H_C; H_D is similarly enantiotopic with H_C and H_B. The sets $H_{A,B}$ and $H_{C,D}$ may be called equivalent (or homotopic) sets of enantiotopic hydrogen atoms. The unlabeled hydrogen atoms at position 3, constitutionally distinct (see Section 3.4) from those at C(2, 4), are homotopic with respect to each other. Enantiotopic ligands need not be attached to the same atom, as seen in the case of *meso*-tartaric acid **32**, and also the just-mentioned pair H_A, H_C (or H_B, H_D) in cyclobutanone.

Symmetry elements of the second kind other than σ may generate enantiotopic ligands. Thus compound **42** in Figure 8.16 (G and Ɔ are enantiomorphic, i.e. mirror-image ligands) has a center of symmetry only; H_A and H_B, which are symmetry related by this center, are enantiotopic. *meso*-Tartaric acid (Fig. 8.16, **32**) in what is probably its most stable conformation also has a center rather than a plane of symmetry, so that its enantiotopic methine hydrogen atoms (H_A, H_B) are related by the i operation. Similarly, the four tertiary hydrogen atoms in **43**, Figure 8.16, are interrelated by the lone symmetry element S_4 in that molecule, and thus H_A is enantiotopic with H_B and H_D and the same is true of H_C. However, since the molecule also has a (simple) C_2 axis, H_A and H_C are homotopic, as are H_B and H_D. It might be noted here and in compound **34** (Fig. 8.13) that, while there can never be more than two enantiomers, a single ligand can have more than one enantiotopic partner.

Enantiotopic faces (Fig. 8.15) are also related by a symmetry plane (e.g., the plane of the double bond in **40**). The faces must not be interchangeable by operation of a symmetry axis, otherwise they are homotopic rather than enantiotopic.

The two faces in question may relate to a molecular plane other than the face of a double bond. Thus the two faces of ethyl methyl sulfide (H_5C_2–S–CH_3) are enantiotopic (Mislow and Raban, 1967), inasmuch as addition of oxygen to one or the other by peracid will give rise to two different, enantiomeric sulfoxides, (R)- and (S)-$CH_3SOC_2H_5$. Alternatively, we may consider this case as one of two enantiotopic ligands, namely, the two unshared pairs on sulfur. Again the two faces of the benzene ring in 1,2,4-trimethylbenzene **44** are enantiotopic: Addition of a chromium tricarbonyl ligand to one or other of the two faces gives enantiomeric coordination compounds **45a,b** (Fig. 8.17).

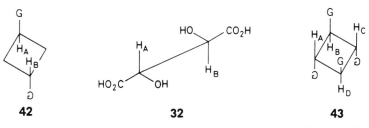

Figure 8.16. Enantiotopic ligands in molecules with a center or alternating axis of symmetry.

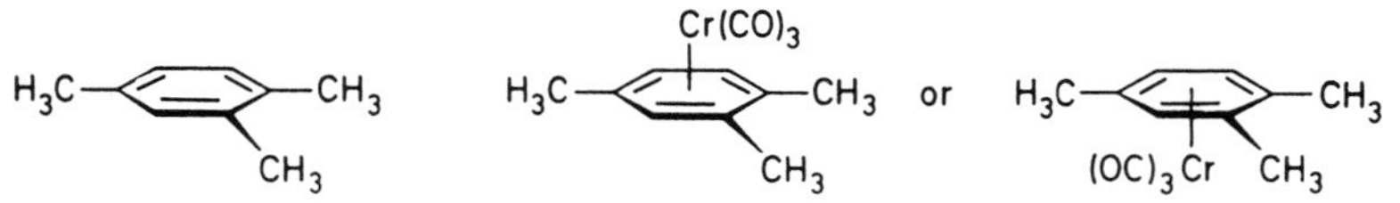

Figure 8.17. Enantiotopic faces in substituted benzenes.

So far we have discussed groups that are enantiotopic by internal comparison. Groups may also be enantiotopic by external comparison, that is, groups in two different molecules are enantiotopic if they are related by reflection symmetry. Clearly, this can be so only if the two molecules themselves are enantiomeric: Corresponding groups in enantiomeric molecules (e.g., the methyl groups in D- and L-alanine) are enantiotopic.

Just as enantiomeric molecules cannot be distinguished in achiral environments, neither can enantiotopic ligands. Such ligands can, however, be distinguished by NMR spectroscopy in nonracemic chiral media (Pirkle et al., 1969, 1971; Kainosho, Pirkle, et al., 1972; cf. Pirkle and Hoover, 1982) or in the presence of chiral shift reagents (discussed in Chapter 6), in synthetic transformations involving either chiral reagents or other types of chiral environment, asymmetric syntheses (see Chapter 12) and, above all, in enzymatic reactions, since the enzyme catalysts are chiral (cf. Section 8-5). It is because of these potential distinctions between enantiotopic ligands and faces that it is important to be able to recognize them.

c. Diastereotopic Ligands and Faces

The earlier mentioned criteria may also be employed to recognize diastereotopic ligands, that is, ligands that are located in a stereochemically distinct but nonmirror-image environment.

Substitution–Addition Criterion

Figure 8.18 shows a number of cases where replacement of first one and then another of two homomorphic ligands by a different achiral test ligand (see also pp. 473 and 475) gives rise to diastereomeric products.

Such ligands are called diastereotopic and are generally distinct both spectroscopically and chemically. Their NMR signals will generally be different (cf. Section 8-4) and their reactivity will, in general, be unequal.

The case of 46 (Fig. 8.18) is straightforward: since C(2) in 2-bromobutane is chiral, H_A and H_B cannot be enantiotopic and the replacement criterion discloses that they are diastereotopic rather than homotopic. The examples of cyclobutanol 48 and 4-tert-butyl-1,1-difluorocyclohexane 52 show (cf. Section 8-1) that presence of a chiral center is not required for the existence of diastereotopic nuclei. The ligands H_A and H_B in 48 and F_A and F_B in 52 are diastereotopic because they are cis and trans, respectively, to the hydroxyl group at C(1) in 48 or the tert-butyl group at C(4) in 52. It might be noted that, after replacement, C(3) in 49 or C(1) in 53 is not a chiral center but is a stereogenic center; the corresponding atoms in

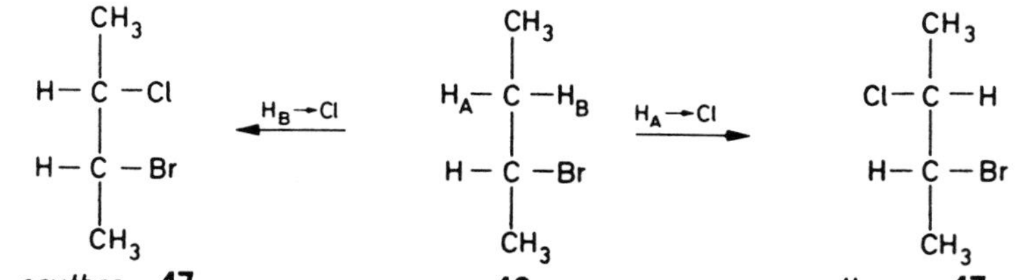

erythro - **47** **46** *threo* - **47**

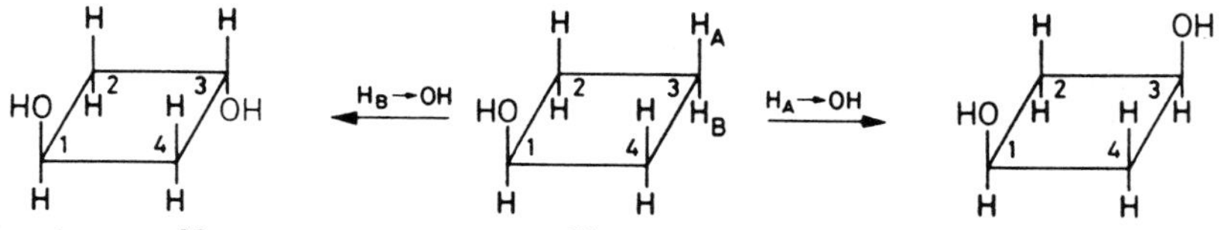

trans - **49** **48** *cis* - **49**

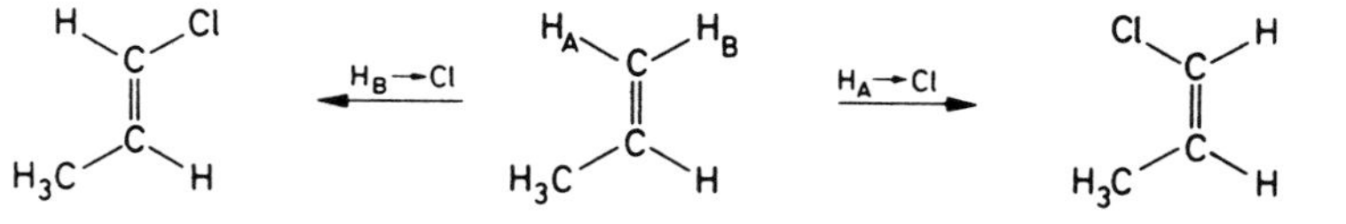

trans -(*E*)- **51** **50** *cis* -(*Z*)- **51**

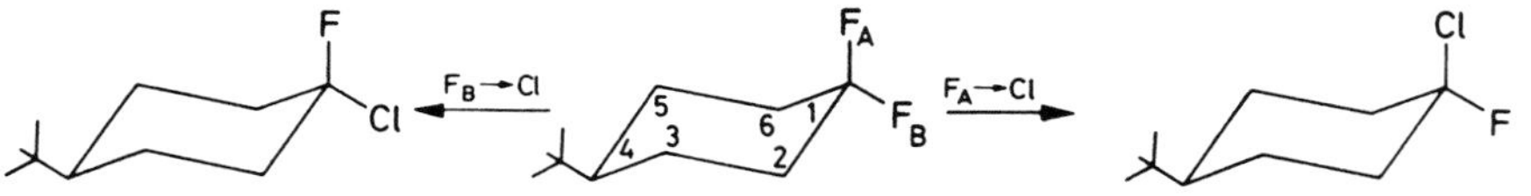

53a **52** **53b**

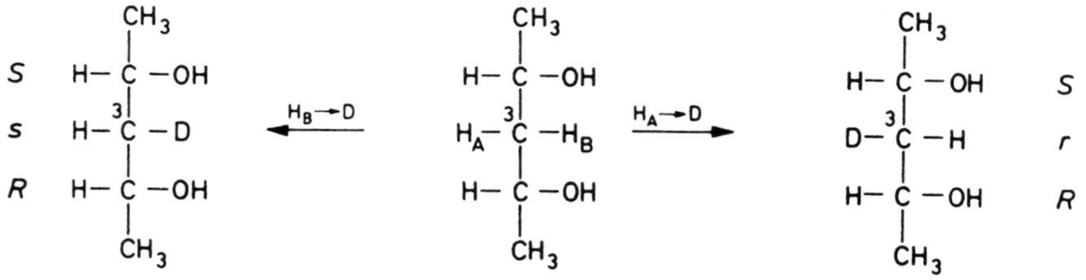

55a **54** **55b**

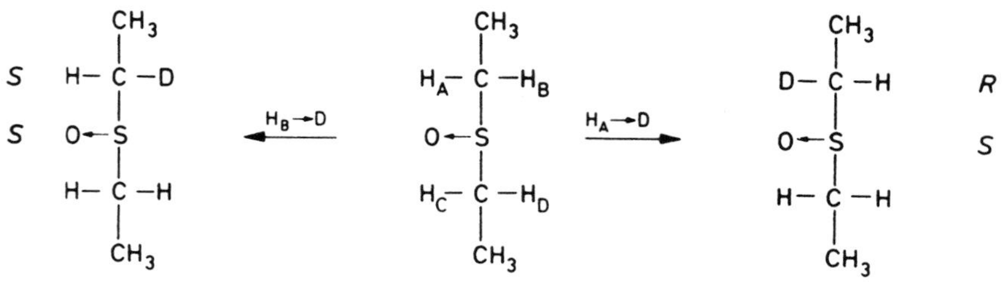

57a **56** **57b**

Figure 8.18. Diastereotopic ligands.

48 and **52** are prostereogenic. In the case of propene (**50**) replacement of H_A and H_B generates a cis–trans pair of (diastereomeric) alkenes, again making H_A and H_B diastereotopic. (One is cis to the methyl group at the distal carbon atom, the other is trans.) The case of *meso*-2,4-pentanediol (**54**) is of note because the products of replacement of H_A and H_B (**55**) are diastereomeric meso forms in which C(3) is pseudoasymmetric (cf. Chapter 3). In diethyl sulfoxide (**56**), H_A and H_B are also diastereotopic as are H_C and H_D; on the other hand, H_A is enantiotopic with H_C and H_B with H_D. Since the molecule as a whole is achiral, the existence of enantiotopic atoms is possible. The situation in **56** is entirely

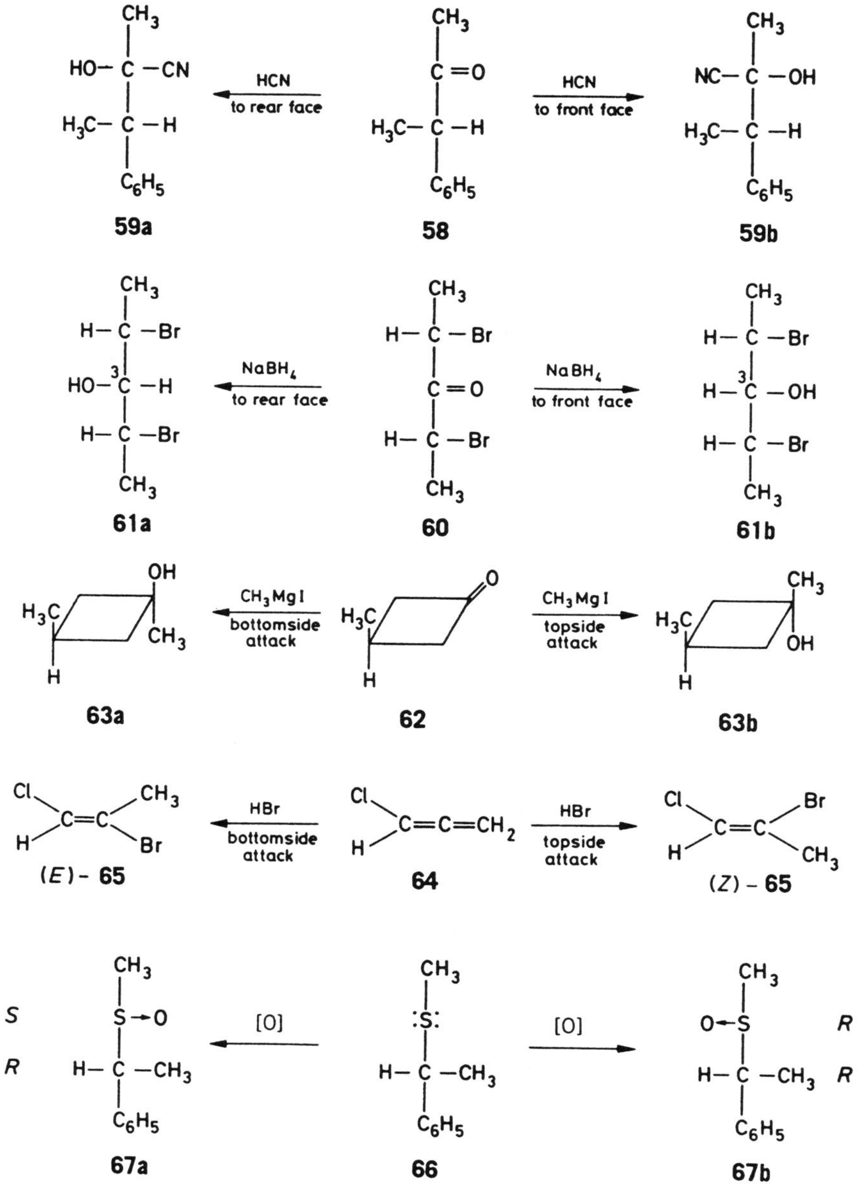

Figure 8.19. Diastereotopic faces of double bonds.

analogous to that in citric acid (Fig. 8.4, **14**) in which we now recognize H_A to be diastereotopic to H_B (as H_C is to H_D), whereas H_A and H_C (or H_B and H_D) are enantiotopic. Citric acid and diethyl sulfoxide contain two enantiotopic pairs of diastereotopic ligands (or two diastereotopic pairs of enantiotopic ligands).

> In passing we may note that the hydrogen atoms attached at C(2) and C(4) in cyclobutanol (Fig. 8.18, **48**) also form enantiotopic pairs of diastereotopic hydrogen atoms. In the trans diol (*trans*-**49**), on the other hand, the corresponding hydrogen atoms form an enantiotopic pair [C(2) vs. C(4)] of geminally homotopic hydrogen atoms whereas in the *cis*-1,3-diol (*cis*-**49**) they form a geminally diastereotopic pair of homotopic [C(2) vs. C(4)] hydrogen atoms. These facts may become clearer when symmetry criteria are applied (see below).

The addition criterion may similarly be applied to recognize diastereotopic faces. Methyl α-phenethyl ketone (**58**, Fig. 8.19) has a chiral center, thus HCN addition gives rise to diastereomers **59a** and **59b**; hence the faces of the carbonyl carbon are diastereotopic. This case is of importance in conjunction with Cram's rule (Chapters 5 and 12). Compounds **60**, **62**, and **64** also display diastereotopic faces even though the products **61**, **63**, and **65** are not chiral; these are cases of prostereogenicity but not prochirality. α-Phenethyl methyl sulfide **66** displays diastereotopic sides of a molecular plane not due to a double bond and may alternatively be considered a case of diastereotopic ligands (unshared pairs on sulfur); attachment of oxygen to one or other of the diastereotopic pairs gives diastereomeric sulfoxides.

Symmetry Criterion

The symmetry criteria of diastereotopic ligands or faces are simple: such ligands or faces must not be related either by a symmetry element of the first kind (axis) or by one of the second kind (plane, center, or alternating axis). The reader should become convinced that the even-numbered molecules depicted in Figures 8.18 and 8.19 (middle column) are either devoid of such symmetry elements or that, when such elements (e.g., σ) are present, their operation does not serve to interchange the ligands or faces designated as being diastereotopic.

> By way of generalization (Kaloustian and Kaloustian, 1975) it might be pointed out that **54** in Figure 8.18 and **60** in Figure 8.19 correspond to the general type **A** in Figure 8-20 [diastereotopic ligands (R) or faces (C=X)]. Type **B** has homotopic ligands or faces as does type **C**. In contrast, the ligands or faces in **D** and **E** are diastereotopic. Those in **F** are enantiotopic [G and Ɔ represent enantiomorphic (i.e., mirror-image) ligands and X represents an achiral ligand].

> We should also understand that diastereotopic ligands or faces must be in constitutionally equivalent environments, such as C(2) and C(4) in *n*-pentane. If this is not the case (e.g., as between C(2) and C(3) in *n*-pentane), one speaks of constitutional heterotopicity (cf. Fig. 8.21 and Section 8-3.d).

Ligands may be diastereotopic by external as well as by internal comparison. Corresponding ligands in diastereomers are diastereotopic under any circum-

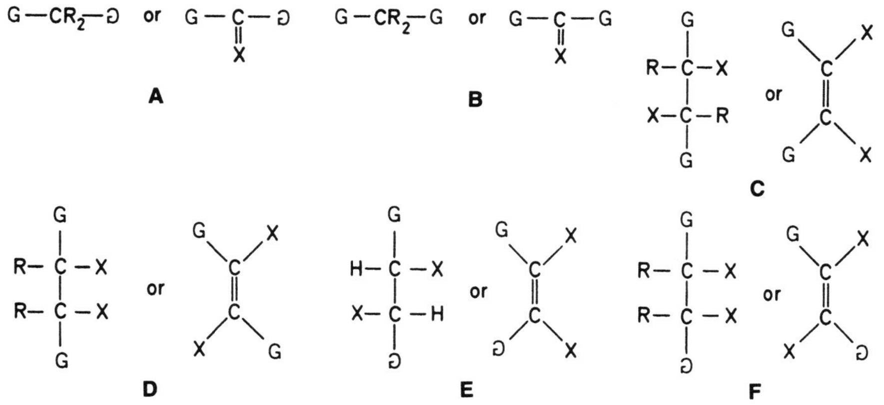

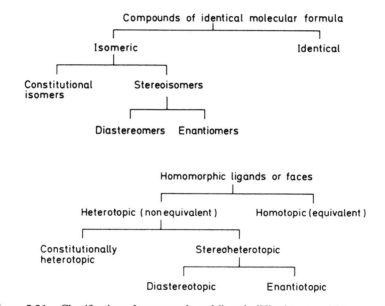

Figure 8.20. Topicity of ligands and faces in molecules with paired chiral substituents.

Figure 8.21. Classification of compounds and ligands (Hirschman and Hanson, 1971a).

stances; corresponding ligands in enantiomers are diastereotopic when viewed in a chiral environment (e.g., a chiral solvent; cf. Pirkle and Hoover, 1982; see also Section 6-5.c, p. 231).

It has been pointed out, however (Reisse, Mislow, et al., 1978) that there is no fundamental difference between internally and externally heterotopic nuclei. For example, when one compares ^{13}C NMR signals of enantiotopic or diastereotopic methyl groups, e.g., in an appropriately placed isopropyl moiety, $(CH_3)_2CH$, one generally creates the fiction that the two ^{13}C labeled methyl groups are in the same molecule (i.e. internally stereoheterotopic). In fact, however, if the spectrum is recorded at the natural abundance of 1.1%, the chance that two adjacent methyl groups in one and the same molecule are both ^{13}C labeled is only about 1 in 10,000;

such species are at too low a concentration to be seen in the spectrum. The actual comparison is therefore between enantiomeric or diastereomeric *molecules* [e.g., $(^{13}CH_3)_aCHX(CH_3)_b$ and $(CH_3)_aCHX(^{13}CH_3)_b$], (where X is an achiral or a chiral group, as the case may be, and the superscript 13 marks the methyl group that contains the isotope); thus it is actually a comparison between externally enantiotopic or diastereotopic groups. Nonetheless, as will be seen in Section 8-4.b, the distinction between internally and externally heterotopic ligands is useful from the experimental point of view.

d. Concepts and Nomenclature

Concepts

First it may be well to review the criteria for homotopic, enantiotopic,, and diastereotopic ligands or faces (Mislow and Raban, 1967). Ligands, or faces, are equivalent or homotopic when they can be brought into coincidence by operation of a proper (C_n) symmetry axis. If this condition is not fulfilled but the ligands or faces can be brought into coincidence by operation of an improper (S_n) axis of symmetry, including a plane σ or center i of symmetry, the ligands or faces are enantiotopic. If neither symmetry operation C_n nor S_n brings the ligands or faces into coincidence, they are diastereotopic or (see below) constitutionally heterotopic.

It is illuminating to make a comparison between isomeric and nonisomeric compounds on the one hand, and homotopic or heterotopic ligands or faces on the other. There is logic to such a comparison since it was explained earlier that stereoisomers are generated by appropriate replacement of heterotopic ligands or addition to heterotopic faces. Figure 8.21 displays such a comparison (Hirschmann and Hanson, 1971a). It is convenient, in conjunction with the diagram of homotopic and heterotopic ligands (Fig. 8.21) to introduce an additional term: If homomorphic ligands (e.g., the hydrogen atoms in a methylene group) occur in constitutionally distinct portions of a molecule, we call them constitutionally heterotopic. Examples would be the methylene hydrogen atoms at C(2) and those at C(3) in cyclobutanol (Fig. 8.18, **48**).

Constitutionally heterotopic ligands are in principle always distinguishable, just as constitutional isomers are. (The same is true of diastereotopic ligands.) Diastereotopic and enantiotopic ligands or faces may be lumped together under the term "stereoheterotopic" just as diastereomers and enantiomers are both called stereoisomers.

Nomenclature

Just as it is convenient to distinguish enantiomers and diastereomers by appropriate descriptors (R, S, E, Z, etc.) it is desirable to provide descriptors for stereoheterotopic ligands or faces. The basic nomenclature to this end has been provided by Hanson (1966; Hirschmann and Hanson, 1971b) and is closely related to the nomenclature of stereoisomers.

If, in a prostereogenic assembly (e.g., a prochiral center Caabc) a hypothetical priority, in the sense of the sequence rules (cf. Chapter 5), is given to one of the identical ligands a over the other a', that ligand a will be called "*pro-R*" if the

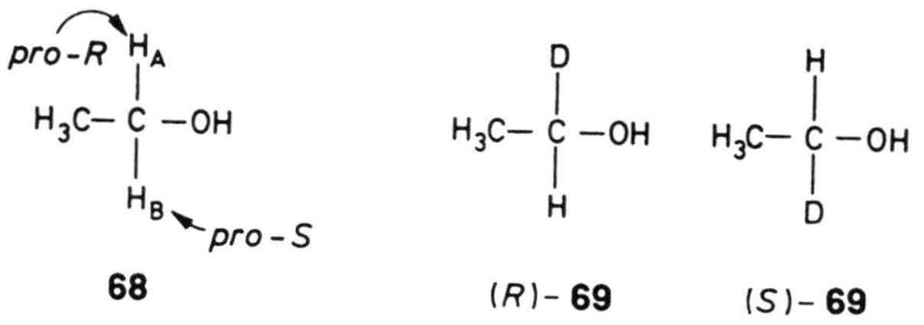

Figure 8.22. Ethanol, (R)-ethanol-1-d and (S)-ethanol-1-d.

newly created "chiral center" Caa'bc (sequence a > a') has the R configuration, but it will be called "*pro-S*" if the newly created "chiral center" has the S configuration. Let us take ethanol (Fig. 8.22, **68**) as an example. The hydrogen atoms H_A and H_B are enantiotopic. If preference is given to H_A over H_B in the sequence rule, the sequence is OH, CH_3, H_A, H_B and the (hypothetical) configurational symbol for **68** would be R, hence H_A is *pro-R*; by default, H_B is *pro-S*. The answer would have come out the same if H_B had been given precedence over H_A; in that case the sequence would have been OH, CH_3, H_B, H_A and the hypothetical configurational symbol for **68** is then S, hence H_B is *pro-S*. It might be noted (cf. Fig. 8.22) that the same result would have been obtained by replacing first one hydrogen and then the other by deuterium since deuterium has priority over hydrogen (Chapter 5); replacement of H_A by D gives (R)-ethanol-1-d [(R)-**69**], and hence H_A is *pro-R*; similarly, replacement of H_B by deuterium gives (S)-ethanol-1-d [(S)-**69**], and hence H_B is *pro-S*.

> This alternative of replacing the atom to be stereochemically labeled by a heavy isotope rather than giving it hypothetical precedence cannot be used when the heavy isotope is already present at the prostereogenic center. Thus the enantiotopic methyl hydrogen atoms of acetic-2-d acid (CH_2DCO_2H) can obviously not be assigned as *pro-R* and *pro-S* by replacing them with deuterium. In this case it is necessary to elevate one or the other of the hydrogen atoms in question to a priority above its counterpart but below D.

In a formula, the "*pro-R* group X" is sometimes written as X_R (and similarly X_S for the *pro-S* group). It is important, however, to read X_R as "the *pro-R* group X" and not as the "R group X," since heterotopicity or prochirality, not chirality, is implied. Indeed, Figure 8.23 shows a case (type of CGGXY) where both CH_3CHOH (G) ligands have the S-configuration but the upper one is *pro-R*, whereas the lower one is *pro-S*.

> In the original structure, the central atom C(3) is achiral (cf. Chapter 3). When priority is given to the upper ligand (indicated in Fig. 8.23 by replacement of CH_3 by $^{13}CH_3$), however, C(3) becomes chiral, and since its configuration is then R, the upper CH_3CHOH ligand is *pro-R* (Eliel, 1971).

Just as chiral centers can be labeled R or S not only in enantiomers but also in many diastereomers, so the designations *pro-R* and *pro-S* are not confined to enantiotopic ligands but may also be used for a number of diastereotopic ones

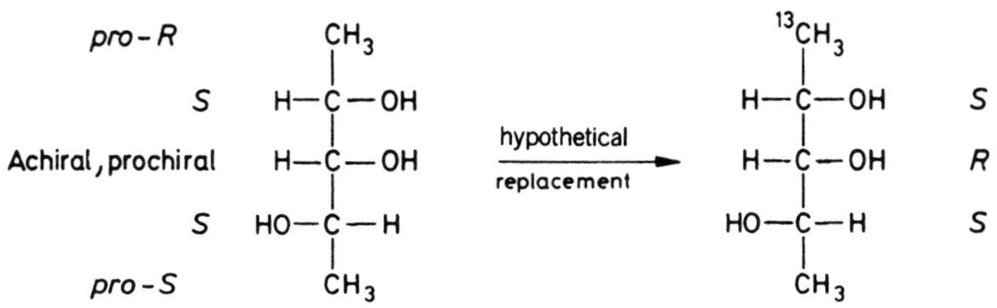

```
     pro-R              CH3                        13CH3
                         |                           |
           S      H—C—OH                     H—C—OH          S
                         |           hypothetical     |
  Achiral,prochiral  H—C—OH      ———————————→   H—C—OH          R
                         |           replacement      |
           S      HO—C—H                      HO—C—H          S
                         |                           |
     pro-S              CH3                         CH3
```

Figure 8.23. Molecule in which the ligand of S-configuration is *pro-R*.

(for exceptions, see below). Thus, for example, the labeling in Figure 8.13 is such that H_A (compounds **30, 32, 34**, but not **36**) or $^1CH_3$ (compound **38**) is the *pro-R* group; the reader should verify this proposition. The same is true for compounds **46** and **56** in Figure 8.18. Compounds **48, 50, 52**, and **54** in Figure 8.18 cannot be labeled in this manner since replacement of the diastereotopic ligands does not produce chiral products. In **54** the substitution gives rise to a pseudoasymmetric center which, in the compound on the left, is s, in the compound on the right r. Hence, H_A is called *pro-r* and H_B *pro-s* (Hirschmann and Hanson, 1971a). In **50**, replacement of H_A and H_B gives rise to Z and E alkenes, respectively; in this case H_A should be called "*pro-Z*" and H_B"*pro-E*" (symbols H_r, H_s, H_E, H_Z may be used). [For nomenclature purposes, the replacement of H by Cl is not appropriate; one should replace H by D or better (see above) "elevated H". In the case of **50** (Fig. 8.18) it happens to make no difference.] In compounds **48** and **52**, the terms "*pro-cis*" for H_A and F_A and "*pro-trans*" for H_B and F_B (symbols H_{cis}, H_{trans}, etc.) respectively are appropriate.

Hanson (1966) also devised a specification of heterotopic faces. Thus if one looks at the chirality in two dimensions (2D) (Fig. 8.14) and the sequence is clockwise, **b**, one calls it *Re*, if the sequence is counterclockwise, **a**, one calls it *Si*, these being the first two letters of *Rectus* and *Sinister*, respectively. [Hanson originally used *re* and *si*, but since prochirality, not propseudoasymmetry, is involved, the use of capital letters is more appropriate (Prelog and Helmchen, 1972) and has now been accepted (cf. Hanson, 1976)]. Thus the face of acetaldehyde turned toward the reader in Figure 8.15 is *Si* (O, CH_3, H are in counterclockwise order) and the corresponding front faces in Figure 8.19 are *Re* for **58** and *re* for **60** (here the use of the lower-case symbol is appropriate since addition to the two faces generates achiral diastereomers: the stereogenic center generated is achirotopic, pseudoasymmetric). In **62** and in **64** (C=CH$_2$ bond) one might call the top face *ci* and the bottom face *tr*, these being the first two letters of cis and trans, respectively (but see below). In **66** (Fig. 8.19) the uninvolved lone pair must be inserted as a phantom ligand; when this is done the right face of the molecule becomes *Si*, the left one *Re*. (As already mentioned, one may alternatively consider both lone pairs as heterotopic ligands, in which case the right pair is *pro-S*, the left *pro-R*, since elevation of the right pair over the left gives a hypothetical S configuration, and vice versa.)

With the recent additions to sterochemical nomenclature, it becomes possible to name the faces of compound **62** in *re/si* terminology also (Prelog and Helmchen, 1982). The cyclobutanone is opened with a double complementation of C(1)

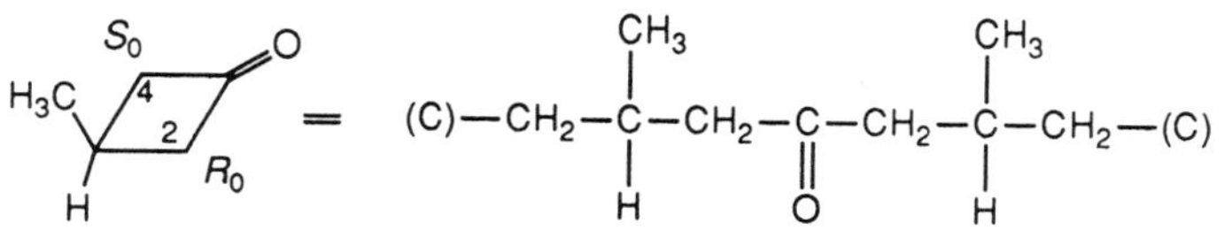

Figure 8.24. Assignment of descriptors to prostereogenic faces in achiral cylanones.

yielding a hypothetical structure in which the left branch has auxiliary descriptor R_0, the right S_0 (Fig. 8.24). The top face is thus *re* and the bottom face is *si*.

We have already mentioned (cf. Fig. 8.3) that heterotopicity may also be found to exist in cases where replacement of one of two homomorphic ligands gives rise to molecules of axial or planar chirality. Compounds **10** (Fig. 8.3) and **36** (Fig. 8.13) are examples of axial prochirality giving rise to enantiotopic ligands; compound **12** in Figure 8.3 is an example of planar prochirality giving rise to such ligands. Figure 8.25 shows examples of axial prochirality giving rise to diastereotopic ligands (Martin et al., 1965), namely, **70** (Martin et al., 1971) and **71** (Beaulieu et al., 1980), and of planar prochirality (if it may be considered as such) giving rise to such ligands, as in **72** (Slocum and Stonemark, 1971).

Although systematic nomenclature is generally to be preferred, there are some instances (e.g., in steroids) where a local system of nomenclature is still generally used. Thus in 3-cholestanone (Fig. 8.26, **73**) the hydrogen atoms above the plane of the paper, which itself represents a projection of the three-dimensional (3D) molecule, are called β and those below the plane α (cf. Chapter 11). Since the geminal hydrogen atoms at each methylene carbon form a diastereotopic pair, it is clear that diastereotopic hydrogen atoms in such pairs

$$C_2H_5CHBr-CO-C(CH_3)=C=CH_2$$

70

$$(CH_3)_2CHCR=C=CR'R''$$

71

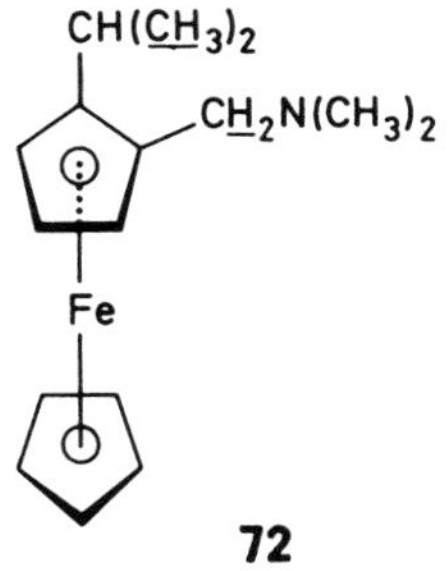

72

Figure 8.25. Prochiral axes and planes.

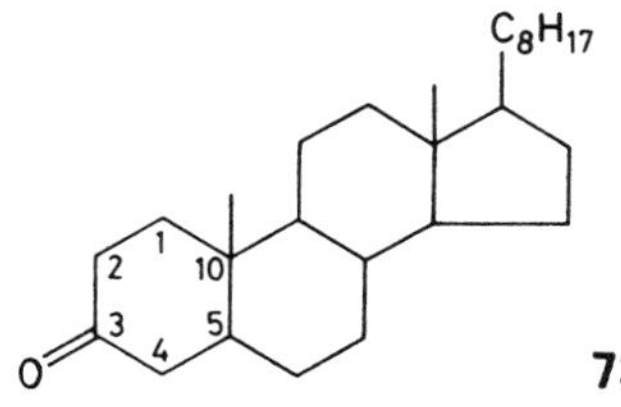

73 **Figure 8.26.** 3-Cholestanone.

may be distinguished by calling them H_α and H_β and this is commonly done. There is obviously no 1:1 connection of such common with systematic nomenclature; for example, the β-hydrogen atoms at C(2) is *pro-S* but that at C(4) is *pro-R*. Not surprisingly this lack of correlation parallels that between α/β and R/S (Chapter 11) when one looks at chiral centers in steroids (as in 2- and 4-cholestanol). The α and β designation may also be used for heterotopic faces; thus the front (*Si*) face of the keto function at C(3) is β and the rear (*Re*) face is α.

Sometimes, especially in enzymology, it is convenient to speak of the face of a molecule quite apart from any particular heterotopic face or set of ligands. For example, one might like to express the fact that a steroid is attached to an enzyme receptor on the α face without making reference to any particular C=O or CH_2 moiety. The *pro-R/pro-S* or *Re/Si* nomenclature is not generally applicable to such cases (for an exception, see Fig. 8.17). The α/β face nomenclature, which applies to steroids as explained above, has been generalized (Rose et al., 1980) to apply to all kinds of rings using the following rules:

1. If the compound is monocyclic, or if the rings are not fused, the faces are designated as α if progression around each ring from the lowest to the next higher number atom by the shortest route is *clockwise*; the faces are designated as β if such progression is *counterclockwise*. The ring is to be examined as a planar regular polygon (i.e. disregarding conformation). If multiple numbering systems are used, the precedence is $1 \rightarrow 2 > 1a \rightarrow 2a > 1'$.

2. If the compound is made up of "ortho-fused" rings only, face designations of the entire system of rings are derived from the ring containing the lowest numbered unshared atoms, as specified in standard numbering for the compound.

Under rule 2, the A ring is pace setting for the entire steroid system (Fig. 8.26) and under rule 1 the front face of this ring is β since the numbering proceeds counterclockwise. Similarly, the front face of heme (Fig. 8.27) is α; oxygen in hemoglobin and myoglobin is known to bind to the β face (Kendrew et al., 1960; Perutz et al., 1960). For further details the original publication should be consulted (Rose et al., 1980).

Alternative Nomenclature

Prelog and Helmchen (1972) presented a view of prochirality that is based on chirality in 2D. This has led them to an alternative system of nomenclature for heterotopic ligands to be discussed now.

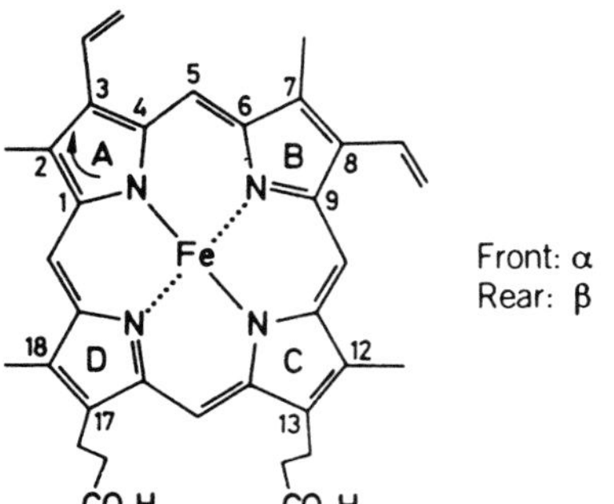

Figure 8.27. Faces of heme.

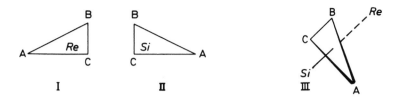

Figure 8.28. Two-dimensional chirality.

A scalene triangle, or any triangle with three differently labeled vertices A, B, C exists in 2D mirror images, as shown in Figure 8.28 (see also Fig. 8.14). These mirror-image representations cannot be made to coincide as long as they are maintained in the plane of the paper. As we have already seen, such 2D chiral representations may be taken to depict heterotopic faces. If the sequence is A > B > C, the front face of triangle **I** is *Re* and that of triangle **II** *Si*.

If one now proceeds into 3D space, one may say that the plane of say, triangle **I** divides all space into two halves, one (*Re* space) in front of and the other (*Si* space) behind triangle **I**. On the right of Figure 8.28, (**III**) a sideways view of the triangle is given with the *Re* space to the right and the *Si* space to the left. Three-dimensional prochirality may now be considered in the following terms: A ligand A′ identical with one already present (say A) is made to form a bond to either the *Re* or the *Si* side of the triangle ABC. The ligand thus "sees" mirror-image representations of ABC depending on the side where it is placed. The ligand may then be labeled A'_{Re} or A'_{Si}, depending on the half-space (*Re* or *Si*) in which it is located, as shown in Figure 8.29, **IV–VII**. Ligands A, B, C, and A′ may alternatively be considered to be attached to a tetrahedral center (Fig. 8.29, **VIII–XI**) in a conventional representation of enantiotopic ligands: the A′-ligand in **VIII** is A'_{Si}, since it is on the *Si* side of the ABC face, the one in **IX** is A'_{Re}, since it is on the *Re* side of the ABC face.

The representation in Figure 8.29 has several advantages. It closely correlates the heterotopic faces of a plane with the heterotopic ligands at a prochiral center

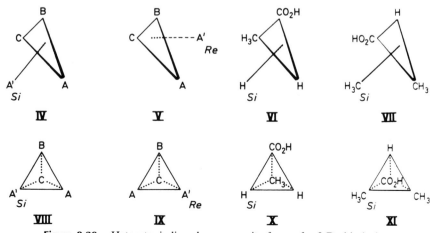

Figure 8.29. Heterotopic ligands on opposite faces of a 2-D chiral plane.

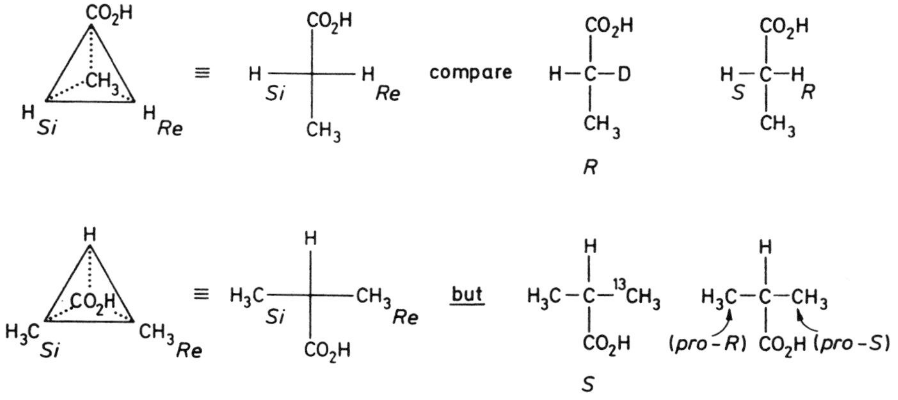

Figure 8.30. Alternative nomenclature for prochiral ligands.

formed by addition to that plane. Even the symbols correspond. And the symbolism is readily applicable to addition to various types of planes, such as ABC, ABG, AGD, etc. (for an application, see Battersby et al., 1980a,b).

There is, however, a countervailing disadvantage: The symbolism implied in Figure 8.30 leads to a loss of the relationship of symbols as between prochirality and chirality. As pointed out earlier, since all stable isotopes in practical use are heavier than the ordinary nuclides, replacing a *pro-R* ligand by an isotopically labeled one will give a molecule of *R*-configuration and analogously for *pro-S*; this relationship is particularly useful in the consideration of enzymatic reactions (Section 8-5). Such a relationship no longer necessarily holds in the alternative nomenclature as shown in Figure 8.30.

In conclusion of this section on nomenclature, it should be pointed out that the concept of chirotopicity (Mislow and Siegel, 1984) introduced in Chapter 3 (p. 53) despite the similarity in terminology, bears no direct relationship to heterotopicity. Whereas, since enantiotopic ligands are in a chiral environment (cf. p. 475), all enantiotopic ligands must be chirotopic, but no such 1:1 correlation applies to homotopic or diastereotopic ligands.

8-4. HETEROTOPICITY AND NUCLEAR MAGNETIC RESONANCE

a. General Principles. Anisochrony

Nuclei that are diastereotopic will, in principle, differ in chemical shift (cf. the reviews by Siddell and Stewart, 1969 and by Jennings, 1975), that is, they will be "anisochronous" (cf. Mislow and Raban, 1967, p. 23; the term was coined by Binsch following the use of "isochronous", for chemical-shift equivalent, by Abragam, 1961). Although such chemical shift differences are often seen, sometimes they are so small that the signals can be resolved only at quite high fields, or not at all. In the latter situation one speaks of "accidental isochrony," meaning that while the nuclei are in principle anisochronous, they are not, in fact, resolved.

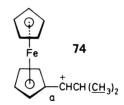

74

Figure 8.31. Example of diastereotopic methyl groups.

Anisochrony for diastereotopic ligands is seen with a number of different nuclei. We have already mentioned that $CH_2BrC(CH_3)BrCO_2CH_3$ displays different signals for the diastereotopic protons (italicized) (Nair and Roberts, 1957) and that $CF_2BrCHBrC_6H_5$ displays different resonances for the diastereotopic fluorine nuclei (Drysdale and Phillips, 1957). The diastereotopic methyl groups in the ferrocenyl cation **74** (Fig. 8.31) are distinct both in their ^1H and ^{13}C signals (Sokolov, Reutov, et al., 1973).

The immediate cause for anisochrony is, of course, the unequal magnetic field sensed by the diastereotopic nuclei. It follows that, as the source of the diastereotopic environment is removed further and further from the test nuclei, the anisochrony is expected to diminish. This prediction has been tested (Whitesides, Roberts, et al., 1964) with the results shown in Table 8.1; the anomaly seen in entry 4 may be due to the molecule "coiling back on itself" so that the methyl groups sense the shielding or deshielding effect of the benzene ring differentially. Table 8.1 also shows a solvent effect; it is thus desirable, in looking for potential anisochronies, to record the spectrum in several different solvents, for example CCl_4, $CDCl_3$, benzene-d_6, or pyridine-d_5 (cf. Schiemenz and Rast, 1971; see also Martin and Martin, 1966; Jennings, 1975). Another way of enhancing (or manifesting) anisochronies is to use lanthanide shift reagents (Schiemenz and Rast, 1971).

Table 8.2 summarizes the shift differences (both ^1H and ^{13}C) between the

TABLE 8.1. Observed Anisochrony of Diastereotopic CH_3 Protons in $(CH_3)_2CH–X–CH(CH_3)C_6H_5$

		Shift Difference	
Entry	X	in CCl_4 (ppm)	in C_6H_6 (ppm)
1	None	0.182	0.133
2	O	0.067	0.013
3	OCH_2	0.005	0.008
4	OCH_2CH_2	0.042	0.030
5	OCH_2CH_2O	0.000	0.013
6	$OCH_2CH_2OCH_2$	0.000	0.000

TABLE 8.2 Chemical Shift Differences (ppm) of Diastereotopic Groups in Compounds **75** and **76**

R	CO_2H	CO_2CH_3	$CO_2C_6H_5$	COCl	$COCH_3$	$CONH_2$	CH_2NH_2	CH_2OH	CH_2OAc	CN
75, ^1H	0.39	0.35	0.41	0.38	0.31	0.27	0.29	0.27	0.20	−0.37
76, ^1H	0.07	0.06	0.00	0.06	0.00	0.06	0.03	0.04	0.00	0.11
75, ^{13}C	1.84	1.90	1.79	0.91	1.14	0.94	−0.80	−0.54	−0.14	0.71
76, ^{13}C	0.85	0.85	0.80	0.50	0.74	0.84	0.42	0.51	0.33	1.67

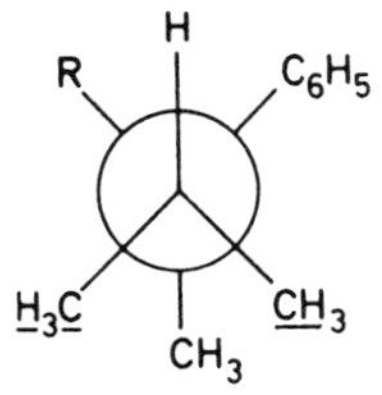

Figure 8.32. Preferred conformations of $(CH_3)_2CHCR(CH_3)C_6H_5$ (**75**) and $(CH_3)_2CHCR(CH_3)C_6H_{11}$ (**76**).

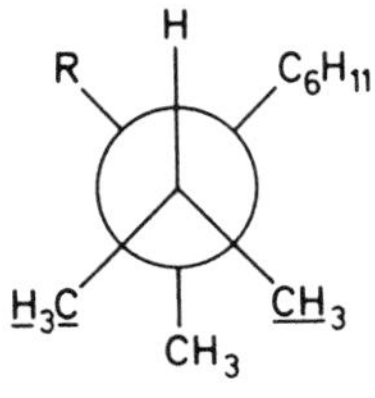

diastereotopic methyl groups of the compounds (Devriese, Mislow, et al., 1976) shown in Figure 8.32. (Arguments are adduced in the paper that the conformation shown is by far the preferred one, at least for R = COX). It is immediately obvious that these differences in shift between diastereotopic protons are much larger for the phenyl than for the cyclohexyl compound; presumably because of the much larger differential shielding of the methyl protons by the phenyl ring. On the other hand, the ^1H shift differences (except in the case of the nitrile) are not strongly dependent on the nature of R. (Neither of these statements is true for the ^{13}C shifts of the phenyl compounds which, in some instances, are opposite to the ^1H shifts.) It is clear, also, that the anisochronies of the diastereotopic methyl groups are much larger in ^{13}C than in ^1H resonance; presumably because of the generally larger shift effects in ^{13}C NMR spectra due to their largely paramagnetic origin. Carbon-13 NMR is thus generally a better probe for diastereotopic ligands than ^1H when both types of nuclei are present in these ligands.

Anisochrony due to axial prochirality is exemplified by the diastereotopic methylene protons in $H_2C=C=C(CH_3)COCHBrR$ (cf. Fig. 8.25); the chemical shift differences (depending on R) may be as high as 0.13 ppm (Martin et al., 1971). Also related to axial chirality are several cases of anisochronous methyl groups in isopropyl moieties that are diastereotopic through being part of a chiral allene of the type $(CH_3)_2CHCR=C=CR'R''$ (Beaulieu et al., 1980; Martin et al., 1966, 1967; Musierowicz and Wroblewski, 1980). These cases resemble that shown in Figure 8.31, where a prochiral center $[(CH_3)_2CH-C\cdots]$ is attached to a chiral ferrocenyl moiety; it should be noted that the ferrocenylmethylcarbenium ion fragment is chiral only if rotation about the $C-C^+$ bond (marked a in Fig. 8.31) is slow on the NMR time scale (Sokolov, Reutov, et al., 1973). Planar prochirality is exemplified by the complexation of dimenthyl maleate and dimenthyl fumarate with iron tetracarbonyl (Fig. 8.33) (Schurig, 1977), which gives rise to two diastereomers in the case of the fumarate (whose olefin faces are diastereotopic) but only a single enantiomer in the case of the maleate (whose faces are homotopic). Nevertheless, both complexes display diastereotopic olefinic protons: the maleate complex because the two ethylenic protons are internally diastereotopic, the fumarate complex because, although the two protons within one molecule are homotopic by virtue of the existence of a C_2 axis, the olefinic protons of the diastereomeric molecules are externally diastereotopic. One may ask, then, whether the two cases are distinguishable, and indeed they are. The two protons in the maleate complex **77** will necessarily be equally intense but, being anisochronous, will split each other and thus give rise to an AB signal. The two diastereomers in the fumarate complex **78** are not necessarily formed in equal amounts and, therefore, their signals are likely to be unequal in intensity. Moreover, since the protons within one molecule are

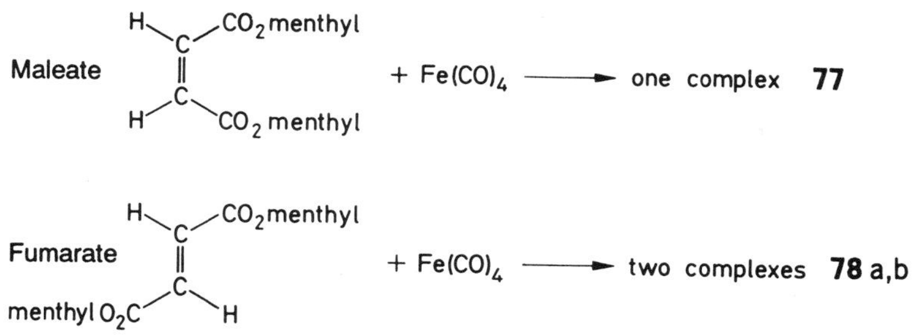

Figure 8.33. Complexation of dimenthyl maleate and dimenthyl fumarate with iron tetracarbonyl.

homotopic, they do not split each other and one thus sees two (probably unequally intense) singlets.

Since NMR is a scalar probe, enantiotopic nuclei are isochronous (i.e., they have the same chemical shift) in achiral media. Such nuclei, however, become diastereotopic in chiral media and thus in principle (though often not in practice) anisochronous. Among many examples (cf. Pirkle and Hoover, 1982) are the enantiotopic methyl protons of dimethyl sulfoxide (DMSO), CH_3SOCH_3, which are shifted with respect to each other by 0.02 ppm (Kainosho, Pirkle, et al., 1972) in nonracemic solvent $C_6H_5CHOHCF_3$. Surprisingly, the ^{13}C signals of the two methyl groups are not resolved under these conditions; this is an exception to the rule that ^{13}C signals of diastereotopic methyl groups generally show larger relative shifts than their ^1H signals (Devriese et al., 1976; Wilson and Stothers, 1974, p. 17). Similarly, the methyl protons of DMSO (Goering et al., 1974) are anisochronous in the presence of chiral lanthanide shift reagents (see Chapter 6) as are the enantiotopic carbinol protons of alcohols RCH_2OH (Fraser et al., 1972).

NMR shift differences between groups that are enantiotopic by external comparison (i.e., in enantiomers) may likewise be induced by either chiral (nonracemic) solvents (cf. Pirkle and Hoover, 1982; Weisman, 1983) or chiral shift reagents (Fraser et al., 1972; Fraser, 1983). Use of this phenomenon for the determination of enantiomeric excess (ee) has already been discussed in Section 6-5.c (p. 231).

The detection of diastereotopic nuclei by NMR is possible only if the diastereotopic nature of such nuclei is maintained on the time scale of the NMR experiment. Thus the equatorial and axial fluorine atoms in 1,1-difluorocyclohex-ane (Fig. 8.34), though diastereotopic, give rise to a single NMR signal because

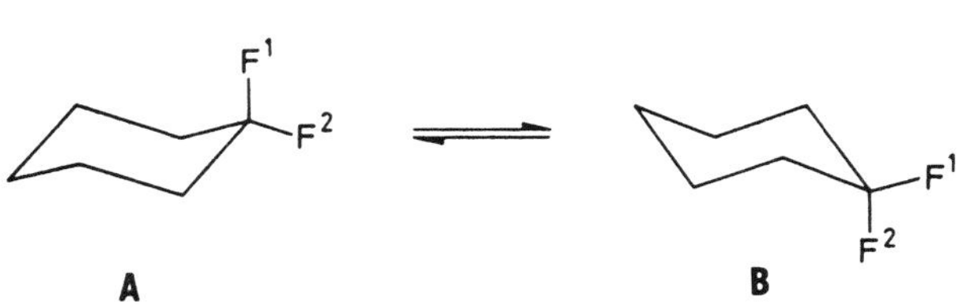

Figure 8.34. 1,1-Difluorocyclohexane.

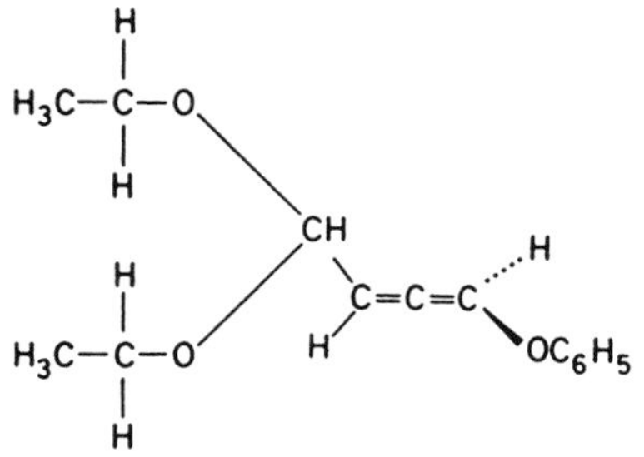

Figure 8.35. Multiple nonequivalence.

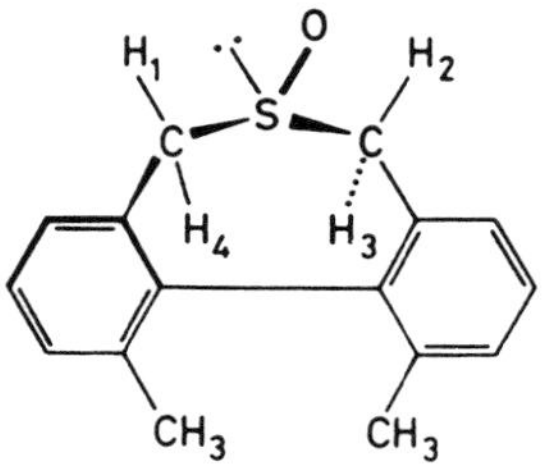

the rate of interchange of these nuclei by ring reversal at room temperature (ca. $100,000 \, s^{-1}$) is much greater than the shift between the fluorine nuclei (884 Hz at 56.4 MHz or $884 \, s^{-1}$; cf. Chapter 11; see also Binsch, 1968, p. 158). However, the fluorine atoms F^1 and F^2 become anisochronous below $-46°C$ when interconversion between the two chair forms (Fig. 8.34, **A, B**) is slowed to a rate less than the separation of the fluorine signals. This situation will be further discussed in Section 8-4.d and in Chapter 11.

In conclusion of this Section we mention the phenomenon of "multiple nonequivalence" (Jennings, 1975; Martin and Martin, 1966), which may occur when there are several nuclei in a molecule that are diastereotopic to each other. Two cases are depicted in Figure 8.35. In the allenic molecule **79** (Martin et al., 1966), the two ethoxy groups are diastereotopic because of the allenic chirality, and within each ethoxy group the methylene protons are similarly diastereotopic since they are not related by either a C_2 or a σ; as a result, all four methylene protons are anisochronous and there will be two sets of AB signals. The same is true of the four methylene protons in the biphenylic sulfoxide **80** (Fraser and Schuber, 1970; Fraser et al., 1972). The symmetry plane that might normally make H_1 enantiotopic with H_2 (and H_3 with H_4) is absent because of the noncoplanarity of the biphenyl ring system. Multiple nonequivalence is also seen in malonates: in the methylene proton signals of $R^*CH(CO_2CHHR')_2$ (Brink and Schjånberg, 1980) and in the isopropyl methyl signals of $R^*CH[CO_2CH(CH_3)_2]_2$, where the asterisk marks a chiral substituent (Biernecki et al., 1972).

b. NMR in Assignment of Configuration and of Descriptors of Prostereoisomerism

Determination of Configuration

This section describes the use of the concept of heterotopicity in assignment of stereochemical configuration (cf. Gaudemer, 1977; see also Chapter 5), usually relative (especially meso vs. chiral rather than absolute configuration), as well as assignment of the appropriate symbols to heterotopic ligands (i.e., experimental recognition as to which ligand is *pro-R* and which is *pro-S* at a prochiral center). Recognition of heterotopic faces as *Re* and *Si* is usually obvious from the

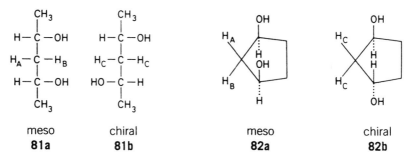

meso   chiral   meso   chiral
81a   **81b**   **82a**   **82b**

Figure 8.36. Distinction of active and meso forms by NMR.

configuration of the addition products thereto and will not be discussed here; examples are found in Section 8.5.b.

In favorable circumstances, chiral and meso stereoisomers may be distinguished directly; an acyclic and a cyclic example are shown in Figure 8.36 (**81, 82**). In both cases the methylene protons H_C in the chiral species (**81b, 82b**) are related by a C_2 axis and are therefore homotopic and isochronous, whereas the corresponding protons H_A and H_B in the meso forms (**81a, 82a**) are not related by either C_2 or σ and thus are diastereotopic and anisochronous. The situation is not altered when the racemic mixture rather than an individual enantiomer is compared with the meso isomer: The (internally homotopic) methylene protons of the two enantiomers are externally enantiotopic and so remain isochronous.

> Cases where the stereogenic centers are further removed from the prostereogenic one [e.g., in compounds of the type $C_6H_5-CHX-CH_2-C(CH_3)_2-CH_2-CHX-C_6H_5$] have also been investigated in both acyclic (Giardina et al., 1979) and cyclic (Giardina et al., 1980) systems.

When no suitable probes for distinction of meso forms and racemic pairs are present in the molecule, they may sometimes be introduced by combination with appropriate reagents. An example (Hill and Chan, 1965) is depicted in Figure 8.37. Benzylation of the amines **83** and **84** gives the *N*-benzyl derivatives **85** and **86**. In **85**, which is derived from the meso isomer **83**, H_A and H_B are enantiotopic, and hence isochronous; the protons constitute a single (A_2) signal. In contrast, in **86**, which is derived from the chiral isomer (whether or not racemic), the benzylic protons are diastereotopic, and hence anisochronous, and constitute an AB system.

Use of this methodology is risky when only one stereoisomer is available. If the benzyl derivative displays a single signal, it may be because one deals with a species of type **85** or because accidental isochrony is encountered in a species of type **86**. If the latter is the case, the method fails even when both stereoisomers are available.

An alternative approach is shown in Figure 8.38 (Kost and Raban, 1972; Raban et al., 1975). The amine (**87** or **88**) is converted into its 2,4-dinitrobenzenesulfenyl derivative. In this species the N=S bond has considerable double-bond character and rotation around it is slow on the NMR time scale at room temperature;

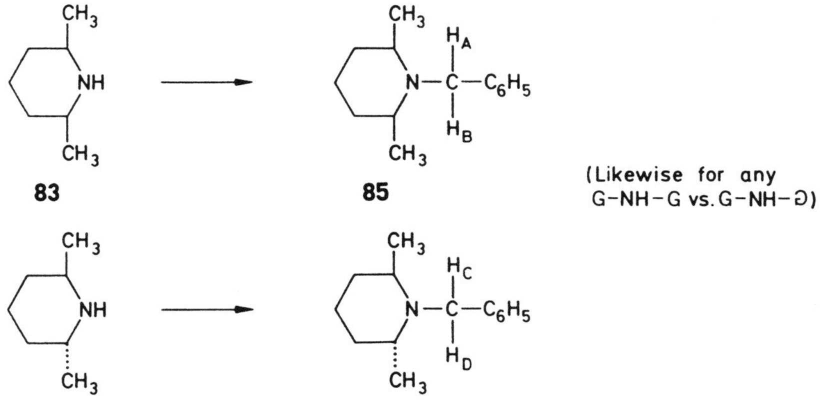

Figure 8.37. Achiral probe to distinguish racemic and meso forms.

moreover, the structure is such that the $N-S-C_{Ar}$ plane is perpendicular to the $C-N-C$ plane; that is, the species resembles an allene and may display axial chirality (or stereogenicity; Chapter 14). Derivatization of the meso isomer **87** thus gives rise to two meso forms, one with the dinitrophenyl group up, the other with the group down (Fig. 8.38, **89a,b**). In each isomer, the methyl groups are internally enantiotopic and thus appear as a sole doublet. However, the methyl groups of **89a** and **89b** are externally diastereotopic and are therefore anisochronous; two methyl doublets will thus be generated from the meso form and are expected to be of unequal intensity, since the two diastereomers are normally not formed in equal amounts. In contrast, the racemic isomer **88** will give rise to a single product **90**. In this compound the methyl groups are internally diastereotopic and hence are also anisochronous, but their signals will be equal in intensity. The meso and racemic isomers can thus be distinguished by signal intensity measurements on the products **89** and **90**; and this measurement is generally possible even if only one isomer is available to begin with. A slight uncertainty is introduced by the possibility that **89a** and **89b**, though diastereomers, might accidentally be formed in equal amounts.

The interplay of external and internal diastereotopicity may sometimes foil attempts to distinguish racemic from meso isomers (Eliel, 1982). However, this difficulty can be alleviated in cases where internally diastereotopic nuclei couple with each other (of course externally diastereotopic ones cannot do so). An example (Kainosho, Pirkle, et al., 1972) is the distinction of meso and racemic

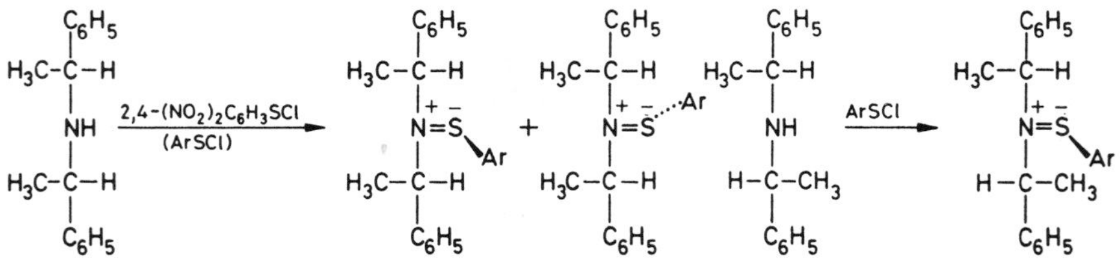

Figure 8.38. 2,4-Dinitrobenzenesulfenyl chloride as probe for racemic and meso forms.

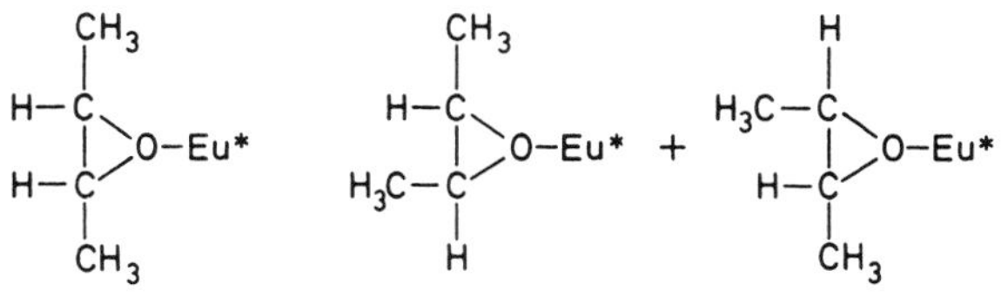

Figure 8.39. Distinction of meso- and racemic 2,3-epoxybutane by chiral shift reagent.

2-butylene oxide (*cis*- and *trans*-2,3-dimethyloxirane, **91, 92**) by means of a chiral shift reagent (Eu,* Fig. 8.39). Upon complexation with such a reagent, the internally enantiotopic C–H protons of the meso isomer **91** become internally diastereotopic and thus anisochronous. The corresponding internally homotopic protons of the chiral isomer **92** remain homotopic and isochronous. But, if **92** is a racemate, the two enantiomers are converted into diastereomers by complexation with the chiral shift reagent and the C–H protons in *rac*-**92** thus become diastereotopic and anisochronous as well. So far the situation appears stalemated. However, in *rac*-**92** the C–H protons in each enantiomer are isochronous and hence do not display coupling. Thus, in the methyl-decoupled proton spectrum, they appear as two singlets. In contrast, the methine protons of the meso form, being anisochronous, do couple and appear as an AB system when the methyl protons are decoupled; the two cases are thus distinguishable.

Use of diastereotopic probes for determination of absolute (as distinct from relative, e.g., meso vs. racemic) configuration is rare. An example relating to chiral amine oxides is shown in Figure 8.40 (Pirkle et al., 1971; Pirkle and Hoover, 1982; Weisman, 1983). The solute–solvent complex shown, composed of the (*S*)-amine oxide and (*S*)-phenyltrifluoromethylcarbinol, has the ethyl group in the amine oxide placed in the shielding region of the phenyl moiety of the solvent; the ethyl protons in this diastereomeric complex will therefore resonate upfield of the ethyl protons of the diastereomeric complex from *S* solvent and *R* amine oxide. The reverse will be true of the protons of the methyl groups in the amine oxide. It is not necessary to have both enantiomers in hand to apply this method. Since the racemate displays two sets of signals (of the two diastereomeric complexes), it suffices to compare the racemate with one enantiomer. This method of configurational assignment suffers, of course, from the usual limitation (cf. Chapter 5) of being dependent on the correctness of the model on which it is based. (See also p. 256.)

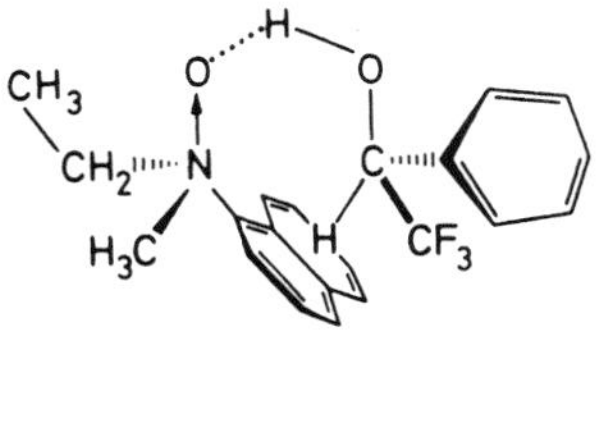

Figure 8.40. Determination of absolute configuration through NMR with a chiral solvating agent. Reprinted with permission from Pirkle, W.H., Muntz, R.L., and Paul, I.C., *J. Am. Chem. Soc.* (1971), **93**, 2818. Copyright © American Chemical Society.

Assignment of Descriptors to Heterotopic Ligands

So far in this section we have discussed the use of stereoheterotopic probes in configurational assignment. We now come to the problem of assigning the stereochemical placement (*pro-R* or *pro-S*) of the stereoheterotopic groups themselves. One way of doing this involves replacement of the prochiral by a chiral center, for example, to replace $RR'CH_2$ by $RR'CHD$ or $RR'C(CH_3)_2$ by $RR'C^{12}CH_3{}^{13}CH_3$ or by $RR'C(CH_3)CD_3$. The groups at the chiral center may then be distinguished and the configuration of that center determined by any of the classical methods described in Chapter 5. Finally, these groups are correlated (usually by NMR) with the corresponding groups at the prochiral center.

If the groups in question are enantiotopic, the correlation of the chiral with the prochiral center is, in most cases, effected through enzymatic reactions (cf. Section 8-5). For example, if an enzyme abstracts the deuterium rather than the hydrogen atoms from (R)-$RR'CHD$ it will abstract the *pro-R* rather than the *pro-S* hydrogen in $RR'CH_2$. Another approach would be to observe an enantiomer of $RR'CHD$ of known configuration by NMR in a chiral solvent or in the presence of a chiral shift reagent. If, under these circumstances, the position of the CHD proton in one enantiomer is different from that of the corresponding proton in the other (or of the other corresponding proton in the racemate) then the position of this proton, say in (S)-$RR'CHD$ (Fig. 8.41, **93**) will correspond, save for small isotope effects, to the position of the *pro-R* proton in $RR'CHH$, **94**. A case of this type (except that it involves covalent bond formation) has been described by Raban and Mislow (Raban and Mislow, 1966) and is shown in Figure 8.42. It was found that in the (R)-O-methylmandelate of (S)-$(+)$-2-propanol-1,1,1-d_3 (**95**) of known configuration (Mislow et al., 1960), the (sole) proton doublet of the CH_3 group (A) of the alcohol corresponds to the higher field doublet A of the corresponding (R)-O-methylmandelate of unlabeled 2-propanol (**96**). The lower field CH_3 doublet B in the unlabeled material dis-

Figure 8.41. Assignment of enantiotopic nuclei in chiral environment.

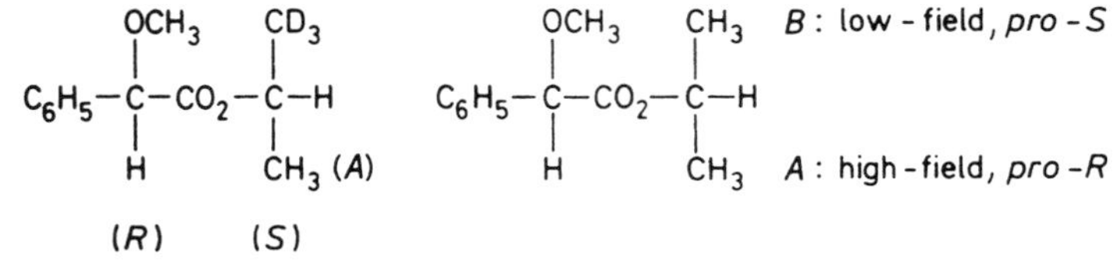

Figure 8.42. Prochirality assignment of C methyl groups in isopropyl O-methylmandelate

appears in the trideuterated species. It may thus be concluded that the lower field signal is due to the *pro-S* methyl group *B* and the higher field one to the *pro-R* group *A*.

We have already mentioned that the enantiotopic protons of benzyl alcohol do, in fact, give distinct chemical shifts in the presence of chiral shift reagents (Fraser et al., 1972). A similar effect can be achieved by "doping" an achiral shift reagent with a chiral complexer (Reuben, 1980). Thus an aqueous solution of sodium or lithium α-hydroxyisobutyrate, $(CH_3)_2COHCO_2^-$ M^+, in the presence of $EuCl_3$ (achiral shift reagent) and L-lactate, $CH_3CHOHCO_2^-$ M^+, or D-mandelate, $C_6H_5CHOHCO_2^-$ M^+, will display two sets of methyl protons (for the two enantiotopic groups) due to the formation of mixed complexes (RCO_2) $(R'CO_2)_2Eu$ ($R = C_6H_5$, $R' = CH_3$, or vice versa) (see p. 193). Moreover, assignments of chirality or heterotopicity can be made by comparing the sense of the shift produced by a given chiral complexer in an unknown situation with that for a compound of known configuration. So far, the method seems to be confined to α-hydroxy acids.

The configurational assignment of the isotopically labeled analogue required in all these cases may, of course, be achieved by synthesis from a chiral precursor. A case in point, but relating to diastereotopic nuclei, is shown in Figure 8.43 (Hill et al., 1973; Aberhart and Lin, 1973, 1974). Valine (**97**) has diastereotopic methyl groups resonating at 1.38 and 1.43 ppm (proton spectrum). In connection with an enzymatic transformation of the molecule, it became of importance to determine which group was which. The methyl groups were introduced stereospecifically starting from methyl (*S*)-lactate via (*S*)-(+)-2-propanol-d_3 and the two diastereomers ultimately obtained were separated by enzymatic resolution (at the conventional chiral center). Corresponding assignments with ^{13}C labeled methyl groups have also been described (Baldwin et al., 1973; Kluender et al., 1973).

By an analogous type of reasoning it has been ascertained (Battersby and Staunton, 1974), which of the two diastereotopic methyl groups at C(12) in Vitamin B_{12} (Fig. 8.44) is derived from methionine: It is the *pro-R* group.

Gerlach and Zagalak, (1973) have suggested that the absolute configuration of primary alcohols (RCHDOH) can be determined by the relative chemical shift of the carbinyl protons in the corresponding ($-$)-camphanate esters in the presence of the (achiral) shift reagent $Eu(dpm)_3$ where dpm = dipivaloylmethane; if this approach is general, it would also serve to identify the *pro-R* and *pro-S* hydrogen atoms in RCH_2OH.

An assignment of the chemical shifts of the diastereotopic oxygen nuclei in

$$\alpha\text{-phenethyl sulfone, } C_6H_5CH(CH_3)\overset{\overset{\displaystyle O}{\|}}{\underset{\underset{\displaystyle O}{\|}}{S}}C_6H_5, \text{ has been effected by means of } {}^{17}O$$

NMR spectroscopy (Kobayashi et al., 1981). The sulfones were obtained from the corresponding ^{17}O-labeled, diastereomerically pure (R, R^*)- and (R, S^*)-α-phenethyl phenyl sulfoxides by oxidation with unlabeled *m*-chloroperbenzoic acid. This manner of preparation defines the relative configuration of the α-

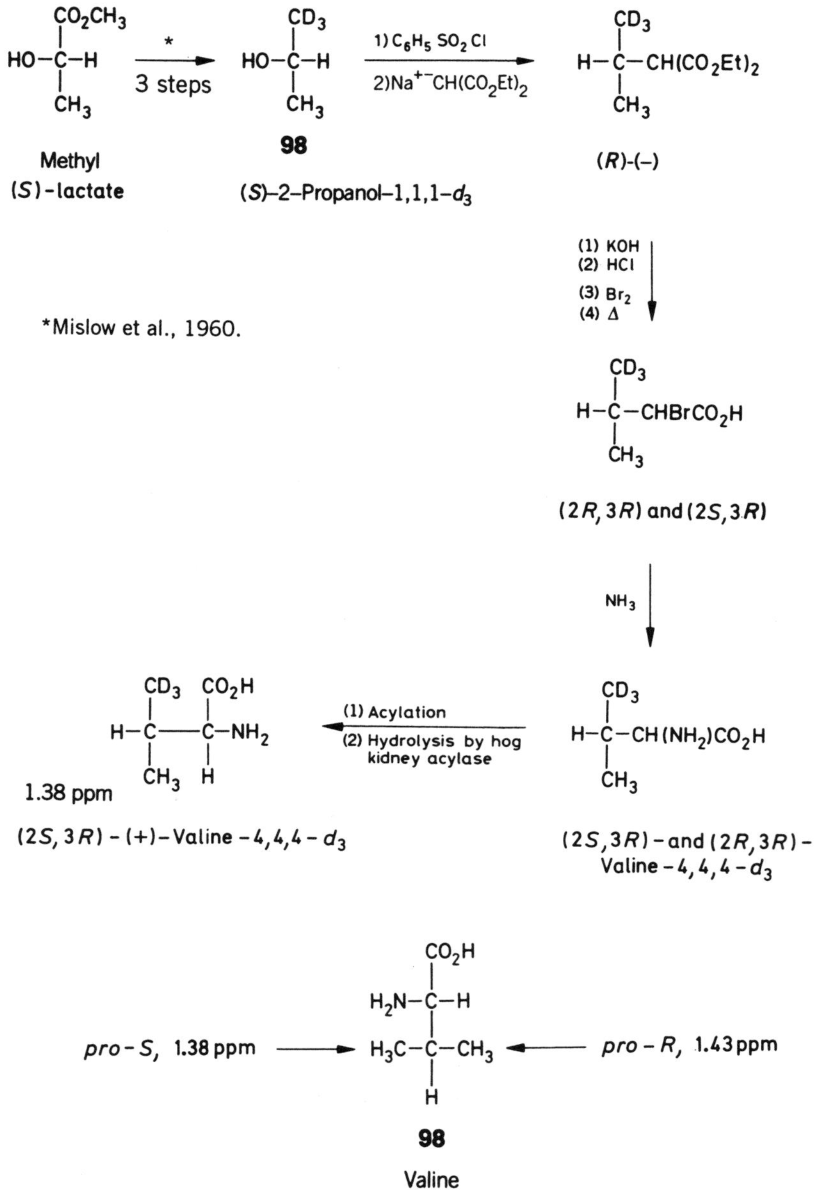

Figure 8.43. Assignment of diastereotopic methyl groups in L-valine.

phenethyl moiety and the labeled sulfone group on the likely assumption that the oxidation proceeds with retention of configuration; the difference in chemical shift between the two ^{17}O nuclei amounts to 4–6 ppm, depending somewhat on the solvent. Even larger differences are found in the corresponding α-phenylpropyl analogue (6–10 ppm). By inference, it follows which of the two diastereotopic oxygen atoms (*pro-R* or *pro-S*) in a randomly labeled α-phenethyl or α-phenylpropyl phenyl sulfone corresponds to which signal, although, in fact, the

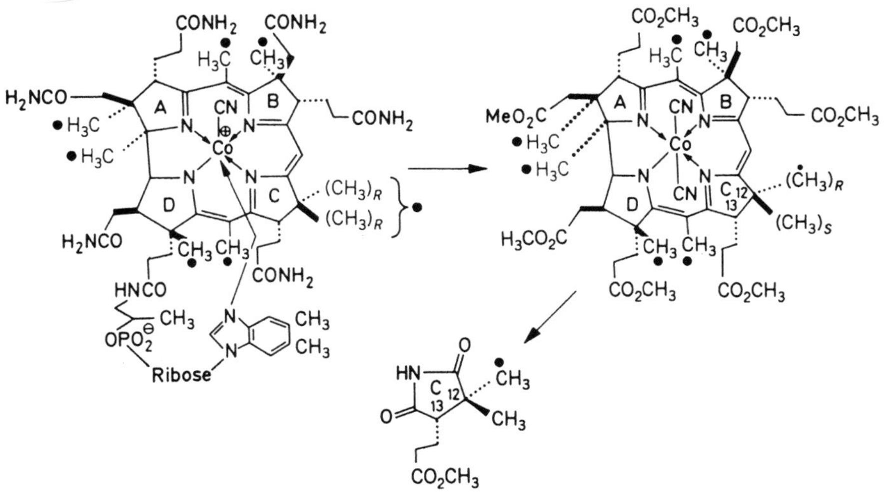

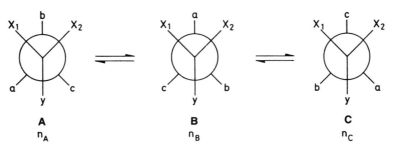

Figure 8.44. Origin of methyl groups in vitamin B_{12}. (The dotted methyl groups are derived from methionine.) Reprinted with permission from Battersby, A.R. and Staunton, J. *Tetrahedron* (1974), **30**, 1714; copyright © Pergamon, Oxford, UK.

signals could not be resolved when the level of ^{17}O was at the natural abundance (see, however, Powers, Pedersen, and Evans 1991).

c. Origin of Anisochrony

The early history of the anisochrony of diastereotopic groups is turbid because there was uncertainty as to whether the cause for the anisochrony was conformational, intrinsic, or both. The problem was finally analyzed clearly by Gutowsky (1962) whose treatment we present here (Fig. 8.45). The compound chosen for illustration is CxxyCabc in which the x nuclei are diastereotopic and anisochronous. For simplicity's sake we shall consider only the three staggered conformations shown (for one enantiomer) in Figure 8.45, assuming that the populations of all other conformations are negligible. [Strictly speaking this is not correct, at room temperature, since there will be a Boltzmann distribution of molecules among all possible conformations (cf. Tabacik, 1968). However, except in cases of very shallow energy wells, the assumption is quite adequate.] The chemical shift of x_1 in conformers **A**, **B**, and **C** may be denoted as $\delta_{a/b}$, $\delta_{a/c}$, and $\delta_{b/c}$,

Figure 8.45. Anisochronous nuclei x in mobile systems.

respectively, according to the groups at the adjacent carbon that are gauche to x_1. If n_a, n_b, and n_c are the mole fractions of **A**, **B**, and **C**, respectively, it follows that the average chemical shift of nucleus x_1 is

$$\delta_1 = n_A \delta_{a/b} + n_B \delta_{a/c} + n_C \delta_{b/c} \tag{8.1}$$

By the same token, the average chemical shift of x_2 is

$$\delta_2 = n_A \delta_{b/c} + n_B \delta_{a/b} + n_C \delta_{a/c} \tag{8.2}$$

Inspection of Eqs. 8.1 and 8.2 immediately discloses that since, ordinarily, $n_A \neq n_B \neq n_C$, $\delta_1 \neq \delta_2$, that is, x_1 and x_2 are anisochronous. Contrary to some misstatements in the literature, this conclusion is independent of the rate of rotation of the $CxyCabc$ system about the carbon–carbon bond; this rotation is assumed, throughout, to be fast on the NMR scale (see below for what happens in the limit of slow rotation).

It might thus appear, at this point, that the anisochrony is due to the unequal population of the three conformations depicted in Figure 8.45. Let us therefore consider the case (hypothetical or otherwise), where $n_A = n_B = n_C$ ($= \frac{1}{3}$). Inspection of Eqs. 8.1 and 8.2 might, at first glance, imply that, in that case, δ_1 and δ_2 are equal. But, on more careful consideration, it turns out that this is a fallacy spawned by the inadequacy of the notation. The assumption leading to this fallacious result is that $\delta_{a/b}$ in Eq. 8.1 is the same as $\delta_{a/b}$ in Eq. 8.2 (and likewise for $\delta_{b/c}$, $\delta_{a/c}$). In fact, however, the neighborhood of x_1 in **A** is not the same as the neighborhood of x_2 in **B**. For example, in the former case, passing beyond a from x_1 one reaches y. In the latter case **B**, proceeding from x_2 beyond a one reaches x_1. Hence the environment a/b of x_1 and the shift $\delta_{a/b}$ in **A** is not the same as the environment a/b of x_2 and the shift $\delta_{a/b}$ in **B**: There is an intrinsic difference so that even if $n_A = n_B = n_C$, $\delta_1 \neq \delta_2$. The conclusion, then, is that both the conformation population difference and the intrinsic difference in chemical shift within each conformer contribute to the observed anisochrony of diastereotopic nuclei in conformationally mobile systems.

Reisse, Mislow et al. (1978) have pointed out that the distinction between population difference and intrinsic difference is artificial: Nuclei are either symmetry related (i.e., interchanged by operation of a symmetry element), in which case they are homotopic or enantiotopic and thus isochronous, or they are not so related, in which case they are diastereotopic or constitutionally heterotopic and therefore anisochronous. While this is certainly correct, we believe that the distinction between population and intrinsic difference is at least pedagogically, and probably in some situations even heuristically, useful.

That intrinsic difference indeed contributes to anisochrony was first pointed out by Raban (1966) on the basis of data for $BrCF_2CHClBr$ provided by Newmark and Sederholm (1965; see also Binsch, 1973; Norris and Binsch, 1973), which are shown in Table 8.3. Conformers **A**, **B**, and **C** correspond to the diagrams in Figure 8.45 with $x_1 = F_1$, $x_2 = F_2$, $y = Br$, $a = H$, $b = Br$, and $c = Cl$. The gauche neighbors of each fluorine nucleus are indicated in the parentheses in

TABLE 8.3 Low-Temperature-Fluorine NMR Data for
$F_2BrC–CHClBr^a$

Conformer	ν_{F_2}	ν_{F_2}
A	2268.4 (H/Br)	3298.8 (Br/Cl)
B	2584.3 (H/Cl)	2628.2 (H/Br)
C	3374.8 (Br/Cl)	2467.4 (H/Cl)
Average^b	2742.5	2798.1
Found^c	2631.3	2819.1

^{a} In CF_2Cl_2 at 123 K; shifts in hertz from CF_2Cl_2 at 56.4 MHz.
^{b} Calculated value.
^{c} At 303 K, experimental values.

Table 8.3. It is immediately obvious that not only are F_1 and F_2 anisochronous in each conformer [more so in **A** and **C** than in **B**; evidently the halogen environments are similar to each other (Br or Cl) but quite different from that of hydrogen] but also there is a substantial difference between nuclei in apparently similar environments (see above). The difference $\nu_{F_2} - \nu_{F_1}$ is 359.8 Hz for the H/Br, −116.9 Hz for the H/Cl, and −76.0 Hz for the Br/Cl environments, with the average intrinsic difference being 55.6 Hz. This difference is a substantial fraction (nearly one-third) of the total observed shift difference of F_1 and F_2 at room temperature (187.8 Hz); the difference between the two numbers (132.2 Hz) is conventionally (Raban, 1966) ascribed to the population difference of the three conformers which, in the above case, happens to be relatively small ($n_A = 0.413$, $n_B = 0.313$, and $n_C = 0.274$) (Newmark and Sederholm, 1965).

It should be noted that, since the H/Br difference is opposite in sign to the H/Cl and Br/Cl differences, the average intrinsic shift difference is considerably smaller than the absolute values of the differences in the individual conformers.

Attempts were also made to establish the existence of an intrinsic shift difference by measuring the shift difference between diastereotopic nuclei as a function of temperature. It was argued that, as the temperature increases, the population difference between conformers should vanish and the residual chemical shift difference (presumably extrapolated to infinite temperature) should be an indicator of the intrinsic difference. However, it has been pointed out (Jennings, 1975) that this assumption is fraught with complications: Since shifts in individual conformers change with temperature, the ratio of conformers does not necessarily converge to unity at high temperature (it will fail to do so when the conformers differ in entropy). Moreover, the accessible temperature range is often too small for comfort. Thus if the individual differences in shift are opposite in sign in different conformers (see above), increase in population of a conformer whose individual shift has the same sign as the population shift may well lead to an initial increase in overall shift difference as the temperature is raised rather than a decrease toward the averaged intrinsic value.

Another elegant way of demonstrating intrinsic nonequivalence, but in a system at room temperature, was suggested by Mislow and Raban (1967) and reduced to practice by Binsch and Franzen (1969, 1973) and subsequently by McKenna et al. (1970). Two of the molecules studied, a bicyclic trisulfoxide **99** (Binsch and Franzen) and a quinuclidine derivative **100** (McKenna et al.) are

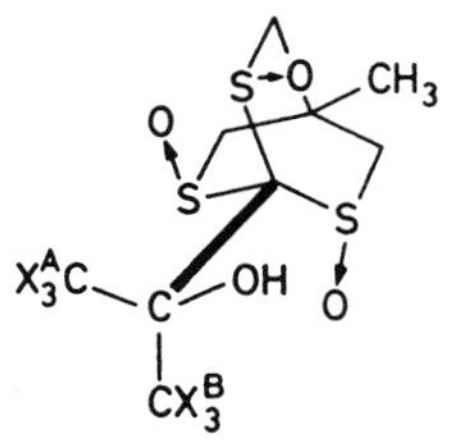

Molecules **99**, 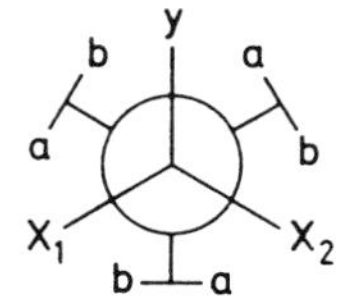 **102**, 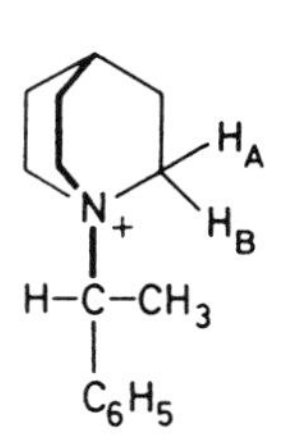 **100**, 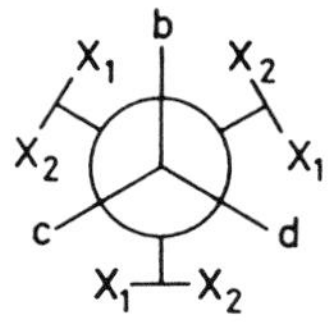 **101**

X = H, δ_{AB} = 0.038 ppm (in C_5H_5N)

X = F, δ_{AB} = 0.282 ppm

δ_{AB} = 0.095 ppm

(decoupled spectrum)

Figure 8.46. Molecules displaying intrinsically anisochronous nuclei.

shown in Figure 8.46. In both cases the presence of a threefold symmetry axis in one of the ligands (the bicyclic trisulfoxide moiety in **99**, the quinuclidine moiety in **100**) assures that the three conformers possible by virtue of rotation about the C–C or N–C bond indicated in heavy type are equally populated. The difference in chemical shifts of the CH_3 of CF_3 groups in **99** and the H_A and H_B methylene protons in **100** must therefore be intrinsic in nature. Additional examples of type **100** (general formula **101**; the general type formula for **99** is **102**) have been adduced (Franzen and Binsch, 1973; Morris et al., 1973; Gielen et al., 1974).

> If, in a molecule of type **102**, x = y, the three ligands at the front carbon are, of course, homotopic because of the C_3 symmetry. However, since the molecule as a whole is chiral (D_3), the x ligands will be chirotopic (p. 53). This has actually been demonstrated by an X-ray analysis of the electron distribution around the chromium tricarbonyl complex of hexakis(dimethylsilyl)benzene, $Cr(CO_3) \cdot C[Si(CH_3)_2H]_6$, which shows a cyclic distortion (Mislow, 1986). The same conclusion is reached by an ab initio calculation of the electron distribution in the (chiral) D_3 conformation of ethane and C_3 conformation of CF_3CH_3 (Gutierrez, Mislow, et al., 1985).

d. Conformationally Mobile Systems

This section deals briefly with the problem of averaging of heterotopic nuclei. In general, the symmetry properties of a given species are dependent on the time scale of observation in that the symmetry of structures averaged by site or ligand exchanges may be higher than the symmetry in the absence of such exchanges (Section 4-5). For the present purpose it is significant that structures lacking C_n or S_n axes may acquire such axes as a result of averaging. It follows that diastereotopic nuclei may become enantiotopic, on the average, through operation of S_n, or they may become homotopic through development of a C_n; in other words, averaging may convert anisochronous nuclei into isochronous ones.

To explore the full potential of what is sometimes called "dynamic NMR" or "DNMR" (Binsch, 1968), that is, NMR studies involving site and ligand exchange, is beyond the scope of this chapter and the reader is referred to numerous reviews (Siddall and Steward, 1969; Binsch, 1968; Lehn, 1970; Kessler, 1970;

Lambert, 1971; Orville-Thomas, 1974; Jackman and Cotton, 1975; Jennings, 1975; Steigel, 1978; Roberts, 1979; Binsch and Kessler, 1980). Only a few examples of the application of this technique can be given here, for example, in the study of ring inversion, rotation about single bonds, and inversion at nitrogen.

An example of ring inversion has already been presented: 1,1-difluorocyclohexane (Fig. 8.34). At rooom temperature the two chair forms average, and the average symmetry is that of a planar molecule, C_{2v} (Leonard, Hammond, and Simmons, 1975), in which the fluorine atoms are related by the C_2 axis and hence are equivalent. Thus the room temperature spectrum of 1,1-difluorocyclohexane displays a single (except for proton splitting) chemical shift for the two fluorine atoms, as shown in Figure 8.47 (Roberts, 1963; Spassov, Roberts, et al., 1967). In contrast, at $-110°C$ the spectrum shows the expected AB pattern for the diastereotopic fluorine atoms expected from the individual structures shown in Figure 8.47. As the temperature is gradually raised, the two doublets broaden and merge into two broad, touching peaks, which on further warming eventually coalesce into one broad single peak at what is called the "coalescence temperature" (in this case $-46°C$, at 56.4 MHz). The spectrum just above coalescence is also shown in Figure 8.47; the single peak sharpens as the temperature is raised further.

An alternative point of view is to recognize that F^1 in structure **A** (Fig. 8.34) is externally homotopic with F^2 in structure **B** (and, similarly, F^2 in **A** and F^1 in **B**). Thus F^2 in **B** has the same shift as F^1 in **A** (and, likewise, for F^1 in **B** and F^2 in **A**); it follows that when the interchange of **A** and **B** is rapid on the NMR scale, all four fluorine nuclei have the same shift, that is, become isochronous and do not couple with each other.

A simple formula for determining the rate of site exchange between two equally populated sites (Pople et al., 1959; Binsch, 1968, 1975) is

$$k_{coal} = \tfrac{1}{2}\pi\Delta\nu\sqrt{2} = 2.221\Delta\nu \tag{8.3}$$

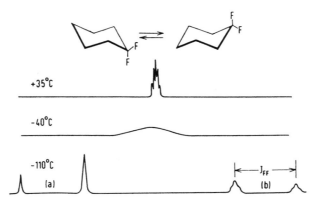

Figure 8.47. The ^{19}F NMR signals, at 56.4 MHz, of 1,1-difluorocyclohexane at various temperatures; signals of (a) equatorial and (b) axial fluorine atoms. Reprinted with permission from Roberts, J.D., *Angew. Chem. Int. Ed. Engl.* (1963), **2**, 58; copyright © VCH, Weinheim, Germany.

where $\Delta\nu$ is the chemical shift difference (in hertz) between the two exchanging nuclei at a temperature well below coalescence. This equation is valid only for a (noncoupled) single site-exchanging nucleus, for example, the proton in cyclohexane-d_{11} (observed with deuterium decoupling). In the case of geminal exchanging nuclei, as in 1,1-difluorocyclohexane (proton decoupled), Eq. 8.4 (Pople et al., 1959; Kurland et al., 1964; Binsch, 1975) should be used,

$$k_{coal} = \tfrac{1}{2}\pi\sqrt{2}\sqrt{(\nu_1 - \nu_2)^2 + 6J^2} \qquad (8.4)$$

J being the coupling constant of the two nuclei in question. Both ν and J are measured at temperatures well below the coalescence temperature. For 1,1-difluorocyclohexane (see above) $\nu_1 = 1522$ Hz, $\nu_2 = 638$ Hz, $J = 237$ Hz, and thus $k = 2349$ s^{-1} at the coalescence temperature of $-46°$C. From this information, in turn, one can calculate the free energy of activation for the site exchange from the Eyring equation (cf. Binsch, 1968, 1975; see also p. 54)

$$k = \kappa(k_B T/h)e^{\Delta G^{\ddagger}/RT} = \kappa(k_B T/h)e^{\Delta H^{\ddagger}/RT}e^{\Delta S^{\ddagger}/R} \qquad (8.5)$$

where k is the rate constant for site exchange, κ is the transmission coefficient (usually taken as unity), k_B is Boltzmann's constant, h is Planck's constant, T is the coalescence temperature, $\Delta G^{\ddagger}$ is the free energy of activation, $\Delta H^{\ddagger}$ is the enthalpy of activation, and $\Delta S^{\ddagger}$ is the entropy of activation. From the data of 1,1-difluorocyclohexane, $\Delta G^{\ddagger} = 9.7$ kcal mol^{-1} (40.6 kJ mol^{-1}) (Spassov, Roberts, et al., 1967).

There has been a controversy (Binsch, 1968; Kost, Raban, et al., 1971) regarding the accuracy of Eqs. 8.3 and 8.4, that is, the validity of inferring rate constants from coalescence temperatures. It now appears that, provided no additional coupling is present, and provided $\Delta\nu > 3$ Hz (Eq. 8.3) and $(\nu_1 - \nu_2) > J$ (Eq. 8.4), the simple coalescence method gives rate constants within 25% of those obtained by more detailed line shape analysis (see below). This leads to an error in $\Delta G^{\ddagger}$ no greater than that produced by the uncertainty of the temperature measurement ($\pm 2°$C) in the NMR probe.

In exchanges between two unequally populated sites, for example, in 4-chloro-1-protiocyclohexane-d_{10} or 4-chloro-1,1-difluorocyclohexane, the rate constants calculated from Eqs. 8.3 and 8.4, respectively, are the average of the forward and reverse rate constants, which we shall designate as k_A and k_B, respectively. These constants are, of course, no longer equal when the populations of A and B in $A \underset{k_B}{\overset{k_A}{\rightleftharpoons}} B$ are unequal at equilibrium. In this case an approximation formula, Eq. 8.6

$$k_A = (1 + \Delta n)k \qquad \text{and} \qquad k_B = (1 - \Delta n)k \qquad (8.6)$$

(Shanan-Atidi and Bar-Eli, 1970), may be used; k may be taken as the average rate constant calculated by Eq. 8.3 or 8.4 and Δn is the difference in mole fractions of the two species A and B at equilibrium (i.e., $\Delta_n = n_B - n_A$, assuming B is the predominant species).

This treatment is somewhat different from that in the original reference. Unfortunately, Eq. 8.3 no longer holds strictly when $k_A \neq k_B$, but the approximate treatment given here gives errors of less than 25% for mole fractions between 0.2 and 0.8. Outside of this region, or for more accurate results inside the region, the original graphic treatment should be used (Shanan-Atidi and Bar-Eli, 1970).

A more general method of measuring site exchange rates is the method of line shape analysis (Pople, 1959; Binsch, 1968). In this method one compares the shape of the broadened lines some 10 or 20°C above and below the coalescence temperature, as well as in the fast and slow exchange limit, with the line shape computed by means of formulas that include the rate of exchange. This method permits one to determine k over a range of temperatures and thus, through a plot of $\Delta G^{\ddagger}$ versus $1/T$, one can arrive at the values of $\Delta H^{\ddagger}$ and $\Delta S^{\ddagger}$ (though the accuracy of determining $\Delta H^{\ddagger}$ and $\Delta S^{\ddagger}$ is often low). The method is applicable to relatively complex spin systems, not just to singlet or AB exchange. It is considered to be the method of choice in determination of rate constants by NMR. A typical comparison of experimental and computed line shapes, referring to the site exchange in furfural, is shown in Figure 8.48 (Dahlqvist and Forsén, 1965).

Because each chemical shift difference between exchanging sites, as well as each spin coupling constant, gives rise to a coalescence of its own when $\Delta \nu$ or $J \approx k$, where k is the rate of site exchange, a system having many such parameters will, because of the presence of a multitude of "internal clocks" be more sensitive in the response of its NMR spectrum to temperature changes. Thus, within the limits of feasibility of computer treatment, the more shifts and coupled spins, the better.

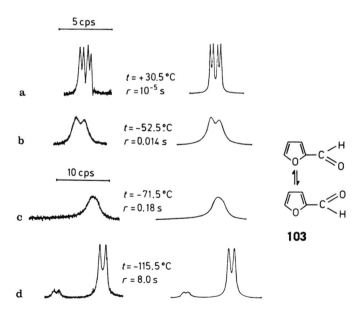

Figure 8.48. Experimental and calculated DNMR spectra for the aldehyde proton of 2-furaldehyde. Reprinted with permission from Dahlqvist, K.-I. and Forsén, S., *J. Phys. Chem.* (1965), **69**, 4068. Copyright © American Chemical Society, Washington, DC.

It is convenient to have terms for structures such as those in Figures 8.35 and 8.48, **103**, which differ only in the position of designated nuclei, and for the process of exchange of such heterotopic nuclei. The term "topomers" has been proposed for the interconverting structures, "topomerization" being the process of interchange (Binsch et al., 1971). An older term, "degenerate isomerization," seems inappropriate since the two structures shown in Figure 8.34 are homomers not isomers. "Automerization" has also been used (Balaban and Farcasiu, 1967); it properly denotes the identity of the two interconverting structures but does not address itself to the significance of the process of their interconversion.

We have already referred to studies of rotational isomerism by DNMR. Many barriers to rotation are below 5 kcal mol^{-1} (21 kJ mol^{-1}) and cannot be investigated by NMR, but there are a number of cases where the barrier is high enough for NMR study (cf. Sternhell, 1975). Frequently, such cases relate to partial double bonds (Kalinowski and Kessler, 1973; Jackman, 1975). Even homotopic ligands, such as the fluorine nuclei in a CF$_3$ group or the methyl groups (^{1}H or ^{13}C signals) in a *tert*-butyl group, may become diastereotopic in the slow exchange limit; for examples see Figure 8.49, **104** (Weigert and Mahler, 1972) and **105** (Bushweller et al., 1972).

We conclude this section with a discussion of inversion of amines of the type NR$_1$R$_2$R$_3$ (cf. Lehn, 1970; Lambert, 1971). In general this process is too rapid to be studied except in special circumstances (Bushweller et al., 1975). Nevertheless, Saunders and Yamada (1963) were able to determine the very high rate of

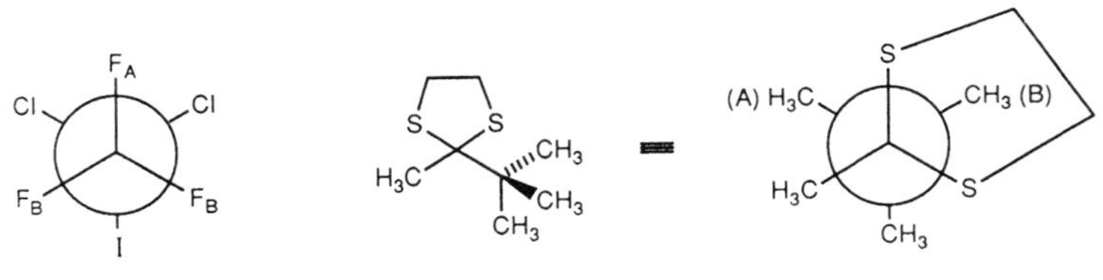

104 **105**

Figure 8.49. Diastereotopic fluorines in CF$_3$ and methyls in C(CH$_3$)$_3$.

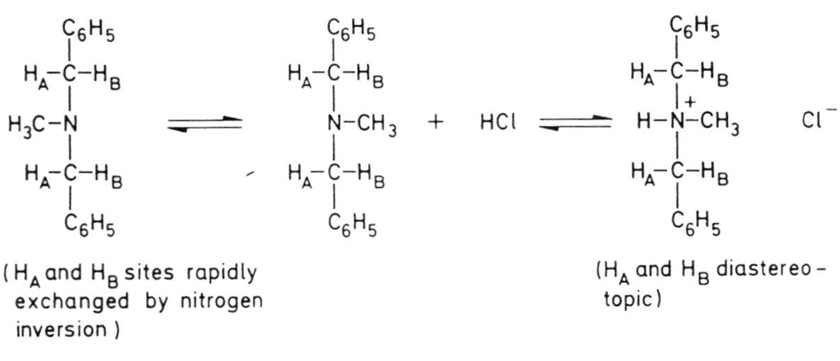

(H$_A$ and H$_B$ sites rapidly exchanged by nitrogen inversion)

(H$_A$ and H$_B$ diastereo- topic)

106

Figure 8.50. Inversion of dibenzylmethylamine.

inversion of dibenzylmethylamine (Fig. 8.50, **106**; $k = 2 \times 10^5$ s^{-1} at 25°C) by the elegant trick of partially neutralizing the amine with hydrochloric acid. Since the hydrochloride cannot invert, the benzylic protons in it are diastereotopic, and are hence anisochronous. Only the small amount of free amine in equilibrium with the salt at a given pH (the measurements were carried out on the acid side) inverts at the rate indicated and it can be easily shown (Saunders and Yamada, 1963) that $k_{obs} = k[\text{amine}]/[\text{salt} + \text{amine}]$, where k_{obs} is the observed rate of site exchange of the diastereotopic protons at a given pH, k is the rate constant for the amine inversion to be determined, and the quantity in the fraction can be ascertained from the measurement of pH and the known basicity of the amine.

e. Spin Coupling Nonequivalence (Anisogamy)

Diastereotopic nuclei that are anisochronous (i.e., differ in chemical shift) may also differ in coupling constants with respect to a third nucleus. Such nonequivalence with respect to spin coupling may be called "anisogamy" and the nuclei are "anisogamous" (term coined by Laszlo; cf. Binsch, Eliel and Kessler, 1971, footnote 14) as well as anisochronous. Thus the ^{13}C–^{1}H spin coupling constants of the diastereotopic methylene protons in $CH_3CH(OCH_2CH_3)_2$ are 139.6 and 141.0 Hz (Rattet, Goldstein, et al., 1967) and the diastereotopic methyl protons in $C_6H_5P[CH(CH_3)_2]_2$ have ^{31}P–^{1}H coupling constants of 11.0 and 14.7 Hz (McFarlane, 1968).

However, even isochronous nuclei may be anisogamous. Thus, whereas the two aromatic protons in 1,3-dibromo-2,5-difluorobenzene (Fig. 8.51, **107**) are both isochronous and isogamous and thus give rise to a single resonance, the two protons H_A and H_C ortho to bromine in p-chlorobromobenzene (Fig. 8.51, **108**) are isochronous but not isogamous; the same is true of the two protons (H_B and H_D) ortho to chlorine. Inspection of **108** shows that the coupling of H_A to H_B differs from that of H_C to H_B; therefore $J_{AB} \neq J_{CB}$, that is, H_A and H_C are anisogamous. The system is of the AA'BB' type (cf. Pasto and Johnson, 1979, p. 223). While this subject is discussed in all textbooks on NMR, it is worthwhile to state here the symmetry criterion for isogamy (or anisogamy) (Mislow, personal communication to ELE): If in a spin system of the type $A_2B_2 \ldots$ (or AA'BB'. . .) substitution of one of the B nuclei by a different nucleus Z leads to a system A_2BZ in which the remaining homomorphic nuclei A are no longer interconvertible by a symmetry operation (C_n or S_n, including σ) then B and B' are anisogamous. But if the A nuclei remain interconvertible by a symmetry operation, then the B nuclei are isogamous. This test should be applied to the

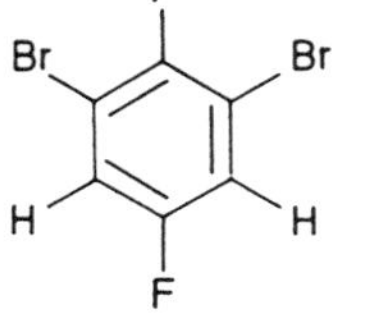

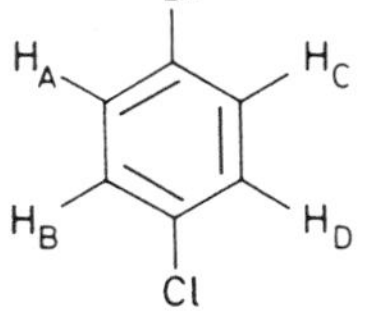

107   **108**   **Figure 8.51.** Isogamous and anisogamous nuclei.

conformation of highest symmetry. It does not apply to cases where the A and B nuclei are themselves symmetry equivalent, such as $ClCH_2CH_2Cl$ or p-dichlorobenzene. By this criterion the protons in CH_2F_2 and $CH_2=C=CF_2$ are isogamous since they are enantiotopic, and hence isochronous in CH_2FBr and $CH_2=C=CFBr$. But in $CHF=C=CHF$ or cis- or $trans$-$CHF=CHF$ the protons are anisogamous, for whereas they are related by a C_2 axis in the compounds shown, and hence are isochronous, they are no longer so related in $CHF=C=CHBr$ or cis- or $trans$-$CHF=CHBr$. The same is true for the methylene protons in $BrCH_2CH_2CO_2H$. These protons become diastereotopic in $BrCH_2CHClCO_2H$ or $BrCHClCH_2CO_2H$.

> It is of interest to note that the protons (or fluorine nuclei) in CH_2F_2 should become anisogamous in a chiral solvent, since the enantiotopic protons in CH_2FBr would become diastereotopic in such a solvent; however, attempts to demonstrate such anisogamy have not so far been successful (Mislow, personal communication to ELE).

8-5. HETEROTOPIC LIGANDS AND FACES IN ENZYME-CATALYZED REACTIONS

a. Heterotopicity and Stereoelective Synthesis

In Section 8-3 we saw that replacement of stereoheterotopic groups or addition to stereoheterotopic faces gave rise to stereoisomers. The rates of such replacements of one or the other of two ligands or additions to one or the other of two faces are frequently not the same. In particular, replacements of diastereotopic ligands or additions to diastereotopic faces usually proceed at different rates because the transition states for such replacements or additions are diastereomeric and therefore are unequal in energy. Thus the reactions shown in Figures 8.18 and 8.19 not only give rise to diastereomeric products, depending on which ligand or face is involved, but they give these products in unequal, sometimes quite unequal amounts, that is, they display diastereoselectivity (Izumi, 1971) or "diastereodifferentiation" (Izumi and Tai, 1977). Replacement of enantiotopic ligands or addition to enantiotopic faces gives rise to enantiomeric products, but here replacement of the two ligands or addition to the two faces ordinarily occurs at the same rate, because the pertinent transition states are enantiomeric and therefore are equal in energy. This situation changes, however, when the reagent (or other entity participating in the transition state, such as the solvent or a catalyst) is chiral. In that circumstance, the two transition states will, once again, become diastereomeric and the two enantiomeric products will be formed at unequal rates and in unequal amounts: the reaction will be enantioselective (Izumi, 1971) or "enantiodifferentiating" (Izumi and Tai, 1977). In this case, where prochiral starting materials give rise to nonracemic chiral products, one speaks of enantioselective (or, less appropriately, asymmetric) syntheses, discussed in detail in Chapter 12. Here we shall deal only with a few applications in enzyme chemistry.

b. Heterotopicity and Enzyme-Catalyzed Reactions

The answer to the question as to which of two heterotopic ligands or faces is involved in an enzyme-catalyzed reaction depends on, and conversely may be used to elucidate, the fit between the substrate and the active site of the enzyme. (However, it is only one of several techniques used to fathom this relationship.) The literature in this area is extensive and only the principles involved and one or two representative examples can be presented here; for more detailed and extensive information, the reader is referred to pertinent reviews (Arigoni and Eliel, 1969; Verbit, 1970; Bentley, 1970, 1976; Hanson, 1972, 1976; Simon and Kraus, 1976; Jones, 1976; Hill, 1978; Young, 1978, 1994; Cane, 1980; Eliel, 1982; Jones, 1985, 1986) and books (Bentley, 1970; Alworth, 1972; Frey, 1986).

We start the discussion with a classical experiment related to the stereochemistry of the oxidation of ethanol and reduction of acetaldehyde mediated by the enzyme yeast alcohol dehydrogenase in the presence of the oxidized (NAD^+) and reduced (NADH) forms, respectively, of the coenzyme nicotinamide adenine dinucleotide (Fig. 8.52). The stereochemically significant feature of this reaction stems from the fact that the methylene hydrogen atoms in CH_3CH_2OH and the faces of the carbonyl in $CH_3CH=O$ are enantiotopic. The question thus arises as to which of the CH_2 hydrogen atoms is removed in the oxidation and to which of the C=O faces the hydrogen attaches itself in the reduction as mediated by the enzyme and coenzyme.

Loewus, Westheimer, Vennesland (1953) found that reduction of ethanal-1-d with NADH in the presence of yeast alcohol dehydrogenase gave ethanol-1-d which, upon enzymatic reoxidation by NAD^+, returned ethanal-1-d without loss of deuterium. There is thus a "'stereochemical memory effect" involved in this reaction: the H and D of the CH_3CHDOH do not get scrambled but the same H that is introduced in the reduction is the one removed in the oxidation. The reason, as we now know, is that the two methylene hydrogen atoms are not identical, but distinguishable by bearing an enantiotopic relationship.

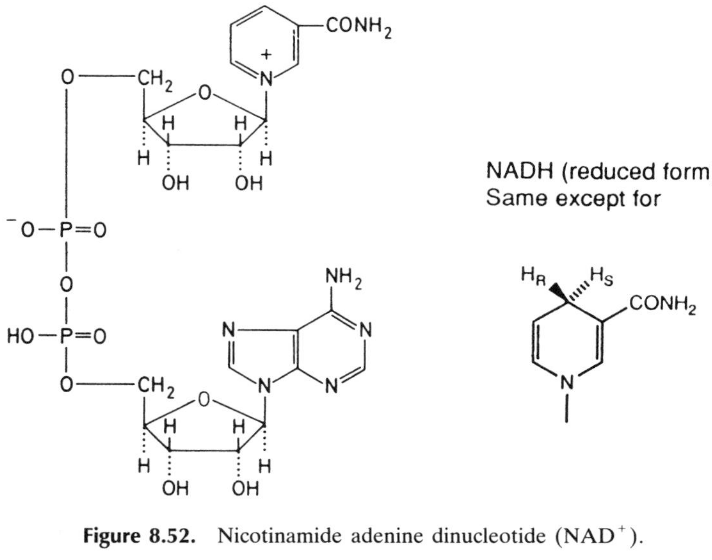

NADH (reduced form):
Same except for

Figure 8.52. Nicotinamide adenine dinucleotide (NAD^+).

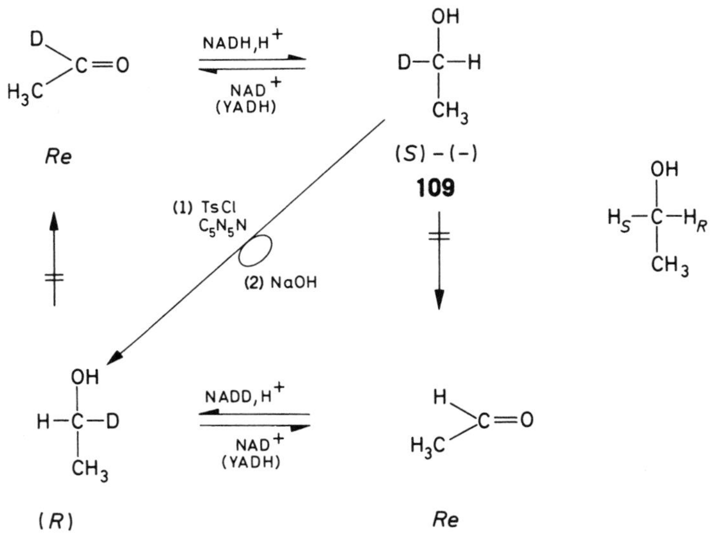

Figure 8.53. Oxidation of ethanol and reduction of acetaldehyde by NAD^+-NADH in the presence of yeast alcohol dehydrogenase (YADH).

When the configuration of the ethanol-1-d is inverted by conversion to the tosylate followed by treatment with hydroxide and the inverted ethanol-1-d is then oxidized with yeast alcohol dehydrogenase and NAD^+, the deuterium atom, which has taken the stereochemical position of the original hydrogen, is now removed and the product is largely unlabeled $CH_3CH=O$. The sequence of events is summarized in Figure 8.53.

Later experiments on a larger scale (Levy, Vennesland, et al., 1957) established that the ethanol-1-d from $CH_3CD=O$ and NADH (Fig. 8.53, **109**) is levorotatory, $[\alpha]_D^{28} - 0.28 \pm 0.03$ (neat) and this finding, coupled with the elucidation of configuration of $(-)$-ethanol-1-d as S (cf. Fig. 5.37) leads to the stereochemical picture summarized in Figure 8.53. It follows that the hydrogen transferred from the NADH in the enzymatic reduction attaches itself to the Re face of the aldehyde and that this hydrogen thus becomes H_R in the ethanol; it is H_R (the pro-R carbinol hydrogen), in turn, which is abstracted by NAD^+ in the oxidative step.

> Of course, in the enzymatic oxidation of unlabeled ethanol one cannot operationally discern that the hydrogen abstracted is pro-R. Reactions of this type have been called "stereochemically cryptic" (Hanson and Rose, 1975).

It is clear that ethanol (and acetaldehyde) must fit into the active site of YADH in such a way as to conform to these stereochemical findings. A model for the reduction of a very similar substrate, pyruvic acid, which is reduced by NADH in the presence of liver alcohol dehydrogenase (LADH) to (S)-lactic acid, is shown in Figure 8.54 (Adams et al., 1973; Vennesland, 1974). Here one can discern Ogston's picture of the three-point contact (cf. Fig. 8.8), one contact being established by the salt bond pyruvate–arginine-H^+, the second by the hydrogen bond (histidine) $N–H \cdots O=C$ (pyruvate), and the third one involving

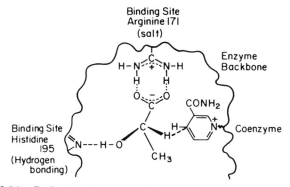

Figure 8.54. Reduction of pyruvate by NADH in the presence of LADH.

delivery of the hydrogen of NADH (bound to the enzyme) to the *Re* face of the C=O of pyruvate. Thus only (*S*)-lactate [not (*R*)-lactate] is formed in the reduction. Similarly, the model explains why, in the reverse reaction, the enzyme is substrate stereoselective for (*S*)-lactate: (*R*)-Lactate, if locked into the enzyme cavity, would have CH_3 rather than C–H juxtaposed with the NAD^+ and could thus not be oxidized.

The reduction of acetaldehyde is probably similar though the absence of the COO^- group requires the contact at the third site to be established in a different manner. It is not certain that covalent or ionic bonding is actually involved in this contact; the shape of the enzyme cavity itself, and the attendant hydrophilic and hydrophobic interactions between certain parts of the enzyme and parts of the substrate, may contribute to the required orientation of the substrate.

The study of the stereochemistry of ethanol oxidation and acetaldehyde reduction and the information relating to the topography of the enzyme derived from this study are typical of a large number of other investigations of this type. For example, the transfer of the hydrogen to and from the coenzyme involves a stereochemical problem of its own (Fig. 8.55): In the reductive step, is it H_R or H_S of the dihydronicotinamide moiety that is transferred from the coenzyme to the substrate; correspondingly, in the oxidation step, is the hydrogen abstracted from the substrate added to the *Re* or *Si* face [or, using Rose's nomenclature, (p. 486) the β or α face] of the pyridinium moiety of the coenzyme? This question was answered as summarized in Figure 8.55 (Cornforth et al., 1962; see also Oppenheimer et al., 1986).

Deuterium was transferred from a dideuterated alcohol RCD_2OH to NAD^+ in the presence of LADH. This transfer creates a chiral center at C(4) in the NAD^2H formed. Degradation of this material in the manner shown (Fig. 8.55, top half) yielded (*R*)-(−)-succinic-*d* acid, which is recognized by its known optical rotatory dispersion (ORD) spectrum. It follows that the configuration of the NAD^2H formed was *R* and it is therefore H_R that is transferred from (and to) the alcohol; the attachment of hydride to NAD^+ thus occurs from the *Re* (β) face. [This result is not general; that is, it does not apply to all oxidation–reduction reactions mediated by NAD^+–NADH (cf. You, Kaplan et al., 1978.)] To confirm this finding and to avoid any remote possibility that the β,β-dimethylallyl-d_2 alcohol used as the source of deuterium would behave differently from ethanol,

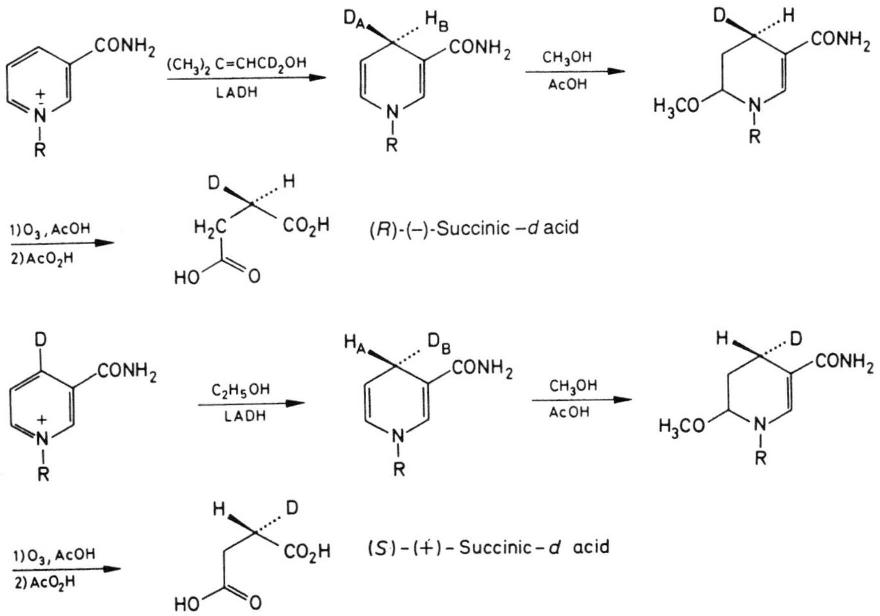

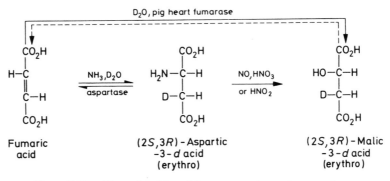

Figure 8.55. Prostereoisomerism of hydrogen transferred to the C(4) of NADH from alcohol in the presence of LADH.

the experiment was repeated with NAD-4-d^+ and ethanol (shown in Fig. 8.55, bottom part). In this case, of course, the ultimate degradation product is (S)-$(+)$-succinic-d-acid.

Next we take up the stereochemistry of an enzymatic addition to a C=C double bond: the hydration of fumaric to (S)-malic acid (England and Colowick, 1956) and the amination of fumaric to (S)-aspartic acid (Krasna, 1958). Both of these reactions are of industrial importance (cf. Chibata, 1982). They are summarized in Figure 8.56. The absolute configurations of both $(-)$-malic and $(-)$-aspartic acids are well known and *erythro*- and *threo*-malic-3-d acids have been identified by NMR spectroscopy (being diastereomers, they differ in NMR spectrum) and their configuration has been unambiguously assigned by a synthesis of controlled stereochemistry (Fig. 8.57; Gawron and Fondy, 1959; Anet, 1960;

Figure 8.56. Stereochemistry of fumarase and aspartase reactions.

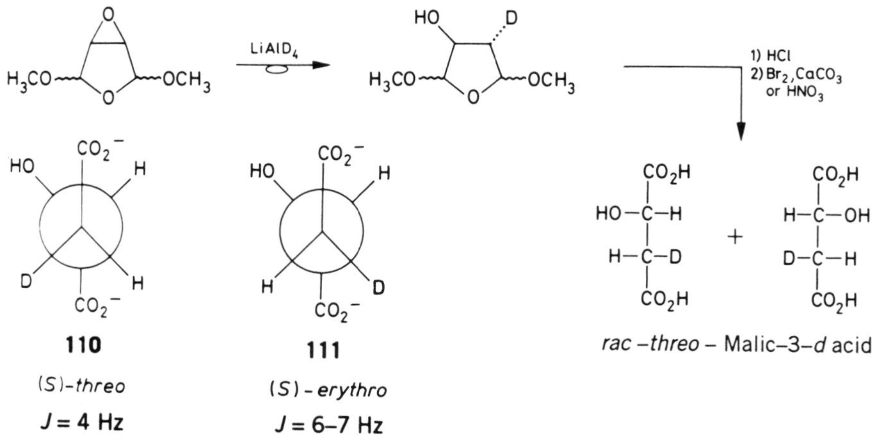

Figure 8.57. Synthesis of racemic *threo*-malic-3*d* acid.

Gawron et al., 1961). In the dianion of the threo acid (**110**) in D_2O solution, the carboxylate groups are predominantly anti (because of electrostatic repulsion) and it follows that the hydrogen atoms are gauche; the coupling constant of these protons is therefore small ($J = 4$ Hz). In contrast the erythro isomer **111** obtained by biosynthesis (see below) has its hydrogen atoms predominantly anti to each other and their coupling constant is thus larger ($J = 6$–7 Hz).

Though addition of D_2O to fumaric acid is reversible, it yields only mono-deuterated malic acid and involves recovery of undeuterated fumaric acid. This indicates that the addition and elimination steps are stereospecific and that their stereochemical course (syn or anti) is the same. The formation of the erythro isomer of (*S*)-($-$)-malic-3-*d* acid in the fumarase mediated D_2O addition (Fig. 8.56) relates the absolute stereochemistry of C(3) to that at C(2) and proves that the configuration at C(3) is *R*. Since the (*S*)-($-$)-aspartic-3-*d* acid formed in aspartase mediated ammonia addition to fumaric acid (Fig. 8.56) is converted to the same (2*S*,3*R*)-($-$)-malic-3-*d* acid by nitrous acid deamination, and since the latter reaction does not affect the configuration at C(3), the aspartic acid must also be 3*R*. (The fact that the steric course of the nitrous acid deamination at C(2) involves retention was already well known in the literature.) Then it follows that the hydrogen added (or abstracted, in the reverse reaction) at C(3) in the conversion of fumaric to malic or aspartic acid is the *pro-R* hydrogen and that addition to the fumaric acid of the proton from either water or ammonia proceeds from the *Re* face at C(3). The *Re* face is the front face of the double bond in Fig. 8.56.) On the other hand, since the configuration at C(2) in both cases is *S*, the addition at C(2) in both cases must be from the rear face in Figure 8.56 (i.e., from the *Si* face). The overall picture, then, is one of anti addition giving the 2*S*,3*R* isomer (Fig. 8.58). An analogous steric course is observed in the addition of water to maleic, citraconic (α-methylmaleic), and mesaconic (α-methyl-fumaric) acids (cf. Arigoni and Eliel, 1969).

We finally return to the subject of citric acid. We posed the question in Section 8.2 as to whether the four hydrogen atoms in citric acid (Fig. 8.4) were distinguishable and we have already seen that H_A and H_B (or H_C and H_D) are

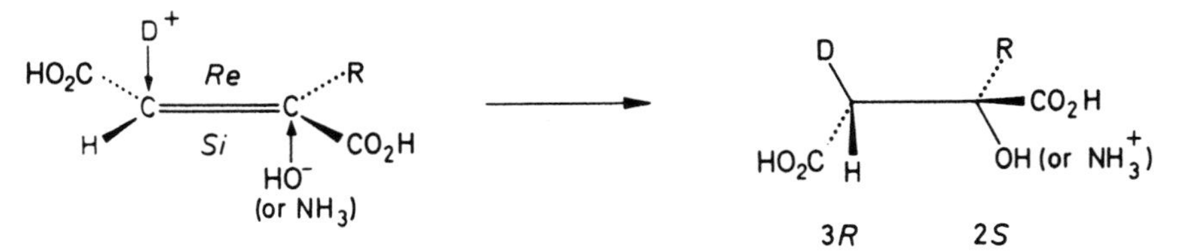

Figure 8.58. Steric course of addition of D_2O and ND_3 (or NH_3/D_2O) to fumaric acid.

diastereotopic, and therefore, at least in principle, are distinguishable by ^1H NMR spectroscopy (cf. Kozarich et al., 1986). These hydrogen atoms will also be distinguishable by their reactivity in nonenzymatic reactions. However, not only are enzymatic reactions likely to be more selective than other types of reactions as between diastereotopic groups (because of the already mentioned multiple interaction of the substrate with the enzyme and its active site) but, in addition, since the enzyme is chiral, it also "sees" the enantiopic ligands H_A and H_C (H_B and H_D) in a diastereotopic environment and thus can distinguish between them. It is therefore to be expected that the dehydration of citric acid mediated by aconitase will affect only one of the four hydrogen atoms and indeed this has been found to be so (England and Colowick, 1957): When citric acid is equilibrated with isocitric acid via aconitic acid in the presence of aconitase (Fig. 8.59), but with D_2O instead of H_2O, only one of the four methylene positions acquires deuterium. Hence, only one of the four hydrogen atoms is eliminated in the dehydration step to aconitic acid and replaced in the reverse reaction.

The problem of identifying which hydrogen atom is eliminated can be factorized into two parts:

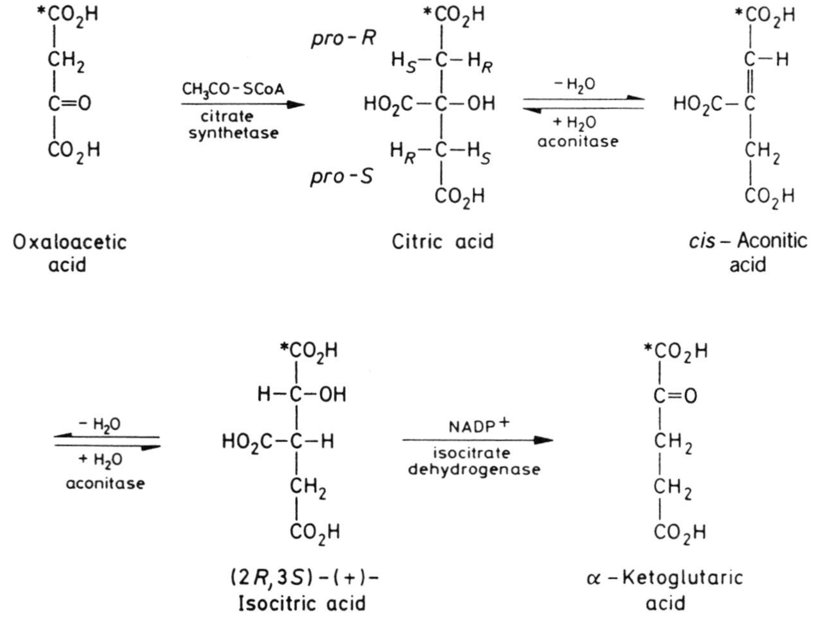

Figure 8.59. Part of tricarboxylic acid cycle.

1. Does the hydrogen come from the *pro-R* or the *pro-S* CH_2CO_2H branch of the citric acid? It is known from the carbon labeling experiments already discussed in Section 8-2 (cf. Fig. 8.6) that the aconitase-active branch is also the one derived from the oxaloacetic acid (cf. the asterisks, used to identify labeled carbon atoms, in Fig. 8.59).

2. Within the branch specified is the hydrogen abstracted H_R or H_S?

The first problem was solved (Hanson and Rose, 1963) through synthesis (followed by enzymatic degradation) of tritiated citric acid stereospecifically labeled in the *pro-R* branch, as indicated in Figure 8.60. The starting material for this synthesis is the naturally occurring 5-dehydroshikimic acid of known configuration. Enzymatic hydration of the double bond in this acid with tritiated water gives the corresponding dihydro compound whose configuration, at C(1), is *R*.

The configuration is *R* because the carbonyl-substituted segment of the ring has precedence over the hydroxyl-substituted segment. The configurational symbol is independent of the tritium labeling; it would be *R* in the corresponding H analogue as well, as required by the sequence rules (Chapter 5). However, in the quinic-6-*t* acid, the configuration is *R* because the applicable sequence rule here demands that the labeled branch precede the unlabeled one, there being no other material difference in the branches.

We turn now to the question of whether the hydrogen removed within the *pro-R* branch is *pro-R* or *pro-S*. An earlier performed sequence of reactions (Englard, 1960), shown in Figure 8.61, had already led to that information. Both of the (2*S*,3*R*)- and (2*S*,3*S*)-malic-3-*d* acids were synthesized by the method discussed earlier (Fig. 8.56) or an extension thereof. These acids were then

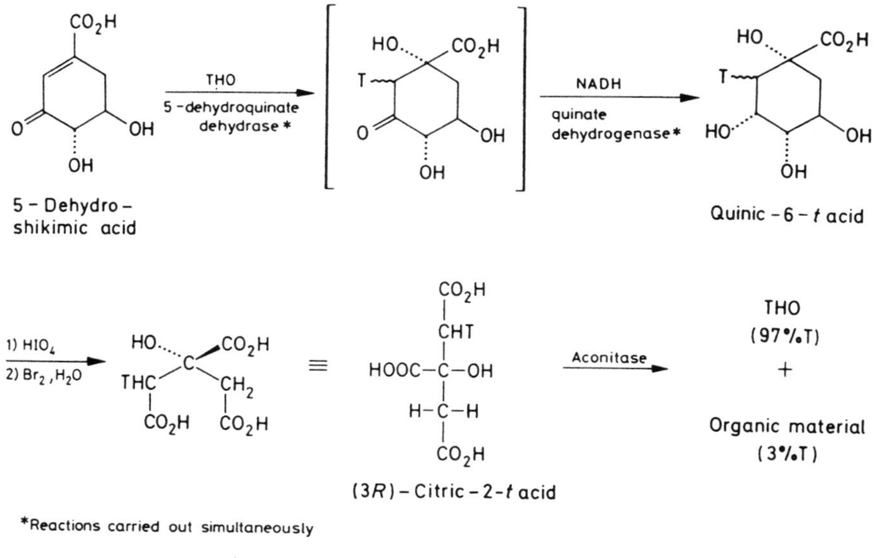

*Reactions carried out simultaneously

Figure 8.60. Stereospecific synthesis of (3*R*)-citric-2*t* acid.

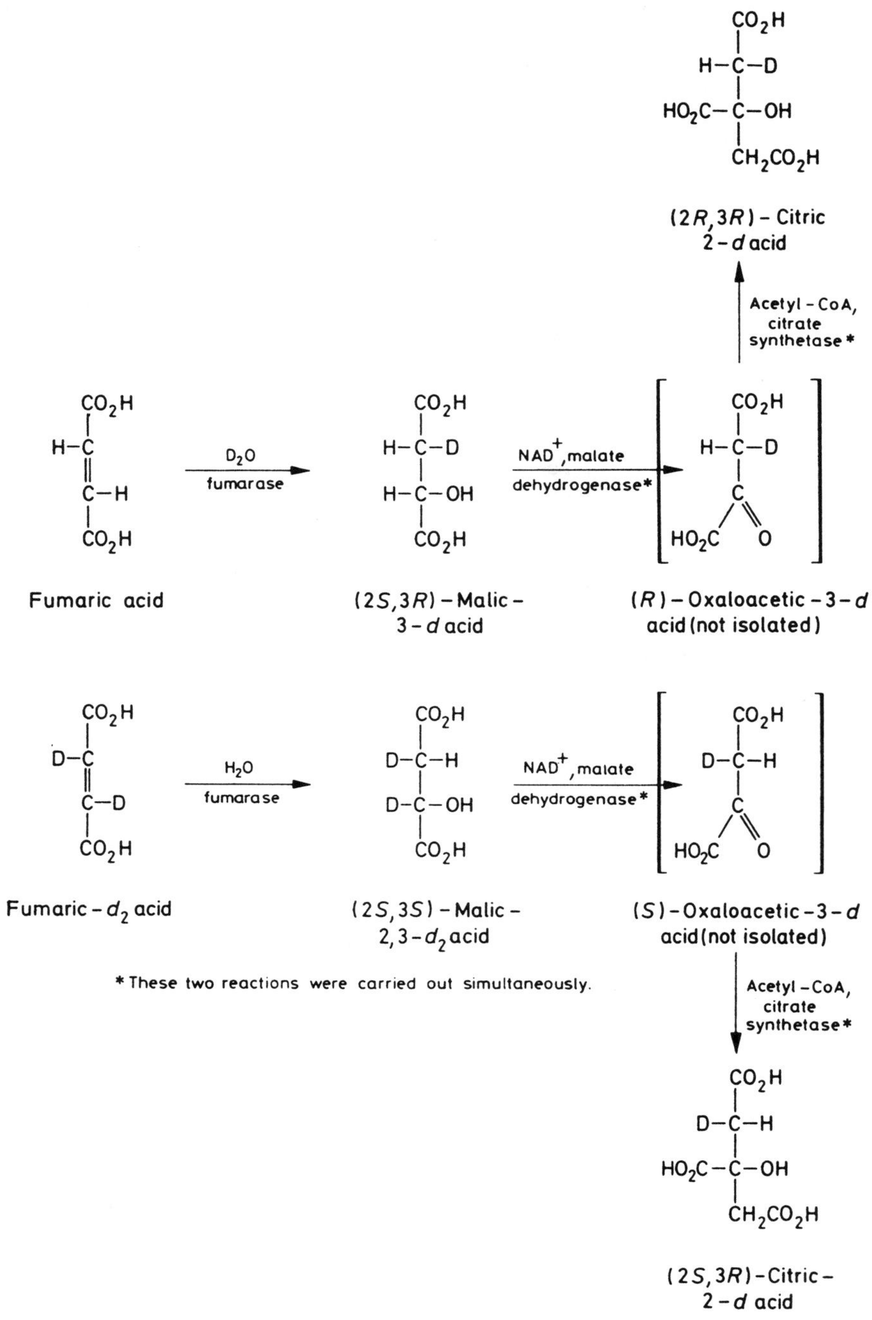

Figure 8.61. Stereospecific labeling of citric acid at C(2).

oxidized enzymatically to oxaloacetic-3-d acid (3R or 3S, depending on the precursor), and the oxaloacetate was condensed in situ with acetyl-CoA by means of citrate synthetase to give citric-2-d acid. Each of the two citric-2-d acids (Fig. 8.61) was now incubated with aconitase. Since it was already known (Evans and Slotin, 1941) from carbon labeling studies (cf. Fig. 8.6) that the oxaloacetate derived branch is the one affected by aconitase, success in this incubation was guaranteed. Indeed, the (2R,3R)-citric-2-d acid lost all of its deuterium in the dehydration to aconitic acid, whereas the 2S,3R isomer retained at least 80% of it, that is, the *pro-R* hydrogen at C(2) is lost. Thus, it may be concluded from the two experiments taken together that the *pro-R* hydrogen in the *pro-R* CH_2CO_2H branch is the one labilized by aconitase.

The conclusion is summarized in Figure 8.62, which shows that, given the known 2R,3S configuration of isocitric acid (cf. Arigoni and Eliel, 1969), the addition of water to aconitic acid to give either citric or isocitric acid proceeds in antiperiplanar fashion to the *Re* face at C(2) and to the *Re* face at C(3) in *cis*-aconitic acid. Finally, the addition of acetyl-coenzyme A (AcCoA) to oxalo-acetic acid proceeds from the *Si* side of the carbonyl function.

> *Si* addition does not occur with all citrate synthetases; enzymes from some sources lead to addition to the *Re* face (Gottschalk and Barker, 1966, 1967). This finding that enzymes from different sources promote one and the same reaction in stereochemically distinct fashion is by no means unique.
>
> Working backward from the results, it may also be concluded that the configuration of the quinic-6-t acid (Fig. 8.60) at C(6) is R and that the enzymatic addition of THO to 5-dehydroshikimic acid is syn.

Many additional examples of the elucidation of prostereoisomerism in biochemical reactions could be given, for example, the elegant elucidation by Cornforth and co-workers (Cornforth, 1969, 1973, 1976; cf. Arigoni and Eliel,

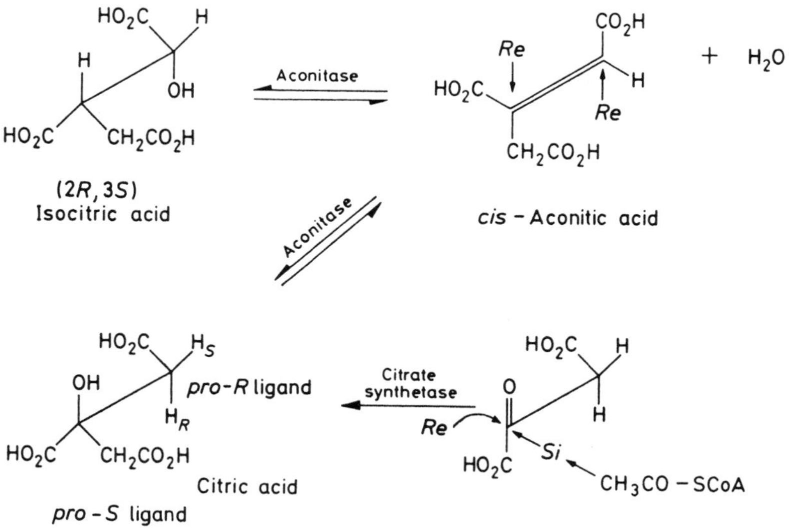

Figure 8.62. Stereochemistry of citric acid cycle.

1969; Cane, 1980) of the biosynthesis of squalene, which was recognized by the Nobel prize in chemistry in 1975, or the more recent studies of the enzymatic decarboxylation of tyrosine (Battersby et al., 1980a), histidine (Battersby et al., 1980b), and 5-hydroxytryptophan (Battersby et al., 1990), as well as the elucidation of the stereochemistry of pyrrolizidine alkaloid biosynthesis (Kunec and Robins, 1985). The stereochemistry of metabolic reactions of amino acids has been reviewed (Young, 1994).

Finally, it should be mentioned that isotopes of carbon (^{12}C, ^{13}C) and of oxygen (^{16}O, ^{18}O) have been used, in addition to hydrogen isotopes, in the elucidation of enzyme stereochemistry (cf. Eliel, 1982, pp. 57–59).

8-6. PRO^2-CHIRAL CENTERS: CHIRAL METHYL, PHOSPHATE, AND SULFATE GROUPS

a. Chiral Methyl Groups

(For reviews, see Floss et al., 1984, 1986; Floss and Lee, 1993)

The earlier-described (pp. 516–517) elucidation of the stereochemical course of the AcCoA-oxaloacetate reaction is incomplete in one respect: There is no indication whether reaction of the methyl group of CH_3CO–CoA proceeds with retention or inversion. Posing the question in this way may be momentarily puzzling, but had the condensation, instead of involving acetate, involved the metabolic poison fluoroacetate ($FCH_2CO_2^-$), the methylene carbon in this compound would be prochiral and one could ascertain whether the *pro-R* or *pro-S* hydrogen were involved in the carbanion formation by working with chiral $FCHTCO_2^-$. Also, by determining the configuration of the fluorocitrate formed, one could find out whether the condensation proceeds with retention or inversion (cf. Alworth, 1972, p. 108). In principle the same considerations should apply to deuterioacetate ($DCH_2CO_2^-$) in which the two hydrogen atoms have become enantiotopic. Stereospecific labeling with tritium would then require chiral $CHDTCO_2H$. This section deals with the synthesis of this material and its use in enzymatic reactions (Floss and Tsai, 1979; Eliel, 1982; Floss et al., 1984). The latter subject is complicated because, unlike in the fluoroacetate case, hydrogen abstraction yielding the intermediate carbanion may occur with any of the three isotopes.

Centers of the type $\underline{C}H_3CO_2H$ or $RO\underline{P}O_3^{2-}$ or $RO\underline{S}O_3^-$ have been called "pro-prochiral" or pro^2-chiral by Floss (Floss and Tsai, 1979, Floss et al., 1984; Mislow and Siegel, 1984). This term may be interpreted in a purely formal way (replacement of one of the ligands at a pro^2-chiral center by another, different from it and from the fourth ligand, yields a prochiral center) or it may be interpreted in terms of a desymmetrization procedure (Mislow and Siegel, 1984; cf. Section 4-4); the term pro-pro-prochiral or pro^3-chiral has also been used for centers such as $\underline{C}H_4$ or $\underline{P}O_4^{3-}$ or $\underline{S}O_4^{2-}$.

As will be shown below, distinguishing homomorphic ligands at pro^2-chiral centers is generally more complicated than distinguishing such ligands at prochiral or

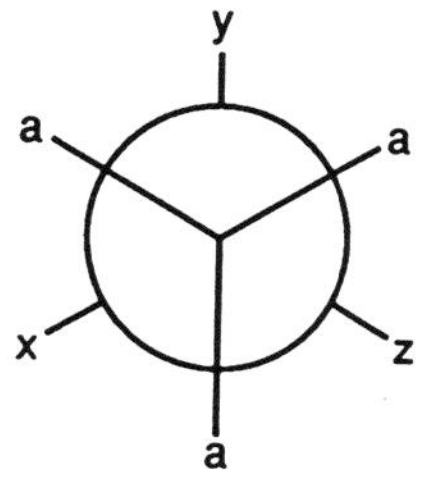

Figure 8.63. Distinguishable ligands in Ca_3Cxyz.

prostereogenic centers. Nonetheless, ligands at pro²-chiral centers that are chirotopic (type $a_3C-Cxyz$) are in principle distinguishable by classical methods, since their environment differs provided rotation about the pertinent C–C bond is slow. Examples of the type shown in Figure 8.63 have been adduced by Ōki (Suzuki, Ōki, et al., 1973; see also Ōki, 1983, p. 53ff).

The synthesis of $CHDTCO_2H$ is, in principle, straightforward and was accomplished in two laboratories in 1969 (Cornforth et al., 1969, 1970; Lüthy, Arigoni, et al., 1969). A modified version of the Cornforth synthesis (Lenz, Cornforth, et al., 1971a) is shown in Figure 8.64. Most of the steps are self-evident. The doubly labeled phenylmethylcarbinol **112** is, of course, a race-mate in the first instance; but the relative configuration of the two chiral centers is defined as $1RS,2RS$ because of the inversion in the epoxide ring opening (cf. Chapter 5); that is, only one of the two possible diastereomers is obtained. The resolution at the methyl group thus occurs automatically when the carbinol is resolved by classical methods.

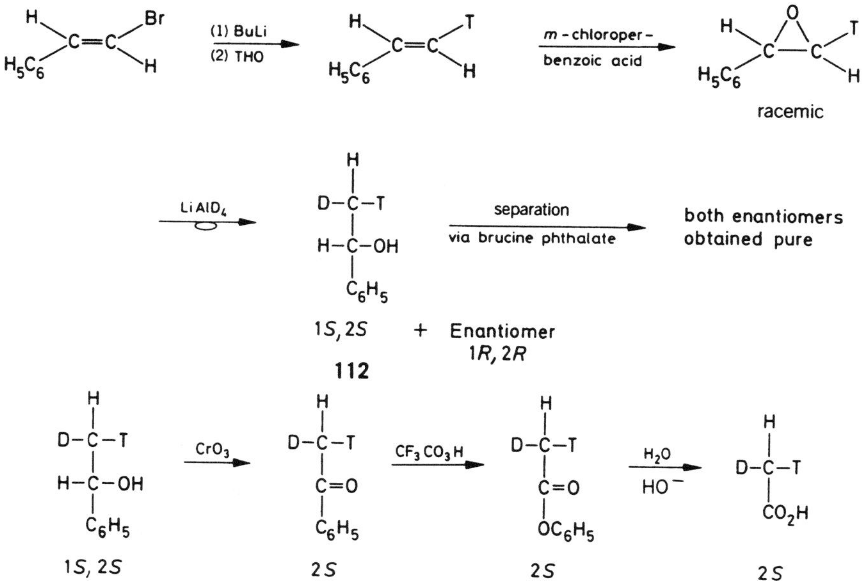

Figure 8.64. Synthesis of $(2S)$-$CHDTCO_2H$. The $2R$ enantiomer is obtained from the enantiomeric $1R, 2R$ precursor in analogous manner.

The method originally described by Arigoni's group (1969) involved generation of the chiral center by enzymatic means, but later an elegant, purely chemical method for the synthesis of chiral $CHDTCO_2H$ with high specific radioactivity was described by the Zurich investigators (Townsend, Arigoni, et al., 1975). Other methods for the synthesis of chiral CHDTX compounds have since been developed (cf. Floss et al., 1984).

Whereas synthesis of chiral, nonracemic $CHDTCO_2H$ of known configuration involves only classical principles, establishment of the configuration, enantiomeric purity, or even the chiral nature of this acid (or any other CHDTX species) cannot readily be effected by classical methods. Quite apart from the question as to whether a chiral center of type CHDTX would display measurable rotation, the $CHDTCO_2H$ is diluted at least 1000-fold by the carrier CH_2DCO_2H, since the tritium is normally present at the tracer level. Thus even if the pure material had a small rotation, it would not be detectable at the existing dilution. Recognition of configuration thus presents a special problem whenever the chiral acetic acid is obtained as a reaction product. The problem was solved (Cornforth et al., 1969, 1970; Lüthy, Arigoni, et al., 1969) by condensing the acetic acid (as CoA ester) with glyoxylate to yield malate and then diagnosing whether the malate formed had the tritium in the *pro-R* or *pro-S* position, by the method already outlined in Figure 8.56. The reaction sequence utilized is shown in Figure 8.65.

Since rapid rotation of the methyl group in $CHDTCO_2H$ averages the position of H, D, and T, vitiating direct experimental configurational distinction, the question as to which is abstracted in the formation of the carbanion or carbanionoid intermediate in the aldol condensation to give malate is resolved by making use of the isotope effect. If the isotope effect is normal [as implied by the fact (Cornforth et al., 1969, 1970) that the malate contains more than two-thirds of the tritium of the acetate] ease of abstraction will be H > D > T. Fortunately, tritium abstraction may be disregarded, for when it occurs, the resulting malate will be nonradioactive and will become comingled with the carrier material as far as counting radioactivity is concerned.

If hydrogen is abstracted preferentially over deuterium, then retention of configuration in the condensation means that (R)-AcCoA gives $(3R)$-malate-3-d,t

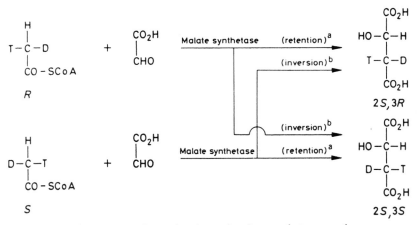

Figure 8.65. Stereochemistry of malate synthetase reaction.

and (S)-AcCoA $(3S)$-malate-3-d,t (Fig. 8.65, arrows labeled a). But if inversion is the course, the (R)-AcCoA gives $(3S)$-malate-3-d,t and (S)-AcCoA $(3R)$-malate-3-d,t (Fig. 8.65, arrows labeled b).

To diagnose the situation, the malates formed were treated with fumarase. The one formed from (R)-acetyl-CoA became equilibrated with tritiated fumaric acid and retained most of its tritium, whereas that from the (S)-acetyl-CoA lost most of its tritium in the reversible dehydration yielding unlabeled fumaric acid (and, then, unlabeled malic acid). Reference to Figure 8.56 shows that the former malate was thus $3S$ and the latter was $3R$, that is, the condensation proceeds with *inversion* of configuration.

The extent of retention of tritium is $79 \pm 2\%$ in the former case and $21 \pm 2\%$ in the latter (Lenz, Cornforth, et al., 1971a; Lenz, Eggerer, et al., 1971b; Lenz and Eggerer, 1976); complete retention (or complete loss) of tritium is, of course, not to be expected, since the isotope effect for extrusion of deuterium rather than protium in the condensation shown in Figure 8.65 is not infinite. The percent tritium retention in the process of dehydration and water exchange (Fig. 8.56) is now generally called the "F value" (cf. Floss et al., 1984); thus enantiomerically pure $CHDTCO_2H$ gives an F value of either 79 (R configuration) or 21 (S configuration); values below 79 or above 21 (i.e., closer to 50), indicate that the acid is not enantiomerically pure. The F-value determination thus takes the place of the classical determinations of enantiomeric purity (cf. Chapter 6) used for conventional chiral compounds of the type CHabc and CHDab. Other methods for determining the configuration and enantiomeric purity of $CHDTCO_2H$ have been described (cf. Floss and Tsai, 1979; Floss, 1984).

Surprisingly, it is also possible to ascertain the configuration of CHDTX*, where X* is an appropriate chiral group, directly by ^{3}H NMR spectroscopy even when rotation about the CHDT–X* bond is fast. Altman et al. (1978), had already used tritium at a high level of incorporation (ca. 20%) to ascertain the configuration of CHTX*Y by ^{3}H NMR. Following earlier experiments with CH_2DX^* (Anet and Kopelevich, 1989), Anet, Floss, et al. (1989) found that the tritium resonances in the two diastereomers of CHDTX*, where X* is 2-methylpiperidyl, are separated by 0.014 ppm (6.9 Hz at 500 MHz), sufficient to ascertain stereoisomeric purity and (knowing the position of the appropriate resonances from model experiments) configuration of the CHDTX* moiety since that of X* is known. The appropriate sequence of reactions is shown in Figure 8.66.

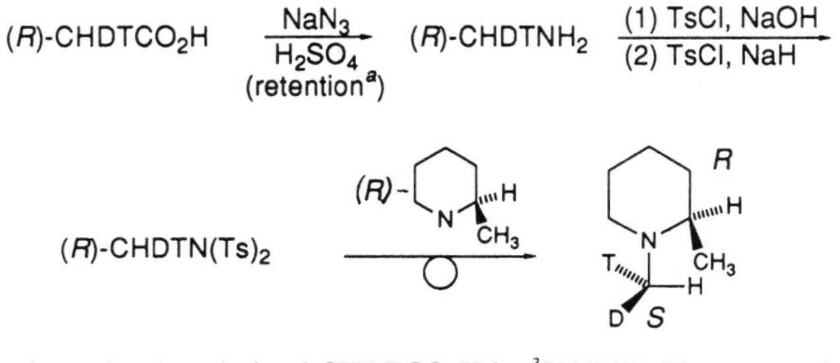

Figure 8.66. Configurational analysis of $CHDTCO_2H$ by ^{3}H NMR. The superscript a indicates that some racemization as well as loss of label occurs in this step. Ts = p-$H_3CC_6H_4SO_2$-.

The ability to analyze $CHDTCO_2H$ stereochemically was put to use in the elucidation (Eggerer et al., 1970; Rétey, Arigoni, et al., 1970) of the stereochemistry of the citric acid condensation to which we have already alluded. Before presenting one of the determinations of the stereochemistry of the citrate synthetase reaction, however, we must develop that of the reverse process, the citrate lyase reaction (Eggerer et al., 1970). The key to this determination is the synthesis of citric acid stereospecifically labeled at the methylene group in the *pro-S* branch [the carbon(1,2) branch of the (2R,3S)-citric-2-*t* acid (Fig. 8.67, **113**)]. At first sight this seems to be difficult to accomplish, at least enzymatically, because that branch is the AcCoA derived one (cf. Fig. 8.61). This difficulty was circumvented by using *Re*-citrate synthetase (rather than the usual *Si*-citrate synthetase, see p. 517) to prepare citric acid (see Fig. 8.67). The stereospecifically

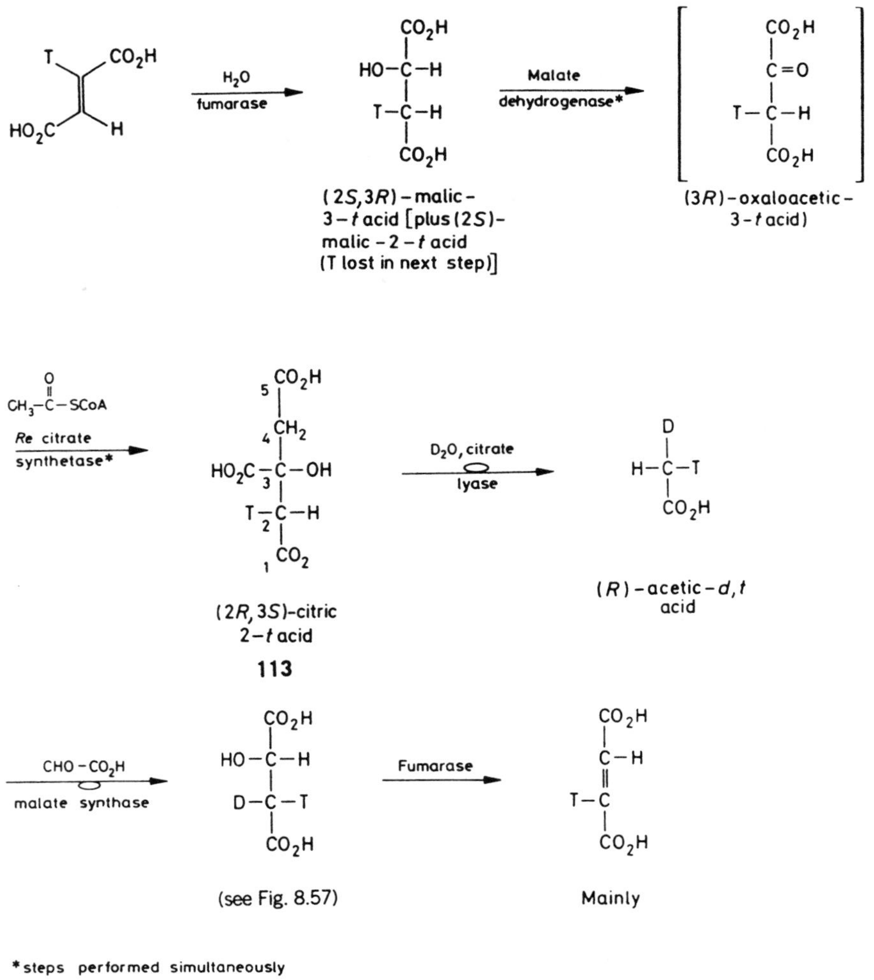

*steps performed simultaneously

Figure 8.67. Stereochemistry of citrate lyase reaction.

labeled $(2R,3S)$-citric-2-t acid **113** was then subjected to the citrate lyase reaction and the acetic-d,t acid formed was diagnosed to be R by the malate synthetase–fumarase sequence shown earlier (Fig. 8.65). It follows that the citrate lyase reaction proceeds with *inversion* of configuration, as shown in Figure 8.67. It should be noted that this result is independent of any assumptions on isotope effects, for while such effects are involved in the transformation XCHDT→XCYDT (rather than XCHYT), they are *not* involved in the reverse process. The diagnosis of, say, the R-configuration of any sample of acetic-d,t acid is independent of the stereochemical course of the malate synthetase reaction. It only makes use of the fact that the course of this reaction is the same as it is when one starts with known (R) acid whose stereochemistry is unequivocally determined by its synthesis.

With the stereochemistry of the citrate lyase reaction determined, that of the Si-citrate synthetase (the common enzyme) was established as shown in Figure 8.68. Condensation of (R)-acetic-d,t acid (configuration known by synthesis) with oxaloacetate gives what turns out to be mainly $(2S,3R)$-citric-2-d,t acid **114**. When this acid is then cleaved with citrate lyase, the major product is (R)-acetic-d,t acid, as established by the malate synthetase–fumarase diagnosis. It follows that both the Si-citrate synthetase and citrate lyase reactions must involve the same stereochemical course. Since that of the lyase reaction is inversion (see above), that of the Si synthetase reaction must be inversion also. And since the overall stereochemical result shown in Figure 8.68 is not dependent on the magnitude of the isotope effect, neither is the demonstration of inversion in the Si-citrate synthetase reaction.

> Because the H/D isotope effect is not infinite, there will also be formed some $(2R,3R)$-citric-2-t acid in which the deuterium was abstracted from the AcCoA. This material will give CH_2TCO_2H in the citrate lyase reaction which, being achiral, behaves like racemic $CHDTCO_2H$ in the subsequent steps. The lower the isotope effect, the less the preservation of enantiomeric purity in the overall reaction. However, even if there were an inverse isotope effect, the stereochemical outcome would not be altered; that is, one would not obtain any excess of (S)-acetic-d,t acid in the end, rather the R acid would be isotopically diluted by much CH_2TCO_2H.

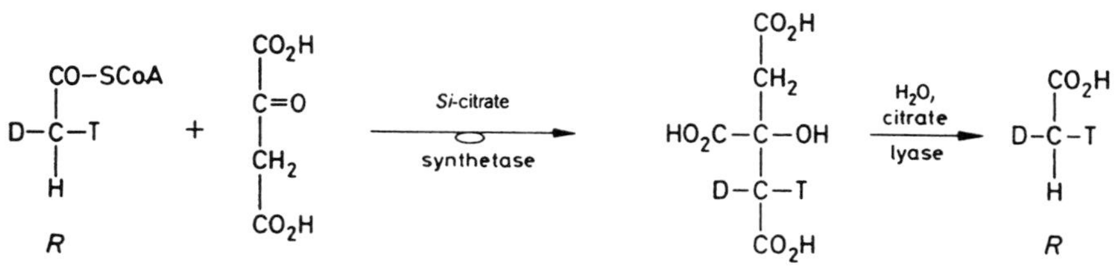

114

(predominant
product)

Figure 8.68. Stereochemistry of Si-citrate synthetase reaction.

Because of the existence of a detailed review (Floss et al., 1984) we shall not summarize the by now quite extensive literature on chiral methyl groups in detail. Suffice it to point out that there are three ways in which chiral methyl groups have been used in stereochemical studies, mostly in bioorganic chemistry. (There is no reason why the CHDTX group could not also be used in classical mechanistic organic studies, except possibly that the practitioners of such studies may be unfamiliar with the enzymatic methodology required in the diagnosis of configuration.) The first instance, exemplified by the malate synthetase (Fig. 8.65) and citrate synthetase (Fig. 8.68) reactions, involves a transformation of the type CHDTX→CDTXY. It is in this transformation that intervention of an isotope effect is required for observation of the stereochemistry, as explained earlier (see below).

Ordinarily one assumes that the isotope effect is normal, meaning that H is replaced or abstracted faster than D. If this assumption is wrong, that is, if there is an inverse isotope effect, then the answer obtained (retention or inversion of configuration) is the opposite of the correct one. Fortunately, such instances are rare and can be guarded against in various ways. One approach is to check the isotope effect independently (Lenz and Eggerer, 1976). A second approach (Sedgwick and Cornforth, 1977) is applicable when the hydrogen isotopes in the CDTXY (or CHTXY) product are in an exchangeable position. Let us take the case depicted in Figure 8.65, starting with (*R*)-CHDTCO–SCoA as a (hypothetical) example and let us assume that the reaction goes with inversion of configuration. Then, if a normal isotope effect is involved, the predominant product will be the (2*S*,3*S*)-3-*d,t* isomer **115**, but if an inverse isotope effect were involved, the predominant product would be (2*S*,3*R*)-3-*h,t* **116** as shown in Figure 8.69. Were the reaction to go with retention, the opposite result would be expected (Fig. 8.69, **117** and **118**). As already mentioned, the configuration at C(3) is determined in that the 3*R* isomer (**116** or **117**) loses tritium with fumarase and the 3*S* isomer (**115** or **118**) retains it. The test for the isotope effect involved a partial exchange of the enolizable T, D, or H at C(3) by H₂O and base. Material that loses tritium becomes unlabeled and is lost to further observation. Of the material that retains tritium, that which contains a CHT group will be racemized by exchange of H; that which contains a CDT group will also be racemized, but more slowly (relative to tritium loss) because base catalyzed enolization is known to have a sizeable *normal* primary isotope effect. Thus, in the case where the reaction course

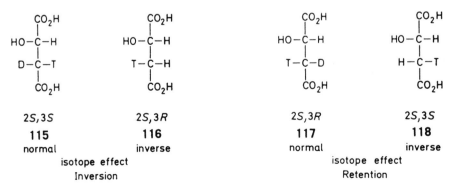

Figure 8.69. Predominant labeled malates.

was inversion (Fig. 8.69, **115**, **116**) the 3*R* isomer (**116**) will be racemized faster than the 3*S* isomer (**115**) and the relative amount of tritium retention upon subsequent fumarase treatment will initially increase. This result is independent of the isotope effect (direct or inverse) operative during the condensation step (Fig. 8.65). But if the original reaction involved retention (Fig. 8.69, **117**, **118**), the 3*S* isomer (**118**) will be racemized faster than the 3*R* isomer (**117**) and the relative amount of tritium *lost* upon fumarase mediated exchange will increase following the base water treatment, again regardless of the initial isotope effect in the glyoxylate CHDTCOSCoA condensation. Thus the effect of the H/D exchange upon the shift in outcome of the final fumarase mediated exchange is in the same direction, whether the isotope effect is normal or inverse, and depends only on the stereochemistry of the malate synthetase condensation. (The actual example related to fatty acid biosynthesis; the case of malate synthetase discussed here is hypothetical.) In fact, what the exchange experiment tells us is which of the two stereoisomers at C(3) has H next to T and which has D next to T; it is interesting that, even if the acetate–glyoxylate condensation (Fig. 8.65) involved no isotope effect at all ($k_H/k_D = 1$), the exchange test could reveal the stereochemistry of the condensation. A third way of evading the isotope problem is to look at both the forward and reverse reactions; the stereochemistry of the latter (see below) can be studied without reference to an isotope effect and that of the former then follows. The citrate lyase–citrate synthetase case (Fig. 8.68) illustrates this approach. A fourth approach involves recognition of the stereochemistry of the product by ^{3}H NMR (Altman et al., 1978). In that case, not only will the chemical shift of a CXYHT group differ from that of the stereochemically analogous CXYDT group (because of an isotope shift), but, in addition, CXYHT but not CXYDT will show a tritium doublet in the proton-coupled spectrum. The two cases are thus readily distinguished and the relative intensity of the signals will reveal if the isotope effect is normal or inverse. An application is in the biosynthesis of cycloartenol (Fig. 8.70); the cyclization of methyl to methylene proceeds with retention of configuration. This result has been independently confirmed by Blättler and Arigoni (Blättler, 1978) using different methodology in an enzyme preparation from *Zea mays*.

A second type of reaction whose stereochemistry was elucidated by the use of chiral methyl groups is of the CDTXY→CHDTX type, exemplified by the citrate lyase reaction (Fig. 8.67).

A third type of reaction studied by chiral methyl groups is of the methyl transfer type: CHDTX→CHDTY. As one might expect, these reactions often involve inversion but not invariably so. An example is the catechol-*O*-methyltransferase (COMT) reaction shown in Figure 8.71 (Woodard, Floss, et al.,

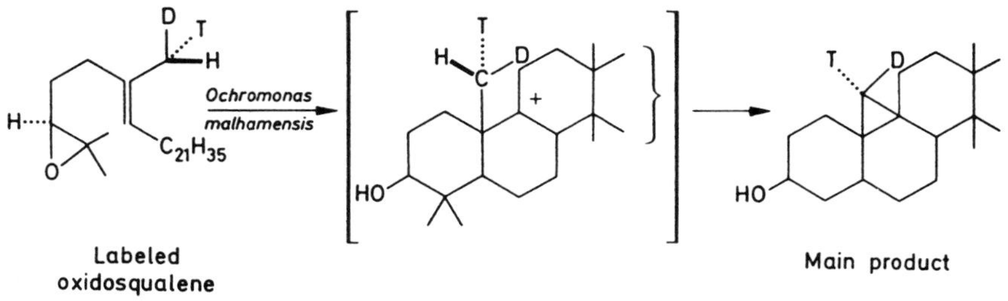

Figure 8.70. Stereochemistry of cycloartenol biosynthesis.

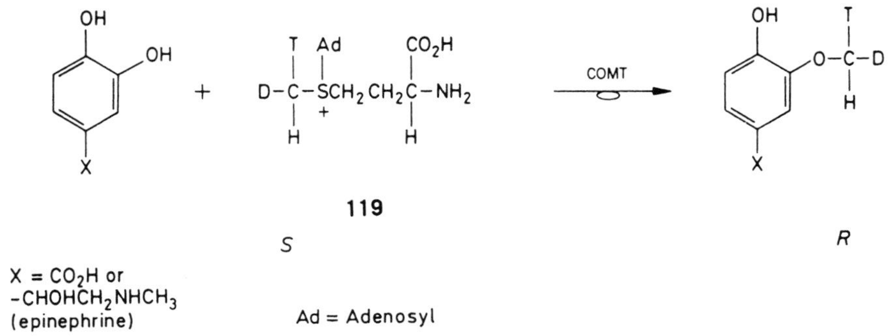

119

X = CO₂H or
-CHOHCH₂NHCH₃
(epinephrine)

Ad = Adenosyl

Figure 8.71. Stereochemistry of catechol-*o*-methyltransferase mediated reaction.

1980). The important methyl transfer reagent *S*-adenosylmethionine with a stereochemically labeled methyl group (**119**) is synthesized from chiral acetate via a Schmidt reaction followed by tosylation to *N*-tosylmethylamine CHDTNHTs, which is converted to the ditosylate in which the N(Ts)₂ group (where Ts = tosyl) becomes a leaving group (cf. Fig. 8.66). Reaction with the S sodium salt of homocysteine gives methionine(CHDT), which is converted to the adenosyl derivative by means of ATP. Analysis of the final product involves oxidation of the benzene ring with ceric ammonium nitrate to CHDTOH, followed by sulfonylation, treatment with cyanide, and hydrolysis to CHDTCO₂H, which is analyzed as described earlier. Both the preparation and the analysis involve one inversion step; this is, of course, taken into account in the consideration of the stereochemistry of the COMT catalyzed reaction which involves inversion. A variety of reactions of the three types have been tabulated (Floss and Tsai, 1979).

A number of additional examples of use of the CHDT group in methyl transfer (Kobayashi, Floss, et al., 1987; Frenzel, Floss, et al., 1990; Vanoni, Floss, et al., 1990; Ho, Floss, et al., 1991), in elucidating the stereochemistry of the Claisen rearrangement of chorismate to prephenate (Knowles, 1984; Grimshaw, Knowles, et al., 1984; Asano, Floss, et al., 1985) and in other applications (Oths, Floss, et al., 1990) are on record.

b. Chiral Phosphate Groups

(For reviews see Eliel, 1982; Lowe, 1983, 1986; Gerlt et al., 1983; Floss et al., 1984)

One other element that is conveniently available in three isotopic forms is oxygen. These isotopes have been used to synthesize (Abbott, Knowles, et al., 1978; Cullis and Lowe, 1978) chiral phosphates of the type $ROP^{16}O^{17}O^{18}O^{2-}$ conveniently written as $ROPO\ominus\bullet^{2-}$. There is one obvious disadvantage and one advantage of $ROPO\ominus\bullet^{2-}$ over CXHDT: the disadvantage is that isotope effects would be very small and probably not practically useful to discriminate among the oxygen (as opposed to the hydrogen) isotopes; the advantage is that the PO bonds are much easier to transform chemically in stereochemically defined ways than the

CH bonds and that ^{17}O (50% enriched) is much easier to use in high concentration than the radioactive T (^{18}O, like D, is available in over 99% purity).

Triply labeled phosphate (Jones, Knowles et al., 1978; Abbott, Knowles, et al., 1978) has been employed to elucidate the steric course of a number of phosphoryl-transfer reactions; the topic has been reviewed (Knowles, 1980; Lowe, 1981; Floss et al., 1984; Gerlt, 1984, 1992), and here we shall present only one example, concerned with the stereochemistry of cyclization of adenosine diphosphate (ADP) to cyclic adenosine monophosphate (cAMP) (Jarvest, Lowe, et al., 1980; Cullis, Lowe, et al., 1981) and the reverse reaction (Jarvest and Lowe, 1980; Coderre, Gerlt, et al., 1981).

The pathway shown in Figure 8.72 represents several syntheses of chirally P-labeled phosphate esters with different R groups, including AMP, carried out by Lowe (1981). Assignment of configuration of the P-labeled AMP depends on the proper assignment of configuration of the cyclic phosphate precursor **122**, R = CH$_3$ (Cullis, Lowe, et al., 1981), which was originally made incorrectly (cf. Jarvest, Lowe, et al., 1980) but has since been corrected (Lowe, 1981; Cullis, Lowe, et al., 1981; Coderre, Gerlt, et al., 1981). The compound (R)-benzoin was prepared from (R)-mandelic acid and was ^{18}O labeled by conversion to the ethylene ketal followed by cleavage with H$_2^{18}$O and acid. Reduction of the resulting compound **120** to chirally labeled hydrobenzoin **121** was followed by conversion to a single cyclic ^{17}O labeled phosphate **122** of established (Cullis, Lowe, et al., 1981) relative position of the =O and OR groups (cis or trans to the phenyl substituents in the five-membered ring); the absolute configuration at phosphorus is thus S, as shown in Figure 8.72, both before (**122**) and after hydrogenolysis (**123**).

When R in **123** is 5'-adenosyl, the product is P-chirally labeled AMP, adenosine-5'[(S)-^{16}O^{17}O^{18}O] phosphate **124**. Its cyclization to chirally ^{16}O, ^{18}O labeled cAMP is shown in Figure 8.73. The NMR analysis of this material is based on the discovery of the very useful effects of ^{17}O (quenching) and ^{18}O (isotope

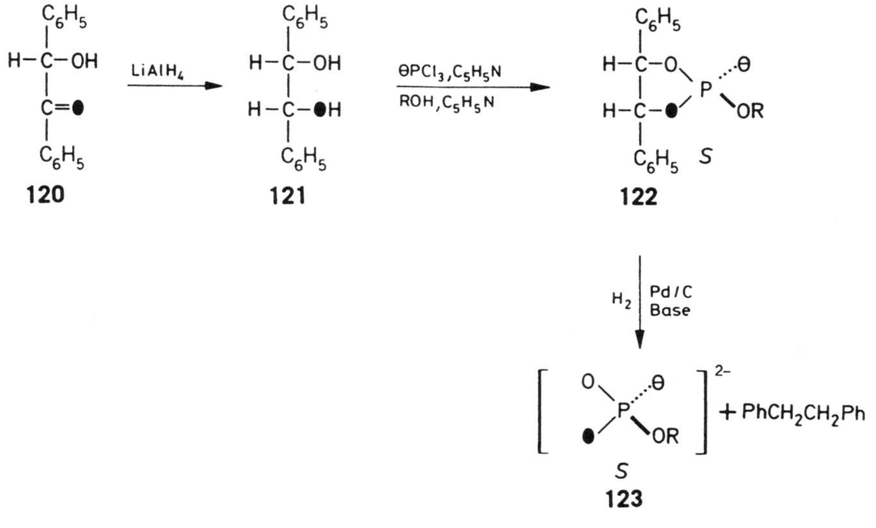

Figure 8.72. Synthesis of chirally labeled phosphates.

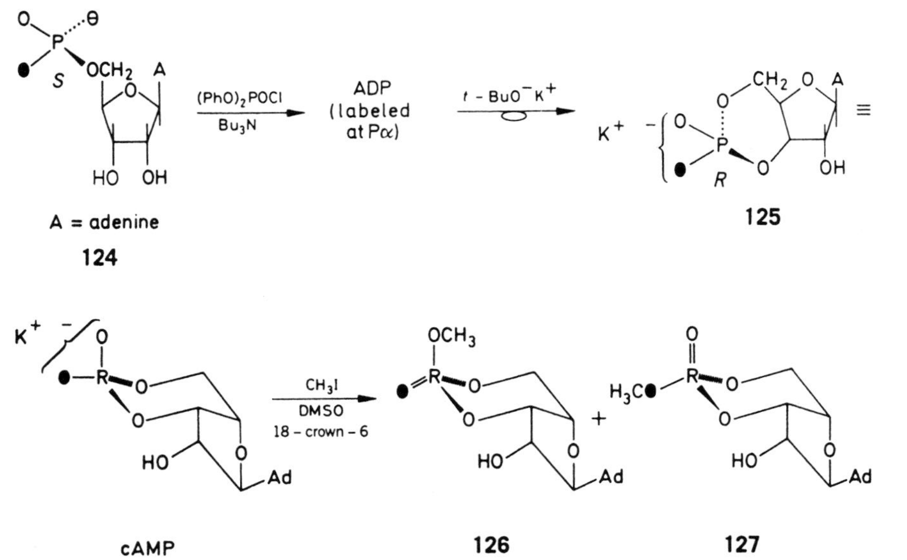

Figure 8.73. Cyclization of stereospecifically PO labeled AMP to cAMP.

shift) on the shift of adjacent ^{31}P nuclei (Cohn and Hu, 1978; Lowe and Sproat, 1978; Lowe et al. 1979; Gerlt et al., 1980; Gerlt and Coderre, 1980). Of the three products that can be formed by displacement of any one of the three oxygen atoms on phosphorus (the isotope effect is negligible in this case), one will contain ^{16}O and ^{17}O, one ^{17}O and ^{18}O, and the third one (**125**) ^{16}O and ^{18}O.

Any species that contains the $P^{17}O$ ($P\ominus$) moiety will have its ^{31}P resonance quenched by the large quadrupole moment of the adjacent ^{17}O nucleus (Tsai, 1979; Lowe et al., 1979). The only cAMP species shown in Figure 8.73 is therefore **125**, formed by displacement of ^{17}O, since other species will retain the $P\ominus$ moiety and will not be seen in the ^{31}P NMR analysis used (Jarvest, Lowe, et al., 1980). The course depicted in Figure 8.73 is the actual one of inversion at phosphorus: If retention occurred, the ^{16}O and ^{18}O nuclei would be interchanged. Thus, in the final methylation product, the ^{18}O is necessarily equatorial, either as $\bullet{=}P$ (**126**) or as $CH_3\bullet P$ (**127**), depending on whether methylation occurs equatorially or axially. Indeed two families of ^{31}P signals are seen for the two diastereomeric species. Since an ^{18}O nucleus induces an isotope shift at an adjacent ^{31}P (Gerlt and Coderre, 1980) and since, moreover, this shift is different for a doubly bonded $\bullet$ than for a singly bonded one, species **126** will show the isotope shift typical for $\bullet{=}P$, whereas **127** will show the shift for $\bullet{-}P$. If the reaction had proceeded with retention, the labeled oxygen atom (^{18}O) would be axial, in which case species **127** would have displayed the $\bullet{=}P$ isotope shift and **126** the $\bullet{-}P$ isotope shift; the two alternatives can be readily distinguished. By this methodology the course of the cyclization was shown to be inversion (Jarvest, Lowe, et al., 1980; Cullis, Lowe, et al., 1981).

The stereochemistry of the reverse reaction, opening of cAMP to AMP by beef heart cyclic cytidine monophosphate (cCMP) phosphodiesterase, was found by Gerlt and co-workers to involve inversion in the deoxyadenosine series

(Coderre, Gerlt, et al., 1981). The same point was demonstrated in the adenosine series (Jarvest, Lowe, et al., 1980; Cullis, Lowe, et al., 1981) by opening the stereospecifically labeled cAMP, formed as shown in Figure 8.73, with $H_2{}^{17}O$ to form stereospecifically labeled AMP (Fig. 8.73, **124**). That this material had indeed the same stereochemistry as **124** was established by recyclizing it as shown in Figure 8.73. Once again, compounds **126** and **127** were obtained predominantly (a considerable amount of isotope dilution occurs in the experiment) and it is therefore clear that the sequence of ring opening followed by ring closure (cAMP → AMP → cAMP) involves overall retention of configuration. Thus, since the second step involves inversion, the first step must do so also.

Inversion of configuration at phosphorus occurs in all the kinase mediated reactions studied so far (Blättler and Knowles, 1979; Lowe, 1981). In contrast, Knowles' group has shown that alkaline phosphatase (Jones, Knowles, et al., 1978) and phosphoglycerate mutases (Blättler and Knowles, 1980) induce retention of configuration, presumably as a result of a two-step process, each step involving inversion, although an "adjacent mechanism" with pseudorotation at phosphorus is an alternative possibility. The stereochemistry of a number of phosphoryl-transfer reactions has been tabulated (Floss et al., 1984; see also Hibler, Gerlt, et al., 1986; Rétey, 1986; Freeman, Knowles, et al., 1989; Seidel, Knowles, et al., 1990 for later examples).

c. Chiral Sulfate Groups

Enantiomeric sulfate esters of the type $(ROS^{16}O^{17}O^{18}O)^-$ have also been synthesized (Lowe and Salamone, 1984). Although these compounds are conceptually similar to phosphates of the type $(ROP^{16}O^{17}O^{18}O)^{2-}$ (see above), their stereochemical analysis (Lowe and Parratt, 1985, 1988) has been effected in a quite different way, namely, by IR spectroscopy. The principle of the analytical method is that in an anancomeric (conformationally strongly biased, cf. Chapter 11) six-membered cyclic sulfite (Fig. 8.74) both the symmetric and the antisymmetric O–S–O stretching frequencies depend on the nature (^{16}O, ^{17}O, or ^{18}O) of both the axial and the equatorial oxygen isotope of the exocyclic sulfate oxygen. For example, interchange of axial and equatorial ^{16}O and ^{17}O involves a shift of $6\,\mathrm{cm}^{-1}$, clearly resolvable, in both the symmetric and antisymmetric stretching frequencies; a corresponding interchange of ^{17}O and ^{18}O changes the frequency by $7\,\mathrm{cm}^{-1}$ in the symmetric and $5\,\mathrm{cm}^{-1}$ in the antisymmetric stretch.

Figure 8.74 indicates the synthesis of an acyclic sulfate of known configuration and its reconversion, in turn, to the cyclic sulfate that can then be analyzed by IR. Since neither the ring-opening nor the ring-closing step involves breakage of any of the sulfur–oxygen bonds, both steps are bound to proceed with retention of configuration. Thus the only stereochemical assumption in the scheme shown in Figure 8.74 is that the ruthenium tetroxide oxidation proceeds with retention of configuration at the sulfoxide chiral center, and this had earlier been demonstrated by a combination of ^{17}O spectroscopy and a lanthanide shift technique (Lowe and Salamone, 1983).

An application to the elucidation of the stereochemistry at sulfur of the

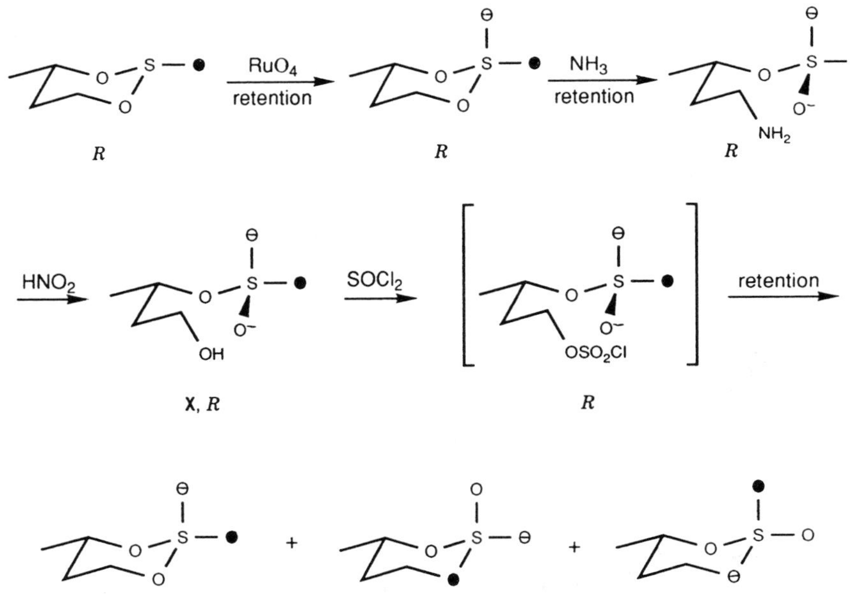

Figure 8.74. Synthesis and stereochemical analysis of $ROS^{16}O^{17}O^{18}O^-$.

azidolysis of epiandrosterone phenyl sulfate is shown in Figure 8.75 (Chai, Lowe, et al., 1991). Sulfite **A** was obtained by treating epiandrosterone with phenol and S●Cl$_2$ and its configuration at sulfur was elucidated by X-ray crystallography (cf. Section 5-5.a). Oxidation with RuO$_4$ (with retention of configuration, see above) gave sulfate **B**, which was azidolyzed with Bu$_4$N$^+$N$_3^-$ to androsteryl azide and PhOS●⊖O$^-$Bu$_4$N$^+$. The configuration of the latter was elucidated as shown in Figure 8.74 with intermediate X being produced by transesterification of the phenyl sulfate with (3R)-1-benzyloxy-3-butanol followed by hydrolytic removal of the benzoate protective group. (See, also, Chai, Lowe et al., 1992.)

d. Pro^3-chiral Centers: The Chiral Thiophosphate Group

Whereas a prochiral center requires one substitution to become a chiral center and a pro^2-chiral center (Floss et al., 1984) requires two such substitutions (see above), a pro^3-chiral center requires three such substitutions. Thus the carbon in methane CH$_4$ or the phosphorus in phosphate PO$_4^{3-}$ are pro^3-chiral. Since there are only three hydrogen isotopes, it is not possible to simulate the pro^3-chiral center in methane by an isotopically substituted chiral center. Strictly speaking, this is true also of phosphate, but, because of the very similar behavior of phosphates and thiophosphates in enzymatic reactions, it is possible to use sulfur as a surrogate for oxygen and thus construct a "pseudo-pro^3-chiral center" of the type SP^{16}O^{17}O^{18}O^{3-}. Such a chiral thiophosphate has been synthesized by Lowe and co-workers (Arnold and Lowe, 1986; Arnold, Lowe et al., 1987; Lowe and Potter, 1989) using methodology similar to that in the synthesis of ROPO⊖●$^{2-}$ (Fig. 8.72) but using PSCl$_3$ in lieu of P⊖Cl$_3$, hydrolyzing the resulting phosphoryl chloride with H$_2$⊖ and base, separating chromatographically at the salt stage and

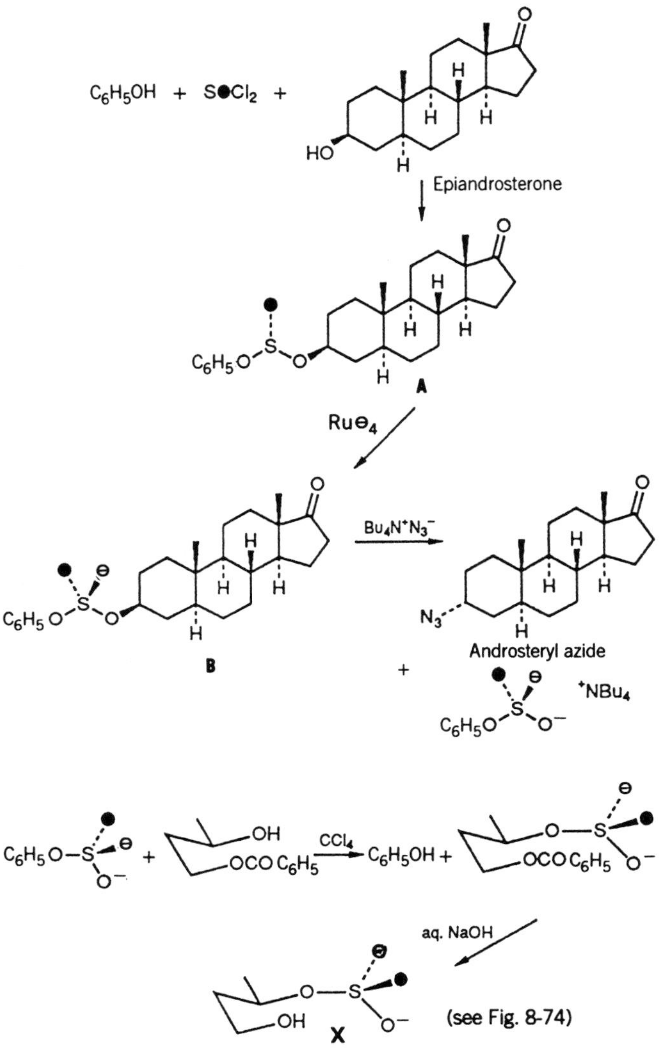

Figure 8.75. Stereochemistry at sulfur in azidolysis of 3-epiandrosterone phenyl sulfate. Reprinted with permission from Chai, C.L.L., Hepburn, T.W., and Lowe, G., *J. Chem. Soc. Chem. Commun.* (1991), 1403. Copyright © Royal Society of Chemistry, London, UK.

then cleaving the dioxolane by reduction with sodium and liquid ammonia to give $SPOO\ominus\bullet^{3-}$ of known configuration (Fig. 8.76). The problem of configurational analysis was solved by alkylating the thiosphosphate at sulfur with enantio-merically pure $C_6H_5CHOHCH_2I$, cyclizing to an oxathiolane with Ph_2PCl and esterifying with diazomethane. The product is then analyzed by the observed ^{18}O induced isotope shift of the ^{31}P signal, somewhat similarly as shown in Figure 8.74, except that the place of the dioxane ring is taken by a (five-membered) oxathiolane ring. Once performed with $SPOO\ominus\bullet^{3-}$ of known configuration, this type of analysis can be used to ascertain the sense of chirality of a doubly labeled

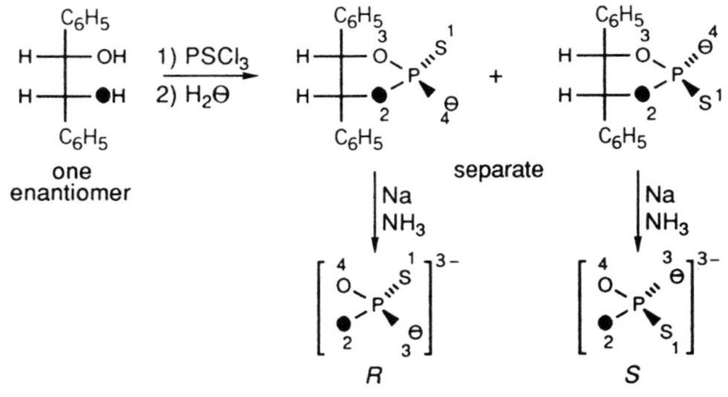

(Stereochemical analysis in same paper)

Figure 8.76. Synthesis of $[SPO\ominus\bullet]^{3-}$. Numbers refer to CIP Priority.

thiophosphate of unknown configuration. Several applications of the technique to ascertain the steric course of enzymatic hydrolysis of phosphates: $R-O-PSO\bullet^{2-} + \ominus H^- \to R-OH + \ominus PSO\bullet^{3-}$ are on record (Harnett and Lowe, 1987; Bethel and Lowe, 1988a,b; Dixon and Lowe, 1989).

REFERENCES

Abbott, S. J., Jones, S. R., Weinman, S. A., and Knowles, J. R. (1978), *J. Am. Chem. Soc.*, **100**, 2558.

Aberhart, D. J. and Lin, L. J. (1973), *J. Am. Chem. Soc.*, **95**, 7859.

Aberhart, D. J. and Lin, L. J. (1974), *J. Chem. Soc. Perkin 1*, 2320.

Abragam, A. (1961), *The Principles of Nuclear Magnetism*, Oxford University Press, p. 480.

Adams, M. J., Rossman, M. G., Kaplan, N. O., et al. (1973), *Proc. Natl. Acad. Sci. USA*, **70**, 1968.

Altman, L. J. et al. (1978), *J. Am. Chem. Soc.*, **100**, 3235.

Alworth, W. L. (1972), *Stereochemistry and its Applications in Biochemistry*, Wiley, New York.

Anet, F. A. L. (1960), *J. Am. Chem. Soc.*, **82**, 994.

Anet, F. A. L. and Kopelevich, M. (1989), *J. Am. Chem. Soc.*, **111**, 3429.

Anet, F. A. L. and Park, J. (1992), *J. Am. Chem. Soc.*, **114**, 411.

Anet, F. A. L., O'Leary, D. J., Beale, J. M., and Floss, H. G. (1989), *J. Am. Chem. Soc.*, **111**, 8935.

Arigoni, D. and Eliel, E. L. (1969), "Chirality due to the Presence of Hydrogen Isotopes at Non-cyclic positions," *Top. Stereochem.*, **4**, 127.

Arnold, J. R. P., Bethell, R. C., and Lowe, G. (1987), *Bioorg. Chem.*, **15**, 250.

Arnold, J. R. P. and Lowe, G. (1986), *J. Chem. Soc. Chem. Commun.*, 865.

Asano, Y., Lee, J. J., Shieh, T. L., Spreafico, F., Kowal, C., and Floss, H. G. (1985), *J. Am. Chem. Soc.*, **107**, 4314.

Balaban, A. T. and Farcasiu, D. (1967), *J. Am. Chem. Soc.*, **89**, 1958.

Baldwin, J. E., Loliger, J., Rastetter, W., Neuss, N., Huckstep, L. L., and De La Higuera, N. (1973), *J. Am. Chem. Soc.*, **95**, 3796.

Battersby, A. R. and Staunton, J. (1974), *Tetrahedron*, **30**, 1707.

Battersby, A. R., Chrystal, E. J. T. and Staunton, J. (1980a), *J. Chem. Soc. Perkin 1*, 31.

Battersby, A. R., Nicoletti, M., Staunton, J., and Vleggaar, R. (1980b), *J. Chem. Soc. Perkin 1*, 43.

Battersby, A. R., Scott, A., and Staunton, J. (1990), *Tetrahedron*, **46**, 4685.

Beaulieu, P. L., Morisset, V. M., and Garratt, D. G. (1980), *Can. J. Chem.*, **58**, 928.

Bentley, R. (1969), *Molecular Asymmetry in Biology*, Vol. 1, Academic, New York.

Bentley, R. (1970), *Molecular Asymmetry in Biology*, Vol. 2, Academic, New York.

Bentley, R. (1976), "The Use of Biochemical Methods for Determination of Configuration," in Jones, J. B., Sih, C. J., and Perlman, D., Eds., *Applications of Biochemical Systems in Organic Chemistry*, Vol. 10, Part 1, Wiley, New York, p. 403.

Bentley, R. (1978), *Nature (London)*, **276**, 673.

Bergman, M. (1934), *Science*, **79**, 439.

Bethell, R. C. and Lowe, G. (1986), *Eur. J. Biochem.*, **174**, 387.

Bethell, R. C. and Lowe, G. (1988a), *Eur. J. Biochem.*, **174**, 387.

Bethell, R. C. and Lowe, G. (1988b), *Biochemistry*, **27**, 1125.

Biernacki, W., Dabrowski, J., and Ejchart, A. (1972), *Org. Magn. Reson.*, **4**, 443.

Binsch, G. (1968), "The Study of Intramolecular Rate Processes by Dynamic Nuclear Magnetic Resonance," *Top. Stereochem.*, **3**, 97.

Binsch, G. (1973), *J. Am. Chem. Soc.*, **95**, 190.

Binsch, G. (1975), "Band-Shape Analysis," in Jackman, L. M. and Cotton, F. A., Eds., *Dynamic Nuclear Magnetic Resonance Spectroscopy*, Academic, New York, p. 45.

Binsch, G. and Franzen, G. R. (1969), *J. Am. Chem. Soc.*, **91**, 3999.

Binsch, G. and Kessler, H. (1980), *Angew. Chem. Int. Ed. Engl.*, **19**, 411.

Binsch, G., Eliel, E. L., and Kessler, H. (1971), *Angew. Chem. Int. Ed. Engl.*, **10**, 570.

Blättler, W. A. (1978), Ph.D. Dissertation, No. 6127, ETH, Zurich, Switzerland.

Blättler, W. A. and Knowles, J. R. (1979), *J. Am. Chem. Soc.*, **101**, 510.

Blättler, W. A. and Knowles, J. R. (1980), *Biochemistry*, **19**, 738.

Brink, M. and Schjånberg, E. (1980), *J. Prakt. Chem.*, **322**, 685.

Bublitz, C. and Kennedy, E. P. (1954), *J. Biol. Chem.*, **211**, 951.

Bushweller, C. H., Anderson, W. G., Stevenson, P. E., and O'Neil, J. W. (1975), *J. Am. Chem. Soc.*, **97**, 4338.

Bushweller, C. H., Rao, G. U., Anderson, W. G., and Stevenson, P. E. (1972), *J. Am. Chem. Soc.*, **94**, 4743.

Cane, D. E. (1980), "The Stereochemistry of Allylic Pyrophosphate Metabolism," *Tetrahedron*, 36, 1109.

Chai, C. L. L., Hepburn, T. W., and Lowe, G. (1991), *J. Chem. Soc. Chem. Commun.*, 1403; see also Chai, C. L. L., Humphreys, V., Prout, K., and Lowe, G. (1991), ibid., 1597.

Chai, C. L. L., Loughlin, W. A., Lowe, G. (1992), *Biochem. J.*, **287**, 805.

Chibata, S. (1982), "Applications of Immobilized Enzymes for Asymmetric Reactions," in Eliel, E. L. and Otsuka, S., Eds., *Asymmetric Reactions and Processes in Chemistry*, ACS Symposium Series 185, Washington, DC, American Chemical Society, p. 195.

Coderre, J. A., Mehdi, S., and Gerlt, J. A. (1981), *J. Am. Chem. Soc.*, **103**, 1872.

Cohn, M. and Hu, A. (1978), *Proc. Natl. Acad. Sci. USA*, **75**, 200.

Cornforth, J. W. (1969), *Q. Rev. Chem. Soc.*, **23**, 125.

Cornforth, J. W. (1973), *Chem. Soc. Rev.*, **2**, 1.

Cornforth, J. W. (1976), *J. Mol. Catalysis*, **1**, 145.

Cornforth, J. W., Redmond, J. W., Eggerer, H., Buckel, W., and Gutschow, C. (1969), *Nature (London)*, **221**, 1212.

Cornforth, J. W., Redmond, J. W., Eggerer, H., Buckel, W., and Gutschow, C. (1970), *Eur. J. Biochem.*, **14**, 1.

Cornforth, J. W., Ryback, G., Popjak, G., Donninger, C., and Schroepfer, G. (1962), *Biochem. Biophys. Res. Commun.*, **9**, 371.

Cullis, P. M., Jarvest, R. L., Lowe, G., and Potter, B. V. L. (1981), *J. Chem. Soc. Chem. Commun.*, 245.

Cullis, P. M. and Lowe, G. (1978), *J. Chem. Soc. Chem. Commun.*, 512.

Dahlqvist, K.-I. and Forsén, S. (1965), *J. Phys. Chem.*, **69**, 4062.

Devriese, G., Ottinger, R. Zimmerman, D., Reisse, J., and Mislow, K. (1976), *Bull. Soc. Chim. Belg.*, **85**, 167.

Dixon, R. M. and Lowe, G. (1989), *J. Biol. Chem.*, **264**, 2069.

Drysdale, J. J. and Phillips, W. D. (1957), *J. Am. Chem. Soc.*, **79**, 319.

Easson, L. H. and Stedman, E. (1933), *Biochem. J.*, **27**, 1257.

Eggerer, H., Buckel, W., Lenz, H., Wunderwald, P., Gottschalk, G., Cornforth, J. W., Donninger, C., Mallaby, R., and Redmond, J. W. (1970), *Nature (London)*, **226**, 517.

Eliel, E. L. (1971), *J. Chem. Educ.*, **48**, 163.

Eliel, E. L. (1980), *J. Chem. Educ.*, **57**, 52.

Eliel, E. L. (1982), "Prostereoisomerism (Prochirality)," *Top. Curr. Chem.*, **105**, 1.

England, S. (1960), *J. Biol. Chem.*, **235**, 1510.

England, S. and Colowick, S. P. (1956), *J. Biol. Chem.*, **221**, 1019.

England, S. and Colowick, S. P. (1957), *J. Biol. Chem.*, **226**, 1047.

Evans, E. A. and Slotin, L. (1941), *J. Biol. Chem.*, **141**, 439.

Floss, H. G. (1986), "Games with Chiral Methyl Groups," see Frey, (1986) reference, p. 71.

Floss, H. G. and Lee, S. (1993), "Chiral Methyl Groups: Small is Beautiful," *Acc. Chem. Res.*, **26**, 116.

Floss, H. G. and Tsai, M.-D. (1979), "Chiral Methyl Groups," *Adv. Enzymol.*, **50**, 243.

Floss, H. G., Tsai, M.-D., and Woodard, R. W. (1984), "Stereochemistry of Biological Reactions at Proprochiral Centers," *Top. Stereochem.*, **15**, 253.

Franzen, G. R. and Binsch, G. (1973), *J. Am. Chem. Soc.*, **95**, 175.

Fraser, R. R., "Nuclear Magnetic Resonance Analysis Using Chiral Shift Reagents," in Morrison, J. D. Ed., *Asymmetric Synthesis*, Vol. 1, Academic, New York, 1983, p. 173.

Fraser, R. R. and Schuber, F. J. (1970), *Can. J. Chem.*, **48**, 633.

Fraser, R. R., Schuber, F. J., and Wigfield, Y. Y. (1972), *J. Am. Chem. Soc.*, **94**, 8795.

Freeman, S., Seidl, H. M., Schwalbe, C. H., and Knowles, J. R. (1989), *J. Am. Chem. Soc.*, **111**, 9233.

Frenzel, T., Zhou, P., and Floss, H. G. (1990), *Arch. Biochem. Biophys.*, **278**, 35.

Frey, P. A., Ed. (1986), *Mechanism of Enzymatic Reactions. Stereochemistry*, Elsevier, New York.

Gaudemer, A. (1977), "Determination of Configurations by NMR Spectroscopy," in Kagan, H. B., Ed., *Stereochemistry, Fundamentals and Methods*, Vol. 1, Thieme, Stuttgart, Germany, p. 73.

Gawron, O. and Fondy, T. P. (1959), *J. Am. Chem. Soc.*, **81**, 6333.

Gawron, O., Glaid, A. J., and Fondy, T. P. (1961), *J. Am. Chem. Soc.*, **83**, 3634.

Gerlach, H. and Zagalak, B. (1973), *J. Chem. Soc. Chem. Commun.*, 274.

Gerlt, J. (1992), "Phosphate Ester Hydrolysis," in Sigman, D.S., Ed., *The Enzymes*, 3rd ed., Vol. 20, Academic, San Diego, CA, p. 95.

Gerlt, J. A. (1984), "Use of Chiral [^{16}O, ^{17}O, ^{18}O] Phosphate Esters to Determine the Stereochemical Course of Enzymatic Phosphoryl Transfer Reactions," in Gorenstein, D. G., Ed., *Phosphorus-31 NMR, Principles and Applications*, Academic, New York, p. 199.

Gerlt, J. A. and Coderre, J. A. (1980), *J. Am. Chem. Soc.*, **102**, 4531.

Gerlt, J. A., Coderre, J. A., and Mehdi, S. (1983), "Oxygen Chiral Phosphate Esters," *Adv. Enzymol.*, **55**, 291.

Gerlt, J. A., Coderre, J. A., and Wolin, M. S. (1980), *J. Biol. Chem.*, **255**, 331.

Giardina', D., Ballini, R., Cingolani, G. M., Melchiorre, C., Pietroni, B. R., Carotti, A., and Casini, G. (1980), *Tetrahedron*, **36**, 3565.

Giardina', D., Ballini, R., Cingolani, G. M., Pietroni, B. R., Carotti, A., and Casini, G. (1979), *Tetrahedron*, **35**, 249.

Gielen, M., Close, V., and de Poorter, B. (1974), *Bull. Soc. Chim. Belges*, **83**, 339.

Goering, H. L., Eikenberry, J. N., Koermer, G. S., and Lattimer, C. J. (1974), *J. Am. Chem. Soc.*, **96**, 1493.

Gottschalk, G. and Barker, H. A. (1966), *Biochemistry*, **5**, 1125.

Gottschalk, G. and Barker, H. A. (1967), *Biochemistry*, **6**, 1027.

Grimshaw, C. E., Sogo, S. G., Copley, S. D., and Knowles, J. R. (1984), *J. Am. Chem. Soc.*, **106**, 2699.

Gutierrez, A., Jackson, J. E., and Mislow, K. (1985), *J. Am. Chem. Soc.*, **107**, 2880.

Gutowsky, H. S. (1962), *J. Chem. Phys.*, **37**, 2196.

Hanson, K. R. (1966), *J. Am. Chem. Soc.*, **88**, 2731.

Hanson, K. R. (1972), "Enzyme Symmetry and Enzyme Stereospecificity," *Annu. Rev. Plant Physiol. Plant Mol. Biol.*, **23**, 335.

Hanson, K. R. (1976), "Concepts and Perspectives in Enzyme Stereochemistry," *Annu. Rev. Biochem.*, **45**, 307.

Hanson, K. R. and Rose, I. A. (1963), *Proc. Natl. Acad. Sci. USA*, **50**, 981.

Hanson, K. R. and Rose, I. A. (1975), "Interpretations of Enzyme Reaction Stereospecificity," *Acc. Chem. Res.*, **8**, 1.

Harnett, S. P. and Lowe, G. (1987), *J. Chem. Soc. Chem. Commun.*, 1416.

Hibler, D. W., Stolowich, N. J., Mehdi, S., Gerlt, J. A., "Staphylococcal Nuclease: Stereochemical and Genetic Probes of the Mechanism of the Hydrolysis Reaction," see Frey (1986) reference, p. 101.

Hill, R. K. (1978), "Enzymatic Stereospecificity at Prochiral Centers of Amino Acids," in van Tamelen, E. E., Ed., *Bioorganic Chemistry*, Vol. 2, Academic, New York, p. 111.

Hill, R. K. and Chan, T.-H. (1965), *Tetrahedron*, **21**, 2015.

Hill, R. K., Yan, S., and Arfin, S. M. (1973), *J. Am. Chem. Soc.*, **95**, 7857.

Hirschmann, H. (1964), "Newer Aspects of Enzymatic Stereochemistry," in Florkin, M. and Stotz, G. H., Eds., *Comprehensive Biochemistry*, Vol. 12, Elsevier, New York, p. 236.

Hirschmann, H. and Hanson, K. R. (1971a), *Eur. J. Biochem.*, **22**, 301.

Hirschmann, H. and Hanson, K. R. (1974), *Tetrahedron*, **30**, 3649.

Hirschmann, H. and Hanson, K. R. (1971b), *J. Org. Chem.*, **36**, 3293.

Ho, D. K., Wu, J. C., Santi, D. V., and Floss, H. G. (1991), *Arch. Biochem. Biophys.*, **284**, 264.

Izumi, Y., (1971), *Angew. Chem. Int. Ed. Engl.*, **10**, 871.

Izumi, Y. and Tai, A. (1977), *Stereo-Differentiating Reactions the nature of asymmetric reactions*, Academic, New York.

Jackman, L. M. (1975), "Rotation about Partial Double Bonds in Organic Molecules," in Jackman, L. M. and Cotton, F. A., Eds., *Dynamic Nuclear Magnetic Resonance Spectroscopy*, Academic, New York, p. 203.

Jackman, L. M. and Cotton, F. A., Eds. (1975), *Dynamic Nuclear Magnetic Resonance Spectroscopy*, Academic, New York.

Jarvest, R. L. and Lowe, G. (1980), *J. Chem. Soc. Chem. Commun.*, 1145.

Jarvest, R. L., Lowe, G., and Potter, B. V. L. (1980), *J. Chem. Soc. Chem. Commun.*, 1142.

Jennings, W. B. (1975), "Chemical Shift Nonequivalence in Prochiral Groups," *Chem. Rev.*, **75**, 307.

Jennings, W. B., Al-Showiman, S., Tolley, M. S., and Boyd, D. R. (1977), *Org. Magn. Reson.*, **9**, 151.

Jones, J. B. (1976), "Stereochemical Considerations and Terminologies of Biochemical Importance," in Jones, J. B., Sih, C. J., and Perlman, D., Eds., *Applications of Biochemical Systems in Organic Chemistry*, Vol. 10, Part 1, Wiley, New York, p. 479.

Jones, J. B. (1985), "Enzymes as Chiral Catalysts," in Morrison, J. D., Ed., *Asymmetric Synthesis*, Vol. 5, Academic, New York, p. 309.

Jones, J. B. (1986), "Enzymes in Organic Synthesis," *Tetrahedron*, 42, 3351.

Jones, S. R., Kindman, L. A., and Knowles, J. R. (1978), *Nature (London)*, **275**, 564.

Kainosho, M., Ajisaka, K., Pirkle, W. H., and Beare, S. D. (1972), *J. Am. Chem. Soc.*, **94**, 5924.

Kalinowski, H.-O. and Kessler, H. (1973), "Fast Isomerization about Double Bonds," *Top. Stereochem.*, **7**, 295.

Kaloustian, S. A. and Kaloustian, M. K. (1975), *J. Chem. Educ.*, **52**, 56.

Kendrew, J. C., Dickerson, R. E., Strandberg, B. E., Hart, R. G., Davies, D. R., Phillips, D. C., and Shore, V. C. (1960), *Nature (London)*, **185**, 422.

Kessler, H. (1970), "Detection of Hindered Rotation and Inversion by NMR Spectroscopy," *Angew. Chem. Int. Ed. Engl.*, **9**, 219.

Kluender, H., Bradley, C. H., Sih, C. J., Fawcett, P., and Abraham, E. P. (1973), *J. Am. Chem. Soc.*, **95**, 6149.

Knowles, J. R. (1980), "Enzyme-Catalyzed Phosphoryl Transfer Reactions," *Annu. Rev. Biochem.*, **49**, 877.

Knowles, J.R. (1984), "Enzyme Catalyses: Lessons for Stereochemistry," *Pure Appl. Chem.*, **56**, 1005.

Kobayashi, K., Sugawara, T., and Iwamura, H. (1981), *J. Chem. Soc. Chem. Commun.*, 479.

Kobayashi, M., Frenzel, T., Lee, J. P., Zenk, M. H., and Floss, H. G. (1987), *J. Am. Chem. Soc.*, **109**, 6184.

Kost, D., Carlson, E. H., and Raban, M. (1971), *J. Chem. Soc. Chem. Commun.*, 656.

Kost, D. and Raban, M. (1972), *J. Am. Chem. Soc.*, **94**, 2533.

Kozarich, J. W., Chari, R. V. J., Ngai, K.-L., and Ornston, L. N. (1986), "Stereochemistry of Muconate Cycloisomerases," see Frey (1986) reference, p. 233.

Krasna, A. I. (1958), *J. Biol. Chem.*, **233**, 1010.

Kunec, E. K. and Robins, D. J. (1985), *J. Chem. Soc. Chem. Commun.*, 1450.

Kurland, R. J., Rubin, M. B., and Wise, W. B. (1964), *J. Chem. Phys.*, **40**, 2426.

Lambert, J. B. (1971), "Pyramidal Atomic Inversion," *Top. Stereochem.*, **6**, 19.

Lehn, J. M. (1970), "Nitrogen Inversion," *Top. Curr. Chem.*, **15**, 311.

Lenz, H., Buckel, W., Wunderwald, P., Biedermann, G., Buschmeier, V., Eggerer, H., Cornforth, J. W., Redmond, J. W., and Mallaby, R. (1971a), *Eur. J. Biochem.*, **24**, 207.

Lenz, H. and Eggerer, H. (1976), *Eur. J. Biochem.*, **65**, 237.

Lenz, H., Wunderwald, P., Buschmeier, V., and Eggerer, H. (1971b), *Z. Physiol. Chem.*, **352**, 517.

Leonard, J. E., Hammond, G. S., and Simmons, H. E. (1975), *J. Am. Chem. Soc.*, **97**, 5052.

Levy, H. R., Loewus, F. A., and Vennesland, B. (1957), *J. Am. Chem. Soc.*, **79**, 2949.

Loewus, F. A., Westheimer, F. H., and Vennesland, B. (1953), *J. Am. Chem. Soc.*, **75**, 5018.

Lowe, G., Cullis, P. M., Jarvest, R. L., Potter, B. V. L., and Sproat, B. S. (1981), *Philos. Trans. R. Soc. London* **B293**, 75.

Lowe, G. (1983), "Chiral ^{16}O, ^{17}O, ^{18}O Phosphate Esters," *Acc. Chem. Res.*, **16**, 244.

Lowe, G. (1986), "The Mechanism of Activation of Amino Acids by Aminoacyl-*t*RNA Synthetases," see Frey (1986) reference, p. 59.

Lowe, G. and Parratt, M. J. (1985), *J. Chem. Soc. Chem. Commun.*, 1073, 1075; id. (1988), *Bioorg. Chem.*, **16**, 283.

Lowe, G. and Potter, B. V. L. (1989), *J. Labelled Compd. Radiopharm.*, **27**, 63.

Lowe, G., Potter, B. V. L., Sproat, B. S., and Hull, W. E. (1979), *J. Chem. Soc. Chem. Commun.* 733.

Lowe, G. and Salamone, S. J. (1983), *J. Chem. Soc. Chem. Commun.*, 1392.

Lowe, G. and Salamone, S. J. (1984), *J. Chem. Soc. Chem. Commun.*, 466.

Lowe, G. and Sproat, B. S. (1978), *J. Chem. Soc. Chem. Commun.*, 565.

Lüthy, J., Rétey, J., and Arigoni, D. (1969), *Nature (London)*, **221**, 1213.

Marckwald, W. (1904), *Ber. Dtsch. Chem. Ges.*, **37**, 349.

Martin, M. L., Mantione, R., and Martin, G. J. (1965), *Tetrahedron Lett.*, 3185.

Martin, M. L., Mantione, R., and Martin, G. J. (1966), *Tetrahedron Lett.*, 3873.

Martin, M. L., Mantione, R., and Martin, G. J. (1967), *Tetrahedron Lett.*, 4809.

Martin, M. L. and Martin, G. J. (1966), *Bull. Soc. Chim. Fr.*, 2117.

Martin, M. L., Martin, G. J., and Couffignal, R. (1971), *J. Chem. Soc. (B)*, 1282.

McFarlane, W. (1968), *J. Chem. Soc. Chem. Commun.*, 229.

McKenna, J., McKenna, J. M., and Wesby, B. A. (1970), *J. Chem. Soc. Chem. Commun.*, 867.

Mislow, K., personal communication to ELE.

Mislow, K., *Introduction to Stereochemistry* (1965), Benjamin, New York, p. 73.

Mislow, K. (1986), "Stereoisomerism and Conformational Directionality," *Chimia*, **40**, 395.

Mislow, K., O'Brien, R. E., and Schaefer, H. (1960), *J. Am. Chem. Soc.*, **82**, 5512.

Mislow, K. and Raban, M. (1967), "Stereoisomeric Relationships of Groups in Molecules," *Top. Stereochem.*, **1**, 1.

Mislow, K. and Siegel, J. (1984), *J. Am. Chem. Soc.*, **106**, 3319.

Morris, D. G., Murray, A. M., Mullock, E. B., Plews, R. M., and Thorpe, J. E. (1973), *Tetrahedron Lett.*, 3179.

Musierowicz, S. and Wroblewski, A. E., (1980), *Tetrahedron*, **36**, 1375.

Nair, P. M. and Roberts, J. D. (1957), *J. Am. Chem. Soc.*, **79**, 4565.

Newmark, R. A. and Sederholm, C. H. (1965), *J. Chem. Phys.*, **43**, 602.

Norris, R. D. and Binsch, G. (1973), *J. Am. Chem. Soc.*, **95**, 192.

Ogston, A. G. (1948), *Nature (London)*, **162**, 963.

Oppenheimer, N. J., Marschner, T. M., Malver, O., and Kam, B. L., "Stereochemical Aspects of Coenzyme-Dehydrogenase Interactions," see Frey (1986) reference, p. 15.

Ōki, M., (1983), "Recent Advances in Atropisomerism," *Top. Stereochem.*, **14**, 1.

Orville-Thomas, W. J., Ed. (1974), *Internal Rotation in Molecules*, Wiley, New York.

Oths, P. J., Mayer, R. M., and Floss, H. G. (1990), *Carbohydr. Res.*, **198**, 91.

Pasto, D. J. and Johnson, C. R. (1979), *Laboratory Text for Organic Chemistry*, Prentice-Hall, Englewood Cliffs, NJ.

Perutz, M. F., Rossmann, M. G., Cullis, A. F., Muirhead, H., Will, G., and North, A. C. T. (1960), *Nature (London)*, **185**, 416.

Pirkle, W. H., Beare, S. D., and Muntz, R. L. (1969), *J. Am. Chem. Soc.*, **91**, 4575.

Pirkle, W. H. and Hoover, D. J. (1982), "NMR Chiral Solvating Agents," *Top. Stereochem.*, **13**, 263.

Pirkle, W. H., Muntz, R. L., and Paul, I. C. (1971), *J. Am. Chem. Soc.*, **93**, 2817.

Pople, J. A., Schneider, W. G., and Bernstein, H. J. (1959), *High-Resolution Nuclear Magnetic Resonance*, McGraw-Hill, New York.

Powers, T. A., Pedersen, L. G., and Evans, S. A. (1991), *Phosphorus Sulfur Silicon*, **59**, 205.

Prelog, V. and Helmchen, G. (1972), *Helv. Chim. Acta*, **55**, 2581.

Prelog, V. and Helmchen, G. (1982), *Angew. Chem. Int. Ed. Engl.*, **21**, 567.

Raban, M. (1966), *Tetrahedron Lett.*, 3105.

Raban, M., Lauderback, S. K., and Kost, D. (1975), *J. Am. Chem. Soc.*, **97**, 5178.

Raban, M. and Mislow, K (1966), *Tetrahedron Lett.*, 3961.

Rattet, L. S., Mandell, L., and Goldstein, J. H. (1967), *J. Am. Chem. Soc.*, **89**, 2253.

Reisse, J., Ottinger, R., Bickart, P, and Mislow, K. (1978), *J. Am. Chem. Soc.*, **100**, 911.

Rétey, J., Lüthy, J., and Arigoni, D. (1970), *Nature (London)*, **226**, 519.

Rétey, J. (1986), "Enzymatic Stereospecificity as a Probe for the Occurrence of Radical Intermediates," see Frey (1986) reference, p. 217.

Reuben, J. (1980), *J. Am. Chem. Soc.*, **102**, 2232.

Roberts, J. D. (1963), "Some Illustrative Applications of Nuclear Magnetic Resonance Spectroscopy to Organic Chemistry," *Angew. Chem. Int. Ed. Engl.*, **2**, 53.

Roberts, J. D. (1979), "Aspects of the Determination of Equilibration Rates by NMR Spectroscopy," *Pure Appl. Chem.*, **51**, 1037.

Rose, I. A., Hanson, K. R., Wilkinson, K. D., and Wimmer, M. J. (1980), *Proc. Natl. Acad. Sci. USA*, **77**, 2439.

Saunders, M. and Yamada, F. (1963), *J. Am. Chem. Soc.*, **85**, 1882.

Schiemenz, G. P. and Rast, H. (1971), *Tetrahedron Lett.*, 4685.

Schurig, V. (1977), *Tetrahedron Lett.*, 3977.

Schwartz, P. and Carter, H. E. (1954), *Proc. Natl. Acad. Sci. USA.*, **40**, 499.

Seidel, H. M., Freeman, S., Schwalbe, C. H. and Knowles, J. R. (1990), *J. Am. Chem. Soc.*, **112**, 8149.

Sedgwick, B. and Cornforth, J. W. (1977), *Eur. J. Biochem.*, **75**, 465.

Shanan-Atidi, H. and Bar-Eli, K. H. (1970), *J. Phys. Chem.*, **74**, 961.

Siddall, T. H. and Stewart, W. E. (1969), "Magnetic Non-Equivalence Related to Symmetry Considerations and Restricted Molecular Motion," *Proc. Nucl. Mag. Reson. Spectrosc.*, **5**, 33.

Simon, H. and Kraus, A. (1976), "Hydrogen Isotope Transfer in Biological Processes," in Buncel, E. and Lee, C. C., Eds., *Isotopes in Organic Chemistry*, Vol. 2, Elsevier, Amsterdam, The Netherlands, p. 153.

Slocum, D. W. and Stonemark, F. (1971), *Tetrahedron Lett.*, 3291.

Sokolov, V. I., Petrovskii, P. V., and Reutov, O. A. (1973), *J. Organometal. Chem.*, **59**, C27.

Spassov, S. L., Griffith, D. L., Glazer, E. S., Nagarajan, K., and Roberts, J. D. (1967), *J. Am. Chem. Soc.*, **89**, 88.

Steigel, A. (1978), "Mechanistic Study of Rearrangement and Exchange Reactions by Dynamic NMR Spectroscopy," in Diehl, P., Fluck, E., and Kosfeld, R., Eds., *Dynamic NMR Spectroscopy, NMR 15*, Springer, Heidelberg, Germany, p. 1.

Sternhell, S. (1975), "Rotation about Single Bonds in Organic Molecules," in Jackman, L. M. and Cotton, F. A., Eds., *Dynamic Nuclear Magnetic Resonance Spectroscopy*, Academic, New York, p. 163.

Suzuki, F., Ōki, M., and Nakanishi, H. (1973), *Bull. Chem. Soc. Jpn.*, **46**, 2858.

Tabacik, V. (1968), *Tetrahedron Lett.*, 561.

Townsend, C. A., Scholl, T., and Arigoni, D. (1975), *J. Chem. Soc. Chem. Commun.*, 921.

Tsai, M.-D. (1979), *Biochemistry*, **18**, 1468.

Vanoni, M. A., Lee, S., Floss, H. G., and Matthews, R. G. (1990), *J. Am. Chem. Soc.*, **112**, 3987.

Vennesland, B. (1974), "Stereospecificity in Biology," *Top. Curr. Chem.*, **48**, 39.

Verbit, L. (1970), "Optically Active Deuterium Compounds", *Prog. Phys. Org. Chem.*, **7**, 51.

Waugh, J. S. and Cotton, F. A. (1961), *J. Phys. Chem.*, **65**, 562.

Weigert, F. J. and Mahler, W. (1972), *J. Am. Chem. Soc.*, **94**, 5314.

Weisman, G. R. (1983), "Nuclear Magnetic Resonance Analysis Using Chiral Solvating Reagents," in Morrison, J. D., Ed., *Asymmetric Synthesis*, Vol. 1, Academic, New York, p. 153.

Whitesides, G. M., Holtz, D., and Roberts, J. D. (1964), *J. Am. Chem. Soc.*, **86**, 2628.

Wilson, N. K. and Stothers, J. B. (1974), "Stereochemical Aspects of ^{13}C NMR Spectroscopy," *Top. Stereochem.*, **8**, 1, especially, p. 17.

Wood, H. G., Werkman, C. H., Hemingway, A., and Nier, A. O. (1942), *J. Biol. Chem.*, **142**, 31.

Woodard, R. W., Tsai, M.-D., Floss, H. G., Crooks, P. A., and Coward, J. K. (1980), *J. Biol. Chem.*, **255**, 9124.

You, K.-S., Arnold, L. J., Allison, W. S., and Kaplan, N. O. (1978), *Trends Biochem. Sci.*, **3**, 265.

Young, D. W. (1978), "Stereospecific Synthesis of Tritium Labelled Organic Compounds using Chemical and Biological Methods," in Buncel, E. and Lee, C. C., Eds., *Isotopes in Organic Chemistry*, Vol. 4, Elsevier, Amsterdam, The Netherlands, p. 177.

Young, D. W. (1994), "Stereochemistry of Metabolic Reactions of Amino Acids," *Top. Stereochem.*, **21**, 381.

9

Stereochemistry of Alkenes

9-1. STRUCTURE OF ALKENES. NATURE OF CIS–TRANS ISOMERISM

a. General

In most alkenes (olefins), the two double-bonded carbon atoms and the four additional ligands attached to them are coplanar (Fig. 9.1). The generally accepted orbital description involves sp^2 hybridized carbon atoms.

The carbon atoms are linked to each other and to the attached ligands (a, b and c, d, respectively) by sp^2 hybridized σ bonds, and they are further linked to a π bond formed by lateral overlap of the remaining p orbitals of the two carbon atoms. Whereas the C–C σ bond strength is about 83 kcal mol^{-1} (347 kJ mol^{-1}), the strength of the π bond, with its less favorable lateral overlap, is only 62 kcal mol^{-1} [259 kJ mol^{-1}; the two numbers add up to the generally accepted total energy of a C=C double bond, 145 kcal mol^{-1} (607 kJ mol^{-1})].

The activation barrier to the thermal isomerization of 2-butene (Fig. 9.1, $Z \rightleftharpoons E$, a = c = CH$_3$, b = d = H), $E_a = 62 \pm 1$ kcal mol^{-1} (259 ± 5 kJ mol^{-1}) (Rabinovitch and Michel, 1959; Cundall and Palmer, 1961; Jeffers and Shaub, 1969; see also Jeffers, 1974) provides a direct measure of the strength of the π bond, since, in the process of rotation of the Z to the E isomer, the p orbitals of the two olefinic carbon atoms become perpendicular, with no overlap. Thereby the π bond is completely broken in the transition state. Earlier measurements of the rotational barrier (Kistiakowsky and Smith, 1936; Anderson et al., 1958) gave lower (sometimes much lower) barriers that were not compatible with the known C–C and C=C bond energies and also led to unreasonably large preexponential A factors in the Arrhenius equation, $k = A\,e^{-E_a/RT}$. The correct A factor (log $A = 13.5 \pm 0.3$) is actually quite normal. The barrier in CHD=CHD, 65 kcal mol^{-1} (272 kJ mol^{-1}) (Douglas, Rabinovitch et al., 1955) is similar to the 2-butene value.

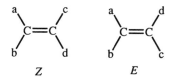

Figure 9.1. *Z*- and *E*-alkenes. Cahn–Ingold–
Prelog priority is a > b and c > d.

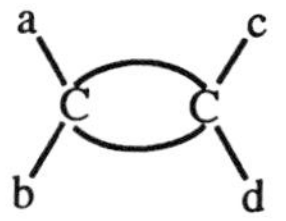

Figure 9.2. "Banana bonds."

A second picture proposed for a double bond is the so-called τ or "banana" bond
in which both bonds linking the double-bonded atoms are bent, much as they
appear in a ball-and-spring molecular model (Fig. 9.2), where the τ bonds lie in a
plane perpendicular to that of the six atoms (Pauling, 1958; cf. Walters, 1966). This
picture is no longer widely accepted, however (but see Sternberg, Haberditzl, et
al., 1982; Wintner, 1987).

The C=C bond length in unstrained, unconjugated ethenes ranges from
133.5–135 pm (1.335 to 1.35 Å) (cf. Luef and Keese, 1991) but is lengthened in
conjugated ethenes and other alkenes in which the C=C bond is weakened (Section
9-1.d). Although elementary textbooks often state that the $\begin{smallmatrix}R\\\\R'\end{smallmatrix}$C=C bond angles
are 120°, this is only approximately true; since a moiety of this type cannot possibly
have local C_{3h} symmetry, there is no reason why the three bond angles (R–C=C,
R'–C=C, and R–C–R') should be equal. In fact, in ethene (ethylene) itself the
H–C–H angle has been measured by electron diffraction as 116.6° (Kuchitsu, 1966)
or 117.8° (Duncan et al., 1972). In propene, the C=C–C angle is 124.3° and the
C=C–H angle is 119° (Lide and Christensen, 1961; Tokue, Kuchitsu, et al., 1973)
and in *cis*-2-butene the C–C=C angle is 125.8° (Kondo, Morino, et al., 1970). The
CH_3–C–CH_3 angle in isobutylene (2-methylpropene) is 115.3° (Scharpen and
Laurie, 1963; Tokue, Kuchitsu, et al., 1974). It appears, thus, that the R–C–R'
angle in the above moiety is generally smaller than 120° and the R–C=C angle is
larger.

cis–trans Isomerism (sometimes called geometric isomerism, a term we shall
not use in this book) is a type of diastereomerism: The cis and trans isomers are
(with rare exceptions, see below) not mirror image stereoisomers. The necessary
and sufficient condition for cis–trans isomers to exist is that one substituent at
each end of the double bond differ from the other; referring to Figure 9.1 this
means a ≠ b and c ≠ d. There are no restrictions as to the identity of a, b with c,
d; thus abC=Cab displays cis–trans isomerism. The other conditions for cis–trans
isomerism are implicit: One condition is that the torsion angles a–C–C–c,
a–C–C–d, b–C–C–c, and b–C–C–d be near 0° or near 180° (i.e., that the alkene be
planar or near-planar; deviations of these torsion angles from 0° or 180° by a few
degrees are common) and the second condition is that the barrier for interconversion of the cis–trans isomers be high enough for these isomers to be distinguishable entities. Both of these conditions are generally fulfilled with alkenes;
as indicated above, the rotational barriers in alkenes are very much higher than
those in alkanes [e.g., 3.6 kcal mol^{-1} (15.1 kJ mol^{-1}) for the central C–C bond in
butane]. Later we shall discuss exceptional cases where the barriers are low
and/or the alkenes are appreciably nonplanar.

b. Nomenclature

The two arrangements shown in Figure 9.1 are called "*Z*" (from German *zusammen* meaning together) and "*E*" (from German *entgegen* meaning opposite) depending on whether the atoms of highest priority in the Cahn–Ingold–Prelog sequence (cf. Section 5-2), which are assumed to be a and c in Fig. 9.1, are on the same side or on opposite sides (Blackwood et al., 1968; Cross and Klyne, 1976; see also Beilstein, preface to each volume). Examples are shown in Figure 9.3, including cases of cis–trans isomerism of C=N and C=O double bonds or partial double bonds. When the descriptor (*E* or *Z*) is part of a name, it is placed in parentheses in front of the name; thus structures **A** and **B** in Figure 9.3 are called (*Z*)-1-bromo-1,2-dichloroethene and (*E*)-(ethanal oxime), respectively. The descriptors *E* and *Z* are always italicized and, in the case of the oxime, since the descriptor refers to the entire name, the two parts of the name are enclosed in parentheses. Figure 9.4 indicates situations where locants must be used in conjunction with the descriptors.

The unaccustomed use of descriptors derived from German (rather than Latin or Greek) names may cause difficulties in memorization to those not familiar with the German language. An additional problem arises because *Z* is a zigzag or transoid letter, whereas *E* might be considered a curly or cisoid letter, so the letters do not mean what they visually imply. Even though two wrongs do not make a right, this "double defect" has served as a mnemonic to many students.

The question may be raised as to what happens when the x–C=C–y torsion angle is not 0° or 180°, that is, when the alkene is nonplanar. In that case there is still cis–trans isomerism as long as there is a sizeable perpendicular (90°) barrier. The isomer with a torsion angle of $\omega < 90°$ between the highest priority groups is called

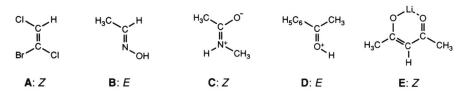

A: *Z*   **B**: *E*   **C**: *Z*   **D**: *E*   **E**: *Z*

Figure 9.3. Examples of *E–Z* nomenclature.

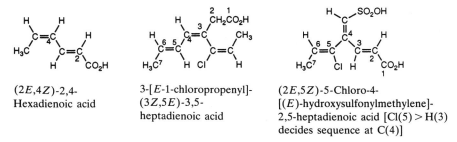

(2*E*,4*Z*)-2,4-Hexadienoic acid

3-[*E*-1-chloropropenyl]-(3*Z*,5*E*)-3,5-heptadienoic acid

(2*E*,5*Z*)-5-Chloro-4-[(*E*)-hydroxysulfonylmethylene]-2,5-heptadienoic acid [Cl(5) > H(3) decides sequence at C(4)]

Figure 9.4. Additional cases of *Z–E* nomenclature.

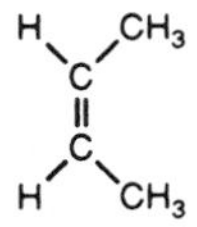

| **F** | **G** | **H** | **I** |

cis-2-butene
Z

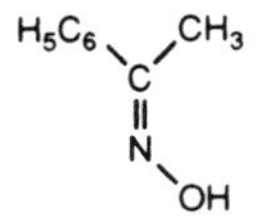

syn-methyl
(or anti-phenyl)
acetophenone oxime
E

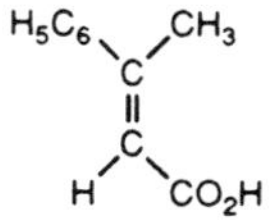

β-Methyl-trans-
cinnamic acid
or β-phenyl-cis-
crotonic acid
(E)-3-phenyl-2-
butenoic acid

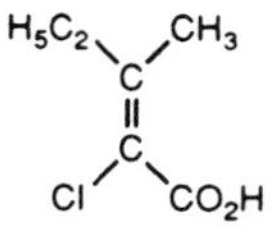

(Z)-2-chloro-3-
methyl-2-pentenoic
acid

Figure 9.5. Old and new nomenclature.

Z and the other ($\omega > 90°$) E. Presumably, there will now also be a barrier in the plane of the alkene. If that barrier is less than RT, the alkene may yet be considered planar (cf. Fig. 2.4). Cases for higher in-plane barriers will be discussed in Section 9-1.d and in Chapter 14.

Before 1968, the prefixes cis and trans were used for alkenes and the prefixes syn and anti for oximes and other aldehyde and ketone derivatives. As shown in Figure 9.5 the old nomenclature ranges from straightforward (as in the butene **F**) to cumbersome (as in the acetophenone oxime **G**) to confusing (as in the 3-phenylbutenoic acid **H**) to being inapplicable (as in many tetrasubstituted ethylenes, such as **I**). The E–Z nomenclature can always be applied and is always unequivocal. It is important to realize that in cases where cis and trans (or syn and anti) can be used, Z does not always correspond to cis or syn (and E does not necessarily correspond to trans or anti). Thus compound **A** in Figure 9.3 is trans but Z; acetaldoxime **B** is (by convention) syn but E.

Although E and Z are always applicable and unequivocal, they do (like R and S) present some practical problems. Thus (Fig. 9.6), while there is no problem in distinguishing the E and Z enolates of 3-phenyl-2-butanone **A** or **B**, problems arise with the enolates of methyl 2-phenylpropanoate **C**. When one writes the free enolate (no attached M^+), since $CH_3O > O:^-$, the structure shown in Figure 9.6, **C** is E. The same is true if one writes a covalent lithium enolate, since O–CH$_3$ has precedence over O–Li. However, the situation is reversed with the covalent sodium, potassium, or cesium enolates; since O–Na (or O–K or O–Cs) precedes O–CH$_3$, the structure is now Z. The problem is vexing, since the E or Z designation depends, in the case of the Na, K, or Cs enolates, on the somewhat

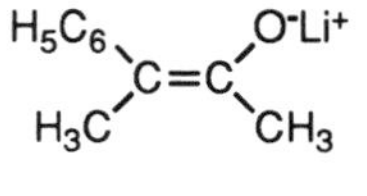

A, Z (cis) 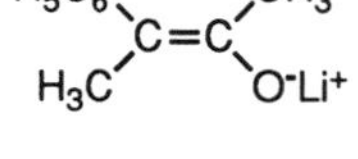   **B,** E (trans)

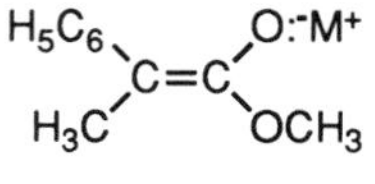

C, "cis" 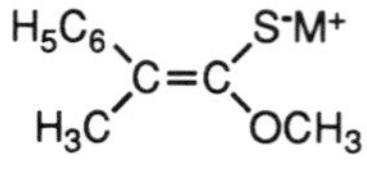   **D,** Z (cis)

Figure 9.6. Enolates.

arbitrary decision as to whether they are written as covalent or ionic species. In contrast, the thioenolate **D** is always Z since S has precedence over O, regardless of what is attached. Because enolates are of great synthetic importance and their stereochemistry is very important in asymmetric synthesis (cf. Evans et al., 1982; Heathcock, 1982; Masamune et al., 1982; see also Chapter 12) it is highly desirable to devise a solution to the dilemma. The best way to do this here as in other situations (e.g., in the steroids; cf. Fig. 11.120) is to use parochial (i.e., local) nomenclature. The descriptors E and Z should *not* be used in such cases, since they must not be redefined, but cis and trans can be defined arbitrarily to suit the situation. Thus it is recommended that cis and trans be defined to mean that the alkoxide or thioalkoxide is on the same side as the higher priority group at the other end of the double bond, and trans be defined to mean that it is on the opposite side. For aldehyde and ketone enolates, cis and trans would then correspond to E and Z, but for ester enolates (cf. Fig. 9.6) this is not necessarily the case.

c. Cumulenes

As van't Hoff (1877) already recognized, cis–trans isomerism exists not only in monoenes but also in polyenes of the cumulene type with an odd number of double bonds [$ab(C=)_n Ccd$, where n is odd]. (When n is even, as in allenes, etc., there is enantiomerism; this will be discussed in Chapter 14.) This isomerism is due to the fact that successive planes of π bonds are orthogonal to each other, as shown in Figure 9.7. The cis–trans isomers in a butatriene (Fig. 9.8) were first observed by Kuhn and Blum (1959). The barrier between the diastereo-mers is relatively low: 31.0 kcal mol^{-1} (129.7 kJ mol^{-1}) for CH$_3$CH=C=C=CHCH$_3$ (Roth and Exner, 1976) and only 27.0 kcal mol^{-1} (113.0 kJ mol^{-1}) for t-Bu(C$_6$H$_5$)C=C=C=C(C$_6$H$_5$)t-Bu (Bertsch, Jochims, et al., 1977). These are $\Delta H^{\ddagger}$ values. The values of $\Delta G^{\ddagger}$ for the two cases, 31.6 kcal mol^{-1} (132.2 kJ mol^{-1}) and 30.0 kcal mol^{-1} (125.5 kJ mol^{-1}) (see also Kuhn et al., 1966), are closer and are in agreement with the values for C$_6$H$_5$CH$_2$(CH$_3$)$_2$C(C$_6$H$_5$)C=C=C= C(C$_6$H$_5$)C(CH$_3$)$_2$CH$_2$C$_6$H$_5$, 29.9 kcal mol^{-1} (125.1 kJ mol^{-1}) (Bertsch, Jochims, et al., 1977). The t-Bu/C$_6$H$_5$ compound has a remarkably negative activation entropy for isomerization ($\Delta S^{\ddagger} = -7.65$ cal mol^{-1} K^{-1}or -32.0 J mol^{-1} K^{-1}). The low barriers are presumably due to zwitterionic or biradical resonance (cf.

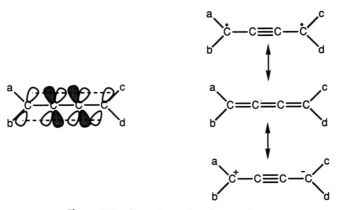

Figure 9.7. Stereoisomerism in cumulenes.

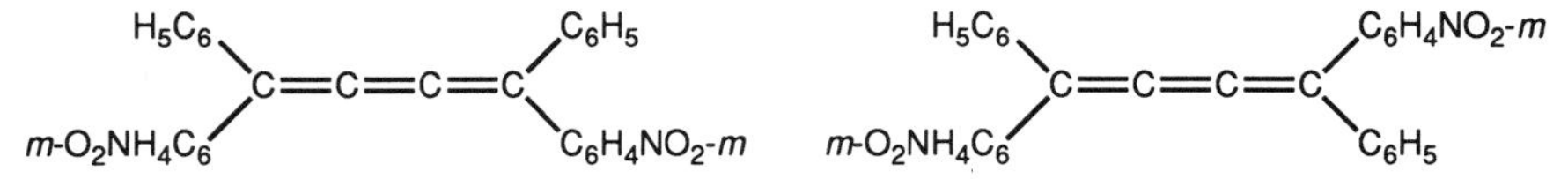

Figure 9.8. The cis–trans isomerism in a butatriene.

Fig. 9.7). It is therefore not surprising that the cis–trans isomers interconvert readily, either photochemically (in diffuse daylight; see the later discussion regarding photochemical cis–trans isomerization of alkenes) or thermally, at 160°C. In the case of the substituted hexapentaenes, [5]cumulenes, the barrier is considerably lower: $19.1 \, \text{kcal mol}^{-1}$ ($79.9 \, \text{kJ mol}^{-1}$) for

$$C_6H_5CH_2(CH_3)_2C(C_6H_5)C=C=C=C=C=C(C_6H_5)C(CH_3)_2CH_2C_6H_5$$

and $19.1 \, \text{kcal mol}^{-1}$ ($79.9 \, \text{kJ mol}^{-1}$) for $t\text{-BuC}_6H_5C=C=C=C=C=CC_6H_5t\text{-Bu}$ (Bertsch and Jochims, 1977). These barriers are too low for the cis and trans isomers to be isolated, but since the equilibrium constant in such isomers is near unity (Roth and Exner, 1976) (inasmuch as the substituents at the two ends are very distant from each other and influence each other little, especially when they are nonpolar), it is relatively easy to measure the barrier by coalescence in the NMR spectrum.

The higher cumulenes with an odd number of double bonds, octaheptaenes ([7]cumulenes), and decanonaenes ([9]cumulenes) polymerize with extreme ease and are only fleetingly stable even in dilute solution (Bohlmann and Kieslich, 1954; see also Rauss-Godineau, Cadiot et al., 1966). Though their UV spectra have been recorded, nothing is known about the cis–trans isomerism in these compounds.

d. Alkenes with Low Rotational Barriers; Nonplanar Alkenes

There are two ways in which the barriers in alkenes can be lowered: by raising the ground-state energy or by lowering the transition state energy, or by a combination of the two (Kalinowski and Kessler, 1973; Sandström, 1983). Steric factors sometimes destabilize the ground state; two examples are shown in Figure 9.9. In the substituted ethylene **A** (Shvo, 1968), as the size of the R group increases from hydrogen to *tert*-butyl, the rotational free energy barrier drops from about $27.7 \, \text{kcal mol}^{-1}$ ($116 \, \text{kJ mol}^{-1}$) to $18.3 \, \text{kcal mol}^{-1}$ ($76.6 \, \text{kJ mol}^{-1}$). In the fulvene **B**, the simple change of the R group from hydrogen to methyl lowers the barrier by $5.7 \, \text{kcal mol}^{-1}$ ($23.8 \, \text{kJ mol}^{-1}$) (Downing, Ollis, et al., 1969); this effect is probably not *entirely* steric.

It will be noted that even in compound **A**, R=H (Fig. 9.9), the barrier is about $35 \, \text{kcal mol}^{-1}$ ($146 \, \text{kJ mol}^{-1}$) lower than that in 2-butene. The relatively mild repulsion of the methoxy and cis-located carbomethoxy group can in no way account for this difference. Rather, compounds **A** are representatives of push–pull or capto–dative alkenes (Sandström, 1983; cf. Viehe et al., 1985) in which there is extensive delocalization of the π electrons, as shown in Figure 9.9, **A'**. One might consider that the central C=C bond has considerable single-bond

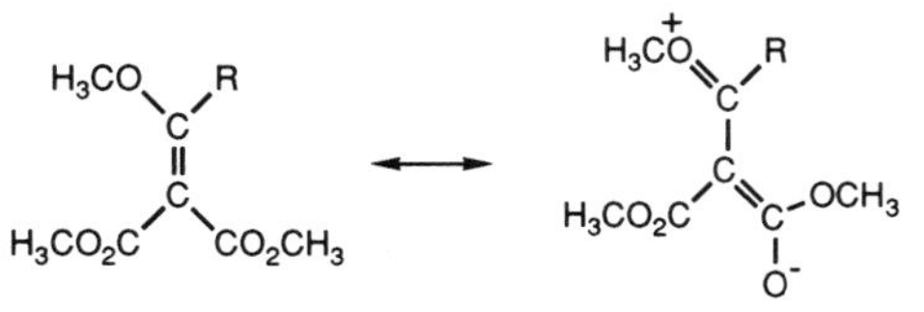

Figure 9.9. Steric effect on barriers in alkenes.

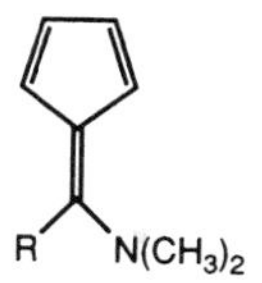

R (in **A**)	$\Delta G^{\ddagger}$ (kcal mol^{-1})	(kJ mol^{-1})	R (in **B**)	$\Delta G^{\ddagger}$ (kcal mol^{-1})	(kJ mol^{-1})
H	27.7	116.0	H	22.1	92.5
CH$_3$	25.7	107.5	CH$_3$	16.4	68.6
C$_2$H$_5$	24.7	103.3			
i-C$_3$H$_7$	23.3	97.5			
t-C$_4$H$_9$	18.3	76.6			

character and that the barrier is therefore unusually low. Perhaps a better way of looking at the situation (Sandström, 1983) is to say that what is shown as canonical form **A'** in Figure 9.9 actually represents the structure of the transition state in which, since the torsion angle at the central bond is now 90°, there is no double-bond character left in that bond. The structure written for **A'** now implies a high degree of stabilization of the zwitterionic transition state with a concomitant sizeable lowering of the activation energy for rotation. Numerous other cases of low rotational barriers in push–pull ethylenes are known (Kalinowski and Kessler, 1973; Sandström, 1983), some of which are shown in Figure 2.15 (double-bonded structures).

Delocalization of a π bond by resonance does not necessarily *require* a push–pull situation. Thus the barrier in stilbene, C$_6$H$_5$CH=CHC$_6$H$_5$ (cis → trans) is 42.8 kcal mol^{-1} (179 kJ mol^{-1}), about 20 kcal mol^{-1} (83.7 kJ mol^{-1}) less than that in 2-butene (Kistiakowsky and Smith, 1934; cf. Schmiegel, Cowan, et al., 1968). Steric strain in the ground state can account for only a small part of this difference, inasmuch as the strained *cis*-stilbene is only 3.7–4.2 kcal mol^{-1} (15.5–17.6 kJ mol^{-1}) less stable than the presumably unstrained *trans*-stilbene (Fischer et al., 1968; Saltiel et al., 1987). There must, therefore, also be stabilization of the transition state, presumably (see Section 9-3.b) as a low-lying resonance-delocalized triplet biradical, H$_5$C$_6$ĊHĊHC$_6$H$_5$ formed by intersystem crossing (Saltiel and Charlton, 1980).

In some alkenes, steric crowding in the planar state may be so severe that the latter is no longer an energy minimum but becomes an energy barrier with the ground state becoming twisted (cf. Luef and Keese, 1991). In this context the compounds in Figure 9.10 are of interest. The drop in rotational barrier from stilbene (42.8 kcal mol^{-1}, 179 kJ mol^{-1}) to compound **A** (21.1 kcal mol^{-1}, 88.3 kJ mol^{-1}) (Rieker and Kessler, 1969) to compound **B** ("small barrier", cf. Kalinowski and Kessler, 1973) may be explained by increasing stabilization of the triplet biradical transition state. However, the "negative" barrier in **C**, that is, the fact that the stable ground state of **C** is nonplanar and that the planar conforma-

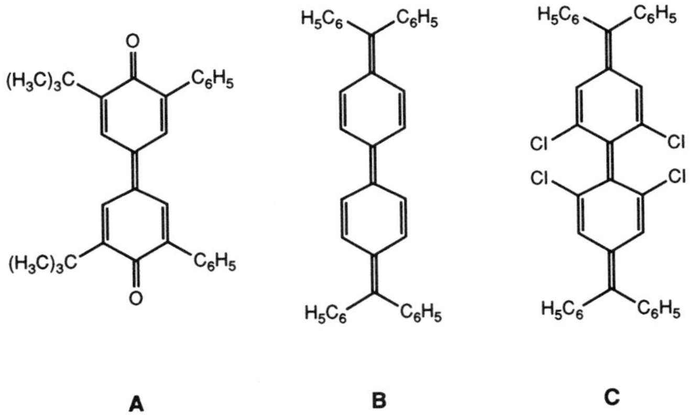

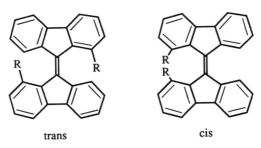

Figure 9.10. Alkenes with low or "negative" barriers.

tion represents the transition state for rotation (Müller and Neuhoff, 1939; Müller and Tietz, 1941) requires a different explanation; the difference between **B** and **C** is presumably due to the steric interaction of the four chlorine substituents in the planar conformation of **C**. This situation resembles that in o,o'-tetrasubstituted biphenyls to be discussed in Chapter 14.

A similar situation is seen in bifluorenylidenes (Fig. 9.11; cf. Favini, Simonetta et al., 1982). Both the parent compound and the isopropyl 1,1'-dicarboxylate are twisted in the ground state, the angle of twist (i.e., angle between the fluorene planes) being 41–43° in the parent compound and 50–54° in the more crowded diester (Bailey and Hull, 1978). (The range of twist angles results because there are eight molecules in the unit cell of the crystal, not all with exactly the same twist angle.) An NMR study (Gault, Ollis, and Sutherland, 1970) indicates that in some of these compounds (various esters were studied) there are two, very similar, energy barriers. This finding has been interpreted (Agranat et al., 1972) as meaning that the energy minima are separated by two energy barriers: one, due to steric factors, at $\omega = 0°$ corresponding to a planar structure, and another, at $\omega = 90°$, corresponding to a structure with minimum π-orbital overlap (Fig. 9.12). The perpendicular barrier leads to interconversions of the cis and trans isomers, which can be seen in the NMR spectrum regardless of the nature of the ester groups, since the cis and trans structures are diastereomeric and thus have different spectra. Moreover, because of the chirality of the twisted molecules, the

trans   cis

Figure 9.11. Bifluorenylidenes, where $R = H$, CO_2CH_3, or $CO_2CH(CH_3)_2$.

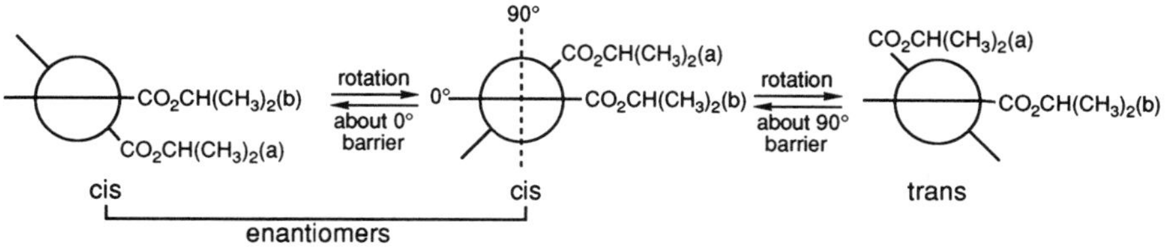

Figure 9.12. Barriers in bifluorenylidenes.

$CH(CH_3)_2$ groups in the isopropyl ester are not related by a symmetry plane; therefore, although the two isopropyl groups as a whole are homotopic, and hence isochronous by virtue of C_2 symmetry, the geminal methyl groups in each are diastereotopic with respect to one another, and hence are anisochronous. Interconversion of enantiomers (leading to exchange or "topomerization" of these methyl groups; cf. p. 505) will occur through rotation through the 0° plane and will thus provide a measure of the 0° barrier. Fortuitously the 90° and 0° barriers are of nearly the same height (22.7 kcal mol^{-1} or 95.0 kJ mol^{-1}).

That the molecule assumes a near −45° torsion angle in the (chiral) ground state is due to a compromise between favorable conjugation or π-orbital overlap, which is at a maximum in the planar conformation, and unfavorable steric effects, which are at a minimum in the 90° conformation. At $\omega = 45°$ steric factors are substantially reduced but appreciable conjugative or π overlap, which falls off as the cosine of the interplanar angle, remains.

A rather different situation is seen in the bianthrones shown in Figure 9.13. X-ray crystallography (Harnik and Schmidt, 1954) shows that these molecules, in addition to being twisted, have a folded (wing-shaped) structure and interpretation of chemical shifts in the NMR spectrum (Agranat and Tapuhi, 1979) indicates that this conformation remains in solution. Under these circumstances, cis–trans isomerization of 2,2′-disubstituted bianthrones (Fig. 9.13) involves a somewhat com-

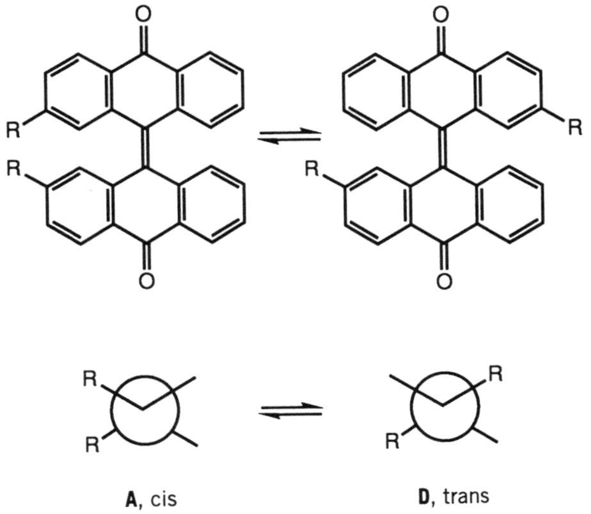

A, cis   **D**, trans

Figure 9.13. Bianthrones and their folded conformations.

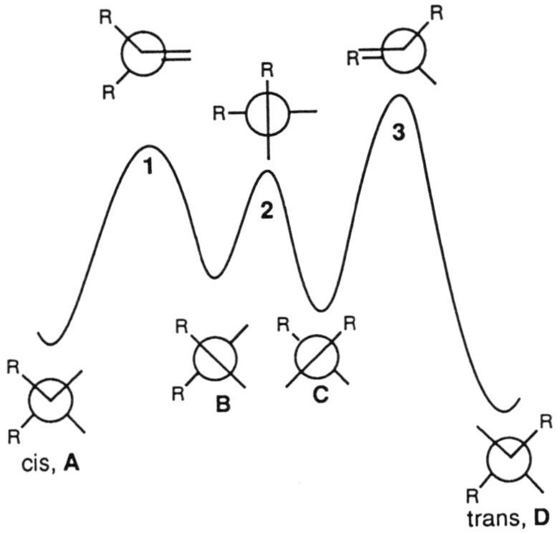

Figure 9.14. Path for cis–trans isomerization and thermochromism of bianthrones.

plex reaction path, as shown in Figure 9.14. Although the trans isomer **D** is slightly more stable than the cis isomer **A** (the twisting and folding minimizes the energy difference for 2,2′-substituted bianthrones, though it is large for 1,1′-substituted ones), the transition state leading away from **D** (3) is higher than that for the cis isomer **A** (1) because it involves R/H rather than H/H eclipsing. The perpendicular transition state (2) appears to be lower than either of the folded ones (1, 3). The activation barriers are low, of the order of 20–23 kcal mol^{-1} (83.7–96.2 kJ mol^{-1}). The twisted conformations (**B** or **C**) are above the folded ones (**A** or **D**) by only 1.3–4.3 kcal mol^{-1} (5.4–18.0 kJ mol^{-1}) and thus are appreciably populated at temperatures not much above room temperature. These states have been associated (Tapuhi, Agranat, et al., 1979) with the phenomenon of thermochromism in bianthrones. Meyer (1909) discovered that bianthrone turns green when heated; the near identity for the activation energy of this phenomenon with the activation energy for cis–trans isomerization leads to the surmise that thermochromism is due to the twisted states **B** and **C** becoming more populated. In accord with this hypothesis, thermochromism is not observed in bianthrones with substituents in 1 and 1′ more bulky than F. Referring to Figure 9.14, such compounds would exist nearly exclusively as trans isomers and the barrier would be too high to permit passage to a twisted form. See also Stezowski et al. (1993); Feringa et al. (1993).

Remarkably, the polycyclic species shown in Figure 9.15 also is chiral, by virtue of a twist of the terminal benzene moieties induced by the half-chair

Figure 9.15. Chiral bridged diphenylanthracene.

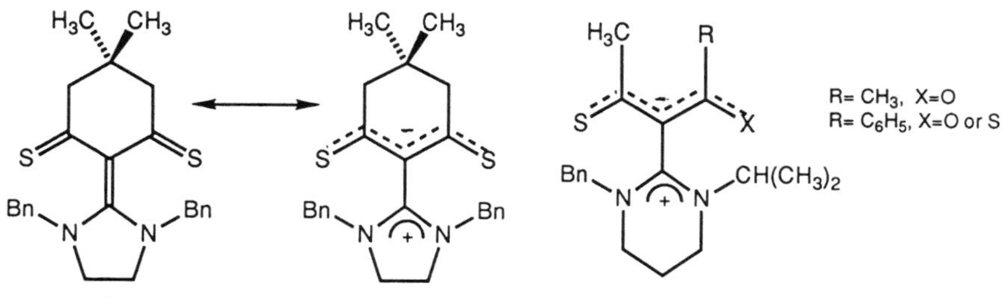

Figure 9.16. Twisted push–pull ethylenes.

conformation of the seven-membered rings. X-ray structure determination revealed that the two CH_2CH_2 bridges in the seven-membered rings are syn, so the symmetry point group of the molecule is C_2 (not C_i). Thus the molecule is chiral and can, in fact, be resolved by chromatography on swollen microcrystalline triacetylcellulose (cf. Section 7-3.d). The stiffness of the CH_2CH_2 bridges is indicated by the ^1H NMR spectrum that indicates an ABCD system. Both from exchange broadening of this spin system and from measurement of the rate of racemization by circular dichroism (CD) (Chapter 13), the barrier to "flipping" of the seven-membered rings was found to be $23.0–23.2 \, \text{kcal mol}^{-1}$ ($96.2–97.1 \, \text{kJ mol}^{-1}$) (Agranat, Sandström, Dale, et al., 1987).

Push–pull ethylenes can also have twisted ground states. An example (Sandström, Venkatesan, et al., 1985) is the dithiodiketone **A** shown in Figure 9.16. The angle of twist (81°) and very long inter-ring bond distance [148 pm (1.48 Å)] suggest that this molecule had better be looked at as existing largely in canonical form **A'**. Related twisted push–pull ethylenes, shown in the general formula **B** in Figure 9.16 (Khan, Sandström, et al., 1987), are actually chiral, have been resolved, and racemize with an activation energy of $29.9–30.3 \, \text{kcal mol}^{-1}$ ($125–127 \, \text{kJ mol}^{-1}$).

We conclude this section with mention of one of the challenges in the area of twisted ethylenes, namely, tetra-*tert*-butylethylene (Fig. 9.17, **C**). It has been calculated (Favini, Simonetta, et al., 1981) that this molecule should exist in two different (diastereomeric) twisted conformations, one, with a twist of 45° and a strain energy of $82.3 \, \text{kcal mol}^{-1}$ ($344 \, \text{kJ mol}^{-1}$), the other with a twist of 13° and a strain energy of $86.3 \, \text{kcal mol}^{-1}$ ($361 \, \text{kJ mol}^{-1}$). Although the molecule has not as yet been synthesized (cf. Dannheim et al., 1987), cyclic analogues have been prepared (Krebs et al., 1984a). Two examples are shown in Figure 9.17, **A**, $n = 1$ or 2; the compound with $n = 2$ shows little of the ordinary reactivity of an alkene (Krebs et al., 1984b).

An even closer approach to the as yet unknown tetra-*tert*-butylethylene (Krebs et al., 1986) is shown in structures **B** and **C** in Figure 9.17. The tetraaldehyde **B** has a long [136 pm (1.36 Å)] central double bond and is twisted by an angle of 28.6°, as evidenced by X-ray diffraction analysis; the CH_3–C–CHO bond angle is compressed to $96 \pm 1°$ (see also Deuter, Irngartinger, Lüttke, et al., 1985). The C=C Raman stretching frequency is at an unusually low $1461 \, \text{cm}^{-1}$, the UV spectrum is anomalous also and rotation about the alkyl bonds is hindered by a barrier of $12.2 \, \text{kcal mol}^{-1}$ ($51.0 \, \text{kJ mol}^{-1}$) as shown by low-temperature NMR

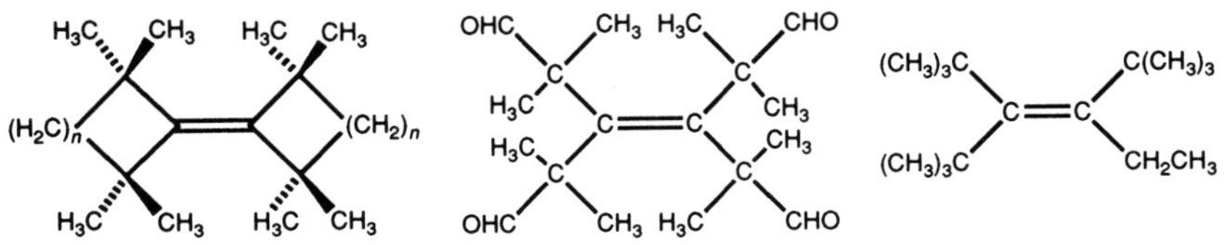

Figure 9.17. Highly strained ethylenes.

spectroscopy. Unfortunately, various efforts to reduce the aldehyde groups terminated in formation of a very stable, double six-membered hemiacetal [cf. Fig. 9.17, structure **A**, CH_2–O–CHOH in place of $(CH_2)_n$] or other cyclic species. A more extensive discussion of strained and nonplanar alkenes may be found elsewhere (Luef and Keese, 1991).

e. The C=N and N=N Double Bonds

The cis–trans isomerism about C=N double bonds is of importance in oximes, imines, hydrazones, and so on, and that about N=N double bonds is of interest in azo compounds (Kalinowski and Kessler, 1973; Jackman, 1975). Partial double bonds also are found in amides and thioamides (pertinent canonical structure

$$R-\overset{|}{\underset{O^-}{C}}=\overset{+}{N}R_2 \text{ and } R-\overset{|}{\underset{S^-}{C}}=\overset{+}{N}R_2),\ \text{esters } (R-\overset{|}{\underset{O^-}{C}}=\overset{+}{O}R'),\ \text{enolates } \left(\overset{\diagup}{\diagdown}C=\overset{|}{\underset{}{C}}-O^- \right),\ \text{and so}$$

on (see also Figure 2.16).

Barriers about various aldehyde and ketone derivatives with amines (imines and hydrazones) are shown in Table 9.1; included are barriers about N=N double

TABLE 9.1.   Barriers to E–Z Isomerization or Topomerization about C=N and N=N Bonds[a]

Compound	$\Delta G^{\ddagger}$ (kcal mol^{-1})	$\Delta G^{\ddagger}$ (kJ mol^{-1})
$(CH_3)_2C=NC_6H_5$[b]	20.3	84.9
$(CH_3)_2C=NCN$[b]	18.9	79.1
$(CH_3)_2C=N-N(CH_3)C_6H_5$[b]	21.1[c]	88.3[c]
$(CF_3)_2C=NC_6H_5$[b]	15.45	64.6
$p\text{-}ClC_6H_4(C_6H_5)C=NCH_3$	25[c]	105[d]
$C_6H_5CH=NC_6H_5$	16.5[d,e]	69.0[d,e]
$C_6H_5N=NC_6H_5$	23.7	99.2
F–N=N–F	35.2	147
$(CH_3)_2\overset{+}{N}=N-\overset{-}{O}$[b]	23.3	97.5
$p\text{-}ClC_6H_4(C_6H_5)C=N-O-CH_3$	>39	>163

[a] Data, taken from Kalinowski and Kessler (1973), refer to isomerization unless otherwise indicated.
[b] Topomerization, cf. Chapter 8.
[c] In hexachlorobutadiene. A much higher barrier was found in diphenyl ether.
[d] E_a.
[e] Data from Anderson and Wettermark (1965).

bonds. Some of these barriers which were determined by NMR spectroscopy involve topomerization rather than isomerization; nonetheless it is clear that they are, in most cases, low enough as to make isolation of stable cis–trans isomers in these series difficult. Notable exceptions are the oximes and oxime ethers (Meisenheimer and Theilacker, 1933) whose E–Z (or syn–anti; cf. Section 9-1.b) isomers are quite stable with barriers above 39 kcal mol^{-1} (163 kJ mol^{-1}). Here, as in the case of F–N=N–F (Table 9.1), the presence of electron-withdrawing substituents on nitrogen greatly increases the barriers for reasons we shall explore shortly. Most of the other (minor) trends in Table 9.1 can be readily explained on the basis of resonance effects (delocalization of the π electrons in C=N lowers the barrier). The low barrier in the hexafluoroacetone phenylimine is an exception; it appears that the normal $\overset{+}{C}$–$\overset{-}{N}$ polarization of the imine bond here is reversed with the negatively charged carbon being stabilized by the electron-withdrawing CF$_3$ groups: $F_3C \leftarrow \overset{\overset{\displaystyle CF_3}{\uparrow}}{\underset{-}{C}} - \overset{+}{N} - C_6H_5$; accordingly electron-donating substituents in the aryl group lower the barrier still further (Hall, Roberts et al., 1971).

A question that arises in cis–trans isomerization about a C=N bond is whether the process is a true rotation or an inversion (flipping) of the nitrogen substituent via an sp-hybridized transition state (Fig. 9.18). A number of arguments have been made in favor of the inversion mechanism (Kalinowski and Kessler, 1973) of which the most convincing rests on the following experiment (Kessler and Leibfritz, 1971): In compounds **A–C**, Figure 9.19, the X and X' groups are diastereotopic by virtue of cis–trans (E–Z) isomerism. In addition, however, because of overcrowding in the planar conformation, the N=CXX' group is forced out of the plane of the benzene ring.

> This overcrowding has long been known in styrenes of the type shown in Figure 9.19, **E**, which can be resolved and are not easily racemized (Adams and Miller, 1940; Adams et al., 1941). Similarly constituted amides (Fig. 9.19, **F**) are also nonplanar, hence chiral and have been resolved (van Lier et al., 1983).

The resulting non-planarity makes molecules **A–C** chiral and hence the methyl groups of individual isopropyl substituents become diastereotopic (cf. Chapter 8). As solutions of **A–C** (Figure 9.19) are warmed, coalescence of X and X' (barrier E_a) and coalescence of the isopropyl methyls (barrier E'_a) occurs with virtually the

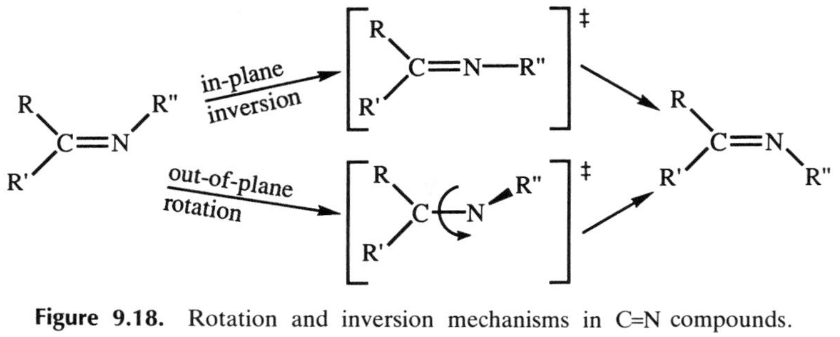

Figure 9.18. Rotation and inversion mechanisms in C=N compounds.

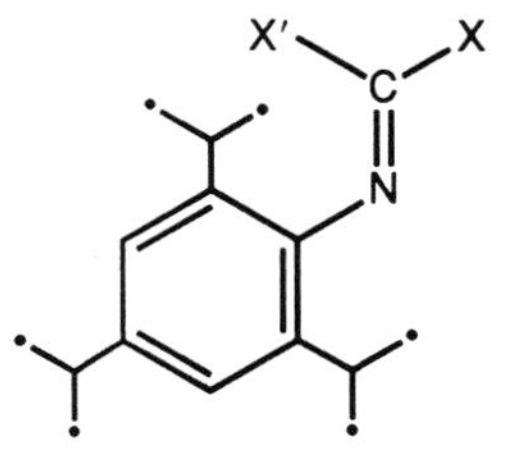

Figure 9.19. Evidence for inversion over rotation mechanism in anils. The dots identify methyl groups that become diastereotopic in the slow exchange limit.

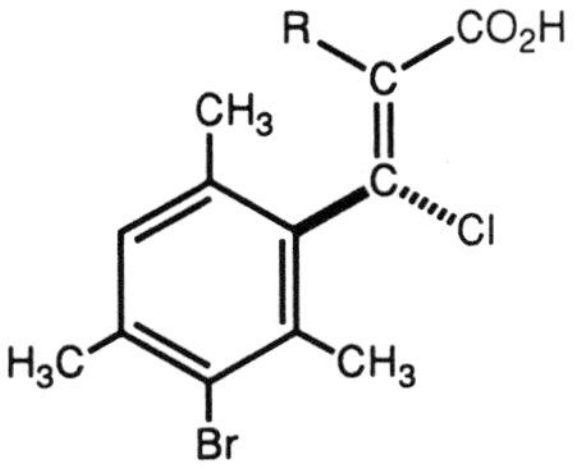

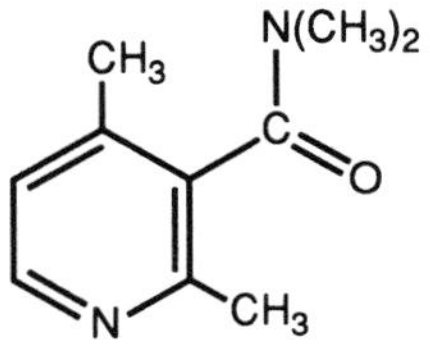

Compound	X	X'	E_a (kcal mol^{-1})	(kJ mol^{-1})	E'_a (kcal mol^{-1})	(kJ mol^{-1})
A	OCH_3	OCH_3	13.4	56.1	13.1	54.8
B	SCH_3	SCH_3	13.2	55.2	13.0	54.4
C	$N(CH_3)_2$	$N(CH_3)_2$	11.4	47.7	11.6	48.5
D	$N(CH_3)_2$	$NHCH_2C_6H_5$	11.7	49.0	23.5	98.3

same activation energies, indicated in this figure. The implication is that the two processes (cis–trans isomerization and racemization) occur via the same transition state, as shown in Figure 9.20; this transition state must involve nitrogen inversion. To eliminate the (unlikely) possibility that two separate processes, C–N single bond rotation around the Ar–N bond (leading to racemization) and purported C=N double-bond rotation (leading to cis–trans isomerization) might fortuitously occur with the same low activation energy, compound **D** (Fig. 9.19) was studied as well. In this case *two* barriers were observed, the lower one being

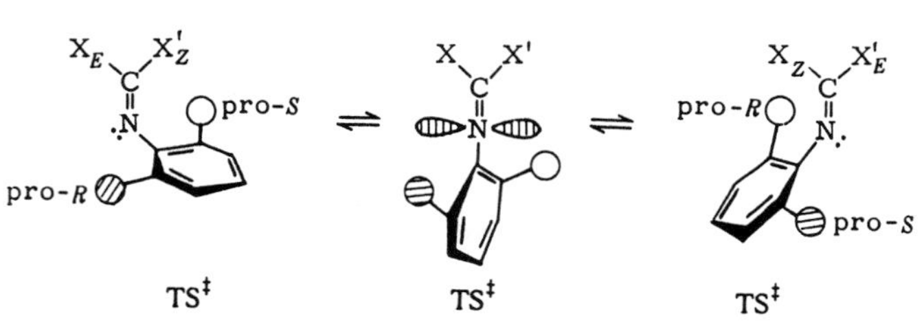

Figure 9.20. Mechanism of cis–trans (E–Z) interchange. (The open and hatched circles denote stereoheterotopic isopropyl groups.) Reprinted with permission from Kessler, H. and Leibfritz, D. (1971), *Chem. Ber.*, **104**, 137, copyright © Verlag Chemie, Weinheim, Germany, 1971.

about the same as in **C**, the higher one being considerably higher. Contemplation of Figure 9.20 reveals that in the case of compound **D**, since $X \neq X'$, nitrogen inversion, while producing cis–trans isomerization, does not lead to enantio-merization (racemization); it takes an *additional* process, Ar–N rotation, to do that. Evidently the activation energy for this process is considerably *higher* than that for exchange. Therefore the simultaneous topomerization of the isopropyl methyl groups and of the E–Z methyl groups on nitrogen in compound **C** is *not* due to the fortuitous coincidence of two barriers but does imply a process that produces both exchanges simultaneously. Nitrogen inversion (Fig. 9.20) does this; C=N rotation does not in that it does not lead to racemization.

We have seen in Chapter 2 that the attachment of one or more electronega-tive atoms to a nitrogen in an amine considerably increases the inversion barrier. This result comes about because the amount of p character in an N–X bond increases with increasing electronegativity of X; as a result the Y–N–X or X–N–X bond angle (Y is any atom other than X) is reduced from its normal, near-tetrahedral value to close to 90° and the transition state for nitrogen inversion, which is planar with about 120° valency angles, becomes more strained. A similar situation is found with respect to the inversion barrier in C=N–X or N=N–X. The transition state for this inversion is linear and thus has sp character, as distinct from the sp^2 ground state. Thus the more p character resides in the N–X bond, the less favorable is the transition from the sp^2 ground to the sp transition state. The high barriers for FN–NF and for the oximes and oxime ether (C=NOR; cf. Table 9.1) can be thus explained if one assumes that in these compounds, also, the barrier involves inversion rather than rotation.

$$\overset{\displaystyle X}{\overset{\displaystyle \|}{}}$$

As mentioned in Chapter 2, amides and thioamides, $R'\text{–}\overset{\overset{\textstyle X}{\|}}{C}\text{–NR}_2$ (X = O or S), though generally written with C=X double and C–N single bonds, in fact have partial C=N double bonds by virtue of the contribution of the $R'C{=}\overset{+}{N}R_2$ canonical

$$\underset{\displaystyle X^-}{\overset{\displaystyle |}{}}$$

form. Barriers for a representative sample of such compounds are shown in Table 9.2 (see also Ōki, 1983); they are of the same order of magnitude as those for the imine-type compounds shown in Table 9.1. It is clear from Table 9.2 that substituents on C=O that are electron donating lower the barrier (e.g., CH_3), whereas groups that are electron withdrawing (CF_3 vs. CH_3, CN vs. H) raise it, presumably because of their unfavorable or favorable effect, respectively, on the electron donation from N to CO ($\overset{+}{N}{=}C{-}O^-$); by the same token, electron-withdrawing groups on nitrogen (CH_2Cl vs. CH_3) lower the barrier. Groups that have p- or π-orbital overlap with the carbon (NH_2, C_6H_5, $CH_2{=}CH$) also lower the barrier because they diminish the C=N double-bond character through competing electron donation to carbon. Thioamides have a higher barrier than amides presumably because the canonical form with C=S double bond is less important than that with a C=O double bond; correspondingly, the canonical form with the $\overset{-}{S}{-}C{=}\overset{+}{N}$ double bond is more important for the thioamide than for the amide. This barrier makes the transition state (C–N single bond) harder to reach. The 2,4,6-tri-*tert*-butylbenzamide shown in Table 9.2 has a barrier high

TABLE 9.2. Barriers in Amides, Thioamides, and Related Compounds[a]

Compound	$\Delta G^{\ddagger}$	
	(kcal mol^{-1})	(kJ mol^{-1})
$HCONH_2$	17.8	74.5
CH_3CONH_2	16.7	69.9
$C_6H_5CONH_2$	15.7	65.7
$HCON(CH_3)_2$	20.9	87.4
$HCON(CH_2Cl)_2^b$	18.3	76.6
$CH_3CON(CH_3)_2$	17.3 (15.6[c])	72.4 (65.3[c])
$CH_3CON(CH_2Cl)_2^b$	16.5	69.0
$CF_3CON(CH_3)_2$	18.6 (16.5[c])	77.8 (69.0[c])
$CH_2=CHCON(CH_3)_2$	16.7	69.9
$C_6H_5CON(CH_3)_2$	15.5	64.9
$C_6H_5CON(CH_2Cl)_2^b$	14.6	61.1
$NC-CON(CH_3)_2$	21.4	89.5
$2,4,6$-tri-$tert$-$BuC_6H_2CON(CH_3)CH_2C_6H_5$	30.3, 32.0[d]	127, 134[c,d]
$HCSN(CH_3)_2$	24.1	101
$CH_3CSN(CH_3)_2$	21.8 (18.0[c])	91.2 (75.3[c])
$C_6H_5CSN(CH_3)_2$	18.4	77.0
$H_2N-CS-N(CH_3)_2$	13.9	58.2

[a] Data taken from Jackman (1975). Since the solvents vary from one case to the other, this source should be consulted for more detailed information. The solvent is clearly important: Not only do barriers vary appreciably from solvent to solvent, but they are appreciably lower (see values in parentheses) in the gas phase (Feigl, 1983).
[b] Vereshchagin et al. (1983).
[c] Gas-phase value (Feigl, 1983).
[d] From one and the other of the two diastereomers, respectively.

enough to permit isolation (and thermal isomerization) of the two diastereomers (Staab and Lauer, 1968), which differ in free energy by 1.66 kcal mol^{-1} (6.95 kJ mol^{-1}). This benzamide, although undoubtedly not planar, must be sufficiently close to planarity in the ground state that the *tert*-butyl groups provide substantial hindrance to C=N rotation. In another amide, L-alanyl–L-proline, cis–trans isomers are sufficiently stable to appear as separate peaks in a reverse-phase liquid chromatogram (Melander, Horvath, et al., 1982); cis–trans isomerism in proline-containing polypeptides is an important determining factor in protein structure.

Returning, for the moment, to the push–pull ethylenes discussed earlier, we note (see Fig. 9.21) that such compounds have not only diminished C=C bond rotation barriers but also enhanced barriers to rotation about the single bonds of

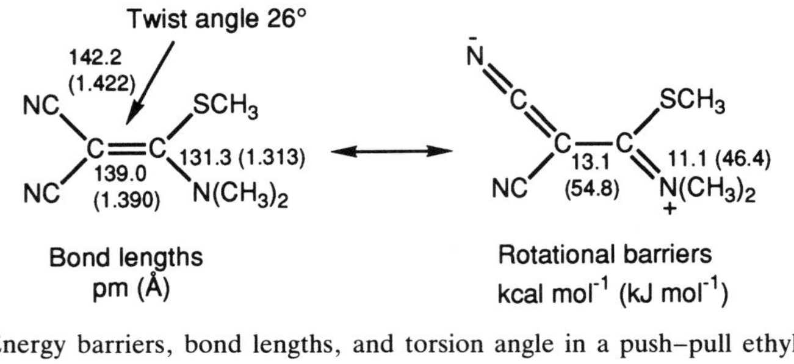

Figure 9.21. Energy barriers, bond lengths, and torsion angle in a push–pull ethylene. Data taken from Sandström, 1983.

the attached donor and acceptor atoms (in the case shown, only the C–N barrier is measurable). Along with this goes a C=C distance considerably longer than the normal 133 pm (1.33 Å) and foreshortened bonds to the donor and acceptor atoms [normal C–C bond length is 153 pm (1.53 Å), C–N, 147 pm (1.47 Å)]. Perhaps surprisingly, in view of the small size of the CN substituents, the molecule shown in Figure 9.21 is also twisted by 26° around the C–C bond.

9-2. DETERMINATION OF CONFIGURATION OF CIS–TRANS ISOMERS

The configuration of cis–trans isomers (E or Z) can be determined by either physical or chemical methods. Chemical methods were the first to be applied; these methods often rest on a very firm basis, and the assignments made on chemical grounds have served as underpinnings for the later developed physical (mainly spectroscopic) methods. However, they are now largely of historical interest; physical methods are simpler and easier to employ, and are therefore used almost exclusively in contemporary chemistry. Our discussion of chemical methods will therefore be limited to but a few cases illustrating the principles involved; a number of additional examples have been assembled elsewhere (Eliel, 1962, Section 12-2; Brewster, 1972).

a. Chemical Methods

These methods are essentially of three types: absolute, correlative not affecting the configuration of the double bond, and mechanistic.

Absolute methods rest on the fact that isomers in which functional or reactive groups are located cis to each other can sometimes be converted to cyclic lactones, anhydrides, amides, and so on, whereas the corresponding trans isomers cannot. Alternatively, the cis, but not the trans, isomers can be synthesized from small ring alkenes. Thus maleic, (Z)-butenedioic, acid is converted, by gentle heating, into its cyclic anhydride (Fig. 9.22) from which the acid can be regenerated by hydration; on the basis of this long-known fact van't Hoff (1875) assigned the cis or Z configuration to maleic acid. The trans isomer, fumaric or (E)-butenedioic acid, is converted into the same anhydride only at a much higher temperature, presumably as a result of thermal isomerization (cf. Section 9-4). The formation of maleic acid by oxidation of benzene or p-quinone (Fig. 9.22) suports the Z assignment. Similar principles have been applied to configurational assignments of oximes (cf. Eliel, 1962).

As in the case of determination of configuration in chiral molecules (Section 5-5), once the configuration of a few alkenes was known with certainty, that of others could be determined by correlative methods. The reliability of such methods is limited on the one hand by the possibility of cis–trans isomerization in the course of the transformations used in the correlations and, on the other hand, in cases where reactions at an olefinic carbon are involved, by potential uncertainty concerning the steric course of such reactions.

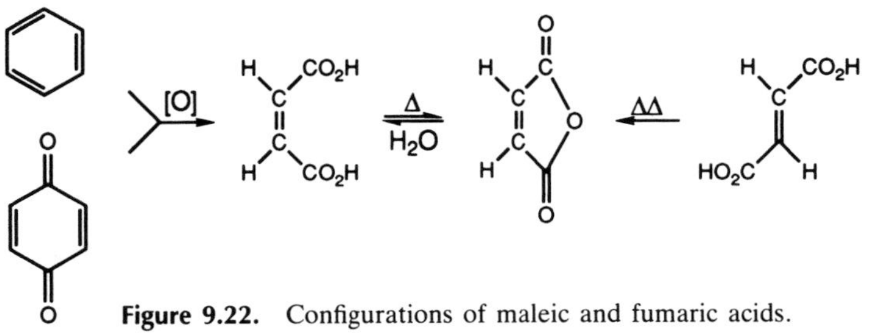

Figure 9.22. Configurations of maleic and fumaric acids.

An example not involving reaction at the olefinic carbon atoms is shown in Fig. 9.23, the correlation of the configuration of the higher melting (trans) crotonic acid with fumaric acid via the relay of trichlorocrotonic acid (von Auwers and Wissebach, 1923). Although the assignment of the E configuration to the crotonic acid melting at 72°C is now known to be correct, the argument embodied in Figure 9.23 is somewhat weak, since isomerization $E \rightarrow Z$ in the course of the transformation was not convincingly excluded. It is preferable in such correlations to make them for both the E and the Z isomer to assure that there is no stereoconvergence, or at least to check (by physical means) that a configurationally pure starting material gives a configurationally pure product. The assumption underlying the latter check is that configurational change, if it occurred, would not be complete but would lead to a cis–trans mixture; this assumption is more certain if the correlation is effected with the less stable (cis?) isomer.

Correlation involving a nucleophilic displacement (with retention) of an olefinic halide of known configuration (Corey and Posner, 1967) is shown in Figure 9.24. Earlier correlations of this type involving electrophilic displacement with retention of vinyllithium compounds, which, in turn, were generated from the corresponding halides by treatment with lithium metal or butyllithium that also involves retention, are due to the pioneering work of Curtin (Curtin and

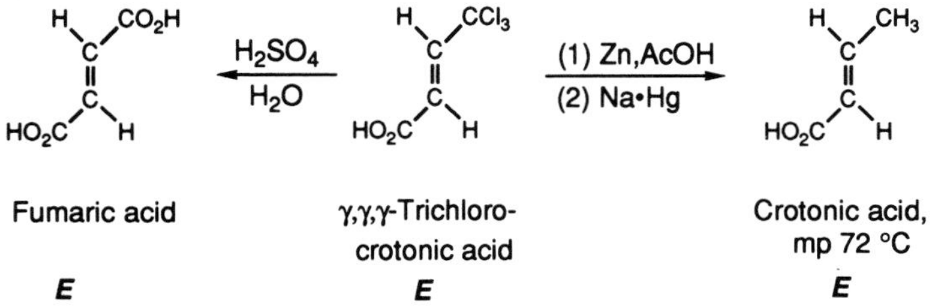

Figure 9.23. Configurational correlation of crotonic acid (mp 72°C) with fumaric acid.

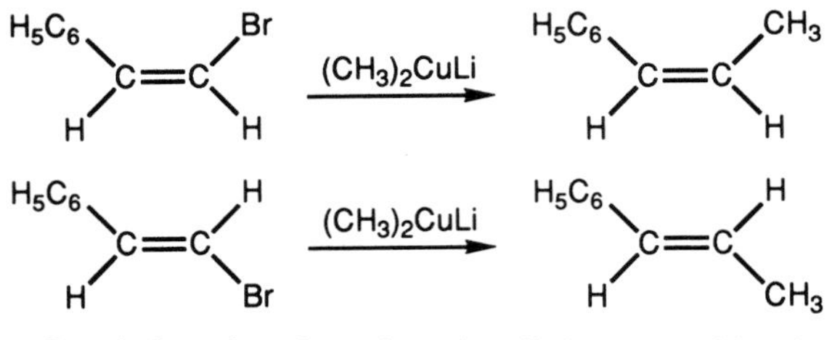

Figure 9.24. Correlation of configuration of β-alkylstyrenes with β-bromostyrene.

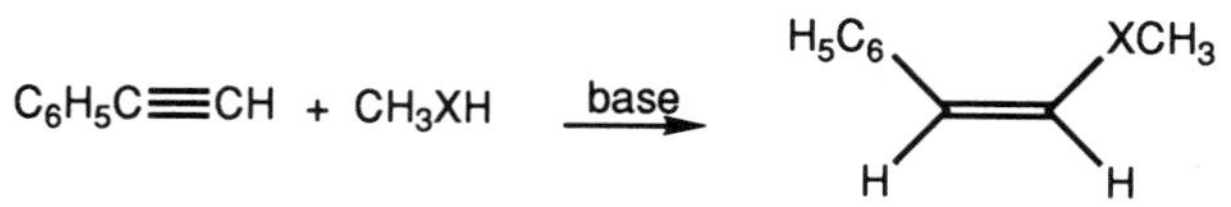

$$C_6H_5C{\equiv}CH \ + \ CH_3XH \ \xrightarrow{\text{base}}$$

Figure 9.25. Nucleophilic addition to acetylenes, where $X = O$ or S.

Harris, 1951; Curtin et al., 1955). Typical electrophiles are aldehydes, acid chlorides, carbon dioxide, alkyl halides and protons.

A third way of assigning E or Z configuration to an alkene is by a directed stereospecific synthesis. The terms "stereospecific" and "stereoselective" are defined in Chapter 12 (see also page 650), "directed" means that the stereochemical outcome of the reaction chosen can be confidently predicted; the principle involved is the same as that discussed in Section 5-5.f for configurational assignment at chiral centers. Although there are many classical examples of this method, only a few of which are mentioned below, its contemporary significance is mainly in diastereoselective synthesis; a number of examples are given in Section 12-2.a.

There are three approaches to this type of assignment: addition to alkynes (acetylenes), synthesis from saturated compounds of known configuration, and "other" methods, of which the Wittig reaction is probably the most important.

Of the three approaches, addition to alkynes is the simplest, since alkynes have no stereochemistry of their own and the stereochemical outcome (formation of cis or trans alkene) depends only on the stereochemical course (syn or anti) of the addition. Thus, in contrast to electrophilic additions (Fahey, 1968) nucleophilic additions, exemplified in Figure 9.25, proceed reliably in anti fashion (cf. Eliel, 1962; Winterfeldt, 1969) giving rise to Z enol (or thioenol) ethers. It is believed that the incoming electron pair of the nucleophile and the p electrons of the π bond tend to stay as far away from each other as possible; subsequent protonation of the anion with retention of configuration then leads to anti addition (Miller, 1956).

In contrast, the addition of an alkylcoppermagnesium halide to a terminal alkyne (Fig. 9.26) is syn (presumably this is a molecular addition of $R'Cu$); when followed by alkylation this is a method for preparing trisubstituted alkenes of defined configuration (Normant et al., 1972).

A similar dichotomy is found in reduction of alkynes to alkenes: addition of alkylboron reagents leads to cis alkenes (as does catalytic hydrogenation) but lithium–ammonia reductions (presumably involving addition of solvated electrons) gives trans alkenes; details are given in Section 12-2.b. Other stereochemically defined syntheses of alkenes from alkynes are also discussed in Section 12-2.b.

We pass now to reactions where alkene configuration is mechanistically linked to the configuration (assumed known) of a saturated precursor (see also Section

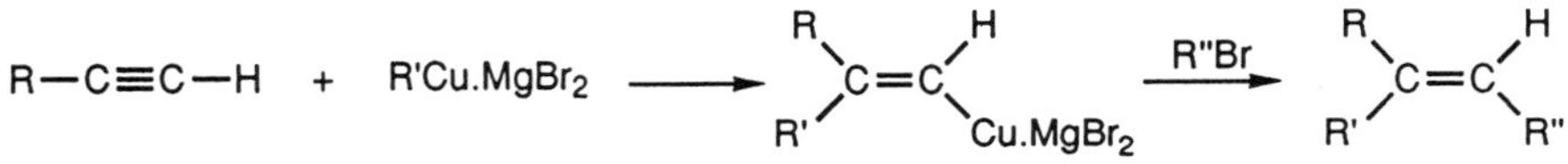

$$R{-}C{\equiv}C{-}H \ + \ R'Cu.MgBr_2 \ \longrightarrow \ \underset{R'}{\overset{R}{>}}C{=}C\underset{Cu.MgBr_2}{\overset{H}{<}} \ \xrightarrow{R''Br} \ \underset{R'}{\overset{R}{>}}C{=}C\underset{R''}{\overset{H}{<}}$$

Figure 9.26. Addition of organocopper reagent to an alkyne, followed by alkylation.

12-2.b). Only such reactions are useful whose stereochemical course is known "with confidence." Ionic E2 eliminations and pyrolytic elimination reactions (Saunders and Cockerill, 1973) best fill this bill. Bimolecular ionic elimination reactions in unconstrained (i.e., noncyclic) systems almost invariably proceed in antiperiplanar fashion (the few exceptions seem to be confined to elimination of quaternary ammonium salts with hindered bases, such as *tert*-alkoxides). The necessary condition for success of the method, then, is that the configuration of the saturated starting material be known. A classical example (Pfeiffer, 1904) is shown in Figure 9.27; the configuration of the precursor "dibromostilbenes" (1,2-dibromo-1,2-diphenylethanes) is readily inferred from their dipole moments (Weissberger, 1945): since the phenyl groups tend to be antiperiplanar, the dipole moment of the meso isomer (antiperiplanar bromines) will be much lower than that of the chiral isomers (gauche bromines). If it is assumed that the steric course of both KOH and sodium thiophenolate (PhS⁻Na⁺) elimination is anti, the configurations of the stilbenes (1,2-diphenylethenes) and α-bromostilbenes shown in Figure 9.27 follow; moreover it follows that the reduction of the bromostilbenes to stilbenes with zinc and ethanol proceeds with retention of configuration.

> The steric course of the reaction of the dibromide with thiophenolate (cf. Otto, 1895) is the same as that with iodide (Mathai and Miller, 1970) and the mechanism (anti elimination) is probably also similar, except that PhSBr, formed in the initial bromine abstraction, reacts with another molecule of PhS⁻ to give PhSSPh.

Additional examples of stereoselective formation of alkenes are given in section 12-2.a.

Perhaps surprisingly, substantial stereoselectivity, and thereby prediction of the configuration of the product, may also be attained in the Wittig reaction (cf. Schlosser, 1970; Vedejs and Peterson, 1994) in which the alkene is, so to speak, put together from its two halves, one in the form of alkylidenephosphorane, $RCH=P(C_6H_5)_3$ (or, in general, $RCH=PR_3$), and the other in the form of an aldehyde or ketone, O=CRR′. When the second reagent is an aliphatic aldehyde,

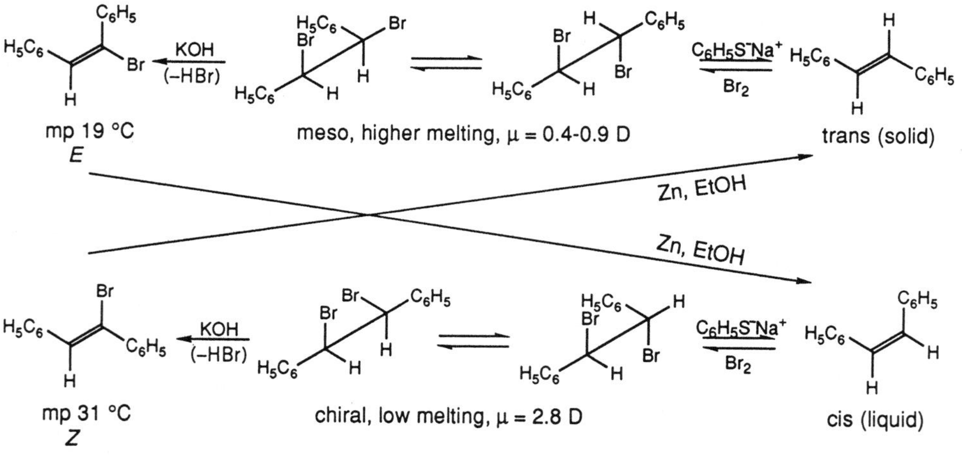

Figure 9.27. Interconversion of stilbene dibromides, stilbenes, and bromostilbenes.

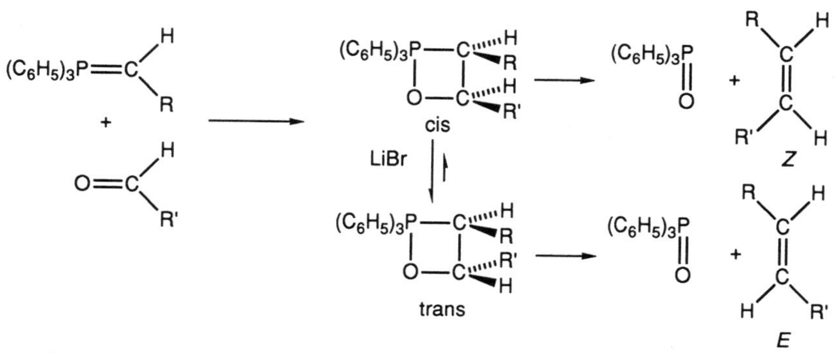

Figure 9.28. Wittig reaction leading to Z alkene (in the absence of Li salt).

O=CHR and the reaction is carried out under carefully circumscribed conditions (for details see Section 12-2.b), the cis or Z alkene will generally be the near-exclusive product, formed via a *cis*-oxaphosphetane intermediate (Fig. 9.28).

The explanations for the stereochemical outcome of the Wittig reaction (cf. Maryanoff et al., 1986; Vedejs and Marth, 1988) are complex and many factors intervene. Under salt-free conditions, the oxaphosphetane intermediate (Fig. 9.28) seems to be formed directly by a [2 + 2] cycloaddition of the aldehyde to the alkylidenephosphorane. The kinetically preferred stereoisomer, for steric reasons (Schlosser and Schaub, 1982; Vedejs and Marth, 1988) is the cis isomer that decomposes to the cis alkene, provided this decomposition is faster than reversal to the aldehyde and alkylidene phosphorane precursors. With "stabilized" (R=aryl) oxaphosphetanes or with $RCH=PBu_3$ [as distinct from $RC=P(C_6H_5)_3$] precursors, decomposition to the alkene is slow and reversal may compete; reversal is also furthered in the presence of lithium salts [possibly via betaine ion pairs, $(C_6H_5)_3\overset{+}{P}-CHR-CHR-\overset{-}{O}\overset{+}{Li}$] and is concentration dependent (Maryanoff
$\overset{}{Br^-}$
and Reitz, 1986). Under such circumstances (assuming equilibration of the intermediate oxaphosphetanes to be complete) Curtin–Hammett kinetics (cf. Section 10-5) will apply and the lower energy transition state from the oxaphosphetane to the alkene product will be traversed; this may well be the one in which the alkyl R groups are trans to each other (Fig. 9.28). It is not clear, however (cf. Vedejs and Marth, 1988), that all Wittig reactions producing predominantly trans (E) alkenes do involve reversion of intermediates to starting materials (see also Section 12-2.b and Vedejs and Peterson, 1994).

Interestingly, when the Wittig reagent is derived from dibenzophosphole (Fig. 9.29), trans alkenes are obtained in better than 6:1 predominance in a kinetically controlled reaction even from aliphatic aldehydes (Vedejs and Marth, 1987). In

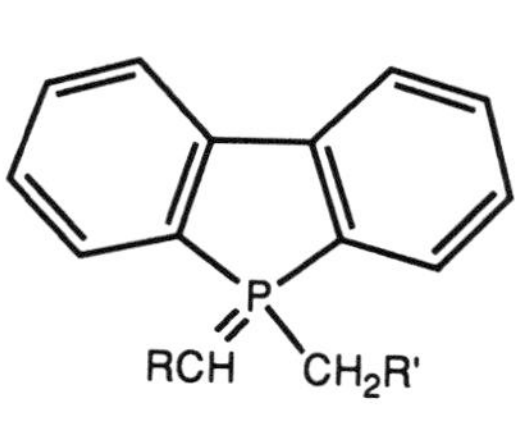

Figure 9.29. Dibenzophosphole derived alkylidenephosphorane.

contrast, triarylphosphine derived Wittig reagents in which the aryl groups bear
o-methyl or *o,o'*-difluoro substituents lead to an enhancement in the predominance of cis alkenes in cases where stereoselectivity is otherwise not satisfactory
(Schaub, Schlosser, et al., 1986).

So far in this section we have discussed reactions in which alkenes are formed
from precursors by processes of known stereochemical course, the configurations
of the products thus being assigned. In principle one can, of course, proceed in
reverse by letting an alkene of unknown configuration undergo a reaction (e.g.,
electrophilic addition) whose steric course is well understood and then analyzing
the stereochemistry of the product formed. Evidently, this also requires reactions
that are stereospecific, that is, in which there is a unique relationship between the
configuration of the starting alkene and the configuration of the addition product.
Such reactions (a trivial example is the addition of bromine to the stilbenes, Fig.
9.27) can be used for configurational assignments in alkenes provided that the
configuration of the saturated compound formed in the reaction is known. This
very constraining proviso is the same as that for the reverse process, stereospecific
formation of an alkene from a saturated precursor, which was discussed earlier.
There is, however, another possibility, namely, to use reaction *rate* rather than
reaction product in configurational assignment. This device (exploitation of
kinetics) can be used both in alkene-forming reactions and in alkene-consuming
reactions. Two examples must suffice to illustrate the principle (see also Curtin,
1954).

In the dehydrohalogenation of 2-bromobutane, the 2-butene isomers are
formed in a ratio *E : Z* of about 6 : 1 (Lucas et al., 1925). The reaction is shown in
Figure 9.30. On the basis of the following argument, one can predict that the
major product formed is the *E* (trans) isomer: Formation of the trans and cis
isomers involves transition states with antiperiplanar Br and diastereotopic hydro-
gen atoms H_1 or H_2. Transition state $A^\ddagger$ (elimination of H_2) is preferred over $B^\ddagger$,
since it lies between starting conformation **A** and product P_A, whereas $B^\ddagger$ lies

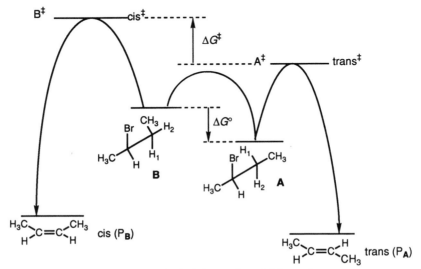

Figure 9.30.   Dehydrohalogenation of 2-bromobutane.

between **B** and P_B, and since the energy level of **A** is below that of **B** (absence of CH_3/CH_3 gauche interaction) and that of P_A lies below that of P_B (trans product more stable than cis; cf. Section 9-4). Therefore, barring the most unlikely possibility of a double crossing of the energy surfaces, transition state $B^{\ddagger}$ lies above $A^{\ddagger}$ (Fig. 9.30) and the major product must thus be the trans isomer.

A slightly different situation (in that the starting states are not in equilibrium) is shown in Figure 9.31, which refers to the dehydrochlorination of the meso and chiral 1,2-dichloro-1,2-diphenylethanes (stilbene dichlorides; Pfeiffer, 1912). At 200°C only one of the two diastereomers is dehydrohalogenated with pyridine. Since the Z (trans) product shown in Figure 9.31 is, for steric reasons (see also Section 9-4) much more stable than the E (cis) product, the transition state leading to it should also be of lower energy. The overall stability of the starting materials is probably not different enough to account for the large difference in reactivity. However, in the chiral isomer the reactive conformation (Cl and H antiperiplanar) is also likely to be the most populated one, because, in it, the bulky phenyl substituents are antiperiplanar. In the corresponding conformer of the meso isomer (Fig. 9.31) the phenyl groups are gauche and the reactive conformer is thus sparsely populated. According to the Winstein–Holness equation (Eq. 10.21, p. 650) this will lead to a depression of the observed rate constant. Thus one can infer from the reaction rates alone and without any other prior stereochemical knowledge that the reactive isomer is the chiral one and that its dehydrohalogenation product is the Z alkene.

Before concluding this section, we must take up one more rather important case where the constitution of a reaction product is used to assign configuration to its precursor, namely, the Beckmann rearrangement of oximes (Donaruma and Heldt, 1960; McCarty, 1970). We have already mentioned (Section 5-5.f) that the migrating group in the Beckmann rearrangement retains its configuration. However, even when the migrating group is achiral, there is an important stereochemical aspect to the rearrangement: The group that migrates is the group trans to the OH moiety of the oxime (Fig. 9.32). This fact allows one to infer the stereochemistry of the oxime from the nature of the amide formed from it. Thus acid

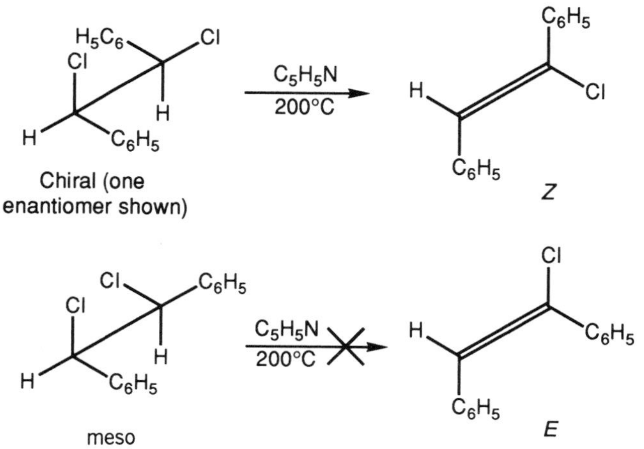

Figure 9.31. Dehydrochlorination of the stilbene dichlorides with pyridine.

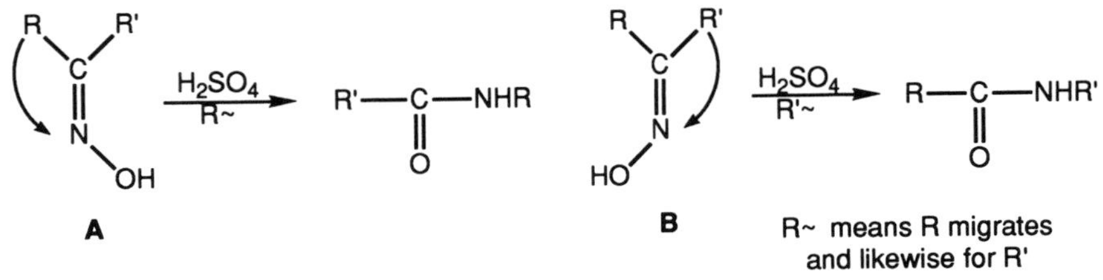

Figure 9.32. Course of the Beckmann rearrangement.

treatment of the (E)-(acetophenone oxime) (Fig. 9.32, **A**, where $R = C_6H_5$, $R' = CH_3$) would lead to acetanilide ($CH_3CONHC_6H_5$) through phenyl migration, whereas the corresponding Z oxime **B** would give N-methylbenzamide ($C_6H_5CONHCH_3$) by methyl migration.

> It is worth mentioning (cf. Blatt, 1933) that Hantzsch (1891), who discovered the stereospecificity of the Beckmann rearrangement, assumed, without proof, that it was the cis group that migrated and assigned configurations of oximes accordingly. As a result, all oxime configurations assigned between 1891 and 1921 are erroneous! It was only after Meisenheimer's unequivocal assignment, in 1921 (q.v.), of the configuration of benzyl β-monoxime (Fig. 9.32, **A**, where $R = C_6H_5$, $R' = C_6H_5CO$) which, in the Beckmann rearrangement, gives the anilide of benzoylformic acid ($C_6H_5COCONHC_6H_5$) by phenyl migration, whereas the isomeric α-oxime **B** gives dibenzoylimide ($C_6H_5CONHCOC_6H_5$) by benzoyl migration that the true nature of the rearrangement (anti migration) was recognized.

b. Physical Methods

Since cis–trans isomers are diastereomers, they will generally differ in physical properties. If the differences are well understood in terms of configuration (E or Z), they can be used for configurational assignment. Only the more easily measurable and often clearcut differences in properties: dipole moments, boiling point, density, refractive index, UV–vis spectra, vibrational (IR–Raman) spectra, NMR spectra, both ^{1}H and ^{13}C, and (though less often useful) mass spectra will be discussed here. Among other techniques, the earlier mentioned X-ray and electron diffraction methods, as well as microwave spectroscopy, also lead to configurational assignment, though the latter two only in the case of relatively small molecules (Landolt-Börnstein, 1976).

Dipole Moments

As shown in Figure 9.33, the relation of dipole moment (cf. Minkin, 1977) to configuration is quite direct. If in a 1,2-disubstituted alkene (XCH=CHY) X and Y are both electron donating or both electron withdrawing, the dipole moment of the cis isomer will generally be sizeable, whereas that of the trans isomer will be small or zero (i.e. $\mu_{cis} > \mu_{trans}$). If, on the other hand, X is electron donating and Y is electron withdrawing, or vice versa, $\mu_{trans} > \mu_{cis}$. In trisubstituted alkenes XCH=CYZ the situation is less clear cut, though if Z is alkyl and X and Y are halogens or other strongly electron withdrawing groups, the disposition (cis or

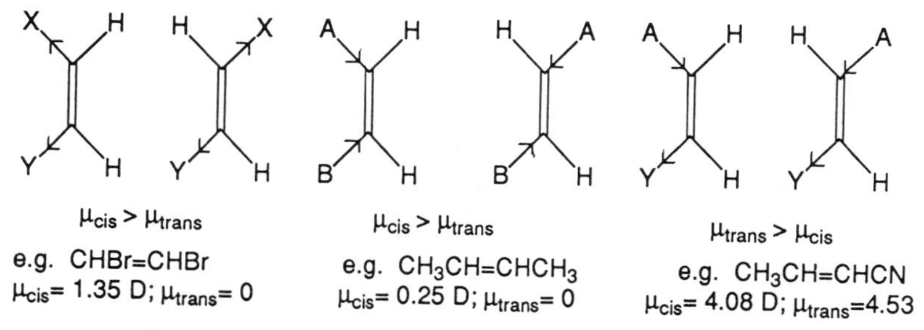

Figure 9.33. Dipole moment and cis–trans configuration.

trans) of X and Y will be decisive. Some salient dipole moments are shown in Table 9.3, column 2. Except for the case of cyclooctene, configuration can be inferred from dipole moment where known, though in some cases the differences are perhaps uncomfortably small. In the case of 1-chloro-2-iodoethylene the original, seemingly unreasonable order of dipole moments (cf. Eliel, 1962) was

TABLE 9.3. Dipole Moments and the Dipole Rule[a]

Compound		$\mu(D)$[b]	bp (°C; 760 mm)	n^{20}	$D^{20}(\text{g mL}^{-1})$
CHCl=CHCl	cis	1.84	60.3	1.4486	1.2835
	trans	0	47.4	1.4454	1.2583
CHCl=CHI	cis	1.27[c,d]	116–117	1.5829	2.2080 (15°C)
	trans	0.55[c]	113–114	1.5715	2.1048 (15°C)
$CH_3CH=CHCl$	cis	1.64[e]	32.8	1.4060	0.9347
	trans	1.97[e]	37.4	1.4058	0.935
$CH_3CCl=CHCl$	cis	2.20[f]	93	1.4549	1.1870 (25°C)
	trans	0.84[f]	76	1.4498	1.1704 (25°C)
$CH_3CH=CHCH_3$	cis	0.25[e]	3.7	1.3931 (−25°C)	0.6213
	trans	0	0.9	1.3848 (−25°C)	0.6044
$(CH_3)_3CCH=CHC(CH_3)_3$	cis	nr[g]	143	1.4266	0.7439
	trans	nr[g]	125.0	1.4115	0.7167
$CH_3CH=CHCN$	cis	4.08[e]	108	1.4182	0.8244
	trans	4.53[e]	122	1.4216	0.8239
$EtO_2CCH=CHCO_2Et$	cis	2.59[h]	223	1.4413	1.067
	trans	2.40[h]	218	1.4411	1.052
Cyclooctene	cis	0.43[i]	74–75[j]	1.4682 (25°C)	0.8443
	trans	0.82[i]	75[k]	1.4741 (25°C)	0.8483
Cyclodecene	cis	0.44[i]	194–195[l]	1.4858	0.8770
	trans	0.15[i]	194[l]	1.4821	0.8672

[a] Data from Eliel (1962), Beilstein, 3rd and 4th supplement, and McClellan (1963, 1974, 1989) unless otherwise indicated.
[b] In benzene at 25°C unless otherwise indicated.
[c] Data of Errera (1928) as corrected (see footnote d).
[d] Henderson and Gajjar, 1971.
[e] In the gas phase, Kondo, Morino, et al. (1970).
[f] At 30°C.
[g] Not reported.
[h] In carbon tetrachloride.
[i] In heptane.
[j] At 84 mm.
[k] At 78 mm.
[l] At 740 mm.

later reversed (Henderson and Gajjar, 1971). In *trans*-cyclooctene, the normal C_{2h} symmetry of *trans*–XCH=CHX is reduced to at most C_2 (in the prevalent twist conformation, see Chapter 14); therefore (cf. Section 4-6.b) the compound can have a dipole moment, whereas compounds of C_{2h} symmetry cannot. That the moment is so large may be related to the highly twisted nature of the double bond (torsion angle between 136° and 157°; cf. Sandström, 1983, p. 166 and Chapter 14).

Density, Refractive Index, Boiling Point, and Melting Point

Equipment to measure dipole moments, while not complex, is perhaps not widely available. Fortunately, the "dipole rule" (van Arkel, 1932, 1933) predicts that the isomer of higher dipole will have the lower molar volume (presumably because of the enhanced tendency of dipolar molecules to self-associate through electrostatic attraction of the oppositely charged ends), and hence the higher density, refractive index, and boiling point, as well as retention volume in gas chromatography (GC) on a nonpolar column. As shown in Table 9.3, columns 3–5, the rule is generally obeyed and so the simple measurements just mentioned can be used for configurational assignments, provided both diastereomers are available and provided that the measured quantities differ appreciably for the two.

Melting points and solubility have also been used as criteria of configuration: In general, the trans isomer in XCH=CHX is higher melting and less soluble than the cis isomer (Eliel, 1962). However, this criterion is more empirical, and therefore less reliable than those given in the previous paragraph.

Acid Strength

Bjerrum's law (Bjerrum, 1923, cf. Section 11-2.c) $\Delta pK_a = pK_2 - pK_1 = 0.60 + 2.3.N.e^2/RT.\varepsilon.r$ is useful in the assignment of configuration of unsaturated dicarboxylic acids. Here ΔpK_a is the difference between the two pK_a values of such an acid, N is Avogrado's number, e is the electronic charge, R is the gas constant, T is the absolute temperature, ε is the effective dielectric constant of the medium, and r is the distance between the acid functions. Since r is generally smaller for a cis dicarboxylic acid than for its trans isomer, ΔpK_a will be larger, a fact that may be used for configurational assignment. For example, ΔpK_a in maleic acid is 4.19, whereas in fumaric acid it is only 1.36 (Fig. 9.34). For monocarboxylic acids (Fig. 9.34) the difference in pK_a between E and Z isomers is much smaller though still palpable, with the Z acid being the stronger one. This is probably due to steric inhibition of resonance in the undissociated acid,

$$RHC=CH-C\diagdown^{\diagup O}_{OH} \leftrightarrow RH\overset{+}{C}-CH=C\diagdown^{\diagup O^-}_{OH}$$

. This type of resonance is less important in the anion than in the acid itself, since in the anion it would place two negative charges on geminal oxygen atoms. Thus its net effect is to stabilize the anion less than the acid, resulting in a weakening of the latter (Fig. 9.35). For such resonance to be at its maximum, the conjugated C=C–C=O system must be coplanar. In the cis isomer, the β substituent interferes with coplanarity; therefore resonance is somewhat inhibited sterically, and since resonance was originally acid

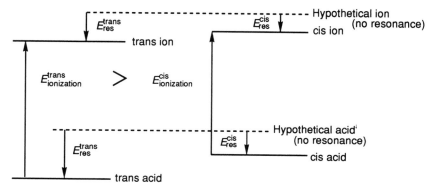

pK_a(1) 2.04	pK_a(1) 3.02	R = CH$_3$	4.42	4.70
pK_a(2) 6.23	pK_a(2) 4.38	C$_6$H$_5$	3.93	4.50
		Cl	3.45	3.79
		Br	3.32	3.71
		I	3.42	3.74

Data from Sarjeant and Dempsey (1979) except those for maleic and fumaric acid, which are from Kortüm et al., (1960) and Lowe and Smith (1974).

Figure 9.34. pK_a values of E and Z β-substituted propenoic acids.

weakening, the cis acid becomes stronger, since the energy required to ionize it is less (Fig. 9.35).

It should be noted (Fig. 9.34) that the *overall* effect of an α, β double bond is acid strengthening (e.g., vis-a-vis acetic acid, pK_a 4.7) presumably because of the inductive effect (electron withdrawing) of the vinyl substituent. The β-haloacrylic acids are stronger than the other acrylic acids because of the electron-withdrawing effect of the halogens; surprisingly, the pK_a values and pK_a cis–trans differences for the Cl, Br, and I compounds are very similar.

Ultraviolet–Visible Spectra

Simple alkenes absorb at about 180 nm (trans isomers) or 183 nm (cis isomers); the absorption is too far in the UV to be routinely useful. However, when the double bond is conjugated with another double bond, an aromatic ring, or a carbonyl function, the region of the absorption becomes readily accessible in the UV (cf. Timmons, 1972) and, as shown in Table 9.4, there are often appreciable differences between cis and trans isomers, both in absorption maximum and in the

Figure 9.35. Resonance weakening of conjugated carboxylic acids. (It is assumed that effects other than resonance affect the acids and ions equally and thus cancel in the difference.)

TABLE 9.4.   Absorption Maxima and Extinction Coefficients for cis–trans Isomers[a]

Compound	λ_{max} (nm)	ε_{max} L mol^{-1} cm^{-1}
trans-Stilbene	295	29,000
cis-Stilbene	280	10,500
(*E*)-α-Methylstilbene[b,c]	272	21,000
(*Z*)-α-Methylstilbene[b,c]	267	9,340
trans-Phenylbutadiene[d]	280	29,500
cis-1-Phenylbutadiene[d]	265	18,200
(*E*)-1-Phenylpropene	249	16,600[e]
(*Z*)-1-Phenylpropene	241	12,400[e]
All-*trans*-β-Carotene[c]	453	158,000
15,15'-*cis*-β-Carotene[c]	340, 453	51,000, 93,000
trans-Cinnamic acid	273	21,000
cis-Cinnamic acid	264	9,500
trans-Cinnamonitrile	269	19,900[e]
cis-Cinnamonitrile	271	19,000[e]
(*E*)-2-Butenoic acid	205	14,000
(*Z*)-2-Butenoic acid	205	13,500
Dimethyl fumarate	208	33,100
Dimethyl maleate	192	21,900
Dimethyl (*E,E*)-2,4-hexadienoate	259, 265	36,700, 34,000
Dimethyl (*E,Z*)-2,4-hexadienoate	260, 266	29,800, 28,900
Dimethyl (*Z,Z*)-2,4-hexadienoate	259, 265	26,400, 23,800
(*E*)-3-penten-2-one	214, 325	12,300, 39
(*Z*)-3-penten-2-one	221, 312	8,700, 41
trans-Benzalacetophenone	227, 298	12,000, 23,600[f]
cis-Benzalacetophenone	248, 290	14,000, 9,000[f]
trans-Azobenzene[c]	318	22,600
cis-Azobenzene[c]	282	5,200

[a] Data from Stern and Timmons (1970) in ethanol, unless otherwise indicated.
[b] Suzuki (1952).
[c] In hydrocarbon solvent.
[d] Grummitt and Christoph (1951).
[e] Fueno et al. (1972).
[f] Dolter and Curran (1960).

molar absorption coefficient. In most cases the trans isomer (or isomer with the larger groups trans) absorbs at longer wavelength λ, and more intensely (i.e., with greater molar absorption coefficient, ε) than the cis isomer. However, it would seem dangerous to rely on this criterion routinely (even when both isomers are available, which is an absolute requirement), for, as Table 9.4 shows, there are just enough exceptions to cause concern. At the very least one should not apply the UV criterion unless both λ_{max} and ε_{max} show the same trend and unless there are no internal contradictions for compounds showing more than one absorption band (e.g., the 3-penten-2-ones and benzalacetophenones in Table 9.4) in the accessible region of the UV.

Since the effect of configuration on UV absorption is not clear-cut and since both steric and electronic factors may play a role in both the ground and excited state (stabilization of the ground state tending to shift λ_{max} to shorter wavelengths, whereas stabilization of the excited state tends to shift it to longer wavelength), a general explanation of the observed trends is not warranted. For example, in the stilbenes (cf. p. 545) it is clear that steric factors (interference of

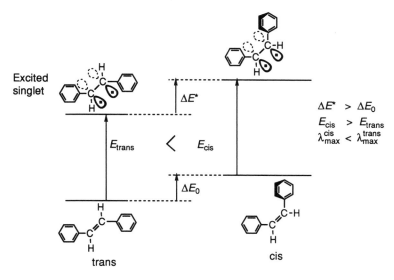

Figure 9.36. Ground and vertically excited states for the stilbenes.

the ortho-hydrogen atoms) force one of the benzene rings out of the plane in the cis isomer (by 43°: Traetteberg et al., 1975), whereas the trans isomer is coplanar. There is thus more π-electron overlap in the trans isomer and, in fact, the trans isomer is more stable. That it nevertheless absorbs at longer wavelength indicates that the energy difference in the excited state must be even greater (Fig. 9.36) to make the energy gap E smaller for the trans isomer (since $E = hc/\lambda$). A plausible explanation is as follows: The singlet excited state is a diradical, and since the Franck–Condon principle demands that it have the same geometry as the ground state, stabilization of the radical on the side of the twisted benzene ring by benzylic resonance is sterically impeded, thus explaining the much higher energy level of the singlet excited state of the cis isomer.

An additional complication in conjugated dienes and enones is that these compounds may exist in s-trans or s-cis conformations (cf. Figs. 10.23 and 10.24). In general, as shown in Figure 9.37, s-cis dienes and enones tend to have considerably lower extinction coefficients than s-trans conformers (cf. Eliel, 1962). Thus the main reason for the much lower extinction coefficient of (Z)-3-penten-2-one as compared to its E isomer (Table 9.4) may be that the E isomer exists largely in the s-trans conformation, whereas steric factors force the Z isomer into the s-cis conformation (Fig. 9.38).

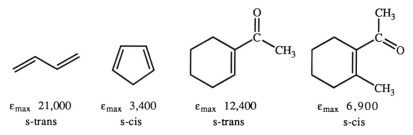

ε_{max} 21,000	ε_{max} 3,400	ε_{max} 12,400	ε_{max} 6,900
s-trans	s-cis	s-trans	s-cis

Figure 9.37. Extinction coefficients of s-cis and s-trans dienes and enones.

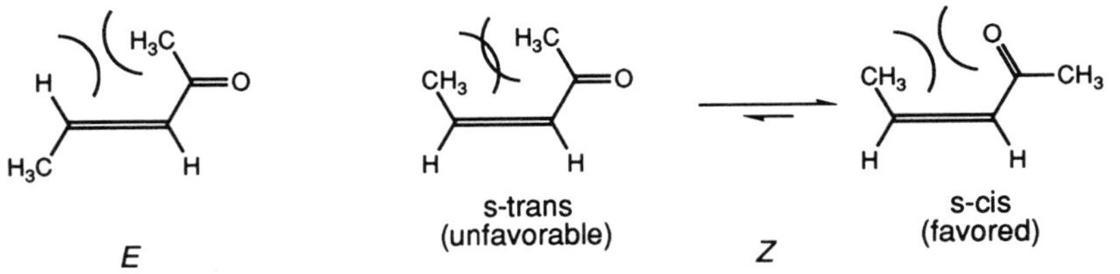

Figure 9.38. Preferred conformations of (*E*)- and (*Z*)-3-penten-2-ones.

Vibrational (Infrared–Raman) Spectra

Differences in IR and Raman spectra (Golfier, 1977; Bellamy, 1975; Eliel, 1962) of cis–trans isomers are found both in the C=C stretching region (ca. 1650 cm^{-1}) and the =C–H out-of-plane vibration region (970–690 cm^{-1}). For a molecular vibration to give rise to IR absorption, it must produce a change in dipole moment of the molecule; no such condition applies to Raman absorption, which, however, requires that the vibration produce a change in polarizability. Thus *trans*-1,2-dichloroethylene, (*E*)-CHCl=CHCl, shows no IR absorption due to C=C stretching, since the dipole moment is zero and remains so during the stretching motion. However, there is a strong Raman absorption at 1577 cm^{-1} corresponding to this vibration. *cis*-Dichloroethylene, (*Z*)-CHCl=CHCl, shows a strong IR C=C stretching vibration at 1590 cm^{-1}, since the molecule has a dipole moment that is altered in the course of the stretching. Similar differences appear in the IR spectra of fumaric and maleic acids, (*E*)- and (*Z*)-HO$_2$CCH=CHCO$_2$H and of *trans*- and *cis*-3-hexenes, (*E*)- and (*Z*)-C$_2$H$_5$CH=CHC$_2$H$_5$. The differences are less marked when the substitution at the olefinic double bond is not symmetrical, for in that case even the trans isomer has a small dipole moment that changes during the C=C stretching. Thus *trans*-2-hexene, (*E*)-CH$_3$CH=CHC$_3$H$_7$, shows a C=C stretching frequency at 1670 cm^{-1}, though much less intense than that of the cis (*Z*) isomer at 1656 cm^{-1} and the *Z* and *E* isomers of 1,2-dichloropropene, ClCH=CClCH$_3$, both show absorptions in the C=C stretching region (at 1614 and 1615 cm^{-1}, respectively), the absorption of the *Z* isomer being considerably more intense (Bernstein and Powling, 1951). As one might expect, this criterion fails in an alkene with an electron-donating group at one end and an electron-withdrawing group at the other (cf. Fig. 9.33); in fact, there is little difference in C=C stretching frequency in the isomeric crotonic acids, CH$_3$CH=CHCO$_2$H, and the situation is actually reversed (trans more strongly absorbing than cis) in the corresponding esters (Allan, Whiting, et al., 1955). In cases where both isomers display the C=C stretching frequency, that of the trans isomer is often at higher wavenumber (1665–1675 cm^{-1}) than that of the cis (1650–1660 cm^{-1}) (Golfier, 1977).

The C–H out-of-plane vibration may also be useful in assigning configuration in alkenes of the type RCH=CHR′. The trans isomers tend to display this mode near 965 cm^{-1} (halogens and oxygens attached to the olefinic carbon lower this number by 35 cm^{-1} per atom). Unfortunately, the C–H out-of-plane vibration in cis alkenes varies over a wide range (650–750 cm^{-1}) and is much less reliable than that in the trans isomer (cf. Golfier, 1977).

Infrared spectroscopy is also useful to distinguish s-cis from s-trans α,β-unsaturated ketones (cf. Fig. 10.24). The s-cis compounds show a higher intensity of the C=C stretching vibration (near 1625 cm^{-1}) and a lower intensity of the C=O stretching mode (near 1700 cm^{-1}) than the s-trans compounds. As a result, the intensity ratio of the two bands (C=O over C=C) is less (0.7–2.5) for the s-cis than for the s-trans compounds (6–9).

Nuclear Magnetic Resonance Spectroscopy

NMR (Gaudemer, 1977; Watts and Goldstein, 1970) is by far the most useful and versatile technique for distinguishing cis and trans isomers. Both ^{1}H and ^{13}C NMR can be used and both chemical shifts and coupling constants are useful for inferring configuration. Moreover, the method is not generally confined to disubstituted alkenes (RCH=CHR′), though some of its variants are.

The chemical shift of the olefinic proton in RCH=CR′R″, where R″ may be H, can be used for configuration assignment by use of a formula first proposed by Pascual, Simon, et al. (1966) and perfected by Matter, Simon, Sternhell, et al., (1969). The formula depends on the fact that the chemical shift increments due to the substituents R, R′, and R″ are additive and, moreover, for R′ and R″ depend on whether these groups are cis or trans to the proton in question: referring to Figure 9.39, $\delta_{C=CH} = 5.25 + Z_{gem} + Z_{cis} + Z_{trans}$, where Z_{gem}, Z_{cis}, and Z_{trans} are parameters characteristic of substituents R_{gem}, R_{cis}, and R_{trans} (Fig. 9.39), given in Table 9.5. The precision of the formula has been assessed through a statistical treatment (Matter, Simon, Sternhell, et al., 1969), which indicated that 81% of the 4298 cases considered yielded calculated shifts within 0.20 ppm of those actually found and 94% yielded shifts within 0.30 ppm of the experimental. Since the calculated differences in shift between a cis and trans disubstituted alkene will be the differences between Z_{cis} and Z_{trans} in Table 9.5, one can easily figure out which substituents allow one to compute the result at the 81 or 94% level of confidence and make a configurational prediction accordingly. By way of illustration, let us, however, choose two trisubstituted alkenes (since methods of configuration assignment are more limited in this series than in disubstituted alkenes) namely, $CH_3CH=CBrCH_3$ and $CH_3CH=CClCO_2H$. For the 2-bromo-2-butenes, the calculated shifts are $5.25 + 0.45 - 0.28 + 0.45 = 5.87$ ppm for the E (cis) isomer and $5.25 + 0.45 - 0.25 + 0.55 = 6.00$ ppm for the Z (trans). The difference of 0.13 ppm is insufficient to allow the Z or E configuration to be assigned to individual isomers at a reasonable level of confidence; although, if shifts for both stereoisomers are available, it may be tentatively concluded that the isomer with the downfield olefinic proton has the Z configuration. In contrast, for the 2-chloro-2-butenoic acids, the calculated vinyl proton shift for the E isomer is $5.25 + 0.45 + 0.32 + 0.18 = 6.20$ ppm and that for the Z $5.25 + 0.45 + 0.98 + 0.13 = 6.81$ ppm. The difference of 0.61 ppm is sufficient to make the assignment from the NMR data with virtually complete confidence even if only one isomer is available. (For a shift difference of 0.50 ppm, the confidence level of assignment is 99.7%. We are assuming here that the shift actually measured is close to one of the two calculated ones.)

In alkenes of type RCH=CHR″ or RCH=CFR′ or RCF=CFR′, proton–proton, proton–fluorine, or fluorine–fluorine coupling constants can be used to assign configuration. Table 9.6 shows the pertinent coupling constants for cis and trans disposed nuclei, respectively (corresponding to cis–trans arrangements of the R and R′ groups). The proton–proton constants follow the Karplus equation (cf. Fig. 10.37) with the cis coupling constant (torsion angle 0°) being smaller than the

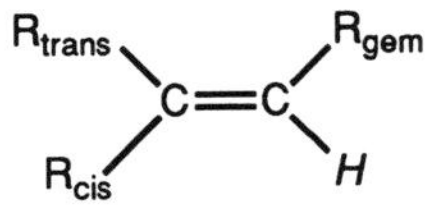

Figure 9.39. Diagram for Table 9.5.

TABLE 9.5. Parameters for Calculation of Proton Shift in the Alkene[a] Shown in Figure 9.39

Substituent R	Z_i for R (ppm)		
	Z_{gem}	Z_{cis}	Z_{trans}
–H	0	0	0
–Alkyl	0.45	−0.22	−0.28
–Alkyl-Ring	0.69	−0.25	−0.28
–CH$_2$O	0.64	−0.01	−0.02
–CH$_2$S	0.71	−0.13	−0.22
–CH$_2$X; X = F, Cl, Br	0.70	0.11	−0.04
–CH$_2$N	0.58	−0.10	−0.08
–C=C isolated	1.00	−0.09	−0.23
–C=C conjugated	1.24	0.02	−0.05
–C≡N	0.27	0.75	0.55
–C≡C	0.47	0.38	0.12
–C=O isolated	1.10	1.12	0.87
–C=O conjugated	1.06	0.91	0.74
–CO$_2$H isolated	0.97	1.41	0.71
–CO$_2$H conjugated	0.80	0.98	0.32
–CO$_2$R isolated	0.80	1.18	0.55
–CO$_2$R conjugated	0.78	1.01	0.46
H –C=O	1.02	0.95	1.17
N –C=O	1.37	0.98	0.46
Cl –C=O	1.11	1.46	1.01
–OR, R: aliphatic	1.22	−1.07	−1.21
–OR, R: conjugated	1.21	−0.60	−1.00
–OCOR	2.11	−0.35	−0.64
–CH$_2$–C=O; –CH$_2$–C≡N	0.69	−0.08	−0.06
–CH$_2$-Aromatic ring	1.05	−0.29	−0.32
–Cl	1.08	0.18	0.13
–Br	1.07	0.45	0.55
–I	1.14	0.81	0.88
–N–R, R: aliphatic	0.80	−1.26	−1.21
–N–R, R: conjugated	1.17	−0.53	−0.99
–N–C=O	2.08	−0.57	−0.72
–Aromatic	1.38	0.36	−0.07
–Aromatic o-subst	1.65	0.19	0.09
–SR	1.11	−0.29	−0.13
–SO$_2$	1.55	1.16	0.93

[a] From Matter, Simon, Sternhell, et al. (1969). Adapted, with permission, from Matter, E.V. et al., *Tetrahedron* (1969), **25**, 693/4. Copyright © 1969 Pergamon Press, Headington Hill, Oxford, UK.

TABLE 9.6. Proton–Proton, Proton–Fluorine, and Fluorine–Fluorine Coupling Constants[a]

Coupled Nuclei	$^3J_{cis}$ (Hz)	$^3J_{trans}$ (Hz)
H, H	+4 to +12	+12 to +19
H, F[b]	−4 to +20	+10 to +50
F, F[b]	+15 to +35	−115 to −134

[a] In part, from Phillips (1972), where X, Y = H or F.
[b] See also Emsley et al. (1976).

trans (torsion angle 180°). There is little overlap of range. For the proton–fluorine couplings, although the range overlaps somewhat, J_{trans} is said to be always larger than J_{cis} and the same is true for the absolute values of the fluorine–fluorine vicinal couplings (Emsley et al., 1976). Some examples are shown in Figure 9.40. A complication arises in that for a symmetrical alkene (RCH=CHR or RCF=CFR) the proton (or fluorine) nuclei are chemical shift equivalent so that the coupling constant cannot be observed directly. However, it is possible to obtain the coupling constant indirectly from the ^{13}C satellite spectrum (since $X–^{13}C–^{12}C–X$, where X = F or H, gives rise to an ABX pattern) or, in the case of protons, from the H–D coupling constant in the spectrum of the deuterated analogue, RCH=CDR ($J_{HH} = 6.49 J_{HD}$).

Since neither IR spectra nor proton coupling constants in NMR are useful for configurational assignment in trisubstituted alkenes (RR′C=CHR″) and since proton chemical shift criteria may also fail in some such cases (see above), it is fortunate that ^{13}C NMR spectroscopy is a particularly valuable tool for configurational assignment in this instance. The only condition is that either R or R′ (or both) must have a carbon atom at the point of attachment to the C=C moiety. The principle involved is that the carbon nucleus cis to the R″ group will be upfield shifted, through a "γ compression effect," relative to the same carbon nucleus positioned cis to H. Both the principle and examples are shown in Figure 9.41

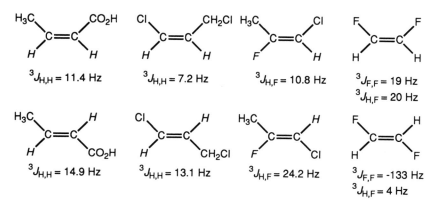

Figure 9.40. Examples of H/H, H/F and F/F vicinal couplings.

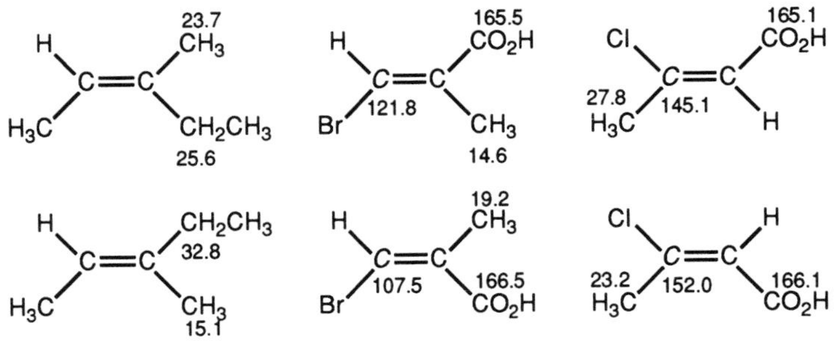

Figure 9.41. The ^{13}C shifts (ppm) of cis–trans isomers.

(data from Breitmaier et al., 1976). It is seen that the CO_2H group does not manifest this upfield compression effect in its shift and that substantial (but apparently not as yet systematized) differences in shift at the β-carbon atom in acrylic acids are observed in some cases.

Recent improvements in NMR techniques, notably 2D NMR and double-quantum coherence experiments, have facilitated the determination of $^3J_{C/H}$ and even $^3J_{C/C}$ coupling constants (cf. Marshall, 1983). Since the Karplus relationship applies to such couplings just as it does to proton–proton couplings (see above), $J_{trans} > J_{cis}$. Examples (from Marshall, 1983) are shown in Figure 9.42.

Spectroscopic techniques as applied to strained and twisted alkenes have been discussed by Luef and Keese (1991).

We conclude this section on the use of NMR in cis–trans configurational assignment with a discussion of two less often used techniques, lanthanide induced shifts (LIS) and the nuclear Overhauser effect (NOE).

Lanthanide induced shifts have already been presented in Section 2-5; in fact, Figure 2.21 shows the distinction of E and Z isomers of oximes by LIS. In principle, any cis–trans pair of alkenes of the type RR′C=CR″X, where X is a group that complexes the lanthanide (such as C=O, SO, etc.) should be amenable to configurational assignment by LIS, since the LIS of the R (or R′) group cis to

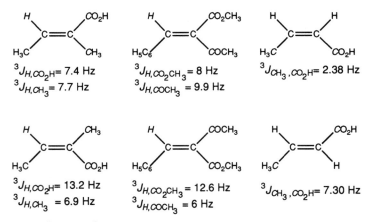

Figure 9.42. Use of $^3J_{C/H}$ and $^3J_{C/C}$ in configurational assignments of cis–trans isomers (The italicized C's represent ^{13}C enriched ligands).

X should be larger than that for the group trans to X (the effect of the lanthanide is transmitted through-space rather than through-bond). An example may be found in the review by Hofer (1976, p. 157).

The NOE (cf. Saunders and Bell, 1970; Noggle and Schirmer, 1971; Bell and Saunders, 1973; Neuhaus, 1989) implies that irradiation of a nucleus (e.g., a proton) in any molecule will facilitate the relaxation of a near-by nucleus (^{13}C, proton, etc.) and thereby enhance its signal, since the more rapidly a nucleus that has been excited to its upper spin state by absorption of energy returns to the ground state, the more energy it can absorb. The NOE, also, is a through-space effect that drops off with the sixth power of internuclear distance. Thus it will generally be important only between cis located nuclei, not between those located trans to each other. Indeed, the first application of the NOE in organic chemistry (Anet and Bourn, 1965) involved assignment of the methyl proton signals in β,β-dimethylacrylic acid, $(CH_3)_2C{=}CHCO_2H$. Irradiation of the protons in one of the methyl groups leads to a substantial enhancement in the signal of the olefinic ($=CH$) proton, whereas irradiation of the other methyl group produced, if anything, a slight diminution in the olefinic proton signal (see also Bell and Saunders, 1970). It was concluded that the methyl group whose irradiation produced the enhancement was located cis to the olefinic hydrogen, its proton signal thus being assigned. In this case, there is only one species with two diastereotopic methyl groups, but determination of the NOE has also been used to assign configuration to the two isomers of citral (E)- and (Z)-$(CH_3)_2C{=}CHCH_2CH_2C^{\beta}(CH_3){=}C^{\alpha}HCHO$ (Ohtsuru et al., 1967). In one isomer there is no enhancement of the α-hydrogen when the β methyl group is irradiated, whereas in the other case there is an 18% enhancement. The isomer showing this enhancement must have $CH_3(\beta)$ cis to $H(\alpha)$ and is therefore the Z isomer (chain cis to CHO), whereas the isomer not displaying the NOE is E. It is important to recognize that in conformationally mobile systems, such as N,N-dimethylformamide (DMF), $HCON(CH_3)_2$, the NOE (unlike the chemical shift or coupling constant) is *not* averaged and the equalization of NOEs may occur at lower temperatures than coalescence of chemical shifts (Bell and Saunders, 1970).

The value of the NOE technique has been enhanced by the possibility of carrying it out for several nuclei at a time in a two-dimensional nuclear Overhauser and exchange spectroscopy (2D NOESY) experiment (cf. Sanders and Hunter, 1993).

Mass Spectrometry

Mass spectrometry (cf. Splitter and Tureček, 1994) is not a major tool for distinguishing stereoisomers since mass spectra of stereoisomers are often very similar. Thus it is not useful in distinguishing unfunctionalized olefinic cis–trans isomers; however, it is occasionally effective in the distinction of stereoisomers containing C=C–C=O moieties (cf. Green, 1976). Thus maleic and citraconic acids, (Z)–$HO_2CCR{=}CHCO_2H$ (R = H or CH_3), show a sizeable $M - 44$ ion (loss of CO_2 from the parent ion), whereas in the corresponding E isomers (fumaric and mesaconic acids) the $M - 44$ peak is small or absent (Benoit et al., 1969).

Chromatography

Although chromatographic methods do not serve to identify cis–trans isomers, they are very useful in analyzing cis–trans mixtures once the relative retention times (or elution orders) of the stereoisomers have been established with authenticated samples. Capillary GC with an Apiezon stationary phase has been used to this end (Lipsky et al., 1959), but, in general, stationary phases containing silver nitrate ($AgNO_3$), which complexes with alkenes, are more efficient (Bednas and Russell, 1958). Gas–liquid chromatography (GLC) employing liquid crystals has also been used (Lester, 1978). In addition, silver nitrate has been employed in liquid–liquid partition chromatography (Heath et al., 1977), including preparative applications (Evershed et al., 1982) and in reverse phase liquid chromatography where the $AgNO_3$ is dissolved in the mobile phase (Vonach and Schomburg, 1978).

A nonchromatographic separation method for cis–trans isomers involves inclusion of the trans isomer in a urea crystal; the cis isomer does not form an inclusion compound (Leadbetter and Plimmer, 1979).

9-3. INTERCONVERSION OF CIS–TRANS ISOMERS: POSITION OF EQUILIBRIUM AND METHODS OF ISOMERIZATION

a. Position of cis–trans Equilibria

When such equilibria (Fig. 9.43) can be established experimentally (see below) the equilibrium constant K (usually obtained chromatographically or spectroscopically from the composition of the equilibrated products) leads directly to the free energy difference between isomers, $\Delta G^0 = -RT \ln K$. If the measurements are carried out at more than one temperature, ΔH^0 can also be obtained as the slope of the plot of $\ln K$ versus $1/T$. Some of the energy values mentioned in this section have been obtained in this fashion, notably those for the 1,2-dihaloethylenes, CHX=CHY. Unfortunately, the method has several drawbacks. One is that cis–trans isomerization is often not clean but is accompanied by double-bond migration. As long as the latter process is not excessive at equilibrium, it is still possible to obtain the equilibrium ratio of the desired cis–trans isomers. Clearly, however, the formation of a number of positional isomers would be a serious problem in equilibration of higher alkenes. With acid catalysts there is also the danger of skeletal rearrangements. In addition, if the equilibrium is quite one-sided, it becomes difficult to measure. An example is the equilibration of the stilbenes (Fig. 9.43, R, R″ = C_6H_5, R′, R‴ = H) by means of iodine activated by visible light (cf. the following section; Fischer et al., 1968; see also Saltiel et al., 1987); the equilibrium mixture contains only 0.09–0.21% of *cis*-stilbene at room temperature, corresponding to $\Delta G^0_{300} = 3.7$–$4.2\ \text{kcal mol}^{-1}$

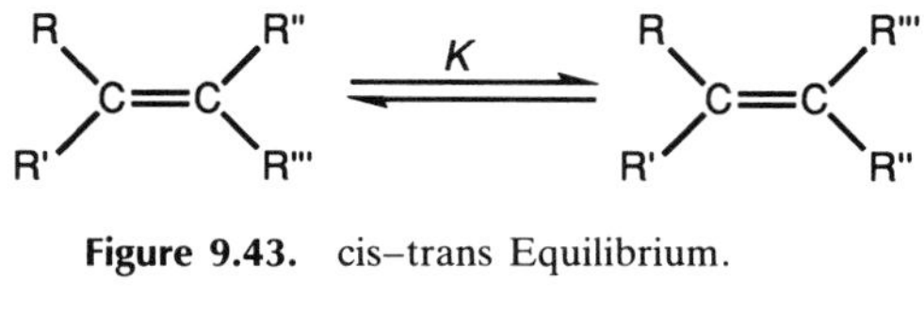

Figure 9.43. cis–trans Equilibrium.

$(15.5–17.6\,\text{kJ mol}^{-1})$. From the variation of K with temperature, $\Delta H^0 = 2.3\,\text{kcal mol}^{-1}$ $(9.6\,\text{kJ mol}^{-1})$ was computed, which agreed with that calculated from heats of combustion (see below; cf. Coops and Hoijtink, 1950; Eliel and Brunet, 1986). However, this value was later corrected to $4.6\,\text{kcal mol}^{-1}$ $(19.2\,\text{kJ mol}^{-1})$ (Saltiel et al., 1987). The reason for the sizeable difference in the ground-state energies for the stilbenes stems mainly from the nonplanarity of *cis*-stilbene (due to steric congestion) which, in turn, leads to steric inhibition of resonance (Fig. 9.36). If the planarity of the *trans*-stilbene framework is also interfered with, as in the α-methylstilbenes or the 2,4,6-trimethylstilbenes, the free energy difference between the Z and E isomers is reduced to $0.83\,\text{kcal mol}^{-1}$ $(3.47\,\text{kJ mol}^{-1})$ in the former case and $1.54\,\text{kcal mol}^{-1}$ $(6.44\,\text{kJ mol}^{-1})$ in the latter (Fischer et al., 1968). In this connection, the finding that the free energy difference between the stilbenes is less than the enthalpy difference is also of interest. Since $\Delta G^0 = \Delta H^0 - T\Delta S^0$, this implies a larger entropy for the cis isomer (by $1.05\,\text{cal mol}^{-1}\,\text{K}^{-1}$ or $4.4\,\text{J mol}^{-1}\,\text{K}^{-1}$). Presumably the essentially planar structure of *trans*-stilbene makes for a rather rigid structure, whereas in *cis*-stilbene at least one of the phenyl rings can librate (oscillate) through a rather wide arc.

Another method of determining ΔH^0 for the equilibrium shown in Figure 9.43 is to determine the heats of combustion (or of formation) of the two isomers and take their difference. The general principle of this method is shown in Figure 9.44. The cis–trans isomers are converted to a common product and the enthalpy change for the process involved is determined. In the case of heats of combustion, the common products are CO_2 and H_2O (formed in the same molar amount for the two stereoisomers), whereas in the case of heats of formation, which must be determined indirectly, they would be hydrogen and carbon in their standard states. Since the products are identical for the two starting isomers, the enthalpy difference found $(\Delta H_{\text{cis}} - \Delta H_{\text{trans}})$ is equal to ΔH^0 between the isomers (cf. Fig. 9.44). Some pertinent data for alkenes are shown in Table 9.7. In disubstituted alkenes, the trans isomer is more stable than the cis. (In the trisubstituted alkenes the enthalpy differences are small or nil.) For unhindered cases the enthalpy difference is close to $1\,\text{kcal mol}^{-1}$ $(4.2\,\text{kJ mol}^{-1})$. Interestingly, here also ΔG^0 is smaller than ΔH^0 because the cis isomer has a slightly larger entropy (for an explanation, see Eliel, 1962, p. 339). When the group at one end of the double bond is *tert*-butyl, it will interact sterically with a cis substituent at the other end

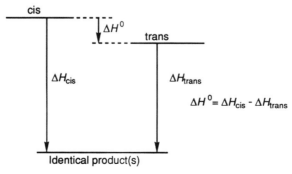

Figure 9.44. Energy picture for conversion of cis–trans isomers to a common product.

TABLE 9.7. Differences in Enthalpies of Combustion (or Formation) and in Free Energies for cis (Z) and trans (E) Isomers[a,b]

Compounds	ΔH^0		ΔG^0	
	(kcal mol^{-1})	(kJ mol^{-1})	(kcal mol^{-1})	(kJ mol^{-1})
$CH_3CH{=}CHCH_3$	0.86[c]	3.6[c]	0.5	2.12
$C_2H_5C{=}CHC_2H_5$	1.19	5.0	1.22	5.1
$(CH_3)_2CHCH{=}CHCH_3$	0.96	4.0	0.60	2.5
$(CH_3)_3CCH{=}CHCH_3$	3.85	16.1	e	
$(CH_3)_3CCH{=}CHC_2H_5$	4.42	18.5	e	
$(CH_3)_3CCH{=}CHC(CH_3)_3^{d}$	10.5	44.0	e	
$C_6H_5CH{=}CHCH_3$	0.98	4.1	0.62	2.6
$CH_3CH_2(CH_3)C{=}CHCH_3$	0.12	0.5	0.07	0.3
$CH_3CH_2(CH_3)C{=}CHC_2H_5$	0.76	3.2	e	
$C_6H_5CH{=}CHC_6H_5^{f}$	4.59	19.2	3.7–4.2	15.5–17.6

[a] At 25°C; gas phase.
[b] From TRC Thermodynamic Tables, Hydrocarbons, Vol. VIII (1983, 1988) unless otherwise indicated: pp. p-2600, p-2630, p-2650, p-2651, p-2672, p-4490, Thermodynamic Research Center, College Station TX.
[c] Golden, Benson, et al. (1964) report $\Delta H^0 = 1.2$ kcal mol^{-1} (5.0 kJ mol^{-1}), $\Delta S^0 = 1.2$ cal mol^{-1} deg^{-1} (5.0 J mol^{-1} deg^{-1}) at 400°C by direct equilibration. The ΔG^0 values in the tables are computed as $\Delta G^0 = \Delta H^0 - T\Delta S^0$, the entropy difference, ΔS^0, being obtained experimentally.
[d] Liquid phase; Rockenfeller and Rossini, 1961.
[e] Not available.
[f] See text.

(more so if that substituent is ethyl than when it is methyl) and the cis–trans energy difference increases. Further examples of this type will be discussed below.

It is important to indicate whether differences in heat of combustion refer to the gas or liquid phase; in the latter case they include the negative heats of vaporization of the two isomers, which may not be the same.

In the case of the stilbenes (see above) the heat of combustion of the trans isomer was determined in the solid (crystalline) state and that of the cis isomer in the liquid (Coops and Hoijtink, 1950). Thus the data were not comparable until they could be converted to the gas phase by measurement of heats of sublimation (Burgess et al., 1980) and of vaporization (Brackman and Plesch, 1952), respectively (cf. Eliel and Brunet, 1986; but see Saltiel et al., 1987).

Yet another means of determining heats of isomerization involves calorimetric measurement of heats of hydrogenation. Since the hydrogenation product of a pair of cis–trans alkenes is one and the same alkane, the energy diagram in Figure 9.44 applies here as well, except that the vertical arrows now represent heats of hydrogenation and ΔH^0 is computed as the difference in heats of hydrogenation of the cis and trans isomers. The advantage of this method over that of determining heats of combustion is that heats of hydrogenation are much smaller, of the order of 30 kcal mol^{-1} (126 kJ mol^{-1}), and can generally be measured to ± 0.1 kcal mol^{-1} (0.4 kJ mol^{-1}) without much difficulty; the minimum precision in ΔH^0, which may amount to no more than 1 kcal mol^{-1} (4.2 kJ mol^{-1}) (cf. Table 9.7) will then be of the same magnitude, (i.e., ca. $\pm 10\%$).

In contrast, heats of combustion tend to amount to well over 1000 kcal mol^{-1} (4200 kJ mol^{-1}) and it is difficult to measure them accurately enough to get meaningful differences of the order of a kcal mol^{-1} (4.2 kJ mol^{-1}) or so.

In fact, determining the enthalpy difference between isomers through differences in heats of combustion has been likened to determining the weight of a naval captain by first determining the weight of an ocean liner with the captain on it and then without the captain and taking the difference! Nonetheless, meaningful data have been obtained through heat of combustion measurements, thanks to the very painstaking experimenters in the field, by working with very pure compounds and using great care in the calorimetry.

Heats of hydrogenation and their differences are shown in Table 9.8 (cf. Jensen, 1976). Here, again, phase must be considered: Data measured in acetic acid include differences in heat of solution. Considering this complication, the data agree quite well with those in Table 9.7 where comparison is possible (entries 1–3, 5, 9) except in the case of the stilbenes (entry 9) where the heat of hydrogenation difference seems to be too large. In unconjugated alkenes large differences in heats of hydrogenation (and therefore stability) are seen if the group at one end of the double bond or both is *tert*-butyl (entries 3 and 5); the case with isopropyl at both ends (entry 4) shows only a marginally enhanced difference and that with two neopentyl groups (entry 6) shows none. Conjugated systems (entries 7–9) show larger differences than unconjugated ones, presumably because of steric inhibition of resonance stabilization in the cis isomer as discussed earlier (p. 564).

The scheme embodied in Figure 9.44 is, in principle, not limited to heats of combustion (or formation) or heats of hydrogenation; calorimetry of any reaction of cis and trans alkenes giving a common product can be used. An example is the addition of HBr to *cis*- and *trans*-2-butene to give, in both cases, 2-bromobutane; ΔH^0 for this reaction is $-18.34 \, \text{kcal mol}^{-1}$ ($-76.7 \, \text{kJ mol}^{-1}$) for the cis isomer and $-17.34 \, \text{kcal mol}^{-1}$ ($-72.6 \, \text{kJ mol}^{-1}$) for the trans. One may conclude that the trans isomer is $1.00 \, \text{kcal mol}^{-1}$ ($4.18 \, \text{kJ mol}^{-1}$) lower in enthalpy than the trans, in excellent agreement with the earlier mentioned value (Lacher et al., 1952).

An exception to the rule that trans isomers are more stable than cis isomers occurs with the 1,2-dihaloethylenes. This apparent anomaly was originally found

TABLE 9.8. Heats of Hydrogenation for cis–trans Isomers and Their Differences[a]

Entry	Compounds	$\Delta H_{hyd.}$ (kcal mol^{-1})			$\Delta H_{hyd.}$ (kJ mol^{-1})		
		cis or Z	trans or E	$\Delta\Delta H^0$	cis or Z	trans or E	$\Delta\Delta H^0$
1	$CH_3CH=CHCH_3$	28.6[b]	27.6[b]	1.0	120[b]	116[b]	4.2
2	$(CH_3)_2CHCH=CHCH_3$	27.3	26.4	0.9	114	110	3.8
3	$(CH_3)_3CCH=CHCH_3$	30.8	26.5	4.3	129	111	18
4	$(CH_3)_2CHCH=CHCH(CH_3)_2$	28.7	26.8	1.9	120	112	7.9
5	$(CH_3)_3CCH=CHC(CH_3)_3$	36.2	26.9	9.3	151	112	39
6	$(CH_3)_3CCH_2CH=CHCH_2C(CH_3)_3$	26.9	26.0	0.9	113	109	3.8
7	$H_5C_2O_2CCH=CHCO_2C_2H_5$	33.2	29.0	4.2	139	121	18
8	$C_6H_5CH=CHCO_2CH_3$	27.8	19.5	8.3	116	81	35
9	$C_6H_5CH=CHC_6H_5$	25.8	20.1	5.7	108	84	24

[a] From Jensen (1976).
[b] These are gas-phase values from Kistiakowsky et al. (1935). Most of the other values were obtained in acetic acid solvent by the group of Turner (e.g., Turner et al., 1958).

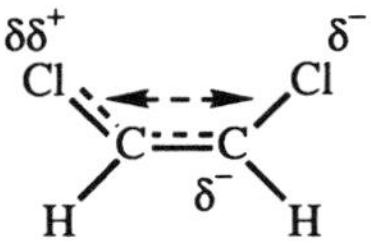

$$\overset{\delta\delta^+}{Cl} \cdots \cdots \overset{\delta^-}{Cl}$$

$$\underset{H}{\overset{}{}}C \!=\! C\underset{H}{\overset{}{}}$$

(Cl⁺- - - -Cl⁻ attraction)

Figure 9.45. Reason for the high stability of *cis*-1,2-dichloroethylene.

in 1,2-dichloroethylene (Pitzer and Hollenberg, 1954) and was ascribed to the halogens being both donors and acceptors of electronic charge (Fig. 9.45). In the meantime it has been found that the cis isomer is more stable than the trans in most 1,2-dihaloethylenes (Viehe, 1960; Viehe et al., 1964) the only exceptions being the bromo-iodo and diodo compounds (Viehe and Franchimont, 1963), where the steric interference of the halogens in the cis isomer is apparently large enough to make it less stable than the trans. Pertinent data, mostly obtained by thermal equilibration of the cis–trans isomers, are shown in Table 9.9. In the case of the dichloro compound, heat of combustion data are also available (Smith et al., 1953); they point to the cis isomer being lower by $0.25\,\text{kcal mol}^{-1}$ ($1.05\,\text{kJ mol}^{-1}$) in enthalpy.

b. Methods of Equilibration

There are two conceptually distinct ways of interconverting cis and trans isomers. One is to use thermal, catalytic, or photochemical means. These will lead to an equilibrium mixture or a photostationary state. Unless one isomer predominates greatly over the other at equilibrium (e.g., in the case of the stilbenes, cf. p. 574, where equilibration produces virtually pure trans isomer) such procedures give rise to mixtures and, while of mechanistic interest (see above), are thus of limited synthetic usefulness. A different approach is to convert an *E* isomer into the *Z* or vice versa in directed fashion. Such interconversion is valuable synthetically in cases where the most readily accessible isomer is not the one wanted as a

TABLE 9.9. Relative stability of *cis*- and *trans*-1,2-Dihaloethylenes[a]

Compound	(trans ⇌ cis)	
	ΔG^0	
	(kcal mol⁻¹)	(kJ mol⁻¹)
FCH=CHF	−0.5	−2.1
FCH=CHCl	−0.8	−3.3
FCH=CHBr	−0.8	−3.3
FCH=CHI	−0.77	−3.21
ClCH=CHCl[b]	−0.5	−2.1
BrCH=CHBr	−0.1	−0.4
BrCH=CHI	+0.2	+0.8
ICH=CHI	+0.7	+2.9
BrCH=CH–CH₃[c]	−0.85	−3.55

[a] From Viehe (1960); Viehe and Franchimont (1963), and Viehe et al. (1964).
[b] Pitzer and Hollenberg (1954).
[c] Skell and Allen (1958) (gas phase). For the liquid phase, see Harwell and Hatch (1955).

synthetic target. Both approaches have been reviewed by Sonnet (1980). We shall first deal with equilibration processes.

Thermal Equilibration

Barriers to rotation have already been discussed in Section 9-1. In simple alkenes, such as 2-butene, the barrier (62 kcal mol^{-1}, 259 kJ mol^{-1}, cf. p. 539) is too high to make thermal isomerization feasible in the absence of a catalysts (cf. Golden, Benson, et al., 1964). Even a somewhat lower barrier of 40 kcal mol^{-1} (167 kJ mol^{-1}) implies a half-life ($t_{1/2}$) of over 58 h at 200°C and though such compounds as fumaric acid have been isomerized "thermally" as mentioned earlier, these must, in fact, be acid-autocatalyzed reactions. In push–pull ethylenes (Fig. 9.9) thermal isomerization may, of course, be facile or may even occur spontaneously. An example of thermal isomerization is the conversion of 15,15′-cis-β-carotene to the all-trans isomer by heating at 80°C (Fig. 9.46; Isler et al., 1956).

Catalyzed Equilibration

A wide variety of catalysts has been employed to bring about cis–trans interconversion (cf. Crombie, 1952; Eliel, 1962; Sonnet, 1980), among them free radicals and free radical generators, such as oxides of nitrogen, halogens in the presence of light, iodine, thioglycolic acid, and diphenyl disulfide in the presence of light (e.g., Rokach et al., 1981). Also employed have been acids, such as halogen acids, sulfuric acid, and boron trifluoride; alkali metals, such as sodium; and hydrogenation–dehydrogenation catalysts and reagents, such as selenium and platinum. Radicals $X^{\cdot}$ probably add reversibly to the alkene $RR'C=CR''R'''$ to give $RR'CX-\dot{C}R''R'''$ radicals in which the central bond is now single and rotation around it therefore facile. Subsequent departure of $X^{\cdot}$ may then give either the original alkene or its stereoisomer. Light is often required to produce the radicals ($I^{\cdot}$ from I_2, $Br^{\cdot}$ from Br_2, and $C_6H_5S^{\cdot}$ from $C_6H_5S-SC_6H_5$); however, a photochemical isomerization process (cf. the following Section) is not involved, inasmuch as the light is not absorbed by the double bond and there is no triplet species formed to produce photosensitization (alkene triplet formation; see below). Moreover the product composition in such cases is different from that produced by triplet excitation (Sonnet, 1980).

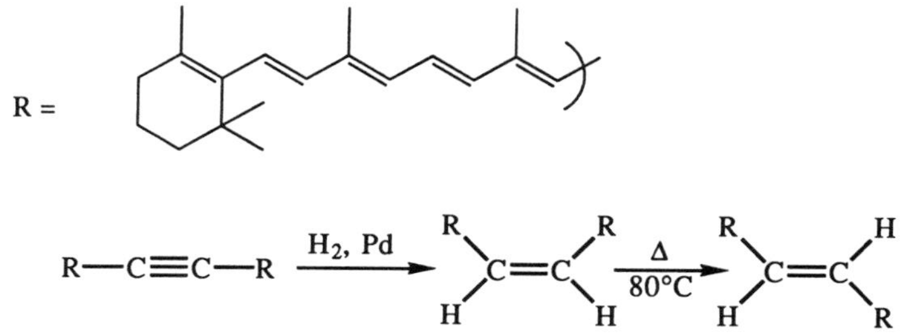

Figure 9.46.   Carotene isomerization.

An interesting experiment (Noyes et al., 1945) involves the isomerization of (Z)-1,2-diiodoethylene to E using radioactive iodine as a catalyst. Both isomerization and radioiodine incorporation into the diiodide occur, but the incorporation of iodine (exchange) is much faster than the isomerization. This result indicates that addition and loss of an iodine atom (which leads to exchange when the atom lost is not the same one that was added) is faster than rotation about a C–C single bond in the intermediate radical (which is presumably required for cis–trans isomerization) and proceeds with retention of configuration.

Acid catalysis may similarly involve addition of the Lewis acid or proton to the double bond generating a cation that may then rotate, similarly as the radical, and subsequently lose the Lewis acid or proton to return the alkene—either the original one, or the isomerized one. The α,β-unsaturated carbonyl compounds are particularly easily isomerized by acid, presumably through involvement of an enolic resonance hybrid: $H{-}\overset{+}{O}{=}C{-}C{=}C \leftrightarrow H{-}O{-}C{=}C{-}\overset{+}{C}$ in which the original C=C double bond is weakened and rotation around it thus facilitated. However, in some instances, although acid catalysis appears to be operating, this is actually not the case. Thus the isomerization of (Z)-stilbene to E by HBr requires oxygen and is stopped by a radical inhibitor, such as hydroquinone (Kharasch et al., 1937). Moreover, HCl does not bring about isomerization. Thus it appears that the catalyst is a bromine atom formed from the HBr, not the acid itself.

Nucleophilic catalysis of cis–trans isomerization has been observed in the case of diethyl maleate and fumarate, (E)- and (Z)-$EtO_2CCH{=}CHCO_2Et$, which are interconverted by secondary amines (Clemo and Graham, 1930). This process presumably involves a reversible 1,4 addition of the amine to the unsaturated ester.

The general mechanism of catalyzed isomerization is shown in Figure 9.47, where the asterisk denotes a positive or negative charge or unshared electron. For isomerization to take place, it is essential that A* add to the alkene (e.g., nucleophiles do not catalyze isomerization of alkenes unless the latter bear electron-withdrawing groups) but that the addition be readily reversible. Species that add exoergically to alkenes (i.e., for which $\Delta G^0_{addition} \ll 0$) will generally add irreversibly rather than induce isomerization, thus I_2 is a much better catalyst for alkene isomerization than Cl_2 which will add to the alkene instead.

Photochemical Isomerization

The cis–trans isomerization of alkenes occurs under irradiation with light that they absorb (Cowan and Drisko, 1976; Saltiel and Charlton, 1980). Absorption of light leads to a singlet excited state whose stable geometry is one in which the planes defined by the three atoms attached to each olefinic carbon are at right angles to each other (Fig. 9.48; for an introduction to photochemistry, see Turro,

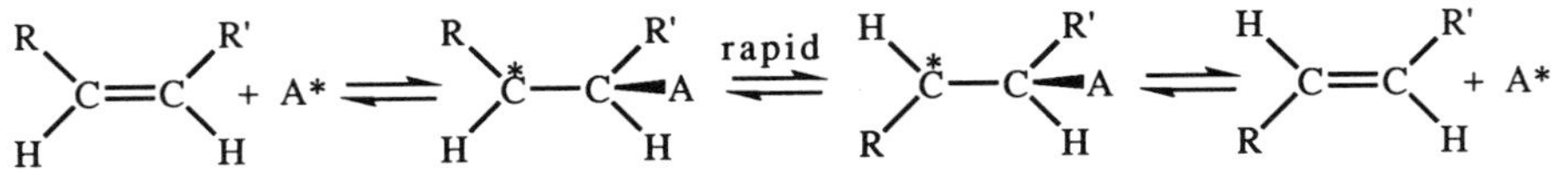

Figure 9.47. Cis-trans Isomerization. A superscript * represents a cation, radical, or anion.

Figure 9.48. Excited state of alkenes abC=Ccd.

1978.) With this geometry the mutual repulsion of the unpaired electrons is minimized and their interaction with the σ-electrons of the bonds to the adjacent carbon (hyperconjugation) is maximized. The initial vertically excited singlet thus relaxes to the twisted minimum and from there to the twisted ground state, which subsequently partitions itself between cis and trans products (Fig. 9.49). The process is difficult to study with simple alkenes because of the very short wavelength of absorption and the intervention of side reactions, such as $2\pi_s + 2\pi_s$ cyclodimerization (Yamazaki, Cvetanović, et al., 1976); however, it has been accomplished with (E)-1,2-di-*tert*-butylethylene (Kropp and Tise, 1981). The ratio of trans/cis isomer in various solvents ranged from 5–8; however, there were other products and it is not certain that a photostationary state was reached. Though an E/Z ratio of 5–8 is much less than the equilibrium ratio ($\Delta\Delta H^0 =$ 9.3 kcal mol^{-1}, 38.9 kJ mol^{-1}, cf. Table 9.8; $\Delta\Delta G^0$ must be similar), it does show that the twisted state relaxes preferentially to the trans isomer, presumably for steric reasons. A somewhat different picture applies to stilbene. Upon irradiation at 313 nm, somewhat on the side of the 295 nm λ_{max} of *trans*-stilbene and that at 280 nm of *cis*-stilbene (cf. Table 9.4), a stationary state is attained in which the cis/trans ratio is about 10:1. There are two reasons for the one-sidedness of the photostationary state in this case. One is that the twisted (90°) conformation relaxes to *cis*-stilbene about 2.5 times as fast as it relaxes to *trans*-stilbene. A look at Figure 9.36 suggests why this may be so: First, for the twisted conformation to relax to *trans*-stilbene, one of the benzylidene (C_6H_5–CH) groups must turn 90° as a whole, whereas relaxation to *cis*-stilbene requires only a swiveling of one of the –CH– moieties with the rings remaining at a large interplanar angle. Second,

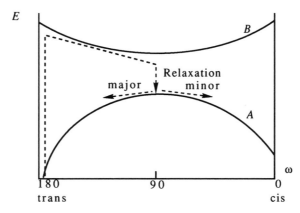

Figure 9.49. Energy profile of ground state and singlet excited states of alkenes. E is the potential energy, A is the ground-state energy as a function of torsion angle, B is the excited state energy as a function of torsion angle, and the dotted line represents the reaction path for trans excitation.

since *trans*-stilbene absorbs considerably more of the light than *cis*-stilbene (see the ε_{max} values in Table 9.4—the fact that irradiation at 313 nm is closer to the *trans*-stilbene than the *cis*-stilbene λ_{max} is also pertinent) the trans–cis interconversion is favored over cis–trans on energetic grounds.

$E \rightarrow Z$ isomerization can also be brought about by IR lasers (Lewis and Weitz, 1985). Photochemical isomerization can moreover be effected via triplet states. Although such states can be attained by intersystem crossing of excited singlet states produced by direct irradiation, it is more straightforward to produce them by photosensitization (triplet energy transfer). A typical photosensitizer is benzophenone ($C_6H_5COC_6H_5$) which, by irradiation, is excited from the ground-state B_0 to a long-lived singlet B_s^* which readily undergoes intersystem crossing to the lower lying triplet B_t^*. This triplet can then transfer energy to some other molecule, provided the triplet state of that molecule lies energetically below, or at least not much above, that of the benzophenone (photosensitizer) triplet. If the acceptor triplet state lies above the photosensitizer triplet state in energy, the energy deficiency must be supplied in the form of thermal activation energy. For stilbene, ground-state S_0, the process may be written

$$S_0 + B_t^* \rightarrow S_t^* + B_0$$

It produces the stilbene triplet S_t^* at the expense of the benzophenone triplet.

In the case of sensitized (triplet) photochemical isomerization of aliphatic alkenes (Kropp, 1979) one may distinguish two possibilities. One is exemplified by the isomerization of *cis*-2-pentene to a cis–trans mixture using benzene as a photosensitizer. Here the photostationary state corresponds to a trans/cis ratio of near unity suggesting that the twisted ground state (formed in this case by intersystem crossing of the triplet and ground-state singlet energy curves; cf. Fig. 9.50 which, however, relates to the stilbenes) relaxes to cis and trans alkene at about the same rate.

A different result is seen when the sensitizer is acetophenone or benzophenone; in that case the photostationary state corresponds to a trans/cis ratio of about 5.5. This value is close to the equilibrium ratio [cf. Table 9.7; a ΔG^0 of

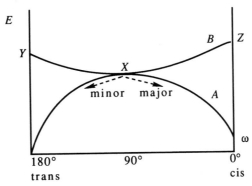

Figure 9.50. Energy curves for sensitized (triplet) isomerization of the stilbenes. *A* represents the ground state energy and *B* represents the triplet energy as a function of torsion angle ω. *Y* is the excited *trans*-stilbene triplet, *Z* is the excited *cis*-stilbene triplet, and *X* is the point of intersystem crossing. See, however, Saltiel et al., 1987.

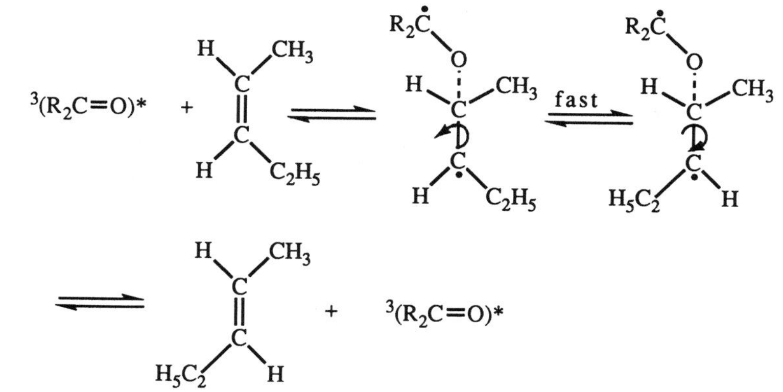

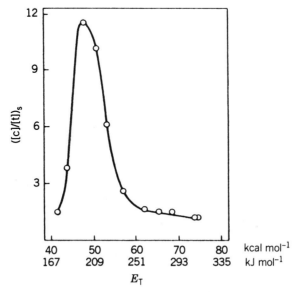

Figure 9.51. Schenck mechanism for cis–trans isomerization; $^3(R_2C=O)^*$ represents an excited triplet; the intermediates are also triplet states.

0.5 kcal mol^{-1} (2.09 kJ mol^{-1}) corresponds to a K of 2.3, whereas $\Delta G^0 = 1.0$ kcal mol^{-1} (4.2 kJ mol^{-1}) would correspond to a K of 5.4 at 25°C] suggesting that the process is actually one of chemical equilibration. This finding has been explained (Saltiel and Charlton, 1980) by the so-called Schenck or relay mechanism (Schenck and Steinmetz, 1962) in which the sensitizer triplet actually adds to the alkene, as shown in Figure 9.51. This addition is akin to a radical-catalyzed ground-state isomerization.

A rather unusual phenomenon is seen in the photosensitized equilibration of the stilbenes (Saltiel et al., 1973). The dependence of the photostationary state (*cis*-stilbene/*trans*-stilbene ratio) on wavelength is shown in Figure 9.52. The

Figure 9.52. Stilbene photostationary compositions as a function of donor triplet energy (Saltiel et al., 1973). Adapted, with permission, from *Organic Photochemistry*, Vol. 3, Chapman, O., ed., p. 12. Copyright © Marcel Dekker, Inc., New York, 1971.

ground-state energies of the *cis*- and *trans*-stilbene triplets are 57 and 49 kcal mol^{-1} (238 and 205 kJ mol^{-1}), respectively, above the ground-state singlets (Saltiel et al., 1973). If the sensitizer that transfers energy has more than about 60 kcal mol^{-1} (251 kJ mol^{-1}) triplet energy [e.g., with benzophenone, triplet energy 69 kcal mol^{-1} (289 kJ mol^{-1}) or with anthraquinone, triplet energy 62 kcal mol^{-1} (259 kJ mol^{-1})], the cis/trans ratio at the photostationary state ([c]/[t])$_s$, is about 1.5, not too far from the 2.5 cis/trans partition ratio of the ground-state twisted conformation mentioned earlier (p. 581). With such sensitizers, both *cis*- and *trans*-stilbene are excited to the triplet state by energy transfer. When the sensitizer triplet energy is between 60 and 49 kcal mol^{-1} (251 and 205 kJ mol^{-1}), the ([c]/[t])$_s$ ratio rises up to above 10 [benzil, 54 kcal mol^{-1} (226 kJ mol^{-1}), ratio 11.5; pyrene, 49 kcal mol^{-1} (205 kJ mol^{-1}), ratio 10] presumably because energy is now transferred only to the trans, not to the cis isomer. However, the curve (Fig. 9.52) drops off only rather gradually for sensitizer triplet energies below 49 kcal mol^{-1} (205 kJ mol^{-1}), even though at these low energies one would have expected little energy transfer to occur to either *cis*- or *trans*-stilbene. For example, with 1,2-benzanthracene, triplet energy 47.2 kcal mol^{-1} (197 kJ mol^{-1}) the ratio ([c]/[t])$_s$ still ranges from 4 to 10, depending on the stilbene concentration; this observation has been ascribed to "nonvertical transitions" from the ground state directly to the twisted geometry (Hammond et al., 1964; Herkstroeter and Hammond, 1966; see also Turro 1978, p. 346; Fouvard, Gorman, et al., 1993). Only when the sensitizer triplet energy drops below 42 kcal mol^{-1} (176 kJ mol^{-1}) (e.g., with eosin) does the cis/trans ratio approach the thermal equilibrium ratio [([c]/[t])$_s$ = 0.002; the calculated equilibrium ratio from Table 9.7 is 0.001–0.002]; in this case presumably the Schenck mechanism (see above) takes over. The sensitizer triplet, being unable to transfer its (insufficient) energy to stilbene now acts as a radical catalyst for cis–trans isomerization.

c. Directed cis–trans Interconversion

Unless equilibria between cis–trans isomers are very one-sided, equilibration methods are of limited synthetic usefulness, since they ordinarily require subsequent diastereomer separation. It should be noted, however, that trans alkenes are often substantially more stable than their cis isomers (Table 9.7), so that quite trans-rich alkene mixtures can frequently be generated by chemical equilibration or by thermodynamically controlled syntheses. An example is the already mentioned synthesis of the all-trans carotene by heating of the 15,15'-cis isomer (Fig. 9.46). In contrast, photostationary states (see above) are frequently rich in cis isomers that can thus be produced from their more stable trans diastereomers either by irradiation at a wavelength where the trans isomer absorbs much more strongly than the cis or by photosensitization with a sensitizer whose triplet energy is below that of the cis but above that of the trans triplet. An example of the latter type is the photochemical conversions of *trans*- to *cis*-β-ionol (Fig. 9.53), which is nearly quantitative with photosensitizers of triplet energies below 65 kcal mol^{-1} (272 kJ mol^{-1}), such as 2-acetonaphthone, E_t, 59 kcal mol^{-1} (247 kJ mol^{-1}), whereas sensitizers of much higher energy [e.g., acetone, E_t, 78 kcal mol^{-1} (326 kJ mol^{-1})] produce a cis–trans mixture containing only about 65% of the cis because of indiscriminate pumping of both stereoisomers to the triplet excited state (Ramamurthy and Liu, 1976).

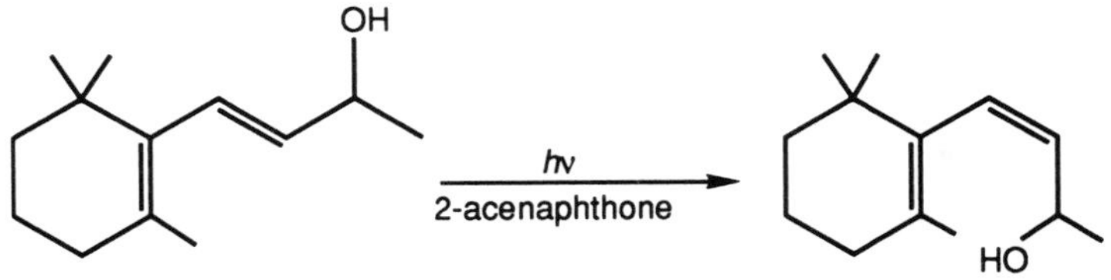

Figure 9.53. Photochemical conversion of *trans-* to *cis-β*-ionol.

A *directed* interconversion of a trans alkene to its cis isomer can be effected in principle by an anti addition followed by syn elimination, by a syn addition followed by an anti elimination, by a syn elimination (to an alkyne) followed by an anti addition, or by an anti elimination followed by a syn addition. Anti addition–anti elimination (or syn addition–syn elimination) or their corresponding elimination–addition sequences, in contrast, normally return alkene of unchanged configuration. However, when the elements added are different from those eliminated, such may not be the case and the sequence may serve for ultimate trans–cis interconversion. An example (Hoff, Greenlee, and Boord, 1951) is shown in Figure 9.54 (see also Fig. 9.27). Here the addition is of Cl_2 (anti) but the elimination (also anti) is of HCl; the original constitution is regenerated by reducing the chloroalkene to an alkene with retention of configuration; the overall result is trans–cis interconversion.

A sequence of anti addition followed by apparent syn elimination (Sonnet and Oliver, 1976a) is shown in Figure 9.55 $[R' = (CH_3)_2CH(CH_2)_4–; R = –(CH_2)_9CH_3]$. Evidently the direct anti elimination of BrCl by iodide is so slow that S_N2 displacement of Br by I (with inversion of configuration) occurs instead (Hine and Brader, 1955); subsequent anti elimination of ICl leads to the cis alkene in high purity, the sequence of S_N2 inversion plus anti elimination being formally equivalent to syn elimination. Similar apparent syn elimination had been observed earlier in the iodide induced conversion of *meso*-CHDBrCHDBr to

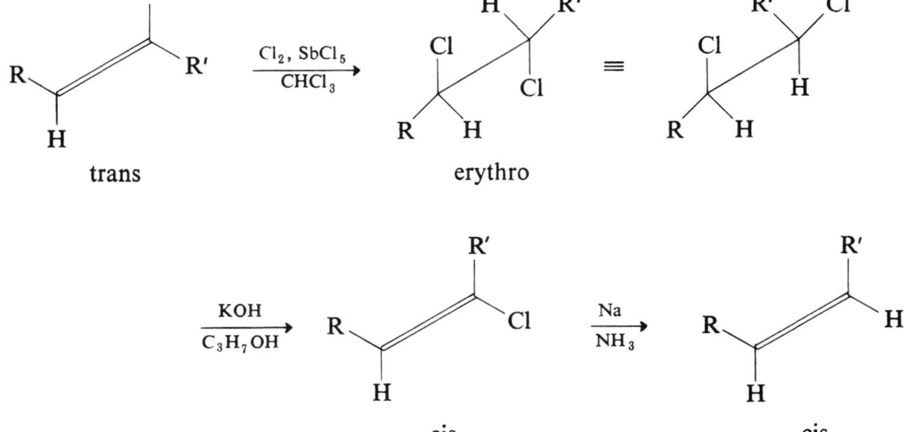

Figure 9.54. Conversion of trans to cis alkenes.

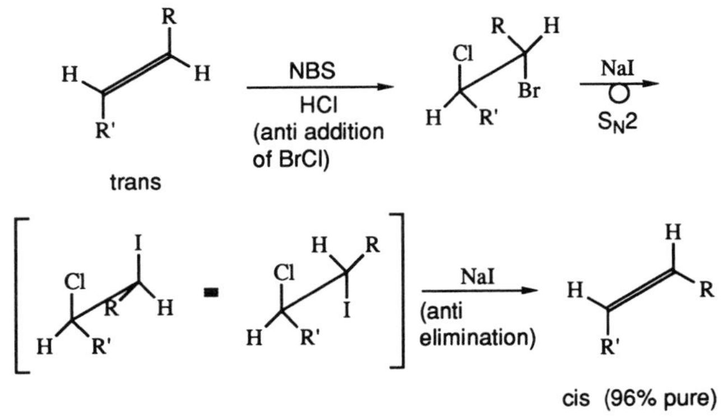

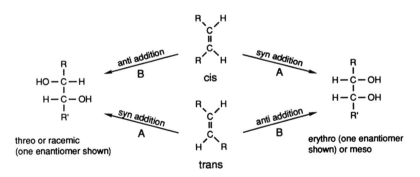

Figure 9.55. A trans–cis interconversion. NBS = *N*-Bromosuccinemide.

cis–CHD=CHD (Schubert et al., 1955). These results are in contrast to the normal anti elimination of nonterminal dibromides with iodide (Eliel, 1962, p. 141).

An apparently attractive cis–trans interconversion sequence would involve conversion of the alkene into a 1,2-diol by syn addition followed by an anti elimination. The attractiveness of this sequence rests on the fact that conversions of alkenes to diols by either syn- or anti-stereospecific addition is facile (Gunstone, 1960). The cis alkenes yield erythro (or meso if R = R′) diols by syn addition or threo (or chiral) glycols by anti addition, respectively, whereas trans alkenes yield threo (or chiral) diols by syn addition and erythro (or meso) diols by anti addition. These transformations are summarized in Figure 9.56, which also contains a listing of reagents with appropriate references.

In principle it might appear that conversion of the glycols to sulfonates

Figure 9.56. Hydroxylation of alkenes.

A KMnO$_4$: Kekulé and Anschütz (1880, 1881); Boëseken (1928); Ogino and Mochizuki (1979).

   OsO$_4$: Hofmann et al. (1913); Schröder (1980).

   CH$_3$CO$_2$Ag, I$_2$, CH$_3$CO$_2$H–H$_2$O (wet Prévost reaction): Woodward and Brutcher (1958); Gunstone and Morris (1957).

B Peracids (RCO$_3$H) followed by hydrolysis: Swern (1953); House (1972, Chapter 6).

   RCO$_2$Ag, I$_2$ (dry Prévost reaction) followed by hydrolysis: Wilson (1957), p. 350; House (1972), p. 438; Küppers (1975).

   Epoxidation followed by hydrolysis: Swern (1953); House (1972, Chapter 6).

followed by stereospecific anti (or syn, depending on the stereochemistry of the bis-sulfonate) elimination would thus provide a path for cis–trans (or trans–cis) interconversion. Unfortunately, *vic*-disulfonates, $-\underset{\underset{RSO_2O}{|}}{C}-\underset{\underset{OSO_2R}{|}}{C}-$ do not undergo direct

elimination with iodide or zinc and the substitution–elimination sequence with iodide is beset with complications (Angyal and Young, 1961; Angyal et al., 1979). Indirect methods to bring about syn elimination of the elements of hydrogen peroxide from a diol are shown in Figure 9.57. The "Corey–Winter reaction" (Corey and Winter, 1963) involves a thionecarbonate intermediate from which sulfur is abstracted by means of trimethyl phosphite; the resulting carbene decomposes to carbon dioxide and alkene. In the other method (Hines, Whitman et al., 1968) the diol is converted to a 2-phenyl-1,3-dioxolane (acetal) which, upon treatment with butyllithium, yields an unstable anion that decomposes to an alkene plus benzoate. The stereoselectivity of both methods is high and both have been used to prepare *trans*-cyclooctene, in the case of the thionecarbonate method in optically active form (Corey and Schulman, 1968). A somewhat less selective method involving phosphate intermediates has also been described (Marshall and Lewellyn, 1977).

A brief discussion of the mechanism of these addition reactions is in order. The stereospecificity of addition generally rests on the formation of cyclic intermediates; depending on the nature of these intermediates it may be anti or syn. Addition of peracids, like that of halogens, proceeds via a three-membered cyclic intermediate: a halonium ion in the case of halogen addition and an epoxide in the case of peracid addition (Fig. 9.58). The epoxide normally opens with inversion of configuration so that the overall course of the reaction (syn addition followed by inversion during ring opening) is anti addition (Fig. 9.58).

Cases are known, however, where epoxide ring opening proceeds with retention of configuration (cf. Eliel, 1962, pp. 102, 356; Parker and Isaacs, 1959; Gorzynski Smith, 1984) in which case the overall sequence would also involve retention.

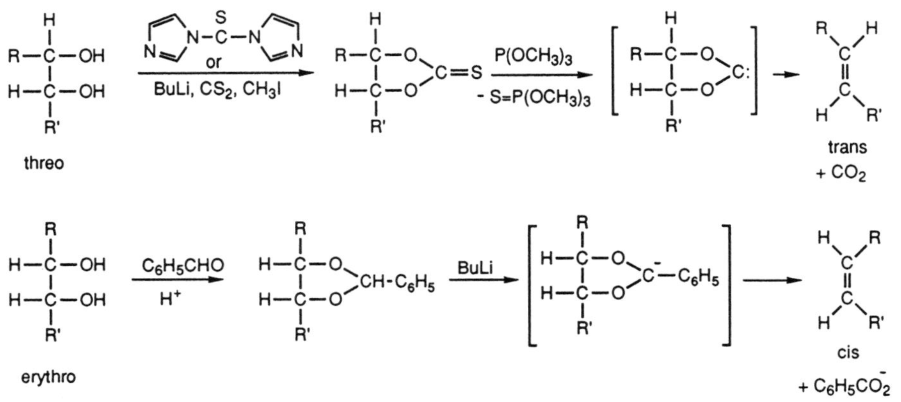

Figure 9.57. A syn elimination from diols.

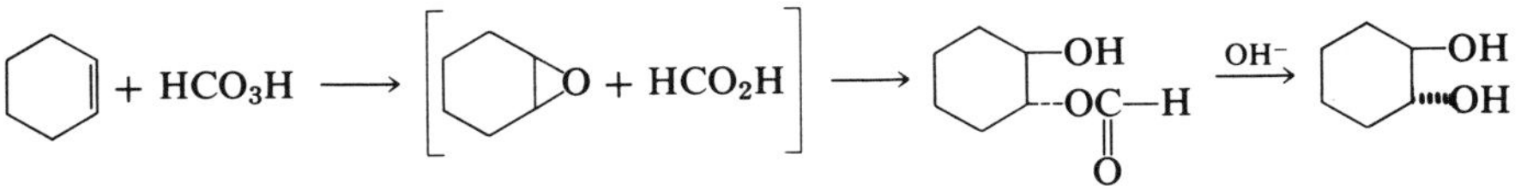

Figure 9.58. A trans hydroxylation.

The Prévost (iodine–silver acetate or benzoate) reaction is more complex in that it involves syn addition of iodine to form an iodonium ion, followed by inversion to form an iodoacetate or iodobenzoate, followed by a second inversion with neighboring group participation (cf. Capon and McManus, 1976) to give a five-membered cyclic intermediate that is then opened by a third inversion to give a diester (Winstein and Buckles, 1942a). Since the syn addition is followed by an odd number of inversions, the overall steric course of addition of the two alkanoate moieties is anti (Fig. 9.59, the example is that of a cyclic alkene).

An interesting change in stereochemistry occurs when the reaction is carried out in wet instead of dry solvent (Winstein and Buckles, 1942b). In that case the cyclic intermediate, instead of being opened with inversion by the alkanoate ion, is hydrolyzed with retention by hydrolytic fission of one of the R–C–O bonds in the acylonium ion. The product is now a monoester rather than a diester; more importantly, since there are now only two inversions following the original syn halonium addition, the overall result is syn addition. This sequence is the basis of Woodward's hydroxylation (Woodward and Brutcher, 1958; see also Djerassi et al., 1955) shown in Figure 9.60. Since the approach to the top face of the steroid shown in Figure 9.60 is impeded by the angular methyl groups, the initial iodonium addition occurs from the bottom (α) side with the result that the final cis diol (2β, 3β) has the opposite configuration at C(2) and C(3) from the cis diol formed by initial α addition of osmium tetroxide.

Other syn hydroxylations proceed more simply, that is, by a single syn addition followed by hydrolysis that does not affect the C–O bonds formed in the initial process. Osmium tetroxide and permanganate oxidations (Fig. 9.61) are of

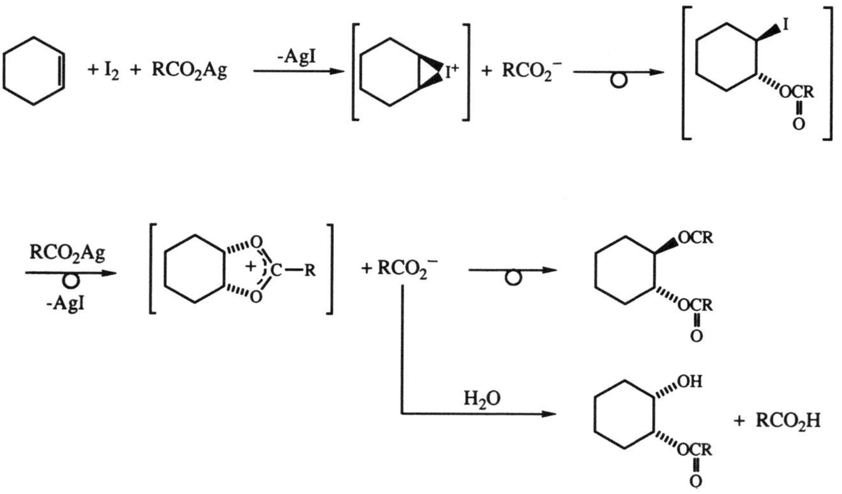

Figure 9.59. Dry and wet Prévost reactions.

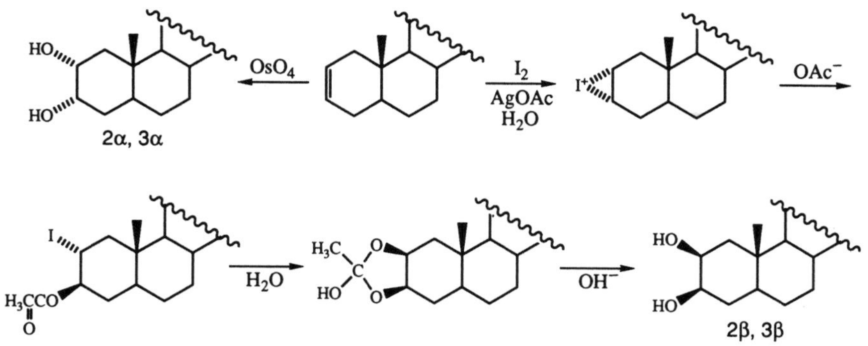

Figure 9.60. Conversion of steroidal alkene to glycol by iodine–silver–wet acetic acid.

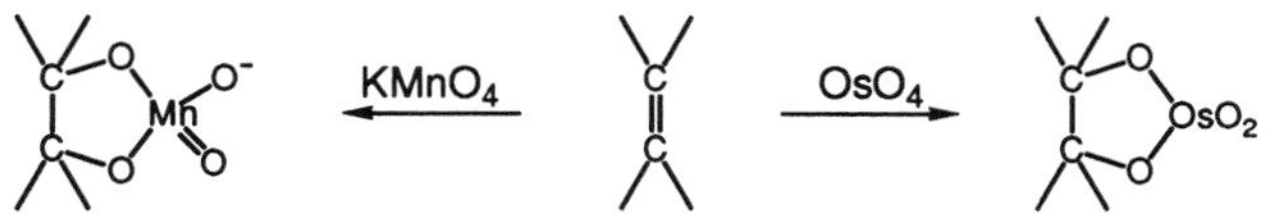

Figure 9.61. Intermediates in oxidation of alkenes to glycols.

this type. The osmate ester is an isolable intermediate (cf. Rotermund, 1975), whereas the intermediacy of the hypomanganate(V) ester has been inferred from spectrophotometric and stop-flow kinetic studies (Lee and Browridge, 1973; Wiberg et al., 1973) of the oxidation of RCH=CHCO$_2$H (R = C$_6$H$_5$ or CH$_3$).

Two additional methods for converting cis to trans alkenes or vice versa are shown in Figure 9.62 (Vedejs and Fuchs, 1973; Vedejs et al., 1973) and Figure 9.63 (Reetz and Plachky, 1976; Dervan and Shippey, 1976; for mechanism see Hudrlik et al., 1975).

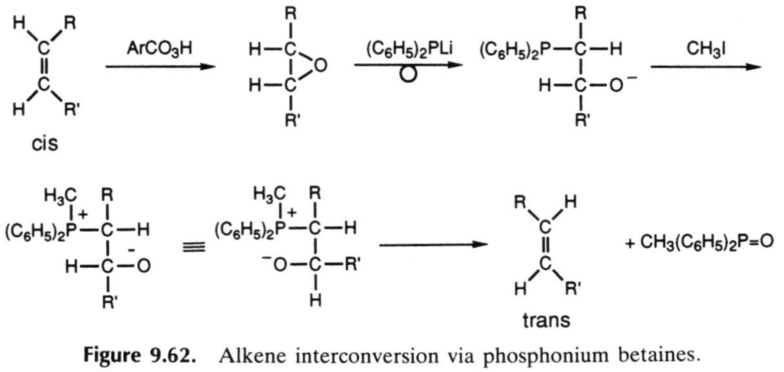

Figure 9.62. Alkene interconversion via phosphonium betaines.

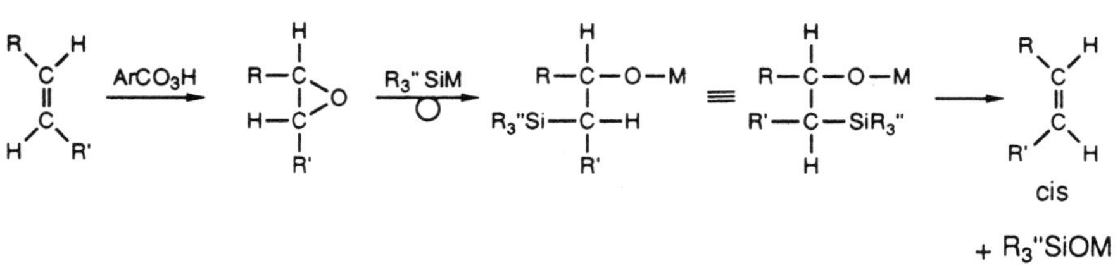

Figure 9.63. Anti deoxygenation of epoxides with trialkylsilyl anions.

A number of additional cis–trans interconversion schemes have been reviewed by Sonnet (1980); for stereoselective syntheses of *E*- and *Z*-alkenes, see also Arora and Ugi (1972).

REFERENCES

Adams, R., Anderson, A. W., and Miller, M. W. (1941), *J. Am. Chem. Soc.*, **63**, 1589.

Adams, R. and Miller, M. W. (1940), *J. Am. Chem. Soc.*, **62**, 53.

Agranat, I., Rabinovitz, M., Weitzen-Dagan, A., and Gosnay, I. (1972), *J. Chem. Soc. Chem. Commun.*, 732.

Agranat, I., Suissa, M. R., Cohen, S., Isaksson, R., Sandström, J., Dale, J., and Grace, D. (1987), *J. Chem. Soc. Chem. Commun.*, 381.

Agranat, I. and Tapuhi, Y. (1979), *J. Org. Chem.*, **44**, 1941.

Allan, J. L. H., Meakins, G. D., and Whiting, M. C. (1955), *J. Chem. Soc.*, 1874.

Anderson, D. G. and Wettermark, G. (1965), *J. Am. Chem. Soc.*, **87**, 1433.

Anderson, W. F., Bell, J. A., Diamond, J. M., and Wilson, K. R. (1958), *J. Am. Chem. Soc.*, **80**, 2384.

Anet, F. A. L. and Bourn, A. J. R. (1965), *J. Am. Chem. Soc.*, **87**, 5250.

Angyal, S. J., Nicholls, R. G., and Pinhey, J. T. (1979), *Aust. J. Chem.*, **32**, 2433.

Angyal, S. J. and Young, R. J. (1961), *Aust. J. Chem.*, **14**, 8.

Arora, A. S. and Ugi, I. K. (1972), "Stereoselektive Olefin–Synthesen," in Müller, E., ed., *Methoden der Organischen Chemie*, Vol. V/lb, Houben-Weyl, 4th ed., Thieme, Stuttgart, Germany, p. 728ff.

Bailey, N. A. and Hull, S. E. (1978), *Acta Cryst.*, **B34**, 3289.

Bednas, M. E. and Russell, D. S. (1958), *Can. J. Chem.*, **36**, 1272.

Beilstein, *Handbuch der Chemie* 4th Suppl., (1978), *Stereochemical Conventions*, Vol. 1, Pt. I, p. X, and in all subsequent volumes.

Bell, R. A. and Saunders, J. K. (1970), *Can. J. Chem.*, **48**, 1114.

Bell, R. A. and Saunders, J. K. (1973), "Some Chemical Applications of the Nuclear Overhauser Effect," *Top. Stereochem.*, **7**, 1.

Bellamy, L. J. (1975), *The Infrared Spectra of Complex Molecules*, 3rd ed., Vol. I., Chapman & Hall, London.

Bellamy, L. J. (1980), *Advances in Infrared Group Frequencies*, 2nd ed., Chapman & Hall, London.

Benoit, F., Holmes, J. L., and Isaacs, N. S. (1969), *Org. Mass Spectrom.*, **2**, 591.

Bernstein, H. J. and Powling, J. (1951), *J. Am. Chem. Soc.*, **73**, 1843.

Bertsch, K. and Jochims, J. C. (1977), *Tetrahedron Lett.*, 4379.

Bertsch, K., Karich, G., and Jochims, J. C. (1977), *Chem. Ber.*, **110**, 3304.

Bjerrum, N. (1923), *Z. Phys. Chem.*, **106**, 219.

Blackwood, J. E., Gladys, C. L., Loening, K. L., Petrarca, A. E., and Rush, J. E. (1968), *J. Am. Chem. Soc.*, **90**, 509.

Blatt, A. H. (1933), "The Beckmann Rearrangement," *Chem. Rev.*, **12**, 215.

Boëseken, J. (1928), *Recl. Trav. Chim. Pays-Bas*, **47**, 683.

Bohlmann, F. and Kieslich, K. (1954), *Chem. Ber.*, **87**, 1363.

Brackman, D. S. and Plesch, P. H. (1952), *J. Chem. Soc.*, 2188.

Breitmaier, E., Hass, G., and Voelter, W. (1979), *Atlas of Carbon-13 NMR Data*, IFI/Plenum Data Co., New York.

Brewster, J. H. (1972), "Assignment of Stereochemical Configurations by Chemical Methods," in Bentley, K. W. and Kirby, G. W., Eds., *Techniques of Chemistry*, Vol. IV, Pt. 3, 2nd ed., Wiley-Interscience, New York.

Burgess, J., Kemmitt, R. D. W., Morton, N., Mortimer, C. T., and Wilkinson, M. P. (1980), *J. Organomet. Chem.*, **191**, 477.

Capon, B. and McManus, S. P. (1976), *Neighboring Group Participation*, Plenum, New York.

Clemo, G. R. and Graham, S. B. (1930), *J. Chem. Soc.*, 213.

Coops, J. and Hoijtink, G. J. (1950), *Recl. Trav. Chim. Pays-Bas*, **69**, 358.

Corey, E. J. and Posner, G. H. (1967), *J. Am. Chem. Soc.*, **89**, 3911.

Corey, E. J. and Shulman, J. I. (1968), *Tetrahedron Lett.*, 3655.

Corey, E. J. and Winter, R. A. E. (1963), *J. Am. Chem. Soc.*, **85**, 2677.

Cowan, D. O. and Drisko, R. L. (1976), *Elements of Organic Photochemistry*, Plenum, New York, Chap. 9.

Crombie, L. (1952), "Geometrical Isomerism about Carbon–Carbon Double Bonds," *Q. Rev. Chem. Soc.*, **6**, 101.

Cross, L. C. and Klyne, W. (1976), "Rules for the Nomenclature of Organic Chemistry," *Pure Appl. Chem.*, **45**, 11; see also (1970), *J. Org. Chem.*, **35**, 2849.

Cundall, R. B. and Palmer, T. F. (1961), *Trans. Faraday Soc.*, **57**, 1936.

Curtin, D. Y. (1954), "Stereochemical Control of Organic Reactions Differences in Behavior of Diastereoisomers," *Rec. Chem. Prog.*, **15**, 111.

Curtin, D. Y. and Harris, E. E. (1951), *J. Am. Chem. Soc.*, **73**, 2716.

Curtin, D. Y., Johnson, H. W., Jr., and Steiner, E. G. (1955), *J. Am. Chem. Soc.*, **77**, 4566.

Dannheim, J., Grahn, W., Hopf, H., and Parrodi, C. (1987), *Chem. Ber.*, **120**, 871.

Dervan, P. B. and Shippey, M. A. (1976), *J. Am. Chem. Soc.*, **98**, 1265.

Deuter, J., Rodewald, H., Irngartinger, H., Loerzer, T., and Lüttke, W. (1985), *Tetrahedron Lett.*, **26**, 1031.

Djerassi, C., High, L. B., Grossnickle, T. T., Ehrlich, R., Moore, J. A., and Scott, R. B. (1955), *Chem. Ind. (London)*, 474.

Dolter, R. J. and Curran, C. (1960), *J. Am. Chem. Soc.*, **82**, 4153.

Donaruma, I. G. and Heldt, W. Z. (1960), *Org. React.*, **11**, 1.

Douglas, J. E., Rabinovitch, B. S., and Looney, F. S. (1955), *J. Chem. Phys.*, **23**, 315.

Downing, A. P., Ollis, W. D., and Sutherland, I. O. (1969), *J. Chem. Soc. B.*, 111.

Duncan, J. L., Wright, I. J., and Van Lerberghe, D. (1972), *J. Mol. Spectrosc.*, **42**, 463.

Eliel, E. L. (1962), "Geometrical Isomerism of Olefins," in *Stereochemistry of Carbon Compounds*, McGraw-Hill, New York, chapter 12.

Eliel, E. L. and Brunet, E. (1986), *J. Org. Chem.*, **51**, 1902.

Emsley, J. W., Phillips, L., and Wray, V. (1976), "Fluorine Coupling Constants," *Prog. Nucl. Magn. Reson. Spectrosc.*, **10**, 83.

Errera, J. (1928), *Phys. Z.*, **29**, 689.

Evans, D. A., Nelson, J. V., and Taber, T. R. (1982), "Stereoselective Aldol Condensations," *Top. Stereochem.*, **13**, 1.

Evershed, R. P., Morgan, E. D., and Thompson, L. D. (1982), *J. Chromatogr.*, **237**, 350.

Fahey, R. C. (1968), "The Stereochemistry of Electrophilic Additions to Olefins and Acetylenes," *Top. Stereochem.*, **3**, 237.

Favini, G., Simonetta, M., Sottocornola, M., and Todeschini, R. (1982), *J. Comp. Chem.*, **3**, 178.

Favini, G., Simonetta, M., and Todeschini, R. (1981), *J. Comp. Chem.*, **2**, 149.

Feigl, M. (1983), *J. Phys. Chem.*, **87**, 3054.

Feringa, B. L., Jager, W. F., and de Lange, B. (1993), *J. Chem. Soc. Chem. Commun.*, 288.

Fischer, G., Muszkat, K. A., and Fischer, E. (1968), *J. Chem. Soc. B.*, 1156.

Forvard, P. J., Gorman, A. A., and Hamblett, I. (1993), *J. Chem. Soc. Chem. Commun.*, 250.

Fueno, T., Yamaguchi, K., and Naka, Y. (1972), *Bull. Chem. Soc. Jpn.*, **45**, 3294.

Gaudemer, A. (1977), "Determination of Configurations by NMR Spectroscopy," in Kagan, H. B., Ed., *Stereochemistry, Fundamentals and Methods*, Vol. 1., Thieme, Stuttgart, Germany, Chap. 2.

Gault, I. R., Ollis, W. D., and Sutherland, I. O. (1970), *J. Chem. Soc. Chem. Commun.*, 269.

Golden, D. M., Egger, K. W., and Benson, S. W. (1964), *J. Am. Chem. Soc.*, **86**, 5416.

Golfier, M. (1977), "Determination of Configuration by Infrared Spectroscopy," in Kagan, H. B., Ed., *Stereochemistry, Fundamentals and Methods*, Vol. 1., Thieme, Stuttgart, Germany, Chap. 1.

Gorzynski Smith, J. (1984), "Synthetically Useful Reactions of Epoxides," *Synthesis*, 629.

Green, M. (1976), "Mass Spectrometry and the Stereochemistry of Organic Molecules," *Top. Stereochem.*, **9**, 35.

Grummitt, O. and Christoph, F. J. (1951), *J. Am. Chem. Soc.*, **73**, 3479.

Gunstone, F. D. (1960), "Hydroxylation Methods," in Raphael, R. A., Taylor, E. C., and Wynberg, H., Eds., *Advances in Organic Chemistry, Methods and Results*, Vol. 1, Interscience, New York.

Gunstone, F. D. and Morris, L. J. (1957), *J. Chem. Soc.*, 487.

Hall, G. E., Middleton, W. J., and Roberts, J. D. (1971), *J. Am. Chem. Soc.*, **93**, 4778.

Hammond, G. S., Saltiel, J., Lamola, A. A., Turro, N. J., Bradshaw, J. S., Cowan, D. O., Counsell, R. C., Vogt, V., and Dalton, C. (1964), *J. Am. Chem. Soc.*, **86**, 3197.

Hantzsch, A. (1891), *Ber. Dtsch. Chem. Ges.*, **24**, 13, 51.

Harnik, E. and Schmidt, G. M. J. (1954), *J. Chem. Soc.*, 3295.

Harwell, K. E. and Hatch, L. F. (1955), *J. Am. Chem. Soc.*, **77**, 1682.

Heath, R. R., Tumlinson, J. H., and Doolittle, R. E. (1977), *J. Chromatogr. Sci.*, **15**, 10.

Heathcock, C. H. (1982), "Acyclic Stereoselection via the Aldol Condensation," in Eliel, E. L. and Otsuka, S. Eds., *Asymmetric Reactions and Processes in Chemistry*, ACS Symposium Series 185, American Chemical Society, Washington, DC.

Henderson, G. and Gajjar, A. (1971), *J. Org. Chem.*, **36**, 3834.

Herkstroeter, W. G. and Hammond, G. S. (1966), *J. Am. Chem. Soc.*, **88**. 4769.

Hine, J. and Brader, W. H. (1955), *J. Am. Chem. Soc.*, **77**, 361.

Hines, J. N., Peagram, M. J., Whitman, G. H., and Wright, M. (1968), *J. Chem. Soc. Chem. Commun.*, 1593.

Hofer, O. (1976), "The Lanthanide Induced Shift Technique: Applications in Conformational Analysis," *Top. Stereochem.*, **9**, 111.

Hoff, M. C., Greenlee, K. W., and Boord, C. E. (1951), *J. Am. Chem. Soc.*, **73**, 3329.

Hofmann, K. A., Ehrhart, O., and Schneider, O. (1913), *Ber. Dtsch. Chem. Ges.*, **46**, 1657.

House, H. O. (1972), *Modern Synthetic Reactions*, 2nd ed., Benjamin, Menlo Park, CA.

Hudrlik, P. F., Peterson, D., and Rona, R. J. (1975), *J. Org. Chem.*, **40**, 2263.

Isler, O., Lindlar, H., Montavon, M., Rüegg, R., and Zeller, P. (1956), *Helv. Chem. Acta.*, **39**, 249.

Jackman, L. M. (1975), "Rotation about Partial Double Bonds in Organic Molecules," in Jackman, L. M. and Cotton, F. A., Eds., *Dynamic Nuclear Magnetic Resonance Spectroscopy*, Academic, New York, p. 203.

Jeffers, P. M. (1974), *J. Phys. Chem.*, **78**, 1469.

Jeffers, P. M. and Shaub, W. (1969), *J. Am. Chem. Soc.*, **91**, 7706.

Jensen, J. L. (1976), "Heats of Hydrogenation: A Brief Summary," *Progr. Phys. Org. Chem.*, **12**, 189.

Kalinowski, H.-O. and Kessler, H. (1973), "Fast Isomerizations about Double Bonds," *Top. Stereochem.*, **7**, 295.

Kekulé, A. and Anschütz, R. (1880), *Ber. Dtsch. Chem. Ges.*, **13**, 2150; (1881), **14**, 713.

Kessler, H. and Leibfritz, D. (1971), *Chem. Ber.*, **104**, 2143.

Khan, A. Z.-Q., Isaksson, R., and Sandström, J. (1987), *J. Chem. Soc. Perkin 2*, 491.

Kharasch, M. S., Mansfield, J. V., and Mayo, F. R. (1937), *J. Am. Chem. Soc.*, **59**, 1155.

Kistiakowsky, G. B., Ruhoff, J. R., Smith, H. A., and Vaughan, W. E. (1935), *J. Am. Chem. Soc.*, **57**, 876.

Kistiakowsky, G. B. and Smith, W. R. (1934), *J. Am. Chem. Soc.*, **56**, 638.

Kistiakowsky, G. B. and Smith, W. R. (1936), *J. Am. Chem. Soc.*, **58**, 766.

Kondo, S., Sakurai, Y., Hirota, E., and Morino, Y. (1970), *J. Mol. Spectrosc.*, **34**, 231.

Kortüm, G., Vogel, W., and Andrussow, K. (1960), "Dissociation Constants of Organic Acids in Aqueous Solution," *Pure Appl. Chem.*, **1**, 187.

Krebs, A., Kaletta, B., Nickel, W.-U., Ruger, W., and Tikwe, L. (1986), *Tetrahedron*, **42**, 1693.

Krebs, A., Ruger, W., Nickel, W.-U., Wilke, M., and Burkert, U. (1984b), *Chem. Ber.*, **117**, 310.

Krebs, A., Ruger, W., Ziegenhagen, B., Hebold, M., Hardtke, I., Muller, R., Schutz, M., Wietzke, M., and Wilke, M. (1984a), *Chem. Ber.*, **117**, 277.

Kropp, P. J. (1979), "Photochemistry of Alkenes in Solution," in Padwa, A., Ed., *Organic Photochemistry*, Vol. 4, Marcel Dekker, New York, p. 1.

Kropp, P. J. and Tise, F. P. (1981), *J. Am. Chem. Soc.*, **103**, 7293.

Kuchitsu, K. (1966), *J. Chem. Phys.*, **44**, 906.

Kuhn, R. and Blum, D. (1959), *Chem. Ber.*, **92**, 1483.

Kuhn, R., Schulz, B., and Jochims, J. C. (1966), *Angew. Chem. Int. Ed. Engl.*, **5**, 420.

Küppers, H. (1975), "Prévost Reaktion" in Houben-Weyl, Vol. IV/lb, Müller, E., Ed., Thieme, Stuttgart, Germany, pp. 948–952.

Lacher, J. R., Billings, T. J., Campion, D. E., Lea, K. R., and Park, J. D. (1952), *J. Am. Chem. Soc.*, **74**, 5291.

Landolt-Börnstein (1976), *Structure Data of Free Polyatomic Molecules*, Vol. 7, Hellwege, K.-H. and Hellwege, A. M., Eds., Springer, New York.

Leadbetter, G. and Plimmer, J. R. (1979), *J. Chem. Ecol.*, **5**, 101.

Lee, D. G. and Brownridge, J. R. (1973), *J. Am. Chem. Soc.*, **95**, 3033.

Lester, R. (1978), *J. Chromatogr.*, **156**, 55.

Lewis, F. D. and Weitz, E. (1985), "Selective Isomerization of Alkenes, Dienes, and Trienes with Infrared Lasers," *Acc. Chem. Res.*, **18**, 188.

Lide, D. R. and Christensen, D. (1961), *J. Chem. Phys.*, **35**, 1374.

Lipsky, S. R., Lovelock, J. E., and Landowne, R. A. (1959), *J. Am. Chem. Soc.*, **81**, 1010.

Lowe, B. M. and Smith, D. G. (1974), *J. Chem. Soc. Faraday Trans. 1*, 362.

Lucas, H. J., Simpson, T. P., and Carter, J. M. (1925), *J. Am. Chem. Soc.*, **47**, 1462.

Luef, W., Keese, R. (1991), "Strained Olefins: Structure and Reactivity of Nonplanar Carbon–Carbon Double Bonds," *Top. Stereochem.*, **20**, 231.

Marshall, J. L. (1983), *Carbon–Carbon and Carbon–Proton NMR Couplings*, Verlag Chemie, Deerfield Beach, FL.

Marshall, J. A. and Lewellyn, M. E. (1977), *J. Org. Chem.*, **42**, 1311.

Maryanoff, B. E. and Reitz, A. B. (1986), *Phosphorus Sulfur*, **27**, 167.

Maryanoff, B. E., Reitz, A. B., Mutter, M. S., Inners, R. R., Almond, H. R., Whittle, R. R., and Olofson, R. A. (1986), *J. Am. Chem. Soc.*, **108**, 7664.

Masamune, S., Kaiho, T., and Garvey, D. S. (1982), *J. Am. Chem. Soc.*, **104**, 5521.

Mathai, I. M. and Miller, S. I. (1970), *J. Org. Chem.*, **35**, 3416.

Matter, U. E., Pascual, C., Pretsch, E., Pross, A., Simon, W., and Sternhell, S. (1969), *Tetrahedron*, **25**, 691.

McCarty, C. G. (1970), "syn–anti Isomerizations and Rearrangements," in Patai, S., Ed., *The Chemistry of the Carbon-Nitrogen Double Bond*, Wiley-Interscience, New York, Chap. 9.

McClellan, A. L. (1963; 1974, 1989), *Tables of Experimental Dipole Moments*. Vol. 1. Freeman, San Francisco; Vols. 2, 3, Rahara Enterprises, El Cerrito, CA.

Meisenheimer, J. (1921), *Ber. Dtsch. Chem. Ges.* **54**, 3206.

Meisenheimer, J. and Theilacker, W. (1933), in Freudenberg, K., Ed., *Stereochemie*, Franz Deuticke, Leipzig, Germany.

Melander, W. R., Jacobson, J., and Horvath, C. (1982), *J. Chromatogr.*, **234**, 269.

Meyer, H. (1909), *Ber. Dtsch. Chem. Ges.*, **42**, 143.

Miller, S. I. (1956), *J. Am. Chem. Soc.*, **78**, 6091.

Minkin, V. I. (1977), "Dipole Moments and Stereochemistry of Organic Compounds. Selected Applications," in Kagan, H. B., Ed., *Stereochemistry, Fundamentals and Methods*, Vol. 2, Thieme, Stutgart, Germany.

Müller, E. and Neuhoff, H. (1939), *Ber. Dtsch. Chem. Ges.*, **72**, 2063.

Müller, E. and Tietz, E. (1941), *Ber. Dtsch. Chem. Ges.*, **74B**, 807.

Neuhaus, D. (1989), *The Nuclear Overhauser Effect in Structural Conformational Analysis*, VCH Publishers, New York.

Noggle, J. H. and Schirmer, R. E. (1971), *The Nuclear Overhauser Effect. Chemical Applications*, Academic, New York.

Normant, J. F., Cahiez, C., Chuit, C., Alexakis, A., and Villieras, J. (1972), *J. Organomet. Chem.*, **40**, C49.

Noyes, R. M., Dickinson, R. G., and Schomaker, V. (1945), *J. Am. Chem. Soc.*, **67**, 1319.

Ogino, T. and Mochizuki, K. (1979), *Chem. Lett.*, 443.

Ohtsuru, M., Teraoka, M., Tori, K., and Takeda, K. (1967), *J. Chem. Soc. B*, 1033.

Ōki, M. (1983), "Recent Advances in Atropisomerism," *Top. Stereochem.*, **14**, 1.

Otto, R. (1895), *J. Prakt. Chem.*, **51**, 285.

Parker, R. E. and Isaacs, N. S. (1959), "Mechanisms of Epoxide Reactions," *Chem. Rev.*, **59**, 737.

Pascual, C., Meier, J., and Simon, W. (1966), *Helv. Chim. Acta*, **49**, 164.

Pauling, L. (1958), "Theoretical Organic Chemistry," in *The Kekule Symposium*, Butterworths, London, p. 1.

Pfeiffer, P. (1904), *Z. Phys. Chem.*, **48**, 40.

Pfeiffer, P. (1912), *Ber. Dtsch. Chem. Ges.*, **45**, 1819.

Phillips, L. (1972), "Applications of ^{19}F Nuclear Magnetic Resonance," in Bentley, K. W. and Kirby, G. W., Eds., *Elucidation of Organic Structures by Physical and Chemical Methods*, Vol. IV, Pt. 1 of *Techniques of Chemistry*, 2nd ed., Wiley-Interscience, New York, p. 323.

Pitzer, K. S. and Hollenberg, J. L. (1954), *J. Am. Chem. Soc.*, **76**, 1493.

Rabinovitch, B. S. and Michel, K.-W. (1959), *J. Am. Chem. Soc.*, **81**, 5065.

Ramamurthy, V. and Liu, R. S. H. (1976), *J. Am. Chem. Soc.*, **98**, 2935.

Rauss-Godineau, J., Chodkiewicz, W., and Cadiot, P. (1966), *Bull. Soc. Chim. Fr.*, 2877.

Reetz, M. T. and Plachky, M. (1976), *Synthesis*, 199.

Rieker, A. and Kessler, H. (1969), *Chem. Ber.*, **102**, 2147.

Rockenfeller, J. D. and Rossini, F. D. (1961), *J. Phys. Chem.*, **65**, 267.

Rokach, J., Young, R. N., and Kakushima, M. (1981), *Tetrahedron Lett.*, **22**, 979.

Rotermund, G. W. (1975), "Osmium (VIII)-oxid als Oxidationsmittel," in Houben-Weyl, Vol. IV/lb, Müller, E., Ed., Thieme, Stuttgart, Germany, pp. 860–864.

Roth, W. R. and Exner, H.-D. (1976), *Chem. Ber.*, **109**, 1158.

Saltiel, J. and Charlton, J. L. (1980), "cis–trans Isomerization of Olefins, in de Mayo, P., Ed., *Rearrangement in Ground and Excited States*, Vol. 3, Academic, New York, pp. 25–89.

Saltiel, J., D'Agostino, J., Megarity, E. D., Metts, L., Neuberger, K. R., Wrighton, M., and Zafiriou, O. C. (1973), "The cis–trans Photoisomerization of Olefins," in Chapman, O. L., Ed., *Organic Photochemistry*, Vol. 3, Marcel Dekker, New York, p. 1.

Saltiel, J., Ganapathy, S., and Werking, C. (1987), *J. Phys. Chem.*, **91**, 2755.

Sanders, J.K.M. and Hunter, B.K. (1993), *Modern NMR Spectroscopy, A Guide for Chemists*, 2nd ed., Oxford University Press, New York, Chapter 6.

Sandström, J. (1983), "Static and Dynamic Stereo-chemistry of Push–Pull and Strained Ethylenes," *Top. Stereochem.*, **14**, 83.

Sandström, J., Stenvall, K., Sen, N., and Venkatesan, K. (1985), *J. Chem. Soc. Perkin 2*, 1939.

Saunders, J. K. and Bell, R. A. (1970), *Can. J. Chem.*, **48**, 512.

Saunders, W. H. and Cockerill, A. F. (1973), *Mechanisms of Elimination Reactions*, Wiley, New York.

Scharpen, L. H. and Laurie, V. W. (1963), *J. Chem. Phys.*, **39**, 1732.

Schaub, B., Jeganathan, S., and Schlosser, M. (1986), *Chimia*, **40**, 246.

Schenck, G. O. and Steinmetz, R. (1962), *Bull. Soc. Chim. Belg.*, **71**, 781.

Schlosser, M. (1970), "The Stereochemistry of the Wittig Reaction," *Top. Stereochem.*, **5**, 1.

Schlosser, M. and Schaub, B. (1982), *J. Am. Chem. Soc.*, **104**, 5821.

Schmiegel, W. W., Litt, F. A., and Cowan, D. O. (1968), *J. Org. Chem.*, **33**, 3334.

Schröder, M. (1980), "Osmium Tetraoxide Cis Hydroxylation of Unsaturated Substrates," *Chem. Rev.*, **80**, 187.

Schubert, W. M., Steadly, H., and Rabinovitch, B. S. (1955), *J. Am. Chem. Soc.*, **77**, 5755.

Serjeant, E. P. and Dempsey, B. (1979), *Ionization Constants of Organic Acids in Aqueous Solution*, Pergamon, New York.

Shvo, Y. (1968), *Tetrahedron Lett.*, 5923.

Skell, P. S. and Allen, R. G. (1958), *J. Am. Chem. Soc.*, **80**, 5997.

Smith, L., Bjellerup, L., Krook, S., and Westermark, H. (1953), *Acta Chem. Scand.*, **7**, 65.

Sonnet, P. E. (1980), "Olefin Inversion," *Tetrahedron*, **36**, 557.

Sonnet, P. E. and Oliver, J. E. (1976a), *J. Org. Chem.*, **41**, 3284.

Splitter, J.S. and Tureček, F. (1994). *Application of Mass Spectrometry to Organic Stereochemistry*, VCH, New York.

Staab, H. A. and Lauer, D. (1968), *Chem. Ber.*, **101**, 864.

Stern, E. S. and Timmons, C. J. (1970), *Electronic Absorption Spectroscopy in Organic Chemistry*, Edward Arnold, London.

Sternberg, U., Bleiber, A., Haberditzl, W., and Lochmann, R. (1982), *Mol. Phys.*, **47**, 1159.

Stezowski, J. J., Biedermann, P. U., Hildenbrand, T., Dorsch, J. A., Eckardt, C. J., and Agranat, I. (1993), *J. Chem. Soc. Chem. Commun.*, 213.

Suzuki, H. (1952), *Bull. Chem. Soc. Jpn.*, **25**, 145.

Swern, D. (1953), "Epoxidation and Hydroxylation of Ethylenic Compounds with Organic Peracids," *Org. React.*, **7**, 378.

Tapuhi, Y., Kalisky, O., and Agranat, I. (1979), *J. Org. Chem.*, **44**, 1949.

Timmons, C. J. (1972), "Applications of Electronic Absorption Spectroscopy," in Bentley, K. W. and Kirby, G. W., Eds., *Elucidation of Organic Stuctures by Physical and Chemical Methods*, Vol. IV, Pt. 1 of *Techniques of Chemistry*, 2nd ed., Wiley-Interscience, New York, p. 57.

Tokue, I., Fukuyama, T., and Kuchitsu, K. (1973), *J. Mol. Struct.*, **17**, 207.

Tokue, I., Fukuyama, T., and Kuchitsu, K. (1974), *J. Mol. Struct.*, **23**, 33.

Traetteberg, M., Frantsen, E.B., Mijlhoff, F.C., and Hoekstra, A. (1975), *J. Mol. Struct.*, **26**, 57.

Turner, R.B., Nettleton, D.E., and Perelman, M. (1958), *J. Am. Chem. Soc.*, **80**, 1430.

Turro, N. J. (1978), *Modern Molecular Photochemistry*, Benjamin/Cummings, Menlo Park, CA.

van Arkel, A. E. (1932), *Recl. Trav. Chim. Pays-Bas*, **51**, 1081; id. (1933) *ibid.*, **52**, 1013.

van Lier, P. M., Meulendijks, G. H. W. M., and Buck, H. M. (1983), *Recl. Trav. Chim.* Pays-Bas, **102**, 337.

van't Hoff, J. H. (1875), letter to Buys Ballot; cited in Brewster, 1972 (q.v.).

van't Hoff, J. H. (1877), *Die Lagerung der Atome im Raume*, Viehweg & Sohn, Braunschweig, Germany, p. 14.

Vedejs, E. and Fuchs, P. L. (1973), *J. Am. Chem. Soc.*, **95**, 822.

Vedejs, E. and Marth, C. F. (1987), *Tetrahedron Lett.*, **30**, 3445.

Vedejs, E. and Marth, C. F. (1988), *J. Am. Chem. Soc.*, **110**, 3948.

Vedejs, E. and Peterson, M. J. (1994), "Stereochemistry and Mechanism of the Wittig Reaction," *Top. Stereochem.*, **21**, 1.

Vedejs, E., Snoble, K. A., and Fuchs, P. L. (1973), *J. Org. Chem.*, **38**, 1178.

Vereshchagin, A. N., Nasyrov, D. M., Klochkov, V. V., Polushin, V. B., and Aganov, A. V. (1983), *Bull. Acad. Sci. USSR, Div. Chem. Sci.*, **32**, 1512. (Engl. Trans., p. 1370).

Viehe, H. G. (1960), *Chem. Ber.*, **93**, 1697.

Viehe, H. G., Dale, J., and Franchimont, E. (1964), *Chem. Ber.*, **97**, 244.

Viehe, H. G. and Franchimont, E. (1963), *Chem. Ber.*, **96**, 3153.

Viehe, H. G., Janousek, Z., Merenyi, R., and Stella, L. (1985), "The Captodative Effect," *Acc. Chem. Res.*, **18**, 148.

Vonach, B. and Schomburg, G. (1978), *J. Chromatogr.*, **149**, 417.

von Auwers, K., and Wissebach, H. (1923), *Ber. Dtsch. Chem. Ges.*, **56**, 715.

Walters, E. A. (1966), *J. Chem. Educ.*, **43**, 134.

Watts, V. S. and Goldstein, J. H. (1970), "Nuclear Magnetic Resonance Spectra of Alkenes" in Zabicky, J., Ed., *The Chemistry of Alkenes*, Vol. 2, Wiley-Interscience, New York, p. 1ff.

Weissberger, A. (1945), *J. Am. Chem. Soc.*, **67**, 778.

Wiberg, K. B., Deutsch, C. J., and Roček, J. (1973), *J. Am. Chem. Soc.*, **95**, 3034.

Wilson, C. V. (1957), *Org. React.*, **9**, 350.

Winstein, S. and Buckles, R. E. (1942a), *J. Am. Chem. Soc.*, **64**, 2780.

Winstein, S. and Buckles, R. E. (1942b), *J. Am. Chem. Soc.*, **64**, 2787.

Winterfeldt, E. (1969), "Ionic Additions to Acetylenes," in Viehe, H. G., Ed., *Chemistry of Acetylenes*, Marcel Dekker, New York, Chapter 4.

Wintner, C. E. (1987), *J. Chem. Educ.*, **64**, 587.

Woodward, R. B. and Brutcher, F. V. (1958), *J. Am. Chem. Soc.*, **80**, 209.

Yamazaki, H., Cvetanović, R. J., and Irwin, R. S. (1976), *J. Am. Chem. Soc.*, **98**, 2198.

10

Conformation of Acyclic Molecules

10-1. CONFORMATION OF ETHANE, BUTANE, AND OTHER SIMPLE SATURATED ACYCLIC MOLECULES

a. Alkanes

As indicated in Section 2-4, the term "conformation" relates to the different spatial arrangements of the atoms in a molecule that arise through rotation about the bonds linking such atoms. Different conformations thus differ in torsion angles about one or more bonds. Since it requires four atoms to define a torsion angle, a molecule must have at least four atoms to display conformational variability. Thus a tetraatomic molecule in which the valency angles differ from 180° and in which the atoms are linked in a row (as in H–O–O–H) will display an infinite number of conformations (Fig. 10.1), since the torsion angle ω around the central O–O bond can vary continuously between 0° and + and −180° (cf. Section 2-4 and Table 2.2).

> If even one of the valency angles is 180°, three of the four atoms are collinear and rotation around the central bonds is equivalent to rotation of the molecule as a whole and does not qualify as a change in conformation. Moreover, the four atoms must be linked in a row, A–B–C–D; an array of the type $\overset{\text{C}}{\underset{\text{D}}{\diagdown}}$A–B (whether pyramidal, as in ammonia or planar, as in boron trifluoride) displays no torsion angles and no conformational variability. (Although the bond angles in a pyramidal molecule may be varied by a vibrational motion, this is not usually considered a conformational change; but see p. 20.)

Most of these conformations are unstable (like the extreme or intermediate positions in the swing of a pendulum); stable conformations that are located at

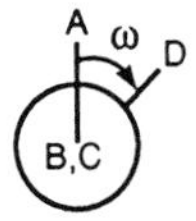

A ω D

B,C $180° > \omega > -180°$

Figure 10.1. Conformations of A–B–C–D.

energy minima are called "conformational isomers" or "conformers." Thus, while ethane ($H_3C–CH_3$) has an infinite number of conformations by virtue of rotation about the C–C bond, there are only three minima, (as shown in Figure 10.2), that is, ethane has only three conformers. Since they are indistinguishable, they are "degenerate."

Conformational isomers, even if nondegenerate as in the case of the butanes (see below), are generally not isolable because of the small energy barriers that separate them (cf. Section 2-1). It was therefore assumed, in the early days of stereochemistry, that rotation about single bonds was "free."

The first surmise that this might not be true appears in work by Bischoff (1890, 1891; cf. Bykov, 1975) concerned with the reactivity of substituted succinic acids; in the 1891 paper the staggered and eclipsed conformations of ethane are shown with the surmise that the staggered form is the more stable. (Bischoff's formulas are exemplified in Fig. 5.21.) Little attention seems to have been paid to Bischoff's work or to conformational analysis in acyclic systems in general until the 1920s when Hermans (1924) explained the difference in ease of acetal formation of racemic and *meso*-hydrobenzoin ($C_6H_5CHOHCHOHC_6H_5$) in conformational terms. (Hermans' formulas are exemplified in Fig. 5.21 also.) The field became more active in the 1930s (Weissberger and Sängewald, 1930; Wolf, 1930; see also Mizushima, 1954; Eliel et al., 1965) and the idea that rotation around the C–C bond in ethane and related compounds was not free had hardened by the middle of the decade (Kohlrausch, 1932; Mizushima et al., 1934; Teller and Topley, 1935; Smith and Vaughan, 1935; for an historical perspective, see Long, 1985). In 1936, Kemp and Pitzer (q.v.) not only asserted, on the basis of lack of agreement of the calculated (spectroscopic) and experimental (calorimetric) entropy of ethane, that rotation was not free but calculated the

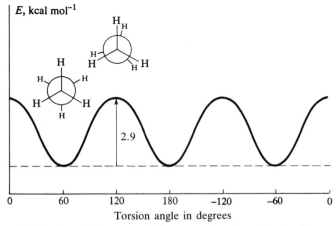

Figure 10.2. Potential energy of ethane as a function of the torsion angle.

barrier to be $3.15 \, \text{kcal mol}^{-1}$ ($13.2 \, \text{kJ mol}^{-1}$), later revised to $2.88 \, \text{kcal mol}^{-1}$ ($12.05 \, \text{kJ mol}^{-1}$, Pitzer, 1951), which is practically identical with the most recent experimental values of $2.89-2.93 \, \text{kcal mol}^{-1}$ ($12.09-12.26 \, \text{kJ mol}^{-1}$) (Weiss and Leroi, 1968; Hirota et al., 1979; Fantoni et al., 1986). The potential energy curve of ethane is shown in Figure 10.2.

The torsional potential (sometimes called *Pitzer potential*) is approximated by the equation $E = \frac{1}{2}E_0(1 + \cos 3\omega)$ (Kemp and Pitzer, 1936) where ω is the torsion angle (taken as zero for the eclipsed conformation) and E_0 is the energy barrier. Three indistinguishable minima, corresponding to staggered conformations, appear at $60°$, $180°$, and $-60°$ (see Section 2-4 for the sign convention of torsion angles) with maxima of $E_0 = 2.9 \, \text{kcal mol}^{-1}$ ($12.1 \, \text{kJ mol}^{-1}$) at $0°$, $120°$, and $-120°$. Because of the occurrence of three maxima and three minima in the potential function (Fig. 10.2) this is sometimes called a "V_3 potential" (see also Section 2-6). (The symbols V and E for energy are used interchangeably.)

The torsional barrier in ethane (reviewed by Pitzer, 1983) is *not* mainly due to steric causes, since the hydrogen atoms of the methyl groups are barely within van der Waals distance (cf. Section 2-6). Thus steric (van der Waals) repulsion accounts for less than 10% of the experimental barrier; electrostatic interactions of the weakly polarized C–H bonds are not of importance, either. The major contribution to the barrier has been ascribed to the unfavorable overlap interaction between the bond orbitals in the eclipsed conformation (Pauli's exclusion principle; Sovers, Pitzer, et al., 1968), but other calculations suggest it results from a favorable interaction (bonding–antibonding orbitals) in the staggered conformation (Bader et al., 1990). Recent calculated values of the barrier (Kirtman and Palke, 1977; Luke, Pople, et al., 1986; Dunlap and Cook, 1986) range from $2.7-3.07 \, \text{kcal mol}^{-1}$ ($11.3-12.8 \, \text{kJ mol}^{-1}$) and thus bracket the experimental value.

The conformational situation in propane is very similar to that of ethane: There are three indistinguishable energy minima (staggered conformations) and three indistinguishable barriers (eclipsed conformations) in the potential energy curve. However, since the barrier now involves the eclipsing of a methyl group (rather than a hydrogen atom) with a hydrogen atom, it is somewhat higher than in ethane, about $3.4 \, \text{kcal mol}^{-1}$ ($14.2 \, \text{kJ mol}^{-1}$) (cf. Lowe, 1968; Grant et al., 1970; Owen, 1974; Durig et al., 1977). Presumably the additional $0.5 \, \text{kcal mol}^{-1}$ ($2.1 \, \text{kJ mol}^{-1}$) is due to CH_3/H van der Waals repulsive interaction. Accordingly, the barrier in 2-methylpropane, $(CH_3)_2CH–CH_3$ amounts to $3.9 \, \text{kcal mol}^{-1}$ ($16.3 \, \text{kJ mol}^{-1}$) (Lowe, 1968; Durig et al., 1970) since it involves two CH_3/H interactions. On the whole, the barriers in $CH_3–CXYZ$ are remarkably constant [between 2.9 and $3.7 \, \text{kcal mol}^{-1}$ (12.1 and $15.5 \, \text{kJ mol}^{-1}$)], regardless of the nature of X, Y, and Z. It takes large substituents on *both* carbon atoms [as in $CCl_3–CCl_3$, barrier $10.8 \, \text{kcal mol}^{-1}$ ($45.2 \, \text{kJ mol}^{-1}$); polar factors probably also play a role here] for van der Waals repulsion to contribute to the barrier in a major way (Lowe, 1968; Allen and Fewster, 1974). Exceptions are fluoroethanes $CFXYCXYZ$ (X, Y, Z halogens; see p. 501).

A different conformational situation is encountered in molecules of the type $XCH_2–CH_2Y$, such as butane (X = Y = CH_3). The potential curve for butane is shown in Figure 10.3. There are now three different energy minima, two corresponding to the enantiomeric (and therefore equienergetic) gauche forms of

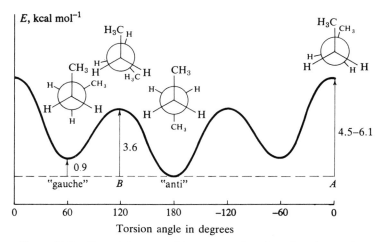

Figure 10.3. Potential energy of butane as a function of torsion angle.

C_2 symmetry and one corresponding to the achiral anti form of C_{2v} symmetry, which is of lower energy because it is devoid of the repulsive CH_3/CH_3 van der Waals interactions of the gauche forms. There are also three barriers between the anti and gauche conformers, two low ones involving CH_3/H eclipsing only (these two barriers are enantiomeric, and therefore equienergetic) and one high one involving CH_3/CH_3 eclipsing.

> In butane there are, of course, three C–C bonds about which rotation is possible. Two of these, CH_3–CH_2CH_2–CH_3 (in which the "rotor" or "top" is the methyl group) present a picture very similar to that in propane, which was discussed earlier. It is the rotation about the third bond (CH_3CH_2–CH_2CH_3) with which we are concerned here. In what follows we shall discuss rotation about specified bonds in a number of other molecules; this does not imply that rotation around other bonds fails to occur, but rather that we are not focusing on it.

The energy difference between the gauche and anti conformers of butane has been determined numerous times (for summaries see Lowe, 1968; Wiberg and Murcko, 1988) and it has become clear that the resulting value is quite phase dependent, being of the order of 0.89–0.97 kcal mol^{-1} (3.7–4.1 kJ mol^{-1}) in the gas phase and 0.54–0.57 kcal mol^{-1} (2.3–2.4 kJ mol^{-1}) in the liquid. This phase dependence may be taken as a manifestation of the von Auwers–Skita or conformational rule (Allinger, 1957; cf. Eliel et al., 1965), which states that the isomer of higher enthalpy has the lower molecular volume, and therefore the higher density, refractive index, boiling point, and heat of vaporization. In the case of the butane conformers, the gauche isomer is more compact, therefore it has the lower molecular volume and greater *inter*molecular van der Waals interactions (as a result of a high surface/volume ratio). Since the latter are predominantly attractive rather than repulsive, the heat of vaporization is larger for the gauche conformer with the resulting situation shown in Figure 10.4. Clearly, the gauche–anti enthalpy difference is greater in the gas than in the liquid (or solution). There is also a corresponding dependence on pressure (Chandler, 1978; Taniguchi, 1985) and on solvent shape (Jorgensen, 1981).

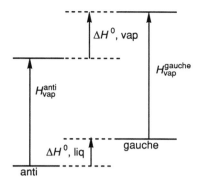

Figure 10.4. Enthalpy difference of butane conformers in the liquid and in the vapor state.

If one wants to estimate the population of gauche and anti conformers in butane, one has to keep in mind that the equilibrium constant $K = \%$anti/$\%$gauche depends on ΔG^0 ($K = e^{-\Delta G^0/RT}$), rather than on ΔH^0, so that the entropy difference between the two conformers also comes into play ($\Delta G^0 = \Delta H^0 - T\Delta S^0$). Although, the rotational and vibrational entropies of the two conformers are not the same (the two structures differ in moments of inertia and in normal vibrations), the difference from this source is apt to be small (but see Reisse, 1971); moreover, there is no difference in translational entropy. However, there are two additional sources of entropy differences, entropy of symmetry and entropy of mixing, which often differ substantially between conformers. The entropy of symmetry is given as $S_{\mathrm{sym}} = -R \ln \sigma$, where σ is the symmetry number characteristic of the symmetry point group of the conformer in question (cf. Section 4-6.c). The entropy of mixing of i components is $S_{\mathrm{mix}} = -R \, \Sigma_i \, n_i \ln n_i$. Thus, for two components $S_{\mathrm{mix}} = -R(n_1 \ln n_1 + n_2 \ln n_2)$. In the case of butane, both the anti and gauche conformers ($\mathbf{C_{2v}}$ and $\mathbf{C_2}$, respectively) have $\sigma = 2$, hence $S_{\mathrm{sym}} = -R \ln 2$ will cancel out in the difference between the two. On the other hand, there is no entropy of mixing for the achiral anti isomer, but the chiral gauche isomer has an entropy of mixing of $-R(0.5 \ln 0.5 + 0.5 \ln 0.5) = -R \ln 0.5 = R \ln 2$. This is the entropy of mixing of any pair of enantiomers. Thus the entropy of the gauche isomers exceeds that of the anti by $R \ln 2$ or 1.38 cal mol^{-1} K^{-1} (5.76 J mol^{-1} K^{-1}). Hence,

$$\Delta G^0 = \Delta H^0 - 1.38T \ \text{(cal mol}^{-1}\text{)} \ \text{(anti} \rightleftharpoons \text{gauche)}$$

or

$$= \Delta H^0 - 5.76T \ \text{(J mol}^{-1}\text{)}$$

For gaseous butane, taking $\Delta H^0 = 900$ cal mol^{-1} (3800 J mol^{-1}), ΔG^0 at room temperature (298.16 K) is $900 - 410 = 490$ cal mol^{-1} (2050 J mol^{-1}), hence $K = 2.3$, corresponding to 30% gauche (g) and 70% anti (a) isomer. The corresponding percentage in the liquid phase [$\Delta H^0 = 550$ cal mol^{-1} (2300 J mol^{-1})] is 44% gauche and 56% anti ($n_{\mathrm{g}} = 0.44$, $n_{\mathrm{a}} = 0.56$).

An alternative, and perhaps conceptually simpler, way of looking at the equilibrium in butane (and similar simple equilibria) is in terms of statistics. It is clear from Figure 10.3 that the single anti conformer of butane is in equilibrium with *two* gauche isomers denoted as

$$g^+ \ (\omega = 60°) \qquad \text{and} \qquad g^- \ (\omega = -60°)$$

Thus

$$g^+ \rightleftharpoons a \rightleftharpoons g^- \qquad \text{a being the anti conformer .}$$

If one assumes

$$\Delta G^0 \approx \Delta H^0$$

(i.e., $\Delta S^0 = 0$) for the two *individual* equilibria, then

$$g^+/a = g^-/a = e^{-900/RT}$$

in the gas phase. At room temperature (298.16 K)

$$g^+/a = g^-/a = 0.22$$

Hence,

$$g/a = g^+/a + g^-/a = 0.44$$

or $a/g = 2.3$, in agreement with the value calculated using the entropy of mixing.

Whereas the torsion angle in the anti form of butane is exactly 180° that in the gauche form is larger than the 60° expected for a perfectly staggered conformation (Bradford, Bartell et al., 1977; Compton et al., 1980; Heenan and Bartell, 1983). This result obtains because the van der Waals repulsion of the methyl groups in the gauche form makes these groups rotate away from each other, even at the expense of increasing the torsional energy. Energy minimization (cf. Section 2-6) occurs near a 65° torsion angle.

Figure 10.3 displays two barriers, one due to the CH_3/CH_3 eclipsing at $\omega = 0°$ and one due to twofold CH_3/H eclipsing at $\omega = \pm120°$. As expected, the value of the latter (Lowe, 1968; Compton et al., 1980; Stidham and Durig, 1986) of about $3.6 \pm 0.2 \text{ kcal mol}^{-1}$ $(15.1 \pm 0.8 \text{ kJ mol}^{-1})$ is close to that in isobutane (see above). It must be noted that this is the energy difference between the lower lying anti conformer and the top of the barrier; the difference between the higher energy gauche isomers and the top is about $0.9 \text{ kcal mol}^{-1}$ $(3.8 \text{ kJ mol}^{-1})$ less (in the gas phase).

The height of the barrier always depends on whether it is measured from the side of the more stable or that of the less stable conformer. The barrier height above the less stable conformer is always less, namely, by the ground-state energy difference of the conformers (cf. Fig. 10.3).

The higher (CH_3/CH_3) barrier, derived from spectroscopic experiments, (Stidham and Durig, 1986) amounts to 4.5–4.9 kcal mol^{-1} (18.8–20.5 kJ mol^{-1});

recent values calculated (Allinger et al., 1990) by either molecular mechanics or ab initio methods appear to agree quite well with the experimental ones [but see Veillard, 1974; Raghavachari, 1984; Wiberg and Murcko, 1988, who calculated a barrier in excess of 6 kcal mol^{-1} (25 kJ mol^{-1}) by ab initio methods].

The gauche-anti enthalpy difference and the rotational barrier in *n*-pentane seem to be similar to corresponding values in butane (Bartell and Kohl, 1963; Lowe, 1968; Darsey and Rao, 1981; Oyama and Shiokawa, 1983; Wiberg and Murcko, 1988) though there is one discrepant report of a larger ground-state energy difference (Hartge and Schrumpf, 1981). Nonetheless, the conformational situation in *n*-pentane is more complex than that in butane, since rotation around *two* CH$_2$–CH$_2$ bonds is now possible, giving rise to nine staggered conformations that are depicted in Figure 10.5. Conformations about the C(2, 3) and C(3, 4) bonds are denoted as anti (a) or gauche (g); for the gauche conformations it is also specified whether the torsion angle is near +60° (g$^+$) or near −60° (g$^-$). Conformations above the horizontal line are indistinguishable from (degenerate with) those below, and conformers to the right of the vertical line are mirror images of those on the left. Thus ag$^+$ and g$^+$a are interconvertible by a rotation of the molecule as a whole (C_2) and the same is true of ag$^-$ and g$^-$a and of g$^+$g$^-$ and g$^-$g$^+$; there are only six distinct conformers. Moreover, g$^+$a (or ag$^+$) is the mirror image of ag$^-$ (or g$^-$a) and these conformers are thus equal in energy. The same is true of g$^+$g$^+$ and g$^-$g$^-$. This leaves four conformations of different enthalpy, the low-lying aa, the intermediate ag, the high-lying (gg)$^\pm$ and the extremely unstable g$^+$g$^-$. (Regarding the exclusion of the very crowded g$^+$g$^-$ conformation, see Pitzer, 1940; Thompson and Sweeney, 1960; Bushweller et al., 1981.) For an assumed gauche–anti enthalpy difference in the liquid phase of 0.55 kcal mol^{-1} (2.3 kJ mol^{-1}) and taking into account that the ga conformers have a statistical weight of 4 and the gg$^\pm$ of 2, the calculated conformer distribution at 298.16 K in liquid pentane is aa, 34.6%; ag, 54.6%; gg, 10.8%; g$^+$g$^-$, less than 0.4%.

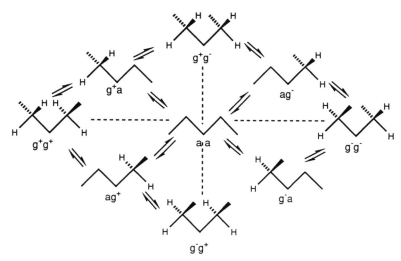

Figure 10.5. Pentane conformations.

The statistical factor may, alternatively, be expressed as an entropy of mixing and an entropy of symmetry as discussed for butane. The aa and gg conformers have a C_2 symmetry axis, hence $\sigma = 2$ and $S_{sym} = -R \ln 2$. The aa conformer is achiral, but the other two are chiral, hence they have an entropy of mixing of $R \ln 2$. It follows that the entropy contribution due to symmetry and chirality is $-R \ln 2$ for aa, $+R \ln 2$ for ag, and $+R \ln 2 - R \ln 2 = 0$ for gg. Since, in general, $S = R \ln W$, where $W =$ probability or statistical weight, the two ways of handling the statistics of conformer distribution are equivalent. It will also be noted that we have taken the population of the g^+g^- conformers as zero because of the severe van der Waals repulsive interaction of the methyl groups in these conformers [estimated interaction energy 3.7 kcal mol^{-1} (15.5 kJ mol^{-1}); cf. Chapter 11]. In higher alkanes, such inaccessible conformations become more numerous; in polymer chains (e.g., polyethylene) they correspond to the "excluded volume" of the polymer, that is, the chain cannot fold back upon itself.

Situations such as that in pentane, where two independent variables [here the C(2, 3) and C(3, 4) torsion angles (ω)] affect the enthalpy of the system are quite common, for example in polypeptides. It is convenient to depict the situation in the form of a contour diagram (Fig. 10.6) in which the abscissa and ordinate represent $\omega_{2,3}$ and $\omega_{3,4}$, respectively, and the contours represent energy (the dots at the centers of the contours are energy minima). The deepest minimum occurs exactly at $\omega_{2,3} = \omega_{3,4} = 180°$ (aa conformer). Transition of barriers (high-energy contours) leads to conformers of higher energy, where $\omega_{2,3} = 65°$, $\omega_{3,4} = 180°$, or vice versa; these are the ag conformers. From these conformers, by further rotation over a barrier, one goes to the still higher lying g^+g^+ and g^-g^- conformers at torsion angles of $+65°$, $+65°$ or $-65°$, $-65°$. Direct transition to these minima from the aa conformer is not favored in that it leads over very high ground (shaded). Minima are not found at $+60°$, $-60°$ (g^+g^-) or $-60°$, $+60°$ (g^-g^+) but there are some high shallow minima on both sides of the g^+g^- conformation representing somewhat deformed g^+g^- conformers between which

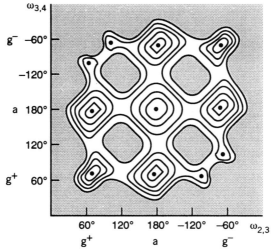

Figure 10.6. Energy contour diagram of *n*-pentane. [Reprinted with permission from Dale, J. (1978), "Stereochemistry and Conformational Analysis." Copyright © Universitetsforlaget, Oslo, Norway, p. 98.]

the g^+g^- conformation constitutes a saddle point. The energy of this conformation has been calculated (Wiberg and Murcko, 1988) to lie over 3.3 kcal mol^{-1} (13.8 kJ mol^{-1}) above that of the aa conformer.

> The conformational situation in higher alkanes evidently becomes increasingly complex (Dale, 1978). *n*-Heptane already has 13 different accessible conformers even counting enantiomers only once and omitting "excluded volume" conformers. In the case of polymers, such as polyethylene, the conformational problem has to be treated statistically, rather than by dealing with each different conformation individually (cf. Flory, 1953, 1969).

Of the branched alkanes, we mention here only 2,3-dimethylbutane (Fig. 10.7) because of the rather unusual finding (Verma, Bernstein, et al., 1974; Bartell and Boates, 1976; Lunazzi et al., 1977; Heinrich and Lüttke, 1977; Ritter et al., 1978) that the anti and gauche conformations are nearly equal in enthalpy (and that the gauche conformation therefore predominates by nearly a factor of 2 at equilibrium, since it is a pair of enantiomers and has an entropy advantage of $R \ln 2$, or, statistically, since there are two gauche conformations but only one anti).

The explanation for the perhaps unexpectedly high stability of the gauche form relative to the anti in 2,3-dimethylbutane and related hydrocarbons of the RR′CH–CHRR′ type (Boyd, 1975; Hounshell, Mislow, et al., 1978; Rüchardt and Beckhaus, 1980; Hellmann, Rüchardt, et al., 1982; see also Eliel, 1985) seems to be as follows: In butane, the CH_3–C–H bond angle is close to tetrahedral. However, in 2,3-dimethylbutane, the CH_3–C–CH_3 angle opens up to near 114° (Bartell and Boates, 1976; Rüchardt and Beckhaus, 1980). As a result, the ordinary Newman projections (Fig. 10.7, **A** and **G**) no longer apply; rather the anti conformer in 2,3-dimethylbutane tends toward a CH_3/CH_3 eclipsed conformation (Fig. 10.7, **A′**) and the gauche toward a CH_3/CH_3 perpendicular one (Fig. 10.7, **G′**). This deformation will enhance van der Waals repulsion in the anti conformer (though the enlargement of the bond angle in that conformer may actually be resisted) but will diminish it in the gauche isomer, which is thereby stabilized. The experimental barrier to rotation in 2,3-dimethylbutane, 4.3 kcal mol^{-1} (18.0 kJ mol^{-1}) (Lunazzi et al., 1977) is remarkably small if considered to represent simultaneous eclipsing of CH_3/CH_3 plus twice CH_3/H; presumably the angle deformations indicated in Figure 10.7 imply (cf. Hounshell, Mislow et al., 1978) that the eclipsing of the various ligands that have to pass each

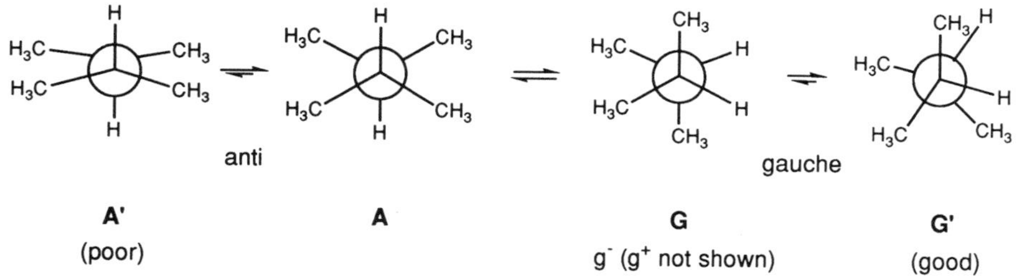

Figure 10.7. Conformational isomers of 2,3-dimethylbutane.

other is *not* simultaneous, the barrier thus being lowered. Similar considerations apply to other moleculess of the RR'CHCHRR' type (cf. Ōsawa et al., 1979; see also Section 14-5.d). An extensive review of the conformations of larger branched hydrocarbons has been published by Anderson (1992); see also Berg and Sandström (1989).

b. Saturated Acyclic Molecules with Polar Substituents or Chains. The Anomeric Effect

The molecules discussed so far are hydrocarbons, devoid of polar groups; therefore electrostatic interaction between groups (cf. Eq. 2.1, Chapter 2, p. 33) is nonexistent and solvation energy (except for phase changes on dissolution, as in the case of butane discussed earlier) is negligible. The situation changes in molecules with polar groups. Such molecules possess substantial dipoles (both local and overall). Since dipole interactions, which are responsible for the electrostatic term V_E in Eq. 2.1, vary from one conformer to another, they affect the conformational energy differences. The magnitude of these interactions may be solvent dependent. Moreover, differences in overall dipole moment between conformers may lead to differences in solvation energy (V_s in Eq. 2.1) between them.

Before entering upon these matters, we shall discuss briefly the history of the conformational analysis of the 1,2-dihaloethanes, XCH_2CH_2X (X = Cl or Br), since these molecules played an important part in the development of the subject (Mizushima, 1954; for a historical retrospect, see Morino, 1985). It was early recognized, on the basis of dipole and Raman spectral measurements (Mizushima et al., 1934), that 1,2-dihaloethanes (cf. Fig. 10.8) could not exist as pure anti isomers: Their dipole moments differ from zero and their Raman spectra, which show a centrosymmetric (anti) structure in the solid, display a number of additional lines in the liquid state. Although it was recognized that both observations could be explained by librational motion (torsional vibrations) in the fluid state, it was considered more likely that a second conformation with a finite dipole moment and a distinct Raman spectrum contributed in the liquid state. Perhaps the best evidence for the latter hypothesis is that molecules of the type CCl_3CHCl_2 or CCl_3CH_2Cl, which can have only one staggered conformation, do not show additional Raman lines on melting.

The question remained as to the nature of the second conformer: Did it correspond to the gauche or syn conformation? This problem was solved elegantly by Neu and Gwinn (1950; see also Bose et al., 1989). The four hydrogen atoms in the anti as well as the syn conformer of $BrCH_2CH_2Br$ (Fig. 10.9) are either homotopic or enantiotopic (cf. Chapter 8), and therefore spectrally indistinguish-

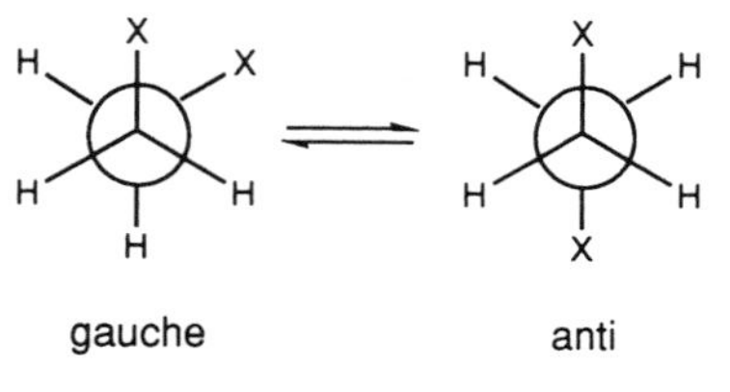

gauche   anti   **Figure 10.8.** Conformers of 1,2-dihaloethanes.

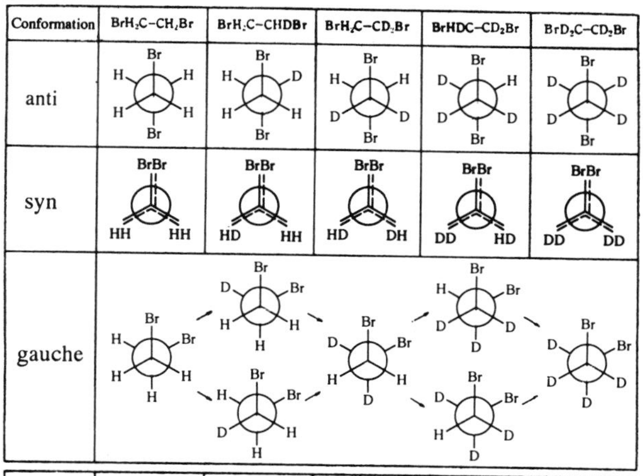

Conformation	BrH_2C-CH_2Br	$BrH_2C-CHDBr$	BrH_2C-CD_2Br	$BrHDC-CD_2Br$	BrD_2C-CD_2Br
anti	*(Newman projection)*	*(Newman projection)*	*(Newman projection)*	*(Newman projection)*	*(Newman projection)*
syn	*(Newman projection)*	*(Newman projection)*	*(Newman projection)*	*(Newman projection)*	*(Newman projection)*
gauche	*(Newman projections)*	*(Newman projections)*	*(Newman projections)*	*(Newman projections)*	*(Newman projections)*
trans	188(10)	188(10)	188(5)	187(10)	188(10)
gauche	355(2) <	342(1) / 352(1) >	341(1) <	321(1) / 333(1) >	316.5(2)
gauche	552(5) <	532(4) / 546(4) >	528(3) <	518.5(4) / 523(2) >	517.5(6)
gauche	581(1) <	565.5(2) / 580(3) >	551(1) <	540(2) >	526.5(3)
trans	660(15)	640(15)	639(6)	617(10)	607(15)

Figure 10.9. Conformations and salient Raman absorption frequencies (cm^{-1}) in 1,2-dibromoethane and deuterated analogues. [Adapted with permission from Neu, J.T. and Gwinn, W.D. (1950) *J. Chem. Phys.* **18**, 1643. Copyright © American Institute of Physics.]

able. In contrast, the geminal hydrogen atoms in the gauche conformer are diastereotopic and in principle distinguishable. While it is not possible to see this distinction in the Raman spectrum of $BrCH_2CH_2Br$, it was manifested in the Raman spectra of $BrCHDCH_2Br$ and $BrCHDCD_2Br$. As seen in Figure 10.9, these molecules (in contrast to $BrCH_2CH_2Br$, $BrCH_2CD_2Br$, and $BrCD_2CD_2Br$) exist in two diastereomeric gauche conformations and, indeed, as a result, a splitting of the Raman absorption bands at 355, 552, and 581 cm^{-1} is observed (Fig. 10.9). No splitting is seen in corresponding bands ascribed to the anti conformer and, most importantly, no splitting would be expected if the second conformer was syn, since no diastereoisomerism is possible in the syn conformer (Fig. 10.9). It follows, then, that the second conformer is gauche and that the splitting observed with some species but not with others is due to the diastereomerism shown in Figure 10.9.

The anti-gauche enthalpy difference in gaseous 1,2-dichloroethane (Fig. 10.8, X = Cl) is in the 0.9–1.3 kcal mol^{-1} (3.8–5.4 kJ mol^{-1}) range (Mizushima, 1954; Lowe, 1968; Tanabe, 1972; Allen and Fewster, 1974; Kveseth, 1975; Felder and Günthard, 1980; Dosěn-Mićović, Allinger et al., 1983) and determinations for 1,2-dibromoethane (Fig. 10.8, X = Br) range from 1.4 to 1.8 kcal mol^{-1} (5.9–7.5 kJ mol^{-1}) (Mizushima, 1954; Lowe, 1968; Tanabe, 1972; Abraham and Bretschneider, 1974; Park et al., 1974; Fujiyama and Kakimoto, 1976; Seki and Choi, 1982, 1983; see also Hammarström et al., 1988). Both differences are appreciably larger than those for butane (see above), not for steric reasons but because of the strong dipole–dipole repulsion of the C–X dipoles in the gauche conformation (cf. Fig. 10.8). This dipole effect, which can be simulated by molecular mechanics calculations (Dosěn-Mićović, Allinger et al., 1983; Dosěn-Mićović and Zigman, 1985) is less important in polar solvents with a consequent increase in the population of the gauche conformation (cf. Table 10.1). This results in part from a diminution of the coulombic interaction of the dipoles in the more polar solvent; a more important factor is that the conformer of higher dipole moment gains more energy by solvation in polar solvents. The latter solvent effect, also, can be simulated by calculation (Abraham and Bretschneider, 1974; Dosěn-Mićović, Allinger et al., 1983).

TABLE 10.1. Solvent Dependence of Conformational (Enthalpy) Differences[a] of 1,2-Dichloro- and 1,2-Dibromoethane[b]

Solvent	ε[c]	ClCH$_2$CH$_2$Cl		BrCH$_2$CH$_2$Br	
		Experimental[d]	Calculated[e]	Experimental[f]	Calculated[f]
Vapor	1.5[d]	0.9–1.3 (3.8–5.4)	0.91[d] (3.8)	1.4–1.8 (5.9–7.4)	NA
c-C$_6$H$_{12}$	2.0	0.91 (3.8)	0.86, 0.97 (3.6, 4.1)	NA	NA
Cl$_4$	2.2	NA	NA	1.31 (5.5)	1.40 (5.6)
CCl$_2$=CCl$_2$	2.5	0.89 (3.7)	0.82, 0.92 (3.4, 3.8)	1.24 (5.2)	1.38 (5.9)
CS$_2$	2.6	0.83 (3.5)	0.80, 0.87 (3.3, 3.6)	1.11 (4.6)	1.33 (5.6)
Et$_2$O	4.3	0.69 (2.9)	0.69, 0.68 (2.9, 2.8)	NA	NA
EtOAc	6.0	0.42 (1.8)	0.61, 0.57 (2.6, 2.4)	NA	NA
C$_6$H$_6$ or C$_6$D$_6$	2.3	0.60 (2.5)	0.56[g], 0.92[h] (2.3, 3.8)	0.69 (2.9)	1.38[h] (5.8)[h]
Liquid (neat)	i	0.31[i] (1.3)	0.48, 0.43 (2.0, 1.8)	0.86[i] (3.6)[i]	1.10 (4.6)
Mesityl oxide	15.0[d]	0.47 (2.0)	0.35, 0.33 (1.5, 1.4)	NA	NA
Acetone	20.7	0.18 (0.75)	0.25, 0.26 (1.05, 1.1)	NA	NA
CH$_3$CN	36.2	0.15 (0.63)	0.04, 0.12 (0.17, 0.50)	0.66 (2.8)	0.57 (2.4)

[a] ΔH^0 in kcal mol^{-1}, values in parentheses are in kJ mol^{-1}. Data not available are indicated as NA.
[b] See Figure 10.9.
[c] Dielectric constant, rounded off to one decimal, from Gordon and Ford (1972) unless otherwise noted.
[d] Dosěn-Mićović, Allinger et al., 1983. See also Hammarström et al., 1988.
[e] When there are two figures, the first refers to Dosěn-Mićović and Zigman (1985) and the second to Abraham and Bretschneider (1974).
[f] Abraham and Bretschneider (1974).
[g] This value was computed assuming a dielectric constant of benzene of 7.5; see text.
[h] This value was computed using the real dielectric constant of benzene of 2.3.
[i] 10.1 for ClCH$_2$CH$_2$Cl, 4.8 for BrCH$_2$CH$_2$Br: Gordon and Ford (1972).
[j] These values are corrected for the variation of ΔH^0 with temperature (resulting, in turn, from the variation of solvent dielectric ε with temperature) and therefore differ from the raw values of 0.0 kcal mol^{-1} (0.0 kJ mol^{-1}) for ClCH$_2$CH$_2$Cl and 0.74 kcal mol^{-1} (3.1 kJ mol^{-1}) for BrCH$_2$CH$_2$Br reported by Lowe (1968). The actual range of values for the dibromide is large, 0.65–1.3 kcal mol^{-1} (2.7–5.4 kJ mol^{-1}): See Takagi et al., (1983); Tanabe et al. (1976).

One value for the anti–gauche enthalpy difference in gaseous $ClCH_2CH_2Cl$, obtained by electron diffraction, is as high as 2.2 kcal mol^{-1} (9.2 kJ mol^{-1}) (Fernholt and Kveseth, 1978). Electron diffraction seems to be less accurate than some other methods for determination of conformational populations; discrepant values have also been obtained for butane (Bartell and Kohl, 1963), for 1,2-dichloroethane (Kveseth, 1974), and for 1,2-difluoroethane (see below; Friesen and Hedberg, 1980).

The situation in 1,2-difluoroethane (Fig. 10.8, X = F), surprisingly, is quite different from that in the dichloro and dibromo analogues in that the gauche form is preferred even in the gas phase (not just in terms of free energy, where it is favored by the statistical factor of 2, but in enthalpy). The ΔH^0 preference seems to amount to 0.6–0.9 kcal mol^{-1} (2.5–3.8 kJ mol^{-1}) (Abraham and Kemp, 1971; Radom, Pople, et al., 1973; Fernholt and Kveseth, 1980; Huber–Wälchli and Günthard, 1981; Hirano et al., 1986; Dixon and Smart, 1988; see also Hammerström et al., 1988; Durig et al., 1992) although there are some discrepant values (Lowe, 1968; see also above). The reason for this preference for the gauche form, despite the repulsive dipole–dipole interaction (a so-called V_1 potential, since there is only one maximum and one minimum of energy in the course of a 360° rotation about the C–C bond; see p. 614), probably lies in a combination of a relatively small van der Waals repulsion (affecting the normal V_3 potential) thanks to the small size of fluorine, plus the intervention of a V_2 potential, which displays energy minima when the C–F bonds are at 90° (or −90°) to each other and maxima when they are at 0° or 180° torsion angles (Radom et al., 1973; Bartell, 1977; Allinger et al., 1977; but see Wiberg and Murcko, 1987). The V_2 (twofold) potential has been pictorially described (Allinger et al., 1977) as being due to a hyperconjugative interaction of the type $FCH_2–CH_2F \leftrightarrow F^-CH_2=CHF\ H^+ \leftrightarrow H^+CHF=CH_2F^-$, which is triggered by the high electron demand of the fluorine. In order to involve both fluorine atoms simultaneously in this interaction, the two C–F bonds must be orthogonal. The combination of this V_2 potential, which is optimal at 90°, and the normal V_3 potential, which is optimal at 60° or 180°, gives rise to the near-gauche energy minimum; the actual F–C–C–F torsion angle is 71° (Takeo, Morini, et al., 1986). Smaller V_2 potentials may occur in other molecules, such as butane. An alternative explanation of the preferred gauche conformation of FCH_2CH_2F is in terms of the so-called "gauche effect" (Wolfe, 1972) which implies that a chain segment A–B–C–D will prefer the gauche conformation when A and D either are very electronegative relative to B and C or are unshared electron pairs (see also below). The two explanations are related in their origin; both imply a stabilization of the gauche conformer. A different explanation, involving destabilization of the anti conformer, has been given by Wiberg et al. (1990). These authors postulate that the bonds in FCH_2CH_2F are bent appreciably; this leads to diminution of the carbon–carbon (sigma) bond overlap, and hence weakening of the C–C bond, more so in the anti than in the gauche conformer.

In a more polar medium, such as the pure liquid (dielectric constant $\varepsilon = 34.4$) the gauche conformer of FCH_2CH_2F is, as expected, even more strongly preferred, to the near exclusion of the anti; $\Delta G^0 = 2.0$–2.6 kcal mol^{-1} (8.4–10.9 kJ mol^{-1}) (Abraham and Kemp, 1971; Harris et al., 1977). The parameter

ΔS^0 is $1.36\,\text{cal mol}^{-1}\,\text{K}^{-1}$ ($5.69\,\text{J mol}^{-1}\,\text{K}^{-1}$) favoring the gauche isomer (Felder and Günthard, 1984), very close to the calculated $R\ln 2$.

A gauche preference similar to that in 1,2-difluoroethane is found in liquid succinonitrile ($NCCH_2$–CH_2CN), the β-halopropionitriles (XCH_2–CH_2CN) (Park et al., 1974), in 1,2-dimethoxyethane (glyme, CH_3O–CH_2–CH_2–OCH_3), and probably also in CH_3O–CH_2CH_2–X ($X =$ halogen) (Matsuura et al., 1977, but see Hoppilliard and Solgadi, 1980); in the latter, the preferred conformation is anti–gauche (ag) about the C–O–C–C and O–C–C–X moieties, respectively (Ohsaku and Imamura, 1978; Ogawa et al., 1977; Anderson and Karlström, 1985; see also Astrup, 1979; Eliel, 1969, 1970). Polyoxyethylene (OCH_2CH_2)$_n$ also prefers the gauche conformation about the C–C bond and therefore, in contrast to polyethylene, assumes a helical conformation overall (Abe and Mark, 1976; Ohsaku and Imamura, 1978; cf. Eliel, 1969, 1970). On the other hand, CH_3S–CH_2–CH_2–SCH_3 resembles $BrCH_2$–CH_2Br in preferring the anti conformation about the C–C bond. The preferred overall conformation is gag', where g and g' represent gauche conformations with opposite torsion angles, the gauche conformation about the C–S–C–C bond being preferred (see below; Ogawa et al., 1977). A similar situation pertains for $CH_3SCH_2CH_2X$ ($X =$ Cl or Br) (Matsuura et al., 1977).

An interesting situation occurs in succinic acid (Fig. 10.8, $X = CO_2H$). At low pH (free acid) the gauche conformation predominates by considerably more than the statistical ratio of 2 [$n_g = 0.79$–0.84, $\Delta H^0 = 0.37$–$0.57\,\text{kcal mol}^{-1}$ (1.5–$2.4\,\text{kJ mol}^{-1}$)]. This finding indicates an attractive interaction, probably of the dipolar $O^- \cdots C^+$ type, since the observations were made in protic solvent that would inhibit intramolecular hydrogen bonding, which is another obvious cause for gauche preference. However, at high pH where the dominant species is the dianion, the anti configuration predominates [$n_g = 0.33$–0.47, $\Delta H^0 = -0.83$– $-0.48\,\text{kcal mol}^{-1}$ (-3.5– $-2.0\,\text{kJ mol}^{-1}$)], presumably because the charged CO_2^- groups now repel each other (Nunes, Gil, et al., 1981; Lit, Roberts, et al., 1993).

We pass, now, to l-propanol and the 1-halopropanes, $CH_3CH_2CH_2X$ ($X =$ OH, F, Cl, Br, or I). In all these molecules the gauche–anti energy difference is small and in the first four the gauche conformation is enthalpically slightly preferred over the anti in the vapor state (Lowe, 1968; Tanabe and Säeki, 1975; Steinmetz et al., 1977; Abraham and Stölevik, 1981; Saran and Chatterjee, 1980; Burkert, 1980; Meyer, 1983; Durig et al., 1984; Yamanouchi, Kuchitsu, et al., 1984). Stabilization of the gauche conformation would seem to be due to an attractive electrostatic interaction between X (negative) and the opposite (positive) end of the carbon chain (Szasz, 1955). An extreme case of this type of attraction is found in acetylcholine, $XCH_2CH_2\overset{+}{N}(CH_3)_3Y^-$ ($X =$ OAc), choline ($X =$ OH) (Terui et al., 1974), and fluorocholine ($X =$ F) (Birdsall et al., 1980), which exist almost exclusively in the gauche conformation; this is not true of chlorocholine ($X =$ Cl) (Terui et al., 1974) nor of thio- and selenocholine ($X =$ SH or SeH) (Birdsall et al., 1980), where steric factors in the gauche conformers are more important.

Yet another potential cause for predominance of the gauche conformation (in addition to the V_2 potential, see above) is found in ethylene glycol, in its monoether and in the haloethanols X–CH_2CH_2OH ($X =$ OH, OCH_3, F, Cl, or Br) (Snyder, 1966; Lowe, 1968; Pachler and Wessels, 1970; Maleknia, Schwartz et al., 1980; van Duin et al., 1986), where hydrogen bonding contributes to stabilization of the gauche conformer (Tichý, 1965).

Next we consider molecules of the type $H_3C-X-CH_2CH_3$, where a heteroatom is part of the chain. In butane ($X = CH_2$) the lesser stability of the gauche conformer (Fig. 10.3) is due to the steric interaction of the terminal methyl groups, which, in turn, depends on the CH_3/CH_3 distance. When $X = O$ (ethyl methyl ether), because of the foreshortening of the C–O [143 pm (1.43 Å), as compared to C–C, 153 pm (1.53 Å)] bonds, the methyl groups approach each other more closely than in butane, and therefore the gauche conformer is even less stable relative to the anti: $\Delta G^0 = 1.1-1.5 \, kcal \, mol^{-1}$ (4.6–6.3 kJ mol^{-1}) (Kitagawa and Miyazawa, 1968; Oyanagi and Kuchitsu, 1978; Durig and Compton, 1978); the same situation is encountered, if to a lesser extent, in ethyl-methylamine [$X = NH$; C–N, 147 pm (1.47 Å)] (Durig and Compton, 1979). The steric repulsion of the methyl groups in $CH_3OCH_2CH_3$ also leads to an increase of the central torsion angle to about 85° (Oyanagi and Kuchitsu, 1978; Jorgensen and Ibrahim, 1981). On the other hand, in ethyl methyl sulfide ($X = S$), the longer C–S bond distance [181 pm (1.81 Å)] engenders a lesser CH_3/CH_3 repulsion and the difference between the gauche and anti conformers is close to zero (Sakakibara, Shimanouchi, et al., 1977; Durig et al., 1979).

We end this rather brief discussion of the conformation of saturated acyclic molecules by considering the conformation of molecules of the type CH_3O-CH_2X (X = halogen, OR, or SR). Such species are important because of their relation to sugars, glycosides, and sugar halides (cf. Chapter 11). In dimethoxymethane ($CH_3OCH_2OCH_3$) one might anticipate, on steric grounds, that gauche conformations would be rather unstable because of the short C–O bond distances (see above). In fact, however, the molecule exists largely in the gg conformation (Astrup, 1971, 1973). From polymer studies of polyoxymethylene, $\{CH_2O\}_n$, which has a helical conformation (cf. Eliel, 1969, 1970; Miyasaka et al., 1981), it had been estimated (Abe and Mark, 1976; Abe, 1976) that the gauche preference is 1.1 kcal mol^{-1} (4.6 kJ mol^{-1}), but calculations for $CH_3OCH_2OCH_3$ referring to the gas phase suggest a value of 2.2 kcal mol^{-1} (9.2 kJ mol^{-1}) (Jeffrey, Pople, et al., 1978) or 2.5 kcal mol^{-1} (10.5 kJ mol^{-1}) (Abe et al., 1990). The discrepancy may well be due to differences in the medium, since calculation also indicates that solvent effects are very important in this type of equilibrium (Tvaroska and Bleha, 1980). In dimethoxymethane, as in pentane (Fig. 10.5), there are, in all, four different conformations of minimum energy engendered by rotation about two C–O bonds: $g^{\pm}g^{\pm}$, ag, aa, and g^+g^-. The energy contour map shown in Figure 10.10 shows the lowest minima at gg, higher minima at ag, and a still higher and shallow minimum at aa; the situation is very different from that in pentane (Fig. 10.6). As mentioned earlier (see Fig. 10.6) there are no minima at g^-g^+, for steric reasons.

The difference between the enthalpy difference (gauche − anti) one would expect on "purely steric" grounds and that actually found is called the "anomeric effect" (Kirby, 1983; Tvaroska and Bleha, 1985; Juaristi and Cuevas, 1992; Graczyk and Mikołajczyk, 1994) or "generalized anomeric effect" (Lemieux, 1971; Eliel, 1972); the origin of the term (which comes from sugar chemistry), its history, and its extensive application in cyclic systems will be discussed in Chapter 11.

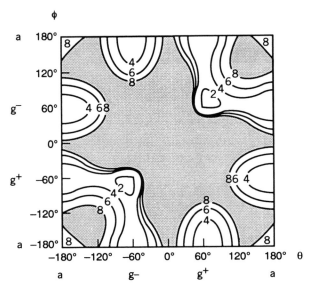

Figure 10.10. Energy contour diagram in kilocalories per mole for dimethoxymethane (CH_3OCH_2-OCH_3). [Reprinted with permission from Jeffrey, G.A., Pople, J.A., Binkley, J.S., and Vishveshwara, S. (1978) *J. Am. Chem. Soc.*, **100**, 373. Copyright © 1978 American Chemical Society.]

To estimate the magnitude of the anomeric effect in dimethoxymethane, one might use *n*-butane (similar to *n*-pentane; see pp. 603–605) as a reference compound; in *n*-butane, the anti conformer is favored by about 0.9 kcal mol^{-1} (3.8 kJ mol^{-1}; gas phase, see above), whereas in $CH_3OCH_2OCH_3$ the gauche conformer is favored by 1.1 kcal mol^{-1} (4.6 kJ mol^{-1}); one might therefore argue that the magnitude of the anomeric effect for CH_3O is $1.1 - (-0.9) = 2.0$ kcal mol^{-1} (8.4 kJ mol^{-1}). [Estimates from cyclic systems range from 0.9 to 1.5 kcal mol^{-1} (3.8–6.3 kJ mol^{-1}); Kirby, 1983.] However, *n*-butane is not a good case for comparison, since, on the one hand, the CH_3/O distance in $CH_3OCH_2OCH_3$ is shorter than the CH_3/CH_3 distance in butane and, on the other hand, O is smaller than CH_2. This point is emphasized by the finding that the C–O–C and O–C–O bond angles in dimethoxymethane are opened to 114.3° and 114.6°, respectively, presumably to minimize CH_3–O interactions (Astrup, 1973). A better estimate of the anomeric effect would come from the difference of the experimental $\Delta H_{\text{trans–gauche}}$ and that calculated by molecular mechanics on purely steric grounds (Tvaroska and Bleha, 1985; for another interesting approach, see Franck, 1983.)

Phenomenologically, the generalized anomeric effect implies that the preferred conformation of a R–O–CH_2–X fragment (where X is an electronegative group, such as halogen, OH, OCOR', OR', SR', or NR'R'') is gauche or, in other words, that an unshared pair of electrons on oxygen will be antiperiplanar to X, as shown in Figure 10.11. A more general statement about the anomeric and related stereoelectronic effects (Kirby, 1983; Epiotis et al., 1977) is "There is a stereoelectric preference for conformations in which the best donor lone pair or bond is antiperiplanar to the best acceptor bond." Originally (Edward, 1955), the effect was ascribed to dipole–dipole repulsion of the C–X dipole and the dipole of the C–O–C moiety (see Fig. 10.11; the C–O–C dipole is directed along the

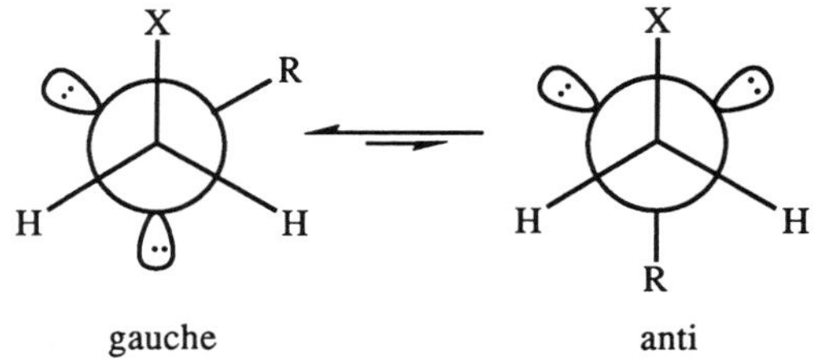

gauche   anti

Figure 10.11. Anomeric preferences.

bisector of the lone pair axes; the lone pairs are considered to be sp^3 hybridized). The solvent dependence of the effect supports this idea; moreover, conformations in which the lone pairs on O and X (which may be considered the generators of the dipoles) are parallel are particularly unstable (Eliel, 1969). Later, it was pointed out (Romers, Havinga, et al., 1969) that the calculated electrostatic repulsion is inadequate to account for the magnitude of the anomeric effect. At the same time, attention was drawn to certain anomalies in bond lengths in molecules displaying the anomeric effect. These anomalies are shown for CH_3-O-CH_2-X (X = F or Cl: Hayashi and Kato, 1980; X = OCH_3, Astrup, 1973) in Figure 10.12; the normal bond distances (cf. Chapter 2) are 137.9 pm (1.379 Å) for C–F, 176.7 pm (1.767 Å) for C–Cl, and 142.6 pm (1.426 Å) for C–O. The data in Figure 10.12 suggest that, whereas the CH_3–O bond length is normal, the O–CH_2 bond is considerably foreshortened; more so when X = halogen than when X = OCH_3. The C–X bond is markedly lengthened when X = Cl, less so when X = F; but it is shortened when X = OCH_3 (in which case, for reason of symmetry, the two CH_2–O bonds must be of equal length). The suggestion was made (Romers, Havinga, et al., 1969) that at least part of the anomeric effect is due to double-bond–no-bond resonance of the type $R-O-CH_2-X \leftrightarrow R-\overset{+}{O}=CH_2 \ \overset{-}{X}$ or its molecular orbital equivalent, which implies overlap of the n orbital of oxygen with the σ^* orbital of the C–X bond (Fig. 10.13). For this overlap to be maximal, one of the lone pairs on oxygen must be antiperiplanar to the C–X bond, as shown in Figure 10.11. When X (Figs. 10.12 and 10.13) is halogen, the O–CH_2 bond order is increased and the bond therefore shortened, whereas the CH_2X bond order is decreased and the bond thereby lengthened as shown. But when X = OCH_3 the hyperconjugative resonance can work either way:

	CH_3——O——CH_2——X		
X = F	142.4	136.2	138.5
	(1.424)	(1.362)	(1.385)
X = Cl	142.1	136.2	182.2
	(1.421)	(1.362)	(1.822)
X = OCH_3	143.2	138.2	138.2
	(1.432)	(1.382)	(1.382)

Figure 10.12. Bond lengths (in picometers; parenthesized values in angströms) in compounds displaying anomeric effect.

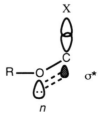

Figure 10.13. Anomeric effect: molecular orbital interpretation.

$$CH_3-\overset{+}{O}=CH_2\overset{-}{O}-CH_3 \leftrightarrow CH_3-O-CH_2-O-CH_3 \leftrightarrow CH_3-\overset{-}{O}-CH_2=\overset{+}{O}-CH_3$$

and the effect on the CH_2-O bond length is equivocal; in actual fact, the bond shortening due to partial double-bond character is more important than the bond lengthening resulting from partial no-bond character.

Detailed ab initio quantum mechanical calculations (e.g., Jeffrey and Yates, 1979) fully support the above picture for X = halogen. The conformational energy V_ω as a function of torsion angle, say in CH_3-O-CH_2-F (Fig. 10.14), may be dissected into a onefold V_1, twofold V_2, and threefold V_3 potential barrier. The onefold barrier (which, by definition, has a single maximum and a single minimum in a 360° rotation) comprises steric (CH_3-F interaction) and dipole factors. The threefold barrier describes the Pitzer potential, which is similar to that found in ethane (Fig. 10.2). The twofold barrier describes, in the main, the effect of the earlier mentioned $n-\sigma^*$ orbital overlap. (That the minimum occurs at $\omega = 90°$ rather than 60° as might be implied in Figure 10.11 results from the use, in that scheme, of an unhybridized p orbital for the unshared pair on oxygen involved in the overlap.) It is significant that the effect of the V_1 potential in stabilizing the gauche conformer is relatively small: The advantage of the 60° over the 180° conformation on the basis of V_1 alone is only about 1.5 kcal mol^{-1} (6.3 kJ mol^{-1}). The overall effect on V_ω (see Fig. 10.14) is much larger; and since V_3 has the same minimum value at 60° and 180°, this must be due to the effect of V_2, which originates in the $n-\sigma^*$ overlap. In fact, in CH_3OCH_2Cl, because of the rather severe CH_3/Cl steric interaction, V_1 quite strongly *dis*favors the synperiplanar and favors the antiperiplanar conformation; that the gauche conformer is nevertheless more stable than the anti is

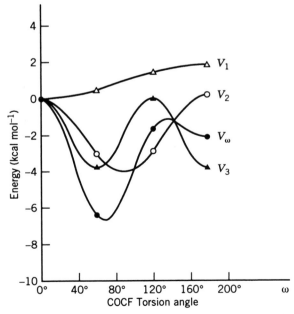

Figure 10.14. Contributing potentials V_1, V_2, and V_3 to torsional potential V_ω of CH_3OCH_2F. [Reprinted with permission from Jeffrey, G.A., and Yates, J.H. (1979), *J. Am. Chem. Soc.*, **102**, 820. Copyright © 1979 American Chemical Society.]

obviously due to the large 90° minimum in V_2. This case shows quite clearly that the anomeric effect is not due mainly to polar causes.

Because of this finding, the above-mentioned strong solvent dependence of the anomeric effect is perhaps surprising. It is possible that solvent simply affects the dipolar V_1 component of the anomeric effect; an alternative or additional explanation (Bailey and Eliel, 1974) is that polar solvents weaken the anomeric effect by interacting with the lone pairs (n electrons) on the ether oxygen thereby interfering with the $n-\sigma^*$ resonance, which is an essential part of the effect. (An analogy exists in the well-known solvent effect on the $n-\pi^*$ transition in ketones: Polar solvents, by stabilizing the ground state vis-a-vis the excited state, shift the carbonyl absorption due to this transition to higher energies, that is, shorter wavelengths.)

10-2. CONFORMATION OF UNSATURATED ACYCLIC AND MISCELLANEOUS COMPOUNDS

a. Unsaturated Acyclic Compounds

Ethylene is planar and, because of the high barrier to rotation about the double bond (cf. Chapter 9), exists in a single conformation. Propylene, on the other hand, can undergo conformational changes by rotation about the $H_3C-CH=CH_2$ single bond. The two achiral conformations (in which the plane of the double bond is the symmetry plane) are shown in Figure 10.15. Interestingly, the most stable conformation is that in which the double bond is eclipsed with one of the C–H bonds of the methyl group (cis or eclipsed, Fig. 10.15; Herschbach and Krisher, 1958). The barrier to rotation in propylene is nearly $2.0\,\text{kcal mol}^{-1}$ ($8.4\,\text{kJ mol}^{-1}$) (Lowe, 1968) and it would appear that the $H_2C=C/C-H$ staggered (synclinal or *bisecting*) conformation corresponds to the energy maximum. Since the H/H eclipsing that occurs in this conformation only accounts for about $1\,\text{kcal mol}^{-1}$ ($4.2\,\text{kJ mol}^{-1}$), i.e., one-half of the total barrier (cf. the earlier discussion on ethane, Fig. 10.2), there seems to be an additional destabilization, of about the same magnitude, due to unfavorable interaction of the other two C–H bonds with the C=C π orbitals in this conformation (cf. Wiberg and Martin, 1985). An alternative would be a corresponding attractive interaction in the eclipsed conformation.

In l-butene (Fig. 10.16) $R_t = R_c = H$ there are two possible eclipsed conformations, one "cis", with the $C=CH_2$ group eclipsed with CH_3, and the other

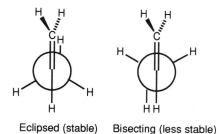

Eclipsed (stable)   Bisecting (less stable)

Figure 10.15. Conformations of propylene.

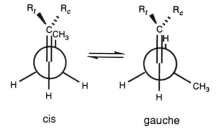

cis   gauche

Figure 10.16. Stable conformers of l-butene, where $R_t = R_c = H$.

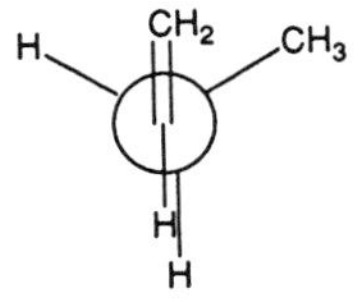

Figure 10.17. Barrier conformation in 1-butene.

"gauche" (really anticlinal), with the $C=CH_2$ group eclipsed with H. Different investigators have found either the cis or the gauche conformer to be lower in enthalpy, but the difference is small, 0.53 kcal mol^{-1} (2.2 kJ mol^{-1}) or less (Van Hemelrijk et al., 1980; Durig and Compton, 1980 and references cited therein). In some substituted propylenes of the type $R–CH_2–CH=CH_2$, where R is a small electronegative substituent, such as F, CH_3O, or CN, the cis conformer clearly predominates (Karabatsos and Fenoglio, 1970). The barrier to interconversion of the l-butene conformers is 1.74 kcal mol^{-1} (7.3 kJ mol^{-1}) above the more stable gauche form. The conformation at the energy maximum is bisecting (Fig. 10.17) and its conformational energy indeed agrees quite well with that of the bisecting conformation in propylene [1.98 kcal mol^{-1} (8.3 kJ mol^{-1}); Fig. 10.15].

Substituents at C(1) in propene, as in $CH_3–CH=CHX$, lower the barrier to methyl rotation if they are cis (Z) to the methyl group but not if they are trans (E) (Table 10.2). Since there is no obvious reason why the bisecting conformation (Fig. 10.17) should be stabilized by substitution at the far end of the double bond, the lowering of the barrier is most likely due to destabilization of the eclipsed form. Contemplation of Figure 10.16 suggests that such destabilization by van der Waals repulsion between a hydrogen of the methyl group and the X substituent should be important only when $R_c = X$ (Z isomer) but not when $R_t = X$ (E isomer).

The situation in aldehydes is similar to that in alkenes. Thus acetaldehyde is more stable in the eclipsed conformation (actually the stable conformation has the methyl group rotated about 9° away from perfect eclipsing) than in the bisecting one by 1.17 kcal mol^{-1} (4.9 kJ mol^{-1}) (Fig. 10.18; Kilb, Lin, and Wilson, 1957). That the energy advantage of the cis over the bisecting form is smaller in acetaldehyde than in propylene may again strike one as unexpected, since O is smaller than CH_2 and, in addition, one might have expected an electrostatic attraction between O and eclipsed H due to hyperconjugation: $O=C–CH_2H \leftrightarrow {}^-O–C=CH_2H^+$. However, the dominating factor seems to be the repulsive $C–H/C=X(\pi)$ interaction in the barrier conformation that is less when $X = O$ than when $X = CH_2$ because of the lesser π character of the (polar)

TABLE 10.2^a. Barriers in $H_3C–CH=CHX$

		Barrier	
X	E or Z	(kcal mol^{-1})	(kJ mol^{-1})
Cl	E	2.17	9.1
Cl	Z	0.62	2.6
CH_3	E	1.95	8.2
CH_3	Z	0.75	3.1

^{a} For a more detailed table, see Bastiansen et al., 1971, Table 4.

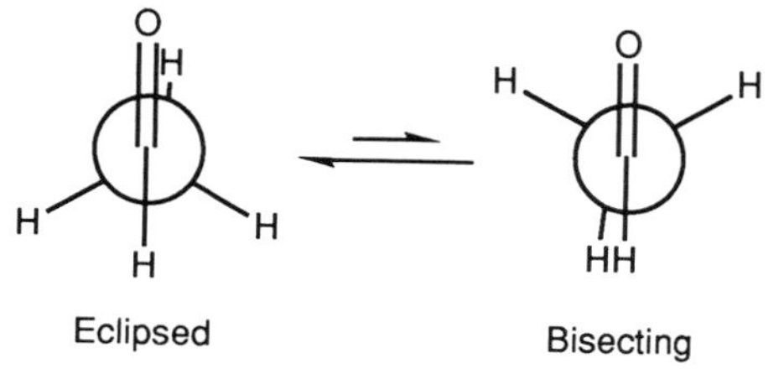

Figure 10.18. Conformations of acetaldehyde.

C=O ↔ C–O bond. In propionaldehyde (in contrast to l-butene) the conformation in which CH_3 is eclipsed with O is preferred [by $0.7–1.2 \text{ kcal mol}^{-1}$ (2.9–5.0 kJ mol^{-1}); Abraham and Pople, 1960; Butcher and Wilson, 1964; Frankiss and Kynaston, 1972; Chapput et al., 1974; Durig et al., 1980; cf. Fig. 10.19, **A**]. That this energy difference represents a net stabilization of the cis conformer is confirmed by the value of the rotational barrier, $2.3 \text{ kcal mol}^{-1}$ (9.6 kJ mol^{-1}) (Butcher and Wilson, 1964), which is $1.1 \text{ kcal mol}^{-1}$ (4.6 kJ mol^{-1}) above that in acetaldehyde, presumably because the cis form in propionaldehyde is stabilized by that amount of energy. That steric factors can vitiate this stabilization is demonstrated in that *tert*-butylacetaldehyde ($(CH_3)_3C–CH_2–CHO$, (Fig. 10.19, **A**, *t*-Bu instead of CH_3) prefers the gauche conformation over the cis by $0.25 \text{ kcal mol}^{-1}$ (1.05 kJ mol^{-1}) (cf. Karabatsos and Fenoglio, 1970).

The preferred conformation of acetone corresponds to that of acetaldehyde in that one hydrogen in each methyl group is eclipsed with the carbonyl oxygen; similarly diethyl ketone resembles propionaldehyde in having as its preferred conformation the one with both terminal methyl groups similarly eclipsed (Karabatsos and Fenoglio, 1970). In 2-butanone (Fig. 10.19, **B** and **C**) the most populated conformation (**B**, $\omega = 0°$) corresponds to that in propionaldehyde; however, an interesting new feature is a long, flat region in the potential energy curve (as calculated by molecular mechanics) near $\omega = 120°$ (**C**). It appears that in the $\omega = 120° - 90°$ region, relief of nonbonded interaction between the terminal methyl groups very nearly balances the energy increase incurred by the loss of the (optimal) eclipsing between one of the methylene hydrogen atoms and the carbonyl oxygen (Bowen, Allinger, et al., 1987).

Imino derivatives of aldehydes and ketones, for example, $R'CH_2–CH=NR$ (aldimines, R = alkyl; oximes, R = OH; hydrazones, R = NHR'') resemble the parent aldehydes $R'CH_2–CH=O$ in that the nitrogen is eclipsed with either H or R'; but the preference of R' eclipsing (cis) over H eclipsing (gauche; see Fig. 10.19 for the aldehyde) is less in the nitrogen compounds and seems to come into

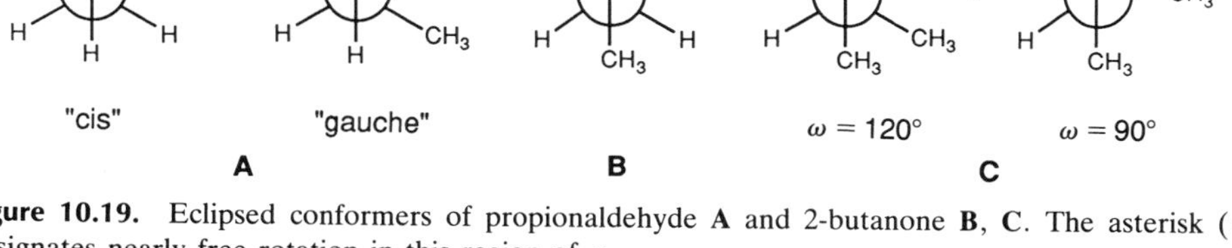

Figure 10.19. Eclipsed conformers of propionaldehyde **A** and 2-butanone **B**, **C**. The asterisk (*) designates nearly free rotation in this region of ω.

play only for aldimines ($R = CH_3$); for oximes and hydrazones the gauche conformation is preferred over the cis (Karabatsos and Fenoglio, 1970). In compounds such as acetaldoxime ($CH_3CH=NOH$) a situation is encountered that is similar to that in the alkenes of the type $CH_3CH=CHX$ discussed above: in the E isomer, the barrier (difference between eclipsed and bisecting conformation, cf. Fig. 10.18) is normal [1.84 kcal mol^{-1} (7.7 kJ mol^{-1}) in (E)–$CH_3CH=NOH$] but in the Z isomer it is much reduced [0.38 kcal mol^{-1} (1.6 kJ mol^{-1}) in (Z)–$CH_3CH=NOH$] (Schwendeman and Rogowski, 1969).

We turn now to acid derivatives: acid halides, RCOX, esters, RCO_2R', and amides $RCONH_2$. The situation in acid halides is very similar to that in aldehydes in that the oxygen will eclipse the carbon chain (cis conformation Fig. 10.19, halogen instead of H at the lowermost position; Stiefvater and Wilson, 1969; Karabatsos and Fenoglio, 1970). The situation with esters is more interesting, since, in principle, these esters might exist in either cis or trans conformations (Fig. 10.20). It has been shown that even for methyl (Curl, 1959) and ethyl (Riveros and Wilson, 1967) formate (Fig. 10.20, $R = H$, $R' = CH_3$ or CH_3CH_2), where the competition is between CH_2/O and CH_2/H eclipsing, the conformer with CH_2 trans to R (Fig. 10.20, Z) wins out. The hydrogen atoms of the methyl group are staggered with the O–C bond, with a barrier to methyl rotation of 1.19 kcal mol^{-1} (5.0 kJ mol^{-1}) in methyl formate; the corresponding lowest barrier for the methylene group in ethyl formate is 1.10 kcal mol^{-1} (4.6 kJ mol^{-1}). Apart from that, methyl formate is planar; but ethyl formate has two almost equally stable conformations, one planar zigzag and another, only 0.19 kcal mol^{-1} (0.79 kJ mol^{-1}) above the first, in which the terminal methyl group is nearly at right angles (95°) to the plane of the rest of the molecule (Riveros and Wilson, 1967).

> Since the alkoxy C–O bond has considerable double-bond character (see below) we prefer to call the major conformer (Fig. 10.20) conformer Z and the other E, the carbonyl oxygen being fiducial at the other end of the O–C partial double bond. This nomenclature is unequivocal, whereas in the cis–trans system it may not be clear to the uninitiated that the R group of the acid is considered fiducial, unless this is specifically stated. (There has, in fact, been considerable confusion on this point in the formates, where $R = H$.) Beilstein uses e and z for this case.

The E–Z energy difference in methyl formate (Fig. 10.20, $R = H$, $R' = CH_3$) is so large as to be difficult to measure, there being only a negligible amount of the E isomer at room temperature. Nonetheless, an estimate of ΔH_{E-Z} has been obtained (Blom and Günthard, 1981) by rapidly cooling hot beams ($T = 286, 662,$ and 803 K) of the ester in an argon matrix so as to "freeze in" the high temperature equilibrium and then analyzing the stable mixture by IR spectroscopy, at liquid helium temperatures (ca. 4 K). (For a discussion of this

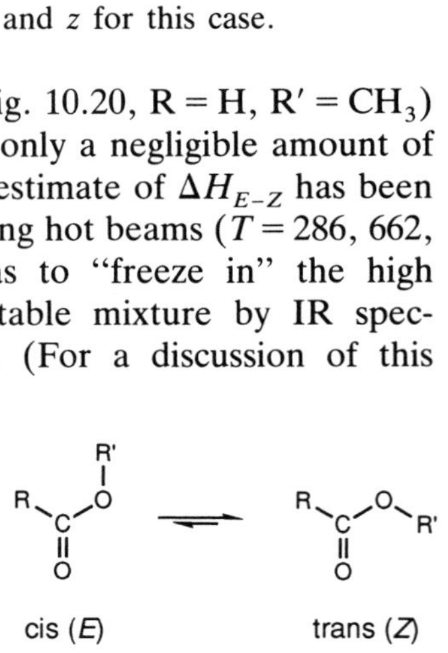

Figure 10.20.   Ester conformations.   cis (E)   trans (Z)

"matrix-isolation IR spectroscopy", see Tasumi and Nakata, 1985; p. 621 and Section 11-4.a.) From the temperature dependence of the absorbance ratio (cf. Section 10-4.d), ΔH^0 was found to be 4.75 ± 0.19 kcal mol^{-1} (19.9 ± 0.8 kJ mol^{-1}) favoring the Z conformer (see also Ruschin and Bauer, 1980). These values are much larger than those found earlier by the ultrasonic absorption method (cf. Wyn-Jones and Pethrick, 1970), which had given results that are internally inconsistent as well as clearly too small. On the other hand, ab initio (Radom, Pople, et al., 1972) and force field calculations (Allinger and Chang, 1977) lead to ΔH^0 values of the right order of magnitude in predicting that the Z conformer in esters is 5–6 kcal mol^{-1} (21–25 kJ mol^{-1}) more stable than the E with a barrier to rotation of 10–13 kcal mol^{-1} (42–54 kJ mol^{-1}) (see also Jones and Owen, 1973; Wiberg and Laidig, 1987).

The geometry of methyl formate (Curl, 1959) is shown in Figure 10.21. The C–O–C angle (114.8°) is considerably larger than that in ethers (111.5°), whereas the H–C=O angle is remarkably small. The acyl oxygen single bond, 133.4 pm (1.334 Å) is considerably shorter than the alkyl oxygen bond, 143.7 pm (1.437 Å). This result suggests considerable overlap of one of the p orbitals on the alkyl oxygen with the C=O bond (Fig. 10.21 *a* shows the corresponding resonance form; see, however, Wiberg and Laidig, 1987). As a result, the H–C–O–C framework is planar and the barrier to rotation about the C–O "single" bond (see above) is high. Surprisingly, the C=O bond is not lengthened; its length of 120.0 pm (1.200 Å) is actually slightly shorter than the 122.0 pm (1.22 Å) usual in aldehydes and ketones. The reasons why the Z conformer of methyl formate is more stable than E are complex (cf. Wiberg and Laidig, 1987). The following factors may contribute: (a) H/CH$_3$ steric repulsion in the E conformer (Fig. 10.20). This factor is probably small in formates but important in methyl esters of higher acids where the repulsion is of the R/CH$_3$ type. (b) van der Waals attraction between CH$_3$ and =O in the Z conformer. (c) Electrostatic attraction between O$^{\delta-}$ and H$_3$C$^{\delta+}$ in the Z conformer. (This factor, equivalent to a dipole–dipole repulsion in the E isomer, is probably the most important one: Wennerström, Forsén, et al., 1972.) (d) n–σ^* overlap in the Z conformer (Fig. 10.21; Kirby, 1983): One of the unshared pairs on the alkyl oxygen is in the R–C–O–R′ plane of the ester function and antiperiplanar to the C=O bond. While this pair is not involved in the p–π overlap, which contributes to the planarity of the ester function, it is properly disposed for overlap with the σ^* orbital of the C=O bond, thus stabilizing the Z conformation in a manner similar to that involved in the anomeric effect discussed earlier.

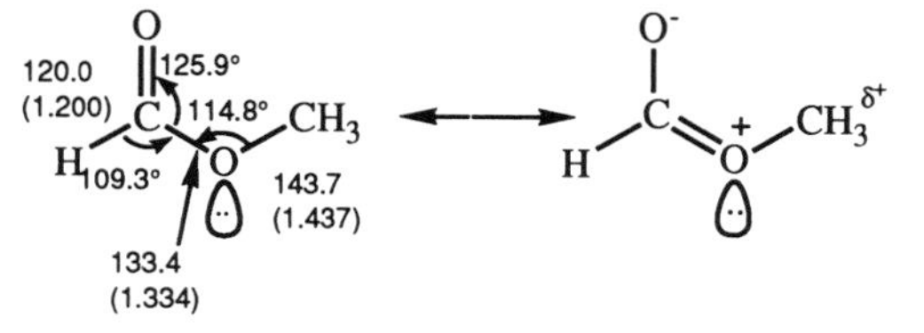

(a)   *(b)*

Figure 10.21. Geometry of methyl formate (bond lengths in picometers; values in parentheses are in angströms).

Acids, such as HCO_2H (Kwei and Curl, 1960) and CH_3CO_2H (cf. Radom, Pople, et al., 1972; Allinger and Chang, 1977, Wiberg and Laidig, 1987), also have a planar Z conformation with a substantial barrier to rotation. The $O=C/$ $C-H$ eclipsed conformation of the methyl group in acetic acid is preferred over the staggered by 0.48 kcal mol^{-1} (2.0 kJ mol^{-1}), similarly as in acetaldehyde (Fig. 10.18) except that the extent of the preference is less, presumably because the $C=O$ and $C-O$ bonds are more similar than $C=O$ and $C-H$. In contrast, the $H-C/C-O$ (single bond) eclipsed conformation in methyl acetate (corresponding to the bisecting conformation in Fig. 10.18) seems to be preferred over $O=C/C-H$ (double bond) by 0.31 kcal mol^{-1} (1.3 kJ mol^{-1}) (Williams et al., 1971). The barrier to rotation around the ester CH_3-O in this compound is 1.20 kcal mol^{-1} (5.0 kJ mol^{-1}), similar to that in methanol. The preference for the Z conformer (now reinforced, in comparison to methyl formate, by CH_3/CH_3 repulsion in the E conformation) is of the order of 8.5 kcal mol^{-1} (35.6 kJ mol^{-1}) (Blom and Günthard, 1981).

From the conformational aspect, amides are probably the most significant acyl derivatives, inasmuch as they serve as prototypes for the important class of peptides. Both formamide (Hirota et al., 1974; Kitano and Kuchitsu, 1974a) and acetamide (Kitano and Kuchitsu, 1973) are planar or nearly planar, as a result of

$$\text{extensive resonance of the } R-\overset{\overset{\displaystyle O^-}{|}}{C}=\overset{+}{N}H_2 \text{ type}$$

extensive resonance of the $R-C=\overset{+}{N}H_2$ type (cf. Section 2-4), with barriers to rotation in the 18–22 kcal mol^{-1} (75–92 kJ mol^{-1}) range (cf. Yoder and Gardner, 1981). The structures of N-methylformamide (Kitano and Kuchitsu, 1974b) and N-methylacetamide (Kitano, Kuchitsu et al., 1973) as determined by electron diffraction (gas phase) are shown in Figure 10.22. The extensive foreshortening of the $OC-NHCH_3$ bond suggests that this linkage has much double-bond character (the bond lengths in N-methylacetamide are similar to those in N-methylformamide). This point is underscored by the nearly 120° $C-N-C$ angle. The barrier

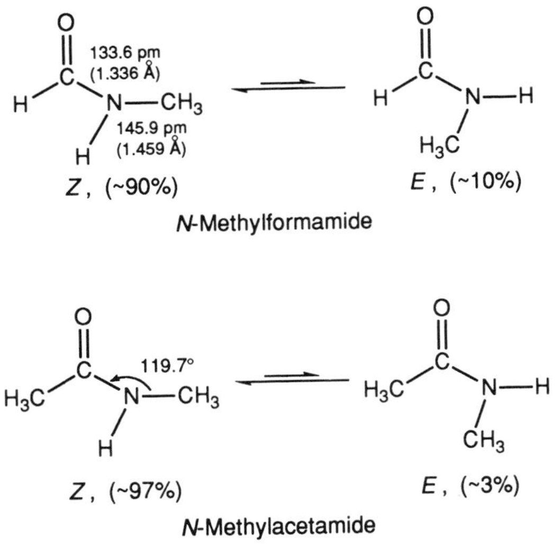

Figure 10.22. Structures of N-methylformamide and N-methylacetamide.

to OC–NC rotation is $20.6 \, \text{kcal mol}^{-1}$ ($86 \, \text{kJ mol}^{-1}$) for $HCONHCH_3$ and $21.3 \, \text{kcal mol}^{-1}$ ($89 \, \text{kJ mol}^{-1}$) for $CH_3CONHCH_3$ [these are free energies of activation; the enthalpies of activation are about $2 \, \text{kcal mol}^{-1}$ ($8 \, \text{kJ mol}^{-1}$) higher, the entropy of activation being positive] (Drakenberg and Forsén, 1971). The Z conformation is more stable than the E by 1.4–$1.6 \, \text{kcal mol}^{-1}$ (5.9–$6.7 \, \text{kJ mol}^{-1}$) in the case of the formamide and by 2.1–$2.5 \, \text{kcal mol}^{-1}$ (8.8–$10.5 \, \text{kJ mol}^{-1}$) in the case of the acetamide (cf. Drakenberg and Forsén, 1971) corresponding to the conformer composition (at room temperature) shown in Figure 10.22. Ab initio quantum mechanical calculations (Perricaudet and Pullman, 1973) are in agreement with the experimental findings. The explanation for the preferred Z conformation for esters in terms of n–σ^* overlap cannot apply to amides, since there are no free electron pairs in the amide plane (the pair involved with C=O is at right angles to that plane). Therefore the marked preference for the Z conformation in amides must be due to a combination of steric (attractive in Z, repulsive in E) and charge interaction factors. Because of strong hydrogen bonding, the conformational energies of N-alkylamides and peptides might be expected to be considerably affected by solvation and self-association.

We next consider unsaturated systems with conjugated double bonds, the simplest of which is 1,3-butadiene (Fig. 10.23). The most stable conformation of this molecule is the antiperiplanar or so-called s-trans conformer ($\omega = 180°$; s refers to the fact that one is dealing with conformation about single bonds rather than an alkene configuration), which has a maximum of conjugation and a minimum of steric interaction. There is a second, higher energy conformer that may be either gauche ($\omega = 60°$) or s-cis (synperiplanar, $\omega = 0°$), the gauche form suffering from some steric interaction and less than optimal orbital overlap, whereas the s-cis form has much better overlap but also greater steric repulsions. Calculations (Tai and Allinger, 1976; de Maré, 1981; Momicchioli et al., 1982) suggest either an s-cis or a gauche conformer 1.5–$2.6 \, \text{kcal mol}^{-1}$ (6.3–$10.9 \, \text{kJ mol}^{-1}$) above the s-trans with a barrier to conformational inversion of 5.5–$7.3 \, \text{kcal mol}^{-1}$ (23.0–$30.5 \, \text{kJ mol}^{-1}$), the decision as to whether this is an s-cis or a gauche conformer being made difficult by the shallow nature of the potential function in the 0–$60°$ region. Despite some doubts that had been raised as to the reality of this higher energy conformer (Bock et al., 1979) it has actually been isolated by a matrix isolation technique (Squillacote, Chapman, Anet, et al., 1979; see also Huber-Wälchli, 1978). A mixture of 1,3-butadiene and argon at

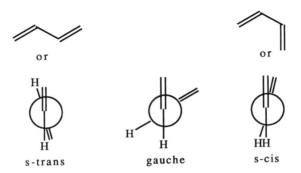

Figure 10.23. Conformations of 1,3-butadiene.

400–900°C was rapidly cooled to -243°C; the UV spectrum of the matrix so obtained displays a band not seen when the same experiment is carried out with vapor cooled from room temperature to -243°C. The nature of the UV absorption suggests that the band is due to the planar s-cis rather than the gauche isomer (but see below) and the rate of its disappearance when the matrix is warmed to -213°C leads to an activation energy of 3.9 kcal mol^{-1} (16.3 kJ mol^{-1}) for the s-cis $\rightleftharpoons$ s-trans interconversion. Raman spectroscopic studies (Carreira, 1975; Durig et al., 1975) imply a barrier in the opposite direction of 6.6–7.2 kcal mol^{-1} (27.7–30.1 kJ mol^{-1}) and a ground-state energy difference of 2.5–3.2 kcal mol^{-1} (10.5–13.4 kJ mol^{-1}) (see also Huber-Wälchli, 1978; Mui and Grunwald, 1982). Overall, these data are compatible with an s-cis conformer lying 2.9 ± 0.4 kcal mol^{-1} (12.1 ± 1.7 kJ mol^{-1}) above the s-trans with a barrier from the s-trans side of 6.8 ± 0.4 kcal mol^{-1} (28.5 ± 1.7 kJ mol^{-1}). Unfortunately, electron diffraction data (Kveseth, Kohl, et al., 1980) are reported not to be compatible with a second conformer of 1,3-butadiene, which lies less than 3.5 kcal mol^{-1} (14.6 kJ mol^{-1}) above the s-trans form; in light of all the other experimental and calculational results one must, however, question this conclusion. Calculations do suggest that the s-cis form is actually a saddle point between two nearby high minima (cf. Furukawa et al., 1983; Bock et al., 1985); Breulet, Schaefer, et al., (1984) and De Mare (1984) calculate a torsion angle of about 38° for these minima. The rotational barrier in styrene, 3.13–3.27 kcal mol^{-1} (13.1–13.7 kJ mol^{-1}) (Hollas and Ridley, 1980; Bock et al., 1985) is considerably smaller than that in 1,3-butadiene, presumably because of less effective π orbital overlap in the (thereby less stabilized) ground state.

Conformational preferences for other conjugated unsaturated molecules have been tabulated (Bastiansen et al., 1971, Table 5; Wilson, 1971; Kiss and Lukovits, 1979); we mention here only three examples: acrolein, acrylic acid, and methyl acrylate. Acrolein (propenal, Fig. 10.24, X = H) seems to exist largely in the sterically less crowded s-trans conformation with an energy difference between it and the less stable s-cis conformer of 1.7 kcal mol^{-1} (7.1 kJ mol^{-1}) and an energy barrier of 5.0–6.4 kcal mol^{-1} (20.9–26.8 kJ mol^{-1}) (Courtieu et al., 1974; Carreira, 1976; Blom and Bauder, 1982). The structure of both conformers has been determined accurately by microwave spectroscopy: Both structures are planar (Blom, Bauder, et al., 1982). In the case of acrylic acid (propenoic acid, Fig. 10.24, X = OH) and methyl acrylate (Fig. 10.24, X = OCH$_3$) there is evidently little steric difference between the s-cis and s-trans conformations; accordingly the energy difference is also small, 0.17 kcal mol^{-1} (0.71 kJ mol^{-1}) in acrylic acid (Bolton et al., 1974) (with a barrier of 3.8 kcal mol^{-1} or 16 kJ mol^{-1}) and 0.31 kcal mol^{-1} (1.3 kJ mol^{-1}) in methyl acrylate (George et al., 1972). Here, also, it is not clear whether the less stable conformer is s-cis or gauche. A similar situation pertains in acryloyl (propenoyl) chloride (Hagen and Hedberg, 1984).

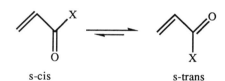

Figure 10.24. Conformation of conjugated unsaturated compounds.   s-cis   s-trans

Figure 10.25. Ground state in *p*-substituted benzaldehydes.

That the s-trans conformer is favored so little in these compounds suggests that it has little stereoelectronic advantage and that its substantial predominance in the acrolein case is due to steric reasons, that is, an unfavorable $=CH_2/O=$ interaction in the s-cis or gauche conformation (see also Section 9-2.b, p. 567). The fact that the barrier (from the s-trans side) in acrylic acid is lower than that in acrolein is in accord with the assumption of a sterically more congested ground state ($=CH_2/$ OH eclipsed) in the acid.

In benzaldehyde, as in acrolein, the planar conjugated conformation is preferred; measurements of the barrier to rotation out of the plane have given results of 7.6–7.9 kcal mol^{-1} (31.8–33.1 kJ mol^{-1}). Electron-donating para substituents increase the barrier, as would be expected from resonance considerations (Fig. 10.25; cf. Drakenberg et al., 1980) to 9.0–9.4 kcal mol^{-1} (37.7–39.3 kJ mol^{-1}) in the *p*-CH$_3$O and 9.9–10.8 kcal mol^{-1} (41.4–45.2 kJ mol^{-1}) in the *p*-(CH$_3$)$_2$N compound. It must be pointed out, however, that the absolute value for the rotational barrier in benzaldehyde is somewhat uncertain (cf. Anderson et al., 1984) and values as low as 4.6 kcal mol^{-1} (19.2 kJ mol^{-1}) have been reported (Fately et al., 1965); the value may be lower in the vapor phase than in the liquid. The barrier in furfural (Figs. 8.48 and 10.26) is somewhat higher, 10.9 kcal mol^{-1} (45.6 kJ mol^{-1}); here as in *o*- and *m*-substituted benzaldehydes, two stable planar conformations are possible, with the one in which the ring and carbonyl dipoles are opposed (**B**) being preferred in the vapor phase but the other (**A**) in polar solvents (cf. Abraham and Bretschneider, 1974; Abraham et al., 1982).

Conjugation also occurs with a cyclopropane ring. The barrier in methylcyclopropane, 2.86 kcal mol (12.0 kJ mol^{-1}), is quite similar to the ethane barrier, but that in cyclopropanecarboxaldehyde [Fig. 10.27; $V_2 = 4.39$ kcal mol^{-1} (18.4 kJ mol^{-1}); Volltrauer and Schwendeman, 1971] is closer to that in acrolein than to that in propionaldehyde (see above). The two stable conformations,

A   B   **Figure 10.26.** Conformers of furfural.

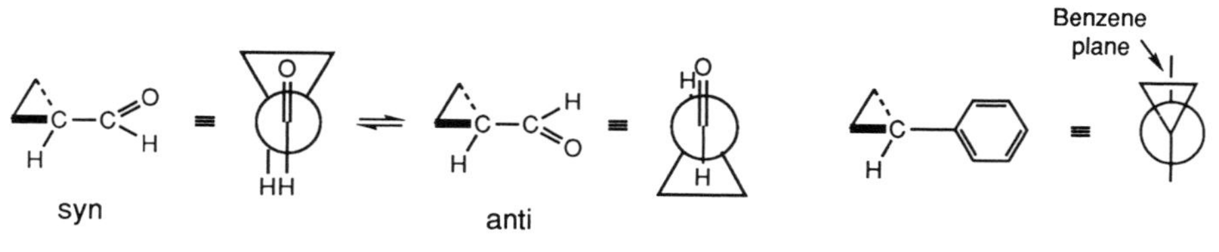

syn   anti

Figure 10.27. Cyclopropanecarboxaldehyde and cyclopropylbenzene.

nearly equal in energy, have the aldehyde hydrogen either synperiplanar or antiperiplanar to the cyclopropyl hydrogen on the adjacent carbon. Similarly, in cyclopropylbenzene (Fig. 10.27) the tertiary C–H bond is eclipsed with the phenyl plane, the barrier to rotation being of the order of 2.0 kcal mol^{-1} (8.4 kJ mol^{-1}) (Parr and Schaefer, 1977). The explanation may be found in the Walsh picture of cyclopropyl bonding (Walsh, 1949; cf. Van-Catledge et al., 1982), which ascribes near sp^2 character to the cyclopropyl hydrogen atoms and places p orbitals in the plane of the cyclopropyl ring (contributing to the ring C–C bonds) and at right angles to the plane of the H–C–H or (exo-C)–C–H bonds. The stable conformations of the molecules shown in Figure 10.27 are such that this p-orbital plane extends in the direction of the filled C=O or aromatic orbitals for maximum overlap, thus explaining both the height and the mainly twofold nature of the barrier.

b. Alkylbenzenes

We now turn to hydrocarbons with aromatic substituents. One might perhaps have expected the barrier to methyl rotation in toluene to be similar to that in propene. However, this is not the case: Methyl rotation in toluene is nearly free, the barrier being only 14 cal mol^{-1} (59 J mol^{-1}) (Pitzer and Scott, 1943; Rudolph et al., 1967; cf. Lambert et al., 1981). Unlike the twofold V_2 and threefold V_3 barriers encountered so far, that in toluene (Fig. 10.28) is sixfold, as a result of the presence of the local C_2 axis in the plane of the aromatic ring. Sixfold barriers, in general, tend to be very low; for example, that in nitromethane, which also has a local C_2 axis along the C–N bond of the nitro group, is about 6 cal mol^{-1} (25 J mol^{-1}) (cf. Lowe, 1968).

The barriers in o-xylene, in higher alkylbenzenes, and in other substituted aromatics are often of the V_3 type (compounds of type $ArCX_3$, such as t-butylbenzene, are exceptions) and the barriers are therefore substantially higher than in toluene (e.g. 1.49 kcal mol^{-1}, 6.23 kJ mol^{-1} in o-xylene, Rudolph et al., 1973), though not high enough to study by low-temperature NMR spectroscopy. As we shall see in Section 10-4.a, whether spectral transitions from distinct conformations of a molecule can be seen separately or whether experimental data from (mole fraction weighted) average conformations are obtained, depends on the relation between the rate of interconversion of the various conformations and the time scale of the observation. It is therefore fortunate that supersonic jet mass resolved excitation spectroscopy (MRES) has been applied to the study of a number of substituted aromatics (Seeman, Bernstein et al., 1989). MRES uses optical absorption processes ($S_1 \leftarrow S_0$ transitions), which occur much more rapidly than nuclear motion (Franck–Condon principle). In addition, the expansion of

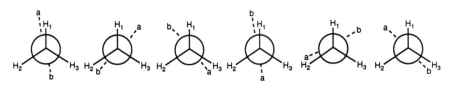

Figure 10.28.   Toluene conformations. (a---b stands for an "oriented" benzene ring plane.)

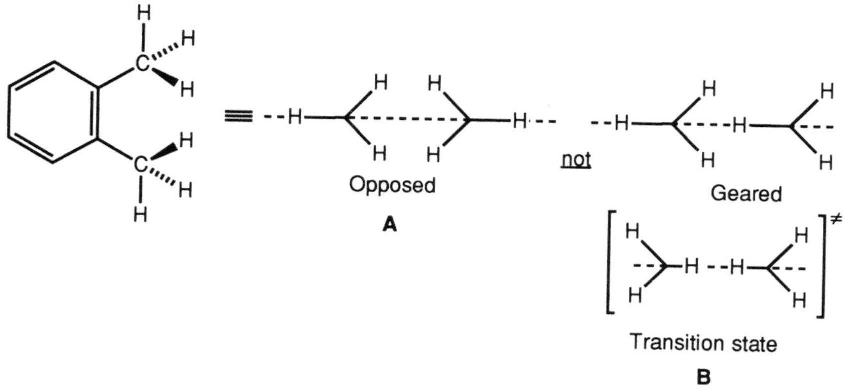

Figure 10.29. Stable conformation of o-xylene.

the molecules through the supersonic jet nozzle leads to very substantial cooling (molecules at near zero K); thus the resolution of the experiment is excellent [< 3 cal mol^{-1} (12.5 J mol^{-1})]. In principle, each conformation of a molecule has its own spectral properties, and the spectroscopic transitions can be seen for each component of a multicomponent (multiconformational) system.

MRES was used to distinguish between the "opposed" and "geared" conformations of o-xylene (Fig. 10.29; Seeman, Bernstein et al., 1992). For o-xylene-α-d_1, spectra for two species were observed; for o-xylene-α,α'-d_2, spectra for four species were observed. These observations are consistent only with the isotopomers possible for the "opposed" conformation **A** (**B**, which would also be compatible with the spectral data, is calculated to represent an energy maximum rather than a stable conformer).

MRES has been applied to other aromatic substituents. Consider ethylbenzene (Breen, Bernstein, and Seeman, 1987; Fig. 10.30). In principle, the molecule may be most stable in the "gauche" **A**, "perpendicular" **B**, or "eclipsed" **C** conformation. An assignment was reached on the basis of the MRES spectra. For ethylbenzene itself, a spectrum consistent with a single conformation was observed, indicating that only one of the **A–C** conformations is present. For 1,3-diethylbenzene, conformation **B** would give rise to two conformers, one (meso) with both methyl groups up (or down, the two possibilities are identical) or the other (chiral) with one methyl group up, the other down (a diastereomeric situation). Conformation **C**, on the other hand, can give rise to three conformers, one with both methyl groups pointing inward, a second with both pointing

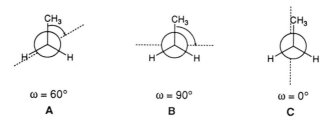

Figure 10.30. Ethylbenzene conformations (benzene plane dashed).

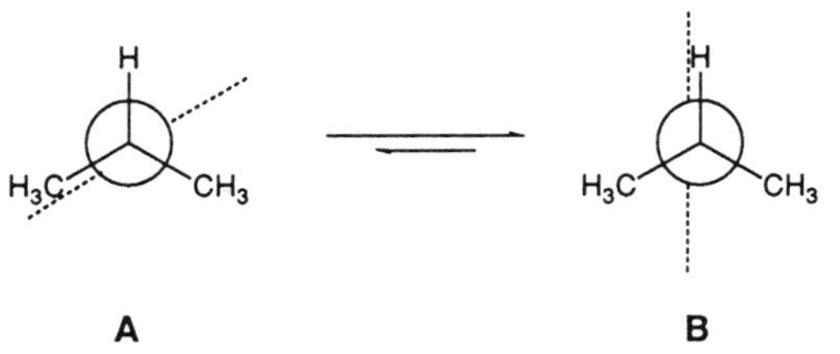

Figure 10.31. Isopropylbenzene conformations (benzene plane dashed).

outward, and the third with one in and the other out. Conformation **A** of 1,3-diethylbenzene can exist in six stereoisomeric forms. As the MRES showed two conformations, the perpendicular conformer **B** is in accordance with the spectral evidence for ethylbenzenes.

That **B** is the most stable conformation is also borne out by calculation (Kříž and Jakeš, 1972; Umeyama and Nakagawa, 1979) although the potential curve in the $\omega = 60°$–$90°$ region seems to be quite shallow (as might be expected by analogy with toluene). Conformation **C** seems to correspond to the energy maximum [barrier 1.16 kcal mol^{-1} (4.85 kJ mol^{-1}); Miller and Scott, 1978]. The conformations of benzyl alcohol (Im, Bernstein, Seeman et al., 1991) and of benzylamine (Li, Bernstein, Seeman et al., 1991) are analogous to that of ethylbenzene (OH or NH$_2$ perpendicular to the plane of the ring).

Isopropylbenzene (Fig. 10.31) is found, by arguments similar to those used for ethylbenzene, to be most stable in the double gauche conformation **B** by MRES (Seeman et al., 1989). The eclipsed conformer **A** is found, by microwave spectroscopy, to be 0.8 kcal mol^{-1} (3.3 kJ mol^{-1}) higher than **B** (Štokr et al., 1972). In both ethylbenzene and isopropylbenzene the conformation with a methyl group eclipsed with the benzene ring is the least stable of those shown, in contrast to the situation in 1-butene (see above).

Other substituted aromatics of particular interest are propylbenzene, anisole (methoxybenzene), and styrene (vinylbenzene). Propylbenzene might, in principle, exist as an eclipsed conformer (Fig. 10.32, **A**) stabilized by C$_6$H$_5$/CH$_3$ attraction (cf. Hirota et al., 1983) or as staggered anti (**B$_i$**) or gauche (**B$_{ii}$**, **B$_{iii}$**) conformers. The observation of two conformers for *n*-propylbenzene (Breen, Bernstein, Seeman, et al., 1987) is not conclusive; they might be either anti and gauche or anti and eclipsed, since the two gauche conformers (**B$_{ii}$**, **B$_{iii}$**) are

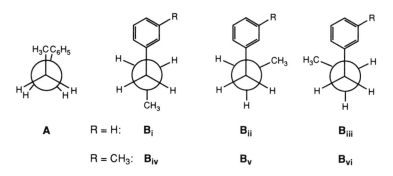

Figure 10.32. Stable conformations of propylbenzene.

enantiomeric. The answer (Breen, Bernstein, Seeman et al., 1987) comes from the study of the MRES spectrum of m-methyl-n-propylbenzene, which shows *three* species to be present that must be the anti ($\mathbf{B}_{iv}$) and the two diastereomeric gauche ($\mathbf{B}_v$, $\mathbf{B}_{vi}$) conformations shown in Figure 10.32. (It must be recalled that, on the basis of earlier evidence (see above), rotation about the $C_6H_5–C_\alpha$ bond is slow on the time scale of the MRES experiment and that the $C_{ortho}–C_{ipso}–C_\alpha–C_\beta$ torsion angle is 90°.)

The situation in anisole is different from that in alkylbenzenes in that the methyl group is in the plane of the aromatic ring (presumably because this favors overlap of the p electrons on oxygen with the π electrons of the ring; Radom, Pople, et al., 1972). Thus p-dimethoxybenzene (in contrast to p-diethylbenzene) shows the presence of *two* conformers (Fig. 10.33, **A** and **B**) both in the vapor phase (by MRES, Breen, Bernstein, Seeman, et al., 1989) and in the liquid phase (by Raman spectroscopy, Tylli et al., 1985), whereas the crystalline solid is entirely in conformation **A** (Goodwin, Robertson, et al., 1950).

Styrene, like anisole, has a planar conformation for similar reasons (optimal $\pi–\pi$ electron overlap). This fact has been confirmed by MRES (Grassian, Bernstein, Seeman, et al., 1989) through studying of p-ethylstyrene (one conformer), m-methylstyrene (two conformers), and p-methyl-*trans*-β-methylstyrene (two conformers). In contrast, data for α-methylstyrene (Grassian, Bernstein, Seeman et al., 1990) suggest that in this molecule the alkene plane is inclined by about 30° with respect to the benzene plane, presumably in order to avoid steric interaction of the α-methyl group with one of the ortho hydrogen atoms. This subject will be returned to in Chapter 14, where it will be shown that replacing the ortho hydrogen atoms by methyl groups increases the steric hindrance sufficiently to make isolation of the (nonplanar) enantiomers possible.

Proceeding now to triply bonded species, it is clear that the barrier in methylacetylene, $H_3C–C{\equiv}C–H$ must be zero since the molecule has $C_{\infty v}$ symmetry, and therefore any arrangement of the methyl top is equivalent to any other. The situation is not necessarily the same in dimethylacetylene ($H_3C–C{\equiv}C–CH_3$) where the rotations of the two methyl groups need not be independent of each other; nevertheless the barrier in this and similar molecules is very small, less than $30\ \mathrm{cal\ mol^{-1}}$ ($125\ \mathrm{J\ mol^{-1}}$) (cf. Lowe, 1968).

c. Miscellaneous Compounds

The barrier in methanol (Table 10.3) is only about one-third of that in ethane and that in methylamine is about two-thirds. These results are most easily rationalized

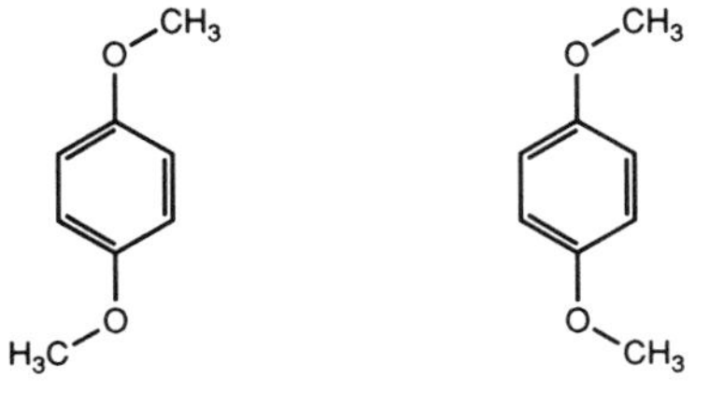

A   **B**

Figure 10.33. Stable conformations of p-dimethoxybenzene.

TABLE 10.3. Rotational Barriers

Compound	Barrier (kcal mol⁻¹)	(kJ mol⁻¹)	Source
H_3C-CH_3	2.91	12.2	(see p. 599)
H_3C-OH	1.07	4.48	Lees and Baker, 1968
H_3C-NH_2	1.96	8.20	Takagi and Kojima, 1971
$H_3C-C_2H_5$	3.4	14.2	(see p. 600)
H_3C-OCH_3	2.7	11.3	Lowe, 1968
$H_3C-OC_2H_5$	2.61	10.9	Durig and Compton, 1978
$H_3C-NHCH_3$	3.62	15.1	Lowe, 1968
$H_3C-NHC_2H_5$	3.12	13.1	Durig and Compton, 1979
$H_3C-SC_2H_5$	2.05	8.57	Durig et al., 1979
$HO-OH$	1.1	4.6	Lowe, 1968
H_2N-NH_2	3.15	13.2	Lowe, 1968

by assuming that there is no bond eclipsing (Pitzer) strain for the opposition of a bond with a lone pair; thus in methanol there is only one set of H/H eclipsed bonds and in methylamine there are two, compared to three in ethane. In contrast, the barrier to methyl rotation (Table 10.3) in dimethyl ether, in ethyl methyl ether (H_2C-O rotor), in dimethylamine and in ethylmethylamine (H_3C-N rotor) are not much lower than the propane barrier; at first sight, this might seem to contradict the assumption that the eclipsing of hydrogen atoms with lone pairs does not cost energy. However, it must be remembered that in ethers and amines, because of the relatively short C–N and C–O distances (relative to C–C), the $H-C-X-C$ (X = O or NR) distance in the eclipsed form is considerably less than in propane (X = CH_2). There may thus be a rather important steric component to the barrier in this H/lone pair eclipsing energy, which fortuitously makes the barriers in the amines and ethers similar to those in propane. The barrier in ethyl methyl sulfide (long C–S bond!) is, in fact, lower than that in the corresponding ether and amine.

The last two molecules to be discussed here (see also Table 10.3) are hydrogen peroxide and hydrazine. In hydrogen peroxide (H–O–O–H) the dihedral angle between the H–O–O planes is near 120° (Oelfke and Gordy, 1969; Bair and Goddard, 1982). In hydrazine, the preferred conformation has an :–N–N–: torsion angle of 91° (Kohata, Kuchitsu, et al., 1982). On the other hand, dialkyl peroxides, such as CH_3O-OCH_3 (Kimura and Osafune, 1975; Rademacher and Elling, 1979) and t-BuO–OBu-t (Käss et al., 1977), have torsion angles of 166–180°, that is the alkyl groups are antiperiplanar. The reason generally advanced for the anticlinal (cf. Fig. 2.13) preference in H_2O_2 and H_2NNH_2 is that in the antiperiplanar conformation there would be an unfavorable orbital interaction of the antiparallel unshared electron pairs (see also Fink and Allen, 1967; Wolfe, 1972). However, this does not explain the variability of the torsion angle nor the fact that RO–OR does, in fact, seem to be nearly antiperiplanar when R = alkyl [when R = $(CH_3)_3Si$ the angle is 143.5°, Käss et al., 1977]. An attractive interaction between the (partially positive) hydrogen atoms and the lone pairs may be a contributing factor, at least in H_2O_2 (Bair and Goddard, 1982) but does not lead to the expected eclipsed conformation in H_2N-NH_2.

There are many molecules other than the few representative ones mentioned here whose conformation has been determined; extensive tabulations have been made by Lowe (1968), Wilson (1971), and George and Goodfield (1980).

10-3. DIASTEREOMER EQUILIBRIA IN ACYCLIC SYSTEMS

Information concerning equilibria between acyclic diastereomers is still quite limited. In principle, if one knows the conformational makeup of each diastereomer, it is possible to calculate the enthalpy and entropy difference between them, and hence the position of equilibrium. We shall illustrate this point with 2,3-dichloro- and 2,3-dibromobutane for which both the underlying conformational energies and some experimental thermochemical results are known.

> We shall assume that the bond energies for stereoisomers are the same and that the energy differences may be entirely ascribed to differences in conformational energy. This statement may be taken as a definition of conformational energy: It is the total energy (or better enthalpy) of formation of conformers minus the (constant) sum of their bond energies. It must be recognized that in some instances, such as s-cis and s-trans conformers of conjugated dienes, the conformational energy so defined may include a major electronic contribution.

We calculate the conformational energies of the individual conformers from known gauche interaction energies, either by (rather inaccurate) summation or (better) by a force field calculation that includes energy minimization. The conformational enthalpy of a given configurational isomer that exists in conformations **A**, **B**, and **C** will be

$$H_{\text{confo}} = n_{\text{A}} \cdot H_{\text{A}} + n_{\text{B}} \cdot H_{\text{B}} + n_{\text{C}} \cdot H_{\text{C}}$$

or, in general,

$$H_{\text{confo}} = \sum_i n_i H_i \tag{10.1}$$

where n_{A}, n_{B}, n_{C}, and so on, stand for the mole fractions of the individual conformers and H_{A}, H_{B}, H_{C}, and so on stand for their conformational enthalpies. The mole fractions of the individual conformers can be computed from their enthalpies (assuming the value of the partition function is unity); for example,

$$\frac{n_{\text{A}}}{n_{\text{B}}} = e^{(H_{\text{B}} - H_{\text{A}})/RT} . \tag{10.2}$$

The two equations can be combined (Schneider, 1972) in the form

$$H_{\text{confo}} = \frac{[H_{\text{A}} + H_{\text{B}} \cdot e^{(H_{\text{A}} - H_{\text{B}})/RT} + H_{\text{C}} \cdot e^{(H_{\text{A}} - H_{\text{C}})/RT}]}{[1 + e^{(H_{\text{A}} - H_{\text{B}})/RT} + e^{(H_{\text{A}} - H_{\text{C}})/RT}]} . \tag{10.3}$$

To obtain the entropy difference between stereoisomers (disregarding differences in vibrational and rotational entropy, which are generally small and, especially in the case of the vibrational entropy, often difficult to assess), one calculates the entropy of mixing for each stereoisomer by the formula

$$S_{mix} = -R(n_A \ln n_A + n_B \ln n_B + n_C \ln n_C)$$

or, in general,

$$S_{mix} = -R \sum_i n_i \ln n_i \qquad (10.4)$$

To this must be added differences in entropy due to symmetry $(-R \ln \sigma)$ (σ is the symmetry number; see Chapter 4) and due to the existence of pairs of enantiomers in racemates $(R \ln 2)$.

> Neglect of differences in vibrational and rotational entropies introduces inaccuracies into this treatment (cf. Reisse, 1971). Further inaccuracies are introduced by assuming that all molecules are in the lowest vibrational energy state and that there are no differences in zero-point energy between conformers. The treatment breaks down when one or more of the isomers to be compared have shallow potential curves, that is, if a number of adjacent conformers are separated by low potential barriers. It works best when the potential wells are deep.

In the case of the 2,3-dihalobutanes (cf. Eliel, 1985), one recognizes that the meso isomer has three conformations, the symmetrical one **A** and the enantiomeric (and hence equienergetic) chiral conformations **B** and **C** (Fig. 10.34), whereas the chiral isomer (only one enantiomer shown) has three different conformations, **D**, **E**, and **F**. We shall assume the following interactions: CH_3/CH_3, gas, 0.9 kcal mol^{-1} (3.8 kJ mol^{-1}); liquid, 0.6 kcal mol^{-1} (2.5 kJ mol^{-1}) (cf. Section 10-1); Br/Br, gas, 1.6 kcal mol^{-1} (6.7 kJ mol^{-1}); liquid, 0.9 kcal mol^{-1} (3.8 kJ mol^{-1}); CS$_2$ solution, 1.1 kcal mol^{-1} (4.6 kJ mol^{-1}); Cl/Cl, gas, 1.2 kcal mol^{-1} (5.0 kJ mol^{-1}); liquid, 0.3 kcal mol^{-1} (1.25 kJ mol^{-1}); CS$_2$ solution, 0.9 kcal/mol^{-1} (3.8 kJ mol^{-1}) (Abraham and Bretschneider, 1974; the value for the "liquid" is, strictly speaking, the value for 1,2-dichloroethane solutions for X = Cl and 1,2-dibromoethane solutions for X = Br, since the values

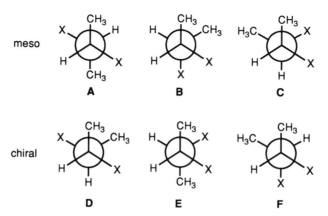

Figure 10.34. Conformers of meso and chiral 2,3-dihalobutanes.

are taken from the conformational ΔH^0 for 1,2-dichloro- and 1,2-dibromo-ethanes as pure liquids); CH_3/Cl, gas, $-0.5\,kcal\,mol^{-1}$ ($-2.1\,kJ\,mol^{-1}$); liquid or solution, $-0.2\,kcal\,mol^{-1}$ ($-0.8\,kJ\,mol^{-1}$); CH_3/Br, gas, $-0.3\,kcal\,mol^{-1}$ ($-1.25\,kJ\,mol^{-1}$); liquid or solution, $0.0\,kcal\,mol^{-1}$ ($kJ\,mol^{-1}$) (these values, though based on experimental data, involve some uncertainty). Based on these values, one calculates the conformational enthalpies of the various conformers for $X = Cl$ and $X = Br$ in the gas, pure liquid, and CS_2 solution ($\varepsilon = 2.6$) given in Table 10.4; from these, in turn, the mole fractions of the various conformers indicated in Table 10.4 are calculated by means of Eq. 10.2. (see also Chia, Huang et al., 1973.)

A vibrational analysis of the 2,3-dichlorobutanes (Jing and Krimm, 1983) suggests crystalline *meso*-2,3-dichlorobutane to exist entirely in the anti conformation **A** (Fig. 10.34) with the gauche conformers **B** and **C** appearing in the liquid. In the chiral isomer, conformer **E** represents the crystalline solid with substantial amounts of conformer **D** appearing on melting. These data support the calculated results for n_x in Table 10.4 as far as the liquid phase is concerned.

From these data, using Eq. 10.1, one may calculate the values of the total H_{confo} for the meso and chiral dichlorides and dibromides and by Eq. 10.4 one may calculate the corresponding conformational entropies. These values are summarized in Table 10.5, as are the corresponding enthalpy and entropy differences between the diastereomers.

The experimental values reported by Rozhnov et al., (1975) shown in the last two columns of Table 10.5 are in reasonable agreement with calculation except for the gas-phase value for the dibromide (Rozhnov and Nesterova, 1973); the same is true for the values reported by Schneider (1972) in CCl_4. Agreement of the calculated with the experimental data might, at first sight, seem surprising, since the racemic isomers are favored by an entropy of mixing of $R\ln 2$. However, this entropy advantage is offset by the entropy of symmetry of this isomer, which has a C_2 axis (hence $\sigma = 2$ and $S_{sym} = -R\ln 2$); thus the net entropy difference between the racemic and meso isomers due to entropy of mixing of the racemic pair and entropy of symmetry vanishes.

It must, however, be conceded that the calculated differences are not very sensitive to the conformational interactions taken as the basis of the calculations. For example Schneider (1972), using different interaction values, calculates a somewhat different conformational composition in the liquid dibromide ($n_D = 0.40$, $n_E = 0.48$, $n_F = 0.12$; cf. Table 10.4); yet his calculated ΔH^0 of $-0.57\,kcal\,mol^{-1}$ ($2.4\,kJ\,mol^{-1}$) is virtually the same as the one in Table 10.5.

On the other hand, the actual conformational composition (Fig. 10.34) is very sensitive to the input conformational enthalpies; it is therefore perhaps not so surprising that the data in Table 10.4 agree rather poorly with experimental data obtained by NMR (Bothner-By and Naar-Colin, 1962; Anet, 1962) and IR spectroscopy (Iimura et al., 1969; Iimura, 1969); the latter suggests, in CS_2 solution, $n_A = 0.7$, $n_B = n_C = 0.15$, $n_D = 0.52$, $n_E = 0.26$, $n_F = 0.22$ for the dichloride; corresponding values for the dibromide are $n_A = 0.75$, $n_B = n_C = 0.125$, $n_D = 0.64$, $n_E = 0.23$, $n_F = 0.13$. In particular, the mole fractions of **B** and **C** and of **F** are appreciably larger than calculated. One might think that the discrepancy is

TABLE 10.4. Conformational Enthalpies[a] (H) and Mole Fractions (n_x) of Conformers **A–F**,[b] in meso and chiral 2,3-Dichloro and 2,3-Dibromobutane in Gas, Liquid, and CS_2 Solution

	A Cl	A Br	B Cl	B Br	C Cl	C Br	D Cl	D Br	E Cl	E Br	F Cl	F Br
H_{confo}												
Gas	-1.0(-4.2)	-0.6(-2.5)	1.6	2.2	1.6(6.7)	2.2(9.2)	-0.1(-0.4)	0.3(1.25)	0.2(0.8)	1.0(4.2)	2.1(8.8)	2.5(10.5)
Liquid	-0.4(-1.7)	0.0(0.0)	0.7	1.5	0.7(2.9)	1.5(6.3)	0.2(0.8)	0.6(2.5)	-0.1(-0.4)	0.9(3.8)	2.1(3.8)	1.5(6.3)
CS_2	-0.4(-1.7)	0.0(0.0)	1.3	1.7	1.3(5.4)	1.7(7.1)	0.2(0.8)	0.6(2.5)	0.5(2.1)	1.1(4.6)	1.5(6.3)	1.7(7.1)
c	0[d](0)[d]	0[d](0)[d]	1.55	1.51	1.55(6.5)	1.57(6.6)	0[d](0)[d]	0[d](0)[d]	0.29(1.2)	1.07(4.5)	1.35(5.6)	1.8(7.5)
n_x												
Gas	0.976	0.982	0.012	0.009	0.012	0.009	0.615	0.751	0.370	0.231	0.015	0.018
Liquid	0.762	0.862	0.119	0.069	0.119	0.069	0.337	0.549	0.559	0.331	0.104	0.120
CS_2	0.898	0.898	0.051	0.051	0.051	0.051	0.583	0.631	0.352	0.271	0.065	0.098
c	0.872	0.876	0.064	0.062	0.064	0.062	0.597	0.826	0.366	0.136	0.037	0.038

[a] Units are in kilocalories per mole (kilojoules per mole).
[b] See Figure 10.34.
[c] From force field calculations in medium of dielectric constant ε = 1.5; chloride: Meyer, Allinger, et al. (1980); bromide: Meyer and Ohmichi (1981).
[d] Arbitrarily assumed.

TABLE 10.5. Conformational Enthalpies[a] and Entropies[b] and Their Differences for meso and racemic 2,3-Dichloro- and 2,3-Dibromobutanes in Gas, Liquid, and CS_2 Solution[c]

	meso Cl	meso Br	racemic Cl	racemic Br	ΔH^0 Calculated Cl	ΔH^0 Calculated Br	ΔH^0 Experimental Cl	ΔH^0 Experimental Br
ΔH_{confo}								
Gas	-0.94(-3.9)	-0.55(-2.3)	0.04(0.17)	0.50(2.09)	-0.98(-4.10)	-1.05(-4.39)		-0.30[d](-1.26)[d]
Liquid	-0.14(-0.59)	0.21(0.88)	0.11(0.46)	0.80(3.35)	-0.25(-1.05)	-0.59(-2.47)	-0.27(-1.13)	-0.43(-1.80)
CS_2	-0.23(-0.96)	0.17(0.71)	0.39(1.63)	0.84(3.51)	-0.62(-2.59)	-0.67(-2.80)		-0.57[e](-2.38)

	meso Cl	meso Br	racemic Cl	racemic Br	ΔS^0 Calculated Cl	ΔS^0 Calculated Br	ΔS^0 Experimental Cl	ΔS^0 Experimental Br
ΔS_{confo}								
Gas	-0.26(-1.09)	-0.20(-0.84)	-1.45(-6.07)	-1.24(-5.19)	-1.19(-4.98)	-1.04(-4.35)		-0.56(-2.34)
Liquid	-1.42(-5.94)	-0.99(-4.14)	-1.84(-7.70)	-1.89(-7.91)	-0.42(-1.76)	-0.90(-3.77)	-0.69(-2.89)	
CS_2	-0.79(-3.31)	-0.79(-3.31)	-1.71(-7.15)	-1.73(-7.24)	-0.92(-3.85)	-0.94(-3.93)		-0.70[e](-2.93)[e]

ΔH^0 (racemic ⇌ meso) $\qquad$ ΔS^0 (racemic ⇌ meso)

[a] Values are given in kilocalories per mole (kilojoules per mole).
[b] Values are given in calories per mole per kelvin (joules per mole per kelvin).
[c] Data from Rozhnov et al., (1975) except as noted.
[d] Rozhnov and Nesterova (1973).
[e] Schneider (1972), in CCl_4.

due to the crudeness of the calculation, but data obtained by (presumably more accurate) force field calculations (albeit for a dielectric constant of 1.5), shown in the last line of Table 10.4, provide no improvement for the meso isomers and, if anything, give worse agreement with the IR data in the case of the racemate. A further pall on the calculation is thrown by the low-temperature ^{13}C NMR results of Schneider et al. (1979), which suggest the presence of 44% **E** for the dichloride and 18% for the dibromide at 133 K, whereas the data in Table 10.4, calculated for that temperature in CS_2 solution, would predict 24.2% dichloride and 12.9% dibromide in conformation **E**.

Among the relatively few diastereomers whose equilibria have been studied are the 2,3-dimethylsuccinic acids (Figure 10.34, $X = CO_2H$). In contrast to the 2,3-dihalobutanes, the dicarboxylic acids display a bias in favor of the racemic pair: $K_{rac/meso} = 1.9$ in 1,4-dioxane and 1.1 in water (Eberson, 1959). In the 2,3-dihalobutanes, the sizeable enthalpic advantage of the meso isomer at equilibrium rested on the dominance of conformer **A** (Fig. 10.34) and on the relative greater stability of **A** over, say, **D**, which has gauche methyl groups, and **E**, which has gauche halogens, that make up the chiral isomer. In the succinic acids, as we have already seen (p. 610; see also ul Hasan, 1980), the gauche conformer (carboxyl groups gauche) is preferred over the anti, at least at low pH. Therefore the meso isomer of the dimethyl homologue should be dominated by conformers **B** and **C** (Fig. 10.34) and the chiral isomer by **E** and possibly **F**. But conformer **E** should be more stable than **B** or **C** since its methyl groups are positioned anti rather than gauche. Thus the preference for the racemic pair over the meso isomer at equilibrium is perhaps not unexpected, despite the fact that the meso compound here gains from entropy of mixing, since the now populous conformers **B** and **C**, being mirror images, must contribute equally. With this reasoning one would predict that in basic solution the salt of the meso acid should be more stable, since at high pH conformers **A** and **D** (Fig. 10.34, $X = CO_2^-$) would be expected to predominate (cf. p. 610), with **A** being more stable than **D**.

Equilibria in 2,4-disubstituted pentanes ($CH_3CHXCH_2CHXCH_3$) have been studied to provide data for the understanding of stereoregularity (cf. Bovey, 1969; Farina, 1986) in polymers of the type $\{CH_2-CHX\}_n$, and we conclude this section with a discussion of the corresponding chlorides (chiral and *meso*-2,4-dichloropentanes) and bromides. The nine (3×3) conformers of the meso isomer (cf. Fig. 10.5) may be grouped into three achiral conformations (C_s) and three pairs of enantiomers (C_1). In the chiral isomer there is a degeneracy that restricts the number of conformers to six, three of which are C_2 and three C_1. However, most of these conformations have either two methyl groups or two chlorine atoms or a methyl and a chlorine in g^+g^- positions, and are therefore of high energy (cf. p. 603). After they are eliminated, only four conformations, two for the chiral and a pair of enantiomers for the meso isomer remain; these are shown in Figure 10.35. Conformer **A** of the chiral isomer stands out in that it has no butane–gauche interactions; it should therefore be the most stable. Next comes the enantiomer pair **B** of the "meso" isomer that has one butane–gauche interaction, and last conformer **C** of the chiral isomer that has two such interactions; thus one would expect the chiral isomer to have the lowest enthalpy and possibly also the lowest free energy, though the latter is less certain.

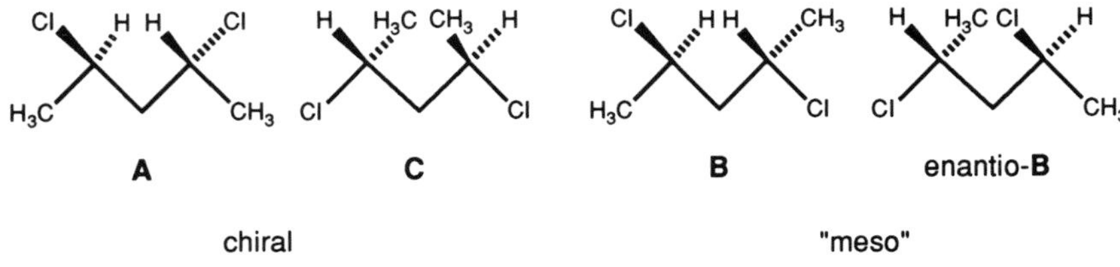

A   **C**   **B**   enantio-**B**

chiral   "meso"

Figure 10.35. Significant conformations for $CH_3CHClCH_2CHClCH_3$.

The so-called "meso" isomer has the entropy advantage since its two equally populated conformers in fact constitute a racemic pair with an entropy of mixing of $R \ln 2$; this offsets the entropy of mixing of the "real" racemic pair, and the latter then loses out, entropywise, since its entropy of symmetry, $-R \ln 2$ ($\sigma = 2$ for both C_2 conformers, **A** and **C**) is not offset by the entropy of mixing of **A** with the (considerably less populous) **C**.

Force field calculations for the dichloride (Boyd and Kesner, 1981) suggest conformational energies of $-0.86 \, \text{kcal mol}^{-1}$ ($-3.6 \, \text{kJ mol}^{-1}$) for **A**, $-0.57 \, \text{kcal mol}^{-1}$ ($-2.4 \, \text{kJ mol}^{-1}$) for **B**, and $+0.36 \, \text{kcal mol}^{-1}$ ($1.5 \, \text{kJ mol}^{-1}$) for **C**. From this, by a calculation of the type explained earlier for the 2,3-dihalobutanes, the enthalpy of the chiral isomer **A** plus **C** is computed, at 343 K, to be $0.69 \, \text{kcal mol}^{-1}$ ($2.9 \, \text{kJ mol}^{-1}$) and the enthalpy difference between the diastereomers as $-0.115 \, \text{kcal mol}^{-1}$ ($-0.48 \, \text{kJ mol}^{-1}$). However, the racemic pair has an entropy disadvantage of $-0.57 \, \text{cal mol}^{-1} \text{K}^{-1}$ ($-2.4 \, \text{J mol}^{-1} \text{K}^{-1}$) at 343 K (see above, $-1.38 \, \text{cal mol}^{-1} \text{K}^{-1}$ entropy of symmetry offset by $0.81 \, \text{cal mol}^{-1} \text{K}^{-1}$ entropy of mixing) and hence, at 343 K, its free energy exceeds that of the meso isomer by $0.57 \times 343 - 115$ or $80 \, \text{cal mol}^{-1}$ ($335 \, \text{J mol}^{-1}$) corresponding to 53% meso and 47% racemic isomer at equilibrium at 343 K. Unfortunately this is not in agreement with the experimental data; Flory and Williams (1969), by equilibrating the chlorides at 70°C (343 K) with LiCl in dimethyl sulfoxide (DMSO), obtained a mixture of 63.6% racemic and 36.4% meso isomer. The discrepancy is even worse than it appears here, since force field calculation for a high-dielectric solvent (as DMSO) suggests that the meso isomer, with its higher dipole moment, should be favored even more. Evidently, some of the parameters used in the calculation are not adequate, but the interesting qualitative point remains that in the 2,4-dihalopentane case, unlike with the 2,3-dihalobutanes, the racemic pair is more stable than the meso isomer and that this can be rationalized by contemplating the relative stability of conformers **A** and **B** in Figure 10.35.

10-4. PHYSICAL AND SPECTRAL PROPERTIES OF DIASTEREOMERS AND CONFORMERS

a. General

Before entering into the subject matter of the physical and spectral properties of acyclic diastereomers, we must recall the discussion concerning isomers given in

Chapter 3. With acyclic compounds, in comparing diastereomers, one is actually comparing conformer mixtures. If, as frequently happens, the time scale of the experiment is slow relative to that of conformational change, no individual conformers are detected and the measurement yields only the ensemble average for all conformers. The thermochemical and thermodynamic measurements described in the previous section are of this type, as are virtually all chemical experiments, plus measurements of such bulk properties as vapor pressure, boiling point, density, refractive index, dipole moment, viscosity, optical rotation, optical rotatory dispersion, and many others. But if the time scale of the experiment is short relative to that of conformer interconversion, individual conformers will be detected. Many of the measurements discussed in Sections 10-1 and 10-2 are of this type, notably measurements of microwave spectra, IR spectra, Raman spectra, UV–vis spectra, and MRES, but excluding electron diffraction. The NMR spectral measurements are on the edge; depending on the frequency separation of the signals observed and the exact rate of the conformational exchange, they may either involve individual conformers or ensemble averages.

From the equation $\nu = c/\lambda$, where c is the velocity of light (3×10^{10} cm s^{-1}), ν is the frequency or time scale of the measurement, and λ is the wavelength, one may infer that microwave spectroscopy [$\lambda \sim 0.3$ mm (0.03 cm)] has a time scale of 10^{12} s^{-1}, IR spectroscopy [$\lambda \sim 3$ μm (0.0003 cm for a C–H stretch)] has a time scale of 10^{14} s^{-1} (actually a 10^{13}–10^{14} range), and UV–vis [centered on $\lambda = 300$ nm (3×10^{-5} cm)] has a time scale of about 10^{15} s^{-1}. In NMR spectroscopy, for a signal separation of 100 Hz the time scale is 10^{2} s^{-1}. This might be compared to the rate of interconversion of acyclic conformers; since (Section 3-1) $k = 2.084 \times 10^{10} T e^{-\Delta G^{\ddagger}/1.987T}$ ($\Delta G^{\ddagger}$ in cal mol^{-1}) or $k = 2.084 \times 10^{10} T e^{-\Delta G^{\ddagger}/8.134T}$ ($\Delta G^{\ddagger}$ in J mol^{-1}), for a barrier of 3 kcal mol^{-1} (12.5 kJ mol^{-1}), k = 3.9×10^{10} s^{-1} at 25°C or 5.9×10^{8} s^{-1} at −100°C. For a barrier of 10 kcal mol^{-1} (42 kJ mol^{-1}), $k = 2.9 \times 10^{5}$ s^{-1} at 25°C or 0.86 s^{-1} at −100°C. It is thus clear that even rotations as fast as that in ethane are slow on the IR or microwave time scale, but that most rotations around "true" single bonds (cf. Fig. 2.14) are fast on the NMR time scale at room temperature and that only those bonds in acyclic systems that have relatively high barriers [around 8–10 kcal mol^{-1} (33–42 kJ mol^{-1})] give rise to rotational isomers that can be seen individually in NMR spectra around −100°C (a temperature that can be routinely reached in standard experimentation).

It might appear from this discussion that microwave spectroscopy would fail for barriers lower than about 1 kcal mol^{-1} (4.2 kJ mol^{-1}). In fact, however, this is not the case. With very low barriers, line splitting due to quantum mechanical tunneling occurs in the microwave spectrum and the barrier height can be deduced from this splitting.

b. Dipole Moments

In Section 4-6.b we have seen that molecules in symmetry point groups $\mathbf{C_n}$ (including $\mathbf{C_1}$), $\mathbf{C_s}$, and $\mathbf{C_{nv}}$ can (and normally will) have dipole moments. If the compounds are hydrocarbons, the dipole moment will ordinarily be small, but for molecules with polar substituents (halogens, OR, SR, NR$_2$, etc.) the dipole moment is often sizeable.

Dipole moments depend on molecular geometry. This fact is particularly obvious in molecules that have two (or perhaps three) polar substituents, such as the 1,2-dihaloethanes. From Figure 10-8, p. 606, it is obvious that the anti conformer of XCH_2CH_2X (X = halogen) belongs to point group C_{2h} and must therefore have zero dipole moment, whereas the moment of the gauche conformers (C_2) is sizeable.

There are two major methods of measuring dipole moments that give different results. One is the classical method of measuring dielectric constant along with density or refractive index in solution as a function of concentration. This method gives an ensemble average dipole moment μ, which is given by the equation,

$$\mu = \left(\sum_i n_i \mu_i^2 \right)^{1/2} \qquad (10.5)$$

or, for the particular case under consideration

$$\mu = (n_g \mu_g^2 + n_a \mu_a^2)^{1/2} \qquad (10.6)$$

where n_g and n_a are the mole fractions of the gauche and anti conformers, respectively, and μ_g and μ_a are their respective dipole moments. (It must be noted that the dipole moment of the ensemble average is *not* the weighted average of the individual conformer dipoles, but that it is the squares of the individual dipole moments that are averaged to give the square of the ensemble dipole moment. This is so because the actual averaging is of the polarizations, which are proportional to the squares of the dipole moments.) The other method is by microwave spectroscopy, which yields the dipole moments of the individual conformers whose spectra are being observed (microwave spectra can, in fact, be obtained only of molecules that have dipole moments).

From the ensemble dipole moment one can calculate conformer composition when there are only two contributing conformations (Eq. 10.6), given that the sum of the mole fractions must add up to unity. (In the case of the 1,2-dihaloethanes, $n_a + 2n_g = 1$.) However, this requires that the dipole moments of the individual conformers be known. There are three ways of getting such moments: from calculation, from model compounds, and from microwave spectra. Calculation relies on vector addition of standard bond dipoles. While this is easy to do with modern computers, it does not take into account the effect of one bond dipole on another (induction effect) and may therefore be unreliable (e.g., the C–Cl bond dipole in 1,2-dichloroethane is not exactly the same as that in chloroethane). The second method, in which dipole moments of individual conformers are taken from model compounds, has not been much used with acyclic molecules but is common with cyclic molecules and will be discussed in Chapter 11. It suffers both from questions regarding the adequacy of the model (is its geometry the same as that of the conformer to be studied?) and questions of the innocuousness of whatever extra substituent is used to stabilize the desired conformation in the model compound (does it contribute to the overall dipole moment, and if so, can the contribution be adequately taken into account?). Thus the most reliable method would seem to be to measure the dipole moments of the individual contributing conformers by microwave spectroscopy and then to measure the ensemble dipole moment by classical methods. To the extent that both measurements are accurate the result obtained by Eq. 10.6 should be unequivocal. For example, in the case of FCH_2CH_2F, the dipole moment of the gauche form, μ_g, is 2.67 D (Butcher et al., 1971) that of the anti conformer, μ_{anti}, may be taken to be zero, and the overall dipole moment of the substance, μ, is 2.5 D at 296.9 K (Goodwin and Morrison, 1992) whence, from Eq. 10.6, $n_g = 0.88$ and $\Delta H^0 = 0.75\,kcal\,mol^{-1}$ $(3.1\,kJ\,mol^{-1})$ (cf. p. 610). However, the method has rarely been used, perhaps because it requires measurement of the overall (ensemble) dipoles in the gas phase (since the microwave measurements of the component dipole moments are in the gas phase), something that is not easy to do, especially when, as will generally be the case, the moments vary with temperature.

Even the assumption of zero dipole moment for the anti conformer is precarious. It is correct if the anti isomer occupies a steep energy well; if the well is shallow and the molecule thus librates appreciably within the well (i.e., carries out to-and-fro torsional motions), the dipole moment will not average out to zero, since it is the squares of the individual moments that must be summed over the libration.

Despite all these problems, some of the earliest conformational computations were made on the basis of dipole moments (Mizushima, 1954; cf. Smyth, 1974). For example, the dipole moment of $ClCH_2CH_2Cl$ was measured as 1.37 D in dilute hexane solution; assuming $\mu_g = 2.57$ D and $\mu_a = 0$, one may calculate $n_g = 0.28$, $n_g = 0.72$ using Eq. 10.6, where $\Delta H^0 = 0.97\,kcal\,mol^{-1}$ $(4.1\,kJ\,mol^{-1})$, in fair agreement with the value tabulated in Table 10.1 obtained by different methods. The dipole moment increases with increasing temperature (because the contribution of the higher dipole and less stable gauche conformer increases) and also increases in higher dielectric solvents, for the same reason. (The repulsive electrostatic forces within the gauche isomer are diminished in higher dielectric media; moreover, the polar gauche conformation is stabilized by solvation; cf. Table 10.1 and Eq. 2.1.)

The question of whether diastereomers can be distinguished by dipole moment measurement is obviously more complex, since each diastereomer is an ensemble of conformers and its dipole moment depends both on those of the contributing conformers and on the conformational composition (Eq. 10.5). Let us take the case of the 2,3-dichlorobutanes as an example. As shown in Figure 10.34, X = Cl, each stereoisomer is made up of three conformers. Of these, **A** and **D** should have low or zero dipole moments, and the others relatively high ones. One might therefore surmise that the dipole moment of the meso isomer **A/B/C** is lower, because conformer **A** contributes so heavily here; whereas that of the chiral isomer, where conformer **E** competes with **D** (cf. the discussion on p. 631), should be higher. The data in Table 10.6 provide but tenuous support for this qualitative prediction: Only in the case of $(CH_3)_2CHCHXCHXCH_3$ is the dipole moment for the threo isomer sustantially larger than that of the erythro. In the other two cases the difference is either very small (in the case of the dichlorides) or even in the wrong direction (in the case of the dibromides). In the case of the *tert*-butyl compound, this anomaly has been traced to the meso isomer being largely in the gauche conformation, apparently because the CH_3/CH_3 g^+g^- interaction in this conformation is less severe than the corresponding CH_3/Br

TABLE 10.6. Dipole Moments[a] of RCHXCHXCH$_3$

	Chloride		Bromide	
	erythro	threo	erythro	threo
R	or meso	or racemic	or meso	or racemic
CH$_3$[b]	1.63[b]	1.79[b]	1.41[b]; 1.73[c]	1.32[b]; 1.62[c]
C$_2$H$_5$[c]			1.67	1.9
(CH$_3$)$_2$CH[d]	0.95	2.40	0.85	2.21
(CH$_3$)$_3$C[d]	2.43	2.51	2.66	2.50

[a] Values are in debye units.
[b] Chia et al. (1973), in CCl$_4$.
[c] Winstein and Wood (1940), neat.
[d] Kingsbury and Best (1967), in cyclohexane.

interaction in the anti conformation (see Section 10-4.d; Kingsbury and Best, 1967). There is, however, no obvious explanation for the dimethyl compounds whose conformational situation is unexceptional (cf. Table 10.4); one must assume here that the dipole moments of the individual conformers are substantially different from what is assumed, either because of mutual induction or because of libration (see p. 637) or, most probably (cf. the discussion concerning 2,3-dimethylbutane, p. 605), because the torsion angles deviate appreciably from the ideal 60°. With calculated dipole moments for **A**, **B/C**, **D**, **E**, and **F** (Fig. 10.34) of 0, 2.94, 0.70, 2.39, and 2.78 D (Meyer, Allinger, et al., 1980), the computed dipole moments for meso and chiral CH$_3$CHClCHClCH$_3$ in CCl$_4$ are 1.65 D and 1.67 D. The calculated data, while close to the experimental, barely reproduce the experimental trend ($\mu_{rac} > \mu_{meso}$). In CH$_3$CHBrCHBrCH$_3$, where the trend is reversed, calculations (Meyer, 1981; Meyer and Ohmichi, 1981) are good for the meso isomer ($\mu_{calcd} = 1.68$ D) but in the wrong direction for the racemic one ($\mu_{calcd} = 1.93$ D). Surprisingly, these calculated data are in good agreement with the experimental ones for C$_2$H$_5$CHBrCHBrCH$_3$!

In summary, it would appear that predictions of dipole moments of diastereomers are hazardous; even qualitative predictions as to which of two diastereomers has the higher dipole moment may be in error, although this picture may improve with better calculations now becoming available.

c. Boiling Point, Refractive Index, and Density

These three quantities are generally related in that they all depend on molar volume: the lower the molar volume, the higher the density, refractive index, and boiling point. (No corresponding correlation applies to melting points.) In non-polar molecules molar volume may be correlated with enthalpy (cf. p. 600), but since most examples relate to cyclic structures (but see Fig. 10.4), we shall defer discussion to Chapter 11. In polar molecules the correlation is generally with dipole moment: Among diastereomers, the one with the highest dipole moment has the highest boiling point, density, and refractive index (van Arkel or dipole rule; cf. Eliel, 1962), presumably because a high dipole moment leads to a high degree of molecular association and thereby to an increase in density and lowering of vapor pressure. It is also true that the higher dipole isomer is apt to have the longer retention time in chromatography on alumina (Ward et al., 1962).

TABLE 10.7. Dipole Moments, Densities, Refractive Indices, and Boiling Points of Diastereomers[a]

Compound	μ (D)		D^{25} (g cm^{-3})		η_D^{25}		bp (°C/mmHg)	
	meso or erythro	racemic or threo	meso or erythro	racemic or threo	meso or erythro	racemic or threo	meso or erythro	racemic or threo
$CH_3CHClCHClCH_3$	1.63	1.79	1.1023	1.1063	1.4386	1.4409	115.9/760	119.5/760
$CH_3CHBrCHBrCH_3$	1.41	1.32	1.7820	1.7864	1.5093	1.5126	103.4/160	107.5/160
$CH_3CH(OAc)CH(OAc)CH_3{}^b$	2.35	1.95	1.0213	1.0244	1.4121	1.4134	76.3/10	82/10
$CH_3CH(OCH_3)CH(OCH_3CH_3{}^b$	c	c	0.8435	0.8464	1.3890	1.3905	108/750	109.5/750
$CH_3CH(OCH_3)CH(OH)CH_3$	c	c	0.9122	0.9032	1.4107	1.4074	132.4/748	126.5/752
$CH_3CHOHCHOHCH_3$	c	c	0.9988	0.9873	1.4372	1.4311	89.3/16	86.0/16

[a] Data from Beilsteins Handbuch, 3rd and 4th supplement.
[b] Data for meso isomer compared with those for chiral isomer. In several other instances, data for enantiomers, or for enantiomers and racemate, are identical.
[c] Not available.

Unfortunately, in practice, the dipole rule is not very reliable (cf. Table 10.7). Thus, for example, the curious inversion of dipole moments in the 2,3-dibromobutanes discussed earlier (Table 10.6) does not reflect itself in the rest of the data shown in Table 10.7. Rather, in the case of the first four entries, the density, refractive index, and boiling point data agree with the *calculated* order of dipole moments, suggesting that the anomaly is in the dipole data and not in the rest. In the last two cases, the order is inverted in that the racemic (threo) isomer has the lower constants and the meso (erythro) isomer the higher. This probably reflects the fact (see next section) that *intra*molecular hydrogen bonding is more pronounced in the chiral or threo compound and *inter*molecular hydrogen bonding, which increases density and boiling point, is therefore more important in the erythro or meso isomer. Concomitant with the inversion of the order of boiling points in the $CH_3CHOHCHOHCH_3$ diastereomers relative to $CH_3CH(OAc)$-$CH(OAc)CH_3$, there is also a reversal of the order of gas chromatographic (GC) retention times (Nasybullina et al., 1973). Similarly, with the 2,3-dichlorobutanes and 2,4-dichloropentanes, in the former case the meso isomer has a shorter retention time on highly polar nitrile columns than the racemic one, which corresponds to the lower boiling point of the meso isomer (cf. Table 10.7), whereas in the latter case the chiral isomer is retained less strongly than the meso (Pausch, 1975).

d. Infrared Spectra

Of the various techniques mentioned in Sections 10-1 and 10-2 for the determination of conformer population, IR and NMR spectroscopy are the most accessible. The use of these techniques is therefore described in some detail.

For the reasons discussed earlier, IR spectra of conformationally heterogeneous substances will usually display absorption bands due to individual conformers. The intensity I of these absorption bands will be proportional to the conformer population (Park et al., 1974); thus for anti (a) and gauche (g) conformers, $I_a = \alpha_a c_a \ell$ and $I_g = \alpha_g c_g \ell$, where the α values are integrated absorption coefficients, the c values are concentrations, and ℓ is the cell length. It follows that the equilibrium constant is

$$K = c_a/c_g = \alpha_g I_a/\alpha_a I_g \qquad (10.7)$$

The ratio I_a/I_g can be measured from the spectrum, provided (and this may cause a problem) that absorption bands can be unequivocally assigned to one conformer or the other. However, the ratio of absorption coefficients, α_a/α_g is generally not known and usually differs from unity.

> It is possible to obtain this ratio from measurement of model compounds of fixed conformation (Pickering and Price, 1958) but this method is not generally satisfactory.

To circumvent this difficulty, the ratio I_a/I_g is measured at two or more temperatures. Then, according to the van't Hoff isochore

$$\ln (K_{T_2}/K_{T_1}) = (\Delta H^0/R)(1/T_1 - 1/T_2) \tag{10.8}$$

where ΔH^0 is the enthalpy difference between two conformers, T_1 and T_2 are two temperatures, and K_{T_1} and K_{T_2} are the equilibrium constants at those temperatures. Insertion of Eq. 10.7 into Eq. 10.8 yields

$$\Delta H^0 = [RT_1T_2/(T_2 - T_1)][\ln(I_a/I_g)_{T_2} - \ln(I_a/I_g)_{T_1}] \tag{10.9}$$

If the measurement is carried out at more than two temperatures, $\ln(I_a/I_g)$ is plotted versus $1/T$; the slope of the straight line so obtained gives $\Delta H^0/R$. Thus the IR method in this form yields ΔH^0, not ΔG^0.

Infrared spectroscopy is particularly suitable for probing intramolecular hydrogen bonding (Tichý, 1965) of the type $-O-H \cdots :X$, $-S-H \cdots :X$, $-N-H \cdots :X$, and so on. Thus the presence of an IR absorption peak at 3612 cm^{-1} in addition to a peak at 3644 cm^{-1} (due to free hydroxyl) for ethylene glycol in dilute CCl$_4$ solution indicates the presence of intramolecularly bonded $-O-H \cdots O-H$ as well as free hydroxyl, and thereby points to the existence of at least part of the material in the gauche conformation (cf. Fig. 10.8, X = OH; Kuhn, 1952).

Manifestation of intramolecular hydrogen bonding can be used for configurational as well as conformational assignment. For example, both meso and racemic 2,3-butanediol (Fig. 10.36) display intramolecular hydrogen bonding (Kuhn, 1958). However, the meso isomer must take up a relatively unfavorable conformation, with methyl groups gauche, to achieve the necessary proximity of hydroxyl groups (Fig. 10.36, **A** or enantiomer), whereas the chiral isomer can exist in conformation **B** in which the OH groups are gauche but the methyl groups are anti. It is therefore not surprising that the ratio of bonded to unbonded OH

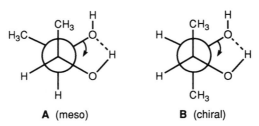

A (meso)   **B** (chiral)

Figure 10.36. Hydrogen-bonded conformations of meso and racemic 2,3-butanediol.

[measured, though not accurately (see above), by the area ratio of the corresponding bands] is higher for the chiral than for the meso isomer. In addition, the difference (in wavenumbers) between intramolecularly bonded and nonbonded OH is larger in the chiral isomer ($3632 - 3583 = 49\,\text{cm}^{-1}$) than in the meso ($3633 - 3591 = 42\,\text{cm}^{-1}$), pointing to a stronger hydrogen bond in the former. This result may be explained by noting (Fig. 10.36) that it is easy to rotate the OH groups together (thus strengthening the bond) in the chiral isomer **B**, whereas the same rotation in the meso isomer **A** requires (unfavorable) increased eclipsing of the methyl groups.

e. NMR Spectroscopy

As explained in Section 10-4.a, the time scale of NMR spectroscopic measurements (Anet and Anet, 1971) may be either fast or slow relative to the time scale of rotation around single bonds. At room temperature, except for bonds having partial double bond character, such as the OC–N bond in amides (cf. Section 10-2.a), rotation around single bonds is generally fast on the NMR time scale, and therefore NMR chemical shifts are averaged over all conformations (Eliel, 1959), as are coupling constants:

$$\delta = \sum_i n_i \delta_i \qquad (10.10)$$

and

$$J = \sum_i n_i J_i \qquad (10.11)$$

where δ and J are the observed chemical shifts and coupling constants, respectively, δ_i and J_i are the shifts and coupling constants of the ith conformer and n_i the corresponding mole fractions. At low temperatures, however, if rotation is "slow" (cf. Section 10-4.a) the conformations may be observed individually. In that case, on the assumption that signal intensity is directly proportional to the number of nuclei (or, in the parlance of IR, Section 10-4.d, that all extinction coefficients are equal; for justification see Booth and Josefowicz, 1976; Eliel et al., 1980; it is essential to compare corresponding nuclei in the conformers in question), equilibrium constants at the temperature of observation are obtained directly from ratios of signal areas.

In fluoroethanes containing other halogens, for example, $CBr_2ClCHBrF$, barriers are indeed high, in the $9\text{--}10\,\text{kcal mol}^{-1}$ ($38\text{--}42\,\text{kJ mol}^{-1}$) range when there are four other halogens (and one hydrogen), and even higher, $13\text{--}15\,\text{kcal mol}^{-1}$ ($54\text{--}63\,\text{kJ mol}^{-1}$) when there are five other halogens (Weigert, Roberts, et al., 1970). Thus it was possible to undertake a complete conformational analysis (i.e., to assess the population of all three rotamers) in a series of fluorinated haloethanes (Norris and Binsch, 1973; see p. 501).

Since normally it requires at least 1% of the minor isomer for its signals to be clearly distinguishable from noise in an NMR spectrum (2–3% is a safer figure), it is not possible to detect, at $-100°C$, minor conformers that are more than

1.5 kcal mol^{-1} (6.3 kJ mol^{-1}) in energy above the major ones. However, a method has been devised (cf. Anet and Basus, 1978; Okazawa and Sorensen, 1978) to quantify minor isomers of this type. When the temperature in such a conformationally heterogeneous system is lowered through the coalescence temperature, the peak of the major isomer will first broaden and then narrow again. Coalescence may be recognized as the temperature of maximum broadening, even if the population of the minor conformer is too small for it to be detected at the slow exchange limit (i.e., well below coalescence). From the line width at coalescence, both the rate of exchange at the coalescence temperature and the equilibrium constant can be calculated if the position of the invisible minor peak can be accurately guessed. Such a guess is often possible with reasonable confidence and accuracy from a knowledge of NMR shift parameters, especially in the case of ^{13}C spectra. Alternately, somewhat less than 1% of a minor isomer can be analyzed by comparing a salient ^{1}H signal with the ^{13}C satellite of the corresponding ^{1}H signal in the major isomer, thereby avoiding dynamic range problems at high signal amplification.

Under circumstances of slow exchange (i.e., well below coalescence), both the chemical shifts for salient nuclei and the coupling constants for appropriate pairs of nuclei can be ascertained for all three conformers of appropriately substituted ethanes and, in principle at least, Eqs. 10.10 and 10.11 can be used to assess the conformational composition at temperatures above coalescence (i.e., where interconversion among conformers is fast on the NMR time scale). However, there are two practical difficulties. One is that chemical shifts are temperature dependent; therefore Eq. 10.10 cannot safely be used because the contributing shifts δ_i at, say, room temperature will not be the same as the actual shifts measured at low temperature. Attempting a temperature extrapolation of the shifts is rarely successful, since the temperature range over which the shifts of individual conformers can be observed is too small, being limited on the upper bound by exchange and coalescence phenomena (cf. Section 8-4.d) and on the lower bound by difficulties due to freezing or excessive viscosity of the solvent, crystallization of the solute, and so on. On the other hand, coupling constants are nearly temperature independent and therefore Eq. 10.11 should be useful to determine room temperature conformer populations. Unfortunately, there is another difficulty: If there are three conformers, Eq. 10.11 takes the form

$$J = n_1 J_1 + n_2 J_2 + n_3 J_3 \tag{10.12}$$

and in addition

$$n_1 + n_2 + n_3 = 1 \tag{10.13}$$

Now, Eqs. 10.12 and 10.13 are a set of two equations in three unknowns (the n values), and therefore have no unique solution unless two different sets of coupling constants can be measured (yielding an extra equation) or unless $n_2 = n_3$. The latter situation will occur for the meso isomer in Figure 10.34 ($n_B = n_C$) and the former for compounds of the type XCH$_2$–CHYZ, which display two different vicinal couplings, since the methylene protons are diastereotopic and display distinct chemical shifts and different coupling constants

TABLE 10.8.   Relation of Coalescence Temperature (T_c) to Barrier $(\Delta G^{\ddagger})^a$ for Signals Separated by 100 and 500 Hz, Respectively

	$\Delta G^{\ddagger}$			
	$\delta = 100$ Hz		$\delta = 500$ Hz	
T_c(K)	(kcal mol^{-1})	(kJ mol^{-1})	(kcal mol^{-1})	(kJ mol^{-1})
100	4.6	19.1	4.2	17.8
110	5.0	21.1	4.7	19.6
120	5.5	23.1	5.1	21.5
130	6.0	25.1	5.6	23.4
140	6.5	27.1	6.0	25.2
150	7.0	29.1	6.5	27.1
175	8.2	34.2	7.6	31.9
200	9.4	39.3	8.7	36.7
225	10.6	44.5	9.9	41.5

^{a} $\Delta G^{\ddagger} = 1.987 \, T_c \, (23.76 + \ln T_c/k_c)$ cal mol^{-1}, where $k_c = 1/2\sqrt{2}\pi\delta$.

with the methine proton. Application to the chiral isomer in Figure 10.34, for example, is not feasible. An example of a somewhat different case has been given by Höfner, Binsch, et al., (1978).

Table 10.8 gives a correlation between coalescence temperature (cf. Section 8-4) and energy barrier between conformers for two signals originating from two different conformers separated by $\delta = 100$ and 500 Hz. (The former separation would correspond to two protons separated by 0.5 ppm in a 200-MHz instrument or two carbon nuclei separated by 2 ppm in the same instrument, the latter to two protons separated by 1 ppm in a 500-MHz instrument or two carbon nuclei separated in such an instrument by 4 ppm.) The data apply strictly to two conformers of equal population only but may be used as an approximation for other cases where the populations do not differ greatly.

Temperatures around $-100°C$ (ca. 175 K) can be reached routinely and coalescence can thus be observed for signals of conformers separated by barriers of about 8 kcal mol^{-1} (33.5 kJ mol^{-1}). However, since complete signal separation and reliable integration requires going well below the coalescence temperature, the barriers required must actually be somewhat higher or the temperature somewhat lower. A few investigators have been able to reach much lower temperatures, down to 100 K ($-173°C$) (Anet and Yavari, 1977; Bushweller et al., 1982) and have thus been able to measure barriers in the 4.2–5.2 kcal mol^{-1} (17.6–21.8 kJ mol^{-1}) range. However, even this leaves out many ethanoid molecules ($RCH_2–CH_2R'$, $RCHX–CH_2R'$, $RCHX–CHYR'$) in which, as we have seen earlier, barriers tend to be in the 3–4 kcal mol^{-1} (12.6–16.7 kJ mol^{-1}) range. For such molecules, Eqs. 10.12 and 10.13 can be used, but a method other than low-temperature measurement must be employed to obtain the coupling constants of the pure conformers (cf. Thomas, 1968; Abraham and Bretschneider, 1974). Moreover, to measure coupling constants for the degenerate protons in RCHXCHXR, it is necessary either to prepare RCHXCDXR and measure J_{HD} (then $J_{HH} = 6.55 \, J_{HD}$) or to study the AB subspectrum of the ^{13}C satellite (ABX) signal of $R^{13}CHXCHXR$ (Cohen, Sheppard, et al., 1958).

Vicinal $^1H/^1H$ coupling constants ($^3J_{HH}$) as a function of torsion angle ω can be obtained from the Karplus equation (Eq. 10.14; Karplus, 1959, 1963; see also Conroy 1960), which is plotted in Figure 10.37. Because of the uncertainty of the coefficients in the equation as well as the uncertainty concerning the

$$J = A \cos^2\omega - B \cos \omega + C \tag{10.14}$$

exact torsion angle in the acyclic conformers, a set of empirical parameters is often used to solve Eq. 10.12. Even so the variability of J is a problem: typically, J_{gauche} varies between 1.5 and 5 Hz and J_{anti} between 10 and 14 Hz. The parameter J_{syn} is typically around 9–9.5 Hz and $J_{90°}$ near zero. In a more recent paper (Haasnoot, Altona, et al., 1980) the Karplus equation is given in the form $J = 7.76 \cos^2\omega - 1.10 \cos \omega + 1.40$. This gives values of $J_0 = 8.08$, $J_{60°} = 2.79$, $J_{90°} = 1.40$, $J_{180} = 10.26$ Hz; the values for $J_{0°}$ and $J_{180°}$ appear somewhat small. It is important to recognize that the coupling constants in a segment $H^1-CX-CY-H^2$ depend not only on the torsion angle but also on several other factors (Karplus, 1963) of which the electronegativity of X and Y (Glick and Bothner-By, 1956, cf. Schaefer et al., 1965) and the torsion angles between Y and H^1 and X and H^2 (Booth, 1965) are the most important. Taking these factors into account, an empirical relationships (Eq. 10.15) has been developed by Haasnoot, de Leeuw, and Altona (1980), which appears to be considerably more accurate and more general than the simple Karplus relationship, though at the expense of introducing several additional parameters.

$$J = A \cos^2\omega + B \cos \omega + \sum_i \Delta\chi_i[D + E \cos^2(\zeta_i\omega + F|\Delta\chi_i|)] \tag{10.15}$$

The first two terms in Eq. 10.15 are Karplus terms; the constant term C in the Karplus equation (10.14) is taken to be zero. The parameter χ_i is the difference in Huggins electronegativity (Huggins, 1953) between the ith substituent and hydro-

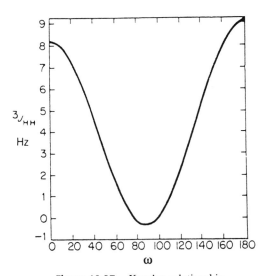

Figure 10.37. Karplus relationship.

gen and ζ_i is either $+1$ or -1: it is $+1$ if the X/H^2 or Y/H^1 torsion angle (looking along the C–C bond) is $+60°$ and -1 if it is $-60°$. The parameters for general use are $A = 13.70$, $B = -0.73$, $D = 0.56$, $E = -2.47$, $F = 16.9°$. An additional parameter G is needed if the substituents X and Y are not single atoms but polyatomic groups (–U–V); in that case $\Delta\chi_i = \Delta\chi_i^U - G \sum_j \Delta\chi_j^V$, the subscript j referring to all the substituents attached to atom U (V in –U–V, V + V in $-UV_2$, V + W in

$$U \begin{matrix} \nearrow W \\ \searrow V \end{matrix}$$

, and so on) with $G = 0.14$. Equation 10.15 correlates 315 coupling constants from the literature with a correlation coefficient of 0.992 and a root mean square deviation of 0.479 Hz (Haasnoot et al., 1980); somewhat better correlation and accuracy are obtained by using separate coefficients for fragments of the $RR'CH–CH_2R''$ and $RR'CH–CHR''R'''$ type. (The correlation is not as good for $RCH_2–CH_2R'$.)

> Several applications of this equation have been made with good results, including an application in an acyclic case (Brunet et al., 1983) to the diastereomers of $C_6H_5CHOX–CH_2SOCH_3$

Coupling other than proton–proton may also be useful in conformational analysis, for example, $^1H/^{19}F$ and $^1H/^{31}P$, as well as $^1H/^2H$ coupling constants in specifically deuterium labeled compounds (see above concerning their utility). Of particular interest, because of their ubiquitousness, are $^1H/^{13}C$ and $^{13}C/^{13}C$ couplings (Marshall, 1983). Unfortunately, the measurement of proton–carbon coupling constants of nonadjacent nuclei requires ^{13}C enriched substrates, but it can be very beneficial in solving certain conformational problems, such as the above mentioned one of analyzing the conformations of RCHXCHYR' systems. It may also be useful in the conformational analysis of $RCHXCH_2R'$ systems where uncertainty in the assignment of the diastereotopic CH_2 protons may lead to two different solutions, based on proton coupling constants alone. Simultaneous study of $^1H/^{13}C$ coupling will tell which solution is the correct one.

In aliphatic sytems, carbon–proton coupling constants of the type $^{13}C–C–C–H$ follow a Karplus relationship with $^3J_{HC} = 0.52 - 0.7\ ^3J_{HH}$, obviously useful in assessing conformation (Marshall, 1983). Measurement of $^2J_{CH}$ may also be informative; this coupling constant depends on the torsion angle in a system $X–^{13}C–C–H$.

Carbon-13 NMR spectroscopy has been used less extensively than ^1H NMR spectroscopy in the conformational analysis of acyclic compounds. Low-temperature NMR, in those cases where barriers are high (Table 10.8), should be at least as straightforward for ^{13}C as for ^1H. In averaged room temperature spectra, the Karplus relationship gives the edge to proton spectroscopy, although application of double quantum coherence spectroscopy (Bax, Freeman, et al., 1980a,b) now makes it possible to ascertain $^{13}C/^{13}C$ coupling constants (cf. Marshall 1983) of vicinal carbon nuclei and, from them, deduce information as to the conformation of the carbon chain (e.g., Phillipi et al., 1982).

> It is implicit in the discussion in Section 8-4 that diastereomers, which contain externally diastereotopic nuclei, will display differences in the NMR signals of such

nuclei. In fact, ^{13}C NMR is a sensitive probe for distinguishing diastereomers and for analysis of diastereomer mixtures; cases where *each* pair of corresponding nuclei in a set of diastereomers is accidentally isochronous are rare, although often only a few of the corresponding pairs will, in fact, display palpable anisochrony.

In the assessment of chemical shift differences in the ^{13}C spectra of diastereomers the upfield shifting γ effect (cf. Duddeck, 1986) is particularly revealing. It must be recalled, however, that in an ensemble averaged spectrum at room temperature the observed effect depends on the conformational composition of the sample; that is, it is conformation that dictates the nature of spectral differences, not configuration as such. Some data from the literature (Ernst and Trowitzsch, 1974; Levy et al., 1980; Schneider and Lonsdorfer, 1981; Grenier-Loustalot and Grenier, 1982), summarized in Table 10.9, will illustrate this point.

In most of the molecules of the $CH_3CHXCHXCH_3$ type, the methyl shift of the chiral isomer is upfield of that of the meso isomer. The reason (cf. Ernst and Trowitzsch, 1974) is implicit in Figure 10.34: The meso isomer is dominated by conformer **A**, whereas the chiral isomer contains substantial amounts (cf. Table 10.4) of conformer **D** in which there is a shielding CH_3/CH_3 gauche interaction. The compound $CH_3CHOHCHOHCH_3$ in carbon tetrachloride provides a plausible exception: As mentioned earlier, hydrogen bonding leads to a favoring of the OH/OH gauche conformer: **B-C** for the meso isomer and **E** or **F** for the chiral one. Since both **B** and **C** but only **F** (and not **E**) have CH_3/CH_3 gauche interactions, the upfield methyl shift in the meso isomer is explained. Again plausibly, in DMSO where intramolecular hydrogen bonding is disrupted, the normal order (chiral upfield of meso) is restored. The only other exception in Table 10.9 is ethylene glycol diacetate; this compound is apparently dominated by an attractive gauche effect of the type discussed in Section 10-1.b for FCH_2CH_2F. This is also evidenced by the larger coupling constant in the chiral isomer, which suggests that conformer **F** of the chiral isomer contributes more than conformer **A**

TABLE 10.9. The ^{13}C NMR Spectra of Diastereomer Pairs

Entry No.	Compound	Solvent	CH_3[a]		CHX[a]		$J_{HH}(Hz)$[b]	
			meso	chiral	meso	chiral	meso	chiral
1	$CH_3CHOHCHOHCH_3$[c]	CCl_4	17.0	19.4	70.7	72.3		
2	$CH_3CHOHCHOHCH_3$[c]	DMSO	19.9	19.2	71.6	71.4		
3	$CH_3CH(OAc)CH(OAc)CH_3$[d]	$CFCl_3$	15.1	16.0	71.0	71.1	3.58	5.08
4	$CH_3CH(OCH_3)CH(OCH_3)CH_3$[c]	[e]	15.3	13.7	80.3	78.9		
5	$CH_3CHClCHClCH_3$[c]	neat	21.9	19.8	61.3	60.2	7.39	3.45
6	$CH_3CHBrCHBrCH_3$[c]	neat	25.2	20.5	53.7	52.1	8.81	3.11
7	$HO_2CCH(CH_5)CH(CH_3)CO_2H$[d]	CH_3OH	15.6	13.9	43.8	42.8		
8	$NaO_2CCH(CH_3)CH(CH_3)CO_2Na$[d]	D_2O	18.0	14.5	48.5	46.4		
9	$CH_3CHClCH_2CHClCH_3$[f]	neat	24.7	25.5	54.5	55.6		
10	$CH_3CHBrCH_2CHBrCH_3$[g]	$CDCl_3$	25.4	26.4	46.8	49.4		

[a] Chemical shifts in parts per million from tetramethylsilane (TMS).
[b] Proton (CHX–CHX) coupling constants, Bothner-By and Naar-Colin (1962).
[c] Levy et al. (1980).
[d] Schneider and Lonsdorfer (1981).
[e] Solvent $CH_3CHOHCH(OCH_3)CH_3$.
[f] Carman, Goldstein, et al. (1971).
[g] Cais and Brown (1980).

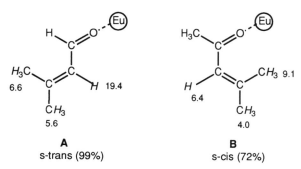

Figure 10.38. Use of lanthanide shift reagents in conformational assignments. The figures indicate the lanthanide induced shifts (shift in presence of lanthanide minus shift in absence of lanthanide) in parts per million of the nearby italicized protons.

of the meso. In $CH_3CHClCH_2CHClCH_3$, Figure 10.35 indicated more gauche interaction in the meso isomer than in the chiral isomer (predominant conformation **A**); therefore the methyl groups in the meso isomer, both in the dichloride and dibromide, are more shielded.

The relative shifts of the CHX carbon atoms [C(2) and C(3)] run parallel to those of the terminal methyl groups. The reason for that is not evident, but the phenomenon itself is common in ^{13}C NMR spectroscopy and will be encountered again in cyclic compounds (Chapter 11).

The ^{1}H NMR spectra also frequently display appreciable chemical shift differences between diastereomers (for the 2,3-dihalobutanes and 2,4-dihalopentanes, see Bothner-By and Naar-Colin, 1962; Satoh, 1964; Doskočilová, 1964; Trost et al., 1971), which are often useful for the quantitative analysis of mixtures. However, they are more difficult to rationalize than the relative magnitudes of vicinal proton coupling constants or the relative chemical shifts in ^{13}C NMR spectra. These latter parameters are therefore more useful in making configurational and conformational assignments.

In concluding this section on NMR as a conformational probe, the use of lanthanide shift reagents to this end must be mentioned (see also pp. 28 and 238). Complexation of 3-methyl-2-butenal (Fig. 10.38, **A**; the lanthanide induced shifts of salient protons in parts per million are indicated) with Eu(fod)$_3$, where fod is 6,6,7,7,8,8,8-heptafluoro-2,2-dimethyl-3,5-octanedionate, indicates that this compound exists largely in the (normal) s-trans conformation; quantitative treatment suggests the percentage to be about 99% (Montaudo et al., 1973). In contrast, in mesityl oxide (4-methyl-3-penten-2-one, Fig. 10.38, **B**), where steric factors militate against the s-trans conformation, about 72% of the molecules are in the s-cis form. The shifts, in parts per million, shown in Fig. 10.38 are, of course, averaged shifts and it must be noted that the indicated percentage of s-cis conformer refers to the Eu(fod)$_3$ complexes, not necessarily to the free carbonyl compounds.

10-5. CONFORMATION AND REACTIVITY: THE WINSTEIN–HOLNESS EQUATION AND THE CURTIN–HAMMETT PRINCIPLE

Although, from a practical point of view, one might want to focus on the different reactivity, in a given reaction, of configurational isomers, the primary relationship of reactivity in conformationally mobile systems is to conformation. Both steric and stereoelectronic factors on reactivity and product composition depend primarily on conformational factors in such systems, in contrast to the situation in (rigid) alkenes (Chapter 9) where reactivity and product composition correlate directly with configuration.

Since the stereochemical aspects of many basic chemical reactions, such as elimination, addition, or substitution, are dealt with in most elementary textbooks, and, to some extent, elsewhere in this text, we shall not take them up here in detail. We must, however, deal with the fact that both reaction rates and reaction products may depend on the conformational composition of starting materials, as well as the conformation of preferred transition states. The dependence of reaction rate on conformational composition is most simply described by the Winstein–Holness equation (Winstein and Holness, 1955; Eliel and Lukach, 1957; Seeman, 1983, 1986) and the related dependence of product composition is governed by the Curtin–Hammett principle (Curtin, 1954; Seeman, 1983, 1986).

Both the Winstein–Holness equation and the Curtin–Hammett principle refer to the kinetic scheme shown in Eq. 10.16. While the scheme is general and has been so treated (Seeman, 1983), the Curtin–Hammett principle and the Winstein–Holness equation

$$C \overset{k_C}{\leftarrow} A \underset{k_B}{\overset{k_A}{\rightleftharpoons}} B \overset{k_D}{\rightarrow} D \tag{10.16}$$

apply to situations in which A and B are in rapid equilibrium (i.e., k_A, $k_B \gg k_C$, k_D), such as when A and B are tautomers or conformational isomers; it is this last situation with which we shall deal in what follows. Products C and D may be different substances, or they may be conformers of the same substance, or (in rare cases) they may even be identical. The energy diagram for the situation where C and D are different is shown in Figure 10.39. For the general kinetic treatment, three different situations must be distinguished (Seeman, 1983).

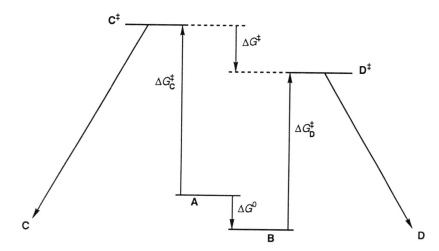

Figure 10.39. Energy diagram for Winstein–Holness and Curtin–Hammett kinetics.

1. Case 1: k_A, $k_B \gg k_C$, k_D. (In practice, this condition may be considered fulfilled when k_A, k_B are at least 10 times as large as the larger of k_C and k_D, provided also the **A–B** equilibrium is not highly one-sided, i.e., k_A and k_B are of the same order of magnitude). In this case the conformational equilibrium $A \rightleftharpoons B$ is maintained throughout the reaction: $[B]/[A] = K = k_A/k_B$. Both the Winstein–Holness equation and the Curtin–Hammett treatment (see below) apply to this case and to this case only.

2. Case 2: k_C, $k_D \gg k_A$, k_B. In this case the **A/B** ratio will not change during the reaction and the product ratio **C/D** will simply reflect the conformer ratio **A/B**. This case is called one of "kinetic quenching" since the $A \rightleftharpoons B$ interconversion is stopped; that is, **A** and **B** are quenched, by the reaction, in their equilibrium concentrations.

3. Case 3: k_C, k_D are of the same order of magnitude as k_A, k_B. In this case, equilibrium is not maintained during the reaction and a kinetic treatment more general than that given below must be applied (Seeman and Farone, 1978); the Winstein–Holness equation and the Curtin–Hammett principle in their usual form do not apply.

We proceed, now, to derive the Winstein–Holness equation for case 1 above. The equation refers to the total empirical rate of reaction of an equilibrating system of the type shown in Eq. 10.16. If the reaction under consideration is first order or pseudo first order, we may write the overall rate as

$$\text{Rate} = \frac{d[A+B]}{dt} = k_{WH}[A+B] \tag{10.17}$$

where $[A+B]$ is the stoichiometric concentration of the substrate and k_{WH} (the Winstein–Holness rate constant) is the experimentally observed rate constant. (We may consider **A** and **B** to be two conformers of one and the same substance; we shall see later that one need not confine oneself to two reacting conformations.) But, from Eq. 10.16,

$$\text{Rate} = \frac{d[A]}{dt} + \frac{d[B]}{dt} = k_C[A] + k_D[B] \tag{10.18}$$

and

$$[B]/[A] = K \tag{10.19}$$

Equating Eqs. 10.17 and 10.18 and replacing $[B]$ by $K[A]$ from Eq. 10.19, one obtains:

$$k_C[A] + k_D K[A] = k_{WH}\{[A] + K[A]\}$$

or, dividing by $[A]$,

$$k_C + k_D K = k_{WH}(1 + K)$$

whence

$$k_{WH} = (k_C + k_D K)/(1 + K) \tag{10.20}$$

Equation 10.20 is the Winstein–Holness equation as derived by Eliel and Ro (1956; see also Eliel and Lukach, 1957). It relates the observed rate constant (k_{WH}) to the individual rate constants (k_C, k_D) at which the conformers **A** and **B** react and to the equilibrium constant K between them. To transform Eq. 10.20 into the form originally given by Winstein and Holness (1955), we set $K = n_B/n_A$, where n_A and n_B are the mole fractions of **A** and **B** at equilibrium, disregarding other components ($n_A + n_B = 1$). Then

$$k_{WH} = (k_C + n_B k_D/n_A)/(1 + n_B/n_A)$$

Multiplying the numerator and denominator on the right-hand side by n_A gives

$$k_{WH} = (n_A k_C + n_B k_D)/(n_A + n_B)$$

or, since $n_A + n_B = 1$,

$$k_{WH} = n_A k_C + n_B k_D \tag{10.21}$$

Equation 10.21 states that the observed rate constant is the weighted average of the rate constants at which the individual conformational isomers (or other rapidly interconverting isomers) react. This is true, however, only if the rates of interconversion of **A** and **B** are fast relative to their transformation into **C** and **D**, that is, only if equilibrium between **A** and **B** is maintained throughout the reaction.

Equation 10.21 can be generalized for any number of contributing conformations:

$$k_{WH} = \sum_i n_i k_i \tag{10.22}$$

where n_i is the population (mole fraction) of the ith conformer and k_i is the rate constant for its conversion to product or products. In other words, the observed rate constant for a conformationally heterogeneous system, under conditions of fast conformational interchange, is the weighted average of all the individual rate constants for the contributing conformers (weighted by the respective mole fractions). This result is already familiar from the averaging of enthalpies, polarizations, chemical shifts, coupling constants, and so on, which were discussed earlier.

By way of an example of Winstein–Holness kinetics we shall consider the debromination of meso and chiral 2,3-dibromobutane with iodide ion (Young et al., 1939). These reactions are quite highly stereospecific* in that the meso dibromide gives largely (96%) *trans*-2-butene, whereas the chiral dibromide gives largely (91%) *cis*-2-butene (Winstein, Young et al., 1939). These reactions have been interpreted in terms of an E2 mechanism in which the groups to be eliminated must, for maximum reactivity, be antiperiplanar (Fig. 10.40) in the

---

* We call a reaction *stereospecific* if stereoisomeric starting materials give stereoisomeric products, that is, if **A** yields **B** then the stereoisomer **A'** of **A** yields a stereoisomer **B'** of **B** (see also Chapter 12).

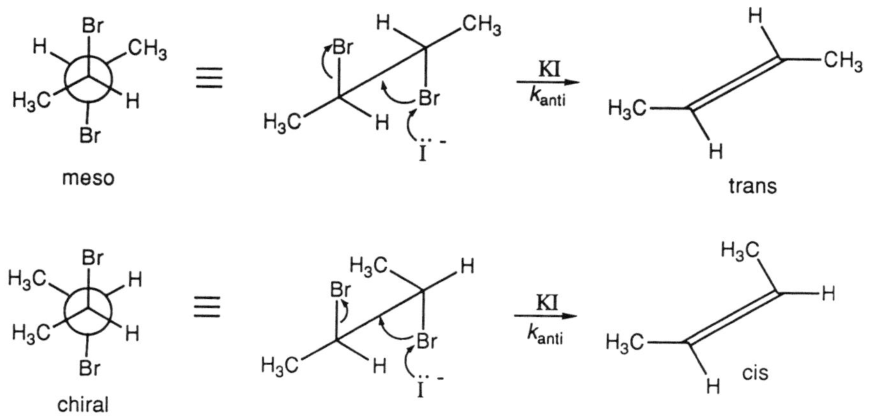

Figure 10.40. Elimination Reaction of $CH_3CHBrCHBrCH_3$ with KI. See also Figure 10.34.

starting material. Referring to Figure 10.34, this means that the meso dibromide needs to be in conformation **A** and the chiral dibromide needs to be in conformation **D**. If the intrinsic reaction rates for iodide-promoted debromination of **A** and **D** are the same, which is not certain, and if it is assumed that no elimination occurs from the other conformations, one has, for the meso isomer (cf. Eq. 10.22)

$$k_{meso} = n_A k_{anti}$$

and for the racemic one

$$k_{rac} = n_D k_{anti}$$

on the assumption that $k_{gauche} = 0$ for conformers **B**, **C**, **E**, and **F**. Here k_{meso} and k_{rac} are the observed reaction rates (k_{WH}) and k_{anti} is the intrinsic rate of reaction for antiperiplanar bromine atoms. It follows, then, that $k_{meso}/k_{rac} = n_A/n_D$, a ratio which, from Table 10.4, is in the 1.3–1.6 region [depending on solvent dielectric; the actual reaction was carried out in acetone, $\varepsilon = 20.7$, which is closest to the pure liquid (Table 10.4) in dielectric]. The experimentally found ratio at 59.7° was 1.93, in reasonable agreement considering the uncertainties of some of the assumptions (for a more detailed treatment of halogen eliminations, see Saunders and Cockerill, 1973).

Additional applications and examples of the Winstein–Holness equation will be discussed in Chapter 11. Whereas this equation deals with reaction *rates*, the conceptually related Curtin–Hammett principle (Curtin, 1954; Seeman, 1983) deals with reaction *products* derived from two starting materials where k_A, $k_B \gg k_C$, k_D, that is, **A** and **B** (Eq. 10.16) remain in equilibrium throughout the reaction. Let us direct attention once more to Eq. 10.16 and Figure 10.39 (kinetics), but this time focusing on the product ratio **D/C**. How will this ratio depend on k_A and k_B (or the equilibrium constant $K = k_A/k_B$)? And how will it depend on the rate constants k_C and k_D? In terms of the energy diagram, these questions relate to the role of the ground-state energy difference ΔG^0, the activation energies $\Delta G_C^\ddagger$ and $\Delta G_D^\ddagger$, and the transition state energy difference $\Delta G^\ddagger$.

In the derivation of the Curtin–Hammett principle, we shall again assume condition 1 above, that is, k_A, $k_B \gg k_C$, k_D. In that case equilibrium is maintained, that is, $[B]/[A] = K$ (Eq. 10.16), $dC/dt = k_C[A]$, and $dD/dt = k_D[B]$, the reaction, as before, being taken to be first order or pseudo first order. It follows that $dD/dC = k_D[B]/k_C[A] = (k_D/k_C)\ ([B]/[A]) = k_D K/k_C$ and since, with the assumption made, the right-hand side is constant, one can integrate to

$$[D]/[C] = k_D K/k_C \tag{10.23}$$

which equation gives the product ratio at the completion of reaction (or at any intermediate point), assuming the initial concentrations of C and D are zero.

Indeed, that $[D]/[C]$ is constant during the course of the reaction is an experimental verification of the Curtin–Hammett assumption.

For a treatment of the Curtin–Hammett and Winstein–Holness kinetics involving second-order reactions, that is, $C \leftarrow R + A \rightleftharpoons B + R \rightarrow D$ (R = second reagent), see Seeman et al. (1980).

Equation 10.23 states that one can calculate the product ratio if one knows the individual rate constants k_C and k_D for the contributing conformers (e.g., from model studies) and the equilibrium constant K between the conformers. Conversely, if one measures the product ratio D/C and knows K, one can calculate the ratio of rate constants, k_D/k_C.

An application of Curtin–Hammett kinetics (to a cyclic substrate) is the determination of the ratio of rate constants k_C and k_D (Fig. 10.41) in the N-oxidation of N-methy-4-tert-butylpiperidine by hydrogen peroxide in acetone (Shvo and Kaufman, 1972). The product ratio D/C is 95:5 in favor of the axial N-oxide D. The ratio of equatorial to axial N-methyl compound (B/A) is found, by quenching kinetics (Crowley, Robinson, et al., 1977) to be about 60.

The determination of the equilibrium constant K is interesting in itself, since it involves the unusual kinetic case 2 (p. 649), where the rate of reaction of conformers A and B, in this case with strong acid, is much faster than the rate of equilibration of A and B. The latter can be estimated from the barrier of nitrogen inversion in piperidine [6.1 kcal mol^{-1} (25.5 kJ mol^{-1}), Anet and Yavari, 1977] to be about 10^8 s^{-1} at room temperature, whereas the rate of protonation, assuming it to be diffusion controlled, is about 10^{10} s^{-1}. By kinetic quenching of N-methyl-4-tert-butylpiperidine (Fig. 10.41, A and B) with acid under conditions where the result was not affected by mixing control (cf. Rys, 1976), Crowley, Robinson, et al.

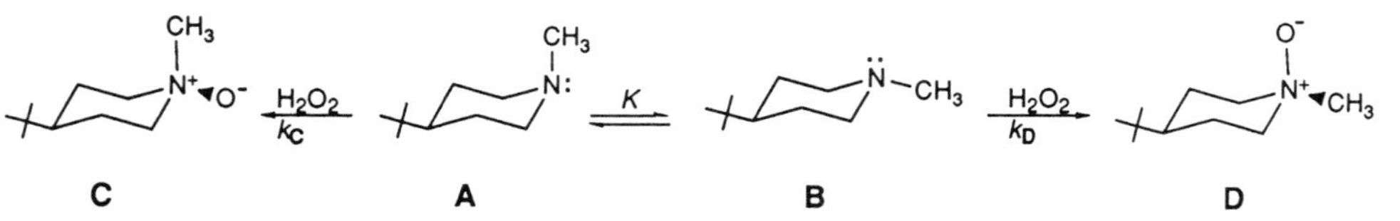

Figure 10.41. Nitrogen Oxidation of N-methyl-4-tert-butylpiperidine.

(1977) determined the conformational energy difference between **A** and **B** to range from 2.4 to 3.15 kcal mol^{-1} (10.1–13.2 kJ mol^{-1}), depending on solvent and phase. The lower value refers to the most hydrogen-bonding solvent chloroform and we have assumed that a similar value applies in aqueous acetone, thus $K \sim 60$.

Applying Eq. 10.23, and assuming pseudo-first-order kinetics, (since H_2O_2 is in large excess) one obtains $95:5 = 60k_D/k_C$ or $k_D/k_C = 0.3$. This result is reasonable, since N oxidation into the more crowded axial position should be slower than equatorial oxidation.

Returning, now, to the derivation of the Curtin–Hammett principle,[*] we note that Eq. 10.23 can be expanded in terms of the free energy and activation free energy parameters displayed in Figure 10.39

$$K = [\mathbf{B}]/[\mathbf{A}] = e^{-\Delta G^0/RT}$$

$$k_C = e^{-\Delta G_C^{\ddagger}/RT}$$

$$k_D = e^{-\Delta G_D^{\ddagger}/RT}$$

Inserting in Eq. 10.23

$$[\mathbf{D}]/[\mathbf{C}] = e^{-\Delta G^0/RT} + e^{-\Delta G_D^{\ddagger}/RT}/e^{-\Delta G_C^{\ddagger}/RT}$$

or

$$[\mathbf{D}]/[\mathbf{C}] = e^{(-\Delta G^0 + \Delta G_C^{\ddagger} - \Delta G_D^{\ddagger})/RT}$$

But, from Figure 10.39, $-\Delta G^0 + \Delta G_C^{\ddagger} - \Delta G_D^{\ddagger} = -\Delta G^{\ddagger}$, where $\Delta G^{\ddagger}$ is the difference in activation energies for the formation of the two products.

It follows that

$$[\mathbf{D}]/[\mathbf{C}] = e^{-\Delta G^{\ddagger}/RT} \tag{10.24}$$

(It should be noted that ΔG^0 and $\Delta G^{\ddagger}$ in Figure 10.39 are vectors pointing downward, that is, ΔG^0 and $\Delta G^{\ddagger}$ are negative; the corresponding upward pointing vectors $\Delta G_C^{\ddagger}$ and $\Delta G_D^{\ddagger}$ are positive, as are $-\Delta G^0$ and $-\Delta G^{\ddagger}$.)

Equation 10.24 seems to imply that the product ratio depends only on the free energy difference of the transition states and not, per se, on the energy difference of the ground states. Curtin (1954) phrased the Curtin–Hammett principle in approximately the following form (see also Gold, 1979): "The relative amounts of products formed from two pertinent conformers are completely independent of the relative populations of the conformers and depend only on the difference in free energy of the transition states, provided the rates of reaction are slower than the rates of conformational interconversion." The principle was interpreted in this way for many years and served mainly as a warning *not* to interpret product ratios

---

[*] Curtin has stated (Curtin, 1954) that the underlying concepts were "pointed out by Professor L.P. Hammett in 1950" and therefore the principle bears Hammett's name as well as Curtin's.

([**D**]/[**C**]) in terms of ground-state conformational equilibrium constants K. However, this interpretation does not allow one to extract quantitative insight, since the difference in energy levels of the transition states ($\Delta G^{\ddagger}$) is not an experimentally determinable quantity. A restated form of the principle (cf. Gold, 1983) is more satisfactory: "*The Curtin–Hammett principle implies that in a chemical reaction that yields one product from one conformer and a different product from another conformer (and provided these two conformers are rapidly interconvertible relative to the rate of product formation, whereas the products do not interconvert), the product composition is not solely dependent on the relative proportions of the conformers in the substrate; it is controlled by the difference in standard Gibbs free energies of the representative transition states. It is also true that the product composition is related to the relative concentrations of the conformers, that is, the conformational equilibrium constant, and the respective rate constants of their reactions; these parameters are, however, often unknown.*"

Yet another way of stating the principle is to say that the product ratio under Curtin–Hammett conditions (conformer equilibration rate faster than rate of product formation) must take into account not only the equilibrium constant of the starting conformers but also the rates of their respective reactions; the sum of the corresponding three free energy terms (ΔG^0, $\Delta G_{\mathrm{C}}^{\ddagger}$, and $\Delta G_{\mathrm{D}}^{\ddagger}$) is expressed in the difference in activation energies of the transition states ($\Delta G^{\ddagger}$), which is directly related to the product ratio.

From the deeper insight gained by a thorough analysis of the scheme in Figure 10.39 it became clear (Seeman, 1983) that maximum information about the kinetic behavior of a system is gained when one focuses not only on the product ratio **C**/**D** but also on the equilibrium constant K and the individual rate constants k_{C} and k_{D}. In fact, k_{C} and k_{D} can be formulated by Eqs. 10.25 and 10.26, derived from Eqs. 10.20 and 10.23 and setting $P = [\mathbf{D}]/[\mathbf{C}]$. The treatment embodied in Figure 10.41 amounts to just that. Another example, due to Seeman

$$k_{\mathrm{C}} = k_{\mathrm{WH}}[(K+1)/(P+1)] \tag{10.25}$$

$$k_{\mathrm{D}} = k_{\mathrm{WH}}[(K+1)/K][P/(P+1)] \tag{10.26}$$

et al., 1980, is shown in Figure 10.42. In this example, additionally, the total rate constant, k_{WH}, was determined and thus, using Eq. 10.21 and knowing $\mathbf{D}/\mathbf{C} = P$ and K, k_{C} and k_{D} could be determined individually by a combination of the Winstein–Holness and Curtin–Hammett kinetic schemes.

The reaction studied was the quaternization of 1-methyl-2-arylpyrrolidines

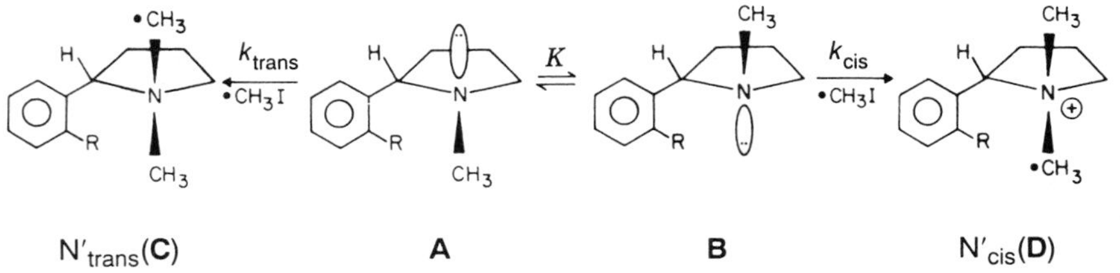

Figure 10.42. Quaternization of *N*-methyl-2-arylpyrrolidines. $k_{\mathrm{cis}} = k_{\mathrm{D}}$; $k_{\mathrm{trans}} = k_{\mathrm{C}}$. The dot indicates ^{13}C.

(Fig. 10.42) in which the 2-aryl group was phenyl or an ortho-substituted phenyl, with the substituent being CH_3, Et, i-Pr, or t-Bu. The (unlabeled) substrates were methylated with $^{13}CH_3I$ and the product ratio (D/C) of the isotopic diastereomers was determined by ^{13}C NMR spectroscopy. The overall rate constant k_{WH} was determined by conductometric titration of the methylation product (quaternary salt) formed and the equilibrium constant K was determined by acid quenching of the starting amines (A, B) and analysis of the product ratio of the ammonium salts so formed by NMR spectroscopy. The results are summarized in Table 10.10.

The following findings from such a complete kinetic analysis are of interest: (a) k_{trans} is about 10–20 times as large as k_{cis}. The aryl group clearly hinders approach from the side of the pyrrolidine ring it occupies. The ratio k_{trans}/k_{cis} increases from about 10 to about 20 when an ortho alkyl group smaller than $tert$-butyl is introduced; the factor of 2 may suggest that the ring turns with the o-alkyl group away from the site of approach of the methyl (the unsubstituted aryl ring, because of its twofold symmetry axis along the aryl–pyrrolidine bond, has twice as many favorable orientations as the substituted ones). With $tert$-butyl the ratio increases to about 15, thus further hindrance clearly comes into play. (Seeman et al., have pointed out that the uncertainty in the measurement of the large K causes greater uncertainty in the value of k_{trans} than in that of k_{cis}; however, the relative values should be reliable in both columns.) (b) $K = [B]/[A]$ increases (though not sharply) as the size of the ortho substituent increases. This is reasonable: The larger the group, the less comfortable it is in being located cis to the N-methyl substituent. Again this ratio for the substituted aryl groups to the unsubstituted one appears to be not much in excess of two, perhaps for reasons similar to those just discussed. (c) The product ratio is nearly constant for the first four compounds; it varies, but only by a factor of 6, for the $tert$-butyl substituted species. This type of constancy in quaternizations is often observed and, in terms of Eq. 10.23, reflects a compensating factor: As the 2-substituent becomes bulkier, K in the expression $[D]/[C] = k_{cis} K/k_{trans}$ becomes larger but k_{cis}/k_{trans} becomes smaller.

We shall return to further examples of the Winstein–Holness equation in Chapter 11.

TABLE 10.10. Experimental Data and Calculated Rate Constant for Quaternization of 1-Methyl-2-arylpyrrolidines[a] with Methyl-^{13}C Iodide[b]

Ortho Substituent R	D/C[c]	k_{WH}[c] $\times 10^4$	K[c]	k_{cis}[d] $\times 10^4$	k_{trans}[d] $\times 10^3$
H	1.7	30.6	17	20	20
CH_3	1.4	7.61	>30	4.6	9.8
C_2H_5	1.3	6.17	>30	3.6	8.0
$(CH_3)_2CH$	1.3	5.32	>30	3.0	6.9
$(CH_3)_3C$	0.28	1.25	>40	0.28	4.0

[a] See Figure 10.42.
[b] Seeman et al. (1980).
[c] Determined experimentally.
[d] $k_{cis} = k_D$; $k_{trans} = k_C$. $k_D/k_C = [D]/K[C]$ from Eq. 10.23, with this ratio known, k_C and k_D can be calculated from Eq. 10.20.

REFERENCES

Abe, A. (1976), *J. Am. Chem. Soc.*, **98**, 6477.

Abe, A., Inomata, K., Tanisawa, E., and Ando, I. (1990), *J. Mol. Struct.*, **238**, 315.

Abe, A. and Mark, J. E. (1976), *J. Am. Chem. Soc.*, **98**, 6468.

Abraham, R. J. and Bretschneider, E. (1974), "Medium Effects on Rotational and Conformational Equilibria," in Orville-Thomas, W. J., Ed., *Internal Rotation in Molecules*, Wiley, New York., p. 481.

Abraham, R. J. and Kemp, R. H. (1971), *J. Chem. Soc. B*, 1240.

Abraham, R. J., Chadwick, D. J., and Sancassan, F. (1982), *Tetrahedron*, **38**, 1485.

Abraham, R. J. and Pople, J. A. (1960), *Mol. Phys.*, **3**, 609.

Abraham, R. J. and Stölevik, R. (1981), *Chem. Phys. Lett.*, **77**, 181.

Allen, B. and Fewster, S. (1974), "Torsional vibrations and rotational isomerism," in Orville-Thomas, W. J., Ed., *Internal Rotation in Molecules*, Wiley, New York, p. 255.

Allinger, N. L. (1957), *J. Am. Chem. Soc.*, **79**, 3443.

Allinger, N. L. and Chang, S. H. M. (1977), *Tetrahedron*, **33**, 1561.

Allinger, N. L., Grew, R. S., Yates, B. F., and Schaefer, H. F. (1990), *J. Am. Chem. Soc.*, **112**, 114.

Allinger, N. L., Hindmann, D., and Hönig, H. (1977), *J. Am. Chem. Soc.*, **99**, 3282.

Anderson, J. E. (1992), "Conformational analysis of acyclic and alicyclic saturated hydrocarbons," in Patai, S. and Rappoport, Z., Eds., *The Chemistry of Alkanes and Cycloalkanes*, Wiley, New York, p. 95.

Andersson, M. and Karlström, G. (1985), *J. Phys. Chem.*, **89**, 4957.

Andersson, S., Carter, R. E., and Drakenberg, T. (1984), *Acta Chem. Scand.*, **B38**, 579.

Anet, F. A. L. (1962), *J. Am. Chem. Soc.*, **84**, 747.

Anet, F. A. L. and Anet, R. (1971), "Configuration and Conformation by NMR," in Nachod, F. C. and Zuckerman, J. J., Eds., *Determination of Organic Structures by Physical Methods*, Vol. 3, Academic, New York, p. 343.

Anet, F. A. L. and Basus, V. J. (1978), *J. Magn. Reson.*, **32**, 339.

Anet, F. A. L. and Yavari, I. (1977), *J. Am. Chem. Soc.*, **99**, 2794, 6752.

Astrup, E. E. (1971), *Acta Chem. Scand.*, **25**, 1494.

Astrup, E. E. (1973), *Acta Chem. Scand.*, **27**, 3271.

Astrup, E. E. (1979), *Acta Chem. Scand.*, **A33**, 655.

Bader, R. F. W., Cheeseman, J. R., Laidig, K. E., Wiberg, K. B., and Breneman, C. (1990), *J. Am. Chem. Soc.*, **112**, 6530.

Bailey, W. F. and Eliel, E. L. (1974), *J. Am. Chem. Soc.*, **96**, 1798.

Bair, R. A. and Goddard, W. A. (1982), *J. Am. Chem. Soc.*, **104**, 2719.

Bartell, L. S. (1977), *J. Am. Chem. Soc.*, **99**, 3279.

Bartell, L. S. and Boates, T. L. (1976), *J. Mol. Struct.*, **32**, 379.

Bartell, L. S. and Kohl, D. A. (1963), *J. Chem. Phys.*, **39**, 3097.

Bastiansen, O., Seip, H. M., and Boggs, J. E. (1971), "Conformational Equilibria in the Gas Phase," in Dunitz, J. D. and Ibers, J. A., Eds., *Perspectives in Structural Chemistry*, Vol. 4, Wiley, New York, p. 60.

Bax, A., Freeman, R., and Kempsell, S. P. (1980a), *J. Magn. Reson.*, **41**, 349.

Bax, A., Freeman, R., and Kempsell, S. P. (1980b), *J. Am. Chem. Soc.*, **102**, 4849.

Berg, U. and Sandström, J. (1989), "Stereochemistry of Alkyl Groups," *Adv. Phys. Org. Chem.*, **25**, 1.

Birdsall, N. J. M., Partington, P., Datta, N., Mondal, P., and Pauling, P. J. (1980), *J. Chem. Soc. Perkin 2*, 1415.

Bischoff, C. A. (1890), *Ber. Dtsch. Chem. Ges.*, **23**, 623.

Bischoff, C. A. (1891), *Ber. Dtsch. Chem. Ges.*, **24**, 1085.

Blom, C. E. and Bauder, A. (1982), *Chem. Phys. Lett.*, **88**, 55.

Blom, C. E., Grassi, G., and Bauder, A. (1984), *J. Am. Chem. Soc.*, **106**, 7427.

Blom, C. E. and Günthard, H. H. (1981), *Chem. Phys. Lett.*, **84**, 267.

Bock, C. W., George, P., Trachtman, M., and Zanger, M. (1979), *J. Chem. Soc. Perkin 2*, 26.

Bock, C. W., Trachtman, M., and George, P. (1985), *Chem. Phys.*, **93**, 431.

Bolton, K., Lister, D. G., and Sheridan, J. (1974), *J. Chem. Soc. Faraday Trans. 2*, **70**, 113.

Booth, H. (1965), *Tetrahedron Lett.*, 411.

Booth, H. and Jozefowicz, M. L. (1976), *J. Chem. Soc. Perkin 2*, 895.

Bose, P. K., Henderson, D. O., Ewig, C. S., and Polavarapu, P. L. (1989), *J. Phys. Chem.*, **93**, 5070.

Bothner-By, A. A. and Naar-Colin, C. (1962), *J. Am. Chem. Soc.*, **84**, 743.

Bovey, F. A. (1969), *Polymer Conformation and Configuration*, Academic, New York.

Bowen, J. P., Pathiaseril, A., Profeta, S., and Allinger, N. L. (1987), *J. Org. Chem.*, **52**, 5162.

Boyd, R. H. (1975), *J. Am. Chem. Soc.*, **97**, 5353.

Boyd, R. H. and Kesner, L. (1981), *J. Polymer Sci. Polymer Phys. Ed.*, **19**, 375.

Bradford, W. F., Fitzwater, S., and Bartell, L. S. (1977), *J. Mol. Struct.*, **38**, 185.

Breen, P. J., Bernstein, E. R., and Seeman, J. I. (1987), *J. Chem. Phys.*, **87**, 3269.

Breen, P. J., Bernstein, E. R., Secor, H. V., and Seeman, J. I., (1989), *J. Am. Chem. Soc.*, **111**, 1958.

Breen, P. J., Warren, J. A., Bernstein, E. R., and Seeman, J. I. (1987), *J. Chem. Phys.*, **87**, 1927.

Breulet, J., Lee, T. J., and Schaefer, H. F. (1984), *J. Am. Chem. Soc.*, **106**, 6250.

Brunet, E., García-Ruano, J. L., Hoyos, M. A., Rodríguez, J. H., Prados, P., and Alcudia, F. (1983), *Org. Magn. Reson.*, **21**, 643.

Burkert, U. (1980), *J. Comp. Chem.*, **1**, 285.

Bushweller, C. H., Fleischman, S. H., Grady, G. L., McGoff, P., Rithner, C. D., Whalon, M. R., Brennan, J. G., Marcantonio, R. P., and Domingue, R. P. (1982), *J. Am. Chem. Soc.*, **104**, 6224.

Bushweller, C. H., Whalon, M. R., and Laurenzi, B. J. (1981), *Tetrahedron Lett.*, **22**, 2945.

Butcher, S. S., Cohen, R. A. and Rounds, T. C. (1971), *J. Chem. Phys.*, **54**, 4123.

Butcher, S. S. and Wilson, E. B., (1964), *J. Chem. Phys.*, **40**, 1671.

Bykov, G. V. (1975), "The Conceptual Premises of Conformational Analysis in the Work of C. A. Bischoff", in Ramsay, O. B., ed., *van't Hoff-Le Bel Centennial*, American Chemical Society Symposium Series, 12, American Chemical Society, Washington, DC, p. 114.

Cais, R. E. and Brown, W. L. (1980), *Macromolecules*, **13**, 801.

Carman, C. J., Tapley, A. R., and Goldstein, J. H. (1971), *J. Am. Chem. Soc.*, **93**, 2864.

Carreira, L. A. (1975), *J. Chem. Phys.*, **62**, 3851.

Carreira, L. A. (1976), *J. Phys. Chem.*, **80**, 1149.

Chandler, D. (1978), *Discuss, Faraday, Soc.*, **66**, 184.

Chapput, A., Roussel, B., and Fleury, G. (1974), *J. Raman Spectrosc.*, **2**, 117.

Chia, L. H. L., Huang, E., and Huang, H.-H. (1973), *J. Chem. Soc. Perkin 2*, 766.

Cohen, A. D., Sheppard, N., and Turner, J. J. (1958), *Proc. Chem. Soc. London*, 118.

Compton, D. A. C., Montero, S., and Murphy, W. F. (1980), *J. Phys. Chem.*, **84**, 3587.

Conroy, H. (1960), "Nuclear Magnetic Resonance in Organic Structural Elucidation," *Adv. Org. Chem.*, **2**, 265.

Courtieu, J., Gounelle, Y., Gonord, P., and Kan, S. K. (1974), *Org. Magn. Reson.*, **6**, 151.

Crowley, P. J., Robinson, M. J. T., and Ward, M. G. (1977), *Tetrahedron*, **33**, 915.

Curl, R. F. (1959), *J. Chem. Phys.*, **30**, 1529.

Curtin, D. Y. (1954), "Stereochemical Control of Organic Reactions Differences in Behavior of Diastereomers," *Rec. Chem. Prog.*, **15**, 111.

Dale, J. (1978), *Stereochemistry and Conformational Analysis*, Verlag Chemie, New York, NY.

Darsey, J. A. and Rao, B. K. (1981), *Macromolecules*, **14**, 1575.

De Maré, G. R. (1981), "Rotational Barriers in Vinyl Compounds," in Czismadia, I. G. and Daudel, R., Eds., *Computational Theoretical Organic Chemistry NATO Adv. Study Inst.*, Ser. C, Reidel, Boston, MA; Vol. 67, p. 335.

De Maré, G. R. (1984), *J. Mol. Struct.*, **107**, 127.

Dixon, D. A. and Smart, B. E. (1988), *J. Phys. Chem.*, **92**, 2729.

Dosěn-Mićović, L., Jeremić, D., and Allinger, N. L. (1983), *J. Am. Chem. Soc.*, **105**, 1723.

Dosên-Mićović, L. and Zigman, V. (1985), *J. Chem. Soc. Perkin 2*, 625.

Doskočilová, D. (1964), *J. Polymer Sci. B*, **2**, 421.

Drakenberg, T. and Forsén, S. (1971), *J. Chem. Soc. D*, 1404.

Drakenberg, T., Sommer, J., and Jost, R. (1980), *J. Chem. Soc. Perkin 2*, 363.

Duddeck, H. (1986), "Substituent Effects on ^{13}C Chemical Shifts in Aliphatic Molecular Systems. Dependence on Constitution and Stereochemistry," *Top. Stereochem.*, **16**, 219.

Dunlap, B. I. and Cook, M. (1986), *Int. J. Quantum Chem.*, **29**, 767.

Durig, J. R., Bucy, W. E., and Cole, A. R. H. (1975), *Can. J. Phys.*, **53**, 1832.

Durig, J. R. and Compton, D. A. C. (1978), *J. Chem. Phys.*, **69**, 4713.

Durig, J. R. and Compton, D. A. C. (1979), *J. Phys. Chem.*, **83**, 2873.

Durig, J. R. and Compton, D. A. C. (1980), *J. Phys. Chem.*, **84**, 773.

Durig, J. R., Compton, D. A. C., and Jalilian, M.-R. (1979), *J. Phys. Chem.*, **83**, 511.

Durig, J. R., Compton, D. A. C., and McArver, A. Q. (1980), *J. Chem. Phys.*, **73**, 719.

Durig, J. R., Craven, S. M., and Bragin, J. (1970), *J. Chem. Phys.*, **53**, 38.

Durig, J. R., Godbey, S. E., and Sullivan, J. F. (1984), *J. Chem. Phys.*, **80**, 5983.

Durig, J. R., Groner, P., and Griffin, M. G. (1977), *J. Chem. Phys.*, **66**, 3061.

Durig, J. R., Liu, J., Little, T. S., and Kolasinsky, V. F. (1992), *J. Phys. Chem.*, **96**, 8224.

Eberson, L. (1959), *Acta Chem. Scand.*, **13**, 203.

Edward, J. T. (1955), *Chem. Ind. (London)*, 1102.

Eliel, E. L. (1959), *Chem. Ind. (London)*, 568.

Eliel, E. L. (1962), *Stereochemistry of Carbon Compounds*, McGraw-Hill, New York, p. 217.

Eliel, E. L. (1969), "The Rabbit Ear Effect. Polar interactions of heteroatoms," *Kem. Tidskr.*, **81**, 6/7, 22.

Eliel, E. L. (1970), "Conformational Analysis in Saturated Heterocycles," *Acc. Chem. Res.*, **3**, 1.

Eliel, E. L. (1972), "Conformational Analysis in Heterocyclic Systems: Recent Results and Applications," *Angew. Chem. Int. Ed. Engl.*, **11**, 739.

Eliel, E. L. (1985), *J. Mol. Struct.*, **126**, 385.

Eliel, E. L., Allinger, N. L., Angyal, S. J., and Morrison, G. A. (1965), *Conformational Analysis*, Interscience-Wiley, New York, (reprinted 1981 by Am. Chem. Soc., Washington, DC).

Eliel, E. L., Kandasamy, D., Yen, C.-y., and Hargrave, K. D. (1980), *J. Am. Chem. Soc.*, **102**, 3698.

Eliel, E. L. and Lukach, C. A. (1957), *J. Am. Chem. Soc.*, **79**, 5986.

Eliel, E. L. and Ro, R. S. (1956), *Chem. Ind. London*, 251.

Epiotis, N. D., Yates, R. L., Larson, J. R., Kirmaier, C. R., and Bernardi, F. (1977), *J. Am. Chem. Soc.*, **99**, 8379.

Ernst, L. and Trowitzsch, W. (1974), *Chem. Ber.*, **107**, 3771.

Fantoni, R., van Helroort, K., Knippers, W., and Reuss, J. (1986), *Chem. Phys.*, **110**, 1.

Farina, M. (1987), "The Stereochemistry of Linear Macromolecules," *Top. Stereochem.*, **17**, 1.

Fateley, W. G., Harris, R. K. Miller, F. A., and Witkowski, R. E. (1965), *Spectrochim. Acta*, **21**, 231.

Felder, P. and Günthard, H. H. (1980), *Spectrochim. Acta*, **36A**, 223.

Felder, P. and Günthard, H. H. (1984), *Chem. Phys.*, **85**, 1.

Fernholt, L. and Kveseth, K. (1978), *Acta Chem. Scand.*, **A32**, 63.

Fernholt, L. and Kveseth, K. (1980), *Acta Chem. Scand.*, **A34**, 163.

Fink, W. H. and Allen, L. C. (1967), *J. Chem. Phys.*, **46**, 2261, 2276.

Flory, P. J. (1953), *Principles of Polymer Chemistry*, Cornell University Press, Ithaca, NY.

Flory, P. J. (1969), *Statistical Mechanics of Chain Molecules*, Interscience-Wiley, New York.

Flory, P. J. and Williams, A. D. (1969), *J. Am. Chem. Soc.*, **91**, 3118.

Franck, R. W. (1983), *Tetrahedron*, **39**, 3251.

Frankiss, S. C. and Kynaston, W. (1972), *Spectrochim. Acta.*, **A28**, 2149.

Friesen, D. and Hedberg, K. (1980), *J. Am. Chem. Soc.*, **102**, 3987.

Fujiyama, T. and Kakimoto, M. (1976), *Bull. Chem. Soc. Jpn.*, **49**, 2346.

Furukawa, Y., Takeuchi, H., Harada, I., and Tasumi, M. (1983), *Bull. Chem. Soc. Jpn.*, **56**, 392.

George, W. O. and Goodfield, J. E. (1980), "Vibrational Spectra at Variable Temperature and the Determination of Energies between Conformers" in Durig, J. R., Ed., *Analytical Applications of FT-IR to Molecular and Biological Systems, NATO Advanced Study Series*, Ser. C, Vol. 57, Reidel, Boston, MA, p. 293.

George, W. O., Hassid, D. V., and Maddams, W. F. (1972), *J. Chem. Soc. Perkin 2*, 400.

Glick, R. E. and Bothner-By, A. A. (1956), *J. Chem. Phys.*, **25**, 362.

Gold, V. (1979), *Pure Appl. Chem.*, **51**, 1725.

Gold, V. (1983), *Pure Appl. Chem.*, **55**, 1281.

Goodwin, A. R. H. and Morrison, G. (1992), *J. Chem. Phys.*, **96**, 5521.

Goodwin, T. H., Przybylska, M., and Robertson, J. M. (1950), *Acta Cryst.*, **3**, 279.

Gordon, A. J. and Ford, R. A. (1972), *The Chemist's Companion*, Wiley, New York, p. 4.

Graczyk, P. P. and Mikołajczyk, M. (1994), "Anomeric Effect: Origin and Consequences," *Top. Stereochem.*, **21**, 159.

Grant, D. M., Pugmire, R. J., Livingston, R. C., Strong, K. A., McMurry, H. L., and Brugger, R. M. (1970), *J. Chem. Phys.*, **52**, 4424.

Grassian, V. H., Bernstein, E. R., Secor, H. V., and Seeman, J. I. (1989), *J. Phys. Chem.*, **93**, 3470.

Grassian, V. H., Bernstein, E. R., Secor, H. V., and Seeman, J. I. (1990), *J. Phys. Chem.*, **94**, 6691.

Grenier-Loustalot, M. F. and Grenier, P. (1982), *Eur. Polymer J.*, **18**, 493.

Haasnoot, C. A. G., de Leeuw, F. A. A. M., and Altona, C. (1980), *Tetrahedron*, **36**, 2783.

Hagen, K. and Hedberg, K. (1984), *J. Am. Chem. Soc.*, **106**, 6150.

Hammarström, L.-G., Liljefors, T., and Gasteiger, J. (1988), *J. Comp. Chem.*, **9**, 424.

Harris, W. C., Holtzclaw, J. R., and Kalasinsky, V. F. (1977), *J. Chem. Phys.*, **67**, 3330.

Hartge, U. and Schrumpf, G. (1981), *J. Chem. Res. (S)*, 189.

Hayashi, M. and Kato, H. (1980), *Bull. Chem. Soc. Jpn.*, **53**, 2701.

Heenan, R. K. and Bartell, L. S. (1983), *J. Chem. Phys.* **78**, 1270.

Heinrich, F. and Lüttke, W. (1977), *Chem. Ber.*, **110**, 1246.

Hellmann, G., Hellmann, S., Beckhaus, H.-D., and Rüchardt, C. (1982), *Chem. Ber.*, **115**, 3364.

Hermans, P. H. (1924), *Z. Phys. Chem.*, **113**, 337.

Herschbach, D. R. and Krisher, L. C. (1958), *J. Chem. Phys.*, **28**, 728.

Hirano, T., Nonoyama, S., Miyajima, T., Kurita, Y., Kawamuru, T., and Sato, H. (1986), *J. Chem. Soc. Chem. Commun.*, 606.

Hirota, E., Saito, S., and Endo, Y. (1979), *J. Chem. Phys.*, **71**, 1183.

Hirota, M., Sekiya, T., Abe, K., Tashiro, H., Karatsu, M., Nishio, M., and Osawa, E. (1983), *Tetrahedron*, **39**, 3091.

Hirota, E., Sugisaki, R., Nielsen, C. J., and Sorensen, G. O. (1974), *J. Mol. Spectroscop.*, **49**, 251.

Höfner, D., Tamir, I., and Binsch, G. (1978), *Org. Magn. Reson.*, **11**, 172.

Hollas, J. M. and Ridley, T. (1980), *Chem. Phys. Lett.*, **75**, 94.

Hoppilliard, Y. and Solgadi, D. (1980), *Tetrahedron*, **36**, 377.

Hounshell, W. D., Dougherty, D. A., and Mislow, K. (1978), *J. Am. Chem. Soc.*, **100**, 3149.

Huber-Wälchli, P. (1978), *Ber. Bunsen-Ges. Phys. Chem.*, **82**, 10.

Huber-Wälchli, P. and Günthard, H. H. (1981), *Spectrochim. Acta.*, **37A**, 285.

Huggins, M. L. (1953), *J. Am. Chem. Soc.*, **75**, 4123.

Iimura, K. (1969), *Bull. Chem. Soc. Jpn.*, **42**, 3135.

Iimura, K., Kawakami, N., and Takeda, M. (1969), *Bull. Chem. Soc. Jpn.*, **42**, 2091.

Im, H.-S., Bernstein, E.R., Secor, H. V., and Seeman, J. I. (1991) *J. Am. Chem. Soc.*, **113**, 4422.

Jeffrey, G. A., Pople, J. A., Binkley, J. S., and Vishveshwara, S. (1978), *J. Am. Chem. Soc.*, **100**, 373.

Jeffrey, G. A. and Yates, J. H. (1979), *J. Am. Chem. Soc.*, **101**, 820.

Jing, X. and Krimm, S. (1983), *Spectrochim. Acta*, **39A**, 251.

Jones, G. I. L. and Owen, N. L. (1973), *J. Mol. Struct.*, **18**, 1.

Jorgensen, W. L. (1981), *J. Am. Chem. Soc.*, **103**, 4721.

Jorgensen, W. L. and Ibrahim, M. (1981), *J. Am. Chem. Soc.*, **103**, 3976.

Juaristi, E. and Cuevas, G. (1992), "Recent Studies of the Anomeric Effect," *Tetrahedron*, **48**, 5019.

Karabatsos, G. J. and Fenoglio, D. J. (1970), "Rotational Isomerism about sp^2–sp^3 Carbon–Carbon Single Bonds," *Top. Stereochem.*, **5**, 167.

Karplus, M. (1959), *J. Chem. Phys.*, **30**, 11.

Karplus, M. (1963), *J. Am. Chem. Soc.*, **85**, 2870.

Käss, D., Oberhammer, H., Brandes, D., and Blaschette, A. (1977), *J. Mol. Struct.*, **40**, 65.

Kemp, J. D. and Pitzer, K. S. (1936), *J. Chem. Phys.*, **4**, 749.

Kilb, R. W., Lin, C. C., and Wilson, E. B. (1957), *J. Chem. Phys.*, **26**, 1695.

Kimura, K. and Osafune, K. (1975), *Bull. Chem. Soc. Jpn.*, **48**, 2421.

Kingsbury, C. A. and Best, D. C. (1967), *J. Org. Chem.*, **32**, 6.

Kirby, A. J. (1983), *The Anomeric Effect and Related Stereoelectronic Effects at Oxygen*, Springer, New York.

Kirtman, B. and Palke, W. E. (1977), *J. Chem. Phys.*, **67**, 5980.

Kiss, A. I. and Lukovits, I. (1979), *Chem. Phys. Lett.*, **65**, 169.

Kitagawa, T. and Miyazawa, T. (1968), *Bull. Chem. Soc. Jpn.*, **41**, 1976.

Kitano, M. and Kuchitsu, K. (1973), *Bull. Chem. Soc. Jpn.*, **46**, 3048.

Kitano, M. and Kuchitsu, K. (1974a), *Bull. Chem. Soc. Jpn.*, **47**, 67.

Kitano, M. and Kuchitsu, K. (1974b), *Bull. Chem. Soc. Jpn.*, **47**, 631.

Kitano, M., Fukuyama, T., and Kuchitsu, K. (1973), *Bull. Chem. Soc. Jpn.*, **46**, 384.

Kohata, K., Fukuyama, T., and Kuchitsu, K. (1982), *J. Phys. Chem.*, **86**, 602.

Kohlrausch, K. W. F. (1932), *Z. Phys. Chem.*, **B18**, 61.

Kříž, J. and Jakeš, J. (1972), *J. Mol. Struct.*, **12**, 367.

Kuhn, L. P. (1952), *J. Am. Chem. Soc.*, **74**, 2492.

Kuhn, L. P. (1958), *J. Am. Chem. Soc.*, **80**, 5950.

Kveseth, K. (1974), *Acta Chem. Scand.*, **A28**, 482.

Kveseth, K. (1975), *Acta Chem. Scand.*, **A29**, 307.

Kveseth, K., Seip, R., and Kohl, D. A. (1980), *Acta Chem. Scand.*, **A34**, 31.

Kwei, G. H. and Curl, R. F. (1960), *J. Chem. Phys.*, **32**, 1592.

Lambert, J. B., Nienhuis, R. J., and Finzel, R. B. (1981), *J. Phys. Chem.*, **85**, 1170.

Lees, R. M. and Baker, J. G. (1968), *J. Chem. Phys.*, **48**, 5299.

Lemieux, R. U. (1971), *Pure Appl. Chem.*, **25**, 527.

Levy, G. C., Pehk, T., and Lippmaa, E. (1980), *Org. Magn. Reson.*, **14**, 214.

Li, S., Bernstein, E. R., Secor, H. V., and Seeman, J. I. (1991), *Tetrahedron Lett.*, **32**, 3945.

Lit, E. S., Mallon, F. K., Tsai, H. Y., and Roberts, J. D. (1993), *J. Am. Chem. Soc.*, **115**, 9563.

Long, D. A. (1985), *J. Mol. Struct.*, **126**, 9.

Lowe, J. P. (1968), "Barriers to Internal Rotation about Single Bonds," *Prog. Phys. Org. Chem.*, **6**, 1.

Luke, B. T., Pople, J. A., Krogh-Jesperson, M. B., Apeloig, Y., Chandrasekhar, J., and Schleyer, P. v. R. (1986), *J. Am. Chem. Soc.*, **108**, 260.

Lunazzi, L., Macciantelli, D., Bernardi, F., and Ingold, K. U. (1977), *J. Am. Chem. Soc.*, **99**, 4573.

Maleknia, S., Friedman, B. R., Abedi, N., and Schwartz, M. (1980), *Spectrosc. Lett.*, **13**, 777.

Marshall, J. L. (1983), *Carbon–Carbon and Carbon–Proton NMR Couplings: Applications to Organic Stereochemistry and Conformational Analysis*, Verlag Chemie, Deerfield Beach, FL.

Matsuura, H., Miyauchi, N., Murata, H., and Sakakibara, M. (1977), *Bull. Soc. Chim. Jpn.*, **52**, 344.

Meyer, A. Y. and Ohmichi, N. (1981), *J. Mol. Struct.*, **73**, 145.

Meyer, A. Y. (1981), *J. Comp. Chem.*, **2**, 384.

Meyer, A. Y. (1983), *J. Mol. Struct.*, **94**, 95.

Meyer, A. Y., Allinger, N. L. and Yuh, Y. (1980), *Isr. J. Chem.*, **20**, 57.

Meyer, A. Y., and Ohmichi, N. (1981), *J. Mol. Struct.*, **73**, 145.

Miller, A. and Scott, D. W. (1978), *J. Chem. Phys.*, **68**, 1317.

Miyasaka, T., Kinai, Y., and Imamura, Y. (1981), *Makromol. Chem.*, **182**, 3533.

Mizushima, S.-I. (1954), *Structure of Molecules and Internal Rotation*, Academic, New York.

Mizushima, S., Morino, Y., and Higasi, K. (1934), *Sci. Pap. Inst. Phys. Chem. Res. Tokyo*, **25**, 159.

Momicchioli, F., Baraldi, I., and Bruni, M. C. (1982), *Chem. Phys.*, **70**, 161.

Montaudo, G., Librando, V., Caccamese, S., and Maravigna, P. (1973), *J. Am. Chem. Soc.*, **95**, 6365.

Morino, Y. (1985), *J. Mol. Struct.*, **126**, 1.

Mui, P. W. and Grunwald, E. (1982), *J. Am. Chem. Soc.*, **104**, 6562.

Nasybullina, R. K., Mar'yakhin, R. Kh., and Vigdergauz, M. S. (1973), *Bull. Akad. Nauk, USSR, Div. Chem. Sci.*, **22**, 800, Engl. Transl. 778.

Neu, J. T. and Gwinn, W. D. (1950), *J. Chem. Phys.*, **18**, 1642.

Norris, R. D. and Binsch, G. (1973), *J. Am. Chem. Soc.*, **95**, 182.

Nunes, M. T., Gil, V. M. S., and Ascenso, J. (1981), *Tetrahedron*, **37**, 611.

Oelfke, W. C. and Gordy, W. (1969), *J. Chem. Phys.*, **51**, 5336.

Ogawa, Y., Ohta, M. Sakakibara, M., Matsuura, H., Harada, I., and Shimanouchi, T. (1977), *Bull. Chem. Soc. Jpn.*, **50**, 650.

Ohsaku, M. and Imamura, A. (1978), *Macromolecules*, **11**, 970.

Okazawa, N. and Sorensen, T. S. (1978), *Can. J. Chem.*, **56**, 2737.

Ōsawa, E., Shirahama, H., and Matsumoto, T. (1979), *J. Am. Chem. Soc.*, **101**, 4824.

Owen, N. L. (1974), "Studies of Internal Rotation by Microwave Spectroscopy," in Orville-Thomas, W., Ed., *Internal Rotation in Molecules*, Wiley, New York.

Oyama, T. and Shiokawa, K. (1983), *Polym. J.*, **15**, 207.

Oyanayi, K. and Kuchitsu, K. (1978), *Bull. Chem. Soc. Jpn.*, **51**, 2237.

Pachler, K. G. R. and Wessels, P. L. (1970), *J. Mol. Struct.*, **6**, 471.

Park, P. J. D., Pethrick, R. A., and Thomas, B. N. (1974), "Infrared and Raman band intensities and conformational change," in Orville-Thomas, W. J., Ed., *Internal Rotation in Molecules*, Wiley, New York, p. 57.

Parr, W. J. E. and Schaefer, T. (1977), *J. Am. Chem. Soc.*, **99**, 1033.

Pausch, J. B. (1975), *Chromatographia*, **8**, 80.

Perricaudet, M. and Pullman, A. (1973), *Int. J. Peptide Protein Res.*, **5**, 99.

Phillipi, M. A., Wiersema, R. J., Brainard, J. R., and London, R. E., (1982), *J. Am. Chem. Soc.*, **104**, 7333.

Pickering, R. A. and Price, C. C. (1958), *J. Am. Chem. Soc.*, **80**, 4931.

Pitzer, K. S. (1940), "Chemical Equilibria, Free Energies and Heat Contents for Gaseous Hydrocarbons," *Chem. Rev.*, **27**, 39.

Pitzer, K. S. (1951), "Potential Energies for Rotations about Single Bonds," *Discuss. Faraday Soc.*, **10**, 66.

Pitzer, K. S. and Scott, D. W. (1943), *J. Am. Chem. Soc.*, **65**, 803.

Pitzer, R. M. (1983), "The Barrier to Internal Rotation in Ethane," *Acc. Chem. Res.*, **16**, 207.

Rademacher, P. and Elling, W. (1979), *Justus Liebigs Ann. Chem.*, 1473.

Radom, L., Lathan, W. A., Hehre, W. J., and Pople, J. A. (1972), *Austr. J. Chem.*, **25**, 1601.

Radom, L., Latham, W. A., Hehre, W. J., and Pople, J. A. (1973), *J. Am. Chem. Soc.*, **95**, 693.

Raghavachari, K. (1984), *J. Chem. Phys.*, **81**, 1383.

Reisse, J. (1971), "Quantitative Conformational Analysis of Cyclohexane Systems," in Chiordoglu, G., Ed., *Conformational Analysis*, Academic, New York, p. 219.

Ritter, W., Hull, W., and Cantow, H.-J. (1978), *Tetrahedron Lett.*, 3093.

Riveros, J. M. and Wilson, E. B. (1967), *J. Chem. Phys.*, **46**, 4605.

Romers, C., Altona, C., Buys, H. R., and Havinga, E. (1969), "Geometry and Conformational Properties of Some Five- and Six-membered Heterocyclic Compounds Containing Oxygen or Sulfur," *Top. Stereochem.*, **4**, 39.

Rozhnov, A. M. and Nesterova, T. N. (1973), *Russian J. Phys. Chem.*, **47**, 2455, Engl. Transl. 1390.

Rozhnov, A. M., Nesterova, T. N., and Alenin, V. I. (1975), *J. Struct. Chem. USSR*, **16**, 545, Engl. Transl. 512.

Rüchardt, C. and Beckhaus, H.-D. (1980), "Towards and Understanding of the Carbon–Carbon Bond," *Angew. Chem. Int. Ed. Engl.*, **19**, 429.

Rudolph, H. D., Dreizler, H., Jaeschke, H., and Wendling, P. (1967), *Z. Naturforsch.*, **A22**, 940.

Rudolph, H. D., Walzer, K., and Krutzik, I. (1973), *J. Mol. Spectrosc.*, **47**, 314.

Ruschin, S. and Bauer, S. H. (1980), *J. Phys. Chem.*, **84**, 3061.

Rys, P. (1976), "Disguised Chemical Selectivities," *Acc. Chem. Res.*, **9**, 345.

Sakakibara, M., Matsuura, H., Harada, I., and Shimanouchi, T. (1977), *Bull. Chem. Soc. Jpn.*, **50**, 111.

Saran, A. and Chatterjee, C. L. (1980), *J. Mol. Struct.*, **65**, 185.

Satoh, S. (1964), *J. Polymer Sci. A*, **2**, 5221.

Saunders, W. H. and Cockerill, A. F. (1973), *Mechanisms of Elimination Reactions*, Wiley, New York, pp. 332–376.

Schaefer, T., Hruska, F., and Kotowycz, G. (1965), *Can. J. Chem.*, **43**, 75.

Schneider, H.-J. (1972), *Justus Liebigs Ann. Chem.*, **761**, 150.

Schneider, H.-J., Becker, G., Freitag, W., and Hoppen, V. (1979), *J. Chem. Res. (S)*, 14; (M), 421.

Schneider, H.-J. and Lonsdorfer, M. (1981), *Org. Magn. Reson.*, **16**, 133.

Schwendeman, R. H. and Rogowski, R. S. (1969), *J. Chem. Phys.*, **50**, 397.

Seeman, J. I. (1983), "Effect of Conformational Change on Reactivity in Organic Chemistry. Evaluations, Applications, and Extensions of Curtin–Hammett/Winstein–Holness Kinetics," *Chem. Rev.*, **83**, 83.

Seeman, J. I. (1986), *J. Chem. Educ.*, **63**, 42.

Seeman, J. I. and Farone, W. A. (1978), *J. Org. Chem.*, **43**, 1854.

Seeman, J. I., Grassian, V. H., and Bernstein, E. R. (1988), *J. Am. Chem., Soc.*, **110**, 8542.

Seeman, J. I., Sanders, E. B., and Farone, W. A. (1980), *Tetrahedron*, **36**, 1173.

Seeman, J. I., Secor, H. V., Breen, P. J., Grassian, V. H., and Bernstein, E. R. (1989), *J. Am. Chem. Soc.*, **111**, 3140.

Seeman, J. I., Secor, H. V., Disselkamp, R., and Bernstein, E. R. (1992), *J. Chem. Soc. Chem. Commun.*, 713.

Seeman, J. I., Secor, H. V., Hartung, H., and Galzerano, R. (1980), *J. Am. Chem. Soc.*, **102**, 7741.

Seki, W. and Choi, P. K. (1982), *Seisan Kenkyu*, **34**, 437; *Chem. Abstr.*, **98**, 125043w (1983).

Shvo, Y. and Kaufman, E. D. (1972), *Tetrahedron*, **28**, 573.

Smith, H. A. and Vaughan, W. E. (1935), *J. Chem. Phys.*, **3**, 341.

Smyth, C. P. (1974), "Dipole moment, dielectric loss and intramolecular rotation," in Orville-Thomas, W. J., Ed., *Internal Rotation in Molecules*, Wiley, New York, p. 29.

Synder, E. I. (1966), *J. Am. Chem. Soc.*, **88**, 1165.

Sovers, O. J., Kern, C. W., Pitzer, R. M., and Karplus, M. (1968), *J. Chem. Phys.*, **49**, 2592.

Squillacote, M. E., Sheridan, R. S., Chapman, O. L., and Anet, F. A. L. (1979), *J. Am. Chem. Soc.*, **101**, 3657.

Steinmetz, W. E., Hickernell, F., Mun, I. K., and Scharpen, L. H. (1977), *J. Mol. Spectrosc.*, **68**, 173.

Stidham, H. D. and Durig, J. R. (1986), *Spectrochim. Acta*, **42A**, 105.

Stiefvater, O. L. and Wilson, E. B. (1969), *J. Chem. Phys.*, **50**, 5385.

Štokr, J., Pivcová, H., Schneider, B., and Dirlikov, S. (1972), *J. Mol. Struct.*, **12**, 45.

Szasz, G. J. (1955), *J. Chem. Phys.*, **23**, 2449.

Tai, J. C. and Allinger, N. L. (1976), *J. Am. Chem. Soc.*, **98**, 7928.

Takagi, K., Choi, P.-K., and Seki, W. (1983), *J. Chem. Phys.*, **79**, 964.

Takagi, K. and Kojima, T. (1971), *J. Phys. Soc. Jpn.*, **30**, 1145.

Takeo, H., Matsumura, C., and Morino, Y. (1986), *J. Chem. Phys.*, **84**, 4205.

Tanabe, K. (1972), *Spectrochimica Acta*, **A28**, 407.

Tanabe, K. and Saëki, S. (1975), *J. Mol. Struct.*, **27**, 79.

Tanabe, K., Hiraishi, J., and Tamura, T. (1976), *J. Mol. Struct.*, **33**, 19.

Taniguchi, Y. (1985), *J. Mol. Struct.*, **126**, 241.

Tasumi, M. and Nakata, M. (1985), *J. Mol. Struct.*, **126**, 111.

Teller, E. and Topley, B. (1935), *J. Chem. Soc.*, 876.

Terui, Y., Ueyama, M., Satoh, S., and Tori, K. (1974), *Tetrahedron*, **30**, 1465.

Thomas, W. A. (1968), "Nuclear Magnetic Resonance Spectroscopy in Conformational Analysis," *Annu. Rep. NMR Spectros.*, **1**, 43.

Thompson, H. B. and Sweeney, C. C. (1960), *J. Phys. Chem.*, **64**, 221.

Tichý, M. (1965), "The Determination of Intramolecular Hydrogen Bonding by Infrared Spectroscopy and Its Applications in Stereochemistry," in Raphael, R. A., Taylor, E. C., Wynberg, H., Eds., *Advances in Organic Chemistry, Methods and Results*, Vol. 5, Wiley, New York, p. 115.

Trost, B. M., Schinski, W. L., Chen, F., and Mantz, I. B. (1971), *J. Am. Chem. Soc.*, **93**, 676.

Tvaroska, I. and Bleha, T. (1980), *Coll. Czech. Chem. Commun.*, **45**, 1883.

Tvaroska, I. and Bleha, T. (1985), *Chem. Pap.*, **39**, 805.

Tylli, H., Konschin, H., and Fogerström, B. (1985), *J. Mol. Struct.*, **128**, 297.

ul Hasan, M. (1980), *Org. Magn. Reson.*, **14**, 309.

Umeyama, H. and Nakagawa, S. (1979), *Chem. Pharm. Bull.*, **27**, 2227.

Van-Catledge, F. A., Boerth, D. W., and Kao, J. (1982), *J. Org. Chem.*, **47**, 4096.

van Duin, M., Baas, J. M. A., and van de Graaf, B. (1986), *J. Org. Chem.*, **51**, 1298.

Van Hemelrijk, D., Van den Enden, L., Geise, H. J., Sellers, H. L., and Schäfer, L. (1980), *J. Am. Chem. Soc.*, **102**, 2189.

Veillard, A. (1974), "*Ab initio* Calculations of Barrier Heights," in Orville-Thomas, W., Ed., *Internal Rotation in Molecules*, Wiley, New York.

Verma, A. L., Murphy, W. F., and Bernstein, H. J. (1974), *J. Chem. Phys.*, **60**, 1540.

Volltrauer, H. N. and Schwendeman, R. H. (1971), *J. Chem. Phys.*, **54**, 260.

Walsh, A. D. (1949), *Trans. Faraday Soc.*, **45**, 179.

Ward, E. R., Poesch, W. H. Higgens, D., and Heard, D. D. (1962), *J. Chem. Soc.*, 2374.

Weigert, F. J., Winstead, M. B., Garrels, J. I., and Roberts, J. D. (1970), *J. Am. Chem. Soc.*, **92**, 7359.

Weiss, S. and Leroi, G. E. (1968), *J. Chem. Phys.*, **48**, 962.

Weissberger, A. and Sängewald, R. (1930), *Z. Phys. Chem.*, **B9**, 133.

Wennerström, H., Forsén, S., and Roos, B. (1972), *J. Phys. Chem.*, **76**, 2430.

Wiberg, K. B. and Laidig, K. E. (1987), *J. Am. Chem. Soc.*, **109**, 5935.

Wiberg, K. B. and Martin, E. (1985), *J. Am. Chem. Soc.*, **107**, 5035.

Wiberg, K. B. and Murcko, M. A. (1987), *J. Phys. Chem.*, **91**, 3616.

Wiberg, K. B. and Murcko, M. A. (1988), *J. Am. Chem. Soc.*, **110**, 8029.

Wiberg, K. B., Murcko, M. A., Laidig, K. E., and MacDougall, P. J. (1990), *J. Phys. Chem.*, **94**, 6956.

Williams, G., Owen, N. L., and Sheridan, J. (1971), *Trans. Faraday Soc.*, **67**, 922.

Wilson, E. B., (1971), "Conformational Studies on Small Molecules," *Chem. Soc. Rev.*, **1**, 293.

Winstein, S. and Holness, N. J.(1955), *J. Am. Chem. Soc.*, **77**, 5562.

Winstein, S., Pressman, D., and Young, W. G. (1939), *J. Am. Chem. Soc.*, **61**, 1645.

Winstein, S. and Wood, R. E. (1940), *J. Am. Chem. Soc.*, **62**, 548.

Wolf, K. L. (1930), *Trans. Faraday Soc.*, **26**, 315.

Wolfe, S. (1972), "The Gauche Effect. Some Stereochemical Consequences of Adjacent Electron Pairs and Polar Bonds," *Acc. Chem. Res.*, **5**, 102.

Wyn-Jones, E. and Pethrick, R. A. (1970), "The Use of Ultrasonic Absorption and Vibrational Spectroscopy to Determine Energies Associated with Conformational Change," *Top. Stereochem.*, **5**, 205.

Yamanouchi, K., Sugi, M., Takeo, H., Matsumura, C., and Kuchitsu, K. (1984), *J. Phys. Chem.*, **88**, 2315.

Yoder, C. H. and Gardner, R. D. (1981), *J. Org. Chem.*, **46**, 64.

Young, W. G., Pressman, D., and Coryell, C. D. (1939), *J. Am. Chem. Soc.*, **61**, 1640.

11

Configuration and Conformation of Cyclic Molecules

11-1. STEREOISOMERISM AND CONFIGURATIONAL NOMENCLATURE OF RING COMPOUNDS

2,2-Dimethylcyclopropanecarboxylic acid (Fig. 11.1, **A**), a derivative of the smallest cyclane (cyclopropane) has a chiral center at C(1) and exists as a pair of enantiomers; its stereoisomerism in no way differs from that of acyclic chiral molecules. The same might be said of cyclopropane-1,2-dicarboxylic acid, which has three stereoisomers: a meso form (Fig. 11.1, **B**) and a pair of enantiomers (**C, D**) diastereomeric with the meso form; the situation is the same as in tartaric acid. However, an additional feature results from the rigidity of the cyclic framework: in the meso diacid **B** the carboxyl groups are on the same side of the ring, whereas in the pair of enantiomers **C** and **D** they are on opposite sides. Therefore one may call the meso form cis and the two (isometric) chiral isomers trans. The Cahn–Ingold–Prelog (CIP) system may, of course, be applied to cyclanes; thus enantiomer **A** is R, the meso form **B** is $1R,2S$ (equivalent to $1S,2R$), and the enantiomers **C** and **D** are $1S,2S$ and $1R,2R$, respectively. The E–Z descriptors (Chapter 9) should not be used for cyclanes.

The CIP system is always unequivocal; nonetheless, when one deals with diastereomers (whether chiral or meso forms) the cis–trans nomenclature is easier to grasp. Unfortunately it has a built-in ambiguity analogous to the earlier discussed problem with the same nomenclature in alkenes (p. 542). Is compound **A** in Figure 11.2 to be called *cis*- or *trans*-2-hydroxy-2-phenylcyclopropanecarboxylic acid? The rule here (Cross and Klyne, 1976) is that the "fiducial" (reference) substituent is to be marked by the prefix *r*- and that the positions of other substituents relative to it are to be denoted as *c*- (cis) or *t*- (trans). Since compound **A** is a cyclopropanecarboxylic acid, the carbon to which the CO_2H group is attached is the reference carbon and the compound is *t*-2-hydroxy-2-

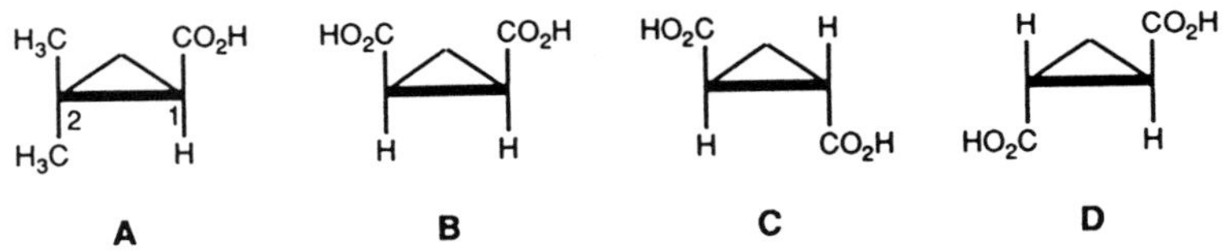

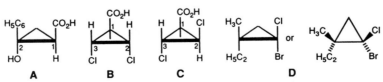

Figure 11.1. Stereoisomerism in cyclopropanes.

phenylcyclopropane-*r*-1-carboxylic acid or *t*-2-hydroxy-2-phenyl-*r*-1-cyclo-propanecarboxylic acid. (The symbol *c* for the phenyl group may be omitted since it necessarily follows from that of the geminal hydroxyl group being *t*.) It should be noted that even if the name and numbering are assigned incorrectly, for example, if the compound were named *r*-1-hydroxy-1-phenyl-*t*-2-carboxyclo-propane, use of the *r*- symbol ensures that the correct stereochemistry can be derived unequivocally from the name given.

The *r* prefix is also useful when there are more than two stereogenic centers in the ring, as in Figure 11.2 **B**. Confusion might result here because the chlorine substituents are trans to the carboxyl group but cis to each other. However, the name *t*-2,*t*-3-dichlorocyclopropane-*r*-1-carboxylic acid is unequivocal: the refer-ence point is the carboxylic acid group and both chlorine substituents are trans to it. In stereoisomer **C** in Figure 11-2, there is a question as to which way around the ring should be numbered: The rule used here is that cis precedes trans, so the compound is properly called *c*-2,*t*-3-dichlorocyclopropane-*r*-1-carboxylic acid (or *c*-2,*t*-3-dichloro-*r*-1-cyclopropanecarboxylic acid). A compound in which there might be a question both about the numbering system and about the reference group is **D** (Fig. 11.2), since there is no suffix (such as -carboxylic acid, -carboxaldehyde, -carbinol, or -ol) in its name; in such cases the reference carbon is the one substituted with the highest priority (CIP) group (here Br), thus the proper name is *r*-1-bromo-1-chloro-*c*-2-ethyl-2-methylcyclopropane.

If it is not known whether a substituent in a cyclane is cis or trans relative to other substituents, this is indicated by a wiggly line (e.g. as in Fig. 11-7, p. 673) from the ring to the substituent (in lieu of a solid line or wedge for a substituent located in front of or above the plane of the ring and a dotted, cross-hatched, or dashed line for one behind or below that plane as in Fig. 11-2, **D**). The symbol used for such a substituent is ξ (Greek xi) in lieu of *c* or *t*.

In a four-membered ring (Fig. 11.3) a new feature emerges: While the stereoisomerism in 1,2-disubstituted cyclobutanes (e.g. Fig. 11.3 **A**) is analogous to that in 1,2-disubstituted cyclopropanes, a different situation arises in 1,3-disubstituted cyclopropanes (Fig. 11.3 **B**, **C**). Although these compounds exist as cis (**B**) and trans (**C**) diastereomers, both **B** and **C** are achiral; C(1) and C(3) are

Figure 11.2. Trisubstituted and tetrasubstituted cyclopropanes.

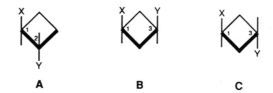

Figure 11.3. 1,2- and 1,3-Disubstituted cyclobutanes.

stereogenic but not chirotopic (p. 53), they are not chiral centers. One should *not* call such diastereomers meso forms, in as much as the set does not contain any chiral isomers.

Nonetheless, the case of, say, the 1,3-cyclobutanediols is similar to that of the trihydroxyglutaric acids (Chapter 5); C(1) and C(3) in such cyclobutanes (**B,C**, $X = Y = OH$) might even be called pseudoasymmetric centers in the sense that the two ligands ($-CH_2CHOHCH_2-$) attached to each of these centers are enantiomorphic. In fact, Prelog and Helmchen (1982) devised a procedure, shown in Figure 11.4, for assigning pseudoasymmetry descriptors to C(1) and C(3) in, say, *cis*- and *trans*-1,3-cyclobutanediol. By the rules explained in Section 5-2 (Fig. 5.7) the C(2), C(3), C(4) and C(4), C(3), C(2) segments are detached from the chiral center at their distal ends (the chiral center still appears as a phantom carbon). Auxiliary chirality descriptors (marked by the suffix zero) may then be assigned to C(3) in the two segments, which are now substituents of C(1). These symbols will now define a priority for the two segments and thus allow assignment of a pseudoasymmetry descriptor (*r* or *s*) at C(1). [The descriptors bear the suffix *n* (new) because they arise from the 1982 modification of the CIP system and were not part of the 1966 publication of Cahn et al.] The symbol at C(1) for the cis isomer (Fig. 11.4) turns out to be s_n (the readers should convince themselves that this conclusion remains the same if the molecule is turned upside down and then unraveled). The

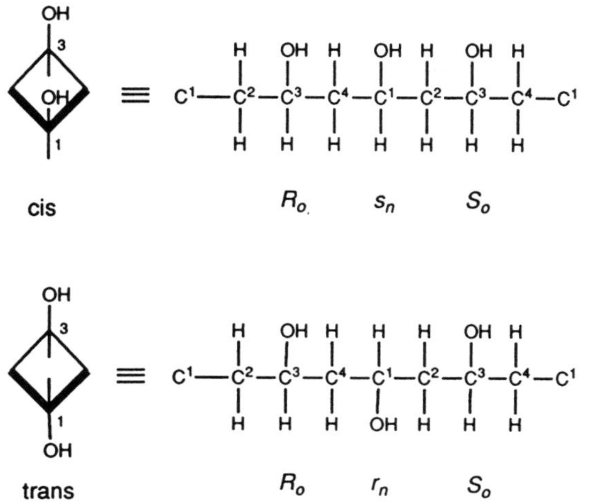

Figure 11.4. Hierarchical digraphs and pseudoasymmetry descriptors at C(1)[and C(3)] in *cis*- and *trans*-1,3-cyclobutanediol.

descriptor at C(3) should be determined independently, but it is clear that, since the molecule has a C_2 axis interchanging C(1) and C(3), it must be the same as that at C(1). The cis isomer is therefore $1s_n, 3s_n$. For the trans isomer (Fig. 11.4) the same procedure leads to the descriptor r_n for C(1); again, because C(1) and C(3) are related by a C_2 symmetry axis, that at C(3) must be the same; hence the trans isomer is $1r_n, 3r_n$. It is interesting that changing the configuration of either C(1) or C(3) (but not both) will change the descriptor of *both* centers, the one that has been inverted and the one that has not.

cis–trans Isomerism without accompanying chirality occurs in all saturated carboxylic rings of n members when n is even and the substituents are at positions 1 and $1 + (n/2)$.

Before one can determine cis–trans descriptors for ring substituents, one must decide whether these substituents are, in fact, on the same side (cis) or on opposite sides (trans) of the ring. This task is trivial with cyclopropanes where cis substituents are synperiplanar and trans substituents are anticlinal (cf. Fig. 2.13). But already with cyclobutane (which, as will be discussed later, is wing shaped) the situation is not quite so clear and with substituted cyclohexanes, in which the ring is usually chair shaped, there are obvious problems (see Fig. 11.5). The trans relationship in **A** (diaxially 1,4-disubstituted) is obvious but the relationship in **B** (diequatorially 1,4-disubstituted) is less so. Indeed, the torsion angle in a 1,2-diequatorially trans-disubstituted cyclohexane **C** is about 60°, just as in the (equatorially–axially) cis-disubstituted stereoisomer **D**. To spot the trans relationship in **B** and **C** one might focus on the tertiary hydrogen atoms (rather than on the X and Y substituents), or one might say that groups are cis if both are uppermost (or lowermost), whereas they are trans if one is uppermost and the other lowermost. Construction of a model of cyclooctane discloses additional difficulties: Here both trans and cis substituents may be synperiplanar, and cis as well as trans substituents may be anticlinal. The IUPAC (International Union of Pure and Applied Chemistry) definition (Cross and Klyne, 1976) specifies that "rings are to be considered in their most extended form; reentrant angles are not permitted," but there is the difficulty that, while planar graphs are easily written in the "most extended form," molecular models are often not easily forced into that form (the reader might experiment with a model of cyclooctane) and real molecules are, as we shall see later, not likely to exist in such a conformation at all (cf. also Anet, 1990).

Parochial (local) systems of nomenclature have been used with sugars, steroids, and certain other polycyclic systems and will be introduced as we deal with these systems.

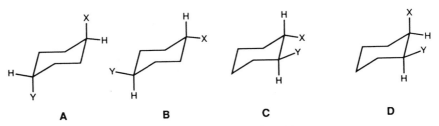

Figure 11.5. 1,4-Disubstituted cyclohexane.

11-2. DETERMINATION OF CONFIGURATION OF SUBSTITUTED RING COMPOUNDS

a. Introduction

Just as with acyclic molecules, methods of determining relative (cis or trans) configuration frequently also depend on conformation. Most notably this is true of spectral (e.g., NMR) methods, whose consideration will therefore be postponed to later parts of this chapter (e.g. Section 11-4.d). Cyclopropane derivatives, in which the ring must necessarily be planar, since three points define a plane, will best serve to illustrate methods that are not dependent on conformational considerations. Only methods for assigning the (relative) configuration of diastereomers will be discussed here; determination of the absolute configuration of enantiomers in cyclic systems is no different, in its principles, from similar determination in acyclic systems (Section 5-3).

Six methods for the determination of cis–trans configuration in cyclic molecules may be recognized: (a) testing resolvability, (b) determining the number of isomers obtained in certain chemical transformations, (c) establishing the ease of bridge formation, (d) drawing conclusions from physical, including spectral, data, (e) drawing conclusions from considerations of reaction mechanism, and (f) correlating one compound chemically or physically with another.

b. Symmetry-Based Methods

The first two methods (a and b) depend on symmetry considerations (see also Section 5-5.c and Chapter 8) and are therefore unequivocal. Among the cyclopropane-1,2-dicarboxylic acids (Fig. 11.1, **B–D**) only the trans isomer can be resolved, since the cis (meso) isomer has a symmetry plane; successful resolution will therefore lead to assignment of (trans) configuration. The converse is not true: Nonresolvability does not guarantee that one is dealing with the meso isomer, since resolution may have failed for technical reasons. This point is particularly vexing if only the achiral meso isomer is at hand: Positive proof of its configuration by the resolvability criterion is evidently impossible. A useful trick here is to convert the achiral compound into a chiral one, resolve the latter, and then reconvert it to its achiral precursor. Thus conversion of the meso acid (Fig. 11.1, **B**) into a monomethyl ester will make it chiral; of course the monomethyl ester of the chiral acid (**C, D**) is chiral also. If one can resolve the monomethyl ester, then, regardless of which diacid was at hand, the configurational problem can be solved: The resolved monomethyl ester of the meso acid **B** will give an optically inactive acid upon saponification, whereas the resolved ester of the chiral acid **C** or **D** will give an optically active acid when saponified. Clearly, this method is applicable only if one diastereomer is chiral and the other is meso (in which case it is not limited to ring compounds: The configuration of the 2,3-dimethylsuccinic acids can be determined analogously). In the general case, where the two substituents on the ring are different (e.g., Fig. 11.3, **A**, X ≠ Y), other approaches must be used.

A number of symmetry-based methods of distinguishing meso and chiral isomers on the basis of the spectroscopically observable heterotopicity (or lack thereof) of appropriate nuclei, either in the parent compound itself or in a derivative, have been described in Section 8-4.b. Similar *chemical* methods have been known for many years; for example, Wislicenus (1901), one of the pioneers of stereochemistry, determined the configuration of the 2,5-dimethylcyclopentane-1,1-dicarboxylic acids by decarboxylating them to the corresponding monocarboxylic acids (Fig. 11.6): The meso (cis) acid, which has diastereotopic CO_2H groups, gives rise to two different meso products, whereas the racemic trans acid, in which the CO_2H groups are homotopic, yields a single racemic product (Fig. 11.6, only one enantiomer shown). In similar fashion, one could distinguish the cis and trans isomers of 2,4-dimethylcyclobutanone by hydride reduction: The cis (meso) isomer should yield two diastereomeric meso forms of 2,4-dimethylcyclobutanol, whereas the chiral (trans) isomer, whether resolved or not, should yield only a single product (enantiomer or racemate, as the case may be). Once again a *caveat* is needed: Formation of two diastereomeric products is unequivocal but a single product (or racemate) may arise from a large mechanistic bias even in situations where two products are, in principle, possible. Therefore, the method is unsafe if only the chiral isomer is at hand (it must necessarily give a single product or racemate) and it obviously fails if both isomers are at hand but both give single products.

Thus, while symmetry-based methods are in principle unequivocal, they may yet meet with practical difficulties. The situation is even less safe, however, in the case of methods discussed below, which are based on chemical reactivity, physical properties, or mechanistic considerations. Since such methods are usually based on spatial proximity, or on a definitive orientation in space of salient substituents, they are basically indicators of conformation rather than configuration. Thus they can be used for the determination of configuration only if conformational factors are well understood.

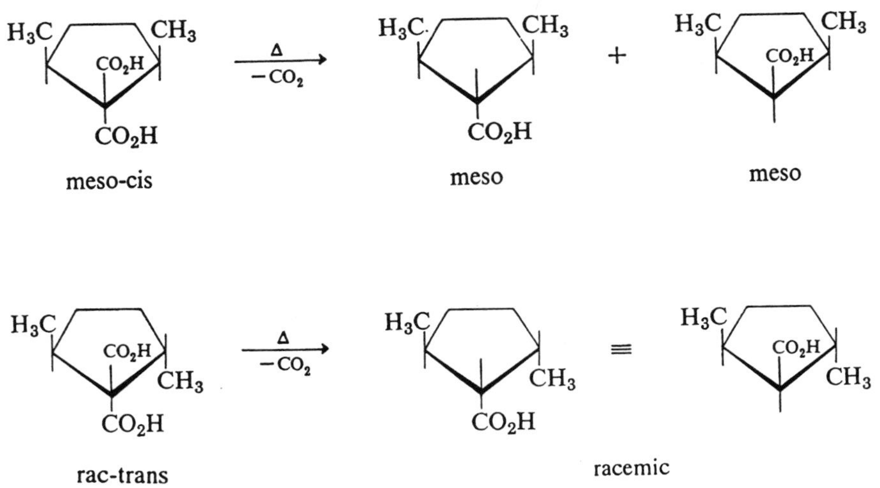

Figure 11.6. Proof of configuration of 2,5-dimethylcyclopentane-1,1-dicarboxylic acids.

c. Methods Based on Physical and Chemical Properties

In *cis*- and *trans*-cyclopropane-1,2-dicarboxylic acids (Fig. 11.1, **B** and **C** or **B** and **D**) the carboxyl groups are close to each other in the cis isomer but not in the trans. Therefore one might expect the cis but not the trans isomer to form a cyclic anhydride; such is, in fact, the case. Also one might expect the cis isomer, because of its eclipsed carboxyl groups, to be less stable; indeed, the cis isomer is converted, by heating with mineral acid, to the trans isomer. Furthermore, passing from chemical to physical properties, one would expect the cis isomer to be more acidic in its first ionization constant (since the monoanion may be stabilized by a through-space effect, either electrostatic or hydrogen bonding, with the free CO_2H), and this is the case: $pK_1^{cis} = 3.56$; $pK_1^{trans} = 3.80$ (McCoy and Nachtigall, 1963). In addition, one might expect the acidity of the second carboxyl group to be higher for the trans acid than for the cis, since double ionization in the latter case leads to CO_2^- groups that are proximal in space and highly repulsive. Again the expectation is fulfilled: $pK_2^{cis} = 6.65$; $pK_2^{trans} = 5.08$ (McCoy and Nachtigall, 1963). The ratio K_1/K_2 is 1210 for the cis isomer and 19.4 for the trans. This ratio is a much more reliable indicator of proximity than are K_1 and K_2 individually; there is a relationship, called Bjerrum's law (Bjerrum, 1923), which correlates the distance of the carboxyl groups with the K_1/K_2 ratio:

$$\ln(K_1/4K_2) = N\,e^2/\varepsilon\,r\,RT \qquad \text{Bjerrum's law}$$

where N is Avogadro's number, e is the charge of an electron, ε is the effective dielectric constant of the medium, R is the gas constant, T is the absolute temperature, and r is the distance between the carboxylic acid groups.

As already explained in Chapter 2 in connection with force field calculations, ε is best taken at a value intermediate between that of the solvent and solute (Kirkwood and Westheimer, 1938; Westheimer and Shookoff, 1939; McCoy, 1965). The reader may recognize that in the formulation of Bjerrum's law the free energy difference between the first and second ionization processes, corrected by a statistical factor of 4, is set equal to the electrostatic interaction energy of the two carboxylate charges as given by Coulomb's law. The statistical factor of 4 comes from the fact that there are two possible monoanions but only one dicarboxylic acid and one dianion; thus the dissociation constant K_1 of the dicarboxylic acid is favored by a statistical factor of 2 but that, K_2, of the monoanion is disfavored by a factor of 2; the factor of 4 in the denominator compensates both for the statistically enhanced K_1 and the similarly diminished K_2.

Rings larger than three membered are, in general, not planar (see individual discussions in Sections 11-5, b–d) and so the criteria just discussed (anhydride or other ring formation from the cis isomer, greater stability of the trans isomer, larger K_1/K_2 ratio for the cis isomer) tend to become less certain as ring size increases from 4 to 7 and more. Thus Perkin, Jr. (1894) found that both in the cyclobutane-1,2-dicarboxylic and the cyclopentane-1,2-dicarboxylic acids, the cis but not the trans isomer forms an anhydride on heating or mild treatment with acetyl chloride, that the anhydride is readily reconverted to the cis acid on treatment with water, and that the cis acid is converted to the trans acid on heating with mineral acid. However, Baeyer (1890) encountered a less clear-cut situation with cyclohexane-1,2-dicarboxylic (hexahydrophthalic) acid: The trans isomer is indeed more stable than the cis (from which it is formed by treatment with hot mineral acid) but *both* acids now form (different) anhydrides, though that from the cis acid is formed more easily, whereas that formed from the trans acid is less stable (apparently more strained) and converted to the cis anhydride on prolonged treatment with acetyl chloride. Greater stability of diesters of the trans over those of the cis acid is found throughout the series C_3-C_8 (Seigler and Bloomfield, 1973; Sicher et al., 1961); again the trend is most marked for the cyclopropanedicarboxylates—the only cycloalkane-1,2-dicarboxylates in which the cis isomer has its functional groups completely eclipsed whereas those in the trans isomer are at a torsion angle of 120°. By the same token, the ratio K_1/K_2 (Table 11.1) for the cis–trans acids shows much more pronounced differences in the cyclopropane series than for the larger ring diacids. Perhaps somewhat unexpectedly, this ratio diverges again in the seven- and eight-membered rings, with K_1/K_2 being extremely large for the cis diacids. It would thus appear that conformational factors lead back to eclipsing of the cis carboxyl groups in such rings.

Among physical properties, dipole moments may be useful in assignment of configuration, as shown in Figure 11.7; in general $\mu_{cis} > \mu_{trans}$ in 1,2-disubstituted 3-, 4-, 5-, and six-membered rings. In 1,3-disubstituted six-membered rings, this generalization is likely to fail; for example, for the 1,3-dibromo compound $\mu_{cis} = 2.19$ D, $\mu_{trans} = 2.17$ D (cf. McClellan, 1974); it probably cannot be relied on either for rings larger than six membered.

The relation does hold in 1,2-dibromocycloheptane, $\mu_{cis} = 3.04$ D, $\mu_{trans} = 2.09$ D, and (barely!) in 1,2-dibromocyclooctane, $\mu_{cis} = 3.07$ D, $\mu_{trans} = 2.90$ D but breaks down in 1,2-dichlorocyclooctane, $\mu_{cis} = 2.34$ D, $\mu_{trans} = 3.05$ D (cf. McClellan, 1974).

TABLE 11.1. Dissociation Constants of Cycloalkane-1,2-dicarboxylic Acids.

1,2-Dicarboxylic Acid Derived from		pK_1	pK_2	K_1/K_2	Reference
Cyclopropane	cis	3.56	6.65	1210	a
	trans	3.80	5.08	19.4	a
Cyclobutane	cis	4.16	6.23	130	b
	trans	3.94	5.55	41	b
Cyclopentane	cis	4.42	6.57	138	c
	trans	4.14	5.99	70	c
Cyclohexane	cis	4.44	6.89	282	d
	trans	4.30	6.06	58	d
Cycloheptane	cis	3.87	7.60	5370	d
	trans	4.30	6.18	76	d
Cyclooctane	cis	3.99	7.34	2240	d
	trans	4.37	6.24	74	d

[a] McCoy and Nachtigall (1963).
[b] Bloomfield and Fuchs (1970).
[c] Inoue, Nakanishi, et al. (1965).
[d] Sicher et al. (1961); see also Delben and Crescenci (1978).

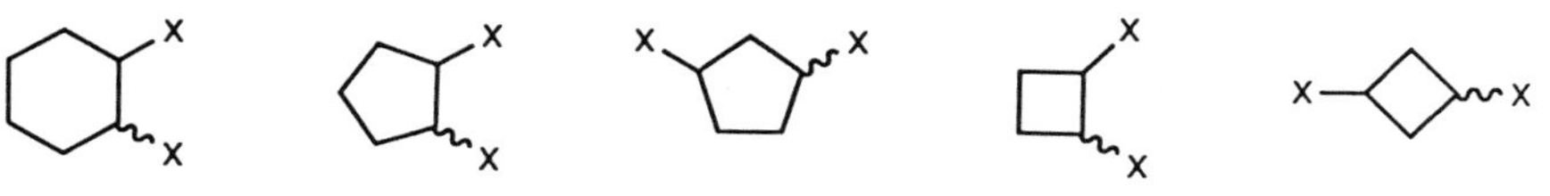

X	cis	trans	cis	trans	cis	trans	cis	trans	cis	trans
Cl	3.10^a	2.63^a		1.64^a	2.76^a	1.44^a	2.98^c	2.39^c	2.00^c	1.20^c
Br	3.12^a	2.15^a	2.92^b	1.56^a			2.91^a	2.39^a	2.02^a	1.10^a
CN					4.70^a	2.87^a			4.26^d	1.23^d

[a] Data from McClellan (1963, 1974, 1989).
[b] From Fuchs (1978).
[c] Lemarié et al. (1977); solvent not specified.
[d] Lautenschlaeger and Wright (1963); data refer to the 2,2,4,4-tetramethyl compound.

Figure 11.7. Dipole moments (in debye units) of dihalocycloalkanes (in benzene at 25–30°C).

Intramolecular hydrogen bonding in vicinal diols, already discussed for acyclic diastereomers in Chapter 10, is also useful for configurational assignment in the cyclic analogues, as shown in Table 11.2. In the case of cyclobutane-1,2-diol and cyclopentane-1,2-diol, consideration of models indicates that only the cis but not the trans isomer is capable of intramolecular hydrogen bonding and assignment of configuration can readily be made on this basis. In the case of the cyclohexane-, cycloheptane-, and cyclooctane-1,2-diols the situation is not so extreme: Both the cis and the trans diol are capable of forming intramolecular hydrogen bonds, but the bond in the cis diol is stronger. The cis isomer therefore shows the greater difference in IR stretching frequency between unbonded and intramolecularly bonded OH.

Another useful criterion for ascertaining the stereochemistry of 1,2-diols is lead tetraacetate cleavage (Table 11.2; Criegee et al., 1956). Apparently, the intermediate in this reaction is a cyclic lead(IV) compound that is formed more

TABLE 11.2. Hydroxyl Stretching Frequencies ν in the Infrared[a] and Lead Tetraacetate Cleavage Rates of Cycloalkane-1,2-diols[b]

Cycloalkane-1,2-diol derived from		ν_{free}	ν_{bonded}	$\Delta\nu$	$k_{Pb(OAc)_4}$	k_{cis}/k_{trans}
Cyclobutane	cis	3640	3580	60^c	>10,000	>1500
	trans	3610		d	12.5	
Cyclopentane	cis	3633	3572	61	>10,000	>1000
	trans	3620		d	12.8	
Cyclohexane	cis	3626	3588	38	5.04	22.5
	trans	3633	3600	33	0.224	
Cycloheptane	cis	3632	3588	44	7.7	6.4
	trans	3626	3589	37	1.2	
Cyclooctane	cis	3635	3584	51		
	trans	3631	3588	43		

[a] In reciprocal centimeters (cm^{-1}). Kuhn (1952, 1954) unless otherwise noted.
[b] Criegee et al. (1956); 1 mol^{-1} min^{-1} at 20°C.
[c] Barnier and Conia (1976).
[d] No intramolecular hydrogen bond formed.

readily if the hydroxyl groups can easily attain a small torsion angle. Therefore cis diols are cleaved more rapidly than trans diols; the difference is enormously large in four- and five-membered rings, for the reasons already discussed; it persists in the larger cycles up to C-8, even though in much less pronounced fashion.

> However, these generalizations break down for rings larger than eight membered. Thus, *trans*-1,2-diols are oxidized faster than their cis isomer in 9-, 10-, and 12-membered rings (cf. Prelog and Speck, 1955) and intramolecular hydrogen bonding is stronger in the trans than in the cis diol in the 10- and 12-membered ring (Kuhn, 1954; see also Tichý, 1965). Presumably in rings of this size, groups located 1,2-trans to each other can approach each other more easily than corresponding groups disposed cis (cf. Section 11-5.d).

Both ^{1}H and ^{13}C NMR spectroscopy are extremely useful in configurational assignment of cyclic compounds, but since the interpretation of NMR spectra is strongly conformation dependent, one can make but few across-the-board generalizations. In cyclopropanes, as one would be led to believe by the Karplus relationship (Fig. 10.37), protons that are located cis to each other (torsion angle 0°) have a larger coupling constant than those located trans (torsion angle 120°) in a corresponding chemical environment (cf. Brügel, 1967). This fact can be used for configurational assignment, for example, in 1-chloro-2-ethoxycyclopropanes (Barlet et al., 1981). However, the ranges of both J_{cis} and J_{trans} are quite wide and overlapping, so that it is unsafe, in the absence of models, to make assignments in

TABLE 11.3. The ^{13}C Resonances of Dimethylcycloalkanes.[a]

Compound		C_1	C_β^b	C_4	C_5	CH_3	Reference
Cyclopropane	*cis*-1,2	9.8	13.6			13.0	c
	trans-1,2	14.2	14.6			19.0	c
Cyclobutane	*cis*-1,2	32.2	26.6			15.4	d
	trans-1,2	39.2	26.8			20.5	d
	cis-1,3	26.9	38.5			22.5	d
	trans-1,3	26.1	36.4			22.0	d
Cyclopentane	*cis*-1,2	37.9	33.5	23.5		15.4	e
	trans-1,2	43.0	35.3	23.6		19.0	e
	cis-1,3	35.7	45.3	34.6[h]		21.4[h]	e
	trans-1,3	33.8	43.4	35.5[h]		21.7[h]	e
Cyclohexane	*cis*-1,2	34.6	31.7	23.9		15.9	f
	trans-1,2	39.7	36.2	27.0		20.4	f
	cis-1,3	33.0	44.9	35.6	26.6	23.0	f
	trans-1,3	27.2	41.6	34.1	20.9	20.7	f
Cycloheptane	*cis*-1,2	37.5	34.0	26.6	29.2	17.9	g
	trans-1,2	41.3	35.8	26.7	29.7	22.6	g
	cis-1,3	34.2	46.9	37.4[h]	26.5[h]	24.9	g
	trans-1,3	31.1	44.7	37.5[h]	29.1[h]	24.3	g

[a] In parts per million from tetramethylsilane (TMS).
[b] The C(3) atom in a 1,2-disubstituted compound and the C(2) atom in a 1,3-disubstituted compound.
[c] Monti et al. (1975); see also Chukovskaya et al. (1981).
[d] Eliel and Pietrusiewicz (1980).
[e] Christl, Roberts, et al. (1971).
[f] Dalling and Grant (1967).
[g] Christl and Roberts (1972).
[h] These entries represent reversals from the expected order.

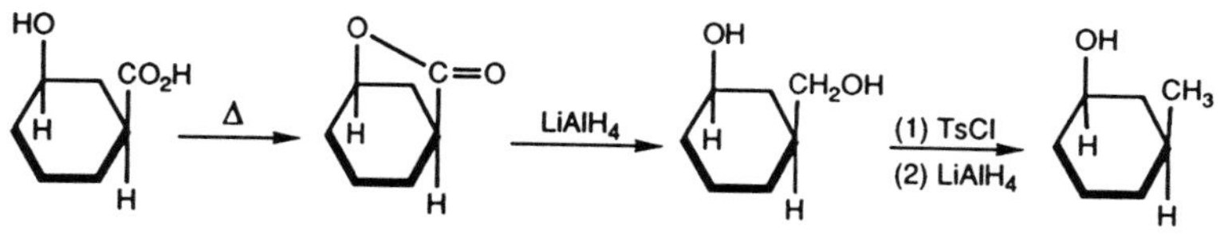

Figure 11.8. Correlation of *cis*-3-hydroxycyclohexanecarboxylic acid with *cis*-3-methylcyclohexanol.

an individual compound where cis and trans couplings cannot be compared. In all larger rings, even four-membered ones (Lemarie et al., 1977), conformational factors conspire to make simple generalizations impossible; coupling constants in such systems will be taken up later in this chapter (e.g. Section 11-4.d) in the context of conformation.

On the other hand, ^{13}C NMR spectra of disubstituted cyclanes (Table 11.3) permit the following generalizations (at least up to and including seven-membered rings): In the 1,2-disubstituted series, the resonances of the cis isomers are upfield of corresponding compounds for the trans isomers, whereas the reverse is true for 1,3-disubstituted compounds.

In summary, physical and spectral properties are often suggestive of relative configuration in rings but they are rarely conclusive.

d. Correlation Methods

In some instances it is convenient to establish the configuration of cyclic molecules by correlative methods, similar to those described in connection with the determination of absolute configuration in Section 5-5.b. For example, the configurations of *cis*- and *trans*-3-hydroxycyclohexanecarboxylic acids are readily established in as much as the cis acid (Fig. 11.8) forms a five-membered lactone but the trans acid does not. Chemical correlation of these acids with the 3-methylcyclohexanols (shown for the cis isomer in Fig. 11.8) establishes the configuration of the latter (e.g., Goering and Serres, 1952). One must, of course, make sure that there is no epimerization (change of configuration at one or other of the chiral centers) in the course of the transformation used for the correlation. In the case depicted in Figure 11.8, this was assured by carrying out the correlation with both the cis and the trans acid; but even if only one of the stereoisomers is available for correlation, the absence of epimerization (leading to formation of a second diastereomer) can nowadays be monitored by chromatographic or spectroscopic (especially ^{13}C NMR) methods.

Conclusions concerning configuration deduced from evidence of reaction mechanism usually involve consideration of conformation and so will be discussed later (e.g. Section 11-4.e).

11-3. STABILITY OF CYCLIC MOLECULES

a. Strain

It was von Baeyer (1885) who first pointed out that the construction of a small ring compound involves strain (*Spannung*). For example, in cyclopropane the angle between the carbon–carbon bonds must, for geometrical reasons, be 60°. (This is the angle between lines connecting the nuclei, i.e., the internuclear, not the interorbital angle, see Section 11-5.a). Since the "normal" bond angle at carbon is 109°28′ (tetrahedral), there is a deviation of 49°28′ from the norm. Baeyer apportioned this strain equally between the two ring bonds flanking the 60° angle and thus called the strain 24°44′ (one-half of 49°28′). In a similar fashion, he reported strain values for other carbocyclic rings as shown in Figure 11.9. This picture, and the very concept of "angle strain," was evidently based on the Kekulé models used at the time of Baeyer (Ramsay, 1975).

As discussed in Chapter 2, strain introduced in a molecule in any fashion tends to be minimized by becoming distributed among several modes, such as bond strain, angle strain, torsional strain, and van der Waals compression. Thus the strain in cyclanes is actually not purely angle strain; and so it becomes desirable to define strain in an entirely different manner, in terms of energy. Strain is the excess of observed over "calculated" heat of formation (or more conveniently from the practical point of view, heat of combustion). The question, then, becomes what is meant by the "calculated" enthalpy value. For a cycloalkane, this is conveniently, though somewhat arbitrarily, taken as the heat of formation or heat of combustion of a CH_2 group multiplied by the number of carbon atoms in the ring. (For a more detailed discussion concerned with the calculation of "strain" see Liebman and Greenberg, 1976; Greenberg and Liebman, 1978; Burkert and Allinger, 1982, p. 185; Wiberg, 1986, 1987; see also below.) The value for a CH_2 group, in turn, is obtained by taking the difference between a large straight-chain hydrocarbon, $CH_3(CH_2)_nCH_3$, where $n > 5$, and its next lower homologue. This difference is quite constant, 157.44 kcal mol^{-1} (658.73 kJ mol^{-1}) for the heat of combustion or 4.93 kcal mol^{-1} (20.6 kJ mol^{-1}) for the heat of formation in the vapor state [Schleyer et al., 1970; Liebman and Van Vechten, 1987; a slightly larger value, 4.95 kcal mol^{-1} (20.7 kJ mol^{-1}) is given by Pihlaja, 1987; the liquid-phase value is 6.09 kcal mol^{-1} (25.5 kJ mol^{-1}), Wiberg, 1987]. The strain thus calculated (Table 11.4), expressed either as total strain or as strain per CH_2 group, is high for cyclopropane, drops to near zero as one proceeds to cyclohexane, increases again to a maximum in the cyclooctane to cycloundecane region and then drops again, reaching values near zero from 14-membered rings on up. The strain in 3-, 4-, and 5-membered rings indeed runs parallel with Baeyer's prediction (Fig. 11.9) but the 6-membered ring is nearly strainless. The Kekulé models that Baeyer used apparently concealed the fact that a puckered (chair or boat shaped) form of cyclohexane can be constructed that is free from angle strain. (The bond angles in a nonplanar ring are always smaller

24°44′	9°44′	0°44′	−5°16′	"angle strain"
9.17	6.58	1.24	0.02	strain per CH_2 group[a] in kcal mol^{-1}
38.4	27.5	5.19	0.09	strain per CH_2 group[a] in kJ mol^{-1}

[a] See Table 8.4.

Figure 11.9. "Angle strain" (Baeyer) and actual strain per methylene group.

TABLE 11.4. Heat of Combustion and Ring Strain for Cyclanes.

Size of Ring (n)	Heat of Combustion[a]		Total Strain		Strain per CH_2[d]	
	(kcal mol^{-1})	(kJ mol^{-1})	(kcal mol^{-1})[b]	(kJ mol^{-1})[c]	(kcal mol^{-1})	(kJ mol^{-1})
3	499.83	2091.3	27.5	115.1	9.17	38.4
4	656.07	2745.0	26.3	110.1	6.58	27.5
5	793.40	3319.6	6.2	26.0	1.24	5.19
6	944.77	3952.9	0.1	0.5	0.02	0.09
7	1108.3	4637.3	6.2	26.2	0.89	3.74
8	1269.2	5310.3	9.7	40.5	1.21	5.06
9	1429.6	5981.3	12.6	52.7	1.40	5.86
10	1586.8	6639.1	12.4	51.8	1.24	5.18
11	1743.1	7293.3	11.3	47.3	1.02	4.30
12	1893.4	7921.9	4.1	17.2	0.34	1.43
13	2051.9	8585.0	5.2	21.5	0.40	1.66
14	2206.1[e]	9230.9[e]	1.9	8.0	0.14	0.57
15	2363.5	9888.7	1.9	7.8	0.13	0.51
16	2521.0	10547.7	2.0	8.0	0.12	0.50
17	2673.2	11184.5	−3.3	−13.9	−0.19	−0.82

[a] Gas phase values. From *TRC Thermodynamic Tables* (1991), p. n-1960.
[b] Heat of combustion minus 157.44n.
[c] Heat of combustion minus 658.7n.
[d] Total strain divided by n.
[e] Corrected values: see Chickos, J. S., Hesse, D. G., Panshin, S. Y., Rogers, D. W., Saunders, M., Uffer, P. M., and Liebman, J. F. (1992), *J. Org. Chem.*, **57**, 1897.

than those in a planar one; thus puckering reduces the angle in cyclohexane from 120° to near tetrahedral.) No more than 5 years after the publication of Baeyer's paper, however, Sachse (1890, 1892) had the insight to recognize that nonplanar cyclohexane can be strainless, or at least has no angle strain. We shall return to this point later.

Schleyer et al. (1970) in connection with a calculation of the strain energy in adamantane, pointed out that the −4.93 kcal mol^{-1} (−20.6 kJ mol^{-1}) increment for the heat of formation of CH_2 is actually inappropriate, since it is derived from heat of formation data of straight-chain hydrocarbons that are mixtures of all-anti (zigzag) and partly gauche conformations. They suggest that a better standard of comparison would be the heat of formation difference between two successive homologous straight-chain hydrocarbons *in the zigzag (all-anti) conformation*; they compute this increment to be −5.13 kcal per CH_2 or 21.5 kJ per CH_2. Using this value, the (theoretical) heats of formation of hypothetical unstrained n-membered cyclanes are calculated to be $0.20 \times n$ kcal mol^{-1} (or $0.84 \times n$ kJ mol^{-1}) lower than estimated earlier and the calculated strain energies are thus $0.20 \times n$ kcal mol^{-1} ($0.84 \times n$ kJ mol^{-1}) higher. In the case of cyclohexane, this leads to a strain energy of 1.31 kcal mol^{-1} (5.48 kJ mol^{-1}) rather than the much smaller value given in Table 11.4. We shall see later that the finding of a small amount of strain energy in cyclohexane is not unreasonable. In any case, this discussion emphasizes that the quantitative assessment of strain is somewhat arbitrary in that it depends on the standard of comparison (i.e., on what is considered an "unstrained" molecule; cf. Greenberg and Liebman, 1978; Kozina, Mastryukov, et al., 1982; Wiberg, 1986). Thus the different values of strain energies given in different sources (see also Schleyer et al., 1970; Wiberg, 1983; Wiberg et al., 1987) do not reflect different experimental values of heats of formation or combustion, but rather different reference states for the pertinent "unstrained" structures.

The data in Table 11.4 suggest that ring compounds might be divided into four families. The three- and four-membered rings are obviously highly strained; they are classified as "small rings". There is relatively little strain in the five-, six-, and seven-membered rings (so-called "common rings"), which abound in natural as well as synthetic products. Strain increases again in the C_8-C_{11} family; rings of this size are named "medium rings" and the larger, nearly strainless ones (C_{12} and larger) are called "large rings." (For the nomenclature, see Brown et al., 1951, footnote 21.)

One might question why the medium rings are so strained, even though it is easy to construct puckered models for them that are free of angle strain (Baeyer strain). The answer, in light of what is now known about molecular mechanics (Section 2-6) is as follows: Models of medium rings that are free of angle strain tend to have large numbers of pairs of eclipsed hydrogen atoms on adjacent CH_2 groups; moreover, with rings in the C_8-C_{11} range, such hydrogen atoms will tend to "bump" each other across the ring, leading to so-called "transannular strain" (Prelog, 1956); this strain arises from the van der Waals compression (non-bonded energy) term V_{nb} in Eq. 2.1. Energy minimization (Section 2-6) will, of course, tend to minimize the total strain; therefore angle deformation will occur, with some development of angle strain that is more than compensated for by a concomitant reduction in eclipsing (Pitzer strain) and transannular (van der Waals) strain. The picture of a puckered ring with normal bond angles suggested by mechanical molecular models is therefore incorrect (just as Baeyer's picture of planar rings was incorrect); in fact, bond angles in cyclononanes and cyclodecanes are as large as 124° (Prelog, 1960; Sicher, 1962). Models tend to lead to overestimation of angle and underestimation of torsional strain and, in the case of nonspace-filling models, van der Waals strain. Rings of specific sizes will be discussed later in Sections 11-4 and 11-5.

b. Ease of Cyclization as a Function of Ring Size

The relative thermodynamic instability of medium-size rings has made their derivatives relatively difficult to obtain, the first successful general synthesis (acyloin reaction) having been described only in the 1940s by Prelog and by Stoll (cf. Bloomfield et al., 1976) although, since then, numerous preparations have been developed. Large rings, despite their low strain, present synthetic difficulties as well. It was recognized many years ago (Ruzicka et al., 1926) that a complicating factor is the difficulty of getting the ends of a long chain to approach each other: The conformational entropy of a chain compound is greater than that of a ring (but see DeTar and Luthra, 1980). A competing reaction in ring closure tends to be dimerization or oligomerization, since, if in $X-(CH_2)_n-Y$ the functional groups X and Y are capable of interacting to form a ring, there is also the possibility of similar interaction between two molecules to form a dimer. These competing reactions are not much of a problem in small and common ring compounds where the loss of translational entropy in a dimerization or oligomerization reaction tends to be much more severe than the loss of conformational entropy in cyclization. But this situation is reversed in medium and especially in large rings, where the possibility of rotation about a large number of bonds leads to a high conformational entropy in the open-chain precursor as well as the linear dimer or oligomer, which is largely lost in the cyclic product. Cyclizations leading to such rings must therefore frequently be carried out in high dilution, where bimolecular reactions tend to be suppressed (but, unfortunately, operating conditions make for long reaction times) (Ruggli, 1912; Ziegler, 1955). Here again a number of new synthetic methods have been developed, principally with the impetus of synthesizing large-ring lactones that occur in a number of natural products known as macrolides (e.g., Nicolaou, 1977; Trost and Verhoeven, 1980).

Although a great deal of synthetic and qualitative information on medium and large rings is available, quantitative data on ring closure are more limited (Winnik, 1981; Illuminati and Mandolini, 1981). For example, the entropies of formation of cyclanes are available only up to C_8; from these data, the following entropies of ring closure (data in $cal\,mol^{-1}\,K^{-1}$, data in parentheses in $J\,mol^{-1}\,K^{-1}$) can be computed: cyclopropane -7.7 (-32.2), cyclobutane -10.9 (-45.6), cyclopentane -13.3 (-55.6), cyclohexane -21.2 (-88.7), cycloheptane -19.8 (-82.8), and cyclooctane -19.0 (-79.5) (Winnik, 1981). The sharp drop for cyclohexane formation reflects the stiffness (or deep energy well) of this ring system compared to the flexibility of cyclopentane, cycloheptane, and cyclooctane rings, but it is not clear from the data if the negative entropy of cyclization has come to a plateau with the eight-membered ring or whether it drops further for larger rings.

In addition to some fragmentary data on the cyclization of ω-bromoalkylamines, $Br(CH_2)_{n-1}NH_2$ (cf. Eliel, 1956; DeTar and Brooks, 1978; DeTar and Luthra, 1980), a few sets of fairly complete kinetic parameters have been reported by Illuminati and Mandolini (1981) of which perhaps the most representative are concerned with the cyclization of salts of ω-bromoacids, $Br(CH_2)_{n-2}CO_2^-$. Figure 11.10 displays rate constants, activation enthalpies ($\Delta H^{\ddagger}$), and activation entropies ($\Delta S^{\ddagger}$) for this series as a function of ring size. The rate constants show a maximum at $n = 5$ and a minimum at $n = 8$, though all the medium-sized ring lactones form rather slowly. That formation of the three-membered ring is so slow is atypical and is undoubtedly due to the presence, in the lactone, of an sp^2 hybridized carbon atom that cannot be readily accommodated in a three-membered ring (cf. discussion on cyclopropanone on p. 755). That the slowest rate is observed at C_8 rather than at C_9 or C_{10} probably reflects, in part, the fact that lactones of up to eight-membered rings must be largely or entirely in the normally disfavored cis (E) conformation (cf. Fig. 10.20), whereas nine-membered rings are mainly in the favorable trans (Z) conformation, and larger rings are entirely so (Huisgen and Ott, 1959). Nonetheless, the rate data do display the expected sharp drop in cyclization velocity beyond the five-membered ring (effect of entropy, particularly severe for $n = 6$, and of ring strain in the case of the medium rings) with a subsequent recovery as one proceeds to the large rings that are nearly strain free.

Rates of ring closure and corresponding activation enthalpies and entropies for closure of three- to five-membered rings are shown in Table 11.5 (Benedetti and Stirling, 1983). These values refer to the cyclization rates of anions derived from ω-halodisulfones. The specific rate is very high for three- and five-membered rings and much lower for four-membered ones. However, the high rate of ring closure in the five-membered ring is due to a relatively low activation enthalpy (reflecting the low amount of strain in the five-membered ring), whereas the high

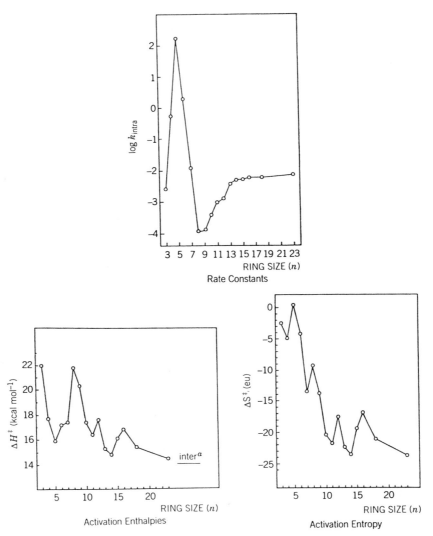

Figure 11.10.   Rate and activation parameters for cyclization of $Br(CH_2)_{n-1}CO_2^-$. The line marked "inter" indicates $\Delta H^{\ddagger}$ for the corresponding intermolecular reaction. Adapted with permission from Illuminati, G. and Mandolini, L. (1981), *Acc. Chem. Res.*, **14**, 96, 99, 100. Copyright © American Chemical Society, Washington, DC, 1981.

TABLE 11.5.   Rates of Ring Closure of $(C_6H_5SO_2)_2\bar{C}(CH_2)_{n-1}Cl$.

n	3	4	5
$k(s^{-1})$	9.05×10^{-1}	6.05×10^{-6}	1.49×10^{-2}
$\Delta H^{\ddagger\, a}$	20.5 (85.7)	21.8 (91.2)	16.3 (68.2)
$\Delta S^{\ddagger\, b}$	+10.0 (+42)	−9.3 (−39)	−12.2 (−51)

[a] In kilocalories per mole (kcal mol^{-1}); data in parentheses are in kilojoules per mole (kJ mol^{-1}).
[b] In calories per mole per degree kelvin (cal mol^{-1} K^{-1}); data in parentheses are in joules per mole per degree kelvin (J mol^{-1} K^{-1}).

rate for the three-membered ring (which is quite strained) is due to a very favorable activation entropy. The two ends of the ring are always favorably disposed for ring formation, but, of course, the C–C–C bond angle needs to be deformed, which causes the unfavorable activation enthalpy. The four-membered ring loses out on both grounds (its ends are apart in the more stable anti conformation of the open-chain precursor, hence unfavorable entropy of activation; the ring is strained, hence unfavorable enthalpy). It is perhaps somewhat surprising that the enthalpy of activation for its formation is higher than that for the three-membered ring, but the difference is small.

Ease of ring closure is usually in the order three-, five-membered greater than four- or six-membered rings, in part for the reasons just discussed. Since the high rate of ring closure of three- and five-membered rings is based on different factors, predictions as to which closure will occur more rapidly in a given case are unsafe. Six-membered rings usually cyclize less rapidly than five-membered rings, but a firm prediction on this point is risky since (Table 11.4) the strain in the six-membered ring is appreciably less; however, this is usually more than outweighed by the much greater loss of entropy (p. 679) in the formation of the six-membered ring. A factor that must be considered is thermodynamic versus kinetic control; since six-membered rings tend to be more stable than five-membered rings, their formation may be favored thermodynamically, even if not kinetically. Moreover, since enthalpy and entropy factors work in opposite directions in the 6:5 equilibrium, such equilibria might be expected to be strongly temperature dependent: Since $K = e^{-\Delta H^0/RT} e^{\Delta S^0/R}$, as ΔH^0 favors the six-membered ring and ΔS^0 favors the five-membered ring, it is clear that lowering the temperature should shift the equilibrium toward the six-membered ring. An example is the acetalization of glycerol with isobutyraldehyde; the reaction scheme is shown in Figure 11.11 and the energy profile is shown in Figure 11.12.

The activation energy is less for formation of the five-membered ring, hence this ring is formed faster (i.e., it is the product of kinetic control: $k_5 > k_6$). However, the reaction is reversible and the six-membered ring is slightly more stable ($K < 1$), so after some time it will become the major product (product of thermodynamic control; both five- and six-membered ring products will exist as a mixture of cis and trans isomers). In the equilibrium shown in Figure 11.11 (bottom), since enthalpy favors the nearly strain-free six-membered ring (whereas entropy favors the more flexible five-membered ring), lowering the temperature

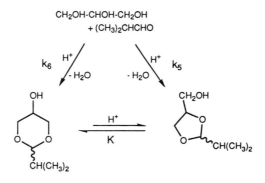

Figure 11.11.   Reaction of glycerol with isobutyraldehyde.

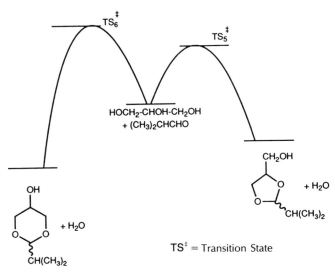

Figure 11.12. Reaction profile for the reaction of glycerol with isobutyraldehyde.

should shift the equilibrium toward the six-membered ring as was indeed found to be the case. Higher yields of the six-membered ring product were obtained when the dioxane–dioxolane mixture was allowed to equilibrate in the refrigerator at 2–5°C after the acetalization and water removal were complete. [Acid-catalyzed equilibration does not require water but proceeds via an oxycarbenium ion intermediate (Abraham, Eliel, et al., 1972)].

c. Ease of Ring Closure as a Function of the Ring Atoms and Substituents. The Thorpe–Ingold Effect

One of the major sources of strain in medium rings is the transannular repulsion of the hydrogen atoms of the CH_2 groups. Therefore replacement of such groups by heteroatoms (O, S, or NH) or sp^2 hybridized carbon atoms (C=O, –CH=CH–) would be expected to reduce the strain. There is a dearth of thermochemical information in this regard, but it is known that the presence of several (not just one) such elements facilitates ring closure. Thus Illuminati and Mandolini (1981) pointed out that the rates of cyclization to phenolic ethers of the compounds shown in Figure 11.13 do not show the customary minimum in the region of the medium rings, but rather display a nearly monotonic decrease with ring size, coming to a plateau with the large rings.

Another interesting manifestation of the effect of changing the nature of the groups in the ring is the so-called "Thorpe–Ingold effect" or "gem dialkyl effect"

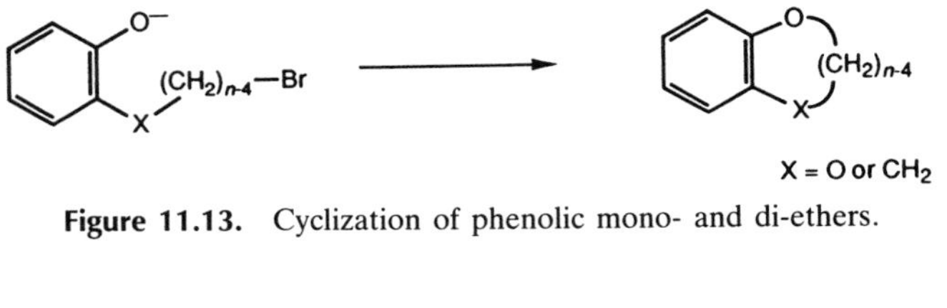

Figure 11.13. Cyclization of phenolic mono- and di-ethers.

(cf. Hammond, 1956). In the formation of small rings, there must occur a substantial reduction in C–C–C bond angle (see above). However, the reductions in angle postulated by Baeyer (Fig. 11.9) for these rings are actually not quite correct, since the normal bond angle in propane is $112.5°$, not $109.5°$ (cf. Section 2-5); thus the strain is larger than calculated by Baeyer. This situation changes when the fragment that cyclizes, $-CH_2CR_2CH_2-$ or $-CH_2CR_2CH_2CH_2-$, bears geminal methyl ($R = CH_3$) or higher alkyl groups instead of hydrogen ($R = H$). Here the carbon that is geminally substituted resembles the central carbon in neopentane at which the bond angle (for symmetry reasons) is reduced to the tetrahedral angle, $109°28'$. This decreases the angle deformation (i.e., strain) incurred upon cyclization; in other words, the formation of rings bearing gem-dialkyl groups should be easier than it would be in the absence of such groups. Although the effect might appear small, it does, in fact, manifest itself, as shown in Table 11.6 for halohydrin cyclization (Nilsson and Smith, 1933); it is also seen in other contexts. For example, whereas $C_6H_5CH_2SCH_2CH_2CH_2OTs$ ($Ts = p$-$C_6H_4SO_2$) methanolyzes without rearrangement and at a rate comparable to that of the sulfur-free analogue, $C_6H_5CH_2SC(CH_3)_2CH_2CH_2OTs$ solvolyzes at an accelerated rate and with complete rearrangement to $C_6H_5CH_2SCH_2$-$CH_2C(CH_3)_2OCH_3$, indicating extensive sulfur participation via a cyclic (four-membered) thietanonium salt intermediate (Fig. 11.14a; Eliel and Knox, 1985).

The effect persists in six-membered and sometimes also in five-membered rings (although eclipsing of geminal groups in the latter is a problem). For example, the ratio of rates of lactone formation (cf. Fig. 11.10) from $Br(CH_2)_{n-3}CR_2CH_2-CO_2^-$, $R = CH_3$, to the rates when $R = H$ (k_{gem}/k_H) is 38.5 for $n = 6$, 6.62 for $n = 9$, 1.13 for $n = 10$, 0.61 for $n = 11$, and 1.22 for $n = 16$ (Illuminati and Mandolini, 1981). The sizeable effect in the formation of the six-membered ring seems at first sight surprising, considering that there is virtually no change of bond angle. A different explanation of the gem-dialkyl effect in such rings has been proposed by Allinger and Zalkow (1960; see also Kirby, 1980; DeTar and Luthra, 1980; Jager, Kirby, et al., 1984; Sternbach et al., 1985). The gem-dialkyl groups in open chains substantially increase the number of gauche interactions, thus increasing enthalpy. When rings are formed, since some of the gauche interactions are now intraannular, and therefore attenuated, the unfavorable enthalpy effect of gem disubstitution is diminished. At the same time, branching diminishes the mobility, and hence the entropy of chains, but has much less effect in this regard in rings, which are already more restricted in their

TABLE 11.6. Relative Rates of Ring Closures of Chlorohydrins.

Compound	Relative Rate
$HOCH_2CH_2Cl$	1
$HOCH_2CHClCH_3$	5.5
$CH_3CHOHCH_2Cl$	21
$HOCH_2CCl(CH_3)_2$	248
$(CH_3)_2COHCH_2Cl$	252
$(CH_3)_2COHCHClCH_3$	1360
$CH_3CHOHCCl(CH_3)_2$	2040
$(CH_3)_2COHCCl(CH_3)_2$	11,600

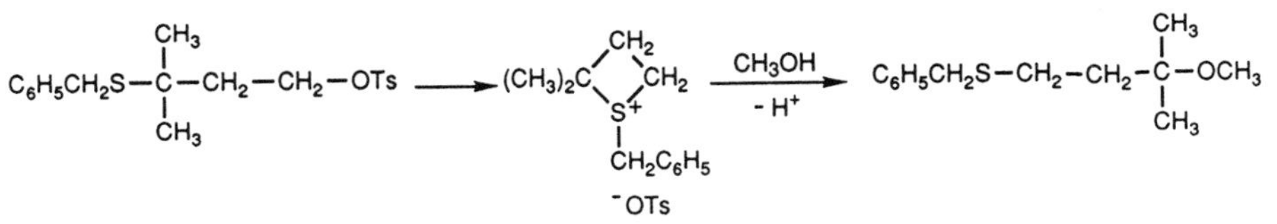

CH3
C6H5CH2S–C–CH2–CH2–OTs
CH3
(CH3)2C
CH2
CH2
S+
CH2C6H5
-OTs
CH3OH
- H+
C6H5CH2S–CH2–CH2–C–OCH3
CH3
CH3

Figure 11.14*a*. Thietanonium salt intermediate.

motions. Thus the entropy change in ring closure is made more favorable (less negative) by alkyl substitution. For rings 10-membered and larger these effects are evidently no longer important. In large rings, the conformational situation resembles that in open chains. In medium rings other, unfavorable, effects of gem-dialkyl substitution may come into play, as we shall see later (Section 11-5.d).

d. Baldwin's Rules

In 1970, Tenud, Eschenmoser, et al. (q.v.) observed that a compound of type $\bar{Y}(CH_2)_4\overset{+}{X}-CH_3$ would not undergo an intramolecular S_N2 displacement [to give $H_3C-Y(CH_2)_4X$] because of the virtual impossibility of approaching a collinear arrangement of the nucleophile Y^- and the leaving group X^+ (cf. Fig. 5.43). In a generalization and expansion of this observation, Baldwin in 1976 enunciated rules of ring closure bearing his name (Baldwin, 1976, 1978; Baldwin et al., 1976). For the purpose of classifying ring closures, Baldwin distinguishes "exo" and "endo" cases, as shown in Figure 11.14*b*. In endo ring closures, the nucleophile at one end of the cyclizing chain attaches itself to the other end of that chain, whereas in exo ring closures, attachment is at the penultimate atom and the terminal (end) atom remains outside of the newly formed ring. Also, the reactions are classified as "tet," "trig," or "dig" depending on whether ring closure involves displacement at a tetrahedral atom or addition to a trigonal (sp^2 hybridized) or digonal (*sp* hybridized) moiety. Finally, the size of the ring to be formed is indicated by an appropriate leading numeral.

(CH2)n
X−
Y
exo-tet
endo-tet
exo-trig
endo-trig
exo-dig
endo-dig

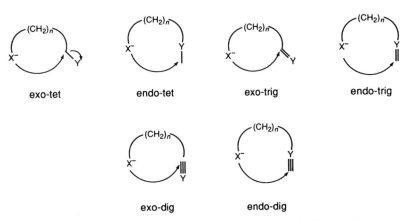

Figure 11.14*b*. Examples of exo- and endo-tet, -trig, and -dig ring closures.

The Eschenmoser case indicates, then, that a 6-endo-tet reaction is disfavored (note that this would not actually be a ring closure) and this is true for other n-endo-tet reactions, where $n \gtrsim 6$ because of the great difficulty in achieving the appropriate collinearity of the incoming and outgoing groups (see above). On the other hand, the n-exo-tet reactions are favorable even for small n values (3–7).

The situation for trig reactions is similar, though for different reasons. According to the Bürgi–Dunitz trajectory (Bürgi, Dunitz, Lehn et al., 1974; see also p. 737), approach of a nucleophile to a double bond does not occur orthogonally but at an angle of about 109°. This angle of approach again makes the exo approach more favorable than the endo, and it is therefore not surprising that, whereas endo-trig reactions are disfavored for rings five-membered and smaller (but not necessarily in formation of six- and seven-membered rings), exo-trig reactions are favored (Baldwin et al., 1976; Baldwin and Reiss, 1977). An example is shown in Figure 11.14c, **A**.

In digonal (sp) systems, however, the angle of nucleophilic approach appears to be less than 90°, rather than 109° (sp^2) or 180° (sp^3) (Baldwin, 1976). Therefore, in this case the exo-dig approach is disfavored for small rings (three- or four-membered) though it seems to be possible for larger rings; the endo-dig process is generally favored.

Baldwin's rules have been applied to a variety of reactions: Conjugate additions of oxygen nucleophiles (Baldwin et al., 1977), endocyclic alkylations of ketone enolates (Baldwin and Kruse, 1977), and intramolecular aldol condensations (Baldwin and Lusch, 1982). However, the rules probably do not apply to nonconcerted reactions, such as the second step in the ring closure of dioxolanes (Fig. 11.14c, **B**), where an oxycarbenium ion intermediate is involved (Baldwin, 1978).

It is of considerable interest (though a detailed discussion is beyond the scope of this book) that a normally disfavored endo–tet ring closure to a six-membered ring may be brought about by the use of an appropriately constituted catalytic antibody (Janda, Lerner, et al., 1993; see also Danishefsky, 1993). In the absence of such a special catalyst, which is produced by immunological methods (cf. p. 411), the reaction occurs by the normally preferred exo–tet pathway to give a five-membered ring.

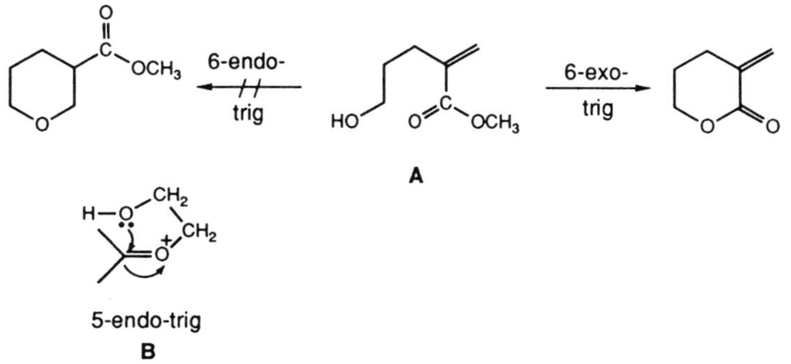

Figure 11.14c. Limitations of Baldwin's rules.

11-4. CONFORMATIONAL ASPECTS OF THE CHEMISTRY OF SIX-MEMBERED RING COMPOUNDS

a. Cyclohexane

Three-membered rings are of necessity flat, but all other rings (from four-membered rings on up) are nonplanar, and thereby present important conformational aspects affecting physical and chemical properties. We consider the six-membered ring first, not only because of its wide occurrence in natural, as well as purely synthetic substances, but also because its conformation is easier to study than that of either smaller or larger rings. This is so because the well-known chair conformer of cyclohexane lies in a deep energy valley, such that chemical changes at the periphery of the ring are unlikely to change the conformation of the ring itself [except (see below) possibly to invert the chair into an alternate chair]. Indeed, it is because of the rigidity of the chair conformer that the entropy of cyclohexane is so much lower than one would calculate on the basis of constant entropy increments in the homologous series of cyclanes (cf. p. 679).

> The other common rings (five- and seven-membered) and to a lesser extent also four-membered rings and rings larger than seven-membered are much more flexible than cyclohexane and their conformation may easily be changed upon substitution on the ring, making conformational study more complex.

As we have already seen, Baeyer considered cyclohexane a strained planar molecule, but Sachse, nearly 100 years ago, recognized that it might be chair or boat shaped (Fig. 11.15) and unstrained (Sachse, 1890, 1892). Sachse died (in 1893) soon after publication of his pioneering papers and his ideas died with him, only to be resurrected a quarter century later by Mohr (1918). The difficulty with Sachse's proposal was that, in light of what was known in 1890, it had to be interpreted in terms of a *rigid* chair model, which should have given rise to two different monosubstituted cyclohexyl derivatives, such as hexahydrobenzoic acid: one (to use present terminology) equatorially substituted, the other axially substituted. For now obvious reasons (to be detailed below) such isomers were, however, never found, and in a widely used textbook of the time Aschan (1905) wrote, "The non-existence of two forms of hexahydrobenzoic acid makes Sachse's notions untenable." It was recognized only many years later that rotation around single bonds (cyclohexane inversion, which interconverts the two forms of hexa-

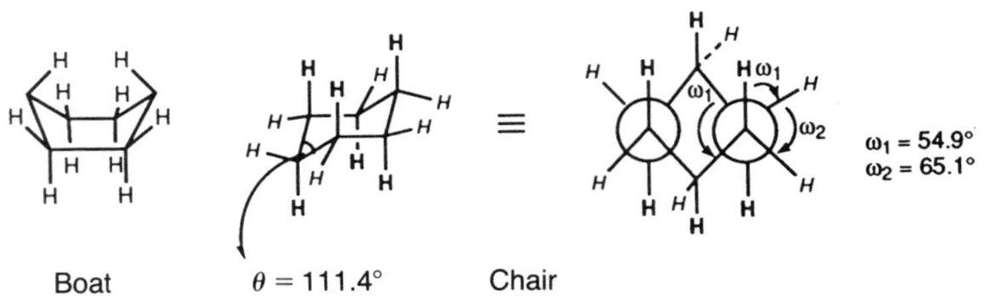

Boat $\theta = 111.4°$ Chair

$\omega_1 = 54.9°$
$\omega_2 = 65.1°$

Figure 11.15. Cyclohexane. **H**, axial hydrogen atoms; *H*, equatorial hydrogen atoms in chair conformer.

hydrobenzoic acid, being a process of this type) could be rapid though not instantaneous. In fact, a well-known textbook of a much later time (Shriner et al., 1938) errs in the opposite direction: "It seems probable that there is an equilibrium between the two forms and that the two models vibrate from one to the other so rapidly that the net average is a planar molecule." However, as pointed out in Section 2-1, the axially and equatorially monosubstituted cyclohexanes are spectroscopically distinguishable molecules; in fact, in the case of chlorocyclohexane, the equatorial isomer has been obtained in pure crystalline form and the axial isomer has been enriched in the mother liquor of crystallization; the two conformers are quite stable at $-150°C$ (Jensen and Bushweller, 1969a).

The idea that cyclohexane was, in fact, a chair-shaped molecule gained much ground based on a variety of physical and chemical experiments carried out in the 1920s, 1930s, and 1940s, but it was not until the appearance of a pioneering paper by Barton (1950) that the physical and chemical consequences of the chair conformations were fully realized. It is with these consequences that this section is mainly concerned; the detailed history of earlier findings that led up to Barton's insight has been dealt with elsewhere (Eliel et al., 1965; Ramsay, 1973; Russell, 1975; Eliel, 1975).

Prominent among the approaches to establishing the chair shape of the cyclohexane ring has been Hassel's work using electron diffraction (cf. Hassel, 1943). Application of this technique to cyclohexane itself (Davis and Hassel, 1963; Alekseev and Kitaigorodskii, 1963; Geise et al., 1971; Bastiansen, Kuchitsu, et al., 1973; Ewbank, Schäfer, et al., 1976) established the molecule to be a slightly flattened chair with bond angles (1971 determination) of $111.4 \pm 0.2°$ and torsion (C–C–C) angles in the ring of 54.9 ± 0.4 (see also Dommen, Bauder, et al., 1990). The bond angle is larger than tetrahedral but smaller than the "normal" (i.e., optimal) bond angle in propane, $112.4°$; the torsion angle deviates from the optimum of $60°$. One might interpret this deviation from optimal values as follows: In propane, bond angle deformation and torsion are independent, so the bond angle can be expanded beyond the tetrahedral with the torsion angle still remaining at or near the $60°$ optimum. In cyclohexane, because of the constraint of the ring (cf. Dunitz, 1970), this is not possible: If the bond angle expands beyond the tetrahedral, the intraannular torsion angle must necessarily decrease below $60°$. Such a decrease causes torsional strain and will be resisted; the total strain is minimized when a compromise is reached with the bond angle being slightly "too small" ($111.4°$ instead of $112.4°$) and the torsion angle slightly "too small" also ($54.9°$ instead of $60°$). Some strain (angle + torsional) remains; it is therefore reasonable that cyclohexane is not an entirely strain-free molecule (cf. the discussion on p. 677).

There is a general relationship between bond angles θ and adjacent torsion angles ω in cyclohexane (cf. Romers, Havinga, et al., 1969): $\cos \omega = -\cos \theta/(1 + \cos \theta)$; the relationship also applies in other rings, except that ω and θ do not then refer to individual angles but to the average torsion and bond angles in the ring. Another useful relation is that for a bond angle near the tetrahedral, a variation in bond angle of $1°$ causes a variation in torsion angle of $2.5°$.

Closing the intraannular torsion angle to $54.9°$ brings with it a *decrease* in the (external) H–C–C–H torsion angle of cis located hydrogen atoms from $60°$ to

54.9° and of trans diaxially located hydrogen atoms from 180° to 174.9° and an *in*crease for the corresponding trans diequatorially located hydrogen atoms from −60° to −65.1° (cf. Fig. 11.15). In view of the Karplus relationship (Fig. 10.37) this has an important impact on vicinal coupling constants in cyclohexane, which will be discussed later (cf. Wohl, 1964).

Cyclohexane has two geometrically different sets of hydrogen atoms, six extending up and down along the S_6 axis of the molecule (whose symmetry point group is $\mathbf{D_{3d}}$), called axial (a) hydrogen atoms and six alternating about an "equatorial" plane at right angles to this axis (not a symmetry plane!), which are called equatorial (e) (Barton et al., 1953, 1954; Fig. 11.15). The axial hydrogen atoms are homotopic relative to each other since they can be brought into coincidence by operation of the symmetry axes (C_3, C_2) of the molecule; the same is true of the equatorial hydrogen atoms relative to each other. However, the axial set is diastereotopic with the equatorial set, the two types being related neither by a symmetry axis nor by a symmetry plane.

Sachse already recognized that a cyclohexane ring could invert (*snap over* as Hassel later called it or *flip*) from one chair form to another (Fig. 11.16). Although he did not explicitly appreciate the ease of this process, he called this motion "version." Today the terms "ring inversion" or "ring reversal" are preferred. The ring reversal produces an evident change: The set of axial hydrogen atoms becomes equatorial and the equatorial set becomes axial; that is, the two diastereotopic sets exchange places: topomerization occurs (cf. Binsch et al., 1971 and Chapter 8). As explained in Section 8-4.d, processes of this type can be detected and the barrier to inversion measured, by means of dynamic (variable-temperature) NMR. The barrier to ring reversal in cyclohexane has been determined a number of times by this method (cf. Anet and Anet, 1975; Sandström, 1982, p. 102) the most accurate results coming from determination in cyclohexane-d_{11}: The "best overall values" estimated for the kinetic parameters are $\Delta G^{\ddagger} = 10.25$ kcal mol^{-1} (42.9 kJ mol^{-1}) at −50 to −60°C, $\Delta H^{\ddagger} = 10.7$ kcal mol^{-1} (44.8 kJ mol^{-1}), $\Delta S^{\ddagger} = 2.2$ cal mol^{-1} K^{-1} (9.2 J mol^{-1} K^{-1}), that is, $\Delta G^{\ddagger} = 10.1$ kcal mol^{-1} (42.3 kJ mol^{-1}) at 25°C; a later determination (Aydin and Günther, 1981) gave a very similar value for $\Delta G^{\ddagger}$: 42.7 kJ mol^{-1} (10.2 kcal mol^{-1}) at −75°C. However, another careful study (Höfner, Binsch et al., 1978) gave slightly different values: $\Delta G^{\ddagger} = 10.5$ kcal mol^{-1} (43.9 kJ mol^{-1}) at −50°C [10.1 kcal mol^{-1} (42.4 kJ mol^{-1}) at 25°C], $\Delta H^{\ddagger} = 11.5$ kcal mol^{-1} (48.2 kJ mol^{-1}), $\Delta S^{\ddagger} = 4.6$ cal mol^{-1} K^{-1} (19.2 J mol^{-1} K^{-1}); the difficulty of obtaining data of this type (especially $\Delta H^{\ddagger}$ and $\Delta S^{\ddagger}$) with high accuracy is evident. Calculations of the barrier by force field methods (cf. Anet and Anet, 1975; van de Graaf et al., 1980) span the same range of $\Delta H^{\ddagger}$ as the experimental values. The activated complex for the ring reversal is probably close to the half-chair

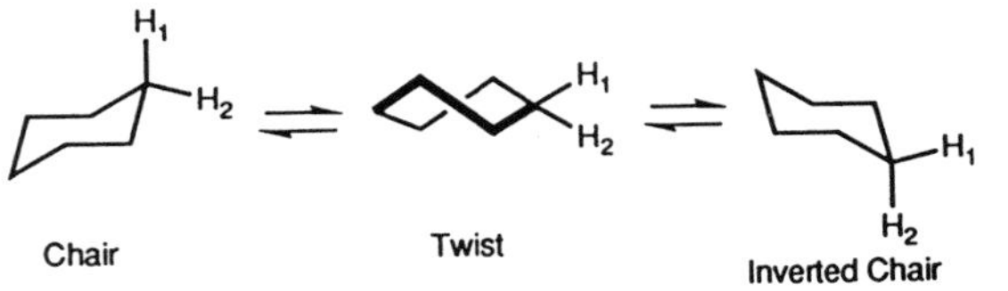

Figure 11.16. Cyclohexane inversion (reversal).

depicted in Figure 11.17 but is actually quite flexible, as evidenced by the fairly large positive entropy of activation.

There are probably three factors contributing to the sizeable positive $\Delta S^{\ddagger}$. (1) The transition state is much less symmetrical than the starting cyclohexane, which has a symmetry number $\sigma = 6$ and hence an entropy of symmetry of $-R \ln 6$. (2) The activated complex has a number of nearly equienergetic states and therefore a sizeable entropy of mixing, including the entropy that results because the half-chair is a racemic mixture. (3) The half-chair has a number of low-lying vibrational modes that contribute to its vibrational entropy (this point is related to the previous one).

As shown in Figure 11.16, the inversion of the cyclohexane chair (see also Anderson, 1974) is not a one-step process. This inversion first produces an intermediate which, in a second step equienergetic with the first, leads to the inverted chair. Manipulation of models might suggest that the intermediate is Sachse's boat form, but force field calculations make this highly unlikely; rather, the intermediate is almost certainly the twist form depicted in Figures 11.16 and 11.17 (Kellie and Riddell, 1974). This conformer, which is obtained from the boat by slight deformations, is less stable than the chair conformer by 4.7–6.2 kcal mol^{-1} (19.7–26.0 kJ/mol^{-1}) in ΔH^{0} according to various indirect experimental determinations; most of the force field calculations also fall into this range (cf. Kellie and Riddell, 1974; Allinger, 1977; van de Graaf et al., 1980; see also Burkert and Allinger, 1982). The calculations also suggest that the true boat form is about 1–1.5 kcal mol^{-1} (4.2–6.3 kJ mol^{-1}) in energy above the twist form; it is apparently the energy maximum in the interconversion of two distinct twist conformers.

Until 1975, the twist form of cyclohexane itself (as distinct from that of certain substituted cyclohexanes, such as the *trans*-1,3- or *cis*-1,4-*di-tert*-butyl derivatives to be discussed later, or of cyclohexanes forming part of fused ring systems) existed

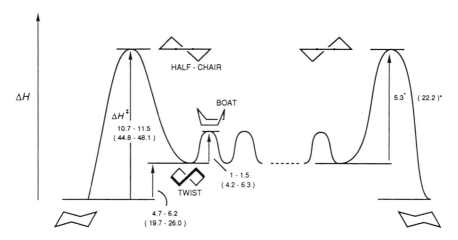

Figure 11.17. Energy profile [$\Delta H^{\ddagger}$ kcal mol^{-1} (kJ mol^{-1})] for cyclohexane ring reversal. The asterisk (*) indicates that $\Delta S^{\ddagger} = 0$ is assumed.

only on paper. In that year, Squillacote, Chapman, Anet, et al., (1975) succeeded in capturing the twist form by a matrix isolation technique, which consisted of rapidly cooling to 20 K (−253°C) a mixture of cyclohexane vapor and argon previously heated to about 800°C (see also p. 621). The IR spectrum of the solid matrix indicated that it contained about 25% of a second entity besides the chair conformer. The rate of disappearance of this entity (presumably the twist form) on slight heating suggested a twist-to-chair activation energy of $\Delta G^{\ddagger} = 5.3$ kcal mol^{-1} (22.2 kJ mol^{-1}). With a chair-to-twist activation enthalpy (see above) of 10.7–11.5 kcal mol^{-1}, 44.8–48.1 kJ mol^{-1}, this places the twist 5.4–6.2 kcal mol^{-1} (22.6–25.9 kJ mol^{-1}) in enthalpy above the chair conformation, on the perhaps dubious assumption that $\Delta S^{\ddagger} = 0$ for the twist-to-chair process. Also, if the 25% twist form at 800°C is taken to be an equilibrium value, $\Delta G^{0}_{\text{chair}\rightarrow\text{twist}}$ at 800°C is about 2.3 kcal mol^{-1} (9.6 kJ mol^{-1}), which requires $\Delta S^{0}_{\text{chair}\rightarrow\text{twist}}$ to be about 3–4.5 cal mol^{-1} K^{-1} (12.6–18.8 J mol^{-1} K^{-1}). This value is only slightly smaller than that of 4.9 cal mol^{-1} K^{-1} (20.5 J mol^{-1} K^{-1}) reported (Allinger and Freiberg, 1960) from the equilibration of cis- and trans-1,3-di-tert-butylcyclohexanes; in any case the entropy of the twist is substantially above that of the chair, again because of lower symmetry, chirality (it is a racemate), and low-lying vibrational states. Even if $\Delta H^{0}_{\text{chair}\rightarrow\text{twist}}$ is as low as 4.7 kcal mol^{-1} (19.7 kJ mol^{-1}) and $\Delta S^{0}_{\text{chair}\rightarrow\text{twist}}$ as high as 4.9 cal mol^{-1} K^{-1} (20.5 J mol^{-1} K^{-1}) one calculates the fraction of twist conformer in cyclohexane at room temperature to be no more than 0.4%.

b. Monosubstituted Cyclohexanes

In monosubstituted cyclohexanes, the topomerization process shown in Figure 11.16 becomes an isomerization process (Fig. 11.18) involving the interconversion of two diastereomers (Jensen and Bushweller, 1971). Like the inversion of cyclohexane itself, this process is very rapid, its rate in most cases being close to that in cyclohexane (ca. 2×10^{5} s^{-1} at room temperature). This result explains why temperatures near −150°C were required to isolate conformational isomers of the type shown in Figure 11.18 and why the notion of Sachse's contemporaries that such isomers might be observable under ambient conditions was erroneous. In fact, no conformational isomers of monosubstituted cyclohexanes have ever been isolated at room temperature, several claims to the contrary having been subsequently disproved (but see p. 358). Nonetheless the equilibrium depicted in Figure 11.18 is very real, with most monosubstituted cyclohexanes being mixtures of two conformers with the equatorial one generally predominating (Hassel, 1943). This phenomenon can be readily demonstrated by IR spectroscopy at room temperature [e.g., cyclohexyl bromide (Fig. 11.18, X = Br) shows a C–Br stretching vibration for the equatorial conformer at 685 cm^{-1} and for the axial conformer at 658 cm^{-1} (Larnaudie, 1954)] and by NMR spectroscopy at temperatures below about −50°C (Berlin and Jensen, 1960; Reeves and Strømme, 1960), under which conditions distinct signals of nuclei due to the two conformers are seen. In fact, as

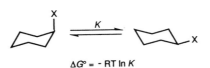

$$\Delta G^{\circ} = -\text{RT} \ln K$$

Figure 11.18. Conformational inversion of monosubstituted cyclohexane.

already described in Chapter 10, both methods lend themselves to a *quantitative* determination of the equilibrium depicted in Figure 11.18: IR spectroscopy (cf. Fishman et al., 1981) for determination of ΔH^0 by observation of the change in band area with temperature (cf. Eq. 10.9), NMR spectroscopy through direct measurement of the intensity of the signals due to the two species shown in Figure 11.18. The ratio of these signal intensities gives K, and hence ΔG^0. Either ^{1}H or ^{13}C (or, in the case of fluorinated compounds, ^{19}F) NMR spectroscopy may be used for this purpose. Application of ^{1}H NMR spectroscopy is limited because the signals of the two conformers, with the possible exception of that for the proton geminal to X (CHX), are often broad and poorly resolved both within one conformer and between the two. ^{13}C NMR spectroscopy (Anet et al., 1971) is therefore much to be preferred; whereas at room temperature $C_6H_{11}X$ (Fig. 11.18) generally shows four distinct ^{13}C signals [C(2,6) and C(3,5) being enantiotopic, and hence isochronous], at low temperature ($-80°C$ or lower) eight signals (for the two individual conformers) are often seen. This finding allows one to compute four different values for K, the degree of whose agreement reflects the accuracy of the determination. (It has already been implied in Chapter 10, p. 641, that differences in nuclear Overhauser effects (NOE) and relaxation times tend not to affect the determination of K as long as corresponding nuclei in the two conformers are compared.)

The NMR method usually yields ΔG^0, in the -80 to $-100°C$ region, since area measurements of signals must be made well below the coalescence temperature. It does not, however, lend itself to determination of ΔH^0 and ΔS^0, which requires accurate measurement of K over a sizeable range of temperature. Unfortunately, the accessible temperature range is quite limited, at the upper bound by incipient coalescence, at the lower by crystallization of the solute or solvent or both. The ΔG^0 values measured at low temperature are therefore often reported as if they were the same at 25°C, which implies $\Delta S^0 = 0$ and ΔH^0 independent of temperature. Neither of these assumptions is likely to be correct. If X lacks the local C_{nv} symmetry of a methyl group (C_{3v}) or of a halogen or C≡N group ($C_{\infty v}$), there may be entropy-of-mixing differences between axial and equatorial conformers. But even if such is not the case, vibrational and rotational entropies will not be the same for conformational isomers (Reisse, 1971). Thus, while ΔS^0 for X = CN was found to be zero (Höfner, Binsch, et al., 1978) that for X = Cl is 0.32 cal mol^{-1} K^{-1} (1.34 J mol^{-1} K^{-1}) and for X = Br 0.06 cal mol^{-1} K^{-1} (0.25 J mol^{-1} K^{-1}), in both cases favoring the equatorial conformer. For CD$_3$O, where one might have expected entropy of mixing to favor the equatorial conformer (which has two favorable and one somewhat less favorable conformations by virtue of rotation of CH$_3$O about the C—O axis, whereas the axial conformer can only have two low-energy rotamers, with the CH$_3$ group pointing outside the ring), the *axial* conformer is actually favored by 0.42 cal mol^{-1} K^{-1} (1.76 J mol^{-1} K^{-1}). The best that can be said here is that entropy effects (and presumably also effects of temperature-variable enthalpy) are small. In the cases investigated, these entropy effects were much less than 1 cal mol^{-1} K^{-1} (4.2 J mol^{-1} K^{-1}), and therefore caused less than a 0.1 kcal mol^{-1} (0.42 kJ mol^{-1}) difference in ΔG^0 between the measurement range of -80 to $-100°$ and room temperature. Attempts to obtain ΔG^0 at room temperature to a greater accuracy than 0.1 kcal mol^{-1} (0.42 kJ mol^{-1}) require measurements of unusual care (Reisse, Oth, et al., 1969; Eliel and Gilbert, 1969; Höfner, Binsch, et al., 1978) and may even so be foiled by extraneous factors, such as solvent effects or self-association, not to mention methodological problems.

In general it would be preferable to measure ΔG^0 at 25°C, which is close to the temperature at which one frequently deals with physical, spectral, and pharmacological properties, as well as synthesis, reaction mechanism, and so on. For the reasons already mentioned, accurate extrapolations from low temperatures are rarely feasible. In principle, however, conformational composition can be measured at any temperature by use of the Winstein–Holness equation (Eq. 10.22) which, applied to a cyclohexyl system (Fig. 11.18) with two conformations and generalized to any appropriate property P, becomes

$$P = n_E P_E + n_A P_A \tag{11.1}$$

or

$$K = n_E/n_A = (P_A - P)/(P - P_A) \tag{11.2}$$

where n denotes mole fraction and the subscripts E and A refer to equatorial and axial conformers, respectively. The first application going beyond kinetics (Winstein and Holness, 1955) and dipole measurements (where $P = \mu^2$ is polarization, μ being the dipole moment) was, in fact, in NMR (Eliel, 1959). Application to chemical shifts, δ, and coupling constants, J, gives $\delta = n_E \delta_E + n_A \delta_A$ and $J = n_E J_E + n_A J_A$; that is, both observed chemical shifts and observed coupling constants in an equilibrating system, such as shown in Figure 11.18, are the weighted averages of the corresponding constants in the individual conformers. Using the form of Eq. 11.2 (cf. p. 641), this becomes

$$K = (\delta_A - \delta)/(\delta - \delta_E) = (J_A - J)/(J - J_E) \tag{11.3}$$

Unfortunately, the room temperature shifts of the pure conformers δ_A and δ_E are not readily accessible; because of sizeable variations of chemical shift with temperature, they can*not* be taken to be equal to the low-temperature shifts. The situation is much more favorable with coupling constants which are very nearly temperature independent and can thus be used to estimate K at room temperature from J, if J_E and J_A are known from low-temperature measurement (Höfner, Binsch, et al., 1978). In the past, this method has been limited to proton spectra, but with the increasing availability of ^{1}H–^{13}C coupling constants (Marshall, 1983), the method may find extension to ^{13}C NMR spectroscopy.

More popular, but less reliable, have been methods to determine ΔG^0 through the use of model compounds. These are compounds confined to a single conformation in which the substituent X is either strictly equatorial or strictly axial. Such confinement may be brought about either by "conformational locking," such as in the *trans*-decalin-2-ols (Fig. 11.19 **A**, R or R' = OH), or by "conformational biassing," such as in the *cis*- and *trans*-4-*tert*-butylcyclohexanols (Fig. 11.19, **B**, R or R' = OH, respectively). Conformational locking was present in many of the compounds cited in Barton's 1950 classical paper (q.v.) dealing with physical and chemical properties of axially and equatorially substituted cyclohexyl compounds. However, it has not been used much in quantitative work, perhaps because of concern that the double substitution only two and three

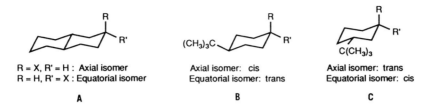

R = X, R' = H : Axial isomer
R = H, R' = X : Equatorial isomer

A

Axial isomer: cis
Equatorial isomer: trans

B

Axial isomer: trans
Equatorial isomer: cis

C

Figure 11.19. Conformationally locked (**A**) and conformationally biassed (*anancomeric*: **B** and **C**) compounds.

positions away from the carbon bearing the substituent X might affect reactivity. Prototypes of the biassed model (Winstein and Holness, 1955) are the 4-*tert*-butylsubstituted compounds (Fig. 11.19, **B**); compounds of this type have also been called "anancomeric" (Anteunis, 1971) meaning "fixed in one conformation" (the word is derived from Greek *anankein* meaning to fix by some fate or law). Models of this type can be used to ascertain appropriate axial or equatorial parameters, such as chemical shifts (of various nuclei), coupling constants, dipole moments, pK_a values of acids, and extinction coefficients for IR determinations (for a review, see Eliel et al., 1965).

One must, of course, ask whether such models (Fig. 11.19, **B** and **C**) are adequate; specifically, whether the introduction of a "holding group," such as 4-*tert*-butyl, to make a molecule anancomeric will alter (falsify) the chemical shifts, coupling constants, dipole moments, pK_a values, and so on, of the equatorial or axial conformer of cyclohexyl-X devoid of such a group. The holding group can have two kinds of adverse effects: it may deform the molecule structurally thus altering many of its properties, or it may affect the particular property to be measured in other ways, for example, by inductive effects or through-space interactions. With respect to the first point, crystallographic study (James and McConnell, 1971; Johnson et al., 1972a; also neutron diffraction: James and Moore, 1975) of cyclohexyl *p*-toluenesulfonate (which crystallizes with the toluenesulfonate group equatorial) and its 4-*trans-tert*-butyl homologue (Fig. 11.19, **B**, R' = OTs, R = H) discloses no obvious significant differences except possibly an unusually long C–OS bond (151 pm, 1.51 Å) in the unsubstituted compound (the corresponding bond length in the 4-*tert*-butyl homologue is a more normal 147 pm (1.47 Å). *cis*-4-*tert*-Butylcyclohexyl tosylate (Fig. 11.19, R = OTs, R' = H) has also been investigated (Johnson et al., 1972b; James and Grainger, 1972); it is dimorphous (i.e., crystallizes in two different forms) but the molecules in the two types of crystal do not differ greatly. As expected, the axial substituent causes some flattening of the ring. Unfortunately, the axial conformer of cyclohexyl tosylate itself cannot be obtained crystalline, so investigation of the effect of the 4-*tert*-butyl substituent on the geometry of the axial OTs group is not possible. However, as far as they go, the crystallographic studies imply no major effect of a 4-*tert*-butyl substituent on molecular shape, though the situation seems to be different for 3-*tert*-butyl substitution (James, 1973) and for the more flexible cyclohexanones (Abraham et al., 1980, 1983). The conclusion with respect to the 4-*tert*-butyl substituent is also supported by force field calculations (Altona and Sundaralingam, 1970).

It has, in fact, been repeatedly found that 3-*tert*-butyl substituents (Fig. 11.19, **C**) as well as *cis*-3,5-dimethyl substituents in cyclohexanes are unsuitable in model studies, either NMR (Eliel and Martin, 1968a) or kinetic or equilibrium studies

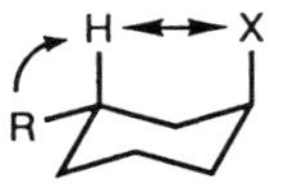

Figure 11.20. "Allinger buttressing."

(Eliel et al., 1966; Eliel and Biros, 1966). The reason has been pointed out by Allinger et al. (1967; see also Burkert and Allinger, 1982): An equatorial alkyl group buttresses the hydrogen geminal with it and thus prevents it from bending outward. If the alkyl group is at position 3 or 5 relative to an axial substituent to be studied, this "lack of give" will increase the synaxial H/X repulsion and thus increase the conformational energy of X as well as cause other changes (Fig. 11.20).

Even though the 4-*tert*-butyl group seems to cause little distortion in model compounds, use of such models in quantitative NMR studies cannot be recommended since it is apt to lead to inaccurate results (e.g., Reisse, 1971). Probably the major reason is a long-range effect of the *tert*-butyl substituent on the chemical shifts. (There are no extensive studies involving coupling constants.) This perturbation is serious in ^{1}H NMR because the whole range of observations (difference between equatorial and axial protons) is usually within less than 1 ppm, and therefore discrepancies of even a few hundredths of a part per million are fatal to accurate determinations. In ^{13}C NMR, the shift differences between equatorial and axial conformers are usually greater, but, unfortunately, this is compensated by the shifts being more sensitive to long-range (δ and ε) effects (cf. Jordan and Thorne, 1986). Fortunately, substituent effects on ^{13}C NMR shifts tend to be additive (see p. 717), and can thus be corrected for on the basis of shifts of suitable reference compounds.

Two methods other than low-temperature NMR studies have been used widely to determine conformational equilibria (Fig. 11.18). One is chemical equilibration of conformationally locked models (Eliel and Ro, 1957a; Fig. 11.21), based on Barton's recognition (1950) that equatorially substituted isomers are generally more stable than axially substituted ones. Since this equilibration is one of cis–trans diastereomers, it requires chemical intervention and one obvious condition for the method to succeed is that equilibrium must be established cleanly, that is, without appreciable side reactions. The second condition is that the equilibria in Figures 11.21 and 11.18 correspond, which again depends on the innocuousness of the holding group. In this situation, experience suggests that a 4-*tert*-butyl substituent does seem to be satisfactory.

Another method consists in the application of the Winstein–Holness equation (Section 10-5) to reaction rates. Using the form of this equation analogous to Eq. 11.2 one has $K = (k_a - k)/(k - k_e)$ (Eliel and Ro, 1956), where K is the conformational equilibrium constant (Fig. 11.18), k is the Winstein–Holness rate constant

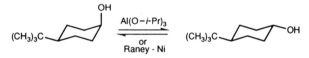

Figure 11.21. Chemical equilibration to determine conformational energies.

(cf. Section 10-5) – in this case the rate constant for the reaction to be investigated in the mobile system (Fig. 11.18) – and k_a and k_e are the rate constants for the same reaction with purely equatorial and purely axial conformers. In this form, despite some questions having been raised to the contrary (see Seeman, 1983 for discussion; see also Section 10-5), the equation is unconditionally correct, provided the rate of interconversion of the axial and equatorial conformers (Fig. 11.18) is fast compared to the rate of the reaction being studied. Since we have already seen that the conformer equilibration proceeds at a rate of about $2 \times 10^5 \, s^{-1}$ at room temperature, this condition will be fulfilled for all but a few extremely fast reactions, such as protonations, nitrous acid deaminations, and carbene additions of amines. Unfortunately, there is no way of determining k_a and k_e directly. These rates must therefore be determined on model compounds. For example, to use acetylation rates in determination of the cyclohexanol conformational equilibrium (Fig. 11.18, X = OH); Eliel and Lukach, 1957) one must use *cis*- and *trans*-4-*tert*-butylcyclohexanol as models to determine acetylation rate constants k_a and k_e, respectively. In most cases the factors mentioned earlier (ring deformation and long-range effects) will vitiate this attempt (cf. Eliel and Biros, 1966) and the kinetic method is therefore not to be recommended, although there may be special situations (especially when the reaction site is somewhat remote from the ring) in which it may succeed (McKenna, 1974).

"Conformational energies," $-\Delta G^0$ or "A values" (cf. Winstein and Holness, 1955; Fig. 11.18) for a variety of common substituents are summarized in Table 11.7. Since almost all ΔG^0 values are negative, we find it convenient to tabulate $-\Delta G^0$. More detailed tabulations have been compiled by Hirsch (1967), Jensen and Bushweller (1971), and Bushweller (in press) (see also Schneider and Hoppen, 1978). Perusal of the halogen values indicates that ΔG^0 is not solely a function of substituent size. As the halogens get larger (i.e., their van der Waals radius increases), the C–X bond becomes longer (i.e., the bond length increases also), and thus X becomes more distant from the carbon and hydrogen atoms at C(3,5) including, especially, the synaxial hydrogen atoms, which are mainly responsible for crowding of the axial substituent. This result leads to a form of compensation that is reinforced because the substituent with the longer bond also benefits more from the outward bending caused by the flattening of the cyclohexane ring (lever principle). The fact that the larger atoms (in the lower part of the periodic table) are also the softer or more polarizable ones, and therefore the atoms for which the attractive part of the van der Waals potential (London force) is more important (cf. Section 2-6) may be a contributing factor to their apparently low conformational energy values. It should, however, be noted that the drop in $-\Delta G^0$ from Cl to Br is due to entropy, not enthalpy effects (see above). A similar continuous decrease in conformational energy is evident in the series $(CH_3)_3C$, $(CH_3)_3Si$, $(CH_3)_3Ge$, $(CH_3)_3Sn$, and $(CH_3)_3Pb$.

Conformational energies for OX [X = H, CH_3, Ac, Ts, $Si(CH_3)_3$, or $C(CH_3)_3$] vary little with X, presumably because the group X can be turned so as to point away from the ring when OX is axial. At worst this will decrease the number of rotamers (rotational conformers) for the axial OX, and thereby produce a slight drop in its entropy. The OH group itself shows a large solvent effect, $-\Delta G^0_{OH}$ being appreciably larger in hydrogen-bond forming solvents (e.g., isopropyl alcohol) than in those not forming hydrogen bonds (e.g., cyclohexane).

TABLE 11.7. Conformational Energies.

Group[a,b]	Conformational Energies (kcal mol^{-1})	(kJ mol^{-1})	t(°C)	References
D	0.006	0.025	25	c
T	0.011	0.046	-88	d
F*	0.25–0.42	1.05–1.75	-86 to -93, -25	e–j
Cl^a*	0.53–0.64	2.22–2.68	-80 to -93, 25–27	e–h, j–m
Br*	0.48–0.67	2.01–2.80	-81, 25–27	e–h, j–m
I*	0.47–0.61	1.97–2.55	-78, -93, 25	e–h, j
OH(C$_6$H$_{12}$)*	0.60	2.51	25	e, n, o
OH(CS$_2$)	1.04^{p,q}	4.35	-83	f, j
OH(CH$_3$CHOHCH$_3$)	0.95	3.97	25	n
OCD$_3$*, OCH$_3$	0.55, 0.58, 0.63, 0.75	2.30, 2.43, 2.64, 3.14	-82, -93	e, f, j, k
OC(CH$_3$)$_3$	0.75	3.14	36	r
OC$_6$H$_5$	0.65	2.72	-93	s
OC$_6$H$_4$NO$_2$-p*	0.62	2.59	-93	s
OC$_6$H$_4$Cl-p*	0.65	2.72	-93	s
OC$_6$H$_4$OCH$_3$-p*	0.70	2.92	-93	s
OCHO	0.27, 0.60^p	1.13, 2.51^p	25, -80 to -93	e, f, j, t
OCOCH$_3$*	0.68, 0.71, 0.79, 0.87	2.85, 2.97, 3.31, 3.64	25, -90 ± 3	e, f, j, u
OCOCF$_3$	0.68, 0.56	2.85, 2.34	25, -88 to -93	f, j, t
OCOC$_6$H$_5$*	0.5^p	2.09^p	-92 ± 1	f, u
OCONHC$_6$H$_5$*	0.77	3.22	-91	u
OSO$_2$C$_6$H$_4$CH$_3$-p	0.50^p	2.09^p	-80 to -83	e, f, j
OSO$_2$CH$_3$	0.56	2.34	-88	e, j
ONO$_2$	0.59, 0.62^p	2.47, 2.59	25, -101	t, v
OSi(CH$_3$)$_3$	0.74	3.10	-103	f
SH	1.21^p	5.06^p	-80	e, f, j
SCD$_3$, SCH$_3$	1.04^p	4.35^p	-79 to -100	e, j, w
SC$_6$H$_5$	1.10–1.24	4.60–5.19	-80	x
SOCH$_3$	1.20	5.02	-90 to -100	w
SO$_2$CH$_3$	2.50	10.5	-90 to -100	w
SCN	1.23	5.15	-79	e, j
SeC$_6$H$_5$	1.0–1.2	4.2–5.0	-50	y
SeOC$_6$H$_5$	1.25	5.23	-60	z
TeC$_6$H$_5$	0.9	3.7	-30	z
NH$_2$ (toluene-d_8; CFCl$_3$)	1.23, 1.47^p	5.15, 6.15^p	-80 to -100	f, aa, bb
NH$_2$(CH$_3$OCH$_2$CH$_2$OH/H$_2$O)	1.7	7.1	20	cc
NH$_3$$^+$	1.7–2.0	7.1–8.4	20–25	cc, kkk
NHCH$_3$(CFCl$_3$–CDCl$_3$)	1.29	5.40	-80	bb
N(CH$_3$)$_2$(CFCl$_3$–CDCl$_3$)	1.53	6.40	-90	bb
N(CH$_3$)$_2$(CH$_3$OCH$_2$CH$_2$OH/H$_2$O)	2.1	8.8	20	cc
NH(CH$_3$)$_2$$^+$	2.4	10.0	20	cc
NHCOC$_6$H$_5$	1.6	6.7	-90	dd
NC	0.20^p	0.84^p	-80 to -93	e, f, j
NCO	0.44, 0.51	1.84, 2.13	-70 to -80	e, j, dd
N$_3$*	0.45–0.62	1.88–2.59	-183, -93	f, ee
NCS	0.25^p	1.05^p	-79 to -93	e, f, j
N=CHCH(CH$_3$)$_2$	0.75	3.14	32	dd
N=C=NC$_6$H$_{11}$	0.96	4.02	-80	j
NO$_2$*	1.1^p	4.8^p	-80 to -90, 25	e, f
PH$_2$	1.6	6.7	-90, 27	ff, gg
P(CH$_3$)$_2$	1.5, 1.6	6.3, 6.7	-90, 27	ff
P(C$_6$H$_5$)$_2$	1.8	7.5	37	hh
PCl$_2$	1.9, 2.0	7.9, 8.4	-90, 27	ff
P(OCH$_3$)$_2$	1.9; 1.5	7.9; 6.3	-90, 27	ff
O=P(C$_6$H$_5$)$_2$	2.46	10.3	-80	ii
S=P(C$_6$H$_5$)$_2$	3.13	13.1	-102	jj
CHO	0.56–0.73, 0.8	2.34–3.05, 3.35	25	kk, ll
COCH$_3$	1.02, 1.21, 1.52	4.27, 5.06, 6.36	-100, 25	mm, nn
CO$_2$H	1.4	5.9	25	nn
CO$_2$$^-$	2.0	8.4	25	nn

TABLE 11.7. (*Continued*)

Group[a,b]	Conformational Energies (kcal mol⁻¹)	(kJ mol⁻¹)	t(°C)	References
CO_2CH_3	1.2–1.3	5.0–5.4	25, −78	e, j, nn, oo
CO_2Et	1.1–1.2	4.6–5.0	25	j
COF	1.4–1.7	5.9–7.1	25	pp
$COCl^*$	1.3	5.4	25	nn
CN^*	0.2	0.84	−79 to −95	e, f, j, k
$C{\equiv}CH$	0.41–0.52	1.71–2.18	−91	e, f, j
$CH{=}CH_2$	1.49, 1.68	6.23, 7.0	−100	kk, qq
$CH{=}C{=}CH_2$	1.53	6.40	−80	rr
CH_3^*	1.74	7.28	27	ss
CD_3	$0.0115^{q,tt}$	$0.048^{q,tt}$	25–27	uu, vv
$CH_2CH_3^*$	1.79	7.49	27	ss
$CH(CH_3)_2^*$	2.21	9.25	27	ss
$C(CH_3)_3$	4.7; 4.9	19.7; 20.5	−120	ww, xx
CH_2Br	1.79	7.49	27	yy
CH_2OH	1.76	7.36	27	yy
CH_2OCH	1.72	7.20	27	yy
CH_2CN	1.77	7.41	27	yy
$CH_2Si(CH_3)_3$	1.65	6.90	27	yy
$CH_2Sn(CH_3)_3$	1.79	7.49	27	yy
$CH_2Pb(CH_3)_3$	1.81	7.57	27	yy
CH_2HgOAc	2.05	8.57	27	yy
CF_3	2.4–2.5	10.0–10.5	27	zz
C_6H_5	2.8^p	11.71^p	−100, 700	qq, aaa
$CH_2C_6H_5$	1.68	7.03	−71	bbb
C_6H_{11}	2.2	9.2	36	ccc
SiH_3	1.45; 1.52	6.07; 6.36	−85, 75	ddd, eee
$Si(CH_3)_3$	2.5	10.5	33	fff
$SiCl_3$	0.61	2.55	−80	e, j
$Ge(CH_3)_3$	2.1	8.8	−70	ggg
$Ge(C_6H_5)_3$	2.90	12.1	not given	ggg
$Sn(CH_3)_3$	1.0^p	4.2	−69 to −90	ggg, hhh, iii
$Sn(i\text{-}Pr)_3$	1.10	4.6	not given	ggg
$Sn(CH_3)_2C_6H_5$	1.08	4.5	not given	ggg
$SnCH_3(C_6H_5)_2$	1.20	5.02	not given	ggg
$Sn(C_6H_5)_3$	1.44	6.0	not given	ggg
$Pb(CH_3)_3$	0.67	2.80	−69	ggg
$HgOAc$	0, −0.3	0, −1.3	−79, −90	e, j, jjj
$HgCl$	−0.25	−1.05	−90	jjj
$HgBr$	0^p	0	−79	e, j, jjj
$MgBr\ (Et_2O)$	0.78	3.26	−75	j
$MgC_6H_{11}\ (Et_2O)$	0.53	2.22	−82	j

[a] Starred values mean that ΔH^0 and ΔS^0 are available in the original reference. [b] The solvent is in parentheses in cases where large solvent dependence is observed. [c] Anet and Kopelevich, 1986. [d] Anet et al., 1990. [e] Jensen et al. 1969. [f] Schneider and Hoppen, 1978. [g] Bugay, Bushweller et al., 1989. [h] Subbotin and Sergeyev, 1975. [i] Chu and True, 1985. [j] Jensen and Bushweller, 1971. [k] Höfner, Binsch et al., 1978. [l] Shen and Peloquin, 1988. [m] Considerably smaller ΔH^0 values were reported by Gardiner and Walker, 1987, by Gardiner et al. 1987, and by Bugay, Bushweller et al., 1989. [n] Eliel and Gilbert, 1969. [o] See also Aycard, 1989. [p] Averaged value; all values given are within experimental error of each other. [q] The alcohol may have been self-associated (oligomeric). [r] Senderowitz, Fuchs et al., 1989. [s] Kirby and Williams, 1992. [t] Allan, Reeves et al., 1963. [u] Jordan and Thorne, 1986. [v] Klochkov et al., 1989. [w] Eliel and Kandasamay, 1976. [x] Subbotin, Zefirov et al., 1978. [y] Duddeck et al., 1985. [z] Duddeck et al., 1991. [aa] Buchanan and Webb, 1983. [bb] Booth and Josefowicz, 1976. [cc] Sicher et al., 1963. [dd] Herlinger and Naegele, 1968. [ee] Sülzle, Klaeboe et al., 1988. [ff] Gordon and Quin, 1976. [gg] Pai and Kalasinsky, 1990. [hh] Juaristi and Aguilar, 1991. [ii] Juaristi et al., 1986. [jj] Juaristi et al., 1987. [kk] Buchanan, 1982. [ll] Buchanan and McCarville, 1972. [mm] Buchanan et al., 1984. [nn] Eliel and Reese, 1968. [oo] Booth et al., 1992. [pp] Della and Rizvi, 1974. [qq] Eliel and Manoharan, 1981. [rr] Gatial et al., 1990, 1991. [ss] Booth and Everett, 1980a. [tt] Difference between CH_3 and CD_3; CD_3 has lesser equatorial preference than CH_3. [uu] Baldry and Robinson, 1977. [vv] Booth and Everett, 1980b, c. [ww] Calculated value: van de Graaf, Webster, et al., 1978. [xx] Manoharan and Eliel, 1984a. [yy] Kitching et al. 1981. [zz] Della, 1967. [aaa] Squillacote and Neth, 1987. [bbb] Juaristi et al., 1991. [ccc] Reisse et al., 1964. [ddd] Penman, Kitching et al., 1989. [eee] Shen et al., 1992. [fff] Kitching et al., 1982b. [ggg] Kitching et al., 1982a. [hhh] Moder, Jensen et al., 1980. [iii] Kitching et al., 1976. [jjj] Anet et al., 1974. [kkk] Eliel et al., 1962.

The value for OH obtained by low-temperature NMR in CS_2 seems to be out of line; it must be strongly suspected that solutions about 0.2 M in cyclohexanol (concentrations appropriate for an NMR experiment) at −80°C are subject to extensive oligomerization of the solute by intermolecular hydrogen bonding, so that the measured $-\Delta G^0$ is not that of the monomeric alcohol.

The value of $-\Delta G^0$ for SH is slightly larger than that for OH. That for SCH_3 is somewhat smaller; that of $SOCH_3$ is similar to SCH_3, but that of SO_2CH_3 is considerably larger. The former three groups presumably confront the ring with their lone electron pairs, which are evidently not greatly repulsive, but SO_2CH_3 must confront the ring with an O or CH_3 moiety (probably the former), which is sterically much more demanding. The progression of $-\Delta G^0$ in the series NH_2, $NHCH_3$, $N(CH_3)_2$ is also slight, for the same reasons adduced for OH versus OCH_3, but $(CH_3)_3\overset{+}{N}$, as expected, has a very large $-\Delta G^0$ value (too large to be measured). The values for PR_2 are of the same magnitude as those for NR_2; the diminution in $-\Delta G^0$ seen for Group 14(IVA) and Group 17(VIIA) elements as one goes down the periodic table is not evident in Group 15(VA) and Group 16(VIA), though in the latter groups there are data for only the first two members. Linear substituents, such as −NC, −NCO, N_3, CN, and C≡CH have expectedly small conformational energy values and those of planar groups, such as COR, CO_2R, CH=CH$_2$ are intermediate between those of linear and those of tetrahedral groups, such as CH_3. The vinyl group has the largest conformational energy in this series; apparently, when it is axial, one of its β (methylene) hydrogen atoms interferes seriously with one of the equatorial ring hydrogen atoms of the cyclohexane.

The sp^2 hybridized groups orient themselves so as to confront the ring with their flat sides, in other words the plane of the substituent is perpendicular or nearly perpendicular to the bisector plane of the cyclohexane ring. In the case of axial phenyl, this rotational conformation, though optimal, imposes steric crowding of the ortho hydrogen atoms (o-H) of the phenyl with *both* adjacent equatorial hydrogen atoms (e-H) of the cyclohexane chair (Fig. 11.22a); this explains the high conformational energy of the phenyl group (Table 11.7; Allinger and Tribble, 1971). Equatorial phenyl, in contrast, is most stable in the bisector plane of the cyclohexane chair (Fig. 11.22b) where the unfavorable o-H/e-H interaction is avoided (see also Section 2-6, Fig. 2.26).

The conformational energy of methyl in methylcyclohexane (a key datum; cf. Anet et al., 1971) has been determined with great accuracy by low-temperature ^{13}C NMR spectroscopy (Booth and Everett, 1976, 1980a). Since the contribution

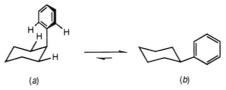

(a)   (b)

Figure 11.22. Equatorial and axial conformers of phenylcyclohexane; $\Delta G^0 = -2.87 \text{ kcal mol}^{-1}$ (12.0 kJ mol^{-1}).

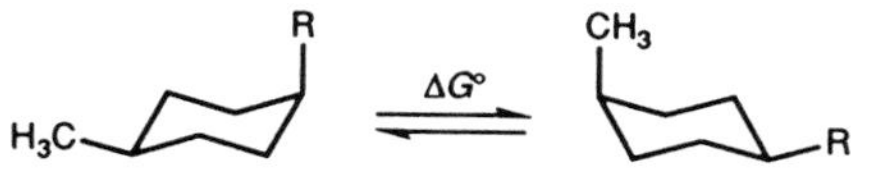

Figure 11.23. Counterpoise method.

of the axial conformer in the temperature range of the experiment (140–195 K) is only about 1%, it was necessary to work with ^{13}C enriched material to see the methyl peak of the minor conformer. With the value for methyl in hand, values for ethyl and isopropyl were determined (Booth and Everett, 1980a) by use of the so-called "counterpoise method" (Eliel and Kandasamy, 1976; see also Eliel et al., 1963) employing *cis*-1-alkyl-4-methylcyclohexanes as objectives of low-temperature NMR investigation (Fig. 11.23). Making the reasonable assumption that the conformational energies of CH_3 and R are additive, one has $\Delta G^0 = \Delta G_R - \Delta G_{CH_3}$, where ΔG^0 is the free energy change for the process shown in Figure 11.23; hence $\Delta G_R = \Delta G^0 + \Delta G_{CH_3}$. By carrying out the equilibration over a range of temperature, it was possible to determine both ΔH^0 and ΔS^0 as well as ΔG^0. The results (axial→equatorial conformer) are shown in Table 11.8 (Booth and Everett, 1980a), along with values calculated by molecular mechanics (Allinger et al. 1968a); the agreement between experimental and calculated values is good. It is of interest that $-\Delta H^0$ *de*creases in the series CH_3, C_2H_5, *i*-Pr whereas ΔS^0 *in*creases. The reason for this finding may be gleaned from Figure 11.24. In ethylcyclohexane the axial conformer has two rotamers (**A** and its mirror image); the third rotamer, in which the methyl group points into the ring, is of very high energy [cf. the $-\Delta G^0$ value of $(CH_3)_3C-$, where one methyl must point into the ring, see Table 11.7]. The equatorial conformer, in contrast, has three populated rotamers: **B** and its mirror image, and **C**. Therefore the equatorial conformer will

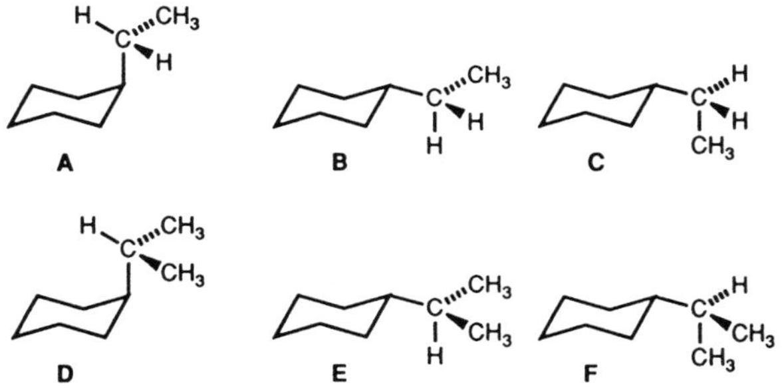

Figure 11.24. Rotamers of ethylcyclohexane and isopropylcyclohexane.

TABLE 11.8. Conformational Thermodynamic Parameters for Alkyl Groups.

Alkyl	$-\Delta H^{0\ a}$		$-\Delta S^{0\ b}$		$-\Delta G_{25}^{0\ a}$	
	Found	Calculated	Found	Calculated	Found	Calculated
CH_3	1.75 (7.32)	1.77 (7.41)	−0.03 (0.13)	0 (0)	1.74 (7.28)	1.77 (7.41)
C_2H_5	1.60 (6.69)	1.69 (7.07)	0.64 (2.68)	0.61 (2.55)	1.79 (7.49)	1.87 (7.82)
$(CH_3)_2CH$	1.52 (6.36)	1.40 (5.86)	2.31 (9.67)	2.18 (9.12)	2.21 (9.25)	2.05 (8.58)

[a] In kilocalories per mole (kcal mol^{-1}); values in parentheses are in kilojoules per mole (kJ mol^{-1}).
[b] In calories per mole per degree kelvin (cal mol^{-1} K^{-1}); values in parentheses are in joules per mole per degrees kelvin (J mol^{-1} K^{-1}).

have a larger entropy of mixing. On the other hand, since the enthalpy of each conformer is the weighted average of the enthalpy of the rotamers, the enthalpy of the equatorial conformer is enhanced by the contribution of the high-enthalpy rotamer **C** in which the terminal methyl of the CH_2CH_3 group is subject to *two* butane–gauche interactions with the ring. (In **A** and **B** there is only one such interaction.) The difference in enthalpy between axial and equatorial conformers for ethyl is thus somewhat diminished relative to the difference for methyl. The same argument applies, a fortiori, to the isopropyl group in which the axial conformer exists as the single rotamer **D**, whereas the equatorial conformer still has three rotamers: The lowest-energy rotamer **E** and the slightly higher energy rotamer **F** and its mirror image (see also Anderson, 1992, p. 118).

> The validity of the frequently used counterpoise method rests on the assumption that the reorientation (axial–equatorial) of one substituent does not affect the ease of reorientation of the other. To the extent that axial and equatorial substituents deform a cyclohexane ring unequally (cf. p. 693), this assumption is probably not strictly correct, but for pairs of small substituents (e.g., CH_3 and C_2H_5, CH_3 and SCH_3) it may be an adequate approximation.

Thermodynamic parameters (ΔH^0 and ΔS^0) have been determined for some of the other substituents in Table 11.7; ΔG^0 values in such cases have been starred; ΔH^0 and ΔS^0 may be found in the original references.

The difference between CO_2^- and CO_2H and between NH_3^+ and NH_2 is of note; in both cases $-\Delta G^0$ for the ion is considerably larger than that for the uncharged species. One way of explaining this is to say that the axial group, when ionic, is swelled by solvation, and therefore more subject to steric repulsion than the neutral ligand. Another complementary explanation implies that the axial substituent, because of crowding, is less readily solvated than the equatorial one, and therefore benefits less from the diminution in free energy than any charged species experiences when it is solvated. (In the case of NH_3^+ vs. NH_2, the steric effect of the extra hydrogen may, of course, also contribute to the larger $-\Delta G^0$ of the former.) The true answer probably lies somewhere in between.

A similar argument applies (if less strongly) to an uncharged but polar substituent and, in general, one might expect conformational energies in such cases to be solvent dependent. However, little solvent dependence of ΔG^0 was found with cyclohexyl fluoride, chloride, and bromide (Eliel and Martin, 1968b).

c. Disubstituted and Polysubstituted Cyclohexanes

1,2-, 1,3- and 1,4-Disubstituted cyclohexanes each exist as cis and trans isomers (Section 11-1). When the two substituents are identical, the cis-1,2 and cis-1,3 isomers are meso forms, whereas the corresponding trans isomers are chiral. In the 1,4-disubstituted series, both the cis and the trans isomers are achiral, regardless of whether the substituents are the same or not.

When one considers conformational factors, the situation becomes somewhat more complex. The 1,4-dimethylcyclohexanes are shown in Figure 11.25. The cis isomer exists as an equimolar mixture of two indistinguishable conformers. Its steric energy (cf. Section 2-6, p. 33) is that of the axial methyl group or

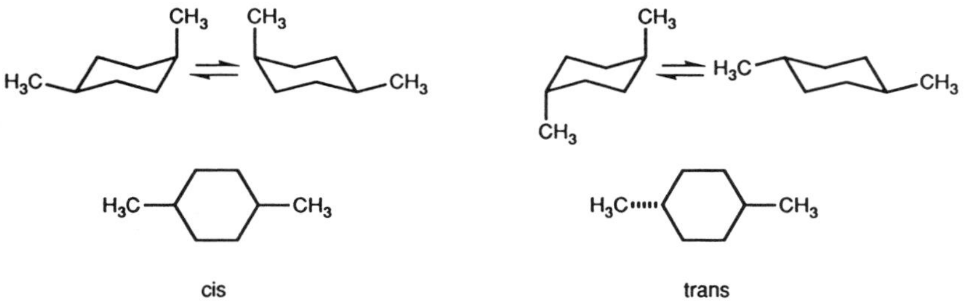

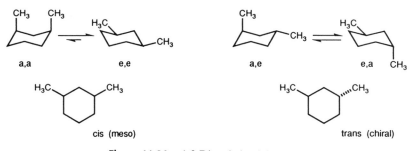

Figure 11.25. 1,4-Dimethylcyclohexanes.

1.74 kcal mol^{-1} (7.28 kJ mol^{-1}). It has no entropy of symmetry (symmetry point group C_s; $\sigma = 1$) and no entropy of mixing, since the two conformers are superposable. The trans isomer consists of two conformers, the predominant e,e and the much less abundant a,a whose energy level is 2×1.74 or 3.48 kcal mol^{-1} (14.56 kJ mol^{-1}) above the e,e, since it has two axial methyl groups. The Boltzmann distribution thus corresponds to 99.7% e,e conformer and 0.3% a,a at 25°C (the amount of the diaxial conformer is less at lower temperatures and more at elevated ones). Use of Eq. 10.1 thus leads to an overall conformational enthalpy of $0.997 \times 0 + 0.003 \times 3.48$ or 0.01 kcal mol^{-1} (0.04 kJ mol^{-1}). The entropy of symmetry is $-R \ln 2$ (symmetry point group C_{2h}; $\sigma = 2$) or -1.38 cal mol^{-1} K^{-1} (-5.76 J mol^{-1} K^{-1}) and the entropy of mixing of the two conformers is $-R(0.997 \times \ln 0.997 + 0.003 \times \ln 0.003)$ or 0.04 cal mol^{-1} K^{-1} (0.17 J mol^{-1} K^{-1}). One thus calculates an enthalpy difference between the two diastereomers of 1.73 $(1.74 - 0.01)$ kcal mol^{-1} (7.24 kJ mol^{-1}) and an entropy difference of $-0.04 + 1.38$ or 1.34 mol^{-1} K^{-1} (5.59 J mol^{-1} K^{-1}), the enthalpy favoring the trans isomer and the entropy the cis. The experimental data, along with the calculations, are shown in Table 11.9. It must be kept in mind that the 1.74 kcal mol^{-1} (7.28 kJ mol^{-1}) value for the conformational energy of methyl is a liquid-phase value and that, as already explained in the case of the butane conformers (p. 601), gas-phase enthalpy differences differ from liquid-phase ones because of differences in heats of vaporization. In the case of enthalpy differences between dimethylcyclohexane diastereomers, such differences can be determined experimentally, and therefore ΔH^0 values are available for both liquid and vapor (Table 11.9).

The situation in 1,3-dimethylcyclohexane (Fig. 11.26) is simpler because both diastereomers exist in single conformations: the trans isomer because the two

Figure 11.26. 1,3-Dimethylcyclohexane.

TABLE 11.9. Enthalpy, Entropy, and Free Energy Differences between Diastereomeric Dimethylcyclohexanes.[a]

Energy or Entropy Difference		Dimethylcyclohexane					
		1,2(liq)	1,2(vp)	1,3(liq)	1,3(vap)	1,4(liq)	1,4(vap)
$-\Delta\Delta H^0$	calcd[b]	1.71		1.74		1.73	
	found[b]	1.5	1.9	1.7	1.94	1.63	1.89
	calcd[c]	7.17		7.28		7.24	
	found[c]	6.4	7.8	7.2	8.1	6.8	7.9
$-\Delta\Delta S^0$	calcd[d]		1.27		1.38		1.34
	found[d]	0.22	0.74	0.88	1.05	0.74	1.10
	calcd[e]		5.31		5.77		5.59
	found[e]	0.9	3.1	3.7	4.4	3.1	4.6
$-\Delta\Delta G^0_{25}$	calcd[b]						
	found[b]	1.48 1.46[f]	1.65	1.46 1.47[f]	1.63	1.41 1.43[f]	1.58
	calcd[c]						
	found[c]	6.2 6.11[f]	6.9	6.1 6.15[f]	6.8	5.9 5.98[f]	6.6

[a] Experimental values for $\Delta\Delta H^0$, $\Delta\Delta S^0$, and $\Delta\Delta G^0$ from *TRC Thermodynamic Tables* (1991), p. n-2100 (differences in enthalpies, entropies, and free energies of formation: 1,2 and 1,4: cis ⇌ trans; 1,3: trans ⇌ cis).
[b] In kilocalories per mole (kcal mol^{-1}).
[c] In kilojoules per mole (kJ mol^{-1}).
[d] In gibbs, that is, calories per mole per degree kelvin (cal mol^{-1} K^{-1}).
[e] In joules per mole per degree kelvin (J mol^{-1} K^{-1}).
[f] Experimental values for $\Delta\Delta G^0$ at 298 K from Allinger et al. (1968b). Since these latter values were obtained by equilibration in the 480–600 K range (ΔG_{25} being computed from the experimental values of $\Delta\Delta H^0$ and $\Delta\Delta S^0$ with corrections for the presence of some twist form at the elevated temperature), they are probably in between liquid and vapor data, since, while most of the material was in the liquid form, some undoubtedly existed as a vapor.

possible conformations are superposable; the cis isomer because the conformation with two synaxial methyl groups is so unstable as to contribute negligibly (i.e., the compound is anancomeric, cf. p. 693). The enthalpy difference (1.74 kcal mol^{-1}, 7.28 kJ mol^{-1}) is thus that of the extra axial methyl group in the trans isomer; the entropy difference of $R \ln 2$ does not result from the fact that the trans isomer has two conformations (see above; the two conformations are superposable, hence there is no entropy of mixing from that source) but from the fact that it exists as a pair of enantiomers (only one enantiomer is shown in Fig. 11.26).

A more complex situation arises in 1,2-dimethylcyclohexane (Fig. 11.27). The

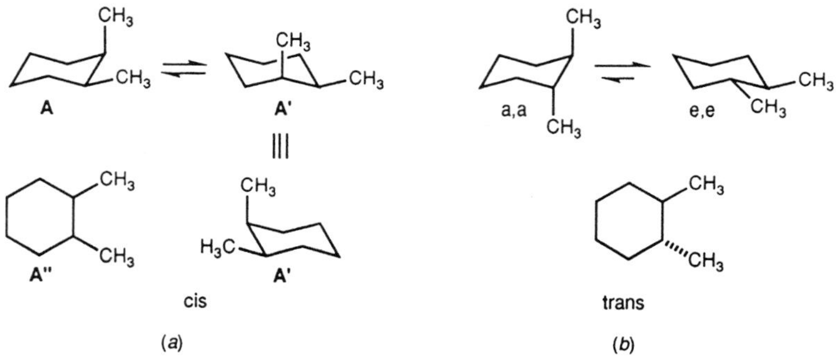

Figure 11.27.   1,2-Dimethylcyclohexane.

(more stable) trans isomer (Fig. 11.27b) has two conformers, the more stable e,e and the less stable a,a. The steric energy of the a,a conformer amounts to 3.48 kcal mol^{-1} (14.56 kJ mol^{-1}) because it has two axial methyl groups. However, the e,e conformer also has nonzero steric energy since its two methyl groups are gauche to each other. One could take its steric energy as equal to that of the gauche form of butane (Section 10-1), but, in fact, a better value is that determined experimentally (Manoharan and Eliel, 1983; Booth and Grindley, 1983) as 0.73–0.90 kcal mol^{-1} (3.05–3.77 kJ mol^{-1}), inter alia by direct determination of the a,a $\rightleftharpoons$ e,e equilibrium (Fig. 11.27b). Using 0.74 kcal mol^{-1} (3.10 kJ mol^{-1}) for this interaction yields a difference of 2.74 kcal mol^{-1} (11.5 kJ mol^{-1}) between the e,e and a,a trans conformers, leading, at room temperature, to a Boltzmann distribution of 99% e,e, 1% a,a. This yields a steric enthalpy of 0.77 kcal mol^{-1} (0.74 × 0.99 + 3.48 × 0.01) or 3.21 kJ mol^{-1} for the ensemble; the steric entropy is composed of an entropy of mixing of the two conformers of 0.11 cal mol^{-1} K^{-1} (0.46 J mol^{-1} K^{-1}), $-R(0.99 \ln 0.99 + 0.01 \ln 0.01)$, an entropy of symmetry of $-R \ln 2$ (symmetry point group C$_2$; $\sigma = 2$), and an entropy of mixing for the enantiomer pair of $R \ln 2$, total 0.11 cal mol^{-1} K^{-1} (0.46 J mol^{-1} K^{-1}).

The cis isomer (Fig. 11.27a) presents a puzzling aspect: In the planar representation **A″** it has a symmetry plane and appears to be a meso form. In contrast, the chair representation (Fig. 11.27, **A** and **A′**) is chiral in either (equilibrating) conformation. However, if one takes the **A′** chair and turns it 120° around a vertical axis, one realizes that it is the mirror image of the **A** chair; in other words, conformational inversion converts one enantiomer of cis-1,2-dimethylcyclohexane into the other. The compound is therefore a racemate in which the enantiomers interconvert rapidly at room temperature, and therefore cannot be individually isolated. In principle, the mixture could be resolved at very low temperatures (ca. −150°C), but this has never been reported because of its obvious technical difficulties. At room temperature, cis-1,2-dimethylcyclohexane is an example of a molecule whose averaged symmetry (**C$_{2v}$**) is higher than the symmetry of the contributing conformers (**C$_1$**). This situation has been discussed in general in Section 4-5; in the specific case of cis-1,2-disubstituted cyclohexanes (identical substituents) it has been shown by group theory that the planar representation properly represents the average symmetry of the two rapidly interconverting chairs (Leonard, Hammond, and Simmons, 1975). On the assumption (probably inaccurate!) that the gauche interaction of adjacent equatorial and axial methyl groups is the same as that of two equatorial methyl groups, the steric enthalpy of cis-1,2-dimethylcyclohexane is 1.74 + 0.74 = 2.48 kcal mol^{-1} (10.4 kJ mol^{-1}) and the steric entropy is the entropy of mixing of the two enantiomeric conformers (not superposable in this case) or $R \ln 2$ (1.38 cal mol^{-1} K^{-1}, 5.76 J mol^{-1} K^{-1}); the cis–trans differences in Table 11.9 are derived from these numbers. The agreement between calculated and experimental $\Delta\Delta G^0$ values in Table 11.9 is quite good, especially if one takes into account that the high temperature at which the experiment had to be carried out may produce a variety of complications.

Inspection of Table 11.9 leads to the conclusion that the agreement between calculated and experimental $\Delta\Delta H^0$ and $\Delta\Delta S^0$ is excellent in the conformationally homogeneous 1,3-dimethylcyclohexanes, provided one considers experimental data for gas-phase entropy. (Liquid–phase entropy is evidently affected by factors other than conformational ones.) Agreement in the 1,4 series is almost as good, especially considering that the experimental values are small differences between large numbers, and therefore affected by sizeable experimental errors. The trends among the 1,4, 1,3, and 1,2 isomers are also as expected, but the actual values in the 1,2 series disagree with the calculated beyond the limits of experimental error, with the experimental values for $-\Delta\Delta H^0$ and $-\Delta\Delta S^0$ being less than the calculated. The cause of this discrepancy is not clear; in fact, one might have expected the opposite for the following reason: Since the torsion angle between cis-1,2 substituents is only 55° but that between trans-1,2 substituents is 65°, one might have expected the vicinal interaction between two methyl groups located cis to each other to be higher than the 0.74 kcal mol^{-1} (3.10 kJ mol^{-1}) value measured for such groups located trans. In fact, for CH_3 and OH it has been found (Sicher and Tichý, 1967) that the $(CH_3)_e/OH_e$ gauche–trans interaction of 0.38 kcal mol^{-1} (1.59 kJ mol^{-1}) is less than either the $(CH_3)_e/OH_a$ [0.66 kcal mol^{-1} (2.76 kJ mol^{-1})] or the $(CH_3)_a/OH_e$ [0.83 kcal mol^{-1} (3.47 kJ mol^{-1})] gauche–cis interaction. The greater gauche–cis interaction should destabilize the cis-1,2-disubstituted isomer relative to the trans.

We have already seen that conformational energies of substituents located 1,4 to each other are close to being additive (Fig. 11.23); this has been confirmed in several cases (e.g., Eliel and Kandasamy, 1976; Manoharan, Eliel, et al., 1983), where $-\Delta G^0$ values have been determined both by direct low-temperature equilibration (Fig. 11.18) and by the counterpoise method (Fig. 11.23). Additivity is not as good for substituents located 1,3 to each other because of buttressing problems (Fig. 11.20). Perhaps not unexpectedly, additivity breaks down altogether for most vicinal 1,2 and geminal 1,1 substituents. With respect to 1,2-(vicinally) disubstituted cyclohexanes, the dimethylcyclohexanes are a case in point, but a more salient example is that of 2-isopropylcyclohexanols (Stolow, 1964), such as menthol and its stereoisomers (2-isopropyl-5-methylcyclohexanols).

We have already seen (Fig. 11.24) that the preferred conformation of an axial isopropyl group **D** is that in which the methine hydrogen $[(CH_3)_2CH]$ confronts the ring, whereas the most stable equatorial rotamer is **E**, with the least number of gauche interactions. Figure 11.28 implies that introduction of an equatorial hydroxyl group into the conformer with an axial isopropyl group introduces a severe OH/CH_3 interaction of the synaxial or g^+g^- type. No such interaction arises from introduction of an axial vicinal hydroxyl into equatorial isopropylcyclohexane. Therefore it is clear that the conformer with an equatorial isopropyl group must be favored by far more than the difference of the $-\Delta G^0$ values (Table 11.7) of isopropyl (2.21 kcal mol^{-1}, 9.25 kJ mol^{-1}) and hydroxyl (0.60 kcal mol^{-1}, 2.51 kJ mol^{-1}) and evidence for this prediction has been adduced (Stolow, 1964).

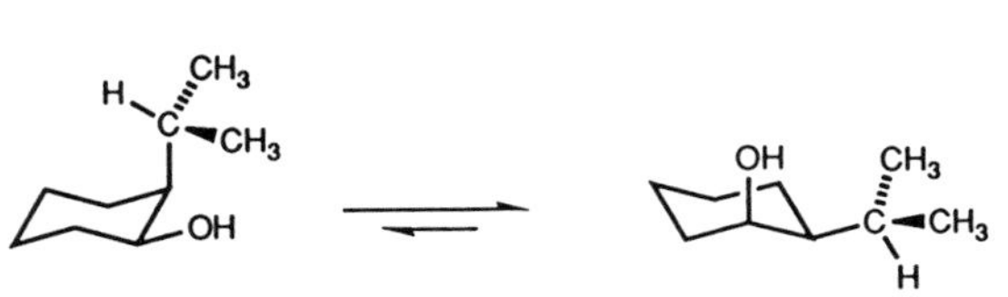

Figure 11.28. Conformations of *cis*-2-isopropylcyclohexanol.

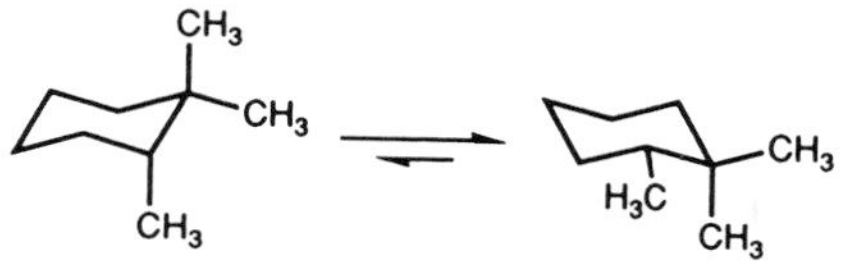

Figure 11.29. Conformational equilibrium in 1,1,2-trimethylcyclohexane.

Another case of vicinal nonadditivity is that seen in 1,1,2-trisubstituted cyclohexanes (Mursakulov et al., 1980), including 1,1,2-trimethylcyclohexane (Fig. 11.29). One might have expected the equatorial–axial free energy difference of the methyl group ($1.74\,kcal\,mol^{-1}$, $7.28\,kJ\,mol^{-1}$) to be diminished by a methyl–methyl gauche interaction ($0.74\,kcal\,mol^{-1}$, $3.10\,kJ\,mol^{-1}$) on the presumption that the equatorial conformer (Fig. 11.29, **E**) has two such interactions but the axial **A** conformer has only one. In fact, however, the equilibrium shown in Figure 11.29, determined by a counterpoise method (Eliel and Chandrasekaran, 1982), corresponds to $\Delta G^{0} = -1.53\,kcal\,mol^{-1}$ ($-6.40\,kJ\,mol^{-1}$), implying that the extra gauche interaction in the equatorial conformer is worth only 0.21 ($1.74 - 1.53$) $kcal\,mol^{-1}$ ($0.88\,kJ\,mol^{-1}$). The reason for this is probably found in a deviation of bond angles similar to that seen in 2,3-dimethylbutane (Fig. 10.7). Other examples of vicinal nonadditivity have been reported (e.g., Chernov et al., 1982; Hodgson, Eliel, et al., 1985).

Another set of compounds of conformational interest is constituted by the di-*tert*-butylcyclohexanes. The cis-1,3 and trans-1,4 isomers have the normal diequatorial chair conformation. The trans-1,3 and cis-1,4 isomers might be expected to have one equatorial and one axial *tert*-butyl group. However (see Table 11.7), the conformational energy of *tert*-butyl, $4.8\,kcal\,mol^{-1}$ ($20.1\,kJ\,mol^{-1}$), is close to the energy difference between the cyclohexane chair and twist forms (Fig. 11.17). It is therefore not surprising that cis-1,4-di-*tert*-butylcyclohexane exists as an equilibrium mixture of a chair conformation with axial *tert*-butyl and twist conformations (in which the *tert*-butyl groups are essentially equatorial; van de Graaf, Wepster, et al., 1974). Experimentally, this has been demonstrated by a combination of electron diffraction, molecular mechanics, and vibrational investigations (Schubert, Schäfer, et al., 1973). The trans-1,3 isomer also exists as a chair–twist equilibrium mixture (Allinger et al., 1968b; Remijnse, Wepster, et al., 1974; Loomes and Robinson, 1977) with the twist conformer having the lower enthalpy. In the 1,2 series, the main factor is the mutual repulsive interaction of the *tert*-butyl groups themselves which causes the trans isomer to exist as a distorted chair with two axial *tert*-butyl groups in the crystal, whereas in solution this conformer appears to coexist with a nonchair conformer (van de Graaf, Wepster, et al., 1974, 1978). The cis isomer appears to exist largely as a chair with axial *tert*-butyl (van de Graaf, Wepster, et al., 1974).

Additivity of conformational energies in geminally disubstituted cyclohexanes (for a tabulation see Eliel and Enanoza, 1972; see also Jordan and Thorne, 1986; Carr, Robinson, et al., 1987) tends to break down for reasons similar to those operative in vicinally disubstituted ones: one substituent interferes with the otherwise optimal rotational conformation of the other in one of the two possible conformations (but not both; or at least the interference is not the same in the two ring-inverted conformations). An extreme example is 1-methyl-1-phenylcyclohexane (Allinger and Tribble, 1971; Hodgson, Eliel, et al., 1985), Figure 11.30 (cf. Fig. 2.26). The conformational energies are $2.87\,kcal\,mol^{-1}$ ($12.0\,kJ\,mol^{-1}$) for

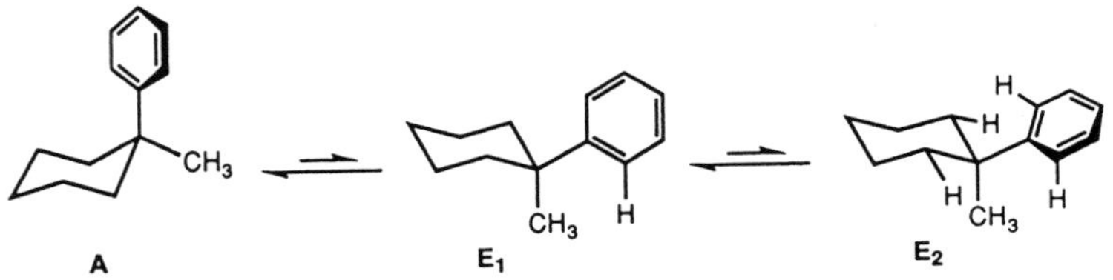

Figure 11.30. Conformational equilibrium in 1-methyl-1-phenylcyclohexane.

phenyl and $1.74 \, \text{kcal mol}^{-1}$ ($7.28 \, \text{kJ mol}^{-1}$) for methyl, so one would expect equatorial phenyl – axial methyl to be preferred by $2.87 - 1.74 = 1.13 \, \text{kcal mol}^{-1}$ ($4.73 \, \text{kJ mol}^{-1}$) if the energies were additive. In fact, however, *axial* phenyl – equatorial methyl is preferred by $0.32 \, \text{kcal mol}^{-1}$ ($1.34 \, \text{kJ mol}^{-1}$), that is, there is a deviation from additivity of $1.45 \, \text{kcal mol}^{-1}$ ($6.07 \, \text{kJ mol}^{-1}$) (Eliel and Manoharan, 1981). The reason becomes clear when one compares Figures 11.22 and 11.30. Introduction of an equatorial methyl group in axial phenylcyclohexane causes no complication: phenylcyclohexane and 1-methyl-1-phenylcyclohexane with axial phenyl groups have corresponding conformations. The same is not true when one introduces an axial methyl group into an equatorial phenylcyclohexane: Either a serious steric interaction between the methyl group and one of the ortho hydrogen atoms of the phenyl substituent arises (Fig. 11.30, $\mathbf{E}_1$) or the phenyl has to turn, in which case its ortho hydrogen atoms will clash with the equatorial hydrogen atoms at C(2,6) in the cyclohexane (Fig. 11.30, $\mathbf{E}_2$). In either case, the **E** conformer will suffer from repulsive interactions over and above those caused by a simple presence of the axial methyl group and these interactions are apparently large enough [$1.45 \, \text{kcal mol}^{-1}$ ($6.07 \, \text{kJ mol}^{-1}$)] to force the molecule into the (now more stable) **A** conformation. The difference in potential energy between $\mathbf{E}_1$ and $\mathbf{E}_2$ is apparently small (Allinger, unpublished MM2 calculations; cf. Hodgson, Eliel, et al., 1985).

Large steric interactions occur when two groups are placed synaxially (Fig. 11.31). The synaxial interaction energies are tabulated in Table 11.10 (Corey and Feiner, 1980); most of these interactions are so large that counterpoise methods had to be used to determine them.

Limitations of space prevent us from discussing tri- and tetrasubstituted cyclohexanes in detail. 1,2,3-Trimethylcyclohexane has a chiral isomer and two meso forms; the 1,2,4 isomer has four (chiral) pairs of enantiomers and the 1,3,5 isomer has only two achiral diastereomers. It is instructive to draw the two conformations of each isomer and calculate the relative thermodynamic stability of each stereoisomer both in this series and in the three (1,2,3,4-; 1,2,3,5-; 1,2,4,5-) constitutionally isomeric sets of stereoisomers in the tetramethylcyclohexane series. The 1,2,4,5-tetramethylcyclohexanes have actually been equilibrated; the position of equilibrium is close to that calculated from the anticipated values of ΔH^0 and ΔS^0 (Werner, Mühlstadt, et al., 1970).

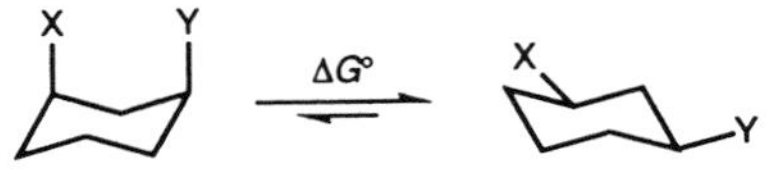

Figure 11.31. The synaxial interaction. $\Delta G^0 = \Delta G^{\text{xy}}_{\text{synaxial}} + \frac{1}{2}(\Delta G_{\text{x}} + \Delta G_{\text{y}})$.

TABLE 11.10. Synaxial Interactions[a].

Groups	CH$_3$/CH$_3$	OH/OH	CH$_3$/OH	OAc/OAc	Cl/Cl	CH$_3$/F	CH$_3$/Br	CN/CN	CH$_3$/CN	CH$_3$/C$_6$H$_5$
$-\Delta G^0$ (kcal mol^{-1})[b]	3.7	1.9	2.4[c]	2.0	5.5	0.4	2.2	3.0	2.7	3.4[d]
$-\Delta G^0$ (kJ mol^{-1})[b]	15.5	7.95	10.0[c]	8.4	23.0	1.67	9.2	12.5	11.1	14.2[d]

Groups	CH$_3$/CO$_2$Et	CH$_3$/CO$_2^-$	CO$_2$CH$_3$/CO$_2$CH$_3$	CO$_2$H/CO$_2$H	CO$_2^-$/CO$_2^-$	CO$_2$H/NH$_3^+$	CO$_2^-$/NH$_3^+$
$-\Delta G^0$ (kcal mol^{-1})[b]	2.8– 3.2	3.4	1.5[e]	1.1	~4.2	0.5	−1.8
$-\Delta G^0$ (kJ mol^{-1})[b]	11.7–13.4	14.1	6.3[e]	4.6	~17.6	2.1	−7.5

[a] From Corey and Feiner (1980) unless otherwise indicated.
[b] Synaxial interaction.
[c] Eliel and Haubenstock (1961).
[d] Manoharan and Eliel (1984b).
[e] Revised value.

Among hexasubstituted cyclohexanes, the inositols (cf. Hudlicky and Cebulak, 1993) and the 1,2,3,4,5,6-hexachlorocyclohexanes (Fig. 11.32) are important. All the inositols are known, either as natural or as synthetic products (Angyal, 1957). Six of the eight diastereomeric hexachlorocyclohexanes are also known; a seventh (θ) has been prepared but not obtained pure (Kolka, Orloff, et al., 1954). The all-cis(ι) isomer (with three synaxial chlorines in the chair form)

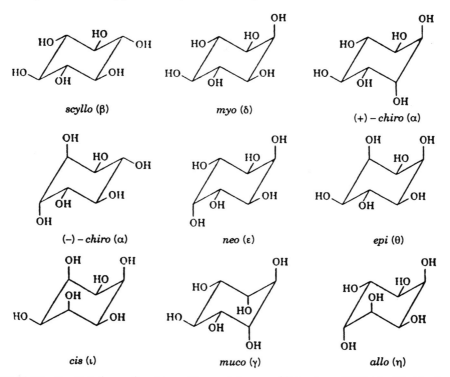

Figure 11.32. Inositols (named) and hexachlorocyclohexanes (Cl in place of OH; identified by Greek letters). Reproduced with permission from Eliel, E.L., Allinger, N.L., Angyal, S.J., and Morrison, G.A., *Conformational Analysis.* Copyright © American Chemical Society, Washington, DC, 1981, p. 353.

has not been synthesized, possibly because of its extreme conformational instability (de La Mare et al., 1983). Of the eight diastereomers, seven are achiral; the eighth (*chiro*-inositol, α-hexachlorocyclohexane) is chiral. The symmetry planes are easy to spot except in the allo or η isomers in which individual chair forms are chiral, but chair inversion leads to the enantiomer, much as in *cis*-1,2-dimethylcyclohexane (see p. 703). α-Hexachlorocyclohexane has been obtained optically active (though not enantiomerically pure) by what was reported to be partial asymmetric destruction of the racemate through incomplete dehydrohalogenation with the optically active alkaloid brucine (Cristol, 1949; see, however, pp. 358 and 405). The γ-isomer (Lindane or hexachloran) is an insecticide, though it is no longer used in the United States.

The all-cis hexamethylcyclohexane is known; its inversion barrier has the unusually high value of 17.3 kcal mol^{-1}(72.4 kJ mol^{-1}), which is explained by the need of vicinal methyl groups to pass by each other in every one of the possible conversion pathways (Werner, Mühlstadt, et al., 1970). All-trans hexaethylcyclohexane has been shown, by X-ray crystallography, to exist in the expected all-equatorial conformation (Immirzi and Torti, 1968); however, the terminal methyl groups in the substituents are gauche to both ring methylenes rather than gauche to only one of them, as in ethylcyclohexane itself. The reason, as shown by force field calculations (Golan, Biali, et al., 1990) is the need to minimize steric and/or torsional interactions between the substituents. In *trans*-hexaisopropylcyclohexane, where such interactions can no longer be avoided in any of the rotational conformations of the isopropyl substituents, the stable conformer is actually the hexaaxial one, as evidenced by both X-ray crystallography and NMR spectroscopy in solution (Golan, Biali, et al., 1990)! Calculations suggest the gear-meshed hexaequatorial conformer to be destabilized by 14.3 kcal mol^{-1} (59.8 kJ mol^{-1}) and a more stable conformer in the equatorial set has a nonchair cyclohexane ring but is still calculated to be 5.6 kcal mol^{-1} (23.4 kJ mol^{-1}) less stable than the hexaaxial conformer. The latter is slightly twisted from the most symmetrical $\mathbf{D_{3d}}$ into an $\mathbf{S_6}$ conformation.

Earlier in this section and also in Section 10-3 we have shown a method for calculating configurational equilibria based on conformational energies by computing conformational enthalpy and entropy differences and then calculating free energy differences and equilibrium constants. There is, however, a shortcut method applicable if all conformational equilibrium constants are available (or can be calculated) at the same temperature. We use 4-methylcyclohexanol (Fig. 11.33) as an example. Both diastereomers exist as mixtures of two conformers.

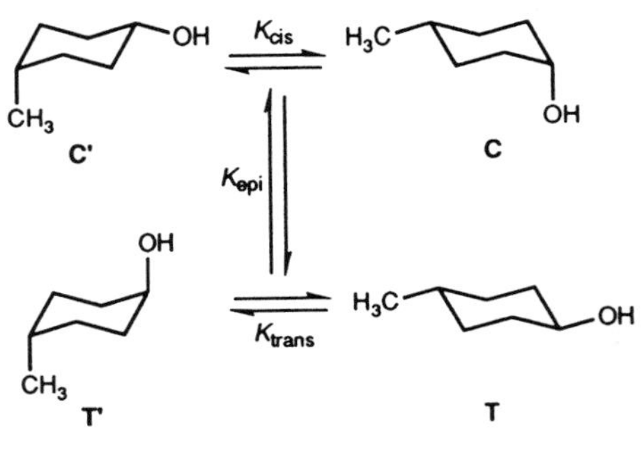

Figure 11.33. 4-Methylcyclohexanol configurational equilibrium.

Let us call K_{epi} the epimerization equilibrium constant (cis $\rightleftharpoons$ trans) and K_{trans} and K_{cis} the conformational equilibrium constants shown in Figure 11.33. Then,

$$K_{epi} = (\mathbf{T} + \mathbf{T'})/(\mathbf{C} + \mathbf{C'}) = \frac{\mathbf{T}(1 + 1/K_{trans})}{\mathbf{C}(1 + 1/K_{cis})}$$

since $K_{trans} = \mathbf{T}/\mathbf{T'}$ and $K_{cis} = \mathbf{C}/\mathbf{C'}$.

If we multiply the numerator and denominator by $K_{cis} K_{trans}$ we obtain

$$K_{epi} = \left(\frac{\mathbf{T}}{\mathbf{C}}\right) \frac{K_{cis}(K_{trans} + 1)}{K_{trans}(K_{cis} + 1)}$$

Moreover, $\mathbf{T}/\mathbf{C} = K_{OH}$, $K_{trans} = K_{OH} K_{CH_3}$ and $K_{cis} = K_{CH_3}/K_{OH}$

Hence,

$$K_{epi} = K_{OH} \frac{(K_{CH_3} K_{OH} + 1)K_{CH_3}/K_{OH}}{K_{CH_3} K_{OH}(1 + K_{CH_3}/K_{OH})}$$

or

$$K_{epi} = (K_{CH_3} K_{OH} + 1)/(K_{OH} + K_{CH_3}) \tag{11.4}$$

If we take $K_{OH} = 4$, $K_{CH_3} = 20$ (from Table 11.7, choosing a value for K_{OH} within the solvent range given) $K_{epi} = 81/24 = 3.38$, that is, the cis–trans equilibrium for 4-methylcyclohexanol is less extreme than the axial–equatorial equilibrium for the cyclohexanol parent, since conformation $\mathbf{C'}$ contributes substantially to the cis isomer but $\mathbf{T'}$ contributes very little to the trans. In fact, as a first approximation, one can assume the population of $\mathbf{T'}$ to be negligible, in which case

$$K_{epi} = \mathbf{T}/(\mathbf{C} + \mathbf{C'}) \quad \text{or} \quad 1/K_{epi} = (\mathbf{C} + \mathbf{C'})/\mathbf{T} = \mathbf{C}/\mathbf{T} + \mathbf{C'}/\mathbf{T} = 1/K_{OH} + 1/K_{CH_3}$$

that is,

$$K_{epi} = K_{CH_3} K_{OH}/(K_{CH_3} + K_{OH}) \tag{11.5}$$

In the present case, this would yield $4 \times 20/24$ or $K_{epi} = 3.33$. The difference between this and the more accurate value is probably within limits of experimental error. Either Eq. 11.4 or 11.5 (which, of course, can be adapted to any two substituents) can be generalized and is rather simple to use.

d. Conformation and Physical Properties in Cyclohexane Derivatives

Many of the differences in physical and spectral properties between conformational and configurational isomers in cyclohexane have been discussed in detail elsewhere (e.g., Eliel et al., 1965, Chapter 3). The most important properties in this category are NMR spectral properties (relating to both ^{1}H and ^{13}C spectra) and chiroptical properties [optical rotatory dispersion (ORD) and circular dichroism (CD)]. Most of this section will therefore be concerned with NMR spectra; chiroptical properties are dealt with in Chapter 13.

Before dealing with NMR spectroscopy, we mention briefly some other salient properties of cyclohexane stereoisomers. Relative boiling points, refractive indices, and densities of stereoisomers can often be predicted on the basis of the conformational rule, a modification of the classical von Auwers–Skita rule (Allinger, 1957): "The isomer with the smaller molar volume has the greater heat content." Since smaller molar volume implies greater density, refractive index, and boiling point, another way to state the rule is to say the isomer of greater enthalpy (*not* free energy) has the higher boiling point, refractive index, density, and also heat of vaporization (Trouton's rule). This rule, incidentally, is the reason why in Table 11.9, $-\Delta\Delta H^0_{vap} > -\Delta\Delta H^0_{liq}$ (cf. Fig. 10.4). The pertinent physical properties of the dimethylcyclohexanes are given in Table 11.11 (for enthalpy data see Table 11.9).

The conformational rule applies only to nonpolar compounds (Kellie and Riddell, 1975); for molecules with dipole moments the dipole (Van Arkel) rule (cf. p. 638) applies, but with less reliability. In alkylcyclohexanols the conformational rule applies as far as refractive index and density is concerned, but the equatorial isomers, which have the lower enthalpy, engage in more extensive hydrogen bonding because the hydroxyl group is more accessible when equatorial. Therefore the equatorial isomers have the higher boiling points (Eliel and Haber, 1958), emerge more slowly from polar gas chromatography columns (Komers and Kochlofl, 1963), and have the higher adsorption affinity (i.e., emerge later in column chromatography on alumina: Winstein and Holness, 1955). These chromatographic properties are thus often very useful for separation of cyclohexane diastereomers.

In connection with Table 11.7, we mentioned that because of solvation or ion pairing, ionic ligands (such as NH_3^+ and CO_2^-) are more bulky than the corresponding neutral ligands (NH_2 and CO_2H), and therefore less comfortable in axial positions. It is thus not surprising that axial amines and axial carboxylic acids are weaker than their equatorial counterparts; for example, the pK_a, 5.55, of *cis*-4-*tert*-butylcyclohexanecarboxylic acid (Fig. 11.34, X = CO_2H), is higher than that of the trans acid with pK_a 5.10, and the cis amine (X = NH_2) with pK_b 3.50 is weaker than the trans amine with pK_b 3.40 [Edward et al., 1976; data in H_2O extrapolated from aqueous dimethyl sulfoxide (DMSO)].

Infrared stretching frequencies of axial and equatorial C–X bonds were at one time popular for qualitative and even quantitative conformational analysis. For

TABLE 11.11. Boiling Points, Refractive Indexes, and Densities of Dimethylcyclohexanes.

Isomer	Major Conformation	bp(°C)	n_D^{25}	d_4^{25}
cis-1,2	e,a	129.7	1.4336	0.7922
trans-1,2[a]	e,e	123.4	1.4247	0.7720
cis-1,3[a]	e,e	120.1	1.4206	0.7620
trans-1,3	e,a	124.5	1.4284	0.7806
cis-1,4	e,a	124.3	1.4273	0.7787
trans-1,4[a]	e,e	119.4	1.4185	0.7584

[a] The isomer of lower heat content in each diastereomeric pair.

Figure 11.34. The *cis*- and *trans*-4-*tert*-butylcyclohexanecarboxylic acids $(X = CO_2H)$ and butyl-cyclohexylamines $(X = NH_2)$.

example, for halocyclohexanes, the equatorial C–X stretching frequency is higher than the axial (C–F, 1053 vs. 1020; C–Cl, 742 vs. 688; C–Br, 687 vs. 658; C–I, 654 vs. 638 cm^{-1}). Corresponding differences have been seen for C–D: equatorial range 2155–2162 and 2171–2177 cm^{-1} (the bands are doubled; see Corey et al., 1954), axial range 2114–2138 and 2139–2164 cm^{-1} (cf. Eliel et al., 1965). However, because of the considerable danger of misassigning such stretching frequencies and arriving at false conclusions as a result, they must be considered as being of limited usefulness in conformational analysis, especially now when better NMR methods are available.

Both IR and UV spectroscopy are useful in distinguishing axial from equatorial α-haloketones (Fig. 11.35, **C, D**). In the IR, equatorial halogen (by depressing C–O polarization and thereby strengthening the C=O bond), causes a sizeable shift of the carbonyl stretching frequency to a higher value; axial halogen (nearly at right angles to the C=O stretch) causes a much lesser shift (Cummins and Page, 1957; Allinger and Blatter, 1962). In the UV spectrum, the situation is the opposite: There is only a small, usually hypsochromic shift due to equatorial halogen whereas the axial halogen, which is aligned with and stabilizes the π^* orbital of the carbonyl, causes a substantial bathochromic shift. The UV and IR shifts of the four halogen atoms in axial and equatorial conformations are summarized in Table 11.12.

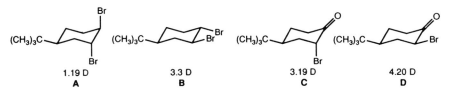

Figure 11.35. Dipole moments of diastereomers. Compounds **A** and **B** were measured in CCl$_4$, compounds **C** and **D** in heptane.

TABLE 11.12. Effect of Equatorial and Axial α-Halogen on UV Absorption Maximum and IR Stretching Frequency of the Carbonyl Function (Relative to Cyclohexanone).

Halogen	IR ($\Delta\nu$, cm^{-1})		UV ($\Delta\lambda$, nm)[ab]	
	Axial	Equatorial	Axial	Equatorial
F	18	27	10	−1
Cl	2–9	18–25	14	−6
Br	0–3	15–27	20	−4
I	−4	8	b	b

[a] Data in cyclohexane as given by Cantacuzène et al. (1972).
[b] Not available.

Dipole moments will sometimes distinguish stereoisomers in an obvious way, as shown in Figure 11.35 **A**, **B** (cf. Abraham and Bretschneider, 1974). However, both in the 1,2-dibromocyclohexane and the 2-bromocyclohexanone (**C, D**) series, the dipole moment of the diaxial or axial conformer is higher than calculated. Moreover, in the equatorial bromoketone, participation of boat forms is suspected. Use of dipole moments for quantitative conformational analysis is therefore somewhat risky.

Nuclear magnetic resonance spectroscopy is undoubtedly the most powerful method for the elucidation of stereochemistry of cyclohexyl derivatives. The most important generalities in ^{1}H NMR spectroscopy (Lemieux et al., 1958) are that axial protons generally resonate upfield of equatorial protons and that, because of the operation of the Karplus relationship (Fig. 10.37) and the fact that axial protons generally have other axial protons antiperiplanar to them, axial protons show larger splitting (or bandwidth, if the splitting is not resolved) than equatorial protons. Thus, since the torsion angles (cf. Fig. 11.15) are $\omega_{aa} = 175°$, $\omega_{ea} = 55°$ and $\omega_{ee} = 65°$, J_{aa} (9–13 Hz) $> J_{ea}$ (3–5 Hz) $> J_{ee}$ (2–4 Hz).

These ranges exclude protons that are antiperiplanar to electronegative atoms or groups, such as the halogens or OR. If one of the two coupled protons is antiperiplanar to such a group, the coupling constant is diminished by 1–2 Hz below the normal value. For example, in the low-temperature spectrum of the axial conformer of cyclohexyl-d_8 methyl-d_3 ether (Fig. 11.36), $J_{ea} < J_{ee}$ because the axial proton at C(2) (but not the equatorial one) is antiperiplanar to electronegative oxygen (Höfner, Binsch, et al., 1978). J_{ea} in the equatorial conformer is much larger.

In the case of equatorial or axial protons for which the splitting is poorly resolved, width at half height W may be taken as a qualitative conformational criterion: Equatorial protons generally have $W < 12$ Hz, while axial protons have $W > 15$ Hz.

The differences in chemical shifts between axial and equatorial protons in cyclohexane have been rationalized in terms of the diamagnetic anisotropy of the C(2)–C(3) bonds (Fig. 11.37). If, in oversimplified fashion, one ascribes the shielding effect of the C–C bond to a magnetic point dipole located at the

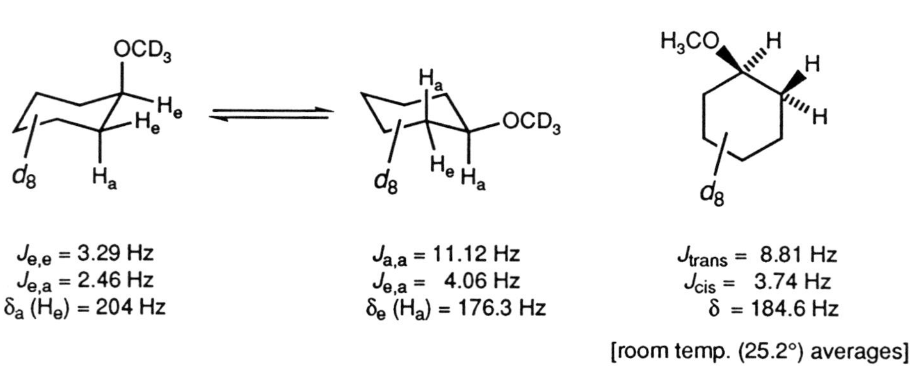

$J_{e,e} = 3.29$ Hz	$J_{a,a} = 11.12$ Hz	$J_{trans} = 8.81$ Hz
$J_{e,a} = 2.46$ Hz	$J_{e,a} = 4.06$ Hz	$J_{cis} = 3.74$ Hz
$\delta_a (H_e) = 204$ Hz	$\delta_e (H_a) = 176.3$ Hz	$\delta = 184.6$ Hz

[room temp. (25.2°) averages]

Figure 11.36. Conformational equilibrium in cyclohexyl-d_8 methyl-d_3 ether; δ at 60 MHz.

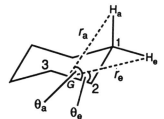

Figure 11.37. Shielding of axial and equatorial protons by C(2)–C(3) bond.

electrical center of gravity of the bond causing the shielding, the shielding (σ) may be expressed by the McConnell equation as

$$\sigma = \Delta\chi(1 - 3\cos^2\theta)/3r^3 \qquad (11.6)$$

where θ is the angle between the bond causing the shielding and a line drawn from the electric center of gravity (G) of this bond to the shielded proton; r is the distance of the proton from G and χ is the diamagnetic anisotropy of the shielding bond (C–C in Fig. 11.37); (McConnell, 1957). Although, because of the inadequacy of the point dipole assumption on a molecular scale (cf. Jackman and Sternhell, 1969), Eq. 11.6 fails quantitatively, it does describe the angular dependence of the shielding: if $\theta < 54.8°$, Eq. 11.6 gives a negative value for σ (assuming positive χ), that is, there is deshielding (this will be the case for the equatorial proton). But for the axial proton, $\theta > 54.8°$, so the effect of the C(2)–C(3) bonds on this proton is one of shielding. Thus a C–C bond will deshield a proton antiperiplanar to it but will shield a proton gauche to it. The effect of a vicinal methyl group can be explained in the same way, as shown in Figure 11.38 (Eliel et al., 1962): The equatorial carbinol proton in **A** is shielded by the methyl groups gauche to it in **B** and **C**; the axial carbinol proton in **D**

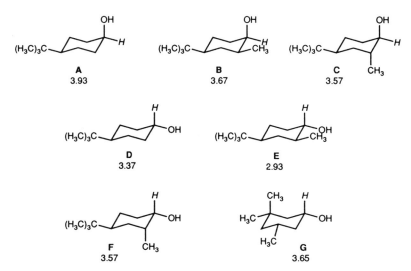

Figure 11.38. Shielding and deshielding effects of vicinal and synaxial methyl substituents (shift in parts per million). (**A** and **D** are reference compounds.)

(upfield of the equatorial one) is shielded by the methyl group gauche to it in **E** but deshielded by the methyl group anti to it in **F**. Compound **G** is included to illustrate another effect, namely, a deshielding caused by van der Waals compression of the axial carbinol proton by the synaxial methyl group.

Everything said so far applies to conformationally homogeneous systems. When rapid conformational inversion occurs, both chemical shifts and coupling constants become weighted averages of those of the contributing conformations (cf. Fig. 11.36).

An application to configurational assignment of 4-*tert*-butylcyclohexyl *p*-toluenesulfonates and 4-*tert*-butylcyclohexyl phenyl thioethers (Eliel and Gianni, 1962) is shown in Figure 11.39. The broad (*b*) upfield signals are due to the axial hydrogen atoms next to sulfur or oxygen, respectively in the trans isomers, whereas the narrow (*n*), downfield signals originate from the corresponding equatorial protons in the cis isomers. The assignment is in agreement with that made earlier on chemical grounds (cf. Fig. 5.44).

Difficulties arise when one wants to obtain *quantitative* conformational information from coupling constants, as pointed out in Section 10-4.e. As explained there, the coupling constants depend not only on torsion angle (Fig. 10.37) but also on the electronegativity of adjacent atoms (cf. Eq. 10.15). Moreover, in mobile systems, such as that shown in Figure 11.40, only average coupling constants can be extracted from the room temperature spectrum:

$$J_{H_1H_4} = J_{cis} = \tfrac{1}{2}(J_{ea} + J_{ae}) \cong J_{ae} \quad \text{and} \quad J_{H_2H_4} = J_{trans} = \tfrac{1}{2}(J_{ee} + J_{aa}) \,.$$

Lambert (1971) pointed out that one can turn this apparent disadvantage to an advantage by basing quantitative conclusions on the ratio $R = J_{trans}/J_{cis}$, this ratio being called the "*R* value" (or perhaps better "Lambert *R* value" to avoid confusion with the *R* factor in X-ray structure analysis, Section 2-5). The ratio can be determined even from the spectra of mobile systems (deuterated species or satellite spectra may have to be used, as discussed earlier). Moreover, the effects of electronegativity tend to cancel out in the J_{trans}/J_{cis} ratio, since electronegativity affects all of the individual couplings (J_{ee}, J_{ea}, J_{aa}), although this effect is strongest

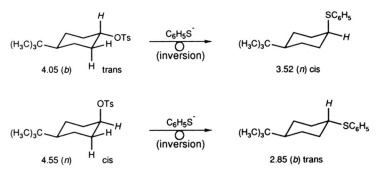

Figure 11.39. The ¹H NMR signals of starting 4-*tert*-butylcyclohexyl tosylates and product 4-*tert*-butylphenyl thioethers in thiophenolate displacement reaction. Shifts are in parts per million; *b*, broad; *n*, narrow signal.

Figure 11.40. Coupling in mobile systems.

for protons antiperiplanar to hetero atoms. For "normal" (i.e., slightly flattened) chairs, such as cyclohexane, R ranges between 1.8 and 2.2; morpholine is a case in point ($R = 2.2$). For flattened chairs R is smaller; cyclohexane-1,4-dione with $R = 1.29$ is an example. On the other hand, in 1,4-dithiane, $R = 3.9$; the chair is puckered by the small C–S–C bond angle. Buys (1969), by setting $^3J_{HH} = A \cos^2\omega$, that is, by neglecting all but the first term in the Karplus equation, arrived at an approximate relationship

$$\cos \omega = \left(\frac{3}{2 + 4R} \right)^{1/2} \tag{11.7}$$

Table 11.13 gives a tabular correlation of torsion angle ω and R. In cases where comparisons between torsion angles determined by the R-value method and determined by X-ray crystallography in solid samples was possible, the values generally agree within two degrees (Lambert, 1971).

For conformationally locked systems, R may be determined by measuring J_{ee}, J_{ea}, J_{ae}, and J_{aa} individually. In conformationally mobile systems, the same can sometimes be done at low temperatures, but it is generally more convenient to measure the average J_{cis} and J_{trans}. If the two equilibrating conformations are not equivalent, the R value and torsion angle are the weighted averages for the two equilibrating conformations.

Carbon-13 NMR spectroscopy (Stothers, 1972; Duddeck, 1986; Morin and Grant, 1989) is at least as useful in identifying configuration and conformation as ^1H NMR. Commonly, each peak in the broad-band decoupled ^{13}C spectrum is sharp and well resolved so that a number of signals can be used for configurational assignment. (There is, of course, a concomitant loss of coupling information; we shall return to this point later.) Diastereomers, with very few exceptions, will differ in the position of at least some of their signals, so that the differences can be used for identification as well as assignment of configuration and quantitative

TABLE 11.13. Correlation of Torsion Angle ω in the X–C–C–X Segment with the R Value.

R	$\omega°$	R	$\omega°$	R	$\omega°$
1.0	45	1.8	55	2.6	60.5
1.1	47	1.9	56	2.7	61
1.2	48.5	2.0	57	2.8	61.5
1.3	50	2.1	57.5	2.9	62
1.4	51	2.2	58	3.1	63
1.5	52	2.3	59	3.4	64
1.6	53	2.4	59.5	3.7	65
1.7	54	2.5	60	4.0	66

TABLE 11.14.   Methyl Substitution Parameters.

α_e	β_e	γ_e	δ_e	ε_e	α_a	β_a	γ_a	δ_a	ε_a
+6.0	+9.0	−0.3	−0.5	−0.4	+1.4	+5.4	−6.4	+0.2	−0.1

analysis. Most useful in signal assignment and configurational and conformational identification are the Grant parameters (Dalling and Grant, 1967, 1972), which indicate the effect of a methyl substituent (either equatorial or axial) on the chemical shifts of the ring carbon atoms in cyclohexane. These parameters are summarized in Table 11.14 (see also Vierhapper and Willer, 1977a); they are to be added to the chemical shift of the appropriate ring carbon in the absence of the substituent. For cyclohexane itself the basic shift [downfield from tetramethyl-silane (TMS) in CDCl$_3$ solvent] is 27.3 ppm. The α-carbon atom is the one to which the substituent (methyl) is attached, the β-carbon atom is next to the α, and so on, as shown in Figure 11.41. [The most remote carbon within the ring is δ. The ε parameter refers to the effect of a ring substituent on the attached carbon of another (equatorial) substituent located 1,4 to it.] By way of an example, in methylcyclohexane (equatorial CH$_3$), the calculated parts per million shift at C(1) is 33.3 ppm (27.3 + α_e) and that at C(2) is 36.3 ppm (27.3 + β_e); the experimental values are 33.0 and 35.6 ppm (Vierhapper and Willer, 1977a). The agreement is better than appears at first sight: If one takes into account that methylcyclohexane is 95% equatorial and 5% axial at 25°C, and adds the approximately weighted shifts for the two conformers (cf. Eq. 11.1) one calculates 33.1 ppm for C(1), 36.1 ppm for C(2).

The α and β effects are generally sizeable and downfield shifting, with the equatorial substitutent having a larger effect than the corresponding axial one and the β effect for a given orientation being larger than the α effect. The (axial) α_a effect is, in fact, small and in some heterocyclic systems even negative (Eliel and Pietrusiewicz, 1979). There has been much controversy about the origins of these effects, including that of the *up*field shifting γ_a effect that was originally ascribed (Grant and Cheney, 1967) to steric compression: The C–H bonding electrons in the ring methylene group are polarized, by the proximal gauche (axial) methyl substituent, in such a way that they recede from the hydrogen atoms to the carbon; the hydrogen (proton) signals are thereby shifted downfield and the carbon (^{13}C) signals upfield. The reciprocal effect is also seen: Axial methyl groups resonate upfield (at 17.5–19 ppm) of equatorial methyl groups (ca. 23 ppm). (The downfield shift on the proton resonance caused by an axial substituent may be seen in compound **G** in Fig. 11.38.) However, the steric origin of the gauche or axial shift

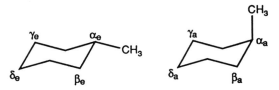

Figure 11.41.   Denomination of substituent effects (parameters) in a (hypothetical) conformationally homogeneous species. The substituent effect (ppm) is equal to the observed shift −27.3 (cyclohexane shift).

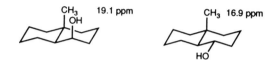

Figure 11.42. The δ-compression effect.

has been questioned (e.g., Gorenstein, 1977, who ascribes the shift to bond and torsion angle effects) and in highly compressed situations, such as that depicted in Figure 11.42, a downfield shift of the compressed methyl is actually found (Grover, Stothers, et al., 1973, 1976). The γ_e shifts due to carbon and the shifts at more remote positions (δ, ε) are quite small.

One attractive feature of the shift parameters is that, except for disturbances that usually occur when substituents are geminal or vicinal, they tend to be additive. Thus in *cis*-1,3-dimethylcyclohexane (Fig. 11.43) the observed shift at C(2) is 45.3 ppm, and is found to be (perhaps fortuitously) exactly that calculated for a cyclohexane (basic shift 27.3 ppm) with two equatorial β_e effects of 9.0 ppm each.

Shift parameters for a large number of groups other than methyl have been tabulated (Schneider and Hoppen, 1978) and are shown in Table 11.15. Except for I and SH, the α effects of heteroatom-linked groups are much larger than

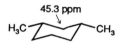

Figure 11.43. Additivity of parameters for C(2) in *cis*-1,3-dimethylcyclohexane.

TABLE 11.15. Miscellaneous Substituent Shifts.[a]

Substituent	α_e	α_a	β_e	β_a	γ_e	γ_a	δ_e	δ_a
F	+64.5	+61.1	+ 5.6	+3.1	−3.4	−7.2	−2.5	−2.0
Cl	+32.7	+32.3	+10.5	+7.2	−0.5	−6.9	−1.9	−0.9
Br	+25.0	+27.5	+11.3	+8.1	+0.7	−6.3	−2.0	−1.1
I	+ 2.1	+ 9.5	+13.8	+9.5	+2.4	−4.5	−2.4	−0.8
OH	+44.1	+38.9	+ 8.5	+6.0	−2.3	−6.9	−1.5	−0.6
OCH$_3$	+52.9	+47.7	+ 5.2	+3.1	−2.0	−6.3	−0.7	−0.1
OSi(CH$_3$)$_3$	+43.5	+39.1	+ 9.0	+6.1	−2.3	−7.2	−2.0	−2.0
OCOCH$_3$	+46.5	+42.3	+ 4.8	+3.2	−2.3	−6.1	−1.5	−1.1
OCOCF$_3$	+51.8	+48.1	+ 4.2	+2.8	−2.4	−6.3	−1.6	−1.2
OTs	+55.5	+52.2	+ 5.5	+3.9	−2.2	−6.7	−2.0	−1.4
SH	+11.1	+ 8.9	+10.7	+6.1	−0.6	−7.6	−2.2	−1.3
NH$_2$	+23.9	+18.1	+10.0	+6.5	−1.6	−7.2	−1.3	−0.3
NHCH$_3$	+32.1	+26.8	+ 6.3	+3.2	−1.8	−6.6	−0.7	−0.1
N(CH$_3$)$_2$	+37.1	+33.7	+ 1.7	+2.6	−1.1	−6.2	−0.6	+0.3
N$_3$	+32.5	+29.8	+ 4.5	+2.0	−2.5	−6.9	−2.5	−1.8
NO$_2$	+58.0	+53.9	+ 4.0	+1.7	−2.4	−5.6	−2.0	−1.1
−C≡CH	+ 1.7	+ 1.0	+ 5.1	+3.0	−1.8	−5.8	−2.1	−1.3
−CN	+ 0.7	− 0.6	+ 2.2	−0.4	−2.6	−5.1	−2.6	−2.0
−NC	+24.9	+23.3	+ 6.7	+3.5	−2.6	−6.9	−1.8	−1.8

[a] Data from Schneider and Hoppen (1978).

those for methyl. On the other hand, the β effects are of the same order of magnitude, ranging from 1.7 to 13.8 ppm for β_e and from -0.4 to 9.5 ppm for β_a. In general, $\alpha_e > \alpha_a$ and $\beta_e > \beta_a$, except for bromine and iodine in the former case and $N(CH_3)_2$ in the latter. (Note that the global α effect for a substituent $-X-Y$ includes the β effect of Y and the global β effect of such a combination includes the γ effect of Y.) All γ_a effects are upfield shifting and of a magnitude similar to that for methyl (Table 11.14). However, most of the substituents also have sizeable negative γ_e effects (Eliel, Grant, et al., 1975). Since the effect is small or even reversed for substituents below the first row of the periodic table (Cl, Br, I, and SH), it was originally thought to be hyperconjugative in origin, especially since it was also noted that the γ_{anti} or γ_e effect would disappear when the substituent causing it was attached to a bridgehead carbon. However, it is now known (e.g., Schneider and Hoppen, 1978) that any high degree of substitution (especially axial substitution) on the two carbon atoms linking the substituent to

the γ_{anti} carbon nucleus observed, that is, at C(2) and C(3) in

$$
\begin{array}{c} 2 \quad \bullet \\ \diagup \diagdown \diagup \\ X \quad 3 \end{array}
$$

will lead to a disappearance or reversal of the effect that thus does not seem to be of hyperconjugative origin (see also Duddeck, 1986). Also notable in Table 11.14 are the almost uniformly upfield shifting delta effects that are much more important for hetero than for methyl substituents. In most instances $|\delta_e| > |\delta_a|$.

The information from ^{13}C spectra discussed so far is based on broad-band decoupled spectra, that is, it is devoid of coupling information. Modern instrumentation and techniques, however, allow one to determine both $^{13}C-^1H$ and $^{13}C-^{13}C$ coupling constants; information of this type promises to become useful in conformational analysis (Marshall, 1983). Thus $^3J_{C-H}$, the vicinal coupling constant in an $H-C-C-C$ segment, follows a Karplus relationship depending on the torsion angle of the central C–C bond: $^3J_{C-H} = 4.26 - 1.00\cos\omega + 3.56\cos 2\omega$, diminishing from a value of 6.82 Hz at $\omega = 0°$ to 0.7 Hz at $\omega = 90°$ (the value at 60° is 2.00 Hz) and then increasing again to a value of 8.81 Hz at 180° (Wasylishen and Schaefer, 1973). The $^{13}C-^{13}C$ vicinal coupling constants for the 1,4-dimethylcyclohexanes (Booth and Everett, 1980b) are shown in Figure 11.44. For the cis isomer both the (individual) low-temperature values and the (averaged) room temperature values are given. The substantial differences in $^3J_{CH_3(e)/C(3)}$ in the cis and the trans isomers suggest that effects of ring deformation on these coupling constants may be appreciable. The direct couplings, $^1J_{C-C}$ also differ as shown in Figure 11.44. It is of incidental interest that the equilibrium constant for the cis labeled compound (Fig. 11.44) is indistinguishable from unity, that is, displays no measurable ^{13}C isotope effect. In contrast, a palpable deuterium isotope effect is seen in *trans-1-*

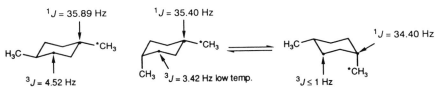

Figure 11.44. $^{13}C-^{13}C$ Coupling constants in dimethylcyclohexanes (*CH_3 is the isotopically enriched methyl group). At room temperature $^3J_{av} = 2.29$ Hz.

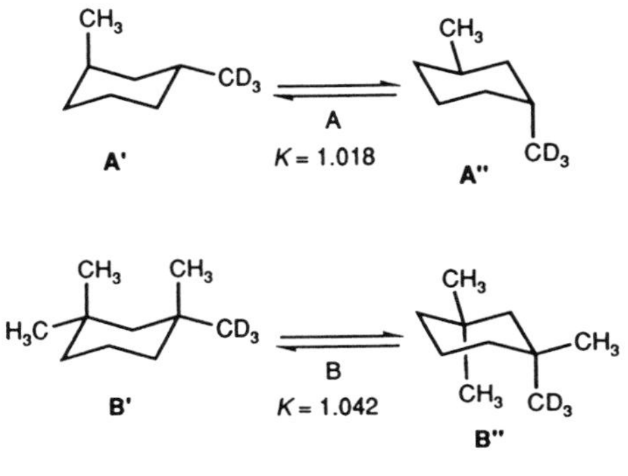

CH₃

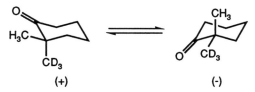

Figure 11.45. Conformational equilibrium involving isotopically labeled species.

trideuteriomethyl-3-methylcyclohexane (Fig. 11.45, **A**), the conformer with axial CD_3 predominating by 46 ± 10 J mol^{-1} (11.0 ± 2.4 cal mol^{-1}), perhaps because C–D has a lesser zero-point vibration than C–H, making the CD_3 group effectively smaller (Baldry and Robinson, 1977; see also Booth and Everett, 1980c). A similar result is seen in [1-methyl-d_3],1,3,3-trimethylcyclohexane (Fig. 11.45, **B**; Anet, Saunders, et al., 1980). The ^{13}C spectrum of this compound shows three resonances in the methyl region: two sharp ones, separated by 0.184 ppm for the geminal CH_3 groups and one broad one in between the other two. [The signal for CD_3 is not seen because of long relaxation time and lack of NOE; the signal for $CH_3(1)$ is broadened by γ-deuterium coupling and does not coincide with either of the other two signals because of an upfield γ-deuterium shift.] The equilibrium constant can be calculated from the shift difference between $(CH_3)_a$ and $(CH_3)_e$ in a conformationally locked situation (**B'** or **B''**) at low temperature (-100°C), which is found to be 9.03 ppm, and the observed shift difference at room temperature; it is $K = \mathbf{B''}/\mathbf{B'} = (9.03 + 0.184)/(9.03 - 0.184) = 1.042$, hence $\Delta G^0 = 24$ cal mol^{-1} (100 J mol^{-1}). The fact that the methyl group at position 1 corresponds to the *low*-field methyl at 3 (the isotope effect, as mentioned above, is *upfield* shifting), that is, to equatorial methyl, indicates that CD_3 prefers the axial position.

In view of these observations, it is curious that the opposite conclusion was arrived at in 2-methyl,[2-methyl-d_3]-cyclohexanone (Fig. 11.46) on the basis of CD study (cf. Chapter 13: $\Delta H^0 = -9.5$ cal mol^{-1} or -39.7 J mol^{-1}) (Barth and Djerassi, 1981; see also Lee, Djerassi et al., 1980; these papers should be consulted for further details).

Nuclear magnetic resonance determinations in cyclohexane-d (Aydin and Günther, 1981; Anet and Kopelevich, 1986) in contrast, showed *equatorial* deuterium to be preferred. The earlier experiment apparently gave too large an equatorial preference; according to the later one, ΔG^0 for the process shown in Figure 11.47, *a*, is 12.6 cal mol^{-1} (52.7 J mol^{-1}) or 6.3 cal mol^{-1} (26.4 J mol^{-1}) for one deuterium. This result was arrived at by observing that the difference in proton

Figure 11.46. Conformational equilibrium in 2-methyl-2-[methyl-d_3]-cyclohexanone (the signs refer to the sign of the Cotton effect; see Chap. 13).

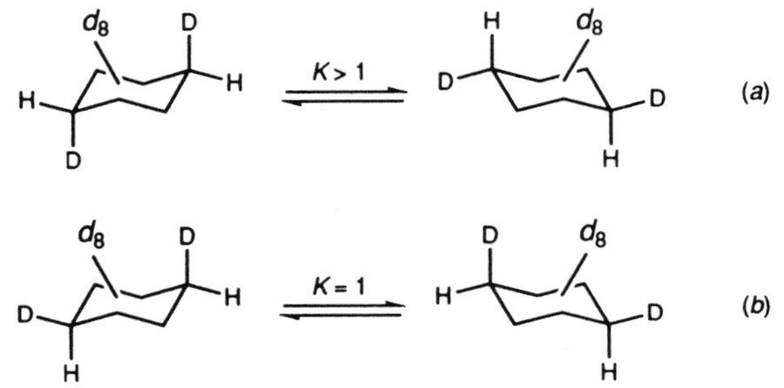

Figure 11.47. Axial–equatorial H/D preference in cyclohexane.

shift between system a and system b (for which $K = 1$) was 0.51 Hz at 200 MHz, whereas the difference between axial and equatorial hydrogen atoms in cyclohexane is known to be 95.6 Hz. It may be shown that $K = (95.6 - 2 \times 0.51)/(95.6 + 2 \times 0.51) = 0.979$, hence $\Delta G^{0} = 12.6$ cal mol^{-1} (52.7 J mol^{-1}). The axial preference of hydrogen in system a follows from the observation that the proton signal in system a is *up*field of that in b (cf. p. 712). A direct determination of the conformational equilibrium in cyclohexane-d (Fig. 11.18, X = D) by low-temperature NMR spectroscopy (Anet and O'Leary, 1989) gave a value of 8.2 ± 0.9 cal mol^{-1} (34.3 ± 3.8 J mol^{-1}) and a similar determination for cyclohexane-t indicated a value of 11.2 ± 0.5 cal mol^{-1} (46.9 ± 2.1 J mol^{-1}) (Anet et al., 1990).

e. Conformation and Reactivity in Cyclohexanes

We mentioned earlier that Barton recognized the greater stability of equatorially as compared with axially substituted cyclohexanes. But probably the most important insight gained from his pioneering work on conformational analysis (Barton, 1950) concerns the effects of conformation on chemical reactivity. It is convenient to divide these effects into two kinds, steric and stereoelectronic effects (even though it is realized that all chemical effects ultimately relate mainly to bonding and nonbonding electrons: The main contributors to steric, or van der Waals, effects are the electron clouds of nonbonding atoms).

By "steric effects" we mean effects due to close approach of two groups in a molecule (or between molecules) such that appreciable van der Waals forces (either attractive, at relative long distances, or repulsive, at short distances) are called into play. Such effects may occur in the ground state of a molecule, or in the transition state for a given reaction, or both. For the sake of simplicity, we shall, in the discussion that follows, assume that the effects are repulsive. While it must be kept in mind that van der Waals interactions can also be attractive, the attractive potential is always quite small (cf. Fig. 2.23), whereas the repulsive potential can become quite large if the nonbonded distances are sufficiently short.

Two kinds of situations are depicted in Figure 11.48. In the first, more familiar situation (a) the repulsion is substantial in the transition state (TS) and small or absent in the ground state (GS). Compared to a reference case, the activation energy in this situation is increased, because the energy level of the transition state is elevated more than that of the ground state: The reaction is slowed down relative to the reference case. This situation is termed one of "steric

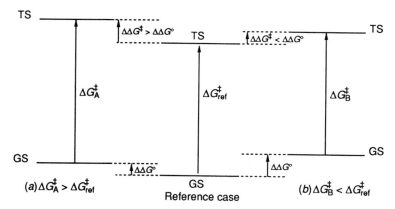

Figure 11.48. Steric hindrance and steric assistance.

hindrance." In the other situation (*b*) the steric repulsion is more important in the ground state than in the transition state. As a result, the energy level of the ground state is elevated more than that of the transition state; thus the activation energy for the reaction is decreased and the reaction is accelerated relative to the reference case. This situation is called one of "steric assistance."

We shall limit ourselves to one example each of these two situations (for additional examples, see Eliel et al., 1965). Saponification of anancomeric ethyl cyclohexanecarboxylates (Fig. 11.49; Eliel et al., 1961; Allinger et al., 1962) illustrates steric hindrance. The carbonyl group in the ground state is sp^2 hybridized; the rate-determining transition state, in which an HO^- moiety becomes attached to the CO_2Et group, is sp^3 hybridized. As we have already seen in Table 11.7, conformational energies of sp^3 hybridized groups are generally larger than those of sp^2 hybridized ones: the latter (in contrast to the former) can escape crowding by turning their flat sides to the ring. In the present case there is the additional factor that the ground state is neutral (at least as far as the organic moiety is concerned) but the transition state is negatively charged, and therefore more solvated. As already pointed out in connection with the conformational

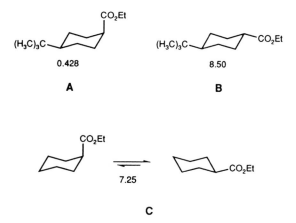

Figure 11.49. Specific saponification rates of ethyl cyclohexanecarboxylates ($L \, mol^{-1} \, s^{-1} \times 10^4$).

energy of CO_2^- versus CO_2H, solvation leads to additional bias against the axial position. The saponification rate of the *cis*-4-*tert*-butyl (axial-CO_2Et, **A**) isomer is therefore about 20 times less than that of the trans (equatorial-CO_2Et, **B**) isomer. [The saponification rate for the unsubstituted, conformationally heterogenous compound **C** is intermediate; cf. Eq. 11.1: $k_c = n_e k_e + n_a k_a$, where n_e and n_a refer to the mole fractions of the equatorial and axial conformers, respectively, and k_e and k_a refer to their specific saponification rates, approximately equal (see above) to k_B and k_A.]

A (presumed) case of steric assistance is shown in Figure 11.50; it concerns the oxidation rates of anancomeric cyclohexanols (Richer, Eliel et al., 1961). It has been pointed out (Schreiber and Eschenmoser, 1955; Eliel et al., 1966) that the rates parallel the degree of crowding of the hydroxyl group, or, perhaps more concisely, the degree of strain relief that occurs when the sp^3 hybridized alcohol (or the corresponding chromate, which is an intermediate in the oxidation) is converted, in the rate-determining step, to the sp^2 hybridized ketone, with resulting relief of synaxial strain. Thus **A** reacts faster than **B** (relief of two synaxial OH/H interactions); **C** also reacts faster than **B** (relief of a synaxial CH_3/H interaction), and **D** reacts much faster than any of the others (relief of a severe CH_3/OH synaxial interaction, cf. Table 11.10). Again the rate of the conformationally mobile species **E** is intermediate.

> The interpretation of the relative oxidation rates of cyclohexanols strictly in terms of relief of steric strain is probably an oversimplification (cf. Kwart and Nickle, 1973, 1976); certainly polar factors also play a role (Šipoš, Sicher, et al., 1962).

We turn now to "stereoelectronic effects" (cf. Deslongchamps, 1983) though at this point we shall confine the discussion to cyclohexane systems. Deslongchamps defines stereoelectronic effects as effects on reactivity of the spatial disposition of particular electron pairs, bonded or nonbonded. In many cases these are electron pairs in bonds that are formed, broken, or otherwise dislocated in the reaction in question; in other examples they are unshared electrons on exocyclic or endocyclic hetero atoms (cf. the discussion of the anomeric effect, Section 10-1.b and p. 749).

A well-known example is the S_N2 displacement reaction; it has long been

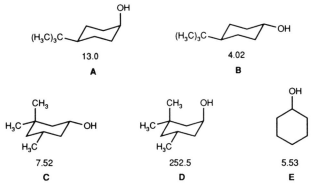

Figure 11.50. Oxidation rates of cyclohexanols (L mol^{-1} s^{-1} $\times 10^3$).

known (cf. Chapter 5) that this reaction proceeds with inversion of configuration, that is, that the incoming nucleophile enters from the back of the leaving one. The stereoelectronic situation is that the σ^* orbital of the bond joining the reacting carbon to the leaving group is the one that gives rise to the new σ bond to the incoming nucleophile. The S_N2 reaction of cyclohexane p-toluenesulfonates with thiophenolate is depicted in Figure 11.39. Clearly, the reaction proceeds with inversion. Moreover, the axial (cis) tosylate ($k_a = 3.61 \times 10^{-4}$ L mol^{-1} s^{-1}) reacts about 31 times faster than the trans (equatorial) tosylate ($k_e = 0.116 \times 10^{-4}$ L mol^{-1} s^{-1}) at 25°C (Eliel and Ro, 1957b). This finding would appear to be due to a combination of steric and stereoelectronic effects: There is ground-state compression in the axial compound, giving rise to steric assistance [but since, according to Table 11.7 $-\Delta G_{OTs} = 0.50$ kcal mol^{-1} (2.10 J mol^{-1}), ground-state compression can only give rise to a rate factor of 2.3 at 25°C] and there is also steric hindrance to backside attack (from the axial side) in the equatorial compound, that is, the stereoelectronic requirement of backside attack produces an adverse steric factor.

A number of bimolecular elimination (E2) reactions in cyclohexyl systems have the stereoelectronic requirement that the ligands to be eliminated be antiperiplanar. This requirement may be an example of the principle of "least motion": After departure of the ligands to be eliminated, two antiperiplanar, that is, antiparallel, orbitals are optimally disposed to overlap and form a π bond. The next best orientation, from the viewpoint of optimal orbital overlap, is a synperiplanar one (which does not normally occur in cyclohexyl systems) whereas a synclinal orientation of the ligands to be eliminated is unfavorable. This condition leads to the conclusion that E2 eliminations in cyclohexanoid systems should be trans–diaxial, but not trans–diequatorial or cis–equatorial–axial, since the former array is antiperiplanar, whereas the latter two are synclinal. Numerous examples are found in the literature (cf. Eliel et al., 1965; Saunders and Cockerill, 1973) of which only a few can be detailed here.

Of the cis- and $trans$-4-$tert$-butylcyclohexyl p-toluenesulfonates (Fig. 11.39), only the cis isomer undergoes bimolecular elimination with ethoxide in ethanol; the trans isomer is constricted to E1 (and S_N) pathways. Here, again, in the cis isomer OTs and H are antiperiplanar (diaxial), whereas in the trans isomer they are synclinal, either e,e or e,a. Cyclohexyl tosylate itself, though predominantly in the conformation with equatorial OTs, does undergo bimolecular elimination with ethoxide at a rate of 0.26 that of the cis-4-$tert$-butyl isomer (Winstein and Holness, 1955) and this result can be readily explained in terms of the Winstein–Holness equation (Section 10-5): $k = n_e k_e + n_a k_a$; if $k_e = 0$, $k = n_a k_a$. According to Table 11.7 the conformational energy of the tosylate group is 0.50 kcal mol^{-1} (2.1 kJ mol^{-1}) from which one may calculate $n_a = 0.326$, at 75.2°C, in reasonable agreement with the experimentally found 0.26.

Among the benzene hexachloride (hexachlorocyclohexane) isomers, the β isomer undergoes HCl elimination with base (eventually to give trichlorobenzene) more slowly by several powers of 10 than all the others, the activation energy for this isomer being about 11–12 kcal mol^{-1} (46–50 kJ mol^{-1}) higher. Contemplation of Figure 11.32 (Cl instead of OH) leads to the conclusion that the β is the only isomer that does not have an axial chlorine next to an axial hydrogen in the most

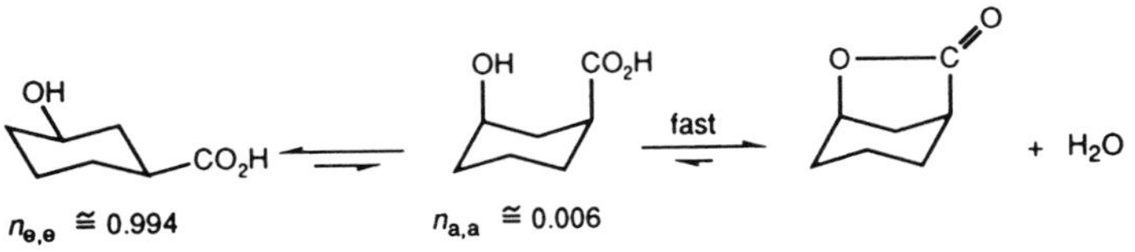

Figure 11.51. Lactonization of *cis*-3-hydroxycyclohexanecarboxylic acid.

stable conformation (or, for that matter, even in the highly unstable hexaaxial alternate conformation); therefore the rate-determining elimination of the first HCl molecule is slowed down by a factor of 7000 or more.

In general, a substrate often reacts in a minor conformation, provided the reaction rate for that conformation is reasonably large. This statement applies to reactions of biochemical and pharmacological import (enzyme-substrate and drug-receptor interactions) just as well as to simpler chemical ones. For example (Fig. 11.51), although the diaxial mole fraction of *cis*-3-hydroxycyclohexanecarboxylic acid is very small [estimated ΔG^0 ca. 3 kcal mol^{-1} (12.6 kJ mol^{-1}) corresponding to 0.6% diaxial conformer at room temperature], the acid lactonizes readily: Once the carboxyl and hydroxyl group are juxtaposed in the synaxial disposition, the rate of lactonization is, for entropic reasons, very high and the equilibrium very favorable to the lactone.

A comparison of diaxial and axial–equatorial elimination is displayed in Figure 11.52: reaction of *cis*- and *trans*-1,2-dibromocyclohexane with iodide to give cyclohexene. That the trans dibromide reacts only 11.5 times as fast as the cis is neither an indication of relatively rapid cis (e,a) elimination, nor does it mean (cf. Fig. 2.24) that the trans dibromide exists mainly in the diequatorial conformation, which is not favorably disposed toward elimination. The most likely interpretation based on kinetic studies is that the reaction in the cis isomer is a (rate-determining) bimolecular substitution of Br by I followed by rapid diaxial elimination. The rate factor of 11.5 thus reflects the somewhat higher rate of diaxial elimination over substitution.

Ring formation, rearrangement, neighboring group participation, and fragmentation constitute another (interrelated) set of reactions with the stereoelectronic requirement that the groups involved must be antiperiplanar. Normally,

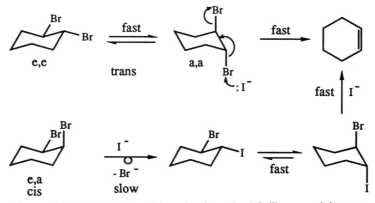

Figure 11.52. Elimination of bromine from the 1,2-dibromocyclohexanes.

this involves diaxial disposition; however, since an element of the ring may be antiperiplanar to an equatorial leaving group, ring contractions can involve equatorial substituents. This situation applies also to fragmentation reactions (cf. Grob and Schiess, 1967; Grob, 1969). The reactions of bromohydrins (derived from cyclohexanes) with base and with silver ions illustrate these principles. These reactions are summarized in Figure 11.53 for the four diastereomeric 2-bromo-4-phenylcyclohexanols (Curtin and Harder, 1960). [The phenyl group, $-\Delta G^0_{\text{confo}} = 2.9\,\text{kcal mol}^{-1}$ ($12.0\,\text{kJ mol}^{-1}$) serves to bias the conformational equilibria of the various stereoisomers although the bias is clearly not compelling.] Case A exemplifies epoxide ring formation: The entering (OH or O$^-$) and leaving (Br) groups are antiperiplanar (a,a). Case B exemplifies ketone formation by either hydride shift or enolate formation (HBr elimination): The hydrogen involved is antiperiplanar to the leaving bromine. In Case C, where the proper stereoelectronic situation is not attained in the most stable starting conformation, the molecule apparently reacts in the alternate conformation, even though this involves synaxial phenyl and bromine. The fourth diastereomer (Case D) is particularly interesting. With Ag$_2$O, departure of the equatorial bromine is induced and the ring bond antiperiplanar to the departing equatorial bromine [C(1)/C(6)] migrates to produce a ring contraction to *cis*-3-phenylcyclopentanecarboxyaldehyde. With base (hydroxide) the driving force for ring contraction in the secondary halide seems to be insufficient (although such a contraction can occur with tertiary halide) and the molecule reacts in the alternate conformation to produce an epoxide (Case E).

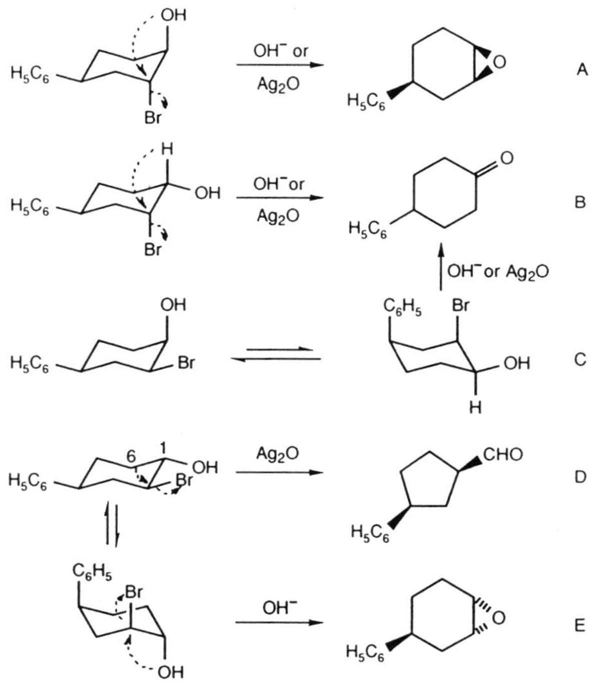

Figure 11.53. Reactions of 2-bromo-4-phenylcyclohexanols with base and silver oxide.

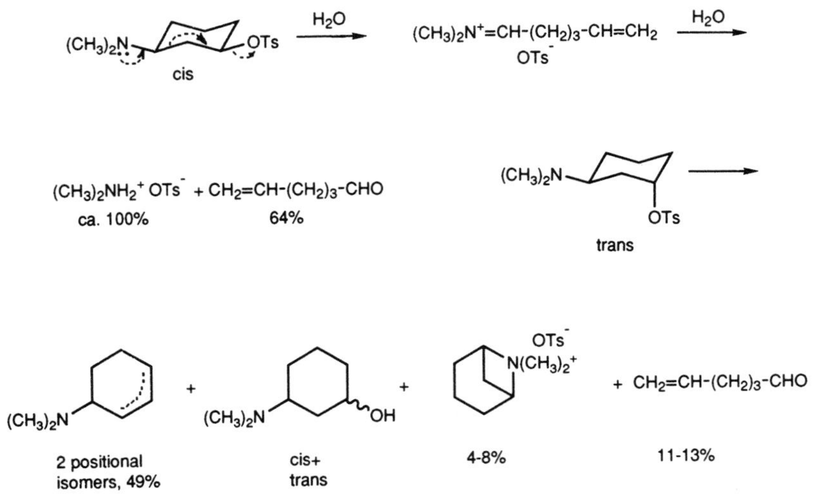

Figure 11.54. Solvolysis and fragmentation of 3-dimethylaminocycloxyl tosylate.

An example of the stereoelectronic requirement for concerted fragmentation is shown in Figure 11.54 (cf. Grob, 1969). The *cis*-3-dimethylaminocyclohexyl *p*-toluenesulfonate, upon treatment with triethylamine in 80% ethanol, fragments rapidly to yield the hydrolysis product of the imminium salt shown in Figure 11.54; the reaction is 39 times faster than the solvolysis of *cis*-3-isopropylcyclohexyl tosylate. In contrast the trans isomer, which is not properly disposed stereoelectronically for fragmentation, gives mainly solvolysis and elimination products; it reacts at about one-eighth the rate of the cis isomer and somewhat more slowly than *trans*-3-isopropylcyclohexyl tosylate (Burckhardt, Grob, et al., 1967).

Two other reactions in which stereoelectronic factors are crucial are electrophilic addition to alkenes and epoxide ring opening; these reactions will be discussed in the next section. Some additional examples for the operation of stereoelectronic effects will be illustrated later in the steroid system (Section 11-5.a).

f. *sp*² Hybridized Cyclohexyl Systems

Various such systems have been discussed in a collection of reviews (Rabideau, 1989a).

Cyclohexene

Cyclohexene has long been assumed to exist in the half-chair form depicted in Figure 11.55, **A** (cf. Anet, 1989) and this has been confirmed by both microwave spectroscopy (Scharpen et al., 1968; Ogata and Kozima, 1969) and electron diffraction (Chiang and Bauer, 1969; Naumov et al., 1970; Geise and Buys, 1970). The torsion angles in cyclohexene are shown in Figure 11.55, **B** (see also Auf der Heyde and Lüttke, 1978). Carbon atoms 3 and 6 lie in the plane of the double bond. Although this plane does not exactly bisect the H–C–H angles at

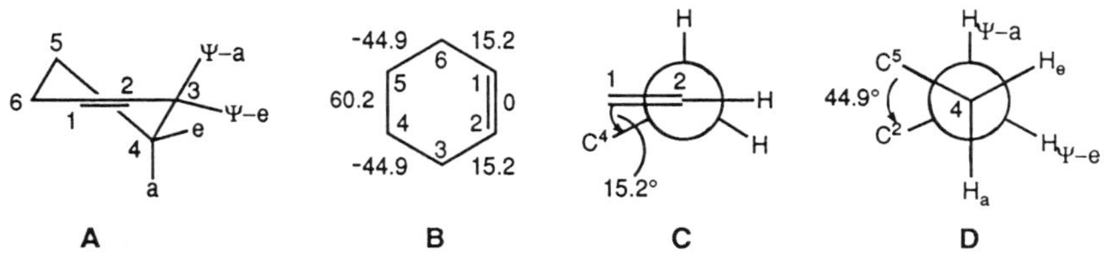

Figure 11.55. Cyclohexene half-chair and its torsion angle.

C(3) and C(6) (this would require $\omega_{2,3} = \omega_{1,6}$ to be 0^0 (see Fig. 11.55, **C**; the actual value is around 15^0), the methylene hydrogen atoms at these positions are quite far from being in truly axial or equatorial positions; such ligands are called "pseudoaxial" ($\Psi-a$) and "pseudoequatorial" ($\Psi-e$). On the other hand, $\omega_{4,5}$ is very close to $60°$ and the hydrogen atoms at these positions are therefore truly axial or equatorial. (For a summary of torsion angles in cyclohexene and cyclohexadienes, see Auf der Heyde and Lüttke, 1978.)

The inversion barrier in cyclohexene has been determined as $5.3 \, kcal \, mol^{-1}$ ($22.2 \, kJ \, mol^{-1}$) by Anet and Haq (1965) and Jensen and Bushweller (1969b) using low-temperature NMR of deuterated analogues. Molecular mechanics (Allinger, 1977; Burkert and Allinger, 1982) and ab initio quantum mechanical calculations (Burke, 1985) are in agreement with a barrier in this range; the transition state corresponds to the boat form (Dashevsky and Lugovsky, 1972; Burke, 1985; see also Bucourt, 1974; Anet and Yavari, 1978; Lipkowitz, 1989).

In 4-substituted cyclohexenes (Lambert et al., 1987) the equatorial conformer is more stable than the axial one, though $-\Delta G^0$ is less than in correspondingly substituted cyclohexanes, presumably because there is only one synaxial X/H interaction in the axially substituted cyclohexene, whereas there are two such interactions in axially substituted cyclohexane. The conformational free energy values are compiled in Table 11.16; except for OH, where intramolecular hydrogen bonding may stabilize the axial conformer (cf. Hanaya et al., 1979), these values are close to one-half the values in cyclohexane (Table 11.7). This finding supports the hypothesis that the steric interaction with the olefinic carbon C(2) is small or absent. Some of the variability of the data in Table 11.16, which were obtained in different solvents, may be due to solvent dependence. Polar solvents stabilize the axial conformer when the substituent is itself polar (Zefirov et al., 1985).

A different situation arises for 3-substituted (allylically substituted) cyclohexenes. Although a 3-methyl group prefers the pseudoequatorial position by $0.97 \, kcal \, mol^{-1}$ ($4.1 \, kJ \, mol^{-1}$) (Senda and Imaizumi, 1974), electronegative groups, such as OH, OCH_3, OAc, Cl, and Br are predominantly pseudoaxial (cf. Lessard, Saunders, et al., 1977) with the preference for OH being $0.45 \, kcal \, mol^{-1}$ ($1.9 \, kJ \, mol^{-1}$) (Senda and Imaizumi, 1974) and that for Cl $0.13 \, kcal \, mol^{-1}$ ($0.54 \, kJ \, mol^{-1}$) by electron diffraction (Lu, Chiang, et al., 1980). Values of $0.64 \, kcal \, mol^{-1}$ ($2.68 \, kJ \, mol^{-1}$) for Cl and $0.70 \, kcal \, mol^{-1}$ ($2.93 \, kJ \, mol^{-1}$) for Br have also been reported based on IR (Sakashita, 1960). An effect akin to the anomeric effect (Section 10-1.b), that is, double-bond/no-bond resonance, may be responsible for the greater stability of the pseudoaxial groups (cf. Fig. 11.56).

TABLE 11.16. Conformational Energies in 4-Substituted Cyclohexenes.

Substituent	CH$_3$	C$_6$H$_5$	CO$_2$H	CHO	CO$_2$CH$_3$	COC$_6$H$_5$	CN	NO$_2$	F	Cl	Br	I	OH	OSi(CH$_3$)$_3$
$-\Delta G^0$ (kcal mol^{-1})	ca. 1; 0.86	0.99	1.0	0.45	0.85	0.45	0.1; 0.15	0.25	ca. 0	0.13; 0.2; 0.31	0.1; 0.27; 0.4	0; 0.16	0; 0.22	0.31
$-\Delta G^0$ (kJ mol^{-1})	ca. 4.2; 3.6	4.1	4.2	1.9	3.6	1.9	0.4; 0.63	1.05	0	0.54; 0.8; 1.3	0.4; 1.13; 1.7	−0.1; 0.67	0; 0.92	1.3
References	a, b	a	c	c	c	c	c, d	c	a	a, d–f	a, d, g	c, d	d, h	d

[a] Jensen and Bushweller (1969); solvent CD$_2$=CDCl.
[b] Fernandez Gomez, Pentin, et al. (1977a) (neat liquid).
[c] Zefirov et al. (1969).
[d] Lambert and Marko (1985); solvent CF$_2$Cl$_2$.
[e] Fernandes Gomes et al. (1973) (neat liquid).
[f] Lu, Chiang, et al. (1980).
[g] Fernandez Gomez, Pentin, et al. (1977b) (neat liquid).
[h] Hanaya et al. (1979). Solvent CCl$_4$.

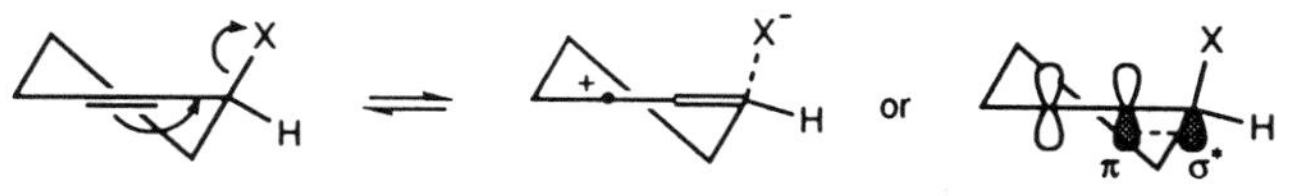

Figure 11.56. Cyclohexenes with electronegative substituents at C(3).

The ΔG^0 values for conformational equilibria in di- and trisubstituted cyclohexenes have been tabulated by Anet (1989).

Since the torsion angle between pseudoaxial and axial protons is less than 180° and that between pseudoequatorial and equatorial ones is more than 60° (Fig. 11.55, **D**), it is not surprising that $^3J_{H_{\psi-a}/H_a}$ (11.0 Hz) and $^3J_{H_{\psi-e}/H_e}$ (1.0 Hz) (Lessard, Saunder, et al., 1977) are smaller than $^3J_{H_a/H_a}$ and $^3J_{H_e/H_e}$ in cyclohexanes (Section 11.4.d). It is less obvious why $^3J_{H_{\psi-a}/H_e}$ and $^3J_{H_{\psi-e}/H_a}$ (1.5 Hz) are also smaller than J_{H_a/H_e} in cyclohexanes; this cannot be a torsion angle effect. In *cis*- and *trans*-4-cyclohexene-1,2-diols the difference in OH stretching frequencies in the IR between bonded and unbonded OH ($\Delta\nu_{OH}$; cf. Section 11-4.d), 32 versus 35 cm^{-1} (Tichý, 1965) is less than in the corresponding cyclohexanediols (Table 11.2) because the torsion angles between substituents at what, in cyclohexene, are 4,5-positions are closer to the ideal 60° than in cyclohexane (cf. Fig. 11.15); the reversal of the values ($\Delta\nu_{cis} < \Delta\nu_{trans}$) might suggest that the trans OH groups stay at ω ca. 60°, whereas the cis groups seem to be forced apart somewhat (ω ca. 65°).

An important reaction of cyclohexenes is electrophilic addition. In the case of bromine addition, which normally involves a bromonium ion intermediate, this addition is generally antiperiplanar. It would appear that in cyclohexenes there are two possible modes of antiperiplanar attack (cf. Fig. 11.57) leading to diaxially or diequatorially 1,2-disubstituted cyclohexanes. However, a careful study of models (cf. Fig. 11.57) shows that antiperiplanar attack from the sides of the proximal axial hydrogen atoms gives rise to a diaxially disubstituted chair, whereas attack from the opposite sides gives rise to a diaxially disubstituted twist form, which must then subsequently invert to the diequatorially disubstituted chair. Since the twist form is over 5 kcal mol^{-1} (21 kJ mol^{-1}) less stable than the chair, the corresponding transition state leading to it is also quite destabilized and it is therefore not surprising that, under conditions of kinetic control, the predominant or exclusive product is that of diaxial addition (Valls, 1961; Valls and Toromanoff, 1961). In fact, it has been shown (Barili, Bellucci, et al., 1972) that bromine addition to an anancomeric cyclohexene, in this case 4-*tert*-butylcyclohexene, gives the diaxial dibromide (cf. Fig. 11.35) in at least 94:6 preference over the diequatorial. However, the same is not true for 3-*tert*-butylcyclohexene, which gives predominantly the product of diequatorial addition. Models show that the otherwise preferred mode of addition is impeded, at C(2), by the steric bulk

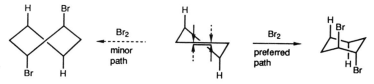

Figure 11.57. Diaxial electrophilic addition to cyclohexenes.

of the *tert*-butyl group. Either this factor here overwhelms the normal preference for diaxial attack, or the molecule actually reacts in the alternate conformation with pseudoaxial *tert*-butyl.

Not all electrophilic additions are antiperiplanar; depending on mechanisms, they may, alternatively, by synperiplanar or nonstereoselective (cf. Fahey, 1968).

Cyclohexene Oxide

Epoxidation of cyclohexenes, for example, with peracids, gives rise to cyclohexene oxides. These molecules, also, exist as half-chairs (Ottar, 1947; Naumov and Bezzubov, 1967). Epoxidation of cyclohexene must necessarily be cis, since the trans bonds in oxirane (ethylene oxide) cannot be spanned by a $(CH_2)_4$ chain. However, trans epoxides, as well as cis, exist in larger rings (cf. Section 11-5.a). When there is an additional substituent in the ring, epoxidation may be cis or trans to the substituent (Fig. 11.58). In 4-*tert*-butylcyclohexene, attack on the two faces of the double bond is nearly random, the cis–trans product ratio being 3:2 (cf. Berti, 1973) but in the 3-*tert*-butyl compound attack on the trans side is favored by a factor of 9:1. Evidently, just as in the case of bromine addition discussed above, the pseudoequatorial *tert*-butyl group blocks the approach from the face cis to it. High stereoselectivity is also observed when there is a hydroxyl group in the allylic position (Henbest and Wilson, 1957). Thus in the 3-*tert*-butylcyclohexene series (Fig. 11.58), when X = OH instead of H, the percentage of cis isomer increases from 90 to 100% (Bellucci et al., 1978) presumably because, in addition to the steric effect of the 3-*tert*-butyl group, there is now a complexing effect of the hydroxyl with the epoxidation reagent that directs the incoming oxygen to the side cis to hydroxyl.

Ring opening of epoxides, either in acidic or in basic medium, usually proceeds with inversion of configuration (cf. Wohl, 1974) although there are some exceptions, for example, in phenyl-substituted oxiranes (cf. Eliel, 1962, p. 102). In addition, however, opening of cyclohexene oxides generally proceeds in such fashion that the diaxial rather than the diequatorial reaction product is obtained (Fürst–Plattner rule, 1951). Thus reduction of *cis*-4-*tert*-butylcyclohexene oxide (Fig. 11.59, **B**) with lithium aluminum deuteride leads to *c*-4-*t*-butylcyclohexan-*r*-ol-*t*-2-*d* (**C**) (Lamaty et al., 1966).

> It is of interest that, even in the enzymatic hydrolytic ring opening of this epoxide with rabbit liver microsomal epoxy hydrolase, the ring opening is entirely trans–diaxial so that the *only* product from either the cis or the trans substituted epoxide is *c*-4-*t*-butyl-*r*-1,*t*-2-cyclohexane-1,2-diol (Fig. 11.59, **A**). However, the trans substituted epoxide gives predominantly the (−)-diol, whereas the cis isomer gives the (+)-diol (Bellucci et al., 1980). Since the epoxide ring opening is inherently stereospecific (i.e., each enantiomer of, say, the cis epoxide gives one enantiomer

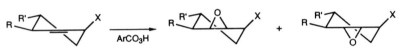

Ar = aromatic group, usually phenyl or m-chlorophenyl

Figure 11.58. Epoxidation of cyclohexenes.

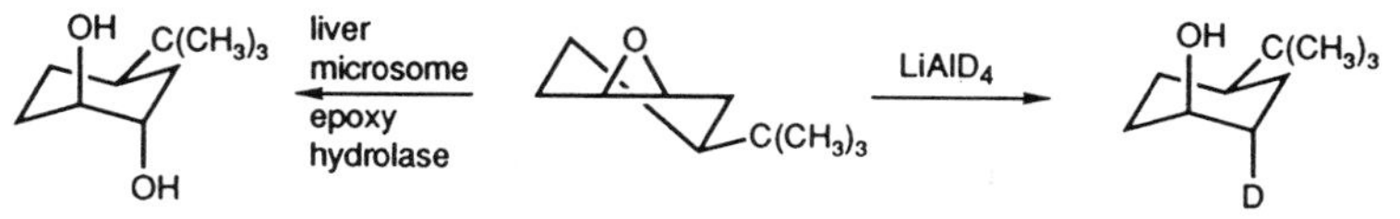

Figure 11.59. Ring opening of cis-4-tert-butylcyclohexene oxide with lithium aluminum deuteride and with rabbit liver microsomal epoxy hydrolase.

of the diol) the formation of optically active products must be due to kinetic resolution. In fact, the optical activity of the product diol is highest at the beginning and drops as the enzymatic hydration proceeds. From the known configuration and enantiomeric purity of the product, it was concluded that the 1S,2R,4S (trans) epoxide was opened much faster than its 1R,2S,4R enantiomer and that the epimeric 1S,2R,4R (cis) epoxide is somewhat faster hydrolyzed than its 1R,2S,4S enantiomer. Moreover, it was evident from the rate of the reactions that the hydrolysis of the 1S,2R,4S epoxide was faster than that of its 1S,2R,4R diastereomer and that the 1R,2S,4S isomer was the slowest reacting of the set. In the 3-tert-butyl series, where steric factors are superimposed upon stereoelectronic ones, as explained earlier, the situation is more complex in that one of the diastereomeric epoxides (cis) gives the diaxial diol and the other epoxide (trans) gives the diequatorial diol (Bellucci et al., 1982).

Cyclohexadienes

In the cyclohexadiene series, the conjugated 1,3 isomer (cf. Rabideau and Syguta, 1989) is nonplanar, as evidenced by microwave (Traetteberg, 1968), electron diffraction (Oberhammer and Bauer, 1969), and NMR (Auf der Heyde and Lüttke, 1978) techniques (see also Lipkowitz, 1989). In contrast, the 1,4 isomer 1,4-dihydrobenzene (cf. Rabideau, 1989b) is planar (Carreira et al., 1973; Auf der Heyde and Lüttke, 1978; Rabert et al., 1982; low-temperature X-ray evidence: Jeffrey, 1988), although its energy minimum is quite shallow and the ground state may be deplanarized by substituents at C(1) (Grossel et al., 1978; Rabideau et al., 1982; Lipkowitz, 1989). We shall see more of these two molecules later: of 1,3-cyclohexadiene as an inherently dissymmetric chromophore of interest in ORD and CD studies (Chapter 13), and of the 1,4 isomer in connection with its dibenzo derivative 9,10-dihydroanthracene (cf. Rabideau, 1978), which is nonplanar and presents interesting conformational features (p. 783).

Cyclohexanone

We pass, now, to a consideration of cyclohexanone (cf. Lambert, 1989). The geometry and conformation of this molecule are summarized in a paper by Dillon and Geise (1980). The most stable conformation of cyclohexanone is the chair; the C–C(O)–C angle is somewhat less than usual (115° with a corresponding expansion of the O–C–C angle to 122°) but the chair is nevertheless appreciably flattened at the carbonyl site, values for the H_{eq}–C–C–O torsion angle ranging from 3.3° (Allinger et al., 1972) to 12.7° (Huet et al., 1976; see also Alonso, 1981). The structure of 4,4-diphenylcyclohexanone has been determined by X-ray diffraction (Lambert et al., 1969); the ring torsion angles are $\omega_{1,2}$ and $\omega_{6,1}$

(average) $= 42°$, $\omega_{2,3}$, and $\omega_{5,6} = 52°$, $\omega_{3,4}$ and $\omega_{4,5} = 59°$, that is, the ring is flattened at the carbonyl end and puckered at C(4) (the latter possibly because of the phenyl substituents). The twist form of cyclohexanone is calculated to be only 2.72 kcal mol^{-1} (11.4 kJ mol^{-1}) above the chair and the activation energy for chair inversion is found (Anet et al., 1973) to be 4.0 kcal mol^{-1} (16.7 kJ mol^{-1}); these values are considerably lower than corresponding values for cyclohexane (Section 11-4.a). The strain in cyclohexanone exceeds the strain in cyclohexane by about 3 kcal mol^{-1} (12.5 kJ mol^{-1}) (Ibrahim, 1990). Cyclohexanone and cyclobutanone (and presumably also cyclopropanone, for which calculations are not available) are the only cyclanones for which this is the case; cyclopentanone and medium ring ketones are less strained than the corresponding hydrocarbons and the strain in simple aliphatic ketones is near zero (Allinger et al., 1972). Angle strain is clearly responsible for the excess strain in the small-ring ketones (deformation of the near-120° angle corresponding to sp^2 hybridization of the carbonyl to 90° or 60° is more difficult than corresponding deformation of the near-tetrahedral 109.5° angle in the hydrocarbon – see also Section 11-5.e). The excess strain in cyclohexanone has at one time been ascribed to eclipsing in the ketone as contrasted to perfect staggering in the hydrocarbon cyclohexane (Brown et al., 1954). However, since ketones naturally prefer carbonyl-eclipsed conformations (Section 10-2.a) this explanation does not seem correct. Allinger et al. (1972) ascribe the strain in cyclohexanone to the fact that the eclipsing here is between C=O and H, whereas in aliphatic ketones the preferred eclipsing is between C=O and C (cf. Chapter 10). In any case, the relief of strain occurring in addition reactions of cyclohexanone makes such reactions particularly facile both with respect to rate and with respect to equilibrium. Thus the reduction of cyclohexanone with sodium borohydride is 355 times as fast as that of di-n hexyl ketone and the equilibrium constant for HCN addition (ketone + HCN $\rightleftharpoons$ cyanohydrin) is 70 times as large for cyclohexanone as for di-n-octyl ketone (see also Table 11.22, p. 770).

Because of the facile epimerization of 2-substituted cyclohexanones through enolate or enol formation, equilibria in this series have been easy to study, with some interesting results that are summarized in Table 11.17. Comparison of the values for the alkyl groups in Table 11.17 with those in Table 11.7 [CH$_3$,

TABLE 11.17. Equilibria ($-\Delta G^0$ values) in 2-Substituted Cyclohexanones[a].

X	CH$_3$	Et	i-Pr	t-Bu	F	Cl	Br$^-$
$-\Delta G^0$ (kcal mol^{-1})	1.56	1.09	0.59	1.62	0.05[b]	-0.68[b]	-1.03[b]
					0.94[c]	0.12[c]	-0.53[c]
$-\Delta G^0$ (kJ mol^{-1})	6.53	4.56	2.47	6.78	0.20[b]	-2.86[b]	-4.30[b]
					3.93[c]	0.50[c]	-2.22[c]

[a] From Eliel et al. (1965), pp. 114, 465. See also Pan and Stothers (1967); Došen-Mićović and Allinger (1978); Abraham and Griffiths (1981).
[b] In cyclohexane and other hydrocarbon solvents.
[c] In CDCl$_3$, Basso, Rittner, Lambert, et al., 1993.

1.74 kcal mol^{-1} (7.28 kJ mol^{-1}); Et, 1.79 kcal mol^{-1} (7.49 kJ mol^{-1}); i-Pr, 2.21 kcal mol^{-1} (9.25 kJ mol^{-1})] suggests that the methyl value is nearly normal but that the values for ethyl and isopropyl next to the ketone function are much lower than in saturated systems. The main reason for this occurrence seems to be that the terminal methyl group of ethyl (or one of the terminal methyl groups of the isopropyl substituent) in the axial conformer is unimpeded when it points toward the carbonyl function of the cyclohexanone, whereas it is subject to a butane–gauche interaction in the corresponding axially substituted cyclohexane (cf. Fig. 11.24). In addition, in the 2-isopropylcyclohexanone there is a destabilization of the equatorial conformer by interaction of one of the terminal methyl groups with the carbonyl oxygen. The situation with the *tert*-butyl compound is more complex, since it appears to exist predominantly in the twist form. The $-\Delta G^0$ values for the halogens are not only diminished from their cyclohexane values [F, 0.36 kcal mol^{-1} (1.51 kJ mol^{-1}); Cl, 0.55 kcal mol^{-1} (2.30 kJ mol^{-1}); Br, 0.49 kcal mol^{-1} (2.05 kJ mol^{-1}); cf. Table 11.7] but they are also strongly solvent dependent; in the case of Cl in the low-dielectric solvent heptane, and for Br in all solvents studied, the axial conformer is actually preferred. The solvent dependence suggests that a dipole interaction is involved at least in part: for equatorial halogen the $^+$CO$^-$ and $^+$CX$^-$ dipoles are nearly parallel and thus repulsive, whereas for the axial conformer they are nearly perpendicular and thus noninteracting (cf. Fig. 13.27, **C**). However, the difference in $-\Delta G^0$ between halocyclohexane and 2-halocyclohexanone varies greatly with the halogen, being (in heptane solvent) 0.31 kcal mol^{-1} (1.30 kJ mol^{-1}) for F, 1.23 kcal mol^{-1} (5.15 kJ mol^{-1}) for Cl, and 1.52 kcal mol^{-1} (6.36 kJ mol^{-1}) for Br. This is so despite the fact that the variation in dipole moment for the three C–X bonds is slight (C$_6$H$_{11}$F, 1.94 D; C$_6$H$_{11}$Cl, 2.21 D; and C$_6$H$_{11}$Br, 2.27 D) and the span of solvent effects (ΔG^0 in heptane minus ΔG^0 in dioxane) does not vary much either for the three halogens [0.89 kcal mol^{-1} (3.73 kJ mol^{-1}) for F; 0.80 kcal mol^{-1} (3.36 kJ mol^{-1}) for Cl; and 0.50 kcal mol^{-1} (2.08 kJ mol^{-1}) for Br]. There is evidently another factor that comes into play, probably the steric interaction of the equatorial halogen with the carbonyl oxygen in the equatorial conformer (Meyer, Allinger, et al., 1980; Abraham and Griffith, 1981). This interaction is attractive for F and increasingly repulsive for Cl and Br. Yet another potential factor is the anomeric effect (σ^*–π orbital overlap) shown in Figure 11.60 (Corey and Burke, 1955). This type of overlap can occur only for the axial, not the equatorial halogen (only for the former is the σ^* orbital nearly parallel to the π orbital) and it is more important for the less electronegative bromine than for the more electronegative fluorine, since charge is transferred from the halogen to the carbonyl group. However, the calculations account for the observed equilibria (Table 11.17) without invocation of such an effect.

Figure 11.60. Possible stabilization of axial α-haloketones by anomeric effect.

Figure 11.61. 3-Alkylketone effect. $\Delta H^0 = 1.36$ kcal mol^{-1} (5.69 kJ mol^{-1}) versus 1.73 kcal mol^{-1} (7.24 kJ mol^{-1}) in the corresponding hydrocarbon.

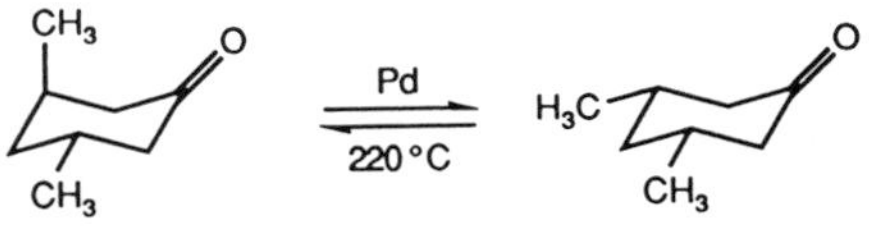

Just as there is a "2-alkylketone effect", that is, a trend toward axial disposition of an alkyl group in position 2 in a cyclohexanone (Table 11.17), there is a corresponding "3-alkylketone effect" (Fig. 11.61), which has to do with the fact that an alkyl group in position 3 to a carbonyl, when axial, encounters only one synaxial hydrogen rather than two; the equatorial–axial enthalpy difference is thereby diminished. Thus ΔH^0 for the equilibrium shown in Figure 11.61 is -1.36 kcal mol^{-1} (-5.69 kJ mol^{-1}), whereas the corresponding value in 1,3-dimethylcyclohexane (Table 11.9) is -1.73 kcal mol^{-1} (-7.24 kJ mol^{-1}) (Allinger and Freiberg, 1962).

It was early recognized (Corey and Sneen, 1956) that in enolization of a ketone, abstraction of the axial α hydrogen should be preferred over that of the equatorial hydrogen for stereoelectronic reasons similar to those implied in Figure 11.60 (but with X = H). Similarly, by the principle of microscopic reversibility, reprotonation of the enolate should take place from the axial direction, and similar stereoelectronic considerations should apply to alkylation and halogenation of enolates and enols. However, surveys of a number of studies along these lines (Toullec, 1982; Reetz, 1982) suggest that the stereoselectivity both in the abstraction and addition steps is disappointing, with the axial preference not exceeding a ratio of 5.5:1 and usually being appreciably less. The difficulty seems to lie not in the stereoelectronic principle but in the flexibility of cyclohexanones (Fraser and Champagne, 1978), which mars the clearcut distinction between C–H bonds nearly perpendicular and nearly parallel to the C=O bond. Indeed, when the molecule is rigidified, as in 2-norbornanone (Fig. 11.62) a ratio of 715:1 for abstraction of the exo over the endo proton is observed (Tidwell, 1970). This case may appear puzzling, since the exo hydrogen extends nearly parallel to the carbonyl group, whereas the endo hydrogen is nearly perpendicular. However, calculations employing the principle of least motion (Tee, Yates, et al., 1974) do lead to the prediction that the exo proton should be abstracted preferentially.

Equally as complex as the stereochemistry of α-hydrogen abstraction and its reversal is the stereochemistry of addition of nucleophiles to cyclohexanones. Both the addition of organometallics (Ashby and Laemmle, 1975) and the addition of metal hydrides (Boone and Ashby, 1979; Wigfield, 1979) have been extensively studied; equilibria in cyanohydrin formation have also been elucidated (Wheeler and Zabicky, 1958; see also pp. 770 and 880 and Eliel et al., 1965, p. 116).

In nucleophilic additions to conformationally locked cyclohexanones, either equatorial attack (to give the axial alcohol) or axial attack (to give the equatorial

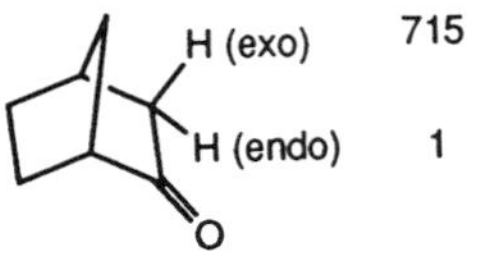

Figure 11.62. Relative rates of exo and endo deprotonation of norcamphor.

alcohol) may occur (Fig. 11.63). The observed facts are as follows: Nucleophiles of small steric requirement add to unhindered cyclohexanones from the axial side to give equatorial alcohol **B**. In contrast, bulky nucleophiles add from the equatorial side to give axial alcohols **A**; predominant equatorial attack is also observed in ketones where the axial side is screened, for example, by synaxial substituents (Fig. 11.63, **C** and **D**). Thus reduction of the sterically open 4-*tert*-butylcyclohexanone with the unhindered LiAlH$_4$ in tetrahydrofuran (THF) gives 90% equatorial alcohol, whereas corresponding reduction of 3,3,5-trimethylcy-clohexanone gives only 23% equatorial alcohol, hydride approach from the axial side being sterically hindered. The somewhat more bulky LiAlH(*t*-BuO)$_3$ still gives 90% equatorial alcohol with the former ketone but only 4–12% with the latter. To reduce 4-*tert*-butylcyclohexanone to the axial alcohol requires a very bulky hydride, such as L-Selectride^{TM}, LiBH(*sec*-Bu)$_3$, which gives 93% axial alcohol at room temperature and 96.5% at −78°C (Brown and Krishnamurthy, 1972). (3,3,5-Trimethylcyclohexanone with the same reagent gives 99.8% axial alcohol.) Even more stereoselective is lithium trisiamylborohydride, LiBH[CH$_3$CHCH(CH$_3$)$_2$]$_3$, which gives almost exclusively the axial alcohol even with the unhindered 4-*tert*-butylcyclohexanone (Krishnamurthy and Brown, 1976). To get pure equatorial alcohol (**B**, Nu = H), one equilibrates the aluminum complex of the axial alcohol or alcohol mixture using AlHCl$_2$ and excess ketone (Eliel et al., 1967).

Steric factors are also important in the reaction of cyclohexanones with organometallics. Thus the following percentages of equatorial attack (to give **A**, Fig. 11.63) on 4-*tert*-butylcyclohexanone are reported (cf. Ashby and Laemmle, 1975): HC≡CNa, 12% (i.e., 88% axial attack); CH$_3$Li, 65%; CH$_3$MgI, 53%; C$_2$H$_5$MgI, 71%; (CH$_3$)$_2$CHMgBr, 82%; and (CH$_3$)$_3$CMgCl, 100%; that is, the more bulky the reagent, the more equatorial and the less axial attack and vice versa. A high degree of equatorial attack (82–99%) is observed with organoiron reagents (Reetz and Stancher, 1993; see also Reetz et al., 1992). An interesting situation arises with (CH$_3$)$_3$Al: when the reagent is used in a 1:1 ratio to ketone, methylation is mostly equatorial (via a four-centered transition state), whereas with a 2:1 ratio it is mostly axial (via a six-membered ring transition state). With 3,3,5-trimethylcyclohexanone, all the reagents [save (CH$_3$)$_3$Al in excess] give

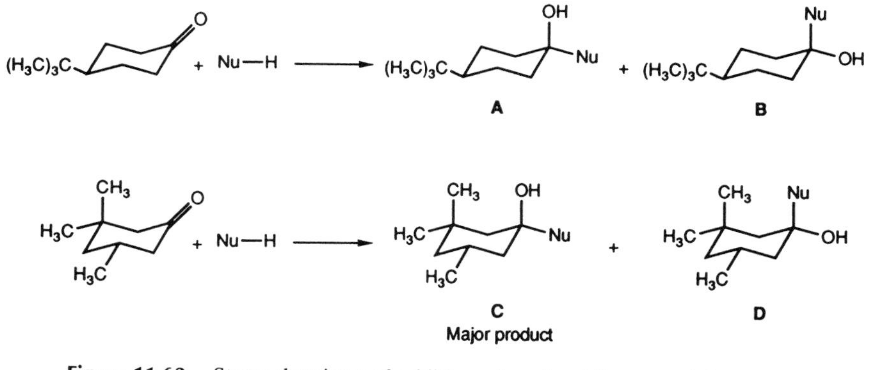

Figure 11.63. Stereochemistry of addition of nucleophiles to cyclohexanones.

exclusively equatorial attack with formation of the axial alcohol **C**. High
stereoselectivity in reactions with both hydrides and organometallics is also seen
with camphor and norcamphor (Fig. 11.64): Camphor reacts nearly exclusively
from the endo side (the exo side is screened by the geminal methyl groups), but
the opposite is true for norcamphor in which the nucleophile approaches from the
less hindered exo side (see also p. 789).

There seems to be general agreement that equatorial attack with bulky reagents or
hindered ketones is due to steric interference to approach from the axial side
("steric approach control", Dauben et al., 1956). The reason why small nucleophiles add to unhindered cyclohexanones from the axial side is less clear.
Originally, it was believed that this was due to "product development control," the
product with the bulky developing alkoxyaluminum or similar moiety in the axial
position being less stable than that where the OH, or rather O–metal, group is
equatorial. However, this hypothesis seems unlikely in as much as the transition
state in the very fast hydride reductions must come early on the reaction coordinate
and must therefore resemble starting material rather than product (Hammond
postulate: Hammond, 1955; see, however, Wigfield, 1979); also the amount of
equatorial isomer formed often exceeds that present at equilibrium. Moreover the
rates of reduction of 4-*tert*-butylcyclohexanone and 3,3,5-trimethylcyclohexanone
with various hydrides from the equatorial side (i.e., to give the axial alcohols: Fig.
11.63, **A** and **C**; Nu = H) are nearly the same (Klein et al., 1968; Eliel and Senda,
1970), despite the fact that in the case of the *tert*-butyl compound, the developing
O–Al= moiety is pushed against a synaxial hydrogen, whereas in the trimethyl
compound it is pushed against a synaxial methyl. Reasons other than product
development thus appear to be responsible for the preferential equatorial attack in
the case of unhindered ketones. Ashby and Laemmle (1975) and Wigfield (1979)
reviewed a number of explanations, such as eclipsing of the incipient incoming
nucleophile-carbonyl carbon bond with the adjacent equatorial hydrogen (Chérest,
Felkin, et al., 1968; Chérest, 1980; Wu, Houk and Paddon-Row, 1992), which
interferes with equatorial attack, and a favorable π–σ^* overlap of the carbonyl
with the axial hydrogen atoms at C(2,6), which promotes axial attack (Anh and
Eisenstein, 1977). Another potential cause is overlap of the σ^* orbital of the
forming Nu–CO bond with the orbital of the vicinal antiperiplanar hydrogen
(Cieplak, 1981). A difficulty in the explanations invoking orbital overlap lies in
gauging the exact geometry of the C=O, C(2)–H_a and C(2)–C(3) bonds and the
relative ability of C–H versus C–C σ or σ^* orbitals to participate in the postulated
interactions (cf. Rozeboom and Houk, 1982).

The steric approach – product stability control hypothesis has been revived by
Rei (1979; see also Rei, 1983; Fang, Sun, and Rei, 1986) who has proposed to put
it on a quantitative basis by equating the difference in activation energy $\Delta\Delta G^{\ddagger}$
between formation of axial and equatorial alcohols with a linear combination of
product stability (Δn) and steric strain ($\Delta\sigma$) terms: $\Delta\Delta G^{\ddagger} = a\Delta_n + b\Delta_\sigma$. For the

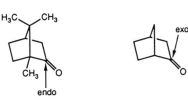

Figure 11.64. Nucleophilic approach to camphor and
norcamphor.

Camphor   Norcamphor

addition of methyllithium, only steric strain is assumed to be controlling, hence $\Delta_n = 0$ and $\Delta\Delta G^{\ddagger} = b\Delta_\sigma$. Moreover, for methyllithium addition b is set equal to 1, Δ_σ being thus defined as $\Delta\Delta G^{\ddagger}_{CH_3Li}$. The parameter Δ_n is equal to ΔG^0_{OH}, the free energy difference between the epimeric secondary cyclohexanol products. For $LiAlH_4$ reduction, $a = 1.4$ and for $LiAlH(OCH_3)_3$ reduction, $a = 0.4$; for both reductions $b = 1$ (as for CH_3Li addition). Thus, for example, for $LiAlH_4$ reduction $\Delta\Delta G^{\ddagger}_H = \Delta\Delta G^{\ddagger}_{CH_3Li} + 1.4\Delta G^0_{OH}$.

It must be mentioned here that the trajectory of the incoming nucleophile (Nu) is not perpendicular to the C=O σ bond but at an angle of 109° (Nu–C–O), in the π plane (Bürgi, Dunitz, et al., 1973; Bürgi, Lehn, et al., 1974); this matter will be treated in more detail in Chapter 12 (p. 877).

Methylenecyclohexane

The structure of methylenecyclohexane (cf. Lambert, 1989) has not been determined experimentally, but force field calculations (Anet and Yavari, 1978; Dosên-Micóvić, Allinger, et al., 1991) suggest that it is quite similar to that of cyclohexanone. The barrier to inversion, however, is considerably higher, 7.7 ± 0.5 kcal mol^{-1} (32.2 ± 2.1 kJ mol^{-1}) (Jensen and Beck, 1968; Gerig, 1968) 8.4 kcal mol^{-1} (35.1 kJ mol^{-1}) (Gerig and Rimmerman, 1970) or 9.0 ± 0.6 kcal mol^{-1} (37.7 ± 2.5 kJ mol^{-1}) (Lessard, Saunders, et al., 1977). About two-thirds of this barrier derives from angle strain in the transition state (Bernard, St.-Jacques, et al., 1974; Anet and Yavari, 1978). This finding lends credence to the hypothesis that the much higher inversion barrier in methylenecyclohexane compared to cyclohexanone results from the greater torsional potential of the X=C–C bonds when X = CH_2 compared to X = O. A similar situation is seen in the isobutylene versus acetone barriers [2.2 kcal mol^{-1} (9.2 kJ mol^{-1}) vs. 0.78 kcal mol^{-1} (3.26 kJ mol^{-1})]. Methylenecyclohexane is thermodynamically unstable with respect to 1-methylcyclohexene; the equilibrium, which is easily established in acid, lies far on the side of the endocyclic alkene. The enthalpy difference of 1.71–1.74 kcal mol^{-1} (7.2–7.3 kJ mol^{-1}) (Yursha et al., 1974; Peereboom, Baas, et al., 1982) is somewhat larger than that between 2-methyl-1-pentene and 2-methyl-2-pentene (1.34 kcal mol^{-1}, 5.6 kJ mol^{-1}); in addition, the entropy is higher for methylcyclohexene so that the free energy advantage of the latter at 25°C is 2.84–2.87 kcal mol^{-1} (11.9–12.0 kJ mol^{-1}). The conformational energy for a methyl group in the 3-position of methylenecyclohexane, 0.8 kcal mol^{-1} (3.3 kJ mol^{-1}) (Lambert and Clikeman, 1976), on the other hand, is substantially less, not only than the value in methylcyclohexane (1.74 kcal mol^{-1}, 7.28 kJ mol^{-1}), but also than the value in 3-methylcyclohexanone (1.36 kcal mol^{-1}, 5.69 kJ mol^{-1}); again it is not obvious why. For polar substituents in the 3-position, however, the conformational energies tend to be larger than in corresponding cyclohexanes in nonpolar solvents but smaller in polar solvents (for a summary, see Lambert, 1989). Evidently, there is some polar interaction even though the dipole moment of methylenecyclohexane is only 0.62 D (Li, 1984; see, however, Dosên-Micóvić, Allinger, et al., 1991).

The conformational energy of a methyl group at C(2) in methylenecyclohexane is also relatively small, 1.0 kcal mol^{-1} (4.2 kJ mol^{-1}) (Lessard, Saunders, et al., 1977). Polar substituents at C(2), such as CH_3O, in this case actually prefer the axial position by 0.4 kcal mol^{-1} (1.7 kJ mol^{-1}). The reason may, in part, be similar to that shown for 3-substituted cyclohexenes in Figure 11.56 (hyperconjugative orbital overlap). However, there is also a steric factor that is quite important in cyclohexenes and very important in methylenecyclohexanes, especially when the terminal substituents on the alkene (R in Fig. 11.65) are larger than hydrogen. This factor is called $A^{(1,2)}$ strain in a 1,6-disubstituted cyclohexene and $A^{(1,3)}$ strain in a 2-substituted methylenecyclohexane (Johnson, 1968; Lambert, 1989); the two types of strain are shown in Figure 11.65. In both instances the strain leads to a disfavoring of the pseudoequatorial position.

The $A^{(1,3)}$ strain is of particular interest in enamines and other ketone derivatives, where it has found useful synthetic applications. Examples are the synthesis of *trans*-1,3-dimethylcyclohexane (Johnson and Whitehead, 1964) and *trans*-2,6-dimethylcyclohexanone (Corey and Enders, 1976) shown in Figure 11.66. The pyrrolidine enamine shows both regio- and stereoselectivity in the following way: In the enamine itself, after alkylation the double bond will be on the unsubstituted side in order to avoid a cis interaction between the 2-alkyl group and the enaminic nitrogen. Similarly, $A^{(1,3)}$ strain forces the 2-alkyl substituent, whether from the incoming nucleophile (CH_3I) or already present in the ketone, into the axial position. Of course it requires a mild method of hydrolysis, as in the case of the *trans*-2,6-dimethylcyclohexanone or avoidance of hydrolysis altogether, as in the case of *trans*-1,3-dimethylcyclohexane to prevent epimerization of the initial product to the more stable diequatorial epimer. In the case of the alkylation of the 2-methylcyclohexanone N',N'-dimethylhydrazone, avoidance of $A^{(1,3)}$ strain leads to formation of the anti or E isomer of the hydrazone (cf. Chapter 12), which then (again to avoid $A^{(1,3)}$ strain) must be alkylated axially. (Geminal dialkylation is presumably shunned because the tertiary CH_3CH site is kinetically disfavored in proton abstraction because of lesser acidity.) In the third example, methyl groups at C(2) and C(6) were both introduced axially, even though this causes a severe synaxial interaction; unfortunately partial epimerization during hydrolysis leads to a mixture of only 60% 2a,6a, and 40% 2e,6a dimethyl ketone. One surprising and as yet not well-understood feature is that the first alkylation even in the imine occurs syn to the N-alkyl substituent (Fraser and Dhawan, 1976).

The $A^{(1,3)}$ strain also seems to be at the root of the stereoselective protonation (to give cis product) of the anions of 2-phenyl-1-acyl-, -1-aroyl- and -1-nitrocyclohexane (e.g. Fig. 11.67) observed by Zimmerman (cf. Zimmerman, 1963). While these reactions were originally interpreted as preferentially equatorial protonations of enolates (or aci forms) with equatorial phenyl groups, it has been shown (cf.

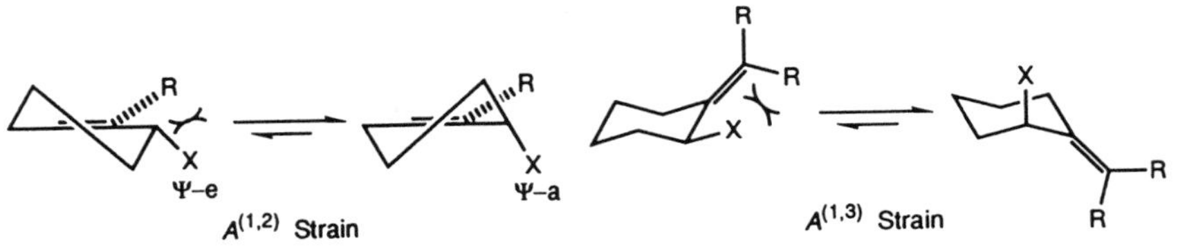

Figure 11.65. Examples of $A^{(1,2)}$ and $A^{(1,3)}$ strain.

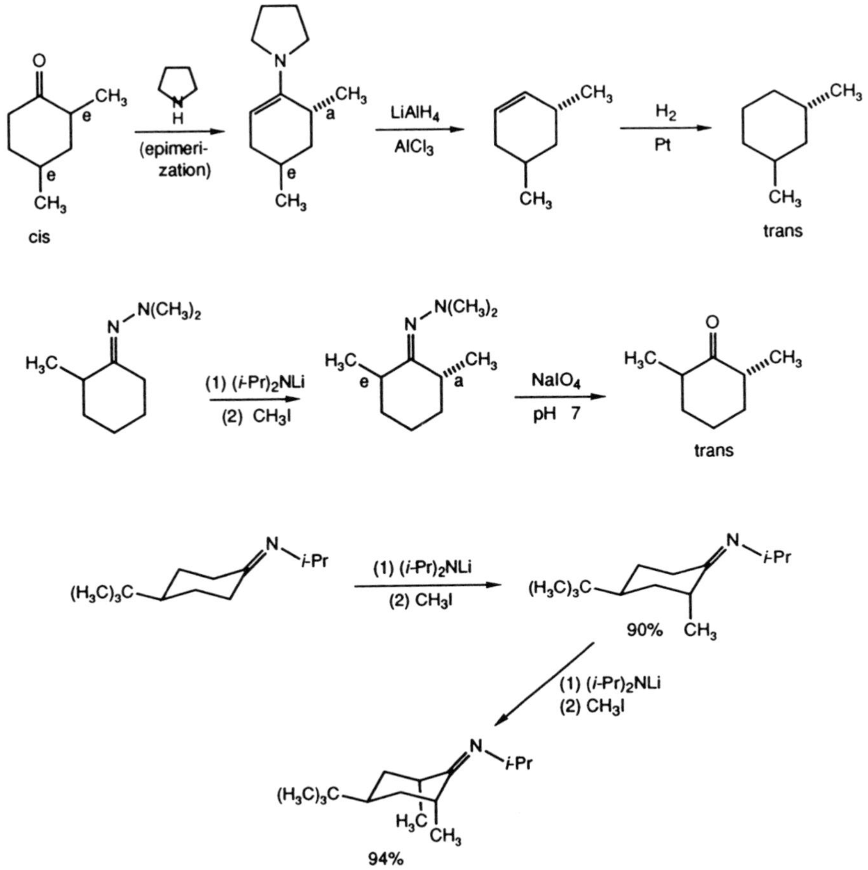

Figure 11.66. Synthetic applications of $A^{(1,3)}$ strain.

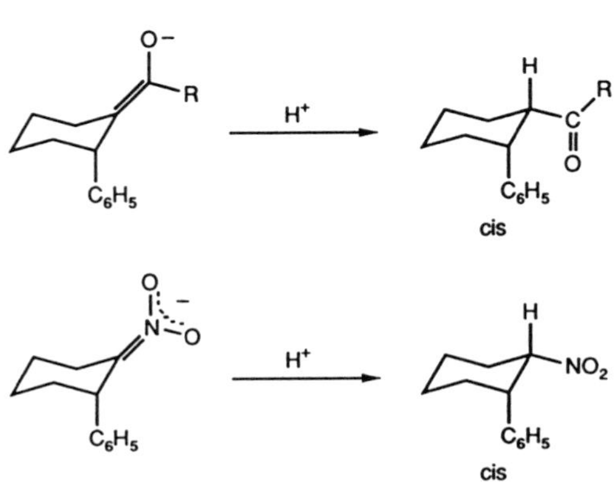

Figure 11.67. Protonation of 2-phenyl substituted cyclohexyl anions with potential $A^{(1,3)}$ strain.

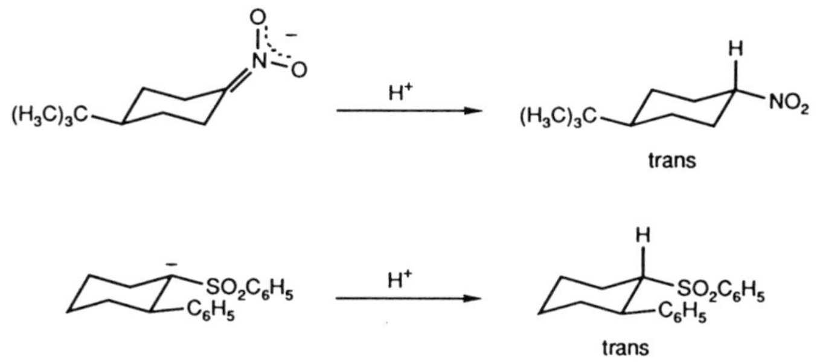

Figure 11.68. Protonation of anions with no potential $A^{(1,3)}$ strain.

Johnson, 1968) that, in fact, they involve axial protonation of the conformer with an axial 2-phenyl substituent, the equatorial conformation of this substituent being destabilized by $A^{(1,3)}$ strain. Protonation occurs from the side away from the 2-phenyl substituent. When there is no such substituent, as in the protonation of the anion of 1-nitro-4-*tert*-butylcyclohexane (Bordwell and Vestling, 1967) or when the carbanion is not delocalized, as in the anion of 2-phenylcyclohexyl phenyl sulfone the more stable diequatorial trans product is formed preferentially (Fig. 11.68).

An interesting application of $A^{(1,2)}$ strain is seen in the synthesis of solenopsin A (Fig. 11.69; Maruoka, Yamamoto, et al., 1983). Ordinary hydride reduction of the precursor imine (a tetrahydropyridine) gives largely the undesired cis (diequatorial) product. However, when $LiAlH_4$ reduction is carried out in the presence of an excess of $Al(CH_3)_3$, complexation of this Lewis acid to the ring nitrogen engenders $A^{(1,2)}$ strain and presumably forces the *n*-undecyl group into the pseudoaxial position; reduction of the C=N bond leading to equatorial CH_3 as before then produces axial–equatorial, that is, trans stereochemistry.

g. Six-Membered Saturated Heterocycles

In a book on conformational analysis published in 1965 (Eliel et al.), the conformational analysis of saturated heterocycles was dealt with in 12 pages. The thinking at that time was that the conformational analysis of saturated heterocycles was, with some minor modifications, very similar to that of cyclohexane. In the intervening years, however, attention has been focused on the differences and

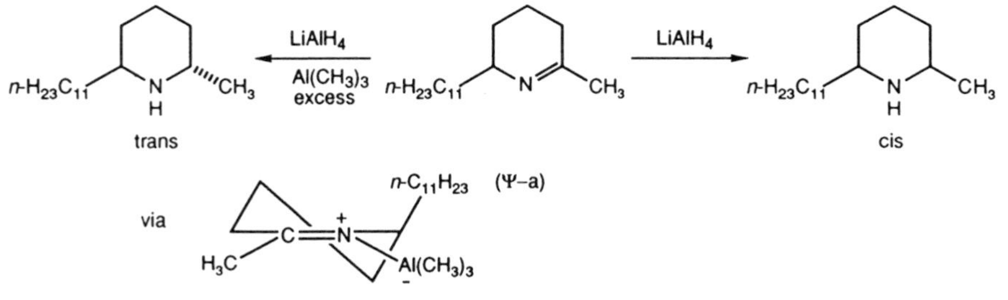

Figure 11.69. Synthesis of solenopsin A.

TABLE 11.18. Inversion Barriers in $C_5H_{10}X$.

X:	CH$_2$	O	S	SO	SO$_2$	Se	SeO	SeO$_2$	Te	NH	NCH$_3$
$\Delta G^{\ddagger}$ (kcal mol^{-1})	10.25	10.3	9.4	10.1	10.3	8.2	8.3	6.7	7.3	10.1	11.9
$\Delta G^{\ddagger}$ (kJ mol^{-1})	42.9	43.1	39.3	42.3	43.1	34.3	34.7	28.0	30.5	42.3	49.8
t(°C); ref.	-60^a	-61^b	-81^b	-70^b	-63^b	-105^b	-102^b	-133^b	-119^b	-63^c	-29^c

[a] See p. 688.
[b] Lambert et al. (1973).
[c] Lambert et al. (1967). $\Delta G^{\ddagger}$ values calculated from chemical shifts and coalescence temperatures.

by 1980 the subject was worthy of a 152-page monograph (Riddell, 1980; see also Armarego, 1977; Lambert and Featherman, 1975; and for nitrogen heterocycles, Delpuech, 1992). Only a brief account of the subject can be given here.

The simple saturated heterocycles all exist as chairs in their most stable conformations, with the inversion barriers given in Table 11.18.

The values for the barrier $\Delta G^{\ddagger}$ are given at the coalescence temperature. Since $\Delta S^{\ddagger}$ is generally positive for the inversion of a six-membered chair (cf. p. 688), $\Delta H^{\ddagger}$ is likely to be larger than $\Delta G^{\ddagger}_{coal}$ by about 1 kcal mol^{-1} (4.2 kJ mol^{-1}) and $\Delta G^{\ddagger}_{25}$ is smaller than $\Delta G^{\ddagger}_{coal}$ by about 0.5 kcal mol^{-1} (2.1 kJ mol^{-1}); the exact differences depend, of course, on $\Delta S^{\ddagger}$ and T_{coal}.

It is seen from Table 11.18 that the inversion barriers for cyclohexane, piperidine, and oxane (tetrahydropyran) are essentially identical; as one goes down the periodic table (O, S, Se, Te) the barriers become lower, probably because the torsional potentials diminish (Lambert et al., 1973). Substitution on the heteroatoms seems to raise the barrier slightly: $S < SO < SO_2$; $NH < NCH_3$; $Se < SeO$ (SeO_2 is out of line), perhaps because the torsional barriers increase. Chair-twist energy differences seem not to be known for the simplest (mono-hetero) systems; a comparison between cyclohexane and the dihetero systems 1,3-dioxane and 1,3-dithiane (Eliel, 1970; see also Kellie and Riddell, 1974) is given in Table 11.19. The twist form in 1,3-dioxane is of higher energy, relative to the chair, than that in cyclohexane, but the opposite is true for 1,3-dithiane. These differences have been explained in terms of molecular dimensions; since the bond lengths are $C-O < C-C < C-S$, steric interactions of carbon and hydrogen atoms across the ring in the twist form should be most serious in 1,3-dioxane and least so in 1,3-dithiane. A contributing factor may be the somewhat lesser torsional (Pitzer) potential in 1,3-dithiane. A ring with four sulfur atoms, 3,3,6,6-tetramethyl-1,2,4,5-tetrathiane (Fig. 7.37), in fact exists in a readily measurable chair–twist equilibrium (Bushweller et al., 1970).

TABLE 11.19. Chair–Twist (c–t) Energy and Entropy Differences in

Compound	X	ΔG^0_{c-t} [a]	ΔH^0_{c-t} [a]	ΔS^0_{c-t} [b]
Cyclohexane[c]	CH$_2$	4.9 (20.5)	5.9 (24.7)	3.5 (14.6)
1,3-Dioxane	O	5.7 (23.8)	7.1 (29.7)[d]	4.8 (20.1)
1,3-Dithiane[e]	S	2.9 (12.1)	4.3 (17.9)	4.7 (19.5)

[a] In kilocalories per mol (kcal mol^{-1}) at 25°C; values in parentheses are in kilojoules per mole (kJ mol^{-1}).
[b] In gibbs or calories per degree per mole (cal deg^{-1} mol^{-1}; values in parentheses are in joules per degree per mole (J deg^{-1} mol^{-1}).
[c] See also Kellie and Riddell (1974).
[d] The true value may be even higher; cf. Kellie and Riddell (1974).
[e] Pihlaja and Nikander (1977).

Conformational equilibria of methyl groups in six-membered saturated heterocycles are tabulated in Table 11.20, which lists the ground-state free energy difference between a ring with an axial methyl substituent at the position indicated and one with the corresponding equatorial substituent.

Little variation in $-\Delta G^0$ between different heterocycles and cyclohexane is seen at the 4 position; equatorial and axial methyl groups at C(4) are evidently "cyclohexane-like" except for possible small variations in the exact ring shape, as one would expect. Rather larger variations are seen at C(2) and C(3), for reasons that will become obvious from Figure 11.70. In the 3-substituted systems, one of the CH_3/H synaxial interactions (in axially substituted cyclohexanes) is replaced by a $CH_3/$: (lone pair) interaction that is known (see below) to be less severe, leading to a drop in $-\Delta G^0$, albeit not a very large one (the difference is especially small for piperidine and N-methylpiperidine). The exception occurs in the N-methylpiperidine salt where there is actually a substantial *rise* in $-\Delta G^0$. One might, perhaps, explain this by the replacement of the lone pair by hydrogen, but this explanation appears inadequate, since the $-\Delta G^0$ value in the N-3-dimethyl-piperidinium salt is even larger than in methylcyclohexane. A contributory factor here, most probably, is solvation that leads to a further "swelling" of the axial proton, just as it contributed (Table 11.7) to the larger size of ammonium and carboxylate ions compared to the uncharged amine and carboxylic acid functions.

In the 2-substituted systems, the nitrogen and especially oxygen heterocycles display a larger $-\Delta G^0$ for methyl than does cyclohexane, whereas the value in 2-methylthiane is smaller. This, like the earlier mentioned situation with respect to chair–twist differences (Table 11.19), appears to be a consequence of changes in molecular dimensions: Since bond lengths are in the order $C–O < C–N <$

TABLE 11.20. Conformational Free Energies[a] $(-\Delta G^0)$ of Methyl Substituents in Saturated Heterocycles

Group X	Position				Reference
	1	2	3	4	
CH_2	1.74 (7.28)	1.74 (7.28)	1.74 (7.28)	1.74 (7.28)	b
O		2.86 (12.0)	1.43 (5.98)	1.95 (8.16)	c
S	$[0.28\,(1.17)]^d$	1.42 (5.94)	1.40 (5.86)	1.80 (7.53)	e
NH	$[0.36\,(1.51)]^f$	2.5 (10.5)	1.6 (6.7)	1.9 (7.95)	g
NCH_3	$[3.0\ \ (12.6)]^h$	1.7 (7.1)	1.6 (6.7)	1.8 (7.5)	g
NCH_3H^+	2.1 (8.8)	1.4 (5.9)	2.2 (9.2)	1.6 (6.7)	g

[a] In kilocalories per mole (kcal mol^{-1}); values in parentheses are in kilojoules per mole (kJ mol^{-1}).
[b] Booth and Everett (1980a).
[c] Eliel et al. (1982).
[d] S-Methylsulfonium salt; Eliel and Willer (1977).
[e] Willer and Eliel (1977).
[f] This is the value for the N–H (equatorial vs. axial H) equilibrium: Anet and Yavari (1977); Vierhapper and Eliel (1979).
[g] Eliel et al. (1980).
[h] Crowley, Robinson et al. (1977).

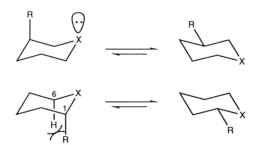

Figure 11.70. Conformational inversion in 3- and 2-substituted heterocyclohexanes.

C–C < C–S, the distance between an axial methyl at C(2) and the synaxial hydrogen at C(6) increases as one passes from oxane to piperidine to cyclohexane to thiane and $-\Delta G^0_{2\text{-}CH_3}$ decreases accordingly (cf. Fig. 11.70). The 2-methyl groups in N-methylpiperidine and its salt are clearly exceptions from the norm, for reasons not yet understood; there appears to be greater steric interaction between N–CH$_3$(e) and equatorial 2-methyl than between N–CH$_3$(e) and axial 2-methyl.

The N-methyl group in N-methylpiperidine has a very large $-\Delta G^0$ value (Crowley, Robinson, et al., 1977). Evidently, the fact that the barrier to nitrogen inversion in N-methylpiperidine is quite low [6.0 kcal mol^{-1} (25.1 kJ mol^{-1}) from the axial side; 8.7 kcal mol^{-1} (36.4 kJ mol^{-1}) from the equatorial side: Katritzky et al., 1979] does *not* reflect itself in a correspondingly low bending force potential of the axial N-methyl group. Rather, this group is quite "stiff" and unable to bend away from the rather close-by (because of the short N–C bond) synaxial hydrogen atoms at C(3,5). The 3.0 kcal mol^{-1} (12.6 kJ mol^{-1}) value was measured in toluene (a slightly higher value obtains in the gas phase), whereas in chloroform the value is lowered to 2.4 kcal mol^{-1} (10.0 kJ mol^{-1}) (Crowley et al., 1977) and upon protonation (Table 11.20) it is further lowered to 2.1 kcal mol^{-1} (8.8 kJ mol^{-1}).

> Changes in configurational (or conformational) equilibrium upon protonation of nitrogenous bases, illustrated here by the cases of N-methyl- and 3-methyl-piperidine, are found repeatedly in the literature (e.g., Eliel et al., 1984). Axial protonation is impeded not only by the synaxial hydrogen atoms (or larger groups) but also by the difficulty of accommodating solvation or pairing with the counterion on the axial side. Hydrogen bonding to solvent, which leads to a lowering of free energy, is preferred from the equatorial side and thus leads to a shift of the N substituent to the axial position; since hydrogen bonding is enthalpically favorable but entropically unfavorable, the effect is particularly marked at low temperatures (Manoharan, Eliel, and Carroll, 1983).
>
> In piperidine itself the N-inversion barrier is 6.1 kcal mol^{-1} (25.5 kJ mol^{-1}) (Anet and Yavari, 1977); values for the preference of equatorial over axial NH range from 0.36 (Anet and Yavari, 1977) to 0.74 kcal mol^{-1} (1.51–3.10 kJ mol^{-1}) (Parkin, Costain, et al., 1981; see also Scott, 1971). That equatorial NH predominates, a fact that was once in doubt (cf. Lambert and Featherman, 1975), seems, however, to be established unequivocally (cf. Scott, 1971; Blackburne, Katritzky, et al., 1975; Anet and Yavari, 1977; Vierhapper and Eliel, 1979; Parkin, Costain, et al., 1981).

We shall deal only briefly with rings having more than one hetero atom. Study of the 1,3-dioxane and 1,3-dithiane systems (Fig. 11.71), preceded that of the simpler monoheterosubstituted species by several years, for whereas the oxane, thiane, and piperidine systems in general have to be studied by low-temperature ^{13}C NMR (cf. Eliel and Pietrusiewicz, 1979; but see below), the conformational equilibria in the disubstituted systems shown in Figure 11.71 could be approached through configurational changes by acid-catalyzed equilibration, as shown. Since these equilibria, in general, present no surprises and have been extensively reviewed (Eliel, 1972; Gittins et al., 1974; Riddell, 1980) we shall refer here to only one case: 2-methyl-5-*tert*-butyl-1,3-dioxane [Fig. 11.71, X = O, R = CH$_3$, R' = C(CH$_3$)$_3$]. This case is remarkable because NMR spectra unequivocally show (Eliel and Knoeber, 1966, 1968; Jones, Grant, et al., 1971) that the preferred conformer has the 2-methyl group in the equatorial position and the 5-*tert*-butyl one in the axial; moreover, $-\Delta G^0$ is only 1.4 kcal mol^{-1} (5.9 kJ mol^{-1}) (see also Riddell and Robinson, 1967). Clearly, the steric compression of the axial 5-*tert*-butyl group by the 1,3-oxygen atoms with their lone pairs is very small compared to the compression by synaxial hydrogen atoms in axial *tert*-butylcyclohexane (cf. Table 11.7).

The *N*-substituted 1,3-diazane system (Fig. 11.71, X = N–CH$_3$) is of interest because cis–trans equilibrium is established spontaneously at room temperature though equilibration is slow on the NMR time scale: Both isomers can be seen spectrally (Kopp, 1972). Similar observations have been made in oxazanes (Bernath et al., 1984)

In cyclohexyl systems (cf. Table 11.7) the equatorial position of a substituent is almost invariably preferred, the lowest observed conformational energy value (for HgX) being about zero. As we shall see presently, this is not necessarily true in heterocyclic systems where there are many cases of axial preference. A case in point is thiane sulfoxide (Fig. 11.72) in which axial SO is preferred by 0.18 kcal mol^{-1} (0.73 kJ mol^{-1}) (Lambert and Keske, 1966); even in the corresponding sulfonium salt the equatorial preference is only 0.28 kcal mol^{-1} (1.15 kJ mol^{-1}) (cf. Table 11.20). It has been alleged (Allinger and Kao, 1976) that the axial preference in SO is due to a destabilization of the equatorial conformer by four O/H gauche interactions (there are only two such interactions plus two apparently lesser synaxial interactions in the axial conformer, cf. Fig. 11.72) but this may be merely a consequence of the force field used in the calculation; with a smaller van der Waals radius for hydrogen the outcome might have been an attractive O/H (synaxial) van der Waals interaction in the axial conformation (cf. Fig. 2.23).

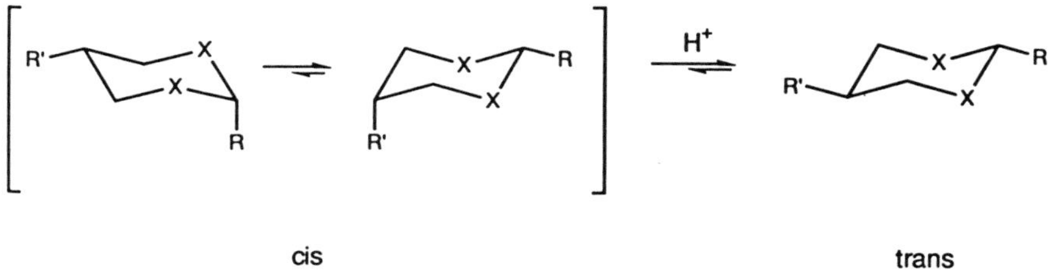

cis   trans

Figure 11.71.   Equilibria in 1,3-dioxanes and 1,3-dithianes, where X = O or S.

Figure 11.72. Thiane sulfoxide equilibrium; $\Delta G^0 = +0.18$ kcal mol^{-1} ($+0.75$ kJ mol^{-1}).

The most important conformational differences between heterocyclic and carbocyclic systems are due to polar factors. In cyclohexanes, polar factors only come into play when there are at least *two* polar substituents in the ring (e.g., see Fig. 11.35). In heterocyclic systems, on the other hand, at least one polar group (and sometimes more than one) is already in the ring and the parent compounds usually have dipole moments (except when, as in the case of 1,4-dioxane, there is a compensation of dipoles). Thus a single polar exocyclic substituent will give rise to dipolar interactions (either attractive or repulsive), which depend on the location of the substituent relative to the endocyclic heteroatom or heteroatoms, as well as on the nature of these atoms. However, as we shall see presently, dipolar interactions are not the only ones operative; other, stereoelectronic, factors are often called into play as well.

Table 11.21 summarizes $-\Delta G^0$ values (Kaloustian, Eliel, et al., 1976; cf. Riddell, 1980) at the 5-position in 1,3-dioxane, including the effect of changing solvent. It is clear that in a number of cases the axial isomer is preferred, whereas in other cases (Cl, Br, or I) the equatorial preference is even larger than in correspondingly substituted cyclohexanes.

To interpret the purely polar factors affecting the conformational preferences in Table 11.21, we start with Eq. 2.1 (p. 33) and draw attention to the last two energy terms in that equation: the electrostatic term V_E and the solvation term V_S. The energy V_E may be either attractive or repulsive, depending on whether the dipoles within the molecules are aligned in antiparallel or parallel orientation. Moreover, in view of Coulomb's law, this term will diminish in absolute value as the solvent dielectric constant increases. The solvation energy V_S is always stabilizing; it increases as the solvent dielectric increases. The overall effect of solvation in the case of a repulsive V_E term is clear: Since V_E decreases with overall solvent dielectric and the $-V_S$ term increases, a conformation with a repulsive electrostatic interaction will become more stable (or less unstable) as the dielectric constant of the solvent increases. The effect is not as clear when the electrostatic interaction is attractive; increase of solvent dielectric will decrease the attraction (negative Coulombic V_E term) but will also increase solvation ($-V_S$ term); the two effects oppose each other and the overall result may be either stabilization or destabilization, but the effect is apt not to be large.

The direction of the resultant ring as well as substituent dipoles is indicated in the structure at the head of Table 11.21, where X is an electronegative substituent. The expectation then is that the axial conformation will be destabilized and the equatorial conformation slightly stabilized by dipole-dipole interaction. An increase in solvent dielectric should therefore shift the equilibrium toward the axial side. Where data are available in two solvents (e.g., the low-dielectric diethyl ether and high-dielectric acetonitrile), the prediction is borne out in all cases but the last two, which are complicated by hydrogen bonding. Generally, $-\Delta G^0$ becomes either more negative or less positive in high-dielectric solvents, that is, the equilibrium

TABLE 11.21. Conformational Equilibria in 1,3-Dioxanes with Polar Substituents at C(5).

X	Solvent	ε^a	$-\Delta G^{0\ b}$ (kcal mol^{-1})	(kJ mol^{-1})	$-\Delta G^{0\ b,c}_{ref}$ (kcal mol^{-1})	(kJ mol^{-1})
F	Et$_2$O	4.34	−0.62	−2.59	0.36	1.51
	CH$_2$CN	37.5	−1.23	−5.15		
Cl	Et$_2$O	4.34	1.20	5.02	0.55	2.30
	CH$_3$CN	37.5	0.25	1.05		
Br	Et$_2$O	4.34	1.44	6.02	0.49	2.05
	CH$_3$CN	37.5	0.68	2.85		
I	CHCl$_3$	4.81	1.43	5.98	0.49	2.05
OCH$_3$	Et$_2$O	4.34	0.83	3.47	0.6	2.5
	CH$_3$CN	37.5	−0.01	−0.04		
CN	Et$_2$O	4.34	0.21	0.88	0.20	0.84
	CH$_3$CN	37.5	−0.55	−2.30		
CO$_3$CH$_3$	Et$_2$O	4.34	0.82	3.43	1.27	5.31
	CH$_3$CN	37.5	0.22	0.92		
NO$_2$	CCl$_4$	2.24	−0.63	−2.64	1.14	4.77
	CH$_2$Cl$_2$	9.08	−0.89	−3.72		
SCH$_3$	C$_6$H$_{12}$	2.02	1.82	7.61	1.04	4.35
	CH$_3$CN	37.5	1.13	4.73		
OSCH$_3$	C$_6$H$_6$	2.28	−0.74	−3.10	1.20	5.02
O$_2$SCH$_3$	C$_6$H$_6$	2.28	−1.07	−4.48	2.50	10.5
S(CH$_3$)$_2^+$	TFA^d	8.4	−2.0	−8.4	e	e
SC(CH$_3$)$_3$	CHCl$_3$	4.81	1.90	7.95^f	e	e
OSC(CH$_3$)$_3$	CHCl$_3$	4.81	−0.10	−0.42^f	e	e
O$_2$SC(CH$_3$)$_3$	CHCl$_3$	4.81	−1.19	−4.98^f	e	e
N(CH$_3$)$_3^+$	HCO$_2$H	58.5	−1.9	−7.95	e	e
OH	C$_6$H$_{12}$	2.02	−0.92	−3.85	0.60	2.51
	(CH$_3$OCH$_2$)$_2$	7.2	−0.27	−1.13	0.74	3.10^g
CH$_2$OH	CCl$_4$	2.24	−0.27	−1.13	1.76	7.36^h
	(CH$_3$OCH$_2$)$_2$	7.2	0.01	0.04		

^{a} Dielectric constant (ε) of solvent at 20 or 25°C.
^{b} Standard free energy change for the process shown at top of this table.
^{c} Reference values for cyclohexyl–X; from Table 11.7 unless otherwise indicated.
^{d} Trifluoroacetic acid.
^{e} Not available.
^{f} Juaristi et al. (1987b).
^{g} Eliel and Gilbert (1969).
^{h} Kitching et al. (1981).

shifts from the equatorial toward the axial conformer. The absolute values of $-\Delta G^0$ are not so easy to interpret, however. Since unfavorable polar factors in the axial conformer come on top of unfavorable steric factors, one would expect the equatorial conformer to be favored, and probably more so than in the reference cyclohexyl compound. [There is a slight uncertainty in that we have already noted the steric preference of nonpolar substituents to be less at C(5) in 1,3-dioxane than in cyclohexane.] This prediction is borne out for Cl, Br, I, OCH$_3$, and SCH$_3$, and may also hold for CN and CO$_2$CH$_3$ if one allows for the diminished steric factor in the dioxane derivative. However, the prediction fails for OH, CH$_2$OH, F, NO$_2$,

CH_3SO, CH_3SO_2, $(CH_3)_2S^+$, and $N(CH_3)_3^+$. In all these cases the axial conformer is the preferred one (in the case of CH_2OH in glyme, $\Delta G^0 \approx 0$).

The situation for OH is easy to account for. The axial isomer shows intramolecular hydrogen bonding in carbon tetrachloride solution; interestingly, this seems to be one of the rare cases of a "bifurcated" hydrogen bond in solution (Jochims and Kobayashi, 1976), that is, the O–H bond points not to one or the other of the ring oxygen atoms, but rests midway in between the two (i.e., in the bisector plane of the chair-shaped ring). Such bonding leads to stabilization of the axial conformer. In a solvent, such as 1,2-dimethoxyethane, which is capable, as an acceptor, of forming an *inter*molecular hydrogen bond with the solute, and thereby partially disrupting the intramolecular hydrogen bond of the latter, the axial preference of the OH groups is diminished. At first sight it might appear that analogous arguments would apply to CH_2OH. Indeed it is likely that hydrogen-bonding to a solvent accounts for the faint equatorial preference of the group in dimethoxyethane; however, the axial preference in carbon tetrachloride is not due to intramolecular hydrogen bonding, since IR studies (Eliel and Banks, 1972) show that no such bonding occurs.

An interesting effect of *intra*- versus *inter*molecular hydrogen bonding is seen in *cis*-3-hydroxythiane oxide (Fig. 11.73). In concentrated solution (2.8 M) in a nonpolar solvent (CD_2Cl_2) the equilibrium lies largely on the side of **E** ($\Delta G^0 = -1.0$ kcal mol^{-1}, -4.2 kJ mol^{-1}), which is favored both sterically and by *inter*molecular hydrogen bonding. But in dilute solution (2.3×10^{-3} M) *inter*molecular hydrogen bonding is no longer favorable and *intra*molecular hydrogen bonding now leads to a favoring of conformer **A** by more than 1.3 kcal mol^{-1} (5.4 kJ mol^{-1}) (Brunet and Eliel, 1986).

Either "reversed" dipole interaction (leading to attraction in the axial conformation) or an attraction of opposite charges (see below) accounts for the strong axial preference of $\overset{+}{S}(CH_3)_2$ and $\overset{+}{N}(CH_3)_3$, but in the case of NO_2, $SOCH_3$, SO_2CH_3, and CH_2OH, all electron-withdrawing groups, the better interpretation is in terms of opposite point charges. Thus in the sulfoxide (Fig. 11.74), the dominant factor appears to be $S^+ \cdots {}^-O$(ring) attraction in the axial conformer. The $S^+ \cdots C(2)^+$ and $O^- \cdots O^-$ repulsions are less important because of the greater distance involved and the $S^+ \cdots {}^-O$(ring) attraction in the equatorial is less important than in the axial conformer, again because of the greater distance. Effects of this type (point-charge interactions) will explain all the remaining cases but the fluorine one.

An alternative explanation of the axial preference of $SOCH_3$, SO_2CH_3, and NO_2 in terms of interaction of the n orbitals on oxygen with the empty d orbitals on sulfur or the π^* orbitals of the nitro group has been suggested (Eliel, 1972) but does not appear necessary. The only substituent whose axial preference requires a stereoelectronic explanation is fluorine. In this case it would seem that the salient factor is an attractive O/F gauche effect similar to that encountered in XCH_2CH_2X (X = F or OCH_3), where the gauche conformation is more stable than the anti (Section 10-1.b). Such an effect may also account for the faint axial preference of

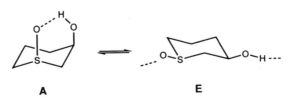

A   **E**

Figure 11.73. Conformational equilibrium in 3-hydroxythiane oxide.

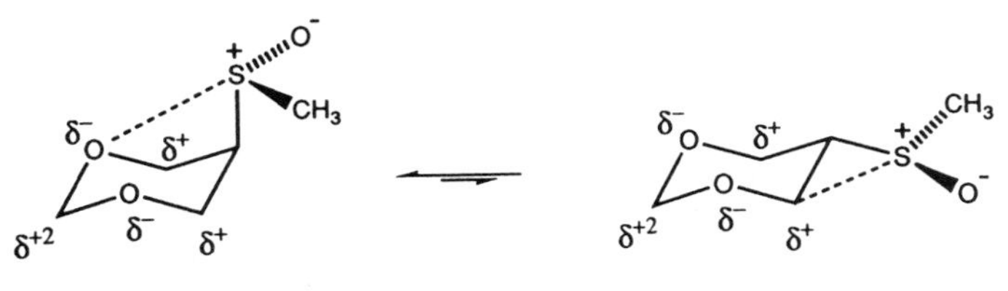

Figure 11.74. Point-charge interactions.

CH$_3$O in 5-methoxy-1,3-dioxane (Table 11.21) in the high-dielectric acetonitrile solvent, where dipole repulsion is largely abolished. The gauche effect (Section 10-1.b) can be repulsive as well as attractive (Zefirov, 1977; Eliel and Juaristi, 1979). The repulsive effect comes to the fore with elements in the lower part of the periodic system where the repulsive interaction of the unshared electron pairs (coming on top of a sizeable normal van der Waals repulsion of the large nuclei) overwhelms the electron–nuclear attraction. This effect is seen for S/O as well as S/S interactions and may be in part responsible for the large equatorial preference for SCH$_3$ shown in Table 11.21. (Even in acetonitrile, this preference is greater than in methylthiocyclohexane, even though on pure steric grounds one might have expected it to be less.)

Perhaps the most important polar factor in the conformational analysis of saturated heterocyclic systems is the anomeric effect (cf. Kirby, 1983; Deslongchamps, 1983; Juaristi and Cuevas, 1992; Graczyk and Mikołajczyk, 1994; Thatcher, 1993). Since recognition of this effect has its origin in carbohydrate chemistry, a brief excursion into the stereochemistry of the aldopyranoses is in order here (see also Stoddart, 1971; Eliel et al., 1965, Chapter 6).

Representative stereoformulas of aldohexoses were shown on Fig. 3.23. The sugars were depicted as open-chain polyhydroxyaldehydes which are, however, in equilibrium with the cyclic hemiacetals shown for D-glucose in Figure 11.75, equilibrium being largely on the side of the latter. And, as also shown in Figure 11.75, the actual shape of the hemiacetals is that of an (oxane) chair, the more stable of the two chair forms being that with the CH$_2$OH group (and, as it

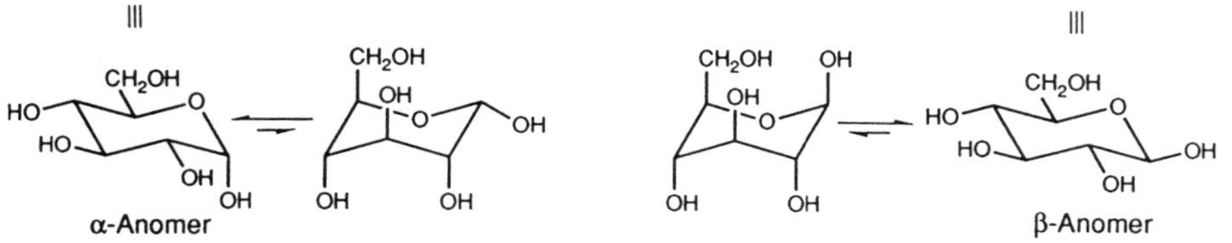

Figure 11.75. Open-chain and various cyclic formulas for D-glucose.

happens, also the hydroxyl groups) equatorial. While it is recognized that cyclization can also lead to five-membered (THF) rings, as in the furanose forms of sugars, only the pyranose forms (six membered) will be discussed here.

We have seen earlier that the aldohexoses, having four chiral centers of the CHOH type, exist as 2^4 or 16 stereoisomers (8 diastereomeric pairs of enantiomers). While, in the modern definition of the term, any pair of diastereomeric aldohexoses differing only in the configuration of one chiral center may be called a pair of epimers, in classic sugar chemistry the term "epimers" referred to pairs of sugars specifically differing in configuration at C(2), such as glucose and mannose (Fig. 3.23). Contemplation of Figure 11.75 indicates that cyclization to a six-membered ring (hemiacetal formation) in any one of the 16 stereoisomeric aldohexoses will lead to two possible stereoisomeric pyranoses differing now in configuration at C(1). Such diastereomers are called "anomers" and C(1) is sometimes called the "anomeric carbon" or "anomeric center." Since equilibrium between the hemiacetal and open-chain forms of sugars is readily established, and since ring closure can lead to either anomer, the anomers are in equilibrium in solution. It is, however, possible to purify them by crystallization; thus crystallization of glucose from water below 50°C in the ordinary manner or from ethanol produces α-D-glucose, $[\alpha]_D^{20} + 112.2$ (H$_2$O), whereas crystallization from pyridine or by vacuum evaporation of the glucose syrup at 115°C produces the β anomer, $[\alpha]_D^{20} + 17.5$. When either anomer is dissolved in water, equilibrium is established in a matter of hours (the exact half-life is dependent on pH) and the initial rotation gradually changes to the equilibrium value of $[\alpha]_D^{20} = +52.7$; this phenomenon is called "mutarotation."

Anomeric sugars are given the symbols α or β, as shown in Figure 11.75. When a D sugar (pyranose) is written in such a way that the six-membered ring is oriented with the ring oxygen at the rear and the anomeric carbon on the right (so-called *Haworth formula* as in Fig. 11.75), the α form is the one with the anomeric hydroxyl (or other functional group) below the plane of the ring, whereas the β form has it above that plane. The opposite (α: OH up; β: OH down) applies in the L series: the mirror image of the α-D isomer is the α-L isomer, not the β-L isomer. In both cases the hydroxyl or other anomeric group in the β series is on the same side as the CH$_2$OH group in aldohexopyranoses. In the major chair conformer of most aldohexoses (Fig. 11.75) the α anomer has an axial OH and the β has an equatorial OH.

From the initial and equilibrium values of the rotations of the anomers of D-glucose one can compute the equilibrium mixture to be made up of about 37% of the axial α anomer and 64% of the equatorial β, hence $K = 1.69$ favoring the equatorial isomer and $\Delta G_{25}^0 = -0.31$ kcal mol^{-1} (-1.30 kJ mol^{-1}) (solvent water). This value should be compared with the conformational energy of hydroxyl in cyclohexane (Table 11.7): $\Delta G^0 = -0.95$ kcal mol^{-1} (-3.97 kJ mol^{-1}) in a hydroxylic solvent; evidently the preference for the β (equatorial) anomer of glucose is considerably less than anticipated. The discrepancy is even larger than this calculation might imply, since, as we have seen earlier, substituents at the 2 position in oxanes ordinarily have *larger* conformational energies than corresponding substituents in cyclohexane (see also Franck, 1983). Moreover, in the corresponding acetal, methyl D-glucoside [CH$_3$O instead of HO at C(1)] the α (axial) isomer predominates over the equatorial at equilibrium by about 2:1. This

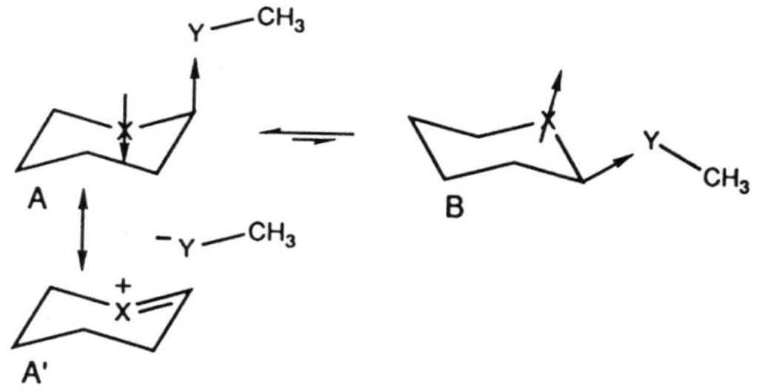

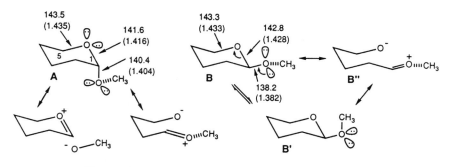

Figure 11.76. Anomeric effect. For X = Y = O see de Hoog, Havinga, et al. (1969); for X = O, Y = S see de Hoog and Havinga (1970). X = S, Y = O, X = Y = S are from Kirby, (1983).

tendency for the axial anomer to predominate was first explained by Edward (1955) and has been termed "anomeric effect" by Lemieux (Lemieux and Chu, 1958; Lemieux, 1964; Lemieux and Koto, 1974). The effect is not limited to sugars but may be seen in simple 2-alkoxytetrahydropyrans as well as the corresponding 2-alkylthio analogues and analogues with sulfur in the ring (2-alkoxy- and 2-alkylthiothians), as shown in Figure 11.76 (cf. Kirby, 1983).

The origin of the anomeric effect, which has long been a matter of controversy, has now been well established by both experiment and quantum mechanical calculations (some of which were alluded to in Chapter 10). It is clear from Figure 11.76 that the axial isomer **A** has a lower dipole moment than the equatorial isomers **B**; therefore it should be favored by the electrostatic term V_E in Eq. 2.1, but, as explained earlier, that advantage should be less in solvents of increasing dielectric constant, both because V_E decreases (Coulomb's law) and because the solvation term V_S becomes more favorable to the higher dipole conformation. There is, however, a second factor, as explained earlier (see Fig. 10.13), namely, the overlap of the p orbital on X with the σ^* orbital of the C(1)–Y bond, shown as the equivalent double-bond/no-bond resonance in Figure 11.76, **A'**. This kind of overlap is not possible for the equatorial CH_3Y group.

However, the reverse motion of electrons, in which the exocyclic CH_3Y is the donor and the endocyclic X the acceptor of electrons, is possible in both isomers, as shown in Figure 11.77 (antiperiplanar pairs and bonds in heavy print). This "reverse" electron donation is called the "exo-anomeric effect" (cf. Kirby, 1983; Lemieux et al., 1979; Wolfe et al., 1979; Praly and Lemieux, 1987; Box, 1990).

Figure 11.77. The C–O bond distances (Å) in glycosides.

Figure 11.77 also shows the C–O bond distances in glycosides (Jeffrey, 1979). The O(endo)–C(5) distance may serve as the standard of comparison. The O(endo)–C(1) distance is appreciably shorter in the axial glycosides than in the equatorial ones because only the axial glycoside can partake of the electron overlap shown in Figure 11.77, **A ↔ A′**, which leads to a foreshortening of the O(endo)–C(1) bond. However, both the axial and equatorial glycosides are capable of the reverse electron motion shown in Figure 11.77; **A ↔ A″, B ↔ B″,** or **B′ ↔ B″**; therefore the C(1)–O(exo) bond is foreshortened in both. This effect is more important in the equatorial glycoside because it represents the only orbital overlap possible there, whereas in the axial glycoside the exocyclic OCH₃ can be either donor **A″** or acceptor **A′**; the foreshortening of the exocyclic C(1)–O(exo) bond is therefore less pronounced in the latter.

> For the kind of overlap shown in Figure 11.77, **A″** and **B″**, to occur, one of the unshared pairs on the exocyclic oxygen (shown in bold in **A, B** and **B′**) must be antiperiplanar to the C(1)–O(endo) bond. In the axial isomer (and granting that the methoxyl methyl group can, for steric reasons, not point into the ring) this requires the exocyclic methyl group to be gauche to the ring oxygen as shown. In the equatorial isomer, it must also be gauche to the ring oxygen, but there are two possible conformations (**B** and **B′**) of which the first one is preferred on steric grounds. In either case there is a definitive preference for the conformation in which the methoxy methyl is gauche to the ring oxygen rather than the ring CH₂(2) group.

A nice illustration of the counterplay of the anomeric and the exo-anomeric effect is provided by compounds containing an $\overset{\diagdown}{\diagup}N–\overset{|}{C}–O–$ moiety. Since nitrogen is the better electron donor but oxygen the better electron acceptor one predicts that by using the valence bond picture in Figure 11.76, $\overset{\diagdown}{\diagup}\overset{+}{N}=C\ \overset{-}{:O}–$ is preferred over $\overset{\diagdown}{\diagup}\overset{-}{N}: \overset{|}{C}=\overset{+}{O}–$. Accordingly, a compound of structure **1** or **2** (Fig. 11.78) should show a strong endo-anomeric but weak exo-anomeric effect, leading to strong favoring of the axial **A** isomer. Indeed, this is seen (Pinto and Wolfe, 1982) in the alkaloid nojirimycin (**2**) for which the OH$_a$/OH$_e$ ratio is four times that for D-glucose (O instead of NH in ring). The contrary behavior is seen in the 2-aminotetrahydropyrans (2-aminooxans) **3** and **4** in which the equatorial isomer **E′**

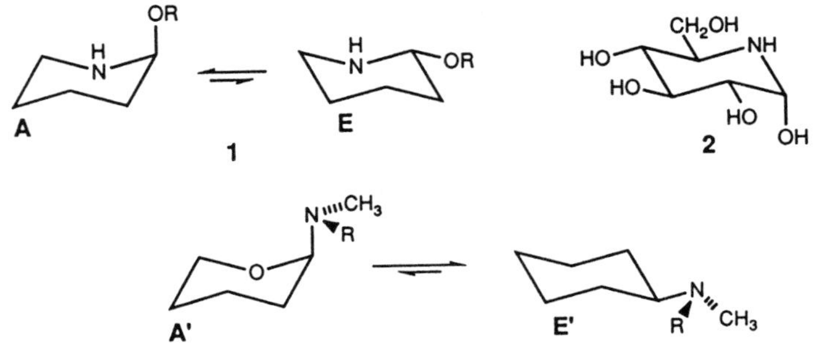

3 R = H
4 R = CH₃

Figure 11.78. O/N anomeric effect.

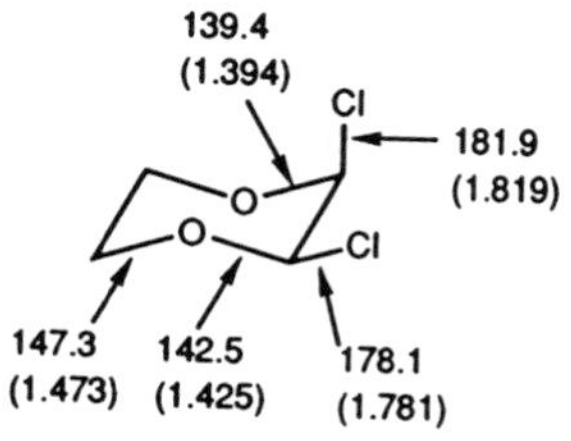

Figure 11.79. Crystal structure of *cis*-2,3-dichloro-1,4-dioxane; distances in pm and (in parentheses) in Å.

is actually favored since the exo-anomeric effect now outweighs the endo-anomeric effect (**3**: Booth and Khedair, 1985; **4**: de Hoog, 1974).

A simpler situation is encountered when the group at C(1) in a sugar, or C(2) in a tetrahydropyran or 1,4-dioxan, is a halogen, which is a good acceptor but a poor donor for electrons. In that case there is a strong preference for the axial conformation of the halogen and a substantial shortening of the endocyclic C–O bond next to it. An example (cf. Kirby, 1983) is shown in Figure 11.79. The fact that the anomeric effect is more important for Br than for Cl (cf. Kirby, 1983) and for SCH_3 than for OCH_3 (cf. Fig. 11.76) is a manifestation of the lower energy level of the σ^* orbital of the C–X bond as X moves down the periodic table. (For studies of the anomeric effect in second- and third-row elements see, for example, Pinto et al., 1988; Juaristi et al., 1989).

The existence of the exo-anomeric effect and the analogous effect in compounds, such as dimethoxymethane ($CH_3OCH_2OCH_3$) described in Chapter 10 indicates the generality of the anomeric effect first seen in hexoses and the corresponding glycosides. The term "generalized anomeric effect" (Lemieux, 1971; Eliel, 1972) has been coined to describe the phenomenon in all its aspects. An example of the generalized anomeric effect is seen in the tendency of *N*-methyl-1,3-oxazanes, *N*,*N'*-dimethyl-1,3-diazanes and *N*,*N'*, *N''*-trimethyl-1,3,5-triazanes to have the (or one of the) *N*-methyl groups axial so as to have the corresponding unshared pair antiperiplanar to a C–O or C–N bond (Fig. 11.80; cf. Kirby, 1983).

We conclude this section on saturated six-membered heterocyclic rings by a consideration of *N*,*N'*-dimethylhexahydropyridazine (*N*,*N'*-dimethyl-1,2-diazane, Figure 11.81). The properties of this system are considerably affected by the following factors: (a) Tendency of the lone pairs in hydrazine to be near 90° to each other (cf. Chapter 10). This tendency leads to a general puckering of the chair and to a substantial destabilization of the diequatorial ($NCH_3–N'CH_3$) conformation (or configuration), which is only $0.23 \text{ kcal mol}^{-1}$ (0.96 kJ mol^{-1}) more stable than the equatorial–axial form (the diaxial isomer is not observed and must be of high energy). (b) Difficulty of the $N–CH_3$ groups to pass by each other. This difficulty leads to a high barrier ($12.6 \text{ kcal mol}^{-1}$, 52.7 kJ mol^{-1}) for the nitrogen inversion process a,e to e,e (process 3 or 3*). In contrast, the a,a to

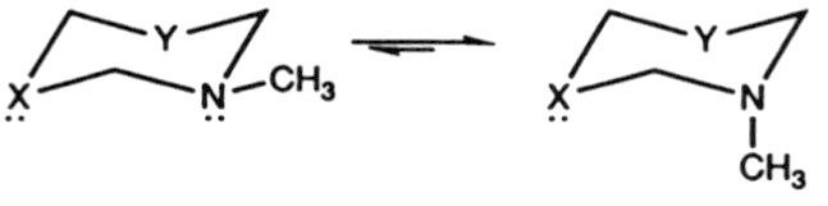

Figure 11.80. Manifestations of the generalized anomeric effect with X = O, Y = CH_2; X = NCH_3(e), Y = CH_2; or X = Y = NCH_3(e).

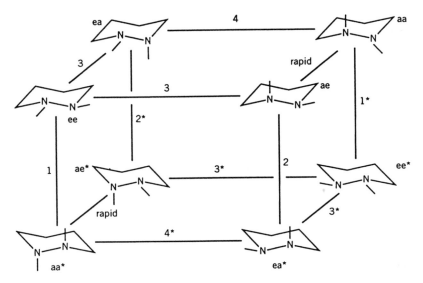

Figure 11.81. Ring and nitrogen inversion of N,N'-dimethylhaxahydropyridazine. (From Riddell, F.G., *The Conformational Analysis of Heterocyclic Compounds.* Copyright (1980) Academic Press. By permission of the publisher.)

e,a barrier (process 4 or 4*), which does not require passing of the N-methyl groups, is considerably lower, 7.5 kcal mol^{-1} (31.4 kJ mol^{-1}). The ring inversion or ring reversal barriers are (process 1 or 1*), 10.2 kcal mol^{-1} (42.7 kJ mol^{-1}); (process 2 or 2*), 11.5 kcal mol^{-1} (48.1 kJ mol^{-1}), the process that requires simultaneous passage of the N-methyl groups being somewhat higher in energy.

11-5. CHEMISTRY OF RING COMPOUNDS OTHER THAN SIX-MEMBERED ONES

a. Three-Membered Rings

Three-membered rings are necessarily planar, and therefore quite highly strained (cf. Fig. 11.9), even though the interorbital angle in cyclopropane is considerably larger than the internuclear angle of 60° inasmuch as the orbitals do not overlap end-over-end (see Liebman and Greenberg, 1989 for a survey of heats of formation of three-membered ring systems).

> The total strain in cyclopropane is only slightly greater than that in cyclobutane (cf. Table 11.4), which implies that angle strain (Fig. 11.9) cannot exclusively account for the difference. There must be an electronic (orbital) stabilization of cyclopropane that partly offsets its greater apparent strain (see Chapter 2 in Greenberg and Liebman, 1978; see also Cremer and Gauss, 1986). Concerning the bond lengths in cyclopropane and cyclobutane, see Hargittai and Hargittai, 1988.

Introduction of double bonds (either endocyclic or exocyclic) into the cyclopropane ring increases the strain further, as expected; the strain in cyclopropene

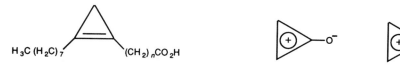

Figure 11.82. Sterculic and malvalic acids.

Figure 11.83. Aromatic character of cyclopropenone and the cyclopropenium cation.

is 52 kcal mol^{-1} (218 kJ mol^{-1}) (Wiberg et al., 1983a) and that in methylenecyclopropane 41.0 kcal mol^{-1} (171.5 kJ mol^{-1}); the isomerization of methylcyclopropene to methylenecyclopropane is exothermic by 10.3 kcal mol^{-1} (43.1 kJ mol^{-1}) (Wiberg and Fenoglio, 1968). Even though cyclopropene avoids the H/H bond eclipsing that exists in cyclopropane and (to a lesser extent) in methylenecyclopropane, angle strain is maximized when two sp^2 hybridized carbon atoms are in the ring.

Despite the high strain in cyclopropene, two cyclopropene derivatives, sterculic acid (Fig. 11.82, $n = 7$) and malvalic acid (Fig. 11.82, $n = 6$), occur in nature in seed oils (Greenberg and Harris, 1982). Derivatives of cyclopropane also are found in nature, e.g., the pyrethrins.

Cyclopropanone (cf. De Boer, 1977) is also very strained and quite unstable; it easily adds water or alcohols to give geminal diols or hemiketals. Its C=O stretching frequency is unusually high at 1813 cm^{-1}, suggesting a high degree of s character in the C=O bond resulting from the C–C bonds in the ring having a high degree of p character to minimize angle deformation. (The normal angle between p bonds is 90°, that between sp^3 hybridized bonds is 109°28′.) On the other hand, cyclopropenone is quite stable (its preparation has been described in *Organic Syntheses*, Breslow et al., 1977) as is the cyclopropenium cation (Fig. 11.83); both of these systems are aromatic (cf. Breslow, 1971). In contrast, anions derived from cyclopropane, which tend to be highly pyramidalized (Walborsky and Impastato, 1959) are "antiaromatic" (cf. Breslow, 1973).

b. Four-Membered Rings

A four-membered ring can be either planar or puckered. The planar ring has minimal angle strain (cf. Fig. 11.9) but will have maximal torsional (Pitzer) strain because of its eight pairs of eclipsing hydrogen atoms. The torsional strain can be reduced by puckering (at the expense of some increase in angle strain) and this is exactly what happens in cyclobutane (cf. Moriarty, 1974, from which much of the following information is taken). Cyclobutane is therefore a wing-shaped molecule with an "angle of pucker" ϕ (cf. Fig. 11.84) of 28° (Egawa, Kuchitsu, et al., 1987)

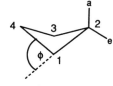

Figure 11.84. Geometry of cyclobutane.

and a barrier to ring inversion of 1.45 kcal mol^{-1} (6.06 kJ mol^{-1}) (Egawa, Kuchitsu, et al., 1987); the angle between the planes defined by atoms 1,2,3 and 3,4,1 is $180 - \phi$.

The barrier in cyclobutane is high enough for the molecule to possess a clear double energy minimum (cf. Fig. 2.4), that is, two (superposable) conformations can be defined; the same is true for a number of heteracyclobutanes. Thus the barrier in thietane (trimethylene sulfide) amounts to 0.75 kcal mol^{-1} (3.1 kJ mol^{-1}), that in trimethylene selenide to 1.07 kcal mol^{-1} (4.5 kJ mol^{-1}), that in trimethylene imine (azetidine) to 1.26 kcal mol^{-1} (5.3 kJ mol^{-1}), and that in silacyclobutane [trimethylene silane, $(CH_2)_3SiH_2$] to 1.26 kcal mol^{-1} (5.3 kJ mol^{-1}). In azetidine, the conformer with equatorial NH is 0.27 kcal mol^{-1} (1.13 kJ mol^{-1}) more stable than that with axial NH (Carreira and Lord, 1969). In contrast, oxetane (trimethylene oxide) has a barrier of only 0.1 kcal mol^{-1} (0.4 kJ mol^{-1}) and may be considered effectively planar since a single molecular vibration will carry the molecule over the barrier. A similar situation occurs in 1,3-dithiacyclobutane which, while showing puckering in the vibrational spectrum, crystallizes in the average planar conformation (according to X-ray structure analysis) (Block et al., 1982). In 1,2,4-trisubstituted azetidines, whereas the 2,4-cis isomers are puckered (see below), the corresponding trans isomers are planar (Kingsbury et al., 1982).

As a result of the puckering in cyclobutane, monosubstituted derivatives, such as cyclobutyl bromide, can exist as axial or equatorial conformers. As in cyclohexyl compounds the equatorial conformer is, for steric reasons, the more stable one. In the case of bromocyclobutane the energy difference between the two conformers, about 1 kcal mol^{-1} (4.2 kJ mol^{-1}) (Durig et al., 1989; see also Durig et al., 1990 for chlorocyclobutane; Durig et al., 1987 for methylcyclobutane, Powell, Klaboe, et al., 1989 for cyanocyclobutane) appears to be greater than the corresponding difference [0.49 kcal mol^{-1} (2.05 kJ mol^{-1})] in bromocyclohexane. The axial conformer is less puckered ($\phi_a = 14°$) than the equatorial one ($\phi_e = 20°$) (Durig et al., 1989; but see Gatial, Klaboe, et al., 1989).

In 1,2-disubstituted cyclobutanes the trans isomer is diequatorial and more stable than the equatorial–axial cis isomer; the equilibrium ratio (trans–cis) of 9 for the cyclobutane-1,2-dicarboxylic acids is nearly the same as for the cyclohexane-1,2-dicarboxylic acids and the ratio of dissociation constants (Table 11.1) is also similar. X-ray structure determination of the trans diacid has shown it to be the diequatorial conformer. In 1,3-disubstituted cyclobutanes the cis (e,e) isomer is predictably the more stable; for example, in the 1,3-dibromocyclobutanes the trans (a,e) isomer is less stable than the cis by 0.58 kcal mol^{-1} (2.43 kJ mol^{-1}); electron diffraction study shows the ring to be puckered in both stereoisomers, which indeed have the expected e,e and a,e conformations. Dipole moments are in agreement with this assessment but suggest that the rings in the trans (e,a) isomers are somewhat flatter than in the cis (e,e); the difference is particularly marked in 1,3-diiodocyclobutane for which ϕ (cf. Fig. 11.84) is 48° for the cis and 24° for the trans isomer.

1,2,3,4-Tetrasubstituted cyclobutanes are available from photochemical (2 + 2) dimerization of 1,2-disubstituted alkenes. Thus dimerization of cinnamic acid gives rise to the truxillic and truxinic acids shown in Figure 11.85, which also occur in

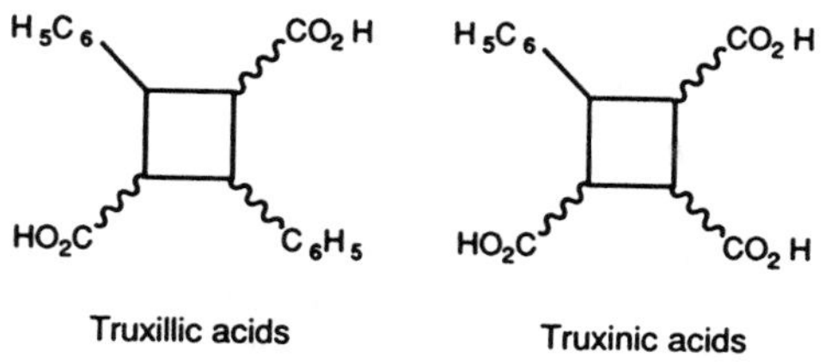

Figure 11.85. Truxillic and truxinic acids.

nature (in coca leaves). The readers should convince themselves that there are five diastereomeric truxillic acids, none of them chiral, whereas the set of truxinic acids comprises four pairs of enantiomers and two meso forms (cf. Moriarty, 1974; Uff, 1967; Green and Rejto, 1974). Interestingly, photodimerization of *trans*-cinnamic acid, which is polymorphic (i.e., it exists in more than one crystalline form), occurs only in the solid state and then with high regio- and stereospecificity in that the α crystalline modification gives rise only to α-truxillic acid, the β form gives rise only to β-truxinic acid, and the γ form does not dimerize at all (Schmidt, 1964). This reaction has been called a "topochemical reaction" (Addadi, Lahav, Leiserowitz et al., 1986; see also Green et al., 1986) in that the reaction depends on the place (Greek *topos*), which the functional group occupies in the crystal, as shown in Figure 11.86 (see also p. 163).

The strain in cyclobutene (29 kcal mol^{-1}, 121 kJ mol^{-1}) and in methylene-cyclobutane (27.9 kcal mol^{-1}, 116.7 kJ mol^{-1})* (Schleyer et al., 1970; Finke et al., 1981; Wiberg et al., 1983a; see also Kozina et al., 1982) exceeds that in cyclobutane by less than 3 kcal mol^{-1} (12.6 kJ mol^{-1}) in contradistinction to the situation in the corresponding three-membered rings mentioned earlier. Although the C=C–C angles in cyclobutene are reduced from 124.3° (in propene) to 94.0°

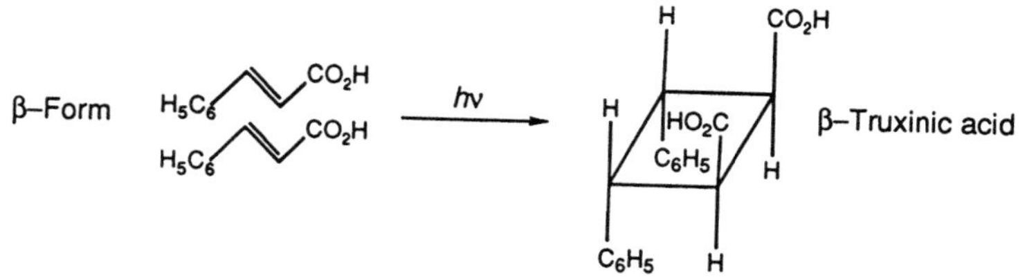

Figure 11.86. Topochemical photodimerization of the α and β forms of cinnamic acid.

*This value diminishes by an additional 0.85 kcal mol^{-1} (3.6 kJ mol^{-1}) if one uses the value of 29.15 kcal mol^{-1} (121.95 kJ mol^{-1}) of Finke et al. (1981) for the heat of formation of methyl-enecyclobutane instead of the value of 30.0 kcal mol^{-1} (125.5 kJ mol^{-1}) used by Schleyer et al. (1970).

(cf. Hargittai and Hargittai, 1988) the increase in angle strain (less than in the three-membered ring!) seems to be nearly offset by a decrease in torsional strain due to the reduction in H–H eclipsing. In contrast, cyclobutadiene is a highly strained and quite unstable nonaromatic molecule that can be isolated only at very low temperatures in a matrix (cf. Bally and Masamune, 1980; Greenberg and Liebman, 1978), or interestingly, at room temperature as an inclusion compound in a "hemicarcerand" (Cram et al., 1991; see also p. 794).

c. Five-Membered Rings

Five-membered rings are common in natural products: carbocyclic rings in the D ring of steroids (Altona et al., 1968; Duax et al., 1976) and in prostaglandins; oxygen containing rings in furanose sugars, nucleosides, nucleotides (Altona and Sundaralingham, 1972), and nucleic acids (Altona, 1982); nitrogen containing rings in the amino acids proline and hydroxyproline (Haasnoot, Altona, et al., 1981), and in alkaloids, such as nicotine (Pitner et al., 1978), among others. A carbocyclic planar five-membered ring would have valence angles of 108° and thus be nearly free of angle strain. On the other hand, the torsional (Pitzer) strain in such a conformation would be sizeable [estimated at 10 kcal mol^{-1} (42 kJ mol^{-1}) for 10 pairs of H–H eclipsings]. To minimize this strain, cyclopentane becomes nonplanar (cf. Aston et al., 1941), thus reducing the sum of residual torsional plus angular (Baeyer) strain to about 60% of the value in a planar ring (cf. Table 11.4). The most stable conformations of cyclopentane are the envelope (or C_2, naming it after its symmetry point group, cf. Section 4-3) and the half-chair (or C_s) conformations shown in Figure 11.87. For cyclopentane itself the difference in energy between these conformations is slight, the envelope form being the more stable by about 0.5 kcal mol^{-1} (2.1 kJ mol^{-1}) (Pitzer and Donath, 1959). The barriers between the two conformations are also very low (the molecule never passes through the high-energy planar form) and it is thus not surprising that cyclopentane is in a rapid "conformational flux" among various C_2 and C_s as well as intermediate conformations. It was early recognized by Kilpatrick, Pitzer, et al. (1947) that this conformation change can be brought about by successive oscillatory motions of the five carbon atoms of cyclopentane in a direction perpendicular to the plane of the ring. The apparent effect of this motion is to create a "bulge" (out-of-plane atom), which appears to rotate around the ring, even though there is in fact no motion of the atoms in that direction. This process has been called "pseudorotation." In cyclopentane itself it is so rapid that it is probably better to consider it a molecular vibration rather than a conformational change. However, in substituted cyclopentanes the barrier is higher, for example 3.40 kcal mol^{-1} (14.2 kJ mol^{-1}) in methylcyclopentane in which the most stable conformation (by

Figure 11.87. Envelope (C_s) and half-chair (C_2) conformations of cyclopentane derivatives.

0.9 kcal mol^{-1} or 3.8 kJ mol^{-1}) is that with the methyl group "equatorial" at the flap of the envelope (cf. Fuchs, 1978 for these and many of the following data) (Fig. 11.87). In cyclopentanone the barrier is 1.15 kcal mol^{-1} (4.81 kJ mol^{-1}) (above the less stable conformation) but the preferred conformation, by 2.4 kcal mol^{-1} (10.0 kJ mol^{-1}) is the half-chair with the carbonyl group in the least puckered region (Figs. 11.87 and 11.88).

The implications in the conformational analysis of cyclopentanes are twofold. One stems from the low barrier which, in general, makes it impossible to "freeze out," by low-temperature NMR spectroscopy, individual conformers in substituted cyclopentanes in the manner described earlier for cyclohexanes. In this regard cyclopentane resembles cyclobutane. However, the pseudorotation causes an additional complication, in that in an unsymmetrically substituted cyclopentane (say cyclopentane-d) there are a total of 20 conformational minima (10 half-chairs and 10 envelopes) on the pseudorotational circuit, plus an infinite number of additional conformations of nearly the same energy as the unique ones just mentioned. Thus, since there is little preference for any particular conformation of the cyclopentane framework itself, substituted cyclopentanes will take up conformations such that the interactions of the substituents with the framework (plus, if there is more than one substituent, the interactions of the substituents with each other) are minimized. As a result, the conformation of any one substituted cyclopentane (or heteracyclopentane) is likely to be different from that of any other. In this regard, cyclopentane is totally different from cyclohexane in which the chair is located in a deep energy valley or mold, the next higher energy minimum (corresponding to the twist form) being about 5 kcal mol^{-1} (20.9 kJ mol^{-1}) above the chair with the energy barrier to interconversion being more than 10 kcal mol^{-1} (42 kJ mol^{-1}) (Fig. 11.17). It takes two very large substituents, for example, two *tert*-butyl groups located such that one would have to be axial in the chair conformation, to force the cyclohexane ring out of its "natural" chair conformation: In contrast, there is no "natural" conformation of the five-membered ring!

In the course of a pseudorotation circuit (i.e., one 360° turn of the puckering) one encounters 10 envelope forms (5 with one of the carbon atoms in turn pointing up and 5 with one of the carbon atoms pointing down) and 10 half-chair forms lying in between 2 envelope forms on the circuit. (There are five pairs of adjacent carbon atoms that may be located at the site of maximum pucker of the half-chair: 1,2; 2,3; 3,4; 4,5 and 5,1, and the puckering may be either up or down,

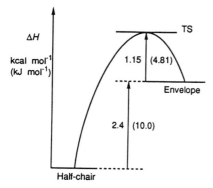

Figure 11.88. Energy diagram for cyclopentanone.

since the half-chair is chiral. Thus there are 10 nonidentical half-chairs.) The puckering may thus be described by two parameters: an angle ψ that describes how far puckering has moved around the five-membered ring starting at an arbitrarily defined origin and a "puckering amplitude" q that indicates how far a given atom moves above and below the average plane of the ring. [The parameter ψ is twice the "phase angle" defined in the original paper by Kilpatrick, Pitzer, et al. (1947).] In cyclopentane itself q is 43 pm (0.43 Å) (Adams, Geise, and Bartell, 1970). If one starts the pseudorotation circuit with a $\mathbf{C_s}$ form, one will encounter a $\mathbf{C_2}$ form at $\psi = 18°$, another $\mathbf{C_s}$ form at $\psi = 36°$, a $\mathbf{C_2}$ form again at $\psi = 54°$ and so on, with $\mathbf{C_s}$ and $\mathbf{C_2}$ forms alternating at 18° intervals. All 20 forms (10 $\mathbf{C_s}$, 10 $\mathbf{C_2}$) thus make an appearance as ψ changes by 360° (i.e., as the puckering moves once around the ring; cf. Fuchs, 1978). (For a review of the calculational aspects of this problem, see Abillon, 1982.)

As ψ changes, both the torsion angles and the bond angles in the five-membered ring change as well. For the envelope ($\mathbf{C_s}$) form, the minimum torsion angle ($\omega_{3,4}$ in Fig. 11.87) is 0°, and the maximum ($\omega_{1,2}$) is about 45°, at the flap, with a value of about 28° for the intermediate torsion angle ($\omega_{2,3}$). These are calculated values and vary somewhat in different calculations. For the half-chair ($\mathbf{C_2}$), the minimum value $\omega_{1,2}$ (in which the four-carbon segment includes the three coplanar atoms) is about 15°, the maximum value ($\omega_{3,4}$) is about 45°, and the intermediate value ($\omega_{2,3}$, Fig. 11.87) is about 37°. The exocyclic torsion angles (e.g., between two hydrogen atoms or between two adjacent hydroxyl groups) vary parallel with the internal angles and thus also range from 0° to about 45°. These are the angles, of course, that determine the magnitude of the vicinal coupling constant in ^{1}H NMR spectroscopy or the strength of the hydrogen bond in, say, vicinal diols. Bond angles evidently also change with puckering, though not very much, the variation of bond angles being between 101° and 106.5° (Hendrickson, 1961).

For the reasons already indicated, the conformations of monosubstituted cyclopentanes (e.g., chlorocyclopentane: Hilderbrandt and Shen, 1982) are much less predictable than those of the corresponding cyclohexanes. Oversimplifications may have been made in the literature by assuming that the ring is either an envelope or a half-chair when, in fact, it may assume a conformation somewhere in between, depending on the substituent. On the other hand, clear-cut experimental data are available for configurational (as distinct from conformational) equilibria in disubstituted cyclopentanes. For 1,2-disubstituted cyclopentanes, the trans isomer is more stable than the cis, the energy difference for the dimethyl compounds, 1.73–1.94 kcal mol^{-1} (7.24–8.12 kJ mol^{-1}), being about the same as for the dimethylcyclohexanes (cf. Table 11.9). However, the agreement may be fortuitous; whereas in the cyclohexane series the major difference is between two equatorial versus one equatorial and one axial methyl (the mutual gauche interaction of the methyl groups nearly cancels), the major factor destabilizing the cis isomer in the cyclopentane series may be the direct interaction of the substituents that are probably at a torsion angle (CH_3–C–C–CH_3) of no more than 50°, since an angle larger than that would involve a serious increase in energy of the cyclopentane framework (Hendrickson, 1961). Greater stability for the trans isomer was also found in the 1,2-dicarbomethoxy- and 1,2-diphenyl-cyclopentanes.

In the conformational study of 1,2- and 1,3-dihalo- and 1,3-dicyanocyclopentanes, studies of dipole moments have been informative. In the 1,2 series, there is a relation between dipole moments and vicinal (H,H) coupling constants, both of which are conformation dependent (Altona, Havinga, et al., 1967). The relatively low dipole moments for the trans-1,2 isomers suggest that these compounds exist as half-chairs with the halogen atoms antiperiplanar in the most puckered region of the half-chair, that is, antidiaxial. However, the moment in the trans-dichloro compound is larger than that in r-1, t-2-dichloro-c-4-tert-butylcyclohexane (diaxial chlorines) (1.70 D vs. 1.21 D). Presumably, the degree of puckering needed to achieve a torsion angle near 180° between the halogens in the trans-1,2-dichlorocyclopentane cannot be achieved and/or other conformers than the one mentioned contribute appreciably.

The configuration of the 1,3-dimethylcyclopentanes (as that of the 1,3-dimethylcyclohexanes) was originally incorrectly assigned; in both series the cis isomer is the more stable. In the cyclohexane series the reason is that one methyl group must be axial in the trans isomer. In the cyclopentane series, the reason is not quite so clear; it was at one time thought that the cis-1,3 isomer existed as an envelope form with equatorial methyl groups on the carbon atoms at the side of the flap, but this is not certain. Even if it is true, the trans isomer would probably not exist in the corresponding equatorial–axial conformation; more likely it has a different conformation altogether. In any case, the gas phase free energy difference ($0.5 \, \text{kcal mol}^{-1}$, $2.1 \, \text{kJ mol}^{-1}$) is considerably less than that ($1.73 \, \text{kcal mol}^{-1}$, $7.2 \, \text{kJ mol}^{-1}$) for 1,4-dimethylcyclohexane (*TRC Thermodynamic Tables*, 1991), confirming that the conformational differences in the five-membered ring are not as clear-cut as in the six-membered one. Indeed, for other 1,3 substituents, the cis–trans energy difference may be near zero (CO_2H/CO_2H or t-Bu/OH) or may actually favor the trans isomer slightly (CN/CN, CO_2R/CO_2R, or CH_3/OH). By the same token, differences in reactivity between cis- and trans-1,3 disubstituted cyclopentanes tend to be slight, with specific rate ratios ranging from 0.8 to 2 (Fuchs, 1978).

A few studies of heterocyclic five-membered rings have been reviewed by Fuchs (1978) and by Riddell (1980, 1992). Both oxygen and nitrogen containing rings, like cyclopentane itself, seem to pseudorotate freely, but there is an appreciable barrier in tetrahydrothiophene ($2.2–2.8 \, \text{kcal mol}^{-1}$, $9.2–11.7 \, \text{kJ mol}^{-1}$) in which the half-chair form with the sulfur in the middle of the unpuckered C–S–C fragment is the preferred conformer. Equilibrium and NMR studies have been carried out in 2,4 disubstituted and 2,4,5 trisubstituted 1,3-dioxolanes (Fig. 11.89) (Willy, Binsch and Eliel, 1970). The values for ΔG^0 are nearly the same (ca. $0.5 \, \text{kcal mol}^{-1}$, $2.1 \, \text{kJ mol}^{-1}$) for all members of the 2,4 disubstituted series (R″ = H), regardless of the nature of the R and R′. In the

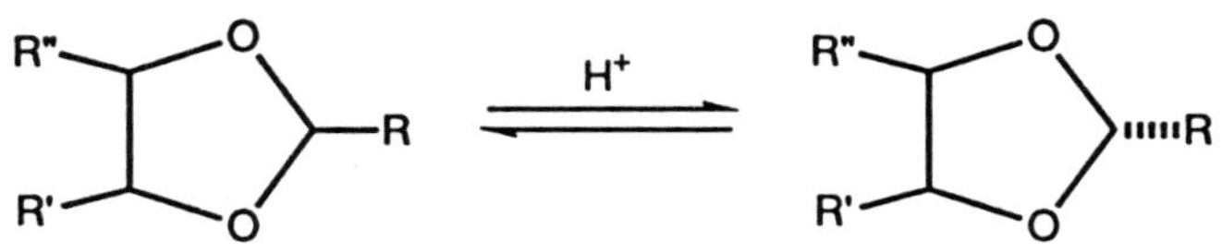

Figure 11.89. Equilibria in 2,4 disubstituted and 2,4,5 trisubstituted 1,3-dioxolanes, where R″ = H or R″ = R′.

trisubstituted series ($R'' = R'$), constancy of ΔG^0 holds for small substituents, but changes, involving a reversal of the position of equilibrium from one favoring the cis to one favoring the trans isomer, occur when both R and R' are isopropyl or *tert*-butyl. It appears that, because of the easy adjustment of the ring conformation to the steric demands of the substituents, appreciable changes in the energy levels of the cis and trans substituted isomers occur only when the substituents are very large and numerous.

The strain in cyclopentene is nearly the same as that in cyclopentane (Cox and Pilcher, 1970) and that in cyclopentadiene is actually considerably less (Wiberg et al., 1983; Kozina et al., 1982). Evidently, the diminution in eclipsing strain more than compensates for the increase in angle strain in these compounds. The strain in methylenecyclopentane is somewhat less than that in cyclopentene (Liebman and Greenberg, 1976), yet the transformation of methylenecyclopentane to methylcyclopentene is exothermic, by $3.5 \, \text{kcal mol}^{-1}$ ($14.6 \, \text{kJ mol}^{-1}$) (cf. Allinger et al., 1982), that is, more so than the corresponding process in the six-membered ring. The explanation may be that in methylenecyclohexane (but not in methylenecyclopentane) the favored C=C/H eclipsed conformation (cf. Section 10-2.a) is achieved.

The conformational situation in cyclopentanone has already been discussed. Addition reactions to cyclopentanone are less favorable than those to cyclohexanone since they lead to an increase in bond eclipsing. Accordingly, the equilibrium constant for addition of HCN to cyclopentanone to give the cyanohydrin is only 3.33 as against 70 in cyclohexanone (Prelog and Kobelt, 1949) and the reduction of cyclohexanone with sodium borohydride is 23 times as fast as the corresponding reduction of cyclopentanone (Brown and Ichikawa, 1957; see also Section 11-5.e).

d. Rings Larger Than Six-Membered

Several excellent reviews of the conformation of seven-membered, medium-sized, and larger rings are available (Sicher, 1962; Anet and Anet, 1975; Dale, 1976; Anet, 1988). By way of introduction to this subject, let us consider that cyclopentane has basically but a single, pseudorotating conformation. In cyclohexane there are two families of conformations: the chair, which happens to be rigid and thus constitutes a single-membered family (for an explanation see Dunitz, 1970) and the twist–boat family. As we proceed to higher cyclanes, we shall find an increasing number of conformational families, each with several members. Within each family there are several conformations that are easily interconverted by a process of "pseudorotation" (see above) which involve a very low barrier, generally inaccessible by NMR study. Between families, however, the barriers approach the $10.3 \, \text{kcal mol}^{-1}$ ($43.1 \, \text{kJ mol}^{-1}$) value in cyclohexane and the separation into families can therefore often be studied by low-temperature NMR, with ^{1}H and ^{13}C spectra yielding complementary results.

In the case of cycloheptane and cyclooctane the number of possible conformations possessing a symmetry element is of manageable size (cf. Figs. 11.91 and 11.93) and common names have been attached to these conformations as shown. However, for larger rings both this form of representation and the naming become too complex and either the torsion angle notation of Bucourt (1974) or the wedge notation of Stoddart and Szarek (1968; cf. Dale, 1976) are best to use (Fig. 11.90). In the torsion angle notation, the symbol + against a bond indicates that the torsion angle, with that bond (2–3) in the middle of the four-atom (1–2–3–4) segment, is positive, that is, the sequence 1–2–3–4 describes a right-handed helix. The opposite arrangement (left-handed helix) is described by the symbol −. In the wedge notation, $1 \triangleleft 2$ means that atom 2 is in front of atom 1 or that atom 1 is behind atom 2; a combination of such wedges, as shown in Figure 11.90, will reveal the conformation.

The principal methods for conformational study in seven-membered and larger rings have been NMR, force field calculations, and to a lesser extent, study of vibrational spectra, electron diffraction, and X-ray study of derivatives.

The conformational situation in cycloheptane (cf. Tochtermann, 1970; Dillen and Geise, 1979) was first unraveled by the classical study of Hendrickson (1961; cf. also Hendrickson, 1967), which represents the first application of the computer to molecular mechanics calculation. There are two families of conformations in cycloheptane, shown in Figure 11.91. One comprises the chair and twist–chair, the other the boat and twist–boat. The situation is thus quite similar to that in cyclohexane, except that the chair is part of a family of flexible conformations and, because of the severe eclipsing at the "flat" end, lies about $2.16 \, \text{kcal mol}^{-1}$ $(9.0 \, \text{kJ mol}^{-1})$ above the twist–chair into which it can be readily pseudorotated and which represents the most stable conformation of cycloheptane. In the second family, comprising the boat and twist–boat, the twist–boat is more stable [by $0.53 \, \text{kcal mol}^{-1}$ $(2.22 \, \text{kJ mol}^{-1})$] than the true boat because of the more severe

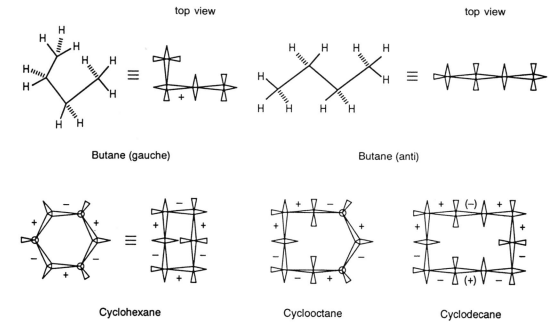

Figure 11.90. Examples of torsion angle notation (Bucourt) and wedge notation (Dale). From Allinger, N. L. and Eliel, E. L. (1976) *Top. Stereochem.* Vol. 9, pp. 204, 205. Copyright © John Wiley & Sons, with permission of the publisher.

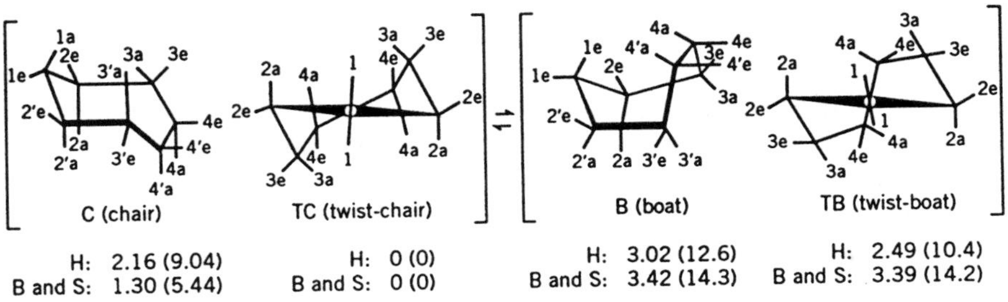

H: 2.16 (9.04)   H: 0 (0)   H: 3.02 (12.6)   H: 2.49 (10.4)
B and S: 1.30 (5.44)   B and S: 0 (0)   B and S: 3.42 (14.3)   B and S: 3.39 (14.2)

Figure 11.91. Conformations of cycloheptane. Relative potential energy values are in kilocalories per mole (kcal mol^{-1}) and (in parentheses) in kilojoules per mol (kJ mol^{-1}). H refers to values given by Hendrickson (1961) and B and S to values given by Bocian and Strauss (1977). Adapted, with permission, from Hendrickson, J. B. (1961), *J. Am. Chem. Soc.*, **83**, 4543, copyright © Am. Chem. Soc., Washington, DC.

eclipsing in the latter, similarly as in cyclohexane. The difference between the twist–boat and the twist–chair is 2.49 kcal mol^{-1} (10.4 kJ mol^{-1}), that is, only about one-half the twist–chair difference in cyclohexane: whereas in cyclohexane the chair is nearly strain-free, in cycloheptane the twist–chair cannot completely escape eclipsing interactions. The chair/twist–chair and boat/twist–boat families in cycloheptane, like their cyclohexane analogues, can be interconverted only by passing over a relatively high barrier of about 8.5 kcal mol^{-1} (35.6 kJ mol^{-1}). The early calculations by Hendrickson have been fully corroborated by later ones and by spectroscopic studies by Bocian and Strauss (1977; see also Elser and Strauss, 1983; Ivanov and Ōsawa, 1984). Figure 11.91 indicates the energy levels of the various conformations (including the transition state) from both the Hendrickson and Bocian and Strauss investigations; however, electron diffraction gives a smaller energy difference (0.9 kcal mol^{-1}, 3.76 kJ mol^{-1}) between the pre-dominating twist–chair and chair conformations (Dillen and Geise, 1979).

The cycloheptene system seems to be largely in the chair form, the C=C moiety taking the place of one of the ring carbon atoms in the cyclohexane chair (Dale, 1978, p. 194; Ermoleeva, Allinger, et al., 1989). Benzocycloheptene derivatives (Ménard and St.-Jacques, 1983) and heterocyclic analogues (e.g., Désilets and St.-Jacques, 1987, 1988) have been studied especially in the laboratories of St.-Jacques with interesting conformational results.

We shall consider only one other compound in the seven-membered ring series: cycloheptatriene. We have already seen (Fig. 3.17) that there is a valence bond tautomerism between this compound and the isomeric norcaradiene (see also Fig. 11.92) with the position of equilibrium depending on the substitutent(s) at the saturated carbon. In addition, however, the boat-shaped (Traetteberg, 1964; Butcher, 1965) cycloheptatriene moiety can undergo a conformational inversion (methylene flip) with a barrier of 6.1 kcal mol^{-1} (25.5 kJ mol^{-1}) (Anet,

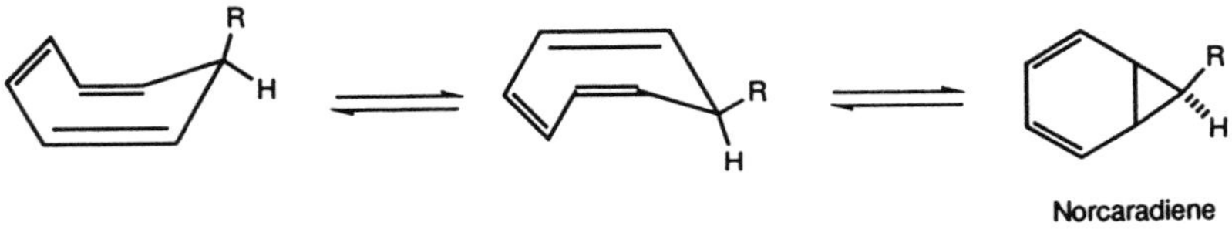

Norcaradiene

Figure 11.92. Conformational equilibrium in 7-substituted cycloheptatrienes.

1964; Jensen and Smith, 1964). In the case of a 7-substituted 1,3,5-cyclohepta-triene, this involves two different conformers that may be called axial and equatorial. Studies involving a variety of substituents (cf. Tochtermann, 1970) suggest that the equatorial conformer is more stable, except when there is also a substituent at C(1) that causes eclipsing. In that case, for example, in 1-methyl-7-*tert*-butylcycloheptatriene, the axial conformer is the more stable (see also Lipkowitz, 1989).

Cyclooctane exhibits no fewer than 10 symmetrical conformations (Fig. 11.93); (cf. Anet, 1974; Anet, 1988, p. 51), which fall into four families. Studies of cyclooctane (Anet and Basus, 1973; Pakes, Strauss, et al., 1981; Dorofeeva, Allinger et al., 1985), as well as X-ray study of a number of derivatives (cf. Anet, 1974), indicate that the most stable family **I** is that of the boat–chair (BC) and twist–boat–chair (TBC) in which the boat–chair represents the energy minimum.

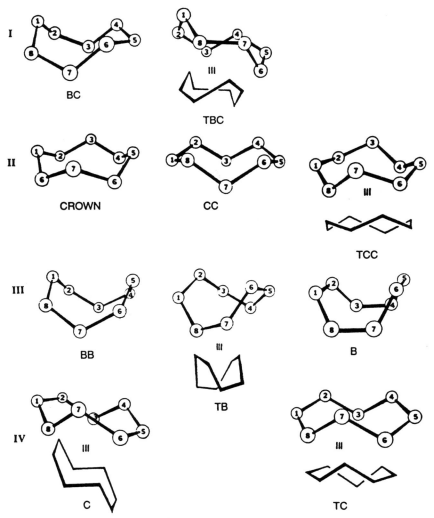

Figure 11.93. Cyclooctane conformations. Adapted, with permission, from Anet, F. A. L. (1974), *Top. Curr. Chem.*, **45**, 178. Copyright Springer Verlag, Heidelberg, Germany.

Interconversion of this family into the one of next-higher energy, the crown/ chair–chair (CC)/twist–chair–chair (TCC) family **II**, requires an activation energy calculated to be 11.4 kcal mol^{-1} (47.7 kJ mol^{-1}), that is, somewhat higher than the chair–twist barrier in cyclohexane. The crown family lies only slightly above the boat–chair family [ca. 1.0–1.6 kcal mol^{-1} (4.2–6.7 kJ mol^{-1})] and it has been estimated that cyclooctane contains about 6% of the crown conformer at room temperature (Anet and Basus, 1973; see also Dorofeeva, Allinger, et al., 1990a). The main problem in the crown form, in addition to high symmetry, and hence low entropy, is eclipsing strain, which is reduced in heteracyclooctanes; thus 1,3,5,7-tetraoxcyclooctane is most stable in the crown form. The third family **III** of cyclooctane conformations comprises the boat (B), twist–boat (TB), and boat–boat (BB) conformations. It is easily seen that these conformations suffer from severe eclipsing strain; they do not appear to be appreciably populated in cyclooctane, even though the barrier between this family and the boat–chair family is calculated to be only 9.4 kcal mol^{-1} (39.3 kJ mol^{-1}). The remaining conformations **IV**, chair (C), and twist–chair (TC), are calculated to be very high in energy [over 8 kcal mol^{-1} (33.5 kJ mol^{-1}) above the boat–chair] and are not viable conformations of cyclooctane (the same applies to the boat member of the boat/twist–boat/boat–boat family).

The conformations of cyclooctadienes and cyclooctatrienes have been discussed by Anet (1988). Cyclooctatetraene (Fig. 11.94; Paquette, 1975, 1993) with its eight ($4n$ with $n = 2$) π electrons is not aromatic, and therefore derives no advantage from being planar; in fact, the molecule is tub shaped (Bastiansen et al., 1957), and thereby relatively unstrained. Similar to the cyclohexane chair, the cyclooctatetraene tub can undergo a ring inversion ($\mathbf{A} \rightleftharpoons \mathbf{B}$) to an alternate tub; the energy barrier to this process in variously substituted cyclooctatetraenes is 14.7 kcal mol^{-1} (61.5 kJ mol^{-1}) (Anet et al., 1964), whereas the energy barrier for double-bond migration ($\mathbf{B} \rightleftharpoons \mathbf{C}$) is appreciably higher, somewhat over 17 kcal mol^{-1} (71 kJ mol^{-1}). The latter process, unlike the former which allows the double bonds to remain localized, may involve an antiaromatic 8π-electron transition state **D** (Fig. 11.94; see also p. 429).

Both cyclononane and cyclodecane (cf. Dunitz, 1971; Anet, 1988, pp. 53–54) present interesting stereochemical aspects. The four minimum-energy conformations of cyclononane, twist–boat–chair (TBC or [333]) (the latter referring to the fact that this conformation has three "straight" segments of three bonds each), twist–chair–boat (TCB or [225]), twist–chair–chair (TCC or [144]), and twist–boat–boat (TBB or [234]) are depicted in Figure 11.95, along with their calculated potential energies. From the calculated energies, it is clear that cyclononane should exist largely as the [333] form, especially at low temperatures, and this is

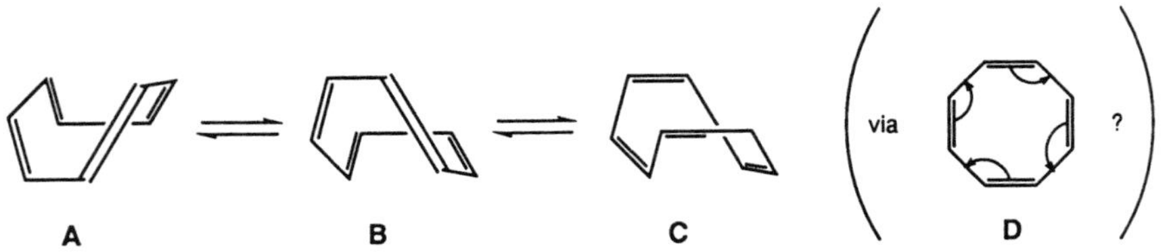

Figure 11.94. Ring inversion $\mathbf{A} \rightleftharpoons \mathbf{B}$ and bond migration ($\mathbf{B} \rightleftharpoons \mathbf{C}$) in cyclooctatetraene.

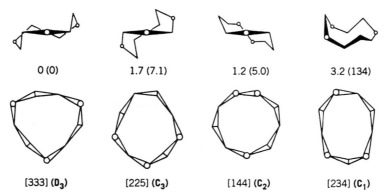

0 (0) 1.7 (7.1) 1.2 (5.0) 3.2 (134)

[333] (**D₃**) [225] (**C₃**) [144] (**C₂**) [234] (**C₁**)

Figure 11.95. Minimum-energy conformations of cyclononane. Relative potential energies are in kilocalories per mole (kcal mol⁻¹); values in parentheses are in kilojoules per mole (kJ mol⁻¹). Adapted, with permission, from Glass, R. S. (1988), *Conformational Analysis of Medium-Sized Heterocycles*, p. 53. Copyright © VCH, New York.

borne out by NMR study (Anet and Wagner, 1971). It was therefore rather surprising that several cyclononane derivatives, such as cyclononylamine hydrobromide (Bryan and Dunitz, 1960), crystallize in the [225] conformation. Detailed NMR reinvestigation of cyclononane (Anet and Krane, 1980) not only disclosed existence of two of the minor conformers, [225] and [144], but also indicated that these conformers have a substantially higher entropy than the symmetrical [333] conformer, something that was not evident from the calculations. Thus it was estimated that, at room temperature, cyclononane consists of 40% of the [333], 50% of the [225], and 10% of the [144] conformer, that is, the [225] conformer is predominant at room temperature. This may explain why the [225] conformer is found in crystals of cyclononane derivatives; it also points up one of the limitations of force field calculations prior to the advent of MM3 (see Section 2-6; see also Dorofeeva, Allinger, et al., 1990b): they estimate potential energy, which is closely related to enthalpy, but in cases where entropy differences are sizeable not to free energy. It should be noted also that, since entropy contributions are small at low temperature, calculations may be borne out by low-temperature NMR experiments; yet, unless a careful determination of temperature dependence of conformational population is made, the composition at room temperature may be appreciably different from both the measured and the calculated values.

Cyclodecane derivatives have provided two interesting conformational findings. The lowest energy conformation calculated is "rectangular" [2323], it fits into a diamond lattice, and may be called boat–chair–boat (BCB) (Fig. 11.96). Most cyclodecane derivatives that have been studied by X-ray diffraction fall into this conformation, with a notable exception to be mentioned later (Dunitz, 1971).

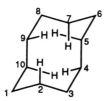

Figure 11.96. Boat–chair–boat (BCB) conformation of cyclodecane.

Indeed, when the electron diffraction pattern of cyclodecane itself was studied (Hilderbrandt et al., 1973), the experimental intensity and radial distribution curves agreed very well with those calculated for the BCB conformer. However, the bond lengths, bond angles, and torsion angles found agreed poorly with those from X-ray studies of various cyclodecane derivatives. It was therefore decided to combine the electron diffraction (ED) studies with force field calculations (this is a known method for enhancing the accuracy of electron diffraction studies). When this was done, it was found that the BCB conformation alone no longer gave a good fit for the ED data, but that a better fit was obtained by assuming cyclodecane to be a conformer mixture of about one-half the BCB conformer, one-third the TBC conformer, and the remainder about equal parts of TBCC and BCC conformers (see also Anet et al., 1972). A low-temperature crystal structure of cyclodecane has been published without comment as to its conformation (Shenhav and Schaeffer, 1981).

Contemplation of the BCB conformation shown in Figure 11.96 discloses six hydrogen atoms pointing to the interior of the ring. Since the ring is fluxional (with a barrier to interchange of all hydrogen atoms of about $7.5\,kcal\,mol^{-1}$ or $31.4\,kJ\,mol^{-1}$) one cannot assign any particular hydrogen atoms to the internal positions (except possibly in NMR spectroscopy at very low temperature). However, for steric reasons it will not be possible in the BCB conformation to place groups larger than hydrogen (e.g., methyl groups) in the inside positions. Let us thus consider a $1,1,n,n$-tetramethylcyclodecane. Using the picture in Figure 11.96, it is clear that one can construct relatively unstrained BCB conformations (with exocyclic methyl groups), if the methyl groups are 1,1,3,3 or 3,3,6,6 (the same as 1,1,4,4 because of pseudorotation) or 1,1,6,6. However, it is not possible to have such conformations when the methyl groups are 1,1,2,2 or 1,1,5,5. It is for this reason that interest in the crystal structure of 1,1,5,5-tetramethylcyclodecane-8-carboxylic acid (cf. Dunitz, 1971) arose. Solving this structure provided an interesting instance in which X-ray diffraction was combined with molecular mechanics calculations (for another example, see Rubin et al., 1984).

The crystal structure of this acid (more properly called 3,3,7,7-cyclodecanecarboxylic acid) as originally determined gave rise to considerable concern, since there appeared two very short C–C bonds [one as short as 133 pm (1.33 Å)] and three very large C–C–C angles (one as large as 136°). Something obviously was wrong. The mystery was solved when X-ray crystallography was combined with force field calculations (Dunitz, Lifson, et al., 1967). The calculation disclosed that the acid could exist in two conformations, TBC and TBCC, which differ in energy by only about $1\,kcal\,mol^{-1}$ ($4.2\,kJ\,mol^{-1}$). When it was then assumed that these two conformers would form a mixture in the crystal in a 4:1 ratio (corresponding approximately to the calculated energy difference) and that the crystal was statically disordered as a result, the X-ray data could be completely accounted for. (It might also be noted that the TBC/TBCC ratio of 4 is very close to the ratio of 4.4 found for these two conformations in cyclodecane itself—see above—albeit at a higher temperature, 130°C).

Only limited information is available for 11-, 13-, and 15-membered rings (Dale, 1976; Anet and Rawdah, 1978a). The 12-membered ring has been studied by X-ray diffraction (Dunitz and Shearer, 1960) and by NMR spectroscopy (Anet et al., 1972; Anet and Rawdah, 1978b; Anet, 1988, p. 55) and by a combination

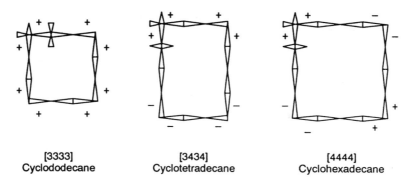

[3333]
Cyclododecane

[3434]
Cyclotetradecane

[4444]
Cyclohexadecane

Figure 11.97. Most stable conformers of cyclododecane, cyclotetradecane, and cyclohexadecane.

of electron diffraction and molecular mechanics (Atavin, Allinger, et al., 1989); the stable conformation is the "square" [3333] one ($\mathbf{D}_4$, Fig. 11.97), and the barrier for site exchange is $7.3\,\text{kcal}\,\text{mol}^{-1}$ ($30.5\,\text{kJ}\,\text{mol}^{-1}$). For monosubstituted cyclododecanes, there is a conformational equilibrium between "corner substituted" and "noncorner substituted" conformers (Schneider and Thomas, 1976). For the larger even-membered rings the most stable conformation corresponds to the diamond lattice type: "rectangular" [3434] in cyclotetradecane (Fig. 11.97; Anet et al., 1972) and "square" [4444] in cyclohexadecane (Fig. 11.97; Anet and Cheng, 1975) with a barrier to site exchange of $6.7\,\text{kcal}\,\text{mol}^{-1}$ ($28.0\,\text{kJ}\,\text{mol}^{-1}$). A cyclotridecane derivative has been investigated by a combination of X-ray diffraction and molecular mechanics (Rubin et al., 1984).

The conformation of rings larger than six-membered containing oxygen, nitrogen, sulfur, and phosphorus has been reviewed (Glass, 1988) as has the conformation and stereochemistry of phosphorus-containing rings of all sizes (Gallagher, 1987).

e. The Concept of I Strain

We have seen that angle strain is important in three- and four-membered rings and that it is larger in rings with unsaturated (sp^2 hybridized) atoms than in saturated (sp^3 hybridized) rings, because the "normal" angle is larger in the former situation than in the latter (120° vs. 109.5°), and therefore the angle deformation incurred in forming the ring is also larger (cf. Schneider et al., 1983). It follows (Brown et al., 1951, 1954; see also Eliel, 1962, pp. 265ff) that in equilibria between saturated and unsaturated species in these rings, the saturated ones should be more favored than in acyclic analogues, where such strain does not come into play. Similar considerations apply to rates, in as much as they are related to energy differences between ground and transition states: If the ground state is sp^2 and the transition state more nearly sp^3 hybridized, the reaction should be more favored than in acyclic analogues, and vice versa. In summary, in three- and four-membered rings, reactions proceeding from sp^2 to sp^3 hybridized states should be favored and those proceeding from sp^3 to sp^2 should be disfavored. This prediction is borne out as shown in lines 2 and 3 of Table 11.22.

TABLE 11.22. Predictions and Observations on I Strain.

Ring Size	Prediction for $sp^3 \rightarrow sp^2$	Reaction			Prediction for $sp^2 \rightarrow sp^3$	Reaction	
		A^a	B^b	C^c		D^d	E^e
Acyclic		1.00	1.00	1.00		1.00	1.00
3	Difficult		0.00002	0	Facile		
4	Difficult	0.97	8.50	0.0075	Facile		581
5	Facile	43.7	10.5	1.2	Difficult	3.33	15.4
6	Difficult	0.35	0.75	0.015	Facile	70.	355
7	Facile	38.0	19.0	0.79	Difficult	0.54	2.25
8	Facile	100	144	0.14	Difficult	0.081	0.172
9	Facile	15.4	129	0.036	Difficult	0.041	0.070
10	Facile	6.22	286	0.050	Difficult	small	0.0291
11	Facile	4.21	30.8		Difficult	0.063	0.0518
12	Facile		2.44	0.0080	Difficult	0.226	0.401
13	Facile	1.00	2.63		Difficult	0.269	0.427
14^f		1.00	0.99	0.020		1.17	

[a] Relative rates of solvolysis of methyl cycloalkyl chlorides in 80% ethanol (Brown et al., 1951; Brown and Borkowski, 1952).

[b] Relative rates of acetolysis of cycloalkyl tosylates (Heck and Prelog, 1955; Brown and Ham, 1956; see also Schneider and Thomas, 1980.

[c] Relative rates of reaction of cycloalkyl bromides with lithium or potassium iodide (Fierens and Verschelden, 1952; Schotsmans, Vierens, et al., 1959).

[d] Equilibrium constants for cyanohydrin formation (Prelog and Kobelt, 1949; Wheeler and Granell de Rodriguez, 1964).

[e] Relative rates of reduction of cyclanones with sodium borohydride (Brown and Ichikawa, 1957).

[f] The 14-membered and larger rings have negligble I strain. In fact, where rates or equilibrium constants have been measured, they are within less than a factor of 2 of those of the acyclic analogues.

A somewhat different rationale pertains to common and medium-sized rings (cf. Schneider et al., 1983). In these rings, angle strain plays but a minor part, but eclipsing strain is important in all but the six-membered ring, which is nearly perfectly staggered in the chair conformation. Since eclipsing is more serious for saturated than for unsaturated ring carbon atoms, the $sp^3 \rightarrow sp^2$ change will be favored and the $sp^2 \rightarrow sp^3$ change disfavored for five-membered and medium-sized rings. The six-membered ring is the one exception. Since it is already perfectly staggered in the sp^3 configuration, conversion to sp^2 will be disfavored; conversely, change from sp^2 to sp^3 is favored. Examples for rings of a variety of sizes are shown in Table 11.22; the predictions are fulfilled in most cases. The reaction of cycloalkyl bromides with iodide is a notable exception; this is an S_N2 reaction and steric hindrance in the transition state due to other causes may overwhelm any predictions based solely on I strain.

The term "I strain" (for internal strain) has been proposed by Brown et al. (1951) for the strain energy changes in rings caused by changes from sp^2 to sp^3 hybridization or vice versa. The concept has stood up under scrutiny from molecular mechanics calculations (Schneider and Thomas, 1980; Schneider et al., 1983) except possibly in the case of cyclododecane derivatives, where there are complications due to the earlier mentioned existence of two conformers. Exceptions may also occur when other factors are more important than I strain, for example, steric hindrance to rearward attack, as mentioned above for cycloalkyl halides, or neighboring group participation (with anchimeric assistance), which leads to a high solvolysis rate in cyclobutyl halides and tosylates (cf. Table 11.22; Roberts and Chambers, 1951).

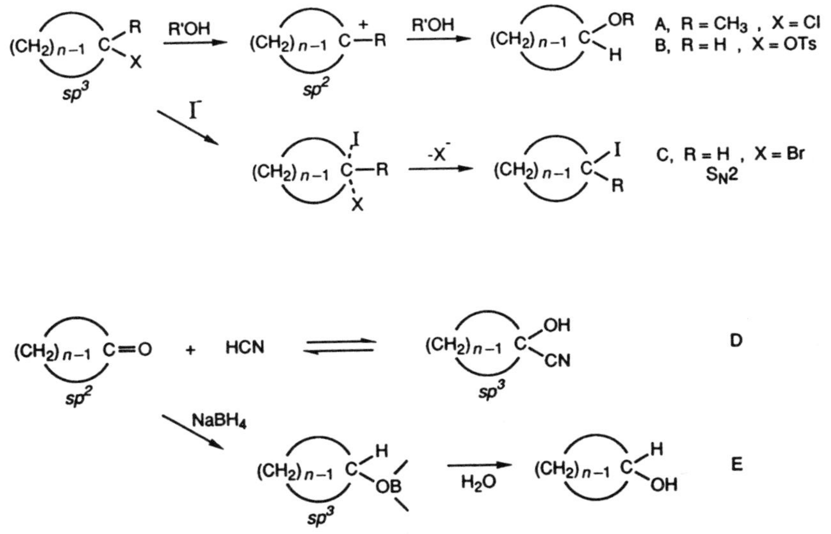

Figure 11.98. Reactions used to illustrate I strain.

Figure 11.98 illustrates the reactions included in Table 11.22 with an indication of the appropriate change in hybridization between starting material and product (equilibrium controlled reactions) or starting material and transition state (rate controlled reactions) (see also Schneider et al., 1983.)

The I-strain concept has also been applied to radical and carbanion reactions and the relative stability of exocyclic and endocyclic double bonds in ring systems, but the results are less complete and often less convincing than those indicated above (cf. Eliel, 1962 and the earlier discussions on pp. 737 and 762 concerning the relative stability of methylenecyclopententane–methylcyclopentene and methylenecyclohexane–methylcyclohexene).

11-6. STEREOCHEMISTRY OF FUSED, BRIDGED, AND CAGED RING SYSTEMS

The chemistry of ring systems of the above types has been extensively studied (e.g. Liebman and Greenberg, 1976; Greenberg and Liebman, 1978; Olah, 1990; Marchand, 1992; Naemura, 1992) and only a few representative cases will be discussed here. We shall pick examples either because they are of particular interest (high strain, extreme cases, high symmetry, etc.) or because the ring systems in question are particularly widespread in occurrence.

a. Fused Rings

The general formula for fused rings is given in Figure 11.99. The smallest, and most strained example is bicyclo[1.1.0]butane, where $m = n = 3$ (Fig. 11.100, **A**), with a strain of $66.5 \, \text{kcal mol}^{-1}$ ($278 \, \text{kJ mol}^{-1}$). [The strain energy here and elsewhere is taken from Liebman and Greenberg (1976) unless otherwise indi-

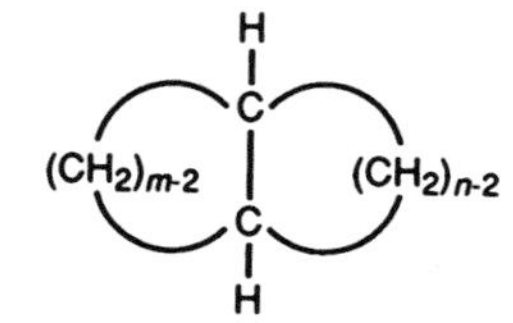

Figure 11.99. Fused rings.

A   **B**

Figure 11.100. Bicyclo[1.1.0]butane and bicyclo[2.1.0]pentane.

cated]. Its structure has been determined by Cox, Wiberg, et al. (1969); its angle of pucker, 58°, is much larger than that of cyclobutane and its C–C bonds are unusually short (149.8 pm, 1.498 Å).

The C–C bond dissociation energy in ethane is 84 kcal mol^{-1} (351.5 kJ mol^{-1}) and that in other unstrained hydrocarbons is similar; thus, if the strain energy in a system exceeds that value and can be relieved by the breaking of a single bond, the molecule might exist as a diradical. Bicyclo[1.1.0]butane is, however, not a candidate for this to happen: Apart from the fact that the strain is not large enough, breaking of a single bond would yield either a cyclopropyl or a cyclobutyl diradical, which would retain a considerable fraction of the strain, since it still contains small, strained rings.

The strain in bicyclo[1.1.0]butane (Fig. 11.100, **A**) exceeds the sum of the strains in two cyclopropane moieties (Table 11.4) by about 11.5 kcal mol^{-1} (48 kJ mol^{-1}). In contrast, the strain in the next higher fused system, bicyclo[2.1.0]pentane (Fig. 11.100, **B**), 57.3 kcal mol^{-1} (240 kJ mol^{-1}) exceeds that of the sum of cyclopropane and cyclobutane by less than 4 kcal mol^{-1} (16.7 kJ mol^{-1}) and the strain in higher bicyclo[n.1.0]alkanes is close to that of the sum of the strain in the two fused rings.

The hydrocarbons shown in Figure 11.100 are cis fused. What is the smallest bicyclic hydrocarbon of the type shown in Figure 11.99 with trans fused rings? *trans*-Bicyclo[5.1.0]octane (Fig. 11.99, $m = 7$, $n = 3$) has been prepared (Kirmse and Hase, 1968; Wiberg and de Meijere, 1969) as have ketone and alcohol derivatives of *trans*-bicyclo[4.1.0]heptane (Fig. 11.99, $m = 6$, $n = 3$), the smallest known trans fused bicyclic hydrocarbon with a three-membered ring (Paukstelis and Kao, 1972; for calculations on even smaller trans fused systems, see Svyatkin et al., 1988). In the [5.1.0] structure, the trans isomer is about 9 kcal mol^{-1} (37.6 kJ mol^{-1}) less stable than the cis (Pirkle and Lunsford, 1972) but this difference vanishes in the [6.1.0] compound (Wiberg et al., 1984). *trans*-Bicyclo[3.2.0]heptane (Fig. 11.99, $m = 5$, $n = 4$) (Mann, 1966) and its ketone precursor (Meinwald et al., 1964) are also known. The structures of the next higher homologues, *cis*- and *trans*-bicyclo[4.2.0]octane (Fig. 11.99, $m = 6$, $n = 4$) have been determined by electron diffraction (Spelbos et al., 1977). The angle of pucker ϕ of the four-membered ring (Fig. 11.84) in the cis isomer (23°) is within the normal range and the cyclohexane ring bears most of the strain, with its endocyclic torsion angle at the ring juncture reduced from the normal 55° to 32.8°. In the trans isomer, the pucker of the four-ring (45°) is abnormally high; at the same time the cyclohexane ring is also exceedingly puckered with an endocyclic torsion angle of 69.8° (so as to reduce the exocyclic e,e torsion angle). Only

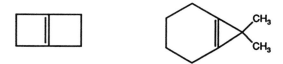

Figure 11.101. Bicyclo[2.2.0]hex-1(4)-ene and 7,7-dimethylbicyclo[4.1.0]hept-1(6)-ene.

calculations are available for *trans*-bicyclo[2.2.0]hexane (Fig. 11.99, $m = n = 4$) (Wiberg and Wendoloski, 1982). The bridged alkenes bicyclo[2.2.0]hex-1(4)-ene (Wiberg et al., 1971) and 7,7-dimethylbicyclo[4.1.0]hept-1(6)-ene (Szeimies et al., 1977; Fig. 11.101) have also been synthesized, but though theoretical work has been done on their lower homologues (Wagner, Schleyer, Pople, et al., 1978; Wiberg et al., 1983), bicyclo[3.1.0]hex-1(5)-ene (Wiberg and Bonneville, 1982) and systems containing the bicyclo[2.1.0]pent-1(4)-ene (Harnish, Szeimies, et al., 1979), and bicyclo[1.1.0]but-1(3)-ene (Szeimies-Seebach and Szeimies, 1978) exist at best as fleeting intermediates.

A molecule that has aroused considerable interest in connection with the possibility of synthesizing a structure with a planar carbon atom is fenestrane (also called windowpane), tetracyclo[3.3.1.0^{3,9}.0^{7,9}]nonane (Fig. 11.102; Venepalli and Agosta, 1987). Both cis and trans isomers can be envisaged; neither has as yet been synthesized. Unfortunately, while calculations agree that the structure would not be planar (**D$_{4h}$**), they do not agree on the minimum energy structure for the cis isomer: **S$_4$** (Minkin et al., 1980) or **C$_{4v}$** (Würthwein, Schleyer, et al., 1981). The **S$_4$** structure would, in fact, have an appreciably flattened central carbon atom, whereas the **C$_{4v}$** structure has an equally unusual pyramidal central carbon with all bonds pointing in the same direction. In any case, the structure is expected to be extremely strained, with strain energy as high as 177.5 kcal mol^{-1} (743 kJ mol^{-1}) (Wiberg and Wendoloski, 1982). Nonetheless, the lower (Wiberg et al., 1980) and higher (Dauben and Walker, 1982; Rao, Agosta, et al., 1985) homologues shown in Figure 11.103 have been prepared; the tricyclic system is no more strained than [2.2.0]bicyclohexane and the tetracyclic systems seem to be quite stable also (see also Krohn, 1991, and Keese, 1992).

We pass now to the more common fused ring systems. The cis and trans isomers of hydrindane (Fig. 11.104) have long been known. The trans (racemic) isomer has a lower heat of combustion than the cis (meso) but the difference is only 1.065 kcal mol^{-1} (4.46 kJ mol^{-1}) (Finke et al., 1972), appreciably less than that (1.74 kcal mol^{-1}, 7.28 kJ mol^{-1}) between *cis*- and *trans*-1,2-dimethylcyclohex-

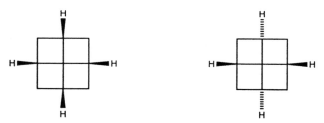

Figure 11.102. Fenestrane.

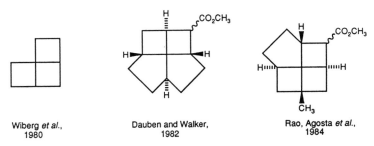

Dauben and Walker,
1982   Rao, Agosta *et al.*,
1984

Figure 11.103. Fenestrane homologues.

ane. Presumably the smaller difference is due to the strain of ring fusion in the trans isomer being larger than in the cis. This is not surprising; it can be seen in models and is a consequence of the normal maximum value of the endocyclic torsion angle in a five-membered ring being about 45° (p. 760), whereas the normal exocyclic 1,2-trans torsion angle in cyclohexane is of the order of 64° (p. 686). There must therefore be a sizeable and energetically unfavorable puckering of the six-membered ring and perhaps of the five-membered ring as well. Electron diffraction data of *trans*-hydrindane (Van den Eden and Geise, 1981) indicate that the endocyclic torsion angle in the six-membered ring at the ring juncture is indeed increased from the normal 55° to 61.1° with a corresponding decrease of the exocyclic (e,e) angle. There is less problem in the cis isomer; although the maximum desirable torsion angle in cyclopentane (45°) is appreciably less than the exocyclic cis torsion angle in cyclohexane (56°), flattening of the six-membered ring to reduce the latter angle is relatively facile (calculated geometry: Van den Eden and Geise, 1981).

> At one time, before the puckered nature of the five-membered ring was fully appreciated, it was thought that the most favorable conformation for *cis*-hydrindane would be that of a cyclohexane boat fused to a nearly flat cyclopentane, but this contention was disproved in 2-oxahydrindane (Eliel and Pillar, 1955).

Entropy favors the cis isomer in the hydrindanes as it does in the 1,2-dimethylcyclohexanes (cf. Table 11.9 and explanation thereto), but the difference is somewhat larger ($2.04 \text{ cal mol}^{-1} \text{ K}^{-1}$, $8.54 \text{ J mol}^{-1} \text{ K}^{-1}$) (Finke et al., 1972); the resulting ΔG^0 at 25°C is $0.50 \text{ kcal mol}^{-1}$ (2.09 kJ mol^{-1}) in favor of the trans isomer. Because of the countertrend of ΔH^0 (favoring trans) and ΔS^0 (favoring cis) there is a cross-over of ΔG^0 at about 200°C; above that temperature the cis isomer is more stable (Allinger and Coke, 1960).

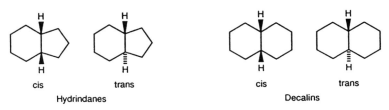

cis · Hydrindanes · trans   cis · Decalins · trans

Figure 11.104. Hydrindanes and decalins.

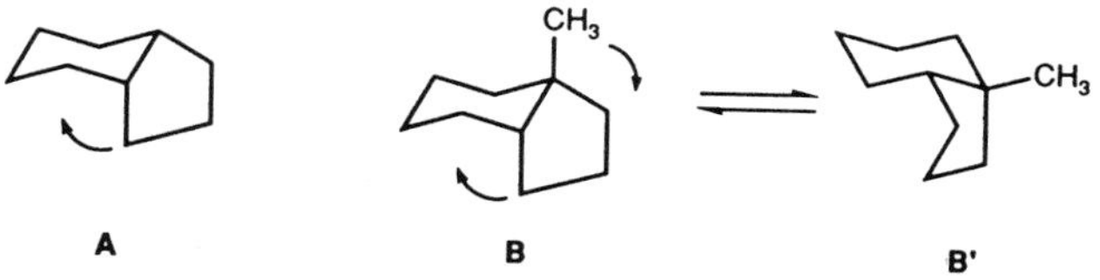

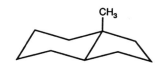

Figure 11.105.   Distortion in *cis*-hydrindane and 3a-methyl-*cis*-hydrindane.

Part of the enthalpy disadvantage of the cis isomer is due to its methylene group being pushed into the ring (Fig. 11.105, **A**). The adverse effect of this is diminished in certain hydrindane derivatives, such as 1-hydrindanone (axial C=O instead of axial CH$_2$), wherein the cis isomer becomes the more stable; the same is true of 4-hydrindanone (Lo Cicero, Dana, et al., 1981).

The situation in 3a-methylhydrindane (cf. Fig. 11.105, **B** and **B'**) is of interest, since this system (but in the trans-fused stereochemistry, cf. Fig. 11.106) exists in the C/D rings of steroids (see below). It would appear that in this system the cis–trans equilibrium should lie on the cis side, inasmuch as the cis isomer can have its methyl group equatorial (Fig. 11.105, **B'**), whereas the trans isomer (Fig. 11.106) must have it axial. Indeed, the cis isomer is favored at equilibrium by a factor of 11.5 at 200°C (Sokolova and Petrov, 1977; see also Lo Cicero, Dana, et al., 1981 for the 4-keto derivative), corresponding to $\Delta G^0 = 2.3\,\text{kcal mol}^{-1}$ (9.6 kJ mol^{-1}) at that temperature. The large magnitude of this free energy difference may come as somewhat of a surprise; it is greater than the axial–equatorial difference for methyl, 1.74 kcal mol^{-1} (7.28 kJ mol^{-1}) (ΔG^0 for the hydrindane framework is near zero at 200°C, see above). Perhaps the large ΔG^0 in part reflects the strain relief in the CH$_3$-axial conformer of the cis isomer (Fig. 11.105, **B**; note the outward bending of the axial CH$_3$), which may contribute appreciably at equilibrium and thus raise the entropy of the cis ensemble. Unfortunately, the conformational equilibrium in 3a-methyl-*cis*-hydrindane (Fig. 8.105, **B ⇌ B'**) has not been determined, presumably because the inversion barrier is quite low. There are, however, some data in the 2-thia- and 2-oxa analogues (Willer, 1981): In 2-thia-3a-methyl-*cis*-hydrindane the conformer with equatorial methyl predominates by only 0.48 kcal mol^{-1} (2.0 kJ mol^{-1}) and in the 2-oxa analogue the conformer with *axial* methyl is actually preferred by about 0.3 kcal mol^{-1} (1.3 kJ mol^{-1}).

The low inversion barrier in *cis*-hydrindane presumably reflects the strain energy in the ground state (Fig. 11.105, **A**), which is partly relieved in the transition state. Measurements in the 148–180 K region (Schneider and Nguyen-Ba, 1982) give $\Delta H^{\ddagger} = 8.8\,\text{kcal mol}^{-1}$ (37 kJ mol^{-1}) and a remarkably large $\Delta S^{\ddagger}$ of 6.7 cal mol^{-1} K^{-1} (28 J mol^{-1} K^{-1}) from which one calculates $\Delta G^{\ddagger}_{140\,\text{K}} = 7.9\,\text{kcal mol}^{-1}$ (33 kJ mol^{-1}); an earlier, more approximate determination gave $\Delta G^{\ddagger}_{140\,\text{K}}$ ca. 6.4 kcal mol^{-1} (27 kJ mol^{-1}) (Moniz and Dixon, 1961). From the more recent data one calculates $\Delta G^0_{298} = 6.9\,\text{kcal mol}^{-1}$ (29 kJ mol^{-1}) at room temperature.

Figure 11.106.   3a-Methyl-*trans*-hydrindane.

cis   trans

Figure 11.107.   The structures of *cis*- and *trans*-bicyclo[3.3.0]nonane.

The lower homologue of hydrindane, bicyclo[3.3.0]nonane (Fig. 11.107), was obtained as cis and trans isomers as early as 1936 (Barrett and Linstead). The considerations mentioned in connection with *trans*-hydrindane apply, a fortiori, to the trans isomer of this lower homologue: The maximum normal pucker in cyclopentane (45°) is ill adjusted to the exocyclic trans torsion angle in cyclopentane (75°) and considerable distortion of both rings must result. Indeed the strain in this isomer is $6.4 \, \text{kcal mol}^{-1}$ ($26.8 \, \text{kJ mol}^{-1}$) [Chang, Boyd, et al., 1970; a very similar difference, $6.0 \, \text{kcal mol}^{-1}$ ($25.1 \, \text{kJ mol}^{-1}$) was measured by Barrett and Linstead in 1936]. Surprisingly, the trans isomers of hetero analogues of *trans*-bicyclo[3.3.0]nonane, with O or S in lieu of $CH_2(2)$ (Owen and Peto, 1955) or O and NR in lieu of CH_2 (1 and 3) (Barkworth and Crabb, 1981) are readily prepared by ordinary synthetic methods; it is not known whether these systems are less strained or whether they can be readily prepared despite being strained.

Decalin (Fig. 11.104) is a molecule of historical importance because it served to confirm, experimentally, the Sachse–Mohr theory of the chair (and boat) shaped six-membered ring (cf. Section 11-4.a). Two planar six-membered rings (as pictured by Baeyer) can only be fused cis; Mohr in 1918 predicted, in contrast, that decalin with puckered six-membered rings should exist as both cis and trans isomers. This hypothesis was confirmed through the isolation of *trans*-decalin by Hückel in 1925 (q.v.). At that time it was believed that *cis*-decalin existed as a double boat, but it is now quite certain, and has, in fact, been demonstrated by electron diffraction measurements (Van den Enden, Geise, et al., 1978) that both isomers exist as double chairs, as undistorted as cyclohexane itself (Fig. 11.108).

> Strictly speaking, of course, the existence of *trans*-decalin can be explained by any number of nonplanar structures of which the chair and boat are only two examples. For example, in the 5/5 system depicted in Fig. 11.107, the rings are clearly not chair or boat shaped, and yet a trans fused structure is possible.

The framework of *trans*-decalin is locked: It can invert to a chair–boat or even boat–boat system of considerably higher energy, but it cannot invert to an

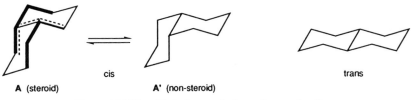

A (steroid)   cis   A' (non-steroid)   trans

Figure 11.108.   Chair forms of *cis*- and *trans*-decalin.

alternative chair–chair system (as cyclohexane can), since this would require spanning two axial positions on opposite sides of the chair with four methylene groups, a geometric impossibility. The *trans*-decalin system, like the anancomeric 4-*tert*-butylcyclohexyl system discussed earlier, thus serves as a conformational anchor; we shall see evidence of this later. In contrast, *cis*-decalin has two interconverting chair–chair combinations (both chairs have to be inverted to effect the interconversion). Just as in *cis*-1,2-dimethylcyclohexane, these two conformers (**A**, **A′**, Fig. 11.108) are not identical but enantiomeric.

If one considers each ring in decalin as being a substituent of the other ring, one notes that in *trans*-decalin all pertinent methylene substituents are equatorial, and therefore antiperiplanar. (There are gauche methylene interactions, but they are all within one ring, as in cyclohexane itself and in *cis*-decalin, therefore they cancel in the computation of the overall enthalpy difference.) In contrast, there are three extraannular gauche interactions in *cis*-decalin. They are indicated in Figure 11.108, **A** either by heavy bonds or by an extra dotted line. It would appear at first that there are four such interactions, since one methylene in each ring is axial to the other, giving rise to two gauche interactions, or four in the two rings. However, if one marks the pertinent interactions, as done in Figure 11.108, one notices that one of them (dotted) is common to the two axial methylenes, so there remain only three. If each such interaction is worth 0.87 kcal mol^{-1} (3.64 kJ mol^{-1}) in the liquid phase (one-half the axial methylcyclohexane interaction of 1.74 kcal mol^{-1} or 7.28 kJ mol^{-1}; cf. Table 11.7) the total should be 2.61 kcal mol^{-1} (10.9 kJ mol^{-1}); the experimental value (heat of isomerization: Allinger and Coke, 1959; Schucker, 1981; difference in heats of combustion: Speros and Rossini, 1960), 2.7 kcal mol^{-1} (11.3 kJ mol^{-1}), is in excellent agreement. Similarly good agreement is found in the gas phase: calculated (on the basis of the data in Table 11.9), 2.91 kcal mol^{-1} (12.2 kJ mol^{-1}); found, 3.1 kcal mol^{-1} (13.0 kJ mol^{-1}).

The symmetry point groups of *cis*- and *trans*-decalin are **C$_2$** and **C$_{2h}$**, respectively; both have a symmetry number of 2 and an entropy of symmetry of $-R \ln 2$. However, the cis isomer exists as a pair of enantiomers (albeit not isolable at room temperature since ring inversion converts one enantiomer into the other), and therefore has an entropy of mixing of $R \ln 2$; the trans isomer is meso, so the entropy advantage of the cis should be $R \ln 2$ or 1.38 cal mol^{-1} K^{-1} (5.76 J mol^{-1} K^{-1}). The experimental value (Allinger and Coke, 1959; Schucker, 1981) of 0.55–0.60 cal mol^{-1} K^{-1} (2.30–2.51 J mol^{-1} K^{-1}) is appreciably less than that, thus there must be additional factors affecting the entropy difference between the cis and trans isomers. The barrier to ring inversion $\Delta G^{\ddagger}$ in *cis*-decalin is 12.3–12.6 kcal mol^{-1} (51.5–52.7 kJ mol^{-1}) at room temperature (Dalling et al., 1971; Mann, 1976), appreciably higher than that in cyclohexane.

As in cyclohexane, there is uncertainty about the dissection of this value into activation enthalpy [13.6 kcal mol^{-1} (56.9 kJ mol^{-1}) in one determination, 12.4 kcal mol^{-1} (51.9 kJ mol^{-1}) in the other] and entropy [3.5 cal mol^{-1} K^{-1} (14.6 J mol^{-1} K^{-1}) vs. 0.2 cal mol^{-1} K^{-1} (0.8 J mol^{-1} K^{-1})]. Calculations (Baas et al., 1981) yield a value of 12.0 kcal mol^{-1} (50.2 kJ mol^{-1}) for $\Delta H^{\ddagger}$ and suggest that the inversion path involves the following transformations: chair–chair (CC) → chair–twist (CT) → twist–twist (TT) with the highest barrier occurring between CT and TT. The TT conformer pseudorotates (over a low barrier) to an

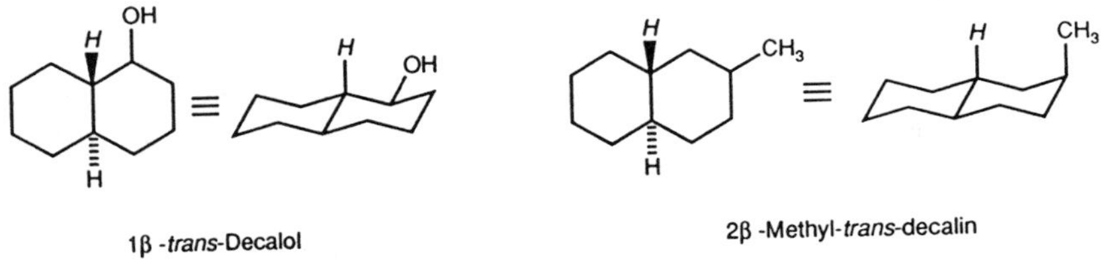

1β -*trans*-Decalol   2β -Methyl-*trans*-decalin

Figure 11.109. Monosubstituted *trans*-decalins.

alternate TT conformation that is then reconverted to the CT and CC by a (mirror imaged) reversal of the path by which it was formed.

Although one must be careful in drawing inferences from manipulations of models (because of the incorrect balance of angle and torsional strain in the latter), mechanical inversion of the two rings in a *cis*-decalin model does correctly pinpoint the high-energy region of the barrier.

Since *trans*-decalin is a conformationally rigid chair–chair system, substituents in *trans*-decalin will occupy well-defined axial or equatorial positions (cf. Fig. 11.19). Examples are shown in Figure 11.109: In 1-hydroxy-*trans*-decalin the hydroxyl substituent is equatorial, whereas in 2-methyl-*trans*-decalin, the methyl substituent is axial.

The parochial descriptors α and β are used to denote relative configuration in the decalin system. The descriptor β is used for a ligand on the same side as the proximal ring-junction hydrogen, and α is used for a ligand on the opposite side. (The proximal hydrogen atoms are italicized in Figs. 11.109 and 11.110.)

Since the *cis*-decalin system is mobile, substituents in it can occupy either the equatorial or the axial position, and since the framework equilibrium is unbiased (1:1, inasmuch as the two conformers are enantiomeric, see above), the preferred conformation will be that with the substituent equatorial. Thus equilibrium **A** in Figure 11.110 will be shifted to the right (equatorial OH) to the extent of the ΔG^0 value of the OH group (Table 11.7), whereas equilibrium **B** will be shifted to the right to the extent of a synaxial $NH_2/CH_2/H$ interaction (cf. Table 11.10).

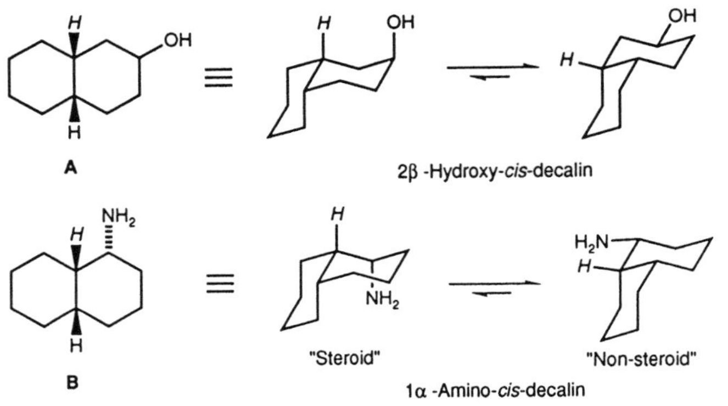

A   2β -Hydroxy-*cis*-decalin

"Steroid"   "Non-steroid"

B   1α -Amino-*cis*-decalin

Figure 11.110. Monosubstituted *cis*-decalins.

For reasons of easier representation, the inverted chair is conventionally rotated through a 60° angle; the reader should become convinced, by the use of models, that the chair conformations depicted on the right in Figure 11.110 are, in fact, obtained from those shown in the middle by an inversion of both rings. Because of analogies with the A/B ring system in cis-fused steroids, to be discussed below (p. 785), the conformations in the middle of Figure 11.110 are sometimes called "steroid" and those on the right "non-steroid" (see also Fig. 11.108).

In hetera-*cis*-decalins (Fig. 11.111), such as *cis*-decahydroquinolines (X = NH or NCH$_3$; Booth and Griffith, 1973; Vierhapper and Eliel, 1977; see also Crabb, 1992) and *cis*-1-thiadecalins (X = S, Vierhapper and Willer, 1977b) the situation is slightly more complex, since the two inverted chair forms are no longer enantiomeric, and therefore no longer of equal energy. In the parent compounds (Fig. 11.111, **C**) the conformation on the right is preferred because it is better to have a methylene group synaxial to a lone pair of electrons than to have it synaxial to a hydrogen (cf. Table 11.20 and Fig. 11.70). The equilibrium constant K is 1.38 for X = S, 2.33–2.45 for X = NCH$_3$, and 10–14.4 for X = NH. (The large difference between X = NH and X = NCH$_3$ is mysterious as to its origin but corresponds to the similar difference between 2-methylpiperidine and N,2-dimethylpiperidine shown in Table 11.20.) When a substituent, such as a methyl group, is introduced in the ring, it may reinforce or oppose the equilibrium shown in Figure 11.111, **C**. Example **D** represents a case where the substituent opposes the equilibrium of the framework and wins out; the equilibrium is shifted to the left (equatorial CH$_3$ preferred over axial). The equilibrium constant K' is 0.20 for X = S, 0.69 for X = NH and for X = NCH$_3$.

The cis–trans energy difference in 4a-methyldecalin (Fig. 11.112) is considerably less than in decalin itself: Addition of the angular methyl substituent increases the conformational energy of the trans isomer by 4 CH$_3$/H synaxial interactions (roughly equivalent to butane–gauche interactions), whereas only two such interactions are added in the cis isomer. Since the cis isomer already has three butane–gauche interactions (see above), addition of the methyl group increases this number to five, one more than in the trans isomer. The difference in enthalpy should

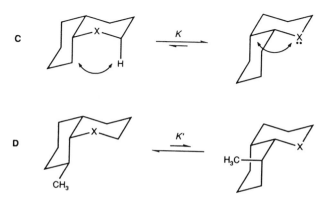

Figure 11.111. Conformational equilibria in 1-hetera-*cis*-decalins and substituted 1-hetera-*cis*-decalins.

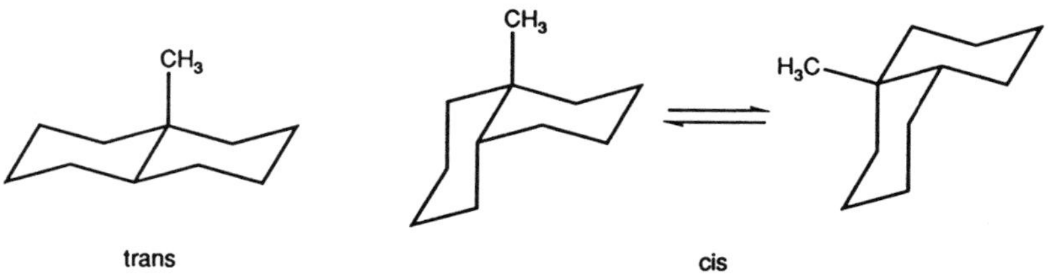

Figure 11.112.   The 4a-methyldecalins.

therefore be of the order of one butane–gauche interaction or $0.6 \, \text{kcal mol}^{-1}$ ($2.5 \, \text{kJ mol}^{-1}$). Experimentally, *cis*-4a-methyldecalin exceeds the trans isomer in enthalpy by $0.55 \pm 0.28 \, \text{kcal mol}^{-1}$ ($2.30 \pm 1.17 \, \text{kJ mol}^{-1}$) as measured by the temperature dependence of the isomerization equilibrium (Allinger and Coke, 1961) or $1.39 \pm 0.64 \, \text{kcal mol}^{-1}$ ($5.82 \pm 2.68 \, \text{kJ mol}^{-1}$) as measured by the difference in heats of combustion (Dauben, Rossini et al., 1960). The two determinations overlap with each other within the reported standard deviations and at least the former is close to the calculated value.

In perhydroazulene (Fig. 11.99, $m = 7$, $n = 5$), a bicyclic system isomeric with decalin, the difference between cis and trans isomers is quite small (Allinger and Zalkow, 1961) and this is probably generally true when one of the fused rings is seven membered or larger and the other is larger than four membered. In the 4-keto derivative of perhydroazulene, the trans isomer is appreciably more stable ($K = 6.75$; House et al., 1983).

Among fused systems containing three rings, the perhydrophenanthrenes and perhydroanthracenes have been most extensively studied. The perhydrophenanthrenes (Fig. 11.113) constitute an ABBA system and there are four pairs of enantiomers and two meso forms (cf. the hexaric acids, Fig. 3.28). The compounds were synthesized and their stereochemistry elucidated in an elegant piece of work many years ago (Linstead et al., 1942, Linstead and Whetstone, 1950), which has been summarized elsewhere (Eliel, 1962, pp. 282ff) and even today presents a challenging exercise in stereochemical reasoning.

The relative stabilities of the six isomers were assigned by Johnson (1953) on the basis of conformational arguments: **A** has no axial substituents, **B** and **C** have one each, **D** has two, **E** has a pair of synaxial substituents (marked by heavy lines) and the configuration of **F** is such that the central ring must be in the twist form. (If it were a chair, the junctions to the outer rings would be e, e, a, a, and, as we have already discussed, the a,a fusion in the *trans*-decalin moiety is sterically prohibited.) The twist structure has been verified in a derivative by X-ray structure analysis (Allinger et al., 1984).

Force field calculations (Allinger et al., 1971) yield the following predicted conformational enthalpies: **A**, 0; **B**, $2.44 \, \text{kcal mol}^{-1}$ ($10.2 \, \text{kJ mol}^{-1}$); **C**, $2.57 \, \text{kcal mol}^{-1}$ ($10.75 \, \text{kJ mol}^{-1}$); **D**, $4.01 \, \text{kcal mol}^{-1}$ ($16.8 \, \text{kJ mol}^{-1}$); **E**, $9.01 \, \text{kcal mol}^{-1}$ ($37.7 \, \text{kJ mol}^{-1}$); **F**, $7.03 \, \text{kcal mol}^{-1}$ ($29.4 \, \text{kJ mol}^{-1}$). Experimental equilibrium data (Honig and Allinger, 1985) indicate the following relative free energies at 348°C: **A**, 0; **B**, $2.25 \, \text{kcal mol}^{-1}$ ($9.4 \, \text{kJ mol}^{-1}$); **C**, $2.66 \, \text{kcal mol}^{-1}$ ($11.1 \, \text{kJ mol}^{-1}$); **D**, $4.60 \, \text{kcal mol}^{-1}$ ($19.2 \, \text{kJ mol}^{-1}$); **E**, $7.43 \, \text{kcal mol}^{-1}$

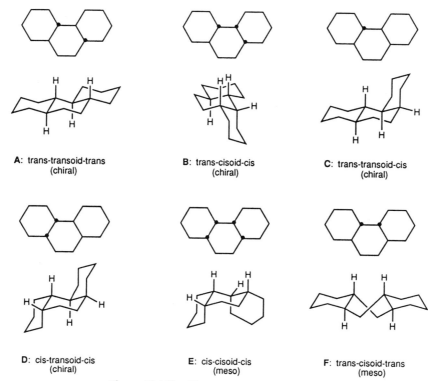

Figure 11.113. The perhydrophenanthrenes.

$(31.1 \, \mathrm{kJ \, mol^{-1}})$; **F**, 8.98 kcal mol^{-1} $(37.6 \, \mathrm{kJ \, mol^{-1}})$. In view of (small) differences in entropy of mixing and entropy of symmetry between isomers, the approximate agreement of calculated $(\Delta\Delta H^0)$ and experimental $(\Delta\Delta G^0)$ energies must be considered satisfactory except for the highly strained isomers **E** and **F**, where the relative calculated energies are the reverse of the experimental.

A note on nomenclature and notation is in order here. A dot at a ring juncture indicates an upward or forward hydrogen, the absence of a dot indicates a downward or backward hydrogen. The descriptors "trans" and "cis" are used for the decalin-like ring junctures (cf. Figs. 11.104 and 11.108), whereas the descriptors "cisoid" and "transoid" are used for two adjacent ring junctures to the central ring involving the two peripheral rings. (In the older literature, "syn" and "anti" were used instead of cisoid and transoid.)

The perhydroanthracenes are shown in Figure 11.114. There are three meso isomers and two enantiomer pairs (the cis–transoid–cis isomer has a center of symmetry). The relative stability of the isomers, originally predicted by Johnson (1953) has been both calculated by molecular mechanics and determined experimentally (Allinger and Wuesthoff, 1971); the calculated and experimental data are summarized in Table 11.23. The instability of isomer **E** is due to the enforced boat (not twist!) conformation of the central ring, that of isomer **D** results from the presence of the synaxial methylene interaction. It is interesting that an alternate conformer of **D** with the central ring as a boat is of only slightly higher energy than the chair and contributes to the extent of 13% at 271°C.

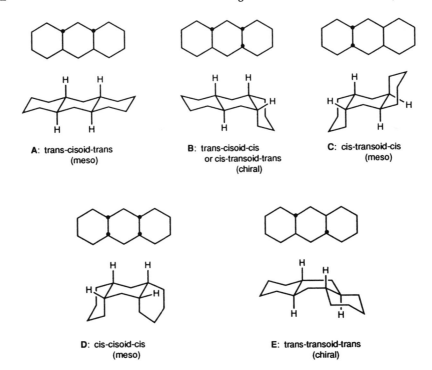

Figure 11.114. The perhydroanthracenes.

We now digress briefly into a highly unsaturated system, 9,10-dihydroanth-racene (Fig. 11.115, **C**; Rabideau, 1989b, p. 110; Morin and Grant, 1989, p. 162). We had mentioned earlier that 1,4-dihydrobenzene (Fig. 11.115, **A**) is a planar molecule with a rather shallow energy minimum. Introduction of bulky sub-stituents at C(1) deplanarizes both this molecule (Grossel et al., 1978; Rabideau et al., 1982) and the related 1,4-dihydronaphthalene (Fig. 11.115, **B**; Rabideau, 1978; Rabert et al., 1982). The substituents will occupy a pseudoaxial position, presumably because of $A^{(1,2)}$ strain (Fig. 11.65) involving the peri position in the

TABLE 11.23. Calculated and Experimental Conformational Enthalpies and Entropies of Perhydroanthracenes[a].

Isomer	ΔH^0		ΔS^0	
	Calcd. (kcal mol^{-1})[b]	Exptl. (kcal mol^{-1})[b]	Calcd. (cal mol^{-1} K^{-1})[c]	Exptl. (cal mol^{-1} K^{-1})[c]
A	0 (0)	0 (0)	0 (0)	0 (0)
B	2.62 (11.0)	2.76 (11.5)	+2.8 (+11.7)	+2.1 (+8.8)
C	5.56 (23.3)	5.58 (23.3)	+1.4 (+5.9)	+0.3 (+1.3)
D	8.13 (34.0)[d]	8.74 (36.6)	+2.2 (+9.2)[d]	+4.0 (+16.7)
E	5.86 (24.5)	4.15 (17.4)	0 (0)	−1.6 (−6.7)

[a] Relative to isomer **A**.
[b] Values in parentheses are in kilojoules per mole (kJ mol^{-1}).
[c] Values in parentheses are in joules per mole per degree kelvin (J mol^{-1} K^{-1}).
[d] Includes contribution of 13% boat form.

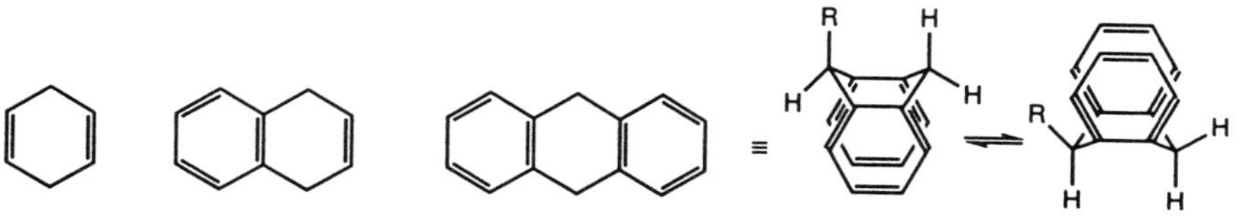

Figure 11.115. Structures of 1,4-dihydrobenzene, 1,4-dihydronaphthalene, and 9,10-dihydroanthracene.

dihydronaphthalene. When there are two peri positions, as in 9,10-dihydroanthracene, the nonplanar geometry (Fig. 11.115, **C′**) is assumed even by the parent molecule, R = H (at least in the solid state: Ferrier and Iball, 1954; Herbstein et al., 1986). Conformational analysis of 9-substituted and 9,10-disubstituted derivatives (cf. Rabideau, 1978) is somewhat difficult, since the inversion barrier in the nonplanar system is quite low and the conformers cannot be "frozen out" by NMR.

An exception is 9-dibromomethylene-10,10-dimethyl-9,10-dihydroanthracene (Fig. 11.116) in which the barrier to ring inversion (and methyl exchange) is as high as 18 kcal mol^{-1} (75 kJ mol^{-1}) (Curtin et al., 1964). In this system $A^{(1,3)}$ is superposed upon $A^{(1,2)}$ strain in the planar transition state for ring inversion, causing the very high barrier.

Nonetheless, indirect methods, such as NOE determinations or chemical shifts taken from model compounds, have been used to assess the conformations of 9-substituted 9,10-dihydroanthracenes and in almost all cases it is found that the substituent is predominantly pseudoaxial (Rabideau, 1978, 1989b). The situation here is quite different from that in cyclohexane: There are no synaxial hydrogen atoms and the only interference for a pseudoaxial substituent comes from the transannular axial hydrogen at C(10). This interference can be minimized by some flattening of the ring. In contrast, in the planar or pseudoequatorial conformation, there would be severe steric interaction of the substituent with the peri hydrogen atoms of the benzene rings. The preference for the pseudoaxial position ranges from very high for isopropyl and *tert*-butyl, to substantial (75–87%) for methyl, and to no preference for phenyl; in the case of phenyl it is clear that interference of the equatorial substituent with the peri hydrogen atoms can be avoided if the phenyl substituent is at right angles to the plane of the fused

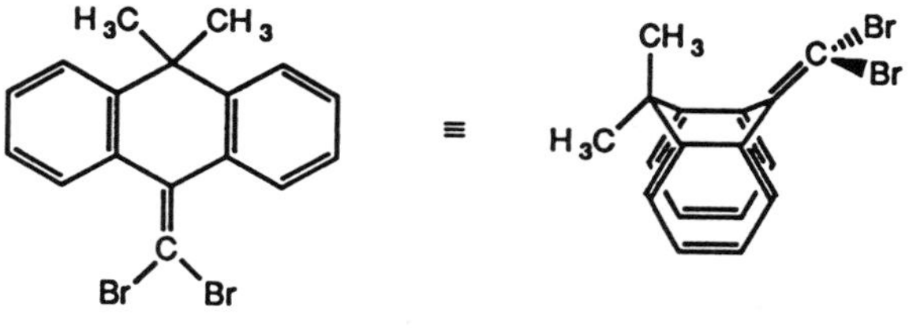

Figure 11.116. Structure of 9-dibromomethylene-10,10-dimethyl-9,10-dihydroanthracene.

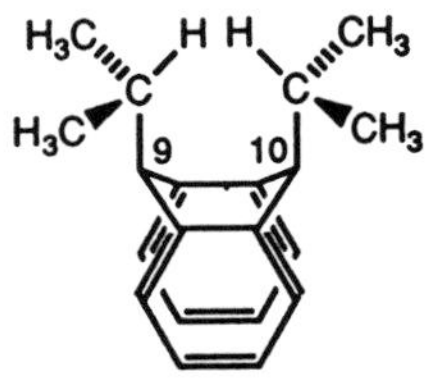

Figure 11.117.   Structure of *cis*-9,10-diisopropyl-9,10-dihydroanthracene.

phenyl moieties of the dihydroanthracene, that is, in a conformation similar to that which it assumes in equatorially substituted phenylcyclohexane (Fig. 11.22).

Even in a 9,10-cis disubstituted derivative as in *cis*-9,10-diisopropyl-9,10-dihydroanthracene (Fig. 11.117), the two substituents are diaxial. Their mutual interaction is diminished by flattening of the central ring and by the isopropyl groups being oriented with their hydrogen atoms inward (Fig. 11.117; Zieger et al., 1969). As a result, the coupling constant between the C(9) protons and the isopropyl protons is as high as 9.5 Hz. However, in 9,9,10,10-tetrasubstituted 9,10-dihydroanthracenes, the ring system is planar (cf. Rabideau, 1989b). Solid state NMR experiments (Dalling, Grant, et al., 1981) generally support the above findings.

A detailed discussion of the conformation of thioxanthenes [S instead of CH_2 at C(10) of 9,10-dihydroanthracenes] and other hetera analogues of the latter is available (Evans, 1989).

A similar system, that is easier to study because inversion is easily stopped on the NMR time scale [barriers of 12.7–15.5 kcal mol^{-1} (53.1–64.9 kJ mol^{-1})], is the 7,12-dihydropleiadene system (Fig. 11.118; Lansbury, 1969). This system also displays a preference for pseudoaxial substituents but less so than the 9,10-dihydroanthracene system: whereas isopropyl, phenyl, carbomethoxy, methoxyl, and chlorine substituents prefer the pseudoaxial position, methyl and hydroxyl substituents are predominantly pseudoequatorial, and for ethyl the ratio is nearly 1:1. The rationale of this order of preference does not seem to be entirely clear; neither steric nor polar factors seem to be wholly responsible.

We return now to the saturated fused ring systems and in particular to steroids, a group of compounds of wide natural occurrence. The constitution and relative as well as absolute configuration of these compounds has been determined both by chemical methods (cf. Fieser and Fieser, 1959; Klyne and Buckingham, 1978) and by X-ray crystallography (Fawcett and Trotter, 1966; cf. Duax et al., 1976). Virtually all natural steroids fall into two stereochemical categories: the A/B trans (trans–transoid–trans–transoid–trans) and A/B cis (cis–transoid–trans–transoid–trans) series as shown in Figure 11.119. The con-

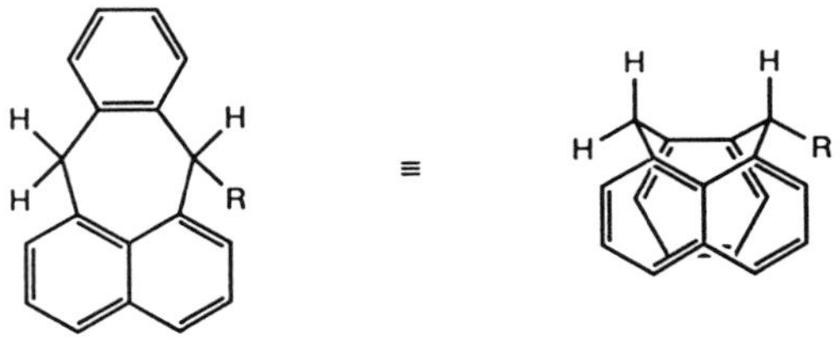

Figure 11.118.   Structure of 7,12-dihydropleiadene.

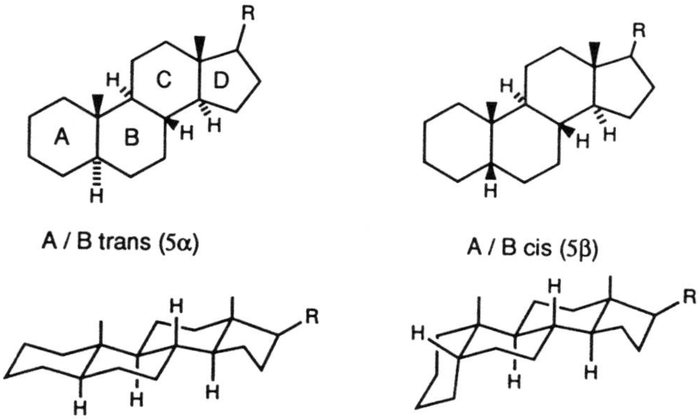

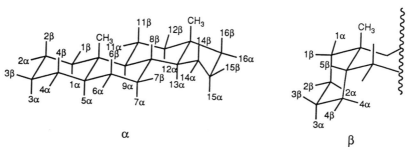

A / B trans (5α)   A / B cis (5β)

Figure 11.119. Steroids.

formational formulas shown in Figure 11.120 were proposed by Barton in his classical 1950 paper and account for many if not most physical and chemical properties of steroids. The conformational principles are the same as those earlier enunciated (Section 11-4.a) for simpler cyclohexanes. Before discussing them, we need to explain the parochial stereochemical nomenclature in steroids: Positions on the same side as the angular methyl groups are denoted as β, those on the opposite side are denoted as α. Figure 11.120 shows these positions for the A/B trans fused system (left) and, for the A ring only, in the A/B cis fused system (right; the descriptors in rings B, C, and D are the same in the two systems). There is no 1:1 relationship between α and β and axial and equatorial, but Figure 11.120 allows one to infer this relationship for each position of the A/B trans or A/B cis fused steroid nucleus.

Numerous examples of the conformational rationale of the reactivity and stability of stereoisomers in steroids as well as triterpenoids and alkaloids have been given elsewhere (Eliel et al., 1965, Chapter 5). Equatorial substituents are more stable than axial substituents and they react more rapidly in reactions governed by steric hindrance, such as acetylation of alcohols and saponification of esters; they react more slowly in reactions governed by steric assistance, such as chromic acid oxidation (cf. Fig. 11.48). Addition and elimination reactions proceed via the diaxial mode; thus bromine addition to 2-cholestene (A/B trans, double bond in 2,3) gives initially the diaxial $2\beta,3\alpha$-dibromocholestane which, on heating, is partially converted to the diequatorial $2\alpha,3\beta$ isomer (thermodynamic

Figure 11.120. Stereochemical descriptors in steroids.

control; both isomers are present at equilibrium because of the earlier discussed counterplay of steric and polar factors). Iodide-induced elimination of the $2\beta,3\alpha$-dibromide (diaxial) is very much faster than that of the $2\alpha,3\beta$-dibromide (diequatorial), and so on. The axial $2\beta,4\beta,6\beta$, and 8β positions are more hindered than other axial positions by the presence of synaxial methyl groups; the 11β position, which has two synaxial methyl groups, is the most hindered of all. The five-membered ring is less puckered than the six-membered ring, thus the axial and equatorial nature of the substituents in this ring is not as definitive.

The two angular methyl groups inhibit approach of reagents from the top (β) side. As a result, additions to double bonds (e.g., epoxidations and carbene additions), which in simple cyclohexenes have an equal chance of occurring on either face of the double bond, in steroids usually proceed by α attack.

We conclude this section with a flow-sheet showing the synthesis of four diastereoisomeric 5α-cholestane-2,3-diols (Fig. 11.121, A/B trans) (Henbest and Smith, 1957; Shoppee et al., 1957), which illustrates a number of the above stereochemical principles (see also Fig. 9.59). Peracid oxidation is initiated by epoxidation on the less hindered face followed by diaxial ring opening (Fürst–Plattner rule) to give the 2β-acetoxy-3α-hydroxy compound, which is then hydrolyzed to the $2\beta,3\alpha$ (trans) diol. Oxidation of 5α-2-cholestene with OsO_4 (molecular addition) proceeds from the less hindered α side to lead to the $2\alpha,3\alpha$ (cis) diol. A

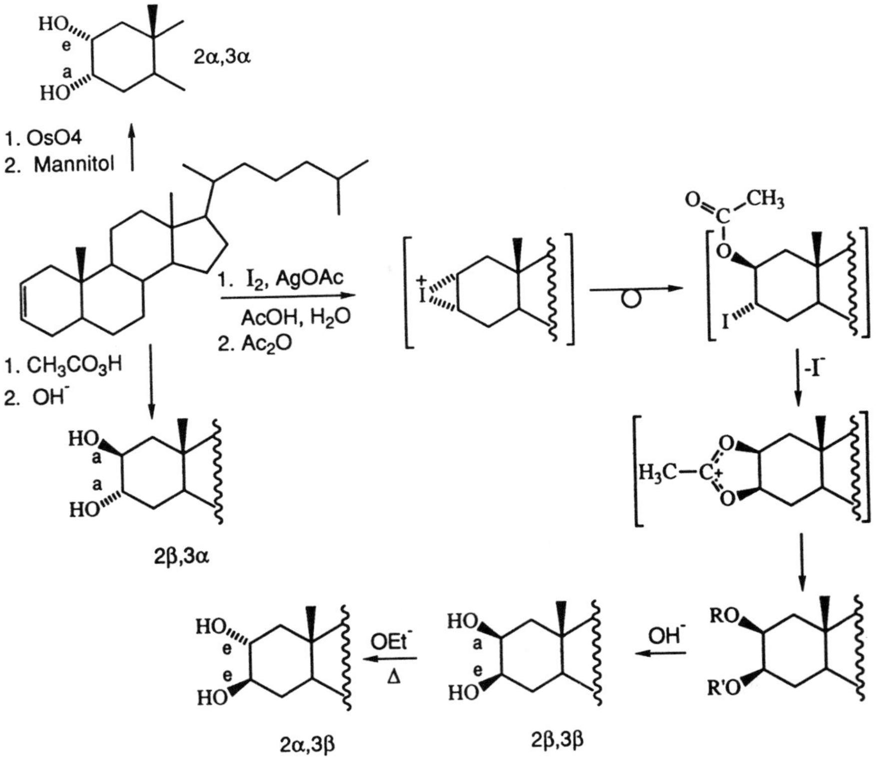

Figure 11.121. Synthesis of 5α-cholestane-2,3-diols. R = H, R′ = Ac before acetylation and R = R′ = Ac isolated after acetylation.

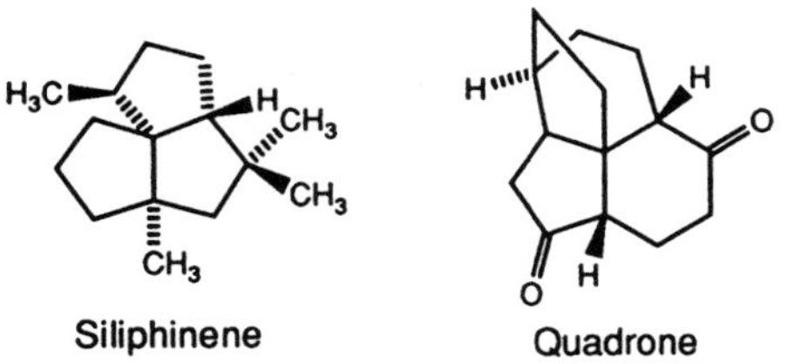

Siliphinene

Quadrone

Figure 11.122. Structures of siliphinene and quadrone.

more complex series of transformations is required to obtain the other cis ($2\beta,3\beta$) diol, although the synthesis is simple in practice. The alkene is treated with iodine and silver acetate in moist acetic acid. The first step is formation of an iodonium ion from the less hindered α side followed by diaxial ring opening to give the 2β-acetoxy-3α-iodo compound, which is not, however, isolated but cyclizes to an α orthoester intermediate [inversion at C(3)], which is immediately opened by water to give one or other of the monoacetates of the $2\beta,3\beta$-diol. Saponification then yields the free $2\beta,3\beta$-diol. The fourth isomer, the $2\alpha,3\beta$-diol (trans), is diequatorial, and therefore thermodynamically the most stable isomer; it can be prepared by equilibrating the $2\beta,3\beta$- or $2\alpha,3\alpha$-diol with sodium ethoxide–ethanol at elevated temperature. All four diastereomers can thus be prepared.

There are many natural products that display multiple fused rings with common bridgeheads. Merely by way of example the natural products siliphinene, with three rings fused to a common vertex (a so-called "angular triquinane", cf. Paquette, 1979, 1984a) and quadrone with four such rings are shown in Figure 11.122. The general question as to how many rings can have a common vertex has been addressed by Gund and Gund (1981) along with a consideration of the strains involved as well as of points of nomenclature. By way of an example, the four bonds emanating from a common vertex can be spanned by six rings (if we number the bonds 1, 2, 3, and 4 the rings can span bonds 1 and 2; 1 and 3; 1 and 4; 2 and 3; 2 and 4; and 3 and 4). Six is therefore the maximum number of rings that can meet in one vertex.

b. Bridged Rings

Figure 11.123 shows the general formula of simply bridged rings; m, n, and $o \neq 0$ (contrast the fused ring system shown in Fig. 11.99). The smallest possible bridged ring system, bicyclo[1.1.1]pentane (Fig. 11.124, **A**) has been synthesized (Wiberg and Connor, 1966) and is reasonably stable. Its structure has been

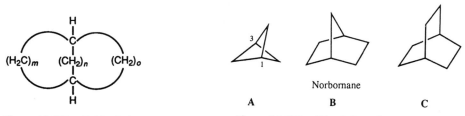

Figure 11.123. Bridged rings.

Norbornane

A   B   C

Figure 11.124. Bicyclo[1.1.1]pentane, norbornane, and bicyclo[2.2.2]octane.

determined by electron diffraction (Chiang and Bauer, 1970; Almenningen et al., 1971) and displays a remarkably short transannular nonbonded C(1)–C(3) distance of 184.5–187.4 pm (1.845–1.874 Å); in fact this is the shortest distance known for nonbonded carbon atoms. There is also a remarkably large H(1)–H(3) coupling in the ^{1}H NMR: $^4J = 18$ Hz, suggesting a strong transannular interaction (cf. Wiberg and Connor, 1966). However, this interaction is not manifest in other properties of the bicyclo[1.1.1]pentane system, such as the effect of 3-substituents on the strength of the 1-carboxylic acid (Applequist et al., 1982).

The strain energy of the bicyclo[1.1.1]pentane system has not been determined experimentally but has been computed by ab initio calculations to be 60–68 kcal mol^{-1} (251–284 kJ mol^{-1}) (Newton and Schulman, 1972; Wiberg and Wendoloski, 1982; Wiberg et al., 1987). The systems with larger m, n, and o (Fig. 11.123) ranging from 1, 1, and 2 to 3, 3, and 3 are also all known (cf. Liebman and Greenberg, 1976; Greenberg and Liebman, 1978) and their strains (cf. Wiberg et al., 1987) have been computed by force field (Allinger et al., 1971; Engler, Schleyer, et al., 1973) or ab initio calculations (Wiberg and Wendoloski, 1982); they are, of course, less strained than the [1.1.1] system. Thus the strain in the very common norbornane system (Fig. 11.124, **B**, $m = n = 2$, $o = 1$) is 17.0 kcal mol^{-1} (71.1 kJ mol^{-1}); the main source of this strain is a serious deformation of the angle at the one-carbon bridge to 93°–96° (Chiang, Wilcox, and Bauer, 1968; Dallinga and Toneman, 1968; Yokozeki and Kuchitsu, 1971) plus the eclipsing of the hydrogen atoms at the two-carbon bridges. Eclipsing must be a major source of strain in bicyclo[2.2.2]octane (Fig. 11.124, **C**). There has been some controversy (cf. Ermer and Dunitz, 1969) whether the molecule is a triple boat, as shown in Figure 11.124 ($\mathbf{D_{3h}}$) or whether it is twisted out of this conformation of highest symmetry into a $\mathbf{D_3}$ shape to relieve eclipsing strain. In the solid state, the compound seems to be $\mathbf{D_{3h}}$; but the electron diffraction pattern in the vapor suggest some twisting (about 10°–12° on either side) with a double-minimum potential curve with a slight hump [ca. 0.1 kcal mol^{-1} (0.4 kJ mol^{-1})] at the $\mathbf{D_{3h}}$ conformation, that is, in the middle (Yokozeki, Kuchitsu, and Morino, 1970); yet, since this hump is much less than RT (cf. Section 2-1) the molecule will rapidly librate from one twist form to the other.

We shall dwell here briefly on the norbornane structure (Fig. 11.124, **B**), both because of its importance in physical organic chemistry and because of its relation to at least one important class of monoterpenes of which camphor $(1R,4R)$-1,7,7-trimethylbicyclo[2.2.1]heptan-2-one (Fig. 11.125) is an important representative. Camphor is, incidentally, one of the few chemicals readily available as both (+) and (−) enantiomers (from different natural sources as well as synthetically, from natural, nonracemic precursors) and as the racemate (by synthesis).

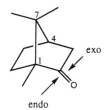

Figure 11.125. (+)-Camphor.

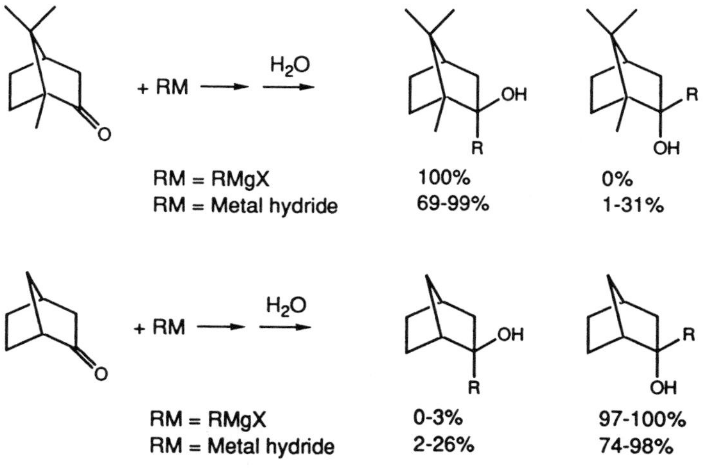

Figure 11.126. Reaction of nucleophiles with camphor and norcamphor.

The two diastereotopic faces of the carbonyl group of camphor (exo and endo, cf. Fig. 11.125) are of quite different accessibility to nucleophilic reagents: Whereas approach to the bottom face is somewhat impeded by the U-shaped cavity of the molecule, that to the top face is strongly hindered by the overhanging methyl group at C(7). Thus Grignard reagents react with camphor exclusively by approach from the endo side (cf. Ashby and Laemmle, 1975) and hydrides attach themselves predominantly to that side, the more so, the more bulky the hydride (cf. Boone and Ashby, 1979). These facts are summarized in Figure 11.126. An entirely different picture is seen with trisnorcamphor, commonly called "norcamphor" (bicyclo[2.2.1]heptan-2-one), also shown in Figure 11.126. Here there is no hindrance from the exo side, but the impedance to approach from the endo side remains, so Grignard reagents in almost all cases approach exclusively from the exo side and hydrides approach mainly from that side.

An analogous situation is seen in norbornene and 7,7-dimethylnorbornene [where studied, bornylene (i.e., 1,7,7-trimethylnorbornene), the unsaturated hydrocarbon corresponding to camphor, behaves just like 7,7-dimethylnorbornene]. The results of epoxidation (Brown et al., 1970) and hydroboration (Brown and Kawakami, 1970; see also Uzarewicz and Uzarewicz, 1976; Brown and Ravindran, 1977) are shown in Figure 11.127. Both epoxidation and hydroboration proceed almost

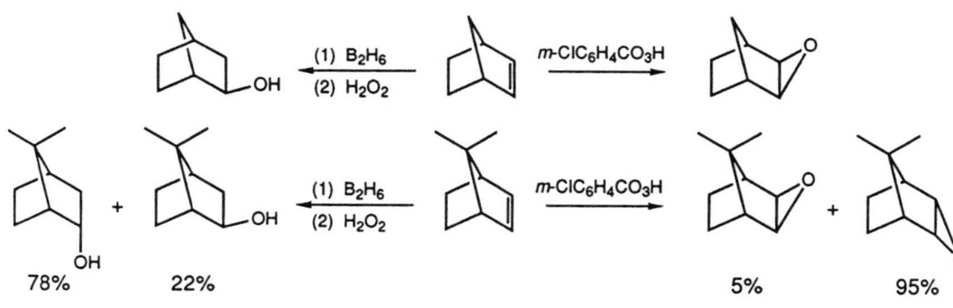

Figure 11.127. Epoxidation and hydroboration of norbornene and 7,7-dimethylnorbornene.

exclusively from the unencumbered exo side in norbornene. In the 7,7-dimethyl derivative attack is predominantly from the endo side but the selectivity is less than in norbornene; presumably the hindrance to approach from the exo side by the geminal methyl group is somewhat offset by hindrance to approach from the endo side by the U-shaped cavity of the molecule. Competitive experiments in epoxidation (Brown et al., 1970) suggest a ratio of about 1:1 for the rate of endo epoxidation of norbornene and its 7,7-dimethyl derivative, whereas exo attack is about 100 times faster than endo for norbornene but 10 times slower for the 7,7-dimethyl analogue. Thus exo attack in norbornene is about 1000 times as fast as for its 7,7-dimethyl homologue.

Bicyclo[2.2.2]octatriene (barrelene; Fig. 11.128) has also been synthesized (Zimmerman and Paufler, 1960); the molecule has $\mathbf{D_{3h}}$ symmetry and its bond lengths and angles are only slightly abormal (Yamamoto et al., 1982).

Among the higher bicycloalkanes we mention here bicyclo[3.3.1]nonane (for a review, see Zefirov and Palyulin, 1991) as an example where primitive model considerations led to an erroneous result. Inspection of a model of the compound (cf. Fig. 11.129) led one of us (Eliel, 1962) to state that the molecule could not exist as a double chair (Fig. 11.129) because of excessive steric compression of the endo hydrogen atoms at C(3) and C(7). However, this compression can be relieved by moderate flattening of the two chairs and it was soon found by X-ray crystallography (Laszlo, 1965; Brown, Sim, et al., 1965) that bicyclo[3.3.1] nonanes do exist in the double-chair conformation. Recent electron diffraction work at different temperatures confirms this conclusion; the ratio of chair–chair to chair–boat conformers at room temperature is 95:5, the chair–chair conformation being favored by 2.5 kcal mol^{-1} (10.5 kJ mol^{-1}) in enthalpy but disfavored by 1.5 cal mol^{-1} K^{-1} (6.3 J mol^{-1} K^{-1}) in entropy (Mastryukov et al., 1981). The transannular H $\cdots$ H distance has been calculated to be a very close 195–196 pm (1.95–1.96 Å) (Jaime, Osawa, et al., 1983; Skancke, 1987) with a resulting strain of 12.3 kcal mol^{-1} (51.5 kJ mol^{-1}) (Warner and Peacock, 1982).

Although, because of limitations of space, we cannot expand on the subject of unsaturated bridged ring systems, we do want to touch on one aspect of such systems, namely, Bredt's rule (Bredt, 1924; see also Fawcett, 1950; Eliel, 1962, p. 298). Bredt's rule states that in a small bridged system one cannot, for reasons of excessive strain, have a double bond at the bridgehead position. Thus, for example, the rule accounts for the fact that bicyclo[2.2.2]octane-2,6-dione (Fig. 11.130a, **A**) lacks the normal acidic properties of a 1,3-diketone: The corresponding enolate would have a bridgehead double bond.

The evident reason for Bredt's rule is that a bridgehead double bond implies that one of the rings containing this bond must be an (E)-cycloalkene. However, as will be seen in Section 14-8.d, (E)-cycloalkenes are isolable with a ring as small

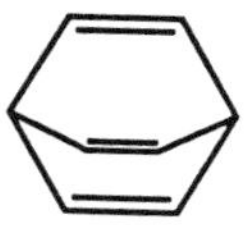

Figure 11.128. Structure of barrelene.

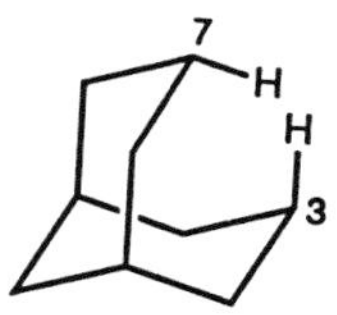

Figure 11.129. Structure of bicyclo[3.3.1]nonane.

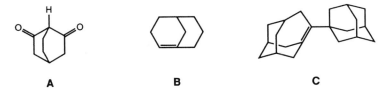

A   **B**   **C**

Figure 11.130a. Bredt's rule and bridgehead alkenes.

as seven membered, and there is good evidence for the fleeting existence of 1-phenyl-*trans*-cyclohexene (cf. Warner, 1989).

It is therefore not surprising that [3.3.1]bicyclo-1-nonene (Fig. 11.130a, **B**) is a stable substance (Marshall and Faubl, 1967; Wiseman, 1967; see also Becker and Pfluger, 1979) as are other bridgehead alkenes (paradoxically called Bredt olefins) containing (E)-cyclooctene rings (Warner, 1989). More surprising is the high stability of 4-(1-adamantyl)homoadamant-3-ene (Fig. 11.130a, **C**) containing an (E)-cycloheptene ring (Sellers, Jones, Schleyer, et al., 1982). Apparently, the attached bulky adamantyl substituent sterically prevents the otherwise facile dimerization of the adamantene moiety. Smaller bicycloalkenes containing (E)-cyclohexene rings are predictably unstable and their existence can, at best, be inferred from the isolation of trapping products (Warner, 1989).

Although camphor and norcamphor (Fig. 11.126) have two chiral centers (the two bridgehead atoms), in both cases there is only one pair of enantiomers rather than the two expected. The reason for this is that the configurations of the bridgehead atoms are not independently variable: In "small" bridged bicyclic compounds, the "bridge" must be cis with respect to the outer ring. In norbornane and bicyclo[2.2.2]octane (Fig. 11.124) the "outer ring" may be considered six membered with the "bridge" being either CH_2 or $(CH_2)_2$; it is clear from models that attempting to bridge a six-membered ring 1,4-trans would incur excessive strain, unless the bridge is quite large, for example, $(CH_2)_5$, in which case the trans (in–out) configuration becomes possible (see below).

When the rings become large enough, however, the restriction that the "bridge" must be cis no longer pertains; in fact there are not just two but *three* possibilities for the bridge now: the normal "out–out" (referring to the location of the bridgehead hydrogen atoms or other substituents), the aforementioned "in–out," and an "in–in" arrangement (see Fig. 11.130b). The "out–out" (cis)

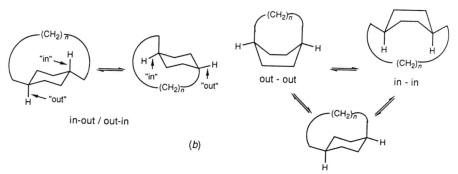

in-out / out-in

(b)

out - out   in - in

Figure 11.130b. In–out, out–in, out–out, and in–in isomers of an [n.2.2]bicyclane.

corresponds to one of the two possible diastereomeric configurations (the other being trans or "in–out"). In contrast, the "in–in" is a conformational isomer of the "out–out"; depending on considerations of strain the two conformers may or may not be interconvertible. There is also conformational isomerism for the "in–out" configurations, the other conformer being "out–in" (Fig. 11.130*b*; Alder, 1990).

These considerations are best appreciated when one considers a cyclohexane 1,4-bridged by a large bridge (Fig. 11.130*b*; the size of the bridge is not specified). The out–in/in–out conversion corresponds to a chair–chair inversion, the out–out/in–in conversion to a boat–boat inversion. Readers who are skeptical about the existence of in–in isomers should convince themselves that a model of in–in bicyclo[3.3.3]undecane can be readily constructed, for example, by severing the bridgehead–bridgehead bond in [3.3.3]propellane (see Section 11-6.c; cf. Alder, 1983).

The first carbocyclic bridged trans-fused system, bicyclo[8.8.8]hexacosane, (in–out) was synthesized by Park and Simmons (1972), along with its cis (out–out) isomer. Other examples have been reported (Krespan, 1980) including one involving in–out/out–in interconversion (Gregory et al., 1977). Indeed, it has been calculated (Alder, 1983) that the in–out isomer of bicyclo[4.4.4]tetradecane (Alder and Sessions, 1979) is more stable (because of a more favorable ring conformation) than the out–out isomer by $50 \, \text{kJ/mol}^{-1}$ ($11.95 \, \text{kcal mol}^{-1}$) and even the in–in isomer, despite H–H interference, is only $9 \, \text{kJ mol}^{-1}$ ($2.15 \, \text{kcal mol}^{-1}$) less stable than the out–out! The smallest known in–out bicyclic ring system appears to be *trans*-bicyclo[5.3.1]undecane, synthesized (Winkler et al., 1986) as the 11-keto derivative (Fig. 11.131). Surprisingly, the compound could not be isomerized to the known (Warnhoff et al., 1967) cis isomer.

However, the most prominent examples of in–out and in–in isomers are those of the bridged bicyclic diamines with nitrogen atoms at the bridgeheads (Fig. 11.132), first synthesized by Simmons and Park (1968; Park and Simmons, 1968a,b; see also Cheney and Lehn, 1972; Smith, Lehn, et al., 1981). These systems have two interesting features: The out–out, out–in, and in–in isomers are in equilibrium through nitrogen inversion, and the H–H repulsion in the in–in isomers of the carbocyclic analogues are replaced by less severe lone pair/lone pair repulsions; there are several other interesting aspects of these systems. One is that the in–in isomer can encapsulate a proton in the center, with resulting high thermodynamic basicity but potentially slow proton addition and removal (Smith, Lehn, et al., 1981; Alder et al., 1979, 1983a). Another feature is that the diprotonated form of a large bridged diamine can encapsulate a counterion, such as chloride, meaning that the ion is no longer readily available in titrable form (Park and Simmons, 1968b). In the [4.4.4] system (Fig. 11.132), which has been found, by X-ray crys-

Figure 11.131.   Structure of *trans*-bicyclo[5.3.1]undecan-11-one.

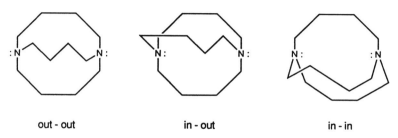

out - out   in - out   in - in

Figure 11.132. Stereoisomers of bridgehead nitrogen bridged bicyclic systems.

tallography, to be most stable in the in–in configuration with $\mathbf{D_3}$ symmetry and a relatively short N–N distance of 281 pm (2.81 Å) (Alder et al., 1983b), pK_1 is 6.5 and pK_2 is −3.2 (Alder et al., 1979). The unusually low pK values (implying unusually low basicity) are explained as being due to the strain involved in protonating from the outside an amine that normally has the lone pair inside (cf. Fig. 11.132, in–in); inside protonation, on the other hand, is very difficult kinetically (see above). Such protonation, when it occurs, *reduces* the N–N distance to 253 pm (2.53 Å)! The [4.4.4] diamine is also very easily oxidized, first to a stable (!) cation radical and then to a dication. Presumably the cation radical and dication contain nitrogen–nitrogen bonds (ordinary in the case of the dication, three-center, two-electron in the case of the cation radical). Crystal structures of a number of protonated bicyclic diamines in the in–in configuration have been recorded (Alder et al., 1988; White et al., 1988; see also Alder, 1989.)

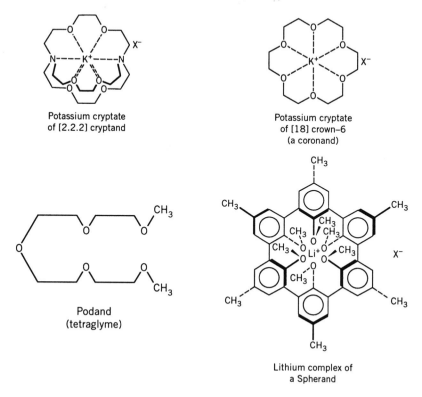

Potassium cryptate
of [2.2.2] cryptand

Potassium cryptate
of [18] crown–6
(a coronand)

Podand
(tetraglyme)

Lithium complex of
a Spherand

Figure 11.133. Cryptands, crown ethers (coronands), podands, and spherands.

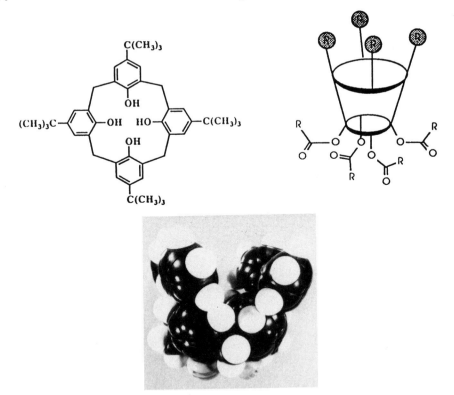

Figure 11.134. Calix[4]arenes.

There are many other fascinating aspects to the chemistry of bridgehead bicyclic diamines, especially those containing oxygen atoms in the bridge (cryptands), which show high selectivity in the formation of metal ion complexes (cryptates) (Lehn, 1978a,b, 1980; Jolley et al., 1982; Hayward, 1983; Parker, 1983; Dietrich, 1984). Detailed discussion of these compounds (Fig. 11.133) and the related crown ethers or coronands (Fig. 11.133; cf. Weber and Vögtle, 1981; Gokel, 1991), podands (Fig. 11.133), spherands (Fig. 11.133; Cram and Trueblood, 1981), hemispherands, carcerands (Cram, 1990) and hemicarcerands (Tanner, Cram, et al., 1990) and calixarenes (Fig. 11.134; Gutsche et al., 1981; Gutsche, 1989; Ungaro and Pochini, 1991; Vicens and Böhmer, 1991) is, unfortunately, beyond the scope of this book (see also Lehn, 1993; Weber, 1993; Schneider and Dürr, 1993 and p. 419).

c. Paddlanes and Propellanes

If one joins the vertices of a bridged bicyclic system with yet another ring, one obtains a tricyclic system (tricyclo[$m.n.o.p^{1,m+2}$]alkane) of the type shown in Figure 11.135, **A**, which has been called "paddlane" (Hahn, Ginsburg, et al., 1973; Greenberg and Liebmann, 1978; Warner et al., 1981). (The heavy dots represent carbon atoms at bridgehead junctures.) While this type of compound has not been systematically studied in general, a great deal of information is available for the special case where $p = 0$, that is, for fused bicyclic systems in

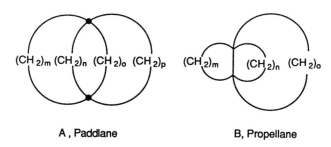

$(CH_2)_m$ $(CH_2)_n$ $(CH_2)_o$ $(CH_2)_p$ $(CH_2)_m$ $(CH_2)_n$ $(CH_2)_o$

A , Paddlane   B, Propellane

Figure 11.135. Paddlanes and propellanes.

which the vertices are spanned by a third ring (Fig. 11.135, **B**). The systematic name for such a system is tricyclo[$m.n.o.0^{1,m+2}$]alkane (the alkane name corresponding to the sum of all the carbon atoms) but the term [$m.n.o$]propellane (Altman, Ginsburg, et al., 1966; Ginsburg, 1972, 1975, 1987; Wiberg, 1989; Tobe, 1992) is now widely used for such systems.

The name is derived from the propeller shape of these molecules, which is most clearly evident in a model of [3.3.3]propellane. The etymologically more appropriate (but more cumbersome) term "propellerane" has also been proposed (Bloomfield and Irelan, 1966, footnote 9; Gassman et al., 1969) but is superseded by usage.

The propellanes in which $m + n + o$ is greater than 8 seem to present no special structural features. For example, [4.4.4]propellane (tricyclo-[$4.4.4.0^{1,6}$]tetradecane, Fig. 11.136) is made up of three nearly undistorted chairs (Ermer, Dunitz, et al., 1971) with a barrier to ring inversion (measured in the 3,3-difluoro compound: Gilboa, Loewenstein, et al., 1969) of 15.7 kcal mol^{-1} (65.7 kJ mol^{-1}), somewhat higher than in *cis*-decalin. The [4.4.3]propellane derivative shown in Figure 11.136 has long been known (Brigl and Herrmann, 1938). A model for [3.3.3]propellane can also readily be constructed, but when $m + n + o$ is 8 or less (as in [3.3.2]propellane whose model shows considerable tension) considerations of strain become important; such propellanes have been called "small propellanes" (Ginsburg, 1972; see also Greenberg and Liebman, 1978, p. 343; Wiberg, 1989) and we shall focus our attention on them. The interest in this class becomes obvious when one compares norbornane with [2.2.1]propellane (Fig. 11.137); in norbornane, the (marked) bridgehead bonds point outward, whereas in the propellane they must point inward to form the bridgehead–bridgehead bond. It is anticipated that this unusual situation will not

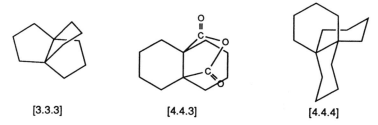

[3.3.3]   [4.4.3]   [4.4.4]

Figure 11.136. Structures of the [3.3.3], [4.4.3], and [4.4.4]propellanes.

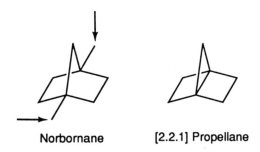

Norborname   [2.2.1] Propellane

Figure 11.137. Norbornane and [2.2.1]propellane.

only lead to considerable strain but also to some unusual physical and chemical features on the part of the bond in question; as we shall soon see, such is indeed the case.

Common methods for synthesizing small propellanes are carbene additions to double-bond bridged fused bicyclic systems or treatment of bridgehead dihalides with alkali metals or alkyllithium (Fig. 11.138). The former method is suitable for the synthesis of [*n.m.*1]propellanes with the exception of [*m.*1.1]propellanes, where the required 1,2-bridged cyclopropene precursors are not readily accessible. Among small propellanes with a total of eight and seven atoms in the bridges, the [6.1.1] and [5.1.1] members, as well as (surprisingly) the [5.2.1] species have not been prepared, perhaps because the challenge to do so was lacking. All other small propellanes are known, though some (see below) only fleetingly so.

The [4.2.2]- (Eaton and Kyi, 1971), [4.3.1]- (Dauben and Laug, 1962), [3.3.2]- (Borden et al., 1970), [4.2.1]- (Warner and LaRose, 1972), [3.3.1]- (Warner et al., 1974), and [3.2.2]propellane (Eaton and Kyi, 1971) appear to be thermally fairly stable substances; the [4.2.2], [3.3.1], and [3.3.2]propellanes are thermally stable up to 160°C; and the [4.2.1] isomer is somewhat less stable (the remaining two have not been tested). On the other hand, in every case where it has been tested, the substances are very sensitive to electrophilic cleavage of the bridgehead–bridgehead bond; thus bromine reverses the reaction shown in Figure 11.138 (bottom line, X = Br). The reactivity at the bridgeheads is so great, in fact,

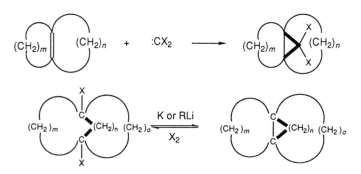

Figure 11.138. Common propellane syntheses.

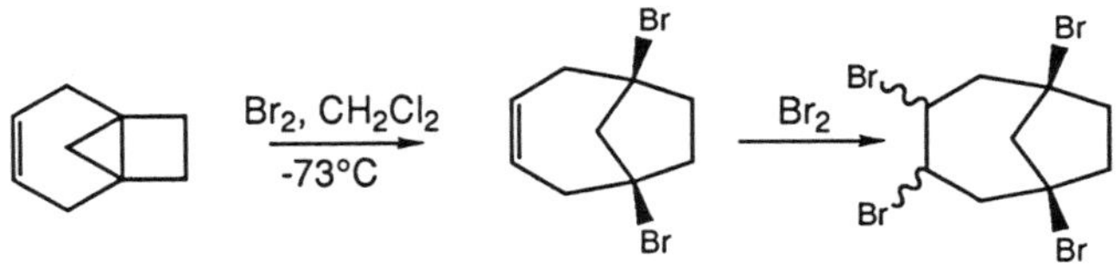

Figure 11.139. Reactivity of propellanes toward bromine.

that the unsaturated propellane shown in Figure 11.139 undergoes ring opening with bromine before the bromine adds to the double bond (Warner and LaRose, 1972).

As expected, stability decreases as the number of atoms in the propellane decreases. Among the propellanes with $m + n + o = 6$ the [4.1.1] isomer (Hamon and Trenerry, 1981) is moderately stable and the [3.2.1] isomer (Gassman et al., 1969; Wiberg and Burgmaier, 1969, 1972; Wiberg et al., 1977) is quite stable in solution, its half-life in diphenyl ether being 20 h at 195°C (the stability of the neat propellanes tends to be less because of the possibility of bimolecular reactions leading to oligomers and polymers). On the other hand, the [2.2.2] isomer is quite unstable; the parent compound has never been isolated (though it has been obtained as a reaction intermediate: Wiberg et al., 1977; see also Prociv, 1982), and the derivative shown in Figure 11.140 (Eaton and Temme, 1973) is stable only below $-30°C$, decomposing even in solution with a half-life of 28 min at room temperature [activation energy ca. 22 kcal mol^{-1} (92 kJ mol^{-1})].

The reverse of the reaction shown in Figure 11.140, photochemical cyclization of appropriate dimethylenecycloalkanes, is a convenient way for synthesizing propellanes (e.g., Cargill et al., 1966; Borden et al., 1970; Bishop and Landers, 1979).

In general, the strain in a propellane somewhat exceeds the sum of the strain in its three rings. As shown in Table 11.4, the strain in cyclopropane and that in cyclobutane are nearly equal (cyclopropane, although it has greater angle strain, benefits from its special type of bonding), but that in cyclopentane and cyclohexane is much less. Therefore, of the above isomers, the one with the three four-membered rings ([2.2.2]) is by far the most strained (see also Table 11.24).

Before proceeding to the four remaining propellanes with $m + n + o < 6$, we present, in Table 11.24, an ab initio calculation by Wiberg (1983; see also Dodziuk, 1984; Wiberg et al., 1987) of the strain in five of the seven smallest propellanes and in the bicycloalkanes, which would be formed by hydrogenolytic

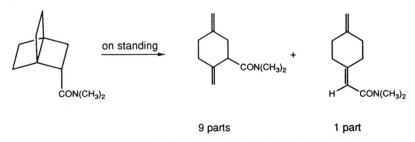

9 parts   1 part

Figure 11.140. *N,N*-Dimethyl[2.2.2]propellane-2-carboxamide and its decomposition.

TABLE 11.24.   Strain in Propellanes and Related
Bicycloalkanes.

Compound	Strain Energy	
	(kcal mol^{-1})	(kJ mol^{-1})
[3.2.1]Propellane	67	280
[2.2.2]Propellane	97	406
[2.2.1]Propellane	109	456
[2.1.1]Propellane	106	444
[1.1.1]Propellane	103	431
Bicyclo[2.2.2]octane	7.4	31
Bicyclo[2.2.1]heptane	14.4	60
Bicyclo[2.1.1]hexane	37.3	156
Bicyclo[1.1.1]pentane	66.6	279

cleavage of these propellanes at the bridgehead–bridgehead bond. First of all it is evident that [2.2.2]propellane has much more strain than [3.2.1]propellane for the reasons mentioned. Moreover, in the opening of the [2.2.2]propellane to the bicyclo[2.2.2]octane diradical, there is a relief of strain of about 90 kcal mol^{-1} (375 kJ mol^{-1}), assuming that the strain in the diradical is about the same as that in the hydrocarbon (this is only approximately correct). Since the carbon–carbon bond strength is only about 84 kcal mol^{-1} (351.5 kJ mol^{-1}) the formation of the diradical from the propellane should be an exothermic reaction! Indeed, Stohrer and Hoffmann (1972) have calculated that there should be two energy minima in [2.2.2]propellane, one for the bonded molecule and the other for the diradical, which are separated by an energy barrier. This energy barrier is probably the activation barrier to decomposition of the propellane (Fig. 11.140) since the further decomposition of the diradical to a 1,4-dimethylenecyclohexane is a symmetry allowed, and therefore probably rapid, process.

Proceeding now to the smaller propellanes, we see that the strain energy in the three is about equal. In fact, the [1.1.1] species with its three cyclopropane rings is slightly more stable than the [2.1.1] homologue (two cyclopropane and one cyclobutane rings), with the [2.2.1] species (two cyclobutane plus one cyclopropane rings) being the least stable of the three. More important, however, is the relative stability of the products of bridgehead–bridgehead bond cleavage: Only for the cleavage of the [2.2.1]propellane to bicyclo[2.2.1]heptane (norbornane) does the exothermicity of the reaction (95 kcal mol^{-1}, 397 kJ mol^{-1}) exceed the strength of a C–C bond, so only in that case should cleavage of the bridgehead–bridgehead bond to a diradical be exothermic. In the other two cases ([2.1.1]propellane to bicyclo[2.1.1]hexane and [1.1.1]propellane to bicyclo[1.1.1]pentane) the exothermicity of the reaction falls short of the strength of a C–C bond, not because the propellanes are so much less strained but because the cleavage products retain so much of the strain. It is thus perhaps not surprising that [1.1.1]propellane (Wiberg and Walker, 1982) is stable enough at room temperature to have its IR spectrum, enthalpy of formation (Wiberg et al., 1985), and molecular structure (by electron diffraction: Hedberg and Hedberg, 1985) determined. The molecule is depicted in Figure 11.141, **A**; of note is the rather long (160 pm, 1.60 Å) central bond.

The [2.2.1] homologue is much less stable than the [1.1.1] homologue (it had

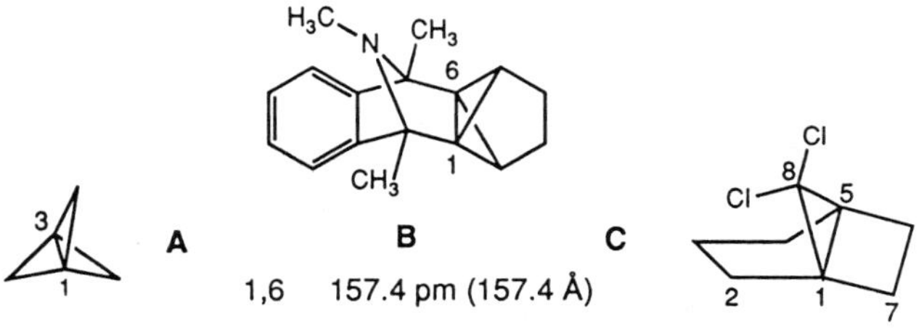

1,6 157.4 pm (157.4 Å)

A		C	
1,2	152 pm (1.52 Å)	1,2	148.3 pm (1.483 Å)
1,3	159-160 pm (1.59-1.60Å)	1,7	152.6 pm (1.526 Å)
∡1-2-3	63°	1,8	145.8 pm (1.458 Å)
∡2-1-4	95°	1,5	157.2 pm (1.572 Å)
∡H-C-H	116°	∡2-1-7	128.9°
		∡2-1-8	120.8°
		∡7-1-8	109.1°

Figure 11.141. Propellane structures.

to be trapped at 29 K and was identified only by IR spectroscopy) (Walker, Wiberg, and Michl, 1982). That [2.1.1]propellane is less stable than [1.1.1]propellane is also clear; it had to be matrix isolated in a solid nitrogen matrix and identified by IR spectroscopy (Wiberg et al., 1983). On the other hand, these experiments give no indication as to whether the relative stability of the [2.2.1] and [2.1.1]propellanes are as predicted.

This leaves for discussion the [3.1.1]propellane, which has been prepared by Gassman and Proehl (1980). As expected, this propellane (whose strain should be similar to that of its [3.2.1] homologue, cf. Table 11.24) is much more robust than its [2.2.1] isomer, being stable in toluene solution at 25°C (though it polymerizes readily in the neat form).

The bridgehead–bridgehead bonds in these propellanes are somewhat longer than normal C–C bonds, 160 pm (1.6 Å) in [1.1.1]propellane **A** (Fig. 11.141), 157.2 pm (1.572 Å) in the [3.2.1]propellane shown in Figure 11.141, **C** (Wiberg et al., 1972) and 157.4 pm (1.574 Å) in a more complex [4.1.1]propellane (Fig. 11.140, **B**; Chakrabarti, Dunitz, et al., 1981). Structure **B**, surprisingly, showed no palpable excess electron density (over that of the unbonded atoms) in the central bond, that is, no evidence for the bond-forming electrons! The determination of structure **C** indicated atoms 1, 2, 7, and 8 to be nearly in the same plane, but slightly bent *into* the center of the molecule, such that all bond vectors emanating from the bridgehead atoms point in the same direction of space! Accordingly, bonds 1,2 and 1,8 (and, to a lesser extent, 1,7) are unusually short, having considerably more *s* character than ordinary sp^3 hybridized single bonds. Consequently, the long bridgehead–bridgehead bond is largely a *p* bond, with but little *s* character (see also Dodziuk, 1984).

Although a number of theoretical studies of small ring paddlanes (e.g., tricyclo[2.2.2.2^{1,4}]decane) are on record (e.g., Würthwein, Schleyer, et al., 1981; Minyaev and Minkin, 1985; Wiberg, 1985) these compounds are enormously strained and only homologues with one relatively large bridge (*n*.2.2.2, with $n = 10$, 12, or 14) have been prepared (Vögtle and Mew, 1978; Wiberg and O'Donnell, 1979; Eaton and Leipzig, 1983).

d. Catenanes, Rotaxanes, Knots, and Möbius Strips

This section (cf. Schill, 1971a,b; Walba, 1985, 1993; Dietrich-Buchecker and Sauvage, 1987; Sauvage, 1993) deals with some molecules of unusual topology. Catenanes (Fig. 11.142) (from Latin *catena* meaning chain; the name was coined by Wasserman, 1960) are molecules containing two or more intertwined rings. Rotaxanes (from Latin *rota* meaning wheel and axis; cf. Schill, 1967a; Schill and Zollenkopf, 1969: Fig. 11.142) are molecules in which a linear molecule is threaded through a cyclic one and prevented from slipping out by large, bulky end-groups. The trefoil knot (one example of the family of knots; cf. Boeckmann and Schill, 1974) is depicted in Figure 11.142; its synthesis via a template method (see p. 804) is depicted in Figure 11.147 (Dietrich-Buchecker and Sauvage, 1989). A Möbius strip molecule is a two-dimensional structure the ends of which are connected after a twist (Walba et al., 1982); a macroscopic analogue is shown in Figure 11.142.

> A reader who has never constructed a Möbius strip may find it instructive to do so. A strip of paper about 25 cm (10 in.) in length and 2.5 cm (1 in.) wide is joined at the ends with glue or tape after twisting the strip once. The result is a Möbius strip with a single twist (or, more correctly, half-twist. Möbius strips with more than one half-twist can also be constructed and have interesting topological properties different from those to be discussed; cf. Schill, 1971a). If the strip is now cut in half along its length, a single (but doubly half-twisted) ring will result. [This operation has actually been performed on the molecular scale (Walba, 1993)!] If this ring is halved again, a catenane-like structure is obtained.

Catenanes and rotaxanes have been synthesized in two different ways: statistical syntheses and directed syntheses. In a statistical synthesis of a catenane, a large ring (say 34-membered or larger) is formed in the presence of another large ring (Fig. 11.143). A finite number of chain molecules will, on a statistical basis, have become threaded through the ring before cyclization. These threaded molecules, upon cyclization of the chain, give rise to a catenane. This method has been used by Wasserman (1960, 1962) as shown in Figure 11.143. However, the yield is very small, $[10^{-4}\%$, based on the ultimate starting material, sebacic (decanedioic) acid: cf. Schill, 1971a]; there is also a difficulty in identifying the product.

> Characterization has been a general problem in the catenane field; as a minimum, it would seem to be essential to demonstrate a molecular ion in the mass spectrum

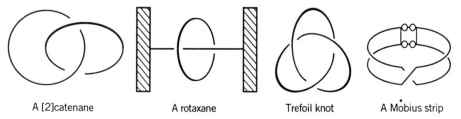

A [2]catenane   A rotaxane   Trefoil knot   A Möbius strip

Figure 11.142. Catenanes, rotaxanes, trefoil knots, and Möbius strips. Adapted, with permission, from Walba, D. M., Richards, R. M., and Haltiwanger, R. C. (1982), *J. Am. Chem. Soc.*, **104**, 3220. Copyright © American Chemical Society, Washington, DC.

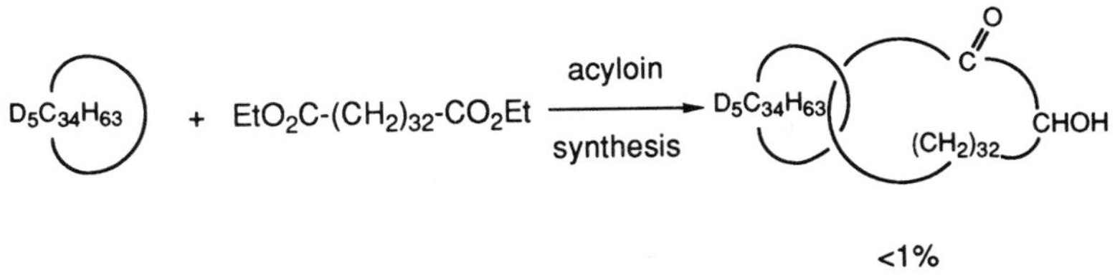

<1%

Figure 11.143. Statistical synthesis of a catenane.

(cf. Vetter and Schill, 1967) so as to prove the existence of a catenane as distinct from two isolated rings; this is especially important in a statistical synthesis.

In the statistical synthesis of rotaxanes (cf. Harrison, 1974), where, in an equilibrium assembly of a large ring and a straight-chain molecule, the latter is capped at both ends with a bulky group so as to "fix" the rotaxane (Fig. 11.144), yields depend on (and increase with) the size of the ring and the length of the chain and may be as high as 11.3% (Schill et al., 1986). Indeed, an ingenious modification of this method constitutes the first published synthesis of a rotaxane (Harrison and Harrison, 1967): the large-ring molecule (a C_{30} acyloin) was connected chemically to a polystyrene resin of the Merrifield (cf. Merrifield, 1965) type, the open-chain molecule (1,10-decanediol) was then threaded and capped [with $(C_6H_5)_3CCl$], and any remaining monomeric material (ditrityl ether of decanediol) washed away. Although the threading is statistically inefficient and only a small fraction of the (bound) ring was converted to a (bound) rotaxane, the fact that the unconverted starting ring compound remains attached to the resin allowed the authors to repeat the threading and capping process 70 times after which the product was chemically detached from the resin and separated into starting material and rotaxane.

It is also possible to increase the chances of threading by chemical affinity. One such effort (Agam et al., 1976) involves the threading of a polyethylene glycol through a large crown ether followed by capping with trityl chloride. The threading is enhanced by the affinity of the crown ether and linear polyether functions and the yield is 15%. In yet another synthesis of similar type, a terminal diamine (1,10-diaminodecane or 1,12-diaminododecane) was threaded in α- or β-cyclodextrin and then capped as a complex with $CoCl(en)^{2+}$ end groups (en = ethylenediamine) (Ogino, 1981). Apparently, threading is favored by the hydrophobic (attractive) interaction between the interior of the cyclodextrin ring and the hydrocarbon chain of the diamine; the yield of rotaxane ranges from 2 to 19%.

The rotaxane prepared from crown ether and polyethylene glycol (see above) was also converted to a catenane (Agam and Zilkha, 1976) in the following way: In lieu of trityl chloride, p-$BrCH_2C_6H_4CCl(C_6H_5)_2$, mono-$p$-bromomethyltrityl chloride, was used as a cap. This reactive cap then allowed cyclization of the

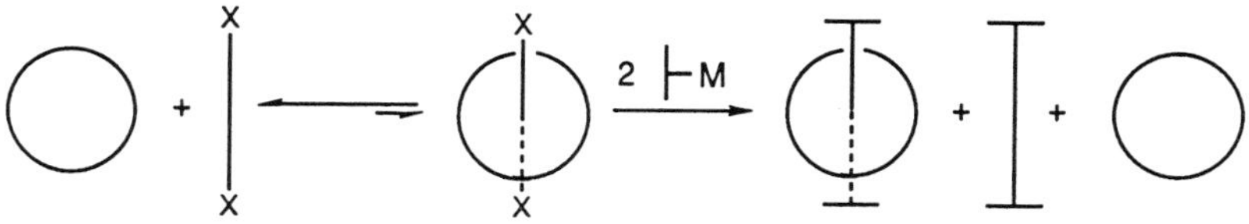

Figure 11.144. Statistical synthesis of a rotaxane.

linear part of the rotaxane by zinc–copper coupling of the benzylic bromide to a bibenzyl, with formation of a catenane in 14% yield.

We proceed now to directed syntheses of catenanes and rotaxanes. In these syntheses the thread, or the second ring, is initially linked covalently to its ring partner, but in a way that scission of certain covalent bonds (designed to be broken readily) will lead to the desired catenane or rotaxane. The final stages of a directed synthesis of a rotaxane of this type (Schill and Zollenkopf, 1969) are depicted in Figure 11.145.

Directed syntheses of [2]catenanes, that is, catenanes with two interlocked rings (Schill and Lüttringhaus, 1964; Schill, 1967b; see also Lüttringhaus and Isele, 1967 for a *semidirected* synthesis) have been reviewed (Schill, 1971a). In most cases, the catenanes are ultimately obtained by cleavage of a carbon–nitrogen bond or bonds (cf. Fig. 11.146), and therefore contain at least one nitrogen in one of the rings. It is therefore of interest that a purely carbocyclic [2]catenane (intercalated 28- and 46-membered rings) has been obtained by cyclization of the chain in a rotaxane (Schill et al., 1986b). Also of interest are syntheses of [3]catenane (Fig. 11.146; see also Schill and Zürcher, 1977), that is, catenanes with three individually interlocked rings (Schill et al., 1981; Rissler, Schill, et al., 1986); the principles involved in directed synthesis of [2]catenanes and [3]catenanes are the same as for the rotaxane synthesis (see above). Interestingly, the synthesis shown in Figure 11.146 gave rise to two "translational" isomers (**A, B**); apparently the hole in the nitrogen heterocycles is too small for the substituted benzenes of the larger ring to slip through, the situation being similar to that in a rotaxane.

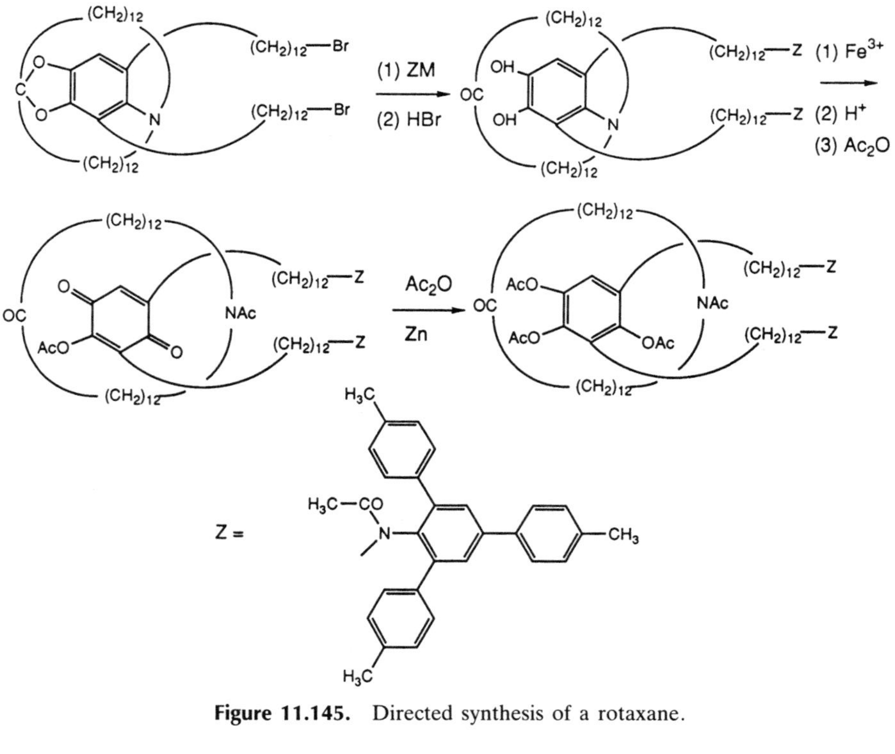

Figure 11.145. Directed synthesis of a rotaxane.

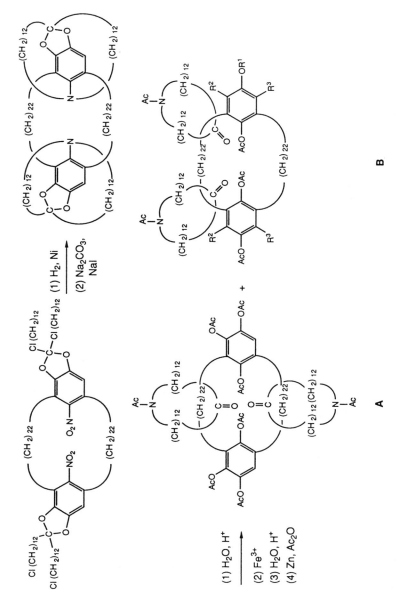

Figure 11.146. Synthesis of a [3]catenane, where $R^2 = CH_3CO_2^-$, $R^3 = H$, or $R^3 = CH_3CO_2^-$, $R^2 = H$.

803

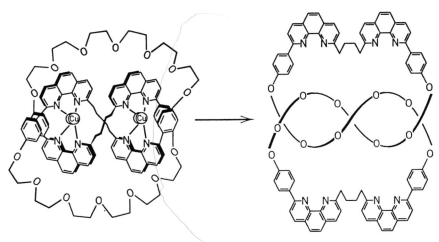

Figure 11.147. Template synthesis of trefoil knot. Reproduced, with permission, from Dietrich-Buchecker, C. O. and Sauvage, J.-P. (1989), *Angew. Chem. Int. Ed. Engl.*, **28**, 190. Copyright VCH Publishers, Weinheim, Germany.

A very interesting synthesis of catenanes involves the use of a metal as a template (e.g., Sauvage and Weiss, 1985). The initial product is a catenane complex (catenate), which may subsequently be demetalated to give the catenane. The sequence is illustrated here by the synthesis of a trefoil knot (Fig. 11.147; Dietrich-Buchecker and Sauvage, 1987, 1989).

A template synthesis of a catenane without intervention of a metal has also been achieved (Ashton, Stoddart, et al., 1989); it relies on complexation, prior to ring closure, of biparaphenylene-34-crown-10 (Fig. 11.148) with a chain precursor containing two 4,4′-bispyridine units that are then capped with *p*-xylyl dibromide; the yield of catenane is an amazing 70%. (See also Ashton, Stoddart, et al., 1991a; Stoddart, 1991; Philip and Stoddart, 1991; Anelli, Stoddart et al., 1992.)

In this case also there are two translational arrangements, but for reasons other than steric: Whereas one of the bipyridine moieties is complexed within the crown ether (shown), the other is wedged between two crown ether moieties in a polymolecular stacked arrangement. Exchange between the two topomers can be observed on the NMR time scale: Depending on which ring moves, the activation energy for the exchange is 12.2 or 14.0 kcal/mol^{-1} (51.0 or 58.6 kJ mol^{-1}).

In a somewhat similar fashion, a "template" synthesis of a rotaxane in 32% yield was accomplished (Anelli, Stoddart, et al., 1991; see also Ashton, Stoddart et al., 1991b, 1992a–c; Stoddart, 1991a,b). The difference between this synthesis and that of the catenane shown in Figure 11.148 is that the "thread" of the rotaxane was introduced as a silyl-capped podand-type (cf. Fig. 11.133) species. Here, also, either of the two phenyl rings of the "thread" may be included in the bisbipyridine-dixylylene ring, and here again a motion of the thread within the ring (interchange of two topomers) can be observed on the NMR time scale with an activation energy of about 13 kcal mol^{-1} (54.4 kJ mol^{-1}). This system (Figure 11.149) has been called a "molecular shuttle" (Anelli, Stoddart, et al., 1991; Philip and Stoddart, 1991; Ballardini, Stoddart et al., 1993, see also Amabilino and Stoddart, 1993).

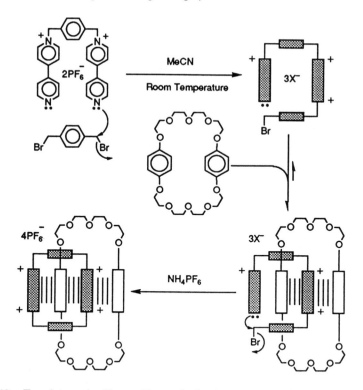

Figure 11.148. Template and self-assembly synthesis of a catenane. The shaded blocks represent 4,4-bispyridine units, the open blocks stand for *p*-phenylene. Adapted with permission from Ashton, P. R. et al. (1989), *Angew. Chem. Int. Ed. Engl.*, **28**, 1396. Copyright VCH Publishers, Weinheim, Germany.

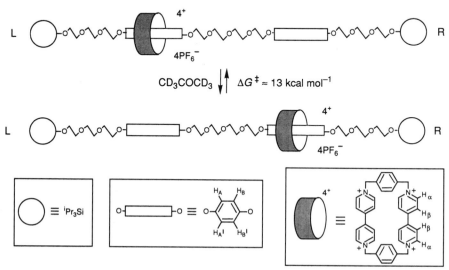

Figure 11.149. "Molecular shuttle." Adapted, with permission, from Anelli, P. L., Spencer, N., and Stoddart, J. F. (1991), *J. Am. Chem. Soc.* **113**, 5132. Copyright © American Chemical Society, Washington, DC.

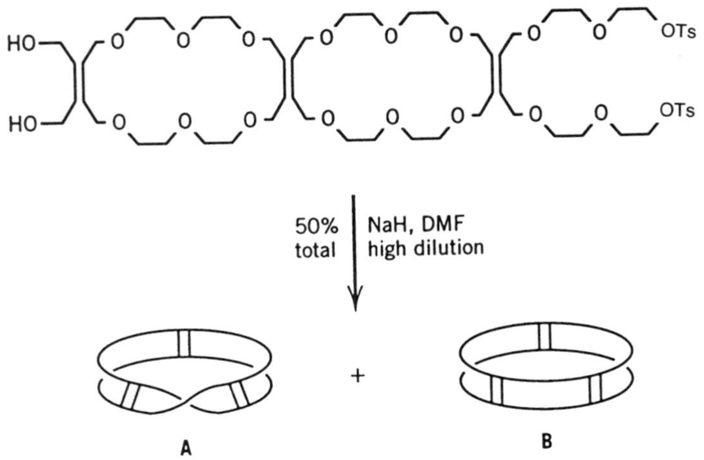

Figure 11.150. Synthesis of a Möbius strip molecule. DMF = N,N-Dimethylformamide. Taken with permission from Walba, D. M., Richards, R. M., and Haltiwanger, R. C. (1982), *J. Am. Chem. Soc.*, **104**, 3220. Copyright © American Chemical Society, Washington, DC.

Catenane structures are also encountered in macromolecules and macroscopic assemblies of various types: in crystals (Duchamp and Marsh, 1969; Herbstein et al., 1981), in polymers (Frisch et al., 1981; see also Maciejwski, 1982; Lipatova, Lipatov, et al., 1982) (rotaxanes), and in DNA, both natural (Hudson and Vinograd, 1967; Clayton and Vinograd, 1967; see also Tabor and Richardson, 1985; Walba, 1985) and biosynthetic (Wang and Schwartz, 1967).

We now come to the synthesis of a Möbius strip (named after the nineteenth century mathematician Möbius). The Möbius (**A**) and cyclinder (**B**) molecules shown schematically in Figure 11.150 were synthesized, from the linear array of crown ethers shown (cf. Walba et al., 1981) in 24 and 22% yields, respectively (Walba et al., 1982, 1993). The cylindrical molecule **B** was crystalline and its structure was confirmed by X-ray crystallography. The Möbius strip molecule **A** is chiral and its chirality could be demonstrated by a doubling of the olefinic ^{13}C signals in the presence of the chiral solvating agent (+)-2,2,2-trifluoro-1-(9-anthryl)-ethanol (cf. p. 232). [It might appear at first sight that molecule **A** contains diastereotopic olefinic carbon atoms even in the absence of a chiral solvating agent. However, this is in fact not the case because the "twist" moves rapidly (on the NMR time scale) around the ring, thereby leading to averaging of the signals of all olefinic (and also of all saturated) carbon atoms.]

The synthesis of the topologically interesting trefoil knot has been shown in Figure 11.147.

e. Cubane, Tetrahedrane, Dodecahedrane, Adamantane, and Buckminsterfullerene

We shall conclude this chapter with an esthetically pleasing subject, namely, the Platonic solids. The Platonic solids are polyhedra of high symmetry: the tetrahedron (**T_d**), the octahedron (**O_h**), the cube (**O_h**), the dodecahedron (**I_h**), and the icosahedron (**I_h**) (see Section 4-3.b). In the octahedron four edges meet in

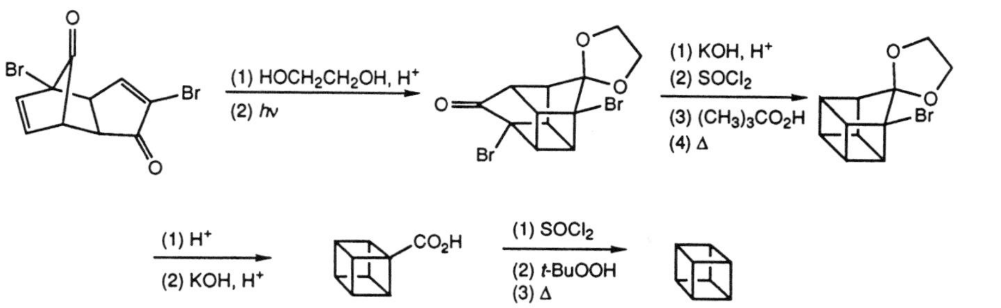

Figure 11.151. Synthesis of cubane.

one vertex and in the icosahedron five, so these figures cannot be constructed from tetracoordinate carbon plus hydrogen; however, the all-carbon compound buckminsterfullerene (C_{60}, see below) has the shape of a truncated icosahedron. Hydrocarbons $(CH)_n$ corresponding to the other three Platonic solids have been synthesized: cubane $(CH)_8$, dodecahedrane $(CH)_{20}$, and tetrahedrane $(CH)_4$, though the last one was synthesized only as a tetra-*tert*-butyl derivative.

Cubane was synthesized in 1964 by Eaton and Cole (q.v.; see also Griffin and Marchand, 1989). The synthesis is schematized in Figure 11.151; the essential features are photochemical dimerization $(2 + 2)$ of a brominated dicyclopentadiene, ring contraction of five- to four-membered rings by Favorskii rearrangement, and free radical decarboxylation. An alternative synthesis (Barborak, Pettit, et al., 1966), shown in Figure 11.152, is interesting not only because it is quite short, but also because the starting material is a cyclobutadiene complex.

Cubane, as expected, displays a single ^1H resonance (at 4.04 ppm, Edward et al., 1976) and a single ^{13}C resonance (at 47.3 ppm, Della et al., 1977). The structure as determined by electron diffraction (Almenningen et al., 1985) shows a bond length (157.5 pm, 1.575 Å) longer than that in cyclobutane. Earlier X-ray data (Fleischer, 1964) indicated a C–C–H angle of 123–127° suggesting a high degree of s character in the C–H bond concomitant with the high degree of p character in the C–C bonds imposed by the 90° bond angle. Surprisingly, the ^{13}C–H coupling constant of 153.8 Hz (Della et al., 1977) suggests only 30.8% s character in the C–H bond if one accepts the relation $\%s = 0.2J^1_{^{13}C-H}$ (Muller and Pritchard, 1959). Perhaps in a structure as highly strained as cubane this relation must be considered suspect. Based on heat of combustion data (Kybett, Margrave, et al., 1966), Engler, Schleyer, et al., (1973) evaluated the strain energy in

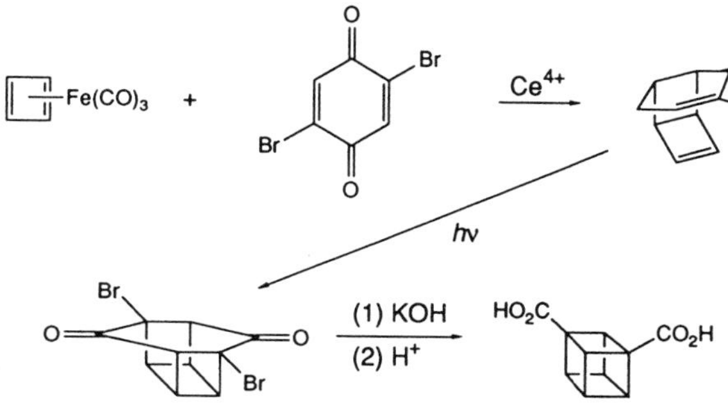

Figure 11.152. Alternative synthesis of a cubane.

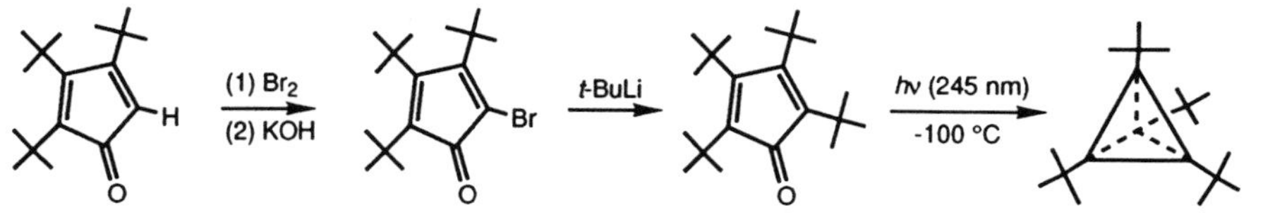

Figure 11.153. Synthesis of tetra-*tert*-butyltetrahedrane.

cubane to be 166 kcal mol^{-1} (695 kJ mol^{-1}). A somewhat smaller value, (154.7 kcal mol^{-1}, 647 kJ mol^{-1}) has been computed by Wiberg et al. (1987). It may or may not be coincidental that these values are close to six times the strain energy in cyclobutane (Table 11.4). (For a review of cubane chemistry, see Higuchi and Ueda, 1992.)

Tetra-*tert*-butyltetrahedrane (Fig. 11.153) was synthesized in 1978 by Maier et al., (1978, 1981a). The rather straightforward synthetic scheme shown in Figure 11.153 conceals both the limited availability of the starting material and the enormous amount of work required in establishing the proper conditions for each step. The material, characterized by X-ray crystallography (Irngartinger et al., 1984) and spectrally (Maier et al., 1981b; e.g., ^{1}H NMR, $\delta = 1.18$ ppm; ^{13}C NMR, $\delta = 9.27, 27.16, 31.78$ ppm; the high-field tetrahedrane carbon, similar to a cyclopropane one, is characteristic; interestingly, the primary and quaternary carbon atoms of the *tert*-butyl group are reversed, the latter being at higher field) is quite stable; it melts at 135°C with decomposition and is thermally converted to tetra-*tert*-butylcyclobutadiene, which can be photochemically reconverted to the tetrahedrane. The activation energy for the interconversion is remarkably high: $\Delta H^{\ddagger} = 25.5$ kcal mol^{-1} (106.7 kJ mol^{-1}) and $\Delta S^{\ddagger} = -10.3$ cal mol^{-1} K^{-1} (-43.1 J mol^{-1} K^{-1}). The strain in tetrahedrane is very large, having been estimated at 130–150 kcal mol^{-1} (544–628 kJ mol^{-1}) (Greenberg and Liebman, 1978; Wiberg et al., 1987). Not only had numerous earlier attempts to synthesize it failed (for a historic review see Greenberg and Liebman, 1978, pp. 85–90; Zefirov et al., 1978) but also attempts to obtain tetrahedranes with substituents less bulky than *tert*-butyl were uniformly unsuccessful (Maier et al., 1981c; 1985). The difference between the parent tetrahedrane and its tetra-*tert*-butyl homologue has been ascribed to a "corset effect" (Maier et al., 1981a): The tetrahedral arrangement of the *tert*-butyl substituents is the one in which these bulky groups keep furthest apart from each other, a fact that opposes rearrangement of tetra-*tert*-butyltetrahedrane to isomeric structures. This hypothesis is supported by calculations (Hounshell and Mislow, 1979), which also suggest that the molecule is more stable in the chiral **T** than in the achiral **T$_d$** symmetry. Unfortunately, the calculated difference is only 2–5 kcal mol^{-1} (8.4–21 kJ mol^{-1}), so that one would expect the enantiomeric **T** structures to interconvert rapidly (even on the NMR time scale) by passing over the low **T$_d$** barrier.

The last of the accessible Platonic solid shaped hydrocarbons to be synthesized was dodecahedrane (Fig. 11.154; Ternansky, Paquette, et al., 1982;

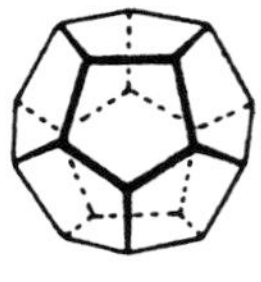

Figure 11.154. Dodecahedrane.

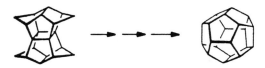

Figure 11.155. Alternative dodecahedrane synthesis. Taken, with permission, from Fessner, W. D., Murty, B. A. R. C., and Prinzbach, H. (1987), *Angew Chem. Int. Ed. Engl.*, **26**, 451. Copyright © VCH Publishers, Weinheim, Germany.

Paquette et al., 1983, 1986; Paquette, 1989; see also Paquette, 1984a,b). (The original synthesis comprises 23 steps and is therefore not reproduced here.) The structure of the compound as determined by X-ray diffraction (Gallucci, Paquette, et al., 1986) shows the expected I_h symmetry, nearly normal C–C bond distances (153.5–154.1 pm, 1.535–1.541 Å) and a C–C–C bond angle of about 108°. It will be interesting to assess the strain in dodecahedrane, because it is one of the few molecules where different force fields yield totally different strain estimates, around $40 \, \text{kcal mol}^{-1}$ ($167 \, \text{kJ mol}^{-1}$), by the Engler–Schleyer (1973) force field (see also Schulman and Disch, 1984) but about twice as much by the Allinger et al., (1971) force field. (For background to the dodecahedrane problem, see Eaton, 1979).

> An alternate dodecahedrane synthesis (Fessner et al., 1987a; Lutz, Prinzbach, et al. 1992) by catalytic isomerization of an accessible "pagodane" precursor (Fessner, Prinzbach, et al., 1987b) is outlined in Figure 11.155.

Numerous other interesting caged hydrocarbons have been synthesized (Scott and Jones, 1972; Greenberg and Liebman, 1978) but we shall mention only one more here: adamantane (Fig. 11.156), not only because of its high symmetry (T_d) and its relation to the diamond lattice, but also because, unlike the hydrocarbons so far discussed in this section, it is a highly stable molecule and constitutes the product of thermodynamically controlled isomerization of many other $C_{10}H_{16}$ hydrocarbons.

Adamantane (for reviews see McKervey, 1975; Fort, 1976; Greenberg and Liebman, 1978, Chapter 4; Ganter, 1992) was first isolated in 1933 from Czechoslovakian petroleum (Landa and Macháček, 1933); its structure was correctly assigned at that time and confirmed some years later by a rational, if lengthy and low-yielding synthesis (Prelog and Seiwerth, 1941). The substance remained a rarity, however, until, in 1957, Schleyer (q.v.) discovered that both hydrogenated *exo*- and *endo*-bicyclopentadienes on treatment with aluminum chloride gave adamantane in about 20% yield. Other $C_{10}H_{16}$ hydrocarbons isomerize more cleanly; for example, twistane (Fig. 4.11) is isomerized to adamantane rapidly in nearly 100% yield on treatment with aluminum chloride (Whitlock and Siefken, 1968).

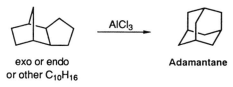

exo or endo
or other $C_{10}H_{16}$ **Adamantane**

Figure 11.156. Adamantane synthesis.

The reader should consult the above reviews for the mechanism involved in the rearrangement and for a discussion of other adamantane analogues. It would appear that the adamantane occurring in petroleum is an isomerization product of other fossil hydrocarbons, produced by adventitious (and possibly very slow acting) acid catalysis.

Although a model of adamantane appears strain-free, the molecule has about 7.6 kcal mol^{-1} (31.8 kJ mol^{-1}) strain. This is more than four times the strain in cyclohexane (cf. p. 677). In adamantane, the framework structure constrains the bond angles to be nearly perfectly tetrahedral. However, as we have seen before, a C–CH$_2$–C bond angle does not want to be exactly tetrahedral; the optimum (in propane) is 112.5° and in the slightly strained cyclohexane the angle is 111.5°. It is the further reduction to 109.5° which causes the additional strain in adamantane, thus demonstrating once again that molecular models can be misleading.

Finally, we must mention the molecule C$_{60}$ which was originally prepared by vaporization of graphite (by a focused pulse laser) into a high-density helium flow. This experiment (Kroto, Curl, Smalley, et al., 1985; cf. Curl and Smalley, 1988; Kroto, 1988) produces a number of large carbon clusters of which C$_{60}$ was discovered, by mass spectrometry, to be by far the most abundant. The authors postulated that the molecule C$_{60}$ is a closed-shell polygon with 60 vertices and 32 faces of which 20 are hexagonal (aromatic) and 12 are pentagonal (Fig. 11.157). Because of the similarity of the structure with a geodesic dome designed by the architect Buckminster Fuller, the compound has been called "buckminsterfullerene"; it has also been called "footballene", which is the Chemical Abstracts name, or "soccerballene." Subsequent to its initial detection in the mass spectrometer, buckminsterfullerene was prepared and isolated by evaporating graphite electrodes into an atmosphere of helium or argon at 50–100 mbar (38–75 mm Hg) followed by extraction of the sootlike product with benzene (Krätschmer, Huffman, et al., 1990; Taylor, Kroto, et al., 1990). This procedure produces mostly C$_{60}$, a mustard-colored crystalline solid, along with the more deeply colored C$_{70}$ (of similar structure) in about 5:1 ratio. Buckminsterfullerene (C$_{60}$) is purified by chromatography (Taylor, Kroto, et al., 1990; see also Hare, Kroto, et al., 1991; Ajie, Huffman, Krätschmer, et al., 1990; Haufler, Curl, Smalley, et al., 1990); or by selective complexation with *p-tert*-butylcalix[8]arene (cf. p. 794; Atwood et al., 1994). The isolation of this new allotrope of carbon has engendered immense interest: A review published in 1991 (Kroto et al.) lists no fewer than 263 references to this substance and in 1992 a complete issue of *Accounts of Chemical Research* was devoted to the subject of fullerenes (see McLafferty, 1992; see also Hammond and Kuck, 1992; Kroto and Walton, 1992).

Figure 11.157. Buckminsterfullerene.

Although X-ray structure determination of C_{60} is complicated by orientational disorder caused by a "rolling" of the near-spherical molecules in the crystal relative to each other (cf. Hawkins et al., 1991a), X-ray structure analysis of the derivative $C_{60}(OsO_4).4$-*tert*-butylpyridine confirms the truncated icosahedral (I_h) structure of the parent molecule (Hawkins et al., 1991b). The bond lengths are 139 pm (1.39 Å) for the six-six-membered-ring fusion and 143 pm (1.43 Å) for the six-five-membered-ring fusion.

The structure of C_{70} has been inferred from its ^{13}C NMR spectrum (Taylor, Kroto, et al., 1990) and appears to have D_{5h} symmetry. Less is known about smaller and larger fullerene C_n species (Kroto, 1990; Parker et al., 1991; Hammond and Kuck, 1992) but at least one of them (C_{78}) seems to contain a diastereomer with chiral D_3 structure (Diederich et al., 1991; see also Taylor, Kroto, et al., 1992; Diederich and Whetten, 1992) and another, C_{76}, has in fact, been resolved (Hawkins and Meyer, 1993).

Work on buckminsterfullerene, its homologues and derivatives continues actively (see, for example Billups and Ciufolini, 1993).

REFERENCES

Abillon, E. (1982), *Biophys. Struct. Mech.*, **9**, 11.

Abraham, R. J., Banks, H. D., Eliel, E. L., Hofer, O., and Kaloustian, M. K. (1972), *J. Am. Chem. Soc.*, **94**, 1913.

Abraham, R. J., Bergen, H. A., and Chadwick, D. J. (1983), *J. Chem. Res. Synop.*, 118.

Abraham, R. J., Bovill, M. J., Chadwick, D. J., Griffiths, L., and Sancassan, F. (1980), *Tetrahedron*, **36**, 279.

Abraham, R. J. and Bretschneider, E. (1974), "Medium effects on rotational and conformational equilibria," in Orville-Thomas, W. J., Ed., *Internal Rotations in Molecules*, Wiley, New York, pp. 481ff.

Abraham, R. J. and Griffiths, L. (1981), *Tetrahedron*, **37**, 575.

Addadi, L., Berkowitch-Yellin, Z., Weissbuch, I., Lahav, M. and Leiserowitz, L. (1986), *Top Stereochem.*, **16**, 65.

Adams, W. J., Geise, H. J., and Bartell, L. S. (1970), *J. Am. Chem. Soc.*, **92**, 5013.

Agam, G., Gravier, D., and Zilkha, A. (1976), *J. Am. Chem. Soc.*, **98**, 5206.

Agam, G. and Zilkha, A. (1976), *J. Am. Chem. Soc.*, **98**, 5214.

Ajie, H., Alvarez, M. M., Anz, S. J., Beck, R. D., Diederich, F., Fostiropoulos, K., Huffman, D. R., Krätschmer, W., Rubin, Y., Schriver, K. E., Sensharma, D., and Whetten, R. L. (1990), *J. Phys. Chem.*, **94**, 8630.

Alder, R. W. (1983), "Medium-Ring Bicyclic Compounds and Intrabridge Chemistry," *Acc. Chem. Res.*, **16**, 321.

Alder, R. W. (1989), "Strain Effects on Amine Basicities," *Chem. Rev.*, **89**, 1215.

Alder, R. W. (1990), "Intrabridgehead Chemistry," *Tetrahedron*, **46**, 683.

Alder, R. W., Casson, A., and Sessions, R. B. (1979), *J. Am. Chem. Soc.*, **101**, 3652.

Alder, R. W., Moss, R. E., and Sessions, R. B. (1983a), *J. Chem. Soc. Chem. Commun.*, 997.

Alder, R. W., Orpen, A. G., and Sessions, R. B. (1983b), *J. Chem. Soc. Chem. Commun.*, 999.

Alder, R. W., Orpen, A. G., and White, J. M. (1988), *Acta. Crystallogr. Sect. C: Cryst. Struct. Commun.*, **C44**, 287.

Alder, R. W. and Sessions, R. B. (1979), *J. Am. Chem. Soc.*, **101**, 3651.

Alekseev, N. V. and Kitaigorodskii, A. I. (1963), *J. Struct. Chem. USSR*, **4**, 163 (Engl. transl., p. 145).

Allan, E. A., Premuzik, E., and Reeves, L. W. (1963), *Can. J. Chem.*, **41**, 204.

Allinger, N. (1957), *J. Am. Chem. Soc.*, **79**, 3443.

Allinger, N. L. (1977), *J. Am. Chem. Soc.*, **99**, 8127.

Allinger, N. L. and Blatter, H. M. (1962), *J. Org. Chem.*, **27**, 1523.

Allinger, N. L. and Coke, J. L. (1959), *J. Am. Chem. Soc.*, **81**, 4080.

Allinger, N. L. and Coke, J. L. (1960), *J. Am. Chem. Soc.*, **82**, 2553.

Allinger, N. L. and Coke, J. L. (1961), *J. Org. Chem.*, **26**, 2096.

Allinger, N. L., Dodziuk, H., Rogers, D. W., and Naik, S. N. (1982), *Tetrahedron*, **38**, 1593.

Allinger, N. L. and Freiberg, L. A. (1960), *J. Am. Chem. Soc.*, **82**, 2393.

Allinger, N. L. and Freiberg, L. A. (1962), *J. Am. Chem. Soc.*, **84**, 2201.

Allinger, N. L., Freiberg, L. A., and Hu, S.-E. (1962), *J. Am. Chem. Soc.*, **84**, 2836.

Allinger, N. L., Gorden, B. J., Tyminski, I. J., and Wuesthoff, M. T. (1971), *J. Org. Chem.*, **36**, 739.

Allinger, N. L., Hirsch, J. A., Miller, M. A., Tyminski, I. J., and Van-Catledge, F. A. (1968a), *J. Am. Chem. Soc.*, **90**, 1199.

Allinger, N. L., Honig, H., Burkert, U., Asolnai, L., and Huttner, G. (1984), *Tetrahedron*, **40**, 3449.

Allinger, N. L., Miller, M. A., Van-Catledge, F. A., and Hirsch, S. A. (1967), *J. Am. Chem. Soc.*, **83**, 4345.

Allinger, N. L. and Kao, J. (1976), *Tetrahedron*, **32**, 529.

Allinger, N. L., Szkrybalo, W., and Van-Catledge, F. A. (1968b), *J. Org. Chem.*, **33**, 784.

Allinger, N. L. and Tribble, M. T. (1971), *Tetrahedron Lett.*, 3259.

Allinger, N. L., Tribble, M. T., and Miller, M. A. (1972), *Tetrahedron*, **28**, 1173.

Allinger, N. L., Tribble, M. T., Miller, M. A., and Wertz, D. H. (1971), *J. Am. Chem. Soc.*, **93**, 1637.

Allinger, N. L. and Wuesthoff, M. T. (1971), *J. Org. Chem.*, **36**, 2051.

Allinger, N. L. and Zalkow, V. (1960), *J. Org. Chem.*, **25**, 701.

Allinger, N. L. and Zalkow, V. B. (1961), *J. Am. Chem. Soc.*, **83**, 1144.

Almenningen, A., Andersen, B., and Nyhus, B. A. (1971), *Acta. Chem. Scand.*, **25**, 1217.

Almenningen, A., Jonvik, T., Martin, H. D., and Urbanek, T. (1985), *J. Mol. Struct.*, **128**, 239.

Alonso, J. (1981), *J. Mol. Struct.*, **73**, 63.

Altman, J., Babad, E., Itzchaki, J., and Ginsburg, D. (1966), *Tetrahedron Suppl. 8*, Part 1, 279.

Altona, C. (1982), "High resolution NMR studies of nucleic acids," NATO Advanced Study Inst. Ser., Ser. A, **45**, 161.

Altona, C., Buys, H. R., Hageman, H. J., and Havinga, E. (1967), *Tetrahedron*, **23**, 2265.

Altona, C., Geise, H. J., and Romers, C. (1968), *Tetrahedron*, **24**, 13.

Altona, C. and Sundaralingam, M. (1970), *Tetrahedron*, **26**, 925.

Altona, C. and Sundaralingam, M. (1972), *J. Am. Chem. Soc.*, **94**, 8205.

Amabilino, D. B. and Stoddart, J. F. (1993), "Self-assembly and macromolecular design," *Pure Appl. Chem.*, **65**, 2351.

Anderson, J. E. (1974), "Interconversion of Six-membered Rings," *Top. Curr. Chem.*, **45**, 139.

Anderson, J. E. (1992), "Conformational analysis of acyclic and alicyclic saturated hydrocarbons," Patai, S. and Rappoport, Z., Eds., *The Chemistry of Alkanes and Cycloalkanes*, Wiley, New York, p. 95.

Anelli, P. L., Spencer, N., and Stoddart, J. F. (1991), *J. Am. Chem. Soc.*, **113**, 5131.

Anelli, P. L., Stoddart, J. F. et al. (1992), *J. Am. Chem. Soc.*, **114**, 193.

Anet, F. A. L. (1964), *J. Am. Chem. Soc.*, **86**, 458.

Anet, F. A. L. (1974), "Dynamics of Eight-membered Rings in the Cyclooctane Class," *Top. Curr. Chem.*, **45**, 169.

Anet, F. A. L. (1988), "Medium-sized Oxygen Heterocycles," see reference to Glass, R. S., Ed. (1988), p. 35.

Anet, F. A. L. (1989), "Conformational Analysis of Cyclohexenes," see reference to Rabideau, P. W., Ed. (1989), p. 1.

Anet, F. A. L. (1990), *Tetrahedron Lett.*, **31**, 2125.

Anet, F. A. L. and Anet, R. (1975), "Conformational Processes in Rings," in Jackman, L. M. and Cotton, F. A., Eds., *Dynamic Nuclear Magnetic Resonance Spectroscopy*, Academic, New York, p. 543ff.

Anet, F. A. L. and Basus, V. J. (1973), *J. Am. Chem. Soc.*, **95**, 4424.

Anet, F. A. L., Basus, V. J., Hewett, A. P. W., and Saunders, M. (1980), *J. Am. Chem. Soc.*, **102**, 3945.

Anet, F. A. L., Bourn, A. J. R., and Lin, Y. S. (1964), *J. Am. Chem. Soc.*, **86**, 3576.

Anet, F. A. L., Bradley, C. H., and Buchanan, G. W. (1971), *J. Am. Chem. Soc.*, **93**, 258.

Anet, F. A. L. and Cheng, A. K. (1975), *J. Am. Chem. Soc.*, **97**, 2420.

Anet, F. A. L., Cheng, A. K., and Wagner, J. J. (1972), *J. Am. Chem. Soc.*, **94**, 9250.

Anet, F. A. L., Chmurny, G. N., and Krane, J. (1973), *J. Am. Chem. Soc.*, **95**, 4423.

Anet, F. A. L. and Haq, M. Z. (1965), *J. Am. Chem. Soc.*, **87**, 3147.

Anet, F. A. L. and Kopelevich, M. (1986), *J. Am. Chem. Soc.*, **108**, 1355.

Anet, F. A. L. and Krane, J. (1980), *Isr. J. Chem.*, **20**, 72.

Anet, F. A. L., Krane, J., Kitching, W., Dodderel, D., and Praeger, D. (1974), *Tetrahedron Lett.*, 3255.

Anet, F. A. L. and O'Leary, D. J. (1989), *Tetrahedron Lett.*, **30**, 1059.

Anet, F. A. L., O'Leary, D. J., and Williams, P. G. (1990), *J. Chem. Soc. Chem. Commun.*, 1427.

Anet, F. A. L. and Rawdah, T. N. (1978a), *J. Am. Chem. Soc.*, **100**, 7810.

Anet, F. A. L. and Rawdah, T. N. (1978b), *J. Am. Chem. Soc.*, **100**, 7166.

Anet, F. A. L. and Wagner, J. J. (1971), *J. Am. Chem. Soc.*, **93**, 5266.

Anet. F. A. L. and Yavari, I. (1977), *J. Am. Chem. Soc.*, **99**, 2794.

Anet. F. A. L. and Yavari, I. (1978), *Tetrahedron*, **34**, 2879.

Angyal, S. J. (1957), *Q. Rev. Chem. Soc.*, **11**, 212.

Anh, N. T. and Eisenstein, O. (1977), *Nouv. J. Chim.*, **1**, 61.

Antenunis, M. J. O. (1971), "Some Results and Limitations in Conformational Studies of Six-Membered Heterocycles," in Chiurdoglu, G. ed., *Conformational Analysis, Scope and Present Limitations*, Academic, New York, p. 32.

Applequist, D. E., Renken, T. L., and Wheeler, J. W. (1982), *J. Org. Chem.*, **47**, 4985.

Armarego, W. L. F. (1977), *Stereochemistry of Hetereocyclic Compounds*, Part 1, Part 2, Wiley-Interscience, New York.

Aschan, O. (1905), *Chemie der Alicyclischen Verbindungen*, Vieweg & Son, Braunschweig, Germany, p. 329.

Ashby, E. C. and Laemmle, J. T. (1975), "Stereochemistry of Organometallic Compound Addition to Ketones," *Chem. Rev.*, **75**, 521.

Ashton, P. R., Bissel, R. A., Górski, R., Philp, D., Spencer, N., Stoddart, J. F., and Tolley, M. S. (1992a), *Synlett*, 919.

Ashton, P. R., Bissel, R. A., Spencer, N., Stoddart, J. F., and Tolley, M. S. (1992b), *Synlett*, 914, 923.

Ashton, P. R., Brown, C. L., Chrystal, E. J. T., Goodnow, T. T., Kaifer, A. E., Parry, K. P., Philp, D., Slawin, A. M. Z., Spencer, N., Stoddart, J. F., and Williams, D. J. (1991a), *J. Chem. Soc. Chem. Commun.*, 634.

Ashton, P. R., Goodnow, T. T., Kaifer, A. E., Reddington, M. V., Slawin, A. M. Z., Spencer, N., Stoddart, J. F., Vicent, C., and Williams, D. J. (1989), *Angew. Chem. Int. Ed. Engl.*, **28**, 1396.

Ashton, P. R., Philp, D., Spencer, N. and Stoddart, J. F. (1991b), *J. Chem. Soc. Chem. Commun.*, 1677.

Ashton, P. R., Philp, D., Spencer, N., and Stoddart, J. F. (1992c), *J. Chem. Soc. Chem. Commun.*, 1124.

Aston, J. G., Schumann, S. C., Fink, H. L., and Doty, P. M. (1941), *J. Am. Chem. Soc.*, **63**, 2029.

Atavin, E. G., Mastryukov, V. S., Allinger, N. L., Almenningen, A., and Seip, R. (1989), *J. Mol. Struct.*, **212**, 87.

Atwood, J. L., Koutsantonis, G. A., and Raston, C. L. (1994), *Nature*, **368**, 229.

Auf der Heyde, W. and Lüttke, W. (1978), *Chem. Ber.*, **111**, 2384.

Aycard, J.-P. (1989), *Spectrosc. Lett.*, **22**, 397.

Aydin, R. and Günther, H. (1981), *Angew. Chem. Int. Ed. Engl.*, **20**, 985.

Baas, J. M. A., van de Graaf, B., Tavernier, D., and Vanheem, P. (1981), *J. Am. Chem. Soc.*, **103**, 5014.

Baeyer, A. von (1885), *Ber. Dtsch. Chem. Ges.*, **18**, 2277.

Baeyer, A. (1890), *Justus Liebigs*, *Ann. Chem.*, **258**, 145.

Balaban, A. T. and Farcasiu, D. (1967), *J. Am. Chem. Soc.*, **89**, 1958.

Balch, A. L., Lee, J. W., Noll, B. C., and Olmstead, M. M. (1993), *J. Chem. Soc. Chem. Commun.*, 56.

Baldry, K. W. and Robinson, M. J. T. (1977), *Tetrahedron*, **33**, 1663.

Baldwin, J. E. (1976), *J. Chem. Soc. Chem. Commun.*, 734, 738.

Baldwin, J. (1978), "Rules for ring closure," Ciba Foundation Symposium 53, *Further Perspectives in Organic Chemistry*, Elsevier, New York, p. 85.

Baldwin, J. E., Cutting, J., Dupont, W., Kruse, L., Silbermann, L., and Thomas, R. C. (1976), *J. Chem. Soc. Chem. Commun.*, 736.

Baldwin, J. E. and Kruse, L. I. (1977), *J. Chem. Soc. Chem. Commun.*, 233.

Baldwin, J. E. and Lusch, M. J. (1982), *Tetrahedron*, **38**, 2939.

Baldwin, J. E. and Reiss, J. A. (1977), *J. Chem. Soc. Chem. Commun.*, 77.

Baldwin, J. E., Thomas, R. C., Kruse, L. I., and Silbermann, L. (1977), *J. Org. Chem.*, **42**, 3846.

Ballardini, R., Balzani, V., Gandolfi, M. T., Prodi, L., Venturi, M., Philp, D., Ricketts, H. G., and Stoddart, J. F. (1993), *Angew. Chem. Int. Ed. Engl.*, **32**, 1301.

Bally, T. and Masamune, S. (1980), *Tetrahedron*, **36**, 343.

Barborak, J. C., Watts, L., and Pettit, R. (1966), *J. Am. Chem. Soc.*, **88**, 1328.

Barili, P. L., Bellucci, G., Marioni, F., Morelli, I., and Scartoni, V. (1972), *J. Org. Chem.*, **37**, 4353.

Barkworth, P. M. R. and Crabb, T. A. (1981), *Org. Magn. Reson.*, **17**, 260.

Barlet, R., Le Goaller, R., and Gey, C. (1981), *Can. J. Chem.*, **59**, 621.

Barnier, J.-P. and Conia, J.-M. (1976), *Bull. Soc. Chim. Fr.*, 281.

Barrett, J. W. and Linstead, R. P. (1936), *J. Chem. Soc.*, 611.

Barth, G. and Djerassi, C. (1981), *Tetrahedron*, **37**, 4123.

Barton, D. H. R. (1950), *Experientia*, **6**, 316.

Barton, D. H. R., Hassel, O., Pitzer, K. S., and Prelog, V. (1953), *Nature (London)*, **172**, 1096.

Barton, D. H. R., Hassel, O., Pitzer, K. S., and Prelog, V. (1954), *Science*, **119**, 49.

Basso, E. A., Kaiser, C., Rittner, R., and Lambert, J. B. (1993), *J. Org. Chem.*, **58**, 7865.

Bastiansen, O., Fernholt, H., Seip, H. M., Kambara, H., and Kuchitsu, K. (1973), *J. Mol. Struct.*, **18**, 163.

Bastiansen, O., Hedberg, L., and Hedberg, K. (1957), *J. Chem. Phys.*, **27**, 1311.

Becker, K. B. and Pfluger, R. W. (1979), *Tetrahedron Lett.* 3713.

Bellucci, G., Berti, G., Bianchini, R., Cetera, P., and Mastriolli, E. (1982), *J. Org. Chem.*, **47**, 3105.

Bellucci, G., Berti, G., Ingrosso, G., and Mastriolli, E. (1980), *J. Org. Chem.*, **45**, 299.

Bellucci, G., Bianchini, R., Ingrosso, G., and Mastriolli, E. (1978), *Gazz. Chim. Ital.*, **108**, 643.

Benedetti, F. and Stirling, C. J. M. (1983), *J. Chem. Soc. Chem. Commun.*, 1374.

Berlin, A. J. and Jensen, F. R. (1960), *Chem. Ind. London*, 998.

Bernard, M., Canuel, L., and St.-Jacques, M. (1974), *J. Am. Chem. Soc.*, **96**, 2929.

Bernath, G., Fülöp, F., Kálmán, A., Argay, G., Sohár, P., and Pelczer, I. (1984), *Tetrahedron*, **40**, 3587.

Berti, G. (1973), "Stereochemistry of Epoxide Synthesis," *Top. Stereochem.*, **7**, 93.

Billups, W. E. and Ciufolini, M. A., Eds. (1993), *Buckminsterfullerenes*, VCH, New York.

Binsch, G., Eliel, E. L., and Kessler, H. (1971), *Angew. Chem. Int. Ed. Engl.*, **10**, 570.

Bishop, R. and Landers, A. E. (1979), *Aust. J. Chem.*, **32**, 2675.

Bjerrum, N. (1923), *Z. Phys. Chem.*, **106**, 219.

Blackburne, I. D., Katritzky, A. R., and Takeuchi, Y. (1975), "Conformation of Piperidine and of Derivatives with Additional Ring Heteroatoms," *Acc. Chem. Res.*, **8**, 300.

Block, E., Corey, E. R., Penn, R. E., Renken, T. L., Sherwin, P. F., Bock, H., Hirabayshi, T., Mohman, S., and Solouki, B. (1982), *J. Am. Chem. Soc.*, **104**, 3119.

Bloomfield, J. J. and Fuchs, R. (1970), *J. Chem. Soc. B.*, 363.

Bloomfield, J. J. and Irelan, J. R. S. (1966), *Tetrahedron Lett.*, 2971.

Bloomfield, J. J., Owsley, D. C., and Nelke, J. M. (1976), *Org. React.*, **23**, 259.

Bocian, D. F. and Strauss, H. L. (1977), *J. Am. Chem. Soc.*, **99**, 2866, 2876.

Boeckmann, J. and Schill, G. (1974), "Knotenstrukturen in der Chemie," *Tetrahedron*, **30**, 1945.

Boone, J. R. and Ashby, E. C. (1979), "Reduction of Cyclic and Bicyclic Ketones by Complex Metal Hydrides," *Top. Stereochem.*, **11**, 53.

Booth, H., Dixon, J. M., and Khedhair, K. A. (1992), *Tetrahedron*, **48**, 6161.

Booth, H. and Everett, J. R. (1976), *J. Chem. Soc. Chem. Commun.*, 278.

Booth, H. and Everett, J. R. (1980a), *J. Chem. Soc. Perkin 2*, 255.

Booth, H. and Everett, J. R. (1980b), *Can. J. Chem.*, **58**, 2709.

Booth, H. and Everett, J. R. (1980c), *Can. J. Chem.*, **58**, 2714.

Booth, H. and Griffiths, D. V. (1973), *J. Chem. Soc. Perkin 2*, 842.

Booth, H. and Grindley, T. B. (1983), *J. Chem. Soc. Chem. Commun.*, 1013.

Booth, H. and Jozefowicz, M. L. (1976), *J. Chem. Soc. Perkin 2*, 895.

Booth, H. and Khedhair, K. A. (1985), *J. Chem. Soc. Commun.*, 467.

Borden, W. T., Reich, I. L., Sharpe, L. A., and Riech, H. J. (1970), *J. Am. Chem. Soc.*, **92**, 3808.

Bordwell, F. G. and Vestling, M. M. (1967), *J. Am. Chem. Soc.*, **89**, 3906.

Box, V. G. S. (1990), "The Role of Lone Pair Interactions in the Chemistry of Monosaccharides. The Anomeric Effect," *Heterocycles*, **31**, 1151.

Bredt, J. (1924), *Justus Liebigs. Ann. Chem.*, **437**, 1.

Breslow, R. (1971), "Quantitative Studies on Aromaticity and Antiaromaticity," *Pure Appl. Chem.*, **28**, 111.

Breslow, R. (1973), "Antiaromaticity," *Acc. Chem. Res.*, **6**, 393.

Breslow, R., Pecoraro, J., and Sugimoto, T. (1977), *Org. Syn.*, **57**, 41; (1988), Coll. Vol. VI, 361.

Brigl, P. and Herrmann, R. (1938), *Ber. Dtsch. Chem. Ges. B*, **71**, 2280.

Brown, H. C. and Borkowski, M. (1952), *J. Am. Chem. Soc.*, **74**, 1894.

Brown, H. C., Brewster, J. H., and Shechter, H. (1954), *J. Am. Chem. Soc.*, **76**, 467.

Brown, H. C. and Ham, G. (1956), *J. Am. Chem. Soc.*, **78**, 2735.

Brown, H. C. and Ichikawa, K. (1957), *Tetrahedron*, **1**, 221.

Brown, H. C. and Kawakami, J. H. (1970), *J. Am. Chem. Soc.*, **92**, 1990.

Brown, H. C. and Krishnamurthy, S. (1972), *J. Am. Chem. Soc.*, **94**, 7159.

Brown, H. C. and Ravindran, N. (1977), *J. Org. Chem.*, **42**, 2533.

Brown, H. C., Fletcher, R. S., and Johannesen, R. B. (1951), *J. Am. Chem. Soc.*, **73**, 212.

Brown, H. C., Kawakami, J. H., and Ikegami, S. (1970), *J. Am. Chem. Soc.*, **92**, 6914.

Brown, W. A. C., Martin, J., and Sim, G. A. (1965), *J. Chem. Soc.*, 1844.

Brügel, W. (1967), *Nuclear Magnetic Resonance and Chemical Structure*, Vol. 1., Academic, New York.

Brunet, E. and Eliel, E. L. (1986), *J. Org. Chem.*, **51**, 677.

Bryan, R. F. and Dunitz, J. D. (1960), *Helv. Chim. Acta.*, **43**, 1.

Buchanan, G. W. (1982), *Can. J. Chem.*, **60**, 2908.

Buchanan, G. W. and McCarville, A. R. (1972), *Can. J. Chem.*, **50**, 1965.

Buchanan, G. W., Preusser, S. H. and Webb, V. L. (1984), *Can. J. Chem.* **62,** 1308.

Buchanan, G. W. and Webb, V. L. (1983), *Tetrahedron Lett.* **24**, 4519.

Bucourt, R. (1974), "The Torsion Angle Concept in Conformational Analysis," *Top. Stereochem.*, **8**, 159.

Bugay, D. E., Bushweller, C. H., Danehy, C. T., Hoogasian, S., Blersch, J. A., and Leenstra, W. R. (1989) *J. Phys. Chem.* **93**, 3908.

Burckhardt, U., Grob, C. A., and Kiefer, H. R. (1967), *Helv. Chim. Acta.*, **50**, 231.

Bürgi, H. B., Dunitz, J. D., and Shefter, E. (1973), *J. Am. Chem. Soc.*, **95**, 5065.

Bürgi, H. B., Dunitz, J. D., Lehn, J. M., and Wipff, G. (1974), *Tetrahedron*, **30**, 1563.

Burke, L. (1985), *Theor. Chim. Acta.*, **68**, 101.

Burkert, U. and Allinger, N. L. (1982), *Molecular Mechanics*, ACS Monograph 177, American Chemical Society, Washington, DC.

Bushweller, C. H. (in press) "Stereodynamics of Cyclohexane and Substituted Cyclohexanes. Substituent A-Values" in Juaristi, E., Ed., *Conformational Studies of Six-Membered Ring Carbocycles and Heterocycles*, VCH, New York.

Bushweller, C. H., Golini, J., Rao, G. U., and O'Neil, J. W. (1970), *J. Am. Chem. Soc.*, **92**, 3055.

Butcher, S. S. (1965), *J. Chem. Phys.*, **42**, 1833.

Buys, H. R. (1969), *Rec. Trav. Chim. Pays-Bas*, **88**, 1003.

Cahn, R. S., Ingold, C. K., and Prelog, V. (1966), "Specification of Molecular Chirality," *Angew. Chem. Int. Ed. Engl.*, **5**, 385.

Cantacuzéne, J., Jantzen, R., and Ricard, D. (1972), *Tetrahedron*, **28**, 717.

Cargill, R. L., Damewood, J. R., and Cooper, M. M. (1966), *J. Am. Chem. Soc.*, **88**, 1330.

Carr, C. A., Robinson, M. J. T., and Tchen, C. D. A. (1987), *Tetrahedron Lett.*, **28**, 897.

Carreira, L. A., Carter, R. O., and Durig, J. R. (1973), *J. Chem. Phys.*, **59**, 812.

Carreira, L. A. and Lord, R. C. (1969), *J. Chem. Phys.*, **51**, 2735.

Chakrabarti, P., Seiler, P., Dunitz, J. D., Schluter, A.-D., and Szeimies, G. (1981), *J. Am. Chem. Soc.*, **103**, 7378.

Chambron, J.-C., Heitz, V., and Sauvage, J.-P. (1992), *J. Chem. Soc. Chem. Commun.*, 1131.

Chang, S.-J., McNally, D., Shary-Tehrany, S., Hickey, M. J. Sr., and Boyd, R. H. (1970), *J. Am. Chem. Soc.*, **92**, 3109.

Cheney, J. and Lehn, J.-M. (1972), *J. Chem. Soc. Chem. Commun.*, 487.

Chérest, M. (1980), *Tetrahedron*, **36**, 1593.

Chérest, M., Felkin, H., and Prudent, N. (1968), *Tetrahedron Lett.*, 2199.

Chernov, P. P., Bazylchik, V. V., and Samitov, Yu. Yu. (1982), *Zh. Org. Chim.*, **18**, 1222 (Engl. transl. p. 1055).

Chiang, J. F. and Bauer, S. H. (1969), *J. Am. Chem. Soc.*, **91**, 1898.

Chiang, J. F. and Bauer, S. H. (1970), *J. Am. Chem. Soc.*, **92**, 1614.

Chiang, J. F., Wilcox, C. F., and Bauer, S. H. (1968), *J. Am. Chem. Soc.*, **90**, 3149.

Christl, M., Reich, H.-J., and Roberts, J. D. (1971), *J. Am. Chem. Soc.*, **83**, 3463.

Christl, M. and Roberts, J. D. (1972), *J. Org. Chem.*, **37**, 3443.

Chu, P.-S. and True, N. S. (1985), *J. Phys. Chem.* **89**, 5613.

Chukovskaya, E. Ts., Dostovalova, V. I., Kamyshova, A. A., and Freidlina, R. Kh. (1981), *Bull. Acad. Sci. USSR Div. Chem. Soc.*, **30**, 1801 (Engl. transl. p. 1470).

Cieplak, A. S. (1981), *J. Am. Chem. Soc.*, **103**, 4540.

Clayton, D. A. and Vinograd, J. (1967), *Nature (London)*, **216**, 652.

Corey, E. J. and Burke, H. J. (1955), *J. Am. Chem. Soc.*, **77**, 5418.

Corey, E. J. and Enders, D. (1976), *Tetrahedron Lett.*, 3.

Corey, E. J. and Feiner, N. F. (1980), *J. Org. Chem.*, **45**, 765.

Corey, E. J. and Sneen, R. A. (1956), *J. Am. Chem. Soc.*, **78**, 6269.

Corey, E. J., Sneen, R. A., Danaher, M. G., Young, R. L. and Rutledge, R. L. (1954), *Chem. Ind. (London)*, 1294.

Cox, K. W., Harmony, M. D., Nelson, G., and Wiberg, K. B. (1969), *J. Chem. Phys.*, **50**, 1976.

Cox, J. D. and Pilcher, G. (1970), *Thermochemistry of Organic and Organometallic Compounds*, Academic, New York.

Coxon, A. C. and Stoddart, J. F. (1977), *J. Chem. Soc. Perkin 1*, 767.

Crabb, T. A. (1992), "Conformational Equilibria in Azabicyclic Systems," Lambert, J. B. and Takeuchi, Y., Eds., *Cyclic Organonitrogen Stereodynamics*, VCH, New York.

Cram, D. J. (1990), *From Design to Discovery*, Am. Chem. Soc., Washington, DC.

Cram, D. J., Tanner, M. E., and Thomas, R. (1991), *Angew. Chem. Int. Ed. Engl.*, **30**, 1024.

Cram, D. J. and Trueblood, K. N. (1981), "Concept, Structure, and Binding In Complexation," *Top. Curr. Chem.*, **98**, 43.

Cremer, D. and Gauss, J. (1986), *J. Am. Chem. Soc.*, **108**, 7467.

Criegee, R., Höger, E., Huber, G., Kruck, P., Marktscheffel, F., and Schellenberger, H. (1956), *Justus Liebigs Ann. Chem.*, **599**, 81.

Cristol, S. J. (1949), *J. Am. Chem. Soc.*, **71**, 1894.

Cross, L. C. and Klyne, W., Collators (1976), "Rules for the Nomenclature of Organic Chemistry; Section E: Stereochemistry," *Pure Appl. Chem.*, **45**, 11.

Crowley, P. J., Robinson, M. J. T., and Ward, M. G. (1977), *Tetrahedron*, **33**, 915.

Cummins, E. G. and Page, J. E. (1957), *J. Chem. Soc.*, 3847.

Curl, R. F. and Smalley, R. E. (1988), *Science*, **242**, 1017.

Curtin, D. Y., Carlson, C. G., and McCarthy, C. G. (1964), *Can. J. Chem.*, **42**, 565.

Curtin, D. Y. and Harder, R. J. (1960), *J. Am. Chem. Soc.*, **82**, 2357.

Dale, J. (1976), "Multistep Conformational Interconversion Mechanisms," *Top. Stereochem.*, **9**, 199.

Dale, J. (1978), *Stereochemistry and Conformational Analysis*, Verlag Chemie, New York.

Dalling, D. K. and Grant, D. M. (1967), *J. Am. Chem. Soc.*, **89**, 6612.

Dalling, D. K. and Grant, D. M. (1972), *J. Am. Chem. Soc.*, **94**, 5318.

Dalling, D. K., Grant, D. M., and Johnson, L. F. (1971), *J. Am. Chem. Soc.*, **93**, 3678.

Dalling, D. K., Zilm, K. W., Grant, D. M., Heeschen, W. A., Horton, W. J., and Pugmire, R. J. (1981), *J. Am. Chem. Soc.*, **103**, 4817.

Dallinga, G. and Toneman, L. H. (1968), *Recl. Trav. Chim. Pays-Bas*, **87**, 795.

Danishefsky, S. (1993), *Science*, **259**, 469.

Dashevsky, V. G. and Lugovsky, A. A. (1972), *J. Mol. Struct.*, **12**, 39.

Dauben, W. G., Fonken, G. J., and Noyce, D. S. (1956), *J. Am. Chem. Soc.*, **78**, 2579.

Dauben, W. G. and Laug, P. (1962), *Tetrahedron Lett.*, 453.

Dauben, W. G. and Walker, D. M. (1982), *Tetrahedron Lett.*, **23**, 711.

Dauben, W. G., Rohr, O., Labbauf, A., and Rossini, F. D. (1960), *J. Phys. Chem.*, **64**, 283.

Davis, M. and Hassel, O. (1963), *Acta. Chem. Scand.*, **17**, 1181.

de Boer, T. J. (1977), *Chimia*, **31**, 483.

de Hoog, A. J. (1974), *Org. Magn. Reson.*, **6**, 233.

de Hoog, A. J., Buys, H. R., Altona, C., and Havinga, E. (1969), *Tetrahedron*, **25**, 3365.

de Hoog, A. J. and Havinga, E. (1970), *Recl. Trav. Chim. Pays-Bas*, **89**, 972.

de la Mare, P. B. D., Hall, D., and Pavitt, N. (1983), *J. Comput. Chem.*, **4**, 114.

Delben, F. and Crescenzi, V. (1978), *J. Solution. Chem.*, **7**, 597.

Della, E. W. (1966), *Tetrahedron Lett.*, 3347.

Della, E. W. (1967), *J. Am. Chem. Soc.*, **89**, 5221.

Della, E. W., Hine, P. T., and Patney, H. K. (1977), *J. Org. Chem.*, **42**, 2940.

Della, E. W. and Rizvi, S. Q. (1974), *Aust. J. Chem.*, **27**, 1059.

Delpuech, J.-J. (1992), "Six-Membered Rings," Lambert, J. B. and Takeuchi, Y., Eds., *Cyclic Organic Stereodynamics*, VCH, New York, p. 169.

Desilets, S. and St.-Jacques, M. (1987), *J. Am. Chem. Soc.*, **109**, 1641.

Desilets, S. and St.-Jacques, M. (1988), *Tetrahedron*, **44**, 7027.

Deslongchamps, P. (1983), *Stereoelectronic Effects in Organic Chemistry*, Pergamon, New York.

DeTar, D. F. and Brooks, W. (1978), *J. Org. Chem.*, **43**, 2245.

DeTar, D. F. and Luthra, N. P. (1980), *J. Am. Chem. Soc.*, **102**, 4505.

Diederich, F. and Whetten, R. L. (1992), "Beyond C_{60}. The Higher Fullerenes," *Acc. Chem. Res.*, **25**, 119.

Diederich, F., Whetten, R. L., Thilgen, C., Ettl, R., Chao, I., and Alvarez, M. M. (1991), *Science*, **254**, 1768.

Dietrich, B. (1984), "Cryptate Complexes," Atwood, J. L., Davies, J. E. D. and MacNicol, D. D., Eds., *Inclusion Compounds*, Vol. 2, Academic, New York, p. 337.

Dietrich-Buchecker, C. O. and Sauvage, J.-P. (1987), "Interlocking of Molecular Threads: From the Statistical Approach to the Templated Synthesis of Catenands," *Chem. Rev.*, **87**, 795.

Dietrich-Buchecker, C. O. and Sauvage, J.-P. (1989), *Angew. Chem. Int. Ed. Engl.*, **28**, 189.

Dillen, J. and Geise, H. J. (1979), *J. Chem. Phys.*, **70**, 425.

Dillen, J. and Geise, H. J. (1980), *J. Mol. Struct.*, **69**, 137.

Dodziuk, H. (1984), *J. Comput. Chem.*, **5**, 571.

Dommen, J., Brupbacher, T., Grassi, G., and Bauder, A. (1990), *J. Am. Chem. Soc.*, **112**, 953.

Dorofeeva, O. V., Mastryukov, V. S., Allinger, N. L., and Almenningen, A. (1985), *J. Phys. Chem.*, **89**, 252.

Dorofeeva, O. V., Mastryukov, V. S., Allinger, N. L., and Almenningen, A. (1990b), *J. Phys. Chem.*, **94**, 8044.

Dorofeeva, O. V., Mastryukov, V. S., Siam, K., Ewbank, J. D., Allinger, N. L., and Schaefer, L. (1990a), *J. Strukt. Chem. (USSR)*, **31**, 167 (Engl. Transl. p. 153).

Došen-Micóvić, L. and Allinger, N. (1978), *Tetrahedron*, **34**, 3385.

Došen-Micóvić, L., Li, S. and Allinger, N. L. (1991), *J. Phys. Org. Chem.*, **4**, 467.

Duax, W. L., Weeks, C. M., and Rohrer, D. C. (1976), "Crystal Structure of Steroids," *Top. Stereochem.*, **9**, 271.

Duchamp, D. J. and Marsh, R. E. (1969), *Acta. Cryst.*, **B25**, 5.

Duddeck, H. (1986), "Substituent Effects on ^{13}C Chemical Shifts in Aliphatic Molecular Systems. Dependence on Constitution and Stereochemistry," *Top. Stereochem.*, **16**, 219.

Duddeck, H., Wagner, P., and Biallas, A. (1991), *Magn. Reson. Chem.*, **29**, 248.

Duddeck, H., Wagner, P. and Gegnor, S. (1985), *Tetrahedron Lett.*, **26**, 1205.

Dunitz, J. D. (1970), *J. Chem. Educ.*, **47**, 488.

Dunitz, J. D. (1971), "Conformation of Medium Rings," *Pure Appl. Chem.*, **25**, 495.

Dunitz, J. D., Eser, H., Bixon, M., and Lifson, S. (1967), *Helv. Chim. Acta.*, **50**, 1572.

Dunitz, J. D. and Shearer, H. M. M. (1960), *Helv. Chim. Acta.*, **43**, 18.

Durig, J. R., Geyer, T. J., Little, T. S., and Kalasinsky, V. F. (1987), *J. Chem. Phys.*, **86**, 545.

Durig, J. R., Lee, M. J., and Little, T. S. (1990), *J. Raman Spectrosc.*, **21**, 529.

Durig, J. R., Little, T. S., and Lee, M. J. (1989), *J. Raman Spectrosc.*, **20**, 757.

Eaton, P. E. (1979), *Tetrahedron*, **35**, 2189.

Eaton, P. E. and Cole, T. W. (1964), *J. Am. Chem. Soc.*, **86**, 3157.

Eaton, P. E. and Kyi, N. (1971), *J. Am. Chem. Soc.*, **93**, 2786.

Eaton, P. E. and Leipzig, B. K. (1983), *J. Am. Chem. Soc.*, **105**, 1656.

Eaton, P. E. and Temme, G. H., III (1973), *J. Am. Chem. Soc.*, **95**, 7508.

Edward, J. T. (1955), *Chem. Ind. (London)*, 1102.

Edward, J. T., Farrell, P. G., Kirchnerova, J., Halle, J.-C., and Schaal, R. (1976), *Can. J. Chem.*, **54**, 1899.

Edward, J. T., Farrell, P. G., and Langford, G. E. (1976), *J. Am. Chem. Soc.*, **98**, 3075.

Egawa, T., Fukuyama, T., Yamamoto, S., Takabayashi, F., Kambara, H., Ueda, T., and Kuchitsu, K. (1987), *J. Chem. Phys.*, **86**, 6018.

Eliel, E. L. (1956), "Substitution at Saturated Carbon Atoms," in Newman, M. S., Ed., *Steric Effects in Organic Chemistry*, Wiley, New York.

Eliel, E. L. (1959), *Chem. Ind. (London)*, 568.

Eliel, E. L. (1962), *Stereochemistry of Carbon Compounds*, McGraw-Hill, New York.

Eliel, E. L. (1970), "Conformational Analysis in Saturated Heterocyclic Compounds," *Acc. Chem. Res.*, **3**, 1.

Eliel, E. L. (1972) "Conformational Analysis in Heterocyclic Systems: Recent Results and Applications," *Angew. Chem. Int. Ed. Engl.*, **11**, 739.

Eliel, E. L. (1975), *J. Chem. Educ.*, **52**, 762.

Eliel, E. L., Allinger, N. L., Angyal, S. J., and Morrison, G. A. (1965), *Conformational Analysis*, Wiley, New York; reprinted (1981) by American Chemical Society, Washington, DC.

Eliel, E. L., Bailey, W. F., Kopp, L. D., Willer, R. L., Grant, D. M., Bertrand, R., Christensen, K. A., Dalling, D. K., Duch, M. W., Wenkert, E., Schell, F. M., and Cochran, D. W. (1975), *J. Am. Chem. Soc.*, **97**, 322.

Eliel, E. L. and Banks, H. D. (1972), *J. Am. Chem. Soc.*, **94**, 171.

Eliel, E. L. and Biros, F. J. (1966), *J. Am. Chem. Soc.*, **88**, 3334.

Eliel, E. L. and Chandrasenkaran, S. (1982), *J. Org. Chem.*, **47**, 4783.

Eliel, E. L., Della, E. W., and Williams, T. H. (1963), *Tetrahedron Lett.*, 831.

Eliel, E. L. and Enanoza, R. M. (1972), *J. Am. Chem. Soc.*, **94**, 8072.

Eliel, E. L. and Gianni, M. H. (1962), *Tetrahedron Lett.*, 97.

Eliel, E. L., Gianni, M. H., Williams, T. H., and Stothers, J. B. (1962), *Tetrahedron Lett.*, 741.

Eliel, E. L. and Gilbert, E. C. (1969), *J. Am. Chem. Soc.*, **91**, 5487.

Eliel, E. L. and Haber, R. G. (1958), *J. Am. Chem. Soc.*, **23**, 2041.

Eliel, E. L., Hargrave, K. D., Pietrusiewicz, K. M., and Manoharan, M. (1982), *J. Am. Chem. Soc.*, **104**, 3635.

Eliel, E. L. and Haubenstock, H. (1961), *J. Org. Chem.*, **26**, 3504.

Eliel, E. L., Haubenstock, H., and Acharya, R. V. (1961), *J. Am. Chem. Soc.*, **83**, 2351.

Eliel, E. L. and Juaristi, E. (1979), "Conformational Interactions in 1,4-Heterobutane Segments," Horton, D. and Szarek, W. A., Eds., *Anomeric Effect: Orgin and Consequences*, ACS Symposium Series 87, American Chemical Society, Washington, DC.

Eliel, E. L. and Kandasamy, D. (1976), *J. Org. Chem.*, **44**, 3899.

Eliel, E. L., Kandasamy, D., Yen, C.-y., and Hargrave, K. D. (1980), *J. Am. Chem. Soc.*, **102**, 3698.

Eliel, E. L. and Knoeber, M. C. (1968), *J. Am. Chem. Soc.*, **90**, 3444.

Eliel, E. L. and Knox, D. E. (1985), *J. Am. Chem. Soc.*, **107**, 2946.

Eliel, E. L. and Lukach, C. A. (1957), *J. Am. Chem. Soc.*, **79**, 5986.

Eliel, E. L. and Manoharan, M. (1981), *J. Org. Chem.*, **46**, 1959.

Eliel, E. L. and Martin, R. J. L. (1968a), *J. Am. Chem. Soc.*, **90**, 682.

Eliel, E. L. and Martin, R. J. L. (1968b), *J. Am. Chem. Soc.*, **90**, 689.

Eliel, E. L., Martin, R. J. L., and Nasipuri, D. (1967), *Org. Syn.*, **47**, 16; Coll. Vol. V, 175.

Eliel, E. L., Morris-Natschke, S., and Kolb, V. M. (1984), *Org. Magn. Reson.*, **22**, 258.

Eliel, E. L. and Pillar, C. (1955), *J. Am. Chem. Soc.*, **77**, 3600.

Eliel, E. L. and Pietrusiewicz, K. M. (1979), "^{13}C NMR of Nonaromatic Heterocyclic Compounds," *Top. C-13 NMR Spectrosc.*, **3**, 171.

Eliel, E. L. and Pietrusiewicz, K. M. (1980), *Org. Magn. Reson.*, **13**, 193.

Eliel, E. L. and Pietrusiewicz, K. M. (1981), *Pol. J. Chem.*, **55**, 1265.

Eliel, E. L. and Reese, M. C. (1968), *J. Am. Chem. Soc.*, **90**, 1560.

Eliel, E. L. and Ro, R. S. (1956), *Chem. Ind. (London)*, 251.

Eliel, E. L. and Ro, R. S. (1957a), *J. Am. Chem. Soc.*, **79**, 5992.

Eliel, E. L. and Ro, R. S. (1957b), *J. Am. Chem. Soc.*, **79**, 5995.

Eliel, E. L., Schroeter, S. H., Brett, T. J., Biros, F. J., and Richer, J.-C. (1966), *J. Am. Chem. Soc.*, **88**, 3327.

Eliel, E. L. and Senda, Y. (1970), *Tetrahedron*, **26**, 2411.

Eliel, E. L. and Willer, R. L. (1977), *J. Am. Chem. Soc.*, **99**, 1936.

Elser, V. and Strauss, H. (1983), *Chem. Phys. Lett.*, **96**, 276.

Engler, E. M., Andose, J. D., and Schleyer, P. v. R. (1973), *J. Am. Chem. Soc.*, **95**, 8005.

Ermer, O. and Dunitz, J. D. (1969), *Helv. Chim. Acta*, **52**, 1861.

Ermer, O., Gerdil, R., and Dunitz, J. D. (1971), *Helv. Chim. Acta*, **54**, 2476.

Ermolaeva, L. I., Mastryukov, V. S., Allinger, N. L., and Almenningen, A. (1989), *J. Mol. Struct.*, **196**, 151.

Evans, S. A. (1989), *Conformational Analysis of Partially Unsaturated Six-Membered Rings Containing Heteroatoms*, see reference to Rabideau, P. W., Ed., (1989a), pp. 169–210.

Ewbank, J. D., Kirsch, G., and Schäfer, L. (1976) *J. Mol. Struct.*, **31**, 39.

Fahey, R. C. (1968), "The Stereochemistry of Electrophilic Additions to Olefins and Acetylenes," *Top. Stereochem.*, **3**, 237.

Fang, J. M., Sun, S. F., and Rei, M.-H. (1987), "Stereochemistry of Hydride Reduction of Cyclic Ketones-Unique Case in the Reactions of *trans*-2-Methyl-4-*tert*-butylcyclohexanone," Kobayashi, M., Ed., *Physical Organic Chemistry 1986*, Elsevier, New York, p. 241.

Fawcett, F. S. (1950), *Chem. Rev.*, **47**, 219.

Fawcett, J. K. and Trotter, J. (1966), *J. Chem. Soc. (B)*, 174.

Fernandez Gomez, F., Lysenkov, V. I., Ul'yanova, O. D., Pentin, Yu. A., and Bardyshev, I. I. (1977a), *Russ. J. Phys. Chem.*, **51**, 2710 (Engl. transl. p. 1585).

Fernandes Gomes, Fr., Ul'yanova, O. D., and Pentin, Yu. A. (1973), *Russ. J. Phys. Chem.*, **47**, 447 (Engl. transl. p. 252).

Fernandez Gomez, F. E., Ul'yanova, O. D., and Pentin, Yu. A. (1977b), *Russ. J. Phys. Chem.*, **51**, 1021 (Engl. transl. p. 610).

Ferrier, W. G. and Iball, J. (1954), *Chem. Ind. London*, 1296.

Fessner, W.-D., Murty, B. A. R. C., and Prinzbach, H. (1987b), *Angew. Chem. Int. Ed. Engl.*, **26**, 451.

Fessner, W.-D., Murty, B. A. R. C., Wörth, J., Hunkler, D., Fritz, H., Prinzbach, H., Roth, W. D., Schleyer, P. v. R., McEwen, A. B., and Maier, W. F. (1987a), *Angew. Chem. Int. Ed. Engl.*, **26**, 452.

Fierens, P. J. C. and Verschelden, P. (1952), *Bull. Soc. Chim. Belg.*, **61**, 427, 609.

Fieser, L. F. and Fieser, M. (1959), *Steroids*, Reinhold, New York.

Finke, H. L., McCullough, J. P., Messerly, J. F., Osborn, A., and Douslin, D. R. (1972), *J. Chem. Thermodyn.*, **4**, 477.

Finke, H. L., Messerly, J. F., and Lee-Bechtold, S. H. (1981), *J. Chem. Thermodyn.*, **13**, 345.

Fishman, A. I., Remizov, A. B., and Stolov, A. A. (1981), *Dokl. Akad. Nauk SSSR, Phys. Chem.*, **260**, 683 (Engl. transl. p. 868).

Fleischer, E. B. (1964), *J. Am. Chem. Soc.*, **86**, 3889.

Fort, R. C. (1976), *Adamantane, The Chemistry of Diamond Molecules*, Marcel Dekker, New York.

Franck, R. W. (1983), *Tetrahedron*, **39**, 3251.

Fraser, R. R., Banville, J., and Dhawan, K. L. (1978), *J. Am. Chem. Soc.*, **100**, 7999.

Fraser, R. R. and Champagne, P. J. (1978), *J. Am. Chem. Soc.*, **100**, 657.

Fraser, R. R. and Dhawan, K. L. (1976), *J. Chem. Soc. Chem. Commun.*, 674.

Frisch, H. L., Frisch, K. C., and Klempner, D. (1981), "Advances in Interpenetrating Polymer Networks," *Pure Appl. Chem.*, **53**, 1557.

Fuchs, B. (1978), "Conformations of Five-membered Rings," *Top. Stereochem.*, **10**, 1.

Fürst, A. and Plattner, P. A. (1951), *Abstr. Papers of the 12th Intern. Congress Pure and Appl. Chem.*, New York, p. 409.

Gallagher, M. J. (1987), "Cyclic Compounds: Conformation and Stereochemistry" Verkade, J. G. and Quin, L. D., Eds., *Phosphorus-31 NMR Spectroscopy in Stereochemical Analysis*, VCH, New York, p. 297.

Gallucci, J. C., Doecke, C. W., and Paquette, L. A. (1986), *J. Am. Chem. Soc.*, **108**, 1343.

Ganter, C. (1992) "A New Approach to Adamantane Rearrangements", see reference to Osawa and Yonemitsu, eds., (1992), p. 293.

Gardiner, D. J. and Walker, N. A. (1987), *J. Mol. Struct.*, **161**, 55.

Gardiner, D. J., Littleton, C. J., and Walker, N. A. (1987), *J. Raman Spectrosc.*, **18**, 9.

Gassman, P. G. and Proehl, G. S. (1980), *J. Am. Chem. Soc.*, **102**, 6862.

Gassman, P. G., Topp, A., and Keller, J. W. (1969), *Tetrahedron Lett.*, 1093.

Gatial, H., Horn, A., Klaeboe, P., Nielsen, C. J., Pedersen, B., Hopf, H., and Mylnek, C. (1990), *J. Mol. Struct.*, **218**, 59; *id.* (1991), *Z. Phys. Chem.*, **170**, 31.

Geise, H. J. and Buys, H. R. (1970), *Recl. Trav. Chim. Pays-Bas*, **89**, 1147.

Geise, H. J., Buys, H. R., and Mijlhoff, F. C. (1971), *J. Mol. Struct.*, **9**, 447.

Gerig, J. T. (1968), *J. Am. Chem. Soc.*, **90**, 1065.

Gerig, J. T. and Rimmerman, R. A. (1970), *J. Am. Chem. Soc.*, **92**, 1219.

Gilboa, H., Altman, J. and Loewenstein, A. (1969), *J. Am. Chem. Soc.*, **91**, 6062.

Ginsburg, D. (1972), "Small Ring Propellanes", *Acc. Chem. Res.*, **5**, 249.

Ginsburg, D. (1975), *Propellanes, Structure and Reactions*, Verlag Chemie, Weinheim, Federal Republic of Germany.

Ginsburg, D. (1987), "Of Propellanes–and Of Spiranes," *Top. Curr. Chem.*, **137**, 1.

Gittins, V. M., Wyn-Jones, E., and White, R. F. M. (1974), "Ring Inversion in some Six-membered Heterocyclic Compounds," Orville-Thomas, W. J., Ed., *Internal Rotation in Molecules*, Wiley, New York, p. 425ff.

Glass, R. S. ed. (1988), *Conformational Analysis of Medium-Sized Heterocycles*, VCH, New York.

Goering, H. L. and Serres, C. (1952), *J. Am. Chem. Soc.*, **74**, 5908.

Golan, O., Goren, Z., and Biali, S. E. (1990), *J. Am. Chem. Soc.*, **112**, 9300.

Gokel, G. W. (1991), *Crown Ethers and Cryptands*, The Royal Society of Chemistry, London, UK.

Gordon, M. D. and Quin, L. D. (1976), *J. Am. Chem. Soc.*, **98**, 15.

Gorenstein, D. G. (1977), *J. Am. Chem. Soc.*, **99**, 2254.

Graczyk, P. P. and Mikołajczyk, M. (1994), "Anomeric Effect: Origin and Consequences," *Top. Stereochem.* **21**, 159.

Grant, D. M. and Cheney, B. V. (1967), *J. Am. Chem. Soc.*, **89**, 5317.

Green, B. S. and Rejto, M. (1974), *J. Org. Chem.*, **39**, 3284.

Green, B. S., Arad-Yellin, R., and Cohen, M. D. (1986), "Stereochemistry and Organic Solid State Reactions," *Top. Stereochem.*, **16**, 131.

Greenberg, A. and Harris, J. (1982), *J. Chem. Educ.*, **59**, 539.

Greenberg, A. and Liebman, J. F. (1978), *Strained Organic Molecules*, Academic, New York.

Gregory, B. J., Haines, A. H., and Karntiang, P. (1977), *J. Chem. Soc. Chem. Commun.*, 918.

Griffin, G. W. and Marchand, A. P. (1989), "Synthesis and Chemistry of Cubanes," *Chem. Rev.*, **89**, 997.

Grob, C. A. (1969), "Mechanisms and Stereochemistry of Heterolytic Fragmentation," *Angew. Chem. Int. Ed. Engl.*, **8**, 535.

Grob, C. A. and Schiess, P. W. (1967), "Heterolytic Fragmentation. A Class of Organic Reactions," *Angew. Chem. Int. Ed. Engl.*, **6**, 1.

Grossel, M. C., Cheetham, A. K., and Newsam, J. M. (1978), *Tetrahedron Lett.*, 5229.

Grover, S. H., Guthrie, J. P., Stothers, J. B., and Tan, C. T. (1973), *J. Magn. Reson.*, **10**, 227.

Gund, P. and Gund, T. H. (1981), *J. Am. Chem. Soc.*, **103**, 4458.

Gunter, M. J. and Johnston, M. R. (1992), *J. Chem. Soc. Chem. Commun.*, 1163.

Gutsche, C. D. (1989), "Calixarenes" (Monographs in Supramolecular Chemistry, No. 1), Royal Society of Chemistry, Cambridge, UK.

Gutsche, C. D., Dhawan, B., No, K. H., and Muthukrishnan, R. (1981), *J. Am. Chem. Soc.*, **103**, 3782.

Haasnoot, C. A. G., de Leeuw, F. A. A. M., de Leeuw, H. P. M., and Altona, C. (1981), *Biopolymers*, **20**, 1211.

Hahn, E. H., Bohm, H., and Ginsburg, D. (1973), *Tetrahedron Lett.*, 507.

Hammond, G. S. (1955), *J. Am. Chem. Soc.*, **77**, 334.

Hammond, G. S. (1956), "Steric Effects on Equilibrated Systems," in Newman, M. S., Ed., *Steric Effects in Organic Chemistry*, Wiley, New York, p. 425ff.

Hammond, G. S. and Kuck, V. J., Eds. (1992), *Fullerenes: Synthesis, Properties and Chemistry of Large Carbon Clusters*, ACS Symposium Series 481, American Chemical Society, Washington, DC.

Hamon, D. P. G. and Trenerry, V. C. (1981), *J. Am. Chem. Soc.*, **103**, 4962.

Hanaya, K., Kudo, H., Gohke, K., and Imaizumi, S. (1979), *Nippon Kakagu Kaishi*, 1666; *Chem. Abstr.*, **91**, 192692a.

Hare, J. P., Kroto, H. W. and Taylor, R. (1991), *Chem. Phys. Lett.*, **177**, 394.

Hargittai, I. and Hargittai, M. (1988), "*Stereochemical Applications of Gas-Phase Electron Diffraction*, VCH, New York.

Harnisch, J., Baumgartel, O., Szeimies, G., van Meerssche, M., Germain, G., and Declercq, J.-P. (1979), *J. Am. Chem. Soc.*, **101**, 3370.

Harrison, I. T. (1974), *J. Chem. Soc. Perkin 1*, 301.

Harrison, I. T. and Harrison, S. (1967), *J. Am. Chem. Soc.*, **89**, 5723.

Hassel, O. (1943), *Tidsskr. Kjemi Bergvesen Met.*, **3**, [5] 32; Engl. Transl. Hedberg, K. (1971) *Top. Stereochem.*, **6**, 11.

Haufler, R. E., Conceicao, J., Chibante, L. P. F., Chai, Y., Byrne, N. E., Flanagan, S., Haley, M. M., O'Brien, S. C., Pan, C., Xiao, Z., Billups, W. E., Cuifolini, M. A., Hauge, R. H., Margrave, J. L., Wilson, L. J., Curl, R. F., and Smalley, R. E. (1990), *J. Phys. Chem.*, **94**, 8634.

Hawkins, J. M., Lewis, T. A., Loren, S. D., Meyer, A., Heath, J. R., Saykally, R. J., and Hollander, F. J. (1991a), *J. Chem. Soc. Chem. Commun.*, 775.

Hawkins, J. M. and Meyer, A. (1993). *Science*, **260**, 1918.

Hawkins, J. M., Meyer, A., Lewis, T. A., Loren, S., and Hollander, F. J. (1991b), *Science*, **252**, 312.

Hayward, R. C. (1983), "Abiotic Receptors," *Chem. Soc. Rev.*, **12**, 285.

Heck, R. and Prelog, V. (1955), *Helv. Chim. Acta.*, **38**, 1541.

Hedberg, L. and Hedberg, K. (1985), *J. Am. Chem. Soc.*, **107**, 7257.

Henbest, H. B. and Smith, M. (1957), *J. Chem. Soc.*, 926.

Henbest, H. B. and Wilson, R. A. L. (1957), *J. Chem. Soc.*, 1958.

Hendrickson, J. B. (1961), *J. Am. Chem. Soc.*, **83**, 4537.

Hendrickson, J. B. (1967), *J. Am. Chem. Soc.*, **89**, 7036, 7043, 7047.

Herbstein, F. H., Kapon, M., and Reisner, G. M. (1981), *Proc. R. Soc. Chem. Ser. A*, **376**, 301.

Herbstein, F. H., Kapon, M., and Reisner, G. M. (1986), *Acta. Cryst. Sect. B*, **42**, 181.

Herlinger, H. and Naegele, W. (1968), *Tetrahedron Lett.*, 4383.

Higuchi, H. and Ueda, I. (1992) "Recent Developments in the Chemistry of Cubane", see reference to Osawa and Yonemitsu, eds., (1992), p. 217.

Hilderbrandt, R. L. and Shen, Q. (1982), *J. Phys. Chem.*, **86**, 587.

Hilderbrandt, R. L., Wieser, J. D., and Montgomery, L. K. (1973), *J. Am. Chem. Soc.*, **95**, 8598.

Hirsch, J. A. (1967), "Table of Conformational Energies, 1967," *Top. Stereochem.*, **1**, 199.

Hodgson, D. J., Rychlewska, U., Eliel, E. L., Manoharan, M., Knox, D. E., and Olefirowicz, E. M. (1985), *J. Org. Chem.*, **50**, 4838.

Höfner, D., Lesko, S. A., and Binsch, G. (1978), *Org. Magn. Reson.*, **11**, 179.

Honig, H. and Allinger, N. L. (1985), *J. Org. Chem.*, **50**, 4630.

Hounshell, W. D. and Mislow, K. (1979), *Tetrahedron Lett.*, 1205.

House, H. O., Gaa, P. C., and VanDerveer, D. (1983), *J. Org. Chem.*, **48**, 1661.

Hückel, W. (1925), *Justus Liebigs Ann. Chem.*, **441**, 1.

Hudlicky, T. and Cebulak, M. (1993), *Cyclitols and Their Derivatives*, VCH, New York.

Hudson, B. and Vinograd, J. (1967), *Nature (London)*, **216**, 647.

Huet, J., Maroni-Barnaud, Y., Anh, N. T., and Seyden-Penne, J. (1976), *Tetrahedron Lett.*, 159.

Huisgen, R. and Ott, H. (1959), *Tetrahedron*, **6**, 253.

Illuminati, G. and Mandolini, L. (1981), "Ring Closure Reactions of Bifunctional Chain Molecules," *Acc. Chem. Res.*, **14**, 95.

Immirzi, A. and Torti, E. (1968), *Atti. Acad. Naz., Lincei, Rend. Cl. Sci. Fis. Mat. Nat.*, **44**, 98; *Chem. Abstr.*, **69**, 90996f.

Inoue, Y., Kurosawa, K., Nakanishi, K., and Obara, H. (1965), *J. Chem. Soc.*, 3339.

Irngartinger, H., Goldmann, A., Jahn, R., Nixdorf, M., Rodewald, H., Maier, G., Malsch, K.-D., and Emrich, R. (1984), *Angew. Chem. Int. Ed. (Engl.)*, **23**, 993.

Ivanov, P. M. and Ōsawa, E. (1984), *J. Comp. Chem.*, **5**, 307.

Jackman, L. M. and Sternhell, S. (1969), *Applications of Nuclear Magnetic Resonance Spectroscopy in Organic Chemistry*, 2nd ed. Pergamon, New York.

Jager, J., Graafland, T., Schenck, H., Kirby, A. J., and Engberts, J. B. F. N. (1984), *J. Am. Chem. Soc.*, **106**, 139.

Jaime, C., Ōsawa, E., Takeuchi, Y., and Camps, P. (1983), *J. Org. Chem.*, **48**, 4514.

James, V. J. (1973), *Cryst. Struct. Commun.*, **2**, 205.

James, V. J. and Grainger, C. T. (1972), *Cryst. Struct. Commun.*, **1**, 111.

James, V. J. and McConnell, J. F. (1971), *Tetrahedron*, **27**, 5475.

James, V. J. and Moore, F. H. (1975), *Acta. Cryst.*, **B31**, 1053.

Janda, K. D., Shevlin, C. G., and Lerner, R. A. (1993), *Science*, **259**, 490.

Jeffrey, G. A. (1979), "The Structural Properties of the Anomeric Center in Pyranoses and Pyranosides," in Szarek, W. A. and Horton, D., Eds., *Anomeric Effect, Origin and Consequences*, ACS Symposium Series 87, American Chemical Society, Washington, DC, p. 50ff.

Jeffrey, G. A. (1988), *J. Am. Chem. Soc.*, **110**, 7218.

Jensen, F. R. and Beck, B. H. (1968), *J. Am. Chem. Soc.*, **90**, 1066.

Jensen, F. R. and Bushweller, C. H. (1969a), *J. Am. Chem. Soc.*, **91**, 3223.

Jensen, F. R. and Bushweller, C. H. (1969b), *J. Am. Chem. Soc.*, **91**, 5774.

Jensen, F. R. and Bushweller, C. H. (1971), "Conformational Preferences in Cyclohexanes and Cyclohexenes," *Adv. Alicycl. Chem.*, **3**, 139.

Jensen, F. R., Bushweller, C. H., and Beck, B. H. (1969), *J. Am. Chem. Soc.*, **91**, 344.

Jensen, F. R. and Smith, L. A. (1964), *J. Am. Chem. Soc.*, **86**, 956.

Jochims, J. C. and Kobayashi, Y. (1976), *Tetrahedron Lett.*, 2065.

Johnson, F. (1968), *Chem. Rev.*, **68**, 375.

Johnson, F. and Whitehead, A. (1964), *Tetrahedron Lett.*, 3825.

Johnson, P. L., Cheer, C. J., Schaefer, J. P., James, V. J., and Moore, F. H. (1972a), *Tetrahedron*, **28**, 2893.

Johnson, P. L., Schaefer, J. P., James, V. J., and McConnell, J. F. (1972b), *Tetrahedron*, **28**, 2901.

Johnson, W. S. (1953), *J. Am. Chem. Soc.*, **75**, 1498.

Jolley, S. T., Bradshaw, J. S., and Izatt, R. M. (1982), "Synthetic Chiral Macrocyclic Crown Ligands: A Short Review," *J. Heterocycl. Chem.*, **19**, 3.

Jones, A. J., Eliel, E. L., Grant, D. M., Knoeber, M. C., and Bailey, W. F. (1971), *J. Am. Chem. Soc.*, **93**, 4772.

Jordan, E. A. and Thorne, M. P. (1986), *Tetrahedron*, **42**, 93.

Juaristi, E. and Aguilar, M. A. (1991), *J. Org. Chem.*, **56**, 5919.

Juaristi, E. and Cuevas, G. (1992), "Recent Studies of the Anomeric Effect," *Tetrahedron*, **48**, 5019.

Juaristi, E., Flores-Vela, A., Labastida, V. and Ordoñez, M. (1989), *J. Phys. Org. Chem.*, **2**, 349.

Juaristi, E., Labastida, V., and Antunez, S. (1991), *J. Org. Chem.*, **56**, 4802.

Juaristi, E., Lopez-Nuñez, N. A., Glass, R. S., Petsom, A., Hutchins, R. O., and Stercho, J. P. (1986), *J. Org. Chem.*, **51**, 1357.

Juaristi, E., Lopez-Nuñez, N. A., Valenzuela, B. A., Valle, L., Toscano, R. A. and Soriano-García, M. S. (1987a), *J. Org. Chem.*, **52**, 5185.

Juaristi, E., Martínez, R., Méndez, R., Toscano, R. A., Soriano-García, M., Eliel, E. L., Petsom, A., and Glass, R. S. (1987b), *J. Org. Chem.*, **52**, 3806.

Kaloustian, M. K., Dennis, N., Mager, S., Evans, S. A., Alcudia, F., and Eliel, E. L. (1976), *J. Am. Chem. Soc.*, **98**, 956.

Katritzky, A. R., Patel, R. C., and Riddell, F. G. (1979), *J. Chem. Soc. Chem. Commun.*, 674.

Keese, R. (1992), *Angew. Chem. Int. Ed. Engl.*, **31**, 344.

Kellie, G. M. and Riddell, F. G. (1974), "Non-chair Conformation of Six-membered Rings," *Top. Stereochem.*, **8**, 225.

Kellie, G. M. and Riddell, F. G. (1975), *J. Chem. Soc. Perkin 2*, 740.

Kilpatrick, J. E., Pitzer, K. S., and Spitzer, R. (1947), *J. Am. Chem. Soc.*, **69**, 2483.

Kingsbury, C. A., Soriano, D. S., Podraza, K. F., and Cromwell, N. H. (1982), *J. Heterocycl. Chem.*, **19**, 89.

Kirby, A. J. (1980), *Adv. Phys. Org. Chem.*, **17**, 208.

Kirby, A. J. (1983), *The Anomeric Effect and Related Stereoelectronic Effects at Oxygen*, Springer, New York.

Kirby, A. J. and Williams, N. H. (1992), *J. Chem. Soc. Chem. Commun.*, 1285.

Kirkwood, J. G. and Westheimer, F. H. (1938), *J. Chem. Phys.*, **6**, 506, 513.

Kirmse, W. and Hase, C. (1968), *Angew. Chem. Int. Ed. Engl.*, **7**, 891.

Kitching, W., Doddrell, D., and Grutzner, J. B. (1976), *J. Organomet. Chem.*, **107**, C5.

Kitching, W., Olszowy, H. A., and Adcock, W. (1981), *Org. Magn. Reson.*, **15**, 230.

Kitching, W., Olszowy, H. A., Drew, G. M., and Adcock, W. (1982b), *J. Org. Chem.*, **47**, 5153.

Kitching, W., Olszowy, H. A., and Harvey, K. (1982a), *J. Org. Chem.*, **47**, 1893.

Klein, J., Dunkelblum, E., Eliel, E. L., and Senda, Y. (1968), *Tetrahedron Lett.*, 6127.

Klochkov, V. V., Chernov, P. P., Agafonov, M. N., Chichirov, A. A., Aganov, A. V., and Kargin, Y. M. (1989), *J. Gen. Chem. USSR*, **59**, 1693 (Engl. transl. p. 1506).

Klyne, W. and Buckingham, J. (1978), *Atlas of Stereochemistry*, 2nd., Vol. 1, Chapman & Hall, London, p. 121.

Kolka, A. J., Orloff, H. D., and Griffing, M. E. (1954), *J. Am. Chem. Soc.*, **76**, 3940.

Komers, R. and Kochloefl, K. (1963), *Coll. Czech. Chem. Commun.*, **28**, 46.

Kopp, L. D. (1973), "Conformational Analysis of Nitrogen Containing Heterocycles," Ph.D. Dissertation, University of Notre Dame, Notre Dame, IN; *Diss. Abstr. Int. B*, **34**, 1425.

Kozina, M. P., Mastryukov, V. S. and Mil'vitskaya, E. M. (1982), "The Strain Energy, Geometrical Structure, and Spin–Spin Coupling Constant of Cyclic Hydrocarbons," *Russ. Chem. Rev.*, **51**, 1337; (Engl. transl. p. 765).

Krätschmer, W., Lamb, L. D., Fostiropoulos, K., and Huffman, D. R. (1990), *Nature (London)*, **347**, 354.

Krespan, C. G. (1980), *J. Org. Chem.*, **45**, 1177.

Krishnamurthy, S. and Brown, H. C. (1976), *J. Am. Chem. Soc.*, **98**, 3383.

Krohn, K. (1991), "Fenestranes – A Look at 'Structural Pathologies'" in Mulzer, J., Altenbach, H.-J., Braun, M., Krohn, K., and Reissig, H.-U., *Organic Synthesis Highlights*, VCH, New York, p. 371.

Kroto, H. (1988), "Space, Stars, C_{60} and Soot," *Science*, **242**, 1139.

Kroto, H. (1990), "Giant Fullerenes," *Chem. Brit.*, **26**, 40.

Kroto, H. W., Allaf, A. W., and Balm, S. P. (1991), "C_{60}: Buckminsterfullerene," *Chem. Rev.*, **91**, 1213.

Kroto, H. W., Heath, J. R., O'Brien, S. C., Curl, R. F., and Smalley, R. E. (1985), *Nature*, (*London*), **318**, 162.

Kroto, H. W. and Walton, D. R. M. "Postfullerene Organic Chemistry," see reference to Osawa, E. and Yonemitsu, O., Eds., (1992), p. 91.

Kuhn, L. P. (1952), *J. Am. Chem. Soc.*, **74**, 2492.

Kuhn, L. P. (1954), *J. Am. Chem. Soc.*, **76**, 4323.

Kwart, H. and Nickle, J. H. (1973), *J. Am. Chem. Soc.*, **95**, 3394.

Kwart, H. and Nickle, J. H. (1976), *J. Am. Chem. Soc.*, **98**, 2881.

Kybett, B. D., Carroll, S., Natalis, P., Bonnell, D. W., Margrave, J. L., and Franklin, J. L. (1966), *J. Am. Chem. Soc.*, **88**, 626.

Lamaty, G., Tapiero, C., and Wylde, R. (1966), *Bull. Soc. Chim. Fr.*, 4010.

Lambert, J. B. (1971), "Structural Chemistry in Solution. The R Value," *Acc. Chem. Res.*, **4**, 87.

Lambert, J. B. (1989), "Conformational Analysis of Six-membered Carbocyclic Rings with Exocyclic Double Bonds," see reference to Rabideau, P. W., Ed. (1989a), pp. 47–63.

Lambert, J. B., Carhart, R. E., and Corfield, P. W. R. (1969), *J. Am. Chem. Soc.*, **91**, 3567.

Lambert, J. B. and Clikeman, R. R. (1976), *J. Am. Chem. Soc.*, **98**, 4203.

Lambert, J. B., Clikeman, R. R., Taba, K. M., Marko, D. E., Bosch, R. J., and Xue, L. (1987), "Steric Interaction of Double Bonds," *Acc. Chem. Res.*, **20**, 454.

Lambert, J. B. and Featherman, S. I. (1975), "Conformational Analyses of Pentamethylene Heterocycles," *Chem. Rev.*, **75**, 611.

Lambert, J. B. and Keske, R. G. (1966), *J. Org. Chem.*, **31**, 3429.

Lambert, J. B., Keske, R. G., Carhart, R. E., and Jovanovich, A. P. (1967), *J. Am. Chem. Soc.*, **89**, 3761.

Lambert, J. B. and Marko, D. E. (1985), *J. Am. Chem. Soc.*, **107**, 7978.

Lambert, J. B., Mixan, C. E., and Johnson, D. H. (1973), *J. Am. Chem. Soc.*, **95**, 4634.

Lambert, J. B. and Taba, K. M. (1981), *J. Am. Chem. Soc.*, **103**, 5828.

Landa, S. and Macháček, V. (1933), *Coll. Czech. Chem. Commun.*, **5**, 1.

Lansbury, P. T. (1969), "Stereochemistry and Transannular Reactions of 7,12-Dihydropleiadenes," *Acc. Chem. Res.*, **2**, 210.

Larnaudie, M. (1954), *J. Phys. Radium*, **15**, 650.

Laszlo, I. (1965), *Recl. Trav. Chim. Pays-Bas*, **84**, 251.

Lautenschlaeger, F. and Wright, G. F. (1963), *Can. J. Chem.*, **41**, 863.

Lee, S.-F., Edgar, M., Pak, C. S., Barth, G., and Djerassi, C. (1980), *J. Am. Chem. Soc.*, **102**, 4784.

Lehn, J.-M. (1978a), "Cryptates: The Chemistry of Macropolycyclic Inclusion Complexes," *Acc. Chem. Res.*, **11**, 49.

Lehn, J.-M. (1978b), "Cryptates: Inclusion Complexes of Macropolycyclic Receptor Molecules," *Pure Appl. Chem.*, **50**, 871.

Lehn, J.-M. (1980), "Cryptate Inclusion Complexes. Effect on Solute–Solute and Solute–Solvent Interactions and on Ionic Reactivity," *Pure Appl. Chem.*, **52**, 2303.

Lehn, J.-M. (1993), "Supramolecular Chemistry", *Science*, **260**, 1762.

Lemarié, B., Braillon, B., Lasne, M.-C., and Thuillier, A. (1977), *J. Chim. Phys.*, **74**, 799.

Lemieux, R. U. (1964), "Rearrangements and Isomerizations in Carbohydrate Chemistry," in de Mayo, P., Ed., Wiley, New York.

Lemieux, R. U. (1971), "Effects of Unshared Pairs of Electrons and Their Solvation on Conformational Equilibria," *Pure Appl. Chem.*, **25**, 527.

Lemieux, R. U. and Chü, N. J. (1958), Abstract Papers American Chemical Society 133rd Meeting, p. 31N.

Lemieux, R. U. and Koto, S. (1974), *Tetrahedron*, **30**, 1933.

Lemieux, R. U., Koto, S., and Voisin, D. (1979), "The Exo-Anomeric Effect," in Szarek, W. A. and Horton, D., Eds., *Anomeric Effect, Origin and Consequences*, ACS Symposium Series 87, American Chemical Society, Washington, DC, p. 17.

Lemieux, R. U., Kullnig, R. K., Bernstein, H. J., and Schneider, W. G. (1958), *J. Am. Chem. Soc.*, **80**, 6098.

Leonard, J. E., Hammond, G. S., and Simmons, H. E. (1975), *J. Am. Chem. Soc.*, **97**, 5052.

Lessard, J., Tan, P. V. M., Martino, R., and Saunders, J. K. (1977), *Can. J. Chem.*, **55**, 1015, 1017.

Li, Y.-S. (1984), *J. Phys. Chem.*, **88**, 4049.

Liebman, J. F. and Greenberg, A. (1976), "A Survey of Strained Organic Molecules," *Chem. Rev.*, **76**, 311.

Liebman, J. F. and Greenberg, A. (1989), "Survey of Heats of Formation of Three-membered Rings Systems," *Chem. Rev.* **89**, 1215.

Liebman, J. F. and Greenberg, A., Eds. (1987), *Molecular Stucture and Energetics*, Vol. 2, VCH, Deerfield Park, FL.

Liebman, J. F. and Van Vechten, D. (1987), "Universality: The Differences and Equivalences of Heats of Formation, Strain Energy and Resonance Energy," see reference to Liebman, J. F. and Greenberg, A., Eds. (1987), p. 315.

Linstead, R. P., Doering, W. E,, Davis, S. B., Levine, P., and Whetstone, R. R. (1942), *J. Am. Chem. Soc.*, **64**, 1985 and papers following.

Linstead, R. P. and Whetstone, R. R. (1950), *J. Chem. Soc.*, 1428.

Lipatova, T. E., Kosyanchuk, L. F., Gomza, Yu. P., Shilov, V. V., and Lipatov, Yu. S. (1982), *Proc. Acad. Sci. USSR (Chem. Sec.)*, **263**, 1379 (Engl. Transl. p. 140).

Lipkowitz, K. B. (1989), "Application of Empirical Force-Field Calculations to the Conformational Analysis of Cyclohexenes, Cyclohexadienes, and Hydroaromatics," see reference to Rabideau, P. W., Ed., (1989a), p. 209.

Lo Cicero, B., Weisbuch, F., and Dana, G. (1981), *J. Org. Chem.*, **46**, 914.

Loomes, D. J. and Robinson, M. J. T. (1977), *Tetrahedron*, **33**, 1149.

Lu, K. C., Chiang, R. L., and Chiang, J. F. (1980), *J. Mol. Struct.*, **64**, 229.

Lüttringhaus, A. and Isele, G. (1967), *Angew. Chem. Int. Ed. Engl.*, **6**, 956.

Lutz, G., Pinkos, R., Murty, B. A. R. C., Spurr, P. R., Fessner, W.-D., Wörth, J., Fritz, H., Knothe, L., and Prinzbach, H. (1992), *Chem. Ber.*, **125**, 1741.

Maciejwski, M. (1982), *J. Macromol. Sci. Chem.*, **A17**, 689.

Maier, G. et al., (1981c), *Chem. Ber.*, **114**, 3906, 3916, 3922, 3935, 3959.

Maier, G., Pfriem, S., Malsch, K.-D., Kalinowski, H.-O., and Dehnicke, K. (1981b), *Chem. Ber.*, **114**, 3988.

Maier, G., Pfriem, S., Schäfer, U., Malsch, K.-D., and Matusch, R. (1981a), *Chem. Ber.*, **114**, 3965.

Maier, G., Pfriem, S., Schäfer, U., and Matusch, R. (1978), *Angew. Chem. Int. Ed. Engl.*, **17**, 520.

Maier, G., Reuter, K. A., Franz, L., and Reisenauer, H. P. (1985), *Tetrahedron Lett.*, **26**, 1845.

Mann, B. E. (1976), *J. Magn. Reson.*, **21**, 17.

Mann, G. (1966), *Z. Chem.*, **6**, 106.

Manoharan, M. and Eliel, E. L. (1983), *Tetrahedron Lett.*, **24**, 453.

Manoharan, M. and Eliel, E. L. (1984a), *Tetrahedron Lett.*, **25**, 3267.

Manoharan, M. and Eliel, E. L. (1984b), *J. Am. Chem. Soc.*, **106**, 367.

Manoharan, M., Eliel, E. L., and Carroll, F. I. (1983), *Tetrahedron Lett.*, **24**, 1855.

Marchand, A. P. (1992), "Policyclic Cage Molecules: Useful Intermediates in Organic Synthesis and an Emerging Class of Substrates for Mechanistic Studies," see reference to Osawa, E. and Yonemitsu, O., Eds., (1992), p. 1.

Marshall, J. L. (1983), *Carbon–Carbon and Carbon–Proton NMR Couplings: Applications to Organic Stereochemistry and Conformational Analysis*, Verlag Chemie, Deerfield Beach, FL.

Marshall, J. A. and Faubl, H. (1967), *J. Am. Chem. Soc.*, **89**, 5965.

Maruoka, K., Miyazaki, T., Ando, M., Matsumura, Y., Sakane, S., Hattori, K., and Yamamato, H. (1983), *J. Am. Chem. Soc.*, **105**, 2831.

Mastryukov, V. A., Popik, M. V., Dorofeeva, O. V., Golubinskii, A. V., Vilkov, L. V., Belikova, N. A., and Allinger, N. L. (1981), *J. Am. Chem. Soc.*, **103**, 1333.

McClellan, A. L. (1963, 1974, 1989), *Tables of Experimental Dipole Moments*, Vol. 1, Freeman, San Francisco, CA; Vols. 2, 3, Rahara Enterprises, El Cerrito, CA.

McConnell, H. M. (1957), *J. Chem. Phys.*, **27**, 226.

McCoy, L. L. (1965), *J. Org. Chem.*, **30**, 3762.

McCoy, L. L. and Nachtigall, G. W. (1963), *J. Am. Chem. Soc.*, **85**, 1321.

McKenna, J. (1974), "Conformational Analysis by the Kinetic Method," *Tetrahedron*, **30**, 1555.

McKervey, M. A. (1974), "Adamantane Rearrangements," *Chem. Soc. Rev.*, **3**, 479.

McLafferty, F. W., Ed., (1992), *Acc. Chem. Res.*, **25**, 97–175.

Meinwald, J., Tufariello, J. J., and Hurst, J. J. (1964), *J. Org. Chem.*, **29**, 2914.

Menard, D. and St.-Jacques, M. (1983), *Tetrahedron*, **39**, 1041.

Merrifield, R. B. (1965), "Automated Synthesis of Peptides," *Science*, **150**, 178.

Meyer, A. Y., Allinger, N. L., and Yuh, Y. (1980), *Isr. J. Chem.*, **20**, 57.

Minkin, V. I., Minyaev, R. M., and Natanzon, V. I. (1980), *J. Org. Chem. USSR*, **16**, 673 (Engl. transl. p. 589).

Minyaev, R. M. and Minkin, V. I. (1985), *J. Org. Chem. USSR*, **21**, 2249 (Engl. transl. p. 2055).

Moder, T. I., Hsu, C. C. K., and Jensen, F. R. (1980), *J. Org. Chem.*, **45**, 1008.

Mohr, E. (1918), *J. Prakt. Chem.*, [2] **98**, 315.

Moniz, W. B. and Dixon, J. A. (1961), *J. Am. Chem. Soc.*, **83**, 1671.

Monti, J. P., Faure, R., and Vincent, E. J. (1975), *Org. Magn. Reson.*, **7**, 637.

Moriarty, R. M. (1974), "Stereochemistry of Cyclobutane and Heterocyclic Analogs," *Top. Stereochem.*, **8**, 270.

Morin, F. G. and Grant, D. M. (1989), "Use of Carbon-13 Nuclear Magnetic Resonance in the Conformational Analysis of Hydroaromatic Compounds," see reference to Rabideau, P. W., Ed., (1989a), pp. 127–167.

Muller, N. and Pritchard, D. E. (1959), *J. Chem. Phys.*, **31**, 1471.

Mursakulov, I. G., Ramazanov, E. A., Guseinov, M. M., Zefirov, N. S., Samoshin, V. V., and Eliel, E. L. (1980), *Tetrahedron*, **36**, 1885.

Naemura, K. (1992), "High-Symmetry Chiral Cage-Shaped Molecules," see reference to Osawa, E. and Yonemitsu, O., Eds., (1992), p. 61.

Naumov, V. A. and Bezzubov, V. M. (1967), *J. Struct. Chem. USSR*, **8**, 530 (Engl. transl. p. 466).

Naumov, V. A., Dashevskii, V. G., and Zaripov, N. M. (1970), *J. Struct., Chem. USSR* **11**, 793 (Engl. transl. p. 736).

Newton, M. D. and Schulman, J. M. (1972), *J. Am. Chem. Soc.*, **94**, 773.

Nicolaou, K. C. (1977), *Tetrahedron*, **33**, 683.

Nilsson, H. and Smith, L. (1933), *Z. Physik. Chem.*, **166A**, 136.

Oberhammer, H. and Bauer, S. H. (1969), *J. Am. Chem. Soc.*, **91**, 10.

Ogata, T. and Kozima, K. (1969), *Bull. Chem. Soc. Jpn.*, **42**, 1263.

Ogino, H. (1981), *J. Am. Chem. Soc.*, **103**, 1303.

Olah, G., Ed. (1990), *Cage Hydrocarbons*, Wiley, New York.

Osawa, E. and Yonemitsu, O. Eds. (1992), *Carbocyclic Cage Compounds*, VCH, New York.

Ottar, B. (1947), *Acta. Chem. Scand.*, **1**, 283.

Owen, L. N. and Peto, A. G. (1955), *J. Chem. Soc.*, 2383.

Pakes, P. W., Rounds, T. C., and Strauss, H. L. (1981), *J. Phys. Chem.*, **85**, 2469, 2476.

Pai, T.-H. and Kalasinsky, V. F. (1990), *J. Raman. Spectrosc.*, **21**, 607.

Pan, Y.-H. and Stothers, J. B. (1967), *Can. J. Chem.*, **45**, 2943.

Paquette, L. A. (1975), "The Renaissance in Cyclooctatetraene Chemistry," *Tetrahedron*, **31**, 2855.

Paquette, L. A. (1979), "The Development of Polyquinane Chemistry," *Top. Curr. Chem.*, **79**, 41.

Paquette, L. A. (1984a), "Recent Synthetic Developments in Polyquinane Chemistry," *Top. Curr. Chem.*, **119**, 1.

Paquette, L. A. (1984b), "Plato's Solid in a Retort: The Dodecahedrane Story," in Lindberg, T., Ed., *Strategies and Tactics in Organic Synthesis*, Academic, New York, p. 175.

Paquette, L. A. (1989), "Dodecahedranes and Allied Spherical Molecules," *Chem. Rev.*, **89**, 1051.

Paquette, L. A. (1993), "The Current View of Dynamic Change within Cyclooctatetraenes," *Acc. Chem. Res.*, **26**, 57.

Paquette, L. A., Miyahara, Y., and Doecke, C. W. (1986), *J. Am. Chem. Soc.*, **108**, 1716.

Paquette, L. A., Ternansky, R. J. Balogh, D. W., and Kentgen, G. (1983), *J. Am. Chem. Soc.*, **105**, 5446.

Park, C. H. and Simmons, H. E. (1968a), *J. Am. Chem. Soc.*, **90**, 2429.

Park, C. H. and Simmons, H. E. (1968b), *J. Am. Chem. Soc.*, **90**, 2431.

Park, C. H. and Simmons, H. E. (1972), *J. Am. Chem. Soc.*, **94**, 7184.

Parker, D. (1983), "Alkali and Alkaline Earth Metal Cryptates," *Adv. Inorg. Chem.*, **27**, 1.

Parker, D. H., Wurz, P., Chatterjee, K., Lykke, K. R., Hunt, J. E., Pellin, M. J., Hemminger, J. C., Gruen, D. M., and Stock, L. M. (1991), *J. Am. Chem. Soc.*, **113**, 7499.

Parkin, J. E., Buckley, P. J., and Costain, C. C. (1981), *J. Mol. Spectrosc.*, **89**, 465.

Paukstelis, J. V. and Kao, J.-l. (1972), *J. Am. Chem. Soc.*, **94**, 4783.

Peereboom, M., van de Graaf, B., and Baas, J. M. A. (1982), *Recl. J. R. Neth. Chem. Soc.*, **101**, 336.

Penman, K. G., Kitching, W., and Adcock, W. (1989), *J. Org. Chem.*, **54**, 5390.

Perkin, W. H., Jr. (1894), *J. Chem. Soc.*, **65**, 572.

Philp, D. and Stoddart, J. F. (1991), "Self-Assembly in Organic Synthesis," *Synlett*, 445.

Pihlaja, K. (1987), "Thermochemical Methods in the Structural Analysis of Organic Compounds," see reference to Liebman, J. F. and Greenberg, A., Eds., (1987) p. 173.

Pihlaja, K. and Nikander, H. (1977), *Acta. Chem. Scand.*, **B31**, 265.

Pinto, B. M., Johnston, B. D., and Nagelkerke, R. (1988), *J. Org. Chem.*, **53**, 5668.

Pinto, B. M., and Wolfe, S. (1982), *Tetrahedron Lett.*, **23**, 3687.

Pirkle, W. H. and Lunsford, W. B. (1972), *J. Am. Chem. Soc.*, **94**, 7201.

Pitner, T. P., Edwards, W. B., Bassfield, R. L., and Whidby, J. F. (1978), *J. Am. Chem. Soc.*, **100**, 246.

Pitzer, K. S. and Donath, W. E. (1959), *J. Am. Chem. Soc.*, **81**, 3213.

Powell, D. L., Gatial, A., Klaeboe, D., Nielsen, C. J., Kondow, A. J., Boettner, W. A., and Mulichak, A. M. (1989), *Acta. Chem. Scand.*, **43**, 441.

Praly, J.-P. and Lemieux, R. U. (1987), *Can. J. Chem.*, **65**, 213.

Prelog, V. (1956), "Bedeutung der vielgliederigen Ringverbindungen für die theoretische organische Chemie", in Todd, A. R., Ed., *Perspectives in Organic Chemistry*, Interscience, New York, p. 96ff.

Prelog, V. (1960), "Some Newer Developments of the Chemistry of the Medium-sized Ring Compounds," *Bull. Soc. Chim. Fr.*, 1433.

Prelog, V. and Helmchen, G. (1982), "Basic Principles of the CIP-System and Proposals for a Revision," *Angew. Chem. Int. Ed. Engl.*, **21**, 567.

Prelog, V. and Kobelt, M. (1949), *Helv. Chim. Acta.*, **32**, 1187.

Prelog, V. and Siewerth, R. (1941), *Ber. Dtsch. Chem. Ges.*, **74B**, 1644, 1769.

Prelog, V. and Speck, M. (1955), *Helv. Chim. Acta.*, **38**, 1786.

Prociv, T. M. (1983), "The Theory and Synthesis of [2.2.2]Propellane," Ph.D. Dissertation, City University of New York, New York; *Diss. Abstr. Int. B*, **44**, 173.

Raber, D. J., Hardee, L. E., Rabideau, P. W., and Lipkowitz, K. B. (1982), *J. Am. Chem. Soc.*, **104**, 2843.

Rabideau, P. W. (1978), "Conformational Analysis of 1,4-Cyclohexadienes, 1,4-Dihydrobenzenes, 1,4-Dihydronaphthalenes and 9,10-Dihydroanthracenes," *Acc. Chem. Res.*, **11**, 141.

Rabideau, P. W., Ed., (1989a), *Conformational Analysis of Cyclohexenes, Cyclohexadienes and Related Hydroaromatic Compounds*. VCH, New York.

Rabideau, P. W. (1989b), "Conformational Analysis of 1,4-Cyclohexadienes and Related Hydroaromatics," see reference to Rabideau, P. W., Ed., (1989a), p. 89.

Rabideau, P. W. and Sygula, A. (1989), "Conformational Analysis of 1,3-Cyclohexadienes and Related Hydroaromatics," see reference to Rabideau, P. W., Ed., (1989a), p. 65.

Rabideau, P. W., Wetzel, D. M., and Paschal, J. W. (1982), *J. Org. Chem.*, **47**, 3993.

Ramsay, O. B. (1973), *Chem. Z.*, **97**, 573.

Ramsay, O. B. (1975), "Molecular Models in the Early Development of Stereochemistry," in Ramsay, O. B., Ed., *van't Hoff–Le Bel Centennial*, ACS Symposium Series 12, American Chemical Society, Washington, DC, p. 74.

Rao, V. B., George, C. F., Wolff, S., and Agosta, W. C. (1985), *J. Am. Chem. Soc.*, **107**, 5732.

Reetz, M. T. (1982), "Lewis-Acid Induced α-Alkylation of Carbonyl Compounds," *Angew. Chem. Int. Ed. Engl.*, **21**, 96.

Reetz, M. T., Harmat, N., and Mahrwald, R. (1992), *Angew. Chem. Int. Ed. Engl.*, **31**, 342.

Reetz, M. T. and Stanchev, S. (1993), *J. Chem. Soc. Chem. Commun.*, 328.

Reeves, L. W. and Strømme, K. O. (1960), *Can. J. Chem.*, **38**, 1241.

Rei, M.-H. (1979), *J. Org. Chem.*, **44**, 2760.

Rei, M.-H. (1983), *J. Org. Chem.*, **48**, 5386.

Reisse, J. (1971), "Quantitative Conformational Analysis of Cyclohexane Systems," in Chiurdoglu, G., Ed., *Conformational Analysis, Scope and Present Limitations*, Academic, New York, p. 219ff.

Reisse, J., Celotti, J. C., Zimmermann, D., and Chiurdoglu, G. (1964), *Tetrahedron Lett.*, 2145.

Reisse, J., Stien, M.-L., Gilles, J.-M., and Oth, J. F. M. (1969), *Tetrahedron Lett.*, 1917.

Remijnse, J. D., van Bekkum, H., and Wepster, B. M. (1974), *Recl. Trav. Chim. Pays-Bas*, **93**, 93.

Richer, J. C., Pilato, L. A., and Eliel, E. L. (1961) *Chem. Ind. London*, 2007.

Riddell, F. G. (1980), *The Conformational Analysis of Heterocyclic Compounds*, Academic, New York.

Riddell, F. G. (1992), "The Stereodynamics of Five-Membered Nitrogen-Containing Rings," in Lambert, J. B. and Takeuchi, Y., Eds., *Cyclic Organonitrogen Stereodynamics*, VCH, New York, p. 159.

Riddell, F. G. and Robinson, M. J. T. (1967), *Tetrahedron*, **23**, 3417.

Rissler, K., Schill, G., Fritz, H., and Vetter, W. (1986), *Chem. Ber.*, **119**, 1374.

Roberts, J. D. and Chambers, V. C. (1951), *J. Am. Chem. Soc.*, **73**, 5034.

Romers, C., Altona, C., Buys, H. R., and Havinga, E. (1969), "Geometry and Conformational Properties of Some Five- and Six-membered Heterocyclic Compounds Containing Oxygen or Sulfur," *Top. Stereochem.*, **4**, 39ff.

Rozeboom, M. D. and Houk, K. N. (1982), *J. Am. Chem. Soc.*, **104**, 1189.

Rubin, B. H., Williamson, M., Takeshita, M., Menger, F. M., Anet, F. A. L., Bacon, B., and Allinger, N. L. (1984), *J. Am. Chem. Soc.*, **106**, 2088.

Ruggli, P. (1912), *Justus Liebigs Ann. Chem.*, **392**, 92.

Russell, C. A. (1975), "The Origins of Conformational Analysis," in Ramsay, O. B., Ed., *van't Hoff–Le Bel Centennial*, ACS Symposium Series 12, American Chemical Society, Washington, DC, p. 159.

Ruzicka, L., Brugger, W., Pfeiffer, M., Schinz, H., and Stoll, M. (1926), *Helv. Chim. Acta*, **9**, 499.

Sachse, H. (1890), *Berichte*, **23**, 1363.

Sachse, H. (1892), *Z. Physik. Chem.*, **10**, 203.

Sakashita, K. (1960), *Nippon Kaguku Zasshi*, **81**, 49.

Sandström, J. (1982), *Dynamic NMR Spectroscopy*, Academic, New York.

Saunders, W. H. and Cockerill, A. F. (1973), *Mechanisms of Elimination Reactions*, Wiley, New York.

Sauvage, J.-P., Ed. (1993) "Topology in Molecular Chemistry," *New J. Chem.*, **17**, 617ff.

Sauvage, J.-P. and Weiss, J. (1985), *J. Am. Chem. Soc.*, **107**, 6108.

Scharpen, L. H., Wollrab, J. E., and Ames, D. P. (1968), *J. Chem. Phys.*, **49**, 2368.

Schill, G. (1967a), *Nachr. Chem. Tech.*, **15**, 149.

Schill, G. (1967b), *Chem. Ber.*, **100**, 2021.

Schill, G. (1971a), *Catenanes, Rotaxanes and Knots*, Academic, New York.

Schill, G. (1971b), in Chiurdoglu, G., Ed., *Conformational Analysis, Scope and Present Limitations*, Academic, New York, pp. 229–239.

Schill, G., Beckmann, W., Schweickert, N., and Fritz, H. (1986), *Chem. Ber.*, **119**, 2647.

Schill, G. and Lüttringhaus, A. (1964), *Angew. Chem. Int. Ed. Engl.*, **3**, 546.

Schill, G., Rissler, K., Fritz, H., and Vetter, W. (1981), *Angew. Chem. Int. Ed. Engl.*, **20**, 187.

Schill, G., Schweickert, N., Fritz, H., and Vetter, W. (1988), *Chem. Ber.*, **121**, 961.

Schill, G. and Zollenkopf, H. (1969), *Justus Liebigs Ann. Chem.*, **721**, 53.

Schill, G. and Zürcher, C. (1977), *Chem. Ber.*, **110**, 2046, 3964.

Schleyer, P. (1957), *J. Am. Chem. Soc.*, **79**, 3292.

Schleyer, P. v. R., Williams, J. E., and Blanchard, K. R., (1970), *J. Am. Chem. Soc.*, **92**, 2377.

Schmidt, G. M. J. (1964), *J. Chem. Soc.*, 2014.

Schneider, H.-J. and Dürr, H. (1991), *Frontiers in Supramolecular Organic Chemistry and Photochemistry*, VCH, New York.

Schneider, H.-J. and Hoppen, V. (1978), *J. Org. Chem.*, **43**, 3866.

Schneider, H.-J. and Nguyen-Ba, N. (1982), *Org. Magn. Reson.*, **18**, 38.

Schneider, H.-J. and Thomas, F. (1976), *Tetrahedron*, **32**, 2005.

Schneider, H.-J., Schmidt, G., and Thomas, F. (1983), *J. Am. Chem. Soc.*, **105**, 3556.

Schneider, H.-J. and Thomas, F. (1980), *J. Am. Chem. Soc.*, **102**, 1424.

Schotsmans, L., Vierens, P. J. C., and Verlie, T. (1959), *Bull. Soc. Chim. Belg.*, **68**, 580.

Schreiber, J. and Eschenmoser, A. (1955), *Helv. Chim. Acta.*, **38**, 1529.

Schucker, R. C. (1981), *J. Chem. Eng. Data*, **26**, 239.

Schubert, W. K., Southern, W. J., and Schäfer, L. (1973), *J. Mol. Struct.*, **16**, 403.

Schulman, J. M. and Disch, R. L. (1984), *J. Am. Chem. Soc.*, **106**, 1202.

Scott, D. W. (1971), *J. Chem. Thermodyn*, **3**, 649.

Scott, L. T. and Jones, M. (1972), "Rearrangement and Interconversions of Compounds of the Formula $(CH)_n$," *Chem. Rev.*, **72**, 181.

Seeman, J. I. (1983), "Effect of Conformational Change on Reactivity in Organic Chemistry. Evaluations, Applications and Extensions of Curtin–Hammett/Winstein–Holness Kinetics," *Chem. Rev.*, **83**, 83.

Seigler, D. S. and Bloomfield, J. J. (1973), *J. Org. Chem.*, **38**, 1375.

Sellers, S. F., Klebach, T. C., Hollowood, F., Jones, M., and Schleyer, P. v. R. (1982), *J. Am. Chem. Soc.*, **104**, 5492.

Senda, Y. and Imaizumi, S. (1974), *Tetrahedron*, **30**, 3813.

Senderowitz, H., Abramson, S., Pinchas, A., Schleifer, L., and Fuchs, B. (1989), *Tetrahedron Lett.*, **30**, 6765.

Shen, Q., Rhodes, S., and Cochran, J. C. (1992), *Organometallics*, **11**, 485.

Shen, Q. and Peloquin, J. M. (1988), *Acta Chem. Scand. A*, **42**, 367.

Shenhav, H. and Schaeffer, R. (1981), *Cryst. Struct. Commun.*, **10**, 1181.

Shoppee, C. W., Jones, D. N., and Summers, G. H. R. (1957), *J. Chem. Soc.*, 3100.

Shriner, R. L., Adams, R., and Marvel, C. S. (1938), "Stereochemistry" in Gilman, H., Ed., *Organic Chemistry*, Wiley, New York, p. 238.

Sicher, J. (1962), "The Stereochemistry of Many-membered Rings," *Prog. Stereochem.*, **3**, 202.

Sicher, J., Jonáš, J., and Tichý, M. (1963), *Tetrahedron Lett.*, 825.

Sicher, J., Šipoš, F., and Jonáš, J. (1961), *Coll. Czech. Chem. Commun.*, **26**, 262.

Sicher, J. and Tichý, M. (1967), *Coll. Czech. Chem. Commun.*, **32**, 3687.

Simmons, H. E. and Park, C. H. (1968), *J. Am. Chem. Soc.*, **90**, 2428.

Šipoš, F., Krupička, J., Tichý, M., and Sicher, J. (1962), *Coll. Czech. Chem. Commun.*, **27**, 2079.

Skancke, P. N. (1987), *THEOCHEM*, **36**, 11.

Smith, P. B., Dye, J. L., Cheney, J., and Lehn, J.-M. (1981), *J. Am. Chem. Soc.*, **103**, 6044.

Sokolova, I. M. and Petrov, A. A. (1977), *Neftekhimiya*, **17**, 498; *Chem. Abstr.*, **87**, 184059b.

Spelbos, A., Mijlhoff, F. C., Bakker, W. H., Baden, R., and Van den Eden, L. (1977), *J. Mol. Struct.*, **38**, 155.

Speros, D. M. and Rossini, F. D. (1960), *J. Phys. Chem.*, **64**, 1723.

Squillacote, M. E. and Neth, J. M. (1987), *J. Am. Chem. Soc.*, **109**, 198.

Squillacote, M., Sheridan, R. S., Chapman, O. L., and Anet, F. A. L. (1975), *J. Am. Chem. Soc.*, **97**, 3244.

Sternbach, D. D., Rossana, D. M., and Onan, K. D. (1985), *Tetrahedron Lett.*, **26**, 591.

Stoddart, J. F. (1971), *Stereochemistry of Carbohydrates*, Wiley-Interscience, New York.

Stoddart, J. F. (1991a), "Making molecules to order," *Chem. Brit.*, **27**, 714.

Stoddart, J. F. (1991b), "Template Directed Synthesis of New Organic Materials" in Schneider, H.-J. and Dürr, H., Eds., *Frontiers in Supramolecular Organic Chemistry and Photochemistry*, VCH, New York, p. 251.

Stoddart, J. F. and Szarek, W. A. (1968), *Can. J. Chem.*, **46**, 3061.

Stohrer, W.-D. and Hoffmann, R. (1972), *J. Am. Chem. Soc.*, **94**, 779.

Stolow, R. D. (1964), *J. Am. Chem. Soc.*, **86**, 2170.

Stothers, J. B. (1972), *Carbon-13 NMR Spectroscopy*, Academic, New York.

Stothers, J. B., Tan, C. T., and Teo, K. C. (1976), *Can. J. Chem.*, **54**, 1211.

Subbotin, O. A., Palyulin, V. A., Kozhushkov, S. I., and Zefirov, N. S. (1978), *J. Org. Chem. USSR* **14**, 209 (Engl. transl. p. 196).

Subbotin, O. A. and Sergeyev, N. M. (1975), *J. Am. Chem. Soc.*, **97**, 1080.

Sulzle, D., Gatial, A., Karlsson, A., Klaeboe, P., and Nielsen, C. J. (1988), *J. Mol. Struct.*, **174**, 207.

Svyatkin, V. A., Ioffe, A. I., and Nefedov, O. M. (1988), *Bull. Acad. Sci. USSR, Div. Chem. Sci.* **37**, 78 (Engl. transl. p. 69).

Szeimies, G., Harnisch, J., and Baumgartel, O. (1977), *J. Am. Chem. Soc.*, **99**, 5183.

Szeimies-Seebach, U. and Szeimies, G. (1978), *J. Am. Chem. Soc.*, **100**, 3966.

Tabor, S. and Richardson, C. C. (1985), *Proc. Natl. Acad. Sci. USA*, **82**, 1074.

Tanner, M. E., Knobler, C. B., and Cram, D. J. (1990), *J. Am. Chem. Soc.*, **112**, 1659.

Taylor, R., Hare, J. P., Abdul-Sada, A. K., and Kroto, H. W. (1990), *J. Chem. Soc. Chem. Commun.*, 1423.

Taylor, R., Langley, G. J., Dennis, T. J. S., Kroto, H. W., and Walton, D. R. M. (1992), *J. Chem. Soc. Chem. Commun.*, 1043.

Tee, O. S., Altmann, J. A., and Yates, K. (1974), *J. Am. Chem. Soc.*, **96**, 3141.

Tenud, L., Farooq, S., Seibl, J., and Eschenmoser, A. (1970), *Helv. Chim. Acta*, **53**, 2059.

Ternansky, R. J., Balogh, D. W., and Paquette, L. A. (1982), *J. Am. Chem. Soc.*, **104**, 4503.

Thatcher, G. R. J., Ed. (1993). *The Anomeric Effect and Associated Stereoelectronic Effects*, ACS Symposium Series 539, American Chemical Society, Washington, DC.

Tichý, M. (1965), "The Determination of Intramolecular Hydrogen Bonding by Infrared Spectroscopy and Its Application in Stereochemistry," in Raphael, R. A., Taylor, E. C., and Wynberg, H., Eds. *Advances on Organic Chemistry, Methods and Results*, Vol. 5, 115.

Tidwell, T. T. (1970), *J. Am. Chem. Soc.*, **92**, 1448.

Tobe, Y. (1992), "Propellanes", see reference to Osawa, E. and Yonemitsu, O., Eds., (1992), p. 125.

Tochtermann, W. (1970), "Konformative Beweglichkeit von Siebenring-Systemen," *Top. Curr. Chem.*, **15**, 378.

Toullec, J. (1982), "Enolization of Simple Carbonyl Compounds," *Adv. Phys. Org. Chem.*, **18**, 1.

Traetteberg, M. (1964), *J. Am. Chem. Soc.*, **86**, 4265.

Traetteberg, M. (1968), *Acta. Chem. Scand.*, **22**, 2305.

TRC Thermodynamic Tables, Hydrocarbons, Vol., VII (1991), Thermodynamic Research Center, The Texax A&M University System, College Station, TX.

Trost, B. M. and Verhoeven, T. R. (1980), *J. Am. Chem. Soc.*, **102**, 4743.

Uff, B. C. (1967), "The Cyclobutane Group," in Coffey, S., Ed., *Rodd's Chemistry of Carbon Compounds*, 2nd ed., Vol. 2, Part A, Elsevier, New York, p. 90.

Ungaro, R. and Pochini, A. (1991), "Flexible and Preorganized Molecular Receptors Based on Calixarenes," in Schneider, H.-J. and Dürr, H., Eds., *Frontiers in Supramolecular Organic Chemistry and Photochemistry*, VCH, New York.

Uzarewicz, I. and Uzarewicz, A. (1976), *Rocz. Chem.*, **50**, 1315.

Valls, J. (1961), *Bull. Soc. Chim. Fr.*, 432.

Valls, J. and Toromanoff, E. (1961), *Bull. Soc. Chim. Fr.*, 758.

van de Graaf, B., Baas, J. M. A., and van Veen, A. (1980), *Rec. J. R. Neth. Chem. Soc.*, **99**, 175.

van de Graaf, B., Baas, J. M. A., and Wepster, B. M. (1978), *Rec. Trav. Chim, Pays-Bas*, **97**, 268.

van de Graaf, B., van Bekkum, H., van Koningsveld, H., Sinnema, A., van Veen, A., Wepster, B. M., and van Wijk, A. M. (1974), *Rec. Trav. Chim. Pays-Bas*, **93**, 135.

Van den Enden, L. and Geise, H. J. (1981), *J. Mol. Struct.*, **74**, 309.

Van den Enden, L., Geise, H. J., and Spelbos, A. (1978), *J. Mol. Struct.*, **44**, 177.

Venepalli, B. R. and Agosta, W. C. (1987), "Fenestranes and the Flattening of Tetrahedral Carbon," *Chem. Rev.*, **87**, 399.

Vetter, W. and Schill, G. (1967), *Tetrahedron*, **23**, 3079.

Vicens, J. and Böhmer, V., Eds. (1991), "Calixarenes: A Versatile Class of Macrocyclic Compounds," Kluwer, Boston, MA.

Vierhapper, F. W. and Eliel, E. L. (1977), *J. Org. Chem.*, **42**, 51.

Vierhapper, F. W. and Eliel, E. L. (1979), *J. Org. Chem.*. **44**, 1081.

Vierhapper, F. W. and Willer, R. L. (1977a), *Org. Magn. Reson.*, **9**, 13.

Vierhapper, F. W. and Willer, R. L. (1977b), *J. Org. Chem.*, **42**, 4024.

Vögtle, F. and Mew, P. K. T. (1978), *Angew. Chem. Int. Ed. Engl.*, **17**, 60.

Wagner, H.-U., Szeimies, G., Chandrasekhar, J., Schleyer, P. v. R., Pople, J. A., and Binkley, J. S. (1978), *J. Am. Chem. Soc.*, **100**, 1210.

Walba, D. M. (1985), "Topological Stereochemistry," *Tetrahedron*, **41**, 3161.

Walba, D. M., Homan, T. C., Richards, R. M., and Haltiwanger, R. C. (1993), *New J. Chem.*, **17**, 661.

Walba, D. M., Richards, R. M., and Haltiwanger, R. C. (1982), *J. Am. Chem. Soc.*, **104**, 3219.

Walba, D. M., Richards, R., Sherwood, S. P., and Haltiwanger, R. C. (1981), *J. Am. Chem. Soc.*, **103**, 6213.

Walborsky, H. M. and Impastato, F. J. (1959), *J. Am. Chem. Soc.*, **81**, 5835.

Walker, F. H., Wiberg, K. B., and Michl, J. (1982), *J. Am. Chem. Soc.*, **104**, 2056.

Wang, J. C. and Schwartz, H. (1967), *Biopolymers*, **5**, 953.

Warner, P. M. (1989), "Strained Bridgehead Double Bonds," *Chem. Rev.*, **89**, 1067.

Warner, P., Chen, B.-L., Bronski, C. A., and Karcher, B. A. (1981), *Tetrahedron Lett.*, **22**, 375.

Warner, P. and LaRose, R. (1972), *Tetrahedron Lett.*, 2141.

Warner, P., LaRose, R., and Schleis, T. (1974), *Tetrahedron Lett.*, 1409.

Warner, P. M. and Peacock, S. (1982), *J. Comput. Chem.*, **3**, 417.

Warnhoff, E., Weng, C., and Tai, W. (1967), *J. Org. Chem.*, **32**, 2664.

Wasserman, E. (1960), *J. Am. Chem. Soc.*, **82**, 4433.

Wasserman, E. (1962), *Sci. Am.*, **207**, No. 5, 94.

Wasylishen, R. and Schaefer, T. (1973), *Can. J. Chem.*, **51**, 961.

Weber, E., Ed. (1993), *Supramolecular Chemistry I – Directed Synthesis and Molecular Recognition*, *Top. Curr. Chem.*, **165**.

Weber, E. and Vögtle, F. (1981), "Crown-type Compounds – An Introductory Overview," *Top. Curr. Chem.*, **98**, 1.

Werner, H., Mann, G., Mühlstadt, M., and Kohler, H.-J. (1970), *Tetrahedron Lett.*, 3563.

Westheimer, F. H. and Shookoff, M. W. (1939), *J. Am. Chem. Soc.*, **61**, 555.

Wheeler, O. H. and Granell de Rodriguez, E. (1964), *J. Org. Chem.*, **29**, 718.

Wheeler, O. H. and Zabicky, J. Z. (1958), *Can. J. Chem.*, **36**, 656.

White, J. M., Alder, R. W., and Orpen, A. G. (1988), *Acta Crystallogr.*, *Sect. C: Cryst. Struct. Commun.*, **C44**, 662, 664, 1465, 1467, 1777.

Whitlock, H. W. and Siefken, M. W. (1968), *J. Am. Chem. Soc.*, **90**, 4929.

Wiberg, K. B. (1983), *J. Am. Chem. Soc.*, **105**, 1227.

Wiberg, K. B. (1985), *Tetrahedron Lett.*, **26**, 5967.

Wiberg, K. B. (1986), "The Concept of Strain in Organic Chemistry," *Angew. Chem. Int. Ed. Engl.*, **25**, 312.

Wiberg, K. B. (1987), "Experimental Thermochemistry", see reference to Liebman, J. F. and Greenberg, A., Eds., (1987), p. 151.

Wiberg, K. B. (1989), "Small Ring Propellanes," *Chem. Rev.*, **89**, 975.

Wiberg, K. B., Bader, R. F. W., and Lau, C. D. H. (1987), *J. Am. Chem. Soc.*, **109**, 985, 1001.

Wiberg, K. B. and Bonneville, G. (1982), *Tetrahedron Lett.*, **23**, 5385.

Wiberg, K. B., Bonneville, G., and Dempsey, R. (1983a), "Strain Energies of Small Ring Alkenes," *Isr. J. Chem.*, **23**, 85.

Wiberg, K. B. and Burgmaier, G. J. (1969), *Tetrahedron Lett.*, 317.

Wiberg, K. B. and Burgmaier, G. J. (1972), *J. Am. Chem. Soc.*, **94**, 7396.

Wiberg, K. B., Burgmaier, G. J., Shen, K.-W., LaPlaca, S. J., Hamilton, W. C., and Newton, M. D. (1972), *J. Am. Chem. Soc.*, **94**, 7402.

Wiberg, K. B., Burgmaier, G. J., and Warner, P. (1971), *J. Am. Chem. Soc.*, **93**, 246.

Wiberg, K. B. and Connor, D. S. (1966), *J. Am. Chem. Soc.*, **88**, 4437.

Wiberg, K. B., Dailey, W. P., Walker, F. H., Waddell, S. T., Crocker, L. S., and Newton, M. (1985), *J. Am. Chem. Soc.*, **107**, 7247.

Wiberg, K. B. and de Meijere, A. (1969), *Tetrahedron Lett.*, 519.

Wiberg, K. B. and Fenoglio, R. A. (1968), *J. Am. Chem. Soc.*, **90**, 3395.

Wiberg, K. B., Lupton, E. C., Wasserman, D. J., de Meijere, A., and Kass, S. R. (1984), *J. Am. Chem. Soc.*, **106**, 1740.

Wiberg, K. B., Olli, L. K., Golembski, N., and Adams, R. D. (1980), *J. Am. Chem. Soc.*, **102**, 7467.

Wiberg, K. B. and O'Donnell, M. J. (1979), *J. Am. Chem. Soc.*, **101**, 6660.

Wiberg, K. B., Pratt, W. E., and Bailey, W. F. (1977), *J. Am. Chem. Soc.*, **99**, 2297.

Wiberg, K. B. and Walker, F. H. (1982), *J. Am. Chem. Soc.*, **104**, 5239.

Wiberg, K. B., Walker, F. H., Pratt, W. E., and Michl, J. (1983b), *J. Am. Chem. Soc.*, **105**, 3638.

Wiberg, K. B. and Wendoloski, J. J. (1982), *J. Am. Chem. Soc.*, **104**, 5679.

Wigfield, D. C. (1979), "Stereochemistry and Mechanism of Ketone Reductions By Hydride Reagents," *Tetrahedron*, **35**, 449.

Willer, R. L. (1981), *Org. Magn. Reson.*, **16**, 261.

Willer, R. L. and Eliel, E. L. (1977), *J. Am. Chem. Soc.*, **99**, 1925.

Willy, W. E., Binsch, G., and Eliel, E. L. (1970), *J. Am. Chem. Soc.*, **92**, 5394.

Winkler, J. D., Hey, J. P., and Williard, P. G. (1986), *J. Am. Chem. Soc.*, **108**, 6425.

Winnik, M. A. (1981), "Cyclization and the Conformation of Hydrocarbon Chains," *Chem. Rev.*, **81**, 491.

Winstein, S. and Holness, N. J. (1955), *J. Am. Chem. Soc.*, **77**, 5562.

Wiseman, J. R. (1967), *J. Am. Chem. Soc.*, **89**, 5966.

Wislicenus, J. (1901), *Ber. Dtsch. Chem. Ges.*, **34**, 2565.

Wohl, R. A. (1964), *Chimia*, **18**, 219.

Wohl, R. A. (1974), *Chimia*, **28**, 1.

Wolfe, S., Whangbo, M.-H., and Mitchell, D. J. (1979), *Carbohyd. Res.*, **69**, 1.

Wu, Y.-D., Houk, K. N., and Paddon-Row, M. N. (1992), *Angew. Chem. Int. Ed. Engl.*, **31**, 1019.

Würthwein, E.-U., Chandrasekhar, J., Jemmis, E. D., and Schleyer, P. v. R. (1981), *Tetrahedron Lett.*, **22**, 843.

Yamamoto, S., Nakata, M., Fukuyama, T., Kuchitsu, K., Hasselman, D., and Ermer, O. (1982), *J. Phys. Chem.*, **86**, 529.

Yokozeki, A. and Kuchitsu, K. (1971), *Bull. Chem. Soc. Jpn.*, **44**, 2356.

Yokozeki, A., Kuchitsu, K., and Morino, Y. (1970), *Bull. Chem. Soc. Jpn.*, **43**, 2017.

Yursha, I. A., Kabo, G. Ya., and Andreevskii, D. N. (1974), *Neftekhimiya*, **14**, 688; (1975) *Chem. Abstr.*, **82**, 57332g.

Zefirov, N. S. and Palyulin, V. (1991), "Conformational Analysis of Bicyclo [3.3.1]nonanes and Their Hetero Analogs," *Top. Stereochem.*, **20**, 171.

Zefirov, N. S. (1977), *Tetrahedron*, **33**, 3193.

Zefirov, N. S., Chekulaeva, V. N., and Belozerov, A. I. (1969), *Tetrahedron*, **25**, 1997; *J. Org. Chem. USSR*, **5**, 630, (Engl. transl. p. 619).

Zefirov, N. S., Koz'min, A. S., and Abramenkov, A. V. (1978), "The Problem of Tetrahedrane," *Russ. Chem. Rev.*, **47**, 289 (Engl. transl. p. 163).

Zefirov, N. S., Samoshin, V. V., and Akhmetova, G. M. (1985), *J. Org. Chem. USSR*, **21**, 224 (Engl. transl. p. 203).

Ziegler, K. (1955), "Methoden zur Herstellung und Umwandlung grosser Ringsysteme" in Houben-Weyl, *Methoden der Organischen Chemie*, 4th ed., Vol. 4, Part 2. Thieme, Stuttgart, Federal Republic of Germany, p. 729ff.

Zieger, H. E., Schaeffer, D. J., and Padronaggio, R. M. (1969), *Tetrahedron Lett.*, 5027.

Zimmerman, H. E. (1963), "Base-catalyzed Rearrangements," in de Mayo, P., Ed., *Molecular Rearrangements*, Part 1, Interscience, New York, p. 345.

Zimmerman, H. E. and Paufler, R. M. (1960), *J. Am. Chem. Soc.*, **82**, 1514.

12

Stereoselective Synthesis

12-1. INTRODUCTION

The most important aspect of the synthesis of organic molecules that contain one or more stereogenic elements is usually that of stereochemical control. It is an essential ingredient of good synthesis design and, in addition to affecting the selection of the most appropriate methodology, it will often have an overriding influence on strategic considerations, including the choice of a particular route. Given the possibility of 2^n stereoisomers for n stereogenic elements, it becomes obvious that without such control, the construction of even moderately sized molecules would be hopelessly inefficient, and without repeated separation of stereoisomers, chaos would occur. The quest for acceptable levels of selectivity has accordingly had a major influence on the development of modern synthetic methodology, especially over the last decade (Bartlett, 1980; Morrison, 1983–1985; ApSimon and Collier, 1986; Morrison and Scott, 1984; Nogradi, 1987; O'Donnell, 1988; Williams, 1989; Mori, 1989; Aitken and Kilényi, 1992; Koskinen, 1993).

The introduction of new stereogenic centers into a target molecule is normally achieved by means of two fundamentally distinct processes: most commonly through addition to one or other stereoheterotopic (i.e. enantiotopic or diastereotopic) faces of a double bond, but also by selective modification or replacement (defined as "substitution" in the present context) of stereoheterotopic ligands (cf. Chapter 8). These reactions may be represented by the generalized outline in Figure 12.1.

Stereoselection by means of ligand substitution may also be extended, at least in a conceptual sense, to certain heteroatom containing compounds, e.g., sulfides, and to the elaboration of axially dissymmetric products, e.g., cumulenes and biaryls (cf. Chapter 14) as shown in Fig. 12.2.

Another important group of stereoselective processes based on discrimination between stereoheterotopic ligands are those based on the substitution of groups or ligands in meso substrates. Two generalized examples are indicated in Figure 12.3.

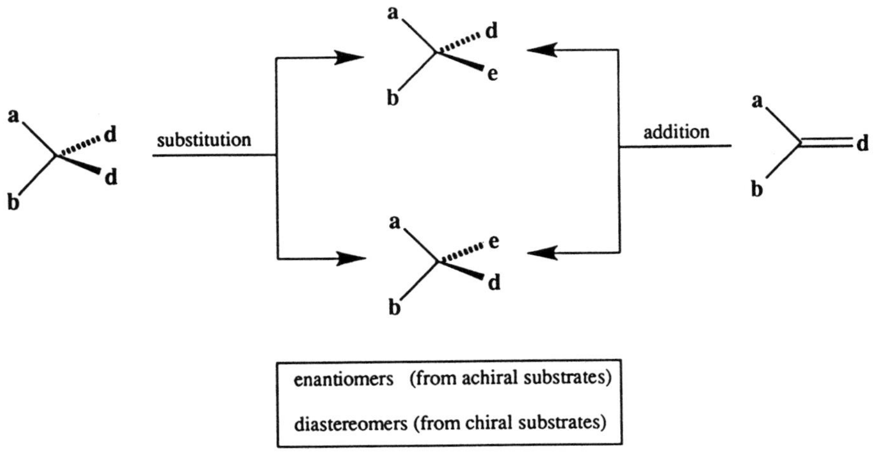

Figure 12.1. Introduction of new stereogenic centers by means of π-facial discrimination or ligand substitution.

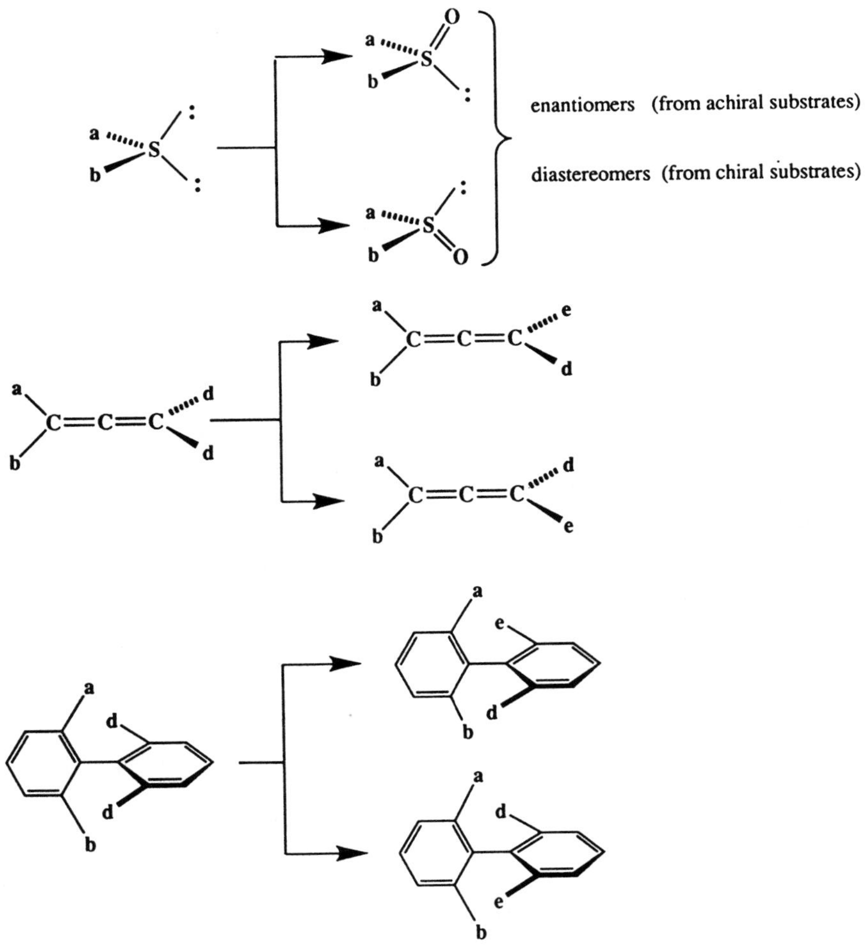

Figure 12.2. Stereoselection by means of ligand substitution in sulfides, cumulenes, and biaryls.

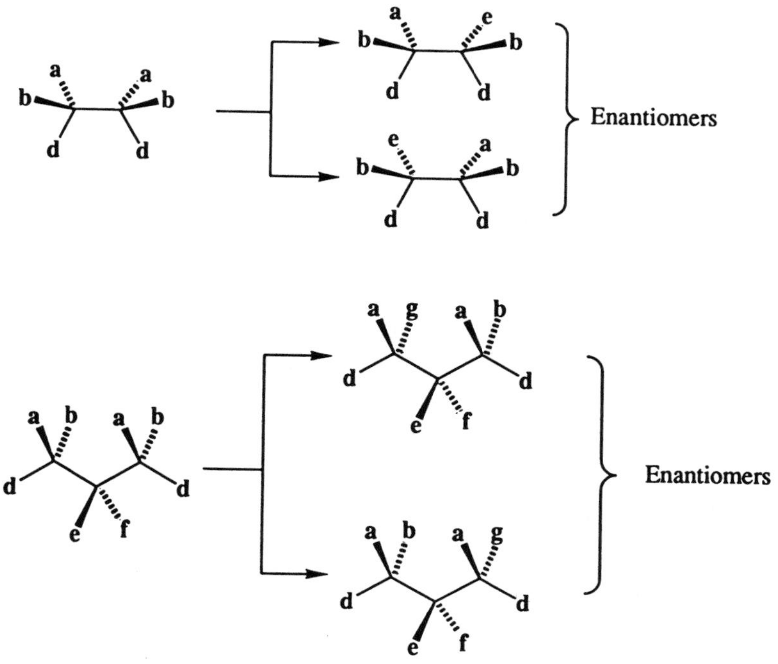

Figure 12.3. Formation of enantiomers from meso substrates.

The stereochemical aspects of elimination reactions (the reverse of the addition process outlined in Fig. 12.1) may also be important in biogenesis and synthesis, especially with respect to the control of regiochemistry and the E/Z geometry of double bonds. The elimination of chiral groups as a means of achieving enantioselection in the formation of chiral alkenes has been reviewed (Goldberg, 1970; Hanessian et al., 1985).

a. Terminology

There appears to be widespread confusion over the terms stereoselective and stereospecific. By convention (Zimmerman et al., 1959; Eliel, 1962), the latter term should be used only when the configuration of the product is related to that of the reactant and when the reaction processes are mechanistically constrained to proceed in a stereochemically defined manner (e.g., S_N2 substitution, hydroboration, and epoxidation of alkenes). A typical example is the trans-addition of bromine to the two isomers of 2-butene, that is, meso-2,3-dibromobutane is formed from the E isomer and ($\pm$)-2,3-dibromobutane is formed, from the Z isomer. The term "stereoselective", however, is used to describe the stereochemical outcome of a reaction when it is possible for more than one stereoisomer to be formed, but one is formed in excess, although its use should desirably be restricted to situations where the proportion of the major stereoisomer is substantially greater than that of the minor one(s) (Seebach et al., 1986). *Even when it appears that only a single stereoisomer is formed, it is not appropriate to use the term stereospecific.* When α-pinene **1** is treated with borane–dimethyl sulfide

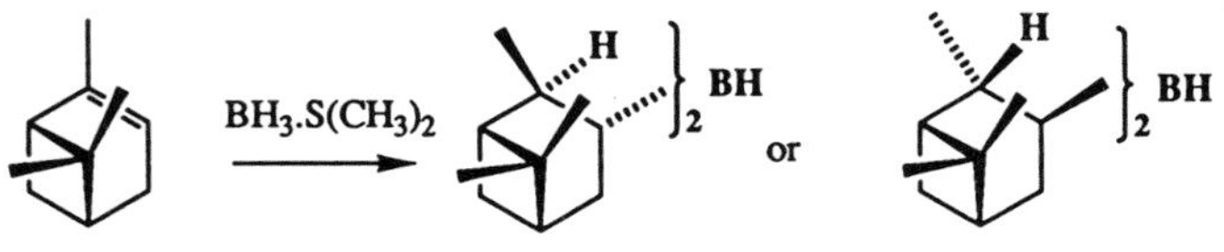

Figure 12.4. Stereoselective addition of a borane to α-pinene.

complex (Brown et al., 1985) (Fig. 12.4), only one stereoisomer (**2**) is obtained. Although the alternative diastereomer **3** cannot be detected among the products, it is nevertheless appropriate to describe the (apparently) exclusive addition of borane to the α face of **1** as *stereoselective*, since it is possible for both diastereomers **2** and **3** to be formed.

b. Stereoselective Synthesis

The task of controlling the elaboration of further stereogenic centers in chiral cyclic or polycyclic molecules is on the whole straightforward, provided that the conformational freedom of the substrate is limited and that preexisting stereo-chemical elements impose a sufficient degree of asymmetry or dissymmetry to the environment of the reaction site(s). The stereochemical outcome is then de-termined by differences in free energy associated with diastereomeric transition states ($\Delta\Delta G^{\ddagger}$) or products ($\Delta\Delta G^{0}$). These differences may usually be estimated in a qualitative sense from inspection of molecular models and the desired outcome determined by an appropriate choice of reagent(s) and reaction sequence.

The conversion of 2-norbornanone **4** into the diastereomeric methyl carbinols **5** and **7** (Fig. 12.5) provides a typical illustration. It has been well established for ketones like **4** that nucleophiles add preferentially to the exo face (cf. Section 12-3.d). Thus, **5** is both the expected and the observed major product from

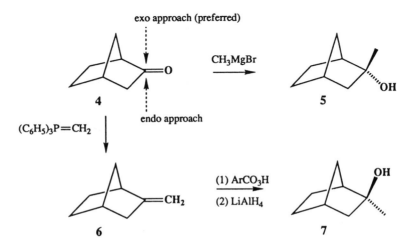

Figure 12.5. Preparation of diastereomeric methyl carbinols from norbornan-2-one.

reaction with a methyl Grignard reagent. In order to obtain the epimer **7**, however, it is necessary to employ an indirect route that reverses the order in which the new C—C and C—O bonds are established (e.g., **4**→**6**→**7**).

Stereoselection in the above sense presents few conceptual difficulties. It has been possible to construct complex targets with reasonable efficiency, and it has normally been unimportant whether the substrate and intermediates were enantiomerically pure or racemic: The outcome has been largely determined by the structure of the substrate. The major difficulties involved with stereodifferentiating processes are associated with conformationally mobile substrates and with the creation of the *first* stereogenic center. A wide range of strategies and new methodology designed to cope with these challenges has emerged over the last decade in response to the need for the production of therapeutically active substances as single enantiomers, and within the context of syntheses of flexible polyketide molecules typified by macrolides and various ionophores. Because syntheses of these molecules lend themselves so readily to convergent designs, there has also been a major driving force to devise preparations that furnish the individual fragments in enantiomerically pure form, since coupling of racemic intermediates normally generates mixtures of diastereomers (cf. pp. 843–845). Yet another important factor in recent developments has been the availability of increasingly powerful computational methods for the estimation of structure (cf. Section 2-6), which allow synthesis to be placed on a more rational basis.

c. Categories of Stereoselective Synthesis

In general, synthetic stereoselective processes fall into three broad categories:

1. Reactions that lead to the selective formation of diastereomers.
2. Reactions that lead to the selective formation of enantiomers.
3. Double stereodifferentiating reactions (in which at least two participants are chiral).

Diastereoselective Synthesis

Three classes of diastereoselective reactions may be recognized, the first of which is concerned with the stereochemically controlled preparation of achiral diastereomeric compounds (or achiral diastereomeric moieties within chiral compounds). These compounds are primarily *E* and *Z* alkenes (cf. Section 12-2.b), but the category also formally includes the preparation of cis–trans diastereomers of achiral cyclic structures (cf. p. 666) and it is important to note that the factors controlling diastereoselection in these latter compounds may be indistinguishable from those associated with chiral substrates. For example, in the reactions of the carbonyl group in cyclohexanones that are "conformationally locked" by a bulky equatorial substituent in either the 3 or 4 position, it is the puckering of the ring that determines the direction of approach by reagents to the diastereotopic π faces of the carbonyl group in each case. The presence or lack of a plane of symmetry has no bearing on the outcome, provided that the other reactant is achiral (cf. Section 12-2.a).

The second, and much more diverse class of diastereoselective reactions, is concerned with the introduction of new stereogenic centers into chiral substrates (cf. Section 12-3). The configurations of the new stereocenters are established in a relative relationship to the preexisting stereocenter(s) in response to differential interactions arising from the reactant approaching along alternative trajectories, for example, avoidance of steric interactions, minimization of torsional strain, and optimization of orbital interactions, (cf. Section 12-3.c). For a racemic substrate the individual enantiomers can be expected to react with achiral reagents at exactly the same rate, but in a stereochemically complementary sense to establish the same *relative* relationship between the new and preexisting stereocenters in each of the enantiomers. In this way the homogeneity of the diastereomeric relationships for the individual enantiomers in the racemic product is maintained. A number of apparent exceptions to this generalization have been reported (Wynberg and Feringa, 1976), but they are almost certainly a consequence of double stereodifferentiation (see below). For example, although the reduction of 1-(*R*)-camphor **8** by one equivalent of lithium aluminum hydride afforded a 90.2:9.8 mixture of isoborneol **9** with borneol **10**, an 88.7:11.3 ratio was obtained with racemic camphor under identical reaction conditions (Fig. 12.6). The differences in outcome could be accounted for by the participation of chiral hydrides of the type $LiAl(OR)_n H_{4-n}$ ($n = 1$, 2, or 3), where R = borneol or isoborneol.

> Measurable differences in physical properties (e.g., by NMR spectroscopy and calorimetry) have been observed between solutions of pure enantiomers and the corresponding racemates, respectively (Horeau and Guetté, 1974; cf. Chapter 6). A transition state based on an aggregate containing both enantiomers clearly must differ from that for a pure enantiomer (i.e., the two transition states will be diastereomeric), but attempts to demonstrate that the impact of these interactions is sufficient to affect chemical reactivity at a detectable level have been inconclusive. For example, in one study, the differences in the rates of ketalization of 2-butanone in (*S*)- and (*R*,*S*)-1,2-propanediol, respectively, were found not to be significant (Wynberg and Lorand, 1981). However, several examples of a nonlinear relationship between the enantiomeric excess (ee) of a reaction product and the ee of the chiral ligand incorporated into a transition metal catalyst of the type ML_2 have been documented (Section 12-4.d).

A third category of diastereoselective processes involves the coupling of two compounds at prostereogenic centers with the formation of two new stereocenters

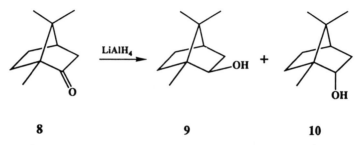

Figure 12.6.   Lithium aluminum hydride reduction of camphor.

as may occur with the aldol and Michael reactions. The focus of concern here is with the stereochemical relationship between these centers (i.e., simple stereoselectivity) and is addressed in greater detail later (Section 12-3.f); cf. also Section 12-1.d, and Oare and Heathcock, 1989).

Enantioselective Synthesis

This category encompasses those processes that are traditionally identified by the term "asymmetric synthesis" (cf. Section 12-4), (Kagan and Fiaud, 1978; Morrison, 1983–1985; Gladysz and Michl, 1992). The substrates are achiral and for most practical purposes, stereoselectivity in the formation of new stereogenic centers (i.e., enantioselection) will require reagents that are chiral, either intrinsically, or as a consequence of chelation, solvation and so on. It has been possible to achieve enantioselective syntheses by adsorption of the reactants onto chiral surfaces or by lattice controlled reactions within chiral crystals or liquid crystal mesophases (Kaftory et al., 1988; cf. Section 6-4.b). One of the more important aspects of enantioselective reactions is the potential for catalyzed processes (in a substoichiometric sense). In spite of rapid progress in recent years, there is still a very restricted choice of reaction types and the level of stereoselectivity is highly substrate dependent. Since our understanding of mechanism is frequently limited and probably speculative in most cases, it is usually more straightforward to rationalize the outcome after the event than to identify the probable outcome beforehand. The most prevalent strategy for overcoming these difficulties and uncertainties has therefore been to attach temporarily a "chiral adjuvant" or "chiral auxiliary" to the achiral substrate. Any new stereocenters are accordingly established in a diastereomeric relationship to those in the auxiliary, which is subsequently cleaved. Although the net result is one of enantioselection, the processes are intrinsically diastereoselective and will accordingly be grouped with those processes and included in Section 12-3. A major advantage for this strategy is the relative ease with which enantiomers may be obtained pure (e.g., through chromatography of diastereomeric intermediates) and the degree of purity established.

For those syntheses that begin with one or mainly one enantiomer of a chiral substrate and continue by means of diastereoselective reactions to elaborate further stereogenic elements with the maintenance of approximately the original enantiomeric purity, it is suggested that the term *enantiospecific* might be applied. The term *enantioselective* should be used to describe a synthesis that involves an enantioselective stage, either a direct enantioselective reaction on an achiral starting material, or a sequence that achieves the selective formation of an enantiomer through the temporary incorporation of a chiral auxiliary, followed by one or more diastereoselective reactions and then cleavage of the auxiliary. These terms would then displace the cumbersome "enantiomerically pure compound (epc) synthesis" (Seebach et al., 1986), and avoids any contention that may arise when a particular substrate or product is not actually "enantiomerically pure."

When a synthesis begins with a racemate or the formation of a racemate and proceeds via a sequence of diastereoselective reactions, then the synthesis should not be described as simply stereoselective, but as *diastereoselective*. We recommend that the description "(nonracemic) chiral synthesis" should not be used and that the term homochiral should only be used as originally defined (Anet et al., 1983); (see p. 215).

An alternative or supplementary nomenclature to the use of terms enantioselective, diastereoselective, and so on, has been advocated by Izumi and Tai (1977) and is based on the term *stereodifferentiation*. It is independent of mechanism and is derived simply from the apparent structural relationship between the substrate and product(s). Six categories are identified and placed into two groups, the first of which is characterized by chirality in the reagent, catalyst, or reaction medium, and the second by chirality in the substrate:

1. Enantio-differentiating reactions: Enantioface-differentiating reactions
   Enantiotopos-differentiating reactions
   Enantiomer-differentiating reactions

2. Diastereo-differentiating reactions: Diastereoface-differentiating reactions
   Diastereotopos-differentiating reactions
   Diastereomer-differentiating reactions

The various kinds of processes, which are illustrated in Figure 12.7, encompass those described earlier (cf. Fig. 12.1), but are more extensive and viewed from a slightly different perspective (note that in one case the term "topos differentiation" equates with heterotopic ligand substitution).

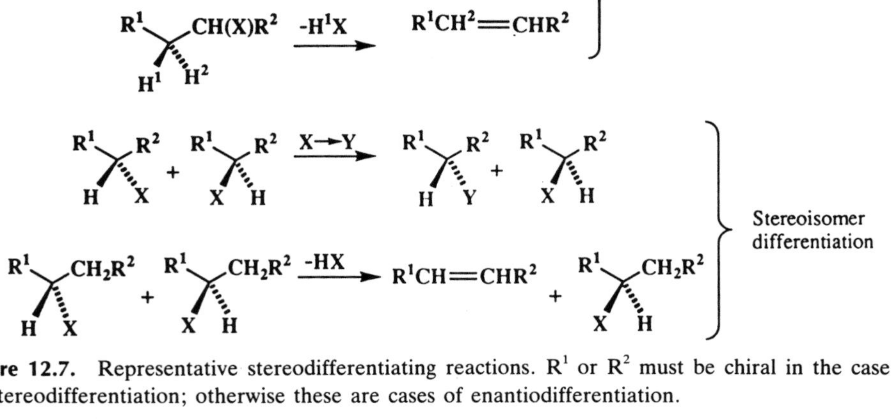

Figure 12.7. Representative stereodifferentiating reactions. R^1 or R^2 must be chiral in the case of diastereodifferentiation; otherwise these are cases of enantiodifferentiation.

A potentially useful alternative to enantioselection may be to employ the strategy of enantioconvergence, whereby a racemate is resolved into its individual enantiomers, which are then both utilized in the synthesis. The objective may be achieved by recycling the undesired enantiomer, or by taking each enantiomer through stereochemically complementary sequences, eventually converging on a common intermediate (cf. Section 12-4.g).

Irrespective of the chosen method, however, enantioselective or enantio-specific syntheses depend on the availability of enantiomerically pure (or enriched) compounds from natural (biological) or industrial sources at a reasonable cost (cf. Scott, 1984, 1989; Blaser, 1992; Santaniello et al., 1992; Collins et al., 1992; Sheldon, 1993) to serve as follows:

1. Resolving agents (i.e., for the formation of diastereomeric salts, esters, or amides).
2. Substrates (e.g., carbohydrates or terpenoids).
3. Catalysts (e.g., enzymes or bases).
4. Chiral auxiliaries attached to achiral substrates.
5. Chiral auxiliaries attached to achiral reagents and catalysts.

Clearly, it is important that in the relevant cases, stereogenic centers in the target molecule are not disturbed while separating it from the chiral auxiliary. It is also desirable that the catalyst or chiral auxiliary can be recovered efficiently in undiminished enantiomeric purity, unless the auxiliary is more readily available or much less expensive than the substrate. Finally, one important class of enantioselective processes is that which utilizes meso substrates, although to date the methodology has been almost exclusively based on the use of enzymes (cf. Sections 12-4.e and f).

Double Stereodifferentiating Reactions

When both participants in a stereodifferentiating reaction are chiral, the outcome will depend to some extent on the relationship between their respective absolute configurations. For example, most enzymes that react rapidly with one enantiomeric form of a chiral substrate will not react at a discernible rate with the other enantiomer, allowing a "kinetic resolution" of the racemic form of the substrate (cf. Chapter 7-6). Kinetic resolutions may also be achieved with nonenzymatic reagents, although the differences in rate are rarely sufficient to achieve high efficiencies. However, relatively small differentials in reactivity between enantiomers may be usefully exploited in enhancing stereoselection over and above that obtained with a single chiral participant, for example, by attaching a chiral auxiliary to an achiral reactant, utilizing a chiral solvent, employing a chiral catalyst in place of an achiral one, or substituting a chiral reactant that is synthetically equivalent to an achiral one. A more extensive discussion is provided in Section 12-5.

d. Convergent Syntheses

When two molecules M^1 and M^2 are connected together there is a range of possible stereochemical outcomes depending on whether or not the participants are chiral

		M^2			
		N^2	N^{2*}	P^2	P^{2*}
M^1	N^1	N^1N^2	N^1N^{2*}	N^1P^2	N^1P^{2*}
	N^{1*}	$N^{1*}N^2$	$N^{1*}N^{2*}$	$N^{1*}P^2$	$N^{1*}P^{2*}$
	P^1	P^1N^2	P^1N^{2*}	P^1P^2	P^1P^{2*}
	P^{1*}	$P^{1*}N^2$	$P^{1*}N^{2*}$	$P^{1*}P^2$	$P^{1*}P^{2*}$

Figure 12.8. Possible stereochemical outcomes from coupling reactions of molecules M^1 and M^2.

and whether or not reaction takes place at a prostereogenic center. Sixteen combinations are possible, as illustrated by the matrix in Figure 12-8, in which the reactants are designated as N (achiral, reacting at a nonprostereogenic center), N* (chiral, reacting at a nonprostereogenic center), P (achiral, reacting at a prostereogenic center), or P* (chiral, reacting at a prostereogenic center).

Assuming the absence of external chiral influences and that the chiral reactants are enantiomerically pure, for combinations (a) N^1N^2, N^1N^{2*}, $N^{1*}N^2$, and $N^{1*}N^{2*}$, no new stereocenters are introduced; (b) N^1P^2 and P^1N^2, a racemate would be formed; (c) P^1P^2, two racemic diastereomers would be formed in unequal amounts; (d) $N^{1*}P^2$, P^1N^{2*}, and N^1P^{2*} and $P^{1*}N^{2*}$, two diastereomers in unequal amounts should be formed in each case; (e) $N^{1*}P^{2*}$ and $P^{1*}N^{2*}$, two diastereomers would also be formed in unequal amounts; (f) P^1P^{2*} and $P^{1*}P^2$, four diastereomers are possible, a situation typified by the mixed aldol reaction (see Section 12-3.f); and (g) $P^{1*}P^{2*}$, up to four diastereomers are also likely to be formed. In categories (e) and (g), double stereodifferentiation is involved, complicating the prediction of diastereomer ratios (cf. Section 12-5). A more detailed treatment is given in a review of the stereochemical aspects of the Michael reaction (Oare and Heathcock, 1989), while combinations involving racemates are considered next.

If two chiral reactants are coupled in a convergent synthesis (disregarding the question of stereoselectivity in the formation of any new stereocenters), and one of the participants is racemic, there is the potential for a kinetic resolution to occur. Thus, for example, (*R*)-M^1 plus (*R*, *S*)-M^2, if the reaction rate $k_{R+R'}$ for (*R*)-M_1 + (*R*)-M^2 is very much greater than the rate $k_{R+S'}$ for (*R*)-M^1 + (S)-M^2, then we could expect a 50% yield of [(*R*)-M^1—(*R*)-M^2] and a residue of 50% (*R*)-M^1 and 50% (*S*)-M^2. If, as is more likely in practice, the rates are not greatly different, only an enantiomeric enrichment will occur, and to optimize the enantiomeric excess of the product it will be necessary to use a significant excess of the racemic reactant.

If both reactants are racemic and the stereogenic centers in the respective participants are remote from each other in the transition state structure, then two racemic diastereomers are likely to be formed in roughly equal amounts. If these stereocenters are close enough to interact, however, there is the potential for mutual kinetic enantioselection (Oare and Heathcock, 1989; see p. 326). That is for the coupling of a racemate composed of enantiomers R^1 and S^1 with a second racemate composed of enantiomers R^2 and S^2, if the rate of coupling (k_{R^1/R^2}) of R^1 with R^2 is significantly greater than the rate of coupling (k_{R^1/S^2}) for enantiomer R^1 with the alternative enantiomer S^2, the reaction should lead to the formation of predominantly one racemic diastereomer. This combination can be dia-grammed as

$$(R^1 + S^1) + (R^2 + S^2) \rightarrow (R^1—R^2 + S^1—S^2)$$

when $k_{R^1/R^2}(\equiv k_{S^1/S^2}) \gg k_{R^1/S^2}(\equiv k_{R^2/S^1})$.

The conjugate addition of ($\pm$)-cuprate **225** to ($\pm$) enone **224** described in Section 12-3.g provides such an example (cf. also Heathcock, 1982).

12-2. DIASTEREOSELECTIVE SYNTHESIS OF ACHIRAL COMPOUNDS

a. Cyclanes

Although diastereoisomerism is most often associated with chiral molecules, chirality is not an essential precondition (cf. Section 3-3). Moreover, the structur-al features that determine a particular stereochemical outcome may not differ significantly between chiral and achiral substrates, as in the chiral ketone **11** and the achiral ketone **12**. Thus, approach of nucleophiles syn to the olefinic bond is sterically favored in both cases and is not influenced by the presence or lack of symmetry elements (Fig. 12.9).

The π faces of a double bond may also be rendered diastereotopic simply as a consequence of the puckering that commonly arises in cyclic systems, especially in six-membered rings (as noted earlier in Section 12-1.c). Nucleophilic addition to an unhindered carbonyl group in a cyclohexanone derivative (Fig. 12.10), for

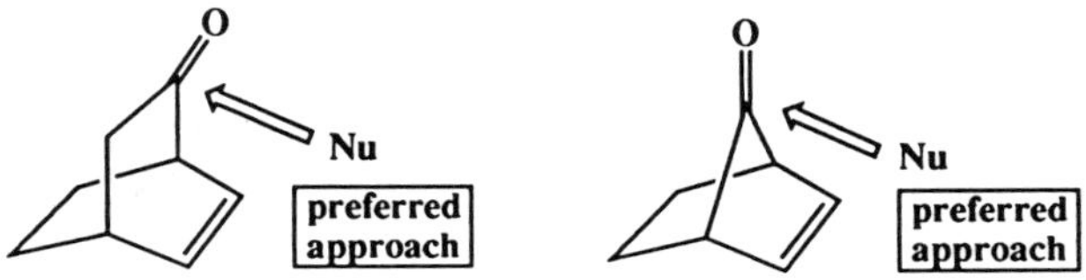

Figure 12.9. Diastereoselective addition of nucleophiles (Nu) to a chiral and an achiral bicyclic ketone.

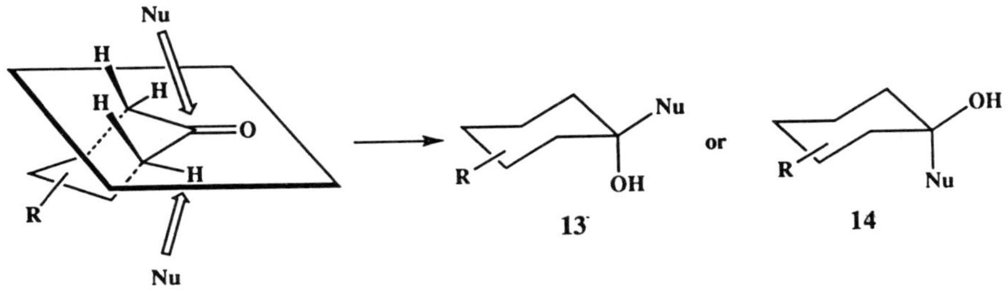

Nu = nucleophile

Figure 12.10. Diastereotopic π faces of the carbonyl group in cyclohexanones.

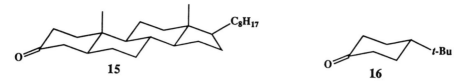

Figure 12.11. Comparison of the topology of the carbonyl group in an achiral and a chiral ketone.

example, may occur along alternative trajectories that result in the new group becoming either an axial or equatorial substituent in the adduct. It can be readily seen that for the methylene groups flanking the carbonyl group, two of the hydrogen substituents lie approximately in the σ plane, while there is a dihedral angle of about 109° between this plane and the remaining pair of hydrogen atoms.

Assuming that conformational freedom is restricted to one particular chair (or possibly boat) arrangement by appropriate substitution, then it could be expected that addition to one or the other π face of the carbonyl group should be more favored and that the outcome will be independent of whether the molecule is chiral or not (except in cases where the reagent or reagent-complex is itself chiral). That is, the topological control elements for the addition of a given nucleophile to cholestan-3-one **15** will not differ significantly from those involved for 4-*tert*-butyl cyclohexanone (Fig. 12.11, **16**). A reagent like lithium aluminum hydride ($LiAlH_4$), for example, will give predominantly the equatorial carbinol in both cases (cf. Morrison and Mosher, 1976, pp. 7–9). The remainder of this section is concerned with the formation of 1,2-di-, tri-, and tetrasubstituted alkenes.

b. Diastereoselective Syntheses of Alkenes

Alkenes have been prepared in a stereocontrolled manner by additions to alkynes, by some kind of elimination process, from pericyclic reactions, and from ring fragmentations. The literature has been comprehensively reviewed (Patai, 1962; Zabicky, 1970; Reucroft and Sammes, 1971; Faulkner, 1971; Henrick, 1977).

Routes from Alkynes

Reductive Methods. The most commonly utilized method for the preparation of 1,2-substituted Z alkenes is based on the catalytic reduction of alkynes that can often be arrested at the alkene stage since alkynes tend to be more strongly adsorbed onto the catalyst than the corresponding alkenes (Marvel and Li, 1973; Rylander, 1979). High chemo- and stereoselectivity has been reported with a highly dispersed nickel-on-graphite catalyst in the presence of ethylenediamine (Savoia, Umani-Ronchi, et al., 1981). The most generally used catalysts, however, are based on palladium on a variety of supports with partial poisoning with lead salts, which are reduced to metallic lead in situ, alone or in conjunction with quinoline, commonly known as "Lindlar catalysts" (Lindlar, 1952); even more selective catalysts have been prepared by the addition of $MnCl_2$ as well (Rajaram, Dev, et al., 1983). Stereoselectivity by catalytic methods is rarely complete,

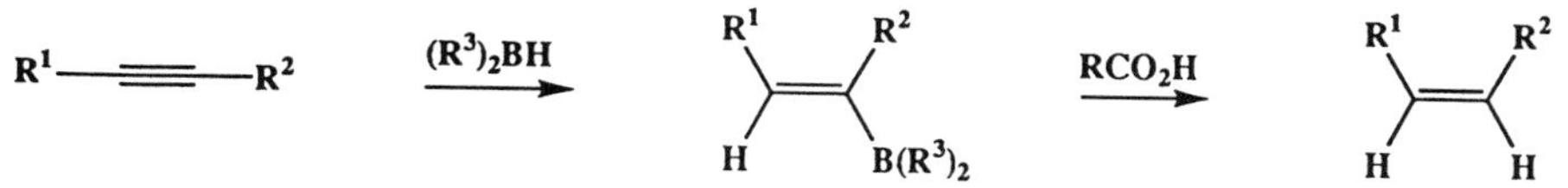

Figure 12.12. Synthesis of a Z alkene by addition of a borane to an alkyne followed by protonolysis.

however, and better results can be obtained through the addition of a borane followed by protonolysis of the vinyl borane adduct **17** (Fig. 12.12; Brown and Zweifel, 1961).

In contrast to catalytic methods, reductions of alkynes by sodium or lithium and alcohol in liquid ammonia normally afford E alkenes (Smith, 1963; Jäger and Viehe, 1977). It is assumed that a 1,2 dianion is formed in which the nonbonding orbitals take up an anti configuration for electrostatic reasons and then protonation takes place with retention of configuration (Greenle and Fernelius, 1942). Cyclodecyne **18**, for example, is reduced by lithium to (E)-cyclodecene **19** (Fig. 12.13; Blomquist et al., 1952) even though the Z isomer is thermodynamically more stable. The latter is in fact formed when sodium is used, although this outcome is attributed to the reaction proceeding via the allene isomer, that is, cyclodeca-1,2-diene (Svoboda, Sicher, et al., 1965).

Solutions of $LiAlH_4$ in diethyl ether or tetrahydrofuran (THF) also afford E alkenes from the reduction of alkynes and this reagent may be especially useful for the reduction of propargylic alcohols that may be prone to hydrogenolysis with other reagents. This reduction was shown by a deuteration experiment to proceed via a cyclic alane, that is, **20 → 21 → 22** (Fig. 12.14; Hochstein and Brown, 1948).

In a subsequent study directed towards the synthesis of (E,E)-farnesol **27**, however, the reaction was found to afford a mixture of β- and γ-substituted products. It was nevertheless possible to achieve regioselective addition either at the β position by employing a 60:1 mixture of $LiAlH_4$ and aluminum chloride ($AlCl_3$), or at the γ-position by using a 1:2 mixture of $LiAlH_4$ with sodium methoxide ($NaOCH_3$). In both cases the intermediate alanes were trapped with

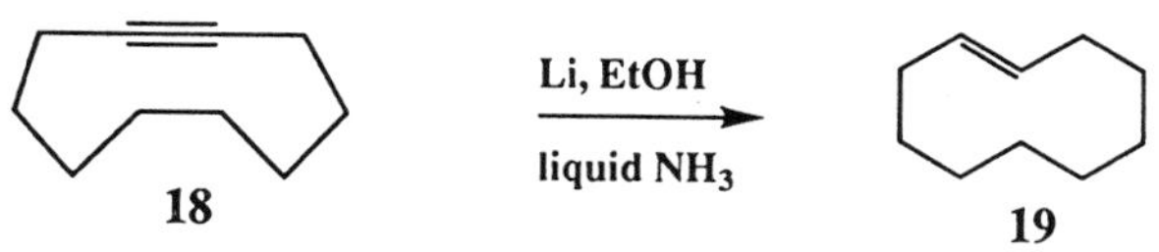

Figure 12.13. Reduction of cyclodecyne to (E)-cyclodecene.

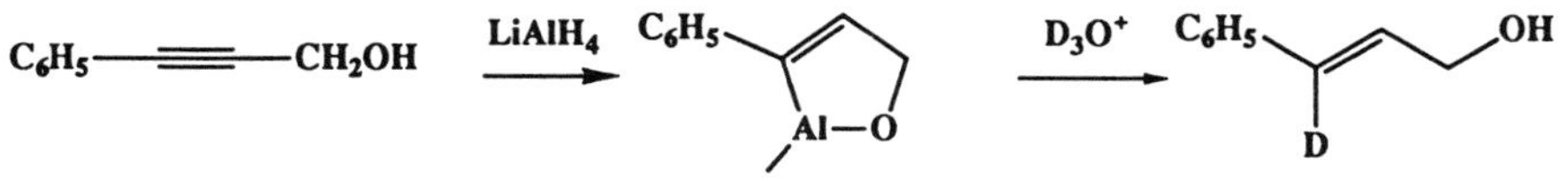

Figure 12.14. Reduction of a propargylic alcohol by lithium aluminum hydride.

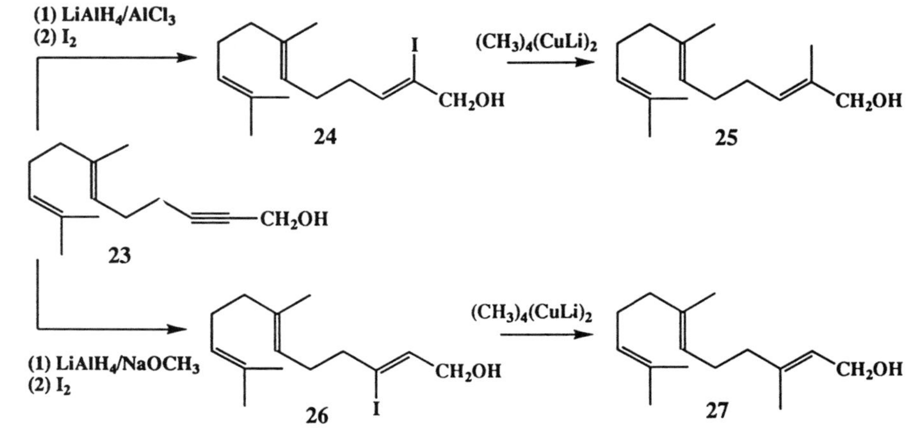

Figure 12.15. Stereocontrolled synthesis of (E,E)-farnesol **27**.

iodine to afford vinyl iodides **24** and **26**, respectively, which were both alkylated subsequently by lithium dimethylcuprate $[\text{LiCu}(\text{CH}_3)_2]$ to afford **25** and **27**, respectively (Fig. 12.15; Corey et al., 1967).

Addition Methods. Although the Friedel–Crafts type of addition of acid chlorides to alkynes to form *trans-β*-vinyl ketones $[(E)\text{-RCOCH}{=}\text{CHCl}]$ is stereoselective (Benson and Pohland, 1964), the steric course of electrophilic additions to alkynes in general is hard to predict (Fahey, 1968), since most reactions proceed stage-wise via vinyl cations (Melloni, Modena, et al., 1981). In contrast, nucleophilic additions proceed reliably in an anti fashion, as in the addition of alkoxides and thioalkoxides to phenyl acetylene to form Z enol and thioenol ethers, respectively (cf. Eliel, 1962; Winterfeldt, 1969).

Alanes and boranes similarly add in a predictable manner. For example E-alkenyl iodides and bromides may be prepared stereoselectivity from simple alkynes by hydroalumination with dialkylalanes, including the commercially available diisobutylalane $[(i\text{-Bu})_2\text{AlH}]$ (Zweifel and Whitney, 1967). Substitution of the alane group by halogen proceeds with retention of configuration, but attempts to extend the methodology to include electrophiles, such as carbon dioxide or formaldehyde, met with only modest success. Following the preparation of an "ate" complex **28** by the addition of methyllithium, however, satisfac-

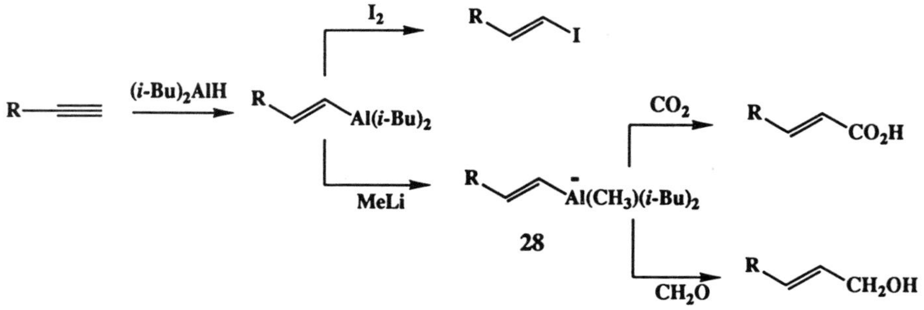

Figure 12.16. Stereocontrolled synthesis of E alkene derivatives via alanes.

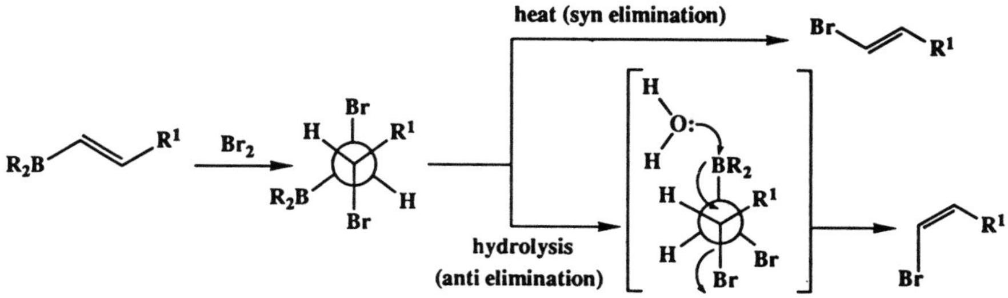

Figure 12.17. Conversion of an *E*-alkenyl borane to *E*- or *Z*-alkenyl bromides.

tory yields of 2-enoic acids and allylic alcohols were obtained, again with complete diastereoselectivity to form the *E* isomers (Zweifel and Steele, 1967; Fig. 12.16).

Alkenyl boranes have been similarly exploited to prepare alkenyl bromides, but instead of direct electrophilic reaction, anti addition of bromine occurs. Thermal elimination then affords the *E* isomer while hydrolysis leads to an inversion of configuration to give the *Z* isomer (Fig. 12.17; Brown et al., 1967).

Alkenyl boranes have been used in a much wider range of applications (Pelter et al., 1988), and are especially useful for transition metal catalyzed coupling processes with alkenyl, alkynyl, and allylic halides to form dienes, polyenes, and, for example, the Suzuki synthesis of dienes mediated by palladium (Miyaura, Suzuki, et al. 1985; Uenishi, Kishi, et al., 1987; Fig. 12.18). If *Z*-alkenyl boranes are required they may be obtained by hydroboration of alkynyl halides followed by treatment with a potassium trialkylborohydride (Negishi et al., 1975) or *tert*-butyllithium (Campbell and Molander, 1978). Both reagents apparently transfer hydride ion to the alkenyl borane **29** followed by rearrangement of the adduct **30** (Fig. 12.19). Again, the scope of syntheses based on alkenyl boranes may be extended by conversion to "ate" complexes, for example, **31 → 32** (Fig. 12.20; Negishi et al., 1973).

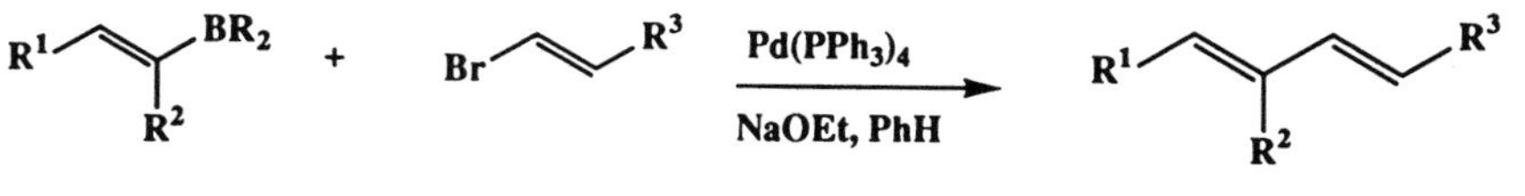

Figure 12.18. Suzuki stereocontrolled synthesis of dienes.

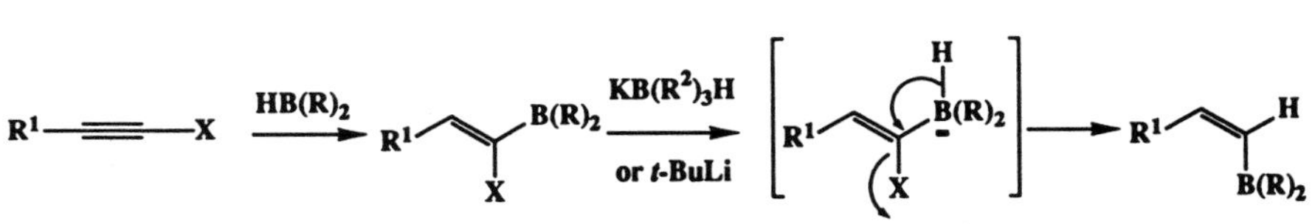

Figure 12.19. Stereocontrolled synthesis of *Z*-alkenyl boranes.

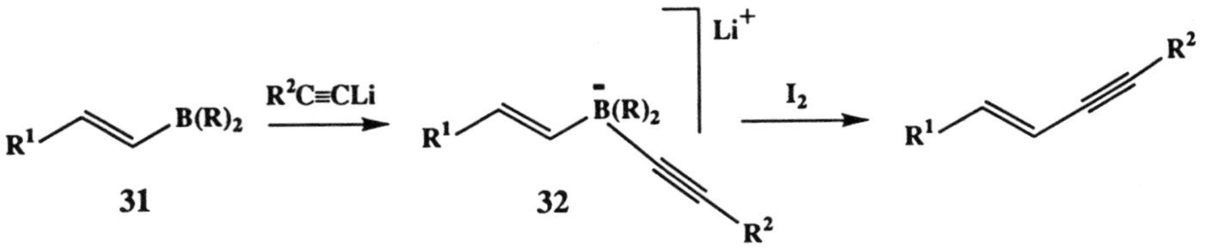

Figure 12.20. Coupling of an alkenyl borane with a lithium acetylide via a boronate intermediate.

Stereoselective Formation of Alkenes from Elimination Processes

In elimination reactions, the geometry of the alkene product may be mechanistically linked in a predictable way to the configuration of the precursor (see also p. 585ff.). It may be safely assumed that in unconstrained systems, bimolecular ionic elimination reactions (e.g., elimination of HX from alkyl halides with strong bases or reductive eliminations of 1,2 halohydrins and their congeners; Fig. 12.21) will almost invariably proceed in an antiperiplanar fashion (Saunders and Cockerill, 1973). In contrast, unimolecular processes, as in the pyrolytic elimination of amine oxides, sulfoxides, selenoxides, esters, and thioesters, will proceed via cyclic transition states in which the participating groups are necessarily syn. The apparently anomalous results obtained from reactions like the Hofmann elimination may be attributed to the participation of cyclic transition states involving ylides.

Cornforth et al. (1959a) designed a new procedure for the stereoselective synthesis of trisubstituted alkenes as a prelude to preparing squalene **33** (Cornforth et al. 1959b). It was reasoned that the reaction of an α-haloaldehyde or α-haloketone with a Grignard reagent or equivalent organolithium reagent should proceed more readily via the conformation in which the C=O and C—Cl dipoles were antiperiplanar, with the reagent approaching anti to the more bulky α-alkyl substituent R^L (although cf. the discussion in Sections 12-3.d and i). The resulting 1,2 halohydrin was then expected to undergo a reductive–elimination reaction with the participating orbitals in an antiperiplanar relationship as illustrated in Figure 12.21.

Although the outcome of the reaction of n-butylmagnesium bromide with 2-chlorobutanal was consistent with the hypothesis, the diastereoselectivity (ca. 70%) was only modest. Improved stereoselectivity was obtained (80–85%) when ethylmagnesium bromide was added to 3-chlorobutanone, however, and even better results were obtained subsequently with similar substrates by the simple expedient of lowering the reaction temperature to −90°C (Brady, Johnson, et al. 1968). Direct reductive elimination of the chlorohydrin intermediates proved not to be feasible, and so these compounds were converted by sodium hydroxide into

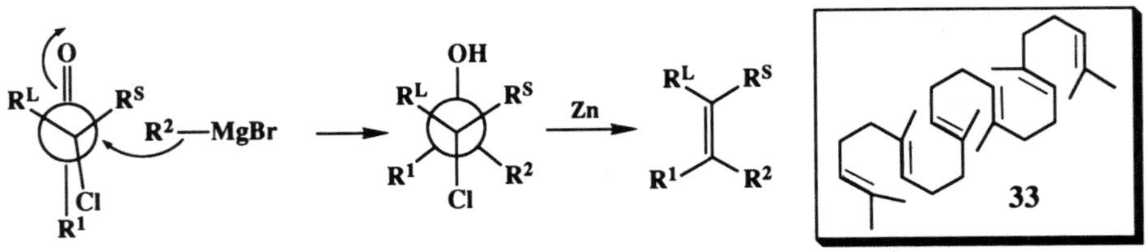

Figure 12.21. Cornforth approach to the stereoselective synthesis of alkenes.

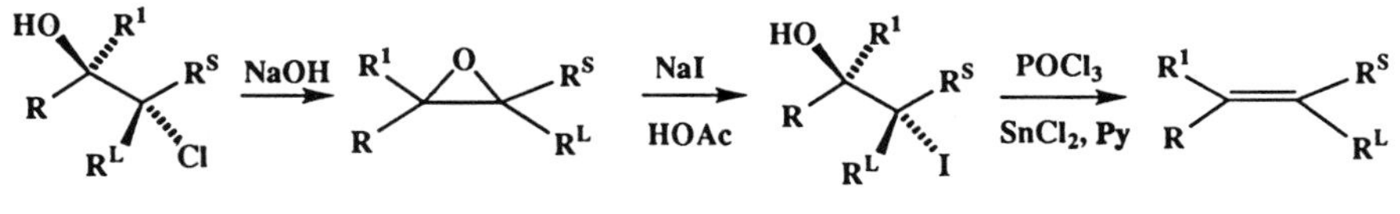

Py = pyridine

Figure 12.22.   Stereocontrolled reduction of a chlorohydrin via an epoxide.

the corresponding epoxides that were, in turn, treated with sodium iodide in acetic acid to form the 1,2 iodohydrins. Reduction to the alkenes was then effected by a mixture of stannous chloride and phosphorus oxychloride in pyridine (Fig. 12.22).

A coupling process to form exclusively E alkenes, which has found more extensive application than the Cornforth method, was introduced by Julia and Paris (1973). It was based on the addition of aryl sulfone carbanions to aldehydes or ketones to form β-hydroxy sulfones. These were converted to their mesylates or tosylates, which subsequently underwent reductive elimination with sodium amalgam in ethanol. A later study (Kocienski, Lythgoe, et al., 1978) introduced a number of refinements and established that in the case of 1,2-substituted alkenes, only the E isomers were formed, irrespective of the stereochemistry of the precursor β-hydroxysulfone. It was therefore concluded that the incipient carbanion derived from reductive cleavage of the sulfone group must have a sufficient lifetime to undergo inversion, allowing the reaction to proceed via the lower energy transition state in which the alkyl groups have a transoid relationship (Fig. 12.23).

One of the most powerful protocols for the regio- and stereoselective formation of an alkene is the Wittig reaction of aldehydes and ketones with phosphoranes (Fig. 12.24) (Maercker, 1965; Maryanoff and Reitz, 1989; Vedejs and Patterson, 1994), although the reaction may be sluggish if the carbonyl group in the substrate is sterically crowded or if the ylide carries too many substituents.

Stereoselectivity is often poor, but it is possible to increase the proportion of either the E or Z isomer through the appropriate choice of reaction conditions (Schlosser, 1970), or by more extensive intervention (see below). To obtain Z

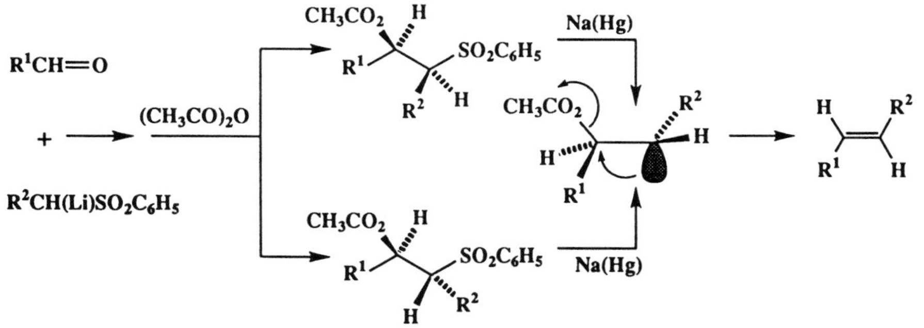

Figure 12.23.   Addition of a sulfone carbanion to an aldehyde with subsequent reduction to an E alkene.

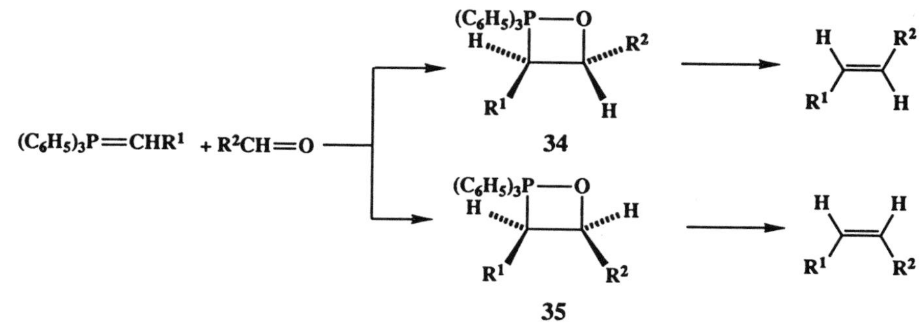

Figure 12.24. The Wittig reaction.

alkenes from aldehydes, for example, the reaction must be carried out under "salt free" conditions (this is readily achieved by the treatment of a suspension of the phosphonium salt in THF with potassium hydride) with an "unstabilized" triphenylphosphine-derived ylide, R^1—CH=P(C_6H_5)$_3$ (Gosney and Rowley, 1979; Bestmann and Zimmerman, 1982). If R^1 = RCO or ROCO, the major product will have the E configuration, while if R^1 is an aryl group, mixtures will result. Ylides derived from phosphinates, $(R^1)_2P(O^-)$=CHR, or trialkyl phosphines (e.g., R^1CH=PEt$_3$) favor E isomers also (Schlosser and Schaub, 1982). Useful examples of ortho-substituted aryl phosphines have been reported (Schaub, Schlosser, et al. 1986; Jeganathan, Schlosser, et al., 1990; Vedejs and Marth, 1988; Vedejs and Patterson, 1994).

In the absence of stabilizing substituents, an effective way of producing E isomers is provided by the method of Schlosser and Christmann (1967) (Fig. 12.25). Lithium salts are added to the reaction mixture and are presumed to stabilize the diastereomeric betaines, which are then deprotonated by butyl lithium, allowing an equilibrium to be established, with the threo isomer **36** predominating. On protonation with *tert*-butyl alcohol and reactivation with potassium *tert*-butoxide (the latter is desirable but not essential), almost pure E alkene is formed.

The mechanism of the Wittig reaction has been the subject of continuing controversy (Johnson, 1966), but recent low-temperature ^{31}P and ^{1}H NMR data support the view that for ylides derived from triphenyl phosphine and that are not

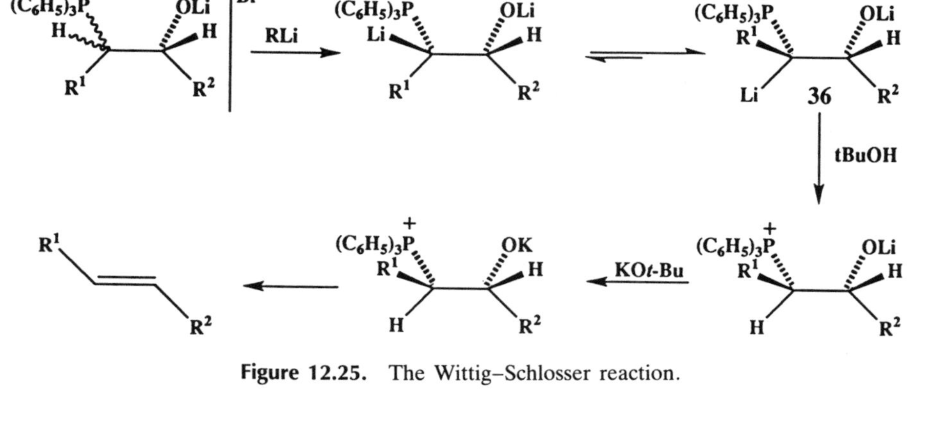

Figure 12.25. The Wittig–Schlosser reaction.

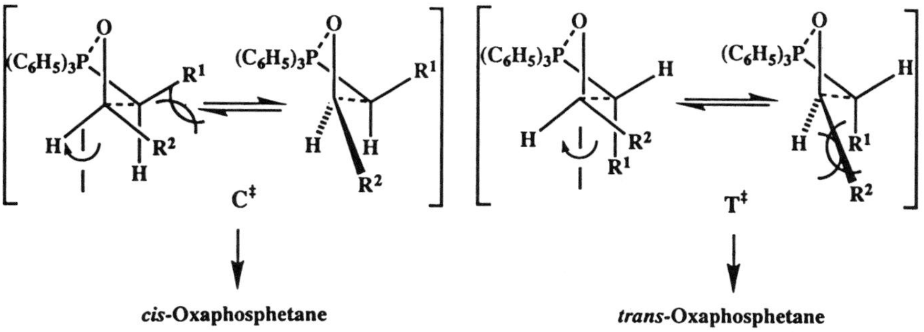

cis-Oxaphosphetane   trans-Oxaphosphetane

Figure 12.26. Transition state structures for the Wittig reaction.

stabilized by electron-withdrawing groups, rapid formation of diastereomeric 1,2-oxaphosphetanes **34** and **35** takes place in the absence of lithium halides (or in dipolar aprotic solvents like dimethyl sulfoxide (DMSO) which coordinate strongly to any lithium cations that may have been formed in the preparation of the ylide). The cis isomer **35** predominates in this mixture as does the Z diastereomer in the alkene mixture that is then formed by extrusion of phosphine oxide (Fig. 12.24).

It is then necessary to explain why formation of the *cis*-oxaphosphetane is favored over the trans isomer and why the proportion of the Z alkene formed from aliphatic aldehydes increases as the steric demand of the alkyl group increases. A rationale has been provided by Vedejs and is based on [2 + 2] cycloadditions with "criss-crossed" transition state structures $C^{\ddagger}$ and $T^{\ddagger}$ in which the plane of the partially rehybridized carbonyl group is tilted towards the ylide (Fig. 12.26). As R^2 becomes larger, nonbonded interactions with the ylide may be relieved by further rotation around the C—O axis in $C^{\ddagger}$, but not in $T^{\ddagger}$, for which the aldehyde hydrogen begins to penetrate the space occupied by R^1 attached to the ylide (for a more detailed discussion and references to alternative hypotheses see Vedejs et al., 1981; Vedejs and Marth, 1988).

When lithium halides are present, the proportion of the E alkene increases and betaine complexes may be observed. It is not clear, however, whether these are formed by a different mechanism or from lithium halide induced opening of the oxaphosphetanes. It is likely that there is a multiplicity of reaction pathways that intervene according to the structure of the reactants, cosolutes, solvent, concentration, and temperature. Moreover, any analysis based on the ratio of stereoisomeric products could be misleading, since it has been shown that the oxaphosphetanes may undergo cis/trans isomerization at rates that are comparable to their conversion to alkenes (Maryanoff et al., 1985, 1986). Further variations on the Wittig reaction, some of which are based on phosphonates, phosphonamides, and sulfinamides have been summarized by Reucroft and Sammes (1971).

Stereoselective Formation of Alkenes from Pericyclic Reactions

Concerted pericyclic reactions are symmetry controlled and in many cases proceed stereospecifically. Where more than one stereochemical outcome is allowed,

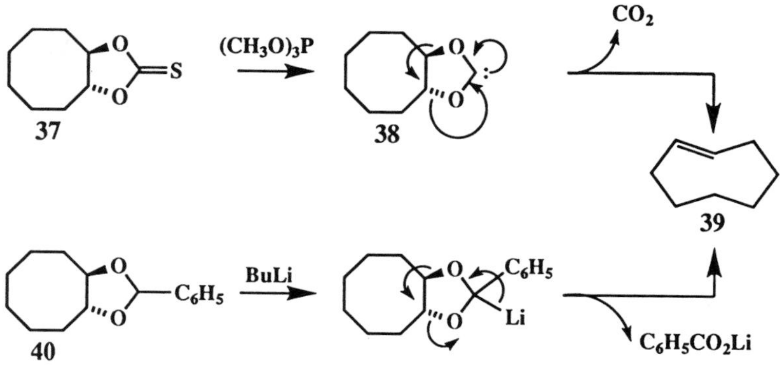

Figure 12.27. Extrusion reactions to form (E)-cyclooctene.

the highly ordered transition states usually ensure a high level of stereoselectivity (Woodward and Hoffmann, 1970; Fleming, 1976).

Cheletropic Reactions. Extrusion processes, that is, cheletropic reactions, proceed stereospecifically to afford alkene products, the geometries of which precisely reflect the configuration of the initial substrates. Thus, treatment of the thionocarbonate **37** with trimethyl phosphite affords carbene **38** and thence (E)-cyclooctene **39** (Fig. 12.27; Corey and Winter, 1963). (E)-Cyclooctene has also been formed by extrusion of benzoate ion from acetal **40** following deprotonation by *n*-butyl lithium (Huisgen, 1963).

N-Nitrosation of aziridines (Carlson and Lee, 1969) and 3-pyrrolines (Lemal and McGregor, 1966) leads to stereospecific deamination (Fig. 12.28).

Sigmatropic Rearrangements. Sigmatropic rearrangements have been used extensively for the stereoselective preparation of alkenes. The Claisen rearrangement of allylic vinyl ethers and related substances has received most attention and a preferred transition state geometry based on a chairlike conformation is normally assumed (Fig. 12.29). Greater than 99% yields of E alkenes are obtained when R^1 is a substituent other than hydrogen. When $R^1 = H$ it has been possible to control the E–Z geometry through the use of bulky methyldiaryloxy-

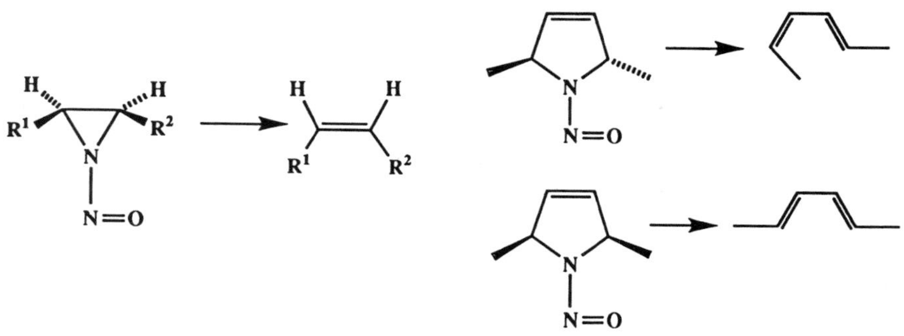

Figure 12.28. Extrusion reactions of *N*-nitroso cyclic amines.

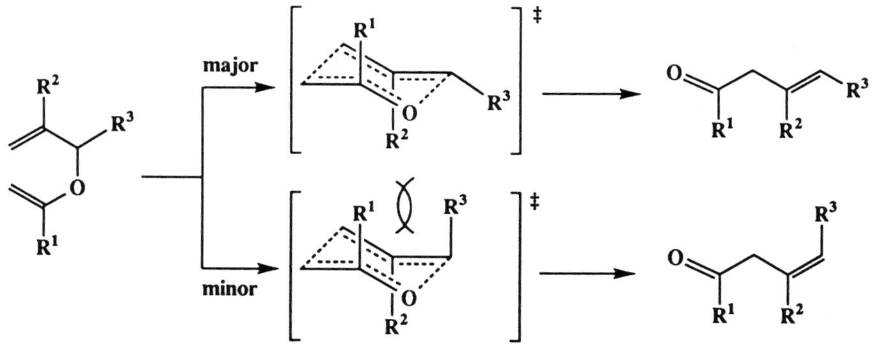

Figure 12.29. The Claisen rearrangement.

aluminum catalysts (Mauroka, Yamamoto, et al., 1988d; Nonoshita, Yamamoto, et al., 1990). A systematic study has led to the development of a semiquantitative predictive model based on the free energy differences between the axial and the corresponding equatorial substituents in a cyclohexane ring (Faulkner and Peterson, 1969; cf. Section 12-3.g).

Synthetic applications abound (Bennett, 1977; Ziegler, 1977 and 1988; Hill, 1984), with the synthesis of squalene **33** outlined in Figure 12.30 (Johnson et al., 1970) providing a spectacular example of the power of the methodology. Further examples of the Claisen rearrangement as well as other sigmatropic rearrangements are addressed in Section 12-3.g.

Electrocyclic Reactions. Electrocyclic reactions have been studied more for mechanistic interest than for their synthetic utility. The general rules for predicting the stereochemistry of products from either ring closures or ring openings (the latter are obviously more pertinent to the present discussion) are very easily derived from the nodal properties of di- and polyene systems: the thermal reaction will be disrotatory for $4n + 2$ ($n = 0, 1, 2, 3, \ldots$) participating π electrons and conrotatory for $4n$ π electrons. In the first excited state the relationships are reversed. There will be two conrotatory and two disrotatory modes for any

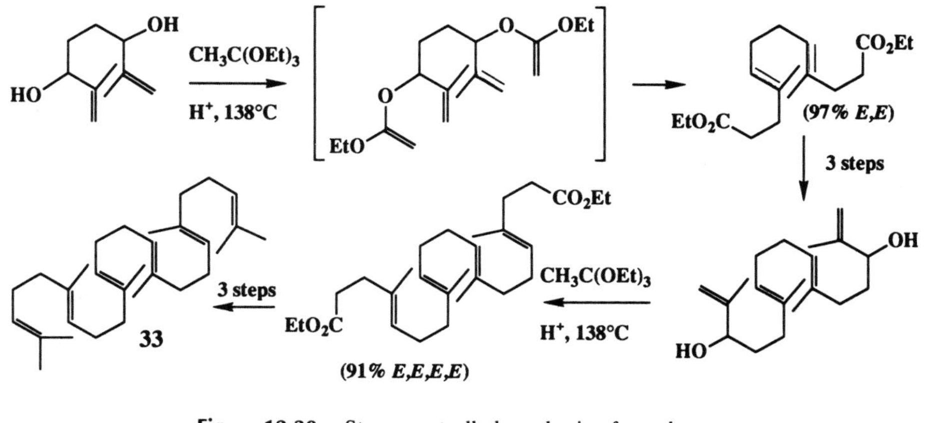

Figure 12.30. Stereocontrolled synthesis of squalene.

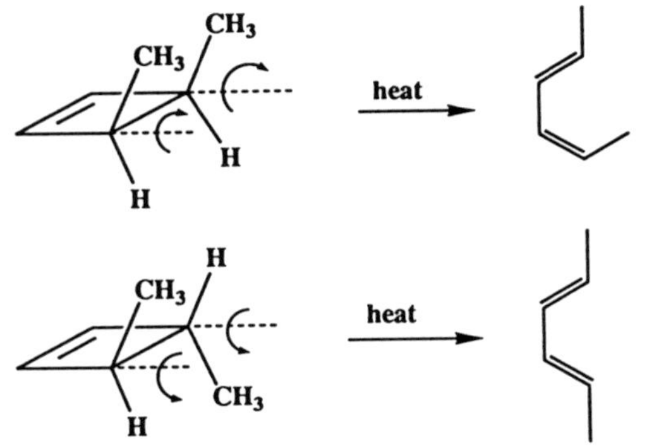

Figure 12.31. Symmetry controlled ring opening of cyclobutenes.

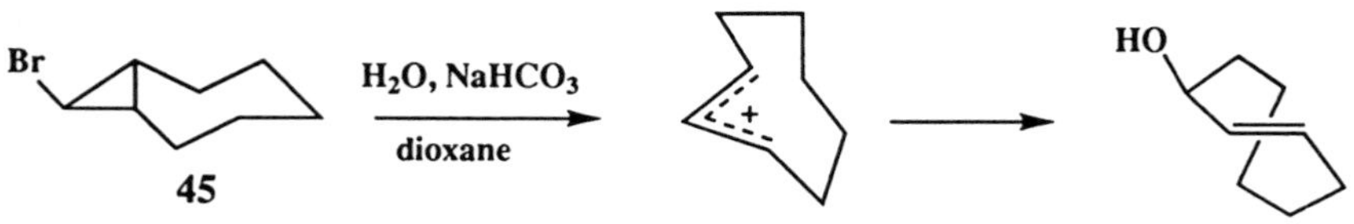

Figure 12.32. Stereocontrolled synthesis of (1*SR*,2*RS*)-(*E*)-cyclooct-2-en-1-ol.

molecule, but it is usually possible to distinguish between each pair on steric grounds. Thus, *cis*-3,4-dimethylcyclobutene **41** furnishes (*E,Z*)-1,4-dimethyl-butadiene **42**, while the trans-isomer **43** forms the (*E,E*)-isomer **44** (Fig. 12.31; Winter, 1965). An elegant confirmation of theory is provided by the synthesis of *rac*-(*E*)-cyclooct-2-en-1-ol **46** from the solvolysis of *exo*-8-bromobicyclo-[5.1.0]octane **45** (Fig. 12.32; Whitham and Wright, 1971). Not only was the *E* alkene formed with greater than 99.5% selectivity, but also only one of the two possible diastereomeric carbinols was formed as well.

Formation of Alkenes from Ring Fragmentations

Treatment of cyclopropyl carbinols with hydrogen bromide (Julia et al., 1960) or the corresponding bromides with zinc bromide (Johnson et al., 1968) provides a useful route to *E*-homoallylic bromides. High levels of stereoselectivity are obtained only with secondary carbinols, however, an outcome that can be rationalized in terms of Newman projections of the two competing transition states **47** and **48** (Fig. 12.33). That is, there will only be a clearcut preference for **47** when $R^1 = H$ and $R^2 = alkyl$.

Fragmentation of larger rings is also possible, provided that the participating bonds have an antiperiplanar relationship (Grob and Schiess, 1967), and this allows access to medium sized cyclic alkenes, the configurations of which are related to the original framework of σ bonds. A range of examples is illustrated in Figure 12.34.

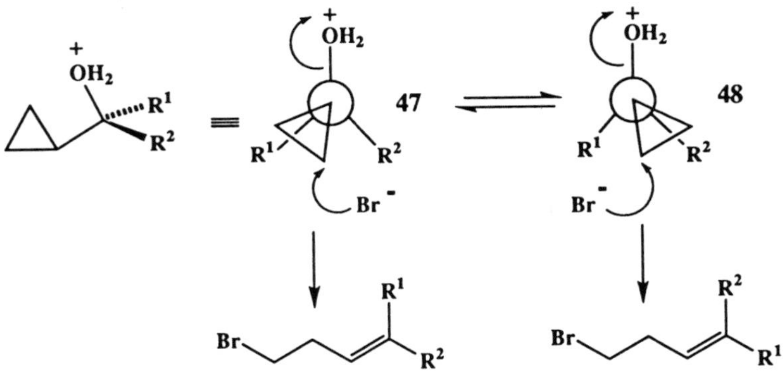

Figure 12.33. Stereocontrolled synthesis of homoallylic halides from α-cyclopropyl alcohols.

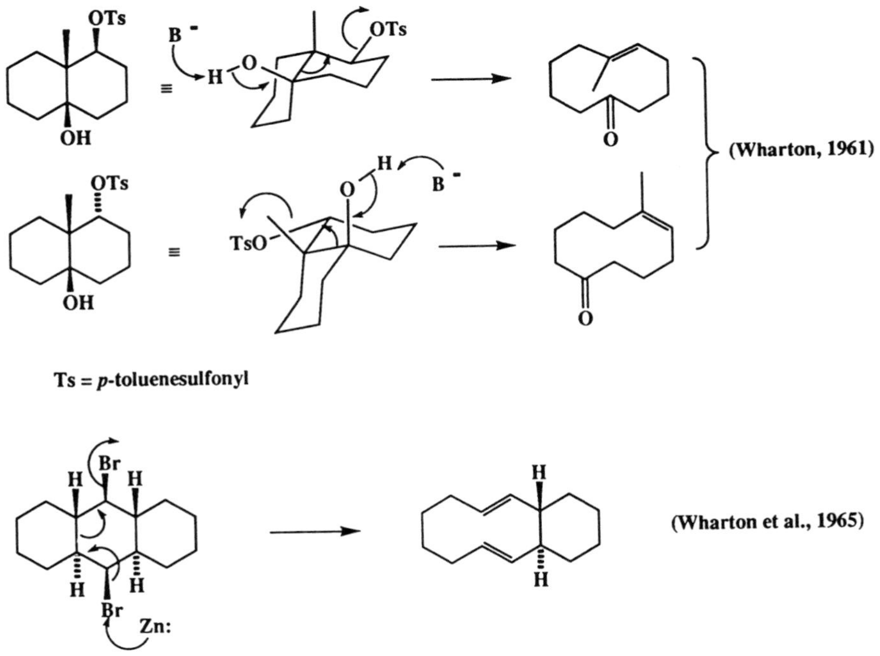

Ts = *p*-toluenesulfonyl

Figure 12.34. Examples of Grob-type fragmentations.

The strategy has also been applied to an ingenious synthesis of ketone **49**, a key precursor to the *Cecropia* juvenile hormone **50** (Fig. 12.35; Zurflüh et al., 1968).

Configurational Inversion of Alkenes

No discussion of the diastereoselective synthesis of alkenes would be complete without reference to methods for the "inversion" of alkene geometry, a subject that has already been addressed in Chapter 9. In brief, equilibrium between *Z* and *E* alkenes may be established thermally and photolytically, and by exposure to free radicals derived from, inter alia, dissociation of iodine, or disulfides. These

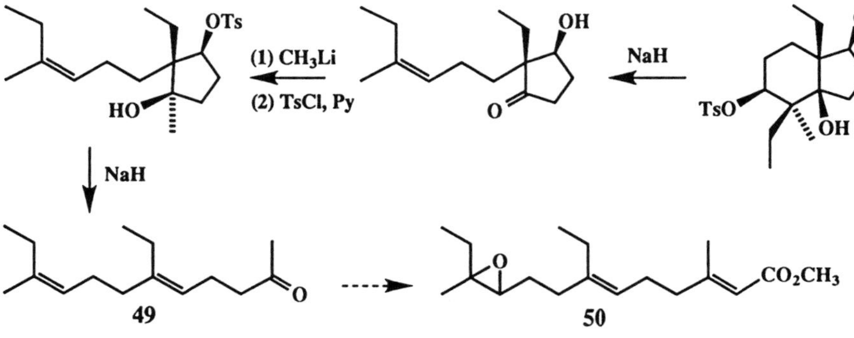

Ts = *p*-toluenesulfonyl

Figure 12.35. Stereocontrolled synthesis of the *Cecropia* juvenile hormone **50**.

conditions usually lead to mixtures of stereo- and positional isomers and naturally favor the thermodynamically more stable isomer. However, the selective photo-sensitized transfer of energy to a geometric isomer with a lower triplet energy in order to enrich the other isomer can be synthetically valuable (Hammond et al., 1964). A more secure approach to configurational inversion involves anti addition followed by syn elimination or vice versa. Alternatively, the addition and elimination steps may be both syn or both anti, but with an intervening S_N2 displacement (Sonnet, 1980; see also Section 9-3.c).

12-3. DIASTEREOSELECTIVE SYNTHESIS

a. Introduction

The more commonly encountered examples of diastereoselective reactions are those that address the task of elaborating additional stereogenic centers in a chiral substrate. The guidelines for determining the outcome are reasonably straight-forward for those substrates in which the preexisting stereogenic elements estab-lish an unambiguous steric or stereoelectronic bias to the environment of the reaction site(s) and a sufficient difference (a minimum of 2 kcal, ca. 8 kJ mol^{-1}, is desirable) between the free energies of the possible diastereomeric products (thermodynamic control) or the transition states leading to them (kinetic control). These cases involve principally cyclic or polycyclic molecules with limited conformational freedom for which a consideration of simple steric and stereoelectronic factors is normally sufficient to define conditions that will lead to a desired result. The less tractable problems are therefore those that are associated with acyclic and conformationally undefined or mobile substrates. A number of strategies designed to cope with these challenges are surveyed briefly in Section 12-3.b.

Surprisingly high levels of stereoselection may in fact be obtained with substrates of undetermined conformation. For example, in a synthesis of 3-deoxyrosaranolide (Fig. 12.36, **51**), 11 kinetically controlled reactions on a 16-membered macrolide

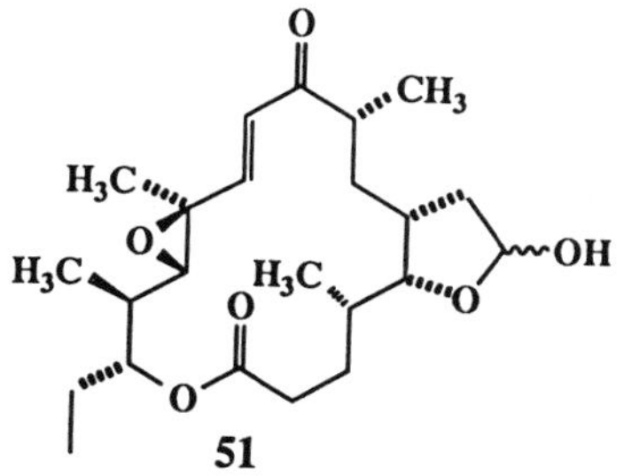

Figure 12.36. 3-Deoxyrosaranolide.

systems were used to establish 7 of the 9 stereocenters. Of these reactions, 8 afforded at least 15:1 ratios of diastereomers, while the remainder gave at least a 5:1 ratio (Still and Novack, 1984). Although the reasons for the observed outcomes were not obvious at the time, there must remain a reasonable prospect that reactions of this type will soon be placed on a rational basis.

A brief treatment of diastereoselective syntheses utilizing naturally occurring chiral substrates is given in Section 12-3.c. The rest of Section 12-3 is organized in terms of reaction types and addresses the mainstream methods for diastereoselective control in synthesis, most of which are concerned with the addition of groups to the diastereotopic faces of double bonds. Procedures that are overall enantioselective, but that involve the use of chiral auxiliaries incorporated into the substrate, will also be treated in this section. The majority of procedures fall naturally into two large categories, that is, nucleophilic (Section 12-3.d) and electrophilic reactions (Section 12.3.e), although characterization in these terms assumes that the description pertains to the reagent and not the substrate (not always a clearcut distinction). The aldol reaction (Section 12-3.f) combines aspects of the first two groupings and is worthy of individual treatment, regardless. Pericyclic processes (Section 12-3.g) afford a rich source of valuable methodology, since the orbital symmetry requirements and geometric constraints tend to lead naturally to high (if not complete) degrees of diastereoselection control. Diastereoselective hydrogenations are addressed in Section 12-3.h, while a final section (Section 12-3.i) is devoted to reactions of free radicals.

b. Strategies for Stereocontrol in Diastereoselective Synthesis

Small Ring Templates

Small rings with limited flexibility and predictable conformations provide relatively rigid templates for the elaboration of new stereogenic centers destined for incorporation into conformationally more mobile targets (Seebach et al. 1986). An illustrative example (Figure 12.37) is provided by the synthetic approach taken by Corey et al. (1978a,b) to erythronolide B **53**. Thus, the desired relative configurations of the stereogenic centers at C(2), C(3), C(4), and C(5) were first established in the carbocyclic ring of intermediate **52** by taking advantage of the thermodynamically more favorable all-trans relationship between the substituents at these centers, that is, they may be all accommodated in equatorial conformations. [The C(6) atom in **53** was elaborated by means of a Baeyer–Villiger oxidation of the 1-oxo-9-oic acid derived from **52**].

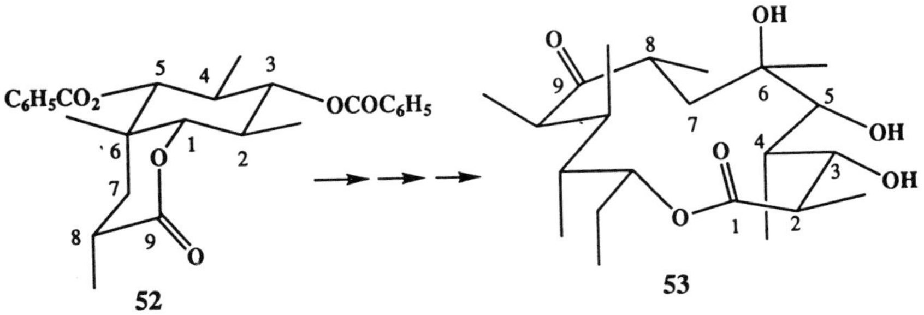

Figure 12.37. Elaboration of stereocenters C(2)–C(5) in the Corey synthesis of erythronolide B.

The more common applications, however, are associated with the construction of a rigid template that is sterically more encumbered on one face. Thus, conversion of threonine into the oxazoline ester **54** ensures that alkylation of the derived enolate occurs predominantly with reaction of the electrophile on the π face opposite to the methyl group (Fig. 12.38; Seebach and Aebi, 1983). With methyl iodide, the diastereofacial selectivity (ds) was 94%.

The expression de (diastereomer excess) is frequently used to relate the diastereomer composition of an intermediate to the "enantiomer excess" (ee) of the target compound. However, we advocate the use of the simple and unambiguous terms "% ds" (% diastereofacial selectivity) or % dr (diastereomer ratio) to define the proportion of a given diastereomer in a mixture (Seebach et al., 1986, pp. 132–133). Given contemporary emphases on the direct estimation of enantiomer composition by chromatographic and NMR methods, there is considerable advantage to be had in using a simple expression of "enantiomer composition" as a supplement to the use of enantiomer excess (Brewster, 1986) as well (cf. p. 1197).

Steric control of alkylation may also be achieved by converting the substrate into a bicyclic derivative with which steric control is exerted by the overall topology of the new system. Thus, the proline derived oxazolidinone enolate **55** undergoes alkylation predominantly on the exo face. This example also illustrates

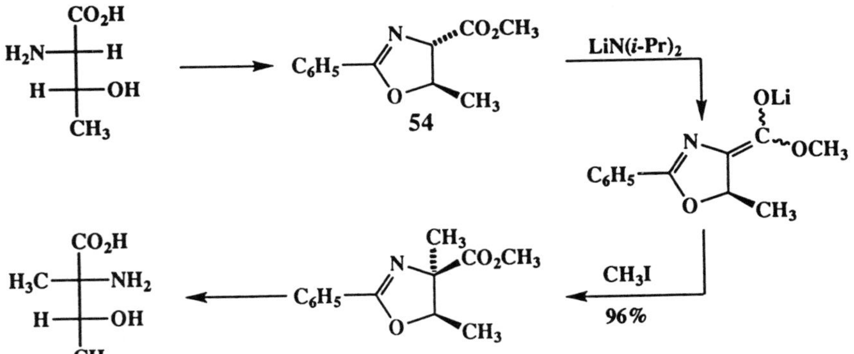

Figure 12.38. Steric control of alkylation in an oxazoline enolate.

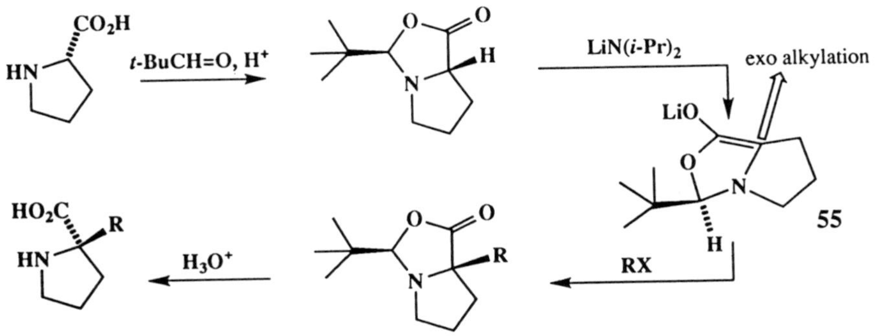

Figure 12.39. Stereoselection controlled by convex topology in an oxazolidinone.

the concept of "self-reproduction of stereogenic centers" (Seebach and Naef, 1981; Seebach et al., 1983), whereby a chiral compound containing a single stereogenic center can undergo stereocontrolled reaction at that center by forming a derivative with a second stereocenter, the configuration of which is determined by that of the original center (Fig. 12.39).

However, the factors controlling formation of the new stereocenter may be quite subtle. Alkylation of enolate **56**, for example, could, a priori, reasonably be expected to occur on the convex (exo face, but in fact affords predominantly endo products in most cases (Fig. 12.40), as does the lithium enolate derived from lactam **57** ($R^1 = CH_3$) (Meyers et al., 1984). The possibility that these results might be due to a stereoelectronic effect associated with the lone pair of the lactam nitrogen atom was explored, but rejected (Meyers and Wallace, 1989). The most promising hypothesis attributes the outcomes to shielding of the exo face by solvation (Durkin and Liotta, 1990).

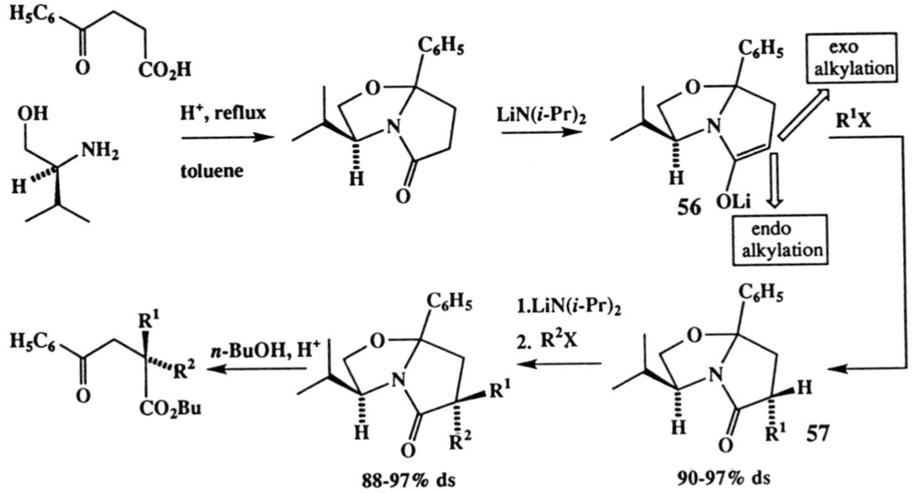

Figure 12.40. The *C* alkylation of lithium enolates derived from a bicyclic lactam.

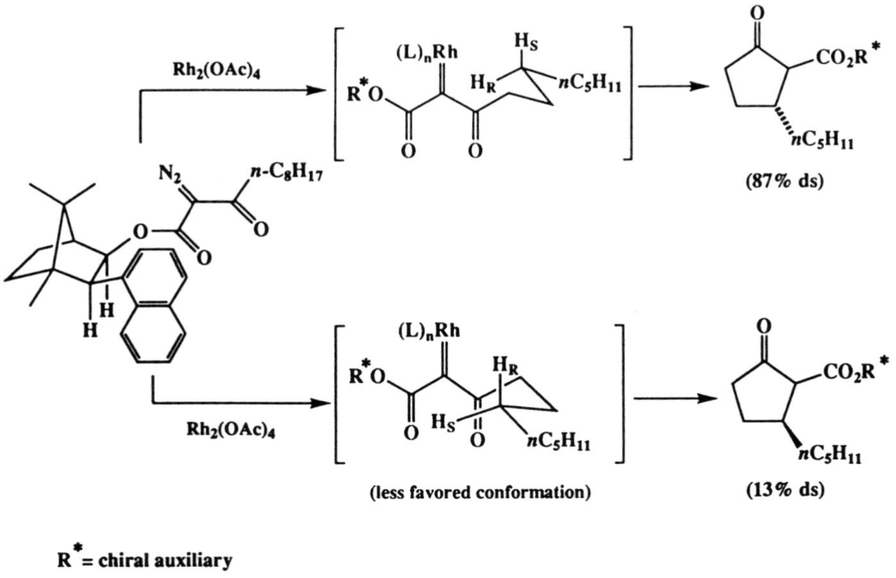

Figure 12.41. Stereoselective intramolecular CH-insertion reaction in an α-diazo-β-ketoester.

"Molecular Walls"

In this strategy for diastereoselection control, a preference between alternative transition state geometries is established through a steric barrier imposed by one or more neighboring substituents. This strategy is best understood by consideration of an example, such as the rhodium catalyzed intramolecular CH insertion reaction of the α-diazo-β-ketoester group illustrated in Figure 12.41 (Taber and Raman, 1983). It can be readily seen that the lower energy conformation is that in which the alkane moiety coils away from the naphthalene substituent in the chiral auxiliary, favoring a transition state in which insertion is into the C-(*pro-R*)-H bond. The use of disubstituted rigid bicyclic molecules in this way has been a successful and popular strategy (Oppolzer, 1987) and further examples are found later in Sections 12-3.d, 12-3.e, and 12-3.h.

Ring-Forming Reactions

The participating functional groups in intramolecular reactions will often be constrained by geometric factors to react in a stereoselective manner. It is apparent, for example, that in the key intramolecular alkylation reaction utilized in a synthesis of the sesquiterpene seychellene (Piers et al., 1969), the electrophilic group can approach only the endo face of the enolate anion **58** (Fig. 12.42).

The more interesting cases are those in which approach to both π faces is possible, however, and for which acceptable levels of stereoselection occur as a consequence of sufficiently marked differences in the free energies of the possible intermediates or products (or activation energies between possible transition states). Thus, the reversible reaction of phosphate **59** with iodine in acetonitrile affords the cis diastereomer **60** in preference to the trans isomer **61** (Bartlett and Jernstedt, 1977; Fig. 12.43).

Ts = *p*-toluenesulfonyl

Figure 12.42. Intramolecular alkylation of an enolate anion.

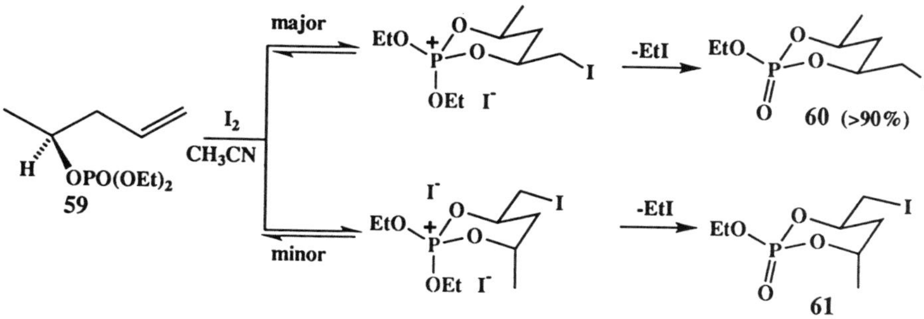

Figure 12.43. Stereoselective formation of a cyclic phosphate under thermodynamic control.

Pericyclic Reactions

The highly ordered transition states associated with pericyclic processes, combined with the restraints imposed by orbital symmetry considerations, ensure that this class of reactions affords excellent and predictable stereochemical outcomes. Cycloadditions have the potential to create up to four new stereogenic centers with predictable relationships (cf. Sections 12-3.g and i), while sigmatropic rearrangements afford, inter alia, controlled 1,3-transfers of stereochemistry in allylic substrates as illustrated in Figure 12.44 (Chan et al., 1976). To the extent that the preferred transition states for [3,3]-sigmatropic rearrangements may be

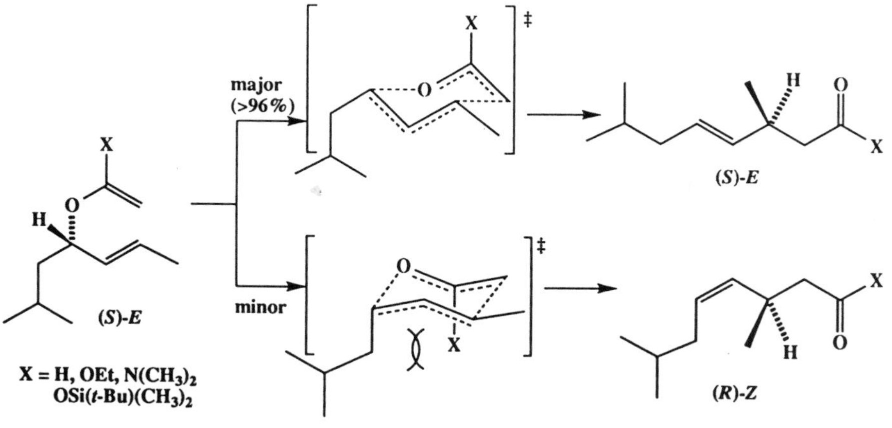

Figure 12.44. 1,3-Transfer of stereochemistry in the Claisen rearrangement.

modeled on chair conformations (cf. Section 12.2.b), the outcome is highly predictable, with good retention of stereochemical integrity. This type of reaction is especially useful for the stereoselective formation of carbon–carbon bonds at the expense of stereogenic centers associated with heteroatoms, since stereocontrol in the construction of such centers may allow a degree of flexibility that is not available in the direct construction of carbon–carbon bonds (e.g., recycling of the undesired stereoisomeric carbinol).

Coordination to Metal Centers

Metal ion chelation may play a crucial organizational role by establishing a fixed stereochemical relationship between stereogenic centers and reaction sites, often reversing the relative diastereoselectivity obtained in the absence of strongly chelating cations, for example, as in the conjugate addition of nucleophiles to the enone sulfoxide 62 (Fig. 12.45). The ground-state nonchelated conformation is at an energy minimum when the carbonyl and sulfinyl oxygen dipoles are oriented in opposite directions, rendering the lower face of the enone function as the less sterically hindered for the depicted enantiomer. In the chelated substrate, however, the upper face is less encumbered (Posner, 1985).

Chelation is especially important in the reactions of enolates and their various analogues derived from imines, hydrazones, oxazolines, and so on, the principal nucleophilic participants in a majority of carbon–carbon bond-forming reactions. The chelation not only ensures π-facial discrimination in a predictable manner, but will often impose a preferred cis or trans geometry. Reaction of the lithium enolates derived from β-hydroxy esters with electrophiles, for example, may be rationally interpreted in terms of reaction on the *Re* face of the chelated cis enolate (Fig. 12.46; Kraus and Taschner, 1977; Fráter, 1979; Seebach and Wasmuth, 1980). In contrast, simple propionate and butyrate esters preferentially

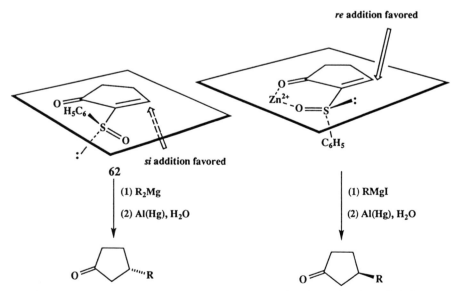

Figure 12.45. Conjugate additions to a vinyl sulfoxide with and without chelation.

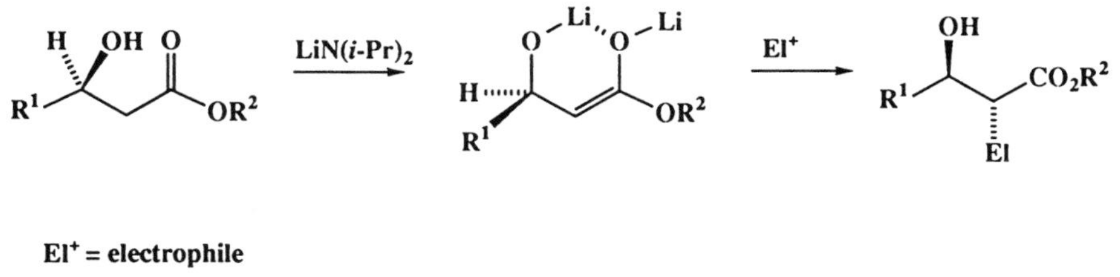

El⁺ = electrophile

Figure 12.46. Stereocontrolled electrophilic reactions of chelated β-hydroxy esters.

form the trans enolates on treatment with lithium diisopropylamide (LDA), although it should be noted that this preference is reversed in the presence of hexamethylphosphoric triamide (HMPA) (cf. Sections 12.3.e and i).

In other applications, enantiotopic substituents or faces may be rendered diastereotopic through ligation of the substrate to metal centers that are stereogenic. Chiral acyl iron complexes, such as **63**, undergo *C* alkylation with unusually high levels of diastereoselectivity. In the case illustrated (Fig. 12.47) the observed stereochemical outcome was rationalized by assuming that the intermediate enolate anion **64** that is formed has the *E* configuration and adopts the conformation in which the negatively charged oxy anion is antiperiplanar to the CO ligand (Baird, Davies, et al. 1983). Numerous more complex cases based on chiral octahedral metal complexes have been reviewed (Sargeson, 1980).

Use of π-Donor Complexes

The formation of π-complexes with transition metals has been used to render the faces of various kinds of aryl and olefinic groups as well as any attached substituents diastereotopic. Thus, H_β in the iron tricarbonyl complex **65** (Fig. 12.48) can be selectively exchanged for deuterium under acid catalysis, while addition of nucleophiles to the cationic complex **66** occurs with "lateral control" (Birch, 1982) and complete stereoselectivity on the opposite face to the metal center, thereby allowing an enantioselective synthesis of gabaculine **67** to be completed (Bandara, Birch, et al., 1984; cf. also Davies, Green, et al., 1978; Lamanna and Brookhart, 1981; Pearson, 1980, 1990; Grée, 1989).

π−π Interactions with Aromatic Substituents

The possibility of π−π interactions has often been implicated in cases of enhanced stereoselection associated with substrates that possess aromatic sub-

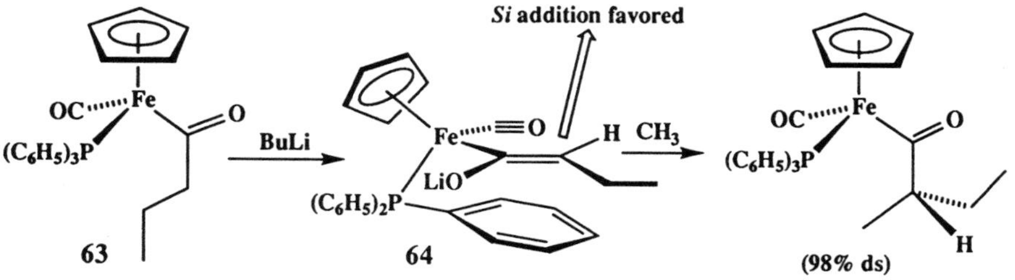

Figure 12.47. Diastereoselective alkylation of a chiral iron acyl complex.

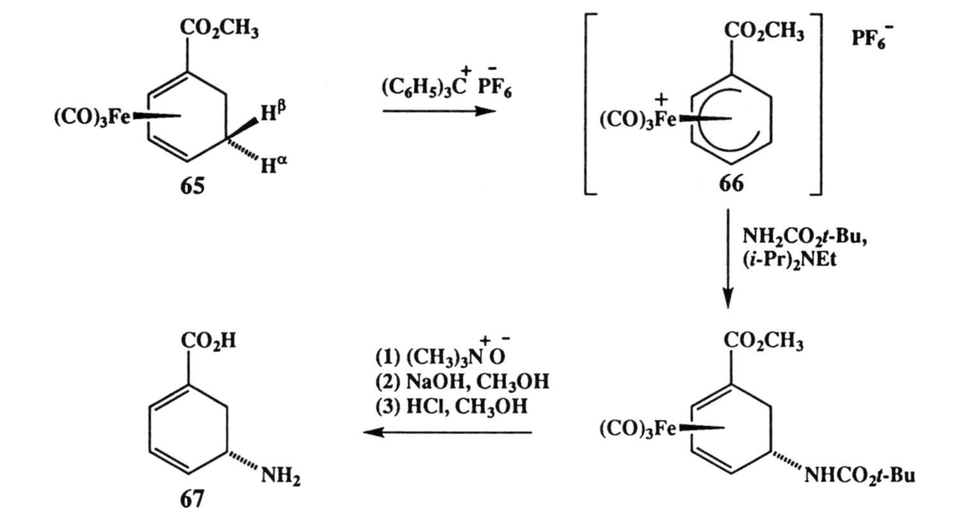

Figure 12.48. Lateral control in the stereoselective synthesis of gabaculine.

stituents within bonding distance of the reaction site (e.g., Trost et al., 1980; Whitesell et al., 1982), but in all of these examples, the precise nature of the interaction remains obscure. However, a recent study comparing the diastereoselectivities in a set of [4 + 2] cycloadditions with a corresponding set of alkylations carried out on closely related substrates (Fig. 12.49) has provided firm evidence for the probable nature of such interactions (Evans et al., 1987a, 1988a). In the series of cycloadditions **68** → **69** + **70** it was observed that for R = benzyl, the ratio **69:70** was considerably higher than for R = isopropyl or cyclohexylmethyl, clearly demonstrating that the diastereofacial bias was dependent on some factor(s) other than a simple steric one. If this had been due to charge transfer between the phenyl ring and the π bond of the electron-deficient dienophile, then a decrease in rate relative to the sterically equivalent nonaromatic analogue would have been expected, whereas an increase (ca. 3:1) was actually observed (Table 12.1). Moreover, the diastereoselectivities for a series of electronically perturbed (p-OCH$_3$, p-Cl, and p-CF$_3$) benzyl derivatives were not significantly different from the parent system.

The study was then extended to C methylation of the group of enolates **71**, which may be regarded as topologically equivalent to the Lewis acid–dienophile complex **74**, and the ratio of diastereomers **72:73** plotted against the ratio **69:70** (Fig. 12.50). The set of substituents R that cannot be involved in π bonding lie on a line (linear correlation $r = 0.997$) representing the simply steric contribution to the diastereofacial selectivities. The benzyl substituted analogues, however, form a distinct cluster with similar diastereoselectivities in the alkylation reaction to that of the sterically comparable cyclohexylmethyl group, but with dramatically enhanced diastereoselectivities in the Diels–Alder reaction. It was therefore concluded that these latter phenomena must originate from an electronic effect, and in view of the minimal variations associated with the π-donor potential of the phenyl ring, must arise from dipole–dipole and van der Waals attractions rather than a charge-transfer process (cf. Lyssikatos and Bednarski, 1990).

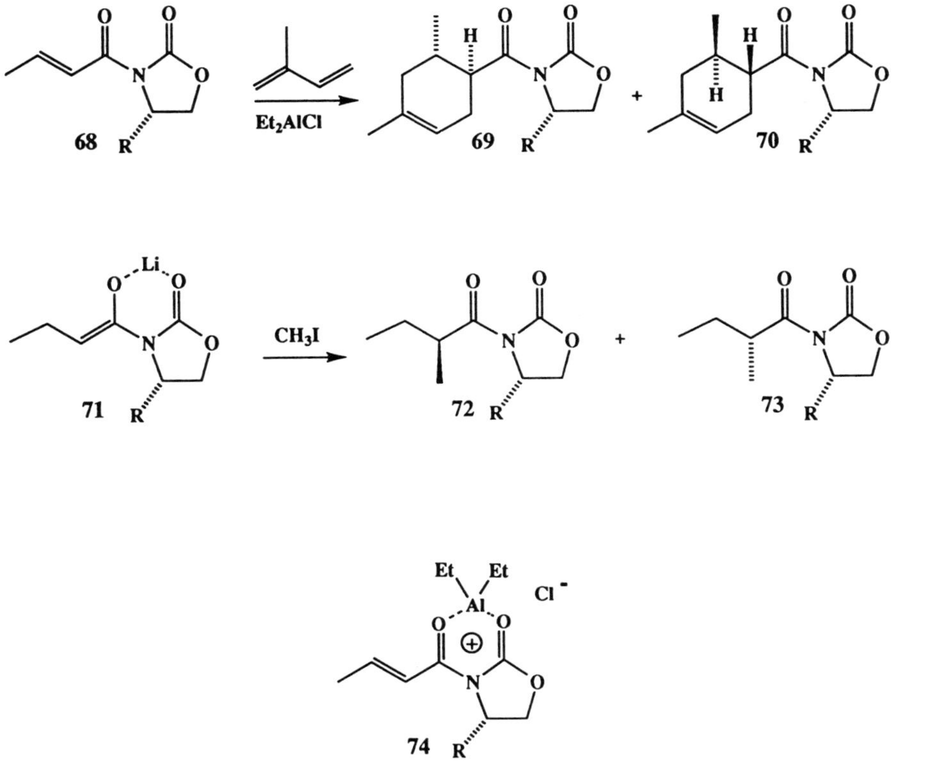

Figure 12.49. Diastereoselectivities in [4 + 2] cycloadditions and alkylations based on oxazolidinone chiral auxiliaries.

TABLE 12.1. Comparisons between the Diastereoselectivities in Alkylation and Cycloaddition Reactions (Evans et al., 1988a).

R	69:70	72:73
C_6H_5	2.06	4.24
CH_3	3.83	6.95
C_2H_5	5.50	9.21
i-Pr	5.34	9.85
$CH_2C_6H_{11}$	9.68	17.4
$CH_2C_6H_5$	20.7	16.7
$CH_2(p\text{-}OCH_3\text{—}C_6H_5)$	23.3	16.9
$CH_2(p\text{-}Cl\text{—}C_6H_5)$	22.8	15.8
$CH_2(p\text{-}CF_3\text{—}C_6H_5)$	21.0	16.8

Directed π-Facial Diastereoselectivity

Unless the respective steric environments of the diastereotopic faces of an olefinic bond are sufficiently distinct, or stereoelectronic influences impart a sufficient bias, good stereoselection can only be achieved with participation by a neighboring heteroatom (Hoveyda et al., 1993). Hydrogen bonding between epoxidizing reagents and allylic or homoallylic alcohols, for example, can lead to enhanced

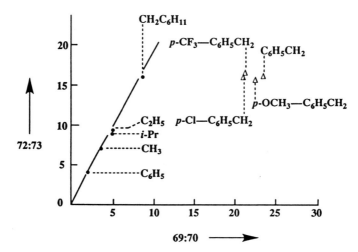

Figure 12.50. Plot of Diels–Alder diastereomer ratios versus enolate C-methylation diastereomer ratios.

rates with associated stereoselectivity (Fig. 12.51; Henbest and Wilson, 1957; Chamberlain et al., 1970).

Ligation between a polar group and a reagent (organometal, transition metal catalyst, or complex metalohydride) has similarly been exploited to effect directed reductions of carbonyl groups and conjugate additions to activated alkenes (Section 12-3.d), epoxidations, cyclopropanations and hydrosilations (Section 12-3.e), and hydrogenations (Section 12-3.h). As in the earlier category of chelation control, the enhanced stereoselection arises from the temporary formation of ring systems that lead to more highly organized reaction complexes with rates of reaction that are greater than those of competing, more direct intermolecular processes.

Chiral Auxiliaries

One of the most reliable strategies for achieving an enantioselective synthesis is the temporary incorporation of a "chiral auxiliary", which is ultimately cleaved following one or more diastereoselective steps, as illustrated in Figure 12.52 (Davies, 1989). The range of examples is as diverse as the subject of stereoselective synthesis itself. Applications of the use of auxiliaries incorporated into the

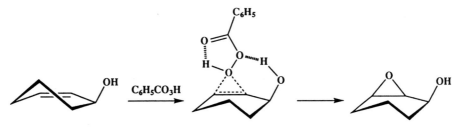

Figure 12.51. Epoxidation of an allylic alcohol directed by hydrogen bonding.

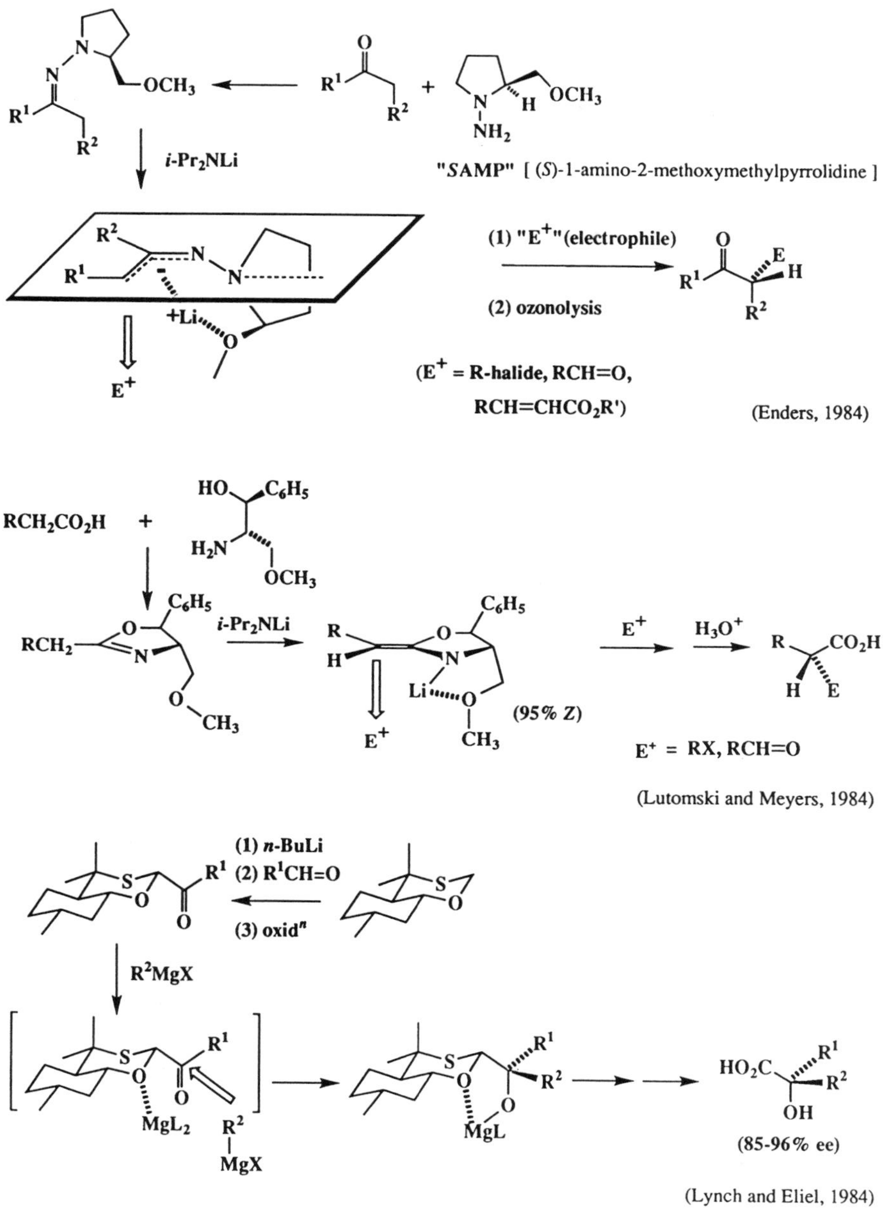

i-Pr2NLi
"SAMP" [(S)-1-amino-2-methoxymethylpyrrolidine]
(1) "E+" (electrophile)
(2) ozonolysis
(E+ = R-halide, RCH=O, RCH=CHCO2R')
(Enders, 1984)
RCH2CO2H +
i-Pr2NLi
(95% Z)
E+ = RX, RCH=O
(Lutomski and Meyers, 1984)
(1) n-BuLi
(2) R1CH=O
(3) oxidn
R2MgX
(85-96% ee)
(Lynch and Eliel, 1984)

Figure 12.52. A selection of examples of enantioselective syntheses utilizing chiral auxiliaries.

substrate are illustrated elsewhere in this section and have been the subject of numerous reviews. The applications include:

- Acetals and miscellaneous heterocycles (Seebach et al., 1986; e.g., Figs. 12.38, 12.39, 12.86, 12.87, 12.100 and 12.102);
- Boronic esters (Matteson, 1989);
- Camphor derivatives (Oppolzer, 1987; e.g., Figs. 12.41, 12.82, and 12.101);

- Dienes and dienophiles for [4 + 2] cycloadditions (Masamune et al., 1983, Paquette, 1984; Helmchen et al., 1986; e.g. Figs. 12.81, 12.132, 12.133, 12.137, and 12.139);
- Enamines and imines (Bergbreiter and Newcomb, 1983; Seebach et al., 1986; e.g., Fig. 12.79);
- Hydrazones (Enders, 1984, 1985; e.g., Fig. 12.52);
- Oxathianes (Eliel, 1983; e.g. Fig. 12.52);
- Oxazolidinones (Evans, 1982; Evans et al., 1981c; Evans et al., 1982c; e.g. Figs. 12.49, 12.127, 12.136, and 12.137);
- Oxazolines (Lutomski and Meyers, 1984); e.g. Figs. 12.40, 12.52, and 12.79);
- Sulfoxides (Posner, 1983, 1985; e.g. Fig. 12.45).

Achiral Auxiliaries

The concept of attaching a sterically demanding ligand to a reagent with a view to improving stereoselectivity is a well-established tactic in synthetic chemistry. Less well developed is the substitution of a group in a substrate for the same purpose, for examples, $(ArS)_3C$— for a methyl group (Stork and Rychnovsky, 1987) or R_3Si for a hydrogen atom (cf. the example illustrated later in Fig. 12.77). Clearly, there are greater limitations when dealing with the substrate rather than the reagent and it is essential that the subsequent conversion to the desired substituent is sufficiently simple and efficient to warrant the more elaborate approach. A further development of this strategem may be to confer chirality on an otherwise achiral intermediate, as illustrated by the enantioselective syntheses of the carbinols 76 and 77 through the utilization of the silyl alcohol 75 as a chiral equivalent of crotyl alcohol and employing the silyl ketene acetal version of the Claisen rearrangement (Fig. 12.53; Ireland and Varney, 1984). Either syn (threo) or anti (erythro) products could be obtained, depending on the stereochemistry of the ester enolate, which could be controlled by the appropriate choice of reagent (for a brief outline of the syn–anti convention see Section 12.3.f).

Intraannular and Extraannular Stereocontrol

In the preceding discussion the important role of cyclic structures and transition states in providing greater control during diastereoselection processes has been repeatedly demonstrated. An especially important aspect of these strategies, however, is the issue of whether the stereogenic elements that provide the basis of the selectivity are "intraannular" (i.e., are contained within the cyclic array), or are "extraannular" (i.e., an absence of cyclic constraints). Evans (1984) classified chiral enolates and their analogues in terms of these two categories, plus a third one: chelate enforced intraannular control (Fig. 12.54).

It can be readily appreciated that in pericyclic reactions, intraannular stereocontrol will also be more effective and predictable (cf. Fig. 12.53). However, the directing stereocenter must be sacrificed as a consequence of rehybridi-

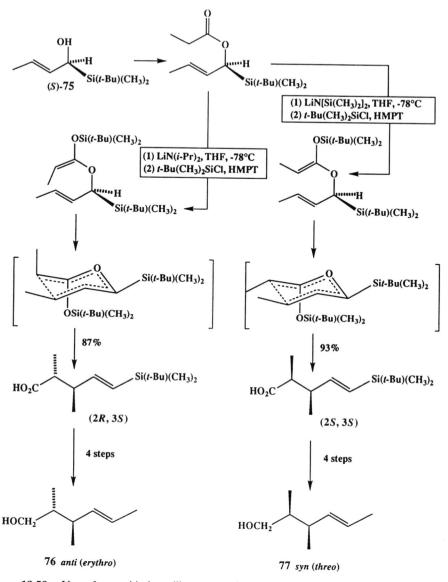

Figure 12.53. Use of an achiral auxiliary to confer chirality on a substrate in the Claisen rearrangement.

zation. Moreover, in some reactions it will be impossible to have a stereogenic center within the ring being formed, as with cycloadditions. The Claisen rearrangement illustrated in Figure 12.55 (Kallmerten and Gould, 1986) demonstrates the potential for extraannular control in these situations (the good levels of diastereoselectivity were attributed to $\pi - \pi$ interaction between the two phenyl rings) and even higher levels of diastereoselectivity have been obtained with the Diels–Alder reaction (cf. Section 12.3.g).

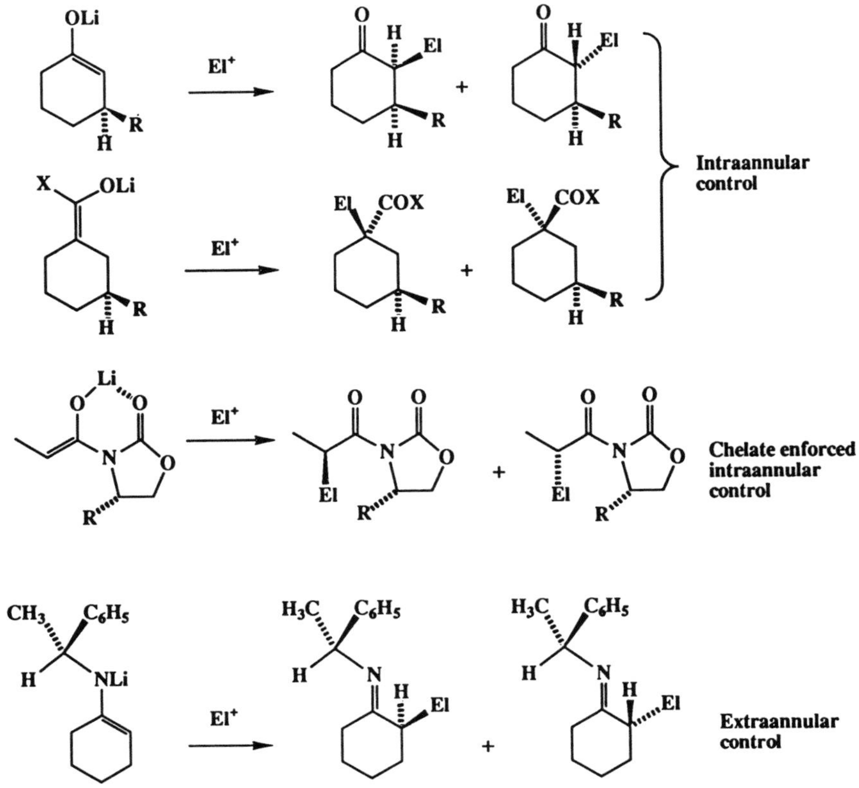

Figure 12.54. Examples of intraannular and extraannular stereocontrolled alkylations (El^+ = electrophile).

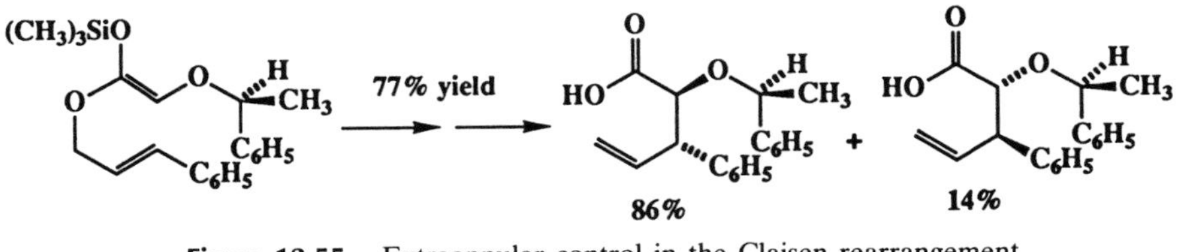

Figure 12.55. Extraannular control in the Claisen rearrangement.

c. Diastereoselective Syntheses Based on Chiral Substrates of Natural Origin

The utilization of chiral substrates obtained from natural sources as substrates and as a source of chiral molecular fragments for incorporation into target structures is a long standing and fruitful practice in the field of organic synthesis. It has nevertheless attained new heights of sophistication over the past two decades or so as a consequence of, inter alia, contributions from Fraser-Reid, Hanessian, Ireland, and their respective co-workers.

The identification (Fig. 12.56) of the naturally occurring terpenoids (−)-carvone **79** and (+)-citronellol **81** as largely intact molecular fragments in the

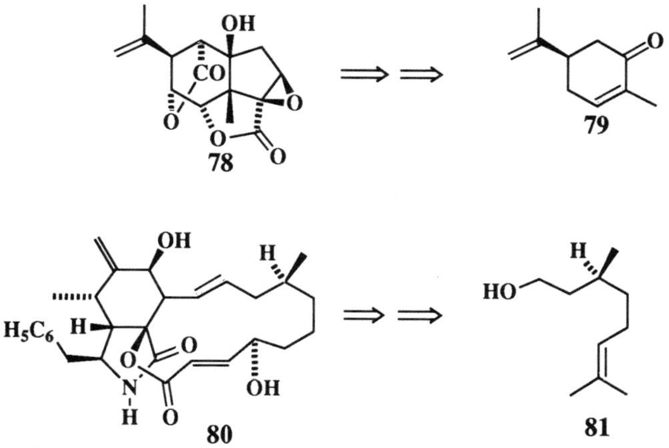

Figure 12.56. Identification of simple terpenoid precursor fragments in the synthesis of complex natural products.

synthetic objectives picrotoxin **78** (Corey and Pearce, 1979) and cytochalasin C **80** (Stork et al., 1978), respectively, is conceptually elementary and requires little more than a good knowledge of the appropriate literature or access to a suitable data base. L-Phenylalanine and (S)-malic acid were also incorporated into **80**. The recognition that application of the second-order Beckmann rearrangement on **83** derived from (−)-camphor **82** might allow access to the steroid intermediate **84** (Fig. 12.57); Stevens and Gaeta, 1977) requires considerably more insight, however.

Hydroxy acids and amino acids have also been valuable sources of chiral "building blocks" (Seebach and Hungerbühler, 1980; Maurer, Takahata and Rapoport, 1984; Coppola and Schuster, 1987; Reetz, 1991), but the most extensive studies have been concerned with the use of sugars, particularly for polyketide derivatives (Vasella, 1980; Hanessian, 1983; Fraser-Reid et al., 1985; Fleet, 1989).

For example, by redrawing the structure of erythronolide A **85** as **89** it may be perceived that the C(1)–C(6) fragment could be obtained from the sugar **87** and the C(9)–C(13) fragment from sugar **88** (Figure 12.58). Once having established this conceptual connection, it is then possible to contemplate the preparation of both these compounds from the common precursor **86**, which may be obtained from D-(+)-glucose (Hanessian and Rancourt, 1977).

The preparation of the precursor **98** to the right-hand one-half of lasalocid A **99** (Fig. 12.59) provides an even more sophisticated example (Ireland et al., 1983). The pivotal step was the enolate based Claisen rearrangement of ester **96** to **97** (cf. Fig. 12.53), thereby uniting the dihydropyran **93** with furan **95**, which

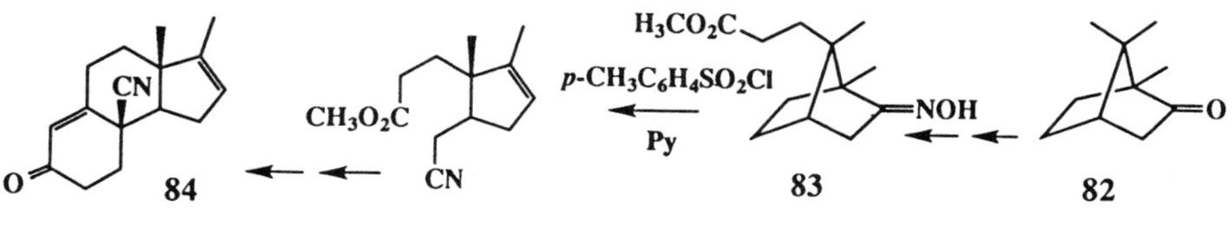

Figure 12.57. Utilization of camphor as a substrate for steroid synthesis.

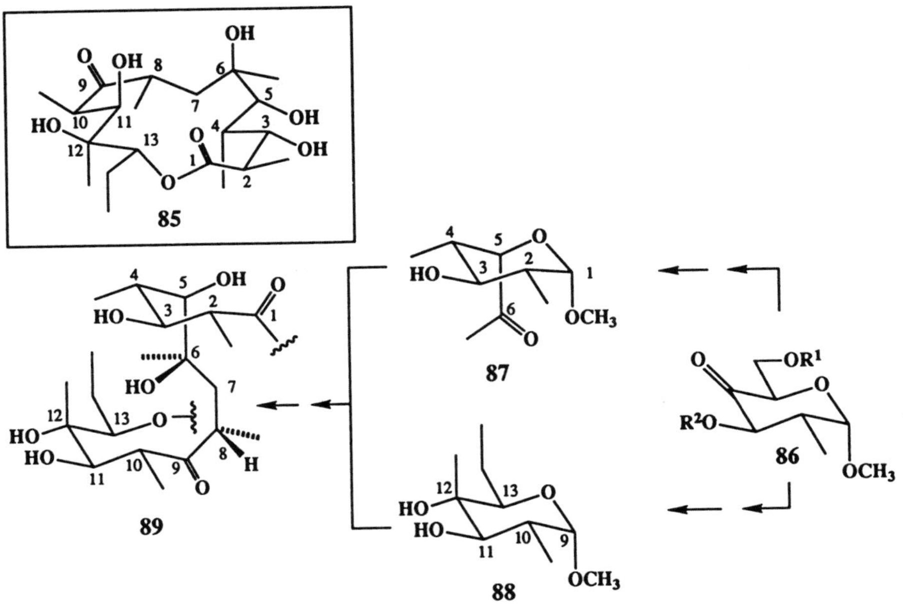

Figure 12.58. Synthesis of erythronolide A **85** from D-glucose.

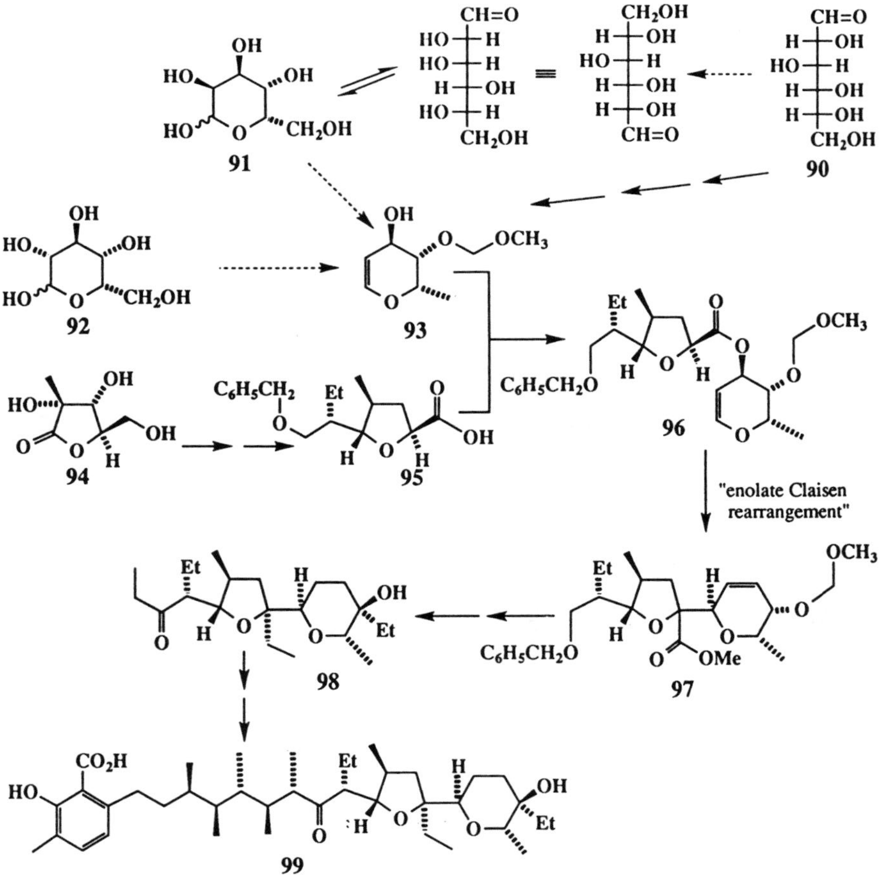

Figure 12.59. Synthesis of the right-hand half of lasalocid A from carbohydrate precursors.

had been obtained by means of a 12 step sequence beginning with the base catalyzed rearrangement of the readily available sugar D-fructose to the "α-saccharinolactone" **94** (Ireland and Wilcox, 1980). The construction of **93** appears at first sight to require access to a rare L-aldohexose, that is, either L-(+)-gulose **91** or L-(+)-idose **92**. However, the realization that the formal swapping of the functional groups at C(1) and C(6) in the open form of D-(+)-glucose **90** would convert it into L-(+)-gulose **91**, allows **90** to be utilized instead as the precursor of **93**. In practice, the preparation of gulose itself was bypassed and the carboxaldehyde function in D-glucose **90** was converted directly into the methyl group corresponding to that in **93**. The recognition of symmetry-based relationships like that between glucose and gulose (cf. Section 5-5.c) has been an important aspect of many other syntheses based on carbohydrates.

Although the treatment above has been confined to the use of naturally occurring chiral molecules, the same principles obviously apply to equivalent materials, which are becoming increasingly available from industrial sources (Scott, 1989).

d. Nucleophilic Additions

This section is concerned with the mainstream synthetic processes involving diastereoselective π-facial additions of nucleophiles to chiral aldehydes, ketones, and alkenes activated by electron-withdrawing groups, largely under kinetic control. For treatment of the more specialized topic of stereoselective nucleophilic substitutions at sp^3 centers, readers are referred to several excellent reviews describing such reactions with O,O and N,O acetals (Seebach et al., 1986) and nucleophilic displacements on allylic compounds (Magid, 1980; Denmark and Marble, 1990).

The search for an understanding of the underlying reasons for π-facial selectivity in the addition of nucleophiles to double bonds, especially the carbonyl group, has generated more models and hypotheses than any other subject in the field of stereoselective synthesis. The proposals, which are not necessarily mutually exclusive, can be summarized as follows:

- Steric repulsion (Prelog, 1953).
- Preference for a staggered conformation in the transition state, that is, minimization of torsional strain (Schleyer, 1967; Chérest, Felkin, and Prudent, 1968; Wu, Houk, and Trost, 1987).
- Electrostatic model (Kahn and Hehre, 1986, 1987; Kahn, Hehre et al., 1986, 1987).
- Dissymmetric π-electron clouds (Inagaki, Fujimoto, and Fukui, 1976; Eisenstein, Klein, and Lefour 1979; Burgess and Liotta, 1981; Burgess et al., 1984).
- Bent-bond or tau-bond model (Vogel, Eschenmoser, et al., 1987; Winter, 1987).
- Preferential attack antiperiplanar to the best electronic donor (Cieplak, 1981; Srivasta and le Noble, 1987; Johnson et al., 1987; Laube and Stilz, 1987).

- Preferential attack antiperiplanar to the best electronic acceptor (Anh and Eisenstein, 1977; Anh, 1980; Kümin, Dunitz, Eschenmoser, et al., 1980; Corey and Boaz, 1985a).
- Principle of least motion (Tee et al., 1974).
- Stereoelectronic control and smallest change in conformation (Toromanoff, 1980).
- Alkene pyramidalization (Houk, 1983).

Additions of Nucleophiles to Aldehydes and Acyclic Ketones

Several models have been developed to predict or rationalize the stereochemical outcome from the kinetically controlled addition of a nucleophile to the carbonyl group of chiral aldehydes and ketones in which a stereogenic center is adjacent to the carbonyl group. These models have been outlined earlier in Section 5-5.g. The Felkin model (Chérest et al., 1968; as refined by Anh and Eisenstein, 1977) appears to give the most satisfactory match with theoretical considerations (Wu and Houk, 1987) and is now the most widely accepted hypothesis. Felkin postulated that torsional strain (Pitzer strain) involving partially formed bonds represents a substantial fraction of the strain between fully formed bonds, even when the degree of bonding in the transition state is quite low. This hypothesis led to the assumption of a preferred transition state based on a conformation in which the bulkiest of the α ligands (L) takes up a perpendicular relationship to the plane of the carbonyl group anti to the incoming nucleophile, and the sterically next most demanding α substituent (M) is placed gauche to the carbonyl function (Fig. 12.60). Supporting experimental evidence was obtained from LiAlH$_4$ reductions on two series of ketones:

1. For R—CO—C*(H)(CH$_3$)cyclohexyl, where R = CH$_3$, Et, i-Pr, and t-Bu.
2. For R—CO—C*(H)(CH$_3$)C$_6$H$_5$, where R = CH$_3$, Et, i-Pr, and t-Bu.

Diastereoselectivity increased as R became bulkier, with the exception of R = t-Bu in the first series, for which nonbonded interactions between the cyclohexyl and t-Bu groups could reasonably be expected to destabilize the otherwise preferred conformation for the transition state (for a more detailed treatment see Eliel, 1983).

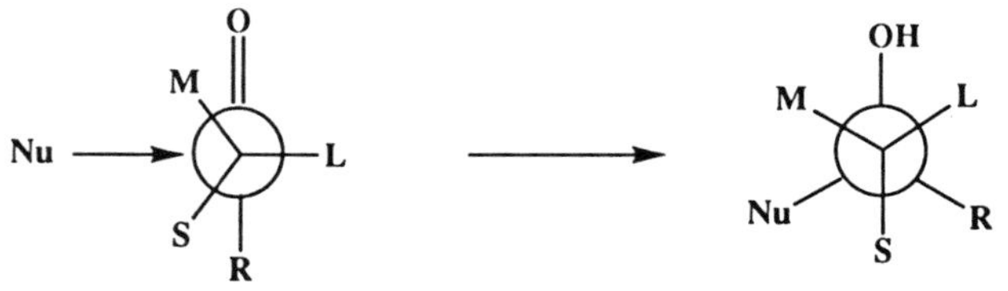

Nu = nucleophile

Figure 12.60 Felkin model for addition of nucleophiles to the carbonyl group in acyclic compounds (L > M > S in order of steric demand).

An evaluation of these models was made by means of ab initio calculations (STO-3G basis set) of hypothetical transition state structures (Anh and Eisenstein, 1977) and it was found that the conformations assumed by Felkin were energetically the most favored, not only for alkyl substituted systems, but also for α-chloro ones as well. In these latter cases the heteroatom takes the place of the large ligand L, thereby minimizing Coulombic repulsion between the electronegative halogen and the incoming nucleophile.

Two further important points were made. By invoking the Bürgi–Dunitz trajectory, that is, approach at an angle of about 109° with respect to the plane of the carbonyl group, (Fig. 12.61), for the incoming nucleophile (Bürgi, Dunitz, et al., 1973; Bürgi, Lehn, et al., 1974; Bürgi et al., 1974), it is no longer necessary to make the somewhat dubious assumption that the medium sized group should be gauche to the oxygen atom in the preferred transition state. Rather, it may be assumed that the two conformations **A** and **B** (Fig. 12.61) are energetically similar, and that differentiation arises out of nonbonded interactions between the nucleophile with either the S or M substituents.

Validation of the refined model arises from a consideration of the molecular orbitals involved, mainly the two-electron stabilizing interaction between the highest occupied molecular orbital (HOMO) of the nucleophile and the lowest unoccupied molecular orbital (LUMO) of the carbonyl group. Although overlap would be maximized by a perpendicular approach, there are destabilizing interactions arising from the out-of-phase overlap with the oxygen atom and from the four-electron interaction with the HOMO (π_{CO}) of the substrate. However, these are reduced by opening up the angle of attack, corresponding to the Bürgi–Dunitz trajectory. A further important aspect raised in this study is the stabilization of the π^* orbital of the carbonyl group (LUMO) by overlap with the σ^* antibonding orbital of the antiperiplanar α substituent, as well as the stabilizing effect of the $n-\sigma^*$ interaction with the electron pair of the nucleophile (Fig. 12.62).

In a very recent study on the reaction of the lithium enolate derived from pinacolone with a series of aldehydes, $RCH(OCH_3)CHO$ and $RCH(C_6H_5)CHO$ (Lodge and Heathcock, 1987a) the authors found that when the bulk of the R group increased beyond a certain point, it was necessary to explain the observed

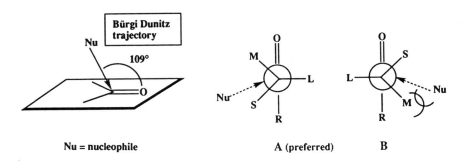

Figure 12.61. The Felkin–Anh model for addition of nucleophiles to the carbonyl group.

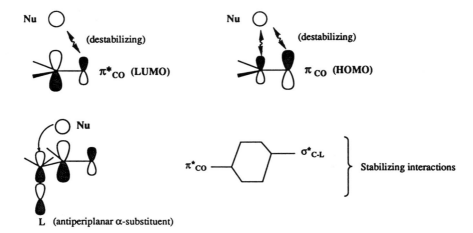

Figure 12.62. Orbital interactions in the Felkin–Anh model for the approach of nucleophiles to the carbonyl group.

trend by considering a second pair of competing reaction pathways. For example, in the four-conformer equilibrium between **W**, **X**, **Y**, and **Z** (Fig. 12.63), reaction may be considered to proceed predominantly through conformer **W** when R = CH$_3$ and Et (*note*: **W** is equivalent to **A** in Fig. 12.61), since this is favored by both the bulk of the phenyl group and the lower σ^* orbital energy of the sp^2—sp^3 bond. When R = *i*-Pr, however, reaction based on conformation **Z** becomes more important, and eventually dominates when R = *t*-Bu, thereby resulting in a reversal of diastereoselectivity.

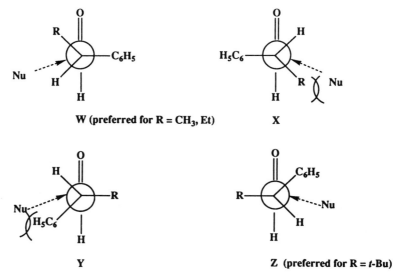

Figure 12.63. Consideration of a four-conformer model for addition of nucleophiles to aldehydes.

Anh and Eisenstein (1977) found that when their calculations were reworked to take into account the effect of coordination of the carbonyl group with a cation (H^+ and Li^+), the outcome was essentially unchanged. Complexation with the cation would be expected to increase the rate of reaction, but not to alter the geometries of the favored transition states. However, it has been suggested (Heathcock and Flippin, 1983) that for the addition of enolate anions to aldehydes, the nucleophile follows a trajectory that is closer to the H substituent than the stereogenic group R* (structure P, Fig. 12.64), thereby attenuating its stereoregulating influence. The very much higher levels of diastereoselection found for the Lewis acid mediated reactions of enol silanes with aldehydes were accordingly rationalized by assuming that the steric demand of the coordinated Lewis acid steered the nucleophile back towards the stereogenic R* substituent (structure Q), thereby accentuating the difference in activation energy between the diastereomeric transition states (cf. Lodge and Heathcock, 1987b; Mori, Bartlett, and Heathcock, 1987).

When an especially bulky Lewis acid is used in conjunction with the nucleophile, a reversal of the normal stereochemical outcome (Maruoka, Yamamoto, et al., 1985) is effected. The most effective reagent was found to be methylaluminum bis(2,6-di-*tert*-butyl-4-methylphenoxide) and the 2,4,6-tri-*tert*-butyl analogue. The transition state structure depicted in Figure 12.65 was invoked to explain the outcome. Parallel results were obtained with cyclic ketones (see below).

By contemporary standards, unsatisfactory levels of diastereoselection are frequently obtained from the addition of nucleophiles to simple aldehydes and ketones, but very much better results are obtained, as might be expected, with substrates possessing polar groups (e.g., OH, OR, or NR_2) proximal to the carbonyl function that allow the formation of stable chelates (Fig. 12.66; cf. Section 12.3.b). When these polar groups equate with the medium size substituents in the open-chain model (Fig. 12.60), the relative stereochemical outcome is the same (Cram and Abd Elhafez, 1952; Eliel, 1983). Otherwise, the complementary diastereomer (the inappropriately called "anti-Cram" product) predominates (Pohland and Sullivan, 1955; Cram and Kopecky, 1959; Angiolini et al., 1969; Andrisano et al., 1970).

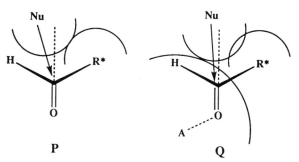

Figure 12.64. Alteration of approach trajectory by Lewis acid complexation.

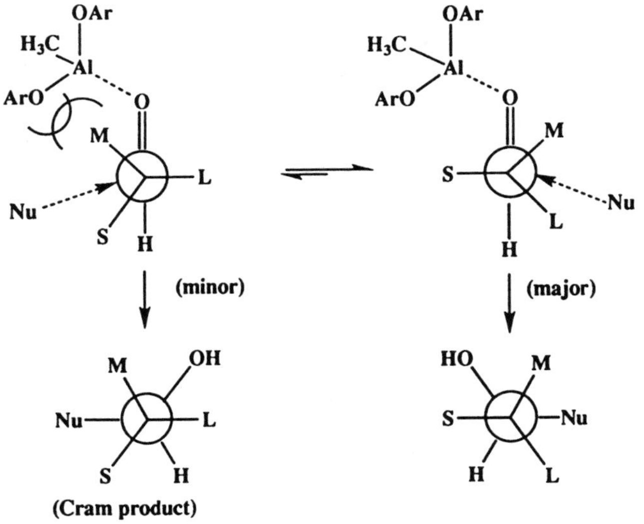

Ar = **2,6-di-*t*-butyl-4-methylphenyl or 2,4,6-tri-*t*-butylphenyl**

Figure 12.65 Addition of nucleophiles to carbonyl groups complexed with bulky Lewis acids.

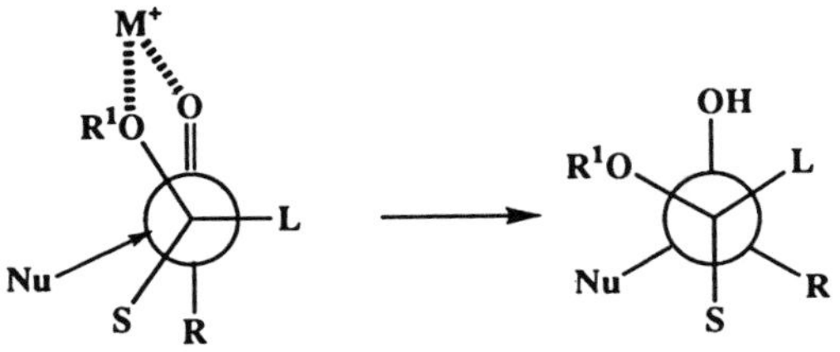

Figure 12.66. Addition of nucleophiles to chelated carbonyl compounds.

Addition of Nucleophiles to Cyclic Ketones

The conformational uncertainties associated with acyclic ketones are considerably reduced when the carbonyl group is contained within a ring. Moreover, as noted earlier, the two π faces of the carbonyl group may be rendered operationally diastereotopic simply as a consequence of puckering, rather than by the proximity of one or more stereogenic centers (cf. Section 12.2.a). Even so, the theoretical basis for the stereochemical outcomes in the irreversible addition of nucleophiles to compounds like the "conformationally locked" cyclohexanones has been the basis of controversy for several decades (Wigfield, 1979; Boone and Ashby, 1979; see also pp. 734–737). There appears to be an intrinsic preference for nucleophiles with a relatively small bulk (e.g., complex metal hydrides, acetylides, and lithiated acetonitrile) to approach along what appears to be the more hindered trajectory, that is, corresponding to "axial" addition. This tendency may be overwhelmed by steric influences arising from nonbonded interactions between the incoming nucleophile and the 3- and 5- axial substituents (cf. Fig. 12.67), however. Early attempts to explain the predominant formation of the more stable equatorial alcohol from the addition of sterically undemanding hydrides or

Axial approach

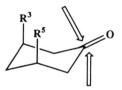

Equatorial approach

Figure 12.67. Comparison of axial and equatorial approaches of nucleophiles to cyclohexanones.

organometallic reagents to unhindered cyclohexanones rested on the rather vague concept of "product development control" involving a productlike transition state (Dauben et al., 1956), which then leads to the preferential formation of the more stable diastereomer.

Felkin was easily able to adapt his model for open-chain carbonyl compounds to the cyclic analogues, however (Chérest and Felkin, 1968). *Both* steric and torsional strain can be minimized when the substrate is acyclic, but it is necessary in cyclohexanones to balance these two factors. Formation of the axial alcohol (equatorial attack) requires a partially eclipsed transition state (Fig. 12.68, $E^{\ddagger}$) involving torsional strain between the nucleophile and the axial α-hydrogen substituents, whereas formation of the equatorial alcohol leads to a staggered transition state $A^{\ddagger}$, which generates steric strain between the nucleophile and the β-axial substituents. Thus, sterically demanding nucleophiles and/or bulky axial β substituents increase the energy of $A^{\ddagger}$ so that formation of the axial alcohols becomes favored.

In an extension of his acyclic model, Anh invoked the importance of the "antiperiplanarity (app) effect". If the portion of the ring containing the carbonyl group is flattened, the axial α-hydrogen substituents take up a conformation that is perpendicular to the plane of the carbonyl group, thereby maximizing the $n-\sigma^*$

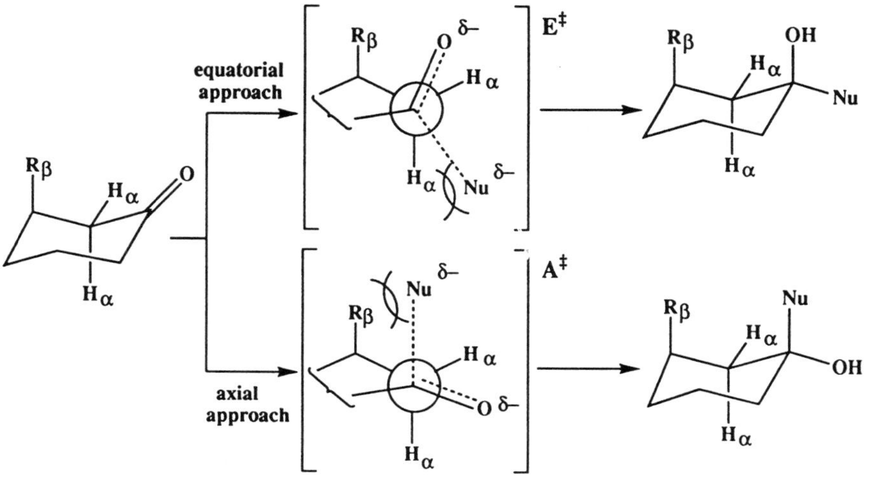

Figure 12.68. Felkin transition states for addition of a nucleophile to a cyclohexanone.

Figure 12.69. Comparison between hydride reduction of flexible and rigid cyclohexanone derivatives.

interaction between the electron pair of the nucleophile and the antibonding orbital of the axial α-hydrogen atoms (app effect); the Bürgi-Dunitz trajectory is angled away from the axial β substituents (Anh, 1980). There appears to be a good correlation between the preference for the formation of equatorial alcohols and the flexibility of the substrate ketones (Fig. 12.69). Reduction of the more flexible bicyclic ketone **100** by LiAlH$_4$, for example, affords greater than 90% of the 2β-alcohol (i.e., predominantly axial attack), whereas the more rigid steroidal ketone **101** is converted into a small excess of the 7α-epimer (equatorial approach).

In one study designed to provide experimental verification of the theoretical estimates of the magnitude of the app effect, the rigid bridged biaryl ketone **102a** and its various α-methyl derivatives **102–e** were subjected to acid-catalyzed exchange with H$_2^{18}$O (Fig. 12.70; Fraser and Stanciulescu, 1987). Evidence was adduced that the n–σ^* interaction between the nucleophile water and an app C—CH$_3$ bond provided a stabilizing energy that was at least 1.9 kcal mol^{-1} greater than that from an app C—H bond.

Yet another hypothesis has been proposed (Cieplak, 1981) in which it is assumed that the carbonyl group has undergone extensive pyramidalization in the transition state, and that the outcome is primarily a consequence of interactions between the vicinal occupied σ orbitals and the σ^* orbital of the incipient bond

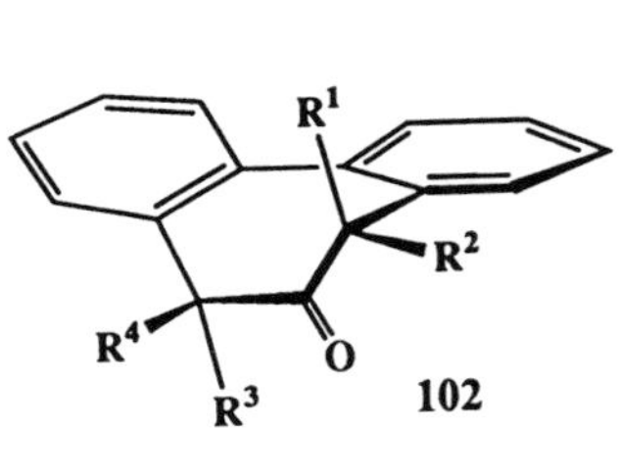

Compound	R^1	R^2	R^3	R^4
102a	H	H	H	H
102b	CH$_3$	H	H	H
102c	H	CH$_3$	H	H
102d	CH$_3$	H	CH$_3$	H
102e	H	CH$_3$	H	CH$_3$

Figure 12.70. Ketones employed in the experimental estimation of the app effect of C—CH$_3$ and C—H.

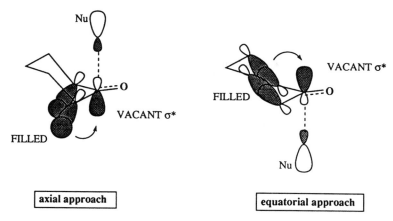

Figure 12.71. Participating orbitals in the Cieplak model.

between the nucleophile and the carbonyl group (Fig. 12.71). On this basis, the preference for axial approach was considered to be due to more effective electron donation from the α—CH bonds relative to that from the C(2)—C(3) and C(5)—C(6) bonds, even though the general view is that a carbon–carbon bond is a better electron–donor (Rozeboom and Houk, 1982; Lodge and Heathcock, 1987a).

The Cieplak hypothesis has nevertheless been invoked to explain the stereochemical trends observed in the addition of hydrides and methyl lithium to 7-norbornanones (Mehta and Khan, 1990) and in the reduction of 5-substituted 2-adamantanones (Cheung, le Noble, et al. 1986). Thus, the depletion of electron density in the C(1)—C(9) and C(3)—(4) bonds by halogens or phenyl groups bearing para electron-withdrawing substituents favors the formation of the (E)-(anti) alcohols **103**. In contrast, the (Z)-(syn) isomers **104** are the major products from the electron-rich p-aminophenyl and p-hydroxyphenyl analogues (Fig. 12.72).

An especially interesting example of hydride approach to what appears to be the more hindered face of a carbonyl group occurs in the reduction of the α-alkyl derivatives of the bicyclooctanones **105** and **106** by LiAlH$_4$ (Fig. 12.73; Brienne et al., 1974). For both series, the proportion of cis alcohol in the product mixture decreased as the bulk of R was increased from methyl to ethyl to isopropyl. Only when R was *tert*-butyl or phenyl was there a predominance of the cis product.

A possible explanation was found in the reaction of LiAl(O*t*-Bu)$_3$H with ketone **105**. Although the bulkier reagent now preferentially approached the less

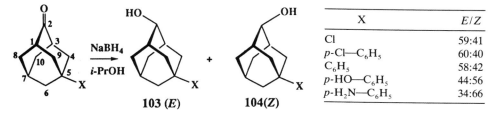

X	E/Z
Cl	59:41
p-Cl—C$_6$H$_5$	60:40
C$_6$H$_5$	58:42
p-HO—C$_6$H$_5$	44:56
p-H$_2$N—C$_6$H$_5$	34:66

Figure 12.72. Hydride reduction of 5-substituted 2-adamantanones.

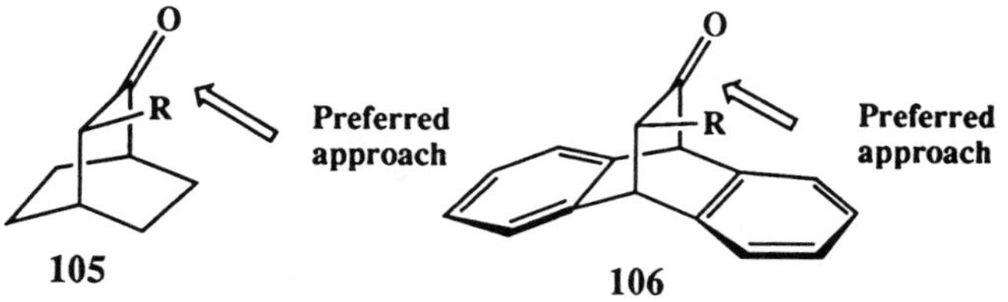

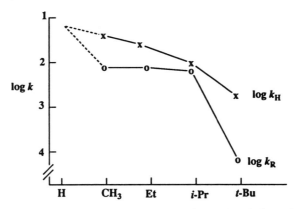

Figure 12.73. Lithium aluminum hydride reduction of bicyclo[2.2.2]octanone derivatives **105** and **106**.

hindered face, the rate of reaction for addition on the same face as the alkyl group remained constant as the substituent was changed from CH_3 to Et to *i*-Pr, while the rate of addition to the *opposite* face fell progressively (Chérest, Felkin, et al., 1977; Fig. 12.74).

This result was explained in terms of an "anisotropy effect" associated with the deactivating inductive effect of the electron-releasing alkyl groups. The anisotropy argument appears to be a variant on similar proposals made by other groups who have suggested stereoselectivity is a result of distortion of the carbonyl π and π^* orbitals; that is, unequal distribution of electron density on the two π faces of the carbonyl group (Klein, 1973, 1974; Burgess et al., 1984). Thus, if one considers the respective interactions between the symmetrical C(2)–C(3) and C(5)–C(6) σ_S and σ_S^* orbitals with the HOMO (π_{CO}) and LUMO (π_{CO}^*) of the carbonyl group in cyclohexanone (Fig. 12.75), it can be seen that the out-of-phase $\sigma - \pi$ interaction induces a higher electron density on the "equatorial face" of the carbonyl group. This hypothesis is consistent with the observation that cyclohexylidene derivatives react preferentially with electrophiles on this face (cf. Section 12-3.e). The $\sigma^* - \pi^*$ interaction, however, leads to higher electron density on the "axial face" of the carbon of the carbonyl group.

Recent work from the Houk group achieved an excellent match between theoretical predictions and experimental results for a wide variety of substrates. In the first of two reports (Wu and Houk, 1987), ab initio molecular orbital calculations (3-21G and 6-31 basis sets) of the activation energies for the addition

Figure 12.74. Plot of the rates of reaction for the syn (k_R) and anti (k_H) addition of hydride to bicyclo[2.2.2]octanone **105** as a function of R (adapted with permission from Chérest et al., 1977).

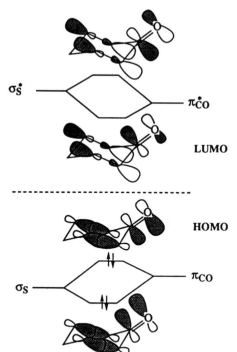

σ_S^* — — π_{CO}^*

LUMO

HOMO

σ_S — — π_{CO}

Figure 12.75. Secondary orbital interactions in cyclohexanone.

of sodium hydride (NaH) to propanal, relative to those for ethanal were completed. Three conformations were considered for the propanal + NaH transition state structure: $\mathbf{B}^{\ddagger}$ with an "inside" methyl group coplanar and syn to the carbonyl group, $\mathbf{C}^{\ddagger}$ with an "outside" methyl group coplanar to the carbonyl group, and $\mathbf{D}^{\ddagger}$ with the methyl group anti to the incoming hydride ion (Fig. 12.76).

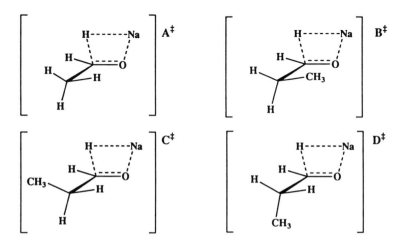

Figure 12.76. Transition state structures for the addition of hydride to ethanal and propanal.

Two further sets of calculations were then conducted based on the "removal" of sodium ion and then NaH itself. It was found that transition state structure $D^{\ddagger}$ was disfavored in all cases, while $B^{\ddagger}$ was favored over $C^{\ddagger}$, presumably for steric reasons. It was reasoned that because the anti-methyl group is a better electron donor than a CH bond, it destabilizes the electron-rich transition state. In the Felkin–Anh model when the α-stereogenic center is secondary, the largest group is anti to the approaching nucleophile. This arises because the two alkyl groups cannot simultaneously achieve their preferred dihedral angles. Thus, the best compromise is for the larger alkyl group to be anti.

When the energetic effects embodied in transition state structures $A^{\ddagger}$–$D^{\ddagger}$ were implemented into the Allinger molecular mechanics force field program (MM2) (Burkert and Allinger, 1982), it was possible to predict ratios of diastereomeric carbinols with a good match with those that have been produced experimentally from the reduction of an extensive range of cyclic and acyclic ketones by LiAlH$_4$. The model differs from the Felkin hypothesis with respect to the minor pathway, however, in that the groups M and L interchange places, rather than S and M (cf. Fig. 12.61). Further experimental evidence based on the reduction of some benzocycloheptenones provided further support for the new computational model (Mukherjee, Houk, et al., 1988), although the validity of using this model system has been questioned (Lin, le Noble, et al, 1988).

Contemporary Guidelines for the Control of Diastereoselection in the Addition of Nucleophiles to Aldehydes and Ketones

Regardless of the theoretical situation, it is not difficult to arrive at a fairly simple set of guidelines that should allow the achievement of a desired stereochemical objective. For acyclic substrates, the most reliable results will be obtained when a neighboring polar group can participate in chelation to provide a more highly organized transition state geometry. Otherwise, if there is a good steric hierarchy of groups connected to a stereogenic center that is attached to the carbonyl group of a ketone and the other flanking substituent has a reasonable steric demand (e.g., phenyl), which ensures a good differentiation between the major and minor Felkin–Anh transition state structures (Fig. 12.61), then acceptable levels of diastereoselection should be obtained. Conversely, the prospects for obtaining good results with aldehydes are generally limited (although the aldol reaction under suitable conditions provides a welcome exception (cf. Section 12-3.f). A constructive solution to this problem has been to employ a trimethylsilyl group as an achiral auxiliary in place of the hydrogen atom attached to the carbonyl group (cf. Section 12-3.b; Nakada, Ohno, et al., 1988). Thus, adduct **108** was obtained with greater than 99% diastereoselectivity from the addition of n-butyl lithium to acylsilane **107**. Protiodesilylation proceeded with retention of configuration to afford **109** in undiminished stereochemical yield (Fig. 12.77).

As noted earlier for unhindered cyclic ketones, there is a preference for nucleophiles with a relatively small steric demand, for example, complex metal hydrides, acetylides, and lithiated acetonitrile (Trost et al., 1987) to approach along what often appears to be the more hindered axial trajectory. Thus, in 2-alkylcyclopentanones and 2- and 4-alkylcyclohexanones, the trans (equatorial) carbinols are the major products from reductions with lithium aluminum hydride.

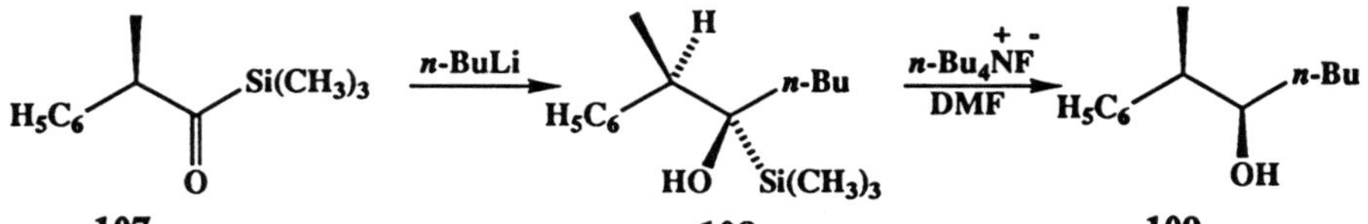

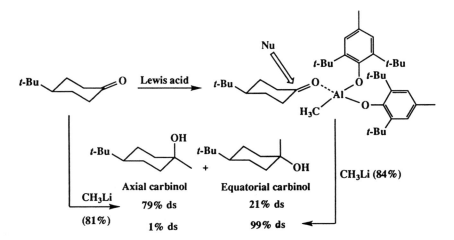

Figure 12.77. Stereocontrolled addition of nucleophiles to acylsilanes.

This tendency is reversed by utilizing sterically demanding reagents, for example, lithium trisiamyl borohydride, which affords greater than 99% of the cis (axial) isomers (Krishnamurthy and Brown, 1976). Steric influences arising from non-bonded interactions between bulky 3- and 5-axial substituents in cyclohexanones and an incoming nucleophile also favor the formation of cis products, regardless of the bulk of the reagent (cf. Fig. 12.67).

If a high yield of an equatorial carbinol is required from either a hindered or non-hindered ketone, it may be best to resort to the Meerwein–Ponndorf–Verley reduction that proceeds under equilibrium conditions (Wilds, 1944) or a reducing metal system, such as lithium and ethanol in liquid ammonia (House, 1972). Samarium(II) iodide, which is a much more selective reducing reagent, may offer advantages, however (cf. Singh, Corey, et al., 1987). If it is not feasible to obtain a particular diastereomer by direct addition to the carbonyl group, it may be necessary to carry out a subsequent inversion (e.g., Mitsunobu, 1981), or use a more circuitous approach (cf. Fig. 12.5). The recently introduced methodology from the Yamamoto group (Maruoka, Yamamoto, et al., 1988b) shows considerable promise for providing high stereochemical yields of equatorial cyclohexanol derivatives. It utilizes very bulky Lewis acids [e.g., methylaluminum bis(2,6-di-*tert*-butyl-4-methylphenoxide)] to complex with and block the more exposed face of a carbonyl group (Fig. 12.78, but cf. Power, Barron, et al., 1990). For more detailed and extensive treatments of the stereocontrolled addition of nucleophiles to aldehydes and ketones see Morrison and Mosher (1976, pp. 84–159) and Nogradi (1987, pp. 131–147 and 160–220).

Figure 12.78. Addition of nucleophiles to a cyclohexanone-bulky Lewis acid complex.

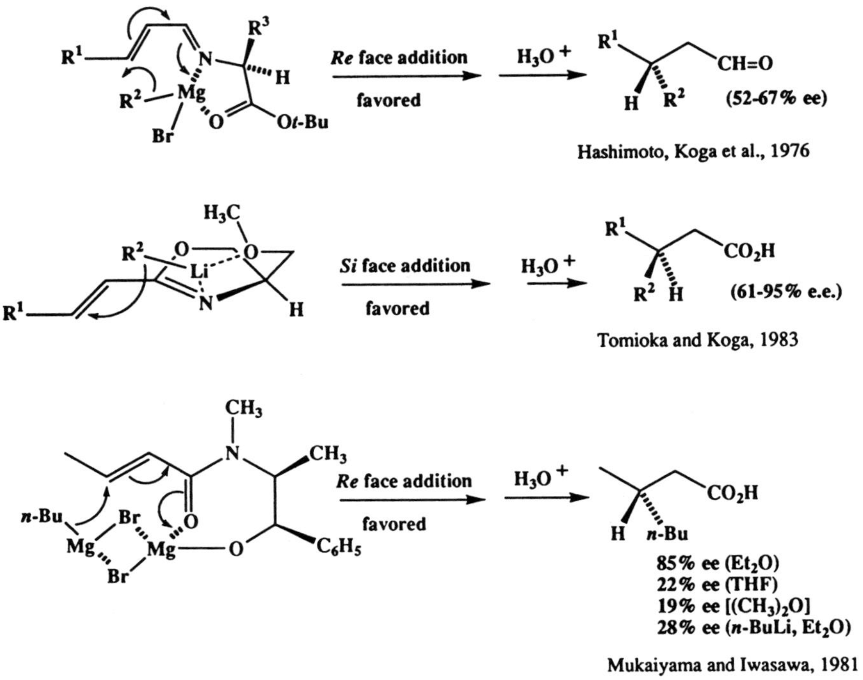

Figure 12.79. Conjugate additions of organometals to chiral α,β-unsaturated imines, oxazolines and amides.

Conjugate Additions to Electron-Deficient Olefinic Bonds

In considering the additions of achiral nucleophiles to olefinic bonds that are activated by an electron-withdrawing group it is possible to identify two categories of diastereoselective reactions, depending on whether the chirality resides in the main body of the olefinic moiety, or in the electron-withdrawing group (Rossiter and Swingle, 1992). Circumstantially, it appears that almost all reported examples in the former category are restricted to cyclic compounds, while those in the latter group pertain to acyclic alkene derivatives. There is also a heavy emphasis on the use of cuprates in order to control the regiochemical problem of 1,4 addition.

Conjugate Additions Mediated by Chiral Electron-Withdrawing Groups. Conjugate additions of Grignard reagents and organolithiums to α,β-unsaturated imines, oxazolines, and amides have been studied in some detail. Intramolecular delivery of the reagent via an intermediate chelate has been invoked to explain the high levels of stereoselectivity obtained (Fig. 12.79; Meyers et al., 1979; Tomioka and Koga, 1983).

In support of this view it was found that in a study of ephedrine derived amides, lower levels of stereoselectivity were obtained if either organolithiums or more strongly coordinating solvents were used (Mukaiyama and Iwasawa, 1981). More recent examples of a similar nature based on a readily obtained camphor-derived sultam auxiliary have been reported (Oppolzer et al., 1987).

Sulfoxides have also served as popular chiral auxiliaries (Posner, 1983, 1985) and an example has been provided earlier (Sections 12-3.d and e). However,

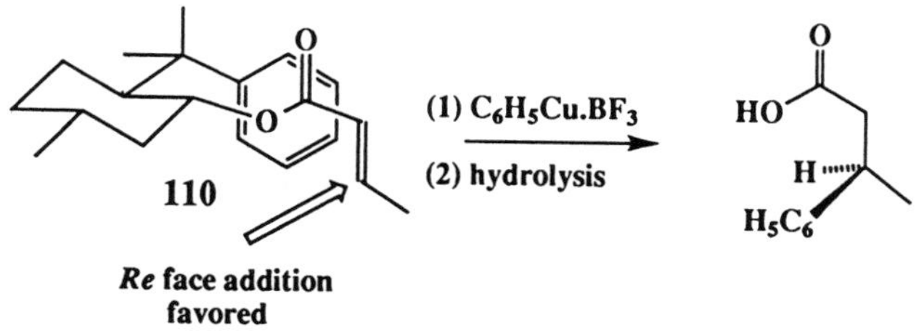

Re face addition
favored

Figure 12.80. Stereocontrolled conjugate addition to 8-phenylmenthyl crotonate.

given the ease of assembly and disassembly, chiral esters are the most appealing group of substrates. One of the earliest examples, conjugate addition of phenyl-magnesium bromide to menthyl crotonate, gave very discouraging levels of diastereoselectivity (Inouye and Walborsky, 1962), but substitution of 8-phenyl-menthol as the chiral auxiliary (Oppolzer and Löher, 1981) established a more encouraging prospect. The configuration of the products may be rationalized by assuming that the α,β-unsaturated esters of secondary alcohols take up a preferred conformation in which the carbonyl group is s-trans (i.e. antiperiplanar to the olefinic bond) and synperiplanar with the alkoxy methine CH bond as depicted for **110** in Fig. 12.80), thereby hindering approach to the Si face of the β-carbon atom of the crotonate moiety.

Although conjugate addition of ammonia and amines to **110** gave poor levels of stereoselection, utilization of the 2-naphthyl analogue **111** gave excellent results, for example, of benzylamine under high pressure (1.5×10^9 pascals, i.e., 15 kbar) afforded the Re addition product almost exclusively (99% ds) (Fig. 12.81). The authors concluded that $\pi-\pi$ interactions (cf. Sections 12-3.b and g) contributed significantly to the greater level of stereoselectivity and cited NMR spectra, which showed that the crotonate methyl group in **111** lay above the plane of the naphthyl ring, since it was shielded by 0.45 ppm relative to the simple menthyl ester; a shift of only 0.15 ppm was observed for **110** (d'Angelo and Maddaluno, 1986).

Numerous further examples of conjugate additions to crotonate esters based on chiral auxiliaries derived from camphor have been reviewed (Oppolzer, 1987). The addition of cuprates to **112** (Fig. 12.82) is typical (Oppolzer and Löher, 1981; Oppolzer and Stevenson, 1986).

Conjugate Additions to Chiral Enones and Related Compounds. On the whole, it appears that the approach of nucleophiles to cyclic enones, and so on, can be rationalized by simple arguments based on steric control that may be

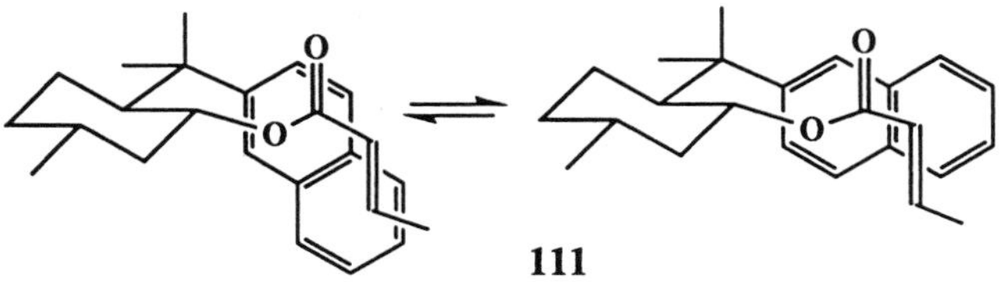

111

Figure 12.81. 8-(2-naphthyl)menthyl crotonate.

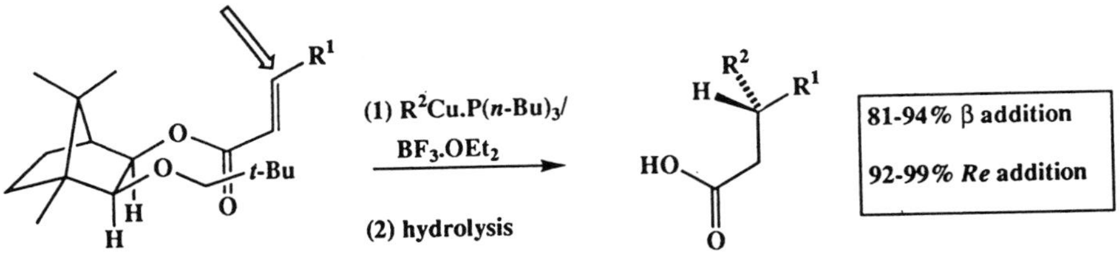

Re face addition favored

(1) $R^2Cu.P(n-Bu)_3/$
   $BF_3.OEt_2$

(2) hydrolysis

| 81-94% β addition |
| 92-99% Re addition |

112

Figure 12.82. Addition of cuprates to α,β-unsaturated esters based on a camphor-derived chiral auxiliary.

imposed either by proximal substituents or the overall topology of the molecule. A requirement to maintain orbital overlap in the transition state when the olefinic bond is endocyclic, however, will often impose stereoelectronic control, leading to introduction of an axial, and therefore more sterically hindered substituent (Marshall and Cohen, 1971). The question of reversibility, and therefore of kinetic versus thermodynamic control, often complicates analysis of the outcomes, however (Nagata et al., 1972). In the case of cuprate additions, the mechanism is uncertain and may vary with both reagent and substrate (Corey and Boaz, 1985a,b; Corey et al., 1989). Trihapto d(Cu)-π_3^* (enone) coordination (Fig. 12.83) has been invoked (Corey and Hannon, 1990) to explain the high level of diastereoselectivity observed for the addition of cuprates to 5-substituted 2-cyclohexenones (House and Fischer, 1968) and related examples (Still and Galynker, 1981).

In the small number of acyclic examples of stereoselective cuprate additions to acyclic substrates, Felkin-type transition states have been invoked to explain the results, for example, the conversion illustrated in Figure 12.84 (Roush and Lesur, 1983).

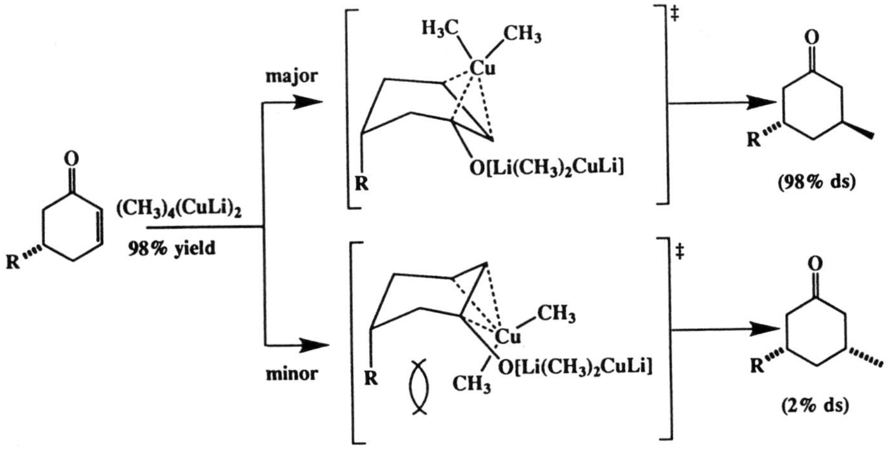

Figure 12.83. Cuprate additions to cyclohexenones.

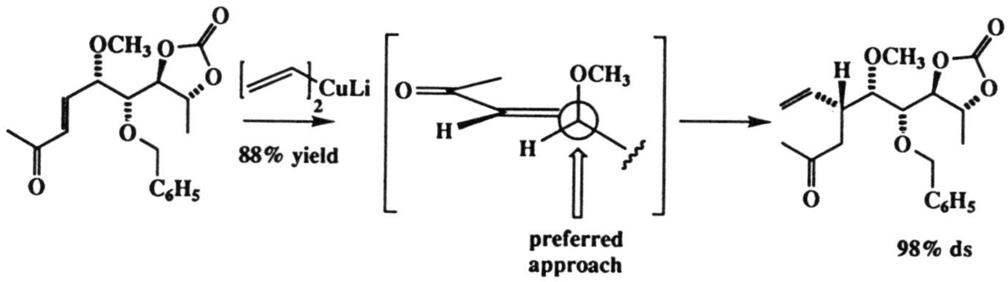

Figure 12.84. Stereocontrolled addition of vinyl cuprate to an acyclic enone.

In further examples, however, it is necessary to seek more subtle explanations. In the predominantly anti addition of cuprates to 5-methoxycyclopent-2-enone **113**, for example, it was shown that the exceptional levels of diastereoselectivity (up to 98% ds) could not be attributed simply to steric factors. Indeed, only modest levels of syn stereoselectivity were obtained with the 5-acetoxy analogue **115**, and although anti selectivity prevailed with the 5-methyl derivative **116**, the preference was marginal. A strong preference for anti addition to the methoxyl was observed even for the 5,5-disubstituted compound **114** (Fig. 12.85; Smith et al., 1988).

Semiempirical calculations using modified neglect of differential overlap (MNDO) for the addition of a diffuse negative charge along the Bürgi-Dunitz trajectory to the enone β-carbon atom, indicated a preference for anti addition of 2.1 kcal (8.8 kJ) at a distance of 270 pm (2.7 Å). Ab initio molecular orbital (3-21G level) calculations for the addition of hydride ion arrived at a similar picture. The authors attributed their results to a basic steric contribution reinforced by a greater electronic bias due to polarization of the enone moiety by the 5-methoxy group. Electrostatic repulsion between the latter and the electron-rich cuprate was also considered as a possible contributing factor. They likened the electronic effect to a vinylogue of the app effect in simple carbonyl derivatives (Smith and Trumper, 1988).

In yet another study of cuprate reactions, completely stereoselective *Si* face additions to the dioxinone **117** were observed (Fig. 12.86; Seebach et al., 1988). It was concluded that neither steric or stereoelectronic considerations could possibly account for the observed outcomes. Crystal structures of related compounds clearly indicated a "sofa" conformation for the substrate with the acetal center C(2) out of the plane defined by the remaining ring members. On this basis it

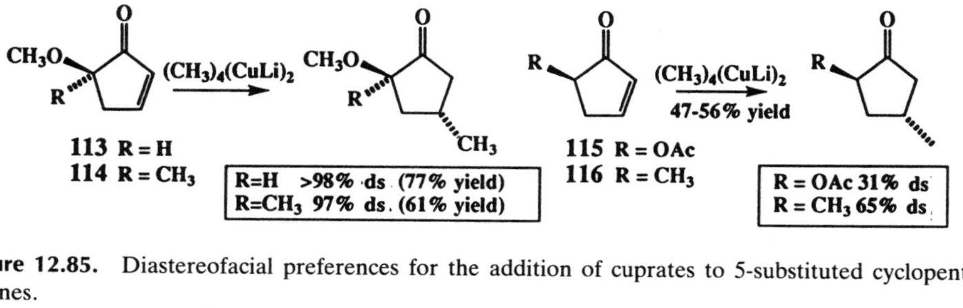

Figure 12.85. Diastereofacial preferences for the addition of cuprates to 5-substituted cyclopent-2-enones.

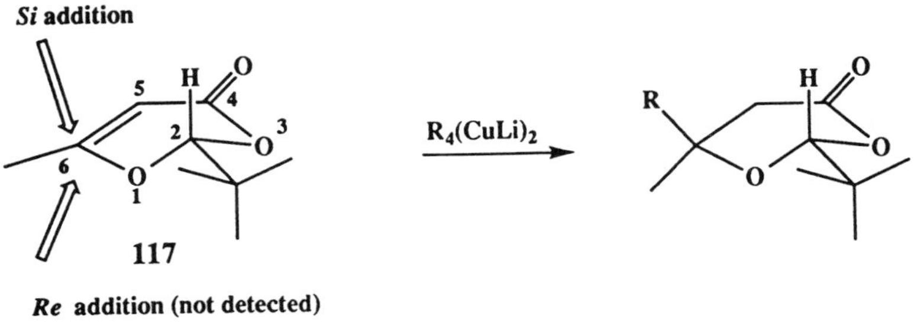

117

Re addition (not detected)

Figure 12.86. Stereocontrolled cuprate additions to dioxinone **117**.

would be expected that *Si* face addition should be hindered by the acetal
hydrogen that has a 1,3-diaxial relationship to the reacting center. Ab initio
molecular orbital calculations on the parent ring system confirmed that the
structures observed in the solid state were an acceptable basis for considering
reactions in solution, and the possibility that the products of addition were the
result of thermodynamic control was discounted.

Noting that the crystal structures of the model systems and numerous related
compounds displayed a small degree of pyramidalization of the sp^2 center C(6)
towards the Si face (i.e., the upper face of **117** as drawn), the authors developed a
rationalization based on an earlier proposal by Houk (1983): "If the pyramidalization of a trigonal center next to a tetrahedral center is caused by a tendency to
minimize torsional strain, approach of the reagent from the direction towards
which pyramidalization has already occurred should reduce torsional strain further, whereas the opposite approach of the reagent would increase strain." In the
dioxinone **117**, O(1) and the 6-methyl group lie slightly below the σ plane defined
by O(3)–C(4)–C(5)–C(6) (Fig. 12.87) so addition to the upper face of C(6)
should, on the basis of the Houk argument, be favored by the resulting
pyramidalization (the dihedral angle is exaggerated to clarify the perspective).

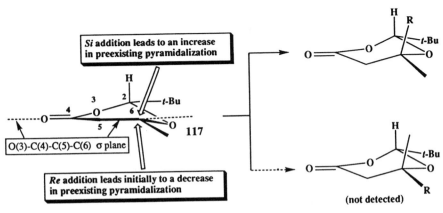

Figure 12.87. Pyramidalization at C(6) in dioxinone **117**.

Directed Nucleophilic Additions

As we shall see in Section 12-3.e, directed addition of electrophiles to alkenes by neighboring ligands in the substrate are commonplace. Conversely, ligand-assisted nucleophilic additions are relatively uncommon. This result should not be unexpected, because nucleophiles are intrinsically electron rich. Nevertheless, recent publications provide illustrations of the considerable value of such an approach, not only for stereochemical control, but for regiochemical control as well. Examples of the alkoxide assisted 1,4 addition of organometals and complex hydrides to enones are provided in Figure 12.88 (Salomon et al., 1984; Solomon, Maryanoff, et al., 1988).

A further illustration is provided in Figure 12.89 with the stereoselective reduction of ketones possessing a proximal hydroxy group of tetramethyl-ammonium triacetoxyborohydride (Evans et al., 1988b). The hydroxyl displaces one of the acetoxy ligands affording a stronger hydride donor than the parent

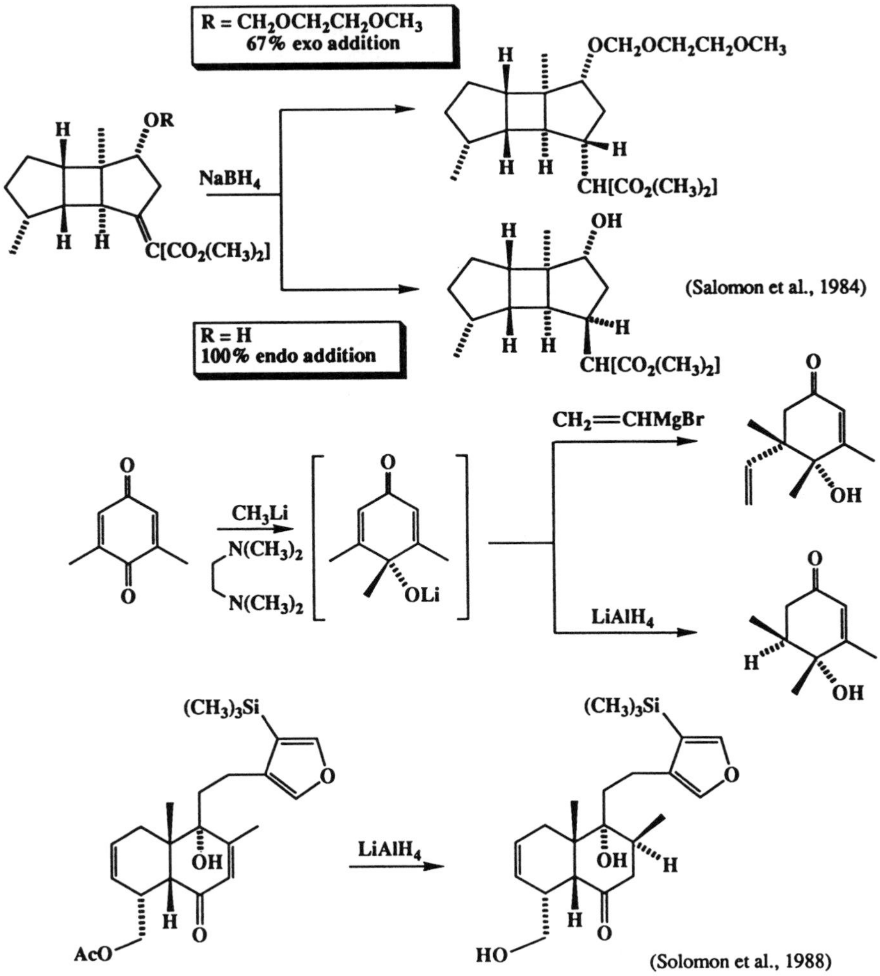

Figure 12.88. Examples of hydroxyl and alkoxide directed conjugate additions.

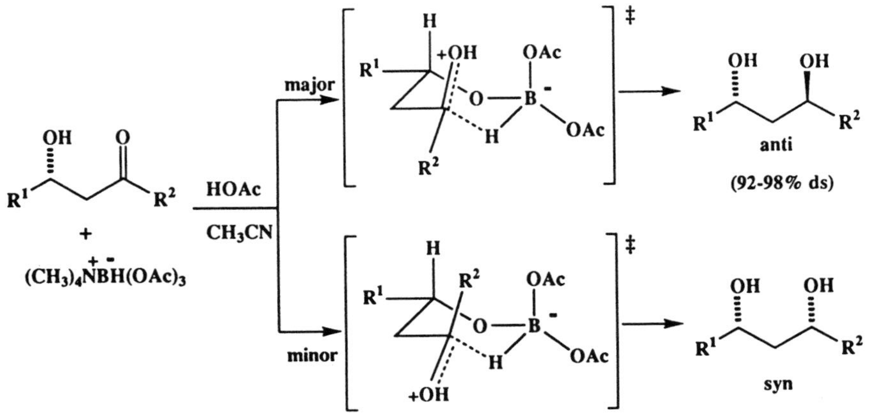

Figure 12.89. Hydroxyl directed reductions in β-hydroxy ketones.

reagent and leads to intramolecular addition of hydride to afford predominantly anti products. The reagent itself is too unreactive to reduce ketones, and therefore intermolecular processes, which are unlikely to be stereoselective, do not compete.

e. Electrophilic Reactions of Alkenes

For the reactions of simple alkenes with a wide range of electrophilic reagents the stereochemical outcomes are on the whole dictated by steric aspects although these may be overridden by stereoelectronic requirements. For relatively rigid cyclic systems and acyclic substrates with clearly preferred conformations it is a straightforward matter to extrapolate from ground-state to transition state structures and predict the direction of approach by reagents. Moreover, reasonable levels of stereoselectivity can be expected. In other situations, however, the participation of neighboring groups may be necessary to produce acceptable results. Electrophilic processes in this section have therefore been grouped into three categories according to whether (a) a simple addition is involved, (b) a neighboring group is coordinated to the reagent, or (c) a second functional group participates in a ring-forming process.

Simple Electrophilic Reactions

Hydroboration. There is surprisingly little variation between the levels of diastereoselectivity obtained from hydroborations with diborane and various monosubstituted and disubstituted boranes. Because donor ligands reduce the electrophilicity of these reagents, there are not the same opportunities for approaches directed by neighboring groups (but see Still and Barrish, 1983). Despite these limitations diastereoselectivity is often good. In a recent synthesis of monensin, for example, intermediate **119** (Fig. 12.90) was obtained as an 8:1 mixture with its alternative diastereomer from hydroboration of the allylic ether

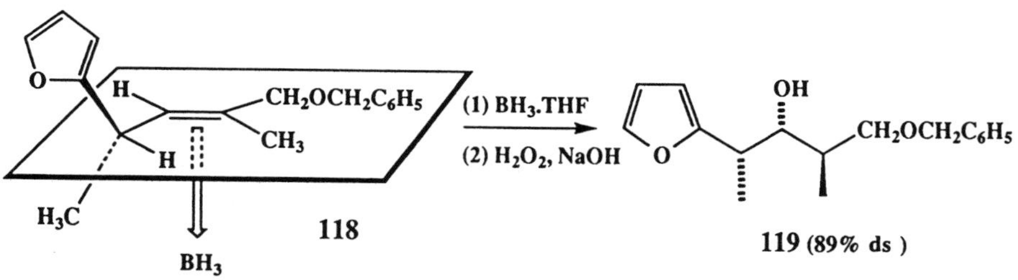

118

119 (89% ds)

Figure 12.90. Stereoselective hydroboration of an intermediate in the total synthesis of monensin.

118 (Schmid et al., 1979). It was concluded that the selectivity could be explained in terms of the indicated conformation, whereby $A^{(1,3)}$-strain (cf. p. 738) is minimized by moving the bulkier substituents out of the plane of the double bond; the reagent then preferentially approaches anti to the aromatic substituent. The reasons for this may involve electronic factors rather than simple steric ones, however (Houk et al., 1984b).

The example above provides an illustration of "1,2-asymmetric induction," but there have been equally good results obtained with stereogenic centers located in a 1,3 relationship to the olefinic bond, for example, ester **120** was converted to an 87:13 mixture of epimers **121** and **122** by disiamylborane (Fig. 12.91; Evans et al., 1982a).

The outcomes for this and similar examples were rationalized in terms of the transition state structures depicted in Figure 12.92. The minor pathway is disfavored by a nonbonded interaction between the medium sized substituent R_M and the vinyl methyl group.

Osmium Tetroxide Hydroxylations. Surprisingly good levels of diastereoselectivity have been obtained in both cyclic and acyclic allylic alcohols and ethers in cases where the results cannot readily be rationalized by simple steric considerations. For cyclic systems the preferred diastereomer is the one in which the preexisting hydroxy function is trans to the newly introduced neighboring hydroxyl group, while for acyclic cases the corresponding relationship is anti

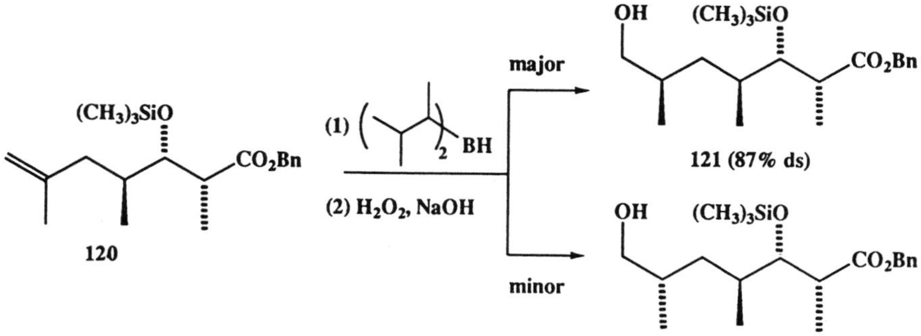

120

121 (87% ds)

122 (13% ds)

Bn = $CH_2C_6H_5$

Figure 12.91. An example of 1,3-asymmetric induction in a hydroboration reaction.

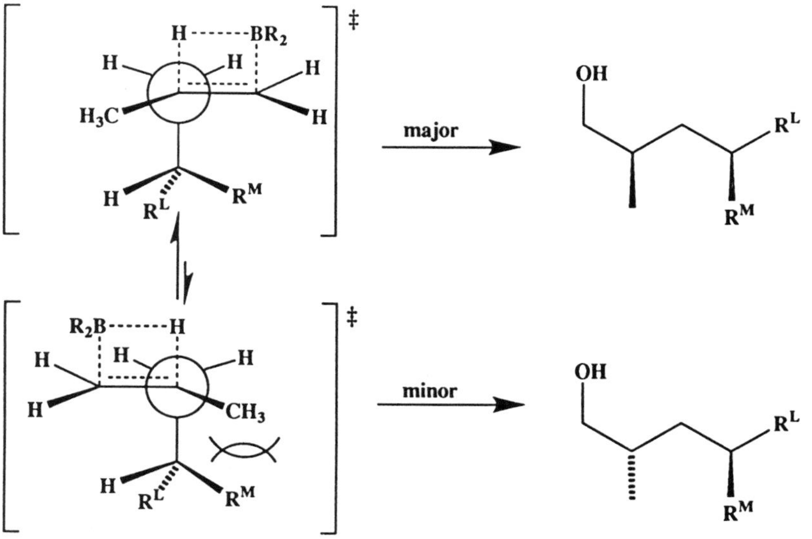

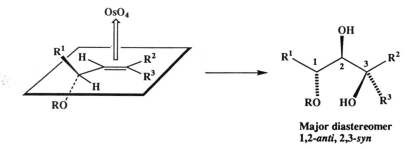

Figure 12.92. Transition state structures for 1,3-asymmetric induction in a hydroboration reaction.

(erythro). A conformation that minimizes $A^{(1,3)}$ strain is depicted in Figure 12.93. Approach by the reagent anti to the oxygen function would then be consistent with the observed outcomes and the level of selectivity (Cha, Kishi, et al., 1983; Christ, Kishi, et al., 1983). This picture is also consistent with results obtained for cyclic systems and the stereoelectronic involvement of the oxygen lone pair appears to be implicated (McGarvey and Williams, 1985; see below) in both cases. This view is reinforced by the observation that the corresponding acyloxy derivatives do not confer any significant control, but the hyperconjugative model has been rejected by the authors of two subsequent studies (Vedejs and McLure, 1986; Evans and Kaldor, 1990; cf. also Stork and Kahn, 1983).

Electrophilic Reactions of Enolates and Related Substrates. The central role played by enolate derivatives and related species, such as enamines and metalated hydrazones in synthesis, especially in the formation of carbon–carbon bonds, has led to the accumulation of an enormous wealth of data (Caine, 1979; Evans, 1984; Seebach et al., 1986). The following treatment has been limited to simple

Figure 12.93. Stereoselective osmylation of an allylic ether.

alkylations, but similar considerations are involved with other electrophiles, for example, brominations (Sinclair, Williams, et al., 1986; Evans et al., 1987b), aminations (Gennari et al., 1986; Trimble and Vederas, 1986; Evans et al., 1986, Evans and Britton, 1987), hydroxylations (Evans et al., 1985a), and conjugate additions (Evans et al., 1991). Aldol reactions are treated in Section 12-3.f.

It appears that the stereochemical outcomes of the reactions of enolate anions and related species depend predominantly on the structural properties of the substrates and is relatively insensitive to the nature of the electrophile, although it is clear that the solvent and counterion can have an important influence. Delocalization of charge in enolate derivatives introduces further complexities that are not encountered in the chemistry of simple alkenes. The stereoelectronic requirement to maintain orbital overlap over three centers favors axial-like trajectories in endocyclic enolates (cf. Fig. 12.97) and may be in conflict with steric factors. This requirement can become a major issue in the electrophilic reactions of enolates derived from carbonyl groups contained within six-membered rings. There are also questions about the location of the counterion and the degree of aggregation. The tetrameric structure **123** (Fig. 12.94) has been established for the THF solvate of the lithium enolates of pinacolone and cyclopentanone by X-ray crystallographic analysis (Amstutz et al., 1981), while detailed NMR studies support the existence of the corresponding tetramer of isobutyrophenone lithium enolate in THF solution and of the dimer **124** in dimethoxyethane (Jackman and Lange, 1977). There is clearly a need to evaluate the degree to which the participation of such aggregates in the reactions of enolates influence the outcomes (Seebach, 1988).

ENOLATE GEOMETRY. As with simple alkenes, the stereochemical outcome from the alkylation of enolates will be related to their cis/trans geometry, and because this is often established in situ, any deficiencies in selectivity at this stage of the process may reduce the ultimate level of stereoselectivity obtained in the formation of the target compound. The range of examples given in Figure 12.95 indicates a number of useful trends in selectivity that correlate the steric properties of both ketone and base.

3-Pentanone has been investigated in the greatest detail and it has been found that treatment with the sterically demanding base lithium tetramethylpiperidide (LTMP) under kinetic control, affords predominantly the trans enolate (trans/

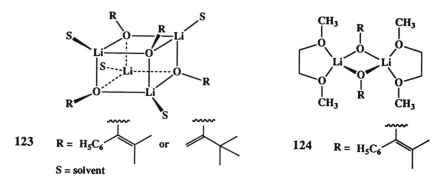

Figure 12.94. Structures of two enolate tetramers and a dimer.

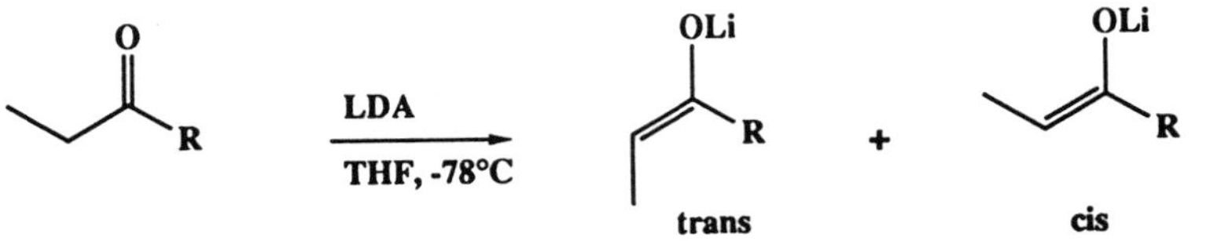

Figure 12.95. Stereoselective enolate formation from the reaction of amide bases with carbonyl derivatives.

TABLE 12.2. Stereochemistry of Enolate Formation in Ethyl Ketones

R	Base	trans Enolate (%)	cis Enolate (%)
—Et	LDA	77	23
—Et	LTMP	86	14
—Et	LTMP (HMPT)	8	92
—Et	LiN[Si(CH$_3$)$_2$C$_6$H$_5$]$_2$	0	100
—OCH$_3$	LDA	95	5
—Ot-Bu	LDA	95	5
—St-Bu	LDA	90	10
—NEt$_2$	LDA	3	97
—C(CH$_3$)$_3$	LDA	2	98
—C$_6$H$_5$	LDA	2	98

cis = 86:14), but that under conditions that promote equilibration (e.g., addition of HMPA), the cis enolate is strongly preferred (cis/trans = 92:8). Formation of the cis enolates is also favored when the disilazide bases LiN[Si(CH$_3$)$_3$]$_2$ and LiN[Si(CH$_3$)$_2$C$_6$H$_5$]$_2$ are employed or when substrates possess bulky α' substituents (Masamune et al. 1982).

Enolization of esters and thioesters by sterically demanding bases (e.g. LDA) under kinetic conditions leads to the preferential formation of the trans enolates, while mainly cis enolates are formed in the presence of HMPT. Dialkylamides, like ketones in which the nonreacting alkyl group is bulky, also afford predominantly the cis enolates under kinetic control.

> We endorse the use of the cis–trans nomenclature advocated by Heathcock et al. (1980) in which the convention describes the stereochemical relationship between the O-metal bond of the enolate and the α substituents and remains consistent for ketones, esters, amides, and so on. Use of this convention obviates the need to corrupt the E/Z system (cf. Evans, 1984, see also Section 9-1.b and Fig. 9.6).

These trends have been rationalized in terms of the six-centered chairlike transition state models depicted in Figure 12.96 (Ireland et al., 1976). Inspection of the developing nonbonded interactions reveals that the transition state structure $\mathbf{C}^{\ddagger}$ leading to the cis enolate is destabilized by the diaxial interaction CH$_3 \leftrightarrow$ L, but also that this may be outweighed by allylic strain (R $\leftrightarrow$ CH$_3$) as R becomes larger, that is, transition state structure $\mathbf{T}^{\ddagger}$ is then less favorable. A more detailed analysis has been provided (Evans, 1984; Heathcock et al., 1980).

ALKYLATION OF ENDOCYCLIC ENOLATES. The stereochemical outcome from the alkylation of endocyclic enolates depends on a balance of several factors, some of

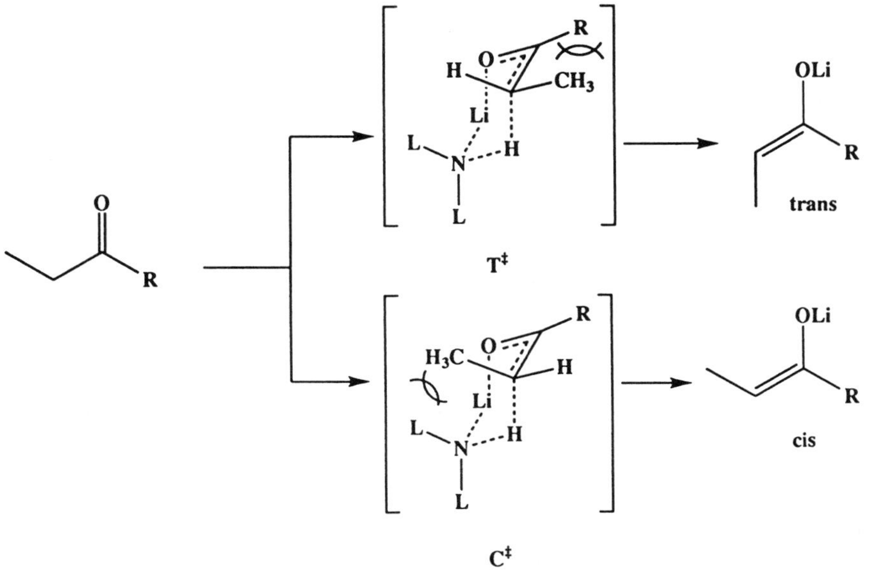

Figure 12.96. Alternative transition state structures for the enolization of carbonyl compounds.

which may be in conflict: substratelike versus productlike transition states, steric interference with electrophile trajectories, torsional strain arising from eclipsing of adjacent groups, and the need to maintain orbital overlap in the transition state structure (stereoelectronic effect). This last aspect leads to the presumption of the requirement of a perpendicular approach to the enolate system (Velluz et al., 1965). For enolate anions contained within six-membered rings, for example, the conformationally locked 4-*tert*-butylcyclohexanone enolate **125**, it is possibe to maintain orbital overlap based on a chairlike transition state structure (Fig. 12.97,

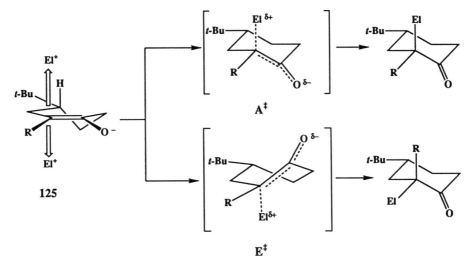

Figure 12.97. Alternative reaction pathways for the reaction of the enolate anion of 4-*tert*-butylcyclohexanone with an electrophile where El$^+$ = electrophile.

$A^{\ddagger}$) when reaction occurs on the upper face (leading to the introduction of an axial substituent, and therefore termed *axial alkylation*). Reaction on the lower face, however, would impose a twist–boatlike transition state structure $E^{\ddagger}$ (resulting in the introduction of an equatorial substituent, i.e., *equatorial alkylation*) if orbital overlap is to be maintained (Corey and Sneen, 1956).

There is little π-facial discrimination in the C alkylation of **125** when it is a simple enolate and it may therefore be concluded that for the more reactive enolates ($R = H$ or CH_3) there is an "early" (i.e., substratelike) transition state (Evans, 1984; cf. Hammond, 1955) in which there is relatively little differentiation between the two possible approaches. The axial alkylation products are more strongly favored in those cases where R is an alkyl group, however. This result could be anticipated from a consideration of the developing eclipsing interaction between the α-alkyl group and the syn β-hydrogen substituent in the boatlike transition state $E^{\ddagger}$ leading to "equatorial alkylation." It has also been suggested (House and Umen, 1973) that $A^{(1,2)}$ strain arising from the eclipsing of the α substituent and the O-metal bond leads to a pyramidalization of the oxygen bearing carbon atom and a consequential predisposition towards axial alkylation (cf. the discussion of nucleophilic addition to dioxinone **117**).

Equatorial alkylation is similarly inhibited by an equatorial β substituent (Coates and Sandefur, 1974) and in many bicyclic systems, such as those depicted in Figure 12.98. Thus, C ethylation of enolate **126** ($R = H$) affords a 95:5 mixture of the epimeric ketones **127** and **128** ($R = H$), respectively, although this preference is reversed when the angular substituent R is a methyl group (Mathews et al., 1970), presumably because axial approach syn to the angular methyl group is so strongly disfavored by steric interactions.

In those cases where the α substituent is an electron-stabilizing group (e.g., CO_2CH_3) the enolate is, as a consequence, less reactive, leading to a more productlike transition state. In the case of the boatlike conformation required for equatorial alkylation, the eclipsing interactions are relatively more severe, especially between substituents attached to C(1) and C(8), and so there is a greater preference for axial alkylation. The β-ketoester **129** therefore affords mainly **130** (i.e., axial alkylation; Fig. 12.99) (Spenser et al., 1968), whereas the C(2) isomer **131** for which a boatlike transition state would not be destabilized to the same extent, undergoes preferential equatorial alkylation (Kuehne, 1970). At first sight it might have been expected that the α-cyano derivative **133** should also have undergone mainly axial alkylation, but equatorial alkylation predominates, form-

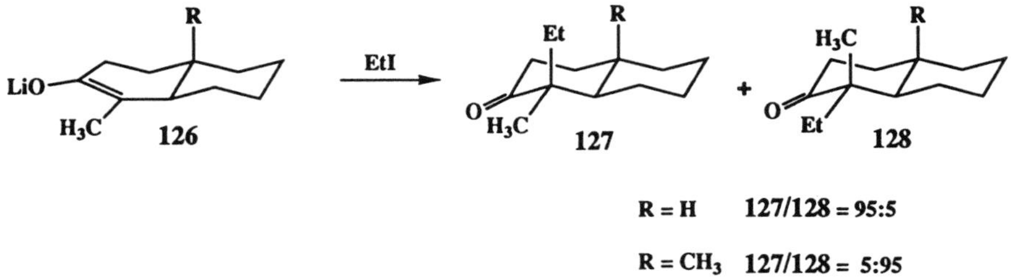

R = H   127/128 = 95:5

R = CH$_3$ 127/128 = 5:95

Figure 12.98. The C alkylation of a bicyclic enolate anion.

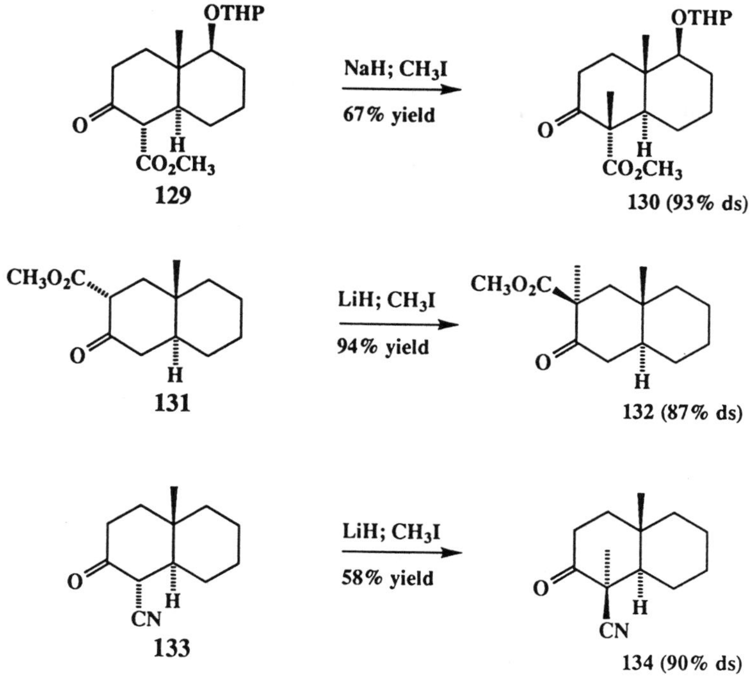

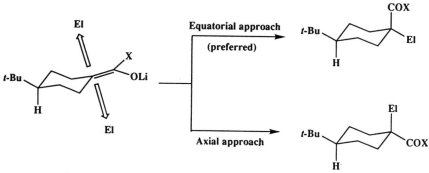

Figure 12.99. Stereoselective *C* alkylations of stabilized enolates.

ing mainly **134** (Kuehne and Nelson, 1970). In this case, however, the eclipsing interaction involving the cyano function in a boatlike transition state would be expected to be significantly reduced, relative to the methoxycarbonyl group.

ALKYLATION OF EXOCYCLIC ENOLATES. When the enolate is located exocyclic to a ring some of the constraints that favor attack of electrophiles along the axial trajectory are removed and so π-facial selectivity appears to depend simply on steric factors. In contrast to the situation that prevails in the superficially similar addition of nucleophiles to unhindered cyclohexanones (Section 12-3.d), alkylation of a conformationally locked cyclohexyl system, results predominantly in the introduction of an equatorial substituent (Fig. 12.100), unless this is sterically

Figure 12.100. Alkylation of a cyclohexylidene enolate anion.

blocked by a proximal substituent (Zimmerman, 1963). Again, the diastereofacial discrimination is poor with the most reactive substrates (e.g., X = OLi), but improves with the more stable enolates. Enolates derived from aldehydes are especially good substrates, affording only the products of equatorial alkylation (Ireland and Mander, 1969). Possibly, the high selectivity in these cases may be largely due to the low steric requirements of the developing aldehyde function as it assumes an axial conformation in the transition state structure.

CHELATE-ENFORCED INTRAANNULAR STEREOCHEMICAL CONTROL. The association between enolate anions and attendant counterions leads naturally to the utilization of chelation to limit conformational freedom and to impose predictable geometries on possible reaction processes. The difference in stereochemical outcomes from the alkylation of the prolinol derivatives **135** and **136** appears to provide a convincing example of chelate enforced stereochemical control (Fig. 12.101). Thus, **137** is obtained as a 92:8 mixture with diastereomer **138** by alkylation of the dilithio derivative, from which it may be inferred that preferential alkylation has taken place on the *Si* face of the chelated enolate **136**. In contrast, the methyl ether **139** afforded a 22:78 mixture of **141** and **142**, respectively, that is, a reversal of π-facial selectivity, indicating that transition state structures based on **140** are now favored (Evans and Takacs, 1980). There is, of course, the presumption in this analysis that *E* enolates, which would engender severe $A^{(1,3)}$ strain (Johnson, 1968) are either not implicated, or intrude only to a limited extent.

Numerous further illustrations of chelate enforced stereoselection are found (or inferred) in the extensive literature describing the utilization of a wide range of chiral auxiliaries (Morrison, 1983–1985; Seebach et al., 1986).

ALKYLATIONS OF ACYCLIC ENOLATES OR CYCLIC ENOLATES WITH EXTRAANNULAR CONTROL ELEMENTS. Acceptable levels of diastereoselection are less likely if bond rotation may occur between the respective parts of the molecule containing the enolate anion and stereogenic centers, although it is possible to construct chiral auxiliaries with a molecular architecture that restricts rotation and shields one of the π faces of the enolate anion, for example, the camphor based system **143**, which is illustrated in Figure 12.102 and in which the urethane moiety inhibits

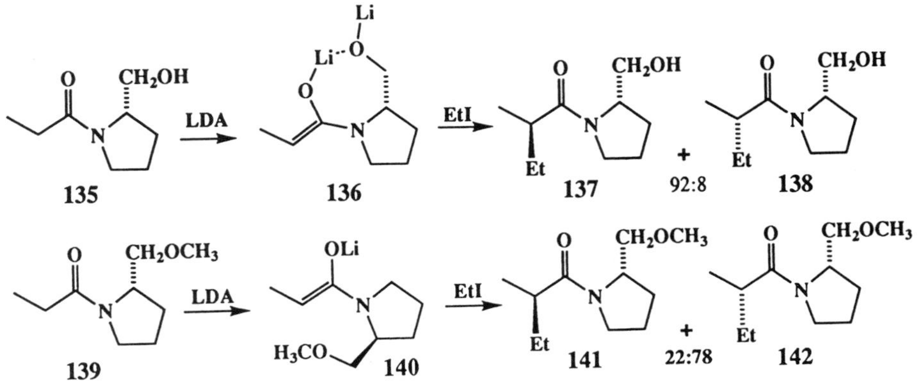

Figure 12.101. An example of π-facial selectivity imposed by chelation.

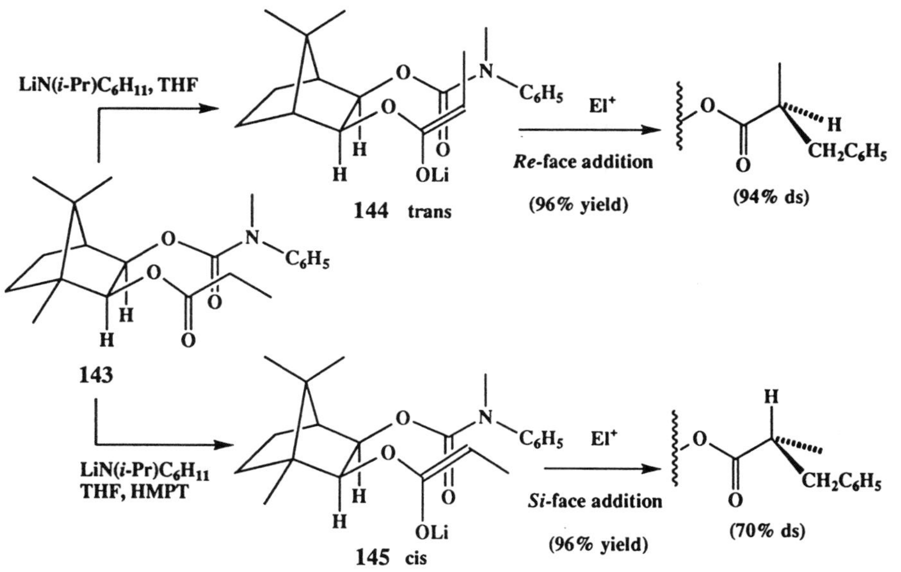

Figure 12.102. Examples of π-facial control in the reactions of propionate enolates attached to a camphor based chiral auxiliary.

approach to the *Re* face of C(2) in the cis enolate **145** or to the *Si* face of C(2) in the trans enolate **144** (Schmierer, Helmchen, et al., 1981).

Such elaborate arrangements are not always necessary, however. Excellent levels of diastereoselection have been observed for the alkylation of enolate **146** (Fig. 12.103) and have been rationalized on the basis of allylic strain considerations. It was assumed that the bulky phenyl and methyl substituents lie, respectively, above and below the plane of the enolate (thereby minimizing nonbonded reactions with the OLi group) and that the electrophile then approached the less encumbered lower *Si* face of the β-carbon atom in **146** (Schöllkopf et al., 1981).

Two further examples that demonstrate the importance of $A^{(1,3)}$-strain (cf. Fig. 11.65) effects are instructive. The *C* methylation of enolate **147** (Fig. 12.104) afforded (to the limits of detection) only the syn diastereomer **149**, a result that can be rationalized in terms of a transition state based on conformer **148** in which the electrophile approaches the π face remote to the very bulky thioacetal

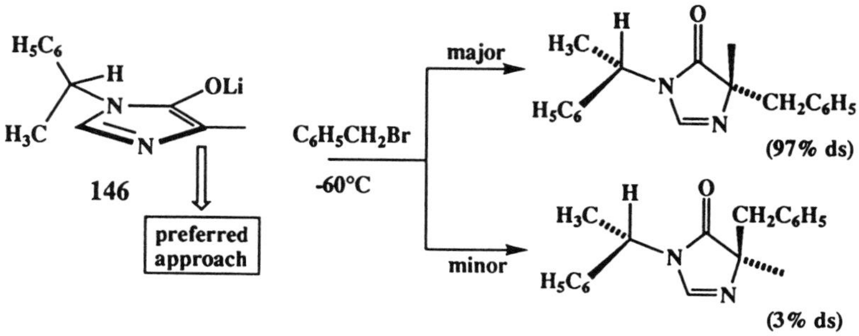

Figure 12.103. Alkylation of an imidazole derived enolate under extraannular stereocontrol.

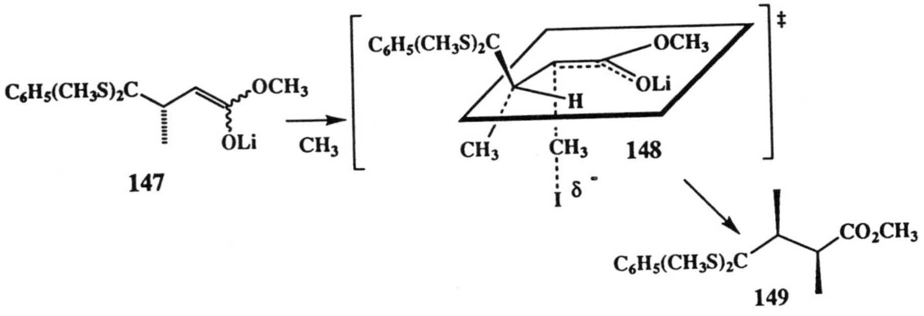

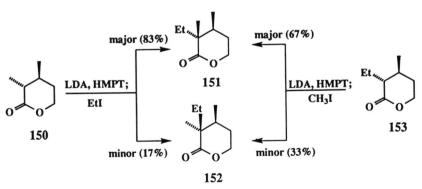

Figure 12.104. Examples of π-facial selectivity based on the minimization of $A^{(1,3)}$ strain.

substituent (Kawasaki et al., 1985). The second example (Tomioka et al., 1988) is of even greater interest. The C ethylation of the enolate anion derived from lactone **150** afforded a 5:1 mixture of diastereomers **151** and **152** (Fig. 12.105) as would be expected from consideration of steric factors. The C methylation of lactone **153** would have been expected on the same steric grounds to have resulted in a stereochemically complementary outcome (i.e., a predominance of **152**), but a 2:1 mixture of **151** and **152**, respectively, was obtained instead. This appears to be a general result when the α substituent to the carbonyl group is neither methyl nor hydrogen (Birch and Subba Rao, 1970; Takahashi et al., 1986). When $A^{(1,3)}$ strain is taken into account, however, it will be realized that the transition state for alkylation should be based on either conformer **154** or **155**. It can be readily envisaged (Fig. 12.106) that the transition state structure **156** based on the latter conformation (i.e., **155**) would be of lower energy, and that if the substituent in the side chain has a greater steric influence than the 4-methyl group on the approach of the electrophile, alkylation syn to the 4-methyl substituent would be more favorable.

An alternative view to these simple sterically based rationalizations has been asserted (McGarvey and Williams, 1985). In a series of β-substituted esters **157** it was found that the formation of diastereomers **158** in preference to **159** (Fig. 12.107) could only be satisfactorily explained if it was assumed that the more

Figure 12.105. Examples of π-facial selectivity in the alkylation of a δ lactone as a consequence of the minimization of $A^{(1,3)}$ strain.

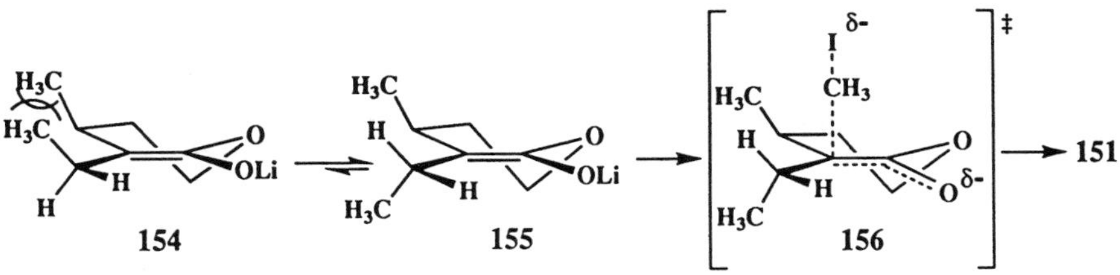

Figure 12.106. Conformations for lithium enolate of lactone **153** and transition state structure leading to the formation of **151**.

favored transition state structure was based on the conformation in which the substituent with the highest energy σ orbital took up a perpendicular relationship with the π system (so that it could participate in a hyperconjugative interaction with the HOMO of the enolate anion), and the less sterically demanding of the remaining substituents was gauche to the enolate system, that is, $\mathbf{A}^{\ddagger}$ (cf. Bernhard, Fleming, et al., 1984).

REACTIONS OF NITROGEN-CONTAINING ANALOGUES OF ENOLATES WITH ELECTROPHILES. The extra valency of nitrogen relative to oxygen provides greater opportunities for the incorporation of chiral centers into the substrate. Enamines or imines derived from prochiral ketones and chiral amines have been especially effective in allowing (overall) enantioselective carbon–carbon bond-forming reactions (Bergbreiter and Newcomb, 1983; Pfau et al., 1985; Seebach et al., 1986; d'Angelo et al., 1988; Revial and Pfau, 1991). A great variety of electrophilic reactions with other nitrogen-containing chiral auxiliaries has been developed and exploited. Several illustrations of the use of such compounds have already been provided (cf. Section 12-3.b) and reviews are available of the synthetically more useful classes of compounds, for example, hydrazones (Enders, 1984), oxazolines (Lutomski and Meyers, 1984), and oxazolidinones (Evans, 1982; Evans et al., 1982c).

Directed Electrophilic Reactions

Participation of a neighboring polar group in electrophilic reaction processes either by hydrogen bonding or by coordination to a metal center, and so on, may

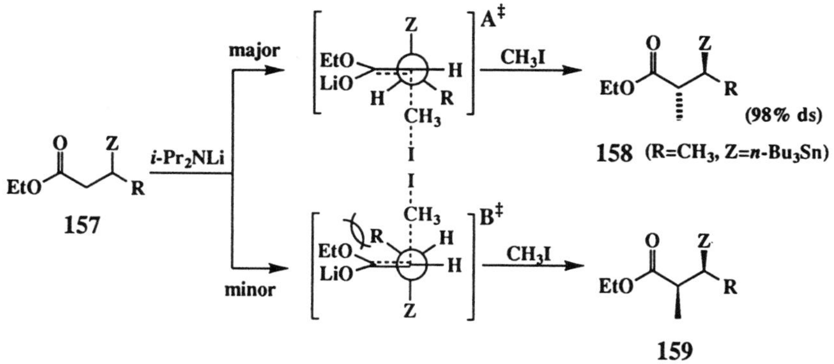

Figure 12.107. Stereocontrolled alkylation of β-substituted esters.

direct a reagent selectively to one or the other π faces of an olefinic bond and/or play an important organizational role in restricting rotational freedom.

Epoxidation. The stereochemical outcome from reactions of simple alkenes with peroxycarboxylic acids to afford epoxides is dictated by steric factors and varies only slightly with the choice of reagent. The solvent can exert a small influence, apparently by increasing the effective bulk of the reagent through solvation. In those cases where the sterically more hindered epoxide is required, it can often be achieved by the addition of hypohalous acids (or their analogues) followed by treatment with base, since it can be expected that in the intermediate 1,2-halohydrin that is formed, the hydroxyl will have been added to the more hindered face of the alkene (cf. Trost and Matsumura, 1977).

Reactions of cyclic allylic and homoallylic alcohols with peroxycarboxylic acids normally afford the syn products (Henbest and Wilson, 1957). These results have been rationalized in terms of a hydrogen-bonded transition state (Section 12-3.b) (Chamberlain et al., 1970). In the case of acyclic substrates, however, analysis of the stereochemical outcome is complex and depends very much on the position and bulk of substituents. In the first comprehensive study on such substrates (Chautemps and Pierre, 1976), four transition state structures were considered (Fig. 12.108) and it was concluded that reactions proceeded predominantly through conformations **B** and **C** (with **B** favored by an increase in the bulk of R), except when the substrate possessed a *cis-β*-alkyl substituent. In these cases, **D** represented the preferred conformation. Subsequent analyses, however, favor conformation **D** for all those cases that do not involve an α-alkyl group (Bartlett, 1980), while dihedral angles (C=C—C—O) of 120° and 150° have been deduced from other investigations (Rossiter, Sharpless, et al., 1979; Itoh, Teranishi, et al., 1979).

A more satisfying analysis based on considerations of A strain has been advanced by Narula (1981, 1983). In this hypothesis it was argued that for the syn transition state $S^{\ddagger}$ there is the possibility of $A^{(1,2)}$ strain between substituents

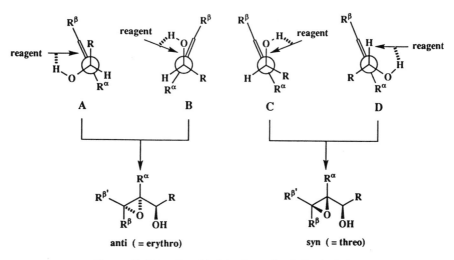

Figure 12.108. Epoxidation of acyclic allylic alcohols.

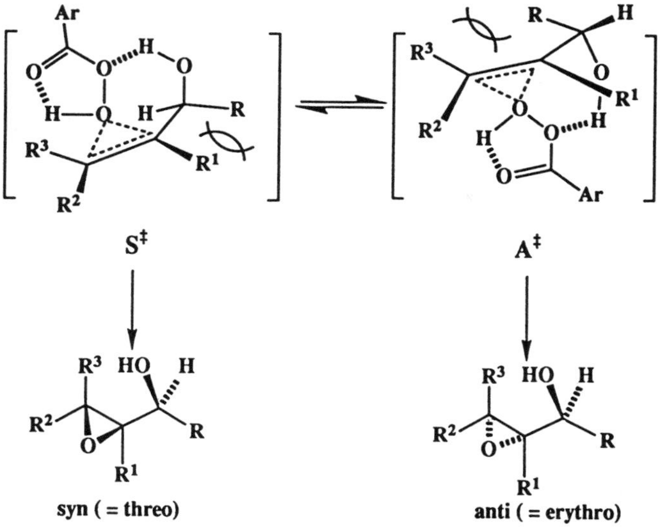

Figure 12.109. Rationalization of diastereoselectivity in epoxidation based on consideration of A strain.

R and R^1 (Fig. 12.109), whereas in the competing anti transition state ($A^‡$) there may be $A^{(1,3)}$ strain (cf. p. 738) between R and R^3. Thus, when both R and R^3 are alkyl groups, $A^‡$ is disfavored and the syn product is formed predominantly (Johnson and Kishi, 1979). When R^1 is especially bulky, however, the anti product is formed exclusively and a trimethylsilyl group has been used as an achiral auxiliary (cf. Section 12-3.b) in this position to regulate stereoselection (Hasan and Kishi, 1980).

Epoxidation by transition metal peroxides provides a valuable alternative method to the use of peroxycarboxylic acids, and although more limited in scope, it is more stereoselective, especially the vanadyl acetoacetate/*tert*-butyl peroxide system (Sharpless and Michaelson, 1973). Syn products are formed consistently from the reactions of cyclic allylic alcohols with peroxycarboxylic aids, except with eight-membered and larger rings, for which the formation of anti products is favored (Itoh, Teranishi, et al., 1979). The most satisfactory method for stereochemical control in many circumstances, however, could be the utilization of the Sharpless–Katsuki tartrate/titanium alkoxide system that has the potential to impose a stereochemical outcome that is independent of the substrate structure (cf. Section 12-5.b).

Effective stereochemical control can also be obtained via halo-lactonization as illustrated for the conversion of **160** into **161** (Fig. 12.110; Bartlett and Myerson, 1978).

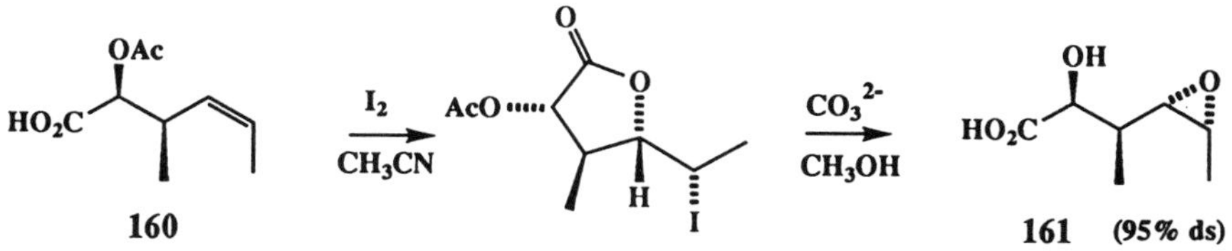

Figure 12.110. Formation of an epoxide via halolactonization.

Cyclopropanation. The stereoselectivity of transition metal catalyzed cyclopropanation reactions with diazo compounds has been reviewed (Maas, 1987). While there is a suggestion of an ylide type of association between the carbenoid and the halogen atom in some allylic halides, diastereoselectivity is generally poor. The reaction of dichlorocarbene with allylic alcohols under phase-transfer conditions, however, proceeds with excellent stereoselectivity (Mohamadi and Still, 1986). This outcome was rationalized by invoking hydrogen bonding between the directing hydroxyl and the filled orbital of the singlet carbene, although no such association was evident in the reactions of allylic alcohols with dichlorocarbene generated from bromodichloromethyl(phenyl)mercury (Seyferth and Mai, 1970).

The only unequivocal cases of directed cyclopropanations are those of allylic and homoallylic alcohols with the Simmons–Smith reagent Zn–Cu/CH_2I_2), a reaction that has been used extensively (Simmons et al., 1973). Syn products are normally formed, even when this involves reaction on the more hindered π face of the alkene bond (cf. Fig. 12.111; Paquette and Cox, 1967). Similar stereochemical outcomes are observed in acyclic substrates as those for equivalent epoxidations and an analogue to the Chautemps–Pierre model has been invoked to explain the results (Ratier, Pereyre, et al., 1978; cf. Figure 12.108). Stereochemical control exerted by allylic chiral ketal auxiliaries has also furnished excellent results and opens up a new avenue for auxiliary-based enantioselective synthesis (Mash et al., 1987; Arai, Yamamoto, et al., 1985). Very recently a samarium iodide based reagent has been introduced that may have advantages over the classical procedure. A model for the transition state has been proposed (Molander and Etter, 1987).

Hydrosilation. Although it may not be feasible to achieve better control over stereoselection in hydroborations by ligation of a neighboring polar group to the borane because of the resulting deactivation, it is possible to achieve equivalent outcomes through the agency of transition metal catalyzed hydrosilation of allylic and homoallylic alcohols with tetramethyldisilazane {[$(CH_3)_2SiH]_2NH$} (Tamao, Ito, et al., 1986, 1988). Excellent regioselectivity and dramatic rate accelerations were also observed. It was found that cyclic homoallylic alcohols afforded exclusively the cis-1,3-diols, but that the diastereomer ratios for acyclic homoallylic alcohols were very substrate dependent. Syn diastereomers were the major products from α-methyl and α-alkoxy secondary allylic alcohols and these outcomes were rationalized on the basis that transition state structure $\mathbf{A}^{\ddagger}$ avoids the nonbonded interaction indicated in the alternative $\mathbf{B}^{\ddagger}$ (Fig. 12.112). An especially useful aspect of this method is that the relative sense of stereochemical

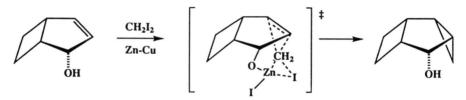

Figure 12.111. Simmons–Smith cyclopropanation of an allylic alcohol.

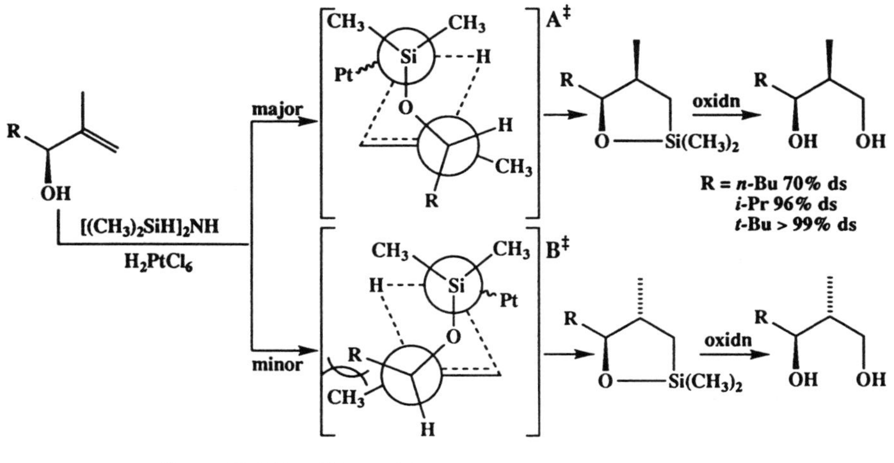

Figure 12.112. Directed hydrosilations of allylic alcohols.

induction is complementary with that obtained through hydroboration (cf. Schmid, Kishi, et al., 1979).

Electrophile Initiated Cyclizations

Heterocycle Formation. Electrophilic reactions involving the intramolecular participation of nucleophilic groups to form heterocycles yield predictable stereoselective outcomes as a consequence of the geometric constraints imposed on the reaction processes (Capon and McManus, 1976). There are several aspects to be considered:

1. The nature of the electrophile, that is, the less reactive electrophiles like iodine (I_2) lead to more extensive bonding by the nucleophilic group with the halonium–π complex in the rate-determining step.
2. Exocyclic versus endocyclic closures: A particular mode may be favored by stereoelectronic factors (Baldwin, 1976).
3. Thermodynamically or kinetically controlled outcomes: These outcomes have a bearing on both the size of the ring and on stereoselectivity.

For example, iodolactonization of acid 162 under "kinetic" control (*N*-iodo-succinimide–chloroform) afforded little discrimination among the products 163–166, a 2:2:3:1 mixture being obtained (Fig. 12.113). There appeared to be partial equilibration when a two-phase system (ether, I_2, NaHCO$_3$ solution) was utilized and so only the more stable γ-lactones 163 and 164 were obtained (as a 3:1 mixture). Under conditions (I_2, acetonitrile), which allowed full equilibrium, however, the thermodynamically most stable product 163 was formed in greater than 95% yield. This last reaction was actually carried out on the acetate of 162 (Myerson, 1980).

In general, "thermodynamic conditions" will lead to good diastereoselection with the bulkier substituents occupying equatorial conformations, either in trans-1,2 or cis-1,3 relationships in six-membered rings. In five-membered rings, the

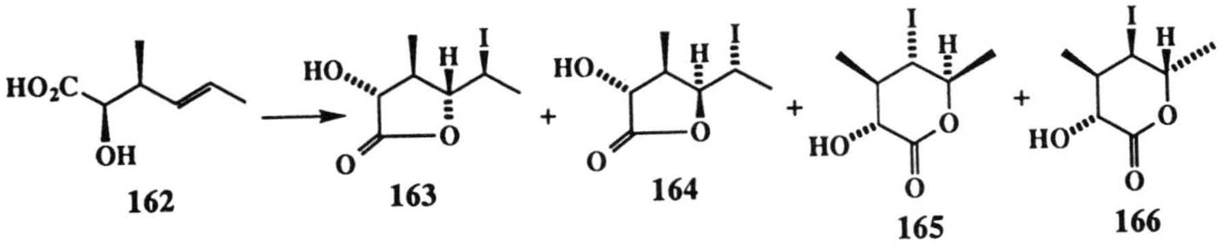

Figure 12.113. Iodolactonization products from the 4-hexenoic acid derivative **162**.

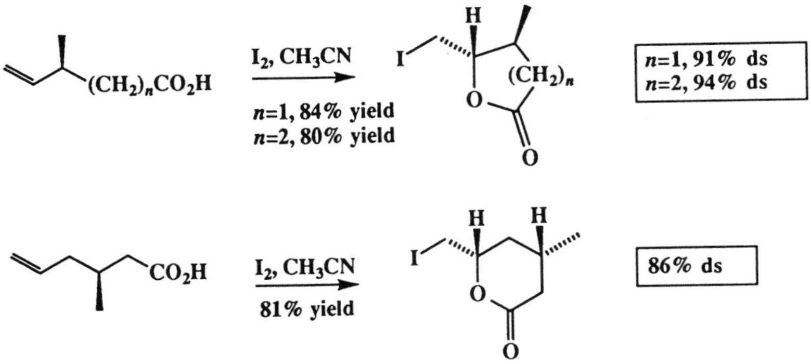

Figure 12.114. Iodolactonizations under thermodynamic control.

substituents need to be in a 1,2 relationship, and then the thermodynamic product will be the trans isomer (cf. Fig. 12.114; Bartlett, 1984a).

Carbocycle Formation. The stereoselective formation of the polycyclic skeleton of triterpenes during their biogenesis from the acyclic polyene squalene has inspired some especially elegant examples of biomimetic synthesis (Johnson, 1976; Bartlett, 1984b) These syntheses are typified by the formation of the cortisone precursor **168** as a mixture of C(17) epimers from tetraene **167** in a single step (Fig. 12.115). The transformation, which proceeds in a yield of 58%, provides an extraordinary example of stereochemical control, producing five new stereogenic centers with the correct configuration (Johnson et al., 1982).

The formation of **168** raises several questions involving mechanism, the degree of concertedness, and the various stereochemical relationships that are involved in such cyclizations, that is, the trans B/C and C/D ring fusions; the anti relationships between C(9) and C(10), and between C(8) and C(14); and the diastereomeric relationship of C(11) to the rest of the molecule. These questions have been addressed in some detail elsewhere (Bartlett, 1984b), but in broad

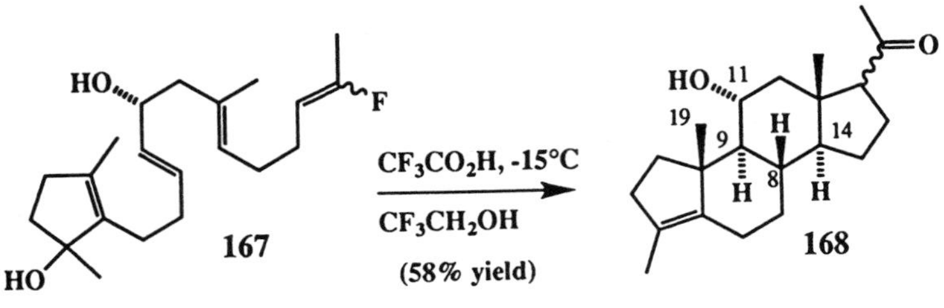

Figure 12.115. Polyene cyclization to form an intermediate for the synthesis of cortisone.

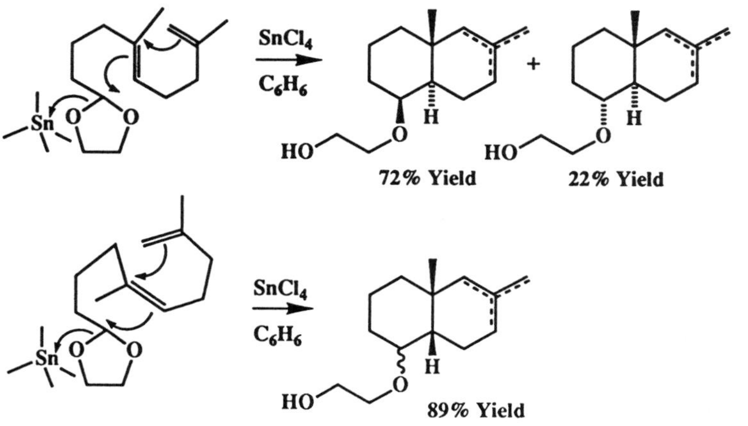

Figure 12.116. Cyclization of diene acetals showing the translation of alkene geometry into the relative configuration of sp^3 centers in the bicyclic products.

terms it is apparent that an all chairlike conformation for the transition state normally prevails (the biosynthesis of protosterol, and hence lanosterol and cholesterol, under enzymic control, is an important exception), and for cyclizations that proceed without the intervention of deprotonation–reprotonation (i.e., without alkene–cation equilibria), the E–Z geometry of olefinic bonds is translated into the stereochemical relationships between substituents along the backbone in the product (Fig. 12.116), thereby implying a considerable degree of concertedness, at least in the early stages of the conversion.

Once a cyclic intermediate has been formed that can establish a sufficient degree of conformational preference, a templatelike influence can be exerted on subsequent events. Cyclization of the cyclohexyl cation **169**, for example, affords predominantly the cis-fused product **170**, whereas the bicyclic cation **172** is converted preferentially into the B,C-trans product **173** (Fig. 12.117).

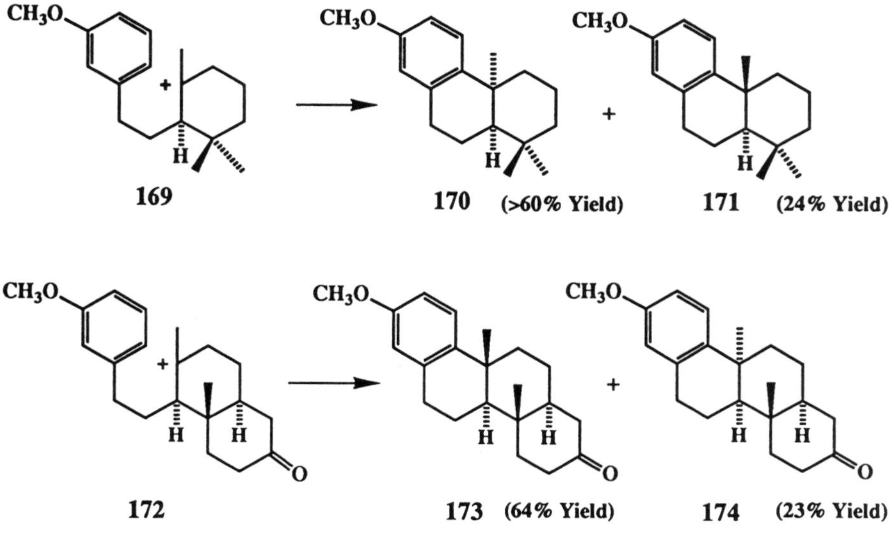

Figure 12.117. Stereoselective cyclization of aryl rings on to cyclohexyl cations.

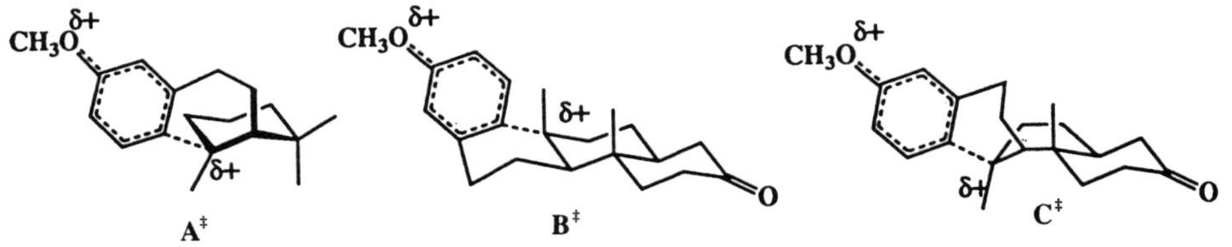

Figure 12.118. Transition states for the cyclization of aryl rings onto cyclohexyl cations.

These outcomes were rationalized in terms of a transition state $A^{\ddagger}$ for the first cyclization that is based on an axial conformation for the reacting side chain, thereby avoiding $A^{(1,2)}$ strain while the second reaction is constrained to react either via transition state $B^{\ddagger}$ in which the side chain is equatorial or through the less favorable boat–chair arrangement $C^{\ddagger}$ in which the side chain is axial (Fig. 12.118; Ireland et al., 1972).

The remaining stereochemical aspects of these cyclizations are concerned with the influence that may be exerted by preexisting stereogenic centers in the substrate. In the absence of the π-facial discrimination afforded by such centers, only racemic products can be obtained. (The desirable option of an effective chiral catalyst has yet to be realized.) It is possible to identify two distinct categories: The chiral element is either involved in the initiating process or it is embedded in the polyene. The latter case is illustrated by the transformation **167 → 168** (Fig. 12.115) and by the conversion of **175** into the 11α-methylprogesterone precursor **176** (Johnson and DuBois, 1976). None of the 11β-epimer could be detected, a result that was rationalized in terms of a destabilization of the alternative transition state $A^{\ddagger}$ because of a diaxial-like nonbonded interaction between the pro-19-methyl group and the pro-11β-methyl substituent (Fig. 12.119). (Note that if any of the 11β-epimer had been formed, the resulting product would be antipodal to **176**.) For analogous compounds with substituents in other positions, for examples, C(7), the axial product may be preferred as a consequence of $A^{(1,2)}$ strain in the transition state leading to the equatorial epimer (Groen and Zeelen, 1978).

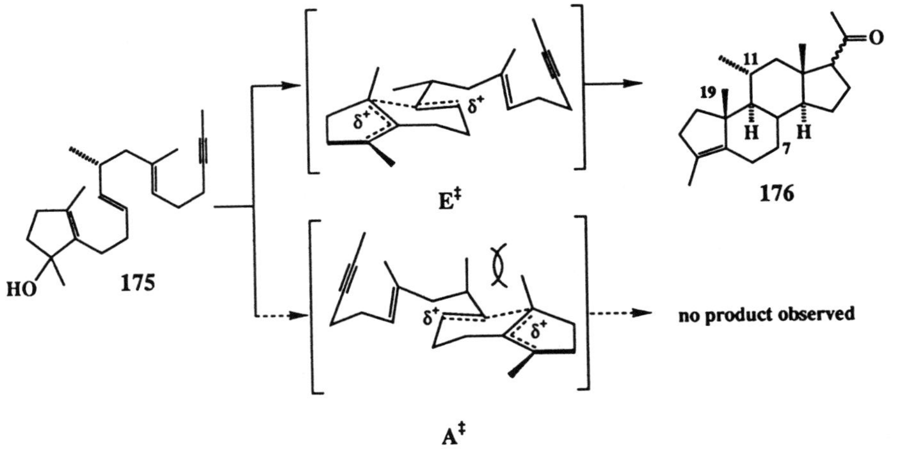

Figure 12.119. Alternative transition state structures for the formation of 11-methyl steroids.

The choice of chiral initiating groups has been reduced more by the limitations of their chemical behavior than by their effectiveness in a stereochemical sense. Epoxides have been extensively investigated (Van Tamelen, 1968), but although nucleophilic attack on the complexed substrates appears to be stereoselective, the chemical yields tend to be very poor. Allylic alcohols provide some of the most effective initiating groups (cf. the conversion of **167** into **168**), but stereochemical information may be lost as a consequence of ionization prior to cyclization. The best prospects appear to lie with the utilization of chiral acetal auxiliaries, but it may be difficult to avoid destruction of the auxiliary group in order to remove it from the parent molecule (Johnson et al., 1976).

f. The Aldol Reaction

Under the impetus provided by the quest for efficient methods to construct polyketide natural products, the stereocontrolled crossed aldol reaction between aldehydes and enolates derived from ketones, and so on, has been the focus of a sustained effort over the past decade (Evans et al., 1981a; Heathcock, 1981; Evans et al., 1982c; Heathcock, 1984a,b; Masamune et al., 1985b; Braun, 1987; Hoffmann, 1987). There are two major stereochemical issues that must be addressed. The first is the relative configuration of the two newly created stereogenic centers, that is, "simple diastereoselectivity" (Heathcock, 1984b), as illustrated in Figure 12.120. The second issue concerns diastereofacial selectivity and arises with (a) the additions of achiral enolates to chiral aldehydes, especially with aldehydes that possess an α-stereogenic center, that is, the stereochemical relationship of this center to the newly formed adjacent carbinol methine (commonly described as the Cram–anti-Cram relationship), and (b) the addition of chiral enolates and their analogues to achiral aldehydes. Reactions of chiral aldehydes with chiral enolates will be treated later in Section 12-5, which deals with double stereodifferentiation.

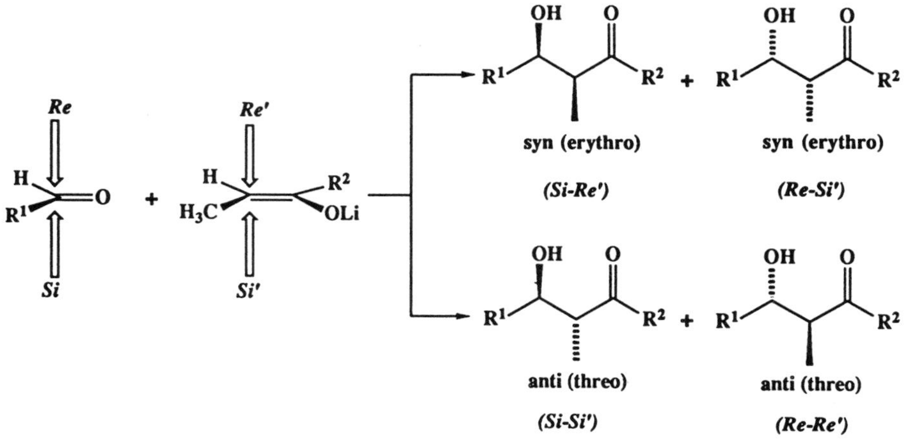

Figure 12.120.   The crossed aldol reaction between achiral reactants.

Addition of Achiral Enolates to Achiral Aldehydes

Both the aldehyde and the enolate have heterotopic faces, so there are two ways in which they can approach each other in a *relative* sense, that is, the *Si* face of the aldehyde can become bonded to either the *Si* face of the enolate to afford the anti (threo) diastereomer, or to the *Re* face to give the syn (erythro) diastereomer. Alternatively, the *Re* face of the aldehyde may react with the *Si* face of the enolate to form the syn product, or the *Re* face to form the anti isomer [Heathcock et al., 1980]. Clearly, the stereochemical outcome is dependent on the cis–trans configuration of the enolate anion, but the factors that control this aspect are now fairly well established (cf. Section 12-3.e).

> The terms syn and anti are derived from the simple and intuitive system of nomenclature advocated by Masamune et al. (1980) to describe the relative configurations of substituents along a chain in a pairwise manner. The terms are used in the sense that the main chain of a molecule is drawn in an extended zigzag and the substituents projected above and below the plane of the main chain. A pair of substituents that both lie on the same side of the plane are described as syn, whereas if they lie on opposite sides, they are termed anti. It is not necessary for the substituents to be adjacent and if more than two stereocenters are involved, it may be necessary to number the stereocenters so as to avoid possible misunderstandings. There can be a degree of arbitrariness involved in the choice of the main chain, however, leading to possible ambiguities. Because there is also widespread use of the erythro–threo nomenclature to specify these kinds of relationships, these terms have also been included, but in this context they are ambiguous and should therefore be abandoned. A discussion of the various conventions is given in Section 5-4 (cf. also Heathcock, 1984b and Brewster, 1986).

It has also been established what factors control the orientation of the reactants in the transition state. Diastereoselection is poor when R^2 is small, irrespective of the geometry of the enolate, but for bulky R^2 groups stereoselection is excellent for both E and Z enolates. The lithium cis enolate **178** from ethyl *tert*-butyl ketone **177** gives almost exclusively the syn (erythro) aldol **179** with benzaldehyde (Heathcock et al., 1980), for example, while the lithium trans-enolate **181** from 2',6'-dimethylphenyl propionate **180** affords predominantly anti (threo) products **182** with a variety of aldehydes (Heathcock et al., 1981a; Fig. 12.121).

Although the need to incorporate a bulky group, such as the *tert*-butyl group, into the enolate appears to be synthetically limiting, it is possible to utilize substituents with similar bulk and degrade them subsequently to more useful groups, for example, **183** serves as a useful syn-selective propanal synthon by virtue of a subsequent oxidative cleavage and has been used for the preparation of **184** in an approach to the total synthesis of erythronolide A (Fig. 12.122; Heathcock et al., 1985).

These outcomes may be rationalized in terms of the transition state models based on closed chairlike structures involving coordination between the two oxygen atoms and the metal center outlined in Figure 12.123 (Zimmerman and Traxler, 1957). If it is accepted that the dominant steric interaction is between R^1 and R^3, it can be readily visualized that the transition state leading from the cis

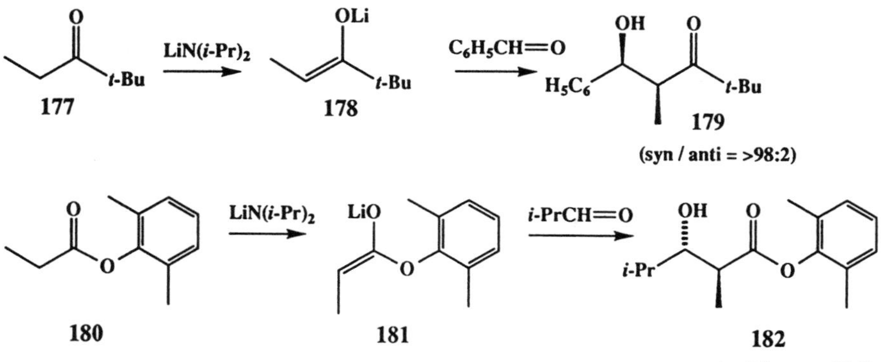

Figure 12.121. Control of simple diastereoselectivity in the aldol reaction.

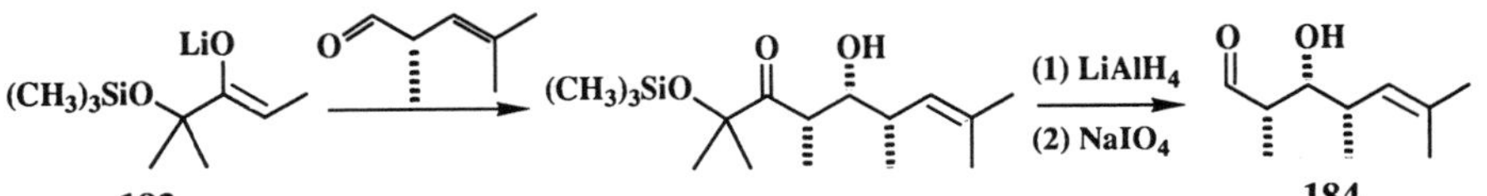

Figure 12.122. Illustration of the use of a syn-selective propanal synthetic equivalent in the aldol reaction.

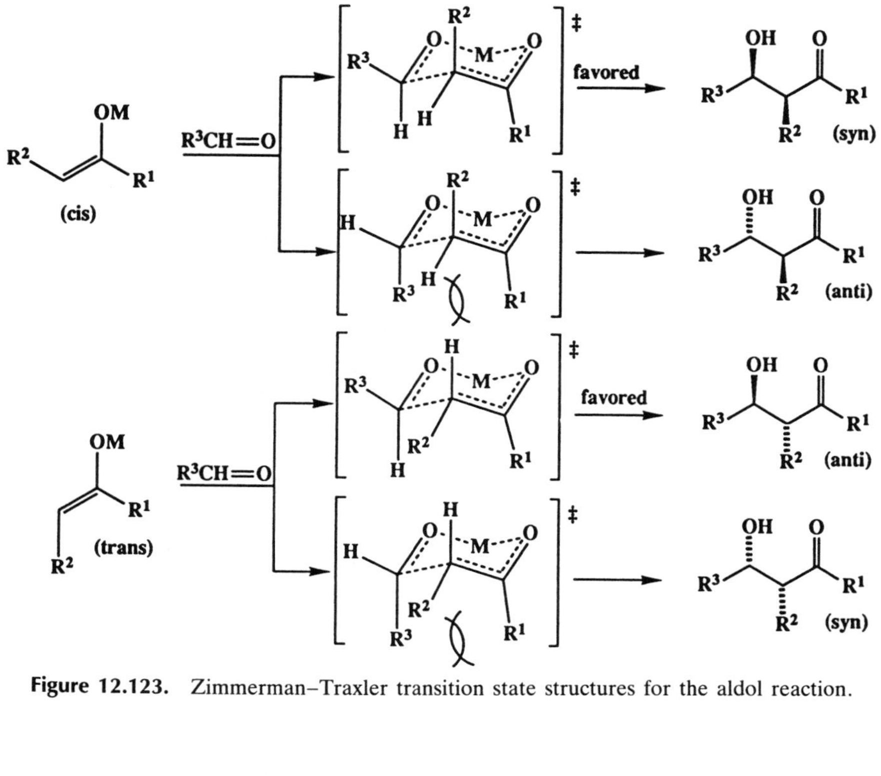

Figure 12.123. Zimmerman–Traxler transition state structures for the aldol reaction.

enolate to the anti aldol product is disfavored, as is the transition state for the trans enolate leading to the syn product. Moreover, $\Delta\Delta G^{\ddagger}$ for both pairs of transition states could be expected to increase as the steric demand of R^1 and R^3 increases. However, these idealized transition states do not account for the observation that cis enolates are significantly more stereoselective than trans enolates when R^1 is not large. It has therefore been suggested that there is a significant skewing of the transition states away from the idealized staggered arrangements implicit in the chairlike structures. This deviation diminishes the torsional angle between R^2 and R^3 during the conversion of the trans enolate to the anti aldol, thereby making this transformation relatively less favorable (Fellmann and Dubois, 1978). An alternative interpretation that considers four possible boat conformations in addition to the four chair arrangements has been suggested (Evans et al., 1982c).

Detailed analyses of the preceding hypotheses and extensions to other systems, including boron derived enolates, which are generally more stereoselective than the Groups 1 (IA) and 2 (IIA) metals (Fenzl and Köster, 1975; Evans et al., 1981a; Van Horne and Masamune, 1979), zirconium enolates, which show a strong preference for syn-selective formation of aldols irrespective of enolate geometry (Evans and McGee, 1980; Yamamoto and Maruyama, 1980), and the Lewis acid catalyzed reactions of enol silanes with aldehydes and acetals (Mukaiyama et al., 1974; Chan et al., 1979; Murata, Noyori, et al., 1980) have been provided by Evans et al. (1982c) and Heathcock (1984a). The theoretical aspects have been summarized by Li, Houk, et al. (1988). Readers should also recall the probable need raised earlier (Section 12-3.e) to consider the degree of aggregation of enolates participating in eletrophilic processes (Seebach et al., 1981; Williard and Salvino, 1985; Horner, Grutzner, et al., 1986; Seebach, 1988).

Aldol Equilibration: Thermodynamic Stereoselection. Aldol products can undergo syn–anti equilibration either by deprotonation–reprotonation, or more commonly by reverse aldolization. It should be noted that the rate of isomerization will be very slow where there is a strong kinetic preference for one of the diastereomers from the aldol reaction. If the syn selectivity is 80:1, for example, the reverse aldol reaction must occur 80 times for each syn aldolate to be converted to an anti aldolate. As might be expected, equilibration is facilitated by structural features that render the enolate less basic (i.e., enolates derived from ketones rather than esters and amides) and introduce a degree of steric crowding into the aldolate products. Those derived from propiophenone, for example, are especially prone to undergo the retrograde aldol process. The rate of equilibration is very sensitive to the nature of the associated cation and the relative rates are consistent with expectations associated with the relative strength of Lewis acidities. Boron enolates are stable, even at elevated temperatures, while potassium derivatives equilibrate with moderate rates even at $-78°C$ (further enhanced by the addition of 18-crown-6). Given this trend, it is reasonable to conclude that tris(diethylamino)sulfonium (TAS) enolates (Noyori et al., 1981) are probably reacting under thermodynamic control.

When the aldol reaction is carried out under reversible conditions, it is generally, but not always, found that the anti isomer predominates. If it is assumed that chelated products are more likely to be thermodynamically more

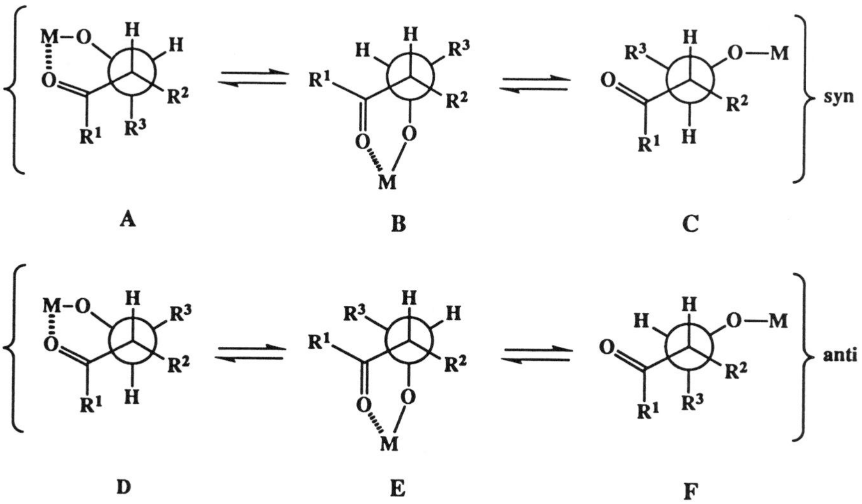

Figure 12.124. Conformations of syn and anti aldolates.

stable than nonchelated ones, this outcome may be rationalized by considering the Newman projections **A–F** for the two possible aldolates (Fig. 12.124). For the anti aldolate it can be seen that gauche interactions in **D** are minimized relative to conformations **A** or **B** of the syn aldolate. However, if the $R^2 \leftrightarrow R^3$ interaction is sufficiently large the anti isomer may take up conformation **E** and the syn isomer adopt the unchelated structure **C**, in which case it is then difficult to evaluate the relative stabilities.

Aldol Reactions of Achiral Enolates with Chiral Aldehydes

The second and more problematical stereochemical issue in the aldol reaction is the question of diastereofacial selectivity when the aldehyde contains a stereogenic center α to the carbonyl group (Cram/anti-Cram selectivity). Up to four diastereomers ($R^2 \neq H$) may be formed as illustrated in Figure 12.125.

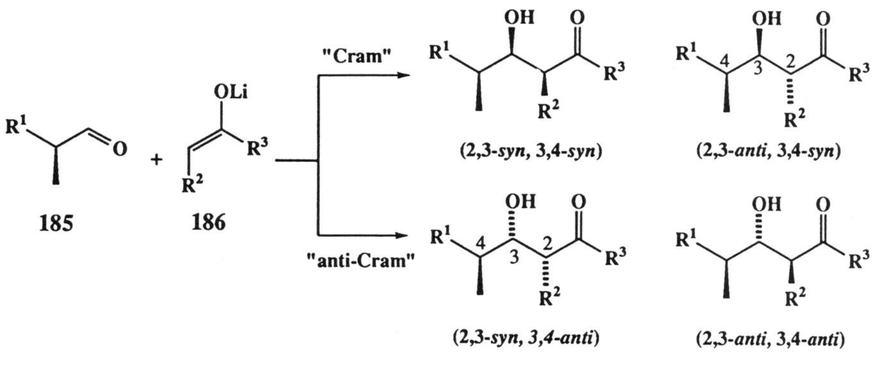

Figure 12.125. Reactions of an achiral enolate with a chiral aldehyde.

Moderately good diastereofacial selectivity with respect to the carbonyl group of 2-methylphenylacetaldehyde (**185**, $R^1 = C_6H_5$) (ca. 4:1 in favor of the "Cram" products) may be obtained with **186** ($R^2 = CH_3$, $R^3 = t\text{-Bu}$) and with **186** ($R^2 = CH_3$, $R^3 = 2',6'$-dimethylphenoxy) (Heathcock et al., 1980; 1981a), but these levels are exceptional; in general the outcomes are very sensitive to the nature of R^1 and difficult to predict (Masamune 1978). Moreover, it is not possible to achieve any significant diastereofacial discrimination for acetyl derived enolates (**186**, $R^2 = H$) (Braun, 1987). It is therefore necessary to incorporate a substituent (e.g., SCH_3), which can be removed subsequently (Evans et al., 1981b,c). The most general solution to both these problems, however, has been to utilize chiral enolates, that is, to resort to double stereodifferentiation (Section 12-5).

Aldol Reactions of Chiral Enolates with Achiral Aldehydes

In contrast to the poor outlook for diastereofacial discrimination described in the previous section, excellent levels of selectivity have been obtained for many chiral enolates derived from ketones, amides, and imides. A wide range of lithium enolates derived from chiral esters has been studied, but the results were on the whole discouraging (Morrison and Mosher, 1976; Dongala, Solladié, et al., 1973;

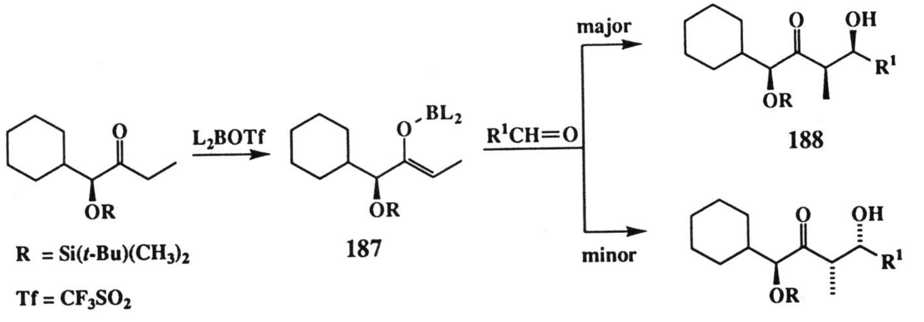

$R = Si(t\text{-Bu})(CH_3)_2$

$Tf = CF_3SO_2$

Figure 12.126. Diastereofacial selectivity in the reactions of an α-silyloxy enolate with aldehydes.

TABLE 12.3. Dependency of Diastereofacial Selectivity on Boryl Ligands.

L_2	R^1	188/189
9-BBN[a]	C_6H_5	14:1
	Et	17:1
	i-Pr	100:1
di-n-Bu	C_6H_5	40:1
	Et	50:1
	i-Pr	100:1
Cyclopentyl	C_6H_5	75:1
	Et	100:1
	i-Pr	No reaction

[a]9-Borabicyclo[3.3.1]nonane = 9-BBN.

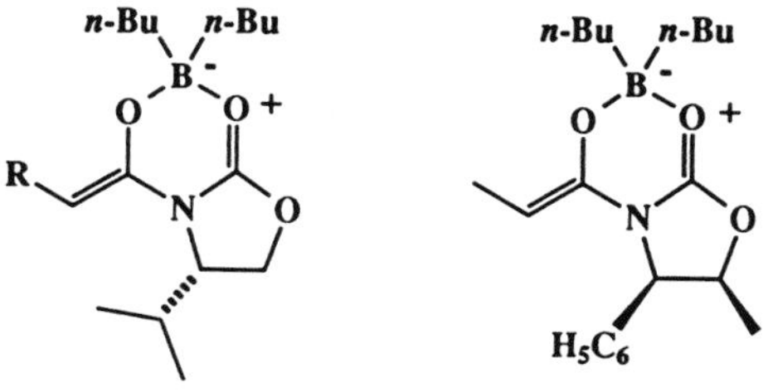

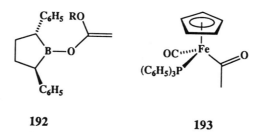

190   **191**

Figure 12.127. Oxazolidinone derived chiral enolates that exhibit high stereofacial selectivity.

Lampe, 1983). A variant on the Mukaiyama reaction utilizing the titanium chloride catalyzed addition of ketene silyl acetals derived from a camphor auxiliary has afforded 93–97% diastereoselection, however (Helmchen et al., 1985). Only moderate success has been obtained with simple ketone enolates (Seebach et al., 1976; Evans and Taber, 1980), but outstanding levels of selectivity have been obtained with zirconium and boron enolates derived from ethyl α-silyloxyalkyl ketones as in the stereoselective formation of **188** (Fig. 12.126; Heathcock et al., 1979; Masamune et al., 1981a,b; Enders and Lohray, 1988; Siegel and Thornton, 1989). For substrates like **187** which may serve as synthetic equivalents to propionate enolates (cf. Fig. 12.122, **183**→**184**), there is a strong dependency of the levels of stereoselectivity on the nature of the boryl ligands as indicated by the ratios of products as summarized in Table 12.3.

New benchmarks for both diastereofacial and syn selectivity have been established with zirconium and boryl enolates obtained from proline amides, and from valinol and ephedrine derived imides, for example, ratios of greater than 99:1 for both kinds of stereoselectivity have been obtained for *cis*-enolates **190** ($R = CH_3$) and **191** (Fig. 12.127; Evans and McGee, 1981; Evans et al., 1981c).

The challenge of obtaining good levels of stereoselectivity with acetyl enolates has been pursued with great vigor and ingenuity. One of the original stratagems of temporarily adding an achiral auxiliary (SCH_3) has already been noted. Satisfactory direct solutions have also been developed, however (Braun, 1987), for example, aldol reactions based on the use of borolane **192** (90% ds) (Reetz, 1988) and the chiral acetyl iron complex **193** (99% ds) illustrated in Figure 12.128 have been reported (Davies et al., 1984; Liebeskind and Welker, 1984; cf. Fig. 12.47).

192   **193**

Figure 12.128. Acetate equivalents for stereocontrolled aldol reactions.

Allyl and Crotyl Metals as Enolate Equivalents in the Aldol Reaction

Finally, note should be made of the additions of various achiral and chiral allyl and crotyl metallic derivatives that have served as masked equivalents of aldehyde enolates, by means of subsequent oxidative cleavage of the alkene bond in the products (Hoffmann, 1982; 1987; Yamamoto and Maruyama, 1982). Numerous silanes (Hayashi et al., 1982, 1983), boranes (Hoffmann and Herold, 1981; Midland and Preston, 1982; Jadhav et al., 1986), stannanes (Suzuki et al., 1985), and chromium derivatives (Lewis and Kishi, 1982; Suzuki et al., 1986) have all been used with good to excellent results as illustrated by the reaction of *B*-allylbis(2-isocaranyl)borane (**194**) with aldehydes (Fig. 12.129). The enantiomer excesses of the homoallylic alcohol intermediates obtained in this way (94–99%) were significantly better than with the analogous bis(4-isocaranyl) and diisopinocampheyl derivatives (Brown et al., 1990).

g. Pericyclic Reactions

Pericyclic reactions have been defined (Woodward and Hoffmann, 1970) as "those [reactions] in which all first-order changes in bonding relationships take place in concert on a closed curve." Four categories of pericyclic processes have been identified: cycloadditions, sigmatropic rearrangements, electrocyclic reactions, and cheletropic reactions, each of which is in principle reversible. The geometric constraints and requirements for the preservation of orbital symmetry in concerted reactions ensure that diastereoselection is very good to excellent, and examples have already been provided in Sections 12-2.b and 12-3.b. The following treatment is largely confined to a brief stereochemical view of the synthetically most useful categories, that is, cycloadditions and sigmatropic rearrangements. For more comprehensive treatments of the Diels–Alder reaction see Paquette (1984); Oppolzer (1984); and Helmchen et al. (1986); for sigmatropic rearrangements see Bennett (1977); Ziegler (1988); and Hill (1984). For more general treatments of pericyclic reactions see Gilchrist and Storr (1979) and Fleming (1976).

Cycloadditions

Cycloadditions involve the reaction of two or more reactants to form a ring in the product without the elimination of any other species (σ bonds are formed but not broken) and are classified according to the number of participating π electrons in

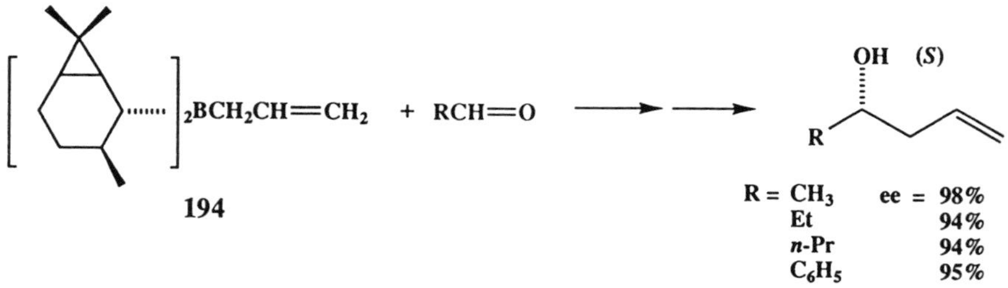

$$[\text{structure}]_2BCH_2CH{=}CH_2 \ + \ RCH{=}O \longrightarrow$$

194

R = CH$_3$	ee = 98%
Et	94%
n-Pr	94%
C$_6$H$_5$	95%

Figure 12.129. Diastereoselective addition of a chiral allylborane to aldehydes.

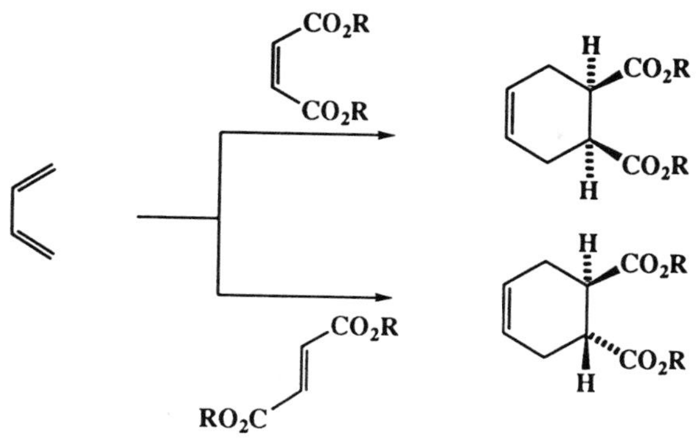

Figure 12.130. Retention of cis and trans geometry in [4 + 2] cycloadditions.

each reacting molecule. Thus, the Diels–Alder reaction, one of the most powerful reactions for assembling organic compounds, is described as a [4 + 2] cycloaddition.

[4 + 2] *Cycloadditions.* As a consequence of the concerted thermally allowed suprafacial–suprafacial addition for this process, cis–trans relationships between substituents in the reactants should be preserved in the products, and so the reactions of butadiene with maleic and fumaric esters, for example, afford cis and trans adducts, respectively (Fig. 12.130). It may be worth noting, however, that a theoretical assessment of the [4 + 2] cycloaddition between ethylene and butadiene at the 4-31G level showed that the synchronous mechanism was energetically favored by only about 2 kcal mol^{-1} (8.5 kJ mol^{-1}) over the nonconcerted mechanism (Bernardi, Robb, et al., 1988), although the addition of dynamic electron correlation should favor the concerted pathway at higher levels of theory (McDouall, Robb, et al., 1987). Nevertheless, less favorable entropy changes for the concerted mechanism may well result in a preference for the asynchronous mechanism when substituents have large steric requirements.

However, it is possible for there to be two relative orientations of the reactants in the transition state as in the reaction of acrylic esters with cyclopentadiene (Fig. 12.131). The endo adduct is assumed to be kinetically favored

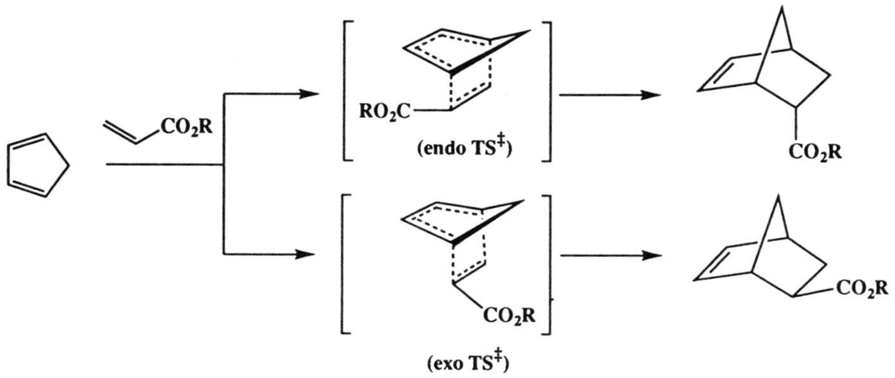

Figure 12.131. Endo and exo pathways for [4 + 2] cycloadditions.

for electronic reasons (secondary orbital interactions) (Ginsburg, 1983), while the exo isomer is thermodynamically preferred because of steric factors.

Given the important role of the Diels–Alder reaction in organic synthesis, it is not surprising to find that some of the earliest attempts at enantioselective syntheses (based on the use of chiral auxiliaries) have centered on this reaction. Since any chiral controlling element must be located externally to the ring being formed (cf. Section 12-3.b) it is necessary to select auxiliaries and reaction conditions with some care with a view to limiting conformational freedom in the transition states. Lewis acid catalysis is almost always essential for the attainment of acceptable levels of stereoselectivity (see below) and $\pi-\pi$ interactions (usually inferred) with an aromatic group in the chiral auxiliary is often a feature of the more controlled reactions (cf. Section 12-3.b).

CHIRAL DIENOPHILES. With the notable exceptions of some sugar-substituted acyclic esters, for example, 195, (Horton and Machinami, 1981) and the furanone 196 (Feringa and de Jong, 1988), diastereoselectivity has been rather poor for uncatalyzed [4 + 2] cycloadditions involving chiral dienophiles. The reaction of (−)-dimenthyl fumarate 197 with butadiene followed by hydrolysis, for example, afforded the (R,R)-(−)-diacid 198 in only a 5.4% ee (Walborsky et al., 1961, 1963; Fig. 12.132).

However, a dramatic increase in diastereoselectivity (interestingly, with the predominant formation of the alternative S,S diastereomer) was observed with Lewis acid catalysis (Walborsky et al., 1963). The best result (78% ee) in this early study was obtained with titanium tetrachloride in toluene, but has subsequently been improved to 90% with the use of diisobutylaluminum chloride (Furuta et al., 1986). These outcomes have been rationalized in terms of a transition state based on the s-trans conformer 199 in which the diene approaches the upper face of the dienophile remote from the isopropyl groups in the menthol auxiliaries (Fig. 12.133). The enhanced diastereoselectivity is attributed to the greater conformational rigidity of the complexed dienophile and to the lower reaction temperatures that are possible with Lewis acid catalysis. Although the present account is focused on the stereochemical aspects of the Diels–Alder reaction, it is worth noting the improved regioselectivity that is obtained with less symmetrical dienes and dienophiles under the influence of Lewis acids.

In chiral acrylates the controlling chiral auxiliary tends to be rather remote

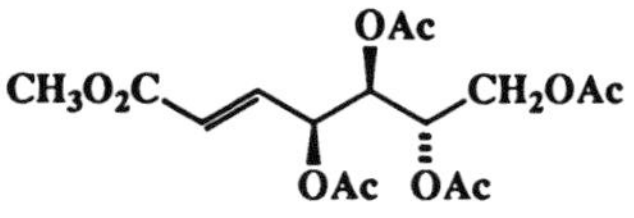

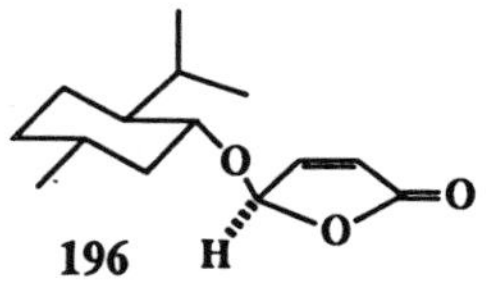

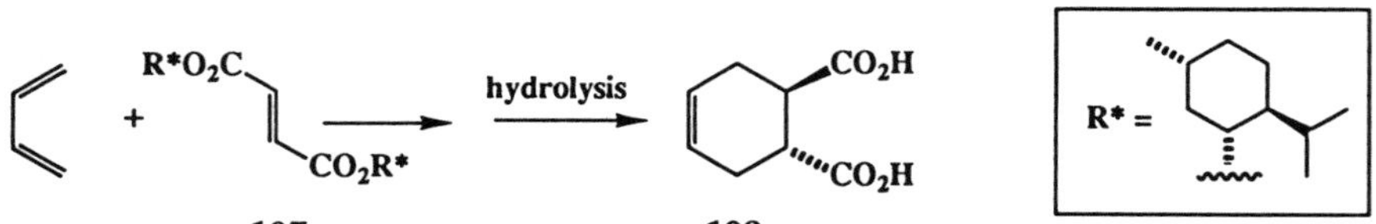

Figure 12.132. A selection of chiral dienophiles utilized in uncatalyzed [4 + 2] cycloadditions.

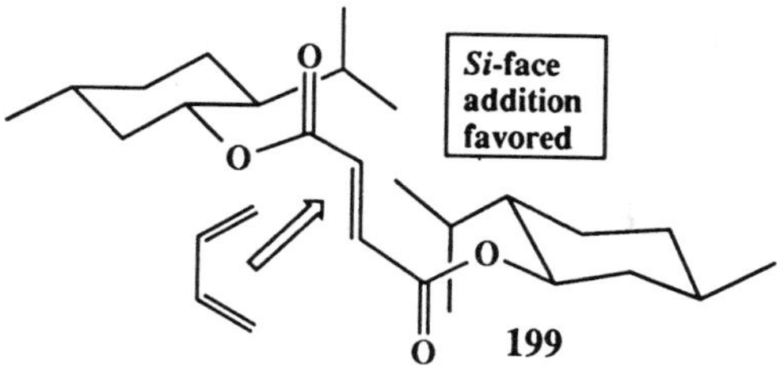

Figure 12.133. The [4 + 2] cycloaddition between butadiene and dimenthyl fumarate.

from the reaction center and menthol is of limited use. The compound 8-phenylmenthol **200** (Fig. 12-134) is considerably superior to menthol itself (Corey and Ensley, 1975; Oppolzer et al., 1980, 1981; Yamauchi and Watanabe, 1988; cf. Fig. 12.80) and excellent results have also been obtained with a variety of 2,3- and 2,10-substituted camphor derivatives in which an acrylate ester derivative is shielded on one side by a neighboring bulky substituent (cf. Fig. 12.82; Oppolzer, 1987). These latter examples illustrate the advantages of creating a concave reaction site (Helmchen and Schmierer, 1981), but other strategies have been just as effective. One of these has been based on the oxazolidinone auxiliaries (cf. Fig. 12.49), which have proven to be so effective in the diastereoselective alkylation and aldol reactions of enolates (Sections 12-3.e, 12-3.f), for example, the [4 + 2] cycloadditions of the crotyl derivative **201** with cyclopentadiene, which forms the endo adduct **202** as the major product (Fig. 12.135).

From a systematic study on the stoichiometry of the reaction it was deduced that the high levels of diastereofacial differentiation were dependent on the formation of bidentate chelate **206** since a dramatic increase in rate (ca. 100-fold), endo/exo ratio, and diastereoselectivity occurred when the molar ratio of the Lewis acid exceeded unity. These outcomes were rationalized in terms of the equilibria outlined in Figure 12.136 (Evans et al., 1988a).

Another approach based on bidentate chelation utilized the α'-hydroxy vinyl ketone **207**, although it should be noted that Lewis acid catalysis is not essential to achieve good levels of diastereoselection, presumably because hydrogen bonding serves a similar organizational role. In this substrate the controlling stereogenic center is positioned more closely to the newly formed stereocenters in the bond-forming processes, as illustrated in Figure 12.137, and **207** may serve as a synthetic equivalent of acrolein by virtue of a subsequent oxidative cleavage (Choy, Masamune, et al., 1983).

CHIRAL DIENES. The utilization of chiral dienes in [4 + 2] cycloadditions has been comparatively neglected (Fig. 12.138). Carbohydrate derived auxiliaries have been employed for the synthesis of di- and trisaccharides, affording good facial diastereoselectivity, but poor endo/exo selectivity (David, Lubineau, et al.,

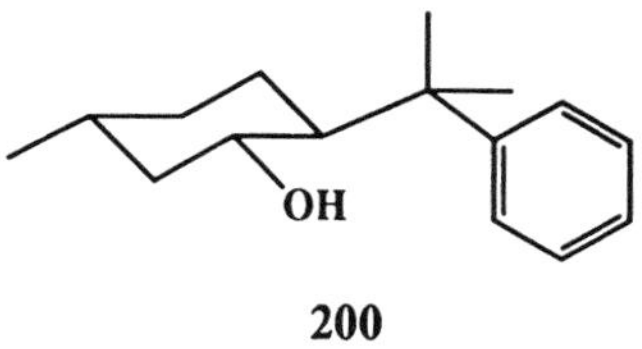

200

Figure 12.134. 8-Phenylmenthol.

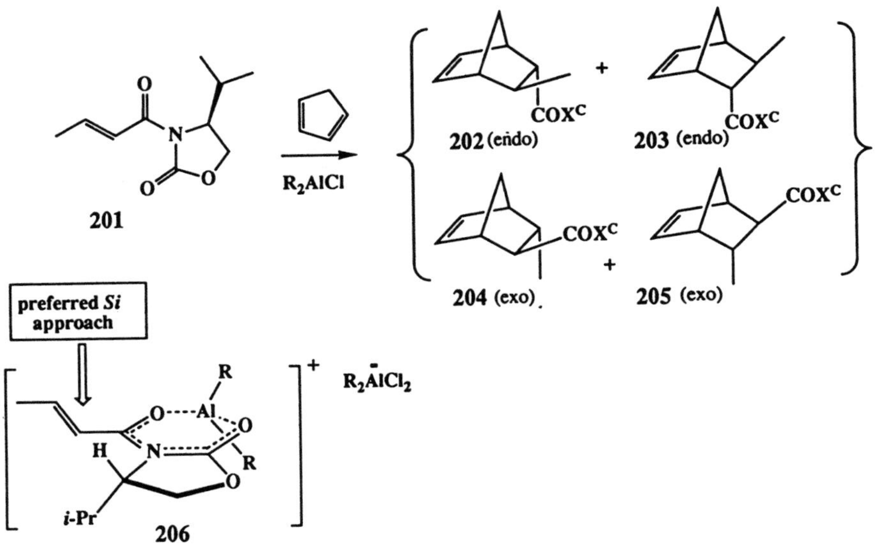

Figure 12.135. The [4 + 2] cycloadditions of an oxazolidinone–Lewis acid complex with cyclopentadiene, where X^C = oxazolidinone auxiliary.

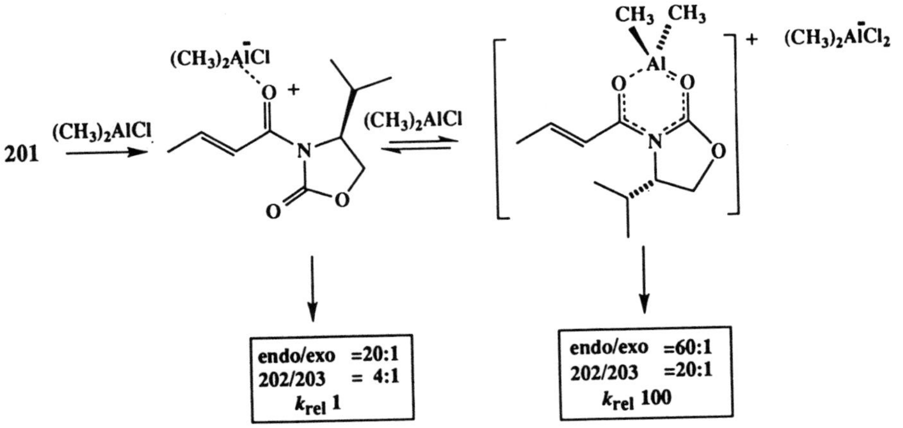

Figure 12.136. Influence of Lewis acid stoichiometry on stereoselectivity in a [4 + 2] cycloaddition.

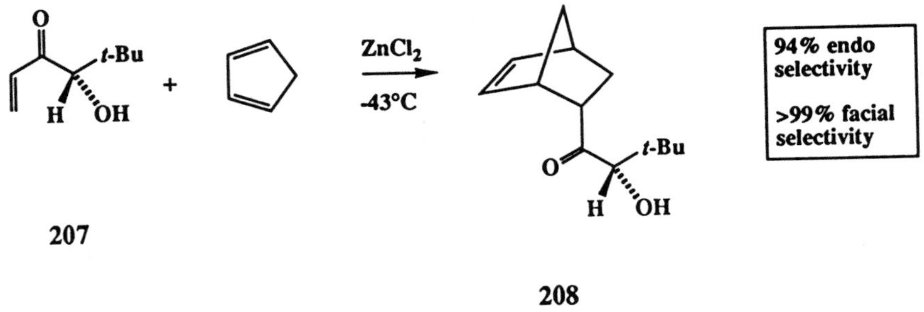

Figure 12.137. The [4 + 2] cycloaddition of an α'-hydroxy vinyl ketone with cyclopentadiene.

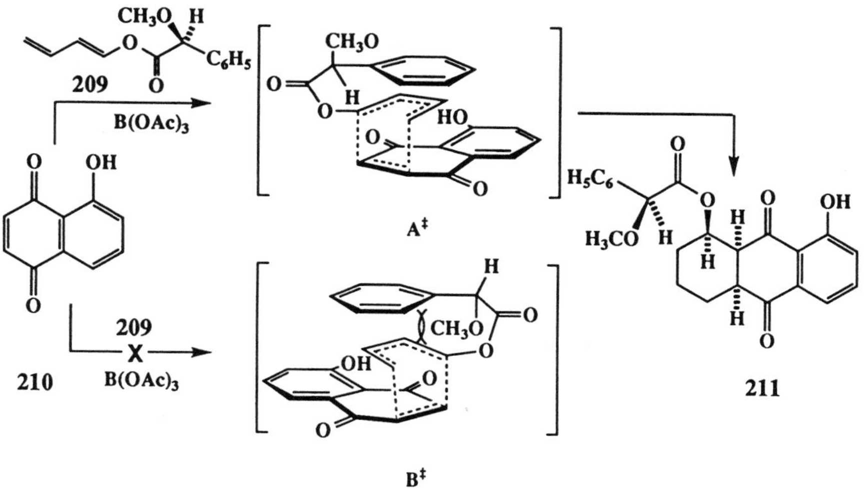

Figure 12.138. Transition state models $\mathbf{A}^{\ddagger}$ and $\mathbf{B}^{\ddagger}$ showing $\pi-\pi$ interactions in a [4 + 2] cycloaddition reaction (the Lewis acid has been omitted from the drawings of $\mathbf{A}^{\ddagger}$ and $\mathbf{B}^{\ddagger}$ for the sake of clarity).

1978). The boron trifluoride catalyzed reaction of the dienyl (S)-O-methylmandelate **209** with acrolein afforded 4:1 diastereoselectivity at $-20°C$ (Trost et al., 1980) and 94:6 at $-78°C$ (Siegel and Thornton, 1989), while reaction with 5-hydroxynaphthoquinone **210** gave only diastereomer **211** (to the limits of detection by ^1H and ^{13}C NMR). The high level of selectivity was rationalized originally by assuming that the alternative transition state structures $\mathbf{A}^{\ddagger}$ and $\mathbf{B}^{\ddagger}$ are preferred over those based on more extended conformations because of stabilizing $\pi-\pi$ interactions between the phenyl ring of the auxiliary and the diene–dienophile complex (Fig. 12.138). The $\mathbf{B}^{\ddagger}$ structure engenders a destabilizing nonbonded interaction between the methoxy group and H(2) of the diene, and hence the reaction proceeds via $\mathbf{A}^{\ddagger}$.

A major difficulty for this hypothesis, however, is the need to invoke the unfavorable Z(s-trans) conformation for the ester group (see p. 618). Moreover, it was reported subsequently that the cyclohexyl analogue of **209** gave a similar ratio of products with acrolein (Masamune et al., 1985a, footnote, p. 12). The most recent view (Siegel and Thornton, 1989) is that the transition state structure is based on the extended conformation **212** (Fig. 12.139) for the diene in which the methoxy and carbonyl groups are eclipsed (calculated to be the most stable arrangement) with the dienophile approaching the face opposite to the bulky phenyl group, although this appears quite remote (cf. also Tucker et al., 1990).

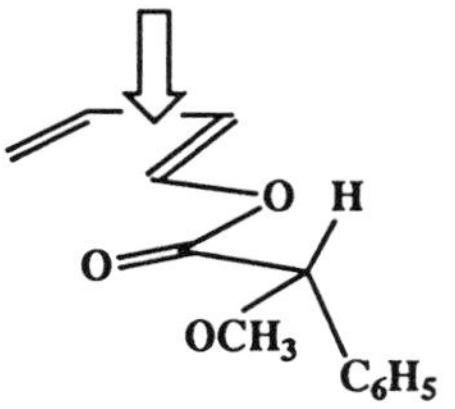

Figure 12.139. Revised transition state structure for the [4 + 2] cycloadditions of diene **209**.

A comprehensive examination of twenty [4 + 2] cycloadditions of chiral dienophiles with stereogenic allylic centers not involving a phenyl substituent has been reported (Tripathy, Franck, and Onan, 1988), but the authors were unable to arrive at a cohesive hypothesis for rationalizing the outcomes. Nor were they able to reconcile the outcomes with theoretical hypotheses (Houk et al., 1984a; Kahn and Hehre, 1987) and other experimental results (Kozikowski and Nieduzak, 1986a, Kozikowski et al., 1986b, 1987). Further studies within the context of a synthetic approach to the biologically important diterpene forskolin indicate that a resolution of the dilemma may be forthcoming, however (Kozikowski et al., 1988).

[3 + 2] *and* [2 + 2] *Cycloadditions.* The stereoselective aspects of [3 + 2] cycloadditions are not expected to differ significantly from those associated with [4 + 2] processes (Huisgen, 1968, but see Firestone, 1968). Reactions of nitrones with chiral groups attached to carbon (DeShong et al., 1984) and nitrogen (Wovkulich and Uskokovic, 1981; Vasella and Voeffray, 1982; Kametani et al., 1985; Iida, Kibayashi, et al., 1986) have been studied, as have the additions of nitrile oxides with chiral alkenes (Kametani et al., 1981; Kozikowski et al., 1983; Curran et al., 1987; Olsson et al., 1988). There are only a few reports of stereoselective [2 + 2] cycloadditions: those of chiral keteniminium salts (Houge, Ghosez, et al., 1982; Saimoto, Ghosez, et al., 1983), of a menthyloxyketene (Fráter et al., 1986) and of the addition of ketenes to chiral alkenes (Greene and Charbonnier, 1985).

Sigmatropic Rearrangements

As a consequence of highly ordered cyclic transition states, sigmatropic rearrangements have provided some of the most effective methodologies for the creation of new stereogenic centers in a controlled and predictable manner. A sigmatropic change of order $[i, j]$ involves the migration of a σ-bond flanked by one or more π bonds to a new position whose termini are $i - 1$ and $j - 1$ atoms removed from the original bonded loci in an uncatalyzed intramolecular process. Thus, the well-known Cope and Claisen rearrangements are sigmatropic rearrangements of order [3,3], while the Wittig rearrangement is an example of a [2, 3]-sigmatropic rearrangement (Fig. 12.140). The most synthetically useful of these sigmatropic processes is the Claisen rearrangement and many variations related inter alia, to esters and amides have evolved (Ziegler, 1988). The "aza-Cope" (Blechert, 1989) and "oxy-Cope" rearrangements also offer considerable utility.

Many examples involve the mutual creation and destruction of stereogenic centers in a 1,3 relationship (often described as 1,3-chirality transfer), which may allow a carbon–carbon bond to be formed at the expense of a carbon–heteroatom bond. This finding can be synthetically advantageous, since it is often possible to effect greater stereochemical control over the formation of stereocenters involving the latter. Also, undesired stereoisomers may, in principle, be recycled.

Claisen Rearrangement. Several examples of acyclic Claisen rearrangements have already been described (Sections 12-2.b, 12-3.b) illustrating the high levels

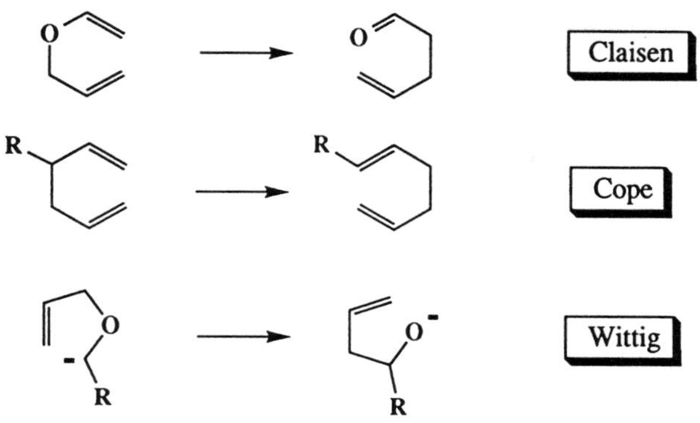

Figure 12.140. Common synthetically useful sigmatropic rearrangements.

of stereoselectivity that are usually obtained with this procedure. Indeed, the selectivity is such that in one model study on the total synthesis of the antineoplastic agent bruceantin **214**, the C(14) stereocenter was constructed by means of the Claisen rearrangement of **213–215**, rather than by the more conventional addition of a substituent to the C ring of a tricyclic structure (Fig. 12.141; Ziegler et al., 1985). The same basic concept of utilizing acyclic stereocontrol to effect the stereoselective synthesis of stereocenters in cyclic compounds had been employed earlier by the same group in a preparation of 11-oxoprogesterone, although on that occasion a Cope rearrangement had been employed (Ziegler and Wang, 1984).

If the olefinic bond of the chiral allylic moiety in the Claisen rearrangement is replaced by an alkynyl bond, the centrodissymmetry of the stereogenic center can be translated into the axial dissymmetry of the allenic product. The (R)-ether **216**, for example, rearranges to the (R)-allenic aldehyde **217** (Fig. 12.142; Evans, Landor, et al., 1965; cf. also Gibbs and Okamura, 1988 and Section 14-2.b).

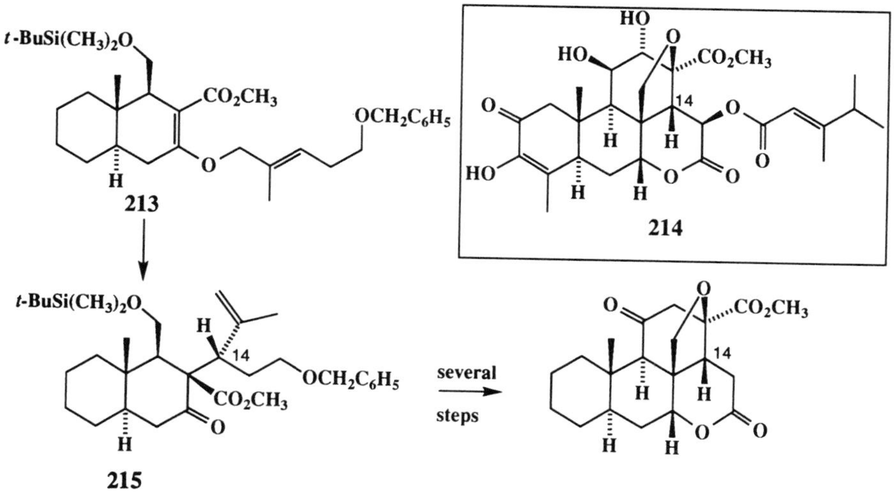

Figure 12.141. Use of an acyclic Claisen rearrangement to control stereochemistry in a cyclic target.

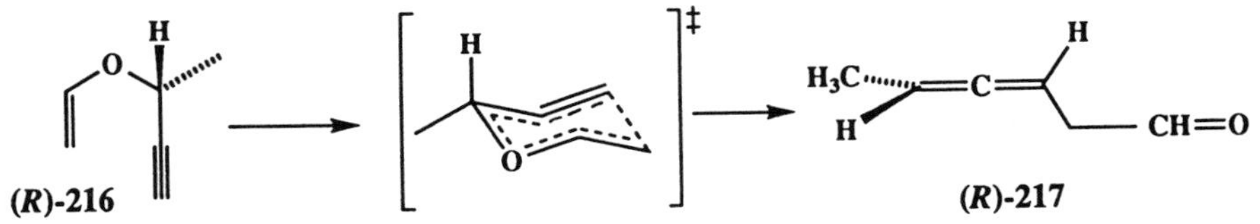

Figure 12.142. Translation of centrodissymmetry into axial dissymmetry during a Claisen rearrangement.

When the allylic or vinylic components are part of a small ring, there is a range of possible stereochemical outcomes as illustrated in Figure 12.143. If the allylic ether moiety is wholly endocyclic, the vinyl group migrates in a suprafacial manner (Eq. 12.1; Church, Ireland, et al., 1966). Otherwise, formation of products is sterically controlled, with the new bond forming on the more exposed face of the ring system (Eq. 12.2; Gibson and Barneis, 1972, except when there are stereoelectronic constraints that favor formation of the axial isomer (Eq. 12.3; Ireland and Varney, 1983). There is little stereoselectivity when the olefinic bonds are exocyclic as in Eqs. 12.4 and 12.5 (House et al., 1975), although this could undoubtedly be improved if the termini of the migrating allyl or vinyl group were substituted (cf. the rearrangement of **223**, Fig. 12.146).

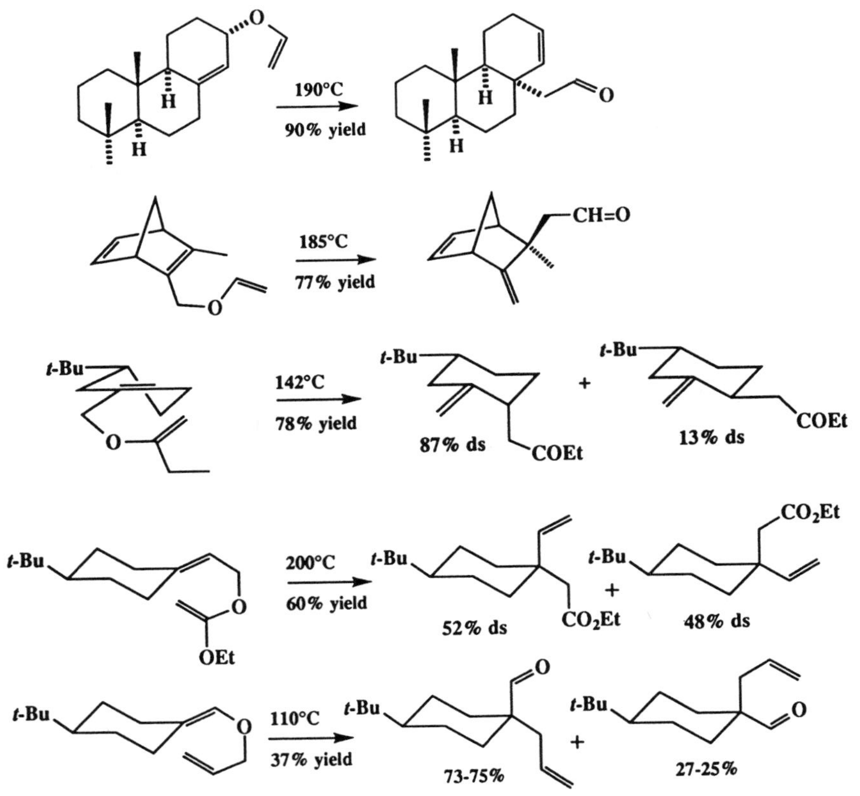

Figure 12.143. Claisen rearrangements of vinyl ethers derived from cyclic allylic alcohols.

Cope Rearrangement. The Cope rearrangement of 1,5-hexadiene derivatives corresponds to the "all carbon" equivalent of the Claisen rearrangement and like these processes the preferred transition state structures are chairlike with substituents located equatorially as far as possible. In a classical stereochemical study (Doering and Roth, 1962) it was found that *meso*-3,4-dimethyl-1,5-hexadiene **218** rearranged to the (*E,Z*)-diene **219** with 99.7% diastereoselectivity, while the racemic isomer **220** afforded a 9:1 mixture of (*E,E*)-**221** and (*Z,Z*)-diene **222**, respectively (Fig. 12.144). It was deduced that the chairlike arrangement was favored over the corresponding boat system by 5.5 kcal mol^{-1} (23 kJ mol^{-1}).

The Cope rearrangement per se has not been used in synthesis as widely as the Claisen rearrangement, possibly because of the higher activation energies required and equilibria that may not lie in the direction of the product. The latter are not a problem with the oxy-Cope variant, however, and enormous rate accelerations may be obtained by conversion of the substrates into potassium alkoxides, enhanced further by the addition of 18-crown-6-ether (Evans and Golob, 1975). Accordingly, there have been numerous applications (Paquette et al., 1990), especially as substrates are in many cases readily assembled by the addition of alkenyl metals to ketones (Fig. 12.145). Another strategem to displace the equilibrium in the desired direction has been to couple the Cope reaction with a sequential in situ (i.e., tandem) rearrangement (Thomas and Ohloff, 1970; Ziegler and Fan, 1981; Ziegler et al., 1982).

[2,3] *Sigmatropic Rearrangements.* The [2,3]-sigmatropic rearrangements of a diverse array of compounds have been observed, including ammonium, oxonium, and sulfonium ylides (Hoffmann, 1979; Doyle et al., 1988), allylic and benzylic ethers, amine oxides, sulfoxides and sulfenates, sulfinates, and phosphites (Hill, 1984; Mikami and Nakai, 1991). Some of the substrates, particularly ylides, have high ground-state energies, and therefore undergo reaction at relatively low temperatures, thereby enhancing prospects for improved stereoselection. The migrating group often carries a bulky substituent that should enhance stereoselection subject to steric control. Rearrangement of ylide **223**, for example, proceeds

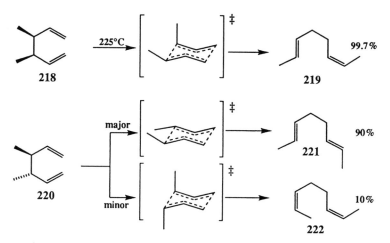

Figure 12.144. Cope rearrangement of 3,4-dimethylhexa-1,5-dienes.

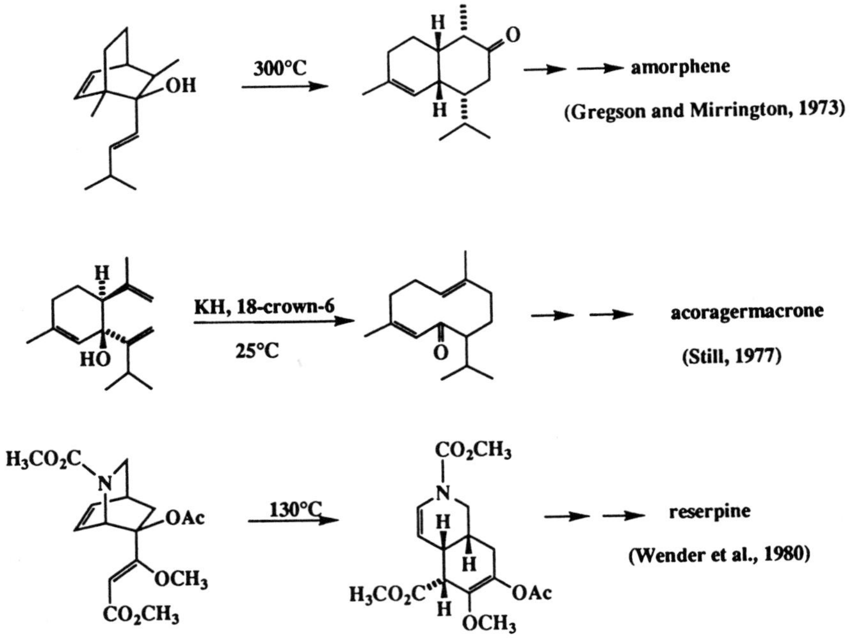

Figure 12.145. Examples of the oxy-Cope rearrangement directed towards the synthesis of natural products.

with significantly improved equatorial selectivity (Fig. 12.146; Andrews and Evans, 1972) relative to the Claisen equivalents (Egs. 12.4 and 12.5 in Fig. 12.143). In the absence of the clearcut conformational differences between chair- and boat-like transition state structures available to the six-centered [3, 3] rearrangements, acyclic examples of [2, 3] processes are less stereoselective, although anti–syn ratios may be improved by utilizing Z alkenes (Rautenstrauch, 1970; Nakai et al., 1981).

As with [3, 3]-sigmatropic rearrangements, 1,3 transfers of chirality are also a feature of [2, 3] processes, but there is rather greater scope and flexibility as illustrated by the especially instructive example outlined in Figure 12.147 (Miller, Stork, et al., 1974). It had been established that the ($\pm$)-enone **224** reacted with the ($\pm$)-(Z)-cuprate **225** in very much higher yield than did the corresponding E-isomer. To obtain the desired target **230**, therefore, the initial ($\pm$)-adduct **226** was converted into sulfenate **227**, which rearranged immediately to sulfoxide **228**. When this was treated with trimethyl phosphite, **230** was obtained by trapping of

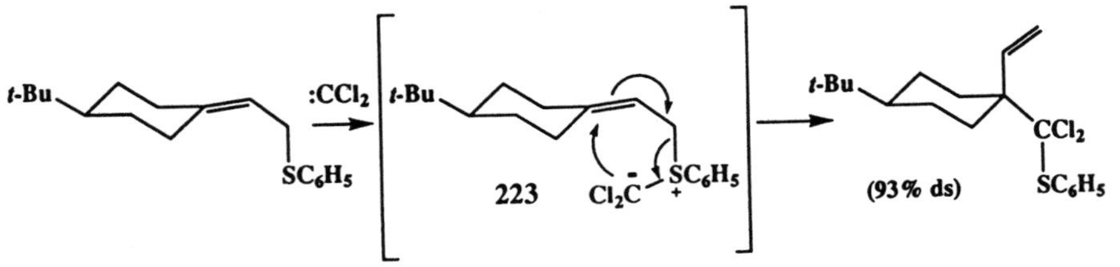

Figure 12.146. The [2,3] sigmatropic rearrangements of a cyclohexylidene ylide.

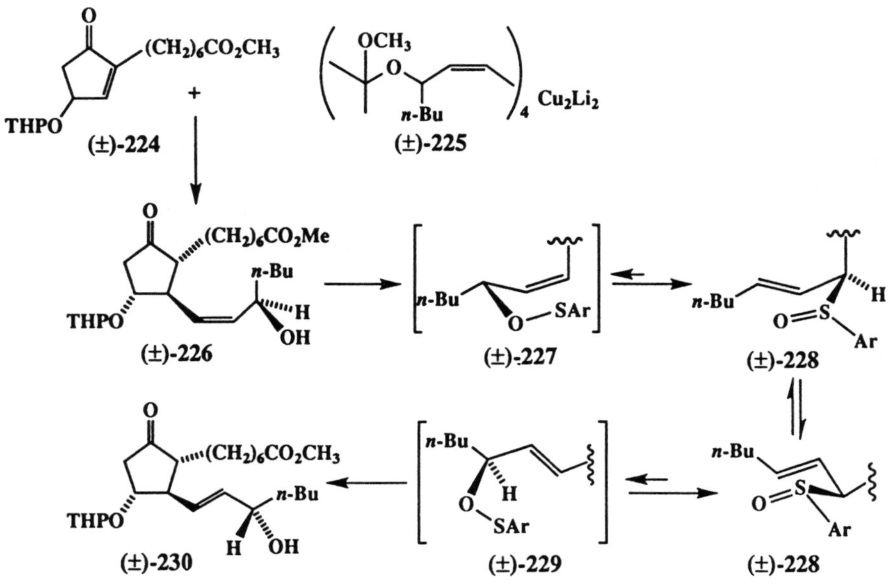

THP = tetrahydropyranyl,
Ar = *p*-CH₃C₆H₄

Figure 12.147. Use of a sulfenate–sulfoxide rearrangement for the stereocontrolled synthesis of a prostaglandin (only one enantiomer is depicted).

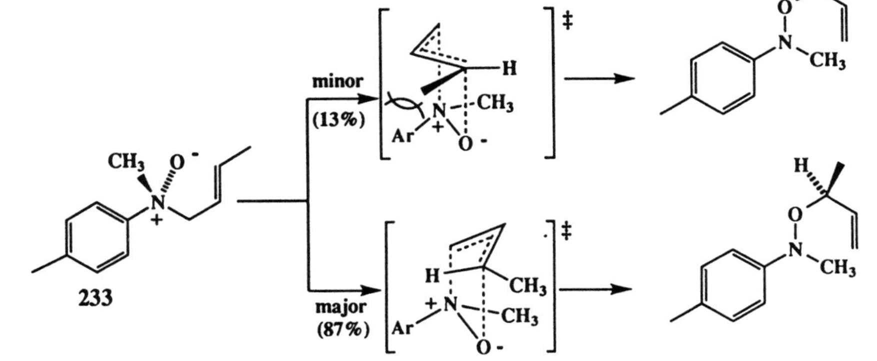

Figure 12.148. Transfer of stereogenicity from a heteroatom to carbon.

the sulfenate **229**. Thus, both the configuration of the alkene bond and the
carbinol center in the side chain had undergone inversion. The coupling of the
two racemates ($\pm$)-**224** and ($\pm$)-**225** to form only one diastereomeric racemate
($\pm$)-**226** is also noteworthy (cf. Section 12-1.d).

The preceding example also highlights a further aspect of [2,3]-sigmatropic
rearrangements, namely, the opportunity to utilize stereogenic heteroatoms as a
source of stereogenic carbon centers (Fig. 12.148). This feature has been studied
in some detail for sulfonium salts (Trost and Hammen, 1973), amine oxides
(Morikawa, Inouye, et al., 1976), and sulfoxides (Bickart, Mislow, et al., 1968;
Hoffmann, 1979).

h. Catalytic Hydrogenations

Hydrogenation over Heterogeneous Catalysts

The stereochemical control of the hydrogenation of olefinic bonds has been a
long-standing, but elusive goal. Although the outcome appears to be governed
largely by steric factors, it is often difficult to predict and is sensitive to catalyst,
solvent, pH, and pressure. A number of early scattered examples involving polar
neighboring groups appeared to provide some prospect of control in heterogeneous catalysis, whereby attractive (haptophilic) interactions between the catalyst surface and groups, such as hydroxyl, would favor syn additions of hydrogen.
One systematic study is summarized in Figure 12.149 (Thompson and Naipawer,
1973).

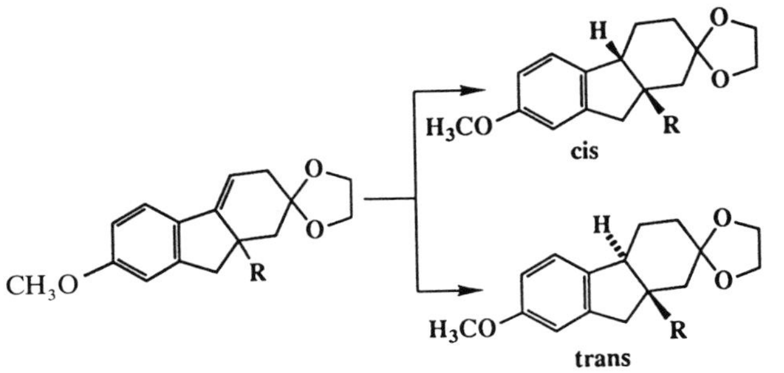

Figure 12.149. Heterogeneous hydrogenation directed by polar functional groups.

TABLE 12.4. Diastereoselectivities in
Hydrogenation Reactions.

R	cis (%)	trans (%)
CH_2OH	95	5
CH=O	93	7
CN	75	25
CH=NOH	65	35
CO_2Na	55	45
CO_2Li	23	77
CO_2H	18	82
CO_2CH_3	15	85
$COCH_3$	14	86
$CONH_2$	10	90

Directed Hydrogenations with Soluble Catalysts

A superior level of control has been realized with allylic and homoallylic carbinols and soluble catalysts, in particular with selected cationic iridium and rhodium catalysts (Thompson and McPherson, 1974; Crabtree et al., 1979; Brown and Naik, 1982). Heterogeneous hydrogenation of hydrindenone **234** over palladium on charcoal, for example, typically afforded only the cis-derivative **235**, whereas reduction over the catalyst $[Ir(COD)Py(PCy_3)]^+PF_6^-$ gave a 96:4 mixture of the *trans*-**236** and *cis*-**235** isomers (Fig. 12.150, Stork and Kahne, 1983). Following careful analysis of the role of pressure, and the nature and proportion of the catalyst (pressures of 1000 psi and/or low ratios of catalyst may be advantageous), very useful levels of diastereoselectivity have been obtained with further substrates (Evans and Morrissey, 1984a).

Extensions of the methodology to acyclic allylic and homoallylic alcohols have also proven to be rewarding in many cases (Evans and Morrissey, 1984b). Hydrogenation of **237** over the cationic rhodium catalyst $[Rh(NBD)(DIPHOS-4)]^+BF_4^-$, for example, afforded a 97:3 mixture of **238** and **239**, while the substrate **240** in which the coordinating hydroxyl is interchanged with the secondary ether group, affords the stereochemically complementary result, that is, a 97:3 mixture of **241** and **242**. The syn-isomer **243** corresponding to **240**, however, was a less satisfactory substrate, giving rise to an 89:11 mixture of epimers **244** and **245** (Fig. 12.151; Evans et al., 1985b).

Analysis of conformational factors allows prediction of both the level and direction of stereoselectivity, for example, in considering the hydrogenation of **240**, conformation **A** should be preferred to conformation **B**, for which there would be severe allylic $A^{(1,3)}$ strain arising from interaction between the alkenyl methyl group and the silyloxy methyl function (Fig. 12.152). Conformation **A** is not so favorable for the syn-diastereomer **243**, however, since the ethyl group would then be *quasi*-axial, thus accounting for the lower level of diastereoselection for this compound.

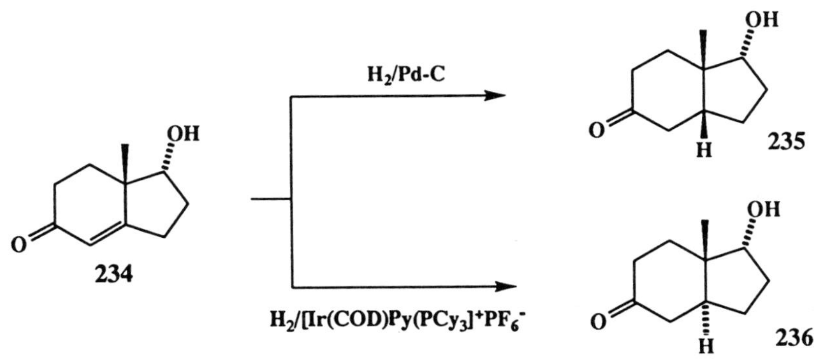

COD = cyclohexadiene
Cy = cyclohexyl

Figure 12.150. Hydroxyl directed hydrogenations over soluble catalysts, where COD = 1,5-cyclooctadiene, Cy = cyclohexyl, and Py = pyridine.

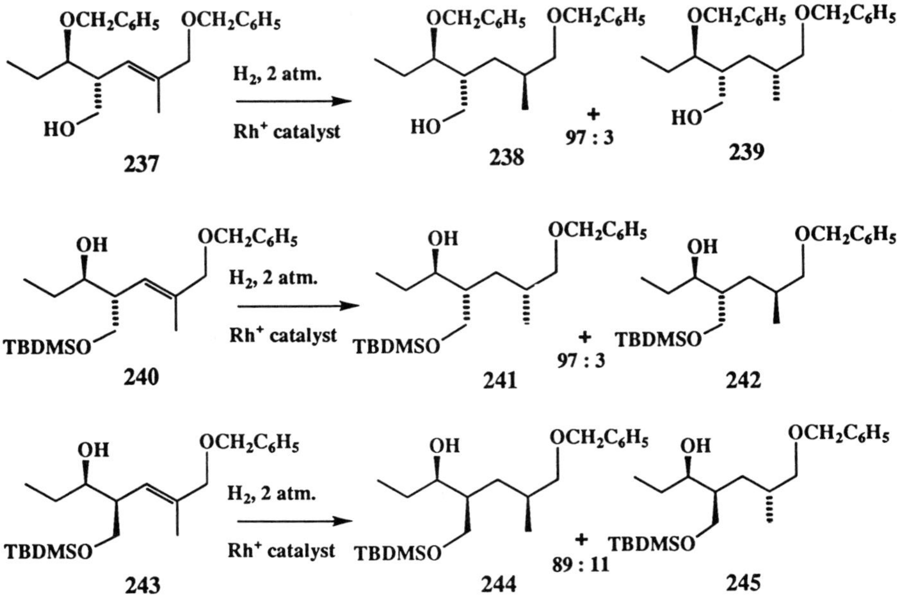

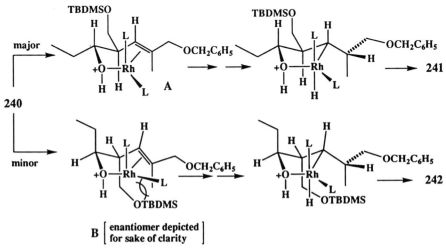

Figure 12.151. Hydroxyl-directed hydrogenations of acyclic alkenes (where nbd = norbornadiene, diphos-4 = 1,4-bis(diphenylphosphino)-butane, and tbdms = *tert*-butyldimethylsilyl.

Improved diastereoselection in the hydrogenation of **243** was achieved by utilizing the BINAP chiral catalysts developed by the Noyori group (Miyashita et al., 1980), that is, by applying the strategy of double stereodifferentiation (cf. Sections 12-1.c and 12-5). Enhanced selection was obtained not only with [Rh(+)-BINAP]$^+$ (**244/245** = 97:3), but also with [Rh(−)-BINAP]$^+$ (**244/245** = 92:8), however.

Figure 12.152. Preferred pathway for the hydrogenation of **240**.

Hydrogenation of Substrates Attached to Chiral Auxiliaries

Hydrogenations of substrates in which the faces of a double bond have been rendered diastereotopic by the attachment of a chiral auxiliary appear to have been directed primarily at the enantioselective preparation of α-amino and α-hydroxy acids. Reductions of N-acyl-α,β-unsaturated amino acids, α-keto esters, oximes, hydrazines, and imines of α-keto acids, amides or esters, and dehydroketopiperazines have been studied (Harada, 1985). A more general approach is provided by a recent example that involves the hydrogenation of the α,β-unsaturated amides **246**. Diastereomer ratios of greater than 9:1 were obtained (Fig. 12.153). From an analysis of the products it was proposed that in the absence of an α substituent, the catalyst was coordinated with the syn oriented sulfonyl and carbonyl groups as well as the lower face (as depicted) of the s-cis olefinic bond, resulting in delivery of hydrogen from the same side of the C(α)-Re face. An α substituent encounters an unfavorable nonbonded interaction in this conformation, however, and low diastereofacial selectivity (reversed in some cases) was observed with such substrates (Oppolzer et al., 1986).

i. Free Radical Reactions

In planning stereoselective syntheses, most chemists have tended to overlook the opportunities afforded by the reactions of free radical intermediates. Certainly the range of synthetic options is fairly narrow and on average, the reactions of free radicals tend to be less stereoselective than those of their polar counterparts. However, the higher reactivity of free radicals may be advantageous, while better chemoselectivity is often possible (cf. Furuta, Yamamoto, et al., 1988a).

Stereochemistry of Free Radical Additions and Atom Transfers

Whereas homolytic substitutions proceed with backside attack (inversion) (Beckwith and Boate, 1986), free radical additions and atom transfers appear to be subject to steric approach control (Fig. 12.154). Thus, the reaction of cyclopentyl bromohydrin **247** with allyl stannane (Keck et al., 1985) affords the trans isomer with 90% ds, although the selectivity is poorer for the cyclohexyl analogue **248** (trans/cis = 4:1) and almost nonexistent for lactone **249** (trans/cis = 1.7:1). The lactam **250** was transformed exclusively into **251**, however, suggesting that more than simple steric factors are involved (cf. Barton et al., 1982).

Stereoselectivity can be expected to be enhanced for addition to β-substituted alkenes. The reduction of **252** by borohydride or stannane in the presence of

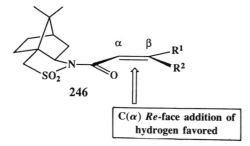

246

C(α) *Re*-face addition of
hydrogen favored

Figure 12.153. Catalytic hydrogenation of α,β-unsaturated substrates attached to a camphor sultam derived chiral auxiliary.

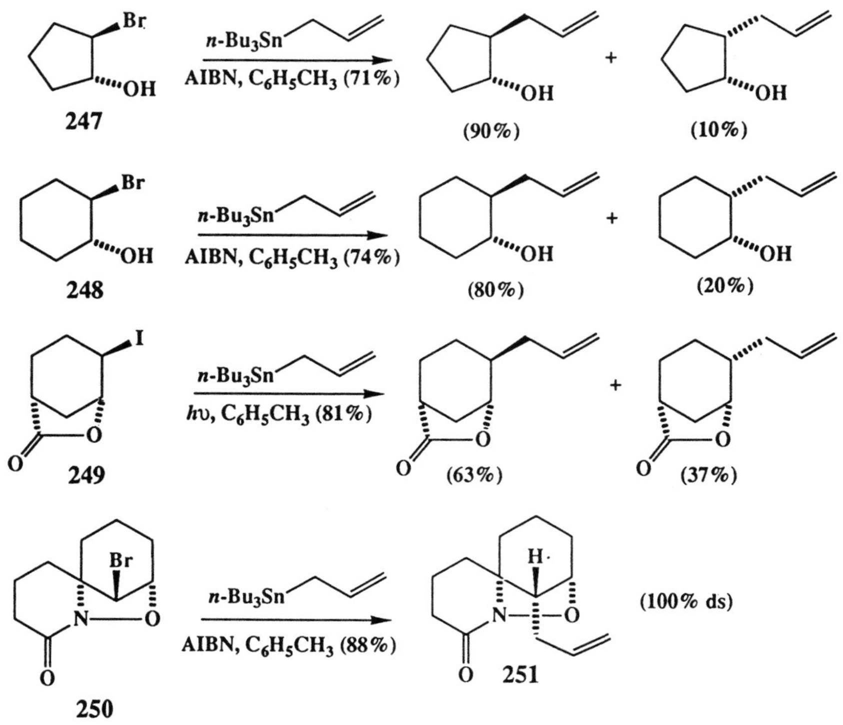

Figure 12.154. Stereoselective free radical additions to monocyclic intermediates, where AIBN = 2,2'-azobisisobutyronitrile.

methyl acrylate, for example, affords a 71:29 mixture of equatorial and axial adducts **253** and **254**, while addition to dimethyl fumarate proceeds with 95% equatorial selectivity to form mainly **255** (Fig. 12.155; Giese and Kroninger, 1984).

The normal preference for approach along the less hindered equatorial trajectory is, at first sight, not observed for the conversion of the glucosyl bromide **256** to the α-C-glucoside **257** (Fig. 12.156).

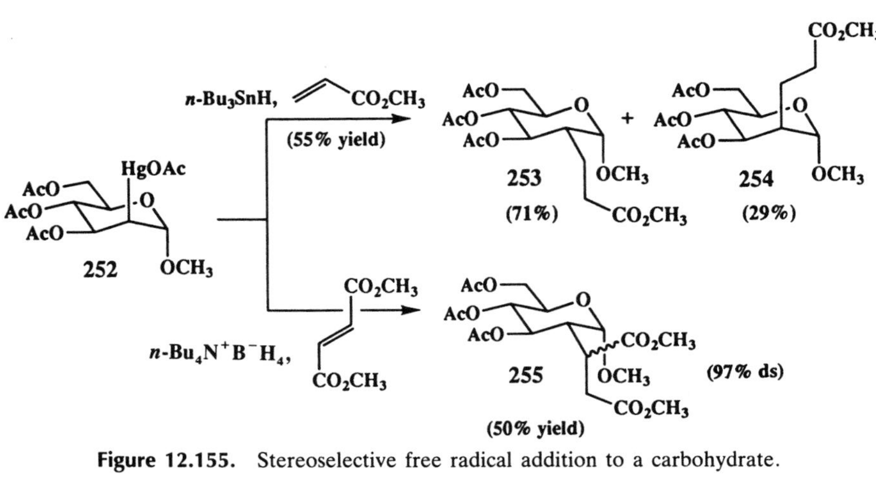

Figure 12.155. Stereoselective free radical addition to a carbohydrate.

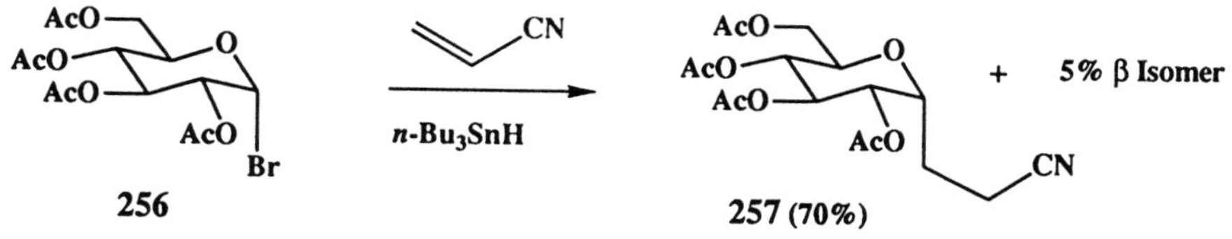

Figure 12.156. Stereoselective formation of an α-C-glucoside.

However, it was shown (Dupuis, Giese, et al., 1984; Korth, Giese, et al., 1986) that the intermediate free radical adopts the boat conformation **258** so as to maintain overlap between the higher energy SOMO (singly occupied molecular orbital) of the alkoxylalkyl radical and the LUMO of the C—O of the neighboring 6-acetoxy group which would consequently have an axial disposition (Fig. 12.157). The major product **257** therefore still arises by preferred equatorial attack.

Additions to bicyclic systems occur predictably on the more exposed face, as in the reaction of the intermediate free radical **260** formed from bromo ether **259** (Stork and Sher, 1983). The decalin system **261** behaved similarly (Fig. 12.158; Stork and Kahn, 1985).

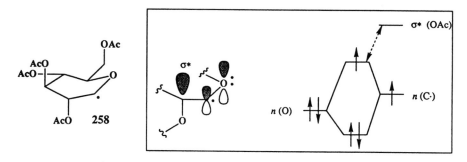

Figure 12.157. Orbital interactions in the glucosyl free radical **258**.

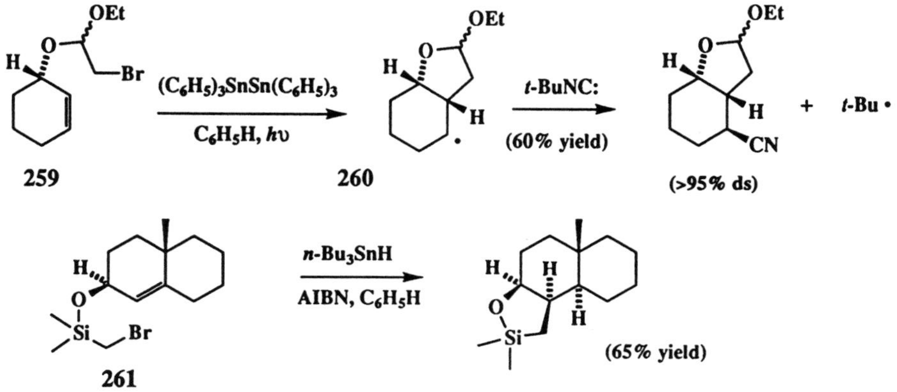

Figure 12.158. Stereoselective additions to bicyclic free radicals.

Stereochemistry of Free Radical Cyclizations

The formation of rings by intramolecular free radical cyclization is largely restricted to five-membered examples, since rates are often too slow for six-membered cases unless the intermediate radical from the cyclization is sufficiently stabilized by a suitable substituent. Thus, cyclization of 5-hexenyl radicals occurs regioselectivity to form five-membered rings unless this is sterically inhibited by substitution at C(5) (Beckwith and Schiesser, 1985). For C(1) and C(3) substituted derivatives, the cis products are favored, while substituents at C(2) and C(4) favor the trans products. These outcomes may be readily rationalized in terms of a chairlike transition state in which the substituents occupy equatorial conformations as depicted for each of the major products in Figure 12.159 (Beckwith et al., 1980). Both 1,5-cis selectivity (Corey et al., 1984) and 1,5-trans selectivity (RajanBabu, 1987) have been observed in more complex systems, however.

Cyclization of an alkenyl side chain onto a cycloalkyl radical normally leads mainly to cis-fused bicyclic compounds (Beckwith et al., 1981) as does the cyclization of radicals in side chains onto cycloalkenes (Fig. 12.160; Beckwith and Roberts, 1986). In the more complex lactones **262** and **263**, however, the newly formed carbocycles are predominantly trans fused (Chuang, Hart, et al., 1988a,b). The formation of the less strained trans-fused hydrindene systems was anticipated by the authors and is consistent with theoretical transition state models (Beckwith and Schiesser, 1985; Curran and Rakiewicz, 1985; Spellmeyer and Houk, 1987). The preference for the trans-fused hydronaphthalene products is not so obvious, however.

A number of excellent reviews providing detailed analyses of these processes and many additional examples are available (Julia, 1974; Beckwith and Ingold, 1980; Beckwith, 1981; Surzur, 1982; Stork, 1983; Hart, 1984; Giese, 1985, 1986, 1989).

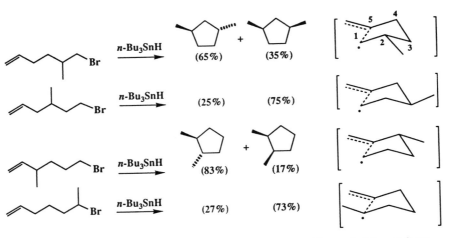

Figure 12.159. Stereochemical preferences in the cyclization of hexen-5-yl free radicals.

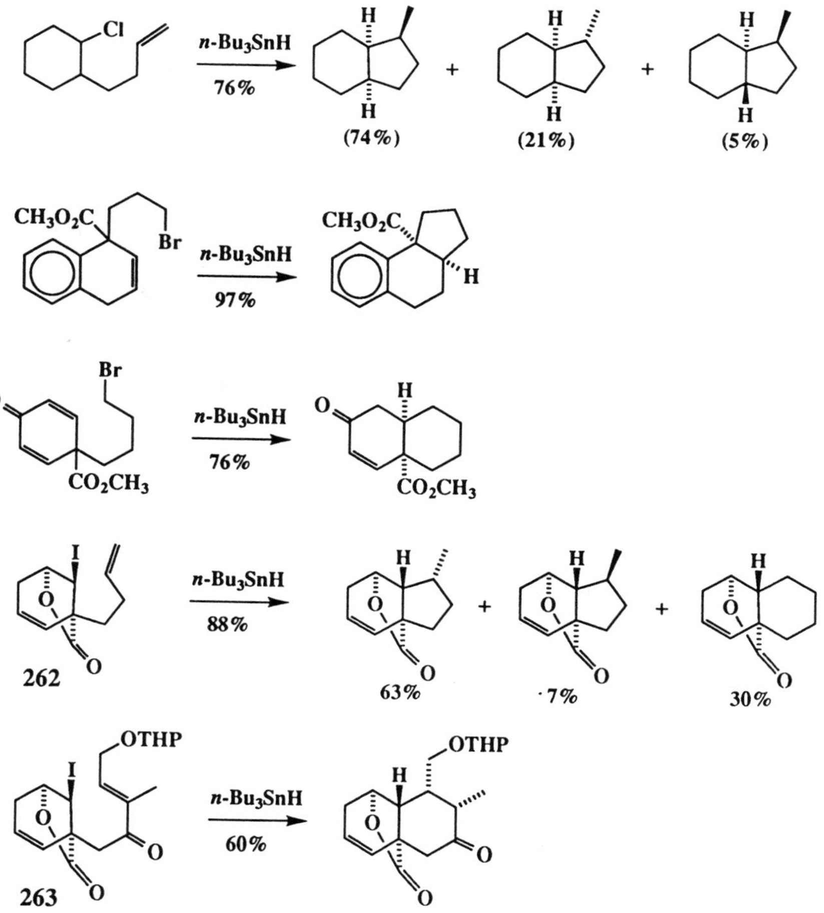

Figure 12.160. Stereochemistry of cyclizations onto preexisting rings, where THP = tetrahydropyranyl.

12-4. ENANTIOSELECTIVE SYNTHESES

a. Introduction

In broad terms, most diastereoselective reactions are "controlled" by the substrate, while enantioselective reactions are under the control of the reagent. From a simple topological perspective there are no essential differences between the reactions of chiral substrates with achiral reagents and those between chiral reagents and achiral substrates. In many of the following examples there are close parallels with the diastereoselective processes outlined earlier: The distinction may be either one of perception or simply one of convenience. The category of aldol reaction described in Section 12-3.f, for example, belongs equally well in Section 12-4.b below. In practical terms, however, there are often important consequences. If the products of a reaction are diastereomeric rather than enantiomeric, separation can usually be achieved more readily and it is possible therefore to accept lower levels of stereoselectivity, although this view may need to be modified as increasingly effective chromatography columns with chiral stationary phases are developed (Pirkle et al., 1986; Allenmark, 1991; cf. p. 253). When enantiometers are formed directly and recrystallization is the only practical method for purification, however, an enantiomeric excess of about 90% may represent the lower limit from which it is possible to obtain enantiomerically pure material without unacceptable losses (but see Section 7-2). More serious problems are the limitations on substrate structure and the uncertainties over the level of stereoselectivity that may be obtained for a specific compound.

The most important aspect of direct enantioselective processes, however, is the potential for "chiral multiplication" (Noyori, 1989) based on catalytic conversions that use substoichiometric quantities of a chiral reagent or ligand. It is, of course, then necessary that the catalyzed process proceeds more rapidly than the uncatalyzed competing reactions and, in those cases where coordination to the catalyst has taken place, that the product is released from it under the reaction conditions.

Until recently, the most effective processes have been dominated by enzymes, but there are now excellent prospects of equally powerful synthetic reagents with far greater flexibility in terms of acceptable substrates and even better levels of stereoselectivity in some cases (Bosnich, 1986; Evans, 1988). Some of the most effective compounds are characterized by C_2 symmetry, that is, derivatives of tartaric acid and 1,1-binaphthyl in particular (Whitesell, 1989). The higher symmetry of the C_2 ligands reduces the number of possible isomers for both the reagent and the reaction complexes. It has also been suggested (Noyori and Takaya, 1985) that the pliability of the binaphthyl ligands is an important property when the structure of the transition state differs significantly from that of the ground state.

The categorization of chiral reactants as reagents rather than as substrates can be quite arbitrary. Thus, many of the reaction processes described in Section 12-3 could be regarded equally well as belonging in this group (e.g., the reactions of various chiral enolate anions, carbanions, and enamines, with achiral partners). The emphasis in this section, however, is on processes for which the chiral ligands of the reagent do not become incorporated into the product or are cleaved during the isolation procedure. The selection of examples examined in Section 12-4.b has been mainly limited to those reactions that have been demonstrated to afford high levels of enantioselection with a reasonably broad selection of substrate structures and mostly involve the use of stoichiometric amounts of the chiral reagent. These cases merge, however, with those for which complexation with the chiral ligand confers greater chemical reactivity on the reagent, allowing the use of relatively small quantities of the auxiliary. The boundary between these latter cases and the more obviously catalytic processes that follow in Section 12-4.c (chiral transition metal complexes, acids, and bases) is somewhat arbitrary. Additional topics to be discussed include "nonlinear relationships" (Section 12-4.d), enzyme-catalyzed reactions (Section 12-4.e), syntheses involving discrimination between enantiotopic groups (Section 12-4.f), and enantioconvergence (Section 12-4.g).

b. Enantioselective Syntheses with Chiral Nonracemic Reagents

This section encompasses the additions of chiral boranes to alkenes, and nucleophilic additions of hydride and organometalic reagents complexed with chiral ligands. The reader is also referred to an interesting review on applications involving chiral bases (Cox and Simpkins, 1991), examples from which have been included in other sections. Apart from the important exceptions noted towards the end of the section, a stoichiometric amount or an excess of the reagent has been used.

Hydroborations with Chiral Boranes

Hydroboration of naturally occurring terpenoids has provided a useful range of chiral boranes (Fig. 12.161). Isopinocampheylborane [IpcBH$_2$] **264**, diisopinocampheylborane [(Ipc)$_2$BH] **265**, and dilongifolylborane **266**, for example, are all readily obtained by the addition of diborane or borane-dimethyl sulfide to the parent isoprenoid (Brown et al., 1981, 1982b). Reasonably priced supplies of the precursor (+)- and (−)-α-pinenes, however, are only available in 90–95% ee. Fortunately, equilibration of diborane with an excess of the pinene for several days at 0°C affords **265** with an ee of greater than 99% (Brown et al., 1981, 1982b; see p. 387). The compound (Ipc)$_2$BH **264** reacts with Z-1,2-disubstituted acyclic alkenes to give (after oxidation) carbinols in 92–98% ees, and with norbornene to afford 2-norbornanol in an 83% ee. Enantioselectivity is poor with 1,1-substituted ethenes, however, and for less reactive alkenes (trisubstituted and E-1,2-disubstituted), displacement of α-pinene competes with hydroboration, resulting in a mixture of hydroborating reagents and lower enantioselectivity. Under these circumstances **264** and **266** may be preferred, although only enantiomeric excesses in the 50–75% range have been obtained (Brown and Jadhav, 1981; Jadhav and Brown, 1981). However, these limitations may be overcome by the utilization of the completely synthetic borane **267**, with 93–99% ees having been obtained with a range of di- and trisubstituted cyclic and acyclic alkenes (Masamune et al., 1985b).

Chiral boranes may, of course, be used for the preparation of numerous products in addition to carbinols (Brown et al., 1986a; Pelter et al., 1988) and have been incorporated into a number of valuable chiral reducing agents (Midland, 1983; see below). Chiral allylic boranes (Hoffmann and Herold, 1981;

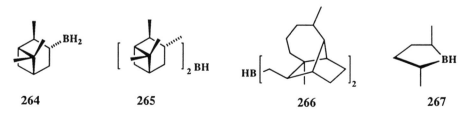

Figure 12.161. A selection of synthetically useful chiral boranes. (*Note*: Boranes are dimeric unless coordinated to donor ligands; cf. Fig. 7.68).

Midland and Preston, 1982; Jadhav, Brown, et al., 1986; Brown et al., 1990), and boronic and borinic esters (Matteson, 1986) have also been utilized in a wide range of enantioselective reactions.

Reductions with Chiral Complex Hydrides

A massive effort has been invested in the search for chiral hydride reagents that might afford good levels of enantioselection in the reduction of ketones, and so on. Much of this has centered on lithium aluminum hydride modified by the attachment of chiral ligands, including diols, amines, diamines, aminocarbinols, and diaminocarbinols (Haubenstock, 1983; Grandbois et al., 1983; Mukaiyama and Asami, 1985; ApSimon and Collier, 1986; Nogradi, 1987; Tomioka, 1990). The attainment of reasonable levels of enantioselectivity has on the whole been limited to a narrow range of substrate structures (aryl alkyl ketones and acetylenic ketones), while the properties of many of the early reagents often varied with time and temperature. The binaphthyl derivative, "BINAL-H" (**268**) (BINAL is the complex formed from equimolar amounts of lithium aluminum hydride, 2,2'-dihydroxy-1,1'-binaphthyl, and ethanol), which is available as either enantiomer (Noyori et al., 1979; Noyori, 1981), and similar compounds obtained from biphenyl (Suda et al., 1984) and 9,9'-phenanthryl derivatives (Yamamoto et al., 1984) provided excellent results, however, and 95–100% enantiomeric excesses have been obtained from the reduction of a reasonable range of ketones (Fig. 12.162). Differentiation between R^1 and R^2 depends on the electronic rather than the steric properties of these substituents and requires that one of them is unsaturated (in the present case this is designated as R^1). Poor enantiomeric excesses are obtained when both are simple alkyl groups (Noyori et al., 1984). The stereochemical outcomes were rationalized by considering four likely transition state structures ($A^\ddagger$–$D^\ddagger$). Two of these ($A^\ddagger$ and $B^\ddagger$) were discounted because of the severe nonbonded interaction between the alkyl group **R** and one of the ortho hydrogen atoms of the binaphthyl residue. The $C^\ddagger$ transition state was then rejected because of the repulsion between the nonbonding orbital of the naphthyloxy oxygen and the π orbital of the axial R^1 substituent. Thus, the favored reaction pathway based on (S)-BINAL proceeds via $D^\ddagger$ to afford the (S)-carbinol.

Hydride transfer from boron in chiral boranes and boranes rendered chiral by complexation with chiral amines has been studied for possible utility in enantioselective reductions, but the outcomes have not been encouraging (Midland, 1983). While hydride transfer from carbon ligands, as in the reactions of "Alpine borane" (Ipc.BBN) **269** (Midland et al., 1977; Midland and Nguyen, 1981; cf. also Midland et al., 1984), has furnished high enantiometric excesses (77–100%) with deuterio aldehydes and acetylenic ketones (Fig. 12.163), sluggish reactivity with other ketones may allow dissociation and, as a consequence, lower levels of enantioselectivity. The problem may be partly ameliorated by performing the reaction at high pressure (Midland and McLoughlin, 1984) or with neat reagent (Brown and Pai, 1982), but diisopinocampheylchloroborane, (Ipc)$_2$BCl, appears to be the reagent of choice for a wide range of ketones, including sterically hindered derivatives (Chandrasekharan, Brown, et al., 1986; Brown et al., 1986b).

Modified borohydrides have been comparatively neglected, but may prove to

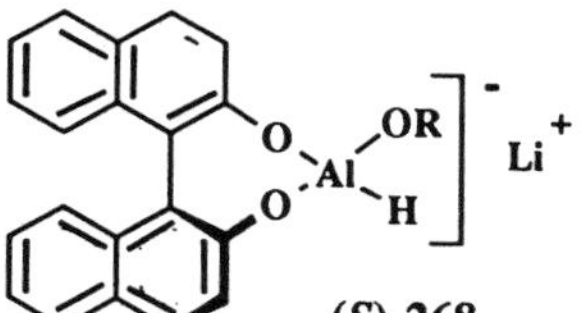

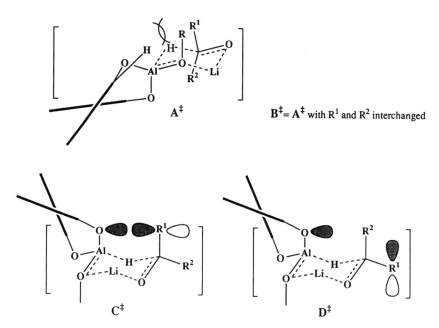

R^1	R^2	R	ee (%)
C$_6$H$_5$	CH$_3$	Et	95
C$_6$H$_5$	Et	Et	98
C$_6$H$_5$	Pr	Et	100
C$_6$H$_5$	iPr	Et	71
n-BuC≡C	CH$_3$	CH$_3$	84
n-BuC≡C	n-C$_5$H$_{12}$	CH$_7$	90

(S)-268

R = CH$_3$, Et

B‡ = A‡ with R^1 and R^2 interchanged

Figure 12.162. An effective reagent for the enantioselective reduction of ketones is BINAL-H **268**.

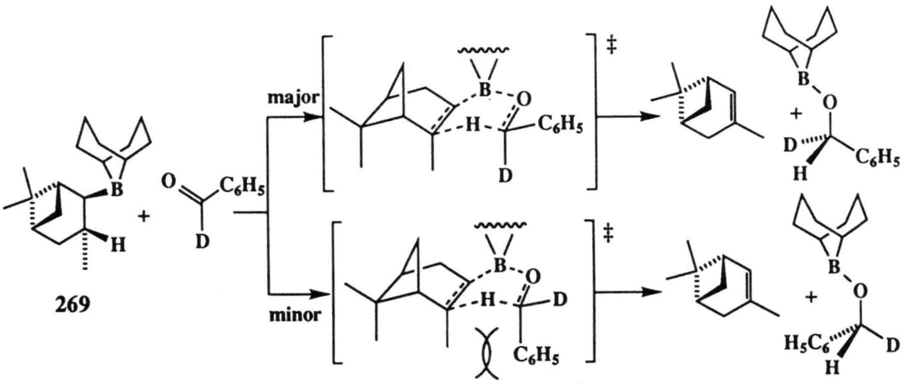

269

major

minor

Figure 12.163. Enantioselective reduction of α-deuterobenzaldehyde with Alpine borane **269**.

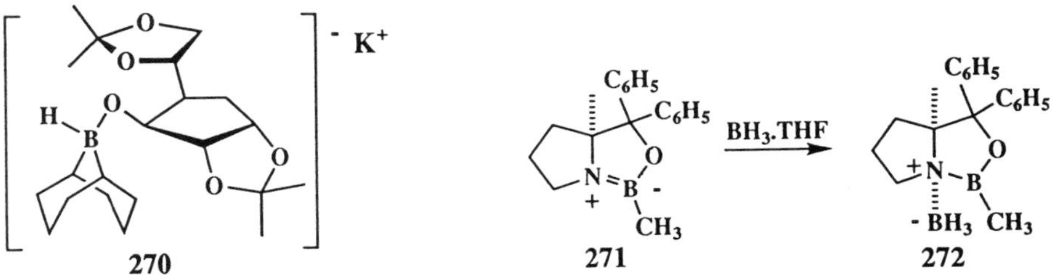

Figure 12.164. Synthetically useful chiral borohydrides.

be the most useful reducing agents of all. The complex hydride **270** has afforded greater than 90% enantiomeric excesses from the reduction of a good range of aromatic ketones and α-keto esters (Brown et al., 1986c,d), while in a development of earlier work (Yamada et al., 1981; Soai et al., 1984; Itsuno et al., 1983, 1987), borohydride **272** has emerged as both a particularly effective and convenient reducing agent (Fig. 12.164; Corey et al., 1987a,b). It is readily prepared from mixing the oxazaborolidine **271** with diborane, and may be transferred in air and stored in closed containers at room temperature. A rationale for the observed stereochemical sense of reduction has been provided (Evans, 1988). Most usefully of all, **271** may be used in substoichiometric quantities (0.05 mol mol^{-1} of ketone $\equiv$ 5 mol%) and affords consistently high enantiomeric excesses over a wide range of structural types.

Enantioselective conjugate reduction of crotonate esters by sodium borohydride has been catalyzed by ligand **273** (1.2 mol %) in conjunction with cobalt(II) chloride (1.0 mol %; Fig. 12.165; Leutenegger, Pfaltz, et al., 1989). Compound **273** has been likened to a "semi-corrin".

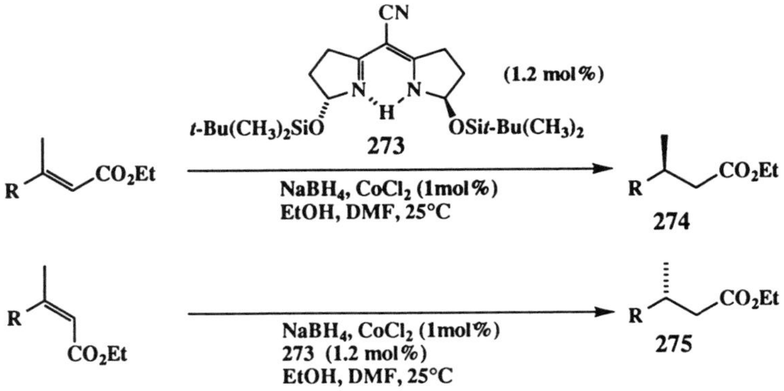

Figure 12.165. Conjugate reduction of crotonate esters catalyzed by a "semi-corrin."

TABLE 12.5. Enantioselectivities in the Conjugate Reduction of Crotonate Esters.

R	% ee (274)	% ee (275)
$C_6H_5CH_2CH_2$	94	94
$(CH_3)_2CH{=}CHCH_2$	94	94
i-Pr	96	90
C_6H_5	81	73

Further categories of chiral reducing agents include chiral organometals (Morrison and Mosher, 1971), alkoxides (Nasipuri and Bhattacharya, 1977; Samaddar, Nasipuri, et al., 1983), and dihydropyridines (Inouye et al., 1983), the last group being inspired by the role of dihydronicotinamide adenine dinucleotide phoshate (NADPH) as an enzyme cofactor. Few of these reagents seem likely to find general application in synthesis, however.

Chiral Organometal Complexes

Many of the chiral auxiliaries that have been developed in the quest for effective chiral hydride reagents have been employed (with minor variations) as ligands to organometals, but with variable outcomes. A selection of the more promising applications is provided in Figure 12.166.

The utilization of substoichiometric quantities of the chiral ligand is an important goal in this area of chemistry as well and is in fact vital in the following example (Fig. 12.167) of the formation of S carbinols by the addition of diethyl and dimethyl zinc to aromatic aldehydes catalyzed by 3-*exo*-dimethylamino-isoborneol **276** [(−)-DAIB], since an excess of dialkyl zinc is necessary to convert the relatively stable complex **277** into a kinetically reactive intermediate as indicated by the abbreviated mechanism. Very good enantiomeric excesses (93–98%) were realized with only 1–2 mol% of ligand, although the levels with aliphatic aldehydes were less satisfactory (61–90% ee) (Kitamura, Noyori, et al., 1986). The kinetics of the process have been thoroughly studied and X-ray crystal

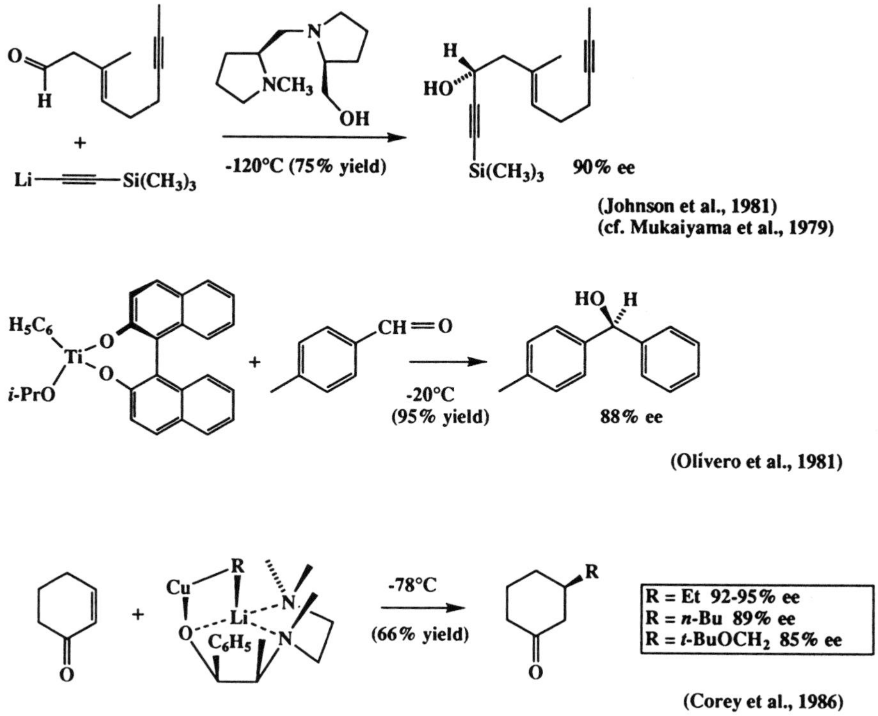

Figure 12.166. Selected examples of enantioselective organometal additions.

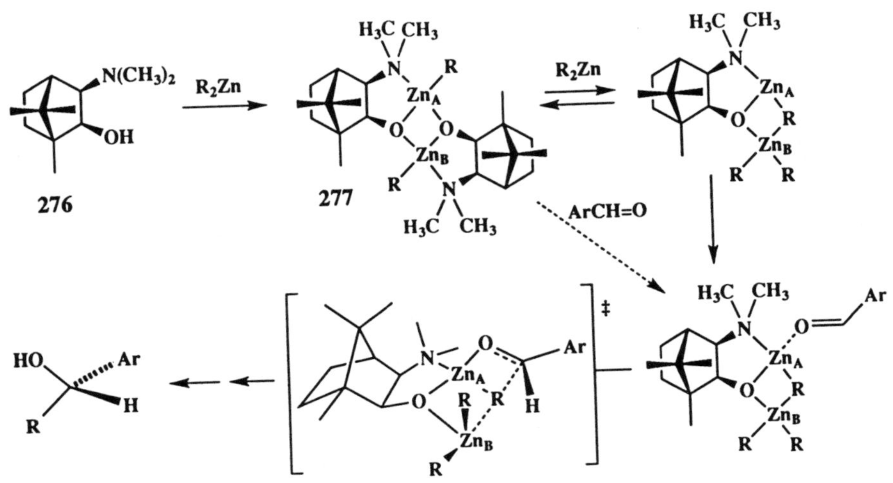

Figure 12.167. Catalyzed additions of diethyl zinc to aldehydes.

structures obtained for some of the intermediates. The favored transition state structure is assumed to have the aryl ring of the aldehyde remote from the proximal alkyl group attached to Zn_B (Kitamura, Noyori, et al., 1989). Similar results were obtained with phenylprolinol ligands (Soai et al., 1987; Soai and Niwa, 1992), while a polymer-bound ephedrine reagent affording 71–83% enantiomeric excesses was also developed by the same group (Soai et al., 1988; cf. Section 12-4.d).

Chiral Enolate Aggregates

The need to take into account the participation of dimeric, tetrameric, and

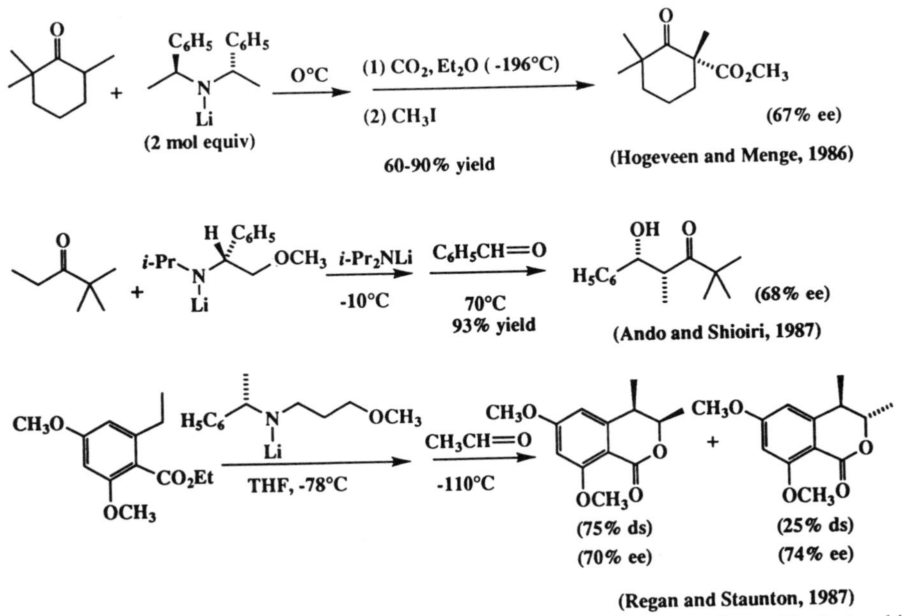

Figure 12.168. Examples of enantioselective achiral enolate–chiral base reactions with electrophiles.

oligomeric aggregates in reaction mechanisms, especially when dealing with enolates has already been noted (Section 12-3.e). Seebach (1988) examined this question in considerable detail and pointed out the likelihood of chiral products being incorporated into an otherwise achiral enolate aggregate. Attempts to exploit this kind of phenomenon with a view to achieving enantioselective reactions of achiral enolates by mixing them with chiral bases (Cox and Simpkins, 1991) are illustrated in Figure 12.168.

c. Enantioselective Reactions with Chiral Nonracemic Catalysts

Processes in this section fall into three natural groups: catalysis by chiral transition metal complexes, chiral bases, and chiral Lewis acids (Martin, 1994).

Catalysis by Chiral Transition Metal Complexes

Utilization of transition metal complexes to mediate enantioselective syntheses has gradually gathered momentum over the past two decades (Bosnich and Fryzuk, 1981; Morrison, 1985; Bosnich, 1986; Brown, 1989; Davies, 1989; Ojima et al., 1989; Ojima, 1993). The first important landmarks were the preparation of chiral variants of the soluble Wilkinson phosphine–rhodium complexes (Knowles and Sabacky, 1968; Horner et al., 1968), which led to highly effective catalysts for the enantioselective synthesis of amino acids by hydrogenation of α-amido-α,β-unsaturated esters (Knowles, 1983). The other turning point was the introduction of tartrate-titanium alkoxide complexes for the epoxidation of allylic carbinols (Katsuki and Sharpless, 1980). These successes have more recently been supplemented by a wider range of processes based on ferrocenyl (Hayashi and Kumada, 1982) and 1,1′-binaphthyl derived ligands (Noyori and Takaya, 1985). One of the most recent reviews of this topic (Brunner, 1988) tabulates a wide array of applications based on approximately 350 chiral ligands, most of them introduced during the period 1984–1986. Much of this development has been empirical, however, and a proper understanding awaits the kind of thoughtful and detailed analysis which, for example, has illuminated the details of the hydrogenation of α-amido-α,β-unsaturated esters and acids over soluble rhodium catalysts (Halpern, 1985; Bosnich, 1985; see below).

Enantioselective Epoxidations of Alkenes. In studies on the enantioselective formation of epoxides from alkenes with chiral peroxycarboxylic acids, the enantiomeric excesses have never exceeded 20% (Ewins, Henbest, et al., 1967; Pirkle and Rinaldi, 1977), presumably because the controlling stereocenters are too remote from the reaction site. An enantiomeric excess of 90% was reported for the poly-(S)-alanine catalyzed epoxidation of chalcone by alkaline peroxide (Julia et al., 1980), but the most successful of the electrophilic nonmetallic reagents (up to 40% ee) have been chiral N-sulfonyloxaziridines (Davis et al., 1983; Davis and Haque, 1986; cf. also Davis et al., 1988). Significant progress has only come by investigating the epoxidation of allylic carbinols mediated by transition metal catalysts, however. The exploration began with a dioxomolybdenum-N-ethylephedrine complex (up to 33% ee; Yamada et al., 1977), progressed to a peroxomolybdenum complex (up to 33% ee; Kagan et al., 1979), and then to vanadium complexes substituted with various hydroxamic acid ligands (50–80% ee; Michaelson, Sharpless, et al., 1977; Sharpless and Verhoeven, 1979). Good enantiomeric excesses have been obtained with manganese salen complexes derived from (R,R)- or (S,S)-*trans*-1,2-cyclohexanediamine and 2-hydroxy-3,5-di-*tert*-butylbenzaldehyde in catalytic amounts (1–8 mol%) with iodosomesitylene as the cooxidant (Zhang, Jacobsen, et al., 1990), or more conveniently with sodium hypochlorite (Zhang and Jacobsen, 1991). However, high enantiomeric excesses have only been obtained with cis-disubstituted alkenes, with especially good results on chromenes (Lee et al., 1991). It was found that when the 5-*tert*-butyl group was replaced by electron-releasing or electron-withdrawing groups that a large variation in enantiomeric excess (22% for NO_2 ranging to 96% for OCH_3) was obtained with 2,2-dimethylchromene (Jacobsen et al., 1991).

In terms of generality, reliability, and easy access to reagents, all of the above studies are overshadowed by one of the most valuable developments in enantioselective synthesis, that is, the utilization of various mixtures of dialkyl tartrates, titanium tetra-alkoxides and *tert*-butyl hydroperoxide (TBHP) to effect the enantioselective formation of epoxides from primary allylic carbinols (Katsuki and Sharpless, 1980). There can be a reasonable expectation of a 90% enantiomeric excess, regardless of the structure of the substrate. The extensive applications of the process to numerous multistep syntheses provide a compelling testimony to its utility and reliability (Rossiter, 1985; Pfenninger, 1986). Numerous group 4 (IVB) and 5 (VB) transition metal alkoxide-diethyl tartrate combinations had been screened with (E)-2-phenylcinnamyl alcohol as the substrate, but the maximum enantiomeric excess observed was 47% and the average less than 20% (Rossiter, 1985). More than 50 ligands were employed in combination with titanium tetraisopropoxide, but none matched the effectiveness of tartrate-derived esters and amides (Finn and Sharpless, 1985). Different ratios of tartrate or tartramide to alkoxide have been studied (Miyano, Sharpless, et al., 1983; Lu, Sharpless, et al., 1984), but the standard reaction employs a ratio of tartrate to titanium alkoxide of 1.2:1 to achieve optimal enantioselectivity (Martin et al., 1981), although the excess of tartrate retards the reaction rate by occupying the catalytic sites on the titanium. Stoichiometric quantities of the catalyst system are often used, although stereochemical yields are usually enhanced when less than one molar equivalent is employed; as little as 5 mol% can be effective. Bulky groups in the tartrate ester, hydroperoxide, and the trans-olefinic substituent all increase the rate of epoxidation and level of enantioselection. A representative reaction is depicted in Figure 12.169 (Hill, Sharpless, et al., 1984).

One of the features of the reaction is that the configuration of the major epoxide enantiomer is entirely dependent on the absolute configuration of the

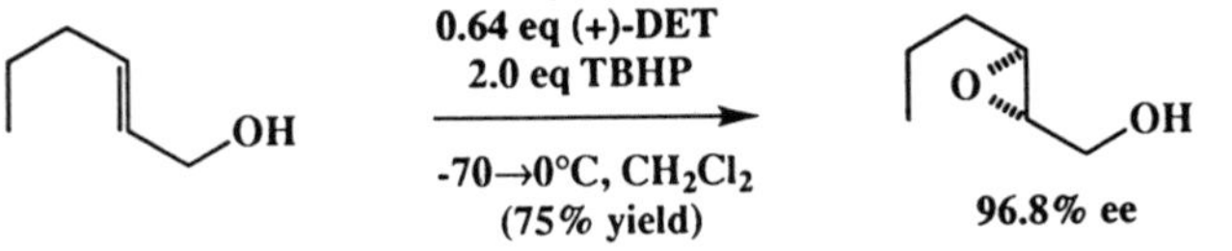

Figure 12.169. Katsuki–Sharpless epoxidation of (E)-2-hexen-1-ol, where DET = diethyl tartrate.

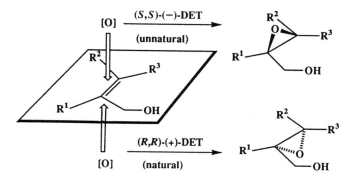

Figure 12.170. Schematic for the prediction of the absolute configuration of epoxides formed by the Katsuki–Sharpless method.

tartrate ester and will be invariably formed according to the schematic outlined in Figure 12.170, regardless of the substituents attached to the olefinic bond. The reagent system has also been very effective for achieving kinetic resolutions of racemic substrates (cf. Section 7-6.a) and in the enantioselective preparation of sulfoxides (Pitchen, Kagan, et al., 1984). However, low chemical yields and enantiomeric excesses have been obtained with homoallylic carbinols, which also react with a reversal of enantiofacial selectivity.

The mechanism of the reaction has not been fully elucidated, however, despite extensive studies and the determination of numerous crystal structures for the complexes involved. The active catalyst was initially believed to be the 10-membered structure **278** (Fig. 12.171; Sharpless et al., 1983) and although later NMR studies gave support to this conclusion (Potvin et al., 1988), a comprehensive analysis (Woodard et al., 1991; Finn and Sharpless, 1991) favors a mechanism based on **279** (cf. also Williams, Sharpless, Lippard, et al., 1984; and Jorgensen, Hoffmann, et al., 1987). The participation of an ion pair based transition state has been suggested (Corey, 1990), but this is apparently inconsistent with the observed kinetic rate expression.

Enantioselective Hydrogenations. The other major success story in catalyzed enantioselective reactions has been the hydrogenation of a range of acrylate derivatives and allylic carbinols with various soluble, chiral rhodium and

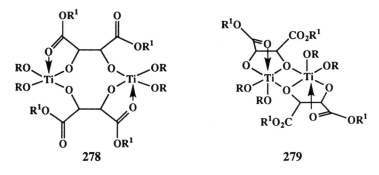

Figure 12.171. Titanium complexes implicated in the Katsuki–Sharpless epoxidation.

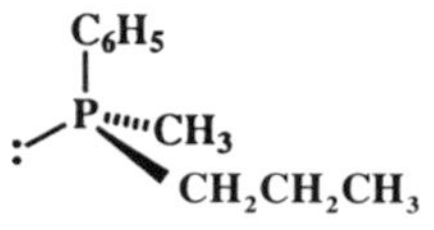

280

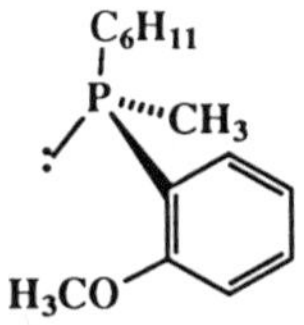

"CAMP"

281

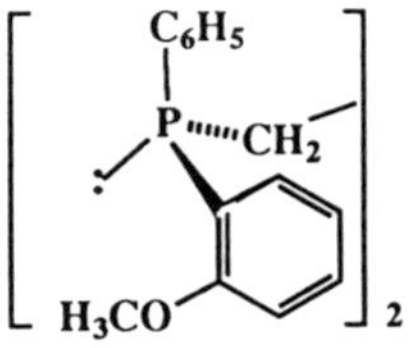

"DIPAMP"

282

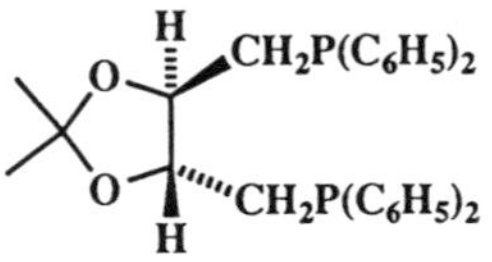

"DIOP"

283

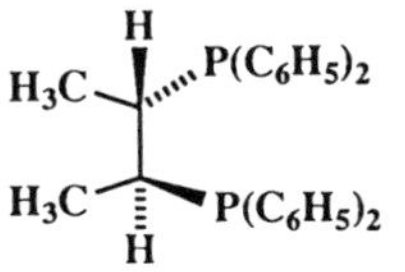

"CHIRAPHOS"

284

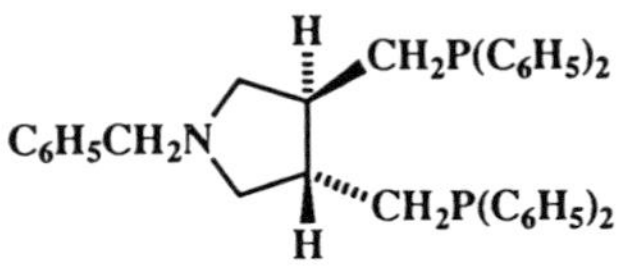

"DEGPHOS"

285

Figure 12.172. Chiral phosphine ligands utilized in enantioselective hydrogenations.

ruthenium complexes. In a search for an effective enantioselective synthesis of amino acids, Knowles and co-workers examined a series of chiral phosphine ligands, beginning with **280**. A breakthrough came with the use of CAMP **281** (80–88% ees; Knowles et al., 1972), although this was eventually displaced by the bis(phosphine) ligand, DIPAMP **282** (95% ee; Fig. 12.172; Vineyard, Knowles, et al., 1977; Knowles, 1983).

At first it appeared that it was necessary to build chiral compounds with stereogenic phosphorus atoms, but it was then discovered that the more easily prepared DIOP **283** (Kagan and Dang, 1972) and CHIRAPHOS **284** (Fryzuk and Bosnich, 1977, 1979, 1981) were also effective ligands. The field has proliferated further with the development of almost 40 distinct ligand types (Kagan, 1985; Koenig, 1985; Ojima et al., 1989), one of the most effective being DEGPHOS **285** (Nagel et al., 1986), while the methodology has provided the basis for inter alia, the commercial production of L-DOPA (3′,4′-dihydroxyphenylalanine) **286** used in the treatment of Parkinson's disease (Fig. 12.173).

A very considerable effort has been expended on the elucidation of the mechanism for this hydrogenation, a difficult task, given the delicate balance of equilibrium and rate constants. In the case of the hydrogenation of (Z)-α-benzamidocinnamic ester catalyzed by $[Rh(DIPAMP)]^+$, it was observed that the

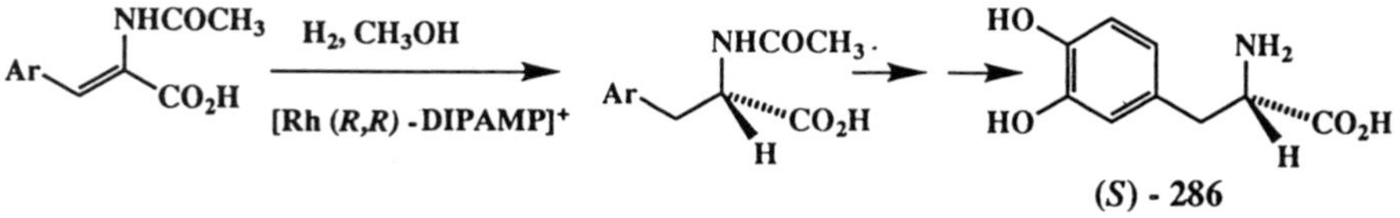

(*S*) - **286**

Figure 12.173. Enantioselective hydrogenation utilized in the synthesis of L-DOPA **286**.

10:1 ratio of the two diastereomeric adducts **287** and **288**, respectively (by NMR spectroscopy), corresponded quantitatively to the enantiomeric ratio of products. This observation led to the conclusion that the stereodifferentiation had occurred at this stage and that the product was formed from the observed major diastereomer (Brown and Chaloner, 1978). However, a detailed kinetic study

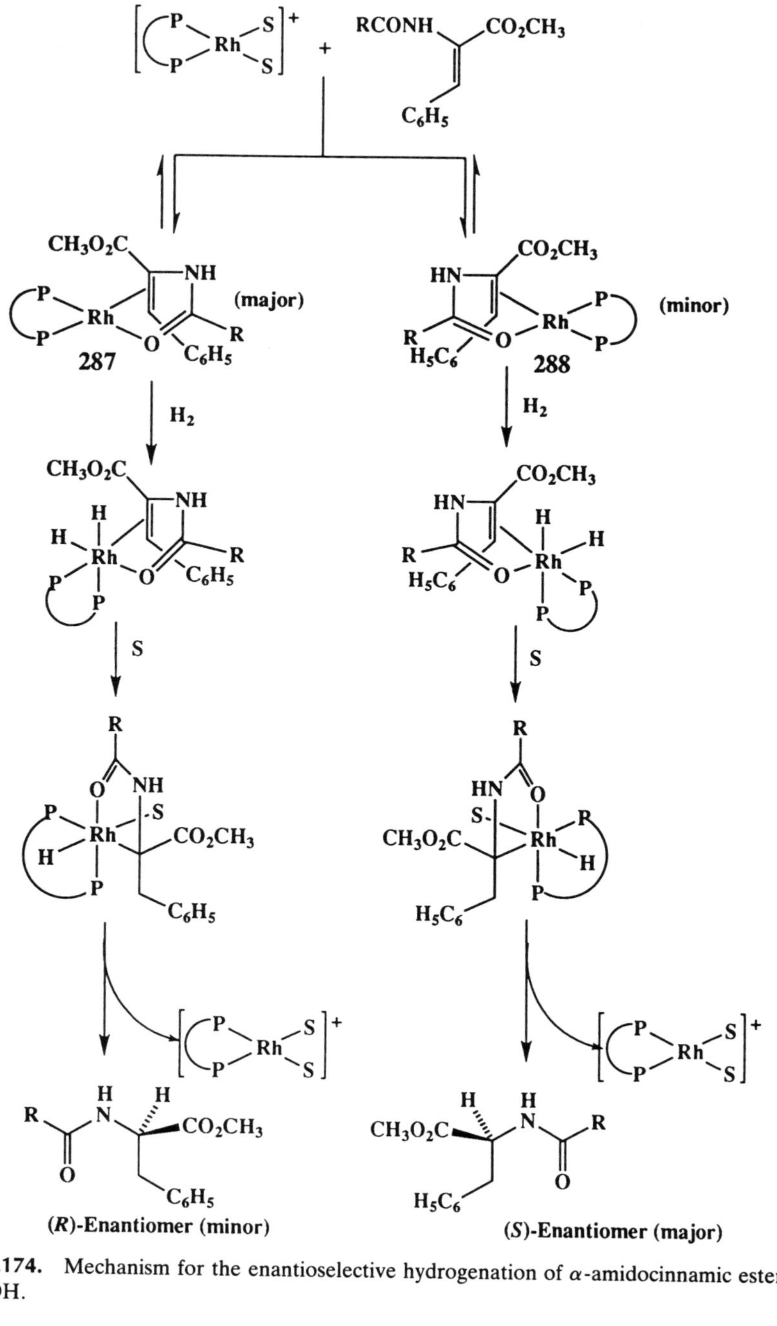

Figure 12.174. Mechanism for the enantioselective hydrogenation of α-amidocinnamic esters, where $S = CH_3OH$.

revealed that the major product in fact arises from the minor diastereomer **288** as a consequence of higher reactivity in the subsequent stages (cf. the Curtin–Hammett principle, Section 10-5) as outlined in Figure 12.174 (Halpern, 1983, 1985). These results emphasize the "treacherous nature of attempts to predict the stability of transition state structures for substrate-catalyst binding and the need to acquire a fairly detailed knowledge of the kinetics of the system" (Bosnich, 1986).

The success of the reaction depends very much on the presence of a carbonyl group in a 1,3 relationship to the olefinic bond that can bind to the catalyst (this is satisfied by an α-amido or α-acyloxy group). It is also necessary that the alkene bond is polarized by an electron-withdrawing group attached in a geminal relationship to the binding substituent, whereas the nature of the β substituents and the geometry of the olefinic bond is largely immaterial. Thus, high enantiomeric excesses have only been obtained from the hydrogenation of acrylic esters and acids possessing a-amido or α-acyloxy substituents, and itaconic acids [$RCH{=}C(CO_2H)CH_2CO_2H$]. In the latter compounds, the carboxymethyl group can obviously take the place of the amido group (Ojima et al., 1978). With the introduction of (aminoalkyl)ferrocenylphosphine (PPFA) complexes, such as **289** (Hayashi et al., 1987), and the BINAP-ruthenium catalyst **290** (Noyori et al., 1986; Takaya, Noyori, et al., 1987; Kitamura, Noyori, et al., 1987; Kawano, Saburi, et al., 1987; Noyori and Takaya, 1990; Fig. 12.175), however, the range of substrates has been broadened considerably to include simple ketones, α,β-unsaturated acids, and allylic and homoallylic carbinols (cf. Section 12-4.g).

Cyclopropanations. One of the earliest applications of transition metal catalysis to enantioselective synthesis (Nozaki et al., 1966) involved the reaction of ethyl diazoacetate and bis[N-(R)-α-phenylethylsalicylaldiminato]copper(II) **291** with styrene, but a discouraging enantiomeric excess of only 6% was obtained. However, extensive testing of numerous other catalysts over the next two decades (Maas, 1987; Fritschi, Pfaltz, et al., 1988) culminated in the development of **292** which affords enantiomeric excesses of 90–98% in the formation of chrysanthemic acid **293** for the production of synthetic pyrethroids (Aratani, 1985). It also provides the basis of a commercial synthesis of the enzyme inhibitor cilastatin **294** (Fig. 12.176), which is used in combination with β-lactam antibiotics (Ojima et al., 1989).

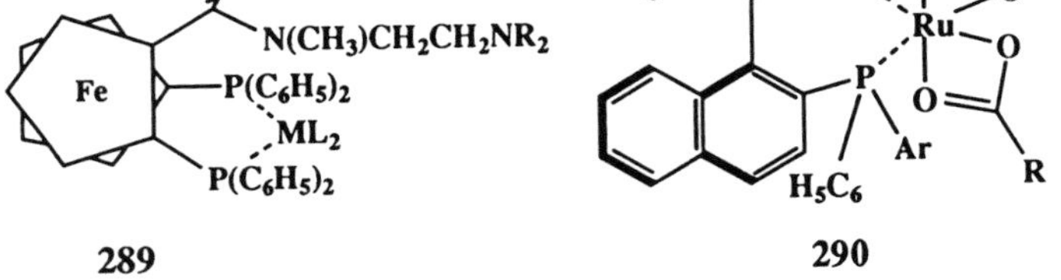

289   290

Figure 12.175. Effective catalysts for the enantioselective reduction of unsaturated carbinols, esters, lactones, and carboxylic acids, and of the carbonyl group in ketones.

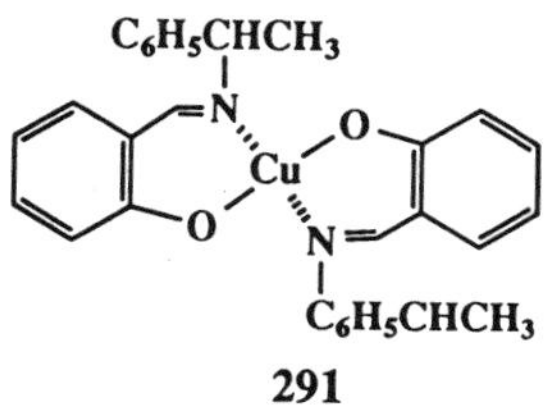

Figure 12.176. Enantioselective cyclopropanation catalysts and products.

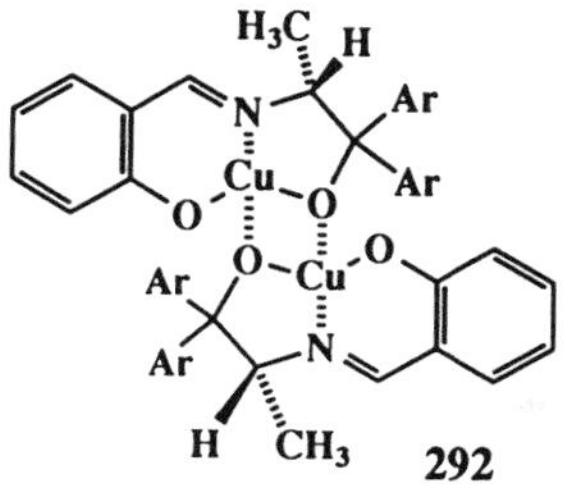

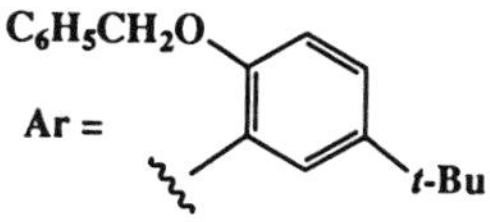

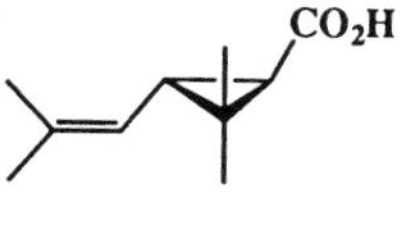

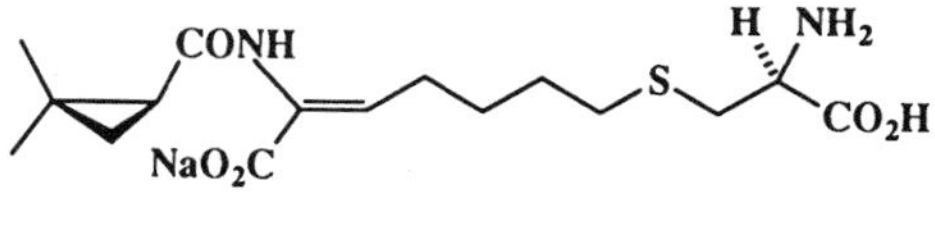

Miscellaneous Reactions. Transition metal catalyzed hydroformylations of alkenes have generally furnished only moderate levels of enantioselectivity (Ojima and Hirai, 1985) and although better results have been obtained from hydrosilations of ketones, keto-esters, keto-amides, imines, and alkenes, enantiomeric excesses above 80% are unusual. However, the reduction of acetophenone by diphenylsilane catalyzed by a thiazolidine–rhodium catalyst is reported to proceed with an enantiomeric excess of 97.6% (Brunner et al., 1984). Cross couplings of simple secondary Grignard reagents with alkenyl halides mediated by PdCl$_2$[(R,S)-PPFA] (cf. **289**) furnished optimal enantiomeric excesses of about 65%, but α-trimethylsilylbenzylmagnesium bromide couples with vinyl bromide to afford an allyl silane in 95% ee (Hayashi et al., 1983). The PPFA ligand has also been used to prepare a gold(I) complex, which catalyzes aldol reactions between aldehydes and methyl isocyanoacetate to form oxazolines with 81–100% trans selectivity and a greater than 95% enantiomeric excess in most cases (Ito et al., 1986). Good enantioselectivity has also been reported for hydrovinylation of cyclohexadiene mediated by a nickel–aluminum–aminophosphine complex (Buono et al., 1985), for palladium catalyzed allylations (Auburn, Bosnich, et al., 1985), and for the alkene isomerization of allylamines to enamines with [Rh-(+) or (−)-(BINAP)]$^+$ complexes (Tani, Otsuka, et al., 1988; Noyori, 1989). This last process provides the basis of the industrial conversion of diethylgeranylamine into citronellal and thence (−)-menthol with a much higher enantiomeric purity than can be obtained from natural sources (Tani, Otsuka, et al., 1984; Otsuka and Tani, 1985; Scott, 1989; Akutagawa, 1992). Rhodium–BINAP has also been used to catalyze the enantioselective addition of catecholborane to alkenes, but the enantiomeric excesses are only modest (up to 70%) (Burgess and Ohlmeyer, 1988; cf. Evans and Fu, 1990).

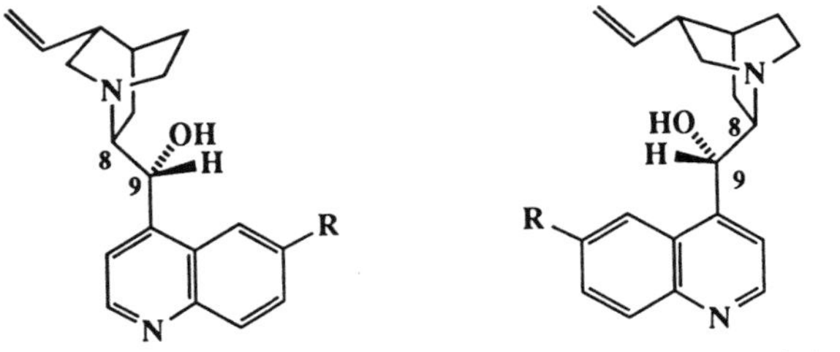

295 (8S,9R) R = OCH₃
296 (8S,9R) R = H

297 (8R,9S) R = OCH₃
298 (8R,9S) R = H

Figure 12.177. Cinchona alkaloids used as chiral catalysts.

Catalysis by Chiral Bases

The utilization of chiral bases for the preparation of enantiomerically pure or enriched compounds has a venerable history. The *Cinchona* alkaloid, quinine **295**, and its derivatives were employed by Pasteur (1853) to effect the resolution of racemic acids. Quinine was also used 80 years ago to catalyze the formation of an optically active cyanohydrin from benzaldehyde (Bredig and Fiske, 1912). The *Cinchona* alkaloids (Fig. 12.177) continued to feature prominently in the field and generally afforded higher enantiomeric excesses than other natural alkaloids (Wynberg, 1986). Quinidine **297**, although diastereomeric with **295**, affords enantiomeric products, since the respective carbinolamine moieties have a "pseudoenantiomeric" relationship. The pair of analogues cinchonidine **296** and cinchonine **298**, are also readily available and have afforded enhanced levels of enantioselectivity relative to **295** and **297**, even though the structural differences appear to be peripheral. Quite good enantiomeric excesses have been obtained with a range of reactions (Fig. 12.178), but the most useful application has been

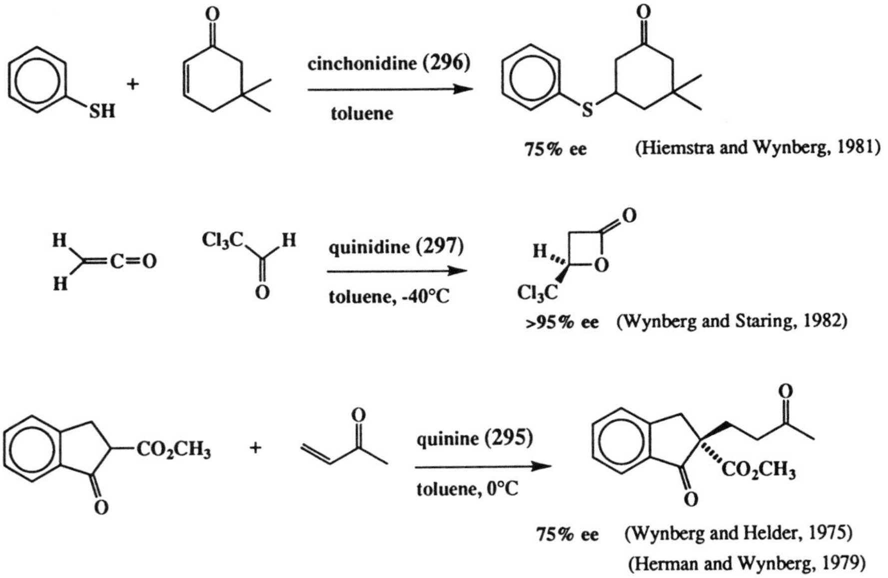

75% ee   (Hiemstra and Wynberg, 1981)

>95% ee (Wynberg and Staring, 1982)

75% ee   (Wynberg and Helder, 1975)
(Herman and Wynberg, 1979)

Figure 12.178. Examples of enantioselective reactions catalyzed by cinchona alkaloids.

in the catalyzed dihydroxylation of alkenes with osmium tetroxide developed by the Sharpless group. Encouraging results were obtained initially with dihydroquinine and dihydroquinine acetates (Hentges and Sharpless, 1980), but then, after an exhaustive search, very much better catalysts with high turnover rates based on dihydroquinidine ethers were introduced (Jacobsen, Sharpless, et al., 1988; Sharpless et al., 1991; Johnson and Sharpless, 1993). The alternative, less hazardous osmium(VI) salt, $K_2OsO_2(OH)_4$ was introduced in place of OsO_4, and an empirical model established that allowed the absolute configuration of the product to be predicted. Most recently, the phthalazine derivative of dihydroquinidine **299** (Fig. 12-179; Sharpless et al., 1992) and the analogous dihydroquinine derivative (in combination with potassium ferricyanide as the cooxidant and a sulfonamide as an accelerant) have proven to be superior catalysts, furnishing high enantiomeric excesses for monosubstituted, geminal and trans disubstituted, and trisubstituted alkenes. For cis disubstituted alkenes, the indolinylcarbamate **300** (although not the quinine analogue) has been found to be reasonably effective (Wang and Sharpless, 1992). The usefulness of the method is enhanced through the development of a simple "one-pot" procedure for the stereospecific conversion of these diols to epoxides via halohydrin ester intermediates (Kolb and Sharpless, 1993) generated with trimethyl orthoacetate and trimethylsilyl chloride; treatment with potassium carbonate then yields the epoxide product with a double inversion (overall retention) of configuration in the two steps. Since these procedures may be applied to isolated alkenes they provide a valuable supplement to the highly useful Sharpless epoxidation that is only applicable to allylic alcohols.

Excellent enantiomeric excesses have been obtained for the osmylation of styrene and E disubstituted alkenes (but not trisubstituted alkenes) with a synthetic C_2 diamine (92–98%), although stoichiometric quantities of reagent and ligand are required (Corey et al., 1989), while in a direct comparison with the Michael addition of the indanone ester to methyl vinyl ketone depicted in Figure 12.178, a 99% ee was obtained when the reaction was catalyzed by potassium *tert*-butoxide complexed with a chiral crown ether (Cram and Sogah, 1981). The average enantiomeric excess was found to be substantially lower for other similar conversions, however.

In addition to serving as an invaluable source of chiral ligands and auxiliaries, amino acids have also been used directly as catalysts (Drauz, Martens, et al., 1982). The most successful application has been to the intramolecular aldol reaction of achiral triones (e.g., **301**), especially within the context of steroid

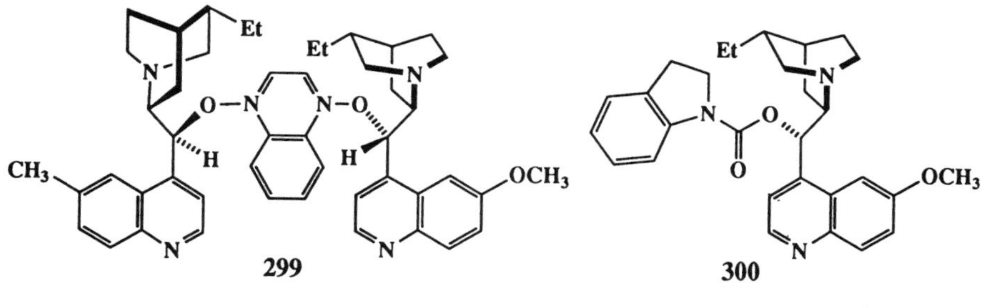

Figure 12.179. New quinidine derivatives for the catalysis of osmylations of alkenes.

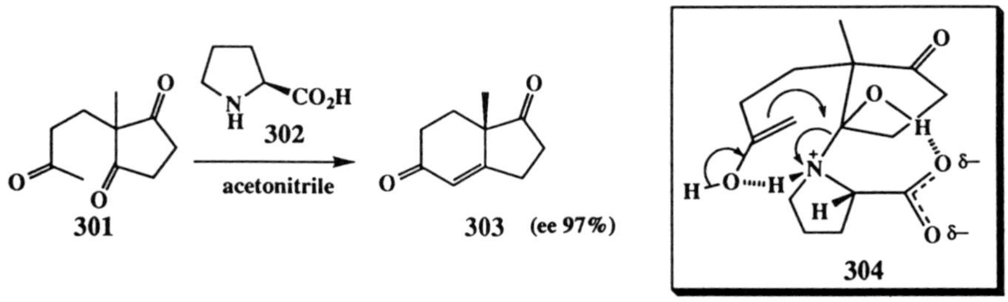

Figure 12.180. Amino acid catalyzed aldol process for the formation of a bicyclic enedione.

synthesis (Eder et al., 1971; Hajos and Parrish, 1974a). Of more than 15 amino acids studied, (*S*)-proline **302** was generally found to give the best results, with the bicyclic enedione **303** being formed in 97% chemical yield and 97% ee (Fig. 12.180; Cohen, 1976). The enantioselectivity dropped significantly when the methyl group in the side chain of **301** was substituted, however.

The original mechanism envisaged for this conversion was based on structure **304** (Hajos and Parrish, 1974b) involving the selective activation of one of the enantiotopic ring carbonyl groups by the proline. However, a later study (Agami et al., 1984) favored the participation of the diastereomeric enamines **305** and **306** (Fig. 12.181). It was observed that only the latter structure (leading to the major enantiomer **303**) could be stabilized by hydrogen bonding (Agami et al., 1986) and to counter the objection that the enamine would thereby be deactivated, participation of a second molecule of proline was invoked, as indicated in structure **307**.

Catalysis by Chiral Lewis Acids

The development of effective chiral Lewis acid catalysts has been a longstanding, but elusive goal. Several investigations have been directed towards enantioselective [4 + 2] cycloadditions (cf. Kagan and Riant, 1992) using terpene-derived auxiliaries, including a series of alkoxyaluminum dichlorides prepared from menthol, isomenthol and borneol (Hashimoto, Koga, et al., 1979), and isopinocampheylboron halides, for example, IpcBCl$_2$ and (Ipc)$_2$BBr (Bir and Kauf-

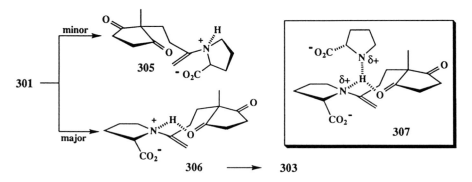

Figure 12.181. Revised mechanism for the enantioselective proline-catalyzed aldol reaction of trione **301**.

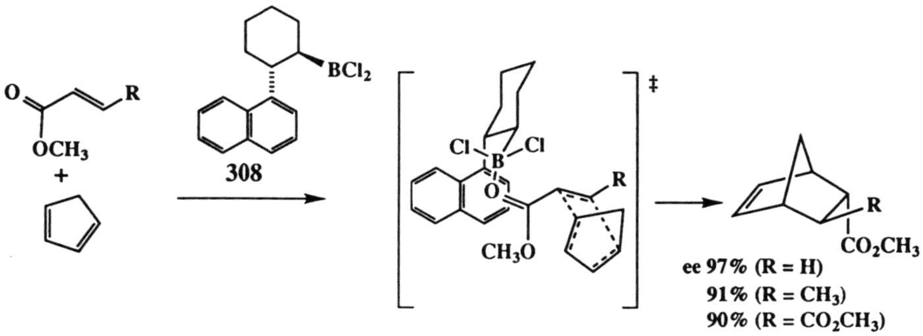

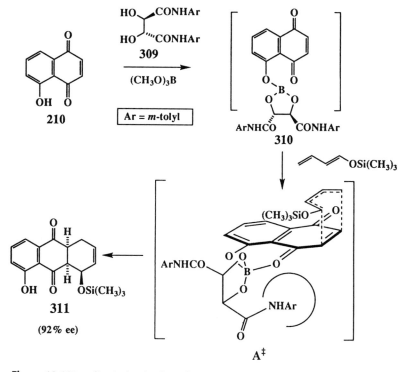

Figure 12.182. Examples of [4 + 2] cycloadditions catalyzed by a chiral borane.

mann, 1987), but the outcomes were discouraging. In sharp contrast, the readily prepared, fully synthetic dichloroborane **308** (resolved as a complex with menthone) catalyzes the addition of methyl acrylate and other dienophiles to cyclopentadiene in 91–97% ee (Fig. 12.182; Hawkins and Loren, 1991). An X-ray structure was determined on the complex between (±)-**308** and methyl crotonate.

Chiral Lewis acids based on ligands with C_2 symmetry of the binaphthyl type and those derived from tartrates have received considerable attention. When juglone **210** was treated with a mixture of tartramide **309** and trimethoxyborane, and the presumed complex **310** allowed to react with 1-trimethylsilyloxy-1,3-butadiene, the cycloadduct **311** was formed with an enantiomeric excess of 92% (Fig. 12.183; Maruoka, Yamamoto, et al., 1986). Tartrate ester derivatives were

Figure 12.183. Catalysis of a [4 + 2] cycloaddition by a C_2 chiral Lewis acid.

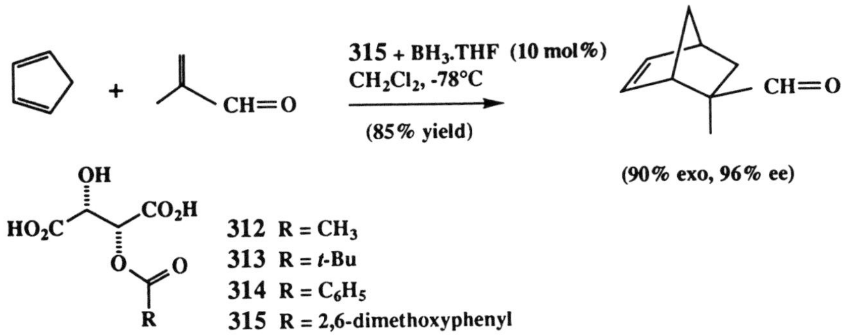

312 R = CH_3
313 R = *t*-Bu
314 R = C_6H_5
315 R = 2,6-dimethoxyphenyl

Figure 12.184. Tartaric acid derived catalysts for the Diels–Alder reaction.

much less effective, while the other tartramides that were examined gave enantiomeric excesses in the 77–87% range. The stereochemical outcome is readily rationalized in terms of the steric shielding of one face of the dienophile by one of the amide groups in the transition state structure **A**‡.

Combinations of the monoacylated tartaric acids **312–315** with the borane tetrahydrofuran complex have been examined as catalysts for the Diels–Alder reaction and 80–97% ees have been obtained with a reasonable range of dienes and dienophiles, an example of which is illustrated in Figure 12.184 (Furuta, Yamamoto, et al., 1988b, 1989). The high enantiomeric excess obtained for an exo aldehyde is unusual.

Reaction of the binaphthyl complex **316** with diene **318** proceeded rapidly (2 min at −78°C) to form **319** with an even better enantiomeric excess (98%) (Fig. 12.185). The need for bulky 2,2′ shielding groups was underlined by finding that an enantiomeric excess of only 70% was obtained from the equivalent reaction with the 2,2′-dimethyl analogue **317** (Kelly et al., 1986).

Several of the above examples involve ligands with C_2 symmetry, a feature that ensures that the boron center does not become stereogenic in the reaction complexes. This aspect has been considered to be important for the avoidance of mixtures of diastereomeric dienophile complexes (Kelly et al., 1986), but the effectiveness of the borane complex derived from **315** shows that the C_2 symmetry is not imperative.

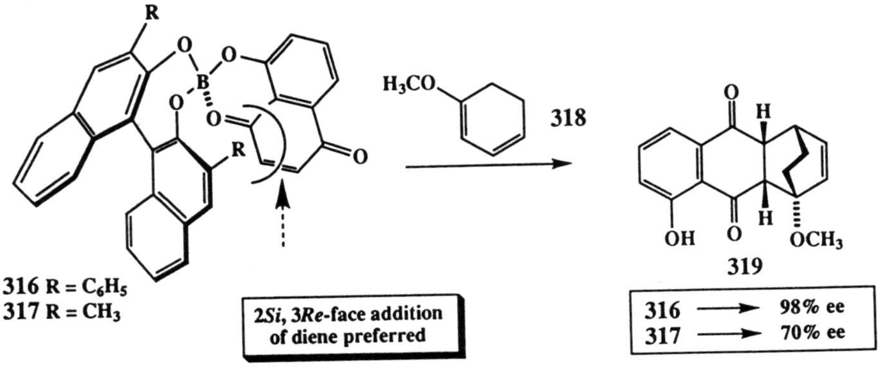

316 R = C_6H_5
317 R = CH_3

Figure 12.185. The [4 + 2] cycloaddition of binaphthyl boryl–juglone complexes.

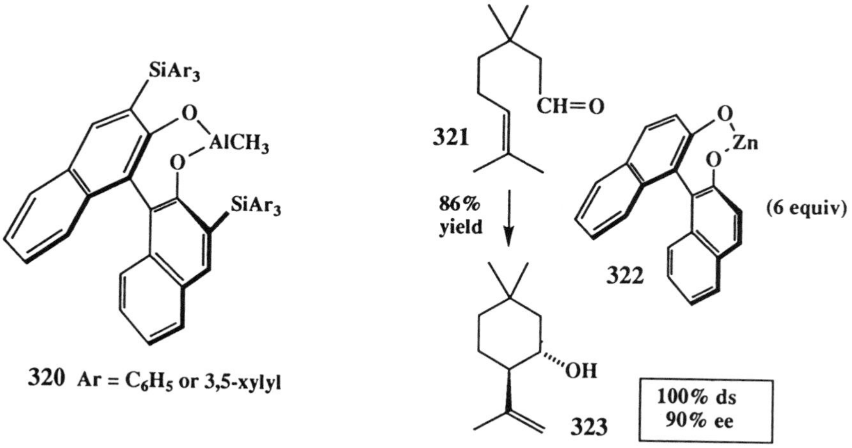

Figure 12.186. Further examples of chiral binaphthyl Lewis acids, where $Ar = C_6H_5$ or 3,5-xylyl.

These advances have not, however, been limited to cycloaddition reactions, nor to boryl complexes. Comparable enantiomeric excesses have also been obtained from [4 + 2] cycloadditions and the ene reactions of several aldehydes catalyzed by the binaphthyl alane **320** in substoichiometric (10 mol%) quantities (Maruoka, Yamamoto, et al., 1988a; 1988c), and from the conversion of aldehyde **321** into carbinol **323** by an intramolecular ene reaction mediated by the binaphthyl zinc complex **322** (Fig. 12.186). In this last case, however, a large excess of complex was required (six molar equivalents) and no enantioselection was observed for the analogue lacking the gem dimethyl group (Sakane, Yamamoto, et al., 1986).

d. Nonlinear Effects in Catalysis

In catalytic reactions for which the reaction complex involves two or more chiral ligands that are not enantiomerically pure, there is a distinct possibility that the enantiomeric excess of the product will not bear a simple relationship to the enantiomeric excess of the catalyst. For example, in the situation where two enantiomeric chiral ligands (L_R or L_S) are attached to a metal center (M), three complexes are possible: $(M)L_RL_R$, $(M)L_SL_S$, and $(M)L_RL_S$. If L_R is in excess and the stability constant for the meso complex $(M)L_RL_S$ is greater than that of the chiral complex, then $(M)L_RL_R$ and $(M)L_RL_S$ will dominate the mixture of catalysts. Now, in those cases where the meso isomer $(M)L_RL_S$ is the more effective catalyst, a lower than expected enantiomeric excess will be obtained. When the meso catalyst is less reactive, however, the enantiomeric excess of the product should be enhanced. These nonlinear relationships were originally demonstrated for the Katsuki–Sharpless oxidation of methyl p-tolyl sulfide and epoxidation of geraniol (Section 12-4.c), and the proline-catalyzed aldol reaction (Section 12-4.c; Puchot, Kagan, et al., 1986). However, the most compelling examples are provided by the addition of diethyl zinc to benzaldehyde catalyzed by a series of amino carbinols, the best of which are **276** and **325** (Fig. 12.187;

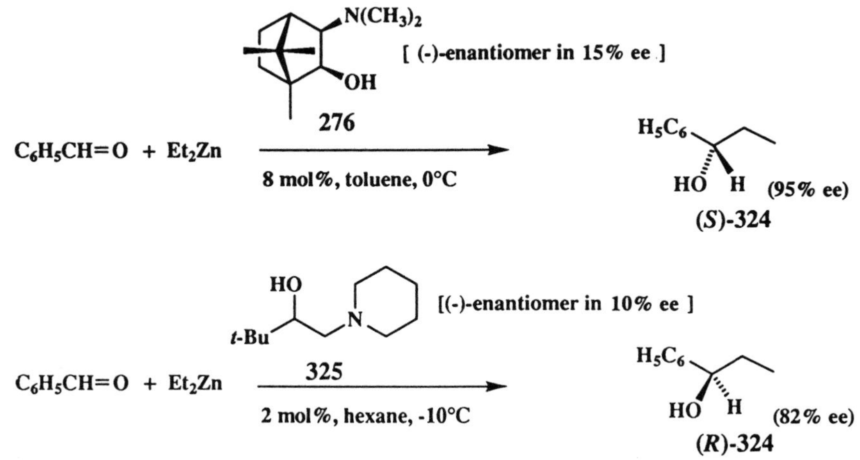

Figure 12.187. Amplified enantiomeric excesses in the catalyzed addition of diethyl zinc to benzaldehyde.

Kitamura, Noyori, et al., 1986; Kitamura, Noyori, et al., 1989; Oguni et al., 1988). In the former case, it was shown that the meso complex corresponding to **277** was very much more stable than **277** (in which the ligands are homochiral) (Fig. 12.167), which therefore reacts selectively with the excess of the dialkyl zinc to give the active catalyst (cf. Section 12-4.b).

e. Enzyme Based Processes

Enzymes provide a paradigm for stereoselective processes (and selective reactions in general, of course) (Santaniello et al., 1992). Apart from cycloadditions and the Cope version of [3, 3]-sigmatropic rearrangements, they catalyze a broad spectrum of standard organic reactions under very mild conditions; the Claisen rearrangement is represented by the chorismate group of rearrangements (Sogo, Knowles, et al., 1984). Enzymes have been classified into six groups:

1. Oxidoreductases, that mediate for example, $CH \rightarrow COH$, $CH(OH) \rightarrow C{=}O$, $CH{-}CH \rightarrow C{=}C$;
2. Transferases, that mediate the transfer of groups such as acyl, sugar, or phosphoryl from one molecule to another;
3. Hydrolases, that catalyze hydrolysis of amides, esters, and glycosides;
4. Lyases (these catalyze additions to π bonds and the reverse processes);
5. Isomerases, that mediate $C{=}C$ migrations, E/Z isomerizations and racemizations;
6. Ligases (or synthetases), that mediate the formation of $C{-}O$, $C{-}S$, $C{-}N$, $C{-}C$, and phosphate ester bonds.

Most of the enzyme catalyzed reactions of interest to organic chemists require coenzymes, such as NAD(P)/H or ATP, but their expense dictates that they too are used in catalytic amounts with an in situ method for their regeneration (Lee and Whitesides, 1985). Soluble enzymes are generally more convenient, but they are not readily recovered and their activity is diminished by exposure to shear, interfacial adsorption, organic solvents, and so on. Immobilization (Chibata, 1982; Chibata et al., 1992) by copolymerization (Pollak, Whitesides, et al., 1980) or containment within dialysis membranes (Bednarski, Whitesides, et al., 1987) allows many of these problems to be avoided, however (Bommarius et al., 1992).

The value of enzymes stems not only from the high (often but not invariably 100%) level of stereoselectivity, but also from their chemoselectivity and their capacity to achieve chemical conversions that remain inaccessible by standard laboratory procedures, at least in practical terms, for example, unactivated $CH \rightarrow COH$. In spite of these advantages, the utilization of enzymatic procedures by synthetic organic chemists has to date not been extensive and has been largely restricted to categories (1) and (2). Not surprisingly, applications tend to occur in areas where standard chemical methodology has been most deficient, namely, to problems that require discrimination between enantiomers and enantiotopic groups. Thus, in addition to simple enantioselective processes, we find numerous examples of the kinetic resolution of racemates and of the conversion of meso compounds into chiral intermediates (cf. Section 1-6.b). A number of illustrations are provided in Figure 12.188, but for a more extensive survey readers are directed towards three excellent reviews (Jones, 1986; Butt and Roberts, 1986; Sih and Wu, 1989).

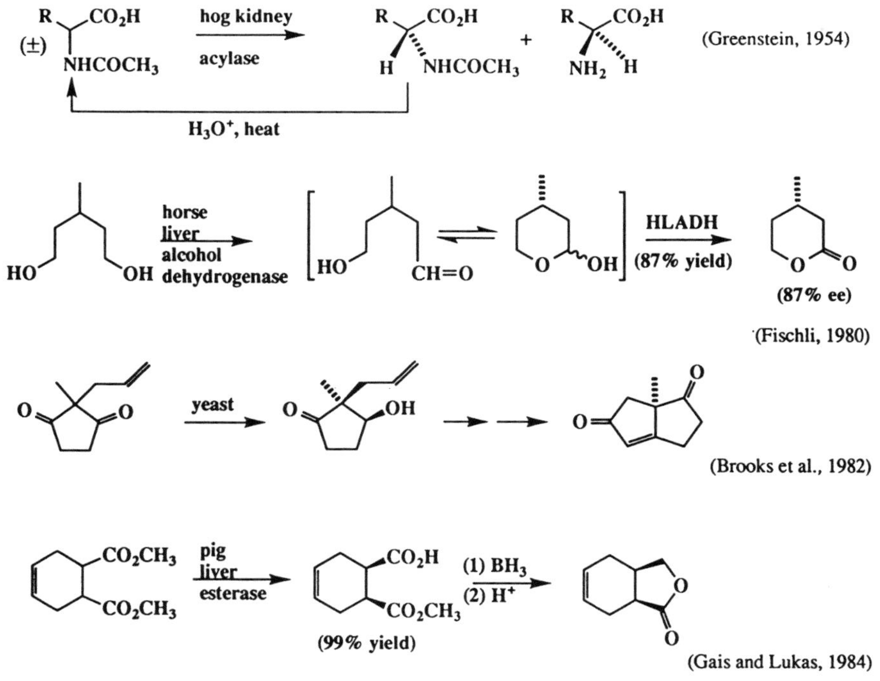

Figure 12.188. Examples of enzyme catalyzed reactions.

f. Enantioselective Synthesis Involving Discrimination between Enantiotopic Groups

As noted in Section 12-4.e, there is a dearth of nonenzymatic methods that might provide chemical discrimination between enantiotopic groups. There has been no systematic assault on such problems, although a number of recent interesting examples indicate an increasing awareness of the challenges involved (Ward, 1990). One specialized example, the amino acid catalyzed aldol reaction of cyclopentanediones and cyclohexanediones, has been outlined in Section 12-4.c. An analogous intramolecular Wittig reaction utilizing chiral phosphine ligands has been utilized to prepare a similar compound (Trost and Curran, 1980), but the enantiomeric excesses (30–40%) obtained were very modest by contemporary standards.

Chiral amide bases have been used, inter alia, to rearrange epoxides to allylic alcohols (Asami, 1984) and deprotonate various ketones, including 4-*tert*-butylcy-clohexanone **16** (Shirai, Koga, et al., 1986), *cis*-2,6-dimethylcyclohexanone **326** (Simpkins, 1986; cf. also Hine et al., 1980) and *cis*-3,5-dimethylcyclohexanone **327**, and the products trapped as enol trimethyl silyl ethers or acetates (Fig.

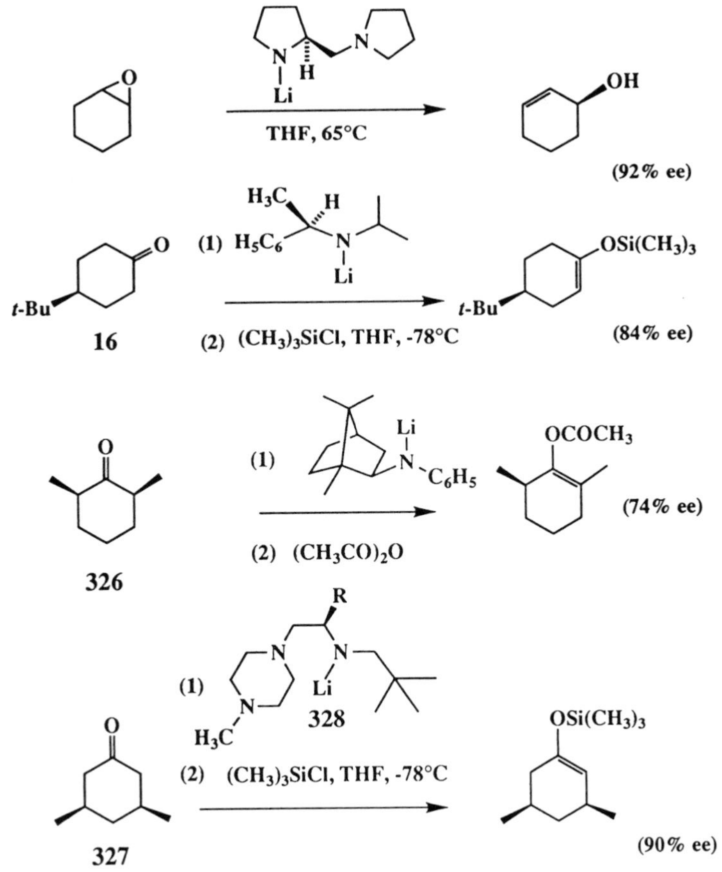

Figure 12.189. Enantioselective deprotonation of cyclohexanone derivatives.

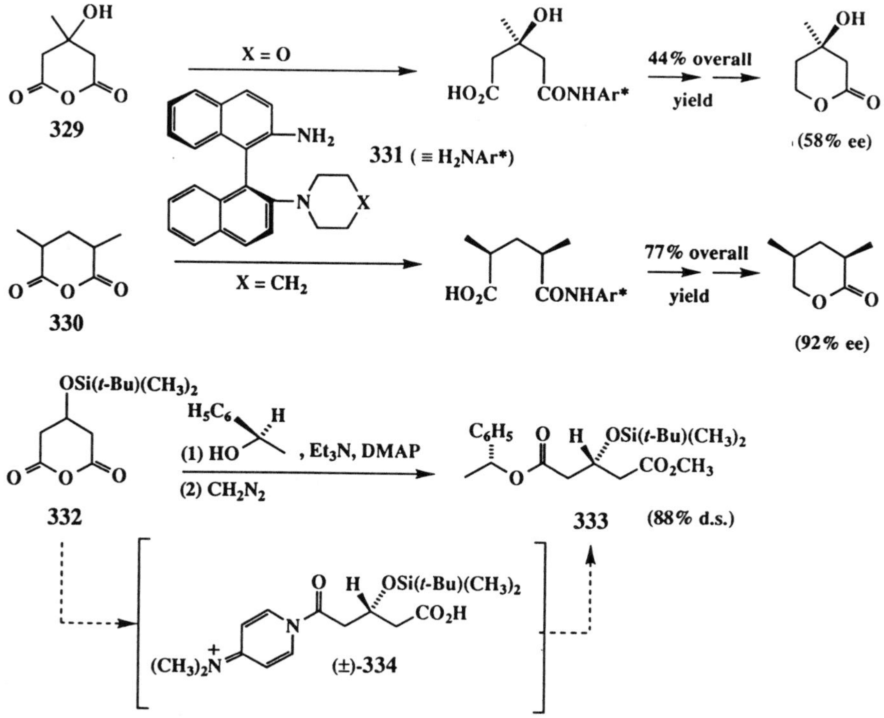

Figure 12.190. Enantiotopic group discrimination in cyclic anhydrides.

12.189). One of the most effective bases was the piperazine derivative **328** (R = C$_6$H$_5$) (Kim, Koga, et al., 1990).

Discrimination between the enantiotopic acyl groups in the prochiral anhydride **329** and the meso anhydride **330** has been achieved with good to excellent selectivity with (S)-(−)-binaphthyl diamines **331** (X = O and CH$_2$) as illustrated in Figure 12.190 (Kawakami, Oda, et al., 1984). In a further example, anhydride **332** was treated with (R)-1-phenylethanol in the presence of 4-N,N'-dimethylaminopyridine (DMAP) to afford ester **333** with 88% ds (Rosen and Heathcock, 1985). In this last case, however, it may be assumed that the DMAP converts **332** into the racemic acylamide **334**, and so the selective formation of **333** is, in effect, a kinetic resolution. Less direct methods that proceed via diastereoselective processes based on the incorporation of a chiral thiazolidine-2-thione (Nagao, Fujita, et al., 1984) or a chiral amine (Mukaiyama et al., 1983) into the substrate have also been reported.

In one final example, the meso-**335** diene has been converted by means of ene reactions into either adduct **336** or **337**, an intermediate in the synthesis of specionin (Fig. 12.191; Whitesell and Allen, 1988).

g. Enantioconvergent Syntheses

The concept of enantioconvergence, whereby both enantiomers of a synthetic intermediate may be utilized in the preparation of a target molecule, has been

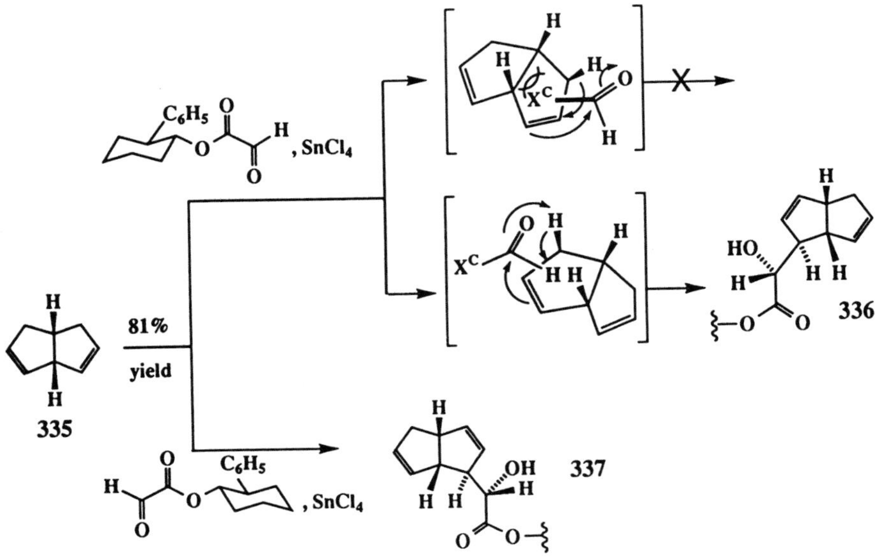

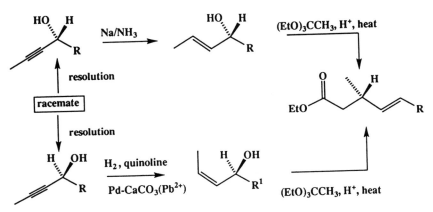

Figure 12.191. Enantioselective ene reactions with chiral glyoxalates.

largely restricted to situations where a racemate is resolved into its component enantiomers and the unwanted enantiomer racemized and recycled. Alternatively, each enantiomer can be taken along different routes, to converge on an appropriate intermediate (cf. Saddler, Fuchs, et al., 1981; Rama Rao et al., 1984). The stereochemically complementary sequences outlined in Figure 12.192 provide an interesting example of the latter strategy (Cohen et al., 1980).

More efficient strategies that avoid the need for formal resolutions, however, are illustrated by the examples outlined below. The first approach (Fig. 12.193) was used to effect an enantioselective synthesis of prostaglandins (Trost et al., 1978) and was based on the attachment of a chiral auxiliary to a racemic substrate, following which the adduct was submitted to a process that effected a configurational inversion of the intermediate under equilibrium conditions. After separation of the major diastereomer, the minor isomer was reequilibrated and recycled.

Figure 12.192. Enantioconvergent synthesis via the Claisen rearrangement.

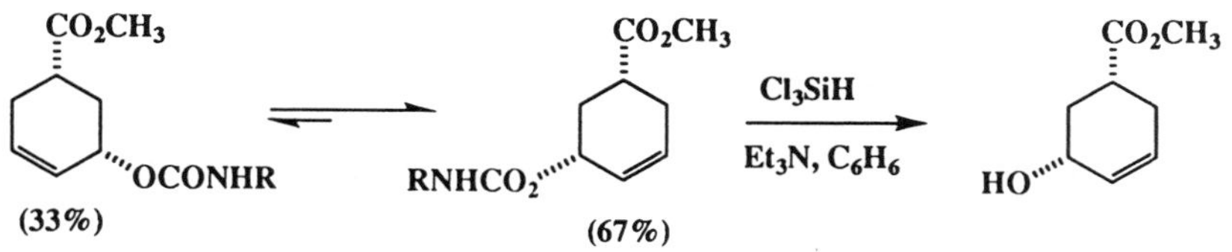

Figure 12.193. Enantioconvergent synthesis of an intermediate for prostaglandin synthesis, where R = (S)-1-α-naphthylethyl.

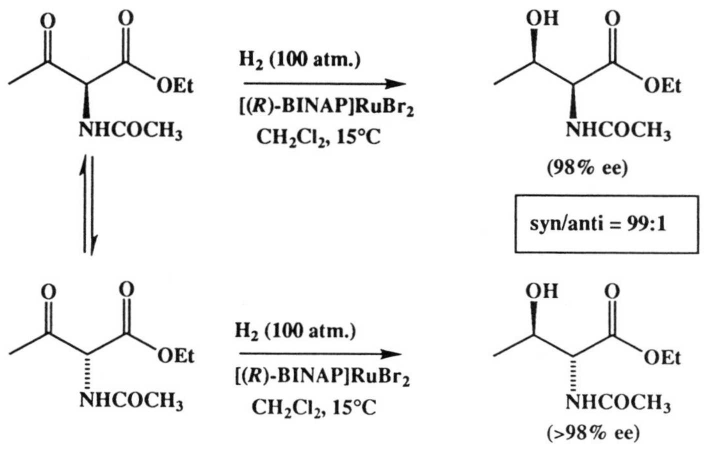

Figure 12.194. Application of dynamic kinetic resolution in the reduction of a racemic β-keto ester.

The very best strategy, however, utilizes kinetic discrimination between rapidly equilibrating enantiomers, whereby one of the enantiomers of a racemate is selectively consumed while in situ concurrent racemization occurs at a faster rate (cf. p. 414). An elegant example is provided by the BINAP–ruthenium(II) catalyzed hydrogenation of β-keto esters to either syn or anti hydroxy esters, an example of which is outlined in Figure 12.194 (Noyori et al., 1989).

12-5. DOUBLE STEREODIFFERENTIATION

a. Introduction

So far, we have dealt almost exclusively with reactions in which only one of the major participants has been chiral, that is, substrate, reagent, or catalyst. The level of stereoselectivity has therefore been independent of the absolute configuration of the chiral reactant. When more than one participant is chiral, however, the outcome will depend to a degree on the relationships between their respective absolute configurations and the proximity of stereogenic elements in the transition state structures. The problem of coupling racemates in a convergent synthesis is normally best avoided, for example. Nevertheless, we have already seen (Section 7-6) that the rate of reaction of a given enzyme or a selected synthetic reagent with one enantiomeric form of a chiral substrate is significantly different from that for the other enantiomer. This difference may then be exploited to achieve a kinetic resolution of the racemic version of the substrate. Although it is rare to find large differences in rates for nonenzymatic processes, the differential in reactivity between enantiomers may still be exploited to enhance stereoselection over and above that obtained with a single chiral participant, for example, by attaching a chiral auxiliary to an achiral reactant, utilizing a chiral solvent, employing a chiral catalyst in place of an achiral one, or substituting a chiral reactant that is synthetically equivalent to an achiral one.

For example, the lanthanide reagents $(+)$-Eu(hfc)$_3$ and $(+)$-Yb(hfc)$_3$ where hfc = 3-(heptafluoropropylhydroxymethylene)-$(+)$-camphorato (Quimpère and Jankowski, 1987; cf. Fig. 6.41), have significantly enhanced the enantiomeric excess obtained from the [4 + 2] cycloaddition of chiral dienes to aromatic aldehydes (Bednarski and Danishefsky, 1986). In a study of the cyclopropanation of alkenes with α-diazoacetic esters catalyzed by a chiral copper salicylaldimine complex, it was found that enantiomeric excesses could be improved by a few percent by substituting menthyl α-diazoacetate for the simple ethyl analogue (Aratani et al., 1982). A similar degree of improvement was observed for the aldol reaction between enolate **338** and aldehyde **339** by conducting the reaction in (S,S)-$(-)$-1,2,3,4-tetramethoxybutane [$(-)$-TMB] instead of THF (Fig. 12.195; Heathcock et al., 1981b), that is, an increase in the product ratio **340/341** from 4.3:1 to 5.0:1 was obtained.

An example of the use of a chiral catalyst to obtain enhanced diastereoselection in the hydrogenation of a chiral substrate has already been noted (Section 12-4c). A further application is provided by the hydrogenation of the premonensin precursor **342** (Fig. 12.196). Reduction over the achiral catalyst [Rh(NBD)(DIPHOS-4)]$^+$BF$_4^-$, for example, afforded an 85:15 mixture of **343** with its C(2) epimer **344**. When [Rh(NBD)$(+)$-BINAP]$^+$.BF$_4^-$ was utilized as the catalyst, however, the selectivity was improved to 98:2 (Evans and DiMare, 1986).

One final example of the opportunities for enhanced selectivity through double stereodifferentiation is provided by the use of the chiral base **328** (R = t-Bu; cf. Fig. 12.189) in place of LDA for the enolization of 3-oxo steroids. It has been well established that the 5α analogues preferentially form the Δ^2-enolates while the 5β isomers tend to form the Δ^3-enolates. Not only is it possible to improve the selectivity of these processes with the appropriate choice of a chiral base (Fig. 12.197; Sobukawa, Koga, et al., 1990), but the regiochemistry of deprotonation may be reversed in each case by selecting the alterna-

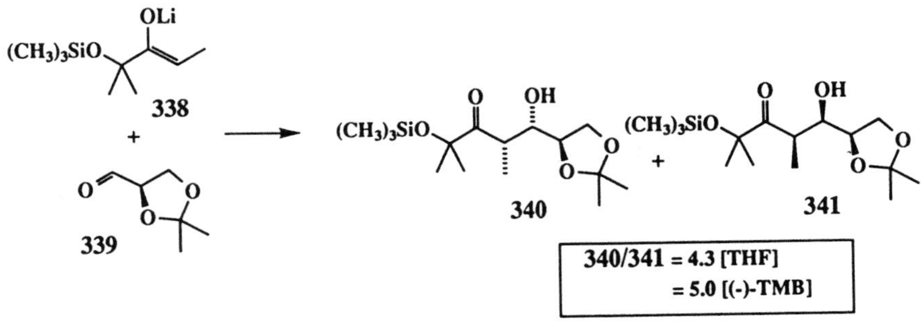

Figure 12.195. Enhancement of diastereoselection in the aldol reaction by the use of a chiral solvent.

342 343 (98%) 344 (2%)

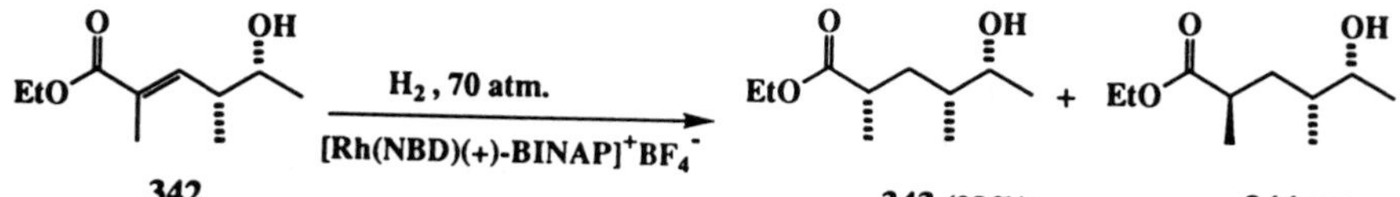

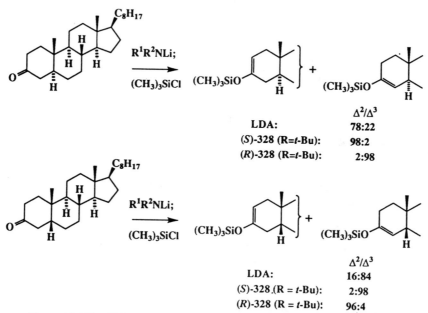

Figure 12.196. Enhanced diastereoselective hydrogenation over a chiral catalyst.

tive enantiomeric base, since the 2β and 4β protons are "pseudoenantiotopic", allowing the possibility of a reagent-imposed outcome (cf. p. 954 and Section 12-5.c).

b. Interactions between Principal Chiral Reactants

If, in situations where two chiral reactants are coupled, stereoselectivity is enhanced relative to a reference reaction in which only one of the participants is chiral, the reactants may be considered to be "matched". Conversely, if the stereoselectivity is diminished, the reactants may be regarded as "mismatched" (Masamune et al., 1985a). For example, the inherent diastereofacial preferences of chiral enolates and aldehydes may reinforce or oppose one another in the crossed aldol reaction. These preferences have been observed for the reaction of the lithium enolate derived from ketone **345** with aldehyde **339** and its *S* enantiomer **348**. As has already been noted (Fig. 12.195) the reaction of **339** with achiral enolate **338** gave a 4.3:1 mixture of **340** and **341**. When **339** reacted with the enolate derived from **345**, however, the stereochemically analogous aldol products **346** and **347** were obtained in a 2:1 ratio (mismatched), whereas the *S* enantiomer **348** afforded **349** with a ds of greater than 97% (matched) (Fig. 12.198; Heathcock et al., 1981b).

	Δ^2/Δ^3
LDA:	78:22
(*S*)-328 (R=*t*-Bu):	98:2
(*R*)-328 (R=*t*-Bu):	2:98

	Δ^2/Δ^3
LDA:	16:84
(*S*)-328 (R = *t*-Bu):	2:98
(*R*)-328 (R = *t*-Bu):	96:4

Figure 12.197. Enhanced regioselectivity in the enolization of 3-oxo steroids.

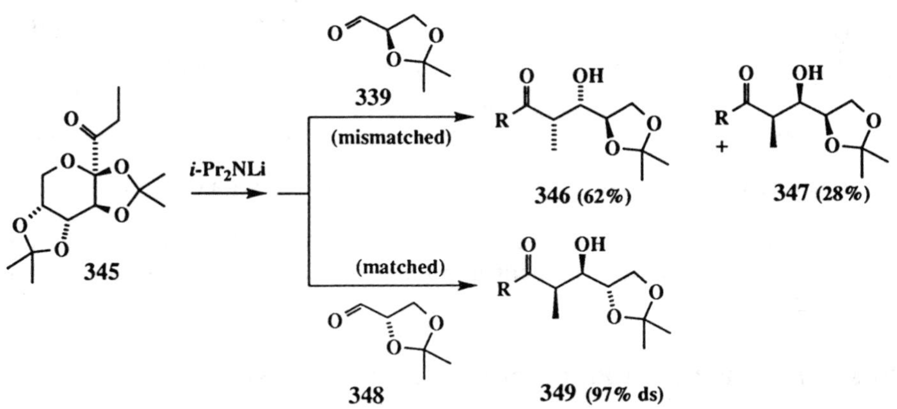

Figure 12.198. Matched and mismatched aldol reactions.

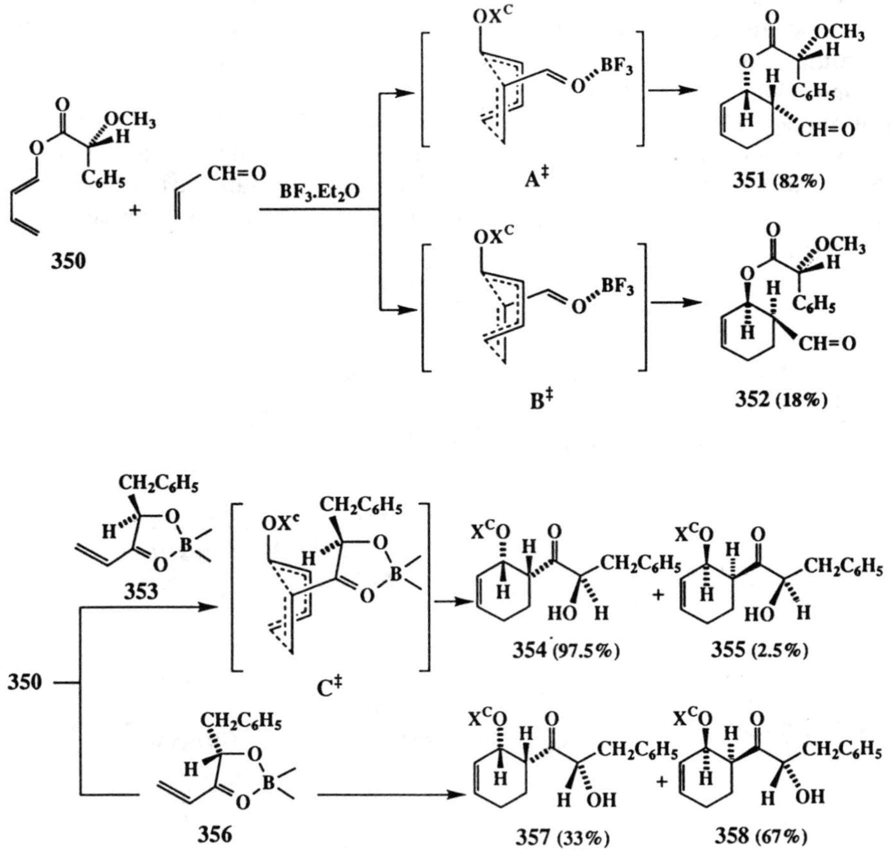

Figure 12.199. Matched and mismatched configurations in a [4 + 2] cycloaddition between a chiral diene and dienophiles.

Processes of this type have been categorized by the terms 'double asymmetric induction' (Horeau et al., 1968) and "double asymmetric synthesis" (Masamune et al., 1985a). Because asymmetry cannot really be qualified, however, the description "double stereodifferentiation" (Izumi and Tai, 1977) is to be preferred (Heathcock and White, 1979).

In a further example, the boron trifluoride etherate catalyzed [4 + 2] cycloaddition of (R)-350 diene with acrolein (cf. Sections 12-3.g and i) gave a 4.5:1 mixture of diastereomeric adducts **351** and **352**. This result is rationalized in terms of the chiral auxiliary X_C attached to the diene shielding the 1-*Si* face of the diene, thereby disfavoring transition state structure **B**‡ (leading to **352**). The reaction of **350** with dienophile (R)-**353**, however, afforded a 40:1 mixture of **354** and **355** (Fig. 12.199) Transition state structure **C**‡ corresponding to **A**‡ would not be expected to be significantly perturbed by the extra substituents in the (R)-dienophile **353**, but in the case of the (S)-dienophile **356** it may be seen that for a transition state structure related to **A**‡, the benzyl group would intrude into the space occupied by the diene, that is, a mismatched situation (Masamune et al., 1985a).

A simple multiplicative model designed to allow a quantitative estimate of the ratio of diastereomeric products from these and related reactions has been proposed (Heathcock, 1984b; Masamune et al., 1985a).

c. Reagent Control

The converse of the strategy, whereby a judicious combination of a chiral substrate with a chiral reagent is employed to enhance diastereoselection, is one based on the utilization of chiral reagents that impose an overwhelming stereochemical preference for a specific sense of chirality in the new stereocenter(s), regardless of the absolute configuration of the substrate. This strategy is nicely illustrated by the Katsuki–Sharpless epoxidation of several allylic alcohols (Sharpless, 1985; cf. p. 399). Many other compelling examples of this approach have been assembled, however (Masamune et al., 1985a), including the aldol reaction illustrated in Figure 12.200 (Evans and Bartroli, 1982). Nevertheless, this strategy

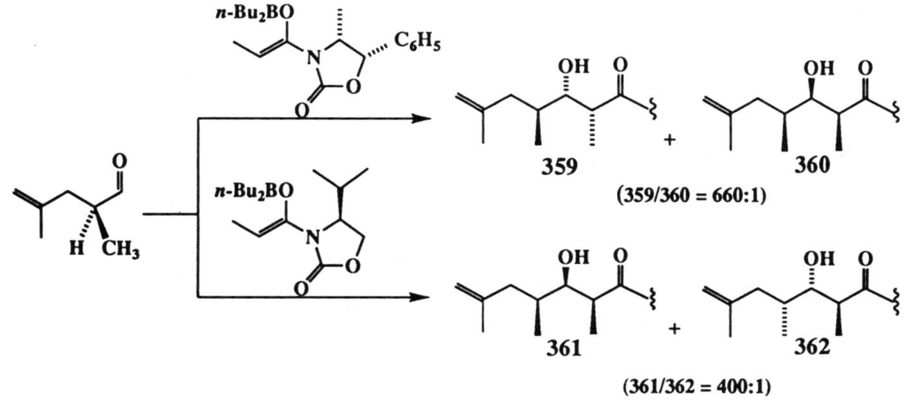

Figure 12.200 Reagent control of stereoselection in the aldol reaction.

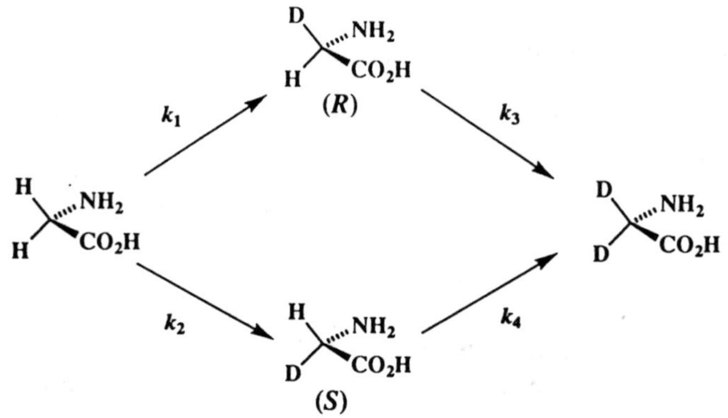

Figure 12.201. Proton–deuteron exchange in glycine.

will only be effective when the substrate itself does not exert too strong an influence on the stereochemical outcome (in the mismatched case, the rate of reaction could become impractically slow).

d. Kinetic Amplification

In stereoselective processes that involve ligand replacement or modification it can be expected that the products will react further with selective consumption of the minor stereoisomer, thereby leading to an enhancement of the initial enantiomeric excess (Bergens and Bosnich, 1987). Consider the case of proton–deuteron exchange for glycine (Fig. 12.201), either by enzyme catalysis (enantioselection)

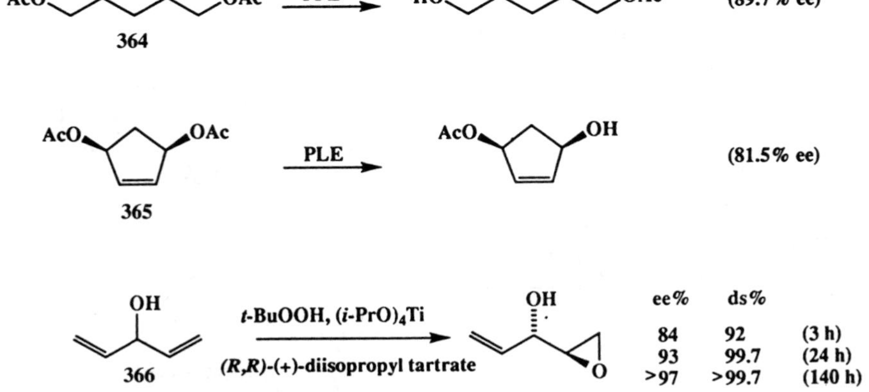

Figure 12.202. Examples of stereoisomer enrichment through kinetic amplification.

or by chemical exchange after attachment to a chiral auxiliary (diastereoselection) (Dokuzovic, Bosnich, et al., 1986). If $k_1 > k_2$, the R enantiomer will be formed in excess. Both enantiomers will continue to react further, however, and the same factors that make $k_1 > k_2$ should also make $k_4 > k_3$. Thus, the minor S enantiomer will be consumed more rapidly, resulting in a "kinetic amplification" of the initial enantioselection thereby affording an excellent enantiomeric excess of the R epimer, even when the differences in rates are quite modest.

Examples of this kinetic amplification are found in the ene reaction of diene **335** described earlier (Section 12-4.f, for hydroboration of diene **363**; Partridge et al., 1973), for the pig liver esterase (PLE) and pig pancreatic lipase (PPL) catalyzed hydrolyses of acetates **364** and **365** (Wang, Sih et al., 1984), and for the Katsuki–Sharpless epoxidation of a series of dialkenyl carbinols, for example, **366** (Schreiber et al., 1987; Fig. 12.202, cf. p. 400). Mathematical analyses have also been provided in the last two cases (cf. also Chen, Sih, et al., 1982; Sih and Wu, 1989).

12-6. CONCLUSION

The aim of this chapter has been to provide an easily digested overview of stereoselective synthesis with sufficient illustrations to make it readily understood. The limitations of space have inevitably resulted in the omission of important topics, although signposts to further examples and more detailed treatments have been provided. A special effort has been made to provide references that supplement and update the many excellent reviews from the early to mid-1980s, including the splendid five-volume series *Asymmetric Synthesis*, edited by Morrison (and in part, by Scott).

Spectacular advances in the levels of stereocontrol have been achieved over recent years, allowing chemists, inter alia, to mount a major assault on what was once considered to be the unchallenged preserve of enzyme-catalyzed processes. Much of this progress has been empirically based, however, and there is a pressing need for more systematic studies substantiated by solid kinetic data. One very important aspect that emerges from the preceding survey is the considerable advantage to be obtained by designing systems in which the reaction rate of a desired stereochemical outcome is accelerated relative to those of competing processes. This approach may also open up opportunities for catalysis by substoichiometric quantities of valuable materials. While the alternative strategy of retarding undesired reactions (usually by steric means) will always be important, by its very nature it requires, as a minimum, stoichiometric quantities of reagents and also tends to lead to cumbersome, "over-engineered" solutions.

REFERENCES

Agami, C., Meynier, F., Puchot, C., Guilhem, J., and Pascard, C. (1984), *Tetrahedron*, **40**, 1031.

Agami, C., Puchot, C., and Sevestre, H. (1986), *Tetrahedron Lett.*, **27**, 1501.

Aitken, R. A. and Kilényi, S. N., Eds. (1992), *Asymmetric Synthesis*, Blackie, London.

Akutagawa, S. (1992), "A Practical Synthesis of (−)-Menthol with the Rh-BINAP Catalyst," in Collins, A. N., Sheldrake, G. N., Crosby, J., Eds., *Chirality in Industry. The Commercial Manufacture of Optically Active Compounds*, Wiley, Chichester, England, p. 313.

Allenmark, G. (1991), *Chromatographic Enantioseparation: Methods and Applications*, 2nd ed. Wiley, New York.

Amstutz, R., Schweizer, W. B., Seebach, D., and Dunitz, J. D. (1981), *Helv. Chim. Acta*, **64**, 2617.

Ando, A. and Shioiri, T. (1987), *J. Chem. Soc. Chem. Commun.*, 1620.

Andrews, G. and Evans, D. A. (1972), *Tetrahedron Lett.*, 5121.

Andrisano, R., Costa Bizzari, P., and Tramontini, M. (1970), *Tetrahedron*, **26**, 3959.

Anet, F. A. L., Miura, S. S., Siegel, J., and Mislow, K. (1983), *J. Am. Chem. Soc.*, **105**, 1419.

Angiolini, L., Costa Bizzarri, P., and Tramontini, M. (1969), *Tetrahedron*, **25**, 4211.

Anh, N. T. (1980), "Regio- and Stereo-Selectivities in Some Nucleophilic Reactions," *Top. Curr. Chem.*, **88**, 145.

Anh, N. T. and Eisenstein, O. (1977), *Nouv. J. Chim.*, **1**, 61.

ApSimon, J. W. and Collier, T. L. (1986), *Tetrahedron*, **42**. 5158.

Arai, I., Mori, A. and Yamamoto, H. (1985), *J. Am. Chem. Soc.*, **107**, 8254.

Aratani, T. (1985), *Pure Appl. Chem.*, **57**, 1839.

Aratani, T., Yoneyoshi, Y., and Nagase, T. (1982), *Tetrahedron Lett.*, **23**, 685.

Asami, M. (1984), *Chem. Lett.*, 829.

Auburn, P. R., Mackenzie, P. B., and Bosnich, B. (1985), *J. Am. Chem. Soc.*, **107**, 2033.

Baird, G. J., Bandy, J. A., Davies, S. G., and Prout, K. (1983), *J. Chem. Soc. Chem. Commun.*, 1202.

Baldwin, J. E. (1976), *J. Chem. Soc. Chem. Commun.*, 734.

Bandara, B. M. R., Birch, A. J., and Kelly, L. F. (1984), *J. Org. Chem.*, **49**, 2496.

Bartlett, P. A. (1980), "Stereocontrol in the Synthesis of Acyclic Systems: Applications to Natural Product Synthesis," *Tetrahedron*, **36**, 3.

Bartlett, P. A. (1984a), "Olefin Cyclization Processes that Form Carbon-Heteroatom Bonds," in Morrison, J. D., Ed., *Asymmetric Synthesis*, Vol. 3, Academic, New York, p. 411.

Bartlett, P. A. (1984b), "Olefin Cyclization Processes that Form Carbon–Carbon Bonds," in Morrison, J. D., Ed., *Asymmetric Synthesis*, Vol. 3, Academic, New York, p. 342.

Bartlett, P. A. and Jernstedt, K. K. (1977), *J. Am. Chem. Soc.*, **99**, 4829.

Bartlett, P. A. and Myerson, J. (1978), *J. Am. Chem. Soc.*, **100**, 3950.

Barton, D. H. R., Hartwig, W., and Motherwell, W. B. (1982), *J. Chem. Soc. Chem. Commun.*, 447.

Beckwith, A. L. J. (1981), "Regio-Selectivity and Stereo-Selectivity in Radical Reactions," *Tetrahedron*, **37**, 3073.

Beckwith, A. L. J. and Boate, D. R. (1986), *J. Chem. Soc. Chem. Commun.*, 189.

Beckwith, A. L. J. and Ingold, K. U. (1990), "Free-Radical Rearrangements," in deMayo, P., Ed., *Rearrangements in Ground and Excited States*, Academic, New York.

Beckwith, A. L. J., Lawrence, T., and Serelis, A. K. (1980), *J. Chem. Soc. Chem. Commun.*, 484.

Beckwith, A. L. J., Phillipou, G., and Serelis, A. K. (1981), *Tetrahedron Lett.*, **22**, 2811.

Beckwith, A. L. J. and Roberts, D. H. (1986), *J. Am. Chem. Soc.*, **108**, 5893.

Beckwith, A. L. J. and Schiesser, C. H. (1985), *Tetrahedron*, **41**, 3925.

Bednarski, M., Chenault, H. K., Simon, E. S., and Whitesides, G. M. (1987), *J. Am. Chem. Soc.*, **109**, 1283.

Bednarski, M. and Danishefsky, S. (1986), *J. Am. Chem. Soc.*, **108**, 7060.

Bennett, G. B. (1977), "The Claisen Rearrangement in Organic Synthesis; 1967 to January 1977," *Synthesis*, 589.

Benson, W. R. and Pohland, A. E. (1964), *J. Org. Chem.*, **29**, 385.

Bergbreiter, D. E. and Newcomb, M. (1983), "Alkylation of Imine and Enamine Salts," in Morrison, J. D. Ed., *Asymmetric Synthesis*, Vol. 2, Academic, New York.

Bergens, S. and Bosnich, B. (1987), *Comm. Inorg. Chem.*, **6**, 85.

Bernardi, F., Bottoni, A., Field, M. J., Guest, M. F., Hillier, I. H., Robb, M. A., and Venturini, A. (1988), *J. Am. Chem. Soc.*, **110**, 3050.

Bernhard, W., Fleming, I., and Waterson, D. (1984), *J. Chem. Soc. Chem. Commun.*, 28.

Bestmann, H.-J. and Zimmerman, R. (1982), "Phosphor-Ylide," in Houben-Weyl, Regitz, M., Ed., "Methoden der Organischen Chemie," 4th ed., Vol. E1. *Organische Phosphorverbindungen I*, Thieme, New York.

Bickart, P., Carson, F. W., Jacobus, J., Miller, E. G., and Mislow, K. (1968), *J. Am. Chem. Soc.*, **90**, 4869.

Bir, G. and Kaufmann, D. (1987), *Tetrahedron Lett.*, **28**, 777.

Birch, A. J. (1982), *Curr. Sci.*, **51**, 155.

Birch, A. J. and Subba Rao, G. S. R. (1970), *Aust. J. Chem.*, **23**, 547.

Blaser, H.-U. (1992), "The Chiral Pool as a Source of Enantioselective Catalysts and Auxiliaries," *Chem. Rev.*, **92**, 935.

Blechert, S. (1989), "The Hetero-Cope Rearrangement in Organic Synthesis," *Synthesis*, 71.

Blomquist, A. T., Liu, L. H., and Bohrer, J. C. (1952), *J. Am. Chem. Soc.*, **74**, 3643.

Bommarius, A. S., Drauz, K., Groeger, U., and Wandrey, C. (1992), "Membrane Bioreactors for the Production of Enantiomerically Pure alpha-Amino Acids," in Collins, A. N., Sheldrake, G. N., Crosby, J., Eds., *Chirality in Industry. The Commercial Manufacture of Optically Active Compounds*, Wiley, Chichester, England, p. 371.

Boone, J. R. and Ashby, E. C. (1979), "Reduction of Cyclic and Bicyclic Ketones by Complex Metal Hydrides," *Top. Stereochem.*, **11**, 53.

Bosnich, B. (1985), *Chem. Scr.*, **25**, NS45. Special Nobel symposium 60 issue.

Bosnich, B., Ed. (1986), *Asymmetric Catalysis*, NATO ASI Series E 103, Nijhoff, Dordrecht, The Netherlands.

Bosnich, B., and Fryzuk, M. D. (1981), "Asymmetric Synthesis Mediated by Transition Metal Complexes," *Top. Stereochem.*, **12**, 119.

Brady, S. F., Ilton, M. A., and Johnson, W. S. (1968), *J. Am. Chem. Soc.*, **90**, 2882.

Braun, M. (1987), "Stereoselective Aldol Reactions with α-Unsubstituted Chiral Enolates," *Angew. Chem. Int. Ed. Engl.*, **26**, 24.

Bredig, G. and Fiske, P. S. (1912), *Biochem. Z.*, **46**, 7.

Brewster, J. H. (1986), *J. Org. Chem.*, **51**, 4751.

Brienne, M. J., Varech, D., and Jacques, J. (1974), *Tetrahedron Lett.*, 1233.

Brooks, D. W., Grothaus, P. G. and Irwin, W. L. (1982), *J. Org. Chem.*, **47**, 2820.

Brown, H. C., Bowman, D. H., Misumi, S., and Unni, M. K. (1967), *J. Am. Chem. Soc.*, **89**, 4532.

Brown, H. C., Chandrasekharan, J., and Ramachandran, P. V. (1986b), *J. Org. Chem.*, **51**, 3394.

Brown, H. C., Cho, B. T., and Park, W. S. (1986c), *J. Org. Chem.*, **51**, 3396.

Brown, H. C., Desai, M. C., and Jadhav, P. K. (1982b), *J. Org. Chem.*, **47**, 5065.

Brown, H. C. and Jadhav, P. K. (1981), *J. Org. Chem.*, **46**, 5047.

Brown, H. C., Jadhav, P. K., and Mandal, A. K. (1981), *Tetrahedron*, **37**, 3547.

Brown, H. C., Jadhav, P. K., and Singram, B. (1986a), "Enantiomerically Pure Compounds via Chiral Boranes," in Sheffold, R., Ed., *Modern Synthetic Methods*, Vol. 4, Springer, Berlin, p. 307.

Brown, H. C. and Pai, G. G. (1982), *J. Org. Chem.*, **47**, 1606.

Brown, H. C., Park, W. S., and Cho, B. T. (1986d), *J. Org. Chem.*, **51**, 1934.

Brown, H. C., Randad, R. S., Bhat, K. S., Zaidlewicz, M., and Racherla, U. S. (1990), *J. Am. Chem. Soc.*, **112**, 2389.

Brown, H. C., Vara Prasad, J. V. N., and Zee, S.-H. (1985), *J. Org. Chem.*, **50**, 1582.

Brown, H. C. and Zweifel, G. (1961), *J. Am. Chem. Soc.*, **83**, 3834.

Brown, J. M. (1989), "Asymmetric Homogeneous Catalysis," *Chem. Br.*, 276.

Brown, J. M. and Chaloner, P. A. (1978), *Tetrahedron Lett.*, 1877.

Brown, J. M. and Naik, R. G. (1982), *J. Chem. Soc. Chem. Commun.*, 348.

Brunner, H. (1988), "Enantioselective Synthesis of Organic Compounds with Optically Active Transition Metal Catalysts in Substoichiometric Quantities," *Top. Stereochem.*, **18**, 129.

Brunner, H., Becker, R., and Riepl, G. (1984), *Organometallics*, **3**, 1354.

Buono, G., Siv, C., Pfeiffer, G., Triantaphylides, C., Denis, P., Mortreux, A., and Petit, F. (1985), *J. Org. Chem.*, **50**, 1781.

Burgess, E. M. and Liotta, C. L. (1981), *J. Org. Chem.*, **46**, 1703.

Burgess, E. M., Liotta, C. L., and Eberhardt, W. H. (1984), *J. Am. Chem. Soc.*, **106**, 4849.

Burgess, K. and Ohlmeyer, M. J. (1988), *J. Org. Chem.*, **53**, 5178.

Bürgi, H. B., Dunitz, J. D., Lehn, J. M., and Wipff, G. (1974), *Tetrahedron*, **30**, 1563.

Bürgi, H. B., Dunitz, J. D., and Shefter, E. J. (1973), *J. Am. Chem. Soc.*, **95**, 5065.

Bürgi, H. B., Lehn, J. M., and Wipff, G. (1974), *J. Am. Chem. Soc.*, **96**, 1956.

Burkert, U. and Allinger, N. L. (1982), *Molecular Mechanics*, American Chemical Society, Washington, DC.

Butt, S. and Roberts, S. M. (1986), "Recent Advances in the use of Enzyme-catalysed Reactions in Organic Research: The Synthesis of Biologically Active Natural Products and Analogues," *Natl. Prod. Rep.*, **3**, 489.

Caine, D. (1979), "Alkylation and Related Reactions of Ketones and Aldehydes via Metal Enolates," in Augustine, R. L., Ed., *Carbon–Carbon Bond Formation*, Marcel Dekker, New York, p. 85.

Campbell, J. B. and Molander, G. A. (1978), *J. Organomet. Chem.*, **156**, 71.

Capon, B. and McManus, S. P. (1976), *Neighboring Group Participation*, Vol. 1, Plenum, New York.

Carlson, R. M. and Lee, S. Y. (1969), *Tetrahedron Lett.*, 4001.

Cha, J. K., Christ, W. J., and Kishi, Y. (1983), *Tetrahedron Lett.*, **24**, 3943.

Chamberlain, P., Roberts, M. L., and Whitham, G. H. (1970), *J. Chem. Soc. (B)*, 1374.

Chan, Ka.-K., Cohen, N., De Noble, J. P., Specian, A. C., and Saucy, G. (1976), *J. Org. Chem.*, **41**, 3497.

Chan, T. H., Aida, T., Lau, P. W. K., Gorys, V., and Harpp, D. N. (1979), *Tetrahedron Lett.*, 4029.

Chandrasekharan, J., Ramachandran, P. V., and Brown, H. C. (1986), *J. Org. Chem.*, **51**, 3395.

Chautemps, P. and Pierre, J.-L. (1976), *Tetrahedron*, **32**, 549.

Chen, C.-S., Fujimoto, Y., Girdaukas, G., and Sih, C. J. (1982), *J. Am. Chem. Soc.*, **104**, 7294.

Chérest, M. and Felkin, H. (1968), *Tetrahedron Lett.*, 2205.

Chérest, M., Felkin, H., and Prudent, N. (1968), *Tetrahedron Lett.*, 2199.

Chérest, M., Felkin, H., Tacheau, P., Jacques, J., and Varech, D. (1977), *J. Chem. Soc. Chem. Commun.*, 372.

Cheung, C. K., Tseng, L. T., Lin, M.-H., Srivastava, S., and le Noble, W. J. (1986), *J. Am. Chem. Soc.*, **108**, 1598.

Chibata, I. (1982), "Applications of Immobilized Enzymes for Asymmetric Reactions," in Eliel, E. L., and Otsuka, S., Eds., *Asymmetric Reactions and Processes in Chemistry*. ACS Symposium Series 185, American Chemical Society, Washington, DC, p. 195.

Chibata, I., Tosa, T., and Shibatani, T. (1992), "The Industrial Production of Optically Active Compounds by Immobilized Biocatalysts," in Collins, A. N., Sheldrake, G. N., Crosby, J., Eds., *Chirality in Industry. The Commercial Manufacture of Optically Active Compounds*, Wiley, Chichester, England, p. 351.

Choy, W., Reed, L. A., and Masamune, S., (1983), *J. Org. Chem.*, **48**, 1137.

Christ, W. J., Cha, J. K., and Kishi, Y. (1983), *Tetrahedron Lett.*, **24**, 3947.

Chuang, C.-P., Galluci, J. C., and Hart, D. J. (1988a), *J. Org. Chem.*, **53**, 3210.

Chuang, C.-P., Galluci, J. C., Hart, D. J., and Hoffman, C. (1988b), *J. Org. Chem.*, **53**, 3218.

Church, R. F., Ireland, R. E., and Marshall, J. A. (1966), *J. Org. Chem.*, **31**, 2526.

Cieplak, A. S. (1981), *J. Am. Chem. Soc.*, **103**, 4540.

Coates, R. M. and Sandefur, L. O. (1974), *J. Org. Chem.*, **39**, 275.

Cohen, N. (1976), "Asymmetric Induction in 19-Nor Steroid Total Synthesis," *Acc. Chem. Res.*, **9**, 412.

Cohen, N., Lopresti, R. J., Neukom, C., and Saucy, G. (1980), *J. Org. Chem.*, **45**, 583.

Collins, A. N., Sheldrake, G. N., Crosby, J. (1992), *Chirality in Industry. The Commercial Manufacture of Optically Active Compounds*, Wiley, Chichester, England.

Coppola, G. M. and Schuster, H. F. (1987), *Asymmetric Synthesis: Construction of Chiral Molecules Using Amino Acids*, Wiley-Interscience, New York.

Corey, E. J. (1990), *J. Org. Chem.*, **55**, 1693.

Corey, E. J., Bakshi, R. K., and Shibata, S. (1987a), *J. Am. Chem. Soc.*, **109**, 5551.

Corey, E. J., Bakshi, R. K., Shibata, S., Chen, C.-P., and Singh, V. K. (1987b), *J. Am. Chem. Soc.*, **109**, 7925.

Corey, E. J. and Boaz, N. W. (1985a), *Tetrahedron Lett.*, **26**, 6015.

Corey, E. J. and Boaz, N. W. (1985b), *Tetrahedron Lett.*, **26**, 6019.

Corey, E. J. and Ensley, H. E. (1975), *J. Am. Chem. Soc.*, **97**, 6908.

Corey, E. J., Jardine, P. D., Virgil, S., Yuen, P.-W., and Connell, R. D. (1989), *J. Am. Chem. Soc.*, **111**, 9243.

Corey, E. J. and Hannon, F. J. (1990), *Tetrahedron Lett.*, **31**, 1393.

Corey, E. J., Hannon, F. J., and Boaz, N. W. (1989), *Tetrahedron*, **45**, 545.

Corey, E. J., Katzenellenbogen, J. A., and Posner, G. H. (1967), *J. Am. Chem. Soc.*, **89**, 4245.

Corey, E. J., Kim, S., Yoo, S., Nicolaou, K. C., Melvin, L. S., Brunelle, D. J., Falck, J. R., Trybulski, E. J., Lett, R., and Sheldrake, P. W. (1978a), *J. Am. Chem. Soc.*, **100**, 4620.

Corey, E. J., Naef, R., and Hannon, F. J. (1986), *J. Am. Chem. Soc.*, **108**, 7114.

Corey, E. J. and Pearce, H. L. (1979), *J. Am. Chem. Soc.*, **101**, 5841.

Corey, E. J., Shimoji, K., and Shih, C. (1984), *J. Am. Chem. Soc.*, **106**, 6425.

Corey, E. J. and Sneen, R. A. (1956), *J. Am. Chem. Soc.*, **78**, 6269.

Corey, E. J., Trybulski, E. J., Melvin, L. S. Jr., Nicolaou, K. C., Secrist, J. A., Lett, R., Sheldrake, P. W., Falck, J. R., Brunelle, D. J., Haslanger, M. F., Kim, S., and Yoo, S. (1978b), *J. Am. Chem. Soc.*, **100**, 4618.

Corey, E. J. and Winter, R. A. E. (1963), *J. Am. Chem. Soc.*, **85**, 2677.

Cornforth, J. W., Cornforth, R. H., and Mathew, K. K. (1959a), *J. Chem. Soc.*, 112.

Cornforth J. W., Cornforth, R. H., and Mathew, K. K. (1959b), *J. Chem. Soc.*, 2539.

Cox, P. J. and Simpkins, N. S. (1991), "Asymmetric Synthesis Using Homochiral Lithium Amide Bases," *Tetrahedron Asymmetry*, **2**, 1.

Crabtree, R. H., Felkin, H., Fellebeen-Khan, T., and Morris, G. E. (1979), *J. Organomet. Chem.*, **168**, 183.

Cram, D. J. and Abd Elhafez, F. A. (1952), *J. Am. Chem. Soc.*, **74**, 5828.

Cram, D. J. and Kopecky, K. R. (1959), *J. Am. Chem. Soc.*, **81**, 2748.

Cram, D. J. and Sogah, G. D. Y. (1981), *J. Chem. Soc., Chem. Commun.*, 625.

Curran, D. P., Kim, B. H., Piyasena, H. P., Loncharich, R. J., and Houk, K. N. (1987), *J. Org. Chem.*, **52**, 2137.

Curran, D. P. and Rakiewicz, D. M. (1985), *Tetrahedron*, **41**, 3943.

d'Angelo, J. and Maddaluno, J. (1986), *J. Am. Chem. Soc.*, **108**, 8112.

d'Angelo, J., Guingant, A., Riche, C., and Chiaroni, A. (1988), *Tetrahedron Lett.*, **29**, 2667.

Dauben, W. G., Fonken, G. S., and Noyce, D. S. (1956), *J. Am. Chem. Soc.*, **78**, 2579.

David, S., Lubineau, A. and Thieffry, A. (1978), *Tetrahedron*, **34**, 299.

Davies, S. G. (1989), "Chiral Auxiliaries," *Chem. Br.*, 268.

Davies, S. G., Dordor, I. M., and Warner, P. (1984), *J. Chem. Soc. Chem. Commun.*, 956.

Davies, S. G., Green, M. L. H., and Mingos, D. M. P. (1978), *Tetrahedron*, **34**, 3047.

Davis, F. A. and Haque, M. S. (1986), *J. Org. Chem.*, **51**, 4085.

Davis, F. A., Harakal, M. E., and Awad, S. B. (1983), *J. Am. Chem. Soc.*, **105**, 3123.

Davis, F. A., Towson, J. C., Weismiller, M. C., Lal, S., and Carroll, P. J. (1988), *J. Am. Chem. Soc.*, **110**, 8477.

Denmark, S. E. and Marble, L. K. (1990), "Auxiliary-Based, Asymmetric S_N2' Reactions: A Case of 1,7-Relative Stereogenesis," *J. Org. Chem.*, **55**, 1984.

DeShong, P., Dicken, C. M., Leginus, J. M., and Whittle, R. R. (1984), *J. Am. Chem. Soc.*, **106**, 5598.

Doering, W. von E., and Roth, W. R. (1962), *Tetrahedron*, **18**, 67.

Dokuzovic, Z., Roberts, N. K., Sawyer, J. F., Whelan, J., and Bosnich, B. (1986), *J. Am. Chem. Soc.*, **108**, 2034.

Dongala, E. B., Dull, D. L., Mioskowski, C., and Solladié, G. (1973), *Tetrahedron Lett.*, 4983.

Doyle, M. P., Bagheri, V., and Harn, N. K. (1988), *Tetrahedron Lett.*, **29**, 5119.

Drauz, K., Kleeman, A., and Martens, J. (1982), "Induction of Asymmetry by Amino Acids," *Angew. Chem. Int. Ed. Engl.*, **21**, 584.

Dupuis, J., Giese, B., Rüegge, D., Fischer, H., Korth, H. G., and Sustmann, R. (1984), *Angew. Chem. Int. Ed. Engl.*, **23**, 896.

Durkin, K. A. and Liotta, D. (1990), *J. Am. Chem. Soc.*, **112**, 8162.

Eder, U., Sauer, G., and Wiechert, R. (1971), *Angew. Chem. Int. Ed. Engl.*, **10**, 496.

Eisenstein, O., Klein, J., and Lefour, J. M. (1979), *Tetrahedron Lett.*, 225.

Eliel, E. L. (1962), *Stereochemistry of Carbon Compounds*, McGraw-Hill, New York.

Eliel, E. L. (1983), "Applications of Cram's Rule: Addition of Achiral Nucleophiles to Chiral Substrates," in Morrison, J. D., Ed., *Asymmetric Synthesis*, Vol. 2, Academic, New York, p. 125.

Enders, D. (1984), "Alkylation of Chiral Hydrazones," in Morrison, J. D., Ed., *Asymmetric Synthesis*, Academic, New York, p. 275.

Enders, D. (1985), *Chem. Scr.*, **25**, NS139. Special Nobel symposium 60 issue.

Enders, D. and Lohray, B. B. (1988), *Angew. Chem. Int. Ed. Engl.*, **27**, 581.

Evans, D. A. (1982), "Studies in Asymmetric Synthesis. The Development of Practical Chiral Enolate Synthons," *Aldrichim. Acta*, **15**, 23.

Evans, D. A. (1984), "Stereoselective Alkylation Reactions of Chiral Metal Enolates" in Morrison, J. D., Ed., *Asymmetric Synthesis*, Vol. 3, Academic, New York, p. 2.

Evans, D. A. (1988), *Science*, **240**, 420.

Evans, D. A. and Bartroli, J. (1982), *Tetrahedron Lett.*, **23**, 807.

Evans, D. A., Bartroli, J., and Godel, T. (1982a), *Tetrahedron Lett.*, **23**, 4577.

Evans, D. A., Bartroli, J., and Shih, T. L. (1981a), *J. Am. Chem. Soc.*, **103**, 2127.

Evans, D. A., Bilodeau, M. T., Somers, T. C., Clardy, J., Cherry, D., and Kato, Y. (1991), *J. Org. Chem.*, **56**, 5750.

Evans, D. A. and Britton, T. C. (1987), *J. Am. Chem. Soc.*, **109**, 6881.

Evans, D. A., Britton, T. C., Dorow, R. L., and Dellaria, J. F. (1986), *J. Am. Chem. Soc.*, **108**, 6395.

Evans, D. A., Chapman, K. T., and Bisaha, J. (1988a), *J. Am. Chem. Soc.*, **110**, 1238.

Evans, D. A., Chapman, K. T., and Carreira, E. M. (1988b), *J. Am. Chem. Soc.*, **110**, 3560.

Evans, D. A., Chapman, K. T., Hung, D. T., and Kawaguchi, A. T. (1987a), *Angew. Chem. Int. Ed. Engl.*, **26**, 1184.

Evans, D. A. and DiMare, M. (1986), *J. Am. Chem. Soc.*, **108**, 2476.

Evans, D. A., Ellman, J. A., and Dorow, R. L. (1987b), *Tetrahedron Lett.*, **28**, 1123.

Evans, D. A. and Fu, G. C. (1990), *J. Org. Chem.*, **55**, 2280.

Evans, D. A. and Golob, A. M. (1975), *J. Am. Chem. Soc.*, **97**, 4765.

Evans, D. A. and Kaldor, S. W. (1990), *J. Org. Chem.*, **55**, 1698.

Evans, D. A. and McGee, L. R. (1980), *Tetrahedron Lett.*, **21**, 3975.

Evans, D. A. and McGee, L. R. (1981), *J. Am. Chem. Soc.*, **103**, 2876.

Evans, D. A. and Morrissey, M. M. (1984a), *J. Am. Chem. Soc.*, **106**, 3866.

Evans, D. A. and Morrissey, M. M. (1984b), *Tetrahedron Lett.*, **25**, 4637.

Evans, D. A., Morrissey, M. M., and Dorow, R. L. (1985a), *J. Am. Chem. Soc.*, **107**, 4346.

Evans, D. A., Morrissey, M. M., and Dow, R. L. (1985b), *Tetrahedron Lett.*, **26**, 6005.

Evans, D. A., Nelson, J. V., and Taber, T. R. (1982c), "Stereoselective Aldol Condensations," *Top. Stereochem.*, **13**, 1.

Evans, D. A., Nelson, J. V., Vogel, E., and Taber, T. R. (1981b), *J. Am. Chem. Soc.*, **103**, 3099.

Evans, D. A. and Taber, T. R. (1980), *Tetrahedron Lett.*, **21**, 4675.

Evans, D. A. and Takacs, J. M. (1980), *Tetrahedron Lett.*, **21**, 4233.

Evans, D. A., Takacs, J. M., McGee, L. R., Ennis, M. D., Mathre, D. J., and Bartroli, J. (1981c), *Pure Appl. Chem.*, **53**, 1109.

Evans, R. J. D., Landor, S. R., and Regan, J. P. (1965), *J. Chem. Soc. Chem. Commun.*, 397.

Ewins, R. C., Henbest, H. B., and McKervey, M. A. (1967), *J. Chem. Soc. Chem. Commun.*, 1085.

Fahey, R. C. (1968), "The Stereochemistry of Electrophilic Additions to Olefins and Acetylenes," *Top. Stereochem.*, **3**, 237.

Faulkner, D. J. (1971), "Stereoselective Synthesis of Trisubstituted Olefins," *Synthesis*, 175.

Faulkner, D. J. and Peterson, M. R. (1969), *Tetrahedron Lett.*, 3243.

Fellmann, P. and Dubois, J. E. (1978), *Tetrahedron*, **34**, 1349.

Fenzl, W. and Köster, R. (1975), *Justus Liebigs Ann. Chem.*, 1322.

Feringa, B. L. and de Jong, J. C. (1988), *J. Org. Chem.*, **53**, 1125.

Finn, M. G. and Sharpless, K. B. (1985), "On the Mechanism of Asymmetric Epoxidation with Titanium–Tartrate Catalysts," in Morrison, J. D., Ed., *Asymmetric Synthesis*, Academic, New York, p. 247.

Finn, M. G. and Sharpless, K. B. (1991), *J. Am. Chem. Soc.*, **113**, 113.

Firestone, R. A. (1968), *J. Org. Chem.*, **33**, 2285.

Fischli, A. (1980), *Modern Synthetic Methods* Vol. 2, Scheffold, R., Ed., Salle-Saurlander, Frankfurt, Germany, p. 269.

Fleet, G. W. J. (1989), "Homochiral Compounds from Sugars," *Chem. Br.*, 287.

Fleming, I. (1976), *Frontier Orbitals and Organic Chemical Reactions*, Wiley-Interscience, Chichester, England.

Fraser, R. R. and Stanciulescu, M. (1987), *J. Am. Chem. Soc.*, **109**, 1580.

Fraser-Reid, B., Tsang, R., and Lowe, D. (1985), *Chem. Scr.*, **25**, NS117. Special Nobel symposium 60 issue.

Fráter, G. (1979), *Helv. Chim. Acta*, **62**, 2825, 2829.

Fráter, G., Müller, U., and Günter, W. (1986), *Helv. Chim. Acta*, **69**, 1858.

Fritschi, H., Leutenegger, U., and Pfaltz, A. (1988), *Helv. Chim. Acta*, **71**, 1553.

Fryzuk, M. D. and Bosnich, B. (1977), *J. Am. Chem. Soc.*, **99**, 6262.

Fryzuk, M. D. and Bosnich, B. (1979), *J. Am. Chem. Soc.*, **101**, 3043.

Furuta, K., Iwanaga, K., and Yamamoto, H. (1986), *Tetrahedron Lett.*, **27**, 4507.

Furuta, K., Miwa, Y., Iwanaga, K., and Yamamoto, H. (1988b), *J. Am. Chem. Soc.*, **110**, 6254.

Furuta, K., Nagata, T., and Yamamoto, H. (1988a), *Tetrahedron Lett.*, **29**, 2215.

Furuta, K., Shimizu, S., Miwa, Y., and Yamamoto, H. (1989), *J. Org. Chem.*, **54**, 1481.

Gais, H.-J. and Lukas, K. L. (1984), *Angew. Chem., Int. Ed. Engl.*, **23**, 142.

Gennari, C., Colombo, L., and Bertolini, G. (1986), *J. Am. Chem. Soc.*, **108**, 6394.

Gibbs, R. A. and Okamura, W. H. (1988), *J. Am. Chem. Soc.*, **110**, 4062.

Gibson, T. and Barneis, Z. J. (1972), *Tetrahedron Lett.*, 2207.

Giese, B. (1985), "Syntheses with Radicals–C–C Bond Formation via Organomercury Compounds," *Angew. Chem. Int. Ed. Engl.*, **24**, 553.

Giese, B. (1986), *Radicals in Organic Synthesis: Formation of Carbon–Carbon Bonds*, Pergamon, Oxford, England.

Giese, B. (1989), "The Stereoselectivity of Intermolecular Free Radical Reactions," *Angew. Chem. Int. Ed. Engl.*, **28**, 969.

Giese, B. and Kroninger, K. (1984), *Tetrahedron Lett.*, **25**, 2743.

Gilchrist, T. L. and Storr, R. C. (1979), *Organic Reactions and Orbital Symmetry*, 2nd ed., Cambridge University Press, London.

Ginsburg, D. (1983), *Tetrahedron*, **39**, 2095.

Gladysz, J. A. and Michl, J., Eds., (1992), "Enantioselective Synthesis," *Chem. Rev.*, **92**, 739.

Goldberg, S. I., (1970), "Asymmetric Selection via Elimination," in Thyagarajan, B. S., Ed., *Selective Organic Transformations*, Vol. 1, Wiley-Interscience, New York, p. 363.

Gosney, I. and Rowley, A. G. (1979), "Transformations via Phosphorus-stabilized Anions. 1. Stereoselective Syntheses of Alkenes via the Wittig Reaction," in Cadogan, J. I. G., Ed., *Organophosphorus Reagents in Organic Synthesis*, Academic, New York, p. 17.

Grandbois, E. R., Howard, S. I., and Morrison, J. D. (1983), "Reactions with Chiral Modifications of Lithium Aluminum Hydride," in Morrison, J. D., Ed., *Asymmetric Synthesis*, Vol. 2, Academic, New York, p. 71.

Grée, R. (1989), "Acyclic Butadiene-Iron Tricarbonyl Complexes in Organic Synthesis," *Synthesis*, 341.

Greene, A. E. and Charbonnier, F. (1985), *Tetrahedron Lett.*, **26**, 5525.

Greenle, K. W. and Fernelius, W. C. (1942), *J. Am. Chem. Soc.*, **64**, 2505.

Greenstein, J. P. (1954), *Adv. Protein Chem.*, **9**, 122.

Gregson, R. P. and Mirrington, R. N. (1973), *J. Chem. Soc. Chem. Commun.*, 598.

Grob, C. A. and Schiess, P. W. (1967), "Heterolytic Fragmentation. A Class of Organic Reactions," *Angew Chem. Int. Ed. Engl.*, **6**, 1.

Groen, M. B. and Zeelen, F. J. (1978), *Recl. Trav. Chim. Pays-Bas*, **97**, 301.

Hajos, Z. G. and Parrish, D. R. (1974a), *J. Org. Chem.*, **39**, 1612.

Hajos, Z. G. and Parrish, D. R. (1974b), *J. Org. Chem.*, **39**, 1615.

Halpern, J. (1983), *Pure Appl. Chem.*, **55**, 1953.

Halpern, J. (1985), "Asymmetric Catalytic Hydrogenation: Mechanism and Origin of Enantioselection," in Morrison, J. D., Eds., *Asymmetric Synthesis*, Vol. 5, Academic, New York, p. 41.

Hammond, G. S. (1955), *J. Am. Chem. Soc.*, **77**, 334.

Hammond, G. S., Saltiel, J., Lamola, A. A., Turro, N. J., Bradshaw, J. S., Cowan, D. O., Counsell, R. C., Vogt, V., and Dalton, C. (1964), *J. Am. Chem. Soc.*, **86**, 3197.

Hanessian, S. (1983), *Total Synthesis of Natural Products: The Chiron Approach*, Pergamon, Oxford, England.

Hanessian, S. and Rancourt, G. (1977), *Pure Appl. Chem.*, **49**, 1201.

Hanessian, S., Delorme, D., Beaudoin, S., and Leblanc, Y. (1985), *Chem. Scr.*, **25**, NS5. Special Nobel symposium 60 issue.

Harada, K. (1985), "Asymmetric Heterogeneous Catalytic Hydrogenation" in Morrison, J. D., Ed., *Asymmetric Synthesis*, Vol. 5, Academic, New York, p. 346.

Hart, D. J. (1984), "Free-Radical Carbon–Carbon Bond Formation in Organic Synthesis," *Science*, **223**, 883.

Hasan, I. and Kishi, Y. (1980), *Tetrahedron Lett.*, **21**, 4229.

Hashimoto, S.-I., Komeshima, N., and Koga, K. (1979), *J. Chem. Soc. Chem. Commun.*, 437.

Hashimoto, S.-I., Yamada, S.-I., and Koga, K. (1976), *J. Am. Chem. Soc.*, **98**, 7452.

Haubenstock, H. (1983), "Asymmetric Reductions with Chiral Complex Aluminum Hydrides and Tricoordinate Aluminum Reagents," *Top. Stereochem.*, **14**, 231.

Hawkins, J. M. and Loren, S. (1991), *J. Am. Chem. Soc.*, **113**, 7794.

Hayashi, T., Kawamura, N., and Ito, Y. (1987), *J. Am. Chem. Soc.*, **109**, 7876.

Hayashi, T., Konishi, M., Ito, H., and Kumada, M. (1982), *J. Am. Chem. Soc.*, **104**, 4962.

Hayashi, T., Konishi, M., and Kumada, M., (1983), *J. Org. Chem.*, **48**, 281.

Hayashi, T. and Kumada, M. (1982), "Asymmetric Synthesis Catalyzed by Transition-Metal Complexes with Functionalized Chiral Ferrocenylphosphine Ligands," *Acc. Chem. Res.*, **15**, 395.

Heathcock, C. H. (1981), *Science*, **214**, 395.

Heathcock, C. H. (1982), "Acyclic Stereoselection via the Aldol Reaction," in Eliel, E. L. and Otsuka, S., Eds., "Asymmetric Reactions and Processes in Chemistry," *ACS Symposium Series*, **185**, American Chemical Society, Washington DC, p. 55.

Heathcock, C. H. (1984a), "Stereoselective Aldol condensation," in Buncel, E. and Durst, T., Eds., *Comprehensive Carbanion Chemistry*, Part B, Elsevier, Amsterdam, The Netherlands, p. 177.

Heathcock, C. H. (1984b), "The Aldol Addition Reaction," in Morrison, J. D., Ed., *Asymmetric Synthesis,*, Vol. 3, Academic, New York, p. 111.

Heathcock, C. H., Buse, C. T., Kleschick, W. A., Pirrung, M. C., Sohn, J. E., and Lampe, J. (1980), *J. Org. Chem.*, **45**, 1066.

Heathcock, C. H. and Flippin, L. A. (1983), *J. Am. Chem. Soc.*, **105**, 1667.

Heathcock, C. H., Hagen, J. P., Young, S. D., Pilli, R., Bia, D.-l., Märki, H.-P., Kees, K., and Badertscher, U. (1985), *Chem. Scr.*, **25**, NS39. Special Nobel symposium 60 issue.

Heathcock, C. H., Pirrung, M. C., Buse, C. T., Hagen, J. P., Young, S. D., and Sohn, J. E. (1979), *J. Am. Chem. Soc.*, **101**, 7077.

Heathcock, C. H., Pirrung, M. C., Montgomery, S. H., and Lampe, J. (1981a), *Tetrahedron*, **37**, 4087.

Heathcock, C. H. and White, C. T. (1979), *J. Am. Chem. Soc.*, **101**, 7076.

Heathcock, C. H., White, C. T., Morrison, J. J., and Van Derveer, D. (1981b), *J. Org. Chem.*, **46**, 1296.

Helmchen, G., Karge, R., and Weetman, J. (1986), "Asymmetric Diels Alder Reactions with Chiral Enoates as Dienophiles," Scheffold, R., Ed., *Modern Synthetic Methods*, Springer, Berlin, p. 261.

Helmchen, G., Leikauf, U., and Taufer-Knöpfel, I. (1985), *Angew. Chem. Int. Ed. Engl.*, **24**, 874.

Helmchen, G. and Schmierer, R. (1981), *Angew. Chem. Int. Ed. Engl.*, **20**, 205.

Henbest, H. B. and Wilson, R. A. L. (1957), *J. Chem. Soc.*, 1958.

Henrick, C. A. (1977), "The Synthesis of Insect Sex Pheromones," *Tetrahedron*, **33**, 1845.

Hentges, S. G. and Sharpless, K. B. (1980), *J. Am. Chem. Soc.*, **102**, 4263.

Herman, K. and Wyberg, H. (1979), *J. Org. Chem.*, **44**, 2238.

Hiemstra, H. and Wynberg, J. H. (1981), *J. Am. Chem. Soc.*, **103**, 417.

Hill, J. G., Sharpless, K. B., Exon, C. M., and Regenye, R. (1984), *Org. Synth.*, **63**, 66.

Hill, R. K. (1984), "Chirality Transfer via Sigmatropic Rearrangements," in Morrison, J. D., Ed., *Asymmetric Synthesis*, Vol. 3, Academic, New York, p. 503.

Hine, J., Li, W.-S., and Ziegler, J. P. (1980), *J. Am. Chem. Soc.*, **102**, 4403.

Hochstein, F. A. and Brown, W. G. (1948), *J. Am. Chem. Soc.*, **70**, 3484.

Hoffmann, R. W. (1979), "Stereochemistry of [2,3]Sigmatropic Rearrangements," *Angew. Chem. Int. Ed. Engl.*, **18**, 563.

Hoffmann, R. W. (1982), "Diastereogenic Addition of Crotylmetal Compounds to Aldehydes," *Angew. Chem. Int. Ed. Engl.*, **21**, 555.

Hoffmann, R. W. (1987), "Stereoselective Synthesis of Building Blocks with Three Consecutive Stereogenic Centers: Important Precursors of Polyketide Natural Products," *Angew. Chem. Int. Ed. Engl.*, **26**, 489.

Hoffmann, R. W. and Herold, T. (1981), *Chem. Ber.*, **114**, 375.

Hogeveen, H. and Menge, W. M. P. B. (1986), *Tetrahedron Lett.*, **27**, 2767.

Horeau, A. and Guetté, J. P. (1974), *Tetrahedron*, **30**, 1923.

Horeau, A., Kagan, H. B., and Vigneron, J.-P. (1968), *Bull. Soc. Chim. Fr.*, 3795.

Horner, J., Vera, M. and Grutzner, J. B. (1986), *J. Org. Chem.*, **51**, 4214.

Horner, L., Siegel, H., and Büthe, H. (1968), *Angew. Chem. Int. Ed. Engl.*, **7**, 942.

Horton, D. and Machinami, T. (1981), *J. Chem. Soc. Chem. Commun.*, 88.

Houge, C., Frisque-Hesbai, A. M., Mockel, A., and Ghosez, L. (1982), *J. Am. Chem. Soc.*, **104**, 2920.

Houk, K. N. (1983), "Theoretical Studies of Alkene Pyramidalizations and Addition Stereoselectivities," in Watson, W. H., Ed., *Stereochemistry and Reactivity of Systems Containing π Electrons*, Chemie International, Deerfield Beach, FL.

Houk, K. N., Moses, S. R., Wu, Y.-D., Rondan, N. G., Jager, V., Schohe, R., and Fronczek, F. R. (1984a), *J. Am. Chem. Soc.*, **106**, 3880.

Houk, K. N., Rondan, N. G., Wu, Y.-D., Metz, J. T., and Paddon-Row, M. N. (1984b), *Tetrahedron*, **40**, 2257.

House, H. O. (1972), *Modern Synthetic Reactions*, Benjamin, W. A., Menlo Park, CA., p. 150.

House, H. O. and Fischer, W. F. (1968), *J. Org. Chem.*, **33**, 949.

House, H. O., Lubinkowski, J., and Good, J. J. (1975), *J. Org. Chem.*, **40**, 86.

House, H. O. and Umen, M. J. (1973), *J. Org. Chem.*, **38**, 1000.

Hoveyda, A. H., Evans, D. A., and Fu, G. C. (1993), "Substrate-Directable Chemical Reactions," *Chem. Rev.*, **93**, 1307.

Huisgen, R. (1963), *Angew. Chem. Int. Ed. Engl.*, **2**, 565.

Huisgen, R. (1968), *J. Org. Chem.*, **33**, 2291.

Iida, H., Kasahara, K., and Kibayashi, C. (1986), *J. Am. Chem. Soc.*, **108**, 4647.

Inagaki, S., Fujimoto, H., and Fukui, K. (1976), *J. Am. Chem. Soc.*, **98**, 4054.

Inouye, Y., Oda, J., and Baba, N. (1983), "Reductions with Chiral Dihydropyridine Reagents," in Morrison, J. D., Ed., *Asymmetric Synthesis*, Vol. 2, Academic, New York, p. 92.

Inouye, Y. and Walborsky, H. M. (1962), *J. Org. Chem.*, **27**, 2706.

Ireland, R. E., Anderson, R. C., Badoud, R., Fitzsimmons, B. J., McGarvey, G. J., Thaisrivongs, S., and Wilcox, C. S. (1983), *J. Am. Chem. Soc.*, **105**, 1988.

Ireland, R. E., Baldwin, S. W., and Welch, S. C. (1972), *J. Am. Chem. Soc.*, **94**, 2056.

Ireland, R. E. and Mander, L. N. (1969), *J. Org. Chem.*, **34**, 689.

Ireland, R. E., Mueller, R. H., and Williard, A. K. (1976), *J. Am. Chem. Soc.*, **98**, 2868.

Ireland, R. E. and Varney, M. D. (1983), *J. Org. Chem.*, **48**, 1829.

Ireland, R. E. and Varney, M. D. (1984), *J. Am. Chem. Soc.*, **106**, 3668.

Ireland, R. E. and Wilcox, C. S. (1980), *J. Org. Chem.*, **45**, 197.

Ito, Y., Sawamura, M., and Hayashi, T. (1986), *J. Am. Chem. Soc.*, **108**, 6405.

Itoh, T., Jitsukawa, K., Kaneda, K., and Teranishi, S. (1979), *J. Am. Chem. Soc.*, **101**, 159.

Itsuno, S., Ito, K., Hirao, A., and Nakahama, S. (1983), *J. Chem. Soc., Chem. Commun.*, 469.

Itsuno, S., Sakurai, Y., Ito, K., Hirao, A., and Nakahama, S. (1987), *Bull. Chem. Soc. Jpn.*, **60**, 395.

Izumi, Y. and Tai, A. (1977), *Stereodifferentiating Reactions*, Academic, New York.

Jackman, L. M. and Lange, B. C. (1977), "Structure and Reactivity of Alkali Metal Enolates," *Tetrahedron*, **33**, 2737.

Jacobsen, E. N., Markó, I., Mungall, S., Schröder, G., and Sharpless, K. B. (1988), *J. Am. Chem. Soc.*, **110**, 1968.

Jacobsen, E. N., Zhang, W., and Guler, M. L. (1991), *J. Am. Chem. Soc.*, **113**, 6703.

Jadhav, P. K., Bhat, K. S., Perumal, P. T., and Brown, H. C. (1986), *J. Org. Chem.*, **51**, 432.

Jadhav, P. K. and Brown, H. C. (1981), *J. Org. Chem.*, **46**, 2988.

Jäger, V. and Viehe, H. G. (1977), "Addition an die C≡C Dreifachbindung ohne Neuknüpfung von C–C Bindungen," in Houben-Weyl, *Methoden der Organischen Chemie*, Vol. V/2a, Thieme, Stuttgart, Federal Republic of Germany, p. 687ff.

Jeganathan, S., Tsukamoto, M., and Schlosser, M. (1990), *Synthesis*, 109.

Johnson, A. W. (1966), *Ylid Chemistry*, Academic, New York.

Johnson, C. R., Tait, B. O., and Cieplak, A. S. (1987), *J. Am. Chem. Soc.*, **109**, 5875.

Johnson, F. (1968), "Allylic Strain in Six-Membered Rings," *Chem. Rev.*, **68**, 375.

Johnson, M. R. and Kishi, Y. (1979), *Tetrahedron Lett.*, 4347.

Johnson, R. A. and Sharpless, K. B. (1993), "Catalytic Asymmetric Dihydroxylations" in Ojima, I., Ed., *Catalytic Asymmetric Synthesis*, VCH, New York.

Johnson, W. S. (1976), "Biomimetic Polyene Cyclizations," *Angew. Chem. Int. Ed. Engl.*, **15**, 9.

Johnson, W. S. and DuBois, G. E. (1976), *J. Am. Chem. Soc.*, **98**, 1038.

Johnson, W. S., Frei, B., and Gopalan, A. S. (1981), *J. Org. Chem.*, **46**, 1512.

Johnson, W. S., Harbert, C. A., Ratcliffe, B. E., and Stipanovic, R. D. (1976), *J. Am. Chem. Soc.*, **98**, 6188.

Johnson, W. S., Li, T., Faulkner, D. J., and Campbell, S. F. (1968), *J. Am. Chem. Soc.*, **90**, 6225.

Johnson, W. S., Lyle, T. A., and Daub, G. W. (1982), *J. Org. Chem.*, **47**, 161.

Johnson, W. S., Werthemann, L., Bartlett, W. R., Brocksom, T. J., Li, T., Faulkner, D. J., and Peterson, M. R. (1970), *J. Am. Chem. Soc.*, **92**, 741.

Jones, J. B. (1986), "Enzymes in Organic Synthesis," *Tetrahedron*, **42**, 3351.

Jorgensen, K. A., Wheeler, R. A., and Hoffmann, R. (1987), *J. Am. Chem. Soc.*, **109**, 3240.

Julia, M. (1974), "Free Radical Cyclizations, XVII. Mechanistic Studies," *Pure Appl. Chem.*, **40**, 553.

Julia, M., Julia, S., and Guegan, R. (1960), *Bull. Soc. Chim. Fr.*, 1072.

Julia, M., Masana, J., and Vega, J. C. (1980), *Angew. Chem. Int. Ed. Engl.*, **19**, 929.

Julia, M. and Paris, J. M. (1973), *Tetrahedron Lett.*, 4833.

Kaftory, M., Yagi, M., Tanaka, K., and Toda, F. (1988), *J. Org. Chem.*, **53**, 4391.

Kagan, H. B. (1985), "Chiral Ligands for Asymmetric Catalysts," in Morrison, J. D., Ed., *Asymmetric Synthesis*, Vol. 5, Academic, New York, p. 1.

Kagan, H. B. and Dang, T. P. (1972), *J. Am. Chem. Soc.*, **94**, 6429.

Kagan, H. B. and Fiaud, J. C. (1978), "New Approaches in Asymmetric Synthesis," *Top. Stereochem.*, **10**, 175.

Kagan, H. B., Mimoun, H., Mark, C., and Schurig, V. (1979), *Angew. Chem. Int. Ed. Engl.*, **18**, 485.

Kagan, H. B. and Riant, O. (1992), "Catalytic Asymmetric Diels–Alder Reactions," *Chem. Rev.*, **92**, 1007.

Kahn, S. D. and Hehre, W. J. (1986), *J. Am. Chem. Soc.*, **108**, 7399.

Kahn, S. D. and Hehre, W. J. (1987), *J. Am. Chem. Soc.*, **109**, 663, 666.

Kahn, S. D., Pau, C. F., Chamberlin, A. R., and Hehre, W. J. (1987), *J. Am. Chem. Soc.*, **109**, 650.

Kahn, S. D., Pau, C. F., and Hehre, W. J. (1986), *J. Am. Chem. Soc.*, **108**, 7396.

Kallmerten, J. and Gould, T. J. (1986), *J. Org. Chem.*, **51**, 1153.

Kametani, T., Nagahara, T., and Honda, T. (1985), *J. Org. Chem.*, **50**, 2327.

Kametani, T., Nagahara, T., and Ihara, M. (1981), *J. Chem. Soc. Perkin Trans. 1*, 3048.

Katsuki, T. and Sharpless, K. B. (1980), *J. Am. Chem. Soc.*, **102**, 5974.

Kawakami, Y., Hiratake, J., Yamamoto, Y., and Oda, J. (1984), *J. Chem. Soc. Chem. Commun.*, 779.

Kawano, H., Ishii, Y., Ikariya, T., Saburi, M., Yoshikawa, S., Uchida, Y., and Kumobayashi, H. (1987), *Tetrahedron Lett.*, **28**, 1905.

Kawasaki, H., Tomioka, K., and Koga, K. (1985), *Tetrahedron Lett.*, **26**, 3031.

Keck, G. E., Enholm, E. J., Yates, J. B., and Wiley, M. R. (1985), *Tetrahedron*, **41**, 4079.

Kelly, T. R., Whiting, A., and Chandrakumar, N. S. (1986), *J. Am. Chem. Soc.*, **108**, 3510.

Kim, H.-D., Shirai, R., Kawasaki, H., Nakajima, M., and Koga, K. (1990), *Heterocycles*, **30**, 307.

Kitamura, M., Nagai, K., and Noyori, R. (1987), *J. Org. Chem.*, **52**, 3174.

Kitamura, M., Okada, S., Suga, S., and Noyori, R. (1989), *J. Am. Chem. Soc.*, **111**, 4028.

Kitamura, M., Suga, S., Kawai, K., and Noyori, R. (1986), *J. Am. Chem. Soc.*, **108**, 6071.

Klein, J. (1973), *Tetrahedron Lett.*, 4307.

Klein, J. (1974), *Tetrahedron*, **30**, 3349.

Knowles, W. S. (1983), "Asymmetric Hydrogenation," *Acc. Chem. Res.*, **16**, 106.

Knowles, W. S. and Sabacky, M. J. (1968), *J. Chem. Soc. Chem. Commun.*, 1445.

Knowles, W. S., Sabacky, M. J., and Vineyard, B. D. (1972), *J. Chem. Soc. Chem. Commun.*, 10.

Kocienski, P. J., Lythgoe, B., and Ruston, S. (1978), *J. Chem. Soc. Perkin Trans. 1*, 829.

Koenig, K. E. (1985), "The Applicability of Asymmetric Homogeneous Catalytic Hydrogenation," in Morrison, J. D., Ed., *Asymmetric Synthesis*, Vol. 5, Academic, New York, p. 71.

Kolb, H. C. and Sharpless, K. B. (1992), *Tetrahedron*, **48**, 10515.

Korth, H. G., Sustmann, R., Dupuis, J., and Giese, B. (1986), *J. Chem. Soc. Perkin Trans 2*, 1453.

Koskinen, A. (1993), *Asymmetric Synthesis of Natural Products*, Wiley, Chichester.

Kozikowski, A. P., Jung, S. H., and Springer, J. P. (1988), *J. Chem. Soc. Chem. Commun.*, 167.

Kozikowski, A. P., Kitagawa, Y., and Springer, J. P. (1983), *J. Chem. Soc. Chem. Commun.*, 1460.

Kozikowski, A. P., Konoike, T., and Nieduzak, T. R. (1986b), *J. Chem. Soc. Chem. Commun.*, 1350.

Kozikowski, A. P. and Nieduzak, T. R. (1986a), *Tetrahedron Lett.*, **27**, 819.

Kozikowski, A. P., Nieduzak, T. R., Konoike, T., and Springer, J. P. (1987), *J. Am. Chem. Soc.*, **109**, 5167.

Kraus, G. A. and Taschner, M. J. (1977), *Tetrahedron Lett.*, 4575.

Krishnamurthy, S. and Brown, H. C. (1976), *J. Am. Chem. Soc.*, **98**, 3384.

Kuehne, M. E. (1970), *J. Org. Chem.*, **35**, 171.

Kuehne, M. E. and Nelson, J. A. (1970), *J. Org. Chem.*, **35**, 161.

Kümin, A., Maverick, E., Seiler, P., Vanier, N., Damm, L., Hobi, R., Dunitz, J. D., and Eschenmoser, A. (1980), *Helv. Chim. Acta*, **63**, 1158.

Lamanna, W. and Brookhart, M. (1981), *J. Am. Chem. Soc.*, **103**, 989.

Lampe, J. (1983), "Stereoselective Aldol Condensation Methodology with Applications to Total Synthesis," Ph.D. Thesis, University of California, Berkeley, CA.

Laube, T. and Stilz, H. U. (1987), *J. Am. Chem. Soc.*, **109**, 5876.

Lee, L. G. and Whitesides, G. M. (1985), *J. Am. Chem. Soc.*, **107**, 6999.

Lee, N. H., Muci, A. R., and Jacobsen, E. N. (1991), *Tetrahedron Lett.*, **32**, 5055.

Lemal, D. M. and McGregor, S. D. (1966), *J. Am. Chem. Soc.*, **88**, 1335.

Leutenegger, U., Madin, A., and Pfaltz, A. (1989), *Angew. Chem. Int. Ed. Engl.*, **28**, 60.

Lewis, M. D. and Kishi, Y. (1982), *Tetrahedron Lett.*, **23**, 2343.

Li, Y., Paddon-Row, M. N., and Houk, K. N. (1988), *J. Am. Chem. Soc.*, **110**, 3684.

Liebeskind, L. S. and Welker, M. E. (1984), *Tetrahedron Lett.*, **25**, 4341.

Lin, M.-H., Silver, J. E., and le Noble, W. J. (1988), *J. Org. Chem.*, **53**, 5155.

Lindlar, H. (1952), *Helv. Chim. Acta*, **39**, 249.

Lodge, E. P. and Heathcock, C. H. (1987a), *J. Am. Chem. Soc.*, **109**, 2819.

Lodge, E. P. and Heathcock, C. H. (1987b), *J. Am. Chem. Soc.*, **109**, 3353.

Lu, L. D.-L., Johnson, R. A., Finn, M. G., and Sharpless, K. B. (1984), *J. Org. Chem.*, **49**, 731.

Lutomski, K. A. and Meyers, A. I. (1984), "Asymmetric Synthesis via Chiral Oxazolines," in Morrison, J. D., Ed., *Asymmetric Synthesis*, Vol. 3, Academic, New York, p. 213.

Lynch, J. E. and Eliel, E. L. (1984), *J. Am. Chem. Soc.*, **106**, 2943.

Lyssikatos, J. P. and Bednarski, M. D. (1990), *Synlett*, 230.

Maas, G. (1987), "Transition-metal Catalyzed Decomposition of Aliphatic Diazo Compounds—New Results and Applications in Organic Synthesis," *Top. Curr. Chem.*, **137**, 75.

Maercker, A. (1965), "The Wittig Reaction," *Org. React.*, **14**, 270.

Magid, R. M. (1980), "Nucleophilic and Organometallic Displacement Reactions of Allylic Compounds: Stereo- and Regiochemistry," *Tetrahedron*, **36**, 1901.

Marshall, J. A. and Cohen, G. M. (1971), *J. Org. Chem.*, **36**, 877.

Martin, S. F., Ed., (1994), "Tetrahedron Symposia-in-Print Number 54: Catalytic Asymmetric Addition Reactions," *Tetrahedron*, **50**, 4235–4574.

Martin, V. S., Woodard, S. S., Katsuki, T., Yamada, Y., Ikeda, M., and Sharpless, K. B. (1981), *J. Am. Chem. Soc.*, **103**, 6237.

Maruoka, K., Hoshino, Y., Shirasaka, T., and Yamamoto, H. (1988a), *Tetrahedron Lett.*, **29**, 3967.

Maruoka, K., Itoh, T., Sakurai, M., Nonoshita, K., and Yamamoto, H. (1988b), *J. Am. Chem. Soc.*, **110**, 3588.

Maruoka, K., Itoh, T., Shirasaka, T., and Yamamoto, H. (1988c), *J. Am. Chem. Soc.*, **110**, 310.

Maruoka, K., Itoh, T., and Yamamoto, H. (1985), *J. Am. Chem. Soc.*, **107**, 4573.

Maruoka, K., Nonoshita, K., Banno, H., and Yamamoto, H. (1988d), *J. Am. Chem. Soc.*, **110**, 7922.

Maruoka, K., Sakurai, M., Fujiwara, J., and Yamamoto, H. (1986), *Tetrahedron Lett.*, **27**, 4895.

Marvel, E. N. and Li, T. (1973), "Catalytic Semihydrogenation of the Triple Bond," *Synthesis*, 457.

Maryanoff, B. E. and Reitz, A. B. (1989), *Chem. Rev.*, **89**, 863.

Maryanoff, B. E., Reitz, A. B., Mutter, M. S., Inners, R. R., and Almond, H. R. (1985), *J. Am. Chem. Soc.*, **107**, 1068.

Maryanoff, B. E., Reitz, A. B., Mutter, M. S., Inners, R. R., Almond, H. R., Jr., Whittle, R. R., and Olofson, R. A. (1986), *J. Am. Chem. Soc.*, **108**, 7664.

Masamune, S. (1978), "Recent Progress in Macrolide Synthesis," *Aldrichim. Acta*, **11**, 23.

Masamune, S., Ali, S. A., Snitman, D. L., and Garvey, D. S. (1980), *Angew. Chem. Int. Ed. Engl.*, **19**, 557.

Masamune, S., Choy, W., Kerdesky, A. J., and Imperiali, B. J. (1981a), *J. Am. Chem. Soc.*, **103**, 1566.

Masamune, S., Choy, W., Peterson, J. S., and Sita, L. R. (1985a), "Double Asymmetric Synthesis and a New Strategy for Stereochemical Control in Organic Synthesis," *Angew. Chem. Int. Ed. Engl.*, **24**, 1.

Masamune, S., Ellingboe, J. W., and Choy, W. (1982), *J. Am. Chem. Soc.*, **104**, 5526.

Masamune, S., Hirama, M., Mori, S., Ali, S. A., and Garvey, D. S. (1981b), *J. Am. Chem. Soc.*, **103**, 1568.

Masamune, S., Kim, B. M., Petersen, J. S., Sato, T., and Veenstra, S. J. (1985b), *J. Am. Chem. Soc.*, **107**, 4549.

Masamune, S., Reed, L. A. III, Davis, J. T., and Choy, W. (1983), *J. Org. Chem.*, **48**, 4441.

Mash, E. A., Nelson, K. A., and Heidt, P. C. (1987), *Tetrahedron Lett.*, **28**, 1865.

Mathews, R. S., Girgenti, S. J., and Folkers, E. A. (1970), *J. Chem. Soc. Chem. Commun.*, 708.

Matteson, D. S. (1986), "The Use of Chiral Organoboranes in Organic Synthesis," *Synthesis*, 973.

Matteson, D. S. (1989), "Boronic Esters in Stereodirected Synthesis," *Tetrahedron*, **45**, 1859.

Maurer, P. J., Takahata, H., and Rapoport, H. (1984), *J. Am. Chem. Soc.*, **106**, 1095.

McDouall, J. J. W., Robb, M. A., Niazi, U., Bernadi, F., and Schlegal, H. B. (1987), *J. Am. Chem. Soc.*, **109**, 4642.

McGarvey, G. J. and Williams, J. M. (1985), *J. Am. Chem. Soc.*, **107**, 1435.

Mehta, G. and Khan, F. A. (1990), *J. Am. Chem. Soc.*, **112**, 6141.

Melloni, G., Modena, G., and Tonellato, U. (1981), "Relative Reactivities of Carbon-Carbon Double and Triple Bonds toward Electrophiles," *Acc. Chem. Res.*, **14**, 227.

Meyers, A. I., Harre, M., and Garland, R. (1984), *J. Am. Chem. Soc.*, **106**, 1146.

Meyers, A. I. and Wallace, R. H. (1989), *J. Org. Chem.*, **54**, 2509.

Meyers, A. I., Smith, R. K., and Whitten, C. E. (1979), *J. Org. Chem.*, **44**, 2250.

Michaelson, R. C., Palermo, R. E., and Sharpless, K. B. (1977), *J. Am. Chem. Soc.*, **99**, 1990.

Midland, M. M. (1983), "Reductions with Chiral Boron Reagents," in Morrison, J. D., Ed., *Asymmetric Synthesis*, Vol. 2, Academic, New York, p. 45.

Midland, M. M. and McLoughlin, J. I. (1984), *J. Org. Chem.*, **49**, 1316.

Midland, M. M. and Nguyen, N. N. (1981), *J. Org. Chem.*, **46**, 4107.

Midland, M. M. and Preston, S. B. (1982), *J. Am. Chem. Soc.*, **104**, 2330.

Midland, M. M., Tramontano, A., Kazubski, A., Graham, R. S., Tsai, D. J. S., and Cardin, D. (1984), *Tetrahedron*, **40**, 1371.

Midland, M. M., Tramontano, A., and Zderic, S. A. (1977), *J. Am. Chem. Soc.*, **99**, 5211.

Mikami, K. and Nakai, T. (1991), "Acyclic Stereocontrol via [2,3]-Wittig Sigmatropic Rearrangement," *Synthesis*, 594.

Miller, J. G., Kurz, W., Untch, K. G., and Stork, G. (1974), *J. Am. Chem. Soc.*, **96**, 6774.

Mitsunobu, O. (1981), "The use of Diethyl Azodicarboxylate and Triphenylphosphine in Synthetic and Transformation on Natural Products," *Synthesis*, 1.

Miyano, S., Lu, L. D., Viti, S. M., and Sharpless, K. B. (1983), *J. Org. Chem.*, **48**, 3608.

Miyashita, A., Yasuda, A., Takaya, H., Toriumi, K., Ito, T., Souchi, T., and Noyori, R. (1980), *J. Am. Chem. Soc.*, **102**, 7932.

Miyaura, N., Yamada, K., Suginome, H., and Suzuki, A. (1985), *J. Am. Chem. Soc.*, **107**, 972.

Mohamadi, F. and Still, W. C. (1986), *Tetrahedron Lett.*, **27**, 893.

Molander, G. A. and Etter, J. B. (1987), *J. Org. Chem.*, **52**, 3944.

Mori, I., Bartlett, P. A., and Heathcock, C. H. (1987), *J. Am. Chem. Soc.*, **109**, 7199.

Mori, K. (1989), "Synthesis of Optically Active Pheromones," *Tetrahedron*, **45**, 3233.

Morikawa, M., Yamamoto, Y., Oda, J., and Inouye, Y. (1976), *J. Org. Chem.*, **41**, 300.

Morrison, J. D., Ed., (1983–1985), *Asymmetric Synthesis*, Vols. 1–3, 5, Academic, New York.

Morrison, J. D. and Mosher, H. S. (1976), *Asymmetric Organic Reactions*, Prentice-Hall, Englewood Cliffs, NJ, 1971; American Chemical Society, Washington, DC, corrected reprint.

Morrison, J. D. and Scott, J. W., Eds., (1984), *Asymmetric Synthesis*, Vol. 4, Academic, New York.

Mukaiyama, T. and Asami, M. (1985), "Chiral Pyrrolidine Diamines as Efficient Ligands in Asymmetric Synthesis," *Top. Curr. Chem.*, **127**, 133.

Mukaiyama, T., Banno, K., and Narasaka, K. (1974), *J. Am. Chem. Soc.*, **96**, 7503.

Mukaiyama, T. and Iwasawa, N. (1981), *Chem. Lett.*, 913.

Mukaiyama, T., Soai, K., Sato, T., Shimizu, H., and Suzuki, K. (1979), *J. Am. Chem. Soc.*, **101**, 1456.

Mukaiyama, T., Yamashita, H., and Asami, M. (1983), *Chem. Lett.*, 385.

Mukherjee, D., Wu, Y.-D., Fronczek, F. R., and Houk, K. N. (1988), *J. Am. Chem. Soc.*, **110**, 3328.

Murata, S., Suzuki, M., and Noyori, R. (1980), *J. Am. Chem. Soc.*, **102**, 3248.

Myerson, J. (1980), "Acyclic Stereocontrol via Iodolactonization Synthesis of (±)-α-Multistriatin," Ph.D. Dissertation, University of California, Berkeley, CA.

Nagao, Y., Inoue, T., Fujita, E., Terada, S., and Shiro, M. (1984), *Tetrahedron*, **40**, 1215.

Nagata, W., Yoshioka, M., and Murakami, M. (1972), *J. Am. Chem. Soc.*, **94**, 4654.

Nagel, U., Kinzel, E., Andrade, J., and Prescher, G. (1986), *Chem. Ber.*, **119**, 3326.

Nakada, M., Urano, Y., Kobayashi, S., and Ohno, M. (1988), *J. Am. Chem. Soc.*, **110**, 4826.

Nakai, T., Mikami, K., Taya, S., and Fujita, Y. (1981), *J. Am. Chem. Soc.*, **103**, 6492.

Narula, A. S. (1981), *Tetrahedron Lett.*, **22**, 2017.

Narula, A. S. (1983), *Tetrahedron Lett.*, **24**, 5421.

Nasipuri, D. and Bhattacharya, P. K. (1977), *J. Chem. Soc., Perkin 1*, 576.

Negishi, E., Lew, G., and Yoshida, T. (1973), *J. Chem. Soc. Chem. Commun.*, 874.

Negishi, E., Williams, R. M., Lew, G., and Yoshida, T. (1975), *J. Organomet. Chem.*, **92**, C4.

Nogradi, M. (1987), *Stereoselective Synthesis*, VCH, Weinheim, Federal Republic of Germany.

Nonoshita, K., Banno, H., Maruoka, K., and Yamamoto (1990), *J. Am. Chem. Soc.*, **112**, 316.

Noyori, R. (1981), *Pure Appl. Chem.*, **53**, 2316.

Noyori, R. (1989), "Centenary Lecture. Chemical Multiplication of Chirality: Science and Applications," *Chem. Soc. Rev.*, **18**, 187.

Noyori, R., Ikeda, T., Ohkuma, T., Widhalm, M., Kitamura, M., Takaya, H., Akutagawa, S., Sayo, N., Saito, T., Taketomi, T., and Kumobayashi, H. (1989), *J. Am. Chem. Soc.*, **111**, 9134.

Noyori, R., Nishida, I., and Sakata, J. (1981), *J. Am. Chem. Soc.*, **103**, 2106.

Noyori, R., Ohta, M., Hsiao, Y., Kitamura, M., Ohta, T., and Takaya, H. (1986), *J. Am. Chem. Soc.*, **108**, 7117.

Noyori, R., and Takaya, H. (1990), "BINAP: An Efficient Chiral Element for Asymmetric Catalysis," *Acc. Chem. Res.*, **23**, 345.

Noyori, R. and Takaya, H. (1985), *Chem. Scr.*, **25**, NS83. Special Nobel symposium 60 issue.

Noyori, R., Tomino, I., and Tanimoto, Y. (1979), *J. Am. Chem. Soc.*, **101**, 3129.

Nayori, R., Tomino, I., Tanimoto, Y., and Nishizawa, M. (1984), *J. Am. Chem. Soc.*, **106**, 6709.

Nozaki, H., Moriuti, S., Takaya, H., and Noyori, R. (1966), *Tetrahedron Lett.*, 5239.

Oare, D. A. and Heathcock, C. H. (1989), "Stereochemistry of the Base-Promoted Michael Addition Reaction," *Top. Stereochem*, **19**, 227.

O'Donnell, M. J., Ed. (1988), "Tetrahedron Symposia-in-Print Number 33: α-Amino Acid Synthesis," *Tetrahedron*, **44**, 5253–5614.

Oguni, N., Matsuda, Y., and Kaneko, T. (1988), *J. Am. Chem. Soc.*, **110**, 7877.

Ojima, I., Ed., (1993), *Catalytic Asymmetric Synthesis*, VCH, New York.

Ojima, I., Clos, N., and Bastos, C. (1989), Recent Advances in Catalytic Asymmetric Reactions Promoted by Transition Metal Complexes," *Tetrahedron*, **45**, 6901.

Ojima, I. and Hirai, K. (1985), "Asymmetric Hydrosilylation and Hydrocarbonylation," in Morrison, J. D., Ed., *Asymmetric Synthesis*, Vol. 5, Academic, New York, p. 104.

Ojima, I., Kogure, T., and Achiwa, K. (1978), *Chem. Lett.*, 567.

Olivero, A. G., Weidman, B., and Seebach, D. (1981), *Helv. Chim. Acta*, **64,** 2485.

Olsson, T., Stern, K., and Sundell, S. (1988), *J. Org. Chem.*, **53**, 2468.

Oppolzer, W. (1984), *Angew. Chem. Int. Ed. Engl.*, **23**, 876.

Oppolzer, W. (1987), "Camphor Derivatives as Chiral Auxiliaries in Asymmetric Synthesis," *Tetrahedron*, **43**, 1969.

Oppolzer, W., Kurth, M., Reichin, D., Chapuis, C., Mohnhaupt, M., and Moffatt, F. (1981), *Helv. Chim. Acta*, **64**, 2802.

Oppolzer, W. and Löher, H. J. (1981), *Helv. Chim. Acta*, **64**, 2808.

Oppolzer, W., Mills, R. J., and Réglier, M. (1986), *Tetrahedron Lett.*, **27**, 183.

Oppolzer, W., Poli, G., Kingma, A. J., Starkemann, C., and Bernardinelli, G. (1987), *Helv. Chim. Acta*, **70**, 2201.

Oppolzer, W., Robbiani, C., and Battig, K. (1980), *Helv. Chim. Acta.*, **63**, 2015.

Oppolzer, W. and Stephenson, T. (1986), *Tetrahedron Lett.*, **27**, 1139.

Otsuka, S. and Tani, K. (1985), "Asymmetric Catalytic Isomerization of Functionalized Olefins," in Morrison, J. D., Ed., *Asymmetric Synthesis*, Vol. 5, Academic, New York, p. 171.

Paquette, L. A. (1984), "Asymmetric Cycloaddition Reactions," in Morrison, J. D., Ed., *Asymmetric Synthesis*, Vol. 3, Academic, New York, p. 455.

Paquette, L. A. and Cox, O. (1967), *J. Am. Chem. Soc.*, **89**, 5633.

Paquette, L. A., Teleha, C. A., Taylor, R. T., Maynard, G. D., Rogers, R. D., Gallucci, J. C., and Springer, J. P. (1990), *J. Am. Chem. Soc.*, **112**, 265.

Partridge, J. J., Chadha, N. K., and Uskokovic, M. R. (1973), *J. Am. Chem. Soc.*, **95**, 532.

Pasteur, L. (1853), *C.R. Acad. Sci.*, **37**, 162.

Patai, S., Ed. (1962), *The Chemistry of Alkenes*, Wiley-Interscience, New York.

Pearson, A. J. (1980), "Tricarbonyl(diene)iron Complexes: Synthetically Useful Properties," *Acc. Chem. Res.*, **13**, 463.

Pearson, A. J. (1990), *Synlett*, 10.

Pelter, A., Smith, K., and Brown, H. C. (1988), *Borane Reagents*, Academic, London.

Pfau, M., Revial, G., Guingant, A., and d'Angelo, J. (1985), *J. Am. Chem. Soc.*, **107**, 273.

Pfenninger, A. (1986), "Asymmetric Epoxidation of Allylic Alcohols: The Sharpless Epoxidation," *Synthesis*, 89.

Piers, E., Britton, R. W., and de Waal, W. (1969), *J. Chem. Soc. Chem. Commun.*, 1969.

Pirkle, W. H., Pochapsky, T. C., Mahler, G. S., Corey, D. E., Reno, D. S., and Alessi, D. M. (1986), *J. Org. Chem.*, **51**, 4991.

Pirkle, W. H. and Rinaldi, P. L. (1977), *J. Org. Chem.*, **42**, 2080.

Pitchen, P., Dunach, E., Deshmulch, M. N., and Kagan, H. B. (1984), *J. Am. Chem. Soc.*, **106**, 8188.

Pohland, A. and Sullivan, H. R. (1955), *J. Am. Chem. Soc.*, **77**, 3400.

Pollak, A., Blumenfeld, H., Wax, M., Baughn, R. L., and Whitesides, G. M. (1980), *J. Am. Chem. Soc.*, **102**, 6324.

Posner, G. H. (1983), "Addition of Organometallic Reagents to Chiral Vinylic Sulfoxides," in Morrison, J. D., Ed., *Asymmetric Synthesis*, Vol. 2, Academic, New York, p. 225.

Posner, G. H. (1985), *Chem. Scr.*, **25**, NS157. Special Nobel symposium 60 issue.

Potvin, P. G., Kwong, P. C. C., and Brook, M. A. (1988), *J. Chem. Soc. Chem. Commun.*, 773.

Power, M. B., Bott, S. G., Atwood, J. L., and Barron, A. R. (1990), *J. Am. Chem. Soc.*, **112**, 3446.

Prelog, V. (1953), *Helv. Chim. Acta*, **36**, 308.

Puchot, C., Samuel, O., Dunach, E., Zhao, S., Agami, C., and Kagan, H. B. (1986), *J. Am. Chem. Soc.*, **108**, 2353.

Quimpère, M. and Jankowski, K. (1987), *J. Chem. Soc. Chem. Commun.*, 676.

RajanBabu, T. V. (1987), *J. Am. Chem. Soc.*, **109**, 609.

Rajaram, J., Narula, A. S., Chawla, H. P. S., and Sukh Dev (1983), *Tetrahedron*, **39**, 2315.

Rama Rao, A. V., Yadav, J. S., Bal Reddy, K., and Mehendale, A. R. (1984), *Tetrahedron*, **40**, 4643.

Ratier, M., Castaing, M., Godet, J.-V., and Pereyre, M. (1978), *J. Chem. Res.*, (S), 179.

Rautenstrauch, V. (1970), *J. Chem. Soc. Chem. Commun.*, 4.

Reetz, M. T. (1988), *Pure Appl. Chem.*, **60**, 1607.

Reetz, M. T. (1991), "New Approaches to the Use of Amino Acids as Chiral Building Blocks in Organic Synthesis," *Angew. Chem. Int. Ed. Engl.*, **30**, 1531.

Regan, A. C. and Staunton, J. (1987), *J. Chem. Soc. Chem. Commun.*, 520.

Reucroft, J. and Sammes, P. G. (1971), "Stereoselective and Stereospecific Olefin Synthesis," *Q. Rev.*, **25**, 135.

Revial, G. and Pfau, M. (1991), *Org. Synth.* **70**, 35.

Rosen, T. and Heathcock, C. H. (1985), *J. Am. Chem. Soc.*, **107**, 3731.

Rossiter, B. E. (1985), "Synthetic Aspects of Asymmetric Epoxidation," in Morrison, J. D., Ed., *Asymmetric Synthesis*, Vol. 5, Academic, New York, p. 194.

Rossiter, B. E. and Swingle, N. M. (1992), "Asymmetric Conjugate Addition," *Chem. Rev.* **92**, 771.

Rossiter, B. E., Verhoeven, T. R., and Sharpless, K. B. (1979), *Tetrahedron Lett.*, 4733.

Roush, W. R. and Lesur, B. M. (1983), *Tetrahedron Lett.*, **24**, 2231.

Rozeboom, M. D. and Houk, K. N. (1982), *J. Am. Chem. Soc.*, **104**, 1189.

Rylander, P. N. (1979), *Catalytic Hydrogenation in Organic Syntheses*, Academic, New York.

Saddler, J. C., Donaldson, R. E., and Fuchs, P. L. (1981), *J. Am. Chem. Soc.*, **103**, 2110.

Saimoto, H., Houge, C., Frisque-Hesbain, A. M., Mockel, A., and Ghosez, L. (1983), *Tetrahedron Lett.*, **24**, 2151.

Sakane, S., Maruoka, K., and Yamamoto, H. (1986), *Tetrahedron*, **42**, 2203.

Salomon, R. G., Sachinvala, N. D., Raychaudhuri, S. R., and Miller, D. B., (1984), *J. Am. Chem. Soc.*, **106**, 2211.

Samaddar, A. K., Konar, S. K., and Nasipuri, D. (1983), *J. Chem. Soc. Perkin 1*, 1449.

Santaniello, E., Ferraboschi, P., Grisenti, P., and Manzocchi, A. (1992), "The Biocatalytic Approach to the Preparation of Enantiomerically Pure Chiral Building Blocks," *Chem. Rev.*, **92**, 1071.

Sargeson, A. M. (1980), "Chirality Induction in Coordination Complexes," in *ACS Symposium Series*, **119**, American Chemical Society, Washington, DC, 1980.

Saunders, W. H. and Cockerill, A. F. (1973), *Mechanisms of Elimination Reactions*, Wiley, New York.

Savoia, D., Tagliavani, E., Trombini, C., and Umani-Ronchi, A. (1981), *J. Org. Chem.*, **46**, 5340.

Schleyer, P. v. R. (1967), *J. Am. Chem. Soc.*, **89**, 701.

Schaub, B., Jeganathan, S., and Schlosser, M. (1986), *Chimia*, **40**, 246.

Schlosser, M. (1970), "The Stereochemistry of the Witting Reaction," *Top. Stereochem.*, **5**, 1.

Schlosser, M. and Christmann, K. F. (1967), *Justus Liebigs Ann. Chem.*, **708**, 1.

Schlosser, M. and Schaub, B. (1982), *J. Am. Chem. Soc.*, **104**, 5821.

Schmid, G., Fukuyama, T., Akasaka, K., and Kishi, Y. (1979), *J. Am. Chem. Soc.*, **101**, 259.

Schmierer, R., Grotemeier, G., Helmchen, G., and Selim, A. (1981), *Angew. Chem. Int. Ed. Engl.* **20**, 207.

Schöllkopf, U., Hausberg, H.-H., Segal, M., Reiter, U., Hoppe, I., Saenger, W., and Lindner, K. (1981), *Justus Liebigs Ann. Chem.*, 439.

Schreiber, S. L., Schreiber, T. S., and Smith, D. B. (1987), *J. Am. Chem. Soc.*, **109**, 1525.

Scott, J. W. (1984), "Readily Available Chiral Carbon Fragments and Their Use in Synthesis," in Morrison, J. D. and Scott, J. W., Eds., *Asymmetric Synthesis*, Vol. 4, Academic, New York, p. 1.

Scott, J. W. (1989), "Enantioselective Synthesis of Non-racemic Chiral Molecules on an Industrial Scale," *Top. Stereochem.*, **19**, 209.

Seebach, D. (1988), "Structure and Reactivity of Lithium Enolates. From Pinacolone to Selective *C*-Alkylations of Peptides. Difficulties and Opportunities Afforded by Complex Structures," *Angew. Chem. Int. Ed. Engl.*, **27**, 1624.

Seebach, D. and Aebi, J. D. (1983), *Tetrahedron Lett.*, **24**, 3311.

Seebach, D., Amstutz, R., and Dunitz, J. D. (1981), *Helv. Chim. Acta*, **64**, 2622.

Seebach, D., Boes, M., Naef, R., and Schweizer, W. B. (1983), *J. Am. Chem. Soc.*, **105**, 5390.

Seebach, D., Ehrig, V., and Teschner, M. (1976), *Justus Liebigs Ann. Chem.*, 1357.

Seebach, D. and Hungerbühler, E. (1980), "Syntheses of Enantiomerically Pure Compounds (EPC-Syntheses). Tartaric Acid, an Ideal Source of Chiral Building Blocks for Synthesis?" in Scheffold, R., Ed., *Modern Synthetic Methods*, Vol. 2, Springer, Berlin, p. 91.

Seebach, D., Imwinkelried, R., and Weber, T. (1986), "EPC Syntheses with C,C Bond Formation via Acetals and Enamines," Scheffold, R., Ed., *Modern Synthetic Methods*, Vol. 4, Springer, Berlin, p. 125.

Seebach, D. and Naef, R. (1981), *Helv. Chim. Acta*, **64**, 2704.

Seebach, D. and Wasmuth, D. (1980), *Helv. Chim. Acta*, **63**, 197.

Seebach, D., Zimmermann, J., Gysel, U., Ziegler, R., and Ha, T.-K. (1988), *J. Am. Chem. Soc.*, **110**, 4763.

Seyferth, D. and Mai, V. A. (1970), *J. Am. Chem. Soc.*, **92**, 7412.

Sharpless, K. B. (1985), *Chem. Scr.*, **25**, NS71. Special Nobel symposium 60 issue.

Sharpless, K. B., Amberg, W., Beller, M., Chen, H., Hartung, J., Kawanami, Y., Lübben, D., Manoury, E., Ogino, Y., Shibata, T., and Ukita, T. (1991), *J. Org. Chem.*, **56**, 4585.

Sharpless, K. B., Amberg, W., Bennani, Y. L., Crispino, G. A., Hartung, J., Jeong, K. S., Kwong, H.-L., Morikawa, K., Wang, Z. M., Xu, D., and Zhang, X.-L. (1992), *J. Org. Chem.*, **57**, 2768.

Sharpless, K. B. and Michaelson, R. C. (1973), *J. Am. Chem. Soc.*, **95**, 6137.

Sharpless, K. B. and Verhoeven, T. R. (1979), "Metal-Catalyzed, Highly Selective Oxygenations of Olefins and Acetylenes with *tert*-Butyl Hydroperoxide. Practical Considerations and Mechanisms," *Aldrichim. Acta*, **12**, 63.

Sharpless, K. B., Woodard, S. S., and Finn, M. G. (1983), *Pure Appl. Chem.*, **55**, 1823.

Sheldon, R. A. (1993), *Chirotechnology. Industrial Synthesis of Optically Active Compounds*, Marcel Dekker, New York.

Shirai, R., Tanaka, M., and Koga, K. (1986), *J. Am. Chem. Soc.*, **108**, 543.

Siegel, C. and Thornton, E. R. (1989), *J. Am. Chem. Soc.*, **111**, 5722.

Sih, C. J. and Wu, S.-H. (1989), "Resolution of Enantiomers via Biocatalysis," *Top. Stereochem.*, **19**, 63.

Simmons, H. E., Cairns, T. L., and Vladuchick, S. A. (1973), "Cyclopropanes from Unsaturated Compounds, Methylene Iodide, and Zinc-Copper Couple," *Org. React.*, **20**, 1.

Simpkins, N. J. (1986), *J. Chem. Soc. Chem. Commun.*, 88.

Sinclair, P. J., Zhai, D., Reibenspies, J., and Williams, R. M. (1986), *J. Am. Chem. Soc.*, **108**, 1103.

Singh, A. K., Bakshi, R. K., and Corey, E. J. (1987), *J. Am. Chem. Soc.*, **109**, 6187.

Smith, A. B., Dunlap, N. K., and Sulikowsk, G. A. (1988), *Tetrahedron Lett.*, **29**, 439.

Smith, A. B. and Trumper, P. K. (1988), *Tetrahedron Lett.*, **29**, 443.

Smith, H. (1963), *Organic Reactions in Liquid Ammonia*, Interscience, New York, p. 213.

Soai, K. and Niwa, S. (1992), "Enantioselective Addition of Organozinc Reagents to Aldehydes," *Chem. Rev.*, **92**, 833.

Soai, K., Niwa, S., and Watanabe, M. (1988), *J. Org. Chem.*, **53**, 928.

Soai, K., Ookawa, A., Kaba, T., and Ogawa, K. (1987), *J. Am. Chem. Soc.*, **109**, 7111.

Soai, K., Oyamada, H., and Yamanoi, T. (1984), *Chem. Lett.*, 251.

Sobukawa, M., Nakajima, M., and Koga, K. (1990), *Tetrahedron Asymmetry*, **1**, 295.

Sogo, S. G., Widlanski, T. S., Hoare, J. H., Grimshaw, C. E., Berchtold, G. A., and Knowles, J. R. (1984), *J. Am. Chem. Soc.*, **106**, 2701.

Solomon, M., Jamison, W. C., Cherry, D. A., Mills, J. E., Shah, R. D., Rodgers, J. D., and Maryanoff, C. A. (1988), *J. Am. Chem. Soc.*, **110**, 3702.

Sonnet, P. E. (1980), "Olefin Inversion," *Tetrahedron*, **36**, 557.

Spellmeyer, D. C. and Houk., K. N. (1987), *J. Org. Chem.*, **52**, 959

Spenser, T. A., Weaver, T. D., Villarica, R. M., Friary, R. J., Posler, J., and Schwartz, M. A. (1968), *J. Org. Chem.*, **33**, 712.

Srivasta, S. and le Noble, W. J. (1987), *J. Am. Chem. Soc.*, **109**, 5874.

Stevens, R. V. and Gaeta, F. C. A. (1977), *J. Am. Chem. Soc.*, **99**, 6105.

Still, W. C. (1977), *J. Am. Chem. Soc.*, **99**, 4186.

Still, W. C. and Barrish, J. C. (1977), *J. Am. Chem. Soc.*, **105**, 2487.

Still, W. C. and Galynker, I. (1981), *Tetrahedron*, **37**, 3981.

Still, W. C. and Novack, V. J. (1984), *J. Am. Chem. Soc.*, **106**, 1148.

Stork, G. (1983), "Vinyl and Beta-Alkoxy Radicals in Organic Synthesis," in Nozaki, H., Ed., *Current Trends in Organic Synthesis*, Pergamon, Oxford, England, p. 359.

Stork, G. and Kahn, M. (1983), *Tetrahedron Lett.*, **24**, 3951.

Stork, G. and Kahn, M. (1985), *J. Am. Chem. Soc.*, **107**, 500.

Stork, G. and Kahne, D. E. (1983), *J. Am. Chem. Soc.*, **105**, 1072.

Stork, G., Nakahara, Y., Nakahara, Y., and Greenlee, W. J. (1978), *J. Am. Chem. Soc.* **100**, 7775.

Stork, G. and Rychnovsky, S. D. (1987), *J. Am. Chem. Soc.*, **109**, 1564.

Stork, G. and Sher, P. M. (1983), *J. Am. Chem. Soc.*, **105**, 6765.

Suda, H., Kanoh, S., Umeda, N., Ikka, M., and Motoi, M. (1984), *Chem. Lett.*, 899.

Surzur, J.-M. (1982), "Radical Cyclizations by Intramolecular Additions," in Abramovitch, R. A., Ed., *Reactive Intermediates*, Vol. 2, Plenum, New York, p. 121.

Suzuki, K., Katayama, E., Tomooka, K., Matsumoto, T., and Tsuchihashi, G. (1985), *Tetrahedron Lett.*, **26**, 3707.

Suzuki, K., Tomooka, K., Katayama, E., Matsumoto, T., and Tsuchihashi, G. (1986), *J. Am. Chem. Soc.*, **108**, 5221.

Svoboda, M., Zavada, J., and Sicher, J. (1965), *Coll. Czech. Chem. Commun.*, **30**, 413.

Taber, D. F. and Raman, K. (1983), *J. Am. Chem. Soc.*, **105**, 5935.

Takahashi, T., Nisar, M., Shimizu, K., and Tsuji, J. (1986), *Tetrahedron Lett.*, **27**, 5103.

Takaya, H., Ohta, T., Sayo, N., Kumobayashi, H., Akutagawa, S., Inoue, S., Kasahara, I., and Noyori, R. (1987), *J. Am. Chem. Soc.*, **109**, 1596.

Tamao, K., Nakagawa, Y., Arai, H., Higuchi, N., and Ito, Y. (1988), *J. Am. Chem. Soc.*, **110**, 3712.

Tamao, K., Nakajima, T., Sumiya, R., Arai, H., Higuchi, N., and Ito, Y. (1986), *J. Am. Chem. Soc.*, **108**, 6090.

Tani, K., Yamagata, T., Akutagawa, S., Kumobayashi, H., Taketomi, T., Takaya, H., Miyashita, A., Noyori, R., and Otsuka, S. (1984), *J. Am. Chem. Soc.*, **106**, 5208.

Tani, K., Yamagata, T., Otsuka, S., Kumobayashi, H., and Akutagawa, S. (1988), *Org. Syn.*, **67**, 33.

Tee, O. S., Altmann, J. A., and Yates, K. (1974), *J. Am. Chem. Soc.*, **96**, 3141.

Thomas, A. F. and Ohloff, G. (1970), *Helv. Chim. Acta*, **53**, 1145.

Thompson, H. W. and McPherson, E. (1974), *J. Am. Chem. Soc.*, **96**, 6232.

Thompson, H. W. and Naipawer, R. E. (1973), *J. Am. Chem. Soc.*, **95**, 6379.

Tomioka, K. (1990), "Asymmetric Synthesis Utilizing External Chiral Ligands," *Synthesis*, 541.

Tomioka, K. and Koga, K. (1983), "Non-catalytic Additions to α, β-Unsaturated Carbonyl Compounds," in Morrison, J. D., Ed., *Asymmetric Synthesis*, Vol. 2, Academic, New York, p. 201.

Tomioka, K., Kawasaki, H., Yasuda, K., and Koga, K. (1988), *J. Am. Chem. Soc.*, **110**, 3597.

Toromanoff, E. (1980), "Dynamic Stereochemistry of the 5-, 6- and 7-Membered Rings Using the Torsion Angle Notation," *Tetrahedron*, **36**, 2809.

Trimble, L. A. and Vederas, J. C. (1986), *J. Am. Chem. Soc.*, **108**, 6397.

Tripathy, R., Franck, R. W., and Onan, K. D. (1988), *J. Am. Chem. Soc.*, **110**, 3257.

Trost, B. M. and Curran, D. P. (1980), *J. Am. Chem. Soc.*, **102**, 5699.

Trost, B. M., Florey, J., and Jebaratnam, D. J. (1987), *J. Am. Chem. Soc.*, **109**, 613.

Trost, B. M. and Hammen, R. F. (1973), *J. Am. Chem. Soc.*, **95**, 962.

Trost, B. M. and Matsumura, Y. (1977), *J. Org. Chem.*, **42**, 2036.

Trost, B. M., O'Krongly, D., and Belletire, J. L. (1980), *J. Am. Chem. Soc.*, **102**, 7595.

Trost, B. M., Timko, J. M., and Stanton, J. L. (1978), *J. Chem. Soc. Chem. Commun.*, 436.

Tucker, J. A., Houk, K. N., and Trost, B. M. (1990), *J. Am. Chem. Soc.*, **112**, 5465.

Uenishi, J.-I., Beau, J.-M., Armstrong, R. W., and Kishi, Y. (1987), *J. Am. Chem. Soc.*, **109**, 4756.

Van Horn, D. E. and Masamune, S. (1979), *Tetrahedron Lett.*, 2229.

Van Tamelen, E. E. (1968), "Bioorganic Chemistry: Sterols and Acyclic Terpene Terminal Epoxides," *Acc. Chem. Res.*, **1**, 111.

Vasella, A. (1980), "Chiral Building Blocks in Enantiomer Synthesis ex Sugars," in Scheffold, R., Ed., *Modern Synthetic Methods*, Vol. 2, Springer, Berlin, p. 173.

Vedejs, E. and Marth, C. F. (1988), *J. Am. Chem. Soc.*, **110**, 3948.

Vedejs, E. and McClure, C. K. (1986), *J. Am. Chem. Soc.*, **108**, 1094.

Vedejs, E. and Peterson, M. J. (1994), "Stereochemistry and Mechanism of Wittig Reaction," *Top. Stereochem.*, **21**, 1.

Vedejs, E., Meier, G. P., and Snoble, K. A. J. (1981), *J. Am. Chem. Soc.*, **103**, 2823.

Velluz, L., Valls, J., and Nomine, G. (1965), *Angew. Chem. Int. Ed. Engl.*, **4**, 181.

Vineyard, B. D., Knowles, W. S., Sabacky, M. J., Bachmann, G. L., and Weinkauff, O. J. (1977), *J. Am. Chem. Soc.*, **99**, 5946.

Vogel, E., Carvatti, P., Franck, P., Aristoff, P., Moody, C., Becker, A. M., Felix, D., and Eschenmoser, A. (1987), *Chem. Lett.*, 219.

Walborsky, H. M., Barash, L., and Davis, T. C. (1961), *J. Org. Chem.*, **26**, 4778.

Walborsky, H. M., Barash, L., and Davis, T. C. (1963), *Tetrahedron*, **19**, 2333.

Wang, L, and Sharpless, K. B. (1992), *J. Am. Chem. Soc.*, **114**, 7568.

Wang, Y.-F., Chen, C.-S., Girdaukas, G., and Sih, C. J. (1984), *J. Am. Chem. Soc.*, **106**, 3695.

Ward, R. S. (1990), "Non-Enzymatic Asymmetric Transformations Involving Symmetrical Bifunctional Compounds," *Chem. Soc. Rev.*, **19**, 1.

Wender, P. A., Schaus, J. M., and White, A. W. (1980), *J. Am. Chem. Soc.*, **102**, 6157.

Wharton, P. S. (1961), *J. Org. Chem.*, **26**, 4781.

Wharton, P. S., Sumi, Y., and Kretchmer, R. A. (1965), *J. Org. Chem.*, **30**, 234.

Whitesell, J. K. (1989), "C_2 Symmetry and Asymmetric Induction," *Chem. Rev.*, **89**, 1581.

Whitesell, J. K. and Allen, D. A. (1988), *J. Am. Chem. Soc.*, **110**, 3585.

Whitesell, J. K., Battacharaya, A., Aguilar, D. A., and Henke, K. (1982), *J. Chem. Soc. Chem. Commun.*, 1989.

Whitham, G. H. and Wright, M. (1971), *J. Chem. Soc. (C)*, 883.

Wigfield, D. C. (1979), *Tetrahedron*, **35**, 449.

Wilds, A. L. (1944), "Reduction with Aluminium Alkoxides (The Meerwein–Ponndorf–Verley Reduction)," *Org. React.*, **2**, 178.

Williams, I. D., Pedersen, S. F., Sharpless, K. B., and Lippard, S. J. (1984), *J. Am. Chem. Soc.*, **106**, 6430.

Williams, R. H. (1989), *Synthesis of Optically Active Amino Acids*, Pergamon, Oxford.

Williard, P. G. and Salvino, J. M. (1985), *Tetrahedron Lett.*, **26**, 3931.

Winter, C. E. (1987), *J. Chem. Educ.*, **64**, 587.

Winter, R. E. K. (1965), *Tetrahedron Lett.*, 1207.

Winterfeldt, E. (1969), "Ionic Additions to Acetylenes," in Viehe, H. G., Ed., *Chemistry of Acetylenes*, Marcel Dekker, New York, Chap. 4.

Woodward, R. B. and Hoffmann, R. (1970), *The Conservation of Orbital Symmetry*, Verlag Chemie, Weinheim, Federal Republic of Germany.

Woodward, S. S., Finn, M. G., and Sharpless, K. B. (1991), *J. Am. Chem. Soc.*, **113**, 106.

Wovkulich, P. M. and Uskokovic, M. R. (1981), *J. Am. Chem. Soc.*, **103**, 3956.

Wu, Y.-D., Houk, K. N. and Trost, B. M. (1987), *J. Am. Chem. Soc.*, **109**, 5560.

Wu, Y.-D. and Houk, K. N. (1987), *J. Am. Chem. Soc.*, **109**, 908.

Wynberg, H. (1986), "Asymmetric Catalysis by Alkaloids," *Top. Stereochem.*, **16**, 87.

Wynberg, H. and Feringa, B. (1976), *Tetrahedron*, **32**, 2831.

Wynberg, H. and Helder, R. (1975), *Tetrahedron Lett.*, 4057.

Wynberg, H. and Lorand, J. P. (1981), *J. Org. Chem.*, **46**, 2538.

Wynberg, H. and Staring, E. G. J. (1982), *J. Am. Chem. Soc.*, **104**, 166.

Yamada, K., Takeda, M., and Iwakuma, T. (1981), *Tetrahedron Lett.*, 3869.

Yamada, S., Mashiko, T., and Terashima, S. (1977), *J. Am. Chem. Soc.*, **99**, 1988.

Yamamoto, K., Fukushima, H., and Nakazaki, M. (1984), *J. Chem. Soc. Chem. Commun.*, 1490.

Yamamoto, Y. and Maruyama, K. (1980), *Tetrahedron Lett.*, **21**, 4607.

Yamamoto, Y. and Maruyama, K. (1982), *Heterocycles*, **18**, 357.

Yamauchi, M. and Watanabe, T. (1988), *J. Chem. Soc. Chem. Commun.*, 27.

Zabicky, J., ed., (1970), *The Chemistry of Alkenes*, Wiley-Interscience, New York.

Zhang, W. and Jacobsen, E. N. (1991), *J. Org. Chem.*, **56**, 2296.

Zhang, W., Loebach, J. L., Wilson, S. R., and Jacobsen, E. N. (1990), *J. Am. Chem. Soc.*, **112**, 2801.

Ziegler, F. E. (1977), "Stereo- and Regiochemistry of the Claisen Rearrangement: Applications to Natural Products Synthesis," *Acc. Chem. Res.*, **10**, 227.

Ziegler, F. E. (1988), "The Thermal, Aliphatic Claisen Rearrangement," *Chem. Rev.*, **88**, 1423.

Ziegler, F. E. and Fang, J.-M. (1981), *J. Org. Chem.*, **46**, 825.

Ziegler, F. E., Fang, J.-M., and Tam, C. C. (1982), *J. Am. Chem. Soc.*, **104**, 7174.

Ziegler, F. E., Klein, S. I., Pati, U. K., and Wang, T.-F. (1985), *J. Am. Chem. Soc.*, **107**, 2730.

Ziegler, F. E. and Wang, T.-F. (1984), *J. Am. Chem. Soc.*, **106**, 718.

Zimmerman, H. E. (1963), "Base Catalyzed Rearrangements," in de Mayo, P., Ed., *Molecular Rearrangements*, Vol. 1, Wiley, New York, p. 345.

Zimmerman, H. E., Singer, L., and Thyagarajan, B. S. (1959), *J. Am. Chem. Soc.*, **81**, 108.

Zimmerman, H. E. and Traxler, M. D. (1957), *J. Am. Chem. Soc.*, **79**, 1920.

Zurflüh, R., Wall, E. N., Siddall, J. B., and Edwards, J. A. (1968), *J. Am. Chem. Soc.*, **90**, 6224.

Zweifel, G. and Steele, R. B. (1967), *J. Am. Chem. Soc.*, **89**, 2754.

Zweifel, G. and Whitney, C. C. (1967), *J. Am. Chem. Soc.*, **89**, 2753.

13

Chiroptical Properties

13-1. INTRODUCTION

"Chiroptical properties" are properties of chiral substances arising from their nondestructive interaction with anisotropic radiation (polarized light), properties that can differentiate between the two enantiomers of a chiral compound. The term, whose use was introduced by Thomson in 1884 [Kelvin, 1904, p. 461; see also Prelog, 1968; Weiss and Dreiding (Weiss, 1968); and Section 1-3], encompasses the classical spectroscopic qualitative and quantitative manifestations of chirality: optical activity and optical rotatory dispersion (ORD), the change of optical rotation with wavelength. Widespread application of another chiroptical technique, circular dichroism (CD), is more recent. Vibrational CD and its counterpart in Raman spectroscopy are relatively new chiroptical techniques presently under active development, and so is the emission of circularly polarized light, circular polarization of emission (CPE).

The foregoing techniques, which are discussed in detail in this chapter, are mainly concerned with "natural optical activity," that is, with inherent properties of nonracemic samples of chiral substances. Properties resulting from optical activity induced in achiral substances or in racemic chiral samples by magnetic and electric fields, the Faraday effect (Caldwell and Eyring, 1972; Thorne, 1972), and the Kerr effect (LeFèvre and LeFèvre, 1972) are not considered to be chiroptical phenomena according to some authorities (Snatzke and Snatzke, 1980). These phenomena are not treated in this book. An alternative analysis treats natural and induced optical activity together as optical birefringence and scattering phenomena (Atkins, 1971; Barron, 1983).

As was pointed out by Thomson in 1884 (Kelvin, 1904, p. 644), Faraday himself understood and made clear the fact that so-called magnetic rotation is not a chiroptical property.

Application of the chiroptical techniques to structural analysis is elaborated in Section 13-4 to 13-7 of this chapter. It will become evident that the application of

TABLE 13.1.   Chiroptical Techniques

Technique	Principle	Accessible Chromophore
Polarimetry and optical rotatory dispersion (ORD)	Refraction	No
Circular dichroism (CD)	Absorption	Yes
Circular polarization of emission (CPE)	Emission	Yes

these techniques is in some cases dependent on the presence of a chromophore in the substance being analyzed; the correlation between the specific chiroptical technique, its underlying principle, and the need for a chromophore is outlined in Table 13.1.

13-2.   OPTICAL ACTIVITY. ANISOTROPIC REFRACTION

a. Origin. Theory

Given the unique place of the phenomenon of optical activity in the methodology and history of stereochemistry, it is well worthwhile to attempt an explanation of its origin. We can do this but rather superficially here; entire books have been devoted to this topic (Caldwell and Eyring, 1971; Charney, 1979; Barron, 1983).

Optical activity (or optical rotatory power) results from the *refraction* of right and left circularly polarized light (cpl) to different extents by chiral molecules (Mislow, 1965, p. 54). The source of the rotation, and hence of ORD as well, is birefringence, that is, unequal slowing down of right (R) and left (L) circularly polarized light ($n_R \neq n_L$, where n is the index of refraction) as the light passes through the sample. In contrast, CD is the consequence of the difference in *absorption* of right and left cpl ($\varepsilon_R \neq \varepsilon_L$, where ε is the molar absorption coefficient).

Let us now consider what happens when a beam of monochromatic polarized radiation passes through a nonracemic sample of a chiral substance. Light is electromagnetic radiation. It has associated with it time-dependent electric and magnetic fields. In ordinary radiation, the electric field associated with the light waves oscillates in all directions perpendicular to the direction of propagation (Fig. 13.1a). Such radiation is called isotropic (or unpolarized). In contrast, if the

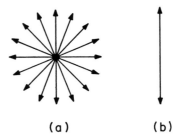

Figure 13.1.   (a) Isotropic and (b) anisotropic (linearly polarized) light beams (electric field only) viewed along the z axis toward the light source. [Adapted with permission from Solomons, T. W. G. (1978), *Organic Chemistry*, pp. 244–245. Copyright © 1978 John Wiley & Sons, Inc.]

(a)   (b)

radiation is filtered so as to remove all oscillations other than in one direction, say in the x,z plane (Fig. 13.1b), then the light is anisotropic and said to be linearly polarized or, less rigorously, plane polarized (cf. Lambert et al., 1987, p. 261).

The relation of the electric field vector **E** and the magnetic field vector **H** of a linearly polarized light beam to the direction of propagation at a given time is shown in Fig. 13.2a. The two fields oscillate at right angles to one another and in phase. A different view (Fig. 13.2b) shows only the magnitude and direction of the electric field vector as a function of time t and at a given distance $z = z_0$ from the light source (Snatzke, 1981). Both relations are cosine functions described by Eq. 13.1

$$E = E_0 \cos(2\pi\nu t - 2\pi z/\lambda) = E_0 \cos \omega(t - z/c_0) \qquad (13.1)$$

where ν is the light frequency, $\lambda = c_0/\nu$ is its wavelength (c_0 is the speed of light in vacuum), E_0 is the maximum amplitude of the wave, and $\omega = 2\pi\nu$.

Circularly polarized light may be described by examining the movement of the electric field vector only. The tip of the electric field vector **E** follows a helical path along the surface of a cylinder that is aligned with the axis of propagation of the light; it is helpful in this connection to think of the helix as being pushed out of the light source in the direction of propagation *but not rotated out*. Figure 13.3a defines a right circularly polarized ray viewed toward the light source and **E** is seen to have traveled toward the observer in a clockwise fashion. Time dependent measurement of the angle is opposite in sense since the observer encounters **E** in the order 6, 5, 4, 3, 2, 1, 0 (Fig. 13.3b; Brewster, 1967; Harada and Nakanishi, 1983, p. 439; Snatzke, 1981).

The definition of the sense of cpl may be a source of some confusion. Helical motion is a combination of translation and rotation occurring at the same time. It is a fact that a static P helix (right handed)—as in a screw—is of the same sense when viewed from either end, "front to back" and, after reversal, "back to front". This fact is of some significance in the designation of absolute configuration of molecules having helical symmetry (cf. Section 14-7). It becomes necessary to stipulate from which end the helix is being viewed the moment the two components of helical motion (translation and rotation) are "decoupled". As the above description points

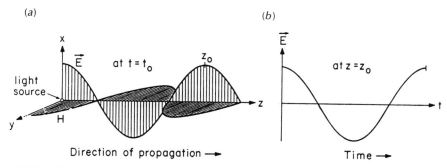

Figure 13.2. Linearly polarized light (a, at a given time and b, at a given place). [(a) Reprinted with permission from Brewster, J. H. (1967), *Top Stereochem.*, **2**, 1. Copyright © 1967 John Wiley & Sons, Inc. and (b) adapted with permission from Snatzke, G., *Chem. Unserer Zeit.*, **15**, 78 (1981).]

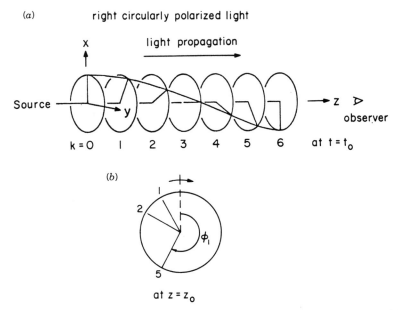

Figure 13.3. Definition of right cpl. (*a*) At $t = t_0$, the electric field vector describes a right-handed helix as viewed toward the light source (z increases as $z = k\ell/12$, $k = 0, 1, 2, \ldots$). [Reprinted with permission from Harada, N. and Nakanishi, K. (1983), *Circular Dichroic Spectroscopy. Exciton Splitting in Organic Stereochemistry*, p. 439. Copyright © 1983 University Science Books, 20 Edgehill Road, Mill Valley, CA.]. (*b*) Right cpl viewed at $z = z_0$ as a function of time with light traveling toward the observer. [Adapted with permission from Snatzke, G., *Chem. Unserer Zeit.*, **15**, 78 (1981).]

out and as indicated in Figure 13.3, analysis of cpl requires that (a) the light ray be viewed *toward* the light source only, and that (b) since the ray is not static, a distinction must be made between an instantaneous view ($t = t_0$, equivalent to the observer traveling with the light wave, Fig. 13.3*a*), and a view at a fixed point ($z = z_0$, equivalent to the observer looking at a slit through which the light source emerges some time after the light is "turned on", Fig. 13.3*b*). These are alternative and complementary, not contradictory, definitions of the sense of cpl.

In addition, when left and right cpl is envisaged by drawing (electric field) vectors in the xy plane (as in Figs. 13.4 and 13.6), it is essential to stipulate the direction in which one is viewing the vectors. This requirement is akin to one needed when observing a transparent clock. In Figure 13.4, when viewed *toward* the light, the --- vector is defined as left cpl. In Figure 13.6, the same convention applies.

Linearly polarized light may be envisaged (represented or conceptualized) mathematically and graphically as a combination of left and right (hence oppositely) coherent rotating beams of cpl. In an isotropic medium, the two components travel at the same velocity, hence in-phase, but in opposite senses (Fig. 13.4). The resultant vector sum, which exhibits the properties of linearly polarized light, is shown as travelling in the x,z plane. The amplitude of the vector sum that decreases and increases as shown in Figure 13.2*a* is double that of each cpl beam.

It is important to understand that cpl is real and not just a mathematical construct; it may be produced by passing linearly polarized light through a

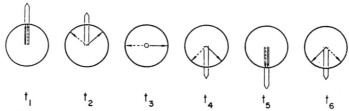

$$t_1 \quad t_2 \quad t_3 \quad t_4 \quad t_5 \quad t_6$$

Figure 13.4. Addition of left (---) and right (—) cpl (of equal frequency, wavelength, and intensity) yields linearly polarized light. Time dependence ($t_1 \rightarrow t_6$) of the view along the light from a given point $z = z_0$; only the electric field vector is shown. [Adapted with permission from Snatzke, G., *Chem. Unserer Zeit.*, **15**, 78 (1981).]

specially cut glass prism called a "Fresnel's rhomb," or by means of the Pockels effect. In the Pockels effect, linearly polarized monochromatic light is subject to one-quarter wave retardation by passage through a biaxial crystal (Fig. 13.5). The Billings cell utilizes an electrooptic modulator that produces alternating left and right cpl in accord with the Pockels effect, whereas all modern CD equipment actually uses elastooptic modulators. In CD (the absorption of cpl by chiral substances, Section 13-3), both right and left cpl pass through a chiral medium and the intensities of the two resultant cpl beams are compared in the detector (Lambert et al., 1976, p. 336).

Let us now consider what happens when linearly polarized light (equivalent to opposite cpl beams of equal intensity) is passed through a sample of a chiral compound containing unequal amounts of the two enantiomers. Let us assume for the moment that the frequency of the radiation is in a region of the spectrum that is free of absorption bands; we will defer discussion of the contrary situation until Section 13-3.

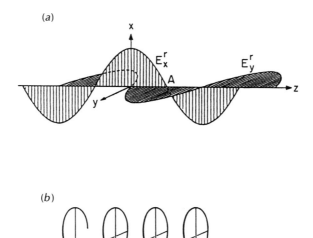

Figure 13.5. (*a*) Right cpl ray (only electric fields are shown); (*b*) The instantaneous electric field is one-quarter wavelength out of phase. [Reprinted with permission from Brewster, J. H. (1967), *Top. Stereochem.*, **2**, 1. Copyright © 1967 John Wiley & Sons, Inc.]

One of the enantiomers interacts with, say, the right cpl beam in such fashion that, within the sample, its velocity is different from that of the corresponding left cpl beam. Both beams are slowed down relative to their (equal) velocities prior to entrance into the sample but to a different extent (anisotropic refraction; the reader should note that the refractive index n for any medium equals c_0/c, where c is the velocity of light in that medium and c_0 is the velocity of light in a vacuum), hence the interaction of the two chiral rays with the chiral sample must have been different. One might say that their interaction was diastereomeric in nature.

If the enantiomer in excess slows down the left cpl beam more than the right beam ($c_L < c_R$, hence $n_L > n_R$), then the sample is defined as being dextrorotatory. In a time-dependent view toward the light source, the linearly polarized light resulting from addition of the two cpl beams appears clockwise rotated relative to the incident beam; α is positive (Fig. 13.6). The difference in light velocity corresponding to a difference in refractive index is given by the Fresnel equation, Eqs. 13.2 or 13.3

$$\alpha = (n_L - n_R)\pi\ell/\lambda_0 \qquad \text{(in rad)} \qquad (13.2)$$

or

$$\alpha = (n_L - n_R)1800\ell/\lambda_0 \qquad \text{(in deg)} \qquad (13.3)$$

where n_L and n_R are the indexes of refraction of the left and right cpl beams in the medium, ℓ is the path length [in centimeters (Eq. 13.2); in decimeters (Eq. 13.3)], and λ_0 is the wavelength in vacuum of the light beam (in centimeters); since $360° = 2\pi$ rads and $1\,\text{dm} = 10\,\text{cm}$, Eqs. 13.2 and 13.3 are equivalent (IUPAC, 1986). When $n_R \neq n_L$, the medium is said to be *circularly birefringent* and to exhibit *optical activity*. As an example of the magnitude of the refractive index difference, consider that at 589 nm (D line of sodium), optically active 2-butanol exhibits a rotation of $\alpha = 11.2°$ without solvent at 20°C (path length 1 dm). We calculate, $n_L - n_R = \Delta n = 11.2(589 \times 10^{-9})/1800(0.1) = 36.6 \times 10^{-9}$. This very small number represents $\Delta n/n = (36.6 \times 10^{-9}/1.3954) \times 100$ or $2.6 \times 10^{-6}\%$ of the value of the isotropic refractive index of 2-butanol.

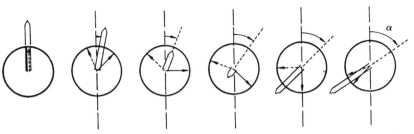

Figure 13.6. The origin of optical activity. Rotation of linearly polarized light by superposition of left (---) and right (—) cpl. Time dependent view toward the light source from a given point $z = z_0$ (left to right). As shown, the rotation is positive (dextrorotation). [Adapted with permission from Snatzke, G., *Chem. Unserer Zeit.*, **15**, 78 (1981).]

A classical and qualitative interpretation of how polarization affects the passage of light through a sample all of whose molecules have the same chirality sense is given in the Feynman lectures (Feynman et al., 1963, p. 33-6). Assume the sample to consist of molecules in the shape of a helix. [The use of helical molecules in this analysis is in keeping with the view that chiral molecules in general may be envisaged as molecular helices (Izumi and Tai, 1977, p. 271; see also Brewster, 1967)]. Suppose that light falls on such a molecule with the beam linearly polarized along the long axis of the molecule. Although the effect of the beam is independent of the orientation of the molecules, for the sake of simplicity, we have chosen the x direction as that both of the polarization and of the long axis of the helix (Fig. 13.7).

The electric field E (cf. Fig. 13.2) exerts a force on charges (essentially on the electrons only since its effect on the heavier atomic nuclei is negligible) in the helical molecules. The up and down motion of the charges generates a current in the direction of the polarization (the x axis). The current in turn generates an electric field that is polarized in the same direction as that of the impinging radiation. Thus the radiation is absorbed and reemitted.

In addition, in a three-dimensional molecule, such as the illustrated helix, electrons driven by the field E_x are constrained to move also in the y direction. Much of the induced electric field generated by the current moving in the y direction produces no radiation since the field arising from current moving in the $+y$ direction is canceled by that from current moving in the $-y$ direction on the opposite side of the helix. However, as the light propagates along the z axis, the fields E_y resulting from the transverse electron motion in the molecule, do not travel together since they are separated by the cross-sectional distance A corresponding to the distance across the spiral. The delay equals A/c (in seconds), where c is the speed of light in the medium. This delay results in a phase difference of $\pi + \nu A/c$, where ν is the frequency of light, such that the E_y fields do not exactly cancel each other.

The upshot of the foregoing analysis is that although the impinging radiation is entirely polarized in the x direction, the emerging radiation has a small component polarized in the y direction. The resultant net polarization is tilted away (rotated) from the x plane in a sense ultimately determined by the handedness of the interacting molecules. That is the origin of the optical rotation. An earlier interpretation due to Stark (1914) bearing some similarity to that of

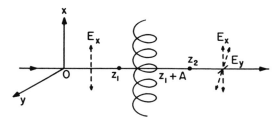

Figure 13.7. Interaction of a beam of light linearly polarized in the x direction with a chiral molecule. [Feynman, R. P., Leighton, R. B., Sands, M. (1963), *The Feynman Lectures on Physics*, p. 33-6. Copyright © 1963 by California Institute of Technology. Adapted with permission of Addison-Wesley Publishing Company, Inc.]

Feynman et al. (1963) is summarized in the book by Lowry (1964, p. 373; see also Kauzmann, 1957, p. 617; Wheland, 1960, p. 204; and Izumi and Tai, 1977, p. 273).

It is evident that replacement of the enantiomerically homogeneous sample in the above experiment by one containing an equal number of *R* and *S* molecules of a given structure, that is, by a racemate, would lead to zero rotation. Each molecule individually rotates the plane of polarization from that of the incident radiation. However, the net rotation is zero since the number of molecules rotating the plane in one sense equals that rotating the plane to the same extent but in the other direction. Intermediate situations obtain for mixtures of chiral molecules containing an excess of one enantiomer. The quantitative treatment of the optical rotation of such mixtures is taken up in Section 13-5.a.

How does the foregoing analysis treat the interaction of *achiral* molecules with polarized radiation? Linearly polarized radiation encounters molecules in chiral conformations or chiral vibrational states such that its plane of polarization is rotated. Virtually all formally achiral molecules are subject to such interaction. However, bulk samples *in the liquid or gaseous states* possess molecules in every conceivable conformation and vibrational state such that the rotation generated by a molecule in one of these conformations or states is canceled by another having the mirror-image conformation or state (see Fig. 2.9 for an example), thus leading to a net zero rotation just as in the case of a racemate. Mislow and Bickart (1976/1977) called the statistical cancellation of local chiral effects *stochastic achirality*.

> Some textbooks speak of the cancellation of rotations due to different *orientations* of achiral molecules with respect to the light beam. However, as has been pointed out above, in isotropic media (gas phase or solution), molecular orientation is irrelevant in the interaction between radiation and matter (Feynman et al., 1963, p. 33-6).

This interpretation of the optical inactivity of achiral molecules is supported in part by the finding that optical activity in the solid state is observed even with certain achiral compounds: Those such as quartz (SiO_2) that crystallize in enantiomorphic space groups (Jacques, Collet, and Wilen, 1981, p. 8), and those that crystallize in chiral conformations, for example, *meso*-tartaric acid (Bootsma and Schoone, 1967).

We conclude this section with a succinct but clear statement by Brewster (Brewster and Prudence, 1973) indicating when one may expect to observe optical activity in general:

> Optical rotation results when the interaction of light with matter produces electric and magnetic moment changes that are not at right angles to one another. Dextrorotation results when these effects are parallel, levorotation when they are antiparallel. In general, such behavior is caused by chiral features of molecular architecture that impose helicity on the motions of electrons: right-handed helical motion gives dextrorotation at long wavelengths and positive Cotton effects at absorption bands.

[This statement is based on the work of Gibbs (1882), Drude (1892), Rosenfeld (1928), Condon (1937), and Condon, Altar, and Eyring (1937)]. We will illustrate and elaborate on this statement, particularly in Section 13-4.a.

b. Optical Rotatory Dispersion

The measurement of specific rotation, $[\alpha]$ (Section 1-3), as a function of wavelength is called optical rotatory dispersion (ORD). In the absence of significant absorption by the analyte, one observes monotonic changes in $[\alpha]$ as a function of wavelength in accord with the Fresnel equation (Eqs. 13.2 and 13.3). More specifically, the absolute value of the rotation increases as the wavelength decreases, that is, as the wavelength tends toward the UV (Fig. 13.8).

A quantitative relation between the wavelength and the molar rotation, $[\Phi]$ (Section 1-3), applicable to transparent spectral regions, was developed by Drude (1900). The Drude equation is an expansion of which the first term is given by $[\Phi] = K/(\lambda^2 - \lambda_0^2)$ where K is a constant, λ is the wavelength of the incident light, and λ_0 is the wavelength of the nearest absorption band (Djerassi, 1960, p. 5). Plots of $1/[\Phi]$ against λ^2 often give straight lines (Lowry, 1964, Chapter 9). The Drude equation remains useful mainly as a means of estimating $[\Phi]_2$ at an inaccessible wavelength λ_2 from a measured $[\Phi]_1$ value at λ_1, both in a transparent region of the spectrum (for the utility of the Drude equation, see also Djerassi, 1960, p. 5).

It has been known for a long time that the monotonic increase in rotation gives way to an anomaly (see below) in the vicinity of an electronic absorption band. It is therefore standard procedure to compare the isotropic (ordinary

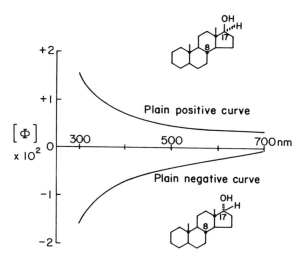

Figure 13.8. Optical rotatory dispersion in transparent regions of the spectrum. [Crabbé, P. (1967) in *Optical Rotatory Dispersion and Circular Dichroism in Organic Chemistry*, Snatzke, G., Ed., p. 2. Copyright © 1967 Heyden & Son. Adapted with permission of John Wiley & Sons, Ltd.]

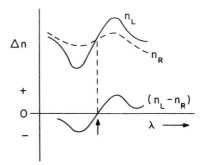

Figure 13.9. Dependence of Δn on wavelength. Anomalous dispersion. [Adapted from Mislow, K. (1965), *Introduction to Stereochemistry*, W. A. Benjamin, New York, p. 85.]

UV/vis) absorption spectrum with the ORD curve (the combined monotonic and anomalous change in optical rotation with wavelength) even though the latter is a measure of refraction and not of absorption.

Figure 13.9 illustrates the fact that anomalous ORD arises from the superposition of two anomalies, namely, those in the indexes of refraction of left and right circularly polarized light beams passing through the sample.

Typical "simple" ORD curves, such as that of D-camphor (Fig. 13.10), are similar in shape to the sum of the refractive index curves $(n_L - n_R)$ shown in Figure 13.9, consistent with the relationship between α and Δn (Eq. 13.3), and the optical null occurs at precisely the same wavelength. The crossover point closely corresponds to the ε_{max} of the UV spectrum provided that the latter does not reflect a superposition of several close lying electronic transitions.

In contrast to a plain curve (Fig. 13.8), an anomalous ORD curve exhibits both a maximum and a minimum, and a point of inflection. (On occasion, one or more of these features may be hidden.) The anomaly is called the Cotton effect (CE) (Cotton, 1895). (A simple ORD curve is one that shows a single CE.) Figure 13.10 shows a single CE curve exhibiting also a change in sign, and illustrates terms that are often used to describe curves without reproducing them. The curve is called *positive* when the rotation first increases as the wavelength decreases; conversely, it is called *negative* when the rotation magnitude first decreases when going towards shorter wavelengths (Djerassi and Klyne, 1957b). The molar amplitude of the curve is given by the following relation (Eq. 13.4):

$$a = \frac{|[\Phi]_1| + |[\Phi]_2|}{100} \tag{13.4}$$

in which $|[\Phi]_1|$ and $|[\Phi]_2|$ are the absolute values of the molar rotations at the first and second extrema (peak and trough, respectively). The wavelength difference between two extrema is called the breadth; the breadth can vary strongly with the wavelength at which the CE occurs.

The wavelength of the CE coincides with the λ_{max} (UV/vis) of an electronic transition (Fig. 13.10), that is, the optical null $[\Phi] = 0$ at 294 nm lies close to, but not precisely at, the same wavelength as the UV λ_{max} (292 nm). The UV spectral band shape is not strictly symmetrical, nor is the shape of the ORD curve. Both curves are influenced slightly by transitions lying at shorter wavelengths. This

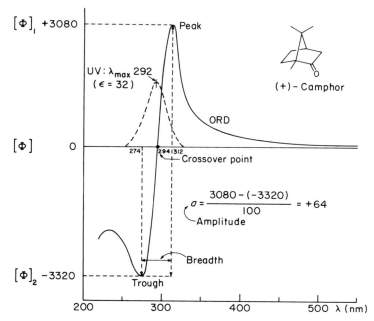

Figure 13.10. Anomalous ORD curve of (1R,4R)-(+)-camphor (—) exhibiting a single positive CE. Nomenclature of ORD curves; the crossover point at 294 nm is an "optical null," $[\Phi] = 0$. The isotropic UV spectrum of camphor (---) is superposed on the ORD curve. [Adapted with permission from Crabbé, P. (1972), *ORD and CD in Chemistry and Biochemistry*, Academic Press, Orlando, FL., p. 6.]

influence is sufficient to cause the difference cited. In addition, $[\Phi] = 0$ does not lie precisely at the midpoint between peak and trough either in intensity or in wavelength.

Figure 13.11a illustrates a less symmetrical single Cotton effect ORD curve in which the sign of the first CE encountered as the wavelength is scanned toward

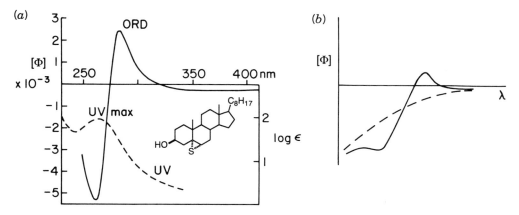

Figure 13.11. (a) Single Cotton effect ORD curve: positive CE with negative rotation in the visible (UV $\lambda_{max} = 264$ nm) [Adapted with permission from Djerassi, C. (1964), *Proc. Chem. Soc., London*, 315. Copyright © Royal Society of Chemistry, Science Park, Milton Road, Cambridge CB4 4WF, UK.]; (b) Shape of an ORD curve that stems from superposition of a positive CE (—) near 264 nm and of a negative (background) CE (--) lying at shorter wavelength. [Adapted with permission from Snatzke, G., *Chem. Unserer Zeit.*, **15**, 78 (1981).]

the UV is opposite to that of $[\alpha]_D$. Figure 13.11 implies that at least one other CE is to be found at shorter wavelengths. The shape of the ORD curve observed is a consequence of superposition of the CE shown and of the "background" curve shown in Figure 13.11b; only the plain part of the negative curve is shown (---) because the corresponding CE is at too short a wavelength to be experimentally manifested.

The sign of the ORD curve reflects the configuration of the chromophore, or of the stereogenic centers that perturb the chromophore, even in the presence of other stereocenters. For a simple example that illustrates the powerful advantage of ORD over the specific rotation, see Figure 13.8. A similar advantage applies to CD (see below). The nearly mirror-image, plain, ORD curves in Figure 13.8 are generated from two diastereomeric sterols whose configuration differs only at the carbinol stereocenter. The configurational relationship between the two stereoisomers can be inferred only from the ORD curve, whereas it could not have been gleaned safely from the fact that the two isomers have oppositely signed $[\alpha]_D$ values. Both ORD and CD serve as the principal spectroscopic probes of absolute configuration of stereocenters in chiral molecules and, in a complementary sense, also of conformation.

More complicated ORD spectra (e.g., Fig. 13.12) exhibit multiple CEs that are close together in wavelength. Such spectra result from electronic transitions of several chirotopic chromophores in a molecule (see below) or from vibrational transitions within the main electronic transition.

Commercially available recording ORD instruments (spectropolarimeters) became available in the mid-1950s and a dramatic increase in the number of publications concerned with chiroptical properties followed in which Djerassi at Wayne State and Stanford Universities was the principal protagonist (Djerassi, 1960, 1964, 1990; Snatzke, 1968). For descriptions of the instrumentation, the

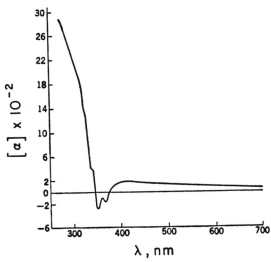

Figure 13.12. Multiple Cotton effect ORD curve of testosterone measured in dioxane (c 0.10) at 25–26°C. [Adapted from Djerassi, C. (1960), *Optical Rotatory Dispersion*, McGraw-Hill, New York, p. 17.]

reader is referred to the reviews by Djerassi (1960, Chapter 3), Woldbye (1967), and Crabbé and Parker (1972). However, as of 1991, commercial ORD instruments have virtually disappeared from the marketplace, being supplanted by CD spectrometers. Polarimeters operable at multiple wavelengths do suffice for some ORD studies; for an example of such usage, see Menger and Boyer (1984). Applications of ORD are discussed in Section 13-4.

13-3. CIRCULAR DICHROISM. ANISOTROPIC ABSORPTION

In addition to the anisotropic refraction of polarized light by chiral matter (circular birefringence), a second chiroptical phenomenon is observed in *nontransparent* regions of the spectrum, namely, CD. The latter phenomenon reflects the anisotropic *absorption* of cpl by chiral samples containing an excess of one enantiomer. Anisotropic absorption, which is also a CE, takes place only in spectral regions in which absorption bands are found in the isotropic UV or visible electronic spectrum (Fig. 13.13). For analogous absorption in the IR region of the spectrum, see Section 13-6.

What is the origin of this absorption? An electronic (or vibrational) transition associated with a chirotopic chromophore in a chiral molecule causes right and left cpl to be absorbed differentially. Provided that the sample contains an excess of one enantiomer, the intensities of the two cpl beams are no longer equal on exiting the sample; the absorbances $A_L \neq A_R$ and $\Delta A = A_L - A_R$ is a measure of the CD. If the molar concentrations are known, then, since $\Delta A = \Delta \varepsilon c \ell$, where c is the concentration in moles per liter (mol L^{-1}) and ℓ is the path length in centimeters (cm), we may write $\varepsilon_L - \varepsilon_R = \Delta \varepsilon$, where ε_L and ε_R are the molar absorption coefficients for left and right cpl, respectively, at the absorption wavelength.

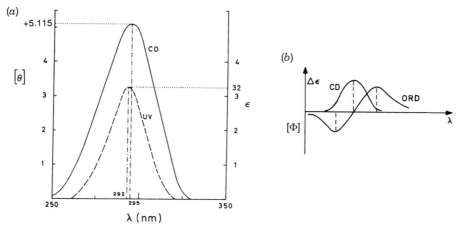

Figure 13.13. (*a*) UV (electronic absorption, EA) and CD (positive CE) spectra of (1*R*,4*R*)-(+)-camphor [Adapted with permission from Crabbé, P. (1972), *ORD and CD in Chemistry and Biochemistry*, Academic Press, Orlando, FL., p. 6.]; (*b*) CD and ORD spectra describing the positive CE of a single electronic (isolated) transition. [Adapted with permission from Snatzke, G., *Chem. Unserer Zeit.*, **15**, 78 (1981).]

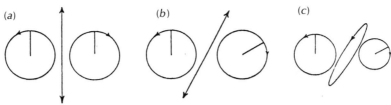

Figure 13.14. Elliptically polarized light. (*a*) Equal velocities of transmission give *no rotation*; (*b*) unequal velocities of transmission give *rotation*; (*c*) unequal velocities and unequal absorptions give *rotation* and elliptical polarization. [Reprinted with permission from Lowry, T. M. (1964), *Optical Rotatory Power*, Dover, New York, p. 152.]

The sign of $\Delta\varepsilon$ defines the sign of the CD (Fig. 13.13*a*, e.g., illustrates a positive CD curve). The signs of the CD curve and that of the corresponding ORD curve in the region of an anomaly are the same (rule of Natanson and Bruhat; Lowry, 1964, p. 427). This correspondence is easily perceived if only one transition is present in a given wavelength range (Fig. 13.13*b*). At a given wavelength, both phenomena, ORD and CD, reflect the interaction of polarized light with the same chirotopic chromophore.

Since the absorbance of left and right cpl by the sample is unequal, $A_L \neq A_R$, and $A = \log(I_0/I)$ (I_0 is the intensity of the impinging light while I is that of the transmitted light), the two cpl components are now of unequal intensity ($I_L \neq I_R$). During its passage through the sample in a region where absorption takes place, the incident linearly polarized light is converted into elliptically polarized light, that is, the resultant electric field vector traces an elliptical path. This conversion is schematically represented in Figure 13.14.

Elliptically polarized light as defined in Figure 13.15 is the most general form

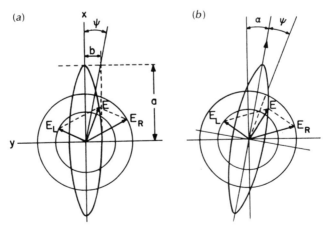

Figure 13.15. Elliptically polarized light (*a*) in a region where $\alpha = 0°$ and (*b*) in a region where α = positive viewed toward the light source. Electric field vectors $\mathbf{E}_R > \mathbf{E}_L$ are both smaller than $\mathbf{E}_0$ (incident cpl); the resultant vector $\mathbf{E}$ traces an elliptical path. The ellipticity angle ψ is given by the geometric construction: arc tangent of minor axis b/major axis a, where $a = \mathbf{E}_R + \mathbf{E}_L$ and $b = \mathbf{E}_R - \mathbf{E}_L$. Since by definition, $\Delta\varepsilon = \varepsilon_L - \varepsilon_R$, ψ is positive if $\varepsilon_L > \varepsilon_R$. [Adapted from Velluz, L., Legrand, M., and Grosjean, M. (1965), *Optical Circular Dichroism*, Verlag Chemie, Weinheim, pp. 22–23.]

of polarized light; linear and circular polarization are special cases of elliptical polarization. The eccentricity of the ellipse $[(a-b)/a]$ is 1 for linearly polarized light ($b=0$), and 0 (zero) for circularly polarized light ($a=b$). The major axis of the ellipse traces the angle of rotation α and the ellipticity ψ is defined by $\tan \psi = b/a$, where b and a are the minor and major axes, respectively, of the ellipse that characterizes the elliptically polarized light (Fig. 13.15).

By analogy with rotation α, one may define a specific ellipticity $[\psi]$ and a molar ellipticity $[\theta]$ (Eqs. 13.5 and 13.6, respectively):

$$[\psi] = \frac{\psi}{c\ell} \quad \text{in } 10^{-1} \text{ deg cm}^2 \text{ g}^{-1} \tag{13.5}$$

$$[\theta] = \frac{[\psi]M}{100} \quad \text{in } 10 \text{ deg cm}^2 \text{ mol}^{-1} \tag{13.6}$$

where the symbols c, ℓ, and M have the same meanings as they do in the definitions of $[\alpha]$ and $[\Phi]$ (Section 1-3; see also p. 1073).

When the ellipticity is small, which is a common occurrence, $\tan \psi \cong \psi$ and the latter is proportional to ΔA (Eqs. 13.7 and 13.8).

$$\psi = 32.982 \, \Delta A = 32.982 \, \Delta \varepsilon c' \ell' \tag{13.7}$$

$$[\theta] = 3298.2 \, \Delta \varepsilon \tag{13.8}$$

In Eq. 13.7, the units of c' are moles per liter (mol L^{-1}) and those of ℓ' are centimeters (cm).

Ellipticities and molar ellipticities are dependent on conditions of measurement. Consequently, the temperature, wavelength, and concentration of the sample should always be specified. While it is possible to measure the ellipticity ψ directly (ellipsometry), this is difficult to do in practice and, due to the incorporation of a photoelastic modulator to generate the cpl (cf. Section 13-2.a), all current commercial CD spectrometers measure ΔA even if they are calibrated in ellipticity (see Eq. 13.7). For descriptions of CD instrumentation, see Velluz et al. (1965, p. 21); Woldbye (1967); Crabbé and Parker (1972); Lambert et al. (1976, p. 333); and Mason (1982, p. 26).

The correspondence of CD maxima with the wavelength of anomalous ORD crossover and the correspondence in sign of the two phenomena in a given sample suggests that it may be possible to calculate a CD curve from the ORD spectrum, and vice versa. This is, in fact, possible through application of the Kronig–Kramers theorem, a general relationship between absorption and refraction (Djerassi, 1960, p. 159; Moffitt and Moscowitz, 1959; Moscowitz, 1961; and Emeis, Oosterhoff, and DeVries, 1967), by means of equations having the following form (Eqs. 13.9 and 13.10):

$$[\Phi(\lambda_0)] = 2099.6 \int_0^\infty \Delta \varepsilon(\lambda) \, \frac{\lambda}{(\lambda_0^2 - \lambda^2)} \, d\lambda \tag{13.9}$$

$$\Delta \varepsilon(\lambda_0) = -\frac{1.9303 \times 10^{-4}}{\lambda_0} \int_0^\infty [\Phi(\lambda)] \, \frac{\lambda^2}{(\lambda_0^2 - \lambda^2)} \, d\lambda \tag{13.10}$$

where $[\Phi(\lambda_0)]$ and $\Delta\varepsilon(\lambda_0)$ are ORD and CD values, respectively, at a specified wavelength λ_0 (IUPAC, 1986).

Less rigorous, but still useful, expressions (Eqs. 13.11 and 13.12) have been proposed that relate the molar amplitude a of an ORD curve (Fig. 13.10) to the intensity of the CD curve, $\Delta\varepsilon$ (Mislow, 1962):

$$a = 40.28\,\Delta\varepsilon \qquad (13.11)$$

and similarly, in view of Eq. 13.8, the ORD molar amplitude is related to the molar ellipticity $[\theta]$ of the CD spectrum (Fig. 13.13).

$$a = 0.0122\,[\theta] \qquad (13.12)$$

These equations were derived for the $n-\pi^*$ carbonyl transition; they should be used with caution with other chromophores (Crabbé, 1967).

Since the two types of measurement, ORD and CD, would appear to provide complementary information, it is fair to inquire into the need for both. Commercial ORD instruments became readily available beginning about 1955 while the first commercial CD spectrometer did not make its appearance until 1960. In the interval since 1960, both types of measurements have been reported, often for the same substance.

Circular dichroism spectra are inherently simpler to interpret; the key point is that CD is nonzero only in the vicinity of an EA band (Figure 13.16). Thus, a typical CD spectrum is less cluttered, bands are better separated, and comparison of anisotropic absorption bands with EA is more straightforward than with ORD. In Figure 13.16, the rotation observed (plain curve) in the region between 350 and

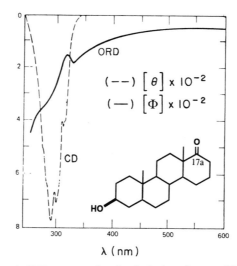

Figure 13.16. The CD and ORD curves of a simple hydroxyketone. The chirotopic chromophore responsible for the CE near 290 nm is the carbonyl group at position 17a. The shoulders in the CD band are due to vibrational fine structure. [Adapted with permission from Crabbé, P. and Parker, A. C. (1972), in Weissberger, A. and Rossiter, B. W., Eds., *Physical Methods of Chemistry*, Part IIIC, *Techniques of Chemistry*, Vol. 1, p. 209. Copyright © 1972 John Wiley & Sons, Inc.]

600 nm is a background or skeletal effect due to CEs below 250 nm. Even the sign of the CE at about 290 nm is nearly masked by the strong skeleton effect. This is not a problem with CD. Therefore, where a choice is possible, CD measurements are preferred over ORD. Consistent with the disappearance of commercial ORD spectrometers from the market, CD has by now essentially replaced ORD as the main chiroptical technique in the study of chiral substances (Scopes, 1975).

However, since ORD of optically active compounds is observable over the entire wavelength range (Fig. 13.16), this technique can provide information about CEs that are outside the range of commercial CD spectrometers. Plain ORD curves, especially in regions where $[\theta] = 0$, can reveal the configuration of a chromophore as illustrated in Figure 13.8. For the same reason, ORD may also be useful in the measurement of chiroptical properties of compounds having low optical activity (Lambert et al., 1976, p. 342) or when the sign of an ORD curve is opposite to that of a small $[\alpha]_D$ as in Fig. 13.11a.

Configurational assignments based on comparisons of specific rotations measured at single wavelengths far removed from CEs can be equivocal since it is not possible to tell if a dispersion (ORD) curve changes sign at lower wavelength. As an example, whereas the $[\alpha]_D$ values of the meta and para isomers of α-(iodophenoxy)-propionic acids are positive, that of the ortho isomer is negative. The configuration of the $(-)$-o isomer is shown to be the same as those of the $(+)$-m and $(+)$-p isomers, since the shapes and signs of the three ORD curves are alike [the ORD curve of the $(-)$-o isomer crosses the zero rotation axis below 350 nm] (Sjöberg, 1959).

13-4. APPLICATIONS OF OPTICAL ROTARY DISPERSION AND CIRCULAR DICHROISM

a. Determination of Configuration and Conformation. Theory

The principal application of CD and ORD is in the assignment of configuration or conformation (Books: cf. Nakanishi, et al., 1994; Purdie and Brittain, 1994). In principle, the theories of optical activity (see below) should permit one to calculate ab initio the magnitude and sign of a CE from the constitution, the relative and absolute configuration of stereocenters, and the conformation of a given compound. Conversely, determination of the configuration through analysis of chiroptical spectra should be possible by application of these theories. In practice, however, we are not at the point where the aforementioned ab initio calculations are possible for any but small molecules, for example, trans-1,2-dimethylcyclopropane (Bohan and Bouman, 1986). In contrast, both procedures are possible by application of semiempirical methods based on three mechanisms, to be described in Section 13-4.c, that account for the generation of "rotational strength." Such calculations are not routinely undertaken; they are, in any event, beyond the purview of this book.

> The semiempirical calculations reproduce the signs and intensities of the measured (experimental) CEs and, in some cases, provide full theoretical CD spectra. In the case of conformationally mobile systems, such calculations can confirm the assignment of the major conformer responsible for the CE. The necessary data, for

example, electric transition moments, **μ** (see below), required for the semiempirical calculations are either obtained from experimental spectra or calculated, again by semiempirical methods; the necessary geometric coordinates are taken from X-ray spectra and/or calculated, for example, by molecular mechanics methods. For examples of such semiempirical calculations, see Rizzo and Schellman (1984); Roschester, Sandström, et al. (1987); Rashidi-Ranjbar, Sandström, et al. (1989); Stevens and Duda (1991); Harada, Khan, et al. (1992).

A "classical" theory, due to DeVoe (1965; see Charney, 1979, p. 268), continues to be applied to the determination of configuration of small and even some large molecules from CD data (Rosini, Giacomelli, and Salvadori, 1984; Rosini, Salvadori, et al., 1985; Pirkle, Salvadori, et al., 1988; Rosini, Zandomeneghi, and Salvadori, 1993). For a recent application of this theory, to 3-(1-naphthyl)-phthalide, see Salvadori, Bertucci, and Rosini (1991).

The following qualitative treatment is intended to be just sufficient to show how the sign of a CE of a specific transition may be related to the absolute configuration of the molecule giving rise to the CD spectrum. With this background, empirical or semiempirical generalizations called sector and helicity rules may be rationalized and applied. Sector rules are widely used in the assignment of configuration by inspection of CD spectra of homologous and analogous compounds bearing the identical functional group. We will see that what is required is a knowledge of the nature of the transition (electronic or vibrational), since only comparable transitions may be so treated, the symmetry properties of the functional group undergoing excitation, and some idea of what structural features give rise to absorption of anisotropic radiation, that is, the origin of the rotational strength of a CD band (see below).

Typical anomalous ORD and electronic CD arise in the same regions of the spectrum in which isotropic radiation excites electrons formally associated with atoms, groups, or moieties in achiral *or* chiral molecules (e.g., C=O, –CH=CH–C=O, C_6H_5). When polarized radiation is used, similar excitation in the identical spectral regions give rise to absorption whose intensity differs for left and right cpl. Because of this relationship between isotropic and anisotropic absorption (cf. Section 13-3), theoretical treatments of chiroptical effects are little concerned with the prediction of wavelengths at which CEs are observed.

In general, any two of the three structural components, constitution, configuration, and conformation, must be known if one is to infer the third from chiroptical spectra (CD or ORD). All three structural components must be known if one is to predict ORD/CD. Hence, as a rule, for chiral molecules possessing torsional degrees of freedom, it is not possible to deduce information about both configuration and conformation simultaneously from chiroptical data. In this connection, the Octant rule (see below) has been called a "one-way rule," that is, one for which prediction of the sign of a CE *from* a known configuration is possible, but for which the converse, assignment of configuration from the CE without further structural information is not (Dugundji, Marquarding, and Ugi, 1976).

The term "absolute conformation" had been suggested to emphasize the fact that the CE signals both the configuration at a chiral center *and* the conformation of the chiral molecule being analyzed (Mislow, Djerassi, et al., 1962; Snatzke, 1979a,c).

For example, the CD of (2S)-2-benzamido-1-propyl benzoate **1** (Fig. 13.17) in the vicinity of 227.5 nm (benzoyl chromophore) depends not only on the configuration at the stereogenic atom C(2) but also on whether conformer *a* or *c* is the dominant one present. The CD of these two conformers would be expected to be opposite in sign since the torsional angle ω between the two benzoyl groups responsible for the CEs (exciton splitting; see Section 13-4.d) reverses in sign in going from *a* to *c*. Parenthetically, conformer *b* does not significantly contribute to the CD since the transition moments (see below) of the two benzoyl groups, being oriented orthogonally, cancel one another (Kawai et al., 1975; Harada and Nakanishi, 1983, p. 159).

While it is important to emphasize this limitation in the analysis of CD spectra, it has been subsequently pointed out that "absolute conformation" is subsumed under the term "absolute configuration" (Hayes, Mislow, et al., 1980; for the definition of the latter term, see Chapter 5). Consequently, we do not use the term absolute conformation in this book.

Let us digress for a moment to recall what is required for electromagnetic radiation, whether isotropic or anisotropic, to be absorbed by matter: (a) the energy $h\nu$ of the impinging photon must correspond to the energy difference ΔE between the pertinent ground- and excited-state orbitals, and (b) the excitation (electronic) must be associated with the migration of charge thus generating a momentary electric dipole that is usually called the *electric transition moment* $\boldsymbol{\mu}$; a similar argument pertains to vibrational spectoscopy. The electric transition moment is related to the area of the UV absorption band by Eq. 13.13

$$D = \mu^2 = 9.188 \times 10^{-39} \int (\varepsilon / \lambda) \, d\lambda \qquad (13.13)$$

where D is called the dipole strength. In practice, the integrated area under the band may be approximated by measuring ε_{max}, λ_{max}, and $\Delta\lambda$ (the width of the band at $\varepsilon_{max}/2$) if the band has a Gaussian shape (Eq. 13.14).

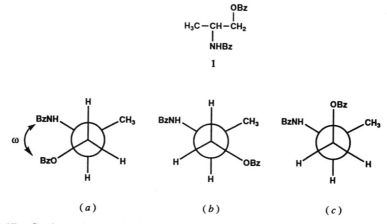

Figure 13.17. Conformations of (2S)-2-benzamido-1-propyl benzoate (Bz = C_6H_5CO) about the C(1)–C(2) bond.

$$\varepsilon_{max}\Delta\lambda/\lambda_{max} \approx \int (\varepsilon/\lambda)\,\Delta\lambda \qquad (13.14)$$

Thus, a measurable "substance property" (absorption intensity) is related to a theoretically calculable "molecular property", namely, the transition moment.

While a precise calculation of $\boldsymbol{\mu}$ is beyond the purview of this book, its magnitude and direction may be estimated by means of a qualitative molecular orbital (MO) method or "recipe" described by Snatzke (1979a,b; 1981). Formal multiplication of the two molecular orbitals between which the transition occurs $\psi_1 \times \psi_2$ gives an electron density (or probability) ψ^2. Where ψ^2 is positive, the charge (developed during the transition) is negative; a positive charge correlates with a negative ψ^2. This sign inversion does not follow from any requirement of physical laws; the inversion is made so as to conform with the deliberate choice of $\Delta\varepsilon = \varepsilon_L - \varepsilon_R$ and not $\varepsilon_R - \varepsilon_L$; q.v. Section 13-3. When the centers of gravity of the charges are properly oriented, the "multipole" (see below) has the character of a dipole. In the latter case, the transition is electrically allowed and the direction of the transition moment may be aligned with a line connecting the centers of gravity of the positive and negative charges (Fig. 13.18). The foregoing argument accounts adequately for the absorption of radiation by chiral as well as achiral molecules. This electric interaction is dominant even when anisotropic radiation interacts with chiral molecules.

Circular dichroism requires not only that charge displacement take place during an electronic transition, but charge rotation as well, that is, the excited electron must be constrained to rotate or to be excited along a helical path (recall the definition of Brewster at the end of Section 13-2.a). The rotation of electric charge creates a magnetic field whose strength and direction may be described vectorially by a *magnetic transition moment* **m**. Note that since no charge rotation takes place during the $\pi-\pi^*$ transition sketched in Figure 13.18, there is no magnetic transition moment associated with this transition ($\mathbf{m} = 0$).

Let us now consider an $n-\pi^*$ transition of a ketone (Fig. 13.19). The relative disposition of the atomic orbitals is now such that there is much less overlap than in the $\pi-\pi^*$ case (Fig. 13.19a). Formal multiplication of the molecular orbitals

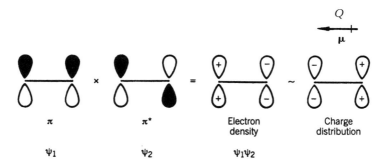

Figure 13.18. Formal multiplication of molecular orbitals for a typical electronic $\pi-\pi^*$ transition (e.g., in alkenes or ketones). Revelation of the direction of the electric transition moment $\boldsymbol{\mu}$ from the dipole character of the charge distribution. [Adapted with permission from Snatzke, G. (1979b), *Pure Appl. Chem.*, **51**, 769.]

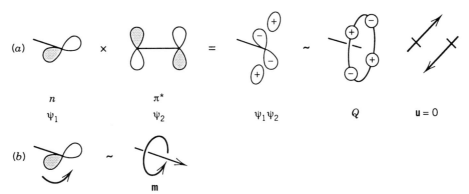

Figure 13.19. Transition moments for $n-\pi^*$ transition. (a) Electric and (b) magnetic transition moments. [Adapted with permission from Snatzke, G., *Angew. Chem. Int. Ed. Engl.*, **18**, 363 (1979a).]

yields an electron probability corresponding to a quadrupole Q that may be seen to comprise two partial dipoles. Since these dipoles compensate each other exactly, there is no electric transition moment associated with the $n-\pi^*$ transition ($\mu = 0$).

However, we see that the transition is associated with the rotation of electronic charge, this being equivalent to the generation of a magnetic transition moment $\mathbf{m}$. Rotation of the applicable orbital lobe through the smaller of the two possible angles (counterclockwise in the example) is carried out so as to superimpose like phase signs. Application of the "right-hand rule" to the rotation of the charge (circular current) reveals the direction of $\mathbf{m}$: If the current flows in the direction of the curved fingers of the right hand, then the outstretched thumb points in the direction of $\mathbf{m}$ (Fig. 13.19b). A cautionary note is in order here, however. The absolute directions of the two transition moments, μ and $\mathbf{m}$, have no physical significance since they derive from the arbitrary selection of signs (or phases) of the atomic orbital lobes chosen in the foregoing analysis. What does matter is the relative orientation of the two transition moments since it is the magnitude of the angle between the two moments that determines the sign of the CE.

For an $n-\sigma^*$ transition, similar analysis reveals that electronic excitation gives rise to a small but finite value of μ and concurrently a finite $\mathbf{m}$ in a direction perpendicular to that of the magnetic transition moment generated in the $n-\pi^*$ transition.

Just as the dipole strength D is related, by theory, to the electric transition moment (Eq. 13.13), so too one may define a rotational strength R in terms of transition moments. It is given by the scalar product of the electric and the magnetic transition moments.

$$R = \mu \cdot \mathbf{m} = |\mu||m| \cos \beta \qquad (13.15)$$

The significance of Eq. 13.15 is that CD arises within an absorption band only when R is finite: The sign of the CE is positive when the angle β between the transition moments ($\mathbf{m}$ and μ) is acute ($0 < \beta < 90°$) or, in the limiting case, parallel, and it is negative if the angle is obtuse ($90° < \beta < 180°$) or, in the limiting

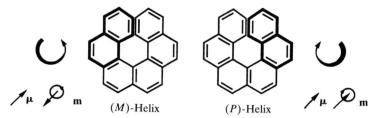

Figure 13.20. Transition moments for hexahelicene. Antiparallel moments for (M) (at left) and parallel moments for (P) helicities (at right) corresponding to $-$ and $+$ Cotton effects, respectively. [Adapted with permission from Nógrády, M. (1981), *Stereochemistry*, p. 74. © 1981 Pergamon Press, Ltd, Headington Hill Hall, Oxford, OX3 0BW, UK.]

case, antiparallel. It follows that when the vectors are perpendicular to each other, the scalar product is zero (cos 90° = 0) and the CE vanishes.

A point that warrants stating explicitly is that although transitions for which μ and/or $\mathbf{m} = 0$ (forbidden transitions) would seem to preclude the measurement of either UV or CD spectra, this is true only to a first approximation. In principle, the ketone $n-\pi^*$ transition is forbidden; however, even in achiral ketones, such as acetone and cyclohexanone, this is not absolutely true. In chiral ketones, even in cases where the obvious stereocenters are far removed from the carbonyl chromophore, the symmetry is sufficiently lifted so that in such instances $\mu \neq 0$. There is a large perturbation in a small quantity, the nominally forbidden μ. This perturbation, together with a small perturbation in the typically large intrinsic magnetic transition moment, can lead to a large value of the rotational strength. Thus, R becomes a very sensitive parameter of structure and of stereochemistry (Charney, 1979, p. 167).

Since both translational and circular electronic displacement are required for CD to be manifest, the electron movement during excitation takes on the character of helical motion. This helical motion is most easily demonstrated in molecules having helical geometries, for example, hexahelicene, wherein the electric transition moment direction is identical for either enantiomer, whereas the magnetic transition moment direction is reversed according to whether the helix is right handed or left handed. In one case, the transition moments are aligned antiparallel thus generating a negative CE, and in the other case, the moments are parallel leading to a positive CE of like magnitude (Fig. 13.20). Prediction of the absolute configuration of the helicenes from chiroptical data requires careful assignment of the transition. The CEs mentioned in the figure refer to the band at about 325 nm, the longest wavelength band of high intensity, presumably corresponding to the $\pi-\pi^*$ transition between the highest occupied molecular orbital (HOMO) and lowest unoccupied molecular orbital (LUMO) (Mason, 1982, p. 69). There is at least one longer wavelength CE having a sign opposite to that at 325 nm (Newman et al., 1967).

While we have not yet so stated explicitly, it may be apparent that R is a measure of the allowedness of a specific CD absorption (a CD band) just as the dipole strength D is a measure of the allowedness of isotropic absorption. The rotational strength of a specific transition, which is a better measure of CD intensity than $\Delta\varepsilon$ alone, or than the amplitude of the ORD curve (Fig. 13.10), is given experimentally by the integrated area under the corresponding CD band (Fig. 13.16).

$$R = 2.297 \times 10^{-39} \int (\Delta\varepsilon/\lambda) \, d\lambda \qquad (13.16)$$

For CD bands having approximately Gaussian shapes, Eq. 13.16 leads to Eq. 13.17

$$R = 2.445 \times 10^{-39}(\Delta\varepsilon_{max}w)/\lambda_{max} \qquad (13.17)$$

where w is the CD width at $\Delta\varepsilon_{max}/2$ (IUPAC, 1986).

A sometimes useful quantity is the g number, the dimensionless ratio of the circular dichroic to isotropic absorbance (Eq. 13.18)

$$g = \Delta\varepsilon/\varepsilon \approx \Delta A/A \qquad (13.18)$$

that was previously called the anisotropy or dissymmetry factor (Kuhn, 1930). The g number is wavelength dependent. Another definition of g stemming from theory (Eq. 13.16 divided by Eq. 13.13) is given by Eq. 13.19 where g^ℓ is the integral value that is constant for each absorption band.

$$g^\ell = 4R/D \qquad (13.19)$$

The magnitude of R is of the order of 10^{-38} cgs units (as a maximum) as is actually found for hexahelicene (Fig. 13.20). A more convenient way of expressing R was suggested by Moffitt, namely, the reduced rotational strength $[R]$ (Eq. 13.20; Moscowitz, 1960)

$$[R] = \frac{100R}{\mu_B \mu_D} \approx 1.078 \times 10^{40}R \qquad (13.20)$$

where μ_B is the Bohr magneton (0.927×10^{-20} cgs) and μ_D is the debye (10^{-18} cgs). Equation 13.21 gives the approximate relation between molar ellipticity and $[R]$ (Gorthey, Vairamani, and Djerassi, 1985):

$$[\theta] = 994[R] \qquad (13.21)$$

b. Classification of Chromophores

The chromophores that are analyzed by means of CD measurements naturally fall into two broad classes as proposed by Moscowitz (1961) on the basis of symmetry considerations (Ciardelli and Salvadori, 1973, p. 13; Snatzke, 1981; Mason, 1982, Chapters 4 and 5):

1. Chromophores that are *inherently achiral* by symmetry, such as the carbonyl and carboxyl groups, the ordinary C=C double bonds (alkenes), and the sulfoxide moiety. Each of these, when considered without substituents, contains at least one mirror plane. Chiral molecules containing inherently achiral (symmetric) chromophores exhibit CEs as a consequence of "chiral perturbations" arising in the chromophore during the electronic

excitation of the latter. These perturbations are exerted by substituents located in the vicinity of the chromophore or by the molecular skeleton itself.

In the preceding statement we have purposely used the language that one finds in most descriptions of such chromophores in the literature. However, as has already been pointed out, since *all* points in a chiral molecule are by definition chirotopic (cf. Section 3-1.a), the notion of an inherently achiral moiety in a chiral molecule is fiction. It might then seem that the proposed bipartite classification of chromophores is invalid. In fact, the classification has an experimental basis (see below); and it may also be retained as a matter of convenience.

2. Chromophores that are *inherently chiral*. This type of chromophore includes compounds, such as the helicenes, in which the entire molecule acts as a single chromophore. Other examples are disulfides, biaryls, enones, cyclic 1,3-dienes (e.g., Chapter 11, p. 731), and strained (twisted) alkenes (for the latter, cf. Section 9-1.d). In all of these, the chirality is built into the chromophore. The rotational strengths R of inherently chiral chromophores (Eqs. 13.16 and 13.17) tend to be very large (see Table 13.2).

When two or more chromophores (identical or not) of either class are in close physical proximity (and have very large values of μ, corresponding to ε of several thousand) yet do not have overlapping orbitals, a third type of CE is observed. For the typical case involving just two such chromophores, two CD bands are

TABLE 13.2. Electronic Transition Magnitudes[a]

	Wavelength $\dfrac{\lambda}{\text{(nm)}}$	UV ε	CD $\Delta\varepsilon$	g Number $g = \dfrac{\Delta\varepsilon}{\varepsilon}$ $(\times 10^3)$	Transition
3-Methylcyclohexanone	298	16	+0.48	30	$n-\pi^*$
	185	1200	+1.0	0.8	$n-\sigma^*$ (3s)
(−)-β-Pinene	200	1.08×10^4	−17.1	2 ⎫	$\pi_x-\pi_x^*$
	181	0.9×10^4	+17.0	2 ⎭	$\pi_x-\pi_y^*$
(+)-Hexahelicene (Fig. 13.20)	325	2.8×10^4	+196	7.0	$\pi-\pi^{*b}$
	244	4.8×10^4	−216	7.7	$\pi-\pi^{*b}$
(+)	247	7×10^4	−245	3 ⎫	
	231	6×10^4	+135	2 ⎭	$\pi-\pi^*$ couplet

[a] Comparison of isotropic (UV) and anisotropic (CD) electronic transition magnitudes in selected chiral molecules. Adapted with permission from Mason, S. F. (1982), *Molecular Optical Activity and the Chiral Discriminations*, Cambridge University Press, Cambridge, UK, p. 49. The data on $(+)_{589}$-hexahelicene (in CHCl$_3$) is taken from Newman et al. (1967) with $\Delta\varepsilon = [\theta]/3300$.

[b] The transitions for hexahelicene are both presumed to be of the $\pi-\pi^*$ type (Moscowitz, 1962).

observed having opposite signs. This phenomenon is called exciton chirality or exciton splitting. The deduction of absolute configurations from such CD spectra is particularly straightforward (Harada and Nakanishi, 1972; cf. Section 13-4.d).

An alternative classification of chromophores, due to Snatzke [Snatzke, 1965, 1967 (p. 208)], was introduced mainly as a way of subdividing the class of inherently achiral chromophores. It is most easily appreciated by examining Figure 13.21. In this example, the molecule is divided into spheres, beginning with the chromophore and moving outward towards the periphery of the skeleton. The principle is that all atoms of all spheres contribute in a nonadditive manner to the CE with the chiral sphere nearest to the chromophore under observation making the largest contribution, even to the point of determining the sign and the magnitude of the CE. For an application of this classification, see Section 13-4.c.

Transitions of inherently chiral chromophores may also be classified on the basis of symmetry and group theoretical considerations. Examination of character tables allow one to identify those point groups having representations that permit electromagnetic transitions having finite rotational strength (Charney, 1979).

The foregoing classification, while based on symmetry considerations, is directly related to the strength of the observed electronic transition. Inherently achiral transitions tend to be weak; they have low rotational strengths. Inherently chiral transitions tend to have high rotational strengths, that is, to produce strong CEs. It has been suggested that high optical activity, for example, an ORD molar amplitude a of the order of 100,000, constitutes strong evidence for the presence of an inherently chiral chromophore in the molecule (Mislow, 1965, p. 65). Circular dichroic spectra exhibiting exciton splitting have the highest amplitudes of the three types of transitions.

It should by now be apparent that rotational strengths, and the more easily observed measures of ORD and CD magnitude (amplitude a, $\Delta \varepsilon$, and molar ellipticity $[\theta]$) of individual transitions, have diagnostic value. It is instructive then to compare the magnitudes of measured CD bands of some specific compounds (Table 13.2). The table clearly reveals that there is a significant gradation in CD intensity ($\Delta \varepsilon$) with differences of the order of 10^2 between an inherently achiral chromophore (the ketone C=O group) and the inherently chiral chromophore comprising the entire skeleton of the helicene. The CD couplet exhibited by the diaminobinaphthyl is attributed to exciton splitting. Regrettably, rotational strength values (R, defined by Eqs. 13.16 and 13.17), cannot be extracted from the tabulated data since information about band widths is not provided. In this connection, it is known that band widths of isotropic and anisotropic absorption bands are not directly comparable even for a given chromophore in the same compound. The CD band widths are typically narrower.

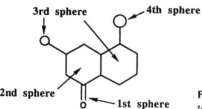

3rd sphere

4th sphere

2nd sphere

1st sphere

Figure 13.21. Definition of spheres in a ketone according to Snatzke. The carbonyl group comprises the first sphere.

Table 13.2 also reveals that all the isotropic transitions are strong and allowed with the exception of the $n-\pi^*$ carbonyl band of the 3-methylcyclohexanone where μ must be close to zero. All the other transitions are said to be electric dipole allowed. For chromophores giving rise to CD, the magnitude of $\mathbf{m}$ is of considerable importance. Recall that, from Eq. 13.15, both μ and $\mathbf{m}$ must be finite for anisotropic absorption to take place.

The magnitude of μ is one debye (10^{-18} cgs units) for an electric dipole allowed transition while $\mathbf{m}$ has a magnitude of the order of one Bohr magneton (0.927×10^{-20} cgs units) for a magnetic dipole allowed transition. Since $R \propto \mu \cdot \mathbf{m}$, the rotational strength is of the order of 10^{-38} cgs units for a transition that is both electric and magnetic dipole allowed. Rotational strength values vary over four orders of magnitude according to the type of chromophore with the inherently achiral type having the smaller R ($\cong 10^{-40}$ cgs). Transitions in hexahelicene are both electric dipole and magnetic dipole allowed and R values for this molecule are of the order of 10^{-38} (Moscowitz, 1960).

The value of g can be calculated from Eqs. 13.13, 13.15 and 13.19:

$$g = \frac{4R}{D} = \frac{4\,|\mu|\,|m|\cos\beta}{\mu^2} \qquad (13.22)$$

hence m/μ provides a measure of the g number. When μ and $\mathbf{m}$ are aligned parallel to one another ($\cos\omega = 0$), $g \cong 10^{-2}$. The value of g also has diagnostic value. Typically, one finds $g = >5 \times 10^{-3}$ for magnetic dipole allowed and electric dipole forbidden transitions, as in the $n-\pi^*$ transition of 3-methylcyclohexanone (Table 13.2). For electric dipole allowed and magnetic dipole forbidden transitions, $g = <5 \times 10^{-3}$ (mostly $\cong 10^{-4}$; Snatzke, 1979a).

The fact that R and g are finite, and hence that CD is observed, even in the case of "forbidden" transitions, is often ascribed to the "stealing" of intensity from allowed transitions (electronic or vibrational). In an alternative interpretation, Moscowitz (1962) pointed out that while the electronic excitations giving rise to the relatively narrow banded CD spectra are usually interpreted as involving only chromophores, the excited chromophoric electrons do, in fact, move over the entire molecule. Accordingly, rotational strength may be induced in the chromophore by a perturbation of the excited electrons in a manner reflecting the chirotopicities of atoms and groups located at some distance to the chromophore. Moreover, it is known that such perturbations have a greater effect on rotational strengths than on dipole strengths. Reiterating for emphasis, circular dichroic spectra are more sensitive to minor structural (i.e., constitutional, configurational, and conformational) changes than are UV spectra.

Let us now apply the Snatzke qualitative procedure to an example to see if we can use it to extract configurational information. This is easiest to do with an inherently chiral chromophore.

The disulfide (–S–S–) chromophore has absorption bands in the vicinity of 300 and 250 nm. It has been known for a long time, on the basis of X-ray data and of theoretical considerations, that in the absence of geometrical constraints, this structural element [torsional barrier 5–15 kcal mol^{-1} (21–63 kJ mol^{-1})] preferentially adopts a skewed conformation with $\omega \cong 90°$. It was pointed out by Beychok (1966), that useful stereochemical information might be extracted from the CD spectra of chiral compounds containing the disulfide chromophore but that evidence relating the "screw sense" of the –S–S– moiety with such spectra was not yet available. Such data were soon provided by Carmack and Neubert (1967) who investigated the CD spectra of compound **2** (Fig. 13.22). On the assumption that the dithiane ring would preferentially adopt a chair conformation, the absolute configuration of compound **2** prepared from (*S,S*)-*trans*-1,2-cyclohexanedicarboxylic acid **B** would be *S,S* as shown in **A**. In **A**, the dithiane is required to adopt an *M* configuration (left-handed helix **C** as viewed along the –S–S– bond); the alternative *P* configuration of –S–S– would involve a twist–boat conformation in the six-membered ring.

The longest wavelength absorption band of **2** at 290 nm is associated with a CD band having $n-\sigma^*$ character. A negative CD band ($[\theta] - 16{,}700$) was found to correspond to the *M* configuration of the disulfide. A second absorption band at 240 nm (CD band at 241 nm), was found to have the opposite sign, that is, a positive CD band was found at 241 nm for the disulfide *M* configuration. These findings were incorporated into an empirical disulfide helicity rule (Charney, 1979, p. 212; Legrand and Rougier, 1977, p. 165). Theoretical calculations that treat the –S–S– moiety as an inherently chiral chromophore have shown that this helicity rule holds only for dihedral angles $\omega < 90°$. At $\omega = 0°$, the CE vanishes and at $\omega > 90°$, the signs are inverted (Linderberg and Michl, 1970).

The above empirical results can be readily rationalized by means of the Snatzke MO approach. The transition responsible for the CD band originates in an MO in which each of the sulfur atomic orbitals (AOs) have n character (Fig. 13.23). First, consider the case in which the interorbital angles are small ($<90°$); combination of the two AOs gives rise to two MOs, n^+ and n^-. The higher energy n^- MO has a nodal plane bisecting the bond between the two sulfur atoms. This MO constitutes the HOMO from which the longest wavelength transition emanates. The unoccupied (virtual) MO that represent the LUMO (transition *target*) has σ^* character. Formal multiplication of the MOs leads to the charge distribution shown in Figure 13.23*b*. The resultant quadrupoles are uncompensated and a net electric transition moment is induced at each sulfur atom. The net resultant moment is a function of the interorbital angle. If the angle is small, the

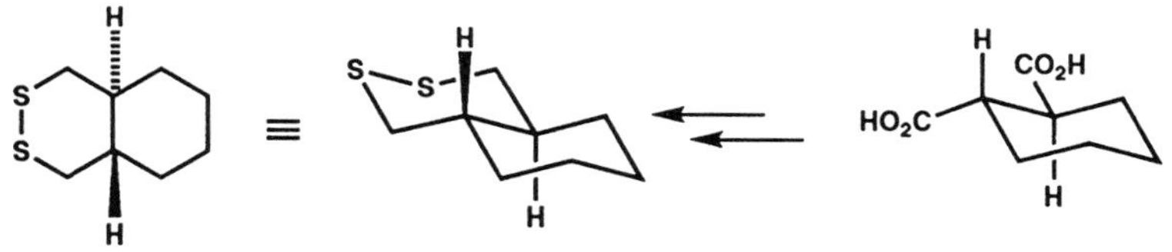

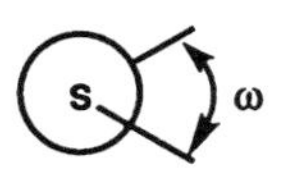

M Configuration

Figure 13.22. Conformation of (9*S*,10*S*)-(−)-*trans*-2,3-dithiadecalin.

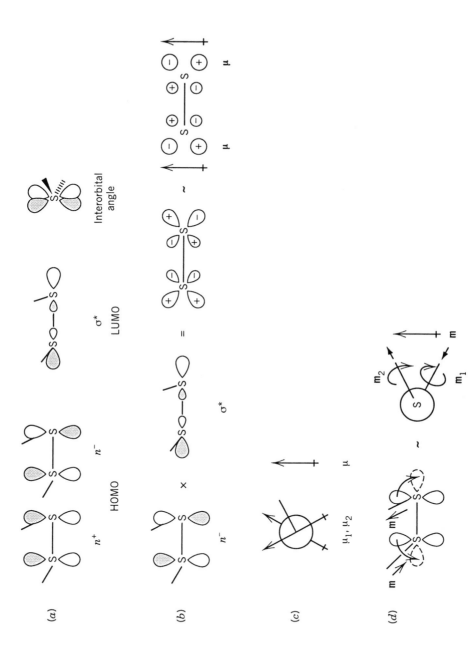

Figure 13.23. Application of the Snatzke MO "recipe" to the disulfide chromophore. A positive long wavelength CE is found for the right-hand or *P* absolute configuration of the –S–S– chromophore (opposite to that of the example in Fig. 13.22) provided that the dihedral angle is less than 90°. [Adapted with permission from Snatzke, G., *Chem. Unserer Zeit.*, **15**, 78 (1981).]

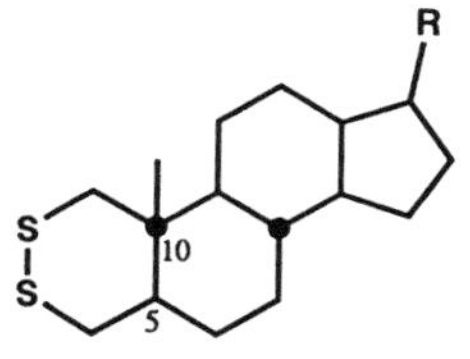

Figure 13.24. A 2,3-dithia-5-α-steroidal disulfide.

sum vector of the two partial moments points up (Fig. 13.23c). Concomitant rotation of charge (shown in Fig. 13.23d as rotation of HOMO to LUMO orbital lobes) gives rise to magnetic transition moments directed approximately along each sulfur to adjacent carbon bond. The sum vector of the two partial magnetic transition moments also points up (Fig. 13.23d). Since both sum vectors are aligned in the same direction (parallel), the long wavelength CE is positive. The transition observed at the shorter wavelength would be expected to have the oppositely signed CE for the same absolute configuration, as is observed (Snatzke, 1979a, 1981).

Two additional comments need to be made here. First, inversion of the phase signs, say in the σ^* MO, would have inverted the orientation of the partial transition moments. However, the relative alignment of the sum vectors would not have changed. Second, we reiterate the following point that has been previously stressed: The just concluded analysis permits one to establish only the absolute configuration of the disulfide moiety. In the 2,3-dithia-5α-steroidal disulfide (Fig. 13.24), the relative configuration of the other stereogenic centers and the conformation of the steroid framework would have to be ascertained independently to establish the absolute configuration of the entire structure.

c. Sector and Helicity Rules

Introduction

Semiempirical rules correlating absolute stereochemistry (conformation or configuration) with chiroptical properties, especially with CD spectra and, in particular, the signs of individual CEs, are of two types: (a) sector rules and (b) helicity (or chirality) rules. The former type pertains to inherently achiral chromophores only while the latter are applicable to inherently chiral ones (IUPAC, 1986; Snatzke, 1979c).

When chirotopic atoms or groups are present (even in nearly symmetrical molecules), these groups are able to perturb the electronic transitions of symmetric chromophores sufficiently to generate chiroptical properties. The designation "Sector rule" stems from the division of 3D space surrounding such symmetrical chromophores into sectors by nodal or symmetry planes as well as by nodal surfaces. In fact, some authors use the term "symmetry rules" to stress the connection (Schellman, 1968). Such rules are designed to assess the contributions of perturbing groups to the sign of the CE according to their location in one or another sector surrounding the chromophore. In order to understand how these rules may be applied, a brief excursion into the history of the theories of optical activity is necessary since it turns out that symmetry rules are related to (one might say, designed to be consistent with) the several "mechanisms" for the generation of rotational strength that have been proposed and developed over the years (Mason, 1982, Chapter 3; Charney, 1979, Chapter 4).

The oldest of the three principal theories is that of Drude (1892) who proposed that optical activity could be explained through the displacement of a charged particle along a helical path thereby generating both electric and magnetic dipole moments. The one-electron model (Condon, Altar, and Eyring, 1937) is an elaboration of this theory taking into account the quantum mechanical treatment of optical activity developed by Rosenfeld (1928).

We saw earlier (Section 13-4.b) that in order for CEs to be observed, both electric and magnetic transition moments associated with a specific transition must be finite and must be so oriented or "coupled" that the transition vectors retain components in the same direction (parallel or antiparallel). However, two of the illustrated cases, the $\pi-\pi^*$ and the $n-\pi^*$ transitions, lack magnetic and electric transition moments, respectively. Among the cases illustrated, the $n-\sigma^*$ transition was the only one shown to have both finite μ and m, but this one has the transition moments oriented perpendicular to one another with the result that here too no rotational strength is developed in the transition. How then do CEs arise in chiral molecules possessing only inherently achiral chromophores, such as C=O or C=C?

In the one-electron theory, transition moments (one electric and one magnetic) in two different transitions of one and the same chromophore are perturbed by the static field, that is, the permanent charge distribution of the rest of the molecule; this is sometimes called the "static-coupling" or "static perturbation" mechanism. This perturbation lowers the symmetry of the functional group and permits "mixing" of the transitions (e.g., $n-\pi^*$ with $\pi-\pi^*$) provided that the moments of the two transitions occur in the same direction (Charney, 1979, p. 105; Schellman, 1968; Mason, 1982, p. 54).

The one-electron theory has been applied principally to transitions that are strongly magnetic dipole allowed, hence those such as the $n-\pi^*$ transition of the carbonyl group that give rise to only weak isotropic absorption bands. Kauzmann, Walter, and Eyring (1940) used this approach to explain the optical activity of 3-methylcyclopentanone, a molecule whose only chromophore is symmetric. It is remarkable that over 20 years elapsed between the formulation of the one-electron theory and the description of the Octant rule (see below) that the former anticipated. As pointed out by Brewster (1967), this fact is a significant example of the difficulties experienced by scientists, especially theoreticians and organic chemists, in communicating with one another. In fairness, however, it would seem that the Octant rule also had to await the development of commercial spectropolarimeters and the accumulation of numerous chiroptical data upon which the empirical generalization is based.

The second of the major theories, the coupled oscillator theory, was developed originally by Oseen (1915), Born (1915, 1918), and Kuhn (1929). It accounts for the optical activity of compounds possessing two or more achiral chromophores (isotropic and oscillators) having strong absorption bands in terms of a multielectron phenomenon. If these chromophores are sufficiently close together, they may be coupled dissymmetrically (dipole–dipole coupling) (Brewster, 1967).

The coupled oscillator theory applies to groups having strong electric transition moments (e.g., in $\pi - \pi^*$ transitions); the electric oscillator transition dipole moment that is largely localized in one part of the molecule "drives" another electric dipole oscillator located elsewhere in the molecule (Weigang, 1979). If properly oriented, interaction or coupling of these moments may give rise to magnetic transition moments and consequently generate rotational strength. With two such chromophores, the CD band is split and becomes *bisignate* (i.e., having two signs); the long wavelength band has one sign while the short wavelength band has the other. The modern theory is due to Kirkwood (1937) who, together with his co-workers Moffitt and Fitts, applied it to the interpretation of the chiroptical properties of macromolecules possessing many chromophores (Moffitt, Fitts, and Kirkwood, 1957; Fitts and Kirkwood, 1956, 1957; Moffitt, 1956a 1956b). The dynamic-coupling mechanism applicable to coupled oscillator theory has been reviewed by Buckingham and Stiles (1974).

Polarizability theories (or models) describe essentially the same phenomenon as does the coupled oscillator theory although the focus of the former is on polarizability or refraction properties (Fitts, 1960; Brewster, 1967; Applequist, 1977).

A particularly fruitful application of the coupled oscillator theory is the exciton splitting method of Moffitt. Interpretation of CD spectra is especially straightforward when exciton splitting is caused by interaction of identical chromophores whose transitions consequently become degenerate. This topic is discussed in Section 13-4.d.

The third theory (electromagnetic, or $\boldsymbol{\mu} - \mathbf{m}$, coupling) is also used to describe the CD of molecules possessing two intrinsically achiral chromophores. In this instance, one of the chromophores has a finite electric transition moment while the other has a nonzero magnetic transition moment. It is particularly applicable to diketopiperazine derivatives of α-amino acids.

Close juxtaposition of the two chromophores allows coupling between the magnetic transition moment of one amide group ($n - \pi^*$ transition) with the electric transition moment of the other amide group ($\pi - \pi^*$ transition). Such coupling between $\boldsymbol{\mu}$ and $\mathbf{m}$, which normally does not occur, takes place via a quadrupole associated with $\mathbf{m}$ (Tinoco, 1962; Schellman, 1968; Charney, 1979, p. 114).

The several theories that we have just briefly described have been combined into one generalized theory by Tinoco (1960, 1962). The equation (Eq. 13.23) for the rotational strength of a given CE developed by Tinoco in connection with the chiroptical properties of polymers contains many terms (T)

$$R = T_1 + T_2 + T_3, \text{ etc.} \tag{13.23}$$

which may be associated with the individual mechanisms for optical activity. The relative magnitudes of the individual terms determine which mechanism is dominant in any given situation. In connection with the verification of empirically established sector rules, the sign and magnitude of the CE may be established by perturbation theory that probes which of the terms in the generalized equation are significant (for an example, see below).

Numerous sector rules have been described for inherently achiral chromophores, carbonyl and C=C (alkenes) in particular. Extensive reviews and summaries of these are given in Crabbé (1972); Legrand and Rougier (1977); Snatzke and Snatzke (1980); Kirk (1986). These are examined in some detail in this section.

Circular dichroism of compounds containing inherently chiral chromophores has also given rise to empirical generalizations (helicity rules) relating the signs of CEs with the sense of helicity of part of a molecule. Cotton effects of helicenes, enones, and peptides, among others, are very large. This effect is characteristic of the presence of both magnetic and electric transition moments. First-order determination of the sign of the CE and even of its magnitude directly from theory is more readily feasible for inherently chiral chromophores than is true for inherently achiral ones.

Saturated Ketones. The Octant Rule

The investigation of chiroptical properties of chiral ketones began around 1954 in the laboratory of Djerassi at Wayne State University when instrumental developments permitted the construction of the first easily usable spectropolarimeter reaching into the UV (to 300 nm). The experimental studies fortunately coincided with the growing understanding of conformational effects that followed the seminal contributions of Barton (1950; cf. Chap. 11). The initial ORD measurements were made on anancomeric (see Glossary) steroidal ketones (Djerassi et al., 1955; Djerassi, 1990, p. 54). Steroids as a class had been thoroughly studied by then and numerous samples of well-characterized compounds were available for the study. Moreover, the rigid steroid backbone was chosen purposely to minimize conformational ambiguities. In addition to the more obvious structural effects, such as the location of the carbonyl group, this permitted the ready assessment of the configurational and conformational effects of numerous substituent groups on the CE of the carbonyl chromophore.

The carbonyl chromophore had been chosen as the "test case" because of its ubiquity in the domain of organic chemistry (Djerassi, 1960, p. 9; Djerassi, 1964). Use of the carbonyl group was an inspired "stroke of luck" in view of two factors: its UV absorption wavelength ($n-\pi^*$ transition) is in a readily accessible spectral region (λ_{max} ca. 300 nm) and its absorption band is so far removed from the next absorption band of higher energy (ca. 190 nm) that there is no danger from overlap of the two nor of confusing the nature of the transition under study (Klyne and Kirk, 1973b). Moreover, the weakness of the $n-\pi^*$ transition avoids experimental difficulties in measuring the rotation right through the absorption band. This measurement was essential since one of the key differences between these studies and the few earlier ones was the emphasis on CE wavelengths, band shapes, and signs.

The very first generalization established as a result of the ORD studies of the Djerassi school was the *axial haloketone rule* (Djerassi and Klyne, 1957a; Djerassi, 1960, p. 120). Just as IR and UV spectral measurements are sensitive to the conformation of halogen substituents in six-membered rings (cf. Section 11-4.d), so too are the ORD parameters: CE intensities and wavelengths (Djerassi, 1960, p. 115). Based on measurements carried out on steroidal ketones, it was observed that equatorial α-halogen (or acetoxy) substituents on either side of the carbonyl group have little effect on the ORD parameters; the sign of the CE at about 300 nm is unchanged in these cases relative to that which is found for the unsubstituted ketone. However, in addition to causing bathochromic shifts, the corresponding axial substituents (a) cause an increase in the amplitude of the CE, and (b) may invert the sign of the CE according to the configuration of the stereogenic atom bearing the halogen atom; this finding was later expanded to include SR, SO_2R, NR_2, and other substituents (Djerassi et al., 1958). The sign of the CE depends on the constitution, the conformation, and the configuration of the haloketone in the vicinity of the carbonyl group. Knowledge of any two of the foregoing factors and of the sign of the CE would suffice to reveal the third.

It was suggested that, in order to predict the sign of the CE, the carbonyl group should be viewed along the O=C bond in the direction of the ring with the carbonyl carbon placed at the "head" of the chair (Fig. 13.25). If the axial α-halogen atom is found to the right of the observer, then a positive CE is found; conversely, if the axial halogen atom is on the left, a negative CE obtains. Axial fluorine substituents have been shown to have effects opposite to those of the other halogens: The S enantiomer (X = F) in Figure 13.25 would exhibit a negative CE.

Applications of the rule include (a) the determination of configuration, e.g., by correlation of CE signs with those of analogous compounds having independently established configurations, assuming that the other two factors, constitution and conformation, are known or can be ascertained, and (b) determination of the constitution of a compound. The determination of the configuration of an 11-bromoketone is illustrated in Fig. 13.26, **B**; for an example of a configurational assignment in the 1-decalone series, see Djerassi and Staunton (1961); Eliel (1962, p. 425). Application (b) is illustrated by the establishment of the location of a halogen atom from the sign of the CE following halogenation of a non-racemic sample of a chiral ketone (Fig. 13.26, **A**; the axial orientation of the halogen—as independently established by UV and/or IR—is implied).

The rule also permits demonstration of conformational mobility as a function

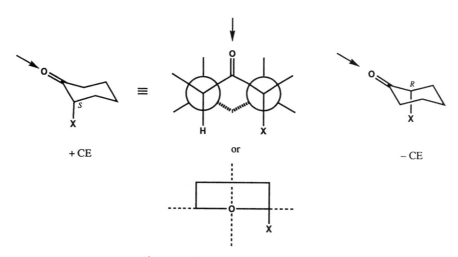

Figure 13.25.   Axial haloketone rule.

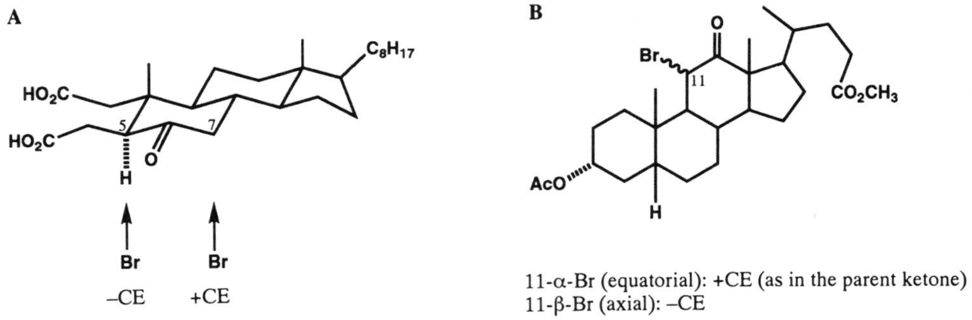

A

–CE   +CE

B

11-α-Br (equatorial): +CE (as in the parent ketone)
11-β-Br (axial): –CE

Figure 13.26. Applications of the axial haloketone rule. (**A**) Position of the halogen substituent. The bromo derivative exhibited a negative CE, hence substitution occurred at C(5); (**B**) determination of the absolute configuration of the 11-bromo substituent in an 11-bromo-12-ketosteroid (Djerassi, 1960, p. 123).

of solvent polarity. On chlorination of (*R*)-(+)-3-methylcyclohexanone, a crystalline 2-chloro-5-methyl derivative is isolated that exhibits a negative CE in octane. With the constitution of the product and the axial position of the halogen having been independently established, the negative CE is consistent only with trans stereochemistry (Fig. 13.27, **A** and **B**). However, a positive CE is observed in methanol. The latter finding is interpreted to mean that the conformational equilibrium has shifted in favor of the higher dipole diequatorial conformer in the presence of a more polar solvent. Presumably, this is due to reduced repulsion between the diequatorially disposed carbonyl and chlorine dipoles in the presence of a solvent of high dielectric moment (Fig. 13.27, **C**) (Djerassi, Geller, and Eisenbraun, 1960; cf. Chapter 11, p. 733).

The axial haloketone rule is also applicable to cyclohexanones that exist in a boat form. Of the two constitutionally related 2-bromo-2-methylcholestan-3-ones (Fig. 13.28), the 2β-bromo-2α-methyl isomer **A** exhibits a strong positive CE as anticipated; the 2α-bromo-2β-methyl diastereomer (shown spectroscopically to have its halogen in an axial orientation) unexpectedly exhibits a negative CE. The results were explained by assuming that in the latter isomer, ring A assumes a

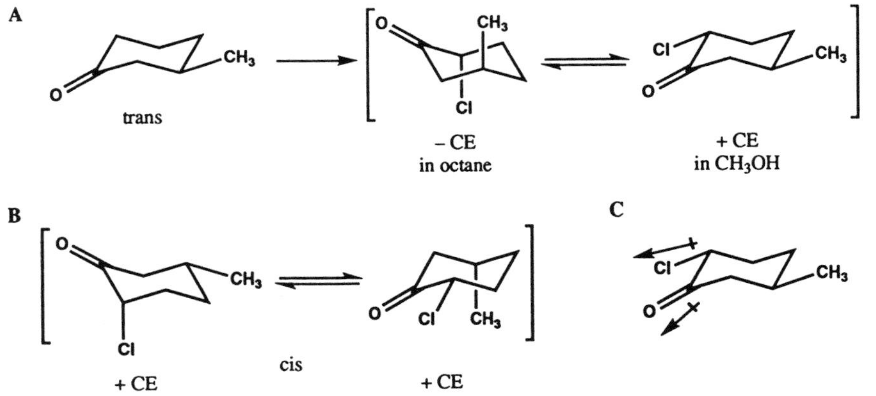

A

trans

– CE
in octane

+ CE
in CH$_3$OH

B

+ CE   cis   + CE

C

Figure 13.27. Conformational mobility in 2-chloro-5-methylcyclohexanones. (**A**) trans isomer; (**B**) the conformational equilibrium of the cis isomer is shown for comparison; (**C**) dipole repulsion in the trans isomer.

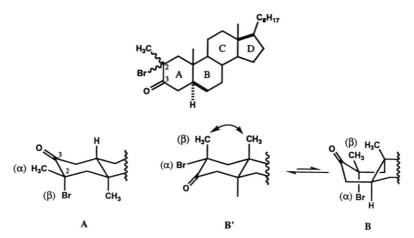

Figure 13.28. Demonstration of a boat form from chiroptical data. In **A**, the steroid A ring has been inverted for ease of comparison with Fig. 13.25.

boat conformation (shown in Fig. 13.28, **B**; Djerassi, Finch, and Mauli, 1959). The destabilization of the chair form **B′** is ascribed to unfavorable axial 1,3-dimethyl interactions and to the unfavorable equatorial orientation of the bromine adjacent to the carbonyl group (cf. Table 11.17, p. 737).

The above discussion of the axial haloketone rule was presented in some detail for two reasons: (a) the clarity of the results beautifully illustrate the power of chiroptical techniques, and (b) the data on which it was based led fairly rapidly to the first of the sector rules, namely, the Octant rule.

The *Octant rule* is an empirical generalization that relates the sign of the CE of the carbonyl chromophore, as measured at about 300 nm in saturated cyclic ketones, to the configuration of chiral centers present in the vicinity of the chromophore (Moffitt et al., 1961). Its most common application is the deduction of configurational or of conformational information from CEs (Reviews: Klyne and Kirk, 1973b; Scopes, 1975; Kirk, 1986).

In spite of being empirical, the rule is firmly based on theoretical studies of the $n-\pi^*$ transition that is responsible for the aforementioned CE observed in chiral ketones. Accordingly, the symmetry properties of this transition provide a framework around which the influence of stereocenters, both near and at some remove from the symmetrical chromophore, may be assessed. A set of Cartesian coordinates is drawn right through a carbonyl group with the origin at the midpoint of the C=O bond and the z axis collinear with the bond as shown in Figure 13.29. The carbonyl group is bisected by x,z and y,z planes that are nodal and symmetry planes of the MOs involved in the electronic transition (cf. Fig. 13.19). The x,y plane (both in character and in location) is merely an approximation to the nodal surfaces associated with (or orbitally determined by) the π^* virtual orbital. Substituent groups attached to the carbonyl carbon atom lie in the y,z plane facing away from the observer.

The coordinate system divides space around the carbonyl group into eight sectors or octants. The principal premise of the rule is that an atom anywhere in the vicinity of the carbonyl group [say at point $P(x,y,z)$] makes a contribution to, and ultimately determines the sign of, the $n-\pi^*$ CE, according to its location, the

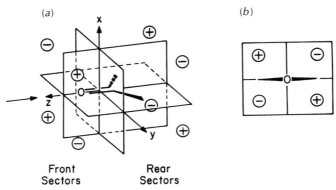

Front Sectors   Rear Sectors

Figure 13.29. Octant rule for saturated ketones. (a) Signs of the sectors in a left-handed Cartesian coordinate system; (b) projection of the rear (−z hemisphere) sectors. [Adapted with permission from Snatzke, G., *Chem. Unserer Zeit.*, **16**, 160 (1982).]

sign being determined by the simple product *xyz* of its coordinates. For example, the contribution of an atom located in a region corresponding to the lower right rear sector (Fig. 13.29b) whose coordinates are −x, +y, −z, is positive in a left-handed coordinate system (see discussion below) and, therefore, would predict a positive CE. The same atoms located in the mirror-image lower left rear sector would induce a negative CE.

A useful mnemonic for learning the signs of the Octant rule sectors is that the *upper*, *front*, *right* sector is positive. Since sector signs alternate in all directions, the signs of all remaining sectors are easily deduced from that of the "UPFRont" (upper, positive, front, right) sector.

The Octant rule was first applied to cyclohexanones whose geometry (bond lengths and angles) were well known and whose conformations were fixed by the ring fusions exhibited in steroids. The cyclohexane skeleton is disposed within the coordinate system as shown in Figure 13.30a, with the α- and α'-carbon atoms in the yz plane, and the remaining carbon atoms projecting upward.

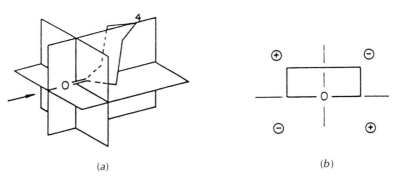

(a)   (b)

Figure 13.30. (a) Stereoprojection of the cyclohexanone ring (chair form) in the octant diagram; (b) projection of cyclohexanone bonds. View facing the carbonyl oxygen with signs of rear octants. [Reprinted with permission from Snatzke, G., *Angew. Chem. Int. Ed. Engl.*, **7**, 14 (1968).]

The article by Moffitt et al. (1961) that announced the Octant rule used a left-handed coordinate system, as in Figure 13.29a. Strictly speaking, therefore, the Octant rule is an inverse octant rule (Snatzke and Snatzke, 1973, p. 189)! This coordinate system and the upward disposition of the cyclohexanone skeleton (Fig. 13.30a) was chosen so as to conform to CE signs of ketones of known absolute configuration. Although the use of a right-hand Cartesian coordinate system has been proposed as a convention (Klyne and Kirk, 1973a), readers should be aware that this proposal has not generally been followed in the literature (e.g., Bouman and Lightner, 1976). All Octant rule diagrams in this chapter are based on a left-handed coordinate system.

Carbon-4 lies in the xz plane. Since, as a first approximation, substituents located on or close to nodal planes make no contribution to the CE (in fact, chiral adamantanones bearing methyl groups lying in symmetry planes do exhibit weak CEs; cf. Lightner and Toan, 1987c), we see that only groups attached to C(2), C(3), C(5), and/or C(6) can make such contributions. Moreover, since equatorial substituents at C(2) and C(6) lie in the yz plane, they have little, if any effect, on the CE. Contributions occurring in more than one ring are additive, and a semiquantitative assessment of such contributions is sometimes possible (see below). Incidentally, it should be evident that the axial haloketone rule is but a special case of the Octant rule (compare Fig. 13.25 with Figs. 13.29 and 13.30).

Carbon atoms, halogen atoms other than fluorine, and sulfur make contributions as predicted above. However, the halogens (again with the exception of fluorine) dominate the CE, completely overriding contributions by alkyl groups, as a consequence of their much higher atomic refraction (Djerassi, 1961). In this connection, it may be noted that fluorine has the lowest atomic refraction of all the halogens; its refraction is *less* than that of hydrogen. Nitrogen- and oxygen-containing groups may exhibit either octant or antioctant (inverse) behavior. The $-N(CH_3)_3^+$ group, for example, exhibits antioctant behavior.

An effect due to a particular substituent is called *consignate* (more accurately sector consignate) if the sign of its contribution to the CD or ORD intensity is the same as the sign of the product of the Cartesian coordinates for the sector in which the substituent is located. A contrary relationship between the sign of the contribution to the CD or ORD and the sign of the sector is termed *dissignate* (or antioctant), for example, fluorine which makes contributions inverse to those of the other halogens (Klyne and Kirk, 1973a). Naturally, con- or dissignate behavior is entirely dependent on the choice of signs of the coordinates, that is, on the coordinate system convention adopted (see above).

Contributions by hydrogen have usually been ignored. This contribution is simply the result of cancellation of effects by hydrogen atoms appearing in oppositely signed sectors. In addition, the atomic refraction of H, hence, the inherent magnitude of its contribution is small. More recent studies have shown that, in some circumstances, contributions particularly from C_α–H make dominant octant-dissignate contributions (Kirk, 1976).

The effect of deuterium as a ketone substituent has also been studied (Lightner et al., 1977; Numan and Wynberg, 1978; Sundaraman and Djerassi, 1978; Sundaraman, Barth, and Djerassi, 1980. Review: Barth and Djerassi,

1981). Deuterium makes a significant dissignate contribution such that the configuration of deuterium substituted ketones may be determined. It has been established that the rotatory strengths of transitions due to conformers in conformationally flexible systems [e.g., (R)-$(2$-$^2H_1)$- and (S)-$(3$-$^2H_1)$-cyclohexanones] are additive. These results require the assumption that axial and equatorial conformers of these ketones are present in nearly equal amounts; steric isotope effects on the equilibrium are below the experimental error limit (Sundaraman, Barth, and Djerassi, 1980; see also Chapter 11).

The compound (+)-3-methylcyclohexanone is known to exhibit a positive CE. Octant rule projections for both axial and equatorial conformers of this compound are shown in Figure 13.31 along with the corresponding configurations at C(3). The C(3) and C(5) carbon atoms make equal and opposite contributions to the CE. Only the positive contribution of the methyl group is unmatched by a negative one. It is evident from the illustration that the preferred equatorial conformation is predicted by the Octant rule if the R configuration obtains (as is indeed known). Arguments based on the additivity of rotatory strengths exclude the possibility that twist conformations play a major role in the conformational equilibrium (Lee, Djerassi, et al., 1980). Conversely, if the preferred conformation (equatorial) had been established independently (e.g., by NMR, cf. Chapter 11), the configuration would have been deduced as being R (Fig. 13.31).

In the case of compounds with reduced conformational mobility, such as 2-methylcyclopentanone, application of the Octant rule suffices to reveal the absolute configuration. The compound (−)-2-methylcyclopentanone exhibits a sizeable negative CE, $[\theta]_{306}$ −4786 (dioxane); accordingly, the methyl substituent must project into the lower left rear octant (projection analogous to that in Fig. 13.30b) and the configuration assigned as $2R$ (Partridge, Chadha, and Uskoković, 1973). An alternative interpretation of the data holds that it is not the contribution of the methyl group that gives rise to the relatively large CE, but rather the helicity of the relatively rigid twisted cyclopentanone skeleton in that conformation that has the methyl group locked in the quasiequatorial position (Snatzke, 1989; Kirk, 1986; see p. 1038).

The Octant rule projections for the three steroidal ketones shown in Figure 13.32 along with the relevant chiroptical data illustrate the fact that semiquantita-

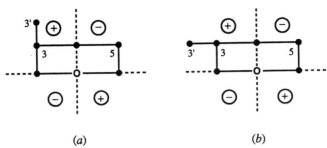

(a) (b)

Figure 13.31.   Octant rule projections for (+)-3-methylcyclohexanone (rear sectors). (a) Projection for the axial conformer (S configuration); (b) projection for the equatorial conformer (R configuration). [Reprinted with permission from Charney, E. (1979), *The Molecular Basis of Optical Activity. Optical Rotatory Dispersion and Circular Dichroism*, p. 176. Copyright © 1979 John Wiley & Sons, Inc.]

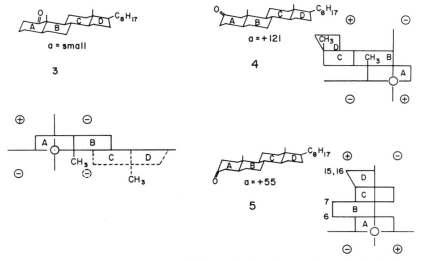

Figure 13.32. Semiquantitative assessment of CE magnitudes. Octant rule projection for isomeric 1-, 2-, and 3-cholestanones (**3–5**, respectively) and experimentally observed CE amplitudes (see Figs. 13.33 and 13.34, respectively). The projection outlined with dashed lines is that in a front octant. [Adapted with permission from Snatzke, G., *Angew. Chem. Int. Ed. Engl.*, **7**, 14 (1968).]

tive assessment of CE magnitudes is often possible: The larger the number of carbon atoms, groups, and/or rings in a given sector, uncompensated by like contributions in sectors of opposite sign, the greater the CE magnitude. The larger positive CE observed for 5α-cholestan-2-one **4** relative to that observed for 5α-cholestan-3-one **5** (Fig. 13.32) is evident, even predicted, from the projections: In the latter, only the methylene C(6) and C(7) atoms found in the positive upper left rear sector make significant contributions to the CE. The angular methyl groups lie in the x,z plane, as do many of the carbon atoms of the C(17) side chain. Neither these nor the remote C(15) and C(16) of **5** make appreciable contributions to the CE, which is less than one-half as large as that of **4**.

The 5α-cholestan-1-one **3** exhibits a negative CE that is much more easily revealed by the CD spectrum (Fig. 13.34) than by the ORD spectrum (Fig. 13.33). The ORD spectrum is more difficult to interpret due to the presence of strong background rotation caused by CEs lying at shorter wavelengths. The latter are not evident in the CD spectrum. The negative CE may be due to uncompensated contributions principally from the front sectors as well as a smaller contribution from rear sectors; the front sectors are indicated in Fig. 13.32 by means of dashed lines..

Figure 13.35 illustrates the utility of empirically determined numerical contributions or increments ($\delta\Delta\varepsilon$) to CE magnitudes (Ripperger, 1977). A cautionary note is in order, however. Such contributions are likely to be accurate only within narrowly defined structural domains. For applications, for example, in the decalone system, see Kirk and Klyne (1974); for a summary, see Kirk (1986, page 793).

There has been some question as to whether ketone $n-\pi^*$ Cotton effects are not more accurately governed by a quadrant rule than by an octant rule, the four sectors of the former rule being the minimum called for by symmetry considera-

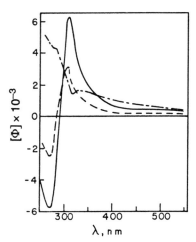

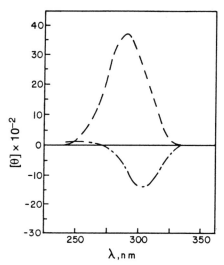

Figure 13.33. The ORD spectra of 5α-cholestan-1-one **3** (– · –), -2-one **4** (—), and -3-one **5** (--) (methanol solution). [Adapted from Djerassi, C. (1960), *Optical Rotatory Dispersion*, McGraw-Hill, New York, p. 42.]

Figure 13.34. The CD spectra of 5α-cholestan-1-one **3** (– · –) and -3-one **5** (--) (methanol solution). [Reprinted with permission from Djerassi, C., Records, R., Bunnenberg, E., Mislow, K., and Moscowitz, A. (1962), *J. Am. Chem. Soc.*, **84**, 4552. Copyright © 1962 American Chemical Society.]

tions (Schellman, 1968). However, we have just seen the example of 1-cholestanone whose CD spectrum can be readily interpreted in terms of an octant, and not of a quadrant, rule whose sectors correspond to the rear hemisphere of the Octant rule. The weight of the evidence, to date, is in favor of an octant rule for saturated ketone $n-\pi^*$ transitions (see below).

Much effort has been expended in identifying the source of the substantial rotatory strengths exhibited by the inherently achiral $n-\pi^*$ transitions of compounds, such as those described above. The question is from where does a small electric dipole transition moment arise that couples with the large magnetic transition moment to generate the CE. Theoretical studies dealing with this issue have taken the form of attempts to "establish" the Octant rule on the basis of one of the theoretical models identified above. A very helpful summary, from the

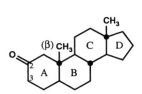

	$\delta\Delta\varepsilon$
Ring B	+ 1.3
Ring C	– 0.15
Ring D	+ 0.35
β-CH$_3$	+ 0.5
(Calcd.)	+ 2.0
Found	+ 1.78

Figure 13.35. Numerical contributions to the CE magnitude. [After Ripperger, 1977.] In this computation, ring D is treated as a six-membered ring.

standpoint of nontheoreticians (including the authors of this book), is given by Snatzke (1979a).

First-order quantum mechanical perturbation theory provides one way to explain (rationalize) from where the rotational strength comes. The Tinoco equation for R (Eq. 13.23; page 1021) includes the following terms: T_1, for an electric transition moment supplied by another chromophore, say $\sigma-\sigma^*$ of a C–C bond; T_2, for an electric transition moment supplied by a different transition, say $\pi-\pi^*$, of the same C=O chromophore; T_3, again for an electric transition moment supplied by yet a higher excited state of the C=O chromophore; and T_4, for a contribution arising from the difference between the permanent dipole moments of the ground and excited states of the $n-\pi^*$ transition, respectively. It turns out that only terms T_1 and T_2 are significant in magnitude.

The signs of T_1 and T_2, determined with respect to perturbing chirotopic atoms (or groups) having residual positive nuclear charge that is incompletely screened by the relevant electrons (the perturbers being located in the upper left rear sector of the octant diagram), were found to be negative and positive, respectively. The foregoing describes the "static coupling" mechanism of the one-electron model (Snatzke, 1979a).

While the term T_1 does "yield" an octant rule, it does so with the wrong sign, hence it is inferred that "interchromophoric" interaction is not the dominant mechanism contributing to the rotational strength of the carbonyl $n-\pi^*$ transition. Conversely, the T_2 term, being of the right sign, is seen to be the remaining and dominant source of the rotational strength. Yet T_2 is consistent with a quadrant and not an octant rule. This analysis is the source of one of the controversies associated with the history of the Octant rule. More recent calculations that take into account the delocalization of nonbonding electrons do lead to an octant rule.

Many theoretical studies have helped to define the limits of the Octant rule. One such theoretical study by Bouman and Lightner (1976) has revealed the curved geometry of the third nodal surface of the Octant rule (Fig. 13.36). In this model, surface B takes the place of plane A of the original Octant rule. As earlier pointed out (see above), the A (x,y) plane is the only one of the three surfaces delineating the sectors that is not determined by the ground state of the carbonyl group. The computed regions of sign change have accounted for a number of results previously characterized as being dissignate (Lightner et al., 1985a,b). From the foregoing, it will be evident that earlier "exceptions" to the Octant rule were responsible for later minor modifications.

The interpretation of CD spectra of compounds containing β-axial substituents in cyclohexanones has proven to be particularly troublesome (Gorthey, Vairamani, and Djerassi, 1985; see also Rodgers, Kalyanam and Lightner, 1982). The contribution of axial β-methyl groups to the (negative) CE is relatively weak and strongly solvent dependent, for example, in the 2-adamantanone derivative **6** shown in Figure 13.37. This contribution was interpreted as being dissignate (Snatzke et al., 1969; Jacobs and Havinga, 1972; Lightner and Wijekoon, 1982). Part of the difficulty may result from the fact that the substituents lie close to the curved nodal surface of the Octant rule, hence the substituents may actually project into a front sector so that its contribution is really consignate (Bouman and Lightner, 1976; Lightner et al., 1985b). Alternatively, the dissignate behavior

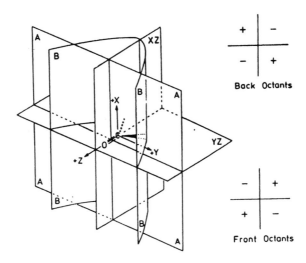

Figure 13.36. Third nodal surface B of the Octant rule as proposed by Bouman and Lightner (1976). Sign change regions computed for the carbonyl $n-\pi^*$ Cotton effect. [Reprinted with permission from Lightner, D. A., Chang, T. C., Hefelfinger, D. T., Jackman, D. E., Wijekoon, W. M. D., and Givens, III, J. W. (1985), *J. Am. Chem. Soc.*, **107**, 7499. Copyright © 1985 American Chemical Society.]

has been interpreted as being due to a specific through-space interaction between orbitals of the β substituent and the carbonyl group (Snatzke, 1979a). Another interpretation is given by Mason (1982, p. 56).

Substituted adamantanones have figured prominently in recent studies of the Octant rule. The structural rigidity of the adamantane skeleton incorporating a cyclohexanone locked in a chair form and the high symmetry of adamantanone (C_{2v}) make such compounds "archetypal" models for probing the limits of the rule (Lightner et al., 1986a). The Octant rule has been shown to apply to thioketones (Lightner et al., 1984). For a rare application of the Octant rule to aldehydes, see Tambuté and Collet (1984).

Computer programs have been developed that permit prediction of the sign, as well as reasonably good relative intensities, of the CD associated with the $n-\pi^*$ transition of rigid and of conformationally flexible ketones, that is, computer modelling of the Octant rule (Wilson and Cui, 1989). For conformationally flexible molecules, the approach relies on molecular mechanics calculations and searches for a set of low-energy conformations. Intensities are parametrized by means of a distance factor formula derived from the concept that the magnitude

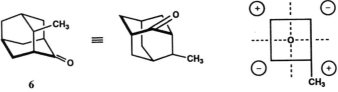

Figure 13.37. Dissignate behavior of axial 4β-methyl-2-adamantanone **6** (Snatzke, Ehrig, and Klein, 1969). At right, the signs of the octant projections are those of the rear sectors.

of the perturbation by a substituent group is inversely proportional to the distance between the chromophore and the substituent (Snatzke, 1979a); in addition, intensities are affected by the distance between the nodal plane and the substituent.

Other Sector Rules

In addition to the Octant rule, a few examples will have to suffice to indicate the scope of sector rules applicable to inherently achiral chromophores. Readers are referred to the most complete summaries of sector and helicity rules available as of this writing, those of Legrand and Rougier (1977); the book by Ciardelli and Salvadori (1973) (see especially the summary in Chapter 1); and that by Crabbé (1972). A compact but useful list may be found in the review by Snatzke and Snatzke (1980). More recent reviews are given in the text where appropriate.

Although the examples that follow are presented in the spirit of empiricism, readers should understand that much of the information that is summarized in these sector rules has been probed by theoretical means. These sector rules are at very least semiempirical and, in some cases, their origin (the relation between CD spectra and stereostructure) is firmly based on modern theories of molecular spectroscopy.

The CD behavior of chiral *alkenes* whose C=C double bond is not significantly distorted out of planarity may be generalized by means of an octant rule (Fig. 13.38; Charney, 1979, p. 202). The relevant transitions are of the $\pi-\pi^*$ type (electric transition moment allowed). There are several possible alkene transitions [two $\pi-\pi^*$ and one $\pi-3s$ (Rydberg)] in the short wavelength region of the spectrum near the limits of typical CD spectrometers. The number of exceptions to consignate behavior in the alkene octant rule is rather larger than one finds with respect to the ketone ($n-\pi^*$) Octant rule. This finding is ascribed at least in part to the difficulty in assigning the alkene absorption bands to specific transitions with confidence. This situation is in contrast to the carbonyl $n-\pi^*$ transition that is far removed from other C=O bands (Mason, 1982, p. 58). After all, a sector rule is specific not only to a given chromophore but to a specific transition of that chromophore (Hansen and Bouman, 1980). Misassignment of the band naturally may lead to an incorrect assignment of the absolute configuration.

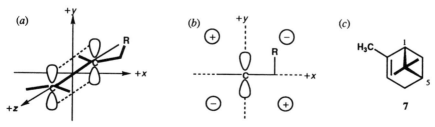

Figure 13.38. (*a*) Sector rule for alkenes for the longest wavelength transition (ca. 210 nm); (*b*) projection of rear sectors ($-z$ hemisphere) showing the position of a substituent R in the upper right rear sector; (*c*) structure of $(-)$-$(1S,5S)$-α-pinene **7**. [Adapted with permission from Hudec, J. and Kirk, D. N., *Tetrahedron*, **32**, 2475 (1976). Copyright © 1976 Pergamon Press, Ltd., Headington Hill Hall, Oxford, OX3 0BW, UK.]

The sign of the CE associated with the longest wavelength $\pi-\pi^*$ ($\pi_x-\pi_x^*$) transition (the longer wavelength band of the CD couplet that is usually observed for alkenes) is especially sensitive to contributions from axial allylic substituents (Yogev, Amar, and Mazur, 1967; cf. Section 13-4.c, page 1043). Application of the above alkene sector rule to (−)-α-pinene **7** is illustrated in Figure 13.38. Compound **7** exhibits a negative CE at about 210 nm (in 3-methylpentane; the band is shifted to ca. 200 nm in the vapor phase; see Table 13.2). The $(CH_3)_2C$ axially oriented bridge determines the sign of the CE by making a significant contribution to one of the upper rear sectors, and hence reveals the absolute configuration to be (−)-(1S,5S) (Hudec and Kirk, 1976; Drake and Mason, 1977).

A bifurcated sector rule (Fig. 13.39) has been proposed for chiral *allenes* in the vicinity of 225 nm ($\pi-\pi^*$ of which there are four one-electron transitions; three or four CD bands are actually seen in the UV region down to ca. 185 nm). The longest wavelength band (possibly $A1 \rightarrow A2$) is expected to exhibit a negative CE with respect to (S)-(+)-1,3-dimethylallene **8** (note the dextrorotation at the sodium D line) as is actually found (ε_{max} 223 nm; $\Delta\varepsilon$ +0.63 in pentane). Figure 13.39c also illustrates application of this sector rule to (R)-(+)-1,2-cyclononadiene **9** that exhibits a positive CE at 236 nm (Crabbé, Mason, et al., 1971; Mason, 1973; see also Chapter 14).

Circular dichroism may be exhibited by the planar *benzene* chromophore (Snatzke, Kajtár, and Snatzke, 1973; Charney, 1979, p. 227). Since the symmetry of simple benzene derivatives differs according to the number and the position of the substituents, many sector rules are needed to describe the chiroptical behavior of benzenic derivatives. Application of the rules proposed by Schellman (1966) makes it possible to determine the number of sectors into which the space surrounding the chromophore must be divided in order to assess the contribution of a perturbing atom or group. Mono- or ortho-disubstitution reduces the D_{6h} symmetry of benzene to C_{2v}, which calls for four sectors (quadrant rule). In

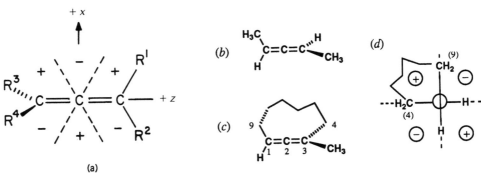

Figure 13.39. (*a*) Bifurcated sector rule for allenes (at the longest wavelength $\pi-\pi^*$ transition). The y coordinate is perpendicular to the plane or the paper (+ is forward). The signs for the sectors are those of the front (+y) hemisphere; the signs are reversed for groups that appear in the −y hemisphere; (*b*) (P)-(+)-1,3-dimethylallene **8**; (*c*) (M)-(+)-1,2-cyclononadiene **9**; (*d*) "quadrant" projection (there are actually 12 sectors) for (R)-(+)-1,2-cyclononadiene. [Adapted with permission from Crabbé, P., Velarde, V., Anderson, H. W., Clarke, S. D., Moore, W. R., Drake, A. F. and Mason, S. F. (1971), *J. Chem. Soc. D*, 1261. Copyright © Royal Society of Chemistry, Science Park, Milton Road, Cambridge CB4 4WF, UK.]

connection with CD spectra of benzene derivatives, it is well to remember that their absorption in the UV is complex. The three principal transitions that are observed are $^1L_b (= {}^1B_{2u}$ in the 260–280-nm region), $^1L_a (= {}^1B_{1u}$ in the 210–230-nm region), and $^1B (= {}^1E_{2u}$ in the 190–200-nm region) in the notation of Platt (1949) (that in parentheses is group theory notation). The caveat of Snatzke, Kajtár, and Snatzke (1973, p. 158) is pertinent: This notation explicitly applies to benzene and is used for benzene derivatives only by analogy. In the case of benzene compounds bearing only one substituent, which contains the chiral center, the CEs associated with the ^1L_b transition (each of these is actually a series of CEs due to transitions between vibrational states) result from "vibronic borrowing" from allowed transitions at shorter wavelengths (Weigang, 1974; Smith, 1989).

For a recent example of the application of Snatzke's rules (Snatzke and Ho, 1971; Snatzke, Kajtár, and Snatzke, 1973) to the determination of configuration of a chiral tetralin, see Speranza et al. (1991).

Application of a four sector (quadrant) rule will be illustrated by means of an example (shown in Fig. 13.40) in which the rule empirically derived by Brewster and Buta (1966) and revised by Smith and Fontana (1991) for conformationally mobile aromatic compounds containing a single stereogenic substituent adjacent to the aromatic ring (one of whose benzylic substituents is hydrogen) is applied to the ^1L_b transition. As is often true, the observed CEs are much more easily interpreted in CD than in ORD spectra.

In conformationally mobile systems, such as that illustrated in Figure 13.40, it is recognized that the observed CD spectrum is the sum of absorptions due to all conformers that are present. The sign of the CE of the ^1L_b transition can be predicted provided that the preferred conformation is known. Here, it is said to be the conformation in which the benzylic hydrogen eclipses (or nearly eclipses) the benzene ring that is preferred and that is the chiroptically determining one (Smith and Fontana, 1991, and references cited therein). The signs of the sectors are consistent with the negative CE observed at 268 nm ($\Delta\varepsilon$ −0.17 in CH_3OH) in the case of (R)-$(+)$-1-phenylethanol (Fig. 13.40; Verbit, 1965). That the alkyl group (in a negative quadrant) is the dominant perturber is consistent with a larger effective transition moment for a C–C bond as compared with a C–O bond in the attachment bonds to the stereocenter. For other groups, also useful in the application of the quadrant rule to compounds having the general formula C_6H_5CHRR', the following order of contributions to the ^1L_b transition obtain: SH, CO_2^-, $C(CH_3)_3 > CH_3 > NH_2$, $^+NH_3$, $^+N(CH_3)_3$, OH, OCH_3, Cl; and $CH_3 > CO_2H > {}^+NH_3$, OH, OCH_3 (Smith and Fontana, 1991). Numerous chiral benzylic alcohols fit this rule, for example, R = $(CH_3)_3X$ and $(C_6H_5)_3X$ (X = C

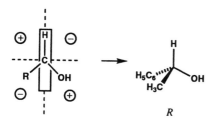

Figure 13.40. Brewster-Buta/Smith-Fontana sector rule (^1L_b transition) for chiral benzylic derivatives; end-on view from the benzylic group end. The plus and minus, respectively, indicate the sign of the ^1L_b Cotton effects induced by the group having the larger effective transition moment in the indicated quadrant. Application to (R)-$(+)$-1-phenylethanol.

and Si) (Biernbaum and Mosher, 1971) and $R = CH_3$, i-Pr, and t-Bu (Salvadori et al., 1984) as does mandelic acid (Colon, Pickard, and Smith, 1991). This sector rule is keyed to the orientation of the electric transition moment; the latter is transverse to the molecular C_2 axis. The rule is applicable also to benzylic amines (Smith and Willis, 1971; Smith and Fontana, 1991).

When additional ring substituents are introduced in the benzene ring, for example, as in α-methyl-(p-chlorobenzyl)amine, bond transition moments are induced in the benzene ring bonds adjacent to the chirotopic group attachment bond (Smith et al., 1983). Enhanced coupling results between the ^1L_b transition of the benzene ring and the chirotopic group. The sign of the ^1L_b Cotton effect may be the same or reversed in comparison to that of the "unsubstituted" compound.

Given that the preferred conformation of the additionally ring-substituted compound is not likely changed significantly with respect to the substituent containing the stereocenter, reversal of the sign of the ^1L_b Cotton effect signals the overshadowing of the vibronic contribution to this CE by an induced contribution of opposite sign (Smith et al., 1983). The sign of the latter contribution depends on the position and the spectroscopic moment of the additional substituent, the spectroscopic moment (Platt, 1951) being an additive property of substituents that is essentially a component of the electric transition dipole moment. The elaboration of the benzene sector rule to account for such sign reversals in the ^1L_b Cotton effect on further substitution has been termed the "*benzene chirality rule*" (Pickard and Smith, 1990). In α-methyl-(p-chlorobenzyl)amine, para substitution with an atom having a positive spectroscopic moment (in this example, chlorine) is responsible for reversing the sign of the CE: $\Delta\varepsilon$ -0.11 at 268 nm and $+0.061$ at 276 nm (both in CH_3OH) for (R)-α-methylbenzylamine and (R)-α-methyl-(p-chlorobenzyl)amine, respectively. In (R)-α-methyl-(m-chlorobenzyl)amine, however, $\Delta\varepsilon$ -0.24 at 274 nm (CH_3OH); both the vibronic and induced contributions have the same negative sign and the sign of the CE is unchanged relative to the unsubstituted amine although increased in magnitude (Smith et al., 1983; Smith, 1989; Pickard and Smith, 1990). For application of the benzene sector and chirality rules to chiral perhydrobenzocycloalkenes, see Lorentzen, Brewster, and Smith (1992).

The absolute configuration of chiral primary amines has been probed through the CE of their cottonogenic (cf. Section 13-4.e, see p. 1055) N-salicylidene Schiff's base derivatives (Fig. 13.54, **20**) at about 250–315 nm (using the *salicylideneamino chirality rule*; Smith, 1983; see Section 13-4.d). For other benzene derivative sector rules, see especially the chapter by Snatzke et al., (1973) and Schoenfelder and Snatzke (1980).

The CEs exhibited by *carboxylic acids* (and some of its derivatives such as esters) are less well understood than those of some other functional groups. Nevertheless, many acids follow the very simple empirical rule illustrated in Figure 13.41 (Listowsky, Avigad, and Englard, 1970). As an example, consider the configuration of $(-)$-2-*exo*-bromonorbornane-1-carboxylic acid **10** for which $\Delta\varepsilon$ -1.13 was found at 215 nm. The carboxyl group preferentially populates a conformation in which the C=O group is synperiplanar to a $C_\alpha–C_\beta$ bond. In the example, dipole repulsion between the CO_2H and Br orients the carboxyl carbonyl toward the methylene bridge C(7) (Fig. 13.41c). In consequence, the

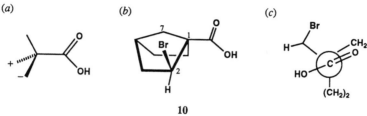

10

Figure 13.41. Empirical sector rule for carboxylic acids at 210 nm (presumably $n-\pi^*$). (*a*) Substituents that project toward the viewer contribute to negative CEs; those away from the viewer contribute to positive CEs; (*b*) (1*R*,2*S*)-2-*exo*-bromonorbornane-1-carboxylic acid **10**; (*c*) preferred conformation of the acid (Newman projection).

enantiomer having a 1*R*,2*S* configuration (Fig. 13.41*b*) would be expected to have a negative CE.

The configuration has also been predicted by application of the Octant rule for saturated ketones (analogous rules are valid for ketones and the corresponding acid). Preferred conformation (*b*) resembles a β-axial bromoketone that exhibits dissignate behavior. Since the 2-bromo substituent is exo relative to the methylene bridge, the configuration of the (−)-acid is 1*R*,2*R*. Analysis by "chiroptical methodology" (to use the authors' terminology) was confirmed by X-ray crystallography on a diastereomeric amide derivative of the acid (Müller, Keese, et al., 1986).

A *benzoate sector rule* is helpful in the determination of configuration of chiral cyclic 2° alcohols (e.g., in the steroid series). Benzoic acid and its esters exhibit a strong intramolecular $\pi-\pi^*$ charge-transfer (CT) band in the CD spectrum at about 225–230 nm. In benzoates, this band dominates the spectrum (Harada and Nakanishi, 1968). The sign of the CE due to the CT band is related to the configuration of the ester by means of an eight-sector rule illustrated in Figure 13.42 (Harada, Ohashi, and Nakanishi, 1968).

Application of the benzoate sector rule is illustrated with the assignment of the absolute configuration of the benzoate of 3-oxoandrostan-17β-ol **11** in which the CE due to the keto group in ring A does not interfere with the interpretation; the intensity of the latter is much smaller than the benzoate CE and its position is

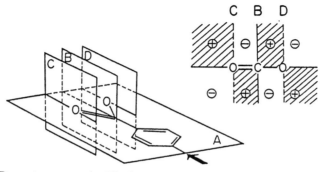

Figure 13.42. Benzoate sector rule. The benzoate group is viewed in the direction of the arrow. [Reproduced with permission from Harada, N., Ohashi, M., and Nakanishi, K. (1968), *J. Am. Chem. Soc.*, **90**, 7349. Copyright © 1968 American Chemical Society.]

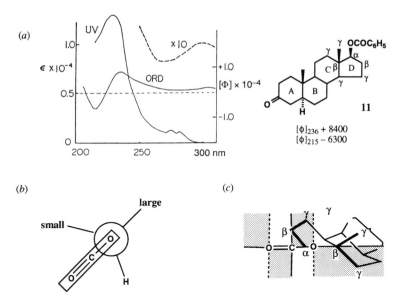

Figure 13.43. (*a*) The UV and ORD spectra of 3-oxoandrostane 17β-benzoate **11** (EtOH). The CE of the keto group is revealed in the expanded scale (10×); (*b*) preferred conformation of C_6H_5-COOR; (*c*) projection of the C and D rings on the sector diagram. [Adapted with permission from Harada, N., Ohashi, M., and Nakanishi, K. (1968), *J. Am. Chem. Soc.*, **90**, 7349. Copyright © 1968 American Chemical Society.]

at a much longer wavelength (Fig. 13.43*a*). In the analysis, it is assumed that the molecule preferentially populates a synperiplanar conformation of the carboxylate group (cf. Section 7-2) and that the benzoyl group is staggered between the geminal hydrogen and the smaller of the two other "groups" on R (Fig. 13.41*b*). Only α,β and β,γ bonds make significant contributions; these can be overridden by even more remote C=C bonds. In the example, the dominant contributions are made in a positive sector; contributions from the C ring and the C(18) methyl group largely compensate each other while the α,β bond in ring D that is hidden in the projection is in a nodal plane (Fig. 13.43*c*; Snatzke and Snatzke, 1973; Legrand and Rougier, 1977, p. 150).

Helicity Rules

The analysis of CEs has been carried out in terms of the two limiting types of chromophores, inherently achiral and inherently chiral. Before we examine the second of these, it is well to remember that the former category is really only a convenient fiction (Deutsche et al., 1969) since, to put it in the language of Mislow and Siegel (1984), all atoms in a chiral molecule are chirotopic. We should then not be surprised to find that some molecular systems have been successfully treated both as containing inherently achiral and inherently chiral chromophores.

Let us consider an example to make this point clear. Cyclopentanones and certain cyclohexanones, constrained into rigid twisted and hence chiral conformations, exhibit CEs of intermediate magnitude. As a first approximation, the sign

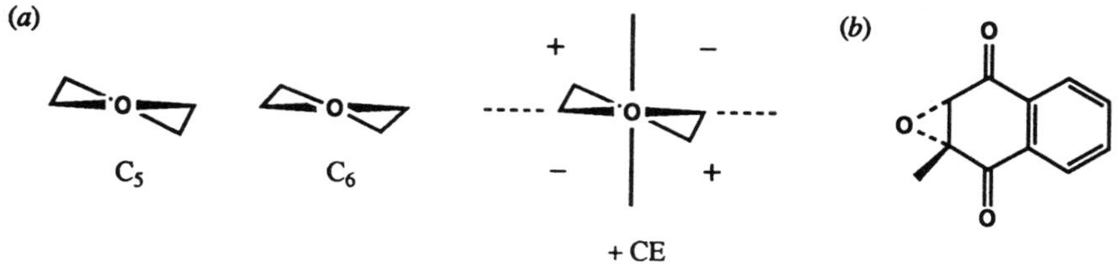

(a)

C₅   C₆

+ CE

(b)

11a

Figure 13.44. Application of the Octant rule to skewed cyclanones. The chirality sense shown gives rise to a positive CE. [Adapted with permission from Kirk, D. N., *Tetrahedron*, **42**, 777 (1986). Copyright © 1986 Pergamon Press, Ltd, Headington Hill Hall, Oxford, OX3 0BW, UK.]

of the CE is given by the Octant rule as if the chromophore were still inherently achiral (Fig. 13.44a; see also Section 13-4.c, p. 1026), though this may be fortuitous (Kirk, 1986).

The rigid twisted cyclanone framework may also be treated as inherently chiral by means of the MO recipe of Snatzke. Figure 13.45 (left) is a schematic representation of the ground-state MO of the $n-\pi^*$ transition. The implication is that the n orbital is combined with the σ_6 HOMO of the cyclohexanone skeleton [$\sigma_{C(2)-C(3)}$ rather than σ_{C-H}, since the latter is of lower energy] *and* with the π MO of the carbonyl group (Snatzke, 1979a). The signs of the lobes [shown only for C(2) though also that for C(6) must be taken into account] are chosen as shown in the figure (with the same stipulation as given on page 1011). Formal multiplication of this "chiral" orbital with the π^* LUMO leads to an electron density for the transition and a charge distribution that reveals the sign and direction of μ. Since both μ and m are aligned in the same sense, the predicted CE for the twisted cyclohexanone having the indicated absolute configuration (same as in Fig. 13.44a) is positive, as experimentally observed (Snatzke, 1979a, 1982). The enantiomeric conformation would have given rise [by choosing an admixture of n, π, and $\sigma_{C(5)-C(6)}$] to μ in the opposite sense; $\Delta\varepsilon$ would have been negative. The mixing of MOs (i.e., transitions) here illustrated gives rise to a helicity rule (rather than to a sector rule—the latter are constructed by mixing excited states) that is illustrated in Figure 13.44a.

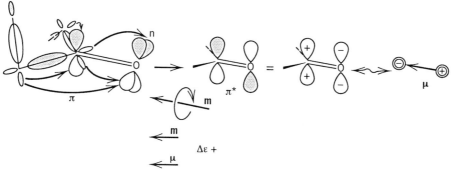

Figure 13.45. Application of the MO recipe of Snatzke to the twisted cyclanone system. [Adapted with permission from Snatzke, G., *Chem. Unserer Zeit.*, **15**, 78 (1981).]

In another example, the absolute configuration of $(-)$-Vitamin K_3 2,3-epoxide (Fig. 13.44b, **11a**) has been deduced from the CD spectrum. The negative CE assigned to the carbonyl $n-\pi^*$ carbonyl transition (340–400 nm) was interpreted as arising from absorption by the $2R,3S$ enantiomer by either a helicity rule (mixing of transitions) or a sector rule (mixing of excited states) (Snatzke, Wynberg, et al., 1980).

Returning to the heuristically useful bipartite treatment of chromophores, generalizations regarding CEs of inherently chiral chromophores are called helicity (and sometimes chirality) rules (Section 13-4.b). Typical examples of such chromophores are unsaturated ketones, dienes, skewed (twisted) alkenes, disulfides [these have already been discussed (p. 1016)], and helicenes.

Often, the interaction between two inherently achiral chromophores present in one molecule suffices to generate the high rotational strength that is the hallmark of inherently chiral chromophores. This type of composite (inherently chiral) chromophore was first observed in chiral β,γ-unsaturated ketones in which the intramolecular C=O and C=C groups are disposed in a twisted array (Mislow et al., 1961).

Shortly thereafter, a β,γ-*unsaturated ketone* helicity rule (Fig. 13.46) for transitions of the $n-\pi^*$ type (ca. 300–310 nm) was described (Moscowitz et al., 1962). The rule effectively states that homoconjugated C=C–CH$_2$–C=O arrays disposed in conformations represented by Figure 13.46a exhibit strongly negative CEs at about 300 nm. These strong CEs are limited to conformations with dihedral angles (ω) between 100° and 120°. Correspondingly large isotropic absorption intensities are also found, especially in nonpolar solvents.

The homoconjugated enone rule was shown to be consignate with the ketone Octant rule with the C=C (considered as a substituent) dominating the sign of the CE (generalized octant rule) (Fig. 13.46c; Mislow et al., 1961).

The helicity rule is also applicable to homoconjugated aldehydes and acid derivatives. In 1961 on the basis of this rule, dehydronorcamphor **12** and

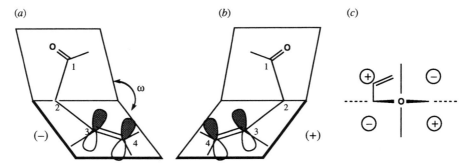

Figure 13.46. (a,b) β,γ-Unsaturated (homoconjugated) ketone helicity rule showing the signs of the CEs generated at about 300 nm [Adapted with permission from Moscowitz, A., Mislow, K., Glass, M. A. W., and Djerassi, C. (1962), *J. Am. Chem. Soc.*, **84**, 1945. Copyright © 1962 American Chemical Society.]; (c) generalized Octant rule (the signs are those of the rear sectors).

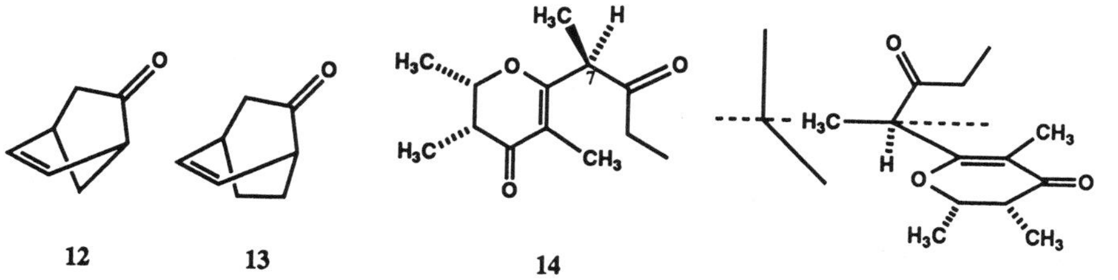

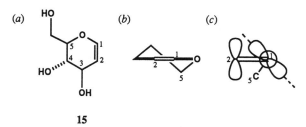

Figure 13.47. Homoconjugated ketones. At right, absolute configuration of C(7) of stegobinone **14** as inferred from the preferred conformation.

bicyclooctenone **13** (Fig. 13.47) were conjectured to have intense positive CEs. The accuracy of this prediction was confirmed in 1962 (Bunnenberg et al., 1962). The absolute configuration of the pheromone stegobinone **14** (derived from the bread beetle) was deduced from the negative sign of the strong CE at 290 nm (Hoffmann et al., 1981; Snatzke, 1982). The absolute configuration at C(7) may be inferred if it is recalled that conformations in which C–C bonds synperiplanar to C=O are energetically preferred to the corresponding (C=O)–C–H conformations; this places the C(7) methyl group rather than the corresponding methine hydrogen in the groove of the helicity rule diagram (Fig. 13.47, right).

A relatively simple semiempirical *quantitative* chirality rule for β,γ enones has shown that the sign and the magnitude of the CE are directly related to the angle between the C=O and C=C bonds (Schippers and Dekkers, 1983b).

The D-glucal **15** (Fig. 13.48a) is a cyclic *enol ether* that incorporates a nonplanar C=C–O moiety locked into a chiral conformation. Figure 13.48b illustrates the case with a negative torsion angle for C(5)–O–C(1)=C(2). Application of the Snatzke MO recipe to this inherently chiral chromophore is shown in Figure 13.49. The MOs of the allyl anion approximate the HOMO and LUMO of the C=C–O chromophore with the terminal AOs twisted to reflect the chirality (left-handed twist). Formal multiplication of HOMO and LUMO reveals the generation of a dipole during excitation, with the partial dipoles oriented obliquely to one another. Translation of charge, for example, from O to C(2) during $\pi°-\pi^-$ excitation generates **μ**. Concomitant rotation of charge along the line connecting C(2)–O generates **m** which, by the right-hand rule, is oriented antiparallel to **μ**. The resulting CE (observed in the vicinity of 195–210 nm) is therefore negative and the absolute configuration is as shown in Figure 13.48c.

15

Figure 13.48. (a) Structure of D-glucal **15**; (b) twisted conformation with M⁻ (left-handed) helicity; (c) Newman projection showing the AOs of the HOMO (oxygen atom at rear).

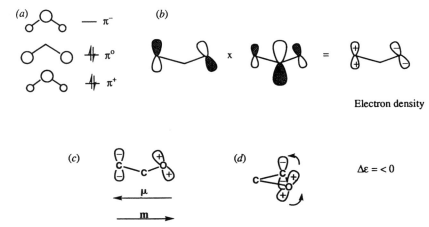

Electron density

$\Delta\varepsilon = < 0$

Figure 13.49. Application of Snatzke MO formalism to enols. (*a*) Allyl anion MOs; (*b*) formal multiplication of HOMO and LUMO; (*c*) charge translation from O to C, and (*d*) concomitant charge rotation. [Adapted with permission from Snatzke, G., *Angew. Chem. Int. Ed. Engl.*, **18**, 363 (1979).]

Assuming that the relative configurations of the OH and CH_2OH groups are known from NMR or other results, the absolute configurations of the chiral centers at C(3), C(4), and C(5) (Fig. 13.48*a*) follow from conformational analysis: L-glucal in this conformation requires all substituents to be axial while D-glucal **15** would have them all equatorial (Snatzke, 1979a).

Conjugated 1,3-dienes absorb in the 220–280-nm range (longest wavelength or π_2–π_3^*) according to substitution and ring size, if any. When the diene is skewed, an inherently chiral chromophore is present. The corresponding absolute configuration is correlated with the sign of the CE by means of the *diene* helicity rule shown in Figure 13.50 (Moscowitz et al., 1961; Weiss, Ziffer, and Charney, 1965). Calculations show that *P* helicity is correlated with a positive CE and *M*

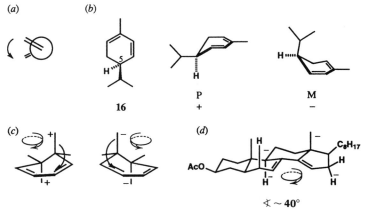

$\sphericalangle \sim 40°$

Figure 13.50. (*a*) Diene helicity rule; (*b*) *P* and *M* absolute configurations of (+)-α-phellandrene **16**; (*c*) reinforcing contributions of allylic pseudoaxial bonds to the CE; (*d*) opposing diene and allylic pseudoaxial contributions: In the example, the CE is negative. [Adapted with permission from Burgstahler, A. W. and Backhurst, R. C. (1970), *J. Am. Chem. Soc.*, **92**, 7601. Copyright © 1970 American Chemical Society.]

helicity gives rise to a negative CE (Charney, 1979, p. 217). Exceptions to the rule are found especially when the double bonds are not homoannular (see, e.g., Koolstra, Jacobs, and Dekkers, 1989).

α-Phellandrene **16** (Fig. 13.50) behaves according to this rule. At room temperature and above, a negative CE is found; however, as the temperature is reduced, the magnitude of the CE decreases, becomes zero, and eventually (down to $-168°$C!) becomes positive. The two conformations consistent with these results are shown in Figure 13.50*b* [the configuration at C(5) as shown in the figure is *S*] with the thermodynamically preferred conformation present at low temperature bearing a pseudoequatorial isopropyl group (Snatzke, Kováts, and Ohloff, 1966). An interesting point is that it is the conformation with the pseudoaxial isopropyl group that dominates the CE at room temperature; this explanation was advanced by Burgstahler et al. in 1961, and confirmed by Snatzke et al. (1966) by means of the just mentioned low-temperature measurements. The greater stability of the pseudoaxial conformation at room temperature may be due to a higher entropy associated with the freer rotation of the pseudoaxial (relative to that of the pseudoequatorial) isopropyl group; the pseudoequatorial isopropyl group in the other conformation may be constrained by the flanking ring hydrogen atoms, and therefore be unable to freely rotate.

It has been found that, in some circumstances, allylic pseudoaxial bonds make contributions to the CE opposite to that predicted by the diene chirality rule (see the alkene sector rule in Section 13-4.c, p. 1034); alternatively, the two effects may reinforce one another (Fig. 13.50*c*). When the two contributions oppose one another, as with the heteroannular cisoid diene shown in Figure 13.50*d*, the pseudoaxial bond contribution wins out (Burgstahler and Backhurst, 1970).

d. Exciton Chirality

When two chromophores are in close spatial proximity to one another and so disposed that a chiral array results, interaction (dynamic coupling) between the individual chromophores gives rise to distinctive CE couplets, often called exciton coupling, in the CD spectrum from which the configuration of the chiral array may be easily deduced. The term exciton [coined by Davydov in connection with his molecular exciton theory (1962)] applies to the nondegenerate excited states of a polychromophoric system.

In 1969, Harada and Nakanishi (q.v.) described a method for determining the configuration of glycols from the sign of the CE of the strong $\pi-\pi^*$ transition of the glycol dibenzoate derivatives. Although considered by them to be an extension of the benzoate sector rule (see above), the new method has little in common with this sector rule (other than the chromophore). The high amplitude of the CEs of such dibenzoates arise mainly from dipole–dipole interaction between the electric transition moments of the two identical but nonoverlapping benzoate chromophores. An example is given in Figure 13.51.

The CD spectrum of compound **17** displays the typical bisignate couplet that characterizes exciton splitting. The two CEs are centered at 307 nm near the UV absorption maximum [310.5 nm, ε 49,000 (EtOH)]. The longer wavelength CE of $(-)$-**17** is negative while the shorter wavelength CE of the couplet is positive. This

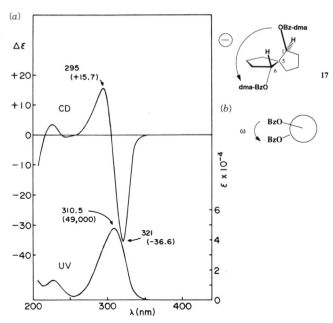

Figure 13.51. (*a*) The UV and CD spectra of the bis(*p*-dimethylaminobenzoate) (dma-BzO) ester of (−)-spiro[4.4]nonane-1,6-diol **17** in ethanol; (*b*) dibenzoate chirality rule. [Reprinted with permission from Harada, N. and Nakanishi, K. (1983), *Circular Dichroic Spectroscopy. Exciton Splitting in Organic Stereochemistry.* Copyright © 1983 University Science Books, 20 Edgehill Road, Mill Valley, CA., p. 150)].

relationship between the two CEs constitutes "negative chirality" as defined by Harada and Nakanishi (1969) and consequently, since the relative configuration of the stereocenters at C(1) and C(6) is known to be 1,5-*cis*-5,6-*trans* (by absence of C_2 symmetry as deduced from the ^{1}H NMR spectrum that shows two different sets of *p*-dimethylaminobenzoate peaks as well as two different methine peaks) (Harada, et al., 1977), the configuration of **17** is 1*S*,5*S*,6*R* (see also p. 1139).

Harada and Nakanishi (1972, 1983) called this type of analysis the *exciton chirality* method. The method itself is historically based on the coupled oscillator method (see above) as extended especially by Moffitt and Tinoco, who applied it to the analysis of chiroptical properties of biopolymers (Section 13-4.g; see e.g., Mason, 1982, p. 88). In 1962, Mason deduced the absolute configuration of the alkaloid calycanthine (Fig. 13.52, **18**) by application of the coupled oscillator method (Mason, 1962; see also Mason, 1982, p. 80). This determination represents the first application of the exciton chirality method to nonpolymeric systems.

The very large magnitude of the characteristic split CEs observed has led to the classification of systems exhibiting exciton splitting as a third type of chromophore, neither inherently chiral nor inherently achiral. Alternatively, dibenzoates, biaryls, and other polychromophoric systems exhibiting exciton splitting have been considered to be inherently chiral chromophores by Mason (1982, p. 72) and by Snatzke (1979) in spite of the lack of electron delocalization between the individual chromophoric units. However, the theoretical analysis of such systems by the coupled oscillator mechanism (applicable also to inherently achiral chromophores) is more rewarding.

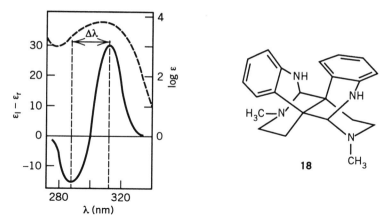

Figure 13.52. The UV (upper) and CD (lower) spectra of calycanthine **18** in ethanol. Exciton splitting. Example of positive chirality. $\Delta\lambda$ represents the "Davydov splitting." [Adapted with permission from Mason, S. F. (1962), *Proc. Chem. Soc., London*, 362. Copyright © Royal Society of Chemistry, Science Park, Milton Road, Cambridge CB4 4WF, UK.]

The calculation of signs and magnitudes of the CD bands in spectra displaying exciton splitting corresponds to the determination of absolute configurations. These calculations seem significantly simpler than those for the other two types of chromophores. Moreover, the exciton chirality method has been claimed to be nonempirical and comparable to the Bijvoet X-ray method (in the sense that reference compounds are not required) in the reliability of the absolute configurations so determined. An incorrect assignment of configuration based on the exciton chirality method, that of Tröger's base (Mason et al., 1967; compare Wilen et al., 1991), may be not an inherent failure of the method so much as a misassignment of transition moment geometries. This interpretation merits further study.

In the exciton chirality method, the electronic transitions of the isolated chromophores cannot be treated individually; to do so would be a violation of the Heisenberg uncertainty principle. Pairing of ground and excited states, as shown in Eq. 13.24,

$$\varphi_{\pm} = 1\sqrt{2}\,\{\phi_{1a} \cdot \phi_{2g} \pm \phi_{1g} \cdot \phi_{2a}\} \tag{13.24}$$

[where ϕ represents orbitals of the isolated chromophore, g stands for ground and a for excited states] splits the excited state energy levels and two nondegenerate excited states are formed. Incidentally, in the UV spectrum (isotropic radiation) the same degenerate states are present but, since no signs are associated with UV spectroscopy, the splitting generally results merely in a doubling of the band intensity.

The energy difference between the common excited states is known as Davydov splitting (Davydov, 1962; see Fig. 13.52). In typical two-chromophore systems (e.g., dibenzoates or calycanthine) the result of this nondegeneracy is the occurrence of a pair of CD bands, one of shorter and one of longer wavelength relative to the absorption wavelength of the monomeric chromophore; the CD is bisignate. The exciton band pairs have approximately equal areas (the bands are said to be conservative) but their intensities ($\Delta\varepsilon$) are generally unequal (Harada and Nakanishi, 1983, pp. 95 and 383). The latter property reveals the operation of a local sum rule (the general sum rule is a requirement that rotational strengths in a chiral system summed over all transitions be equal to zero, $\Sigma R = 0$) (Kuhn, 1930).

When the longer wavelength band of the CD couplet describes a positive CE (first CE) and the adjacent shorter wavelength CE band is negative (second CE), the CD spectrum is said to represent positive chirality (Fig. 13.52). The reverse situation is called negative chirality (Fig. 13.51). In the case of dibenzoate esters, the strong UV absorption found at about 230 nm is due to charge-transfer absorption bands involving the benzene ring and CO_2R. The large electric transition moment of each benzoate group is oriented approximately collinearly with the long axis of the benzoate group (Fig. 13.53a). The conformations of benzoates are well known and are approximated by Figure 13.53b. The individual transition moments μ are aligned approximately parallel to the "alcoholic" C–O bonds. The dihedral angle ω between the C–O bonds thus corresponds roughly to that between the two transition moments.

In the electronic transition giving rise to the CD spectra, the individual electric transition moments couple in parallel (symmetric) and antiparallel (antisymmetric) ways (Figs. 13.53c and d, respectively) in a manner reminiscent of the symmetric and antisymmetric stretching vibrations of CH_2 and CH_3 groups that

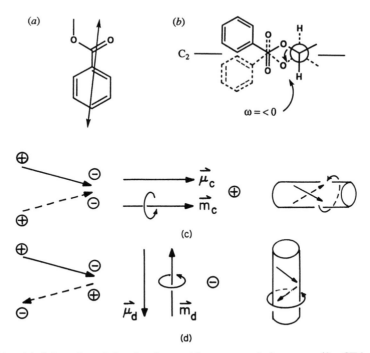

Figure 13.53. (a) Orientation of the electric transition moment in benzoates (for CT band at about 230 nm); (b) preferred configuration of a *vic*-dibenzoate; for the absolute configuration shown, the dihedral angle is negative; (c) and (d) combination of the individual electric transition moments and resultant magnetic transition moment. [Adapted from Snatzke, G., *Angew. Chem. Int. Ed. Engl.*, **18**, 363 (1979).]

lead to doublet bands in infrared spectra. The "total" electric transition moments for the dibenzoate system are oriented along (in case c) and perpendicular (case d) to the chromophoric C_2 axis.

The charge rotation associated with the transition may be visualized by placing the partial moments tangent to a cylinder with the latter aligned along the axis of the combined moment $\boldsymbol{\mu}$. Parallel and antiparallel arrangements of the electric and magnetic transition moments are found and these lead, in turn, to the adjacent bisignate CEs that are characteristic of exciton chirality (the CEs are positive and negative, respectively).

Repulsion between poles of the same sign in the c mode dictates that parallel coupling is of higher energy than the antiparallel mode d, where the poles of unlike sign are closest together. In consequence, the positive CE band associated with the c mode is found at shorter wavelength than the negative CE related to d. For the absolute configuration shown in Figure 13.53b (with negative dihedral angle between the isolated benzoate chromophores), the bisignate couplet is negative overall (negative chirality) as in Figure 13.51. Obviously, the enantiomeric absolute configuration would give rise to a "positive chirality" couplet at the same wavelengths.

The following general requirements apply for exciton chirality to be manifest.

1. Electron delocalization between the two chromophores (equivalent to the presence of an inherently chiral chromophore) should be minimal and the CE couplet should be isolated from other strong CE bands. Substitution in the benzoate ring especially by a p-dimethylamino group has been recommended so as to displace the exciton couplet to longer wavelengths (ca. 311 nm) to achieve the second of these requirements. Analogous displacement of the couplet to 311 nm is achieved by application of the p-methoxycinnamate group (Wiesler, Vázquez and Nakanishi, 1987). For a discussion of other substituents and derivatives applied to this purpose, see Section 13-4.e, Cottonogenic substituents. The p-(dimethylamino)cinnamate group that absorbs at even longer wavelengths (ca. 361 nm) has been applied to the determination of configuration of mitomycin C, an antitumor antibiotic (Verdine and Nakanishi, 1985).

2. The directions of the electric transition moments of the individual chromophores must be known for the chromophore(s) being examined. Chromophores of high symmetry are, consequently, especially desirable. In many cases (e.g., benzoate esters) these moments are well known. In other cases, evaluation of $\boldsymbol{\mu}$ may require the analysis of polarized single crystals or of linear dichroism measurements on solutes oriented anisotropically (e.g., in liquid crystals), or suspensions in stretched polymer films.

3. Exciton chirality is generally limited to chromophores that are isotropically strongly absorbing, especially aromatic compounds and acyclic conjugated systems. The method may nevertheless be more widely applicable than the foregoing statement suggests, for example, to allyl benzoates in which only one of the CEs of the bisignate couplet is evident (Gonnella, Nakanishi, et al., 1982).

4. The amplitude ($\Sigma\Delta\varepsilon$) of split CEs is inversely proportional to the square of the interchromophoric distance (Harada, Chen and Nakanishi, 1975). This

fact would suggest that the interacting chromophores must be fairly close to one another, though not necessarily adjacent. Nevertheless, in spite of weak coupling due to the remoteness of interacting chromophores, the exciton chirality method has been successfully applied to such systems as the dibenzoate of 2,17-dihydroxy-A-nor-5α-steroids (Canceill, Collet, and Jacques, 1982).

5. The attribution of bisignate CE couplets to exciton coupling requires care. If $\Sigma \Delta \varepsilon > 50$ there can be little doubt about the origin of the bands (see Table 13.2 and Fig. 13.51). In the case of weaker CEs, attribution of a bisignate couplet to exciton coupling may tentatively be made on the basis of the following rule of thumb: $\Sigma \Delta \varepsilon$ within the couplet should have at least twice the usual magnitude obtained with an identical but isolated chromophore absorbing in the same region of the spectrum (Snatzke, 1989).

In the case of complex chromophores, for example, in biflavones (atropisomeric bicoumarins), bisignate CEs are not due to "pure" exciton coupling and the exciton chirality method could not be applied. Assignment of configuration was carried out by theoretical calculation of the CD spectrum and comparison of the calculated spectrum with that measured experimentally (Harada, Khan, et al., 1992).

6. For glycol benzoate chromophores, the amplitude of the exciton coupling is maximal when the dihedral angle between the two chromophores $\omega \cong 70°$ (Fig. 13.51b), and is zero when $\omega = 0°$ or $180°$ (Harada and Nakanishi, 1983, pp. 55, 73).

From requirement 6, it follows that the absence of exciton coupling can also be revealing. The absence of expected coupling of the phthalide ^1L_a transition with the ^1B_b transition of the naphthyl ring in **18a** (Fig. 13.54a) at about 230 nm (as evidenced by the absence of split CEs) implies that the two transition dipole moments are aligned parallel to one another. This explanation was confirmed by a calculation using the DeVoe model (Pirkle, Salvadori, et al., 1988).

The exciton chirality method is readily applicable to a wide variety of systems consisting of two nonconjugated (or homoconjugated) chromophores. These range from the highly symmetrical (e.g., 1,1'-binaphthyl; Fig. 13.54b, **19**; Mason, 1982, Chapter 6) to those in which the two chromophores are very different from one another, for example, allyl benzoates (see above), and the p-chlorobenzoate derivatives of cinchonine and cinchonidine (**30** and **32**, respectively, in Fig. 7.19a;

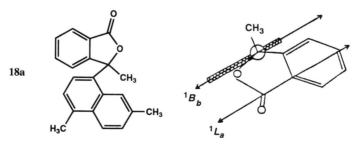

Figure 13.54a. Preferred conformation of (R)-3-(4,7-dimethyl-1-naphthyl)-3-methylphthalide **18a** showing the parallel alignment of the phthalide ^1L_a and the naphthyl ^1B_b transition dipole moments. [Reproduced with permission from Pirkle, W. H., Sowin, T. J., Salvadori, P., and Rosini, C. (1988), *J. Org. Chem.*, **53**, 826. Copyright © 1988 American Chemical Society.]

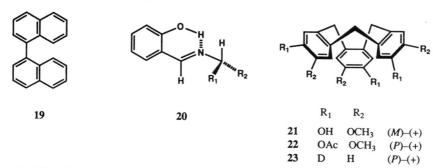

	R_1	R_2	
21	OH	OCH₃	(M)–(+)
22	OAc	OCH₃	(P)–(+)
23	D	H	(P)–(+)

Figure 13.54b. Compounds whose configurations have been determined by the exciton chirality method: 1,1′-binaphthyl **19**; a primary amine $R_1R_2CHNH_2$ as its salicylideneamino derivative **20** (as enolimine tautomer); cyclotribenzylenes **21**, **22**, and **23** with $R_1 \neq R_2$.

Harada and Nakanishi, 1972). Similarly, configurations of 1° amines and some α-amino acids and amino sugars may be predicted by analysis of the CD spectra of their salicylideneamino derivatives (salicylideneamino chirality rule; Smith, 1983). Intramolecularly hydrogen-bonded Schiff's base derivatives prepared, for example, by reaction of an arylalkylamine with salicylaldehyde (Fig. 13.54b, **20**) give rise to CEs attributed to exciton coupling between the salicylideneamino and phenyl chromophores (but no bisignate signals, presumably because the $\pi-\pi^*$ transitions of the phenyl chromophores are at too short a wavelength). For a thorough survey of applications of the exciton chirality method, see the book by Harada and Nakanishi (1983).

Systems consisting of three chromophores, for example, triptycenes as well as tribenzoates of sugars and steroids are also amenable to analysis by the exciton chirality method (Harada and Nakanishi, 1983). The cyclotriveratrylenes (e.g., **21**; Fig. 10.54b, $R_1 = OCH_3$, $R_2 = OH$) give rise to split CD bands from which the configuration may be deduced by application of exciton theory. The CD is extremely sensitive to the nature and position of the substituents, for example, acetylation of the OH groups of **21** gives rise to **22** in which the handedness of the CD couplets is reversed (Collet and Gottarelli, 1981). Even the case of **23** with $R_1 = D$, $R_2 = H$ (Fig. 10.54b) gives rise to a measurable and interpretable CD spectrum (Canceill, Collet, Gottarelli, et al., 1985).

Ring substituents rotate the electric transition dipole moments with respect to the short axes of the rings [for the ^1L_b (longer wavelength) transition at ca. 284 nm]. The sense and extent of this rotation depends on the relative magnitude (R_1 vs. R_2) of the substituent spectroscopic moments (Section 13-4.c, p. 1036). Changes in the relative magnitudes of the spectroscopic moments are responsible for the above-cited sign reversal in the CD couplet. A recent study of the CD spectra of substituted cyclotribenzylenes has demonstrated that analysis based on incorrect assumptions of polarization directions in the chromophores can sometimes lead to wrong configurational assignments (Canceill and Collet, 1986).

Exciton coupling in multichromophoric systems may be analyzed in terms of additivity of pairs of electronic transitions that couple. The CD of **24** is the sum of pairwise interactions, one **B/B**, one **C/C**, and four of the **B/C** type (Fig. 13.55). In fact, the CD spectrum of **24** is fairly accurately reproduced by summation of the CD spectra of the six glucopyranosides exhibiting each of the six above-mentioned pairwise interactions (basis pairs). Three-way or higher interaction

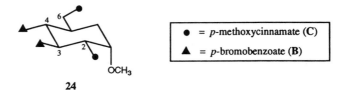

Figure 13.55. Abbreviated structural formula for methyl α-D-glucopyranoside 3,4-di-p-bromo-benzoate 2,6-di-p-methoxycinnamate **24**.

terms need not be considered (Liu and Nakanishi, 1982; Wiesler, Vázquez, and Nakanishi, 1987). This summation is an application of the *principle of pairwise additivity* (Kauzmann et al., 1961). The additivity relation holds also for amino, deoxy, and *N*-acetylated sugars; moreover, the additivity holds even when the sugar is derivatized with different chromophores.

The exciton chirality method has been applied to the structure determination of oligosaccharides, on their "chromophoric" degradation products (Vázquez, Wiesler, and Nakanishi, 1988; Chang, Nakanishi, et al., 1989). The identity the component monosaccharides, the linkage patterns of the latter and their configurations can be ascertained from the CD spectra. Analyses based on the additivity of bisignate CE pairs is facilitated by availability of a growing CD data base of bi-, tri-, and tetrachromophoric sugar derivatives (Wiesler, Nakanishi, et al., 1990).

It must be emphasized that this type of analysis is so successful because there is little ambiguity about the conformations of the sugar derivatives. The complex conformational equilibria that obtain for underivatized sugars give way to single pyranose conformers in the benzoate derivatives exemplified. While the orientation of the anomeric methyl group has little effect on the CD, the principal deviations from pairwise additivity have been ascribed to the stable C(5)–C(6) bond rotamers associated with 6-benzoates. The application of the exciton chirality method to the determination of configuration of acyclic 1,3-polyols has been facilitated by subtraction of CD spectra of appropriate derivatives (difference CD) (Mori et al., 1992).

Exciton coupling has been applied to the establishment of the configuration of cryptophanes incorporating six equivalent chromophores in a $\mathbf{D_3}$ point group (Canceill, Collet, et al., 1987; Collet and Gottarelli, 1989).

Intermolecular exciton coupling effects have also been observed and provide information about the helical organization of hydrogen-bonded oligomers in solution (Sheves, Mazur, et al., 1984).

e. Other Applications. Induced ORD and CD

Introduction

In addition to "structural analyses" described in the preceding sections (and below), CD measurements, in particular, have been applied also to other analytical problems, which are illustrated below. For reviews, see Barrett (1972), Scopes (1975), Legrand and Rougier (1977), Snatzke and Snatzke (1977) and Purdie and Swallows (1989). Qualitative and quantitative analysis of the plant growth regulator abscisic acid (Fig. 13.56, **25**), in plant extracts has been described by measurement of a specific CE in preference to polarimetric measurements at a single (or a few) wavelength (Dörffling, 1967; Cornforth et al., 1966).

Chiral reaction byproducts, particularly when present in very small amounts, are more easily identified and monitored in nonracemic samples by CD than by many other methods, for example, conjugated unsaturated ketones ($\lambda_{max} \approx$ 340 nm) in samples of the corresponding saturated ketones ($\lambda_{max} \approx$ 300 nm). A similar application of CD is the identification of diene **26** in **27** (Fig. 13.56) isolated from frankincense (Snatzke and Vértesy, 1967). The critical micelle concentration of a chiral detergent was determined by means of CD (de Weerd et al., 1984). A CD detector has been applied to analyses and chromatographic resolutions by high-performance liquid chromatography (HPLC) to permit direct determination of the enantiomeric purity (Drake, Gould, and Mason, 1980; see also Salvadori et al., 1984).

Numerous other applications involve the induction of ORD or CD either in achiral or chiral compounds under the influence of temperature, different solvents or environments, or under the influence of a variety of reagents. Optical activity can be induced in racemates by cpl via the photointerconversion of the enantiomers, as well as by the preferential photodestruction of one enantiomer of a racemate (Zandomeneghi, Cavazza, and Pietra, 1984; Cavazza et al., 1991). Chiroptical properties induced in achiral compounds are often called "extrinsic" to distinguish them from those (*intrinsic* properties) of chiral compounds.

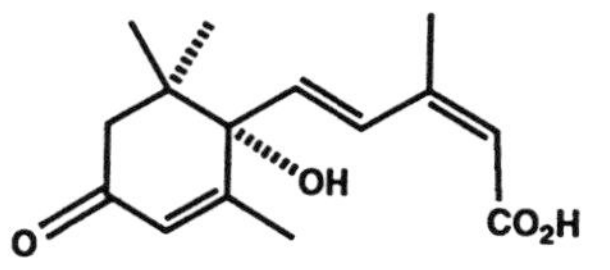

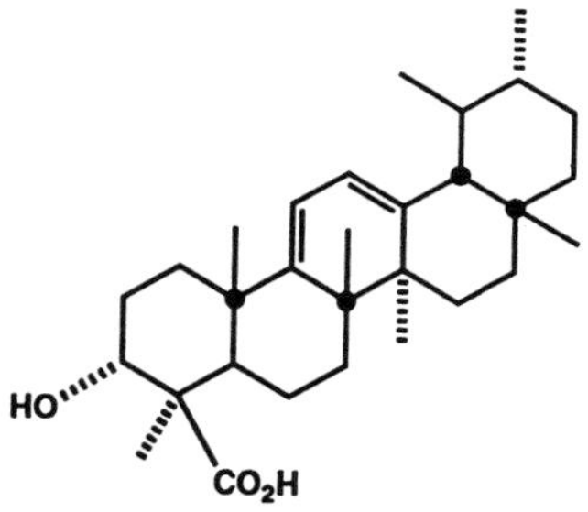

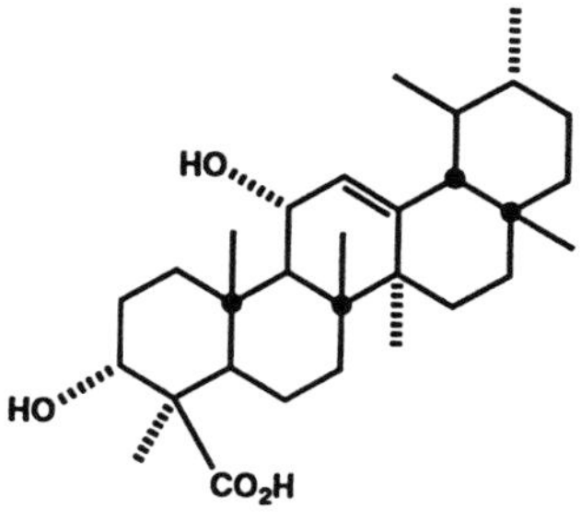

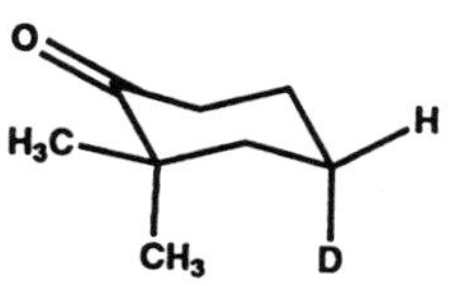

Figure 13.56. Structures **25–28**.

A recent report describes evidence that irradiation of the two enantiomers of a chiral compound with opposite senses of cpl (in a transparent region of the spectrum) induces differential ^{1}H NMR chemical shift changes. No significant heating is observed and the involvement of the reverse Faraday effect appears to be precluded. While the observations (termed laser-enhanced NMR spectroscopy) are thus far unexplained, the potential of this combination of CD and NMR is an exciting prospect (Warren et al., 1992).

Changes in the CD spectra of adducts of guanosine with carcinogenic hydrocarbons induced by pH conveniently permit determination of the pK_a values that reflect the point of attachment of the hydrocarbon on the base (Nakanishi et al., 1977).

Optical rotary dispersion is a sensitive reporter of solvent-induced changes in the position of equilibrium in conformationally mobile systems. In a study aimed at assessing specific solvent effects, the conformational equilibrium of (+)-*trans*-2-chloro-5-methylcyclohexanone (cf. Section 13-2.c, p. 1024) was reexamined in 28 solvents; $[\Phi]_{330}$ values ranged from 680 in dimethyl sulfoxide (DMSO) to 8.3 in cyclohexane (Menger and Boyer, 1984).

Temperature dependence of the CD is generally interpreted as being caused by equilibration between at least two chiral species (Moscowitz, Wellman, and Djerassi, 1963). The method is sufficiently sensitive to estimate the conformational energy of deuterium versus hydrogen. In (4S)-2,2-dimethyl-4-deuteriocyclohexanone (Fig. 13.56, **28**), the gem-dimethyl groups serve as "chiral probes" for the deuterium that resides in the Octant rule x,z symmetry plane. Changes in the rotational strength of the carbonyl $n-\pi^*$ transition as a function of temperature can be related to the conformational equilibrium, leading to the conclusion that axial D is preferred over H with $\Delta H^\circ = -3.3 \, \text{cal mol}^{-1}$ ($-13.8 \, \text{J mol}^{-1}$) (Lee, Barth, and Djerassi, 1978; see, however, p. 719).

The observation of a change in sign of the CE of (S)-4-methyl-3-hexanone (Fig. 13.57, **29**) at about 290 nm when measured in heptane as against the vapor phase (both at 27°C) has led to the suggestion that a change in magnitude and

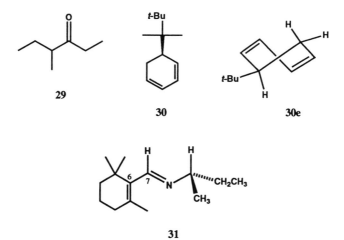

Figure 13.57. Structures **29–31**.

sign of the same CE with temperature (in the hydrocarbon solvent) is not solely attributable to a conformational equilibrium. Strong dependence of the $n-\pi^*$ transition on solute–solute interactions must be responsible for the change in sign of the CE resulting from the aforementioned change in phase (Lardicci, Pino, et al., 1968).

Conformational analysis of $(+)$-$(5R)$-*tert*-butyl-1,3-cyclohexadiene **30**, a compound that incorporates a diene dissymmetric chromophore (Fig. 13.57), by variable temperature CD, leads to the conclusion that the *tert*-butyl group in the pseudoequatorial conformation **30e** is only slightly preferred over the pseudoaxial conformation $[\Delta G^{\circ}_{\text{ax}-\text{eq}}\ 0.40 \pm 0.05\ \text{kcal mol}^{-1}\ (1.67 \pm 0.21\ \text{kJ mol}^{-1})]$ in major contrast to what obtains in *tert*-butylcyclohexane (Lightner and Chappuis, 1981).

Temperature dependence of the CD that is observed in the β-cyclocitral Schiff's base (Fig. 13.57, **31**), is ascribed to s-cis–s-trans equilibration about the C(6)–C(7) bond. Twisting about this bond yields an inherently chiral chromophore (Buss et al., 1984; for another example of temperature dependent CD, see Section 13-4.c, p. 1043). For a review of solvent and temperature effects as probed by CD, see Legrand (1973).

Induced ORD and CD of achiral compounds was apparently first demonstrated on dyes bound to polypeptides in their helical (but not random) conformations (Blout and Stryer, 1959). Studies of this type were suggested as tools for probing the conformations of macromolecules (Stryer and Blout, 1961) (cf. Section 13-4.f). The first clearcut demonstration of extrinsic CD induced in achiral molecules dissolved in nonracemic chiral solvents appears to be that of $Na_2[PtCl_4]_2$ in $(-)$-2,3-butanediol. The metal complex exhibited a negative CD band at about 400 nm (Bosnich, 1966). Bosnich drew attention to the possibility of determining the polarization direction of the transition from the sign of the CD. Induced CD was subsequently observed in the $n-\pi^*$ transition of benzil and benzophenone. Both exhibit negative CD at about 390 and 340 nm, respectively, when dissolved in $(-)$-2,3-butanediol (Bosnich, 1967), and $(-)$-menthol induces CD in the spectra of acetone, cyclohexanone, and nitromethane (Noack, 1969).

Bosnich described two possible mechanisms that might be responsible for the induced CD: (a) the generation of stable chiral conformations in the carbonyl compounds under the influence of hydrogen bonding between solute carbonyl group(s) and solvent hydroxyl groups, or (b) dissymmetric solvation of the achiral solutes possibly involving the aromatic rings. The available evidence did not permit Bosnich to rule out the operation of one or the other of these mechanisms. Bosnich called attention to the relation between the induced chiroptical properties and the operation of an asymmetric transformation of the first kind (see Section 7-3.e). Induced CD may serve as a sensitive probe of the geometry of interaction between the chiral inducer and the achiral chromophore (Bosnich, 1966). It has since been established that "transmission of chirality between freely rotating systems" is also possible. A model for such dispersion-induced CD has been proposed (Schipper, 1981; Schipper and Rodger, 1985).

Induction of chiroptical properties in achiral covalent or ionic molecules may take place (a) by their solution in nonracemic chiral solvents (Hayward and Totty, 1969; Hayward, 1975; and see above), in rigid chiral polymers (Saeva and Olin, 1977), in biopolymers (Hatano, 1986), and in chiral surfactants (Tachibana and Kuihara, 1976); (b) in associated ion pairs (Takenaka et al., 1978); (c) on addition of nonracemic chiral solutes (Sherrington et al., 1982; Okamoto et al., 1984); (d) on covalent derivatization with chiral reagents (Breslow et al., 1973; Lightner et al., 1987a; Nishino et al., 1992); (e) by addition of chiral shift reagents (see the following section); and (f) by inclusion compound formation with chiral host molecules. The geometry of inclusion of hydrocarbons, such as pyrene, into the cavity of β-cyclodextrin could be inferred from the sign of the induced CD with information about the direction of polarization of the transition, as previously suggested by Bosnich (see above; Kobayashi, Osa, et al., 1982). Induction of extrinsic CD in achiral dyes on simple mechanical stirring of solutions has been shown to be an artifact of linear dichroism (Saeva and Olin, 1977).

A well-known and dramatic example of induced CD involves the achiral metabolite bilirubin **32** (Fig. 13.58*a*), a major constituent of bile, that is produced in humans by degradation of heme and is responsible for the pigmentation evident in jaundice. Bilirubin can exhibit induced CD when simply dissolved in α-methylbenzylamine (Blauer, 1983; Lightner et al., 1987b), when examined in the presence of cyclodextrins (Lightner et al., 1985c), in the presence of albumins (Lightner et al., 1986b), on complexation with cinchona alkaloids in CH$_2$Cl$_2$

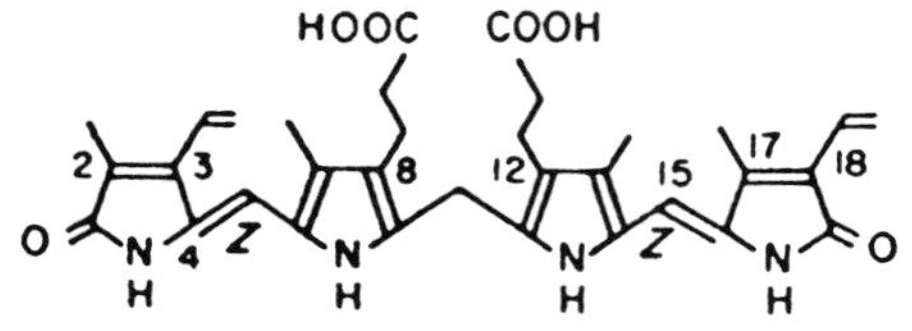

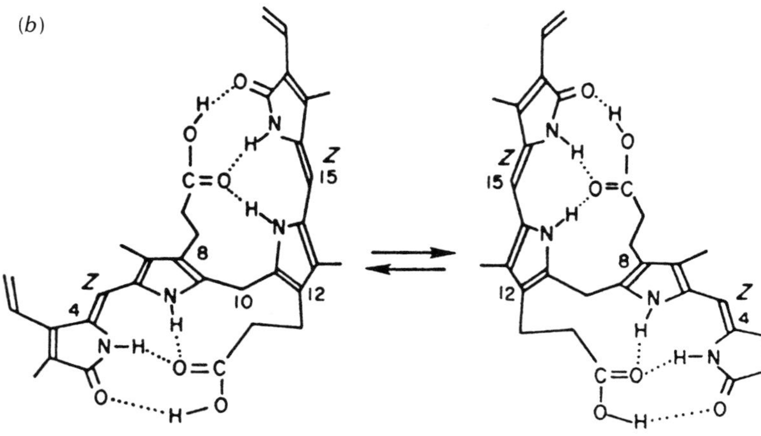

Figure 13.58. (*a*) Structure of (4*Z*,15*Z*)-bilirubin **32**; (*b*) interconverting intramolecularly hydrogen-bonded enantiomeric conformers of **32**. [Reproduced with permission from Lightner, D. A., Gawronski, J. K., and Wijekoon, W. M. D. (1987), *J. Am. Chem. Soc.*, **109**, 6354. Copyright © 1987 American Chemical Society.]

solution (Lightner et al., 1987a), or in covalent derivatives, such as the N-acetyl-L-cysteine adducts to the C(18) vinyl group (Lightner et al., 1988). The nature of the inherently dissymmetric chromophore exhibited by bilirubin and its chiral as well as achiral analogues was first recognized and studied by Moscowitz et al. (1964). Intramolecular hydrogen bonding permits a helical conformation to be easily achieved. Except in covalent derivatives, the induced CD results from an asymmetric transformation of the two interconverting enantiomeric conformations of bilirubin (Fig. 13.58b).

The induction of CD in a racemic helical polymer on dissolution in a nonracemic chiral solvent has been described in Section 7-2.e.

Cottonogenic Substituents

In order to overcome the inability of CD and ORD spectrometers to record Cotton effects in the UV region below about 210 nm, the so-called transparent functional groups were derivatized so that they would reveal themselves through new chromophores that could be studied in easily accessible wavelength regions (Djerassi, 1960, Chapter 15; Djerassi, 1964; Sjöberg, 1967; Scopes, 1975). To emphasize the generation of CEs, such derivatives have been dubbed *cottonogenic* (or chromophoric) (Sjöberg, 1967). For example, simple alcohols do not exhibit CD bands down to 200 nm; and, while they do exhibit plain ORD curves, deductions of configuration from such plain curves are either unfeasible or unsafe (Crabbé, 1972). Similarly, deductions about configuration were difficult to make for many simple aliphatic and alicyclic amines. These functional groups were converted, for example, to nitrites or to hemiphthalates in the case of alcohols and to phthalimides or to isothiocyanates in the case of amines. Thus, for example, 2-amino-1-butanol exhibits three CEs from 200 to 350 nm when derivatized as the isothiocyanate of the methyl carbonate ester (Fig. 13.59, **33**; Halpern et al., 1969).

Extensive lists of such derivatives and discussion of the configurational correlations made possible by examination of their CEs have been given by Crabbé (1965, 1968), Sjöberg (1967), and by Crabbé and Parker (1972). The principal types of derivatives that have been studied are, for alcohols, esters [in the early literature, acetates, xanthates, and nitrites (−O−N=O), and more recently, benzoates and cinnamates]; for amines, dithiocarbamates, N-salicylidene and N-phthaloyl derivatives, dimedone (enamine) derivatives, N-nitroso and N-chloro derivatives; for carboxylic acids, acylthioureas, and thionamides.

Organometallic derivatives have figured prominently in studies of ORD and CD. Cuprammonium solutions of glycols and of amino alcohols exhibit CEs in the

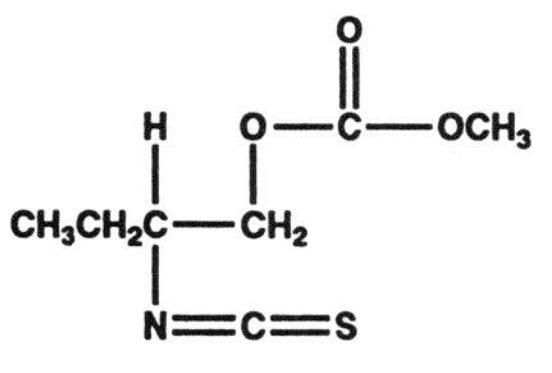

33   **Figure 13.59.** Cottonogenic derivative of 2-amino-1-butanol.

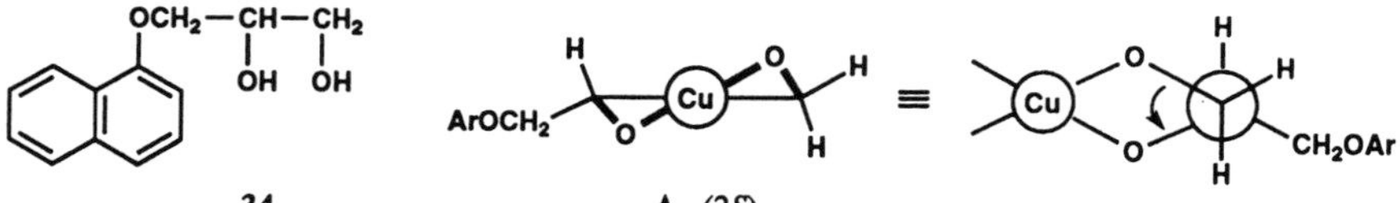

Figure 13.60a. Cupra A complex of (2*S*)-diol **34** in the λ conformation with the aryloxymethylene substituent in the equatorial position. [Reproduced with permission from Nelson, W. L., Wennerstrom, J. E., and Sankar, S. R. (1977), *J. Org. Chem.*, **42**, 1006. Copyright © 1977 American Chemical Society.]

vicinity of 600 nm in the metallic *d–d* transition region as a result of 1:1 in situ complex formation between the copper and a bidentate ligand. This finding has been called the Cupra A effect. The sign of this CE has been correlated with the sense of chirality particularly in the carbohydrate series (Reeves, 1951, 1965). For example, (2*S*)-diol (Fig. 13.60a, **34**) exhibits two CEs when dissolved in Cupra A solution: $[\theta]_{540}$ −50 and $[\theta]_{320}$ +370 both of these being associated with a complex of structure **A** (Fig. 13.60a) having an absolute configuration with a negative chirality sense (Bukhari, Scott, et al., 1970; Nelson et al., 1977). The method may be applicable to α-hydroxy acids as well (Nelson and Bartels, 1982).

The NMR lanthanide shift reagents, such as Pr(DPM)₃ (DPM = dipivaloylmethane) and the nickel complex Ni(acac)₂ (acac = 2,5-hexanedione), induce long wavelength CE's originating in the transitions of the metallic element. These CEs may be empirically linked to the configurations of vicinal glycols and amino alcohols (Dillon and Nakanishi, 1975).

Molybdenum, rhodium, and ruthenium cations form acylate (e.g., acetate) complexes of the general formula [Met₂(O₂CR)₄; Fig. 13.60b, **A**] whose ligands may be rapidly exchanged in situ with (in the case of rhodium, exchange requires heating) or add typically nonabsorbing chiral compounds, for example, carboxylic acids, diols, amino alcohols, peptides, and nucleosides. The chiral complexes

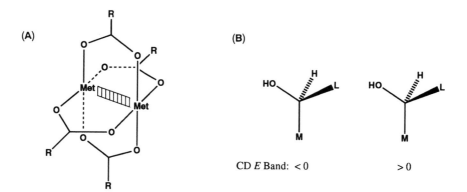

Figure 13.60b. (A) Acylate complexes of type [Met₂(O₂CR)₄], Met = Mo²⁺, Rh²⁺, Ru²⁺, Ru³⁺ (R = CH₃, CF₃, C₃H₇); (B) relation of absolute configuration of 2° alcohol to the sign of the CE at about 350 nm (*E* band) generated by complex formation with [Rh₂(O₂CCF₃)₄]; M and L are medium and large groups, respectively. [Adapted with permission from Gerards, M. and Snatzke, G., *Tetrahedron: Asymmetry*, **1**, 221 (1990). Copyright © 1990 Elsevier Science Ltd, The Boulevard, Langford Lane, Kidlington, OX5 1GB, UK.]

exhibit several CEs, the most reliable for determination of absolute configuration being those in the 300–400-nm range. Since the conformation of the ligands is restricted in the complex, the sign of the CE may be used for the empirical assignment of configuration of the complexing ligands (mostly bidentate) molecules in a reliable way (Snatzke et al., 1981; Frelek, Snatzke, et al., 1985, 1986).

Secondary alcohols, alkenes, epoxides, and ethers rapidly form in situ complexes by binding to the axial position of $[Rh_2(O_2CCF_3)_4]$ (Fig. 13.60b, **A**). The CD spectra of the complexes exhibit up to five CEs in the 300–600-nm range. In the case of 2° alcohols, the configuration can be established from the sign of the CE at about 350 nm (E band) by means of empirical rules (Fig. 13.60b, **B**; Gerards and Snatzke, 1990).

Widespread application of cottonogenic derivatives to the determination of configuration of chiral compounds containing two or more chromophores in close proximity to one another, for example, CD measurement of a chiral 1,2-diol following its conversion to a dibenzoate, is described in Section 13-4.d in connection with the operation of the exciton splitting mechanism.

Liquid Crystal Induced Optical Activity and CD

"Liquid crystals" is a general term that describes intermediate phases (or mesophases) existing between the solid crystalline state and the isotropic liquid state. On warming, liquid crystals are characterized by the stepwise occurrence of reversible transitions between the solid and several possible mesophases. Many compounds possessing rodlike or, less commonly disclike, molecular structures exhibit such behavior. When such compounds are heated, the regular 3D organization of the solid gives way to a two-dimensional (2D) layered structure wherein the rods are aligned with their long axes orthogonal or tilted to the layer planes (smectic phases). On further warming, the smectic phase gives way to a turbid nematic phase in which the rods remain oriented parallel along their long axes (thread like) but the layers disappear. In the case of chiral molecules, the smectic phase on warming forms a twisted nematic (cholesteric) phase that retains local nematic packing. However, the long axes of molecules in individual layers are rotated about an axis that is perpendicular to the layer planes formed into a stack. Cholesteric phases thus have helical structure. The three types of mesophases, smectic, nematic, and cholesteric, all of which are visibly turbid, together are called liquid crystals. On further warming, both nematic and cholesteric phases give way to isotropic liquids (Brown and Crooker, 1983).

Cholesteric liquid crystals, formed exclusively from chiral molecules (e.g., cholesteryl benzoate), are characterized by the helix pitch p of the molecule (the pitch is the periodicity of twist; it is given by the distance corresponding to a 360° turn of the long axis) and by the handedness of the helix (Solladié and Zimmerman, 1984). Cholesteric mesophases exhibit extremely high molar rotations (e.g., in excess of 50,000). Cholesteric phases can apparently induce CD in *achiral* solute molecules (liquid crystal induced circular dichroism, LCICD). The cholesteric liquid crystal does not behave just as a chiral solvent; it can be shown that the induction is caused by the macroscopic helical structure of the mesophase (Saeva and Wysocki, 1971). The CD spectra of anthracene and even substituted benzenes dissolved, for example, in a mixture of cholesteryl nonanoate and cholesteryl chloride, exhibit CEs in the absorption bands of the solute (Saeva et al., 1973).

> The just cited induction has been qualified with the adverb "apparently" (see above). This qualification is necessary because Shindo and Ohmi (1985) called attention to the possible generation of artifacts when measurements are made on cholesteric mesophases as a result of the coupling of nonideal optics and the electronics of CD spectrometers. These artifacts give the appearance of CD induced by achiral molecules. Caution must evidently be exercised in interpreting LCICD spectra.

When nonracemic *chiral* solutes are dissolved in achiral nematic phases, the latter are transformed into cholesteric phases. This observation, made by Friedel in 1922 (q.v.), was ignored for nearly 50 years. Small amounts of virtually any chiral substance, including those that do not form mesophases by themselves, can effect this transformation. One of the most widely used thermotropic liquid crystals (those whose order is altered on heating) for induction of optical activity or CD is *N*-(4-methoxybenzylidene)-4'-butylaniline (MBBA) **35**, Figure 13.61.

The induction of cholesteric phases may be observed with a polarizing microscope in the visible range, and by infrared ORD (Korte et al., 1978). At low concentrations, the induced pitch (p, in μm; methods for the measurement of the pitch are described in Solladié and Zimmerman, 1984) is inversely proportional to the molar concentration of the solute c and to its enantiomeric purity r (Eq. 13.25).

$$1/p = \beta_\mathrm{M}\, r\, c \qquad\qquad (13.25)$$

It has been pointed out that while the sign (handedness) and the magnitude of the helical twisting power β_M may characterize a chiral compound just as well as the specific rotation $[\alpha]$, the origin of the two properties is quite different. The rotation is a measure of the interaction of light with a chromophore in the molecule, while β is dependent on the nature of, and the extent of interaction between, solute and solvent molecules (Gottarelli, Spada, and Solladié, 1986).

As a consequence of their very high rotatory power, induced cholesteric mesophases may serve to detect or to amplify very small optical activities, for example, in compounds whose chirality is due solely to isotopic substitution (Gottarelli et al., 1981), or to observe the rotatory power in a very small amount of a compound (Gaubert, 1939). This induction is the basis of a test for conglomerate behavior described by Penot, Jacques, and Billard (1968). Induction of cholesteric behavior in nematic phases (also termed LCICD) has been characterized as a method of "chirality amplification" (Rinaldi et al., 1982; Gottarelli, Spada, and Solladié, 1986). Chiroptical properties of large magnitude

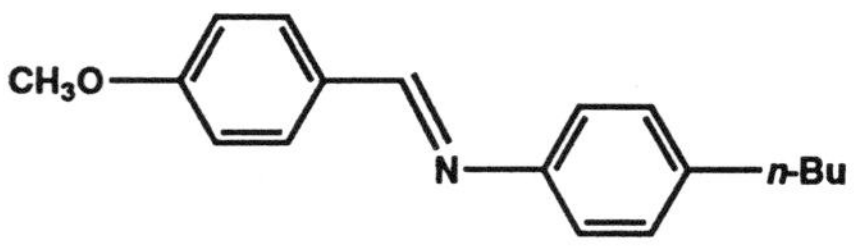

35

Figure 13.61. *N*-(4-Methoxybenzylidene)-4'-butylaniline (MBBA).

are indeed produced when the chiral solute induces the formation of chiral conformations in the achiral liquid crystalline solvent. However, we believe that it is misleading to call the phenomenon "amplification of chirality" since the observed chiroptical properties are due to both the solute and the solvent, and they arise as a result of incompletely understood interactions.

From Eq. 13.25, it follows that the thermal racemization of a chiral compound can be followed as a function of time by measurement of the variation of the pitch of a cholesteric mesophase created by dissolving the chiral compound in a nematic phase (the nematic director).

Numerous studies have been made to relate the helical sense of the induced cholesteric phase to the configuration of the chiral compound responsible for the induction. While the two enantiomers of the guest always induce cholesteric phases of opposite handedness, the induced sense unfortunately is not related to the configuration of the inducing molecule in a simple way.

In general, it can be said that close matching of the structure and conformation of the chiral solutes and the nematic director yields the largest twisting power β_M and the best correlation of the configuration of the solute molecules with the sign of β_M. High twisting power is a sign of strong solute–solvent interaction (Solladié and Zimmerman, 1984; Solladié and Gottarelli, 1987).

f. Fluorescence Detected Circular Dichroism

When a nonracemic sample of a chiral compound is excited with cpl and the resulting fluorescence is monitored, the chiroptical information so obtained is called fluorescence detected circular dichroism (FDCD)(Brittain, 1985). This technique was developed by Tinoco, Jr., and co-workers (Turner, Tinoco, Jr., and Maestre, 1974). The FDCD signal is defined by Eq. 13.26

$$\text{Signal} = \frac{2(F_L - F_R)}{F_L + F_R} = \frac{2(g_F - 2R)}{2 - g_F R} \approx g_F - 2R \tag{13.26}$$

where F_L and F_R are fluorescence intensities excited by left and right cpl, respectively, R is a function of the absorbances for lcpl and rcpl as obtained from CD and absorbance measurements, and g_F is the FDCD g number (dissymmetry factor) $[g_F = \Delta\varepsilon_F/\varepsilon_F]$ for the chromophore being observed (analogous to the g_{abs} of Eq. 13.18 (Section 13-4.a; Lamos, Lobenstine, and Turner, 1986). Measurement of FDCD is made possible by a detector attachment mounted perpendicular both to the exciting light and to the standard detector of a CD spectrometer.

It must be emphasized that, although the measurement is one of luminescence, FDCD does not measure the circular polarization of the emitted radiation, a technique that provides information about the excited state of molecules [cf. circular polarization of emission (CPE), Section 13-7]. Fluorescence detected circular dichroism reveals information about the ground state of molecules just as does the ordinary (sometimes termed transmission) CD; however, it does so with a significant difference.

The FDCD technique is more selective than CD in that it permits measurement of the CD solely of fluorescent chromophores (so-called fluorophores). A mixture of (S)-tryptophan and (R)-cystine, for example, exhibits a CD spectrum whose main features are due to the cystine; in contrast, the FDCD spectrum of the mixture only reveals the presence of the tryptophan since cystine is non-fluorescent (Turner, Tinoco, Jr., and Maestre, 1974).

Since fluorescence is a phenomenon that requires both absorption and emission, g_F is a measure of the chirality of the fluorophore (or, more specifically, of one of its electronic transitions) plus that of any molecular entity that transfers energy to it. The FDCD technique has been applied to proteins, such as adrenocorticotropic hormone (ACTH), monellin, ribonuclease T1 (RNase T1), and human serum albumin (HSA), each of which contain a single, fluorescent tryptophan. In the proteins studied, part of the observed fluorescence is believed to be due to energy transfer from tyrosine. Of the several proteins studied, monellin and RNase T1 have the largest absolute g_F values (at ca. 235 nm) implying that the tryptophans in these protein molecules are "rigidly ordered". Fluorescence detected circular dichroism is, therefore, a selective chiroptical technique that is sensitive to the local environment of the fluorophore (Lobenstine, Schaefer, and Turner, 1981).

The FDCD technique has been shown to be useful in the study of fluorescent drug binding to macromolecules (e.g., homidium, a substituted phenanthridinium ion) bound to nucleic acids, even in in vivo systems (Lamos, Lobenstine, and Turner, 1986) and the technique has been applied to assess the contributions of side-chains and backbone α helix to the CD of poly(L-tryptophan). Noncoincident FDCD and CD spectra were ascribed to exciton coupling between indole chromophores in the polymer side chains (at ca. 230 nm) (Muto et al., 1986). For other applications of FDCD, see Section 13-7.

g. Circular Dichroism of Chiral Polymers

The use of CD spectroscopy, and to a lesser extent ORD, is especially important in the study of macromolecules because it is one of the few spectroscopic techniques that is able to probe the helical secondary structure that characterizes such molecules. In contrast to X-ray diffraction techniques, CD and ORD provide information about the conformations of large molecules in solution. As a consequence, applications of CD to the analysis of biopolymers, in particular, exceed those to virtually any other class of compounds. Numerous books, chapters, and review articles have summarized the many applications of chiroptical properties to the study of biopolymer conformations (Goodman et al., 1970; Scheraga, 1971; Van Holde, 1971; Ciardelli and Salvadori, 1973; Jirgensons, 1973; Freifelder, 1976; Charney, 1979, Chapter 8; Ciardelli et al., 1979; Sélégny, 1979; Johnson, 1985; Urry, 1985; and Woody, 1985). This section provides but a brief overview of the chiroptical properties of biopolymers and of synthetic polymers, a very active research area endowed with a large literature.

As we have already seen, CD is very sensitive to the interaction between neighboring chromophores. In polypeptides and in polynucleotides, for example, interaction between adjacent amide groups and adjacent aromatic nuclei, respectively, during light absorption accounts for most of the intensity of the CD bands.

Since the configurations of the constituent amino acids and nucleotides are known, the CD principally provides information about the conformations (i.e., the secondary structure) of the biopolymers that are built up from these chiral monomers. The key structural elements that are responsible for the wavelengths and intensities of CD bands are the relative orientation of the neighboring chromophores and the distance between them. Since the magnitude of CD bands that arise from interaction between chromophores falls off rapidly as the chromophores are further removed from one another, in polychromophoric systems it is only adjacent absorbing groups that need be considered in first-order analyses of CD spectra (recall that the amplitude of exciton splitting depends approximately on the inverse square of the distance between the chromophores; Section 13-4.d). These chromophores have a greater effect on CD spectra than does the primary structure, that is, the number, kind, and location of stereocenters in the backbone and side chains of chiral macromolecules (Cantor and Schimmel, 1980; Johnson, 1985).

Biopolymers

Let us first consider the case of polypeptides and proteins. As a consequence of the conformational mobility of single bonds flanking the chiral center of amino acids, linear polypeptides spontaneously assume the following principal conformations in solution:

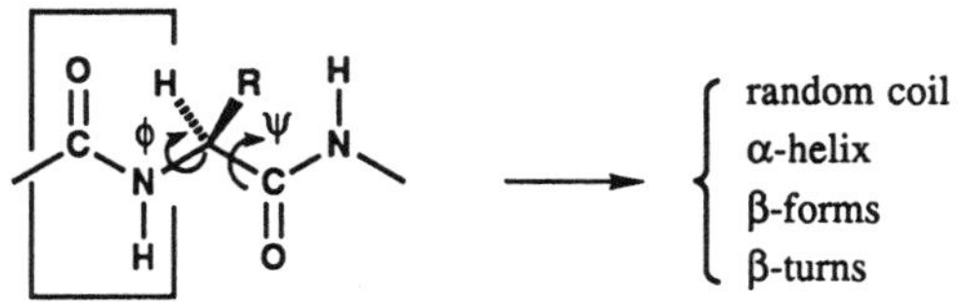

Figure 13.62. Organization of polypeptides into their principal conformational forms. The atoms within the rectangle constitute a rigid planar unit.

These conformations are stabilized especially by *intra*molecular hydrogen bonds in the case of α helices (typically right handed in the case of polymers of L-amino acids) and by *intra*molecular as well as *inter*molecular hydrogen bonds in the case of β forms. When such conformations are not able to form, for example in polymers of a single amino acid having repeating R groups bearing like charges, such as poly(L-glutamate) (at pH 8) or poly(L-lysine) (at pH 7), the polymer molecules assume flexible and partially disordered arrangements called random coils. When the charges are neutralized, for example, at pH 4.5 for poly(L-glutamic acid) (Fig. 13.63a) and at pH 12 for poly(L-lysine) (Fig. 13.63b), conformations approximating α helices form spontaneously.

The CD of polypeptides in the random coil conformation does not approximate the sum of the monomer (dipeptide) contributions to the CD. This finding reflects the fact that polypeptides are partially organized even in the random coil form (Goodman et al., 1970). The term random coil is therefore a misnomer. Designations, such as unordered (or aperiodic) form(s), are to be preferred.

The organization of peptides into stereoregular forms begins when the number of peptide units in the polymer reaches a range of 5–12 (Blout, 1973). As the polypeptide molecules organize themselves by extending the chains and by

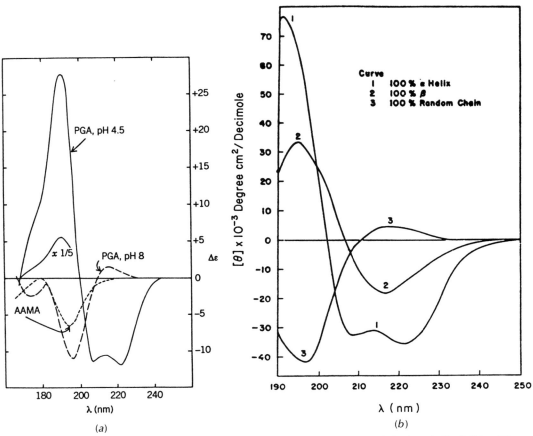

Figure 13.63. (*a*) CD of poly(L-glutamic acid) (—) and poly(L-glutamate) (– –): α helix (PGA, pH 4.5) and random coil (PGA, pH 8). CD of *N*-acetyl-L-alanine-*N*′-methylamide (AAMA, · · · ·) [Reproduced with permission from Johnson, W. C., Jr., and Tinoco, I., Jr. (1972), *J. Am. Chem. Soc.*, **94**, 4389. Copyright © 1972 American Chemical Society]; (*b*) CD of poly(L-lysine): (1) α helix; (2) β form; (3) random coil. [Reproduced with permission from Greenfield, N. and Fasman, G. D. (1969), *Biochemistry*, **8**, 4108. Copyright © 1969 American Chemical Society.]

hydrogen bonding intermolecularly (between NH and C=O groups) with like chains either in parallel or antiparallel ways (β forms), the CD bands increase in intensity and change in sign, the CD spectrum becoming virtually enantiomeric with that of the random coil form (Fig. 13.63*b*).

When intramolecular hydrogen bonding prevails, coiling of the polymer chain into an approximate α helix takes place. Under such conditions, there is a dramatic increase in the intensity of the CE in the vicinity of 190 nm. The peptide backbone itself having become helical, absorption of radiation by numerous properly oriented like chromophores is enhanced giving rise to large CEs by the interaction of the chromophores with near neighbors (exciton coupling; see below and Section 13-4.d). The CD spectroscopy allows the presence of the α helices in proteins to be inferred with greater certainty than that of any other secondary structural feature (Regan and De Grado, 1988).

The CD spectrum of poly(L-alanine), which is a fairly typical homopolypeptide, in the α helix form is shown in Figure 13.64. The experimental (boldface) CD spectrum exhibits three distinctive extrema, at 191, 207, and 221 nm. The spectrum can be deconvoluted into component (theoretical) Gaussian curves (Fig. 13.64) addition of which approximates the experimental spectrum. This analysis reveals the presence of the negative exciton couplet due to splitting of the $\pi - \pi^*$ transition as found in right-hand α helixes of L-polypeptides. The long wavelength negative band at 221 nm is due to the carbonyl $n_\tau \pi^*$ chromophore. In the absence of solvent interactions or of strongly absorbing (chromophoric) side chains, as in phenylalanine, the CD spectra of many polypeptides are essentially identical well into the vacuum UV region (to ca. 140 nm).

The amide absorption is associated with a strong electric transition moment. As pointed out originally by Moffitt (1956a; Moffitt and Yang, 1956), the transition of individual amide groups in the helix gives rise to, that is, splits, into two (helix) transitions whose moments are aligned either parallel or perpendicular to the screw axis (Fig. 13.65; review: Charney, 1979, Chapter 8).

Just as in the CD of glycol dibenzoates (Section 13-4.d), the relative positions (i.e., wavelengths), of the $-$ and $+$ CEs resulting from the coupled electric and

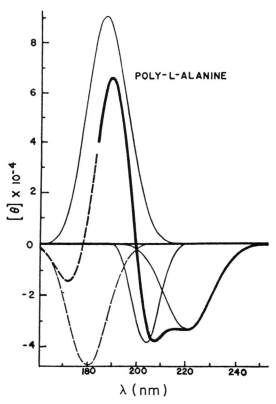

Figure 13.64. Resolved CD spectrum of the pure helical form of poly(L-alanine) in trifluoroethanol–trifluoroacetic acid (98.5 : 1.5 v/v). The bold faced curve represents the experimental data. The 180 nm negative CD band (——) is inferred to facilitate and improve the curve resolution. [Reproduced with permission from Quadrifoglio, F. and Urry, D. W. (1968), *J. Am. Chem. Soc.*, **90**, 2755. Copyright © 1968 American Chemical Society.]

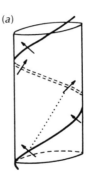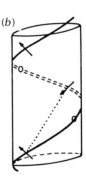

(a)   *(b)*

Figure 13.65. Polarization of $\pi-\pi^*$ electric transition moments of polypeptide amide bonds along a right-handed α helix. (*a*) Polarization is parallel ($\parallel$), and (*b*) polarization is perpendicular ($\perp$) to the helix (or screw) axis. Coupling of each $\boldsymbol{\mu}$ with the associated $\mathbf{m}$ gives rise to $+$ and $-$ CE couplets, respectively. [Reproduced with permission from Moffitt, W. (1956), *Proc. Natl. Acad. Sci. USA*, **42**, 736.]

magnetic transition moments, respectively, are related to the handedness of the α helix: A negative couplet (Fig. 13.51) requires that the polypeptide α helix be right handed (*P* configuration) (Snatzke, 1982; Mason, 1982).

A second major stereochemical feature distinguishable in CD spectra is the so-called β sheet conformation. The latter, exemplified by poly(L-lysine) (at pH 11.1), consists of adjacent molecules aligned side by side either in parallel or in antiparallel senses. Beta sheets exhibit a negative band at about 216 nm, a positive band between 195 and 200 nm, and a negative band near 175 nm (Fig. 13.63*b*, curve 2). The CD spectral features of β forms are subject to more variation than are those of α helix forms. At lower pH, poly(L-lysine) (pH 5.7) becomes unordered, and, although strong CD bands are still evident especially in the vicinity of 200 nm, these are of limited usefulness in conformational analysis (Fig. 13.63*b*, curve 3; Woody, 1985).

Interest in the measurement of chiroptical properties increased beginning in the late 1950s when it became evident that ORD (and later CD) spectra could provide measures of the extent of α helix and β forms present in polypeptides and proteins in solution (Doty, 1957).

Quantitative analysis of the secondary structure of proteins (Review: Yang, 1984) is generally based on treatment of the CD spectrum as a linear combination of ellipticities at specified wavelengths contributed by each conformational form (Greenfield, Davidson, and Fasman, 1967; Greenfield and Fasman, 1969). Such analyses have been refined to include additional conformations that are present in proteins, such as β turns (structural domains in which peptide chains reverse direction; Chang et al., 1978). Equations of the form

$$[\theta]_\lambda = f_H[\theta]_H + f_\beta[\theta]_\beta + f_t[\theta]_t + f_R[\theta]_R \qquad (13.27)$$

for such analyses have now been incorporated in the data handling software of contemporary CD spectrometers. In this equation, $[\theta]_\lambda$ is the mean residue ellipticity at a given wavelength, that is, the ellipticity of the macromolecule per peptide unit. The other $[\theta]$ terms are the corresponding ellipticities for the α-helical H, β-form β, β-turn t, and unordered R forms, respectively. Simultaneous solution of the equation with ellipticity data measured at several wavelengths leads to the f values, f being the fraction of each conformational form present in the protein.

Originally, Fasman used the CD of synthetic homopolymers [e.g., poly(L-lysine)] that may be prepared in each of the three distinct conformational forms: α helix, β pleated, and random coil, as reference spectra for such analyses. In spite of numerous assumptions inherent in this quantitative treatment, for example, lack of interference from side-chain chromophores (Freifelder, 1976), estimates of protein composition calculated in this way are in good agreement with fractional conformations determined by X-ray diffraction (Greenfield and Fasman, 1969; Woody, 1985). Quantitative analysis of this type has been extended into the 205–165-nm wavelength range (Brahms and Brahms, 1980).

In an alternative approach developed by Saxena and Wetlaufer (1971), reference spectra for the several conformational forms were computed from the CD spectra of proteins (lysozyme, myoglobin, and ribonuclease were used in the original treatment) whose 3D structures, including the fraction of each conformational domain present, were established by X-ray diffraction (Fig. 13.66).

The latter approach overcomes two limitations inherent in that of Fasman, namely, (a) reference homopolypeptides used in the analysis are of much greater molecular weight than are the α helix and β form regions of typical globular proteins, and (b) CD spectra of globular proteins incorporate contributions due to the interaction between α-helical and β-form fragments within a given protein. Use of proteins to generate the basis spectra, of necessity, compensates for differences in the lengths of the conformational forms present in proteins, and to some extent, also for the interaction between segments that have different forms (Cantor and Schimmel, 1980).

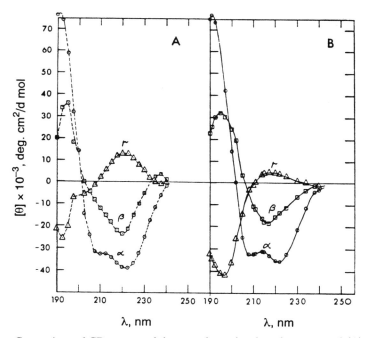

Figure 13.66. Comparison of CD spectra of three conformational modes computed (A) from X-ray diffraction data and CD spectra of lysozyme, myoglobin and ribonucleases, and (B) from CD spectra of the pure conformational forms of poly(L-lysine) (r = random coil). [Reproduced with permission from Saxena, V. P. and Wetlaufer, D. B. (1971), *Proc. Natl. Acad. Sci. USA*, **66**, 969.]

Other macromolecular conformations of polypeptides and proteins having characteristic CD spectra that have been examined closely are the two helical forms (one right handed and the other left handed) of poly(L-proline), and the double and triple helical forms of the proteins myosin (muscle protein) and collagen, respectively. Polypeptides of proline and its derivatives, in which hydrogen bonding is precluded, form stable conformational forms as a result solely of restricted rotation about the polypeptide backbone (Goodman et al., 1970).

Although nucleic acid and polynucleotide CD spectra, for example, those of transfer-ribonucleic acid (tRNA) and deoxyribonucleic acid (DNA), have been analyzed in ways similar to those described for proteins, the process is much more complicated. These polymers form triple, double, as well as single helices because of differences among the strongly absorbing monomer components and because the heterocyclic bases that are present strongly bind to one another. Not only the nature of the bases present (there are at least four distinct chromophores: adenine, guanine, cytosine, and uracil or thymine) but also their sequences significantly affects the CD. In spite of these complications, some impressive computations of CD spectra of polynucleotides have been achieved (Cantor and Schimmel, 1980).

Figure 13.67 illustrates the CD spectrum of the polynucleotide polyadenylic acid (Poly A, a polymer of adenosine 5'-phosphoric acid) in the region in which the planar but chirotopic 6-aminopurine chromophore absorbs. The sign of the CD between 260 and 280 nm is inverted and its ellipticity is increased nearly 10-fold relative to that of the monomer (adenylic acid, Fig. 13.67). Most of the CD intensity can be accounted for by interaction between adjacent (nearest neighbor) bases that are oriented (rotated along the polymer backbone) so that

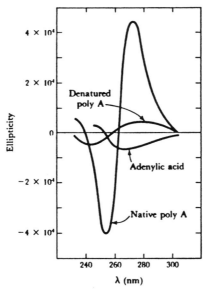

Figure 13.67. The CD spectra of adenylic acid, native polyadenylic acid (poly A) and denatured poly A. [Reprinted with permission from Freifelder, D. (1976), *Physical Biochemistry*, p. 467. Copyright © W. H. Freeman and Company, New York.]

they form a stack. Much of the increased CD intensity is lost on denaturation (Freifelder, 1976).

Even in systems as simple as dinucleoside phosphates (containing a chromophore dimer), e.g., adenosyl adenosine (ApA), the ORD revealed stacking of the bases and a conformation that obviously is the beginning of a right hand helix (Tinoco, Jr., 1979).

A hexanucleotide composed of L-deoxyribose in lieu of the naturally occurring D-deoxyribose (the only chiral structural component of DNA) has been synthesized. This DNA oligomer exhibits a CD spectrum that is the mirror image of the synthetic (natural) D hexamer. The CD spectrum is known to reflect principally the conformation of the hexanucleotide; it exhibits the same sign inversion at 295 nm signaling a conformational transition (in the case of the L hexamer) from left handed to right handed double helical conformation under high salt conditions (the inverse of that of the D hexamer). It is inferred that L-DNA and D-DNA must possess the same conformation and dynamic conformational properties, except for the sense of chirality (Urata, Akagi, et al., 1991).

Circular dichroism spectra provide information about prosthetic groups, that is, defined structural moieties, such as hemes, that are attached to proteins. Analysis of prosthetic groups requires that they absorb in spectral regions that do not interfere with polypeptide backbone absorption bands. For example, reduction of cytochrome and of hemoglobin leads to distinctive ellipticity changes (in ferro- vs. ferri- and in oxy- vs. deoxy-, respectively) in the 240–360-nm region where the heme moiety absorbs.

Heme–heme interaction due to stacking of the large and relatively flat prosthetic groups in proteins is revealed by CD as is the interaction between heteroaromatic moieties in nucleotides. Information about the conformation of macromolecules in biomembranes is also derived from CD spectra. The bulk of the purple membrane in halobacteria is composed of the protein bacteriorhodopsin. The CD spectrum is especially sensitive to the physical state of the membrane, that is, whether its suspension has or has not been sonicated or whether the membrane has been dissolved. Application of correction factors to the CD ellipticities makes it possible to obtain reliable estimates of the amount of α helix present in the membrane (Urry, 1985).

Another very useful application of CD is the detection of *changes* in conformation, for example, on denaturation (visible changes associated with the unfolding of a peptide chain or its reorganization into a new conformation), on reaction of a chiral polymer with a chemical agent (including the solvent and the pH), or on binding of a substrate, inhibitor, or coenzyme to an enzyme.

Synthetic Polymers

While they contain stereocenters, the stereoregular (isotactic and syndiotactic) diastereomeric forms of vinyl polymers, such as polypropylene, ${CH_2-CH(CH_3)}_n$, are considered to be achiral. Several explanations for this view have been advanced. While this subject cannot be examined in detail here, the explanations involve considerations of symmetry elements (the molecules of stereoregular polymers may, in some cases, be considered to be centrosymmetric), and/or the concept of cryptochirality (Section 13-5.a; Goodman, 1967; Farina, 1987).

Poly(methylene-1,3-cyclopentane), an isotactic polymer whose main chain incorporates rings with trans-oriented attachment atoms is chiral (Fig. 13.68). Optically active samples have been obtained by enantioselective cyclopolymerization of 1,5-hexadiene. The magnitude of its rotation, $[\Phi]_D^{20}$ 22.8 ($c = 7.8$, CHCl$_3$) per monomer unit, relative to that of optically active trans-1,3-dimethylcyclopentane, $[\Phi]_D^{20}$ 3.1, suggests that part of the rotation magnitude of the polymer is conformational in origin (Coates and Waymouth, 1991).

In contrast, atactic vinyl homopolymers, even when prepared from achiral monomers, are chiral. Based on statistical considerations, samples of such polymers containing the shorter chains (degree of polymerization, $n < 60$) are "conventionally racemic" (the probability of finding equal numbers of enantiomeric pairs is high) while those having longer chains, that is, having a high degree of polymerization ($n > 70$), would be expected to be cryptochiral, that is, optically inactive as a result of intermolecular compensation of many different enantiopure diastereomers whose rotations are expected to average to zero (Green and Garetz, 1984).

Most stereoregular macromolecules assume helical conformations in the solid (crystal) state (Farina, 1987, p. 46; Vogl and Jaycox, 1987) and chiroptical properties can be, and have been, measured in the solid state even for isotactic polypropylene (cf. Farina, 1987, p. 95). However, in solution, stereoregular polymers behave essentially as random coils (Bovey, 1982). In solution, such polymers do not usually exhibit chiroptical properties associated with the presence of helices: formation of the M and P helical conformations is equally probable and the helices are unstable (such systems are stochastically achiral; see Section 13-2.a). Stereoregular vinyl polymers exemplify the dictum that the application of the term chiral to any system or molecular model "depends on the conditions of measurement" (Mislow and Bickart, 1976/1977).

Only when the side chain R in $(CH_2-CHR)_n$ is very bulky does restricted rotation about CC single bonds of the polymer backbone stabilize helical conformations yielding conformationally rigid polymers even in solution. In any event, the observation of CD depends not only on the presence of an excess of one stable enantiomeric polymer conformation but also on the absorption wavelength of its chromophores. Four types of conformationally rigid synthetic chiral polymers (atropisomeric polymers; cf. Nolte and Drenth, 1987) that are devoid of chiral side-chain groups yet exhibit chiroptical properties are illustrated in Figure 13.69.

Polymerization of (S)-(+)-sec-butyl isocyanide $CH_3CH_2CH(CH_3)N=C$ generates an optically active polymer (Fig. 13.69, **A**, R = sec-butyl). Simple molecular modeling considerations [assuming that the steric fit in this poly (S)-(+)-sec-butyliminomethylene is the controlling parameter in determining the screw sense

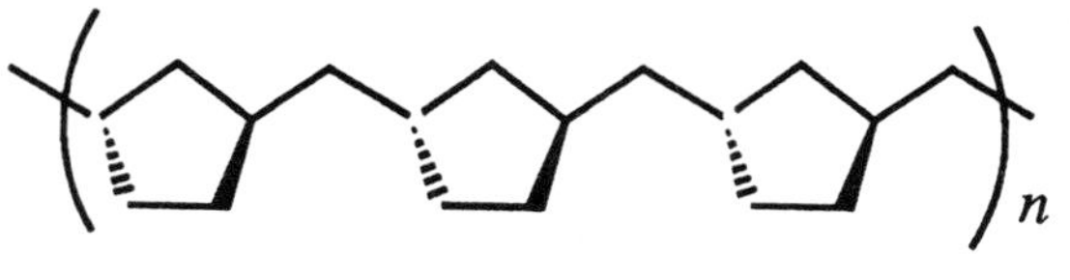

Figure 13.68. Isotactic poly(methylene-1,3-cyclopentane). The nonracemic polymer produced actually contains only about 68% trans rings (Coates and Waymouth, 1991).

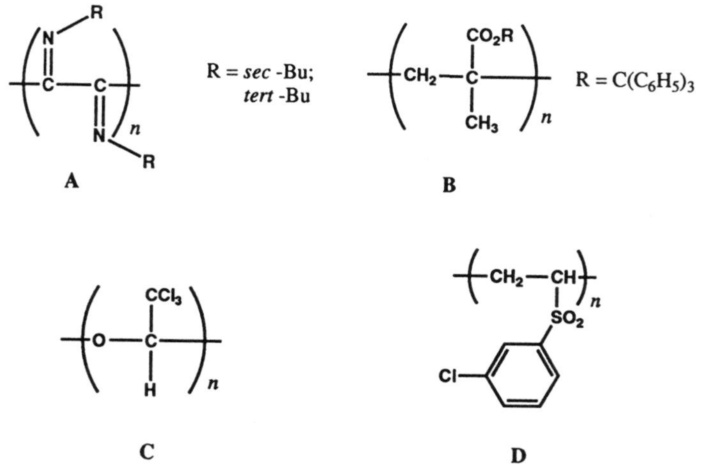

Figure 13.69. Conformationally stable helical polymers. (**A**) poly(*sec-* and *tert*-butyliminomethylene); (**B**) poly(triphenylmethyl methacrylate); (**C**) polychloral; and (**D**) poly(*m*-chlorophenyl vinyl sulfone).

of the polymer helix] led to the conclusion that levorotatory fractions $[\alpha]_D < 0$ are composed mostly of P helixes. Calculation of the sign of the CD band in the vicinity of 300 nm led to the same conclusion (van Beijnen, Drenth, et al., 1976; Drenth and Nolte, 1979).

An analogous polymer (with degree of polymerization, $n \cong 20$, hence actually a large oligomer) devoid of chirotopic atoms in the side chain, poly(*tert*-butyliminomethylene) (Fig. 13.69, **A**, R = *tert*-butyl), could be resolved chromatographically using the optically active *sec*-butyl analogue (see above) as an enantioselective stationary phase. The apparent sole stereochemical element of this remarkable macromolecule is its helicity (though this is an obvious oversimplification since the nitrogen atoms in the imino side chains are stereogenic, alkyl groups having syn–anti relationships about the $>C=N-$ bond; cf. Green et al., 1988b). If one assumes that "parallel screws have a smoother mutual fit," then optically active fractions of the *tert*-butyl polymer consist mainly of helical molecules having a like helical sense. Comparison of the experimental CD spectrum with the calculated spectrum (see above) leads to the conclusion that fractions with $[\alpha]_D < 0$ consist mainly of helical molecules having a P helix sense (van Beijnen, Drenth, et al., 1976). The same conclusion is reached by analysis of the polymerization mechanism (Kamer, Nolte, and Drenth, 1986).

Stable helical conformations have been inferred also for the homopolymer of triphenylmethyl methacrylate (Fig. 13.69, **B**). When obtained by polymerization of triphenylmethyl methacrylate in the presence of (−)-sparteine–butyllithium, one of the helical forms predominates. Applications of the highly levorotatory polymer as an enantioselective stationary phase in HPLC are described in Chapter 7. Hydrolysis of the polymer shown in Figure 13.69, **B** and methylation of the resulting poly(methacrylic acid) yields optically inactive isotactic poly-(methyl methacrylate) (Okamoto, Yuki, et al., 1979). It thus appears that methacrylates with less bulky pendant ester groups are unable to maintain stable helical conformations (Cram and Sogah, 1985; Okamoto, Nakano, and Hatado, 1989).

Two other polymer types whose chirality is due to the presence of stable helical conformations are polychloral (for which $[\alpha]_D > 4000$ has been recorded on some samples) prepared, for example, by polymerization of chloral in the presence of nucleophiles, such as the lithium salt of methyl mandelate (Vogl and Jaycox, 1986, 1987; Vogl et al., 1992; Fig. 13.69, **C**) and poly(aryl vinyl sulfones) (Toda and Mori, 1986; Fig. 13.69, **D**).

Polyisocyanates, $(NR\text{--}CO)_n$, readily form helical conformations both in the solid state and in solution (Bur and Fetters, 1976). Since these helical forms readily undergo enantiomerization, they do not exhibit chiroptical properties. As has already been pointed out (Section 7-2.e), when dissolved in (R)-2-chloro-butane, racemic poly(n-hexyl isocyanate) undergoes an asymmetric transformation of the first kind, that is, the equilibrium between the two enantiomers is displaced and the polymer exhibits a positive CD at about 250 nm (a spectral region in which the chiral solvent is transparent) attributed to an excess of right handed helix (Khatri, Green, et al., 1992).

A large optical rotation is exhibited by poly[(R)-1-deuterio-n-hexyl isocyanate] **36** (Fig. 13.70) $[\alpha]_D^{25} -367$ (CHCl$_3$). It has been suggested that the helical sense, that is, the ratio of enantiomeric polymer helixes in **36**, is strongly influenced by a cooperative conformational equilibrium isotope effect due to a difference in energy of α-D versus α-H involving many deuterium atoms (Green et al., 1988a; Lifson, Green, et al., 1989; Green et al., 1991).

Vinyl polymers containing chiral substituents may, of course, be prepared in optically active forms, as may polymers containing heteroatoms, for example, polypropylene oxide, $(CH_2\text{--}CH(CH_3)\text{--}O)_n$, and/or double bonds in the polymer backbone (Pino, 1973). Space limitations preclude a detailed exposition of the chiroptical properties of such polymers. For a summary of such properties, see Farina (1987, p. 84).

Early efforts to study the chiroptical properties of hydrocarbon polymers derived from optically active alkenes were hampered by the absence of chromophores accessible to existing ORD and CD spectrometers. Pino et al., (1968) conceived of a way of partially circumventing this limitation by synthesizing a copolymer formed from styrene and a chromophore-free optically active monomer, (R)-3,7-dimethyl-1-octene. The CD spectrum exhibits a negative band centered at 262 nm with fine structure typical of the $\pi\text{--}\pi^*$ ^1L_b transition of aromatic chromophores. The benzene side chain incorporated in the copolymer thus acts as a chirality reporter by revealing the chirotopicity of an inherently achiral chromophore. The CD also reveals a 10-fold enhancement of the CD intensity relative to low molecular weight model compounds. This enhancement

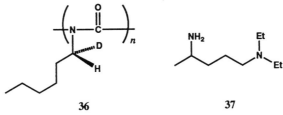

                             36   **37**

Figure 13.70. Poly[(R)-1-deuterio-n-hexyl isocyanate](left) and 4-amino-1-(N,N-diethylamino)pentane (right).

was ascribed to conformational rigidity and to the presence of isotactic helical blocks of a preferred helical sense in the copolymer. A similar example of such a cooperative phenomenon (with a resulting high optical rotation) involving a copolymer consisting of mostly achiral poly(n-hexyl isocyanate) incorporating as little as 0.12 mol% of a nonracemic chiral isocyanate [(*S*)-(−)-2,2-dimethyl-1,3-dioxolane-4-methylene isocyanate] has been described (Green et al., 1989).

> Attachment of achirotopic chromophoric groups such as dehydrophenylalanine and azobenzene to polypeptides [e.g., poly(L-glutamic acid)] generates synthetic macro-molecules whose CD spectra reflect the presence of inherently chiral chromo-phores. This spectral feature results because the pendant side chains serve as chirality reporters of the secondary structure of the peptides (Ciardelli and Pieroni, 1980).

In subsequent studies on dimethyloctene–styrene and similar copolymers, the CD could be measured deeper into the UV spectral region. Other CD bands typical of isolated benzenoid electronic transitions (^1B and ^1L_a) signaling the chirotopicity of the benzene ring were observed. In contrast to model compounds, such as the conformationally restricted 2-phenyl-3,3-dimethylbutane (Salvadori et al., 1972), the copolymer exhibits, in addition, an exciton-like couplet centered at about 190 nm. The latter was attributed to a helical conformation in which benzene rings in the same chain are sufficiently close to couple. The enantiomer exhibiting negative chirality incorporates a right handed helix (Ciardelli et al., 1972). The CD of such helical copolymers can be calculated by means of a classical theory developed by DeVoe (DeVoe, 1969, 1971; see Hug et al., 1974 and Section 13-4.a).

It is only since the 1980s that CD studies have permitted the observation of CEs in the vacuum UV region of hydrocarbon polymers devoid of aromatic groups. The CD spectra of films of poly-(*S*)-4-methyl-1-hexene and poly-(*R*)-3,7-dimethyl-1-octene exhibit a CD band at 158 nm that is ascribed to conformations containing helical segments having a common helix sense (Ciardelli and Salva-dori, 1985).

13-5. APPLICATIONS OF OPTICAL ACTIVITY

a. Polarimetry

The actual measurement of optical activity may be carried out with either manual or photoelectric polarimeters. Manual polarimeters have changed relatively little since the first instruments were developed some 140 years ago (Lowry, 1964, p. 180). Photoelectric polarimeters, the type nowadays commonly found in research laboratories, have greatly reduced the tedium formerly associated with the measurement of optical rotation with manual instruments. Moreover, photoelec-tric polarimeters are much more accurate and sensitive, permitting the rapid and meaningful recording of quite small absolute rotation values α to about $\pm 0.002°$ and, consequently, the use of smaller samples. Polarimeters fitted with microcells may even serve advantageously as detectors in HPLC resolutions (Mannschreck,

Eigelsperger, and Stühler, 1982; Mannschreck, 1992; Pirkle, Salvadori, et al., 1988; Lloyd and Goodall, 1989). A laser-based polarimetric HPLC detector has been shown to be sensitive to as little as 12 ng of sample (Yeung et al., 1980). For the advantages of the use of CD detectors in HPLC, see Salvadori, Bertucci, and Rosini (1991); Mannschreck (1992).

The laser polarimetric detector has been adapted to HPLC analysis not only of optically active samples but also, in a different way, as a universal detector for *achiral*, that is, optically inactive, substances. In this technique, termed "indirect polarimetry," the mobile phase is optically active, containing, for example (−)-2-methyl-1-butanol or (+)-limonene, and the detector output due to the optically active solvent is zeroed. Under these conditions, any optically inactive fraction passing through the detector cell is sensed since the concentration of the optically active solvent is thereby reduced. The response of the detector is universal, like that of a refractive index detector, but is more sensitive than the latter (Bobbitt and Yeung, 1984, 1985; Yeung, 1989). The simultaneous measurement of absorbance and optical rotation during the liquid chromatographic resolution of chiral substances on enantioselective stationary phases make possible the determination of the enantiomer composition in spite of extensive peak overlap (Mannschreck et al., 1980; Mannschreck, Eigelsperger, and Stühler, 1982; compare Drake, Gould, and Mason, 1980).

For a brief discussion of polarimetry and its instrumentation, see the review by Lyle and Lyle (1983); for a more extensive treatment, see Heller and Curmé (1972).

The measurement of optical activity has traditionally been the method of choice to establish the nonracemic character of a sample of a chiral compound and, when quantitatively expressed as a ratio $[\alpha]/[\alpha]_{max}$, of its enantiomeric composition (optical purity). In contemporary practice, chiroptical measurements have to a large extent been replaced by NMR and by chromatographic analyses for the purpose of determining enantiomeric compositions (cf. Chapter 6). Nevertheless, the use of $[\alpha]$ for this and other purposes continues. The reasons are that the measurement is easy to carry out and one may wish to compare experimental values of $[\alpha]$ with those in the literature. While substantial collections of optical rotation data exist, for example those in various handbooks and chemical supplier catalogs, it should not be assumed that values of $[\alpha]$ provided are those of enantiomerically pure compounds. A consistent set of specific rotation data for amino acids including temperature coefficients has been compiled by Itoh (1974).

Optical activity has been used (*a*) to determine if a given unknown substance is chiral or achiral; (*b*) to ascertain the enantiomeric composition of chiral samples, either qualitatively or quantitatively; (*c*) to study equilibria; the mutarotation or change in rotation of equilibrating stereoisomers as a function of time is one such phenomenon (Eliel, 1962; for a recent example, see Arjona, Pérez-Ossorio, et al., 1984); and (*d*) to study reaction mechanisms. Other chiroptical techniques, namely, ORD and CD, have increasingly replaced polarimetry in these applications, especially in the past 20 years. For reviews of applications of polarimetry, see Lowry (1964), Eliel (1962), Legrand and Rougier (1977), and Purdie and Swallows (1989).

Polarimetric methods remain useful for quality control in pharmacology and food-related industries (Lowman, 1979; Chafetz, 1991); there are also numerous applications in forensic, clinical, pharmaceutical, and agricultural chemistry (Purdie and Swallows, 1989). The percentage of sucrose in commercial samples is still being determined by polarimetry (saccharimetry); in the trade this is called "direct polarization". The cost of raw sugar is based on the results of the polarimetric analysis; if the analyte solution is dark, the raw sugar is first clarified by precipitation of the dark side products with basic lead acetate (Cohen, 1988). An example of application d (above) is the methanolysis of the tosylate of (R)-$(+)$-$C_6H_5CH_2SCH_2CH(CH_3)CH_2OH$ that leads to a partially racemized methyl ether. The intervention of a cyclic (symmetrical, and hence achiral) intermediate, via neighboring group participation, was inferred (Eliel and Knox, 1985).

The magnitude of rotation α, in degrees, fundamentally depends on the number of molecules of the sample being traversed by the linearly polarized light as well as on their nature, hence optical activity is not a colligative property. Values of α are affected by many variables, among which are wavelength, solvent, concentration, temperature, and presence of soluble impurities. It must also be mentioned that large molecules, such as proteins, may spontaneously orient themselves in solution, and consequently no longer be isotropic. The measurement of the rotatory power of such substances may then be complicated by the occurrence of linear dichroism (see Heller and Curmé, 1972, p. 67).

As already pointed out in Section 1-3 (q.v.), rotation magnitudes are usually normalized to a quantity called the *specific rotation* $[\alpha]$ that was introduced by Biot in 1835 (Biot, 1835; cf. Lowry, 1964, p. 22), Eq. 13.28,

$$[\alpha] = \alpha / \ell \rho = \alpha / \ell c \tag{13.28}$$

where ℓ is the length of the cell in decimeters, ρ (for undiluted liquids) is the density in grams per milliliter ($g\, mL^{-1}$) and c is the concentration also in grams per milliliter. The units of $[\alpha]$ are $10^{-1}\, deg\, cm^2\, g^{-1}$ (see also Eq. 1.1 and Section 1-3).

Comparison of specific rotations of homologues, and of organic compounds generally, is more significant if a modified Biot equation is used in which the quantity called the *molar rotation* $[\Phi]$ depends on the number of moles of substance traversed by the linearly polarized light, Eq. 13.29.

$$[\Phi] = [\alpha]M / 100 \tag{13.29}$$

where M is the molecular weight. The units of $[\Phi]$ are $10\, deg\, cm^2\, mol^{-1}$ (see also Eq. 1.2) (IUPAC, 1986).

The cumulative effect of the above-mentioned variables on $[\alpha]$ or $[\Phi]$ is potentially very large. A practical consequence is that precise reproduction of published rotation values, from laboratory to laboratory, or even from day to day in the same laboratory, is difficult to achieve (Lyle and Lyle, 1983). This sensitivity to numerous variables (Schurig, 1985) and the absence of major tabulations of critically evaluated absolute rotation data is responsible for the decreasing reliance on optical activity as a measure of enantiomeric composition.

Much preliminary work may be necessary to increase the low accuracy of routine optical rotation measurements. For an outstanding example of a study of some of the variables affecting the rotation of α-methylbenzylamine, mandelic acid, and ephedrine, see the dissertation by Zingg (1981).

The sign of rotation is often the only experimental criterion for the specification of configuration. It is important to stress how frequently and how easily this property may change for a given substance, for example, (R)-2-hydroxy-1,1'-binaphthyl has $[\alpha]_D^{20}$ +4.77 [c = 0.86, tetrahydrofuran (THF)], +13.0 (c = 1.12, THF), and $[\alpha]_D^{20}$ −5.2 (c = 1.03, CH$_3$OH) (Kabuto et al., 1983). Even tartaric acid, one of our configurational standards, does not exhibit an invariant sense of rotation: $[\Phi]_{578}$ −12.9 (24°C), −0.9 (57°C), and +10.8 (94°C) (all c = 10, dioxane); +21.3 (24.7°C, H$_2$O), +6.6 (24°C, EtOH), +0.3 [25.3°C, N,N-dimethylformamide (DMF)], −12.9 (24°C, dioxane), and −14 (25.2°C, Et$_2$O) (all c = 10) [both sets of data measured on (R,R)-tartaric acid] (Hargreaves and Richardson, 1957).

When the sign of rotation shows a strong solvent, concentration, wavelength, or temperature dependence, the association of such sign with a given configurational descriptor is arbitrary. In the case of 4-amino-1-(diethylamino)pentane (Fig. 13.70), **37** [α] is less than 0 in ethanol (365–589 nm) but greater than 0 when measured in the absence of solvent (neat). The configuration of the sample derived from L-glutamic acid was referred to as (R)-(−) rather than (R)-(+) because of the much larger magnitude of the rotation of the neat sample over that in ethanol (Craig et al., 1988). This, and the other examples cited, serve to emphasize the crucial importance of specifying and recording the precise experimental conditions of measurement of optical rotations and of chiroptical properties in general. In particular, confusion can arise when the sign of rotation is related to a given configuration and the solvent is not specified.

Occasionally, specific rotations of samples are very small. When that situation arises, especially in resolutions or in stereoselective syntheses in which strongly rotating reagents are used, exceptional care must be taken to insure that the rotation of the product is not spurious. A small amount of impurity having a large [α] may overwhelm (or at least seriously falsify) the rotation of a sample having a small [α] (for a problem case, see Baldwin et al., 1969; see also Goldberg et al., 1971). Achiral contaminants, particularly solvents, will also affect the optical activity of a sample (Lyle and Lyle, 1983; Schurig, 1985).

Traces of achiral compounds, including solvent residues, normally would be expected to reduce [α] (by dilution of the sample) and hence to artificially lower the optical activity of nonracemic samples (but not the enantiomeric purity of the chiral solute). However, the converse may also be observed, for example, the optical activity of 1-phenylethanol at 589 nm is increased when acetophenone, a possible contaminant, is present in the sample (Yamaguchi and Mosher, 1973). The enhanced optical activity of the alcohol comes about because chiroptical properties are *induced* in the achiral ketone by the alcohol; in the example, the induction is superimposed on and swamps the typical and opposite dilution effect (cf. Section 13-4.e).

The case of low rotation warrants further comment. There are two situations in which no optical rotation is observed with enantiomerically enriched samples:

(a) the experimental device used (by implication this includes the eye) is of insufficient sensitivity, and (b) the specific conditions of measurement are such that α is, in fact, accidentally equal to zero.

In the first situation, the measurement threshold is such that there is no clear signal (rotation) distinguishable from instrumental noise. The condition is one of *operational null*. Progressive dilution of a solution of an optically active compound eventually leads to a sample that is no longer palpably optically active when the operational null threshold is crossed. Such a sample no longer reveals its enantiomeric excess; the sample is said to be *cryptochiral* (Mislow and Bickart, 1976/1977).

> The term cryptochiral is not to be confused with the analogous expression "stereochemically cryptic" that refers to a stereoselective chemical reaction whose stereochemical outcome is hidden (Hanson and Rose, 1975; cf. also Section 8-5).

Notable examples of enantiomerically enriched compounds that are cryptochiral as a consequence of inherently low optical rotation magnitude are shown in Figure 13.71. The cryptochirality condition may conceivably be lifted by measuring a different chiroptical property, for example, vibrational circular dichroism (VCD) (Section 13-6).

The second type of cryptochirality arises when the measurement of rotation accidentally takes place in the vicinity of a change in sign (see below and above). For an example involving a change in concentration of dimethyl α-methylsuccinate, $CH_3O_2CCH(CH_3)CH_2CO_2CH_3$, see Berner and Leonardsen (1939). At a certain concentration, the measured rotation is necessarily zero (crossover point) and the sample is then accidentally cryptochiral. Note that a distinction between stochastic achirality (cf. Section 13-2.a) and cryptochirality cannot be made unless

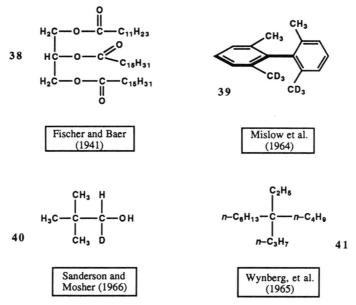

Figure 13.71. Compounds illustrating cryptochirality.

the former be lifted by a change in measuring device or the latter by a change in conditions, the latter being easier to achieve. Other examples of the second type of cryptochirality have been given above and in Section 13-4.

The dependence of optical rotation on the wavelength of the light, ORD, has been discussed in Sections 13-2.b and 13-4.

Effect of Temperature

The effect of temperature on chiroptical properties may be ascribed to the following phenomena (Legrand and Rougier, 1977): (a) changes in density of the solute and/or the solvent that alter the number of molecules being observed; (b) changes in the population of vibrational and rotational energy levels of the chiral solute; (c) displacement of solute–solvent equilibria; (d) displacement of conformational equilibria; and (e) aggregation and microcrystallization of the chiral solute (cf. enantiomer discrimination, Section 6-2).

In general, $[\alpha]$ changes 1–2% per degree Celsius, but larger changes (up to 10% per degree Celsius) are not unknown, for example, $[\alpha]_D$ of aspartic acid, $HO_2CCH(NH_2)CH_2CO_2H$, in water ($c = 0.5\%$) is 4.4 at 20°C, 0 at 75°C, and −1.86 at 90°C. The change in sign at 75°C (temperature of cryptochirality, see above) is noteworthy (Greenstein and Winitz, 1961, p. 78).

An early example of an increase in specific rotation with increasing temperature that was ascribed to a shift in a conformational equilibrium is that of 2-butanol (Horsman and Emeis, 1965). Other examples are discussed in Section 13-4.e).

Strong dependence of the optical rotation on temperature may be found even among hydrocarbons, for example, 3-phenyl-1-butene whose neat rotation, $[\alpha]_D^{22} - 5.91$, for the enantiomerically pure R enantiomer increases linearly 0.18°/°C from 16 to 29°C. Here, it is likely that the temperature exerts a strong conformational bias (Cross and Kellogg, 1987).

Effect of Solvent

The "nonspecific" influence of solvent on the specific rotation may be corrected by calculation of a quantity called the specific rotivity Ω' that includes the refractive index of the solvent n_s (see Heller and Curmé, 1972):

$$\Omega' = [3\alpha]/(n_s^2 + 2) \tag{13.30}$$

Several examples of dramatic changes in the angle of rotation as a function of solvent have been given above. Many instances of changes in sign of $[\alpha]$ have also been recorded for amphoteric substances, such as the amino acids, as the pH is changed (Greenstein and Winitz, 1961, p. 1727). An exceptional example of the effect of solvent on $[\alpha]$ is given in Figure 13.72. Given examples such as these, it is disconcerting how frequently the mention of the solvent is omitted from experimental descriptions of the optical rotation.

Care in choosing the solvent to be used in the measurement of $[\alpha]$ is necessary in view of the several specific types of interaction that are possible between solute and solvent. In general, one recognizes the intervention of hydrogen bonds when oxygen-containing solutes, such as carboxylic acids, alde-

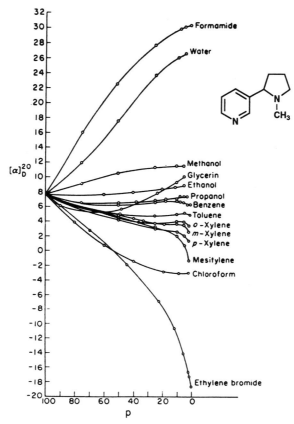

Figure 13.72. Specific rotation of nicotine in various solvents (p = concentration of solute in grams per 100 grams of solution). At $p = 100$, the "bulk" rotation $[\alpha]_{100}$ should be a constant, as observed, and at $p = 0$, $[\alpha]$ should tend to the intrinsic rotation $\{\alpha\}$ (p. 1079). [Adapted with permission from Winther, C. (1907), *Z. Phys. Chem.*, **60**, 621.]

hydes and ketones, and alcohols, are dissolved in hydroxylic solvents; in some cases reactions, such as hemiacetal formation, may occur. In addition, dipole–dipole interactions and changes in conformer populations are important sources of solvent-induced variations in rotation magnitude (Lyle and Lyle, 1983).

The effect of intermolecular solute association of polar solutes on $[\alpha]$ in nonpolar solvents has already been pointed out (see above). Solute–solute association effects may be leveled out or suppressed in polar solvents by competition with (concentration-independent) solute–solvent association. Polar solvents, such as ethanol, may break up solute–solute association leading to a smaller concentration dependence of $[\alpha]$, as is found with nicotine (Fig. 13.72). Such findings illustrate the desirability of using methanol or ethanol as a solvent in polarimetry.

In some instances, hydrogen bonding is known to be responsible for changes in $[\alpha]$ with concentration and/or solvent. Compounds **42** and **43** (Fig. 13.73) exhibit a remarkable solvent dependence of the sign of $[\alpha]_D$ for the *RR/SS* (syn) **42** diastereomer that is not found in the case of the *RS/SR* (anti) diastereomer **43**. The sign of $[\alpha]_D^{20}$ of (4*R*,5*R*)-**42** is (+) in methanol and (−) in chloroform. This

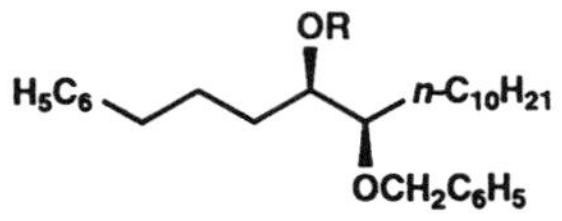

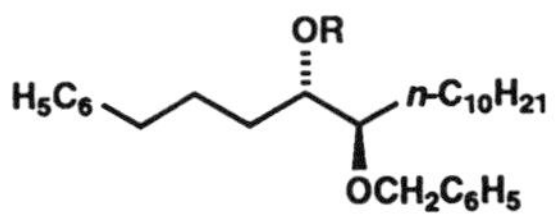

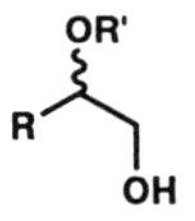

42 (R = H) *R,R* shown
42a (R = Ac)

43 (R = H) *S,R* shown

44

Figure 13.73. Structures **42–44**.

difference has been ascribed to a conformational change: the predominant methanol-solvated ($OH/OCH_2C_6H_5$) anti conformer gives way to a ($OH/OCH_2C_6H_5$) gauche intramolecularly hydrogen-bonded conformation in chloroform. Such sign reversal is not seen in the benzyl ether–acetate derivative **42a**, in the corresponding diol (the latter appears to prefer the gauche conformation regardless of solvent) or in the diol acetonide. Reversal of the sign of $[\alpha]_D$ would seem to be precluded in the predominant zigzag (all-anti) conformation of the molecular skeleton. A similar sign reversal was observed in a series of 2-alkoxy alcohols **44** (Fig. 13.73) presumably for the reasons advanced above. The free diol (S)-1,2-dodecanediol exhibits $[\alpha]_D^{20}$ −10.1 (EtOH) but +0.9 ($CHCl_3$). This suggests that here, too, the intramolecularly hydrogen-bonded conformer prevails in $CHCl_3$ (Ko and Eliel, 1986).

Another example of optical rotation sign reversal is found with compound **45** (Fig. 13.74): $[\alpha]_D$ +14.0 ± 0.6 ($CHCl_3$) and −2.7 ± 0.6 (CH_3OH) (Suga et al., 1985). The molar rotation $[\Phi]_D$ was found to be independent of concentration in 12 solvents of varying polarities. The principal factor responsible for the sign reversal was intramolecular hydrogen bonding between the carbonyl and hydroxyl groups in nonpolar solvents (confirmed by IR and ^{13}C NMR measurements) and its absence in the presence of strongly solvating media (alcohols, CH_3CN, or acetone) as a result of competing intermolecular hydrogen bonding between the solvent and the solute (Suga et al., 1985). A sign reversal of a CE was also noted in the ORD of **46** (Fig. 13.74) in $CHCl_3$ and CH_3OH but this was not reflected in the sign of $[\alpha]_D$. The latter fact points up once again the desirability of carrying out studies of chiroptical properties over a range of wavelengths and preferably into regions that reveal the responsible CEs.

A particularly clear-cut example of a conformational equilibrium that is responsible for changes in chiroptical properties over the range of 210–350 nm is shown by ketone (+)-**47** (Fig. 13.75). As the solvent is changed from cyclohexane to acetonitrile or methanol, the effect of increasing solvent polarity on the dipole–dipole repulsion (as well as solvation effects) between the adjacent permanent dipoles (C=O and C–Br) causes a conformational change: the bromine changes from axial to equatorial and significant changes in the CD ensue

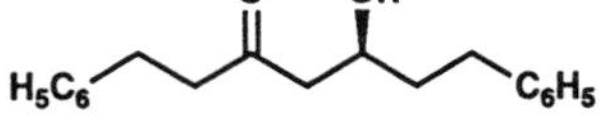

45

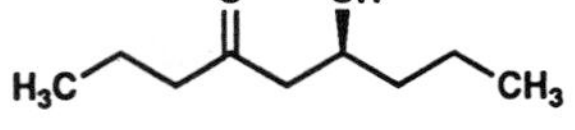

46

Figure 13.74. Structures **45–46**.

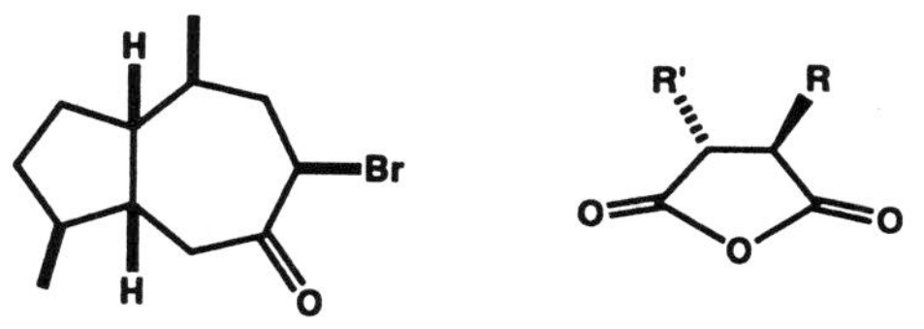

47   **48**

Figure 13.75. Structures **47–48**.

(Kuriyama et al., 1967). Striking changes in $[\alpha]_D$ of propylene oxide, including sign reversal, are observed as the solvent is changed from benzene to water. In this instance, we are cautioned against ascribing the effect of solvent directly to conformational changes (Kumata et al., 1970).

Effect of Concentration

Equation 13.28 suggests that the specific rotation should be independent of concentration. It is not hard to find evidence that this constancy holds only over very narrow concentration ranges and, in some solvents, not at all (the example of nicotine is found in Fig. 13.72; other examples may be found in the book by Lowry, 1964, Chapter VII).

As early as 1838, Biot suggested that the specific rotation followed a linear relationship, such as that of Eq. 13.31, where a and b are constants

$$[\alpha] = a + bc \tag{13.31}$$

and c is the concentration (Lowry, 1964, p. 90). Constant a has been equated with a new quantity called the "intrinsic rotation" $\{\alpha\}$, a true constant corresponding to the specific rotation in a given solvent at infinite dilution; $[\alpha]_{c \to 0} = \{\alpha\}$ (Heller and Curmé, 1972, p. 163). Obviously, $\{\alpha\}$ can only be calculated since experimentally, as the concentration is reduced the rotation must vanish.

The intrinsic rotation is the specific rotation for a system free of solute–solute interactions. However, solute–solvent interactions are maximized in $\{\alpha\}$, which can differ greatly from solvent to solvent, for example, for nicotine (Fig. 13.72). Since, obviously $\alpha = 0°$ at 0% solute, the values at very low concentrations must be extrapolated. Conversely, as the concentration of solute increases, solute–solute interactions become dominant and the effect of the solvent eventually vanishes: $[\alpha]_{c \to 100} = [\alpha]_{\text{neat}}$ tends to a constant value that is identical for all solvents.

A recent report on the specific rotation of (S)-2-phenylpropanal, $CH_3CH(C_6H_5)CH{=}O$, makes it clear that even at relatively low concentrations in benzene ($c = 1\text{–}4$), changes of the order of 1–2% are found in $[\alpha]_D$ as the concentration is doubled (see Table 13.3; Consiglio et al., 1983). The accurate determination of optical purities is thus seen to be dependent on the careful measurement of rotations as well as on comparison of the resulting $[\alpha]$ values with reference $[\alpha]$ values measured in the *same* solvent, at the *same* temperature, and at the *same* concentration (Consiglio et al., 1983). The reader is also

TABLE 13.3. Influence of Dilution on $[\alpha]_D^{25}$ for (S)-2-Phenylpropanal[a,b]

Concentration (g/100 mL^{-1})[c]	$[\alpha]_D^{25}$	$[\alpha]_D^{21}$
Neat	161.8	166.6
46.43	177.9	182.2
18.57	190.5	195.4
9.29	196.6	201.9
7.43	202.7	207.9
3.72	205.8	211.3
1.49	209.1	214.7

[a] Reprinted with permission from Consiglio, G., Pino, P., Flowers, L. I., and Pittman, C. U., Jr. (1983), *J. Chem. Soc. Chem. Commun.*, 612. Copyright © Royal Society of Chemistry, Science Park, Milton Road, Cambridge CB4 4WF, UK.
[b] Optical purity 68%.
[c] Benzene solution.

reminded that $[\alpha]_D$ reflects, but is not necessarily linearly related to, the enantiomeric composition (Horeau effect: Horeau and Guetté, 1974; see Section 6-5.c).

Some very subtle effects are manifested by the rotatory power. For example, $[\alpha]_{436}$ of menthol and of menthol-O-*d* diverge as the solute concentration is increased 100-fold (in cyclohexane). This differential concentration effect reveals that intermolecular hydrogen bonding is subject to a thermodynamic isotope effect (Kolbe and Kolbe, 1982).

In the absence of experimental information, calculations, or obvious structural features, the detailed aspects of chiroptical data including sign reversal of CEs cannot be explained. A detailed study of the chiroptical properties of alkyl substituted succinic anhydrides **48** (R' = H, R = alkyl; Fig. 13.75) revealed a great sensitivity of the CD (CEs at ca. 220 and 240 nm) to solvent polarity, to temperature, and to the size of the alkyl substituent. The results could generally be accommodated by a sector rule (for anhydrides with local C_{2v} symmetry of the chromophore, that is, a planar anhydride group) (Gross, Snatzke, and Wessling, 1979), but the absence of information about conformational equilibria and solvation effects made it impossible to offer explanations for the detailed features of the CD (Sjöberg and Obenius, 1982).

b. Empirical Rules and Correlations. Calculation of Optical Rotation

Ever since simple curiosity about optical activity gave way to its application, efforts have been made to calculate the magnitude and sign of the optical rotation in relation to structure and configuration.

One of the very oldest correlations between structure and rotatory power is that of Walden who observed that the molar rotations of diastereomeric salts in dilute solution are additive properties of the constituent ions (Walden, 1894; Jacques et al., 1981, p. 317). Arithmetic manipulation of the molar rotations of diastereomeric salts, such as those obtained in a resolution, may thus permit one to estimate the enantiomeric purity achieved during a resolution mediated by these salts.

The molar rotations $[\Phi]_D$ of two of the nonenantiomeric diastereomeric salts of α-methylbenzylamine mandelate are -182.3 and $+169.7$ (in H_2O). From these data, the molar rotations of the constituent ions are calculated to be $[\Phi]_D = \frac{1}{2}$ $([-182.3] + [+169.7]) = -6.3$, that is, ± 6.3 and $[\Phi]_D = \frac{1}{2}([-182.3] - [+169.7]) = -176$, that is, ± 176. The latter value is assigned to the mandelate ion [literature values of $[\Phi]_D$ for sodium and potassium mandelates are 182 and 178, respectively (both in H_2O; Ross, et al., 1937)]. Combination of this value with that for ephedrinium hydrochloride, $[\Phi]_D$ 69.8 (H_2O) (Overby and Ingersoll, 1960) yields the molar rotations of the two ephedrine mandelate diastereomers, $[\Phi]_D = 176 + 69.8 \cong 246$ and $176 - 69.8 \cong 106$. The experimental values of these rotations obtained by Zingg, Arnett, et al. (1988) are ± 250 and ± 107, respectively (both in H_2O).

Analogously, molar rotations of inclusion compounds would be expected to be additive properties of the host and guest molar rotations. In both cases, additivity of rotations would not necessarily obtain when strong intermolecular interaction takes place.

In the case of covalent compounds, early correlation attempts also made use of the concept that the rotations of compounds containing several chiral centers might be calculated by adding rotation contributions from each of these centers. This concept, incorporated in van't Hoff's empirical "Principle of Optical Superposition," that individual chiral centers in a chiral compound make independent contributions to the molar rotation (van't Hoff, 1893; Kuhn, 1933, pp. 394, 423) is still successfully being applied in very limited contexts.

The relative configurations of the diastereomeric (R)-O-methylmandelate esters of **49** were assigned by application of the van't Hoff principle. Contributions to the specific rotations from the octalin portions of the ester molecules were estimated from the rotations of ($+$)-dihydromevinolin **50** and lactone **51** (Fig. 13.77) to be approximately 100 ($148.6 - 48.8$) while that of the (R)-mandelate portion was independently known to be strong and positive [(S)-($-$)-methyl O-methylmandelate] has $[\alpha]_D^{25}$ -124 (Barth, Mosher, Djerassi, et al., 1970). Accordingly, the configurational assignments were ± 75 for the octalin portion of the esters and $+55$ for the mandelate (respectively, **B** and **A** in Fig. 13.76; Hecker and Heathcock, 1986). These assignments were confirmed by chemical correlations.

van't Hoff's principle is likely to be invalid when the stereocenters contributing to the molar rotation are close together (so-called "vicinal action" limitation). However, when the stereocenters are separated by several saturated atoms, as in the above example, the principle holds reasonably well (Eliel, 1962, p. 110).

The estimation given above ignores differences in solvent and concentrations of the rotation measurements and it neglects the contribution of the chiral methylbutanoyl group. More seriously, however, we observe that a similar calculation involving molar rotations rather than the specific rotations used by Hecker and Heathcock (1986), while qualitatively still leading to the same configurational assignment, is quantitatively much less persuasive.

Along with van't Hoff's "principle of optical superposition," another empirical rule, Freudenberg's "rule of shift" (also termed Rotational Displacement Rule; Freudenberg, Todd, and Seidler, 1933), is considered quite reliable for

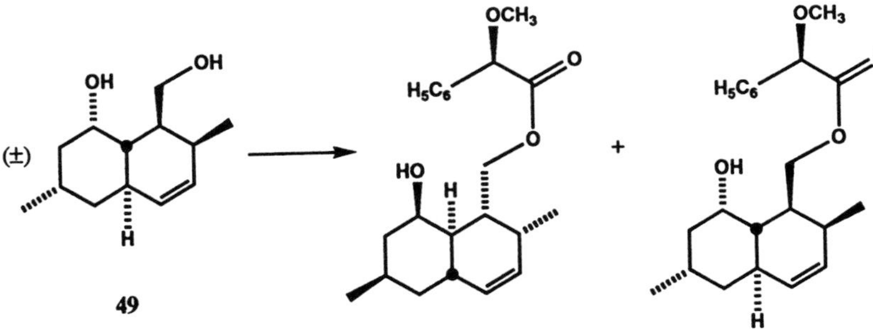

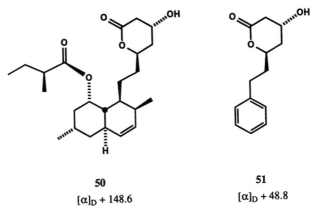

A + B = +130

A − B = −20

A = +55 (due to the mandelate)

B = +75 (due to the octalin)

Figure 13.76. Application of van't Hoff's "principle of optical superposition." Simultaneous solution of the two equations at left gives the values of **A** and **B** shown at right. Assignment of the relative configurations depends on the correct attribution of rotations **A** and **B** (see discussion in the text; Hecker and Heathcock, 1986).

establishing the relative configuration of pairs of compounds (by examining the sign and magnitude of molar rotation changes as these pairs are subjected to like chemical changes, e.g., derivatization). Illustrations are given by Freudenberg (1933), Lowry (1964, pp. 274–278), Eliel (1962, pp. 110–114), Barrett (1972), and Potapov (1979, pp. 200–207). For applications to carbohydrates, see Eliel et al. (1965).

By way of example, the configuration of (+)-phenyltrifluoromethylcarbinol (Fig. 13.78, **52**) was shown to be related to that of (+)-phenylmethylcarbinol **53** and other (+)-phenylalkylcarbinols. Positive rotational shifts on the molar rotations were observed in the transformation of both carbinols to the corresponding acetate, benzoate, and hydrogen phthalate esters (Peters, Feigl, and Mosher, 1968). For other empirical rules relating the signs of optical rotations of certain functional group types (allylic alcohols, amino acids, lactones, nucleosides, sugars) to their configurations, see Snatzke (1974).

Figure 13.77. Rotations of (+)-dihydromevinolin **50** and of lactone **51**.

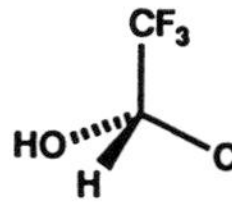

Figure 13.78. Application of Freudenberg's rule of shift to substituted trifluoromethylcarbinols. (S)-$(+)$-**52** is correlated with (R)-$(+)$-**53** since, unlike the other carbinols, the CF_3 group has a higher order of precedence than phenyl in the CIP sequence rule (Sec. 5-2).

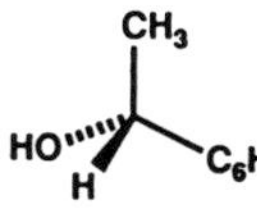

Empirical and semiempirical treatments for the prediction of the magnitude of optical rotations from structural formulas have been developed that are based on polarizability and one-electron theories developed beginning in the 1930s (for a summary, see Charney, 1979, pp. 1–5; 191–195). These theories have also had as a goal the prediction of configuration from the sign of the experimentally measured rotatory power at a single wavelength. A new semiempirical theory that has been applied to the calculation of the optical activity of the saccharides is illustrative of recent developments in this area (Stevens and Sathyanarayana, 1987; Sathyanarayana and Stevens, 1987). Ab initio calculations of chiroptical properties is presently limited to small molecules, such as *trans*-1,2-dimethyl-cyclopropane (Bohan and Bouman, 1986; see Section 13-4.a). While the examination of the several theories of optical rotatory power is not possible in this book, nor even a description of all of the treatments alluded to above, we will illustrate in detail at least the most successful of the empirical ones that is also easily accessible to organic chemists, namely, that of Brewster (1959a,b; 1961; 1967). Brewster's model fits within the framework of the coupled oscillator theory of optical activity (Section 13-4.c). It is still being applied some three decades after its appearance, albeit in modified form.

The reader will recall that optical activity arises because nonracemic samples of chiral compounds are circularly birefringent (Section 13-2). Since the refractive index is related to the polarizability of atoms and groups in molecules, that is, the sensitivity of these molecular constituents to deformation by electrical fields, as well as to their relative positions, it should not surprise us that the rotatory power of chiral molecules also should be related to polarizabilities (Brewster, 1967).

Brewster's original formulation empirically factored the optical rotatory power into two components, as had earlier been proposed by Whiffen (1956); contributions to the rotation result from differences in the polarizability of atoms attached to asymmetric atoms (this was called atomic asymmetry), and from conformational dissymmetry (see below) (Brewster, 1959a). Nowadays we might attribute these contributions to chirotopicity and chiral conformations, respectively.

Since both atomic and conformational dissymmetry describe chiral screw patterns of polarizability, both contribute to the molar rotation in conformationally flexible molecules. Although there is no simple way of assessing its magnitude (see below), the contribution of the atomic asymmetry component is small especially when the polarizabilities of two of the attachment atoms are equal or nearly so (Boter, 1968).

Attempts to calculate the magnitude of rotation of molecules, such as CHBrClF, that are devoid of conformational dissymmetry by summation of terms describing the interaction between pairs of atoms fails. The theoretical treatment based on polarizability theory in which pairwise interaction of groups is invoked predicts zero

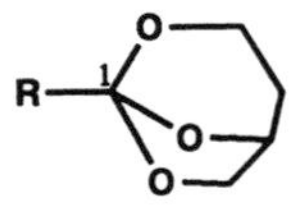

54

Figure 13.79. Rigid 1-(R)-2,7,8-trioxabicyclo[3.2.1]octanes (R = H, CH₃, C₆H₅).

rotation for chiral methanes endowed with substituents that possess $C_{\infty v}$ symmetry (Boys, 1934; Kirkwood, 1937). It is only when the calculations (within the classical dipole interaction theory) are refined to the level of fivefold interactions (Applequist 1973, 1977) that the calculated molar rotation ($[\Phi]_D$ 2–16 deg cm² dmol⁻¹) reaches the correct order of magnitude (the experimentally observed value is $[\Phi]_D$ +1.7 ± 0.5; Canceill, Lacombe, and Collet, 1985; Wilen et al., 1985).

The same type of calculation, wherein atoms interact with one another through the fields of the electric dipole (transition) moments generated by the field of the light wave, has been applied to rigid bicyclic ortho esters **54**, R = H, CH₃, C₆H₅ (Fig. 13.79; Wroblewski, Verkade, et al., 1988). The calculations were shown to be quite sensitive to the precision of bond lengths as well as to the values of the polarizability parameters, but not to the nature of the bridgehead substituent R or, in the case of phenyl, to its rotation about the C(1)–R axis. More recently, the gas phase CD of **54**, R = H and CH₃, has been measured down to 150 nm. No evidence was found for possible exciton splitting resulting from interaction between the three nonidentical oxygen chromophores (Davar, Gedanken, et al., 1993).

Compounds whose molecules are adequately described by the model in Fig. 13.80b where the substituents A–D are atoms or small groups having average cylindrical or conical symmetry (for average symmetry, see Section 4-5), and whose absolute configuration is depicted by Figure 13.80a, are dextrorotatory when the polarizability order of the attached atoms is A > B > C > D. The polarizability order is given by the atomic refractions (Vogel, 1948). In the case of attachment atoms in C≡C and C=C, atomic refractions are calculated by taking one-half the value of the group refraction (the latter are labeled*; a slightly more complicated apportionment is used in the case of CN, C₆H₅, and CO₂H): I (13.954) > Br (8.741) > SH (7.729) > Cl (5.844) > C≡C (3.580 = 7.159*/2) > CN (3.580; 5.459*) > C=C (3.379 = 6.757*/2) > C₆H₅ (3.379 = 6.757*/2) > CO₂H (3.379; 4.680*) > CH₃ (2.591) > NH₂ (2.382) > OH (1.518) > H (1.028) > D (1.004) > F (0.81). For example, on the basis of the rule, (R)-ethanol-1-d, is predicted to be dextrorotatory at 589 nm ($[\Phi]_D > 0$). This is in accord with experiment (Klyne and Buckingham, 1978).

This polarizability order is insufficient, however, in determining the rotation since it turns out that polarizability is affected by the nature of the attachment

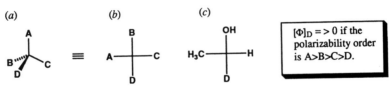

Figure 13.80. (a) Prediction of the sign of rotation in a system exhibiting atomic asymmetry; (b) Fisher projection; and (c) application to ethanol-1-d.

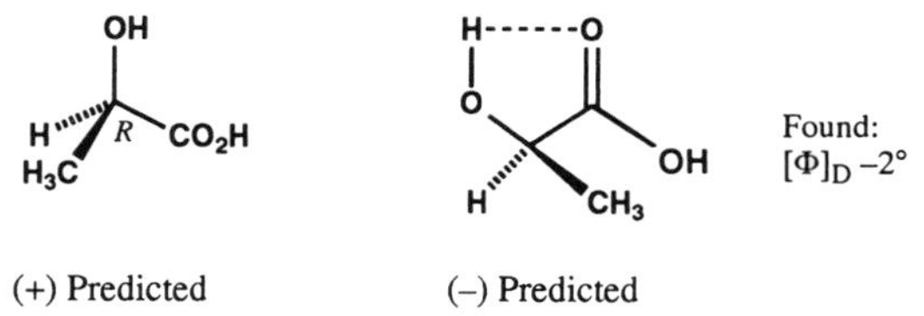

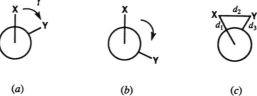

(+) Predicted   (−) Predicted

Figure 13.81. Influence of intramolecular hydrogen bonding on the molar rotation of lactic acid. The observed rotation of lactic acid is $[\Phi]_D - 2$ (c 1.24, water), however, the sodium salt is dextrorotatory.

atoms, for example, NH_2 and OH must be ranked ahead of groups whose attachment atom is carbon ($C\equiv C$, CN, C_6H_5, $C=C$, CO_2H, CH_3) when they are alpha to a phenyl group (Eliel, 1962, p. 403). Moreover, if two of the groups can interact intramolecularly (e.g., by hydrogen bonding) there is a conformational dissymmetry contribution to the rotation (see below). If the atomic asymmetry and the conformational dissymmetry components (the latter was originally called conformational asymmetry) predict the same sense of rotation, then Figure 13.80 still gives the configuration accurately. However, if the two components are predicted to have oppositely signed rotations, then the model may lead to ambiguous results. Lactic acid is an example of the latter situation (Fig. 13.81).

Brewster's empirical "conformational dissymmetry model" has the general form given by Eq. 13.32, where X and Y are properties of the atoms X and Y (Fig. 13.82) and Δ is the contribution of the XY unit (Fig. 13.82) to the molar rotation $[\Phi]_D$.

$$\Delta[\Phi]_D = +kXY \tag{13.32}$$

The model has the following provisions (Brewster, 1959a, 1961):

1. A four atom skew conformational unit X–C–C–Y is the basic structural unit responsible for dextrorotation (at 589 nm) when its absolute configuration has the sense of a right-handed helix as viewed end-on (Newman projection) (Fig. 13.82).

2. The magnitude of the rotatory power of this conformational unit is proportional to the atomic refractions R of the terminal atoms of the unit according to the following empirical expression.

$$\Delta[\Phi]_D = 165\sqrt{R_X}\sqrt{R_Y} \tag{13.33}$$

(a)   (b)   (c)

Figure 13.82. (a,b) Skew conformational dissymmetry unit and helix model of optical activity. The four atom units depicted that are responsible for dextrorotation at 589 nm describe a right-handed helix; (c) bond distances and angles; the dihedral angle γ is shown in (a).

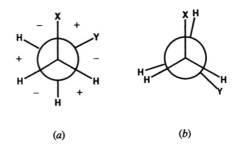

Figure 13.83. (a) Staggered chiral conformation of X–CH$_2$CH$_2$–Y showing signs of contributions from the six conformational units to the molar rotation; (b) eclipsed chiral conformation.

(a)   (b)

Equation 13.33 holds for compounds in which the central atoms of the chiral axis of X–C–C–Y are carbon atoms (Brewster, 1961). However, it has been found that the proportionality constant of Eq. 13.33 is relatively insensitive to replacement of carbon in X–C–C–Y by oxygen or phosphorus (Boter, 1968).

3. The rotatory power of a molecule X–CH$_2$CH$_2$–Y is the sum of the contributions of the six conformational units present whose signs alternate around the central bond (as shown in Fig. 13.83a). It becomes evident that the molar rotation is the sum of interactions between pairs of atoms, in either staggered or eclipsed chiral conformations (Fig. 13.83a and b, respectively). Note that some of the contributions cancel one another.

Brewster found that contributions of the conformational units (Fig. 13.83a) could be factored thereby reducing the number of empirical constants required (the possibility of factoring the contributory rotations demonstrates the connection between Brewster's models and van't Hoff's principle of optical superposition).

$$\sum \Delta[\Phi]_D = k(XY - YH + HH - HH + HH - HX) \tag{13.34}$$

$$= k(XY - YH + HH - HX) \tag{13.35}$$

$$= k(X - H)(Y - H) \qquad \text{after factoring terms} \tag{13.36}$$

For relatively rigid saturated molecules, the molar rotation is simply the sum of the bond rotations $k(X - H)(Y - H)$, where X, Y, and H are the square roots of the atomic refractions listed above, due regard being given to the sign of each bond's rotational contributions. α-Isosparteine, known to exist in the all-chair conformation, is thereby predicted by the rule to have $[\Phi]_D$ +120 for the conformation and configuration shown in Figure 13.84 (Brewster, 1961). The observed molar rotation (for α-isosparteine hydrate) is +129 (EtOH) (Marion et al., 1951); +141 (MeOH) (Leonard and Beyler, 1950).

For conformationally flexible compounds, the rotation is given by Eq. 13.37 in which ϕ_i refers to the rotations of individual conformers and f_i is a function of the conformer populations (Brewster, 1974):

$$[\Phi]_i = \sum \phi_i f_i \tag{13.37}$$

The calculated rotation is thus a population weighted average rotation of all conformations having rotations of significant magnitude.

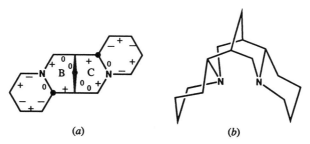

(a)   (b)

Figure 13.84. Calculation of the molar rotation of α-isosparteine (a) showing the signs of the rotatory contributions of each bond in the all-chair conformation (b). There is a net excess of two dissymmetric conformations each contributing $k(\text{C–H})(\text{C–H}) = 60°$ (calculated by substitution of XY and other terms in Eq. 13.34 by Eq. 13.33) to $[\Phi]_D$. [Reprinted with permission from Brewster, J. H., *Tetrahedron*, **13**, 106 (1961). Copyright © 1961 Pergamon Press, Ltd, Headington Hill Hall, Oxford, OX3 0BW, UK.]

For such compounds, the following additional rules are required:

4. Only staggered conformations count.

5. Conformations c, c', and d (Fig. 13.85) do not contribute to the molar rotation; however, the exclusion of all-gauche conformations may not be empirically justified in all cases (cf. Chapter 10 and p. 1089).

6. The simplifying assumption is made that all other "allowed," that is, stable, conformations are considered equally probable.

Application of the conformational dissymmetry model to 2-chlorobutane (only two of three conformations about the C(2)–C(3) bond were considered; see Fig. 13.86; conformation c was omitted on the basis of Rule 5) leads to a value of $[\Phi]_D$ +39.5 that is of the correct sign and order of magnitude (the highest observed rotation is $[\Phi]_D$ +34; Davis and Jensen, 1970):

$$[\Phi]_D = ([\Phi]_a + [\Phi]_b)/2 = [k(\text{Cl} - \text{H})(\text{C} - \text{H}) - k(\text{C} - \text{H})(\text{C} - \text{H})]/2$$

$$[\Phi]_D = (139 - 60)/2 = +39.5$$

The model is quite successful in the calculation of molar rotations of saturated acyclic hydrocarbon molecules in which conformations about more than one bond must be considered. For examples, see Brewster (1959a, 1961). For examples of calculations involving cyclic molecules see Brewster (1959b, 1961) and Eliel (1962).

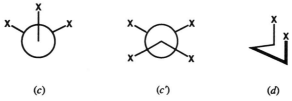

(c)   (c')   (d)

Figure 13.85. Conformations that do not contribute to $[\Phi]_D$ in Brewster's conformational dissymmetry model (for terminal atoms that are larger than H): (c),(c') doubly skewed (g^+g^-) conformations and (d) equivalent to 1,3-diaxial interactions.

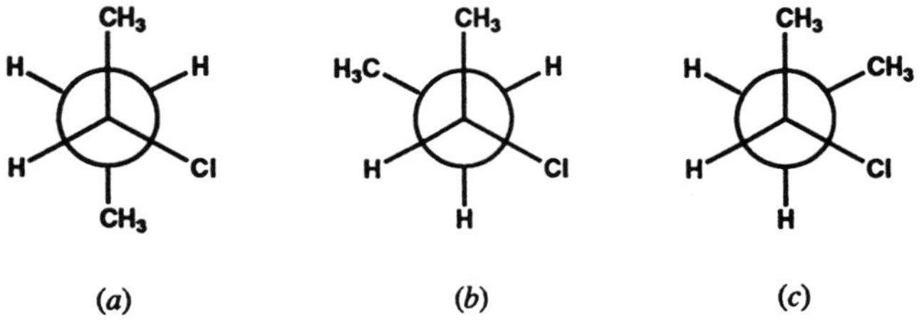

$$[\phi]_D = ([\phi]_a + [\phi]_b)/2 = [k(Cl - H)(C - H) - k(C - H)(C - H)]/2$$
$$[\phi]_D = (139 - 60)/2 = +39.5$$

Figure 13.86. Conformations of 2-chlorobutane about the C(2)–C(3) bond. Evaluation of $[\Phi]_D$ by application of the calculated values of $k(Cl-H)(C-H)$ and $k(C-H)(C-H)$, respectively.

An empirical linear correlation related to Brewster's model is that between the molar rotation and bond refractions. It has been applied to alkyl halides and to organometallic compounds derived from them, for example, for the *sec*-butyl series, $[\Phi]_D$ 3.78 $\Sigma R_D + 8.8$, where ΣR_D is the bond refraction parameter (Davis and Jensen, 1970).

In Brewster's later developed helix model of optical activity, the so-called "uniform helix conductor model" (Brewster, 1967), atomic asymmetry is explicitly neglected; the molar rotation of the conformational unit (the helix rotation) is calculated by incorporating a term describing the geometry of the system including, as suggested by intuition, the dihedral angle γ (Fig. 13.82a)

$$[\Phi]_D = \sum \Delta[\Phi]_D = 652.4 \frac{d_1 d_2 d_3}{2(d_1 + d_2 + d_3)^2} \sin\alpha \sin\beta \sin\gamma \left[\sum \Delta R_D\right] f(n)$$

$$(13.38)$$

where d_1, d_2, and d_3 are bond distances, and α, β, and γ are bond angles ($\angle C-C-X$, $\angle C-C-Y$, and the dihedral angle, respectively) in the skew conformational unit (shown in Fig. 13.82c), R_D are bond and group refractions, and $f(n)$ is a term reflecting the refractive index of the medium (rotivity correction) $f(n) = [(n^2 + 2)/3][(n^2 + 2)/3n]$. Typically, $f(n) = \frac{4}{3}$ (using $n = \sqrt{2}$). When the bond angles α, β are tetrahedral and the dihedral angle γ is 60°, Eq. 13.38 reduces to Eq. 13.39.

$$\Delta[\Phi]_D = 251 \frac{d_1 d_2 d_3}{(d_1 + d_2 + d_3)^2} \left[\sum \Delta R_D\right] f(n) \qquad (13.39)$$

For X = H and Y = OH, for example, $\Delta[\Phi]_D = 225$. Table 13.4 lists calculated helix rotation values for common X and Y groups.

Recalculation of the molar rotations of each of the three 2-chlorobutane conformations using the helical conductor model is shown in Figure 13.87. The six relevant single helix rotations are algebraically added to obtain the molar rotations of each conformation. Since the conformational equilibrium of 2-chlorobutane in the liquid state is known, the fractional contributions of these

TABLE 13.4. Single Helix Rotations $[\Delta\Phi]_D/f(n)^{a,b}$

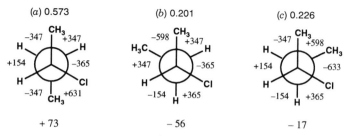

X	Y = H	F	OH	NH$_2$	CH$_3$	Cl	CO$_2$H	SH	Br	I	C$_6$H$_5$
H	154	159	225	297	347	365	405	480	483	698	1082
F	159	168	240	319	376	398	436	528	531	775	1180
OH	225	239	314	395	453	478	575	609	615	860	1271
NH$_2$	297	317	395	477	537	566	600	700	707	957	1363
CH$_3$	347	374	453	537	598	633	661	767	777	1037	1435
Cl	365	396	478	566	633	672	699	813	823	1100	1509
CO$_2$H	405	436	575	600	661	699	727	824	844	1104	1501
SH	480	525	609	700	767	813	848	956	971	1125	1648
Br	483	528	615	707	777	823	844	971	988	1270	1669
I	698	770	860	957	1037	1100	1104	1255	1270	1570	1945
C$_6$H$_5$	1082	1180	1271	1363	1435	1509	1501	1648	1669	1945	2275

[a] Calculated by means of Eq. 13.39 in the case where the pivot atoms are carbon. Adapted with permission from Brewster, J. J. (1967), *Top. Stereochem.*, **2**, 1. Copyright © 1967 John Wiley & Sons, Inc. and from Potapov, V. M. (1979), *Stereochemistry*, Mir, Moscow, p. 301.
[b] Caution should be exercised in using the values for CO$_2$H and C$_6$H$_5$ since these groups lack the conical symmetry of the other groups.

conformations to the molar rotation of the compound can be calculated directly and application of Rules 5 and 6 (admittedly the weakest of the above rules for applying the conformational dissymmetry model; see above) can be avoided. Note that conformation c (Fig. 13.86), whose contribution was excluded in the calculation by the conformational dissymmetry method, makes a nonnegligible contribution to the rotation.

If the conformational populations for a given compound are unknown, they can often be approximated by molecular mechanics (force field) calculations. For an example, see Whitesell et al. (1988).

The molar rotation is then equal to $(+73 \times 0.573) + (-56 \times 0.201) + (-17 \times 0.226)$ multiplied by the rotivity correction (n_D^{20} 1.40 for neat chlorobutane;

Figure 13.87. Calculation of $\Delta[\Phi]_D$ for the three conformations of 2-chlorobutane (as in Fig. 13.86) according to the Brewster uniform conductor model. [Potapov, V. M. (1979), *Stereochemistry*, Mir, Moscow, p. 301.] The mole fractions [(a) 0.573; (b) 0.201; (c) 0.226] are taken from data by Nomura and Koda (1985).

f(n) = 1.24) or $[\Phi]_D$ +33.1 (Nomura and Koda, 1985; see also Potapov, 1979). This calculated molar rotation value is significantly closer to the observed molar rotation than that obtained using the cruder calculation shown on p. 1087. For a discussion of the assumptions in the derivation of the helical conductor model and a critical comparison of this model with the conformational dissymmetry model, see Brewster (1974b).

The Brewster model has been successfully applied to the diastereomeric 2-amino-1,2-diphenylethanols whose conformational equilibrium is known (Stefanovsky et al., 1969). The calculation of the rotations included a correction for intramolecular hydrogen bonding.

When the rotations are small, as is the case when opposing atomic versus conformational effects are present, for example, lactic acid (see above), or when a molecule contains multiple skew conformational units of opposing senses as in cis-1,2-disubstituted cyclopropanes (Bergman, 1969), the model reaches its limits. Brewster also explicitly recognized the limitations imposed by a model for the calculation of rotations at a single wavelength that may be far remote from that of the CE responsible for that rotation. It is remarkable that, in spite of the approximations inherent, this type of calculation is so successful.

Refinement of the Brewster model calculations depends in an important way on the number of conformations considered (the above examples included only staggered ones). Refinements that take into account additional conformations have been described by Brewster (1974a) and by Colle et al. (1981) and applied to the calculation of the molar rotations of flexible chiral alkanes.

When the molar rotation of an enantiomerically pure compound is known, the Brewster model may be used in a converse manner to probe the equilibrium of a conformationally mobile system. An early study of this type is that of α-phellandrene, a nonplanar molecule possessing an inherently dissymmetric chromophore (Ziffer et al., 1962).

The conformational equilibrium of isopropyl methylphosphonofluoridate (Sarin **55**, Fig. 13.88), a compound with a non-carbon stereocenter, has been calculated. In this calculation, the refraction of a lone electron pair was taken to be 0.24 (Boter, 1968). Brewster's original conformational dissymmetry model has recently been applied to the determination of the conformational energy differences $\Delta G°$ of 4-substituted cyclohexenes (Fig. 13.88) with substituents having low (X = Cl or F) and nearly zero (X = CN or D) conformational free energies (Lauricella et al., 1987). If the experimental specific rotation is the weighted average of the rotations of the two conformers present, then

$$\Delta G^0 = G_a^0 - G_e^0 = -RT \ln \frac{[\alpha]_e - [\alpha]_{exp}}{[\alpha]_{exp} - [\alpha]_a} \qquad (13.40)$$

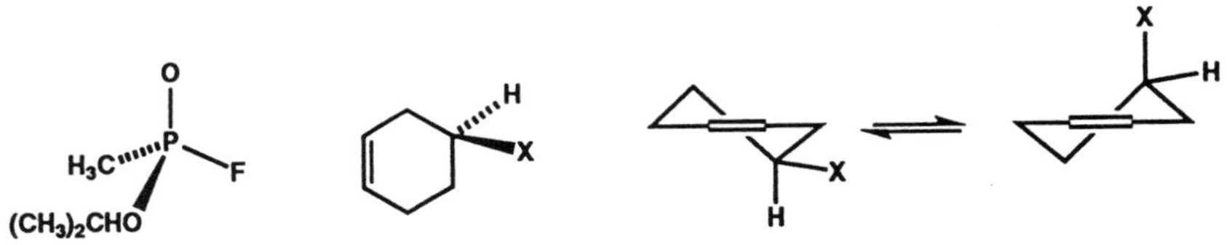

Figure 13.88. Sarin (**55**) and conformations of 4-substituted cyclohexenes (right).

Values of $[\alpha]_a$ and $[\alpha]_e$ were calculated by Brewster's method (Brewster, 1959a). Analysis of the thermal variation of the rotatory power with solvent (rotivity) taken into account permits an independent assessment of ΔG^0, which is concordant with the calculations described above. In particular, it was deduced that the equatorial conformer of 4-deuteriocyclohexene is more stable than the axial one ($\Delta H^0 = H_a^0 - H_e^0 = +10 \, \text{cal mol}^{-1}$).

The configuration of allenes and of alkylidenecycloalkanes (cf. Chapter 14) is predicted by Lowe's rule (Lowe, 1965) at least for structures in which substituents do not contribute to conformational dissymmetry (Fig. 13.89) and their molar rotations are calculable by means of Brewster's helix model (Brewster, 1967).

(S)-1,3-Diphenylallene **56** (Fig. 13.89b with B and Y = H, X and A = C_6H_5) is thereby predicted to have $[\Phi]_D > 0$; this is in accord with experiment (see below). Lowe's rule also correctly predicts the configuration of allene **57** (Fig. 13.89): The R enantiomer is predicted to have $[\Phi]_D < 0$ if the polarizability order is *tert*-butyl > CH_3. The configuration of (−)-**57** has recently been shown to be R (Eliel and Lynch, 1987; see also p. 1129). In the case of the cyclic allene 1,2-cyclononadiene, Lowe's rule fails (see also p. 1131) due to the significant conformational dissymmetry associated with the $(CH_2)_6$ methylene chain (Moore et al., 1971, 1973). A sector rule for allenes has been developed that subsumes both Lowe's and Brewster's treatments and provides an electronic basis for them (Crabbé, Mason, et al., 1971).

Lowe's rule has also been applied to spirans that have similar symmetry properties to simple allenes. However, the rule fails to correctly predict the absolute configuration of the spiran **58** (Fig. 13.90a; Krow and Hill, 1968). An attempt to determine the configuration of Fecht acid, spiro[3.3]heptane-2,6-dicarboxylic acid (Fig. 13.90b, **59**), by means of Lowe's rule leads to the conclusion that the (+) enantiomer has the R configuration. Other correlational models including those of Brewster and of Eyring–Jones (Jones and Eyring, 1961) lead to the same conclusion. These results are in contradiction of the findings resulting from anomalous X-ray crystallographic studies on the barium salt of the acid (for a discussion of these studies see Chapter 14). The failure of Lowe's rule in this instance is due in part to the unusual geometry of this spiran system (Fecht acid is better represented by Fig. 13.90c than by a structure in which the four-membered rings are flat). In addition to the conformational mobility associated with the puckered rings, the failure of Lowe's rule was ascribed to the

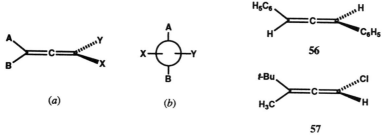

(a)   *(b)*

56

57

Figure 13.89. Absolute configuration of allenes (a) by Lowe's rule. The model shown in (a), corresponding to (b) in Newman notation, is dextrorotatory at 589 nm if the polarizability order is X > Y, A > B (right-hand helix) and levorotatory if the polarizability order of X and Y is reversed.

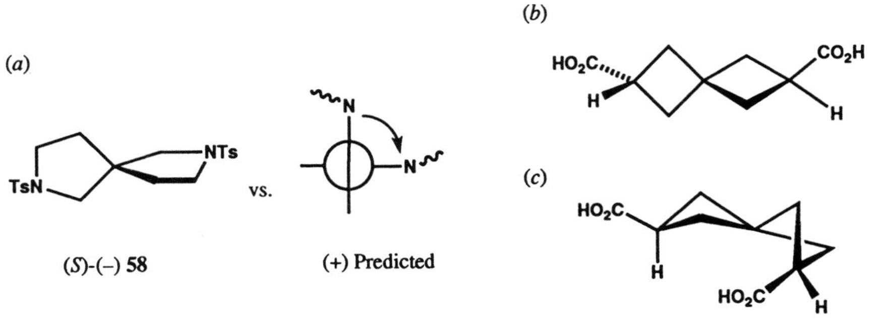

Figure 13.90.   Application of Lowe's rule to spirans.

presence of several chirotopic elements in **59** (the spiro[3.3]heptane skeleton itself, that resulting from disubstitution at positions 2 and 6 and that due to preferred conformations of the 2- and 6-substituents), which are not evident when the spiran is viewed as shown in Fig. 13.90*b* rather than as in Figure 13.90*c*. Moreover, because of the low rotation of the acid ($[\Phi]_{578}$ +8.7, in acetone), Cotton effects of the carboxyl group have disproportionate influence on the rotation in the visible part of the spectrum (Wynberg and Hulshof, 1974; Hulshof, Wynberg, et al., 1976).

A general algebraic theory whose goal is to gain insight into the phenomenon of chirality has been proposed by Ruch (1972). By means of "chirality functions" (χ), Ruch and co-workers (Ruch et al., 1973) demonstrated that the molar rotations of chiral allene derivatives (Fig. 13.91, **60**) may be calculated by manipulation of constant increments of a (pseudoscalar) property associated with individual parts (ligands) of the molecules (Ruch, 1972; cf. also Section 14-2.c). The calculation makes use of a polynomial equation of the type

$$\chi(R_1, R_2, R_3, R_4) = [\lambda(R_1) - \lambda(R_2)][\lambda(R_3) - \lambda(R_4)] \qquad (13.41)$$

in which (R_n) are molar rotations, and $\lambda(R_n)$ are the ligand-specific parameters [$\lambda(R) = 0$, +44.3, +15.4, +7.7, and +12.6 for H, C_6H_5, CO_2H, CH_3, and C_2H_5, respectively]. For compound **56**, $\chi = (+44.3 - 0)(+44.3 - 0) \cong +1960$; this may be compared with the experimental value $[\Phi]_D$ +1958. The ligand parameters themselves (see Table 14.1) may be obtained by manipulation of a set of polynomials analogous to Eq. 13.41, one for each compound used (these not including enantiomer pairs), with the number of compounds being greater than the number of kinds of ligands present in the test case (Ruch et al., 1973; Runge and Kresze, 1977; see also p. 1130).

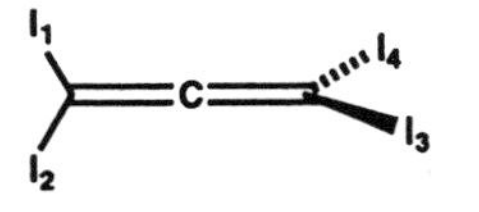

60

Figure 13.91.   Chiral allenes **60** with ligands ℓ_n = H, Me, Et, C_6H_5 and CO_2H whose molar rotations may be calculated by means of chirality functions.

Semiempirical application of chirality functions also yields satisfactory results in the case of chiral methanes (Richter, Richter, and Ruch, 1973). Chirality function analysis of Raman optical activity (ROA) data (Section 13-6) has been helpful in deducing the sense of chirality of conformers that contribute to the overall ROA intensity in the C–X (X = Cl or Br) stretching region of the spectrum (Diem, Burow, et al., 1976).

Equation 13.41 is identical in form to Brewster's Eq. 13.36. Note, however, that the chirality functions do not require a second set of parameters, such as the polarizabilities of the ligands, hence they are very general. Because these functions appear to be applicable only to systems having little or no conformational mobility, they are complementary to the Brewster model.

For chiral molecules not in point group C_1, "chirality observations" may be approximated by mathematical functions consisting of sums of pair functions. This approximation is justified by the theory of pairwise interactions (Kauzmann et al., 1961). Applications of the Octant rule (Section 13-4.c, p. 1022) and of circular dichroic exciton coupling (Section 13-4.d) are thus seen to be approximations to chirality functions (Dugundji, Marquarding, and Ugi, 1976).

An alternative algebraic description of chirality has been provided in terms of simple polynomials deduced from the topological properties of knots tied around chiral centers (Mezey, 1986). An important property of these knots and of the polynomials that describe them is that they are invariant to conformational changes. Consider, for example, that a string tied and knotted around a chiral object contains the necessary information to reproduce the chirality sense of the chiral object even after the object has been removed.

13-6. VIBRATIONAL OPTICAL ACTIVITY

Vibrational optical rotatory dispersion, vibrational circular dichroism (VCD) and Raman optical activity (ROA) are all manifestations of the interaction of polarized IR radiation with chiral substances. The three spectroscopic techniques together are aspects of vibrational optical activity (VOA) (Nafie and Diem, 1979; Mason, 1982a; Freedman and Nafie, 1987).

Few infrared ORD measurements have been reported (Korte and Schrader, 1973, 1975; Korte, 1978). On the other hand, VCD data and, to a lesser extent, ROA data are being reported with increasing frequency. As of 1992, VCD measurements are still being carried out on home-built dispersive or Fourier Transform IR (FTIR) spectrometers modified for VCD (Stinson, 1985; Nafie, 1992).

Vibrational optical activity arises from vibrational transitions in the electronic ground state of chiral molecules. Vibrational circular dichroism spectra are characterized by numerous usually well-resolved bands (CEs) consistent with the presence of many chirotopic vibrational motions that absorb in the IR region of the spectrum; but not all IR bands find their counterpart in VCD spectra. Within a given spectral region, the latter are generally simpler than IR spectra (Nafie, Keiderling, and Stephens, 1976). A typical VCD spectrum is shown in Figure 13.92. Just as in electronic CD, VCD spectra are often recorded with abscissa units of $\Delta\varepsilon$.

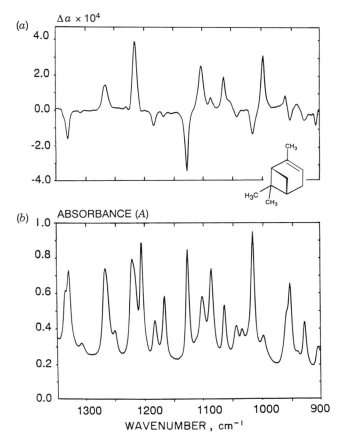

Figure 13.92. Spectra of $(-)$-α-pinene in the mid-IR range: (a) VCD and (b) isotropic IR. Ordinate units are absorbance (A) for IR, and the difference between absorbance for left and right circularly polarized radiation, $\Delta A = A_L - A_R$ for VCD. [Personal communication from L. A. Nafie, 1992. Reproduced by permission.].

Two features of VCD spectra immediately deserve comment. In a nonracemic chiral environment, the familiar multiplicity of IR absorption bands, for example, of methyl and methylene groups [ν_{C-H} both unsymmetrical doublets] is not always observed in the corresponding VCD spectra. The degeneracy of the antisymmetric stretching mode of CH_3 (ν_{as}) is lifted and two bands are observed, for example, in the VCD spectrum of alanine-N-d_3 (Lal, Nafie, et al., 1982). The VCD technique can reveal the chirotopicity of the formally symmetric methyl group in nonracemic samples of chiral molecules. Vibrational optical activity spectra are the most sensitive probes of *local* chirality presently available.

A second significant feature is that, relative to electronic CD bands, VCD signals are reduced by two or more orders of magnitude. There is a similar reduction in intensity of VOA signals

$$\Delta A/A = \Delta\varepsilon/\varepsilon = g \text{ number} \qquad (13.42)$$

relative to the isotropic IR and Raman effects. In contrast to electronic transitions

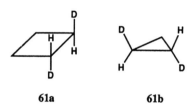

61a   **61b**

Figure 13.93. *trans*-Cyclobutane-1,2-d_2 and *trans*-cyclopropane-1,2-d_2.

in which the rotational strength R is about 10^{-38}–10^{-42} (esu cm)^2, those of VCD bands are only 10^{-43}–10^{-45}. Nevertheless, significant g numbers, of the order of 10^{-3}–10^{-4} are sometimes encountered. Even with g numbers of the order of 10^{-5}, the corresponding peaks may easily be observed.

The first observation of VCD on a liquid sample, $C_6H_5CH(OH)CF_3$, was that of $\nu_{C-H} = 2920$ cm^{-1} with $D = 1.4 \times 10^{-39}$ (esu cm)^2 and $R = +2 \times 10^{-44}$ (esu cm)^2. From Eq. 13.15 (Section 13-4.a), one calculates $g = 5.8 \times 10^{-5}$ (Holzwarth, Mosher, Moscowitz, et al., 1974). Also observed was $\nu_{C-D} = 2204$ cm^{-1} of *tert*-C_4H_9CHDCl, $g = -2 \times 10^{-5}$.

While in electronic spectroscopy, isotopic substitution has little effect on the wavelengths and intensities of the signals (UV) or is revealed indirectly by its effect on other, usually adjacent chromophores (CD) (Section 13-4.c), significant VCD signals due to vibrations involving the isotope itself are observed. Isotopically labeled compounds, such as **61a** (Fig. 13.93), in nonracemic form, exhibit CEs due to C–D bond stretching and bending that are as intense as those associated with corresponding C–H bonds (Cianciosi, Nafie, et al., 1989). Consequently, VCD may be applied to studies such as the measurement of the relative rates of isomerization (cis–trans) to racemization of **61b**, a compound devoid of typical chromophores (Chickos et al., 1986). Vibrational CD may also be applied to the determination of the enantiomer composition of deuterated chiral hydrocarbons by analysis in the ν_{C-D} region (Spencer, Nafie, et al., 1990).

The advantage of VCD relative to electronic CD is strikingly illustrated by the proton exchange of racemic acetoin, $CH_3COCH(OH)CH_3$ (Fig. 13.94, **62**) catalyzed by the enzyme acetolactate decarboxylase. In D_2O, proton exchange at the stereogenic center is observed to stop at 50% incorporation. The product of the exchange reaction (Fig. 13.94) exhibits only a very weak negative CE at 279 nm (associated with the $n-\pi^*$ carbonyl transition) relative to the strong negative CD typical of a chiral ketone, such as that observed in (R)-(−)- acetoin **62a**. This finding is due to the fact that the reaction product is a 50:50 mixture of two compounds having nearly equal and opposite CDs.

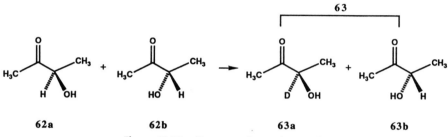

62a   **62b**   **63a**   **63b**

Figure 13.94. Proton exchange in acetoin.

Examination of the ν_{C-D} region by VCD, on the other hand, reveals several negative CEs from 2050 to 2200 cm^{-1}. The product mixture is clearly not racemic, contary to what might be inferred from the electronic CD spectrum [a 50:50 mixture of (S)-acetoin **62b**, and (R)-acetoin-d **63a**, would be expected to have nearly no rotation at any wavelength; however, since the IR spectra of **62** and **63** differ, they can be distinguished in VCD spectra]. Also, the H–D exchange must be stereoselective: only (R)-acetoin undergoes exchange to give acetoin-d **63a** with retention of configuration (**63a** is the so-called isotopomer of **62a**) and the same product exhibits positive VCD in the ν_{C-H} region of the spectrum due to unreacted (S)-acetoin **62b** with an intensity about one-half that of pure **62a** (Drake et al., 1987).

One of the simplest VCD spectra observed is that of ethylene oxide-2,3-d_2 [(S,S)-[2,3-^{2}H$_2$]-oxirane] (Fig. 13.95, **64**). The spectrum of **64** in the high frequency (C–H and C–D stretching) region of the spectrum consists of two bisignate pairs of bands, ν_{C-H} at 3026 and 3004 cm^{-1} and ν_{C-D} at 2254 and 2228 cm^{-1}, of comparable intensity, and both of negative chirality (p. 1046), consistent with the operation of the coupled oscillator mechanism (Section 13-4.c). While this mechanism calls for each of the $(+,-)$ couplets to be conservative (i.e., the $+$ and $-$ bands are expected to be of equal areas), this is not observed (Freedman, Nafie, et al., 1987).

The bias observed in each couplet [$+$band area $> -$band area for ν_{C-H} and $-$band area $> +$band area for ν_{C-D}] was interpreted by Freedman and Nafie as arising from the presence of a ring current generated *during* C–H and C–D stretching (Freedman, Nafie, et al., 1987). Such a ring current gives rise to a large magnetic dipole transition moment. An empirical rule for an oscillator external to the ring and adjacent to a hetero atom contained in the ring permits one to determine the sense of electronic current flow during contraction or elongation of the adjacent C–H or C–D bond (Freedman and Nafie, 1987). The effects on the electric and magnetic dipole transition moments are illustrated in Figure 13.95. Since the relative orientation of the C–H and C–D bonds in each molecule is reversed, the bias (but not the chirality sense at each stereocenter; Section 13-4.d) is reversed as well.

The production of angular electron current density during vibrations, either in existing cyclic structures encompassing delocalizable electron density, or as a

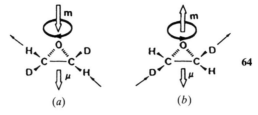

(a)   (b)

Figure 13.95. VCD ring currents and associated electric and magnetic transition moments in oxirane **64**; (a) generation of positive VCD bias in the ν_{CH} region; (b) generation of negative VCD bias in the ν_{CD}. [Reprinted with permission from Freedman, T. B., Paterlini, M. G., Lee, N.-S., Nafie, L. A., Schwab, J. M., and Ray, T. (1987), *J. Am. Chem. Soc.*, **109**, 4727. Copyright © 1987 American Chemical Society.]

result of intramolecular hydrogen bonding, reveals a fundamental distinction between electronic and vibrational CD, namely, that in the latter, electronic and nuclear motion both make contributions to the spectrum. Thus, the neglect of vibrational motion that is usually invoked in interpreting electronic spectra (including chiroptical phenomena) (the Born-Oppenheimer approximation) is not freely applicable in VCD. The coupling of achiral but chirally disposed oscillators and the production of vibrationally generated electronic current density are the two principal mechanisms responsible for the generation of VCD (Freedman and Nafie, 1987).

The principal difficulty in using the stereochemical information contained in VCD spectra is the requirement for theoretical models or empirical rules linking configuration or conformation to the form (mono- or bisignate) and the sense (the net positive or negative integrated intensity over a particular spectral region, i.e., the bias) of the spectral bands. Moreover, the bands in question must have been accurately assigned. An example of an empirical rule may be found in the characteristic C=C=C asymmetric stretching vibration of chiral allenes: (S)-allenes exhibit positive monosignate VCD at about $1950 \, \text{cm}^{-1}$ (Narayanan, Keiderling, et al., 1988). Many theoretical models exist (Freedman and Nafie, 1987) yet the straigtforward extraction of stereochemical information from VCD spectra by nonspecialists remains elusive.

Among the significant results of VCD studies thus far reported are those on amino acids, on phenylethanes, and on carbohydrates. The L-amino acids exhibit a strong positive bias in the $\nu_{C_\alpha-H}$ region of the VCD spectrum. The results appear to be the first experimental observations leading to the proposal for the aforementioned vibrationally generated electronic ring current. An empirical chirality rule was proposed linking the positive bias of the methine $C_\alpha-H$ stretching band to the (S)-α-amino acid (and α-hydroxy acid) configuration (Nafie et al., 1983; Oboodi, Nafie, et al., 1985). The proposed theoretical model is shown in Figure 13.96a.

The most prominent features of the CH stretching region of VCD spectra of phenylethane derivatives have also been interpreted in terms of the vibrationally generated electronic ring current as illustrated in Figure 13.96b for (S)-(−)-1-phenylethanol (Freedman et al., 1985). The sign of a VCD band at about $1200 \, \text{cm}^{-1}$, assigned tentatively to CH deformation (bending), correlates with the

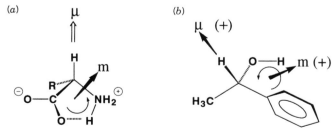

Figure 13.96. Vibrationally generated electronic ring current in the dominant conformation of (a) (S) α-amino acids; and (b) (S) α-phenylethane derivatives (contraction illustrated). The scalar product of μ and m [R = Im($\mu \cdot m$)] for the vibration of the methine $C_\alpha-H$ bond is positive, hence so is the CD bias for either stretching or contraction. [Adapted with permission from Freedman, T. B. and Nafie, L. A. (1987), *Top. Stereochem.*, **17**, 113. Copyright © 1987 John Wiley & Sons, Inc.]

configuration of phenylcarbinols $C_6H_5CR(OH)H$ for, for example, R = D, CH_3, and CO_2H (Polavarapu et al., 1986; Fig. 13.97a).

The VCD spectra of polypeptides, such as poly(γ-benzyl L-glutamate), exhibit bisignate absorption in the mid-IR range (amide C=O and NH stretching) whose sense depends less on the absolute configuration of the component amino acid than on the sense of the α-helix conformation. This explanation was arrived at by comparison with the opposite sense of the couplet observed, for example, in poly(β-benzyl L-aspartate), a polypeptide that forms left-handed α-helices (Lal and Nafie, 1982). The secondary structure of poly(L-lysine) in aqueous solution (random coil, α helix, β sheet, and transitions between them as the pH is changed) is readily revealed by VCD in the amide region of the spectrum (Yasui and Keiderling, 1986).

In the past, CD data on underivatized carbohydrates have been virtually unavailable due to the inaccessibility of the OH chromophore to existing CD instrumentation. With the advent of VOA, this has begun to change. In the first report of VCD spectra of carbohydrates (Marcott, Moscowitz, et al., 1978), the spectra of the α- and β-methyl D-glucosides were shown to be nearly enantiomeric in the 2800–3000-cm^{-1} (ν_{C-H}) region, although the compounds being compared are diastereomers, all of whose stereocenters have the same configuration save for that at the anomeric carbon. This result is a beautiful example of the ability of VCD to probe a single and specific stereocenter in molecules that possess a multiplicity of stereocenters. Though few such studies have been carried out thus far (Freedman and Nafie, 1987), VCD spectra of carbohydrates are likely to be quite useful for the assignment of configuration (e.g., Back and Polavarapu, 1984).

Chiral cyclohexane derivatives possessing the CH_2CH_2CH fragment exhibit a VCD pattern of three bands from 2890 to 2960 cm^{-1}, independent of the

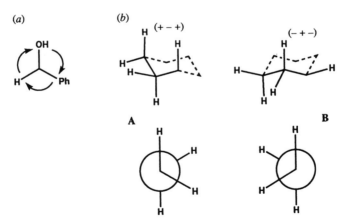

Figure 13.97. (a) Phenylcarbinol configuration from the sense of the VCD band at bout 1200 cm^{-1}. The fourth ligand is in the back of the plane. A clockwise pattern (according to the sequence rule) corresponds to a negative VCD band (Polavarapu, et al., 1986); (b) Inherently dissymmetric CH_2CH_2CH chromophore in cyclohexane derivatives, the corresponding Newman projections, and the signs of the resulting VCD pattern in the ν_{CH} region. [Reprinted with permission from Laux, L., Pultz, V., Abbate, S., Havel, H. A., Overend, J., Moscowitz, A., and Lightner, D. A. (1982), *J. Am. Chem. Soc.*, **104**, 4276. Copyright © 1982 American Chemical Society.]

functional group present, that has been interpreted as an inherently dissymmetric chromophore (Fig. 13.97*b*). While the fragments (in relevant stereoisomers) are not exactly enantiomeric since the methines in the two alternatives are axial and equatorial, respectively, enantiomeric $(+ - +)$ and $(- + -)$ patterns are found in numerous cases with structures as disparate as β-pinene, pulegone, and menthol that are fully in accord with prediction based on known configurations (Laux, Moscowitz, Lightner, et al., 1982). When the configurations are known, as in the cases of $(-)$-menthone and $(+)$-isomenthone, the signs of the three-band VCD patterns $(+ - +$ and $- + -$, respectively) reveal the nature of the dominant conformers present in solution (Fig. 13.98).

A technique complementary to VCD for the measurement of VOA is ROA, a chiroptical technique that is unique to the vibrational spectral range (Barron, 1980, 1983). In ROA, one measures the Raman circular intensity difference (CID), the differential scattering intensity of left and right circularly polarized light by nonracemic chiral matter. The CID is defined as the ratio of the differential scattering to the total scattering

$$\text{CID} = \frac{I_R - I_L}{I_R + I_L} = \frac{\Delta I}{I} \tag{13.43}$$

where I_L and I_R are the scattered intensities in left and right circularly polarized incident light. For ROA, in lieu of a g number (Eq. 13.42), the dimensionless ratio $2(\Delta I)/I$, where $\Delta I = I_R - I_L$ and $I = I_R + I_L$ has been called "chirality number" (Freedman and Nafie, 1987; Mason, 1982, p. 143).

Raman optical activity is obtained by exciting samples with circularly polarized visible radiation (alternating periodically between left and right circular polarization); the vibrational transitions are measured "as small displacements from the frequency of the incident light in the visible spectrum of the scattered light" (Barron, 1983). The fundamental problem of low source intensities that is inherent in VCD measurements is therefore absent in ROA. The first ROA spectra were measured in 1973 on α-methylbenzylamine and 1-phenylethanol (Barron, Bogaard, and Buckingham, 1973).

Raman optical activity may be measured to about 50 cm^{-1}. The low-frequency region of the spectrum is that in which bending vibrations and torsions involving

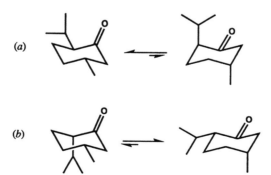

Figure 13.98. Conformational analysis of (*a*) menthone and (*b*) isomenthone from analysis of VCD spectra between 2960 and 2900 cm^{-1}.

large parts of the molecular framework are manifest. One of the advantages of ROA is that, in this spectral region, which contains much stereochemical information, the magnitude of VOA is larger than at the higher ($>2000\,\mathrm{cm}^{-1}$) frequencies. A principal limitation, however, is that ROA is more difficult to interpret than VCD. In any event, fewer routine ROA results have been reported (e.g., Waki et al., 1980).

Raman optical activity is particularly sensitive to molecules whose chirality results from isotopic substitution, for example, $C_6H_5CHD(OH)$ (Barron, 1977), and to molecular conformation (Barron, 1975). Distinct CIDs observed in the ν_{C-X} region of 1-halo-2-methylbutane (X = Br or Cl) were identified as arising from different conformers of these flexible molecules (Diem et al., 1976). Application of ROA to the analysis of peptides and proteins (aqueous solutions) has been significantly improved by recent advances in the geometry of the instrumental optics and in detector sensitivity (Barron et al., 1990).

13-7. CIRCULAR POLARIZATION OF EMISSION. ANISOTROPIC EMISSION

The systematic study of the emission counterpart of CD was initiated by Oosterhoff and his students (Emeis and Oosterhoof, 1967). Although the preferred IUPAC designation for the phenomenon is circular polarization of emitted light, CPE (IUPAC, 1986), it is commonly called circular polarization of luminescence, CPL (Richardson and Riehl, 1977; Richardson, 1978; Brittain, 1985; Riehl and Richardson, 1986). Yet another term equivalent to CPE is circularly polarized fluorescence (CPF). Care must be taken not to confuse CPF with fluorescence detected circular dichroism (FDCD; Section 13-4.f).

Here too, a dissymmetry ratio or g number (g_{em}) analogous to $g_{abs} = \Delta\varepsilon/\varepsilon$ may be defined

$$g_{em} = \frac{2(I_L - I_R)}{I_L + I_R} = \frac{\Delta I}{I} \tag{13.44}$$

where I is the intensity of the emitted left (I_L) or right (I_R) cpl. Photoexcitation of nonracemic samples carried out with unpolarized or linearly polarized light may give rise spontaneously to left and right cpl having different intensities with each enantiomer absorbing and emitting cpl in proportion to its presence in the sample.

The g_{em} is a property of the excited state from which the radiation arises and the associated rotational and dipole strengths of the emitted rays may differ from those of the absorption. The CPE spectrometer is essentially a modified fluorimeter that is able to measure both the total (or mean) fluorescence (the denominator of Eq. 13.44) and the differential fluorescence (the numerator of the equation).

Whereas CD measurements require that the samples contain an excess of one of the enantiomers, this restriction is removed in the case of emission phenomena. Photoexcitation with left or right circularly polarized light makes possible the determination of chiroptical properties on racemic samples (Dekkers, Emeis, and Oosterhoff, 1969). An interesting corollary to the measurement of CPE from racemates is that, on absorption of light, compounds such as **65** and **66** (Fig.

13.99) should give rise to two enantiomeric excited states, R*S and RS*, since the two C=O are enantiotopic and, therefore, would be expected to respond differently to cpl.

On excitation with cpl of a given sense, the concentrations of the enantiomeric excited ketones would no longer necessarily be equal and CPE might be measured. This has been observed (Schippers and Dekkers, 1983a). Moreover, the rate of intramolecular energy transfer between the two carbonyl groups could be assessed. Compound **65** behaves as if the carbonyl groups are isolated. The measured differential polarization, $G_L - G_R = (58 \pm 3) \times 10^{-4}$ (where the circular anisotropy $G_L = \frac{1}{2} g_{abs} g_{em}$ and $G_R = -\frac{1}{2} G_L$ for left and right cpl excitation light, respectively) is nearly identical to that of **67**. In compound **66**, on the other hand, the carbonyl groups are much closer together and $G_L - G_R$ vanishes as a result of very rapid intramolecular energy transfer (Schippers and Dekkers, 1983a). An attempt to apply this type of experiment to the generation, by chemiluminescence, of a locally excited and chiral state of a *meso*-diketone, such as **65**, in high enantiomer purity has thus far been unsuccessful possibly because of the aforementioned rapid intramolecular energy transfer (Meijer and Wynberg, 1979; Meijer, 1982).

From the foregoing, it may be evident that in order for substances to exhibit CPE they must be luminescent. This limitation may actually be turned to good advantage in that CPE is thus seen to be a more selective technique than CD. The CPE measurements of proteins, where only some of the chromophores are luminescent, illustrates this selectivity relative to CD, for example, tryptophan and tyrosine in proteins give rise to CPE effects while the peptide bond or the disulfide bond do not (compare these findings with those made by means of the FDCD technique, Section 13-4.f). The greater sensitivity of fluorescence measurements relative to CD spectroscopy is another strength of the technique (Brittain, 1985).

The CPE spectrum of 1,1'-bianthryl-2,2'-dicarboxylic acid, measured by Gafni and Steinberg (1972) exhibits some of the salient features characterizing the CPE phenomenon. The g_{em} number varies with the wavelength (as is true also of g_{abs} numbers); it is proportional to the enantiomer composition p, $G_{meas} = p g_{em}$; mirror-image CPE spectra are obtained for the two enantiomers; CPE bands are red-shifted relative to the CD bands associated with the same electronic transition; and, significantly, observed g_{em} numbers are appreciably smaller than g_{abs} numbers by factors of 10 or more. Since the latter would be expected to be approximately equal if the transition involved "exactly the same pairs of quantum states," Gafni and Steinberg suggested that the marked difference in the intensity

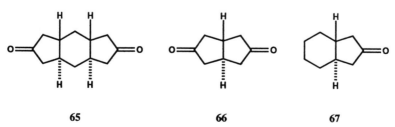

65   **66**   **67**

Figure 13.99. Structures **65–67**.

of the two g numbers undoubtedly is a reflection of changes in conformation taking place during the excitation to the first singlet excited state responsible for the emitted radiation. A similar suggestion was put forward independently by Emeis and Oosterhoff in 1971.

The latter suggestion has come to be generally accepted, and CPE is considered to be a technique that reflects the chirality and the geometry (in particular the conformation) of excited-state molecules in the same way as CD probes the configuration and conformation of ground-state chiral molecules.

Though in principle, CPE from triplet states (phosphorescence) is possible, CPE measurements to date have been derived only from emissions of singlet states (fluorescence), with one significant exception (Brittain, 1985). At room temperature, the CPE of the luminescent D-camphorquinone is mostly fluorescence; at $-150°C$, strong emission centered at 556 nm is ascribed to phosphorescence. Moreover, above 540 nm, the g_{em} number is about one order of magnitude smaller at $-150°C$ than it is at room temperature. This large difference may be due in part to a difference in conformation between the T_1 and the S_1 states (Steinberg, Gafni, and Steinberg, 1981). The ORD of triplet species obtained on photoexcitation of benzoin (Carapellucci et al., 1967) and the CD of bridged biaryls in the triplet state have been observed. In all but one case, compounds examined were configurationally stable in their excited triplet states; no photoracemization (photoatropisomerization) was observed (Tetreau and Lavalette, 1980; Tetreau et al., 1982).

Detailed analysis of CPE spectra and evaluation of the relevant dipole and rotational strengths in absorption (D_a, R_a) and in emission (D_e, R_e) have made it possible to assess changes in the geometry (bond lengths) and in the conformation (bond angles) that occur during excitation to the chiral $n–\pi^*$ singlet excited state. Large differences, even changes in sign, have been observed between g_{abs} and g_{em} in a series of β,γ enones. An example is given in Figure 13.100. Note, in particular, that the CPE signal at about $25,000 \text{ cm}^{-1}$ (ca. 415 nm) due to the carbonyl $n \leftarrow \pi^*$ fluorescence is of opposite sign to that of the corresponding CD (at ca. 300 nm), suggesting that the local configuration of the excited state is "inverted" relative to that of the ground state (Schippers et al., 1983c). It was suggested that the out-of-plane distortions in the (pyramidal) $n–\pi^*$ excited states may be the first step along the reaction coordinate to intramolecular formation of oxetanes that takes place on photoexcitation of flexible β,γ enones (Schippers et al., 1983c).

Thermal decomposition of optically active dioxetanes, such as $(-)$-**68**, is accompanied by beautiful chemiluminescence at about 430 nm that is circularly polarized. The observation of CPE requires that the emitting species be chiral. The CPE $[g_{em} + (3 \pm 1) \times 10^{-3}$ (fluorescence)$]$ can only result from the emission of singlet excited state adamantanone-ketal $(-)$-**69** (Fig. 13.101; Wynberg, Numan and Dekkers, 1977).

The widespread occurrence of bioluminescence in nature has prompted a study of the possible existence of circular polarization of the emitted light. It was observed that the left and right lanterns of firefly larvae that had been fed rac-amphetamine, $C_6H_5CH_2CH(CH_3)NH_2$, emitted yellow-green radiation at about 540 nm of opposite senses (right lanterns $\rightarrow$ +CPE and left lanterns $\rightarrow$ −CPE). The interpretation of this finding is anything but obvious since

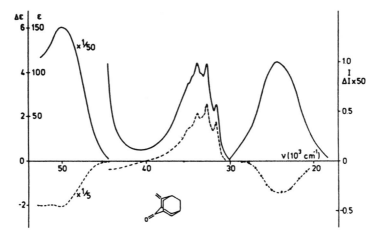

Figure 13.100. Spectra of (1*S*,3*R*)-4-methyleneadamantan-2-one in heptane at room temperature: (—) unpolarized absorption (ε) and fluorescence (*I*); (- - -) CD ($\Delta\varepsilon$) and (· · ·) CPE (ΔI). At left, the UV and CD charge-transfer band (50,000–40,000 cm^{-1}; 2500–250 nm) and the carbonyl $n \to \pi^*$ transition (at about 33,000 cm^{-1}; ca. 300 nm). At right, the $n \leftarrow \pi^*$ fluorescence and the negative circular polarization of the fluorescence. [Reprinted with permission from Schippers, P. H., van der Ploeg, J. P. M., Dekkers, H. P. J. M. (1983), *J. Am. Chem. Soc.*, **105**, 84. Copyright © 1983 American Chemical Society.]

the accepted mechanism of firefly bioluminescence involves the achiral species luciferin. While recognizing that a chiral complex between luciferin and an enzyme, for example, luciferase, might be responsible for the observed CPE, a more likely explanation involves a macroscopic phenomenon. Wynberg et al. (1980) hypthesized that passage of partially linearly polarized light through biopolymers in the lantern membranes (these being linearly birefringent and therefore capable of transforming linearly to elliptically polarized light) was responsible for the observed CPE. Moreover, the biopolymers in the left and right membranes may be oriented as the two stereocenters of a meso structure, and therefore capable of emitting CPE of opposite senses (Wynberg et al., 1980; Meijer, 1982).

Although the application of CPE to the determination of enantiomer purity was foreseen many years ago (Eaton, 1971; Kokke, 1974), a test of the two suggested procedures and of yet a third method involving CPE was not carried out until the early 1980s (Schippers and Dekkers, 1982). Three measurements are

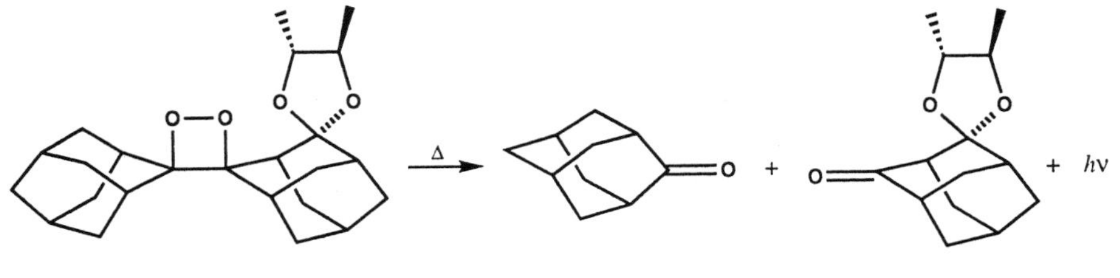

(–)-**68**   (–)-**69**

Figure 13.101. Observation of CPE on thermal decomposition of an optically active dioxetane.

required in each case. Significantly, the method is applicable to compounds devoid of functional groups, for example, compound **70** (Fig. 13.102) whose enantiomer composition may be difficult to measure by other means and it gives the best results with small enantiomeric excess (*p*) values, for example, experimental values of $p = 3.3$ and 3.4, respectively, both ±0.1%, were determined for compound **71** (Fig. 13.102) for the second and third methods mentioned above.

The principal limitations of the method are that the analyte molecules must fluoresce, that *g* numbers not be too small, and that differential quenching of excited-state molecules by ground-state molecules in nonracemic samples (intermolecular energy transfer prior to fluorescence) be minimal.

The basis of the determination rests on the assumption that the enantiomer composition *p* of the ground state and of the excited state of a given sample are equal, and that the measured *g* numbers are directly proportional to the enantiomer compositions, $G^p_{\text{meas}} = pg$, both in absorbance (G_{abs}) and in emission (G_{em}). The latter measurement must, of course, be carried out with unpolarized or linearly polarized light in order to avoid photoselection. A third measurement is required, $G^{p=0}_L = -G^{p=0}_R$, this one on a racemic sample and with left cpl or right cpl. In the procedure of Schippers and Dekkers, combination of three equations gives

$$p = \frac{(pg)(pg_{\text{em}})}{gg_{\text{em}}} = \frac{G^p_{\text{abs}}\, G^p}{2G^{p=0}_L} \tag{13.45}$$

Additional assumptions are that no racemization takes place during the lifetime of the excited state and that the concentrations $c^*_1,\ c^*_r \ll c_1,\ c_r$ so as to avoid complications due to participation of excimers in the fluorescence (see above).

Circular polarization of emission is evidently a useful, if not powerful, technique for the measurement of molecular geometry in excited states. Among other applications, CPE measurements permit one to probe geometric features of inclusion compound formation with fluorescent guest molecules (even achiral ones in which CPE may be induced locally) (Kano, Sisido, et al., 1985) and of fluorescent chromophores, such as naphthyl and anthryl groups, bound to helical polymers (Sisido et al., 1983). A combination of CD, FDCD, and CPE measurements is used to reveal the inclusion of achiral pyrene molecules in the cavity of γ-cyclodextrin with attendant formation of a chiral pyrene dimer, that is, one in which two pyrene molecules are twisted relative to one another and not stacked face-to-face. The chirality sense of the pyrene excimer bound to the 1° hydroxyl side of the cavity was determined to be *S* (Kano et al., 1988).

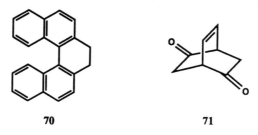

70   **71**

Figure 13.102. Structures **70–71**.

REFERENCES

Applequist, J. (1973), *J. Chem. Phys.*, **58**, 4251.

Applequist, J. (1977), "An Atom Dipole Interaction Model for Molecular Optical Properties," *Acc. Chem. Res.*, **10**, 79.

Arjona, O., Pérez-Ossorio, R., Pérez-Rubalcaba, A., Plumet, J., and Santesmases, M. J. (1984), *J. Org. Chem.*, **49**, 2624.

Atkins, P. W. (1971), "Natural and Induced Optical Activity," *Chem. Br.*, **7**, 244.

Back, D. M. and Polavarapu, P. L. (1984), *Carbohyd. Res.*, **133**, 163.

Baldwin, J. E., Hackler, R. E., and Scott, R. M. (1969), *J. Chem. Soc. D*, 1415.

Barrett, G. C. (1972), "Applications of Optical Rotatory Dispersion and Circular Dichroism," in Bentley, K. W. and Kirby, G. W., Eds., *Techniques of Chemistry*, Vol. IV, 2nd. ed., Part. I, Wiley, New York, Chap. 8.

Barron, D. (1975), *Nature (London)*, **255**, 458.

Barron, L. D. (1977), *J. Chem. Soc. Chem. Commun.*, 305.

Barron, L. D. (1980), "Raman Optical Activity: A New Probe Stereochemistry and Magnetic Structure," *Acc. Chem. Res.*, **13**, 90.

Barron, L. D. (1983), *Molecular Light Scattering and Optical Activity*, Cambridge University Press, Cambridge, UK.

Barron, L. D., Bogaard, M. P., and Buckingham, A. D. (1973), *J. Am. Chem. Soc.*, **95**, 603.

Barron, L. D., Gargaro, A. R., and Wen, Z. Q. (1990), *J. Chem. Soc. Chem. Commun.*, 1034.

Barth, G. and Djerassi, C. (1981), "Circular Dichroism of Molecules with Isotopically Engendered Chirality," *Tetrahedron*, **37**, 4123.

Barth, G., Voelter, W., Mosher, H. S., Bunnenberg, E., and Djerassi, C. (1970), *J. Am. Chem. Soc.*, **92**, 875.

Barton, D. H. R. (1950), *Experientia*, **6**, 316.

Bergman, R. G. (1969), *J. Am. Chem. Soc.*, **91**, 7405.

Berner, E. and Leonardsen, R. (1939), *Justus Liebigs Ann. Chem.*, **538**, 1.

Beychok, S. (1966), "Circular Dichroism of Biological Macromolecules," *Science*, **154**, 1288.

Biernbaum, M. S. and Mosher, H. S. (1971), *J. Org. Chem.*, **36**, 3168.

Biot, J. B. (1835), *Mém, Acad. Roy. Sciences de l'Institut France*, **13**, 116 through Kuhn, W., "Theorie und Grundgesetze der optischen Aktivität," in Freudenberg, K., Ed., *Stereochemie*, Franz Deuticke, Leipzig and Vienna, 1933, p. 318.

Blauer, G. (1983), *Isr. J. Chem.*, **23**, 201.

Blout, E. R. (1973), "Polypeptides and Proteins," Chap. 4.5 in Ciardelli, F. and Salvadori, P., Eds. (1973).

Blout, E. R. and Stryer, L. (1959), *Proc. Natl. Acad. Sci. USA*, **45**, 1591.

Bobbitt, D. R. and Yeung, E. S. (1984), *Anal. Chem.*, **56**, 1577.

Bobbitt, D. R. and Yeung, E. S. (1985), *Anal. Chem.*, **57**, 271.

Bohan, S. and Bouman, T. D. (1986), *J. Am. Chem. Soc.*, **108**, 3261.

Bootsma, G. A. and Schoone, J. C. (1967), *Acta Crystallogr.*, **22**, 522.

Born, M. (1915), *Phys. Z.*, **16**, 251.

Born, M. (1918), *Ann. Phys.* **55**, 177.

Bosnich, B. (1966), *J. Am. Chem. Soc.*, **88**, 2606.

Bosnich, B. (1967), *J. Am. Chem. Soc.*, **89**, 6143.

Boter, H. L. (1968), *Recl. Trav. Chim. Pays-Bas*, **87**, 957.

Bouman, T. D. and Lightner, D. A. (1976), *J. Am. Chem. Soc.*, **98**, 3145.

Bovey, F. A. (1982), *Chain Structure and Conformation of Macromolecules*, Academic, New York, Chap. 7.

Boys, S. F. (1934), *Proc. R. Soc. London, A*, **144**, 655.

Brahms, S. and Brahms, J. (1980), *J. Mol. Biol.*, **138**, 149.

Breslow, R., Baldwin, S., Flechtner, T., Kalicky, P., Liu, S., and Washburn, W. (1973), *J. Am. Chem. Soc.*, **95**, 3251.

Brewster, J. H. (1959a), *J. Am. Chem. Soc.*, **81**, 5475.

Brewster, J. H. (1959b), *J. Am. Chem. Soc.*, **81**, 5483, 5493.

Brewster, J. H. (1961), *Tetrahedron*, **13**, 106.

Brewster, J. H. (1967), "Helix Models of Optical Activity," *Top. Stereochem.*, **2**, 1.

Brewster, J. H. (1974a), *Tetrahedron*, **30**, 1807.

Brewster, J. H. (1974b), "On the Helicity of Variously Twisted Chains of Atoms," *Top. Curr. Chem*, **47**, 39.

Brewster, J. H. and Buta, J. G. (1966), *J. Am. Chem. Soc.*, **88**, 2233.

Brewster, J. H. and Prudence, R. T. (1973), *J. Am. Chem. Soc.*, **95**, 1217.

Brittain, H. G. (1985), "Excited-State Optical Activity," in Schulman, S. G., Ed., *Molecular Luminescence Spectroscopy: Methods and Applications–Part 1*, Wiley, New York, Chap. 6.

Brown, G. H. and Crooker, P. P. (1983), "Liquid Crystals. A Colorful State of Matter," *Chem. Eng. News*, **61**, Jan. 31, 1983, p. 24.

Buckingham, A. D. and Stiles, P. J. (1974), "On the Theory of Natural Optical Activity," *Acc. Chem. Res.*, **7**, 258.

Bukhari, S. T. K., Guthrie, R. D., Scott, A. I., and Wrixon, A. D. (1970), *Tetrahedron*, **26**, 3653.

Bunnenberg, E., Djerassi, C., Mislow, K., and Moscowitz, A. (1962), *J. Am. Chem. Soc.*, **84**, 2823.

Bur, A. J. and Fetters, L. J. (1976), "The Chain Structure, Polymerization and Conformation of Polyisocyanates," *Chem. Rev.*, **76**, 727.

Burgstahler, A. W. and Barkhurst, R. C. (1970), *J. Am. Chem. Soc.*, **92**, 7601.

Burgstahler, A. W., Ziffer, H., and Weiss, U. (1961), *J. Am. Chem. Soc.*, **83**, 4660.

Buss, V., Kolster, K., Wingen, U., and Simon, L. (1984), *J. Am. Chem. Soc.*, **106**, 4621.

Caldwell, D. J. and Eyring, H. (1971), *The Theory of Optical Activity*, Wiley, New York.

Caldwell, D. J. and Eyring, H. (1972), "Theory of Optical Rotation," in Weissberger, A. and Rossiter, B. W., Eds., *Physical Methods of Chemistry*, Part IIIC, *Techniques of Chemistry*, Vol. 1, Wiley, New York, Chap. 1.

Canceill, J. and Collet, A. (1986), *Nouv. J. Chim.*, **10**, 17.

Canceill, J., Collet, A., Gabard, J., Gottarelli, G., and Spada, G. P. (1985), *J. Am. Chem. Soc.*, **107**, 1299.

Canceill, J., Collet, A., Gottarelli, G., and Palmieri, P. (1987), *J. Am. Chem. Soc.*, **109**, 6454.

Canceill, J., Collet, A., and Jacques, J. (1982), *J. Chem. Soc., Perkin Trans. 2*, 83.

Canceill, J., Lacombe, L., and Collet, A. (1985), *J. Am. Chem. Soc.*, **107**, 6993.

Cantor, C. R. and Schimmel, P. R. (1980), *Biophysical Chemistry. Part II: Techniques for the Study of Biological Structure and Function*, Freeman, San Francisco, Chap. 8.

Carapellucci, P. A., Richtol, H. H., and Strong, R. L. (1967), *J. Am. Chem. Soc.*, **89**, 1742.

Carmack, M. and Neubert, L. A. (1967), *J. Am. Chem. Soc.*, **89**, 7134.

Cavazza, M., Festa, C., Veracini, C. A., and Zandomeneghi, M. (1991), *Chirality* **3**, 257.

Chafetz, L. (1991), Optical Rotation and Drug Standards, *Pharm. Technol.*, **15**, 52.

Chang, M., Meyers, H. V., Nakanishi, K., Ojika, M., Park, J. H., Park, M. H., Takeda, R., Vázquez, J. T., and Wiesler, W. T. (1989), "Microscale structure determination of oligosaccharides by the exciton chirality method," *Pure Appl. Chem.*, **61**, 1193.

Chang, T. C., Wu, C.-S. C., and Yang, J. T. (1978), *Anal. Biochem.*, **91**, 13.

Charney, E. (1979), *The Molecular Basis of Optical Activity. Optical Rotatory Dispersion and Circular Dichroism*, Wiley, New York.

Chickos, J. S., Annamalai, A., and Keiderling, T. A. (1986), *J. Am. Chem. Soc.*, **108**, 4398.

Ciancosi, S. J., Spencer, K. M., Freedman, T. B., Nafie, L. A., and Baldwin, J. E. (1989), *J. Am. Chem. Soc.*, **111**, 1913.

Ciardelli, F., et al., (1979). "Circular Dichroism and Optical Rotatory Dispersion in Polymer Structure Analysis," in Hummel, D. O., Ed., *Proceedings of the Fifth European Symposium on Polymer Spectroscopy, 1978*, Verlag Chemie, Weinheim, Germany, pp. 181–216.

Ciardelli, F. and Pieroni, O. (1980), "Chiral Synthetic Macromolecules as Models for Biopolymers," *Chimia*, **34**, 301.

Ciardelli, F. and Salvadori, P., Eds. (1973), *Fundamental Aspects and Recent Developments in Optical Rotatory Dispersion and Circular Dichroism*, Heyden, London.

Ciardelli, F. and Salvadori, P. (1985), "Conformational Analysis of Chiral Polymers in Solution," *Pure Appl. Chem.*, **57**, 931.

Ciardelli, F., Salvadori, P., Carlini, C., and Chiellini, E. (1972), *J. Am. Chem. Soc.*, **94**, 6536.

Coates, G. W. and Waymouth, R. M. (1991), *J. Am. Chem. Soc.*, **113**, 6270.

Cohen, M. A. (Amstar Sugar Corporation) (1988), personal communication to SHW.

Colle, R., Suter, U. W., and Luisi, P. L. (1981), *Tetrahedron*, **37**, 3727.

Collet, A. and Gottarelli, G. (1981), *J. Am. Chem. Soc.*, **103**, 204.

Collet, A. and Gottarelli, G. (1989), *Croat. Chem. Acta*, **62**, 279.

Colon, D. F., Pickard, S. T., and Smith, H. E. (1991), *J. Org. Chem.*, **56**, 2322.

Condon, E. U. (1937), *Rev. Mod. Phys.*, **9**, 432.

Condon, E. U., Altar, W., and Eyring, H. (1937), *J. Chem. Phys.*, **5**, 753.

Consiglio, G., Pino, P., Flowers, L. I., and Pittman, C. U., Jr. (1983), *J. Chem. Soc. Chem. Commun.*, 612.

Cornforth, J. W., Milborrow, B. V., and Ryback, G. (1966), *Nature (London)*, **210**, 627.

Cotton, A. (1895), *Compt. Rend.*, **120**, 989, 1044; (1896), *Ann. Chim. Phys.*, [7] **8**, 347.

Crabbé, P. (1965), *Optical Rotatory Dispersion and Circular Dichroism in Organic Chemistry*, Holden-Day, San Francisco; see especially Chap. 11.

Crabbé, P. (1967), "An Introduction to Optical Rotatory Dispersion and Circular Dichroism in Organic Chemistry," in Snatzke, G., Ed., *Optical Rotatory Dispersion and Circular Dichroism in Organic Chemistry*, Heyden, London, Chap. 1.

Crabbé, P. (1968), *Applications de la Dispersion Rotatoire Optique et du Dichroisme Circulaire Optique en Chimie Organique*, Gauthiers Villars, Paris.

Crabbé, P. (1972), *ORD and CD in Chemistry and Biochemistry: An Introduction*, Academic, New York.

Crabbé, P. and Parker, A. C. (1972), "Optical Rotatory Dispersion and Circular Dichroism," in Weissberger, A. and Rossiter, B. W., Eds., *Physical Methods of Chemistry*, Part IIIC, *Techniques of Chemistry*, Vol. 1, Wiley, New York, Chap. 3.

Crabbé, P., Velarde, E., Anderson, H. W., Clark, S. D., Moore, W. R., Drake, A. F., and Mason, S. F. (1971), *J. Chem. Soc. D*, 1261.

Craig, C. J., Bhargava, H. N., Everhart, E. T., LaBelle, B., Ohnsorge, U., and Webster, R. V. (1988), *J. Org. Chem.*, **53**, 1167.

Cram, D. J. and Sogah, D. Y. (1985), *J. Am. Chem. Soc.*, **107**, 8301.

Cross, G. A. and Kellogg, R. M. (1987), *J. Chem. Soc. Chem. Commun.*, 1746.

Davis, D. D. and Jensen, F. R. (1970), *J. Org. Chem.*, **35**, 3410.

Davar, I., Gedanken, A., Menge, W., and Verkade, J. G. (1993), *J. Chem. Soc. Chem. Commun.*, 262.

Davydov, A. S. (1962), *Theory of Molecular Excitons*, Kasha, M. and Oppenheimer, Jr., M. (translators), McGraw-Hill, New York.

de Weerd, R. J. E. M., van Hal, H. M. P. J., and Buck, H. M. (1984), *J. Org. Chem.* **49**, 3413.

Dekkers, H. P. J. M., Emeis, C. A., and Oosterhoff, L. J. (1969), *J. Am. Chem. Soc.*, **91**, 4589.

Deutsche, C. W., Lightner, D. A., Woody, R. W., and Moscowitz, A. (1969), "Optical Activity," *Ann. Rev. Phys. Chem.*, **20**, 407.

DeVoe, H. (1965), *J. Chem. Phys.*, **43**, 3199.

DeVoe, H. (1969), *Ann. N. Y. Acad. Sci.*, **158**, 298.

DeVoe, H. (1971), *J. Phys. Chem.*, **75**, 1509.

Diem, M., Diem, M. J., Hudgens, B. A., Fry, J. L., and Burow, D. F. (1976), *J. Chem. Soc. Chem. Commun.*, 1028.

Dillon, J. and Nakanishi, K. (1975), *J. Am. Chem. Soc.*, **97**, 5409, 5417.

Djerassi, C. (1960), *Optical Rotatory Dispersion: Applications to Organic Chemistry*, McGraw-Hill, New York.

Djerassi, C. (1961), *Tetrahedron*, **13**, 13.

Djerassi, C. (1964), "Applications of Optical Rotatory Dispersion and Circular Dichroism in Stereochemistry," *Proc. Chem. Soc., London*, 314.

Djerassi, C. (1990), *Steroids Made it Possible*, American Chemical Society, Washington, DC.

Djerassi, C., Finch, N., and Mauli, R. (1959), *J. Am. Chem. Soc.*, **81**, 4997.

Djerassi, C., Foltz, E. W., and Lippman, A. E. (1955), *J. Am. Chem. Soc.*, **77**, 4354.

Djerassi, C., Geller, L. E., and Eisenbraun, E. J. (1960), *J. Org. Chem.*, **25**, 1.

Djerassi, C. and Klyne, W. (1957a), *J. Am. Chem. Soc.*, **79**, 1506.

Djerassi, C. and Klyne, W. (1957b) "Recording and Nomenclature of Optical Rotatory Dispersion," *Proc. Chem. Soc., London*, 55.

Djerassi, C. and Staunton, J. (1961), *J. Am. Chem. Soc.*, **83**, 736.

Djerassi, C., Osiecki, J., Riniker, R., and Riniker, B. (1958), *J. Am. Chem. Soc.*, **80**, 1216.

Djerassi, C., Records, R., Bunnenberg, E., Mislow, K., and Moscowitz, A. (1962), *J. Am. Chem. Soc.*, **84**, 1962.

Dörffling, K. (1967), *Naturwissenschaften*, **54**, 23.

Doty, P. (1957), Proteins, *Sci. Am.*, **197**, Sept. 1957, p. 173.

Drake, A. F., Gould, J. M., and Mason, S. F. (1980), *J. Chromatogr.*, **202**, 239.

Drake, A. F. and Mason, S. F. (1977), *Tetrahedron*, **33**, 937.

Drake, A. F., Siligardi, G., Crout, D. H. G., and Rathbone, D. L. (1987), *J. Chem. Soc. Chem. Commun.*, 1834.

Drenth, W. and Nolte, R. J. M. (1979), "Poly(iminomethylenes): Rigid Rod Helical Polymers," *Acc. Chem. Res.*, **12**, 30.

Drude, P. (1892), *Göttinger Nachrichten*, 366. See Drude, P., *The Theory of Optics*, 1902, Longmans, Green & Co., New York; English translation by Mann, C. R. and Millikan, R. A., Dover Reprint, New York, 1959, pp. 400–407.

Dugundji, J., Marquarding, D., and Ugi, I. (1976), "Chirality and Hyperchirality," *Chem. Scr.*, **9**, 74.

Eaton, S. S. (1971), *Chem. Phys. Lett.*, **8**, 251.

Eliel, E. L. (1962), *Stereochemistry of Carbon Compounds*, McGraw-Hill, New York.

Eliel, E. L., Allinger, N. L., Angyal, S. J., Morrison, G. A. (1965), *Conformational Analysis*, Wiley, New York, pp. 381–394.

Eliel, E. L. and Knox, D. E. (1985), *J. Am. Chem. Soc.*, **107**, 2946.

Eliel, E. L. and Lynch, J. E. (1987), *Tetrahedron Lett.*, **28**, 4813.

Emeis, C. A. and Oosterhoff, L. J. (1967), *Chem. Phys. Lett.*, **1**, 129.

Emeis, C. A. and Oosterhoff, L. J. (1971), *J. Chem. Phys.*, **54**, 4809.

Emeis, C. A., Oosterhoff, L. J., and De Vries, G. (1967), *Proc. R. Soc. London, A*, **297**, 54.

Farina, M. (1987), "The Stereochemistry of Linear Macromolecules," *Top. Stereochem.*, **17**, 1.

Feynman, R. P., Leighton, R. B., and Sands, M. (1963), *The Feynman Lectures on Physics*, Vol. I, Addison-Wesley, Reading, MA.

Fischer, H. O. L. and Baer, E. (1941), "Preparation and Properties of Optically Active Derivatives of Glycerol," *Chem. Rev.*, **29**, 287.

Fitts, D. D. and Kirkwood, J. G. (1956), *Proc. Natl. Acad. Sci. USA*, **42**, 33.

Fitts, D. D. and Kirkwood, J. G. (1957), *Proc. Natl. Acad. Sci. USA*, **43**, 1046.

Freedman, T. B., Balukjian, G. A., and Nafie, L. A. (1985), *J. Am. Chem. Soc.*, **107**, 6213.

Freedman, T. B. and Nafie, L. A. (1987), "Stereochemical Aspects of Vibrational Optical Activity," *Top. Stereochem.*, **17**, 113.

Freedman, T. B., Paterlini, M. G., Lee, N.-S., Nafie, L. A., Schwab, J. M., and Ray, T. (1987), *J. Am. Chem. Soc.*, **109**, 4727.

Freifelder, D. (1976), *Physical Biochemistry: Applications to Biochemistry and Molecular Biology*, Freeman, San Francisco, Chap. 16.

Frelek, J., Konowal, A., Piotrowski, G., Snatzke, G., and Wagner, U. (1986), "Absolute Configuration of Natural Products from Circular Dichroism," in Atta-ur-Rahman and Le Quesne, P. W., Eds., *New Trends in Natural Products Chemistry 1986*; Studies in Organic Chemistry, Vol. 26, Elsevier, Amsterdam, The Netherlands, p. 477.

Frelek, J., Majer, Z., Perkowska, A., Snatzke, G., Vlahov, I., and Wagner, U. (1985), "Absolute configuration of *in situ* transition metal complexes of ligating natural products from circular dichroism," *Pure Appl. Chem.*, **57**, 441.

Freudenberg, K. (1933), "Konfigurative Zusammenhänge optisch aktiver Verbindungen," in Freudenberg, K., Ed., *Stereochemie*, Franz Deuticke, Leipzig and Vienna, 1933, p. 662.

Freudenberg, K., Todd, J., and Seidler, R. (1933), *Justus Liebigs Ann. Chem.*, **501**, 199.

Friedel, G. (1922), *Ann. Phys. (Paris)*, [9] **18**, 273.

Gafni, A. and Steinberg, I. Z. (1972), *Photochem. Photobiol.*, **15**, 93.

Gaubert, P. (1939), *C.R. Hebd. Seances Acad. Sci.*, **208**, 43.

Gerards, M. and Snatzke, G. (1990), *Tetrahedron: Asymmetry*, **1**, 221.

Gibbs, J. W. (1882), *Am. J. Sci.*, **23**, 460.

Goldberg, S. I., Bailey, W. D., and McGregor, M. L. (1971), *J. Org. Chem.*, **36**, 761. See especially Note 16.

Gonnella, N. C., Nakanishi, K., Martin, V. S., and Sharpless, K. B. (1982), *J. Am. Chem. Soc.*, **104**, 3775.

Goodman, M. (1967), "Concepts of Polymer Stereochemistry," *Top. Stereochem.*, **2**, 73.

Goodman, M., Verdini, A. S., Choi, N. S., and Masuda, Y. (1970), "Polypeptide Stereochemistry," *Top. Stereochem.*, **5**, 69.

Gorthey, L. A., Vairamani, M., and Djerassi, C. (1985), *J. Org. Chem.*, **50**, 4173.

Gottarelli, G., Samorí, B., Fuganti, C., and Grasselli, P. (1981), *J. Am. Chem. Soc.*, **103**, 471.

Gottarelli, G., Spada, G. P., and Solladié, G. (1986), "Some Stereochemical Applications of Induced Cholesteric Liquid Crystals," *Nouv. J. Chim.*, **10**, 691.

Green, M. M. and Garetz, B. A. (1984), *Tetrahedron Lett.*, **25**, 2831.

Green, M. M., Andreola, C., Muñoz, B., Reidy, M. P., and Zero, K. (1988a), *J. Am. Chem. Soc.*, **110**, 4063.

Green, M. M., Gross, R. A., Schilling, F. C., Zero, K., and Crosby, C., III (1988b), *Macromolecules*, **21**, 1839.

Green, M. M., Lifson, S., and Teramoto, A. (1991), *Chirality*, **3**, 285.

Green, M. M., Reidy, M. P., Johnson, R. J., Darling, G., O'Leary, D. J., and Willson, G. (1989), *J. Am. Chem. Soc.*, **111**, 6452.

Greenfield, N., Davidson, B., and Fasman, G. D. (1967), *Biochemistry*, **6**, 1630.

Greenfield, N. and Fasman, G. D. (1969), *Biochemistry*, **8**, 4108.

Greenstein, J. P. and Winitz, M. (1961), *Chemistry of the Amino Acids*, Wiley, New York.

Gross, M., Snatzke, G., and Wessling, B. (1979), *Liebigs Ann. Chem.*, 1036.

Halpern, B., Patton, W., and Crabbé, P. (1969), *J. Chem. Soc. (B)*, 1143.

Hansen, A. E. and Bouman, T. D. (1980), "Natural Chiroptical Spectroscopy: Theory and Computations," *Adv. Chem. Phys.*, **44**, 545.

Hanson, K. R. and Rose, I. A. (1975), Interpretation of Enzyme Reaction Stereospecificity, *Acc. Chem. Res.*, **8**, 1.

Harada, N., Chen, S. L., and Nakanishi, K. (1975), *J. Am. Chem. Soc.*, **97**, 5345.

Harada, N. and Nakanishi, K. (1968), *J. Am. Chem. Soc.*, **90**, 7351.

Harada, N. and Nakanishi, K. (1969), *J. Am. Chem. Soc.*, **91**, 3989.

Harada, N. and Nakanishi, K. (1972), "The Exciton Chirality Method and Its Application to Configurational and Conformational Studies of Natural Products," *Acc. Chem. Res.*, **5**, 257.

Harada, N. and Nakanishi, K. (1983), *Circular Dichroic Spectroscopy–Exciton Coupling in Organic Stereochemistry*, University Science Books, Mill Valley, CA.

Harada, N., Ochiai, N., Takada, K., and Uda, H. (1977), *J. Chem. Soc. Chem. Commun.*, 495.

Harada, N., Ohashi, M., and Nakanishi, K. (1968), *J. Am. Chem. Soc.*, **90**, 7349.

Harada, N., Ono, H., Uda, H., Parveen, M., Khan, N. U.-D., Achari, B., and Dutta, P. K. (1992), *J. Am. Chem. Soc.*, **114**, 7687.

Hargreaves, M. K. and Richardson, P. J. (1957), *J. Chem. Soc.*, 2260.

Hatano, M. (1986), *Induced Circular Dichroism in Biopolymer-Dye Systems*, Advances in Polymer Science, No. 77, Okamura, S., Ed., Springer, Berlin.

Hayes, K. S., Nagumo, M., Blount, J. F., and Mislow, K. (1980), *J. Am. Chem. Soc.*, **102**, 2773.

Hayward, L. D. (1975), *Chem. Phys. Lett.* **33**, 53.

Hayward, L. D. and Totty, R. N. (1969), *Chem. Commun.*, 676; *Can. J. Chem.*, **49**, 624 (1971).

Hecker, S. J. and Heathcock, C. H. (1986), *J. Am. Chem. Soc.*, **108**, 4586.

Heller, W. and Curmé, H. G. (1972), "Optical Rotation–Experimental Techniques and Physical Optics," Chap. 2 in Weissberger, A. and Rossiter, B. W., Eds., *Physical Methods of Chemistry*, Part IIIC, *Techniques of Chemistry*, Vol. I, Weissberger, A., Ed., Wiley, New York.

Heller, W. and Fitts, D. D. (1960), "Polarimetry," Chap. XXXIII, in Weissberger, S., Ed., *Physical Methods of Organic Chemistry*, 3rd. ed., Part 3, Wiley, New York, p. 2147.

Hoffmann, R. W., Ladner, W., Steinbach, K., Massa, W., Schmidt, R., and Snatzke, G. (1981), *Chem. Ber.*, **114**, 2786.

Holzwarth, G., Hsu, E. C., Mosher, H. S., Faulkner, T. R., and Moscowitz, A. (1974), *J. Am. Chem. Soc.*, **96**, 251.

Horeau, A. and Guetté, J.-P. (1974), "Interactions diastéréoisomères d'antipodes en phase liquide," *Tetrahedron*, **30**, 1923.

Horsman, G. and Emeis, C. A. (1965), *Tetrahedron Lett.*, 3037.

Hudec, J. and Kirk, D. N. (1976), *Tetrahedron*, **32**, 2475.

Hug, W., Ciardelli, F. and Tinoco, Jr., I. (1974), *J. Am. Chem. Soc.*, **96**, 3407.

Hulshof, L. A., Wynberg, H., van Dijk, B., and de Boer, J. L. (1976), *J. Am. Chem. Soc.*, **98**, 2733.

Itoh, T. (1974), "Quality of Amino Acids" in Kaneko, T., Izumi, Y., Chibata, I., and Itoh, T., Eds., *Synthetic Production and Utilization of Amino Acids*, Kodansha, Tokyo and Wiley, New York, Chap. 5.

IUPAC (1986), "Provisional Recommendations on Chiroptical Techniques, Nomenclature, Symbols, Units." Forthcoming, in *Pure Appl. Chem.*; see *Chem. Internat.*, **8**, (6) 23 (December 1986).

Izumi, Y. and Tai, A. (1977), *Stereo-differentiating Reactions*, Kodansha, Tokyo and Academic, New York.

Jacobs, H. J. C. and Havinga, E. (1972), *Tetrahedron*, **28**, 135.

Jacques, J., Collet, A., and Wilen, S. H. (1981), *Enantiomers, Racemates and Resolutions*, Wiley, New York.

Jirgensons, B. (1973), *Optical Activity of Proteins and Other Macromolecules*, 2nd ed., Springer, New York.

Johnson, Jr., W. C. (1985), "Circular Dichroism and Its Empirical Application to Biopolymers," in Glick, D., Ed., *Methods of Biochemical Analysis*, Vol. 31, Wiley, New York, p. 61.

Johnson, Jr., W. C. and Tinoco, Jr., I. (1972), *J. Am. Chem. Soc.*, **94**, 4389.

Jones, L. L. and Eyring, H. (1961), *Tetrahedron*, **13**, 235; *J. Chem. Educ.*, **38**, 601 (1961).

Kabuto, K., Yasuhara, F., and Yamaguchi, S. (1983), *Bull. Chem. Soc. Jpn.*, **56**, 1263.

Kamer, P. C. J., Nolte, R. J. M., and Drenth, W. (1986), *J. Chem. Soc. Chem. Commun.*, 1789.

Kano, K., Matsumoto, H., Hashimoto, S., Sisido, M., and Imanishi, Y. (1985), *J. Am. Chem. Soc.*, **107**, 6117.

Kano, K., Matsumoto, H., Yoshimura, Y., and Hashimoto, S. (1988), *J. Am. Chem. Soc.*, **110**, 204.

Kauzmann, W. (1957), *Quantum Chemistry*, Academic, New York, p. 617.

Kauzmann, W. J., Walter, J. E., and Eyring, H. (1940), "Theories of Optical Rotatory Power," *Chem. Rev.*, **26**, 339.

Kauzmann, W., Clough, F. B., and Tobias, I. (1961), *Tetrahedron*, **13**, 57.

Kawai, M., Nagai, U., and Katsumi, M. (1975), *Tetrahedron Lett.*, 3165.

Kelvin, Lord (W. Thompson) (1904), *Baltimore Lectures on Molecular Dynamics and the Wave Theory of Light*, Clay, London. The lectures were given in 1884 and 1893 at Johns Hopkins Univ., Baltimore, MD.

Khatri, C. A., Andreola, C., Peterson, N. C., and Green, M. M. (1992), 204th American Chemical Society National Meeting, Washington, DC, August 1992, Polymer Division Preprints.

Kirk, D. N. (1976), *J. Chem. Soc. Perkin 1*, 2171.

Kirk, D. N. (1986), "The Chiroptical Properties of Carbonyl Compounds," *Tetrahedron*, **42**, 777.

Kirk, D. N. and Klyne, W. (1974), *J. Chem. Soc., Perkin 1*, 1076.

Kirkwood, J. G. (1937), *J. Chem. Phys.*, **5**, 479.

Klyne, W. and Buckingham, J. (1978), *Atlas of Stereochemistry*, 2nd ed., Vol. I, Oxford, New York, p. 211.

Klyne, W. and Kirk, D. N. (1973a), *Tetrahedron Lett.*, 1483.

Klyne, W. and Kirk, D. N. (1973b), "The Carbonyl Chromophore: Saturated Ketones," Chap. 3.1 in Ciardelli and Salvadori (1973).

Ko, K.-Y. and Eliel, E. L. (1986), *J. Org. Chem.*, **51**, 5353.

Kobayashi, N., Saito, R., Hino, Y., Ueno, A., and Osa, T. (1982), *J. Chem. Soc. Chem. Commun.*, 706.

Kokke, W. C. M. C. (1974), *J. Am. Chem. Soc.*, **96**, 2627.

Kolbe, A. and Kolbe, A. (1982), *Z. Phys. Chem. (Leipzig)*, **263**, 61.

Koolstra, R. B., Jacobs, H. J. C., and Dekkers, H. P. J. M., (1989), *Croat. Chem. Acta.*, **62**, 115.

Korte, E. H. (1978), *Appl. Spectrosc.*, **32**, 568.

Korte, E. H. and Schrader, B. (1973), *Messtechnik (Braunschweig)*, **81**, 371.

Korte, E. H. and Schrader, B. (1975), *Appl. Spectroc.*, **29**, 389.

Korte, E. H., Schrader, B., and Bualek, S. (1978), *J. Chem. Res. Synopsis*, 236; *Miniprint*, 3001.

Krow, G. and Hill, R. K. (1968), *Chem. Commun.*, 430.

Kuhn, W. (1929), *Z. Phys. Chem.*, **B4**, 14.

Kuhn, W. (1930), *Trans. Faraday Soc.*, **26**, 293.

Kuhn, W. (1933), "Theorie und Grundgesetze der optischen Aktivität," in Freudenberg, K., Ed., *Stereochemie*, Franz Deuticke, Leipzig, Germany, p. 317.

Kumata, Y., Furukawa, J., and Fueno, T. (1970), *Bull. Chem. Soc. Jpn.*, **43**, 3920.

Kuriyama, K., Iwata, T., Moriyama, M., Ishikawa, M., Minato, H., and Takeda, K. (1967), *J. Chem. Soc. C*, 420.

Lal, B. B., Diem, M., Polavarapu, P. L., Oboodi, M., Freedman, T. B., and Nafie, L. A., (1982), *J. Am. Chem. Soc.*, **104**, 3336.

Lal, B. B. and Nafie, L. A. (1982), *Biopolymers*, **21**, 2161.

Lambert, J. B., Shurvell, H. F., Lightner, D. A., and Cooks, R. G. (1987), *Introduction to Organic Spectroscopy*, Macmillian, New York.

Lambert, J. B., Shurvell, H. F., Verbit, L., Cooks, R. G., and Stout, G. H. (1976), *Organic Structural Analysis*, Macmillan, New York, Part. 3.

Lamos, M. L., Lobenstine, E. W., and Turner, D. H. (1986), *J. Am. Chem. Soc.*, **108**, 4278.

Lardicci, L., Salvadori, P., Botteghi, C., and Pino, P. (1968), *Chem. Commun.*, 381.

Lauricella, R., Kéchayan, J., and Bodot, H. (1987), *J. Org. Chem.*, **52**, 1577.

Laux, L., Pultz, V., Abbate, S., Havel, H. A., Overend, J., Moscowitz, A., and Lightner, D. A. (1982), *J. Am. Chem. Soc.*, **104**, 4276.

Lee, S.-F., Barth, G., and Djerassi, C. (1978), *J. Am. Chem. Soc.*, **100**, 8010.

Lee, S.-F., Edgar, M., Pak, C. S., Barth, G., and Djerassi, C. (1980), *J. Am. Chem. Soc.*, **102**, 4784.

LeFèvre, C. G. and LeFèvre, R. J. W. (1972), "The Kerr Effect," in Weissberger, A. and Rossiter, B. W., Eds., *Physical Methods of Chemistry*, Part IIIC, *Techniques of Chemistry*, Vol. 1, Wiley, New York, Chap. 6.

Legrand, M. (1973), "Use of Solvent and Temperature Effects," Chap. 4.2, in Ciardelli and Salvadori (1973).

Legrand, M. and Rougier, M. J. (1977), "Application of the Optical Activity to Stereochemical Determinations," in Kagan, H. B., Ed., *Stereochemistry, Fundamentals and Methods*, Vol. 2, Thieme, Stuttgart, Germany, p. 33.

Leonard, N. J. and Beyler, R. E. (1950), *J. Am. Chem. Soc.*, **72**, 1316.

Lifson, S., Andreola, C., Peterson, N. C., and Green, M. M. (1989), *J. Am. Chem. Soc.*, **111**, 8850.

Lightner, D. A., Bouman, T. D., Wijekoon, W. M. D., and Hansen, A. E. (1984), *J. Am. Chem. Soc.*, **106**, 934.

Lightner, D. A., Bouman, T. D., Wijekoon, W. M. D., and Hansen, A. E. (1986a), *J. Am. Chem. Soc.*, **108**, 4484.

Lightner, D. A., Chang, T. C., Hefelfinger, D. T., Jackman, D. E., Wijekoon, W. M. D., and Givens, III, J. W. (1985b), *J. Am. Chem. Soc.*, **107**, 7499.

Lightner, D. A., Chang, T. C., and Horwitz, J. (1977), *Tetrahedron Lett*, **1977**, 3019; **1978**, 696.

Lightner, D. A. and Chappuis, J. L. (1981), *J. Chem. Soc. Chem. Commun.*, 372.

Lightner, D. A., Crist, B. V., Kalyanam, N., May, L. M., and Jackman, D. E. (1985a), *J. Org. Chem.*, **50**, 3867.

Lightner, D. A., Gawroński, J. K., and Gawrońska, K. (1985c), *J. Am. Chem. Soc.*, **107**, 2456.

Lightner, D. A., Gawroński, J. K., and Wijekoon, W. M. D. (1987b), *J. Am. Chem. Soc.*, **109**, 6354.

Lightner, D. A., McDonagh, A. F., Wijekoon, W. M. D., and Reisinger, M. (1988), *Tetrahedron Lett.*, **29**, 3507.

Lightner, D. A., Reisinger, M., and Landen, G. L. (1986b), *J. Biol. Chem.*, **261**, 6034.

Lightner, D. A., Reisinger, M., and Wijekoon, W. M. D. (1987a), *J. Org. Chem.*, **52**, 5391.

Lightner, D. A. and Toan, V. V. (1987c), *J. Chem. Soc. Chem. Commun.*, 210.

Lightner, D. A. and Wijekoon, W. M. D. (1982), *J. Org. Chem.*, **47**, 306.

Linderberg, J. and Michl, J. (1970), *J. Am. Chem. Soc.*, **92**, 2619.

Listowsky, I., Avigad, G., and England, S. (1970), *J. Org. Chem.*, **35**, 1080.

Liu, H.-w. and Nakanishi, K. (1982), *J. Am. Chem. Soc.*, **104**, 1178.

Lloyd, D. K. and Goodall, D. M. (1989), "Polarimetric Detection in High-Performance Liquid Chromatography," *Chirality*, **1**, 251.

Lobenstine, E. W., Schaefer, W. C., and Turner, D. H. (1981), *J. Am. Chem. Soc.*, **103**, 4936.

Lorentzen, R. J., Brewster, J. H., and Smith, H. E. (1992), *J. Am. Chem. Soc.*, **114**, 2181.

Lowe, G. (1965), *Chem. Commun.*, 411.

Lowman, D. W. (1979), "Bibliography: Methods of Sucrose Analysis," *J. Am. Soc. Sugar Beet Technol.*, **20**, 233.

Lowry, T. M. (1964), *Optical Rotatory Power*, Dover, New York, 1964; this is a reprint of the book originally published in 1935 by Longmans, Green and Co., London.

Lyle, G. G. and Lyle, R. E. (1983), "Polarimetry," in Morrison, J. D., Ed., *Asymmetric Synthesis*, Vol., 1, Academic, New York, Chap. 2.

Mannschreck, A. (1992), "Chiroptical Detection During Liquid Chromatography," *Chirality*, **4**, 163.

Mannschreck, A., Eiglsperger, A., and Stühler, G. (1982), *Chem. Ber.*, **115**, 1568.

Mannschreck, A., Mintas, M., Becher, G., and Stühler, G. (1980), *Angew Chem. Int. Ed. Engl.*, **19**, 469.

Marcott, C., Havel, H. A., Overend, J., and Moscowitz, A. (1978), *J. Am. Chem. Soc.*, **100**, 7088.

Marion, L., Turcotte, F., and Ouellet, J. (1951), *Can. J. Chem.*, **29**, 22.

Mason, S. F. (1962), *Proc. Chem. Soc.*, London, 362.

Mason, S. F. (1973), "The Development of Theories of Optical Activity and of their Applications," Chap. 2.1 in Ciardelli and Salvadori (1973).

Mason, S. F., Ed. (1979), *Optical Activity and Chiral Discrimination*, Reidel, Dordrecht, The Netherlands.

Mason, S. F. (1982), *Molecular Optical Activity and the Chiral Discriminations*, Cambridge University Press, Cambridge, UK.

Mason, S. F. (1982a), *Molecular Optical Activity and the Chiral Discriminations*, Cambridge University Press, Cambridge, UK, Chap. 8.

Mason, S. F., Vane, G. W., Schofield, K., Wells, R. J., and Whitehurst, J. S. (1967), *J. Chem. Soc. B*, 553.

Meijer, E. W. (1982), Ph.D. Dissertation, University of Groningen, Groningen, The Netherlands.

Meijer, E. W. and Wynberg, H. (1979), *Tetrahedron Lett.*, 3997.

Menger, F. M. and Boyer, B. (1984), *J. Org. Chem.*, **49**, 1826.

Mezey, P. G. (1986), *J. Am. Chem. Soc.*, **108**, 3976.

Mislow, K. (1962), *Ann. N. Y. Acad. Sci.*, **93**, 459 and references cited therein.

Mislow, K. (1965), *Introduction to Stereochemistry*, Benjamin, New York.

Mislow, K. and Bickart, P. (1976/1977), "An Epistemological Note on Chirality," *Isr. J. Chem.*, **15**, 1.

Mislow, K., Glass, M. A. W., Moscowitz, A., and Djerassi, C. (1961), *J. Am. Chem. Soc.*, **83**, 2771.

Mislow, K., Glass, M. A. W., O'Brien, R. E., Rutkin, P., Steinberg, D. H., Weiss, J., and Djerassi, C. (1962), *J. Am. Chem. Soc.*, **84**, 1455.

Mislow, K., Graeve, R., Gordon, A. J., and Wahl, G. H., Jr. (1964), *J. Am. Chem. Soc.*, **86**, 1733.

Mislow, K. and Siegel, J. (1984), *J. Am. Chem. Soc.*, **106**, 3319.

Moffitt, W. (1956a), *J. Chem. Phys.*, **25**, 467.

Moffitt, W. (1956b), *Proc. Natl. Acad. Sci. USA*, **42**, 736.

Moffitt, W., Fitts, D. D., and Kirkwood, J. G. (1957), *Proc. Natl. Acad. Sci. USA*, **43**, 723.

Moffitt, W. and Moscowitz, A. (1959), *J. Chem. Phys.*, **30**, 648.

Moffitt, W., Woodward, R. B., Moscowitz, A., Klyne, W., and Djerassi, C. (1961), *J. Am. Chem. Soc.*, **83**, 4013.

Moffitt, W. and Yang, J. T. (1956), *Proc. Natl. Acad. Sci. USA*, **42**, 596.

Moore, W. R., Anderson, H. W., and Clark, S. D. (1973), *J. Am. Chem. Soc.*, **95**, 835.

Moore, W. R., Anderson, H. W., Clark, S. D., and Ozretich, T. M. (1971), *J. Am. Chem. Soc.*, **93**, 4932.

Mori, Y., Kohchi, Y., Suzuki, M., and Furukawa, H. (1992), *J. Am. Chem. Soc.*, **114**, 3557.

Moscowitz, A. (1960), "Theory and Analysis of Rotatory Dispersion Curves," Chap. 12 in Djerassi (1960).

Moscowitz, A. (1961), *Tetrahedron*, **13**, 48.

Moscowitz, A. (1962), "Theoretical Aspects of Optical Activity. Part One: Small Molecules," *Adv. Chem. Phys.*, **4**, 67.

Moscowitz, A., Charney, E., Weiss, U., and Ziffer, H. (1961), *J. Am. Chem. Soc.*, **83**, 4661.

Moscowitz, A., Krueger, W. C., Kay, I. T., Skewes, G., and Bruckenstein, S. (1964), *Proc. Natl. Acad. Sci. USA*, **52**, 1190.

Moscowitz, A., Mislow, K., Glass, M. A. W., and Djerassi, C. (1962), *J. Am. Chem. Soc.*, **84**, 1945.

Moscowitz, A., Wellman, K., and Djerassi, C. (1963), *J. Am. Chem. Soc.*, **85**, 3515.

Müller, S., Keese, R., Engel, P., and Snatzke, G. (1986), *J. Chem. Soc. Chem. Commun.*, 297.

Muto, K., Mochizuki, H., Yoshida, R., Ishii, T., and Handa, T. (1986), *J. Am. Chem. Soc.*, **108**, 6416.

Nafie, L. A. (1992), Personal communication to SHW.

Nafie, L. A. and Diem, M. (1979), "Optical Activity in Vibrational Transitions: Vibrational Circular Dichroism and Raman Optical Activity," *Acc. Chem. Res.*, **12**, 296.

Nafie, L. A., Keiderling, T. A., and Stephens, P. J. (1976), *J. Am. Chem. Soc.*, **98**, 2715.

Nafie, L. A., Oboodi, M. R., and Freedman, T. B. (1983), *J. Am. Chem. Soc.*, **105**, 7449.

Nakanishi, K., Berova, N., and Woody, R. W., Eds., (1994), *Circular Dichroism: Principles and Applications*, VCH, New York and chapters therein.

Nakanishi, K., Kasai, H., Cho, H., Harvey, R. G., Jeffrey, A. M., Jennette, K. W., and Weinstein, I. B. (1977), *J. Am. Chem. Soc.*, **99**, 258.

Narayanan, U., Keiderling, T. A., Elsevier, C. J., Vermeer, P., and Runge, W. (1988), *J. Am. Chem. Soc.*, **110**, 4133.

Nelson, W. L. and Bartels, M. J. (1982), *J. Org. Chem.*, **47**, 1574.

Nelson, W. L., Wennerstrom, J. E., and Sankar, S. R. (1977), *J. Org. Chem.*, **42**, 1006.

Newman, M. S., Darlak, R. S., and Tsai, L. (1967), *J. Am. Chem. Soc.*, **89**, 6191.

Nishino, N., Mihara, H., Hasegawa, R., Yanai, T., and Fujimoto, T. (1992), *J. Chem. Soc. Chem. Commun.*, 692.

Noack, K. (1969), *Helv. Chim. Acta.*, **52**, 2501.

Nolte, R. J. M. and Drenth, W. (1987), "Atropisomeric Polymers," in Fontanille, M. and Guyot, A., Eds., *Recent Advances in Mechanistic and Synthetic Aspects of Polymerization*, Reidel, Dordrecht, The Netherlands, p. 451ff.

Nomura, H. and Koda, S. (1985), *Bull. Chem. Soc. Jpn.*, **58**, 2917.

Numan, H. and Wynberg, H. (1978), *J. Org. Chem.*, **43**, 2232.

Oboodi, M. R., Lal, B. B., Young, D. A., Freedman, T. B., and Nafie, L. A. (1985), *J. Am. Chem. Soc.*, **107**, 1547.

Okamoto, Y., Nakano, T., and Hatada, K. (1989), *Polym. J. (Tokyo)*, **21**, 199.

Okamoto, Y., Suzuki, K., Ohta, K., Hatada, K., and Yuki, H. (1979), *J. Am. Chem. Soc.*, **101**, 4763.

Okamoto, Y., Takeda, T., and Hatada, K. (1984), *Chem. Lett.*, **1984**, 757.

Oseen, C. W. (1915), *Ann. Phys.*, **48**, 1.

Overby, L. R. and Ingersoll, A. W. (1960), *J. Am. Chem. Soc.*, **82**, 2067.

Partridge, J. J., Chadha, N. K., and Uskoković, M. R. (1973), *J. Am. Chem. Soc.*, **95**, 532.

Penot, J. P., Jacques, J., and Billard, J. (1968), *Tetrahedron Lett.*, 4013.

Peters, H. M., Feigl, D. M., and Mosher, H. S. (1968), *J. Org. Chem.*, **33**, 4245.

Pickard, S. T. and Smith, H. E. (1990), *J. Am. Chem. Soc.*, **112**, 5741.

Pino, P. (1973), "Optical Rotatory Dispersion and Circular Dichroism in Conformational Analysis of Synthetic High Polymers," Chap. 4.4 in Ciardelli, F. and Salvadori, P., Eds., (1973).

Pino, P., Carlini, C., Chiellini, E., Ciardelli, F., and Salvadori, P. (1968), *J. Am. Chem. Soc.*, **90**, 5025.

Pirkle, W. H., Sowin, T. J., Salvadori, P., and Rosini, C. (1988), *J. Org. Chem.*, **53**, 826.

Platt, J. R. (1949), *J. Chem. Phys.*, **17**, 484.

Platt, J. R. (1951), *J. Chem. Phys.*, **19**, 263.

Polavarapu, P. L., Fontana, L. P., and Smith, H. E. (1986), *J. Am. Chem. Soc.*, **108**, 94.

Potapov, V. M. (1979), *Stereochemistry*, Mir, Moscow.

Prelog, V. (1968), "Das Asymmetrische Atom, Chiralität and Pseudoasymmetrie," *Proc. K. Ned. Akad. Wet.*, **B71**, 108.

Purdie, N. and Brittain, H. G., Eds. (1994), *Analytical Applications of Circular Dichroism*, Elsevier, Amsterdam and chapters therein.

Purdie, N. and Swallows, K. A. (1989), "Analytical Applications of Polarimetry, Optical Rotatory Dispersion, and Circular Dichroism," *Anal. Chem.*, **61** (2), 77A.

Quadrifoglio, F. and Urry, D. W. (1968), *J. Am. Chem. Soc.*, **90**, 2755.

Rashidi-Ranjbar, P., Man, Y.-M., Sandström, J., and Wong, H. N. C. (1989), *J. Org. Chem.*, **54**, 4888.

Reeves, R. E. (1951), "Cuprammonium–Glycoside Complexes," *Adv. Carbohydrate Chem.*, **6**, 107.

Reeves, R. E. (1965), "Optical Rotations in Cuprammonium Solutions for Configurational and Conformational Studies," in Whistler, R. L., Ed., *Methods in Carbohydrate Chemistry*, Vol. V, Academic, New York, p. 203.

Regan, L. and DeGrado, W. F. (1988), *Science*, **241**, 976.

Richardson, F. S. (1978), "Circular Polarization Differentials in the Luminescence of Chiral Systems," in Mason, S. F. Ed., *Optical Activity and Chiral Discrimination*, Reidel, Dordrecht, The Netherlands, p. 189.

Richardson, F. S. and Riehl, J. P. (1977), "Circularly Polarized Luminescence Spectroscopy," *Chem. Rev.*, **77**, 773.

Richter, W. J., Richter, B., and Ruch, E. (1973), *Angew. Chem. Int. Ed. Engl.*, **12**, 30.

Riehl, J. P. and Richardson, F. S. (1986), "Circularly Polarized Luminescence Spectroscopy," *Chem. Rev.*, **86**, 1.

Rinaldi, P. L., Naidu, M. S. R., and Conaway, W. E. (1982), *J. Org. Chem.*, **47**, 3987.

Ripperger, H. (1977), "Räumliche Struktur der Moleküle und chiroptische Eigenschaften–neuere Ergebnisse," *Z. Chem.*, **17**, 250.

Rizzo, V. and Schellman, J. A. (1984), *Biopolymers.*, **23**, 435.

Rodgers, S. L., Kalyanam, N., and Lightner, D. A. (1982), *J. Chem. Soc. Chem. Commun.*, 1040.

Roschester, J., Berg, U., Pierrot, M., and Sandström, J. (1987), *J. Am. Chem. Soc.*, **109**, 492.

Rosenfeld, L. (1928), *Z. Phys.*, **52**, 161.

Rosini, C., Bertucci, C., Salvadori, P., and Zandomeneghi, M. (1985), *J. Am. Chem. Soc.* **107**, 17.

Rosini, C., Giacomelli, G., and Salvadori, P. (1984), *J. Org. Chem.*, **49**, 3394.

Rosini, C., Zandomeneghi, M., and Salvadori, P. (1993), *Tetrahedron: Asymmetry*, **4**, 545.

Ross, J. M. D., Morrison, T. J., and Johnstone, C. (1937), *J. Chem. Soc.*, 608.

Ruch, E. (1972), "Algebraic Aspects of the Chirality Phenomenon in Chemistry," *Acc. Chem. Res.*, **5**, 49.

Ruch, E., Runge, W., and Kresze, G. (1973), *Angew. Chem. Int. Ed. Engl.*, **12**, 20.

Runge, W. and Kresze, G. (1977), *J. Am. Chem. Soc.*, **99**, 5597.

Saeva, F. D. and Olin, G. R. (1977), *J. Am. Chem. Soc.*, **99**, 4848.

Saeva, F. D., Sharpe, P. E., and Olin, G. R. (1973), *J. Am. Chem. Soc.*, **95**, 7656, 7660.

Saeva, F. D. and Wysocki, J. J. (1971), *J. Am. Chem. Soc.*, **93**, 5928.

Salvadori, P., Bertucci, C., and Rosini, C. (1991), *Chirality*, **3**, 376.

Salvadori, P., Lardicci, L., Menicagli, R., and Bertucci, C. (1972), *J. Am. Chem. Soc.*, **94**, 8598.

Salvadori, P., Rosini, C., and Bertucci, C. (1984), *J. Org. Chem.*, **49**, 5050.

Sanderson, W. A. and Mosher, H. S. (1966), *J. Am. Chem. Soc.*, **88**, 4185.

Sathyanarayana, B. K. and Stevens, E. S. (1987), *J. Org. Chem.*, **52**, 3170.

Saxena, V. P. and Wetlaufer, D. B. (1971), *Proc. Natl. Acad. Sci. USA*, **68**, 969.

Schellman, J. A. (1966), *J. Chem. Phys.*, **44**, 55.

Schellman, J. A. (1968), "Symmetry Rules for Optical Rotation," *Acc. Chem. Res.*, **1**, 144.

Scheraga, H. A. (1971), "Theoretical and Experimental Studies of Conformation of Polypeptides," *Chem. Rev.*, **71**, 195.

Schipper, P. E. (1981), *Chem. Phys.*, **57**, 105.

Schipper, P. E. and Rodger, A. (1985), *J. Am. Chem. Soc.*, **107**, 3459.

Schippers, P. H. and Dekkers, H. P. J. M. (1982), *Tetrahedron*, **38**, 2089.

Schippers, P. H. and Dekkers, H. P. J. M. (1983b), *J. Am. Chem. Soc.*, **105**, 79.

Schippers, P. H. and Dekkers, H. P. J. M. (1983a), *J. Am. Chem. Soc.*, **105**, 145.

Schippers, P. H., van der Ploeg, J. P. M., and Dekkers, H. P. J. M. (1983c), *J. Am. Chem. Soc.*, **105**, 84.

Schoenfelder, W. and Snatzke, G. (1980), *Isr. J. Chem.*, **20**, 142.

Schurig, V. (1985), "Current Methods for Determination of Enantiomeric Compositions (Part 1); Definitions, Polarimetry," *Kontakte (Darmstadt)*, 54.

Scopes, P. M. (1975), "Applications of the Chiroptical Techniques to the Study of Natural Products," *Fortschr. Chem. Org. Naturst.*, **32**, 167.

Sélégny, E., Ed. (1979), *Optically Active Polymers*, Reidel, Dordrecht, The Netherlands.

Sherrington, D. C., Solaro, R., and Chiellini, E. (1982), *J. Chem. Soc. Chem. Commun.*, 1103.

Sheves, M., Kohne, B., Friedman, N., and Mazur, Y. (1984), *J. Am. Chem. Soc.*, **106**, 5000.

Shindo, Y. and Ohmi, Y. (1985), *J. Am. Chem. Soc.*, **107**, 91.

Sisido, M., Egusa, S., Okamoto, A., and Imanishi, Y. (1983), *J. Am. Chem. Soc.*, **105**, 3351.

Sjöberg, B. (1959), through Djerassi, C., *Optical Rotatory Dispersion*, McGraw-Hill, New York, 1960, p. 236.

Sjöberg, B. (1967), "Optical Rotatory Dispersion and Circular Dichroism of Chromophoric Derivatives of Transparent Compounds," in Snatzke, G., Ed., *Optical Rotatory Dispersion and Circular Dichroism in Organic Chemistry*, Heyden, London, Chap. 11.

Sjöberg, B. and Obenius, U. (1982), *Chem. Scr.*, **20**, 62.

Smith, H. E. (1983), "The Salicylidenamino Chirality Rule: A Method for the Establishment of the Absolute Configurations of Chiral Primary Amines by Circular Dichroism," *Chem. Rev.*, **83**, 359.

Smith, H. E. (1989), "Correlation of the Circular Dichroism of Unsubstituted and Ring-Substituted Chiral Phenylcarbinamines with Their Absolute Configurations," *Croat. Chem. Acta*, **62**, 201.

Smith, H. E. and Fontana, L. P. (1991), *J. Org. Chem.*, **56**, 432.

Smith, H. E., Neergaard, J. R., de Paulis, T., and Chen, F.-M. (1983), *J. Am. Chem. Soc.*, **105**, 1578.

Smith, H. E. and Willis, T. C. (1971), *J. Am. Chem. Soc.*, **93**, 2282.

Snatzke, F. and Snatzke, G. (1977), "Molekülstrukturanalyse durch Circulardichroismus," *Fresenius' Z. Anal. Chem.*, **285**, 97.

Snatzke, G. (1965), *Tetrahedron*, **21**, 413.

Snatzke, G., Ed. (1967) *Optical Rotatory Dispersion and Circular Dichroism in Organic Chemistry*, Heyden, London.

Snatzke, G. (1968), "Circular Dichroism and Optical Rotatory Dispersion–Principles and Application to the Investigation of the Stereochemistry of the Natural Products," *Angew. Chem. Int. Ed. Engl.*, **7**, 14.

Snatzke, G. (1974), "Application of Circular Dichroism, Optical Rotatory Dispersion and Polarimetry in Organic Stereochemistry," in Korte, F., Ed., *Methodicum Chimicum*, Vol. 1A, Academic, New York, and Thieme, Stuttgart, Germany, Chap. 5.7.

Snatzke, G. (1979a), "Circular Dichroism and Absolute Conformation: Application of Qualitative MO Theory to Chiroptical Phenomena," *Angew. Chem. Int. Ed. Engl.*, **18**, 363.

Snatzke, G. (1979b), "Semiempirical Rules in Circular Dichroism of Natural Products," *Pure Appl. Chem.*, **51**, 769.

Snatzke, G. (1979c), "Chiroptical Properties of Organic Compounds: Chirality and Sector Rules," in Mason, S. F., Ed., *Optical Activity and Chiral Discrimination*, Reidel, Dordrecht, The Netherlands, p. 25.

Snatzke, G. (1981), "Chiroptische Methoden in der Stereochemie I," *Chem. Unserer Zeit*, **15**, 78.

Snatzke, G. (1982), "Chiroptische Methoden in der Stereochemie II," *Chem. Unserer Zeit*, **16**, 160.

Snatzke, G. (1989), Personal communication to SHW.

Snatzke, G., Ehrig, B., and Klein, H. (1969), *Tetrahedron*, **25**, 5601.

Snatzke, G. and Ho, P. C. (1971), *Tetrahedron*, **27**, 3645.

Snatzke, G., Kajtár, M., and Snatzke, F. (1973), "Aromatic Chromophores," Chap. 3.4. in Ciardelli and Salvadori (1973).

Snatzke, G., Kováts, E., and Ohloff, G. (1966), *Tetrahedron Lett.*, 4551.

Snatzke, G. and Snatzke, F. (1973), "Other Chromophores", Chap. 3.5. in Ciardelli and Salvadori (1973).

Snatzke, G. and Snatzke, F. (1980), "Chiroptische Methoden," in Kienitz, H., Bock, R., Fresenius, W., Huber, W., and Tölg, G., Eds., *Analytiker-Taschenbuch*, Band 1, Springer, Berlin, p. 217.

Snatzke, G. and Vértesy, L. (1967), *Monatsh. Chem.*, **98**, 121.

Snatzke, G., Wagner, U., and Wolff, H. P. (1981), *Tetrahedron*, **37**, 349.

Snatzke, G., Wynberg, H., Feringa, B., Marsman, B. G., Greydanus, B., and Pluim, H. (1980), *J. Org. Chem.*, **45**, 4094.

Solladié, G. and Gottarelli, G. (1987), *Tetrahedron*, **43**, 1425.

Solladié, G. and Zimmerman, R. G. (1984), "Liquid Crystrals: A Tool for Studies on Chirality," *Angew. Chem. Int. Ed. Engl.*, **23**, 348.

Spencer, K. M., Ciancosi, S. J., Baldwin, J. E., Freedman, T. B., and Nafie, L. A. (1990), *Appl. Spectrosc.*, **44**, 235.

Speranza, G., Manitto, P., Pezzuto, D., and Monti, D. (1991), *Chirality*, **3**, 263.

Stark, J. (1914), *Jahrb. Radioakt., Elektronik*, **11**, 194; *Chem. Abstr.*, 1914, **8**, 3740.

Stefanovsky, J. N., Spassov, S. L., Kurtev, B. J., Balla, M., and Ötvös, L. (1969), *Chem. Ber.*, **102**, 717.

Steinberg, N., Gafni, A., and Steinberg, I. Z. (1981), *J. Am. Chem. Soc.*, **103**, 1636.

Stevens, E. S. and Duda, C. A. (1991), *J. Am. Chem. Soc.*, **113**, 8622.

Stevens, E. S. and Sathyanarayana, B. K. (1987), *Carbohydr. Res.*, **166**, 181.

Stinson, S. C. (1985), "Vibrational Optical Activity Expands Bounds of Spectroscopy," *Chem. Eng. News*, **63**, November 11, 1985, 21.

Stryer, L. and Blout, E. R. (1961), *J. Am. Chem. Soc.*, **83**, 1411.

Suga, T., Ohta, S., Aoki, T., and Hirata, T. (1985), *Chem. Lett.*, 1331.

Sundararaman, P. and Djerassi, C. (1978), *Tetrahedron Lett.*, **1978**, 2457; **1979**, 4120.

Sundararaman, P., Barth, G., and Djerassi, C. (1980), *J. Org. Chem.*, **45**, 5231.

Tachibana, T. and Kurihara, K. (1976), *Naturwissenschaften*, **63**, 532.

Takenaka, S., Ako, M., Kotani, T., Matsubara, A., and Tokura, N. (1978), *J. Chem. Soc. Perkin. Trans. 2*, 95.

Tambuté, A. and Collet, A. (1984), *Bull. Soc. Chim. Fr. II*, 77.

Tetreau, C. and Lavalette, D. (1980), *Nouv. J. Chim.*, **4**, 423.

Tétreau, C., Lavalette, D., Cabaret, D., Geraghty, N., and Welvart, Z. (1982), *Nouv. J. Chim.*, **6**, 461.

Thomson, W. see Kelvin (Lord).

Thorne, J. M. (1972), "The Faraday Effect," in Weissberger, A. and Rossiter, B. W., Eds., *Physical Methods of Chemistry*, Part IIIC, *Techniques of Chemistry*, Vol. 1, Wiley, New York, Chap. 5.

Tinoco, Jr., I. (1960), *J. Chem. Phys.*, **33**, 1332.

Tinoco, Jr., I. (1962), "Theoretical Aspects of Optical Activity. Part Two: Polymers," *Adv. Chem. Phys.*, **4**, 113.

Tinoco, Jr., I. (1979), "Circular Dichroism of Polymers: Theory and Practice," in Selégny, E., Ed., *Optically Active Polymers*, Reidel, Dordrecht, The Netherlands, p. 1.

Toda, F. and Mori, K. (1986), *J. Chem. Soc. Chem. Commun.*, 1059.

Turner, D. H., Tinoco, Jr., I., and Maestre, M. (1974), *J. Am. Chem. Soc.*, **96**, 4340.

Urata, H., Shinohara, K., Ogura, E., Ueda, Y., and Akagi, M. (1991), *J. Am. Chem. Soc.*, **113**, 8174.

Urry, D. W. (1985), "Absorption, circular dichroism and optical rotatory dispersion of polypeptides, proteins, prosthetic groups, and biomembranes," in Neuberger, A., and Van Deenen, L. L. M., Eds., *Modern Physical Methods in Biochemistry*, Part A, *New Comprehensive Biochemistry*, Vol. 11A, Neuberger, A., and Van Deenen, L. L. M., Gen. Eds., Elsevier, Amsterdam, The Netherlands, Chap. 4.

van Beijnen, A. J. M., Nolte, R. J. M., Drenth, W., and Hezemans, A. M. F. (1976), *Tetrahedron*, **32**, 2017.

Van Holde, K. E. (1971), *Physical Biochemistry*, Prentice-Hall, Englewood Cliffs, NJ, Chap. 10.

van't Hoff, J. H. (1894), *Die Lagerung der Atome im Raume*, 2nd ed., Vieweg, Brunswick, Germany, p. 119. See also, Guye, P. A. and Gautier, M. *C. R. Hebd. Séances Acad. Sci.*, **119**, 740 (1894).

Vázquez, J. T., Wiesler, W. T., and Nakanishi, K. (1988), *Carbohydr. Res.*, **176**, 175.

Velluz, L., Legrand, M., and Grosjean, M. (1965), *Optical Circular Dichroism*, Verlag Chemie, Weinheim and Academic, New York.

Verbit, L. (1965), *J. Am. Chem. Soc.*, **87**, 1617.

Verdine, G. L. and Nakanishi, K. (1985), *J. Chem. Soc. Chem. Commun.*, 1093.

Vogel, A. I. (1948), *J. Chem. Soc.*, 1833.

Vogl, O. and Jaycox, G. D. (1986), "Macromolecular Asymmetry Can Produce Optical Activity," *CHEMTECH*, **16**, 698.

Vogl, O. and Jaycox, G. D. (1987), "Helical Polymers," *Polymer*, **28**, 2179.

Vogl, O., Jaycox, G. D., Kratky, C., Simonsick, Jr., W. J., and Hatada, K. (1992), "Mapping the Genesis of Helical Structure in Polymers of the Trihaloacetaldehydes," *Acc. Chem. Res.*, **25**, 408.

Waki, H., Higuchi, S., and Tanaka, S. (1980), *Spectrochim. Acta*, **36A**, 659.

Walden, P. (1894), *Z. Phys. Chem.*, **15**, 196.

Warren, W. S., Mayr, S., Goswami, D., and West, A. P., Jr. (1992), *Science*, **255**, 1683.

Weigang, O. E., Jr. and Ong, E. C. (1974), *Tetrahedron*, **30**, 1783.

Weigang, O. E., Jr. (1979), *J. Am. Chem. Soc.*, **101**, 1965.

Weiss, U. (1968), *Experientia*, **24**, 1088.

Weiss, U., Ziffer, H. and Charney, E. (1965), "Optical Activity of Non-Planar Conjugated Dienes. I. Homannular Cisoid Dienes," *Tetrahedron*, **21**, 3105.

Wheland, G. W. (1960), *Advanced Organic Chemistry*, 3rd ed., Wiley, New York.

Whiffen, D. H. (1956), *Chem. Ind. (London)*, 964.

Whitesell, J. K., LaCour, T., Lovell, R. L., Pojman, J., Ryan, P., and Yamada-Nosaka, A. (1988), *J. Am. Chem. Soc.*, **110**, 991.

Wiesler, W. T., Berova, N., Ojika, M., Meyers, H. V., Chang, M., Zhou, P., Lo, L.-C., Niwa, M., Takeda, R, and Nakanishi, K. (1990), *Helv. Chim. Acta*, **73**, 509.

Wiesler, W. T., Vázquez, J. T., and Nakanishi, K. (1987), *J. Am. Chem. Soc.*, **109**, 5586.

Wilen, S. H., Bunding, K. A., Kascheres, C. M., and Weider, M. J. (1985), *J. Am. Chem. Soc.*, **107**, 6997.

Wilen, S. H., Qi, J. Z., and Williard, P. G. (1991), *J. Org. Chem.*, **56**, 485.

Wilson, S. R. and Cui, W. (1989), *J. Org. Chem.*, **54**, 6047.

Winther, C. (1907), *Z. Phys. Chem.*, **60**, 590.

Woldbye, F. (1967), "Instrumentation," Chap. 5 in Snatzke, G., Ed., *Optical Rotatory Dispersion and Circular Dichroism in Organic Chemistry*, Heyden, London.

Woody, R. W. (1985), "Circular Dichroism of Peptides," in Udenfriend, S., and Meienhofer, J., Eds., *The Peptides*, Academic, New York, Vol. 7, Hruby, V. J., Ed., Chap. 2.

Wroblewski, A. E., Applequist, J., Takaya, A., Honzatko, R., Kim, S.-S., Jacobson, R. A., Reitsma, B. H., Yeung, E. S., and Verkade, J. G. (1988), *J. Am. Chem. Soc.*, **110**, 4144.

Wynberg, H., Hekkert, G. L., Houbiers, J. P. M., and Bosch, H. W. (1965), *J. Am. Chem. Soc.*, **87**, 2635.

Wynberg, H. and Hulshof, L. A. (1974), *Tetrahedron*, **30**, 1775.

Wynberg, H., Meijer, E. W. J., Hummelen, J. C., Dekkers, H. P. J. M., Schippers, P. H., and Carlson, A. D. (1980), *Nature (London)*, **286**, 641.

Wynberg, H., Numan, H., and Dekkers, H. P. J. M. (1977), *J. Am. Chem. Soc.*, **99**, 3870.

Yamaguchi, S. and Mosher, H. S. (1973), *J. Org. Chem.*, **38**, 1870.

Yang, J. T., Wu, C.-S., and Martinez, H. M. (1986), "Calculation of Protein Conformation from Circular Dichroism," in Hirs, C. H. W. and Timasheff, S. N., Eds., *Methods in Enzymology*, Vol. 130, Academic, New York, p. 208.

Yasui, S. C. and Keiderling, T. A. (1986), *J. Am. Chem. Soc.*, **108**, 5576.

Yeung, E. S. (1989), "Indirect Detection Methods: Looking for What Is Not There," *Acc. Chem. Res.*, **22**, 125.

Yeung, E. S., Steenhoek, L. E., Woodruff, S. D., and Kuo, J. C. (1980), *Anal. Chem.*, **52**, 1399.

Yogev, A., Amar, D., and Mazur, Y. (1967), *Chem. Commun.*, 339.

Zandomeneghi, M., Cavazza, M., and Pietra, F. (1984), *J. Am. Chem. Soc.*, **106**, 7261.

Ziffer, H., Charney, E., and Weiss, U. (1962), *J. Am. Chem. Soc.*, **84**, 2961.

Zingg, S. P. B. (1981), *Chiral Discrimination in the Structure and Energetics of Ion Pairing*. Ph.D. Dissertation, University of Pittsburgh; *Diss. Abstr. Int. B* (1983), **44**(1) 178.

Zingg, S. P., Arnett, E. M., McPhail, A. T., Bothner-By, A. A., and Gilkerson, W. R. (1988), *J. Am. Chem. Soc.*, **110**, 1565.

14

Chirality in Molecules Devoid of Chiral Centers

14-1. INTRODUCTION. NOMENCLATURE

In Chapter 1 it was pointed out that a necessary and sufficient condition for a molecule to be chiral is that it not be superposable with its mirror image. The presence of a (single, configurationally stable) chiral center in the molecule (central chirality) is a sufficient condition for the existence of chirality but not a necessary one. In this chapter we shall turn our attention to chiral molecules devoid of chiral centers. We shall include some types of molecules (certain spiranes and metallocenes) in which, for nomenclatural purposes, a chiral center may be defined to exist (Cahn, Ingold, and Prelog, 1966) even though these molecules are closely akin to others in which no chiral centers can be discerned.

Classes of molecules to be discussed here (Eliel, 1962; Krow 1970) are allenes; cumulenes with even numbers of double bonds (cf. Chapter 9 for cumulenes with odd numbers of double bonds); alkylidenecycloalkanes; spiranes; the so-called atropisomers (biphenyls and similar compounds in which chirality is due to restricted rotation about a single bond); helicenes, propellerlike structures; and molecules, such as cyclophanes, chiral *trans*-cycloalkenes, ansa compounds, and arene–metal complexes including metallocenes, which are said (Cahn, Ingold, and Prelog, 1966) to contain a "plane of chirality." Also included is the phenomenon of cyclostereoisomerism, even though it does not strictly fit the condition set down in the first paragraph.

Allenes, alkylidenecycloalkanes, biphenyls, and so on, are said to possess a "chiral axis" (Cahn, Ingold, and Prelog, 1956). If we stretch a tetrahedron along its S_4 axis, it is desymmetrized to a framework of $\mathbf{D_{2d}}$ symmetry (Fig. 14.1). With proper substitution, the long axis of this framework constitutes the chiral axis. Because of the intrinsically lower symmetry of the framework shown in Figure 14.1 compared to a tetrahedron, it no longer takes four different substituents to make the framework chiral: A necessary and sufficient condition for chirality is that a ≠ b and c ≠ d. Thus, even when a = c and/or b = d, the framework retains chirality, for example, in abC=C=Cab (see below).

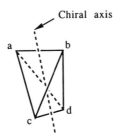

Figure 14.1. Chiral axis.

To specify the sense of chirality (i.e., configuration) of a molecule possessing a chiral axis (*axial chirality*, examples are shown in Fig. 14.2) an additional sequence rule is needed: Near groups precede far groups. The application of this rule to the molecules shown in Figure 14.2 is shown in Figure 14.3. In all cases the molecules in Figure 14.2 are viewed from the left. However, the reader should take note that the same configurational descriptor results when the molecules are viewed from the right, so no specification in this regard is needed. In the case of biphenyl it is important to note that the ring substituents are to be explored from the center on outward, regardless of the rule given above. Thus, in the biphenyl in Figure 14.2, in the right ring the sequence is C—OCH$_3$ > C—H; the chlorine atom is too far out to matter, a decision being made before it is reached in the outward exploration. The fiducial atoms (i.e. those that determine the configurational symbol, cf. p. 665) are the same when the molecule is viewed from the right. The descriptors a*R* and a*S* are sometimes used to distinguish axial chirality from other types, but the use of the a prefix is optional.

Molecules with chiral axes may alternatively be viewed as helices (in this respect they resemble the helicenes to be disussed below) and their configuration may be denoted as *P* or *M*, in a manner similar to that of conformational isomers (Chapter 10; Prelog and Helmchen, 1982). For this designation, only the ligands of highest priority in front and in the back of the framework are considered (ligands 1 and 3 in Figure 14.3). If the turn from the priority front ligand 1 to the priority rear ligand 3 is clockwise, the configuration is *P*, if counterclockwise it is

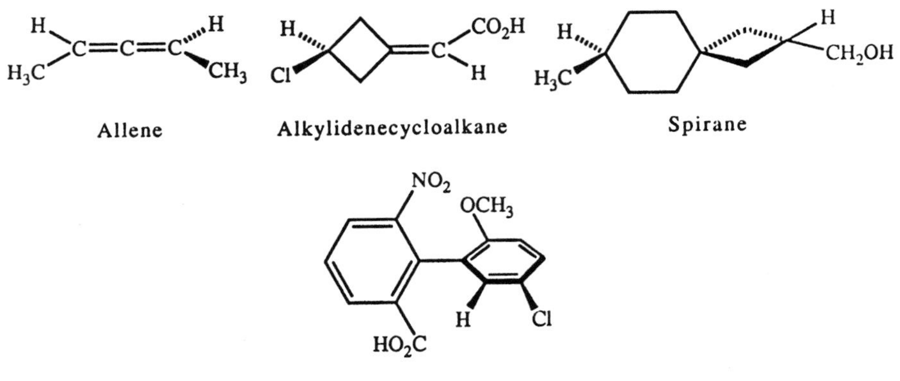

Allene   Alkylidenecycloalkane   Spirane

Biphenyl

Figure 14.2. Molecules with chiral axes.

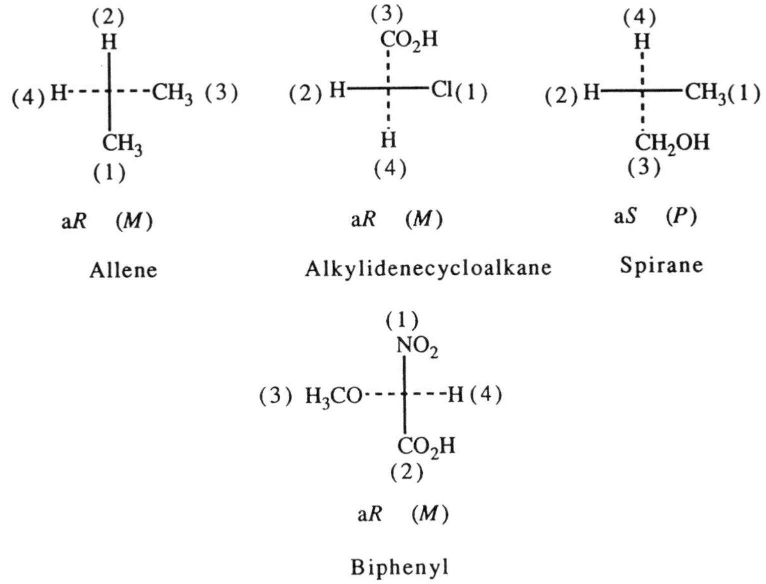

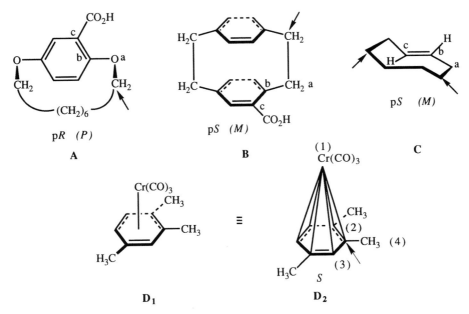

(2)
H
(4) H
CH3 (3)
CH3
(1)
aR (M)
Allene
(3)
CO2H
(2) H
Cl(1)
H
(4)
aR (M)
Alkylidenecycloalkane
(4)
H
(2) H
CH3(1)
CH2OH
(3)
aS (P)
Spirane
(1)
NO2
(3) H3CO
H (4)
CO2H
(2)
aR (M)
Biphenyl

Figure 14.3. Descriptors for molecules with chiral axes.

M. Thus three of the four structures in Figures 14.2 and 14.3 are a*R* (chiral axis nomenclature) or *M* (helix nomenclature); the spirane is a*S* or *P*. (The correspondence of a*R* with *M* and a*S* with *P* is general.)

Figure 14.4 shows molecules with chiral planes. The definition of a chiral plane is less simple and clear-cut than that of a chiral center or axis. It is a plane that contains as many of the atoms of the molecule as possible, but not all; in fact,

CO2H
c
b
O
O a
CH2
CH2
(CH2)6
pR (P)
A
H2C
CH2
H2C
b
CH2 a
c
CO2H
pS (M)
B
c
H
b
H
a
pS (M)
C
(1)
Cr(CO)3
Cr(CO)3
CH3
CH3
H3C
≡
CH3
(2)
CH3 (4)
(3)
H3C
S
D1
D2

Figure 14.4. Molecules with chiral planes.

the chirality is due (and solely due) to the fact that at least one ligand (usually more) is *not* contained in the chiral plane. Thus the chiral plane of the "ansa compound" **A** (in which the alicyclic ring is too small for the aromatic one to swivel through) is the plane of the benzene ring; the same is formally true of the arenechromium tricarbonyl compound **D**; in the paracyclophane **B**, the more highly substituted benzene ring (bottom) is considered the chiral plane and in *trans*-cyclooctene **C** the chiral plane is that of the double bond. To find the descriptor for planar chiral molecules one views the chiral plane from the out-of-plane atom closest to the plane (if there are two or more candidates, one chooses the one closest to the atom of higher precedence according to the sequence rules, cf. Section 5-2). This atom, sometimes called the "pilot atom," is marked with an arrow in Figure 14.4 (for compound **C** there are two equivalent such atoms). Then, if the adjacent three atoms a, b, and c (again chosen by precedence if there is a choice) describe a clockwise array in the chiral plane, the configuration is p*R*, if the array is counterclockwise, the descriptor is p*S*. (The prefix "p" may be used to signal planar chirality.)

Compound **D**, although it would also appear to have a chiral plane, is conventionally treated as having chiral centers by replacing the η_6 π bond by six σ single bonds, as shown in structure **D$_2$**. The (central) chirality is now determined for the atom of highest precedence (the ring carbon marked by an arrow) and the descriptor is thus found to be *S* (Cahn, Ingold, and Prelog, 1966; see also Schlögl, 1967; Klyne and Buckingham, 1978, Vol. 1, p. 222).

Planar chirality, like axial chirality, may alternatively be looked at as a type of helicity (Prelog and Helmchen, 1982). To determine the sense of the helix one uses the pilot atom plus atoms a, b, and c specified as above. It is then seen (Fig. 14.4) that p*R* compounds correspond to *P* and p*S* corresponds to *M*, opposite to the correlation in axial chirality (see above).

14-2. ALLENES

a. Historical. Natural Occurrence

It was already pointed out by van't Hoff (1875) that an appropriately substituted allene should exist in two enantiomeric forms. A simple case is shown in Figure 14.5, **A**; a necessary and sufficient condition for such an allene to be chiral is that a $\neq$ b. The reason for the dissymmetry is that the groups a and b at one end of the system lie in a plane at right angles to those at the other end. If the doubly

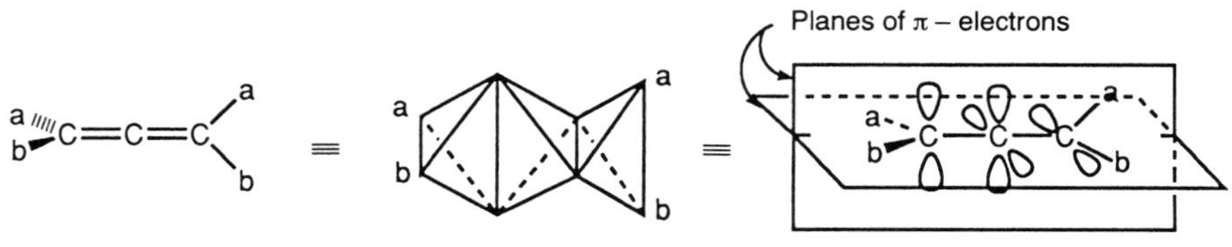

Figure 14.5. Dissymmetric allene.

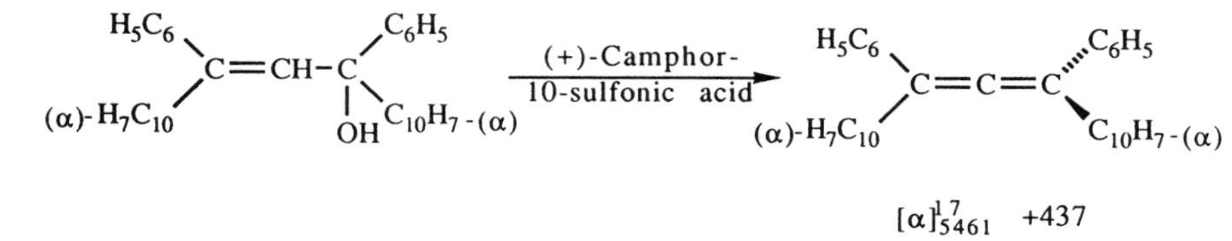

Figure 14.6. Asymmetric synthesis of optically active allene.

bonded carbon atoms are viewed as tetrahedra joined edge to edge, a view that was originally proposed by van't Hoff (see also Chapter 9), the noncoplanarity of the two sets of groups follows directly from the geometry of the system (Fig. 14.5, **B**). If, on the other hand, one views a double bond as being made up of pairs of σ and π electrons, orbital considerations indicate that the two planes of the π bonds attached to the central carbon atom must be orthogonal, and since the a and b groups attached to the trigonal carbon lie in a plane at right angles to the plane of the adjacent π bond, their planes are orthogonal to each other (Fig. 14.5, **C**).

The experimental realization of van't Hoff's prediction proved to be quite difficult, and 60 years elapsed before the first optically active allene was obtained in the laboratory (Maitland and Mills, 1935, 1936). The route chosen was one of asymmetric synthesis: Dehydration of 1,3-diphenyl-1,3-α-naphthyl-2-propen-1-ol with (+)-camphor-10-sulfonic acid gave (+)-1,3-diphenyl-1,3-di-α-naphthylallene (Fig. 14.6) in slight preponderance over its enantiomer [enantiomer excess (ee) ca. 5%]. Fortunately, the optically active allene forms a conglomerate (cf. Chapter 6) and the pure enantiomer could be separated from the racemate by fractional crystallization without excessive difficulty. The material has the high specific rotation $[\alpha]_{546}^{17} + 437$ (benzene), $[\alpha]_D^{20} + 351$ (cyclohexane). Use of (−)-camphor-10-sulfonic acid gave the enantiomer of $[\alpha]_{546}^{17} + 438$ (benzene). Shortly after this asymmetric synthesis was accomplished, the allenic acid shown in Figure 14.7 (R = CH_2CO_2H) was resolved by crystallization of the brucine salt (Kohler et al., 1935). Earlier attempts to resolve the simpler allenic acid shown in Figure 14.7 (R = H) had failed, but a quite similar acid of related type, $CH_3CH{=}C{=}C(n\text{-}C_4H_9)CO_2H$ was finally resolved by means of strychnine in 1951 (Wotiz and Palchak).

In 1952, it was recognized that optically active allenes also occur in nature. In that year Celmer and Solomons (1952, 1953) established the structure of the antibiotic mycomycin, a fungal metabolite, to be that of a chiral allene: $HC{\equiv}C{-}C{\equiv}C{-}CH{=}C{=}CH{-}CH{=}CH{-}CH{=}CH{-}CH_2CO_2H$. Since then a number of other chiral allenes have been found in nature (for tabulations see Rossi and Diversi, 1973, p. 27; Murray, 1977, p. 972; Runge, 1982, p. 595; see also Landor, 1982).

In recent years, numerous optically active allenes have been obtained in a variety of ways (resolution, transformation of chiral precursors, and enantioselec-

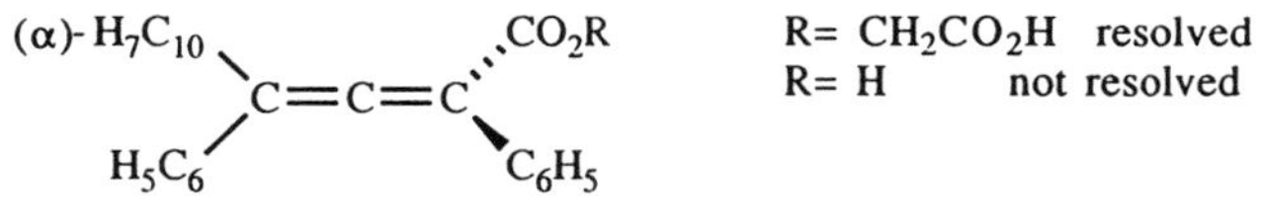

R= CH_2CO_2H resolved
R= H   not resolved

Figure 14.7. Allenic acids used in resolution experiments.

tive synthesis) and only the highlights of allene stereochemistry can be presented here. Fortunately, several detailed reviews are available (Krow, 1970; Rossi and Diversi, 1973; Murray, 1977; Runge, 1980, 1982).

b. Synthesis of Optically Active Allenes

In addition to classical resolution (Runge, 1982, p. 597) there are a few other methods capable of giving allenes of high enantiomeric purity. One of these is kinetic resolution (so-called *asymmetric destruction*, cf. Section 7-6.a) by incomplete reduction with tetra-α-pinanyldiborane (Fig. 14.8; Waters and Caserio, 1968; Moore et al., 1973). The method is simple but the enantiomeric purity of the products is generally low (cf. Runge, 1982, p. 644).

There are a number of schemes of synthesizing chiral allenes from precursors having chiral centers. An example is the reductive rearrangement of tetrahydropyranyl ethers of chiral acetylenic alcohols shown in Figure 14.9 (Olsson and Claesson, 1977), which proceeds in 75–100% optical yield, depending on R (which contained an alcohol function in all cases). The addition is trans, that is, hydride approaches the acetylene from the side opposite to that of the —OTHP (THP = tetrahydropyran) leaving group. A conceptually similar scheme involves the transformation of the methanesulfonate of a chiral acetylenic alcohol to a chiral allenic halide by means of lithium copper halides (Elsevier et al., 1982; for related reactions, see Elsevier et al., 1983; Hayashi, Kumada et al., 1983).

An orthoester Claisen rearrangement, which also proceeds highly stereospecifically, has been used by Mori et al. (1981) in the preparation of an intermediate X toward the synthesis of the sex pheromone produced by the male dried-bean beetle, $CH_3(CH_2)_7CH=C=CH—CH=CHCO_2CH_3$ (Fig. 14.10). This reaction proceeds by suprafacial attack, that is, the enol ether on the bottom side of the molecule (as drawn) attaches itself to the acetylenic triple bond from the same side. A somewhat similar principle is also involved in the thermal rearrangement of the acetal of isobutyraldehyde and (S)-(−)-2-butynol to

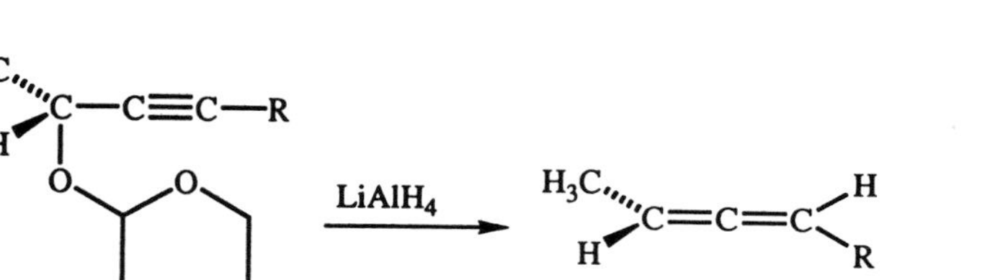

Figure 14.8. Kinetic resolution of chiral allenes.

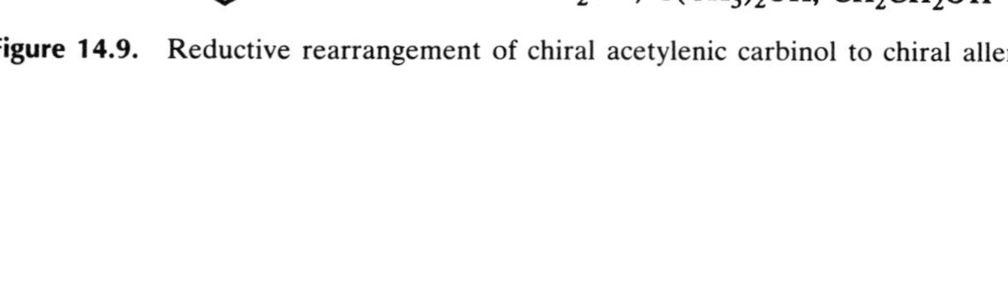

$R = CH_2OH, C(CH_3)_2OH, CH_2CH_2OH$

Figure 14.9. Reductive rearrangement of chiral acetylenic carbinol to chiral allene.

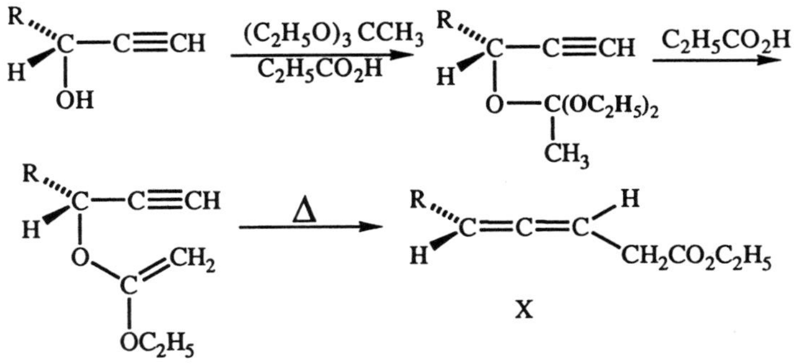

Figure 14.10. Orthoester Claisen rearrangement of chiral acetylenic carbinol to chiral allene.

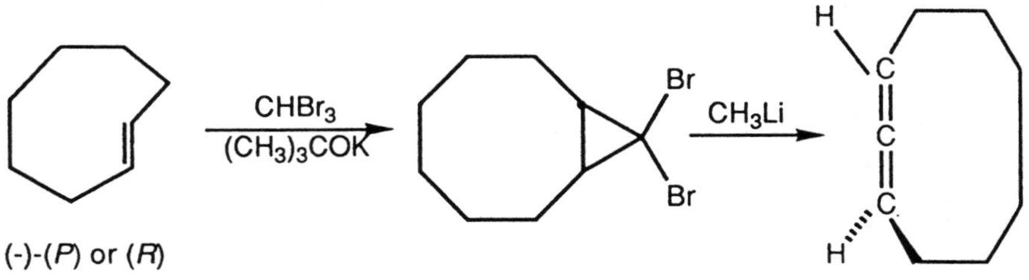

Figure 14.11. Synthesis of (+)-1,2-nonadiene from (−)-*trans*-cyclooctene.

(−)-2,2-dimethyl-3,4-hexadienal, whose configuration was thus shown to be *R* (Jones et al., 1960; cf. Eliel, 1962, p. 315). A somewhat different scheme (Fig. 14.11) is illustrative of the synthesis of allenes from cyclopropane derivatives, presumably via carbene or carbenoid intermediates (Cope et al., 1970). Here a compound with planar chirality (*trans*-cyclooctene) is converted to one with axial chirality.

Additional examples of syntheses of chiral allenes from precursors having chiral centers will be found in the next section.

c. Determination of Configuration and Enantiomeric Purity of Allenes

Experimental Methods

As was discussed in Section 5-3.a, the classical absolute method of determining configuration is by anomalous X-ray scattering (Bijvoet method). There seems to be only one application of this method to allenes, namely, the elucidation of the relative and absolute configuration of the allenic ketone shown in Figure 14.12

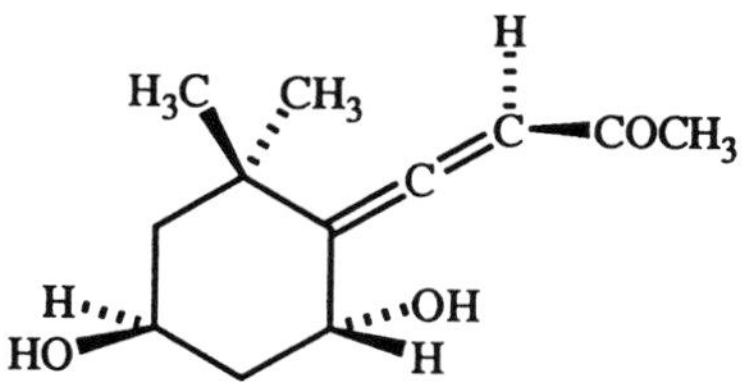

Figure 14.12. Allene whose configuration has been elucidated by X-ray structure analysis.

isolated from ant-repellant secretions of a species of grasshopper (De Ville, Weedon, et al., 1969; Hlubucek, Weedon, et al., 1974). This compound contains both chiral centers and allenic chirality and although it has been correlated with other naturally occurring chiral allenes, the chiral allenic moiety has not been cut out. In fact, neither the Bijvoet method nor correlative X-ray methods have yet been applied to the determination of absolute configuration of a chiral allene devoid of other chiral elements (chiral centers).

There are, however, a number of indirect methods, notably mechanistic correlation of chiral allenes with molecules possessing chiral centers of known configuration, methods based on interpretation of optical rotatory dispersion–circular dichroism (ORD/CD) spectra (cf. Chapter 13) and theoretical methods. Again, only a few examples can be presented; detailed discussions may be found elsewhere (Krow, 1970; Rossi and Diversi, 1973; Runge, 1982; see also Klyne and Buckingham, 1978, Vol. 1, p. 215, Vol. 2, p. 105; Buckingham and Hill, 1986, p. 157).

One of the first determinations (Evans, Landor, et al., 1963; Evans and Landor, 1965) of the configuration of a chloroallene by treatment of an acetylenic carbinol with thionyl chloride has now been shown to be in error (Caporusso et al., 1985, 1986; Elsevier and Mooiweer, 1987; Eliel and Lynch, 1987) since the configuration of the starting material, contrary to earlier indications (see also Eliel, 1960), is the reverse of what was originally alleged on the basis of Prelog's rule (p. 142; Prelog, 1953), which was evidently incorrectly applied. The corrected correlation is shown in Figure 14.13; the (−)-chloroallene has the aR configuration. The correlation of the configuration of the allenic chloride and that of the acetylenic carbinol (Evans, Landor, et al., 1963) rests on the assumption that the reaction of the carbinol with thionyl chloride follows an $S_N i'$ (i.e., internal displacement with allylic rearrangement) mechanism but the conclusion would be the same, if perhaps less firm, if an $S_N 2'$ mechanism (attack of chlorine at the terminal acetylenic carbon with double inversion and displacement of the OSOCl group) were operative.

> The reversal of configuration of the chloroallene shown in Figure 14.13 resolves a number of apparent internal contradictions (cf. Runge, 1982). Examples will be presented in the next Section on the theoretical prediction of configuration.

A firmer correlation of configuration, based on the Diels–Alder reaction of the so-called glutinic acid (pentadienoic acid) with cyclopentadiene (Agosta, 1964), is shown in Figure 14.14. If the (−) acid has the aR configuration (cf. Fig. 14.14, **A**), there are four diastereomers (**B**–**E**) that can be formed in the addition, two (**B** and **C**) by approach from the less hindered side (out-of-plane hydrogen,

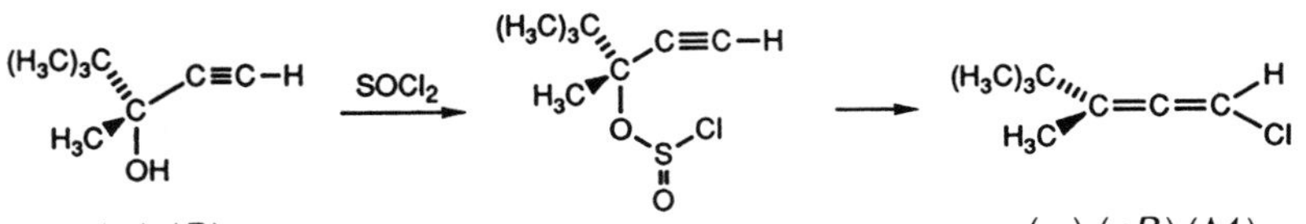

$$(-)-(R) \qquad\qquad\qquad\qquad (-)-(aR)\,(M)$$

Figure 14.13. Configurational correlation of acetylenic carbinol with allenic chloride.

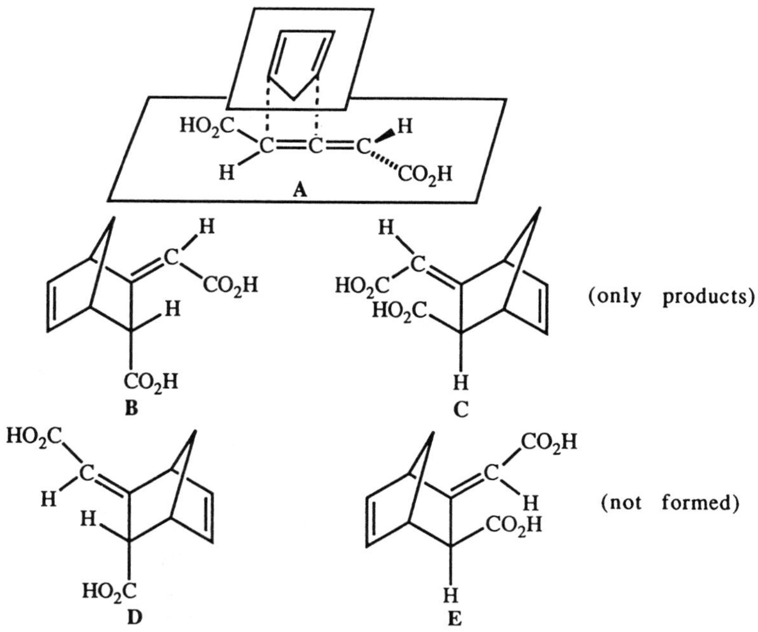

Figure 14.14. Addition of (−)-glutinic acid to cyclopentadiene.

see **A**) and two (**D** and **E**) by approach from the more hindered side (out-of-plane carboxyl). If the (−) acid has the aS configuration, the enantiomers of **B–E** would result. (It is to be noted that the two double bonds of glutinic acid are symmetry equivalent being interconverted by a C_2 operation.)

> According to the Alder–Stein rules, the endo addition product **B** should be the major or exclusive one (Alder and Stein, 1934, 1937), the mode of addition being as shown in **A**, where the double bonds of the cyclopentadiene have maximum overlap with the C=O π bond of the in-plane carboxylate group (see also Eliel, 1962, p. 295). However, many exceptions to the Alder–Stein endo addition rule are now known.

In fact, only two (separable) products (shown to be **B** and **C**; see below) were formed in the addition. Both had proximate carboxyl groups, as demonstrated by cyclic anhydride formation upon treatment with acetic anhydride; this excludes **D** and **E**. The endo configuration of the carboxyl group in **B** was proved by iodolactone formation, whereas the exo configuration of the carboxyl group in **C** (which formed no iodolactone) was demonstrated by hydrogenation to the known cis dicarboxylic acid **F** as shown in Figure 14.15.

With the relative configuration of **B**, and therefore also of **C**, thus firmly established, it remained to elucidate their absolute configuration. This feat was accomplished (cf. Fig. 14.15) by selectively hydrogenating the ring double bond, esterifying, ozonizing the exocyclic double bond, hydrolyzing, and decarboxylating. In this way it was found that (+)-**B** formed from (−)-glutinic acid gave (+)-norcamphor known (Berson et al., 1961) to have the 1S configuration as shown in Figure 14.15. Diastereomer (−)-**C**, also formed from (−)-glutinic acid,

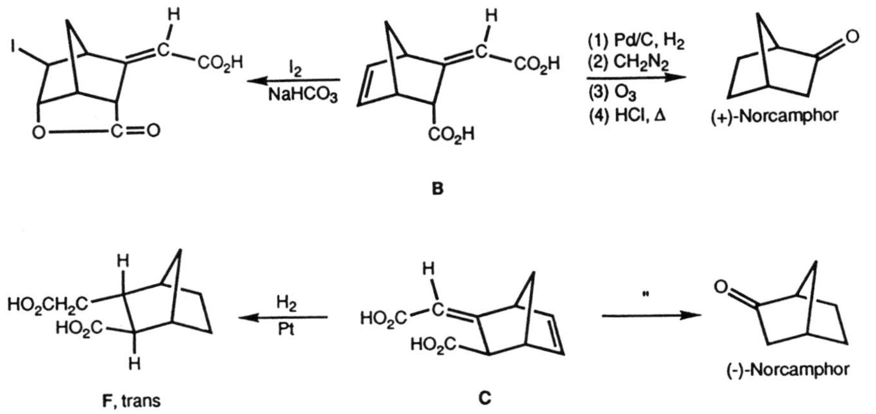

Figure 14.15. Determination of relative and absolute configuration of **B** (Fig. 14.14).

similarly gave (−)-norcamphor, [identified as the (−)-2,4-dinitrophenylhydrazone] as expected. The configuration of (−)-glutinic acid is thus aR, as shown in Figure 14.14, **A**.

Figure 14.16 shows a configurational assignment of a series of resolved phenylallenecarboxylic acids (R = H, CH$_3$, Et, i-Pr, t-Bu) through a bromolactonization procedure followed by oxidation to known α-hydroxy acids (Shingu et al., 1967) using a procedure originally devised by Gianni (1961). It is found that the (+)-phenylallenic acids give rise to (S)-α-hydroxy acids and thus have the aS configuration themselves.

An even simpler scheme (if perhaps a less reliable one, inasmuch as the reaction is not intramolecular, and therefore its stereochemical course is less certain) is shown in Figure 14.17; it involves partial hydrogenation of the allene (Crombie et al., 1975) from the less hindered face (i.e., the one on the side of the C(2) hydrogen rather than the carboxyl group). Formation of the ester of the S acid of known configuration suggests that the starting allene also has the aS configuration.

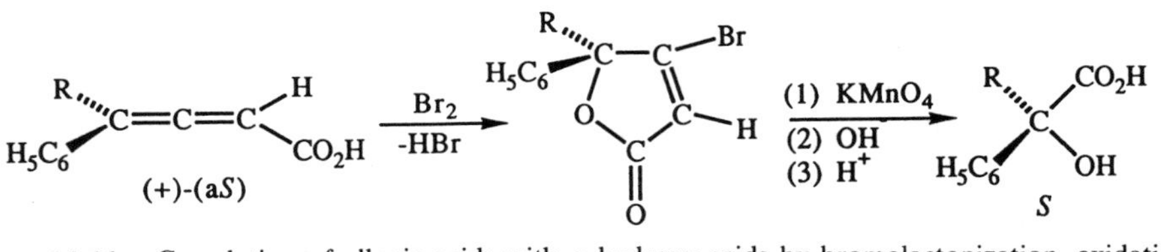

Figure 14.16. Correlation of allenic acids with α-hydroxy acids by bromolactonization–oxidation.

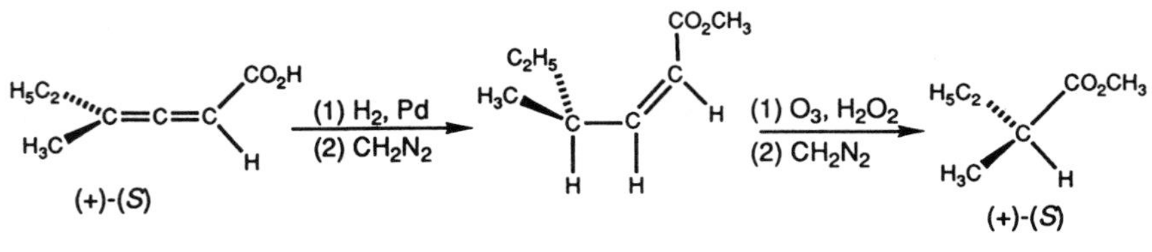

Figure 14.17. Configurational assignment of allene by partial hydrogenation.

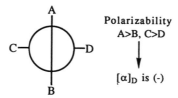

Figure 14.18.   Lowe–Brewster rule (also called "Lowe's rule").

A Priori Methods

The Lowe–Brewster rules (Lowe, 1965; Brewster, 1967; Crabbé, Mason et al., 1971; see also Section 13-5.b), derived by Lowe on empirical grounds and by Brewster on semitheoretical grounds, are summarized in Figure 14.18. If ligand A is more polarizable than B and C than D, the molecule as shown is levorotatory at the sodium D line. The molecules shown in Figures 14.13, 14.14, 14.16, and 14.17 are shown in the proper perspective in Figure 14.19. It is clear that correct predictions are made in all these cases (and a number of others) if reasonable assumptions for relative polarizabilities are made. In the case of Figure 14.13 this requires $(CH_3)_3C > CH_3$ and $Cl > H$.

At one time (Eliel, 1960) it was assumed, on the basis of the observed dextrorotation of (S)-pinacolyl alcohol, $(CH_3)_3CCH(CH_3)OH$, that the relative polarizabilities were $CH_3 > (CH_3)_3C$ and the configuration of the $(-)-(CH_3)_3CCOH(CH_3)C\equiv CH$ (Fig. 14.13) was misassigned on this basis, using Brewster's rules, as explained in Section 13-5.b. Unfortunately, this also makes the wrong aS configuration of the $(-)$-chloroallene (Fig. 14.13) compatible with Brewster's rules. However, it is now clear (Taft et al., 1978; Runge, 1982, p. 589)

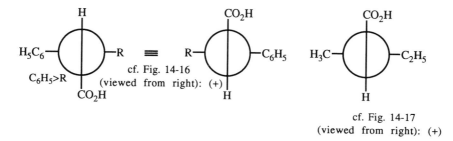

Figure 14.19.   Application of the Lowe–Brewster rule.

that the polarizability order is $(CH_3)_3C > CH_3$. Attempts to justify the originally assigned aS configuration (Evans and Landor, 1965) to the $(-)$-chloroallene (Fig. 14.13) by assuming that chlorine is *less* polarizable than hydrogen are in disagreement with other evidence (Runge, 1982) and are now (see above) recognized to have been misguided.

An alternative way of predicting rotations in allenes is by means of chirality functions (Ruch and Schönhofer, 1968; Ruch, 1972, 1977; see also p. 1092). The full chirality function for an allene $R_1R_2C{=}C{=}CR_3R_4$ is of the form

$$\chi(R_1, R_2, R_3, R_4) = \varepsilon_1[\lambda(R_1) - \lambda(R_2)][\lambda(R_3) - \lambda(R_4)] + \varepsilon_2[\mu(R_1) - \mu(R_2)]$$
$$\cdot [\mu(R_1) - \mu(R_3)][\mu(R_1) - \mu(R_4)][\mu(R_2) - \mu(R_3)]$$
$$\cdot [\mu(R_2) - \mu(R_4)][\mu(R_3) - \mu(R_4)] \qquad (14.1)$$

where $\chi(R_1, R_2, R_3, R_4)$ denotes the molar rotation $[\Phi]_D$. The parameters $\lambda(R)$ and $\mu(R)$ are ligand specific parameters, and $\varepsilon = \pm 1$ are sign factors (Ruch et al., 1973; Runge and Kresze, 1977; Runge, 1980, 1981, 1982; Elsevier, Runge, et al., 1985). It is clear that the first (λ) term depends only on differences between ligands on one and the same allenic carbon (and will be zero when $R_1 = R_2$ or $R_3 = R_4$, in which case there is no chirality), whereas in the second μ term all ligands are compared pairwise; this term will be zero if *any* pair of ligands is identical. Thus if $R_1 = R_2$ or $R_3 = R_4$, $\chi(R_1, R_2, R_3, R_4) = 0$ as it should be (the molecule is achiral) and if $R_1 = R_3$ or $R_1 = R_4$ or $R_2 = R_3$ or $R_2 = R_4$, the second (μ) term is zero and Eq. 14.1 simplifies to

$$\chi(R_1, R_2, R_3, R_4) = [\lambda(R_1) - \lambda(R_2)][\lambda(R_3) - \lambda(R_4)] \qquad (14.2)$$

(it is assumed that the λ values are so signed that $\varepsilon = +1$).

This simplified form of the chirality function (Eq. 14.2) has been used with some success even when all four R substituents are unequal, suggesting that the μ terms in Eq. 14.1 are relatively small. The λ parameters for a variety of ligands at the sodium D line are shown in Table 14.1 (Runge, 1982); as in the Brewster–Lowe treatment (see above) they are related to polarizability. The parameters may be calculated from the measured rotations of chiral allenes of known configuration and enantiomeric purity or they may be computed from NMR chemical shift data or by CNDO (complete neglect of differential overlap) quantum mechanical calculations (Runge, 1980). Runge (1980, p. 124) has made a comparison of calculated and experimental rotations of 29 allenes of known configuration and enantiomeric purity; in all cases the sign of rotation is predicted correctly and in about two-thirds of the cases the calculated magnitude is within 10% of the experimental. There are small changes of rotation with solvent; if these are nonspecific they can be allowed for by using the equation

$$\chi(S) = \chi(EtOH)[n_D^2(S) + 2]/[n_D^2(EtOH) + 2] \qquad (14.3)$$

where $\chi(S)$ is the chirality function in solvent S of refractive index $n_D(S)$ and $\chi(EtOH)$ is the chirality function in ethanol, refractive index $n_D(EtOH)$. Compli-

TABLE 14.1. The $\lambda(R)$ parameters for the calculation of the molar rotation of allenes at the wavelength of the sodium D line (in ethanol).[a]

R	$\lambda(R)$	R	$\lambda(R)$	R	$\lambda(R)$
H	0	CO_2H	+15.4	NCO	+12.4
CH_3	+ 7.7	CO_2CH_3	+16.5	$POCl_2$	+27.5
C_2H_5	+12.6	CO_2Na	+19.3	$PO(OH)_2$	+24.5
n-C_3H_7	+ 7.7	C_6H_5	+44.3	F	−17.3[b]
$(CH_3)_2CH$	+16.5	p-ClC_6H_4	+44.3	Cl	− 2.5[b]
n-C_4H_9	+12.6	p-BrC_6H_4	+44.3	Br	+ 9.2[b]
$(CH_3)_3C$	+18.9	p-$CH_3C_6H_4$	+42.3	CH_3O	−21.1[b]
n-C_5H_{11}	+ 7.7	α-Naphthyl	+82.9	CH_3S	+ 9.2[b]
n-C_6H_{13}	+12.6	$HC\equiv C-C\equiv C$	+37.5	CH_3SO_2	+15.8
n-$C_{11}H_{23}$	+15.4	$CH_3C\equiv C-C\equiv C$	+37.5		
$C(CH_3)_2OH$	+14.7	$HC\equiv C$	+21.8		
CH_2OH	+ 9.5	$H_2C\equiv CH$	+25.6		
CH_2OCOCH_3	+ 9.5	$COCH_3$	+10.4		
CH_2SCH_3	+ 7.7	CN	+19.1		
$CH_2CO_2CH_3$	+ 9.5				
$(CH_2)_2OH$	+12.6				
$(CH_2)_3OH$	+ 9.3				
$(CH_2)_3CO_2CH_3$	+ 9.3				
$(CH_2)_4OH$	+ 7.7				
$(CH_2)_4OCOCH_3$	+ 7.7				

[a]From Landor, S., ed., The Chemistry of the Allenes, Vol. 3, copyright 1982, Academic Press. With permission of the publishers.
[b]Tentative values.

cations will occur when there are specific solute–solvent interactions or solute–solute association (cf. Chapter 13) or when the substituent is not conformationally homogeneous (e.g., long-chain alkyl) and its conformation changes from one case to another.

A case where the rules fail is 1,2-cyclononadiene; the (+) isomer has been shown to be R (Moore et al., 1971), whereas the Lowe–Brewster rules as well as the chirality function treatment predict it to be S (cf. the case of glutinic acid, Fig. 14.14). The failure has been explained (see p. 1091); the correct configuration can be derived from the CD spectrum (Rauk, Mason et al., 1979).

Allenes generally absorb in the UV below 250 nm. Down to about 330 nm (i.e., in the *transparent region*) the ORD curve is plain (cf. Fig. 13.8) and the chirality function at a wavelength λ (in nm) other than the sodium D line may be approximated (Runge, 1980, p. 130) by Eq. 14.4.

$$[\chi(R)]_\lambda = 0.224[\chi(R)]_D[1 + (1.20 \times 10^6/\lambda^2)] \qquad (14.4)$$

Below 250 nm ORD and CD extrema are found which can also be used in the assignment of configuration (Mason and Vane, 1965; Runge, 1980). Vibrational circular dichroism (VCD) has also been used in the configurational correlations of allenes (Narayanan, Keiderling et al., 1987). Especially useful is the VCD of the $C\equiv C\equiv C$ stretch near 1950 cm^{-1}: positive VCD corresponds to the S configuration of the allene (cf. Section 13-6).

Enantiomeric purity of allenes may be crudely estimated from rotations calculated according to Eq. 14.3 or it may be determined by the more usual methods (use of chiral shift reagents in NMR, chromatography or NMR analysis of derivatives with chiral reagents, chromatography on chiral stationary phases, and so on; cf. Section 6-5). However, the application of such methods to allenes so far seems to have been limited (Runge, 1982, p. 592).

d. Cyclic Allenes, Cumulenes, Ketene Imines

1,2-Cyclonoadiene is the smallest stable cyclic allene to be isolated (Moore and Bertelson, 1962; Skattebøl, 1963; see also p. 1175) although 1,2-cyclooctadiene has been detected spectroscopically (Marquis and Gardner, 1966) and both it and 1,2-cycloheptadiene can be trapped as platinum complexes (Visser and Rama-kers, 1972; see also Balci and Jones, 1980); trapping has succeeded even with the very unstable 1,2-cyclohexadiene (Wittig and Fritze, 1968; Wentrup et al., 1983; cf. Greenberg and Liebman, 1978, p. 127; Johnson, 1986, 1989). The cyclic diallene dodeca-1,2,7,8-tetraene (Fig. 14.20) has been obtained both as the meso isomer **A** and in optically active form **B** (Garratt, Sondheimer, et al., 1973), whereas only the meso isomer has been synthesized in the case of the lower homologue deca-1,2,6,7-tetraene (Skattebøl, 1963; cf. Nakazaki, 1984). Of particular interest is the synthesis of the doubly bridged allene **C** (Fig. 14.20) through a dihalocyclopropane synthesis (cf. Fig. 14.11) both in racemic and (by use of a butyllithium-(−)-sparteine combination) in optically active form (Nakazaki et al., 1982; cf. Nakazaki et al., 1984; pp. 19–21).

According to van't Hoff (cf. Fig. 14.5), any properly substituted cumulene with an even number of double bonds, $RR'C=C(=C=C)_n=CRR'$ should be chiral. For allene $n = 0$; cumulenes of this type with $n \geq 2$ seem to be unknown (for cumulenes with odd numbers of double bonds, see Section 9-1.c). In any case such cumulenes, if obtainable, would probably not be configurationally stable, since the barrier to rotation around the $C=C$ double bonds in a cumulene decreases with increasing number of double bonds (Bertsch and Jochims, 1977). Although tetraaryl substituted pentatetraenes were first synthesized in 1964 (Kuhn et al.), resolution of $(CH_3)_3C(C_6H_5)C=C=C=C=C(C_6H_5)C(CH_3)_3$ (Karich and Jochims, 1977) was achieved only in 1977 (Bertsch and Jochims) through chromatography with an enantioselective stationary phase. The optical activity of material purified by crystallization at −80°C was $[\alpha]_D^{22} \pm 336$ (both enantiomers were obtained) and the material racemized in a few hours in n-nonane solution at −15°C with an activation barrier $\Delta G^{\ddagger} = 114.8 \text{ kJ mol}^{-1}$

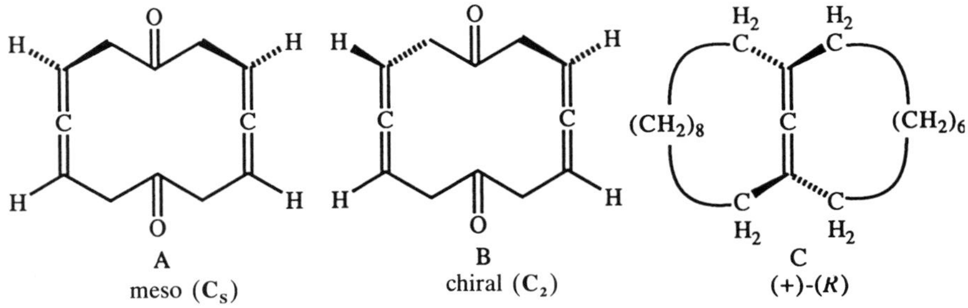

Figure 14.20. Cyclic bis-allenes and a doubly bridged allene.

(27.4 kcal mol^{-1}). This barrier is slightly less than the cis-trans interconversion barrier of the next lower homologue [triene, 122.5 kJ mol^{-1} (29.3 kcal mol^{-1}) in chlorobenzene] but considerably above that of the next higher one [pentaene, 86.9 kJ mol^{-1} (20.8 kcal mol^{-1}) in nitrobenzene]. It is, of course, much lower than the racemization barrier in an allene RCH=C=CHR [R = CH$_3$ or (CH$_3$)$_3$C, 46–47 kcal mol^{-1} (192–197 kJ mol^{-1}), Roth et al., 1974] which, in turn, is appreciably lower than the 62.2 kcal mol^{-1} (260 kJ mol^{-1}) cis–trans interconversion barrier in *trans*-2-butene (see Section 9-1.a).

Among the nitrogen analogues of allenes, the ketene imines RR′C=C=NR″ and their quaternary immonium salts, RR′C=C=$\overset{+}{N}$R″R‴, and the carbodiimides RN=C=NR are of particular interest. Racemization barriers in both cases were determined indirectly, by observing the coalescence of diastereotopic groups in R [R = (CH$_3$)$_2$CH— or C$_6$H$_5$CH$_2$C(CH$_3$)$_2$—]. For the ketene imines (Lambrecht, Jochims, et al., 1981; Jochims, Lambrecht, et al., 1984) barriers generally varied between 37 and 63 kJ mol^{-1} (8.8 and 15.1 kcal mol^{-1}); these species are clearly not resolvable. The barrier in (CH$_3$)$_2$CHN=C=NCH(CH$_3$)$_2$, 6.7 ± 0.2 kcal mol^{-1} (28.0 ± 0.8 kJ mol^{-1}) is even lower (Anet et al., 1970); however, diferrocenylcarbodiimide has been resolved and its absolute configuration determined (Schlögl and Mechtler, 1966). The barriers in ketene immonium salts (Lambrecht, Jochims et al., 1982) are too high to be measured by NMR (>115 kJ mol^{-1}, 27.5 kcal mol^{-1}) and these compounds should be resolvable.

It is of interest that, although ketene imine CH$_2$=C=NH is linear, the ketene CH$_2$=C=C=O is "kinked" with the C(3)—C(2)—C(1) angle 142° and the C(2)—C(1)—O angle −164.5° (cf. Brown, 1987).

14-3. ALKYLIDENECYCLOALKANES

Although optically active alkylidenecycloalkanes, in contrast to allenes, probably do not occur in nature, their resolution in the laboratory preceded that of allenes by over a quarter of a century. Already in 1909, Perkin et al. (q.v.) resolved 4-methylcyclohexylideneacetic acid (Fig. 14.21, **A**). The absolute configuration of this compound was assigned by Gerlach (1966) through correlation with (2*R*)-(−)-isoborneol of known configuration. The key intermediate in this correlation is *cis*-(*R*)-(+)-4-methylcyclohexylacetic-α-*d* acid **C**, obtained from the resolved α-deuterated analogue of 4-methylcyclohexylideneacetic acid, **B**, by catalytic hydrogenation (with Pd/BaSO$_4$), which proceeds cleanly (i.e., without atom or bond migration) and is thus assumed to be syn (cf. Section 12-3.h). The two diastereomeric products **C** and **D** (Fig. 14.21) were separated and their relative configurations (cis or trans) ascertained by comparison with known samples; their optical rotations [α]$_{546}$, +0.44 and +0.65, respectively, were appreciable. To determine the absolute configuration of (+)-**C**, it was converted to the corresponding amine by Curtius degradation, which is known to proceed with retention of configuration (cf. Section 5-5.f, see also Eliel, 1962, p. 120). The (−) amine obtained was then synthesized by an alternative route involving an asymmetric reduction with isobornyloxymagnesium bromide made from (2*R*)-(−)-isoborneol

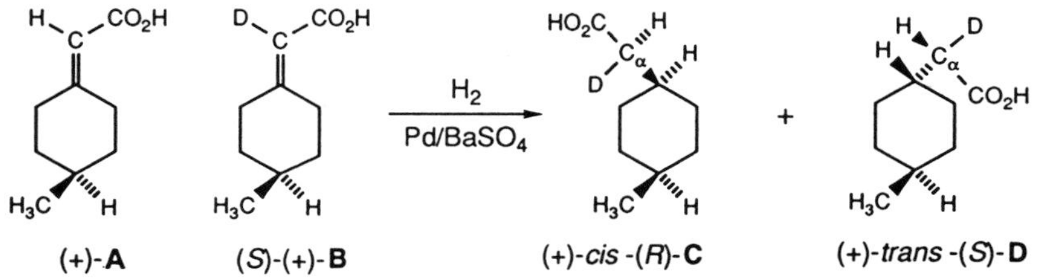

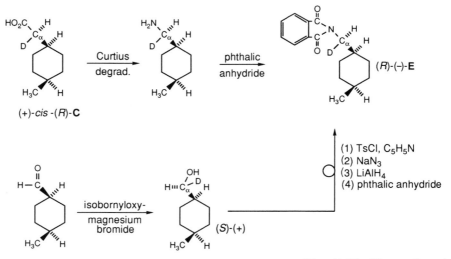

Figure 14.21. Correlation of the configuration of (+)-4-methylcyclohexylideneacetic acid. The configurational descriptors in the products relate to C_α.

(Fig. 14.22). It is known from earlier work (Streitwieser et al., 1959; see also Arigoni and Eliel, 1969, p. 161) that the isobornyloxymagnesium bromide reduction of RCHO leads to (S)-RCHDOH. Since the conversion of the alcohol to the amine involves a single inversion (in the tosylate to azide step), the (−)-cis-phthalimide (Fig. 14.22, **E**) is R and its precursor acid (+)-**C** is also R. Reference to Figure 14.21 then indicates that (+)-**B** must be S or P, and since chiral compounds bearing hydrogen and deuterium in the same position generally differ little in rotation (cf. Arigoni and Eliel, 1969), (+)-**A** is assigned the S or P configuration also. The same assignment is obtained through the trans isomer (+)-**D** which, by further degradation, is assigned the S configuration (Fig. 14.21; Gerlach, 1966).

In the above correlation, an alkylidenecycloalkane is converted to a compound with a chiral center of known (or demonstrable) configuration by a reaction (catalytic hydrogenation) of known mechanism (cf. also Eliel, 1962, p. 350 ff). The opposite course, conversion of a compound with central chirality to a chiral alkylidenecyclohexane, was taken by Brewster and Privett (1966) as shown in Figure 14.23. The compound (R)-(+)-3-methylcyclohexanone (**F**) of known configuration was condensed with benzaldehyde to give one of the corresponding

Figure 14.22. Determination of absolute configuration of **C** (Fig. 14-21). The configurational descriptors refer to C_α.

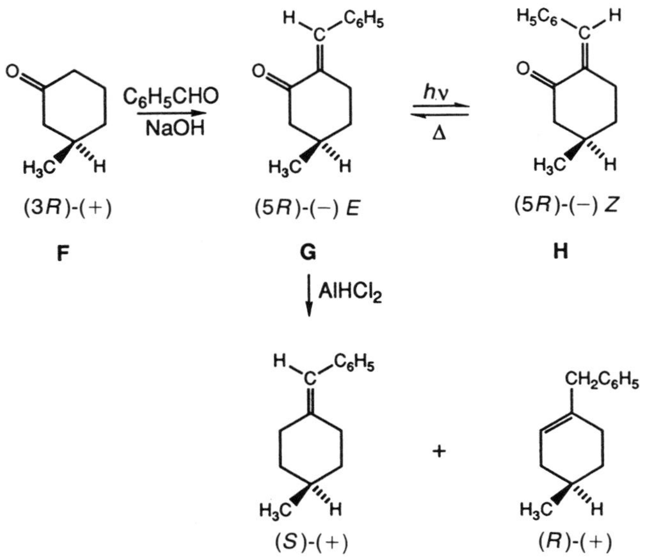

Figure 14.23. Assignment of configuration of (+)-1-benzylidene-4-methylcyclohexane.

(5R)-(−)-2-benzylidene derivatives (**G**). Photochemical isomerization led to a thermally less stable compound (**H**) with a lower λ_{max} and ε_{max} than that of the original isomer. It was concluded that the product of condensation was the E isomer (**G**) and the less stable photoproduct the Z isomer (**H**, Fig. 14.23). Removal of the carbonyl group from the E isomer gave as one of two products (+)-1-benzylidene-4-methylcyclohexane (**I**) whose configuration is thus shown to be S or P (Fig. 14.23). The Lowe–Brewster rules (Fig. 14.18) evidently apply to the alkylidenecyclohexanes shown in Figures 14.21 and 14.23 (assumed polarizabilities $CH_3 > H$, $C_6H_5 > H$, and $CO_2H > H$) but the proviso is added (Brewster and Privett, 1966) that the rotation in the visible must not be controlled by a near-UV Cotton effect (CE) for this to be so.

The Gerlach and Brewster–Privett assignments of compounds **A** and **I** have been correlated by Walborsky's group (Walborsky and Banks, 1981; Walborsky et al., 1982; Duraisamy and Walborsky, 1983) who also established, by correlation, the configuration of a number of additional alkylidenecycloalkanes (Fig. 14.24). The assignment of the dextrorotatory 1-bromomethylidene-4-methylcyclohexane (**J**) as S had already been proposed by both Gerlach (1966) and Brewster and Privett (1966) on the basis of earlier work by Perkin and Pope cited below (cf. Fig. 14.25, which relates to the opposite (R)-(−) enantiomer). It was confirmed by a lithiation–carboxylation sequence (Fig. 14.24), which proceeds with retention of configuration (Walborsky and Banks, 1980); the sequence of reactions shown in Figure 14.25 (**A** to **J**) followed by recarboxylation (Fig. 14.24, **J** to **A**; Walborsky and Banks, 1980) proceeds without loss of enantiomeric purity of **A**. Both the dehydrobromination and decarboxylative debromination studied by Perkin and Pope (1911) are anti eliminations (Fig. 14.25).

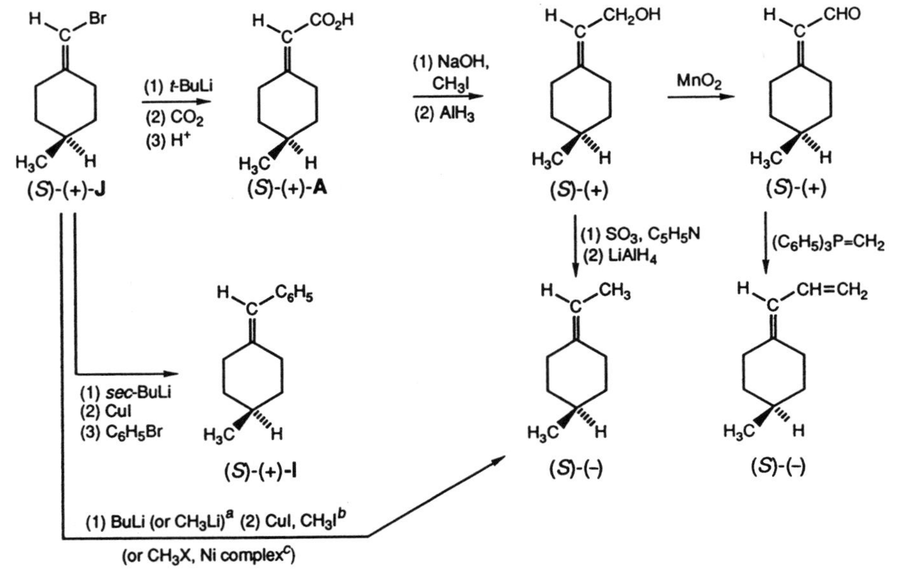

Figure 14.24. Configurational correlations in alkylidenecyclohexanes. [a]Walborsky and Banks (1981). [b]Walborsky, et al. (1982). [c]Solladié and Zimmerman (1984).

An interesting enantioselective synthesis of alkylidenecyclohexanes (Hanessian et al., 1984) is shown in Figure 14.26. The observed stereochemistry [formation of the $(R)-(+)$-alkylidenecyclohexane from the (R,R)-phosphonamide precursor of **A** can be explained by an equatorial approach from the *Si* side of the anionic center of ylid **A** to the cyclohexanone (R-group equatorial) followed by a syn elimination of phosphoramide **B**. The *Si* side of **A** can be seen, in a model, to be the more open side (Hanessian and Beaudoin, 1992).

Alkylidenecyclohexanes have also been enantioselectively synthesized in very high optical yield by high-temperature, based-catalyzed elimination from chiral

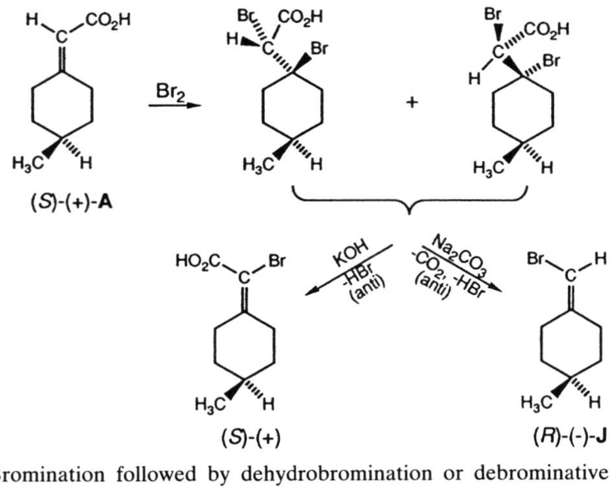

Figure 14.25. Bromination followed by dehydrobromination or debrominative decarboxylation of 4-methylcyclohexylideneacetic acid.

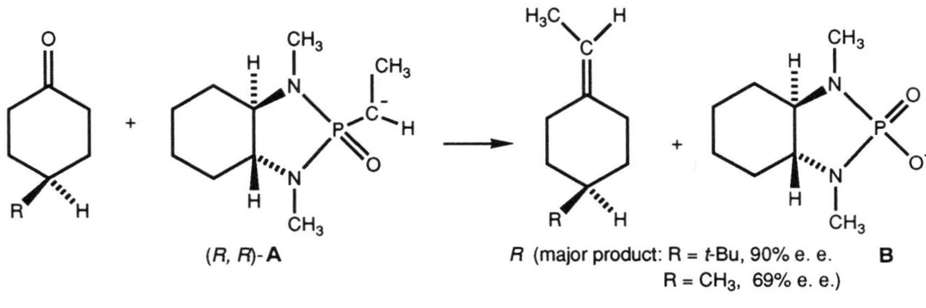

Figure 14.26. Enantioselective synthesis of alkylidenecyclohexane.

sulfoxides, for example, 4-$CH_3C_6H_{10}CHBrSOC_6H_4CH_3$-$p$ (Solladié and Zimmermann, 1984).

The ORD and CD spectra of some chiral conjugated bisalkylidenecyclohexanes have been reported (Walborsky et al., 1982) but estimates of enantiomeric purity of alkylidenecyclohexanes usually rest on comparisons of rotation at the sodium D line. The use of chiral shift reagents for this type of compound has so far not been successful.

Alkylidenecycloalkanes with substituents in the 2 and 3 positions (cf. Duraisamy and Walborsky, 1983; Hanessian et al., 1984) display central rather than axial chirality, the carbon atoms in positions 2 or 3 being stereogenic centers. In addition, there is cis–trans isomerism. However, apparent axial chirality may be found in cis-3,5-disubstituted alkylidenecyclohexanes and similar structures (Fig. 14.27, **A**), and a nitrogen analogue of this type (Fig. 14.27, **B**) has, in fact, been synthesized and resolved (Lyle and Lyle, 1959). The configuration of this compound has been elucidated (Lyle and Pelosi, 1966) as shown in Fig. 14.27. Beckmann degradation (cf. Fig. 9.32) followed by β elimination gave (R)-($-$)-2-methylamino-2-phenylethylamine, whose configuration was established by correlation with the known (R)-($-$)-phenylglycine (Fig. 14.27). The configuration of the ($+$) oxime is therefore Z (equivalent to syn-R).

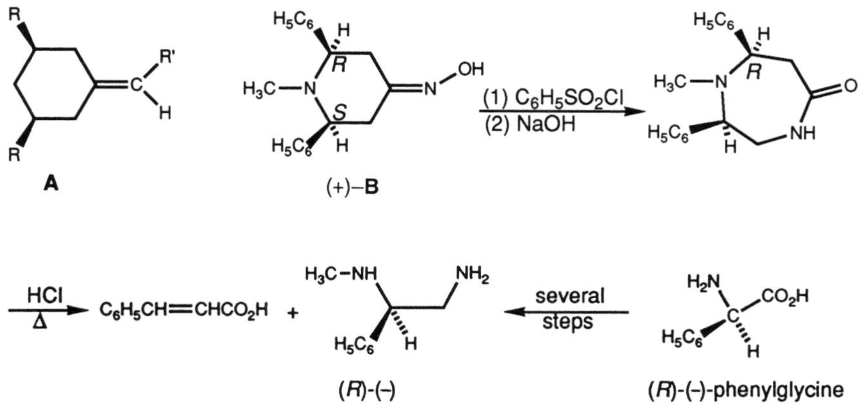

Figure 14.27. Absolute configuration of ($+$)-1-methyl-2,6-diphenyl-4-piperidone oxime **B**.

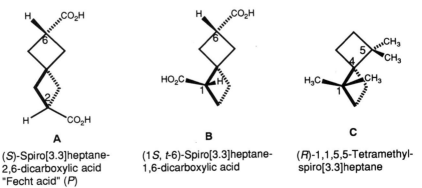

Figure 14.28.   cis–trans Enantiomers.

It should be noted that, at least from the nomenclatural point of view, compounds **A** and **B** in Fig. 14.27 are *not* considered cases of axial chirality (Cahn, Ingold, and Prelog, 1966). To name such compounds one labels the chiral centers in the ring as R or S, as the case may be, and since $R > S$ in the sequence rules (other things being equal), the configuration is Z, the OH of the oxime being on the side of the R substituent on the ring. The type of chirality seen in Figure 14.27, originally called "geometric enantiomorphic isomerism" (Lyle and Lyle, 1959) and later "geometric enantiomerism" (Eliel, 1962, p. 320) should now be called cis–trans enantiomerism. In fact, it need not coincide with apparent axial chirality at all; a more general case is shown in Figure 14.28.

14-4. SPIRANES

The name "spirane," from the Latin *spira* meaning twist or whorl implies that spiranes (cf. Fig. 14.2) are not planar; it is their nonplanarity that gives rise to their chirality.

Among the chiral spiranes (Fig. 14.29) one may discern three types: **A**, which definitely displays axial chirality similar to that of allenes and alkylidenecycloalkanes (see above); **B**, which, like corresponding alkylidenecycloalkanes (see above), displays central rather than axial chirality; and **C**, which conceptually would appear to display axial chirality but, for purposes of nomenclature, is considered to have a chiral center (Cahn, Ingold, and Prelog, 1966). Compound **A** is described as indicated in Figure 14.3, the descriptor is aS or P. Compound **B** has four stereoisomers (2 pairs of enantiomers); C(1) is a chiral center, whereas C(6) displays cis–trans isomerism and the stereoisomer shown is 1S,6-trans. To name **C** one arbitrarily gives one ring preference over the other; the more substituted branch in that ring then has priority 1 and the less substituted has

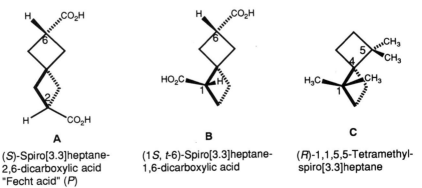

A

(S)-Spiro[3.3]heptane-
2,6-dicarboxylic acid
"Fecht acid" (P)

B

(1S, t-6)-Spiro[3.3]heptane-
1,6-dicarboxylic acid

C

(R)-1,1,5,5-Tetramethyl-
spiro[3.3]heptane

Figure 14.29.   Types of spiranes.

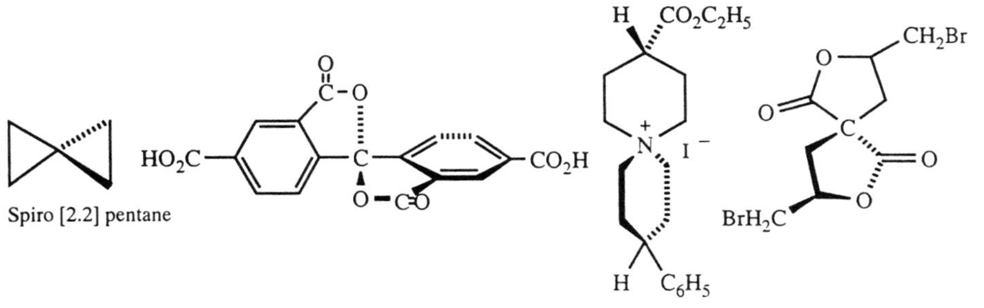

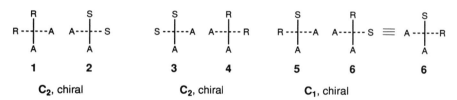

Figure 14.30. Examples of spiranes.

priority 3, whereas the corresponding priorities in the arbitrarily less favored ring are 2 and 4. The configuration is then $4R$; the spiro center $C(4)$ is considered a chiral center.

The most strained saturated spirane, spiro[2.2]pentane (Fig. 14.30, **A**), was apparently first synthesized in 1896 by Gustavson (q.v.) although it was not then recognized as such (cf. Applequist et al., 1958). Its strain of 65 kcal mol^{-1} (272 kJ mol^{-1}) is only about 10 kcal mol^{-1} (42 kJ mol^{-1}) greater than that of two isolated cyclopropane rings (Humphrey and Spitzer, 1950; Fraser and Prosen, 1955; for an interpretation, see Bernett, 1967). Chirality in spiranes, first recognized by Aschan (1902) was demonstrated in 1920 by Mills and Nodder (q.v.) by resolution of a spirodicarboxylic acid (Fig. 14.30, **B**). This compound is of type **C** in Figure 14.29; the central carbon atom can be described as a chiral center. However, a compound of type **A** in Figure 14.29 was resolved 5 years later (Mills and Warren, 1925); it is shown in Figure 14.30, **C**. It is of interest that the spiro center of compound **C** is a quaternary nitrogen rather than a carbon atom. Compound **D** in Figure 14.30 is also interesting; it has a spiro center and two conventional chiral centers; contemplation of models (cf. Fig. 14.31) indicates the existence of three diastereomeric racemates, which have, in fact, been isolated (Leuchs and Giesler, 1912).

Several assignments of absolute configuration of spiranes of type **C** in Figure 14.29 have been accomplished (see also Klyne and Buckingham, 1978, Vol. 2, p. 106; Buckingham and Hill, 1986, pp. 150–151). The first correct one (Gerlach, 1968) is concerned with the dione shown in Figure 14.32 and is based on that of its diol precursor shown in the same figure. The relative configuration of this diol had been established by the absence of intramolecular hydrogen bonding and by reductive correlation with a single one of the two diastereomeric mono-ols. The

Figure 14.31. Combination of spirane and conventional chiral centers.

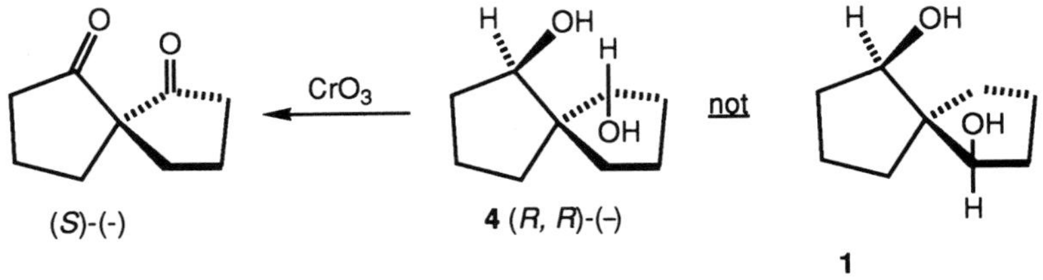

Figure 14.32. Configuration of spiro [4.4] nonane-1,6-dione. Intramolecular hydrogen bonding is possible in isomer **1** and in its enantiomer (**2**, Fig. 14.31).

latter result demands C_2 symmetry, that is, structures **5** and **6** (Fig. 14.31) are excluded. The former result (absence of intramolecular hydrogen bonding) excludes **1** (Fig. 14.31) and its enantiomer **2**. Finally, the absolute configuration of the diol was established by Horeau's rule (cf. Horeau, 1977 and pp. 140–142), to be R,R (Fig. 14.32). Oxidation of this diol then must lead to the (S)-dione (Fig. 14.32, chiral center naming) which, experimentally, was found to be levorotatory.

Brewster and Jones (1969) correlated the levorotatory quaternary ammonium salt **A** (Fig. 14.33) with the tertiary amine **B** (Fig. 14.33) by a Stevens rearrangement; the configuration at the stereogenic carbon in the product is correlated with that of the starting spirane on the assumption that the rearrangement is suprafacial. The configuration of **B** is, in turn, correlated chemically with that of (S)-aspartic acid as shown, i.e. by oxidizing the aromatic rings to carboxyl groups. The configuration of the levorotatory spirane is therefore R (Fig. 14.33).

The third, in principle more direct, correlation of configuration of a spirane of type **C** (Fig. 14.29) by synthesis from a compound with a chiral center of established configuration at first miscarried because of experimental problems. The correct correlation (Overberger, Krow, Hill, et al., 1981) is shown in Figure 14.34. The configuration of the key intermediate, (R)-(−)-**A** was established by conversion to sulfonamide (R)-(+)-**B** whose enantiomer was chemically correlated with (S)-methylethylsuccinic acid. The configuration of the latter had been established (Porath, 1951) by the method of quasiracemates (Section 5-5.e). The compound (R)-(−)-**A** was then converted to a spirane, (+)-**C**, further reduced to

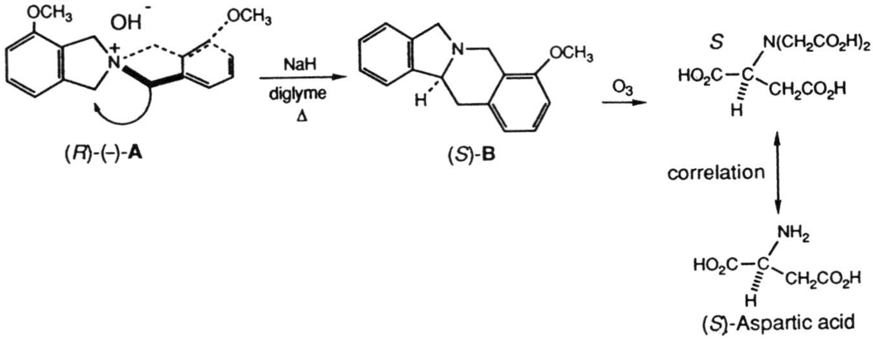

Figure 14.33. Correlation of spiroammonium salt with (S)-aspartic acid.

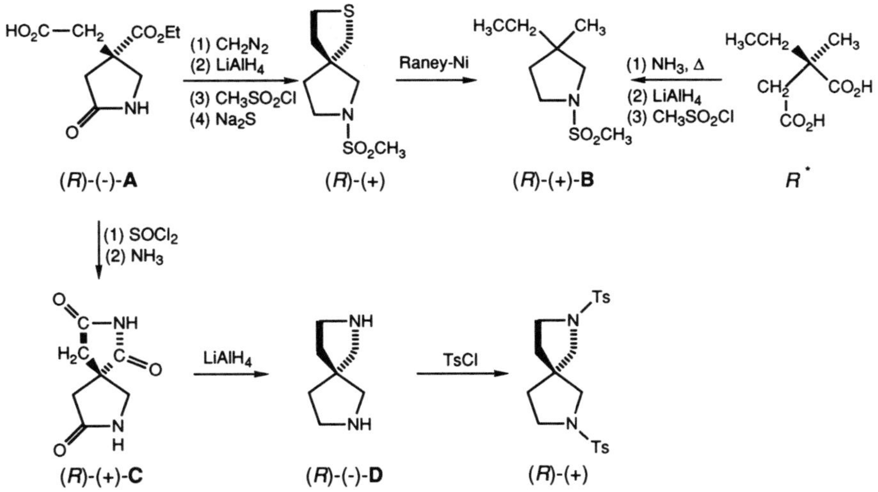

Figure 14.34. Configurational assignment of 2,7-diazaspiro[4,4]nonane. The asterisk (*) signifies that the actual experiment was carried out with the *S* acid, which gave (*S*)-(−)-**B**. The sign of rotation of the acid depends on concentration as well as on configuration.

diazaspirane (−)-**D** by reduction. The correlation shown in Figure 14.34 proves that (+)-**C**, (−)-**D** and its derived dextrorotatory ditosylate have the *R* configuration.

Similar experimental difficulties (initially incorrect results) delayed the correct determination of the absolute configuration of spiro[3.3]heptane-2,6-dicarboxylic acid (Fecht acid; Fig. 14.29, **A**), the only type **A** spirane whose absolute configuration has been determined directly (though other type **A** spiranes have been correlated with it, Wynberg and Houbiers, 1971). Eventually, however, the Bijvoet method (Section 5-3.a), carried out both with the free acid and its barium salt (Hulshof, Wynberg, et al., 1976) showed the a*S* configuration (Fig. 14.29) to correspond to the (+) acid.

Although a number of empirical rules, including the Lowe–Brewster rules (Fig. 14.18) have been proposed for spiranes, it is now clear (Brewster and Prudence, 1973; Hulshof, Wynberg, et al., 1976) that such rules based on long-wavelength rotation are not general. Nonetheless, in a *limited* field (the spirobiindanes shown in Fig. 14.35, **A**) chirality functions have been applied successfully (Neudeck and Schlögl, 1977). Circular dichroism methods have been applied to spiranes (e.g., Hagishita, Nakagawa, et al., 1971) but caution is necessary here also; for example, the three stereoisomeric 3,8-di-*t*-butyl analogues of spiro[4.4]nonane-1,6-dione (Fig. 14.32) with identical configuration at the spiro center show quite different CD spectra (Sumiyoshi, Nakagawa, et al., 1980). Perhaps application of exciton coupling CD (Harada and Nakanishi, 1983, pp. 149, 212, see also Section 13-4.d) will prove more successful.

Circular dichroism methods have also been used for configurational assignment of the so-called [*n*, *n*]-vespirenes (*n* = 6,7,8) shown in Figure 14.35, **B**. These chiral compounds (Haas and Prelog, 1969; Haas et al., 1971) are notable in that they display **D₂** symmetry and that the four ligands attached to the spiro center are structurally identical; chirality is enforced by the methylene bridges that cannot lie integrally in the planes of either of the fluorene ring systems.

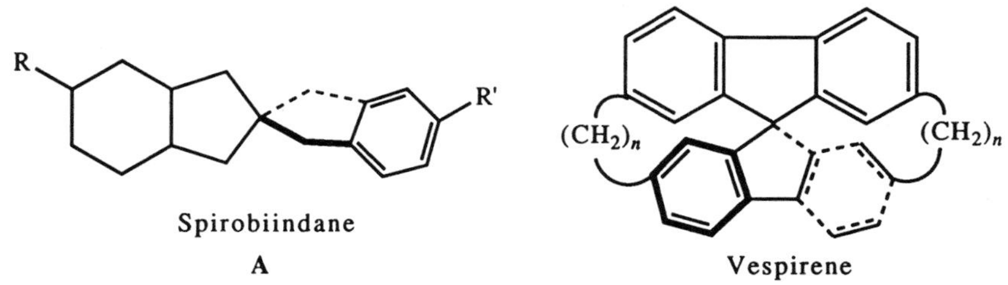

Figure 14.35. Spirobiindanes and vespirenes.

14-5. BIPHENYLS. ATROPISOMERISM

a. Introduction

In the examples given so far, the chiral axis is sustained (and the screw or helical sense of the molecule maintained) either by the "stiffness" (high barrier to rotation) of a double bond (allenes) or by the molecular framework as a whole (spiranes) or by a combination of the two (alkylidenecycloalkanes). We now come to molecules with a chiral axis whose helical sense is maintained through hindered rotation about single bonds, the hindrance in general being due to steric congestion. The classical examples of such molecules are the biphenyls (or biaryls in general) shown in Figure 14.36. If $X \neq Y$ and $U \neq W$ and, moreover, the steric interaction of X–U, X–V, and/or Y–V, Y–U is large enough to make the planar conformation an energy maximum, two nonplanar, axially chiral enantiomers (Fig. 14.36) exist. If the interconversion through the planar conformation is slow enough they may, under suitable circumstances, be isolated (resolved). This type of enantiomerism was first discovered by Christie and Kenner (1922) in the case of 6,6'-dinitro-2,2'-diphenic acid (Fig. 14.36, $X = U = CO_2H$; $Y = V = NO_2$), which they were able to resolve. It was later called (Kuhn, 1933) "atropisomerism" (from Greek *a* meaning not, and *tropos* meaning turn). (For references to the early history, see Eliel, 1962, p. 156.)

Reference to Chapter 10 suggests that atropisomerism is a type of conformational (rotational) isomerism in which the conformational isomers or conformers can be isolated. It is immediately obvious that the term suffers from all the problems discussed in Sections 2-4 and 3-1.b: How slow must the interconversion of the enantiomers be (i.e., how long is their half-life) before one speaks of atropisomerism? At what temperature is this measurement to be made? Does atropisomerism still exist when isolation of stereoisomers becomes difficult or impossible but their existence can be revealed by NMR (or other spectral) study,

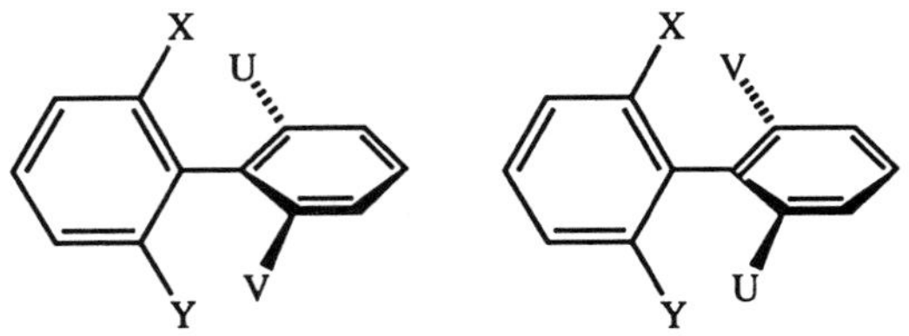

Figure 14.36. Enantiomeric chiral biphenyls.

and so on. Ōki (1983) arbitrarily defined the condition for the existence of atropisomerism as one where the isomers can be isolated and have a half-life $t_{1/2}$ of at least 1000 s (16.7 min). This value still does not define the required free energy barrier, which evidently now depends on temperature; it is 22.3 kcal mol^{-1} (93.3 kJ mol^{-1}) at 300 K, 26.2 kcal mol^{-1} (109.6 kJ mol^{-1}) at 350 K, and 14.7 kcal mol^{-1} (61.5 kJ mol^{-1}) at 200 K. Though this definition is entirely arbitrary, it is convenient and quite essential if the concept of atropisomerism is to be maintained at all.

b. Biphenyls and Other Atropisomers of the *sp²–sp²* Single-Bond Type

General Aspects

Atropisomers are numerous in number and type and only a very brief treatment can be given here. Biphenyl isomerism has been extensively discussed earlier (Adams and Yuan, 1933; Shriner, Adams, and Marvel, 1943; Eliel, 1962, p. 156; Krow, 1970), especially in regard to the structural attributes needed to "restrict" rotation. Half-lives of racemization of numerous biphenyls have been determined with the following general findings:

1. Most tetra-ortho substituted biphenyls (Fig. 14.36, U,V,X,Y ≠ H) are resolvable and quite stable to racemization unless at least two of the groups are fluorine or methoxy.

A nonresolvable, tetra-ortho substituted biphenyl is shown in Figure 14.37, **A** (Adams and Yuan, 1933). It should be noted that although the condition U ≠ V and X ≠ Y is not fulfilled by this molecule, the perpendicular conformation lacks a plane of symmetry because of the meta substituents (Cl ≠ CO₂H).

2. Tri-ortho substituted biphenyls are readily racemized (short $t_{1/2}$ values) when at least one of the groups is small (CH₃O or F), otherwise racemization tends to be slow (but is possible, generally at elevated temperatures).

3. Di-ortho substituted biphenyls are generally resolvable only if the substituents are large. An interesting example is 1,1′-binaphthyl (Fig. 14.37, **B**), originally obtained optically active by deamination of the resolved 4,4′-diamino derivative (Cooke and Harris, 1963). The compound exists in

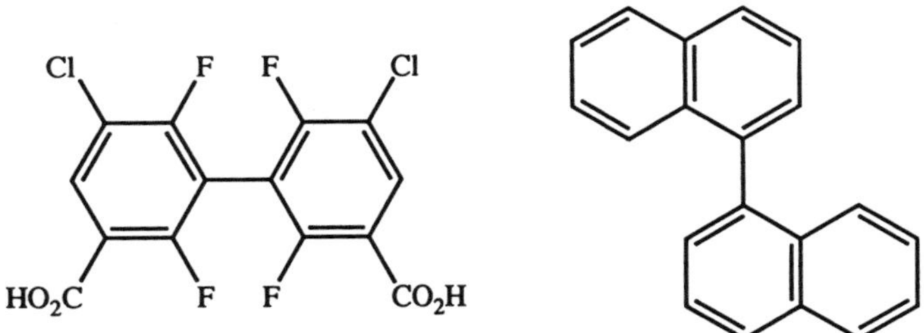

Figure 14.37. A tetra-*o*-fluorosubstituted biphenyl **A** and 1,1′-binaphthyl **B**.

two crystalline modifications, a racemic compound, mp 145°C and a conglomerate (cf. Section 6-3), mp 158°C. The latter is easily resolved either spontaneously or by seeding of the melt or solution (Wilson and Pincock, 1975; see also p. 317); above the melting point the enantiomers are in rapid equilibrium ($t_{1/2} \approx 0.5$ s at 160°C; $\Delta G^{\ddagger} = 23.5$ kcal mol^{-1} (98.3 kJ mol^{-1}).

4. Mono-ortho substituted biphenyls are, in general, not resolvable although the (+)-camphorsulfonate of the arsonium salt shown in Figure 14.38 shows mutarotation (cf. p. 750) suggesting that an interconversion of diastereomers occurs in solution because the two diastereomers are not equally stable (asymmetric transformation of the first kind, cf. Section 7-2.e).

5. Substituents in the meta position tend to enhance racemization barriers by what is known as a "buttressing effect," that is, by preventing the outward bending of an ortho substituent, which would otherwise occur in the transition state (coplanar conformation) for racemization. (This outward bending allows the ortho substituents to slip past each other more readily by energy minimization of the activated complex (cf. Section 2-6).

6. The apparent size of substituents (as gauged by racemization rates of differently ortho-substituted biphenyls) is $I > Br \gg CH_3 > Cl > NO_2 > CO_2H \gg OCH_3 > F > H$. This order roughly parallels van der Waals radii ($I > Br > C > Cl > N > O > F > H$; in polyatomic groups allowance must be made for the outer substituents) and is quite different from that of the ΔG^0 values in cyclohexanes (axial-equatorial equilibrium, Table 11.7) In contrast to synaxial substituents in cyclohexanes, ortho substituents on the two rings in biphenyls point at each other, so their interaction should increase with increasing van der Waals (and bond) radii.

7. Activation barriers to racemization can be calculated quite closely by molecular mechanics (cf. Section 2-6); in fact, calculation of barriers of this type constitutes the first application of what is now called the molecular mechanics or force field method (Westheimer and Mayer, 1946; cf. Westheimer, 1956). Semi-empirical methods have also been applied to the calculation of barrier height, with some success (Kranz, Schleyer, et al., 1993).

8. Diastereomers are found not only in biphenyls with chiral substituents but also in terphenyls. The compound shown in Figure 14.39, **A** is an example; the cis isomer has been resolved, whereas the trans isomer, which was separated from the cis, cannot be resolved because it has a center of symmetry (Knauf, Adams, et al., 1934).

Oxidation of the optically active cis hydroquinone **A** gave an optically active quinone **B**, which was reduced back to optically active **A**. This

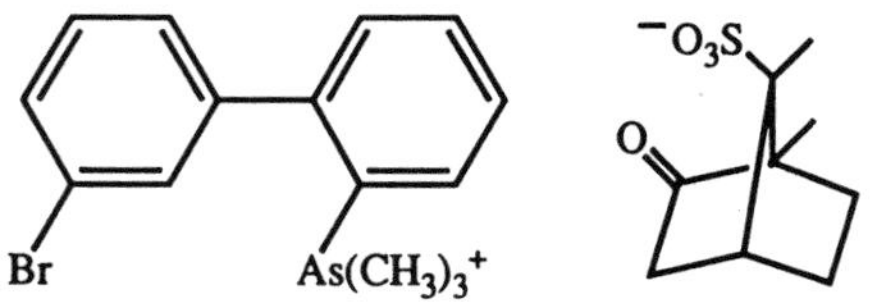

Figure 14.38. Mutarotating mono-orthosubstituted biphenyl.

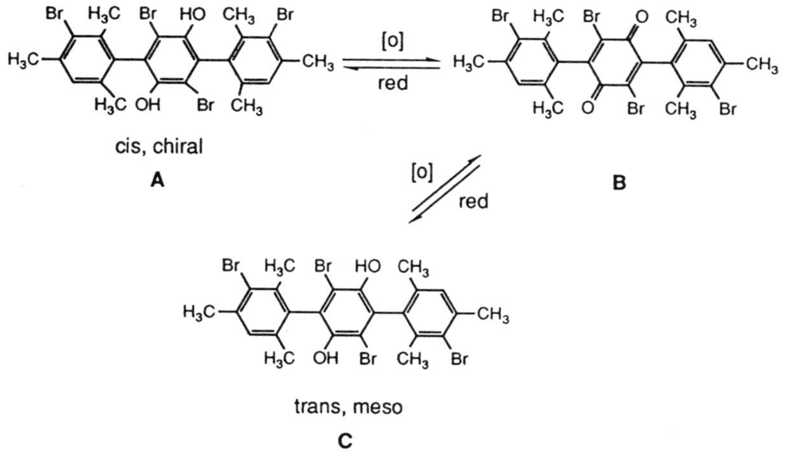

Figure 14.39. cis–trans Isomerism in terphenyls and diphenylquinones.

finding showed that atropisomerism is possible in structures other than biphenyls. Additional cases of resolvable atropisomers are shown in Figure 14.40 (Mills and Dazeley, 1939; Adams and Miller, 1940; Adams et al., 1941). Examples of thioamides were discussed earlier (Section 9-1.e; see also Ōki, 1983).

9. Noncoplanarity may be assisted or enforced by bridging (cf. Hall, 1969). Molecule **A** in Figure 14.41 is resolvable and moderately optically stable (Adams and Kornblum, 1941) even though o,o'-diphenic acid (Fig. 14.36, $X = U = CO_2H$, $Y = V = H$) is not. The ortho bridged **B** and doubly ortho bridged **C** biphenyls (Fig. 14.41) have been synthesized in optically active form by Mislow et al. (1964). The doubly bridged biphenyl enantiomers **C** with $X = CO$ or S are quite stable but can be racemized thermally, whereas the compound with $X = O$ is labile and racemizes with a half-life of only 54 min at 10.1°C. The enantiomers of the corresponding three-atom singly bridged species **B** ($X = O$) are much more stable (see also Iffland and Siegel, 1958). In contrast, the two-atom bridged compound **D** (Mislow and Hopps, 1962) has quite low enantiomeric stability ($t_{1/2} = 108$ min at 28.1°C) contrary to the high stability of tetra-ortho substituted biphenyls in general (see above); it is thus clear that ortho bridging diminishes the enantiomeric stability of ortho-substituted biphenyls, presumably because

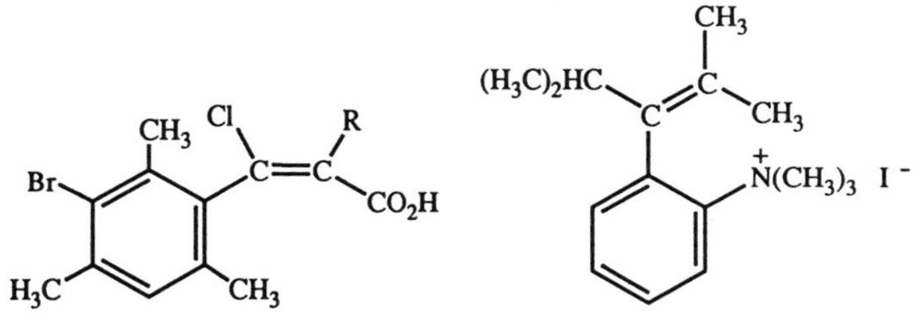

Figure 14.40. Resolvable styrenes, where $R = CH_3$ or H.

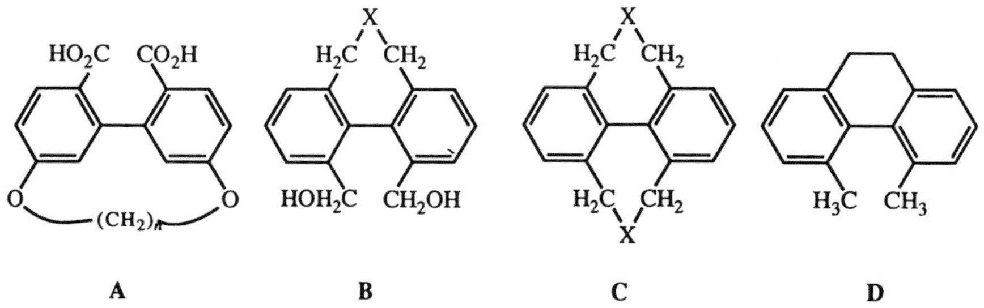

Figure 14.41. Bridged biphenyls, where $n = 8$ or 10.

it diminishes the dominant van der Waals interaction of the ortho substituents. (In compound **D**, Fig. 14.41, the methylene substituents are bonded and have therefore no nonbonded interaction at all, although they may impose a slight torsional preference for nonplanarity.)

The energy profile in unbridged biphenyls is more complex than the discussion so far might imply. Conceptually, one might want to think of a "hindered" biphenyl (Fig. 14.36) as being constituted of two orthogonal phenyl rings; the condition that $X \neq Y$ and $U \neq V$ serves to abolish a plane of symmetry that would otherwise exist in that conformation. In fact, however, in solution the rings in biphenyls are neither coplanar nor orthogonal in the lowest energy conformation. The tendency for nonplanarity, which is imposed by the steric demands of the ortho substituents, is opposed by π-electron overlap, which produces maximum stabilization when the rings are coplanar. Even biphenyl itself is nonplanar in the ground state, the inter-ring torsion angle being 44° in the vapor phase (Almenningen, Bastiansen, et al., 1985), though the rings are coplanar in the crystal phase, perhaps because of packing forces (cf. Brock, 1979, Brock and Minton, 1989). The result is a compromise with the interplanar angle in biphenyls varying from 42° to 90° (cf. Ingraham, 1956; Bastiansen and Samdal, 1985). The energy profile for rotation in ortho substituted biphenyls is shown in Figure 14.42; the maxima (barriers to rotation) occur at 0° and 180° and the regions to the two sides of the 0° maximum correspond to the two enantiomers which may or may not be isolable, depending on the energy difference between the minimum and the lower

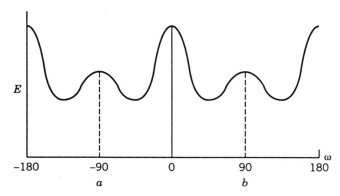

Figure 14.42. Energy profile for rotation in biphenyls.

maximum ($0°$ or $\pm 180°$). The curves to the right and left of the $\omega = 0°$ maximum are mirror images of each other, but these parts of the curves for each enantiomer, in turn, display a secondary maximum at $\pm 90°$ with minima near $\pm 44°$ on both sides. The maximum at $90°$ is due to the complete absence of resonance (π-orbital overlap) stabilization of the biphenyl system at this angle. Nonetheless the curve may be quite flat in the $\pm 90°$ region and in some cases of large steric repulsion between ortho substituents (cf. Ingraham, 1956) there may actually be a minimum in energy at $90°$ instead of a low maximum. For biphenyl itself and biphenyls lacking ortho substituents, the barriers at $0°$ and $90°$ are of comparable magnitude, in the range of 1.4–2.0 kcal mol^{-1} (6–8 kJ mol^{-1}).

The maxima at $0°$ and $\pm 180°$ will, in general, be unequal if (Fig. 14.36) $X \neq Y$ and $U \neq V$, since the X/U + Y/V in-plane interaction will generally be different from X/V + Y/U. Even if $X = Y$ or $U = V$, the barriers may not be equal, since, in that case, a meta substituent is required to maintain chirality, and such a substituent is likely to exert a buttressing effect, enhancing steric interaction on its side. By the same token, the two minima near $+44°$ and $+136°$ (enantiomer region *b* in Fig. 14.42) will not be exactly equal in energy (see below).

The strong conjugation band in the UV spectra of biphenyl, *o*-methylbiphenyl and some ortho, ortho' disubstituted biphenyls in the 240–250-nm region, which disappears upon more extensive ortho substitution or with more bulky ortho substituents (Pickett et al., 1936; O'Shaughnessy and Rodebush, 1940; cf. Hall, 1969), is an indication of residual conjugation (π-orbital overlap); that is, the interplanar angle must be less than $90°$ in these molecules. In *o,o'*-dichlorobiphenyl, the torsion angle is near $70°$ with the two chlorine substituents being syn to each other (Bastiansen, 1950; Rømming et al., 1975), perhaps because of attractive van der Waals forces.

Nonplanarity can sometimes be demonstrated by NMR spectroscopy even when the $0°$ barrier is too low for stereoisomers to be isolated. Thus *o,o'*-diacetoxymethylbiphenyl (Fig. 14.43, **A**) shows an $(AB)_2$ system for the italicized protons (Meyer and Meyer, 1963) at room temperature, whereas at 127°C the pattern collapses to a single line. The CH$_2$ protons would be enantiotopic (cf. Chapter 8) if the biphenyl system were planar (the biphenyl plane, in that case, would be a symmetry plane relating the two hydrogen atoms), but since the system is nonplanar, at room temperature, these protons become diastereotopic and anisochronous (cf. Section 8-4.d). On heating, topomerization (p. 505) occurs

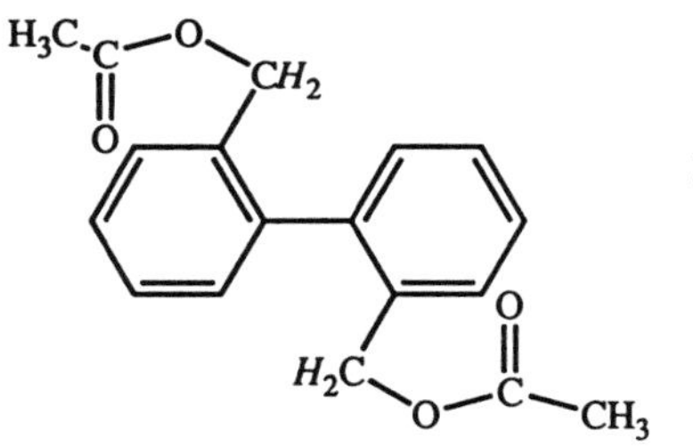

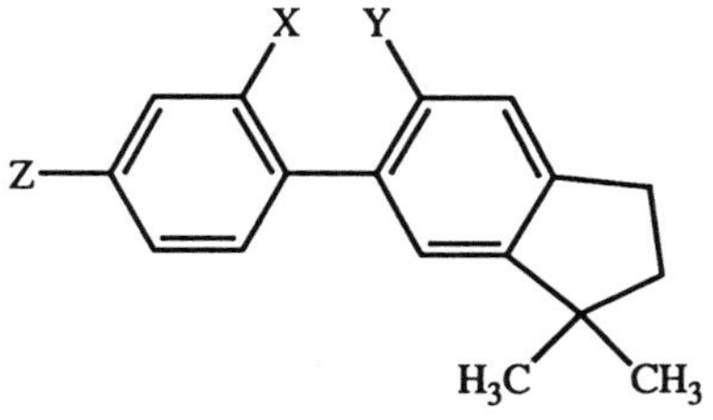

Figure 14.43. Biphenyls in which the rotational barrier has been determined by NMR.

and the protons coalesce. From the coalescence pattern, an energy barrier to biphenyl rotation of 13 kcal mol^{-1} (54.4 kJ mol^{-1}) was estimated; this barrier is evidently too low for resolution to be possible. Clearly, this method can be used to predict if a system might reasonably be resolved (and if so, how easily) before resolution is attempted. The NMR method does not require optically active substances though it does require the presence of NMR active diastereotopic nuclei. Barriers in a number of ortho, ortho'-disubstituted phenylindanes (Fig. 14.43, **B**) have been determined experimentally in this way and have been rationalized theoretically (Bott, Sternhell, et al., 1980). Barriers in a number of other sp^2–sp^2 type systems with hindered rotation about the connecting bond have been evaluated by this method also (Binsch, 1968; Ōki, 1983).

Configuration of Biphenyls and Binaphthyls.

Early assignments of configuration of biphenyls and binaphthyls (Klyne and Buckingham, 1978, Vol. 1, pp. 219–221; Vol. 2, p. 108; Buckingham and Hill, 1986, pp. 152–153) were based on correlations with optically active compounds possessing chiral centers of known configuration (cf. Eliel, 1962). Such correlations (e.g., Newman, Mislow, et al., 1958; Mislow and McGinn, 1958; Berson and Greenbaum, 1958a) involved mechanistic arguments including the application of Prelog's rule (Section 5-5.g; Mislow, Prelog, and Scherrer, 1958). The following example is illustrative: Meerwein–Ponndorf reduction of racemic ketone **A** (Fig. 14.44) with (S)-(+)-pinacolyl alcohol or (S)-(+)-octanol (**B**) was interrupted short of completion (kinetic resolution, cf. Section 7-6). The two faces of the cyclic ketone are homotopic (by virtue of the existence of a C_2 axis in **A**), but the faces in **A** and its enantiomer are (externally) enantiotopic, and therefore their respective interactions with (S)-(+)-**B** give rise to diastereomeric transition

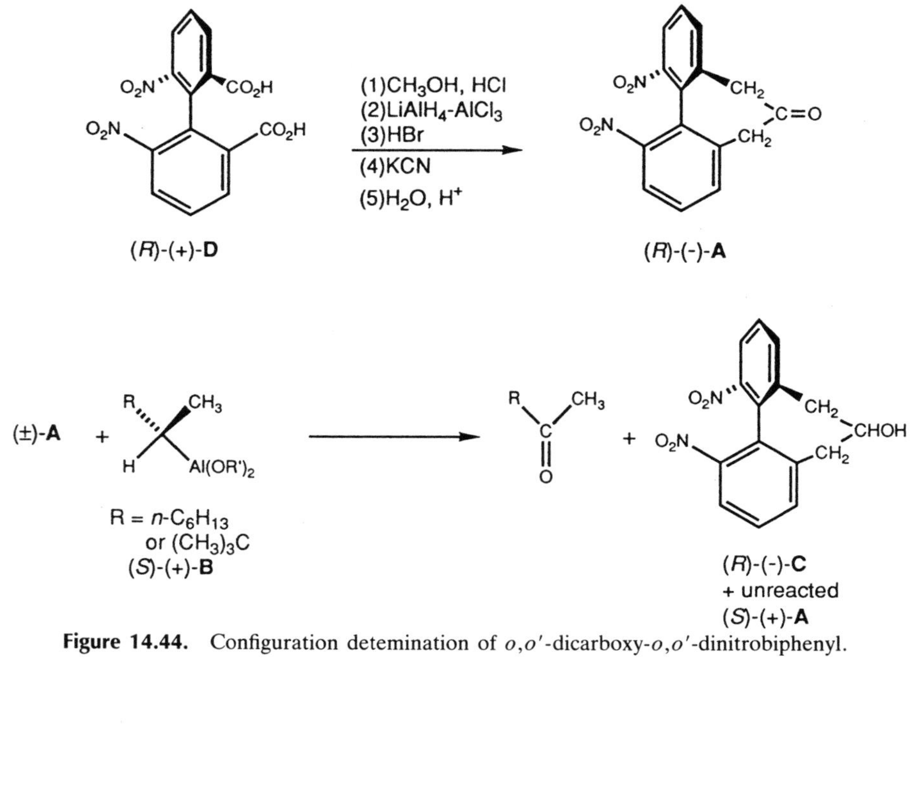

Figure 14.44. Configuration detemination of o,o'-dicarboxy-o,o'-dinitrobiphenyl.

states. The more favorable interaction (from model considerations) is with (R)-**A**, which is therefore reduced [to (R)-**C**] more rapidly than is (S)-**A** to (S)-**C**. When the reaction is interrupted short of completion, the predominant product must therefore be (R)-**C** and the predominant ketone left-over is (S)-**A**. Since the product alcohol is levorotatory and the left-over ketone is dextrorotatory, it follows (provided the model is correct) that $(-)$-**C** is R and $(+)$-**A** is S; consistency was checked by reduction of $(+)$-**A** to $(+)$-**C** by a symmetric reducing agent, $\text{Al}(\text{O}i\text{-Pr})_3$. Since (S)-$(+)$-**A** is obtained by cyclization of $(-)$-6,6'-dinitro-2,2'-diphenic acid **D**, $(-)$-**D** is also shown to be S. A number of other biphenyls were configurationally connected with $(+)$- or $(-)$-**D** by correlative methods (Mislow, 1958).

Unfortunately, application of Prelog's rule to 2-hydroxy-1,1'-binaphthyl (Berson and Greenbaum, 1958b) was later found to give the wrong result (Kabuto, Yamaguchi, et al., 1983).

Meanwhile, these biphenyl configurations have been confirmed by the Bijvoet X-ray method (Section 5-3.a) of a cobalt complex of 2,2'-diamino-6,6'-dimethyl-biphenyl (Pignolet, Horrocks, et al., 1968). The $(+)$ ligand has the R configuration; this is in accord with the above-described configurational assignment of (R)-$(+)$-**D** (Fig. 14.44), which can readily be converted chemically into the dextrorotatory 2,2'-diamino-6,6'-dimethyl analogue (Mislow, 1958; cf. Eliel, 1962, p. 172). The Bijvoet method has also been applied (Akimoto, Yamada, et al., 1968) to $(+)$-2,2'-dihydroxy-1,1'-binaphthyl-3,3'-dicarboxylic acid; the crystal contained a molecule of bromobenzene of solvation which provided the desired heavy atom. The $(+)$ isomer is R and it was chemically correlated with a number of other binaphthyls, including (S)-$(+)$-1,1'-binaphthyl itself and the (S)-$(-)$-1,1'-binaphthyl-2,2'-dicarboxylic acid whose absolute configuration had been earlier determined by Mislow and McGinn (1958). Yamada and Akimoto (1968) also worked out a chemical correlation between the 1,1'-binaphthyl and biphenyl series; all these correlations are internally consistent. Meyers and Wettlaufer (1984) correlated, by oxidation, a centrally chiral 3-substituted 4-(1-naphthyl)-1,4-dihydroquinoline with an axially chiral 3-substituted 4-(1-naphthyl)quinoline; this correlation, like the earlier correlation of $(-)$-thebaine with $(+)$-phenyldihydrothebaine and its congeners (Berson and Greenbaum, 1958a; cf. Eliel, 1962, pp. 170–171), connects axially chiral with centrally chiral compounds by direct interconversion of the latter into the former.

Internal correlation with (R)-$(-)$-6,6'-dimethyl-2,2'-diphenic acid has been used to determine the configuration of the α-phenethylamides of biphenyl-2,2',6,6'-tetracarboxylic acid (Fig. 14.45; Helmchen, Prelog et al., 1973). This compound can exist as diastereomers with interesting symmetry properties, noted at the bottom of Figure 14.45 (see also p. 378).

Atropisomerism about an sp^2–sp^2 single bond should, in principle, be possible in appropriately substituted butadienes (Fig. 14.46, **A**) provided the substituents R and R' are large enough. In fact, fulgenic acid (Fig. 14.46, **B**) was resolved as early as 1957 by Goldschmidt et al. (q.v.); however, it racemized completely in 20 min at room temperature. More recently, barriers to rotation in a variety of hindered butadienes have been measured by NMR spectroscopy (e.g., Köbrich et

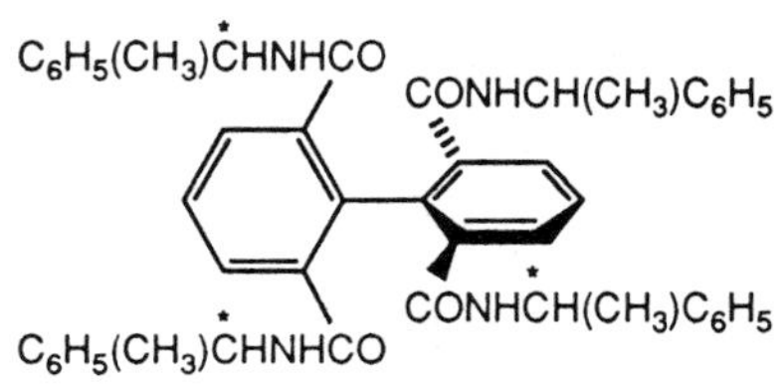

C₆H₅(CH₃)ĊHNHCO

CONHCH(CH₃)C₆H₅

C₆H₅(CH₃)ĊHNHCO

CONHCH(CH₃)C₆H₅

*: RR RR (**D₂**)
*: RR RS (**C₁**)
*: RR SS (**S₄**)
*: RS RS (**C₂**)

Figure 14.45. Symmetry of tetra-α-phenethylamides of biphenyl-2,2′,6,6′-tetracarboxylic acid.

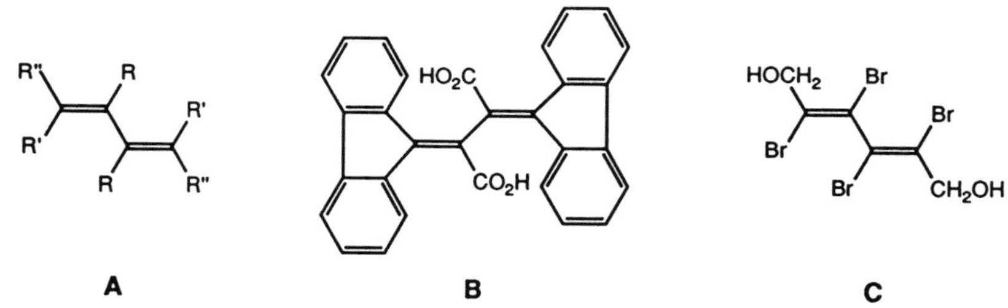

A   B   C

Figure 14.46. Resolvable 1,3-butadienes.

al., 1972; Mannschreck et al., 1974; Becher and Mannschreck, 1983). Compound **C** (Fig. 14.46) has been resolved both by classical methods (Rosner and Köbrich, 1974) and by chromatography on the optically active stationary phase, triacetylcellulose (Becher and Mannschreck, 1981). The racemization barrier in this diene is only 23.7 kcal mol^{-1} (99.2 kJ mol^{-1}) and so it racemizes rapidly in solution on standing.

c. Atropisomerism about sp^2–sp^3 Single Bonds

Whereas atropisomerism about sp^2–sp^2 single bonds involves either a twofold or a fourfold barrier (Fig. 14.42; a substantial energy maximum at 90° will lead to a fourfold barrier; with no or a negligible energy maximum at 90° the sp^2–sp^2 barrier is twofold), atropisomerism about an sp^2–sp^3 single bond is either threefold or sixfold, as implied in Figure 14.47. If X is small the situation will be similar to that in propene (Section 10-2.a), that is, conformer **A** (and corresponding conformers with X eclipsing N or O) will correspond to an energy minimum (since double bonds tend to be eclipsed with single bonds), and conformer **B** and

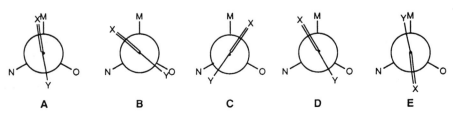

Figure 14.47. sp^2–sp^3 barrier. Only **A**, **B** and **E** are eclipsed conformations.

similar conformers in which Y eclipses M or N (such as **E**) will correspond to energy maxima; thus the barrier is threefold. But if X is large and the barrier is dominated by steric interactions, conformers **C** and **D** and corresponding conformers in which X flanks N or O on either side will correspond to energy minima and **A**, **B** and **E** (plus three other eclipsed conformations) in which X or Y eclipses M, N, or O will correspond to energy maxima and the barrier will be sixfold.

It was stated in Section 10-2.b that sixfold barriers are normally very low, of the order of a few calories per mole. However, this may not be true when the eclipsed conformation involves serious steric congestion.

In practice, the situation tends to be less complex, because if X=C—Y is part of an aromatic system, the C=X and C—Y bond orders are equal (or nearly equal) and conformers **A** and **E** have nearly identical torsional interactions. Assuming that X and Y are of moderate size and M is appreciably smaller than N and O, only conformers **A** and **E** need be considered, plus the lower of the two barriers between them, and the situation then formally resembles that in biphenyls (Fig. 14.42) with two minima and two maxima. However, since conformers **A** and **E** are diastereomeric (not enantiomeric) sp^2-sp^3 atropisomerism involves separation and equilibration of diastereomers, not resolution and racemization of enantiomers.

For nomenclature purposes, the Prelog–Klyne conformational designation shown in Table 2.2 is used. If M and X are the fiducial (determining) groups, **A** in Figure 14.47 is synperiplanar (*sp*) and **E** antiperiplanar (*ap*). Ligand M is always fiducial, by definition, when N = O. But when N ≠ O and N or O happen to be fiducial, **A** is anticlinal (*ac*) and **E** synclinal (*sc*) with X still being fiducial; cf. p. 1120.

One might think that increasing the size of M, as well as N, O, X, or Y in Figure 14.47 would increase the activation energy for rotation, but this is not necessarily so. For example the **A** ⇌ **E** interconversion in Figure 14.47 involves passage of X by N and Y by O (or vice versa) but does not require either X or Y to pass by M. Thus if M is made larger (but not so large as to make **A** or **E** energy maxima) the main effect may be to raise the ground-state energy of **A** and/or **E**, and thereby to *lower* the activation energy, assuming that the transition state energy level is largely unaffected (cf. Fig. 11.48). This does, in fact, happen.

The first instance of an sp^2-sp^3 barrier was reported by Chandross and Sheley (1968) who found nonequivalent ortho methyl groups in 9-mesitylfluorene (Fig. 14.48, **A**) at all temperatures studied. A much lower barrier was found in the 9-chloro compound **B**.

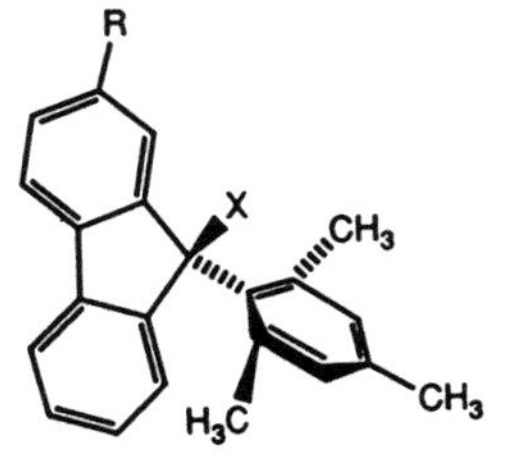

A, X = H, R = H
B, X = Cl, R = H
C, X = OH, R = H
D, X = Cl, R = CH(CH₃)₂

Figure 14.48. 9-Mesitylfluorenes.

TABLE 14.2.   Barriers to Rotation about the Aryl-to-Fluorenyl Bond in 9-Arylfluorenes.

		ΔG^0			
Compound	9-Substituent	Mesityl		2,6-Dimethoxyphenyl	
		(kcal mol^{-1})	$(\text{kJ mol}^{-1})^a$	(kcal mol^{-1})	$(\text{kJ mol}^{-1})^a$
A	H	$>25^b$	$>104^b$ (>190)	20.6	86.2 (145)
C	OH	20.2	84.5 (145)	14.4	60.2 (24)
B	Cl	16.2	67.8 (66)	9.2	38.5 (-81)

[a]Numbers given in parentheses are coalescence temperatures (°C) at which the free energy of activation for rotation was obtained. From Rieker and Kessler (1969).
[b]The actual value is probably about 27 kcal mol^{-1} (113 kJ mol^{-1}), see text.

In the original work it was postulated that the lower barrier for the 9-chloro compound **B** was due to a facile dissociative process yielding a 9-mesitylfluorenyl carbocation and chloride anion. However, this was ingeniously and conclusively disproved by study of the 2-isopropyl homologue **D** in Figure 14.48 (Ford et al., 1975). Topomerization of this compound on heating leads to coalescence of the ortho methyl groups of the mesityl substituent but *not* of the α-methyl groups of the isopropyl substituent. If topomerization involved a (planar) carbocation, it would introduce a symmetry plane in the activated complex that would make the isopropyl methyl groups enantiotopic, and hence isochronous, so they would coalesce. Rotation of the mesityl ring, on the other hand, topomerizes the ortho methyl groups *without* introducing a symmetry plane, and hence the isopropyl methyl groups do not become isochronous (cf. Chapter 8).

Later studies by Rieker and Kessler (1969) indicated the barriers shown in Table 14.2 (mesityl column). The magnitude of the barriers is $\mathbf{A} > \mathbf{C} > \mathbf{B}$ (Fig. 14.48), in other words, as predicted above (p. 1151) the smallest 9-substituent leads to the highest barrier, and vice versa. Replacing the mesityl group by 2,6-dimethoxyphenyl, on the other hand, lowered the barrier considerably (Table 14.2, 2,6-dimethoxyphenyl column); in other words, diminishing the size of X and Y (Fig. 14.47) lowers the barrier.

High barriers can be observed in compounds of the type shown in Figure 14.49, where atropisomers can actually be isolated and their interconversion studied by classical kinetic methods. The barrier in **A** (Nakamura and Ōki, 1974), 27.1 kcal mol^{-1} (113.4 kJ mol^{-1}) is probably close to that of **A** in Figure 14.48. The barrier in **B** (Ford et al., 1975), 33.3 kcal mol^{-1} (139.3 kJ mol^{-1}), is substantially higher than that in **C** (Siddall and Stewart, 1969), 29.9 kcal mol^{-1} (125.1 kJ mol^{-1}), showing that an increase in size in N (or O), Figure 14.47, does

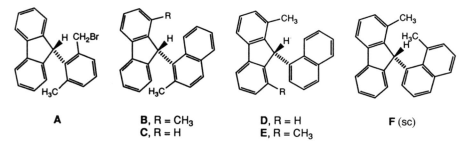

A	**B,** R = CH$_3$	**D,** R = H	**F** (sc)
	C, R = H	**E,** R = CH$_3$	

Figure 14.49.   Structures of 9-Arylfluorenes.

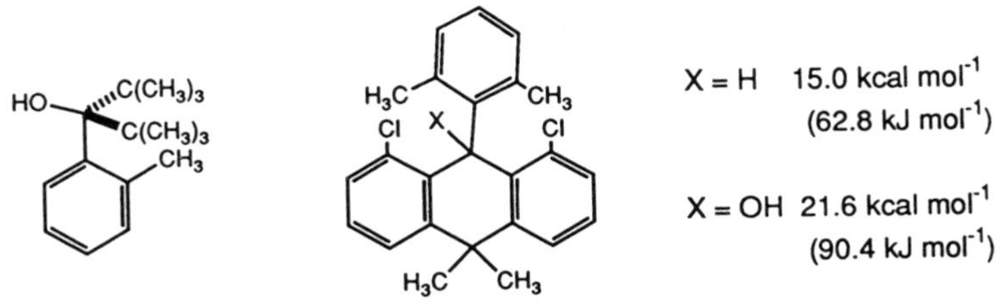

X = H   15.0 kcal mol^{-1}

(62.8 kJ mol^{-1})

X = OH   21.6 kcal mol^{-1}

(90.4 kJ mol^{-1})

A   **B**

Figure 14.50. Other examples of sp^2–sp^3 rotational barriers. In **A**, $\Delta G^0 = 25.6$ kcal mol^{-1} (107 kJ mol^{-1}).

raise the barrier, as predicted. Increasing both N and O in size, on the other hand, does not have a cooperative effect, since the passage of X by N and Y by O (or vice versa) is not synchronous. Indeed, the activation energies in ap–sp interconversion in the monomethyl [Fig. 14.49, **D**, 21.4 kcal mol^{-1} (89.5 kJ mol^{-1}), Kajigaeshi et al., 1979] and dimethyl [Fig. 14.49, **E**; 20.6 kcal mol^{-1} (86.2 kJ mol^{-1}), Mori and Ōki, unpublished results] compounds are nearly the same.

> It must be noted that, when the ap and sp (or sc and ac) isomers are diastereoisomers (e.g., Fig. 14.49, **A**) rather than topomers (e.g., Fig. 14.48), their ground-state energies, and therefore also the activation energies for their interconversion, are unequal. For example (Mori and Ōki, 1981), in Figure 14.49, **F**, the sc isomer (shown) is more stable than the ac isomer by 1.96 kcal mol^{-1} (8.20 kJ mol^{-1}), $K_{ac \rightleftharpoons sc} = 33$ at 30°C; therefore the activation energy (E_a) $sc \rightarrow ac$ being 25.2 kcal mol^{-1} (105 kJ mol^{-1}), that for the $ac \rightarrow sc$ conversion is 23.2 (25.2–1.96) kcal mol^{-1} (97 kJ mol^{-1}).

Other sp^2–sp^3 atropisomers have been discussed by Ōki (1983) or will be presented in the section on molecular propellers (Section 14-6.a). Of particular interest are the isolation of atropisomers of type **A** in Figure 14.50 (Lomas and Dubois, 1976) and the much lower rotational barriers in 9-aryl-9,10-dihydroanthracenes (Fig. 14.50, **B**; Nakamura and Ōki, 1980) as compared to the earlier discussed 9-arylfluorenes. This difference is caused by the nonplanarity of the boat-shaped central ring in **B** (cf. Fig. 11.115), which diminishes the steric interaction between the ortho substituents of the benzene moiety and 1,8-substituents on the dihydroanthracene ring.

d. Atropisomerism about sp^3–sp^3 Bonds

Barriers to rotation in ethanes have been discussed in Chapter 10. In principle, if these barriers are made high enough, and if the structures are appropriately desymmetrized, atropisomers should be isolable. The framework within which atropisomerism of this type has been most successfully demonstrated contains the triptycene (tribenzobicyclo[2.2.2]octatriene, Fig. 14.51) or dibenzobicyclo-[2.2.2]octatriene (Fig. 14.52) structures. Much of this work has been pioneered by Ōki and co-workers and has been reviewed by him (Ōki, 1983, 1993).

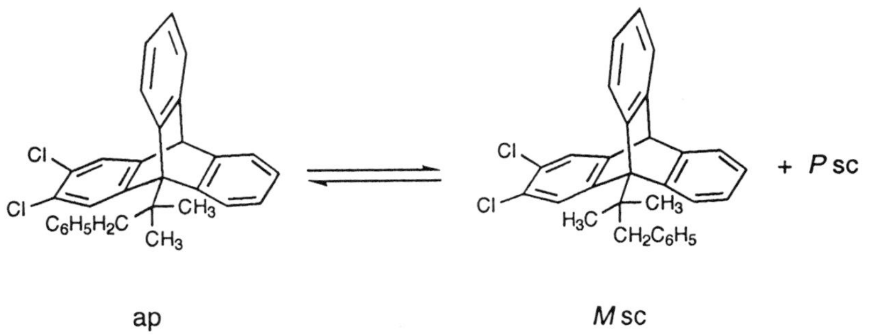

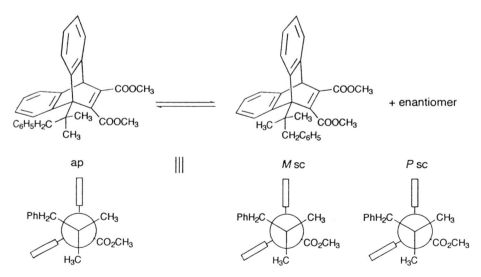

Figure 14.51. Atropisomerism in the 3,4-dichlorotriptycene system.

Thus Yamamoto and Ōki (1972) isolated the *ap* and *sc* diastereomers of the dibenzobicyclo[2.2.2]octadiene structure shown in Figure 14.52. The simplified Newman projections show that the *ap* isomer is a meso form whereas the *sc* exists as *M* and *P* enantiomers; the *sc* isomer was, in fact, resolved (Yamamoto, Ōki, et al., 1975). The *ap* isomer was the initial product of the Diels–Alder synthesis (from a substituted anthracene and dimethyl acetylenedicarboxylate) and was equilibrated thermally to a 3:1 *sc:ap* mixture; the *sc* isomer is evidently slightly favored above the 2:1 statistical ratio. The barrier to isomerization is $E_a = 33.2 \, \text{kcal mol}^{-1}$ ($139 \, \text{kJ mol}^{-1}$).

The triptycene systems shown in Figure 14.51 were prepared by in situ addition of dichlorobenzyne to an appropriately substituted anthracene (*ap*) and benzyne to an appropriately substituted dichloroanthracene (*sp*) respectively (Yamamoto and Ōki, 1975). Equilibration to a statistical (1:2) mixture proceeded on heating with an activation energy (E_a) of $36.6 \, \text{kcal mol}^{-1}$ ($153 \, \text{kJ mol}^{-1}$). A number of similar systems, with different substituents on the aromatic rings and different functional groups in the aliphatic Cabc portion have been examined (Ōki, 1983).

Figure 14.52. Atropisomers in the dibenzobicyclo[2.2.2]octatriene system.

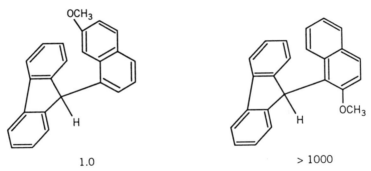

1.0   > 1000

Figure 14.53. Relative rates of lithiation of *ap-* and *sc-*9(2-methoxy-1-naphthyl)fluorene. [Reprinted with permission from M. Ōki (1983), "Recent Advances in Atropisomerism Topics in Stereochemistry," Vol. 14, Wiley, New York, p. 72. Copyright © John Wiley and Sons, Inc.).

Just as interesting as the stereoselective synthesis of one or other of two atropisomers (diastereomers) of the 9-alkyltrypticene type is their differential reactivity. The clearest examples here involve the sp^3–sp^2 type; thus (cf. Fig. 14.53) in replacement of the acidic H(9) in the fluorene moiety by lithium (with butyllithium) the *sp* isomer reacts over 1000 times as fast as the *ap*, presumably because the methoxy group is available for chelation in the former but not in the latter (Nakamura and Ōki, 1975). The anion appears to have the expected *sp* configuration, since on treatment with water it gives the *sp* starting material.

As for simpler compounds, atropisomerism has been demonstrated (Brownstein, Ingold, et al., 1977) in 1,1,2,2-tetra-*tert*-butylethane (Fig. 14.54, **A**) inasmuch as this compound shows two different sets of signals for the *tert*-butyl protons; the stable conformation is a distorted gauche form (cf. Fig. 10.7), as supported by force field calculations (Hounshell, Mislow, et al., 1978; Fitzwater and Bartell, 1976). In the related 1,2-di-*tert*-butyl-1,2-di(1-adamantyl)ethane (Fig. 14.54, **B** and **C**), two atropisomers of the chiral species (as established by crystal structure determination) have actually been isolated (Flamm-Ter Meer, Rüchardt, et al., 1986).

The strain energy of the tetra-*tert*-butyl compound (Flamm-Ter Meer, Rüchardt, et al., 1984) of 66.3 kcal mol^{-1} (277 kJ mol^{-1}) is the largest known in a noncyclic hydrocarbon.

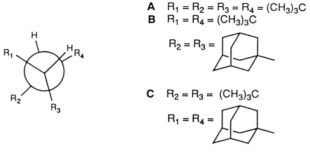

A $R_1 = R_2 = R_3 = R_4 = (CH_3)_3C$
B $R_1 = R_4 = (CH_3)_3C$

$R_2 = R_3 = $

C $R_2 = R_3 = (CH_3)_3C$

$R_1 = R_4 = $

Figure 14.54. Ethanoid atropisomers.

14-6. MOLECULAR PROPELLERS AND GEARS

a. Molecular Propellers

This section deals with a particular kind of atropisomerism involving so-called "molecular propellers" (Mislow et al., 1974; Mislow, 1976, 1989) because of their analogy with the (chiral) propellers (two-, three-, or more bladed) of airplanes or boats. Molecules of this type consist of two or more subunits (the blades) radiating from a central axis of rotation (propeller axis), which may be a single atom or a combination of atoms (e.g., a C_2-ethanoid or C_2-ethenoid or C_2-benzenoid unit). Each blade must be twisted in the same sense. If the blades are identical in structure, this may lead to symmetry as high as D_n, but the term "molecular propeller" is not confined to cases where the subunits (blades) are identical.

> As in an airplane propeller, helicity can be imposed by having planar blades (subunits), which are all tilted in the same sense or by having truly helical blades, all of the same sense of helicity. The latter case has apparently not yet been realized on the molecular scale.

A straightforward, if perhaps not simple example of a three-bladed propeller is the tri-ortho-substituted triarylboron shown in Figure 14.55, **A**. If the three aryl rings are not coplanar, the molecule is chiral. If we assume, for the moment, that the rings are perpendicular to the plane defined by the boron and the three attached carbon atoms of the aryl rings, there are four diastereomeric arrangements, one with all substituents (X, Y and Z) on the same side of the plane and three with one atom (X, Y, or Z) on one side, and the other two on the other side. Since each arrangement exists in two enantiomeric forms, there are four racemic pairs. If we now change the assumption to one where the rings are not orthogonal to the boron plane but tilted, all in the same direction, with respect to it, then the system acquires helicity with either a right-handed or a left-handed pitch and the number of stereoisomers doubles to 8 racemic pairs. Finally, if the central atom is also made chiral, as in the triarylmethane (Fig. 14.55, **B**), the number of stereoisomers is doubled again, to 16 racemic pairs. This number will be reduced if either two or all three of the rings are identical (X = Y or X = Y = Z) or if the rings have local C_2 axes, as in Fig. 14.55, **C**). The number of racemic pairs in these various cases (Mislow et al., 1974) is shown in Table 14.3.

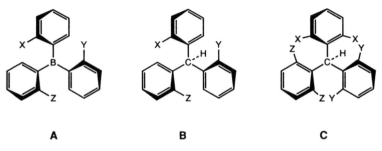

| A | B | C |

Figure 14.55.   Three-bladed propeller molecules.

TABLE 14.3. Number of Racemic Pairs for ArAr'Ar"Z and ArAr'Ar"ZX Systems.

System	Number of Identical Rings	Number of Rings with C_2 Axes			
		0	1	2	3
ArAr'Ar"Z	0	8	4	2	1
	2	4	3	1	1
	3	2	*a*	*a*	1
ArAr'Ar"ZX	0	16	8	4	2
	2	8	4	2	1
	3	4	*a*	*a*	1

*a*The system is achiral and there are no diastereomers.

The interconversion of stereoisomers has been considered in terms of "flip mechanisms" (Kurland, Colter, et al., 1965) by Gust and Mislow (1973), a "flip" being defined as a passage of one or more rings through the plane perpendicular to that of the central atom and its three neighbors ("reference plane"; cf. Fig. 14.55, **A**). In compounds **B** and **C** in Figure 14.55, the "flip" carries a ring through a plane containing the central atom, the attached atom of the ring in question, and the singular atom, which is hydrogen in cases **B** and **C**. One can then discern four kinds of mechanisms, depicted in Figure 14.56, called the zero-ring, one-ring, two-ring, or three-ring flip. (The ring or rings that do not flip pass through the reference plane in the transition state.) Each flip mechanism reverses the helicity and, for molecules of the type shown in Figure 14.55, **A** and **B** (i.e., without local C_2 axes) leads to a different stereoisomer. For each stereoisomer, eight single-step isomerization paths are possible (one zero-ring, one three-ring, three one-ring, and three two-ring flips) and for the 16 isomers of **A** there are $(16 \times 8)/2$ or 64

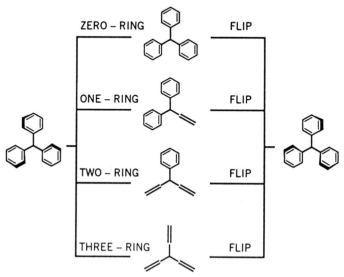

Figure 14.56. Transition states for "flip" mechanisms. [Reprinted with permission from D. Gust and K. Mislow (1973), *J. Am. Chem. Soc.*, **95**, 1535, copyright 1973 American Chemical Society.]

interconversion paths possible (Mislow et al., 1974). The same is true for **B**, since the "flips" do not affect the chiral center, but in **B** only diastereomer interconversion is possible by flipping, whereas in **A** flipping will eventually interconvert enantiomers.

We shall mention here only two examples from several investigations of molecular propellers of the type shown in Figure 14.55 pioneered by Mislow and co-workers. One concerns the tris-1-(2-methylnaphthyl)borane system shown in Figure 14.57, **A** (Blount, Mislow, et al., 1973). According to Table 14.3, this system (Ar$_3$B) should exist as two racemic pairs, one with C_3 symmetry and one with C_1. Low-temperature ^{1}H NMR spectroscopy in fact discloses two diastereomers; but at 85°C the spectrum coalesces to that of a single species. Spectral study of the thermodynamics and kinetics of the system shows that the symmetrical isomer (all methyl groups on the same side) is of lower enthalpy ($\Delta H^0 = 0.61$ kcal mol^{-1}, 2.55 kJ mol^{-1}) but also, as one might expect from its symmetry number of 3 (cf. Chapter 4), of lower entropy (3.1 cal deg^{-1} mol^{-1}, 13.0 J deg^{-1} mol^{-1}). The barrier to diastereomer interconversion is quite low [15.9 or 16.2 kcal mol^{-1} (66.5 or 67.8 kJ mol^{-1}), depending on the starting isomer]. The C_1 isomer can enantiomerize without diastereoisomerization with a barrier of 14.6 kcal mol^{-1} (61.1 kJ mol^{-1}). A detailed analysis of the four flip mechanisms and the kind of NMR coalescences they would produce (Blount, Mislow, et al., 1973) leads to the conclusion that only the one- and two-ring flip mechanisms are compatible with the data; on steric grounds, the two-ring flip mechanism is more likely.

Molecule **B**, Figure 14.57 (Finocchiaro, Mislow et al., 1974) belongs to the general case **B**, Figure 14.55 and, according to Table 14.3, should have 16 racemic pairs. Indeed, at −40°C this compound shows a multitude of signals in the ^{1}H NMR. As the temperature is raised, however, many of these signals coalesce, and at 87°C only two sets of signals in a nearly 1:1 ratio remain; evidently two diastereoisomers are quite stable and not readily interconverted. Indeed two crystalline isomers (albeit not quite pure) can be isolated from the material; they interconvert (on the laboratory time scale) on heating to 122° (with an activation energy of 30.4–30.6 kcal mol^{-1} (127.2–128.0 kJ mol^{-1}) from one side or the other. Analysis of the flip mechanisms discloses that the two-ring flip will *not* interconvert all isomers but will leave two separate families of "residual stereoisomers" (cf. Section 3-1.b). These two families are interconverted one into the

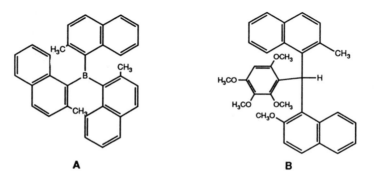

A   **B**

Figure 14.57. Molecular propellers.

other only at much higher temperatures, presumably by a one-ring flip mechanism. [Molecular mechanics calculations show the three-ring and zero-ring flips (Fig. 14.56) to have even higher activation energies.] Evidently the rotation of the rings at 87°C, while quite free according to the two-ring flip mechanism, which interconverts 8 of the 16 racemic pairs, is yet not entirely unrestricted, or else there would be no residual diastereoisomerism. The rotation of the three rings is said to be "correlated" (Finocchiaro, Mislow, et al., 1974). Unlike in the case of biphenyls (cf. Fig. 14.42) the two residual stereoisomers (or stereoisomer sets) can*not* be differentiated by the torsion angles of any *one* of the three rings because in each set members will be found that have the same torsion angles for a given aryl ring. Thus it is the *relationship* between the torsion angles of all three aryl rings, not the individual torsion angles, that defines and differentiates the two residual stereoisomers, hence the term "correlated rotation." [The Prelog–Klyne nomenclature (cf. Table 2.2) cannot be applied in this case.]

A more detailed treatment of the nature of the isomerization processes in molecular propellers is beyond the scope of this book and the reader is referred to the articles and reviews already cited for details (also see Brocas et al., 1983). Nuclear magnetic resonance spectroscopy and molecular mechanics calculations combined with group and graph theory are the main tools used to study stereoisomerism in this class of compounds. Suffice it to mention here that molecular propellers may be represented by chemically quite different structures, such as tri-*o*-thymotide (cf. Fig. 4-7) and a number of inorganic bis- and tris-chelates (e.g., Willem, Gielen, et al., 1985a,b) and that molecular propellers with a variety of axes or hubs have been studied. Examples are hexa- and pentaarylbenzenes (Gust, 1977; Gust and Patton, 1978; Pepermans, Gielen, et al., 1983), tetraarylethanes (Finocchiaro, Mislow, et al., 1976a,b), tri-arylethylenes (enols) (e.g., Biali and Rappoport, 1985), tetramesitylethylene which has been resolved (Gur, Biali, Rappoport et al., 1992) and in which the flip barrier amounts to an amazing $39.6 \text{ kcal mol}^{-1}$ ($165.7 \text{ kJ mol}^{-1}$), and tri-arylamines (Glaser, Mislow, et al., 1980; cf. Hellwinkel et al., 1973, 1975). The molecule involved in the last case (see also Mislow, 1989) is shown in Figure 14.58; it crystallizes as a mixture of two forms that can be separated by hand on the basis of the different morphology of the crystals. The forms are stable in the solid state and are identified as two different diastereoisomers both by X-ray crystallography and by NMR spectroscopy after dissolution in CD_2Cl_2 at $-40°C$. However, on warming in solution to room temperature the two diastereomers equilibrate to a 1:1 mixture with an activation energy of $\Delta G^{\ddagger} = 17.8 \text{ kcal mol}^{-1}$ (74.5 kJ mol^{-1}) at $-21°C$; essentially the same energy barrier is obtained by NMR coalescence experiments at higher temperature. Again the two-ring flip leaves two families of residual diastereomers; the observed energy barrier be-

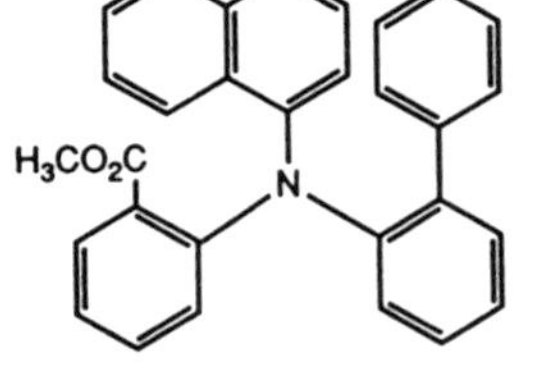

Figure 14.58. A case of diastereoisomerism in a triarylamine.

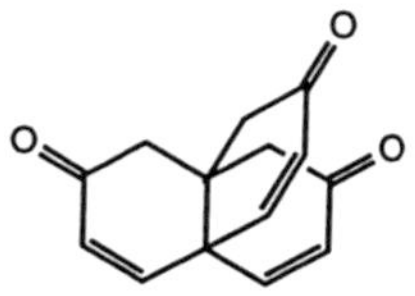

Figure 14.59.   Helical [4.4.4]propellatrienetrione. 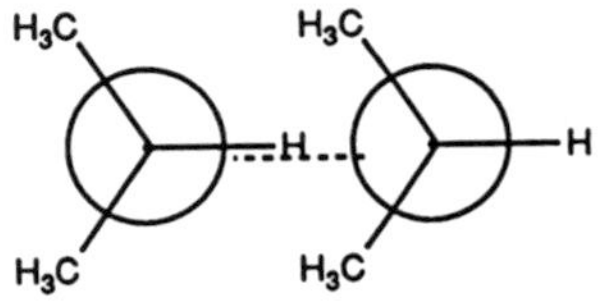   **Figure 14.60.**   Example of static gearing.

tween them corresponds to a one-ring or three-ring flip. At lower temperatures more isomers appear in the NMR spectrum; from the coalescence of their signals the barrier for the two-ring flip may be determined as about $12 \, \text{kcal mol}^{-1}$ $(50 \, \text{kJ mol}^{-1})$.

A borate propeller derived from 2,2′-dihydroxy-1,1′-binaphthyl (Fig. 7.36, **113**) has been found to be a very effective chiral catalyst for Diels–Alder reactions, with exo- and enantio-selectivities of 9:1 or more.

By now it will be quite clear that there is a distinction between propellanes (Section 11-6.c) which, like achiral paddle-wheels, may have planes of symmetry (symmetry point group as high as $\mathbf{D_{nh}}$) and molecular propellers that do not (symmetry point group as high as $\mathbf{D_n}$). Therefore, the synthesis of a propellane that is also a molecular propeller with appreciable conformational stability (Jendralla, Paquette, et al., 1984; Fig. 14.59) is of interest; the barrier to reversal of helicity (inversion of the three rings) is $16.5 \, \text{kcal mol}^{-1}$ $(69.0 \, \text{kJ mol}^{-1})$.

b. Gears

The finding of correlated rotation (preferred two-ring flip) in systems, such as triarylmethanes (Fig. 14.57, **B**) and triarylamines (Fig. 14.58), brings up the question how general such correlated rotation, also called "gearing" or "cog-wheeling," is (Berg, Sandström, et al., 1985). Hounshell, Mislow, et al. (1980) have made a distinction between static and dynamic gear effects. Static gear effects (Fig. 14.60) are seen, for example, in hexaisopropylbenzene (a partly deuterated analogue is shown in Fig. 14.61, **A**; Siegel and Mislow, 1983; Siegel, Mislow, et al., 1986; see also Arnett and Bollinger, 1964) and in tetraiso-propylethylene (Fig. 14.61, **B**; Langler and Tidwell, 1975; Bomse and Morton, 1975; see also Casalone, Simonetta, et al., 1980); in both molecules the isopropyl groups on adjacent carbon atoms are statically geared as shown in Figure 14.60, the symmetry being $\mathbf{C_{6h}}$ in hexaisopropylbenzene and $\mathbf{C_{2h}}$ in tetraisopropyl-ethylene. The geared groups (e.g., isopropyl) cannot readily rotate.

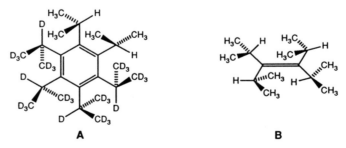

A   **B**

Figure 14.61.   1,2-Diisopropyl-3,4,5,6-(tetraisopropyl-d_7)-benzene **A** and tetraisopropylethylene **B**.

The demonstration of static gearing in tetraisopropylethylene rests on X-ray evidence (Casalone, Simonetta, et al., 1980) and on pairwise anisochrony of corresponding nuclei of isopropyl groups Z on adjacent carbon atoms, with a coalescence barrier of $17 \, \text{kcal mol}^{-1}$ ($71 \, \text{kJ mol}^{-1}$) (Bomse and Morton, 1975).

The C_{6h} symmetry of hexaisopropylbenzene makes the corresponding nuclei of all six isopropyl groups isochronous, but a high barrier to isopropyl rotation [$>22 \, \text{kcal mol}^{-1}$ ($>92 \, \text{kJ mol}^{-1}$)] was ingeniously demonstrated by Siegel and Mislow (1983) through a study of 1,2-diisopropyl-3,4,5,6-(tetraisopropyl-d_7)-benzene (Fig. 14.61, **A**). The deuterated isopropyl groups destroy the C_6 axis and corresponding nuclei on the two adjacent protiated isopropyl groups are now anisochronous by virtue of an isotope effect and cannot be made to coalesce by heating.

A situation somewhat related to static gearing is seen in 1-methylnaphthalene **A** and 9-methylanthracene **B** (Fig. 14.62) and has been probed by measurement of relaxation times in ^{13}C NMR (Levy, 1973). Relaxation in NMR is caused mainly by adjacent magnetic dipoles (magnetic nuclei): so-called "dipole–dipole relaxation." This type of relaxation is most effective when the frequency of tumbling or rotation of the molecule investigated or one of its parts is similar to the Larmor frequency of the nucleus observed. Usually, however, the tumbling or rotation rate is considerably faster than the Larmor frequency and relaxation becomes inefficient, the more so, the faster the rotation. This mismatch results in long relaxation times (T_1). As shown in Figure 14.62, the rotation of the methyl group in **A** is hindered by the peri hydrogen at C(8), which acts like a pin stuck between the cogs of a cogwheel. As a result, relaxation is quite efficient and T_1 is shorter than in **B**. In **B**, the barrier to rotation is lower, since there will always be clashing of the hydrogen atoms of the methyl with one of the peri hydrogen atoms, either at C(8) or at C(1). As a result, while the energy maximum in **B** is no lower than in **A**, the energy minimum is elevated and the barrier is thereby lowered. The lower barrier leads to faster rotation (much faster than the Larmor frequency) and resulting inefficient relaxation, leading to longer T_1.

In contrast to static gearing, dynamic gearing calls for actual correlated rotational motion of two or more parts of a molecule. Whereas examples of static gearing are common in chemistry, bona fide examples of dynamic gearing are rare (cf. Berg, Sandström, et al., 1985). For example, one might think that the isopropyl groups on adjacent olefinic carbon atoms in tetraisopropylethylene (Fig. 14.61, **B**) remain "gear meshed" in the transcourse of the 180° rotation of the two groups that is required for the observed topomerization at elevated temperature. Calculation by molecular mechanics (Ermer, 1983) suggests, however, that this is not the case; rotation actually involves several minima and maxima, the highest of

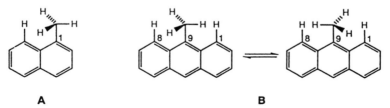

A　　　　　　　　　　　　　　　**B**

Figure 14.62. Methylnaphthalene and 9-methylanthracene.

which corresponds, in calculated energy above the minimum, to the observed activation energy.

Nonetheless, bona fide cases of dynamic gearing (cf. Iwamura and Mislow, 1988; Mislow, 1989) have been found in bis(9-triptycyl)methanes (Guenzi, Mislow et al., 1983; Bürgi, Mislow et al., 1983; Kawada and Iwamura, 1983) and other 9-triptycyl (Tp) compounds of the type Tp_2X ($X = O$, CH_2, CO, CHOH, NH, S, or SiH_2). Only one of the simpler cases of such dynamic gearing will be discussed here. When, in a triptycene, one benzene ring is distinct (by substitution) from the other two, the corresponding bis(9-triptycyl)methane, Tp_2CH_2, will exist as separable meso and racemic species, as shown in Figure 14.63, by virtue of gearing. Correlated disrotatory motion of the triptycene rings is facile enough to make the CH_2 protons (and corresponding carbon atoms in the two triptycene moieties) isochronous in all three isomers even at $-90°C$. However, uncorrelated rotation or correlated conrotation, which would interconvert the diastereomers, evidently does not occur at room temperature, implying that the "notch" between two aryl rings of one triptycene moiety is deep enough to accommodate the aryl ring of the other and not to let it slip out.

The two diastereomers are formed by Diels–Alder addition of 4,5-dimethylbenzyne to bis(9-anthryl)methane and are separated by column chromatography. The fact that the meso isomer has a symmetry plane that is absent in the enantiomers serves to identify the diastereomers: The meso isomer has 12 aromatic ^{13}C NMR signals, whereas the chiral isomers have 18 signals. The energy barrier for gear slippage (i.e., interconversion of the chiral and meso isomers) is $\Delta G^{\ddagger} = 34$ kcal mol^{-1} (142 kJ mol^{-1}); thus temperatures well above 100°C are needed for interconversion. Empirical force field calculations agree with the experimental results in predicting a low barrier to concerted disrotation and a high one for concerted conrotation.

Interestingly, dynamic cogwheeling of the type encountered in Tp_2CH_2, which accounts for the facile topomerization by disrotatory motion coupled with a very high barrier to slippage, is possible only if the number N of "meshed cogwheels" is even. Examples are the ditriptycylmethane shown in Figure 14.63 ($N = 2$) and hexaisopropylbenzene (cf. Fig. 14.61) in which $N = 6$. However, when N is odd, coordinated disrotatory motion of all the cogwheels is mechanically impossible, and rotation then requires gear slippage, with a resulting high-energy barrier. Actual examples (Chance, Mislow, et al., 1990) are seen in tris(9-triptycyl)germanium chloride and tris(9-triptycyl)cyclopropenium perchlorate. X-ray structure analysis shows these molecules to be statically geared but

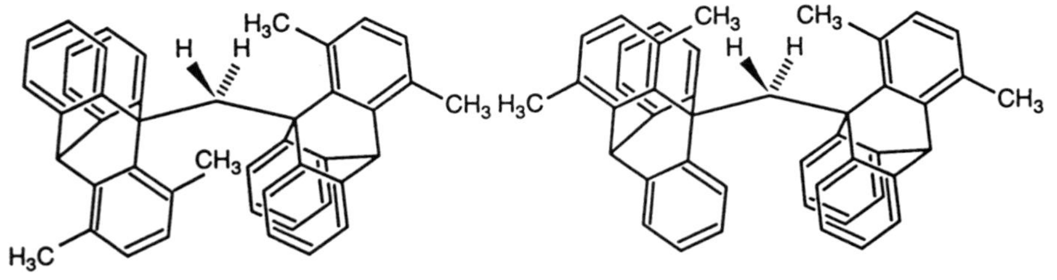

Figure 14.63. Meso and racemic bis(2,3-dimethyl-9-triptycyl)methane.

enantiomerization [following chromatographic resolution on a poly(triphenyl-methyl) methacrylate column (cf. Section 7-3.d and Fig. 6-57)] involves activation barriers of 22.2 kcal mol^{-1} (92.9 kJ mol^{-1}) in the first case and of 23.4 kcal mol^{-1} (97.9 kJ mol^{-1}) in the second; these high barriers are evidence for nonconcerted (gear clashing) cogwheel motion of the triptycyl groups in these compounds.

14-7. HELICENES

In the chiral molecules described so far, helicity was due to some form of restricted rotation about a chiral axis, due to a high bond order, a rigid framework, or steric factors, combined with an appropriate substitution pattern. In this section we consider molecules whose helicity is inherent in the molecular framework (cf. Martin, 1974; Laarhoven and Prinsen, 1984; Meurer and Vögtle, 1985; Oremek et al., 1987).

In 1947, Newman and Hussey (q.v.) succeeded in resolving 4,5,8-trimethyl-phenanthrene-1-acetic acid (Fig. 14.64, **A**) and correctly ascribed its optical activity to nonplanarity enforced by the crowding of the 4,5-methyl substituents. The molecule was of low optical stability and racemized in a matter of minutes, but much more stable molecules of this type (Fig. 14.64, **B**) were synthesized and resolved subsequently (Newman and Wise, 1956). This research culminated in the synthesis of the first optically active helicene, hexahelicene (Fig. 14.64, **C**) (Newman and Lednicer, 1956). Resolution was effected by complexation with α-(2,4,5,7-tetranitro-9-fluorenylideneaminooxy)propionic acid (TAPA, structure Fig. 7.32, **105**; for details, see p. 352). The material has the remarkable rotation $[\alpha]_D^{24}$ −3640 (CHCl$_3$) and begins to racemize only at the melting point of 266°C. Contemplation of models shows that the molecule is helical and that passage through a planar transition state for racemization would meet with extraordinary steric difficulty. That the molecule racemizes at all $[\Delta G_{300}^{\ddagger} = 36.2$ kcal mol^{-1} (151.5 kJ mol^{-1}); $t_{1/2} = 13.4$ min at 221.7°C] must mean that the transition state is in fact not planar but that the two ends of the helix slip across the mean plane of the molecule one after the other (cf. Laarhoven and Prinsen, 1984, p. 93; Oremek et al., 1987). As one passes from hexahelicene to [9]helicene, the activation free energy to racemization increases only modestly to 43.5 kcal mol^{-1} (182 kJ mol^{-1}) (Martin and Marchant, 1974b; see also p. 428).

Higher carbon helicenes, up to [14]helicene (Martin and Baes, 1975), as well as a number of heterohelicenes, especially those containing thiophene units (cf.

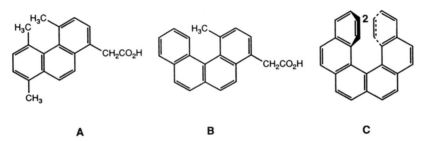

Figure 14.64. Chirality due to "molecular overcrowding."

Wynberg, 1971) up to 15 rings (Yamada et al., 1981) have meanwhile been synthesized, as have double helicenes (Fig. 14.65). Compounds **A** and **B** are [10]helicenes with and without a reversal of helicity at the center and may be considered as two superposed hexahelicenes. Depending on whether the two halves have opposite (P, M) or identical (P, P or M, M) twists, the compound will have a center of symmetry (meso isomer, **A**) or a C_2 axis (active isomer or racemic pair **B**) (Laarhoven and Cuppen, 1973). The compounds were made photochemically from stilbene precursors and separated chromatographically. As might be expected from the analogy with hexahelicenes, the two diastereomers are interconverted on heating above 320°C. Compound **C**, "propellicene" or bi-2,13-pentahelicenylene (Thulin and Wennerström, 1976) results from the fusion of two pentahelicenes (though it is synthesized in a different way) and has **D₂** symmetry. It is both a helicene and a molecular propeller.

The layered nature of the helicene molecule expresses itself in the ^{1}H NMR spectrum. Thus H(2) in hexahelicene (Fig. 14.64, **C**) lies in the shielding region of the aromatic ring at the other end of the helix and its chemical shift ($\delta = 6.65$) is about 1 ppm upfield of the normal phenanthrene resonance; in [13]helicene this proton is shifted to 5.82 ppm. In compound **A**, Fig. 14.65, the H(2) resonates at 6.40 ppm, slightly upfield from the helicene value, but in isomer **B** it is found at 7.12 ppm, perhaps because the two end rings, while shielded by the middle ones, deshield each other (cf. Laarhoven and Prinsen, 1984, pp. 99, 101).

X-ray crystallographic data of hexahelicene (DeRango et al., 1973) and of higher helicenes (cf. Laarhoven and Prinsen, 1984, p. 113) leave no doubt as to the helical nature of these molecules. In hexahelicene, the interplanar angle of the terminal rings is 58.5°. Hexahelicene forms ostensibly chiral crystals and thus appears to be a conglomerate (cf. Section 6.3). However, single crystals of hexahelicene give solutions with only a 2% enantiomeric excess. This disappointing result has been ascribed (Green and Knossow, 1981) to "lamellar twinning" (cf. p. 302) of alternate layers of P- and M-hexahelicenes; the layers are 10–30 μm thick. Nonetheless, once hexahelicene is resolved to the extent of about 20% ee, enantiomerically pure crystals can then be obtained from the solution. Similar problems are not encountered with [7]-, [8]-, and [9]helicene whose

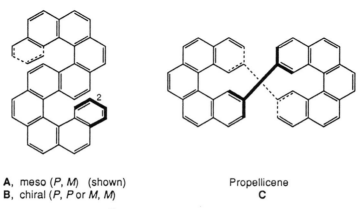

A, meso (P, M) (shown)
B, chiral (P, P or M, M)

Propellicene
C

Figure 14.65. Double helicenes.

conglomerates can be separated by repeated recrystallization after initial seeding with hand-picked crystals (Martin and Marchant, 1974a).

As already mentioned for hexahelicene, the specific rotation of helicenes is remarkably high, reaching a value of $[\alpha]_D^{25}$ 9620 in [13]helicene (Martin and Libert, 1980). As one proceeds to higher helicenes, $[\alpha]_D^{25}$ increases, but with decreasing increments (cf. Laarhoven and Prinsen, 1984, p. 91). Because of their very high rotation, combined with the fact that they are readily synthesized photochemically from stilbene precursors in the presence of oxidants (stilbene → phenanthrene synthesis), helicenes have been prime targets for asymmetric synthesis with circularly polarized light, a technique that had not been noted for success with other types of molecules but did succeed with helicenes (Kagan, Martin, et al., 1971; Moradpour, Kagan, et al., 1971, 1975; Bernstein, Calvin et al., 1972a,b, 1973; see also Buchardt, 1974). Even though optical yields were very small (rarely more than 1%), the high specific rotation of the products left no doubt about the accomplishment of a photochemical asymmetric synthesis. The extent of asymmetric synthesis appears to reach a maximum with octahelicene, drops rapidly with the nona and deca compounds and is nil for [11]–[13]helicenes. Left and right circularly polarized light appropriately yield products of opposite rotation. The possibility that optical activation might be due to preferential asymmetric photochemical destruction (a known process; cf. Kuhn and Knopf, 1930) of one of the helicene enantiomers was excluded by control experiments.

The absolute configuration of (−)-hexahelicene has been determined as M (as shown in Fig. 14.64, **C**) by a Bijvoet X-ray structure determination of the (−)-2-bromo derivative which was then chemically converted to (−)-hexahelicene (Lightner et al., 1972). The absolute X-ray method had earlier been applied to a heterohelicene (Groen, Wynberg et al., 1970) and, in both cases, is in agreement with the best available calculations (hexahelicene: Hug and Wagnière, 1972, based on CD spectra; heterohelicene: Groen and Wynberg 1971) though not with earlier ones, using less refined methodology (Moscowitz, 1961).

A very straightforward chemical correlation of (+)-pentahelicene with (S)-(−)-2,2′-bisbromomethyl-1,1′-binaphthyl of known configuration (Section 14-5.b) (Bestmann and Both, 1972; see also Mazaleyrat and Welvart, 1983) is shown in Figure 14.66; the configuration of (+)-pentahelicene is P. A more complex correlation (Tribout, Martin, et al., 1972; see also Nakazaki et al., 1981) confirms the assignment of the M configuration to levorotatory helicenes by correlation with a cyclophane (cf. Section 14-8.b; for other correlations see Klyne and

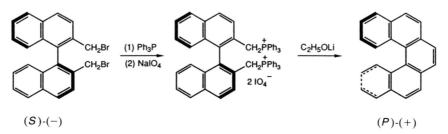

Figure 14.66. Configurational correlation of pentahelicene with binaphthyl.

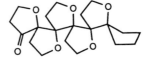

Figure 14.67. Cyclopentylhelixane.

Buckingham, 1978, Vol. 1, p. 221; Vol. 2, p. 110; Buckingham and Hill, 1986, p. 154).

An interesting spirohelicene, named "cyclopentylpentahelixane," shown in Figure 14.67, has been synthesized by the groups of Magnus and Clardy (Gange, Magnus, Clardy, et al., 1980). This molecule is a chiral, helical spirane; it has been called a "primary helical molecule" because the helix is inherent in the structure and not, like in all previous examples, enforced by molecular over-crowding; it has not yet, however, been resolved.

> Helical structures are also of great importance in naturally occurring macro-molecules, for example in the α helices of polypeptides and the double helices of nucleic acids; they also occur in synthetic polymers, such as isotactic polypropylene (cf. Meurer and Vögtle, 1985). A general discussion of such secondary and tertiary structures (conformations) in polymers is beyond the scope of this book; some of their chiroptical manifestations have been discussed in Section 13-4.g.

14-8. MOLECULES WITH PLANAR CHIRALITY

a. Introduction

Among molecules with planar chirality (cf. Schlögl, 1984), examples of which have been shown in Figure 14.4, the cyclophanes are the most important. Other examples are bridged annulenes, *trans*-cyclooctene, and related molecules, which may alternatively be considered to possess axial chirality (cf. Krow, 1970) and metallocenes and other metal complexes of arenes.

b. Cyclophanes

The topic of cyclophanes is very extensive (cf. Vögtle and Neumann 1974; Vögtle and Hohner, 1978; Boekelheide, 1980; Vögtle, 1983a,b; Keehn and Rosenfeld, 1983; Kiggen and Vögtle, 1987; Diederich, 1991; Collet et al., 1993; Vögtle, 1993); only a short presentation of the stereochemical aspects of these molecules can be given here.

The chirality of cyclophanes was discovered by Lüttringhaus and Gralheer

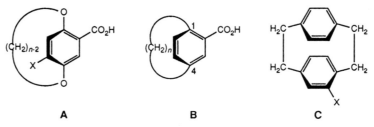

Figure 14.68. Chiral [*n*]cyclophanes.

(1942, 1947) in compounds of type **A** in Fig. 14.68, which, at the time, were called "ansa compounds" (from Latin *ansa* meaning handle) but would now be called 1,*n*-dioxa[*n*]paracyclophanes. The first compound to be resolved was **A**, *n* = 12, X = Br; the two bulky ortho substituents prevent the benzene ring from swiveling through the larger ring formed by the dioxamethylene chain, thus this is a form of atropisomerism. When the bulk of X is reduced by reduction of **A**, X = Br to **A**, *n* = 12, X = H the product becomes nonresolvable. Evidently, rotation of the phenyl ring is now rapid enough for racemization to occur on a time scale faster than that of the experiment. However, when *n* = 10, X = H, compound **A** can be resolved by the classical salt formation with alkaloids (Section 7-3.a) and racemizes only extremely slowly even at 200°C. The compound with *n* = 11, X = H (Lüttringhaus and Eyring, 1957) is intermediate in enantiomer stability; it racemizes at 82.5°C with $t_{1/2} = 30.5$ h $[E_a = 28.4 \, \text{kcal mol}^{-1} \, (119 \, \text{kJ mol}^{-1})]$. The [10]paracyclophanecarboxylic acid (Fig. 14.68, **B**, *n* = 10), in which the bridge is purely carbocyclic, has also been prepared and resolved (Blomquist et al., 1961). The [*n*]paracyclophane with the smallest bridge chemically stable at room temperature is [6]paracyclophane (Kane, Jones, et al., 1974) and its carboxylic acid derivative (Fig. 14.68, **B**, *n* = 6; Tobe et al., 1983). X-ray structure determination of the latter shows the substituted carbon atoms of the benzene ring, C(1) and C(4), to be bent out of the plane of the other four by about 20°; the attached benzylic carbon atoms of the bridge are bent up further by about 20° (relative to the plane of the three nearest aromatic ring carbon atoms) and the bond angles in the bridge are distorted to an average of 126.5°. The even more strained [5]paracyclophane (Jenneskens, Bickelhaupt, et al., 1985) is chemically stable up to 0°C (at which temperature the flipping of its bridge can be studied by dynamic NMR), but decomposes at room temperature.

The configuration of the levorotatory acid **A** in Figure 14.68 (*n* = 11, X = H) has been determined by correlation with a compound of known configuration possessing a chiral center, as shown in Figure 14.69 (Schwartz and Bathija, 1976). The key assumption, which appears eminently reasonable, is that catalytic hydrogenation of the aromatic ring occurs from the side opposite the bridge and that the product **B** is stereochemically stable at the ether linkages. Esterification, beta-elimination, hydrogenation, followed by equilibration (to the cis isomer of the carboxylic acid), and reesterification yields (−)-**C**, which served as a relay. The dextrorotatory enantiomer, (+)-**C**, was prepared in several steps from (+)-*cis*-3-hydroxycyclohexanecarboxylic acid, which is known to be *R* at the carbinol carbon (Klyne and Buckingham, 1978; see also Buckingham and Hill, 1986, p. 154). Hence, (−)-**C** is 3*S* and **B** has the configuration shown in Figure 14.69; it follows that the bridge in (−)-**A** is forward when the CO_2H group is on the right. The configuration of (−)-**A** is therefore *S* or *M*.

The smallest [*m.n*]paracyclophane, [2.2]paracyclophane (Fig. 14.68, **C**, X = H) was first obtained accidentally as a byproduct of the pyrolytic polymerization of xylene to poly-*p*-xylylene (Brown and Farthing, 1949; Farthing, 1953; Brown, 1953). Systematic work on this system is largely due to Cram and co-workers (Cram and Steinberg, 1951; cf. Cram et al., 1974). Since the benzene rings cannot freely turn in this structure, the corresponding acid (Fig. 14.68, **C**, X = CO_2H) is chiral and has been resolved (Cram and Allinger, 1955); the absolute configuration of the levorotatory enantiomer has been determined as (−)-(*R*) or (−)-(*P*) by X-ray diffraction (Bijvoet method; Frank, unpublished results; cf. Tribout et al., 1972).

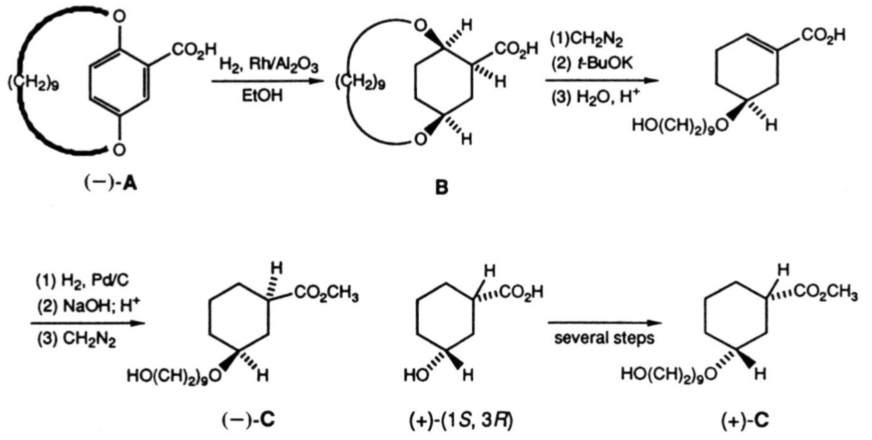

Figure 14.69. Configurational correlation of (−)-1,11-dioxa[11]cyclophanecarboxylic acid.

The configuration of several cyclophanes has also been assigned by kinetic resolution methods (e.g., Horeau's method; cf. Section 5-5.g) and by chiroptical (CD) methods (Chapter 13). These methods are less secure than X-ray methods or chemical correlations (cf. footnote 4 in Schwartz and Bathija, 1976) and will not be discussed here in detail; they have been reviewed by Schlögl (1984). As mentioned earlier (p. 1165) (R)-(−)-[2.2]paracyclophanecarboxylic acid has been elaborated into a helicene (Fig. 14.70), which must be M (the cyclophane moiety blocks formation of the P helix) and is levorotatory; comparison of its ORD curve with that of (−)-hexahelicene suggests that the latter is also M, in accord with the direct X-ray evidence cited earlier (Tribout et al., 1972; Nakazaki et al., 1981).

The carboxylic acid derived from [3.4]paracyclophane has been resolved and racemizes when heated above its melting point (154°C) (Cram et al., 1958), whereas resolution of the homologous [4.4] acid was unsuccessful (Cram and Reeves, 1958). Evidently, as the bridges are made larger, the substituted ring can turn within the cyclophane system.

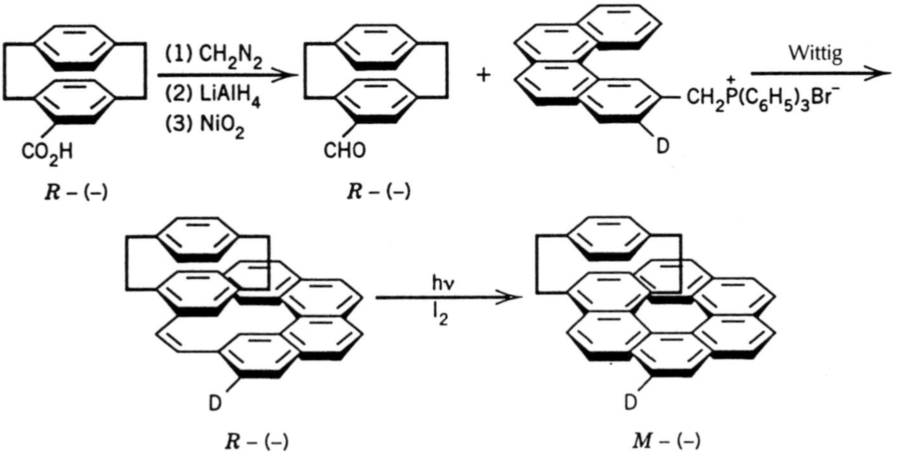

Figure 14.70. Correlation of cyclophane and helicene configurations.

Only brief mention can be made here of other types of cyclophanes (Fig. 14.71): [*m*][*n*]Paracyclophanes **A**; layered [2.2]paracyclophanes **B** also called "[*n*]chochins" (Nakazaki et al., 1977a; Misumi and Otsubo, 1978; Nakazaki, 1984); the highly strained [2.2]paracyclophane-1,9-diene **C** (Dewhirst and Cram, 1958) – whose crystal structure (Coulter and Trueblood, 1963) shows it to be quite distorted, though the benzene rings are less bent than in [6]paracyclophane (see above) – metacyclophanes (e.g., **D**, X = H); metaparacyclophanes (e.g., **E**, X = H) (Semmelhack et al., 1985; Bates et al., 1991); and superphane (**F**) (Sekine, Boekelheide et al., 1979). (The calixarenes, exemplified in Fig. 11.134, are metacyclophanes.)

Compounds **A** in Figure 14.71, where $m = 8$, $n = 8$ or 10, were built up from optically active (+)-[8]paracyclophanecarboxylic acid whose configuration was established to be *S* by a combination of chemical and CD correlations (Nakazaki et al., 1977b). When $m = n = 8$, the compound has $\mathbf{D_2}$ symmetry; the (+) enantiomer is *S* (*M*). The chochin **B** is built up from optically active [2.2]paracyclophane fragments of known configuration; the (−) isomer is *R,R,R,R* or *P,P,P,P*. In the metacyclophanes **D**, the anti, staircase-shaped, or chairlike arrangement of the rings is preferred in the [2.2], [2.3], [2.4], and [3.4] compounds, as inferred from NMR and other evidence (Krois and Lehner, 1982), presumably because van der Waals interaction of the aromatic protons located between the bridges is minimized in this way. However, the syn conformation (Fig. 14.71, **F**) is preferred in [3.3]metacyclophane and substitution on the bridge may change the conformational preference.

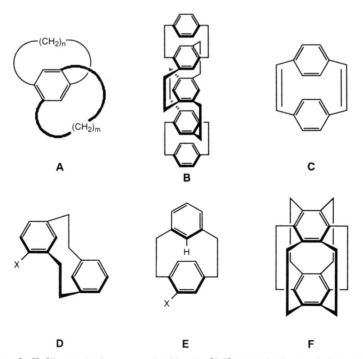

Figure 14.71. [*m*][*n*]Paracyclophanes **A**, chochins **B**, [2.2]paracyclophane-1,9-diene **C**, metacyclophane **D**, metaparacyclophane **E**, and superphane **F**.

The racemization barrier in **D**, $X = CO_2H$ (Fig. 14.71) is $\Delta G^{\ddagger}_{423} =$ 31.5 kcal mol^{-1} (132 kJ mol^{-1}). The racemization, which is insensitive to substituents in the ring, probably corresponds to something akin to a chair inversion (Glotzman, Schlögl, et al., 1974). The barriers in large metacyclophanes are too low to permit resolution of derivatives but have been determined by NMR (cf. Krois and Lehner, 1982): [2.3], 73 kJ mol^{-1} (17.4 kcal mol^{-1}); [3.3], 50 kJ mol^{-1} (11.95 kcal mol^{-1}); [2.4], 60 kJ mol^{-1} (14.3 kcal mol^{-1}); [3.4] < 38 kJ mol^{-1} (<9.1 kcal mol^{-1}).

The metaparacyclophanecarboxylic acid **E** (Fig. 14.71), $X = CO_2H$ has been resolved (Hefelfinger and Cram, 1970) and is configurationally highly stable, indicating that the ring-substituted benzene cannot swivel (rotate) with any ease (see also Vögtle, 1969). On the other hand, various metaparacyclophanes, including the parent compound (Fig. 14.71, **E**, $X = H$) display temperature-dependent ^1H NMR spectra, presumably because of "rocking" of the meta-substituted benzene ring (Vögtle, 1969; Akabori et al., 1969; Hefelfinger and Cram, 1970) with an activation energy of 20–21 kcal mol^{-1} (84–88 kJ mol^{-1}). This motion leads to diastereomerization (in one case, $X = CHO$, equilibration of the diastereomers on the laboratory time scale was observed at $-13.5°C$; Hefelfinger and Cram, 1970) but not to racemization. As one might expect, the rocking barrier is increased when the hydrogen located between the meta bridges (shown in Fig. 14.71, **E**) is replaced by fluorine (Vögtle, 1969).

Since the (+)-acid **E**, $X = CO_2H$ can be converted chemically to the (−)-methyl homologue (**E**, $X = CH_3$) which, in turn, may be obtained by rearrangement of a paracyclophane of known configuration by $HAlCl_4$, the dextrorotatory acid has the R (P) configuration (Delton, Cram, et al., 1971).

In conclusion of this section we mention the synthesis of "superphane" (Fig. 14.71, **F**), a molecule which, like benzene, has $\mathbf{D_{6h}}$ symmetry, by Sekine, Boekelheide, et al., (1979), see also Kriggen and Vögtle (1987).

c. Annulenes

Annulenes are cyclic molecules with alternate single and double bonds (thus benzene may be considered [6]annulene and cyclooctatetraene is [8]annulene). According to Hückel's rule, annulenes are aromatic if they contain $2n + 2$ (as distinct from $2n$, where n is any integer or zero) double bonds, provided also that they are planar or nearly planar. Much of the pioneering work in this field was done by Sondheimer's group (e.g., 1972; see also Garratt, 1979) who obtained the smallest stable simple annulenes, [14]annulene (aromatic) and [16]annulene (antiaromatic) in crystalline form. While this work is outside the scope of this book, we shall briefly discuss the bridged annulenes shown in Figure 14.72, whose chemistry was largely elucidated by Vogel (cf. 1980, 1982), because these compounds may display planar chirality, loosely speaking.

In actual fact the "chiral plane" described by the 10-membered ring (Fig. 14.72, **A**, $X = CH$) is not a plane at all; the molecule is wing shaped with the two halves bent away from the CH_2 group; the deviation from planarity is 35° for $X = C—CO_2H$ (Dobler and Dunitz, 1965). The resonance energy of this "[10]annulene" is 17.2 kcal mol^{-1} (72.0 kJ mol^{-1}), about 40% of that of naphthalene (Roth et al., 1983).

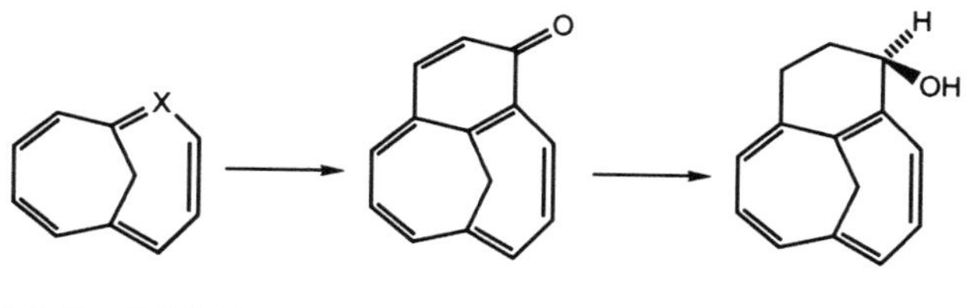

(+)-**A**, X = C CO₂H   **B**   (+)-**C**

Figure 14.72. Bridged [10]annulenes and configurational assignment.

The parent molecule **A**, X = CH, has C_{2v} symmetry but when X = C—CO₂H or N, the molecules become chiral and have been resolved, in the case of the acid by crystallization of the (+)- and (−)-α-methylbenzylamine salts (Kuffner and Schlögl, 1972) and in the case of the amine by medium-pressure chromatography on the chiral stationary phase microcrystalline triacetylcellulose (Schlögl et al., 1983; cf. p. 259) (the amine is insufficiently basic for resolution by means of chiral acids). The configuration of the acid (X = C—CO₂H) was tentatively assigned as (+)-(S) by kinetic resolution of the anhydride with (−)-α-methylbenzylamine. Support for this assignment comes from chemical transformation of **A** (X = C—CO₂H) to ketone **B** which, on hydride reduction (attack from the less hindered exo side) gives mainly the endo or cis alcohol **C** whose relative configuration was confirmed by chromatography and ^{1}H NMR spectroscopy. The absolute configuration at the carbinol carbon was established to be R by Horeau's method (kinetic resolution with (+)-α-phenylbutyric anhydride; cf. Section 5-5.g) and the planar chirality in (+)-**C**, and hence (+)-**A** (X = C—CO₂H) is therefore S (Kuffner and Schlögl, 1972).

In a bridged [10]annulene it appears impossible to effect racemization by forcing the bridge through the "plane" of the conjugated ring system. The situation is different in bridged [14]annulenes (Fig. 14.73), which can exist as cis and trans isomers. In the case of X = C=O, both the cis and trans isomers are quite stable; the cis isomer survives even at 500°C under vacuum flash pyrolysis conditions (cf. Vogel, 1982)! When X = O, however, passage of the bridges through the plane becomes possible. Because of the much greater stability of the (aromatic) cis isomer, the direct thermal cis $\rightleftharpoons$ trans equilibration cannot be observed; in fact, the trans isomer is unknown. However, the process has been demonstrated indirectly in the dibromo derivative shown in Figure 14.73 (Vogel, Schlögl, et al., 1984), which is chiral and can be resolved on microcrystalline triacetylcellulose (cf. Chapter 7). The compound racemizes at 150°C with an activation barrier of 31.8 kcal mol⁻¹ (133 kJ mol⁻¹). Calculation confirms that the racemization involves reversible conversion of the cis isomer to the trans (i.e.,

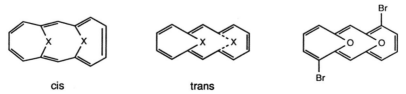

cis   trans

Figure 14.73. Bridged [14]annulenes.

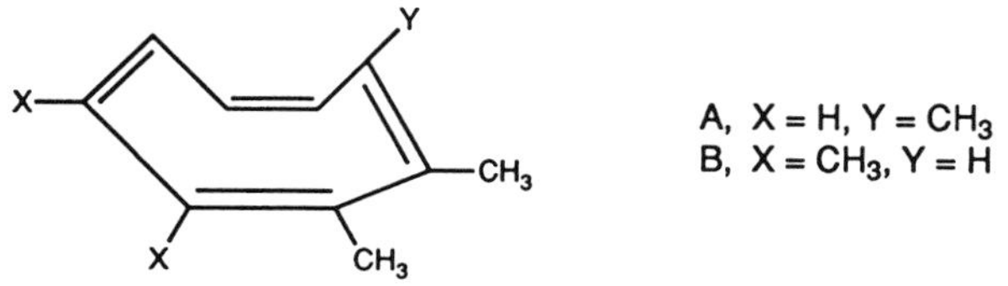

A, X = H, Y = CH₃
B, X = CH₃, Y = H

Figure 14.74. Chiral substituted cyclooctatetraenes.

stepwise passage of the two oxygen bridges through the aromatic plane) rather than an unlikely concerted passage.

The [8]annulene cyclooctatetraene is tub shaped with an inversion barrier of about $13.7 \, \text{kcal mol}^{-1}$ ($57.3 \, \text{kJ mol}^{-1}$) (Anet, 1962). However, the barrier to inversion can be enhanced by appropriate substitution, which also introduces chirality (Mislow and Perlmutter, 1962). Thus the tri- and tetramethyl substituted cyclooctatraenes shown in Figure 14.74 have been obtained optically active by a valence bond isomerization of substituted [4.2.0]bicyclohexadienes (Gardlik, Paquette, et al., 1979a) of established configuration; this transformation leads to products of predictable absolute configuration. The barrier to inversion is about $26 \, \text{kcal mol}^{-1}$ ($109 \, \text{kJ mol}^{-1}$) in the trimethyl compound **A** and $31.8 \, \text{kcal mol}^{-1}$ ($133 \, \text{kJ mol}^{-1}$) in the tetramethyl compound **B** (Gardlik, Paquette, et al., 1979b; see also Paquette, 1993 and p. 429).

d. *trans*-Cycloalkenes

When the saturated segment of a *trans*-cycloalkene (Fig. 14.75) is sufficiently short, it is forced out of the plane of the olefinic moiety, which is normally a symmetry plane of the molecule. The *trans*-cycloalkene then becomes chiral (cf. Schlögl, 1984; Nakazaki et al., 1984); its chirality being of the planar type.

> Since the double bond itself becomes twisted (Ermer, 1974; Traetteberg, 1975) when the bridge is short, the chirality has, alternatively, been considered axial (Moscowitz and Mislow, 1962). This point of view is not generally held, nowadays, however (cf. Schlögl, 1984). The optical activity of (E)-cyclooctene is due in part to the twisting and in part to the dissymmetry of the bridge (Levin and Hoffmann, 1972), so both points of view have theoretical justification.

That (E)-cyclononene is, in principle, chiral was recognized by Blomquist et al., in 1952 (q.v.); however, the two enantiomers of this compound are rapidly interconverted and its was not until 10 years later that the configurationally much more stable (E)-cyclooctene was resolved through conversion to and separation of diastereomeric platinum(IV) complexes, *trans*-$C_8H_{14} \cdot PtCl_2(R^*NH_2)$, where

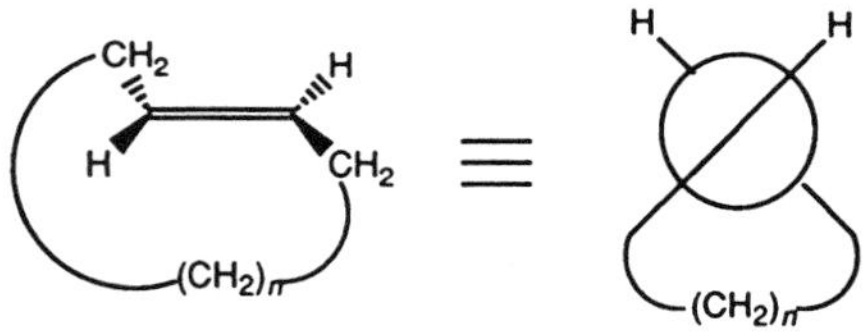

Figure 14.75. *trans*-Cycloalkenes.

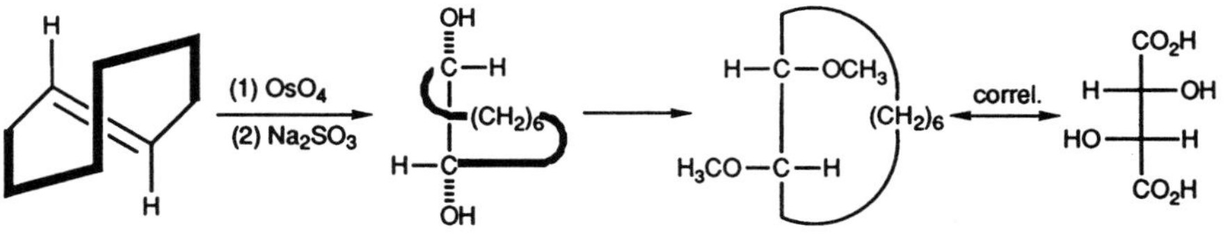

Figure 14.76. Configurational correlation of $(-)$-(E)-cyclooctene with $(+)$-tartaric acid. The osmium tetroxide oxidation takes place from the rear (Si, Si) face, the front face being screened by the hexamethylene bridge.

$R^* = (+)$-$C_6H_5CH_2CH^*(CH_3)$- (Cope et al., 1962a; the asterisk marks an enantiomerically pure ligand; see also p. 351).

The configuration of $(-)$-(E)-cyclooctene was shown (Cope and Mehta, 1964) to be R or P by oxidative (OsO_4) conversion to (S,S)-$(+)$-1,2-cyclooctanediol whose configuration, in turn, was demonstrated by transformation to its dimethyl ether, which was synthesized independently from (R,R)-$(+)$-tartaric acid of known configuration (Fig. 14.76).

The R or P configuration of $(-)$-(E)-cyclooctene was confirmed by X-ray crystallography of the trans-$PtCl_2[(+)$-$H_2NCH(CH_3)C_6H_5]$ complex that correlates the configuration of the levorotatory (E)-alkene with that of (R)-$(+)$-$C_6H_5CH(CH_3)NH_2$ (cf. Section 5-5.a); an absolute determination of the configuration of the same complex by the Bijvoet method (Section 5-3.a) confirmed the R assignment to the $(-)$-alkene (as well as the R configuration of $(+)$-α-methylbenzylamine) (Manor, Shoemaker, et al., 1970; see Nakazaki et al., 1984, p. 3 and Klyne and Buckingham, 1978, Vol. 1, p. 216, for additional correlations). Initial theoretical predictions of the configuration of $(-)$-cyclooctene (Moscowitz and Mislow, 1962) based on the twist of the alkene moiety gave the wrong answer, but later predictions (Levin and Hoffmann, 1972) do agree with experiment.

In larger trans-cycloalkenes, swiveling of the double bond through the polymethylene bridge becomes more facile and the racemization barrier (E_a) drops accordingly: (E)-cyclooctene, 35.6 kcal mol^{-1} (149 kJ mol^{-1}) (Cope and Pawson, 1965); (E)-cyclononene, 20 kcal mol^{-1} (83.7 kJ mol^{-1}), measured at low temperature (Cope et al., 1965); (E)-cyclodecene, 10.7 kcal mol^{-1} (44.8 kJ mol^{-1}), measured by dynamic NMR (Binsch and Roberts, 1965).

Contemplation of a molecular model of (E)-cyclooctene discloses (cf. Binsch and Roberts, 1965) that the mere swiveling of the double bond through the polymethylene bridge, while inverting the configuration of the planar chiral moiety, does not convert the molecule into its enantiomer, rather it interconverts a chair and a crown conformation (Fig. 14.77). A further conformational change is required to change the (distorted) chair to a crown conformer enantiomeric to the original one. X-ray structure determination of an (E)-cyclooctene derivative (allylic 3,5-dinitrobenzoate) (Ermer, 1974), as well as electron diffraction study of the alkene itself (Traetteberg, 1975) indicates the crown conformation to be the stable one. Molecular mechanics calculations (Ermer and Lifson, 1973) support this finding and lead to an energy difference favoring the crown of 3.14 kcal mol^{-1} (13.1 kJ mol^{-1}). Since the barrier to (S)-crown $\rightleftharpoons$ (S)-chair interconversion (step

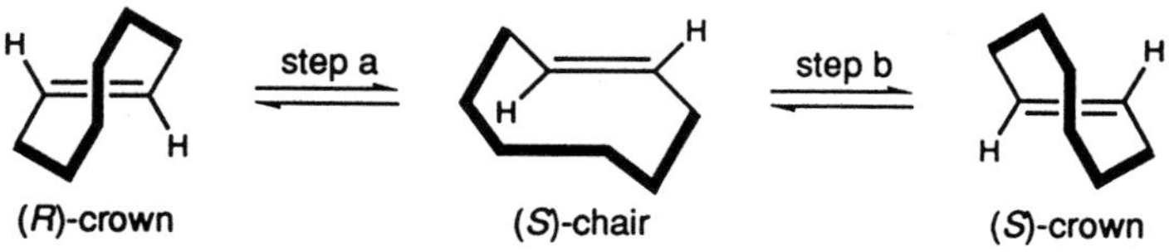

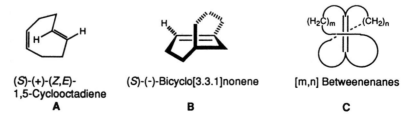

(R)-crown   (S)-chair   (S)-crown

Figure 14.77. Racemization of (E)-cyclooctene. [a]Barrier 35.6 kcal mol^{-1} (149 kJ mol^{-1}). [b]Barrier 10.35 kcal mol^{-1} (43.3 kJ mol^{-1}).

b, Fig. 14.77) is calculated to lie only 10.35 kcal mol^{-1} (43.3 kJ mol^{-1}) above the ground state of the crown (Ermer, 1975), this process cannot be the rate-determining step in the racemization of (R)-(E)-cyclooctene; rather, the 35.6 kcal mol^{-1} (149 kJ mol^{-1}) barrier to racemization (see above) must pertain to the swiveling [(R)-crown $\rightleftharpoons$ (S)-chair] step (Fig. 14.77, step a; cf. Ermer, 1981, pp. 123–165).

(E)-Cycloheptene is surprisingly stable; it can be obtained at $-78°C$ by methyl benzoate sensitized photoisomerization of the cis isomer; the barrier to thermal isomerization to the latter is $\Delta H^{\ddagger} = 17$ kcal mol^{-1} (71 kJ mol^{-1}) and the substance persists for several minutes at $1°C$ (Inoue et al., 1983, 1984; Squillacote et al., 1989; see also Warner, 1989 for a review on *trans*-cycloalkenes).

Replacement of the alkene hydrogen atoms in (E)-cycloalkenes by methyl groups substantially increases the racemization barrier (Marshall et al. 1980), presumably because the swiveling (Fig. 14.77) is now sterically impeded by the larger methyl substituents (as compared to hydrogen) on the double bond. Thus (E)-1,2-dimethylcyclodecene and (E)-1,2-dimethylcycloundecene were obtained optically active and the enantiomers appear to be stable at room temperature. In contrast, attempts to obtain optically active (E)-1,2-dimethylcyclododecene by a kinetic resolution method (treatment with one-half equivalent of monopinanyl-borane) failed, even though the same method was successful with (E)-1,2-dimethylcyclodecene. It was concluded that (E)-1,2-dimethylcyclododecene is not enantiomerically stable at room temperature.

Among other *trans*-cycloalkenes that can be resolved we mention (Z,E)-1,5-cyclooctadiene (Fig. 14.78, **A**), which was first obtained optically active by Cope and co-workers (1962b) by Hofmann elimination of a chiral intermediate (for configuration, by conversion to chiral *trans*-1,2-cyclooctanediol, see Leitich, 1978) and the anti-Bredt compound **B** in Fig. 14.78, $(-)$-bicyclo[3.3.1]-1(2)-nonene (cf. p. 791), which was obtained by Nakazaki et al., (1979) from $(1R, 3S)$-$(-)$-*cis*-3-hydroxycyclohexanecarboxylic acid of known configuration and is thus known to have the S configuration. Compound **B** is seen to be an (E)-cyclooctene with a

(S)-(+)-(Z,E)-   (S)-(-)-Bicyclo[3.3.1]nonene   [m,n] Betweenenanes
1,5-Cyclooctadiene
A   **B**   **C**

Figure 14.78. Miscellaneous chiral *trans*-cycloalkenes.

1,5-methylene bridge, which prevents the swiveling motion (cf. Fig. 14.77) by which unbridged *trans*-cycloalkenes racemize.

An interesting class of cyclic polyunsaturated compounds is represented by the cyclic enediynes containing the —C≡C—CH=CH—C≡C— moiety. Such molecules, which include the natural product calicheamicin, readily cyclize thermally in an exothermic process to reactive benzene 1,4-diradicals in a so-called Bergman cyclization (Jones and Bergman, 1972). The diradicals, in turn, promote DNA cleavage, a fact that has generated interest in their enediyne precursors (Nicolaou and Smith, 1992).

The so-called "betweenenanes" are trans-bridged *trans*-cycloalkenes in which the second bridge impedes swiveling at least when the rings are small; such compounds should thus be resolvable and optically stable. Indeed, Nakazaki et al. (1980) synthesized [8.8]betweenenane (Fig. 14.78, **C**, $m = n = 8$) in optically active form by photochemical asymmetric isomerization of a Z precursor (ketone) in a chiral solvent and later resolved the racemate by chromatography; both procedures gave material of relatively low enantiomeric purity. Marshall and Flynn (1983) later obtained nearly enantiomerically pure [10,10]betweenenane (Fig. 14.78, **C**, $m = n = 10$) by a reaction sequence involving Sharpless oxidation (cf. Section 12-4.c); since the steric course of the Sharpless oxidation is known, the R configuration could be assigned to $(+)$-[10,10]betweenenane.

Finally, we note that the singly and doubly bridged allenes discussed in Section 14-2.d (cf Fig. 14.20) may be considered to display planar as well as axial chirality. However, by convention, the latter takes precedence even though the former may make the greater contribution to the optical rotation (cf. Runge, 1980, p. 129: the molar rotation of (R)-1,2-dipropylallene is -61.1, whereas that of (R)-1,2-cyclononadiene is $+214$).

e. Metallocenes and Related Compounds

Metallocenes (metal "sandwich compounds") and other aryl metal complexes display chirality when the arene is properly substituted (Fig. 14.79). The parent compounds may display average symmetry as high as D_{nh} (e.g., D_{5h} for ferrocene; cf. Fig. 4.20 and discussion on p. 83) since the molecules generally pivot rapidly around the arene–metal axis. Yet an appropriate set of substituents on one of the aromatic rings (1,2,4-trisubstitution pattern for equal substituents and 1,2 for unequal ones) will destroy all symmetry planes and lead to chiral structures (Fig. 14.79; see also Fig. 14.4, **D**). Although, conceptually, this is presumably a case of planar chirality, it is, for purposes of nomenclature, treated in terms of chiral centers (cf. Section 14.1).

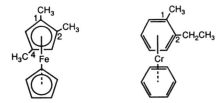

Figure 14.79. Chiral metallocenes.

A detailed disussion of metallocenes is outside the scope of this book and the reader is referred to two reviews by Schlögl (1967, 1970).

14-9. CYCLOSTEREOISOMERISM

The case of the inositols (Fig. 11.32) shows that severe degeneracies result when constitutionally identical stereogenic centers occur in rings. Thus, although inositol has six stereogenic centers, there are only nine stereoisomers: seven meso forms and one pair of enantiomers (i.e., eight diastereomers). A similar degeneracy occurs in cyclic compounds where the stereogenic center is outside the rings, for example, a hexaalkylbenzene with six constitutionally identical chiral substituents (C_6L_6, L being a chiral ligand; cf. Chapter 2), such as hexa-(2-methylbutyl)benzene. The readers may convince themselves that here, also, there are eight diastereomers, one with six L groups, one with five L and one ⅃, three with four L and two ⅃ (the latter can be ortho, meta, or para to each other), and three with three L and three ⅃ (the latter being 1,2,3, 1,2,4, and 1,3,5). However, in this case there are five chiral pairs of enantiomers and only three meso isomers ($C_6L_3⅃_3$).

> The degeneracies themselves are not related to the cyclic structures, although their pattern is. For example, cyclobutane-1,2,3,4-tetrol has, as the reader might work out, four achiral diastereomers with C_{4v}, C_s, C_{2h}, and D_{2d} symmetry. In contrast, $C(CHOHCH_3)_4$, an acyclic compound with four constitutionally identical stereogenic centers, has only three diastereomers: two pairs of enantiomers and one meso isomer (Fig. 3.31). No acyclic carbon analogue to the inositols can be constructed, but an octahedral compound, such as hexa-*sec*-butylaminocobalt CoL_6 would fill the bill: this compound exists as four pairs of enantiomers and two meso isomers. Evidently, the number of stereoisomers is more severely reduced in the acyclic than in the cyclic cases.

In connection with what follows it should be recognized that no ring direction can be defined in the above cases. Thus, while the clockwise sequence in **A** (Fig. 14.80) differs from that of **A′** and corresponds to a counterclockwise sequence in the latter, **A** can be superposed (i.e., is identical with) **A′** by turning over the benzene ring. (The reader should start the cyclic progression at the arrow.)

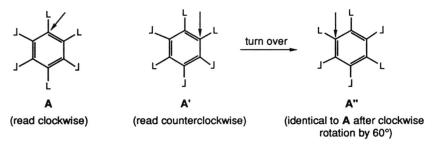

A	**A′**	**A″**
(read clockwise)	(read counterclockwise)	(identical to **A** after clockwise rotation by 60°)

Figure 14.80. Absence of directionality in $C_6L_3⅃_3$.

Similar arguments apply to the inositol case. In general, no directionality exists in $+X^*-a+_{2n}$ (in inositol, case **I**, a is absent) or

$$\left(\begin{array}{c} X^* \\ | \\ \overline{}a-b\overline{} \end{array}\right)_{2n} \quad (C_6L_6, \text{ case } \textbf{II}, \text{ b is absent}).$$

(From now on we confine ourselves to cases where the chiral substituents X^* are evenly balanced between n L and n ⅃ groups.

In 1964, Prelog and Gerlach (q.v.) showed that a new kind of stereoisomerism evolves when the two directions in the ring can be distinguished, as in $+X^*-a-b+_{2n}$ (case **III**) or

$$\left(\begin{array}{c} X^* \\ | \\ \overline{}a-b-c\overline{} \end{array}\right)_{2n} \quad (\text{case } \textbf{IV}),$$

provided there are six or more stereogenic units in the ring. Examples of case **III** (cyclohexaalanyl) and case **IV** (hexa-*N-sec*-butylcyclohexaglycyl) are shown in Figure 14.81. Prelog and Gerlach called this kind of isomerism "cyclostereoisomerism" (see also Cruse, 1966; Bentley, 1969). The two examples in Figure 14.81 suggest that this phenomenon is of particular importance in the stereochemistry of cyclopeptides (Gerlach, Owtschinnikow, and Prelog, 1964; Shemyakin, Ovchinnikov, et al., 1967; Goodman and Chorev, 1979).

Prelog and Gerlach (1964) considered cases **III** and **IV** jointly and they defined three terms, *building blocks*, [e.g., (*R*)- and (*S*)-alanine, represented by filled circles = *R* and open circles = *S* in Fig. 14.82]; *distribution pattern*, referring to the sequence or connectivity of the building blocks in the cycle; and *directionality*, referring to a clockwise or counterclockwise array of the building blocks (monomers) in an otherwise identical distribution pattern.

Figure 14.82 summarizes the situation for $n = 2-5$. When $n = 2$ two ordinary diastereomers exist, both achiral. When $n = 3$ there are three diastereomers. Compounds **1** (C_i) and **4** (S_6) are achiral. However, **2** and its mirror-image **3**,

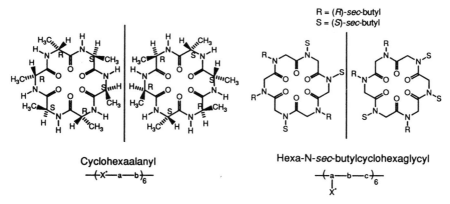

Cyclohexaalanyl

$+x^*-a-b+_6$

R = (*R*)-*sec*-butyl
S = (*S*)-*sec*-butyl

Hexa-N-*sec*-butylcyclohexaglycyl

$+a-b-c+_6$
     |
     x*

Figure 14.81. Cyclostereoisomerism (Prelog and Gerlach, 1964).

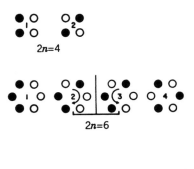

2n=4

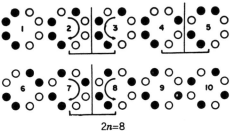

2n=6

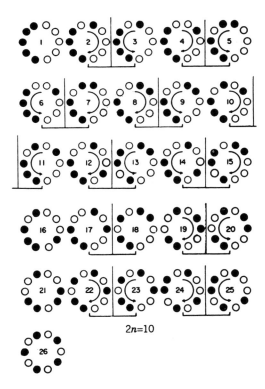

2n=8

2n=10

Figure 14.82. Cyclic distribution patterns for *n* constitutionally identical building blocks, one-half of *R*, one-half of *S* configuration, for *n* = 2–5. [Reprinted with permission from V. Prelog and H. Gerlach (1964) *Helv. Chim. Acta*, **47**, 288, copyright Helv. Chim. Acta.]

while superficially appearing identical, are in fact not, because the directionality in which the sequence of building blocks occurs in the two species is opposite, as indicated by the arrows. Note (cf. Fig. 14.81) that the intercalation of —NH—CO, corresponding to —a—b— in the general formula for case **III**, and of N—CO—CH$_2$, corresponding to a—b—c in the corresponding formula for case **IV** given above, precludes the existence of a vertical symmetry plane. Thus **2** and **3** are enantiomers; Prelog and Gerlach (1964) called them "cycloenantiomers."

For $n = 4$ there are seven diastereomers. Four of them, **1** (C_i), **6** (S_4), **9** (C_i), and **10** (S_8), are achiral. The pair **4/5** is an "ordinary" pair of enantiomers. However, **2** and **3** as well as **7** and **8** have pairwise identical distribution patterns but opposite ring directionality: They are cycloenantiomers.

For $n = 5$ there are 15 diastereomers (Fig. 14.82). Four of these (**1, 16, 21, 26**) have centers of symmetry and are thus achiral. Five pairs (**4/5, 6/7, 12/13, 14/15,** and **17/18**) display ordinary enantiomerism and six others (**2/3, 8/9, 10/11, 19/20, 22/23,** and **24/25**) display cycloenantiomerism in that, in each pair, the partners differ only in the direction in which the building blocks occur in the ring. Careful inspection of the patterns discloses, moreover, that **4** and **6**, **5** and **7**, **12** and **14**, and **13** and **15** have pairwise identical distribution patterns and differ only in directionality. These pairs thus also represent cyclostereoisomers but the partners are now not enantiomeric but diastereomeric. Such pairs are said to represent cyclodiastereomers (Prelog and Gerlach, 1964).

These authors have also given a formula for computing the numbers of stereoisomers (enantiomers being counted as two) for any n for cases **III** and **IV** above; it is equal to the number z of cyclic permutations of $n + n$ identical elements

$$z = \frac{\dbinom{2n}{n} - \dbinom{n}{2} x_m}{2n} + r$$

where r = number of cyclic arrangements with axes of rotation C_m, m = number of equivalent positions, $x_m = 2n/m$ for all C_m present, $x_m = 0$ for all C_m not present. They have also tabulated the results of this computation for n up to 15.

Singh, Mislow et al. (1987) have raised objections to the classification of both cases **III** and **IV** under the common rubric of cyclostereoisomerism. The kernel of their objection is that one should speak of a new element of stereoisomerism (here cyclostereoisomerism) only if the number of stereoisomers is augmented by the existence of such an element. In this respect cases **III** and **IV** are different even though the number and kind of isomers are the same in the two cases. Case **IV**,

$$\left(\begin{array}{c} X^* \\ | \\ \text{a—b—c} \end{array} \right)_{2n}$$

reduces to case **II**

$$\left(\begin{array}{c} X^* \\ | \\ \text{a—b} \end{array} \right)_{2n} , \text{ or } \left(\begin{array}{c} X^* \\ | \\ \text{a} \end{array} \right)_{2n}$$

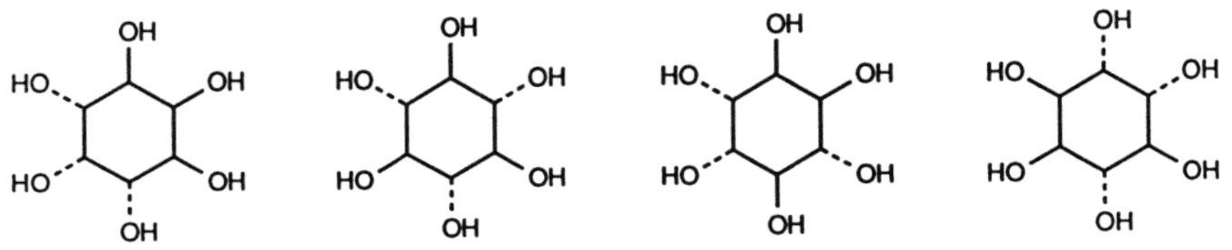

Figure 14.83. Inositol isomers (cf. Fig. 8.32) with 3 OH groups up and 3 down.

when ring directionality is removed. As mentioned earlier, all three $C_6L_3J_3$ isomers are achiral. (Note especially that this is true for the least symmetrical isomer shown in Fig. 14.80.) Thus the introduction of ring directionality does introduce a new element in that it creates chirality in one of the three stereo-isomers. The situation is different when one compares case **III** $+X^*{-}a{-}b\}_{2n}$ with case **I** $+X^*{-}a\}_{2n}$, or $(X^*)_{2n}$, the inositol case. The three inositol isomers in which three OH groups are up (OH) and three are down (OH) are depicted in planar average form in Figure 14.83. Here there are two meso isomers and one enantiomeric pair of chiral isomers, just as in the case of the cyclohexaalanyl depicted in Figure 14.81. In fact, it is clear from Figure 14.83 that the (−)- and (+)-chiro-inositols have the same distribution pattern but differ in the direc-tionality of the arrangement of their CHOH building blocks. Thus one would either have to define these isomers as cycloenantiomers (this has not been proposed) or drop the cyclostereoisomerism label from compounds of type **III**, which includes the cyclopeptide case.

An interesting case of "conformational cycloenantiomerism" has been re-ported by Singh, Mislow, et al. (1987). This case is related to the earlier discussed (p. 1160) situation in hexaisopropylbenzene in which there is static gearing in that all the methine hydrogen atoms point in the same direction. A case similar to that depicted in Figure 14.80, but with superimposed directionality, could thus be created if the isopropyl groups were replaced by suitably configured chiral (e.g., *sec*-butyl) groups. While this proved experimentally unachievable, a related molecule, 1,2-bis(1-bromoethyl)-3,4,5,6-tetraisopropylbenzene (Fig. 14.84, **A**) was synthesized. Here four stereoisomers (*RR, SS, RS* and *SR*) are possible rather than the three (*RR, SS,* and *RS*) that would exist in the absence of gearing that causes directionality. In fact, two pairs of enantiomers (identified by four CHBrCH$_3$ quartets and CHBrCH_3 doublets in the ^{1}H NMR spectrum) were

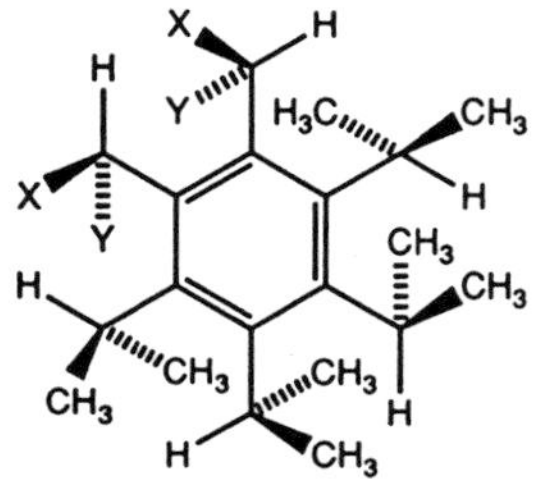

A, X = Br, Y = H
or X = H, Y = Br
B, X = Br, Y = Cl
or X = Cl, Y = Br

Figure 14.84. 1,2-Bis-(1-bromoethyl)-3,4,5,6-tetraisopropylbenzene **A** and 1,2-bis(bromochloro-methyl)-3,4,5,6-tetraisopropylbenzene **B**.

found. One of them was isolated pure; the high temperature NMR study showed that the barrier to interconversion (by reversal of the orientation of the six ring substituents) of the two cycloenantiomers was in excess of $24 \, \text{kcal} \, \text{mol}^{-1}$ ($100 \, \text{kJ} \, \text{mol}^{-1}$), since coalescence of the two diastereotopic CH_3 groups, which become enantiotopic on fast C_{ar}—C_α rotation about the ring-to-isopropyl bond, was not observed up to the decomposition temperature. With so high a barrier the two *RS* enantiomers should, in principle, be resolvable. In a later paper (Biali and Mislow, 1988) referring to the sterically very similar compound 1,2-bis(bromochloromethyl)-3,4,5,6-tetraisopropylbenzene (Fig. 14.84, **B**), the barrier, determined by saturation spin transfer, was found to be $26.8 \, \text{kcal} \, \text{mol}^{-1}$ ($112 \, \text{kJ} \, \text{mol}^{-1}$).

REFERENCES

Adams, R., Anderson, A. W., and Miller, M. W. (1941), *J. Am. Chem. Soc.*, **63**, 1589.

Adams, R. and Kornblum, N. (1941), *J. Am. Chem. Soc.*, **63**, 188.

Adams, R. and Miller, M. W. (1940), *J. Am. Chem. Soc.*, **62**, 53.

Adams, R. and Yuan, H. C. (1933), "The Stereochemistry of Diphenyls and Analogous Compounds," *Chem. Rev.*, **12**, 261.

Agosta, W. C. (1964), *J. Am. Chem. Soc.*, **86**, 2638.

Akabori, S., Hayashi, S., Nawa, M., and Shiomi, S. (1969), *Tetrahedron Lett.*, 3727.

Akimoto, H., Shioiri, T., Iitaka, Y., and Yamada, S. (1968), *Tetrahedron Lett.*, 97.

Alder, K. and Stein, G. (1934), *Justus Liebigs Ann. Chem.* **514**, 1.

Alder, K. and Stein, G. (1937), "Untersuchungen über der Verlauf der Diensynthese," *Angew. Chem.*, **50**, 510.

Almenningen, A., Bastiansen, O., Fernholt, L., Cyvin, B. N., Cyvin, S. J., and Samdal, S. (1985), *J. Mol. Struct.*, **128**, 59.

Anet, F. A. L. (1962), *J. Am. Chem. Soc.*, **84**, 671.

Anet, F. A. L., Jochims, J. C., and Bradley, C. (1970), *J. Am. Chem. Soc.*, **92**, 2557.

Applequist, D. E., Fanta, G. F., and Henrikson, B. W. (1958), *J. Org. Chem.*, **23**, 1715.

Arigoni, D., and Eliel, E. L. (1969), "Chirality due to the Presence of Hydrogen Isotopes at Non-cyclic Positions," *Top. Stereochem.*, **4**, 127.

Arnett, E. M. and Bollinger, J. M. (1964), *J. Am. Chem. Soc.*, **86**, 4729.

Aschan, O. (1902), *Ber. Dtsch. Chem. Ges.*, **35**, 3396.

Balci, M. and Jones, W. M., (1980), *J. Am. Chem. Soc.*, **102**, 7608.

Bastiansen, O. (1950), *Acta Chem. Scand.*, **4**, 926.

Bastiansen, O. and Samdal, S. (1985), *J. Mol. Struct.*, **128**, 115.

Bates, R. B., Gangwar, S., Kane, V. V., Suvannachut, K., and Taylor, S. R. (1991), *J. Org. Chem.*, **56**, 1696.

Becher, G. and Mannschreck, A. (1981), *Chem. Ber.*, **114**, 2365.

Becher, G. and Mannschreck, A. (1983), *Chem. Ber.*, **116**, 264.

Bentley, R. (1969), *Molecular Asymmetry in Biology*, Academic, New York, Vol. 1, pp. 26ff.

Berg, U., Liljefors, T., Roussel, C., and Sandström, J. (1985), "Steric Interplay between Alkyl Groups Bonded to Planar Frameworks," *Acc. Chem. Res.*, **18**, 80.

Bernett, W. A. (1967), "A Unified Theory of Bonding for Cyclopropenes," *J. Chem. Educ.*, **44**, 17.

Bernstein, W. J., Calvin, M., and Buchardt, O. (1972a), *Tetrahedron Lett.*, 2195; (1972b) *J. Am. Chem. Soc.*, **94**, 494; (1973), *J. Am. Chem. Soc.*, **95**, 527.

Berson, J. A. and Greenbaum, M. A. (1958a), *J. Am. Chem. Soc.*, **80**, 445.

Berson, J. A. and Greenbaum, M. A. (1958b), *J. Am. Chem. Soc.*, **80**, 653.

Berson, J. A., Walia, J. S., Remanick, A., Suzuki, S., Reynolds-Warnhoff, P., and Willner, D. (1961), *J. Am. Chem. Soc.*, **83**, 3986.

Bertsch, K. and Jochims, J. C. (1977), *Tetrahedron Lett.*, 4379.

Bestmann, H. J. and Both, W. (1972), *Angew. Chem. Int. Ed. Engl.*, **11**, 296.

Biali, S. E. and Mislow, K. (1988), *J. Org. Chem.*, **53**, 1318.

Biali, S. E. and Rappoport, Z. (1985), *J. Am. Chem. Soc.*, **107**, 1007.

Binsch, G. (1968), "The Study of Intramolecular Rate Processes by Dynamic Nuclear Magnetic Resonance," *Top. Stereochem.*, **3**, 97.

Binsch, G. and Roberts, J. D. (1965), *J. Am. Chem. Soc.*, **87**, 5157.

Blomquist, A. T., Liu, L. H., and Bohrer, J. C. (1952), *J. Am. Chem. Soc.*, **74**, 3643.

Blomquist, A. T., Stahl, R. E., Meinwald, Y. C., and Smith, B. H. (1961), *J. Org. Chem.*, **26**, 1687.

Blount, J. F., Finocchiaro, P., Gust, D., and Mislow, K. (1973), *J. Am. Chem. Soc.*, **95**, 7019.

Boekelheide, V. (1980), "[2]$_n$Cyclophanes: Paracyclophane to Superphane," *Acc. Chem. Res.*, **13**, 65.

Bomse, D. S. and Morton, T. H. (1975), *Tetrahedron Lett.*, 781.

Bott, G., Field, L. D., and Sternhell, S. (1980), *J. Am. Chem. Soc.*, **102**, 5618.

Brewster, J. H. (1967), "Helix Models of Optical Activity," *Top. Stereochem.*, **2**, 1.

Brewster, J. H. and Jones, R. S. (1969), *J. Org. Chem.*, **34**, 354.

Brewster, J. H. and Privett, J. E. (1966), *J. Am. Chem. Soc.*, **88**, 1419.

Brewster, J. H. and Prudence, R. T. (1973), *J. Am. Chem. Soc.*, **95**, 1217.

Brocas, J., Gielen, M., and Witters, R. (1983), "The Permutational Approach to Dynamic Stereochemistry," McGraw-Hill, New York.

Brock, C. P. (1979), *Mol. Cryst. Liq. Cryst.*, **52**, 163.

Brock, C. P. and Minton, R. (1989), *J. Am. Chem. Soc.*, **111**, 4586.

Brown, C. J. (1953), *J. Chem. Soc.*, 3265.

Brown, C. J. and Farthing, A. C. (1949), *Nature (London)*, **164**, 915.

Brown, R. D. (1987), "Understanding 'kinky' molecules," *Chem. Br.*, **23**, 1189.

Brownstein, S., Dunogues, J., Lindsay, D., and Ingold, K. U. (1977), *J. Am. Chem. Soc.*, **99**, 2073.

Buchardt, O. (1974), "Photochemistry with Circularly Polarized Light," *Angew. Chem. Int. Ed. Engl.*, **13**, 179.

Buckingham, J. and Hill, R. A. (1986), *Atlas of Stereochemistry*, 2nd ed., Supplement, Chapman & Hall, New York.

Bürgi, H.-B., Hounshell, W. D., Nachbar, R. B., and Mislow, K. (1983), *J. Am. Chem. Soc.*, **105**, 1427.

Cahn, R. S., Ingold, C. K., and Prelog, V. (1956), "The Specification of Asymmetric Configuration in Organic Chemistry," *Experientia*, **12**, 81.

Cahn, R. S., Ingold, Sir C., and Prelog, V. (1966), "Specification of Molecular Chirality," *Angew. Chem. Int. Ed. Engl.*, **5**, 385.

Caporusso, A. M., Rosini, C., Lardicci, L., Polizzi, C., and Salvadori, P. (1986), *Gazz. Chim. Ital.*, **116**, 467.

Caporusso, A. M., Zoppi, A., Da Settimo, F., and Lardicci, L. (1985), *Gazz. Chim. Ital.*, **115**, 293.

Casalone, G., Pilati, T., and Simonetta, M. (1980), *Tetrahedron Lett.*, **21**, 2345.

Celmer, W. D. and Solomons, I. A. (1952), *J. Am. Chem. Soc.*, **74**, 1870; 2245; (1953), *id., ibid.*, **75**, 1372.

Chance, J. M., Geiger, J. H., Okamoto, Y., Aburatani, R., and Mislow, K. (1990), *J. Am. Chem. Soc.*, **112**, 3540.

Chandross, E. A. and Sheley, C. F. (1968), *J. Am. Chem. Soc.*, **90**, 4345.

Christie, G. H. and Kenner, J. H. (1922), *J. Chem. Soc.*, **121**, 614.

Collet, A., Dutasta, J.-P., Lozach, B., and Canceill, J. (1993), "Cycloveratrylene and Cryptophanes: Their Synthesis and Applications to Host-Guest Chemistry and to the Design of New Materials," *Top. Curr. Chem.*, **165**, 103.

Cooke, A. S. and Harris, M. M. (1963), *J. Chem. Soc.*, 2365.

Cope, A. C., Banholzer, K., Keller, H., Pawson, B. A., Whang, J. J., and Winkler, H. J. S. (1965), *J. Am. Chem. Soc.*, **87**, 3644.

Cope, A. C., Ganellin, C. R., and Johnson, H. W. (1962a), *J. Am. Chem. Soc.*, **84**, 3191.

Cope, A. C., Howell, C. F., and Knowles, A. (1962b), *J. Am. Chem. Soc.*, **84**, 3190.

Cope, A. C. and Mehta, A. S. (1964), *J. Am. Chem. Soc.*, **86**, 5626.

Cope, A. C., Moore, W. R., Bach, R. D., and Winkler, H. J. S. (1970), *J. Am. Chem. Soc.*, **92**, 1243.

Cope, A. C. and Pawson, B. A. (1965), *J. Am. Chem. Soc.*, **87**, 3649.

Coulter, C. L. and Trueblood, K. N. (1963), *Acta Cryst.*, **16**, 667.

Crabbé, P., Velarde, E., Anderson, H. W., Clark, S. D., Moore, W. R., Drake, A. F., and Mason, S. F. (1971), *J. Chem. Soc. Chem. Commun.*, 1261.

Cram, D. J. and Allinger, N. L. (1955), *J. Am. Chem. Soc.*, **77**, 6289.

Cram, D. J., Hornby, R. B., Truesdale, E. A., Reich, H. J., Delton, M. H., and Cram, J. M. (1974), *Tetrahedron*, **30**, 1757.

Cram, D. J. and Reeves, R. A. (1958), *J. Am. Chem. Soc.*, **80**, 3094.

Cram, D. J. and Steinberg, H. (1951), *J. Am. Chem. Soc.*, **73**, 5691.

Cram, D. J., Wechter, W. J., and Kierstead, R. W. (1958), *J. Am. Chem. Soc.*, **80**, 3126.

Crombie, L., Jenkins, P. A., and Robin, J. (1975), *J. Chem. Soc. Perkin* **1**, 1090.

Cruse, R. (1966), in Eliel, E. L., *Stereochemie der Kohlenstoffverbindungen*, German transl. by Lüttringhaus, A. and Cruse, R., Chemie, Weinheim, Germany, p. 215.

Delton, M. H., Gilman, R. E., and Cram, D. J. (1971), *J. Am. Chem. Soc.*, **93**, 2329.

DeRango, C., Tsoucaris, G., Declerq, J. P. Germain, G., and Putzeys, J. P. (1973), *Cryst. Struct. Commun.*, **2**, 189.

DeVille, T. E., Hursthouse, M. B., Russell, S. W., and Weedon, B. C. L. (1969), *J. Chem. Soc. D Chem. Commun.*, 754, 1311.

Dewhirst, K. C. and Cram, D. J. (1958), *J. Am. Chem. Soc.*, **80**, 3115.

Diederich, F. (1991), *Cyclophanes*, The Royal Society of Chemistry, Cambridge, UK.

Dobler, M. and Dunitz, J. D. (1965), *Helv. Chim. Acta*, **48**, 1429.

Duraisamy, M. and Walborsky, H. M. (1983), *J. Am. Chem. Soc.*, **105**, 3252.

Eliel, E. L. (1960), *Tetrahedron Lett.*, No. 8, 16.

Eliel, E. L. (1962), *Stereochemistry of Carbon Compounds*, McGraw-Hill, New York.

Eliel, E. L. and Lynch, J. E. (1987), *Tetrahedron Lett.*, **28**, 4813.

Elsevier, C. J., Meijer, J., Tadema, G., Stehouwer, P. M., Bos, H. J. T., and Vermeer, P. (1982), *J. Org. Chem.*, **47**, 2194.

Elsevier, C. J. and Mooiweer, H. H. (1987), *J. Org. Chem.*, **52**, 1536.

Elsevier, C. J., Stehouwer, P. M., Westmijze, H., and Vermeer, P. (1983), *J. Org. Chem.*, **48**, 1103.

Elsevier, C. J., Vermeer, P., and Runge, W. (1985), "Optical Rotations of Open-Chain Allenes Revisited," *Isr. J. Chem.*, **26**, 174.

Ermer, O. (1974), *Angew. Chem. Int. Ed. Engl.*, **13**, 604.

Ermer, O. (1975), *Tetrahedron*, **31**, 1849.

Ermer, O. (1981), *Aspekte von Kraftfeldrechnungen*, Wolfgang Baur, Munich, Germany.

Ermer, O. (1983), *Angew. Chem. Int. Ed. Engl.*, **22**, 998.

Ermer, O. and Lifson, S. (1973), *J. Am. Chem. Soc.*, **95**, 4121.

Evans, R. J. D. and Landor, S. (1965), *J. Chem. Soc.*, 2553.

Evans, R. J. D., Landor, S. R., and Taylor-Smith, R. (1963), *J. Chem. Soc.*, 1506.

Farthing, A. C. (1953), *J. Chem. Soc.*, 3261.

Finocchiaro, P., Gust, D., Hounshell, W. D., Hummel, J. P., Maravigna, P., and Mislow, K. (1976a), *J. Am. Chem. Soc.*, **98**, 4945.

Finocchiaro, P., Gust, D., and Mislow, K. (1974), *J. Am. Chem. Soc.*, **96**, 3198, 3205.

Finocchiaro, P., Hounshell, W. D., and Mislow, K. (1976b), *J. Am. Chem. Soc.*, **98**, 4952.

Fitzwater, S. and Bartell, L. S. (1976), *J. Am. Chem. Soc.*, **98**, 5107.

Flamm-Ter Meer, M. A., Beckhaus, H.-D., Peters, K., v. Schnering, H.-G., Fritz, H., and Rüchardt, C. (1986), *Chem. Ber.*, **119**, 1492.

Flamm-Ter Meer, M. A., Beckhaus, H.-D., and Rüchardt, C. (1984), *Thermochim. Acta*, **80**, 81.

Ford, W. T., Thompson, T. B., Snoble, K. A. J., and Timko, J. M. (1975), *J. Am. Chem. Soc.*, **97**, 95.

Frank, G. W. (unpublished results); see reference to Tribout, J. et al., 1972, p. 2842, footnote *b*.

Fraser, F. M. and Prosen, E. J. (1955), *J. Res. Natl. Bur. Stand.*, **54**, 143.

Gange, D., Magnus, P., Bass, L., Arnold, E. V., and Clardy, J. (1980), *J. Am. Chem. Soc.*, **102**, 2134.

Gardlick, J. M., Johnson, L. K., Paquette, L. A., Solheim, B. A., Springer, J. P., and Clardy, J. (1979a), *J. Am. Chem. Soc.*, **101**, 1615.

Gardlick, J. M., Paquette, L. A., and Gleiter, R. (1979b), *J. Am. Chem. Soc.*, **101**, 1617.

Garratt, P. J. (1979), "Annulenes and Related Systems," in Stoddart, J. F., Ed., "*Comprehensive Organic Chemistry*," Vol. 1, Pergamon, New York, p. 361.

Garratt, P. J., Nicolaou, K. C., and Sondheimer, F. (1973), *J. Am. Chem. Soc.*, **95**, 4582.

Gerlach, H. (1966), *Helv. Chim. Acta*, **49**, 1291.

Gerlach, H. (1968), *Helv. Chim. Acta*, **51**, 1587.

Gerlach, H., Owtschinnikow, J. A., and Prelog, V. (1964), *Helv. Chim. Acta*, **47**, 2294.

Gianni, M. H. (1961), *Diss. Abstr.*, **21**, 2474.

Glaser, R., Blount, J. F., and Mislow, K. (1980), *J. Am. Chem. Soc.*, **102**, 2777.

Glotzmann, C., Langer, E., Lehner, H., and Schlögl, K. (1974), *Monatsh. Chem.*, **105**, 907.

Goldschmidt, S., Riedle, R., and Reichardt, A. (1957), *Justus Liebigs Ann. Chem.*, **604**, 121.

Goodman, M. and Chorev, M. (1979), "On the Concept of Linear Modified Retro-Peptide Structures," *Acc. Chem. Res.*, **12**, 1.

Green, B. S. and Knossow, M. (1981), *Science*, **214**, 795.

Greenberg, A. and Liebman, J. F. (1978), "Strained Organic Molecules," Academic, New York, p. 127.

Groen, M. B., Stulen, G., Visser, G. J., and Wynberg, H. (1970), *J. Am. Chem. Soc.*, **92**, 7218.

Groen, M. B. and Wynberg, H. (1971), *J. Am. Chem. Soc.*, **93**, 2968.

Guenzi, A., Johnson, C. A., Cozzi, F., and Mislow, K. (1983), *J. Am. Chem. Soc.*, **105**, 1438.

Gur, E., Kaida, Y., Okamoto, Y., Biali, S. E., and Rappoport, Z. (1992), *J. Org. Chem.*, **57**, 3689.

Gust, D. (1977), *J. Am. Chem. Soc.*, **99**, 6980.

Gust, D. and Mislow, K. (1973), *J. Am. Chem. Soc.*, **95**, 1535.

Gust, D. and Patton, A. (1978), *J. Am. Chem. Soc.*, **100**, 8175.

Gustavson, G. (1896), *J. Prakt. Chem.*, [2] **54**, 97.

Haas, G., Hulbert, P. B., Klyne, W., Prelog, V., and Snatzke, G. (1971), *Helv. Chim. Acta*, **54**, 491.

Haas, G. and Prelog, V. (1969), *Helv. Chim. Acta*, **52**, 1202.

Hagishita, S., Kuriyama, K., Shingu, K., and Nakagawa, M. (1971), *Bull. Chem. Soc. Jpn.*, **44**, 2177.

Hall, D. M. (1969), "The Stereochemistry of 2,2'-Bridged Biphenyls," *Prog. Stereochem.*, **4**, 1.

Hanessian, S. and Beaudoin, S. (1992), *Tetrahedron Lett.*, **50**, 7659.

Hanessian, S., Delorme, D., Beaudoin, S., and Leblanc, Y. (1984), *J. Am. Chem. Soc.*, **106**, 5754.

Harada, N. and Nakanishi, K. (1983), "Circular Dichroic Spectroscopy—Exciton Coupling in Organic Stereochemistry," University Science Books, Mill Valley, CA.

Hayashi, T., Okamoto, Y., and Kumada, M. (1983), *Tetrahedron Lett.*, **24**, 807.

Hefelfinger, D. T. and Cram, D. J. (1970), *J. Am. Chem. Soc.*, **92**, 1073.

Hellwinkel, D., Melan, M., and Degel, C. R. (1973), *Tetrahedron*, **29**, 1895.

Hellwinkel, D., Melan, M., Egan, W., and Degel, C. R. (1975), *Chem. Ber.*, **108**, 2219.

Helmchen, G., Haas, G., and Prelog, V. (1973), *Helv. Chim. Acta*, **56**, 2255.

Hlubucek, J. R., Hora, J., Russell, S. W., Toube, T. P., and Weedon, B. C. L. (1974), *J. Chem. Soc. Perkin 1*, 848.

Horeau, A. (1977), "Determination of the Configuration of Secondary Alcohols by Partial Resolution," in Kagan, H. B., Ed., *Stereochemistry, Fundamentals and Methods*, Vol. 3, Thieme, Stuttgart, Germany, p. 51.

Hounshell, W. D., Dougherty, D. A., and Mislow, K. (1978), *J. Am. Chem. Soc.*, **100**, 3149.

Hounshell, W. D., Iroff, L. D., Iverson, D. J., Wroczynski, R. J., and Mislow, K. (1980), "Is the Effective Size of an Alkyl Group a Gauge of Dynamic Gearing?" *Isr. J. Chem.*, **20**, 65.

Hug. W. and Wagnière, G. (1972), *Tetrahedron*, **28**, 1241.

Hulshof, L. A., Wynberg, H., van Dijk, B., and de Boer, J. L. (1976), *J. Am. Chem. Soc.*, **98**, 2733.

Humphrey, G. L. and Spitzer, R. (1950), *J. Chem. Phys.*, **18**, 902.

Iffland, D. C. and Siegel, H. (1958), *J. Am. Chem. Soc.*, **80**, 1947.

Ingraham, L. L. (1956), "Steric Effects on Certain Physical Properties," Newman, M. S., Ed., *Steric Effects in Organic Chemistry*, Wiley, New York.

Inoue, Y., Ueoka, T. and Hakushi, T. (1984), *J. Chem. Soc. Perkin 2*, 2053.

Inoue, Y., Ueoka, T., Kuroda, T., and Hakushi, T. (1983), *J. Chem. Soc. Perkin 2*, 983.

Iwamura, H. and Mislow, K. (1988), "Stereochemical Consequences of Dynamic Gearing," *Acc. Chem. Res.*, **21**, 175.

Jendralla, H., Doecke, C. W., and Paquette, L. A. (1984), *J. Chem. Soc. Chem. Commun.*, 942.

Jenneskens, L. W., de Kanter, F. J. J., Kraakman, P. A., Turkenburg, L. A. M., Koolhaas, W. E., de Wolf, W. H., and Bickelhaupt, F. (1985), *J. Am. Chem. Soc.*, **107**, 3716.

Jochims, J. C., Lambrecht, J., Burkert, U., Zsolnai, L., and Huttner, G. (1984), *Tetrahedron*, **40**, 893.

Johnson, R. P. (1986) "Structural Limitations in Cyclic Alkenes, Alkynes and Cumulenes," in Liebman, J. F. and Greenberg, A., *Molecular Structure and Energetics*, Vol. 3, VCH, Weinheim, Federal Republic of Germany, p. 85.

Johnson, R. P. (1989), "Strained Cyclic Cumulenes," *Chem. Rev.*, **89**, 1111.

Jones, E. R. H., Loder, J. O., and Whiting, M. C. (1960), *Proc. Chem. Soc.*, 180.

Jones, R. R. and Bergman, R. G. (1972), *J. Am. Chem. Soc.*, **94**, 660.

Kabuto, K., Yasuhara, F., and Yamaguchi, S. (1983), *Bull. Chem. Soc. Jpn.*, **56**, 1263.

Kagan, H., Moradpour, A., Nicoud, J. F., Balavoine, G., Martin, R. H., and Cosyn, J. P. (1971), *Tetrahedron Lett.*, 2479.

Kajigaeshi, S., Fujisaki, S., Kadoya, N., Kondo, M., and Ueda, K. (1979), *Nippon Kagaku Kaishi*, 239; *Chem. Abstr.*, **90**, 203375g.

Kane, V. V., Wolf, A. D., and Jones, M. (1974), *J. Am. Chem. Soc.*, **96**, 2643.

Karich, G. and Jochims, J. C. (1977), *Chem. Ber.*, **110**, 2680.

Kawada, Y. and Iwamura, H. (1983), *J. Am. Chem. Soc.*, **105**, 1449.

Keehn, P. M. and Rosenfeld, S. M., Eds. (1983), "Cyclophanes," Academic, New York, 2 Volumes.

Kiggen, W. and Vögtle, F. (1987), "More than Twofold Bridged Phanes," in Izatt, R. M. and Christensen, J. J., Eds., *Synthesis of Macrocycles*, Wiley, New York, p. 309.

Klyne, W. and Buckingham, J. (1978), "Atlas of Stereochemistry," 2nd ed., Vols. 1 and 2, Chapman & Hall, London.

Knauf, A. E., Shildneck, P. R., and Adams, R. (1934), *J. Am. Chem. Soc.*, **56**, 2109.

Köbrich, G., Mannschreck, A. Misra, R. A., Rissmann, G., Rosner, M., and Zundorf, W. (1972), *Chem. Ber.*, **105**, 3794.

Kohler, E. P., Walker, J. T., and Tishler, M. (1935), *J. Am. Chem. Soc.*, **57**, 1743.

Kranz, M., Clark, T., and Schleyer, P. v. R. (1993), *J. Org. Chem.*, **58**, 3317.

Krois, D. and Lehner, H. (1982), *Tetrahedron*, **38**, 3319.

Krow, G. (1970), "The Determination of Absolute Configuration of Planar and Axially Dissymmetric Molecules," *Top. Stereochem.*, **5**, 31.

Kuffner, U. and Schlögl, K. (1972), *Monatsh. Chem.*, **103**, 1320.

Kuhn, R. (1933), "Molekulare Asymmetrie," in Freudenberg, K. Ed., *Stereochemie*, Franz Deutike, Leipzig, Germany, p. 803.

Kuhn, R., Fischer, H., and Fischer, H. (1964), *Chem. Ber.*, **97**, 1760.

Kuhn, W. and Knopf, E. (1930), *Z. Phys. Chem.*, **7B**, 292.

Kurland, R. J., Schuster, I. I., and Colter, A. K. (1965), *J. Am. Chem. Soc.*, **87**, 2279.

Laarhoven, W. H. and Cuppen, Th. H. J. M. (1973), *Recl. Trav. Chim. Pays-Bas*, **92**, 553.

Laarhoven, W. H. and Prinsen, W. J. C. (1984), "Carbohelicenes and Heterohelicenes," *Top. Curr. Chem.*, **125**, 63.

Lambrecht, J., Gambke, B., von Seyerl, J., Huttner, G., Kollmannsberger-von Nell, G., Herzberger, S., and Jochims, J. C. (1981), *Chem. Ber.*, **114**, 3751.

Lambrecht, J., Zsolnai, L., Huttner, G., and Jochims, J. C. (1982), *Chem. Ber.*, **115**, 172.

Landor, S. R. (1982), "Naturally Occurring Allenes," in Landor, S. R., Ed., *The Chemistry of the Allenes*, Vol. 3, Academic, New York, p. 679.

Langler, R. F. and Tidwell, T. T. (1975), *Tetrahedron Lett.*, 777.

Leitich, J. (1978), *Tetrahedron Lett.*, 3589.

Leuchs, H. and Gieseler, E. (1912), *Ber. Dtsch., Chem. Ges.* **45**, 2114.

Levin, C. C. and Hoffman, R. (1972), *J. Am. Chem. Soc.*, **94**, 3446.

Levy, G. C. (1973), "Carbon-13 Spin-Lattice Relaxation Studies and Their Application to Organic Chemical Problems," *Acc. Chem. Res.*, **6**, 161.

Lightner, D. A., Hefelfinger, D. T., Powers, T. W., Frank, G. W., and Trueblood, K. N. (1972), *J. Am. Chem. Soc.*, **94**, 3492.

Lomas, J. S. and Dubois, J.-E. (1976), *J. Org. Chem.*, **41**, 3033.

Lowe, G. (1965), *Chem. Commun.*, 411.

Lüttringhaus, A. and Eyring, G. (1957), *Justus Liebigs Ann. Chem.*, **604**, 111.

Lüttringhaus, A. and Gralheer, H. (1942), *Justus Liebigs Ann. Chem.*, **550**, 67; (1947) *id., ibid.*, **557**, 108, 112.

Lyle, R. E. and Lyle, G. G. (1959), *J. Org. Chem.*, **24**, 1679.

Lyle, G. G. and Pelosi, E. T. (1966), *J. Am. Chem. Soc.*, **88**, 5276.

Maitland, P. and Mills, W. H. (1935), *Nature (London)*, **135**, 994; (1936), *id., J. Chem. Soc.*, 987.

Mannschreck, A., Jonas, V., Bödecker, H.-O., Elbe, H.-L., and Köbrich, G. (1974), *Tetrahedron Lett.*, 2153.

Manor, P. C., Shoemaker, D. P., and Parkes, A. S. (1970), *J. Am. Chem. Soc.*, **92**, 5260.

Marquis, E. T. and Gardner, P. D. (1966), *Tetrahedron Lett.*, 2793.

Marshall, J. A. and Flynn, K. E. (1983), *J. Am. Chem. Soc.*, **105**, 3360.

Marshall, J. A., Konicek, T. R., and Flynn, K. E. (1980), *J. Am. Chem. Soc.*, **102**, 3287.

Martin, R. H. (1974), "The Helicenes," *Angew. Chem. Int. Ed. Engl.*, 649.

Martin, R. H. and Baes, M. (1975), *Tetrahedron*, **31**, 2135.

Martin, R. H. and Libert, V. (1980), *J. Chem. Res., (Synop.)*, 130; *J. Chem. Res., Miniprint*, 1940.

Martin, R. H. and Marchant, M.-J. (1974a), *Tetrahedron*, **30**, 343.

Martin, R. H. and Marchant, M.-J. (1974b), *Tetrahedron*, **30**, 347.

Mason, S. F. and Vane, G. W. (1965), *Tetrahedron Lett.*, 1593.

Mazaleyrat, J.-P. and Welvart, Z. (1983), *Nouv. J. Chem.*, **7**, 491.

Meurer, K. P. and Vögtle, F. (1985), "Helical Molecules in Organic Chemistry," *Top. Curr. Chem.*, **127**, 1.

Meyer, W. L. and Meyer, R. B. (1963), *J. Am. Chem. Soc.*, **85**, 2170.

Meyers, A. I. and Wettlaufer, D. G. (1984), *J. Am. Chem. Soc.*, **106**, 1135.

Mills, W. H. and Dazeley, G. H. (1939), *J. Chem. Soc.*, 460.

Mills, W. H. and Nodder, C. R. (1920), *J. Chem. Soc.*, **117**, 1407.

Mills, W. H. and Warren, E. H. (1925), *J. Chem. Soc.*, **127**, 2507.

Mislow, K. (1958), "Die Absolute Konfiguration der atropisomeren Diaryl-Verbindungen," *Angew. Chem.*, **70**, 683.

Mislow, K. (1976), "Stereochemical Consequences of Correlated Rotation in Molecular Propellers," *Acc. Chem. Res.*, **9**, 26.

Mislow, K. (1989), *Chemtracts-Org. Chem.*, **2**, 151.

Mislow, K., Gust, D., Finocchiaro, P., and Boettcher, R. J. (1974), "Stereochemical Correspondence Among Molecular Propellers," *Top. Curr. Chem.*, **47**, 1.

Mislow, K., Glass, M. A. W., Hopps, H. B., Simon, E., and Wahl, G. H. (1964), *J. Am. Chem. Soc.*, **86**, 1710.

Mislow, K. and Hopps, H. B. (1962), *J. Am. Chem. Soc.*, **84**, 3018.

Mislow, K. and McGinn, F. A. (1958), *J. Am. Chem. Soc.*, **80**, 6036.

Mislow, K., Prelog, V., and Scherrer, H. (1958), *Helv. Chim. Acta*, **41**, 1410.

Mislow, K. and Perlmutter, H. D. (1962), *J. Am. Chem. Soc.*, **84**, 3591.

Misumi, S. and Otsubo, T. (1978), "Chemistry of Multilayered Cyclophanes," *Acc. Chem. Res.*, **11**, 251.

Moore, W. R., Anderson, H. W., and Clark, S. D. (1973), *J. Am. Chem. Soc.*, **95**, 835.

Moore, W. R. and Bertelson, R. C. (1962), *J. Org. Chem.*, **27**, 4182.

Moore, W. R., Anderson, H. W., Clark, S. D., and Ozretich, M. (1971), *J. Am. Chem. Soc.*, **93**, 4932.

Moradpour, A., Kagan, H., Baes, M., Morren, G., and Martin, R. H. (1975), *Tetrahedron*, **31**, 2139.

Moradpour, A., Nicoud, J. F., Balavoine, G., Kagan, H., and Tsourcaris, G. (1971), *J. Am. Chem. Soc.*, **93**, 2353.

Mori, T. and Ōki, M. (1981), *Bull. Chem. Soc. Jpn.*, **54**, 1199.

Mori, T. and Ōki, M., unpublished results, cited in Ōki, M. (1983), p. 38.

Mori, K., Nukada, T., and Ebata, T. (1981), *Tetrahedron*, **37**, 1343.

Moscowitz, A. J. (1961), *Tetrahedron*, **13**, 48.

Moscowitz, A. and Mislow, K. (1962), *J. Am. Chem. Soc.*, **84**, 4605.

Murray, M. (1977), "Methoden zur Herstellung und Umwandlung von Allenen bezw. Kumulenen," in Houben-Weyl, 4th ed., Vol. V/2a, Thieme, Stuttgart, Germany, p. 963.

Nakamura, M. and Ōki, M. (1974), *Tetrahedron Lett.*, 505.

Nakamura, M. and Ōki, M. (1975), *Chem. Lett.*, 671.

Nakamura, M. and Ōki, M. (1980), *Bull. Soc. Chem. Jpn.*, **53**, 2977.

Nakazaki, M. (1984), "The Synthesis and Stereochemistry of Chiral Organic Molecules with High Symmetry," *Top. Stereochem.*, **15**, 199.

Nakazaki, M., Naemura, K., and Nakahara, S. (1979), *J. Org. Chem.*, **44**, 2438.

Nakazaki, M., Yamamoto, K., Ito, M., and Tanaka, S. (1977b), *J. Org. Chem.*, **42**, 3468.

Nakazaki, M., Yamamoto, K., and Maeda, M. (1980), *J. Org. Chem.*, **45**, 3229.

Nakazaki, M., Yamamoto, K., and Maeda, M. (1981) *J. Org. Chem.*, **46**, 1985.

Nakazaki, M., Yamamoto, K., and Naemura, K. (1984), "Stereochemistry of Twisted Double Bond Systems," *Top. Curr. Chem.*, **125**, 1.

Nakazaki, M., Yamamoto, K., Maeda, M., Sato, O., and Tsutsui, T. (1982), *J. Org. Chem.*, **47**, 1435.

Nakazaki, M., Yamamoto, K., Tanaka, S., and Kametani, H. (1977a), *J. Org. Chem.*, **42**, 287.

Narayanan, U., Keiderling, T. A., Elsevier, C. J., Vermeer, P., and Runge, W. (1987), *J. Am. Chem. Soc.*, **110**, 4133.

Neudeck, H. and Schlögl, K. (1977), *Chem. Ber.*, **110**, 2624.

Newman, M. S. and Hussey, A. S. (1947), *J. Am. Chem. Soc.*, **69**, 3023.

Newman, M. S. and Lednicer, D. (1956), *J. Am. Chem. Soc.*, **78**, 4765.

Newman, M. S. and Wise, R. M. (1956), *J. Am. Chem. Soc.*, **78**, 450.

Newman, P., Rutkin, P., and Mislow, K. (1958), *J. Am. Chem. Soc.*, **80**, 465.

Nicolaou, K. C. and Smith, A. L. (1992), "Molecular Design, Chemical Synthesis and Biological Action of Enediynes," *Acc. Chem. Res.*, **25**, 497.

O'Shaughnessy, M. T. and Rodebush, W. H. (1940), *J. Am. Chem. Soc.*, **62**, 2906.

Ōki, M. (1983), "Recent Advances in Atropisomerism," *Top. Stereochem.*, **14**, 1.

Ōki, M. (1993), *The Chemistry of Rotational Isomers*, Springer, New York.

Olsson, L.-I. and Claesson, A. (1977), *Acta Chem. Scand.*, **B31**, 614.

Oremek, G., Seiffert, U., and Janecka, A. (1987), "Synthese und Eigenschaften von Helicenen," *Chem.-Ztg.*, **111**, 69.

Overberger, C. G., Wang, D. W., Hill, R. K., Krow, G. R., and Ladner, D. W. (1981), *J. Org. Chem.*, **46**, 2757.

Paquette, L. A. (1993), "The Current View of Dynamic Change within Cyclooctatetraenes," *Acc. Chem. Res.*, **26**, 57.

Pepermans, H., Gielen, M., and Hoogzand, C. (1983), *Bull. Soc. Chim. Belg.*, **92**, 465.

Perkin, W. H. and Pope, W. J. (1911), *J. Chem. Soc.*, **99**, 1510.

Perkin, W. H., Pope, W. J., and Wallach, O. (1909), *Justus Liebigs*, *Ann.*, *Chem.*, **371**, 180; *id.*, *J. Chem. Soc.*, **95**, 1789.

Pickett, L. W., Walter, G. F., and France, H. (1936), *J. Am. Chem. Soc.*, **58**. 2296.

Pignolet, L. H., Taylor, R. P., and Horrocks, W. DeW. (1968), *Chem. Commun.*, 1443.

Porath, J. (1951), *Ark. Kemi*, **3**, 163.

Prelog, V. (1953), *Helv. Chim. Acta*, **36**, 308.

Prelog, V. and Helmchen, G. (1982), "Basic Principles of the CIP-System and Proposals for a Revision," *Angew. Chem. Int. Ed. Engl.*, **21**, 567.

Prelog, V. and Gerlach, H. (1964), *Helv. Chim. Acta*, **47**, 2288.

Rauk, A., Drake, A. F., and Mason, S. F. (1979), *J. Am. Chem. Soc.*, **101**, 2284.

Rieker, A. and Kessler, H. (1969), *Tetrahedron Lett.*, 1227.

Rømming, C., Seip, H. M., and Aanesen Øymo, I.-M. (1975), *Acta Chem. Scand.*, **A28**, 507.

Rosner, M. and Köbrich, G. (1974), *Angew. Chem. Int. Ed. Engl.*, **13**, 741.

Rossi, R. and Diversi, P. (1973), "Synthesis and Absolute Configurations and Optical Purity of Chiral Allenes," *Synthesis*, 25.

Roth, W. R., Böhm, M., Lennartz, H.-W., and Vogel, E. (1983), *Angew. Chem. Int. Ed. Engl.*, **22**, 1007.

Roth, W. R., Ruf, G., and Ford, P. W. (1974), *Chem. Ber.*, **107**, 48.

Ruch, E. (1972), "Algebraic Aspects of the Chirality Phenomenon in Chemistry," *Acc. Chem. Res.*, **5**, 49.

Ruch, E. (1977), "Chiral Derivatives of Achiral Molecules: Standard Classes and the Problem of Right–Left Classification," *Angew. Chem. Int. Ed. Engl.*, **16**, 65.

Ruch, E., Runge, W., and Kresze, G. (1973), "Semiempirical Use of Chirality Functions for the Description of Optical Activity of Allene Derivatives in the Transparent Region," *Angew. Chem. Int. Ed. Engl.*, **12**, 20.

Ruch, E. and Schönhofer, A. (1968), *Theor. Chim. Acta*, **10**, 91.

Runge, W. (1980), "Chirality and chiroptical properties," in Patai, S., Ed., *The Chemistry of Ketenes, Allenes and Related Compounds*, part 1, Wiley, New York, p. 99. See also p. 45.

Runge, W. (1981), "Substituent Effects in Allenes and Cumulenes," *Prog. Phys. Org. Chem.*, **13**, 315.

Runge, W. (1982), "Stereochemistry of Allenes," in Landor, S. R., Ed., *The Chemistry of the Allenes*, Vol. 3, Academic, New York, p. 579.

Runge, W. and Kresze, G. (1977), *J. Am. Chem. Soc.*, **99**, 5597.

Schlögl, K. (1984), "Planar Chiral Molecular Structures," *Top. Curr. Chem.*, **125**, 27.

Schlögl, K. (1967), "Stereochemistry of Metallocenes," *Top. Stereochem.*, **1**, 39.

Schlögl, K. (1970), "Configurational and Conformational Studies in the Metallocene Field," *Pure Appl. Chem.*, **23**, 413.

Schlögl, K. and Mechtler, H. (1966), *Angew. Chem. Int. Ed. Engl.*, **5**, 596.

Schlögl, K., Widhalm, M., Vogel, E., and Schwamborn, M. (1983), *Monatsh. Chem.*, **114**, 605.

Schwartz, L. H. and Bathija, B. L. (1976), *J. Am. Chem. Soc.*, **98**, 5344.

Sekine, Y., Brown, M., and Boekelheide, V. (1979), *J. Am. Chem. Soc.*, **101**, 3126.

Semmelhack, M. F., Harrison, J. J., Young, D. C., Gutiérrez, A., Rafii, S., and Clardy, J. (1985), *J. Am. Chem. Soc.*, **107**, 7508.

Shemyakin, M. M., Ovchinnikov, Yu. A., Ivanov, V. T., and Ryabura, I. D. (1967), *Experientia*, **23**, 326.

Shingu, K., Hagishita, S., and Nakagawa, M. (1967), *Tetrahedron Lett.*, 4371.

Shriner, R. L., Adams, R., and Marvel, C. S. (1943), "Stereoisomerism," in Gilman, H., Ed., *Organic Chemistry*, 2nd ed. Vol. 1, Wiley, New York, p. 343ff.

Siddall, T. H. and Stewart, W. E. (1969), *J. Org. Chem.*, **34**, 233.

Siegel, J., Gutierrez, A., Schweizer, W. B., Ermer, O., and Mislow, K. (1986), *J. Am. Chem. Soc.*, **108**, 1569.

Siegel, J. and Mislow, K. (1983), *J. Am. Chem. Soc.*, **105**, 7763.

Singh, M. D., Siegel, J., Biali, S. E., and Mislow, K. (1987), *J. Am. Chem. Soc.*, **109**, 3397.

Skattebøl, L. (1963), *Acta Chem. Scand.*, **17**, 1683.

Solladié, G. and Zimmerman, R. G. (1984), *Tetrahedron Lett.*, **25**, 5769.

Sondheimer, F. (1972), "The Annulenes," *Acc. Chem. Res.*, **5**, 81.

Squillacote, M., Bergman, A., and De Felippis, J. (1989), *Tetrahedron Lett.*, **30**, 6805.

Streitwieser, A., Wolfe, J. R., and Schaeffer, W. D. (1959), *Tetrahedron*, **6**, 338.

Sumiyoshi, M., Kuritani, H., Shingu, K., and Nakagawa, M. (1980), *Tetrahedron Lett.*, **21**, 1243.

Taft, R. W., Taagepera, M., Abboud, J. L. M., Wolf, J. F., DeFrees, D. J., Hehre, W. J., Bartmess, J. E., and McIver, R. T. (1978), *J. Am. Chem. Soc.*, **100**, 7765.

Thulin, B. and Wennerstrom, O. (1976), *Acta Chem. Scand.*, **B30**, 688.

Tobe, Y., Kakiuchi, K., Odaira, Y., Mosaki, T., Kai, Y., and Kasai, N. (1983), *J. Am. Chem. Soc.*, **105**, 1376.

Traetteberg, M. (1975), *Acta Chem. Scand.*, **B29**, 29.

Tribout, J., Martin, R. H., Doyle, M., and Wynberg, H. (1972), *Tetrahedron Lett.*, 2839.

van't Hoff, J. (1875), *La Chimie dans l'Espace*, Bazendijk, Rotterdam, The Netherlands, p. 29.

Visser, J. P. and Ramakers, J. E. (1972), *J. Chem. Soc. Chem. Commun.*, 178.

Vogel, E. (1980), "Bridged Anulenes," *Isr. J. Chem.*, **20**, 215.

Vogel, E. (1982), "Recent Advances in the Chemistry of Bridged Anulenes," *Pure Appl. Chem.*, **54**, 1015.

Vogel, E., Tückmantel, W., Schlögl, K., Widhalm, M., Kraka, E., and Cremer, D. (1984), *Tetrahedron Lett.*, **25**, 4925.

Vögtle, F. (1969), *Chem. Ber.*, **102**, 3077.

Vögtle, F., Ed., (1983a), "Cyclophanes I" *Top. Curr. Chem.*, **113**.

Vögtle, F., Ed., (1983b), "Cyclophanes II" *Top. Curr. Chem.*, **115**.

Vögtle, F. (1993), *Cyclophane Chemistry: Synthesis, Structure and Reactions*, Wiley, New York.

Vögtle, F. and Hohner, G. (1978), "Stereochemistry of Multibridged, Multilayered and Multistepped Aromatic Compounds. Transannular Steric and Electronic Effects," *Top. Curr. Chem.*, **74**, 1.

Vögtle, F. and Neumann, P. (1974), "[2.2] Paracyclophanes, Structure and Dynamics," *Top. Curr. Chem.*, **48**, 67.

Walborsky, H. M. and Banks, R. B. (1981), *J. Org. Chem.*, **46**, 5074.

Walborsky, H. M., Banks, R. B., Banks, M. L. A., and Duraisamy, M. (1982), *Organometallics*, **1**, 667.

Walborsky, H. M. and Banks, R. B. (1980), *Bull. Soc. Chim. Belg.*, **89**, 849.

Warner, P. M. (1989), "Strained Bridgehead Double Bonds," *Chem. Rev.*, **89**, 1067.

Waters, W. L. and Caserio, M. C. (1968), *Tetrahedron Lett.*, 5233.

Wentrup, C., Gross, G., Maquestiau, A., and Flammang, R. (1983), *Angew. Chem. Int. Ed. Engl.*, **22**, 542.

Westheimer, F. H. (1956), "Calculation of the Magnitude of Steric Effects," in Newman, M. S., Ed., *Steric Effects in Organic Chemistry*, Wiley, New York, p. 523.

Westheimer, F. H. and Mayer, J. E. (1946), *J. Chem. Phys.*, **14**, 733.

Willem, R., Gielen, M., Pepermans, H., Brocas, J., Fastenakel, D., and Finocchiaro, P. (1985a), *J. Am. Chem. Soc.*, **107**, 1146.

Willem, R., Gielen, M., Pepermans, H., Hallenga, K., Recca, A., and Finocchiaro, P. (1985b), *J. Am. Chem. Soc.*, **107**, 1153.

Wilson, K. R. and Pincock, R. E. (1975), *J. Am. Chem. Soc.*, **97**, 1474.

Wittig, G. and Fritze, P. (1968), *Justus Liebigs Ann. Chem.*, **711**, 82.

Wotiz, J. H. and Palchak, R. J. (1951), *J. Am. Chem. Soc.*, **73**, 1971.

Wynberg, H. (1971), "Some Observations on the Chemical, Photochemical, and Spectral Properties of Thiophenes," *Acc. Chem. Res.*, **4**, 65.

Wynberg, H. and Houbiers, J. P. M. (1971), *J. Org. Chem.*, **36**, 834.

Yamada, S. and Akimoto, H. (1968), *Tetrahedron Lett.*, 3967.

Yamada, K.-I., Ogashiwa, S., Tanaka, H., Nakagawa, H., and Kawazura, H. (1981), *Chem. Lett.*, 343.

Yamamoto, G., Nakamura, M., and Ōki, M. (1975), *Bull. Chem. Soc. Jpn.*, **48**, 2592.

Yamamoto, G. and Ōki, M. (1972), *Chem. Lett.*, 45.

Yamamoto, G. and Ōki, M. (1975), *Bull. Chem. Soc. Jpn.*, **48**, 3686.

Glossary*

Absolute configuration. The spatial arrangement of the atoms in a chiral molecule that distinguishes it from its mirror image, and its stereochemical description (R or S, M or P). See also Sense of chirality and Chapter 5.

Achiral. An entity, such as a molecule, is achiral if it is superposable (q.v.) with its mirror image.

Achirotopic. Antonym of Chirotopic (q.v.).

Alternating axis of symmetry of order n (S_n). An axis in a molecule, model, or object, such that rotation of the entity about such axis by an angle $360°/n$ followed by reflection across a plane at right angles to the axis produces a superposable entity. Also called rotation–reflection axis.

Anancomeric. Fixed in a single conformation either by geometric constraints, as (axial) 2β-chloro-*trans*-decalin, or because of an overwhelmingly one-sided conformational equilibrium, as *cis*-4-*tert*-butylcyclohexanol.

Angle strain. Excess enthalpy of a molecule, caused by bond angle deformations, over and above that of a corresponding molecule possessing "normal" bond angles. Also called "Baeyer strain." See Chapter 2.

Anisometric. Antonym of Isometric (q.v.).

Anomeric effect. Originally defined as the thermodynamic preference for the axial conformation of an aglycone, for example, an alkoxy group at C(1) (the anomeric carbon) in a glycopyranoside. Now generalized to imply the preference for the synclinal (gauche) conformation of a fragment Y–C–X–C, where X and Y are heteroatoms, at least one of which is O, N, or F. In some instances, for example, when $X = O$ and $Y = \overset{+}{N}(CH_3)_3$ in a glycopyranoside, the antiperiplanar (anti) conformation (Y equatorial in this case) may be preferred because of what is called the "reverse anomeric effect." See Chapters 10 and 11.

---

* The authors acknowledge, with thanks, having received several preliminary drafts of the *Basic Terminology of Stereochemistry* document (as yet unpublished) from the Joint Working Party on Stereochemical Terminology of IUPAC. While we have benefited from these drafts, we must stress that the glossary here presented neither duplicates the *Basic Terminology of Stereochemistry*, nor does it have IUPAC approval. Moreover, this glossary deviates in a number of definitions from those proposed by IUPAC in minor or major ways.

Anomers. Diastereomers in glycosides, sugar hemiacetals, or related cyclic forms of sugars and sugar derivatives differing in configuration at C(1) in aldoses or at C(2) in 2-ketoses (the so-called *anomeric center*).

Anti. See Antiperiplanar. Now also used to denote the relative configuration of any two stereogenic centers (see stereogenic element) in a chain: If the chain is written in a planar zigzag conformation and the ligands at the stereogenic center are on opposite sides of the plane, the relative configuration is called anti; if they are on the same side of the plane, the configuration is syn. Formerly also used to describe configuration of oximes, hydrazones, and similar compounds, now obsolete in that sense, for which see *E, Z*.

Anticlinal (*ac*). In X–A–B–Y, X and Y are anticlinal if the torsion angle (q.v.) about the A–B bond is between $+90°$ and $+150°$ or between $-90°$ and $-150°$. See Chapter 2 and Figure 2.13.

Antiperiplanar (*ap*). In X–A–B–Y, X and Y are antiperiplanar (or anti) if the torsion angle about the A–B bond is between $+150°$ and $-150°$. See Table 2.2.

Antipodes. See Optical antipodes.

Asymmetric. Lacking all elements of symmetry (other than the identity E or C_1); belonging to the symmetry point group $\mathbf{C_1}$.

Asymmetric carbon atom. A carbon atom with four different substituents, Cabde. The term, originally coined by van't Hoff, may also be applied to other tetra-hedral atoms (e.g. $\overset{+}{N}abcd$). See also Chiral center and Stereogenic element.

Asymmetric center. Incorrect synonym for Chiral center (q.v.).

Asymmetric destruction. See Kinetic resolution.

Asymmetric induction. A term referring to the extent of excess of one enantiomer over the other achieved in an asymmetric synthesis (q.v.). See Chapter 12.

Asymmetric synthesis. De novo synthesis of a chiral substance from an achiral precursor such that one enantiomer predominates over the other. Because of lack of agreement on how to extend the definition to substances where molecules already contain at least one chiral element (q.v.) and where the synthesis introduces a new chiral element, it is preferable to replace this term by stereoselective synthesis, enantioselective synthesis (q.v.), or diastereoselective synthesis (q.v.), as the case may be. See Chapter 12.

Asymmetric transformation. The transformation of a mixture (usually 50:50) of stereoisomers into a single stereoisomer, or into a different mixture of stereo-isomers, by an equilibrium process. The term is limited to mixtures which, at least at the end of the transformation, are enantiomerically enriched or enantio-merically pure.

An asymmetric transformation of the first kind (also inappropriately called first-order asymmetric transformation) involves equilibration without concomitant separation. In the case of enantiomers, such enrichment requires a chiral (non-racemic) solvent or other chiral influence.

In an asymmetric transformation of the second kind (also misnamed second-order asymmetric transformation), following equilibration, one stereoisomer crystallizes from solution and may thus be obtained enantiomerically or diastereo-merically pure or nearly so in a yield approaching 100% (based on the original mixture). Also called crystallization-induced asymmetric transformation. The term may be extended to equilibration on a chromatographic column with concomitant separation—see Chapter 7.

Atropisomers. Stereoisomers resulting from restricted rotation about single bonds where the rotational barrier is high enough to permit isolation of the isomeric species. See Chapter 14.

Automerization. See Topomerization.

Axial bonds, axial ligands. Bonds directed along the S_6 or C_3 axis of cyclohexane in the chair conformation. In general, bonds perpendicular to a plane containing, or nearly containing, the majority of atoms in a cyclic molecule, and ligands attached to such bonds. See Chapter 11 and Figure 11.15.

Axial chirality. Chirality stemming from the nonplanar arrangement of four groups about an axis, called a chiral axis; as, for example, in allenes abC=C=Cab. See Chapter 14.

Axis of chirality. Also called chiral axis. See Axial chirality.

Axis of symmetry. See Symmetry axis.

Baeyer strain. See Angle strain.

Biot's law. See Specific rotation.

Bisecting conformation. In a compound $R_3C-C(X)=Y$, the conformation in which Y is antiperiplanar (q.v.) to one of the R groups and the torsion angles R–C–C=Y to the other two R groups are equal or nearly equal. See Figure 10.15.

Boat. The C_{2v} conformation of cyclohexane and similar conformations of higher cycloalkanes and their hetera analogues. See Figure 11.15.

Bowsprit position. Position of bonds (and substituents) attached to the two out-of-plane carbon atoms of the cyclohexane boat and pointing in a direction nearly parallel to the plane of the other four atoms. The positions geminal to the bowsprit positions (b) are called flagpole positions (f).

CD. See Circular dichroism.

CIP system. Abbreviation for the 'Cahn–Ingold–Prelog' system. A system of rules for the assignment of descriptors (R, S, M, P, r, s, m, p, E, Z) for stereoisomers. See Chapters 5 and 9.

Center of symmetry *i*. A point such that, if a line is drawn from any element in a molecule, model, or object to this point and then extended an equal distance beyond the point, an identical element is found. Equivalent to twofold alternating axis of symmetry (S_2).

Center of chirality. See Chiral center.

Chair. The (rigid) $\mathbf{D_{3d}}$ conformation (Fig. 11.15) of cyclohexane and similar conformation of its derivatives and hetera analogues. See Chapter 11. Also applied to similar conformations of larger rings.

Chiral. Not superposable (q.v.) with its mirror image, as applied to molecules, conformations, as well as macroscopic objects, such as crystals. The term has been extended to samples of substances whose molecules are chiral, even if the macroscopic assembly of such molecules is racemic (q.v.). See also Nonracemic.

Chiral axis. See Axial chirality.

Chiral center. In a tetrahedral (Xabcd) or trigonal pyramidal (Xabc) structure, the atom (X) to which four (or three, respectively) different ligands abc(d) are attached and to which a CIP (q.v.) chirality descriptor R or S can be assigned. Reflection of the molecule reverses the sense of chirality (q.v.) and changes the descriptor.

Chiral element. Chiral center, axis, or plane (q.v.).

Chiral plane. A planar unit connected to an adjacent part of the structure by bonds that result in restricted torsion, so that the plane cannot lie in a symmetry plane. For example, with *trans*-cyclooctene the chiral plane includes the double-bond carbon atoms and all four atoms attached to the double bond; with a monosubstituted paracyclophane the chiral plane includes the monosubstituted benzene ring with its three hydrogen atoms and the three other atoms linked to the ring (i.e., the substituent and the two chains linking the two benzene rings). See Section 14-8.

Chiral recognition. A term used by some for the discrimination between enantiomers, or between enantiotopic ligands (q.v.), achieved by appropriately structured reagents or catalysts, either natural, such as enzymes, or synthetic. A preferred and broader term is stereoisomer discrimination.

Chirality (Handedness). The property of being chiral (q.v.). Chirality in a molecule or other entity implies absence of an S_n axis (see alternating axis of symmetry) including a plane ($\sigma = S_1$) or center ($i = S_2$) of symmetry.

Chirality rule. A rule that defines the descriptor as R or S for a tetrahedral atom Xabcd once the priority of a, b, c, and d is established. See also CIP system and Sequence rules; cf. Chapter 5.

For an entirely different use of the term (as a synonym for helicity rule) see Chapter 13.

Chirality sense. See Sense of chirality. Also see Absolute configuration.

Chiroptical (or chiroptic). A term referring to optical techniques (polarimetry, optical rotatory dispersion, circular dichroism, circular polarization of emission, cf. Chapter 13) that can differentiate between the two enantiomers of a chiral compound and that can be used to characterize chiral (nonracemic) substances.

Chirotopic. The description, in a molecule or structure, of any point (including a material point, such as an atom) that resides in a (local) chiral environment. (The molecule or structure as a whole need not be chiral.) A point or atom located on a plane or at a center of symmetry, or at the point where an alternating axis of symmetry (q.v.) intersects its reflection plane is "achirotopic."

Circular dichroism (CD). The differential absorption of left and right circularly polarized radiation by a nonracemic sample. Circular dichroism is characterized by a signed absorption band in the UV, vis, or IR region of the spectrum. See also Cotton effect (q.v.) and Chapter 13.

Circularly polarized light. Electromagnetic radiation filtered so that the tip of its electric field vector describes a helix along whose axis the radiation is propagated.

cis (c). A stereochemical term for the relationship between ligands located on the same side of a double bond or of a ring structure in a conformation (real or hypothetical) devoid of reentrant angles. In the case of alkenes only, *E* or *Z* (q.v.) are preferred as descriptors in conjunction with a chemical name.

cisoid. See s-cis. The term "cisoid", which is obsolete as a descriptor for polyenes, is still used to denote the configuration (cis) between the terminal rings in fused cyclic systems (cf. Chapter 11).

cis–trans Isomers. Diastereomers of alkenes and cyclanes having at least two nongeminal substituents. Use of the older term "geometric(al) isomers" is discouraged. See Chapter 9 for use of *E*, *Z* descriptors and Chapter 11 for use of reference atoms (denoted *r*) in polysubstituted cyclanes.

Configuration. The spatial array of atoms that distinguishes stereoisomers (isomers of the same constitution) other than distinctions due to differences in conformation (q.v.). See Chapters 2 and 5. See also Sense of chirality, Absolute configuration, and Relative configuration.

Configurational isomers. Stereoisomers that differ in configuration.

Conformation. The spatial array of atoms in a molecule of given constitution and configuration. Conformation of such molecules can be changed by (rapid) rotation around single bonds (and, in the definition of some, by rapid inversion at trigonal pyramidal centers) without, in general, affecting the constitution and configuration. See Chapters 2, 10, and 11.

Conformational analysis. The interpretation or prediction of physical (including spectral) and chemical properties and of relative energy content of substances in terms of the conformation or conformations of their molecules.

Conformational energy. The energy (free energy, enthalpy, or potential energy, as defined in each specific instance) of a given conformational isomer (q.v.) over and above that of the corresponding conformational isomer of global minimum energy. For example, the difference in free energy between axial and equatorial chlorocyclohexane (Table 11.7).

Conformational isomer (conformer). One of a set of stereoisomers that differ in conformation (q.v.), that is, in torsion angle or angles. See Conformation. Only structures corresponding to potential energy minima (local or global) qualify.

Conformer. See Conformational isomer.

Conglomerate. An equimolar mechanical mixture of crystals that form a eutectic of two enantiomers. Formerly called "racemic mixture." Relatively few racemates crystallize as conglomerates from solution. See Chapter 6.

Constitution. The description of the nature of the atoms and the connectivity (including bond multiplicity) between them in a molecule, disregarding their spatial array [configuration, conformation (q.v.)]. See Chapter 2.

Constitutional isomers. Isomers differing in constitution (connectivity).

Constitutionally heterotopic ligands. Heterotopic ligands (q.v.) whose separate substitution by a new ligand gives rise to constitutional isomers.

Cotton effect. Manifestation in ORD of the refraction, or in CD (q.v.) of the absorption, of linearly or circularly polarized radiation by a nonracemic chiral sample in the vicinity of a UV–vis absorption band. In optical rotatory dispersion (ORD) a peak followed by a null followed by a trough, or the reverse; in circular dichroism (CD) a maximum or minimum (extremum). Characterized by the wavelength of the null in ORD or of the extremum in CD. See Chapter 13.

D,L. Configurational descriptors for carbohydrates or α-amino acids. See Chapter 5. Their use for other kinds of chiral compounds is obsolete.

d,l. Obsolete characterization of enantiomers by the sign of their optical rotation at a specified wavelength, usually 589 nm (sodium D line emission). Now supplanted by ($+$) or ($-$) for dextrorotatory and levorotatory, respectively. The prefix *dl* is still used for racemates but should be replaced by ($\pm$) or *rac*.

Descriptors. See CIP system.

Diastereomers (diastereoisomers). Stereoisomers not related as mirror images. They usually differ in physical and chemical properties. See Chapter 3.

Diastereomer excess (de). In a diastereoselective reaction (q.v.) producing two (and only two) diastereomers in amounts A and B, de $= 100(|A - B|)/(A + B)$. This definition is useful only if the diastereomers are ultimately converted to enantiomers (see enantiomer excess). Otherwise it is better to report the diastereomer ratio A/B, or proportion A:B.

Diastereoselective reaction or synthesis. A chemical reaction in which a new stereogenic element (q.v.) is introduced in such a way that diastereomers are produced in unequal amounts. The reaction or synthesis is said to display diastereoselectivity. See also Stereoselectivity.

Diastereotopic faces. Faces of a double bond that are not symmetry related.

Addition of a new ligand to one or the other such face gives rise to diastereomers. See Chapter 8.

Diastereotopic ligands. Homomorphic (q.v.) ligands in constitutionally equivalent locations that are not symmetry related (i.e., not interchanged by a C_n or S_n operation – Chapter 4). Replacement of one or other of a set of diastereotopic ligands by a new ligand produces diastereomers. See Chapter 8. See also Heterotopic ligands.

Dihedral angle. The angle between two defined planes. See also Torsion angle.

Dihedral symmetry. An object has dihedral symmetry if it possesses one or more (a total of n) symmetry axes perpendicular to the main C_n symmetry axis. See Chapter 4.

Dissymmetric. Obsolete synonym for Chiral (q.v.). Not equivalent to asymmetric (q.v.), since dissymmetric or chiral entities may possess C_n axes ($n > 1$).

Dynamic nuclear magnetic resonance (DNMR). Nuclear magnetic resonance of a compound or mixture observed as a function of an external variable, such as temperature, used to study rates of fast intra- or intermolecular chemical reactions. See Section 8-4.d.

Dynamic stereochemistry. A term referring to the stereochemical aspects of chemical reactions (as distinct from the "static stereochemistry" of resting molecules).

E (*entgegen*), Z (*zusammen*). Stereochemical descriptors for alkenes or for cumulenes with an odd number of double bonds (and their hetera analogues, such as oximes, hydrazones, and azo compounds) with at least two nongeminal substituents (other than H) at the two ends of the double bonds. *E* (*entgegen*) denotes that the substituents of highest CIP priority (cf. Chapter 5) at each end of the double bond are trans to each other, that is, on opposite sides. See also CIP system, trans. If the pertinent substituents are on the same side (cis to each other) the descriptor is *Z* (*zusammen*). The nomenclature may be used also with respect to partial double bonds such as the C–N bond in N-methylformamide, $OHC-NHCH_3$. *E* and *Z* should *not* be used for substituted cycloalkanes.

Eclipsed conformation. Conformation in which two substituents, X, Y on adjacent atoms A, B are in closest proximity, implying that the torsion angle X–A–B–Y or X–A–B=Y is 0°. See also Synperiplanar, Chapter 10, and Figures 10.2 and 10.15.

Eclipsing strain. See Pitzer strain.

Enantiomer. One of a pair of molecular species that are mirror images of each other and not superposable. Mirror-image stereoisomers.

Enantiomer composition. An expression of the proportion $R:S$ of enantiomers R, S present in a sample of a chiral compound.

Enantiomer excess (ee). In a mixture of a pure enantiomer (R or S) and a racemate (q.v.) (RS) ee is the percent excess of the enantiomer over the racemate.

$$ee = \frac{|R - S|}{R + S} \times 100 = |\%R - \%S|$$

Enantiomeric purity. This term is not clearly defined. It may be used synonymously to enantiomer excess (q.v.) or it may (less commonly) refer to the percentage of the major isomer. In the latter case it is better to refer to enantiomer composition (q.v.) or to enantiomer ratio $\%R/\%S$. See Enantiomer excess and Optical purity.

Enantiomerically enriched (enantioenriched). Having an enantiomer excess of more than 0 but less than 100%.

Enantiomerically pure (enantiopure). Having 100% ee (within the limits of measurement).

Enantiomerization. Conversion of one enantiomer into the other. Usually not applied to racemization (q.v.).

Enantiomorph. One of a pair of chiral objects, such as molecular models, or crystals, that are nonsuperposable mirror images. The term is sometimes incorrectly applied to molecules and used as a synonym for enantiomer (q.v.).

Enantioselective reaction or synthesis. A chemical reaction or synthesis that produces the two enantiomers of a chiral product in unequal amounts. The reaction is said to display enantioselectivity. See also stereoselectivity; Chapter 12.

Enantiotopic ligands and faces. Homomorphic (q.v.) ligands in constitutionally equivalent locations that are related by a symmetry plane (or center or alternating axis of symmetry) but not by a (simple) symmetry axis. Replacement of one or the other enantiotopic ligand by a new ligand produces enantiomers. See also heterotopic.

Enantiotopic faces of a double bond are similarly related by a symmetry plane but not by a C_2 axis; addition to one or the other of such faces of one and the same achiral reagent gives rise to enantiomers. See Chapter 8.

endo, exo. Stereochemical descriptor in a bicyclic system of a substituent on a bridge (*not* bridgehead) that points toward the larger of the two remaining bridges. If the substituent points toward the smaller remaining bridge, its descriptor is exo.

Envelope conformation. The C_s (envelope shaped) conformation of cyclopentane. See Figure 11.87.

Epimerization. Interconversion of diastereomers by change of configuration at one of more than one stereogenic elements (q.v.). Interconversion of epimers (q.v.).

Epimers. Diastereomers differing in configuration at one of two or more stereogenic elements (q.v.). Originally, the term applied to aldoses of opposite configuration at C(2), such as glucose and mannose, but it has now been generalized.

Equatorial bonds, equatorial ligands. Bonds forming a small angle to a plane containing (or nearly containing) the majority of atoms in a cyclic molecule. Contrast to axial bonds (q.v.). Ligands attached to such bonds. See Figure 11.15.

erythro, threo. If the "main chain" of a molecule is drawn vertically in a Fischer projection (q.v.), the erythro isomer has identical or similar substituents at two adjacent (nonidentical) chiral centers on the same side of the chain, whereas in the threo isomer these corresponding substituents are on opposite sides of the chain; see Figure 5.24. For difficulties with this definition, see Chapter 5.

exo. See endo.

Fiducial group. The group that is decisive in the assignment of the stereochemical descriptor (configurational or conformational). Thus in Figure 5.6, the bromine but not the iodine containing ligand is fiducial. In the assignment of conformational descriptors to $CClBr_2CClBr_2$ the Cl rather than the Br is fiducial; in r-1, c-3,t-5-trimethylcyclohexane the substituent at C(1) is fiducial.

Fischer projection. A planar projection formula of a three-dimensional molecular model obtained by drawing the main chain in a vertical arrangement and other groups on either side of that chain such that bonds drawn vertically are considered to be below or behind the projection plane and horizontal bonds above or in front of that plane.

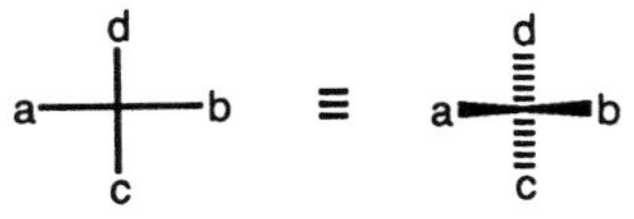

Flagpole position. See Bowsprit position.

Fluxional isomers. See Valence bond isomers.

Flying wedge formula. A way of representing stereoisomers with one or two chiral centers in two dimensions.

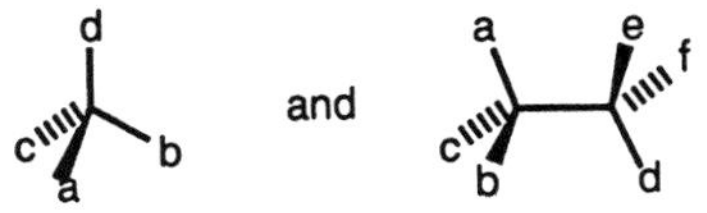

Gauche. In A–B–C–D, ligands A and D are gauche if the torsion angle ω(ABCD) about the B–C bond is near $+60°$ or $-60°$. See also Synclinal.

Gauche effect. In a conformational array 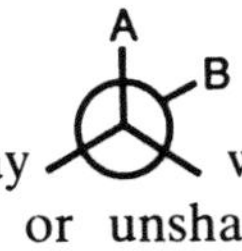 where A and B are second-row electronegative atoms (N, O, or F), or unshared electron pairs, the often observed preference for the gauche conformation of A and B (shown above) is called the "gauche effect."

Geometric isomerism. Obsolete synonym for cis–trans isomerism.

Half-chair conformation. The conformation of a five- or six-membered ring in which three or four atoms, respectively, are in a plane and the remaining two atoms lie on opposite sides of the plane. Its symmetry is C_2. See Figures 11.17 and 11.87.

Helicity. The sense of twist of a helix, propeller, or screw. It is right handed (symbol *P* for plus) if the sense of twist is clockwise as one progresses along the axis of the helix, propeller, or screw away from the observer, left handed (symbol *M* for minus) if the corresponding sense of twist is counterclockwise. See Chapter 14.

Heterochiral. Two isometric (q.v.) molecules are heterochiral if their sense of chirality is opposite, for example, if one is *R* and the other *S*. See also p. 103.

Heterofacial. On opposite sides of a defined plane or face. Thus in a trigonal bipyramid the two apices of the bipyramid are heterofacial with respect to the (triangular) basal plane. See Section 5-5.f.

Heterotopic ligands and faces. Two or more ligands in a molecule that are identical when detached [homomorphic (q.v.)] are heterotopic if replacement of each in turn by a "new" ligand (i.e., one not already present at the sites of attachment of the ligands) gives rise to distinct products. See Constitutionally heterotopic, Stereoheterotopic, Enantiotopic, Diastereotopic ligands and faces. See Chapter 8.

Heterotopic faces of a double bond are faces such that addition of one and the same reagent to one or the other face gives different products.

Homochiral. Isometric molecules (q.v.) are homochiral if they have the same sense of chirality, that is, if they are all *R* or all *S*. The term should *not* be used to describe enantiomerically pure substances. Antonym: Heterochiral (q.v.).

Homofacial. On the same side of a defined plane or face. Antonym: Heterofacial (q.v.). See Section 5-5.b.

Homomorphic ligands. Ligands that are structurally (including configurationally) identical when detached.

Homotopic ligands and faces. Homomorphic (q.v.) ligands in constitutionally and configurationally equivalent positions. Such ligands are interchanged by a C_n operation. Separate replacement of one or the other of such ligands by a new ligand (see heterotopic ligands) gives identical products. Similarly, addition to one or the other of the homotopic faces of a double bond gives identical products. Ligands are considered homotopic when they fulfill the above condition on the time scale of the experiment under consideration (thus the hydrogen substituents on the methyl group in CH_3CO_2H are considered homotopic). Antonym: Heterotopic. See Chapter 8.

Homomers. Molecules that are superposable (q.v.) on each other on the time scale of the experiment (cf. Chapters 2 and 3). Antonym of isomers (q.v.).

Inversion of configuration

1. If X = Y, conversion of a molecule into its enantiomer in a nonequilibrium process. (See Pyramidal inversion for the similar equilibrium process.)
2. If X ≠ Y, see figure: Conversion of CabcX into CabcY by a heterofacial process, that is, approach of Y to the plane that contains the central atom (and is perpendicular to the line of approach) from the side opposite to that from which X recedes. This case is sometimes called "Walden inversion." Antonym: Retention of configuration. See also Relative configuration, Retention of configuration and Section 5-5.f.

Isomers. Chemical species that have the same number and kind of atoms but differ in physical and/or chemical properties because of a difference in structure [constitution and/or configuration and/or conformation (q.v.)]. The time scale of the experiment matters in the distinction of isomers from homomers (q.v.). See Chapters 2 and 3.

Isomerism. Term referring to the existence of isomers and their relationship.

Isometric. Species that are either superposable or mirror images of each other are called isometric, meaning, in the case of molecules, that all distances between corresponding atoms in the species in question are the same. Antonym: Anisometric.

Kinetic resolution. Achievement of partial or complete resolution (q.v.) of a racemate (or of further resolution of a partially resolved mixture) by virtue of unequal reaction rates of the enantiomers with a chiral, nonracemic agent (reagent, catalyst, or enzyme) the reaction not being allowed to proceed to completion.

L, *l*. See **D**, *d*.

ℓ (meaning "like"). Stereodescriptor for diastereomers with two stereogenic (q.v.) centers where both are *R* or both *S* within a given molecule. Antonym *u* (meaning unlike). See Chapter 5.

ℓk (meaning "like"). Stereodescriptor for the approach of an achiral reagent to a double bond in a chiral molecule with diastereotopic faces (q.v.) such that in the *R* isomer preferential approach is to the *Re* face and in the *S* isomer preferential approach is to the *Si* face. Antonym *uℓ* (unlike). Also preferential approach of a reagent of *R* sense of chirality to the *Re* face of a substrate containing a double bond, or of the *S* reagent to the *Si* face. See Chapter 5.

Linearly polarized light. Electromagnetic radiation filtered so that its electric field vector oscillates in one plane, which contains the line along which the radiation is propagated.

London force. The attractive component of a nonbonded interaction (q.v.).

M. See Helicity.

meso. Stereodescriptor for an achiral member of a set of diastereomers that also includes at least one chiral member.

Molar rotation. $[\Phi] = [\alpha]MW/100$ where $[\alpha]$ is the specific rotation (q.v.) and MW is the molecular weight of the substance whose rotation is measured. The alternative term "molecular rotation" is now deemed incorrect (IUPAC) and the symbol $[M]$ (instead of $[\Phi]$) is considered obsolete.

Mutarotation. The change in optical rotation (q.v.) accompanying epimerization [i.e., interconversion of epimers (q.v.)], which comes about because the epimers (diastereomers) differ in specific rotation (q.v.). It requires a labile chiral center, that is, one whose configuration or sense of chirality (q.v.) changes spontaneously or under the influence of a catalyst on a readily observable time scale. In sugar chemistry the term usually refers to the change in optical rotation brought about by change of configuration at the anomeric (hemiacetal) carbon atom.

Neat liquid. The word *neat* is used for an undiluted liquid, a liquid without solvent; synonym *homog*. Used especially in relation to measurement of optical rotation.

Newman projection. A projection formula representing the spatial arrangement of bonds on two adjacent atoms in a molecule, viewed along the bond joining the two atoms (see left formula below). Bonds to these two atoms, which become superposed in the projection, are represented as follows: at the front atom, as spokes drawn to the center of the circle, at the rear atom, as spokes drawn to the periphery of the circle.

Newman   Sawhorse

Nonbonded interactions. Intramolecular through-space interactions (attractive or repulsive, see Fig. 2.23) between atoms in a molecule that are not bonded to each other. Also called (intramolecular) van der Waals interactions. See also London force.

Nonracemic. A term describing a sample of a chiral substance in which molecules of one enantiomer are in excess over those of the other.

Nonsuperposable. Antonym of Superposable (q.v.). Enantiomers, though isometric (q.v.), are nonsuperposable.

Octant rule. Rule predicting the sign of the Cotton effect, in ORD, or the sign of the extremum, in CD, as a function of substitution in a cyclohexanone. See Chapter 13. It has been extended to some other ketones.

Optical activity. Attribute of nonracemic assemblies of chiral molecules or of a chiral crystal by virtue of which these species rotate the plane of polarized light. See Optical rotation.

Optical antipodes. Obsolete term for enantiomers.

Optical isomers. Ill-defined term that should be abandoned. Sometimes used for enantiomers, at other times for any stereoisomers (for example, diastereomers) containing chiral elements.

Optical purity. The absolute value of the ratio of the observed specific rotation of a sample made up only of two enantiomers (i.e., otherwise chemically pure) to the corresponding specific rotation of one pure enantiomer, expressed as a percentage. Usually though not invariably equal to enantiomer excess (q.v.). Since enantiomer excess is now often measured by non-polarimetric methods, use of the title term is becoming obsolete.

Optical resolution. See Resolution.

Optical rotation. The rotation of the plane of plane-polarized light, generally measured in a polarimeter, caused by the presence of either chiral molecules or a chiral crystal in the light path. The angle of rotation α is positive, symbol $(+)$, if the plane is turned clockwise as seen by an observer *towards whom* the light travels, negative, symbol $(-)$, if the plane is turned counterclockwise.

Optical rotatory dispersion (ORD). The change of specific rotation $[\alpha]$ (q.v.) or molar rotation $[\Phi]$ (q.v.) with the wavelength of the light used in the observation. The plot of $[\alpha]$ or $[\Phi]$ versus wavelength λ is often related to the sense of chirality (q.v.) of the substance under observation. See Cotton effect; Chapter 13.

Optical yield. In a chemical reaction, the ratio of the enantiomer excess of the product over the enantiomer excess of the starting material. If the reaction is stereospecific (q.v.) and no racemization occurs, the optical yield is 100%. Although the term appears outmoded, none better has been devised.

Optically labile. Undesirable term for substances whose optical rotation changes with time (either spontaneously or under the influence of a catalyst) either due to racemization or (in less common use) due to epimerization. Better alternatives are "readily racemized" or "readily epimerized."

P. See Helicity.

Perspective formulas. Formulas that represent the (three-dimensional) stereochemical features of molecules in two-dimensional perspective, such as flying wedge (q.v.) and sawhorse (q.v.) formulas. Contrast to projection formulas (q.v.).

Pitzer strain. Excess enthalpy of a molecule, caused by the torsion angles deviating from their optimum values (normally 60° in a saturated molecule), over that of the lowest energy conformation. Also called "eclipsing strain" or "torsional stain."

Planar chirality. Chirality resulting from the arrangement of out-of-plane groups with respect to a reference plane, called the "chiral plane" (q.v.). See Chapter 14.

Plane polarized light. See Linearly polarized light.

Plane of symmetry. See Symmetry plane.

Point group (symmetry point group). The symmetry classification of a molecule, model, or object according to the symmetry elements (axes, planes, or alternating axes of symmetry) it contains. See Chapter 4.

Primary structure. The sequence of the amino acids in a peptide or protein (usually without considering disulfide cross-links, if any). The constitutional formula of a peptide or protein.

Priority. This term, used in connection with the CIP system (q.v.), refers to the ordering of the ligands connected to a stereogenic element (q.v.) with a view to obtaining the stereochemical descriptor for that element. The priority rules (see Chirality rules) are explained in Chapter 5.

Prochirality. A term referring to the existence of stereoheterotopic ligands or faces (q.v.) in a molecule, such that appropriate replacement of one such ligand or addition to one such face in an achiral precursor gives rise to chiral products. A more general term is "prostereoisomerism," since, in some cases, replacement of one or other of two heterotopic ligands or addition to one or other of two heterotopic faces gives rise to achiral diastereomers that contain stereogenic (but not chiral) elements (q.v.). The descriptors *pro-R* or *pro-S* are used for heterotopic ligands, depending on whether replacement of a given ligand by one identical, but arbitrarily assumed to be of higher priority, gives rise to a chiral element with descriptor *R* or *S*, respectively. The descriptors *Re*, *Si* (q.v.) are used for heterotopic faces. See Chapter 8. If the elements or faces are prostereogenic but not prochiral, the descriptors are *pro-r*, *pro-s*, *re*, and *si*.

Projection formulas. Two-dimensional projections of a three-dimensional molecular model (as might be obtained by shining a light through a molecular model and observing the shadow on a plane surface). If the spatial arrangement of bonds is indicated in the projection formulas, they are sometimes called stereochemical formulas. Examples are Fischer projections, Newman projections (q.v.), and formulas with bold and dashed bonds. See also perspective formulas, which are distinct. See Figure 5.21.

Prostereoisomerism. See Prochirality.

Pseudoasymmetric atom. A stereogenic [but achirotopic (q.v.)] atom of whose four distinct ligands two are enantiomorphic, as in $Cab\ell^+\ell^-$. Molecules containing such atoms are achiral, but interchange of two ligands gives rise to diastereomers. The descriptors for pseudoasymmetric atoms are *r* or *s*. See Chapter 3.

Pseudoaxial, pseudoequatorial. Description of a bond or ligand attached to the allylic atoms C(3) and C(6) in a cyclohexene or heteroanalogue. Because of the smaller torsion angle (45°, versus 55° in cyclohexane), such bonds (or ligands), while recognizably similar to axial or equatorial bonds (or ligands) are somewhat differently inclined. Therefore one uses the prefix "pseudo", which is also applied in rings other than six membered, to substituents and bonds that are not truly axial or equatorial because of deviation of ring torsion angles from 60°.

Pseudorotation. The out-of-plane motion of the ring atoms in the facile conformation changes that occur in cyclopentane and that simulate a (hypothetical) rotation of a bulge or pucker around the ring.

The term is also used for the often facile ligand interchange that takes place in trigonal bipyramidal molecules, such as PF_5.

Pyramidal inversion. The change of bond directions to a three-coordinate central atom having a pyramidal (or tripodal) arrangement of bonds in which the central atom (at the apex of the pyramid) appears to move to an equivalent position on the other side of the base of the pyramid. If the central atom is chiral, pyramidal inversion inverts its sense of chirality.

Quasienantiomers. Heterofacially (q.v.), substituted tetrahedral molecules Xabcd and Xabec (e taking the place of d but on the opposite face of the abc plane). Note that these molecules would be enantiomeric if d = e. Thus (R)-2-bromobutane is the quasienantiomer of (S)-2-chlorobutane. See also Quasi-racemate.

Quasi-racemate. A 1:1 mixture of quasienantiomers that may form a compound, a eutectic, or a solid solution (see Section 5-5.e).

Quaternary structure. The structure of a protein consisting of two or more individual subunits (smaller proteins) held together by hydrogen bonds, Coulombic, or London (q.v.) forces.

R (rectus), S (sinister). Stereochemical descriptors in the CIP (Cahn–Ingold–Prelog) system. See Chapter 5. When the descriptors refer to axial chirality (q.v.), they may be modified to aR, aS, if referring to planar chirality (q.v.) to pR, pS. If the chiral atom being described is other than carbon, its atomic symbol is sometimes indicated as a subscript, such as R_P, R_S for R chirality sense at phosphorus or sulfur, respectively. The symbols R^* and S^* may be used for relative configuration. See Chapter 5.

Racemate. A composite (solid, liquid, gaseous, or in solution) of equimolar quantities of two enantiomeric species. It is devoid of optical activity. In the chemical name or formula it may be distinguished from the individual enantiomers by the prefix ($\pm$)- or rac- or by the descriptors RS. In the older literature the term was confined to a racemic compound (q.v.) and the general term used was "racemic modification." See also Conglomerate. Synonym: Racemic mixture.

Racemic compound. A racemate in which the two enantiomers form a crystalline compound (which can be recognized from the melting phase diagram or by X-ray structure analysis: The unit cell contains equal numbers of enantiomeric molecules). Formerly sometimes called "true racemate." See Chapter 6.

Racemic conglomerate. See Conglomerate.

Racemic mixture. In the old literature this term was used as a synonym for conglomerate (q.v.), but nowadays the term is used as a synonym for racemate (preferred term, q.v.) or racemic modification.

Racemic modification. Older term for Racemate (q.v.).

Racemization. The formation of a racemate (q.v.) from a chiral nonracemic starting material.

Re, Si. Descriptors for heterotopic faces as in abX=Y, indicating their two-dimensional chirality (see Chapter 8). When additions to X=Y gives rise to a pseudoasymmetric (q.v.) rather than a chiral center, the symbols are *re* and *si*.

Reflection invariant. Term applied to a structure or substructure (part of a structure) that remains unchanged by reflection, such as an achiral or meso compound.

rel. Prefix used to indicate relative configuration (q.v.). However, the descriptor R^*,S^* is preferred to *rel–R,S*.

Relative configuration.
1. The configuration at any stereogenic element (center, axis or plane) with respect to that of any other stereogenic element in the same molecule. Unlike absolute configuration, relative configuration is reflection invariant. Descriptors for relative configuration in molecules possessing only two stereogenic elements are R^*,R^* (or ℓ), R^*,S^* (or *u*), *cis* (*c*), *trans* (*t*), *E*, and *Z*.
2. Two different chiral molecules (or molecules containing stereogenic centers), such as Xabcd (i) and Xabce (ii) are said to have the same relative configuration when d and e are homofacial (q.v.) to the abc plane, that is, if replacement of d by e converts i into ii. The molecules are said to have opposite relative configurations when d and e are heterofacial (q.v.) to the abc plane, in which case replacement of d by e in i gives Xabec, the enantiomer of ii. See also Inversion of configuration and Retention of configuration.

Resolution. Separation of the enantiomers from a racemate (racemic mixture) with recovery of at least one of the enantiomers.

Retention of configuration. In a chemical reaction, configuration (q.v.) is said to be retained if the product has the same relative configuration (q.v.) as the starting material, that is, if the reaction proceeds homofacially (q.v.). See also Relative configuration and Inversion of configuration.

Ring inversion or ring reversal. The interconversion of the chair forms of cyclohexane or substituted cyclohexanes (ring flipping) accompanied by interchange of the axial and equatorial bonds and substituents. By extension, interconversion of conformers having equivalent ring shapes in any cyclic structure.

Rotamer. One of a set of conformers (q.v.) arising from restricted rotation about one or several single bonds.

Rotation. See Optical rotation.

Rotation–reflection axis. See Alternating axis of symmetry.

Rotational barrier. The difference between a given minimum and an adjacent maximum of potential energy in a conformational change brought about by bond rotation.

s-cis, s-trans. Descriptors that refer to the conformation about a single bond that links two conjugated double bonds. The synperiplanar (q.v.) conformation is called s-cis and the antiperiplanar (q.v.) one is called s-trans.

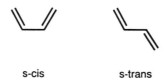

s-cis   s-trans

Sawhorse formula. A perspective drawing indicating the spatial arrangement of all the bonds on two adjacent atoms. The bond between the atoms is represented by a diagonal line, usually from lower left to upper right, the left-hand bottom end representing the atom nearest to the observer and the right-hand top end the atom that is farther away. Two of the remaining bonds to the two atoms are drawn vertically and the other four at $+120°$ or $-120°$ angles to these two (see diagram under Newman projection on p. 1202).

Secondary structure. Parts of protein or polypeptide structure that are shaped by hydrogen bonds between nonadjacent amino acids. Examples are the α-helix (coiled structure) and the β-pleated sheet (folded structure).

Sense of chirality. The handedness, usually expressed as R or S, or P or M, of a chiral molecule. Also applied to macroscopic chiral objects, such as hands or screws (which may be right handed or left handed).

Sequence rules. In the CIP (q.v.) system, rules that establish the priority (q.v.) of ligands to a chiral or stereogenic element.

Specific rotation [α]. The optical rotation (q.v.) of a nonracemic (q.v.) chiral substance in solution at a concentration of $1 \, g \, L^{-1}$ in a 10-cm cell under specified conditions of temperature and wavelength of the polarized light employed in the measurement. In general $[\alpha] = \alpha/\ell c$, where α is the observed rotation in degrees, c is the concentration in grams per liter and ℓ the cell length in decimeters. This formula (Biot's law) also applies to neat liquids (q.v.) with c then standing for density. Concerning the units of $[\alpha]$, see Chapter 13.

Spontaneous resolution. Resolution (q.v.) by spontaneous crystallization of individual enantiomers from a solution or melt of the racemate, possible only if the latter is a conglomerate (q.v.) in the solid state.

Staggered conformation. Conformation of an ethanoid moiety abcX–Ydef in which a, b, and c are at the maximum distance from d, e, f. This requires the torsion angles to be 60°. See Figure 10.2.

Stereochemistry (adjective stereochemical). Chemistry in three dimensions, chemistry with consideration of its three-dimensional aspects, but also used in relation to chemical and physical properties of cis–trans isomers (q.v.) in alkenes.

Stereoconvergent. This term is applied to a reaction in which stereoisomerically differing starting materials yield identical products.

Stereodescriptor (stereochemical descriptor). A prefix specifying configuration or conformation, such as *R*, *S*, *r*, *s*, *P*, *M*, *Re*, *Si*, *E*, *Z*, *cis*, *trans*, *ap*, or *sc*. See also CIP system.

Stereoelectronic control. Control of the outcome of a reaction by stereoelectronic effects (q.v.).

Stereoelectronic effect. An effect determining the properties or reactivity of a species that depends on the orientation of filled or unfilled electron orbitals in space.

Stereogenic element. A focus of stereoisomerism (stereogenic center, axis, or plane) in a molecule such that interchange of two ligands attached to an atom in such a molecule (e.g., a and b in a tetrahedral atom Cabcd or in an alkene abC=Cab or in an allenic species abC=C=Cab) leads to a stereoisomer. If the element is chirotopic (q.v.), for example, if it occurs in a chiral molecule, one speaks of a chiral center, axis, or plane (q.v.), but if the element is achirotopic as in abC=Cab these terms are not appropriate, yet the center, axis, or plane is stereogenic.

Stereoheterotopic. Homomorphic ligands (q.v.) whose separate replacement gives rise to stereoisomers; also faces of a double bond, separate addition to which gives rise to stereoisomers. If the products are enantiomeric, the ligands or faces are enantiotopic (q.v.); if the products are diastereomeric, the ligands or faces are diastereotopic (q.v.).

Stereoisomers. Isomers of identical constitution but differing in the arrangement of their atoms in space. Subclasses are Enantiomers and Diastereomers (q.v.).

Stereomutation. A general term for the conversion of one stereoisomer into another, such as Racemization, Epimerization, or Asymmetric transformation (q.v.).

Stereoselectivity. The preferential formation of one stereoisomer over another in a chemical reaction. If the stereoisomers are enantiomers, one speaks of enantioselectivity [quantified by enantiomer excess (q.v.)]; if they are diastereomers, one speaks of diastereoselectivity (q.v.). The term "enantioselective" may be applied to the ultimate outcome of a sequence of reactions, even if individual steps are diastereoselective.

Stereospecific. A reaction is termed stereospecific if, in such a reaction, starting materials differing only in their configuration are converted to stereoisomerically distinct products. According to this definition, a stereospecific process is necessarily stereoselective, but stereoselectivity does not necessarily imply stereospecificity.

The term may be extended to a process involving a chiral catalyst (including an enzyme) or chiral reagent when the configuration of the product of the reaction depends uniquely on the configuration of the catalyst or reagent., i.e. becomes reversed when a catalyst or reagent of opposite configuration is employed.

The use of the term "stereospecific" merely to mean "highly stereoselective" is discouraged.

Steric energy. The energy, calculated by molecular mechanics, over and above the bond energy, calculated by a bond energy additivity or group increment scheme. Strictly speaking this is potential energy.

Strain energy. The experimentally determined energy (usually enthalpy) of a strained structure (e.g. cyclobutane) over and above that of the corresponding unstrained structure (e.g. cyclotetramethylene) as calculated from group increments.

s-trans. See s-cis.

Structure. The complete arrangement in space of all the atoms in a molecule as determined, for example, by X-ray crystal analysis. It includes constitution, configuration, and conformation (q.v.). The use of the term as a synonym for constitution alone is strongly discouraged.

Structural isomers. An obsolete term for constitutional isomers (q.v.); use now is strongly discouraged. Under the current definition of structure (q.v.) all isomers are structural isomers, so the term is redundant.

Superimposable. See Superposable.

Superposable. Two structures, stereoformulas, or models are said to be superposable if, after suitable translation and rigid rotation, they can be brought into coincidence. The term is often extended to structures that can be brought to coincidence only after internal rotation, or in disregard of internal rotation, about single bonds. Thus two molecules of (S)-2-bromobutane may be said to be superposable regardless of their conformations.

Syn. See Synperiplanar. Formerly also used to describe the configuration of oximes, and so on. Antonym of anti (q.v.).

Synclinal (*sc*). In X–A–B–Y, X and Y are synclinal if the torsion angle about the A–B bond is between 30° and 90° or between −30° and −90°. See also Gauche; Table 2.2.

Synperiplanar (*sp*). In X–A–B–Y, X and Y are synperiplanar if the torsion angle about the A–B bond is between −30° and +30°. See also Eclipsed conformation; Figure 2.13.

Symmetry axis (C_n). A symmetry axis of order n is an axis such that rotation of a molecule, model, or object by an angle of $360°/n$ about such an axis produces a superposable (q.v.) entity.

Symmetry plane. A plane such that if a molecule, model, or any object is reflected across this plane, the reflected entity is superposable (q.v.) with the original one.

Tautomers. Tautomers are readily interconvertible constitutional isomers, but, in contrast to conformational isomers (q.v.) and valence bond isomers (q.v.), in tautomers there is a change of connectivity of a ligand, as in the keto and enol forms of ethyl acetoacetate.

Tertiary structure. The spatial organization, resulting from the folding of helices and sheets over one another, of an entire peptide or protein molecule consisting of a single peptide chain.

threo. See erythro.

Topomerization. A reaction converting a molecule into another that is superposable with the original one but in which two or more atoms or ligands have exchanged place. An example is the pyramidal inversion (q.v.) of $C_6H_5CH_2N(CH_3)_2$, which gives an identical product in which, however, the two methyl groups are interchanged. Also called "automerization" or "degenerate isomerization." The two structures that are interconverted may be called "topomers." See Chapter 8.

Torsion angle. In a nonlinear molecule A–B–C–D, the [dihedral (q.v.)] angle ω between the planes containing A–B–C and B–C–D. If, looking (in either direction) along the B–C axis, the turn from A to D or D to A is clockwise, ω is positive; if the turn is counterclockwise, ω is negative. See Figures 2.2 and 2.13.

Torsional isomers. Stereoisomers that can be interconverted (actually or conceptually) by torsion about a bond axis. Rotamers, atropisomers (q.v.), and cis–trans isomers of alkenes are in this category.

Torsional strain. See Pitzer strain.

trans (t). Descriptor indicating the relationship between ligands on the opposite sides of a double bond or ring. Antonym of cis (q.v.).

Transannular strain. Nonbonded interaction (q.v.) between ligands attached to nonadjacent ring atoms as in a cyclohexane boat (see Bowsprit position) or medium-sized ring.

transoid. Antonym of cisoid (q.v.).

Twist form. The intermediate conformer of D_2 symmetry (cf. Fig. 11.16) between two cyclohexane chairs; also called skew–boat or skew in the older literature. By extension, the term is used in larger rings (cf. Section 11-5.d).

u **(meaning unlike).** Stereodescriptor for diastereomers with two stereogenic centers when one is *R* and the other is *S* or vice versa. See ℓ (meaning "like").

uℓ **(unlike).** Stereodescriptor for approach of an achiral reagent to a double bond in a chiral molecule with diastereotopic faces such that in the *R* isomer preferential approach is to the *Si* face and in the *S* isomer to the *Re* face. Also, preferential approach of a chiral reagent of *R* chirality sense to the *Si* face of a substrate containing a double bond, or of an *S* reagent to the *Re* face. See also ℓk (meaning "like").

Valence bond isomers. Isomers that differ only by the position and order of the bonds between their atoms, which latter may, however, move slightly. Examples are cycloheptatriene and norcaradiene (Fig. 11.92) or cyclooctatriene and bicyclo-[4.2.0]octadiene. Also called fluxional isomers.

van der Waals interactions. See Nonbonded interactions. Also used for interactions (attractive or repulsive) resulting from close approach of two separate molecules.

Walden inversion. See Inversion of configuration.

Z. See *E, Z*.

Index